Dam Construction in China
—A Sixty-year Review

Three Gorges Project

The Three Gorges Project is the largest hydropower station in the world with installed capacity of 22,500MW. The dam is concrete gravity dam with height of 181m and the crest length is 2,335m. It is the highest concrete gravity dam in China, completed in 2010. The project was awarded as the international milestone concrete dam in 2011 by ICOLD, CHINCOLD and USSD.

South-to-North Water Diversion Project

The south-to-north water diversion project, consisting of eastern, middle and western routes, is designed to transfer water from south part of the country mainly the Yangtze River, the country's longest river, to the north. The total capacity of three routes for water diversion is about 44.8 billion m^3 according to planning. It will take 40-50 years for the construction of the project. The whole project will be implemented in phases according to the actual situation.

Dam Construction in China
—A Sixty-year Review

Xiaolangdi Project

Xiaolangdi Project is located at the main stream of Yellow River. It was the highest rockfill dam with clay core in China in the last century with the maximum dam height of 160m. The installed capacity of the power station is 1,800MW. The project was completed in 2001 and was awarded as the international milestone rockfill dam in 2009 by ICOLD, CHINCOLD and CBDB.

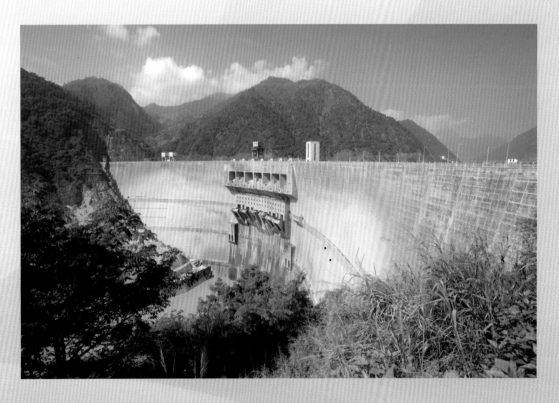

Ertan Project

Ertan Project is located at the lower reaches of the main stream of Yalong River in Sichuan Province. The dam was the highest concrete double-curvature arch dam in Asia and the third highest in the world with maximum height of 240m in the 20th century. The installed capacity of the power station is 3,300MW. The project was completed in 2000 and was awarded as the international milestone concrete dam in 2011 by ICOLD, CHINCOLD and USSD.

Longtan Project

Longtan Project is located on the upper reaches of Hongshui River in Guangxi Zhuang Autonomous Region. It is the highest RCC gravity dam in the world with dam height of 192m at the stage I. The maximum dam height is 216.5m. The installed capacity of the power station is 6,300MW. The project was awarded as the international milestone RCC dam in 2007 by ICOLD, CHINCOLD and SPANCOLD.

Shuibuya Project

Shuibuya Project is located at the middle reaches of Qingjiang River in Hubei Province. It is the highest CFRD in the world with dam height of 233m. The installed capacity of the power station is 1,840MW. The project is in good condition and has brought the remarkable benefits of electricity generation, flood control etc. It was completed in 2009 and was awarded as the international milestone project of rockfill dam at the same year by ICOLD, CHINCOLD and CBDB.

Dam Construction in China
—A Sixty-year Review

Xiaowan Project

Xiaowan Project is located at the middle reaches of Lancang River with maximum dam height of 294.5m and installed capacity of 4,200MW. It was completed in 2012 and was the highest arch dam during the construction. A lot of technical problems related has been investigated and solved.

Jinping-I Project

Jinping-I hydropower station is located at the lower reaches of the main stream of Yalong River. The concrete double-curvature arch dam is 305m high and it is the highest double-curvature arch dam in the world. The installed capacity of the power station is 3,600MW. The project will be completed very soon.

Xiluodu Project

Xiluodu Project is located on the Jinsha River (one branch of Yangtze River) in Sichuan Province. The maximum arch dam height is 285.5m. The total installed capacity is 13,860MW. It is the 2nd largest hydropower station in China and the 3rd in the world. The project was completed in 2015.

Nuozhadu Project

Nuozhadu Project is located at the downstream of Lancang River. The maximum height of the rockfill dam with clay core is 261.5m. The project ranks the third in the world and the first in China within the same dam type that have been built or under construction. The installed capacity of the power station is 5,850MW. The project was completed in 2015.

Guangzhao Project

Guangzhao Project is located at the midstream of Beipan River in Guizhou Province. With the maximum dam height of 200.5m, it is the highest RCC gravity dam that has been built in the world. It is the first RCC dam with the height over 200m in the world. The installed capacity of the power station is 1,040MW. The project was completed in 2007 and was awarded as the international milestone RCC dam in 2012 by ICOLD, CHINCOLD and SPANCOLD.

Hongjiadu Project

Hongjiadu Project is located at the main stream of Wujiang River in Guizhou Province. It is a CFRD with height of 179.5m and the installed capacity of the power station is 600MW. The project was completed in 2005 and was awarded as the international milestone CFRD project in 2011 by ICOLD, CHINCOLD and CBDB.

Jiudianxia Project

Jiudianxia Project is located at the midstream of Taohe River in Gansu Province. The maximum height of the CFRD is 133m and the installed capacity of the power station is 300MW. By the construction of this project, key technologies to construct high CFRD under the conditions of narrow river valley, thick overburden layer foundation and high seismic intensity were studied and new progresses have been made. It was completed in 2008 and was awarded as the international milestone CFRD in 2013 by ICOLD, CHINCOLD and CBDB.

Zipingpu Project

Zipingpu Project is located at the upstream of Minjiang River in Sichuan Province. The maximum height of the CFRD is 156m and the installed capacity of the power station is 760MW. The dam withstood the test of magnitude-8 Wenchuan earthquake in 2008. And its lessons can provide a reference example for the seismic performance of high CFRD in the world. It was completed in 2006 and won special award as the international CFRD in the world in 2009.

Pubugou Project

Pubugou Project is located at the midstream of Dadu River in Sichuan Province. The maximum height of rockfill dam with clay core is 186m and the installed capacity of the hydropower station is 3,600MW. It is the highest rockfill dam on the deep overburden foundation with the maximum thickness of 77.9m in China. The project was completed in 2009 and awarded as the international milestone rockfill dam in 2011 by ICOLD, CHINCOLD and CBDB.

Dam Construction in China
—A Sixty-year Review

Puding Project

Puding Project is located at the upper reaches of Wujiang River in Guizhou Province. It was the highest RCC arch dam in the world with maximum height of 75m during its construction. The total installed capacity of the power station is 75MW.

Guangzhou Pumped-Storage Power Station

Guangzhou Pumped-Storage Power Station is located in Conghua district of Guangzhou city in Guangdong Province. It is the largest pumped-storage hydropower station in the world with the total installed capacity of 2,400MW. The project was completed in 1994.

Dam Construction in China
—A Sixty-year Review

Shisanling Pumped-Storage Power Station

Shisanling Pumped-Storage Power Station is about 30km away from Beijing. The total installed capacity is 800MW. It is the first project in China of which the upper reservoir was completely lined by the reinforced concrete slabs with successful results. The project was completed in 1997.

Tianhuangping Pumped-Storage Power Station

Tianhuangping Pumped-Storage Power Station is located in Anji county of Zhejiang Province. The total installed capacity of the power station is 1,800MW. It is one of the stations completed or under construction at present with the highest water head in China. The project was completed in 1997.

Dujiangyan Project

Dujiangyan Project is the oldest and still well used large irrigation system in the world. It was built 2,200 years ago. Since its operation, the Chengdu Plain has been free of flooding. Now it still plays a crucial role in draining off floodwater, irrigating farms and providing water resources for more than 50 cities in Sichuan Province.

Dam Construction in China
—A Sixty-year Review

Shilongba Project

Shilongba Project located in the suburbs of Kunming city in Yunnan Province is the first hydropower station in China. It was completed in 1912. The total installed capacity is 6,000kW. It has been put into operation for more than 100 years.

SERIES OF CHINESE NATIONAL COMMITTEE ON LARGE DAMS

Dam Construction in China
—A Sixty-year Review

Chief editor   Jia Jinsheng

Abstract

This book summaries systematically the great achievements and successful experiences in dam construction in China over the past 60 years, especially high dam construction practices in recent decades. This book is divided into 3 parts and 26 chapters. It covers the strategic importance of dam construction and hydropower projects in China, the main technical progresses in dam planning, survey, design, construction, operation management and scientific research, as well as introduction of typical high dams at home and abroad. The objectives of this book are to showcase the nation's achievements in dam construction, particularly key projects, new concepts and the latest progress in recent 10 years from an international perspective.

It is a professional and informative book that provides a comprehensive review of dam construction over the past 60 years until 2010 in China, and could be reference to dam engineers.

图书在版编目（CIP）数据

中国大坝建设60年 = Dam Construction in China
—A Sixty-year Review : SERIES OF CHINESE NATIONAL
COMMITTEE ON LARGE DAMS : 英文 / 贾金生主编. -- 北
京 : 中国水利水电出版社, 2015.12
 ISBN 978-7-5170-4026-2

Ⅰ. ①中… Ⅱ. ①贾… Ⅲ. ①大坝－水利建设－成就
－中国－英文 Ⅳ. ①TV64

中国版本图书馆CIP数据核字(2015)第321650号

书　名	SERIES OF CHINESE NATIONAL COMMITTEE ON LARGE DAMS **Dam Construction in China—A Sixty-year Review**
作　者	Chief editor　Jia Jinsheng
出版发行	中国水利水电出版社 （北京市海淀区玉渊潭南路1号D座　100038） 网址：www.waterpub.com.cn E-mail：sales@waterpub.com.cn 电话：(010) 68367658 （发行部）
经　售	北京科水图书销售中心（零售） 电话：(010) 88383994、63202643、68545874 全国各地新华书店和相关出版物销售网点
排　版	中国水利水电出版社微机排版中心
印　刷	北京瑞斯通印务发展有限公司
规　格	210mm×285mm　16开本　63.5印张　3876千字　8插页
版　次	2015年12月第1版　2015年12月第1次印刷
定　价	**180.00美元**

凡购买我社图书，如有缺页、倒页、脱页的，本社发行部负责调换
版权所有·侵权必究

Editorial Committee

Director:

Chen Lei — Minister of Water Resources

Vice Directors:

Jiao Yong — Vice Minister of Water Resources
Kuang Shangfu — President of China Institute of Water Resources and Hydropower Research
Yan Zhiyong — Chairman of the Board of Power Construction Corporation of China
Cao Guangjing — Former Chairman of the Board of China Three Gorges Corporation

Members (*Ordered alphabetically*):

Chen Houqun — Academician, Chinese Academy of Engineering
Chen Zuyu — Academician, Chinese Academy of Science
Cheng Niangao — President, China Huadian Corporation
Feng Junlin — President, HydroChina Kunming Engineering Corporation
Gao Anze — Former Chief Engineer of the Ministry of Water Resources
Gao Bo — Director General, Department of International Cooperation, Science and Technology
Kou Wei — Vice President, China Huaneng Corporation
Liao Yiwei — Deputy Director, Yellow River Conservancy Commission of the Ministry of Water Resources
Lin Chuxue — Vice President, China Three Gorges Corporation
Liu Jinhuan — Assistant General Manager, China Guodian Corporation
Ma Hongqi — Academician, Chinese Academy of Engineering
Niu Xinqiang — Academician, Chinese Academy of Engineering
Qu Bo — Chief Engineer, China Datang Corporation
Shi Lishan — Deputy Director General, Department of New and Renewable Energy, National Energy Administration
Sun Hongshui — President, Power Construction Corporation of China
Sun Jichang — Director General, Department of Construction and Management, the Ministry of Water Resources
Tian Zhongxing — Director General, Bureau of Rural Hydropower and Electrification Development, the Ministry of Water Resources
Wang Hao — Academician, Chinese Academy of Engineering
Wu Guotang — Director General, Department of Safety Supervision, the Ministry of Water Resources
Yang Chun — Deputy Director, Changjiang Water Resources Commission of the Ministry of Water Resources
Yue Xi — Commander, The Hydropower Headquaters of the Chinese Armed Police Force

	President, China An'neng Construction Corporation
Zhang Chaoran	Academician, Chinese Academy of Engineering
Zhang Jianyun	President, Nanjing Hydraulic Research Institute
	Academician, Chinese Academy of Engineering
Zhang Liying	Former Chief Engineer, State Grid Corporation of China
Zhang Xiaolu	Former Vice President, China Power Investment Corporation
Zheng Shouren	Chief Engineer, Changjiang Water Resources Commission of the Ministry of Water Resources
	Academician, Chinese Academy of Engineering
Zhang Ye	Deputy Director, Office of the South-to-North Water Diversion Project Commission of the State Council
Zhou Dabing	Former President, Chinese Society for Hydroelectric Engineering
Zhu Bofang	Academician, Chinese Academy of Engineering

Honorable Chief Editor:

Wang Shucheng	President of Chinese National Committee on Large Dams (CHINCOLD), Former Minister of Water Resources

Advisors:

Lu Youmei	Honorable President of CHINCOLD, Academician of Chinese Academy of Engineering
Pan Jiazheng	Honorable President of CHINCOLD, Academician of Chinese Academy of Engineering, Academician Chinese Academy of Science

Chief Editor:

Jia Jinsheng	Vice President and Secretary General of CHINCOLD, Vice President of China Institute of Water Resources and Hydropower Research

Vice Chief Editors:

Zhou Jianping	Chief Engineer, Power Construction Corporation of China
Liu Zhiming	Vice President, General Institute of Water Resources & Hydropower Planning & Design

Reviewers:

Gao Jizhang	Former President of China Institute of Water Resources and Hydropower Research
Jiang Guocheng	Professor, China Institute of Water Resources and Hydropower Research

Editors:

Zheng Cuiying   Xu Yao   Yuan Yulan   Ye Weimin

Preface I

China has a large population and inadequate water resources, which are distributed unevenly in time and space. Suffering from frequent drought and flood, China has the most complicated water regime and the heaviest tasks for water management in the world. It is an inevitable way for China to build water projects so as to properly regulate the natural water, harness water resources and prevent water disasters.

Since the foundation of the People's Republic of China in 1949, China has staged a large-scale campaign to build water projects. Numerous dam projects are constructed with functions of flood control, irrigation, power generation, navigation, ecological protection, etc. By 2012, 87,800 reservoirs had been completed with a total storage capacity of 716.2 billion m^3; 5,564 dams with the heights over 30 m were completed or under construction; the installed capacity reached 230 GW. China has subsequently become the country with the most numbers of reservoirs and dams, and the largest installed hydropower capacity. Meanwhile, rehabilitation and reinforcement has been carried out in recent years for defective reservoirs and dams which are inadequately maintained or in dangerous state. Within three years, 7,356 large, medium, and certain important small reservoirs were reinforced, 5,400 small-sized-I reservoirs (1 million $m^3 \leqslant$ storage capacity $<$ 10 million m^3) and 40,900 small-sized-II reservoirs (storage capacity $<$ 1 million m^3) were rehabilitated. This is an unprecedented achievement in the world history of dam construction. With the successful construction and scientific regulation of these reservoirs, China have triumphed over many severe flood and drought disasters. Having only 6% of global fresh water resources and 9% of global farmland, China has managed to ensure the safety of water and food supply for 21% of the world population, and maintained strong socio-economic development to become a comprehensive well-off society. The achievements are indeed remarkable.

Evolving through six decades of extensive dam practices, the technologies of dam construction have developed by leaps and bounds. New technology, techniques, materials, and equipment are widely applied. Major breakthroughs are made in areas such as core wall construction for seepage control, foundation treatment, high slope treatment, river diversion, flood discharge and energy dispassion of high dams, etc. The dam height has also risen step by step from 100m to 200m and to 300m now. The currently highest arch dam, concrete faced rockfill dam and roller compacted concrete dam in the world are all located in China. In particular, several mega hydropower projects commanding extremely complicated technologies are successfully built and run stably, such as Three Gorges, Xiaolangdi, Ertan, Shuibuya and Longtan. China has become one of the leading countries in terms of dam construction technology.

That being said, China still has an uphill road to go before the comprehensive modernization can be realized and a well-off society be built. Better socio-economic development raises new and higher requirements for flood control, water and energy supply, food safety and ecology protection. Compounded by global climate change, the uneven distribution of water becomes more prominent. Water shortage is growing, and flood and drought are occurring more frequently. To accelerate water infrastructure construction and enhance the capability of water control and allocation remain significant and urgent. Future emphasis of dam construction, as envisioned by the central government, will be laid on building large and medium reservoirs and developing water supply projects in Southwest China; basin-scale flood control projects, key water supply systems, and lake-river connectivity enhancement; localized small-scale water impoundment, diversion, withdrawal, rainwater harvesting and usage; as well as eliminating safety risks of dams and reservoirs. The scientific and sustainable development of water resources hopefully will serve for sustainable socio-economic development.

Dam Construction in China—A Sixty-year Review draws on authoritative and professional wisdom to render a circumstantial review of how dam construction has evolved in China—its successes and experiences. It introduces the ideas, strategic plans and future perspectives; it narrates the latest achievements in planning, design, construction, operation and management both at home and abroad; and it elaborates on new technologies, techniques, materials, and methodologies. Contributed by Chinese famous dam experts, this book provides a kaleidoscope into the conceptual and technological advancement of dam construction in China and makes valuable reference for the international dam community in promoting sustainable development of dams.

H. E. Chen Lei
Minister of Water Resources, China
May 29, 2012

Preface II

Over the past six decades, water and hydropower sector in China have developed rapidly, and great progresses have been made in dam construction. As of 2010, 87,000 dams are built, among which more than 25,000 dams are over 15m high. The total storage capacity of reservoirs is 71.62 billion m^3, ranking the 4th in the world after the USA, Brazil, and Russia. The installed capacity of hydropower reaches 230GW, ranking top in the world. Particularly, small hydro with installed capacity of less than 50MW reaches 62,000MW. Dams have played an important role in flood control, water supply, food and energy safety, and ecological protection. They provide a strong support for improving the living standards and advancing socio-economic development.

Before 1949, there were only 21 dams (Taiwan included) with the height over 15m. Flood and drought disasters posed constant threat. Although dams were urgently needed. It was difficult that due to weak and strained national economy in progress. Since the foundation of the People's Republic of China in 1949, the water and hydropower sector ushered in a period of rapid growth. The evolvement of dams in China experienced three stages. The first stage started from 1949 to 1978 when China adopted reform and opening-up. The number of dams mushroomed, making China the most active country to build dams in the world. The numbers of dams higher than 15m increased from 21 to 11,760 (Taiwan included), the installed hydropower capacity increased from 540MW to 18,670MW (Taiwan included), and the dams were mainly for flood control and irrigation. Though great achievements had been made, China still fell behind the developed countries due to limited technology and investment. The second stage started from 1978 to the early 21st century. With huge projects like Three Gorges, Xiaolangdi and Ertan being completed, China transformed from a late comer in dam construction to a lead runner in various aspects, and many dams withstood the test of the major flood in 1998 and Wenchuan Earthquake in 2008. Dams constructed in this period were characterized by high quality design, fast construction, high reliability in safety, and sound benefits. The successful construction of the Xiaowan arch dam, Longtan RCC gravity dam, Shuibuya CFRD and Jinping-I arch dam with heights around 300m marked the third stage. During this period, China built the highest dams in the world of three types—arch dam, concrete faced rockfill dam and roller compacted concrete dam. Environment protection and ecological benefits of dams were given more attention. A lead runner by then in various aspects, China began to widely participate in international dam construction and consulting services.

In 2011, the No.1 Document released by the Chinese government articulated its will to speed up the reform and development of water conservancy, water security was elevated as sine qua non for national security, and water infrastructure was prioritized in national infrastructure construction. Though China topped the world in terms of the number and capacity of dams, water ownership per capita, storage capacity per capita and many other indicators remained low due to the large population and uneven distribution of water, etc. Therefore, more dams are needed to ensure a sustainable future. On the other hand, in the fight against climate change, the Chinese government promises to reduce 40%-45% of carbon dioxide emissions per unit of GDP compared with that of 2005 by 2020, and that non-fossil energy will account for about 15% of primary energy consumption. Therefore, the energy structure should be adjusted, hydropower and other renewable energy should be vigorously advanced. As far as China is concerned, hydropower is abundant in reserve, technologically mature, highly cost-effective and flexible in dispatch among all the renewable energy sources. Dam construction in China will witness another golden period coming. It is estimated that by 2020, the installed hydropower capacity in China will reach to 350 - 400GW from 200GW in 2010, which means at least over 100GW of hydropower projects will be approved during the "Twelfth Five Year Plan" period (2011 - 2015).

Most of the future dam projects in China are located in mountain valleys in Southwest China. They are generally huge, face complex geological conditions and high technological difficulty, posing great challenges to construction. Dam-related technical and environmental issues warrant continual research and practices. The major objective of *Dam Construction in China—A Sixty-year Review* in taking stock of experiences in the past six decades, and the last decade in particular, is to guide future dam activities in China in a scientific and man-water harmonious way, and provide reference to the next generations of dam engineers in their quest of new heights. In addition, publishing this book is expected to facilitate China's exchanges with the international community, and advance the joint efforts to a sustainable future of dams worldwide.

Wang Shucheng
President of Chinese National Committee on Large Dams
May 28, 2012

Forword

In order to systematically summarize the main progresses and achievements of dam construction in China over the past 60 years, and the last decade in particular, Chinese National Committee on Large Dams (CHINCOLD), with the support of relevant parties and invited academicians, well-known experts and professionals in China, compiled the book of *Dam Construction in China—A Sixty-year Review*. Adopting an internationally-oriented perspective, the book provides a kaleidoscope into the highlights of plan, design, construction, operation and management of dams as well as research progress in dam engineering in China. It provides valuable reference for future damming activities both at home and abroad, helps guarantee more secure and environment-friendly dams, and offers powerful support for sustainable socio-economic development.

Drawing upon experiences in the last decade in dam construction, and in high dam projects in particular, this book accentuates innovative thoughts and technological breakthroughs, and renders numerous examples of dams with technological progress. It is thus a forward-looking, scientific and practical book, with altogether 26 chapters in three parts that elaborates on the strategic importance and development planning of dam construction in China. The scope of this book covers construction technologies including hydrological calculation and analysis, geological engineering and survey, dam type selection, hydropower structure layout, construction technologies, concrete temperature control, high slope engineering, seismic resistance of high dams, flood discharge and energy dissipation in high dams, underground hydraulic structures, and power generation equipment, etc. And damming technologies involved in pumped storage, cemented sandy gravel dams, dams with the height over 300m, overseas dam projects undertaken by Chinese companies, as well as internationally hot issues of dam operation and management are addressed. In addition, the world top 100 projects in terms of reservoir storage capacity, the world top 100 projects in terms of dam height and representative projects of various dam types are listed for readers' reference.

To ensure the consistency of dam parameters in each chapter, the secretariat of CHINCOLD have checked all the data on dam heights, dam lengths and reservoir storage capacity, etc. according to the statistics of CHINCOLD, and referred to recent survey on Chinese dam safety jointly carried out by the Ministry of Water Resources, State Administration of Work Safety, State Electricity Regulatory Commission, and National Energy Administration.

Dam Construction in China—A Sixty-year Review holds itself up to the standard of being professional and informative, offering systematic and integrated technical documentation and references. It hopes to provide valuable reference for engineers engaged in planning, survey, design, construction, operation, management and scientific research in dam engineering. Compiling and publishing efforts have received guidance and help from the Ministry of Water Resources, National Energy Administration and other authorities. Our sincere thanks also goes to China Institute of Water Resources and Hydropower Research, China Three Gorges Corporation, Hydro China Corporation, Dadu River Hydropower Development Co. Ltd, China Guodian, Changjiang Institute of Survey, Planning, Design and Research, Kunming Investigation Design and Research Institute Chinese Hydropower Consulting Group. The leaders of CHINCOLD and invited academicians, experts, professionals also must be appreciated for their contributions.

This book covers a wide array of contents in multiple disciplines, and disagreements or debates on some of them may be unavoidable. Opinions offered by the readers are welcome.

Secretariat of Chinese National Committee on Large Dams
August, 2012

Contents

Preface I
Preface II
Forword

Part I
Planning and Sustainable Development of Dams and Reservoirs

Chapter 1 Strategic Importance of Dams and Reservoirs 3
  New Leap in New Era—Several Issues on Dam Construction in China .... Wang Shucheng 3
  Social Responsibility of Dam Projects .... Lu Youmei 7
  Achievements, Problems and Prospect of Water Conservancy Construction in China .... Pan Jiazheng 12
  Hydropower Miracle of China .... Zhang Guobao 25
  Large Dams and Harmonious Development—Practice and Exploration in China .... Jiao Yong 34
  Thoughts on Developing Water Projects in a Man-Nature Harmonious Way .... Liu Ning 38
  Regulation of Water and Sediment for the Yellow River Based on Joint Operation of Reservoirs
    and Artificial Intervention .... Li Guoying 45
  The Three Gorges Project and Flood Prevention of the Yangtze River .... Cai Qihua 54
  To Achieve the Full Potential of the Three Gorges Project in Ensuring
    People's Livelihood .... Cao Guangjing 59
  Achievements and Prospects of China's Hydroelectric Power in the Past 30 Years
    after the Reform and Opening-up .... Zhou Dabing 62
  Philosophical Speculations on Disputes in Dam Construction .... Lin Chuxue 71
  A Study on Hydropower Strategic Development for Long and Medium
    Term in China .... Yan Zhiyong, Qian Gangliang 78

Chapter 2 Dam Construction in China and International Comparison
  .... Jia Jinsheng, Yuan Yulan, Wang Yang, Zheng Cuiying, Zhao Chun 83
  2.1 Dams in China before 1900 and International Comparison 83
  2.2 Development of Dams in China in 1900-1949 86
  2.3 Development of Dams in China in 1949-1978 87
  2.4 Development of Dams in China in 1978-2011 89
  2.5 Dam Comparison between China and other Countries in the World 91

Chapter 3 Dams and Environmental Protection .... Xie Xinfang, Sun Zhiyu, Chen Min 99
  3.1 General Overview of Environmental Protection of Dam 99
  3.2 Environmental Management of Dams 100
  3.3 Ecological and Environmental Conservation of Dams 104
  3.4 Water Environmental Protection of Dams 109
  3.5 Environmental Protection of Dams Construction 116
  3.6 Environmental Protection Benefits of Dams 120
  3.7 Problems and Outlook in Connection with Environmental Protection Effort 123

Chapter 4 Dam, Reservoir and Water Resources Allocation in China
.... Liu Zhiming, Li Jianqiang, Li Aihua 125
  4.1 Characteristics of Water Resources in China 125
  4.2 Status of Water Resources Development and Utilization in China 126
  4.3 Exploration and Practice on Dams in China 128
  4.4 General Situation of Water Resources Allocation in China 130
  4.5 China Hydropower's Prospect and Layout 131

Chapter 5 Dams and Hydropower Development .... Qian Gangliang 135
  5.1 Hydropower Resources 135
  5.2 Hydropower Development Planning of Rivers 142
  5.3 Research Subjects 150

Part Ⅱ
Construction Technologies for Dams and Reservoirs

Chapter 6 Hydrological Analysis and Calculation for Dam Construction
.... Yang Baiyin, Wang Zhengfa 169
  6.1 Overview 169
  6.2 Design Runoff 170
  6.3 Design Flood 182
  6.4 Hydrological Forecast 201
  6.5 Engineering Sediment 206

Chapter 7 New Techonologies for Engineering Geology and Survey in Dam Construction
.... Cai Yaojun, Yuan Jianxin, Chen Deji 218
  7.1 New Situations and New Problems of Engineering Survey 218
  7.2 Development of Engineering Survey Technologies under Complex Conditions 226
  7.3 Research and Evaluation on Major Geological Problems 235

Chapter 8 Dam Type and Project Layout .... Shi Ruifang 264
  8.1 General 264
  8.2 Concrete Gravity Dams 272
  8.3 Arch Dams 289
  8.4 Earth and Rockfill dams 305

Chapter 9 Design and Construction Technologies of Concrete Gravity Dams
.... Zheng Shouren, Zhou Jianping 322
  9.1 Development and Current Situation of Gravity Dams 322
  9.2 Research on Engineering Geological Conditions 324
  9.3 Layout of Hydropower Structure and Energy Dissipation Mode 326
  9.4 Shape and Structure Design of Gravity Dams 328
  9.5 Design and Construction Technologies of Foundation Treatment 331
  9.6 Stress and Deformation Analysis for Gravity Dams 334
  9.7 Study on Dynamic Analysis and Seismic Design 336
  9.8 Concrete Design and Temperature Control 340
  9.9 Diversion and River Closure Construction Technology 342
  9.10 Concreting Technology in Dam Construction 344
  9.11 Safety Monitoring of Gravity Dam 346
  9.12 Planning of Gravity Dam Construction and Key Research Subjects 347

Chapter 10 Arch Dam Construction Technique .... Zhu Bofang 352
  10.1 Introduction 352
  10.2 Shape Design and Optimization for Arch Dam 355

10.3　Stress Analysis of Arch Dam ··· 362
　10.4　Anti-sliding Stability of Arch Dam ··· 366
　10.5　Three-dimensional Finite Element Elastic-pastic Analysis of Arch Dam ················ 369
　10.6　Selection of Foundation Surface and Foundation Treatment for Arch Dam ············ 371
　10.7　Empirical Assessment of Shape of Arch Dam ·· 372
　10.8　Masonry Arch Dam ··· 374
　10.9　Conclusion ·· 374

Chapter 11　Rolled Compacted Concrete（RCC）Damming Technology
　·· Zhang Guoxin，Chen Gaixin　377
　11.1　Overview ··· 377
　11.2　RCC Gravity Dam ·· 379
　11.3　RCC Arch Dam ··· 383
　11.4　RCC Seepage Resistance and Drainage System ··· 386
　11.5　RCC Thermal Stress and Temperature Control ··· 388
　11.6　Materials and Mixed Proportion of RCC ··· 391
　11.7　RCC Construction ·· 395
　11.8　Conclusion ·· 401

Chapter 12　Construction of High Rockfill Dams ················ Jiang Guocheng，Xu Zeping　404
　12.1　Introduction ·· 404
　12.2　Earth Core Rockfill Dam ·· 407
　12.3　Concrete Faced Rockfill Dam ··· 424
　12.4　Conclusion ·· 436

Chapter 13　New Progress of Cemented Material Dam
　·· Jia Jinsheng，Zheng Cuiying，Ma Fengling，Du Zhenkun，Zhao Chun　438
　13.1　Principles for Design and Construction of Cemented Material Dam ················· 438
　13.2　Studies on CSGR Dams and its Application ·· 439
　13.3　Studies on RFC Dam and its Application ··· 445
　13.4　Conclusion ·· 446

Chapter 14　Temperature Control Technology for Dam Concrete ············ Zhu Bofang　449
　14.1　Introduction ·· 449
　14.2　Basic Theory and Analytical Method of Thermal Stresses of Mass Concrete ········ 450
　14.3　Thermal Stress of the Concrete Dam Block ··· 451
　14.4　Thermal Stress of the Gravity Dam ·· 454
　14.5　Thermal Stress of the Arch Dam ··· 459
　14.6　Thermal Stress of the Sluice ··· 462
　14.7　Thermal Stress of the Foundation Beam ··· 464
　14.8　Pipe Cooling of Concrete ·· 465
　14.9　Comprehensive Temperature Control and Crack Control Measures ·················· 472
　14.10　Long Time Superfical Thermal Insulation with Comprehensive Temperature Control and End of the History of "No Dam without Crack" ······································· 474
　14.11　Anti-crack Safety Factor ·· 475
　14.12　Conclusion ·· 476

Chapter 15　Slope Engineering in Dam Construction ············ Chen Zuyu，Wang Yujie　479
　15.1　Introduction ·· 479
　15.2　Engineering Geology ·· 479
　15.3　Slope Stability Analysis and Safety Assessment ··· 482
　15.4　Engineering Approaches for Slope Stabilization ··· 484
　15.5　Case Studies of Slope Engineering of China's Hydropower Engineering ············ 488

Chapter 16 Seismic Safety of Dams ········· 500
- 16.1 Seismic Safety of High Concrete Dams ········· Chen Houqun 500
- 16.2 Seismic Safety Evaluation of High Concrete Dams Part I: State of the Art Design and Research ········· Zhang Chuhan, Jin Feng 531
- 16.3 History and Progress of Seismic Analysis and Safety Evaluation of High Dams ········· Lin Gao 549

Chapter 17 Energy Dissipation and High Speed Flow in High Dams
········· Gao Jizhang, Liu Zhiping, Guo Jun 584
- 17.1 Introduction of Technological Development of High Head and Large Discharge Flow and Energy Dissipation ········· 584
- 17.2 Technologies and Facilities of High Head, Large Discharge Flow and Energy Dissipation ········· 588
- 17.3 New Types of Energy Dissipator ········· 599
- 17.4 Issues of High-speed Flow ········· 606
- 17.5 Conclusion ········· 609

Chapter 18 Underground Hydraulic Structures
········· Wang Renkun, Chen Zhonghua, Chen Ping, Cui Wei, Liao Chenggang, Xing Wanbo, Fu Yingru 611
- 18.1 Introduction ········· 611
- 18.2 Development of Underground Hydraulic Structure in China ········· 612
- 18.3 New Development of Underground Powerhouse of China's Hydropower Station ········· 623
- 18.4 New Progress of China Hydraulic Tunnel Construction ········· 634
- 18.5 Prospects of Hydraulic Underground Engineering Construction in China ········· 643

Chapter 19 High Dam Construction in China ········· 644
- 19.1 Construction of High Earth-rock Dam ········· Miao Shuying, Chu Yuexian 644
- 19.2 High Concrete Gravity Dam Construction ········· Dai Zhiqing, Sun Changzhong, Huang Jiaquan 659
- 19.3 Construction of High Concrete Arch Dam ········· Cheng Zhihua 673

Chapter 20 Hydraulic Generating Equipments
········· Zhu Yaoquan, Zhu Yunfeng, Liu Jie, Zhang Liangying 687
- 20.1 Hydraulic-Turbine and Generator Unit ········· 687
- 20.2 The Advanced Technology of Hydraulic Turbine Governor ········· 700
- 20.3 Advanced Excitation Technique of Hydraulic-turbine Generator ········· 704
- 20.4 Computer Monitoring and Control System ········· 706
- 20.5 Key Technology of Hydraulic Turbine Operation ········· 710

Chapter 21 Pumped Storage Power Station ········· Qiu Binru, Lv Mingzhi, Jiang Zhongjian 717
- 21.1 Development of Pumped Storage Power Station in China ········· 717
- 21.2 Layout of Main Structures of Pumped Storage Power Station ········· 720
- 21.3 Impervious Lining of Upper (Lower) Reservoir ········· 726
- 21.4 High Pressure Pipe and Powerhouse ········· 745

Chapter 22 Typical Projects of High Dam ········· 754
- 22.1 Three Gorges Project ········· Sun Zhiyu, Chen Xianming 754
- 22.2 Ertan Hydroelectric Project ········· Zhu Zhonghua, Wang Gang 778
- 22.3 Xiaolangdi Project ········· Yin Baohe 787
- 22.4 Xiaowan Hydropower Project ········· Yu Jianqing 795
- 22.5 Shuibuya Hydropower Project ········· Sun Yi, Chen Runfa, Cai Jinyan 806

Chapter 23 Typical Problems of High Dams around 300m ········· 816
- 23.1 Typical CFRD Projects with the Height around 300m and the Main Challenges ········· Ma Hongqi 816
- 23.2 Typical Projects of 300m Earth Core Rockfill Dam and Main Problems
········· Zhang Jianhua, Yao Fuhai, Xiao Peiwei, Ma Fangping 822

  23.3 Outstanding Issues of High Arch Dams with the Height of 300m Duan Shaohui 833

Chapter 24 Dams in the World Constructed by China Wang Ruihua, Liang Jian 848
  24.1 The International Brand of Dams Built by China 848
  24.2 Earth-rock Dams and Concrete Face Rockfill Dams (CFRD) 849
  24.3 Concrete Dam 860
  24.4 Prospect Forecasting 868

Part Ⅲ
Operation and Management of Dams and Reservoirs

Chapter 25 Dam Operation and Management
  Cai Yuebo, Sheng Jinbao, Yang Zhenghua, Wang Shijun, Wu Suhua 873
  25.1 General 873
  25.2 Dam Operation and Management Technologies 881
  25.3 Dam Management Informatization 913

Chapter 26 Dam Operation and Management
  Zhang Xiuli, Xie Xiaoyi, Du Dejin, Zhao Huacheng, Huang Shiqiang, Xu Chuangui 922
  26.1 Dam Safety Supervision and Administration Organizations 922
  26.2 Dam Safety Administrative Legislations 926
  26.3 Dam Safety Technology Supervision Administration 933
  26.4 Dam Safety Monitoring, Detection and Appraisal Technology 942
  26.5 Reinforcement Technology for Dams 963

Appendix Dams and Reservoirs in China and the World
  Jia Jinsheng, Yuan Yulan, Zhao Chun, Zheng Cuiying 970

Part I

Planning and Sustainable Development of Dams and Reservoirs

Chapter 1

Strategic Importance of Dams and Reservoirs

New Leap in New Era—Several Issues on Dam Construction in China[*]

Wang Shucheng[❶]

In 1992, I wrote an article titled *Great Leap in New Century*, and in the last paragraph of the article, I wrote like this: "By 2012, through a century of unremitting efforts, the national hydropower installation will exceed 100 million kW and achieve the great leap in the new century. We shall not only build the world's largest hydropower station, the Three Gorges Hydropower Station to win the "individual" champion, but also enable the country with the highest hydropower installation (conventional units) over the world to win the "team champion". In August 2010, we held the Chinese Hydropower Centennial Assembly and celebrated the national total hydropower installation exceeding 200 million kW at the same time. Compared with the target of 100 million kW that I assumed in *Great Leap in New Century*, the actual completed hydropower installation has been doubled, which is beyond the expectation of all people. Today, when we are exercising the scientific concept of development I would like to present four points of view with respect to the development prospect of hydropower in China and my envisage on further promotion of healthy and rapid development of dam construction in China.

1 Hydroelectric Power is a Preferred Clean Energy for Combating Climate Change

In order to cope with climate change, the Chinese government has made a solemn commitment of three objectives at the international level, two of which are related to hydropower: one, to cut the CO_2 emission per unit of GDP by 40%–45% by year 2020 off the year 2005 level; and the other, to raise the share of the non-fossil energy to 15 percent of total consumption by 2020. To meet such requirement, nuclear power, hydropower, solar, wind power and other clean energy in China will witness the large-span development. However, it is just at this moment that Japan suffered "3·11" earthquake, and the nuclear pollution accident of Fukushima Nuclear Power Plant had imposed extreme impact upon the nuclear power development throughout the world. I believe that China's nuclear power development planning would also be influenced significantly. At present, the State Council has decided to check and rectify the development and planning of nuclear power. The issues on nuclear pollution and nuclear safety would inevitably cause the attention of the whole society. Under such circumstance, in order to achieve the goal of greenhouse gas emission reduction, the hydropower would play a more important role. Recently, I have repeatedly emphasized a statement at various conferences that: "at present, hydropower is a renewable non-fossil energy with the richest resources, the most mature technology, the most cost effective and the most flexible power dispatching, the most realistic and preferred energy with large-scale development potential, and the sustainable low-carbon energy." I used a lot of attributives and adjectives, which are directed against some social and governmental improper perceptions. Hydropower is unexpectedly ignored during the discussion of clean energy in some important articles published at present. So I have repeatedly stressed that: hydropower is a clean, renewable and low-carbon energy as well as the preferred energy.

According to the national review results for water and energy resources from 2000 to 2003, the theoretical reserve of hydropower resources of the mainland China is 694 million kW, the technically exploitable amount is 542 million kW, and the economically exploitable amount is 402 million kW. When I still held the post of the Ministry of Water Resources, when considering the ecological problems and the comprehensive planning of a river, I proposed that: the upper limit for developing a river can be set as 60%, and the original ecology of certain reaches should be retained. After Fukushima nuclear accident

---

[*] This paper is the keynote report at the opening ceremony of 2011 Annual Meeting of Chinese National Committee on Large Dams on June 16, 2011.

[❶] Wang Shucheng, President of Chinese National Committee on Large Dams and Former Minister of Water Resources, China.

in Japan, I suggest that the maximum limit of hydropower development should be raised to 80%-85% after careful consideration. The impact of the Japanese nuclear pollution accident on the global energy made me realize the heavier burden of hydroelectric power. In this case, the magnitude of hydropower development should be raised, and otherwise China's energy problems would be difficult to be solved. Generally the hydropower development process of China can be conceived as follows: the hydropower installed capacity reached 200 million kW in 2010 representing 37% of the total installation; by 2020, the hydropower installed capacity will reach 300 million kW which will occupy 56% of the total, the hydropower installation will reach 400 million kW by 2030, which will be 74%; the hydropower installation will reach 460 million kW by 2040 which will be 85%, in other words, hydro potential will be fully developed by 2050. Calculated as the above assumptions, 10 million kW of hydropower installed capacity should be newly increased every year for 40 years. The spring of hydropower development is coming, but I don't hope it artificially becomes the summer, and we should go steadily instead of aggressively. In the next 20 - 30 years, I think we should keep a reasonable pace, i.e. the annual newly-added installation be kept at around 10 million kW, and great efforts must be taken for realizing such objective.

2 The Quality and Safety of Dam Projects Should be Always Placed as the Top Priority

The quantity of dams will be expected to increase significantly in future, in terms of quality, project safety first. Most hydropower projects in China will be concentrated in the southwestern region of mountains and canyons in future. We will thus face four challenges for constructing these projects: firstly, the geological condition is complex with most regions at earthquake-prone belt.

Secondly, the project requirements have become increasingly high, and these projects are usually high dams and the power stations with large flow and installation, which are difficult technical wise.

Thirdly, the mountains and narrow construction place with poor access conditions will result in unexpected difficulties for construction.

Fourthly, the time for preliminary preparation is urgent. Most of the power plants under construction now have gone through decades of efforts and long-term repeated and considerable work in geological exploration, preliminary design and scientific experiments. Nevertheless, the preparation time for lot of projects to be started in future is very urgent. Under the circumstances of tight schedule and heavy task, we must place the quality and safety of project first, which is the historical responsibility. For those who engaged in survey should fulfill their own responsibilities. For those who engaged in design should do a solid job at the preliminary stage. For those who engaged in the construction must pay attention to the construction quality. In my opinion, the historic opportunity has enable us to accomplish a lot in our career, however, we must be cautious in fulfilling our responsibility, we should neither be condemned by history and nor be unworthy of the motherland and the people.

3 China Will Pay More Attention to the New Function of Ecological Protection of Dams, and Ensure the Ecological Security of the River Basin

From the perspective of the relationship between man and nature, man and nature depended on each other, in the primitive society, and water resources were used to develop irrigation and power generation in the agricultural society, while in highly developed industrial society, the relationship between man and nature has somewhat become the plundering of resources by human beings. Of course, at current stage, what we are pursuing is the harmony between man and nature. Our hydropower is still in the stage of development and utilization for a long term, and more attention is paid to the development and utilization of water and hydropower resources. This is certainly correct, but we must take pains to avoid over development. For a long period of time both at school and at work after graduation, our understanding on the function of dams is how to fully develop and utilize the resources, such as flood control, power generation, navigation, water supply and irrigation, but we are lack of the awareness about the protection of water resources and the ecological problems. At present, the stress is laid on the harmony between man and nature, so the protection and conservation of water resources is placed into the important agenda. In this case, the ecological function of the dams, I think, is a new function, which must be strengthened during the processes of design, research, construction, operation and management.

What are the ecological functions? For example, as far as the Yellow River is concerned, Sanmenxia and Xiaolangdi projects have the similar locations, but Sanmenxia project failed while Xiaolangdi project has played a great role in the ecological protection for the middle and lower reaches of the Yellow River. Now, the Yellow River has kept continuous stream since 1999, and the Yellow River estuary wetland is well protected. Through these years of water and sediment regulation, the riverbed is cut by 1.2m to 2.0m. The artificial density flow released from Wanjiazhai Reservoir scours the sediment of Xiaolangdi Reservoir. All of these are the ecological functions. After the completion of Daliushu project in Ningxia Autonomous Region, the ecological environment of the section of the Yellow River in Ningxia and Inner Mongolia will also be improved greatly. In case when flow from upstream of the Pearl River is insufficient, the subsequent seawater intrusion will cause a scant supply of water in some areas, especially in Macau. Therefore, we regulate the reservoir 1,000km away to increase the flow so as to reduce the salt and supplement the fresh water, which has played an important role in guaranteeing the security of water supply for Macau. There are many other operations in these years. Baiyangdian Lake, Zhalong wetland and Xianghai wetland are replenished with water, Tahe River and Heihe River begin to flow once again, the sewage arising from the water pollution accident of the Songhuajiang River was diluted by increasing the discharging flow, so as to drain pollutants quickly; the sewage of Taihu Lake was regulated, the Three Gorges Reservoir sacrificed its power generation to increase the discharge in order to reduce the drought loss for the middle and lower reaches of the Yangtze River and the lake, all of these are the outstanding contributions of water conservancy and hydropower projects in the protection of the ecological environment. For dam construction in the future, the concept of ecological protection should be fully considered during design stage. As to the constructed dams, greater attention should be paid to the ecological issues during operation. Ecological protection is also another important function, which should be added in our textbooks in addition to the conventional functions of power generation, flood control, irrigation, tourism, shipping and the like. I hope there are more and more terms, such as ecological water yield, ecological flow, ecological water level, ecological regulation, ecological benefit, ecological safety etc. would frequently emerge in our articles, speeches and work in future, so as to adapt to the requirements of modern society.

4 China Highly Values the Construction of the Pumped Storage Power Stations

Along with the widely utilization of clean energy, the rapid development of distributed energy, the substantial increase of the proportion of long-distance power transmission, the widespread application of large-capacity generator sets, the changes of power loads in modern society has led to higher requirements to peak cutting and valley filling in the power system, and the pumped storage power stations will certainly become an important means for ensuring the safe and stable operation of power network.

(1) To widely utilize clean energy. Nuclear power is generally used as a base load, while the wind power and solar power are unstable. Therefore, with the extensive utilization of clean energy, the smart power grid has imposed higher requirements on the pumped storage power station.

(2) To rapidly develop distributed energy. As an indispensable part of modern power industry, distributed energy should be given more attention with the development of high-voltage, supercritical, large-capacity and large-power grids. Taking wind power generation for example, we tend to build the large wind power station with dozens of thousands or hundreds of thousands kW, and hundreds or thousands of units. However, there are less large wind power stations in Denmark and the Netherlands. And distributed energy occupied a considerable part and widely used in a community, a farm or a business where have been set up two wind turbines for their own use. The diesel engines can be used for power generation for no wind case, and electricity surplus, if any, can be sold to the grid; rather, the power grid will supplement in case of power shortage.

After the separation of transmission and distribution in Chinese power grid, our reform effort will develop the distributed energy in the future. In such case, it is very important to build the pumped storage power stations with different capacities and progressively eliminate the difference between peak and valley.

(3) To substantial increase the proportion of long-distance transmission. Long-distance extra high voltage (EHV) transmission has been debated vigorously among many experts in different industries, even some experts wrote to the State Council and proposed that China should not develop ultra-high voltage and supercritical transmission. In my opinion, the long-distance transmission voltage level in a country is always related to the construction of large-scale hydropower station. When the installed capacity of hydropower increases and the large-capacity power station is built up, it is necessary to resolve the problems about the long-distance EHV transmission. Therefore, the EHV transmission is always developed along with hydropower development. China's hydropower is concentrated in the western and southwestern areas, but the electricity load centers are located in the eastern region. For this reason, China's EHV transmission is bound to emerge and develop. Just now I mentioned the issue of nuclear power, and nuclear safety has always been a very important issue in our country. Before the nuclear accident in Japan, I had put forward that if China's nuclear power is developed up to 400 - 500 million kW in 2050, which means a total of 400 - 500 units. Supposedly, 1 million kW per unit, there will be so many units distributed in China with more than ten units in each province on average, which will result in extremely high risks. Before I left office in the Ministry of Water Resources, I had once told the leaders of the Water Resources Department that we should veto any nuclear power plants supposedly built in the middle and upper reaches of a river. When I visited Ukraine, they introduced the Chernobyl Nuclear Power Plant accident and treatments afterwards, and I was shocked after watching the video. Chernobyl Nuclear Power Plant is located at the upstream of Kiev, and as water source to Kiev, a reservoir is in the middle of them, which is very terrible if nuclear accident happens. Before Fukushima nuclear pollution accident, I had

also put forward a proposal that: nuclear power should be developed in the western regions of Xinjiang, Qinghai, Gansu and Inner Mongolia where no inhabitants is; a nuclear city could have the capacity of 100 million kW or 200 million kW, and 100 - 200 units could be arranged, so that the safety management would be ensured. Even if accident occurs, it would not cause a serious impact. Then someone may say that there is no water and drought is serious. I have to say, the availability of water is relative. In Xinjiang, for example, the average annual amount of water resources for many years is 88 billion m^3 (the average annual flow of the Yellow River over multiple years is 58 billion m^3), and water resource per capita is 5,700m^3 (2,200m^3 per capita nationwide). I don't mean that Xinjiang has rich water resources, but the problem of water is out of question for developing a 100 - 200 million kW nuclear city. Lu Youmei, a member of Chinese Academy of Engineering, proposed that the nuclear power might be placed in caves to ensure the security. The development of hydropower, nuclear power, thermal power, wind power and solar energy in the west of China should need the long-distance EHV transmission, and the pumped storage power station with large capacity can be built in the electricity utilization end for regulating the electricity.

(4) To commonly use of large-capacity units to adapt to fast changes of power use load in modern society. The wide application of large capacity units can bring about a higher requirement on the shift peak load capacity of electric network. The power load changes in modern society, such as the rapid development of the electric vehicles, the rapid increase of the cold and heat loads, also need a means to store the energy and adjust the difference between peak and valley. All of these will promote the development of pumped storage power station.

The construction of pumped storage power station should adapt to local conditions, match with the grid hierarchical scheduling, and coordinate different station sizes. To make the translation cohesive and coherent with the grid planning, the development of pumped storage station should be in close communication and cooperation with the grid planning department, design department and dispatching department. To promote the development of the pumped storage power stations, it should be studied on its systems and a clear relation with power grid.

I have talked about the four perspectives as above: the first is the quantitative development of hydropower; the second is the qualitative improvement of hydropower; the next is to extend hydropower functions; and the last one is the expansion of hydropower development mode.

It is a long way to go for China's hydropower construction. I believe that, the dam construction in China will achieve a new leap in the current era under the joint efforts of the colleagues.

Social Responsibility of Dam Projects*
Lu Youmei[1]

The advancement of human civilization has promoted the development of water conservancy and dam projects. The process of human civilization is fundamentally dependent on the utilization of water, which can be traced back to 5,000 years ago when the earliest water diversion projects and reservoirs appeared on earth. So it is clear that water availability has promoted the development of human civilization, in turn, human civilization has contributed to the development of hydropower towards a path of modern engineering. I appreciate the two philosophical dictum very much, one is "I think, therefore I am" by Descartes, an ancient French philosopher and the other one is "I create, therefore I am" by Professor Li Bocong, a member of the Chinese Society for Dialectics of Nature. The two dictums outlined the role and position of human being in the process of evolution and civilization. It is estimated that there are more than 400,000 dams worldwide today (or 600,000 according to others), of which more than 86,000 are in China and more than 70,000 are in the United States. Most of dams higher than 15m were built up in the last 100 years. The reservoirs formed by the dams can meet the demands for water resources of population growth, industrialization and development of human society.

Since the 21st century, China's economy has experienced a period of sustained rapid growth. As better living environment, safe, stable and sufficient water supply, abundant energy for improving life quality are demanded, there came some negative voices criticizing dams and hydropower projects in the name of ecology and environment. For a certain period of time, a lot of media and public opinions have attacked dam projects with severe critique. Some influential newspapers have blamed dam projects for ecological degradation and environmental deterioration resulting from high-speed economic development, population growth, unplanned development, and excessive waste of resources by saying that "Dams are trouble-making", "The dam age ends", "The world has entered a phase of dam removal, but we are still building dams". Some people are even proud of being a "the most ardent opponents of the dam" and gave endless criticism over dams. In face of the backlash in public opinion, water engineers should ponder over the issues.

P. Mc Cully from USA wrote a book named *Silenced River*, with the sub-title as *Ecology and Politics of Large Dams*. The title in Chinese was translated to *Economics of Dams*. This book was written in prose-style but with political tendency. Here I wouldn't talk much about the content. The purpose of the book is to denote that the era of dams has been over; construction of new dams should be restricted and stopped, existing dams should be removed, thus to restore natural rivers to flow freely. In the book, a lot of biased data and false information were cited to exaggerate negative impacts of dams on ecological environment, and human society development was seen from a static perspective. There was more emotion than rational thinking, and a lack of scientific attitudes. In fact, the book has misled public opinions and fomented the "anti-dam" tendency in China.

It is well-known that water is the basis of all life. Up to now, it has been not found another planet with so abundant water resources as on earth. There is a total of 1.396 billion km^3 of water on earth, in which 0.2 million km^3 can be directly utilized by human, only accounting for 0.015% of the total water amount on earth. Water cycles with the energy of the sun and the period of the earth revolution can provide the perpetual source for terrestrial life on earth. However, it is the natural law that water resources which can be utilized by human beings are not evenly distributed in time and space. By this law, human beings take effective measures, among which water projects are the most important. With water projects, people can get stable, safe, sufficient and high-quality water. These are also the social responsibilities of dam projects.

Hydraulic engineering consists of dam, reservoir, dike, water-diversion irrigation and urban water supply projects, hydroelectric projects, channel regulation works, navigation engineering and others. I wouldn't talk much about the functions of such structures but rationally analyze their negative and positive impacts on ecological environment.

Dam construction can regulate the instable rivers' flows, control flood and mitigate disaster, attain clean renewable hydropower, and improve navigation and water supply conditions. At the same time, it also can change the natural flow, submerge lands, and change the original environment and ecology to some extent. Ecology is the relationship among plants, animals including man, and the rest of the environment. The relationship among the living organisms in the natural world is constantly changing, the ecological balance actually exists for a relatively short time span and the imbalance is absolute,

---

  * This paper was published in Issue 5, 2005, *Construction of China's Three Gorges*.

  [1] Lu Youmei, Academician of the Chinese Academy of Engineering and Honorary President of Chinese National Committee on Large Dams.

only this imbalance can generate the power of nature into a balanced direction, to promote the biological evolution rules including "survival of the fittest" and "natural selection", which is the basis for Darwin evolution.

The ecological state is completely dependent on the environment. Both the extinction of dinosaurs and the rise of humankind depend on the circumstances, divided into natural environment and artificial environment. In fact, man is a part of the natural world, and artificial environment should also be considered as a component of natural environment. Man, as highly emotional creature with high intelligence, for its own survival and development, has started to use natural resources to change the environment since primitive times. With such efforts, human beings have evolved into today's civilized men. Humans no longer live in primitive and purely natural environment but in an artificialized environment.

Building hydropower projects on rivers and thus forming reservoirs has changed the environment and the ecology to some extent. We should make scientific analysis and propose appropriate countermeasures.

1 Impacts on Aquatic Life

Dam construction can block the channels for aquatic life and change the living environment for fish especially for migratory fish, whose the channel of breeding and spawning are blocked, and lead to the extinction of some fishes. We should scientifically study on the species and populations, quantities and living habits of aquatic life in the natural river, and try to keep the dam away from the breeding and spawning areas of fishes while making plans for dam construction and at the same time explore new breeding areas for fish. In the process of dam construction, we should consider the possibility of building fishway to protect fish passages, apply artificial propagation technology to protect the reproduction of some fish species by establishing fish conservation center and building up gene pool for biodiversity protection for some endangered fishes. To ensure people have edible fish, enhanced fish species should be put into the reservoir to increase economic value.

The above measures have many examples of success, for example, a series of fish ways built on the cascade hydropower on the Columbia River, fish ways newly built in the Itaipu Hydropower Station of Brazil, the Chinese Sturgeon Research Institute for the Three Gorges Project of the Yangtze River, and fish stocking at the Xin'anjiang Reservoir (Thousand island Lake of China), all of which are worth of being popularized. It can be seen that the sharp decreased fish number in the inland rivers of China is mainly caused by water pollution, excessive fishing and interference by navigation and human activities, which is not the case all by dams.

2 Losses from Reservoir Inundation

Building reservoirs submerges land but forms surface area at the same time. China is a continental country, with a lake area of only 71,787km^2 in the land area of 9,600,000km^2, accounting for 0.0075% of the total land area. It indicates that the water storage capacity of China's mainland is very low. To exchange some land area for surface area can generally improve China's ecological environment. However, submersion of precious fertile plains and lush forest vegetation should certainly be reduced and even avoided in dam construction and basin planning. In fact, hydropower projects are mostly in the upstream of the rivers of western China where most reservoirs are gorge-shaped and the submerged lands are mostly barren slopes. From the perspective of flood control and disaster mitigation, the essence of flood disaster is scrambling for land area between humans and water. Human beings should not reclaim land from lakes without limit; or the water area can be occupied. And it should be rationally made way for land by for example building reservoirs and flood diversion area to hold excessive precipitation. Taking the Three Gorges Reservoir as an example, 638km^2 of land area was converted into surface area, in which 238km^2 was cultivated areas, 49km^2 was economic forests and 351km^2 was barren rock slopes. But in exchange, 40,000km^2 of fertile Jianghan Plain was well protected. In other words, it gained the high-quality plain with a small amount of poor land to reserve the land storage capacity for excessive floods.

3 Reservoir Resettlement

In a half century, around 16 million people have been resettled due to dam construction in China. It is reported that at present about 10 million people are living below the poverty line. Reservoir resettlement is a social problem. Before the Reform and Opening up in 1978, resettlement compensation was very low, which made most resettled people live near the poverty line and they were unable to take the advantage of the opportunities of economic development; With China's economic development and China's economy entering period of socialist market economy, the arrangement of reservoir migration has experienced a change process from the arrangement migration to development migration, which is a course of combination of the resettlement with poverty alleviation and economic development. And this translation can be traced back to the 1980s. Governments at all levels have supported the built reservoir migrations. To new reservoirs migrations, the governments have raised compensation standards with clear policy support after dam construction and strengthened policy implementation in the migration districts, which can generally improve the life quality of the immigrants. It can be seen that the Three Gorges project has paved a successful and feasible way of reservoir migrants. Most reservoirs in China were built in mountainous areas where land resources are scarce. To relocate people reasonably with the poverty-reduction of mountain inhabitants to, live a modern life, reservoir immigrants, in this sense, has become the progress of a nation. Although it leaves something to be desired, I believe that reservoir resettlement demonstrates the progress of a nation. With social progress and economic development, the natural resources and social wealth will be reasonably distributed and the migrant benefits will be fairly ensured.

4 *Reservoir Sedimentation*

All rivers have sedimentation problems, which are caused by rock-soil bed erosion due to heavy rain. Rainfall washes out upstream sediments, which silts up downstream and forms alluvial plain at the river mouth to the sea, which is a natural law with the consequence of the elevation of downstream riverbed, the decrease of flood discharge capacity and flood disasters. When a reservoir is built in the upper reaches, sediments are deposited in the reservoir, especially coarse gravels in the terminal of the reservoir, which will cause block river channel and increase flood level upstream. The most typical case is the Sanmenxia Dam on the Yellow River, which is a most valuable but tragic lesson for water engineers. But we should not give up dam construction because of a slight risk. Instead, when planning a reservoir and dam project, we should conduct a thorough investigation of the river sediment beforehand, collect enough materials and carry out scientific experiment. Furthermore, we must make magnitude analysis and design the regulation mode of reservoir in a scientific way, so as to avoid adverse consequences. Take the Three Gorges Reservoir as an example, there have always been some voices of doubt that the Three Gorges Reservoir may follow the old path of what happened at the Sanmenxia Reservoir. By detailed analysis on sediment data of many years, large scale sediment model test are carried out to simulate reservoir operation mode to draw out the results of sedimentation magnitude analysis, and adopted the operation mode of "storing clear and releasing muddy". Through reasonable regulation of the reservoir, the storage capacity can remain over 80% for long, and the Sanmenxia phenomenon will never happen. The Three Gorges Reservoir has started impounding since June 1, 2003, the incoming sediment to the reservoir decreased each year. Because reservoirs on the main stream and tributaries upstream have been built and some sediment has been stored in the scattered reservoirs upstream with vegetation protection and random distribution of heavy rain, the sedimentation at the Three Gorges Reservoir was much less than estimated values, which can only be drawn by thorough monitoring for a long time.

As some sediment silted up in the reservoir, the discharged water from the dam became clearer. This may scour the downstream channel for a period of time, thus lowering the water level of the downstream with some adverse effects. Therefore, protection measures should be taken at some important sections. The river channel can finally reach scouring and depositing equilibrium and become stable.

In short, we should take a scientific attitude towards sedimentation problems in reservoirs together with integrated measures to eventually reach scouring and depositing equilibrium and avoid disastrous sedimentation.

5 *Water Pollution*

As the old saying goes, "Running water is never stale and a door-hinge is never worm-eaten". When a reservoir is formed, the flow will slow down, which causes water pollution within the reservoir; and discharging sewage and waste directly into the reservoir will ferment and decay, resulting in the increase of nitrogen and phosphorus to contaminate water. Water pollution appears particularly severe in multi-year regulating storage reservoirs in the tropical regions. And this does not normally happen in other types of regulating (annual or seasonal) storage reservoirs due to its faster renewal period of water bodies. Hydropower station itself does not discharge sewage and waste, however, which is from domestic wastes from surrounding towns, industrial wastes and agricultural non-point source pollution. Only controlling the source head can the pollution treat, and rivers should not be the drainage channels for disposal of waste. Pollution control should be accelerated even in regions if there is no reservoir on rivers. Only in this way river environment can be protected in real terms. Building reservoir has improved the development of garbage and sewage treatment works in the reservoir area. It should further improve and execute the management system in strict accordance with the law, boost the law enforcement to make the water quality reach the national standard as early as possible in rivers, reservoir and lakes.

Essentially, water shortage is one-time activity of excessive use of clean water and too much sewage discharge. However, water molecules do not change a little in the nature. Therefore, we should vigorously promote water conservation, develop sewage treatment technology and equipment, enhance water recycling, and improve water use efficiency.

6 *Earthquake and Geological Disasters*

Reservoirs increase the water level over the river bed, which adds the water weight on the rock surface. 10-meter increase on rock face of every square meter is equivalent to a weight increase of 10t; 100-meter increase in water level is equivalent to 100t in weight. The weight change altered the stress situation of the rock foundation. During the stress regulation process of the rock foundation, the geological structure will be slightly deformed, resulting in reservoir induced earthquake called reservoir induced earthquake. According to the statistics of International Commission on Large Dams (ICOLD), the probability of reservoir induced earthquake around the world is 0.2%. Earthquake happens frequently in China. According to statistics, the average probability of earthquake induced by reservoirs with capacity over 100 million m^3 is 5%. The probability of earthquakes depends on the geological structure of the dam area and the reservoir area. The intensity of reservoir-induced earthquake is generally low. An earthquake on magnitude scale of 6.1 in Xinfengjiang Reservoir happened during impounding time, ranking No. 4 in reservoir-induced mega earthquakes of the world and is much more severe than others. As in the case of the Three Gorges Reservoir, the largest scale of earthquakes is 3.0 since the water storage level reached 135m, which occurred only once. If the water level is stable, reservoir-induced earthquakes will be gradually decreased.

In an area where natural earthquakes frequently happen, sufficient safety for anti-seismic load in dam design should be considered based on the basic seismic intensity of the region provided by the State Seismological Bureau.

With regard to the safety and stability of reservoir banks, comprehensive survey should be conducted before impounding, and necessary reinforcement must be made to certain sections. Measures must be taken to ensure that resettlement areas are free from possible landslides and collapse to avoid geological disasters. The Phase I and Phase II reinforcing projects have been finished for the Three Gorges Project. And RMB 4 billion from the Three Gorges Project Construction Fund was used for the reinforcement projects. Now the project has entered in its Phase III and the government will continue to increase its investment.

7 *Dam Project Safety*

Dam project safety has always been an important issue of great concern. The disaster of dam collapse is not rare since ancient times.

The dam failures of Shimantan Reservoir and Banqiao Reservoir of China due to heavy rain in August 1975 brought about casualties of 86,000; the Vajont Dam in Italy resulted in disaster due to massive rock slide at dam abutments. Those distressing major accidents remained fresh in the memories of dam engineers.

In addition, secondary disasters caused by dam failure are more serious than other engineering damage. Hydraulic engineers must clearly recognize that and should bear their social responsibility.

Dam failure is caused either by human or nature. Among the human factors, the most important ones are the destruction of war and unexpected factors in modern society. Therefore, the corresponding countermeasures and predetermined precepts must be taken into account in engineering design, such as the possibility of quickly emptying the reservoir, and sufficient resistance to blast structurally, and under no circumstances should we avoid secondary disasters. Natural factors, such as floods or earthquakes which exceeded the preplanned criteria for the dam projects, may cause dam bursts, threatening lives and property. Hence, it is necessary to continuously recognize nature and explore the laws of nature for human. To avoid possible secondary disasters, we always prefer to be conservative in design and construction of dam rather than lower standards for low costs.

8 *Historical Sites and Cultural Relics*

Human beings have been living around water since ancient times. Reservoirs building may submerge some cultural relics. For some heritages that are on the ground, it is easier to be protected either in terms of in situ conservation or ex situ conservation; For some buried in the ground, it should be carefully excavated and well protected after archeologist's identification, such as the relocation of the old temple Aswan Reservoir in Egypt, the relocation of Zhangfei Temple, Quyuan Temple and protection of the underwater relics of Baiheliang Ridge, which can protect not only historical heritages but also the national culture.

But not least, the above list represents the major aspect of reservoirs effects on environment. I believe that the ecological impacts caused by dam constructions can be reduced or avoided with scientific means and methods. In engineering practice, it should be constantly deepen understanding of objective things and propose correspondent measures. However, all those negative effects were insufficient reason to give up building dam.

Water project has essentially improved the lost ecological balance to create a good environment. Water scarce and the shortage of energy and electric power in the north of China are the puzzles and constraints of China's sustainable economic development. It is always conflicting between water scarcities with frequent flow cutoff and increasing water consumption caused by the economic development and people's life quality improvement. In fact, the problem of the Yellow River itself is the expression of ecological imbalance and can only be resolved by building large-scale water diversion projects. The Yangtze River has abundant water resources, which is a top option of water transfer from south to north. The engineering measures are made up of a series of complicated water projects, which are the most important ecological responses.

From the perspectives of energy resources and energy demand on the continuous economic development in China, there is also imbalance between supply and demand. Electricity, as a kind of secondary energy, is directly related to the safety and stability of the whole society. The life today cannot be without electricity even for a moment. In China, the primary energy for electricity production is mainly dependent on coal. China has abundant coal resources, 1.96 billion tons of coal was produced in 2004, nearly 1 billion tons of which was used for power generation. Coal is fossil fuel and a nonrenewable energy when the more coals are used, the lesser their reserve are. From the environmental effects perspective, burning 1 billion tons of coal produces about 2 billion tons of CO_2, along with SO_2 and NO and other harmful gases that can cause acid rain which will become our country and even global environment pollution. However, the present state of China's resources shows that coal power is the main use of resources. There are also fairly rich waterpower resources in China, with developable total reserves of 400 million kW; if the hydropower operate 4,000 hours per year, 1,600 billion kW · h of electricity can be generated, which is equivalent to that generated from 640 million tons of coal. Hydro power can be generated electricity by using the water potential energy without any production of waste gas, water and solid, and become a clean and renewable energy, which is a very simple principle. Up to now, China has developed 100 million kW of hydropower, and there is still 300 million kW to be developed. In order to protect our environment, it should be consumed as less coals as possible and developed more hydropower as an important energy policy. In addition, *Kyoto Protocol* and other important international protocols also advocate more hydropower development. As China's energy policy, it is the exact correct way to encourage the development of renewable hydropower and comply with the international trend of energy development. Dam is the key project for a hydroelectric station, of which the most important is the ecological responses of dam from this point.

We protect all living things, natural landscapes, and cultural relics for the sustainable development of mankind. Today's environmental protection is for tomorrow's ecology, which embodies the people-oriented principle. The ecological effects of hydropower are huge, especially for dam project. Scientific planning, meticulous design and construction and standard operation are extremely important to reduce or avoid the negative ecological effects of hydropower projects.

China's economy is now in a period of sustainable development. Faced with the harsh international environment, the fact is that a country lagging behind means exposure to invasion. Therefore, there's no reason we would restrain ourselves to slow down the speed of development. We have no power to inhibit people from getting clean water, to make our world brighter, to make mountainous areas remain in poverty and backwardness, or to stop people in poor living standard getting richer rather than remaining "original living state" for the well-off to visit and comment on and also no right for river to free flow without thinking about people's treasure and security. Even though there is still a lot of tough work to do about rivers and lakes, it is the social responsibility of water engineers.

There are both successful and unsuccessful experiences for dam engineers. We need to learn from lessons, listen to various opinions, establish a platform for communicating with dam opponents, and effectively carry out our work to benefit people. To blindly oppose the dam-building is ignorant and it should not interfere hydropower in a normal, healthy and rapid development with such wrong idea as pseudoscience and pseudo-environmental-protection ideas.

We should follow the principle of putting people first, maintain the strategic development in a sustainable path and guide action by applying the Scientific Outlook on Development. And we firmly believe that China and its people will have a better future.

Achievements, Problems and Prospect of Water Conservancy Construction in China*

Pan Jiazheng❶

1 Achievements of Water Conservancy Construction in China over the Past 50 Years

Water conservancy construction in China, with more than 4,000 years of history, can date back to river-harnessing by Da-Yu during the Xia Dynasty. It has also showed that China have long historical culture and an unevenly distribution of water resources in time and space can lead to the construction of water conservancy projects to early withstand natural disasters for the development and survival for the Chinese people. China is located in the southeastern part of Eurasia, adjacent to the Pacific. The Northwest frontier of China situates in the hinterland of Eurasia. The terrain of China is high in the west but low in the east with a marked continental monsoonal climate characterized by great variety, which leads to extremely unevenly precipitation distribution in time and space and even the frequent floods and droughts. Spatially, such unevenness is manifested as dry in the west and north, and humid in the east and south; temporally, this unevenness is manifested as tremendous inter-annual changes in the total annual precipitation (with more than even 10 years of consecutive dry years or wet years), and the precipitation concentrations within a year in three or four months of the flood season.

Due to the above mentioned features, floods and droughts has become the main natural disasters in China besides geological disasters (e.g. earthquakes) and biological disasters (e.g. locust plague), and none of the river basins in the country could be exempted. According to the statistics, the Yellow River basin witnessed a total of 110 great floods and 95 serious droughts during the 7th–20th century, and such type of great flood and serious drought should occur on average only once every 6.8 years. During serious droughts, the land became utterly desolated and littered with the corpses of the starved, and even cannibalism was witnessed. In case of dike breach in great floods, houses were destroyed, tens of thousands of people died and millions more suffered deadly flooding, even rivers and ecosystem could be destroyed. The Yellow River is the abundant sand river with a world-famous "suspended river" in the lower reach, which makes it more difficult to tackle. The Huaihe River basin has always been one of the hardest-hit and poverty-stricken areas in China, proven by a saying "There are big floods in heavy rain, small floods in light rain and drought when it does not rain". In the affluent Yangtze River basin, the increasingly exacerbated disasters are even more astonishing due to the largely destroyed areas, and huge affected population as well as severe consequences; in case of severe droughts and floods, the affected area may add up to tens of millions of mu (6,667km^2), and the conflicted population up to 10 million; the direct and indirect death toll could amount to hundreds of thousands and even one million. The severe disasters in other areas, such as the Haihe River Plain, Liaohe River Plain, Songhuajiang River Plain, and Pearl River Plain, and northwestern drought areas are also registered. Floods and droughts have become the most serious disasters of the Chinese people. Although the governments in different phases of history adopted various measures including setting specific positions, deploying dedicated officials, building water projects and allotting goods and funds for disaster relief, and numerous people with lofty ideals devoted their whole life to disaster control, such situations had not been effectively changed due to the political, economic, and technological limits.

The People's Republic of China, founded in 1949, fundamentally reversed this disastrous situation, pushing China into a period of water project construction with the largest scale and the most significant benefit.

1.1 Preventing Floods and Alleviating Disasters to Safeguard People's Life and Property and Social Stability by Building Dams and Reservoirs with Attention to both Flood Discharge and Storage

There were totally 40,000km of dikes in China before 1949 spreading mainly in the lake areas of the Yangtze River and the lower reach of the Yellow River. Those dikes were characterized with small size, insufficient capacity and poor quality. As a result, dike breach often occurred in case of relatively heavy floods. After 50 years' development, 260,000km of dikes for rivers and lakes and 7,900km of seawalls have been built or reinforced all over China. Moreover, 85,000 small, medium and large reservoirs have been built; a flood prevention system has been formed for the seven main rivers of China, consisting of dikes, river regulation, reservoirs, as well as flood detention and diversion areas, which have been basically con-

---

* The paper was published in No. 2, Volume 4, *Engineering Sciences*, 2002.

❶ Pan Jiazheng, Academician of Chinese Academy of Engineering, Academician of Chinese Academy of Sciences, and Honorary President of Chinese National Committee on Large Dams.

trolled floods. In combination with the non-engineering protection system such as flood forecasting, flood prevention scheduling, floodplain area management, emergency rescue and disaster relief, this flood prevention system has protected 33 million hm^2 of farmland, over 600 cities, major industries and mines as well as transportation, reducing loss from floods.

Over the past 50 years, the Huayuankou section of the Yellow River witnessed 12 floods, each of which had a peak discharge over 10,000m^3/s, but dike breach never happened there. Among the 12 floods, the flood in 1958 has been the largest since the observed water level was available. This creates a historical record of the Yellow River with no flood disaster for 50 consecutive years. In 1954, a rare flood over the whole basin occurred in both the Yangtze River and the Huaihe River for about 71 years. In 1957, the Songhua River experienced a flood which was the second largest since 1898. In 1963, the Haihe River suffered from a 50-year flood. All of these floods were conquered without devastating consequence. In the past decade, with the constant development of flood prevention projects, flood control and disaster alleviation measures have been demonstrating a more and more important benefit. For instance, in the 1991 basin-wide big flood of the Huaihe River, with the combined operation of the 51 large reservoirs in the basin and the use of flood retention and diversion areas in the lower basin, the whole basin with all the cities, railways, industries and mines in the whole basin were successfully protected. In 1998, both the Yangtze River and the Songhua River suffered basin-wide major floods when the river embankment played significant roles, and 1,335 large and medium sized reservoirs in relevant provinces and cities were put into operation of flood retaining and peak regulation, which eventually helped the river basin out of the difficulty. (Notes: the Yangtze River flooding in 1998 was similar to that in 1931. Both were the basin-wide devastating floods, but respective consequences were incomparable.)

1.2 Boosting Food Production to Solve the Clothing and Food Problems of 1.2 billion People by Irrigation Development

Over the past 50 years, China has built over 5,600 irrigation districts, covering an area of over 667hm^2, and 4 million pumped wells, as well as a large number of reservoirs, ponds and cisterns. The effective irrigation area of the whole country has increased from 16 million hm^2 in 1949 to 53.33 million hm^2; water logging has been resolved for a total area of 20 million hm^2, accounting for 82% of the areas prone to the waterlogged areas; 5.33 million hm^2 of saline-alkali land has been improved, accounting for 71% of the total areas; presently the water-saving irrigation area has reached 15.2 million hm^2, where modern technology has been applied, such as spray irrigation, drop irrigation and micro-irrigation; the national layout of the farmland irrigation has basically taken form, and the current agricultural water consumption is about 410 billon m^3/a.

Such irrigation and drainage projects for farmland have greatly enhanced the capability to mitigate droughts and water loggings. For example, historically the Yellow River Plain, Huaihe River Plain and Haihe River Plain were areas suffering most from floods and droughts where "heavy rains cause great floods, small rains cause small floods, and droughts occur when it does not rain". However, they have now become grain bases. Rice, wheat and cotton yields continue to rise, and the grain yields in the whole country have reached and even exceeded 500 million tons. China has managed to feed 22% of the world's population with only 7% of the world's arable land, answering the question "Who can feed the Chinese". It is predicted that China will still be able to feed itself even when the population peaks reach about 1.6 billion in the 21st century.

1.3 Improving People's Livelihood and Promoting Urbanization by Ensuring Industrial and Urban Water Consumption

Over the past 50 years, China has built a large number of water supply and delivery projects to meet the demands of industrial and mining enterprises and cities, including Beijing, Tianjin, Dalian, Qingdao, Xi'an, Shenzhen, Hong Kong and other cities with water shortage. Meanwhile, township and rural water supply utilities have made tremendous progress. At present, the national industrial and urban water consumption is about 150 billion m^3 per year, including about 110 billion m^3 of industrial water consumption and above 30 billion m^3 of urban water consumption. Annual water supply for agriculture, industry, and cities totals 560 billion m^3 (only about 100 billion m^3 in 1949), and the overall per capita water consumption is about 460m^3 per year, more than doubling that of 1949, which has created favorable conditions for the country's economic development and social stability.

1.4 Becoming an Important Part of China's Energy Structure with the Rapid Development of Hydropower

China has abundant hydropower resources, with a theoretical reserve of 676GW and exploitable resources of 378GW, and its annual electricity generation numbers 1,920TW·h, ranking first in the world. In 1949, the national total installed capacity of hydropower was only 360MW, and the electricity generation 1.2TW·h, which was mainly generated from the Fengman Hydropower project in northeast China which was built during the Japanese occupation. With 50 years of efforts, the national total installed hydropower capacity has reached 79.35GW, and the annual electricity generation has amounted to 243.1TW • h, accounting for 24.8% and 17.8% of the country's total, ranking the 2nd and the 3rd in the world. Among them, large-scale (above 250MW) and small-scale (below 25MW) hydropower projects respectively take up 1/3 and more, with the rest being medium-sized ones. The installed capacity of the medium and large-sized hydropower projects under construction have reached 46.2GW; the installed capacity of Three Gorges Hydropower Project reached 18.2GW, which is the largest hydropower station in the world and will be put into operation in 2003. China has become the world's largest producer of hydropower.

Among China's hydropower stations, many of them are famous projects above 1,000MW-class, including the Three Gorges, Ertan, Xiaolangdi, Liujiaxia, Shuikou, Longyangxia, Yantan, Lijiaxia, Geheyan, Tianshengqiao I and II, Wuqiangxi, Wanjiazhai, Dachaoshan, Gezhouba, and Baishan, etc. The construction of pumped storage power stations started relatively late, but developed at a high speed. Some pumped-storage power stations with large installed capacity and high water head have been built, such as Guangzhou, Tianhuangping, and Ming Tombs Reservoir. Among them, Guangzhou's pumped storage station has an installed capacity of 2,400MW, ranking first in the world.

Hydropower development not only provides an important source of energy, but also effectively eases the problem of pollution caused by coal combustion (243 TW • h of hydropower each year is equivalent to a decrease of raw coal combustion of 120 million tons). At the same time, it promotes the development of shipping, tourism, fisheries and so on. Large hydropower stations also bear comprehensive functions such as flood mitigation, irrigation, water supply, etc.

1.5 Other Achievements

Over the past 50 years, 780,000km² of soil erosion areas have been improved across the country, including 166,000km² in the Yellow River Basin, which has protected the land resources and reduced the sediment deposition in river and reservoirs. Since the 1970s, the sediment flowing into the Yellow River has decreased by 300 million tons annually.

Inland river navigation has been developed by adopting measures like dredging silt, exploding reef, and building ship locks. At present, inland navigable distance has reached 100,000km, among which the Yangtze River is known as the golden waterway.

Since the policy of reform and opening-up was adopted, a great deal of reform has been carried out in legislation, administration, financing and so on. The Water Law was promulgated in 1988, followed by *Law of the People's Republic of China on Water and Soil Conservation*, *Flood Control Law of the People's Republic of China*, and *Law of the People's Republic of China on Prevention and Control of Water Pollution*, and *Regulations on River Management*, as well as the water license system. With all these laws and regulation, China has gradually stepped on the track of harnessing water according to law. Investment in water projects has been relatively concentrated. In the past, China practiced a single mode characterized by money from the central government and labor from farmers. Now it has developed into a new pattern with joint efforts of all walks of life, including central and local governments, collectives and individuals. The funding sources have been diversified. In addition to fiscal appropriation, such means as policy preferential loans, bank loans, and special funds for water projects and foreign investment have all been leveraged to solve the fund shortage all of which have greatly boosted the construction of water projects. Water resources planning, allocation, and supervision and administration systems have been strengthened, and the reform of operational mechanism is also being deepened. For water resources, use with payment has been practiced and water tariff policy has been adjusted, and a reasonable cost is levied, so as to conditionally transform to the market economic mode. In this context, some water-related enterprises have embarked on a virtuous circle of self-improvement and development.

With the large-scale development of water projects, the corresponding science and technology have developed rapidly, and the competence of the water professionals constantly have enhanced, which in turn promotes the development of water project construction again. At present, in water resources and hydropower sector, there are six major river basin commissions (all with investigation and designing institutes), planning and designing institutes at national level and various designing institutes (consultancy companies). There are such institutions at provincial level (autonomous regions, and municipalities directly under the central government), and even some prefectures. In addition, there are numerous construction teams (engineering companies) throughout the country. As for scientific research, there are three professional national research institutes and a large number of local scientific research centers. There are three special universities for water resources and hydropower. Some famous universities (e.g. Tsinghua University) have set up water resources and hydropower departments, which have cultivated a new generation of talents. Both systems have formed a team with strong dedication and sophisticated techniques, advanced exploration technology, powerful design and research strength, as well as significant electromechanical and metal structure manufacturing capability. Now, China is able to manufacture 320MW Francis turbine units, and propeller units with a diameter of over 10m, and to subcontract the manufacturing of 550MW and 700MW turbine units, 500kV HV electrical equipment and the world's largest pressure pipe, lock gate, and common construction equipment. Over the past five decades, people in China's water sector have built a large number of well-known large hydropower projects, including extra-large reservoirs and hydropower stations, ultra-long water conveyance tunnels, inter-basin water transfer projects, ultra-large irrigation and water supply projects, etc. As the world's largest water project, the Three Gorges project is now under construction, and more projects are in the designing stage to be constructed. Research institutions have been organized to tackle the academic problems in water sciences; the research in many areas has reached the international advanced level, such as high slope analysis and treatment, high dam design and construction, high arch dam anti-seismic performance, sediment movement and river dynamics, dam construction in Karst areas, high dam large-flux flood discharge and energy dissipation. Take dams for instance, China has built a double curvature arch dam (Ertan) of 240m in height, and concrete faced rockfill dam (Tianshengqiao) of 178m high. Some dams have been initiated or under construction, such as the Three Gorges gravity dam (178m), Xiaowan arch dam (294.5m), Longtan RCC dam (192m for stage I and 216.5m for the stage II), and Shuibuya concrete faced rockfill dam (233m), all of them ranking highest in the world list. That's why some foreign experts came up with the conclusion after the on-site visit that Chinese engineers can build any dams on any rivers as needed.

It is difficult to list all the achievements of water resources development in China over the past five decades in the limited length of this paper. Based on the above brief introduction, any person without prejudice shall admit that Chinese people have indeed achieved unprecedented achievements in water project construction, which has ended people's sorrow and suffering from the frequent disasters, marking this 50 years the most magnificent and fruitful period in terms of water project construction.

2 Existing Problems

Over the past 50 years, China has reached achievements of world renown, but has also made a detour in the course of exploration. This is caused by both unresolved historical problems and new contradictions arising from errors in the development. By unraveling the superficies, we could explore the in-depth factors.

2.1 Flood Mitigation

Although there had never been catastrophic flood disasters over the past 50 years, the flood control standards of large rivers are still relatively low failing to reach the expected standards. Thus, people have yet been exempted from the flood threat. It has not formed a complete and scientific flood control system. In case of the excess flood standard, China still lacks the relevant resolutions.

Before the Three Gorges Project, Xiaolangdi Project and the corresponding supporting works were completed, the flood mitigation standards of the Yangtze River and Yellow River were insufficient to resist 100-year floods, letting alone the Huaihe River, Haihe River, Pearl River, Liaohe River, and Songhua River where the dikes can only resist 10 – 20 year floods. In case of great floods, no comprehensive countermeasures are available for those rivers.

Flood mitigation requires comprehensive measures. For example, in the middle and lower reaches of the Yangtze River, in case of catastrophic floods, it is necessary to rely simultaneously on the protection of river embankments, retaining capacity of the reservoirs, such as the Three Gorges Reservoir, and operation of relevant flood diversion areas to avoid the devastating disasters. Currently, the current work has been weighted too heavily towards the construction of dam and dike and ignored the flood diversion. Hundreds of thousands of people dwelled in the planned flood diversion areas, where is the grain and cotton production bases. Therefore it is hard to make the decision to divert flood to such areas. Meanwhile, the transmission of information about flood diversion, the organization of retreatment or evacuation of the tens of thousands of people, as well as the compensation for people after the flood, all of these are very complicated issues which yet fail to catch due attention and be resolved as the dam construction.

The reality even totally goes against "the planned flood diversion". Because of the soaring population, in many places, people are constantly reclaiming lakes, low-lying lands, wastelands and beaches for cultivation, immoderately developing the land that is necessary for floods discharge or storage, and building all kinds of "flood mitigation projects" regardless of necessary priority. The consequence is that the dikes nationwide have added up to 260,000km; while flood level has also been constantly raised, leading to a vicious circle of flood embankment building and flood level rising. There are many reasons for the rising of flood level; however, it is no doubt that the major cause lies in the numerous enclosed areas for cultivation and failure to carry out flood diversion timely. Due to the heightened embankments, the flood mitigation burdens and risks are continuously increasing. During flood seasons, there are always a large number of servicemen and common people engaged in flood fighting. In 1998, a million people including military forces joined the battle against flood. In case of dike breach, the consequences could be unimaginable.

It is more complicated for the Yellow River, as about 1 billion tons of sediment flows into the sea every year; such problem has not been solved properly up to now. During the past 50 years, riverbed has been constantly elevated due to aggradations. From 1986 to 1997, the average annual sedimentation was 250 million tons, and most of the sedimentation concentrated in the main channels, and the previous bankfull discharge reached 6,000m^3/s, which has decreased to 300m^3/s currently, causing high water level and large scale inundation even in small floods. In August 1996, the peak discharge was only 7,600m^3/s at the Huayuankou, while its water level was even 0.91m higher than that of the great flood in 1958 with a peak discharge of 22,300m^3/s. At present, drought is threatening in the Yellow River basin. Entering the 21st century, if the hydrological cycle turns to be abundant, it remains a question whether the Yellow River can maintain safe. Although the Xiaolangdi Reservoir was built, due to startling annual sediment yield of the Yellow River, it is still a complex and difficult problem on how to maintain the long term function of Xiaolangdi Reservoir in silt arresting, flood regulation, and sedimentation reduction. Besides, there are three reservoir sites on the main streams of the Yellow River, whose development shall be seriously considered.

All these situations are thought-provoking. In fact, it is impossible to "completely eliminate floods". People should not only control floods appropriately, but also adapt to floods actively and coordinate the relations between human and floods. It seems that flood shall be mitigated, instead of being prevented and controlled. People can't occupy lands for flood storage in an uncontrolled manner, especially can't pass one's miseries to its neighbors or to protect the interest of the minority while sacrificing that of the majority. In addition to the construction of reservoirs and embankments, flood protection works should be focused on the construction and operation of flood diversion and detention areas. Flood mitigation is a comprehensive project, in which attention should be given to not only the structural measures but also the nonstructural measures, especially the social insurance. In a word, the strategy of flood mitigation shall be shifted from the engineering construction to a comprehensive flood mitigation and disaster relief system, so as to achieve the harmony between man and floods.

2.2 Irrigation and Water Supply

As for water consumption for industrial, agricultural and urban purposes, the outstanding problem is the serious shortage of water resources and startling low efficiency of water utilization.

China is a country with water shortage. The total volume of water resources is about 2.8 trillion m^3, while the per capita volume is only 2,200m^3, ranking the 121st in the world, which will decrease to 1,700m^3 in the middle of the 21st century. The water resources are rather unevenly distributed in China. For example, the per capita level is only 700m^3 for the northern China, and 500m^3 for the Yellow River, Huaihe, and Haihe plains (only 358m^3 for the Haihe River Basin), which is below 1,000m^3, the internationally recognized threshold of water shortage and even 500m^3, the threshold of serious water shortage. Over the past 50 years, with constant increase of population and the continuous development of industry and agriculture and urbanization, the extent of exploitation and utilization of water resources have been subsequently enhanced. In particular in the North of China, the utilization rates of water resource in the Yellow River, Huaihe River, and Haihe River have reached 67%, 59%, and even 90% respectively, much higher than reasonable level. The excessive exploitation of water resources has caused a series of problems, such as the drying-up of lakes, the river flow cutoff, the groundwater overexploitation and ecological deterioration in the estuaries and arid areas.

What makes the situation seriousness is that such overexploitation is in coexistence with the inefficient use (even waste) of water resources. The water utilization coefficient of agricultural irrigation is 0.3 - 0.4 in China, compared with 0.7 - 0.8 in the advanced countries, representing a 30 - 50 year gap. Water consumption per unit of industrial output value in China is 5 - 10 times that in advanced countries. Moreover, industrial water recycling rate is 30% to 40%, while the advanced countries can reach from 75% to 85%. In most cities, leakage rate of water pipelines reaches at least 20%. These phenomena also exist in areas with serious water shortage. In Huaihe River Basin, for example, industrial water recycling rate is only 30%, and that of rural township enterprises is as low as 15%; wild flooding is very common in the irrigation areas of Xinjiang, Ningxia, and Inner Mongolia, in northwestern China with agricultural water utilization rate of only 40%. In the Yellow River, Huaihe River and Haihe River Basin, grain production per cubic meter of water is only 1kg, while the number of Israel is 2.3kg. The living standard of people in the driest cities of Hebei Province is still very low, yet the per capita water consumption is as high as 216m^3, even more than that of Seoul, Madrid, and Amsterdam.

Such phenomena are quite thought-provoking. Chinese people have always taken water as a kind of free and inexhaustible resource, knowing only about its utilization and development, yet little about its protection and conservation, and emphasizing only the forms but not the actual results. Such situation was even more common in the era of planned economy. This cannot meet completely the sustainable development require.

2.3 Aggravated Pollution and Degraded Water Quality

China's industrialization has not been completed yet, while environmental pollution and ecological deterioration have been rather serious. In addition to atmospheric pollution, water environment pollution is the most severe. In fact, all rivers, lakes and even seas of China have been polluted in general and with a deteriorating trend.

According to the incomplete statistics, each year a total of 62.4 billion m³ of sewage is discharged into rivers, lakes, reservoirs, or directly used for irrigation, with the vast majority untreated at all or poorly treated. Among the approximately 100,000km of river sections under evaluation, 47% of the sections demonstrate a pollution of Category Ⅳ or worse. For basins like the Liaohe River, Yellow River, Haihe River and Huaihe River, the ratio of sewage to surface runoff can reach 1 : 6. Over 75% of the total lake area in China, have been seriously contaminated. It is indicated from a survey on 118 cities in the country, that 64% of the groundwater of the cities have been seriously polluted, and 33% mildly polluted.

In the context of the unreasonable industrial structure and extensive development mode, chemical industry, papermaking, and mining and smelting enterprises have become important pollution sources. With the rise of township enterprises, numerous point source pollutions have emerged. Meanwhile, the farmland using a large quantity of sewage and fertilizer become the massive non-point pollution source. Some reservoirs and lakes become the pool with full concentration of pollutants. In cities, due to much lower sewage treatment rate (in 1997, the urban water consumption reached 50.8 billion m³, while only 4.74 billion m³ was treated), a large amount of sewage was directly discharged and did not curb its development trend in a short term.

In addition to water environment pollution, water-related ecological environment damage and deterioration are also serious, including the shrinkage of natural oases in the northwestern arid areas of China, the absence of flow in inland rivers at lower reaches, the disappearance of terminal lakes, degradation of pasturing grassland, deforestation, expanding of desertification, and intensified sand storms and soil erosion in the Loess Plateau.

Without curbing or reversing such situation, water pollution and ecological environment deterioration will not only affect the sustainable development of China's economy, but also cause great harm for the health and survival of the Chinese people.

China has undergone backwardness, poverty and weakness over the past 200 years and suffered from the foreign bullying, which almost cause the dying out of national states. After the founding of New China in 1949, people have an urgent need to accelerate their economic development to lift our country from the state of "poverty and blankness" as soon as possible. These requirements and sentiments are understandable and justifiable, which nonetheless led to the insufficient attention on ecological and environmental issues during this process. The contents about environmental impact were not included in the planning or design documents of early projects, China stepped on the road of development first and treatment later. Perhaps it was inevitable in the early period of New China. However, with the constant expansion of construction scale and growing national strength, it is necessary to give higher priority to the ecological and environmental protection issues, which will definitely reduce some rather tricky problems for now.

2.4 Other Issues

Besides the above mentioned three categories of problems, many other problems could also be pointed out as follows:

(1) The construction scale of hydropower engineering are massive, but most of them are inefficient, poorly management with a lack of supporting infrastructures which can lead to aging without repair and being scrapped without any regarding it.

(2) Most projects mainly rely on government investment with poor economic profits, and even without no operation fund, and then they have been the government's burden, not to mention self-development and self-optimizing to form a virtuous circle mechanism.

(3) Some projects are of low safety level and poor quality, especially the projects built during the "Great Leap Forward" (in 1958) and "Cultural Revolution" (from 1966 to 1976). Accidents of some structures, such as dams and reservoirs, will cause grave losses of life and property in the downstream areas.

(4) Not enough attention is paid to migration settlement work during the dam construction. Only economic compensations were made without considering any comfortable housing and subsequent development. Even worse, economic compensations were not provided sufficiently for emigrants in some projects, leading to the misery of the emigrants and social instability.

(5) The environmental impact was neglected during the construction of some hydropower projects. Efforts to mitigate those negative impacts were not sufficient, either.

(6) Over the past 50 years, the navigation conditions of inland rivers have been improved significantly. However, no enough attention was paid to this outstanding transportation means. For instance, volume transported through the Yangtze River is less than that of the Tennessee River, a tributary of the Mississippi River in USA. Moreover, the relevant sectors didn't coordinate well each other. In some cases, dam construction affects the navigation; while in others, the built large

navigation structures fail to deliver any performance.

(7) Despite the great progress in hydropower development, only 12% of the available water energy was tapped. The current policies (taxation, interest, and electricity price) are not conducive to promoting the development of such clean energy. The underexploited hydropower mainly concentrates in the southwest part of China with more complicated conditions, and speeding up hydropower construction in the future is itself threatened by a series of challenges. It can be concluded from the previous description that the problems and mistakes are very serious, which obviously shall not be imputed to any specific project, department or engineer. Instead, such problems involve people's awareness and some a country's behavior. Of course, ideological understanding of leadership and the governing party plays a decisive role, for that it is the basis of any guideline, policy or measure. In essence, the problem lies in whether people have a scientific understanding about the objective things and their laws without any prejudice, stubbornness or impulsiveness. Such problems certainly affect the overall situation of national construction, far beyond the scope of the water projects. However, the consequences of wrong understanding and policy can be perceived clearly from the contradictions in water project construction.

The author summarizes the causes of such problems into three categories as follows.

(1) The first is awareness. Since the founding of the People's Republic of China, the belief of "Man definitely triumphs over the nature" has prevailed for long term. People don't understand the principle that sustainable development can only be achieved by adapting to the nature and living in harmony with it. Such wrong awareness is mainly reflected in the following aspects:

1) One-sided emphasis on "remaking nature" runs against objective laws, putting people in a passive position. Chinese people's survival and development certainly rely on development and utilization of the nature, and it is inevitable to encounter setbacks in such a process. It is no ground for blame to encourage the people to hold an optimistic attitude of "Man definitely triumphs over the nature". However, we can adapt to the natural environment, and step on a road to sustainable development only when we can understand the real situations, make scientific planning taking into consideration of the overall situation and long-term interests, and implement with correct methods. It is wrong to unilaterally promote "struggle philosophy".

Take flood mitigation for instance still, we should recognize that human beings cannot completely eliminate floods and conquer the nature. Human beings must appropriately control floods and actively adapt to floods. Mastering the technology of embankment construction does not mean that we can endlessly reclaim lakes and beaches or occupy flood channels, because that would lead to a vicious circle.

2) We should not merely ask from the nature without letting the nature restore. There are abundant natural resources, but not infinite, water in particular, which is often considered as a kind of free resource endowed by the nature. The Chinese would like to describe China as a country of "vast territory and abundant resources". In fact, many resources are insufficient, including water, in terms of per capita volume. This naturally makes controlling population growth and practicing economy the basic national policy for long term. The "new population theory" proposed by Mr. Ma Yinchu in the 1950s pointed out the essence of the problem, which however was criticized as "Malthusism". Consequently, Chinese population was out of control for a period of time, resulting in permanent burden; and the traditional practice of economy was given up. Inefficient and even devastating "development and utilization" of water, forest, mineral, fish resources, among others have caused startling wastes, leading to rather severe consequences.

3) We should not merely focus on development while neglecting protection. Once the ecological environment is destroyed, it is difficult to restore. As for water resources, the most obvious consequence is reckless and uncontrolled exploitation and utilization of water resources, as well as overall water environment pollution. Some people even proposed to use up water, and the so-called water consumption is only limited to industrial, agricultural and domestic use, but excludes the eco-environmental water requirement. Various disasters take place in many places, including dry rivers and lakes and huge funnels formed due to sharp decline of groundwater level. The situation in some places even reached a condition where "all rivers are dry and all water is polluted". Meanwhile, there is massive soil erosion, constant degrading of natural forests, shrinking fisheries, and extinction of some species, all of which are hard to restore.

Due to impact of the idea that "man definitely triumphs over the nature" and "remake nature", an atmosphere was formed in which, development, utilization and engineering measures are partially stressed while environmental protection, conservation and non-engineering measures are neglected. This is a kind of practice that "seeks short-term gains but impairs long-term development", which is like "eating the grains of the forefathers while leaving nothing for the descendants".

(2) Working style was once subjective and not practical. The Communist Party of China (CPC) believes in materialism, and continues a good tradition of seeking truth from facts. Unfortunately, due to the interference of "leftism" route, it was deviated from the correct path, falling into subjective idealism and leading to some tragic results.

1) Emphasizing politics, overlooking and even opposing to science and technology, and substituting scientific verification and democratic decision-making with subjective imagination: After the founding of the People's Republic of China, there has been recurrent political movements, stressing that politics tops everything else, thereby mistakenly criticizing intellectuals or even denying the role of science and technology, all of which were even more extreme during the "Great Leap Forward" and "Cultural Revolution". With hasty decisions, many projects were started in a rush. Later on, most projects were suspended, some projects were even delayed by more than 10 years, some became dangerous dams and reservoirs due to poor quality, and some failed to play their role after completing.

2) Emphasizing the superficial appearance and data, and paying no attention to the real effect, or even practicing frauds. The party and governmental officials are public servants, whose mission is to serve the people. In the early years after the founding of the People's Republic of China, honesty and pragmatism of the officials were highly praised by people. However, due to the impact of the ultra-left thoughts, they gradually deviated from the track of pragmatism and turn to pursue things that are "big but false and empty" instead of actual effect. On one hand, new projects were put into construction successively; on the other hand, many old projects were no longer used. Later on, some officials practiced frauds and took water projects as a means to show "political achievements", seek "promotion", or cope with superior leadership. Therefore, the phenomenon, such as so-called "cadres make data and data make cadres". That's why tree planting campaign are carried out every year, yet still no trees are seen anywhere. In each winter and spring, there will be campaigns to build irrigation and drainage projects, but many of them are a mere formality.

3) Craving for greatness and success, seeking instant benefits, and lacking long-term perspective. Cadres at all levels only focused on immediate interests and engaged in short-term construction. As incumbent officials, they all launched projects for their "achievements", which caused chaotic development and construction, and resulted in resource waste, improper layout, and environmental deterioration.

The above-mentioned subjective and idealistic practices formed the irregular climate of stressing formality, quantity, main project, and phony politics instead of actual effect, quality, supporting facilities, and science & technology.

(3) Traditional and historical factors and planned economy system affected the construction of water projects.

1) Impact of localism: In fact, China never achieved unification since the fall of Qing Dynasty in the early 20th century until the founding of the People's Republic of China. Even after 1949, localism was not seriously criticized or eliminated in a very long period of time. In particular, water issues often involves the interests among adjacent provinces and regions, the upper and lower reaches of a river and various sectors, weaving a complex network of relationship. Some locality only pursues the interest of the small region even at the expense of the interest of its neighbors. Such behaviors affected the unity between regions even leading to fighting with weapons. Some contradictions remained unresolved for a long time; for water resources planning involving several provinces it is often very difficult to make decisions and implement. Therefore, water projects with the most significant comprehensive benefits are often hard to build.

2) Impact of the planned economy: China adopted strict planned economy during the 30 years after the founding of the People's Republic of China. Major water projects were invested and operated by the state, while for irrigation and drainage projects, farmers were involved in the hydropower construction as the workforce. Under such a model, no attention was paid to cost, construction period, efficiency & benefits, or water economy; so the projects benefiting the people failed to gain support from them. After the projects were completed, many failed to gain returns; some even became burdens on the government. Free use of water or unreasonably low price can cause wasting water. Such situations have been preliminarily reversed until the adoption of the reform and opening-up policy.

The above factors led to the one-side emphasis on local benefits, technology and construction instead of long-term benefits, economy, and management during water conservancy construction.

In conclusion, the past mistakes can be summarized as "10 focuses and 10 neglects". To learn lessons from the mistakes, we must realize the shifts in mind, styles, policies, and measures from "focus on development instead of environmental protection" to reasonable development under the premise of environmental protection; from "focus on utilization instead of saving" to reasonable utilization of resources under the premise of saving resources; from formalism to practicality; from partial focus on main projects to equal attention to both main projects and supporting facilities; from paying more attention the quantity than the quality to giving top priority to quality and a veto in quality; from biased focus on construction not management to more emphasis on modernized management to improve; from the exclusive focus on technology to more attention to the adjustment of economic mechanism, bringing water project construction into the socialist market-oriented economy model; from focus on locality and short term interests to the consideration of overall situations and long term interests; from partial focus on engineering measures' to paying attention to both engineering and social measures, making them complementary to each other; and from pursuing phony politics to upholding the belief that science & technology is the primary productivity.

3 Outlook of Water Conservancy Development in China in the 21st Century

The 21st century is an important period for China. In the first half of the century, China will realize the third step of its strategic development plan transforming from well-off society into a moderately developed country, increasing per capita GDP from the current US＄800 to US＄5,000 - 6,000 (comparable value), eliminating poverty, backwardness and ignorance, ending a historic foreign bullying, fully achieving modernization and completing national rejuvenation.

In this great historic period, water conservancy development undertakes an arduous task and faces severe challenges. Briefly speaking, the new generation of hydraulic engineers needs to finish the tasks as follows:

3.1 Properly Mitigating Floods of the Major Rivers

The flood control standards for the major rivers and relevant cities shall be raised to appropriate levels (to be able to resist 100-year events in general), and countermeasures shall be available in case of floods beyond standards so as to avoid the devastating disasters.

With 50 years of construction and development, China has built a considerable economic foundation, and people's living standards have been greatly improved. In face of catastrophic floods like those in the past two centuries, the major rivers suffered from unimaginable losses and serious consequences, and even the national development plan could be disturbed. Therefore, it is the top priority to prevent such disasters.

From the engineering perspective, the comprehensive flood mitigation measures still rely on discharge, storage and diversion, with discharge as the main measure. As for discharge, embankments of rivers and lakes should continue to be heightened and reinforced, and the potential dangers and illegal obstacles in the flood discharge routes shall be eliminated. Some enclosed areas for cultivation shall be leveled off, so as to restore flood discharge routes. We must learn how to coexist with floods, and ensure embankment safety under certain flood level.

For the main streams and tributaries of rivers, necessary reservoirs shall continue to be built with the function for other uses. During flood season, these reservoirs shall be involved in the scientific joint regulation so as to mitigate flood peaks.

River bank-plains and enclosed areas for cultivation can be still available for agricultural purposes, but they must be abandoned when flood has reached a certain level according to the planning, and shall not be enclosed or blocked without permission. And the requisite flood diversion areas shall continue to be built and improved, so that they can be used as a program in case of catastrophic floods. In this context, it is necessary to ensure that people in the affected areas can evacuate safely and obtain reasonable compensations later on.

In addition to the above measures, non-structural measures shall also be emphasized, such as applying modern science and technology to improve forecast accuracy and efficiency, extending forecast period, optimizing regulation plans, organizing capable and mobile flood mitigation teams, setting up the unified and authoritative institution for flood mitigation dispatch, and implementing social insurance system for flood mitigation, so as to prevent the frequent dispatch of large number of people for embankment protection during flood period.

Efforts shall be made to accomplish these missions within 15-20 years, and set up a scientific, safe and reasonable system for flood mitigation system.

3.2 Soundly Governing the Yellow River

Among the major rivers in China, the Yellow River demonstrates special complexities, and its governance is a major issue for Water Conservancy Development in the 21st century.

Upon its completion and operation, the Xiaolangdi Dam can play tremendous roles in flood regulation, sediment retention, (downstream) silting reduction, and channel sluicing for about two or three decades. However, both its storage capacity and sediment retention period are limited after all, thus this rare opportunity must be seized to carry out scientific regulation, tracking and monitoring, researches, and to make a correct deployment for the management of the Yellow River. The major task is to handle the sediment into the Yellow River and the suspended river bed in the lower reaches.

As for a fundamental measure for the issues, we must adhere to the soil and water conservation in the upper and middle reaches, rationally return farmlands to forests (grasslands), and enclose hillsides to carry out forestation. The relevant investment shall be increased, and its work shall be deepened. The effective sediment retention projects shall be built on the tributaries to reduce sediment into the Yellow River. According to the preliminary plans of the water authorities, newly controlled soil and water loss area will reach 145,000km² by 2015, and increase by another 242,000km² by 2030, with annual average reduction of sediment inflow in the Yellow River by 400 million tons, 600 million tons and 800 million tons respectively in 2015, 2030 and 2050.

For the remaining sediment inflow into the Yellow River, scientific regulation shall be conducted to transport it all the way into the sea; the related measures shall be adopted to appropriately reduce the elevation of the suspended reach, such as discharging artificial flood, and deepening estuary to generate backward erosion. The suction and excavation can be used as supplementary measures to pile sediment into embankments or divert to both banks for utilization, so as to secure the long-term stability of the Yellow River.

If China can bring the Yellow River under control in the 21st century, it will become a world-leading achievement.

3.3 Reasonably Allocating Water Resources with the Implementation of South-to-North Water Diversion Project

Many regions of China are suffering severe water shortages, such as northern China, especially the Yellow River, Huaihe River, and Haihe River basins, which limit economic development and improvement of people's living standards, and destruct ecological environment. For these regions, based on water conservation and reasonable development, allocation and utilization of water resources to establish a water saving society, giving equal emphasis to both use decrease and supply increase and both development and protection while prioritizing use decrease and protection. Under this premise, the inter-basin water diversion of appropriate scales can be adopted.

The most important water diversion project is the so-called South-to-North Water Diversion Project which is designed to divert water from the Yangtze River rich in water resources to the northern part of China. After decades of study, the water authorities proposed the plan of diverting water from the Yangtze River to the northern part of China through the east, middle and west routes respectively, and each route has respective water supply areas while being complementary to each other, which constitute a reasonable overall layout. After the project is implemented, water shortage can be eased in the northern part of China, and a basic balance can be gradually realized between water supply and demand.

For the eastern route, water will be diverted from the north bank of the Yangtze River near Yangzhou City, Jiangsu Province, which mostly utilizes Beijing-Hangzhou Grand Canal and parallel river channels as trunk for water diversion, and diverts water to Tianjin by crossing the Yellow River near Weishan in Shandong Province. At the same time, there is a branch canal supplying water to Jiaodong Peninsula. Because of the high terrain of the Yellow River, the southern section (660km) of the diversion project needs to uplift water to a certain elevation, while water in the northern section (490km) can flow naturally. The eastern route is to supply water for the eastern parts of the Yellow River Plain, Huaihe Plain, and Haihe Plain, as well as Shandong Peninsula, which mainly supply domestic water and industrial water for Tianjin and other cities, as well as meet requirements for agriculture and navigation along the route. Taking 2020 as the design average year, pumping rate from the river is 1,000m³/s, and diversion rate to Tianjin 180m³/s, and annual water supply volume can reach 15.4 billion m³, and up to 1,400m³/s and 250m³/s respectively in the forward level year. There is no major technical problem for the east route. However, major issues still lie in how to prevent water diversion from pollution discharge along the route and how to reduce cost and coordinate with the requirements for water demands of local authorities, so as to achieve unified management and optimal benefits.

The middle route supplies water to five provinces (municipality directly under the Central Government), including Beijing, Tianjin, Hebei, Henan and Hubei. The water is mainly targeted to satisfy the demand for domestic and industrial use of the riparian cities, with a certain consideration for agricultural and other uses. The transmission line of the middle route is 1,246km in length (482km in the south of the Yellow River, and 764km in the north of the Yellow River) with water intake from Danjiangkou Reservoir of Hanjiang River (supplemented from the Yangtze River in the forward level year). Water will be transmitted along the newly built main channel with an annual average transmission amount of 13.0 – 14.5 billion m³. The middle route project will enable high-quality reservoir water to flow by gravity to Beijing, Tianjin, and North China which suffers from water scarcity. However, this project will involve a huge investment and a large number of resettled people, and it is necessary to solve the problems related to the balance between water storage and diversion, engineering risks, and impact on the water source areas. The project operation and management and water price are also very complicated.

Both eastern and middle routes have planned to complete in 15-20 years, and they are expected to divert water across the Yellow River by 2010. Two routes can divert 30 billion m³ of water northward, which can not only meet the need in the northern part of China, but also improve the deteriorated ecological environment, adjust water consumption in the middle and lower reaches of the Yellow River, and increase water consumption and sediment sluicing flux in the upper reach, bearing great significance.

For the western route project, dams will be built to store water on the Tongtian River section of the Yangtze River main stream, the Yalong River, tributary of the Yangtze River, and the source area of the Dadu River, and the water will be diverted into the upper reaches of the Yellow River through Bayanhar Mountain via a tunnel. The project is designed to mainly solve the water shortage of the Yellow River and the northwestern region, and mainly supply water for urban domestic and industrial water uses, with certain consideration of the water user in agriculture, forestry and animal husbandry. If this route can directly divert water to the upper reaches of the Yellow River, it will be far-reaching significance to change the ecological landscape in some parts of the Northwest and promote the comprehensive management of the Yellow River. The project intends to be implemented by stages, and water diversion amount will gradually increase from 4 billion m³ to 17 billion m³.

As elevation of the upper Yellow River is higher than that of tributaries and main streams of the Yangtze River, high dams need to be built on the water intake sections to divert water by means of long tunnels (from one-hundred and tens of kilometers to hundreds of kilometers in length). The project area is located in the cold alpine area with insufficient oxygen, so the project construction will be extremely difficult, and its investment will be huge. The preliminary planning and study are underway, and efforts have been made to start this project before 2020, and complete it in the first half of the 21st century.

In addition, some people suggest diverting water from farther Jinsha River, Lancang River, Nujiang River, and even the Yarlung Zangbo River to the northern part of China. Preliminary analysis suggests that such plans are unrealistic in the foreseeable decades, and also unnecessary. Therefore such plans do not need to be taken into account at present.

Besides the South-to-North Water Diversion Project, it is also necessary to build some inter-basin water transfer projects in Northeast China, Xinjiang, Gansu, and Shaanxi provinces, etc., which shall be achieved in early 21st century.

As a large-scale water diversion project across main basins, the South-to-North Water Diversion Project is a major fundamental and strategic project which can rationally allocate the water resources and underpin the sustainable development of the country. Therefore, it concerns the long-term interests of future generations. Therefore, over the past decades, many people in the water ministries have devoted themselves to the project. However, the previous studies were done mainly by the water ministries, focusing on demand, technology, funding, and so on. Therefore, it is necessary to comprehensively study the far-reaching problems, such as the ecological environment impact (favorable and unfavorable); engineering risks; re-distribution and coordination of interests among localities and authorities; water economy and management & operation modes; policy measures. The study requires joint efforts from such relevant departments from water, environmental protection, and urban construction to secure a smooth implementation of the project.

Recently, the State Council have pointed out that the South-to-North Water Diversion Project shall give priority to water conservation, pollution control and ecological protection, and then to water diversion, water transmission and water consumption. It is our belief that this great project will be launched and will demonstrate benefits in the 21st century, so long as we follow the correct direction and fulfill our work thoroughly.

3.4 Paying Special Attention to Water Conservation and Pollution Control to Protect and Improve Ecological Environment

Development in the past usually was achieved at the cost of resources wastes and the destruction of ecological environment. Such practice must not continue in the 21st century. The future water projects shall give top priority to water conservation, pollution control, and ecological environment protection and improvement.

China's annual water supply volume has grown from about 100 billion m³ in 1949 to about 560 billion m³ at the end of the 20th century. In the 21st century, population, economy, urbanization will continue to increase and develop, and water consumption will increase accordingly. According to experts' analysis and forecast, water consumption for industry and agriculture, and urban and rural domestic water will reach 730 billion m³ even with stringent control. Coupled with water consumption for ecological environment, such figure could reach 800 billion m³. But the total available water resources all over the country is about 800 – 950 billion m³, close to the limit, hence the situation is grim. The only way is to practice economy, and build a water-saving society.

Among all sectors, agriculture is the biggest water consumer. Therefore, it is necessary to transform the extensive growth mode of agriculture into water-efficient modern irrigation and dry farming. We should use water in a scientific manner, control irrigation water, and increase of crop yield (grain yield increases from 500 million tons to 700 million tons) to meet 1.6 billion people's needs for agricultural products without increasing agricultural water demand. A variety of measures shall be adopted to enhance water use efficiency, so that average grain yield of per cubic meter of water can grow from 1.1kg to 1.5 – 1.8kg. To this end, we shall fully develop water-saving irrigation, focusing on improving surface irrigation, and combining with wells and canals to prevent leakage and reduce evaporation, and appropriately develop sprinkler irrigation techniques. The State shall invest in the supporting facilities, renovation and extension projects in the irrigation area (in the past, the state only invested in projects that can increase water supply and key projects as well). The key areas for water-saving transformation include the Yellow River Plain, Huaihe Plain, Haihe Plain, Ningxia, Inner Mongolia and Xinjiang provinces.

For the industrial system, water-saving reform shall be taken as an important task. For urban water, leakage shall be reduced, water-saving apparatuses shall be developed, and people's awareness of water conservation shall be improved, making sure that demand is determined by supply, and development scale is determined by available volume of water resources. Water quota system should be adopted. The government shall develop water quota standards for different areas, industries and cities. In case of exceeding the standards, heavy penalties will be imposed, or water supply will be suspended. Utmost efforts shall be made to achieve zero growth in water consumption as soon as possible, so as to achieve a balance between supply and demand, as well as sustainable development.

Equally important as water conservation is the overall pollution control and eco-environment protection and restoration. Efforts shall be made in the following aspects:

(1) Focus on prevention, control from the sources and promote clean production: It is necessary to adjust industrial structure, reform production processes, transform technology, strengthen production management, advocate green technology & products, and eliminate pollution in the production process. Wastewater to be discharged must be treated to meet discharge standards, and the total pollutant volume must be control.

(2) Speed up the construction of wastewater treatment plants to reuse of urban wastewater: The nationwide wastewater treatment rate in 2010, 2030, and 2050 is planned to respectively reach over 40%, 60%, and 80%, and 50%, 80%, and 95% for severely polluted areas and important cities, which is an arduous task.

(3) Launching non-point source pollution control including sewage from farmland and township enterprises: waste water and residue shall be recycled through rational use of chemical fertilizers and pesticides, in combination with the construction of "ecological agriculture", so as to implement the clean production in the township enterprises and minimize non-point source pollutions. While strengthening land pollution abatement, pollutants discharged into the sea shall be reduced, and maritime pollution sources shall be controlled, so as to strengthen marine environmental pollution control.

(4) Fully promoting the ecological construction of soil and water conservation: The following measures shall be promoted according to the local conditions, including returning farmland to forests (grassland), enclosing hills for greening, planting trees and grass, sand control, hillsides transformation, and groundwater recharge, so as to increase coverage rate of forests and grasslands as far as possible, reduce water and soil erosion, curb the expanding trend of desertification, maintain river flow, abundant water of lakes, clear water and blue sky, so as to realize coexistence between human and nature.

In the past years, we owe much to the ecological environment. In the new century, China shall take advantage of new situations to launch a nationwide initiative to save water, mitigate pollution, and relieve ecological environment, so as to make the environment change notably and rapidly, and enter a virtuous circle in sustainable development by adapting to the nature. This requires the improvement of people's awareness, the formulation of appropriate policy, the implementation of measures, increasing investment and strict supervision with persistence, only by which, can we finally reap benefits. Such a task is arduous, and the cost is also high, which sometimes might seemingly affect development pace. However, if we fail to do so, China will see no future.

3.5 Vigorously Developing Hydropower Resources to Provide Clean Energy and Improve the Environment

As mentioned above, China ranks the first worldwide in terms of its hydropower resources. Over the past 50 years, the country's hydropower has witnessed rapid development, and the national installed capacity has reached 79.35GW, ranking 2nd in the world, but it is still in a low degree of development. In the new century, hydropower will receive more attention, especially the hydropower resources in western China which will gain an unprecedented development.

In 1989, the Ministry of Energy planned to build 12 large hydropower bases with a total capacity of 215GW, among them, 9 bases (Jinsha River, Yalong River, Dadu River, Wujiang River, the upstream of the Yangtze River, Hongshui River, Lancang River, the upstream and midstream of the Yellow River) are located in the western China with a capacity of 181GW. With sufficient development of the western hydropower, it can supply power to eastern China through national network connection, which is the so-called West-to-East Power Transmission. Like South-to-North Water Diversion project, this is one of the major policies of China and a great project that will be completed in the new century.

The western power is to be transmitted eastward basically along three routes: north, middle and south route. For the north route, cascade reservoirs development will be carried out in the upper and middle reaches of the Yellow River (and thermal power in Ningxia, West Inner Mongolia and Northern Shaanxi), and the power will be connected to the North China Power Grid. The key hydropower stations include Gongboxia, Laxiwa, Heishanxia, or Daliushu hydropower stations. For the middle route, besides the Three Gorges Project under construction, Shuibuya, Xiluodu, Xiangjiaba, Pubugou, and Jinping cascade reservoirs will also be built to transmit power to Central China and North China Power Grids. For the south route, Hongjiadu, Goupitan, Xiaowan, Nuozhadu, Jinghong, Longtan, and other hydropower projects will be built on the Wujiang River, Lancang River, or Hongshui River which can transmit power to Guangdong Province (and Thailand). At present, the state has approved a number of key projects including Gongboxia, Longtan, Xiaowan, Shuibuya, and Hongjiadu projects, as well as some large water pumping storage power station in the eastern part of China. It is hopeful to enable the national installed capacity of hydropower to reach 120GW, 180GW, and 240GW in 2010, 2020 and 2030 respectively. At that time, China becomes the world's largest country in terms of hydropower capacity.

For hydropower development, lessons must be learned to reduce errors, with particular attention to the following issues:

(1) Fulfill migrant settlement work to make sure that migrants can migrate, stabilize and become rich. The western hydropower projects are mostly located in barren areas, so they cause relatively less migrants and inundation losses, which are major advantages, but special attention shall be paid to migration.

(2) Focus on the side effects arising from hydropower development, such as environmental capacity, sediment, soil erosion, landscape changes, and rare species. Try the best to eliminate, mitigate these impacts or make compensations for them.

(3) Maximize the overall benefits of hydropower stations as much as possible to lead to the optimal and most comprehensive benefits.

In the 21st century, the task of water project construction in China will be extremely heavy and arduous. The above mentioned are just some listed key tasks. As practice makes perfect, it can be ascertained that, as in the 20th century, with the progress of these grand projects and the improvement of science and technology and management, the water related disciplines will be flourished, contributing to the people all over the world.

So long as we fully recognize our achievements, soberly learn the lessons, stick to the forward direction, carry forward the fine traditions, and continue to move forward pragmatically and unswervingly, we will definitely be able to accomplish our historic missions, and achieve water development on the right track, so as to make due contributions to the national prosperity and rejuvenation of the Chinese nation.

Hydropower Miracle of China[*]
Zhang Guobao[●]

Sponsored by Yunnan private investment, construction of the first Chinese hydropower station, Shilongba Hydropower Station, was commenced on the Tanglang River at the outlet of Dianchi Lake, in the suburb of Kunming city on August 21 (July 17 on traditional Chinese calendar) one hundred years ago, starting the Chinese history of hydropower development by themselves. The project lasted for 21 months at the cost of over 500,000 silver dollars. The station was constructed in accordance with the international bidding procedures and became one of the earliest joint stock enterprises with global bidding. Meantime, it also constructed the longest (34km) power transmission and transformation line with the highest voltage (23kV) in China. After dozens of years of construction and development, a batch of large, medium and small hydropower stations sprung up over the main streams and branches of large rivers throughout China. The ages using pine torch and oil lamp had gone forever because hydropower stations were constructed in many places of China, spurring the economic and social development. From the small Shilongba Station in the early 20th century to the grand Three Gorges Project on the Yangtze River in the 21st century, China's hydropower has gone through a century of resplendent history, with more than 45,000 hydropower stations in various sizes completed and the total installed capacity over 200GW, ranking the first place in the world. Significant changes have taken place in China over a century after braving the wind and rain and developing from the poor and weak hard times to the peacefully developing age with stability and unity. At the same time, it's international status increasingly improved. The fast-growing hydropower cause is a miniature of the process that Chinese people have become independent in the world. It developed from scratch and was then expanded and strengthened. Since 2004, the total installed capacity of hydropower of China has been ranking number one in the world, and the technological level of hydropower construction has reached the world leading level. All these significant development and glorious achievements benefit from the leadership of the Chinese communist party, the advantageous socialist system with concentrated efforts on important issue, the persistent reform and opening-up policy, as well as the selfless dedications of generations of Chinese hydropower sector.

1 Brilliant History

Human began to utilize hydropower with the exploration and use of water resources. More than 2,000 years ago, our ancestors had developed implements like mill wheel and grinder, utilizing water energy to develop production and meet life needs. It was the initial example that ancient people started to utilize water energy. Hydropower generation was born in the industrial revolution period. In 1878, the first hydropower station in the world was constructed in France, which was of epoch-making significance in the process of water energy development and utilization. This qualitative leap from water conservation to hydropower represented a big step forward of human from agricultural society to industrial society. China started using the hydropower 100 years ago, and has rapidly become a big and strong country of hydropower. In the early stage of construction and operation of Shilongba Hydropower Station, two German experts were employed. A German couple Von Karl and Mosig, who worked in this station early 20th century, published an article on Siemens Journal (Vol. 7 Issue 1) in 1927, which was titled *Yunnan Province: First Hydropower Station in China*. It said, "In this remote inland, isolated from world fashion and western culture, some people have introduced western technology to their own homeland. This is what excellent intellectuals and pioneers are doing…"

Over the past century, many Chinese have dedicated themselves to China's hydropower cause for the great rejuvenation of the country and the nation. In 1910, some private individuals with the ideal of "Revitalizing China through Industry" sponsored to build Shilongba Hydropower Station on Tanglang River in Yunnan. It produced electric energy with 480kW capacity for citizens in Kunming city, enlightened the modern industry, and was the first hydropower station in China, which is still operated now. This China-made hydropower station commenced a hard but glorious journey of China's hydropower cause. Although it's only more than two decades later than the first hydropower station in France, China's hydropower developed much more slowly than the powerful countries in Europe and America due to repeated warfare in which people could hardly survive, and construction of hydropower was difficult. Till new China was founded in 1949, the total installed capacity of national hydropower was only 360MW, and annual electricity production was 1,200GW·h, which was mainly attrib-

---

[*] This paper comes from the speech of the author at the 100th Anniversary of China's Hydropower Development on Aug. 26, 2010.

[●] Zhang Guobao, Former Director of National Energy Administration (NEA) and the Chairman of National Energy Expert Consulting Committee.

uted to the Fengman Hydropower Station built during the occupation of the northeast China by the Japanese army.

Hydropower can develop only after a nation prospers. After the founding of PRC, full-scale reconstruction was launched, and a fast-developing era was ushered for the hydropower cause. The leadership of the communist party and the establishment of socialist system laid a solid political foundation for the hydropower development of China.

History shall never forget that the first generation of party and state leadership, with Mao Zedong and Zhou Enlai at the core, attached great importance to water conservation and hydropower cause. Chairman Mao wrote in his poem, "Walls of stone will be sure to stand in the west river, to hold back Wushan mountain's clouds and rain till a smooth lake rises in the great narrow gorges. The mountain goddess will be not affected if she is still there, will marvel at a world so changed". In December 1957, Premier Zhou wrote an inscription for the National Electric Power Conference, saying "fight for the full use of 540 million kW hydropower resources and the ambitious goal of constructing the Three Gorges Dam". All the leaders and workers on the hydropower front took the bull by the horn and unveiled the new page of China's hydropower development. Soon after the founding of new China, the first hydropower station "designed and constructed independently with homemade equipments", Xin'an River Hydropower Station was completed. During the periods of the "First Five-Year Plan" and "Third-line construction", two climaxes of hydropower construction were achieved. A batch of medium and small hydropower stations and some large stations such as Sanmenxia, Liujiaxia, Danjiangkou, Wujiangdu and Gezhouba dams were built, providing premium electric power for national construction. The identification and planning of water energy resource, trained and expanded hydropower team, and accumulated and developed technology and equipment had created favorable conditions for the rapid development of hydropower in China.

History shall never forget that comrade Deng Xiaoping led China to the reform and opening-up path, which brought innovation and vitality to hydropower development. Because of the vast investment in hydropower construction and the long construction period, it is an important subject for China to seek fund support and provide guidance for system reform. The use of foreign fund and financing introduced a new model and concept for hydropower construction and development. Lubuge Hydropower Station was a pioneering attempt of project management system and construction model, and had formed "Lubuge Impact" on hydropower management. As a result, owner responsibility system, bidding invitation contract system, and construction supervision system were implemented in successive five stations with capacity over GW-class, namely Guangzhou Pumped Storage Power Station, Yantan, Manwan, Shuikou, and Geheyan dams, which were later named "Five Gold Flowers" for its recognized achievements in project duration, quality, and cost. According to the proposal of establishing socialism market economy system and the *Company Law* of the People's Republic of China presented at the 3rd Plenary Session of the 14th CPC Central Committee, basin cascade development reorganization was exercised for the companies located at Qingjiang River, Wuling, Wujiang River, and the upper reaches of the Yellow River, and the fundamental transformation of hydropower constructing companies was achieved from owner responsibility system to legal person responsibility system, greatly promoting the development of hydropower. With the implementation of opening-up policy, the global advanced technology, equipment and managerial experience were introduced into China. A lot of hydropower workers visited Itaipu Dam in Brazil and Aswan High Dam in Egypt, opening their eyes, and enhancing China's hydropower technological level.

History shall never forget that the third generation of collective leadership, with Jiang Zemin at the core, constructed the Three Gorges Project with remarkable resourcefulness, turning the grand blueprint of the Three Gorges dream into reality. China's campaign of developing its western region has created abundant opportunities for hydropower development. During 1992 – 1999, hydropower production had exceeded 3GW capacity annually for consecutive 7 years, and the hydropower construction of China achieved the climax once again. By the end of 1999, the national hydropower installed capacity reached 72.79 million kW, and the annual electricity production reached 212.9 billion kW·h, the world rankings of which were number two and number four respectively.

History shall never forget that the Party Central Committee with Hu Jintao as the General Secretary continued to lead us to success. After entering the new century, a batch of major hydropower stations such as Longtan, Xiaowan, Jinghong, Pubugou, and Laxiwa dams were successively commissioned and put into operation as the strategies of developing west China and "west-to-east electricity transmission" were implemented. Enormous progress had been achieved in the aspects such as the installed capacity and electricity output, the design and construction, and the equipment manufacturing and operation management. In 2004 and 2005, China's installed capacity and annual electricity output of hydropower ranked first in the world in succession. During 2003 – 2009, the installed capacity put into operation in China increased by more than 10GW capacity annually and reached 23.69GW in 2009 at a record level. By the end of 2009, the national hydropower installed capacity reached 196.29GW, and the annual electricity output 571.7TW·h, which were respectively 545 times and 476 times those in the early years after the new China was founded. Currently, as the No. 4 generating unit of Xiaowan Hydropower Station was commissioned, installed capacity of hydropower in China has exceeded 200GW. With the rapid development of hydropower, represented by the Three Gorges Project, China is among the top ones in the world in technologies of hydropower. China's Hydropower is embracing a new era of large-scale, large unit, high voltage, automated and information age. China, a weak, small and under-developed country in hydropower with low installed capacity and power output as well as outdated technologies and equipment manufacturing techniques, has grown into a world-recognized large and strong country of hydropower. China not only ranks first in the world in terms of installed capacity, but also is the country with the largest scale of hydropower projects under construction and the fastest development in the world. China is also making its mark in the international hydropower market. China not only built the Merowe Hydroelectric Station in Sudan and Bakun Hydroelectric Station in Malaysia but also supplied the equipment. Hydropower has become one of China's competitive industries in the world market. The dream of generations of Chinese has come true, and Chinese hydropower workers are now all over the world. The German couple once working in Shilongba had predicted a century ago that China would become a large hydropower nation in the world a century later! Now this prophecy has become a reality.

2 Great Achievements

China's hydropower is world famous. As a leader around the globe, it has achieved great results and attracted the world's attention.

2.1 Hydropower Resources are Ascertained

Hydropower stations are mostly located in mountains and valleys, with sparse population and tough exploration and construction conditions. In order to determine the theoretical hydropower potential and provide basis for hydropower development, the state has organized several general surveys and double-checks of hydropower resources. Generations of new and old hydropower engineers had overcome the difficulties and collected more precise and complete data. Many hydro-geologists had lived in mountains and forests since graduating from college and accumulated valuable hydrological data of hydropower resources year after year, and, living a lonely life, they made great contributions to the establishment and complementation of the hydropower resources survey data of China.

After new China was founded in 1949, the Ministry of Fuel Industry and the Ministry of Water Conservancy and Electric Power conducted three times of surveys on national hydropower resources in 1955, 1958, and 1977-1980 respectively, and basically figured out the condition of hydropower resources in China. However, with economic and social development and the establishment of market economy system, changes have taken place in terms of the technological level, economy and society, and production conditions. In order to better develop and utilize the hydropower resources, National Development and Reform Commission mobilized more than 1,000 engineers throughout the country during 2000 – 2003 to review the aggregate amount of national hydropower resources. The review covered 3,886 domestic rivers of the mainland with the installed capacity of 10MW and above, and identified the theoretical hydropower potential, technologically feasible installed capacity, and economically feasible installed capacity, which were 694.4GW, 541.6GW, and 401.79GW respectively. Information of the hydropower resources in Taiwan Province was known, too. In 2006, more than 50 engineers from China Renewable Energy Engineering Institute explored the Yarlung Zangbo River valley on foot, and documented in detail the hydropower resources of the great gorge of Yarlung Zangbo River for the first time. The review result serves as important references for hydropower development.

China, with vast territories and numerous mountains and rivers, has affluent hydropower resources and a large number of sites suitable for pumped-storage hydropower stations. According to the preliminary study and general survey result of 23 provinces and municipalities nationwide, more than 250 premium pumped storage hydroelectricity sites have been identified, with the total installation capacity up to 310GW, superior engineering technological and economic indicators, and high development value, which are able to meet the need of optimal energy structure and safe operation of power grid.

2.2 Scale of Hydropower Development has Continuously Gone up to a New Level

As a clean and renewable energy source, hydropower plays an extremely important role in energy structure in the world. Since the People's Republic of China founded in 1949, the Party and the Government have paid much attention and attached great importance to hydropower planning and construction. However, due to complicated historical reasons and low productivity level, the hydropower construction in China was slow before the national strategy of reform and opening-up. After the 3rd Plenary Session of the 11th CPC Central Committee, the state readjusted the development strategy of hydropower resources, and constantly expanded the construction of hydropower. Construction of key national projects was commenced one by one. Construction of hydropower was accelerated, the average annual new production capacity was increased, and clean energy was provided constantly during the thriving socialist economic development. After entering the 21st century, construction of hydropower stations came to a new development stage, related policies were constantly improved while the development was enhanced continuously, and a new concept for developing the hydropower effectively was proposed. In September 2004, the total hydropower installed capacity exceeded 100GW, and 200GW in 2010, making

China the world's largest hydropower nation. It only took us about ten years to nearly quadruple the total installed capacity of the last 50 years since the founding of new China, and the utiliztion rate of the hydropower resource was increased to 34% from the technically feasible ratio of 10% before reform and opening-up. Now there are nearly 5,200 dams with height over 30m already built or under construction, including more than 140 dams with height over 100m. These dams and hydropower stations embody the outstanding wisdom and creativity of the Chinese people. The rapid growth of hydropower plays an important role in improving and optimizing the national energy structure, reducing emissions of greenhouse gases, and boosting the fast and sound development of national economy and society.

2.3 Technologies of Hydropower Station and Dam Construction have Reached the International Advanced Level

Science and technology is the first productive force. With science and technology as the lead, electricity development as the main battlefield, national key projects as the basis and transformation of scientific and technological achievements into project as the starting point, hydropower development and technological progress has been constantly promoted, breaking the technological, financial and environmental constraints and advancing dam construction technology. So far, many key technological breakthroughs have been made, including 200m super high roller compacted concrete gravity dam technology, 200m super high concrete face rock-fill dam technology, 300m super high arch dam, 100m super high roller compacted concrete arch dam technology, dam foundation treatment technology, and technological innovation of energy dissipater of high-velocity flow. These remarkable dam construction technologies enable the hydropower industry to be on top of China's modern scientific and technological achievements. China has become one of advanced or even leading countries in dam construction in terms of the quantity and scale, the technological complexity and the technological innovation in the world, significantly contributing to the power science and technology of China. Hydropower projects in China stood the test of Wenchuan earthquake happened in 2008, and none collapsed.

2.4 Localization Rate of Hydropower Equipment has been Raised Significantly

Since the first hydro turbine was developed in 1927, China's hydropower equipment industry has been developing for more than 80 years. Before the People's Repubilc of China was founded in 1949, hydropower equipment industry was very backward. There were not any specialized hydropower equipment plants, the output was low, capacity of the hydropower units was small, the technological level was low and the cost was high. From 1949 to the adoption of reform and opening up policy, hydropower equipment industry developed considerably, and Francis turbine, Kaplan turbine, and impulse turbine were developed independently, which basically met the needs of national economic development. After the adoption of reform and opening-up policy, with the development of an innovative nation, the innovative capability was steadily improved, and a series of remarkable achievements in scientific research were made, which effectively boosted the development and the international competitiveness of high-tech industry. Hydropower cause of China flourished, hydropower equipment developed into higher level, a batch of 300MW units were put into operation, e. g. Liujiaxia, Longyangxia, Yantan dams, and Guangzhou pumped-storage power stations; Lijiaxia (unit capacity of 400MW), Ertan (550MW), especially more than 30 units at Three Gorgers Project and Longtan dam (both 700MW) were commissioned successfully; stations under construction or to be constructed, such as Xiluodu, Xiangjiaba, Laxiwa, Baihetan, and Wudongde projects, shall generally adopt units of 700MW to 1GW. Through the practice of "technological transfer, digestive absorption and independent innovation" of projects such as the Three Gorges Project, it only took China's large hydropower equipment manufacturing industry seven years to realize a big leap that would have taken other countries by several decades, indicating that China had entered the age of large hydro generating units that have been independently designed, manufactured, and installed. Now, China has built 32 units of 500MW and above, and 109 more units are under construction, achieving the installed capacity of units, built or being built, up to 90.75GW, both among the largest in the world.

Meanwhile, with the introduction of R&D and design technology of pumped-storage hydropower units through unified bidding and combined technology and trade for three pumped-storage power stations, i. e. Baoquan, Huizhou and Bailianhe, as well as Heimifeng and Pushihe later on, the pumped-storage group equipments are increasingly localized. Xianju Pumped-Storage Power Station in Zhejiang Province is designed fully independently with its equipment supplied locally. The pumped-storage power stations have entered the age of large capacity, high head and high lift, and raised the manufacturing technology of unit equipments, which greatly promotes the construction and development of pumped-storage power stations of China.

2.5 Development of Major Hydropower Stations has Promoted the Continuous Innovation of Large Capacity, Long Distance, and High Voltage Power Transmission Technologies

With the continuous development of national economy, the operating voltage level of power grid system is constantly enhanced to cater for the needs of large capacity and long distance power transmission. China's reform and opening-up policy further accelerates the rapid development of the power industry, leading power grid construction to the age of extra-high voltage power transmission. The construction of high dams, large reservoirs and large stations promote the prosperity of large and high-voltage power grid. Direct Current (DC) high voltage power transmission lines, such as the lines for Fengman Hydropower Station (220kV), Liujiaxia Hydropower Station (330kV), Gezhouba Hydropower Station (500kV), Gongboxia Hydropower Station (750kV), and Xiangjiaba Hydropower Station (±800kV) are put into operation in succession. China has continuously increased power grid construction technology, and been a leading country in terms of Alternating Current (AC) and DC power transmission technology in the world.

More than 70% of China's coal and hydropower resources are located in the relatively underdeveloped western regions, while the economically developed but energy-hungry eastern part has a higher power load. In order to change such an unbalanced energy structure, the state implements the west-to-east electricity transmission strategy to provide large amounts of clean power for the east and promote regional interconnection. At present, the power transmission capacity of northern, central and southern channels of the west-to-east electricity transmission project has reached 60GW. In order to further meet the needs of hydropower development and power transmission of west China, the state has launched 1,000kV AC and ±800kV DC power transmission demonstration projects, which are proceeding smoothly. As the backbone grid of China, they span and connect the power grids of each region, and will play an active role in realizing inter-regional and inter-basin power transmission, the mutual supplementation between hydro and thermal power resources, and the optimal allocation of national energy resources.

2.6 Hydropower Development Brings Enormous Comprehensive Benefits to National Economic and Social Development

While providing premium power for economic and social development, hydropower development of China also plays an important role in the comprehensive utilization of water resource, boosting energy-saving and emission-reduction, improving atmospheric environment, promoting "China Western Development" campaign, developing regional economy, building new socialist countryside, flood control, navigation, irrigation, water supply, aquaculture and the like. The Three Gorges Project alone can raise the flood control standard of Jingjiang Section in the middle reaches of the Yangtze River from the once-in-a-decade occurrence frequency to once-in-a-century, greatly reducing the flood threat to people's lives and property and protecting 15 million people living in the Jianghan Plain and Dongting Lake area. Over the past century, the cumulative electricity generation of hydropower was 7,299TW, equivalent to 2.7 billion tons of standard coal, reducing emission of about 7 billion tons of carbon dioxide. Therefore, the rapid development of hydropower is a great contributor to the fast and sound development of the national economy and society.

Small hydropower stations are an important element of the hydropower construction. They have advantages in quantity, extent, and resources, and play an important role in solving power supply in remote areas. A series of projects, such as small hydropower development, rural electrification with hydropower, and substituting electricity for firewood, have addressed the power supply problems of 300 million farmers living in one third of all counties or cities in half of China's territory. The implementation of China Western Development and the west-to-east electricity transmission strategies not only accelerate the hydropower development of the western regions, but also spurs local infrastructure construction such as highway transportation, promotes the development of regional industry, agriculture, tourism, commerce, and farming, and drives the development of new countryside and townships. It plays a great role in expanding employment, boosting regional economic development, and transforming local resource advantage into economic advantage.

Today, under the guidance of Scientific Outlook on Development, China's hydropower becomes more appealing as a clean, renewable energy source. Livelihood-oriented, green, ecological, and harmonic hydropower, now with a new look, is playing an important role in China's energy development and structural adjustment.

3 Successful Experience

A smooth lake rises in the great narrow gorges, and a goddess, if there is any, has not been affected. This project will marvel at a world so changed. Over the past 100 years, several generations of Chinese hydropower people have won world recognition with their hardworking in hydropower construction, bravely creating and promoting the hydropower development of China. To summarize the Chinese glorious achievements on hydropower is to leave precious wealth for future generations.

(1) Keeping pace with the times, hydropower is developed in compliance with national energy strategic development direction. In the early days of power industry in China, hydropower was a primary industry, followed by thermal power. Then hydropower and thermal power were developed equally, and the former was promoted to suit local conditions. After reform and opening-up, the emphasis was first placed on both hydropower and thermal power and then on hydropower develop-

ment. During China's 9th and 10th Five-Year plan period, priority was given to thermal power while hydropower was developed vigorously, nuclear moderately and power grid synchronously. During the 11th Five-Year, while protecting the ecosystem, hydropower was developed in an orderly manner. From the very beginning, hydropower is always deemed as an important element of national energy strategy. In the new circumstance of adapting to climatic change and under the conditions of protecting ecology and solving relocation issues, it is very important to develop hydropower vigorously. During 2000 – 2009, China's GDP increased from RMB 9.9 trillion to RMB 33.5 trillion, the aggregate installed capacity of power generation increased from 319GW to 874GW, in which installed capacity of hydropower increased from 79.35GW to 197GW. The Scientific Outlook on Development shall be implemented for developing hydropower. Guided by the concept of national energy strategy development, hydropower and the national economy shall be developed synchronously, so as to make major contributions to building a well-off society in an overall manner.

(2) Boosting reform, hydropower has been developed in a diversified manner. Guided by the general policy of reform and opening-up and the market-oriented policy, by the end of the 1970s, China had started the reform of hydropower system and realized the transformation from the planned economy model to the market economy model by constant improvement. The former is dominated by self-management system, while the latter was developed from the transitional models after reform and opening-up, such as investment responsibility system, estimate contracting model and the like, and formed gradually three systems, i.e. owner responsible system, bidding contract system and construction supervision system, after the impact of Lubuge bidding contract system, followed by project legal person responsible system established according to the modern enterprise system. Now, China's hydropower has been developing in a diversified way instead of a single form, and the hydropower development is energetic.

(3) Scientific planning promotes the sustainable development of hydropower. Energy security is an integral part of national security. Energy resource reserve and assessment are closely related to the planning and general objective of national economic and social development. With close attention and full support of the CPC Central Committee and the State Council, China carried out four surveys and reviews of hydropower resources, and figured out the theoretical hydropower resources nationwide. Through scientific planning, careful design and construction, and strict quality management, hydropower construction on large rivers were developed preferentially, and 13 major hydropower bases were planned gradually, including Jinsha River, Yalong River, Dadu River, Wujiang River, the upper reaches of the Yangtze River, Nanpan-Hongshui River, Lancang River, the lower and middle reaches of the Yellow River, western Hunan Province, Nujiang River, northeast China, Fujian-Zhejiang-Jiangxi region, optimizing the allocation of resources, boosting the rational development and utilization of hydropower resources and promoting the sustainable development of hydropower industry.

(4) Environmental protection and relocation of local people are emphasized for realizing the harmony among hydropower development, man and nature which is the precondition for hydropower construction, and also the scientific concept of developing hydropower in China in a rational and orderly manner at present. This has been widely accepted by all sectors of society and carried out by Chinese hydropower workers to realize all-win among hydropower development, environmental protection and life improvement of relocated people. For years, the relocation work and measures for hydropower development has been improved constantly. The guidelines for relocation have experienced a change process from the arrangement migration to development migration, that is, giving the relocated people not only financial compensation but also ensuring their life and work for a long time by the late-stage supporting policies implementation in a scientific and reasonable manner, and to achieve the goal of "move out, settle down and gradually become rich". At the same time, such development resettlement shall be combined with the campaign of building new socialist countryside. Especially after entering the 21st century, on the basis of the well-arranged resettlement, more attention is paid to the late-stage supporting policy. While fully understanding and implementing the relocation policies and regulations, the developer shall develop characteristic regional and local late-stage supporting policy to help the relocated people get rich and boost economic and social development of that region. For years, hydropower workers had attached great importance to environmental protection continuously, including planning and design, construction work, site cleanup and vegetation restoration after completion. In the report to the 17th National Congress of CPC, the concept of ecological civilization development was proposed, raising up requirements of environmental protection on the hydropower construction in the new era. As is required by the CPC Central Committee and the State Council to "developing the hydropower systematically on the basis of ecological protection", the environment shall be protected in the process of hydropower development while hydropower developed in the process of environmental protection to achieve the harmonious man-nature coexistence in hydropower development.

(5) Adhering to the reform and opening up policy, the hydropower workers should not rest on their laurels and refuse to make progress while learning from predecessors and others. Western culture and technology helped China develop its own hydropower cause, including Shilongba Hydropower Station. Self-reliance and independent innovation do not mean isolation from the world. Since reform and opening-up, China has imported foreign advanced technology and equipments to narrow the gap with foreign countries. Under the new circumstance, China shall avoid arrogance and absorb the advanced experi-

ence from others.

(6) Independent scientific and technological innovation, as well as technological progress of dam construction and equipment localization, shall be promoted. Led by the strategic objective of "developing an innovative nation", and relying on key projects, China has made phenomenal progress in dam construction technology as well as equipment manufacturing technology. After years of practice, the level of equipment modernization, and the design, construction, manufacturing, installation, operation and management of hydropower construction, have reached the international advanced level. As advanced equipment and core technology were introduced into China, the leading role of key projects for hydropower technological innovation and major technology equipment innovation is fully exerted, so as to realize the equipment localization. The construction period of GW-class hydropower station is shortened from eight years to five to six years, reducing construction cost, increasing economic benefits and enhancing market competitiveness. Hydropower construction quality is considerably improved. In Wenchuan earthquake on May 12, 2008, buildings in many cities collapsed, but all the hydropower projects have survived with zero collapse and secondary disaster did not happen. The high, thin roller compacted concrete arc dam of Shapai Station, located in the epicenter, remained sound. Zipingpu Hydropower Station resumed operation on the 5th day after the earthquake. The quality of the stations and dams withstood the severe tests.

4 Current Situation and Prospects

Today, the global climate change, ecological damage and energy shortage are having a deep impact on the survival and development of human society. To reduce the fossil energy consumption, it is the shared mission of all humans to develop clean energy vigorously, prevent global warming and save the earth. In 2009, the Chinese government promised to "make efforts in increasing non-fossil energy to 15% of total primary energy consumption and reducing by 40%-45% carbon dioxide emission per unit of GDP from that in 2005 by 2020". In China, hydropower is the second major conventional energy resource only after coal, and is the cleanest and most economical type of energy with the most defined resource and the most mature technology among the renewable and non-fossil energy resources so far. With increasing pressure and responsibility of cutting carbon dioxide emission, hydropower can play a more and more important role in reducing carbon dioxide emission and developing low carbon economy. Hydropower resource of China ranks first in the world, and the technologically feasible capacity is 542GW. Calculating by 185GW of conventional hydropower at present, the utilization ratio of hydropower is only around 34%, far lower than the average level (60%-70%) of developed countries. With the accelerated construction of nuclear power plants and the fast development of wind energy, the peak-valley difference in the operation of regional power grids is increasing, calling for pumped storage power stations to eliminate such difference and ensure safe operation. This is also the new requirement and opportunity for hydropower development. According to the latest planning, in order to realize the energy-saving and emission-reduction objective in 2020, the installed capacity of hydropower must reach 380GW, including 330GW of conventional hydropower and 50GW of pumped-storage power generation. Only when we are determined to develop and utilize hydropower in an orderly manner can the energy structure of China be improved, the commitment of energy-saving, emission-reduction and non-fossil energy development for 2020 be achieved, and the sound and sustainable development of national economy and social cause be realized.

In the 21st century, the supply of safe, reliable and clean energy has become an important guarantee for the sound and sustainable development of both the economy and the society. Standing at the new starting point of the development of new energy and renewable energy, the hydropower people shall get a clear understanding of the situation, seize the opportunity, address the environmental protection and relocation challenges, seek unity in thinking and work uniformly to enforce the national policies and laws for the ecological environmental protection, relocation of local people, farmland protection and new socialist countryside development, assume social responsibility actively, and strive for developing hydropower into livelihood-oriented, green, ecological and harmonious projects, so as to benefit the human society and future generations.

(1) To implement scientific outlook on development to maintain social harmony with harmonious hydropower. According to the general requirements of "building socialist harmonious society" proposed at the 16th National Congress of CPC, hydropower people shall further learn and implement the scientific outlook on development, change their minds, improve the cognition level, develop ecological hydropower instead of simple engineering hydropower, and construct social projects instead of pure technological projects. More attention shall be paid to the interests of relocated people and ecological environmental protection, hydropower development be combined organically with comprehensive utilization of water resource, development of ecological engineering and regional economic development so as to ensure the sustainable development of China's economy and energy while maintaining the development of harmonious socialist society with the harmony of hydropower.

(2) To put forward the concept of ecological civilization and promote ecological engineering of hydropower development. While developing hydropower in a rational and orderly way, ensuring power supply and realizing the medium – and

long-term hydropower development strategies of the state, the new concept of "ecological civilization development" proposed in the report of the 17th National Congress of CPC must be taken as the center, the fundamental principle be followed of "protecting the environment in the process of hydropower development while developing hydropower in the process of environmental protection", environmental protection be enhanced in the process of hydropower construction, measures of environmental protection be improved, and hydropower development be combined in a real sense with environmental protection and ecological development to realize the integrated development of basin-wide water resources, beautiful ecological environment and the coordinated development of man and nature with scientific and orderly utilization.

(3) To attach great importance to relocation of local people and social stability, and construct livelihood-oriented projects in the process of hydropower development. People's livelihood are one of the important issues addressed in the report of the 17th National Congress of CPC. Relocation is most important for hydropower development, which can proceed smoothly only when stability is achieved among the relocated people, and which can prosper only after they get rich. The principles of people first and harmonious development shall be further adhered to, hydropower developed in an orderly manner on the basis of environmental protection and well-arranged relocation, the work be enhanced in relocating local people and maintaining social stability, the late-stage support policy be studied and innovated, and the benefits be combined between hydropower and the relocated people, so as to win support of hydropower development from the relocated people, promote the construction of new socialist countryside, boost the coordinated development of local economy and society, maintain social stability and consolidation, and focus on developing livelihood-oriented projects in hydropower development.

(4) To strengthen the guiding of public opinion and create a sound social environment for hydropower development. Currently, there are some public opinions opposing to dam and hydropower construction. The advantages and disadvantages of hydropower shall be analyzed, pursuing the advantages while avoiding the disadvantaged, problems corrected seriously, public supervision accepted and the public guided to understand and judge the hydropower objectively at the same time. It is an important task for the hydropower construction under the new situation to work on public opinion and create the sound social environment. The scientific outlook on development shall be taken as the guideline to conduct scientific, systematic, comprehensive and active publicity of China's hydropower construction, with a focus on the outstanding social and economic benefits of the major, jumbo backbone hydropower projects, in order to answer questions about hydropower development, improve the public image of hydropower, create good public opinion and social environment, and prop up sustainable economic and social development with the healthy and sustainable development of hydropower.

(5) To deepen basin-wide cascade development and unified regulation and realize the overall benefits of hydropower development. The experience of basin development shall be carefully summarized, the policy of basin-wide cascade development shall be upheld, and basin cascade regulation with unified and scientific control shall be realized. The goal is to increase and maintain the economic and social benefits and ensure the safe operation of power grid. While addressing the interests and conflicts between power generation and water conservation, among different sectors, between upstream and downstream, among neighboring provinces and between station and power grid, efforts shall be accelerated to study and promote basin cascade unified control so as to realize the integral, unified control of cascade power station clusters of all major basins in the country, maximizing the overall economic and social benefits of hydropower stations and better serving China's economic and social development.

(6) To improve the hydropower construction systems and mechanisms constantly and promote the healthy and sustainable hydropower development. Faced with the new situation and new tasks, it is important to focus on such systems and mechanisms that affect the further development of hydropower. Rules and regulations shall be established and improved for hydropower development. *The Regulations on Management of Hydropower Development* shall be formulated to clarify the requirements related to hydropower development and further regulate hydropower development. Procedures and measures for managing hydropower construction shall be improved to clarify the requirements for different stages such as pre-project, checking for approval, and construction, while the checking and approval systems shall be established that conform to the requirements of investment system reform and the reality of hydropower construction. Study shall be strengthened on the formation mechanism of hydropower tariff and fair pricing mechanism while actively promoting equal feed-in tariff and encouraging price competition for both hydropower and thermal power in accordance with the objective and direction of electricity price reform and the requirements of socialist market economy. Policies related to hydropower development shall be studied and improved to better combine hydropower development with the prosperity of relocated people and the local economic and social development, so as to stimulate the local economy and benefit the local people from the hydropower development.

Reviewing the Chinese hydropower development with the past 100-year history, we are inspired and proud of its brilliant achievements; looking into the future development, China's hydropower development and construction will be more healthy and orderly with a bright future. There's still a long way for Chinese hydropower people to go, requiring more intelligence, wisdom and innovation. We shall hold high the great banner of Chinese characteristic socialism and implement the spirits of the 17th National Congress of CPC and the Third and Fourth Plenary Session of the 17th CPC Central Committee and the national energy development policies under the guidance of the Scientific Outlook on Development. We shall commit ourselves to the cause of new and renewable energy development such as hydropower, developing and innovating hydropower science and technology and contribute to boosting the sound and sustainable development of hydropower, realizing the energy-saving and emission-reduction objectives and promoting the sustainable economic and social development!

Large Dams and Harmonious Development
—Practice and Exploration in China[*]
Jiao Yong[❶]

More than 2,000 years ago, the Chinese nation already began to build water conservancy projects, including the world-famous Dujiangyan Project, and to realize the purposes of mitigating the disasters and acquiring benefits from water projects. In China, like many other countries in the world, along with the social progress and development, our understanding on the functions and roles of dams and reservoirs has gone through a development process from aiming at single function to diversified functions, and from focusing on socio-economic function to incorporating the ecological function into the targets of major project construction and operation. Now, dam construction has become an indispensable part of China's socio-economic development. Its influence and position in the coordinated development of economy, society and environment are increasingly remarkable.

1 *The Three Gorges Project on the Yangtze River—Practices on Enhancing the Integrated and Coordinated Development of the River Basin*

The Three Gorges Project is a successful example of enhancing the integrated and coordinated development of the whole river basin through relying on the powerful function of a huge hydro project.

Firstly, the Three Gorges Project has achieved the greatest importance in flood control function in the world. After completion, it will directly protect the fertile Jianghan Plain and important cities like Wuhan by depleting a 100 – year flood to a 10 – year flood through the huge reservoir. In case of a 1,000 – year flood event, the downstream major cities and regions can also be protected from suffering catastrophic disasters by jointly operating the flood diversion and detention projects. Secondly, the Three Gorges Hydropower Station has the world's largest installed capacity of 22,500MW, providing enormous clean energy to Central China, East China, South China and Chongqing city and thereby promoting the sustainable development of economy and society in the vast region of the country. This will mean a reduction of burning 34 million tons of coal, and consequently a reduction of emission of 74.8 million tons of carbon dioxide and 680 thousand tons of sulfur dioxide. Thirdly, the Three Gorges Project has the world's largest navigation ability, which will greatly improve the navigation capacity of the mainstream of the Yangtze River, famous for the "Golden Waterway". The annual shipment volume can be increased up to 300 million tons and the transportation cost reduced by 35%–37%. Fourthly, the Three Gorges Project with its huge storage capacity can supply water to the middle and lower reaches during extremely dry seasons to mitigate the water demand conflict between socio-economy and ecosystem. The Yangtze River runs across vast region from West to East China, and the Three Gorges Project having huge functions in flood control, power generation, navigation and water supply has ensured the safety of the vast region along the middle and lower reaches, thereby playing a significant role in promoting the coordinated development of West, Central and East China.

2 *Xiaolangdi Reservoir on the Yellow River—Practices on Restoring the River Ecology*

The water quantity of the Yellow River accounts for 2.5% of the national total, but it provides water supply for 8.7% of the total population, 12% of the total irrigation area and 7% of the total economic output. The contradiction between water supply and demand is more and more serious. The Yellow River is scarce in water resources but plentiful in sediment amount, causing serious sedimentation and becoming a world famous suspended river. As a result, the economic growth of the river basin area maintained at the sacrifice of damaging ecosystem. In the 1980s and 1990s of the 20th century, the Yellow River suffered from downstream river desiccation almost every year. In 1997, the river desiccation at the downstream reaches continued for 226 days, resulting in serious shrinking of the river and essential loss of river ecological function; this has aroused worldwide concern. After the completion of Xiaolangdi Hydro Project on the Yellow river, the Chinese government has set a high requirement for the reservoir operation, targeting at "neither river desiccation nor river bed aggradation". The Project should not only play the roles of flood control, power generation, water supply and irrigation, but also play an important role of restoring the ecological function of the river channel. Since 1999, unified water regulation

---

[*] This paper is the keynote presentation at the 23rd Congress of ICOLD on 24th May, 2009.

[❶] Jiao Yong, Vice Minister of Water Resources and Vice President of Chinese National Committee on Large Dams, Professor Senior Engineer.

has been conducted for the Yellow River. Experiments of water and sediment regulation have been carried out on the Yellow River by artificially creating large river flow from Xiaolangdi Reservoir. Such experimental practice has effectively flushed away the riverbed sediment in the lower river channel and also has resulted in an increase of the minimum floodway capacity in the lower river channel from the originally less than $1,800 m^3/s$ to $4,000 m^3/s$. This has led to: improving the reliability of safety flood discharge in the lower river channel, ensuring no desiccation of the Yellow River for nine consecutive years, recovering the ecological function of the river, and improving the ecosystem of wetland at the Yellow River Delta area. In addition, the capability for ensuring water supply to domestic and production use has also increased by a big margin.

3 Xin'anjiang Reservoir and Ertan Hydropower Station—Practices on Building Eco-environmentally Friendly Projects

The harmonious development of dams and river basins has been given wide concern by the international dam community. The important tasks now facing the dam constructors are how to minimize the adverse impacts of project on environment while building environmentally friendly hydro projects, and how to create harmony between water and human-being and improve the ecosystem while displaying the function of reservoir in creating new environment.

Xin'anjiang River is one of the important rivers flowing through Anhui and Zhejiang provinces in south China. The precipitation is abundant but uneven in spatial and time distribution, causing frequent flooding in rainy season on the one hand and water scarcity in dry season on the other. For this reason, in the 1960s the central government decided to build Xin'anjiang Reservoir to solve the issues of flood control, power generation and water supply. The completion of this reservoir does not damage the river ecosystem. On the contrary, a huge artificial lake has been formed and it was named Thousand-islands Lake after the 1,078 islands have been created in the reservoir. The reservoir storage capacity is 21.6 billion m^3, and the installed capacity of hydropower station is 662MW, capable of playing significant functions of power generation and flood control. Since the completion of the project, the lake has maintained very good water quality for several decades and has become a famous National 5A-Class tourist resort. Currently, for the purpose of coordinating the relations between development and protection of Thousand-islands Lake, Xin'anjiang River Basin becomes one of the three pilot basins selected by the government for joint investment of development and joint share of benefits, and the work is proceeding actively.

Ertan Hydropower Station is located on the lower reach of the Yalong River in Sichuan Province, Southwest China. When completed in the 20th century, it was China's largest hydropower stations in terms of total installed capacity of 3,300MW. The 220m high concrete double-curvature arch dam ranks the third in the world. During the various stages of construction, we persistently stick to the target of coordination and harmony between ecological environment and social-economic development in the river basin. After the commissioning, the hydropower station has not only played important roles in power generation, flood control, navigation and tourism, but also significantly supported the local economic and social development. Remarkable eco-environmental benefits have been achieved and great improvement in the regional ecological environment has been obtained as compared to the pre-construction period. The World Bank has recommended this hydropower project to other developing countries as an outstanding example using the World Bank's loan. Since this hydropower project has realized the harmonious development of the river basin and the region, and has been constantly providing environmental protection and improving the basin ecology, the project won the highest China's environmental protection award "The National Environmentally Friendly Project" in 2006, and become a milestone for the water resources and hydropower projects.

4 Zipingpu Hydro Project—Practices on Constructing and Operating of Safety Dams and Reservoirs

Zipingpu Hydro Project is located on the upper Mingjiang River. The project construction commenced in 2000 and operation started in 2006. The Project with a storage capacity of 1.112 billion m^3 is mainly for flood control, water supply and power generation. The safety was given the highest priority during construction and operation. The Wenchuan Strong Earthquake of May 12, 2008 with a magnitude of 8 in Richter scales shocked the whole world. Zipingpu Project at only 17km away from the epicenter suffered great damages. After the earthquake, a number of large horizontal and vertical deformations were found in the concrete face rockfill dam and the power station was shutdown. However, the dam had withstood the serious test of strong earthquake, becoming the first higher than 100m CFRD experiencing Magnitude-8 strong earthquake and maintaining safety in the world.

After the earthquake, a series of emergency measures were adopted to ensure the safe operation of the station. Firstly, about 10 minutes after the earthquake, the operation staff of the power station opened the idling units to provide emergency water supply to the lower reaches. Secondly, the bottom outlet holes and sediment flushing holes were put under emergency repair and they began to release water in 28 hours after the earthquake to maintain no rising of reservoir water level for guaranteeing the safety of the dam. Thirdly, the power station was under emergency repair and started commissioning again in 5 days after the earthquake for supplying electric power to the earthquake disastrous areas. Fourthly, the face slab and seepage control facilities of the dam were timely repaired to maintain the overall stability and safety of the dam.

The safety of Zipingpu Dam assured the confidence of more than 10 million people living downstream and guaranteed the water and electricity supply during the post-disaster period. A rescue transportation route on the reservoir has been formed, which had played an irreplaceable role in timely shipping rescue team and materials into and evacuating people from the earthquake-stricken areas.

The lesson learned from Wenchuan strong earthquake in the dam construction is that: large dams, especially dams over 100m high, should be handled by optimized design, careful construction, stringent quality control, strict management so as to ensure safety in all cases. Only the safety of the dams can reassure the people to have confidence and to eliminate panic, and this is imperative in recovering the normal livelihood and in supporting the post-disaster production.

The post-disaster reconstruction work of Zipingpu Project is still ongoing and will be completed by September 2010.

5 Unified Regulation of Reservoirs in the Pearl River Basin—Practices on Integrating the Management of Water Resources of a River Basin

The Pearl River Delta Region, neighboring Hong Kong and Macao, is the most densely populated and economically developed region in China. During the rapid economic and social development of the Pearl River Delta Region, such various factors as upstream inflow reduction, water consumption increase and climate change, have contributed to the decrease in the estuary runoff year after year, resulting in the rise of seawater level, the continual decrease of water flow in the river channels and estuaries, and the frequent and serious tidal intrusion. All these lead to severe problems of water supply security and estuarine ecology. However, the water crisis in this Delta Region cannot be solved by the regulation of local water resources on its own, and it must be settled by integrating the management of the water resources of the whole river basin.

In order to ensure the water supply security in the Pearl River Delta Region, the National Headquarters on Flood Control and Drought Relief implemented the urgent measure of water transfer to increase the river flow for preventing tidal water intrusion in the estuarine areas. This is realized by joint operation of eight upstream reservoirs of the whole river basin. The water transfer involves the West River and North River tributaries totaling 1,336km, flowing across Guizhou Province, Guangxi Zhuang Autoromous Region and Guangdong Province. The water transfer quantity amounted to 843 million m^3. All these have ensured the drinking water safety for 15 million people in the Pearl River Delta Region and Macau Special Administrative Region. The enterprises affected by the tidal intrusion can quickly recover their normal operation. In addition, 230 million m^3 of water body in the Pearl River Delta network has been replaced; hence the water environment has been remarkably improved.

In recent years, China has targeted the protection of ecology, implemented scientific and rational regulation of water projects, and restored the river ecosystem. There are many other successful case studies in maintaining the healthy rivers, including emergency water transfer to Taihu Lake, the Tarim River, the Heihe River, and the Zhalong Wetland, etc.

6 Further Thoughts

From the case studies I have just mentioned, we can clearly see the multi-functional roles of dams in the socio-economic development of human being. With the continual growth of population and economic development, the human being has found not much space for the selection of habitation. As we all know, the regions along the rivers are the best places for human being to live and to develop. However, with the increase of population growth and total economic output, the threat from flood disasters is getting more and more severe, whereas the pollution caused by human activities is also getting more and more intensive. Facing the increasing size of population and economy, the rivers that can satisfy the demand of human being through their natural regulation are becoming less and less. Most rivers need construction of dams to achieve artificial regulation and this is where the reservoirs can play such a function. Of course, reservoirs will bring some negative impacts. The most obvious one is the impact on ecosystem. Therefore, this necessitates us to establish a new thought, to use advanced scientific and technological methodology, and to achieve optimized operation and management of reservoirs. All these are to realize harmony between human and nature, to promote harmony between human and human, to resolve the conflicts of water demand among the regions. By so doing, we can maximize the benefits of human being from the reservoirs and at the same time we can minimize the adverse impact of reservoirs, for enabling the rivers to create forever benefits for us, generation after generation.

During dam construction and operation, China and many other countries in the world will encounter many problems and difficulties. We are very much concerned over several key issues and are now conducting researches on solving these issues. They include, for instance: The construction of dams at the upper river reaches will cause changes in the relation between the river regimes of the middle and lower reaches, and also changes in the relation between rivers and lakes located downstream. The reservoir impoundment may possibly trigger geological disasters in the reservoir areas. The zone where the reservoir water level fluctuates may affect the effective protection of the ecological environment. The reservoir operation modes and joint operation of all cascade hydropower stations have to aim at protecting ecological environment on a dimension of a whole river basin. The policy of compensation has to be decided upon to tackle the river basin ecology affected by the socio-economic development carried out within the river basin.

Like Switzerland and other countries, for the sake of basin health, China has also promoted the development of "green hydropower" and started relevant work at the giant Nuozhadu Hydropower Station on the Lancang River. Besides, China has practiced very stringent measures to protect the water resources of the headwaters of the three major rivers of China.

Currently in China, there are a large number of world-class dams with significant scale and functions under construction or under planning. We are fully aware that China still has a long way to go to achieve the harmonious development of dams with the development of society and economy. However, we are fully confident and are looking forward to conducting more extensive and sincere cooperation in future with all the other countries in the world.

Thoughts on Developing Water Projects in a Man-Nature Harmonious Way[*]
Liu Ning[1]

1 New Ideas for Water Project Construction

Since the founding of new China, water conservation and hydropower have experienced fast track development and made remarkable achievements. In recent years especially, the Chinese government has substantially increased input to the water cause, and China has consequently embraced a period of water project construction with an unprecedented scale, which has greatly promoted water conservation development. Generally speaking, current water engineering practices exhibit the following salient features:

(1) Flood control projects are built to guarantee the security of economic and social development.

(2) Water projects aimed at satisfying people's needs are extensively carried out.

(3) Water projects in rural areas are augmented as a support of new rural development.

(4) A water-saving society is further boosted.

(5) Hydropower construction is accelerated.

(6) Much attention has been paid to the construction of water diversion projects.

Modern water projects in China combine traditional water conservation concept with innovative ideas. Decades of experience, rich theoretical basis and modern technologies are closely linked to meet the needs of sustainable development. While tremendous achievements are made, there emerge some unusually tricky, difficult issues. At present, we need to further strengthen our studies on dam construction technologies including stress and stability of high dam-rock foundation system, high dam seismic dynamics and anti-seismic measure, high dam concrete material, high dam hydraulics and hydromechanics. More studies are needed on ecological evaluation and appraisal of water projects, compensation technologies and mechanisms of river ecological impact assessment index system and assessment method, water project ecological protection guidelines, dam environmental protection standard and certification methods. Some of the projects are super large in size and impose great technological challenges, which in turn require more project experience, more advanced experimental and testing technology and algorithms. At the same time, given the restrains of terrain, geology, earthquake, water resource availability, social and economic development status, new technological breakthroughs are needed, and decisions on setting the objective of water projects need to be made in a more scientific manner. The ultimate objective is that water projects will have to be constructed, utilized and managed in such a way as to achieve the synergy created by human-water harmony. There's no doubt that China has acquired advanced technologies of hydropower construction under various complex conditions, and achieved unprecedented results in flood control and disaster reduction, hydrology and water resource, water ecology and environment, rural water conservation, hydraulic structures and materials, soil and rock mechanics, project construction, information technology, and so on, which have greatly boosted hydro science-technology progress in water development.

These achievements also manifest that all engineering activities are done with a certain engineering philosophy as guidance. Apart from technological progress and economic benefit, water projects must lay emphasize on their environmental benefit, and respect social morality, ethics, justice and equality. A rightful engineering philosophy must be based on respecting objective natural, economic and social laws. Implementing the Scientific Approach to Development and developing a proper engineering philosophy, it should be applied to the planning, design, construction, operation, management and maintenance of water projects. It is crucial to the sound and sustainable development of water resources and hydropower.

After disinterested analyses of the pros and cons of water project construction, some intellectuals home and abroad have brought up a lot of new concepts and ideas that aim to make social benefits and environmental soundness compatible in water projects construction. Since the 1960s, a significant amount of studies have been carried out by developed countries on the impact on terrestrial ecosystem and species diversity, the impact of runoff variation on aquatic ecosystems and species diversity, the impact of closure and storage on the fishery ecosystems upstream, reservoir area, downstream, and the river

---

[*] This article was published on China Water Resources, Issue 4, 2007.

[1] Liu Ning, Vice Minister of Water Resources in China, Professor Senior Engineer.

mouth, damming-induced secondary environment effect, assessment and prediction of water projects impact on the ecosystem, and measures for reducing their impact, etc.

Internationally, the academic discussions and cooperation surrounding water project construction are becoming more in-depth and frequent. In 1973, "Damming Impact On the Environment" was for the first time introduced as a special issue at the 11th International Conference on Large Dams, and remains as an important discussion in subsequent meetings. Such concepts or theories as "green hydropower" and "green dam" are put forward, stressing that water project construction and utilization shall go hand in hand with ecological and environmental protection so as to realize the twofold benefits for both ecological protection and economic development.

Some countries have developed certification procedures and criteria to reduce the negative impact of water project, such as Low Impact Hydropower Certification of the U. S., and Green Hydropower Certification of the Switzerland. It is intended to establish objective, scientific and fair eco-environmental verification criteria and market incentives through comprehensive assessment and effective management of hydropower impact on the eco-environment, so as to minimize the adverse impact of dams on eco-environment. The United States was the first to release the National Environmental Policy Act in 1970, specifying environmental impact assessment of large construction projects. In 1996, the Federal Energy Regulatory Commission (FERC) demanded new reservoir operation programs to mitigate eco-environmental impact during its review and approval process of reservoir operation. The programs included increasing minimum discharge volume, expanding or improving fish way, periodic high-flow discharge, and ecological protection measures of land-based ecosystems. Some European countries and Japan also promulgated laws or measures to raise environmental protection requirement in water project construction and to minimize eco-environmental impact. In 1985, the World Bank stipulated that all the dam projects directly or indirectly financed by the Bank must conduct environmental impact analysis.

After several decades of construction and reflections, like many other countries, China has made breakthroughs in the related research. In 1982, the Ministry of Water Resources issued *Provisions on Several Issues Concerning Environmental Impact Assessment of Water Conservation Project*; the central government promulgated the *Environmental Protection Law* in 1989; subsequently *Specifications on Environmental Impact Assessment of Water Projects*, *Specifications on Environmental Impact Assessment of River Basin Planning*, and *Technical Guidelines for Environmental Impact Assessment* were issued. In 2003, the *Law on Environmental Impact Assessment of P. R. C* was enacted, which stipulated environmental impact assessment must be carried out during river basin development planning. In newly revised *Water Law of the People's Republic of China*, water resource conservation and protection were emphasized and explicit provisions were made on water resource planning. In recent years, according to the requirements of sustainable development of water resources, China has attached great importance to developing water projects that are aimed at a harmonious coexistence between man and nature. In practice, we have actively made our efforts to further improve project construction, management and regulation, and conducted a series of work, e. g. regulating project construction approval procedure, reform on water management system, and water eco-environmental protection. Remarkable achievements have been made in using water projects regulation to improve the eco-environment. For example, Yangtze – Taihu Water Diversion pilot project launched in 2002 has considerably improved the water environment of the Taihu Basin. Water authorities in Jiangsu Province improved its water environment through rational regulation of water project and river networks. Water ecology was restored through integrated regulation of water projects in the Haihe River Basin. Meanwhile, many creative achievements on ecological water demand, river ecosystem health indicator system, eco-environment impact assessment method of water conservation project, cumulative ecological effects of the river basin cascade development, eco-water engineering, and river restoration and compensation measures.

2 Current Positive and Negative Views on the Water Project Construction

2.1 Main Positive Views

Due to climate and geographical influence, the precipitation in China is unevenly distributed both spatially and temporally. For the purpose of survival and development, building water conservation projects to bring benefits and minimize water disasters have been an essential statecraft throughout China's history. In the 21st century, socio-economic development renders a greater demand of water and is more dependent on water engineering. It's believed that water and hydropower construction must be further strengthened so as to meet the needs of flood control, water supply, food production, electricity generation, and navigation, etc. To summarize, positive views of the water cause include:

(1) Flood control project can effectively reduce losses incurred by floods and waterlog. Flood is considered to be the top natural disaster in the world in terms of frequency, economic loss and mortalities. In China, the area directly threatened by flood and waterlog is about 800,000km^2, where the GDP accounts for 70% of national total. In six years of the 1990s, the major river basins in China suffered heavy floods. Flood control projects such as reservoirs and dikes can effectively regulate

and control flood, and reduce loss of life and property. Taking the benefit of flood control of reservoirs in 2003 for instance, reservoirs and dams throughout the nation stored 44.7 billion m³ of water, spared 110 million people and 3.35 million hm² of land from being affected, prevented 445 cities above the county level from flooding and generated direct economic benefit of up to 113 billion RMB.

(2) Irrigation and water supply projects can guarantee water security for industry, farming and residential use. China has about 56 million hm² of irrigated area, and can supply 645.9 billion m³ of water annually. More than 100 large and medium-sized cities depend mainly or completely on reservoirs. Water security has been substantially improved. However, water resources per capita are low; precipitation is unevenly distributed spatially and temporally, being heavy in summer and few in winter, more in the east and the south and less in the west and the north. Economic growth and domestic water use are thus seriously constrained. Statistics show that since the founding of new China, the annual average drought affected farmland exceeds 21 million hm², and annual average grain loss reaches more than 14 million tons. Irrigation and water supply projects can optimize the allocation of water resource spatially and temporally, ensuring good and stable water supply for industries, farmland and residents.

(3) Hydropower produces rich and clean energy. Since the reform and opening-up, hydropower in China has developed rapidly. The installed capacity in 2004 exceeded 100 million kW, ranking the world top. By the end of 2006, the installed capacity amounted to 128.57 million kW, accounting for 20.67% of the national total; the electricity generated was 416.7 billion kW·h, accounting for 14.70% of the total. Hydropower, as a relatively low-cost energy, helps reduce our reliance on fossil energy, and makes a premium renewable energy source. As a major basic industry in a hydro-rich country, hydropower plays a positive role in utilizing and protecting land resources, improving eco-environment, and realizing socio-economic sustainable development.

(4) Water projects have augmented navigation. Using natural and artificial rivers channels, inland river navigation has such advantages as large transport capacity, low cost, low energy consumption and less pollution, and is hence prioritized in modern transportation development. China has abundant inland navigation capabilities; there are 123,000km of navigable waterways, over 8,600km of which have a shipping capability exceeding 1,000 tons. Artificial canals connect once isolated waters, thus expanding shipping range and capacity. Reservoirs can raise the water level and widen channels in the reservoir area, reducing flow velocity, increasing water depth and curvature radius, minimizing some rapids and shoals unfavorable to navigation. Similarly, river training can control and stabilize river regime, and serves as an important means for navigation security and development.

(5) Water project promotes the prosperity of aquaculture and tourism. Reservoirs and other water projects can form large expanses of water, which is particularly favorable to aquaculture. Many reservoirs have become major aquatic produce hubs. Reservoir areas boast nice environment and fresh air, constituting ideal places for recreation. For example, Three Gorges Reservoir and Xin'anjiang Reservoir have become famous tourist resorts.

(6) Water projects are conducive to local economic development. Water projects can not only generate enormous social and economic benefit from flood control, power generation, irrigation, navigation, water supply, aquiculture and tourism, but also provide plenty of job opportunities to promote local economic growth and welfare of the people. In many regions rich with hydro resources, the local appeal for faster hydropower development from both the government and citizens is very strong; they expect hydropower development could help spur on the local economy and lift them from poverty.

2.2 Main Negative Views

As the awareness of environmental protection increases, people pay unprecedented attention to the eco-environmental impact of large water conservation and hydropower projects, and many begin to reflect upon the advantages and disadvantages of water projects. Since the 1980s, heated debate regarding dam projects has been raised across the world. In recent years, with the construction of the Three Gorges Project and South-to-North Water Diversion Project, the accelerated hydropower development in the southwest, and negative impact from certain water projects such as Sanmenxia Water Control Project, domestic debate and discussion on water projects have become ever intensified. Quite a number of experts and scholars call for addressing the negative impact of dam projects on eco-environment, climate, and resettlement, and strongly advocate preserving the original ecology of river. They oppose to developing hydropower and water project as it is believed that the pros outweigh the cons. The negative views can be generalized as follows.

(1) Water projects construction could cause deterioration of the river ecosystem. River is the carrier of material flow, energy flow, and information flow of the ecosystem. River continuity refers to not only hydrological continuity of water flow, but also the continuity of nutrient transportation, biotic community, and information flow. The construction of water and hydropower projects, such as river dredging, river channel training, damming and large-scale water diversion and son on, can block the natural river channel, obstruct material and energy flow of the basin, and change watershed habitat and pri-

mary productivity of the basin, which will in turn bring irreversible effect to biotic community structure, function and biodiversity, and cause aquatic ecosystem diversity deterioration and even massive species extinction. Once these negative impacts become excessive, it will then cause river variation, and is likely to affect the basic function and sustainable utilization of the basin, and jeopardize the living environment for human and other lives.

(2) Water projects could have an influence on climate, geology, and soil, etc. After a reservoir is built, the original land becomes water body or wetland. Consequently, the amount of rainfall and the spatial and temporal distribution of rainfall are changed, and the average temperature will go slightly higher and result in local climate change. Water projects may induce geological disasters like earthquake, bank collapse, landslide, etc. Water body weight causes increased lubrication between earth crust stress and fault, which is prone to induce earthquake. According to statistics, reservoir-induced earthquake occur in 66 high dams in the world. Water level rises after reservoir impoundment, and reduced shearing resistance of bank slope soil mass make it vulnerable to collapse, landslide, and destabilization, bank collapse, and landslide of risky rock mass. In addition, impoundment causes land inundation, swamping, and salinization in the reservoir area.

(3) Water projects could change hydrology, sediment, and water quality. Water projects change the flow downstream, causing significant drop of water level and even drying-up. The groundwater level of the surrounding area drops too, and natural lakes or ponds downstream become dry because of water supply cutoff. Sedimentation occurs in the estuary because of reduced water flow, which leads to salinization or sea water encroachment. Change of hydrologic conditions has an adverse impact on flood control, water supply, and navigation. The rising water temperature in reservoir and lowered flow velocity reduce water-air interface exchange rate and the migration and diffusion capability of pollutants, so that self-clean capability of the water body is weakened, and that ditches and branches of reservoirs are more vulnerable to water pollution and worsened water quality.

(4) Water projects result in a large number of immigrants. Water projects, especially reservoirs have to flood much land and inevitably lots of people have to relocate. The relocated people's living space will be changed considerably, which might trigger social instability. According to statistics, since the founding of new China, there are more than 18 million reservoir-caused immigrants, ranking top among all nations and regions in the world. The Three Gorges Project flooded 632 km² of land area, and resulted in up to 1.1 million immigrants.

(5) Water projects have an impact on cultural relics and biological species. China has a long and rich history, having numerous cultural relics and historical sites. Reservoirs and water diversion projects may flood or destroy some relics and sites, and cause heavy losses to historical research and preservation of cultural relics, especially the Three Gorges Project and South-to-North Water Diversion Project. Reservoir impoundment inundate virgin forests, diversion by culvert cause drying-up of riverbed, large-scale project construction usually destroys surface vegetation, divide and encroach on wildlife habitat, threaten their survival, and aggravate extinction of species.

(6) Defective and dangerous projects impact the safety of people's life and property. Due to historical reasons, many water projects suffer from potential quality issues and unsafe factors. Also, since the engineering conditions and the environment have undergone great changes over the long years, a number of water projects are found to suffer from aging and defects. According to statistics in 1999, dangerous reservoirs accounted for 36.3% of the total, and many were located upstream of the cities (towns), posing great threats to the urban area. There are 543 large and medium hazardous reservoirs located in towns at or above the county level, and 178 in cities above the county level, involving 146 million people and 8.8 million hm² of arable lands. The security of many major traffic arteries, factories, mines, enterprises and military communication infrastructure are threatened by such dangerous projects, which have high stakes in people's lives and property. It has therefore drawn extensive attention from the public.

3 Thoughts on Man-Nature Harmony-Oriented Water Projects

3.1 Put into Perspective the Role of Water Projects

Water projects are indispensable and irreplaceable infrastructure and public utilities. By making rational use of natural resources, water projects make it possible to realize coordinated development of man, nature, and society. It is necessary to establish a correct and scientific perception about water project. Water projects are an important leverage that support and push scientific development, serve as one of the important means to evaluate and implement the scientific development, and constitute a vital cornerstone of a harmonious society. Being an important component of the engineering system, water projects share common features of an engineering project, but also serve their unique functions. Compared with other kinds of infrastructure, water projects are more closely related to ecology and nature, and need be guided by a more scientific and coordinated development manner. Therefore, water projects should be built in consideration of both the value of rivers and that of human development need. To realize socio-economic sustainable development, it is imperative to shift from unrestrained claim for nature to harmonious coexistence with nature, from wasting water resources, polluting the environment to conserving resources and protecting the environment.

(1) Water projects should be the bridge between human and water toward harmonious coexistence. The relationship between man and nature, and between man and water is essential during the development of human society. We cannot pay attention only to the development and utilization of water resources while ignoring the damage to the eco-environment. Similarly, we should not fall into absolute nature protectionism. Water projects manifest man's proactive endeavors to transform nature, by which man directly interact with water. The Dujiangyan Irrigation Project built 2,260 years ago is considered as an excellent example of man-water harmonious development. On the one hand, the project makes good use of the Minjiang River, irrigating significant areas, and on the other hand, it can preserve the river's natural function and environment. However, there are cases in which water projects jeopardize the man-water relationship. Such projects destroy the water eco-environment and consequently have an adverse impact on human sustainable development.

Under the background of the foundetmental values transformation on the economic development in China, which has already shifted from only pursuing economic growth to developing economy and society in a comprehensive, coordinated, sustainable manner, the perspectives on water project construction has changed significantly, with a special emphasis on restraining irrational human activities to explore disciplined development mode on the basis of respecting for the nature lawsand. Project construction undergoes a rational transition from "man can conquer nature" to "man-nature harmony and man-society harmony". Guided by such a concept, today's water projects should combine the development of economic society with natural ecological protection, and should be built in line with the eco-environment carrying capacity. Only when immediate interests are combined with long-term interests, water projects can help achieve harmonious coexistence between human and water.

(2) Water projects are an important measure to realize sustainable utilization of water. With the rapid socio-economic development and the improvement in living standards, water imbalance between supply and demand will become increasingly prominent. In order to build a water-saving society and mitigate sharp imbalance between water supply and demand, we should adhere to the principle of sustainable utilization of water, and focus on water conservation, protection and optimized allocation. It is important to note that the emphasis on transforming from project based development to resource conservation orientated development is by no means to weaken project construction. In fact, sustainable utilization and optimized allocation of water resource must be based on sound engineering methods and measures. Guided by the Scientific Outlook on Development and the concept of man-water harmony, the construction of water storage projects, such as reservoir is essential to water resource regulation and sustainable utilization. Developing canal system and field water-saving works is an important measure to reduce waste and improve utilization efficiency. Developing water diversion works and other allocation projects is an important method to realize rational allocation of water resource, which could balance areas with abundant and ones with water shortages, and between southern and northern areas. It should be noted that water diversion must ensure that "less water, efficiency and compensation". At the same time, the principle of "saving water first, water diversion second; pollution control before water supply; and environment protection first, water use second" should be abided by. The ultimate purpose is to realize coordinated, sustainable development of both diverting-out and diverting-in areas.

(3) Water projects are the foundation of scientific development to achieve social progress. The progress and development of human society rely on engineering projects. As the key to turn knowledge into actual productivity, project engineering is a direct productive force, a hub of innovations, and inarguably has tremendous social, economic, environmental, and cultural benefits. The North-South Railway Artery and Hatton Project of the U. S., China's ancient Dujiangyan Irrigation Project, and modern Three Gorges Project, South-to-North Water Diversion Project, West-to-East Natural Gas Transmission Project, Qinghai-Tibet railway project are all fine examples. At the 16th National Congress of the Communist Party of China, it is proposed that the goal of building a well-off society in a comprehensive way shall be achieved by 2020, and China, with one fifth of the global population, shall usher in a richer, more civilized society. It is the most difficult but spectacular process of social progress in the history of mankind. Building a well-off society has multiple implications, including thousands of engineering projects, large and small, to be planned, designed and built throughout China. That is to say, a well-off society depends on various engineering activities, especially the construction of large and mega engineering projects. China's water resource characteristics determine that many large-scale, comprehensive water conservation projects are needed. It is the energy structure and development trend that warrant the requirement of developing hydropower in the next 15 to 20 years.

(4) Water projects are a collective reflection and achievement of human civilization. Civilization is the advanced form of cultural development, and the prosperity and development of human civilization is closely related to water. The decline of some early civilization has something to do with water waste and abuse. Water project construction is the important activity of mankind to utilize and transform nature. The history of water project construction is consistent with that of civilization, and the engineering concept and technical level reflect the stage of human civilization and its progress. The water projects such as

the Three Gorges Project, South-to-North Water Diversion Project and so on, have created many world tops and rewritten the history, which not only represent the advanced level in modern water project construction. They symbolize contemporary civilization and progress, and mark the great rejuvenation of the nation. Focusing on water ecological protection in water engineering to ensure sustainable development indicates a certain stage of civilization development. Only by stressing cultural taste, water environment can meet the needs of spiritual and cultural life. Only by injecting cultural connotation, water landscape can exhibit water personality and charm. Only by playing with aesthetic functions, water project could be more picturesque, harmonious, and dynamic. Water cultural landscape shall be constructed in combination with architecture, tourism, transportation, environmental protection, and greening, so as to symbolize modern civilizations, as beautiful as glittering pearls. One of the important tasks of water conservation work in the future is to build a picturesque water environment, improve the quality of living environment, and provide support for building a harmonious society.

3.2 Identify and Evaluate Water Project Objectives

Whatever the projects are, their objectives must be rational, economic, and conforming to public interests. Depending on the time, location and affected people, the objectives and focuses may vary. Either positive or negative views regarding water projects, or newly emerged eco-environmental protection opinions e.g. green hydropower, green dam are all different understandings and pursuits of the objectives of water projects.

In view of its public benefits, the necessity of water project is well recognized by the society. As a developing country, China places development in the first place. According to the status quo and economic requirements, vigorous development of water projects are still the major task in future so as to ensure flood control, water supply, drinking water security and energy demand. The key is whether water project construction can conform to requirements of human-nature harmonious coexistence, and whether it can benefit the future generations. When identifying project objectives, it's necessary to place public interest and eco-environmental protection in a strategic, restrictive position, and perform lawful, democratic and scientific decision-making.

Advocation of protection only without considering the necessity of development, and of ecosystem restoration only and totally repudiate the necessity of water projects are in essence contrary to the development of human society. Without project construction, there will be no development and progress of human society, not to mention eco-environmental protection. Water projects are irreplaceable means and requirements of mankind along the course when man try to adapt to nature, seek for man-nature harmony, and achieve social progress. Taking "dam removal" in the U.S. for example, the dams being removed were less than 10m high on average, and 20m on top. They were removed because their initial purpose had been changed, e.g. dams for industrial water supply such as textile and papermaking were decommissioned because of shutdown of textile and paper mills. Developing water projects is still an important task internationally and the age of massive dam removal believed by some people has not arrived yet.

In recent years, many new concepts, views, theories and practices proposed and conducted by some experts are significant for sound and sustainable development of water projects. For instance, developing green hydropower and green dam, establishing river health assessment system, and conducting scientific regulation, ecological regulation of hydropower project have fully considered the negative impacts of water project construction, especially for the impacts on eco-environment. It should be managed to avoid or mitigate such impacts, which serve as valuable reference for water project construction and management in China. However, most of these concepts and opinions are, for the time being, limited to theoretical discussion, and have a long way to go before being applied. There still exist difficulties in actual operation.

Building water projects not only involves project and economic objectives, but also eco-environment and social objectives. Decision-making is decisive to the success of a project. Project decision-making includes feasibility analysis and demonstration, scenario selection and optimization, and it also includes various decision-making behaviors during project implementation. For example, Sanmenxia Water Control Project on the Yellow River is in its current dilemma mainly due to cognitive shortcomings of decision-maker and constructors, insufficient knowledge or neglect of the eco-environment, flood control, and resettlement. The Aswan Dam in the Nile neglected salinization protection and estuary erosion of the downstream during construction, inviting worldwide criticism for a long time. In order to build water projects in compliance with the requirements of the time and sustainable utilization, scientific and overall verification of the project must be conducted, in line with the overall socio-economic plan. To meet the requirement of harmonious coexistence, decision-making should be done following a democratic and scientific procedure. The selection, construction, and operation of projects should truly reflect coordination of ecological, economic and social benefit. Therefore, the planning of a water project should be made in connection with river basin planning, and water resource planning. Moreover, reasonable project verification and approval procedures must be established, and eco-environmental protection must be emphasized and implemented during project construction, so as to ensure sound development of water conservancy and hydropower.

3.3 Boost Man-water Harmony Oriented Water Project Construction

Water project planning shall conform to not only the development requirements and natural conditions, but also scientific demonstration, democratic consultation and legally feasible plan. This kind of planning is conducive to the fundamental interests of the nation and the people, to the sustainable socio-economic development, to the continuation of resource and environment, and to integrated basin planning. Project engineering is therefore a major measure to realize planning objectives.

Putting people's interests first, water project development should be based on multiple-objective verification and optimization in a lawful, democratic and scientific manner, and strict implementation of basic construction programs and relevant policies and laws, making project the foundation and engine for a harmonious society.

Project construction is a fundamental task crucial for generations. In this regard, quality shall always come first so as to perform functions without hidden risks. On that premise, rational analysis of the project cost-benefit shall be conducted to acquire maximum comprehensive benefits.

Project application shall fullly play the project's function, realize cascade regulation and project cluster regulation to ensure maximum comprehensive benefits.

Project management shall ensure safe, reliable, and durable operation, and shall have regular maintenance and reinforcement to reduce risk and minimize public concerns.

It is said that engineers cannot be blamed for a disharmonious project or an imperfect project since the decisions are made mainly by politicians and entrepreneurs other than engineers. But as engineers are destined to constantly explore within the interactions of human society and the nature, they must have the attitude of "full responsibility", uphold the Scientific Outlook on Development, and make their best efforts for planning, design, construction, operation, monitoring, management and maintenance of water projects.

Regulation of Water and Sediment for the Yellow River Based on Joint Operation of Reservoirs and Artificial Intervention*

Li Guoying[1]

1 Research Background

The Yellow River flows through the Loess Plateau, the largest loess area in the world, because of the rare rainfall and water shortage in the river basin. According to statistics, the average annual natural river runoff of the Yellow River is only 58 billion m^3; however, the sediment load is up to 1.6 billion tons. The multi-year average measured sediment concentration is $35kg/m^3$, therefore, both sediment discharge and sediment concentration of the Yellow River are the largest in the world. Less water and more sediment, and its unharmonious relationship constitute the basic characteristics of water and sediment of the Yellow River. The water-sediment regulation means that based on the premises of fully considering sediment transport capacity of the downstream channel of the Yellow River and sediment carrying capacities at different flow levels, the water and sediment can be effectively regulated by utilizing the regulation storage of reservoirs, so as to duly store or discharge and regulate the natural water and sediment process, thereby coordinating the relationship between water and sediment as far as possible, and reaching the purposes of sediment flushing through water delivery, alleviation of river channel shrinkage, and recovery and maintenance of medium-sized flowing channel.

Based on the above principles, two water-sediment regulating tests for the Yellow River had been conducted in 2002 and 2003 respectively.

The first test in May to June, 2002, the inflow from the upstream and midstream of the Yellow River was more abundant than the stream inflow in the corresponding period when compared to recent years. At the end of June, the water level of Xiaolangdi Reservoir reached 236.09 m with 4.341 billion m^3 of water volume in the reservoir, and 1.421 billion m^3 of water volume above flood control level of 225m, the total water volume above flood control level in the reservoir plus forecasted inflow in the next few days would be enough for regulation test of water and sediment. Considering the flood carrying capacity of main channel of part downstream river reaches was less than $3,000m^3/s$, the final testing scheme was determined as: controlling the flow discharge at the Huayuankou Station as $2,600m^3/s$ no less than 10 days, and the average sediment concentration no more than $20kg/m^3$, corresponding flow discharge at the Aishan Station as about $2,300m^3/s$, and at the Lijin Station as about $2,000m^3/s$. The test was implemented from 9:00a.m., 4 July to 9:00a.m., 15 July, 2002, with the flow discharge over $2,600m^3/s$ at the Huayuankou Station for 10.3 days and the average sediment concentration of $13.3kg/m^3$, while the flow discharge over $2,300m^3/s$ at the Aishan Station maintained for 9.9 days and $2,000m^3/s$ at the Lijin Station for 9.9 days. During the test, the sediment erosion of the downstream river channel was 36.2 million tons, the water volume into the sea from Lijin was 2.335 billion m^3, and sediment into the sea was 50.5 million tons.

Second test from 25 August to the beginning of November, 2003, there was a rarely seen autumn flood in the history of the Yellow River. Until 8:00a.m., September 5, 2003, the water volume in Xiaolangdi Reservoir was 5.37 billion m^3, which was only 620 million m^3 less than the corresponding volume of flood control level of 248 m for the post flood season after 11 September. Based on the forecasting that there would be a large precipitation after 5 September, in Shanxi and Shaanxi provinces, the Jing River and Wei River basins, and the region between Sanmenxia-Huayuankou, test plan was made. The flow regulation, taking the discharge between Xiaolangdi and Huayuankou as the base flow, maintained the discharge at Huayuankou Station about $2,400m^3/s$; the sediment concentration regulation taking the sediment concentration of the Yi River, Luo River and Qin River as base and considering the added sediment from the main stream between Xiaolangdi and Huayuankou, maintained the average sediment concentration at Huayuankou Station as $30kg/m^3$. The test was carried out from 18:30 of 6 September to 18:30 of 18 September, the measured average flow and sediment concentration at Huayuankou Station were $2,390m^3/s$ and $31.1kg/m^3$ respectively. During the test, a total of 45.6 million tons of sediment was flushed up from the downstream river channel. The water and sediment into the sea from Lijin were 2.719 billion m^3 and 120.7 million tons respectively.

From 19 June to 13 July, 2004, the third regulation test of water and sediment was performed on the Yellow River. The

---

* This paper was published in *Journal of Hydraulic Engineering*, Issue 12, Volume 37 of 2006.

[1] Li Guoying, Former Vice Minister of Water Resources, Professor Senior Engineer.

test covered more areas than the previous two, which included the 2,000km long of the midstream and downstream reaches and three reservoirs of Wanjiazhai, Sanmenxia, and Xiaolangdi. The test fully utilized the water power from the outflows above flood control level of the three reservoirs. The reservoir sedimentation reduction was realized by artificial density current, as well as the artificial agitating and dredging measure in river channel, so as to increase the sediment carrying capacity, realize the sedimentation reduction in the river channel, and particularly, enlarge the flood carrying capacity in the main channel at the bottle necks (with low discharge capacity) of downstream reaches. This test was quite different from the previous two concerning the conditions and approach of the regulation tests in July of 2002 and September of 2003[1,2], which created a new model for regulation of water and sediment of the Yellow River through emptying the water storage of reservoirs before flood season without taking the inflow from upstream.

2 Reservoirs Impoundment on the Main Streams of the Yellow River before Flood Season in 2004

Before this regulation test of water and sediment in 2004 (as to 8:00, 19 June), the total water storage of three reservoirs of Xiaolangdi, Sanmenxia, and Wanjiazhai was 6.816 billion m^3, and the water volume above flood control level was 3.859 billion m^3. According to the flood control requirements, the reservoir water levels should be maintained below the flood control level before the flood season, therefore, at least 3.859 billion m^3 of water must be discharged to the downstream. For rivers with clear water, when the flood season is coming, the reservoirs can sluice the discharge until the water level is lower than the flood control level in accordance with the safety discharge for the downstream. However, for the Yellow River, due to the existence of a large amount of sediment, the sediment discharge issue should be fully considered.

3 Sedimentation in Reservoirs and River Channel

3.1 Downstream River Channel

Analysis on the measured data of the river cross-section before flood season in 2004 showed that the bankfull discharges of the Lower Yellow River were: about 4,000m^3/s above Huayuankou Station, about 3,500m^3/s from Huayuankou to Jiahetan, about 3,000m^3/s from Jiahetan to Gaocun, about 2,500m^3/s from Gaocun to Aishan, and about 3,000m^3/s below Aishan. The bankfull discharge at some cross-sections were less than 2,600m^3/s (the critical flow for sediment erosion and deposition at the Lower Yellow River), particularly for Xingmiao-Yanglou and Yingtang-Guonali, the figures were less than 2,400m^3/s, while the bankfull discharges of cross-sections at Xumatou and Leikou were only 2,260m^3/s and 2,390m^3/s respectively. Xumatou and Leikou are two obvious block reaches, and the lengths of Xumatou reach (Xingmiao-Yanglou) and Leikou reach (Yingtang-Guonali) are 20km and 10km respectively (see Fig. 1). If the discharge capacity in the main channel of the two block reaches could be increased, the bankfull discharge of the Lower Yellow River could be increased as a whole.

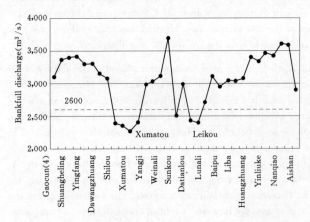

Fig. 1 Variation of Bankfull Discharge along the Lower Yellow River before Regulation Test of Water and Sediment

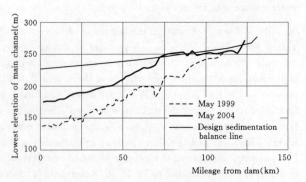

Fig. 2 Longitudinal Profile of Xiaolangdi Reservoir before Regulation Test of Water and Sediment

3.2 Xiaolangdi Reservoir

Since October 1999, the beginning of impounding and until May 2004, Xiaolangdi Reservoir had deposited sediments the reached 1.394 billion m^3 in the reservoir area and the designed volume was 2.614 billion to 3.443 billion m^3. But the operating water level of Xiaolangdi Reservoir was high in autumn flood season of 2003, and the sedimentation delta at the reach 70

-93km above the dam exceeded the design sedimentation equilibrium profile of 38.5 million m³ (see Fig. 2).

According to the underwater sediment measurement before the flood season of 2004, the closer to the dam, the deposited sediment grain size would be finer. The sediment median size from cross-section 18 (29.35km from the dam) to front section of dam was 0.006 – 0.010mm, while the figure was 0.117mm for cross-section 52 (105.85km from the dam) near the end of backwater of the reservoir. The sedimentation delta of reservoir tail part was middle and coarse sediment with median size of 0.036 – 0.073mm (see Fig. 3).

3.3 Sanmenxia Reservoir

From the post flood season of 2003 to prior flood season of 2004, the sedimentation of Sanmenxia Reservoir was about 88 million m³ below the elevation of 318 m. There was about 6.00 million tons fine particle sediment in the funnel of dam area (with median size of 0.008 mm), which could be desilted at the initial stage of erosion of the reservoir. There was about 20 million m³ sediment in the main channel of the near dam area (with median size of 0.012mm), the desilting could be done for discharge at low water level (see Fig. 4).

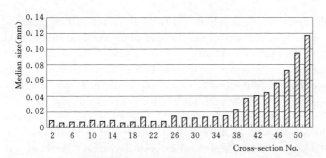

Fig. 3 Median Size of Each Cross-section of Xiaolangdi Reservoir before Regulation Test of Water and Sediment

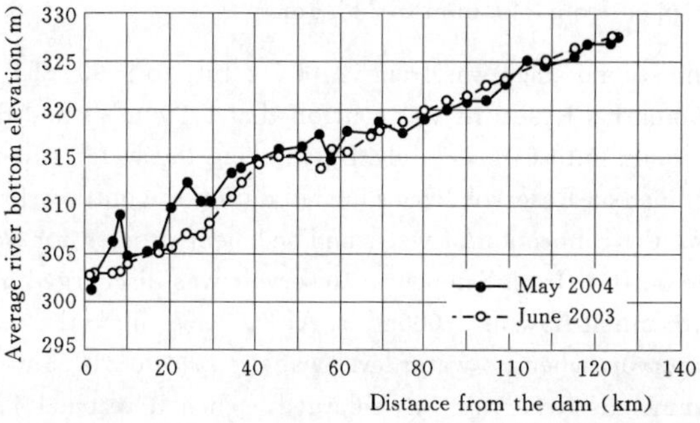

Fig. 4 Longitudinal Profile of Sedimentation in Sanmenxia Reservoir

4 Design Idea and Expected Objectives on Regulation Plan of Water and Sediment

4.1 Design Idea

Utilize the water storage of reservoir, completely and tactfully use nature forces, accurately schedule three hydraulic projects of Wanjiazhai, Sanmenxia, and Xiaolangdi and adopt artificial intervention measure, so that artificial density current in Xiaolangdi Reservoir can be created to adjust the sedimentation shape of the reservoir tail reach and increase the desilting sediment amount from Xiaolangdi Reservoir. Meanwhile, by utilizing the extra sediment carrying capacity in the downstream, riverbed sediment can be intervened at the "secondary perched river" and the block reaches with most severe sedimentation in the main channel of the Lower Yellow River, thereby strengthening the flood carrying capacity of main channel.

4.2 Expected Objectives

This regulation test of water and sediment has set up four objectives as follows.

(1) Adjust the sediment deposition shape in Xiaolangdi reservoir tail part to desilt out of the reservoir. Through the erosion of outflow from Sanmenxia and Wanjiazhai reservoirs, and artificial sediment intervention measure, the sedimentation part occupying the long term available storage of Xiaolangdi Reservoir tail part would be eliminated. In addition, in order to reduce the reservoir sedimentation, the desilting out of reservoir could be conducted as much as possible by utilizing the density current of reservoir.

(2) Increase the flood carrying capacity of two block reaches in the downstream. Through the erosion of water flow and artificial sediment intervention in the block reaches, the unfavorable pro-phase "hump" sedimentation shape would be eliminated, thereby enlarging the cross section of the main channel.

(3) Through joint operation of reservoirs on the main stream, create artificial flood peak with "harmonious" water and sediment relationship, and thereby allow the whole main channel of the Lower Yellow River to erode and carry sediment to the sea.

(4) Through the realization of creating artificial density current, evolution of water and sediment in reservoirs and river channels and their spatial connection, and adding sediment by artificial intervention to balance the water and sediment, deepen the knowledge about the water and sediment movement law of the Yellow River and can be translated into theory, which can guide the river governing practice in the future.

5 Test Process and its Effects

The test was divided into two stages: The first stage is to boost the flood carrying capacity of the main channel in the downstream reaches by utilizing the outflow of Xiaolangdi Reservoir and artificial intervention; the second stage is to regulate the sedimentation shape of tail area of Xiaolangdi Reservoir, create artificial density current and realize desilting out of the reservoir by joint operation of reservoirs on the main stream and by artificial intervention.

The first stage was from 9:00, 19 June to 0:00, 29 June. During this period, the water level of Wanjiazhai Reservoir was controlled at about 977m, the water level of Sanmenxia Reservoir was kept not exceeding 318m, while clear water was released from Xiaolangdi Reservoir at a controlled flow of 2,600m^3/s at the Huayuankou Station until the water level was dropped from 249.06m to 236.6m.

The second stage was from 12:00, 2 July to 8:00, 13 July. From 12:00, 2 July to 15:00, 5 July, the outflow from the Wanjiazhai Reservoir was controlled at 1,200m^3/s and the water level of reservoir was dropped below the flood control level of 966m until 6:00, 7 July. By applying the water balance between inflow and outflow of the reservoir, the outflow from Sanmenxia Reservoir was adjusted during the outflow from Wanjiazhai Reservoir on the way to Sanmenxia Reservoir, realizing the connection of water and sediment process for Wanjiazhai and Sanmenxia. From 15:00, 5 July to 13:00, 10 July, the outflow from Sanmenxia Reservoir was discharged in accordance with the principle of "from small increasing to large" with initial flow of 2,000m^3/s. At 8:00, 7 July, the outflow from Wanjiazhai Reservoir of 1,200m^3/s reached Sanmenxia Reservoir when its water level was dropped to 310.3m (the design water level was 310m). After that, the outflow from Sanmenxia Reservoir was enlarged, when it reached 4,500m^3/s, it would be utilized as open discharge, therefore, the density current of Xiaolangdi Reservoir was pushed out of the reservoir by utilizing the force caused by outflow from Wanjiazhai and Sanmenxia reservoirs. With the increase of the main channel bankfull discharge of block section of downstream river driven by flow scour and artificial intervention, the outflow from Xiaolangdi Reservoir was controlled at no higher than 2,800m^3/s at Huayuankou Station from 21:00, 3 July; then the outflow was gradually increased from 2,550m^3/s to 2,750m^3/s. And when the water level was dropped to the flood control level of 225 m at 8:00, 13 July, the test was completed.

5.1 Increasing of Flood Carrying Capacity of the Main Channel in the Downstream Reaches by Utilizing the Outflow from Xiaolangdi Reservoir and Artificial Intervention

(1) Timing of discharging clear water from Xiaolangdi Reservoir. In accordance with the requirements of flood prevention, the water level of Xiaolangdi, Sanmenxia, and Wanjiazhai reservoirs must be dropped below the flood control level. Considering the discharges from the main streams and tributaries, as well as district water diversion, if the flow at Huayuankou was controlled at 2,600m^3/s, it will take more than 20 days to regulate water and sediment, which required that Xiaolangdi Reservoir should start discharging before 20 June. At the same time, the wheat harvest at beach area of downstream of the Yellow River usually ends at around 15 June. Through overall consideration, it is determined that the regulation shall start from 19 June.

(2) Flow discharge. In accordance with the water and sediment relationship for non-sedimentation of flood in downstream reaches, if the whole line erosion in the downstream reaches was realized, the flow at Huayuankou Station should be no lower than 2,600m^3/s. According to the measurement data of cross-section before flood in 2004, the bankfull discharge of block area of river reaches near Xumatou and Leikou was only about 2,300m^3/s. Considering the flood attenuation and energy loss during the process, the flow of Huayuankou was set at 2,600m^3/s at the beginning of the test.

(3) Intervention river reaches. Two principles were followed for selection of intervention river reaches. The one principle is that the river reaches with most developed "secondary perched river"; The other one is that the river reaches with the smallest bankfull discharge. In accordance with the measurement data of cross-section before flood, the "secondary perched river" of river reaches near Xumatou and Leikou was the most severe, the beach edge was 1–3m higher than beach lands at both banks, the traverse gradient was 5‰–22‰, which was the place with the smallest bankfull discharge in of the Lower Yellow River (with bankfull discharge of only 2,260–2,390m^3/s). Therefore, the two block sections at Xumatou and Leikou reaches were selected as the intervention river reaches.

(4) Intervention Measures. In accordance with the sediment flushing effects, the cost, and the field test results, it was determined that the sediment pumping and scattering, as well as underwater jet flow were adopted as the intervention meas-

ures.

The arrangement at Xumatou Reach: 11 sediment intervention work platforms, including two 80 tons automatic barges, one 200 tons binary automatic barge, two 120-typed dredgers, and one civil boat. Each boat was equipped with jet flow equipment and diving slurry pump.

The arrangement at Leikou Reach: 15 sediment intervention work platforms, including 2 automatic barges, 2 mobile pressure-bearing boats, and 11 work platforms combined with floating bridges and floating bodies. Each boat was equipped with jet flow equipment.

(5) Intervention command. The field intervention headquarters were established in Xumatou and Leikou, which were responsible for command, organization, equipment, logistical support and security work for respective work zone under the unified command of general headquarter for the regulation test.

(6) Intervention technology. In order to scientifically allocate the sediment adding amount and sediment grain composition of sediment intervention area, the front station and feedback station to monitoring water and sediment factors were established. The upstream front station of sediment intervention area was established in hydrometric station of Gaocun Village, 76km away from sediment intervention spot of Xumatou. The sediment adding timing, sediment adding amount and sediment grain composition of sediment the intervention area were controlled by water and sediment collocation at front station. If the sediment concentration of flow or sediment with certain grain size at front station of Gaocun was lower than the control value (equilibrium value), the controlling sediment intervention area would be supplemented with sediment in accordance with its difference. The feedback station was established in Aishan hydrometric station, which was at 51km of downstream of Leikou sediment intervention station, and was designed to control the relationship of water and sediment entering the Aishan to Linjin reach. If the sediment concentration at Aishan Station was higher than the control level (equilibrium value), the sediment intervention amount would be stopped or decreased, so as to realize the objective of erosion in the whole line of the Lower Yellow River.

Considering the influence of sediment grain size to erosion and deposition of river channel, the data of 250 floods in 1960-2002 were analyzed, and thereby the critical sediment concentration formula under different water and sediment conditions of Aishan Station was established:

$$S = k Q^\alpha P_*^\beta \qquad (1)$$

In the formula, S is the calculated critical sediment concentration at Aishan Station, kg/m^3; Q is the Aishan flow discharge, m^3/s; P_* is weight of sediment smaller than 0.05mm at Aishan Station; k, α, β are undetermined parameters, which could be determined by measured data.

The formula (1) determinatively reflected the discipline of the thicker the grain size, the smaller the critical sedimentation concentration.

In accordance with the calculation of formula (1), if the deposition reach between Aishan-Lijin was not happened, when the flow at Aishan Station was $2,400m^3/s$, $2,500m^3/s$ and $2,600m^3/s$ respectively and the median diameter of suspended sediment was about 0.025mm, the maximal allowable carrying sedimentation concentration would be $27.3kg/m^3$, $30.0kg/m^3$ and $32.9kg/m^3$ respectively; when the flow at Aishan Station was $2,600m^3/s$, and the median diameter of sediment was 0.045mm, the critical sediment concentration at Aishan Station would be $18kg/m^3$.

Meanwhile, the controlling parameters for intervention sediment adding under different water and sediment factors at the front station of Gaocun were:

$$S = K \frac{(TQ)^{2.432}}{m} \qquad (2)$$

In the formula, S is the calculated sedimentation concentration at Gaocun Station, kg/m^3; Q is the flow at Gaocun Station, m^3/s; m is the evolution coefficient of flow, the value of m depended on the measured sedimentation concentration at Gaocun, the K value was 1.012×10^{-7}.

In accordance with the formula (2), the controlling sedimentation concentration S could be calculated by the measured discharge flow and sedimentation concentration with actual measurement at Gaocun Station. If the measured value was lower than the control value, the intervention sediment adding was allowed near Sunkou, otherwise, which should be stopped. Through calculation, if the flow at Gaocun Station was between $2,400m^3/s$ and $2,600m^3/s$, the sedimentation concentration lower than $23-29kg/m^3$, the intervention sediment adding was allowed, with the adding amount determined by the calculation of water and sediment.

(7) Intervention time. In accordance with the water and sediment process at front station of Gaocun and feedback station of

Aishan, this intervention was divided into two stages: the first stage was from 12:00, 22 June to 8:00, 30 June, 188 hours in total; the second stage was from 7:00, 7 July to 6:00, 13 July, 143 hours in total. The two stages cost 331 hours altogether.

(8) Intervention effects: Sediment intervention can cause the increase of the downstream sediment concentration in sediment intervention area and the flood carrying capacity of block section.

In accordance with the quantity and performance of equipment invested to the two river reaches, the calculated sediment engaged intervention was 1,641,300m³ without considering the erosion quantity increased by intervention.

Before the sediment intervention, the bankfull discharge at Xumatou was 2,260m³/s, and at Leikou was 2,390m³/s. In the process of test, when the flow of 2,900m³/s at Gaocun was not over the both banks, therefore, the bankfull discharge in the river reaches with sediment intervention was increased to 2,900m³/s. The bankfull discharge increased by 510 – 640 m³/s after the sediment intervention.

Based on the cross-section data of Xumatou measured before and after the test, it was found that the sediment intervention river reach was obviously eroded, and the average erosion of four cross-sections within the sediment intervention reaches was 135m². In order to analyze the sediment intervention effects, contrast calculation by utilizing the mathematical model was conducted. If the sediment intervention was not performed, the average erosion area of Gaocun-Sunkou reach would be 75m², it was thus clear that the area of cross-section of Xumatou river reach was completely enlarged by 60m² due to sediment intervention (see Fig. 5). Through the calculation of mathematical model, this sediment intervention test made the river channel below Gaocun erode an additional 410,000m³, approximately accounting for 25% of the total amount of sediment intervention, and about 1/4 sediment can be transported for a long distance. The added value of bankfull discharge in the river reaches with sediment intervention was 200 – 300m³/s greater than the average added value of other river reaches of the Lower Yellow River.

5.2 Regulation of Sedimentation Shape of Tail Area of Xiaolangdi Reservoir, Creation of Artificial Density Current and Realization of Desilting out of Reservoir by Joint Operation of Reservoirs on Main Stream and Artificial Intervention

(1) The sediment intervention and effects of Xiaolangdi Reservoir tail section reach. 8 boats were put into sediment intervention in reservoir area, including 4 intervention boats, and 4 boats for measurement and logistical support. The intervention boats were loaded with high-pressure jet flow equipment, which were adaptive for underwater work at a depth of 1 – 10 m and with a jet flow speed of 23m/s. The river reach of sediment intervention work was performed in the reach of cross-section 34 – 40 (57.00 – 69.34km away from the dam) of reservoir area, which was completed in two stages from 19 June to 29 June, and from 3 July to 10 July, see Fig. 5. The accumulated work for sediment intervention last 886 hours in total with 60 (times) measurements of cross-section in the reservoir area, and collection of 80 sediment samples of riverbed material.

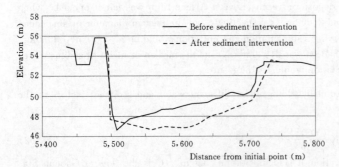

Fig. 5   Contrast of Xumatou Cross-section before and after Sediment Intervention

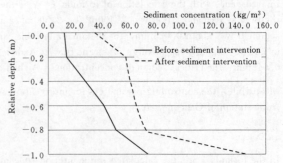

Fig. 6   Contrast of Sediment Concentration of Cross-section along Vertical Line of Reservoir Tail Section before and after Intervention

The functions of reservoir tail intervention on artificial creation of density current are in three aspects: first, being favorable for concentrating water flow above the plunging point and increasing flow velocity; second, improving the longitudinal shape of river channel, reducing the energy loss of water flow, and being good for direct plunging of water flow; third, loosening the riverbed sediment above the plunging point, and making it easy for the sediment being eroded and carried by water flow. Through intervention, the sediment on the riverbed can be eroded, thereby increasing the sediment concentration of water flow. Fig. 6 shows the test observation for change of sediment concentration along vertical line before and after the intervention, and the maximal increase could be more than 30kg/m³.

(2) Creation of artificial density current and erosion of the delta deposit in Xiaolangdi Reservoir tail area by joint operation of Sanmenxia and Wanjiazhai reservoirs. The so-called density current referred to the relative movement generated by density difference of two mixable fluids with similar densities. In reservoirs on sediment laden rivers, when the water flow carrying sediment met the clear water of reservoir area, since the density of the former was larger than the latter, the water flow carrying sediment would plunge into the bottom of clear water to continuously flow forward under appropriate conditions, and it was called the density current of reservoir.

1) Formation condition and movement law of density current. The formation of density current was related to inflow, sediment concentration, sediment grain composition, cross-section characteristics of plunging point, and reservoir river bed slope etc. The sign for generation of density flow was that the surface of reservoir clear water had obvious plunging point, and the sediment condition at plunging point was the formation condition of density current.

The density current was propelled to the dam after being formed, which needed stable dynamics that was sufficient to overcoming the on-way resistance, namely, the continuous follow-up density current should be supplemented, once the follow-up density flow was stopped, the moving density current in the front would stop soon and disappear gradually.

Based on the data of density current naturally generated in Xiaolangdi Reservoir, the critical condition for formation of density current and continuous movement of Xiaolangdi Reservoir can be analyzed and concluded, namely: under the condition of meeting with the duration of flood and the fine sediment weight in to the reservoir should be about 50%, one of the following conditions should also be satisfied: ① the inflow was greater than 2,000m³/s and the sediment concentration was greater than 40kg/m³; ② the inflow was greater than 500m³/s and the sediment concentration was greater than 220kg/m³; ③ when the discharge was 500 - 2,000m³/s, the corresponding sediment concentration should be met by the following:

$$S \geqslant 280 - 0.12Q \tag{3}$$

In the formula, S is the inflow sediment concentration, kg/m³; Q is the inflow, m³/s.

2) Creation of artificial density current. ①Determination of outflow from Sanmenxia Reservoir. The discharge objective of Sanmenxia Reservoir was to create an artificial density current in Xiaolangdi Reservoir. Four conditions should be considered for the selection of discharge flow and time of Sanmenxia Reservoir: Firstly, the erosion of sedimentation delta of Xiaolangdi Reservoir needed larger amount of outflow from Sanmenxia Reservoir; Secondly, there was no flood with high sediment concentration in midstream, and the water flow of Sanmenxia Reservoir had certain sediment concentration; Thirdly, when flow from Wanjiazhai Reservoir to Sanmenxia Reservoir, the water level of Sanmenxia could not be too high, which should be at about 310 m, otherwise the effects of sediment sluicing from Sanmenxia Reservoir would not be obvious; Finally, the water level of Xiaolangdi Reservoir should not be too high when discharging Sanmenxia Reservoir, otherwise, the energy of discharge flow of Sanmenxia Reservoir would be eliminated, and if the water level of Xiaolangdi Reservoir was too high, the effects of artificial sediment intervention would not be obvious. Through overall considerations, it was determined that the discharge timing of Sanmenxia Reservoir was 15:00, 5 July with outflow of 2,000m³/s. When the flow of Wanjiazhai Reservoir reached Sanmenxia at 8:00, 8 July, the discharge of Sanmenxia Reservoir was increased. ②Regulation of sediment concentration of water flow. There were two sediment sources that formed artificial density current of Xiaolangdi Reservoir, the first sediment source was the sedimentation delta of Xiaolangdi Reservoir tail part, which should be eroded with large quantity of clear water discharged from Sanmenxia Reservoir and supplemented with artificial intervention measure to make it enter into the water flow; the other sediment source was the fine sediment in deposited in the main channel of Sanmenxia Reservoir, which should be eroded and drained by the outflow from Wanjiazhai Reservoir when Sanmenxia Reservoir was at low water level. ③Connection of outflows from Wanjiazhai Reservoir and Sanmenxia Reservoir. The objective for connection was that the outflow from Wanjiazhai Reservoir evolved to Sanmenxia Reservoir when the water level of Sanmenxia dropped to and below 310m, the furthest eroding the sediment of Sanmenxia Reservoir, and providing the density current of Xiaolangdi Reservoir with continuous water dynamics and sufficient fine sediment source. It was that Wanjiazhai Reservoir was discharged at 12:00, 2 July, which was in advance of discharging time of Sanmenxia Reservoir of 15:00, 5 July, with discharge flow of 1,200m³/s.

3) Test effects. ①Improvement of sedimentation shape of Xiaolangdi Reservoir tail part. Through natural erosion by water flow and intervention of Xiaolangdi Reservoir, the sedimentation delta summit of Xiaolangdi Reservoir tail part moved downstream from 70km to 47km above the dam, the delta surface elevation was dropped about 4m after the intervention, thereby making the sedimentation delta erode sediment of 132.9 million m³ (cross-sections 37 - 53 of reservoir area, and 62.49 - 110.27 km from the dam), completely eroding and eliminating the deposited sediment of 38.5 million m³ above the design deposit equilibrium profile, and rationally regulating the deposit formation of the reservoir tail part (see Fig. 7). ②Creation of artificial density current and desilting out of reservoir. The creation of artificial density current was divided in-

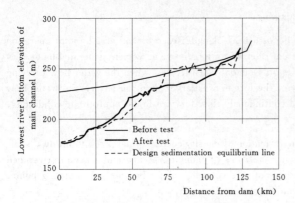

Fig. 7 Change on Profile Shape of Xiaolangdi Reservoir Sedimentation before and after Test

to two stages. The first stage was from 15:00, 5 July, Sanmenxia Reservoir was discharged with the discharge of 2,000m³/s, and the sedimentation delta of the reservoir was strongly eroded. At 18:30, 5 July, the density current plunged into the cross-section 34 (about 57km from the dam) and propelled toward the dam. The second stage was from 8:00, 7 July, after the connection of discharges from Wanjiazhai and Sanmenxia reservoirs, the deposited sedimentation in Sanmenxia Reservoir was eroded, and the outflow from Sanmenxia Reservoir was increased. At 14:10 of the same day, the peak discharge was 5,130m³/s, and at 14:00, the desilting started, and the sediment concentration reached 446 kg/m³ at 20:00. The flood with high sediment concentration continuously eroded the sedimentation delta of Xiaolangdi Reservoir, and formed the follow-up dynamics of density current to promote the density current to move towards the Dam. At 13:50, 8 July, the density current was desilted out of the reservoir, and the average sediment concentration of desilting tunnel flow was about 70kg/m³. At 2:00, 9 July, the sediment peak of density current was out of the reservoir, the maximal sediment concentration of density current was 126kg/m³, and the continuous duration was about 80 hours. In the movement of density current, the sediment out of Xiaolangdi Reservoir was 4.4 million t. The maximal velocity, thickness, and sediment concentration of the density current at Cross-section 1 (1.3km from the dam) were 0.54m/s, 2.98m and 742kg/m³ respectively. The sediment grain was relatively fine, 90% was smaller than 0.025mm, and the median grain size was about 0.005 – 0.009 mm.

5.3 Water and Sediment Amount into Downstream and Change of River Channel Erosion and Deposition

The test started from 9:00, 19 June to 8:00, 13 July, 2004 with the duration of 24 days. If deducting the 5 days from 0:00, 29 June to 21:00, 3 July with small discharge were deducted, the actual test duration was 19 days. During the entire test duration, the discharging water from Xiaolangdi Reservoir was 4.68 billion m³ with the sediment of 4.4 million tons; the water through Huayuankou Station was 4.757 billion m³ with the sediment of 21.1 million tons and average sediment of 4.43kg/m³; the water at Lijin Station flowing into the sea was 4.801 billion m³ with sediment of 69.7 million tons and average sediment of 14.52kg/m³.

During the test, the all river reaches upstream Lijin Station was eroded, the total erosion between Xiaolangdi and Lijin was 66.5 million tons, including the erosion of 16.9 million tons between Xiaolangdi and Huayuankou, 14.7 million tons between Huayuankou and Gaocun, 19.8 million tons between Gaocun-Aishan, and 15.1 million tons between Aishan and Lijin.

6 Understanding and Experience

(1) The main problem of the Yellow River is the imbalance of water and sediment, and the sedimentation problem shall be considered for reservoir discharging before the flood season. Due to the imbalance of water and sediment in both spatial and temporal distributions, the water used for sediment transport is usually not sufficient and cannot satisfy the transport project. After the construction of water projects such as Xiaolangdi Reservoir, a certain amount of water can be stored in the non-flood season in normal years, the water and sediment relationship can be regulated by utilizing the projects before flood season, so as to avoid the empty running of clear water, the abandoned water can be used to transport sediment, eliminate the imbalance of water and sediment of the Yellow River from the aspects of time and space, and avoid the insufficiency of water used for sediment transportation, thereby realizing the "harmonious" relationship of water and sediment.

(2) The double sedimentation reduction effects can be obtained through fully utilizing the flow energy with artificial intervention. In the third regulation test of water and sediment for the Yellow River, the used water was mainly stored in reservoirs, which should be abandoned before the flood season. The abandoned water was taken as the main force for downstream erosion, and if artificial dredging was added at the river block section, the flood carrying capacity in entire downstream reaches can be improved. The sediment eroded from river bed in this test was 66.5 million tons. If the sediment was dug up by manual work for RMB 10/t, it would cost RMB 665 million in total. In addition, the expense would increase if considering the fees for transport and stack.

(3) The application mode for water and sediment regulation has been developed. Three regulation tests of water and sediment conducted for the Yellow River have basically covered different regulation types providing solid foundation for transferring the regulation tests to practical applications in the future. For example, the first test was performed aiming at medium

and small floods on the upstream of Xiaolangdi, regulating the unharmonious water and sediment relationship into harmonious one to enter the downstream river; The second test was performed aiming at the upstream muddy water and downstream clear water of Xiaolangdi, jointly operating the water and sediment of Xiaolangdi, Luhun, and Guxian reservoirs to realize the spatial "connection" of water and sediment at Huayuankou, and evolve to the downstream river through forming the "harmonious" relationship of water and sediment after mixing the clear water and muddy water; As for the third test, as there was no flood in the main stream of the Yellow River, the water was mainly stored in reservoirs at the end of last flood season, and create artificial density current in Xiaolangdi Reservoir through operating Wanjiazhai, Sanmenxia, and Xiaolangdi reservoirs, which was supplemented of artificial sediment intervention in the sedimentation delta area and downstream block, so as to make the abandoned water of reservoir turn into density current, and sediment could be transported into sea by riverbed intervention. Therefore, it was a new regulation mode of water and sediment different from the previous two tests.

(4) It is imperative to construct a complete water and sediment control system for the Yellow River. In this regulation test, the joint operation of reservoirs on the main streams has produced huge effects, alleviated the deposit in Xiaolangdi Reservoir, and eroded the sediment of downstream river channel, which cannot be realized only depending on the operation of Xiaolangdi Reservoir. The double effects can only be realized with the joint operation of these three reservoirs. Although Wanjiazhai and Sanmenxia reservoirs were involved in the joint operation, the defects of these two reservoirs were obviously revealed: Firstly, the storage capacity of the two reservoirs is relatively small, and the water stored is quite limited, so they cannot provide sustained dynamics to maintain the continuance and stability of artificial density current in Xiaolangdi Reservoir, otherwise, more sediment will be discharged from Xiaolangdi Reservoir; Secondly, Wanjiazhai Reservoir is too far from Sanmenxia Reservoir, making it difficult to guarantee the accuracy of joint operation. Therefore, if new reservoirs could be constructed at Guxian or Qikou on the north main stream of the middle stream of the Yellow River, a perfect water and sediment control system on the middle stream of the Yellow River can be constituted, thereby realizing the regulation of sediment for years, and achieving double objectives of sedimentation reduction for both reservoirs and river channel.

(5) The regulation of water and sediment is one of most effective measures to maintain the healthy lives of the Yellow River. The ultimate objective of management of the Yellow River is to maintain the healthy operation of the Yellow River, once this objective is determined, all measures managing the Yellow River should serve this objective. In recent ten years, the Yellow River has seen the sustained silting sedimentation in the main trough channel due to small flow and unharmonious water and sediment, or flood disasters due to overabundant flow and large area of flood plain, and or running of clear water into the sea which caused water and energy waste. In the long run, it will be impossible to create and maintain a healthy Lower Yellow River. However, the continuous deterioration trend of river channel in the Lower Yellow River can be suppressed through water and sediment regulation of "harmonious" amount of flow, sediment concentration, and sediment grain composition, and in this way the Yellow River will be able to recover with a healthy life and be virtuously maintained in the end.

References

[1] Li Guoying. First Regulation of Water and Sediment for the Yellow River. *Science*, 2003, (1): 41-43.
[2] Li Guoying. Theory and Practice of Temporal and Spatial Dispatch of Water and Sediment in Middle-Lower Reaches of the Yellow River. *Journal of Hydraulic Engineering*, 2004, (8): 1-7.

The Three Gorges Project and Flood Prevention of the Yangtze River

Cai Qihua[1]

The Three Gorges Project, which has attracted the worldwide attention, is a pivotal project for harnessing and protecting the Yangtze River. The construction of the Three Gorges Project was officially approved at the 5th Plenary Session of the 7th National People's Congress in April 1992; the Three Gorges Project formally started construction in December 1994; the damming of the river was successfully carried out in November 1997; the damming of diversion channels was made in November 2002; its water storage level reached 156m after the floods in 2006; and the experimental water storage level reached 172m after the floods in 2008; the target of the normal water level of 175m was realized for the first time after the floods in 2010, and then the TGP has comprehensively displayed the reservoir's multiple benefits including flood prevention, power generation, navigation, and ecology etc. As the backbone of the flood protection for the Yangtze River, the flood prevention standard of the Jingjiang Reach rise to a 100-year flood from a 10-year flood after its completion, which can considerably mitigate the flood threat to the middle and lower Yangtze River.

I

Nature always holds surprises for humanity with its uncanny workmanship. The Himalaya Mountains was created 30-40 million years ago as a result of the orogenesis, and the Three Gorges formed during the miraculous Earth revolution has given birth to the brand-new Yangtze River in its entirety, together with the strength inherent in the nationality which was soon to come into being. This strength burst into life in the 20th century through tens of millions of years of accumulation and development. By taking the advantages of the unique terrain and location of the Three Gorges, the contemporary "Da Yu" has miraculously built up the Three Gorges Project by combining the human intelligence with the magic of nature, which has signalled a new level for governing and developing the Yangtze River and started the new progress of the Yangtze River which benefits the Chinese nation. At present, the raging floodwater have been controlled effectively; the galloping river water has now transformed into the sparks for illuminating the gorgeous territory of China; the magnificent picture of thousands of ships setting sail is now unfolding on the boundless river; the grand Three Gorges Dam and the pretty Goddess landscape have greatly combined and complemented each other.

The Three Gorges Project is the key to ensure a safe flood protection system for the Yangtze River. After the Three Gorges Project comes into operation with the designed normal water level of 175m, it has a flood controling storage capacity of 22.15 billion m^3, which boosts the flood protection capability of the Jingjiang Reach, Hanjiang River Basin, and Dongting Lake greatly. Without using the flood diversion-retention areas, the flood protection standard of the Jingjiang Reach can be improved from a 10-year flood to a 100-year flood; when it suffered a flood with a return period greater than 100 years to 1,000 years, or even like the record flood occuring in 1870, the flow at Zhicheng can be controlled within 80,000m^3/s by the regulation of the Three Gorges Reservoir; after using the planned flood diversion-retention areas, the water level in Shashi District can be guaranteed not to lower than 45m, which mitigates or avoids the devastating disasters around the Jingjiang Reach. The Three Gorges Project is irreplaceable for its role of regulating the extraordinary flood at the upper reaches of the Yangtze River. During the flood season in 2010, the Three Gorges Dam successfully carried out the flood retention scheduling for five times, with the amount of more than 23 billion m^3. Among them, the maximum peak discharge reached 70,000m^3/s since the completion of the Dam; the Three Gorges Reservoir cut the peak discharge at the lower reaches by roughly 30,000m^3/s under its control of discharging, with a peak cutting rate of 40% and an impoundment amount of 7.3 billion m^3, thereby reducing a water level by 0.45-2.55m along the main lines of the middle reaches of the Yangtze River. As a result of this, by keeping the water levels of the middle and lower reaches, in particular the reaches at Shashi and Wuhan, lower than the warning water level, the embankment of the main lines at the middle and lower reaches has avoided the risks, which effectively relieves the flood protection pressure for the middle and lower Yangtze River. Besides, the Three Gorges Project is conducive to improving the flood control of the Dongting Lake area. As a result of a huge amount of flood and silt emptying into Dongting Lake was reduced, the life of the Dongting Lake can be prolonged, creating favorable conditions for the Dongting Lake to accept the flood from the four rivers—Xiangjiang River, Zijiang River, Yuanjiang River, and Lijiang River, or flood from the Yangtze River. The Three Gorges Reservoir can also make compensative regulation for the flood water of the middle reaches; should be a flood akin to the 1954 one occur, the flood diversion amount to

---

[1] Cai Qihua, Former Vice Minister of Water Resources, Professor Senior Engineer.

the middle and lower reaches can be reduced from 49.2 billion m³ at present to 39.8 billion m³ or 33.6 billion m³, mitigating the loss in the flood diversion-retention areas.

The Three Gorges Project is pivotal for developing and harnessing the hydropower resources of the Yangtze River reasonably. The Three Gorges Hydropower Station, with the total installed capacity of 22.50 million kW, and the average annual electricity generation of 88.2 billion kW·h, has such merits as stability in power generation, durability, and favorable adjustability. It can be applied to solving the problem of the electricity consumption for Central China and Eastern China for a certain period at the beginning of the 21st century, and lessening the transportation pressure by reducing the transport amount of power generation-oriented coal from the coal production bases in the northern part of China. By dint of its superior geographical location, the Three Gorges Hydropower Station is capable of creating conditions for the "west-to-east electricity transmission" project, achieving the capacity exchange of the hydropower station and thermal power station in the thermal power-oriented electrical power system in the northern part of China, with a view to complementing the hydroelectric power and thermal power, to a greater extent, taking full advantage of the hydropower resources. The Three Gorges Hydropower Station can also be regulated in the hydropower station system to achieve the benefit compensation in the overall system.

The Three Gorges Project is the hub for connecting the Yangtze River waterway. After its completion, more than 600km waterway in the reservoir area, which was originally characterized by many river shoals, rapids, and narrow channels, becomes the good deep channel, with a view to fundamentally improving the waterway of the Chuanjiang River, realizing the direct connection between Shanghai and Chongqing via Wuhan for 10,000-ton fleets of vessels, cementing the East-West relationship, promoting the economic development in the western regions of the Yangtze River, and boosting the competitiveness of the waterway transportation in the integrated transportation system. The channelization of the Three Gorges Reservoir for the 1,200km-course of network of rivers in the main streams and tributaries in the reservoir area has irreplaceable value for improving the waterways fundamentally, which will facilitate the construction of the water network system in the Yangtze River greatly. As for the middle and lower reaches, the Three Gorges Project can deepen the navigable depth in the waterway during the dry season.

The Three Gorges Project is the treasure-house for regulating the water resources of the Yangtze River. After the completion of the Three Gorges Project, an additional discharge volume of 1,000 - 2,000m³/s can be achieved during the dry season, compared with that under the natural conditions, which creates the favorable conditions for increasing the water transfer amount in the Danjiangkou Reservoir concerning the central route of the South-to-North Water Diversion Project. The increase of flow during the dry season plays a considerable role in guaranteeing the eastern route of the South-to-North Water Diversion Project. In the long run, the water can be diverted from the Three Gorges Reservoir or the area downstream of the Dam, should a greater quantity of water diversion from the Yangtze River be needed in the northern part of China.

The Three Gorges Project is beneficial for the ecological improvement in the middle and lower Yangtze River. The project, on completion, would mitigate the risks of devastating disasters inflicted in the Jingjiang Reach area and lessen the ecological environment problems caused by the flood protection and flood diversion greatly. The water regulation during the dry season is beneficial for the ecological environment of the lakes. The increase in water during the dry season boosts the assimilative capacity in the area downstream of the Dam, improves and stabilizes the water quality in the area downstream of the Dam, and reduces the salinity of the Yangtze River, thus having the distinct functions in terms of diluting the salt tide at the river mouth, shortening the duration of the salt tide, improving the water quality, etc. The hydropower generated by the Three Gorges Dam, as a kind of clean energy, is equivalent to the energy produced by burning 50 million tons of raw coal annually, without the adverse impact on the ecological environment by thermal power station of the same scale. After the operation of the Three Gorges Reservoir, the discharge of clear water under the action of the deposition of sediments and the biochemical action will improve the water quality.

The Three Gorges Project is conducive to promoting the economic and social development of the area. Although it has caused loss due to its reservoir inundation, millions of migrants have passed the secluded production and lifestyle and gained new development opportunities. The investment environment and the infrastructure conditions in the Three Gorges area have been improved; the injection of a wealth of compensatory funds and other investments have created favorable conditions for the development of regional economy, infused new vitality for promoting the economic structural adjustment and the progress of urbanization, and promoted the local development of culture, education, and science and technology.

II

The Three Gorges Project is the result of arduous exploration and practice by generations of Chinese for nearly a century. From the magical ideas of the utilization of water resources of the Three Gorges and the improvement of navigation proposed by Sun Yat-sen in the *International Development of China* in 1919 to the lofty ideas of "a smooth lake appears in the high gorges" by Chairman

Mao Zedong; from the attention to the development of gigantic hydropower to the guarantee of rehabilitation for the people; from the economic and social functions to the benefit of ecological environment, we have constantly deepened and improved the understanding of the Three Gorges Project. During the historical process, the Changjiang Water Resources Commission (hereinafter referred to as "Changjiang Commission") was honored to become the overall organization for planning and designing the Three Gorges Project and played its unique role. In a sense, the Three Gorges Project has become a yardstick for assessing the prowess of Changjiang Commission, and is the representation of the spirit of Changjiang Commission of "solidarity, dedication, science and innovation". The members of Changjiang Commission are proud of all these.

Changjiang Commission has conducted a huge amount of survey, planning, design and scientific research regarding the Three Gorges Project since the 1950s. In 1986, the State Council organized experts and scholars for the further proof of the project nationwide and proposed 14 special proof reports, upon which the *Feasibility Study Report of the Three Gorges Project* was rewritten by the Changjiang Commission. The said report has passed the review. The resolution of "the Construction of the Three Gorges Project", passed by the National People's Congress in April 1992, has confirmed that it would be implemented in accordance with the schemes of the normal pool level of 175m and the principles of first-level development, one-off construction, stage-by-stage water storage, and continuous migration. Subsequently, the Changjiang Commission has initiated a new round of design for the Three Gorges Project once again.

The design for the Three Gorges Project consists of five phases: the feasibility research, the preliminary design, the technical design of individual projects, the bidding design and the detailed construction drawings; and the Changjiang Commission assumes overall responsibility for this. The preliminary design report consists of three main parts, namely the pivot project, the handling of reservoir inundation and migrant resettlement and the power transmission and transformation project. The preliminary design (the pivot project) was prepared and reported to a higher body in December 1992 and was reviewed and approved by the Three Gorges Project Construction Committee of the State Council in July 1993. Subsequently, the technical design of individual projects, the bidding design and the design for the detailed construction drawings concerning partial works were carried out. The scale of the Three Gorges Project is ever so huge, and its technology is complicated, with some individual projects amounting to a large-scale project. Consequently, the Changjiang Commission has conducted the technology design for the eight individual projects—the dam, the power plants, the dual-lane five-level navigation lock, the vertical ship lift, the mechanical and electrical products (including the head-end convertor station), the phase-II upstream transversal cofferdam, the safety monitoring, the fluctuating backwater area waterway and the harbor improvement (including the influence and countermeasures of the waterway undercutting in the area downstream of the Dam) in accordance with the requirements of the Three Gorges Project Construction Committee of the State Council in August 1993. During this process, more than 200 special design researches have been conducted. At present, apart from the completion of the first phase of the fluctuating backwater area waterway and the harbor improvement (including the influence and countermeasures of the waterway undercutting in the area downstream of the Dam), the other seven individual technical designs have been successfully completed. During the stage of the design of construction drawings, the Changjiang Commission has kept an open mind about the suggestions and opinions from the experts and the participating parties and provided powerful technology support for the smooth operations of the Three Gorges Project construction through careful organization, close cooperation, continuous improvement and optimized design in accordance with the design documents.

During the arduous and glorious design process of the Three Gorges Project, the Changjiang Commission has adopted the up-to-date science achievements and the state-of-the-art engineering technology with its innovative mind fully in addressing a great number of technological problems in the design in terms of the flood discharge, the energy dissipation, the form of the water inlet for power plant, the penstock, the desilting, the hydraulics of filling-emptying system of the ship lock, the deep-cut high slope, the miter gate, the headstock gear, the river closure, and the deep-water cofferdam, etc. These techniques have filled the gaps in China and created successively world miracles. Take the arrangement of the pivot layout for example, based on the researches and experiments on a number of schemes and structure forms, and in light of the features, such as the great flow rate for flood discharge and diversion through the dam, the onerous tasks of flood prevention and desilting, the large variation of the upstream water level and so on, the barrage adopts the form of concrete gravity dam, and the flood discharge structure is put on the river main channel with both deep outlet and surface outlet, the downstream power plants are arranged on the both banks. According to the river regime of the dam site, and considering the hydraulic conditions at the entrances and exits for pilotage upstream and downstream and the prevention of sediment accumulation, the navigation structure is placed on the left bank. The design for the dam has researched and solved the problems such as the selection of configuration of the three-level flood discharge outlets under different operating conditions, and the anti-cavitation and anti-silt abrasion under high-velocity flow, as well as the major problems of the joint flood discharge through deep outlets and surface outlets, the hydraulic connection and energy dissipation and erosion prevention under the joint flood discharge through deep outlets and bottom outlets, etc. For another example, the unit with high-capacity and the fast working flow rate for the Three Gorges Power Plant adopts the single-hole and small-hole water inlet, which is a breakthrough in the design technology of the water inlet for the hydropower station in China; the penstock of the power station adopts the penstock on downstream surface of dam, which not only solved the joint stress of the steel lining and the concrete construction, but also facilitated the construction; by taking advantages of the successful experience from the Gezhouba Dam and the scientific experiments for the Three Gorges Project, the disperse desilting method is capable of preventing the silt from clogging the water inlet of the power station, and the coarse sand from entering the machines, thus guaranteeing the safe operations of the power plant. Furthermore, the dual-lane five-level continuous navigation locks are the world's largest with the highest water head and greatest levels. Since it's the continuous lock, the hydraulics for the filling and emptying system is extremely complicated; for a great number of locks and valves operating under the high water head, it requires the mutual coordination and stable stream flow, and the structural design and the automatic control system design are of crucial importance. The hydraulics problem for the locks with high water head was solved through the researches on the filling-emptying system of lock arrangement and the valve equipment by the ventilation and quick opening method; the combination of flood prevention and discharge reduces the underground water level of the side slopes and provides the anchorage technology including pre-stressed anchor cable and anchor rod, which solves the problems such as the stability of high slope for locks and the change of lock head impacting upon the application of the miter gate; the bottom pintle adopts the self-lubricating structure in order to make allowances for the high miter gate; the lock chamber adopts the high-strength anchor rod and the new structure of thin-lining wall; the long-span miter gate with high water head and the hydraulic headstock gear are top-flight products globally.

The resettlement for millions of migrants from the Three Gorges Reservoir is a worldwide difficult problem during the planning and design of the project. Throughout the world history of migrant resettlement from the reservoir, a project with millions of migrants is unprecedented and will also be rare in the future. Most of the cities and counties in the reservoir area were largely located in mountain areas in the eastern Sichuan Province and the western Hubei Province. Due to the low economic development level, the rapid population growth and the limited natural ecological environment capacity, migrant resettlement remained an enormous problem. Therefore, the key to the success of the Three Gorges Project is, in a sense, the migrant resettlement. The engineering and technical staffs of the Changjiang Commission involved in the reservoir migrant resettlement have devoted themselves to the migrant resettlement of the Three Gorges Reservoir since the 1950s. The report: *Planning Report for the Reservoir Inundation and Migrant Resettlement for the Three Gorges Project*—serving as a guideline for the resettlement of millions of migrants from the Three Gorges, was creatively formulated after years of effort and practice. This report has implemented the governmental policy of "the development-oriented migration" for the Three Gorges Project and has blazed a trail in the key technical problems such as the environment capacity for migrants, the planning and design, the compensation and development, the system of investment responsibility, the fund limited planning, the construction of mountain cities, etc.

During each phase of the design of the Three Gorges Project, a great number of Chinese experts, especially a bunch of renowned experts, have actively participated in terms of guidance, assistance and review, and some domestic research institutes, colleges and universities have all offered close cooperation. Several generations of the members in Changjiang Commission have put all of their painstaking efforts into the planning and design of the Three Gorges Project during the day and night. A bunch of pre-eminent experts, under the leadership of the academician Zheng Shouren, racked their brains and worked assiduously and carefully; they are the backbone of the Three Gorges Project and the incarnation of the spirit of the Changjiang Commission. All these will be carved on the historical monument of the Three Gorges Project forever.

III

The human beings have continuously gained understanding of the objective world, mastered the laws of nature, adapted to natural changes and transformed the natural conditions in order to achieve sustainable development during their survival and development, on the basis of the process "from practice and understanding to further practice and understanding". In the 21st century, people have taken seriously the relation between the human beings and the river basin from the wider perspective of the global environment and sustainable development, and made efforts to achieve the harmonious coexistence of humanity and environment. The construction of the Three Gorges Project provides us with a new method and a new platform for establishing the compatible relationship between the people and the Yangtze River. At present, our task is that in the process of endeavoring to provide various technical services for the 3rd stage of the Three Gorges Project, we should have the enthusiasm and sufficient energy to understand the new change and influence on the ecosystem of the Yangtze River and to actively study and solve the related major problems after the completion of the Three Gorges Project; we should ensure the sustainable utilization of the Three Gorges Project and the water resources of the Yangtze River by regulating scientifically and reasonably and taking necessary measures, so as to promote the sustainable development of economy and society. To this end, we must balance the following relationships through further research.

1. Balance of the benefits among the economy, society and ecology

In essence, the Three Gorges Project is a major ecological environment project, and the electricity generation shall subject to the flood prevention and the ecology. As a result, the relationship among the flood prevention, power generation, navigation and ecological environment shall be coordinated fully through the studies on the scheduling schemes, and the scheduling and operation mode for the Three Gorges Project shall be scientifically and reasonably decided by the in-depth researches on the computing method of flood regulation for the Three Gorges Reservoir, the sediment accumulation for the reservoir, the scheduling method of flood prevention (including the compensation for the Jingjiang or Chenglingji reaches) and the scheduling mode for dual or multiple water level limitation for flood season etc. In the past, the research on the comprehensive benefits of the project mainly focused on the flood prevention, power generation, navigation, etc. Though a wealth of work has been made regarding the environmental impact assessment of the project, the studies on the ecological benefits are still insufficient. Next, further studies shall be made on how to bring into play the overall role of the Three Gorges Project and implement the sustainable development strategies better.

2. *The relationship between the Three Gorges Reservoir and the integrated flood prevention system of the Yangtze River*

In relation to obvious contradictions between the flood discharge capacity and the flood features like the high flood peak and great flood volume in the middle and lower reaches of the Yangtze River, the flood control capacity of the Three Gorges Reservoir is still insufficient. The flood prevention at the middle and lower reaches of the Yangtze River mainly relies on the integrated flood prevention system of the Yangtze River including the project. To this end, the orientation, classification, adjustment and construction of the flood diversion-retention areas at the middle and lower reaches of the Yangtze River must be carried out further, and the different planning ideas shall be taken for different flood diversion-retention areas. In the order of priority and with greater input, the construction of the flood prevention engineering and safety facilities shall be carried out in phases and in batches, so that the flood diversion-retention areas can play a proper role in the flood control and disaster reduction. Meanwhile, the research shall be conducted on the flood storage capabilities of the reservoir in the main streams and tributaries upstream, and the flood control capacity of the cascade reservoirs on the Jinsha River shall be planned.

3. *The relationship between the water level and the discharge of the main streams at the middle and lower reaches of the Yangtze River, and the relationship between river and lake*

The operation of the Three Gorges Reservoir will change the flow and water level of the main streams at the middle and lower reaches of the Yangtze River. During the flood seasons, the flow on the Jingjiang and Luoshan sections will increase under the conditions of the same water level due to the scour of river channel, and the flood diversion amount will decrease; however, the flood control pressure at the river section downstream of the Chenglingji may increase, and the distribution of the water diversion amount shall be adjusted, which requires further research. The amount of the water and sediment into the Dongting Lake through the three-inlet water diversion will decrease considerably due to the discharge of the clear water, the riverbed scour of Jingjiang Reach and the falling of the water level, the relationship between river and lake will change, and the status of silting in the Dongting Lake will change accordingly. This increases the regulation and storage capacity of Dongting Lake and lessens the flood control pressure. At the same time, it proposes new requirement for the flood channel adjustment and the river channel improvement of the river system of the Dongting Lake.

4. *The relationship between the sediment movement in river and the riverbed evolution*

The regulation of water and sediment by the Three Gorges Project has impact on the storage capacity of the reservoir, waterway, harbor and backwater, etc., and the discharge of clear water results in the scour and transformation of the river bed. In order to prevent or to reduce the adverse impact as far as possible and ensure the smooth operation of the Three Gorges Project, efforts shall be made to keep abreast of these changes, and to conduct observational analysis and scientific experimental study on the information of hydrology, sediment and river channel.

IV

There is no end to this process. Friedrich V. Engels points out that "With every day passing we are acquiring a better understanding of the laws of natural and getting to perceive the more immediate or the more remote consequences of our interference with the traditional course of nature" in the *Dialectics of Nature*. As for the grand immense Three Gorges Project, we shall constantly make efforts to gain an understanding and exploration in terms of the width and depth of its influence. Meanwhile, due to the complexity of the Yangtze River, it is a deepening process for understanding and governing the Yangtze River. The members of the Changjiang Commission shall, as always, keep improving the design capabilities and optimizing the scheduling and operation on the basis of the spirit of being highly responsible for the people, the Country and the history, with the purposes of satisfying the needs of the people and the history, and establishing the great cause of the Chinese nation jointly.

To Achieve the Full Potential of the Three Gorges Project in Ensuring People's Livelihood*

Cao Guangjing[1]

Water conservancy is a public cause that benefits people's livelihood. Large-scale water conservancy infrastructure plays significant role in terms of flood control and drought alleviation, reasonable allocation and efficient utilization of water resources, as well as water resources protection, etc. To develop water conservancy for people's livelihood, it should be given full play to ensure the effect of large-scale water conservancy infrastructure to guarantee people's livelihood. As a backbone project for harnessing and developing the Yangtze River, the Three Gorges Project (TGP) is able to provide tremendous comprehensive benefits in terms of flood control, power generation, navigation, water resources allocation and comprehensive utilization, as well as ecological environment protection, etc. which make the Three Gorges Project a fundamental water conservancy project for people's livelihood. As the owner of TGP, China Three Gorges Corporation (hereinafter referred to as "CTG"), under the guidance of "Scientific Outlook on Development", scientifically manages and carefully operates the project to give full play to TGP's role in ensuring people's livelihood, and to facilitate sustainable economic and social development.

1 Making Social Benefits in the First Place and Giving Full Play to the Public Welfare Functions of Three Gorges Project in Terms of Flood Control, Drought Relief and Disaster Mitigation, Allocation & Utilization of Water Resources, and Navigation

Giving full play to the effect of flood-control facilities and improving the flood control and disaster relief capability of the Yangtze River basin: Since ancient times, the flood control capability of the midstream and downstream of the Yangtze River have been severely inadequate; frequent disasters have brought people on both banks great sufferings. The Three Gorges Project marks the initial formation of flood control system of the Yangtze River. The huge reservoir capacity of the Three Gorges Project for flood control has improved the flood control standard of the Jingjiang section at the midstream and downstream Yangtze River from 10-year flood to 100-year flood. Every year before flood, CTG draws up regulation and operation plan for the flood season and submits it to related national authorities for approval, formulates contingency plan for flood control and organizes drills, as well as elaborately inspects and maintains the structures and facilities, such as the dam, power stations, and ship locks, etc. During the flood season, CTG strictly abides by the regulation and operation plan for flood season, executes the regulation instructions from the State Flood Control & Drought Relief Headquarters and Yangtze River Flood Control & Drought Relief Headquarters, decreases the reservoir level to flood control limit operation level, and makes the dam strong shield for flood control at the midstream and downstream of the Yangtze River through precise monitoring, accurate forecasting, scientific implementation of reservoir regulation, as well as effective peak cutting and shifting. The functions of the backbone project should be fully achieved, and the capability for water resources allocation and comprehensive utilization should be enhanced. The natural inflow of the Yangtze River has shown obvious seasonal characteristics with unevenly spatial and temporal distribution, the domestic, production and shipping water demand on the middle and lower reaches are secured through flood storage and water supply of the Three Gorges Reservoir, as well as optimization and regulation of spatial and temporal distribution of water resources. We should give full play to the effect of golden waterway of the Yangtze River and promote the prosperous development of the shipping industry of the Yangtze River. The Yangtze River is the main artery for China's inland water transportation, and is economic tie connecting the east, middle, and west parts of China. After the reservoir impoundment, the navigation conditions of the Yangtze River have been greatly improved and 10,000-tonage fleets can directly reach Chongqing from Shanghai. Since impoundment, the average annual growth rate of Yangtze River cargo volume crossing the dam has reached 13%, among which the total volume of cargo through the dam reached 87.94 million tons in 2010, more than eight times of average annual volume before impoundment, which have greatly improved the economic and social development along the Yangtze River.

---

* This paper was published in the *Truth Seeking*, Vol. 20, 2011.

[1] Cao Guangjing, Former Chairman of China Three Gorges Corporation, Professor Senior Engineer.

2 Insisting on Elaborate Operation and Scientific Regulation to Provide High-quality, Safe, and Economical Clean Energy for the Economic and Social Development

To enhance water resources development is an effective measure for increasing energy supply, improving energy structure, and protecting the ecological environment. Since the completion of the Three Gorges Project, the annual electricity production has reached 88.2TW · h. The continuous clean electric energy supply has not only provided strong support to the national economic development, but also made significant contribution to realizing the objectives of energy conservation and emission reduction in China, as well as alleviation of the global greenhouse effect. CTG has been comprehensively carrying out lean production to improve the power producing level, upholding the lean production management philosophy, constantly improving the operation and maintenance ability for the Three Gorges Hydropower Station and mega turbine generating units, optimizing reservoir regulation, improving water resources utilization effectiveness by effectively utilizing flood resources under the condition of ensuring the flood control safety. Since the first batch of generating units were put into production, 500TW · h of electricity was accumulatively produced, which greatly stimulated the development of the national economy. Safe production is comprehensively promoted and safe operation of the grid is ensured. As a backbone power source of "west-to-east power transmission" and "double-way supply between the north and south", the Three Gorges Project, at the center of the national electricity grid, has very large installed capacity. Therefore, its safe operation is crucial to grid safety. CTG has always given great attention to safe electricity production, attached great importance to technical innovation, strengthened equipment maintenance, put safe enterprise standards into practice, and continuously improved safe operation level. After the Three Gorges Hydropower Station was put into operation, the proportion of hydropower installed capacity has been significantly improved in China's total electricity installed capacity, which plays an important role in the optimization of Chinese energy structure and energy conservation and emission reduction. According to the thermal power emission standard from the State Electricity Regulatory Commission, after the Three Gorges Hydropower Station was put into production, the generated electric energy is equivalent to saving 167 million tons of the standard coal with emission reduction of 380 million tons of CO_2 and 4.5 million tons of SO_2.

3 Putting People First to Promote Migrants' Stability and Prosperity and Economic and Social Development in the Reservoir Area

People's livelihood is the most important for water conservancy project construction. CTG has actively worked with China's related authorities and local governments to do well the migration work of the Three Gorges Project, fulfilling social responsibilities of the enterprise, practicing the hydropower development principal of "build a hydropower station to stimulate the local economy, improve the environment, and bring benefits to the local people". Through more than a decade of migrants' resettlement, millions of migrants have rebuilt their homes, and the reservoir area has taken on a brand-new look. In order to ensure successful resettlement of migrants and realize the objectives of "moving out, stabilizing, and gradually becoming prosperous", CTG has earnestly implemented the national migration policies and regulations, and timely appropriated the migrants' compensation funds and follow-up support funds. By the end of 2010, the appropriated compensation funds to migrants from Three Gorges area accumulated to nearly RMB 84 billion, and the follow-up support funds for post-resettlement drawn from the revenue from power production of the Three Gorges Project were about RMB 3.1 billion. In recent years, CTG has also actively carried out designated poverty alleviation through various activities, such as talent assistance, industrial support, and goods providing to the needing people; CTG cooperated with the local governments to help improve water facilities, supported the construction of ecological industrial parks for migrants from the Three Gorges Project area, and provided the migrants with employment opportunities through various channels. The construction of the Three Gorges Project has effectively driven the economic and social development of the reservoir area. From 1994 to 2010, the GDP growth of the reservoir area was higher than the average level of China; education, sanitation, and culture undertakings have also made important progresses.

4 Giving Equal Emphasis on Protection and Development to Maintain Ecological Balance and Health of the Yangtze River Basin

The Three Gorges Project is essentially a project improving the ecology of the Yangtze River. During the construction and operation of the Three Gorges Project, the principle of "developing with protection and promoting protection through development" has been followed to strive to reduce the adverse effects to ecological environment, paying attention to the harmony between people and water, maintaining the ecological balance of the basin, as well as keeping healthy rivers. The Three Gorges reservoir area is important water source and significant ecological barrier for the Yangtze River basin, which is crucial to flood control safety and ecological safety in the downstream of the Yangtze River. During the stages of demonstration, planning, design, construction, and operation of Three Gorges Project, the related parties have paid great attention to water quality protection, geological disaster prevention and control, as well as ecological environment monitoring of reservoir area. In project construction, the water quality protection of the reservoir has been strengthened through measures of water pollution control, ship pollution treatment in the reservoir area, reservoir bottom dredging, reservoir floating objects cleaning and monitoring, etc. The state has successively invested a special fund of over RMB 12 billion to regulate the geological disasters in reservoir area. At present, according to various monitoring results, the ecological environment of the Three Gorges Reservoir area is generally in accordance with the prediction in demonstration and feasibility study phase, and partial ecological environment have been improved. The environmental restoration of dam area has been strengthened and ecological demonstration base has been built. The environmental protection of Three Gorges Dam area is the window and important part of environmental protection of Three Gorges Project. During project construction, CTG earnestly implemented various measures in environmental impact assessment documents, controlled the adverse environmental impacts on the construction area with measures of pollution control and prevention, water & soil conservation, environmental greening, population health protection, as well as environmental monitoring, etc. After the Three Gorges Project entered operation stage, CTG has increased investment and planned to invest RMB 480 million by 2012 for water and soil conservation of the dam area, ecological restoration and environmental greening, as well as construction of ecological demonstration base. The protection of rare animals and plants has been strengthened, so as to construct new aquatic ecosystem. The Three Gorges Project has brought certain impacts on the biodiversity. In order to alleviate the adverse impacts, measures have been adopted to protect the terrestrial rare plants and rare fish, such as Chinese sturgeon. The measures include: protection zone and section building, generic resources protecting, field ex-situ conservation, artificial reproduction and release, as well as special monitoring and study, etc. After the Three Gorges Reservoir was formed, the aquatic ecosystem suitable for the characteristics of new environment has been under building.

Achievements and Prospects of China's Hydroelectric Power in the Past 30 Years after the Reform and Opening-up*

Zhou Dabing[1]

Water is the vital source for life, and electricity is the lifeblood of social development. Changing flood disasters into water benefits, converting water current into electric current, and making full use of hydropower, a clean renewable energy, is an important issue for the sustainable development of the global economy, the only way to optimize the allocation of resources, and an inevitable choice for scientific development. The theoretical reserves of China's hydropower resources rank first in the world. Before reform and opening-up, however, less than 10% of national hydropower resources were exploited, and the per capita electricity consumption was equivalent to only 1/3 of the world average. Floods and electricity shortage were the two factors restricting the development of the national economy and the increase of people's living standards. The Third Plenary Session of the Eleventh Central Committee of the Chinese Communist Party convened in December 1978 marked China's stepping fully into the new historical period of reform and opening-up, and started the spring of hydropower development. In only 30 years, Chinese people working in hydropower sector carried forward the hardworking, intelligent, diligent and forging-ahead spirit of entrepreneurship, and build a monument for the leapfrog development of hydroelectric power in China.

1 Outstanding Achievements of China's Hydropower in the past 30 Years after Reform and Opening-up

The 30 years after reform and opening-up have witnessed the constant exploration and progress of China's hydropower cause. In the 30 years, under the correct leadership of the CPC Central Committee and the State Council, China's hydropower has progressed with each passing day and made remarkable achievements.

1.1 Hydropower Scale has Been Continuously Increasing

Hydropower as a clean renewable energy plays a very important role in the world energy structure. Since the founding of the new China in 1949, the Communist Party and the government have been concerned with and paid much attention to the planning and construction of hydropower. But because of various historical reasons and the relative backward productive force in the early days after the founding of the new China, hydropower development in China was relatively slow before reform and opening-up. After the Third Plenary Session of the Eleventh Central Committee of the Party, the Country re-adjusted the strategy on hydropower development, and continuously made efforts in basic hydropower construction. A number of key projects were launched, providing constant clean energy for the socialist economic development in full swing.

After reform and opening up, the pace of China's hydropower construction has been accelerated remarkably, and the average annual added installed capacity increased year by year. Especially after 1993, the annual commissioned installed capacity has been more than 3GW. In September 2004, China's total installed capacity of hydropower exceeded 100GW, making China a country with the largest installed capacity of hydropower throughout the world. As of the end of 2007, China's total installed hydropower capacity reached 145GW, maintaining its first rank in the world. It takes only 8 years to double the total installed hydropower capacity accumulated in the 50 years since the founding of new China, while the utilization rate of hydroelectric energy has also increased to 27% from less than 10% of the technically feasible reserve before reform and opening up. The rapid growth of hydropower has made important contributions to the national economy and social development.

1.2 Technologies of Hydropower and Dam Construction Take a Leading Position in the World

After reform and opening-up, China's hydropower and dam construction technology has made a comprehensive breakthrough with science and technology as the guidance, power construction as the battleground, and the major national projects as the footing.

(1) Construction technology of ultra-high roller compacted concrete (RCC) gravity dams higher than 200 meters.

---

* This article was published in *Hydroelectric Power*, Vol. 12, 2008.

[1] Zhou Dabing, Honorable President of China Society for Hydropower Engineering, Vice President of Chinese National Committee on Large Dams.

So far China has built 66 high RCC gravity dams and 35 dams more under construction. Among them, Longtan Hydropower Station with a maximum height of 216.5 meters is the world's largest and highest RCC gravity dam.

(2) Construction technology of ultra-high concrete face rockfill dams (CFRD) higher than 200 meters.

At present, among the CFRDs in China, 160 dams are at or higher than 30 meters, and 40 at or higher than 200 meters. Shuibuya Hydropower Station, with a height of 233 meters, is the world's highest CFRD.

(3) Construction technology of ultra-high arch dams higher than 300 meters and ultra-high RCC arch dams higher than 100 meters.

Design theory and construction technology of these two kinds of dams have reached the world-class level. Ertan (240m), Xiaowan (292m) and Jinping I (305m) Power Stations, for example, have elevated the arch dam construction technology of China toward the pinnacle of dam construction throughout the world.

(4) Technology of dam foundation engineering.

China has mastered the treatment technology for 150-meter-deep cut-off wall foundation in geologically complex conditions. The maximum depth of the foundation grouting hole has reached 206m, and the high slope excavation and reinforcement technology has reached the 700-meter level.

(5) Technological innovation of energy dissipator for high-speed water flow.

With the accumulation of experience and independent innovation, China has made distinctively innovative fruits of the hydraulic structure layout and the systematic studies of the form and structure of energy dissipator, which ensure the safety of the operation of the hydropower project.

(6) Construction technology of magnesium oxide micro-expansion concrete dam.

This Chinese pioneering technology with independent intellectual property rights is hailed as a major innovation and breakthrough in international dam construction technology.

It is just these high-level dam construction technologies that elevate China's hydropower development to the peak of the achievements of modern science and technology. And China is among the top countries of dam construction technology in terms of amount, size, technical difficulty and technological innovation.

1.3 Localization Level of Large and Giant Units has Significantly Increased

Over the past 30 years, hydropower electromechanical equipment in China has been significantly improved. Following the operation of a batch of units with single-unit capacity of 300MW, such as Liujiaxia, Longyangxia and Yantan hydropower stations, as well as Guangzhou Pumped Storage Power Station. Lijiaxia Hydropower Station has the single-unit capacity of 400MW, and Ertan 550MW. Either the Three Gorges or the Longtan Hydropower Station has over 30 units, each with the single-unit capacity of 700MW, and has been successfully put into operation. For Xiluodu, Xiangjiaba, Laxiwa, Baihetan, Wudongde and other hydropower stations, the singe-unit capacity has increased from 700MW toward 800MW, or even 1GW. Among the 26 generating units on the left and right banks of the Three Gorges Project, 8 are fully-nationalized with independent intellectual property rights. Through technology transfer, digestion and absorption, and independent innovation, it take only 7 years for China's manufacturing industry of large hydropower electromechanical equipment to achieve a 30-year leap, marking a new era of giant hydraulic generating sets designed, manufactured and installed locally in China. At present, China has built 32 units with singe-unit capacity of 500MW and above, and 109 more under construction, both ranking top in the world.

The localization process of the pumped storage units is constantly accelerated. Construction of pumped storage power stations in China was started in the late 1960s and entered into large-scale construction stage in the late 1980s after reform and opening up. By the end of 2007, 17 pumped-storage power stations had been built with a total installed capacity of 8.945 GW, and 11 are under construction with a total installed capacity of 12.16GW. Guangzhou Pumped Storage Power Station I and II, which have been put into production with the total installed capacity of 2.4GW and a water head of 535 meters, is the world's largest pumped storage power station. Xilongchi Pumped Storage Power Station, which is still under construction, has a total installed capacity of 1.2GW with a nominal water head of 640 meters. Huizhou Pumped Storage Power Station has a total installed capacity of 2.4GW with a nominal water head of 517.4 meters. As a result, construction of pumped storage power stations in China has entered the high-capacity, high water head and high lift era.

1.4 Construction of Large Hydropower Stations has Promoted the Improvement of Large-capacity, Long-distance and High-voltage Transmission Technology

After the founding of new China, with continuous development of the national economy, the grid system operating voltage level has also been improved in order to meet the needs of high-capacity and long-distance power transmission. After the country entered a new era of reform and opening-up, which has further promoted the rapid development of China's heavy industry, the power grid construction soon stepped into an era of extra-high-voltage (EHV) transmission. Over the past 30 years, the construction of high dams, large reservoirs, and large power stations has promoted the vigorous development of large power grids and high voltage transmission lines. Large power grids such as those for Fengman Power Station (220kV), Liujiaxia (330kV), Gezhouba (500kV), Gongboxia (750kV) and Xiangjiaba (800kV), etc. have been built and put into operation, pushing forward China's power grid construction technology towards the world leading level in terms of direct current (DC) transmission technology. Currently, in order to achieve the great development in the western part of China and electricity transmission from the west to the east, as well as to meet the needs of power transmission for hydropower development in southwest China, the state has initiated the construction of 1,000kV alternative current (AC) and ±800kV DC transmission demonstration projects which will serve as the backbone network of the national grid to bridge major regional grids, and main energy channels for inter-regional and inter-basin power transmission, hydro-thermal supplementation and regulation, and national energy resource optimization configuration.

1.5 Hydropower Development has Brought Huge Benefits to National Economic and Social Development

Over the past 30 years, the CPC Central Committee and the State Council have unswervingly pushed forward reforms, promoted the opening-up policies and made tremendous achievements in socialist modernization, while China's hydropower cause has also born fruitful results. In addition to the huge direct economic benefits, hydropower generation has also brought immeasurable economic efficiencies in such areas as building a new socialist countryside, regional economic development, improving the atmospheric environment, reducing river siltation, improving water quality, protecting power grid operation in a safe and economical way, as well as flood control, transportation, irrigation, water supply and aquaculture.

Power generation and environmental benefits: by the end of 2007, China's total installed hydropower capacity had reached 145GW, and the annual electricity output 486.7TW·h, replacing the thermal power generated by 243 million tons of raw coals and reducing carbon dioxide emissions by 487 million tons.

Flood control: Taking the Three Gorges Project an example, the huge reservoir storage capacity will be able to effectively regulate floods from upper reaches of the Yangtze River and raise the flood control standard from ten-year return period to a hundred year in the Jinjiang Section of the middle reach, greatly reducing the threat of floods to people's lives and property, and ensuring the safety of 15 million people living in the Jianghan Plain and Dongting Lake area in flood season.

Transportation efficiency: Navigation conditions in the Three Gorges Reservoir have been greatly improved after the reservoir impounding, and the shipping amount in the Three Gorges was increased from the historic record of 18 million tons dramatically to more than 60 million tons in 2007. A golden age of navigation has been ushered in for water traffic.

Relocation benefits: As a typical example of hydropower-incurred relocation, Wujiang Development Company actively cooperates with the power plant designer and the local government to prepare for the relocation plan and compensation investment budget and mobilize local government and the relocated people to ensure that the relocation work be completed in a complete and steady manner, that the relocated people get rich gradually, and that the reservoir area keep stable in a long term. As of the end of 2006, a population of 68,785 people had been relocated and RMB 2.438 billion of fund allocated for the hydropower stations built by Wujiang Corporation, including Hongjiadu, Suofengying, Goupitan, Silin and Dahuashui power stations. Relocation and resettlement had been carried out smoothly, ensuring the construction of the power plants on schedule.

Reduction of river siltation and improvement of water quality. Due to the enormous flood-regulating function of the Three Gorges Reservoir, sand content of discharged water is decreased significantly, the sediment erosion distance downstream lengthened, and erosion volume increased, thereby reducing sediment deposition, and greatly improving the water quality.

Ensuring safe and economical grid operation. Hydropower has the features of quick start-up, peak and frequency adjusting, economical operation, emergency standby and black start (pumped storage), and thus plays a key role in power structure optimization, quality ensuring of power supply, resource conservation promotion, green and environmental protection, power system security, and improvement of the economy of power system operation.

Promotion of the construction of new socialist countryside. At present, installed capacity of small hydropower in China has reached about 55GW. Development of rural small hydropower and replacing firewood by electricity can not only improve the ecological environment of the vast rural areas and accelerate the realization of power supply in all villages, but also increase the grid coverage area, reduce the production and living cost of the broad masses of peasants, improve the level of economic income, and speed up the construction of a new socialist countryside.

2 Successful Experience of China's Hydropower Development in the 30 Years after the Reform and Opening-up

2.1 Following the National Energy Strategic Development Direction and Advancing with the Times, in the Hydropower Development

The electrical energy is one of the important pillars of the national economy and social development. Since the disposable fossil fuels such as oil and coal have limited proven reserves in China, and carbon dioxide and other greenhouse gas emissions generated in coal combustion will cause more and more severe global warming problems, hydropower as a kind of clean, green and renewable energy, will have a more prominent position in the electric energy structure. After China's reform and opening up policy, the state has continued to increase efforts in adjusting the power structure, and always attached great importance to the development of hydropower. The 17th CPC Congress report pointed out the construction of a resource-saving and environment-friendly society must be highlighted in the development strategy of industrialization and modernization. *China's Energy Conditions and Policies White Paper* explicitly requested, under the protection of the ecological condition and properly solving the relocation problem, hydropower should be vigorously developed. During a short period of 8 years from 2000 to 2007, China's GDP grew from RMB 9.9 trillion to RMB 24.95 trillion, and national power generation capacity grew from 319GW to 713GW, in which the hydropower installed capacity grew from 79.35 GW to 145 GW. The speed of the development of electric power industry is consistent with the economic growth, and it has played an important role in supporting the development of the national economy.

In over 30 years of reform and opening up, China's hydropower development has insisted on implementing the scientific development outlook, and the hydropower development is consistent with the national economic development under the guidance of the energy strategy of CPC Central Committee and the State Council, making important contributions to speeding up the process of building a moderately prosperous society in all aspects.

2.2 A Diversified Development Pattern of Hydropower Development Through a Series of Reforms

Under the guidance of the general policy of the country's reform and opening up and the market economy-oriented reform ideas, China's hydropower system reform was initiated in the late 1970s, generally experienced four stages in the past 30 years: ①Hydropower design and construction management system was transferred from the local governments during the Cultural Revolution to the central government. After the restructuring of hydropower enterprises, it was made clear that hydropower projects be contracted and constructed by the construction enterprises. ②Lubuge Impact triggered by Yunnan Lubuge Power Plant that introduced the World Bank loans and carried forward the international bidding, leading to the implementation of the hydropower construction market competition mechanism. Afterwards, the nationwide large-scale hydropower projects began to implement market-oriented reform, and the hydropower reform was fully conducted in the national hydropower industry. The construction of basin companies of the relevant rivers steps on an important stage. ③The deepening period of hydropower system reform and the adjusting period of the hydropower development. On the one hand, the initiation of the Three Gorges Project and the building of Ertan, Xiaolangdi and other large projects greatly facilitated the reform and management capacity of major rivers and large projects; on the other hand, due to the overheating of investment in fixed assets, the state had adopted some adjustment measures. Consequently, the number of hydropower projects entered an adjustment period, so did the hydropower development. ④With the initiation and implementation of the power system reform, implementation of scientific development outlook proposed by CPC Central Committee and strong electric energy demand of national economic development, the hydropower system reform and development have entered a new stage with both opportunities and challenges coexisted.

After 30 years of development, China's hydropower has stepped on the diversified development pattern from single national financial allocation before reform and opening up, to fund-raising for electricity and multi-channel power generation, making the hydropower development more flexible and orderly, and further accelerating the process of China's hydropower development.

2.3 Promoting the Sustainable Development of Hydropower with Scientific Planning

Energy security is one of the foundations of national security, while reserves and assessment of energy resources involve planning and overall goal of national economy and social development. Three surveys on hydropower resources in China were conducted in the 20th century. The first and second were implemented respectively in the 1940s and the 1950s. Limited by the depth and breadth of these two surveys, the results were of less accuracy and reliability. The truly large-scale national hydropower resources survey is the third during 1979-1980, through which the situation of China's hydropower resources was essentially clarified. But with the development of the country and the establishment of the market economy, changes in economic, social and productive conditions of hydropower construction occurred. For better development and utilization of hydropower resources in China, during 2000-2003 the state organized the fourth survey, mobilizing more than 1,000 engineering and technical personnel to spend more than three years making new check and assessment on 3,886 rivers in China's mainland territory, each of which has the installed capacity of 10MW or more. The results showed that theoretically hydroelectric reserves of the mainland was 694.4GW, technically feasible reserve 541.6GW, and economically feasible reserve 401.79GW, ranking first in the world in all aspects.

During the more than 30 years of reform and opening-up, China's hydropower development, under the care and attention of the CPC Central Committee and the State Council, has made scientific planning and layout, careful design and construction, high quality and strict management, and given priorities to hydropower development of major rivers, formed 13 major hydropower bases, implemented the policy of optimal allocation of resources, stimulated the rational development and utilization of hydropower resources in China, and promoted sustainable development of hydropower industry.

2.4 Promoting Orderly Development of Major River Basins by the Establishment of River Basin Development Companies

In the 1980s, considering the distinct basin characteristics in terms of hydropower resources distribution in China and the need to make the most of the economic and social benefits of water resources through basin regulation, the state put forward the "basin, cascade and expansion" development principle of hydropower; in combination with national hydropower development planning and construction of 13 major hydropower bases, a number of basin development companies have been established, and some of them have gradually developed into diversified investment companies jointly invested by multiple parties. Basin development companies follow the market principles and implement integrated management on basin development and operation, so as to expand and amplify the investment. Some river basin companies have reached agreement with power grids to implement unified water flow and power regulation for basin cascade stations, and strive to optimize the allocation of hydropower resources, in order to make the most rational use. In the structure of most basin development companies, the local companies on behalf of the interests of the local governments are involved so as to integrate the relocation of local people, environmental protection and local economic development during the hydropower development process, and promote the harmony and unity between hydropower development and regional economic as well as social development.

The resources reserve of the 13 major hydropower bases exceeds a half of the total national amount, and the hydropower installed capacity accounts for more than 60% of the total, so the basin development companies have promoted the development and construction of hydropower base. At present, about 30% of hydropower resources in the 13 major hydropower bases have been developed, which has played a great role in promoting the development of the west regions and implement west-to-east power transmission.

2.5 Attaching Great Importance to Environmental Protection and Relocation of Local People, and Realizing the Harmony between Human and Nature in Hydropower Development

Vigorously developing hydropower on the premise of protecting ecological environment and solving properly the relocation of local people is the scientific philosophy to promote hydropower development in a reasonable and orderly manner, which has been responded extensively by all sectors of the community, agreed by Chinese hydropower practitioners, and implemented seriously during the project implementation process.

Relocation and resettlement of local people is a very difficult task in hydropower development and construction. In the past 30 years, relocation caused by hydropower development has advanced constantly in trials, and relevant initiatives are also adjusted accordingly. The policy is adjusted from the past resettlement-oriented to development-oriented relocation , which means, in addition to the financial compensation for the relocated people, more importantly the follow-up supportive policies are introduced to ensure the long-term production and life of migrants in a scientific and reasonable manner, realizing the objective of "successful resettlement, stabilization, and gradual well-off" . Meanwhile, the development-oriented resettlement of local people and the building of a new socialist countryside are organically combined. Especially after entering the 21st century, on the basis of doing well in relocation, more attention is paid in the process of hydropower development to implementing follow-up supportive policies on resettlement; all development owners, while fully understanding and implementing the national policies and regulations on resettlement in hydropower development, have innovated relocation follow-up support policies with regional and local characteristics to help the relocated people get rich through their hard work and promote economic and social development of the region where power stations are situated.

In the process of hydropower development, the laws of nature have to be complied to realize the scientific and rational development; the development activities and natural protection must complement each other in order to achieve the harmony between man and nature; otherwise the irrational development of hydropower resources and the changes of river morphology without demonstration may cause new natural disasters. In the past 30 years, water and electricity workers have further strengthened the concept of environmental protection; whether in the planning and design stage or the construction period, or the site clearance and vegetation recovery period after the completion and dismantling of the project, great importance is attached to environmental protection. The report of the 17th CPC National Congress proposed the concept of ecological civilization building. Hydropower construction in the new era has put forward higher requirements for the environmental protection, and the requirements of the CPC Central Committee and the State Council should be achieved: "to develop hydropower on the basis of environmental protection", strive to achieve win-win in development and environmental protection, and truly achieve the harmony between human and nature during hydropower development.

2.6 Accelerating Localization Process of Hydropower Generating Units by Independent Innovation

During the localization process of China's hydropower generating units, both advanced equipment and core technologies are introduced simultaneously, and the driving force of national major projects upon the hydropower technology innovation and major technical equipment innovation is fully exploited. Through introduction, digestion, absorption and re-innovation, the successful path on the equipment localization has been found. After reform and opening up, in the context of development of large hydropower bases, construction of many large-scale hydropower stations, and high market demand for large hydropower equipment, China had gradually mastered the rules of the market; relying on national major projects, China has attracted world-class manufacturers with strong market demand; while purchasing advanced equipment, China has introduced key technologies for digestion, absorption and re-innovation. Therefore, China's hydropower equipment design, manufacturing, installation and operation level are increased substantially, and a great leap is realized. Through the implementation of independent innovation, on the one hand, technological transformation of the old unit is increased, transition and innovation from the old unit to the new modern units achieved, continuing to play and improve the economic benefits of the old unit; on the other hand, foreign advanced equipment manufacturing technology experience is learned, upgrading and updating of the localization of the large units sped up, and the level of automation and control of large units increased. A batch of 300MW, 400MW, 500MW to 700MW units have researched and developed, and now 800MW and even 1GW higher-level units are under development. Manufacturing, processing and installation capacity of China's huge hydropower units have reached world leading level. At the same time, localization of large pumped storage units is stepped up, more high-quality and reliable electricity produced, the grid scheduling made more flexible and efficient, the grid operation made more economical, safe and reliable, and the economic and social development better served.

2.7 Giving Priority to Science and Technology to Elevate Modernization of Hydropower Construction Equipment

In the 30 years of the reform and opening-up, the level of domestically produced large-scale hydropower units have reached the international first-class, a variety of dam engineering techniques are leading the world, and hydropower construction equipment modernization has increased continuously. The vertical ship lift, designed and constructed independently in the Three Gorges Project, has the largest lift tonnage, the highest lift height and the most difficult technology in the world. The two-way five-stage ship lock of the Three Gorges is a large ship lock with the largest water head and the most continuous grades. Its herringbone gate height and single gate weight rank number one in the world. Under the guidance of the strategic goal of building an innovative country proposed by CPC Central Committee and the State Council, and relying on major projects, China's technologies on hydropower equipment manufacturing and building construction equipment modernization stride rapidly, and have reached the international advanced level, so that China's hydropower design, construction, manufacturing, installation, operation, and management level has increased significantly. Hydropower construction period is shortened enormously, million-kilowatt hydropower construction period is shortened from seven years in the past to 4-5 years, greatly reducing the cost of construction, improving the economic benefits of hydropower, and enhancing the market competitiveness of hydropower. The quality of hydropower projects is the significantly improved. During Wenchuan earthquake on 12 May 2008, many urban buildings collapsed in the earthquake, but no hydropower stations in the earthquake region suffered significant danger or collapsed, or caused any secondary disasters. The RCC arch dam of Shapai Hydropower Station in the epicenter was in sound condition; Zipingpu Hydropower Station was recovered for power generation in 5 days after the earthquake, which showed that the quality of China's hydropower dam construction had withstood the toughest test.

2.8 Actively Developing Small Hydropower, and Vigorously Promoting the Building of New Socialist Countryside

China's reform and opening up has injected vigor for rural electrification. During a grassroots research in 1982, Mr. Deng Xiaoping fully affirmed the practice of "setting up small hydropower stations by self-reliance to deal with the local power supply" created by the rural local governments and farmers, which has opened up a road of rural electrification in China through the development of small hydropower stations. Before the reform and opening up, national small hydropower installed capacity was less than 7GW. Now, more than 45,000 small hydropower stations have been built with the total installed capacity of about 55GW, accounting for 38% of the installed capacity of hydropower nationwide. Since the small hydropower stations in rural areas are numerous and widely distributed, basic power supply function is far greater than its share of 5% in the national power generation capacity. At present, 1/2 of the land, 1/3 of the counties and 1/4 of the population around the country rely on small rural hydropower stations for power supply. China's small hydropower construction has formed a rural electrification road with Chinese characteristics, and China has become a veritable kingdom of small hydropower.

The rapid development of small hydropower stations in rural areas has improved rural electrification, driven economic and social development in rural areas, improved the production and living conditions of farmers, played a key role in solving power supply in rural areas, promoted poverty reduction, achieved prosperity of the economic and social development in countryside, and made a significant contribution to comprehensively promotion of the building of a new socialist countryside. In the new era, the unique role of rural hydropower with small hydropower stations as the mainstay in mountain ecological construction and environmental protection, energy conservation is widely acclaimed by all sectors of the community. Especially in the southern snow and ice storms in the first half of 2008, and the Wenchuan earthquake on May 12, 2008, small hydropower stations, owing to the advantages of distributed energy, had played a huge role in ensuring network security and disaster mitigation.

3 Enlightenment and Prospects of China's Hydroelectric Power in the 30 Years after the Reform and Opening-up

In 30 years, after the successful power system reform and the introduction of competition mechanism, China's electric power cause has witnessed a rapid development. However, during the process of introducing the electric power market competition mechanism, some extreme environmental thoughts and forces at home and abroad had formed a social trend, slandering the hydropower development by exploiting the intense competition among major power companies, and their indifference and inability to timely respond to false propaganda in the society. As a result, the hydropower development, which should have been developed in the competition actually, is blocked in different degrees. Therefore, China's hydropower development is seriously lagging behind thermal power and hydropower proportion seriously imbalanced, the energy structure constantly worsened, and the development of the electric power industry suffering unprecedented hardship.

The theoretical reserves of water energy in China is more than 6,000TW · h. According to the current survey and design level, the technically feasible amount is 2,470TW · h. If fully developed, the energy equivalent of 1.0-1.3 billion tons of raw coal can be provided. But by the end of 2007, the annual hydropower generating capacity in China is only 486.7TW · h, roughly equivalent to 243 million tons of raw coal. Under the condition that the hydropower development level in China is much lower than developed countries, the domestic criticism over China's excessive hydropower development has risen one after another. Under the circumstance that the current annual new installed power generating capacity is more than 100GW, only 2.7GW of large and medium-sized hydropower projects were approved by the state in 2007, with the ratio less than 3% of the total growth of the entire power industry. If so, the energy development policy of giving priority to the development of hydropower can not be realized.

Countries in different development stages have different advantages. For all developing countries where the hydropower development technology, fund and environment are restricted, hydropower development has both opportunities and challenges. China's hydropower energy reserves rank first throughout the world, which determines that the role of nuclear power in China is unlikely to exceed hydropower. Wind power and solar power, relative to the huge growth of energy demand in China, have much greater symbolic meaning than the actual meaning in fact. Only when we are determined to develop and exploit hydropower as soon as possible, can China's energy structure be fundamentally improved and the sustainable development of the national economy and social development be achieved.

3.1 Implementing the Scientific Development Concept, and Promoting the Process of a Harmonious Society through Hydropower Development

The 16th CPC National Congress proposed building a socialist harmonious society. Hydraulic engineers should further study and implement the scientific development outlook, raise awareness, attach great importance to the development of hydropower, ensure the sustainable development of China's economic construction and energy, and promote the process of building a socialist harmonious society through hydropower development. Hydropower is an important energy resource, and its development and utilization of is an important measure to increase energy supply, improve the energy structure, ensure energy security, reduce greenhouse gas emissions, protect the ecological environment, resolve the climate change, and achieve sustainable development.

Giving priority to the development of hydropower is an important principle of China's energy development. *Medium and Long-term Development Plan of Renewable Energy* clearly puts forward the target of 300 GW of newly installed capacity of hydropower development in 2020, so hydropower construction task is still arduous. Facing the new situation and new requirements based on hydropower construction, hydropower development must get guided by the scientific development outlook, effectively change the view of development from a pure engineering hydropower to ecological hydropower, transform from a purely technical engineering to social engineering, pay more attention to the interests of migrants and ecological environmental protection, combine hydropower development and comprehensive utilization of water resources, ecological engineering construction and regional economic development, and promote sustainable economic and social development.

3.2 Implementing National Concept of Building up Ecological Civilization and Building the Ecological Engineering of Hydropower Development

To develop hydropower in a reasonable and orderly manner, ensure electricity supply, and realize hydropower development goals in the Eleventh Five-Year and 2020, we must closely follow the new concept of building ecological civilization proposed by the 17th CPC National Congress, follow the fundamental principle of simultaneous development and protection, further increase environmental protection efforts in the construction of the power stations, improve environmental protection measures, combine truly hydropower development and comprehensive utilization of water resources, ecological construction and regional economic development, achieve comprehensive development of river basin water resources, thus making hydropower construction conductive to environmental protection, form a good ecological environment, and promote coordinated development of man and nature with scientific and orderly manner.

Practices have proved that the built hydropower stations have played important roles in promoting ecological development. According to domestic and international experiences, scientific increase of hydropower development will be more conducive to the protection of ecological environment. For example, our previous cascade hydropower development was not in large scale; more run-of-river hydropower plants were built, but it is short of regulation by leading reservoirs, so not only the quality of power generation can not be ensured but floods and low water shut-off can not be effectively dealt with. For a typical example, before Xiaolangdi Reservoir on Yellow River was built, drying up of the Yellow River was almost impossible to avoid. Only after the Xiaolangdi Reservoir was built and the whole basin has been rationally regulated, was the drying up of the Yellow River completely solved. Due to the obvious contradiction of spatial and temporal distribution of water resources, insufficient lack of development will inevitably lead to inadequate protection. To increase hydropower development efforts in a scientific and orderly manner is also an important means of ecological protection.

3.3 Properly Addressing Relocation Resettlement, and Building People's Well-being Projects of Hydropower Development

Improving people's well-being is an important focus of the 17th CPC Congress report. Hydropower construction inevitably results in relocation of local people, which has become an important factor restricting the construction of hydropower. We should further adhere to the people-oriented concept and the harmonious society construction, develop hydropower orderly on the basis of good environmental and resettlement work, earnestly strengthen the relocation resettlement and stability in hydropower construction work, combine the hydropower construction and interests of the migrants better, promote the socialism new countryside construction, promote local economic and social development, and focus on the building people's well-being projects of hydropower development.

Relocation of local people is the priority of hydropower development. Only if society is stable and the relocated people get rich can hydropower construction be smooth and prosperous. In recent years, the state unveiled a new *Resettlement Regulations* and follow-up support policies to create conditions for properly addressing the relocation problems left over. But with regard to the present situation, the old and new relocation problems are still quite prominent, and the following tasks should be focused: firstly, make full use of existing relocation follow-up support policies, adjust measures to local conditions, take multiple channels to resolve the issues left over, and ensure that living standards of the relocated people are gradually improved and the reservoir areas stable; secondly, strengthen the preliminary relocation work of new power plants, conduct resettlement-related work synchronized with the main part of the project with the working efforts no less than the requirements of the main project, and earnestly implement state policy related to relocation issues; thirdly, strengthen research work on resettlement. In the framework of existing policy, according to the actual situation of each reservoir area, we need to make innovative working ideas, effectively solve issues, and realize long-term stability of resettlement; fourthly, in combination with the needs of local economic development, support local economic development, give full play to the advantage of local resources, make reasonable distribution of the efficiency of hydropower development, and make hydropower construction as a booster of economic development. Only if the local economy is developed and powerful can all the people's production and life issues (including the relocated people) be resolved fundamentally.

3.4 Improving Modern Corporate System and Building of River Basin Developers and Maximizing Benefits of Power Stations

Concerning the current status of several investment and development bodies in the same basin, if a tacit understanding on the power plant operation scheduling can not be reached, the capacity and efficiency of the cascade hydropower stations will not be maximized. To improve the basin comprehensive hydropower benefits, the important thing is the implementation of a basin-wide joint scheduling as well as unified scheduling and optimized scheduling. In order to maximize hydropower construction, it is recommended that: if there is more than one investment and development body in the same basin, we can follow the philosophy of modern corporate system, that is, several investment subjects can, through consultation and restructuring, establish and complete the corporate governance structure with a diversified investment, stock cooperation and regulated operation, establish modern enterprises for cooperation in basin development, realize basin-wide cascade joint scheduling, give better play of benefits of cascade hydropower stations on the same river, and achieve win-win situations.

3.5 Improving Earthquake Resistance Capacity of the Power Stations in Seismic Region and Promoting Sustainable Hydropower Development

After the Wenchuan earthquake on May 12, 2008, the public paid enormous attention to the safety of hydropower development in the southwestern region and questioned whether the southwestern seismic belt is suitable for construction of hydropower stations. If devastating earthquakes occur in the earthquake belt where reservoirs are built, how can we ensure the local people living nearby will not suffer secondary disasters? In addition, the public concerns for reservoir-induced earthquake also forms resistance on the hydropower development in some areas. How to make full use of modern scientific and technological means to ensure the seismic safety of dam construction and eliminate concerns of the public, are the first issue to face in the future hydropower development. On the one hand, the relevant national departments should strengthen the publicity and popularization of the knowledge of the earthquake; on the other hand, the construction of buildings through developed technology is the most effective method to resist and defeat the earthquake disasters.

3.6 Intensifying Publicity of Hydropower Policies to Create a Good Social Environment for Development of Hydropower

At the end of the 20th century, there was a tendency in the international community to deny the large-scale hydropower development. After repeated research and debate among experts around the world, especially in the new context of global reduction of greenhouse gas, the action to support hydropower development in developing countries was passed unanimously in the International Sustainable Development Summit held in 2002. In October 2004, the Beijing International Conference on Hydropower and Sustainable Development stressed that hydropower is an important renewable energy, and has an important role in achieving sustainable development. However, due to various reasons, there are still widespread opinions against hydropower construction at home and abroad, imposing a negative impact on the public understanding of hydropower. How to work well on the public opinion and publicity on hydropower development and create a good social environment are important tasks of hydropower construction in the new circumstance. Hydropower development should strengthen the publicity efforts, including the basic knowledge of hydropower, the latest progress and trends of hydropower, the achievements and beneficial aspects of the environment, and report problems and shortcomings in hydropower construction objectively to form good atmosphere to promote hydropower development in a healthy and orderly manner.

In the course of reform and development in the past 30 years, China's hydropower is prospering along with the socialist modernization. Standing at a new historical starting point, hydropower development is promising and have shouldered with heavy responsibilities. We should get more closely united around the CPC Central Committee, get guided by the scientific development outlook, conscientiously implement the country's hydropower development policies, seize the golden period for the development of electric power industry as an opportunity to forge ahead and press on, provide a steady stream of clean energy to build a socialist innovative country, and make greater contributions to building a moderately prosperous society in all respects and a socialist harmonious community.

Philosophical Speculations on Disputes in Dam Construction*

Lin Chuxue[❶]

In recent years, the fierce debate on the advantages and disadvantages of domestic dam construction reflects, to some extent, the opposition and argument between different viewpoints of the western post-modern society ideological trend and the traditional mainstream philosophy thought and their increasingly important influence on sectors of engineering, social sciences, media and government departments in China. It has realistic and far-reaching significance to analyze and explore the philosophical root of those disputes from the position and viewpoint of Marxist materialist dialectics, absorb nutrition from scientific and rational spirit and modern humanism thoughts, refine the basic concept and paradigm of engineering philosophy which conform to the modernized development stage of our country, implement the scientific concept of development, correctly formulate and implement the industrial policy of water resources utilization and energy development in our country, and improve the construction of water conservancy and hydropower projects in our country.

1 The Dispute on Dam Project Roots in Differences between the View of Nature and the Theory of Value

Engels pointed out that "we are faced with two big changes in this world which are the reconciliation between people and nature and reconciliation of people themselves (Page 603 of Volume 1 in *Karl Marx and Frederick Engels*). This is also the triple relationship to be coordinated for building harmonious society today. On the micro level, the balance between people's conscience and cupidity; on the middle level, coordination among people including coordination among different nationalities, countries and civilizations; on the macro level, the harmony between man and nature, which refers to general long term harmony between human and their habitats where they were born and raised. The most basic and most ancient relation between man and nature is the relation between man and water, and the important means for human to adjust such relationship in an active way is to build kinds of water conservancy projects.

Water is the mother of life, the change of water distribution in terms of space and time on the earth, has cultivated the splendid human civilization, but extinguished some ancient civilization at the same time. The progress in science, technology and management contributes to growing capacity of water conservancy project engineering, which can adjust the space and time distribution of water partially. The construction of various water conservancy projects meets the demand on water resources by agriculture, population growth, industrialization, urbanization and the modern economic and social development. The modern water conservancy projects often have functions in flood control, power generation, navigation, recreation, etc. Today, hundreds of thousands of dams in different scales, as core facilities of modern water conservancy engineering, are standing over main streams and tributaries of rivers in the world to supply water for domestic and industrial use, electric power and transportation convenience.

Science and technology have been criticized during their development, so have the dam projects, although they have brought great benefits to human beings. The advantages and disadvantages of dam construction have been disputed for a long time, sometimes loud, sometimes weak. Under these circumstances, the western developed countries have built thousands of dams, thus completing a water conservancy infrastructure system and basically completed water resources development. In recent years, fierce debates on the ecological and social influence of large-scale water conservancy and hydropower engineering have been also carried out in China and become white-hot sometimes.

The reasons for serious difference in dam construction opinions are complicated, but generally from several aspects, such as: difference in understanding of basic facts and actual data, different in skills for using the data for professional analysis, different views due to different aspects of different disciplines, imbalance of profit sharing among different stakeholders, in addition, technological, economic and social influences caused by a failure case of dam construction or mismanagement in certain aspects, etc. From philosophical roots, the most basic difference roots in different views of nature and theories of value which reflect the collision of different philosophical thoughts.

In early period, the most famous dispute case for dam construction is the 7-year long great debate on the reservoir project of

---

* This paper was published in *Construction of China's Three Gorges*, Issue 6, 2006.

❶ Lin Chuxue, Vice President of China Three Gorges Corporation and Vice President of Chinese National Committee on Large Dams, Professor Senior Engineer.

the American Hetch Hetchy Valley in the early 20th century. The reservoir was key project for remote water supply to San Francisco. The Hetch Hetchy Valley is located inside the scenic Yosemite National Park. The protest movement sponsored by John Muir (the founder of "Sierra Club" which is the World's earliest and one of the largest nature conservancy organizations at present, and the advocator of natural reserve) had drawn the attentions and quite wide response of the whole nation. After judging and weighing the environmental and social benefits and losses, the special committee of U. S. Congress has passed the reservoir proposal, and the construction started after approved by the President Theodore Roosevelt. Another supporter for the construction of this project was Gifford • Pinchot (the pioneer of environmental protection movement and famous leader of progressivism, the first director of Forest Service of America and conservationist). The "wise use" principle by Pinchot clearly pointed out that the ultimate goal and value basis for human to protect the natural resources is "the greatest good of the greatest number in the long run", which has become the classic guide for resources protectionism of the traditional environmental protection movement and has formed the early ideological basis of sustainable development theory as well. The protection and utilization policy of the national forest and land, rivers and other resources based on this idea played a decisive role in reversing the difficult situation of American natural resources protection. But as "wise use" principle demonstrated humanistic thought and rational subject. According to the "post-modern" view of history, "materialized" and "dwarfed" the nature. Pinchot's resource conservation idea was labeled as "anthropocentrism" and was lambasted.

Muir, who was hailed as a prophet of environmental protection and was the "Western Taoist", had advocated "the spirit of wilderness", traveled the beautiful mountains and rivers of North America and illustrated his naturalistic philosophy with large amount of wonderful travel notes. The stunning and beautiful words "have moved the whole nation" . Muir was against the "wise use" principle and believed that the beauty of wilderness (representing the nature) was independent of human. Muir's contemporary and later proponents inherited and developed his thought, formed "the land ethics", "deep ecology" and other theories of eco-centrism, adhered to the idea of the "inner" value of nature itself and the equal rights and proposed to protect the nature for nature instead of human. These ecological ethic thoughts echo the appeal by postmodern philosophy to negate the science rationality and re-understand and re-assess the relationship between human and the world. With eco-centrism as the theoretical weapon, the international anti-dam movement organization shouted out the slogan of "let the rivers run freely" and made action plans. Domestic anti-dam responders also proposed "retaining the original ecology" which is a radical environmental protection claim with a strong "postmodern" color and condemned the dam projects both in speech and writing, so as to influence the public opinion and the governmental decision-making on large-scale construction project of water conservancy and hydropower projects.

In the Case of Hetch Hetchy Valley Reservoir, the dispute between Pinchot and Muir marks the policy division between the resources protectionism and the nature reserves ideas deriving from the internal environmental protection front due to the difference of natural values. Their ideas have been followed by different environmental protection organizations or theoretical schools, and influenced the trend of American environmental protection policy constantly.

2 Discriminating and Analyzing the Philosophy of Advantages and Disadvantages of Dam Project by Marxist Philosophy

Marxist philosophy has realized the scientific unity of materialism and dialectics, the study of the conception of nature, the methodology and the humanistic spirit in Marxist philosophy can enlighten and guide us to understand the philosophy in dam engineering. Three aspects should be taken to understand the ideas: the practice theory and the subjectivity theory analyzing man's subjective initiative are the starting point and standpoint for researching on engineering problems; epistemology and dialectics analyzing contradiction movement are the method to discriminate and analyze dam engineering disputes; Marxist view of nature as well as the values of people-oriented throughout provide the basic dimensions and specifications for building a harmonious project.

"All social life is about practice" (Page 56 of Volume 1 in *Karl Marx and Frederick Engels*, which must reveal the problem of main body of practice. In Marxist materialism, "firstly, existence should be regarded as the main practice body and the object to be transformed; it is believed that people's practical activities constantly change the world around, and the free nature constantly changes into humanized nature" (Page 26 of *the Basic Questions of Marxism-Leninism*).

Today, monthly, the newborn baby can have self awareness under enlightenment of parents, and annually, the young man can be familiar with the environment where they grow up with the help of education and science. When did hominid have self awareness is still a difficult question for research in the history of philosophy. Philosophical historians tell us that it is a great leap of human development that human had self consciousness and were able to distinguish the object and subject. As the object had deeper understanding of the subject, human realized of being able to adapt to or use or even partly change the nature and make it more suitable for human existence, i. e. , the generation of human subjective initiative which represents greater leap of human development and reveals a new chapter in the history of the relation between man and nature. The progress of science and technology amplifies the human subjective motivation, changes the weak position of human in front of the natural forces to a certain extent and helps the human society to approach much nearer to the free kingdom from the realm of necessity. During the development of western philosophy over more than two thousand years, from the subject-object relation firstly found by the ancient Greek Socrates Plato, to the subjectivity philosophy of "ego cogito ergo sum" by Descartes and the Marxist philosophy of dialectical materialism have all analyzed and witnessed the development and evolution course of subject-object relation between man and nature.

Human beings have interfered with the nature with their subjective initiative and change the the evolution rhythm of the nature. Since the industrial revolution, science and technology have advanced rapidly, human's ability and the scale to intervene with the nature have shown unprecedented growth, at the same time, some side effects have emerged gradually. Just like the incisive analysis by Engels in *the Dialectics of Nature*, "we shouldn't be reveling in our victory over the nature. The nature will take revenge on us for each of such victory." The deterioration of ecology and the exhaustion of resources result in the imbalance of the natural ecological environment for the survival of humans. Modern science and technology achievements have been appropriated for the force contest between people or countries to scramble for interests, with potential threat to destroy the entire earth and the human. People have to face with and be questioned by the ultimate question of "to be or not to be" again. From philosophers to engineers, from politicians to artists, science, philosophy and even the human responsibility was re-pondered seriously. This thought itself is a part of people's subjective initiative. Man's subjective initiative not only reflected in understanding and changing nature, but also displayed in correcting the mistakes of humans themselves. "Transform the objective world and our own subjective world – change our own understanding ability and reform the relation between the subjective world and the objective world", and "this is the theory of the unity of knowledge and practice in the dialectical materialism" (Page 297 of Volume 1 in *On Practice* Selected Works of Mao Zedong, People's Publishing House, in 1991). Through reflection, the human thinking on sustainable development is clearer and clearer and a global consensus forms gradually. The Scientific Development View is the representation of such global consensus shown in the guiding principle of economic and social development.

However, various disadvantages of society, economy, culture and environment at present are attributed to human's conquest over the natural forces after mastering the science and technology in the post-modern social ideological trend. Some philosophy schools use the reflection to criticize the scientific rationality, trying to subvert the conception of nature of the subject-object relationship of the traditional mainstream philosophy thoroughly, criticize it as anthropocentrism, uplift the flags of eco-centrism and advocate equal rights between human and living things that human beings should not and are not able to intervene and change nature. These reflective philosophies constituted the ideological basis of the contemporary radical environmentalism, and "land ethics", "deep ecology", etc. which have influence in the academia and the public are their academic proposition. Mr. Engels has once criticized Feuerbach's the old materialism view of nature as "worshiping the nature passively and being in ecstasies over worshiping the magnificence and universal of the nature" (Page 360 of Volume 42 in *Karl Marx and Frederick Engels*), through which we can also realize the negative and one-sided eco-centrism.

Philosophers often quote Einstein's words that "How much does fish know the water where it swims in all its life?" Human's perception of the world is always limited. Solely from the perspective of epistemology, humans are not really the measure of all things. Most of the philosophy schools questioning scientific rationality use it as breakthrough to launch fierce attack. Religion, agnosticism, obscurantism, mysticism and even pseudoscience, also look for living space here. Materialist dialectics tells us that the progress of science, as the commander of human's cognitive knowledge of the universe, is always the constant process of negating negation and approaching from finite toward infinite. Scientific theory is an open system, which shows the truthfulness during constant enrichment and update by putting forward problems and assumptions through real evidence or falsification. Lenin said that "the nature of human thinking can provide and is providing us with the absolute truth composed of the total relative truth. Every state of the scientific development can add new momentum to the sum of such absolute truth" (Page 135 of Volume 18 in *The Collected Works of Lenin*). Science today has made great achievement in terms of meeting the basic human survival needs, but still has a long way to go to solve the problems for humans' sustainable development. However, it neither hides the great role of science and technology projects in the process of the human recognizing and transforming the nature nor shows that the shame and negation of science can achieve the harmony of man and nature. On the contrary, only under the guide of Scientific Development View and relying on the progress of science and technology and the enhancement of engineering capability, should various problems including the ecological and resource problems that the world faces be solved finally, so as to create the green civilization of mankind.

"The philosophers interpret the world in different ways, but the problem lies in how to change the world" (Page 6 of Volume 3 in *Karl Marx and Frederick Engels*). The general values of believers of eco-centrism are vague and ambiguous, and its worldview that breaks away from the practice is also pessimistic and negative. They deny the rationality and feasibility of changing the nature by science, technology and engineering, but try to bypass the science to search for salvation, which leave in serious paradox. On the one hand, they put forward staggering eschatology prophecy to render the fear of science and technology; on the other hand, they couldn't find solution apart from scientific rationality and engineering practice, and many people's thoughts can only wander among literature, religion, mythology, and metaphysics for a long time.

Different values fiercely contradict with each other on dam construction. Those who are against dam construction and advocate "let rivers run freely" and "retaining the original ecology" emphasize "the intrinsic value" of the natural rivers, but lose the fundamental measure of human survival benefit. Their so called "the nature understood abstractly in isolation and fixed as being separated from people, is nothing at all for people" (Page 178 of Volume 42 in *Karl Marx and Frederick Engels*). As a typical process of humanized nature, dam projects embody humanitarian efforts for people to improve their own safety conditions and living quality. Since ancient times, the overflowing rivers and floods in summer has threatened the safety of life and property of the people along the river every year. The floods threat has been reduced greatly by launching water-conservancy projects, and the peaceful years have increased gradually. Our country is in the period of important strategic opportunities of the economic and social development and is facing with the severe challenges from resource and environmental problems. Developing and using the natural resources more scientifically and more reasonably and exchanging the maximum economic development with minimum ecological price are the standards for measuring the gains and losses of engineering construction including dam projects construction.

3 The Disputes on Dam Construction should be Carried out in Healthy Context

The People's Republic of China is the country that has built the largest number of dams in the world. Water and hydropower projects have played an important role in agricultural irrigation, industrial and urban water supply, flood control and disaster reduction at river basins, rural power supply and so on. Although China has comparative advantage for top one water resources reserves in the world, the development of large and middle scale hydropower engineering went a tortuous path. China has experienced difficult periods, such as "Lesson from Soviet Union", "Struggles between Water and Fire", the capital and technical constraints, etc. (Preface of *Don't let river go for null* by Li Rui, China Electric Power Press, 2002). With the reform and opening-up, the hydropower development entered into the relatively smooth stage, and by 2004, China's hydropower installation surpassed the United States and became the world's largest.

Hydroelectric power is renewable clean energy, the undeveloped economically-developable hydropower resources in our country is nearly 300 million kW at present. Once it was developed, coal consumption equivalent to 500 million tons per year would be saved and emissions equivalent to about 1.5 billion tons of carbon dioxide would be reduced, which would make significant contribution to overcome the greenhouse gases issues, the biggest threats to our common habitat. From the perspectives of utilizing water resources, the construction of water conservancy projects and the development of hydropower themselves are the ecological engineering to improve human survival environment and deliver good social, economic and ecological benefits. On an overall basis, the positive benefits are greater than the negative. It is the responsibility of China's engineering field, especially hydropower workers to develop the rich water power resources of rivers our country to serve the overall benefits of the national social and economic development.

There were also a few failure cases of dam engineering in China that brought profound lessons. Furthermore, the negative effects of mismanagement in terms of migration and environment in the past have become conspicuous gradually over the time. The issue of developing water conservancy hydropower resources scientifically and orderly draws high attention from academic circles, engineer and public opinion. Several articles, such as *Adopt a New Engineering Concept and Promote the New Development of Productivity* by Xu Kuangdi, the President of the Chinese Academy of Engineering, *On Harmonious Coexistence of Human and Nature and on Dams and Ecology* by Wang Shucheng, the former Minister of the Ministry of Water Resources, *Should We Build Dams* by Lu Youmei, the academician of the Chinese Academy of Engineering, *the Water Conservancy Engineering and Ecological Economy* by Tengteng, the former Vice President of Chinese Academy of Social Sciences, *Promoting Harmony between the Water and Electricity Engineering and Environment* by Qu Geping, the original Director of Resources and Environment Committee of National People's Congress, have dialectically analyzed the important position of dam construction in relationship between man and nature, affirmed the positive effects of dam construction on social and economic development as well as ecological protection in China and also pointed out the practical significance of improving the dam project construction management and controlling its negative influences on ecology and society. These discussions have had great instructive effects on considering the basic proposition of engineering philosophy and guiding the disputes of dam construction to develop to a positive side.

Different from the disputes in the past, the main force of opponents in the debate on dam construction today includes scholars, experts, environmentalists and media reporters due to their strong opposition to the modern development model based on science and technology. The words from them are harsh and some extreme ideas are trying to the social and economic value of dam projects.

Many social science scholars at home and abroad have incisively analyzed the idealism essence and the negative harm of social advocates of the philosophical thought of the post-modern social ideological trend and pointed out that "the realization of such social advocates can only make us homeless at most in the rubble of modernism", which was published in *Modernization Cannot Reject Modernity* by Wang Xiaolin(2006). Some phenomena in the anti-dam disputes are particularly noteworthy.

Firstly, equalizing anti-dam with environmental protection, i. e. , to put the construction of dam and reservoir for flood control and power generation on the opposite to the environmental protection, to erase the positive effect of ecological improvement of the dam projects, to amplify the negative effects so as to denigrate dam project, especially the hydropower project. For the publicity, they deliberately filter out the repeated affirmations by the international society, especially the official organizations and specialized agencies responsible for preparing sustainable development policy (such as the United Nations, the World Bank, WEC and the energy departments of governments of various countries) in all kinds of official documents of that water conservancy and hydropower engineering are encouraged and developed as the facilities for comprehensive utilization of water resources and clean renewable energy. Such activities of labeling their environmental thought orientation with anti-dam opinion is supported and followed by some public, which is related with the one-sided propaganda of blocking and distorting mainstream voice by some mainstream media. This tendency is obvious in academia as well. Deep ecology and other light green (or translated as "soft green") theories have been introduced as a classics for environmental protection and have accordingly been well-respected, while other green theories including Marxism natural view and even the viewpoints, policies and practices of conservative environment which are dominant in the field of environmental protection are mostly regarded as lagged-behind, rejected or attacked. The glory of environmental protection can only be enjoyed by the radical ideas of environmental protection exclusively.

Secondly, the active thought of "science studies" has been fully demonstrated during the dam controversy and the later debate on "the fear of nature" . In addition to degenerating and vilifying science and technology violently as in the post-modern social ideological trend, some anti-dam people always behave as savior to guide the public unconsciously and believe that their values of eco-centrism are more advanced than those of anthropocentrism, their own environmental protection ideas are lofty with sacred missions and occupy the moral high ground which mean that they are competent to be the moral tutor of scientists and engineers and are able to transcend expertise field and practices and use "the poetic romance" to build "scientific facts" and "theory of science and technology" to point out the right way to experts in the industry. In fact, they are exactly the same as the superficial and ridiculous followers of science studies during the American Sokal Affair at the end of the 20th century. For example, there is still a popular anti-dam idea with Chinese characteristics, namely, the low-dam theory, which advocates low dams over the rivers of lowland European countries to be used for the hydropower development for the mountainous rivers in southwest China. Some anti-bam followers said it is not that we are against all kinds of dams. Why don't we learn from Europeans to build low dams? The ancients are much wiser than you, for example, Dujiangyan Dam is not a high dam. The Dujiangyan Dam is the crystallization of the wisdom of the Chinese nation, the marvel of ancient projects over thousands of years, impressing the world and appraised by many engineers at present for its perfectness. However, we should not use the case of Dujiangyan Dam that channels water with dam structure to disprove that modern water conservancy dam projects and high dams are stupid. The history of science and technology tells us that high dam project is product of modern hydraulic and dam engineering and the result of integrating the scientific calculation, engineering materials and construction technology improvement. There was no high dam in the ancient times because the ancients were not able to build high dams over the main stream, and today, the selection of low or high dam projects are decided by the terrain and the situations instead of smart wisdom. Most of the southwestern rivers in our country with rich water resources have V-shaped river ways among canyons, and the water volume changes greatly with seasons. These two conditions have determined that low dams are inapplicable (the only option is run-of-river type of hydropower station, which may easily result in cutoff at sections, which can be avoided by ecological environmental protection with efforts). To those who understand basic knowledge about dam engineering and know about the hydrology, it is easy to understand. Facts have proven that such "pseudo-technology theory" has confused the public.

The rational reflection and criticism on science do not mean to overthrow the scientific theory itself; the public participation in democratic decision on national large-scale public project doesn't mean that non-professionals are eligible to transcend professionals to make decisions on science and engineering. Hayek, who believed that scientism was the ideological basis of planned economy, pointed not at the science and the scientists but at those who believed that the science and technology could solve all problems in the world while criticizing the scientific rationality sharply. When anti-dam dispute occurs in the context with insufficient scientific and rational development and less rigorous academic atmosphere, the domestic followers of postmodern trend go further beyond the line than their foreign teachers in substituting "vision" for science, interpreting values as scientific knowledge, quoting academic views as scientific rules, and neglecting fact, logic and operability, etc. For example, dam-opponents always prove how their thoughts of environmental protection and water conservancy are advanced by quoting the beautiful environment of European and American countries, however, the hydropower development degree there is quite high, which very well demonstrates that the development of hydropower and maintaining beautiful mountains and rivers could coexist, doesn't it? For another example, the anti-dam publicity says that the mainstream of the Rhine River is beautiful and clean because there is no dam. But they don't know that the Rhine River was known as European sewer with serious environmental pollution in the 1960s, which demonstrates that the river pollution is not directly associated with dam construction, isn't it? There are numerous similar examples in dam disputes. The most appalling case is the theory of so-called "wooden barrel effect" which attributes the severe drought in the Sichuan Basin to the Three Gorges Dam.

The case which can reflect the academic atmosphere of anti-dam extremists at most should be the false publicity of American dam removal and the mistranslated name of the book *Silenced Rivers*. In recent years, several hundreds of old non-functional waste dams over the tributaries and brooks in America have been removed, which are averagely about 6m high and 65m long as investigated, and most of them are small unknown dams unrelated to hydropower. Some dam-opponents hype that the dam removal age has come in developed countries. While publishing articles in the anti-dam column of the magazine of *Chinese National Geography*, for the purpose of arousing people's anger and confusing the public, they deliberately delete the data of dam length and width from the list and add images of magnificent dams such as Hoover, Glen, etc. The book *Silenced River* with the subtitle as *The Economy and Politics of Large Dams*, written by the director of American IRN (International Rivers Network), the core of world anti-dam movement, is the bible for global dam opponents. This book describes in prose style that dam and the hydropower as the root of all evil and the most stupid thing that human beings have ever done. All social evils and economic crises, such as human rights, pollution, corruption, poverty, pestilence and waste, are connected with dams. This book does not contain any theory of dam engineering or economics, but as translated into Chinese and introduced to China, its name was intentionally changed to *the Economics of Dam*, with the purpose of giving the book an academic title, changing the idiosyncratic views into academic thus facilitating marketing the claims against dams. It is eye-popping that as the book of Chinese version is reprinted, its English name was fabricated as *The Large Dams Economics* and the book was designated as higher education textbooks. The gimmicks reveal the arbitrary nature and utilitarian pursuit of anti-dam scholars. These gimmicks have deteriorated the healthy context of debate on dams and other problems seriously, interfered with the debate justice, distorted clarification and problem solving.

4 The Dam Projects in the New Century Call for the Integration of Scientific Spirit and Humanistic Thought

Science, technology and engineering are the major means for human beings to change nature in modern time. Science focuses on the discovering, technology focuses on invention, and the engineering focuses on creation. Engineering is the reconstruction activities that human beings make based on their understanding of the natural law (namely science) and tools and techniques (namely technology) to seek advantages and avoid disadvantages of nature (namely behavior purpose distinguished by values). The functions and effects of engineering on nature and human society will depend on the cognitive ability, the technical level and the correctness of values. Due to the practicality and sociality of engineering and the long term influence of economic, social and ecological consequences of large-scale projects, the engineering practice is always indispensable from the direction of Scientific Development Concept. The Chinese enlightenment leaders in China had realized apart from "Mr. Science", "Mr. Democracy" was also needed to save China as early as during the May 4th Movement. However, since the war of science and metaphysics, science and philosophy gradually went far apart in our country due to various historical reasons, and a certain gap has formed between engineering and philosophy. In recent years, the domestic scholars have firstly carried out researches on the philosophy of engineering and made efforts to "build a bridge between engineering and philosophy", which is a good attempt for introducing the speculation of philosophers and sages on the universal problems of the world into planning, implementation and post evaluation of a project.

The engineering construction contains many philosophical questions. In the analysis of debates over dam construction theory, more opposing and argumentative aspects of different views in the modern western philosophy ideological trend are seen. When we explore with the vision of materialist dialectics, we would see the complementation and integration side of such philosophy thoughts, especially the scientific rationalism and the modern humanism, including many reasonable parts existing in the postmodern criticism consciousness, which can be entirely used as our spirit nutrition to establish the construction concept for constructing the dam project of the new century. "The natural science includes human science in the future, just as the human science includes natural science" (Page 128 of Volume 42 in *Karl Marx and Frederick Engels*). The truth, the goodness and the beauty belong to the research category of scientific humanism. The main mission for science is to identify truth, the values are to judge the goodness, and the artistic thought is to evaluate the beauty. The mission for engineering philosophy research is to summarize these aspects together and enable our engineers with stronger sense of social responsibility and global consciousness, collaborate with scientists, politicians, financiers, entrepreneurs and artists to undertake the arduous task of river governance and dam construction jointly.

The pros and cons of constructing dam over the river should be analyzed objectively and realistically based on the reality of the economic and social development level of China. We should face up to the great benefits as well as a variety of negative influences brought by the projects of water conservancy and hydropower projects. The academic ideas should be widely open and encourage diverse thinking in order to obtain constructive opinions, especially absorb reasonable composition from the opposite opinions to improve the construction of water conservancy and hydropower projects continuously and take all kinds of engineering and non-engineering measures to minimize various negative effects. The government's policy guidance and implementation standards should be more clear for scientific planning, sufficient argumentation, careful decision-making, acting by law, making consideration as it stands and grasping the principal contradiction, and they should also guide the developers to surpass the development mode of traditional water conservancy and hydropower engineering by legislation and policy and use enough resources and efforts to protect the public interests of migrants and improve the ecological environment of the reservoir area.

Today, many people are too embarrassed to sing the beautiful passages and lyrics of "awakening the sleeping mountain and changing the appearance of the rivers" in the song of *My Motherland*, which used to touch numerous builders. In fact, the real issue lies in the purpose and the way of "awakening" and "changing". Our country is still a developing country in the modernization stage, which should set development strategy and the environmental policy according to the reality instead of accepting the post-modern advocacies of deep ecology, which are radical even for the West. The engineering of China directly shoulders the responsibility of the construction of national major projects, from the sky to the sea, cross mountains and rivers, for building bridges, paving roads, building factories and opening mines. It is a glorious mission to do as much as possible to make our country stronger and revive. Under the guidance of Scientific Development Concept, the water conservancy and hydropower industry will surely accomplish its mission for river regulation and hydropower development in our country very well by using the successful experience and failure lesson accumulated from water conservancy and hydropower construction project since the founding of the People's Republic of China. When with the outstanding engineers, builders with advanced engineering philosophical concept solved the two subjects of immigrant and ecology through long-term planning, scientific decision-making, meticulous design, orderly development and people-oriented harmony construction, and the water resources and hydropower sector should fulfill its mission of river regulation and hydropower development.

A Study on Hydropower Strategic Development for Long and Medium Term in China

Yan Zhiyong[1], Qian Gangliang[2]

1 Introduction

Hydropower resources in China are abundant, with reserve ranking first in the world, and constitute a very important part of conventional energy sources. After the founding of new China in 1949, hydropower in China saw a rapid development despite of its poor basis in every aspect. Especially since reform and opening-up, and along with rapid economic and social development and technology advancement, hydropower development has made remarkable and brilliant achievements, and played a vital role in the development of national economy, society and environment. Referring to the experiences of developed countries, in order to adjust the energy structure, achieve energy's low-carbon development mode, and guarantee the country's sustainable development and global climatic change adaptation, we still have to make great efforts to accelerate hydropower, such clean renewable energy, in China in the future under the premise of sound protection of eco-environment and proper arrangement of resettled people.

2 Overview of Hydropower Resource

According to the *Results of Survey of Hydropower Resources of People's Republic of China* (*in 2003*) issued by the National Development and Reform Commission in 2005, the technically exploitable installed capacity of the hydropower in Mainland China was 541.64GW, and the annual generation was 2.474 trillion kW · h. Among them, the economically exploitable installed capacity was 401.79GW, and the annual power generation was 1.7534 trillion kW · h. According to the 2003 national review of hydropower resources, the technically exploitable installed capacity of small hydro ranging from 500kW to 50MW reached 65.21GW. In 2009, the Ministry of Water Resources issued the findings of the national survey on rural hydropower resources. According to its results, the technically exploitable installed capacity of hydroelectric projects ranging from 100kW to 50MW (inclusive) reached 128GW, with annual power generation of 535 billion kW · h. According to the result of the national review of hydropower resources, in combination with the assessment of the national survey of rural hydropower resources issued by the Ministry of Water Resources, the technically exploitable installed capacity of hydropower stations of 100kW and above can reach 600GW in total, and annual power generation of 2.7197 trillion kW · h. With economic and social development and technological progress, the installed capacity of both the technically economically exploitable hydropower resource in China will be further increased.

The hydropower resources in 12 provinces (autonomous regions and municipalities directly under the central government) in Western China approximately account for more than 80% of the country's total, especially the five provinces (autonomous regions and municipalities directly under the central government) in the southwestern area including Yunnan, Guizhou, Sichuan, Chongqing and Tibet, which account for 2/3. The hydropower resources are mainly concentrated in the hydroelectric energy bases including Jinsha River, Yalong River, Dadu River, Lancang River, Wujiang River, the upper reaches of the Yangtze River, Hongshui River of Nanpan River, the upper reaches of the Yellow River, Western Hunan, Fujian-Zhejiang-Jiangxi, Northeast China, main stream in the north of the Yellow River, Nujiang River, etc., and the total installed capacity reaches 280GW which accounts for 52% of the country's total of technically exploitable potentials. Especially in western China, the total installed capacity of the main streams of the middle and lower reaches of the Jinsha River reaches nearly 60GW, the total installed capacity of the main stream of the upper reaches (from Yibin to Yichang) of the Yangtze River exceeds 30GW, while that of the Yalong River, Dadu River, the upper reaches of the Yellow River, Lancang River, and Nujiang River exceeds 20GW each, and that of Wujiang River and Hongshui River of Nanpan River exceeds 10GW each. The concentration of hydropower resources in these rivers provides favorable conditions for the cascade and rolling development of the river basin, the building of large hydropower energy source bases, and the utilization of the large capacity of hydropower resources to implement the program of "West-to-East Power Transmission".

---

* This article was published on *China Engineering Sciences*, Issue 6, 2011.

[1] Yan Zhiyong, Chairman of PowerChina Group Co. Ltd. and Vice President of Chinese National Committee on Large Dams, Professor Senior Engineer.

[2] Qian Gangliang, Vice Chief Engineer of HydroChina Corporation, Professor Senior Engineer.

3 Status and Role of Hydropower Resources

The composition of the total residual exploitable reserves of conventional energy sources (among them, hydropower resources are renewable energy resources, and which are calculated against the assumption of the technically exploitable deposit for 100 years) is as follows: raw coal 61.6%, hydropower 35.4%, crude oil 1.4%, and natural gas 1.6%. Hydropower resources, ranking the second only after coal, have a very important strategic position. In terms of power generation, based on the technically exploitable volume of hydropower resources, each year 1.143 billion tons of raw coal can be replaced by hydropower, and that means 114.3 billion tons for 100 years. Therefore, to develop hydropower resources and hydropower generation is an effective approach to adjust the energy, structure, develop low-carbon energy sources, carry out energy conservation and emission reduction, and protect ecological environment.

Meanwhile, in addition to power generation, hydropower projects also have multiple benefits such as flood mitigation, irrigation, water supply, navigation, tourism, etc. With the development of hydropower, our investigation, design and construction technologies of hydropower engineering, manufacturing of large hydraulic turbine generator units, and long distance power transmission technology, etc. have reached the world level. The development of abundant hydropower resources in western region is an important part of Western Development, and the implementation of "West-to-East Power Transmission" is beneficial to the optimized configuration of energy resources in China and the economic development in western region. Therefore, the hydropower development is of great significance for the sustainable economic and social development in China.

4 Hydropower Development History and Current Status

When the People's Republic of China was founded in 1949, the installed hydropower capacity of China was only 163MW. In 1980 when China just started the reform and opening-up, the total installed capacity of the country was 65.87GW, among which hydropower numbered 20.32GW, accounting for 30.8%. In the decade from 1981 to 1990, the installed capacity of hydropower increased by 15.73GW, reaching 36.05GW in 1990. From 1991 to 2000, the installed capacity of hydropower increased by 43.3GW, reaching 79.35GW in 2000. At the turn of the 21st century, hydropower in China had entered an unprecedented development stage. In 2000, the installed capacity of hydropower in China exceeded that in Canada, which made China become the second largest hydroelectric power country in the world. In 2004, the Gongboxia Hydropower Station on the Yellow River was put into operation, marking that the installed hydropower capacity of China topped 100GW, and China surpassed the U.S. to become the No.1 hydropower country in the world. In 2010, with the operation of Xiaowan Hydropower Station on Lancang River, the installed capacity of hydropower in China broke through 200GW mark, and China has been No.1 in the world in terms of hydropower for seven consecutive years.

In the 11th Five-Year Plan period from 2006 to 2010, the installed capacity of hydropower increased by 96.88GW, which exceeded the total of that in the first 50 years after the founding of New China. By the end of 2010, the installed capacity of hydropower in China reached 213.4GW, accounting for 22.2% of the total installed capacity nationwide; the annual power generation was 731 billion kW·h, accounting for 16.3% of the national total. In 2010, the composition of installed capacity of hydropower was: 196.4GW from conventional hydropower stations (out of which, 138GW of large and medium-size hydropower stations, and 58.4GW from small hydropower stations), and 17GW from pumped storage power stations. In 2010, the installed capacity and the power generation volume of conventional hydropower stations respectively accounted for 36.3% and 29.6% of that of the technically exploitable installed capacity and annual power generation from hydropower resource in China as a whole.

5 Inspiration of Hydropower Development from Other Countries

Hydropower development of western developed countries is 30 years earlier than China in average. China is far behind those western countries whose hydropower development has got mature. At present, the average development level of hydropower in developed countries is over 60%. Among them, about 82% hydropower resources have been developed in the U.S., 84% in Japan, 65% in Canada, 73% in Germany, and over 80% in France, Norway and Switzerland, while the development and utilization level of hydropower resources in China is less than 30%. In terms of the development stage of hydropower, China is also much later than western countries. For example, it was the peak period for the construction of dams and hydropower projects in the U.S. in the 1960s and the 1970s. The development of hydropower in Norway started at the end of the 19th century, and the large-scale development started after the World War II, and peak period for the development of hydropower in Norway was in the 1960s. The average annual growth of yearly installed capacity exceeded 10%. In the 1980s, the growth rate gradually slowed down (3%-4%); during the late 1990s, installed capacity was little increased. Compared to the developed countries, China started its hydropower development relatively late, and now China is experiencing a very critical development stage.

The development course of hydropower abroad have showed that since hydropower provides clean renewable power to meet the demand of industrial and agricultural development as well as production and living of residents, it has been the first choice at the early stage of economic development in every country, thus giving priority to the hydropower development is a common choice of every country. For example, hydropower in U.S. played an important and irreplaceable role in stimulating national economic development during the Westward Movement and the Great Depression and in providing power for the production of strategic materials during the World War II. All countries have been giving gradually increased attention to the protection of ecological environment in hydropower development, such as doing more research, taking various measures, protecting the biological diversity in river ecosystems, further optimizing the operation and regulation of reservoirs, strengthening ecological management, compensating the ecological environment in rivers, alleviating environmental impact, strengthening science and technology communication and enhancing the public sense of the entire river basins.

Some principles adopted in the hydropower development of some international rivers could be taken as reference for that of trans-boundary river of China. These principles include benefit sharing, rational investment, sharing and compensation mechanism, sound management institution (run by business bodies instead of government department or independent law persons), equal involvement in management. The measures of eco-environment protection in other country's hydropower development and the mechanism of interest sharing with resettled residents, etc. are worthy of reference for China, including the increasingly improved laws and regulations for the eco-environment protection during hydropower development, the gradually implemented certification system for the environmental impact of hydropower, the continuously strengthened measures for eco-environment protection, the sharing of taxes on hydropower by the resettled residents, the development funds for resettlement, long-standing compensation and favorable power rate entitled to resettled residents, and so on.

6 *Challenges for Hydropower Development*

Atmospheric pollution in China is mainly caused by coal-burning. 85% of SO_2 and CO_2 emission and 70% of smoke and dust are generated by coal burning. CO_2 is the main gas which leads to greenhouse effect. The global climatic change has become a common threat to mankind. It has become an international agreement to take measures to reduce the greenhouse gas emission, deal with climatic change and alleviate the impact of climatic change. As one of the main countries of CO_2 emission, China will inevitably face huge pressure from other countries. Climate change may become a major uncertain factor which influences the economic development of China in the future. Chinese government has specified two restrictive targets: one is the proportion of non-fossil energy shall reach 15% by 2020, and the other is the emission of CO_2 per unit of GDP shall be reduced by 40%–45% compared to that in 2005. In order to reach these two goals, it is required to accelerate the development of renewable energy sources and new energy sources.

China is one of the main countries for the production and consumption of energy sources with coal as the main energy source. The intensive exploitation and consumption of coal has become the main cause of environmental pollution, and has exerted serious impact on the implementation of the sustainable development strategy. To optimize energy structure is an important task for the sustainable development of energy. At present, the coal consumption in China accounts for as high as about 70% of the national total energy consumption, while the proportion of renewable energy sources such as hydropower, wind power, etc. is less than 7%. Among China's coal consumption each year, nearly 50% is used for power generation. China's petroleum and natural gas resources are limited, while the external dependence of petroleum has exceeded 50%. In order to ensure the national petroleum security, it is necessary to control such dependence under 60%, which will be very hard. Therefore, there is quite a problem about energy security, and it is hard to realize a coordinated economic, social and environmental development by solely relying on fossil energy.

In addition to the great efforts to promote energy conservation and improve energy efficiency, the approaches to effectively increase and guarantee energy supply mainly include the following: firstly, to give priority to domestic resources, greatly develop and utilize non-fossil energy such as hydropower, nuclear energy, wind power and so on, and to increase the supply of localized clean energy sources; secondly, to expand international cooperation, sufficiently and effectively utilize international resources as important supplement. To greatly develop renewable energy sources such as hydropower, wind power, etc. can not only provide clean electric energy and service for economic and social development, but also effectively optimize the energy supply structure in China. Among the available renewable energy sources (hydropower, wind power, photovoltaic power generation, biomass energy, geothermal energy, ocean energy, etc.) in China, hydropower features the greatest development value, the largest economy of scale and the most mature technology for a long time. Therefore, to fully tap hydropower is the key for the development of renewable energy sources in the future.

7 Target and Focus of Hydropower Development

For a quite long period of time at present and in the future, the guiding ideology for the development of hydropower is: to make meeting the requirement of energy conservation and emission reduction and local economic and social development in western China as our basic starting point; and being environment-friendly as the basic principle; and social harmony as the basic prerequisite; and science & technology innovation, system innovation and administration innovation as the support; and enhancing the development of key river basins, ensuring continuous development of hydropower bases and realizing "the electricity transmission from the west to the east" as focal point; strengthen the hydropower development in hydropower bases, especially in Tibet, intensify international exchange and cooperation in order to realize a sound and active hydropower development.

By the end of 2010, among the total installed hydropower capacity of China, the conventional hydropower stations reached 196.4GW, and the hydropower stations under construction reached 50GW. By the end of 2015, the installed capacity of conventional hydropower stations will reach 260GW, of which, the large and medium-sized hydropower stations will reach 192GW, and the small hydropower stations will reach 68GW. During the period from 2011 to 2015, 100GW of installed capacity will have to be increased in order to meet the target of commissioning of conventional hydropower projects with installed capacity of 350GW by 2020.

By 2020, hydropower will account for 8.1% of the whole energy consumption in China, and non-fossil energy sources will account for more than 50%. By 2020, the installed capacity of hydropower in China will reach 350GW, of which the large and medium-sized hydropower stations will reach 270GW, and small hydropower stations 80GW. At that time, the hydropower development in the mainstream of Yalong River and Dadu River will be completed for their main parts.

By 2030, with an additional installed capacity of 80GW generated by large and medium-sized hydropower projects to be built, and plus that in 2020, the installed hydropower capacity of China will reach 430GW, among which, the large and medium-size hydropower stations will represent 340GW, and the small-size hydropower stations will represent 90GW. By then, the hydropower on the Jinsha River of the upper reaches of the Lancang River will be put into operation; the construction of planned cascade hydropower projects on the Nujiang River of Tibet will be started on full swing; efforts will be made to start the construction of 1 – 2 cascade hydropower stations on the downstream of the Yarlung Zangbo River. The development of hydropower in the mainstream of the major rivers such as the Lancang River, Jinsha River, Yalong River, Dadu River, etc. will be basically finished.

By 2050, with an additional installed capacity of 80GW generated by large and medium-sized hydropower projects to be built, and plus that in 2030, the total installed hydropower capacity of China will reach 510GW, among which, the large and medium-sized hydropower stations will account for 410GW, and the small hydropower stations will reach 100GW. By then, the development of hydropower in the mainstream of the major rivers such as the Lancang River, Jinsha River, Nujiang River, Yalong River, Dadu River, etc. will be basically completed (See Table 1).

Table 1 Strategic Goal for Hydropower Development in China   Unit: GW

Year	2015	2020	2030	2050
Goal	260	350	430	510
Large and medium-sized hydropower stations	192	270	340	410
Small hydropower stations	68	80	90	100

8 Development of Pumped Storage Power Stations

Since pumped storage power stations are an important standby power source in electric power system for peak regulation, frequency regulation, condenser operation and emergency accidents, its development has attracted attention in every country worldwide. The 1950s marks the beginning of the development of pumped-storage power stations. After the World War II, economy of all countries, especially European and American countries, was restored and put back on the track of rapid development, the peak-to-valley difference gradually increased. Besides, the conventional hydropower resources were approaching their limits, so the construction of pumped-storage power stations were started in many countries to take the peak loads and regulate frequencies in the electric power system. From the 1970s to the 1980s, pumped-storage power stations saw the fastest development abroad. In that period, the pumped-storage power stations were introduced to other countries of the world from the industrially developed countries in Europe, America, Japan, etc. In the late 1990s, the construction of pumped-storage power stations in Europe, America, and Japan slowed down, while the moderately-developed countries and developing countries in Asian regions enjoyed a fast economic development in the same period, which boosted the power demand and increased the electric peak-to-valley difference, which is why most of the pumped-storage power stations in this period were constructed in Asia.

The construction of pumped-storage power stations in China started in the late 1960s with a relatively small scale. In the 1980s, with the construction of a group of large pumped-storage power stations in the eastern regions such as Guangdong province, East China, North China, etc., the development of pumped-storage power stations was promoted to a new height. By the end of 2010, the installed capacity of the existing pumped-storage power stations reached 17GW.

Both the existing pumped-storage power stations and those under construction in China are mainly distributed in the regions where thermal power is playing the leading role, including South China, Central China, North China, East China, Northeast China, etc. to solve the problem of peak regulation in the power grid, while few in the inland provinces. Due to the uneven distribution of energy resources and economic development, when planning and constructing pumped-storage power stations, consideration shall not only be given to the factor of economic development and power source structure, but also to the requirements of trans-regional, large-scale and long-distance power transmission, the optimization of power source structure and the development of intelligent power grids as well as the factor of intensive development of new energy and renewable energy sources such as wind power, nuclear power, etc. According to the planning, by 2020, the total installed capacity of pumped-storage power stations in China will reach 80GW, which will be distributed in the areas where there are large peak-to-valley difference in the load of electric power systems, the peak capacity is relatively inadequate, the power receiving and the source regions where wind power has been intensively developed and concentrated, and where nuclear power is developed.

9 The Guarantee Measures on Hydropower Development

To further improve and optimize the supporting mechanism in hydropower development; specify the strategic position of hydropower, and establish a long-acting and steady development mechanism. To further establish the electric market for hydropower, further improve and optimize the pricing system of power tariff, continuously promote the administration mechanism for construction of the cascade hydropower on rivers. Establish the investment sharing mechanism for the multipurpose hydroelectric projects, and set up the compensation mechanism for cascade hydropower projects.

The eco-environmental protection should be highly paid attention and environmental protection measures should be practically implemented. To establish a scientific evaluation system, make overall planning and coordinate the relations between hydroelectric development and the eco-environment protection. The relevant government departments shall jointly organize and carry out the study on the bearing capacity of environments in river basins of hydropower development and the zoning of functional areas of water energy resources so as to reach agreements on scale, and layout and time sequence of development. Meanwhile, corresponding protection measures shall be proposed to reach the win-win situation for development and protection. To establish a practical and feasible compensation mechanism for eco-environment, enhance the study on protective measures of relevant eco-environment, take effective further measures, and minimize or even avoid negative impact of hydropower development on eco-environment.

It should always be highlighted that the residents affected by the construction of hydropower projects are properly resettled, legal rights and interests of project affected residents are ensured, and affected residents get rid of poverty and become better off. To improve and optimize the administrative system of affected residents, study and establish necessary supporting policies and regulations for affected residents, and supplement and revise relevant regulations. To be innovative on resettlement arrangement, implement the development oriented measures for affected residents, establish a mechanism to support development, and make efforts to solve the pending problems of affected residents of the existing projects, and establish robust system and mechanism so that the hydropower development could boost the economic and social development of the project region.

To strengthen science & technology innovation, speed up technical progress in hydropower construction, further improve the technical level of hydropower development, improve the designing and manufacturing capacity of water turbines with high water head and large capacity, and solve the technical and equipment problems for long-distance and large capacity power transmission projects in plateaus, frigid zones.

All the countries in the world are faced with the problems of climatic change, energy conservation and emission reduction, renewable energy development, which have to be solved properly. Hydropower is a kind of renewable energy resource with mature technology. It is expected that the strategic study will provide a very good basis for the decision-makers, and offer a solution to the tough problems in energy development, and promote the sustainable development of energy in China.

Chapter 2

Dam Construction in China and International Comparison

Jia Jinsheng[1], Yuan Yulan[2], Wang Yang[3], Zheng Cuiying[4], Zhao Chun[5]

Floods and droughts occur frequently in China. In the past 2,000 years, 1,092 floods and 1,056 droughts occurred nationwide, that means each year at least one flood or one drought occurred which was heavily serious. Drought in North China in 1920 made more than 500 thousand people starve to death. The heavy flood of the Yangtze River in 1931 killed 145 thousand people. Floods and droughts have always threatened the Chinese people. However, because of the large lag behind the developed countries in economy and technology in modern time of Chinese history, there were very few dams and shortage of water storage facilities in China before 1949, which made it impossible to handle the natural disasters. After the foundation of new China, the Party and the nation have paid high attention to water conservation and hydropower, and dam constructions has ushered in a new period of great development. Especially for the past 30 years since the reform and opening-up, China's hydropower has made astounding advances in every respect, which can provide the important guarantee for flood control, irrigation, water supply, energy and ecological security, etc. In order to follow the trend of hydropower development, this article reviewed the information of the dams higher than 15m in other countries before 1900 and that in China before 1949, and compared per capita reservoir storage capacity, hydropower exploitation level, dam type and dam height with other countries. The results have showed that, with respect to the per capita reservoir storage capacity and economic and social development indices issued by the UN, the dam construction in China is basically consistent with the level of social and economic development. Furthermore, the dam construction technology in China has been shifted from catching up with the advanced international level to leading the world in some aspects.

2.1 Dams in China before 1900 and International Comparison

It does not clearly record the first dam construction about when and where in world history, but it is universally recognized that China, India, Iran and Egypt are the countries in which dams constructed earliest in the world. Before 2000 B. C. China had begun to construct dike for flood control. At that time, learning from the "natural dike", people gradually formed the concept that water can be retained by soil. Some legends of China not only indicated the long history of dam development in China, but also showed the complexity of water control and the seriousness of the disaster by dam failure.

Historical records show that dam failure probability is astonishing, particularly for the early dams. Although lots of dams were built in the early period throughout the world, only a limited amount of them existed till today. Table 2.1-1 and Table 2.1-2 summarizes the large dams higher than 15m built in the ancient times in some countries according to the limited records, most of them collapsed due to the defects in construction and management, and only a few dams existed till today and played their roles in result of being rehabilitated and reinforced frequently.

---

[1] Jia Jinsheng, Vice President of China Institute of Water Resources and Hydropower Research, Vice President and Secretary General of Chinese National Committee on Large Dams, Professor Senior Engineer.

[2] Yuan Yulan, Office Director, Secretariat of Chinese National Committee on Large Dams.

[3] Wang Yang, Secretariat of Chinese National Committee on Large Dams, Engineer.

[4] Zheng Cuiying, China Institute of Water Resources and Hydropower Research, Professor Senior Engineer.

[5] Zhao Chun, China Institute of Water Resources and Hydropower Research, Professor Senior Engineer.

Table 2.1-1 The Table of Dam Built Higher than 15m before 1900

No.	Country	Number of dams higher than 15m	The maximum dam height (m)
1	Sri Lanka	47	25
2	USA	28	56
3	Iran	15	55
4	Japan	15	32
5	Spain	14	50
6	India	7	54
7	France	6	60
8	Turkey	2	16
9	China	1	48
10	Iraq	1	17
11	Yemen	1	37
12	Syria	1	18
13	Mexico	1	23.5
14	Germany	1	22
15	Egypt	1	15

During the Spring and Autumn and the Warring States Period of China (770 B.C.-221 B.C.), the dike construction had been started in the Yellow River, and it was known that dam construction should be carried out in the dry seasons. The breach blocking engineering promoted the technical progress in river closure. The vertical blocking and horizontal blocking were developed. The "Cifang", a water-retaining structure made by binding the willows, was invented to block the breaches. Huzi Breach Blocking and Dongjun Breach Blocking were the typical projects at that time.

In China, the weirs originated from the farmland irrigation project, which had two main types. One was built to cut off the upstream water, then the water gates were built on dam to regulate water; and the other one was overflow dam built on river to raise water level. The first type of dam was with long length, irregular shape, small width and low height, forming plain reservoir with large inundated area, such as, Que Bei (now called AnfengTang) built in 598 B.C., Hongxi Bei (pond) built in Western Han Dynasty, Liumen E (literally, six-gate weir) built in 34 B.C., and Jianhu Lake built in 140 A.D. The second type dam was built in the dry season by dump filling into the riverbed, and the dam's slope was protected by stone. For example, weir at the canal head of Qisi-Yulou Irrigation Zone built in about 600 B.C., Zhibo Ditch Dam built in 453 B.C., weir at the canal head of Zhangshui Twelve-Canals built in about 422 B.C., Dujiangyan built in about 255 B.C., weir at the canal head of Zheng State Canal built in about 246 B.C., and weir of Lingqu Canal built in about 219 B.C., etc.

In 514 A.D., during the Northern and Southern Dynasties period, the construction of Fushan Weir was started, but it was not completed due to the failed river closure, and was finished after the second river closure. It was earth dam and with two spillways at the both sides. According to some literature, the dam was about 48m in height; however it was thought to be from 30m to 32m in height by other literature. The total reservoir storage capacity was approximately over 10 billion m^3. Only 4 months after its completion, the dam collapsed due to the flood, which resulted in serious disaster.

Tashan Weir (lower than 5m) in Tang Dynasty was built by laying large boulder strip for the first time, which still exists now. For Mulan Bei (pond) built in the Northern Song Dynasty, the construction was carried out by inserting 32 stone columns into the bottom of the riverbed, then reinforced columns by pouring the melted iron at their foundation and surrounding the columns with heavy stone blocks, each with the weight above 500kg. Large strip stones were connected by the iron ingots. This weir still exists today.

Table 2.1-2 The List of Dam Higher than 30m in the World Built before 1900 According to Incomplete Statistics (in an Order of the Dam Height)

No.	Current country	Years	Dam name	Dam height (m)	River (catchment area) or geographic location	Remarks
1	France	1866	Gouffre d'enfer Dam	60	Le Furan River	Cement-laid rubble gravity arch dam with 252m of the average arch radius
2	France	1866	Furens Dam	56	—	Gravity dam
3	USA	1872	Bowman Dam	56	—	
4	Iran	—	Farfan Dam	55	Esfahan City	
5	India	1887-1897	Periyar Dam	54	—	Gravity dam with materials of lime slurry, sand and crushed stones
6	Iran		Korit Dam	54	Yazd	
7	USA	1886	Lower Otay Dam	51	—	
8	Spain	1802	Puentes	50	Rio Guadalentin	Dam with rubbles
9	USA	1893	Lake Hemet Dam	49	—	
10	China	516	Fushan Weir	About 48	Huaihe River, east of Wuhe County, Anhui Province	Earth dam (Some inferred that the dam height was only 30-32m)
11	USA	Phase I in 1875	Lower san leandro Dam	Phase I 35, Phase II added up to 47	East San Francisco	Earth-rock dam
12	France	1870	Ban Dam	46.3	—	Gravity dam
13	Spain	1594	Tibi Dam	46	Near Alicant City	Gravity dam
14	Spain	1880	Hijar Dam	43	—	Gravity dam
15	France	1854	Zola Dam	42	—	Cement-laid rubbles with stone blocks. The average arch radius was 51m
16	Spain	1580-1594	Alicante (Tibi) Dam	41	Rio Monegre	Masonry dam, which was the highest masonry dam in that period
17	India	1892	Tansa Dam	41		Mainly serving Mumbai City
18	India	1869-1879	Poona Dam	40		Cement-laid stone dam
19	Yemen	1000 B.C.	Marib Dam	37	320km away from south of Aden City	Earth-rock dam
20	France	1667-1675	St. Ferreol Dam	36	Laudot River	Earth-rock dam
21	Spain	1786	Val de Infierno Dam	35.5	Tributary of Rio Guadalentin River	Gravity dam
22	USA	1870	San Andreas Dam	35	—	—
23	India	1500	Mudduk Masur Dam	33	Madras City	Earth dam
24	France	1811	Couzon Dam	33	Near St. Etienne	Earth dam, with a masonry wall in the middle
25	USA	1888	Walnut Dam	33	—	—
26	USA	1893	Austin Dam	33	—	—
27	Japan	1128	Daimonike Dam	32	Near Nara City	Earth dam
28	USA	1866	Pilarcitos Dam	32	—	—
29	Spain	17th century	Relleu Dam	28, added up to 3.85	Rio Amadorio	Arch dam with rubbles and stone blocks
30	USA	1890	Castlewood Dam	31		
31	India	322 B.C. - 298 B.C.	Sudarsana Dam	about 30	Kathiawar Peninsula	For agricultural irrigation

Gaojia Weir (higher than 5m) was built in the Northern Song Dynasty, and was systematically repaired in the Ming and

Qing dynasties with the slope protection of cement-laid stone masonry to be preserved till today. And its reservoir is called today as Hongze Lake with a very large reservoir capacity.

Shen and Zheng (2000) listed the details of the weirs. Based on it as well as Table 2.1-1 and Table 2.1-2, the following conclusions can be drawn:

(1) There is a long history of dam construction in China, but there are limited number of large dams according to the International Commission on Large Dams (ICOLD) definition, which is the dam higher than 15m, or 5m to 15m high with the reservoir storage capacity larger than 3 million m^3. The largest dams built in ancient China include Anfengtang (dam height 6.5m, reservoir storage capacity 97 million m^3), Fushan weir (dam height 48m), Gaojia Weir (dam height above 5m and large reservoir capacity); Gukow River Dam in Shanxi Province built in 240 B.C. recorded (Robert, 1980) was 30m in height. And there are other large dams, but due to the limited references, they are not described in details.

(2) The dam height is the significant symbol of the dam's level of development. Before 1000 A.D., though there were many dams, but only three dams higher than 30m. The first one is Marib Dam in Yemen, 37m high, built in about 1000 B.C.. The second one is Sudarsana Dam 30m high in India, built in 322 B.C., and the last one is Fushan weir in China, 48m high, built in 516 A.D..

(3) The maximum dam height built before 1900 was only 60m, which was a masonry arch dam built in 1860 in France.

(4) Science and technology development not only has promoted the construction of high dams, but also has laid the foundation for guarantee for safe dam operation of. Dam collapse happened in most of dams built in the ancient time with low dam height due to the finite level of science and technology.

2.2 Development of Dams in China in 1900-1949

In early 20th century, the modern technology of dam construction was introduced into China by those who came back to China after studying overseas. However, because of lack of fund, there were only a few dams constructed and the general level of technology was still lagging behind the world.

In 1910-1912, the first hydropower station was completed in China which is Shilongba Hydropower Station in Yunnan Province. It is a diversion hydropower station with a low head, which is still in operation now.

In three northeastern provinces of China, when occupied by Japan, several dams were built. In 1936-1942, the Naodehai concrete gravity dam was built with the height of 32m in the middle reaches of the Liuhe River in Liaoning Province. In 1937-1943, Shuifeng concrete gravity dam was built with the height of 106m on the Yalujiang River, and is now commonly owned by China and North Korea. Fengman gravity dam, started in September 1937, was situated on the main stream of the 2nd Songhua River. Due to poor construction quality, this dam has been rehabilitated for years and now the decision has been made to dismantle it for reconstruction.

Compared with western developed countries and the former USSR in this period, the level of dam construction in China obviously lagged behind. In 1936, the famous Hoover Concrete Gravity Arch Dam was completed in USA with the dam height up to 221.4m. According to the statistics of the International Commission on Large Dams (ICOLD), there were 5,196 dams with the height above 15m in the world in 1950, only 21 of which were in China (including Taiwan and Hong Kong). See Table 2.2-1 for details.

During this period, the hydropower development was quite slow. Until 1949, the total installed capacity of the hydropower stations was 360MW, with the annual power generation 1,200GW·h (excluding Taiwan Province). China ranked No. 20 by the total installed capacity, and No. 21 by the power generation capacity at that time.

Table 2.2-1 Dams with the Height above 15m Completed and under Construction during 1900-1949
(Order in the Year of Starting Construction)

No.	Dam name	Beginning year	Year of completion	River	Province	Dam type	Dam height (m)	Dam length (m)	Total reservoir capacity (million m^3)
1	Wangjiadian	1914	1917	—	—	Masonry dam	29.09	—	—
2	Wushantou	1920	1930	Guantian River	Taiwan	ER	56	1273	154.16
3	Chengmen Reservoir	1923	1936	—	Hong Kong	—	35.05	—	13

Chapter 2 Dam Construction in China and International Comparison

Countinue

No.	Dam name	Beginning year	Year of completion	River	Province	Dam type	Dam height (m)	Dam length (m)	Total reservoir capacity (million m³)
4	Xishi	1926	1927	Xishi River	Taiwan	PG	30	102.4	0.56
5	Wujie	1927	1934	Wanda River	Taiwan	PG	57.6	86.5	14
6	Shangli	1927	—	—	—	Masonry arch dam	27.3	—	—
7	Riyuetan	1934	1937	Wucheng River	Taiwan	TE	30.3	364	171.62
8	Jianshanpi	1936	1938	Guichong River	Taiwan	TE	30	255.6	8.11
9	Shuifeng	1937	1943	Yalu River	Liaoning	PG	106	900	14,666
10	Fengman	1937	1953	Songjiang River	Jilin	PG	91.7	1,080	10,800
11	Luliao	1937	1939	Luliao River	Taiwan	TE	30	270.4	3.783
12	Naodehai	1938	1942	Liuhe River	Liaoning	PG	44.5	167	222.7
13	Xipan	1940	1954	Liwu River	Taiwan	PG	30	129.6	0.354
14	Lantan	1942	1944	Bazhang River	Taiwan	TE	34	546	9.72
15	Yuanyangchi	1943	1947	Taolai River	Gansu	Dam with gravelly soil core wall	37.8	240	110
16	Erlongshan	1943	1966	Liaohe River	Jilin	TE	31.2	410	1,762
17	Agongdian	1946	1953	Agongdian River	Taiwan	TE	31	2,380	45
18	Haitang	1947	1977	Xunjiang River	Guangxi	TE	36	96	0.07
19	Daxishan	—	1934	Malan River	Liaoning	TE	34.59	570	22.99
20	Mingqin	—	1944	Qiaoergou River	Hubei	TE	33.6	117	0.624
21	Longwangtang	—	1925	Longweigou River	Liaoning	PG	39.7	156.4	19.05

Note: TE, earth dam; ER, rockfill dam; PG, gravity dam.

2.3 Development of Dams in China in 1949-1978

Since the founding of New China, the dam technology developed quickly. In the 1950s, China had constructed almost all main types of dams in the world at that time, such as multi-arch dam, massive head dam, deck buttress dam, concrete gravity dam, concrete arch dam, rockfill dam with clay core or inclined core, directed blasting rockfill dam and so on. It can be seen that he types and heights of the dams built and under construction with the height above 30m in 1950-1978 in Table 2.3-1.

Table 2.3-1 Distribution of the Types and Heights of the Dams Completed and under Construction during 1950-1978

Dam type	The number of various Dam height				Total
	30 – 60m	60 – 100m	100 – 150m	>150m	
Gravity dam	235	42	9	1	287
Arch dam	299	44	4	3	350
Earth dam	2,531	54	5	0	2,590
Buttress dam	18	1	2	0	21
Rockfill dam	149	11	3	0	163
Concrete faced rockfill dam	19	4	1	0	24
Mixed dam	200	3	0	0	203
Others	13	0	0	0	13
Total	3,464	159	24	4	3,651

There were 3,651 dams with the height above 30m in the starting or completing stage during 1949-1978, including 287 gravity dams, 350 arch dams, and 2,777 earth (rock) dams. There were 28 dams with the height above 100m.

Concrete dam was the main type of high dam in this period. In Mainland China, the highest concrete dam was Wujiangdu Gravity Arch Dam, 165m high, and Deji Arch Dam in Tainwan was 181m in height.

1. Concrete dam

During 1949-1978, most of the high dams built in China were concrete dams. There were 28 dams higher than 100m, in which concrete dams accounted for 60.71%. The highest concrete gravity dam built in 1958-1974 was Liujiaxia Dam, with the dam height of 147m; the highest concrete arch dam was Wujiangdu Concrete Gravity Arch Dam started in 1970, with the dam height of 165m; the highest concrete buttress dam was Hunan zhen Dam built during 1958-1980, with the dam height of 129m. Deji Concrete Dam built in Taiwan during 1969-1974 was 181m in height.

2. Earth (Rock) dam

In the 1950s, modern technology started to be used for the construction of earth (Rock) dams in China. Since the construction could be manually carried out with the support of a small amount of light-duty mechanical, earth (rock) dam often was the first choice of dam type. The dam height was generally less than 50m. The dam type was mostly homogenous earth dam or earth-core sand-gravel dam, and some dams were made as hydraulic filling rockfill dams, rockfill dams of dumping soil into water and directional blasting rockfill dams with low demand of mechanical devices and applicable for the dam construction by the public. The rockfill dam with dumping fill generally had the height of no more than 50m. The highest dam was Nangudong Earth-Rock Dam completed during 1958-1961 with the height of 78.5m. For the rockfill dams by applying the technology of directional blasting, the highest one is Yiyi Dam with the height of 85.5m in Wuding County, Yunnan Province.

In 1965, the vibratory roller was introduced into the rockfill dam construction in U.S.A., and the thin-layered rolling replaced the dumping fill. That signified that the rockfill dam construction has stepped into the modern period. The new technology speeded up the development of two types of dam, that is, concrete-faced rockfill dam and rockfill dam with clay core. Subsequently, vibratory-roller dam construction technology was introduced into China. The typical project was Bikou dam with clay Core completed in 1976 with the height of 101m. During the later construction phase of the project, the 13.5t vibratory roller was introduced and made as well as rolling the gravel materials of the dam shells and rockfill material.

In the early 1970s, the dam with asphalt concrete core (inclined core) were started to construct in China. The cores were generally formed by pouring the materials. Danghe River dam with asphalt concrete core in Gansu Province was completed in 1974. The height of the core is 58.5m. Due to the limitation of construction machines, tools and high quality bitumen materials, this type of dam wasn't popularly used.

3. Stone masonry dam

With the development of modern dam construction technology, fewer and fewer stone masonry dams were built. But during 1949-1978, more than 300 stone masonry dams higher than 15m were built in China, including Qunying Gravity Dam built during 1966-1971 with the dam height of 103.5m.

4. Typical projects

Foziling Multi-Arch Concrete Dam built during 1952-1954 was the first concrete multi-arch dam independently designed and constructed by Chinese, with the dam height of 74.4m, which was then enhanced to 75.9m.

Xianghongdian Dam built during 1956-1958 was the first concrete gravity dam of China self-designed and self-constructed, which was situated on the upper reach of the Pihe River, the tributary of the Huaihe River. The dam was made as concrete gravity arch dam with the fixed circle center and equal radius. The dam has the height of 87.5m, top arch is 361m long, top arch thickness is 6m, and bottom arch thickness is 39m. The stress of the dam body calculated by trial load method, and verified with the 1:100 model test.

Xin'anjiang Hydropower Station constructed during 1957-1965 was the first large hydropower station independently designed and constructed by Chinese. It is concrete gravity dam with wide-joint space within its body. The height is 105m with length of 465.4m with installed power generating capacity of 850MW.

Gezhouba Hydropower Station started in December 1970 was the first large Hydo-Project on the main stream of the Yangtze River. The river closure was made in January 1981. The dam has the maximum height of 53.8m, and the crest length is 2,606.5m and installed capacity 2,715MW. The scale of the project and its technical complexity were very representative, which have laid technological foundation for the construction of the Three Gorges Dam.

During 1949-1978, China was in the planning economy period with economic difficulties. Thus dam construction needed to overcome many difficulties, such as lack of technical information, poor construction machines and so on. Overall, Dam construction in China was in the process of independence and self-reliance. Technically, it was focused on digestion, absorption and catching-up, and made a great process in various aspects to cultivate a large number of talents for the late development.

2.4 Development of Dams in China in 1978-2011

After the reform and opening up in 1978, great changes have taken place on the construction system of water resources and hydropower project. River harnessing and hydropower exploitation have stepped into a new period of development. Dam construction has accomplished in setting up a pattern of import, digestion, absorption, and re-innovation and the leaping and bounding development has been attained. During this period, the typical characteristics of the dam construction in China are: the construction techniques of roller compacted concrete dam, concrete faced rockfill dam, rockfill dam with cores and arch dam are constantly improved after being introduced, and then applied in large scale. Signified by the Three Gorges gravity dam, Ertan arch dam and Xiaolangdi dam with inclined core, China has reached the international advanced level in many aspects of the dam construction.

During the construction of high dams, concrete dam still plays a significant role, but the earth (rock) dam, particularly concrete faced rockfill dam, has achieved rapid development. Table 2.4-1 has showed the distribution of the types and heights of the dams completed and under construction with the height above 30m.

Table 2.4-1 Distribution of Types and Heights of the Dams Completed and under Construction with the Height above 30m during 1978-2011

Dam type	The dam numbers of various height					Total
	30-60m	60-100m	100-150m	150-200m	>200m	
Gravity dam	292	46	10	3	0	351
Roller compacted concrete gravity dam	38	36	21	3	2	100
Arch dam	503	76	11	6	7	603
Roller compacted concrete arch dam	9	21	15	1	0	46
Earth dam	856	45	0	0	0	901
Buttress dam	11	1	0	0	0	12
Rockfill dam	135	33	17	2	2	189
Concrete-faced rockfill dam	97	108	46	14	3	268
Mixed dam	103	4	0	0	0	107
Gate dam	18	0	0	0	0	18
Others	30	6	1	0	0	37
Total	2,092	376	121	29	14	2,632

1. Development of concrete gravity dam (including RCC dam)

The Three Gorges Dam is representative of the gravity dam in China is the largest concrete volume in the world till now with the height of 181m. As an international milestone project, the Three Gorges Dam has achieved prominent success in the respects of construction quality, technological innovation, environmental protection, as well as economic and social benefits, and tremendous influence at the international level. Since the 1980s, China has started to develop the technology of RCC (Roller Compacted Concrete) dam. The Kengkou RCC Gravity Dam was completed in 1986 with the dam height of 56.8m. Then RCC dam technology has been widely used in the construction of gravity dams, with the dam height quickly above 100m. Since the completion of Longtan RCC Dam with the designed height of 216.5m (192m in Phase I at present) and Guangzhao RCC Dam with the dam height of 200.5m, China has the largest number of RCC dams in the world and the dam height firstly over 200m.

2. Arch dam

Longyangxia Hydropower Station, started in July 1978, was a gravity arch dam with the dam height of 178m. It represented the top level of dam construction of China in the 1980s, and was the highest dam in China at that time. Ertan Concrete Arch Dam, completed in 2000, was the milestone project for the arch dam construction in China, and also the milestone project worldwide. Ertan has the highest dam height of 240m, which marked the first breakthrough 200m of the dam in China.

After Ertan Concrete Arch Dam, a series of super high arch dam projects have been started, including Laxiwa Double-curvature Arch Dam with the height of 250m, Xiluodu Double-curvature Arch Dam with the height of 285.5m, Xiaowan Double-curvature Arch Dam with the height of 294.5m, and Jinping I Double-curvature Arch Dam with the dam height of 305m. Xiaowan, Xiluodu and Jinping dams all have the height above the former highest Inguri Arch Dam (271.5m) in the world. At present, Xiaowan and Laxiwa arch dams have been completed and put into operation. Arch dam higher than 300m means that the high arch dam technology in China has shifted from catching up with the world level to being in the leading position in some aspects in the world.

Since the RCC dam technology was successfully applied into the gravity dam construction, the construction of RCC arch dams was started in the 1990s in China. The first one was Puding Arch Dam built in 1991-1993, which was followed by Shapai RCC Double-curvature Arch Dam (132m), Yunlong River III RCC Arch Dam (135m), and Dahuashui RCC Double-curvature Arch Dam (134.5m). Under the condition of full reservoir, Shapai Arch Dam successfully withstood the test of Wenchuan Earthquake of Sichuan Province in 2008. Qinglong Dam (139.7m) under construction is the highest RCC arch dam in China.

3. Earth (rock) dam (concrete faced rockfill dam)

Since the 1980s, modern concrete faced rockfill dam technology has been introduced into China. Xibeikou Concrete Faced Rockfill Dam, started in 1985, was the first one in China with the dam height of 95m. Then this type of dam has been widely applied. According to incomplete statistics, about 75% of high earth (rock) dams with the height above 100m completed or under construction during 1985-2009 in China were concrete faced rockfill dams. Tianshengqiao I Concrete Faced Rockfill Dam with the height of 178m was completed in 2000; Shuibuya Concrete Faced Rockfill Dam with the height of 233m was completed in 2009. Shuibuya Dam is the highest concrete-faced rockfill dam either in the world.

Xiaolangdi Rockfill Dam with inclined core is the typical example of core dam, which was started in 1991 with the dam height of 160m and maximum depth of riverbed cover layer is over 80m. The anti-seepage of the riverbed cover was achieved with concrete wall. It was the highest core rockfill dam in China in that period, realizing the leap of the height for core rockfill dam from 100m high lever before the reform and opening up in 1978 to 150m high level. In August 2009, Pubugou Core Rockfill Dam was filled to the top elevation with the maximum height of 186.0m, which is the highest core rockfill dam completed in China. It marked the height of the core rockfill dam reach the level of 200m high. Nuozhadu Core Rockfill Dam was started in 2006 with the dam height of 261.5m, and Shuangjiangkou Core Rockfill Dam with the dam height up to 314 m is ready to start construction, which will be the highest dam in China.

4. Roller compacted asphalt concrete dam

There are two typical roller compacted asphalt concrete dams, one is Maopingxi Roller Compacted Asphalt Concrete Dam with the height of 104m started in 1994, and the other is Yele Asphalt Concrete Core Rockfill Dam with the height of 124.5m started in 2001. The construction technology for roller compacted asphalt concrete has been used in other projects, such as the surface lining of upper reservoirs of Tianhuangping, Zhanghewan, Xilongchi and Baoquan Pumped-storage power stations, and this technology has acquired good effects.

5. Stone masonry dam

With the technology development of concrete dam and earth (rock) dam, and increase of the labor cost, the advantage of the stone masonry dam has been constantly decreased. Since 1978, the construction of stone masonry dams has significantly slowed down. During this period, the typical stone masonry dam projects include Shiziguan Stone Masonry Gravity Dam started in 1999, and Xiahuikeng masonry Double-curvature arch dam with Cement-laid Rock started in 1998.

During 1978-2011, the dam construction in China made rapid progresses. During this period, milestone projects such as the Three Gorges Project, Ertan Project and Xiaolangdi Project were constructed which are well known in the world. Longtan Roller Compacted Concrete Gravity Dam, Jinping I Arch Dam and Shuibuya Concrete Faced Rockfill Dam and Shuangjiangkou Core Rockfill Dam were the highest dams in their respective dam types in the world. So the dam construction in China has led the world's development in many aspects. In addition, some types of dam constructed before 1980 have gradually become history. For example, directional blasting dam, dam with dumping soil into water, hollow gravity dam, multi-arch dam, buttress dam and wide-joint gravity dam have no longer been used basically, and the number of the stone masonry dams is also significantly reduced.

2.5 Dam Comparison between China and other Countries in the World

2.5.1 Dam Construction in China Compared with that in the World

In 2010, there were 5,564 dams with the height above 30m completed or under construction in China (13,629 dams in the same period in the world, in which the dams in China accounted for 40%), including one dam with the height above 300m, 13 dams with the height from 200m to 300m, 30 dams with the height from 150m to 200m, 141 dams with the height from 100m to 150m, and 5,379 dams with the height from 30m to 100m.

There was large number of dams in some provinces: 685 dams in Hunan Province, 610 dams in Yunnan Province, 474 dams in Zhejiang Province, 470 dams in Hubei Province, 403 dams in Guangdong Province, 346 dams in Fujian Province, 308 dams in Sichuan Province and 306 dams in Guangxi Zhuang Autonomous Region. And the number of the dams completed or under construction in Guizhou, Shaanxi and Jiangxi Provinces was also above 200.

There were 870 arch dams (including 47 roller compacted concrete arch dam), 679 gravity dams (including 103 roller compacted concrete gravity dams), 577 rockfill dams (including 254 concrete faced rockfill dams), 3,077 earth dams and 361 dams of other types in the 30m height dams that have be built and been building according to the type of dam. The distributions of dam type with the height of above 30m can be seen in Fig. 2.5-1.

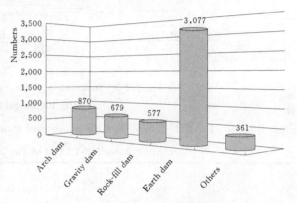

Fig. 2.5-1 Distribution of the Dam Types with the Height above 30m in 2010

In 2010, there were 353 dams under construction with the height above 60m in the world, mostly in Asia (270 dams), accounting for 76%, including 109 dams in China, and the others in 20 other countries, including Iran, Japan and India, etc. 30 dams were in Europe, accounting for 8%, and were mainly situated in 10 countries, including Greece, Italy and Spain, etc. 20 dams were in South America, accounting for 6%, and were mainly in 7 South American countries, including Brazil, Chile and Columbia, etc. 15 dams were distributed in each of central-north America and Africa, mainly in 6 countries of Africa, including Algeria, Ethiopia and Morocco, etc.; and mainly in 8 countries of Central and North America, including Mexico, Dominica, Canada and USA, etc. 3 dams were in Australia (shown in Fig. 2.5-2).

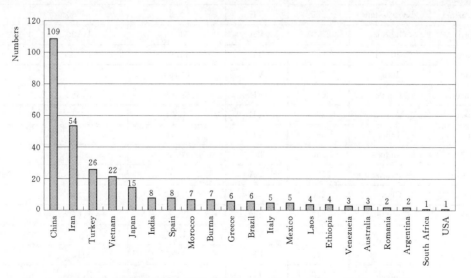

Fig. 2.5-2 Distribution of the Dams under Construction with the Height above 60m in the Main Countries in 2010

Part I Planning and Sustainable Development of Dams and Reservoirs

The rockfill dam is most popular dam type. There were 141 rockfill dams, accounting for 40% in the total, in which concrete faced rockfill dams are most widely constructed, accounting for 50% in the rockfill dams. Gravity dam is secondly popular dam type, and the number is 116 (in which roller compacted concrete gravity dams account for 66% in gravity dams), accounting for 33% in the total. There were 39 earth dams, accounting for 11% in the total. There were 35 arch dams, accounting for 10% in the total (in which roller compacted concrete arch dam accounted for 40% in the arch dams). There were 22 dams of other types and mixed dams (shown in Fig. 2.5-3).

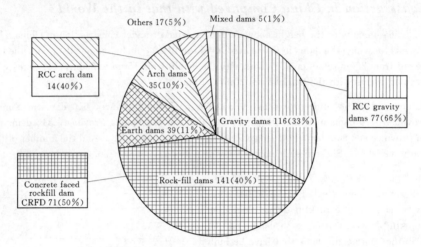

Fig. 2.5-3 Distribution of the Types of the Dams under Construction with the Height above 60m in 2010

From Table 2.5-1 to Table 2.5-7 show the comparison among the high dams.

Table 2.5-1 List of Top 20 Highest Dams in the World

No.	Dam name	Country	Dam height (m)	Dam type	Total reservoir volume (10^9 m^3)	Installed capacity (MW)	Year of completion
1	Jinping I	China	305	VA	7.988	3,600	2014
2	Nurek	Tajikistan	300	TE	10.5	2,700	1980
3	Xiaowan	China	294.5	VA	15	4,200	2012
4	Xiluoduo	China	285.5	VA	12.67	13,860	2015
5	Grande Dixence	Switzerland	285	PG	0.4	2,069	1962
6	Kambarazin-1	Kyrgyzstan	275	TE	3.6	1,900	1996
7	Inguri	Georgia	271.5	VA	1.1	1,320	1980
8	Boruca	Costa Rica	267	TE	14.96	1,400	1990
9	Vajont	Italy	262	VA	0.169		1961
10	Nuozhadu	China	261.5	ER	23.703	5,850	2015
11	Chicoasen	Mexico	261	TE	1.68	2,430	1981
12	Tehri	India	261	TE	3.54	1,000	1990
13	Alvaro Obregon	India	260	PG	0.4	86.4	1952
14	Mauvoisin	Switzerland	250.5	VA	0.212	114	1991 (added)
15	Laxiwa	China	250	VA	1.079	4,200	2010
16	Guavio	Columbia	247	TE	0.9	1,600	1992
17	Deriner	Turkey	247	VA	1.97	670	2004
18	Alberto Lleras C.	Columbia	243	ER	0.97	1,150	1989
19	Mica	Canada	243	TE	25	2,104	1972
20	Gibe III	Ethiopia	243	RCC	14	1,870	2013

Note: TE, earth dam; VA, arch dam; ER, rockfill dam; PG, gravity dam; RCC, roller compacted concrete dams.

Chapter 2 Dam Construction in China and International Comparison

Table 2.5-2 Top 10 Highest Gravity Dams in the World

No.	Dam name	Country	Maximum dam height (m)	Total reservoir volume (10^9 m³)	Installed capacity (MW)
1	Grande Dixence Dam	Switzerland	285	0.4	2,069
2	Alvara Obregon	Mexico	260	0.4	86.4
3	Kishau	India	236	1.81	30
4	TaSang	Myanmar	227.5	1.81	30
5	Bhakra	India	226	9.62	1,325
6	Dworshak	USA	219	4.28	400
7	Longtan	China	192 (phase I) 216.5 (phase II)	18.8 (phase I) 29.92 (phase II)	4,900 (phase I) 6,300 (phase II)
8	Toktogul	Kyrgyzstan	215	19.5	1,200
9	Keban	Turkey	210	31.0	1,330
10	Lakhwar	India	204	0.58	300

Table 2.5-3 Top 10 Highest Arch Dams in the World

No.	Dam name	Country	Maximum dam height (m)	Total reservoir volume (10^9 m³)	Installed capacity (MW)
1	Jinping I	China	305	7.988	3,600
2	Xiaowan	China	294.5	15	4,200
3	Xiluodu	China	285.5	12.67	13,860
4	Inguri	Georgia	271.5	1.1	1,320
5	Vaiont Dam	Italy	262	0.17	
6	Mauvoisin Dam	Switzerland	250.5	0.21	114
7	Laxiwa	China	250	1.079	4,200
8	Deriner	Turkey	247	1.97	670
9	Sayano-Shushenskaya	Russia	242	31.3	6,800
10	Ertan	China	240	6.1	3,300

Table 2.5-4 Top 10 Highest Earth (Rock) Dams in the World

No.	Dam name	Country	Maximum dam height (m)	Total reservoir Volume (10^9 m³)	Installed capacity (MW)
1	Nurek	Tajikistan	300	10.5	2,700
2	Kambarazin-1	Kyrgyzstan	275	3.6	1,900
3	Boruca	Costa Rica	267	14.96	1,400
4	Nuozhadu	China	261.5	23.703	5,850
5	Chicoasen	Mexico	261	1.68	2,430
6	Tehri	India	261	3.54	1,000
7	Guavio	Columbia	247	0.9	1,600
8	Alberto Lleras C.	Columbia	243	0.97	1,150
9	Mica	Canada	243	25.0	2,104
10	Changhe	China	240	1.075	2,600

Part I Planning and Sustainable Development of Dams and Reservoirs

Table 2.5-5 The Countries with Large Number of Concrete Faced Rockfill Dams and Roller Compacted Concrete Dams

Country	Number of CFRD higher than 30m completed or under construction	Dam height with the highest CFRD (m)	Country	Number of RCC dams higher than 30m completed or under construction	Dam height with the highest RCC (m)
China	254	233	China	150	216.5
Spain	23	108	Japan	49	156
USA	17	150	Brazil	35	95
North Korea	17	125	USA	30	97
Romania	11	110	Vietnam	19	139
Mexico	11	210	Morocco	18	120
Brazil	10	200	Spain	17	120
Turkey	9	135	South Africa	12	85
Chile	8	115	Mexico	11	132
Columbia	7	190	Australia	10	82

Table 2.5-6 Top 10 Highest CFRDs Completed or under Construction in the World

No.	Dam name	Country	Maximum dam height (m)	Total reservoir volume (10^9 m^3)	Year of completion
1	Shuibuya	China	233.2	4.58	2009
2	Tasang	Myanar	227.5	—	Under construction
3	Houziyan	China	223.5	0.706	Under construction
4	Jiangpinghe	China	221	1.424	Under construction
5	El Platanal	Peru	221	—	Under construction
6	La Yesca	Mexico	210	—	Under construction
7	Bakun	Malaysia	205	—	2003
8	Campos Novos	Brazil	202	1.65	2006
9	Karahnjukar	Iceland	193	—	2008
10	Sogamoso	Columbia	190	—	2005

Table 2.5-7 Top 10 Highest Roller Compacted Concrete Dams Completed or under Construction in the World

No.	Dam name	Country	Maximum dam height (m)	Total reservoir volume (10^9 m^3)	Year of completion
1	Gibe III	Ethiopia	243	14.0	2013
2	Longtan[①]	China	216.5	29.92	Phase I completed
3	Portugues	Puerto Rico	210	—	Under construction
4	Guangzhao	China	200.5	3.245	2007
5	Miel I	Columbia	188	0.57	2002
6	Nam Thuen 1	Laos	177	—	Under construction
7	Guandi	China	168	0.783	Under construction
8	Pervari	Turkey	165	—	Under construction
9	Murun	China	160.5	0.2793	Under construction
10	Wanjiakouzi	Malaysia	160	—	Under construction

① Longtan Phase I has the maximum dam height of 192m.

2.5.2 Coordination between the Construction of Dam and Hydropower and Economic-social Development

Dam and hydropower construction is closely related to economic-social development. In order to better explain this correlation, such indicators as reservoir storage capacity per capita, development technology rate of hydropower potential and human development index (HDI), were introduced here and related data were collected and analyzed about those indictors in about 100 countries. HDI is a parameter reflecting the GDP, life expectancy and education level, which is between 0 and 1. The figure closer to 1 indicates the higher human development level. The countries with HDI above 0.8 mostly are the developed countries, for example, Norway (0.943), USA (0.910); the countries with HDI between 0.7 and 0.8 are mostly the relatively developed countries, such as Russia (0.755), Brazil (0.718); and the countries with HDI between 0.5 and 0.7 are mostly the developing countries in Asia, Africa and Latin America, for example, China (0.687), Egypt (0.644); the countries with HDI less than 0.5 are mostly the underdeveloped countries, such as Nigeria (0.459) and so on.

In 2008, Professor Luis Berga presented the concept that the development of reservoirs and dams is closely related to the development of economy and society. Based on this, we have made further detailed analysis. Fig. 2.5-4 compares the relevant relationship between per capita reservoir storage capacity in different types of countries and the human development index. The result shows that most developed countries have a good foundation in the guarantee of water safety and enough water storage facilities to handle various kinds of changes; while developing countries still shoulder tough development tasks due to the restriction in fund, technology and talent. As a whole, the construction level of reservoir and dam of a country is generally in harmony with the development level of human. This coincides with "the distribution of global water infrastructure is in inverse relation with the distribution of global water risks". It needs to be pointed out that there are also some exceptional cases. For example, the human development index of Mozambique is very low (HDI = 0.322), while per capita reservoir storage capacity reaches 2,727m^3. The human development indexes of Israel (HDI = 0.888) and Switzerland (HDI = 0.903) are relatively high, while the per capita reservoir storage capacity is 27m^3 and 440m^3 respectively. The reason is that Mozambique has a sparse population with large amount of water resources; while Israel is a country with drought and low precipitation, and the water resources are very limited; while Switzerland owns a large number of natural lakes.

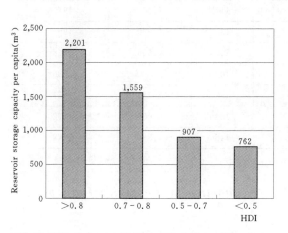

Fig. 2.5-4 Relevant Relationship between Reservoir Storage Capacity per Capita and HDI in 2007

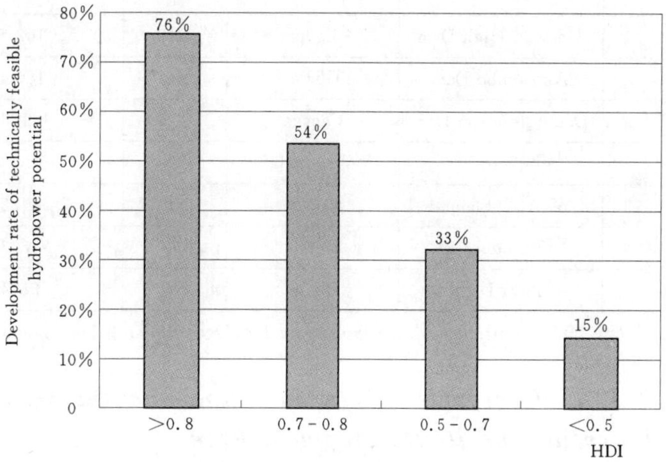

Fig. 2.5-5 Relevant Relationship between Development Rate of Technically Feasible Hydropower Potential and HDI in 2007

The development level of hydropower is also closely related to HDI. Here comparison will be made in terms of developed rate of technically feasible hydropower potential. With the rapid development of economy and technical progress, development technology rate of hydropower potential has a relatively large change. Fig. 2.5-5 compares relevant relationships between average development level of technically feasible hydropower potential and the human development index in various countries. The results have shown that developed countries enjoy high development level of hydropower, while developing countries and underdeveloped countries still undertake tough task for hydropower development. For example, the human development index in USA is 0.910, and the development rate of technically feasible hydropower potential is approximately 67%; the human development index of China is 0.687, and the development rate of technically feasible hydropower potential is 27%. The development levels of hydropower of USA and China are generally in harmony with the level of national economic and social development. This is the similar case with the other countries. However, there may be quite a few countries exceptions. For example, the development rate of technically feasible hydropower potential in Australia (HDI = 0.929) is only 29%, and the development rate of technically feasible hydropower potential in Norway (HDI=0.943) is 41%. The reason is that Australia is rich in coal resources, so the development urgency for hydropower is not high. Hydropower has accounted for more than 95% in national electric power supply in Norway, which has met the requirements.

2.5.3 International Comparison of Total Reservoir Storage Capacity

Annual river runoff in the world is about 55 trillion m^3, and total amount of available water resource is about 9 trillion m^3. Till 2007, total reservoir storage capacity in the world was nearly 7 trillion m^3, in which the dam reservoir accounted for 98%. Effective storage capacity of all reservoirs in the world was about 4 trillion m^3.

In China, the annual runoff of river is 2.8 trillion m^3, accounting for 5% that of the world, ranking No. 6. Brazil has 8 trillion m^3, Russia has 4.3 trillion m^3, USA has 3 trillion m^3, Canada has 2.9 trillion m^3, and Indonesia has 2.84 trillion m^3, these countries rank in the first five positions respectively. By the end of 2008, China has constructed 87,151 reservoirs with total reservoir storage capacity of 706.4 billion m^3 (excluding Hong Kong, Macau and Taiwan regions), ranking No. 4 in the world according to National Water Development of 2009 Statistical Bulletin.

To utilize water resources more efficiently, total reservoir storage capacity of some countries is more than the river runoff, for example, in Turkey, the river runoff is 186.0 billion m^3, and in 2007, this country's total reservoir storage capacity can reach 198.0 billion m^3.

The Three Gorges Project has the largest reservoir storage capacity in China, that is 45 billion m^3, ranking No. 26 in the world. Table 2.5-8 showed the top 10 reservoirs according to the reservoir capacity in the world.

Table 2.5-8 Top 10 Dams with the Largest Storage Capacity in the World

No.	Dam name	Country	Dam type	Total storage capacity (10^9 m^3)	Dam height (m)	Installed capacity (MW)	Year of completion
1	Owen Falls Reservoir	Uganda	PG	204.8	31	180	1954
2	Kariba	Zambia/Zimbabwe	VA	180.6	128	1,500	1976
3	Bratsk	Russia	PG	169.0	125	4,500	1964
4	Haswan High Dam	Egypt	ER	162.0	111	2,100	1970
5	Akosombo Dam	Ghana	ER	150.0	134	1,020	1965
6	Daniel Johnson Dam	Canada	VA	141.85	214	2,656	1968
7	Guri	Venezuela	PG	135.0	162	10,235	1986
8	W. A. C. Bennett	Canada	TE	74.3	183	2,730	1967
9	Krasnoyarsk	Russia	PG	73.3	124	6,000	1972
10	Zeya Dam	Russia	PG	68.4	115	1,330	1978

Note: TE, earth dam; VA, arch dam; ER, rockfill dam; PG, gravity dam.

2.5.4 Comparison between China and the World in Economically Feasible Hydropower Potential and its Development Rate

There is not a unified formula in calculating the rate of hydropower development. The technically exploitable amount and the economically exploitable amount are likely to change with the development of technology and economy and energy demand. Therefore, the exploitation level announced by various countries may change frequently. For comparison, in this paper, hydropower generating capacity issued by various countries is used as the parameter for statistical calculation. Considering the statistical difficulty of the designed power generation of hydropower stations and the variation of the hydro electricity power amount in dry year and wet year, when calculating the development rate of technically and economically feasible potential, the maximum power generation announced by various countries recently (after 1998) is used to be divided by the recently released value of technically and economically potential. Since the power generation through pumped storage power plant is included in the hydropower generation released by various countries, and it is hard to make strict classification of economically feasible potential and technically feasible potential, the developing rate of economically feasible potential in some countries may higher than 95% and even higher than 100%. In this case, it will be uniformly stated that the development rate of economically feasible potential is more than 95%.

In 2010, the total hydropower generation in the world was 3.4097 trillion kW·h, accounting for 39% of the economically exploitable amount (8.7099 trillion kW·h), the development rate of economically exploitable amount in most of the developed countries is above 80%, and in most developing countries is below 30%, such as 13% in Africa, 34% in Asia (including Turkey and Russia), 49% in Australia and Oceania, 45% in South America, 66% in Centre and North America, 72% in Europe (excluding Turkey and Russia). The economically exploitable hydropower potential in China is 1.7530 trillion kW·h, accounting for 20.1% of the world and ranking No.1 in the world.

In 2010, total hydropower generation in China was 662.2 billion kW·h, accounting for 19.42% of the world, 16% of total power generation in China, and the development rate of economically exploitable hydropower potential can reach 38%. The average development rate of the economically exploitable hydropower potential in developed countries has been above 60%. There are some countries whose actual development level of economically exploitable hydropower potential have been above 95%, such as Germany, Switzerland, Spain, Italy. And the economically exploitable hydropower potential in USA is up to 82%, Japan is 90%. The development rate of technically exploitable hydropower potential is 74% in Germany, 92% in Switzerland, 67% in USA, 67% in Spain, 86% in Italy, 73% in Japan, but 27% in China.

According to the statistic data in 2010, there are 19 countries whose hydropower generation is above 90% of the total power generation, including Norway, Paraguay, Albania and so on; in 57 countries above 50% (included), including Brazil, Canada, Switzerland, Austria, and so on; 64 countries above 40%, including most of South American countries. Total hydropower generation in the world accounts for 16% of total power generation in the world. Total installed hydropower capacity is 937,000MW, in which above 185,180MW is under construction.

The installed capacity of the Three Gorges Project is 22,500MW, ranking No.1 in the world. Table 2.5-9 shows the top 20 hydropower installed capacity in the world.

Table 2.5-9 Top 20 Hydropower Installed Capacity in the World

No.	Dam name	Country	Dam type	Dam height (m)	Total reservoir volume ($10^9 m^3$)	Installed capacity (MW)
1	The Three Gorges	China	PG	181	45.05	22,500
2	Xiluodu	China	VA	285.5	12.67	13,860
3	Itaipu	Brazil/Paraguay	PG	196	29.0	14,000
4	Guri	Venezuela	PG	162	135.0	10,000
5	Tucurui	Brazil	ER	95	45.54	8,370
6	Sayano-Shushenskaya	Russia	VA	245	31.3	6,400
7	Xiangjiaba	China	PG	162	5.163	6,400
8	Krasnoyarsk	Russia	PG	124	73.3	6,000
9	Longtan (Guangxi)	China	RCCPG	192 (phase I) 216.5 (phase II)	18.8 (phase I) 29.92 (phase II)	4,900 (phase I) 6,300 (phase II)
10	Nuozhadu	China	ER	261.5	23.7	5,850
11	Jinping II	China	BM	34	0.0192	4,800
12	Bratsk	Russia	PG	125	169.0	4,500
13	Ust-Ilim	Russia	PG	105	59.4	4,500
14	Xiaowan	China	VA	294.5	15.0	4,200
15	Laxiwa	China	VA	250	1.079	4,200
16	Jinping I	China	VA	305	7.988	3,600
17	Pubugou	China	ER	186	5.337	3,600
18	Ilha Solteira	Brazil	PG/TE	74	21.17	3,444
19	Ertan	China	VA	240	6.1	3,300
20	Goupitan	China	VA	232.5	6.454	3,000

Note: TE, earth dam; VA, arch dam; ER, rock fill dam; PG, gravity dam; BM, gate dam; RCC, roller compacted concrete dams; RCCPG, roller compacted concrete gravity dams.

References

[1] Shen Chonggang, Zheng Liandi. History of Dam Construction in China // Pan Jiazheng, He Jing. *Dam Construction for 50 Years in China*. China Water Conservancy and Hydropower Press, 2000.
[2] Robert B. Jansen. *Dams and Public Safety*. United States Government Pringting Office. 1980.
[3] China Society for Hydropower Engineering. *Electric Power for 60 Years in China*. Beijing: China Electric Power Press, 2009.
[4] China Society for Hydropower Engineering. *Almanac of China's Water Power*, Beijing: China Electric Power Press, 1984.
[5] Berga, L. Dams for Sustainable Development. Proceedings of the High Level International Forum on Water Resources and Hydropower, 2008.
[6] The United Nations Development Programme (UNDP). *Human Development Report* 2007/2008. 2008.
[7] Word atlas & industry guide 2008, *The International Journal on Hydropower and Dams*. London, UK. 2009.
[8] Jia Jinsheng, Ma Jing, Zheng Cuiying. Dam and hydropower development in the changing world, *Hydropower Generation*, 2011 (4).
[9] Jia Jinsheng, Xu Yao. Investigation on hydropower development for one hundred years in China compared with international hydropower development, *Hydropower Generation*, 2012 (1).
[10] Jia Jinsheng, Yuan Yulan, Li Tiejie et al. Hydropower stations completed or under construction with the dam height above 100m and the installed capacity above 500 thousand kW in China (till 2003). *Hydropower Generation*, 2004 (12).
[11] Jia Jinsheng, Yuan Yulan, Zheng Cuiying, et al. Brief Introduction to the statistics, technological progress and concerned problems of the dams in 2008 in China. *Hydropower Generation*, 2008.
[12] Jia Jinsheng, Yuan Yulan, Zheng Cuiying, et al. Brief Introduction to the statistics, technological progress and concerned problems of the dams in China. *Hydropower Generation*, 2010 (1).

Chapter 3

Dams and Environmental Protection

Xie Xinfang[1], Sun Zhiyu[2], Chen Min[3]

3.1 General Overview of Environmental Protection of Dam

With an area of about 9.6 million square kilometers, China has numerous water systems and rivers with a long history; including the Yangtze River, Yellow River, Pearl River, Huaihe River, Haihe River, Songhua and Liaohe River, Taihu lake and part of the international rivers shared with other countries, such as the Yarlung Zangbo River, Lancang River, Nu jiang River, Yalu River, Tumen River, Heilongjiang River, Irtysh River, Yili River and Aksu River. Since 1949, hydropower has been rationally developed in accordance with the principle of all-round consideration, comprehensive utilization, positioning on local conditions and combination of the large-medium-small rivers in China.

Hydropower has played, and will continue to play an important role in the continued, healthy development of the Chinese national economy and in improving the people's living standards. With the rapid economic development of China, it is an urgent task to deal with electrical energy's shortage. As clean reproducible energy, hydropower is vital for national socio-economic development. However, the rate of hydropower harness in China is relative low, compared with most of other developed nations. Therefore, speeding up hydropower development is a long term program for the demand of national economic reconstruction.

As one of the major form of hydropower exploitation, dam has multiple socio-economic benefits in terms of flood control, generation, irrigation, water supply and navigation; however, dam construction has simultaneously come with serious negative consequences for the environment, such as water level rise and land inundation. Because of the reservoir impoundment and its occupying lands, part of local people are forced to leave their homeland where they have been living for generations in form of moving backward or over a long distance to other places, which affect the production, living and mentality of the people involved. Dam construction can also produce "three-waste" pollution and the reservoir operation has impacts on water temperature, water quality, terrestrial biodiversity, aquatic ecosystem as well as cultural relics and landscapes. Though the extent and elements of these impacts vary from project to project, it is an unavoidable fact that any specific project has some impact on environment.

To overcome the shortcomings of dams and solve the environmental problems in the construction and operation of hydropower, an positive way is to, in addition to improving the scientific verification and design quality of dam developments, strictly control environmental pollution and ecological destruction throughout the dam construction, practically highlight environmental protection for resettlement, and provide the optimal solution for alleviating negative environmental impact. It will contribute to the coordinated development of economy and environmental protection by enhancing the environmental science and technology for dam development in China and preventing environmental problems both from the start and during the process. Environmental hazards resulting from dam construction can be reduced to the minimum extent and a concerted coexistence of economic, social and environmental benefits will not be a dream as long as positive and effective environmental protection measures are taken, e.g. Environment Impact Assessment (EIA) work and the feasibility study on dam construction in view of environmental protection for pre-project; environmental management throughout the whole processes of project construction and environmental pollution control during the construction stage; environmental protection in immigrant resettlement regions; measures to maintain stable production for the migrants and virtuous circle of the ecological environment in the resettlement area. Over the past years, along with the rapid progress of dam construction in China, envi-

---

[1] Xie Xinfang, Vice Chief Engineer of the Yellow River Engineering Consulting Co., Ltd. Professor Senior engineer.
[2] Sun Zhiyu, Director of Department of Science-technology and Environmental Protection, China Three Gorges Corporation.
[3] Chen Min, Officer of Department of Science-technology and Environmental Protection, China Three Gorges Corporation.

ronmental protection work, as an important part of the whole project, has been widely recognized by both the government and the public, and it has made some progress.

After considerable achievements in the environmental protection work for individual dam project, China has opened a new page of watershed planning on environmental protection. A basin is a highly relevant and relatively integral region where natural elements are so closely correlated that the change of one element will affect the other elements in the whole region. In addition to the impact of each individual project on ecological environment, the implementation of basin planning and the development of basin cascade reservoirs also cause environmental changes in the whole basin. Therefore, a more detailed analysis on multi-projects environmental impact and accommodating the ecological protection of the upper, middle and lower reaches of the basin will have great significance for the coordinated development of the basin's or the regional economy, society and environment.

The environmental impact assessment of cascade hydropower have been well studied and piloted in China since the 1980s. For example, of Guangdong Province Water Resources Department carried out EIA of cascade development in Dongjiang Basin Plan; Water Resources Department of Xinjiang Uyghur Autonomous Region conducted watershed EIA in Ye'erqiang Basin Plan with Xinjiang Survey & Design Institute of Water Resources and Hydropower. These works have gained experience for the subsequent EIA on basins.

To improve the Planning Environmental Impact Assessment (PEIA) quickly and orderly, China has made related policy for it, which is an important milestone in China's EIA history. Since then, Technical Guide of PEIA and its specifications have been successively issued. As part of the river basin planning, more and more planning EIA programs have been initiated nationwide. PEIA programs have already undertaken in basin comprehensive planning of the Yangtze River basin, Yellow River basin, Huaihe River basin and Jingou River basin located in Xinjiang Uyghur Autonomous Region. And PEIA of hydropower development are also conducted in the main streams of the Lancang River, Nujiang River and Jinsha River.

To meet the demand of rapid development for China's economy, hydropower must be developed in different levels. But, at the same time, environmental problems are increasingly standing out, which put forward new demand for environmental protection effort in basin planning for a long time. By the implementation of basin PEIA, the projects arrangement have been constantly optimized, advanced science and technologies have been explored and effective environmental protection measures have been taken, all of which will certainly minimize the adverse environmental impacts from basin planning and finally build a harmonious environment with beautiful waters, picturesque landscapes, sound ecosystem and healthy rivers.

3.2 Environmental Management of Dams

Substantial developments have been witnessed in China in dam environmental protection since the Environmental Protection Law (for trial implementation) was promulgated in September 1979. Over the past three decades, China, where there had been no records of dam environmental protection before, has laid a considerable cornerstone in dam environmental protection, gained many experiences and made significant achievements in the environmental management measures for dam construction by arduous cultivation, continuous exploration and gradual improvement.

3.2.1 Setting Environmental Management Procedures

Environmental management is an important way to balance human social and economic development with environmental protection. It has two different meanings. Broadly speaking, environmental management mainly refers to regulate human socio-economic activities in the range of the environmental capacity by utilizing technical, economic, legal, educational and administrative means, based on environmental science theories. Narrowly, it means that environmental managers control the pollution and destructive impacts, which were caused by the social and economic development, to realize the optimal economic, social and environmental performance.

The development of hydropower aims at combining the promotion of benefits with the elimination disasters and promoting comprehensive utilization of water resources, with the purpose of improving people's production and living conditions. However, the environmental pollution and ecological destruction associated with hydropower should be received more attention. Especially, it is obviously shown that the number and magnitude of hydropower development are increasing in face of adverse impacts on environment. It is well proven that environmental problem is still left or exposed from some of hydropower, to a large extent resulting from the inappropriate management of the projects such as improper planning arrangement and economic development without the constraint of environmental protection in a pursuit of the maximal economic profits. Some other environmental problems also exist which, though not directly resulting from the improper management, could have been effectively solved if environmental management had been strengthened. From this standpoint, regarding enhancement of environmental management as the working guideline for environmental protection while increasing investment in environmental protection by means of advanced technology and strictly enforcing the environmental management procedure (Fig. 3.2-1), which have great realistic significance for improving the entire environmental protection work.

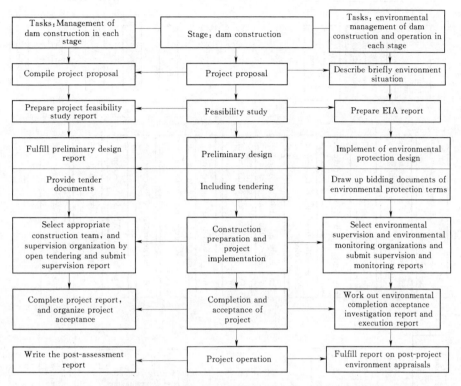

Fig. 3.2-1 Environmental Management Procedure for Dam Projects in China

3.2.2 Establishing Environmental Protection Laws

Long before, China had the laws and regulations concerning about the environmental protection. Historical records have showed that wildlife and plant protection clauses dates back to the Qin Dynasty and Tang Dynasty. Since 1949, particularly over the past three decades, China has promulgated dozens of laws, acts, provisions, regulations and methods concerning environmental protection, which serve as powerful basis for the nation's environmental protection work in its transition from the man-based rule to the law-based rule.

Legal references applied in dam construction for environmental protection can be found in nearly any type of laws including the constitution at the top level and the local regulations and methods if not the exact basis, at least legal reasons.

Environmental Protection Law of the People's Republic of China is a sub-law of the nation's constitution as well as the basic law of environmental protection. The legal framework for environmental protection is roughly classified as comprehensive environmental protection laws, pollution and other public nuisance control laws, natural resources protection laws, ecological environment protection law, environmental standards and the environmental protection regulations from other departments. The more details of the legal framework for environmental protection work are shown in Fig. 3.2-2 below.

3.2.3 Promulgating Environmental Protection Guidelines

Ever since the 1980s, EIA system implemented on dam construction in China has played a positive role in the environmental protection of hydropower projects. With the society's increasing awareness of the environmental protection and rigid requirement for environmental protection, the Ministry of Environmental Protection has published a series of technical guidelines to further regulate EIA work according to environmental protection standards, such as the *Technical Guidelines for Environmental Impact Assessment General Principles* (HJ 2.1—2011), *Atmospheric Environment* (HJ 2.2—2008), *Surface Water Environment* (HJ/T 2.3—93), *Acoustic Environment* (HJ/T 2.4—2009), *Non-Pollution EIA* (HJ/T 19—2011) and *Hydroelectric Engineering* (HJ/T 88—2003), which gradually made the EIA work of hydroelectric engineering advancing along scientific and standard road. The issuance and execution of the above-mentioned Guidelines clearly define

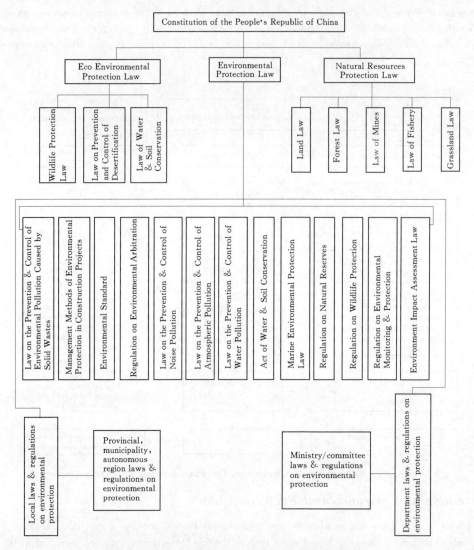

Fig. 3.2-2 Legal System for Environmental Protection

the EIA procedure, content and methods of construction project such as hydroelectric engineering.

Whether EIA is for the overall evaluation or for environmental elements, a three-step procedure of EIA work is followed: the preparatory stage, regular working stage and compiling report stage.

Preparatory stage: Firstly, an EIA starts with initial analysis of project and investigation of environmental baselines. Based on its results, the next step is to screen the proposed project on the criteria of identifying significant environmental impacts and establish the action-class for each individual impact according to the project features, size, output and environment parameters. Then, according to the content of action-class, an EIA action-outline should be prepared which directs the subsequent EIA process, and also provide criteria for reviewing the environmental impact reports (EIR).

Regular working stage: With the detailed study of project and baseline environmental conditions, the core step in EIA is to determine environmental indicators, to predict and evaluate the holistic effects of environment or individual effect of environmental elements with public involvement.

Compiling report stage: From the environmental protection point of view, the final step of an EIA is a separate decision-making procedure from any permission given by the conclusion on mitigation measures, which are proposed on the basis of the above possible impacts, and then to complete the preparation of a report. Turning to the EIA process itself, the content of EIA is composed of these environmental elements such as water, atmosphere, noise, ecology, population health, natural landscape, as well as cultural relics. Evaluation indicators, methods and depth of each element have a type determined by evaluation grade, which in return is made by the project scale, the impact scope and degree on its surroundings, pollution sources. coverage of its impact on the ambient environment, subjective and source intensity of the impacts.

EIA methods include mathematical simulation, physical model, analogy and contrast method, characteristic indexes, and so on. The evaluating methodology of each element is used as corresponding model or evaluation method in a given situation according to the assessment grade. Generally speaking, the higher assessment grade is, the more comprehensive investigation is and the more accurate prediction is.

3.2.4 Perfecting EIA System to Raise the Executable Rate

Ever since its initiation, China's dam EIA has undergone an experimental stage and an overall implementation stage. Until the 1980s, EIA execution had remained at a low level in the established large and medium-sized hydropower projects. In 1986, *Management Guidelines on the Environmental Protection of Construction Projects* was jointly issued by the Environmental Protection Committee under the State Council, State Development Planning Commission and State Economic & Trade Commission, which clearly stipulates that "any construction project will have an impact on the environment, the report about the environmental impacts must be strictly submitted and approved." and "the Environmental Impact Report (EIR) or Environmental Impact Form (EIF) shall be completed in the stage of project feasibility study". The law of the People's Republic of China on Environmental Impact Assessment ("EIA Law" for short) which comes into force in 2003, further stipulated the project-level EIA work and explicitly required that the "the State shall implement classified EIA management for construction projects according to the effect degree on environment. Projects, which are likely to cause a range of significant adverse environmental impacts, need to finish an Environmental Impact Report and comprehensively evaluate impact; while projects which are likely cause to a slight environmental impacts need to fill an Environmental Impact Form with environmental impact analysis or specific assessment. Enacting and implementing of environmental protection laws, regulations, norms and standards have played a substantial role in improving the EIA execution rate for hydropower in China. To date, EIA of large and medium-sized hydropower projects under construction have been fully implemented.

3.2.5 Expanding the Scope of Environmental Protection

For a long time, the scope of dam environmental protection has been limited to the preparation of an environmental impact report during the feasibility study stage. It is mainly because of the different departments for the investigation, planning, construction and management of projects, the lack of a perfect system of law and incomplete understanding of the EIA work. Today, with the comprehensive implementation of environmental protection work for hydropower projects, particularly with the introduction of international financial organization's loans programs, the public is more aware of environmental protection. Besides professionals in environmental protection for hydropower projects, other professionals, managers and leadership are also more aware of environmental protection. This has continuously expanded the scopes of the environmental protection work.

For the completed dams such as the world-renowned Three Gorges Dam in the Yangtze River and Xiaolangdi Dam in the Yellow River, environmental protection works have been carried out throughout the whole project process including the pre-planning, project proposal, feasibility study, design, tendering, construction and completion stages. After completing the preliminary environmental impact assessment in the pre-planning of the project, the EIA report at the feasibility study stage and environmental protection design at the preliminary design stage and after obtaining approval from the superior environmental protection authority, and environmental protection provisions are also incorporated into the tendering documents of the project in the tendering design stage of the project according to the International Federation of Consulting Engineers (FIDIC for short in French) terms, project characteristics and environmental protection requirements. The tendering can provide sufficient basis for the planning and execution of environmental protection.

During the design stage of the project, the EIA staff work out an implementing programs for environmental protection, which can identify the overall objective and individual goals of environmental protection, defined the mode, method and content of individual tasks.

In the implementing stage of the project, a special environmental administration is set up to perform environmental supervision, environmental monitoring, disease control and other work in connection with environmental protection.

In the completion stage of the project construction, special inspection and acceptance of environmental protection for large and medium-sized construction projects are part of the work involved in the completion work. In December 2001, to standardize the completion acceptance work of environmental protection for hydropower projects, the State Environmental Protection Administration promulgated *Management Procedures for the Inspection and Acceptance of Environmental Protection of Completed Construction Project*, stipulating that a construction project should satisfy itself with nine prerequisites including obtaining compulsory environmental protection review and approval, and making available all necessary technical data and environmental protection documentation before they can proceed to the completion acceptance of environmental protection. The management procedures also specified that the acceptance should cover all environmental protection facilities in relation to the construction project concerned, including works, equipment, devices and monitoring means established or supplied for controlling pollution and protecting environment and all the infrastructures of ecological protection; and other environmental protection measures taken as specified in environmental impact report, environmental impact form and the design documents of the items concerned. Upon completion of the construction project, the construction organization should submit to completion acceptance of environmental protection with the competent department of environmental protection administration for approval.

3.2.6 Promoting Planning EIA Work

China's EIA law of promulgation firstly has documented and broadened EIA coverage from the original project EIA to the Planning EIA (PEIA), which specifies that EIA should be developed in land utilization related plans, the integrated plans of areas, drainage basin, and sea areas and relevant special plans for industry, agriculture, animal husbandry, forestry, energy, water conservancy, transportation, urban construction, tourism, natural resources development to prevent from the source environmental pollution and ecological damage so called "source controlling". This marks a new stage for China's EIA system.

To carry out the EIA Law, to guide the implementation of PEIA and to promote scientific and standard PEIA, the State Environmental Protection Administration organized the related experts and issued *the Technical Guideline for Plan Environmental Impact Assessment* (Technical Guideline for PEIA), which stipulates the concept of PEIA and environmentally feasible recommendations. PEIA is a process in which potential environmental impact from planning implementation should be analyzed, predicted and assessed during the planning preparation stage, and countermeasures should be proposed for preventing or alleviating negative environmental impacts. Environmentally feasible recommendations should be proposed in line with the objectives of planning and environment. As different types of planning vary significantly in the content and its method, it is difficult to provide a unified method or content for PEIA, thus making it even more difficult to prepare a technical guideline. As a result, the Technical Guideline only set out the general principle, work procedure, method and content of PEIA, which proves to be a positive guide and instructor for further enriching and improving the theory and technical method for PEIA.

To standardize the EIA of river basin planning and to provide a unified basic principle, content and technical requirement for the EIA work, standard for *Environmental Impact Assessment of River Basin Planning* (SL 45—2006) was promulgated by the Ministry of Water Resources in 2006. PEIA is generally required to predict environmental impact of planning implementation considering the basin environmental system and its each of environmental element characteristics and to assess according to the water resources distribution, planning layout, size, implemented sequence and the project for basin development and harnessing in the near future. The direct impact, indirect impact and cumulative impact from the proposed planning should be predicted by implementing the strategy of sustainable development, boosting the social, economic and environmental sustainability, and taking into account the association and correlation of the projects.

As part of river basin planning, the environmental impact assessment in basin planning can be implemented throughout the entire process of river basin planning. Environmental impact assessment in basin planning includes planning analysis, environmental baseline investigation and assessment, identification of environmental impact and objective of environmental protection, environmental impact prediction and assessment, scenario comparison, the countermeasures of environmental protection, environmental monitoring and follow-up assessment plans and public involvement.

3.3 Ecological and Environmental Conservation of Dams

When a dam is built, the water level is raised and the underlying surface is turned from land surface into water surface. This leads to negative impacts on the species and habitat of the original terrestrial life in the reservoir areas. Therefore, to conduct ecological environmental impact assessment of dam construction and provide countermeasures is an important way for protecting ecological environment.

3.3.1 Impact on Terrestrial Plants and its Mitigation Measures

3.3.1.1 Impact on Terrestrial Plants

1. Reservoir inundation

Water storage in reservoir directly causes inundation of plants, and biological individuals lose their growth environment in the reservoir areas. The extent of impact is irreversible. Reservoir inundation leads to destruction and fragmentation of habits and has a negative impact on terrestrial plants. In topographically complicated mountain with rich biological diversity, dam inundation deprives the habitats of numerous plants and reduces the species communities in the reservoir area, while habitat fragmentation will change the composition of plant populations, and generation of many habitat boundaries directly prevents the dispersion and resettlement of species within the population. The impact is particularly significant on wind-pollinated plants, reducing the potential chance of for generating new populations. Furthermore, the light, wind and temperature in the new boundaries are different from those inside the former habitat, which has an effect on the survival and component composition of the species. The difference of inundation areas, topographical conditions and the intensities of human activities have changed the distribution of plant species. Generally speaking, greater impact is generated on the plant resources in the local area where a larger area is inundated in the reservoir and human activities are more frequent.

After an inundation of the reservoir, the water forms many branches, and the original forest community is artificially separated, resulting in habitat loss and fragmentation. The most serious impact of reservoir inundation on terrestrial plants is that on rare, endangered and endemic species.

2. Hydro-fluctuation

In the hydro-fluctuation of reservoir, how much of an effect on inundation vegetation depends on the magnitude of the project and reservoir operation. Habitat of exposure and inundation alternations will have an adverse effect on the survival and growth of some vegetation.

3. Construction activities

Foundation excavation and the construction of temporary road for dam projects can have permanent or temporary influence on plants. The worst impact on plants from land occupation for project construction and construction personnel activities leads to the loss of rare, endangered species and economically valuable species. The influence of human activities on vegetation during the construction usually results from improper management such as deforestation and expansion of construction boundaries.

3.3.1.2 Protective Measures for Terrestrial Plants

Protection of terrestrial plants needs to be more sharply focuses on rare, endangered and endemic plant species, ancient and famous trees suffered from reservoir inundation and project construction.

In-situ protection and ex-situ protection are to be adopted respectively to protect biodiversity according to the features of the terrestrial plants concerned and the adverse effects caused by construction works. The in-situ protection should be mostly considered ecological characters, quantities, distributions and growing conditions of protection subjects within the scope of protection to propose measures such as geological evasion, in-situ enclosure, public listing, and the establishment of protective area so on. While the ex-situ protection is taken to consider the characters of protected plants to put forward the measures of transplanting, introduction and breeding, the conservation of species germ plasm resources, and setting up the seed bank. The objective of the two measures is to protect the natural ecological system with vegetation restoration measures by use of the original plant resources. And more efforts should be made to strengthen environmental management and regulate the behavior of construction personnel during the construction period.

3.3.2 Impact on Aquatic Ecosystem and its Mitigation Measures

Dam construction has a great impact on aquatic species and its habitat. Fish has a most direct relationship with humans in aquatic ecosystems so the impacts of dam construction on fish and its measures were only analyzed emphatically and some mitigation measures were put forward based on the impact analysis in the following studies.

3.3.2.1 Fish

Dam cause fragmentation of aquatic habitat, obstruct the passage of migratory fish, and loss of fish genetic diversity in the upper and lower reaches. The alteration of flow regimes such as water depth and stream flow caused the changes of natural aquatic habitat and the changes in community composition, resulting in negative impact on rare and endemic species.

Fish spawning habitat is crucial for the protection of fish species and population. The impacts of dam construction on fish spawning habitat and fish reproduction have the following several points:

(1) Fish migration passage barriers can block the access to essential habitats. The blockage of dam reduces the river flow, deprives the survival of fish floating eggs by incapability of floating enough distance because of channel reduction or loss and slows the river's current in the reservoir area leading to fish eggs for sinking not to hatch.

(2) Water impoundment of the reservoir directly inundates fish spawning habitats.

(3) During the spawning season, flow regulation can drop down the flood peak in the downstream of the dam. The reduction of the discharge in the channel and the low temperature of the discharged water have an important influence on fish spawning habitat in the downstream and the ability of fish to reproduce.

3.3.2.2 Protection Measures for Fish

(1) When a dam goes through the fish passage for rare, endemic and important economic fishes, measures must be taken to ensure fishes passing. The blockage and low-head dam is well suited for the construction of permanent fish passage such as fish ways, fish ladders and fish locks. As for the high dams, fish lift should be introduced with fish pump, dam-passing ships for fishing and net over fishing of passing-dam facilities.

(2) The construction and operation of the dam cause the reduction of fishery resources, and the measures of artificial propagation and releasing should be implemented to avoid the reduction of fishery resources. It is necessary to set up an aquatic ecosystem monitoring system to long-term monitor the results of fish propagation and releasing.

(3) Dam constructions have an adverse impact on the destruction and the loss of wintering habitat, feeding habitat and spawning habitat (so called "three habitats") and other important habitats. And an appropriate river reach should be selected to recreate the corresponding aquatic habitats.

(4) Beside the artificial propagation and releasing, setting up fish reserves and prohibited fishing areas in the appropriate river reach is another measure to prevent the decline in rare and endemic fish resources causing the instability of fish populations.

3.3.3 Impact Assessment on Sensitive Protection Objectives and its Mitigation Measures

3.3.3.1 Impact Assessment on Wetland and its Protection

Wetland is a unique ecological system featuring rich biological diversity and multiple important functions. Wetland is not only the growth areas to numerous hydrophytes and the habitats of many wading birds and fowls, but also the spawning ground and feeding ground of various fish, shrimps and shellfish. Besides, it also has multiple ecological environment functions like regulating river flood peak discharges, controlling the flood water, storing water to supplement groundwater, improving the local microclimate, containing wastes and purifying water quality.

Water storage in reservoirs may either inundate parts of wetlands in the overflow land or reduce the water inflow going into the wetland and thereby break the water balance of the wetland. While water projects try to control flood by altering flood rules, many wetlands are nothing but the products of floods which make up and store water when a flood peak comes and then slowly release and discharge water after the flood peak. The wetland ecosystem varies with the alteration of flood rules.

In protecting the wetland's ecological environment, an important measure is to address Ecological Water Requirement (EWR), in addition to preventive measures and compensation practices. Ecological water requirement is the major limiting factors of the wetland ecosystem. The ecological water requirement bases on maintaining the minimum area of wetland habitat, and must leave free to prevent the risk from competitive water demand in the continuous drought years. Water needs or the mininum area of wetland habitat are closely related to the ecological environmental functions in the wetland, including the water supply function, the survival of wading creatures and sustaining of an economy producing function of the wetland.

3.3.3.2 Impact Assessment on Natural Reserves and its Protection

Nature reserves may be divided into three parts: the core zone, buffering zone and experimental zone. The core area is the intact natural ecosystems and the area where the rare and endangered animals or plants are in obviously centralized distribution within natural reserves. To assess the impact of dam construction on nature reserves, it should be firstly to identify the property, function and the level of protection of the natural reserves and then to scientifically analyze its actual impact on the natural reserve according to applicable regulations, provisions and requirements for the natural reserves.

To effectively play its role of biodiversity protection in the nature reserves, it must formulate the following strategy by prohibiting from entering the core area in the nature reserve, maintaining as natural as possible outside the core area, keeping the minimal critical size, restoring the natural habitat corridor and strengthening the outer protection zones.

3.3.3.3 Impact Assessment on Wildlife Habitats and its Protection

Threats to the wildlife from dam construction typically come from the fragmentation of the wildlife habitat due to reservoir inundation and land occupation for the project, resulting in shrinkage and even occupation of the habitat. Encroaching on the habitat shall inevitably lead to reduction of usable area of the habitat and thereby have an impact on rare (including native) and endangered species and especially the important species of economic, scientific and cultural values. The dam breaks up the originally linked habitats into many small segments and thereby leads to habitat fragmentation, the alteration of the ecological conditions, shrinkage of the coverage of the feeding, mating and avoiding predators for some animals, destroying genetic information transfer and eventually making the species fragile and extinct.

To mitigate the impact of dam construction on wildlife habitats, it is the key to select the competitive schemes during the addressing stage and propose the optimum scheme to avoid the important habitats for environmental protection and socio-economic development purposes. And it should strengthen environmental protection and management, regulate the behavior of construction personnel, scientifically and strictly implement environmental protection measures, minimize the occupation of animal habitats and lower the degree of habitat fragmentation to the largest extent during the project construction period.

3.3.4 Impact Assessment on the Eco-environment of the Three Gorges Project and its Protection Measures

3.3.4.1 Terrestrial Plants Protection

The Three Gorges reservoir area lies in the north of China's subtropical zone. Due to the subtropical monsoon climate, the zonal vegetation is basically of evergreen broad-leaved forest represented by *castanopsis fagaceae* and *nanmu phoebe*. Because of its high mountains, the Three Gorges reservoir area has obvious vertical vegetation zonation. Evergreen broad-leaved forest is widely distributed below 1,300m, coniferous and broad-leave mixed forest about 1,300 – 2,200m and subalpine coniferous forest above 2,200m. The dominant type of vegetation is secondary, which is in the succession and transition of vegetation below 1,700m. Because of human's intense reclamation, little is left of the primeval forest below 1,000m, instead, where is widely distributed of *pinus massoniana* forests, *cypress* forests and sparse forests, shrubs, grass and farmlands.

The impact on species from the Three Gorges Project construction involves 120 families, 358 genera and 550 species. Perhaps the most significant species in quantities are mainly *Gramineae*, *Compositae*, *Euphorbiaceae* and *Rosaceae*. The two genus of *Sapindaceae* in the reservoir area may be completely inundated. Although the impacts of the Three Gorges Project on rare and endemic species are not as serious as extinction of the endangered species, it can submerge the living origin of some species, which threatens the quantity of plants as an example of the *Adiantum reniforme var. sinense*. In the eleventh part of *Preliminary Design Report of the Three Gorges Hydroelectric Project in the Yangtze River (Key Project)* about Environmental Protection is prepared to reexamine and the results showed that an endemic plant of area, *Myricaria laxiflora* in the Three Gorges reservoirs also distributed on the river beach in the reservoir area at 80 – 130m. It is discovered that there are wild *Myricaria laxiflora* communities in the three beach wetlands within the Yangtze River's main stream from about 100km in the downstream of the dam between December 2007 and February 2008 which suggests that *Myricaria laxiflora* is not endemic to the Three Gorges reservoir area.

The EIA Report of the Three Gorges Project in the Yangtze River has presented the measures, such as the establishment of natural reserves and biodiversity protection which had rechecked about the impacts on terrestrial plants, and carried out a detailed measures of environmental protection by means of establishing natural reserves and reserve places in the eleventh part of *Preliminary Design Report of the Three Gorges Hydroelectric Project in the Yangtze River (Key Project)* about Environmental Protection.

In project implementation stage, the Three Gorges Project launched to the projects to protect plant biodiversity in Dalaoling of Yichang city, as well as a protection projects and scientific research of terrestrial plants with an emphasis on *Adiantum reniforme var. sinense* and *Myricaria laxiflora* by means of multiple alternatives to choose from the establishment of natural reserves and reserve place, germplasm conservation and ex-situ conservation. So far, the work of ex-situ conservation has been performed for 30 rare plant species and 73 dominant plants. Fig. 3.3-1 showed that the detailed experiment about in-situ and ex-situ conservation of wild plant in the Three Gorges Project. Besides, it also initiated the construction and preservation of the conserving broad-leaved evergreen forests in the Longmen River of Xingshan according to the regionally representative terrestrial ecosystem.

In addition, a cultivation base for rare and endemic plant species in the Three Gorges Project has been established in the dam construction area to investigate the survival and growth of endemic and rare plants and introduce, domesticate and reproduce the species with the purpose to study and protect the endemic and rare plants in the Three Gorges area. So far, preliminary development has been achieved in plant asexual reproduction in the Three Gorges reservoir area and some have been successfully cultivated.

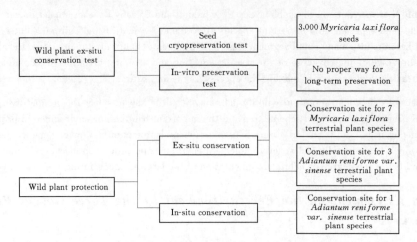

Fig. 3.3-1   Detailed Experiment about In-situ and Ex-situ Conservation of
Wild Plant in the Three Gorges Project

3.3.4.2 Aquatic life Protection

The impact of the Three Gorges Project on the rare and endangered aquatic animals and Four Chinese Carp resources is the key part of the impact of the Three Gorges Project on aquatic life. In the area affected by the Three Gorges Project, there are 6 rare and endangered aquatic animals, of which the Yangtze River dolphin (*Lipotes vexillifer*), Chinese paddlefish (*Psephurus gladius*), Chinese sturgeon (*Acipenser sinensis*) and Dabry's sturgeon (*Acipenser dabryanus*, also called Yangtze River sturgeon) are first class national protected animals while the finless porpoise (*Neophocaena phocaenoides asiaeorientalis*) and Chinese sucker (*Myxocyprinus asiaticus*) are second-grade national protected animals. The total resources of four major Chinese carps declined.

According to the environmental impact predictive results in the EIA report, protective measures are taken including establishing natural reserves, artificial proliferation and releasing, defining the closed-fishing periods and catch limits, research on eco-environmental protection and its monitoring systems.

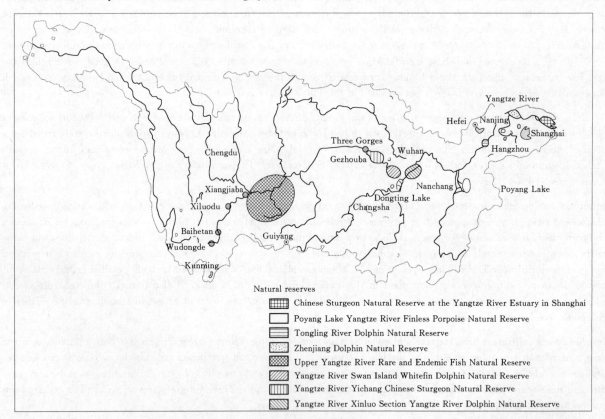

Fig. 3.3-2   Distributing Sketch of Aquatic/Wild Animal Natural Reserves in the Yangtze River MainStream

(1) Establishing natural reserves: There are many established reserves and habitats of rare and endemic species in the Yangtze River such as the Chinese sturgeon natural reserve in Yichang, Hubei Province; Chinese sturgeon natural reserve at the Yangtze River estuary; rare and endemic fish natural reserve in the upper reaches of the Yangtze River; Yangtze River dolphin natural reserve in Swan Island of the Yangtze River and a national natural reserve of Yangtze River dolphin at Xinluo section of the Yangtze River (Fig. 3. 3-2).

(2) Artificial proliferation and releasing: Proliferation and releasing of the Chinese sturgeon started in 1984 and was extended after 2005 from Chinese sturgeon to the Dabry's sturgeon, Chinese sucker and a variety of economic fish.

(3) Defining closed-fishing periods and catch limits: Based on the ecological habits of Yangtze fish and other data base with reproduction, management and the fishery social conditions, a comprehensive closed fishing was fully implemented in the Yangtze River since 2003. It was carried out from February 1 to April 30 every year in the upstream of the Gezhou Dam and during April 1 to June 30 for the downstream of the Gezhou Dam in the Yangtze River.

(4) Research on eco-environmental protection and its monitoring: It was carried out that the researches on eco-environmental protection of related rare animals and ecological water requirement in the middle and lower reaches of the Yangtze River such as series researches on the artificial propagation and ecological protection of rare fish species like the Chinese sturgeon, Dabry's sturgeon, Yangtze River dolphin and finless porpoise; previous research on ecological operation scheme for the Three Gorges Project specific to the natural propagation requirement of four major Chinese carps, research on the environmental change trends of the Yangtze River dolphin (*finless porpoise*) natural reserve and other habitats after the operation of the Three Gorges Reservoir and its protection measures, and preliminary research on the impact of the operation of the Three Gorges Project on the relationship among the Yangtze River, the Dongting Lake, Poyang Lake and the ecological environment in the lake area[4]. In Oct. 2009, a fully artificial propagation of the Chinese sturgeon achieved success. In the ecological environmental monitoring system for the Three Gorges Project, rare and endemic fish, fishery resources and fishery environment are continuously monitored as an important part of the program and the monitoring data shall be vital for scientific research and implementation of other measures.

3.4 Water Environmental Protection of Dams

After the completion of the hydropower, the water reservoir area becomes larger, water depth increases, current velocity becomes slower, the self-clean capacity obviously decreases but water quality goes towards a negative direction. The thermal changes in the water body of the reservoir will stratify the water temperature in the reservoir, which in turn directly affects the water quality, aquaculture and farmland irrigation. As a result, it is an important part of dam water environmental protections to analyze the water environment changes in connection with dam construction and take proper measures to mitigate its negative impacts.

3.4.1 *Impact Assessment on Water Quality and its Measures*

3.4.1.1 Water Quality

The alteration of flow regimes causes the change of water quality. Possible latent effect of dam on water quality was severe including reduction of flow velocity and obvious clarification effects which helps to curb dissoluble mineral matters, cut down the turbidity and biochemical oxygen demand (BOD), increase nutrient concentration and enlarge the environmental capacity. At the same time, however, as the dilution and self-purification ability of per unit water body in the reservoir decreases and a clear thermal stratification is formed, the water in the reservoir would form a density shield which turns the cold water layer into an anaerobic micro-organism layer. The indissoluble solid matters in reservoir would deposit and enrich on the bottom of reservoir.

From the water quality property of the reservoir, dam construction would have remarkable changes in the organic matters, heavy metals and nutritional status in water quality.

1. Organic matter

The river without dams maintains a certain velocity. Of particular concern are the dilution and mixture of organic matter because of the turbulent diffusion. Dam construction can lead to intensified deposition and decrease of dilution and mixture abilities, and less solar radiation to absorb. The seasonal temperature stratification and pumping of the reservoir can have an effect on the purification function of the organic matter in the water body.

Generally speaking, in case that no significant change is expected for the water quality inflow into the reservoir, the BOD_5

concentration in the water body of the reservoir would be lower than that of no reservoir, and the BOD_5 concentration of the water outflow would be lower than that of the water inflow. But there is an inverse condition at the bay and tail of the reservoir, where heavier organic pollution happens in the water areas. This is particularly serious at the tail of the reservoir or along the bank side subject to urban pollution, where a bank-side pollution zone is likely to take shape.

In a stratified reservoir, the concentration of dissolved oxygen (DO) has a remarkable change in the vertical depth of water body. As lower velocity and weaker aeration happens after dam construction, the concentration of dissolved oxygen reduces. On the other hand, however, as expanding water surface and increasing air velocity with dam construction have an advantage to the airborne oxygen into the water body. Besides, when the flow velocity reduces, the transparency, more photosynthesis and the dissolved oxygen on the top of the reservoir would increase for the aquatic plants. However, the photosynthesis would reduce and even deprive of oxygen, and the concentration of the dissolved oxygen shows a decrease trend with the increase of the water depth.

2. Heavy metals

Researches have indicated that suspended substances are the principle carriers of heavy metal pollutants. The heavy metals move in the water with the suspended substances and then deposit at the water bottom by flocculation and become sediments under different hydraulic actions and physical changes.

The river is flowing very fast with the strong capacity of adsorption and deposition and diffusion displacement, so the water body is highly capable of diluting and self-purification and heavy metals in the river are quickly adsorbed by the suspended substances. After dam constructed, the water velocity slows down and the settlement action becomes stronger, which is helpful for the settlement of the heavy metals in the water body. When the pH value reflects on alkalescency, it is more helpful for the flocculation and settling of heavy metals and the adsorbed heavy metals deposits and accumulates at the bottom of the reservoir.

3. Eutrophication

Eutrophication of reservoirs is closely related to human pollution. The discharge of industrial wastewater and the rain-wash of the chemical fertilizers and livestock manure into the reservoir cause the increase of nutrients load in the reservoir, which gradually turns the water body in the reservoir from poor nutrition to rich nutrition.

The eutrophication reservoir is the deterioration of water quality due to the enrichment of nitrogen and phosphorus, represented by the fast growth and fecundity of aquatic life and abnormal proliferation of the algae. However, reservoir eutrophication is by no means an irreversible process and can lighten the nourishment-enrichment in waters by controlling nutrient discharge from human activities and taking effective measures to nutrient removal in the reservoir waters.

4. Gas supersaturation

When the dam is releasing water, a large amount of gas is involved into the flow and thereby develops a high aerated flow because of high head water and quick stream speed. When the water flow carrying high-concentration air bubbles flows to downstream of the dam, large amounts of gases enter and are dissolved into water to supersaturating flows because of increase in pressure. The dissolved gas supersaturation can cause gas bubble disease to some fish, especially juvenile fish.

3.4.1.2 Mitigation Measures

To ameliorate the negative impacts of dams on the water quality, the following measures are mainly taken:

(1) To implement total emission control of the pollution sources in the vicinity of the reservoir area. Pollution sources should be subject to removal, rectification, discharge reduction or even discharge prohibition if they fail to meet the goal of pollution amount control.

(2) To improve the dissolved oxygen content at the bottom of the reservoir by properly utilizing reservoir regulation mode.

(3) To take some measures of gas supersaturation. These measures are as followed: adjust release structures; moderately extend the discharge time and reduce the discharge ensured flood control against danger; discharge in rational combinations of multiple facilities.

3.4.2 Impact Assessment on Water Temperature and its Measures

3.4.2.1 Water Temperature Stratification in Reservoirs

Due to dam construction, water depth in the reservoir area could increase and the rate of exchange could slow down, thus changing the heat exchange at the water-gas interface and heat transfer process inside the water body and leading to changes in effect of the water temperature in the reservoir water. Proving by a lot of the measured data, water temperature effect of multi-year regulating storage reservoir is represented by the water thermal stratification in vertical direction. At the beginning of air temperature rise in early spring, the frequently strong winds may weaker this stratification which becomes an important mechanism for raising the water temperature in the bottom layers. In early summer, water thermal stratification deeply increases, a stable warm water layer is developed in the top layers, which stops the vertical mixture through the full depth and develops a thermocline between the top warm water layers and bottom cold water layers. In late summer and early autumn, net heat flux turns from temperature rise to temperature fall, the surface water temperature gradually decreases and the convection from gravity gradually makes the temperature gradient of the thermocline reduce. In late autumn, under the joint effects of high winds and gravity convection, an isothermal level is reached through the full water depth.

Water temperature stratification in reservoirs imposes a series of impacts on water quality in the reservoir, water temperature of the released flow and its physiochemical properties.

3.4.2.2 Identification of Water Temperature Structures in Reservoirs

Water temperature in reservoirs can be classified into stratified type and mixed type according to vertical distribution of water temperature in the reservoirs.

(1) Mixed type: The type of water temperature is characterized that the water temperature in the reservoir is relatively evenly distributed at any time of the year, the temperature gradient is very small, the bottom water temperature varies with the surface water temperature and significant heat exchange takes place between the water body and the reservoir bottom. In general, reservoirs featuring shallow water and low regulation capacity are mostly mixed water temperature reservoirs.

(2) Stratified type: It normally refers to the case where, as the flow velocity slows down inside the reservoir and the water exchanges less frequently, the reservoir with high regulation capacity usually has a stratified temperature structure. The stratified temperature structure of a reservoir can be roughly divided into the top layer, middle layer and bottom layer. The top layer is the surface temperature layer, the middle layer is the thermocline layer and the bottom layer is the stagnation layer. During the period of the temperature rise, as the surface water temperature of the reservoir is noticeably higher than the middle and bottom layers, temperature stratification takes place. The temperature gradient in the thermocline is about 1 - 1.5℃/m. The water temperatures in the bottom layer can slightly vary in vertical direction.

There are several methods of identifying the type of water temperature in reservoirs. These include the indices α and β, of reservoir water substitution timing and density Froude number et al, of which α, β method is the simplest and most useful.

Index α is used to indicate whether the reservoir is stratified while β is used to indicate the impact of reservoir inflow flood in a flood season on the water temperature structure.

The criteria of index α and β are as follows:

$$\alpha = \frac{\text{Mean annual runoff}}{\text{Total storage capacity}} \tag{3.4-1}$$

$$\beta = \frac{\text{Flood volume one time}}{\text{Total storage capacity}} \tag{3.4-2}$$

When $\alpha \leqslant 10$, the water temperature of reservoir belongs to stratified type.

When $10 < \alpha < 20$, the water temperature of reservoir belongs to transitional type.

When $\alpha \geqslant 20$, the water temperature of reservoir belongs to mixed type.

For a stratified reservoir, when a flood with $\beta \geqslant 1$ takes place, a temporary mixture of water temperature will take place;

When a flood with $\beta \leqslant 0.5$ takes place, little impact will be imposed on the distribution structure of water temperature in the reservoir.

When a flood with $0.5 < \alpha < 1$ takes place, the stratification of water temperature in the reservoir is between the two above.

3.4.2.3 Analysis of Environmental Impact from Water Temperature

Low temperature water discharged from reservoirs has some effect on aquatic life, fish and farmland irrigation on the watercourse. Fish propagation requires certain water temperature. For example, four major Chinese carps in China when water temperature is above 18℃. Therefore, the low temperature of the reservoir outflow will affect the fish spawning within a certain distance downstream of the dam, and may also postpone the fish spawning. Besides, the releasing of water with low temperature may also degrade the metabolic ability of fish and slow down its growth. The low temperature and the consequently slow growth of food organisms have a direct effect on fish life including its growth, fattening and wintering. Furthermore, low temperature water has a "cold damage" to crops and lead to crop yield dropping or the total loss. Where the crop in the irrigated area is a hemophilic and hygrophilous rice, the releasing of water with low temperature will affect the rice yield.

3.4.2.4　Measures to Reduce the Effects of Low Water Temperature

To mitigate the negative impacts of the releasing of water with low temperature to maintain favorable circulation of ecosystem in the lower reaches, based on operation experiences in built hydropower, the following mitigation measures should be taken:

(1) To use multiple stratified water intake facilities at the design stage of a hydropower, such as shaft well, culvert horizontal-tube, and multiple intake structures at different elevations.

(2) To improve the vertical structure of water temperature in the reservoir area by properly utilizing reservoir regulation mode.

(3) To take measures to regulate the farmland temperature such as extending the water transmission channels and providing additional regulating ponds to increase the releasing water temperature.

(4) To adjust the crop planting structure and switch to the irrigation practices in the irrigated area.

3.4.3　*Water Quality Protection for the Three Gorges Reservoir Area in the Yangtze River*

According to the method of individual indicator assessment in *Environmental Quality Standards for Surface Water* (GB 3838—88), the analysis on about 10,000 monitoring data on water quality in 1984-1991, showed that all other monitoring factors of reservoir water were better than class Ⅲ except for Escherichia coli, oil pollutant and total mercury, and the water quality was sound at the river cross sections in the reservoir area to meet with the multi-purpose water demand[3].

After the reservoir is built, the water velocity in the reservoir decreases, the abilities of the reoxygenation, dilution and diffusion degrade and the river side pollution becomes more serious, particularly in the Chongqing and Wanxian river reaches. Dam construction impounds the nutrients of nitrogen and phosphorous, and contributes to the improved biological productivity in the reservoir water. However, negative impact is also expected, e.g. intensified runoff pollution in the farmlands, heavy pollution of total phosphorus and eutrophication tendency in some of the tributaries and reservoir bays[3].

According to the above predictions, proposed in the *Environmental Impact Assessment Report on the Three Gorges Project in the Yangtze River*, the main countermeasures are to eliminate and control major pollution sources within a time limit, to formulate and gradually implement water pollution prevention under the principle of total amount control, to implement cleanup of the reservoir bottom and solid waste along the river banks and to establish monitoring system of water quality so on.

During the project implementation, Chinese government has taken the measures of policy adjustment, project implement and scientific researches and made great effort in the water quality protection of the Three Gorges Project from pollution prevention, source control, water body restoration, monitoring and researches.

To protect the water quality in the reservoir area, the Chinese government has implemented migrants' settlements instead of the farmland-orientated resettlement, and adjusted the industrial structure in the reservoir area, which prohibited rebuilding small enterprises featuring heavy pollution, poor economic profits and difficult control to relieve the pressure for pollution control in reservoir area. In 2001, the government approved implementation of *The Water Pollution Prevention Plan for the Three Gorges Reservoir and Upstream* (2001 - 2010), as amended in 2008, covering city's sewage treatment, household wastes and disposing of hazardous waste, industrial pollution prevention, ecological environmental conservation and ambulatory ship pollution control specific to the Three Gorges reservoir area. Relevant protection measures of water environment should be continuously implemented as planned, which include strict cleanup of the bottom of the reservoir before water storage, control of ship pollutants and removal of floating substances. Moreover, hydrological and water quality monitoring is included as an important topic in the construction and continuous operation of the ecological and environmental monitoring system for the Three Gorges Project. The water environmental protection research projects for the reservoir area is approved and adjusted in time according to water qualities. These research projects mainly include a series of research on the water bloom prevention, investigation about polluting source in the reservoir area, emergency of water pollution incidents in the Three Gorges reservoir and management system researches on the Three Gorges reservoir. Besides, after the Three Gorges Project came into the operation stage from the construction stage, a series of the demonstration pilot projects

Chapter 3 Dams and Environmental Protection

have been carried out by the State Council Three Gorges Project Construction Committee. These projects mainly include the environmental management in the riparian zone, emergency management and long-term prevention of water bloom in the tributaries, pollution retention in rural and urban areas, the drinking water insurance system in the tributaries and construction of ecological defense along the reservoir banks.

Since 1996, water quality has been monitored within the affected areas of the Three Gorges Project. The water quality monitoring sections of the Three Gorges reservoir and its main stream and distributaries upstream are shown in Fig. 3.4-1 below. The monitoring results indicated that the water quality in the Three Gorges reservoir has been fairly stable and there was no tendency of deterioration since the monitoring system operated.

It is noticeable that since 2003, there have been changes in the sampling and assessment standards for water monitoring though these changes are not significant. From the average annual water quality variation in each individual indicator at Cuntan Station, which is the inflow section of the Three Gorges reservoir in the main stream of the Yangtze River, and the sections in the reservoir area in 1996-2009, the results showed that the water quality on the dam-front section has been typically subject to the inflow water quality and water pollution in the vicinity of the reservoir in nearly 10 years. After water has been stored, the water quality on the main stream of the Three Gorge Reservoir remains fairly stable and is very sound to reach the standard class Ⅲ (Fig. 3.4-2).

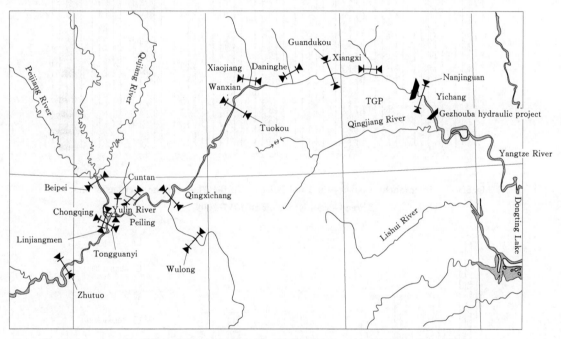

Fig. 3.4-1 Location Map of Water Quality Monitoring Sections in the Three Gorges Reservoir Area

Note: The Zhutuo, Tongguanyi and Cuntan sections are the water quality monitoring sections for water inflows into the Three Gorges reservoir which are not subject to water storage, while Qingxichang, Tuokou and Guandukou sections are the reservoir sections.

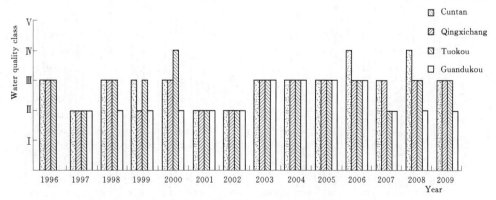

Fig. 3.4-2 Water Quality Classifications in the Three Gorges Reservoir

Note: The water quality assessment of cross section in 1996-2002 was employed the National *Environmental Quality Standard for Surface Water* (GB 3838—88), and that in 2003-2009 was the National *Environmental Quality Standard for Surface Water* (GB 3838—2002).

As shown from Fig. 3. 4-3 to Fig. 3. 4-6 above, the data source is from *Ecological and Environmental Monitoring Network for the Three Gorges Project*, the concentrations of principal nutrients such as the total nitrogen (TN) and total phosphorus (TP) in the main stream and primary tributaries are generally high before water storage and remain basically unchanged after water storage, suggesting that water storage does not lead to remarkable rise in the main nutrient level in the main stream and primary tributaries of the Three Gorges Reservoir, and the changes are basically consistent with those on the inflow section.

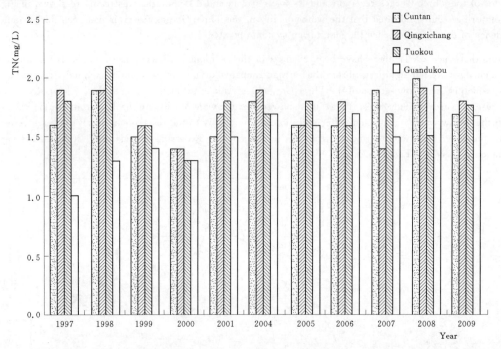

Fig. 3. 4-3 Interannual Variation in Total Nitrogen in the Main Stream Sections of the Three Gorges Reservoir in 1997-2009

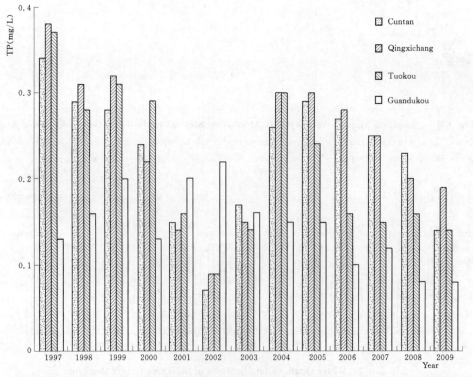

Fig. 3. 4-4 Interannual Variation in Total Phosphorus in the Main Stream Sections of the Three Gorges Reservoir in 1997-2009

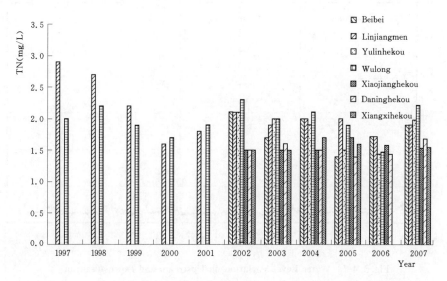

Fig. 3.4-5 Interannual Variation in Total Nitrogen in Main Tributaries of the
Three Gorges Reservoir in 1997-2007

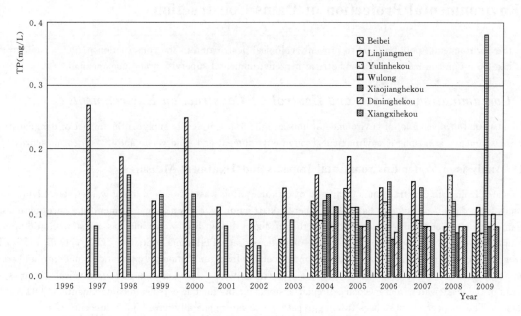

Fig. 3.4-6 Interannual Variation in Total Phosphorus in Main Tributaries of the
Three Gorges Reservoir in 1996-2009

Through the above analysis, it can be concluded that there is no significant difference between the water quality classification and nutrient saline levels of the water bodies in the main stream and tributaries of the reservoir before and after the Three Gorges Project. This is close related to the efforts of the Chinese government in water pollution prevention in the Three Gorges Reservoir area and upstream. Besides, as the Three Gorges Reservoir is a classic canyon river-type reservoir lined up by tall and rolling mountains at both sides, with normal water level of 175m and average surface width of about 1,100m, which increases about double compared with that of the natural conditions, it still keeps some features of stream flow even after water storage. Furthermore, as the Three Gorges Reservoir is the seasonal flow regulation with frequent water exchanges, and its regulating storage is roughly equivalent to 3.7‰[3] of the annual runoff at the dam site, it has limited effects on the water environment capacity in the stretch of river built the reservoir. In addition, the reservoir regulation mode also determines its impact on the reservoir water quality. The high levels of water in the reservoir is only about 6 months, that is annually from late October till early May of the next year, as shown in Fig. 3.4-7. Water flow was low during the winter of 2006-2009, the Three Gorges Dam would start to increase its water release and provide water to the downstream area.

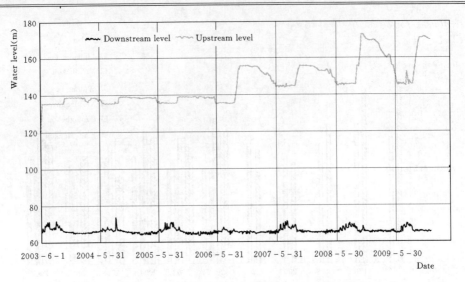

Fig. 3.4-7 Water Level Variation on Upstream and Downstream of
the Three Gorges Dam in 2003-2009

3.5 Environmental Protection of Dams Construction

To reduce the environmental pollution and to prevent ecological destruction in the construction period, it is extremely important to take pollution control measures and strengthen environmental supervision and management.

3.5.1 Contamination Analysis and Control of Construction Environment

As dam construction involves a lot of environmental impacting factors, it deeply analyzes the impact of dam construction on the water environment, atmospheric environment, sound environment and solid waste and its control measures.

3.5.1.1 Analysis of Water Environmental Impacts and Treatment Measures

The waste water from dam construction is composed of manufacturing wastewater and living wastewater. Living wastewater mostly comes from daily canteens and bathrooms where construction workers and management staff are more concentrated. The contaminant concentrations is generally less than ordinary living dirty water and has a characteristic of centralized discharge time and significant coefficient of variation of water storage. Pollutants contained COD_{Cr}, BOD_5, SS, TP, ammonia nitrogen, etc. And manufacturing wastewater mostly stems from foundation pit excavation; underground powerhouse excavation, aggregate processing systems, concrete mixing systems, car washing and mechanical repair systems et al. Pollutants in the wastewater from mechanical repair systems mainly include petroleum contaminants, COD_{Cr} and suspended substances. The suspended substances from aggregate wastewater generally have higher concentration.

To protect the water environment in and around the construction area, manufacturing wastewater and living wastewater are treated according to the requirement of water environment functions in the project area.

1. Wastewater treatment for aggregate processing system

In the treatment of wastewater from aggregate processing systems, a treatment scheme should be determined first based on characteristics of the suspended matter in the wastewater. Wastewater with high-concentration suspended matter may be handled by natural sedimentation or coagulation sedimentation. The sewage treatment flow by natural sedimentation is:

Natural sedimentation process: high suspension wastewater flows out of the screening plant into the sedimentation pond and deposits naturally without any flocculants. The supernatant liquid is used in circulation.

Coagulation sedimentation process: After flowing out of the screening plant, firstly wastewater is stripped of coarse sand in the settled sand treatment unit to the coagulated sedimentation unit. Small-sized suspended substances can be removed quickly and effectively due to dosing of the flocculants. Coagulated sedimentation performs well when most of suspended substances from the aggregate processing wastewater are inorganic particles with sound sedimentation performance, and wastewater treatment is highly demanding and construction land is very limited.

2. Wastewater treatment for mechanical repair system

If oil content is high in the wastewater, flocculated sedimentation can be firstly used to strip suspended substances, COD_{Cr} and some petroleum in the wastewater, then deliver the outflow from the sedimentation pond into an oil-water separator assembly to treat the oil wastewater further. If the petroleum concentration in the wastewater is lower than 30mg/L, the water outflow from the sedimentation pond can be delivered into a small oil separator. Oil-water separator is quite effective with the high oil-recovery and oil-removal efficiency, but if so, the investment is great, and high requirement for maintenance technique. And oil separator had simple structure, costs lower and implements management more conveniently, but need regular cleanup of the tank.

3. Living wastewater treatment

According to the characteristics of dispersed distribution, small discharge, low pollutant concentration and short running time in living wastewater, the process equipment has been adopted such as sewage treatment complete sets of equipment, and urban sewage purification methane tank or biofilter. For water areas subject to stricter discharge standards, domestic sewage treatment assemblies with perfect processing effect are used.

3.5.1.2 Analysis of Atmospheric Environmental Impacts and Treatment Measures

Waste gases from the dam construction typically consist of transport dust, surface dust and waste gases from machines and vehicles. Atmospheric pollutants in project construction are generally emitted without control and the pollution sources appear in areas or lines. In the process of prediction, the dusts from aggregate processing systems and concrete mixing systems are taken as area source, while the dust pollution source from the transport system is taken as linear source. Besides, it also should be analyzed the impact of mechanical fuels on the environmental air.

While atmospheric environmental protection for the characteristics of the decentralized air pollution sources in the construction area are hard to end control by centralized treatment, such measures should be adopted as optimization of the construction technological process, the utilization of advanced construction equipment, the protection of the construction area and externally sensitive areas to mitigate the air pollution from blasted excavation dust, dust from aggregate processing and concrete processing systems, transport dust and fuel waste gases.

3.5.1.3 Analysis of Acoustic Environmental Impacts and Treatment Measures

Noises produced in the construction activities of the dam mainly include moving noises from transport vehicles, noises from fixed and continuous boring and construction machinery as well as short and fixed-time blasting. Acoustic environmental impact analysis mainly addresses the impact of fixed, traffic movement or blasting noises on resident points, schools, hospitals and other sensitive subjects.

Road signs are mandatory at environmentally sensitive area to restrict vehicle speed during the night and avoid honking. It should be strengthen the regular maintenance of road and vehicles to control noise sources. Construction materials should be selected to having the satisfactory sound absorption, elimination and insulation properties inside the construction living quarters. For residential areas with higher exposure to traffic noises, a sound-proof wall should be built at one side of the road nearby the residential area. Blasting time should be strictly controlled and timely operation to avoid between 22:00 p.m. and 7:00 a.m. at the next day. Advanced blasting technology should be used to mitigate blasting noises.

3.5.1.4 Analysis of Environmental Impacts from Solid Wastes and Treatment Measures

Solid wastes from dam construction include abandoned materials excavated and domestic garbage. The slag of the project can lead to consequent soil and water erosion during the foundation excavation, cofferdam removing, road-building, ground-flattening, auxiliary plants of construction and excavation of quarries.

A lot of crew participate the construction and domestic and operation areas are decentralized. Domestic garbage from construction workers can not only pollute the air, destroy the surrounding natural sceneries to contaminate soil and water, but also provide a congregated place for flies, mosquitoes, bacteria and mice that spread epidemics and threaten people's health.

Production spoils from construction operations are generally disposed in the designed. As domestic garbage has a very complex composition, it should be proposed integrated disposal scheme for solid waste according to specific ingredients, characteristics, quantity and the number of years disposal waste as well as site conditions. So far, garbage is disposed by sanitary landfill, burning, compost and comprehensive treatment. It's proved that sanitary landfill, burning or composting is not suitable for disposing construction domestic garbage alone. At present, a satisfactory way to carry garbage is to select one suitable arrangement from transporting garbage to the nearest rubbish disposal yard, completing sets of equipment of household garbage and building a rubbish landfill in the construction area.

3.5.2 Strengthening in Construction Environmental Supervision and Management

3.5.2.1 Supervising Construction Environment

Environmental supervision refers to the construction supervising units should be entrusted by the project owners to supervise and administer works regarding the environmental protection of project construction in an objective, independent, impartial manner by specified method of supervision according to the contract and related regulations.

Practices have proven that the organization of environmental supervision is an important way to oversee the implementation of the environmental protection measures during the engineering construction and to prevent environmental pollution and ecological destruction. Environmental supervision is a new mode of environmental management. The environmental supervision work in relation to water projects includes:

(1) Managing the environmental protection in dam construction in a unified way according to related laws of environmental protection and environmental protection regulations for engineering.

(2) Supervising the contractor's enforcement of the articles related to environmental protection in the dam construction and interpreting this articles; and proposing treatment comments and reports on major environmental problems and authorizing the chief supervision engineer of the project to urge the concerned department to take corrective actions within a given time.

Discovering and keeping hold of environmental problems in engineering construction and giving monitoring orders for some environmental indices; studying the monitoring result and providing improvement plans for environmental protection.

(3) Attending inspection meetings of the construction organization design, construction technical solution and construction progress schedule proposed by the contractor and putting forward suggestions to environmental protection; reviewing the construction material and equipment lists as well as its environmental indices.

(4) Coordinating the relationship between the owner and the contractor, handling contract breaches related to environmental protection; dealing with bidirectional claims related to environmental protection in an impartial way according to the contract provisions and following the claim procedure.

(5) Recording environmental problem at the site and its processing results each day in table form of the environmental supervision, submitting a monthly report to the owner each month, sorting the environmental supervision files according to the accumulated data and submitting an environmental supervision evaluation report every quarter.

(6) Participating in the acceptance of the different stages, subprojects, individual project and the completion of the construction companies, and ordering cleanup and site restoration for completed projects to comply with the related provisions about environmental protection. As environmental supervision is in close combination with engineering construction, shorting the time lag in discovering problems and solving them to lessen the economic losses, transmitting from post management to process management, from mandatory environmental management to mandatory and instructive management, and from passive control to positive prevention and process control.

3.5.2.2 Establishing Environmental Management System to Identify Responsibility

To improve the overall management efficiency of projects and guarantee successful implementation of environmental protection measures in dam construction, environmental administrations have been established recently at different levels in China during dam construction, and a series of measures such as clear responsibilities have been taken to improve their decision and execution ability. An example is the environmental management system established during the implementation of the Xiaolangdi project of the Yellow River, which consists of leading institution, organizing agency, implementation institution, assisting organization and advisory body (Fig. 3.5-1). The agencies are closely related and coordinated with one another and independent individually, with clear division.

The leading institution of environmental management mainly consists of leaders from related departments of construction, management and local government and experts. It is the highest decisive board of environmental management and functions to solve major issues about environmental protection by holding regular meetings.

The organizing agency of environmental management division which, under the direct leadership of the leading agency, drafts environmental protection methods and detailed provisions suited to the specific conditions of the project concerned, prepares the master plan and annual plans of the engineering environmental protection, organizes the overall implementation of these plans, estimates the annual expenditure for environmental protection with the financial department in their planning and management of environmental protection outlays, promotes environmental protection propaganda work, organizes nec-

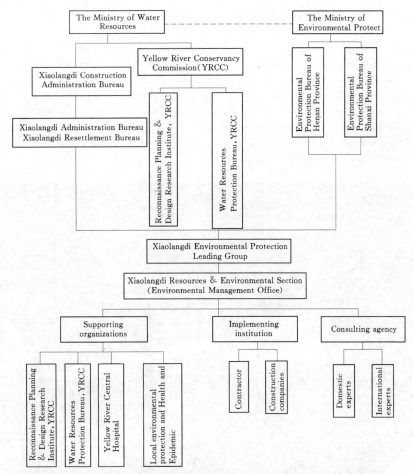

Fig. 3.5 - 1 Environmental Management System of Xiaolangdi Project

essary universal education and improves the environment awareness of people concerned.

The implementation institution of engineering environmental management consists of contractors, subcontractors and relevant resident companies of individual bid sections, each of which undertakes environmental protection work independently, performs the specific environmental protection work within the area and receives the inspection and supervision from the organization.

The assisting organization for engineering environmental management consists of the design institutes, monitoring units and the departments of disaster control which undertake the environmental protection task of a specific aspect either independently or jointly as authorized by the owner according to their technical advantages and without changing their administrative subordination.

The consulting agency for engineering environmental management is a consultant team that is employed by the owner and consists of international and domestic experts in environment and resettlement. The team members can be alternated or recomposed to certain specific environmental requirements, assume responsibilities directly for the leader function and conduct regular or random consultation activities. The task of the consultant team is to identify and solve environmental protection problems during the dam construction and submit an environmental consultation and evaluation report by means of consultation and evaluation of difficult problems in the environmental protection work.

As implementing rigorous management, supervision and various measures in environmental protection, the Xiaolangdi Project of the Yellow River successfully passed the environmental acceptance in 2002 and awarded *Top* 100 *National Environmental Protection Projects* with highly-rated praise of the World Bank experts as good example of construction projects environmental management in developing countries.

3.5.3 *Construction Environmental Protection for the Three Gorges Project of the Yangtze River*

The construction area of the Three Gorges Project is the main area subject to apparent environmental impact during the construction stage of the Three Gorges Project. To implement the environmental measures set out in EIA Report carefully, it was executed that an integrated mechanism for unified management by China Three Gorges Corporation (formerly China Three Gorges Project Development Corporation) in the environmental protection of the construction area, and individual management by each construction company, actively followed the latest environmental protection information in China, employed professional environmental supervisors and water and soil conservation experts, taken all-dimensioned protective and control actions with respect to engineering, techniques, management and investment to conduct a lot of work including pollution control and prevention, water and soil conservation, environmental afforestation, population health protection and environmental monitoring in accordance with the approved *Implementation Plan for the Environmental Protection in the Construction Area of the Three Gorges Project*. So far, regional environmental impact is basically controlled and the environmental quality is recovering and improving (Fig. 3.5-2).

Fig. 3.5-2 The Three Gorges Dam and Local Environment

Environmental monitoring results have indicated that water quality in the Yangtze River mainstream within the construction area and over all the sections in the near-bank is excellent; the water quality over the upstream reference section, the downstream control section and in the near-bank shall comply with class Ⅱ-Ⅲ standard; the ambient air quality in the construction area has been sound as a whole and its various indicators have complies with class 1-2 standard; the ambient noise in the construction area have also complied with related standards; the controlled ratio of water and soil loss and the percentage of dammed slag or ashes in soil conservation area have met national industry standards.

3.6 Environmental Protection Benefits of Dams

Construction of dams not only brings about multiple social and economic benefits including flood control, power generation, irrigation, water supply and navigation, but also produces favorable regulation effects for the resources, environment and ecosystem of the basin, and plays a multifunctional role in social, economic and environmental sustainability.

3.6.1 *Flood Prevention and Disaster Alleviation*

Dams can intercept and store flood. Utilizing the regulating capacity of the reservoir and storing the upstream water inflow help to reduce the erosion and destruction caused by floods to the downstream, alleviate the environmental impact from floods and maintain the safety of life and properties. An example is the Three Gorges Project on China's Yangtze River for which flood control is the greatest ecological and environmental benefit. The mainstream and tributary networks of the Yangtze River contain numerous lakes and frequent flooding hazards have always occurred. History shows that flood hazards occurred seven times in 1911-1949. The flood in 1954 inundated 3.2 million hm^2 arable lands, killed 33,000 people and blocked the Beijing-Guangzhou railway up to 100 days, despite the soldiers and civilians struggled to rescue. In 1998, the whole Yangtze River basin suffered a severe flood event, which inundated 240,000 hm^2 arable lands and killed 1,526 people. Through the flood-relief efforts for about 6 million civilians and soldiers, the safeties of Shashi and Wuhan had been protected[5] (See Fig. 3. 6-1).

Fig. 3. 6-1　Flood Diversion in Xiaohekou Town of Shishou in 1998

With flood control storage of 22. 15 billion m³, the Three Gorges Reservoir can control the upstream floods effectively, cut the peak discharges in the middle reaches and improve the flood control capacity. Since the Three Gorges Reservoir was built, the flood control capacity of the middle reach and the Yangtze River embankment has been upgraded from the original 10-year return period to the present 100-year. This helps prevent destructive disasters along the banks of the Jingjiang River and safeguard the 20 million people, 1. 533 million hm² arable lands and the towns in the Jianghan Plain. When a 1,000-year flood or a devastating flood of its kind in 1870, the reservoir, as supported by flood diversion in Jingjiang River and operation of other flood diversion and storage works, shall guarantee flood safety in Jingjiang River section, degrade the threat of the flood of Yangtze River to the Dongting Lake area, prevent the flood to inundate 126,000km² of area, 75 million people, 6 million hm² arable lands and large and medium-sized towns in the plain area along the middle and lower reaches of the Yangtze River, and mitigate the environmental deterioration and post-disaster epidemic due to the flood hazard and diversion actions[5].

In June 2006, the roller compacted concrete (RCC) cofferdam was successfully dismantled and the Three Gorges Dam started full-scale impoundment. In early November 2008, the Three Gorges Reservoir water level tentatively rose to 172. 8 m and the Three Gorges Project had the key function of regular flood control. In late July 2007, the Three Gorges Project successfully conducted its first initiative flood control, with total flood storage of 1. 043 billion m³ and drop down the flood peak of nearly 5,100m³/s. In late July 2008, in regulating the flood peak discharge of 52,500m³/s, the Three Gorges Project limited the downstream discharge to 48,000m³/s, with total flood storage of 1. 334 billion m³. In 2009, the Three Gorges Project cut the maximum flood peak to 15,000m³/s and made significant contribution to alleviating the flood control pressure in the middle reaches of the Yangtze River.

3. 6. 2　Low-carbon Clean Energy Production

As renewable energy, speed-up developing and utilizing clean hydropower power, lowering the percentage of thermal generation and limiting the consumption of nonrenewable raw coal energy can not only alleviate the scarce problem of energy, but also can contribute to the reduction greenhouse gas emissions like CO_2, and open up a broader environmental space for the social and economic sustainable development of China.

China has formulated and implemented *the National Climate Change Program*, which explicitly puts forward legally binding targets at the national scale to reduce the energy consumption per unit of GDP, main pollutant emission, and increase the forest coverage and the proportion of renewable energy in 2005-2010. As an important part of China's effort to push the nation's energy structure toward clean and low carbon, hydropower is expected to account for over 50% of the emission reduction task of the energy industry in 2010. According to statistics, from the grid-connected operation of the first generator unit of the Three Gorges Hydropower Station on July 10, 2003 till December 31,2009, the Three Gorges power plant generated about 368 billion kW·h, by an amount equivalent to 0. 32 billion ton of CO_2 emission reduction.

The energy-saving emission reduction of the Three Gorges Project is also present in the increase freight volume and improved navigation conditions. As the most energy-efficient mode of transportation, water transportation can save a lot of energy and reduce CO_2 emission. Water storage of the Three Gorges Project can improve the navigational conditions of the Yangtze River from Yichang to Chongqing, reduce the transport cost by 35%-37%, cut down 46% per unit GDP energy consumption, and also significantly improve the navigation conditions in the reach downstream of Yichang in the dry season by replenishing water. Ever since the navigation test in 2003, about 0.36 billion ton of cargo has been carried through the Three Gorges dam area, exceeding the total cargo transported through the Gezhouba ship gate over the 20 years without the Three Gorges Project impoundment.

3.6.3 *Recovery of the River Ecological Functions*

Dam construction is playing an increasingly important role in restoring river ecological functions and improving the estuarine wetland's ecological environment. The Yellow River is the second longest river in China, about 8.7% of the nation's total population, 12% of total irrigated area in the whole country, but 2.5% of the nation's gross of water resources are in the Yellow River Basin. The contradictions between supply and demand of water resources are obvious in this region.

The Yellow River faces a problem of less water and more sediment with serious depositions, and has been the most renowned suspended river in the world. Due to water scarcity, the Yellow River Basin had to maintain its economic growth at the expense of ecological environment. In the 1980s and 1990s, the lower reach of the Yellow River frequently dried up nearly every year. In 1997, the downstream of main stream had dried out for 226 days with severe channel shrinkage and almost the loss of river functions which has caught the attention of the word. After the Xiaolangdi Project was completed and put into operation, the Chinese government targeted at preventing the Yellow River from running dry and the riverbed from heightening as the main purpose of operating the reservoir and put forward higher requirements for the regulating operation to it. That is, the reservoir should not only have functions of flood control, generate power, supply water and irrigation, but also play an important role in restoring the ecological functions of river channel. From 1999, the Yellow River started to implement the unified regulation for water, conducted a series of large-scale experiments on the water-sediment regulation to artificially created considerable discharge and effectively scour the downstream riverbed, which increased the minimum flood discharge capacity of the main channels in the lower reaches of the Yellow River from the original less than $1,800 m^3/s$ up to $4,000 m^3/s$, significantly improving the downstream flood safety, keeping the downstream river from running for the 9th consecutive year. Now, river channel has fully recovered its functions and the wetland ecological environment in the Yellow River delta has been effectively improved. In addition to recovering the ecological function of the downstream river, it also greatly protect the normal producing and living of the people in the downstream area.

3.6.4 *Water Supplying and Regulation Functions*

With the regulation and storage functions of reservoirs, the occurrence of heavy draughts or extreme water shortage during the dry season in the lower reaches of the dam can have a serious influence on water demand for industrial and agricultural production and for ecological environment. To prevent this accident, the water volume in the river can be regulated to supply water to the downstream area and mitigate its negative impact.

In the initial impoundment of the Three Gorges Reservoir, the regulating water has been implemented in the dry season to improve the water quality in the downstream and water transport, control salt at the estuary and raise the utilization rate of the consumption of industrial and agricultural water and other ecological water.

In 2006-2007, the Three Gorges Reservoir supplied water to the downstream area for 80 days totaling water volume over 4 billion m^3; in 2007-2008, the reservoir supplied an accumulated water 2.25 billion m^3; From 1 Jan. 2008 to 11 Feb. 2009, the reservoir supplied water of an 2.5 billion m^3; in 2009, the reservoir supplied water for 125 days totaling 7.164 billion m^3.

In 2006, the serious event of a 100-year drought happened in Sichuan. To ease the drought in the downstream area, Zipingpu Reservoir, located in the upstream of the Minjiang River, utilized its storage water to provide water of an accumulated 1.6 billion m^3 to the downstream area, with an additional supply of 0.35 billion m^3 compared with the natural water inflow. It is an effective way to abate the drought in the Dujiangyan irrigated area and the water stress in Chengdu, and maintained the enough water for spring irrigation, production and living in the rural and urban area.

In mid-March 2006, water environmental pollution event occurred in Qingbai River's Guanghan section. Within a short time, Zipingpu Reservoir dispatched over 4 million m^3 of water to the downstream area in an emergency, which diluted the pollutants in this river section and provided the safety of water use for the production of industry and agriculture and for people's life in the local and downstream area.

3.7　Problems and Outlook in Connection with Environmental Protection Effort

(1) Reservoirs are creating new environment. The development of human civilization is inevitably accompanied by reconstruction of the natural environment. In the process of developing and utilizing water resources, dam, it is necessary to build more and more water projects. It is believed that dam construction can lead to various impacts on the local natural and social environment, either positive or negative or even harmful. However, with the development of society and continually progress of science and technology, people begin to realize and take effective measures of scientific management and optimized operation of hydroelectric stations to expand all positive impacts of dam construction, to avoid or limit various harmful impacts to the furthest extent, to fully utilize the reservoirs to create new environment, to achieve man-water harmony and improve the ecological environment.

(2) Environmental protection should penetrate the whole process of dam construction and operation. With the development of China's national drive of water resources and hydropower construction, environmental protection work have been highly recognized and expanded for each individual stage of dam construction. However, not the same importance has been attached to environmental protection work for all individual stages of dam construction. A fair amount of environmental protection work is limited to the environmental impact assessment during the feasibility study stage and the environmental design in the preliminary design stage whereas no sufficient attention is paid to the environmental protection in other stages. Along with further development of environmental protection in the coming years, environmental protection through the whole process of dam construction and operation shall be given attention. Horizontal expansion of dam environmental protection work is typically reflected in the resettlement project. Dam construction gives rise to resettlement, and resettlement is closely correlated and interactive with the environment. As a result, thorough and meticulous environmental protection work is mandatory in the resettlement stage of planning, siting, removing and resettlement to achieve mutual coordination and promotion between migrant resettlement and environment conversation.

(3) Environmental work should implement dynamic management system. Of the many approaches for China in its endeavor to intensify environmental protection work in the future, relying on technological advancement could be one of the most important ways. For the dam environmental protection work of China, applying new technology, new method and advanced equipment is the trend of the times. And in the near future, a separate EIA database is expected to be established for each single large hydropower project and will, in line with the dynamic changes of the project, continuously replenish important new information to verify the predicted environmental impact and the recommended countermeasures, make adjustment to the practical conditions and perform dynamic environmental management.

(4) The method of Planning Environmental Impact Assessment (PEIA) is growing mature. China does not have much a long history in implementing PEIA as compared with project EIA. The EIA technique for basin planning and multi-reservoir joint operation scheme is still on the elementary stage. The purpose of environmental impact analysis and assessment of basin planning is to make systematic studies on the mutual impacts between multiple projects within the same basin, cross-impacts between different environmental elements, cumulative impacts of cascade hydropower projects on the environment and impacts on the ecological corridors in the upper, middle and lower reaches, and to optimize the design plan at the standpoint of environmental protection, to adjust the dam arrangement. To achieve the optimal economic, social and environmental benefit of the hydropower projects, in-depth studies and continual improvement are still necessary for the analysis and method. It is sure that the technique and method for PEIA will become more and more mature with the gradual consummation of PEIA systems, thorough and extensive work and enough experiences of PEIA.

(5) Channel wetland environment should be further improved. With the incessant in-depth of the environmental technology for dam construction, the negative impact of dam construction on the channel ecological environment and the wetland along the way should be mitigated. So far, the major attention is paid to flood control, power generation, irrigation and water supply in reservoir operations in China, whereas little is addressed to the ecological environment protection of the channel and wetland in the downstream of the dam and the estuary. The development of run-of-river hydropower station and its peak-regulating operation can usually give rise to water-reducing sections and even water-blocking in the downstream of the dams, and destroy immensely the surviving environment of the aquatic life, particularly fish, in the downstream of the dams, severely affect the biological species and their habitats in some of the water-reducing or water-blocking sections, and directly threaten the ecological safety in the downstream of the dams. As large areas of tidal flat wetland are distributed in the lower reaches or at the estuary of damming rivers, the altered hydrological regime and sediment conditions have certain effects on the structure and functions of the wetland. However, it has been believed that channel ecology and tidal flat wetland environment can be improved with the increasing people's environmental awareness and enhancing science and technology.

(6) Sustainable development should be persisted. A transformation is taking place in which the dam goal is becoming multi-objective combination of disaster elimination and benefit generation from single objective of eliminating disasters in China, it has been recognized that China should promote sustainable development. In building reservoir dams and conquering nature, man should first obey to the "Law of Nature", establish a new man-nature relationship, turn from simply using the nature to meet our demand to respecting nature and protecting nature, and establish harmonious coordination between man and nature. Whether in basin planning or regional scheme, or in the process of planning, design and implementation of hydropower projects, it should be considered both current benefits and the benefits of our next generations. Only by combining development constructions and the preservation of the natural world can nature bring benefits to humanity "forever".

References

[1] Department of Environment Evaluation of Chinese Academy of Sciences, Research Institute for Yangtze River Water Resources Protection. *Environmental Impact Assessment Report on the Three Gorges Hydroelectric Project of the Yangtze River (simplified)*. Beijing: Science Press, 1996.

[2] Lu Youmei. Decision and Practice of the Three Gorges Project, *China Engineering Sciences*, 2003, 5 (6): 1-6.

[3] Three Gorges Project Construction Committee Office of the State Council: Structure of the Monitoring System. http://www.tgenviron.org/sysintro/sysintro_structure2.html.

[4] China Three Gorges Project Corporation. *Annual Environmental Protection Report*. 2007-2008.

[5] Wang Rushu. Flood Control is the Maximal Ecological Environmental Performance of the Three Gorges Project, *Three Gorges Forum*, 2010, 222: 3-9.

[6] Three Gorges Project Construction Committee Office of the State Council. *Implementation Program for the Ecology and Environment Monitoring System of the Three Gorges Project (Revision)*, 1996.

[7] An Shenyi, Liu Langui, Zhao Shenshan, et al. *Hydropower Engineering in China: Environmental Protection in Resettlement*. Beijing: China Electric Power Press, 2000.

[8] Xie Xinfang, Shang Yuming, Zhang hongan, et al. *Environmental Protection Practice in Xiaolangdi Project of the Yellow River*. Zhengzhou: Yellow River Conservancy Press, 2001.

Chapter 4

Dam, Reservoir and Water Resources Allocation in China

Liu Zhiming[1], Li Jianqiang[2], Li Aihua[3]

Water resources are of the basic natural resources and strategic economic resources. Featured with unevenly spatial-temporal distribution, insufficient self-regulation capacity of rivers, and distributional mismatching between water resources and population as well as arable lands, etc, it is necessary for China to build the water resources allocation projects such as dams to regulate and control spatial-temporal distribution of water resources, optimize water resources allocation and promote regional water resources carrying capacity. Due to insufficient nature of water resources endowment, inadequate storage capacity of water resources, together with constantly increasing population, continuous improvement of people's living standard and the sustained and rapid socio-economic development in China, which propose higher demand on quantity, process and security degree of water supply, water supply and demand contradictions have become increasingly prominent. In order to develop and utilize water resources in a rational manner and in good order, it is necessary to introduce an overall and systematic analysis and evaluation on water resources situation in China, to formulate the integrated water resources plan at national and river basin levels, and to constitute overall layout of key projects for water resources allocation at river basin and regional levels. Above measures will provide water safeguards for national social and economic development. Dam construction will also be effective measures to realize rational allocation of water resources, and to treat and mitigate problems and challenges in China.

4.1 Characteristics of Water Resources in China

China is located in the eastern part of Euro-Asian continent and to the west of the Pacific Ocean. Due to wide span in both latitude and longitude, China has various climate features in different regions. China is characterized a complicated topography characterized by a physical feature with the high west of mountain areas and low east, showing the interlacing distribution of mountains, basins and plains with the complex landforms. The obvious differences of water cycle in different parts of China is determined by the particular natural geography conditions in China which lead to the unevenly distribution of water resources in time and space and the distribution mismatching among water resources, population, productivity and land resources and others.

4.1.1 Inadequate Water Resources per Capita and Great Variation of Water Resources at the Interannual and Intra-annual Scale

China's total water resources are rich, around 2.8 trillion m^3. However, water resources per capita and water resources per hectare of cultivated land are far less than world average. The available water resources per capita are less than 2,100m^3, only 30% of world average. The available water resources per hectare of cultivated land are 22,500m^3, about 50% of the global average.

Affected by the monsoon climate, China is one of the medium and low latitude countries which have highly concentrated precipitation and runoff within a year with large inter-annual variation. In most regions, precipitation and water resources distribute unevenly within a year. Around 60% to 80% of annual precipitation and river runoff are concentrated in flood sea-

---

[1] Liu Zhiming, Vice President, General Institute of Water Resources and Hydropower Planning and Design, Professor Senior Engineer.
[2] Li Jianqiang, Director, General Institute of Water Resources and Hydropower Planning and Design, Professor Senior Engineer.
[3] Li Aihua, General Institute of Water Resources and Hydropower Planning and Design, Engineer.

sons, and the percentage in wet years is even higher, especially in northern regions where annual runoff is often constituted by one or several precipitation process. Meanwhile, inter-annual precipitation and runoff in China vary greatly. In southern China the maximum annual precipitation is 2 to 4 times more than the minimum one, while the number in northern China could reach 3 to 6 times. The maximum annual river runoff could be 10 to 15 times more than the minimum one with consecutive wet (dry) years. If this happens frequently, the water supply and demand contradictory have become increasingly prominent in dry years, especially in successive dry years. For example, the Songhua River Basin, Haihe River Basin and lower reach of the Yellow River once confronted successive drought of more than 7 years, where the annual water resources in the driest year are only 41% to 67% of the mean annual water resources and the total water resources of the successive years reduced by 350.1 billion m^3, 68.1 billion m^3 and 76.3 billion m^3 than that of the annual average. The southern region also confronted successive drought of 3 to 6 years, which led to water scarcity within a local area or even in the whole river basin. In addition, 58% of groundwater is recharged by precipitation infiltration in China's plain areas. The groundwater recharge variety is influenced by precipitation and surface water recharge, particularly in a dry year and successive dry years, the groundwater recharge reduces remarkably. Combined with water stress and the increasing exploit strength, water table is declining dramatically.

Obvious inter-annual variance and concentrated distribution of water resources within a year result in frequent flood and drought disasters in China, as well as the mismatch between natural flow regime and water demand process. For example, industrial and urban domestic water supply often requires a stable process, while the peak of agricultural water demand often coincides with dry period. In order to mitigate flood and drought disasters effectively and regulate natural runoff process to meet the water consumption demand, it is necessary to build dams to artificially regulate nature water resources in most areas of China.

4.1.2 *The Mismatch between Water Resources Distribution and Social and Economic Development Pattern*

Generally, the southern and eastern China yields more water resources than the northern and western China, while the mountainous areas yield more water than the plain areas. The distribution of water resources does not significantly match with social and economic development pattern.

The northern China, with 46% of the country's population, 60% of the country's arable lands and 45% of the country's GDP, yields only 19% of the country's total water resources, featured with water shortage and scarcity. Particularly, the Yellow River, Huaihe River and Haihe River Basins are the areas with the most intensive contradiction between water supply and demand in China. With only 7% of the country's total water resources, the Huang-Huai-Hai River Basins take up 36% of the country's arable lands, and water resources per capita basis are less than 450 m^3. The southern China has relatively abundant water resources. With 54% of the country's population, 40% of the country's arable lands and 55% of the country's GDP, the southern China yields 81% of the country's total water resources. However, 25% of water resources in the south are distributed in the Southwest Rivers. With only 3% of the country's arable lands, the Southwest Rivers accounts for 20% of the country's water resources, which is 1.1 times of water resources in northern China. Because of deep down-cutting of rivers, low elevation of water flows while high elevation of land, it is difficult to utilize and develop water resources in the Southwest Rivers Basins. Due to China's specific climate and geomorphology, most rivers flow from the west to the east. Usually most of water resources are from the upper and middle region, while large population and major water utilization concentrate in the middle and downstream of the river basin. At the same time, majority of water resources in China is from river runoff. 93% of river runoff is concentrated in mountainous areas while the plain areas only contribute a little. However, the plain areas are always densely populated with cities and irrigation districts, which have the increasing need for water. Generally, water requirements of socio-economic development and eco-environment improvement in the plain areas can only be met through rational distribution of water resources at river basin and regional level because water sources of the plain area are mainly from external areas.

The mismatch between the distribution of water resources and social and economic development patterns makes the regions differ from one to another in terms of water resources conditions. And this creates acute contradiction between water supply and demand. It is impossible to solve the problem of uneven spatial distribution of water resources only by the river's natural storage and regulation capacity. To solve this issue, it is necessary to introduce rational water resources allocation through dam construction as well as the implementation of inter-basin water diversion projects.

4.2 Status of Water Resources Development and Utilization in China

By 2008, China has completed construction of more than 86,000 reservoirs with a total storage capacity of 692.4 billion m³. As the important infrastructure to national economy, reservoirs play an enormous role in guarantying social and economic development. However, compared to the demand of national economy development, people's living standard promotion and the eco-environment improvement, the capacity of water resource security still lags behind. The capacity of regulating, controlling and storing water resources are insufficient, water shortage is a long existing problem, and drought occurs from time to time.

4.2.1 *Water Resources Development and Utilization*

In 2008, national total water supply was 591 billion m³. Surface water supply was 479.6 billion m³, about 81% of total water supply. Among the surface water supply, water supplied by reservoirs only account for 34%.

According to local water supply and water resources data from year 2000 to 2008, the exploitation ratio for water resources was 21% in China, while the exploitation ratio for surface water resources was 18%. In the northern China, the average exploitation ratio for water resources was 50%, while the average exploitation ratio for surface water resources was 38%. The exploitation ratios of the Haihe River Basin, the Yellow River Basin and the Liaohe River Basin were 76%, 63% and 59% respectively. In the southern China, the exploitation ratios for water resources and surface water resources were both 14%.

4.2.2 *Water Shortage*

Currently, China's water shortage could be demonstrated in two aspects. The first one is water shortage resulted from insufficient water supply, which has direct impact on people's living quality and reasonable water use of normal socio-economic activities. The second one is overdraft and over-exploitation of groundwater to squeeze instream ecological water requirements is used to make up insufficient water supply and to ensure socio-economic development.

Based on the analysis on the balance between water demand and supply in various river basins and regions, mean annual water shortage resulted from insufficient water supply is 18.9 billion m³, in which, 2.2 billion m³ is for urban domestic and industrial water demand and 16.7 billion m³ is for agricultural water demand. With the current utilization level, mean annual water shortage gap caused by groundwater overexploitation and misappropriation of instream ecological water requirements is 34.7 billion m³. Mean annual amount of groundwater overexploitation is 21.5 billion m³, and Mean annual amount of misappropriation of instream ecological water requirements is 13.2 billion m³. Mean annual water shortage in China is shown in Table 4.2-1.

Table 4.2-1 Mean Annual Water Shortage in China   Unit: billion m³

Zone	Water shortage due to insufficient water supply	Groundwater overexploitation	Squeezed instream ecological water requirements	Total water shortage
Nation	18.9	21.5	13.2	53.6
Songhua River Zone	1.0	2.5	—	3.5
Liaohe River Zone	1.2	1.6	1.3	4.1
Haihe River Zone	1.0	9.2	2.2	12.4
Yellow River Zone	4.8	2.2	2.6	9.5
Huaihe River Zone	3.4	3.3	2.7	9.4
Yangtze River Zone	2.8	0.5	—	3.3
Southeast Rivers Zone	0.8	0.1	—	0.9
Pearl River Zone	1.6	0.2	—	1.8
Southwest Rivers Zone	0.7	—	—	0.7
Northwest Rivers Zone	1.7	2.0	4.5	8.2
Northern China	13.0	20.7	13.2	46.9
Southern China	5.9	0.8	—	6.7

Among the total of mean annual water shortage in China, water shortage in northern China is 46.9 billion m³, with 13 billion m³ caused by insufficient water supply and 33.9 billion m³ resulted from misappropriation of instream eco-environmental requirements. Water shortage in northern China is in forms of lack of water resources and irrational development and utilization of water resources. Total water shortage of the Yellow River, the Huaihe River, the Haihe River and the Liaohe River Zones, all located in northern China, takes up 66% of national water shortage. Water shortage in southern China is 6.7 billion m³, is mainly displayed by engineering water shortage and some of the areas still exist in resources water shortage.

From current situation of water resources development and utilization as well as water shortage, it can be seen that infrastructure construction for water resources development and utilization in China lags behind the demand of social and economic development. Key projects for water resources allocation in many river basins and regions have not been built, appropriate allocation pattern of water resources has not been formulated, and water supply system is not completed. Featured by small-size and scattered distribution, water infrastructure with existing capacity of water supply, regulation and storage could not meet demand of socio-economic development and eco-environmental restoration. Among all facilities of surface water supply, water supply capacity from water diversion, pumping and transfer projects accounts for 66% of total surface water supply capacity. The remaining 34% is from reservoirs and dams, which is mainly contributed by the medium and small sized reservoirs and ponds, accounting for 68% of the total amount. However, medium and small sized reservoirs and ponds are often with low storage capacity, minimal control effectiveness, leading to relatively low guarantee degree of water supply. Besides, water sources of many cities and important industrial regions are fragile and there are no alterative water sources, which results in low guarantee degree of water supply, and it is difficult to supply water during drought period. It is also difficult for many irrigation districts to fully realize the benefits because of water sources variation and shortage as well as irrational utilization.

4.3 Exploration and Practice on Dams in China

Reservoirs have multiple functions of flood control, power generation, irrigation, water supply, navigation, aquaculture, recreation and eco-environment improvement, etc. Large reservoir is an integrated project that can regulate spatial and temporal distribution of water resources as well as optimize water resources allocation. They can not only bring tremendous social and economic benefits in terms of flood control, irrigation, water supply and emergency response, but also generate positive regulating impacts on resources, environment and ecosystem of river basins. Reservoir construction takes a very important position in the process of China's economic development. As early as two thousand years ago, China constructed worldwide well-known water projects, such as the Dujiang Weir, and recognized comprehensive effects of promoting benefits and mitigating disasters by constructing reservoirs. After years of construction, operation and management of reservoirs, a lot of valuable experiences and theories have been accumulated, which lead to objective and scientific cognition on socio-economic benefits and impacts on eco-environment of reservoirs. Nowadays, the construction of the Three Gorges Reservoir in the Yangtze River and the Xiaolangdi Reservoir in the Yellow River and their comprehensive benefits have placed increasingly obvious impacts on coordinated development of economy, society and environment in China. These are helpful and useful explorations and practices on construction of reservoirs in China.

4.3.1 *Practice on Improving Basin's Capacity on Overall Coordinated Development—the Three Gorges Project*

By making use of strong functions of large scale water project, the Three Gorges Project has promoted the basin's capacity on overall coordinated development. The Three Gorges Project owns the most important role of flood control, the largest power generation capacity, and the largest ship-lock in the world. After the completion of the project, it can directly protect the fertile and rich Jianghan Plain and major cities (Wuhan City, etc.) in the lower Yangtze River against serious flood damages and losses. The generated clean energy equals to the energy generated by using 34 million tons of standard coals, which means reducing CO_2 emission of 74.8 million tons and 680 thousand tons of SO_2. The navigable condition of the mainstream of the Yangtze River (called "the Gold Water Course") is greatly improved while the transportation cost is reduced by 35% to 37%. Besides, due to its huge storage capacity, the Three Gorges Reservoir could complement water to the middle and lower reaches in the driest season so as to ease the contradiction between water supply and ecological protection. The Yangtze River spans from the west to the east of China. The construction of the Three Gorges Project can secure the safety of the middle and lower reaches because of its enormous functions on flood control, power generation, navigation and water supply, so it plays an important role in promoting the coordinated development between the east, central and west regions in China.

4.3.2 *Practice on Restoring River Ecological Function—the Xiaolangdi Reservoir in the Lower Stream of the Yellow River*

The Yellow River, with 2.5% of China's total water resources, supplies water for 8.7% of total population, 12% of total irrigation areas and 7% of total economic yield in China. The contradiction of water supply and demand is very acute. The Yellow River has the characteristics of scarce water with surplus sediment and serious sediment deposition, and it is the most famous suspending river in the world. The river's functions are almost lost. The main objectives of the Xiaolangdi Reservoir are to prevent dry-up of the river course and rise-up of the riverbed after its completion. The reservoir needs to play a role not only in flood control, power generation and irrigation, but also needs to benefit a lot for restoring ecological functions of the river course. Since 1999, the Yellow River has implemented the integrated water amount regulation and the large-scale testing of water and sediment regulation. Through that, downstream riverbed was scoured effectively, which improves the reliability of flood drainage greatly. The downstream of the Yellow River experienced no drying up for successively 9 years since then, the river functions were rehabilitated, and the wetland eco-environment of the Yellow River delta was improved effectively. At the same time, the guarantee capacity on normal production activities and daily life of the downstream areas was also greatly enhanced.

4.3.3 Practice on Construction of Eco-environment Friendly Projects — the Xin'anjiang Reservoir and Ertan Hydropower Station

Harmonious development between dams and river basins or region is a common concern in the international dam society. It is an important issue for dam construction to clarify how to minimize negative impacts through an environment friendly construction manner, as well as to promote harmony between mankind and water and improve eco-environment by making full advantages of the dam's functions on recreating new environment.

The Xin'anjiang River is an important river in southern China, which spans over Anhui and Zhejiang Provinces. Rainfall in the river basin is abundant, but unevenly distributed in time and space. Inundation often occurs in flood season while in dry season water supply is difficult. The Xin'anjiang Reservoir was built to solve problems of flood control, power generation and water supply in the river basin. Construction of the reservoir did not damage the ecology of the Xin'an River Basin. Instead, a huge artificial lake called the Thousand-Islands Lake was formed. With a total storage capacity of 21.6 billion m^3, the Xin'anjiang Reservoir has significant benefits in terms of power generation and flood control. Furthermore, the reservoir area has become a well-known national 5A-class tourist resort because of maintenance of good water quality and beautiful scenery since it was built for decades.

Located at the downstream of the Yalong River, the Ertan Hydropower Station was the largest one built in the 20th century in China. In each phase of the construction process, the objective was always stuck to the aim of harmonious development between eco-environment and economy as well as society. After its operation, the hydropower station not only supports the local socio-economic development through achieving enormous benefits of power generation, flood control, navigation and tourism, but also has significant eco-environmental benefits. Regional eco-environmental condition was improved obviously compared to before the construction of the hydropower. In 2006, this project was given the award of the National Environment Friendly Project, which is the highest-level award for environmental protection of construction project in China.

4.3.4 Practice on Safe Construction and Operation of Reservoirs—the Zipingpu Water Conservancy Project

Located at the upstream of the Minjiang River, the Zipingpu Water Conservancy Project was constructed for main functions of flood control, water supply and power generation with a total storage capacity of 1.112 billion m^3. The project was started in 2000 and put into operation in 2006. In the whole process of construction and operation, the project always put safety in a dominant position. The reservoir was only 17 km away from the epicenter of 2008 Wenchuan Earthquake. After the earthquake, the dam experienced relatively large displacement both vertically and horizontally, the power station units were stopped running, and both the internal and external systems were turned off. However, the dam itself withstood the severe shock. It becomes the first concrete face rock dam with height of more than 100m that remained safety after the 8 magnitude earthquake in the world. Safe operation of the Zipingpu Reservoir ensured emotional stability of more than 10 million people, water supply security and power supply in the downstream areas in extraordinary times. It also turned into a waterway of rescue channel for life saving, which played a remarkable role in rescue and disaster relief for the quake-hit area.

4.3.5 Practice on Integrated Basin Management of Water Resources—Joint Dispatching of Cascade Reservoir in the Pearl River Basin

Adjacent to Hong Kong and Macau, the Pearl River Delta is one of the most populated and economically developed regions in China. In the process of rapid social and economic development, a number of negative phenomena emerged because of continuously decreasing run-off from the upstream, growing water supply for social and economic development and sea level rising caused by climate change. Water flow in the river course and the estuary continuously decreased, and salt-water intrusion occurred more frequently and seriously which lead to serious exceeding of drinking water quality standard such as in Macao and Zhuhai Cities. The issues of water supply security and estuarine ecology have become more and more severe. To ensure water supply security of the Pearl River Delta regions, integrated water resources management and regulation have been adopted in the whole Pearl River Basin. Eight reservoirs in the upstream are jointly regulated, and emergency water transfer for preventing salt water flowing further inland by supplement of fresh water is implemented. The length of water transfer route is 1,336 km, including Guizhou and Guangdong Provinces and Guangxi Autonomous Region, as well as the Xijiang River and Beijiang River, with delivered water quantity of 843 million m^3. This has ensured drinking water security of 15 million people in the Pearl River Delta area and Macau Special Administrative Region. The economic order of enterprises affected by salt water is also recovered quickly. At the same time, 230 million m^3 of water are replaced in river network of the Pearl River Delta regions, which improve water environment significantly.

Besides, focusing on ecology protection, there are many other successful practices on scientific and rational regulation of water projects, river ecology restoration and river health maintenance. For example, emergency water transfer projects of the Tarim River, the Heihe River and the Zhalong Wetland, etc.

4.3.6 Dam Construction is an Inevitable Choice to Solve Water Resources Problems in China

Based on knowledge of the law of nature, dam construction can improve human's survival environment by developing and utilizing water and hydropower resources in a better manner in virtue of power of nature. The above practices preliminarily have illustrated the role of reservoirs in regulating and controlling rivers and their multi-functions in development of human economy and society. Of course, reservoirs might also bring about some negative effects, and one of the most obvious side-effects is on river's eco-environment. To solve the negative impacts, in-depth research is needed with new thinking and application of advanced technologies and methods. Through optimized reservoir management and regulation, harmony between man and nature as well as between human beings could be realized, and water use contradictions among different regions could be eased. The reservoirs could fully benefit the human beings while their negative impacts could be minimized, thus the rivers could benefit our future generations in a sustainable way.

China's specific geographical location, complicated geomorphology and monsoon climate make it very difficult in developing and utilizing water resources. In-superior water endowment, together with the continuously increasing demand of rapid socio-economic development and eco-environmental protection, poses a very severe water resources situation. Compared to other countries with superior natural conditions, China's water resources problems are prominent. The security of drinking water, urban water supply, grain and aquatic eco-environment have been influenced seriously. Currently, China is constructing and will construct a bunch of world-class dams with large size and multiple functions. This must be implemented through scientific planning and deep research on the basis of transforming the development mode and water use manner, reinforcing water saving and drainage reduction as well as pollution control. By dams construction with large regulation and storage capacity, the goal is to realize harmonious development between reservoirs and economy and society, realize development and protection at the same time, and support sustainable development of economy and society through sustainable utilization of water resources.

4.4 General Situation of Water Resources Allocation in China

4.4.1 General Requirements of Water Resources Allocation

In order to change irrational manner of water resource development and utilization, realize sustainable utilization of water resources, ensure stable and rapid socio-economic development, maintain good eco-environment and ensure national water security, the general requirements of reasonable water resource allocation are as the following: ①control total amount of water use strictly, and depress over-consumption of water resources; ②manage water use quota rigorously, and promote water use efficiency and effectiveness; ③enhance eco-environment protection, and realize sustainable utilization of water resources; ④regulate and allocate water resources rationally, and improve regional water resource carrying capacity; ⑤improve the water supply security system, and ensure socio-economic development in a rapid and good manner; ⑥implement the most strict water resources management mechanism, and promote social management capacity comprehensively.

4.4.2 Water Resources Allocation in China in the Future

In accordance with the demand of building a resource-conserving and environmental-friendly society, based on the predicted socio-economic development indicators, water supply in the future is forecasted with water use quota and efficiency indicators under the strengthened water saving scheme. In order to return and make up current groundwater overexploitation and misappropriation of instream eco-environmental water use, it is predicted that in 2030 the balance between water supply and demand can only be reached when annual water supply nationwide is 711.3 billion m³. In the northern China, water supply will increase by 66.8 billion m³ to basically ensure that no water shortage will occur in normal years. In the southern China, local water resources will be regarded as the main source for water supply increase, which will be 54.6 billion m³. Water use of future incremental water supply is shown in Fig. 4.4-1.

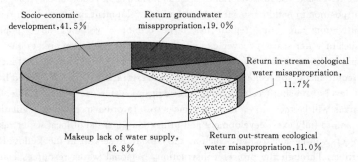

Fig. 4.4-1 Water Usage of Incremental Water Supply in the Future

Among national out-stream water allocation of 711.3 billion m³, local surface water supply is 553.6 billion m³, which is 121.4 billion m³ more than the current number. Local surface water supply is 78% of the total water supply.

4.5 China Hydropower's Prospect and Layout

According to the national and basin's water resources allocation scheme, considering the demand of flood control, power generation, navigation and eco-environment, it is necessary to build large and medium sized key reservoirs to enhance the regulation and control capacity on natural runoff with feasible construction conditions and acceptable eco-environment impacts. Through optimizing water resources allocation and promoting regulation and control water capacity, water source conditions of major regions, river sections, cities and grain production bases can be improved, the guarantee degree of water supply safety can be enhanced, and the rational water demand on socio-economic development and eco-environmental protection can be met.

4.5.1 The Songhua River and the Liaohe River Zones

The Songhua River and the Liaohe River Zones have abundant natural and mineral resources, and are important production base of commodity grain in China. In the future, the irrigated areas will be increased by more than 2.60 million hm². Regional economic and social activities are mainly focused on the hinterland areas and city clusters, including Harbin, Changchun, Shenyang and Dalian, etc. Considering the water demand of urban flood control, agricultural production and economic development in this region, the emphasis of project construction is to form the layout of water resources allocation in the northeast China step by step through construction of the Hadashan and Jinling Reservoirs as well as implementation of the Water Transfer Project of the Dahuofang Reservoir. Thus, water supply can be guaranteed for development of the city clusters, construction of national grain production base, revitalization of northeast old industrial base and eco-environmental protection.

In the Songhua River Zone, it is planned to build more than 40 large and medium sized reservoirs with a total storage capacity of more than 30 billion m³, including the Hadashan, Wendegen and Beiguan Reservoirs. The Hadashan Water Control Project and the water diversion project of the Da'an irrigation district could provide water resources for the grain production base in the Songhuang River and Nenjiang River Plains and cities such as Qiqihar, Daqing and Baicheng, gradually mitigate the contradiction between regional water demand and supply, and improve the eco-environment and water environment. In addition, the emergency response water source project will be constructed for special dry years to ensure water supply for cities in the Songhua River Zone under emergency situation.

In the Liaohe River Zone, it is planned to construct more than 10 large reservoirs with a total storage capacity of more than 8 billion m³, including the Jinling and Dashimen Reservoirs. An auxiliary project, the water transfer project of the Dahuofang Reservoir, will also be implemented. Through the above measures, the regional security system of water supply could be established step by step to ensure water supply for domestic and industrial use of 6 cities (Fushun, Shenyang, Liaoyang, Anshan, Yingkou and Panjin), to ease contradictions between eco-environmental water requirements and agricultural water demand in the Hunhe River and Taizi River Basins. The Jinling and Dashimen Reservoirs will not only secure the regional safety of flood control, but also ensure water supply for industrial and agricultural water use of river basins in the area of Huanghai and Bohai Sea Circle.

4.5.2 The Yellow River, the Huaihe River and the Haihe River Zones

The Yellow River, Huaihe River and Haihe River Zones are major industrial base, energy base and grain production base, which hold very significant position in national social and economic development in China. With relatively developed industry and high agricultural cultivation rate, this region has large economic volume, large population and dense urban system. However, the issues of water scarcity, over-exploitation and water pollution are serious. In the future, the growth rate of water demand should be lower than 0.4%. Incremental water supply, used for rational increase in industrial and domestic demand and water requirements for ecological protection and restoration, could be gained by means of inter-basin water transfer and promoting level of reusing treated sewage. Agricultural water use will be stable. Incremental water demand caused by new irrigation areas will be met via water saving measures. In order to enhance regulation and control capacity of regional water resources, and to fully exert development and utilization potential of local water resources, large-sized water regulation and storage projects will be constructed, such as the Guxian Reservoir in the Yellow River and the Chushandian Reservoir in the Huaihe River. Through the project construction, regional water resources could be regulated and controlled, and river and regional eco-environment could be improved.

Until now a number of projects for water resources development and utilization have been built in the Yellow River Zone, which has played a significant role in treatment and development of the Yellow River. Along with socio-economic development and water use increase, the contradiction between water supply and demand in the Yellow River becomes increasingly acute. In order to meet the continuously increasing water demand in the river basin and adjacent regions, the Guxian Water Control Project in the main stream and the Hekoucun Reservoir in the tributary Qinhe River as well as some other projects are planned in the Yellow River Zone to further enhance runoff regulation and control capacity and to ensure water demand for sediment transportation in flood season and ecological flow in non-flood season of the Yellow River and no drying up in the downstream Yellow River. The Guxian Water Control Project together with the existing four reservoirs, namely the Longyangxia, Liujiaxia, Sanmenxia and Xiaolangdi Reservoirs, will constitute the water and sediment regulation and control system as well as water allocation system of the Yellow River with total storage capacity of 16.4 billion m^3, which have played a strategic role as connecting link in the water and sediment regulation and control system. Its main task is flood control and siltation reduction, supplemented with such comprehensive utilization as power generation, water supply, irrigation, water and sediment regulation, and so on.

Major water storage projects in the Huaihe River Zone include reservoirs that will be built in the mainstream and important tributaries, for regulating and controlling water resources. In the Huaihe River Zone, there are 16 large-sized water supply oriented reservoirs under construction or on the planning process, with a total storage capacity of 8 billion m^3. There are also 29 important medium-sized reservoirs. In the upper reach area, it is planned to build large reservoirs, such as the Chushandian Reservoir and so on. In the middle reach area, it is planned to build large reservoirs, such as the Qianping Reservoir and so on. In the Yi-Shu-Si Rivers area, it is planned to build the Zhuangli Reservoir and so on. In the Shandong Peninsular, it is planned to build the Shuangwangcheng Reservoir and so on to regulate and store water resources.

In the Haihe River Zone, surface water has been overexploited. The purpose of building new reservoirs is to increase water supply in mountain areas and to optimize water allocation. It is planned to build large reservoirs (e.g. the Shuangfengsi Reservoir) with total storage capacity of 500 million m^3.

4.5.3 The Yangtze River and the Southwest Rivers Zones

The Yangtze River and Southwest Rivers Zones own relatively abundant water resources. However, the water resources condition and socio-economic situation vary greatly in different regions. Water shortage due to lack of resources or projects and seasonal water shortage co-exist simultaneously. Future water demand growth is mainly for industrial and domestic water use, while agricultural water use will be reduced by 6.5 billion m^3. By making full use of the advantage of abundant water resources, it is planned to construct a number of reservoirs for regulating and storing water resources in proper regions. Incremental surface water supply is more than 20 billion m^3. The safety of drinking water in urban and rural areas could be secured, social and economic water use could be ensured, water resources can be utilized and protected in a rational way, and water environment condition could be improved.

In the Upper Yangtze River and the Southwest Rivers Zones, social and economic development lags behind relatively. Water resources regulation and control capacity is insufficient since lack of major water resources controlling projects. The water development ratio, utilization level and water supply capacity are relatively low. The guarantee degree of water supply is not high and water shortage often occurs in dry year. The Zones belong to water shortage areas lacking of projects. In addition, water demand will be increased in the Upper Yangtze River and the Southwest Rivers Zones due to Western Development Program and construction of new rural areas. On the basis of exploiting the potential of current facilities, combined with hydropower development, a bunch of controlling regulation and storage projects and a number of medium and small sized water source projects will be built. The emphasis is to solve water shortage caused by lack of projects and water resources in the Yungui Plateau and mountainous areas of the northern part of Sichuan and Chongqing. In the Upper Yangtze River, large reservoirs such as the Xiaozhongdian and Zaodu Reservoirs and medium and small sized reservoirs including the Jiulongtan Reservoir will be built. In the Southwest Rivers Zone, the Dehou Reservoir will be built in the Red River, and the Pangduo and Laluo Reservoirs will be built in the Yarlung Zangbo River.

The Middle Yangtze River area is significant agricultural production base and core region of the economic belt along the Yangtze River, connecting the eastern and western regions. Taking into account the demand of social and economic development in the developed city and town areas in the middle stream, the need of ensuring the national food security as well as the need of water transfer to outside river basins, the emphasis of project construction is to reinforce water saving and pollution prevention and treatment, to enhance auxiliary construction of existing projects and reform current pattern of water saving, as well as to build a bunch of water resources regulation and control projects and river-lake connection projects in a scientific and reasonable way. Through above measures, water resources regulation and control capacity in the mainstream and key tributaries will be promoted. The condition of water supply in dry seasons and dry years and in-stream eco-environmental water use could be improved. Several large reservoirs (e. g. the Xiajiang Reservoir) and a number of medium and small sized reservoirs (e. g. Xiaolongdong Reservoir) will be built to meet the increasing water demand of urban domestic and industrial use.

The Lower Yangtze River is an important economic center in China. Generally, water utilization level is relatively high, water resources are mainly used by city and industry, and water supply requirements for good water quality and high guarantee degree are relatively high. Severe water pollution and Lake Eutrophication in some areas are prominent issues. The measures adopted to solve these issues include enhancing diverting and draining capacity of projects along the Yangtze River, and properly increasing water quantity diverted from the Yangtze River. In the future, the number of new-built dams is relatively small.

4.5.4 *The Pearl River and the Southeast Rivers Zones*

Both economically developed and under developed regions co-exist in the Pearl River and Southeast Rivers Zones. The developed regions are mainly located in the Pearl River Delta and coastal areas, which are featured with developed economy, intensive buildings and dense population. The under developed regions are mainly located in mountainous and hilly areas. Though, water resources condition is generally good, regional differences are prominent and existing problems vary in different regions. In some regions, water scarcity due to lacking projects to control, regulate and distribute water resource is prominent. In the Pearl River Delta and downstream regions of the Qiantang River and Minjiang River, water scarcity due to poor water quality problem is a dominant issue, meanwhile, water pollution, salt water intrusion in the estuary, and eco-environmental problems are acute in the areas. In coastal areas (such as eastern Zhejiang Province, southern Fujian Province, eastern and western Guangdong Province, southern Guangxi Province, southeastern Yunnan Province, central Guizhou Province and Hainan Island), the guarantee degree of water supply is not high, and water shortage is serious in dry years or dry seasons. In the future, on the basis of strict water saving and wastewater drainage reduction, a number of dams will be built to rationally regulate and allocate water resources, with considering combined demands of regional flood control, salt water intrusion control, water supply and power generation. The water supply will be increased by more than 10 billion m^3, and the Pearl River Zone will take up about 3 billion m^3, which are mainly for the Nanpan River, Beipan River, Hongliu River and coast-borne rivers to secure water supply safety in key areas, to improve natural runoff process of some rivers, and to meet demand of water flowing into the sea for preventing salt intrusion in the estuary of the Pearl River. In the Southeast Rivers Zone, the water supply will be increased by more than 8 billion m^3, which are mainly for the Minjiang River and the Qiantang River to meet urban and industrial water use, to solve water shortage problems in coastal areas and some sea inlands, and to improve eco-environmental condition of some cities and major ecological regions.

In the Pearl River Zone, it is planned to implement the Rundian Project and the Ziqian Project by stages (Rundian Project means Moisten Yunnan Province and Ziqian Project means Moisten Guizhou Province). 77 large-and medium-sized reservoirs will be built in Rundian Project and 68 medium sized reservoirs will be built in Ziqian Project. Besides, in the short-term, the Datengxia and Wacun Reservoirs in the Xijiang River, and the Hongling Reservoir in Hainan Province will be built, which will play critical roles in rational allocation of water resources in river basins. Thus, water resources regulation and allocation system of the Pearl River Basin will be preliminarily formulated with major reservoirs (e. g. the Longtan and

Datengxia Reservoirs in the Xijiang River, the Feilaixia Reservoir in the Beijiang River, the Xinfeng Reservoir in the Dongjiang River). In the long term, in order to secure regional water supply safety by 2030, a bunch of large and medium sized reservoirs (e.g. the Maiwan Reservoir in Hainan Province) will be constructed. Then, a good water resources regulation and allocation system will be formed. In the southeast rivers zone, a number of large and medium sized reservoirs (e.g. Huokou Reservoir) will be built to promote regulating and allocating capacity of regional water resources.

4.5.5 The Northwest Rivers Zone

Most areas in the northwest are arid with serious water scarcity, and natural ecology and environment are very fragile. The northwest region is an important ecological protection barrier in China, and is also the key area of Western Development. Although, this region is of typical oasis economy, water utilization rate is high in most regions and contradiction between water supply and demand becomes increasingly prominent due to population growth and rapid socio-economic development. With the prerequisite of protecting ecology and environment rigorously, the economic structure will be adjusted and the manner of economic development will be changed in accordance with regional water resources carrying capacity. The growth rate of water demand will be controlled lower than 0.26% till 2030. A bunch of large and medium sized reservoirs will be built (e.g. the Zhengyixia and Huangzangshi Reservoirs in the Heihe River, the Aertashi Reservoir in the Ye'erqiang River, the Kalabeili Reservoir in the mountainous part of the Kezi River in southern Xinjiang Uigur Autonomous Region, the Kensiwate Reservoir in the Ma'nasi River) to further enhance runoff regulation and control capacity of the northwest rivers, to rationally regulate and allocate regional water resources, to reduce overexploitation of groundwater, and to return in-stream eco-environmental water use misappropriation. Consequently, in-stream runoff could be basically ensured, eco-environment of lower reaches could be realized gradually, and water resources for the Western Development could be guaranteed.

For above water allocation projects such as large and medium sized reservoirs proposed to be built in the Integrated Water Resources Plan, a five-year construction plan needs to be formulated based on social and economic development condition, and scientific demonstration should be carried out. It is necessary to identify the priorities, deeply carry out the preliminary work, and strengthen the environmental impact assessment, especially for augmentation and evaluation of eco-environmental impacts of inter-basin water transfer projects. In addition, the construction scale, scheme and time should be scientifically demonstrated.

References

[1] *Summary report on investigation and assessment of water resources and its development utilization in China*. General Institute of Water Resources and Hydropower Planning and Design, 2005.
[2] Jiao Yong. Dams, reservoirs and harmonious development—China's exploration and practice. *China Water Resources*, 2008.12: 1-3.
[3] *National integrated water resources plan*. National Development and Reform Commission, Ministry of Water Resources, 2010.
[4] *Integrated water resources plan of the Songhua River and Liaohe River Basins*. Songliao Water Resources Commission, MWR, 2009.
[5] *Integrated water resources plan of the Yellow River Basin and the Northwest Rivers Zone*. Yellow River Water Conservancy Commission, MWR, 2009.
[6] *Integrated water resources plan of the Huaihe River Basin and Shangdong Peninsula*. Huaihe River Water Resources Commission, MWR, 2009.
[7] *Integrated water resources plan of the Haihe River Basin*. Haihe River Water Resources Commission, MWR, 2009.
[8] *Integrated water resources plan of the Yangtze River Basin and Southwest Rivers Zone*. Yangtze River Water Resources Commission, MWR, 2009.
[9] *Integrated water resources plan of the Pearl River Basin*. Pearl River Water Resources Commission, MWR, 2009.
[10] *Integrated water resources plan of the Southeast Rivers Basin*. Taihu Basin Authority, MWR, 2009.

Chapter 5

Dams and Hydropower Development

Qian Gangliang[1]

5.1 Hydropower Resources

5.1.1 Hydropower Resources Re-investigation

From 1977 till 1980, the third national hyropower resources census was organized by the former Water Resources and Power Ministry with the former Ministry of Power Industry, putting forward the theoretical reserves and potential development capacity of the hydropower resources across China. In the following two decades, with the gradual development of China's economy, science & technology and the deepening of the preliminary work and hydropower development, there was an increasingly significant disparity between the results of the third census and the actual reserves of hydropower resources. In addition, it became necessary to study and evaluate hydropower resources with the international concept of economically exploitable amount across China, and to propose the economically exploitable hydropower resources. These works could provide important criteria to study and draw up national and local plans for energy and electric power development. In order to make a thorough investigation of hydropower resources and to fill in the gaps of economically exploitable hydropower resources, the former State Development Planning Commission decided to carry out a three-year re-investigation on the hydropower resources in China since 2001, covering all the provinces, autonomous regions and municipalities of the China Mainland and the regions of Taiwan, Hong Kong and Macau.

Accomplished in 2003, the *Re-investigation Report of Hydropower Resources of the People's Republic of China* (2003) basically checked up on China's hydropower resources and their potential for development, and obtained the re-investigation results of hydropower resources in different provinces (autonomous regions and municipalities) and various river basins.

5.1.1.1 Gross Hydropower Resources

In China, there are a total of 3,886 rivers with theoretical water resource reserves of no less than 10MW. Table 5.1-1 has shown the summary of re-investigation results of national hydropower resources (2003) and Table 5.1-2 for the summary of the exploitable amount of national hydropower resources by scale. The gross hydropower resources in China, including theoretical reserves, technically exploitable amount and economically exploitable amount, rank the first place in the world.

Table 5.1-1　Summary of Re-investigation Results of National Hydropower Resources (2003)

No.	Item		Unit	Amount
1	Theoretical reserves	Annual energy generation	TW·h	6,082.9
		Average power	MW	694,400
2	Technically exploitable amount	Number of hydropower stations	—	13,286+28/2
		Installed capacity	MW	541,640
		Annual energy output	TW·h	2,474
3	Economically exploitable amount	Number of hydropower stations	—	11,653+27/2
		Installed capacity	MW	401,795
		Annual energy output	TW·h	1,753.4

Note: 1. The scope of statistics in this table only covers the rivers with theoretical reserves of 10MW or above and hydropower stations with single installed capacity of 0.5MW or above on these rivers, excluding those in the regions of Hong Kong, Macau and Taiwan.

2. The symbol of "+28/2" in this table indicates that there are another 28 hydropower stations in the bordering international rivers, the same as the followings.

---

[1] Qian Gangliang, Deputy Chief Engineer in HydroChina Corporation, Professor Senior Engineer.

Part I Planning and Sustainable Development of Dams and Reservoirs

Table 5.1-2 Summary of the Exploitable Amount of National Hydropower Resources by Size

No.	Item			Unit	Amount
1	Large-sized hydropower stations (300MW or above)	Technically exploitable amount	Number of hydropower stations		263+10/2
			Installed capacity	MW	388,700
			Annual energy output	TW·h	1,792
		Economically exploitable amount	Number of hydropower stations		175+10/2
			Installed capacity	MW	276,082
			Annual energy output	TW·h	1205
2	Medium-sized hydropower stations (50MW to 300MW)	Technically exploitable amount	Number of hydropower stations		785+10/2
			Installed capacity	MW	87,730
			Annual energy output	TW·h	392.7
		Economically exploitable amount	Number of hydropower stations		631+9/2
			Installed capacity	MW	68,252
			Annual energy output	TW·h	293.3
3	Small-sized hydropower stations (0.5 MW to 50MW)	Technically exploitable amount	Number of hydropower stations		12,238+8/2
			Installed capacity	MW	65,210
			Annual energy output	TW·h	289.3
		Economically exploitable amount	Number of hydropower stations		10,847+8/2
			Installed capacity	MW	57,461
			Annual energy output	TW·h	255.1

Note: The statistics scope in this table only covers the rivers with theoretical reserves of 10MW or above and the single hydropower station with installed capacity of 0.5MW or above in these rivers, excluding those in the regions of Hong Kong, Macau and Taiwan.

5.1.1.2 Comparison with the Results of 1980

The results of the 3rd national hydropower resources census completed in 1980 excluded Hong Kong, Macau and Taiwan, and economically exploitable amount was not proposed, so in order to unify the comparison basis, this comparative analysis is limited to the theoretical reserves, technically exploitable amount, the exploited and exploiting amount of the re-investigation results in 2003 and the census results in 1980, excluding the amount of resources in Hong Kong, Macau and Taiwan (see Table 5.1-3).

Table 5.1-3 Comparison with Re-investigation Results in 2003 and Census Results in 1980

Item		Re-investigation in 2003	Census in 1980	Increment	Increase percentage (%)
Theoretical reserves	Number of quantified rivers	3,886	3,019	867	28.7
	Annual energy generation (TW·h)	6,082.9	5,700	382.9	6.7
	Average power (MW)	694,400	650,000	44,400	6.7
Technically exploitable amount	Number of hydropower stations	13,286+28/2	11,000	2,286+28/2	20.9
	Installed capacity (MW)	541,640	370,000	171,640	46.4
	Annual energy output (TW·h)	2,474	1,900	574	30.2
Exploited and exploiting amount (with single station installed capacity of 10MW or above)	Number of hydropower stations	827+4/2	214+3/2	613+1/2	284.6
	Installed capacity (MW)	119,975	25,610.5	94,364.5	368.5
	Annual energy output (TW·h)	477.8	108.575	369.225	340.1

Note: In the exploited and exploiting amount in the above table, the census value in 1980 refers to the hydropower stations completed in 1978 and those approved for construction with at least one unit being put into operation; the re-investigation value in 2003 refers to the hydropower stations completed and being constructed by the end of 2001.

According to the comparison of the re-investigation results in 2003 and the census results in 1980, theoretical reserves had a slight increase, and technically exploitable amount, exploited and exploiting amount increased by a large margin. The increase in the theoretical reserves was mainly caused by the enlarged re-investigation coverage and a larger number of rivers. The number of rivers involved in re-investigation in 2003 was 867 more than the ones in 1980. The primary reason for the increase in the technically exploitable installed capacity and annual energy output is that since 1980, a great deal of surveying, planning and design work had been carried out on many rivers, there were quite a few adjustments and complements to river cascade development schemes, and a growing number of hydropower stations with correspondingly more technically exploitable amount were constructed. Furthermore, the installed capacity of the established and planned hydropower stations grew substantially, leading to a considerable increase of the total exploitable amount of hydropower resources.

5.1.1.3 Characteristics of Hydropower Resources

(1) The extremely uneven regional distribution of hydropower resources demands for the from-West-to-East power transmission strategy in hydropower development. Since China has a vast territory with distinct differences in terrain and rainfall, there is an uneven regional distribution of hydropower resources, which are intensively distributed in the west and scattered in the east. According to the statistics of technically exploitable installed capacity, 12 provinces (autonomous regions and municipalities) in west China, i.e. Yunnan, Guizhou, Sichuan, Chongqing, Shaanxi, Gansu, Ningxia, Qinghai, Xinjiang, Tibet, Guangxi and Inner Mongolia boast 81.46% of the total hydropower resources in China, 66.70% of which are located in the southwest provinces, including Yunnan, Guizhou, Sichuan, Chongqing and Tibet; the central regions, such as Heilongjiang, Jilin, Shanxi, Henan, Hubei, Hunan, Anhui and Jiangxi, possess the second largest amount of hydropower resources, which accounts for 13.66%; 11 provinces (municipalities) in the east including Liaoning, Beijing, Tianjin, Hebei, Shandong, Jiangsu, Zhejiang, Shanghai, Guangdong, Fujian and Hainan have only 4.88% of the total amount. China's economy is relatively developed in the eastern areas and backward in the west, therefore, the development of the western hydropower resources shall take into account the eastern market and implement the from-West-to-East power transmission strategy in hydropower development, in addition to the demand of the western power market.

(2) The uneven temporal distribution of hydropower resources requires reservoirs regulation. Located in southeast Eurasia and close to the Pacific Ocean, China is affected by typical monsoon climate, and the runoff in intra-annual and interannual change in most rivers is unevenly distributed, with the runoff demonstrating a tremendous difference between the wet and dry seasons. Hence, China needs to build reservoirs with outstanding regulatory capacity to regulate the river runoff and improve the overall quality of hydropower generation to better meet the demands of the power market.

(3) Hydropower resources are concentrated in large rivers, so it is convenient to set up hydropower bases to implement strategic development plans. There are abundant hydropower resources in the Jinsha River, Yalong River, Dadu River, Lancang River, Wujiang River, the Upper Yangtze River, Nanpan River (part of Hongshui River), the Upper Yellow River, western Hunan, Fujian-Zhejiang-Jiangxi, Northeast China, the North Mainstream of the Yellow River and Nujiang River hydropower bases, whose total installed capacity accounts for approximately 50.9% of the national technically exploitable amount. The gross installed capacity of the Middle-and-Lower Jinsha River comes at 58,580MW and that of the Upper Yangtze River is as much as 33,197MW. The installed capacities of the Yalong River and Dadu River in the upstream tributaries of the Yangtze River, the Upper Yellow River, Lancang River and Nujiang River all exceed 20,000MW for each, and those of the Wujiang River and Nanpan River (part of Hongshui River) are more than 10,000MW for each. Hydropower resources are concentrated in these rivers, which can facilitate the construction of large-scale hydropower bases and give full play to the economies of scale of hydropower resources.

(4) The installed capacity of large-sized hydropower stations takes up a large proportion of installed capacity and there are many medium and small sized hydropower stations with wide location. Among the 13,314 technically exploitable hydropower stations (among them 28 stations are on international boundary rivers), there are 273 large-sized hydropower stations (among them 10 stations on rivers forming international boundaries) with an installed capacity of 300MW or above for each, but the ratio of their technically exploitable installed capacity and annual energy output account for 71.76% and 72.43%, respectively. There are only 116 extra-large stations (among them 5 stations on international boundary rivers) with an installed capacity of 1,000MW or above, but their total installed capacity and annual energy out can reach 315,594MW and 1,457.907TW·h and the proportions of both exceed 50%. An overwhelming majority of extra-large hydropower stations are distributed in the southwest region, and there are a total of 12,246 small-sized hydropower stations (among them 8 stations on rivers forming international boundaries) distributed across the country, accounting for 92.1% of the total.

5.1.1.4 The Distribution of Hydropower Resources in Basins or Regions

There are numerous rivers in China, with more than 50,000 covering an area of over 100km^2 and 1,500 rivers above 1,000km^2, respectively. According to the flow direction, China's rivers can be classified into two categories—inland rivers with no connection with the ocean and outflow rivers flowing to the sea, and the watershed area of the later makes up about two-third of the Chinese territory. Most rivers in China flow from west to east or from southeast to the Pacific Ocean, including but not limited to the Heilong River, Liaohe River, Haihe River, Yellow River, Huaihe River, Yangtze River, Zhujiang River, Lancang River, Qiantang River and Minjiang River, etc.; the rivers such as the Nujiang River, Yarlung Zangbo River, and so on, flow southward out of China's territory to the Indian Ocean; and the Irtysh River in northwest Xinjiang flows into the Arctic Ocean through Kazakhstan and Russia. Based upon the amount of hydropower resources and geographic location of various river basins, the analytic statistics of re-investigation on hydropower resources in 2003 divided the country's rivers into 10 regions, which are the Yangtze River Basin, Yellow River Basin, Zhujiang River Basin, Haihe River Basin, Huaihe River Basin, northeast rivers, southeast coastal rivers, southwest international rivers, Yarlung Zangbo River and other Tibetan rivers, and inflow rivers in the north and other rivers in Xinjiang. According to the priority order of the technically exploitable amount of hydropower resources, the top three are the Yangtze River Basin with 256.27GW, followed by the Yarlung Zangbo River with 67.85GW and the Yellow River Basin with 37.34GW (see Fig. 5.1-1, Fig. 5.1-2 and Fig. 5.1-3).

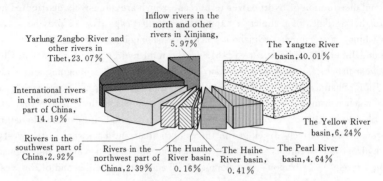

Fig. 5.1-1 Distribution of Theoretial Reserves of National Hydropower Resources in Various River Basins

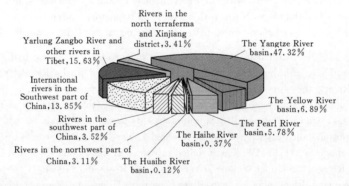

Fig. 5.1-2 Distribution of Technically Exploitable Installed Capacity of National Hydropower Resources in Various River Basins

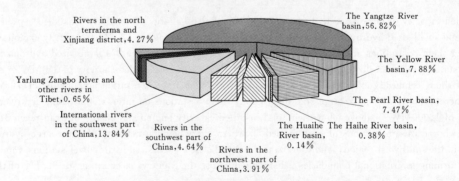

Fig. 5.1-3 Distribution of Economically Exploitable Installed Capacity of National Hydropower Resources in Various River Basins

The distribution of theoretical reserves of China's hydropower resources is extremely unbalanced, with 70.6% centralized in the southwest, and only 2% in north China, 1.9% in the northeast and 4% in the eastern developed areas (see in Fig. 5.1-4, Fig. 5.1-5 and Fig. 5.1-6). The same situation happens to the technically exploitable capacity and economically exploitable capacity, with the southwest region taking up 66.7% and 58.9%, and northern China merely 1.6% and 1.9%. Fig. 5.1-7 and Fig. 5.1-8 have shown provincial (or municipal) hydropower distribution in a decending sequences.

Table 5.1-4 and Table 5.1-5 show the summaries of the re-investigation results of national hydropower resources by river basin and provinces (autonomous regions and municipalities).

Chapter 5 Dams and Hydropower Development

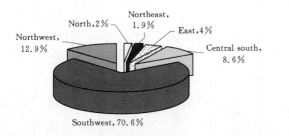

Fig. 5.1-4 Regional Distribution of Theoretical Reserves (Annual Electricity Generation) of Hydropower Resources

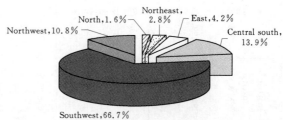

Fig. 5.1-5 Regional Distribution of Technically Exploitable Amount (Installed Capacity) of Hydropower Resources

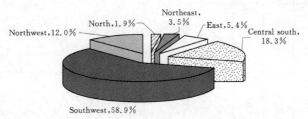

Fig. 5.1-6 Regional Distribution of Economically Exploitable Amount (Installed Capacity) of Hydropower Resources

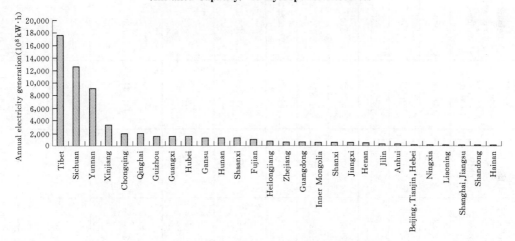

Fig. 5.1-7 The Ordination of Theoretical Reserves of Hydropower Resources in Various Provinces (Autonomous Regions and Municipalities) of China

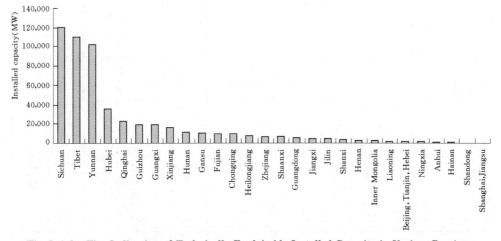

Fig. 5.1-8 The Ordination of Technically Exploitable Installed Capacity in Various Provinces (Autonomous Regions and Municipalities) of China

Part I Planning and Sustainable Development of Dams and Reservoirs

Table 5.1-4 Summary of Re-investigation Results of National Hydropower Resources (By River Basin)

No.	River basin	Theoretical reserves		Technically exploitable amount			Economically exploitable amount			Exploited and exploiting amount		
		Annual electricity generation (TW·h)	Average power (MW)	Number of hydropower stations	Installed capacity (MW)	Annual energy output (TW·h)	Number of hydropower stations	Installed capacity (MW)	Annual energy output (TW·h)	Number of hydropower stations	Installed capacity (MW)	Annual energy output (TW·h)
1	Yangtze River Basin	2,433.598	277,808.0	5,748	256,272.9	1,187.899	4,968	228,318.7	1,049.834	2,441	69,727.1	292.496
2	Yellow River Basin	379.413	43,312.1	535	37,342.5	136.096	482	31,647.8	111.139	238	12,030.4	46.479
3	Pearl River Basin	282.394	32,236.7	1,757	31,288.0	135.375	1,538	30,021.0	129.768	957	18,100.7	78.578
4	Haihe River Basin	24.794	2,830.3	295	2,029.5	4.763	210	1,510.0	3.501	123	803.4	1.950
5	Huaihe River Basin	9.800	1,118.5	185	656.0	1.864	135	556.5	1.592	75	310.3	0.958
6	Northeast rivers	145.480	16,607.4	644+26/2	16,820.8	46.523	510+26/2	15,729.1	43.382	196+4/2	6,396.8	15.174
7	Southeast coastal rivers	177.611	20,275.3	2,558+1/2	19,074.9	59.339	2,532+1/2	18,648.3	58.135	1,388	11,653.7	36.308
8	Southwest international rivers	863.007	98,516.8	609+1/2	75,014.8	373.182	532	55,594.4	268.436	313	9,322.7	44.277
9	Yarlung Zangbo River and Other Tibetan rivers	1,403.482	160,214.8	243	84,663.6	448.311	130	2,595.5	11.969	52	346.6	1.155
10	Inflow rivers in the North and other rivers in Xinjiang	363.357	41,479.1	712	18,471.6	80.586	616	17,174.0	75.639	270	2,290.2	8.510
	Total	6,082.9	694,400	13,286+28/2	541,640	2,474.0	11,653+27/2	401,795	1,753.4	6,053+4/2	130,980	525.9

Note: The scope of statistics in this table are rivers with theoretical reserves of 10MW or above and with a single station installed capacity of 0.5MW or above on these rivers, excluding those in Hong Kong, Macao and Taiwan.

Chapter 5 Dams and Hydropower Development

Table 5.1-5 Summary of Re-investigation Results of National Hydropower Resources (By Province, Autonomous Region and Municipality)

No.	Province, autonomous region and municipality	Theoretical reserves		Technically exploitable amount			Economically exploitable amount			Exploited and exploiting amount			Development degree (Installed capacity) (%)
		Annual electric generation (TW·h)	Average power (MW)	Number of hydropower stations	Installed capacity (MW)	Annual energy output (TW·h)	Number of hydropower stations	Installed capacity (MW)	Annual energy output (TW·h)	Number of hydropower stations	Installed capacity (MW)	Annual energy output (TW·h)	
1	Beijing, Tianjin and Hebei	19.926	2,274.6	179+1/2	1,751.3	3.712	104+1/2	1,252.5	2.527	65+1/2	686.8	1.493	39.2
2	Shanxi	49.361	5,634.8	169+7/2	4,020.4	12.055	149+7/2	3,973.8	11.896	59+3/2	943.3	2.861	23.5
3	Inner Mongolia	50.914	5,812.2	103+10/2	2,624.5	7.345	82+10/2	2,567.3	7.191	24+2/2	736.9	1.901	28.1
4	Liaoning	17.792	2,031.0	200+3/2	1,767.3	6.039	171+3/2	1,728.9	5.914	68+2/2	1,360.3	4.726	77
5	Jilin	30.130	3,439.6	188+14/2	5,115.5	11.793	157+14/2	5,042.3	11.547	71+2/2	3,882.9	7.922	75.9
6	Heilongjiang	66.420	7,582.2	166+11/2	8,161.9	23.831	111+11/2	7,226.9	21.179	41+1/2	971.8	2.046	11.9
7	Shanghai and Jiangsu	15.220	1,738.1	22	57.9	0.173	14	22.4	0.067	14	22.4	0.067	38.7
8	Zhejiang	53.759	6,136.8	1,070+4/2	6,643.8	16.146	1,067+4/2	6,613.2	16.072	651	4,423.7	10.501	66.6
9	Anhui	27.350	3,122.0	157	1,074.0	3.000	144	996.0	2.730	65	613.0	1.670	57.1
10	Fujian	94.101	10,742.0	1,031+4/2	9,979.7	35.302	1,027+4/2	9,697.7	34.497	482	5,808.1	21.330	58.2
11	Jiangxi	42.557	4,858.1	533	5,162.9	17.096	448	4,161.9	13.774	255	2,036.9	6.053	39.5
12	Shandong	10.248	1,170.1	50	64.2	0.159	37	50.8	0.132	21	34.9	0.087	54.4
13	Henan	41.232	4,706.6	214+3/2	2,880.6	9.695	137+3/2	2,726.4	9.140	83+2/2	2,331.1	7.682	80.9
14	Hubei	150.712	17,204.5	704+2/2	35,540.5	138.631	649+2/2	35,355.9	138.045	326	31,475.7	125.720	88.6
15	Hunan	116.251	13,270.4	967+2/2	12,020.9	48.619	769+2/2	11,349.8	45.758	448+1/2	6,856.8	28.219	57
16	Guangdong	53.160	6,068.5	1,051	5,401.4	19.814	970	4,878.8	17.779	672	3,435.3	12.404	63.6
17	Hainan	7.378	842.2	85	760.5	2.103	82	710.5	1.994	54	452.9	1.315	59.6
18	Guangxi	154.537	17,641.2	821+5/2	18,913.8	80.884	759+5/2	18,575.0	79.498	410+3/2	12,333.7	52.511	65.2
19	Sichuan	1,257.189	143,514.7	1,992+27/2	120,040.0	612.159	1,836+19/2	103,270.7	523.289	756+2/2	16,301.2	86.083	13.6
20	Chongqing	201.167	22,964.3	421	9,808.4	44.578	323	8,195.9	37.804	194	1,556.7	6.730	15.9
21	Guizhou	158.437	18,086.4	574+27/2	19,487.9	77.799	448+26/2	18,980.7	75.242	194+7/2	10,189.8	37.000	52.3
22	Yunnan	914.421	104,386.0	769+27/2	101,939.1	491.881	729+25/2	97,950.4	471.283	470+3/2	11,257.6	53.903	11
23	Tibet	1,763.898	201,358.2	333+5/2	110,004.4	575.969	191	8,350.4	37.625	82	456.5	1.757	0.4
24	Shaanxi	111.856	12,768.9	343+6/2	6,623.8	22.216	312+6/2	6,501.6	21.722	134+1/2	1,901.1	6.877	28.7
25	Gansu	130.416	14,887.3	310+11/2	10,625.4	44.434	195+10/2	9,009.0	37.043	85+2/2	3,692.4	16.431	34.8
26	Qinghai	191.614	21,873.8	229+12/2	23,140.4	91.344	170+8/2	15,479.1	55.462	96+2/2	5,082.8	20.640	22
27	Ningxia	18.419	2,102.6	10+1/2	1,458.4	5.894	10+1/2	1,458.4	5.894	5	428.9	1.952	29.4
28	Xinjiang	334.446	38,178.7	518	16,564.9	71.259	495	15,670.5	68.281	213	1,706.7	5.999	10.3
	Total	6,082.9	694,400	13,286+28/2	541,640	2,474	11,653+27/2	401,795	1,753.4	6,053+4/2	130,980	525.9	24.2

Note: The scope of statistics in this table are rivers with theoretical reserves of 10MW or above and with a single station installed capacity of 0.5MW or above on these rivers, excluding those in Hong Kong, Macao and Taiwan.

5.1.2 *Investigation on the Hydropower Resources of the Lower Yarlung Zangbo River*

Originated from the Chemayungdung Glacier at the northern slope of Himalayas, the Yarlung Zangbo River exits at the Pasighat in Linzhi and then into the India. With a source elevation of 5,590m, the Yarlung Zangbo River is 2,057km in domestic length, 5,435m in total drop, and 240,480km^2 in drainage area in China. It is the third largest river in China, just second to the Yangtze River and the Pearl River.

Rechecked by investigations, from the Paizhen Town in the mainstream of the lower Yarlung Zangbo River to the estuary of the Lugong River, the watercourse distance approximates 257km, the natural drop is 2,335m and the average gradient is 9.1‰, with 69,620MW of technically exploitable hydropower resources and 347.4TW·h of annual energy output. It has the country's highest enrichment, which is rarely seen in the world, so it is suitable to build an extra-large follow-up energy base for the from-West-to-East power transmission.

The lower Yarlung Zangbo River has complex natural conditions and favorable ecological environment. By means of the rational layout of the hydropower development programs and the adoption of relevant engineering measures, they are blessed with the topographic and geological conditions to build dams (floodgates), diversion tunnels and powerhouse (underground powerhouse), and meanwhile, mitigate the negative impacts upon the ecological environment. The socioeconomic development of the lower Yarlung Zangbo River is backward, with few and concentrated inhabitants and cultivated lands. Therefore, except for Daduka regulating reservoirs, the construction of other tentative hydropower stations has little to do with inundated farmlands and resettlement.

The mainstream of the Yarlung Zangbo River from the Lugong River estuary to the Pasighat is 237km in length, with a drainage area of 12,579km^2, an annual average runoff of around 47.1 billion m^3, a natural drop of 410m and an amount of 15,000MW of technically exploitable hydropower resources.

5.2 Hydropower Development Planning of Rivers

After the founding of New China, with view to ensuring the right direction of the hydropower development in China, it was put forward to comprehensively study the experience of the Soviet Union in hydropower development and in 1953, the first batch of experts paid a visit to China. A year later, the *Technical and Economic Survey Regulations* was promulgated, with a systematic introduction to the experience of the Soviet Union in planning for river hydropower development, including the following emphases: ①the importance of basic data, particularly the hydrological data and geological prospecting data; ②conduct a complete investigation on rivers, understand the comprehensive utilization requirement of various sectors of national economy and earnestly handle the integrated utilization in river planning; ③carry out the survey and investigation on the entire river, formulate the cascade development program based on the whole river and select the first-stage projects from a great number of potential ones; and ④distinguish the long-term and short-goal of river development. These planning thoughts and experience had played a vital role in directing the planning for the Yellow River and other rivers, and exerted a profound influence on the subsequent river hydropower planning in China.

The improvement and development of the Yellow River was started soon after the founding of the People's Republic of China. In October 1954, *the Technical and Economic Report on the Compressive Utilization Planning for the Yellow River* was compiled and submitted to the Central Committee of CPC and state leaders for approval after being examined by the Party leadership group of the former State Development Planning Commission (SDPC) and the former State Infrastructure Commission in April 1955. A month later, the Political Bureau of the CPC Central Committee debriefed this report, and in July that year, the Second Session of the First National People's Congress discussed and passed this report, which is the first comprehensive utilization planning for large rivers adopted by the National People's Congress. The said Report is composed of two parts: long-term planning and first-stage plan. In addition to the comprehensive utilization planning for the Yellow River, similar efforts were made toward the Yangtze River, Zhujiang River, Heilongjiang River, Hanjiang River and Yuanshui River as well as medium and small rivers such as Longxi River, Shangyou River, Gutian River, Yili River, Maotiao River, Xin'an River, Zishui River and Huijiang River. Concerning these rivers, some completed the planning report, some proposed preliminary planning report or key planning points. Though most of the planning outcomes were not officially examined or approved, they played a critical role in guiding further planning as well as the hydropower construction of the aforesaid rivers.

From the 1960s to the 1970s, just like the other economic development in China, the hydropower planning and construction was disturbed and destroyed by the "leftism", the basic laws of hydropower construction were ignored, and the role of hydropower planning was denied, with the work at a standstill in fact. At the late 1970s, with the transfer of the national key tasks, the hydropower planning for rivers were gradually put on the agenda. In order to offer guidance, *the Several Provi-*

sions on the River Hydropower Planning was formulated in 1976, and was printed and implemented in the *Circular of the Several Opinions on Strengthening Electric Power Planning*, No. 47 Document of the Ministry of Water Resources and Power in 1978.

In 1981, the General Administration Bureau of Water Resources and Hydropower convened a preliminary meeting, and in terms of planning, it raised *the Several Opinions on Strengthening Hydropower Planning* in light of the conditions then, and issued official documents, whose guiding ideology and specific contents played a certain role in enhancing the recognition of river hydropower planning and reinforcing such planning at that time and over a specified period. Tremendous planning achievements with higher quality had been made during this period, including the *Report on the Comprehensive Utilization Planning for the Hongshui River* in 1980, *Report on the Development Planning for the Section from Dukou to Yibin in the Jinsha River* in 1981, *Report on the Cascade Planning for the Section from Longyang Gorge to Qingtong Gorge in the Upper Yellow River* in 1983, *Report on the Mainstream Planning for the Dadu River* in 1984, *Development Planning for the Middle-and-Lower Lancang River* in 1986, *Report on the Mainstream Planning for the Wujiang River* in 1987, *Mainstream Planning for the Yuanshui River* in 1988 and *Report on the Mainstream (from Kala to Estuary) Hydropower Planning for the Yalong River* in 1992.

With the purpose of making medium-sized hydropower the backbone power source of local regions and strong backup of large power grids, since the 1980s, the state has continuously arranged and compiled a package of medium-sized river plans. For instance, it has successively been completed its hydropower planning for Longxi River, Futun River, Shaxi River, Dazhang River, Jianxi River, Muyang River, etc. Fujian Province, and it has been completed in succession its hydropower planning for Nanya River, Mabian River, Wasi River, Baoxing River, Huoxi River and Meigu River in Sichuan Province. The competent departments, various provinces and autonomous regions have examined and approved these plans on a timely basis, which has steered the preliminary work and construction of medium-sized hydropower, and created conditions for attaining the strategic goal of "speeding up constructing medium-sized hydropower stations with capacity of 10GW within the 11 years from 1990 to 2000" explicitly indicated in the *Outline of the Interim Development Plan for the Energy Industry in China* formulated by the former Ministry of Energy in 1990.

In order to further unify the guiding principle and technical requirement of river hydropower planning, the Water Resources and Hydropower General Construction Corporation arranged the drafting of the *Specifications on Compiling Hydropower Planning for Rivers as early* as 1984. After much investigation, the compiling institue completed the draft version of this specification for opinion soliciting in 1987. In January 1988, the *Water Law of the People's Republic of China* was issued, and revisions of this specification were made on this law. After modifications and improvements again and again, the *Specifications on Compiling Hydropower Planning for Rivers* (Draft for Approval) was finalized in October 1994. On July 13, 1995, the Ministry of Power Industry released the *Specifications on Compiling Hydropower Planning for Rivers* (DL/T 5042—1995), as the power industry standard, which was implemented as of January 1, 1996.

After 50 years' practices in river hydropower planning, we have fully realized: ①river hydropower planning should take precedence in the preliminary hydropower work; ②river hydropower planning report shall go through the procedures for examination and approval; ③it is imperative to actively implement the principle of comprehensive utilization, seek truth from facts and coordinate river hydropower planning with comprehensive planning; ④geological prospecting is the basis of river hydropower planning; ⑤reservoir resettlement shall be taken seriously; ⑥the reservoir inundation of trans-provincial rivers shall be properly dealt with and hydropower resources be rationally distributed; ⑦the law of reservoir sediment accumulation shall be emphasized; ⑧the planning for controlling reservoir project shall be highlighted; ⑨approved river plans also call for supplement and amendment; and ⑩ river hydropower planning is in need of rolling planning.

Local socioeconomic development and ecological environment conservation have gradually turned to the main considerations in hydropower planning. The *Environmental Protection Law of the People's Republic of China* enforced in 2002 particularly stipulates that environmental impact assessment on hydropower planning shall be carried out simultaneously. In light of the new situation and new demand, the revision of the *Specifications on Compiling Hydropower Planning for Rivers* was organized by the Hydropower and Water Resources Planning and Design General Institute in 2007, which was accomplished in the early of 2010 and implemented since December 2010. Since 2000, the hydropower planning for the Middle Jinsha River, the mainstream adjusted hydropower planning for the Dadu River, the hydropower planning for the Middle Yalong River and the hydropower planning for the the Upper Lancang River (section from Gushui to Miaowei) have been finished. These completed, examined and approved reports have provided strategic guidance for hydropower planning. According to the opinions on planning layout and hydropower development, various cascades have conducted pre-feasibility studies, prospecting and design in the feasibility study phase one after another, advancing the hydropower development of sections in an orderly rapid pace. After 2000, a series of river plans have been mapped out in line with the early-stage arrangements of National Development and Reform Commission (NDRC). The hydropower planning has been conducted in the river reaches including the Upper reach of Jinsha River, Lancang River (within Tibet), Nujiang River and

Yellow River (section from Hukou to Erduo) and the Middle Yarlung Zangbo River. Remaining to be approved, the said plans have paid more attention to reservoir inundation and ecological environment protection, the combination with local socioeconomic development while stressing hydropower engineering. The hydropower planning for main rivers (sections) after 2000 is as follows.

5.2.1 Hydropower Planning for the Middle Jinsha River

The Jinsha River is the upstream of the Yangtze River, and is 2,326km long along its mainstream from Zhimenda to Yibin, with a drop of 3,279m. The section from Shigu to the estuary of the Yalong River is the Middle Jinsha River and has a length of about 564km, and a drop of nearly 838m. The said reach is mainly located within Lijiang City and Diqing Zang Autonomous Prefecture of Yunan Province.

In September 1990, the State Council approved the *Brief Report on the Comprehensive Utilization Planning for the Yangtze River Basin (Revised Version of 1990)* (hereinafter referred to as the "Planning"). According to the Planning, the major improvement and development tasks of the section from Shigu to Yibin are power generation, navigation, flood control, wood drifting and soil and water conservation, etc., and a five-cascades-development scenario is recommended for the middle Jinsha River, i.e. Tiger Leaping Gorge, Hongmenkou, Zili, Pichang and Guanyinyan. In the light of the approval requirement of the State Council and the arrangement of the SDPC, the *Hydropower Planning for the Middle Jinsha River* was completed in December 1999. Based upon the tasks put forward by the Planning, the *Hydropower Planning for the Middle Jinsha River* proposed that the development and improvement tasks of the Middle Jinsha River are power generation, water supply, flood control, soil and water conservation, navigation and tourism on the basis of the specific conditions of the said section as well as the comprehensive demonstration and analysis.

In consideration of the complicated topographical, geological, social and environmental conditions in the Tiger Leaping Gorge section, the location of controlling upstream reservoir turned out to be the key problem in planning. While the *Hydropower Planning for the Middle Jinsha River* was implemented, a monographic study on the cascade development scheme of the Tiger Leaping Gorge section was in process, followed by the *Research Report on the Cascade Development Scheme Design of the Tiger Leaping Gorge Section* as a part of the *Hydropower Planning for the Middle Jinsha River*. Based on regional geological studies and tentative screening of dam sites and considering the topographical and geomorphic conditions and distribution features of hydropower resource, the Report presented Qizong and Longpan as the controlling upstream reservoir site options, and recommended the two-cascade (Longpan and Liangjiaren) development scheme for the Tiger Leaping Gorge section through comprehensive technical and economic comparisons.

According to the natural, social and economic conditions of the Middle Jinsha River, the research results of development way for the Tiger Leaping Gorge reach, the comprehensive technical and economic comparisons of the six cascade development programs concerning the midstream section, the development scheme recommended for the Middle Jinsha River was "One-controlling Reservoir and Eight-Cascades": Longpan, Liangjiaren, Liyuan, Ahai, Jin'anqiao, Longkaikou, Ludila and Guanyinyan. The former State Planning Commission presided over the examination of the *Hydropower Planning for the Middle Jinsha River* in April 2002, and approved and released the examination document in January, 2003.

In line with the comprehensive analysis of the outcomes of the Planning and the *Hydropower Planning for the Middle Jinsha River*, the following conclusions can be made:

(1) River planning shall be revised with the times. River planning is an important basis for hydropower resource development, and along with the social progress and economic growth, the changes of the utilization requirements of the power system and other departments on water resources as well as the market conditions, particularly the enhancement of public awareness of reservoir resettlement and environmental conservation, it is necessary to adjust and re-examine the original planning.

(2) More emphasis is placed on reservoir inundation. In a bid to achieve integrated balance from various aspects, such as minimizing the overall inundation losses from cascade development, the *Hydropower Planning for the Middle Jinsha River* adjusted the five-cascade development scheme suggested in the Planning to an eight-cascade one. After adjustment, the total inundation population decreased from 71 people per 10MW to 55 people per 10MW, and the cultivated farmland inundated from 9 hectares per 10MW to 4.67 hectares per 10MW, which showed a sharp fall in inundation losses.

(3) The role of the leading reservoir is taken seriously. Both the Planning and the *Hydropower Planning for the Middle Jinsha River* regard the Longpan Reservoir as the leading one for the cascade development in the mainstream of Jinsha River. By virtue of its storage, the cascade runoff has been well regulated. Water head is mainly used by downstream cascades. With the purpose of reducing reservoir inundation, engineering difficulty and construction costs, all cascades under Longpan universally adopt low dam development scheme.

5.2.2 Adjusted Hydropower Planning for the Dadu River

The Dadu River is the largest tributary of the Minjiang River system in the upper reaches of the Yangtze River. In order to develop and exploit the hydropower resources of the Dadu River, a series of census, survey, resurvey, planning and designing work have been carried out since the 1950s, with *the Dadu River Census Report*, *Resurvey Report of Mainstreams and Tributaries of the Dadu River*, *Water Resources and its Exploitation in the Dadu River Basin*, etc. compiled. In accordance with the arrangement of former Ministry of Water Resources and Power, *Planning Report of the Dadu River Mainstream* was proposed in June 1983, which suggested that the exploitation of the Dadu River mainstream should be centered on power generation while giving due considerations to wood drifting, flood control, shipping and irrigation and formed a 17-cascade development scheme including Dusong, Dagangshan, Pubugou, Gongzui (high dam) as main cascades. From the upstream to the downstream, the cascade hydropower stations successively consist of Dusong, Manai, Danba, Jijiaheba, Houziyan, Changheba, Lengzhuguan, Luding, Yingliangbao, Dagangshan, Longtoushi, Laoyingyan, Pubugou, Shenxigou, Zhentouba, Gongzui (high dam) and Tongjiezi, among which the regulating reservoirs of Dusong and Pubugou are controlling projects, with Pubugou being recommended as the development and construction project following Tongjiezi Hydropower Station.

The hydropower planning of the Dadu River mainstream has been reexamined and regulated since 2001, in order to promote the progress of preliminary work and development and settle on key cascade for research after the Pubugou Hydropower Station. The *Adjusted Report on the Hydropower Planning for the Mainstream Dadu River* was finished in July 2003, passed review in November 2003, and was approved by Sichuan Provincial Government in September 2004.

In consideration of complicated geological structure in part of the Dadu River reaches, large population and arable land along the river, deep overburden in the river bed and the impact of Chengdu-Kunming railway line, the following aspects of relationships must be handled well in cascade arrangement and development scheme of the Dadu River, which are the relationship between demand and likelihood, short-term and long-term, compensation benefits and reservoir inundation as well as residents relocation, cascade connection and environment protection.

Based on the original planning research, the adjusted report recommends a 22-cascade development plan. From the upstream to the downstream they are successively: Xiaerga, Bala, Dawei, Busigou, Shuangjiangkou, Jinchuan, Badi, Danba, Houziyan, Changheba, Huangjinping, Luding, Yingliangbao, Dagangshan, Longtoushi, Laoyingyan, Pubugou, Shenxigou, Zhentouba, Shaping, Gongzui (low dam) and Tongjiezi, with Xiaerga as the upstream controlling reservoir and Shuangjiangkou as main reservoir in the upstream, and Pubugou as controlling reservoir in the midstream. With a large storage capacity and competence of annual regulation, Shuangjiangkou Reservoir can benefit the downstream cascade hydropower stations. Annual regulation of the Dudu River's runoff can be realized by joint operation of Shuangjiangkou, Xiaerga and Pubugou reservoirs.

The cascade development plan recommended by the new scheme excels the original one in the following aspects: the overall planning of the cascade location in mainstream will benefit from extending planning scope, setting the Xiaerga Reservoir as the upstream controlling one that reduces the burden of the rest reservoirs; it saves the inundating Jinchuan County and vast tract of arable lands alongside by setting Shuangjiangkou Reservoir as the controlling reservoir in upstream instead of Dusong Reservoir; development of the river courses between Danba and Changheba are adjusted from cascade three to two, which makes the cascade location more reasonable; cascade adjustment from Lengzhuguan to Huangjinping in Guzan Town saves Guzan Town, the educational and cultural area in Ganzi Prefecture, from being inundated; the development of Yingliangbao, which has poor regional and engineering geological conditions, is adjusted from dam type to diversion-conduit type, making it more feasible to exploit resources in this region; the original Gongzui (high dam) plan is adjusted to 2-cascade development plan with Gongzui (low dam) and Shaping, which saves Ebian County from being inundated.

According to the new scheme, the total coverage of population living in inundated area is 36,776, with 2,289 hectares inundated arable land, 574km inundated highways, and 11km inundated railway, excluding Gongzui, Tongjiezi and Pubugou; no counties or big market towns will be inundated. Compared with the original one, there are 84,591 fewer people, 1,891 hectares fewer arable land, 2 fewer counties, 1 fewer town at county level, 1 fewer market town and 39km fewer railway being inundated. It can significantly reduce the losses of reservoir inundation and protect scarce land resources in the river basin. Negative impacts on the region's social environment have been notably diminished due to less inundation coverage. The new scheme sets apart a 50km long natural river course in upstream, midstream and downstream respectively, a good coordination between exploitation of hydropower resources and environment protection; meanwhile, it benefits the reproduction of the existing aquatic creatures.

5.2.3 Hydropower Planning in the Middle of Yalong River

Yalong River is the largest tributary of the Jinsha River. The hydropower planning for the Middle Yalong River ranges from

Heping Township (located in Xinlong County, backwater area of the Lianghekou hydropower station reservoir) to Kala Township (located in Muli County, backwater area of the Jinpin I hydropower station reservoir). The river planning reach is 385km long with a 980m natural drop. Initiated from 2002 and completed in June 2006, the said planning has been examined and approved by Sichuan Provincial Government.

The following aspects should be given due considerations in hydropower planning work of the Middle Yalong River. Firstly, balance should be maintained between high dam reservoir and residents living in inundated area. Thanks to a relatively smaller population and fewer arable lands in the middle Yalong River, fewer inundation losses will be brought by dam construction. However, most reclaimable land has already been developed, reserve arable land in this area is inadequate, which leads to a small relocation capacity and poses a big challenge to relocation. The inundation sensitive area in midstream is Yajiang County, of which the inundation condition has been investigated at the beginning of the planning. It turns out that 604 people are living in the inundated area if the normal reservoir water level in Yagen Reservoir reaches the elevation of 2,602m. Such level is determined in view of the small affected population. Secondly, balance the reationship between resources development and environment. In the beginning, it needs to be found out that whether there are natural reserves, scenic spots, important cultural relics, mineral resources, religious facilities in the planning area and efforts should be made to keep clear of them. Planning of development and environment must be coordinated and carried out correspondingly in order to get the planning approved smoothly. Thirdly, integrated exploitation requirement should be taken into account in development planning. Fourthly, dam sites with good topographical and geological conditions should be chosen as cascade development. There are many fracture structures in the midstream, possibly leading to numbers of and large scale of landslides. Disadvantages should be avoided in planning and cascade development located in better-conditioned river courses. Fifthly, due attention should be paid to choosing upstream controlling reservoir. Setting such reservoir in this reach plays a key role in increasing dry season output and power generation of numerous hydropower stations located in downstream Yalong River, Jinsha River, as well as Three Gorges and Gezhouba hydropower stations. The Lianghekou dam site, connecting the three rivers, forms a multi-annual regulation reservoir, through which due emphasis are paid to both upstream and downstream, exploitation of resources are optimized, overall benefits are maximized. Sixthly, make reasonable use of water head, pay attention to cascade connection and raise the scale of the cascade hydropower stations moderately. Emphasis should be attached to reducing the reservoirs inundation and protecting environment and the topographical and geological conditions of the dam sites; meanwhile, cascades should be connected to the fullest extent and precious water resources should be exploited reasonably. The hydropower stations should be constructed in as large scale as possible in respect that these stations in midstream are located in the national hydropower bases, which undertake the important task of from-West-to-East power transmission.

In accordance with planning principles mentioned above, especially new concepts in terms of land acquisition, resident relocation and environment protection, the development pattern and planning for the midstream has been studied repeatedly. A six-cascade development scheme including Lianghekou, Yagen, Lenggu, Mengdigou, Yangfanggou and Kala, with an installed capacity of 11,238MW, finally excels by technological and economic comparison with the other three hydropower cascade development schemes. The development in midstream usually adopts the dam type of damming, in view of the plenty water quantity and a slow slope of river averaged at 2.55‰. However, the river course between Lenggu and Menggushan features a steep slop, which is as high as 8.24‰ with a 105m drop in roughly 13km long river course, and a large curve, which resembles a "Ω" style river bend in downstream Menggushan with a 38m drop by curve cutoff. In addition, there is a high intensity of earthquake in this river course, with lots of and large-scaled landslides along the river. In the light of factors mentioned above, the development of Lenggu cascade combines lower barrage and water diverting tunnel, which utilizes the advantageous river geography and at the same time is more quake-proof as well as avoiding large landslides.

5.2.4 Hydropower Planning for the Section from Gushui to Miaowei in the Upper Reach of Lancang River

At the end of 2000, hydropower planning for the Upper Lancang River from Gushui (including the reservoir area) to Miaowei was conducted. This section reaches 1,036m in regards to water head and is as long as 489 km, with about 34km-long section in Tibet, 37km on Tibet-Yunnan border and 418km in Yunnan Province.

The development tasks of the planning section mainly lay in power generation, but also included coordinating water supply, irrigation, environmental protection, and soil and water conservation as well as promoting the development of tourism, navigation in reservoir area and aquaculture. After technical and economic comparison of the five cascade schemes, it was recommended to develop the "One-controlling-reservoir and Seven-cascade" program in the said section with the total installed capacity reaching 7,810MW. Four hydropower stations, Miaowei, Lidi, Huangdeng and Gushui, should be the projects preferentially getting started.

The planning attached great importance to environmental protection and ecological balance, fully weighed the compensation benefits of reservoirs as well as reservoir inundation and resident relocation, appropriately dealt with the relationship between hydropower development and tourism, reasonably handled the issue of connection with the cascade stations in upper and lower reaches, paid close attention to the selection of storage reservoirs, and pursued reasonable utilization of water power resources, as a result of which, the "One-controlling-reservoir and Seven-cascade" program put forward in the planning performed quite well in coordinating the relationships between different parties.

Coordinate the relationship between the World Natural Heritage Site of "Three Parallel Rivers", and hydropower development. The Guonian Hydropower Station, which was proposed at the beginning of the planning, is in the Baimang-Meri Snow Mountain Area (the core area), one of the World Natural Heritage Sites, while all the other cascade projects are not. The Planning Environmental Impacts Assessment suggested that this station should not be developed, and Wunonglong Reservoir, which lies in its lower reaches, should maintain pool water level without reaching the World Natural Heritage Site, "Three Parallel Rivers". To this end, the normal pool water level of Wunonglong Reservoir decreased from 1,943m to 1,906m in the recommended development program for the said section. Besides, the Guonian cascade was also listed as an unfavorably developed cascade, and the unexploited water head and natural watercourse reached 171m and 50km respectively.

Fully balance the compensation benefits of reservoirs and inundation losses. Reservoir inundation in the planning area mainly occurs in the section from Wunonglong to Tuoba, along which stand four townships and towns, which are Yezhi, Kangpu, Baijixun and Zhonglu, as well as plenty of farmland. Tuoba cascade has excellent reservoir-forming conditions, so a large storage capacity would be achieved as a result of a high dam scheme. But vast fields in the four townships and towns mentioned above would be submerged. So out of the consideration of protecting valuable cultivated land and alleviating resident relocation difficulties, Tuoba low dam scheme with 65m in water head unutilized was recommended. A natural water course with a length of 37km from Kangpu to Badi, Yezhi - "Granary of Diqing Prefecture", and the township government locations of Kangpu, Yezhi, and Zhonglu were reserved, which meant a lot to the economic development of Diqing Prefecture and Weixi County.

Explore reasonable connection between upstream and downstream cascades. In Gushui high dam scheme, the upstream connecting cascade is the Guxue cascade in Tibet, while in Gushui low dam scheme, it is the Biyonggong cascade on the Tibet-Yunnan boundary reach of Lancang River. To properly handle matters concerning reservoir inundation of trans-provincial rivers and reasonable allocation of water power resources so as to achieve the goal of reasonably utilizing water power resources, a comparison was made between one-cascade development of Gushui high dam program and two-cascade development of Gushui low dam and Biyonggong cascade. Although Gushui low dam development scheme is better than the high dam from the aspect of construction workload and duration, the high dam scheme excels in distribution of key hydraulic structures and economy. In addition, the high dam scheme could provide substantial regulation and compensation benefits for the said section and would play an important role in improving the electric energy quality of the cascade hydropower station on the said section. So Gushui high dam scheme is recommended in the planning report.

Pay close attention to the setting and selection of storage reservoirs. There are two multi-year regulating storage reservoirs in the middle and lower reaches of the Lancang River, Xiaowan and Nuozhadu, which yet can not control the runoff of the section from Gushui to Miaowei in the upstream. So it is necessary to set up a main reservoir in the said section on account of cascade compensation benefits and power transmission requirement. After analysis of the reservoir-forming conditions of each cascade, three cascades were primarily considered as suitable to construct controlling reservoirs, i.e. Gushui, Tuoba, and Huangdeng. With a view to the poor geological and not suitable conditions for Huangdeng high dam scheme and in consideration of the only 170m of water head which could be utilized in the section below Huangdeng, accounting for 16% of the planning section, Tuoba high dam, Gushui high dam and Tuoba high dam plus Gushui high dam were proposed for making a comparison. The comparison revealed that although the two-reservoir scheme could provide the power with better quality, the hydrogeological conditions in Tuoba high dam scheme are quite complicated with deep unloading zonation and great difficulties in handling. Moreover, Tuoba high dam scheme would submerge 43% of the high yield farmland in the five townships and towns in Weixi County, Diqing Prefecture -Yezhi, Kangpu, Baijixun, Pantiange, and Zhonglu, which would result in the loss of substantial inundation and the difficulties in resident relocation. So after analysis of various indicators such as dam and reservoir construction conditions, reservoir inundation, compensation benefits, investment, etc., it is recommended to set up only one storage reservoir in Gushui in the planning section.

5.2.5 *Hydropower Planning on Other Major Rivers*

Under the arrangement of SDPC, the *Hydropower Planning for the Middle and Lower Reaches of the Nujiang River* was worked out in 2003 and was reviewed by NDRC on August 13, 2003. On September 30, 2003, NDRC issued the Letter of the General Office of the National Development and Reform Commission on Seeking for Opinions on the Evaluation Comments on Hydropower Planning for the Middle and Lower Reaches of the Nujiang River (Draft) (the General Office of NDRC [2003] No. 987) to ask for the comments of various organizations. On December 14, 2004, NDRC and the former State Environmental Protection Administration (SEPA) made a joint review of *the Environmental Impact Assessment Report of the Hydropower Planning for the Middle and Lower Reaches of the Nujiang River*. For a variety of reasons, the hydropower development in the lower Nujiang River faced a great number of controversies, and thus the planning has not been approved. According to the Hydropower Planning for the Middle and Lower Reaches of the Nujiang River, it was recommended to adopt the "Two-controlling-reservoir and Thirteen-cascade" development scheme in the middle and lower reaches of the Nujiang River with a total installed capacity and average annual energy generation amounting to 21,720MW and 103.42TW · h respectively.

In accordance with the preparatory hydropower work schedules of NDRC, hydropower planning for the upper reach of Jinsha River was carried out in 2004; planning for the middle reach of Yarlung Zangbo River and for the upper reach of Lancang River (the section in Tibet) in 2005; planning for the upper reach of Nujiang River in 2006; planning for the section from Hukou to Erduo in the Upper Yellow River in 2007. Until now, the hydropower planning reports and environmental impact assessment reports of these rivers have been completed, but have not yet been reviewed.

In line with the planning, the 13-cascade development scheme with Gangtuo as the upstream controlling reservoir was recommended for the upper reach of Jinsha River, in which the total installed capacity and average annual energy generation would reach 13,920MW and 64.23TW · h respectively; the 11-cascade development scheme was recommended for the Middle Yarlung Zangbo River, with the total installed capacity and average annual energy generation amounting to 4,450MW and 20.537TW · h respectively; the 7-cascade development scheme was recommended for the Upper Lancang River (the section in Tibet), with 5,850MW of total installed capacity and 29.57TW · h of average annual energy generation; the 12-cascade development scheme was recommended for the Upper Nujiang River, with the total installed capacity and average annual energy generation amounting to 14,630MW and 75.23TW · h respectively; the 10-cascade development scheme was recommended for the section from Hukou to Erduo in the Upper Yellow River, with 4,283MW of total installed capacity and 18.866TW · h of average annual energy generation.

Almost all these rivers (sections) are located in Tibetan area in Western China which is characterized by high altitude, poor traffic and bad climatic conditions, low living standard, and lagging economic development. People in this region have strong religious belief and different habits and customs from those living in other hydropower planning areas in China. So this region is equipped with distinct ethnic characteristics. In the process of hydropower planning for various rivers, we should consistently adhere to the policy of "Unified planning, Integrative utilization, Protecting ecosystems, Seeking for benefits, and Comprehensive Consideration", take into consideration the local economic and social development reality as well as the public demands, and draw up the planning program in the light of the principle of reasonable and harmonious development.

1. Keep consistent with power development planning and other industrial planning

All these rivers are located in the region with low-level economic and social development, so in order to guarantee local power supply for economic development and with a view to the national demands for power development, it is necessary to fully consider domestic demands and power transmission requirements while mapping out the cascade scale. For example, the hydropower resources in the Middle Yarlung Zangbo River have been defined as supplying power for the power grid of central Tibet. Out of the consideration of the power demands of central Tibet power grid, medium-sized hydropower station should be taken as the main choice while determining the development size. In the hydropower planning for the Upper Lancang River, it is scheduled to construct several medium-sized hydropower stations with the capacity of 120MW in the section near to Changdu. But as the power source of Changdu and for the purpose of meeting the demands of long-distance power transmission, all other hydropower stations in this section are equipped with the capacity of over 1,000MW. Since the reaches of the Yellow River above Maqu is characterized by poor natural conditions, undeveloped traffic, and lower living standards, and some counties have not yet been included into a large power grid, it is scheduled to construct six cascade hydropower stations in the reaches above Maqu in the hydropower planning, i.e. Tageer, Guancang, Saina, Mentang, Tajike I, and Tajike II. The construction of these stations and their supporting power grids will satisfy the local power demands step by step. In view of the lagging local economic and social development, it is necessary to make the hydropower planning combined with local economic and social development, take eco-environmental protection, helping settlers to shake off poverty and set out on a road to prosperity, and promoting economic and social development as the development tasks of hydropower planning, and carry out these tasks in future hydropower development. Besides, hydropower planning should also be coordinated with the special planning of other industries. For example, in the hydropower planning for the Middle Yarlung Zangbo River, the planning section would involve the railway from Lhasa to Shigatse and the road from Lhasa to Nyingchi, so the hydropower planning department specially coordinated with relevant departments of railways and roads.

2. Set up the upstream controlling reservoirs with excellent regulation capacity

According to the water resources distribution characteristics and the uneven annual runoff distribution of the planning reaches, setting up an upstream controlling reservoir in the planning section, which is used to regulate the runoff, will play a favorable role in improving the power quality of each cascade. For instance, there is a great gap between the runoffs of the Upper Jinsha River in wet season and dry season, and the average annual runoff in dry season occupies only 10.4% to 13.3% of the annual runoff, while the basic tendency is the lower the reach is, the higher ratio it gets. So in accordance with the runoff characteristic of the said section, a storage reservoir was set up in Gangtuo with the storage capacity amounting to 3.2 billion m^3, and through the compensation effect of the main reservoir, the electric energy generation of the cascade in level period showed a 52% increase comparing to that in last year when the main reservoir has not yet been built. For another example, in the planning for the Upper Nujiang River, the Reyu River section and the Tongka River section with slight slope possess the topographic conditions for large storage capacity, and meanwhile, there are no important inundation constraints in the reservoir area, so the initial choice was to constructannual storage reservoirs in two dam sites-Reyu and Tongka. Reyu dominates 97% of the total water head and 60% of the annual runoff of the planning section, and Tongka dominates 76% and 87% respectively. The aggregate regulated storage capacity of the two reservoirs comes to 4.5 billion m^3, and with the regulation of the two reservoirs, utilization of water resources and quality of electric energy would be remarkably improved. There are much more such examples as setting up a upstream controlling reservoir which is Ningmute Reservoir in the section from Hukou to Erduo in the Upper Yellow River, setting up Rumei Reservoir in the Upper Lancang River, etc.. Thanks to the regulation of these reservoirs, the electric generation of the cascades in lower reaches in dry season has been ameliorated a lot.

3. Pay close attention to reservoir inundation and ethnic culture protection

These rivers are situated in the regions inhabited by the minority nationalities in China, such as Qinghai, Sichuan, Tibet and Yunnan, where there are deficient land reserve resources, limited resettlement capacity and the difficulties in immigrants resettlement. In hydropower planning, it should be avoided to the greatest extent that reservoir inundations have an influence on the section with good living environment and rich farmland. For those areas with a certain amount of residents and intensive farmland as well as the religious facilities along the planning section, cascade stations should be laid out with them as the controlling factors.

In the hydropower planning for the Upper Jinsha River, the scheme of setting up a large reservoir in E'nan was abandoned for the purpose of avoiding inundation impacts on Luoxu Town as well as the religious facilities and open valley land near the town; the scheme of setting up a reservoir in Yanbi was abandoned for avoiding inundation impacts on the National Highway 317 as well as the intensive residential areas and farmland along the section from Gangtuo to Yanbi; the program of integrating Boluo with Yebatan was abandoned for avoiding inundation impacts on Boluo Township, Boluo Temple as well as the residents and farmland in high elevation regions along the section. Through reasonable distribution of hydropower planning in the Upper Lancang River (the section in Tibet), it steered away from the section with intensive residents and farmland such as the 10km-long section below Cege with quite intensive villages, Kagong Township which is 85km away from Changdu County seat, and Zhuka Township of Rumei Town, Mangkang County.

In the initial planning for the upper reach of Nujiang River, the topographical and geological conditions of the Xinrong dam site are quite suitable for high dam construction, but E'xi Township in the reservoir area is a comparatively large township with intensive population and farmland and a fairly big temple. So in order to avoid the impacts on E'xi Township and the temple and alleviate the reservoir inundation at the best, the planning abandoned the development of a 26km-long section near E'xi Township and left about 35m of water head unemployed but kept about 3,700 residents and 733.3 hectares of land from inundation. The Zhonglinka Township section in the middle of the planning section is another example with intensive population and farmland. The planning also abandoned the development of a section with the length of 33km and left about 65m of water head unemployed but kept about 3,500 residents and 233.3 hectares of land from inundation.

Guancang and Mentang were originally selected to be dam sites in the planning for the Upper Yellow River, as they possessed the topographical and geological conditions of constructing high dam and this project would be quite economical if it follows the one-cascade high dam development plan. But the reservoir inundation loss would be great by which the Mashijia Temple and the Ganglong Township government in Gande County as well as the Mentang Township in Jiuzhizhi County would be submerged. For the purpose of decreasing loss, the Guancang cascade was changed into two cascades, Tageer and Guancang, and the Mentang cascade was changed into Saina and Mentang.

For these regions, ethnic and religious culture protection is also an issue demanding for much attention in the hydropower planning. Hydropower planning should always hold the principle of fully respecting local religious belief and avoiding the impacts on major ethnic and religious facilities to the greatest extent. In the process of hydropower planning for the Middle Yarlung Zangbo River, Bayu dam site was found to possess the topographical conditions of constructing a high dam and a storage reservoir on a certain scale was actually needed for the Middle Yarlung Zangbo River. But for the sake of not affecting the Shannan Prefecture Samye Monastery and the vast broad valley, the planning abandoned the scheme of setting up a high dam in Bayu cascade and lowered the normal pool water level; in order to avoid the impacts on the Chongkang Villa in Zilong Township, Lang County, the planning reduced the size of the Langzhen cascade. Similar situation also happened to the hydropower planning for the Upper Jinsha River—for the purpose of not impacting the important religious facility Boluo Temple near Boluo County seat in Changdu region, the planning replaced the one-cascade development program with a two-cascade development scheme (Boluo and Yebatan).

4. *Attach great importance to eco-environmental protection, and try to avoid or reduce the impacts on environment-sensitive objects*

The hydropower planning should adhere to the idea of harmonious development, i.e. "development in the protection and protection in the development", make the hydropower development in harmony with the economy, society and eco-environment of related basins, try to avoid or reduce the inundation impacts on nature reserves and scenic resorts, and take environmental impacts as an important controlling factor of drawing up and selecting cascade programs.

In the hydropower planning for the upper reach of Jinsha River, the scheme of having the normal pool water level of Suwalong cascade, which is back to the tailwater of Lawa cascade, was gave up for avoiding the inundation impacts on the Zhubalong Nature Reserve; similarly, in order to avoid the inundation impacts on Rhinopithecus Bieti National Nature Reserve, another abandoned scheme is integrated Xulong cascade into Changbo cascade with the backwater to Suwalong cascade. While giving precedence to environment-sensitive objects, the hydropower planning for the Upper Jinsha River reserved a natural watercourse with the length of 73km. To avoid the impacts on the eco-function reserve and the Qomolangma National Nature Reserve at the source of the Yarlung Zangbo River as well as the Grus Nigricollis Nature Reserve, the GyaidE'xiu National Forest Park, and some important towns and facilities in the middle reaches as a controlling factor, the cascade development scheme recommended by the hydropower planning for the Middle Yarlung Zangbo River performed quite well in coordinating the relationship between hydropower development and environmental protection. Besides, this planning also reserved three natural watercourses, i.e. the sections from Lizi Village to Nuda Village, from Kare to Woka estuary, and from Lang Town dam site to Paizhen Town, the total length of which amount to 1,158.3km. The section under the hydropower cascade development only takes up 15.4% of the whole planning section in the Middle Yarlung Zangbo River which comes up to 1,369km. From the perspective of paying close attention to eco-environmental protection, a 640km long natural watercourse is also reserved in the hydropower planning for the section from Hukou to Erduo in the Upper Yellow River, which occupies 51% of the total length of the planning section.

In hydropower planning, it is necessary to fully consider the possible results on water reduced section by diversion-type and mixed-type development, as well as natural characteristics of rivers so as to draw up a development program in a reasonable manner. For those sections with slight slope and great runoff, dam-type development should be adopted; while for those sections with steep slope, diversion-type or mixed-type development program is economically better than the dam-type, but out of the consideration of water reduced sections, dam-type development is still recommended if there are little difference between them from the aspect of economical efficiency.

5. *Pay close attention to project safety, and construct dams in sections with sound topographical and geographical conditions*

The Ganzi-Yushu fault zone in the Upper Jinsha River, the Jinsha River fault zone, and most of the Yarlung Zangbo River deep fault zone are passing by along the sections under planning. The regional geology there is not quite good and in some sections, harmful geo-physical phenomena occur, so the selected cascade dam sites should be away from or avoid active fault zone to the fullest extent so as to reduce the impacts of earthquake on the project safety. Some sections even have to decrease the scale of the cascade project. In view of the complicated geological conditions of the planning section and domestic advanced technologies of hydropower development, the hydropower planning for the Upper Nujiang River takes the principle that the height of the largest dam in each cascade station is not supposed to exceed 300m as one of the preconditions of drawing up the cascade program, for the sake of lowering the project risks.

5.3 Research Subjects

With the reform and opening-up, in accordance with such principles as separating government functions from enterprise management, regarding the province as an entity, interconnecting power grids, dispatching centrally and collecting fund for electricity construction, as well as adjusting measures to local and grid conditions, collecting fund for electricity construction through multiple channels and enterprises has been practiced in power industry. With the transition of our economic system from the planned economy to the market economy and deepening of the reforms in planning, investment, finance, fiscal revenue, etc. , a series of major reforms has been launched in power industry, and a system of holding the legal person responsible for projects has been gradually implemented and project capital system has been established. But many problems have also occurred in the development of power construction. One of important aspects is structural imbalance. Influenced by such factors as fund, market and building cycle, hydropower development has become increasingly difficult and the proportion of hydropower in electricity system has witnessed a decline, which is not only extremely unfavorable to rationally readjusting our energy structure to optimize the allocation of energy resources and improve the environment, but also contrary to the policy of developing the hydropower vigorously. Determining which policy to adopt requires in-depth investigation and study so that hydropower development planning can be put forward and referenced by the relevant decision-making bodies of the country.

In order to promote the development of hydropower and study relevant measures and policies, special studies regarding hydropower development macroscopic planning and hydropower policies had been carried out, and problems existing in the hydropower development had been investigated during the periods of the Ninth Five-Year Plan and the Tenth Five-Year Plan. Focusing on the main problems and integrating the power industry system reform, these studies and investigations aimed to discuss the solutions and measures to promote hydropower development. As a result, a series of accomplishments were achieved and some were even adopted by SDPC and former State Electric Power Corporation. The major research subjects are the existing problems in hydropower development and countermeasures; policies and measures to strongly develop hydropower; policies to promote hydropower development and inter-province power investment; monographic study on macroscopic planning of hydropower development; electricity generation end market's two-system power price and pumped-storage power stations' economic evaluation methods; pilot study on inter-basin and trans-regional hydropower compensation benefit; compensation benefit of Longtan Hydropower Station for Three Gorges Power Station; China's from-West-to-East power transmission planning for the next 30 years; hydropower cascade compensation benefit and distribution policy; review and summary of social benefits of the existing hydropower stations—analysis on social benefits of the existing hydropower stations in the Upper Yellow River, etc.

In 2002, with the introduction of electric system reform plan, a power market featuring separation of the operations of power plants and power grids and bidding for access to power grids has gradually taken shape. As hydropower is related to the nation's energy policy and optimal allocation of energy resources, the adjustment of power structure, from-West-to-East power transmission strategy, comprehensive utilization and optimal allocation water resources, resource development and eco-environmental protection, reservoir inundation and relocation policies, as well as investment scale, economic and fiscal policies, the development of hydropower is inseparable from national macro-control and policy support. Especially in times of further reform of China's electric power system, state macro-control can assure the orderly, coordinated and sustainable hydropower development. With the deepening of electric power system reform, hydropower development is facing some new problems. A series of research subjects carried out after 2,000 have provided suggestions for the integrated departments and decision-making departments and played an important role in the further hydropower development.

5.3.1 *Research on the Hydropower Development and Transmission Planning in Sichuan Province*

With its theoretical reserves ranking second and technically exploitable installed capacity ranking first nationwide, Sichuan is rich in hydropower resources, which occupies an important strategic position among China's energy resources. In line with the strategic arrangement of the development of the west region and from-West-to-East power transmission, the hydropower in Sichuan should be developed with great efforts. To further enhance the directive function of the planning, recommend Sichuan's time arrangement for hydropower development in the Eleventh Five-Year Plan period and 2020, market consumption and power outward transmission scale in every target year, include the planning into national construction plans and make Sichuan's hydropower resources development more reasonable, orderly and sustainable, and Hydropower and Water Resources Planning and Design General Institute carried out a study on the hydropower development and market consumption planning of Sichuan province in 2004. The main research achievements are as follows.

(1) Speeding up hydropower development in Sichuan, making proper use of hydropower resources and implementing the strategy of power transmission from Sichuan can make hydropower generating capacity of Sichuan replace the consumption of power generation fuel in energy-deficient areas, save precious energy resources and optimize energy structure. It is in line with the strategy of sustainable energy development of our country and can contribute to sustainable economic and social development. Accelerating the development of hydropower resources in Sichuan province, implementing from-West-to-East power transmission strategy and complementing with the Northwest China Grid in terms of thermal power and hydropower

is in compliance with the strategy of optimal allocation of energy resources to promote the networking in the whole nation.

(2) Combining long-and short-term development in an orderly manner to make sure that the installed hydropower capacity in Sichuan in 2015 and 2020 will reach 52,980MW and 61,200MW respectively (excluding the hydropower in the mainstream of Jinsha River).

(3) Making overall planning for east and west regions, combining station-to-grid with grid-to-grid and dovetailing supply with demand features to achieve a target transmission scale of 25,000MW from Sichuan Province before 2020. Giving consideration to power supply-demand relationship in Sichuan, it is recommended the target scale of Sichuan power outward transmission be 25,000MW which will be accomplished between 2015 and 2020 (see Table 5.3-1).

Table 5.3-1 Regional Capacity and Generation Distribution of Outward-transmited Sichuan Hydropower in the Various Design Level Years

Design level year	Transmission end region	Transmission capacity (MW)	Annual generation (TW·h)	Dry season (from Dec. to next Apr.) energy output (TW·h)
2010	Chongqing	4,300	21.761	4.317
	Four Provinces in Central China	5,700	28.533	5.258
	Northwest Grid (Interchanged)	750MW/1,500MW (5h at Peak/other time intervals)		
	Subtotal	10,000	50.294	9.575
2015	Chongqing	7,200	41.078	11.345
	Four Provinces in Central China	9,620	51.564	13.309
	East China Grid	7,400	35.849	7.320
	Northwest China Grid (Interchanged)	1,500MW/3,000MW (5h at Peak/other time intervals)		
	Subtotal	24,220	128.490	31.974
2020	Chongqing	7,200	44.083	14.390
	Four Provinces in Central China	10,400	57.745	17.150
	East China Grid	7,400	38.670	11.258
	Northwest China Grid (Interchanged)	1,500MW/3,000MW (5h at Peak/other time intervals)		
	Subtotal	25,000	140.498	42.798

(4) As Sichuan's hydropower resources occupy an important strategic position in national energy resources, it shall be integrated into the equilibrium process of national primary energy resources. Energetic efforts shall be made to include the plan of Sichuan's hydropower development sequence and its power outward transmission plan for every target year into national construction plans in order to make the hydropower development more rational, orderly and sustainable so that it can make contributions to the balance of national energy resources.

5.3.2 Research on the Hydropower Development and Transmission Planning in Yunnan Province

Ranking third nationwide with both theoretical reserves and technically exploitable installed capacity, Yunnan is rich in hydropower resources, which occupies an important strategic position among China's energy resources. In line with the strategic arrangement of the development of the west region and from-West-to-East power transmission, the hydropower in Yunnan should be developed with great efforts. To actively support and guide the orderly hydropower development and optimal allocation of energy resources in Yunnan, to achieve sustainable development, and to implement relevant policies of the development of the west region issued by the State Council, it is urgently required to conduct in-depth research on the hydropower development and outward transmission planning of Yunnan Province to present a hydropower development plan that can not only gratify its own economic development, but also ensure a long-term stable hydropower transmission scale. In addition, it is also imperative to integrate Yunnan hydropower resources into the equilibrium process of national energy resources and propose proper outward transmission scale, transmission direction and transmission mode. Therefore, HYDROCHINA CO. established the Research on the Hydropower Development and Power Outward Transmission Planning of Yunnan Province as its technical program. The main research achievements are as follows.

(1) Speeding up hydropower development in Yunnan, making proper use of water resources and implementing the strategy of power outward transmission from Yunnan can make hydropower generating capacity of Yunnan replace power generation fuel consumption in energy-deficient areas, save precious energy resources and optimize energy structure. It is in line with the strategy of sustainable energy development of our country and can contribute to sustainable economic and social development.

(2) Combining long-and short-term development in an orderly manner to make sure that the installed hydropower capacity in Yunnan in 2015 and 2020 will reach 52,110MW and 83,330MW respectively.

(3) Make overall planning for East and West regions, combining station-to-grid with grid-to-grid, and dovetail supply with demand features to achieve a target transmission scale of 60,000MW from Yunnan province before 2020 (see Table 5.3-2).

Table 5.3-2 Combination of Major Hydropower Stations and Power Outward-transimssion in Yunnan

Item	Transmission capacity (MW)	Target Year
Northwest Kunming direct current channels	5,000 (Xiaowan 2,100, Other 2,900)	2010
Nuozhadu, Small and middle-sized hydropower stations, and Daying River fourth cascade hydropower station	5,700 (Nuozhadu 3,000, the total of Small and middle-sized hydropower stations and Daying River fourth cascade hydropower station 2,700)	2015
Guanyinyan, Ludila and Longkaikou	5,400 (Yunnan's share)	2015
Xiluodu and Xiangjiaba	9,300 (Yunnan's share)	2015
Baihetan	6,000 (Yunnan's share)	2020
Longpan, Ahai and Liyuan	6,000 (Installed capacity 8,600, 2,600 in Yunnan)	2020
Maji, Wunonglong and Tuoba	6,300	2020
Saige, Yabiluo, Miaowei, Huangdeng and Lushui	8,000	2015
Nuozhadu and Jinghong	3,000	2015
Other	5,000	

(4) Based on the research and analysis of Tibet power outward transmission, the primary energy, market demand and power development of five countries of Indo-China Peninsula (i.e. Viet Nam, Lao PDR, Cambodia, Thailand and Myanmar), it is proposed that the transmission target scale may remain long-term sustainability after connecting power transmission from Tibet and introducing power from Laos and Burma.

5.3.3 Research on the Hydropower Development and Transmission Planning in Southeast Tibet

With both its theoretical reserves and technically exploitable installed capacity ranking first nationwide, Tibet Autonomous Region is rich in hydropower resources, which occupy an important strategic position among China's energy resources. In line with the strategic arrangement of the development of the west region and from-West-to-East power transmission, it should be made great efforts to develop the hydropower in Tibet Autonomous Region. To implement relevant requirements on the Development of the West Region and economic and social development of Tibet Autonomous Region made by the Central Committee of CPC and the State Council, in-depth research regarding the hydropower development in Tibet Autonomous Region and outward transmission planning should be conducted. This research project, which started in 2007, mainly includes preliminary program study on hydropower energy base development in southeast Tibet as well as outward transmission planning, analysis of the hydropower base's transmission capacity in southeast Tibet in every target year, analysis of the power transmissing target market and power price bearing capability, etc. The research aims to clearly define the hydropower energy base development in southeast Tibet, guide its orderly development and smooth advancement, and provide basis for the State's formulation of energy strategic planning.

The hydropower development in Tibet Autonomous Region should first consider gratifying its own power development, and then study Tibeten hydropower outward transmission. Taking Tibet's relatively small power demands into consideration, power generated from hydropower stations along the main streams of rivers and major tributaries within Tibet would be mainly outward transmitted. Principles combining connection and long-distance direct outward transmission should be followed. Connection mainly makes use of the existing Sichuan and Yunnan power transmission channels, which makes it possible for hydropower in southeast Tibet to continue and maintain the outward transmission scale of Sichuan power and Yunnan power. To facilitate social and economic development of Tibet Autonomous Region and realize the transformation from resources advantage to economic advantage, some cascades can be selectively banded and directly outward transmitted over a long distance. The first hydro generating unit in the lower Yarlung Zangbo River will start to operate in around 2036. Due to factors such as the great installed scale along this section and broad power supply scope, the sustainability of Sichuan power and Yunnan power outward transmission scale after 2036 can be ensured by developing the downstream cascade of the Yarlung Zangbo River. Before 2035, the upper reaches of the Jinsha River and Nujiang River are used to continue and maintain the target competence of Sichuan power and Yunnan power outward transmission. The main research achievements are as follows.

1. Conception of hydropower development and construction in southeast Tibet

To facilitate the sound and smooth socio-economic development of Tibet Autonomous Region, the Fifth Tibet Work Conference listed six hydropower projects with a total installed capacity of 12,870MW, i.e. Yebatan, Lawa and Suwalong along the Upper Jinsha River, Rumei and Gushui along the Upper Lancang River and Songta hydropower station along the Upper Nujiang River, all of which aim to start construction during the 12th Five-Year Plan period. With regard to other hydropower projects of hydropower energy base in southeast Tibet, conception of development and construction is drafted on the basis of the combination of connection and long-distance direct outward transmission, considering the preliminary work progress and comprehensive analysis of the rivers' geographical and transportation conditions.

The three cascades, Benzilan and Xulong on the Sichuan-Yunnan boundary reach and Changbo on the Sichuan-Tibet boundary reach, in the planning for the Upper Jinsha River, boast an aggregate installed capacity of 5,160MW. Therefore, long-distance power outward transmission can be considered. The group transmission time for the three cascades is envisioned to be from 2020 to 2025. Cascades such as Boluo, Batang and Gangtuo can be developed at the proper time depending on the progress of Sichuan power outward transmission and used as the supplementary power source. The development time is decided by when Sichuan power needs to be transmitted outside.

Since the preliminary work for the cascades in the Upper Lancang River is solid and fast, and the three cascades' scale is moderate, direct long-distance group outward transmission is appropriate, that is, the three cascades Banda, Rumei and Guxue would be directly transmitted over a long distance with a scale of 5,510MW. The transmission time is tentatively selected from 2020 to 2025.

Located in Tibet's border with Yunnan Province, the preliminary work of Songta cascade in the Upper Nujiang River is well grounded. The moderate scale of Songta cascade and E'mi cascade that joins with Songta upstream makes it suitable for direct long-distance outward group transmission with the scale of 5,600MW. According to the geographical location of the Upper Nujiang River cascade, Luola and Angqu cascades near Yunnan Province can be connected with outward transmission of Yunan electricity with the scale of 3,050MW. Cascades of Tongka, Kaxi, Nujiangqiao and Lalong can be connected with outward transmission of Sichuan electricity with the scale of 6,350MW. Duration of outward transmission is up to the connection time required by Yunnan electricity and Sichuan electricity, and connection power supply with relatively favorable exportation conditions can be developed with priority.

Till 2050, the eastward transmission scale of southeast Tibet hydropower will reach 104,090MW (hereinto Tibet Autonomous Region outward transmission capacity is 95,650MW), among which eastward transmission of the Lower Yarlung Zangbo River is 68,800MW, eastward transmission of the Upper Nujiang River is 17,000MW, eastward transmission of upstream Lancang River is 5,510MW, and that of the Upper Jinsha River is 12,780MW (hereinto the eastward transmission capacity of Tibet Autonomous Region is 4,340MW).

2. Transmission end market of southeast Tibet hydropower

According to preliminary analysis of consumption capacity of receiving end market for southeast Tibet hydropower, East China, four provinces in east-central China (i.e. Provinces of Henan, Hubei, Hunan and Jiangxi), Guangxi and Guangdong boast certain market space. In particular, thanks to the relatively developed social and economic status in East China and its thermal power grid, substitution rate of southeast Tibet hydropower to East China grid is relatively high with broad market space.

In order to improve the overall consumption economy of the southeast Tibet hydropower outward transmission, it is suitable for the southeast Tibet hydropower, as connection power supply for southwest hydropower, to be consumpted nearby with Sichuan, Yunnan, Chongqing, and southwest Guizhou as priorities, for a steady and sustainable outward transmission of southwest hydropower. If southwest region can not consume southeast Tibet hydropower transmitted, it can be transmitted to electricity shortage area, such as four provinces in east-central China, Guangdong and Guangxi as substitute for coal-generated power.

3. The contributions of southeast Tibet hydropower base development to social and economic development of Tibet Autonomous Region

The development of southeast Tibet hydropower base can not only contribute to energy saving, emission reduction and environmental protection, but also to the rapid development of Tibetan economy and social stability. After the full operation of southeast Tibet hydropower base, raw coal will be saved by nearly 150 million tons annually, the substitution for thermal power will be equivalent to reducing carbon dioxide emission by 240 million tons annually, reducing sulfur dioxide emission by 2.3 million tons annually, reducing dust release by 48 million tons annually, and waste residue of 12 million tons annually, which will be positive and effective to the improvement of national air quality.

5.3.4 Research on the Western Hydropower Development and Eastward Transmission Planning

Rich in hydropower resources, the west region of our country includes 12 provinces (autonomous regions and municipalities), i.e. Yunnan, Guizhou, Sichuan, Chongqing, Shaanxi, Gansu, Ningxia, Qinghai, Xinjiang, Tibet, Guangxi and Inner Mongolia. Its technically exploitable installed capacity is more than 480GW, accounting for more than 80% of the country's total. Its power demand, however, is relatively low. By the end of 2007, the installed capacity was no more than 100GW, and utilization rate of hydropower development was less than 20%. To utilize our nation's energy resources in a rational and orderly manner requires speeding up the development of clean and renewable hydropower resources. According to China's energy development strategy, large hydropower bases in the west region will be built in an orderly manner with its total installed capacity standing at 300GW in 2050.

HYDROCHINA CO. carried out researches on the hydropower development and outward transmission planning in Sichuan, Yunnan and Tibet respectively, in 2006, and raised the subjects again to comprehensively study such issues as development time arrangement, electricity market, supply and demand relations and the continuation of from-West-to-East power transmission in order to coordinate the relationship between the hydropower development in the whole west region and the eastward transmission planning, give guidance on the planning of west region's hydropower development and optimal allocation of energy resources, and provide accurate reference data for establishing a proper power grid structure of our nation.

The research work shall regard the Scientific Outlook on Development as its guideline, scientifically analyze the status of China's socio-economic development and industrial restructuring, study the reasonable trend of delivery ends and receiving ends' electricity demand in the west region, study the appropriate hydropower development sequence of the west region; conduct in-depth research on hydropower delivery and receiving markets in the west region, and analyze power market space and marginal demand features; coordinate the work of power transmission from west to east and the continuation of eastward transmission, put forward its target scale; follow the reasonable flow of China's primary energy and the objective law of electricity development, give positive consideration to the new technology of power grid, bring forward the suggestion on from-West-to-East power transmission; put forward recommendations on the distribution proportion of generating capacity in service areas based on the relevant regions' actual situation of power grid and the requirement on optimal allocation of energy resources; calculate reasonable electricity price for power transmission and analyze its competitive advantages based on hydropower consumption plan of the west region and research findings of power transmission planning. The main research results are as follows.

1. Speeding up hydropower development in the west region is of great significance

The technically exploitable installed capacity of China can be 583,900MW, in which the west region accounts for more than 80%. Speeding up hydropower development of the west region and eastward transmission and taking measures to utilize energy resources in a rational and orderly manner, implementing energy conservation and environmental protection and developing the western economy are integrated parts of the western region development strategy.

2. The total installed capacity of western hydropower can reach 383,000MW

The total installed capacity of western hydropower in 2020 and 2050 will reach 236,000MW and 383,000MW, respectively, and its hydropower exploitable rates are respectively 53.7% and 87.2% of the technically exploitable amount (see Table 5.3-3).

Table 5.3-3   Summary of Research Results of Western Hydropower Development   Unit: MW

Year Regions	2015	2020	2030	2040	2050
Jinsha River Boundary	11,800	39,100	58,620	58,620	58,620
Sichuan Province	57,075	76,450	85,230	90,000	94,000
Yunnan Province	49,660	66,540	77,280	77,280	77,280
Guizhou Province	18,100	18,700	18,700	18,700	18,700
Tibet Autonomous Regison	1,400	7,000	22,160	55,010	98,810
Northwest Regions	23,590	28,220	35,780	35,780	35,780
Total	161,625	236,010	297,770	335,390	383,190

Note: The total installed capacity of Sichuan, Yunnan and Tibet excludes the hydropower of provincial boundary reaches of the Jinsha River.

3. The total target scale of from-West-to-East power transmission is from 135,000MW to 160,000MW

The total target scale of from-West-to-East hydropower transmission is from 135,000MW to 160,000MW, the main reason for the range of variation is the uncertainty of Tibet hydropower development, and the other reason is the uncertainty of the power transmission scheme of the middle and lower reaches of the Jinsha River; the total target scale of 135,000MW will be reached in around 2030 (see Table 5.3-4). In addition, the target scale of electricity input from five countries of Indo-China Peninsula is estimated to be 20,000MW and electricity transmission from Yunnan to Thailand is about 3,000MW.

Table 5.3-4 Summary of Analysis Results of From-West-to-East Hydropower Transmission Target Scale

Item	Target Development Scale (MW)	Target Transmission Scale (MW)
Total Scale	350,000 – 380,000	135,000 – 160,000
In which: Jinsha River Boundary	58,620	42,000 – 48,600
Sichuan Province	94,000	25,000
Yunnan Province	77,280	38,000
Guizhou Province	18,700	3,600
Tibet Autonomous Regsion	78,800 – 98,800	70,000 – 90,000
Northwest Regions	35,780	4,000

4. Western hydropower shall be mainly transmitted to East China, Central China and Guangdong Province

Based on the coordination of the marginal demand characteristics of each receiving end and power allocation among various receiving ends in consideration of prioritizing the acceptance of the southwest hydropower, the power supply through from-West-to-East transmission can be well consumed. The space of East China power market is huge and is suitable to western hydropower characteristics.

5. The conception of western hydropower transmission planning

In order to achieve optimal allocation of power resources in a wider scope, meet the needs of large-capacity and long-distance from-West-to-East transmission, and effectively mitigate the shortage of transmission corridor resources, the hydropower transmission from the west to the east shall be based on ±800kV and ±1,000kV DC UHV station-to-grid transmission and AC UHV grid-to-grid transmission, and in some cases, the comparatively mature ±500kV DC UHV transmission in light of transmission distance and transmission capacity (see Table 5.3-5).

Table 5.3-5 Summary of Consumption Planning from-West-to-East Hydropower Transmission   Unit: MW

Item	Target development scale	Eastward power transmission target scale	Chongqing	Central China	East China	Guangdong Province	Guangxi Province	Sichuan Province	Yunnan Province	North China (Shandong Province)
I. Total	350,000 – 380,000	135,000 – 160,000	14,690	37,060 – 41,260	47,700 – 64,800	35,542 – 41,842	2,348	25,000	30,000	4,000
II. Jinsha River Boundary										
1. Scenario I	58,620	42,000		13,200	22,500	6,300		4,350	4,350	
Cascades on the Upper Jinsha River		9,000		9,000						
Xiluodu and Xiangjiaba		18,600		6,000	6,300	6,300		5% respectively in dry season		
Baihetan and Wudongde		14,400		7,200	7,200			4,350	4,350	
2. Scenario II (Recommended)	58,620	48,600		9,000	39,600	0		0	5,100	
Cascades on the Upper Jinsha River		9,000		9,000						
Guanyinyan		3,000		3,000						
Xiluodu and Xiangjiaba		18,600		6,000	12,600			5% respectively in dry season		
Baihetan and Wudongde		18,000			18,000				5,100	

Chapter 5 Dams and Hydropower Development

Countinued

Item	Target development scale	Eastward power transmission target scale	Chongqing	Central China	East China	Guangdong Province	Guangxi Province	Sichuan Province	Yunnan Province	North China (Shandong Province)
III. Sichuan	94,000	25,000	6,500	11,300	7,200					
1. Station-to-grid		16,100	2,900	6,000	7,200					
Ertan		900	900							
Xichang Platform for the Yalong River		9,200	2,000		7,200					
Dadu Ya'an Platform		6,000		6,000						
2. Grid-to-grid		8,900	3,600	5,300						
IV. Yunan	77,280	38,000		6,000		32,000				
Longpan, Ahai and Liyuan		6,000		6,000						
500kV AC Grid Transmission		2,900				2,900				
Xiaowan Platform		5,000				5,000				
Nuozhadu Platform		5,700				5,700				
Ludila and Longkaikou		3,900				3,900				
Maji, Wunonglong and Tuoba		6,200				6,200				
Saige, Yabiluo, Miaowei, Huangdeng and Lushui		8,300				8,300				
V. Guizhou	18,700	7,840	190	1,760	0	3,542	2,348		505	
Cascades on Qingshui River		1,570		1,570						
Cascades on Boundary River		1,463					1,463			
Other Power Stations on Boundary River		0							505	
Coordinated Power Grid Transmission		4,808	190	190	0	3,542	886			
VI. Tibet	78,800 – 98,800	70,000 – 90,000	8,000	9,000	18,000			25,000	10,000 – 30,000	
VII. Northwest	35,780	4,000							exchange 3,000	4,000
Other: Five Countries of Indo-China Peninsula	20,000	20,000				0 – 20,000			20,000 – 0	

6. Eastern and central areas shall give priority to the acceptance and consumption of western hydropower

By calculating the factory power price from West region (including increment tax) and analysing the transmission price, to-grid price, grid price showed thermal power pole of receiving end in the east region and marginal price, the results can be seen in Table 5. 3-6.

Table 5. 3-6 Analysis of Bidding Ability of Power Transmission from Southeast to East Unit: RMB/(kW · h)

Item	Thermal power on-grid thermal power benchmarking price of receiving ends	Marginal on-grid marginal price of receiving ends grid	On-grid price of western hydropower to gird price
I. Chongqing	0.3793	0.4771 – 0.5455	Sichuan Electricity: 0.3 – 0.6 Tibet Electricity: 0.662 – 0.999
II. Four Provinces of Central China	0.3942 – 0.4405 (average about 0.42)	0.4771 – 0.5455	Sichuan Electricity: 0.35 – 0.63 Tibet Electricity: 0.716 – 1.067
III. East China	0.3980 – 0.4657	0.5113 – 0.5797	Sichuan Electricity: 0.40 – 0.69 Tibet Electricity: 0.758 – 1.117
IV. Guangdong	0.5302	0.5455 – 0.6482	Yunnan Electricity: 0.42 – 0.66 Five countries of Indo-China Peninsula: 0.362 – 0.543

The on-grid price of Yunnan hydropower and Sichuan hydropower has a competitive edge in the power grid market of receiving ends; compared with renewable energy such as wind power, the receiving ends of eastward hydropower transmission from Tibet should be able to stand the expected price of Tibet hydropower. The price of from-West-to-East hydropower transmission shall be set a bit higher than marginal on-grid price of eastern grid and not lower than the corresponding price of western grid, and the resulting benefits could be used for advancing the economic and social development of the west region so as to fully implement the western region development strategy.

5.3.5 Study on the Involvement of Southern Hydropower in the Competition and Trading in Southern Power Market

Founded in March 2003, the State Electricity Regulatory Commission has promoted the development of power market in China, with *Program for Building Southern Power Market*, which would be implemented under initiation, development and maturity stages, being approved. In November 2005, the southern power market entered the first stage of the simulation run with those first participants including 5 power grid companies and 14 thermal power providers and 39 thermal power generators totaling 13,540MW. Hydropower was not involved in this market competition. It was necessary to conduct research as soon as possible on the plan and rules of southern power market competition and trading involved by the hydropower in South China, in order to comprehensively optimize the distribution of the resources by the market mechanism on a larger scale, improve power price system, adjust power investment by the market, facilitate the development and utilization of western hydroelectric resources and the implementation of from-West-to-East power transmission strategy, and provide better services for the coordinated development in the south. In 2005, HYDROCHINA CO. conducted the *Study on the Involvement of Southern Hydropower in the Competition and Trading of Southern Power Market*.

This project has three sub-subjects, including competition and trade planning study, competition and trading module study, and ancillary service market. Firstly, collect a large amount of basic information for study and gather a variety of viewpoints from power grid companies and some power generation groups. Then analyze the characteristics of hydropower and the potential issues regarding competition and trading. Finally, put forward the feasible general framework and market mode of competition and trading involved by hydropower. The report is composed of three parts, i.e. southern power market overview; study on the involvement of southern hydropower in the southern primary power market; study on the involvement of southern hydropower in the southern ancillary service market. The main study results are as follows.

1. Study on the involvement of southern hydropower in the southern primary power market

(1) The main problems faced by southern hydropower in competing in the power market. The key problems to be resolved are: uncertainty of the coming water; the huge diversity of reservoir regulation performance; price difference between participating power stations and imbalanced regional economic development; coordination of cascade hydropower competition and trading and dispatching coordination between cascades; electric quantity brought by hydropower surplus water.

(2) The first phase overall framework of market competition with south regional hydropower involvement.

In terms of "proper start-up and stable operation", the partial electric quantity transferred from a batch of hydropower plants are taken into market competition firstly, then gradually increases the competing plants. A unified hydropower and thermal power transaction platform is to be established.

(3) The market mode and transaction means in the first phase.

Market mode: the long-term aim of the South China power market is to establish a unified market. In order to make sure that the market goes smoothly after the hydropower is involved into the market competition, the following market modes are used by one platform of the hydropower and thermal power into, single electricity price, involving part of plants and electric quantity to the competition.

The main body in the competition: the principle of the main body of the hydropower competition in the first phase is the hydropower plants with the installed capacity no less than 100MW. The pumped-storage power plants or hydropower station run by the smart grid enterprises for meeting the emergency and peak load regulating are not included into the competition. In the starting phase, the first batch of hydroelectric plants participating in the testing operation is shortlisted according to the principles that the plants are operated by different electricity groups from different provinces, being in conformity with the involved thermal capacity. "Easy first" and "Pilot project first" principles need to be considered. The suitable plants are easier to be selected by the comprehensive analysis and more hydropower stations will be gradually involved with the development of the market.

Transaction categories: after the hydropower has participated in the market competition in the first phase, the available dealing types in the near future include: annual contract transaction of non-competition, annual competitive trade, monthly competitive trade, day-ahead competitive trade and generation rights trade.

The proportion of competition electricity: through research, the electric quantity for annual contract of non-competition would be better set at 50% of the average annual power generation of each plant with high extent of assurance. The competitive electricity is 50% of the average power generation by all the competitive plants. After each plant signs the contract of non-competition electricity according to the 50% of the average power capacity, the remaining electricity could be participate in different types of trading competition according to the distinct characteristics of the plant, the water level status of reservoirs and the possible forecasting inflow.

Competitive marketing unit of the hydropower: at the initial stage of hydropower's involvement in the power market competition in South China, only single hydropower station, regarded as one single marketing unit, is allowed to take part in the competition by principle. However, association of the whole cascade or grouping association of the cascade would also be allowed for competition without resulting in market power (one single united power generation company features the installed capacity less than 20%).

Competitive trading means of the hydropower: the trading means of each category are drawn up. Several measures are taken to thrash out the issues in terms of disparity between hydropower and thermal power prices, the price differences among the competition involved by hydropower plants and the price differences caused by the imbalance of economic development in different regions. These measures are: in the first phase, to choose hydropower stations whose electricity prices are close to those of the thermal power, then to select proportional power stations for competition and dealing, and to adjust the long-term contract electricity price in the future to thoroughly solve the issue of fair play in market competition.

2. Study on the involvement of south regional hydropower in the southern auxiliary service market

The auxiliary service of Southern Power Grid in China consists of basic auxiliary service and fee-based auxiliary service. Basic auxiliary service is offered free by market members for the sake of safe and smooth operation of the power grid, including the primary frequency regulation of the generating set, base peak load regulating and base reactive power regulation. Fee-based auxiliary service includes automatic generation control (AGC), paid peak load shaving, backup, paid reactive power regulation and black start.

The cost analysis principle is used to analyze the cost of fee-based auxiliary service, and the indemnifying measure of the fee-based auxiliary service in the first phase is put forward. Fee-based peak load shaving, paid reactive power regulation and black start are based on the principle of "dispatching first and then settling accounts on the basis of the compensation standard." AGC and backup service are grounded on the competitive price, "dispatching first and then settling accounts", in an effort to guarantee the safety and smoothness of the power grid.

The research has showed that under the condition of the current electric power system and electricity price mechanism, it is not yet ripe for the hydropower plant to participate in the market competition. The reasonable price of those plants available for the competition should be equal to that of thermal power stations; otherwise the fair play of the competitive dealing could not be reached in the process of benefit distribution. In the research, the principle and detailed rules of competitive dealing involved by the hydropower in the South China is formulated, but not yet put into action according to the requirement of rational usage of water resource and benefit allocation. Nevertheless, they can provide a valuable reference for the establishment of a genuine electricity market in the future.

5.3.6 Study on the Promotion of Hydropower Development to Local Economic Development

Hydropower resources are very important energy resources in China, accounting for 40% of the country's total remaining exploitable conventional energy resources. Therefore, the development of these hydropower resources is an important measure to meet our growing energy needs, reduce greenhouse gas emissions, adjust our power structure and make comprehensive use of water resources. However, due to historical and policy reasons, hydropower development, the vital interests of resettlement and local economic and social development are not well combined, which results in the situation that there are still many problems regarding reservoir resettlement and the economical and social development in the reservoir area is still lagging behind. In order to further play the positive role of hydropower development in poverty elimination of settlers, and local economic and social development as well as to promote the sound development of hydropower and the building of a harmonious society, Hydropower and Water Resources Planning and Design General Institute organized a policy research on how to use hydropower development to promote local economic and social development. The main research achievements are as follows.

China's hydropower resources are mainly in the southwest and northwest with weak economic foundation and low economic development level. Since the founding of new China, especially since the reform and opening up, China's hydropower construction has achieved remarkable growth. It has not only become able to ensure power supply, but also promoted local economic and social development. However, it has not fully transformed China's resource advantages into economic advantages. According to the requirement of the Scientific Outlook on Development and the actual situation of hydropower development, it is necessary to adjust the current hydropower development policies so as to further promote the economic and social development of reservoir areas, improve the benefits sharing mechanism of stakeholders and finally further ensure sustainable hydropower development.

1. Make appropriate adjustments to fiscal policy and sharing mechanism of hydropower development

Western provinces (autonomous regions and municipalities) such as Sichuan, Yunnan, Guizhou, Qinghai and Tibet are rich in hydropower resources. Hydropower resources development is an important part of the western development and thus it is hoped that through exploiting of these resources, the resource advantages of these regions could be turned into economic advantages, and hydropower resources development could become an industry that can promote economic development. Sustainable development of hydropower and the economic and social development of these provinces (autonomous regions and municipalities) are mutually reinforcing. As for the allocation of value added tax (VAT) and corporate income tax between the central government and provincial (autonomous regions and municipalities) government, the proportion of the central government could be reduced and that of the provinces (autonomous regions and municipalities) which have resources could be increased according to the characteristics of the hydropower industry. For example, the allocation ratio of hydroelectric generation's VAT and corporate income tax between central government and provinces (autonomous regions and municipalities) could be adjusted to each half of 50%. As for the VAT and corporate income tax retained for local government, it should be allocated between constructions sites and company registration places should be decided based on the actual situation of the construction sites, according to the principles of favoring the former and the specific ratio. If there are more than one county (city) involved in the construction, the distribution proportion among these counties (cities) should be determined based on resettlement people, inundated area, contribution of these places to the project and other factors. It is suggested that relevant government departments should make a unified water resources fee standard for hydropower projects and the distribution and use of that fee should also favor local counties (cities) where these projects are. Specific distribution proportion and purposes of the fee should be made clear. Meanwhile, it should be conducive to the development of local economic and social development as well as resettlement.

2. Make a paid development mechanism for hydropower resources

It should charge hydropower resources development compensation fee according to the principle of paying for use of hydropower resources. The levy of this fee is to reflect the resource advantages in these places, and transform resource advantages into economic advantages so as to further promote local economic and social development and improve sustainable development capability of reservoir area and resettlement area, although the ownership of hydropower resources belongs to the country. Therefore, the compensation fee should be given to local government as a special expense without taking into account the distribution proportion of the central government. The levy of that fee could significantly promote development of local economy, increase local revenue as well as balance regional economic development. According to the characteristics of hydropower, the levy of that fee should be based on electric capacity and quantity, RMB 100/kW (validated capacity) for the former and 5% of annual income from electricity sales for the latter. Moreover, the levied fee should be distributed among provinces (autonomous regions and municipalities), prefectures (regions and municipalities) and counties (cities), with counties where projects and reservoirs are being favored. These standards should be applied in the projects of new construction while finished projects should only pay quantity fee according to the above standard. The charge of compensation fee should stop once the power station goes out of use. Considering that it may take a long time to finish the process from proposing the levying of that fee to its implementation, a "hydropower resettlement support and development fund" should be established recently so as to improve sustainable development of reservoir and resettlement areas. The on-grid price could be increased by RMB 0.02/(kW·h) and that money could be used to set up the fund which is also conducive to local economic and social development as well as improving sustainable development of reservoir and resettlement areas. The charged compensation fee of hydropower resources development according to the relevant standards will also be mainly used to support that fund.

3. Implement the repayment mechanism of regulating reservoir generation compensation benefits

The reimbursement of generation compensation benefits of cascade hydropower station involves the adjustment of economic relationship among many companies and between upstream and downstream regions. Therefore, there should be a proper and reasonable distribution method for the reimbursement based on laws and regulations and following the principle of openness, fairness and impartiality, and the method should be constantly revised and improved throughout the operation of cascade hydropower stations so as to maintain and adjust the interests of the reservoir (including the power generation companies and regulatory surrounding areas of the reservoir) as well as ensure the long-term sound operation and development of cascade hydropower stations group.

Regulating reservoirs should have the ability to adjust seasonal electricity capacity of wet season and dry season which means that these reservoirs should at least be able to adjust seasonal performance. The scope of compensation benefits should be determined by the ratio between regulation capacity of the reservoir and annual runoff of downstream reservoirs. Moreover, the difference in annual sales revenue between beneficiary stations should be calculated based on whether they have regulating reservoirs stations, which is compensation of regulating reservoirs to benefited stations. In order to make the implementation and promotion easier, it is recommended that downstream beneficiary stations should give 50% of their incremental in electricity sales revenue to their upstream regulating reservoirs, if they have. Moreover, a joint operation and management agency for cascade hydropower stations should also established to be in charge of scheduling and optimization of all reservoirs in the watershed, supervising and implementing the benefits compensation mechanism so as to ensure effective implementation of the mechanism. Meanwhile, the management and paying methods of compensation benefits should also be set up, making clear the ways to calculate, collect, distribute of compensation benefits as well as liabilities for breach of contract.

4. Share comprehensive utilization investment of hydropower projects

The allocation of comprehensive utilization investment of hydropower projects is related to economic relations among multiple departments, enterprises, and regions' governments in the upstream and downstream river and residents. Therefore, there should also be a proper and reasonable allocation method for the reimbursement based on laws and regulations and following the principle of openness, fairness and impartiality. According to the principle of "Who derive the benefit, who take the responsibility", relevant departments or beneficiary regions (persons) should bear not only their specific project investment, but also the investment apportion of utilities system according to the benefits they receive. This is a comparatively fair and reasonable method. With the continuous improvement of China's hydropower engineering construction system, the cost allocation policy should be implemented comprehensively in hydropower projects that undertake the task of comprehensive utilization. In addition, the generating efficiency losses of hydropower stations during their operation caused by comprehensive utilization should also be properly compensated.

5. Improve the price mechanism of hydropower projects

After a series of reform in the electricity pricing mechanism, currently China mainly uses benchmark electricity price. Now on average, the hydropower on-grid benchmarking price is RMB 0.08-0.12/(kW·h) lower than that of thermal power. Thus following the principle of same quality, same price in one grid, the price of hydropower could be increased. Hydropower generation capacity in 2007 accounts for about 15% of the national total. So an appropriate increase in the hydropower benchmarking price will not exert much impact on sale-side price. However, further research is needed due to the fact that electricity price has a huge impact on national economy and the increase should be implemented step by step. It is recommended that hydropower benchmarking price could be increased by RMB 0.02/(kW·h) before 2009 so as to resolve the issue of the hydropower resettlement support and development fund. After that, the gap between the hydropower and thermal power benchmarking price levels could be gradually reduced as the market changes and by the year 2012, that gap should be further narrowed, making the price of hydropower close to that of thermal power and the principle of same quality, same price in one grid should be fully implemented by the year 2015.

After the increase of hydropower benchmarking price and the principle of same quality, same price in one grid, some part of the profits should be used as compensation for hydropower resources development (hydropower resettlement support and development fund) on condition that hydropower companies have reasonable profits. In addition, some hydropower may have some excess profits, which may pay segment excess profits tax. This tax could be used as hydropower resources development adjustment fund to give support to hydropower stations with poor long-term development conditions but viable economic evaluation so as to achieve sustainable development of hydropower resources and make contributions to national energy security.

5.3.7 Study on the Medium and Long-term Development Strategy of Hydropower in China (in the Planning Level Year of 2030 and 2050)

Sustainable energy development is an important part of sustainable economic and social development. As China's economy witnesses steady and rapid development, its problem in energy supply and demand and ecological environmental pressures become increasingly prominent, which restricts the sound development of our economy and society and becomes one of the major bottlenecks of sustainable development, to some extent. Over the years, relevant authorities have conducted a large number of studies on energy development and put forward some ideas about medium and long-term development. Moreover, some consensuses over China's energy development before 2020 have been reached. However, more in-depth studies and discussions on the development of China's energy supply and demand, and the development potential, structure, development strategy and technology development roadmaps of a variety of energies after 2020 are still needed. In order to meet our goal of building a socialist harmonious society and promoting sustainable energy development, the Chinese Academy of Engineering organized relevant units to carry out consultation studies over China's medium and long-term energy development strategy in 2008. Among the said studies, HYDROCHINA CO. is mainly responsible for the renewable energy studies and the compiling of the report *Study on the Medium and Long-term Development Strategy of Hydropower in China*, working together with related organizations.

Development plans and objectives of China's hydropower from 2030 to 2050 as well as their implementation roadmaps and main problems and difficulties have been made clear after the project research, which can serve as a guide to China's hydropower development and is conducive to sound and orderly development of China's hydropower. The main research achievements are as follows.

1. *Planning objectives*

It is planned that China's hydropower installed capacity should reach 340GW and annual power output should reach 1,405TW · h in 2020, among which, large-sized and medium-sized hydropower should account for 265GW and small hydropower 75GW. The total size of the eastern region should reach 30GW, accounting for about 9% of the total size of China, the central region 65GW, about 19% and the west region 245GW, about 72%.

It is planned that China's hydropower installed capacity should reach 430GW and annual power output should reach 1,853TW · h in 2030, among which, large-sized and medium-sized hydropower should account for 337GW and small hydropower 93GW. The total size of the eastern region should reach 32GW, accounting for about 7% of the total size of China, the central region 68GW, about 16% and the west region 330GW, about 77%.

It is planned that China's hydropower installed capacity should reach 510GW and annual power output should reach 2,290TW · h in 2050, among which, large-sized and medium-sized hydropower should account for 410GW and small hydropower 100GW. The total size of the eastern region should reach 33GW, accounting for about 6% of the total size of China, the central region 70GW, about 14% and the west region 407GW, about 80%.

According to the hydropower resources development in China's neighbors, the Southeast countries can transmit power of about 20GW to 30GW to China through Yunnan Province to Guangdong Province and Guangxi Autonomous Region.

Considering current progress and development trends as well as background, the strategic goal of China's hydropower development can be achieved.

2. *Development arrangement*

China's hydropower resources are mainly distributed in the rivers (or river sections) with thirteen large hydropower bases and the Yarlung Zangbo River. Considering from the current situation of China's hydropower development, the hydropower development in the Yalong River, Dadu River, Lancang River, Nujiang River, Jinsha River, and Yellow River in the Western and the Yarlung Zangbo River is still very low. Specifically, there is still no hydropower development in the upper Jinsha River, the upper Lancang River, the Nujiang River and the Yarlung Zangbo River. Therefore these above rivers (sections) will be the future important regions for hydropower development in China, which are mainly located in provinces such as Sichuan, Yunnan, Tibet and Qinghai Provinces.

In the period before 2020 there will be rapid development in the thirteen large hydropower bases. The focuses are the mainstreams of the Yalong River, Dadu River, Nujiang River, Yellow River, Lancang River and Jinsha River. Among them, the mainstream of the Jinsha River will generate 39GW, the mainstream of the Dadu River 22GW, the mainstream of the Yalong River 18GW and the mainstream of the Lancang River 18GW.

The focuses of China's large-sized and medium-sized hydropower development from 2020 to 2030 will be in the lower-and-middle Nujiang River, the upper Jinsha River (mainly along the boundary section between Sichuan and Tibet). The hydropower development objectives in the Yalong River and Dadu River will mainly be the ones with relatively poor economic indicators and late preliminary preparation. From 2020 to 2030, China's newly installed capacity will be gradually located in the Tibet Autonomous Region. By the end of 2030, the mainstream hydropower development of the Nujiang River, Lancang River, Yalong River, Dadu River and Jinsha River will be largely completed. Among them, the Jinsha River will generate 33GW and the Nujiang River is 13GW.

The focuses of China's large- and medium-hydropower development from 2030 to 2050 will be in the Upper Nujiang River and the lower mainstream of the Yarlung Zangbo River. The upperstream of the Nujiang River will generate 11GW and the lower stream of the Yarlung Zangbo River will generate 60GW. These stations will mainly be located in the Tibet Autonomous Region and the total installed capacity will be 73GW.

3. Speeding up the development of pumped-storage power plants

The construction of China's pumped-storage power plants did not begin until the late 1960s and its scale was very small. In the 1980s, the construction of many pumped-storage power plants in east region such as Guangdong Province, East China and North China pushed the country's pumped-storage power plants development to a new height. By the end of 2010, the installed capacity of China's pumped-storage power plants had reached 17GW.

China's pumped-storage power plants, both built and under construction ones are mainly located in areas such as South, Central, North, East and Northeast China where thermal power is the majority, to meet the peak regulating requirement in these regions while there are not many such plants in inland provinces. As China's energy resources distribution and economic development are imbalanced, in addition to economic development and power structure, factors such as cross-regional, large-scale, long-distance power transmission, power structure optimization and strong and smart grid development, large scale development of new energy such as wind power and nuclear power and renewable energy should also be taken into consideration in the planning and building of pumped-storage power plants. It is planned that the total scale of China's pumped-storage power plants will reach 80GW in 2020, located in areas where there is large peak and off-peak difference and load regulating capacity is low. Moreover, these areas also include the receiving and delivering ends of large-scale wind power development and areas where there is nuclear power.

4. Measures to secure hydropower development

To further improve China's hydropower development supporting mechanisms. The strategic position of hydropower and a long-term stable hydropower development mechanism should be established. The hydropower market shall be further established. Besides, the electricity pricing system and hydropower cascade development management mechanism should be further improved. Moreover, the investment sharing mechanism for comprehensive utilization of hydropower project should be reestablished and hydropower cascade compensation mechanism should be set up.

To attach great importance to ecological and environmental protection, and effectively implement environmental protection measures. It is necessary to establish a scientific evaluation system to coordinate the relationship between hydropower development and ecological environment protection. Relevant government departments should organize and carry out the studies on watershed environmental carrying capacity of hydropower development and zoning of hydropower resources function areas so as to reach consensus over scale, arrangement and sequence of development and put forward protection measures and finally achieve a win-win situation between development and protection. Moreover, a practical ecological compensation mechanism should be established. The studies on hydropower development-related ecological environment protection measures should be further strengthened to mitigate negative impacts of hydropower development on ecological environment with effective measures.

We should place high importance on properly arranging the migrants, protecting their legitimate rights and interests, and helping them become rich in a prominent position. To improve the resettlement management system, it is necessary to formulate supporting policies of the current resettlement regulation, as well as supplement and amend relevant regulations; make innovation in ways of resettlement, properly implement migrant development measures, set up supporting mechanisms for their development, and make efforts to resolve remaining problems of built hydropower projects; and moreover, establish and improve systems and mechanisms so as to achieve the goal of use hydropower development to promote local economic and social development.

We should strengthen scientific and technological innovation, accelerate technological progress in hydropower development, further improve technology and skills in hydropower development, enhance designing and manufacturing capacity of high water head, large-capacity hydro-generating units, as well as solve technical and equipment problems faced by large-capacity, long-distance power transmission on high plateaus and in alpine areas.

5.3.8 Research on the Policy of Same Network and Same Price for Hydropower and Thermal Power

Commissioned by National Energy Administration, Hydropower and Water Resources Planning and Design General Institute organized and carried out the "Policy Study of same network and same price for Hydropower and Thermal Power" in 2009, which discussed the executive conditions of the existing hydropower and thermal power on-grid prices from following several aspects such as hydropower station regulation performance, resources exploitation conditions, difficulty of migrant resettlement, development duties, and differences between provinces, and put forward the features of our current hydropower on-grid price and main existing problems based on the executive conditions of the existing hydropower and thermal power on-grid prices which were collected and investigated from 31 provinces (autonomous regions, municipalities). This research raised power pricing mechanism and methods appropriate to China by means of analyzing current hydropower and thermal power on-grid pricing mechanism at home and abroad; analyzed main factors affecting hydropower price and its influence on power price change through looking into the rational composition of hydropower and thermal power on-grid price; put forward the rationality of same network and same price for hydropower and thermal power at last by way of analyzing China's current condition of hydropower and thermal power on-grid price and rational composition. The research put forward the price-setting standard and methods for same network and same price. For the first time, it systematically ascertained the actual executive condition of nationwide hydropower on-grid price, revealed the problems faced by China's hydropower on-grid price on a deeper level, and furnished China's implementation of the policy of same network and same price for hydropower and thermal power with foundations. The main research achievements are as follows.

1. Current hydropower on-grid price is lower than thermal power on-grid price

Seeing from the investigation results of hydropower and thermal power price in China's major provinces, a comparatively big difference exists between current hydropower on-grid price and thermal power on-grid price. Hydropower on-grid price is generally lower than thermal power on-grid price. In provinces abundant in hydropower resources such as Sichuan, Yunnan and Guizhou, hydropower on-grid benchmarking price (or on-grid power price of hydropower recently being invested) is about RMB 0.1/(kW·h) lower than thermal power on-grid benchmarking price.

2. On-grid power price of newly-built hydropower stations should be increased

Calculating the on-grid power price of hydropower and thermal power stations to be built per 30 years' operating period on the basis of 8% of internal capital gains and taking the increase of uncertainty caused by investment in hydropower stations into consideration, the on-grid prices between hydropower and thermal power have been very close (see Table 5.3-7).

Table 5.3-7 Measurement and Calculation Results of On-grid Power Prices of Hydropower and Thermal Power Stations to Be Built

Province	Hydropower weighted average price (RMB/(kW·h))	Hydropower average price after increasing hydropower investment (RMB/(kW·h))	Thermal power on-grid price (RMB/(kW·h))
Sichuan	0.2832 - 0.3683	0.3399 - 0.4303	0.3550
Yunnan	0.2435 - 0.3098	0.2893 - 0.3704	0.3230
Guizhou	0.2573 - 0.2764	0.3079 - 0.3300	0.3385

Viewing from the measurement and calculation results of on-grid power prices for hydropower and thermal power stations to be built in the three provinces of Sichuan, Yunnan and Guizhou, the hydropower weighted average on-grid price of reservoirs with regulation capacity is higher, and price of some reservoirs even surpasses thermal power on-grid price. Moreover, with the development of economy and society, the implicit costs (environmental cost, resettlement costs) of hydropower development might increase by a large margin. The on-grid price for hydropower stations to be built is higher in provinces with higher hydropower resources development. For example, the to-be-built hydropower on-grid prices in provinces like Shaanxi, Hunan, Hubei, Zhejiang are respectively RMB 0.3825/(kW·h), RMB 0.4396/(kW·h), RMB 0.6091/(kW·h) and RMB 0.8362/(kW·h), which are prominently higher than thermal power on-grid prices.

Viewing from the measurement and calculation results of to-be-built hydropower on-grid price, current hydropower benchmarking on-grid price is apparently lower. Power price that is too low is unfavorable to not only the development of the power industry but also to the normal development of the national economy. Therefore, starting from the perspective of facilitating the sound and sustainable development of the hydropower industry, the on-grid power price of to-be-built hydropower projects should be increased.

3. New situation and new environment pose new requirements to hydropower development, and the external costs for hydropower development are increasing progressively

In recent years, as the national economy grows rapidly and hydropower development steps up, the external environment hydropower faces great changes. In particular, the increasing prominent problems, for instance, coordination between hydropower development and protection of ecological environment and resettlement caused by hydropower development, have posed new challenges to hydropower development. A series of new regulations, policies and measures have also raised new requirements for hydropower development. In addition, social responsibilities hydropower development bears become more and more obvious. Its development obligations, which directly lead to the increase of hydropower electricity generation cost, are not only restricted to electricity generation, but also protecting ecological environment, enhancing migrants' life quality and promoting local economic and social development.

4. Pricing mechanism of same network and same price for hydropower and thermal power

a. Continue to carry out one-system electricity quantity and electricity price at present

The *Interim Measures on On-grid Power Price Management* issued in 2005 in China definitely regulated that the on-grid power price is defined according to the rational gains per consideration of social average cost before bidding for access to power grids, and the two-system power price will be carried out after the bidding. The implementation of two-system power price can better reflect hydropower stations' characteristic of flexible operation as well as the needs of the power market. However, the current condition is mature to carry out two-system power price. One-system power pricing dominated by government keeps on being carried out.

b. Carry out the policy of same network and same price per provinces (autonomous regions and municipalities)

Since the economic development in different provinces (autonomous regions and municipalities) in China varies widely, so does the gap between hydropower and thermal power benchmarking on-grid power prices in different regions. Therefore, the policy of same network and same price cannot be carried out nationwide. In accordance with the relevant requirements on power price reform in China, the power price reform should be in concert with actual conditions such as local conditions of energy resources, power structure, economic development level, and measures should be adjusted to local conditions. Considering the hydropower within the same province (autonomous regions and municipality) bears same features, the policy of same network and same price in current stage is implemented per provinces (autonomous regions and municipalities).

c. Different levels of the same network and same price in different provinces (autonomous regions and municipalities)

Due to the inconsistency of hydropower resources development in different provinces (autonomous regions and municipalities) and the wide gap between development conditions of the remaining resources, the implementation of the one plant and one price policy has its rationality. If such a policy is being carried out, different hydropower and thermal power having "same network and same price" pricing programs are recommended since hydropower stations in different provinces (autonomous regions and municipalities) generate different electricity quantity within its own boundary. It is inappropriate to measure Sichuan, Yunnan with the same power price level since they have rich remaining hydropower resources and varied types of power stations such as run-off power stations and reservoirs power stations. It is recommended that Sichuan takes 8% storage coefficient as the boundary, Yunnan takes 6% storage coefficient as the boundary, and hydropower stations with different regulation performances implement two power prices. It is also recommended that provinces such as Guizhou, Qinghai, Chongqing, Gansu execute the current hydropower benchmarking on-grid power price in line with their remaining resources. It is appropriate for provinces with limited remaining hydropower resources and larger development difficulties and economic costs such as Xinjiang, Zhejiang, Hunan, Hubei to carry out the one plant and one price policy (see Table 5.3-8).

Table 5.3-8 Recommended Hydropower Pricing Program for Different Provinces in China

Province	Current coal thermal power benchmarking on-grid price (RMB/(kW·h))	Pricing program recommended by this Research
Sichuan	0.3837	① First-grade power price program: RMB 0.3837/(kW·h) ② Stepped power price program: storage coefficient $\beta \geq 8\%$, power price calculated by RMB 0.4555/(kW·h); storage coefficient $\beta < 8\%$, power price calculated by RMB 0.3503/(kW·h)
Yunnan	0.3230	① First-grade power price program: RMB 0.3230/(kW·h) ② Stepped power price program: storage coefficient $\beta \geq 6\%$, power price calculated by RMB 0.360/(kW·h); storage coefficient $\beta < 6\%$, power price calculated by RMB 0.3035/(kW·h)
Guizhou	0.3281	First-grade power price program: RMB 0.3281/(kW·h)
Qinghai	0.2790	First-grade power price program: RMB 0.2790/(kW·h)
Chongqing	0.3793	First-grade power price program: RMB 0.3793/(kW·h)
Gansu	0.2765	First-grade power price program: RMB 0.3041/(kW·h)

Countinued

Province	Current coal thermal power benchmarking on-grid price (RMB/(kW·h))	Pricing program recommended by this Research
Xinjiang		Define the on-grid power price of power stations by satisfying capital fund FIRR=8%
Hunan	0.4405	Define the on-grid power price of power stations by satisfying capital fund FIRR=8%
Hubei	0.4250	Define the on-grid power price of power stations by satisfying capital fund FIRR=8%
Zhejiang	0.4657	Define the on-grid power price of power stations by satisfying capital fund FIRR=8%
Other Provinces		Define the on-grid power price of power stations by satisfying capital fund FIRR=8%
Trans-provincial transmission capacity		Defined by coal thermal power benchmarking on-grid price subtracting power transmission and distribution prices

As for the hydropower stations power quantity of from-West-to-East power transmission, the landing power price after subtracting transmission power price should be considered to adopt the policy of same network and same price as thermal power price of the receiving power network. As for the big clients who directly purchase power quantity, the price can be negotiated by both parties.

d. Further Conduct research on and resolve the resource differences after implementing the policy of same network and same price

From the perspective of the market economy, the power prices as end product should be close. The state has proposed benchmarking power price policy, namely, from regional point of view, to define one benchmarking power price according to the pattern of social average cost plus certain profits. However, due to hydropower resources development condition, the on-grid prices of hydropower stations in different types vary widely, even the price from the same type also have a wide gap. Implementing the same power price will inevitably result in irrational gains level in some power stations which gains exorbitant profits exceeding their rational gains, especially for strategic projects in the cascade hydropower stations. Only for the compensation to the controlling hydropower station, can its development smoothly propel. Further research needs to be carried out as to the quantization, management and usage of exorbitant profit.

Part II

Construction Technologies for Dams and Reservoirs

Chapter 6

Hydrological Analysis and Calculation for Dam Construction

Yang Baiyin[1], Wang Zhengfa[2]

6.1 Overview

Hydrology is the scientific study the spatio-temporal distribution and dynamics of water on the earth's surface. Closely connected with geosphere, atmosphere and biosphere studies, hydrology belongs to geoscience from a scientific system. It serves the development of water and hydropower resources, and contributes to flood control, which makes hydrology an integral part and foundation of water and hydropower science. As an important branch of hydrology, engineering hydrology has the features of both basic science and engineering application by applying the basic hydrological theories to practical engineering issues. Additionally, it provides hydrological analysis for engineering plan, design, construction, operation and management. Engineering hydrology mainly consists of hydrologic analysis and calculation and hydrologic forecast. It covers water circulation and runoff generation, hydrological information collection and compilation, hydrological statistical theory, design runoff analysis and calculation as well as runoff stochastic simulation, design flood analysis and calculation, PMP (Probable Maximum Precipitation) and PMF (Probable Maximum Flood) analyses and calculations, watershed runoff yield and flow concentration calculations, hydrologic forecast, river sediment measurement and engineering sediment, etc.

Now with growing awareness of the significance of hydrology and water resources, engineering hydrology has evolved from the pure service for hydropower project, as in the past into an important specialized basic discipline serving hydropower project, water engineering, water supply and drainage engineering, channel ports and coastal engineering, agricultural water conservancy engineering, environmental engineering and civil engineering, etc.

Based on the principles, methods, specifications of engineering hydrology, the hydrological design for hydropower stations is to calculate design runoff, design flood, stage-discharge relations of a certain design section, engineering sediment, meteorological elements, etc, according to the features of hydropower project itself by basic information such as precipitation, discharge, water level, sediment under the help of automatic hydrological data acquisition system and hydrologic forecast. The accuracy and reliability of the results of hydrological design play a considerably fundamental role in hydropower station engineering design, construction, operation and management, which make a strong impact on the economical efficiency and safety of the hydropower station engineering and determine whether a hydropower station works well.

After 60 years of hydropower development in China, engineering hydrology has developed rapidly and a team of highly skilled scientific and technological personnel for hydrology have been cultivated to undertake various hydrological analyses and calculations for the hydropower engineering construction and to participate in domestic and international academic exchange activities concerning engineering hydrology. Since the reform and opening-up launched three decades ago, especially since 2000, the hydropower construction in China has been developed on a rapid and in-depth basis and has brought new tasks to hydrologists, such as improvement of the parameter estimation method used in frequency analyses, design flood calculations and flood control standard selection for cascade reservoirs, reservoir inflow flood calculations, analyses of long-term consecutive low flow period for multi-year regulating reservoirs, study on Probable Maximum Precipitation (PMP) and Probable Maximum Flood (PMF), paleo-flood studies, Prediction in Ungauged Basins (PUB), reservoir sedimentation in heavy sediment-laden rivers, reservoir ice prevention, human activities impacts on river flow regimes, hydrological simulation technology, fuzzy mathematics, neural networks, wavelet analyses, GIS applications in engineering hydrology, etc. Nowadays, such tasks have made great progresses, given good solutions to some engineering hydrological problems in hydropower station, and laid a solid foundation for further in-depth studies. Meanwhile, a good many achievements have

---

[1] Yang Baiyin, Planning Director of HydroChina Corporation, Professor Senior Engineer.
[2] Wang Zhengfa, PowerChina Xibei Engineering Corporation Ltd. Professor Senior Engineer.

been made in scientific researches, a lot of hydrological scientific papers have been issued and a number of engineering hydrological monographs have been published. These monographs include those representative publications closely connected with engineering hydrology for hydropower projects in various periods, such as *Flood Survey and Calculation* (1957), *the Principles and Methods of Hydrological Statistics* (1964), *Flood Forecast Methodology* (1955), *Hydrological Forecast Methodology* (1993), *Calculation of Storm and Flood in Small Watershed* (1985), *Hydrological Analysis and Calculation* (1989), *Probable Maximum Precipitation and Flood* (1983), *Stochastic Hydrology* (1988), *Historical Extraordinary Floods in China* (1988), *Sediment Manual* (1992), *Stochastic Analysis of Hydrological and Water Resources* (1993), *Design Flood Calculation Manual for Hydropower Engineering* (1995), *Engineering Hydrology for Hydropower Stations* (1995), *Hydrological Studies in the Three Gorges Project* (1997), *Storm and Flood in the Yellow River Basin* (1997), *Calculation Principles and Methods for Probable Maximum Precipitation and Flood* (1998), *Hydropower Engineering in China-Engineering Hydrology* (2000), *Rationality Assessment of Design Flood* (2002), *Analysis and Study on Hydrological and Water Resources* (2003), *Atlas of Statistical Precipitation Parameters in China* (2006) as well as some engineering hydrological textbook published by the concerned colleges and universities.

6.2 Design Runoff

6.2.1 Introduction

The purpose of design runoff calculation is to analyze river runoff characteristics, volumes as well as the law of their intra-annual, and inter-annual variation, to define the rational scale of the construction of water resources and hydropower projects, to meet the people's demands for water and to offer accurate and reliable design data. The design runoff calculation mainly consists of the analysis and calculation of annual runoff as well as its spatial and temporal distribution, the analysis and calculation of low water flow in dry season and runoff stochastic simulation.

The analysis and calculation for design runoff are an important part of the hydropower station design. Hydropower stations with different regulating capacity have different requirements for design runoff. For example, the daily, weekly or ten-day design minimum flow is required for a run-of-river hydropower station; the annual runoff and the runoff in dry season for an annual regulating hydropower station; and the length and extent of consecutive low flow years is critical for a multi-year regulating hydropower station.

Runoff designs provide the basic design data for hydropower planning, engineering sediment, water conservation and environmental protection, etc. The design annual runoff, which is the key basis for identifying reservoir regulating capacity, normal water level, installed capacity, annual power production, firm capacity and other hydropower indicators, is closely related to the economic efficiency of hydropower station, and ultimately exerts indirect influence on those decisions on development of the hydropower station.

6.2.2 Runoff and its Characteristics

Runoff means the rainfall or snowmelt water which flows aboveground or underground by gravity. The runoff may be classified as rainfall runoff and snowmelt runoff by different water sources and as surface runoff and groundwater runoff by flow mode. And the surface runoff may be sub-classified as overland flow and river channel flow.

The water flow which passes through the river outlet section within one year is the annual runoff of the watershed upstream of this section. The water balance equation in a watershed shows that the annual runoff in the watershed is affected mainly by climate, surface and human activities.

Due to the abovementioned factors, intra-annual and interannual as well as spatial distributions of runoff vary a lot. According to the analysis of those discharge data observed in hydrological stations, runoff has the following characteristics.

6.2.2.1 Intra-annual Variation of Runoff

In China, river flow is mainly from the atmospheric precipitation with the glacier melt water as an additional recharge source for some rivers in the northwest and the southwest. Therefore, runoff recharge changes with climate and has great impacts on the intra-annual runoff distribution. Most parts of China are located in the North Temperate and subtropical monsoon climate zones, featuring obvious seasonal changes which follow the course of nature of high flow in summer, low flow in winter and transition in spring and autumn. Wet season and dry season are also called flood season and non-flood season respectively. The flood season and non-flood season start and end at different time in different regions with the difference between the maximum daily average flow and the minimum daily average flow several times or even dozens of times higher than the minimum.

For the majority of rivers in China, the intra-annual runoff distribution is mainly dependent on seasonal changes in precipitation.

(1) In winter (December to February), the runoff accounts for 4% to 6% of annual runoff for most rivers. This figure is 6% to 10% for rivers in the south, up to 25% for rivers in the northeast of Taiwan, less than 1% for some rivers in the north of the Amur River Basin.

(2) In spring (March to May), the runoff increases gradually as temperature rises. The runoff in spring accounts for 6% to 10% of annual runoff in the north, most of the southwest and the Sichuan Basin, 20% on the Altai Mountains in the northwest of Xinjiang, and up to 40% in the Dongting Lake Basin between the Yangtze River and the Nanling Mountains.

(3) Summer (June to August), which generally refers to the 4 consecutive months with the maximum runoff, is known as the flood season. The runoff in the flood season accounts for 60% to 90% of the annual runoff in the north China Plain and the costal regions of Liaoning and up to 60% in the most regions to the south of the Yangtze River and to the east of the Yunnan-Guizhou Plateau.

(4) In autumn (September to November), the runoff accounts for 20% to 30% of annual runoff in most parts of China with up to 50% in Hainan Island which is the region with the maximum runoff in autumn in China and less than 10% in arid and semi-arid regions with the minimum runoff. Table 6.2-1 shows annual runoff distribution in different seasons observed in the hydrological stations in some regions of China.

Table 6.2-1   Annual Runoff Distribution in Different Seasons Observed at Some Hydrological Stations in China

River name	Station	Number of years	Percentage in annual runoff (%)			
			Winter	Spring	Summer (4 consecutive months with the maximum runoff)	Autumn
Yarkant River	Kaqun	45	6.25	7.03	79.9	18.5
Irtysh River	Fuyun	50	2.73	24.8	67.5	11.4
Yarlung Zangbo River	Nugesha	24	8.30	8.50	73.6	32.6
Lancang River	Jiuzhou	51	7.93	14.6	62.8	29.7
Xijiang River	Wuzhou	37	6.70	18.9	64.2	21.4
Choshui River (Taiwan)	Chi-chi	26	9.20	17.1	62.2	26.9
Songhua River	Kiamusze	29	3.90	16.8	63.6	33.7
Hunhe River	Shenyang	42	3.30	16.5	71.0	21.0
Yellow River	Lanzhou	81	8.11	15.4	57.7	34.2
Heihe River	Yingluo Gorge	65	6.90	14.2	68.0	25.0
Weihe River	Xianyang	46	19.1	26.9	61.4	31.4
Huaihe River	Dapoling	28	6.20	20.7	64.8	17.4
Jialing River	Tingzikou	55	7.82	14.5	61.9	33.4

6.2.2.2 Regional Runoff Distribution

The regional runoff distribution is dependent on latitude, landform, weather conditions, etc. The coastal regions in the southeast of China have higher precipitation and more runoff. Since the climate in China varies on the both sides of the Nanling Mountains and the Qinling Mountains, the runoff is subject to certain geographic and vertical variations as well as some special variations in some areas. According to the national runoff depth isoline, China may be divided into five zones: humid zone, semi-humid zone, transition zone, semi-arid zone and arid zone. With more than 800mm of annual runoff depth, the humid zone mainly covers most coastal regions in the southeast of China, mountainous area in Hunan, Jiangxi and South Guangxi, Southeast Tibet, Southwest Yunnan, etc. With 800mm to 200mm of annual runoff, the semi-humid zone covers mountainous regions in the east of Northeast China, regions in the lower reaches of the Yangtze River and to the south of the Huaihe River and the Qinling Mountains, most regions in Yunnan and Guangxi Provinces, etc. With 200mm to 50mm of annual runoff, the transition zone covers some regions in the Songnen Plain, the Sanjiang Plain, the plain in the lower reaches of the Liaohe River, most regions in the North China Plain, Shanxi and most of Shaanxi, central part of the Qinghai-Tibet Plateau, regions in the west of Xinjiang, etc. Respectively with 50 to 10mm and less than 10mm of annual runoff depth, the arid zone and semi-arid zone mainly cover those plateau areas in Northeast China, Northwest China and Inner Mongolia, etc.

6.2.2.3 Inter-annual Variation of Runoff

The coefficient of variation for annual runoff (C_V) and the extremal ratio (the ratio of maximum to minimum, K) are used to show inter-annual variation of runoff. The values of C_V for the rivers in China are significantly subject to a geographic distribution rule, increasing gradually from Southeast to Northwest China. In general, the geographic distribution rule to which the values of annual runoffs in all medium-sized watershed areas in China are subject is that the value of C_V is lower than 0.5 for those regions to the south of the hilly areas along the Yangtze River and the Huaihe River as well as the Qinling Mountains, 0.60 to 0.80 for most regions in the Huaihe River Basin, about 1.0 for the North China Plain, lower than 0.5 for those mountainous regions in Northeast China and inland river basins, and higher than 0.8 for plains and basins. Table 6.2-2 shows the average annual discharge as well as the values of K and C_V observed at some hydrological stations for some rivers in China.

Table 6.2 – 2 Average Annual Discharge and Values of K & C_V Observed at Hydrological Stations for Some Rivers in China

River	Station	Control area (km^2)	Number of years	Annual average discharge ($10^8 m^3$)		K	C_V
				Maximum	Minimum		
Yarkant River	Kaqun	50,248	45	95.6	44.8	2.13	0.17
Irtysh River	Koktokay	5,005	51	25.1	7.22	3.48	0.32
Niyang River	Gengzhang	15,581	32	196	112	1.75	0.16
Songhua River	Harbin	390,526	78	847	123	6.90	0.43
Taizi River	Shenyang	8,082	40	50.1	10.8	4.60	0.41
Chaobai River	Suzhuang	17,595	58	64.7	3.35	19.31	0.74
Luanhe River	Luanxian	44,100	50	128	16.1	8.00	0.54
Yongding River	Guanting Reservoir	43,402	47	32.2	7.16	4.50	0.35
Huaihe River	Sanhe Sluice	158,160	53	944	64.1	14.80	0.62
Yellow River	Shanxian	688,421	61	823	242	3.40	0.25
Heihe River	Yingluo Gorge	10,009	65	23.5	10.9	2.16	0.17
Xijiang River	Wuzhou	329,705	37	3,470	1,070	3.20	0.20
Jinsha River	Pingshan	458,592	40	1,950	1,070	1.80	0.16
Jialing River	Tingzikou	61,089	55	350	89.9	3.89	0.33
Yangtze River	Hankou	1,488,036	113	10,130	4,531	2.20	0.14
Yangtze River	Yichang	1,005,501	100	6,037	3,345	1.80	0.12

In China, the inter-annual variation in annual runoff is also characterized by the occurrence of consecutive wet and dry years which often last for 3 to 5 consecutive years or even more than 10 years. According to the statistics of observed dataset, consecutive dry years happened to many rivers in China (see Table 6.2-3).

Table 6.2 – 3 Consecutive Dry Years in Large Basins of China

River	Control section	Average annual discharge Q_N (m^3/s)	Consecutive dry years		
			Starting and ending time	Years	Average annual discharge Q_N (m^3/s)
Songhua River	Harbin	1,190	1916 – 1927	12	678
Yellow River	Sanmen Gorge	1,350	1922 – 1932	11	994
Yalu River	Yunfeng	278	1939 – 1950	12	236

Countinued

River	Control section	Average annual discharge Q_N (m³/s)	Consecutive dry years		
			Starting and ending time	Years	Average annual discharge Q_N (m³/s)
Nenjiang River	Buxi	338	1967–1977	11	258
Huaihe River	Bengbu	788	1970–1979	10	703
Yangtze River	Hankou	23,400	1955–1963	9	21,322
Yuanjiang River	Wuqiangxi	2,060	1955–1966	12	1,771
Xin'an River	Xin'an River	338	1956–1968	13	265
Yujiang River	Xijin	1,620	1952–1962	11	1,314
Minjiang River	Zhuqi	1,750	1963–1972	10	1,388
Xiushui River	Tuolin	254	1956–1965	10	210
Hanjiang River	Ankang	608	1965–1973	9	525
Minjiang River	Gaochang	2,840	1969–1979	11	2,570

6.2.3 Analysis and Calculation of Design Average Annual Runoff

Before the analysis and calculation of the runoff, the reliability, conformity and representativeness of the observed runoff data in the design representative station shall be checked. If the influence of the human activities seems obvious before and after the data acquisition, the conditions of the underlying surface of the watershed change noticeably and thus alter the intra-annual or inter-annual distribution of runoff, the restoration calculation for runoff must be conducted to ensure the conformity of the data series.

6.2.3.1 Statistical Calculation of Annual Runoff

When observed data for a long term are available at the design representative station, direct statistical calculation may be made. Otherwise, before making the statistical calculation, the adjacent reference station observed data for a long term shall be selected to perform correlation analysis, interpolate and extend the data series of design representative station. According to the design requirements of hydropower stations, the statistical calculation may respectively calculate the average annual discharge, count the historic monthly maximum and minimum discharge and calculate the runoff modulus by calendar year and hydrological year. The average annual runoff series of the dam site can be obtained by area ratio method, area ratio with the correction of precipitation, or a linear interpolation of the areas between the upstream hydrological site and the downstream one.

6.2.3.2 Interpolation & Extension of Runoff Series and Its Representativeness Analysis

The main routine analysis on the runoff series include correlation analysis, series representation analysis, hydrological period analysis, etc., which are respectively described below.

1. Correlation analysis and interpolation & extension

It's a common situation in hydrological analysis that the observed data are rare or insufficient for a certain phenomenon while being abundant for another related phenomenon. In this case, the short series can be interpolated and extended through regression and correlation analysis and calculation. Before the correlation analysis and calculation, it must be analyzed whether the phenomena are correlated. Otherwise, the correlation analysis is meaningless. During engineering hydrology design, the bivariate correlation analysis with different curve types available for selection is most commonly used.

The parameters of the correlation equation are estimated using the least square method. For the linear function, the parameters can be directly estimated and, for nonlinear function, it must be firstly converted into linear function by variable substitution before parameter estimation. The line type is selected based on the principle of the minimum mean square deviation. The linear function is the most commonly used type in engineering hydrological analysis and calculation.

While interpolating the hydrological data using the regression equation, the correlation coefficients and correlation equation must receive statistical test to determine whether the created regression equation is reasonable and applicable. The test methods include correlation test, t test and F test.

2. Representation analysis

According to specifications of engineering hydrology, representation analysis of the runoff series is required while analyzing the frequency of average annual runoff.

Representation analysis generally refers to the representation of runoff series. The annual runoff series with conformity is the sample, and the similarity to the overall statistical characteristics of that of the sample is called representation. The higher the similarity is, the better the series representation is, and the higher the accuracy of the result of frequency analysis is. Otherwise the accuracy is lower. However, the overall condition is still unknown and direct comparison is unavailable. The judgment is mainly based on the designer's knowledge. The common analysis methods of runoff representation are as follows.

(1) Periodicity analysis method. The runoff series of n years features the runoff value of each year fluctuating around the mean value, wet-year and dry-year groups alternately occurring. For the runoff series of n years, the test shall focus on whether it includes the wet, normal and dry periods and whether the wet and dry periods are approximately distributed in symmetry. If so, the representation is good. Otherwise the representation is poor. Generally speaking, longer runoff series means better representation. However, exceptions may exist. If the number of the wet periods is larger than that of dry periods, the average annual runoff is higher than the reality. Otherwise it's lower. A better representation can be obtained by removing one of the wet or dry periods. When doing so, careful and thorough analysis must be made. The periodicity of the annual runoff series can be analyzed using the annual runoff sequence chart and integral indifference curve or spectrum of annual runoff modulus.

(2) Comparative analysis method for statistical parameters of long and short series. According to the three grouping methods by various lengths (forward sequence and backward sequence) and setting length (slide), use the moment method to calculate series mean, C_V, C_S and mean and the sampling error of estimated value of C_V for the designer to check the variation of the statistical parameters and analyze the series representation.

With the accumulative increment of the series length, both the calculated mean and coefficient of variation tend to stabilize and the sampling errors of the mean and C_V estimation are gradually decreasing, then the series is of good representation. Otherwise, it has low representation.

(3) Fractal theory analysis method. In recent years, some practical achievements have been made in studying the hydrological phenomena using the fractal theory, chaos theory and wavelet analysis method.

The fractal theory, chaos theory etc. are the important parts of the modern nonlinear science and complexity study. According to the fractal theory, many objects are of geometrical or statistical self similarity, both partially and as a whole and with unlimited nesting. Such a self similarity structure in form, function, information, time, space etc. and with infinite hierarchies among the entirety and its ingredients creates numerous complex objects and phenomena and forms a type of highly difficult issues with the so called "Scaleless Property". The representation analysis of runoff series is one of the issues.

There is still lack of a recognized good method of the representation analysis of runoff series. It's generally considered that longer series is of better representation which, however, isn't always the case. It's indeed of great difficulty to accurately predict and determine the representative period of the annual runoff data of long series (sample length $\geqslant 50$ years). In most cases, the annual runoff series in certain length (sample length $\geqslant 30$ years) are all periodic, in other words, with time self similarity. Therefore, the part (sub-series) with time self similarity to its entirety ascertained by analyzing the annual runoff series using relevant methods of fractal theory is representative.

3. Hydrological periodicity analysis

The periodicity analysis of runoff series is the main point of representation analysis. The conventional methods to determine the primary period of stochastic sequence are the power spectrum method and periodogram method of R. B. Blackman-J. W. Tukey (1959). They are both of obvious flaws, including the influence of the truncation order of autocorrelation function and the selection of the window function on spectrum estimation. To overcome the insufficiency in the conventional estimation method, other new spectrum estimation methods are proposed, including the most broadly adopted maximum entropy spectrum method.

6.2.3.3 Calculation of Design Annual Runoff

Design annual runoff refers to the average annual runoff and runoff series that comply with the design standard. The annual runoff design standard is generally expressed by reliability or frequency. Design annual runoff is an important hydrological parameter in the planning and design of the hydroelectric generation, irrigation, water supply, navigation projects etc. The calculation contents include ①the average annual discharge; ②the annual runoff and runoff series in conformity with the specified design reliability. The annual runoff and runoff series reflect the intra-annual and inter-annual variation of runoff in a certain watershed, and represent the possible runoff process might occur during the future operation of the project. The design reliability of annual runoff is determined according to the requirements of the water consumer. Depending on the hydrological data condition of the project location, the design annual runoff is calculated using different methods, generally included as the follows.

(1) Frequency calculation method. This method is applicable to the situations with long term observed discharge records (more than 30 years, interpolating and extending data included) at the project site or nearby locations. Firstly check the reliability, conformity and representativeness of the observed data. If the observed data are influenced by human activities and the runoff is of obvious change, restoring calculation is required to restore them to the natural condition without the influence of human activity to achieve the conformity in the data series. The main items of the restoring calculation include the industrial, agricultural and living water consumption, the impoundage and discharge of the water storage work, water diversion and dam-break water flows, inter-basin water transfer, etc. The series representation analysis is made on the restored runoff series and the representation is evaluated using the previous representation analysis methods. If the series representation is inadequate, the statistical parameters shall be properly adjusted by extending the series or referring to regional comprehensive results. Then the statistical parameters are determined by statistical calculation to obtain the annual runoff at different reliabilities.

(2) Isogram method. The method is applicable to medium-sized and small projects in ungauged basin. As there is generally a certain internal connection between the hydrological phenomena and geographical factors, the hydrological parameters mostly distribute following the geographical rules that can be used to conduct hydrological calculation in the region lacking hydrological data. The presently compiled hydrological manuals of each province and city (district) of China include a set of isograms of the statistical parameters of annual runoff. By looking up the statistical parameters of the project location, the annual runoff at different reliabilities can be obtained.

The runoff frequency analysis method is to determine a probability distribution function, preliminarily estimate the statistical parameters and determine the design runoff parameters of design representative station through performing curve fitting, comprehensive coordination of regional parameter and rationality analysis using the mathematical statistics tools.

For the majority of rivers in China, the type of the runoff frequency curve shall be Pearson-Ⅲ type or P-Ⅲ to give it the usual abbreviation. Other types may be adopted based on analysis and argumentation in special conditions.

The density curve $f(x)$ of P-Ⅲ type curve is:

$$f(x)=\frac{\beta^{\alpha}}{\Gamma(\alpha)}(x-a_0)^{\alpha-1}e^{-\beta(1-a_0)} \qquad (6.2-1)$$

where α, β and a_0 are parameters.

The three parameters, α and β and a_0 of P-Ⅲ type curve are related to the mean value, deviation coefficient and coefficient of skewness as follows:

$$\alpha=\frac{4}{C_S^2} \qquad (6.2-2)$$

$$\beta=\frac{2}{\overline{X}C_V C_S} \qquad (6.2-3)$$

$$a_0=\overline{X}\left(1-\frac{2C_V}{C_S}\right) \qquad (6.2-4)$$

Since the minimum of many hydrological data are greater than zero, the following condition must be met:

$$C_S \geqslant 2C_V \qquad (6.2-5)$$

The statistical parameters of P-Ⅲ type frequency curve are expressed using the mean value X, deviation coefficient C_V and coefficient of skewness C_S. The statistical parameters are often preliminarily estimated by the moment method or other parameter estimation methods and determined with the curve fitting method.

In recent years, China conducted a lot of researches on the parameter estimation of P-Ⅲ type curve and proposed the new methods as the weight function method, single weight function method, double weight function method, probability weight moment method and linear moment method, etc.

After determining the statistical parameters, calculate the design flood at each frequency according to the table of coefficient of deviation Φ_p or table of modulus rate coefficient K_p for P-Ⅲ type curve and using the following formula:

$$X_p = \overline{X}(1+\Phi_p C_V) \qquad (6.2-6)$$

or

$$X_p = K_p \overline{X} \qquad (6.2-7)$$

The mathematical expectation formula shall be used for calculating the empirical frequency for the curve fitting of runoff frequency in China.

The frequency curve of average annual discharge at the dam site of the Yellow River Longyangxia Hydropower Station is shown in Fig. 6.2-1.

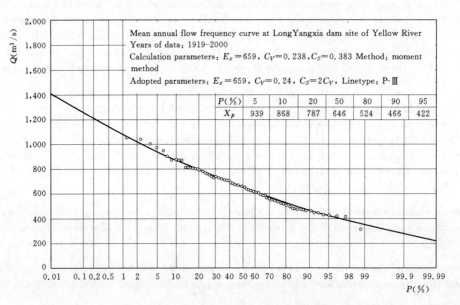

Fig. 6.2-1 Frequency Curve of Average Annual Discharge at Dam Site of
Yellow River Longyangxia Hydropower Station

Comprehensive balancing shall be made on the statistical parameters of the runoff series after frequency analysis according to the regional distribution rules of the parameters. Specifically for the cascade reservoirs, the runoff frequency curves of the upper reach and lower reach should not intersect, so as to avoid unbalancing the water volume between the upper reach and lower reach in the design conditions. Generally, with the increase of the watershed area, the mean value increases and C_V value gradually decreases. If abnormality exists, its rationality shall be analyzed and proven according to the specific characteristics of hydrological, meteorological and natural geographical conditions. Otherwise, proper correction shall be made.

6.2.3.4 Temporal Distribution of Design Annual Runoff

The distribution of the annual runoff at different reliability in a year is the runoff variation in one year with the time. If the runoff data of longer time are available in the location of a hydropower station project, usually the distribution in the year of longer series is taken as the calculation basis of the runoff regulation of reservoir. If the design representative years of runoff at different reliability are required, the typical year method is used. The method is to select the year close to the runoff flow at design reliability as the typical year of the reliability and to control and enlarge the wet and dry periods using the annual runoff flows of different reliabilities to obtain the representative year of the design runoff at different reliabilities of the design project. In the region without observed data, the distribution of the annual runoff flow in the year can be estimated based on the distribution percentage in each month of the year that is provided by the reference station or the hydrological manual, with proper correction made when necessary.

6.2.4 Analysis and Calculation of Low Flow in Non-flood Season

The low flow in non-flood season is also called the minimum flow, which is a special form of river runoff. The low flow is different from but closely related to drought. Generally, drought doesn't necessarily occur each year while low flow does. Depending on the project requirements, the design period of low flow can be expressed by instant, a day, ten days, a month, a year or longer time. The low flow period with duration over one year is called the consecutive low flow period.

The low flow is usually characterized by its scale, duration and probability of occurrence. The low flow is the important factor of determining the firm capacity and the regulating capacity of reservoir.

6.2.4.1 Analysis of Low Flow Characteristics

China has a marked continental monsoonal climate characterized by hot rainy season. Due to the water vapor accompanied by the southwest monsoon from the Indian Ocean and the southeast monsoon from the Pacific Ocean etc., the precipitation pattern of China has the following features: it's rainy in the southeast and in mountainous region while dry in the northwest and on the plains. As a result, the five precipitation distribution zones in the country have been created, as shown in Table 6.2-4.

Table 6.2-4 Characteristics of Precipitation Distribution Zones of China and Minimum Monthly Flow Modulus

Precipitation Zonality	Mean annual precipitation (mm)	Annual rainy days (d)	Annual runoff depth (mm)	Minimum monthly flow modulus $[dm^3/(s \cdot km^2)]$	Months of occurrence
Ample rain zone	>1,600	>160	>900	16.00 - 1.50	December (or April, March)
Rainy zone	1,600 - 800	120 - 160	900 - 200	13.80 - 0.10	December (or January, February, March, April)
Semi-humid zone	800 - 400	80 - 100	200 - 50	3.64 - 0.00	December (or May, June)
Semi-arid zone	400 - 200	60 - 80	50 - 10	2.90 - 0.00	January, February
Arid zone	<200	<60	<10	0.73 - 0.00	January, February

The precipitation in each zone has important influence on the occurrence of the low flow runoff. In basically the same natural geographical conditions, the amount of precipitation directly determines the low flow. The differences in the precipitation distribution between the south and north China clearly reflect that the low flow runoff is larger in the south than in the north. Meanwhile, the distribution of the minimum flow in China reveals a certain regional characteristics. The distribution trend of the minimum monthly flow modulus is shown to be approximately conformable to the zone distribution of the annual precipitation. Taking the upper and middle reaches of the Yellow River for example, the distribution trend of the annual precipitation is basically similar to that of the modulus isoclines of the mean annual minimum monthly flow. As the distribution characteristic across the country, they basically gradually decrease from the rainy zone along the southeast coastline to the dry area in the northwest inland. This fully reflects the variation of humidity conditions from south to north and the regional variation.

The runoff in dry seasons of the rivers in China accounts for approximately 5%-25% of the annual runoff, and the low flow varies greatly among the years. There are seasonal rivers in the northwest regions, which run dry in the dry seasons. Also, only a few large rivers run dry in the dry seasons due to the influence of human activities. Most rivers have consecutive low flow period of more than two years.

In analyzing the low flow characteristics, the following results of the low flow analysis should be provided partially or completely, depending on the data conditions and the design requirements of projects.

(1) The general rules and abnormal changes of the low flow in a year include the normal year (or average annual), dry year, low flow period in extraordinary dry year, average flow and its proportion in three or two consecutive driest months, driest monthly runoff flow and minimum daily average flow; the occurrence month of dry and zero flow in low flow period and the duration.

(2) The general rules and abnormal changes of the low flow among the years include dry year, the extraordinary dry year and the occurrence periodicity rule of the period of consecutive dry years; the occurrence of particularly extraordinary dry year and the possibility of its future recurrence.

During the hydrological design of the Three Gorges Hydropower Station, Changjiang Water Resources Commission analyzed the runoff region composition and inter-annual variation of Yichang station during the low flow period (January-March) according to the data of 1951-1983 at Yichang station, see Table 6.2-5.

Table 6.2-5 Statistics of Runoff Region Composition and Inter-annual Variation at Yichang Station in Low Flow Period (January-March)

River	Name of hydrological station	Watershed area (km²)	Region composition (%)	Average annual runoff (100 million m³)	Runoff modulus (10,000 m³/km²)	Proportion in Yichang runoff (%)	Proportion in inter-annual variation of Yichang runoff (%)	C_V	Extremal ratio of runoff
Jinsha River	Pingshan	485,099	48.2	112.8	2.33	36.0	34-45	0.09	1.50
Minjiang River	Gaochang	135,378	13.5	59.6	4.40	19.0	17-22	1.4	—
Tuojiang River	Lijiawan	23,283	2.3	6.1	2.62	1.9	1-3	0.34	3.78
Jialing River	Beibei	156,142	15.5	33.6	2.15	10.7	8-16	0.18	2.44
Wujiang River	Wulong	83,035	8.3	38.1	4.59	12.2	8-18	0.26	2.66
Wujiang River	Qujian	122,564	12.2	63.5	5.18	20.2	12-25	0.23	2.50
Yangtze River	Yichang	1,005,501	100.0	313.7	3.17	100	—	0.11	1.81

6.2.4.2 Frequency Analysis of Low Flow

The frequency analysis of low flow is the important method of designing the low flow runoff. P-Ⅲ type is also adopted for the frequency distribution of the low flow in China. In special cases, other types of distribution function may also be used if proven.

1. Conventional analysis methods of low flow frequency

When the long series observed runoff data are available for the design representative station, the average flow within the shortest period of the year can be selected to constitute the sample series according to the sampling rule of annual minimum. The minimum flow of statistical period shorter than 12 months is called the consecutive hydrological drought event in the year and its frequency analysis is called the analysis of frequency of yearly low flow. The analysis method is the same as the ordinary frequency analysis method. However, attention shall be paid to the following points:

(1) Determine the statistical periods according to project requirements, which are generally 1d, 3d, 5d, 7d, 15d and 30d.

(2) Count the average minimum flow (or runoff, the same below) in the selected period each year as the sample series.

(3) Divide the years by hydrological year or water year instead of calendar year.

(4) Rank the minimum flow from the smallest to the biggest and calculate the deficient empirical frequency $P\ (X \leqslant x_i)$.

(5) Inflection point may exist in the parts below 20% and above 90% in the frequency curve of the annual low flow. The inflection point of frequency below 20% may be considered to be exhausted for surface water and phreatic water. The rivers are mainly recharged by the deep phreatic water. The inflection point of frequency above 90% may result from the large low flow in particular years and the recharged by surface water.

(6) If the sample series is long, the relation between the exceedance probability p and non-exceedance q may be approximately expressed by $q = 1 - p$.

The frequency curve of annual low flow at Shangsi station on Qingshui River of Bailongjiang River is shown in Fig. 6.2-2.

2. Frequency Analysis of low flow series with zero flow

China is vast in territory and the conditions of low flow are varied. Small and medium-sized rivers in the northwest run dry frequently. The water recourses of large rivers, such as the Yellow River basin, become tenser each day due to the influence of human activities, and zero flow also occurs frequently in the dry season in the lower reach at the estuary. For the frequency analysis of the low flow series with zero flow, the Chinese hydrologists have conducted many researches and proposed some practi-

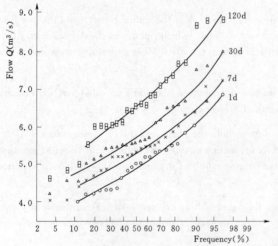

Fig. 6.2-2 Frequency Curve of Annual Low Water Flow at Shangsi Station on Qingshui River of Bailongjiang River (1957-1986)

cal methods such as the simple and practical frequency ratio method proposed by Jin Guangyan.

Let there be k years of non-zero series in the low water flow series of n years, which are taken as independent and stochastic continuous series to make frequency analysis and obtain the frequency curve. Then, convert the design frequency of a part of the series into the frequency of the whole series through the following formula:

$$P_{design} = (k/n) P_{non\text{-}zero} \qquad (6.2-8)$$

where P_{design} is the design frequency of the whole series and $P_{non\text{-}zero}$ is the corresponding frequency of the non-zero series.

6.2.4.3 Calculation of design low flow

Design low flow is the low water flow in conformity with the design standard. When observed runoff data are available for the design representative station, the design low flow of the design representative station can be calculated directly using the frequency analysis method. Otherwise, the design low flow is often estimated using the isogram method or hydrologic analogy method. If there are cascade reservoirs or multiple reservoirs with large regulation and storage effects built in the upstream design section, the fact that the regulation effect of those reservoirs obviously changes the natural hydrological condition and its influence on the design low flow of the downstream design section shall be considered when calculating the design low flow.

6.2.5 *Analysis of Multi-year Consecutive Low Flow Period*

The analysis and calculation of design annual runoff is based on a year period. The observation station with more observed data is only of decades of data. Those data meet the design requirements of a hydropower station with annual regulating capacity. However, they are too short for large hydropower station with multi-year regulating reservoir. If there is long but only one consecutive low flow period in the series, the adopted long consecutive low flow period will largely influence the engineering design of the hydropower station. The occurrence probability of the low flow period and the probability of its average flow (or water flow), as well as how the low flow period will be considered in the planning and design of the hydropower station, shall all be carefully analyzed.

Since the 1990s, the Chinese hydrologists have conducted a lot of researches on the multi-year consecutive low flow period and made some important research achievements. Shi Fucheng, et al. analyzed and studied the consecutive low flow period of 11 years from 1922 to 1932 of the Yellow River in 1991. Also in 1991, Wang Weidi, Sun Hanxian, et al. analyzed the upstream consecutive low flow period of the Yellow River, analyzed the recurrence period of the low flow period of 11 consecutive years from 1922 to 1932 in the upper reach of the Yellow River and studied the frequency analysis of the consecutive low flow period. In 1998, Yang Baiyin made a review and summarization on the studies of consecutive low flow period in the upstream of Yellow River, and applied the achievement to the operation of Longyangxia Multi-year Regulating Reservoir on the Yellow River, which had got good results. In 1995, based on the basic principles of the probability theory and turn theory, Wang Zhengfa carried out further theoretical study on the probability analysis and determination of recurrence period of multi-year consecutive low flow. He studied the probability distribution $R_{i,k}(n)$ of the multi-year consecutive low flow with the turn length of k years and provided the solving equation. In 2010, Gu Ying, Zhang Shifa, et al. studied the characteristics of the consecutive dry years in China and their variation trends and put forward corresponding strategies.

6.2.5.1 Determination of Recurrence Period and Frequency Analysis of Multi-year Consecutive Low Flow Period

Use the measured or randomly generated flow data to analyze the number of occurrences of the multi-year consecutive low flow period with the length of k years in the series and determine its recurrence period, for example, as determined by Wang Weidi, Shi Fucheng, et al (1991). through analysis, the recurrence period of the consecutive low flow period of 11 years from 1922 to 1932 in the upper reach of the Yellow River is 80-200 years. With level of severity of the low flow in the consecutive low flow period taken into consideration, the recurrence period will be approximately 1,000 years.

The frequency analysis of multi-year consecutive low flow period is usually limited by the series length and can't be performed effectively. In the latest 30 years, with the advancement in engineering technology and the engineering hydrological practice, some experience was accumulated for reference.

If the length of the consecutive low flow years is small (for example, 2-3 years), the runoff series not less than 50 years can be adopted. If the length is more than 4 years, it's recommended to adopt the extra-long runoff series generated using the stochastic simulation technology.

Count the average flows Q_k of the consecutive low flow period with the turn length of k in the runoff series of N years and rank the flows from small to large, whereby obtain the series part of the multi-year consecutive low flow period. The empirical frequency P and recurrence period T of the low flow period with the duration of k years are still calculated by the mathematical expectation formula, i. e. :

$$P = \frac{M}{N+1} \qquad (6.2-9)$$

$$T = \frac{N+1}{M} \qquad (6.2-10)$$

where M is the serial number of Q_k in the series composed of the consecutive low flow periods of k years that are ranked from the smallest to the biggest. N is the total number of years of the series.

Formulas (6.2-9) and (6.2-10) refer to the frequency and recurrence period of the average flow Q_k in the condition of k years of consecutive low flow periods within N years. When converted to annual probability and recurrence period, the following are obtained:

$$P_k = kP \qquad (6.2-11)$$

$$T_k = \frac{T}{k} \qquad (6.2-12)$$

By plotting the graph of relation between the average discharges Q_k and the corresponding P and k in the consecutive low flow periods of k years, a group of empirical frequency curve can be obtained, as shown in Fig. 6.2-3.

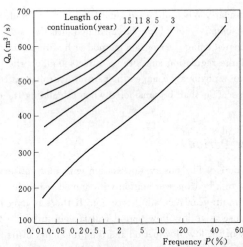

Fig. 6.2-3 **Empirical Frequency Curve of Average Flow in Multi-year Consecutive Low Flow Period at Longyangxia Hydropower Station of Yellow River**

The probability distribution function of the multi-year consecutive low flow period is yet to be clearly specified in China. The frequency curve can be generated referring to the conventional method to use P-III distribution for curve fitting.

6.2.5.2 Analysis of Frequency of Multi-year Consecutive Low Flow Period

Wang Zhengfa (1995) assumed that the water inflow of a river is an infinitely continuous Bernoulli event and, based on the basic principles of probability theory and theory of rounds, studied the probability distribution $R_{i,k}(n)$ of the multi-year consecutive low flow of k years and provided the following solving equation of $R_{i,k}(n)$:

$$\begin{cases} R_{1,k}(n) = f_1 R_{0,k}(n-1) + f_2 R_{0,k}(n-2) + \cdots + f_{n-1} R_{0,k}(1) \\ R_{2,k}(n) = f_1 R_{1,k}(n-1) + f_2 R_{1,k}(n-2) + \cdots + f_{n-1} R_{1,k}(1) \\ \vdots \\ R_{I,k}(n) = f_1 R_{I-1,k}(n-1) + f_2 R_{I-1,k}(n-2) + \cdots + f_{n-1} R_{I-1,k}(1) \end{cases} \qquad (6.2-13)$$

The calculation of the factor f_i ($i=1, 2, \cdots, n$) in formula (6.2-13) has the following recurrence relation:

$$\begin{cases} f_1 u_0 = u_1 \\ f_1 u_1 + f_2 u_0 = u_2 \\ \vdots \\ f_1 u_{n-1} + f_2 u_{n-2} + \cdots + f_n u_0 = u_n \end{cases} \qquad (6.2-14)$$

The calculation of the factor u_i ($i=1, 2, \cdots, n$) in formula (6.2-14) has the following recursion relation:

$$\begin{cases} u_n + u_{n-1} q + \cdots + u_{n-k+1} q^{k-1} = q^k \\ u_1 = u_2 = \cdots = u_{k-1} = 0, u_0 = 1 \end{cases} \qquad (6.2-15)$$

To calculate $R_{i,k}(n)$ by recurrence using formula (6.2-15), the nonoccurrence probability $R_{0,k}(n)$ of the multi-year consecutive low flow period with the turn length of k years in the runoff series of n years is to be calculated. The solving formula of $R_{0,k}(n)$ is as follows:

$$R_{0,k}(n) = 1, n < k \qquad (6.2-16)$$

$$R_{0,k}(n) = 1 - p^k, n = k \qquad (6.2-17)$$

$$R_{0,k}(n)=1-g(n), n>k \qquad (6.2-18)$$

where $g(n)$ is the probability of at least one occurrence of the multi-year consecutive low flow period with the turn length of k years in the year with the order of n, which can be calculated by recursion using the following formula:

$$g(n+1)=g(n)+p^k q[1-g(n-k)], n>k \qquad (6.2-19)$$

$$g(n)=0, n<k \qquad (6.2-20)$$

$$g(k)=p^k, n=k \qquad (6.2-21)$$

In practice, if $n<k$ or $n=k$, calculate $R_{0,k}(n)$ directly using formulas (6.2-16) and (6.2-17). If $n>k$, firstly use formula (6.2-19) to solve $g(n)$ by recursion and then calculate $R_{0,k}(n)$ using formula (6.2-18).

6.2.6 Stochastic Simulation of Runoff

The variation of runoff with time has both certainty and uncertainty. The variation of the runoff in wet period and dry period during a year, as well as the inter-annual difference between wet year and dry year, are the representation of the certainty of the runoff variation. However, the quantity of runoff in a future period can't be accurately predicted, which is the representation of the uncertainty of runoff variation.

During the 1950s-1970s, mainly the probability distribution function of the previously observed runoff series or runoff was adopted in China to describe the future variation of the runoff. Sometimes, the Markov process was also used to describe the future variation of runoff, which considers that a certain correlation only exists between the discharges within adjacent periods.

According to the analysis of the observed runoff series of the main rivers of China, for the rivers on the north of the Yellow River (including the Yellow River), there is obvious alternation between the wet year and dry year groups. The attenuation of the autocorrelation function is slower, and the autocorrelation functions between the original series and the series with periodic items filtered are clearly different from each other. The situation is contradictory, however, for the rivers on the south of the Yellow River, where there is no obvious transition between wet year and dry year groups, the autocorrelation function quickly attenuates and there is barely any difference between the autocorrelation functions of the original series and the series with periodic items filtered. This indicates that the variation of runoff neither repeats the historic process nor obeys the purely stochastic probability distribution function.

The improvement in stochastic simulation technology of runoff brings the study of runoff description into a new development stage and provides a new approach of taking the certainty and uncertainty of the runoff variation into overall consideration. It utilizes some basic information of runoff generation mechanism and fully uses and analyzes the runoff data obtained. Therefore, more reasonable runoff variation rules can be obtained through the stochastic simulation of runoff. Therefore, the stochastic simulation of runoff has drawn general attention from hydrologists since it comes out.

The study of the stochastic simulation of runoff in China started in the early 1980s. During the present simulation of annual and monthly runoff series, the commonly used models are mainly the auto-regression model and disaggregation model. The auto-regression model could better reflect various spatiotemporal dependence relations of the runoff series and is a theoretically and methodologically matured stochastic simulation of runoff. Its mathematical expression is:

$$x_t = f(x_{t-1}, x_{t-2}, \cdots, x_{t-p}) + \varepsilon_t \qquad (6.2-22)$$

where x is runoff, t is time, p is the autoregressive process of order, $f(*)$ is a deterministic function describing the deterministic part of the runoff variation, ε_t is a random variable describing the uncertain part of the runoff variation.

A special case of formula (6.2-22) is that the linear auto-regression model with the process order of p is:

$$x_t = a_1 x_{t-1} + a_2 x_{t-2} + \cdots + a_p x_{t-p} + \varepsilon_t \qquad (6.2-23)$$

where $a_1, a_2, \cdots, a_p$ are the auto-regression coefficient.

The linear auto-regression model simulates the characteristics of runoff using linear auto-regression equation (6.2-23) and directly generates runoff series in time sequence. In China, the commonly used auto-regression models include Markov Model, Thomas-Fiering Model and multivariate auto-regression model etc. and most models that have been used so far are of first-order.

The basic idea of the solution set model is to, firstly study the stochastic simulation of the high aggregation level variable (annual runoff, for example) and then break the high aggregation level variable into low aggregation level variable (monthly runoff, daily runoff, for example) using certain methods. The commonly used disaggregation models include the correlation disaggregation model and typical disaggregation model. The correlation disaggregation model is based on the assumption that the high aggregation level variable is formed by adding up the low aggregation level variables which can be obtained by disaggregating the high aggregation level variable. The basic formula is:

$$Y = AX + B\varepsilon \qquad (6.2-24)$$

where Y is the high aggregation level variable, X is the low aggregation level variable, A reflects the correlation between the high and low aggregation level variables, B reflects the correlation between the low aggregation level variables and ε is a random variable.

The concept of the typical disaggregation is basically identical to that of correlation disaggregation, i. e., that the low aggregation level variables compose and can be obtained by disaggregating the high aggregation level variable. It differs from the correlation disaggregation in that the disaggregation is obtained by the measured typical process instead of the correlation relation. The main advantage of the typical disaggregation model is that the sum of the components exactly equates to the total from which they are disaggregated, which meets the water balance requirements.

The simulation series of each component reflects its own statistical characteristics and variation principle, while the amount series reflects the statistical characteristics and variation principle of the amount.

In the stochastic model of runoff, in addition to the auto-regression model and disaggregation model above, there are also the shot noise model, fractional Gaussian noise model, threshold model, conversion model and single-variable, multivariate, seasonal, out-of-season, normal, non-normal models etc. Therefore, it's still an important question for hydrologists to select from numerous stochastic simulation models the one suitable for the characteristics of runoff variation of a certain watershed and to create better stochastic simulation model of runoff.

6.3 Design Flood

6.3.1 Introduction

Flood prevention capacity during the operation period of a dam is one of indices for measuring its safety. Besides, some reservoirs are responsible for the flood control of some areas or cities downstream of them. Thus, there appears a question on design criteria for flood control safety of dams and areas or cities downstream of reservoirs. Flood control safety of dams and protection areas downstream of reservoirs is related to the flood standard which is adopted as the design basis of dams. The higher the flood standard adopted, the lower the risk of hydro-structures damaged and protection areas submerged during the operation period. However, the engineering investment will increase accordingly. Conversely, if the standard of design flood is too low, the engineering investment will reduce naturally. However, the accident rate of the project will increase accordingly. Once accident happens, the downstream area will suffer huge losses. As a result, what flood standard shall be selected as the design basis? The flood standard shall be determined on the basis of analyzing flood control safety risk, flood control benefit, consequences of accidents, investment and other relations through integrated economic-risk analysis and taking into consideration such factors as possible personal casualty, economic loss, as well as political, social and environmental impacts. Flood control standard has reflected economic, technical, political, social, environmental and other factors of a nation integrally through engineering, which is related to the project safety and the safety of human life and property, industrial and mining enterprises and facilities as well as ecological environment. Moreover, it will directly affect normal functioning of project benefit, project cost and construction speed. Its determination is an important link showcasing the design complying with laws of nature and laws of economics and demonstrating the economic policy and technical policy of a nation. In principle, the flood control standard is balancing between flood control safety and economy. Therefore, it shall be appropriate to economic strength of a nation. The flood control shall be neither overdue nor insufficient. And the flood control standard shall solve the conflict between safety and economy, society as well as environment properly.

For over 60 years of development of China hydropower, researches at different depths have been carried out for flood control standard in different periods according to economic development of the nation, and appropriate codes have been issued and implemented in the development of Chinese hydropower stations, which has ensured safe operation of large-scale hydropower stations and functioning of normal economic benefits.

In China, design flood is used to represent flood control standard. The result of design flood calculation is ultimately shown in the flood control standard. As important design parameters for hydropower projects, design flood calculation result plays an important role in the engineering scale, flood control, and structure layout. It will decide whether the project is successful or not. In this sense, it's very necessary to research on design flood calculation in actual work.

Design flood, referring to flood complying with the design criteria for flood control, is the general term for various standard flood used for the design of flood control safety for hydropower engineering. It includes not only floods which are displayed in frequency and probable maximum flood, but also floods used for the design of normal operation condition for permanent hydro-structures (design flood) and floods for emergency operation condition (check flood) as well as floods adopted for temporary structures in the construction period. Its analysis and calculation generally includes peak flow, frequency analysis of period flow and the preparation of design flood hydrograph. Essentially, the determination of design flood has estimated flood which may occur to the dam in the future operation period. The dam will usually operate for dozens of years, one hundred year or even longer. Therefore, the estimation is obviously an extra-long term forecast.

Due to the impact of changes in natural and unnatural factors, the occurrence of a flood features randomness. People's knowledge level on its variation principle is quite low, somehow semi-experience and semi-theory. In addition to scarce flood information, the analysis and calculation of design flood are mainly based on preliminary knowledge on statistical law of storm and flood data. It's really hard to forecast design flood which comes around once in a millennium or 10,000 years based on 30 – 50 years' information. In this sense, you can imagine its uncertainty. Actually, everything is changing whenever and wherever possible. The true design flood which comes around in many years doesn't have a true value. And the estimated design flood result is only an approximate value to "True value" in the imagination. With the increment of storm and flood data, the design flood result estimation is changing. In addition to different experience of designers and methods adopted, the estimation may become much more uncertain. In this sense, hydrologists should, on the basis of current storm and flood data, adopt many analysis and calculation methods to carry out design flood calculation for hydropower projects, reasonably analyze on results, determine adopted design flood results in a comprehensive way, try their best to minimize the uncertainty of results adopted and balance economy and safety complying with current specifications and codes.

Over the past 60 years, especially over the past 30 years of the reform and opening-up, dam construction in China has been following a very hard and challenging path. We have developed from learning, imitation, referring foreign experience to self-dependent innovation with respect to design flood calculation. After cascade development has been formed on various major and middle rivers, we have worked on to a new level with respect to design flood calculation of cascade reservoirs, gained a wealth of experience and obtained fruitful results.

6.3.2 Design Flood Calculation Method and Development History in China

The calculation path and method on design flood for hydropower projects is developing and improving gradually with the accumulation of storm and flood data, the boosting of engineering construction and operation experience and people's deepening knowledge on flood rules. For over 60 years since the establishment of the People's Republic of China in 1949, Chinese hydrologists, proceeding from the reality, have been researching and exploring the design flood calculation path and method which satisfy the requirements of China's project with persistence, establishing a set of comparatively complete analysis and calculation method system on design flood gradually. The development process of Chinese design flood calculation method can roughly be divided into the following six periods.

6.3.2.1 Early Decades of the Founding of the People's Republic of China

In the early 1950s, new China had just been established and everything was waiting to be improved. There were few hydrometrical stations, precipitation gauging stations and meteorological stations. The data could only be dated back for quite a short time and those of some hydrometrical stations were incomplete. Due to the limit of data, there was no general standard and requirement on calculation method of design flood. Therefore, methods adopted for design flood calculation for each project were different from each other. There were mainly two methods:

(1) Mostly, a disastrous flood that occurred in the past or properly added flood would be used as design flood. As the history of storm data was generally longer than that of flood flow data at that time, the design flood was calculated indirectly by adopting the relationship between storm and flood relationship.

(2) For a few projects, there were long periods of flood flow data. Therefore, flood frequency analysis method was used directly to calculate the design flood. Moreover, in river basins where some projects were located, there were quite long periods of rainfall data. Therefore, frequency analysis method was used to calculate the design storm first, which would then be converted into design flood.

Due to the scarce data and engineering experience, the design flood calculation value during this period was generally quite small.

6.3.2.2 From the Middle of the 1950s to the Early 1960s

From the middle of the 1950s, frequency analysis method was used more and more widely in China to calculate the design flood. By the early 1960s, most hydropower projects used flood appropriate to a certain frequency (or return period) as the design basis. During this period, Chinese hydrologists, starting from the national conditions of China, carried out historical flood investigation and textual research in a comprehensive way and accumulated very precious material on various major rivers in China. However, there were different opinions on how to use historical flood data in the frequency calculation. Due to the urgent development need of hydropower projects, extensive research had been carried out on design flood calculation method during this period. Take Hydrology Research Department of China Institute of Water Resources and Hydropower Research as an example, in-depth research had been carried out on various links of frequency analysis and calculation method and specific method for incorporating historical flood into frequency analysis and calculation, and research findings had been applied to detailed projects, achieving many important results.

The importance of incorporating historical flood into frequency calculation to the stability of design flood results was further recognized by people. However, hydrologists also felt urgently that a comprehensive technical standard and quite complete design flood calculation method was required to normalize various links of national design flood calculation so as to make results much more stable and reliable.

Therefore, from the early 1960s, *the Regulation for Calculating Design Flood of Water Resources and Hydropower Projects* (Draft) started to be prepared under the organization of former Sinohydro Bureau and Hydrology Research Department of China Institute of Water Resources and Hydropower Research. It was formally proposed in 1964. In the following decades, the regulation was referred and executed by various design companies, achieving favorable effect, presenting quite reasonable design flood calculation result in many projects and stabilizing design data in a quite small variability.

6.3.2.3 From the Late 1960s to the Early 1970s

In the late 1960s, due to the impact of the "Cultural Revolution" left-wing ideas, the frequency analysis method was once criticized. Many projects rejected to use frequency flood as the flood standard. Instead, historical flood addition method was preferred, which had actually lowered flood standard and reduced design flood data, causing unfavorable effect and putting flood control safety of many projects under risk.

6.3.2.4 From the Middle of the 1970s to the Early 1980s

In August 1975, extraordinary rainstorm and flood occurred on the upper basin of the Huaihe River, leading to the dam break of Banqiao and Shimantan reservoirs and resulting in extraordinarily serious calamities. Thus, it was stipulated in China that probable maximum precipitation and flood should be used as flood standard for emergency application for important projects which may cause serious result in the event of a hazard. Therefore, the preparatory work of probable maximum precipitation isocline map and the corresponding storm-runoff table was carried out nationwide. And by the early 1980s, each province (district) had prepared storm-runoff table in succession. So, it is justifiable to say that the national popularity of probable maximum precipitation flood calculation by hydro meteorology had gained a wealth of experience and obtained fruitful results.

In 1979, *the Regulation for Calculating Design Flood of Water Resources and Hydropower Projects* (Trial) (SDL 22—79) was jointly issued by the former Ministry of Water Resources and Ministry of Power. By then, China eventually had a unified method and technical standard on design flood calculation. The Regulation had thoroughly summarized the experience and lessons of Chinese design flood calculation, underlining the design flood calculation principle of "multiple methods, comprehensive analysis and reasonable selection".

6.3.2.5 From the Middle of the 1980s to the Late 1990s

With the implementation of reform and opening-up policy in China, the national economy has developed rapidly, and so did hydropower projects. Especially, the policy of underlining the development of a batch of "hydropower-rich projects" was proposed, forming a consecutive cascade development pattern for a river or its section and presenting new problems for design flood calculation. For example, how to calculate flood probability distribution of downstream section after the storage of upstream reservoir? How to use the regulation of the upstream reservoir to lower design flood of downstream projects in the construction period? Moreover, people discovered that the confluence conditions of reservoir zones had changed obviously after the construction of some reservoirs. The reservoir inflow flood was obviously different from the dam site flood. Therefore, here arose the problem of when to take the change into consideration. To solve these problems, in 1983, the former Ministry of Water Resources and Electrical Industry Northwest Engineering Corporation and Nanjing Institute of Hydrology and Water Resources, organized by the Hydropower and Water Resources Planning and Design General Institute, carried out in-depth study on design flood calculation method for cascade reservoirs. Bureau of Hydrology (BOH) Changjiang Water Resources Commission and the former Ministry of Water Resources and Electrical Industry Central Southern Geotechnical Design Institute had carried out in-depth study on calculation method for the reservoir inflow flood. All of them had achieved favorable results. During this period, stochastic simulation technology had been widely used in flood control planning, design flood calculation and the design of flood control dispatching of reservoirs. Various single station and multiple station model and stochastic generation technology had developed substantially and were applied to some engineering design and operation. In the 1980s, great achievement has been made in paleo-flood research, which played a certain role in determining the return period of historical flood correctly and selecting design flood result reasonably. In some large-scale projects, flood control risk analysis has been started and some models and methods for calculating flood risks have been proposed. These research results have been tried in the comparison of design schemes for flood structures respectively.

6.3.2.6 From the Early 2000 to 2010

Since 2000, Chinese hydropower construction has entered into high-speed and in-depth development stage. Many rivers have been covered with cascade development pattern. While summarizing previous experience, Chinese hydrologists have further carried out in-depth research on characteristics and laws of Chinese storm statistical parameters, flood frequency distribution, flood frequency analysis peak-over-threshold (POT) series, palaeoflood as well as design flood calculation based on reliability theory according to newly added storm and flood data. Central Southern Geotechnical Design Institute and Northwest Engineering Corporation, organized by Hydropower and Water Resources Planning and Design General Institute of China Hydropower Engineering Consulting Group, has carried out systematic research on design flood calculation of cascade reservoir group and the selection of flood control standard for cascade reservoir group, achieving significant progress. In the last decade, China has carried out abundant stochastic modeling of hydrologic time series and achieved many new results. During the flood control dispatching research of reservoirs, stochastic simulation is underlined domestically to determine design flood control storage capacity and other appropriate water parameters. Application research has been conducted for stochastic simulation models (mainly linear disturbing model, typical disaggregation model, multi-station autoregressive model and BP network sequence model) applied in design flood calculation for cascade reservoirs. Chinese hydrologists also try to incorporate Copula function into stochastic simulation for flood process, making flood process simulation take into consideration the relationship between flood peak and flood volume and keeping the statistical characteristics of observed flood peak and flood volume. Impacted by global climate change, water disasters occur more and more frequently. Chinese hydrologists have started to carry out preliminary research on the impact of climate change on hydrological extreme events.

To sum up, since the issuance and implementation of *Regulation for Calculating Design Flood of Water Resources and Hydropower Projects* (Trial)(SDL 22—79), a unified technical standard has been formed for various engineering designs, and the design flood calculation result becomes much more stable. During this period, in the traditional frequency calculation method, plenty of researches have been carried out for parameter estimation method and great achievements have been made. From the late 1980s to the beginning of 1990s, Bureau of Hydrology (BOH) Changjiang Water Resources Commission and other units, organized by Hydropower and Water Resources Planning and Design General Institute, modified design flood calculation specification issued in 1979. In 1993, the Ministry of Water Resources and Ministry of Energy issued new *Regulation for Calculating Design Flood of Water Resources and Hydropower Projects* (Trial) (SL 44—93), increased the composition of relevant design flood areas and design flood calculation method for drought, karst and glacier areas. In 1995, *Manual for Calculating Design Flood of Water Resources and Hydropower Projects* corresponding to the 1993 Regulation was prepared and published. In 2006, Hydropower and Water Resources Planning and Design General Institute organized experts to modify the 1993 Regulation. In 2010, Hydropower and Water Resources Planning and Design General Institute of China Hydropower Engineering Consulting Group organized experts to modify the *Regulation for Calculating Design Flood of Water Resources and Hydropower Projects* (Provisional)(SL 44—93) which was jointly issued by the Ministry of Water Resources and Ministry of Energy. It would publish separate specification for calculating the design flood of hydropower engineering, increasing design flood affected by reservoir storage and design flood for pumped storage power stations and tidal power stations. By then, a set of comparatively complete system had been formed for calculation contents and methods for design flood in China.

6.3.3 China's Major Experience in Design Flood Calculation

Over the past 60 years, Chinese hydrologists have explored new problems through practices, carried out in-depth analysis on rules of Chinese storm and flood, and made research and improved design flood calculation method continually while tracing, referring to and learning the latest theory on design flood calculation from many countries around the world. As a result, a set of design flood calculation method system complying with natural conditions of Chinese rivers has been formed and a wealth of experiences and lessons has been learned. China's major experiences in design flood calculation are listed in the following points:

(1) Correct calculation method and results can be obtained only by adhering to the practical and realistic attitude.

(2) Attach great importance to basic data.

(3) Attach great importance to investigation and textual research of historical storm and flood.

(4) Implementation of the principle of "multiple methods, comprehensive analysis and reasonable selection" carefully.

(5) Attach great importance to rationality test of calculation results.

6.3.4 Flood Frequency Analysis

Although Yan Kai had introduced frequency analysis method into China in the early 1930s, Chen Chunting has conducted preliminary analysis on flood frequency curve for five major rivers in China. It was after the 1950s that Chinese hydrologists used frequency analysis method to determine design flood. In 1951, Liu Guangwen has proposed basic principles and problems on frequency analysis of design flood in China. In 1958, Jin Guangyan prepared *Principles and Methods of Hydrologic Statistics*, describing frequency analysis method in a systematic way. Probability theory and mathematical statistics are the theoretical basis for frequency analysis. However, many hydrological features have been incorporated into frequency analysis in China to suite the characteristics of hydrology samples.

There are three basic problems concerning the frequency analysis calculation method: sampling, selection of probability distribution function and estimation of statistical parameters. In the 1950s, China started to develop frequency analysis for hydrology. After 60 years of development, recent 30 years especially, due to the information expansion and popularization of computer technology, in-depth research has been carried out for flood data sampling method, type of frequency curve, empirical frequency formula, and parameter estimation method, abundant research findings have been obtained and a wealth of engineering experience has been gained.

6.3.4.1 Sampling

In order to present quite reliable foundation for flood frequency analysis, the review of flood data and flood series (sample series) should be completed to check their consistency, representation and reliability and to determine the flood series of adoption.

Flood is a natural phenomenon. Therefore, it's both deterministic and random. And, its data can be analyzed by probability theory. For quite a long time, hydrological science and technological researchers usually classified flood frequency analysis as extremal statistics. Therefore, sampling statistics shall be carried out for flood data based on the sampling principle. Generally, peak flow and flood volume in a period are used to calculate the design flood. As a digital characteristic describing a flood process, one maximum value shall be selected each year according to the principle of independent sample selection. As for single peak flood process, the peak flow can be obtained directly from the flow hydrograph. Generally, maximum flow at regular interval W_t is collected, i.e., take duration t (for example $t=$1d, 3d, 5d, 7d, 15d and 30d), move back and forth so as to obtain the maximum flood volume. Generally, the base flow is not deducted during the determination of flood characteristics. Multi-peak flood will not be divided either.

The selection of flood flow statistical interval shall be determined according to duration and shape feature of design section flood process, flood control interval of reservoirs as well as flood control requirements of downstream rivers, etc. Meanwhile, to facilitate statistics, the selected interval shall be consistent with interval adopted in hydrological statistical data issued by hydrological test departments.

So far, few flood observation series used in engineering design in China can date back 100 years, in which some have a history of tens of years (some series have a history of less than 30 years). Therefore, information expansion is always one of the keys to improve design flood calculation precision and promote flood frequency analysis technology.

China has quite a long history and its cultural relics and document literature are quite abundant, which have provided favorable conditions for investigating and verifying historical storm and flood. Since 1955, Chinese hydrologists have conducted substantial historical flood investigation and verification nationwide, collecting expensive data and carrying out planned collection, analysis, preparation and research on investigated historical flood data. By 1981, over 20,000 historical extraordinary flood data on almost 6,000 river reaches in China had been collected. Moreover, 92 historical extraordinary floods since 1482 had been selected to prepare the *Chinese Historical Flood*, which was edited and published in 1992.

The research and application of historical flood plays an important role in flood frequency calculation: firstly, it is beneficial for determining empirical frequency of extraordinary flood in the measurement period in a more reasonable way; secondly, it can increase flood data other than actual measurement. All of them are beneficial for improving flood frequency calculating precision.

China's paleo-flood study has started since the 1980s by the utilization of the method of water level of slack water deposits. The most obvious achievement of paleo-flood study is its application to design flood for the Three Gorges Hydropower

Station in the Yangtze River, the Xiaolangdi Hydropower station in the Yellow River, and the Zangmu Hydropower Station in the Yarlung Zangbo River in Tibet Autonomous Region. Paleo-flood was used to rebuild historical flood information, which had lengthed the time series of historical flood record substantially.

Chinese hydrologists initially applied the historical flood data to flood frequency calculation, which is undoubtedly a major breakthrough and conducive to improving the reliability of design flood results.

According to China's experiences, historical flood and paleo-flood shall be taken into consideration in the process of sampling flood data if conditions permit so as to increase the representation of flood series and reduce expanded extent. Meanwhile, the uncertainty of historical flood and paleo-flood estimation results shall also be taken into consideration.

Recently, Chinese hydrologists have realized that: in the frequency analysis, one maximum value is taken by the method of annual maximum sampling, which cannot fully take advantage of useful information in the measured series. Therefore, some series (POT or PDS) frequency analysis research have been developed and certain achievements have been made.

6.3.4.2 Type of Frequency Curve

Frequency analysis provides hydrological design value of probability meaning to various hydropower engineering, so as to determine the scale, investment and benefits of projects. Hydrologists should not only calculate the design value of return period in the length range of data series but also make expanded calculations, so as to calculate the design value whose return period is far longer than series period. In fact, frequency curve is a model which reveals data distribution and statistical rule, or an extended or interpolating frequency analysis tool.

China has started to study the type of hydrological frequency curve since the 1960s. Generally, the flood frequency analysis would mainly involve the tail end performance of frequency curve. In this way, the type of frequency curve can be divided into the following two types. One is thin tail distribution. All exceeding probability curves decrease progressively and exponentially at the tail end like the normal distribution, P-Ⅲ distribution, Gumble distribution, etc. The other is thick tail distribution. All exceeding probability curves decrease progressively at the tail end based on the rule of power function like the logarithmic normal distribution, logarithmic P-Ⅲ distribution, Wakeby distribution, etc. As the decrement of power function is much slower compared with the exponential function, the thick tail distribution is preferable to thin tail distribution for larger values. Currently, no physical factors can be used to argue what kind of probability distribution should be obeyed by flood variation. Since the 1960s, Chen Zhikai, Wang Jiaqi and other people have utilized hydrological data collected from major rivers in China, applied statistical assumption and inspection methods, and carried out substantial analysis and comparison on logarithmic normal distribution, P-Ⅲ distribution, K-M distribution, Gumble distribution, Wakeby distribution, Weibun distribution and exponential P distribution. Afterwards, it was pointed out that in Chinese three-parameter Γ distribution curve (always known as Pearson Ⅰ curve—P-Ⅲ curve) can better fit storm, flood and other series in the most parts of China. Therefore, in the previous design flood calculation specification, it was stipulated that "Pearson Ⅲ curve is generally used as the type of frequency curve. Other types may be adopted after analysis and argumentation under special circumstances". For this purpose, the line type used for flood frequency analysis in most rivers in China use P-Ⅲ distribution. Substantial research has been carried out for other types with many results obtained. However, they are seldom used in the engineering design.

Recently, Chinese hydrologists have also carried out hydrological frequency analysis and research with uncertain type selection, achieving some results.

6.3.4.3 Empirical Frequency Formula

In the hydrological frequency analysis, curve fitting method is always used to calculate parameters and design values of frequency curves. Therefore, the empirical frequency formula at point position shall be determined.

By far, there are tens of hydrological empirical formulae with different precisions. The occurrence frequency estimation of each element in the sample is based upon order statistics. Digital characteristics of probability distribution of order statistics can be used as estimation of occurrence frequency of each element in the sample. If the mathematic expectation of probability distribution of order statistics is taken, the mathematic expectation formula of estimated empirical frequency is:

$$p_m = \frac{m}{n+1} \tag{6.3-1}$$

If mid value of probability distribution of order statistics is taken, the mid value formula of estimated empirical frequency is:

$$p_m = \frac{m-0.3}{n+0.4} \tag{6.3-2}$$

In the aforementioned two formulae, n refers to sample size, m refers to the serial number of sample element in the order of size, and p_m refers to the empirical frequency numbered as m as per size in the sample.

Chinese practices demonstrate that it's better to use formula (6.3-1) to estimate the empirical frequency of sample elements.

Chinese hydrologists have carried out substantial comparison research on empirical frequency calculation formula and achieved abundant results. The most prominent contribution is the proposition of empirical frequency calculation formula for discontinuous flood series.

Generally, flood series composed of measured and interpolation extended data is deemed to obtain by independently random sampling from general distribution. If, after paleo-flood study and historical flood verification, in the flood series obtained by random sampling, no extraordinary flood value should be dealt with independently and various flood value are arranged in order as per size, the sample is deemed as continuous series; Otherwise, if after paleo-flood study and historical flood investigation and verification, the measured and investigated extraordinary flood value should be arranged in a long period, this sample is deemed discontinuous sample.

It was stipulated in regulations for calculating design flood issued in 1979 and other following specifications in China that expectation formula shall be used for calculating empirical frequency. Different formulae should be adopted for calculating empirical frequency based on continuous series and discontinuous series.

As for continuous series of n items, the empirical frequency shall be calculated with mathematic expectation formula:

$$p_m = \frac{m}{n+1}, m=1,2,\cdots,n \tag{6.3-3}$$

As for discontinuous series, in 1963, Qian Tie has proposed the following formula to calculate the empirical frequency of discontinuous series. And the empirical frequency formula for n extraordinary flood is as follows:

The frequency of n extraordinary flood is:

$$p_M = \frac{M}{N+1}, M=1,2,\cdots,a \tag{6.3-4}$$

The empirical frequency of n-l continuous floods is:

$$p_m = \frac{a}{N+1} + \left(1 - \frac{a}{N+1}\right)\frac{m-l}{n-l+1}, m=l+1,\cdots,n \tag{6.3-5}$$

or

$$p_m = \frac{m}{n+1}, m=1,2,\cdots,n \tag{6.3-6}$$

where a refers to quantity of extraordinary flood of continuous sequence in year N, N refers to historical flood investigation and verification period, n refers to items of observed flood series, l refers to items of flood in the observed flood series which has been collected out as extra-large value treatment, M refers to the serial number of a extraordinary flood in the order of size, m refers to the serial number of n measured flood in the order of size, p_M refers to the empirical frequency of item M in historical flood and p_m refers to the empirical frequency of m item in the measured series.

If there is investigation flood in further N' year beyond N year, historical flood in N year and historical flood and observed flood in N' year can form a discontinuous series. And various empirical frequencies can be estimated based on the aforementioned formula.

6.3.4.4 Parameters Estimation Method

As for parameter estimation for frequency calculation, one sample function $\hat{\theta}(X_1, X_2, \cdots, X_n)$ is formed to estimate the general parameter θ. And $\hat{\theta}$ is the estimation value of θ. $(X_1, X_2, \cdots, X_n)$ refers to a random sample collected from the total with a capacity of n. When a set of observation value $(x_1, x_2, \cdots, x_n)$ is set for $(X_1, X_2, \cdots, X_n)$, $\hat{\theta}(x_1, x_2, \cdots, x_n)$ can be obtained and used as an estimation value of general parameter θ. In the mathematical statistics, moment method, maximum likelihood method and other methods are generally used to construct this type of sample function. The sampling error of estimation parameter by moment method increases with the increment of order of moments. When maximum likelihood method is used to estimate parameters, the type of distribution function of known general is a premise. Therefore, in the flood frequency calculation, both moment method and maximum likelihood method are not practical enough. Since the 1950s, Chinese hydrologists have dedicated themselves to developing curve fitting method to estimate parameters.

As for P-Ⅲ curve, three parameters shall be estimated, i. e., mean value $\overline{X}$, deviation coefficient C_V, and coefficient of skewness C_S. In the frequency analysis, the estimated frequency curve shall fit properly with empirical dots and be provided with favorable statistical characteristics. According to many years of practical experiences in China and current technical level on frequency analysis, the statistical parameters for estimating frequency curve can proceed as the following three steps:

(1) Preliminary estimate of parameters: moment method and other parameter estimation methods can be used for the preliminary estimate of statistical parameters. Major parameter estimation methods include moment method, probability weighted moment method, maximum likelihood method, weighted function method, single weighted function method, double weighted function method, method of L-moments, etc.

(2) Using curve fitting method to adjust statistical parameter from the preliminary estimation: during adjustment, target functions shall be selected to get statistical parameters or empirical curve fitting method may be used. When empirical curve fitting method is used, efforts shall be made to fit all dots. Otherwise, much more reliable heavy flood dots shall be considered. When curve fitting method is used to adjust statistical parameters of preliminary estimation, the rules can be used including the minimum of sum square variation, sum of the minimum of absolute value of deviation squares sumand the minimum of sum relative square variation, etc.

(3) Parameter coordination: statistical parameters determined through curve fitting method shall be comprehensively re-determined according to flood peak discharge, statistical parameters for design flood and design values at different intervals and results of stations in the upper and lower reaches, branches and adjacent rivers.

6.3.4.5 Curve Fitting Method

There are many methods to calculate P-III distribution statistical parameters, for example, moment method, probability weighted moment method, maximum likelihood method, weighted function method, single weighted function method, double weighted function method, and method of L-moments, but, by far, no parameter estimation method can be accepted by both hydrologists and designers in China. Curve fitting method is always used to determine theoretical frequency curve in the engineering design.

The curve fitting method is used based on a certain curve fitting principle to calculate the statistical parameters of theoretical frequency curve best fitting the empirical dots. Firstly, the occurrence frequency of each element in the sample is estimated to obtain empirical frequency dots. Then, whether fitting empirical frequency dots is the best one or not is used to select the type of frequency curve and to determine corresponding statistical parameters. As a parameter estimation method, the sample function of the curve fitting method is formed through the estimation of empirical frequency of sample elements and best fitting of empirical frequency dots. Different empirical frequency formulae and best fitting principle correspond to different parameter estimation methods. Key technologies relevant to the curve fitting method are shown as follows: Firstly, how to estimate the empirical frequency? Secondly, which type of best fitting principle shall be used? Thirdly, how to select reasonable hydrological frequency curve type?

Currently, two types of curve fitting methods are adopted in China actually: ①optimum curve fitting method, i. e., firstly select objective function for curve fitting (i. e., curve fitting principle), and then get corresponding optimum statistical parameters; ②empirical curve fitting method.

1. Optimum curve fitting method

Since the 1970s, according to a certain curve fitting principle (i. e., objective function), Chinese hydrologists have developed researches using computer technology curve fitting to obtain statistical parameters.

(1) Principle of sum of squares, also known as least square estimation method (LSCE), that is to say, sum of squares of differences between ordinates of empirical dots and frequency curve of the same frequency (deviation or residual) shall be the minimized, i. e.,

$$\text{Min} \quad S_1(\overline{X}, C_V, C_S) = \sum_{i=1}^{n+a-1} [x_i - f(p_i, \overline{X}, C_V, C_S)]^2 \tag{6.3-7}$$

(2) Principle of sum of absolute deviation, that is to say, the sum of absolute values of differences between ordinates of empirical dots and frequency curve of the same frequency (deviation or residual) shall be minimized, i. e.,

$$\text{Min} \quad S_2(\overline{X}, C_V, C_S) = \sum_{i=1}^{n+a-1} |x_i - f(p_i, \overline{X}, C_V, C_S)| \tag{6.3-8}$$

(3) Principle of sum of relative squares. The error of flood estimation is related to it. However, its relative error is comparatively stable. Therefore, the sum of relative squares minimum can better fit the assumption of least square estimation. Then, the curve fitting principle can be written as:

$$\text{Min} \quad S_3(\overline{X}, C_V, C_S) = \sum_{i=1}^{n+a-1} \left[\frac{x_i - f(p_i, \overline{X}, C_V, C_S)}{f_i(\overline{X}, C_V, C_S)} \right]^2 \qquad (6.3-9)$$

2. Empirical curve fitting method

If the error difference of various flood (historical flood and paleo-flood especially) is really huge and optimum curve fitting method doesn't produce satisfactory theoretical frequency curve, empirical curve fitting method should be used. The empirical curve fitting method plots various values of hydrological series and corresponding empirical frequency dots on the frequency paper, uses ocular estimation and adjusts statistical parameters continuously as per the fitting between theoretical frequency curve and measured dots until a satisfactory theoretical frequency curve is obtained.

The empirical curve fitting method is simple and flexible, capable of reflecting the experience of designers. However, the designers' knowledge and experiences vary from person to person. It's difficult to avoid the subjectivity, which may result in high uncertainty in design results. Moreover, to facilitate curve fitting, empirical C_S/C_V value also lacks foundation. As for curve fitting for flood frequency, it is necessary to try to observe trend of dot group to make the curve pass through the dot group center. If dots are irregular, upper and middle dots may be underlined and efforts shall be made to make the curve adjacent to dots with higher precision. As for extraordinary flood, its possible error range shall be analyzed. It is not appropriate to pass through the extraordinary flood mechanically so that frequency curve breaks away from the dot group.

6.3.4.6 Reasonable Analysis of Flood Frequency Calculation Results

According to the experiences of Chinese hydrologists, in the flood frequency calculation, besides making the best of historical flood and paleo-flood data, the precision of measured or interpolation data, representatiion of series and rationality analysis of statistical parameters are also very important.

Generally, the statistical parameters are, to some extent, of hydrological explanations. For example, the mean value is a characteristic value measuring the flood level in the river basin, C_V is a characteristic value displaying the annual variation of flood in the river basin, and C_S is a characteristic value reflecting the probability of large and small floods in the basin. As flood characteristic value is mainly dominated by climatic conditions and can be displayed in function of geological longitude and latitude, the isoclines of flood characteristic value can be drawn on the map. The geological principle displayed in the contour map can be used to carry out necessary analysis for flood statistical parameters in a certain region or necessary modification. Ye Yongyi, et al. have already proposed this method in the 1950s and called it the regional coordination of statistical parameters.

Moreover, the statistical parameters of the same flood in the upper and lower reaches of a river and the statistical parameters of the flood volumes in different periods are to some extent related. Chinese experts have already discussed these relationships by taking into consideration actual work and used them for rationality analysis of flood statistical parameters.

6.3.4.7 Safe Adjustment Value

For a variety of reasons, design flood X_p complying with some design criteria for flood control standard p obtained from frequency calculation has error. Jin Guangyan made researches to estimate statistical parameters of consecutive series with moment method and to calculate the sample error of X_p. As a result, the following formula calculating the X_p sampling error is obtained:

$$\sigma_{X_p} = \frac{\overline{X} C_V}{\sqrt{n}} B \qquad (6.3-10)$$

where σ_{X_p} is sampling error of X_p, $\overline{X}$ is mean value, C_V is coefficient of deviation mean, n is sample capacity, and B is the function of deviation coefficient C_S and frequency p.

Tan Weiyan, et al. used the curve fitting principle of sum of absolute deviation minimum to obtain the calculation chart of the function through statistical tests. Chinese experts add σ_{X_p} as safe adjustment value to the X_p obtained to ensure the flood control safety of dams. However, obviously, this practice has certain defects.

Qualitative judgment can only be made after rationality analysis of precision of original data, representation of series, degree of investigation and verification of historical flood, precision of paleo-flood study results as well as statistical parameters and design value. If it's lower, safe adjustment value shall be added on the calibrated standard flood design value for safety. The safe adjustment value can be determined by comprehensive analysis for possible amplitude of lower results and referring to error calculation results of mean squares.

In China, the specification of design flood says that "as for large-scale projects or important medium-sized projects, sampling error shall be calculated for calibrated standard design flood calculated by frequency analysis method. After comprehensive analysis and inspection, safe adjustment value shall be added if the results are slightly lower. However, generally, the safe adjustment value shall not exceed 20% of the calculated value."

6.3.5 Calculation of Design Flood

6.3.5.1 Calculation of Natural Design Flood

According to the calculation specification of design flood in China, the following three methods can be used to calculate the natural design flood based on data available.

(1) When 30 years of observed as well as interpolated and extended flood data are available and the survey of historic flood has been carried out at the dam site or in the places just upstream and downstream of the dam, the design flood can be calculated using the frequency analysis method.

(2) If there are more than 30 years of observed as well as interpolated and extended storm data in the project location which are also correlated to the storm flood, the design storm can be calculated using the frequency analysis method and then the design flood calculated.

(3) Due to the lack of flood and storm data of the watershed in which the project is located, the observed or surveyed storm and flood data in the vicinity can be utilized to make regional comprehensive analysis and estimate design flood.

The calculation methods of natural design flood above can be summarized in two categories. The first category is called the "direct method", which is to calculate the design flood via discharge data, and the second category is called the "indirect method", which is to calculate the design flood via precipitation data. As the basic data on which the first method is based is the measured flow process of relatively less intermediate procedures. It's the mostly adopted method in the calculation of project design. Here mainly the "direct method" is introduced.

The main contents of calculating the natural design flood based on the flow data include three factors of the design data, respectively, the analysis and calculation of the design flood peak discharge, flood volume in design period and design flood hydrograph. The frequency analysis method is used in calculating both the design flood peak discharge and flood volume in design period. Therefore, the calculation of the design flood peak discharge is taken as an example for introducing the detail. Firstly, sample the measured flood data according to the meanings and independent sampling rules of the flood control safety standard to obtain the sample series of flood peak discharge and analyze its reliability, conformity and representation. Then, analyze and determine the historic flood and its textual research period. Finally, use the frequency analysis method to analyze and calculate the design flood peak discharge and determine the results to be adopted based on rationality analysis.

As specified by the present *Regulation for Calculating Design Food of Water Resources and Hydropower Projects* (SL 44—2006), P-III type curve shall be generally used as the type of frequency curve.

The relation and estimation method of the three parameters, α, β and a_0, and the mean value, deviation coefficient and coefficient of skewness of P-III type curve are the same as that of runoff design, which are shown in Section 6.2.3.3 above.

To calculate the natural design flood of the cascade reservoirs, firstly select the representative station of hydrological design of each stage of the power station and respectively calculate the design flood of each station. Then, check the rationality of the design result and properly select the design flood result of each hydrological station.

According to the catchment area of each representative station and the dam site of each cascade reservoir, properly make the station-dam conversion to obtain the result of natural design flood of each cascade reservoir.

6.3.5.2 Design Flood Hydrograph

Corresponding to flood control safety standard of a dam, the flood control capacity is related to such factors as the reservoir inflow flood hydrograph, the safety discharge of the downstream river channel, the pattern, scale and control and operating methods of the flood discharge structures of the reservoir. With the safety discharge of the downstream river channel and the pattern, scale and control and operating methods of the flood discharge structures of the reservoir already determined, the flood control storage capacity of the reservoir obviously only depends on the reservoir inflow flood hydrograph.

$$V = f_1(Q_1, Q_2, \cdots, Q_n) \qquad (6.3-11)$$

or

$$V = f_2(W_1, W_2, \cdots, W_n) \qquad (6.3-12)$$

where V is the flood control storage capacity of the reservoir, Q_1, Q_2, $\cdots$, Q_n are respectively the inflows at the times of t_1, t_2, $\cdots$, t_n, W_1, W_2, $\cdots$, W_n are respectively the maximum inflows with the durations of T_1, T_2, $\cdots$, T_n.

Therefore, theoretically, the design flood hydrograph shall be a flood hydrograph complying with the following relation:

$$P = \iint \cdots \iint \varphi_1(Q_1, Q_2, \cdots, Q_n) dQ_1 dQ_2 \cdots dQ_n \qquad (6.3-13)$$
$$\Omega: f_1(Q_1, Q_2, \cdots, Q_n) \geqslant V_p$$

or

$$P = \iint \cdots \iint \varphi_2(W_1, W_2, \cdots, W_n) dW_1 dW_2 \cdots dW_n \qquad (6.3-14)$$
$$\Omega: f_2(W_1, W_2, \cdots, W_n) \geqslant V_p$$

where P is the standard of the flood control safety design of the dam, V_p is the flood control storage capacity essential for complying with the standard of the flood control safety design of the dam, Ω is the domain of integration; $\varphi_1(Q_1, Q_2, \cdots, Q_n)$ or $\varphi_2(W_1, W_2, \cdots, W_n)$ is the probability density function of the flood hydrograph used as the stochastic process.

As early as the 1950s, China started using the same-multiple enlarging method and the same-frequency control enlarging method in different periods to estimate the flood hydrograph. Let the flood control storage capacity be the function of the maximum flood flow W_T in the period T, formulas (6.3-11) and (6.3-13) can be respectively simplified as

$$V = f_2(W_T) \qquad (6.3-15)$$

and

$$P = \int \varphi_2(W_T) dW_T \qquad (6.3-16)$$
$$\Omega: f_2(W_T) \geqslant V_p$$

where T is called the design period, which depends on the regulation water supply capacity of the reservoir. The bigger the flood control storage capacity of the reservoir is, the stronger the flood control capacity of the reservoir is and the bigger the value of T is. Otherwise it's smaller.

To calculate the design flood hydrograph according to the rules revealed by formulas (6.3-15) and (6.3-16), firstly the flood volume in the design period compliant with the design standard of flood control safety P must be calculated. Then, the typical flood hydrograph is enlarged based on the same multiple. The design flood hydrograph is usually a rare flood hydrograph. Therefore, the large flood hydrograph based on reliable data, of good representation and more unfavorable to the flood control safety must be selected as the typical flood hydrograph. Presently in China, a set of reasonable rules for selecting the typical flood hydrograph have been proposed. Different from the same-multiple enlarging method that uses the flood volume of only one period for control, the same-frequency control enlarging method of different periods firstly select several control periods of different lengths to obtain the flood volumes of different control periods compliant with the design standard of flood control safety and then perform the same-frequency enlarging in different periods of the typical flood hydrograph. The same-frequency control enlarging method in different periods is still based on formulas (6.3-15) and (6.3-16). However, it depends less on the selection of the typical flood hydrograph than the same-multiple enlarging method. Therefore, the same-frequency control enlarging method of different periods has become the most commonly used method to calculate the design flood hydrograph.

The design flood hydrograph calculated for the flood control safety of the dam shall be the inflow flood hydrograph which differs from the design flood hydrograph of the cross section of dam site. The inflow flood hydrograph is generally of sharper peak than the flood hydrograph of the cross section of the dam site and earlier peak time. Therefore, the flood control storage capacity of the dam calculated according to the flood hydrograph of the cross section of the dam site is generally smaller than the flood control storage capacity calculated according to the inflow flood hydrograph. At present, many handling methods have been proposed for this issue.

6.3.5.3 Calculation of Design Flood Influenced by Regulation of Upstream Reservoirs

In recent years, due to the development pattern of cascade reservoirs on the stem streams and tributaries that are gradually formed on the rivers rich of hydropower resources, the regulation influence of the upstream cascade reservoir groups is more and more required to be considered in the engineering design stage of the water resources and hydropower projects.

During the flood control safety design for the reservoir project itself, if there are cascade reservoirs or a group of reservoirs already built or to be built in the near term of large regulation and storage effects at the upstream of the project, the flood regulation effect of those reservoirs on the design flood of the downstream design cross-section shall be considered. Compared to the natural flood process of the downstream design cross section of reservoir, the discharge flow process after the flood regulation of the upstream cascade reservoir is generally of less peak flows and short-time flood flows and delayed peak time that also vary with different scales of the natural flood and the shapes of the flood hydrograph. The discharge flows process of the upstream cascade reservoir combines with the district flood process and forms the flood process of the design cross section of the downstream reservoir with the flood regulation influence of the upstream reservoir. If the district peak appears behind the natural flood peak at the design cross section of the upstream cascade reservoir, the flood regulation of the upstream cascade reservoir may increase the chance that the discharge flow of the upstream cascade reservoir encounters the district peak flow. Otherwise, the chances may be decreased. If the upstream reservoir adopts the flood regulation method of certain fixed discharge flow for the flood below certain flood control standard (generally called the "simplified flood-releasing control" method). In this case, the maximum discharge flow of the upstream reservoir lasts very long and largely increases the changes of encountering the peak flow of the downstream district flood, which is unfavorable to the flood control safety of the downstream reservoir.

The flood regulation effect of the upstream cascade reservoir changes the peak flow, periodic flood flow and hydrograph shape of the natural flood at the design cross section of the downstream reservoir and therefore changes the flood probability distribution at the design section of the downstream reservoir. To calculate the design flood at the design section of the downstream reservoir, the most direct method is to make flood regulation simulation of the measured flood flow data year by year according to the flood regulation rule of the cascade reservoirs, calculate the flood hydrograph at the design cross section of the downstream reservoir and count the characteristic value series of the flood at the design section of the downstream reservoir influenced by the flood regulation of the upstream reservoir. However, there are practical difficulties in calculating the frequency according to the flood series influenced by the flood regulation of the reservoir. On one hand, such series is difficult to be fit using any known type of frequency curve to achieve extension. On the other hand, it's hard to fit the empirical frequency points plotted according to the series using a smooth curve and the extension trend is uncertain. Especially, the discharge flows of certain reservoirs are subject to sudden change about the flood at certain frequency. The empirical frequency curve, even without large extension amplitude, may be of great error. Therefore, during the practical application, the approximation methods in certain generalization conditions are always adopted, mainly including the zone composition method, frequency combination method and stochastic simulation method.

1. Zone composition method

The flood process of the design cross section influenced by the flood regulation of the upstream reservoir is formed by the combination of inter-zone flood process and flood process after the flood regulation of the upstream reservoir. Therefore, to calculate the design flood influenced by the flood regulation of the upstream reservoir, firstly the zone composition analysis of the flood shall be made according to the data of large floods that have occurred actually upstream of the design section, so as to study out the flood zone composition method unfavorable to the flood control of the project.

The design flood zone composition patterns generally to be considered include: ①the upper and lower reaches are at the same frequency and the interval is corresponding; ②the interval and the lower reach are at the same frequency and the upstream is corresponding; ③typical year composition.

Whether the proposed design flood district composition is reasonable in the design condition can be judged by analyzing whether the composition complies with the flood composition rules in each subzone upstream of the design control cross section of the reservoir.

If there is built reservoir of regulation and storage capacities in the upstream of the present reservoir, firstly analyze and calculate the natural design flood hydrograph at the design section of each cascade reservoir. Then, according to the design flood zone composition pattern studied out, calculate the flood regulation of the upstream cascade reservoir and obtain its discharge flow process. Finally, combine the discharge flow process of the upstream cascade reservoir with the district flood to obtain the design flood result of the present reservoir stage taking into consideration the influence by the regulation and storage of the upstream cascade reservoir.

2. Frequency combination method

The frequency combination method is to, taking the flood flow in each subzone upstream of the design section as the combination variable and through the frequency combination calculation and the flood regulation calculation of the upstream reservoir, directly calculate the flood frequency curve and design value of the downstream design cross section influenced by the regulation and storage of the upstream reservoir. By applying the method to the flood frequency in each subzone upstream of the design cross section, the calculation result is more reliable and the flood peak relation is better. It's also more suitable for the condition of significant flood regulation effect of reservoir.

Depending on different processing methods, the frequency combination method is further divided into the numerical integration and scatter sum methods.

If the zone flood frequency combination method is used, generally the periodic flood flow imposing major regulation effect on the project in each subzone is taken as the combination variable. The subzones shouldn't be too many.

During practical application, generally the frequency combination calculation is carried out only on the mutually independent combination variables. Therefore, independency check is required on the combination variable. If the variables are found not independent from each other, independency check shall be made. If they are not independent, they shall be converted into independent stochastic frequency combination for calculation.

3. Stochastic simulation method

As the hydrological variable is randomized, the mutual encountering combination of flood in each subzone upstream of the design dam site of reservoir also has obvious randomness.

The stochastic simulation method takes the flood process in each subzone as a stochastic process and selects proper model according to the engineering requirements of the design reservoir, the watershed characteristics and data condition. Then, make the stochastic simulation of the flood hydrographs of multiple stations and generate the synchronous flood hydrographs of multiple stations of sufficiently long series. Then, according to the flood regulation rules of the cascade reservoirs, perform flood regulation simulation year by year and calculate the flood hydrograph at the design cross section of the downstream reservoir. Then, count the characteristic value series of the flood at the design cross section of the downstream reservoir taking into consideration the flood regulation of the upstream reservoir, obtain the probability distribution of the flood characteristic values at the cross section downstream of the reservoir and calculate its design flood.

If the flood stochastic simulation method is used, proper model shall be selected and the statistical characteristics and rationality of the simulation result shall be checked.

The zone composition method, frequency combination method and stochastic simulation method have their own advantages and disadvantages.

The zone composition method features clear concept and simple calculation and is more applicable to engineering design. However, it's of limitations because the flood process in each subzone is randomized and varied, and the flood zone composition method studied out is hard and impossible to cover all encountering combinations.

The frequency combination method can cover all zone compositions of flood and the corresponding occurrence probabilities and better reflect the flood regulation effect of the flood at the frequencies of the reservoir. It's also intuitive without requiring the simplification of the flood regulation rule of the reservoir and is applicable to various conditions, especially the condition in which the flood regulation rule of the reservoir is complex and the combined flood control and regulation is adopted for the cascade reservoir. However, it's of high requirements for data condition. Furthermore, the calculation works increase exponentially with the increase in the number of cascade reservoirs.

The stochastic simulation method uses the randomly generated synchronous flood hydrographs of multiple stations of sufficiently long series to directly calculate the flood regulation and obtain the probability distribution of the flood characteristic values at the cross section downstream of the reservoir. It necessitates neither the simplification of flood regulation function nor the processing of complex flood combination encountering issue. Meanwhile, it has the advantage of simplifying the problem by processing the flood characteristic values (peak, flow) and hydrograph in a combined manner and reducing certain processing procedures. Its accuracy mainly depends on whether the model established is proper and reflects the objective rule of the flood in the design watershed. However, the flood hydrograph is generated using the existing flood data and information, while the future variation of the flood in each subzone is known. The flood hydrograph obtained through stochastic simulation doesn't cover the information of the future variation of flood and is merely the simple repetition of the characteristics of the historic flood process, which is inadequate to guide the engineering design.

To sum up the above analysis, although the zone composition method has certain limitation, it's already broadly adopted in engineering design. Additionally, it's considerably safe to select the combination unfavorable to the flood control safety of the project.

In recent years, National Hydropower and Water Resources Planning and Design General Institute, CHECC organized the experts to carry out the special research on the calculation of the design flood of cascade reservoir group. They carried out deep research on calculating the design flood of the serial, parallel and mixed cascade reservoir groups and comprehensive study on various method recommended by the Chinese specifications of the calculation of design flood based on the summarization of many years' research achievements. The research enriched the knowledge of the calculation method of design flood of the cascade reservoir group and put forward some innovative ideas. It's put forward that, during the same-frequency zone composition calculation of the cascade reservoir group, the same-frequency composition method and typical flood composition method shall be coupled in application; for the cascade reservoir group of large difference in the control periods of flood regulation, the multi-period typical flood composition method should be adopted. For corresponding flood division in the same-frequency composition method, the new method of "redistribution method of typical process node" shall be used. The zone composition method is the basic method of calculating the design flood of the cascade reservoir group with priority given to the typical flood zone composition method. For the frequency combination method, the summation shall be made per zone, per level and discretely while calculating the design flood of the cascade reservoir and the annual maximum sampling method shall be used to sample the flood characteristic value of each subzone. It's recommended in calculating the design flood of the cascade reservoir group to adopt the multi-station steady auto-regression model for the stochastic simulation method in the design flood calculation of the cascade reservoir group and the recursion algorithm of parameter matrix for parameter estimation. They made beneficial attempt on the method of adding the sample series to the historic flood information and provided reference and technical support for calculating the design flood of the cascade reservoir group in the future.

6.3.6 Risk-based Flood Control Safety Design of Dam

6.3.6.1 An Overview

At present, the flood control safety design of reservoirs and dams for hydropower station in China is based on the design flood and check flood. However, as the calculation of design flood is influenced by many factors, the calculation result is to some extent uncertain. Therefore, the flood control safety of the dam designed in this manner may not be truly safe but with certain imperfections. As it neither takes into consideration that the dam may collapse when the reservoir level exceeds the check flood level nor the uncertainty that exists in estimating the design flood, it can't provide the concept of the actual safety level of the reservoir dam. With the formation of the cascade development pattern on most rivers, while conducting the flood control safety design of the hydropower project itself, if there are cascade reservoirs or reservoir groups upstream of large regulation and storage effects that are already or to be built in short time, the flood regulation effect of those reservoirs will change the natural flood condition and directly influence the design flood of the downstream design cross section. On the other hand, in the cascade reservoir group, the flood control safety standard of reservoir dam selected based on a single flood control standard or specification of reservoir may not be able to guarantee the flood control safety of the reservoir dam to be designed.

In consideration of the imperfections in the flood control safety design of the reservoir dam, the Chinese hydrological workers started the study on the risk-based flood control safety design of dam in 1980. In 1984, Wu Shiwei, Zhang Sijun, et al. counted and analyzed the data series of measured annual maximum dam front water level of the 82 reservoirs already built in China using the Kol-mogorov-Smirnov method and concluded that the density function of the annual maximum dam front water level can be fit using normal distribution or logarithmic normal distribution. In 1989, Xu Zuxin and Guo Zizhong put forward the risk calculation mode in the hydraulic design of open spillway and for the first time used the JC method in calculating the discharge risk. In 1989, Zheng Guanping and Wang Mulan probed into the calculation method of the reliability of the flood discharge capacity of the spillway and comprehensively studied the influence of the uncertain hydrological and hydraulic factors on the flood discharge capacity of the spillway. In 1990, Jiang Shuhai used the JC method to calculate and analyze the cavitation-free probability design of the discharge components. In 1991, based on the systematic study of the effect of hydraulic uncertainty on the risk analysis of the flood control and discharge system, Jin Ming concluded that the influence of the hydraulic uncertainty model mainly depends on its coefficient of variation, and it's more reliable and practical to calculate the risk using the simple probability distribution model of bearing capacity and the complete probability method. In 1992, by using the manning formula as the hydraulic calculation model, using the FOSM method to make parameter estimation and combining the MC method and optimal selection to obtain the optimum probability model of the discharge capacity, Chu Xiangyuan thought that the normal distribution is the best distribution of the hydraulic discharge capacity. In 1993, Jiang Shuhai analyzed and proved the Wiener process characteristics of the flood storage capacity of reservoir, deduced the flood regulation calculation equation with random input item and random initial conditions and solved the probability density distribution of the reservoir water level process closely related to the flood discharge rate. In 1995, Zhu Yuansheng and

Wang Daoxi studied the safety design of reservoir and the risk issue of dam collapse and thought that generally in China that the flood with the recurrence period of 1,000 - 10,000 years was taken as the design standard, while abroad mostly the probable maximum flood is taken as the standard. They concluded that the design standard shall be changed according to the place and time. In 1996, Yang Baiyin and Wang Ruichen, et al. studied the flood discharge risk modes of a single reservoir and cascade ones. They thought that not only the flood exceeding all standards, but the flood at different frequencies would lead to overtopping. Besides taking flood as a control condition of the occurrence of risk, the situation in which the maximum dam front water level Z after regulation and storage exceeds the dam crest elevation Z_d may also be taken as a condition of the occurrence of the accident to introduce the JC method into the calculation of the analysis of the flood discharge risk of reservoir. In 1999, they also put forward the risk analysis mode and calculation method to the scheme with single reservoir after flood regulation. In 1996, Wang Changxin, Wang Huimin, Cheng Fenglan, et al. compared the JC method and MC method for calculating the risk of flood discharge and energy dissipation based on the analysis of the characteristics of P-III type distribution probability density function and obtained similar calculation results. However, the calculation efficiency of the JC method is much higher than that of the MC method. In 1997, Xie Chongbao, Yuan Hongyuan, et al. fully analyzed the uncertainty of the relation between the hydrology, hydraulics and water level and reservoir capacity in calculating the flood control risk of reservoir, probed into its distribution and parameter determination method and studied the application issue of full risk model for flood control of reservoir. In 1998, in preparation for the risk map of check flood and dam-break flood of the large/medium-sized reservoirs of Jiangsu, Xu Xiangyang, Yu Xiaozhen, et al. firstly carried out the inundation calculation of the two flood types, then estimated the flood inundation loss and finally plotted the flood rick map using the large-scale topographic map. In 1998, Fu Xiang and Ji Changming established the risk analysis model of the flood limit level of large reservoir using the systematic analysis method and, by taking the Three Gorges Reservoir as the study object and according to the scheduling and operation principles, calculated the relation between different flood limit levels and the maximum flood risk. In 1998, considering that the dam safety analysis doesn't include many random and fuzzy and uncertain factors in a quantitative manner, which leads to one-sidedness and limitation to a certain extent, Jiang Shuhai established random and fuzzy risk analysis model to achieve the quantification of the flood control capacity of dam. In 1999, he used the probability tree method to discuss the formation of the overtopping accident layer by layer and sequentially and provide corresponding quantitative flood control risk. In 1999, Xiong Ming pointed out that the assumption that the flood control safety standard equates to the flood design standard lacks necessary basis, put forward the principle, method and applicable condition of calculating the risk of flood control safety of dam and thought that the dam front water level series didn't comply with the general theoretical distribution function. In 2000, Wang Bende, Liang Guoha, et al. established the real-time risk scheduling model for the flood control of reservoir, considered the objectives of downstream flood control benefit and reservoir risk and put forward the principle of determining the allowable limit risk. In 2000, Wanjun, Chen Huiyuan, et al. put forward the three types of unexpected events for the impoundment and operation of the reservoir in the flood season, including the inundation of the upstream reservoir area, the flooding downstream and the threat of dam failure, respectively calculated their risk losses and provided the basis for selecting the proper interception and storage level scheme. In 2002, during the risk analysis of the flood control safety of dam, Mei Yadong used the stochastic simulation method to simulate the possible future flood regimen the reservoir may face in the future. Besides the consideration of the uncertainty of the hydraulic condition, reservoir capacity and boundary condition of water level and the initial condition of the starting regulation level, the consideration of the hydraulic uncertainty is more adequate. In 2005, based on the successive study of the dam front level distribution by Wu Shiwei, Ding Jing and Deng Yuren, Xiao Yi put forward the calculation mode of flood control risk based on the measured water level series. In 1989 by Liu Guangwen and in 2005 by Huang Jinchi, based on the relation between the design flood and risk, it's considered that the risk level (flood control standard) represented by the design flood is the risk presently acceptable in China. Therefore, the flood control safety risk of the dam is the probability of occurrence of standard-exceeding flood.

It's a development trend that the flood control design based on risk analysis will improve the traditional one based on flood frequency.

6.3.6.2 Relations Among Risk, Reliability and Flood Frequency

Water load is one of the main loads which the dam bears and can be measured according to the reservoir level formed by the inflow. Assume X as the water load of the dam and Y as the bearing capacity of the dam. When $X>Y$, the dam is unsafe and may collapse. Otherwise the dam will be safe. Under ordinary conditions, both X and Y are random variables. According to the probability theory, the probability of the event $\{X>Y\}$ shall be:

$$P\{X>Y\} = \iint_{\Omega: X>Y} f(x,y)\mathrm{d}x\mathrm{d}y \qquad (6.3-17)$$

where $f(x,y)$ is the joint density function of X and Y. The probability $P\{X>Y\}$ is the possibility of dam failure during its operation, which is called the risk expressed by R. Generally, X and Y are independent of each other. Therefore, formula (6.3-17) can be simplified as:

$$R = P\{X > Y\} = \iint_{\Omega: X>Y} f_x(x) f_y(y) \mathrm{d}x \mathrm{d}y = \int_0^\infty \int_0^x f_x(x) f_y(y) \mathrm{d}x \mathrm{d}y \qquad (6.3-18)$$

where $f_x(x)$ and $f_y(y)$ are the density functions of X and Y.

Thus it can be seen, with the probability density function $f_x(x)$ and $f_y(y)$ reflecting the water level in front of dam and the bearing capacity of dam determined, the risk can be calculated using formula (6.3-18).

It should be pointed out that the risk-based flood control safety design of dam is more reasonable. The flood control safety design of dam based on flood frequency is actually a special instance of formula (6.3-17). In fact, if Y is a deterministic constant y, the formula is changed into the following:

$$R = P\{X > Y\} = \int_y^\infty f_x(x) \mathrm{d}x \qquad (6.3-19)$$

Apparently, the risk is the flood frequency when Y is the deterministic constant y.

The traditional flood control safety design selects the design flood frequency and calibration flood frequency as the flood control standard of the project according to specification and calculates the flood at corresponding frequency as the basis of the design of the hydro-structure. This is a simplification adopted mainly to facilitate the engineering design.

The risk shown in formula (6.3-19) is actually the annual probability that the reservoir dam fails due to standard-exceeding flood in the operation period. Let the flood control design standard of the reservoir dam be p and the service life of the project be n years, the calculation of the flood control safety risk can be expressed in detail as follows:

$$R_{f,\text{Year}} = p \qquad (6.3-20)$$

$$R_f = 1 - (1-p)^n \qquad (6.3-21)$$

Thus the reliability is

$$\overline{R} = 1 - R_f = (1-p)^n \qquad (6.3-22)$$

The risk-based flood control safety design of dam shall be the inverse problem of the problem above, that is, firstly calculating the comprehensive risk with consideration of the influences of the uncertain factors in hydrology, hydraulics and engineering structure etc. and then determining the discharge capacity of the reservoir dam whereby it's ensured that the risk of the reservoir dam failure is the minimum and acceptable during the service life of the project.

6.3.6.3 Calculation of Flood Control Risk of Single Reservoir

Actually, the flood control safety risks of reservoir dams are not merely related to flood frequency. In traditional flood control safety design, only the risks resulting from standard-exceeding floods during the useful life period of the project are taken into account, while the risks of hydraulic, engineering structure, operation management, etc. are not taken into account. Therefore, traditional safety designs are not all-sided. Yang Baiyin (1998) and Wang Ruichen, et al (2000) deeply studied the risks of flood discharge for a single reservoir and cascade reservoirs based on the flood control task of reservoirs, which improves the risk-based flood control safety design of dams in China.

As for the flood control design of a reservoir, the flood control tasks of the reservoir itself and of its downstream protected subjects need to be taken into account. Considering these two tasks, the contents covered by the flood control design are analyzing the frequency of the coming flood from the area upstream of the reservoir, adopting proper flood control standard, comparing and selecting flood discharging facility proposals and determining applicable flood discharge capability and flood control storage capacity. It is certain that these factors related to flood control are uncertain more or less. If only the objective uncertainty of floods alone is taken into account, the reservoir will only be subjected to the risks caused by standard-exceeding floods during the reservoir's useful life period. This is the risk caused by the recommended flood control standard by current Chinese flood control standards and specifications. As a matter of fact, besides the objective uncertainty of floods, reservoir level, trial value of the discharge coefficient of flood discharging facilities and flood discharging outlet sizes are subjected to construction error uncertainty, which results in the hydraulic uncertainty of reservoir discharge capability. In this case, a flood under the design standard usually goes beyond the flood control storage capacity and thus causes overflowing because of the under-discharge capability of flood discharging facilities. As a result of this, the reservoir fails to control flood. That is to say, flood discharging facilities are also subjected to some risks due to the above-mentioned uncertain factors. Therefore, the probability of such flood control failure caused by uncertain hydrology and hydraulic factors is defined as the comprehensive flood discharging risks to which flood discharging structures are subjected, and such comprehensive flood discharging risks is represented by R.

Due to different flood control tasks, reservoirs can be classified into those with flood regulation storage capacity and those without flood regulation storage capacity. The design discharge capability of reservoirs without flood regulation storage capacity is usually equal to or beyond the maximum peak flow of the check flood. If the discharge capability of a reservoir dam is lower than the flood peak flow from the upstream, an overflowing event occurs. The flood regulation storage capacity of reservoirs with flood regulation function is closely related to the discharge capability. Due to the function of the flood regulation storage capacity, the required reservoir discharge capability may be lower than the inflow flood peak flow more or less. Only when the flood occurring upstream of a dam makes the discharge capability during the check flood level lower than the discharge capability required according to a flood regulation principle and the post-flood regulation reservoir level higher than the check flood level, an overflowing event may occur.

Let the peak discharge of the reservoir inflow be X, the corresponding flood regulation storage capacity V, the corresponding maximum discharge Q_m, the reservoir capacity at the check flood level V_p, the corresponding discharge capability Q_c. Based on the above-mentioned reservoir types, the flood discharge risks (annual risks) of reservoirs are defined as follows:

(1) In case of reservoirs without flood regulation storage capacity, overflowing events occur when reservoir inflow go beyond the discharge capability. In this case, the flood discharge risk is:

$$R = P\{Q_c < X\} \tag{6.3-23}$$

(2) In case of reservoirs with flood regulation storage capacity, overflowing events occur when the flood regulation storage capacity required by the upstream incoming floods is larger than the check storage capacity. In this case, the flood discharge risk is:

$$R = P\{V_p < V\} \tag{6.3-24}$$

At a determined regulation-start level, the discharge capability of the reservoir q_c is in a single-valued correspondence relationship with the reservoir capacity, namely when $V_1 > V_2$, $q_{c1} > q_{c2}$. So, formula (6.3-24) can be represented as follows:

$$R = P\{V_p < V\} = P\{Q_c < Q_m\} \tag{6.3-25}$$

It can be seen from the two results above that $Q_m = X$ when the reservoir has no flood regulation storage capacity. Virtually, this is an exception to the reservoirs with flood regulation storage capacity.

Because Q_c and Q_m are respectively affected by uncertain hydraulic and hydrological factors, they are random variables. Therefore, formula (6.3-25) may be further presented as follows:

$$R = P(Q_c < Q_m) = \sum_{q_m=0}^{\infty} [P\{Q_c < Q_m = q_m\} P\{Q_m = q_m\}] \tag{6.3-26}$$

According to the analysis, we know that both Q_c and Q_m are continuous random variables. The probability density functions and distribution functions of Q_c and Q_m are respectively represented as $f_{Q_c}(\cdot)$, $f_{Q_m}(\cdot)$, $f_{Q_c}(\cdot)$ and $f_{Q_m}(\cdot)$. Because Q_c and Q_m are random variables caused by different reasons, they are independent of each other. Therefore, formula (6.3-26) can be represented as the following integral form.

$$R = P\{Q_c < Q_m\} = \int_0^{\infty} \left[\int_0^{q_m} f_{Q_c}(q_c) dq_c \right] f_{Q_m}(q_m) dq_m = \int_0^{\infty} [F_{Q_c}(q_m)] f_{Q_m}(q_m) dq_m \tag{6.3-27}$$

It can be seen from formula (6.3-27) that if only the uncertainty of floods is taken into account, and the uncertainty of the discharge capacity is not taken into account, namely $F_{Q_c}(q_m) = 1$, the flood discharge risk of the reservoir is the single over-check standard flood risk. Actually, at the same reservoir level, the discharge capacity is a random variable rather than a constant, affected by various uncertain factors. Therefore, usually, it is inappropriate to determine reservoir flood control safety only according to the flood standard. So, formula (6.3-27) is the analysis mode for the comprehensive risks that are brought by uncertain floods and uncertain discharge capability to reservoir flood discharging facilities. Therefore, it can be concluded that in case of various floods, the annual probability of non-overtopping is the reliability of flood discharging structure $\overline{R}$, which is defined as follows:

$$\overline{R} = 1 - R \tag{6.3-28}$$

6.3.6.4 Calculation of Flood Discharge Risk of Cascade Reservoirs

Cascade reservoirs include two or more upstream and downstream reservoirs that are constructed in a river (or a river reach) and are in hydraulic connection with each other. Regarding the flood control design of cascade reservoirs, because cascade reservoirs consist of double reservoir system or multiple reservoir system, independent operation and joint compensation operation systems, and the systems with and without flood regulation function, the uncertain factors affecting the flood discharge safety of cascade reservoirs are much more complicated than those of the single reservoir. The flood discharge safety of cascade reservoirs is affected not only by the natural inflow water of each hydropower station reservoir upstream of the studied hydropower station, but also by the uncertain factors of interval floods and flood area. Therefore, to analyze flood discharge risks of cascade reservoirs, all the above-mentioned uncertain factors should be taken into account. In point of fact, the above-mentioned uncertain factors can fall into two categories: the one is the uncertainty of floods and reservoir group operation, and the other is the hydraulic uncertainty affecting each cascade reservoir's discharge capability. So, in order to study the flood discharge risk mode of cascade reservoirs, the characteristics of design floods of cascade reservoirs must be analyzed at first. In 1986, Wang Ruichen, Song Dedun, et al. studied the flood probability distribution of the downstream design section affected by the upstream reservoir flood regulation in case of cascade reservoirs, and proposed a whole set of calculation method, which is briefly stated as follows: in case of cascade reservoirs consisting of an upstream reservoir (A) and a downstream reservoir (B), assume that X_A, Y and Q_A respectively represents the reservoir (A)'s inflow from the upstream, interval and discharge flow, and X_B represents the inflow flood of the reservoir B (a study subject). Obviously, they are random variables, and their distribution functions are represented as $F_{X_A}(\cdot)$, $F_Y(\cdot)$, $F_{Q_A}(\cdot)$ and $F_{X_B}(\cdot)$ respectively. Assume that $g_A(\cdot)$ is the flood regulation function of reservoir A, the inflow flood distribution of reservoir B is shown as follows:

$$F_{X_B}(x_B) = P\{X_B = Q_A + Y \leqslant x_B\}$$
$$= \int_{y_0}^{x_B - q_{A_0}} \int_{q_{A_0}}^{x_B - y} f_{Q_A, Y}(q_A, y) \mathrm{d}q_A \mathrm{d}y$$
$$= \int_{y_0}^{x_B - q_{A_0}} f_Y(y) F_{Q_A/A}(x_B - Y) \mathrm{d}y \qquad (6.3-29)$$

where $F_{Q_A/Y}(\cdot)$ represents the probability distribution function of Q_A when $Y = y$ and q_{A_0} and y_0 respectively represents the sample space lower bounds of Q_A and Y.

Because $F_{Q_A}(\cdot)$ represents the set of all the X_A meeting $g_A(x_A) \leqslant q_A$,

$$F_{Q_A/Y}(q_A) = P\{X_A \leqslant g_A^{-1}(q_A)/Y = y\} \qquad (6.3-30)$$

If Q_A and y are independent of each other, the above formula is changed to the following one:

$$F_{Q_A/Y}(q_A) = F_{Q_A}(q_A) = F_{X_A}[g^{-1}(q_A)] \qquad (6.3-31)$$

So, the inflow flood probability distribution of reservoir B can be obtained through formulas (6.3-29), (6.3-30) or (6.3-31). The discharge flow distribution function of reservoir B can be obtained through formula (6.3-31) in the same way so long as the operation function $g_B(x_B)$ of reservoir B has been figured out. Formula (6.3-31) is also applicable for the single reservoir. When reservoir A and B are operated jointly, $g_{AB}(x_A, y)$ is the corresponding joint operation function. Q_B represents the discharge flow of reservoir B, and the distribution function of Q_B can be figured out by the following formula:

$$F_{Q_B}(Q_B) = \iint_{g_{AB}(+) \leqslant q_B} f_{X_A, Y}(x_A, y) \mathrm{d}x_A \mathrm{d}y \qquad (6.3-32)$$

Yang Baiyin (1996) proposed the flood discharge risk mode of cascade reservoirs by the above distribution function of inflow for cascade reservoirs.

Let the natural inflow from the most upstream section of cascade reservoirs be X, the inflow from each interval Y_1, Y_2, $\cdots$, Y_{n-1}, the operation function of cascade reservoirs $g_{A_1 \cdots A_n}(x, y_1, y_2, \cdots, y_{n-1})$, the study subject A_n, and the discharge capability Q_c. According to the definition of flood discharge risk, the overflowing probability caused by under-discharge of facilities when floods occur is equal to the risks to which flood discharging structures are subjected to. Because the influence of uncertain hydraulic factors of the discharging equipment upstream of cascade reservoirs on discharge risks is much smaller than that of floods, and because taking the uncertain hydraulic factors of cascade reservoirs into account makes risk calculation very complicated and it is very hard to propose a risk mode analysis formula, usually only the hydraulic uncertainty of the discharge capability of the study subject is taken into account. Therefore, the flood discharge risk mode of cascade reservoirs is defined as follows:

$$R = P\{Q_c < Q_m[X, Y_1, \cdots, Y_{n-1}, g_{A_1 \cdots A_n}(x, y_1, \cdots, y_{n-1})]\} \tag{6.3-33}$$

where $Q_m(\cdot)$ represents the maximum discharge flow that the reservoir A_n is required to discharge when a flood X occurs, and the inflow from intervals are Y_1, Y_2, $\cdots$, Y_{n-1}.

According to the design flood features of cascade reservoirs, formula (6.3-33) can be changed into the following one:

$$\begin{aligned} R &= \int_0^\infty \cdots \int_0^\infty \int_0^{g_{A_1 \cdots A_n} x, y_1, \cdots, y_{n-1}} f_{Q_c}(q_c) f_{X, Y_1, \cdots, Y_{n-1}}(x, y_1, \cdots, y_{n-1}) dq_c dx dy_1 \cdots dy_{n-1} \\ &= \int_0^\infty \cdots \int_0^\infty F_{Q_c}[g_{A_1 \cdots A_n}(x, y_1, \cdots, y_{n-1})] f_{X, Y_1, \cdots, Y_{n-1}}(x, y_1, \cdots, y_{n-1}) dx dy_1 \cdots dy_{n-1} \end{aligned} \tag{6.3-34}$$

If X, Y_1, Y_2, $\cdots$, Y_{n-1} are independent of each other, formula (6.3-34) can be changed into the following one:

$$R = \int_0^\infty \cdots \int_0^\infty F_{Q_c}[g_{A_1 \cdots A_n}(x, y_1, \cdots, y_{n-1})] f_X(x) f_{Y_1}(y_1) \cdots f_{Y_{n-1}}(y_{n-1}) dx dy_1 \cdots dy_{n-1} \tag{6.3-35}$$

In this case, formulas (6.3-34) or (6.3-35) is the flood discharge risk analysis mode that we need. The corresponding reliability of such mode is

$$\overline{R} = 1 - R \tag{6.3-36}$$

6.3.7 Probable Maximum Precipitation and Probable Maximum Flood

As early as the 1940s, America began to study probable maximum precipitation (PMP) and probable maximum flood (PMF), and proposed that PMP and PMF should be taken as the standard for flood control safety design of important dams. China began to study PMP and PMF in 1958. PMP and PMF were firstly employed for the design flood study of the Three Gorges Project. In 1963, Zhan Daojiang wrote a paper to describe systemically PMP calculation methods in China. In August 1975, an unusual extraordinary storm occurred at the upper basin of the Huaihe River. Afterwards, universal importance is attached to the study of PMP and PMF. It's thought that PMP and PMF should be taken as the dam protection standards of important hydropower projects, and "Isocline Map of Possible 24h Point Rainfall in China" was prepared in 1977 based on a lot of studies. *The Specifications for Calculating Hydropower Engineering Design Floods* (SL 44—93) issued in 1993 states specifically that "To meet the requirements of engineering design, probable maximum precipitation can be estimated by the hydro-meteorological method, and then probable maximum flood can be figured out".

PMP means the theoretical probable maximum precipitation during a certain duration of a stage in one year at a specific geographic location and in the given torrential rain area. After years of study, Chinese scholars thought that the practical method for calculating the design watershed PMP should include three steps: the first is to maximize the water vapor of the actually measured extraordinary torrential rain, namely maximize water vapor; the second is to assume that the maximized torrential rain occurs in the design area; and the third is to get the envelop curve of the duration-area-depth relationship of the maximized torrential rain. Like this, the PMP duration-area-depth relationship in the design area is gotten. Based on such a relationship, the average PMP of a design watershed of the design area can be gotten.

According to the above, we can obviously see that theoretically, any method for calculating PMP needs to solve two key problems: one is how to establish the torrential rain mode, and the other is how to maximize the torrential rainfall. To solve the first problem, the emphasis of study should be on rainstorm transposition and storm combination. The study of storm transposition should mainly focuses on weather condition consistency, orographic rain correction, transposition direction, etc., while the study of storm combination should mainly focuses on the rationality and possibility. To solve the second problem, China currently has adopted the water vapor magnifying method and the water vapor efficiency magnifying method.

PMF is calculated according to the PMP watershed runoff generation and flow concentration. For the study on watershed runoff generation and flow concentration theory and calculation methods, refer to the runoff generation and flow concentration theory in Section 5.4 of this chapter.

6.3.8 Paleo-flood Study

During a flood period, the organic substances carried by the flood, like spore and pollen, vegetation, etc. are floating on the water surface. In the short advection period when the highest river level is achieved, such organic substances carried by water may deposit on bank sides and become slack water deposits. Usually, the grotto, terminals of branches, and gorge mesa on the river bank sides can keep such slack water deposits up to now. The elevation at which slack water deposits exist is equal to the highest level of the corresponding flood. So, the flood peak flow can be figured out according to the elevation at which slack water deposits exist. The depositing year of slack water deposits can be dated with ^{14}C isotope assay, and thus the recurrence period of the flood peak flow can be figured out. The data of extraordinary floods gotten in this way is different from that gotten by investigating the historical floods. Because the former involves the principles of sedimentology and stratum chronology, the floods measured according to sedimentology and stratum chronology principles are called paleo-floods.

Chinese hydrologists began to study paleo-floods based on the abundant results of flow frequency calculation, historical flood application and PMF study, etc. Therefore, Chinese paleo-flood study has a very significant background, and is the focus of the hydrology field all the time. From 1985 to today, paleo-flood study has been applied to the reinforcement of Xianghongdian Reservoir on the Huaihe River, reinforcement of Huangbizhuang Reservoir on the Haihe River, the Three Gorges Project on the Yangtze River, Xiaolangdi Water Project of the Yellow River, Zangmu Water Project on the Yarlung Zangbo River, and other important water projects. For example, in the paleo-flood study of the Three Gorges, Zhan Daojiang, et al. spent five years, namely from 1990 to 1994, to totally take 92 samples of paleo-flood deposits in the section of the Three Gorges Project. They analyzed and detected such deposits, and learned that in the past 2,500 years, a total of 4 extraordinary floods whose peak flow exceeds 100,000m³/s occurred in this section, which increases the flood recurrence period (840 years) previously gotten in 1870 through historical floods investigation to 2,500 years, and makes great contributions to settling the long-term pending technical problems of design flood calculation of the Three Gorges Project.

6.4 Hydrological Forecast

6.4.1 *Introduction*

During dam construction period and dam operation period, hydrological forecast plays a role in ensuring safety during flood, and ensuring flood control and power generation. Therefore, hydrological forecast is of great importance.

A practical hydrological forecast should contain forecast value and lead time meeting accuracy requirements. According to the length of lead time, hydrological forecast can be classified into short-term flood forecast and medium-and long-term hydrological forecast. The flood forecasts provided according to the actually measured torrential rain course or the short-term quantitative torrential rain forecast results of a meteorological department authority are called as short-term flood forecasts, which are based on the runoff generation and flow concentration theory. The lead time of medium-and long-term hydrological forecasts ranges from 10 days to 1 year usually. Usually, the methods for medium-and long-term hydrological forecasts are based on meteorology or mathematical statistics theory.

6.4.2 *Runoff Generation and Flow Concentration Theory*

The runoff generation and flow concentration theory aims at researching the physical mechanisms of rainfall-runoff under various climatic and underlying surface conditions, the hydrodynamics law of water flow concentration in various media, and the calculation methods of runoff generation and flow concentration. It is the theoretical base for studying deterministic hydrologic models and short-term flood forecast methods.

China's contributions to the runoff generation and flow concentration theory are mainly made after the 1950s. In the runoff generation aspect, Zhao Renjun, et al. (1984) analyzed lots of hydrologic data under various climatic and underlying surface conditions in China, and then in the 1960s got a significant conclusion that the runoff yield under saturated storage dominates the wet regions, and the runoff yield under excess infiltration dominates the arid regions, and such conclusion was verified by the observed data of many Chinese runoff test stations. Later on, they proposed the watershed storage capacity curve and then the watershed infiltration capacity distribution curve in the 1970s, and such curves became the theoretical basis for studying Xin'anjiang Two Water Resources Model and Shaanbei Model. In the 1970s, Chinese scholars perfected the concept of hierarchical calculation of watershed evapotranspiration according to soil evapotranspiration law, and such calculation developed into the major calculation method for confirming the watershed evapotranspiration. In the middle of the 1970s, the study on the runoff generation mechanism in the world made a new breakthrough, whose symbol is the discovery of subsurface runoff and saturated surface runoff, and the proposal of regressive runoff concept. Based on this, Yu Weizhong, et al. proposed the interface runoff generation law in 1980, Wen Kang, et al. proposed the uniform runoff generation model under various natural conditions in 1982, and Rui Xiaofang, et al. proposed the uniformities and reciprocal transformation conditions of runoff generation mechanism in 1991. All of these studies were tried to theoretically and practically uniform the physical conditions for the generation of various runoff components. The discovered generation mechanism of subsurface runoff and saturated surface runoff became the theoretical basis for Zhao Renjun to study Xin'anjiang Three-component Model later. In 1988, Li Changxing and Shen Jin applied the orient theory of the soil water dynamics to treating the area changes of the runoff yield under excess infiltration, and proposed a new method for calculating runoff yield under excess infiltration. In 1992, Gu Weizu utilized environmental isotope technology to experimentally analyze the runoff components of a flood hydrograph, and pointed out that the surface runoff may not originate from the rainfall, and some of the subsurface runoff and groundwater runoff must not come from the rainfall. Therefore, Gu Weizu thought that rainfall and runoff are not in a strict one-to-one correspondence relationship. This is a challenge against the current runoff generation theory, and is worthy of attention.

In the aspect of flood routing, at the end of the 1950s, Chinese scholars introduced the characteristic river length concept from the Soviet Union, explained the physical meanings of the parameters of Muskingum Method previously introduced from America, and got the theoretical formulas for calculating the parameters of Muskingum Method, that is:

$$x = \frac{1}{2} - \frac{l}{2L} \qquad (6.4-1)$$

$$K = \frac{L}{C} \qquad (6.4-2)$$

where x and K represent the flow proportion factor and channel storage coefficient of Muskingum method respectively, l the characteristic river length, L the calculus river reach length and C flood wave speed.

Based on these, Zhao Renjun proposed the reach-by-reach continuous calculus for Muskingum Method in 1962, which shot the irrational negative flow phenomenon occurring during the long river reach calculation by Muskingum Method, and raised the accuracy of long river reach flood routing calculation. The flow concentration coefficient of constantly completed reach-by-reach continuous calculus of Muskingum Method is:

$$\begin{cases} P_m(n) = C_2^m & n = 0 \\ P_m(n) = \sum_{j=1}^{m} B_j C_1^{m-j} C_2^{n-j} A^j & n > 0, n-j \geqslant 0 \end{cases} \qquad (6.4-3)$$

where

$$A = C_0 + C_1 C_2$$

$$B_j = \frac{m! \ (n-1)!}{j! \ (j-1)! \ (m-1)! \ (n-1)!}$$

$$C_0 = \frac{K'x' + 0.5\Delta t}{K'(1-x') + 0.5\Delta t}$$

$$C_1 = \frac{0.5\Delta t - K'x'}{K'(1-x') + 0.5\Delta t}$$

$$C_2 = \frac{K'(1-x') - 0.5\Delta t}{K'(1-x') + 0.5\Delta t}$$

where $P_m(n)$ represents flow concentration coefficient, m serial number of sub-reach, n time period serial number, x' and K' respectively represents flow proportion factor and channel storage coefficient of sub-reach, Δt the calculation period, and the meanings of other symbols are the same as the above-mentioned meanings.

The river network flood routing develops fast recently. At present, there are two classes of methods for river network flood routing: one is river network unit partition method and the other is also to use Preissmann 4-point implicit finite differential numerical scheme with a weight coefficient in order to calculate river network flood movement.

Tan Weiyan, Hu Siyi, et al. (1990) had deeply studied dynamic wave calculus. They analyzed the various schemes of St. Venant equations theoretically, summarized the way and proper principle of establishing the conservation difference scheme, and explored how to apply the high-performance schemes in aerodynamics field that meet requirement for conservation, counterblast, positivity, high discontinuous resolution, etc. to computing the numerical values of St. Venant equations. In practice, they adopted one dimensional and two dimensional flow models, followed the modeling concept of combining hydraulics with hydrology methods, and successfully developed the flow simulation model for the huge and complicated flood control system of Dongting Lake in the middle reach of the Yangtze River. Unstructured grid two dimensional unsteady high-performance finite volume schemes was adopted for the lake part of the model for the purpose of adapting to the complicated boundary geometry and keeping water balance. One dimensional unsteady flow conservation explicit scheme was adopted for the river network part in order to avoid complicated implicit scheme matrix algorithm, which at the same time was beneficial to the coupling with the two dimensional model, and the explicit connection to various complicated connection relationships. The calculated results have shown that this model is successful.

Another major feature of flood routing study reflects the modern technology influence and the crossing of various study fields. In 1993, Rui Xiaofang took the single river as single input-single output system for flood routing, and adopted the

ARMA model and optimal control theory of the time series analysis method to establish the method of flood routing under free lower boundary condition. In 1995, Zhong Denghua, et al. introduced artificial neutral network method for flood routing, which can solve both single output and multiple output problems, and both linear and non-linear problems.

In watershed flow concentration theory aspect, at the beginning of the 1950s, Chinese scholars proposed the Huaihe River synthesis unit hydrograph method, disclosing the nonlinearity nature of the unit hydrograph. This made contributions to the development of the synthesis unit hydrograph method proposed by Snyder in 1938 in China. In 1958, Hydrologic Research Department, China Institute of Water Resources and Hydropower Research verified the non-linearity existence of unit hydrograph through indoor simulated tests. In 1962, Zhao Renjun developed the origin flow concentration theory proposed by the hydrologists of the Soviet Union. Afterward, some experts developed the origin flow concentration theory into the watershed flow concentration calculation method that can realize the combination of the hydrology method and the hydraulic method. The instantaneous unit hydrograph proposed by Nash was introduced in the 1960s for the first time, which made positive contributions to Chinese scholars' study on conceptual watershed flow concentration calculation method. Chinese scholars not only conquered some theoretical defects of Nash method by combining Nash method and area-time curve, but also expanded Nash method to non-linear conditions, which expanded the application range of Nash method. In the 1970s, Ye Shouze, Xiajun, et al. introduced systematical analysis method to watershed flow concentration study, and made achievements in the aspect of identifying watershed flow concentration system. In addition, Shen Jin, Shen Bing, et al. disclosed the slope flow concentration mechanism through indoor tests, and did a lot of quantitative analysis and calculation study work by virtue of the finite element analysis method. In 1989, Rui Xiaofang derived the instantaneous unit hydrograph formula of the groundwater flow concentration covered by homogeneous soil from the groundwater dynamics equation. The landform instantaneous unit hydrograph theory created by Rodriguze, Iturbe, et al. at the end of the 1970s represents one of the most important developments of watershed flow concentration theories in recent years. People got the probability theory interpretation of the watershed instantaneous unit hydrograph through disclosing the statistical mechanism of watershed flow concentration, and theoretically derived the watershed instantaneous unit hydrograph formula expressed by landform, topography and hydraulic parameters. This is the new attempt of solving watershed flow concentration problems with "particle" view. In 1988, Wen Kang, et al. derived the general formula for the landform instantaneous unit hydrograph in any grade of watershed

$$U(t) = -\sum_{j=1}^{\Omega} \left\{ \lambda_j \left[\sum_{i=1}^{j} \theta_{i,\Omega}(0) A_{ij} \right] e^{-\lambda_j t} \right\} \qquad (6.4-4)$$

where

$$A_{ij} = \frac{B_{ij}(\Omega)}{\prod_{a=i}^{\Omega}(-\lambda_j)(\lambda_a - \lambda_j)} (a \neq j)$$

$$\theta_{i,\Omega}(0) = \frac{P_B^{\Omega-1}}{R_A^{\Omega-1}} \left(R_A^{j-1} - \sum_{i=1}^{j-1} R_A^{i-1} R_B^{j-i} P_{ij} \right)$$

$$P_{ij} = (R_B - 2) R_B^{\Omega-j-i} \frac{\prod_{k=i}^{j-i-1}(R_B^{\Omega-i-k} - 1)}{\prod_{k=i}^{j-1}(2R_B^{\Omega-i-k} - 1)} + \frac{2}{R_B} \delta_{i+1,j}$$

$$\delta_{i+1,j} = \begin{cases} 1, j = i+1 \\ 0, j \neq i+1 \end{cases}$$

where Ω represents watershed grade, λ_j represents the reciprocal of the average flow concentration time of grade j river, R_B and R_A respectively represents watershed bifurcation ratio and area ratio, and $B_{ij}(\Omega)$ is realized by subroutine.

In recent years, based on this way, some researchers have explored the non-linearity mechanism of watershed flow concentration, and some have proposed the view that watershed flow concentration is controlled by geomorphologic diffusion and hydrodynamic dispersion. Such original studies make people full of confidence in the prospect of the instantaneous landform unit hydrograph theory.

It should be noted that the crossing and reciprocal influences of various disciplines, the combination of experimental measures and theoretical analysis, and the complementation of deterministic and stochastic methods have made great contributions to the development of modern runoff generation and flow concentration theories. Therefore, in the future, we shall further explore watershed runoff generation theory by combining climatology, soil water dynamics and geology, and explore river flood routing theories and watershed flow concentration theories by combining geomorphology, topography, hydrodynamics, and statistical physics.

6.4.3 *Watershed Hydrological Model*

Watershed hydrological model is a logical structure consisting of a mathematical expression to describe the watershed rainfall and runoff formation. The model input is rainfall, evapotranspiration capacity, model parameter and initial condition. The model output is evapotranspiration discharge, soil moisture content or watershed outlet section discharge curve, etc. The emergence of the watershed hydrological model is inseparable from the computer invention and extensive application. Stanford model, the first watershed hydrological model in the world, emerged in the 1960s. Afterwards, the model study has become popular worldwide. There are hundreds of watershed hydrological models proposed successively, but few of them are proven by practice to be of better application value, American Sacramento river forecast center model (SAC model for short), Japan National Disaster Prevention Research Center tank model (tank model for short) and Italian constrained linear system model (CLS model for short) etc are typical of watershed hydrological models. On one hand, China has, since the1970s, actively introduced foreign helpful watershed hydrological models, and on the other hand, been committed to the development of new watershed hydrological models.

In the early 1970s, Zhao Renjun, et al. successfully developed the Xin'anjiang two -component hydrological model applicable to humid or semi-humid areas giving priority to runoff yield under saturated storage, based on domestic study results in the runoff generation and flow concentration theory for many years and foreign valuable experience in the development of watershed hydrological models. In the early 1980s, they timely absorbed new discovery of foreign hillside hydrology on runoff generation mechanism, and improved the Xin'anjiang two water sources model, thereby causing the development of the Xin'anjiang three-component model, including watershed runoff generation calculation, watershed evapotranspiration calculation, runoff component classification, surface runoff calculation and ground water runoff flow concentration calculation subsystems. The two water sources model differs from the three water sources model in the classification of runoff components only: the former falls into two runoff components, i.e. surface and subsurface runoff according to the Horton theory, while the latter falls into three runoff components, that is, surface, subsurface and groundwater runoff according to the hillside hydrology theory. Later, they have adopted objective optimization technique of parameters in the model to make beneficial exploration to the regional laws of model parameters. Practices have proved that, with the Xin'anjiang model, higher accuracy is achievable in the China's vast humid and local semi-humid areas. So far, the model has been actively recommended by World Meteorological Organization (WMO).

In China's arid loess plateau region, Zhao Renjun has successfully developed Shaanbei model based on many years' study, the model calculates the single point excess infiltration surface runoff according to the infiltration curve, and runoff yield area changes of excess infiltration surface runoff may be taken into account with the infiltration capacity distribution curve, and the flow concentration computing method may be configured itself.

In 1981, Wang Juemou, Zhang Ruifang, et al. introduced the international CLS model from Italy, and introduced the China's calculation method of runoff yield under saturated storage into CLS model while keeping the existing runoff generation computing method so as to further improve the model accuracy. In 1985, they have increased a structure of the classification of runoff components by the Xin'anjiang three-component model and a structure of the calculation of excess infiltration surface runoff by Shaanbei model to finally develop an improved model called Synthetic Constrained Linear System model (hereinafter referred to as SCLS model), which has been applied well in China Gezhouba Hydropower Station, Sanmenxia Reservoir and Fengman Hydropower Station, and incorporated into the WMO Hydrological Operation Management System (HOMS) software to be presented to all countries worldwide in 1986.

It should be noted that the middle of the 1970s-1980s is the flourishing development period of watershed hydrological models, which are popular because they are mathematical models, and compared with the physical model, characterized by the following obvious advantages. First, all conditions of the mathematical model can be given directly from the prototype measurement data, not subject to scales, that is, the mathematical model has no similarity rules. Second, boundary conditions and other conditions of the mathematical model may be subject to strict control or changes as required anytime. Third, the mathematical model is of strong generality, and different practical problems can be solved as long as a kind of proper application software is developed. Fourth, the mathematical model is characterized by perfect anti-interference property, and repeated simulation can yield the same results in case of invariable conditions. Fifth, the mathematical model has low development cost and operation management fees. However, since the late 1980s, the watershed hydrological models are in a slow development stage, with neither breakthrough nor new influential model. Why? There are the following limits in the current watershed hydrological model study. Firstly, hydrology is so complex that we fail to fully describe any of its sub-processes with physical law and mathematical and physics equations as yet. Therefore, the watershed hydrological models are, in many structural links, still dependant on conceptual element simulation or empirical function to make most of parameters contained in the watershed hydrological model which lack explicit physical significance. Secondly, the model parameters determined by optimization methods are greatly dependent on the actually measured rainfall and runoff data, and in general, a watershed hydrological model includes more than two parameters to be calculated by means of the actually measured rainfall and runoff data. Model parameters calculated with the optimization method can describe the model fitting degree only, instead of revealing the physical significance of parameters. Thirdly, the input of watershed hydrological model is the rainfall process of all points in the watershed, while its output is the flow or water level process through the outlet section. Therefore, its input is discrete and its output is integrated, with which the existing watershed hydrological model structure is not matched. Fourthly, it is the effective calculation method issues. The watershed hydrological model always simulates the continuous process of watershed rainfall runoff formation in a discrete way. Besides rational structures and clear physical significance of parameters, the success of the watershed hydrological model is also determined by proper discrete format and effective calculation method. Most of the existing watershed hydrological models adopt the explicit format, and almost fail to carefully study the stability of numerical calculation, which is an issue to be noticed.

6.4.4 *Short-term Flood Forecast*

In China, there are typically three basic short-term flood forecast issues: the first is river reach flood forecast, i. e. to forecast the lower reach water regime based on the upper reach water regime of a river, with its theoretical basis being the movement rules of flood wave in the river reach, and the travel time of flood wave in the river reach is the theoretical lead time of flood forecast in the river reach; the second is watershed rainfall runoff forecast, that is, to forecast the watershed outlet flow duration curve based on a spatiotemporal distributed rainfall in the watershed, with its theoretical basis being the watershed rainfall runoff formation rules, the watershed flow concentration time is the theoretical lead time of watershed rainfall runoff forecast, which is traditionally divided into two stages, i. e. the watershed runoff generation and flow concentration forecast; the third is "river-watershed" flood forecast, a watershed may be always divided into several unconnected sub-watersheds according to the natural water division, and these sub-watersheds are connected by channel. Each sub-watershed shall be first subject to watershed rainfall runoff forecast, its contribution to the total outlet section flood of watershed shall be calculated through flood routing, and be superposed to forecast the flood duration curve of total outlet section flood. This is the basic train of thought for "river-watershed" forecast, with its theoretical basis being the combination of watershed rainfall runoff formation rules and river flood wave movement rules and with the summation of the watershed flow concentration time of watersheds and flood wave travel time in the river reach as the theoretical lead time of "river-watershed" flood forecast.

In terms of river reach flood forecast, classical corresponding water level and flow methods have been improved substantially in combination with physical mechanism of river flood wave movement to make them applicable to water level and flow forecast of multi-tributary rivers, heavily silt-carrying rivers and tidal reaches. Flood routing methods have also been improved and innovated to some extent under the guidance of hydraulics principles and other modern science and technology achievements, which not only puts forward some forecast methods of water level or flow duration applicable to river reaches without subject to backwater jacking influence, but also proposes certain forecast methods of water level or flow duration of reaches able to consider the impact of the backwater jacking and tidal reaches. Beneficial exploration has been made for river flood forecast methods with flood storage areas.

In terms of watershed rainfall runoff generation forecast, traditional experience rainfall runoff correlation chart has been improved continuously under the instruction of the constantly developed rainfall runoff theory to make it reach better forecast accuracy as long as the influencing factors are selected properly in humid, arid, semiarid or semi-humid and semiarid regions. In addition, it further puts forward the watershed storage curve suitable for runoff generation forecast in humid regions and infiltration curve method and initial loss & continuing infiltration method applicable to runoff generation forecast in arid regions.

In terms of watershed flow concentration forecasting, traditional unit hydrograph method and contour method of equal travel time are improved, and practical methods to process the watershed flow concentration nonlinearity with separate rainfall unit hydrograph, separate rainstorm central position unit hydrograph and variation isochrone are proposed. Much exploration has been made for the application of Nash instantaneous unit hydrograph and geomorphologic instantaneous unit hydrograph to the watershed flow concentration forecast.

In terms of "river-watershed" flood forecast, with this method, not only the rainfall runoff forecast problems in large watershed or watershed with great impacts of human activities have been solved well, but also the rainfall runoff forecast problems in the tidal waters have been solved with a combined method of hydrology and hydraulics.

6.4.5 *Medium and Long-term Hydrological Forecast*

With the increase of forecast period, hydrological uncertainty is increasingly playing a leading role. Therefore, methods used for medium and long-term hydrological forecast are significantly different from those for short-term flood forecast. In

China, existing methods for medium and long-term hydrological forecast mainly fall into three categories. The first is synoptic method, which, based on the concept that long-term runoff changes shall be closely related to the evolution of the macroscopic synoptic process, uses the future evolution rules of earlier atmospheric circulation features to make long-term forecast for hydrologic features. The second is statistics method, which, according to the analysis on a great many long-term historical hydrological and meteorological data, makes medium-and long-term hydrological forecast by establishing a statistical relation between the forecast objects and forecast factors. Depending on features of factors to be selected for formulating forecast schemes, this method falls into single element method and multifactor synthesis method. The former takes the sequential variation rules of the forecast object itself as the forecast basis, while the latter takes the analysis of statistical correlation between the forecast objects and influencing factors as the forecast basis. The third is geophysics-based physical genetic analysis method, which tries to take the relation between universe physical geographical factor changes and long-term changes of hydrologic features in some regions on the basis of proposed methods for medium and long-term hydrological forecast.

China's medium and long-term hydrological forecast dates back to the early 1960s, and statistical methods ever used include multiple regression, stepwise regression, threshold regression, time series analysis, harmonic analysis, power spectrum analysis, cluster analysis, natural orthogonal function decomposition, etc. in which harmonic analysis is a common method to analyze the long wave and ultra-long wave activities, the power spectrum analysis is an objective method to reveal the periodic changes of hydrologic features, and the orthogonal function decomposition is characterized by rapid convergence. Multiple regression, stepwise regression, threshold regression, time series analysis and cluster analysis are methods of the most frequent and widespread application.

It should be noted that the forecast accuracy is, due to the existence of chaos, bound to be greatly reduced with the increase in forecast period. Therefore, it is rather difficult to improve the medium and long-term hydrological forecast accuracy.

6.5 Engineering Sediment

6.5.1 Sediment Problems on Hydropower Projects and Study Contents

In the process of hydropower station construction, sediment problems tend to key technical ones in the engineering design. The proper solutions of projects with prominent sediment problems are important conditions to determine the engineering success.

In the early 1950s, limited by insufficient recognition of reservoir sediment, China failed to make in-depth analysis and study on reservoir sediment problems in the design and take effective sediment prevention and releasing measures for some water conservancy and hydropower projects, causing that the completed project failed to operate normally due to impacts of sediment. For example, on the Yellow River Sanmenxia Hydro-junction Project the severity and impacts of reservoir sedimentation were underestimated in the design, and the operation mode of high dam, large reservoir, impoundment and sediment trapping was adopted. As a result, after the project was put into operation, the installed capacity was substantially reduced with serious sedimentation, and the power station became a seasonal one. Though some projects were provided with sediment releasing facilities, but sediment discharge capacity was not enough, and they operated according to the condition for rivers with light sediment, therefore the sediment problems emerged ceaselessly and caused great loss.

After the 1960s, China's engineering technicians summarized the past success and failure experiences and lessons in terms of planning, design and operation, constantly studied the reservoir sediment problems on the basis of foreign advanced technology, and yielded excellent effects. Through engineering practice, China's sediment design, study and management level are improved continuously, to put forward a set of engineering development modes and effective reservoir sediment prevention and releasing measures in all conditions of flow and sediment inlet, minimizing the design and operation errors arising from sediment problems. For instance, in projects of Bikou, Longyangxia, Three Gorges, Xiaolangdi and other large hydropower stations constructed later, successful experiences in sediment design and reservoir water and sediment regulation management and operation have been gained.

With the continuous development of domestic water conservancy and hydropower engineering, a series of new requirements have been proposed for dealing with the sediment problems. It is necessary to study existing sediment problems in the engineering construction, comprehend the sediment movement and change patterns and put forward prevention and control measures in order to carry out the engineering construction on a reliable basis.

Reservoir sediment problems mainly include: reservoir sedimentation rises the water level in the reservoir area, resulting in land inundation and influencing residents around the reservoir; the incoming sediment is deposited in the reservoir, which encroaches a part of flood control storage capacity, and affects the downstream flood control effects of the reservoir and the dam safety itself; the sedimentation reduces the regulating performance of the reservoir, and the planned power generation, irrigation benefit indexes of the reservoir will be decreased; local sedimentation in the downstream will raise the elevation of tailwater of power station, affecting the power output; as for the hydropower stations with serious reservoir sedimentation, with the development of sedimentation, sediment discharge from the reservoir increases constantly, the sediment grain size coarsens gradually, water flow carrying a large amount of coarse sand gives rise to wear and tear through the sluicing of water turbines and outlet structures; sedimentation in front of the dam will cause the water inlet of power stations to be clogged by sediment, and result in inflexible gate operation.

Engineering sediment study generally covers river sediment data investigation and analysis, sediment delivery characteristics, calculation of reservoir sediment erosion and deposition, engineering sediment prevention and releasing measures and sediment regulation mode.

6.5.2 Calculation of the Scour and Sediment of Reservoir

6.5.2.1 Sediment Data

Analysis on the problems that the engineering sediment problems are serious or not, sediment prevention and releasing facility should be taken or not in the design, and the facility scales, shall be based on correct estimation of sediment from rivers. The river sediment is divided into suspended load and bed load in terms of grain size and state of motion, and shall be treated separately to calculate the incoming sediment. For the areas with insufficient sediment data, incoming sediment shall be calculated by indirect means.

Incoming suspended load sediment into the reservoir consists of sediment from upstream as well as tributaries, etc. Through analysis and classification of sediment data measured from hydrological stations on main stream and tributaries, the sediment yield series are determined with the interpolation and extension method. On the basis of design sediment yield series, further analysis shall be made on sediment delivery characteristics and temporal distribution features, including sediment delivery variation of the intra-annual and inter-annual period, concentration ratio of sediment delivery, etc. Knowing the sediment delivery characteristics of rivers provides basis for the identification of reservoir operation mode.

With respect to bed load, it is rather difficult to identify the bed load sediment delivery of rivers accurately. Up to now, there has not been a sound solution; the key issue lies in the fact that the bed load measurement method is not perfect enough. Therefore, it is most calculated with flume tests, field survey or empirical formula indirectly. It is proven by practice that samplers in measured data fail to collect all typical sand samplers, with results tending to be smaller; formula computing fails to fully reflect the integrated relation between sediment yield supply and hydraulic conditions, and the computing results are often larger, and all formula considerations are different and the results are in big difference; flume test results fall in between. Thus, when it is required to identify the bed load, analysis must be made by several means to ensure the reasonable identification.

Basic characteristics of sediment are one of basic data to carry out sediment analysis and calculation, including suspended load grain size distribution, bed material size distribution, falling velocity of suspended load, turbid water density, mineral composition of suspended load, etc. To carry out sediment pressure calculation in front of structures or reservoir sedimentation calculation, it is required to know well variation characteristics of dry density of deposited sediment.

6.5.2.2 Judgment of Reservoir Sediment Problems

Severity of engineering sediment problems shall be judged based on the incoming sediment, in combination with reservoir terrain, engineering layout and other features, and subject to itemized analysis and study in terms of sediment problems in the project. In judgment of reservoir sediment problems, the ratio $K(K=V/W_s)$ of reservoir storage capacity V to annual incoming sediment load W_s may be taken as the initial assessment index to judge the severity of reservoir sediment problems, which may fall into the three categories, as shown in Table 6.5-1.

Table 6.5-1 Judgment of Severity of Engineering Sediment Problems

Category	V/W_s	Severity of sediment problems	Recommended treatment measures
Class I	$K \leqslant 30$	With relatively small reservoir storage capacity, large incoming sediment, rapid sedimentation speed, the reservoir may reach sedimentation balance in 10-20 years or a shorter time, with serious sedimentation impacts	To put forward effective deposition prevention and reduction, sediment prevention and releasing measures to ensure the normal operation of reservoirs

Continued

Category	V/W_s	Severity of sediment problems	Recommended treatment measures
Class II	$30<K\leqslant100$	With relatively large reservoir storage capacity and large incoming sediment, or with relatively small reservoir storage capacity and small incoming sediment. Suffering from medium sediment problems, some with small incoming sediment, thanks to small reservoir storage capacity, hazards caused by reservoir sedimentation cannot be ignored	To carefully analyze and study the sedimentation at the end of reservoirs area, effective reservoir storage capacity maintained, and unimpeded shipping, etc. and provide with favorable sediment releasing facilities
Class III	$K>100$	With large reservoir storage capacity, small incoming sediment, and slow sedimentation speed, it takes the reservoir over 100 years to reach sedimentation balance, and its sediment problems are not serious	To properly simplify the computing contents, provided with no sediment releasing facilities

It must be pointed out that the index K can be used as a determination standard for initial assessment only. It may be understood in such a way that, the sediment problems of projects with K value of less than or equal to 30 cannot be ignored; the projects with large K value may also need to take into account potential special sediment problems, for example, the Three Gorges Project on the Yangtze River of large K value, but the impacts of its sediment on shipping may need further study.

6.5.2.3 Calculation of Sediment Scouring and Deposition

Calculation of sediment scouring and deposition mainly aims to forecast the volume and patterns of sediment scouring and deposition in every operation stage of the reservoir as well as the influence on the project, which enables the analysis on the degree of sediment scouring and deposition and on the result incurred thereby when the project is completed and put into operation. Therefore, the calculation is of great importance for determining hydro-project layout, establishing a reasonable operation mode, and taking appropriate sediment prevention measures. At present, three ways are usually used to forecast the scouring and sedimentation patterns, i.e. analogy analysis of similar reservoirs, mathematical modeling and physical modeling of reservoir sediment scouring and deposition.

Calculation of reservoir sediment scouring and deposition is a key task in the entire sediment design. It mainly includes backwater deposition calculation and scouring calculation. The backwater deposition calculation mainly covers calculation of the sedimentation volume of suspended load and bed load, sedimentation position and patterns. The scouring calculation mainly covers headwater erosion and streamwise erosion.

Due to limited data and experience in the earlier period, the common method for calculation of sediment scouring and deposition at home and abroad considered that the incoming sediment was simply laid in the dead storage. In the 1970s, based on the analysis on the measured data for sediment scouring and deposition of the completed project reservoir, Chinese engineering technicians clarified preliminarily the sediment scouring and sedimentation patterns in the reservoir and concluded that the sedimentation patterns were characterized by balance orientation. The analysis showed that the plane form of river channel, sedimentation elevation, variation of storage capacity, and dry density of deposit, etc. tend to be stable when the average releasing sediment from reservoir was 90%-95% in that year; and the sediment transport features of natural river channel was recovered when the reservoir sedimentation was in a balance. Due to different factors such as characteristics of incoming water and sediment, degree of reservoir backwater, operation mode, and landform of reservoir, etc., the sedimentation patterns in reservoir could be divided into delta sedimentation, cone sedimentation, and belt sedimentation. For the large reservoir of high dam with broad river channel, the sedimentation is of typical delta pattern. Cone sedimentation often occurs in the relatively steep river channels with low reservoir backwater and large incoming sediment volume. Belt sedimentation occurs in the reservoir with high backwater, great variation in water level in front of dam, and small incoming sediment volume since the sedimentation was evenly distributed along the reservoir bottom. Therefore, the sedimentation patterns change with different types of reservoirs. On this basis, Chinese researchers proposed a semi-empirical and semi-theoretical sedimentation equilibrium slope method as well as various analogy methods and empirical methods. These analysis and calculation methods had clear physical concepts, which could reflect the sedimentation characteristics of reservoir and control the sedimentation degree in each stage from a broader view. However, the calculation accuracy was limited to some degree.

Since the 1980s, Han Qiwei, et al. proposed mathematical modeling for non-uniform sediment transport; Zhang Qishun, et al. proposed mathematical modeling for river and reservoir scouring and deposition, and Yang Guolu, et al. (1992) proposed SUSBED mathematical model. These models are commonly characterized by solid theoretical basis for sediment calculation, broad calculation scope, and rapid calculation, which allow the calculation of specific sedimentation position, process, elevation, and volume for sediment scouring and deposition in river channels and reservoirs in respect of different river reaches and different periods of time.

Physical model testing is the most intuitive and practical method for estimation of reservoir sedimentation. Therefore, the reservoir project with severe sedimentation shall be subject to physical model testing of sediment. Physical model testing of sediment has developed into a mature technology, so it allows intuitive observation of sediment movement in the river channels or reservoirs, direct learning of different operation modes of reservoirs, sedimentation patterns in front of dam at different operational water levels, development orientation of scouring funnel, layout of various sediment release facilities, volume of releasing sediment and release effect, etc. Whether the project is subject to physical model testing depends on the scale and importance of the project. Generally model testing is conducted in some parts of the project, such as the area in front of dam or the river reaches at the backwater end of reservoir. Important projects shall be subject to both integral model testing and local model testing. For example, the Three Gorges Project on the Yangzte River and Xiaolangdi Project on the Yellow River have been subject to several physical model tests concurrently across China, facilitating the comparison and study on the results. Therefore, the testing has delivered good results and provided the basis for engineering design and operation.

1. Calculation method of reservoir sedimentation pattern

The reservoir sedimentation calculation requires analysis and determination of sedimentation volume, sedimentation position and patterns. Generally the reservoir sedimentation volume is calculated by sediment release ratio method. The sediment release ratio, expressed by η, is the ratio of outflow sediment to incoming sediment into reservoir in the same period. When the sediment release ratio η equals to 1.0, all the incoming sediment is released out of the reservoir and the equilibrium of scouring and deposition in the reservoir area is achieved; when the sediment release ratio $\eta < 1.0$, sedimentation occurs in the reservoir area.

The main factors affecting the sediment release ratio are VQ_1/Q_0^2, ω, and B. V is the storage capacity at the operational water level of reservoir; Q_1 and Q_0 are the inflow and outflow discharges; ω is the settling velocity of sediment particles; and B is the width of reservoir. Northwest Hydraulic Research Institute and Tsinghua University have obtained the empirical relationship between the sediment release ratio η and VQ_1/Q_0^2 upon analysis on the measured data in Sanmenxia, Qingtongxia, Heisonglin, Bajiazui, Guanting, Yanguoxia, and Danjiangkou reservoirs. See Fig. 6.5-1 for details.

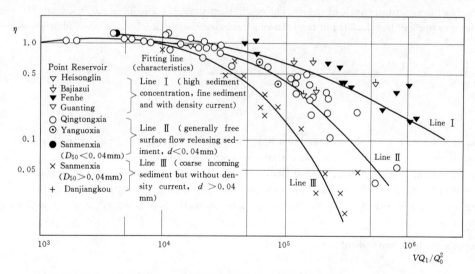

Fig. 6.5-1 **Empirical Relationship between Sediment Release Ratio and VQ_1/Q_0^2 of Some Reservoirs in China**

This empirical relationship between the sediment release ratio η and VQ_1/Q_0^2 is mostly applied for calculation of reservoir sedimentation. The three curves in the Fig. 6.5-1 are divided according to the sediment concentration of river, grain size, and longitudinal slope of reservoir area, etc. For the calculation of reservoir sedimentation, the sedimentation volume shall be calculated upon selection according to the practical condition of the reservoir or interpolation of the sediment release ratio η.

The reservoir sedimentation pattern is divided into delta sedimentation, cone sedimentation, and belt sedimentation, so the reservoir pattern of the designed project shall be determined before the calculation. So far some empirical formulas for discrimination have been developed. The commonly used formula is the discriminant proposed by the Survey and Design Institute of 11th China Water Conservancy and Hydropower Engineering Bureau, as shown in Table 6.5-2.

Table 6.5-2 Discriminant of Reservoir Sedimentation Patterns

Sedimentation pattern	Delta	Belt	Cone
Discrimination condition	$SV/Q > 10^8$ $\Delta H/H_0 < 0.1$	$0.25 \times 10^8 < SV/Q < 10^8$ $0.1 < \Delta H/H_0 < 1$	$SV/Q < 0.25 \times 10^8$ $\Delta H/H_0 > 1$

Note: V is the reservoir storage capacity, in m³; ΔH is the water-level amplitude in front of dam, in m; Q is the discharge, in m³/s; H_0 is the average water depth in front of dam or average water depth above sill elevation of releasing facility in front of dam, in m; $\Delta H/H_0$ is the relative water-level amplitude of reservoir; and S is the sediment concentration, in kg/m³.

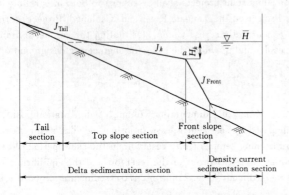

Fig. 6.5-2 Longitudinal Profile of Delta Sedimentation

(1) Calculation of delta sedimentation. The delta sedimentation can be divided into front slope section, top slope section, and tail section, as shown in Fig. 6.5-2. As indicated from the concept of equilibrium, the reservoir sedimentation achieves equilibrium, the water flow complies with uniform flow equation and continuity equation, sediment transport complies with the regime sediment charge formula, and the contour of river channel complies with the hydraulic geometric relation of river. Therefore, the calculation parameters of the delta sedimentation include equilibrium slopes J_k, equilibrium water depth H_k, and delta crest elevation a, etc. The department of River Research in the former Beijing Institute of Water Resources and Hydropower Research had obtained the calculation formula of top slope equilibrium slope by simultaneous solution of manning formula, water flow continuity equation, river sediment-carrying capacity formula, and hydraulic geometric relation formula, etc. on the premise that the fundamental principle of equilibrium is met. The tail slope of the delta sedimentation is the mean value of the original river channel slope and the delta top slope. The front slope is far deeper than the top slope. In the case density current sedimentation does not occur, the front slope is formed mainly by free silting of grains and its value approaches the underwater angle of repose. The equilibrium slope J_k, equilibrium water depth H_k, and stable river width B_k of the top slope can be calculated in the following formula. Calculation formula of equilibrium slope of top slope:

$$J_k = A^* \frac{\rho^{*5/6} d^{5/3} D_{50}^{1/3}}{\left(\frac{Q}{B}\right)^{1/2}}$$

$$\rho^* = K' \frac{v^3}{gh\omega} \tag{6.5-1}$$

where J_k is the equilibrium slope of the sedimentation bed; A^* is the comprehensive coefficient; Q is the average discharge in flood season, in m³/s; ρ^* is the sediment concentration of average bed material load in flood season, in kg/m³; d is the median diameter of bed material load of suspended load, in mm; D_{50} is the median diameter of bed material load, in mm; B is the river width, in m; K' is the sediment-carrying coefficient; v is the cross-sectional average flow velocity, in m/s; ω is the settling velocity of sediment particles, in m/s; g is the acceleration of gravity, in m/s²; h is the average water depth, in m.

Calculation formula of stable water depth:

$$h_k = \left(\frac{nQ}{\xi_1 J_k^{1/2}}\right)^{3/11}$$

$$\xi_1 = \frac{\sqrt{B}}{H} \tag{6.5-2}$$

where h_k is stable water depth, in m; Q is the bed forming discharge, in m³/s; ξ_1 is the river regime coefficient; B is the river width; H is the average water depth.

Calculation formula of stable water width:

$$B_k = \xi_1^2 h_k \tag{6.5-3}$$

The position of the delta can be roughly determined by determination of the slope, sedimentation volume, stable water depth h_k, stable water width B_k, and delta crest elevation a for each part of the delta sedimentation. For actual calculation, the delta crest elevation a is required to move forward and backward further to allow that the volume of the delta sedimentation equals to the sedimentation volume of the delta. In this way, the position of the delta is determined finally.

For the reservoir with density current, after the delta sedimentation is determined, the position of the density current sedimentation could be determined according to the distribution law of the density current sedimentation.

(2) Calculation of cone sedimentation. It is easier than that of delta sedimentation. See Fig. 6.5 – 3 for the sedimentation pattern. The figure shows that the three basic parameters for calculation of the cone sedimentation are the sedimentation elevation in front of dam F, sedimentation slope J_k, and sedimentation end A. On the premise that the sedimentation volume has been determined, the sedimentation position may be determined by trial method.

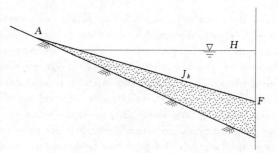

Fig. 6.5 – 3 Profile of Cone Sedimentation

(3) Calculation of belt sedimentation. Belt sedimentation often occurs in the rivers with few incoming sediment into reservoir. Therefore, the sedimentation is not so thick. The error of topographic survey may be greater than the actual sedimentation thickness. For this reason, no sedimentation calculation is performed for this type of reservoir.

2. Mathematical model of reservoir sedimentation

In the calculation method mentioned above, the calculation of the sedimentation volume and the sedimentation pattern is conducted separately. Thus the calculation result can only reflect the consequence of sedimentation rather than the process of sedimentation. This method is only applicable to the calculation of reservoir sedimentation when sedimentation was in balance.

In actual design, in order to compare the different schemes, the sedimentation bed in different sedimentation years has to be calculated. In this way, the calculation of the sedimentation process is required. Quantities of mathematical models come into being accordingly. The representative models are mathematical modeling for river and reservoir scouring and deposition, one-dimensional mathematical modeling for reservoir sedimentation and riverbed process, CRS-1, SUSBED-2, etc. For example, non-equilibrium sediment transport is considered in the mathematical modeling for non-equilibrium sediment transport, allowing the calculation of the suspended sediment concentration and gradation, sediment-carrying capacity and gradation, bed load discharge and gradation, sedimentation area and volume, hierarchical gradation of sedimentation bed load, dry density of deposit, variation of roughness, generation and movement of density current, formation and sediment release of muddy water reservoir, upward extension of reservoir sedimentation at the backwater end, and scouring and deposition in fluctuating backwater area of each section in different periods of time. The model can reflect the whole scouring and deposition process, pattern, and variation of the river channel and reservoir. It has been verified in some rivers or reservoirs that the effect meets the design requirement. At present, it is widely applied in calculation of sediment in project.

When mathematical model is used to solve the engineering sediment problems, the model has to be selected according to the nature or accuracy of the problems to be solved. Different models have their own limitation and applicability. Due to the complexity of the sediment movement, the mathematical model must depend on the experience of the designer and the measured data, so it is empirical to some degree. At present, the fast growing mathematical model is one-dimensional model which is generally used to study the variation of riverbed caused by great change of incoming water and sediment conditions and scouring basis. For the detail riverbed deformation in short river reach, such as sedimentation pattern in front of dam, discharge funnel at bottom orifice, approach sedimentation of navigation channel in dam area, scouring and deposition in fluctuating backwater area with complex boundary condition, etc., the combination of physical model test and mathematical model is advised for solution.

6.5.3 Sediment Release Facilities of Hydropower Station

According to years' study and reservoir operation, in order to reduce or eliminate the sediment hazard to the completed project, sediment release of the reservoir is proved to a very effective way. The sediment release facilities are provided to form a scouring funnel in front of the water intake of the hydropower station, which allows that the area in front of the intake is kept a low sediment concentration. When the reservoir is in operation, these facilities can take multiple measures to release sediment and recover part of storage capacity; meanwhile, they can handle some particular problems and allow for unforeseen circumstances.

6.5.3.1 Scale of Sediment Release and Flow Discharge

The practical experience reveals that when the sediment release facilities are required for the project, the scale of sediment release and flow discharge directly affects the reservoir life, sedimentation pattern in front of dam, and effect of sediment release. Therefore, the finalization the scale of the sediment release and flow discharge in a scientific manner is the key to determine the success of the project.

Since the sediment in the sandy river mostly concentrates in flood season, the reservoir has to be operated at a low water level (sediment releasing level) in order to control the sedimentation. For the reservoir with severe problems in sedimentation, the water level must drop to the sediment releasing level every year or every few years for scouring and sediment release of the reservoir. Therefore, the discharge corresponding to the sediment releasing level is undoubtedly related to sediment release. The discharge shall meet the requirement of sediment release on the premise of remained storage capacity in different stages. For the reservoir with large volume of incoming sediment, all the sediment in the reservoir throughout a year (including flood season and non-flood season) should be released out of the reservoir in order to achieve equilibrium of scouring and deposition within a year. In order to determine the discharge reasonably, a preliminary discharge should be proposed. Later on, this value may be determined through specific calculation of scouring and reservoir sedimentation for various schemes.

From that point of view, the method for preliminary determination of the sediment release and flow discharge is proposed as shown in Table 6.5 – 3, on the basis of the design of the actual project and existing study result in China.

Table 6.5 – 3 Relationship between Water Levels and Corresponding Release Discharge

Reservoir water level	Corresponding elevation	Sediment control	Corresponding release discharge	Effects	Water release structure
Operational water level in flood season	Flood plain	Control sedimentation	Annual average peak flow discharge	Discharge normal flood and flood not overbank	Bottom orifice + middle orifice
Minimum pool level or sediment release water level	River channel and plain	Control scouring	Discharge corresponding to sediment release ratio of 1	Sediment release	Bottom orifice (including sediment releasing orifice)

6.5.3.2 Sediment Release Structure and Layout

The structures of releasing sediment mainly include bottom orifice for sediment release, sluice pipe, flush gallery, and sand-sluicing gate, etc. Due to simple structure and comprehensive effect, the bottom orifice for sediment release is widely used. Since the bottom orifice is generally provided lower than the water intake of the hydropower station, the sediment release is highly efficient. It plays a great role in preventing sediment entering the turbine, dropping the water level rapidly to empty the reservoir, releasing the suspended load and bed load. In addition, it can also discharge part of normal flood. The sluice pipe, which has the lowest sill elevation, is laid out in a dispersive manner (at both sides of each unit intake) and meets the requirement of "low and dispersive" sediment release structure in the sandy river, so a low sediment concentration can be kept in the area in front of the hydropower station after being put into operation. In this way, the sedimentation elevation in front of the hydropower station is lower than the sill of water intake. It not only reduces the cleaning workload of the trash rack and eases the wearing of the water turbine, but also benefits the sediment release of the reservoir. The flush gallery is provided in the front edge of the water intake of the hydropower station, forming a scour pit, which allows the sediment in front of the dam is released to the downstream through the conduit. It can reduce the volume of sediment passing the unit and avoid deposition block of the gate at the water intake when the hydropower station is shut down for service. The sand-sluicing gate is applicable to low water head hydro-project. Since the gate has low elevation and large discharge, it can release both the sediment and flood. The silt-releasing tunnel is applicable to sediment release of the hydro-project with an earth dam. Due to the limit from engineering characteristics, tunnel is major structure of the sediment release.

The reasonability of the design layout of the sediment release structure relates to the effect of sediment release. Therefore, it shall be laid out in such manner that the unit can be always positioned within the scope of the scouring funnel. It can greatly reduce the volume of sediment passing the unit and ease the wearing of the water turbine. For the reservoir that requires sediment regulation, the bottom orifice shall be provided at a low elevation in order to control the elevation of the scour base level and achieve a certain effect of sediment release. Therefore, the design of the sediment release structure is closely related to sediment prevention and release. According to the experiences in layout of the project at home and abroad currently, layout shall adhere to the following principles:

(1) Plane layout. In consideration of the sediment movement and distribution regularity, the sediment release structure shall be laid out in the main flow direction or in a dispersive manner at the both banks of the river. The diversion hydropower station shall be laid out at the convex bank.

(2) Elevation setting. The elevation shall be determined to allow the scouring funnel, which is formed based on the sediment characteristics of the river and the hydraulic conditions, to keep a low sediment concentration in the area in front of the water intake of the unit, and to achieve the most efficient sediment release. For this purpose, the bottom orifice shall be positioned below the unit. The statistics on the relative elevation difference between the sediment release structure and the unit of the hydropower stations, which have been in operation for more than a decade in China, is shown in Table 6.5-4.

Table 6.5-4 Relative Elevation Difference between Sediment Release Structure and Generator Unit of Some Hydropower Stations in China

Hydropower stations	Water intake elevation (m)	Elevation of sediment release structure (m)	Elevation difference between water intake and sediment release structure (m)
Liujiaxia	1,680	Flood discharge gate 1,675	5
		Sand-sluicing gate 1,665	15
		Sluiceway 1,665	15
Bapanxia	1,558	Flood discharge gate 1,555	3
		Flush gallery 1,551.5	6.5
Qingtongxia	1,130.15	Sluice pipe 1,124	6.15
Tianqiao	816	Bottom orifice for sediment release 809.5	6.5
		Flood-discharge & sediment release tunnel 811	5
		Flood discharge gate 811	5
Sanmenxia	287	Bottom orifice of overflow dam 280	7
Gezhouba	39.8 (Main channel)	Sediment releasing orifice 29.2	10.6
	39.7 (Distributery)		10.5
Gongzui	494	Bottom orifice for sediment release 472	22
		Sediment releasing orifice 493.5	0.5

6.5.4 Sediment Regulation Mode

In order to reduce the threat of sediment to the hydropower station, desilting facilities in this project are used reasonably at the beginning of the impoundment of the reservoir, so as to give full play to their roles. Reasonable operation modes are formulated for controlling the silt sedimentation places, the sedimentation elevation and the sedimentation amount in the reservoir, and therefore ensure that the project can create benefits and safely run for a long term. The general principle is to lower the water level in front of the dam to the greatest extent by desilting before the flood season when large amounts of sediment and the water flow are large, and to raise the water level at the end of the flood season for impounding clear water flow. This operation mode is usually called "storing clear and releasing muddy". Presently, the operation modes in China can be classified as follows:

(1) Lowering water level in flood season for desilting: in order to control the elevation of the reservoir sedimentation surface, when sediment inflow is accumulated in flood season, the water level is lowed for the whole flood season or temporarily according to a certain flow range for desilting. This water level is generally higher than the dead water level and lower than the normal water level.

(2) Emptying reservoir to flush silt: this method is to completely empty the reservoir to flush silt once or several times in a year or several years, which is applicable to low-head and low-gate (dam) diversion-type power station with a small stor-

age capacity, and can produce satisfactory desilting results.

References

[1] Wang Weidi, Zhu Yuanshen, Wang Ruichen. *Engineering Hydrology of Hydropower Station*. Nanjing: Hohai University Press, 1995.
[2] Wang Ruichen. *China Hydroelectric Engineering—Engineering Hydrology Volume*. Beijing: China Electric Power Press, 2000.
[3] Shi Fucheng. *Analysis and Study on Continuous 11 Years of Low Flow Period for the Yellow River from 1922 to 1932*. Advances in Water Science, 1991 (4).
[4] Yang Baiyin. *Analysis on Continuous Low Flow Period for Upper Reach of the Yellow River*. Hydroelectric Science and Technology Report, 1989 (1).
[5] Wang Weidi, Sun Hanxian. *Analysis on and Design Inspection for Continuous Low Flow Period for Upper Reach of the Yellow River*. Advances in Water Science, 1991 (4).
[6] Yang Baiyin. *Discussion on Continuous 11 Years of Low Flow Period for Upper Reach of the Yellow River*. Hydrology, 1993 (3).
[7] Wang Zhengfa. *Analysis of the Runoff Series Representativeness of the Upper Yellow River above Longyangxia Hydropower Station*. Northwest Water Power, 2010 (2).
[8] Wang Zhengfa. *Analysis on Hydroperiod of Upper Stream of the Yellow River*. Northwest Water Power, 1998 (2).
[9] Li Xiuyun, Fu Suxing, Song Xianfeng. *Research on Formation Mechanism of Low Water Runoff and Extreme Value of Rivers*. Journal of Desert Research, 1999 (3).
[10] Yang Baiyin. *Summery on Key Technology of Longyangxia—Analysis on Continuous Low Flow Period of Upper Reach of the Yellow River*. Beijing: China Electric Power Press, 1998.
[11] Wang Zhengfa. *The Probability Analysis of a Multiyear Persistent Drought and the Determination of Its Recurrence Duration*. Journal of Shaanxi Water Power, 1995 (1).
[12] Wang Zhengfa. Forecasting and Mitigation of Water-related Disasters Proceedings of Theme C XXIX Congress Beijing, China, *The Contrast Research of Calculating Formula of Non-occurrence Probability of Multi-year Persistent Water Disaster*, September 16-21, 2001. (In English)
[13] Yan Kai. *Frequency of Flood of the Huaihe River*. China Water Resources Monthly, 1933, 5 (2).
[14] Chen Chunting. *Research on Flood Flow Frequency Curve for Chinese Five Rivers*. China Water Resources Monthly, 1947, 14 (6).
[15] Liu Guangwen. *Basic Understanding of Hydrology Analysis Method*. Engineering Construction, 1951 (12).
[16] Jin Guangyan. *Principle and Method of Hydrologic Statistics*. Beijing: Water Conservancy and Electric Power Press, 1959.
[17] Cong Shuzheng, Tan Weiyan. *Statistical Test for Parameter Estimation Method in Hydrologic Frequency Calculation*. Journal of Hydraulic Engineering, 1980 (3).
[18] Chen Zhikai, Wang Jiaqi. *Discussion on Applicability of Pealson Type-III Model and Kritski - Menkel Curve to Flood Design*. Hydrologic Calculation Experience Compilation (Volume 2). Beijing: China Industry Press, 1963.
[19] Song Dedun, Ding Jing. *Probability Weighted Moments Method and Its Application to P-III Distribution*. Journal of Hydraulic Engineering, 1988 (3).
[20] Liu Guangwen. *Pealson Type-III Distribution Parameters Estimate*. Hydrology, 1990 (4) - (5).
[21] Ma Xiufeng. *Weight Function Method for Calculating Hydrological Frequency Parameters*. Hydrology, 1984 (3).
[22] Xia Letian. *One Non-parametric Estimation Method in Hydrological Frequency Calculation—Density Estimation*, China Hydrology Prospecting in 2000. Nanjing: Hohai University Press, 1991.
[23] Hu Mingsi, Luo Chengzheng. *China's Historical Floods*. China Bookstore, 1992.
[24] Ye Yongyi, Chen Zhikai. *Treatment of Historical Data in Flood Frequency Analysis*. Journal of Hydraulic Engineering, 1962 (1).
[25] Qian Tie. *Determination of Empirical Flood Flow Frequency with Historical Flood Data*. Journal of Hydraulic Engineering, 1964 (2).
[26] Ye Yongyi. *Calculating Design Flood according to Hydrological Data*. China Water Resources, 1957 (1) - (2).
[27] Guo Shenglian. *Development and Assessment on Design Flood Research*. Beijing: China WaterPower Press, 2005.
[28] Yang Baiyin. *Discussion on Flood Optimization in Cascade Reservoir Construction*. Hydrology, 2004 (1).
[29] Tan Weiyan, Zhang Dongran. *Figures and Tables Commonly Used in Hydrological Statistics*. Water Resources and Electric Power Press, 1982.

[30] Cong Shuzheng. *Research on Expected Frequency of Design Flood for Three Gorges Project*. Journal of Hohai University, 1988, 16 (3).

[31] Gao Rongsong, Deng Yuren. *Stochastic simulation of Flood Process for Xiangjiaba Hydro-junction on the Jinsha River*. Sichuan Hydrology Journal, 1983.

[32] Wang Zhongming, Ding Jing. *Stochastic simulation of Flood Process for Yichang Station of the Yangtze River*. Minutes of First National Water Resources and Hydropower System Applied Probability and Statistics Scientific Conference. Nanjing: Hohai University Press, 1985.

[33] Deng Yuren. *Discussion on Applicability of Shot Noise Model*. Journal of Chengdu Electronic Technology University, 1986 (4).

[34] Chen Yuanze, Gui De. *Simulation and Adjustment for Flood Process of Two Stations of Shangquan*. Hydrologic Calculation, 1984 (4).

[35] Cong Shuzheng, Chen Yuanfang. *Stochastic Simulation of Flood and Its Theoretical Research*. Report on Research Project Subsidized by National Science Foundation Committee, 1990.

[36] Xiong Ming. *Research on Stochastic Simulation of Daily Flood Process of Cascade reservoir*. Water Resources and Power, 1992 (3).

[37] Hydrological Office, Changjiang Water Resources Committee. *Research on Stochastic Simulation of Flood of the Yangtze River*. Hydrology, 1993 (4).

[38] Wu Shiwei, Zhang Sijun. *Variation and Statistic for Headwater Level of Dam*. Journal of East China University of Water Conservancy, 1984 (4).

[39] Wang Ruichen, Chen Yuanze. *Calculation Method of Flood Probability Distribution for Section of Lower Reach of Cascade Reservoir*. Conference Preceding of Northwest Survey and Design Institute of Ministry of Energy and Ministry of Water Resources. Nanjing: Hohai University Press, 1991.

[40] Jiang Shuhai. *Application of Stochastic Differential Equation to Analysis on Flood Discharge Risk*. Journal of Hydraulic Engineering, 1994 (3).

[41] Yang Baiyin. *Analysis and Research on Reliability and Risk of Fire-fighting Arrangement and Flood Discharge of Cascade Reservoir*. Hydroelectric Power, 1996 (10).

[42] Yang Baiyin. *Analysis and Research on Flood Discharge Risk of Single Reservoir*. Hydrology, 1999 (2).

[43] Huang Qiang, Ni Xiong. *Research on Flood Control Standard of Cascade Reservoir*. Yellow River, 2005 (1).

[44] Chen Jionghong, Guo Shenglian. *Research on Design Flood Calculation Method Based on Reliability Theory*. Yellow River, 2008 (3).

[45] Xiong Lihua, Guo Shenglian. *Research on Reliability for Pealson Type-III Design Flood*. Water Resources and Power, 2002 (2).

[46] Hydro Meteorology Research Group of East China Institute of Water Conservancy. *Calculation on Probable Maximum Rainfall*. Collection of Hydrologic Calculation Experience (Volume 3). Beijing: China Industry Press, 1965.

[47] Zhan Daojiang, Zou Jinshang. *Probable Maximum Rainstorm and Flood*. Beijing: Water Conservancy and Electric Power Press, 1983.

[48] Zhan Daojiang. *New Method of Flood Calculation*. Journal of Hohai University, 1988, 16 (3).

[49] Zhan Daojiang, Xie Yuebo. *Research on Ancient Floods*. Beijing: China WaterPower Press, 2001.

[50] Zhan Daojiang, Xie Yuebo, Yang Yurong. *Research on China's Ancient Floods*// Zhu Guangya, Zhou Guangzhao. China Science and Technology Library-Astronomy and Earth Science. Beijing: Science and Technology Literature Press, 1998.

[51] Xie Yuebo, Yang Yurong, Wang Hui. *Spilling Sediment Index System for River Reach of Three Gorges*. Yangtze River, 1999 (8).

[52] Xie Yuebo, Wang Jingquan, Li Li. *Action of Ancient Floods to Xiaolangdi Design Flood*. Hydrology, 1998 (6).

[53] Zhao Renjun. *Hydrological Simulation of Drainage Basin*. Beijing: Water Conservancy and Electric Power Press, 1984.

[54] Yu Weizhong. *Discussion on Drainage Basin Runoff*. Journal of Hydraulic Engineering, 1982 (8).

[55] Wen Kang. *Mathematical Model for Drainage Basin Runoff Calculation*. Journal of Hydroelectric Engineering, 1982 (8).

[56] Rui Xiaofang. *Formation Principle of Runoff*. Nanjing: Hohai University Press, 1991.

[57] Li Changxing, Shenjin. *Drainage Basin Runoff Model Considering Spatial Variability of Soil Properties*. Journal of Hydraulic Engineering, 1989 (10).

[58] Gu Weizu. *Research on Environmental Isotope Test Response to Rainfall Runoff of Catchment Area*. Advances in Water Science, 1992, 3 (4).

[59] Zhao Renjun. *Calculation Method of Drainage Basin Runoff*. Journal of Hydraulic Engineering, 1962 (2).

[60] Rui Xiaofang. *Method of Kinematic Wave Numerical Diffusion and Flood Routing*. Journal of Hydraulic Engineer-

ing, 1987 (2).

[61] Rui Xiaofang. *Application of Analytic Solutions of Diffusion Wave and Linear Diffusion Model*. Journal of East China Water Conservancy University, 1985, 13 (3).

[62] Cheng Haiyun, Rui Xiaofang. *Analytic Solutions of Linear Diffusion Wave Equation and Its Application to Water Level Forecasting*. Advances in Water Science, 1997, 8 (2).

[63] Yang Baiyin. *Application of Adaptive Kalman Filter to Flood Forecasting of the Taohe River for Northwest Water Power*. Northwest Water Power, 1988 (4).

[64] Cheng Wenhui. *Application of Dual-elimination Method to Calculation of Unsteady Flow for Open Channel*. Journal of East China Water Conservancy University, 1985, 13 (3).

[65] Tan Peiwen, Wang Chuanhai. *Evolution Model for Water Supply of Middle Reach of the Huaihe River*. Advances in Water Science, 1996, 7 (2).

[66] Tan Weiyan, Hu Siyi. *Conservation Scheme for Calculation of One-dimension Unsteady Flow of Natural Rivers*. Advances in Water Science, 1990, 1 (1).

[67] Hu Siyi, Tan Weiyan. *Three Kinds of High Performance Conservation Schemes for Calculation of One-dimension Unsteady Flow*. Advances in Water Science, 1991, 2 (1).

[68] Tan Weiyan, Hu Siyi. *Water Flow Simulation of Flood Control System for Dongting Lake on Middle Reach of the Yangtze River*. Advances in Water Science, 1996, 7 (4).

[69] Rui Xiaofang. *Flood Routing Method Based on Optimal Control Theory of Time Series Analysis*. Journal of Hydraulic Engineering, 1993 (4).

[70] Zhong Denghua. *Time Series Neural Network Model for Hydrologic Forecast*. Journal of Hydraulic Engineering, 1995, (2).

[71] Rui Xiaofang. *Affluxion Production Principle*. Beijing: Water Conservancy and Electric Power Press, 1995.

[72] Ye Shouze, Xiajun. *Identification of Hydrologic System*. Beijing: Water Conservancy and Electric Power Press, 1989.

[73] Shen Qing, etc. *Research on Dynamic Hydrological Experiment*. Xi'an: Shaanxi Science and Technique Press, 1991.

[74] Shen Bing. *Surface Hydrological Finite Element Modeling*. Xi'an: Northwest Polytechnic University Press, 1996.

[75] Rui Xiaofang, Zhu Jianying. *Groundwater Flow Concentration Model and Its Application to Watershed Hydrological Simulation*. Journal of Hydraulic Engineering, 1989 (4).

[76] Wen Kang. *Mathematical Simulation for Process of Surface Runoff*. Beijing: Water Conservancy and Electric Power Press, 1991.

[77] Rui Xiaofang. *Several Comments on Geomorphic Instantaneous Unit Linear Theory*. Advances in Water Science, 1991, 2 (3).

[78] Zhao Renjun. *Objective Optimization Method for Parameters of Watershed Hydrology Model*. Hydrology, 1993 (4).

[79] Wang Juemou, Zhang Ruifang. *Comprehensive Restraint Linear System Model*. Journal of Hydraulic Engineering, 1987 (7).

[80] Yang Baiyin. *Using Grey Dynamic Model for Long-term Hydrologic Forecast*. Northwest Water Power, 1987 (4).

[81] Yang Baiyin. *Fuzzy Clustering Method for Long-term Hydrologic Forecast*. Hydrology, 1989 (2).

[82] Yang Baiyin. *Low Streamflow Forecast Model for Lonyyangxia Hydropower Station on the Yellow River*, Pra of U. S. A - China Sym. on Dro. and Arid Hydrology, 1991, U. S. A. (In English)

[83] Yang Baiyin. *Low Streamflow Forecast Model*. Conferennce Prceeding of Sino-german Bilateral Scientific Conference. 1991.

[84] Rui Xiaofang. *Multiple Issues in Research of Watershed Hydrology Model*. Advances in Water Science, 1997, 8 (1).

[85] Hua Shiqian. *Flood Forecast Method*. Hydrographic Office, Ministry of Water Resources, 1955.

[86] Ge Shouxi, Zhang Ruifang. *Online Realtime Forecasting System for Flood of Middle and Lower Reaches of the Yangtze River*. Paper Anthology of Scientific Conference for National Hydrologic Forecast and Disaster Reduction. Nanjing: Hohai University Press, 1997.

[87] Zhao Weimin. *Online Realtime Forecasting System for Flood of Middle and Lower Reaches of the Yellow River*. Paper Anthology of Scientific Conference for National Hydrologic Forecast and Disaster Reduction. Nanjing: Hohai University Press, 1997.

[88] Le Jiaxiang, Liu Jinping. *Online Realtime Forecasting and Management System for Flood of Main Stream of the Huaihe River*. Paper Anthology of Scientific Conference for National Hydrologic Forecast and Disaster Reduction. Nanjing: Hohai University Press, 1997.

[89] Zhu Hua. *Automatic Prediction System of Flood Condition*. Beijing: Water Conservancy and Electric Power

Press, 1993.
[90] Wei Budong, Peng Gongbing. *Discussion on Influence Manner of Antarctic Sea Ice on Subtropical High of Northwest Pacific Ocean*. Advances in Water Science, 1992, 3 (3).
[91] Zhang Xinping, Fan Zhongxiu. *Influence of Air-Sea Interaction on Flood Season of Upper Reach of the Yangtze River*. Geographical Science, 1993, 13 (3).
[92] Wang Xiaolan, Peng Gongbing. *Relationship between water regime in Flood Season on Upper and Middle Reaches of the Yangtze River and Arctic Sea Ice Extent*. Meteorological Bulletin, 1992, 50 (1).
[93] Rui Xiaofang, Huang Zhenping. *Preliminary Research on the Applicability of Decomposing Model for Long-Term Variation of Estimated Hydrographic Features*. Journal of Hohai University, 1989 (4).
[94] Liu Quanshou, Fang Lerun. *Multi-level Autoregression Stochastic Simulation for Monthly Runoff and Its Application*. Hydrology, 1988 (5).
[95] Wang Wensheng, Ding Jing, Jin Juliang. *Statistical Hydrology*. Version 2. Chengdu: University of Science and Technology of Chengdu Press, 2008.
[96] Sediment Research Institute of China Institute of Water Resources. *Sediment Manual*. Beijing: China Environmental Science Press, 1992.
[97] Han Qiyou. *(One-dimensional) Mathematical Model of Reservoir Sedimentation and Fluvial Process*. Journal of Sediment Research, 1987 (3).
[98] Yang Guolu. *River Mathematical Model*. Wuhan Institute of Water Conservancy and Electric Power, 1992.
[99] Pan Jiazheng, He Jing. *Large Dams in China A Fifty-Year Review*. Beijing: China WaterPower Press, 2000.

Chapter 7

New Techonologies for Engineering Geology and Survey in Dam Construction

Cai Yaojun[1], Yuan Jianxin[2], Chen Deji[3]

7.1 New Situations and New Problems of Engineering Survey

In ten years after the publication of *Large Dams in China*, *A Fifty-Year Review*, an unprecedented and new situation of dam construction in China has appeared as a result of sustained and rapid growth of social economy and continued adjustment of domestic energy structure. Not only the number of dams has been increasing dramatically, but the dam heights are at its unprecedented height. According to statistics, more than 200 dams above 70m have been started to construct or completed since 2000, in which more than ten dams are above 200m in height. The highest concrete faced rockfill dam in the world has been completed with a height of 233m (Shuibuya). The highest double arch dam (305m) of Jinping and the highest core wall rockfill dam (Shuangjiangkou, 320m) in the world are under construction. Most of these large dams are constructed in western China where topographical and geological conditions are extremely complex. New situations and new problems that have never occurred in dam construction history in the world put forward unprecedented challenges to engineering geological survey of dam construction. Some representative projects with prominent engineering geological problems are as following.

7.1.1 *High Seismic Intensity*

Western China is located in the Qinghai-Tibet Plateau and its fringes—the collision part of Indian Plate and Eurasian Plate—where the tectonization and crustal uplift are intense. Meanwhile, it is also in the seismic zone between Central Asia and the Qinghai-Tibet Plateau, which is a high-frequency and high seismic area. For example, in last 300 years, the northern section of Xianshuihe fault zone experienced 9 earthquakes with above 7 magnitudes. Strong earthquakes reoccurred in a rather short period and some violent earthquakes were witnessed in modern times. On May 12, 2008, a 8.1-magnitude earthquake occurred in Longmenshan fault zone, which brought severe casualties and property losses. Therefore, the seismic fortification problems of the dam projects in this area are very prominent, not only the seismic intensity and the seismic peak acceleration are high, but also the influence on the dam projects caused by secondary geological disasters, such as landslides, collapse, unloading rock mass slopes, etc. is severe.

Seismic fortification problems in dam project are represented in three aspects: firstly, the shaking effect of strong earthquake on hydraulic construction. Considerable projects in western area are in Ⅷ seismic intensity zone. Therefore, the seismic fortification parameters are usually set above 0.3g. Table 7.1-1 shows the seismic fortification accelerations adopted by part of the dams in China. As the table shows, the dam with the highest seismic fortification parameter is the arch dam of Dagangshan Hydropower station on Daduhe River, Sichuan Province, where the dynamique horizontal acceleration is up to 0.5575g. It is followed by the concrete facing rockfill dam of Jilintai Hydropower Station in Xinjiang, where the seismic acceleration is 0.462g.

Secondly, the breaking problem of dams and other hydraulic structures caused by seismic active fault. Generally speaking, the dam of water conservancy and the hydropower project are not allowed to span over the active faults because of devastating impact caused by possible damage of the dams. However, the problems of active faults in the tunnels projects such as the water transportation and traffic are relatively easy to deal with. Their dislocation will not cause devastating damage. So

---

[1] Cai Yaojun, Deputy Chief Engineer in Changjiang Institute of Survey, Planning, Design and Research under Changjiang Water Resources Commission, Director of Changjiang Institute of Survey Technology, MWR; Professor Senior Engineer.

[2] Yuan Jianxin, Deputy Chief Engineer in HydroChina Cooperation, Professor Senior Engineer.

[3] Chen Deji, Scientific and Technological Commission, Changjiang Water Resources Commission, Professor Senior Engineer.

the tunnels may span across active faults after the demonstration. So far, there is not still an example that the active fault across the large dams. This situation exists in some small-scale water conservancy projects, such as the subsidiary dam of Heizier Reservoir in Xinjiang, and Xicker reservoir. In large hydroelectric projects, only a few water transfer systems cross the active faults, such as Dafa Hydrpower Station on the Tianwan river, Sichan Province.

Table 7.1 - 1 Ground Motion Parameters of Most Dams in China

No.	Dam Name	Dam Type	Dam height (m)	Design horizontal acceleration (g)	Remarks
1	Dagangshan	Double arch concrete dam	210	0.5575	Under construction
2	Jilintai I	Concrete faced rockfill dam	157	0.462	Completed
3	Yele	Asphalt core rockfill dam	124.5	0.45 (9 degree fortification)	Completed
4	Longpan	Double arch concrete dam	276	0.4076	Pre-construction
5	Baihetan	Double arch concrete dam	289	0.325	Pre-construction
6	Xiluodu	Double arch concrete dam	285.5	0.321	Under construction
7	Xiaowan	Double arch concrete dam	294.5	0.308	Basically completed
8	Wudongde	Double arch concrete dam	265	0.265	Pre-construction
9	Laxiwa	Double arch concrete dam	250	0.23	Completed
10	Jinping I	Double arch concrete dam	305	0.197	Under construction

Thirdly, the adverse effect caused by secondary disasters caused by earthquake. The canyon areas in western China are prone to landslides, collapse and other geological disasters, for which seismic events are important triggering factors. Earthquake damage survey after Wenchuan Earthquake has showed that a large number of secondary geological disasters caused by earthquake not only affect the safety of buildings (such as collapse of the right abutment of Taipingyi plant), but also bring a serious difficulty for the rescue. It is difficult to imagine the damage on the dam and the downstream life and property if Tiangjiashan landslide took place in upstream or near the the dam. Therefore, avoiding the risk of secondary geological disasters is another important issue.

7.1.2 Deep Overburden Layer

The thicknesses of river overburden layers in China are tens to hundred meters (see Table 7.1 - 2), some can reach up to hundreds of meters. In the southwest, the river beds are deep with thick overburden layers. In upper reach of Minjiang River, the overburden layer from Yingxiu plant to Shilipu plant is as thick as 60 - 100m. It is 100 - 250m from Qizong to Longpan in Shigu Basin of Jinsha River; more than 100m in Xinshi Town to Yibin in lower reach of Jinsha River. It is 20 - 134m in mainstream of Dadu River. It reaches 420m in dam site of Yele hydropower station on Nanya River, the tributary of Dadu River, which is the thickest covering we've ever found in the completed and proposed hydroelectric projects. It is 12 - 50m in Yalong River. The overburden layer of upper reaches of Yarlung Zangbu River in Tibet is above 120 meters, and its lower gully zone may reach hundreds of meters. The maximum overburden thickness of its tributaries, Lhasa River and Niyang River, are over 400m and 100m respectively. In Taxkorgan River, the tributary of Yarkand River in Xinjiang, is 147.95m for overburden thickness. In the Three Gorges Project region on the Yangtze River, there has about 90 deep grooves about 35 - 40 meters in depth, whose cumulative length accounts for about 45% of the total length of the river.

There are various causes resulting in the deep overburden, such as the rise and fall (or pause) of new tectonic movement to form accumulation layer, fast erosion and serious collapse, frequent strong earthquake inducing landslide and slide groups, the action of quaternary glaciation and modern glaciation, the landslides and debris flow caused by rainfalls, which are also often controlled by the lithotomic characters and the active faults.

The overburden layers are featured with complex genetic types, loose structures, discontinuous series, violent variation in thicknesses and inhomogeneity of physical-mechanical properties. In the past, they were dealt with by excavation method. However, when the layer is quite deep (more than 40 meters), the excavation is neither economic nor safe for its deep pit and high slope. Now the current method is studying its geological characteristics to take appropriate measures to reasonably use overburden layers as dam foundation. Many large dams are constructed on deep overburden layers in China. For example, Pubugou plant on Dadu River in Sichuan is a 186-meter rockfill dam with clay core constructed on 70 - 80 meters overburden layer. Chahanwusu station on Kaidu River in Xinjiang is a 110m concrete faced rockfill dam, whose plinth is constructed on 40m overburden layer.

Table 7.1-2 Maximum Depth of Overburden Layer of Some Dam Sites in China

River name	Dam site	Covering depth (m)	River name	Dam site	Covering depth (m)
Jinsha River	Lawa	55	Wujiang River	Hongjiadu	8
	Rimian	30		Dongfeng	17.12
	Benzilan	42		Suofengying	25
	Qizong	120		Wujiangdu	21.5
	Tacheng	100.6		Goupitan	8
	Shangjiang	206		Silin	4.5
	Longpan	100		Shatuo	4.62
	Hutiaoxia	40		Pengshui	10
	Liangjiaren	62.8	Lancang River	Gushui	26.8
	Liyuan	15.5		Huangdeng	34
	Ahai	17.34		Xiaowan	30.8
	Jinanqiao	8		Nuozhadu	31.26
	Guanyinyan	24.02		Jinghong	28
	Longkaikou	43.1		Ganlanba	22.5
	Wudongde	72.8		Miaowei	21.2
	Baihetan	54.15		Tuoba	50
	Xiluodu	40	Nujiang River	Lumadeng	28.37
	Xiangjiaba	80		Fugong	40.77
Yalong River	Lianghekou	12.4		Bijiang	33
	Yagen	14.7		Lushui	27.56
	Lenggu	59.5		Liuku	35.16
	Yangfanggou	32.1		Shitouzhai	22.58
	Kala	46.19		Saige	44.3
	Jinping I	47		Guangpo	12.5
	Jinping II	51	Qingjiang River	Shuibuya	21.7
	Guandi	35.8	Nanpan River	Nuozu	15
	Ertan	38.29		Daqiao	22.76
	Tongzilin	36.74		Leidatan	15.09
Dadu River	Shuangjiangkou	67.8		Yunpeng	15.92
	Jinchuan	80		Fenghuanggu	13.3
	Busigou	19.28		Tianshengqiao	25.61
	Xia'erga	13.22		Lubuge	4
	Badi	130	Hongshui River	Longtan	15
	Danba	127.66	Honghe River	Gasajiang I	35
	Houziyan	85.5		Qiaotou	15
	Changheba	79.3		Luozhi	38
	Huangjinping	133.9		Nalan	24.3
	Luding	148.6	Yellow River	Longyangxia	2.68
	Yingliangbao	116		Lijiaxia	18
	Longtoushi	70		Liujiaxia	10
	Laoyingyan	70		Daxia	34.83
	Anshunchang	73		Daliushu	10.63
	Yele	>420		Tianqiao	33
	Pubugou	75		Ganzepo	48
	Shenxigou	55		Sanmenxia	10
	Zhentouba	48		Renjiadui	70
	Shaping II	38		Balihutong	67.5
	Gongzui	70		Wangjiatan	110
	Tongjiezi	73.5			
	Dagangshan	20.9			
	Dawei	30			

Countinue

River name	Dam site	Covering depth (m)	River name	Dam site	Covering depth (m)
Minjiang River	Shilipu	96	Yellow River	Xiaolangdi	80
	Futang	92.5	Yongding River	Guanting	28
	Taipingyi	80		Youzhou	32
	Yingxiu	62		Zhuwo	38
	Zipingpu	31.6		Xiaweidian	37.8
	Yuzui	23.8	Chaobai River	Miyun	44
Bailong River	Shachuanba	35		Shisanling	59.5
	Miaojiaba	35		Huangbizhuang	57
	Haozidian	65		Yuecheng	30
	Xianyaba	73		Nangedong	51
	Bikou	37	Hanjiang River	Ankang	29
	Dujiaba	22		Danjiangkou	22
	Baozhusi	31.2		Xinglong	60
	Zilanba	30			
Taxkorgan River, the tributary of Yarkand River	Xiabandi	147.9	Xin'an River	Xin'an River	10
			Tingjiang River	Mianhuatan	5.13
			Minjiang River	Shuikou	29
Kaidu River in Xinjiang	Chahanwusu	46.7	Lhasa River	Pangduo	>400
			Baishui River, Gansu	Hanpingzui	46.6
Yarkand River	Altash	93.9	Taohe River, Gansu	Jiudianxia	56

Note: The data are from "*Hydroelectric Engineering Geology Manual*", Xinjiang Institute of Water Conservancy and Hydroelectric Survey and Design; Gansu Institute of Water Conservancy and Hydroelectric Survey and Design; Changjiang Institute of Survey, Planning, Design and Research; Shaanxi Institute of Water Conservancy and Hydroelectric Survey and Design.

When the water conservancy and hydroelectric projects are constructed on the overburden layers, the following engineering geological will be considered: load and deformation stability; leakage and seepage stability; sliding stability; sand liquefaction. When overburden layers are used as the foundation of hydraulic structures, it is necessary to identify soil distribution, genetic type, thickness, series structure, physical and mechanical properties, hydraulic properties, hydro-geological properties, to propose physical and mechanical properties such as permeability coefficient of dam soil body, allowable hydraulic seepage gradient, bearing capacity, deformation modulus and strength, and so on, and to evaluate ground subsidence, sliding stability, leakage, seepage deformation and liquefaction.

Xiabandi water conservancy project is typical example of building on the deep overburden layer in recent years in China, which is located in downstream of the Yarkand River, one of the major tributaries of the Taxkorgan River in Xinjiang. The dam is gravel dam with asphalt concrete core, with the 78m of the maximum dam height, 867 million m³ of total reservoir capacity and 150MW of the hydropower installed capacity.

Three proposed dam sitesare primarily located in a 3.0km range. The depth of overburden layer is about 80 ~ 150m with complex genetic causes, tangled accumulation, size disparity and poor homogeneity. The maximum rock block size reaches more than 10.0m. In view of the special geologic condition, it can be mainly from the two aspects to study engineering adaptability, one is selecting the core wall sandy gravel dam which has a low demand to the foundation, and the second is using 100m of concrete core-wall for the foundation watertight which creates a new record of concrete cut-off wall depth in China.

Main engineering geological problems of the deep overburden layers include: ①leakage and seepage stability of the dam foundation; ②liquefaction of the sand layer of the dam foundation; ③strength and stability of soft clay in the dam foundation. These issues have important influence on dam site selection and foundation treatment.

7.1.3 *Complicated Rock Foundation*

Most large-scale dam projects in the western area are located in geotectocline, for example, a number of projects are distributed in Songpan-Ganzi geosynclinal folded system. The sedimentary rocks and their corresponding metamorphic rocks in this area are characterized by the great petrological transformation, the complicated lithological features; the inhomogeneous rock mass strength. Both hard rocks and soft rock are distributed in this area. Soft rock, due to its low intensity, is the study focus of deformation stability in high-dam projects. Meanwhile, the soft rocks evolve into soft interlayer during long-term geological process and become the boundary controlling sliding stability, which greatly and negatively impact the slide-resistance and stability of dam and slope. Affected by complex tectonic conditions, various faults and fractures are developing and the integrity of the rocks differs greatly. Sometimes hard rocks show deformation characters similar to soft rocks due to intense fragmentation. For example, Xiangjiaba hydropower dam site faces with slide-resistance problems. Guanyinyan hydropower dam site on Jinsha River have many serious problems concerning the compression deformation, the leakage and the seepage deformation, the slide stability due to Jurassic calcirudite, sandstone, siltstone and muddy siltstone in the dam foundation.

Because there exists a high geo-stress and the incised valley is rapid in Cenozoic in western China, the geological evolution not only forms unique landscape and topographical features, but also reveals many engineering geological problems that have never been met in the past. Take Pusiluogou Arch Dam site in Jinping Project as example, large-scale cracks are found in the deep part of the marble rocks on the left bank. Accompanied with relaxed tension of marble rocks at various degrees, the deformation performance of mass rock is severely weakened and slide resistance stability of the dam abutment is influenced.

7.1.4 High Geostress

Geostress can often cause various deformation and damages associated with unloading rebound and stress release, which harmfully influence project construction and safety operation. During dam construction, rapid excavation of rock mass and stress lease result in unloading rebound and cracking. For example, Grand Coulee concrete gravity dam in America has granite foundation. During foundation pit excavation, the granite was discovered horizontally split layer after layer. In China, when initial excavation of Xiaowan arch dam foundation was completed, rock mass had very good quality and the excavation control was very satisfied. But it was not long when unloading rebound occurred in granite gneiss, the stripping ever the burst occurred at rock base. The integrity and intensity of dam foundation was severely affected. Enforcement measures such as re-excavation, anchoring and grouting were taken. Laxiwa arch dam also has granite foundation and rather high geostress. It encountered with similar problems during foundation excavation. The harmful influences from high geostress were mediated through modification of dam corps. The release of high ground stress may cause rock burst which seriously threatens construction safety, construction period and structure safety of underground project. The harm of rock burst is one of the main threats in construction of long diversion tunnel of Jinping Ⅱ Hydropower Station. Table 7.1-3 shows geostress measurement of some domestic projects, and it can be seen that large dam projects in the western area can not avoid high ground stress problems. The influences of geostress with different values are also different.

Table 7.1-3 Test Results of Geostress in Some Hydropower Projects

Project name	Lithology	Test depth (m)	σ_1			σ_2			σ_3		
			Value MPa	Direction α (°)	Angle β (°)	Value MPa	Direction α (°)	Angle β (°)	Value MPa	Direction α (°)	Angle β (°)
Three Gorges	Granite	42-130	2.8-12	33-340	3-64	-1.5-5.8	238-349	1-53	-4.1-3.3	54-336	23-81
Xiluodu	Granite	200-625	15-21	257-344	0.5-25	5-17	22-338	14-83	4-13	2-376	1-35
Jinping Ⅰ	Marble	200-400	20-40	100-357	3-64	10-37	15-351	20-81	6-13	8-352	1-34
Jinping Ⅱ	Marble	463-1,843	32-47	120-260	5-75	20-35	18-120	15-80	12-19	30-169	1-30
Guandi	Granite	150-500	20-39	280-340	1-42	10-15	11-60	30-70	4-11	6-72	6-60
Ertan	Syenite, basalt	>200	17-38	13-35	—	7-23	52-160	—	3-19	123-277	—
Shuangjiangkou	Granite	200-570	15-38	31-332	1-48	6-24	50-335	7-58	2-16	11-300	11-65
Houziyan	Limestone	250-570	15-33	250-316	4-54	8-23	9-353	5-68	6-16	11-209	4-47
Dagangshan	Granite	240-500	11-22	18-61	1-39	10-16	53-168	51-88	2-10	279-324	2-22
Xiaowan	Gneiss	>200	16-28	188-311	1-53	11-20	220-276	0-60	6-10	130-278	30-40
Laxiwa	Granite	60-364	8.8-30	9-350	19-55	6-20.6	60-336	6-55	0.8-13	9-330	12-69

7.1.5 Deep Karst

Deep Karst refers to the karst below local erosion base level. It might be ancient karst buried deeply underground through earth's crust descend and have nothing to do with current base level. It may also be formed through deep circulation of groundwater in modern times. From the real conditions of water conservancy and hydropower engineering, the deep karst refers to the karst below the local lowest water drainage base level.

The formation of deep karst of valley mainly related to siphon circulation of groundwater and drainage system of karsts. The main genetic factors are: ①favorable karst hydro-geological structures, such as corrosion layer, permeable fault zone and so on; ②favorable groundwater siphon circulation conditions, such as confined water; ③favorable water chemistry conditions for deep karst development, such as mixing corrosion and infiltration between aqueous solutions with different concentrations, which promotes ground water deep cycle or deep erosion; ④temperature variation effect (often accompanied by favorable hydrochemical conditions) resulting in cold and hot water convection and promoting alternate cycle; ⑤high topographic difference between valley and watershed, which brings a high seepage pressure of groundwater; ⑥deep cycle of the hot springs and mixture with shallow groundwater to form deep karst. Two dam sites at Pengshui project have deep karst at the intersection of hot spring and cold water.

Distribution features of deep karst under the valley are: ①Horizontally, the deep karsts mainly distributes at both sides of the valley in deep cutting gorges. They are connected with karst systems. The karst development under river bed is relatively weak, such as in two dam sites of Pengshui and Wujiangdu station. In the wide valley areas, the deep karsts are well developed under river bed and both banks. It is, probably, because relatively long stability of the crust. When the gorges evolves into the wide valley, the deep karst under river banks may be under river bed, such as Sanjiangkou dam site in Lishui, Hunan Province. ②Vertically, the development of deep karst under the valley is gradually weakened with the increase of depth. The strong karstified zone is normally less than 50 – 60m below water drainage base. The zone 50 – 100m below water drainage base is karstified moderately. The karstification below 100m is greatly weakened.

At Goupitan Hydropower Dam site of Wujiang River, the river bed elevation is about 410 meters. Drilling or excavation revealed 76 karst caves higher than 0.1m below 410m elevation. Horizontally, the deep karst is usually seen near the banks, in which 21 karst caves were revealed on left bank, 40 caves were revealed on right bank; and 15 caves were under the river bed. Vertically, more karst caves are distributed in 360 – 410m elevations, accounting for 75.0% of the total. The lowest elevation for karst distribution is 201.87m. The karst developments under river bed is divided into three zones (Ancient erosion – strong corrosion zone on top of P_{1m}^{2-3} layer is excluded): 0 – 140m under river bed is dissolution pore and karst cave zone; 140 – 230m under river bed is dissolution pore and dissolving fissure zone; 230m below is weak karst zone.

Table 7.1 – 4 shows the drilling-revealed development of deep karst at different levels at dam Site of Pengshui Hydropower Station.

Table 7.1 – 4 Karst Development Elevation Revealed by Drilling at Dam Site

Stratum Code	Left bank				Right bank			
	Karstification lowest elevation (m)		Drilling-revealed karst lowest elevation (m)		Karstification lowest elevation (m)		Drilling-revealed karst lowest elevation (m)	
	Karst cave	Karst fissure	Karst cave	Karst fissure	Karst cave	Karst fissure	Karst cave	Karst fissure
O_{1h}	320.89	213.37	320.89	169.63			270.69	204.86
O_{1f}			328.56	207.97				174.25
O_{1n}^{5}	225.00	208.00	163.00	174.74	205.83	218.00	43.00	143.56
O_{1n}^{4}	259.45	192.25	259.45	163.90			91.85	
O_{1n}^{1+2+3}				122.43				
ϵ_{3m}^{2-2}			323.14	205.13			143.29	152.47
ϵ_{3m}^{2-1}			293.91	208.47				
ϵ_{3m}^{1-2}	263.76	94.26	263.76	94.26				
ϵ_{3m}^{1-1}	47.75	54.00	17.82	2.85			187.62	
$\epsilon_{3g}+f_1$	83.41	87.00	83.41	−9.23				

KW_{17} system on left bank reveals that the lowest elevations of karst cave and karst fissure are respectively 174.8m and

163.90m. KW_{65} system on left bank reveals that the lowest elevation of karst cave is 17.82m, and karst caves are still well-developed at 50-80m elevation. KW_{51} system on right bank reveals that the lowest elevation of karst cave is 43.31m, and the deep karst caves intensely developed above 193m in diversion tunnel section. W84 warm spring system on right bank reveals that the lowest karst cave elevation is 152.47m, lowest karst fissure elevation is 143.29m.

Deep and shallow karsts have apparent inheritance features. Deep karsts on both banks look like inverted siphon tubes, which is to say, the karst development elevation is far lower than the river water level of dry season. But the groundwater is still drained as springs near the river. The outlets are above 215m elevation.

7.1.6 Deep Unloading and Strong Unloading

In western mountain and gorge areas, the unloading not only expresses in the shallow part of valley but also develops into deep sections and forms deep concentrated relaxed-tension crack zone. It is a very common phenomenon in western projects, for example, Jinping I Hydropower Station has a series of deep cracks on left bank. In Mabukan high slope near left bank of Xiangjiaba, some tension cracks about tens of centimeters can be seen as deep as 200m. The studies in recent years have shown that the deep unloading and the high ground stress are closely correlated. The high ground stress stored in western areas will release during rapid cutting process of the valley, as a result, the rock mass will be cracked and relaxed, the low-velocity rocks will be formed. Meanwhile, the measured stress value in relatively complete rock mass in relaxation zone decreases dramatically. For example, the relatively complete marble in the mountain on the left bank of Jinping I, the longitudinal wave velocity usually does not exceed 3,000m/s, where the measured stress value is only about 5MPa. Therefore, it is critical to determine the unloading rochmass, to judge the evolution and development phases of the unloading cracks, to analyze whether the deformation tends to be convergent, to assement whether the cracks go on developing and lead to instability of the mountain, to predict when the mountain will become unstable, to judge whether the mountain stability will be ensured during the project life. These questions will always be considered by engineering and technical personnel, especially the engineering geological staff. There are still some experts suspecting the construction of world's highest dam under geological conditions as complex as Pusiluogou in Jinping. What they are worrying about are issues above.

The similar geologic problem occurs also in the projects of Dagangshan and Baihetan. Consistent understandings are not yet formed on the formation of deep cracks. There are tectonic, seismic and unloading hypothesis.

7.1.7 Geological Disasters near the Dams

The unfavorable geology structures such as collapse, landslides, unstable rock mass and debris flow have a direct or indirect damage to the safety and normal operation of hydraulic structures. Direct damage could directly destroy the dam body, the flood spillway, the inlet and outlet of water diversion and the power generation systems, the diversion tunnel, the discharge (tail water) channels, and directly threaten the human being. Indirect hazards include the surges caused by large slumps in the reservoir, which might damage buildings and have an influence on buildings and residence in the downstream. For example, Vaiont landslide in Italy made the whole project scrapped. In China, the landslide in Tuoxi tangyanguang Reservoir resulted in a great loss of human lives and property. Therefore, since the construction of Longyangxia Hydropower Station, a great deal of works had been done to investigate and study the stability of near-dam reservoir bank, as well as analyze its stability under various conditions. In recent years, the researches have shown that the real threats of near-dam geological desasters are not the revivification of the ancient landslides and old landslides. Those slides have relatively low barycentre and small potential energy because of their previous movements. The revivification caused by the reservoir is basically manifested by the creeping, cracking, tractive slide and deformation. It is uneasy to form an overall, high-speed violent sliding. Although the deformation is great sometimes, the substantial threat to project operation is small. What we should give special attention is those instable slopes which have been already of sliding conditions and mechanisms but the sliding does not occur or the slide has not completed. These slopes have very small safety margin and very high potential energy. Under the action of reservoir water, it may take place the "a straw can crush a camel" critical effect which probably leads to disastrous results. These characteristics had been shown in Vaiont and Tangyangguang slides.

7.1.8 Reservoir Bank Stability

The stability of reservoir banks is gaining more and more attentions along with the increasing concern on environmental effect, resettlement and social stability in reservoir area. Jipazi slide and Xintan slide took place on mainstream of Changjiang in 1982 and 1985 respectively, causing great losses of reservoir navigation and people's lives and property. Since then, the stability of reservoir bank has attracted more and more attention. The awareness of this problem abroad started from Vaiont Slide on Oct. 9, 1963.

The stability of reservoir banks has the following rules:

①Instability events on reservoir banks occurred mainly in the Quaternary sediment slopes. ②The degree of development and the distribution of Landslides, collapse, unstable rock body are obviously determined by lithology, geological structure and slope conditions. ③The collapse of bedrock banks is closely related with lithology of parent rock. The collapse of reservoir banks are related with the rock bodies liable to be weathered, slacking and easy to be fragmented and cleaved under tectonization. ④The development of bedrock bank collapse is obviously determined by geological structure of the reservoir area and the type of geological structure of the slope.

The failure modes are divided into three categories according to the geological condition of the bank and the mechanism for bank deformation and instability.

(1) Erosion-denudation type: under the action of reservoir water, surface water and other exotic forces, surface materials of the slope are gradually moved away and the bank slope surface receded slowly. It mainly occurs in fully and strongly weathered slope and soil slope with gentle terrain. The rebuilding process is generally slow with small scale.

(2) Collapse type: toe or lower part of the slope has been softened or dredged and eroded under long-term action by reservoir water, and upper material without support will result in partial collapse. Generally, this mode occurs in soil slope with sharp terrain or unloading zone of bedrock. It tends to occur suddenly when there is rainfall or abrupt water level change in reservoir.

(3) Overall slide type: under the action of reservoir water and other factors, rock and soil mass of the slope overall slide along one or more weak planes. It is very hazardous in large scale. This category includes revivification of ancient or old slides, arc-shaped slides of soil slope and the slide of chair-like structure.

Normal water level of the Three Gorges Project is 175m. it is a canyon-type reservoir about 600km in length, 200–660m (2,000m at its widest point) in width, 5,927km in total length of the shoreline. The state has organized a number of regulation projects according to stability projections after the Three Gorges Reservoir was impounded. Through engineering and non-engineering regulations, the Three Gorges Reservoir, an area with high incidence of geological disasters, only had a few small slide damages. Relocation of residence had been completed before some slopes were deformed and damaged. The geologic hazard treatment has gotten an evident effect.

7.1.9 *New Problems on Natural Construction Materials in the Western Area*

In western China, especially the southwest, the large rivers are located in strong uplift region since cenozoic era, where the mountains are very high and the river current is torrential. Although sometimes deep overburden layers are distributed on river bed, they are deep under the water and have complex composition with various genetic origins. Therefore, it is difficult to be used as natural construction materials. The dam in western region can only adopt artificial aggregates as raw material.

It is not easy to find qualified stone though western area is mountainous with rocks distributed everywhere. Dam projects there are mainly located in geosynclinals area of the tectonic units. Great lithology and intensity differences, adding to laminated or layered distribution of soft and hard rocks, make it impossible to form large scale qualified quarry. Such as the Lancang River basin, it is basically composed of metamorphic sandy slates. Sand stone and slate are inter-layered. Schistosity, split and cleavage develop in rock body. It is very hard to be processed into qualified concrete aggregates, even not ideal to be used as rockfill. In addition, alkali-silica activity problem wildly exists in sandstones.

Hydroelectric projects in western China are also characterized by high-dam and large reservoirs, which demands natural construction materials with both high quality and adequate reserves for engineering use. The arch dam of Jinping I hydropower station on Yanglong River is 305 meters in height. The highest grade of concrete shall be C40. Considering that aggregate strength shall be 1.5 times higher than concrete strength, aggregate strength shall be 60MPa. Great amounts of marbles are distributed in vicinity of Jinping dam site. However, the crystalline substances of marbles are coarse with strength between 40–50MPa, which cannot meet the needs for construction. 30%–40% of the yields are powder during the process of aggregates, which seriously affect the yield quality. At last, thick sandstone was chosen as aggregates because its strength was qualified. However, alkali-silica activity problem cannot be avoided, it is necessary to use low-alkali cement and mix a certain amount of fly ash. During the construction of Xiangjiaba Hydropower Station, Tianping, a plance 30km away from the dam site, was used as limestone quarry because of alkali activity problems in sandstone and sand gravel. A 30km tunnel was excavated to convey materials to the site. Excavating and transporting costs were increased.

Another common problem existing in western hydraulic projects is insufficient core wall material. On the one hand, the projects are mainly located in upper reach of the rivers which have very little fine-grained materials and rarely have large clay terraces; on the other hand, less precipitation, strong evaporation, large temperature difference, physical weathering rather than chemical weathering in western China, make it difficult to keep thick soil layer in slope area. Sometimes, insufficient core wall materials will become a key factor that restricts the development of this dam type. The lack of soil material, followed by scattered materials yards and insufficient reserves, become very prominent environmental and social problems. For example, Lianghekou dam on Yalong River is as high as 300m. It chose soil core rockfill dam because the project is located in the upstream of Yalong River which is far away from cement and steel producing area and the transportation cost is high. To satisfy 5–6.3 million m^3 impervious core wall materials needed, 12 soil yards located in the upper reaches, the lower reaches, left bank and right bank were investigated. Total reserve is only 2 times than requiring amount. In addition, the exploitation involves replacement of about 2,000 people, which is very difficult to deal with.

However, despite all problems in reserves, quality, exploitation and transportation conditions in the western area, there are still many successful cases that no specific quarry is exploited. Dachaoshan station on Lancang River is a one-hundred-meter concrete gravity dam. It takes advantage of the tuff characteristics and satisfies the requirement for concrete aggregates depending on the excavation of dam foundation and underground project. Gongguoqiao Station on Lancang River is also a one-hundred-meter gravity dam with sandy slate dam site. The planned sandstone quarry was failed because of the objection of local residents. Later on, after restricted material management and fully utilization of excavated sandstone and sandy slate, the requirement for concrete aggregates is almost satisfied. The remnant materials can be bought through the market.

7.2 Development of Engineering Survey Technologies under Complex Conditions

7.2.1 New Technologies for Drilling Deep Overburden Layers

Since the adoption of Reform and Opening-up Policy until now, the exploration technologies have developed rapidly and made many important achievements, including ① miner diamond core drilling technology; ② cord core diamond drilling technology; ③ directional core drilling technology with second drilling; ④ successful development of MS vegetable gum drilling fluid and SD diamond drilling tools; ⑤ large-diameter drilling technology which can drill 60m into medium hard rocks; ⑥ successful development and extensive application of a series of hydro-geological drilling test equipment and technology.

In recent years, the drilling and core technologies tailored to deep and wide overburden layers in western China have been developed. SinoHydro Chendu Engineering Corporation developed DTH drilling with simultaneous casing and coring technology and corresponding drill, drill bits and joints. They also developed elaborated operational procedures. This technology can be used in deep sand and gravel layer or other loose layers. The project quality and progress have been improved. Hydrochina Zhongnan Engineering Corporation is developing new directional drilling and coring technology. It applies self-designed directional and deflecting device to realize the planned bending drills, which can partly substitute horizontal tunnel at river bottom and opposite inclination holes, as well as being used in geotechnical engineering under special conditions.

Changjiang Institute of Survey, Planning, Design and Research has developed some new technologies and techniques when surveying overburden layers in Rimian, Tacheng, Wudongde Hydropower Stations in upper and middle reaches of Jinsha River. These sites are featured with accumulation and mixture of colluviation, alluviation and slides, inhomogeneous material composition and structure, diverse material s and particle sizes.

(1) Optimized drilling program. Firstly, standard penetration, then drilling, followed by TV to make sure not to miss fine soft soil (such as fine sand, silt and clay, etc.).

(2) Using advanced drilling equipment. New equipments, such as ϕ114mm double core barrel with diamond bits, ϕ130m bottom discharge double barrel with diamond bits, are applied. Single barrel or thin-walled electroplated diamond bit is usually used to quickly pierce large stones, pebbles with large size or boulders; in other cases, double barrel drills and supporting diamond bits are used.

(3) Improving drilling operation process. In drilling process, standard operations shall be emphasized. Pressure, rotary speed and water parameters shall be reasonably selected. Footage per run shall be strictly controlled (decreased). High quality mud is used as circulating fluid. The density of mud in drilling process shall be maintained to increase which will increase the powder-carrying ability of mud and keep the hole's bottom cleaning, and increase and consolidate the wall protective role of mud to reduce collapse and blocks and improve the drilling efficiency.

(4) Developing hammer type sampler. Existing overburden layer samplers, such as double barrel inner ring cutting sampler, ϕ108mm undisturbed sand sampler, double barrel advance boot L type sampler, can be used to sample thick fine sand

layer, soil layer or clay gravel layer. To this end, undisturbed sampler with high strength are developed to sampling in pebble layer and compacting layers of ϕ130mm、ϕ110mm single barrel hammer sampler, ϕ130mm and ϕ110mm double barrel hammer samplers. Meanwhile, L-type drill bit is technologically improved. Hammer samplers can successfully get undisturbed gravel (pebble) samples whose diameter is less than 70mm. it can not only basically meet the needs for particle analysis, soil test and other routine laboratory tests required, but also provide reliable basis for lab simulation configuration.

At dam site of Wudongde Hydropower Station, the deep overburden layer has very complex genetic causes, inhomogeneous material composition and structure, prominent stability problems of high cofferdam and deep foundation pit. It is required to identify material composition of overburden layer, structure and spatial distribution characteristics, to accurately evaluate its engineering geological characteristics and provide reliable data for designing cofferdam and foundation pit projects. By optimizing drilling operation process, improving drilling operations and combining exploration coring and visualization technologies, data are obtained to reveal material composition, structure and spatial distribution features of overburden layers. Undisturbed samples as required by test are also obtained. Meanwhile, a large number of complementary and comparative pilot studies were carried out between various types of in-situ testing (heavy dynamic penetration, super-heavy dynamic sounding, pressuremeter test, pumping test and sound test, etc.) and lab tests (particle analysis, soil conventional test, simulation sample test). These studies effectively protected the accuracy and reliability of test results and laid an important foundation for accurately identify and correctly evaluate the engineering geological characteristics of overburden layer.

In Tacheng dam site on Jinsha River, the overburden layer is up to 100m, and Shangjiang dam site on downstream has overburden layer as thick as 206m. During the survey process, Changjiang Institute of Survey and design emphasized on exploring drilling technologies and comprehensive logging technologies besides the application of new drilling equipment. Satisfactory results had been achieved on improving drilling efficiency and obtaining in-situ parameters of overburden layer.

7.2.2 *New Technologies for Geophysical Prospecting*

7.2.2.1 New Development of Geophysical Technologies

(1) Coupling system for acoustic test in drill hole without water. It is technical improvements of traditional acoustic probe. It is attached by specially designed inlet section, lengthened exhaust section and sealed pressurized bladders and is able to be used with acoustic instruments and supporting hydraulic equipment, which changes acoustic test conditions in drill holes without water and make such test no longer a difficulty.

(2) Bolt anchoring density detection. Anchor is a concealed project. Drawing force test used before has been applied less because it cannot reflect grouting anchoring density. It was replaced by acoustic nondestructive testing technologies.

(3) 2D complex structures triangulation network ray tracing CT technology. Triangulation network model was applied for complex external geometry boundary and internal features of tomography to realize triangulation of the model. Triangulation enables free loading of priori geological information, such as faults and cavities. This technology is a major upgrade of elastic wave and electromagnetic tomography and has been successfully applied to the Three Gorges and Shuibuya hydraulic projects.

(4) Additional mass method for rockfill density Inspection in rockfill dam construction. Addition mass method is applied for real-time tracking of filling quality and controlling of construction quality. Inspection information is timely feedback. Re-compaction is required for non-conforming parts. The numbers of pits are reduced and project term is shortened.

(5) CT 3D visualization system. It is a technology integrating data management of electromagnetic wave CT, acoustic CT, seismic wave CT with visualization technology.

(6) Shallow 3D seismic technology. It uses multi-wave, multi-component and distributed network observation technology. Informative data are processed by modern processing and analysis technology to provide sectional, plane or 3D images of underground geological structures. It provide new means for effectively solve problems on complex geological structure and karsts.

(7) Digital Television system of drilling overall-hole wall. It is a new technology developed based on drilling TV technology in recent years. It can form continuous scanned images through the conical mirror or curved mirror reflected images of the drill holes wall and continuous movement of the probe in drill holes.

(8) Multi-channel transient surface wave technology. Many geological problems of the project can be solved by utilizing correlation between dispersion characteristics, propagation velocity and the physical and mechanical properties of rock to realize

high resolution and high precision in shallow geotechnical exploration.

7.2.2.2 Application of New Geophysical Technologies

1. Deep buried and long tunnel exploration

General investigation methods tend to be ineffective when the tunnel is deeply buried and long. Deep geophysical methods based on magneto-telluric exploration provided support for solving such problems. These methods can detect some changing "trends" of important geological phenomena along with depth, such as changes of geological location of intrusive rock an other rock interfaces, soft rock and other rock interfaces, carbonate rocks and other rock interfaces, fault location, orientation and width as well as aquifer location. It can also reveal the distribution of large fault zone, water burst, soft rock deformation and karsts.

The most commonly used magneto-telluric geophysical methods are: magnetotelluric (MT), audio magnetotelluric (AMT) and controlled source audio magnetotellurics (CSAMT). Their probing depth is up to 1,200 - 1,500m. Theories and practical applications of these methods are basically the same. MT and AMT are natural source with difference detecting frequency bands. The lower the frequency, the greater the depth reflected by apparent resistivity. MT acquisition frequency band is usually between 400Hz and 0.00001Hz, which is suitable for deep geophysical survey. AMT acquisition frequency band is up to 1,000 to 10,000Hz. AMT is applicable profound investigation of shallow layers. In terms of CSAMT, it adopts controlled artificial signal source to supplement the randomness of MT and AMT signal sources, weak signals and difficult observations. It is particularly effective in thorough investigation of shallow layers.

A hydraulic project in Qinghai province was planned to introduce the water of Datong River into Baoku River, the upper reach of Beichuan River, a tributary of Huangshui River, in order to solve water shortage problems in Huangshui Basin of Qinghai Province. The sectional length of Dabanshan tunnel in this project is 22,650m. Geophysical Company of Changjiang Design Institute investigated Dabanshan tunnel by AMT method to detect and analyze deep geological phenomena.

The tunnel project was started in October 2004. After 5-year construction on site, it revealed that the magneto-telluric conclusions coincide with real geological conditions which accumulated some detective experiences in exploring deep-buried long tunnel.

In the first stage in west route of South-to-North water diversion project, deep-buried water diversion tunnel was planned to pass through Bayankala Mountain, the first watershed divide between the Yellow River and the Yangtze River. The tunnel section in Aba is about 40km in length, 10m in diameter of hole. The mountain elevation is 3,700 - 4,300m. The tunnel is buried 300 - 800m under the ground. In this section, there are few outcrops so that it is difficult to determine stratigraphic signature. CSAMT exploration can overcome the deficiencies of conventional geological investigation methods. The location of exploration lines is shown in Fig. 7.2 - 1.

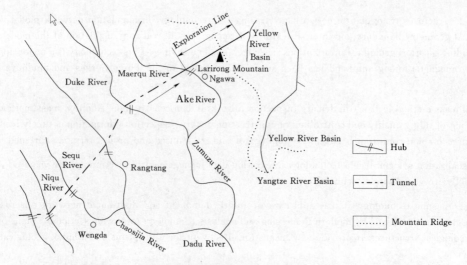

Fig. 7.2 - 1 Location of CSAMT Exploration Line (from Engineering Construction of and Yellow River Conservancy Commission)

High resistivity showed on CSAMT resistivity cross-section diagram means complete rock structure with simple geological conditions. It is good section for construction. A wide range of low-resistivity zone on CSAMT resistivity cross-section diagram was determined by lithology and groundwater conditions rather than tectonic causes. Drilling-hole test verified the geological speculation. The low resistivity is caused by water-rich thin sandstone and slate in medium-thickness stratum. It reflects that the permeability and water-richness of deep rock mass is higher than superficial rock mass. The tunnel has to pass through a long low-resistivity zone. Seepage, small fracture zone might occur, even water burst, and soft-fractured rock mass burst might occur. Because of poor groundwater recharge, the geological disasters are not in large scale. But drainage and geological prediction shall be emphasized in tunnel construction. When banded low-resistivity zone was shown in CSAMT resistivity cross-section diagram, it was inferred to be faults which are hazardous for project construction.

2. Deep overburden exploration

In preliminary water conservancy and hydropower projects, geophysical technology for deep overburden exploration becomes more and more mature. One or more geophysical methods shall be reasonably selected according to mission and purpose of the exploration, the depth, scale and physical properties of surrounding medium of the object to be explored, as well as topographical, geological and geophysical conditions.

When site openness allows, shallow refraction method will be effective to explore depth of overburden and bed rock.

Seismic refraction method, transient Rayleigh wave and electric method can be applied to determine the layers of overburden.

Transient Rayleigh wave method is applicable to overburden having soil coverage and less than 70m in depth.

If grounding conditions are good and the site is open with small undulation, electrical prospecting is applicable when apparent electrical differences exist between overburden and bedrock or overburdens.

Transient electromagnetic method can be used under poor conditions such as in desert, grassland, Gobi, bare rock and frozen ground.

Seismic reflection wave method and controlled source audio magnetotellurics (CSAMT) are optional to explore deep overburden. When topographic, geological and geophysical conditions are complex and there is no known drilling data, a single geophysical method is prone to uncertainty or multiple-solution. A variety of geophysical methods are preferred to be used in main line and sections with complex geological conditions.

Take Tacheng dam site in Jinsha River as example, Controlled-Source Audio magnetotellurics, supplemented by seismic refraction method are selected according to its topographic, geological and geophysical conditions to detect overburden depth. Controlled-Source Audio magnetotellurics result is shown in Fig. 7.2-2. It can be seen from the diagram that the electrical interface is stable with good continuity, corresponding bedrock resistivity (ρ) is $300-1,000\Omega \cdot m$. Geophysical interpretation of the overburden depth is $0-97m$, which is consistent with the drilling results; In 195-670m section, a clear low-resistance band (resistivity $\rho=1-25\Omega \cdot m$) is clearly reflected in the overburden. The center elevation of the low-resistivity object from left to right is $1,850-1,856m$ and the thickness is about 10m. And it is proved to be clay layer.

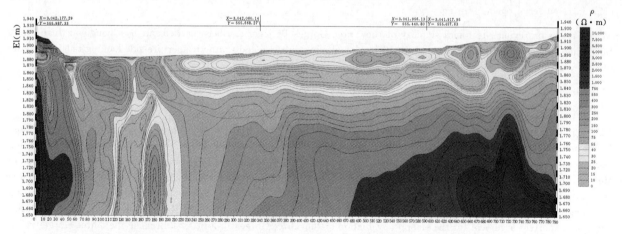

Fig. 7.2-2 Controlled-Source Audio Magnetotellurics Result of Tacheng Dam Site on Jinsha River

3. Non-destructive detection

Non-destructive testing is mainly used in quality inspection of concrete, bolt cable and cutoff wall. Main methods for testing concrete strength are: rebound method, ultrasound and ultrasound-rebound combined method. Rebound method is mainly used for testing the strength of concrete surface or strength of structural concrete with thinner thickness. It is not very effective for testing the strength of deep concrete with large volume. Ultrasound is mainly used for testing the strength of deep concrete with large volume. But its accuracy depends on the correlation of empirical formula. Rebound-ultrasound combined method adopts many physical parameters and is able to more comprehensively reflect the factors revealing concrete strength. It is also able to offset certain factors that influence the correlation of strength and physical quantity. For this reason, it is more accurate and reliable than single non-destructive method. It is a method that is most frequently studied and most extensively applied in testing concrete strength both home and abroad. Detection of internal defects of concrete includes non-compacted area (honeycomb, segregation etc.) and extended depth of crack. Non-destructive methods for concrete defects testing are: ultrasound, pulse echo, ground-penetrating radar, digital Television system of drilling overall-hole wall and borehole TV. The use of ultrasound can be divided into flat measurement method, opposite measurement method, slanting measurement method, CT and drilling method according to actual situation.

Non-destructive testing for blot cable is to detect its length and mortar saturation. If bolt (cable), mortar and wall rock are grouted evenly and densely, the vibration characteristics throughout the bolt (cable) will have no significant differences; when stress wave encounters defects, the vibration will change. Acoustic reflection method is adopted to detect the quality of bolt cable.

Quality inspection of cutoff wall include: depth, uniformity and joints of cutoff wall. This technology has been widely used in dangerous reservoirs and Changjiang embankment project, and also extensively used in cutoff wall detection of large hydropower dam. Cutoff wall tests are divided into ground geophysical method and bore geophysical method. In terms of ground geophysical method, there are high density receptivity method, controllable source audio frequency magneto-telluric prospecting and ground penetrating radar; in terms of bore geophysical method, there are single bore acoustic method, penetrating acoustic method, elastic wave CT and digital Television system of drilling overall-hole wall.

7.2.3 Remote Sensing Technology

Remote sensing technology has been widely used in the investigation of tectonic structure, geological survey, and geological disaster monitoring, migration and environment capacity survey in water conservancy and hydropower projects.

7.2.3.1 Remote Sensing Technology in the Study of Regional Tectonics

Aviation and space remote sensing images are featured with wide viewing angle, fidelity and efficiency. It can accurately figure out the tectonic distribution and activity strength of regional faults and rift basins. Especially the application of GPS, high-resolution satellite data and satellite radar interferometry technology can precisely analyze quantitative data of ground or fault activities, which has been used as strong evidence for evaluating regional tectonics stability. In the project survey of the Three Gorges Project and Danjiangkou, the serious problems of regional tectonics stability had been properly settled through using high-resolution satellite remote sensing, Synthetic Aperture Radar aviation remote sensing and large-scale color infrared aerial remote sensing.

Many studies on regional tectonics stability of the Three Gorges Project have been carried out by Changjiang Institute of Surveying Technology via using color infrared, black and white infrared and synthetic aperture side-looking radars. A series of achievements have been made:

(1) Interpretation diagram and image maps for the Three Gorges area have been prepared to solve environmental problems on regional tectonics and identify the tectonic skeleton of this area.

(2) The north extension problem of Fairy Hill fault. Being about 18km away from the Three Gorges Dam, Fairy Hill fault is not only the largest fault in this area but also a seismogenic structure for moderately strong earthquakes. It plays an important role in evaluating regional stability. One issue that had been debated for a long time is whether this fault has passed through the Yangtze River. Processed images displayed that the fault ended near Fengchuiya, 6km south to the Yangtze River. After interpretation of color infrared aerial photographs and ground surveys and verification by artificial seismic sounding profiles, it was confirmed that this fault did not extent through the Yangtze River in the north. This conclusion provides important argument for the stability evaluation of the Three Gorges Project area. Because the Three Gorges reservoir fault does not penetrate into the reservoir, the anxiety about reservoir-induced earthquake shall be ruled out.

(3) The argument of whether the linear images of Shizikou and Sandouping were faults. The NW direction dense linear images located in southwest of the Three Gorges Dam about 12km were considered to be large regional faults. Changjiang Institute of Survey Technologies took advantage of satellite images, side-looking radar images and color infrared aerial photos and adopted multiple image processing methods and interpretation approaches. Supported by in-situ detection and many exploration methods, a conclusion was made: NNW direction linear image of Shizikou is ridge-groove alternated after Sinian

upright soft and hard rocks alternated stratum was weathered and eroded. Sandouping NNE direction linear image is discontinuous distribution of gullies and ridges rather than faults.

7.2.3.2 Remote Sensing in Surveying Exogenetic Geological Phenomenon

Landslides, collapse and debris flow are exogenetic geological phenomena which are common geo-hazards in reservoir area and dam site during the construction of hydropower projects. They can be clearly reflected in remote sensing images, especially aerial 3D images because of their visual representations of geological micro-topography. Since the 1980s, remote sensing interpretations based on aerial three-dimensional images have become a common tool of investigating geological disasters of water conservancy and hydropower projects. During the development of the Three Gorges project and Jinsha River Cascade hydropower projects, landslides and debris flows distributed in mainstream and tributary of a series of high dam and large reservoir have been successfully identified.

In Wudongde reservoir area, geological disasters such as collapses, landslides and debris flows are well developed. Changjiang Institute of Design made remote sensing interpretation for environmental engineering geology including collapse, landslides and debris flows within a range of 4km from backwater lines at both banks via adopting 1/30,000 black-and-white aerial photograph. The remote sensing interpretation based on 3D air photo reveals that the main deformation and failure modes of reservoir banks in Wudongde mainstream and tributary are collapse, landslides and debris flows. Within the Wudongde reservoir area, 120 collapses, landslides and debris flows with certain scales were extracted and interpreted.

7.2.3.3 Remote Sensing in Investigation of Migration and Environmental Capacity

Water conservancy and hydropower projects result in a lot of migrants. The key issue for resettlement is environmental capacity. In recent years, migrant resettlement plans got a strong technical support from the rapid development of remote sensing technologies and the improvement of multi-level remote sensing data acquisition, data analysis and process and synthetic application of remote sensing data.

Remote sensing images have wide vision and broad perspectives, which overcome the limitation of ground survey. Remote sensing image recognition and interpretation, combined with field surveys, can meet investment requirements for most of submergences. In the Three Gorges Plan and migrant investigation of Wudongde, Tingzikou reservoir areas, remote sensing technologies were extensively used in surveys of land use, houses, roads, population, historical relics and determining administrative boundaries.

Remote sensing technology can demonstrate its unique advantages in environmental capacity survey of resettlement area.

(1) Obtaining information of natural landscape, geological structure, lithology, surface water system, etc.

(2) Obtaining information of land use and water resources distribution, such as distribution and acreage and arable land, garden, woodland, grassland, reclaimable land and land unsuitable to use; the distribution of surface water system, reclaimable acreage and location.

(3) The investigation of geological disasters, including lithology of stratum, rock mass or accumulation, rupture, level, lithology interfaces, slope gradient, aspect, shape and elevation.

(4) The survey of soil and water loss as powerful technical supports for developing soil and water conservation planning and improving living environment of migrants.

(5) Reservoir resettlement planning, such as city and town location selection, environmental capacity of resettlement area, the planning of the rural residential area, the selection of roads and power transmission line and so on.

7.2.4 *Numerical Simulation Technology*

Numerical analysis method is widely used with the need for further research on the engineering geological problems. This method is able to consider anisotropic and non-homogeneous characteristics of geomaterials and their changes over time, complex boundary conditions and complex geological conditions such as discontinuities. These are its prominent advantages. However, the reliability of numerical analysis is still depending on investigations of engineering geological conditions. Therefore, investigation and study of geological conditions and qualitative analysis of engineering geology have played a significant role in numerical simulation analysis.

7.2.4.1 Methods of Numerical Modeling for Engineering Geology

While the Finite-Difference Method (FDM) is the pioneer method of numerical analysis, the Finite Element Method (FEM) became popular in 1950s, and the Boundary Element Method (BEM) appeared in the 1970s. The basic idea of semi-analytical element method was suggested by Y. K. Cheung in the 1968, that is, the finite strip method. It is a solution combining analytic method of mathematical equation and numerical method. By utilizing part of analytical solution to reduce the workload of purely numerical calculation method, it is applicable to solve high-dimensional, infinite and dynamic field problems. Discrete element method was first proposed by Cundall in 1971. It was a promising numerical method with rapid development. Unbounded element method was proposed to solve the problem of "difficulty of determining calculation scope and boundary conditions" encountered by finite element method. It is another effective way to solve problems on rock mechanisms.

In recent years, coupling analysis of numerical methods has made great progresses in order to solve complex engineering geological problems. For example, the couplings of finite element method and boundary element method, finite element method and discrete element method, boundary element method and discrete element method are wildly applied to solve numerical simulation problems under complex conditions.

In recent years, some new numerical methods emerged to deal with complex soil-rock structures: Joint Element (JE) of finite element method, Block Theory (BT), Discontinuous Deformation Analysis (DDA), Fast Lagrangian Analysis of Continua (FLAC), Block Spring Method (BSM), Element Free Galerkin method (EFGM) and Manifold Method (MM). These methods are more targeted at geotechnical engineering problems and the post-processing has also been significantly strengthened.

7.2.4.2 Safety Criteria of Finite Element Method

1. Overload method

The load is multiplied by K times, and K is gradually increased. Each K value is corresponded by one elastic-plastic finite element analysis. When main load that causes geotechnical instability is increased proportionally, a series of results on displacement, stresses and plastic zone distribution will be obtained. Based on the displacements reflecting geotechnical deformations, load-displacement curves can be developed in which you can find the corresponding dramatic increase in load. This is the load causing failure of rock and soil mass. Thus, overloading safety factor of soil and rock mass is obtained. This factor is usually high and difficult to be compared with traditional result of limit equilibrium method.

2. Safety coefficient method

This method can be used to study the safety of specific failure surfaces and obtain different safety factors of them respectively (or slide-resistance safety margin coefficient).

3. Strength reduction method

This method is first proposed by Zienkiewicz in 1975. It is used to extend plastic zone of rock and soil mass by reducing the strength indicators of the materials. When the calculation no longer converges, or the displacement of feature points increases abruptly, it is considered that the rock and soil mass is in the critical state. Its result is comparable to the safety factor obtained by limit equilibrium method. As a result, it gains more and more applications.

7.2.4.3 Numerical Analysis in Water Conservancy and Hydropower Engineering

1. Stability analysis of high slope and slide

Numerical analysis can be applied in determining the stability of high slope and landslide, studying failure mechanism and analyzing environmental sensitivity, which has important engineering significance.

In high slope survey at Wudongde Dam Site, UDEC method was used and concluded that the safety margin of stability was moderate and overall ability was fair. In the stability analysis of natural slope of tailrace tunnel in Pubugou Hydropower Station, nonlinear finite element analysis was used to predict the overall stability of the slope and the distribution of plastic zone. For HP12 landslide on the right bank of Zhaikou Hydropower Project in Xinjiang, strength reduction method was applied to make elastic-plastic finite element analysis and revealed that landslide failure mode under different conditions was mainly revivification of slip mass along slip surface. It provided basis for landslide regulation.

2. Anti-slippery stabiligy analysis

Numerical analysis is able to simulate the non-linear constitutive relations between the dam and its foundations and the complex geological conditions of dam foundation. It also considers the load conditions of construction process, water storage, seepage and earthquake. It can get the data of stress, spatial distribution of deformation region and development of plastic zone. It can also learn about the distribution and scope of failure zone to find the most dangerous part and analyze the function of various reinforcement measures. Therefore, numerical analysis has been widely used in analyzing stability against sliding and engineering analysis of the dam.

In stability analysis of $1^\#-5^\#$ sections of the Three Gorges Gravity Dam, strength reduction method is used and supported by two models of elastic-plastic model and elastic-brittle-plastic model to analyze stability against sliding of their foundations by comprehensively considering plastic zone connection in sliding channel and abrupt displacement changes of feature points. In Gaobazhou Project on Qingjiang River, stability against sliding of dam is analyzed by finite element shear strength and shear stress ratio method.

3. Stability analysis of underground cavern

The conditions for determining stability of surrounding rock are the initial stress filed of the rock mass in cavern area and the changing pattern of surrounding rock stress during the excavation of caverns. They are also support parameters for reasonably selecting bolt (cable) and determining reinforcement planning of surrounding rocks in the cavern.

In Goupitan hydropower project, elastic-plastic rock theory, three-dimensional nonlinear finite element method were applied to study deformations of surrounding rock, stress features and failure development degree during excavating main plant, main change hole, tailrace surge chamber. It guided the selection of support parameters of bolt (cable) and determined the reinforcement plan for surrounding rocks.

4. Seepage control analysis

Seepage numerical analysis can demonstrate the rationality of seepage control design and the possibility for optimization. It was widely applied in projects. In Shuibuya project, numerical method was taken to analyze seepage for underground plant. It provides basis for the reasonable design of seepage control project.

7.2.5 3D Collaborative Design

The study of complex geological issues should be presented as the relationships between topography, stratum and faults in a stereo, clear and intuitive way and quickly select geological analysis sections. The complexity of geological problems and the precision of geological survey and design promote the rapid development of 3D visualization technology and generate the demand for 3D collaborative design.

3D co-design is a new technology developed after computer aided design (CAD). It is one multi-disciplinary and cross-regional 3D CAD software based on 3D solid modeling and network technology. The developing orientation of 3D co-design is towards a platform for geological visualized technology and design, engineering and mechanical technologies to realize data, parameter and model sharing, as well as serving as a platform for conflict test between various designs.

Since the beginning of this century, a number of 3D visualization software has been developed by research institutes, universities and business development companies both at home and abroad. 3D-GVS software is a geological 3D visualization device developed by Changjiang Institute of Survey Technology based on American 3D development platform IDL6.0 (Fig. 7.2-3). It can realize 3D modeling for terrain, stratum, faults, weathering boundaries and karsts, and generate geological cross-section and horizontal-section map on 3D geological models and create DXF format file at the same time. It can also output 3D geological model as interface file (SLT format) of general modeling software for the use of other software.

GOCAD software has very powerful performances of 3D modeling, visualization and geological analysis. It has mesh object (sgrid) and entity object, realizing the transformation from surface modeling to volumetric modeling; it can not only express the geometric objects, but also manage and explain space attributes. Mesh generation followed by numerical modeling extends the application of 3D geological model.

The most typical 3d co-design software is CATIA platform from French Dassault. The versions published after CATIA R17 has more and more powerful mesh creating and modifying functions, particularly the accuracy of terrain modeling; in addition, the secondary development based on CATIA enables software interface, multi-source modeling import, 3D modeling and 2D plotting of topography and geology to meet the demands of geological works.

7.2.6 Reflection of Technical Progress in Survey Technical Standards

With a large number of engineering experiences and profound studies of geological problems, some major geological issues have made important progress or achieved breakthrough. The most typical issues include the minimum standards of active faults, index value of rock and soil mechanics, utilization of rock mass of dam foundation and division of unloading zones.

In Technical Specification for *Geological Investigation of Water Conservancy and Hydropower Engineering* (GB 50487—

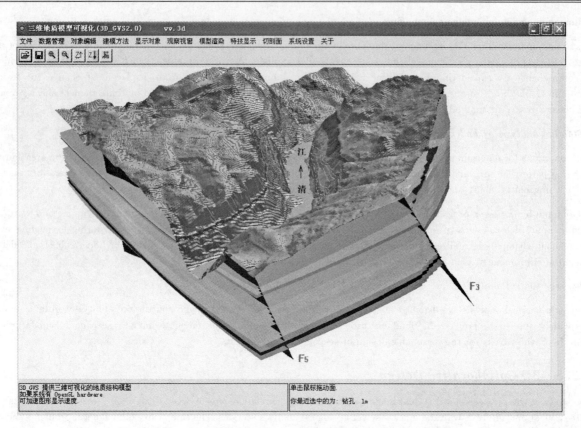

Fig. 7.2-3 3D-GVS Software Interface and 3D Geological Model of
Shuibuya Dam on Qingjiang River

2008), three levels of regional tectonics were amended from original 300km, 20-40km and 8km to 150km, 25km and 8km. The absolute age of active fault was adjusted from 100,000-150,000 years to 100,000 years, reflecting the progress of the study on regional tectonic stability and cohesion of related codes. In Technical Specifications for *Stability of Regional Tectonics of Water Conservancy and Hydraulic Projects* (DL/T 5335—2006), the three levels were adjusted to ≥ 150km, 20-40km, 5km. Throughout the changes and development nearly a decade, on the one hand, conditions of regional tectonic stability are more and more complex in water conservancy and hydropower projects. Seismic intensity in many sites in western China is as high as 8-9 degree (ground motion acceleration is 0.3-0.6g); on the other hand, the projects are more and more demanding on seismic fortification levels. For large projects, besides design of 2‰ 100-year probability of exceedance of ground motion parameter, seismic fortification stability check shall be done according to a 1‰ probability of exceedance in 100 years.

Technical Standard for Geological Investigation of Water Conservancy and Hydropower Engineering (GB 50487—2008) stresses the importance of in-situ test for rock mass strength and deformation properties and cancels the specifications of lab rock shear test in old standard. It reflects the technological progress in recent 10 years. Along with the accumulative experience on expansive-soil dam construction and improving treatment technology of expansive-soil slope, the contents about expansive soil strength parameters should be rheological strength parameters were abolished in the new standard.

In past ten years, with the construction of a large number of water conservancy and hydropower projects, geological problems in dam foundations were constantly emerging. For example, the utilization of "hard, brittle and fragment" rock mass in Zaoshi dam foundation of water conservancy project in Hunan, rock crushing due to geniculate tectonics of Xiangjiaba dam foundation, sliding stability of Wudu dam foundation, determination of rock surfaces for utilizing the dam foundation in high stress area of Xiaowan Hydropower station. These problems and corresponding researches promote the progress of technologies for rock surface determination on dam foundation utilization and treatment.

Weathering zone division is an importation composition of pre-survey of inland water conservancy and hydropower projects. But in the western regions, the division of unloading zone is as importance as weathering zone, even more important. In past decade, geological engineers not only recognize and reveal normal unloading and deep abnormal unloading, but also present the weathering zones division standards for carbonate and non-carbonate rock, and put forward quantitative and qualitative criteria for unloading zone classification.

7.3 Research and Evaluation on Major Geological Problems

7.3.1 Safety Evaluation of High Seismic Density Zone

In preparatory work of water conservancy and hydropower project, seismic resistance is an important part of research contents in planning, pre-feasibility and feasibility studies, preliminary design and even at the operation stage after completion. All large dam projects, without exception, made specific evaluation for regional tectonics stability and seismic safety of the site. The research must be approved by China Earthquake Administration before being used as design basis.

Specialized seismic hazard studies have mainly addressed three aspects: Firstly, seismogeological background at the engineering site. The focus of much research is on stability of regional tectonics, distribution of seismogenic structure, seismogenic ability and shock effects on buildings after earthquake within a 150km range and the study is finally carried out into the motion acceleration in site area, as important for designing dam and other buildings. Secondly, fault activity near the dam site, especially within the 5 – 8km range. The study can answer the fault-resistance questions of hydraulic buildings brought about by seismic action and displacement of active fault. Generally, dam project is not allowed to bridge over known active fault or their contemporaneous faults. Thirdly, analysis and prediction of secondary seismic-geological hazards probably occurring in the dam site area. Corresponding treatments will be taken in engineering construction. The seismic safety of the dam will be guaranteed as long as the three aspects are addressed properly.

7.3.1.1 Xiaowan Hydropower Station

The Xiaowan Dam is located on mainstream of Lancang River in Nanjian county and Fengqing county in Yunnan Province. It is composed of concrete double-arch dam and underground plant on the right bank. Its normal water level is 1,240.0m and the maximum height of the dam is 294.5m. Its total installed capacity is 4,200MW.

From east to west of the periphery of site area, the main regional faults are: Chenghai fault, Honghe fault, Chuanhe fault, Wuliangshan fault, Lancang River fault, Nujiang River fault, Nanting River fault, Changning fault, Kejie fault and Baoshan fault. The site is located in seismic zone of southwest Yunnan. Xianshuihe-East Yunnan seismic zone is in northeast of the site. Seismic activities in these zones are featured with high-frequency and high-intensity. Seismic hazard analysis has showed that the biggest seismic density is Ⅶ degrees, which probably caused by seismic hazard zones near the dam site, such as 7-magnitude Dali – Midu zone, 6-magnitude Yongping – Lushi Zone, and 7-magnitude Yun County Zone. The project is in a Lushi – Majie, a potential 7 magnitude focal region. Another influence on the dam site is Gengma – Yunxian, a potential 7.5 magnitude focal region.

The major regional faults in reservoir and dam area belong to the middle section of Lancang River fault and its secondary faulted structures. The middle section fault is arch connection of north-south Langcang River fault, which is composed of a series of paralleled east-west faults. Earthquake history and modern observation show that seismic activity in Lancang River fault, especially middle and south sections, are weak. Tectonic activities and modern seismic activities were moderate. Considering the complex tectonic settings, a long-term seismic and geological study has been taken for Xiaowan Dam. In Dec. 1982, Yunnan Earthquake Administration proposed "Opinions on Basic seismic Intensity of Xiaowan and Manwan Regions in Yunnan", in which basic intensity was Ⅷ. In the end of 1988, Geological Institute of National Earthquake Administration presented "Comprehensive Study on Seismic Issues of Xiaowan Hydropower Project in Lancang River, Yunnan Province". At the beginning of 1989, this report was approved by National Seismic Intensity Assessment Committee. The basic intensity of project area was Ⅷ degree. 50-year 5% exceeding possibility of bedrock acceleration peak value is 0.172g. 100-year 2% exceeding possibility of bedrock acceleration peak value is 0.289g. 600-year 10% exceeding possibility of bedrock acceleration peak value is 0.308g. The approved ground motion parameters were used as the deign basis for Xiaowan arch dam. After being testified by detailed tectonic geological studies and a variety of dating methods, F_1 and F_{13} faults in project area were proved no significant signs of fault activity since late Pleistocene. They are non-active faults. Being measured with thermoluminescence (TL) and electron spin resonance (ESR) methods, the latest activity of F_7 fault was 60,000 – 90,000 years ago, while scanning electron microscopy (SEM) showed that no activity in past 100,000 years. Some experts suspected that it was an active fault. F_7 fault is 5.6km in length. Its attitude is N80°– 85°W, NE∠74°– 90°. The fracture zone is 18 – 37m in width and composed of breccia, calaclastic rock and gouge. Gouge thickness is 1.5 – 2.5m with strong bonding. 10°– 16° slickensides can be seen on fault face. This fault is about 50m away from dam heel and passes through the discharge hole. In view of the large scale and complex seismic settings of Xiaowan Dam, monitoring projects have been arranged in Xiaowan dam site since 1990, such as microseismic monitoring, strong motion monitoring, 2D and 3D displacement measurement of F_7 fault. 11-years-observation showed that there was no any tectonic activity in this fault.

After Wenchuan earthquake on May 12, 2008, the state required to check the seismic safety of the site since Nov. 2008. The check concluded that basic seismic intensity of dam site is Ⅷ degree; in terms of bedrock horizontal ground motion acceleration, 10% possibility of exceedance in 50 years and 5% possibility of exceedance in 50 years of are 0.166g and 0.205g respectively, 2% possibility of exceedance in 100 years is 0.313g. The check result is almost the same as original design basis except a slight increase of ground motion parameter. The check and long-term monitoring results further clarify that there is no evidence of activity since late Pleistocene in F_1, F_7, F_{13} faults. There is no active fault in this region. Cut slopes in the project region were reinforced.

7.3.1.2 Dagangshan Hydropower Station

Dagangshan Hydropower Station is located in Shimian County, the middle reach of Dadu River in Sichuan Province. It is one of the large hydropower projects recently constructed on mainstream of Dadu River. The project is mainly composed of concrete double-arch dam, underground plant on the left bank and flood discharge system on the right bank. Maximum height of this dam is 210m. The normal water level is 1130m and the installed capacity is 2,600MW.

The project is located at the north end of the Sichuan-Yunnan north-south direction fault, which is the intersection of many south-north, north-west and north-east direction faults. It has complex regional tectonic settings and poor tectonic stability. According to evaluation of seismic safety, the tectonic stability is controlled by Moxi fault 4.5km away from the site. Moxi fault is a regional left-lateral strike-slip active fault, which an important seismogenic fault. It was identified a potential 8.5-magnitude focal area. Further studies have showed that its activity decreases from north to south, especially weak in south of Wandong. Its activity after late Pleistocene is weak. 1260 earthquakes have been recorded within 25km range of the dam site. These earthquakes mainly took place in Detuo fault, Moxi fault, Shimian fault and the regions to the west of the damsite. They were zonal-distributed and converged in main faults. Seismic activities to the east of the damsite were more moderate, indicating that the seismic activities were significantly controlled by the fault structures on the western boundary of the project area.

Evaluation Report on Seismic Safety of Dagangshan Hydropower Station on Dadu River was jointly prepared by Geological Institute of China Earthquake Administration, Geophysics Institute of China Earthquake Administration and Institute of Earthquake Enginneering of Sichuan Earthquake Administration in October, 2004. It was checked by National Earthquake Safety Evaluation Committee and approved by China Earthquake Administration. The report showed that basic seismic intensity in Dagangshan site is Ⅷ. In terms of bedrock horizontal peak acceleration, 10% possibility of exceedance in 50 years is 251.7g and 2% possibility of exceedance in 100 years is 557.5g.

The faults in the dam site are in small scale with short extension. Most of them develop along the dikes and form "dike fault". It is the result of stresses adjustment of different rock in granite tectonic deformation. Tectonic activity researches had been done for many times and in-situ discussions and studies had been carried out by leading seismic experts in China. A conclusion was achieved. Compared with regional active faults 4~4.5km away (Moxi faults and Dadu River faults), the small faults in damsite area have obvious differences in nature from the perspective of tectonic status, occurrence characteristics and scale. In addition, dating results show that they have no activity since late Pleistocene. These factors indicate that those small faults do not genetically relate to Moxi fault 4.5km away and will not generate surface rapture when the Moxi fault is active. There is no active fault in dam site area.

This project has been approved by the state and is under construction. Relevant seismic resistance measures will be implemented gradually with the progress of the project.

7.3.1.3 Jilintai Hydropower Station

Jilintai I Hydropower Station is located in the middle section of Jilintai Gorge and in the middle reach of Kashi River. The hydropower station is composed of concrete face rockfill dam, power tunnel and ground plant on the left bank, and the spillway system on the left bank. The maximum height of the barrage is 157m. The total installed capacity of power plant is 500MW.

The project is located in Kashi River basin with Boluohuoluo Mountain to its north, Alawule Mountain to its south, Habi'erga Mountain to its east and Yili basin to its west. The project lies in Jilintai fault block in the middle of Kashi River which located at western Tianshan latitudinal tectonics zone. The block constitutes a series of east-west folded faults. To its east is north-west fault tectonic zone. To its west is NNE tectonic zone. The east-west tectonic zone is composed of Boluohuoluo compound anticline, Awulale anticline, Gongnaisi depression, Tekemin compound anticline and Zhaosu depression. They are all fault bounded. Main east-west faults:

(1) Boluohuoluo fault develops along the northern slope of Tianshan Mountain. It is formed by three large cross sections. The fault offsets Q_1 stratum about 30m. A 6.5-magnitude earthquake occurred in 1958 in the west section, a 6.3-

magnitude earthquake occurred in 1962 in middle section, and a 6.5-magnitude earthquake occurred in 1955 in east section.

(2) Zhaikoufala fault (South Boluohuoluo fault): it is in the depression boundary between Boluohuoluo hill and Kashi River, 7km away to the north of the dam site. It has a 50 – 100m compressive rupture zone. The fault offsets the second and the third terrace. Granite intrusion is found along the fault. Hot springs are exposed. New tectonic activities are intense. In 1812, the 8-magnitude Nileke earthquake occurred in faults intersection between east extension and northwest rupture of Zhaikoufala fault. Earthquake deformation zone was found in secondary fault to the south of epicenter.

(3) Fault in northern slope of Awulale Mountain: it is the dividing line of Kashi depression and Alawule uplift. The relative height difference is 500 – 1,000m and distributed in east-west direction. It starts from Wulangdaban in the west, extending eastward through Suozimute. The fault plane dips to the south and dip angle is 45°. It offsets the Quaternary strata. It is a recent active fault.

(4) Fault in southern slope of Mt Awulale: this fault is large, serving as the dividing line between Awulale compound anticline and Gongnaisi depression. The neotectonic is intensive, by which the Paleozoic strata were overthrust onto the Q_3 pebble layer. The eastern section went through a 7.25 magnitude earthquake in 1944. The western section (in Kazakhstan) had frequent earthquakes in history, and two 8-magnitude earthquakes occurred in 1889 and 1911.

(5) Qiakebu fault and Mt Tekemin fault are in the southern site. They are modern active boundary faults of tectonic zones.

NW tectonic zone mainly distributes 50km away to the east of the site. It is dozens of kilometers in width and hundreds of kilometers in length, diagonally crossing the EW tectonic zone. Its right-lateral offset is about 0.5 – 1km. but several offsets are about 2km. The right-lateral rate is up to 2.0 – 6.7mm/a. Obvious activities have been found since late Pleistocene. A number of 6.0 – 6.5 magnitude earthquakes has occurred along this zone. In 1812, the 8-magitude Nileke earthquake was occurred at the intersection between Zhaikulafa fault and this one.

The NNE tectonic zone is mainly exposed 30km away from the project. It is to the west of Nileke River and to the east of Yamadu-Mazhaer. This fault is a small-scale one. Left-lateral offset F_1 and F_2 is about 2km. The vertical offset in N_2 basal conglomerate layer is about 80m. In Yamadu where this fault intersects with the EW fault occured a 6.5-magnitude earthquake.

The major impact of seismic activity in project area is from the influence of peripheral strong earthquakes. Since 1716, 17 earthquakes above 6-magnitude have occurred in peripheral area.

In 1999, Xinjiang Earthquake Administration completed "Supplementary Opinions on Seismic Hazard Analysis of Jilintai Hydropower Station in Nileke County, Xinjiang". Further researches showed that, the ground motion horizontal peak acceleration of bedrock is 188.92g for 10% possibility of exceedance in 50 years and 461.97g for 2% possibility of exceedance in 100 years.

Jilintai project is in Jilintai fault block which is constituted by Carboniferous volcanic rocks. F_1 and F_2 faults are its south and north boundaries. Intense tectonic activities occurred in Pliocene with big uplift. It develops NW and NE faults inside. The F_{32} fault is the larger one.

The F_1 fault is 1.5 – 2km to the north of dam site. The fault offsets N_2 strata in Jiasibulake channel. But the trial trench cross Motuo channel reavels that the F_1 fault doesn't affect Q_3 loess layer. In Sep. 1993, Geological Institute of China Earthquake Administration applied TL and ESR methods to test F_1 and other faults. The test results showed that the gouge age is about 196,000±15,550 years and 176,100±52,800 years. The last intense activity of F_1 is in late Pleistocene. The age of un-deformed calcite vein is about 90,000 years. The ideal upper time limit for the F_1 fault last intense activity shall be about 90,000 years.

The F_{32} fault crosses the dam area diagonally and passes through riverbed and toe slab of the dam. It is NW in direction and is the largest fault in this area. Gauge samples of two small faults F_{300} and F_{279} obtained in PD1 adit were tested by TL and ESR methods. The result showed that the latest activity is 128,100±10,200 and 167,900±50,700 years ago. The F_{32} fault and two small faults mentioned above were revealed in PD1. Comparative analysis by scanning electron microscopy and interference microscope methods showed that their activities were in the same period-late Pliocene to Pleistocene.

The F_2 fault is 1.2km away from the south of the dam site. It is a depression band. F_2 fault does not offset the Kashi River gravel strata in the third terrace. C^{14} sample from gravel strata is 320,000-year-old deposit in late Q_3. Un-deformed gypsum sample dating is 372,000±112,000 years (ESR method), which means F_2 fault is inactive in 370,000 years. We can conclude from the relationship between F_1, F_2 and F_{32} and planation surface analysis of Jilintai fault block that the activities of F_1, F_2 and F_{32} were synchronized. Main activities of F_2 occurred in late Pliocene. It has been inactive since late Pleistocene. The block is relatively stable.

The hydraulic structures, including the dam, are not cut through by active fault. The slope of this area is not very high so that earthquake-induced secondary disasters are not serious. The cut slopes have been reinforced. The project is completed in 2005 and under normal operation now.

7.3.1.4 Zipingpu Water Conservancy Project

Zipingpu Hydraulic Project is located in the upper reaches of Minjiang River, the largest water resource regulation project before Minjiang River enters into Chengdu Plain. The project is composed of concrete face rockfill dam, plant at the right bank and spillway system at the right bank. The maximum height of the dam is 156m.

The site of the project is located in the intersection of Longmenshan mountain central fault (Yingxiu-Beichuan fault) and mountain-front fault (Anxian-Guanxian fault). The dam is 8km away from Yingxiu-Beichuan fault and 5km away from Anxian-Guanxian fault. Longmenshan fault zone is formed by a series of nappe structures including two faults mentioned above. It is a famous seismogenic zone in western China. The largest earthquake recorded is no more than 6.5 magnitudes. After 20-year researches and occasional controversies, the seismic intensity of Zipingpu site was recognized be Ⅷ and the first class fortification was applied in the engineering design. On 12 May, 2008, an 8.1-magnitude earthquake occurred in Wenchan. The epicenter is located in Yingxiu – beichuan fault whose activities were improved by the earthquake. The fault upper walls near Yingxiu Town and Bajiaomiao were uplifted 2.5m and 4.5m respectively. However, Anixan – Guanxian fault did not have any activity. The straight distance between the epicenter and the dam is only 17km. Researches after earthquake showed that no coseismic dislocation was found within project area. It is correct to determine that the faults in project area are inactive.

In Wenchuan earthquake, the actual seismic intensity influence undergone by Zipingpu project is Ⅺ which is far more than the original degree of seismic fortification. But the project was not catastrophically affected and started to generate power soon after the earthquake. The reservoir also played an important role as water lifeline during disaster relief. Although some ancillary buildings in the project were damaged, the main buildings had been restored and put into normal use, surviving the flood season that year. This example fully illustrated that the dam project is very capable of withstand seismic shaking effect. The safety is guaranteed to building large dam in regions with high seismic intensity. But the selection of project site must avoid locating above active fault. It is hard to imagine what would happen if the dam spans over Yingxiu-Beichuan fault and undergoes several meters displacement. Therefore, it is an important principle to observe that dam site selection should avoid the affect of active fault. Seismic damage survey showed that all dam (sluice gate) site and plant site were not damaged by co-seismic dislocation of the active fault. They were kept away from the damages caused by active fault and branch fault tectonically associated with the active fault. It proves that the avoidance principle mentioned above is right and able to stand the test of history and catastrophic earthquake.

After the earthquake, most of the artificially-processed slopes in hub area, including Zipingpu project, have not been severely damaged other than some cracks and deformations in certain section. But rolling stones, landslide and collapse are prone to occur at artificial slopes including those outside the boundary of artificial slope, or unprocessed slopes like reservoir bank near the dam, inlet and outlet, power house and switching station. The falling rocks bury outlet and inlet, hit the power houses, waterlines and switching stations to cause damages. At the same time, it may result in extensive surface loose of the mountain, critical slope and unstable slope. These phenomena tell us that reinforced artificial slope with loose rocks getting excavated can effectively guarantees seismic stability of the slopes. The supporting measures of the slope are with strong seismic reliability. Meanwhile, it reminds us more attention shall be paid to secondary disasters potentially existing in hub area. The exploration shall be extended and researches shall be further improved. Large dam project is not only possibly to be constructed in areas with high seismic intensity but also be safe as long as the selection of site avoids the influence of active faults and major geological hazards, the potential geohazards in project area are treated thoroughly and effective prevention measures are taken for areas outside the excavation boundary of the project. On April 14, 2010, a 7.1-magnitude earthquake occurred in Yushu County, Yushu State of Qinghai province. The researches afterwards over damages of hydropower projects have showed that the recognitions above are correct.

7.3.2 Geological Evaluation and Practices of Constructing Dam Projects in High-stress Area

High stress usually causes stability problems of tunnel surrounding rock. In recent years, dam foundation projects in high mountain canyon area in western China are disturbed by high stress. In the following we will discuss the formation mechanism, damage characteristics, hazard degree of dam foundation rocks under high stress, as well as how to avoid or deal with these damages. The engineering practice of Xiaowan Hydropower Station and Laxiwa Hydropower station will be discussed.

7.3.2.1 Xiaowan Hydropower Station

The attitude of rock in the project region of Xiaowan Hydropower Station is N75°– 80°W, NE∠75°– 80°. The axis of the arch dam and the rock strike are nearly orthogonal. The rocks on the foundation surface of the arch dam are: biotite granite gneiss in the river bed, hornblende plagioclase gneiss (100 - 120m thickness) on the dam abutments of both banks, biotite granite gneiss on the top of the right abutment. These two types of gneisses are intercalated with lamellate and lenticular schist generally with the thickness of 0.1 - 0.4m.

Only fault F_{11} on the foundation surface belongs to Level Ⅲ structural plane, with the overall attitude of N85°- 90°W, NE∠75°- 90°, and the width of fault fracture rock zone of about 4m. Affected by F_{11}, the rock mass in the upstream ranging from 10m to 12m are more broken, develop with multiple bedding extrusion surfaces and secondary structural planes.

The small faults with many extrusion surfaces belong to Level Ⅳ fracture structural plane with steep dip angle. The joints belong to Level Ⅴ structural plane.

The dam foundation has 6 larger alteration belts. Its trend is nearly SN and its width varies from 2m to 10m. There is E_8 alteration belt on the left bank, $E_4 + E_5$, E_1, E_9, on the right bank, and E_{10} on the river bed.

The rock mass on the foundation surface belongs to weak pervious zone with permeability rate of $q = 10 - 1$Lu. The bedrock which buried depth below river bed 40 - 50m and inside the two banks (at elevation of 1,130m) about 130m belongs to weak pervious zone, under which is little pervious zone ($q < 1$Lu).

The direction of measured maximum principal stress σ'_1 inclines along the slope surface, which in the shallow slope it is parallel to bank slope and becomes steeper inside. At the horizontal buried depth of 50 - 130m of both banks (with elevation ranging from 1,100m to 1,170m), $\sigma'_1 = 8 - 17$MPa. The direction of the maximum principal stress σ'_1 in shallow river bed changes randomly, and σ'_1 generally is 22 - 35MPa, when local stress concentrates, $\sigma'_1 = 44 - 57$MPa.

The rock mass on the foundation surface are mainly slightly weathered rocks. Only partial parts in dam toe have moderately weathered rocks and unloading rocks. The rock mass on the dam foundation are mainly Class Ⅱ rock mass. According to the percentage of the projected area, Class Ⅱ rocks account for 89.3%, Class Ⅲa rocks for 3.0%, Class Ⅲb rocks for 5.4%, Class Ⅳa rocks for 0.3%, Class Ⅳb rocks for 1.8%, and Class Ⅳb + Ⅳc rocks for 0.2%. Class Ⅲa, Ⅲb and Ⅳa rock mass are mainly distributed at the high elevation of dam foundation and the dam toe. Overall rock mass quality of the dam foundation is good. As the bed rock has high geostatic stress, after the excavation of the dam foundation, the two banks' abutments, especially dam foundation rocks at low-elevation, have shown strong relaxation and become a major engineering geological problem. The forms of rock relaxation are: ①The abutment parts of two banks have had fissure dislocation, opening and expansion. ②The middle and high-elevation dam foundation rocks show an "onion skin" phenomenon, which thin film rock slices are distributed like imbrication, with the crack extending into the slope by general 20 - 30cm. ③Low elevation and the rock mass of dam foundation in the river bed show a "cracked plate" phenomenon, which the rock beam of excavation face at smooth-going place is relatively thin, with the thickness of generally 3 - 20cm; the rock beam is medium thick in upright steep slopes and tunnel entrances. ④Shallow parts of the dam foundation in the river bed bring about differential rebound and creep along new or original horizontal unloading fissures. ⑤Rock burst occurs at some low-elevation rocks. Relaxation of the above rock excavation is largely proportionate to the depth of the excavation. Excavating deeper, the more intensive the relaxation is. So the relaxation at the dam heel is relatively more intensive compared to that at the dam toe. Pre-anchoring or advanced anchoring measures play a certain role in controlling rock relaxation and deformation.

A large number of acoustic tests and observations have been carried out in the rock mass before and after the excavation. The depth of the burst impact on the excavation of most sections of the dam is less than 1.0m. The decay rate of single-hole wave velocity is generally less than 10%, and some acoustic wave holes leak water. The depth of unloading is increasing with time. The biggest increase appears generally 60 to 90 days after the burst, and then slows down 90 to 180 days after the burst and levels off after 180 days.

In order to properly evaluate the impact of relaxation rock mass at dam foundation to take appropriate measures to deal with it, the relaxation rock mass at dam foundation is divided into three zones based on its characteristics of rock mass at foundation surface: relaxation zone, transition zone and normal zone. The zone characteristics are shown in Table 7.3 - 1.

Site deformation modulus tests after dam foundation excavation have showed that the measured deformation modulus of Class Ⅱ and Ⅲa which cover the overwhelming area of the dam foundation sharply reduces due to the unloading. The suggested deformation modulus of Class Ⅱ and Ⅲa of rock mass are 16 - 22GPa and 12 - 16GPa respectively, and after excavation, the block structure of biotite granite gneiss is 13.77GPa and sub-block structure is 10.66GPa.

After the relaxation of rock mass at dam foundation, although the strong relaxation rock mass is removed, excessive excavation may bring new relaxation. Thus cleaning foundation are conducted before pouring concrete, and after foundation cleaning concrete should be poured immediately to load weight, and then conduct consolidation grouting. With the pouring of the dam, the unloading rebound of rock mass at dam foundation is suppressed and the stress is partially restored. Deformation monitoring has showed that: when the concrete in dam section on river bed is 10m thick and the dam section on slope is around 20 – 30m thick, continued rebound of rock mass at dam foundation can be effectively inhibited; when the concrete of the dam is 30 – 40m thick, the rock mass shows a compression trend. Prior to consolidation grouting, the velocity of longitudinal wave of the acoustic wave test hole in the depth of 0 – 2m has reached approximately 5,000m/s. Before the river bed dam section is covered by concrete, the wave velocity is high below the depth of 10m with pie core distribution. After pouring the concrete, pie core appears at 5 – 10m below bedrock surface, indicating that the dam foundation rock mass stress is recovered to some extent. Consolidation grouting of dam foundation and the corresponding inspection, test and deformation monitoring show that after grouting the average velocity of longitudinal wave V_p are: at 0 – 2m section below the dam foundation, $V_p = 5,070 - 5,340$m/s; at 2 – 5m section, $V_p = 5,200 - 5,490$m/s; below 5m, $V_p = 5,230 - 5,540$m/s. The deformation characteristics are basically similar to the intact rock mass, indicating that the deformation characteristics of dam foundation have a considerable degree of recovery.

Table 7.3 – 1 **Characteristics of Relaxation Rock Zoning**

Zoning		Geological characteristics	Reference index			
			Relaxation strength (10^{-3})	Average Wave velocity V_p (km/s)	V_p Decay rate (%)	Permeability rate q (Lu)
Relaxation zone	Strong relaxation zone	The rock mass is basically detached from the parent rock with an obvious displacement. Each part varies hugely in thickness. It has been removed during foundation cleaning				
	Relaxation zone	The structure of rock mass is clearly relaxed and joint fissure opens, which are accompanied by dislocation. The degree of relaxation intensifies over time and the fissures tend to link up together. The extent and depth of rock relaxation are different in space	>1	<4.5	10 to 20, up to 30 in individual cases	>10, some> 100
Transition zone		Open fissures are few and not concentrated, most of them are slightly open, and some hidden fissures appear	1 – 0.1	4.5 – 5.2	<10	<3
Normal zone		Joint fissures are mainly closed or hidden, and occasionally slightly open	<0.1	5.2 – 5.5	<5	<1

Throughout the unloading relaxation characteristics of rock mass at dam foundation and the analysis of basic treatment effect of the Xiaowan Arch Dam, you can get the following understanding: The main adverse effect of relaxation rock mass after excavation is that unloading fissure zone is formed in the shallow surface of rock mass at dam foundation, reducing the deformation modulus and shear strength of rock mass. The development of unloading fissures diminishes with depth with no clear boundaries. According to the measured acoustic data, when the elevation exceeds 975m, the relaxation zone is generally within 5m; when the elevation is less than 975m, the relaxation zone is mainly within the depth of 2m. After the pouring of dam concrete and the consolidation grouting, the acoustic velocity is above 5,000m/s, its deformation modulus value should not be smaller than the recommended value according to the rock mass classification at dam foundation; the unloading fissures in the shallow surface of dam foundation are mainly new developed fissures, the fracture surface is generally flat, straight and rough. After compaction and consolidation grouting, the shear strength along the direction of fissure may have a comparison with that of cemented surface of concrete and moderately-slightly weathered rocks. The average value of 4 on-site concrete/bed rock shear test results is: peak strength $f = 1.68$, $c = 1.8$MPa; the minimum residual strength $f = 0.966$, $c = 0$MPa. After treatment, the shear strength of unloading rock mass should be lower than the peak strength, but should be significantly greater than the minimum residual strength. Shallow relaxation zone under the dam foundation is the control layer, since friction strength accounts for a relatively large proportion of shear strength due to the high dam and the shear strength has safety margin. After a comprehensive treatment, the dam foundation may meet the foundation requirements.

By the end of 2010, the dam water level reached about 1,210m. There was no abnormal change in deformation, stress and seepage of dam foundation in the water storing process of the reservoir. The treatment of unloading relaxation zone of dam foundation has been initially stood the test by impounding water. In the subsequent water storing process, it is necessary to strengthen the observation and analysis on dam foundation.

7.3.2.2 Laxiwa Hydropower Station

Laxiwa Hydropower Station is the second cascade power station adjacent to Longyangxia Hydropower Station at the upstream of Yellow River. It is located in the main stream of Yellow River at the border between Guide County and Guinan County of Qinghai Province. The hydropower project comprises concrete bi-arch dam, plunge pool behind the dam and auxiliary weir, the underground water power generation system on the right bank, etc. The maximum height of bi-arch dam is 250.0m, the elevation of dam crest is 2,460m, the normal water level is 2,452m and the installed plant capacity is 4,200MW (6×700MW).

The normal range of Laxiwa Hydropower Station crustal stress σ_1 is 20–23MPa, in some parts it increases to nearly 30MPa which is a high-stress area. The dam situates in the narrow valley, and the natural stress field is the local stress field formed by the superposition of regional stress filed with the slope self-weight stress field. The dominant direction of maximum principal stress is NE18.67°.

The excavation of abutment on both banks results in unloading and relaxation of the rock mass. The macroscopic features are:

(1) Relaxation and cracking of the original structural fissure. The steep-dipping NWW fissures in the left bank and NEE fissures in the right bank are nearly parallel to or intersected with the spandrel groove slopes at small angle. The original high stress and the overlying loading effect before excavation make these fissures slightly open or close. After the excavation of foundation surface, the rock mass is rebounded and deformed, and all these fissures are open or appear.

(2) Local surface rocks slightly detach, crack and loosen, which appears mainly in the low-elevation site near the river bed, with the slightly detaching thickness of less than 20cm, and the depth of rock cracking, loosening are generally within 3m.

(3) The "onion skin" phenomenon. The thin film rock slices cut by unloading fissure is distributed like imbrication on the intact or blocky slightly weathered and fresh rock surface, with the thickness of thin film rock slice of 0.5–5cm; they are generally newly developed cracks without the lag due to its damage; fissures are mostly closed–open or slightly open, with the maximum opening width no more than 5mm, and some parts filled with rock powder. The cracks do not extend far into the slope, generally within 20–30cm. They are mainly distributed on both sides of river bed foundation surface within a certain elevation, with the upstream side of deeper excavation develops more than the downstream side. Generally, it is more noticeable at low elevation than at high elevation.

The rock mass is tested with the drilling acoustic wave, and the relaxation characteristics of spandrel groove slope on both sides are listed in Table 7.3–2. From the table it can be seen that the thickness of the relaxation zone on the left bank is generally less than 2.6m, it's up to 6.4m at local parts; it's generally less than 3m on the right bank, and up to 7m at local parts.

Table 7.3–2 List of Relaxation Characteristics of Spandrel Groove Slope on Both Sides

Part	Elevation (m)	Relaxation thickness (m)	Wave velocity of relaxation zone (m/s)		Original rock wave velocity (m/s)		Degree of relaxation (%)
			Wave velocity range	Average wave velocity	Wave velocity range	Average wave velocity	
Left bank spandrel groove	2,460–2,410	1.8–2.6	1,510–2,930	2,000	2,780–3,330	3,000	26.3–39.2
	2,410–2,340	1.2–2.0	2,700–5,190	3,500	4,480–5,280	4,700	9.1–23.4
	2,340–2,260	1.6–2.5 Local 6.4	1,710–4,620	3,000	4,100–4,970	4,800	39.2–46.0
	2,260–2,240	1.0–2.4 El. 2240: 3.5	3,570–4,500	4,000	3,950–5,450	4,900	9.2–22
Right bank spandrel groove	2,460–2,370	1.4–1.8	3,080–5,030	4,000	3,160–5,770	4,800	5.2–16.8
	2,370–2,340	2.4–4.8	1,910–5,200	3,300	2,830–5,580	4,300	19.8–27.3
	2,340–2,280	1.4	3,920–4,950	4,000	4,040–5,720	4,900	8.2–19.7
	2,280–2,265	1.8–2.4	1,460–4,880	3,500	1,980–6,000	4,700	24.2–28.8
	2,265–2,240	3.0–7.0	1,550–4,160	2,500	1,830–5,220	4,000	23.5–49.6

Note: Relaxation degree refers to the damage and relaxation of rocks at relaxation zone. It is determined by the percentage of the average longitudinal wave velocity in the rock mass of relaxation zone to the average wave velocity in the original rock (not relaxed relatively).

Thickness of the vast majority of relaxation zone at rock mass of river bed dam foundation is less than 2.2m, with the maximum of 4m; the average velocity of longitudinal wave of rock mass in relaxation zone is less than 3,500m/s, with 2,000 - 3,000m/s prevailing. Borehole camera observation is conducted in the dam foundation of river bed. It is revealed that the fissures generally close or open by about 0.1 - 0.3cm, with most distributed at the elevation of over 2,203m, few at the elevation of over 2,195m.

The maximum relaxation of rock mass at dam foundation is 49.6%, and the minimum is 5.2%; the average relaxation is 28.2% on the left bank and 16.6% on the right bank, and the average relaxation on both banks is 22.4%. Except that the average wave velocity of the original rocks at the elevation of over 2,410m on the left bank is 3,000m/s which is relatively low, the average wave velocity of remaining parts of the original rocks are above 4,200m/s, and wave velocity of the original rocks on both banks are close. The relaxation zone of spandrel groove slope on both banks is dominated by slight-medium relaxation rock mass, without strong relaxation rock mass.

For relaxation time of dam foundation rock mass, on the third day after the excavation, the decay rate of wave velocity of rock mass has reached 60% of that on the 45th day (total decay rate), indicating the stress adjustment is completed mainly in 3 days after the excavation; on the 9th day after the dam foundation excavation, about 80% of the total decay rate is completed, indicating that most of the rock stress excavation adjustments have been completed within 9 days after the blasting and excavation.

Similar to Xiaowan Project, acoustic wave is also used to detect the relaxation on the foundation surface of spandrel groove on both abutments of Laxiwa Hydropower Station. Drilling depth is generally 10m. In addition to the general obvious decrease of 1 - 3m waves and obvious relaxation of rock mass, the rock mass in the slope is slightly affected by the blasting and stress decrease. The wave velocity and integrity coefficient of the rock mass is worse compared to that in the deeper parts. The affected depth is mainly 6 - 7m with the deeper one being 9 - 10m. Therefore, the rock mass of spandrel groove from the slope surface to inner slope is divided into three zones: relaxation zone which is 0 - 3m deep; transition zone which is 3 - 10m deep; normal rock zone which is more than 10m deep.

The wave velocity of foundation surface at dam foundation and the wave retest results of residual exploration tunnels in abutment after excavation are used to calculate the deformation modulus of relaxation zone and transition zone, and then deformation modulus is used to extrapolate the parameters of shear strength.

The calculation results on the left bank have showed that: The values of relaxation zone E_0 are 3.2 - 14.1GPa, but most are 4 - 9GPa, with the average deformation modulus of 6.8GPa. The E_0 of relaxation zone does not have an obvious correlation with the variation of elevation, but in general it is low at high elevation while it is slightly high at low elevation, which corresponds with the basic characteristics of rock mass. The distribution range of deformation modulus in transition zone is 11.3 - 31GPa, with the average value of 17.8GPa. Compared with the previous recommended value of geology, it is evident that low-elevation excavation has a greater impact on the parameters. The calculation value of deformation modulus in normal rock zone is mostly 20 - 31GPa, with the average of 25.1GPa. For elevation of over 2,400m, it is 14 - 22.1GPa, with the average of 18GPa. It is generally in line with the situation that the deformation modulus of the original rocks in the deep parts at the elevation under 2,400m is greater than 25GPa, while above 2,400m is 15GPa. The parameters of sheer resistence are basically consistent with the previous recommended value.

The calculation method on the right bank is the same as that on the left bank. The results show that: The range of deformation modulus value is 1.8 - 20.1GPa, but most of them are 4 - 10GPa. At the elevation of 2,270 - 2,220m, the fissures are more than other parts. In addition, two times of blasting and excavation pose severe damages to the rock mass at the depth of 6 - 7m. The average value of deformation modulus is only 2.9 - 5.9GPa. The secondary clearing and digging are carried out and concrete is backfilled. The deformation modulus of rock mass in transition zone is obviously affected by the excavation, and the average deformation modulus at the elevation of 2,460 - 2,400m is about 18GPa, slightly higher than the previous recommended value of 15GPa (the recommended value of elevation above 2,430m is 10GPa). The average value of deformation modulus at the elevation of 2,400 - 2,270m is 19.6 - 23GPa, consistent with the previous recommended value of 20 - 23GPa. The average value of deformation modulus at the elevation of 2,270 - 2,240m is only 10.1GPa, more than 50% lower than the originally recommended value of 23GPa and the design adopted value of 20GPa on average. This area has undergone secondary excavation. The average value of deformation modulus at the elevation of 2,240 - 2,220m is 16.7GPa. Although it decreases compared with the design adopted value of 20GPa, the decrease is not more than 20%, it remains good Level II rock, no cleaning or digging is executed. The average value of deformation modulus of rock mass in deeper parts is 21 - 26GPa. The total average is about 23.7GPa, also close to the value of greater than 25GPa which determined before excavation, thus it has small impact on the dam deformation.

Calculation and check of deformation modulus of dam foundation in river bed: by using the six relations between acoustic wave velocity and deformation modulus proposed in the preliminary stage and construction period, the deformation modulus is calculated based on the acoustic wave of foundation surface. The E_0 value of slightly weathered – fresh Level I rock mass is greater than 19GPa, and the average value of lower limit is 27GPa; the E_0 value of Level II rock is basically greater than 13GPa, and the average value of lower limit is 16GPa; the E_0 value of Level III 1 rock mass that is weakly weathered and with relative intact lower part is greater than 8GPa, and the average value of lower limit is 10GPa; the E_0 of Level III 2 rock mass is greater than 3GPa and the average value is 4GPa. It is almost the same as the previously recommended value.

The excavation depth of dam foundation of Laxiwa Hydropower Station is relatively deep. The excavation depth of downstream and upstream corners of abutment on the left bank is generally about 30m and 60m respectively, while the excavation depth of downstream and upstream corners of abutment on the right bank is generally about 36m and 80m respectively, and the digging depth of bed rock of river bed is 5 – 18m. Since the dam site is located in high-stress areas, because of the unloading rock mass and release of curstal stress after excavation, relaxation is caused to the rock mass on the abutments of both banks and rock mass at the foundation surface of river bed. From the relaxation characteristics of the rock mass, the relaxation zone is mainly characterized by the phenomena that the fissure of original structure opens, the local surface of rock mass is slightly exfoliated, cracked and loosened, and local "onion skin" appears. The weathered degree and structure of the rock stay unchanged. The major adverse effect is the decrease of deformation modulus and shear strength of the rock mass. The rock mass in transition zone shows no macroscopic relaxation, but mainly the overall unloading rebound of rock mass and the decline of acoustic wave value; the acoustic wave value of normal rock mass zone is not evidently reduced. The thickness of the vast majority of relaxation zones on the dam foundation of river bed is less than 2.2m, with the maximum thickness of 4m, the macroscopic characteristics of relaxation of rock mass is substantially the same as the abutment on both banks, with the weathered degree and structure of the rock mass staying unchanged. Judging macroscopically, the rock mass is at Level I or II, and based on the acoustic wave value the local rock mass is judged as Level III or IV where the acoustic wave value declines significantly. The consolidation grouting of rock mass at dam foundation of river bed produces favourable effects. After grouting, the average acoustic wave value is above 4,980m/s. The rock mass is at Level II or above.

The dam constructionhas been completed, and the first batch of power unit has started generation in May 2009. The observation results have showed that no abnormalities have been found in the working behavior of dam foundation.

7.3.3 *Complex Foundation Treatment*

The treatment for the defects of dam foundation mainly involves two types of problems of inadequate bearing capacity, insufficient deformation resistance, leakage and seepage deformation, etc., which are caused by inadequate strength of rock mass, broken rocks, aggravated weathering, corrosion etc., as well as the sliding stability of dam foundation controlled by rock discontinuities.

7.3.3.1 **Xiangjiaba Hydropower Station**

Situated at Sichuan-Yunnan border of downstream Jinsha River, the Xiangjiaba Hydropower Station is the last cascade of downstream Jinsha River. The project comprises concrete gravity dam, power house at dam toe on the river bed and underground power house on the right bank, flood discharge and energy dissipation systems on the right bank, and other buildings. The maximum height of the river dam is 162.00m.

Xiangjiaba dam site has complex bedrock lithology and lithofacies. It is Triassic Xujiahe Formation medium-huge thick bed sandstone intercalated pelitic siltstone, silty mudstone and coal lines. Influenced by lithofacies change and the Limeiwan Flexure, the rock strata attitude changes greatly, joints and fissures of the rock mass, gently inclined soft interlayers and intetlayer shear zones are relatively developed. Moreover, influenced by cross bedding, the thicknesses of soft interlayer and intetlayer shear zones vary significantly, and the engineering geological conditions are complicated.

Exploration incidates mainly T_3^{2-5}、T_3^{2-3} soft rock zones and 8 weak intercalated layers which are distributed in T_3^{2-6}、T_3^{2-4} rock strata are existed in the dam foundation. These soft discontinuities not only exert influence on the compressive deformation and seepage deformation to the dam, but also form the crucial boundaries for anti-sliding stability of the dam foundation.

The layered discontinuities of dam foundation gently inclined toward the downstream are clear, including interlayer soft rock strata, weak intercalated layers and bedding planes. The development of discontinuities gently inclined to upstream and its combination with the layered discontinuities become the controlling factors for anti-sliding stability. According to the statistics about the joint development characteristics on both banks, reef in the river bed and adit at the river bottom, the number of low angle joints in the aforesaid locations accounts for 5.8%, 3.78% and 24.83% of total joints respectively, among which the number of those dipping towards upstream accounts for 21.8%, 16.7% and 67.9% of low angle discontinuities respectively, indicating that there are indeed some low angle fissures at dam foundation. Therefore, when analyzing the boundary of anti-sliding stability at the dam foundation, mulitple combination methods with the first rupture plane is taken into consideration, for example, the second rupture plane dominated by low angle fissures is directly cut out or extends like steps, etc.; during the analysis, since it's impossible to obtain the connection rate from the statistics of the actual parts at dam foundation, the different connection rates of 10%, 20%, 30%, 40% or even 50% are adopted to make sensitivity stability analysis; meanwhile, by taking into consideration that the steep fracture zone with big width is distributed at the downstream of the flood discharge dam section on the right bank, whose deformation modulus is low and rock stiffness is insufficient, it can be considered as a compressible space to analyze the adverse effect of anti-sliding stability.

The actual excavation of Xiangjiaba Project confirms not only the complexity of deformation at dam foundation and anti-sliding stability, and it also indicates that the severity of the problem is worse than expected. A large area of sandy rock fragments has been found on the left bank of the dam foundation and the degree of broken of dam foundation on the right bank is also more severe than expected. According to the actual conditions, such problems were treated by deep digging and replacement, which increased the amount of excavation and concrete work. From the perspective of engineering safety and reliability, this treatment is necessary.

7.3.3.2 Jinping I Hydropower Station

Jinping I Hydropower Station is an important backbone project in the hydropower development Yalong River. The dam site is located in both Mianning and Muli Counties of Liangshan Yi Autonomous Prefecture in Xichang, Sichuan Province. The project is composed of a concrete bi-arch dam, underground power house on the right bank and other buildings. The maximum height of arch dam is 305m, which is now the tallest in the world.

The bed rock of dam site mainly comprises the Triassic Zagunao metamorphic rocks and also a small amount of post-intrusive lamprophyre veins. The left bank of the project is a typical reverse slope, with the upper part being sandy slate and the lower part being marble. On the whole it is characterized by great depth of unloading, wide opening of unloading fissures and complex unloading types. There is generally no strong unloading on the river bed. The vertical depth of weak unloading is generally 20 – 40m and the corresponding base elevation is 1,560 – 1,580m. The deep unloading on the left bank (deep cracks) is a unique geological phenomenon on the left bank of project region of Jinping I Hydropower Station. The horizontal depth of deep unloading boudary in the marble is generally 150 – 200m. In general, the horizontal depth of deep unloading, the fissure density and opening width in unloading zone become smaller with the reduction of elevation; the horizontal depth of deep unloading boundary in the middle and upper parts of the sandy slate may be up to 200 – 300m, and the deep unloading overlapped with shallow unloading in the high elevation of bank slope. The occurrence of deep unloading problems leads to a serious argument about whether high concrete arch dam should be built in this dam site. After extremely complex argumentations, most experts come to share an identical understanding about deep unloading, and believe that on the basis of original structural plane and under the condition of high ground stress, the Yalong River Valley were rapidly incised downward, stress of the bank slopes releases and the unloading rock mass rebounds, the deep fissures are formed. When selecting dam axis, topography of the dam area and the distribution of the deep unloading are fully taken into account. The dam axis is moved toward upstream as much as possible to keep away from the concentrated region of deep unloading. Meanwhile, the comprehensive foundation treatment is conducted on the unloading relaxation rock mass on the left bank.

The exposed strata of the dam foundation below elevation of 1,820 – 1,850m on the left bank, river bed and right bank is the second section of Zagunao formation marble. The third section sandy slate is distributed above the elevation of 1,820 – 1,850m on the left bank. The lamprophyre vein strips spread to the direction of approximately N60°E on the left bank, the river bed and the right bank, with a general thickness of 2 – 5m.

On the left bank of the dam, besides the faults f_2, f_5, f_8 and f_{42-9} developed, there are many interlayer extrusion and shear belts developing in the sixth layer of the second section marble on the lower bank slope, while there are many minor faults developed in the third section sandy slate on the upper bank slope. Besides the faults f_{13}, f_{14} and f_{18} developed on the right bank of the dam, there are a few interlayer extrusion and shear belts developing in the sixth layer of the marble on the upper bank slope. Both bank slopes of the dam are high-steep. The lower bound of horizontal depth of the strong unloading zone on the right bank is 5 – 10m in general, and that of the weak unloading zone is 20 – 40m. The horizontal depth of the marble strong unloading zone on the left bank is 10 – 20m while that of the weak unloading zone is 50 – 80m in general, and the lower bound of horizontal depth of the deep unloading zone is 150 – 200m in general. The horizontal depth of the strong unloading zone of the sandy slate on the left bank is 50 – 90m while that of the weak unloading zone is 100 – 160m in general, and the lower bound of horizontal depth of the deep unloading zone is 200 – 330m in general, and in most areas the deep unloading overlaps with the shallow weak unloading zone. Bedrock of river bed does not have strong unloading zone generally and the vertical depth of the weak unloading zone is generally 20 – 40m under the bedrock surface.

Chapter 7 New Techonologies for Engineering Geology and Survey in Dam Construction

The rock mass of dam foundation in the river bed is fresh or slightly weathered and weak unloading, in which the interlayer extrusion and shear belts do not develop. Slightly relaxed, relatively high permeability rock mass and river bed stress-concentrated belts are mainly lied about 50 – 60m beneath the bedrock surface. The actual dam foundation excavation is successful without obvious unloading rebound problems.

The bedrock of the dam foundation on the right bank is marble, which is mainly fresh or slightly weathered grade II rock mass with relatively intact and uniform massive structure and high deformation resistance.

Truly complex is the left bank of Jinping dam foundation, which is divided into marble and sandy slate in accordance with lithology. Two aspects of the main treatment are as follows:

(1) Treatment of the rock mass of the dam foundation. As the unloading of sandy slate section of the dam foundation is strong and the unloading depth is deep, it is difficult to meet the requirements for the foundation of high arc dam. Combining with the treatment of fault on the left bank, a 155-meter-high concrete pedestal is set at the height of 1,730m to 1,885m of the dam crest to disperse the concentrated load of the abutment. Bedrock of the pedestal is mainly weakly to slightly weathered with partial fault belts weakly to intensely weathered, which is overall in the weak and deep unloading zones. There are grade III 1、III 2、IV 2 and V 1 rock mass, which can basically meet the requirements of pedestal foundation. In order to homogenize the foundation, an entire consolidation grouting to the foundation is needed.

(2) Consolidation of dam abutment resisting force. The resisting rock mass on the left bank makes the problem of the abutment rock mass deformation stability and stability against sliding extremely prominent because of the faults, the lamprophyre veins and the intense unloading effect. Thus besides setting a large concrete pedestal, it is necessary to effectively consolidate the resisting rock mass on he left bank to meet the requirements of the high arch dam deformation and anti-sliding. For this reason a complicate and comprehensive strengthening project is designed. The main measures include:

1) the entire consolidation grouting and local grouting of the resisting rock mass. The consolidation grouting is carried out in the range between El. 1,885m and El. 1,635m with the altitude difference of 250m. To ensure the grouting effect and the effective range of construction, five layers consolidation grouting galleries are set at the elevations of 1,885m, 1,829m, 1,785m, 1,730m and 1,670m respectively. All grouting works are accomplished in the above galleries, and the accumulated works quantity is approximately 500,000 linear meters.

2) Concrete replacement of fault f_5 and lamprophyre vein. Two layers concrete replacement adits are set at El. 1,730m and El. 1,670m, between which one impermeable curtain inclined shaft is set and three concrete replacement inclined shafts are set to replaced fault f_5 (f_8) with concrete grids. Three concrete replacement adits are set at El. 1,785m, 1,730m and 1,670m respectively; one impermeable curtain inclined shaft is set between El. 1,885m and El. 1,730m; three concrete replacement inclined shafts between the adits of El. 1,785m and 1,730m, and one inclined shafts between El. 1,730m and 1,670m are set to replace lamprophyre vein with concrete grid.

3) Concrete plugs on the left bank. Three concrete plugs are set at El. 1,829m, 1,785m and 1,730m respectively to transmit the arch thrust to the III 1 type rock mass beyond the lamprophyre vein. The dimension of the concrete plugs is 9.0m × 12.0m (width × height).

4) Curtain grouting on the left bank. Five layers impermeable curtain grouting adits are set on the left bank at El. 1,885m, 1,829m, 1,785m, 1,730m and 1,670m respectively. In order to fully seal the possible unloading loosening and rupture range, the curtain grouting adits are as deep as over 500 meters.

5) Drainage of dam foundation and resisting rock mass on the left bank. Four layers drainage adits are set in the resisting parts on the left bank, with three to six drainage adits at each layer along the elevation.

Through the analysis, it can be concluded that the arch dam abutment deformation and stability against sliding can meet the requirements after completing the consolidation on the left bank. The above treatment measures are still under construction at present, thus the final effects of the treatment will be tested after the impoundment of the dam.

7.3.3.3 Wudu Key Water Control Project

Wudu reservoir is located on the main stream of Fu jiang River in Jiangyou, Sichuan Province. It is a large (1) -scale Water conservancy project which gives priority to flood control and irrigation, combining with comprehensive utilization of power generation and industrial, daily and environmental water supply in urban and rural areas, etc. The dam is an RCC gravity dam, with a maximum height of 120.34m and a total capacity of 572 million m³. Its designed normal storage level is 658m with an installed capacity of 150 MW.

Faults $10f_2$, f_{101}, f_{114}, f_{115}, f_{127} and f_{31-1} of the dam foundation rock mass are the control sliding surfaces of the dam foundation's deep stability against sliding. Through the cores identification of $11^{\#}-20^{\#}$ drill holes in dam foundation at the river bed and 24 drill holes of downstream resisting rock mass, combined with the detailed panoramic photographs illustrating materials, the attitude of each long and big gentle inclined fissure is analyzed. The fractures in adjacent drill holes with nearly the same apparent dip are connected together. The analysis thereof confirms that in the different heights of riverbed dam foundation, among those gentle inclined fissures which may form the major sliding surfaces, there are 102 fissures inclined toward upstream and 60 fissures inclined toward downstream respectively.

Finally on the representative longitudinal section of each dam monolith, all the fissure slip surfaces and faults that inclined toward upstream and downstream are marked, different controlling sliding surfaces are formed. The amounts of controlling sliding surface combinations in $11^{\#}-19^{\#}$ dam monolith are 12, 6, 20, 27, 25, 19, 15, 10, and 20 respectively.

For various sliding modes in dam foundation, through survey, design and comprehensive analysis, faults f_{101}, f_{101-1}, f_{115} in $11^{\#}-17^{\#}$ dam monoliths receive partial replacement treatment; in $18^{\#}$ dam monolith the schemes of front cutoff trench, apron and dam toe concrete replacement are adopted, i. e. partial replacement of faults $10f_2$, f_{114}, f_{115}; $19^{\#}$ dam monolith receives large cutoff trench and pressing foot schemes, i. e. deep cutoff trench is set in the dam upstream heel and fault $10f_2$ is replaced, increasing the cutoff trench size, sandy pebble pressing foot behind the dam. The replacement lengths of all the rest dam monoliths are calculated and determined in accordance with the design requirements of stability against sliding. Water storage of this project is started on 29 December, 2010.

7.3.4 High Slope Treatment

Building water conservancy and hydropower project in the high canyon area involves a prominent problem of the high slope. Since the slope is often hundreds of meters high, the deformation and failure mechanism of the slope is complex and difficult to survey, and the adverse influence on the project, the treatment difficulty and the amount of treatment work are great.

Due to high relative elevation on the two banks, there are many high slope stability problems on the dam site of Jinping I Hydropower Project. "Deep crack" is the special geological phenomenon on the left bank of Jinping I Hydropower Project. It is a set of slope deep unloading fracture system formed based on the original structure surface, by the unloading and rifting of high-steep slope of Jinpin under the combined conditions of particular high stress environment and lithologic, accompanied by the valley incising process, strong release of slop stress.

The deep cracks (belts) developed on the left bank differ greatly in different lithology and different parts. From the perspective of lithology, most deep cracks developed in the hard metamorphic sandstone and marble while few in relatively weak slate; from the perspective of spatial distribution, it has the tendency that the developing depth of the deep cracks gradually increases from elevation about 1,660m to the upper valley slope, and becomes strong from the upstream to downstream.

The statistical analysis of deep fissure attitude have showed that there are two groups of dominant directions: ①N40°-70°E, SE∠50°-75°; ②N0°-30°E, SE∠50°-65°, which are basically the same as the dominant directions of the fractures, joints and fissures in the project region.

According to the opening width and fillings' characteristics, deep cracks (belts) are divided into four grades.

Grade I: Generally, it a single empty crack, the biggest opening width is ≥20cm without fillings or with a few rocks and rubbles. The cracks narrow and broaden alternatively along the strike and inclination, and generally most of them develop based on original small faults and have dozens of centimeters dislocation move. The extended length is relatively big, usually above one hundred meters.

Grade II: Generally, it is a relaxation belt with some width. The single biggest opening width is < 20cm, but ≥10cm. It is dense with a wide belt and most cracks are fresh without fillings, which has several to dozens of centimeters dislocation move. The extended length is relatively big, usually dozens of meters.

Grade III: Generally, it is a relaxation rock zone composed of several small cracks developed in parallel. The cumulative opening width is < 10cm but ≥3cm, part of which filled with rocks and rock debris and other parts of which are empty cracks without fillings. And it generally has no dislocation move. The extended length is ranging from several meters to dozens of meters.

Grade IV: It is a fissure relaxation belt. Fissures are densely developed while rock mass is relaxed without typical empty cracks. The cumulative opening width is < 3cm. The extended length is small, generally ranging from several meters to dozens of meters.

Through continuous research and the adjustment of dam axis, the deep cracks developed in the abutment and within the resistance part on the left bank of the project region are mainly small deep cracks of Grades Ⅲ and Ⅳ and the cracks are distributed randomly. Although the influence of the cracks in Grades Ⅰ and Ⅱ is basically avoided, the stability of the slope is influenced by the small cracks to some degree.

The upper part of project region on the left bank is sandy slate while the lower part is marble. Because of differences in lithology, rock mass strength and the original structure, the different lithological parts showing different deformation and failure mechanism and the degree of stability of the slope. On the sandy slate bank slope soft and hard rock strata are distributed alternatively with obvious toppling, ripping deformation. In the 6th layer of the marble the interlayer crushed bedding belts are developed, and form the bottom sliding surface that controls the stability of the bank slopes. Most thick massive marbles form the stable bank slopes because of its hardness and relatively integral structure.

The slope on the left bank is a reverse slope at macroscope, but the inclined steep cracks are well developed due to the epigenetic deformation and some slopes develop into inclined structure.

Slope on the left bank of Jinping Ⅰ Hydropower Project could be divided into three parts according to different deformation modes and boundary: toppling deformation slope of sandy slate with the cable-crane platform as the representative; potentially unstable blocks cut by structure planes on the left bank abutment; plunge pool slope with No. Ⅳ ridge as a representative.

1. Cable-crane platform slope on the left bank

Cable-crane platform slope on the left bank is Ⅱ$^{\#}$, Ⅴ$^{\#}$ ridge clip gully in terms of the landscape, but the terrain integrity is poor. The slope in this area is a reverse slope composed of the hard and soft sand slates alternatively. After excavation of the slope and drainage tunnel, it is found that there are phenomena of inclining, ripping, unloading etc. in the rock mass above El. 2,000m, most of which have a cataclastic or loose structure. Rock mass relaxes and opens obviously under El. 2,000m, the rock mass is in general a cataclastic or slab-cataclastic structure. Meanwhile there are many fractures and small faults in NE direction in the slope, constituting the rear edge cutting surface of the slope. To reinforce the slope, anchor cable reinforcement measures are taken, and steel anchor bunches are applied in addition in step excavation, which can control siope deformation and sliding. At the same time many drainage tunnels are set in the slope, and shotcrete-anchorage is generally used to seal off the slope surface. After the execution of reinforcement measures like anchor cable, anchorage beam, and drainage tunnels, the analysis of actual deformation mechanism and stability monitoring results prove that they are effective and feasible supporting measures. Because of the relaxation and ripping of the slope rock mass, the groubility of slope rock mass is good, and the amount of borehole consolidation grouting in the construction of anchor cable is great.

2. Abutment slope on the left bank

The abutment slope on the left bank and the cable-crane slope are indivisible as a whole. With the excavation of the abutment and spandrel slot, the border of sandy slate and marble is at El. 1,820m. The early exploration of adits indicates the probable sliding modes within the slope include: controlling "large block" sliding composed of borders of fault f_{42-9}, lamprophyre veins, SL_{44-1} cracks etc. ; relatively small block ("medium block") sliding composed of borders of of f_{42-9}, hanging wall of SL_{44-1} and other structure surfaces in parallel, and "small block" sliding composed of random cracks.

Fault f_{42-9} extends long, is large-scale and its fault properties are poor with gravity creeping characteristics which is speculated to result from the creeping of the large blocks composed of borders of it (f_{42-9}) and SL_{44-1}. This block is large in dimension, and located upstream of the left abutment, it will be influenced by the reservoir impoundment in the future, its stability is the critical issue of the abutment on the left bank. During the reinforcement treatment, f_{42-9} is set with three layers concrete anti-shear tunnels, the anti-shear strength is increased to ensure the stability of rock mass under the abutment in the process of excavation. With the excavation of the abutment pedestal, anchor cables are used to deal with the large and small blocks in the large blocks.

It is found through the monitoring that during the excavation of the cable-crane platform slope, the abutment pedestal and spandrel on the left bank, the shallow and deep deformation of mountains on the left bank increases continuously, and the slope deformation magnitude is much greater than the right bank. Especially the deformation in the deep mountains causes experts to worry about whether the deformation would converge. From the perspective of the whole excavation, the deep deformation is closely related to the excavation of left bank. Since the altitude difference of the slope on the left bank is above 1,500m, and the geological conditions of the slope as a whole is unsatisfactory, the deep unloading and relaxation and ripping of rock mass are generally developped, the excavation scope is rather big to meet the requirements of arc dam. Hence the large-scale excavation of the lower part of whole natural slope will cause a readjustment of the stress of the whole lower part of the slope. This adjustment can be divided into several stages to analyze: ①it shall be obvious to make the adjustment when the excavation takes place above El. 1,885m, it involves the removal of a large amount of toppling rock mass of the cable-crane platform slope; ②at El. 1,885m to 1,730m, in order to set the pedestal, the range of the excavation and its influence on displacement are both great; ③when the excavation of spandrel slot takes place at El. 1,730m to 1,580m, the range of excavation and horizontal depth decreases and deformation magnitude will significantly decrease until tending to convergence; ④after the large scale concrete pouring of the pedestal and the arch dam, deep deformation will eventually converge; ⑤after the dam is built and the reservoir is impounded, the stability of the slope will be finally tested under the effect of water load. At the end of 2011, the excavation of the dam foundation and concrete works of the pedestal have been completed. Deep deformation is changing as assumed in accordance with the above steps basically, from the initial deformation of cm magnitude, gradually reducing to the monthly deformation of 2mm, 1mm and 0.5mm. Viewed from this trend, the stability of the slope on the left bank as a whole can be guaranteed, but is expected to be tested finally after the operation of the arch dam.

3. Ⅵ-Ⅳ *Ridge slope*

Ⅵ-Ⅳ exploration lines ridge has been transformed into deformation and ripping rocks. The deep cracks develop in the slope, the slope stability is poor and failure may occur under the influence of the precipitation in the atomization. In terms of the landscape, the slope is two ridges with a cut-shallow gully, which is evident above El. 1,800m. Slope below El. 1,920m is composed of the 6th, 7th and 8th layers of marble of the second section of the Zagunao group with slope of 60°–70°; above El. 1,920m, there are the third section of sandy slate of the Zagunao group with slope of 40°–50°. Attitude of the rock stratum is N10°–35° E, NW∠30°–45°, with the strike basically the same as the natural slope. The slope belongs to the typical reverse inclination slope.

This part is the typical section of deep cracks developed in the left bank of the project region. Relatively large-scale structure planes in the slope include faults f_2, f_5, f_9 and a series of grade Ⅰ and Ⅱ deep cracks with wide openings.

Geological structure analysis of Ⅵ-Ⅳ exploration lines ridge shows that overall strike of fault f_9 (SL_{24}), SL_{13}, SL_{18}, SL_{30} deep cracks development zones is oblique intersecting or in parallel with bank slope, medium or steep dipping out in good connectivity. Thus, these structure planes form the rear edge cut surfaces that control the deep stability of the slope, which is extremely adverse to the slope stability. The deep deformation and failure modes have two types: one is fault f_9 (SL_{24}) as rear edge cut surface, bottom slide face shear the rock mass at the slope toe; the other is deep crack development zone nearly parallel with bank slope forming the rear edge cut surface, bottom slide face shear the rock mass at the slope toe. Among them, in Ⅵ exploration line profile the rear edge cut surface is deep cracks SL_{13} (which extending upward into the sandy slate is relaxed opening unloading fissure XL_1); in Ⅳ exploration line profile is SL_{18} (which extending upward into the sandy slate is relaxed opening unloading fissure XL_2); in A exploration line profile is SL_{30} (SL_{30} extending upward into the sandy slate is relaxed opening unloading fissure XL_3).

In the Ⅳ-Ⅵ exploration lines ridge deformation and ripping rock mass, deep cracks of fault f_9 (SL_{24}), SL_{13}, SL_{18}, SL_{30}, etc., developing above El. 1,700m, form the rear edge cutting boundaries to control stability of bank slope. But because deep cracks under El. 1,700m weaken and even disappear, whose exterior still has 50–100m wide and relatively intact rock mass, in which the gently dipping out structure planes do not develop, and the specific cut surface is not found on the upstream side, which is an important anti-sliding and locking rock mass on the slope and plays a vital role in keeping the overall slope stability. Therefore, it is believed by the macro analysis, judgment and calculation of geological conditions that Ⅵ-Ⅳ lines ridge deformation and ripping rock mass is stable as a whole in natural state, but the shallow strata need to be applied with systematic shotcrete-rockbolt support. The slope stability and water pressure are closely related, and the groundwater level of the natural slope is very low with almost no influence on the stability of the slope; but after flood discharge the rain and fog may provide a great amount of water supply, which can increase water pressure and reduce the degree of stability of the slope. Therefore, the surfaces and slope drainage are important measures to guarantee of slope stability.

Serious problems of the high slope stability were involved in the construction of right abutment of Wujiang Goupitan Hydropower Station, right abutment of Shuibuya Dam on Qingjiang River, both abutments of Xiluodu Hydropower Station on Lancang River etc. Besides, the high slope stability problem of Baihetan Hydropower Station on the Jinsha River and Laxiwa Hydropower Station on the Yellow River is one of their major geological problems each.

7.3.5 *Karst Foundation Treatment*

Karst phenomenon is the most important geological problem that hydraulic and hydro-power engineerings in karst areas confronted with, which not only directly affects the engineering construction, but also has great impact on seepage of dam foundation, foundation deformation and building safety. Due to the complexity of karst hydrogeology, the investigation study of underground karst system is of great difficulty. In places like Guizhou and Guangxi of China, when underground reservoir is constructed, trial impounding is operated to test the karst groundwater recharge and the reliability of underground anti-seepage project.

7.3.5.1 Treatment of Karst Foundation at Goupitan Hydropower Station

Wujiang Goupitan Hydropower Station is located in Yuqing County, Guizhou Province. Dam of the power station is a concrete bi-arch dam, with the maximum height of 232.5m. The hydropower station, with the normal water level of 630m, total reservoir capacity of 6.454 billion m^3 and the installed capacity of 3,000MW, is a key project of the cascade development along the main stream of Wujiang River.

The river valley where Goupitan Hydropower Station is located is a V-shaped symmetrical valley. The rock mass of the dam foundation is the Permian system lower Maokou formation (P_{1m}^1) medium-and thick-bedded micro crystal biodetritus limestone, thin and extremely thin layers of peaty biodetritus limestone are interlayed at local parts. The dam foundation of the river bed is mainly located on the layer of P_{1m}^{1-1} and dam foundations of both banks are gradually located on layers of P_{1m}^{1-2} and P_{1m}^{1-3}. The rock stratum inclines to the upstream (NW) with the angle of 45°–55°, and some angles of 65°. Interlayer shear zone, fault and fissure are major structural feature of this region. The karst is developed in the dam site, which on the left bank is relatively undeveloped, while on the right bank and the river bed, an amount of karst caves have been developed, among which Cave K_{280} on the right bank is a large scale cave.

1. Karst cave

Cave K_{280} on the right bank outcrops at the elevation of 545–565m and runs through the foundation surface of the dam along the river trend. This cave is mainly shaped by wide cracks and inclined shafts, with the height difference of nearly 90m and the volume of 20,000m^3, fully filled with clay, grit and broken rocks. Because of the large scale and wide sphere of influence, Cave K_{280} is the key point of engineering treatment.

2. Weathered-leached zone

There are two weathered-leached zones in the dam foundation, of which the weathered-leached zone (KJ_1) at the bottom of the layer P_{1m}^{1-2} is widely distributed. In about 0.6 - meter-thick stratum at the bottom of layer P_{1m}^{1-2}, the limestones on the side of three layers siliceous belts have been weathered and leached to red clay discontinuously distributed in bed-parallel chain. The thickness per layer is 0.3–5cm, and the cumulative thickness is 8–15cm. The distribution of the clay layer is continuous at local parts. The thickness from the base of the clay layer to the top F_{b113} of layer P_{1m}^{1-1} is around 5m. Within the horizontal depth of around 80m, this layer is continuously distributed with poor property. This layer on the left bank enters foundation surface from the upstream at the elevation of 450m and passes through the foundation surface to enter the arch abutment rock in the downstream at the elevation of 517m. On the right bank, it enters the foundation surface from the upstream at El.505m, and passes through the foundation surface to enter the arch abutment rock in the downstream at El.575m.

Another weathered-leached zone (KM_1) distibutes in the dam section on river bed and near the dam toe below the elevation of 430m on both banks, which developed along the bedding plane with thickness of 0.5–0.8m. Its property on left side of the river bed is poorer than that on the right side. On the right side at the elevation of 410m the horizontal foundation surface it has been almost pinched out. The rock mass in the zone is mainly alveolately weathered and corroded. On the left side of the river bed, there is continuous silt interlayers, with the thickness of 5–10cm, stretching around 30m downward.

3. Interlamination dislocation corroded and intercalated clay

The major interlamination dislocations in the foundation rock of the dam are F_{b82}, F_{b112}, F_{b113} and F_{b86}. Small-scale interlamination dislocation also develops in layer P_{1m}^{1-3}. The interlamination dislocation breccia in layers F_{b82} and F_{b86} is usually well bonding, which is only partially corroded and intercalated clay, and has small influence on the stability of dam foundation. F_{b112} and F_{b113} develop on the top of Layer P_{1m}^{1-1}, with the distance of around 7m. The width of the dislocation belt is generally 0.1–0.4m, which is made of breccias and cataclasites. The two-layer interlamination dislocation corrosion on the right side at the elevation of 520–600m is serious. At the junction of the fracture and NW and NWW, there is a large-scale karst cave K_{280} developed. The corrosion on the left bank F_{b113} is very serious and with continuous mud intercalations. Below the elevation of 540m, F_{b112} corrosion is serious, with the above part just locally corroded. On the river bed dam section, F_{b112} and F_{b113} emerge in the upstream of dam foundation, and gradually enter into the dam foundation on both banks or remain close to the arch abutment.

4. Corroded fracture

The corroded fracture in the rock of the dam foundation is mainly featured high dip angle along the direction of NW and NWW, with medium scale and the length less than 100m. The scale of F_{37} is relatively large, but the breccia is well bonding with slight corrosion. The corroded fracture on the right bank mainly develops at the elevation of 520 – 580m, among which the scale of f_{280} is relatively large, Cave K_{280} develops along the corrosion, and this fault is partially corroded below the elevation of 517m. In the dam section on the river bed, Fault f_{38} and f_{48} develop along the river trend, running through the foundation surface from upstream to downstream and seam-filled karst caves develop along the fault. The corroded fracture mainly develops in Layer P_{1m}^{1-3}, located near the toe of dam and in the rock of downstream arch abutment at the elevation of 580 – 590m and 600 – 615m, amomg which the scale of f_7, is larger, which has been partially corroded to mud-filled gap.

5. Corroded fracture zone

The foundation surface revealed 3 corroded fracture zones, among which the scale of KM_2 on the river bed dam section is relatively large. The other two are located at the toe of dam on the left bank at the elevation of 550 – 560m and near the toe of dam on the right bank at the elevation of 590 – 600m respectively, both of which are small-scaled. KM_2 is located at the groove part in the middle of the river bed, with the width of 3 – 6m and the area of 500m². In the fracture zone, there developed short and small, and even tiny, irregular fissures with the length mostly less than 1m. The fissures are usually corroded, with clay and calcareous filled and partially caves developed in smaller scale, stretching downward to the elevation of around 405m.

The above mentioned geological defects have negative influence on the deformation of arch dam and the anti-sliding stability. Therefore, these defects call for engineering treatments. According to the type and distribution of various defects, shallow and deep treatments are adopted respectively. Shallow treatment mainly includes cleaning and digging, consolidation grouting and devilling on the surface of the revealed karst cave and fractured rock, while deep treatment mainly means to excavate replacing tunnel (well) to backfill concrete and form a force transmission column. For treatment measures, please refer to Table 7.3 – 3. The consolidation grouting of the dam foundation resorts to the combination of normal concrete-cover-weight and non-cover-weight, which means that deep grouting is operated before concrete pouring with bedrock cover-weight, and when the corresponding dam section is poured to a certain height, the cylinder surface will be drilled to conduct surface grouting. This composite grouting method reduces mutual interference between each process, so as to accelerate the construction progress.

Table 7.3 – 3 Treatment Measures on Geological Defects

Type of geological defects	Major treatment measures	Auxiliary measures
Karst cave	Cleaning and digging	Deep replacement, consolidation grouting
Interlamination dislo cation	Deep replacement	Devilling on the surface, consolidation grouting
Fault	Cleaning and digging	Devilling on the surface, consolidation grouting
Weathered – leached zone	Deep replacement	Devilling on the surface, chemical grouting
Corroded fracture zone	Cleaning and digging	Consolidation grouting
Structural fracture zone	Cleaning and digging	Consolidation grouting

Cave K_{280} has a large scale, and most part of it is shallowly buried below the dam foundation, it has great influence on the bearing capacity of the dam foundation, anti-sliding stability, deformation stability and seepage-prevention. Hence, it is the key geological defect to be treated of the dam foundation. The treatment method mainly includes cleaning and digging, in combination with deep replacement.

KJ1 is severely weathered and corroded and continuously filled with mud. Calculations by two-dimension and three-dimension linear, nonlinear finite elements methods indicate that it has great influence on the deformation stability of the arch dam. Furthermore, because of the discrepancy of the distance from mudded intercalation zone on either bank to the arch abutment, the asymmetry of deformation displacement is caused and which calls for engineering treatment. Since KJ1 spreads in the shape of a plane under the dam foundation and in the abutment rock, the excavation scheme is not feasible. Through calculation and analysis, treatment of deep replacement is selected, i.e. to construct a horizontal replacement tunnel at every 15 – 20m in height. The depth of the replacement tunnel is controlled according to the influence magnitude by the stress of the arch dam, and is generally 50 to 100m. Furthermore, replacement inclined shafts are set between horizontal replacement tunnels along interlamination dislocation tendency or based on the direction of inclination, and the replacement tunnels are arranged in Chinese character of "well" shape, so as to enhance the effect of overall power transmission. For KJ1, there are 7 layers of replacement tunnels on the left bank and 8 on the right. Among them, two layers on the right bank fall into the excavation scope of Cave K_{280}. Hence, these two layers at lower elevation will be cancelled. Replacement inclined shafts are arranged at locations with severe corrosions and continuous mudded intercalations according to the characters of weathered-leached zone between two layers of replacement tunnels. There will be 6 shafts on the left bank and 4 on the right bank. Similar deep replacement treatment scheme will be adopted for the interlamination dislocation of F_{b112} and F_{b113}. Calculation indicates that with the deep replacement to corroded and filled mud interlamination dislocation and weathered-leached zone, the stress distribution, displacement distribution and plastic zone of the dam body, arch abutment and the rock mass are significantly improved; stress of the dam body is approximately distributed symmetrically; stress of the arch abutment tends to be even. The effect of the treatment is remarkable. As for mudded intercalation emerging on the dam foundation surface, treatment of devilling on the surface and consolidation grouting are adopted.

7.3.5.2 Karst Foundation Treatment at Pengshui Hydropower Station

Pengshui Hydropower Station is located at the limestone valley river section, 11km upstream of Pengshui County in Chongqing City. The dam is a concrete curved gravity dam, with the maximum height of 116.5m. The bed rock of the dam site is carbonates rock with a small amount of shale. Karst here strongly develops. On the right bank, from the upstream to the downstream, there are four major karst systems: KW_{14}, KW_{51}, W_{84} and KW_{54}, among which KW_{51} and W_{84} karst system develop in bedding to the dam foundation. The excavation indicates that the karst caves develop in large scale with low elevation, which exerts great influence on the construction. Therefore, it is the key part of dam foundation treatment. Karst treatment in dam foundation is of great importance to ensure the safety of the dam.

The rock stratum in the dam site area is inclined to the right bank of upstream, forming an angle of 70°–75° with the current direction of Wujiang River. Middle and upper Ordovician and Cambrian stratum emerges here, including lower Ordovisian Dawan formation (O_{1d}) thick-bedded shale with a small amount of limestone, Honghuayuan formation (O_{1h}) limestone, Fenxiang formation (O_{1f}) limestone intercalated with shale and chert, Nanjinguan formation (O_{1n}) limestone and limy dolomite intercalated with shale, upper Cambrian Maotian formation (ϵ_{3m}) limestone and dolomite and Gengjiadian formation (ϵ_{3g}) dolomite.

The dip angle in the dam site area is 60°–70°. The fault and fissure in the area are more developed. The major development direction of fault is NNE, NW and NWW, with the fault dip angle of over 70° in general, and a few of 56°–65°. NNE group mainly consists of f_1 crossing both banks; NW group (along the river trend) contains f_8 and f_{110} on the right bank, and f_7 and f_5 on the left bank; for NWW group there is f_{36} on the left bank. These fractured tectonites are usually well bonding, but most of them are corroded in strong corrodible layer.

In the stratum from Gengjiadian formation to Dawan formation, there are 59 weak intercalations with poor revealed characters. According to formation condition, composition and property, weak intercalations are divided into three basic types: muddy intercalation (Ⅰ), broken intercalation (Ⅱ) and weathered and corroded mud-filled weak intercalation (Ⅲ). Type Ⅲ intercalations in layer O_{1n}^5 are distributed in dam foundation and power house area and have some impact on buildings. They are usually weathered karst caves. Type Ⅱ intercalation includes 401 and 404 of layer O_{1n}^4 and 093, 082 and 085, etc. of layer ϵ_{3m}.

Since the angle between stratum strike and Wujiang River's flow direction is big and the stratum steeply inclines to the upstream, the karst layer is alternatively distributed with impermeable layer and relative impermeable layer, and there developed 11 karst systems on two banks in symmetry.

Karst systems on the left bank include: W_9, KW_{17}, W_{202}, KW_{65}, hot spring W_{10} and KW_{66} (Shuangbikong ϵ_{2p}).

Karst systems on the right bank include: KW_{14} and WH_{11}, KW_{51}, W_{84}, KW_{54} and KW_{40} (Dalongdong ϵ_{2p}).

KW_{51} and W_{84} have relatively great impact on the dam foundation construction.

1. Features of system KW_{51}

It is located on the right bank, about 10–20m upstream from the dam axis. Its outlet position is the boundary of O_{1n}^{5-2} and O_{1n}^{5-1}. The outlet elevation is 227.4m.

System KW_{51} mainly develops in layer O_{1n}^{5-2}. The karst has a bedding development, mainly develop severely along the big weak intercalations and lithologic interface. For example, along C_2, C_4, C_5 and C_{504} usually developed large-scale karst caves, and some times, the corrosion zone of $C_4 - C_5$ forms 15-meter-high halls. Fault f_3, f_8 and f_{110} on the right bank intersect the C_2, C_4 and C_5 weathered and corroded mud-filled layer zone. The intersection of faults and layer zones usually forms corroded halls. The karst in layer O_{1n}^{4+5} at the off-bank zone is developed strongly. Underground water emerges in the junction of flood plain and slope. Deep karst is in the state of inverted siphon.

The excavation of the dam indicates that karst caves of system KW_{51} mainly exist in 12#, 13# and 14# dam blocks and slopes near dam foundation. Moreover, they developed along the intercalations of C_2, C_4 and C_5. Elevation 235m (dam foundation of 13# dam block) and 14# dam block (El. 255m) – elevation 280m (entrance of P17 adit), along intercalation

C_2, formed karst caves (K_{16}, K_{22}, etc.) with the width of 1.5 – 8m. These caves, filled with clay and detritus with speleothem, has the horizontal depth of 13.4m. The width of the water-eroded groove will gradually narrow down to 0.3 – 0.5m.

Along the intercalation C_4, usually karst caves are formed in the interior dam foundation of 13# dam block (El. 235m) and the dam foundation of 14# dam block (El. 255m) to the slope of elevation 275m. The width of entrances to these karst caves ranges from 0.3m to 2.5m. These caves are filled with yellow mud, gravelly soil and detritus, etc. Cleaning and digging from the inside slope of 14# dam block to 21.5m deep inward, one can find the corroded wide seam narrows down to 0.1 – 0.5m.

Along the intercalation C_5, at the dam foundation of elevation 235 – 255m, on the slope of middle part of 14# dam block, there is a bedding of 1.0m long and 0.3m wide corrosion, and clean and digg 0.2 – 0.8m downward after yellow mud is seen, then one can see moderately weathered rock. There is karst cave with the width of 1 – 1.5m along the intercalation C_5 in the slope with the elevation of 255 – 290m. Cleaning 1 – 7.2m inward, the width of the intercalation corrosion narrows down to 0.2 – 0.5m.

2. Features of system W_{84}

Spring W_{84} is located near the low water of Wujiang River, 10m from the downstream of toe of dam on the right bank, with outlet position in layer C_{3m}^2, spring elevation of 210.81m and flow of 200 – 300L/min. The largest exit point of this hot spring is in the former P16 exploration adit, with the temperature of 40℃ and stable flow. The main channel direction of W_{84} karst system develops almost completely along layer C_{3m}^{2-2}. It mainly develops at the bottom of layer C_{3m}^{2-2}, and only a tiny part on the top of C_{3m}^{2-2}. The main channel of the part facing the river has developed to the top of C_{3m}^{2-2}. The width of the karst cave is 1.5 – 5m, with some parts being wider. From the development elevation to the elevation of 140m, fillings of the karst cave mainly consist of clay, sand inclusion, detritus and speleothem. Some unfilled part is empty.

The excavation of dam foundation indicates that the corrosion zone W_{84} emerges near the dam toe of dam block 8# and 9#. It is a 0.2m wide karst breccia zone, and filled with a small amount of mud. It develops along the bottom of layer C_{3m}^{2-2}. On the right side of 8# dam block, there is a karst cave filled with yellow mud, with the diameter of 0.5m×1.2m and the depth of around 1.1m. The geophysical prospecting test of this zone is abnormal. Running through 9# dam block, W_{84} corrosion zone enters post-dam apron side slope. From the elevation of 188m (at the side of right bank behind 9# dam block) to the apron slope of the elevation of 227m is a corrosion zone with the width of 0.3 – 3.8m. Around the elevation of 200m lies the original water outlet of W_{84}. From the elevation of 227m to the position with the same elevation of C_{3m}^{2-2} at 250m is a 0.1m crevice. Above the elevation of 250m, no corrosion is found.

The existence of karst system KW_{51} and W_{84} destroys the integrity of rock mass, thus result in relative reduction of overall strength of the rock mass. Secondly, as the fillings in karst cave come with low strength, the required engineering strength cannot be met. The construction of power station also blocks the original natural drainage channels of karst water, thus there is a need to drain water, and seepage prevention treatment is required to meet the anti-seepage demand. Of the above two karst systems, the natural drainage channels of KW_{51} karst system run through the impermeable curtain to the dam, so it is of great importance to conduct blocking and ground water drainage for karst channels.

(1) Within the outline range of dam and its outer surrounding of 10m, the "V-shaped" groove carving method is used to excavate the fillings of the revealed corrosion layers (caves) like the C_2, C_4 and C_5 and other local bedding caves at the dam foundation. After the fillings in cave are removed, concrete is backfilled therein. For the karst cave in the side slope of dam foundation, concrete is backfilled after horizontal and downward tracking for complete excavation. After backfill of the cave, consolidation grouting is conducted to the dam foundation, and contact grouting conducted to the side slope.

(2) For the caves on and near the impermeable curtain line, backfilling after excavation and backfill grouting shall be adopted to ensure the formation and safety of the main impermeable curtain. At the same time, the surface blocking is conducted to the surface exit, karst cave and karst fissure of KW_{51} system on the right bank in front of the dam, and the impermeable curtain is made in front of the dam.

(3) Based on the knowledge of the inflow and variation rule of karst system, a drainage tunnel is specially arranged within the mountain at the right bank and behind the impermeable curtain, in order to drain the karst water in both the KW_{51} and W_{84} karst systems. After excavating the KW_{51} water tunnel and the drainage tunnel of power house, all the karst water is exposed in the water tunnel and the drainage tunnel within the mountain. And the revealing positions are mainly in the fault corrosion zones of f_{110} and f_8, and etc. There is no water in corrosion zone when excavating the dam foundation. The lower water tunnel (El. 165m) of W_{84} karst system has hot spring emerged in the corrosion zone of C_{3m}^{2-2} layer. The hot spring water is all poured out from the corrosion zone within lower water tunnel, with the flow of 200 – 300L/min. After this, there

is no groundwater emerged in corrosion zone when excavating the corrosion zone at $\mathcal{C}_{3m}^{2-2}$ layer of dam foundation 8# and 9# to the elevation of 184m. The spring water from the two karst systems is cut-off within the mountain, so the spring water will not flow into dam foundation during the operating period of the power station.

Pengshui Hydropower Station began to store water at the beginning of 2008. From the monitoring of dam and long-term observation of karstic ground water in the right bank and the curtain drainage tunnel, the following monitoring results from various aspects have showed that the treatment of dam karst foundation and blocking and drainage of karstic ground water achieve expected effects.

The horizontal displacement at dam crest is between -4.61mm and 7.25mm, and displacements at the galleries of dam body and dam foundation are small and are all within 1mm. Also, the horizontal displacement of dam body is still normal after the reservoir is filled with water and the dam foundation is stable. The cumulative vertical displacement at dam crest varies between 1.72mm and 6.78mm, and the displacement trend is sinking as a whole. The foundation gallery of dam body is also sinking with the settlement of about 3mm, and the settlement is even, there's no nonuniform settlement.

The drainage hole of grouting adit at El. 239m of right bank is basically characterized by no water or water droplet, and the total leakage of drainage hole in grouting adit at El. 193m of right bank is about 60 – 80L/min, with the maximum leakage of single hole being about 14L/min. When the reservoir water level rises to the normal water level of elevation 293.0m, the uplift pressure coefficient of dam piezometer tube is less than 0.20, which is within the allowable range. This shows that the dam has a sound anti-seepage effect. From storing water to now, the piezometric level on both right and left banks changes very small, and also the rainfall has little influence on piezometric level. The rock mass at the deep part of dam foundation on the two banks features neither obvious seepage nor seepage around the dam. After successfully blocking the karst gallery of KW_{51} system on the right bank, there is neither reservoir water before dam nor karstic ground water flow into dam foundation through the karst channel.

The flow of KW_{51} karst water at main revealing locations in right-bank mountains: the flow in rainy season is 50 – 80L/min and 5 – 30L/min in dry season at f_{110} fault area of drainage tunnel at the top layer upstream of powerhouse. The flow of KW_{51} water tunnel is 8.5 – 12L/min. And also the total flow of the above two locations is basically the same as that of the spring water under original natural state.

In the W_{84} lower water tunnel (El. 165m), the hot spring water is poured out from the corrosion zone within the lower water tunnel, with the flow of 200 – 300L/min and the temperature of 41°– 45°. This revealing position is the lowest point of W_{84} karst system in dam area, after revealed by this revealing position, there is no such spring water emerged again in the karst channel unveiled by various tunnels in dam foundation and right-bank mountain. All these show that the W_{84} lower water tunnel completely cuts off the karst water of this system.

In general, karst water of the two karst systems is fully cut off and drained, and the karst water is completely blocked outside of dam foundation and the desired results are achieved.

7.3.6 Seepage Prevention of Deep and Thick Covering Layer

The deep and thick covering layer is adopted for building high dam. For the dam type, dam using local materials is generally selected, which is to accommodate to possible problems of compression deformation, sand liquefaction, etc. of dam foundation. However, the treatment of leakage and seepage deformation of dam foundation is the key to successful construction of dam. At present, there are two methods of horizontal impervious blanket and vertical anti-seepage, which are mainly used for deep and thick covering layer for seepage prevention treatment in both home and abroad. When the water head is not high and the seepage deformation of base foundation is not prominent, the horizontal impervious blanket can be employed; while when the water head is high and the dam foundation seepage comes with a great threaten to dam safety, the vertical anti-seepage treatment is used more. The record of vertical impervious wall abroad is the Manic Ⅲ Hydropower Station concrete impervious wall in Canada that completed in the 1960s, whose maximum depth is 131m. There are more than 70 impervious walls deeper than 40m at present in China, among which the Yele Hydropower Station comes with maximum depth of 140m. Also, the maximum depth of impervious wall at Xiaolangdi Project on Yellow River is 81.9m, that height of second-stage upstream cofferdam of the Three Gorges is 73.5m, the maximum depth of impervious wall at Xinjiang Xiabandi Project is 102m and also the successful trial depth of impervious wall in Pangduo, Tibet is 145.9m, creating a new world record.

7.3.6.1 Pubugou Hydropower Station

Pubugou Hydropower Station is located on trunk stream of Dadu River in Hanyuan County, Sichuan Province, whose hydraulic structure mainly composed of soil core rockfill dam, left bank water-diversion project, underground powerhouse, spillway on bank and spillway tunnel, etc. The dam is 186m in height. It is a soil core rockfill dam built on covering layer. The bed rock of project region is mainly made of hard epimetamorphic basalt, rhyolite porphyry, rhyolitic tuff and medium-grained to coarse-grained granite.

The general thickness of covering layer of dam foundation in the river bed is 40 – 60m, and the maximum depth can reach 77.9m. From old layer to new layer, it can generally be divided into four layers: the first layer is the (Q_3^2) pebble and boulder, the second is the (Q_4^{1-1}) gravel layer, the third is the (Q_4^{1-2}) pebble and boulder layer and the fourth is the (Q_4^2) pebble and boulder layer. The underneath of the third layer (Q_4^{1-2}) near to the bank has sand lens, and the thicker part below dam comes with upstream sand lens and downstream sand lens, which are respectively made of fine & medium sand and fine sand. The surface of the fourth layer (Q_4^2) has lenticular sand which is discontinuously distributed along the bank.

The main engineering geological problems of dam foundation are:

1. Deformation of dam foundation

The deep and thick covering layer of dam foundation at river bed features many complicate layers with multiple interlayers of sand lens, of which the physic-mechanical properties are greatly different. The thickness of each layer is different and great changes in longitudinal and sectional areas, thus causing the problems of dam foundation deformation and uneven deformation. All these may generate negative influence on the stress distribution and deformation of dam body, core wall and impervious wall. From the perspective of engineering geology, the river bed covering layer within the bearing stratum of dam foundation is mainly composed of coarse granule, having certain resistance to deformation and mechanical strength, in addition, the earth-rock dam features large bottom width and is of flexible structure, and thus all these constitute good flexibility for dam foundation deformation. To reduce to the maximum the influence of uneven settlement deformation of foundation on core wall and impervious wall, the excavation and replacement treatment is conducted for the lenticular sand that is discontinuously distributed along the bank of the fourth (Q_4^2) layer surface and sand lens discovered during the construction. In addition, shallow consolidation grouting is conducted at the core wall foundation, in order to improve the base stress state, and weight loading zone is set downstream of the dam body.

2. Leakage and seepage of dam foundation in river bed

The covering layer of dam foundation in the river bed has features of coarse granule with many boulders, obvious cavitation at local areas, strong permeability, and with no relative aquiclude to be made use of, it is the main channel for dam foundation leakage and even a concentrated seepage may occur. The covering layer lacks 5 – 0.5mm middle granule, therefor has lower anti-seepage failure ability and contact scouring as well as piping failure may occur. Thus, anti-leakage and seepage deformation prevention is one of the critical technical issues for dam foundation treatment. The design uses two paralleled concrete impervious walls, the leakage channel of covering layer of dam foundation is full-section sealed. Also, the bottom of impervious wall is embedded in bed rock for 1.5m, and it will be embedded in bed rock for 5m when encountering fault fracture zone. Furthermore, a horizontal filter is set between the downstream covering layer and the rock-fill.

3. Sand layer liquefaction of dam foundation

The sand lens of covering layer of river bed is mainly distributed near the bank down the river, the thickness is generally less than 2m and the maximum thickness can reach about 13m. The sand layer lens are mainly distributed at the bottom of the third layer (Q_4^{1-2}), which can also be found in the Q_3^2 boulder and pebble layer at the nearshore of the left bank.

Carbon-14 dating shows the age of the sand lens located at the bottom of the third layer is about 7 – 10 thousand years, whose sedimentary era is in the early Holocene. The embedded depth of the sand lens top is 26 – 48m. According to the reconnaissance results, the sand lens below dam foundation may be the layer of liquefied soil. The three-dimensional dynamical analysis conducted according to different technical routes have showed that when the upper and lower sand layer lens encounter earthquake, both the generated pore water pressure and the dynamic shear stress are relatively small and will not result in liquefaction without regard to the weight loading of downstream dam toe. To prevent the adverse effect of liquefaction on dam stability, the weight loading zone at dam toe is built at downstream. In addition, excavation and replacement treatment is conducted for the sand lens discontinuously distributed nearshore on the surface of the fourth layer (Q_4^2) as well as the sand lens discovered during the construction.

Based on a great deal of exploration experiments, mainly inside experiments and field tests, according to the engineering geological conditions of covering layer, as well as the engineering geological analogy analysis conducted in the built projects, the physical and mechanical parameters are selected and determined. The arithmetic mean value of tests is taken as the standard value for the parameter of physical properties of the soil in this project, and the average line is regarded as the representative value of the granulometric composition, the average value of local maximum values of water pumping and injection tests on site as standard value of permeability coefficient. The ϕ value of shear strength parameters uses the effective stress shearing strength parameter, taking the average value of tests or multiplying by 0.75 - 0.85 as standard values. Modulus of compression, allowable bearing capacity and slope ratio are determined according to the geologic analogy, and also the allowable bearing capacity of sand layer is set based on the relationship between standard penetration number and bearing capacity. This project is already completed and the power is generated, and its water level basically reaches the normal water level. The monitoring results show that both the deformation and seepage of dam foundation are normal.

7.3.6.2 Yele Hydropower Station

Yele Hydropower Station is the leading reservoir project established through the cascade development of Nanyahe basin at the tributary of Dadu River. The core wall rockfill dam of asphalt concrete is used as the water retaining structure of the station, with the maximum height of 124.5m. The dam site is at the river valley which is about 1.5km away from the downstream of two-branch river mouth (the confluence of lime Cellar River and Leya River) located at the border of Yele basin. The medium to upper Pleistocene gravel layer and silty loam as well as the gravelly soil, which are all at the bottom of right bank and river bed, constitute the five lithological groups ($Q_2 - Q_3^2$). The groups are the glacio-lacustrine deposits, with the thickness of about 500m, and also feature varying degrees of calcareous and argillaceous cementation and over consolidation. The loose accumulation layer from Upper Pleistocene to Holocene comes with ice water accumulation layer, alluvial layer, diluvial layer and colluvial layer etc. Ice water accumulation form the Ⅲ-Ⅳ terrace composed of tawny gravel. The alluvial sand gravel (alQ_4) of Holocene is distributed in the Ⅰ and Ⅱ terrace, floodplain and river bed. The diluvial layer is commonly seen at the gully exit, which is comprised by gravelly soil and boulder clay. The Xigeda Formation silty fine sand (Q_{1x}) of Early Pleistocene is sporadically distributed in the boundary of the basin, and has the characteristics of half-diagenesis.

The gravel layer, silty loam layer and the gravelly soil with yellow stiff soil layer ($Q_2 - Q_3$) of the Middle to Upper Pleistocene are the main Quaternary layer in the dam site. According to the dating material of sporopollen, ^{14}C and thermoluminescence, we can see that its sedimentary era is almost 600 - 32 thousand years ago. According to the sedimentary rhythm, lithological change and engineering geological characteristics, it can be divided into five lithological groups from top to bottom. The overall attitude of the rocks is stike N20°- 65°E and the inclination angle (5°- 12°), inclining to SW direction.

The fifth lithological group (Q_3^{2-3} Ⅴ) is the silty loam and the silty loam layer intercalated charry plant detritus, whose thickness is about 90 - 107m. The thickness of single layer of silty loam is generally about 15 - 20m. Silty loam softens easily when encountering water, within which are several layers of intercalated charry plant detritus with thickness of about 5 - 15cm. There are also 3 - 8 layers of gravel in the intercalation, and single layer thickness is about 0.8 - 5m and has the characteristics of small particle and relatively poor cementation.

The fourth lithological group (Q_3^{2-2} Ⅳ) is the weakly cementted gravel layer, whose thickness is 65 - 85m. It is of thick-extremely thick layers, with the thickness of single layer of generally about 2 - 10m. Within these layers are several lenticular silty sand layers or the silty sand loam with the thickness of about 0.2 - 3m, which all feature glacio-lacustrine sedimentary characteristics. In most cases, overhanging cliffs would be formed in topography, and plus occurrence of corrosion.

The third lithological group (Q_3^{2-1} Ⅲ) is the interbeded gravel layer and silty loam layer, whose thickness ranges from 46m to 154m. Silty loam is characterized by over consolidation and slight cementation, and both the sedimentary rhythmic layers and thickness gradually increase and thicken from the margin of basin to the basin center.

The second lithological group (Q_3^1 Ⅱ) is the tawny, grey yellow and grayish-green gravelly soil intercalated with stiff soil layer, whose general thickness is about 31 - 54m and the thin part is only about 10 - 22m. It is made of the glacier (water) dammed lake deposit, which mainly originates from the palaeo-glacier Valley of Sancha River. Within the layers are several tawny stiff cohesive soil with slight stratification, and the thickness of single layer is generally about 1.5 - 3.5m and the maximum thickness can reach 5.5 - 9.1m. This lithological group is mainly distributed on the left bank of dam site, both sides of downstream river valley and the channels of Sancha River, which is deeply buried in dam site and the upstream as well as the basin. Its changing trend from basin margin to basin center and the thickness is gradually thinned to be pinched out. There is yellow palaeo colluvium gravel soil layer occasionally emerged near the left bank of basin at dam site.

The first lithological group (Q_2^2 Ⅰ) is the weakly cemented gravel layer, with the maximum thickness >100m. The thickness of the basin margin is only about 15 - 35m, and thin silt layer occasionally emerges, which belongs to the glacio-lacustrine sedimentary layer. This lithological group is deeply buried in the bottom of basin and river valley. The boreholes indicate that along the river valley near to the basin margin at dam site, there is a layer of dark gray - light yellow gravelly soil intercalated with cohesive soil layer (Q_2^1) at the bottom of this lithological group, which directly covers on the quartz diorite and the thickness varies from 28m to 36m.

According to the granulometric composition, the lithological group can be divided into coarse and fine grained soils. The coarse grained soils are mainly composed of gravel, whose content is 43%–63% and the nonuniformity coefficient is 178.57–828.6, the cumulative curve is basically the low and gentle slope type with small sand content rate. The fine grained soils are mainly composed of silt grain, whose content is 50%–63% and the nonuniformity coefficient is 8.3–15.0, it's silt of good gradation. The fine grained soils have the features of uniform soil, small void space and compact structure, and also have been pressed by the preconsolidation pressure. Thus, they belong to the good natural foundation soil.

The five lithological groups at dam site feature compact structure and high density as well as weak water permeability. Among which, the permeability coefficient of silty loam, yellow stiff soil and gravelly soil is $K=3.51\times10^{-5}-5.6\times10^{-9}$ m/s. Thus it belongs to the extremely weak-slightly weak pervious layer, which can be regarded as the relative aquiclude. And the permeability coefficient of gravel layer is $K=5.19\times10^{-3}-1.58\times10^{-4}$ cm/s. That it belongs to the weak pervious layer.

The results of seepage deformation tests show that the silty loam of the third lithological group and the gravelly soil and yellow stiff soil of the second lithological group has high anti-permeability strength, and their damage gradient is above 12.2–13.93. Also, the main damage form is quick condition. As the gravel of both the third and fourth lithological groups has the calcium cementation effect, its anti-permeability strength is also very high and the damage gradient is above 4.25–10.4. The contact scouring test of gravel and silty loam shows that when the water seepage gradient reaches 4.25, there is only local piping failure but not the contact scouring phenomenon within gravel.

The large load test on site is conducted for the gravel soils with stiff clay of the second lithological group of dam foundation and the caesious silty loam of the third group as well as the gravel layer of both the third and fourth lithological groups, consequently the maximum load can reach 2.2MPa, the P–S relation curve does not have any obvious turning point. The settlement of proportional limit load calculated from graphing method is only 0.314–1.122cm, which is not large. The indoor compression test of original samples of yellow stiff cohesive soil of the second lithological group and the silty loam of the third and fifth lithological groups indicate a maximum vertical load of 1.6MPa, so they all belong to low compression soil. When conducting the compression test for the yellow stiff cohesive soil of the second lithological group and the caesious silty loam of the third lithological group, when the maximum pressure applied reaches 3.2MPa, the normal dense compression section of the compression curve still does not appear, the preconsolidation pressure of fine grained soils in the second and third lithological groups worked out by the graphical method is between 4.5–6.0MPa.

Both the silty loam of the third lithological group and the gravel soils with yellow stiff cohesive soil of the second lithological group have the in-situ large shear testing and indoor small triaxial shear testing etc. The large direct shear test results show that the shear strength values of the same soil under different testing conditions are high and the difference is not large. In addition, the friction angle ϕ is between 33.27°–35.94° and the cohesion C value is greater than 0.1MPa. The stress strain curve from the small triaxial shear test has peak value emerged. It generally presents the hump-type softening brittle failure, and also shows the shear failure features with overconsolidated soil.

The standard penetration number of the yellow stiff cohesive soil of the second lithological group is 46–59, and the velocity of transverse wave is 488–499m/s. The standard penetration number of the caesious silty loam of the third lithological group is 56–106, and the velocity of transverse wave is 463–499m/s. During the dynamic triaxial test, the pore water pressure of the soil samples in vibration process generally generates slowly, and there also shows softening brittle failure with high dynamic strength value. As a result, this kind of soil belongs to the over consolidated soil, and has the features of high standard penetration number, high wave velocity and high dynamic strength value.

Seepage prevention and control project of dam foundation uses both the cutoff wall and curtain grouting, and there is the suspended cutoff wall embedded into the relative aquiclude at the river bed. At the dam foundation on the left bank, there is the totally-enclosed cutoff wall with curtain grouting, which is horizontally extended 150m into the mountain. Also, the deep covering layer of dam foundation on the right bank and bank slope employs the suspended cutoff wall with curtain grouting, and is extended into the mountain for 288m. Furthermore, curtain grouting is applied under the cutoff wall on the right dam head. All the anti-seepage bases use the relative impermeable second lithological group as anti-seepage base. Because of the deep and thick covering layer on the right bank and the deep cutoff wall, the section construction method is adopted. The upper wall is built on platform terra, and the lower wall is built through grooving within the gallery. The connection of the two walls employs the embedded or contacting modes, and also the grouting needs to be strengthened. Furthermore, the link up curtain connection method is also used at the locations with relatively long seepage paths. This project has already been finished to generate electricity. In 2010, it passes the project acceptance, and the project is under normal operating condition.

7.3.7 Natural Building Materials of Certain Projects

7.3.7.1 Material Sources of Artificial Concrete Aggregate for Xiangjiaba Hydropower Station

Xiangjiaba Hydropower Station uses about 13 million m³ of concrete and needs 28.6 million tons of aggregates. The demand of aggregate for its main construction, river diversion and temporary works are respectively 24.13 million tons and 4.47 million tons.

During the pre-feasibility study stage, the selection of concrete aggregate source is conducted at 6 stock grounds of natural sandy gravel materials, limestone material and sandstone material. It basically tends to use Shuanghe limestone as main material concrete aggregate, and the excavated materials of underground powerhouse as auxiliary materials. Furthermore, the natural sandy gravel materials are used for temporary works.

After the feasibility study, further explorations and tests are conducted for the natural sandy gravel materials, artificial limestone materials and sandstone materials. Then, the proven aggregates material sources that can be choosen are: natural sandy gravel materials at stock grounds downstream of Jinsha River and downstream of Minjiang River, Shuanghe limestone stock ground, Xintanxi limestone stock ground, Hanwangshan limestone stock ground, Taiping limestone 1# and 2# stock ground, Xintanba sandstone stock ground, Dashipan sandstone stock ground and Longcangxi sandstone stock ground and so on. Although there are many stock grounds available for selection, each stock ground has different problems. Therefore, there is certain difficulty in selecting the stock ground.

1. Natural sandy gravel stock ground

Each of the Jingshajiang stock grounds generally comes with low sand content ratio, whose exploitable yield is only 7%-13%. The fineness modulus of sand is relatively low and is 1.2 - 2.5, while the mud content is high and is 3.9%-8.6%. Thus, the quality of aggregate is relatively poor. The exploitation during flood season is greatly affected by flood, and also the reserves are scattered and the total volume is small, plus the poor mining conditions, these stock grounds cannot meet the large-scale mining requirements for the main works. Thus, it is not suitable to be used as concrete aggregates sources for the main works, but can be used in temporary works.

Minjiang natural sandy gravel stock ground is at the downstream of Minjiang River, and there are 6 stock grounds which are 38 - 54km away from the dam site, with the total reserves of 46.96 million m³. The main stock ground covers large agricultural acreage and has many residents, so the exploitation of stock ground will come with more resettlement problems. The sand gravel of this stock ground is mainly composed of Triassic sandstone that comes with potential alkali-silica active reaction. The content of needle-like coarse aggregate exceeds the standard, with the particle shape being bad, and the sand gravel gradation not balanced. The fineness modulus is very small, thus the sand belongs to the superfine sand, whose quality cannot meet the requirements of relevant codes. However, by mixing with artificial coarse sand to improve the gradation, the quality of the produced concrete can basically meet the requirements of relevant specifications. This stock ground needs to be exploited underwater, and the exploitation needs to be stopped for 5 months because of flood during the flood season each year, the transportation intensity is high, and the transportation distance is very long, both the transportation and exploitation conditions are relatively poor.

2. Sandstone stock ground

Sandstone stock grounds include Xintanba, Longcangxi and Dashipan stock grounds. The lithology of this stock ground is the Triassic Xujiahe formation (T_3^2) sandstone, mainly the grey white thick-extreme thick mid-fine to mid-coarse sandstone. Within the layers are the laminar or lenticular mudstone, siltstone, carbonaceous belt and coal line. The soft stratum averagely accounts for 5.56% of the minable layer. The average saturated compression strength of rock is 115.59MPa. And the rock is hard, with small porosity and water absorption rate. The aggregate has potential alkali-silica active reaction, but if adding 30% of flyash into the cement, the alkali-silica reaction can be obviously controlled, the requirement of non-active indices can be met. The overburden amount to be excavated is large, and the mining condition is poor.

In addition, the 1.35 million m³ of sandstone excavated from underground power house of this project and the Xintanba sandstone belong to the same rock group, which have the same lithology and quality. They can also be used to produce concrete aggregates.

3. Limestone aggregate

The limestone stock grounds in successive operations are Shuanghe stock ground, Xintanxi stock ground, Hanwangshan stock ground and Taiping 1# and 2# stock ground. The Shuanghe limestone stock ground is 14km of linear distance and 45km of highway mileage away from the dam site. The lithology of this stock ground is the Triassic Leikoupo Formation (T_2^1) limestone, and the monaxial saturated compression strength of rock is 40 - 170MPa, with the average strength of

85.6MPa. The alkali activity test results show that this stock ground has the potential alkali-carbonate active reaction, which the expansion values of test specimen highly exceed the standard values, but there is no effective method to control the alkali-carbonic acid active reaction either at home or abroad at present.

The lithology of both Xintanxi and Hangwangshan limestone stock grounds is the middle Triassic Leikoupo Formation (T_2^1) limestone. Similar as the Shuanghe stock ground, they all have the potential alkali-carbonic acid active reaction.

The Taiping limestone stock ground 1# is located within the Huangpingxi gully on the right bank in the reservoir area, which are 56km of highway mileage and 27km of linear distance away from the dam site. Its lithology is the Permian Maokou Formation limestone. Karst is developed at the stock ground and the overburden amount to be excavated is large with poor construction conditions and difficultly avaiable transportation conditions.

The Taiping 2# limestone stock ground is located in the Xintanxi Town in Suijiang County on the right bank, which is about 56km highway mileage from dam site and 31km linear distance. The exposed elevation of limestone is above 1,160m, and the production area is more than 436,000m². Most of the bed rock is exposure to surface and the effective reserves are 38.67 million m³. The lithology of this stock ground is the upper Permian Maokou Formation (P_{1m}) limestone, mainly the grayish white, dark grey medium thickness dense massive fine-crystalline and microcrystalline limestone. There also occasionally have less bioclastic limestone. The average monaxial saturated compression strength of rock is 90.14MPa, the integrity is fine. Karst of this stock ground is not developed. According to the alkali activity test results, there exists no hazard alkali activity reaction. The stripping ratio is only 0.03, and the exploitation conditions are very good. However, this stock ground is far from the dam site, and the available transportation condition is very poor.

Xiangjiaba Hydropower Station is of large scale and very important for the development of Jingsha River. Also, the alkali activity of hydraulic concrete aggregate is very complicate, and there are some uncertainties in the long-term safe operation of concrete. Therefore, considering the long-term and operation safety of project, it is finally determined to use the Taiping 2# limestone stock ground as the concrete aggregate source for the main works. And the materials excavated from the underground chambers and the natural sandy gravel materials from downstream Jingshajiang can be used in the river diversion works and other temporary works. As the Taiping 2# stock ground is far away from the dam site, it is better to use the long-distance tunnel belt conveyor to convey the aggregate to concrete system in dam area.

7.3.7.2 Selection of Earth Materials for Lianghekou Hydropower Station

Lianghekou Hydropower Station is located in Yajiang County of Tibetan Autonomous Prefecture of Garze, which is the most upstream and the leading reservoir project of the hydropower planning in the midstream of Yalong River. The project is made up of core wall rockfill dam, water diversion and power generation system on the left bank and flood discharge and energy dissipation system and etc. The maximum dam height of gravel soil core wall dam is close to 300m, and the project scale is large. For the preliminarily planned impervious core wall, 5 - 6.3 million m³ of earth materials are needed. As the core wall will bear high water head, its requirements for earth materials is very strict.

During the prefeasibility study stage, reconnaissance works of earth materials are conducted in advance, and some exploration work on stock ground of earth materials reaches detailed exploration precision. To study the plan of rockfill dam with clay core wall, the exploration work of 13 stock grounds from upstream and downstream at dam site is carried out. Among them, the Gala stock ground whose reserves reach 5.26 million m³ is the one with most concentrated reserves among all the stock grounds. However, it is also the stock ground that has most concentrated agricultural land and residents, and it is a Tibetan habitation that is impossible to conduct the land allocation and relocation work. By detailed investigation of Pubarong, Guali, Zhili, Yizha, Yazhong and Pingguoyuan that distributed at the upstream of dam site as well as Xidi, Jiaonibao and Baizi and other nine anti-seepage stock grounds of earth materials at downstream, it's indicated that the total geological reserves can reach 12.44 million m³. This can basically meet the requirements of the project. The investigated stock grounds are all earth materials caused by pluvial sediments and slide rock, and the soil forming rock is basically the metamorphosed sandy slate. As the genetic types of earth materials are similar, their grain composition and physic-mechanical properties are similar as well. The clay content of earth materials is within 18.74%-29.40%, and the permeability coefficient can reach $10^{-6} - 10^{-8}$ cm/s. In addition to the fact that the seepage failure gradient is relatively high, the materials feature fine impermeability and resisting ability to seepage deformation. However, most of the material sources are made of very fine granules, whose index of shear strength is very low. Through adding gravel for modification treatment, the mechanical property of earth materials is greatly improved, thus it can basically meet the technical requirements of impervious wall of 300m earth and rockfill dam. As the stock grounds are dispersedly distributed and there are some differences among the physic-mechanical properties of earth materials from different stock grounds, in order to make reasonably division for the core wall filling and control the quality, it's necessary to fully consider various features of stock ground, to reasonable division and design of core wall filling.

7.3.7.3 Artificial Aggregate Selection for Jinping Ⅰ Hydropower Station

For the exploration of artificial concrete aggregate for Jinping Ⅰ Hydropower Station, investigations and tests are conducted to multiple marble stock grounds, sandstone stock grounds and granite stock grounds within 50km of dam site. After comprehensive study and comparison of reserves, quality, exploitation and transportation conditions of each stock ground, the stock ground study gradually focuses on the right bank of Santan upstream of dam site for marble, Dabenliu on the left bank of downstream of dam site for metamorphic sandstone, and downstream of Jiulong River mouth for granite.

The stock ground of marble on the right bank of Santan is located at about 3km from the upstream dam site, whose elevation is within 1,950 - 2,650m. The lithology of this stock ground is grey medium-fine crystalline marble, mainly the light grey-grey fine crystalline marble, intercalated with medium crystalline marble stripe. The total thickness of marble in this stock ground is about 400m, and the calculated reserves between the elevation of 1,950 - 2,370m are about 15.5 million m^3. The original aggregate test has showed that the wet compression strength of gray medium-fine crystalline marble is 51.5 - 86.1MPa, and the wet compression strength of light gray-gray-white fine crystalline marble is 75.3-76.2MPa, and that of the white medium crystalline marble is 54.5MPa. The strength of white medium crystalline marble is relatively low, which is distributed in other marbles in the shape of strips, and it cannot be removed when mining. The test shows that the rock has no alkali active reaction. The tests of aggregates show that both the crush index and freezing loss of coarse aggregate of gray medium-fine crystalline marble and light-gray - gray fine crystalline marble cannot meet the standard requirements, and the finished product ratio of this marble is very small; whose abrasive resistance is very poor and the fall lose is great. In addition, the fineness modulus of artificial sand is very small; sand powder content is high (it can reach 30% - 40%). Furthermore, the solidity also cannot meet requirements. Thus, in terms of rock strength and aggregate property, the marble stock ground of Santan is not suitable to be used as concrete aggregate for the dam.

The stock ground of metamorphic sandstone at Dabenliu is located on the left bank of downstream dam site, which is about 9km from the dam site. The stock ground is mainly composed of thick-extremely thick caesious fine-grained sandstone of metamorphic quartz. The main mining area (Area B) has the useful layer reserves of about 120 million m^3 at the elevation of 1,700 - 2,100m. And the volume of useless layers of lateral sandy slate needed to be stripped and weathered surface layer is about 1,800,000 m^3. There are also useful layer reserves of 120 million m^3 in Area A. The wet compression strength of sandstone is 100.8 - 147MPa. The finished product ratio of coarse and fine aggregates is high, and the main performance index can meet the standard requirements. By using CECS48 rapid method and mortar length method, it's assessed that the aggregate has no alkali-silica active reaction, while the accelerated motar-bar test and concrete prism test show that the aggregate has potential alkali-silica active reaction. Because of the problem of alkali-silica active reaction, the high-quality low alkali cement shall be used for the 300m grading concrete arch dam, and at the same time, some high-quality flyash are mixed therein to restrain the reaction. Meanwhile, because of the steep slope and being close to river, the area is narrow and small, and the mining conditions are relatively poor.

The stock ground of granite at Jiulong River mouth is located at about 51km away from the downstream of dam site. It has bed rock exposure to surface, and the lithology is the white medium-grained biotite monzogranite. The lithology is even and the strength is high. The useful layer gross reserves at the elevation of 1,610 - 2,000m is 13 million m^3, and the volume of the intensely weathered and unloading rock on surface that need to be stripped is about 2.8 million m^3. The strength of original rock on the stock ground can meet the requirement, but the mica content of rock is relatively high. After concrete test, the thermotics index and ultimate tension etc. can not meet requirements. The stock ground is abundant in reserves, and offers fine mining conditions. But the transportation distance is long.

The marble stock ground of Lanba is located on the right bank which is 5km of the upstream of dam site. Its useful layers are distributed in the light gray dense marble and gray mesh marble at the elevation below 1,900m, and the rock is dense and hard. The gross reserves of useful layers at the elevation of 1,700 - 1,900m are only about 965,000 m^3, while the gross reserves of useful layers at the elevation of 1,810 - 1,900m are about 128,000 m^3. The wet compression strength of density marble is 74.9 - 130.2MPa, and the wet compression strength of mesh marble is 77.1 - 94.4MPa. Apart from the facts that the sand powder content of artificial sand of the mesh marble is high and the index of solidity is unqualified, the other main performance indexes of aggregate can meet the requirements of specifications. However, the reserves of this stock ground are far from satisfying the project requirements, and the mining conditions are relatively poor.

Through repeated selection and research, it is finally decided in the artificial aggregate scheme of Jinping Ⅰ Hydropower Station that the metasandstone is adopted as the coarse aggregate of concrete and the marble in Santan as the fine aggregate of concrete. The actual processing after the construction shows that the apparent density, the sturdiness, and the trial value of crushing index of finished coarse aggregates can meet the standard with the oversize content generally up to the standard. However, extra large stones, large stones in the needle shape (columnar) are in large contents. They are strip, flat grains in inferior particle shape, and the stones with the maximum of single sided length over 200mm account for 30%

−50%. The large content of stones in the needle shape will influence the operation of tube machine and forced concrete mixers on the coarse aggregate supplyline as well as the construction workability and anisotropy of concrete. The appearance of the concrete pedestal core indicates that the aggregates of large stones tend to be in a row side to side. Relevant reshaping inspection shall be conducted in large stones, aggregates of large stones (counter-strike, cone crusher) in order to guarantee the constructional quality of the concrete. Reshaping equipment to increase the coarse aggregates shall be conducted on the basis of the inspection to promote the acceptability concerning coarse aggregate in the needle shape of extra large and large stones. Due to a poor abrasion resistance of sandstones, the rock flour produced in the aggregate processing course tends to be absorbed by the aggregate surface, which results in powders on the aggregate surface after the aggregate rescreening and increases the total content of rock flour in the concrete. Therefore, the flushing technology shall be strengthened and improved to control the content of rock flour in the coarse aggregate within 0.3%.

The finished sands from the processing of fine aggregates of marbles on the right bank of Santan have discontinuous grading, a low content of intergrade grains of 0.63−2.5mm, and a high content of rock flour. It has some influence on the production quality, and leads to waste of aggregates, a low rate of finished products of artifical sand, and a low output. From the actual processing condition of the above aggregates, the performance of aggregates for the project isn't satisfactory, but considering the actual condition of Jingping Project, the current scheme of aggregate sources is the only choice.

7.3.8 Research on Reservoir-induced Earthquakes

It is indicated by statistics that the reservoir-induced earthquake is a small probability event, that is to say, only a small amount of reservoirs in thousands of reservoirs may induce an earthquake. According to the statistical standard of International Commission On Large Dame (ICOLD), there were around 50,000 dams with a height over 15m worldwide until 2003, and only over 130 reservoir-induced earthquakes have been reported up to now, Less than 100 of which are widely acknowledged, account for around 2‰ of all the dams higher than 15m. China is one of the countries where many reservoir-induced earthquakes happened. There have been 34 reported reservoir-induced earthquakes, 22 of which have been widely acknowledged. It only accounts for around 1‰ of all the 25,800 dams higher than 15m in China.

The reservoir-induced earthquake is a complicated natural phenomenon. The formation mechanism and occurrence conditions of earthquakes, especially the time of occurrence, location, strength, and prediction, still remain unsolved. But people have acquired some basic knowledge about the activity features and patterns of the reservoir-induced earthquakes.

(1) Regard to the spatial distribution, most of the reservoir-induced earthquakes happen in the reservoir basins and within 3-5km from the bank, seldom beyond 10km.

(2) The origin time of main shock is closely related to the process of reservoir filling. In the early stage of the reservoir filling, there is a coherence between earthquake activities and the rise and fall of the reservoir water level. The climax of strong earthquakes activities usually appears in the season of high water level in the previous storage periods, and with some time delay, and it has certain relevance with the rise speed of the water level and the duration of the high water level.

(3) Changes in the internal and external conditions arising from the reservoir filling are gradually adjusted to a balance as time goes on. Therefore, the frequency and intensity of the reservoir-induced earthquakes decrease obviously as time prolongs. According to the statistics of 55 reservoirs both at home and abroad, there are 37 main shocks happened within a year after the reservoir filling, accounting for 67.3%; 12 earthquakes happened in 2 to 3 years, accounting for 21.8%; 2 happened within 5 years, accounting for 3.6%; and 4 happened after 5 years, accounting for 7.3%.

(4) The vast majority of the reservoir-induced earthquakes are slight shocks and weak earthquakes with the earthquake magnitude below 4 in general. According to statistics, the reservoir-induced earthquakes with the magnitude no higher than 4 account for 70%−80% of the total amount of earthquakes, and those with the magnitude over 6 (6.1−6.5) only account for 3%.

(5) The focus depth is extremely shallow, most of which are 3−5km and even near the surface.

(6) Due to the shallow depth of seismic focus, compared with natural earthquakes, the reservoir-induced earthquakes have a relatively high frequency, ground peak acceleration, and epicentral intensity. But the meizoseismal areas are small, and the intensity decays fast.

(7) Generally the probability of the reservoir-induced earthquakes is only about 0.1% of the total amount of projects. However, as the height of dams and the capacity of reservoirs keep rising, there is an obvious rise in the proportion. The earthquake probability of dams with the height over 100m and the reservoir capacity over 10 billion m^3 is around 30%.

(8) The magnitude of some reservoir-induced earthquakes may exceed that of the most serious earthquake magnitude in history there, or the basic earthquake intensity in that place. Therefore, neither can be used to judge the maximum strength of potential reservoir-induced earthquakes in a region.

Through analysis and research of a large amount of earthquakes, the contributing factors of the reservoir-induced earthquakes can be divided into three types:

1) Structure type. It refers to earthquakes induced by weak parts of certain fracture structures in the reservoir area triggered by the water. The reservoir-induced earthquakes of this type have a relatively strong intensity, a large influence on the water conservancy projects. It is a major type being researched by countries around the world most frequently.

2) Karst (KST) type. It occurs in Karst developed region in carbonatite distribution area. The rising water in reservoirs flows in the Karst caves, and the high hydraulic pressure creates gas explosion, water hammer effect, and large-scale karst collapse in the caves. It is one of the most common reservoir-induced earthquakes. 70% of reservoir-induced earthquakes in China belong to this type. But earthquakes of this type have a low magnitude, more often less than magnitude 2-3, and only around magnitude 4 as the maximum.

3) Microfracture on the surface, also called the surface unloading type. It refers to earthquakes arising from adjustment fractures, displacement, or deformation of rock mass on the surface under the impact of reservoir water. Earthquakes of this type usually occur in hard brittle rock mass or in the so-called insufficient unloading area on the bottom of river valleys. Earthquakes of this type usually have a very small magnitude, less than magnitude 3 and with a short duration. According to statistics in recent years, the induced earthquakes of this type are more common than expected in the past.

Besides, the rising water drowns abandoned mines, and causes slope deformation, etc., which may also cause the vibration of rock mass to create an "earthquake".

The research of the reservoir-induced earthquakes in the Three Gorges Project has gone through the analogy judgment of seismology and geology, multi-factor comprehensive analysis, and special test on key factors, etc. Methods adopted include common research and survey of seismological and geological conditions, multiple special tests, and many ways of numerical simulations as well as analysis calculations. A satisfactory result was acquired through all these methods.

On basis of the above work, methods below are mainly adopted for the prediction of the reservoir-induced earthquakes: ①Analogy prediction through the seismology and geology. ②Prediction through statistics analysis. ③Prediction by numerical analysis. ④Comprehensive evaluation of the reservoir-induced earthquakes.

Since the impoundment of the Three Gorges reservoir, tens of thousands of micro earthquakes occurred in the regions such as Badong and Zigui. The magnitude of most earthquakes there was around magnitude 1, and the maximum around magnitude 4. The locations where the earthquakes occurred and the magnitudes basically coincide with the prediction result.

7.3.9 Near-dam Geological Hazard
7.3.9.1 Laxiwa Hydropower Station

With the achievement of river closure on January 9, 2004, the main construction was commenced on 15 April, 2006. The diversion tunnel closed gate on March 1, 2009 for impoundment of reservoir. During a patrol after reservoir impoundment in May 2009, it was found that the top platform of Guopu deformation mass on the right bank before dam was showing an obvious sign of deformation. The subsided belt at the elevation of 2,950m platform of original slope top was showing obvious tensile crack and increased fissure, and new tensile cracks and fissures and local landslides appeared on the bank slope as well. From August 15, 2009 to December 5, 2009, from the observation point at the front edge of deformation mass platform, the accumulative maximum horizontal displacement reached 3.52m, and the accumulative maximum vertical displacement reached 2.28m. Mean daily vertical displacement reached 28mm, and maximum horizontal displacement reached 43mm, showing a rapid deformation rate.

Guopu deformation mass is located 900-1,700m at right bank in front of the dam, between No. 1 Channel and Huanghua Channel, covering a large range of deformation of the bank slope. For its short distance to the dam and water inlet, its stability will have a vital impact on project safety. Guopu deformation mass was found at early reconnaissance. There is subsided tensile belt at platform of El. 2,950m, which was covered with soil layer on the surface. With several years of simple observation during the reconnaissance, no signs of deformation were found in the rock mass. It is infered that the rock mass was formed at geological time, with downward incision of Yellow River, it's gradually formed with the valley slope free face. Its bottom boundary is at the elevation of around 2,750m, under the control of Hf_{104} gentle dipping structure plane, with an estimated volum of 30 million m³.

Since Guopu deformation mass was found with obvious deformation after impoundment of the reservoir, Northwest Hydro Consulting Engineers (NWH) of China Hydropower Engineering Consulting Group Co., the project owner and designer, has attached great importance to this issue and carried out corresponding reconnaissance, deformation monitoring and analysis. The deformation monitoring and reconnaissance are still under way at present.

Based on the analysis of field reconnaissance and existing deformation monitoring information, current Guopu deformation is characterized by the following features:

(1) Absolute deformation is increasing continuously. By the end of September 19,2010, the maximum synthetic displacement at the front edge of platform (Point K1 at the front edge of No. 2 Ridge) had reached 21.09m, in which the horizontal displacement was 17.76m, and the vertical displacement was 11.37m;

(2) The displacement is large at front edge of high elevation platform, relatively small at the rear edge; the front edge is predominated by horizontal deformation, while the rear edge by vertical deformation;

(3) The displacement vector dig angle of top platform is large ($\angle 40°-50°$), and the deformation displacement vector of 2,400 - 2,500m elevation at bottom bank slope is predominated by approximately horizontal displacement (approximate dig angle $\angle 10°-20°$);

(4) The tensile crack and fissure on top platform are increasing continuously. Vertical crack and fissure also appear, deformation points on the slope surface increase, and the area and scale of collapse are also constantly expanding;

(5) New tensile crack and fissure appear on top of No.1 Ridge, of which the deformation range is on the rise and spreads to the downstream free face. However, no sign of obvious deformation and destruction has yet been found below elevation of 2,800m up to date;

(6) The deformation is closely related with impoundment of reservoir. With the rise of water level, the deformation rate grows basically without delay. When the water level is stable at a certain elevation, the deformation rate tends to be stable; however, the total deformation amount is on the rise constantly, and the development tendency of the slope deformation is facing severe challenge.

The preliminary study suggests that Guopu bank slope underwent overall creeping deformation at geological time, which resulted in relaxation of rock mass. Under the influence of reservoir impoundment and other factors, new overall deformation was generated, whose bottom boundary may lie around the original river level. Meanwhile, due to its low safety or stability, high and steep bank slope, strong degree of rock weathering and unloading on superficial layers, under the effect of gravity, various deformations and failures such as tensile crack, fall and landslide occur on the slope. With low overall safety and stability of the slope, high-steep topography conditions, unloading and relaxing of superficial rock mass, the disintegrated creeping deformation and failure occur under the effect of gravity; however, the possibility of overall failure shall not be eliminated.

According to analysis of existing information, the maximum total deformation is approximate 60 million m³, suppose that the maximum slide is 15 million m³, maximum slide speed 40m/s, the maximum climb height of surge is no more than 40m. At the time of risk analysis, an active research on thorough treatment measures and pre-arranged planning shall be commenced upon a preliminary recognition on formation mechanism, stability and deformation tendency of the deformation mass in light of reconnaissance progress and slope deformation and failure condition.

To avoid the risk of the project brought by deformation of bank slope, at the time of closely monitoring the deformation of bank slope, a corresponding critical slide early warning system and emergency plan shall be researched and formulated, the monitored information is used to guide the impoundment process of reservoir. Controllable impoundment strategy shall be adopted in reservoir impoundment, which divides water level into some steps by 5 - 10m and stays in a step of water level for a certain period, after making analysis on monitored information, the water level of reservoir will be raised to another step. The water level of reservoir is kept stable at approximate 2,430m in early 2011 after gradually passing the water level steps of 2,400m, 2,410m, 2,420m and 2,430m, etc.

7.3.9.2 Suofengying Hydropower Station

In the event that the physical and mechanical parameters required for stability analysis and treatment design of deposit body can not be achieved through test method, the mechanical parameters of corresponding boundaries can be achieved by back analysis on the basis of ascertaining the boundary conditions under a judgment regarding the stability in a natural state in macroscopic view. Regarding comprehensive treatment of the deposit body, the comprehensive treatment measures such as reducing load, supporting and retaining, drainage, protection and etc. are adopted after the safety grade of the slope is determined according to the location in the project and its importance. For instance, if the bottom of No. 1 deposit body is a diversion channel, no overall deformation is permitted, the specific plan is to reduce load on the top of the deposit body, and make use of the diversion channel structure at the bottom, and plus the cast-in-pile pile to execute the anti-slide treatment; establish drainage system within slope to decrease underground water level, make framework and afforest on slope surface for protection purpose, after the aforesaid treatments are completed, the long-term deformation monitoring system could be arranged.

For dangerous rock mass, the project treatment design, based on the stability requirements, adopts corresponding measures according to the possible instability mode: anchoring the unloading crack with reinforce ferroconcrete beam at the top, making treatment of all cracks by channeling and backfilling for the purpose of rendering anti-toppling ability and preventing penetration of surface water; and laying out long-term monitoring facilities between cracks and rocks. The middle of dangerous rock mass is reinforced by pre-stressed anchor cable. Meanwhile, in order to improve the integrity of dangerous rock mass, the pre-stressed anchor cable and bolt system are set to combine various rocks into an integrity and enhance rigidity of the rock mass. Anti-slide piles are set perpendicular to the sliding surface of soft rock at the bottom to prevent dangerous rock mass from causing shearing and sliding failure by sliding along weak intercalated layer, and meanwhile consolidation grouting is carried out in base rock mass to improve the deformation resistance of the soft rock. The long-term monitoring facilities are established for long cracks of interior dangerous rock mass and bottom base.

7.3.9.3 Zaoshi Hydro Project

Shuiyangping-Dengjiazui landslide treatment is predominated by unloading and drainage. Such measures as surface shotcrete support, building drainage ditch, building drainage tunnel and hole in the sliding body etc. are adopted; comprehensive measures are taken to regulate Shuiyangping platform, such as cutting slope & reducing load to the elevation of 165 – 170m, cleaning specially developed part of unloading crack at the front edge of platform, and shotcreting with wire mesh; slope cutting and leveling on a moderate basis, surface drainage, spraying and protection are carried out in at local places of Dengjiazui, and gabion slope protection and dumping rock for foundation protection are primarily carried out to the edge part which is eroded directly by water. After completion of landslide treatment project in the year of 2006, each monitoring index indicates sound stability of landslide.

For 3 years of impoundment of the project, there are no abnormal phenomena occurred to the landslide, which shows that the treatment is effective.

7.3.9.4 Wudongde Hydropower Station

The normal impoundment level of Wudongde Hydropower Station is 975m, with total reservoir storage of 7.405 billion m^3, maximum dam height of 265m, and installed generation capacity of 8,700MW. Jinpingzi landslide is located on the right bank of downstream Wudongde Hydropower Station, approximately 900m away from the dam site at the upstream. The landslide is vital to the feasibility of cascade development.

The elevation of front edge of Jinpingzi landslide is 820 – 850m, the elevation of rear edge of top slope is over 1,900 – 2,200m, and the elevation difference between front and rear edge is over 1,000 – 1,200m. Jinpingzi landslide is divided into five areas, of which Area I, Area II, and Area III are Quaternary deposit area, Area III, as the closest one to the Wudongde dam site and directly affected by the project, is a key focus on stability research; while Area IV is a bed rock outcrop area, and its original position is a key point to determining whether the landslide of 0.6 billion m^3 exists or not. Although Area V is also a bed rock outcrop area, its original position is an important reference to make macroscopic judgment on the stability of Area III, therefore the research on original positions of bed rock of two areas should be given top priority among researches.

Upon reconnaissance analysis and argumentation, and in combination with monitoring information, it is indicated that Area IV and Area V of Jinpingzi Landslide are abundant and thick bed rock, the deposits of Area I and Area III are in a state of overall stability, Area II Abai-Xiongjia well gully landslide is active in local area, however, it is relatively far from upstream dam and its deformation and failure modes are predominated by creeping and debris flow or small debris flow within gully, which results in limited instability scale. Therefore, Jinpingzi landslide will not affect the cascade development of Wudongde Hydropower Station and selection of dam site in nearby Wudongde valley.

Chapter 8

Dam Type and Project Layout*

Shi Ruifang❶

8.1 General

8.1.1 *Dam Construction from 1949 to 1980*

Since the foundation of New China in 1949, around 30 years till 1980, lots of dams had been built up for the purposes of water conservancy and hydropower development, among which a large number of them are over 70m high, and incomplete statistics shows that the total number is 44 as summarized in Table 8.1-1. Besides, some other distinctive dams like stone masonry arch dams, hollow gravity dams and multi-arch dams were constructed as well, though their reservoirs or installed capacities are not so large in scale. The 30-years' achievements on the dam types and project layout patterns opened a way forward for seeking the law of dam construction in China under the guideline of self-dependence, accumulated rich experience and acquired plenty of lessons in the dam engineering cause, and laid a solid foundation for further development in recent 30 years since the Reform and Opening-up Policy.

Table 8.1-1 Dam Higher than 70m Completed before 1980 (in Order of Dam Height)

No.	Name of project	Province (municipality/ region)	Dam type	Max. height (m)	Reservoir storage capacity ($10^8 m^3$)	Installed capacity (MW)	Dam crest length (m)	Dam volume ($10^4 m^3$)	Year of construction
1	Wujiangdu	Guizhou	PG	165	23.0	1250	395	193.2	1970-1982
2	Liujiaxia	Gansu	PG	147	64	1350	204	76.0	1958-1974
3	Yunfeng①	Jilin	PG	113.75	38.95	400	828	274.0	1959-1965
4	Shitouhe	Shaanxi	CSD	114	1.47	54.7	590.0	835	1974-1981
5	Fengtan	Hunan	PG	112.5	17.4	815	488	117	1970-1979
6	Panjiakou	Hebei	PG	107.5	29.3	420	1,039.11	280.0	1975-1985
7	Huanglongtan	Hubei	PG	107	11.7	510	371	98	1969-1976
8	Sanmenxia	Henan	PG	106	96	410	713.2	163	1957-1960
9	Shuifeng①	Liaoning	PG	106.4	146.66	900	899	340	1937-1943
10	Xinanjiang	Zhejiang	PG	105	216.26	850	850	138	1957-1965
11	Xinfengjiang	Guangdong	PG	105	138.96	355	440	91	1958-1977
12	Zhexi	Hunan	PG	104	35.7	947.5	330	65.8	1958-1975
13	Bikou	Gansu	CERD	101.8	5.21	300	297.36	424.1	1969-1976
14	Qunying	Henan	VA	100.5	0.2	1	154.28	18.1	1966-1971
15	Danjiangkou②	Hubei	PG/CERD	97	339.1	900	3,442	292.8	1958-1974
16	Fengshuba	Guangdong	PG	95.3	19.32	200	400	73.1	1970-1974

* Owing to the engineering design evolution and construction in different phases, parameters of some projects (power stations, reservoirs) referenced in this chapter might be different. And more examples can not be included herein due to limitation of the text length.

❶ Shi Ruifang, Former president of Hydrochina Xibei Engineering Corporation, Professor Senior Engineer.

Chapter 8 Dam Type and Project Layout

Continued

No.	Name of project	Province (municipality/region)	Dam type	Max. height (m)	Reservoir storage capacity ($10^8 m^3$)	Installed capacity (MW)	Dam crest length (m)	Dam volume ($10^4 m^3$)	Year of construction
17	Ansha	Fujian	PG	92	7.4	115	168	47.3	1970–1978
18	Fengman③	Jilin	PG	91.7	109.88	554	1,002.5	194	1937–1953
19	Niululing	Guangdong	PG	90.5	7.78	80	341.2	40.7	1976–1985
20	Meishan	Anhui	VA	88.24	22.63	40	443.5	23.6	1954–1956
21	Shimen	Shaanxi	VA	88	1.1	40.5	254.42	20	1969–1973
22	Xianghongdian	Anhui	PG	87.5	26.32	40	367.6	28	1956–1958
23	Gongzui	Sichuan	PG	85	3.73	700	447	74.5	1966–1979
24	Mozitan	Anhui	PG	83.1	3.47	20	331	29.4	1956–1958
25	Shibianyu	Shaanxi	CFRD	82	0.26	3	285	190	1971–1982
26	Zhouwan	Shaanxi	TE	82	0.70	—	171	1362	1970–1976
27	Quanshui	Guangdong	VA	80	0.22	24	140.34	5.7	1972–1976
28	Huanren	Liaoning	PG	75	34.6	246.5	593.3	119.6	1958–1972
29	Nangudong	Henan	ER	78.5	0.775	—	205	222	1958–1960
30	Liuxihe	Guangdong	VA	78	3.78	48	255.52	13	1956–1958
31	Wangwushan	Hunan	VA	77.3	0.694	—	105.47	4	1971–1975
32	Zhaikou	Henan	TE	77.2	1.85	4.8	258	195	1959–1973
33	Chencun	Anhui	VA	76.3	27.06	180	419	70	1958–1971
34	Yaxi I	Zhejiang	VA	75	0.29	6.4	181	13	1970–1977
35	Foziling	Anhui	PG	75.9	4.91	40.1	510	19.5	1952–1954
36	Lishimen	Zhejiang	VA	74.3	1.79	8.25	265.57	12	1973–1979
37	Bajiazui	Gansu	TE	75.6	5.4	26.5	565	400	1958–1962
38	Fengjiashan	Shaanxi	TE	73	3.89	2.5	282	255	1970–1982
39	Nanshan	Zhejiang	TE	72	1.05	3.67	242	271	1965–1973
40	Xinyugong	Henan	VA	71	—	—	83	—	1969–1970
41	Chengbihe	Guangxi	TE	70.4	11.21	30	425	291	1958–1961
42	Liuduzhai	Hunan	TE	70	1.205	2.9	480	236	1974
43	Bianqiangqu	Qinghai	TE	70	0.069	—	270	237	1970–1976
44	Gucheng	Beijing	VA	70	0.085	0.55	93.37	7.2	1973–1979

Note: This table is quoted from *the Summary on Investigation and Design of Water Conservancy and Hydropower Projects*, edited by the General Institute of Water Resources and Hydropower Planning and Design, MOE & MWR, 1993. Missing parameters in the original table have been supplemented or updated.

① Yunfeng and Shuifeng dams are on the rivers shared between China and The Democratic People's Republic of Korea (DPRK);
② Danjiangkou dam has been raised up over 100m high;
③ Fengman power station has been extended with a total capacity over 1,000MW.

(1) A number of large-sized high dam projects with distinctive features and multiple benefits were constructed in combination with river training works. Since the foundation of New China, for mitigating water troubles and utilizing hydropower resources of rivers such as the Huai River and Yellow River, a number of medium or large-sized water conservancy and hydropower dam projects were constructed, and much benefits were gained from these projects in terms of power generation, flood control, irrigation, water supply and etc. Till 1980, 44 dams (Table 8.1-1) higher than 70m were built, of which 15 dams exceed the 100m height, 18 reservoirs retained by dams have the storage capacities over 1 trillion m^3, and 13 hydropower stations have the installed capacities over 250MW. Scattered over the major rivers in China, these projects have

gained obviously multi-benefits and played important roles in mitigating flood hazards and promoting the development of power industry.

(2) As for the dams constructed in the 30 years before the reform and opening-up, the dam types were designed independently and selected as suitable to local conditions, and they all have their own uniquenesses. According to statistics, relevant data of Taiwan Province are not included[1], by the end of 1970s, together about 8,400 reservoirs of all types had been constructed, with a gross storage capacity of about 400 billion m^3, of which 319 reservoirs have the capacity over 100 million m^3; among the constructed dams, 89 dams are higher than 60m, of which the earth-rockfill dams, concrete dams and masonry dams account for 41.6%, 35.9% and 22.5% respectively.

Earth-rockfill dams were widely applied and lots of new types were developed to be adapted to local conditions, for instances, the 101.8m high Bikou loam core earth-rockfill dam in Gansu Province, the 81.3m high Nanshui directional blasting rockfill dam in Guangdong Province, and the 60m high earthfill dam constructed underwater in Fenhe River in Shanxi. With the development of earth rockfill dams, those previously under-evaluated "poor quality materials" such as weathered gravelly soil, red clay, medium-fine sand and excavated rock mucks were extensively used in the dam embankment works, and the variety of local earth and rock materials applied in dam embankments has been greatly extended.

Among the concrete dams, the gravity ones and slotted gravity ones predominate, such as the first over-100m high Xin'anjiang slotted gravity dam designed and constructed solely on China's own forces; the first over-1,000MW Liujiaxia Hydropower Station with a 147m high gravity dam. Besides, a lot of thin dams such as multi-arch dams and buttress dams were built up as well. Much progress has been made in arch dam construction, for instances, the 80m high Quanshui double-curvature arch dam in Guangdong Province, with a thickness-height ratio of 0.11; the 112.5m high Fengtan hollow arch dam in Hunan Province with unique features like the powerhouse contained in the dam body and water overflowing on the dam crest; and the 54.8m high Zhaixiangkou overflow double curvature arch dam on the Maotiao River in Guizhou Province, though the dam is not so high, it is founded on the riverbed gravel stratum.

Masonry dams have a very long history in China, and in the 30 years after founding of new China, many more masonry dams had been constructed. The highest masonry dam is Qunying Dam in Henan Province, with a height of 100.5m. The 48m high Tonghangxi double curvature masonry arch dam in Zhejiang is a masonry dam with the smallest thickness-height ratio (0.11).

(3) Some "head" reservoirs were constructed on the upstream location of rivers or river reaches for regulating runoff and making cascade development. In the 30 years, a series of large-or medium-sized water conservancy and hydropower projects had been constructed on the Lanzhou reach of the Yellow River as well as several other mediam rivers, and cascade development in succession could be made thanks to the storage and regulating capability of the "head" reservoirs. For instances, with the Liujiaxia high dam with large reservoir as the "head" in the upper reach of the Yellow River, Yanguoxia and Bapanxia Hydropower Stations were constructed; with the Huanren high dam with reservoir as the head on the Hunhe River in Liaoning Province, three cascade hydropower stations were constructed in the downstream. In the 30 years after the reform and opening-up, this kind of development mode and reservoir construction have made much progress, for instances, several cascade projects were constructed in the downstream location of Longyangxia Reservoir on the Yellow River, Wangfuzhou Hydropower Station in the downstream location of Danjiangkou Reservoir, Lingjintan Hydropower Station in the downstream location of Wuqiangxi Reservoir, Tongzilin Hydropower Station in the downstream location of Er'tan Reservoir, Hongshi Hydropower Station in the downstream location of Baishan Reservoir, Xixiayuan Hydropower Station in the downstream location of Xiaolangdi Reservoir, etc. Successive development of these cascade projects saved a lot and gained great benefits.

(4) Based on successful experience and lessons, a complete series of codes, specifications and standards well-suited with China's actual conditions have been published and used. During the 30 years after the founding of new China, including the "Great Leap Forward" period in the 1950s and the "Cultural Revolution" period in the 1960s, much successful experience has been gained on the water and hydropower resources development, the types of dam and project layout selection, as well as compensation and resettlement arrangement because of reservoir inundation. However, there are many lessons should be thought over and learned from, namely, the reservoir sedimentation influence on the construction progress, dam foundation consolidation or detrimental cracks treatment, damage from high speed flow to flood and sand sluice structures, and so on. Perception and understanding were furthered on the dam type selection, dam aseismic and reinforcing design, flood design standard as well as resettlement issues. All these practical experiences and lessons promoted the complete amendment and formulation of codes and standards in accordance with China's actual conditions. From the 1970s, the flood design standard and series of specifications on engineering design, construction and management of gravity dams, arch dams,

---

[1] Quoted from *the Large Dams in China*, Editorial Group, China Water Power Press, 1987.

Chapter 8 Dam Type and Project Layout

earth-rockfill dams, hydraulic tunnels, hydraulic concrete structures and reservoir inundation treatment were issued in succession, which played important roles in and laid both theoretical and practical bases for guiding and promoting sustainable development of water resources and hydropower since the 1980s.

8.1.2 Dam Development from 1980 to 2000

From 1980 to the end of 20th century, it was a fast developing period for water conservancy and hydropower development in China. Within 20 years, about 36 dams higher than 75m had been constructed and put into operation, as indicated in Table 8.1-2 (Projects with the dam higher than 75m are not included if its storage capacity is small), of which 30 dams are higher than 100m, the number is 2 times that of the 15 dams constructed and put into operation before 1980. The completed highest arch dam is Er'tan Hydropower Project, with a dam height of 240m and an installed capacity of 3,300MW. The highest earth rockfill dam is Xiaolangdi Multipurpose Project with a dam height of 154m and an installed capacity of 1,800MW. And the highest concrete face rockfill dam (CFRD) is Tianshengqiao I Project with a dam height of 178m and an installed capacity of 1,200MW. They all had been put into service by the end of the 20th century. The Three Gorges Project on the Yangtze River is the world largest hydropower station with a capacity of 22,500MW, and its main dam is the highest gravity dam (181m high) in China in the 20th century. The early installed units were put into operation in 2003. The dams constructed within the twenty years manifested the major achievements and key features in China within the last 50 years and paved a firm base for sustainable development of the dams in the 21st century.

Table 8.1-2 Constructed Dams Higher than 75m from 1980 to 2000 (in Order of Dam Height)

No.	Name of project	Province (municipality/ region)	Dam type	Max. height (m)	Reservoir storage capacity ($10^8 m^3$)	Installed capacity (MW)	Dam crest length (m)	Dam volume ($10^4 m^3$)	Year of commissioning
1	Ertan	Sichuan	VA	240	61	3,300	774.69	414	1998
2	Tianshengqiao I	Guizhou, Guangxi	CFRD	178	102.57	1,200	1,104	1,800	1999
3	Longyangxia	Qinhai	PG	178	247.7	1,280	1,226	157	1986
4	Dongfeng	Guizhou	VA	162	10.25	695	250	48	1994
5	Xiaolangdi	Henan	ER	160	126.5	1,800	1,667	4,900	1999
6	Dongjiang	Hunan	VA	157	92.7	500	438	96	1987
7	Lijiaxia	Qinhai	VA	155	17.5	1,600	414	110	1996
8	Geheyan	Hubei	VA	151	34.0	1,200	674	237	1993
9	Baishan	Jilin	PG	149.5	59.21	1,800	676.5	163	1983
10	Wuluwati	Xinjiang	CFRD	133	3.47	60	365	680	2000
11	Baozhusi	Sichuan	PG	132	25.5	700	524.48	200	1996
12	Manwan	Yunnan	PG	132	10.06	1,670	418	153	1995
13	Shanxi	Zhejiang	CFRD	132.8	18.24	200	418	500	2000
14	Hunanzhen	Zhejiang	PG	129	20.6	320	440	130	1982
15	Ankang	Shaanxi	PG	128	32	800	541.5	321	1989
16	Jinpan	Shaanxi	TE	125	1.72	18.2	405	559	UC
17	Heiquan	Qinhai	CFRD	123.5	1.82	12	438	550	UC
18	Baixi	Zhejiang	CFRD	124.4	1.68	18	398	370	UC
19	Guxian	Henan	PG	125	11.75	60	315	109	1995
20	Baiyun	Hunan	CFRD	120	3.6	54	200	170	1998
21	Gudongkou	Hubei	CFRD	117.6	1.476	45	193	190	UC
22	Dachaoshan	Yunnan	RCCPG	111	9.4	1,350	460.39	130	2000
23	Yantan	Guangxi	RCCPG	110	33.8	1,210	525	171.7	1995
24	Gaotang	Guangdong	CFRD	110.7	1.0	36	288	195	UC

Continued

No.	Name of project	Province (municipality/region)	Dam type	Max. height (m)	Reservoir storage capacity ($10^8 m^3$)	Installed capacity (MW)	Dam crest length (m)	Dam volume ($10^4 m^3$)	Year of commissioning
25	Shitouhe	Shaanxi	TE	105	1.5	16.5	465	855	1982
26	Qiezishan	Yunnan	CFRD	106.1	1.25	16	236	140	UC
27	Lubuge	Yunnan, Guizhou	ERD	103.8	1.22	600	217	222	1992
28	Chaishitan	Yunnan	CFRD	101.8	4.37	60	312.6	217	UC
29	Jinshuitan	Zhejiang	VA	102	13.93	300	350.6	30	1987
30	Shuikou	Fujian	PG	101	26	1,400	783	181	1995
31	Xibeikou	Hubei	CFRD	95	1.96	18.9	222	162	UC
32	Wanjiazhai	Inner Mongolia, Shanxi	PG	105	8.96	1,080	443	156.7	1998
33	Wuqiangxi	Hunan	PG	85.83	42.9	1,200	719.7	147	1997
34	Tongjiezi	Sichuan	PG	82	2.6	600	1,084.59	230	1992
35	Dahua	Guangxi	PG	74.5	8.74	566	1,166	128	1982
36	Puding	Guizhou	RCCVA	75	4.01	75	195.17	13.7	1994

Note: 1. This table is quoted from *Summary on Investigation and Design of Water Conservancy and Hydropower Projects*, Chapter 3 Hydraulic Structures (1993), as well as references or data from pertinent engineering deign institutes. Missing parameters in the previous table have been supplemented or updated.

2. Dams higher than 75m with small installed capacities are not included in this table. Dams under construction marked in the table had been commissioned around 2000.

1. A number of large hydropower stations and high dam projects had been constructed on the major rivers, which greatly increased the gross hydropower installed capacity

As shown in Table 8.1-2, these dams are widely distributed over the large rivers in China, most of them higher than 100m, and 9 of them higher than 150m. The total storage capacities of the reservoirs are generally 2.5 billion m^3 to 3.0 billion m^3, bringing about multiple benefits such as power generation, flood control, navigation, irrigation and water supply. There are 13 hydropower stations as indicated in Table 8.1-2 having the installed capacity over 1,000MW with big single unit capacities of 200MW to 300MW, and most units supplied by Chinese manufacturers. A number of pumped storage projects were rapidly built up along the coastal area, and the upper and lower reservoirs of the commissioned pumped storage power plants in Guangzhou, Shisanling and Tianhuangping are constructed with different types of dams. In the 20-year period from 1980 to 2000, China had made great achievements in the dam construction as well as water conservancy and hydropower development, and it was also a 20-year of development to a higher level on the dam and hydropower science and technology, which drew much attention from the world.

2. A number of large dams and hydropower projects with distinctive features were constructed

(1) The number of dams with a height of 150m to 200m was growing, and the maximum dam height had surpassed 200m. As shown in Table 8.1-2, most of the high dams over 75m are concrete ones. In the 36 high dams, the gravity dams and arch dams account for 55.5%, the earth-rockfill dams and CFRDs account for 41.7%, the only one buttress dam accounts for 2.8%, and the masonry dam is nil. The highest gravity dam is the 181m high Three Gorges Dam under construction. The highest arch dam is the 240m high Ertan Dam. The highest CFRD is the 178m high Tianshengqiao I Dam. The highest core earth-rockfill dam is the 154m high Xiaolangdi Dam, with an embankment volume of 49 million m^3, and an 80m-deep concrete wall in the dam foundation. The only one buttress dam is the exceptionally structured Hu'nanzhen Dam with a maximum height of 129m.

(2) The types of dam obviously varied and dam technologies developed further. During 1980 to 2000, the roller compacted concrete dam technology had been widely applied instead of the conventional block concreting method; the arch dams developed rather fast, and the 75m Puding RCC arch dam was constructed; among the earth-rockfill dams, the CFRDs experienced fastest development and were extensively applied for dams lower than 150m in height. As indicated in Table 8.1-2, the number of completed CFRDs over 100m high is 11, accounting for 1/3 of the similar high dams, and there are more CFRDs under construction or to be constructed.

Dam construction for types that require too much labors changed a lot, and thanks to the dam building technology development and wide application of modern construction equipment and machines, many types adopted before 1980 were gradually not used. For instance, during 1980 to 2000, the sand gravel dam, directional blasting fill dam, earth-filled underwater embankment dam, hollow dam, multi-arch dam, buttress dam and slotted gravity dam basically were not selected any more.

(3) The dam and project layout developed diversely with unique features. There is an obvious trend of arranging underground powerhouses for large-sized hydropower stations. The scales of underground powerhouses for the Er'tan, Lubuge, Xiaolangdi, Tianshengqiao II hydropower stations as well as Guangzhou I & II, Shisanling and Tianhuangping pumped storage power stations are very huge, and their tunneling and cavern excavation dimensions are extraordinary.

Lijiaxia Hydropower Station, with the damsite located at a narrow canyon on the Yellow River, adopted a double-row unit arrangement in the powerhouse at the dam toe, which was the first attempt in China. The powerhouse was installed with 5× 400MW units, the double curvature arch dam is 165m high, and the five penstocks with size of 8.0m in diameter were constructed on the downstream face of dam.

The high dams, large hydropower stations and navigation facilities constructed on the major navigable rivers were laid out in another different pattern. For instance, Shuikou Hydropower Station on the Minjiang River has a 3-stage navigation lock capable for a fleet of two 500t ships to pass by; Wuqiangxi Hydropower Station on the Yuanjiang River has a 3-stage navigation lock to allow a 500t ship fleet sailing; Gezhouba Water Control project has three single-step ship locks for passing 10,000t ship fleets. The Three Gorges Project has two 5-stage locks for passing 10,000t ship fleets and the yearly transporation capacity may amount to 50 million tons.

3. The overall dam construction period was shortened noticeably with faster building speed

Thanks to introducing international advanced technologies and project management mechanism, availability of large construction equipment and transportation vehicles, as well as continuous improvement of dam building technologies, the dam construction speed had been geared up continually. For those projects listed in Table 8.1-2, the yearly concrete placement volume was generally 400,000m^3 to 600,000m^3, and the peak monthly filling intensity for the earth-rockfill dams could amount to 1,000,000m^3. At least two sets of large turbine-generator units could be installed every year. Gezhouba Dam even set a record of commissioning 4 units within one year. Hence, the first unit commissioning schedule for a 1,000MW scale dam and large scale hydropower station could be shortened to 4 to 6 years after the river closure, and the overall construction period would last only 6 to 8 years.

8.1.3 Large Dams from 2001 to 2010

(1) According to the *Briefing on Statistics of Large Dams in China and World Large Dams*[1], by the end of 2005, the total number of dams higher than 30m completed or under-construction in China is 4,860, of which one dam is higher than 300m, 8 dams are between 200m to 300m, 22 dams are between 150m to 200m, 99 dams are between 100m to 150m, 422 dams are between 60m to 100m, and 4,308 dams are between 30-60m. Among the dams higher than 30m, the numbers of earth dams, gravity dams, rockfill dams, arch dams and others are 2865, 545, 391, 729 and 330 respectively. Jinping I dam commenced in 2005 is the highest one, with a maximum height of 305m.

The total number of dams higher than 100m in the world is more than 800. China, Japan and USA are the leading countries based on the number of dams higher than 100m.

From 2000 to 2009[2], the gross hydropower installed capacity of China increased to 182,100MW from 72,790MW. During 2000 to 2009, the total energy output was 3,551.9TW·h, accounting for 52.8% of 6,727.3TW·h, the total hydropower ouput in the last 60 years. Both the installed capacity and yearly energy output ranked first in the world, and the hydropower achieved a stride forward in the new century.

(2) From the beginning of the 21st century to 2010, it is a fast developing 10 years for dams and hydropower stations in China. Till 2010, the gross hydropower capacity in China has amounted to 200,000MW. The twenty-six (26) units of the Three Gorges Project with the capacity of 700MW in each have been completely commissioned (the underground six units would be commissioned soon in succession). The large-sized hydropower projects like Ertan, Xiaolangdi and Tianshengqiao I have been completed. fifty-one (51) large-sized or mega hydropower projects with dams higher than 100m (Table 8.1-3) have been commenced as well, of which 27 are projects with installed capacity exceeding 1,000MW in each, 24

---

[1] Extracted from Reference 6, *Technological Development and Engineering Examples of Large Dams in China*.

[2] Extracted from *Hydropower in China-A Sixty-Year Rever*.

are projects with capacity of 300-1,000MW in each, they were all completed and commissioned in 2010 or in later years respectively. These large-sized or mega projects represent some novel characteristics and developing trends.

Table 8.1-3 Large-sized Hydropower Stations and Dams Higher than 100m Constructed or under Construction from 2001 to 2010

No.	Name of project	Location: Province (Autonomous region) & river	Power plant		Dam		Reservoir		Reservoir inundation	
			Installed capacity (MW)	Annual energy output (10^8 kW·h)	Dam type	Maximum height (head) (m)	Gross storage ($10^8 m^3$)	Mean annual discharge (m^3/s)	Resettlement (persons)	Inundated land (10^4 mu)
1	Three Gorges	Hubei, Yangtze River	22,500	882	PG	181	450.5	14,300	1,130,000	17,160
2	Xiluodu	Sichuan Yunnan, Jinsha River	13,860	571.2	VA	285.5	126.7	4,620	27,738	25,071
3	Xiangjiaba	Sichuan Yunnan, Jinsha River	6,400	308.8	PG	162 (116)	51.63	4,620	78,769	35,900
4	Nuozhadu	Yunnan, Lancang River	5,850	239.12	ER	261.5 (215)	237	1,750	27,400	64,000
5	Longtan (stage I)	Guangxi, Hongshui River	4,900	156.7	RCC PG	192	188 272.7	1,640	73,392	82,924
6	Xiaowan	Yunnan, Lancang River	4,200	190	VA	294.5	150	1,210	32,737	55,678
7	Laxiwa	Qinghai, Yellow River	4,200	102.3	VA	250	10.79	680	962	249
8	Pubugou	Sichuan, Dadu River	3,600	147.9	ER	186	53.37	1,230	61,900	37,800
9	Jinping I	Sichuan, Yalong River	3,600	166.2	VA	305	79.88	1,200	3,163	7,628
10	Jinping II	Sichuan, Yalong River	4,800	209.7	PG	34 (312)	0.19	1,180	—	—
11	Guanyinyan	Yunnan, Jinsha River	3,000	122.4	ER (left bank) /RCCPG (right bank)	159	22.5	1,830	8,853	14,059
12	Guxian	Shaanxi, Shanxi, Yellow River	2,560	71.6	CFRD	186	150.5	717.6	31,790	44,550
13	Jin'anqiao	Yunnan, Jinsha River	2,400	110.43	PG	160	9.13	1,670	1,868	2,478
14	Ludila	Yunnan, Jinsha River	2,160	99.57	PG	140	17.18	1,750	17,655	31,229
15	Goupitan	Guizhou, Wujiang River	3,000	96.82	VA	232.5	64.54	724	25,304	27,199
16	Guandi	Sichuan, Yalong River	2,400	110	PG	168	7.6	1,370	0.03	0.07
17	Shuibuya	Hubei, Qingjiang River	1,840	39.84	CFRD	233.2 (207)	45.8	300	9,800	10,700
18	Gongboxia	Qinghai, Yellow River	1,500	51.4	CFRD	132.2	6.2	717	5,340	7,585
19	Jinghong	Yunnan, Lancang River	1,750	63.62	RCCPG	108	11.4	1,820	1,983	4,750
20	Datengxia	Guangxi, Qianjiang River	1,200	64.34	RCCPG	62 (34)	17.6	4,290	37,046	53,762
21	Pengshui	Sichuan, Wujiang River	1,750	63.51	RCCPG	113.5 (87.4)	14.65	1,280	14,329	13,421
22	Guangzhao	Guizhou, Beipan River	1,040	27.54	RCCPG	200.5	32.45	257	15,800	8,950
23	Sanbanxi	Guizhou, Qingshui River	1,000	24.28	CFRD	185.5	40.9	239	17,100	4,100
24	Jishixia	Qinghai, Yellow River	1,020	33.63	CFRD	103	2.94	720	4,500	2,600
25	Silin	Guizhou, Wujiang River	1,050	40.64	RCCPG	117 (74)	15.93	844	15,123	19,462
26	Shatuo	Guizhou, Wujiang River	1,120	45.52	RCCPG	101	9.21	951	12,511	10,265
27	Tingzikou	Sichuan, Jialing River	1,100	31	RCCPG	116	41.0	660	32,500	41,400

Chapter 8　Dam Type and Project Layout

Continued

No.	Name of project	Location: Province (Autonomous region) & river	Power plant		Dam		Reservoir		Reservoir inundation	
			Installed capacity (MW)	Annual energy output (10^8 kW·h)	Dam type	Maximum height (head) (m)	Gross storage (10^8 m^3)	Mean annual discharge (m^3/s)	Resettlement (persons)	Inundated land (10^4 mu)
28	Dongqing	Guizhou, Beipan River	880	30.26	CFRD	150	9.55	370	1,554	2,573
29	Zipingpu	Sichuan, Minjiang River	760	34.17	CFRD	156	11.2	469	28,275	8,847
30	Shiti	Sichuan, Youshui River	720	13.7	RCCPG	131	31.7	176	55,634	52,812
31	Changzhou	Guangxi, Sunjiang River	630	30.143	RCCPG	56	56	624	915	37,616
32	Tankeng	Zhejiang, Xiaoxi River	604	10.23	CFRD	162	41.9	121	43,010	30,720
33	Suofengying	Guizhou, Wujiang River	600	20.11	RCCPG	115.95	2.01	359	1,307	1,557
34	Hongjiadu	Guizhou, Liuchong River	600	15.59	CFRD	179.5(163)	49.47	149	42,671	29,102
35	Pankou	Hubei, Duhe River	500	10.801	CFRD	114	23.38	167	21,838	35,272
36	Baise	Guangxi, Youjiang River	540	16.9	RCCPG	130	56.6	263	24,716	39,359
37	Jilintai I	Xinjiang, Yili Kashgar River	500	9.38	CFRD	157	25.3	112	2,717	18,392
38	Longshitou	Sichuan, Dadu River	700	25.82	ER	58.5 (50)	1.39	1,010	1,236	7,260
39	Tongzilin	Sichuan, Yalong River	600	25.5	RCCPG	71.3(28.5)	0.91	1,890	220	65
40	Malutang II	Yunnan, Panlong River	240	19.22	CFRD	154	5.46	82.3	1,409	5,890
41	Longkou	Inner Mongolia, Shanxi, Yellow River	420	13.02	PG	51	1.96	182	100	3,991
42	Gelantan	Yunnan, Honghe River Lixian River	450	20.18	RCCPG	113	4.09	408	467	72
43	Yinzidu	Guizhou, Sancha River	360	9.78	CFRD	129.5	5.31	140	5,039	8,590
44	Xunyang	Shaanxi, Hanjiang River	320	8.0	PG	(24)	2.08	699	1,0279	5,072
45	Hechuan	Sichuan, Jialing River	300	13.75	RCCPG	51 (27)	10.70	1,683	1,236	7,260
46	Jiangya	Hunan, Loushui River	300	7.56	RCCPG	131	17.41	132	11,839	14,023
47	Jiudianxia	Gansu, Taohe River	300	9.94	CFRD	133	9.42	122	11,660	11,823
48	Jiemian	Fujian, Longxi River	300	3.6	CFRD	126	18.2	45.2	11,097	10,895
49	Chahanwusu	Xinjiang, Kaidu River	309	10.8	CFRD	110	1.25	99.2	4	839 (meadow)
50	Wanmipo	Hunan, Youshui River	240	7.92	RCCPG	66.5	3.78	299	19,360	12,950
51	Zhouning	Fujian, Muyangxi River	250	6.58	RCCPG	73.4(425)	0.47	20.3	130	350

Note: This table is quoted from *Atlas of Planned Large-and Medium-sized Hydropower Stations in China*, compiled by the General Institute of Water Resources and Hydropower Planning and Design, 1994. The extension projects, pumped storage projects and large dam projects in Taiwan Province are not included in the table. Missing parameters in the original table have been supplemented.

1) Most projects are concentrated on the rivers in the western region, mainly on the mainstreams of the Jinsha River, Yalong River, Dadu River and Wujiang River in the upper reach of the Yangtze, as well as the Lancang River, Beipan River and Hongshui River in Yunnan Province. As for the projects with obviously multi-benefits and superior technical and economic indexes, particularly those governing reservoirs with regulating effects, such as Longtan on the Hongshui River, Xiluodu on the Jinsha River, Xiaowan on the Lancang River and Sanbanxi on the Yuanjiang River mainly developed for power generation and with considerable flood control and other multiple benefits, their investigation and engineering design works had been completed in the last century, construction commenced in the early 21st century, and commissioned around 2010.

2) The installed capacities of hydropower stations associated with the large dams are rather huge. Fifteen hydropower projects respectively have the total capacity over 2,000MW with the single unit capacity of 500MW to 700MW. The construction progress is rather swift, able to commission 2 to 4 units every year. At the same time, the engineering investigation and design for a number of large-sized hydropower projects and dams over the middle and lower reaches of Jinsha River as well as the Yalong River, Dadu River, Nu River, Yellow River and the upper and middle reaches of the Yarlung Zangbo River, would be accomplished within 10 years and ready for preparation or construction, of which 26 projects have the capacity of 2,000MW or above; in view of power head, except for those diversion type hydropower schemes, about 12 dams have the maximum height of 200m or above.

3) The types of high dam show some new developing trends. Table 8.1-3 suggests that the main dams for 51 large hydropower stations can be generally classified as three types: 20 CFRDs (or earth-rockfill dams), accounting for about 40%; 26 gravity dams (RCC gravity dams), accounting for about 50%; 5 double curvature arch dams, accounting for about 7%. The gravity dam and CFRD (or earth rockfill dam) are the main types. Almost all the gravity dams are RCC gravity ones, no any slotted or hollow gravity ones. And almost all the arch dams are with double curvatures.

In the fifteen 200m-scale high dams completed or under construction as indicated in Table 8.1-3, there are 7 CFRDs (or core rockfill dams), accounting for 47%; five double curvature arch dams, accounting for 33%; and three gravity dams, accounting for 20%. Particularly, there are five dams with a height scale of 250m to 300m, they are either double curvature arch dams or core rockfill dams; the highest one is the 305m high Jinping I double curvature arch dam; there is no any CFRD or concrete gravity dam. This new trend implies that, there exist numerous difficulties for the two types of dams (CFRD and gravity dam) to be built in a 250m to 300m height, and scientific research should be strengthened to make further breakthroughs on key dam building technologies.

8.2 Concrete Gravity Dams

8.2.1 General

1. Dam type

The concrete gravity dam is a conventional type of dam to attain stability depending on its own weight, diversified as solid gravity dams, slotted gravity dams, buttress dams, hollow gravity dams, arch gravity dams (mainly depending on the gravity stability besides the arch action) and gravity arch dams (mainly depending on the arch action).

In the 1950s, two slotted gravity dams as Xin'anjiang and Gutian I were constructed; in the 1960s, Yunfeng and Danjiangkou slotted gravity dams as well as Sanmenxia and Liujiaxia solid gravity dams were constructed; in the 1970s, Hu'nanzhen trapezoid gravity dam was constructed; in the early 1980s, Wujiangdu arch gravity dam was constructed. Till the 1980s, more than twenty over 70m high concrete gravities had been built up, of which solid gravity dams and slotted gravity dams dominate, as shown in Table 8.1-1.

In the early period after foundation of new China, due to limited international communication as well as the constraints of finance capability and materials availability, most gravity dams were constructed as slotted gravity dams or buttress dams in order to save three main materials (steel, cement and timber). In the "Cultural Revolution" period, in response to the government's call for technology revolution, and application of preliminary computer technology, a number of slotted gravity dams were developed and constructed, and based on that, it was further innovated as the hollow gravity arch dam. Fengtan Hollow Gravity Arch Dam was the representative project, highly recognized by the world, with its powerhouse set in the dam body and the spillway located on the powerhouse (dam) crest.

Since the 1980s, China put much emphasis on the water conservancy and hydropower construction, summarized and learned lessons about dam construction in other countries, and amended relevant national design codes and specifications. The design principle for checking the stability against sliding of a gravity dam was altered to apply the universal shear-friction formula instead of the previous classic pure friction formula. And the solid gravity dam had shown its advantages. Furthermore, as China is situated in the temperate zone, subject to frequent cold waves in the winter season as well as seasonal change from winter to spring, due to constraints of certain factors like temperature control and dam building technology, the slotted gravity dam was gradually not used and replaced by the solid gravity dam. Solid gravity dams such as Sanmenxia (106m), Liujiaxia (147m), Wujiangdu (165m) and Danjiangkou (97m) were constructed in monolithic blocks with longitudinal joints.

Since the reform and opening-up policy, China has learned a great deal of practical experiences from the international dam engineering companies, and the new dam building technologies such as roller compacted concrete (RCC) dams and CFRDs were introduced. In the late years of the 20th century, the two types of dams developed rather rapidly. Within about 10 years, more than 10 RCC dams had been constructed or under construction (Table 8.1-2), such as Mianhuatan (113m), Yantan (110m), Dachaoshan (111m), Jiangya (131m), Baise (130m) and Gaobazhou (57m). The RCC gravity dam design and building technology fully replaced the conventional design and construction methods for 100m high concrete gravity dams to be constructed in monolithic blocks with longitudinal joints, which was the usual practice before 1980s. The 132m high Baozhusi dam constructed in the middle of 1980s is the last gravity dam placed in monolithic blocks with longitudinal joints. With the design and construction experience accumulations and technology innovations on about 10 RCC gravity dams in the last century, the RCC gravity dam's height was upgraded from 100-150m scale to 200m scale in the first 10 years of the 21st century, and two RCC gravity dams, namely Longtang (216.5m) and Guangzhao (200.5m), were successfully built up and became the highest RCC gravity dams in the world.

2. Project layout

The general layout for a concrete gravity dam would be rather flexible, so that there might be several alternative layout plans for comparison in terms of the flood outlet works, powerhouse location, river diversion, dam concrete construction etc., particularly the dam concrete placing method has most obviously influence to the project layout.

Since the 1980s, with the development of RCC gravity dams, the project layout also has made great progress and variation. For adapting to RCC dam building technology and extending its application scope, speeding up the dam height rising rate for earlier power generation and energy gains, the powerhouse would be located in the banks or underground, and except for the diversion bottom outlet(s) and surface spillway works on the dam crest, there would be no or least outlet structures in the dam body, only necessary bottom outlet(s) would be provided to supply water for downstream riparian users or release a basic environmental flow during the reservoir impounding period. Numerous practical examples have proved that, the project layout schemes made on the above principles are rational both technically and economically.

3. Reservoir emptying facilities

As for reservoir emptying facilities, in the early period after foundation of new China, for a 100m-scale gravity dam, no matter how big the reservoir is, almost no consideration was made to provide a reservoir emptying facility, such as Xin'anjiang (dam height 105m, reservoir storage capacity 21.6 billion m^3), Zhexi (dam height 104m, reservoir storage capacity 3.57 billion m^3) and Xinfengjiang (dam height 105m, reservoir storage capacity 13.9 billion m^3). As for Liujiaxia Hydropower Station (dam height 147m, reservoir storage capacity 6.4 billion m^3), when commencing construction, there was no such design for an emptying facility which was added later during the project construction was postponed.

Based on available data from the international communication, the 100m to 150m high gravity dams were seldom provided with emptying facilities. Fengman Dam (dam height 91.7m, reservoir storage capacity 10.99 billion m^3) on the Songhua River in Northeast China, which was designed by Japanese engineers, had no any emptying facilities, and the current flood release tunnel on the left side was added in the 1970s and the 1980s.

Whether or not providing emptying facilities for reservoirs experienced an investigation and practice. In the 1960s, due to sedimentation in Sanmenxia Reservoir, the water conservancy and hydropower community became aware that the sediment silting issue must be fully considered when building reservoirs on those rivers with high silt content. And large reservoirs with emptying facilities would allow flexible control of the reservoir levels when necessary, which would be favourable to the reservoir's safety and enable to positively avoid unforeseeable risks. For prevention and treatment of unstable rock slopes on the reservoir slope near the dam site due to the reservoir impoundment and operation, provision of emptying facilities would be more flexible and essential. As for the scale of emptying facilities as well as how to give full play to their multiple functions, since 1980 there have been many advances and innovations. For instance, emptying facilities also assume functions like silt release, flood release and diversion during flood season, as well as adding new generation units for peaking need of electricity.

4. Development of underground powerhouse

In the large dams and hydropower projects before the 1980s, it was quite rare to construct an underground powerhouse. Thereafter, geological exploration, rock mechanics and computation science have made great advance, especially the development of underground construction technology, heavy machines, quality inspection and control technology; and under the drive of "Lubuge Effect (new way for management of the dam construction)", the construction of underground powerhouse headrace system and underground caverns/tunnels have gained much development. Even for a dam site with a relatively wide valley and large flood discharge volume, it is usually to choose the underground powerhouse, constructing a high cofferdam to divert flow through diversion tunnels, and close the river course at one time, so as to leave the riverbed space for building the RCC gravity dam, which could relieve the flood release and energy dissipation burden on the riverbed, easing the difficulties of diversion and easy for passing through the flood season, accelerating the dam embankment speed, and shortening the straight construction schedule for the dams. If possible in the preparation period, the underground powerhouse system could be excavated when building the diversion tunnel, which would obviously accelerate the underground powerhouse system construction, so that the first unit (s) commissioning time and the overall construction period could be shortened at least one year compared with an ordinary schedule.

Among the twelve cascade schemes planned on the middle and lower reaches of the Jinsha River as well as other large-sized hydropower projects on the rivers in the western region, those under construction or preparation would mostly have underground powerhouses. Even in some projects where it is possible to arrange a powerhouse at the dam toe on the riverbed, an underground powerhouse would be selected for the purpose of minimizing interference with the RCC dam construction and the river diversion in the flood season, and achieving the goal of earlier power generation. However, it should be clarified that, the engineering geological conditions for these underground powerhouse systems are sound and favourable for arranging underground caverns and tunnels.

8.2.2 Development of Concrete Gravity Dams in the Last 30 Years since the Reform and Opening-up

8.2.2.1 Gravity Dam Projects in 1980-2000

Up to 2010, there are 36 gravity dams over 100m high completed or under construction in China mainland as indicated in Table 8.2-1, of which in the twenty years from 1980 to 2000, the number of concrete gravity dams completed or under construction is 13. These dams have features as follows.

Table 8.2-1　(Arch) Dams Higher than 100m Completed or under Construction Till 2010

No.	Name of project	River	Dam type	Dam height (m)	Concrete volume (10^4 m^3)	Total storage capacity (10^8 m^3)	Installed capacity (MW)	Discharge capacity (m^3/s)	Year of completion
1	Longtan I	Hongshui River	RCCPG	192	574/736	188	4,900	27,692/26,085	2009
2	Guangzhao	Beipan River	RCCPG	200.5	293.71	32.45	1,040	9,857	2009
3	Three Gorges	Yangtze River	PG	181	1610	450.50	22,500	102,500	2010
4	Longyangxia	Yellow River	VA	178	157	247	1,280	6,000	1992
5	Guandi	Yalong River	RCCPG	168	357.023	7.6	2,400	16,300	UC
6	Wujiangdu	Wujiang River	PG	165	193.2	23.00	1,250	21,235	1982/EW
7	Xiangjiaba	Jinsha River	PG	162	1221.01	41.9	6,400	48,600	UC
8	Jin'anqiao	Jinsha River	RCCPG	160	360	9.13	2,400	17,653	UC
9	Geheyan	Qingjiang River	VA	151	268	27	180	24,000	1994
10	Liujiaxia	Yellow River	PG	147	182	64	1,350	7,419	1974
11	Baozhusi	Baolong River	PG	132	200	25.5	700	16,060	2000
12	Manwan	Lancang River	PG	132	153	10.06	1,670	16,805	1995/EW
13	Jiangya	Loushui River	RCCPG	131	137	17.41	300	10,491	2000
14	Baise	Youjiang River	RCCPG	130	260	56.60	540	11,487	2006
15	Ankang	Hanjiang River	RCCPG	128	321	32	800	36,700	1995
16	Hongkou	Hongtongxi River	RCCPG	130	72.7	4.5	200	10,200	2008
17	Guxian	Luohe River	PG	125	156.4	11.75	60	11,436	1995
18	Wudu Reservoir	Pujiang River	RCCPG	120.34	161	5.72	150	7,795	UC
19	Danjiangkou	Hanjiang River	PG	117	292.8	339.1	900	82,300	1973
20	Silin	Wujiang River	RCCPG	117	110	15.93	1,050	32,922	UC
21	Pengshui	Wujiang River	RCCPG	113.5	93.3	14.65	1,750	42,200	2007

Continued

No.	Name of project	River	Dam type	Dam height (m)	Concrete volume (10^4 m^3)	Total storage capacity (10^8 m^3)	Installed capacity (MW)	Discharge capacity (m^3/s)	Year of completion
22	Suofengying	Wujiang River	RCCPG	115.95	62	2.01	600	15,954	2006
23	Yunfeng	Yalv River	PG	113.75	304	38.95	400	24,204	1965
24	Mianhuatan	Tingjiang River	RCCPG	113	61.5	20.35	600	11,490	2001
25	Gelantan	Lixian River	RCCPG	113	117.27	4.09	450	14,000	UC
26	Yantan	Hongshui River	RCCPG	110	62.6	33.8	1,210	32,768	1995/EW
27	Dachaoshan	Lancang River	RCCPG	111	130	9.40	1,350	23,800	2002
28	Jinghong	Lancang River	RCCPG	108	320	11.39	1,750	34,800	UC
29	Panjiakou	Luanhe River	PG	107.5	280	29.30	420	56,200	1985
30	Honglongtan	Duhe River	PG	107	98	11.7	510	14,700	1976/EW
31	Shuifeng	Yalv River	PG	106.4	340	146.66	900	59,500	1971/EW
32	Sanmenxia	Yellow River	PG	106	163	96	410	9,030	1960
33	Shatuo	Wujiang River	RCCPG	101	187.29	9.1	1,120	32,019	UC
34	Wanjiazhai	Yellow River	PG	105	178.85	8.96	1,080	21,200	1965
35	Xin'anjiang	Xin'an River	PG	105	138	216.26	850	13,200	1960
36	Shuikou	Minjiang River	PG	101	180	26.00	1,400	51,640	1996

Note: This table is quoted from Chapter 1 of *Design of Gravity Dams—A Twenty-Year Review*. Those under-construction (UC) projects have been completed around 2010.

1. Most dams are located in the middle and east of China and quick development in the western region

In the twenty years, about ten gravity dams with the height of 100 – 150m were built up in succession in the middle and east of China. To meet the rapid socio-economic development after initiating the reform and opening-up, it was urgent to develop local hydropower potential. Even though they involved relatively more resettlement and land occupation, the projects were built up in succession within the 20 years as long as there have no major technical or ecological problems to hinder their implementation. The typical gravity dams include Yantan on the Hongshui River, Wuqiangxi on the Yuanshui River, Shuikou on the Minjiang River, Baozhusi on the Jialing River and Ankang on the Han River.

As indicated in Table 8.2 – 1, in the twenty years, among those 100m to 150m high dams with large reservoirs in the central, eastern and western parts of China, besides gravity dams, there are many CFRDs, earth rockfill dams or arch dams completed, such as Manwan and Dachaoshan on the Lancang River, Geheyan on the Qingjiang River, Tianshengqiao I and II on the Hongshui River, Baishan on the Songhua River, Xiaolangdi and Lijiaxia on the Yellow River, and Ertan on the Yalong River. And the Three Gorges Project on the Yangtze River was commenced in that period.

2. Almost all gravity dams are solid gravity ones

Since mid-1980s, driven by the "Lubuge Effect", the employer's responsibility system, tendering system and construction supervision system have been widely applied in the water conservancy and hydropower projects execution. Payment on schedule would exempt the construction progress from being affected by finance constraints, and with the development of RCC gravity dams, the solid gravity dam is widely applied. In the twenty years, the gravity dams account for about 50% of the 100m to 150m high dam projects.

3. Layout of gravity dams experienced further development and lots of innovations

With the accumulation of practical experience and development of dam building technologies, many innovations were made for the gravity dam complex layout in the twenty years after initiating the reform and opening-up:

(1) Surface spillways on the gravity dams in canyons were evolved into a layout allowing flow to jump over the powerhouse from the previous manner of flowing over the powerhouse crest. Overflowing on the powerhouse crest has certain vibrating influence on the powerhouse structure that is located at the dam toe. In Wujiangdu Hydropower Station, application of the

release pattern of flow jumping over the powerhouse prevents the powerhouse crest from being impacted by jet flows. Manwan Hydropower Station completed in the early 1990s made some new evolutions. Its surface spillways would allow outflow to jump from the large-differential flip bucket, and by utilizing the high flood level at the dam downstream, a plunge pool was constructed therein, so that the energy of flood flow across the powerhouse crest could be dissipated in the plunge pool. The surface spillway arrangement for Manwan dam is shown in Fig. 8.2-1.

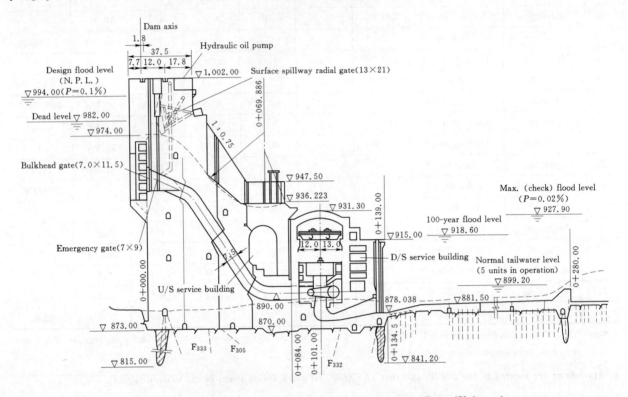

Fig. 8.2-1 Profile of Surface Spillway Section for Manwan Dam (Unit: m)

(2) Great development has been made in the scale of surface spillways and middle outlets on the dam body as well as the energy dissipators. As most of the flood discharge volume would be released through the surface spillways, enlarging the surface spillway size, increasing the unit width discharge capacity and optimizing the length of the flood release front edge could remarkably refine the gravity dam complex arrangement, reduce the overall work quantity and speed up construction. According to the actual topographical and geological conditions at the dam sites, the hydraulic jump or ski-jump energy dissipation manner was adopted for most dams by combining surface spillways with middle outlets. The scale of surface spillways and middle outlets could also reveal the scale of gates and hoists. All these advancements laid a solid foundation for further development and innovation of gravity dams in the 21st century.

There was a clear developing trend to construct deep (bottom) outlets with large openings for releasing flood, flushing silt, ensuring safety in the high water season and regulating the reservoir level. Besides their basic functions of releasing flood, desilting, keeping "clean in front of the power intake" and overcoming flood seasons in cooperation with the diversion works during the construction period, these deep (bottom) outlets in the gravity dam projects also would assume their vital responsibility of assuring the reservoir's operation safety by lowering or controlling the reservoir level when necessary. In those rivers with high sediment concentration, for controlling sediment silting in the reservoirs, maintaining their effective storage capacities and prolonging their service life, large flood release and sand flushing deep (bottom) outlets (or flood release tunnels) would become more essential.

(3) Large capacity turbine generator units were widely applied in the hydropower projects with high gravity dams. Not only for gravity dams, but also for arch dams and earth rockfill dams, units with a single unit capacity over 200MW were widely adopted in the power plants. No matter it is a powerhouse at the dam toe or under ground, applying large capacity units can minimize the width of the generation flow diversion front and shorten the overall length of powerhouse, which would be extremely important for optimizing the project layout. For ease of construction and accelerating the working progress, penstocks for a powerhouse at the dam toe were evolved as pipes laid on the downstream dam slope instead of pipes embedded in the dam body, which has become a consensus trend for building an RCC gravity dam. The inner diameter of penstocks on the downstream dam slope for Three Gorges Project reaches 12.4m, the biggest among the hydropower projects under con-

struction in China.

(4) The diversion works is closely integrated with the gravity dam complex. Open diversion channels and large diversion bottom outlets at the dam bottom have been widely used for construction diversion and passing through the high water season. As required by the diversion criteria and flood standard as well as the discharge capacity, there is a developing trend that the number of diversion bottom outlets is growing, and the outlet size is increasing, as shown in Table 8.2-2. For the Three Gorges Dam initiated in the 1990s, every sluice dam block over the riverbed is provided with an outlet, and there are totally 23 diversion bottom outlets in the dam body, the maximum number among the completed gravity dams.

Table 8.2-2 Parameters of Open Diversion Channels and Bottom Diversion Outlets for Gravity Dams Constructed in China

No.	Name of project	Design discharge (m^3/s)	Section type of open diversion channel	Size of diversion channel (m)		Bottom diversion outlet No. $-W \times H$ (m)	Other purposes of open channel	Year of completion
				Length	Width			
1	Shuikou	32,200	Rectangular	1,170	75	$10-8 \times 15$	Navigation, log passing	1989
2	Wan'an	15,500	Trapezoid	1,530	50	-	Navigation	1990
3	Ankang	4,700	Trapezoid	412	40	$3-8 \times 15$	Navigation	1990
4	Tongjiezi	10,300	Rectangular	590	54	-	Log passing	1992
5	Yantan	15,100	Rectangular	1,110	65	$8-4 \times 10$		1993
6	Feilaixia	15,500	Combound	1,697	300		Navigation	1994
7	Baozhusi	9,570	Rectangular	527	35	$2-5 \times 10$		1996
8	Daxia	5,000	Rectangular	628	40			1996
9	Three Gorges	79,000	Combound	3,410	350	$23-6 \times 9$	Navigation	1997
10	Xin'anjiang					$3-10 \times 13$		1962

4. Wide application of RCC gravity dams

Especially for the 100m high gravity dams, no matter in the East, West or chilly Northeast and Northwest (like Kalasuke), almost all dams are constructed of RCC; even the 150m high gravity dams are partially constructed of RCC. And based on international experiences, China has innovated some RCC dam building technologies with unique Chinese characteristics.

Though the 181m high Three Gorges Dam commenced in the 1990s was still constructed in the normal monolithic concrete pouring method, both the gravity dam retaining wall and the over 100m high concrete cofferdams were constructed of RCC.

5. Typical projects in the 1980s and the 1990s

Two representative projects with 100m to 150m high dams are illustrated below for readers to review the country's gravity dam works standard in the 1980s and the 1990s.

(1) Baozhusi Hydropower Station❶ (Fig. 8.2-2): The project was developed mainly for power generation as well as other purposes like flood control and irrigation. The gravity dam was concreted in monolithic blocks, with the power plant located in the middle of riverbed, and release structures on both sides of the powerhouse. The reservoir is a yearly regulating one with a gross storage capacity of 2.55 billion m^3 and a regulating storage capacity of 1.34 billion m^3. The power plant is equipped with 4 units with a total installed capacity of 700MW, capable of generating electricity 2.2TW·h per year. The river closure was achieved on November 29, 1991, the first unit started power generation in December 1996, and the whole project was completed in 2000.

The river valley at the damsite is relatively narrow. Bedrock is dominated by rigid silty sandstone and silty shale with low strength. The dam has a maximum height of 132m, and a maximum bottom width of 92m. The dam crest elevation is at El. 595.00m, and its basic triangle apex is at El. 600.00m. The upstream dam slope is in vertical, and the downstream dam slope in 1:0.6 and 1:0.65. The gravity dam's cross section is relatively economical compared with those of similar pro-

---

❶ Extracted from *Design of Gravity Dams—A Twenty-Year Review*.

Part II Construction Technologies for Dams and Reservoirs

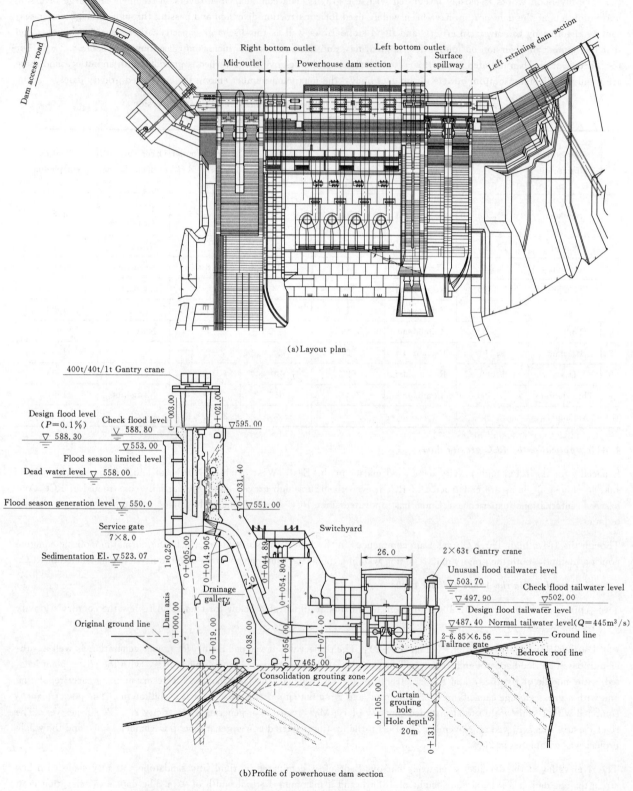

(a) Layout plan

(b) Profile of powerhouse dam section

Fig. 8.2-2 Project Layout of Baozhusi Hydropower Station (Unit: m)

jects in China.

1) Cut-off and drainage measures were adopted to decrease uplift pressure on the dam body. The design assumed that the upstream seepage pressure at the second row curtain is 0. Since the flood duration is not long, for an unusual flood case,

the head applied for computing the buoyancy is counted as the normal tailwater depth plus 0.5m to 1.0m head, and full head is applied at the powerhouse end. Since its commissioning, all the monitored uplift values at the dam base are below the design value, and most uplift parameters vary around 0.1, less than the design-assumed value of 0.2.

2) Combination of the dam and powerhouse enhances the dam's safety and stability against sliding. The aim of combination is for the powerhouse to bear partial force transferred to the dam, so as to satisfy the dam's sliding stability and stressing requirements, optimize the dam body's section design, and decrease the dam body's volume. The dam-powerhouse compound makes that only compressive stress, no tensile stress, would occur in the dam and powerhouse base, and both principal tensile stress and principal compressive stress in the dam body would be smaller than the stressing values in case a joint is set between the dam and the powerhouse, which suggests that the combined action could improve the stressing condition in the dam body and be favourable for the dam structure's safety and stability.

3) Application of grouting along the transverse joints consolidated the dam's integrity. Trapezoid keys were set in the transverse joints within the 90m range of the dam height below El. 550m, grouting was conducted in the joints for consolidating the dam body and improving the dam's stability against sliding. Based on finite element analysis, there exists arching effect at the dam block on the riverbed, which can bear about 9% of the water pressure. Grouting along transverse joints is helpful for the dam's stability, though the grouting effect was regarded as an additional safety margin and had not been taken into account when making the dam body section design.

(2) Dachaoshan Hydropower Station (Fig. 8.2-3): Located in the mid-lower reach of the Lancang River in Yunnan, Dachaoshan Dam has a maximum height of 111m, and the gross storage capacity of the reservoir is 940 million m^3. The total installed capacity is 1,350MW, capable of generating electricity 5.931TW·h per year, and its energy output may reach 7.021TW·h after Xiaowan Hydropower Station being put into operation. The project is composed of an RCC gravity dam and an underground powerhouse at the right bank followed by a long tailrace tunnel. The river closure was achieved in November 1997, the first unit was put into operation in December 2001, and the whole project was completed in October 2003. The project has a unique layout. The works quality is superior though the progress is very fast.

1) The project has a unique layout pattern. The main dam is an RCC one with surface and deep outlet structures constructed in the riverbed section dam body, and energy dissipation depends on the hydraulic jump and ski-jump in the bailer-type stilling basin after the flaring pier gate. The maximum discharge capacity of the outlet works is 23,800m^3/s. By utilizing the available space at the right abutment, the power intake and the underground powerhouse are located at the right bank, so that a compact arrangement including the headrace and flood release structures can be made around the dam axis. The underground powerhouse is equipped with 6 sets of 225MW units. The long tailrace tunnel rightly takes the advantage of the drop of the big bend river course at the dam downstream. A short diversion tunnel is located at the left bank. The upstream cofferdam is an RCC arch dam. The overall project layout and construction arrangement are compact and rational.

2) The surface spillway overflow is in steps ($H \times W = 1m \times 0.7m$) at its straight section, to be formed during the continuous construction of RCC in layers. Under the design and check flood levels, the unit width discharge values at the weir crest are 142m^3/s and 237.8m^3/s, respectively. Such big discharge capacity is quite rare for a stepped type overflow surface.

3) Local materials like excavated materials were fully utilized as much as possible. In the project area, there exists no natural aggregates. All coarse and fine aggregates for concrete were sourced from the underground excavated material-basalt rock. No dedicated quarry was exploited, so extra land acquisition and access road construction for quarry areas were not required.

Because the flyash was unavailable in Yunnan, the admixture in concrete for Dachaoshan dam resorted to a grinded mix of local phosphorite slag and tuff, which achieved good results. The cement use for concrete placed inside the dam body was only 68kg/m^3 to 72kg/m^3, both its mechanical and thermal properties could meet the design requirements.

8.2.2.2 Gravity Dams Construction from 2001 to 2010

1. Developing trend of gravity dams

In the 60-year history of new China, the gravity dams normally account for 40% to 45% of the high dam projects in different periods. The gravity dams are the dominant dam type in the 100m to 150m high dam projects. However, among the nine over-200m-high dams constructed or under construction (Table 8.1-3), there are only two gravity dams (Longtan and Guangzhao), and it is anticipated that, around 2020, in the scale of 2,000MW hydropower stations and 200m high dams, there might be a few gravity dams, but the number would be far less than the numbers of CFRDs and double curvature arch dams.

2. Gravity dams still having obviously advantages for further development

In the first 10 years of the 21st century, the number of 100m to 150m high gravity dams completed or under construction to-

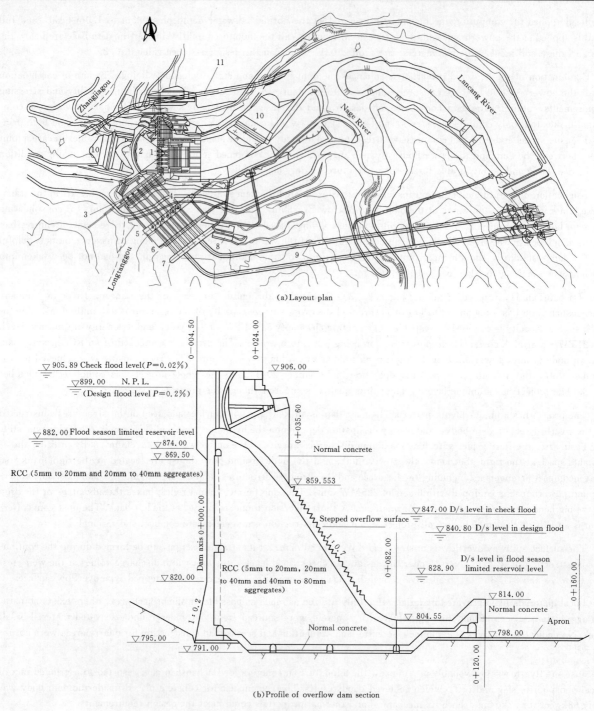

Fig. 8.2-3 Project layout of Dachaoshan Hydropower Station (Unit: m)

1—RCC dam; 2—U/s arch cofferdam; 3—Power intake; 4—Penstock; 5—Underground powerhouse; 6—Main transformer cavern; 7—Tailrace surge chamber; 8—Ground outgoing line yard; 9—Tailrace tunnel; 10—U/s and D/s temporary cofferdams; 11—Diversion tunnel

tals about 15 (Table 8.1-3). More experience has been obtained on the basis of the projects in the 20th century, and its overall advantages remain.

(1) The solid gravity dam is still one of the prime choices. When the dam height is 100m to 150m, the flood discharge is quite large, and geological conditions are relatively sound, the solid concrete gravity dam would still be a favored pick, because it can be constructed with RCC, which would be favourable for the project layout in terms of flood release, river diversion and flood control facilities, and its construction pace can be accelerated for higher investment gains. These advantages can hardly lay with other types of dam.

(2) The rapid development of RCC gravity dams promotes continuous innovations on the general layout. During the 1980s and the 1990s, great progress has been made in RCC gravity dams, almost all 100m to 150m high gravity dams were constructed with RCC, and the RCC dam building technology also encouraged the project layout innovations for dams sited in canyons. For instance, the Dachaoshan RCC gravity dam "made the space for" setting the surface spillways and middle outlets at the overflow dam section, and "shifted" the powerhouse to the underground at the right bank instead of the dam-toe powerhouse option, which not only minimized the power intake and waterway arrangement in the dam body for ease of RCC placement, but also enlarged the flood release front favourable for decreasing the unit width discharge at the flood release dam section, simplifying the energy dissipation, and minimizing erosion to the riverbed after the dam. As the unit width discharge of the surface spillway decreases, the water jet would be directly drained off over the RCC steps of the overflow surface, which would make the construction procedure simpler, the progress faster and favourable for passing through the flood season.

(3) The development of large capacity units also promotes the gravity dams construction at the broad valley dam sites. For large-sized hydropower stations at wide river channels, thanks to application of large capacity units, the number of units could be reduced and the powerhouse could be downsized, which would be more favorable for arranging the powerhouse at the dam toe, or under the ground at either or both flanks to facilitate flood release through the riverbed dam section. This is another key factor that the RCC gravity dams have developed and kept its own advantage for the last 30 years.

3. Typical 200m high gravity dam projects

The 200m high gravity dams constructed in other countries were mostly concreted in conventional monolithic blocks with few exceptions built with RCC. While the Longtan and Guangzhao dams are the typical 200m high gravity dams constructed in China, and also are the highest RCC gravity dams in the world. The two dams commissioned around 2010 have demonstrated the state-of-the-art of concrete gravity dams in China. Three representative projects are briefed below:

(1) Longtan RCC gravity dam (Fig. 8.2-4): Located in Tian'e County of Guangxi, Longtan Hydropower Project is developed mainly for power generation besides multi-purposes including flood control and navigation. The dam was designed as the normal pool level of El. 400.00m, and was constructed to El. 375.00m in the initial stage. The corresponding gross storage capacities in the initial stage and late stage are 16.21 billion m^3 and 27.27 billion m^3 respectively, capable for regulating flow in one year and multiple years respectively, and with total installed capacities of 4,900MW (7×700MW) and 6,300MW (9×700MW) correspondingly. The dam was constructed as the initial stage section (the underwater portion was constructed once for all as per the late stage section). The project was commenced in July 2001, river closed in November 2003, reservoir impoundment initiated in September 2006, first unit commissioned in May 2007, and the whole project completed by the end of 2009. This project set three world records: the highest RCC gravity dam (maximum height 216.5m, dam concrete volume 7.36million m^3), the largest underground powerhouse ($L \times W \times H = 388.5m \times 28.5m \times 74.4m$), and the highest lifting height for the ship lift (179.0m, lifted in two steps of 88.5m and 90.5m).

Longtan Hydropower Project had experienced several decades of investigation, research, design and comparison of numerous dam types and project layout schemes, and absorbed the constructing experience of high concrete gravity dams in the last 30 years since the reform and opening-up. Built up as the highest RCC gravity dam in the world, Longtan Dam is of great significance to China's large dam development cause.

The project is mainly composed of the main dam, outlet structures, powerhouse and navigation facilities. Its salient features are summarized below:

1) The previous preliminary design proposed to construct a conventional concrete gravity dam, and after a great deal of scientific studies and tests, it was finally determined to build an RCC gravity dam, with its full height and whole dam section constructed of RCC.

2) The previous dam-toe powerhouse scheme was altered to an underground one at the left bank, which extended the RCC zone in the dam and facilitated speed construction. The whole project was completed one year ahead of schedule, and gained considerable economical benefits.

3) The maximum flood discharge capacity of the dam is 27,144m^3/s and 26,085m^3/s respectively in the initial stage and late stage. The flood release structures are arranged in the middle of the river channel. There are 7 surface spillways ($W \times H = 15m \times 20m$) and 2 bottom outlets ($W \times H = 5m \times 8m$). Energy of flood is dissipated through the differential flip buckets.

4) The thickness of the rock pillar between the underground powerhouse and the main transformer cavern is 43m, and that between the main transformer cavern and the tailrace surge shaft is 28.4m, having a span ratio of 1.79 and 1.40 respectively with neighboring caverns and tunnels. The tailrace surge shaft is a long corridor throttle type surge chamber to be shared

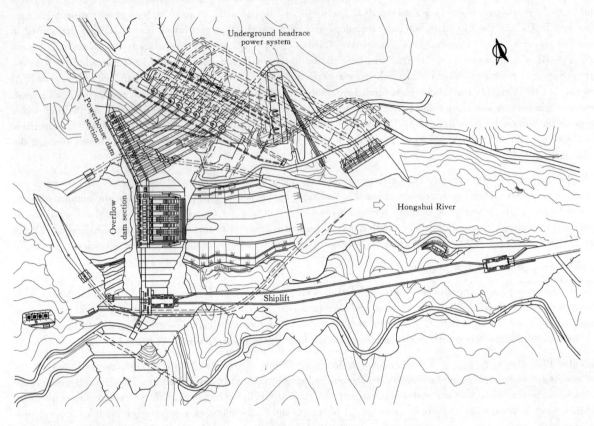

(a) Project layout plan

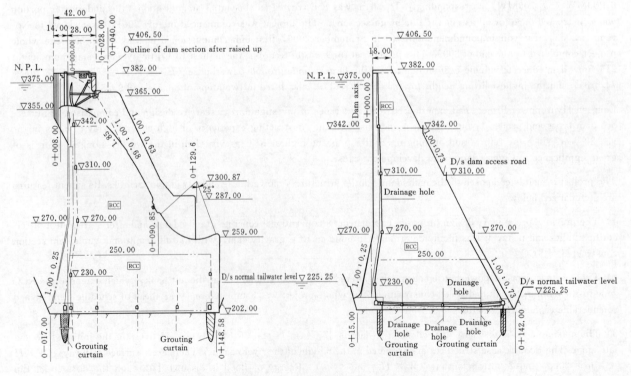

(b) Profile of typical dam section (the broken line indicating the section after being raised, dimension in m)

Fig. 8.2-4 Project Layout of Longtan Hydropower Station

by three units.

5) During the actual construction, the construction equipment worked rather efficiently, so that the dam RCC placement intensity could meet the design target. The daily maximum placement intensity hit 16,800m^3 in one place. RCC construction went on continuously in the high temperature and rainy seasons. In 2005, the maximum monthly placement intensity hit 318,000m^3, and the whole year RCC placement 2,400,000^3, which rightly manifested the fast placement feature of RCC.

(2) Guangzhao RCC gravity dam (Fig. 8.2-5): Located in the middle reach of the Beipan River mainstream, Guangzhao Hydropower Station was developed mainly for power generation besides navigation and other multi-purposes. At its normal pool level of El. 745.00m, the total storage capacity of the reservoir is 3.135 billion m^3, with incomplete multi-year regulating capability. The total installed capacity of the power plant is 1,040MW (4×260MW). The project achieved river closure in October 2004, first unit commissioned in June 2008, and was completed within 5 and half years. With the maximum height of 200.5m, Guangzhao dam is another 200m high RCC gravity dam built in China after Longtan dam. The project is composed of a main dam and one surface powerhouse system at the right bank, leaving space for the navigation infrastructure.

1) The maximum design discharge capacity is 9,857m^3/s. The flood release structures include three surface spillways 16m ×20m ($W \times H$) and one bottom outlet 4m×6.5m ($W \times H$). Outflow energy would be dissipated through ski-jumps out of slit buckets.

2) The total concrete volume is 2.8 million m^3, of which the RCC is 2.42 million m^3, accounting for 85% of the total.

3) The overall layout with the headrace system and the surface powerhouse arranged at the right bank greatly facilitates the RCC dam construction. The headrace and power system is arranged in the pattern of "one tunnel for two units", having two throttle type surge shafts. The layout plan is compact and tight.

4) Below El. 622.50m of the dam, RCC was directly hauled to the designated places by trucks; above El. 622.50m, RCC was placed via slip pipes after the conveyance belts with the assistance of cable hoists. Dam concrete was compacted ascendingly in sloped layers within big lifts. The biggest lift has an area of 20,086m^2. And the maximum RCC placement intensity reached 210,000m^3 per month.

(3) The Three Gorges Dam Project (Fig. 8.2-6): The Three Gorges Project is a mega water conservancy and hydropower scheme constructed for multipurposes mainly including flood control, power generation and navigation. Its main dam is 2,309m long in total along the axis and 181m high to the maximum. The design flood discharge capacity (corresponding to 1,000-year flood) is 98,800m^3/s and the check flood discharge capacity (corresponding to 10,000-year flood plus 10%) is 124,000m^3/s. The reservoir's gross storage capacity is 39.3 billion m^3, with a flood control capacity of 22.1 billion m^3.

1) The flood discharging dam section is situated in the middle of the river channel, and on both sides, are the powerhouse dam sections, non-overflow dam sections and power intake dam section for the right bank underground powerhouse. The total concrete volume of the dam reaches 14.8 million m^3. The project layout has three key features: The flood discharge capacity is rather great with 22 surface spillways, 23 deep outlets, 3 flood anddebris outlets and 7 silt sluice outlets; there are many diversion facilities like the 23 large sized diversion bottom outlets under the flood discharging dam section besides the 350m wide open diversion canal at the left side; the installed capacity of the power plant is enormous, with 26 units installed in the two surface powerhouses on the two sides, and 6 units in the right bank underground powerhouse, and the single unit capacity of all units is 700MW. The over length of the two powerhouse dam sections plus the flood sluice dam section reaches 1,690m, which has exceeded the width of the original river channel (about 1,300m). Hence, part of the surface powerhouses was excavated out from the rock slopes on both sides.

2) The diversion flood design standard is the 100-year flood corresponding to the discharge of 73,800m^3/s. River water was discharged through the open diversion canal on the right bank before the deep outlets in the dam body had been constructed. When the open diversion canal was blocked for building the right bank powerhouse after the dam toe, river water was discharged jointly by the bottom diversion outlets and permanent sluice deep outlets.

3) In order to keep navigation on the Yangtze River during the construction period, ships navigated through the open diversion canal in the low level period and the temporary ship lock, and navigation during the high level period depended on the vertical ship lifts established at the left bank non-overflow dam section. After completing the permanent ship lock construction, the temporary ship lock was restructured as a flood and silt sluice outlet.

4) A double-line 5-step large tonnage ship lock was excavated on the left bank, for passing ship fleets of 10,000 tons, every year capable of transporting 50 million tones altogether in one way. The total length of upstream and downstream channels is 4,835m, and the bottom width of the channels is 180m. The approach channels and entrances are far from the upstream and downstream flood release areas, and the exit is 4.5km away from the dam axis.

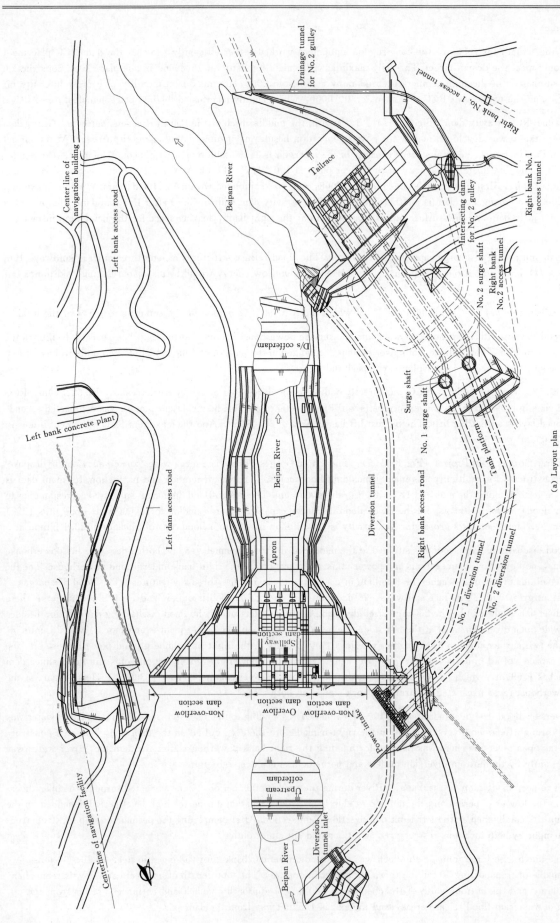

Fig. 8.2-5 (Ⅰ) Project Layout for Guangzhao Hydropower Project (Unit: m)

(a) Layout plan

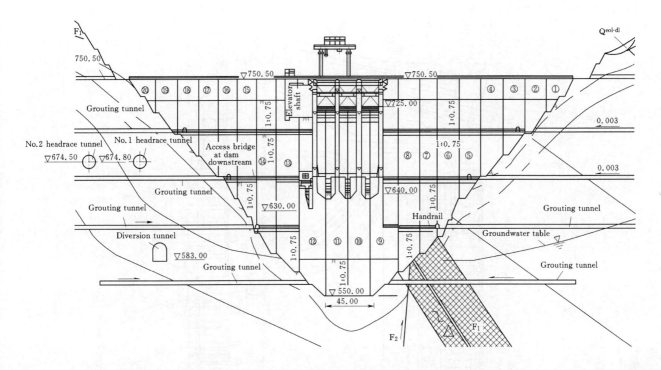

(b) Downstream dam elevation view

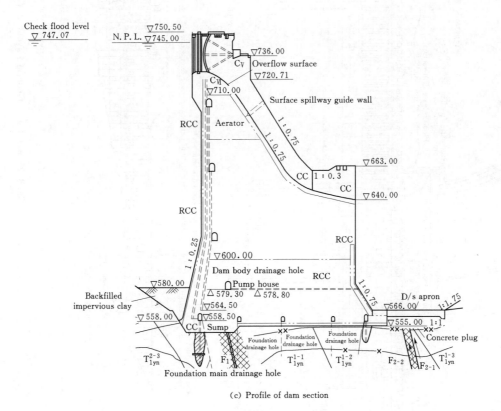

(c) Profile of dam section

Fig. 8.2-5 (Ⅱ) Project Layout for Guangzhao Hydropower Project (Unit: m)

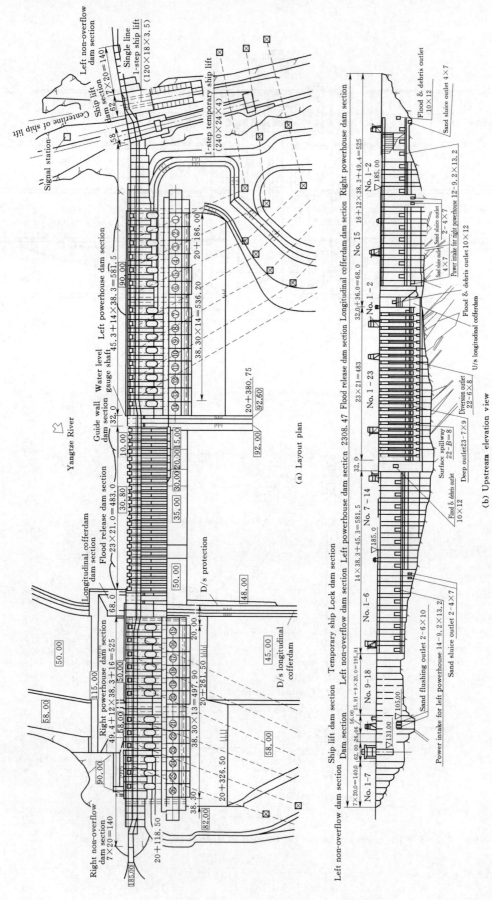

Fig. 8.2-6 (Ⅰ)　General Layout of Three Gorges Project (Unit: m)

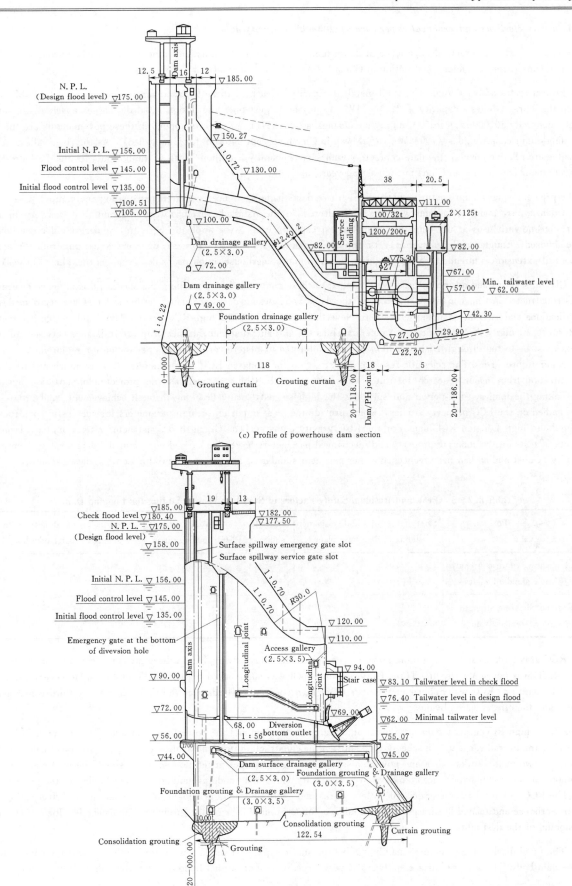

Fig. 8.2-6 (Ⅱ)　**General Layout of Three Gorges Project** (Unit: m)

4. Main characteristics and practical experience of 200m high gravity dams

Main characteristics of practical experience of 200m high gravity dams (mainly including Longtan, Guangzhao and Three Gorges) constructed or under construction in China before 2010 are summarized as follows:

(1) Project scales are very huge. The total installed capacity of each of these hydropower stations exceeds 1,000MW, of which the Three Gorges's capacity is 22,400MW, being the largest one. The design flood discharge capacities are rather great, generally 30,000m³/s to 100,000m³/s (Guangzhao, 9,857m³/s). Units applied in these power plants are the largest single unit capacity ones (700MW to 750MW) in China (except 240MW for Guangzhao). As the river valleys at the damsites are relatively wide, the dam concrete volumes are generally 3 million m³ to 16 million m³, being the largest for concrete dams constructed in China in the past 60 years.

In the previous stages before actual construction, these dams had experienced comprehensive study and comparison on different dam types, and gravity dams were the final selected. As the installed capacities were so gigantic, it was very difficult to arrange powerhouses at the dam toes even biggest capacity units were applied. Therefore, underground powerhouses were choosed (Longtan and Guangzhao), or bankside powerhouses (Three Gorges) were excavated out from the gentle slopes, if extension is intended, it may build an underground powerhouse on either banks (such as the Three Gorges).

(2) Classic sections and design theories were applied to these dams, and by absorbing the successful experience of gravity dams constructed in China in the 20th century, the dams were designed in accordance with the latest gravity dam design specifications and aseismic design specifications. Nevertheless, for the 200m high RCC gravity dams, the design parameters of RCC, particularly the mechanical parameters such as the anti-shearing and the shear rupture resistance between concrete layers must be tested through trial embankment, and subjected to austere verification as per the design requirements on the dam concrete properties. The scientific research and study on the Longtan RCC dam concrete attracted much attention and concentration from major research institutes and universities in China as well as the project administration authorities. Taking the longest time period and spending the highest cost, many breaking-through achievements and progresses were gained on the RCC property studies, which provide the most important design parameters for successful construction of the 200m high Longtan RCC dam. For the RCC gravity dams such as Guangzhao constructed after Longtan, large amounts of tests on the concrete property and mechanical parameters were also carried out, and the dam's safety factors of stability against sliding had been rechecked as per pertinent standards. The safety factors for Longtan dam are shown in Table 8.2 - 3❶.

Table 8.2 - 3 Peak and Residual Safety Factors of Stability Against Sliding for Longtan Dam

Position / Strength value	Safety factor of stability against sliding at dam base plane	Safety factor of stability against sliding at RCC1 bedding plane	Safety factor of stability against sliding at RCC1 dam body
Peak shearing rupture strength f' and c' standard values	3.048	3.301	3.897
Residual shearing strength f_r and c_r standard values	2.094	2.108	2.296

(3) RCC gravity dams and underground powerhouses were simultaneously applied in these projects. Due to a number of reasons, RCC was not applied to the Three Gorges Dam which was commenced in the middle of 1990s. However, after the early 21st century, RCC was applied to the over-100m high concrete cofferdam at the right bank in its third constrcution stage, and the effects were obvious.

For a 200m high RCC gravity dam, the duration of dam construction and the powerhouse construction are the main factors governing the overall construction schedule. To shorten the construction time and achieve power generation as earlier as possible, the two major works, dam and powerhouse, governing the power generation schedule should be well planned. The common and successful experience is to have an RCC gravity dam and an underground powerhouse in view of the project layout. The RCC gravity dam and associated outlets works and energy dissipators should be arranged on the "riverbed", and the powerhouse and related headrace system should go "underground" (or on the ground at the bankside) for earlier commissioning of the first units.

(4) The flood discharging structures mainly involve the surface spillways and a few bottom outlets, seldom having outlets in the dam body. The scale of surface spillways is very large. In case of normal floods, the turbine flow accounts for an im-

---

❶ Quoted from *the Design of Gravity Dams—A Twenty-Year Review*, Table 7.3 - 5.

portant part of the flood discharge. Some deep (bottom) outlets set in the dam body are favourable for flushing sand and releasing basic necessity flows in the initial impoundment period, and if necessary, they can lower and control the reservoir level as well. The flood release and energy dissipation works may adopt the flip buckets or plunge pools according to the dam foundation rock condition and downstream tailwater level variation.

(5) The dam concrete volumes are very huge, and both the monthly placement intensity and the monthly rising height set new records in China. As the two highest RCC gravity dams in China, the dam concrete volumes of Longtan and Guangzhao are $5,740,000m^3$, $7,360,000m^3$ and $2,940,000m^3$, respectively. The yearly concrete placement capacity surpasses several times the yearly placement capacity of $60,000m^3$ to $80,000m^3$ in 1980s and 1990s.

5. Mega dams and large hydropower plants require an ultra-high voltage grid's support

In China's western region, the 200m to 300m high mega dams and large-sized hydropower projects constructed, under-construction or to be constructed, are closely associated with each other. No matter they are gravity dams, arch dams or earth rockfill dams, all these hydropower plants require ultra-high voltage grids' support, in order to transmit the huge hydropower energy to the East of China, thousands of kilometers away, for achieving the strategy of transmitting western electricity to the eastern region. As reported by *the People's Daily* on February 28, 2011, in the article of *Extrahigh Voltage Grid-A key to solve the grid's bottleneck problem*❶:

"The extrahigh voltage (EHV) grid refers to a power grid with the voltage rating of AC 1,000kV, DC ±800kV or above. It features being capable to transmitting electricity by large capacity with low loss in a long distance. It is estimated that, the transmission capability of a 1,000kV AC EHV transmission line can exceed 5,000MW, almost 5 times that of a 500kV AC UHV transmission line. The transmission capability of a ±800kV DC EHV transmission line can exceed 7,000MW, almost 2.4 times that of a ±500kV DC UHV transmission line."

8.3 Arch Dams

8.3.1 General

8.3.1.1 Arch Dam Type Development

China is one of the countries having built most arch dams in the world. Up to 2010 since the founding of new China, about 60 years, there are 35m over 70m high concrete arch dams constructed or under construction. The over-70-m-high dams are dominated by concrete arch dams, and most of the medium and low ones below 70m high are masonry arch dams, as shown in Table 8.3-1.

Masonry dams in China are rather distinct, and the masonry arch dams are the most important type among the masonry dams. Only two above 15m high masonry dams were built up in China before 1949, and 6 in the 1950s, 56 in the 1960s, 591 in the 1970s, and in the 1980s after the reform and opening-up, the number of masonry dams decreased obviously. Though the number of masonry arch dams is large, but the number of dams over 70m high totals only 12. The 100.5m high Qunying masonry gravity arch dam built up in 1971 in Henan is the highest masonry arch dam in the world.

The masonry type arch dam is one exceptional feature in China's dam construction history. At that time, China had a huge number of stonemen in the countryside. They were master craftsmen good at carving and sculpture work. Masonry arch dams are not inferior to those concrete arch dams in the durability and safety. However, since the reform and opening up, peasant workers are migrating to cities, and almost no masonry arch dams has been constructed in China as there is no sufficient stonemen to work on that. It is anticipated, in the near future, the masonry arch dams might turn out to be tangible cultural heritage in China's water conservancy and hydropower development history.

Construction of concrete arch dams was set out after new China's foundation. In the 1970s, the arch dams had rapid development. China's arch dam building technology has gone through much upgrading since the 1980s. The 240m-high Ertan arch dam constructed in the 1990s ranks the third highest in the world among the double curvature arch dams, which symbolizes that China's arch dam construction has reached up to a new level. After entering the 21st century, high arch dams over 200m develop ever faster, the completed and under-construction 250m to 300m high arch dams include Xiaowan, Xiluodu, Jinping Ⅰ, Laxiwa, Goupitan etc., of which Xiaowan and Laxiwa arch dams started impoundment and power generation in the early of 2009, which demonstrates that China's high arch dams construction has stepped up to a new height.

---

❶ *People's Daily* on February 28, 2011, Page 19, by Ran Yongping and Yao Lei.

Part II Construction Technologies for Dams and Reservoirs

As shown in Table 8.3-1 (incomplete statistics), among the 35 concrete arch dams over 70m high, 10 dams constructed between 1980 and 2000, accounting for 28.6%; 25 dams completed during 2001 to 2010, accounting for 71.4%; and 15 hydropower stations among the 35 arch dams have the installed capacity over 300MW, accounting for around 50%.

Table 8.3-1 Over 70m High Concrete Arch Dams Completed or under Construction

No.	Name of project	Province (municipality/ region)	River	Dam type	Dam height (m)	Dam volume ($10^4 m^3$)	Installed capacity (MW)	Reservoir storage capacity ($10^8 m^3$)	Max. discharge (m^3/s)	Year of completion
1	Jinping I	Sichuan	Yalong River	VA	305	540	3,600	79.88		UC
2	Xiaowan	Yunnan	Lancang River	VA	294.3	838	4,200	150	20,745	UC
3	Xiluodu	Sichuan/Yunnan	Jinsha River	VA	285.5	685	13,860	126.7	52,300	UC
4	Laxiwa	Qinghai	Yellow River	VA	250	292	4,200	10.79	6,000	2010
5	Ertan	Sichuan	Yalong River	VA	240	414	3,300	61	23,900	2000
6	Goupitan	Guizhou	Wujiang River	VA	232.5	316	3,000	64.54	35,600	2012
7	Longyangxia	Qinghai	Yellow River	VA	178	157	1,280	274	6,000	1987
8	Dongfeng	Guizhou	Wujiang River	VA	162	48	695	10.25	14,200	1994
9	Dongjiang	Hunan	Leishui River	VA	157	96	500	92.7	24,100	1992
10	Lijiaxia	Qinghai	Yellow River	VA	155	110	1,600	17.5	6,340	1997
11	Geheyan	Hubei	Qingjiang River	VA	151	237	180	17.5	27,800	1993
12	Baishan	Jilin	Songhua River	VA	149.5	163	1,800	59.21	26,200	1986
13	Dahuashui	Guizhou	Qingshui River	RCCVA	134.5	69.8	200	2.76		2008
14	Shapai	Sichuan	Caopo River	RCCVA	132	38.5	36	0.18		2001
15	Fengtan	Hunan	Youshui River	VA	112.5	117	815	17.4	34,800	1979
16	Shimenzi	Xinjiang	Maxi River	RCCVA	110	21	600	0.5		2001
17	Zhaolaihe	Hubei	Zhaolai River	RCCVA	107	22	36	0.7		2005
18	Bailianya	Anhui	Manshui River	RCCVA	104.6	66.9	50	4.6	15,760	2007
19	Jinshuitan	Zhejiang	Oujiang River	VA	102	30	300	13.93	14,900	1988
20	Linhekou	Shaanxi	Lanhe River	RCCVA	96.5	29.3	72	1.49	3,480	2003
21	Bieshan	Henan		VA	94.5	72	0.2	0.01		
22	Shimen	Shaanxi	Baohe River	VA	88	20	40.5	1.1	5,190	1973
23	Xianghongdian	Anhui	Xipi River	VA	87.5	28	40	26.32	618	1961
24	Quanshui	Guangdong	Nashui River	VA	80	5.7	24	0.22	2,340	1976
25	Liuxihe	Guangdong	Lixi River	VA	78	13	48	3.78	2,100	1959
26	Wangwushan	Henan	Tieshan River	VA	77.3	4		0.694	1,770	1978
27	Tongtou	Sichuan	Baoxing River	VA	77		80	0.225		1995
28	Chencun	Anhui	Qingge River	VA	76.3	70	180	27.06	14,400	1971
29	Yaxi I	Zhejiang	Lishui River	VA	75	13	6.4	0.29	3,070	1977
30	Pudding	Guizhou	Wujiang River	VA	75	13.7	75	4.01	6,610	1994
31	Lishimen	Zhejiang	Shifeng River	VA	74.3	12	8.25	1.79	7,316	1979
32	Dashankou	Xinjiang	Kaidu River	VA	70	19.3	80	0.29		1991
33	Tiantangshan	Guangdong	Zengjiang River	VA	70	18	19.5	2.43	3,370	1994
34	Xinyugong	Henan		VA	71					1970
35	Gucheng	Beijing	Guishui River	VA	70	7.2	0.55	0.085	493	1979

Note: Quoted from *Hydropower Engineering in China*: Hydraulic Engineering Volume as well as Bibliography in this Chapter. Arch dams in Taiwan Province are not included.

8.3.1.2 Project Layout Development

The project layout for a concrete arch dam appears to be not so flexible compared to that of a concrete gravity dam, its layout has some particular features. At an arch dam site, the valley would be relatively narrow, even very deep with high mountains and steep slopes on both sides, so the project layout would be very uneasy to arrange, and the construction conditions are rather unsound. Particularly for a high dam and huge reservoir, with a big design flood discharge and large installed capacity, the project layout should give first priority to the arrangement of the outlet and energy dissipation works as well as the powerhouse. The past 30-years development of arch dams manifests some common characters:

1. About 100m high arch dam projects

For 100m high arch dam projects, though the layout patterns are diverse, their general arrangements are relatively simple and concise. Especially for projects with small scale and low flood discharge capacity, in recent 10 years, almost all the project layouts are made in the RCC arch dams with surface spillways and a bank side powerhouse, which would be favourable for shortening the overall construction duration and making better economic gains.

2. High arch dam with the height of 100 - 150m

As for 100m to 150m high arch dam projects, if the design flood is relatively great, generally surface spillways and middle outlets would be set in the dam body, with a ski-jump discharge and energy dissipation arrangement, and the powerhouse would be set at the dam toe on the riverbed. The Jinshuitan Arch Dam Project is a typical example. After entering the 21st century, the RCC arch dam height has exceeded 130m, the dam-toe powerhouse plan is not the preferred scheme any more. The project layout develops to the pattern of an RCC arch dam with surface spillways and an underground powerhouse (or bank side waterway and powerhouse).

3. High arch dams with the height of 100 - 150m

For 150m to 200m high arch dams, the dam bodies could be placed with conventional concrete without setting longitudinal joints, and their project layouts are distinctively made in different patterns due to topographic and geologic restrictions. For instance, though dam-toe powerhouses were adopted in Dongjiang, Geheyan and Lijiaxia Hydropower Stations, their powerhouse arrangements are not alike, and their flood release and energy dissipation structures are disparate. Longyangxia and Baishan dams are gravity type arch dams, with broad foundations, and their dams are still required to be concreted in column blocks. Due to geological constraints at the dam sites, their powerhouses are arranged at the dam toe and under ground, respectively. The above five high arch dams constructed in succession in the 1980s and 1990s have operated safely for many years and accumulated much valuable experience in the project layout planning, and laid foundations and provided lessons for further building over 200m high arch dams.

4. High or above mega arch dams with the height of 200 - 250m

For mega arch dams 200m to 250m or above 250m in height, the completed or under-construction of Xiaowan, Laxiwa, Xiluodu, Jinping I and Goupitan as well as the under-preparation of Wudongde, Baihetan, Dagangshan etc., have following common features. The flood release and energy dissipation works is laid on the riverbed section. Flood will be discharged through the surface spillways on the dam crest and the middle and bottom outlets in the dam body. A secondary dam created a plunge pool will be built up at the downstream of the main dam. And considering the design flood characteristics as well as the requirement for emptying the reservoir or lowering its level, an auxiliary bank side flood release outlet will be provided. The headrace and power generation system is set under ground, so that large underground power tunnels and caverns would be formed on either side or both sides of the dam site. If the diversion flow is too big, several large diversion tunnels should be arranged on the two banks.

This general layout pattern has become a model for mega arch dams. Since the Ertan mega arch dam (240m) completed by the end of the 20th century, almost all mega arch dams constructed in the 21st century adopted such layout, which turns out to be a unique feature of mega arch dams in China.

8.3.2 Development of Concrete Arch Dams in Last 30 Years

8.3.2.1 Arch Dam Development from 1980 to 2000

During the 20-year period from 1980 to 2000, altogether 30 over 100m high dams were constructed, as summarized in Table 8.1-2, of which 8 are concrete arch dams, accounting for, about 26.6% of the total number. Though the percentage of arch dams is not much, far less than those of gravity dams and earth rockfill dams, the arch dam construction technology achieved great progress, with the Er'tan arch dam (Fig. 8.3-1) as the representative one, reaching a new level. The development and features of the arch dams in the twenty years can be summarized as follows:

Part II Construction Technologies for Dams and Reservoirs

1. The arch dam type and overall project layout had continuous development and innovations in various aspects

Both the construction of Longyangxia (dam height 178m) in the 1970s and Baishan (dam height 149.5m) hydropower stations in the 1980s adopted the gravity arch dam scheme separately with a dam-toe powerhouse and an underground one. Both Dongjiang (dam height 157m) and Lijiaxia (dam height 155m) projects completed in 1980s and 1990s have a double curvature thin arch dam with a dam-toe powerhouse and penstocks installed over the downstream dam slope. Details refer to Fig. 8.3-2 to Fig. 8.3-5.

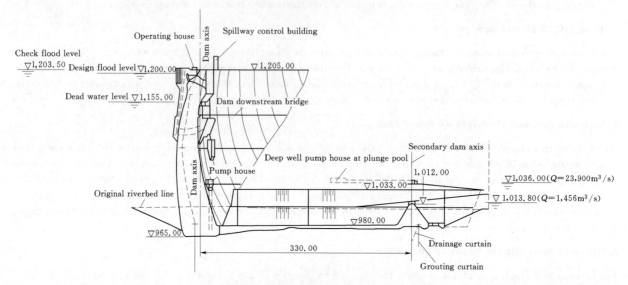

Fig. 8.3-1 Profile of Ertan Arch Dam (Unit: m)

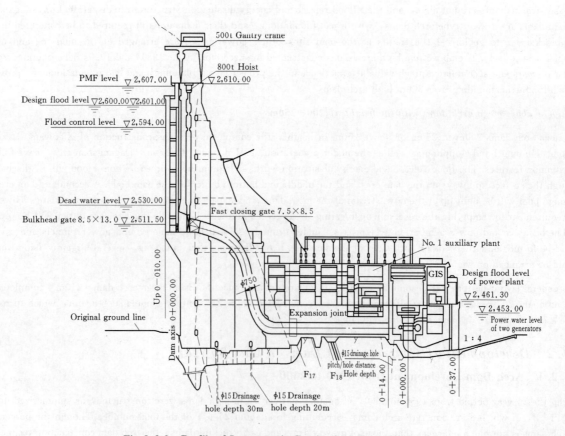

Fig. 8.3-2 Profile of Longyangxia Gravity Arch Dam (Unit: m)

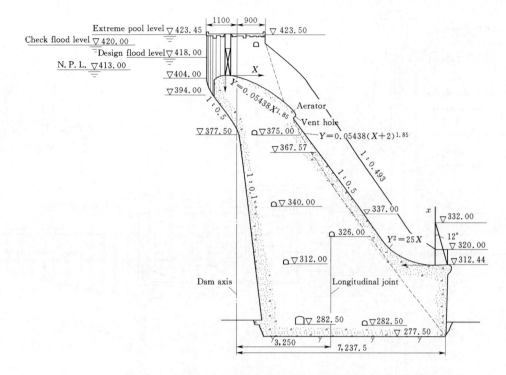

Fig. 8.3-3 Profile of Baishan Gravity Arch Dam (Unit: m)

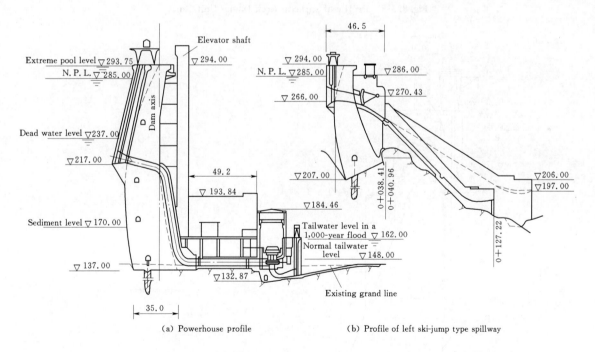

(a) Powerhouse profile (b) Profile of left ski-jump type spillway

Fig. 8.3-4 Profile of Dongjiang Arch Dam (Unit: m)

Geheyan arch dam with the height of 151m had some new things in its arrangement that the lower portion is an arch and the upper portion a gravity section. Fengtan Arch Dam (112.5m high) is more exceptional as its powerhouse is located in the dam body and excessive water overflows the dam crest. Details refer to Fig. 8.3-6 and Fig. 8.3-7.

The above six arch dams completed during the 1980s and the 1990s are the typical projects, and have operated safely for many years. They all have won excellent design or engineering awards.

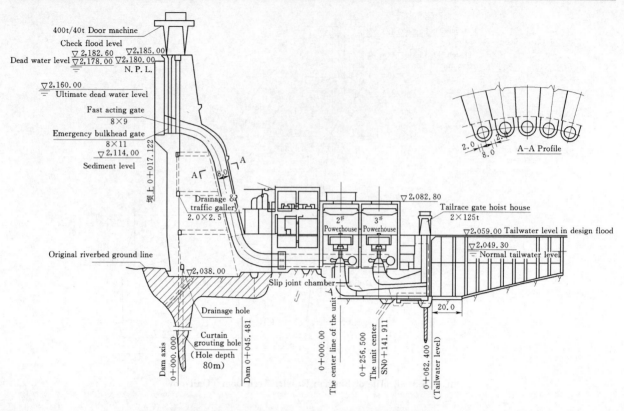

Fig. 8.3-5 Profile of Lijiaxia Arch Dam (Unit: m)

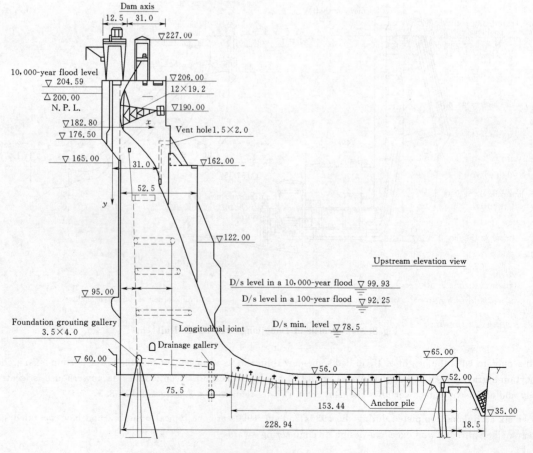

Fig. 8.3-6 Profile of Geheyan Arch Dam (Unit: m)

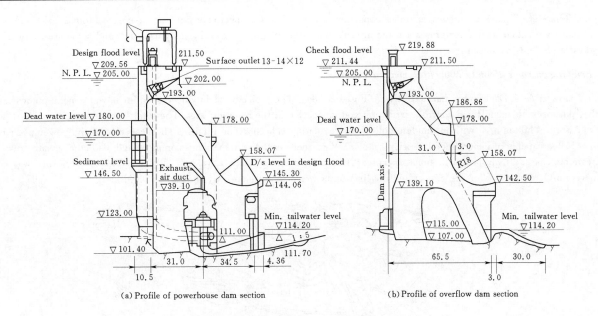

Fig. 8.3-7 Profile of Fengtan Arch Dam (Unit: m)

2. The discharging facility scale and energy dissipation for high arch dams have made great progress

The flood release works for high arch dams widely adopt the hybrid release method combining surface spillways and deep outlets (or surface spillways, middle and bottom outlets) on the dam body, and if the site topographic condition is favourable, an auxiliary bank side flood tunnel or spillway would be arranged for discharging flood. The size of outlets in the dam bodies is very large, basically equivalent to that of openings in the gravity dams. In 1980 – 2000, the outlet works for 100m to 150m high arch dams were diversified in the layouts, though the energy dissipation method of most projects resorted to the trajectory jet. Most 150m to 200m high arch dams adopted the trajectory jet energy dissipation over the downstream riverbed. As for over 200m high arch dams, such as the 240m high Ertan arch dam, because of the high head and great energy of flood release, a secondary dam was constructed to form a plunge pool in the downstream energy dissipation zone.

3. Development of arch dam safety design criteria makes arch dams still competitive

Owing to experience accumulation and key technologies development of the high arch dam constructionin in the 20th century, arch dams have great priority over other dam types, and the arch dam has been equally recognized as a safe, reliable, and economical dam type.

The arch dam is a higher-order hyperstatic structure constrained by rock foundation and has strong monolithic action. Under various external loads, inner forces in the dam body can be adjusted and homogenized through deforming and coordinating actions to make the arch dam structure have higher overloading capability, and the arch dam also can be evaluated in quantative method through numerical simulation analysis, which is one important factor why the arch dam possesses obvious advantage. Even if the geological conditions have some weak points, this aim also can be realized by making an appropriate arch dam shape and taking certain physical measures. The high arch dam construction in Ertan has enriched the experience in arch dam safety design. Much development has been made on the high arch dam strength design criteria, abutment stability safety design criteria, high arch dam seismic fortification design criteria, high arch dam monolithic safety assessment criteria etc., which also laid a solid foundation for building 250m to 300m high arch dams including Laxiwa, Xiaowan, Xiluodu, Jinping I and Goupitan in the early 21st century.

4. Large openings in the high arch dam bodies and associated hydraulic steel structures have great development

With the development of arch dams, many surface and deep outlets are required to be set in the dam body, and such high head large openings are generally steel lined or reinforced concrete structures, which also promotes the development of large-sized high head gates and large-capacity hydraulic hoists. Successful experience has been gained in the design, fabrication, installation and operation of hydraulic steel structures, as essential references for building high arch dams in the 21st century.

8.3.2.2 High Arch Development from 2001 to 2010

The first 10 years of the 21st century is an important development period for high arch dams and over 250m arch dams in

China. There are 5 dams completed or under construction, they were respectively put into operation around 2010. These arch dams with their unique features reach the international state-of-the-art level in many aspects, and some are the world highest arch dams at present.

1. Briefing on some 250m to 300m high mega arch dams

(1) Xiaowan Arch Dam on the Lancang River (Fig. 8.3-8): The project is mainly developed for power generation besides other purposes including flood control, irrigation and navigation in the reservoir area. The total storage capacity of the reservoir is 14.9 billion m^3, with incomplete multi-year regulating capacity. The total installed capacity of the power plant is 4,200MW, capable of generating electricity 18.890TW·h per year. And its commissioning would also compensate downstream six cascade hydropower stations to increase the annual energy output by 2.575TW·h. The river closure was achieved in October 2004 for the project and the first units were commissioned by the end of 2009.

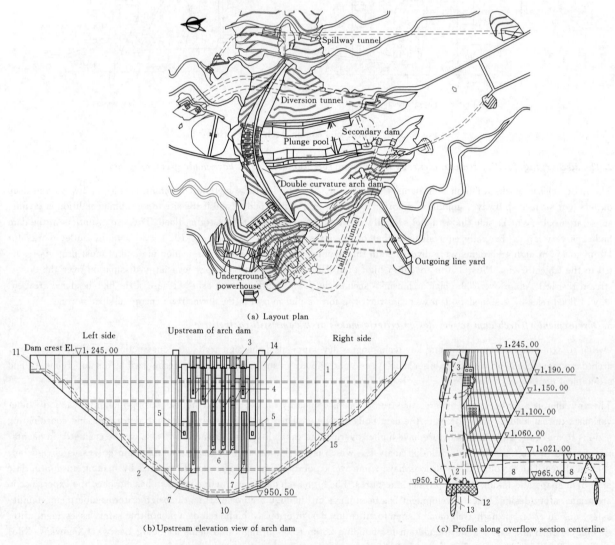

Fig. 8.3-8 Layout Plan of Xiaowan Arch Dam

1—DCAD; 2—Curtain grouting gallery; 3—Surface spillway; 4—Mid-outlet; 5—Bottom outlet; 6—Diversion mid-outlet; 7—Diversion bottom outlet; 8—Plunge pool; 9—Secondary dam; 10—Inducing joint; 11—Thrust pier; 12—Consolidation grouting; 13—Grouting curtain; 14—Elevator shaft; 15—Inspection gallery

The project consists of the main dam and an underground powerhouse on the right bank. The dam is a concrete double curvature arch one with a maximum height of 294.5m, a dam base thickness-dam height ratio of 0.238, a dam crest arc length of near 900m, and a maximum arc end thickness of 72.9m. The headrace power generation system is arranged on the right bank with the underground powerhouse ($L \times W \times H = 323.0m \times 30.6m \times 82.0m$) installed with 4 sets of 700MW units. There are two tailrace tunnels, 900m and 700m long respectively. The release works is composed of 5 surface spillways, 6 middle outlets and one spillway tunnel with the length of 1,542m on the left bank.

1) Xiaowan Arch Dam is the highest arch dam commissioned in China among the 300m high arch dams, and also the highest arch dam under construction in the world. The dam is constructed in a high seismic intensity region with a horizontal PGA of 0.308g. Design of an extra-high arch dam is not an easy task. For instance, the current Chinese code has no specific stipulations on the dam heel cracking problem for high arch dams. In this project, based on numerous analyses and calculations, a physical measure for preventing or mitigating the dam heel cracking problem was proposed, i.e. setting structural inducing joints at the dam heel, providing silting material in front of the dam and seepage-proofing system over the dam surface. For dealing with possible opening of transverse joints at the middle and higher elevations in the event of strong quake, aseismic reinforcement and dampers are provided, and their acting mechanisms and effects have been subjected to numerical analysis and model testing.

2) The flood discharge and power can reach 20,745m^3/s and 46,000MW respectively, so the energy dissipation issue would be a great challenge in the flood release process. Based on the numerous studies, it was determined to apply a joint flood release manner relying on the surface spillways and middle outlets on the dam body as well as one bank side flood release tunnel, and the energy dissipation depends on the outflow trajectory jets to fall into the downstream plunge pool.

3) The scale of caverns and tunnels for the underground headrace power generation system is quite large. The surge shaft in a diameter of 32.0m has a height of 92.0m. The tailrace system stability problem resorts to the cylinder throttle type surge shaft with an upper chamber.

4) Raw materials and temperature control for the arch dam concrete are two key technical problems for the project construction. The maximum concreting block in the mid-lower part of the dam is more than 90m long, so measures for controlling temperature and against cracking are of great significance. Based on study and testing, the control criteria like the suitable minimum outlet temperature of concrete, placement temperature, flow water cooling temperature and duration, arc sealing grouting temperature as well as lift thickness and placement interval time were proposed, and strict quality measures were taken for the concrete raw materials and concrete temperature control.

5) The project is located in a remote and deep canyon with steep bank slopes rising 1,000m high above the river. Particularly, the slopes at the two abutment troughs are more than 1,000m high, and the slopes to be manually treated are 700m high. It is quite rare to treat such high slopes with poor stability for high arch dam projects in China and abroad.

(2) Laxiwa Arch Dam on the Yellow River (Fig. 8.3-9): Laxiwa Arch Dam project is located in the downstream of Longyangxia reservoir on the Yellow River mainstream in Qinghai Province. Its maximum design flood discharge is 6,000m^3/s. The reservoir area is a 40km long canyon with great mountains on both sides, about 600m to 700m high above the riverbed. The main dam is a logarithmic spiral double curvature arch dam with a height of 250m. Its reservoir storage capacity is 1.079 billion m^3, with daily regulating capability. The project is developed mainly for power generation. Its total installed capacity is 4,200MW (6×700MW). The project layout turns out to be a typical arrangement model for those over 250m high arch dams in China, with the dam located on the riverbed, surface and deep outlets set in the dam body for releasing flood, energy of outflow flood to be dissipated in the downstream plunge pool retained by the secondary dam, and an underground powerhouse at the right bank. The project features are as follows:

1) The dam site is a V-shaped valley, with sound topographic and geological conditions, and bedrock is hard and intact massive granite. The arch dam's base thickness-height ratio is about 0.2. The basic seismic intensity in the region is Ⅷ degree, and the seismic fortification intensity is Ⅸ degree.

2) Besides the available regulating capacity of the upstream Longyangxia reservoir, Laxiwa reservoir also has an accident standby storage capacity of 120 million m^3, which would enable Laxiwa power station to become a reliable backup power source in case of the system accident. The voltage rating of the power plant's outgoing line is 750kV. This is the first power station in China adopting the 750kV transmission line.

3) Dam construction is an outstanding issue for building a high arch dam in a narrow and steep valley. The river closure was made in advance. Under the protection of the upstream and downstream cofferdams, mucks resulted from the high slopes treatment, cable crane platform excavation as well as excavation of upper parts of the abutments were contained in the foundation pit, and removed out once for all in a highly efficient manner to win construction time as much as possible. The internal traffic network was arranged in a pattern of interconnecting different-level roads and adits on both banks, and the concrete conveying routes and access to the cable crane platform went through adits. It has demonstrated that this practice is safe, reliable, saving both cost and time.

4) Successful application of a reverse arc plunge pool and providing a cushion layer between the spiral case and its concrete cover: In a narrow and steep canyon, the ground stress at the valley bottom is rather high, building a reverse arc plunge pool could ease adverse impacts on the bank slopes in the energy dissipation area. After consultation with the turbine manu-

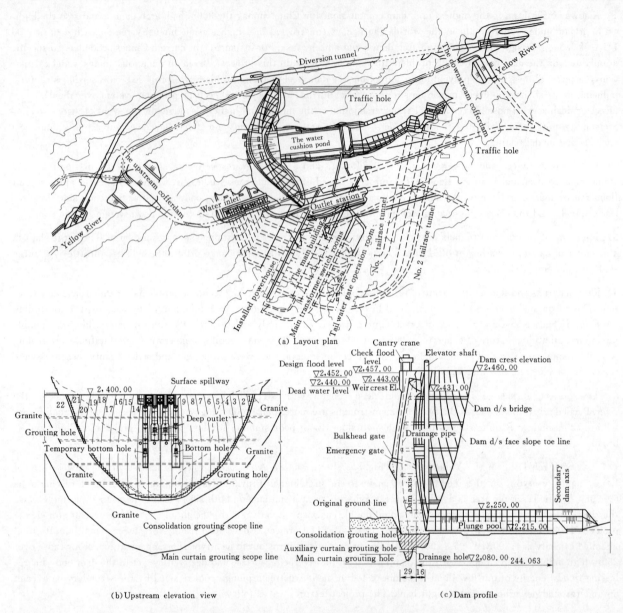

Fig. 8.3-9 Layout Plan of Laxiwa Hydropower Project (Unit: m)

facturer, it was determined to provide a cushion layer between the spiral case and its concrete cover to transfer internal pressure. This structural form is seldom applied to similar projects in China. In this respect, the cushion material and the spiral case structure had been subjected to special case studies to justify their viability.

(3) Xiluodu Arch Dam on the Jinsha River (Fig. 8.3-10): Located in the lower reach of the Jinsha River and the tranboudary river between Yunnan and Sichuan Provinces, and the Xiluodu Project is a huge hydropower station to be developed mainly for power generation besides sediment retention, flood control and improvement of the downstream navigation. Its total installed capacity is 13,860MW, only second to that of Three Gorges Project, to be the second largest one in China. It is one of the largest hydropower stations under construction in China. The catchment area till the dam site is 454,400km^2, accounting for 96% of the total catchment area of the Jinsha River. The total storage capacity of the reservoir is 9.27 billion m^3, with a regulating storage of 6.46 billion m^3, and a flood retention capacity of 4.65 billion m^3. The project is composed of the main dam, headrace power generation system and outlet works. It was commenced in December 2005, river closed in 2008, first units commissioned in 2013, and the overall construction period would last 12 years and 2 months. The project has the following main features:

1) The main dam is a concrete double curvature arch dam with a maximum height of 285.5m. Its length of the dam crest is 714m. The dam bottom thickness is 70m, with a thickness-height ratio of 0.27. Xiluodu Dam, like Xiaowan Arch Dam and Jinping I Arch Dam, also is one of the highest arch dams with the height of 300m under construction in China.

Chapter 8 Dam Type and Project Layout

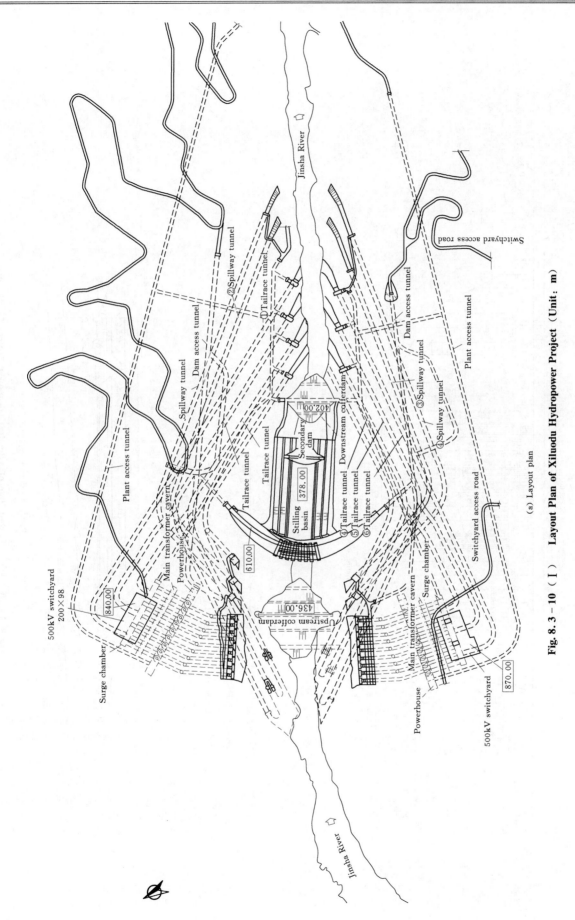

Fig. 8.3-10 (I) Layout Plan of Xiluodu Hydropower Project (Unit: m)

(a) Layout plan

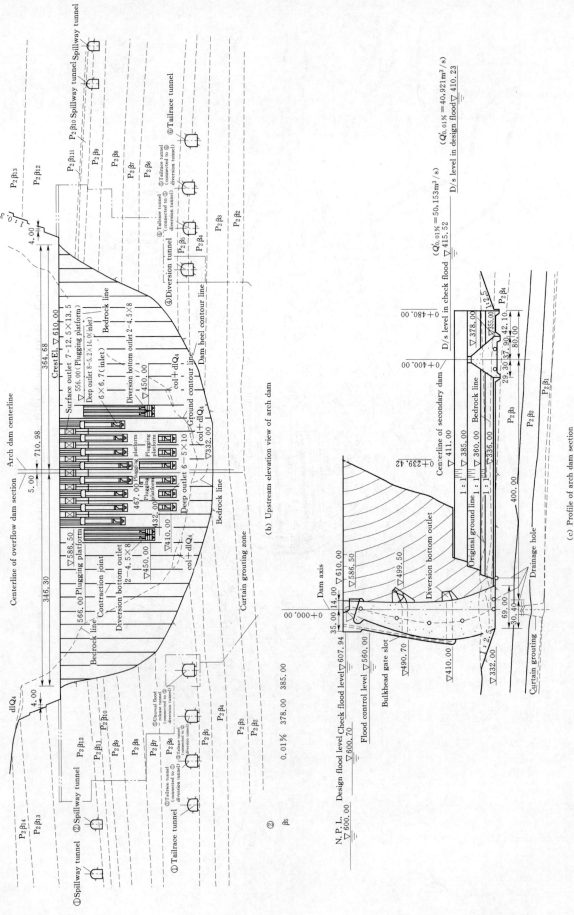

Fig. 8.3 - 10 (Ⅱ) Layout Plan of Xiluodu Hydropower Project (Unit: m)

2) The largest discharge capacity of the outlet works is 52,300m³/s ($P=0.01\%$), jointly discharging flood through 7 surface spillways ($W \times H = 12.5m \times 13.5m$) and 8 deep outlets ($W \times H = 6.0m \times 6.7m$) on the dam, and 4 spillway tunnels on the two banks. Energy of outflow from surface spillways and deep outlets will be dissipated in the plunge pool retained by the downstream secondary dam, and energy of flow out of the flood release tunnels will be dissipated through trajectory jets. It is a rare case both in China and abroad, that the outlet works scale and corresponding discharge capacity are so large in a canyon damsite like this.

3) There are two underground powerhouses, with their headrace systems respectively set at the upstream side of the dam axis on both banks. Both powerhouses will be equipped with 9 sets of 700MW units. Both the horizontal and vertical rock covers over the underground power caverns and tunnels for the two power systems are greater than 300m. According to the rating of rock quality, the tunneling condition is sound. The total volume of underground tunnel and cavern excavations approximates to 15 million m³, which would overpass the total volume of any underground powerhouse constructed both in China and abroad.

4) As the design diversion flow amounts to 32,000m³/s ($P=2\%$), 3 large diversion tunnels are designed respectively on both banks. The upstream cofferdam is a high earth rockfill dam.

(4) Jinping I Arch Dam on the Yalong River (Fig. 8.3-11): Jinping I Project is the regulating reservoir and power station in the lower reach of the Yalong River, followed by downstream cascade hydropower projects, i.e. Jinping II, Guandi, Ertan and Tongzilin. Its main purpose is to generate electricity besides flood control and sediment retention. The catchment area controlled at the dam site is 102,600km². The annual average discharge at the dam site is 1,200m³/s. The storage capacity of the reservoir at the normal high level is 7.765 billion m³, with a regulating storage of 4.91 billion m³ with mean annual regulating capability. The power plant's total installed capacity is 3,600MW. The project is composed of the concrete double curvature arch dam, plunge pool at the main dam downstream, secondary dam, right bank spillway tunnel and underground headrace and power generation system on the right bank. The project preparation was initiated in the end of 2003, commenced in November 2005, river closed in December 2006, estimated to commission first units in 2012, and to be completed in 2014. The project has the following features:

1) The arch dam's maximum height is 305.0m as the highest arch dam in the world. The crown cantilever's top thickness is 16.0m, bottom thickness 63.0m, the thickness-height ratio 0.207, the crown centerline arc length 552.23m, and the arc-height ratio 1.811. The dam body would withstand the total water thrust near 12 million tons under the normal high reservoir level. The dam foundation rock is dominated by thick-layered massive marbles. The dam site enjoys some favorable conditions like a narrow valley and hard and rigid rocks on both banks. However, the topographic and geologic conditions are asymmetric, and the deformation modulus of the two side foundations at the same elevation are rather disparate, which would cause the arch dam stressing and deformation to be asymmetric, and be extremely unfavorable for a high arch dam's stability against deformation. Based on extensive research on mitigating the unfavorable effects of the asymmetric dam foundation and such complex base, large amounts of foundation reinforcement have been made to the works so as to improve the arch dam's overall stability and safety.

2) Flood release and energy dissipation are the two outstanding issues. 4 surface spillways and 5 deep outlets are designed on the dam body, one plunge pool and a secondary dam are set at the downstream of the main dam, besides there is one flood release tunnel set on the right bank.

3) The underground powerhouse will be installed with 6 sets of 600MW units, capable of generating electricity 16.620TW·h per year. The maximum head on the turbine is 240m. The right bank headrace and power generation system is arranged in a classic layout pattern, with other unique features like tailrace surge shafts and every three units sharing one tailrace tunnel.

2. Main features of completed and under-construction 250m to 300m high mega arch dams

(1) All these 250m to 300m high arch dams are located at the grand canyons of middle or upper reaches of the rivers in the western remote mountainous regions scarcely populated. High mountains rise hundreds of meters high above the dams on both banks. The valleys are narrow and take on "V" or "U" shapes. At the dam sites, the ratios of dam crest arc lengths to heights are 1.97 for Laxiwa, 1.811 for Jinping I, 3.16 for Xiaowan, 2.47 for Goupitan, and 2.44 for Xiluodu, respectively. From the topographic point of view, these are good sites for building arch dams. During the construction and operation periods, the stability and safety of high slopes on both sides is one of the priority issues that the developers are highly concerned with.

(2) All the reservoirs are rated as Large-Scale (Grade I), with good regulating capability. Besides their main purposes of power generation, these arch dam-reservoirs have other multi-purposes of flood control, water supply and navigation in the reservoir areas. The total storage capacities of Xiaowan, Xiluodu and Goupitan reservoirs separately are 15 billion m³, 9.27 billion m³

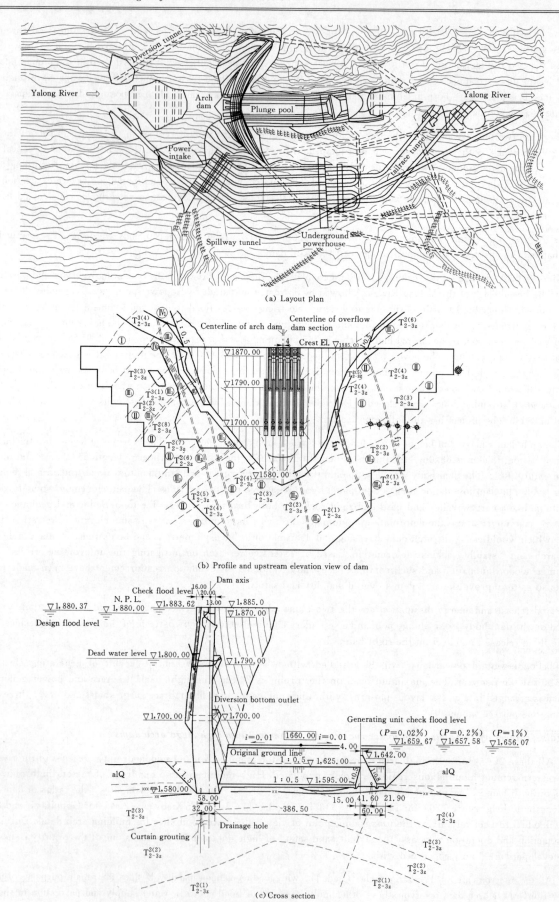

Fig. 8.3-11　General Layout of Jinping Ⅰ Arch Dam Project（Unit：m）

and 6.454 billion m³, and respectively with regulating storage capacities of 9.895 billion m³, 6.46 billion m³ and 3.154 billion m³, which reflects their great regulating capability.

(3) The seismic intensity values at these dam sites are high, and the design flood discharges are huge. The dam sites are situated in the regions with high seismic intensity, and the designed seismic intensity for most dams is IX degree. The performance and safety of arch dams under seismic conditions are issues of great concern for project construction. As the flood discharge capacities of these arch dams are rather great (25,300m³/s for Xiluodu, 20,745m³/s for Xiaowan) and the flood periods last relatively long, the flood release and energy dissipation design for such arch dams become one of the outstanding technical problems affecting the project layouts.

(4) There has formed one typical project layout mode. The layouts of above mentioned arch dams are similar in their basic patterns. The outlet works including associated energy dissipators as well as the powerhouses are the major components affecting the layouts.

The flood release and energy dissipation works are usually arranged in a typical layout pattern, i.e., the surface spillways on the dam crest are the main flood release works with auxiliary facilities such as middle and bottom outlets in the dam body as well as a bank side spillway tunnel, a secondary dam is constructed on the riverbed at the downstream of the main dam to form a plunge pool for energy dissipation, and the flood release tunnel outlet would have a trajectory jet or underflow energy dissipater. The flood discharge capacities of the outlet works for mega arch dams in China are fairly large and rare among those projects constructed in other countries.

All the powerhouses are underground. At those sites suitable for building arch dams, the engineering geological conditions on both banks are generally also suitable to arrange underground powerhouses and headrace power systems. According to the diversion flood standards and corresponding diversion flows, one or several large-diameter diversion tunnels also can be arranged on both banks spaced with the headrace & power generation system.

(5) The underground powerhouses are huge in scale. It should not be over-exaggerating that the scale of such underground powerhouses for these mega arch dam projects is "Huge". They also have vast installed capacities of 4,000MW to 5,000MW or above, and 5 to 6 or more sets of biggest single capacity (700MW to 750MW) units to be equipped in the powerhouses. The under-construction Xiluodu Hydropower Project will install 18 units with a total capacity of 12,600MW.

All these projects become the most largest hydropower underground works in China since 1949, and also the largest hydropower underground works in the present world. Their construction promoted rapid development of many sciences and technologies such as geognostic, rock mechanics, underground water dynamics, engineering design and new construction technologies on the underground big-span tunnels and caverns, and ample theoretical and practical experience has been acquired.

(6) The safety and risks of construction diversion were paid great attention to. The diversion standard usually adopts the upper limits of the current specifications, generally in a 30 years to 50 years flood. The river channel will be closed once for all. Dam construction can be carried out in the foundation pit all the year round. A preventive emergency plan will be prepared for probable over-standard flood during the diversion period, and the standard for overcoming flood season shall be reviewed as per the water retaining capacity of the dam in different stages, so as to assure the dam's safety during the construction period, and to minimize possible risks.

The diversion works is fairly majestic, but it is only a prelude for building a high arch dam. As the designed peak flood diversion flow is very large, and the flood retention and regulating capacity of the cofferdam in the canyon is limited, so a number of large-diameter diversion tunnels are needed. The diversion works for the five mega arch dams are summarized in Table 8.3-2.

Table 8.3-2 Parameters of Diversion Works for Mega Arch Dam Projects

No.	Name of project	Max. dam height (m)	Diversion standard		Diversion tunnel			Height of upstream cofferdam (m)
			Whole year	Diversion flow (m³/s)	No. of tunnels	Diameter (m)	Length (m)	
1	Xiaowan	294.5	30-year flood	10,300	2	16×19	860,980	60.6
2	Laxiwa	250	20-year flood	2,000	1	13×14.5	1,416	44.8
3	Goupitan	232.5	10-year flood	13,500	3	15.6×17.7	—	—
4	Xiluodu	285.5	50-year flood	32,000	6	18×20	—	—
5	Jinping I	305	30-year flood	9,370	2	15×19	1,214	—

Note: Tables 8.3-2, 8.3-3, 8.3-4 and 8.4-2 are provided by Senior Engineer Duan Zhenmei from Hydrochina Xibei Engineering Corporation.

As shown in Table 8.3-2, the diversion scale of Xiluodu is the largest one among the five mega arch dams, with a 50-year design flood diversion discharge of 32,000m³/s, 6 diversion tunnels (18m×20m) are required, respectively set at both banks. The diversion scale for Laxiwa is the least, only one diversion tunnel (13m×14.5m) is satisfactory to the regulating capability of Longyangxia Reservoir in the upstream.

(7) Construction layouts for these projects are rather difficult to make and different from each other as the available spaces are limited. Building 250m to 300m high arch dams in the canyon sites, large-capacity aggregation and concrete producing systems are required to be established near the dam sites, as well as heavy cable machines and gantry or tower cranes for accomplishing high intensity concrete placement task. For cutting the high slopes and excavating the abutments on both banks as well as conveying concrete, construction access roads should be provided at different levels. The difficulty of making the construction layout, the huge quantity of earth rock excavation works, and the uneasiness of arranging internal roads, all these things have never been seen before in the sixty years of hydropower and water conservancy undertakings. However, the construction layout schemes have been successfully made with their own characteristics.

(8) The dam concrete volumes are very great, placement intensity is very high, and construction duration is relatively short. Relevant concreting parameters for the five mega arch dams completed or under-construction are summarized in Table 8.3-3.

Table 8.3-3 represents the current state-of-the-art technology of high arch dams, reflects the highest level of current aggregate and concrete systems' producing capability as well as the dam concrete placement capability, which also is the highest level in the past 60-year high arch dams construction in China.

All these high arch dams adopt the underground powerhouse arrangement. Based on the analysis of the construction schedule, annual increase of the main dam' height instead of the underground powerhouse system is the factor controlling the power generation schedule. The increasing height is such an important comprehensive index that can broadly reflect the dam concrete placement intensity, quality control, temperature control, dam's 2nd stage cooling and grouting in transverse joints, as well as the essential capability of the dam to pass through flood seasons.

Table 8.3-3 Concreting Parameters for High Arch Dams

Name of project	Max. dam height (m)	Powerhouse type	Total volume of dam concrete (10^4 m³)	Max. bottom width of dam (m)	Max. monthly placement intensity (10^4 m³/month)	Max. yearly placement intensity (10^4 m³/year)	Dam's full section monthly rise height (m/month)	Dam foundation concrete temperature control (℃)
Xiaowan	294.5	Underground	838	72	22.2	247.5	8.41/5.77	14
Laxiwa	250	Underground	292	49	12.17	94.98	12.67/5.93	14
Goupitan	232.5	Underground	316	50	—	—	—	—
Xiluodu	285.5	Underground	685	69	—	—	—	15
Jinping I	305	Underground	540	60	16 (19)	190 (200)	8	14

Note: The Max. monthly and yearly placement intensity as well as the dam's full section monthly rise height are the design values, and those figures in brackets are ever-achieved values.

(9) No longitudinal joints were set in the dam body, and the concrete temperature was under strict control. All these high arch dams have no longitudinal joints. The dam base's maximum bottom width is an important factor for working out the concrete placement intensity and temperature control measures. The arch dam designer also tries to control the dam base's bottom width within a reasonable and viable scope. For the dam section to adapt to the canyon topography and geological features, different arch dam outlines with different variable curvatures have been studied through numerous calculation methods and necessary tests, so as to control the dam stresses within the permissible limit.

(10) The power station's electromechanical equipment and power transmission works technology reach the highest level in the sixty years. The characteristics of electromechanical equipment and transmission works for the five completed or under-construction mega arch dams are summarized in Table 8.3-4.

The above table shows that the technical indexes of these power stations' electromechanical equipment and transmission works are the highest ones in the 60-year history of high arch dams construction. All the electromechanical equipment are developed and produced by Chinese companies or joint ventures, and their performances have come up to a new high level since 1949.

Chapter 8 Dam Type and Project Layout

Table 8.3-4 Main Technical Indexes of High Arch Dams and Hydropower Stations and Electromechanical Equipment

Name of project	Installed capacity (MW)	No. of units & single capacity (MW)	Max. design head of unit (m)	Turbine flow under full load operation (m³/s)	Max. diameter of waterway (m)	Diameter of turbine runner (m)	Main transformer capacity (MW)	Output voltage calss (kV)
Xiaowan	4,200	6×700	251	6×360.3	9	6.6	3×260	500
Laxiwa	4,200	6×700	220	6×378.0	9.5	6.9	3×260	750
Goupitan	3,000	5×600	200	5×385.5	9.5	7.0	3×223	500
Xiluodu (*)	13,860	18×770	226	18×423.8	10	8		500
Jinping I	3,600	6×600	240	6×337.4	9	6.55		500

* Parameters in the table are mainly extracted from construction Xiluodu design documentation.

8.4 Earth and Rockfill dams

8.4.1 General

According to incomplete statistics, in the 40 years from 1949 to 1989, China had built up 86,000 large, medium or small dams with various types, of which most are earth-rockfill dams. Among the 2,668 dams higher than 30m, the earth and rockfill dams are 2,174, accounting for 81.5%. But almost all the dams higher than 100m are concrete ones.

Since the 1990s, with the rise of earth-core rockfill dams and concrete faced rockfill dams (CFRDs), the number of completed and under-construction earth-rockfill dams higher than 100m obviously increases, and a few of them are constructed on large rivers. At the beginning of the 21st century, there are 7 earth-rockfill dams (200m to 250m high) completed or under-construction, and the to-be-constructed 4 earth-rockfill dams would exceed the 250m height. Surely, they would become the world highest earth-rockfill dams. Details refer to Table 8.4-1.

Among the over 100m high earth rockfill dams, most of them are CFRDs, especially in those medium-sized water conservancy projects on medium rivers, CFRDs dominate. The constructed or under-construction over 100m high earth rockfill dams for the large-sized hydropower and water conservancy projects with installed capacity over 250MW are listed in Table 8.4-1.

Table 8.4-1 Constructed or Under construction over 100m High Earth and Rockfill Dams and CFRDs in 2006

No.	Name of project	Location	River	Dam type	Dam height (m)	Dam volume (10⁴m³)	Area of face slab (m²)	Reservoir storage capacity (10⁸m³)	Installed capacity (MW)	Year of completion
1	Nuozhadu	Yunnan	Lancang River	ER	261.5	3,360	—	237.03	5,850	UC
2	Shuibuya	Hubei	Qingjiang River	CFRD	233.2	1,526	137,000	45.8	1,840	2009
3	Jiangpinghe	Hubei	Loushui River	CFRD	219	718		13.66	450	UC
4	Pubugou	Sichuan	Dadu River	ER	186.0	2,055	—	53.4	3,600	2009
5	Sanbanxi	Guizhou	Qingshui River	CFRD	185.5	828	84,000	40.94	1,000	2006
6	Hongjiadu	Guizhou	Liuchong River	CFRD	179.5	920	75,100	49.47	600	2005
7	Tianshengqiao I	Guizhou, Guangxi	Nanpan River	CFRD	178.0	1,800	177,000	120.57	1,200	2000
8	Tanking	Zhejiang	Xiaoxi River	CFRD	162.0	980	95,000	41.9	604	2008
9	Zipingpu	Sichuan	Minjiang River	CFRD	156	1,117	108,800	11.12	760	2006
10	Jilintai	Xinjiang	Karshgar River	CFRD	157.0	836	74,000	25.3	500	2005
11	Bashan	Chongqing	Renhe River	CFRD	155.0	—	—	3.1.5	140	2009
12	Xiaolangdi	Henan	Yellow River	ER	160	4,900		126.5	1,800	2001

Part II Construction Technologies for Dams and Reservoirs

Continued

No.	Name of project	Location	River	Dam type	Dam height (m)	Dam volume ($10^4 m^3$)	Area of face slab (m^2)	Reservoir storage capacity ($10^8 m^3$)	Installed capacity (MW)	Year of completion
13	Malutang	Yunnan	Panlong River	CFRD	154.0	800	—	5.46	240	2009
14	Dongqing	Guizhou	Beipan River	CFRD	150	950	—	9.55	880	2010
15	Longshou II	Gansu	Heihe River	CFRD	146.5	253	26,400	0.86	157	2004
16	Wawushan	Sichuan	Zhougong River	CFRD	138.76	350	20,000	5.84	240	2007
17	Jiudianxia	Gansu	Taohe River	CFRD	133.0	385	41,300	9.43	300	2008
18	Jinpeng	Shaanxi	Heihe River	ER	133.0	559	—	2	20	2001
19	Wuluwati	Xinjiang	Kalakashi River	CFRD	133.0	649	75,800	3.47	60	2001
20	Shanxi	Zhejiang	Feiyun River	CFRD	132.8	580	70,000	18.24	200	2001
21	Gongboxia	Qinghai	Yellow River	CFRD	132.2	476	57,500	6.2	1,500	2006
22	Longma	Yunnan	Babian River	CFRD	135			5.9	240	2007
23	Yinzidu	Guizhou	Sancha River	CFRD	129.5	310	37,500	5.31	360	2003
24	Jiemian	Fujian	Youxi River	CFRD	126	340	30,000	18.2	300	2007
25	Baixi	Zhejiang	Baixi River	CFRD	124.4	403	48,400	1.68	18	2001
26	Eping	Hubei	Huiwan River	CFRD	125.6	298	43,000	3.03	114	2006
27	Heiquan	Qinghai	Baoku River	CFRD	123.5	540	79,000	1.82	12	2000
28	Qinshan	Fujian	Muyang River	CFRD	120	248	42,000	2.65	70	2001
29	Baiyun	Hunan	Wushui River	CFRD	120.0	170	14,500	3.6	54	2001
30	Gudongkou I	Hubei	Gufu River	CFRD	117.6	190	28,100	1.476	45	1999
31	Bajiaohe	Hubei	Bajiao River	CFRD	113	192	36,000	0.96	34	2006
32	Sinanjiang	Yunnan	Sinan River	CFRD	115.0	297		2.71	201	UC
33	Shitouhe	Shaanxi	Shitou River	ER	114.0	835	—	1.5	55	1981
34	Gaotang	Guangdong	Baishui River	CFRD	111.3	195	26,400	0.96	36	2000
35	Jinzaoqiao	Fujian	Jinzaoxi River	CFRD	111.3	175		0.95	60	UC
36	Shuanggou	Jilin	Songjiang River	CFRD	110.5	258	37,300	3.88	280	2004
37	Nalan	Yunnan	Tengtiao River	CFRD	109	259	40,800	2.86	150	2005
38	Chahanwusu	Xinjiang	Kaidu River	CFRD	110	410		1.25	309	UC
39	Yutiao	Chongqing	Daxi River	CFRD	106.0	195	18,800	0.95	48	2001
40	Liyutang	Chongqing	Taoxi River	CFRD	103.8	180	25,300	1.02	15	2009
41	Dongba	Guangxi	Xiyang River	CFRD	105.0	316	52,700	3.15	72	UC
42	Maopingzhen	Hubei	Maoping River	ER	104.0	1,180		—	—	2003
43	Lubuge	Yunnan	Huangni River	ER	103.8	396		1.11	600	1989
44	Qiezishan	Yunnan	Supa River	CFRD	106.11	129	22,000	1.25	16	1999
45	Si'anjiang	Guangxi	Si'an River	CFRD	103.4	210	41,200	0.94	12	2003
46	Panshitou	Yunnan	Supa River	CFRD	102	548	73,500	6.08	10	2004
47	Chaishitan	Yunnan	Nanpan River	CFRD	101.8	235	38,200	4.37	60	2000
48	Bikou	Gansu	Bailong River	ER	101.8	397	—	5.21	300	1978
49	Taoshui	Hunan		CFRD	101.5	—	28,700	5.25	69	2008
50	Baishuikeng	Zhejiang	Jiangshan Port	CFRD	101.3	150	—	2.48	40	2003

Chapter 8 Dam Type and Project Layout

Continued

No.	Name of project	Location	River	Dam type	Dam height (m)	Dam volume ($10^4 m^3$)	Area of face slab (m^2)	Reservoir storage capacity ($10^8 m^3$)	Installed capacity (MW)	Year of completion
51	Jishixia	Qinghai	Yellow River	CFRD	103		35,500	2.64	1,000	UC
52	Taian Upper Reservoir	Shandong	Yingtaoyuan gou River	CFRD	99.8	386	—	0.12	1,000	2005
53	Shuanghekou	Guizhou	Beipan River	CFRD	99.5	250		1.95	120	2006

Note: 1. Quoted from the *Technological Development and Engineering Examples of Large Dams in China*, those under-construction (UC) projects have been commissioned in succession around 2010.

2. Some data are extracted from the *Hydropower Engineering in China: Hydraulic Engineering Volume* and over 100m high core wall rockfill dams have been supplemented.

The 60-year history of earth and rockfill dams development manifests following characteristics:

(1) At the earlier stage after the founding of new China, starting from harnessing the Huai River, a number of earth and rockfill dams were constructed. They are homogeneous earth dams or clayey soil core sand gravel dams which are less than 50m in height. The rockfill dams were not developed. Most earth and rockfill dams constructed in this period are water conservancy projects mainly for flood control and/or irrigation purposes, the installed capacities of most hydropower stations are in medium or small scales, and the project layouts are simple and uncomplicated.

(2) From 1958 to 1980, in the period of the "Great Leap Forward" and "Cultural Revolution", the number of high earth rockfill dams became more and more, and a few of them reached the 100m scale. From the dam type point of view, homogeneous earth dams and clay core or inclined core gravel dams were still dominant. On dam building technology, besides the roller compacted earth rockfill dam, the underwater earthfill, hydraulic fill and directional blasting dam type were developed to suit local conditions. The highest underwater earthfill dam in China is the Fenghe Reservoir homogeneous earthfill dam in Shanxi Province, 61.4m high, with an earthfill dam embankment volume of 5.8 million m^3.

Rockfill dams did not develop much due to constraints of construction machines and tools. However, in response to the government's call for saving "three main materials" (steel, cement and timber), the directional blasting dam construction technology was applied into practice from 1958, nearly 50 directional blasting dams were successively built up. The largest one is the 90.0m high Jiyi dam in Yunnan Province though its reservoir storage capacity and installed capacity are relatively small. The directional blasting dam construction technology was not widely applied in large-sized projects. Since 1980, the directional blasting dam construction technology has no further development with increasing environmental concerns.

(3) During 1980 to 2000, with the development of national economy and advances on science and technology, and given the water and hydropower construction enterprises equipped with heavy earth rockfill dam construction machines, the high earth rockfill dams became more competitive when compared with other dam types. Particularly, the CFRDs had obviously development. As shown in Table 8.1-3, among the 51 completed or under-construction over 100m high dams, the number of earth rockfill dams is 19, constructed since the 1980s, accounting for about 40% of the total; and among the 19 earth rockfill dams, the number of CFRDs is 16, accounting for 80% or so.

During 1980 to 2000, the representative projects are the Tianshengqiao I CFRD on the Hongshui River and the Xiaolangdi inclined earth core rockfill dam on the Yellow River. Both are mega projects and large hydropower stations on major rivers, as well as the representatives of the two prevailing types of earth rockfill dams constructed in the late 20th century.

(4) During 2000 to 2010, earth core (or inclined core) rockfill dams and CFRDs gradually have become two prevailing dam types among the roller compacted high earth rockfill dams. The former is mainly applied to dam projects 250m high or above (with the 263m high Nuozhadu dam as the representative) and founded on deep and thick overburden, and the later is applied to dam projects nearly up to 250m high (with the 236m high Shuibuya dam as the representative).

CFRD technology is experiencing fast development. Nowadays, the 150m high CFRDs have concrete plinths and connecting slabs be built on the cut-off wall, which indicates that CFRD has entered a new technical development stage, i.e., to construct CFRDs on deep and thick overburden foundations, and to build high CFRDs with plinths and connecting slabs to be constructed on the deep-embedded cut-wall. For instance, Jiudianxia dam on the Tao River in Gansu and Chahanwusu dam on the Kaidu River in Xinjiang, both have been completed and put into operation.

8.4.2 High Earth-rockfill Dams Construction from 1980 to 2000

After the 1980s, the high earth rockfill dams built up with modern technologies had rapid development, especially CFRDs developed very fast, widely distributed here and there in the whole country. Both the dam height and the project scale were growing. The development features are illustrated below.

8.4.2.1 Rapid Development of High Earth Rockfill Dams

The representative projects built in 1980-2000 are Xiaolangdi inclined core wall rockfill dam and Tianshengqiao I CFRD.

(1) Xiaolangdi Inclined Core Wall Rockfill Dam (Fig. 8.4-1): Xiaolangdi dam is the largest loam inclined core wall rockfill dam built in China at the end of the 20th century, with a maximum height of 160.0m and the total embankment volume of 50.73 billion m^3. It was commenced in 1995, river closed in 1997, initial impoundment in October 1999, and completed in November 2000. The dam crest is 1,667.0m long and 15.0m wide. The dam bottom has a maximum width of 864.0m. The upstream dam slope is in 1 : 2.6 and the downstream dam slope in 1 : 1.75. The riverbed overburden has a maximum depth over 70m, and contains interbedded sand layers and big boulders. A concrete diaphragm wall with the thickness of 1.2m was adopted to control seepage from the reservoir.

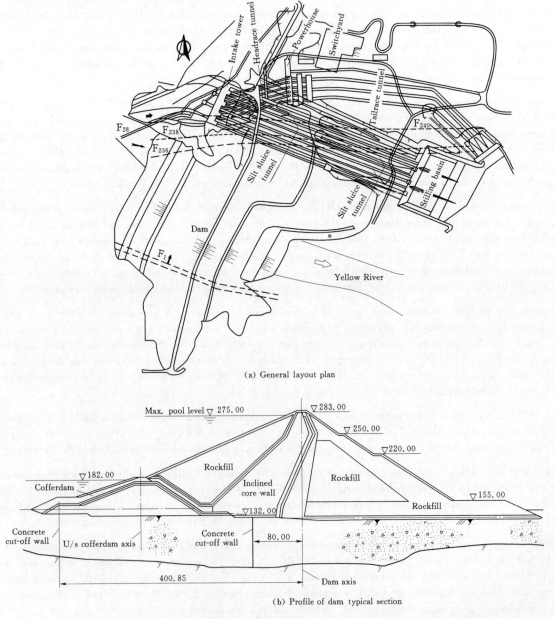

Fig. 8.4-1 General Layout of Xiaolangdi Inclined Core Wall Rockfill Dam (Unit: m)

Located at the outlet of the last canyon on the Yellow River mainstream, Xiaolangdi dam project was developed mainly for flood control, ice flood control and desilting, besides other purposes including water supply, irrigation and power generation. After its completion, the downstream flood control standard was upgraded to the 1,000-year flood from the previous 60-year flood, and the downstream ice flood hazard was basically solved. The total installed capacity of the power plant is 1,800MW.

The normal high level of the reservoir is at El. 275.0m. The total storage capacity of the reservoir is 12.65 billion m³, with a flood control capacity of 4.05 billion m³, a water and sediment regulating capacity of 1.05 billion m³, and a sediment storage capacity of 7.55 billion m³. The 1,000-year design flood peak discharge out of the dam is 40,000m³/s, and the 10,000-year check flood peak discharge is 52,300m³/s. The measured annual average sediment transport is 1.351 billion tons.

(2) Tianshengqiao I CFRD (Fig. 8.4-2): Tianshenqiao I Project is located on the Nanpan River mainstream, an upstream tributary of the Hongshui River. The lake formed by the dam is a regulating reservoir on the Hongshui River, with a gross storage capacity of 10.257 billion m³, a regulating storage of 5.796 billion m³, and having yearly regulating capability. The project is developed mainly for power generation besides flood control and other multi-purposes. The project is composed of the CFRD, spillway, emptying tunnel, left bank headrace system and surface powerhouse. It was commenced in June 1991, river closed in December 1994, initial impoundment in August 1998, first unit commissioned in December 1998, and completed in December 2000.

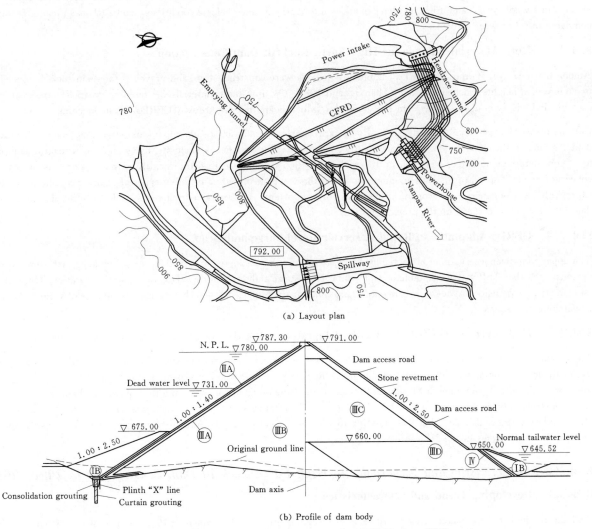

Fig. 8.4-2 General Layout of Tianshengqiao I CFRD Project (Unit: m)

ⅠB—Random material; ⅡA—Transition material; ⅢC—Soft rock material; Ⅳ—Clay material;
ⅠA—Cushion material; ⅢB—Main rockfill zone; ⅢD—Secondary rockfill zone

The CFRD has a maximum height of 178.0m, a crest length of 1,104.0m, a dam filled volume of 18 million m³, and a total face slab concrete volume of 174,000m³. The dam design presented many innovations. After so many years' operation under high water level, the seepage of the dam decreases gradually year after year. For the CFRD deformation control measures, self-healing top waterstops were applied in the joints for the first time. Excavated materials were utilized as fully as possible, and local soft rock materials like sand shale were used appropriately as well.

After the reservoir's regulation, the maximum flood discharge is 21,750m³/s, to be released through the spillway located in an eroded limestone trough at the right bank. The spillway approach channel is 1,187.5m long, with a bottom width of 120m. The total excavation volume is 17.64 million m³, of which 15.2 million m³ was filled at different zones of the dam. The headrace power generation system was arranged on the left bank side. Four sets of 300MW in each unit were installed in the surface powerhouse. The penstocks are 9.6m in diameter. No surge shaft was provided. Two diversion tunnels with a flow passage section of 13.5m×13.5m were arranged on the left bank. The approach channels at entrances and exits of the diversion tunnels, the power intake approach channel as well as the tailrace channel were all excavated out. The excavation volume and the dam fill volume were coordinated and balanced on the whole.

8.4.2.2 CFRDs with Obvious Advantages

Among the 11 over 100m high earth rockfill dams constructed in the late 20th century, it was dominated by 9 CFRDs. These projects encountered various complicated climatic factors and different topographical and geological conditions at the dam sites. Practical experience proves that CFRD could easily adapt to local conditions, construction can be carried out almost all seasons in a whole year, and it has no high demands on topographic and geologic conditions, so CFRD has great advantages compared to other types of dams.

8.4.2.3 Rock Material Balance for High Earth Rockfill Dam Construction

Among the nine over 100m high CFRDs completed, 4 dams were constructed of sand gravel at the main rockfill zone. The maximum dam heights reached 130.8m (Shanxi) and 138m (Wuluwati). Materials for building an CFRD are easily acceptable and widely available at the dam site localities, which is another feature why CFRDs are so welcome.

Current practical experience suggests that: Rock excavation materials are commonly used in the main body of rockfill dam. As long as the excavated materials are qualified and the hauling distance is moderate. For a high earth rockfill dam project, considerations no longer were paid to the rock excavation volume, instead the rockfill material balance issue is more concerned about. This feature shows an important trend that, when comparing different types of structures, the rock excavation volume is not merely an economical factor, more important, the excavated materials could be utilized as a kind of dam fill materials and speed up the project progress.

8.4.2.4 CFRDs Adaptable to Rapid Reservoir Level Surge and Drop

The upper and lower reservoirs of many pumped storage projects have adopted the CFRDs, usually higher than 70m, which demonstrated that CFRDs are adaptable to rapid water level surge and down during the power plant's operation. This important practical experience provides a safe operation example for the high CFRDs on the major rivers in case of various possible dispatching operations.

8.4.2.5 High Earth Rockfill Dams Easily Adaptable to Local Topographic and Geologic Conditions

The completed high earth rockfill dams have shown their own characteristics on the project layouts, and all have made good use of the local topographic and geologic conditions at the dam sites. Likewise, it was far-sighted that the energy dissipation area for the flood release tunnel in Xiaolangdi Project adopted the plunge pool type; and the spillway and approach channels were constructed by large excavation at the right side of Tianshengqiao I CFRD, which was very instructive for seeking short-hauling distance dam building materials for rockfill dams. These projects provide rich experience for construction of high earth rockfill dams in the 21st century.

8.4.3 Characteristics and Experience in High Earth Rockfill Dam Construction in 2001 – 2010

8.4.3.1 Developing Trend and Characteristics

As indicated in Table 8.1-3, in the first 10 years of the 21st century, among the over 100m high dams completed or under-construction, the number of earth rockfill dams is 19, accounting for 40%, of which the CFRDs dominate with a total number of 16, accounting for 84% of the earth rockfill dams.

In Table 8.1-3, among the twenty-two high dams with the height of 150 – 200m and over 250m, there are 12 earth rockfill dams, accounting for 54.5%. This is a developing trend of high dams construction in China in the early 21st century, and by absorbing practical experience in high earth rockfill dams since the 1990s, the development also shows some new characteristics:

(1) As high earth rockfill dams can better adapt to local geological and topographical conditions, they would have obvious advantages when compared with alternative gravity type or arch dams. Especially for the 200m to 250m high earth rockfill dams seated on thick overburden (over 60m to 80m), most of them are core wall rockfill dams such as Nuozhadu (263m) and Pubugou (186m). At present, the completed Xiabandi Bituminous Concrete Core Rockfill Dam on the upper reach of the Yeerqiang River in Xinjiang has a maximum height of 87m and a total storage capacity of 980 million m^3. The overburden at the dam site is about 150m thick, being the thickest among the completed 100m high earth rockfill dams.

(2) For the 200m to 250m high dams founded on relatively thin overburden, based on comparison of different dam types, most of them selected the CFRD scheme with the dam body upstream portion as well as face slabs and plinths placed on bedrock, such as the Shuibuya Dam (233.2m) on the Qinghe River, the highest CFRD completed and put into operation in the early 21st century, and the Sanbanxi Dam (185.5m) and Hongjiadu Dam (182.3m), the nearly 200m high CFRDs completed in the first 10 years of the 21st century, all with their face slabs and plinths placed on bedrock.

The CFRD shows a new developing trend that: the 150m high CFRDs founded on 40m to 60m thick overburden adopt cut-off walls to control seepage, with their plinths and connecting slabs constructed on the cut-off walls, and their dam bodies constructed on overburden. Such examples are the Jiudianxia CFRD on the Taohe River (tributary of the Yellow River) and Chahanwusu CFRD on the Kaidu River in Xinjiang, both have been constructed and impounded.

(3) For the 150m to 200m high earth rockfill dams, their design flood discharges are very large. Flood release generally relies on the bankside chute spillway (or surface spillway followed by flood tunnel) with auxiliary deep (bottom) outlets for flood discharge, sediment flushing and reservoir emptying. The dam downstream energy dissipation is mostly by ski-jumping.

The energy dissipation works for the 200m to 250m high earth rockfill dams completed or under-construction had some development. For minimizing high head flood flow's scouring and damage to the downstream riverbed and bank slopes, most projects such as Suibuya and Nuozhadu learned the experience of Xiaolangdi Project to construct a plunge pool at the dam downstream.

(4) Most projects would have an underground or bank side surface powerhouse, so as to shorten the headrace system alignment and provide no surge shaft if possible. For those over 200m high earth rockfill dams, the underground powerhouse schemes always have advantages over different alternatives for the headrace power generation systems. This trend is obvious, and it has gradually turned out to be a classic project layout pattern.

(5) Large-diameter diversion tunnels are adopted for the river diversion works, and deep (bottom) outlets aid in diversion and passing flood season. Reservoirs would be impounded in stages with the rising up of dam embankment. The construction mode for earlier impoundment and commissioning in stages has become a common practice.

8.4.3.2 Representative Projects

The high earth rockfill dams constructed or under-construction in the first 10 years of the 21st century can be basically classified into the following three modes, and they have become model projects. Besides, there are a few 100m high bituminous concrete core rockfill dams such as Yele, Maopingxi and Xiabandi.

1. CFRDs with front portion or whole body founded on rock

The representative CFRDs are Shuibuya, Hongjiadu and Sanbanxi, with the maximum height over 200m, nearly up to 250m, being the highest CFRDs in the world. The typical project layout pattern for the 250m high CFRDs in China is that, the underground powerhouse would be arranged on one bank side, the spillway on another bank side, energy to be dissipated in the plunge pool, and deep flood outlet tunnels to be provided for sand flushing as well as emptying the reservoir or lowering its level. One of the salient features is that, the rockfill body's front portion or whole body would be founded on rock, and the seepage control system for the dam foundation and main body consists of the reinforced concrete face slabs, plinths constructed on bedrock and consolidation grouting curtains.

(1) Shuibuya CFRD (Fig. 8.4-3): The project is a trunk hydropower station on the Qing River mainstream, and one major power source for peak load and frequency regulation in the Central China Grid. Its total installed capacity is 1,840MW, capable of generating 3.920TW·h electricity per year. The total storage capacity of the reservoir is 4.58 billion m^3, with multi-year storage regulating capability. The project consists of the CFRD, right bank diversion type underground powerhouse, left bank spillway and right bank release outlet tunnel. The project achieved the river closure in October 2002, first unit commissioned in 2007, and completed in 2009. The project has following features.

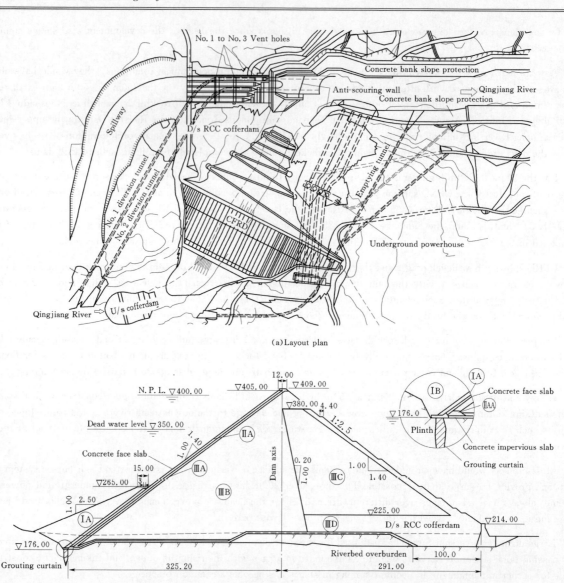

Fig. 8.4－3　General Layout of Shuibuya CFRD Project（Unit：m）

ⅠA—U/s Blanket； ⅡA—Cushion layer zone； ⅡAA—Small zone； ⅢA—Transition zone； ⅠB—U/s Blanket；

ⅢB—Main rockfill zone； ⅢC—Secondary rockfill zone； ⅢD—D/s rockfill zone

1) The dam has a maximum height of 233.2m, being the highest CFRD in the world. The embankment volume is 15.64 million m³. Its crest length is 660.0m. Both the upstream and downstream dam slopes have an overall gradient of 1:1.4. The downstream RCC cofferdam was integrated with the dam body. The total area of the concrete face slab is 137,000m³, and the total length of joints is 12,500m. The dam was designed as per the seismic intensity of Ⅶ.

The riverbed overburden at the dam foundation area is generally 12.0m to 14.4m thick, maximum in 20.0m. The cover within the range of 150m downstream of the plinth has been removed, and the left part remains there subject to heavy tamping treatment for an average settlement of 40cm to 60cm for improving its compactness. The dam filling materials were mainly from the excavation of structure foundations, and the balance resorted to exploitation from quarries. In the excavated limestone materials about 20 million m³, only those with high compressive strength and good compacting property were utilized. To control the overall deformation of the dam body, 25t large vibrating rollers were employed. The porosity of rockfill after compaction was about 20%. According to monitoring data, the maximum vertical displacement of the dam is 225cm, about 1% of the dam height. The overall deformation of the dam is basically controlled within the anticipated value.

2) The design flood discharge for the reservoir is 16,300m³/s, and the check flood discharge 18,320m³/s. The outlet works includes the left bank spillway with 5 surface spillways ($W \times H = 14.0m \times 7.0m$), and the right bank emptying tun-

nel ($W \times H = 6.0\text{m} \times 7.0\text{m}$) which is a pressurized one after a non-pressure section at the beginning, with a maximum discharge capacity of $1,605\text{m}^3/\text{s}$. The outlets of the two diversion tunnels on the left bank are located below the flip bucket of the spillway chute. For the sake of energy dissipation safety in the flood discharge process, a 40m deep anti-scouring wall was constructed adjacent to the left side of the energy dissipation area, and both bank slopes were protected with concrete.

3) Suibuya dam is the highest CFRD exceeding the 200m height in China. The 10-year study before construction and 4-year continuous investigation based on feedback information during construction gained many innovative achievements, made breakthroughs on various aspects like the dam zoning, damfilling materials selection, dam deformation control, acquired successful experience, and provided essential references for later 200m to 250m high CFRDs construction in China.

(2) Hongjiadu CFRD in Fig. 8.4-4: The project is located on the Liuchong River, an upstream tributary of the Wu River in Guizhou. The lake formed by the dam is the only one regulating reservoir among the 11 cascade hydropower stations on the Wu River mainstream, with multi-year storage regulating capability. The total storage capacity of the reservoir is 4.947 billion m^3. The project is developed mainly for power generation, besides other purposes including water supply, ecological improvement, tourism and navigation. The major components of the project include the CFRD, tunnel type spillway, flood release tunnel, headrace system and surface powerhouse at the dam toe. It was commenced in November 2000, river closed in October 2001, initial impoundment in April 2004, first unit commissioned in July 2004, and completed in 2005.

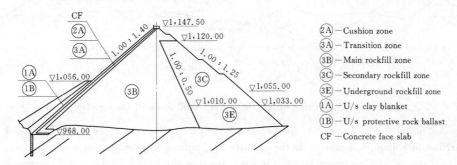

Fig. 8.4-4　Zoning of Hongjiadu CFRD (Unit: m)

The dam site takes on an asymmetric-V shape in the narrow valley. The CFRD has a maximum height of 179.5m, a total crest length of 427.79m and a width-height ratio of 2.38. Both the upstream and downstream dam slopes have a slope gradient of 1:1.4. The dam embankment volume is 9.2 million m^3. The concrete face slab placed in 3 stages has a total area of $75,100\text{m}^3$.

Hongjiadu dam is a 200m high CFRD constructed up by 2005, which was the highest in China at that time. Many innovations were made on the dam building technologies.

1) The dam body deformation control technology for the 200m high CFRDs has innovative breakthroughs. Two quantative control indexes were raised respectively as pre-settling time and pre-settlement. For improving the dam body's compactness, various deformation control measures and methods were applied, such as the dam body overall rising up in parallel, roller compacting in main and secondary rockfill zones to achieve equal compactness, and trimming steep slopes to improve their moduli of deformation.

2) The joint waterstop system for the CFRD had great development, and new waterstop forms and new waterstop materials suitable to the 200m high CFRDs were developed and applied.

3) Important progress have been made in application of the narrow (new) plinth structure and high-steep slope treatment technology for CFRDs in canyons. The new structure of "continuous narrow plinth with an equal width" was applied for the first time, and excavation and "locking" technology (special fortification measures) was applied for the first time at the junctions between the dam and the steep high slopes.

4) CFRD safety monitoring technology was developed and upgraded to a new level.

2. Rockfill dams with Core

Nuozhadu and Pubugou are the two representative rockfill dams with core. The project layout was made in such way that, the underground powerhouse be arranged on one bank side, the spillway and plunge pool on another bank side, and auxiliary deep flood outlet tunnels provided for flushing sand as well as emptying the reservoir or lowering its level. The layout pattern is a classic arrangement mode in China, and basically similar to that of the 200m to 250m high CFRDs.

One of the distinct features of the 200m to 250m high core wall rockfill dams is the seepage control system for the dam body and foundation. The wide and thick soil (or gravel soil) core wall and the concrete diaphragm wall in the dam foundation form a vertical seepage control system, and long-hole curtain grouting was carried out in-depth at both abutments. The concrete diaphragm wall has penetrated through overburden and been built into bedrock, and curtain grouting was also carried out in the rock mass to improve seepage control performance.

(1) Nuozhadu Core Wall Rockfill Dam (Fig. 8.4-5).

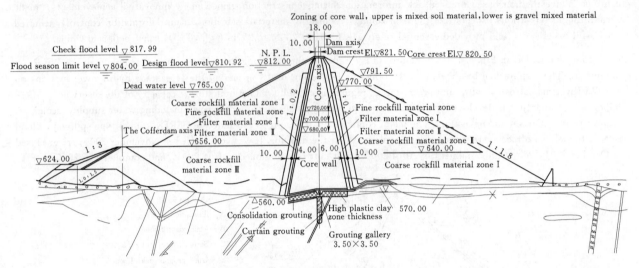

Fig. 8.4-5　Profile of Nuozhadu Core Wall Rockfill Dam Section

The project is located in the mid-lower reach of the Lancang River in Yunnan, developed mainly for power generation, as well as other purposes including flood control for the downstream Jinghong City and for improving navigation conditions in the downstream river course. The catchment area controlled at the dam site is 144,700km^2, and the annual average runoff volume is 54.6 billion m^3. The total storage capacity of the reservoir is 23.7 billion m^3, with a flood retention capacity of 2 billion m^3. The project is composed of the gravel soil core wall rockfill dam, left bank chute spillway and underground headrace power generation system at the left bank. Preparation for the project construction was initiated in February 2004, diversion tunnel works commenced in January 2006, river closed in November 2007, the first units estimated to be commissioned in 2012, and overall completion in 2015. It has the features below:

1) The core wall rockfill dam has a maximum height of 261.5m, ranked the third in the world and the highest one in China among the same type of dams completed or under construction. As the natural soil material is relatively fine, the content of clayey particles is rather low. Though its seepage control performance is sound, its compressibility could not satisfy the core wall demand for a 260m high dam. Hence, certain amount of artificial dams should be added in to improve the core wall's mechanic properties. Based on tests of soil materials with different admixture quantities, it was concluded that the gravel admixture quantity should be within the range of 30% to 40%. And it was finally determined to add 35% of gravel in soil according to core wall settlement calculation, dam slope stability analysis, seepage flow analysis as well as static and dynamic stress analysis. And the zoning of impervious core wall had been studied. The zoning boundary line was finally determined at El. 720.00m, gravel soil applied below El. 720.00m, and mixed earth material without gravel applied above El. 720.00m. The ratio of the later stage settlement at the dam crest to the dam height was controlled to less than 1.0%, so as to meet the specification requirement.

Based on rational zoning of the dam fill materials, quality rockfill material with larger sizes and high strength is applied at the 1/5 height range of the dam body, and seismic fortification measures are taken at the seismic weak locations like the dam crest, which allows the dam slopes to be optimized as 1:1.9 for the upstream slope and 1:1.8 for the downstream slope.

For satisfying the dam foundation's requirements on strength, seepage control and seepage stability, the core wall and filter layer foundation is required to be excavated to the weakly weathered bedrock. Bind concrete with reinforcement is poured at the contact zone between the core wall impervious material/filter layer and bedrock, which also serves as a cover for consolidation and curtain grouting.

2) The design flood discharge is 19,400m^3/s, and the check flood discharge is 31,320m^3/s. The bank side chute spillway has 8 radial gates ($W \times H = 15m \times 20m$), being the largest one among the bank side chute spillways in term of discharge capability. The maximum flow velocity at the spillway is 52m/s. Energy of outflow would be dissipated through trajectory jet

into the plunge pool.

3) In the powerhouse of the underground headrace power generation system, 9 sets of 650MW units with a gross capacity of 5,850MW are installed, capable of generating 23.900TW • h electricity per year in the yearly utilization of 4,088 hours. The large underground power tunnels and caverns encountered many complicated engineering geological conditions, which brought much difficulty to engineering design and construction works.

4) The work quantities are rather huge, with a total earth/rock open excavation volume of 52.61 million m^3, rock tunnel excavation of 6.28 million m^3, and concrete volume of 4.75 millon m^3. The total embankment volume of the core wall rockfill dam is 33.6 million m^3, of which the core wall impervious material is 4.68 million m^3, filter material 2.02 million m^3 and rockfill material 26.9 million m^3. The embankment works features large volume, high construction intensity, various dam fill materials, and complicated construction techniques. For example, the soil material exploited for the core wall needs to be mixed with gravel as per the 35% weight ratio. The monthly average dam embankment volume is 720,000m^3, and the embankment in peak month is 940,000m^3.

5) Construction diversion and flood control standard: In the initial stage, the flood control standard is a 50-year flood corresponding to the discharge of 17,400m^3/s; in the middle stage, the flood standard is a 200-year flood corresponding to the discharge of 22,000m^3/s; in the later stage, the flood standard is a 500-year flood corresponding to the discharge of 25,100m^3/s. The check flood standard is a 1,000-year flood corresponding to the discharge of 27,500m^3/s. Five diversion tunnels with the diameter of 20m were constructed to divert the river water and pass through flood season in combination with permanent outlet works.

6) The maximum height of the core wall rockfill dam is 261.5m, about 100m higher than the highest core wall rockfill dam of Xiaolangdi constructed in the late 20th century, which is a great leap with many technical breakthroughs. Implementation of Nuozhadu dam project accumulates precious experience and lays a technical foundation for building 300m high core wall rockfill dams in the next decade.

(2) Pubugou Core Wall Rockfill Dam (Fig. 8.4-6).

Located in the middle reach of the Dadu River, Pubugou Hydropower Project with a controlling reservoir, is mainly developed for power generation besides flood control and other purposes. The normal pool level of the reservoir is at El. 850.00m. The gross storage capacity of the reservoir is 5.337 billion m^3, including a regulating storage of 3.882 billion m^3. This is an incomplete yearly regulating reservoir. The project is composed of a rockfill dam with gravel soil core, left bank underground headrace power generation system, left bank side spillway, left bank deep flood release tunnel, right bank emptying tunnel, etc. The powerhouse is installed with six sets of 550MW units, 3,600MW in total, capable of generating 14.581TW • h electricity per year on average.

The gravel soil core rockfill dam is 186.0m high to the maximum and 540.5m long along its crest. Its upstream and downstream dam faces are sloped in 1 : 2.0 to 1 : 2.25 and 1 : 1.8 respectively. The dam bottom is about 900m wide. The dam body incorporating the cofferdam has a gross embankment volume of 23.72 million m^3.

The strata at the dam site area are composed of basalt, tuff, rhyolite and etc. Overburden at the riverbed is generally 40m to 60m thick, dominated by coarse particles of boulders, cobbles and gravels. Voids prevail in overburden besides evidence of sand lens. Its permeability is about 0.023 - 0.104cm/s.

Two fully enclosed concrete diaphragms both in the thickness of 1.20m are constructed into the riverbed overburden underneath the dam base for seepage control purpose. The centerline distance between the two diaphragms is 14.0m. Diaphragms penetrate into bedrock by 1.5m, below the dam foundation level of El. 670.00m. The maximum depth of the diaphragms is about 68m. The upstream diaphragm inserts into the core about 10m. A grouting (also serving monitoring purpose) gallery will be constructed over the downstream diaphragm top.

The core wall impervious soil source came from deluvial and pluvial gravel soils about 16km away from the dam site. Based on numerous tests and studies for years, the main indexes of the applied core wall impervious soil materials are prescribed as follows: The maximum particle size is 60mm to 80mm. Gradation of particles should be continuous. Content of particles $>$ 5mm shall not exceed 55%, and that of particles $<$0.075mm shall be more than 15%. The permeability of the soil materials shall be lower than 1×10^{-5}cm/s. The critical gradient against seepage deformation shall be greater than 2.5. The soil plasticity shall be 8 to 20.

3. CFRDs with plinths constructed on overburden

In the first 10 years of the 21st century, for the 100m to 150m high CFRDs, it was studied and developed to build dams on deep and thick overburdens, with plinths constructed on overburden, concrete cut-off walls used for seepage control in

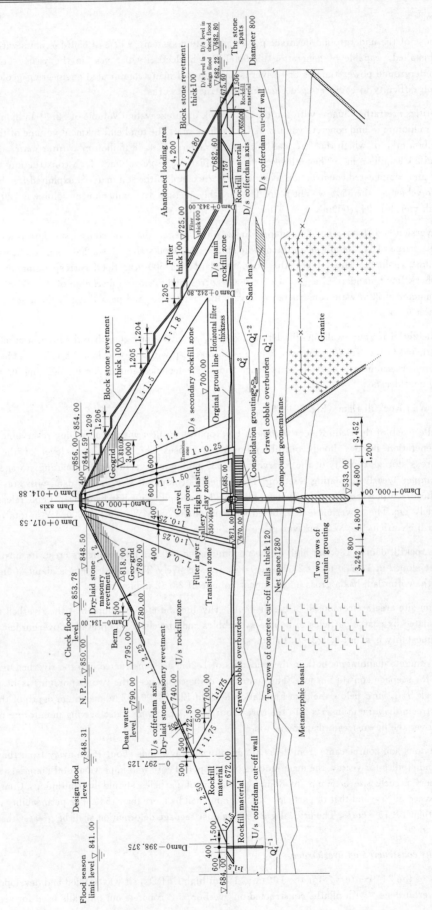

Fig. 8.4-6 Typical Section of Pubugou Gravel Soil Core Rockfill Dam (Unit: elevation in m, dimension in cm)

overburden, and plinths connected with cut-off walls with concrete connecting slabs, thus forming a complete seepage control system. Such dam type along with a vertical seepage control system is developing very fast in recent 10 years, thanks to application of precise exploration and advanced computer technologies as well as the technology basis of constructing cut-off walls in deep and thick overburdens. As for the project layout of such dam type, most adopt a bank side spillway (or surface spillways followed by flood tunnels) and a surface powerhouse on the downstream river shore, which gradually becomes a typical layout pattern. It is expected that this type of dams will have a good development prospect.

(1) Jiudianxia CFRD on the Tao River (Fig. 8.4-7).

Located in the middle reach of the Taohe River in Gansu Province, the reservoir impounded by Jiudianxia CFRD is the water source for the Taohe River Water Diversion Project (annual water supply capacity of 550 million m^3), with multiple purposes including water supply, power generation and flood control. The yearly runoff to the reservoir is 3.825 billion m^3. The total storage capacity of the reservoir is 907 billion m^3, including a regulating storage capacity of 5.71 billion m^3. The total installed capacity of the power plant is 300MW. The main dam was previously designed as a concrete gravity one. However, due to the thick overburden at the riverbed maximum to 56m, the CFRD scheme was selected based on numerous studies and comparison. The project consists of a CFRD, left bank spillway tunnel, right bank headrace power generation system plus surface powerhouse, and an emptying tunnel at the right bank.

The CFRD has a maximum height of 133m and an embankment volume of 3.28 million m^3. The dam was seated on deep and thick overburden while the plinth was put on a relatively compacted sand gravel layer. The plinth is 0.8m thick and 10m wide (in two segments of 6m+4m). A concrete cut-off wall was constructed at the upstream of the plinth to prevent seepage, and the plinth and the cut-off wall were directly connected. The cut-off wall has a thickness of 1.2m, an area of 1,100m^2, about 30m to 35m deep, and 1m penetrating into bedrock.

The design flood for the project is $P=1\%$ corresponding to the flood discharge of 2,560m^3/s, and the check flood is $P=0.02\%$ corresponding to the flood discharge of 4,650m^3/s. Flood discharge relies on two spillway tunnels at the left bank, with an equal opening width of 12m, respectively 770m and 780m long, with a total flood discharge capacity of 3,660m^3/s, and energy of outflow flood would be dissipated through flip bucket. Besides, there has one flood release tunnel at the right bank, in a diameter of 6m, a discharge capacity of 640m^3/s, and energy dissipated through flip bucket. The headrace tunnel at the right bank is about 2,250m long, for utilizing the hydraulic gradient of the river course to increase the power head. The surface powerhouse is equipped with 3 units of 88MW each.

(2) Chahanwusu CFRD on the Kaidu River (Fig. 8.4-8).

Located in the middle and lower reach of the Kaidu River in Xinjiang Automous Region, at the southern foot of Tianshan Mountain, Chahanwus Project was developed to supply power and water to the Tarim Oil Field for recovering oil and building the petrochemical industry base therein. The reservoir has a storage of 125 million m^3, with incomplete yearly storage regulating capability. The project is composed of a CFRD, right bank headrace system plus a surface powerhouse, river diversion tunnel (also as flood tunnel) at the left bank, and spillway tunnels and emptying tunnel at the right bank. The headrace tunnel is about 3.6km long, with a drop about 120m after being straightened at the downstream great bend. The total capacity of the power plant is 309MW.

Having a maximum thickness of 46.7m, the overburden at the riverbed is mainly composed of alluvial sand gravels and cobbles with occasional boulders. The upper and lower layers contain sand gravels and cobbles with boulders as well, respectively 17m to 27m and 11m to 19m in thickness, moderately compact to fairly compact, and the middle layer is a coarse sand layer with gravels, 4.5m to 8.5m in thickness.

Founded on deep and thick overburden, the CFRD has a maximum height of 110m, and an embankment volume of 4.1 million m^3. A concrete cut-off wall was constructed to control seepage, and connected with the face slab and plinth. The cut-off wall is 1.2m thick, 41.8m high, 1.0m penetrating into bedrock. The plinth is 10m long and 0.8m thick, separated into three segments (4m+3m+3m) by two expansion joints.

The project was started in October 2004, and initial impounding stated in October 2007. The first unit was put into operation in December 2007, and rest units commissioned in July 2008.

8.4.3.3 Summary of Main Characteristics

(1) Dams founded on deep and thick overburden: The first step is to investigate the property of deep and thick overburden by utilizing advanced precise exploration techniques. The above mentioned overburdens at these high rockfill dams generally have a thickness of 40m to 50m, even up to 70m to 100m. Those overburdens have good properties, mainly composed of cobbles and gravels, with fair gradation and good compactness, normally in the unit weight of 2t/m^3 or so, permeability

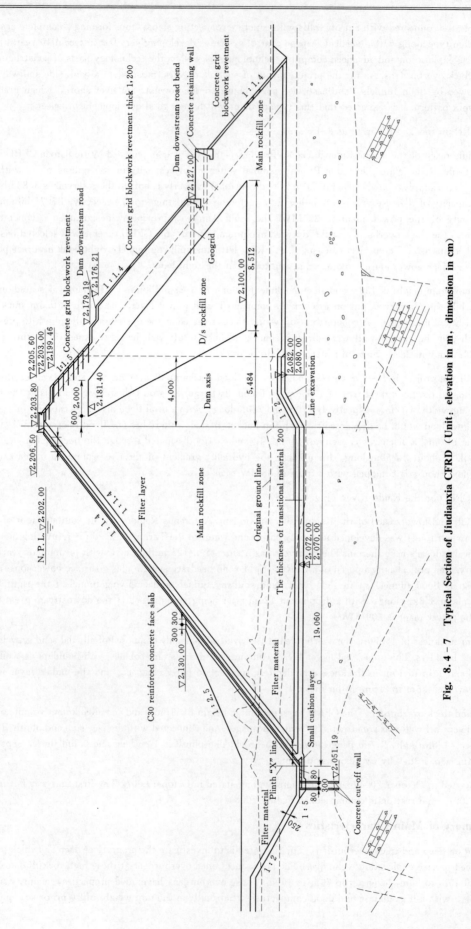

Fig. 8.4-7 Typical Section of Jiudianxia CFRD (Unit: elevation in m, dimension in cm)

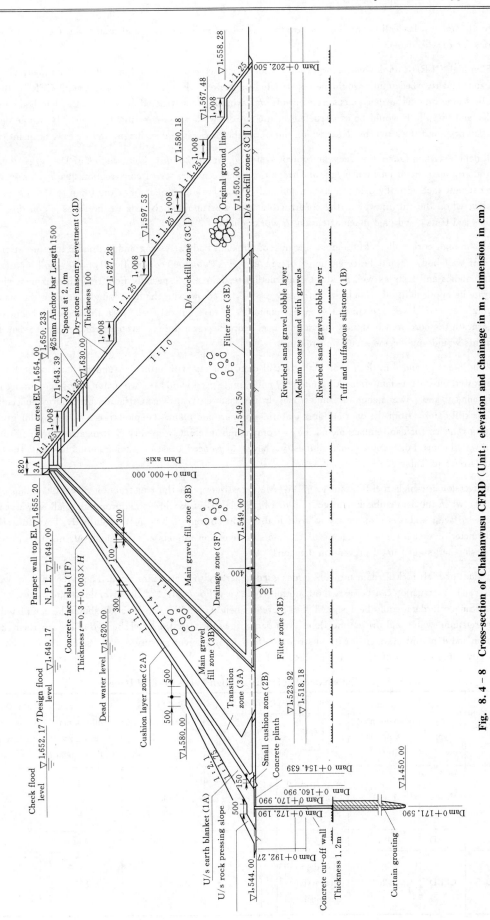

Fig. 8.4-8 Cross-section of Chahanwusu CFRD (Unit: elevation and chainage in m, dimension in cm)

from 10^{-2} cm/s to 10^{-3} cm/s, basically containing no continuous silty fine sand layer and mud layer. Hence, they can be used as foundations for rockfill dams.

For a 200m to 250m high CFRD, its downstream portion can be constructed on overburden, while all its upstream cushion layer, transition zone and riverbed plinth would be constructed on bedrock. For a 100m to 150m high CFRD, the whole dam body would be seated on overburden, a concrete cut-off wall would be constructed underneath the dam body and penetrate into bedrock, and the plinth would be constructed on the cut-off wall. For a core rockfill dam, the whole dam body would be founded on deep and thick overburden, which can noticeably lower the dam height and save work quantities.

(2) High rockfill dam body and foundation seepage control system: For the 200m to 250m high CFRDs, great importance was attached to the arrangement of cushion layers and transition zones, assurance of construction quality, as well as the connection of the concrete face with the dam foundation and bedrocks at both abutments. For the 100m to 150m high CFRDs with plinths constructed on the cut-off walls, the constructors took a scrupulous attitude to the seepage control systems, made lots of studies and tests, and paid much attention to works quality.

For the 200m to 250m high core wall rockfill dams, because of the dam body stressing and hydraulic fracture strength requirements, the core wall body should be constructed of gravel soil (Pubugou) or artificial gravelly soil (Nuozhadu), based on particular studies and tests as well as strict construction techniques, especially field trial embankment should be made to demonstrate its applicability. It should be extremely careful for connecting the dam foundation with the overburden as well as arranging the vertical seepage control system. Generally, 1 to 2 cut-off walls in a thickness of 1.2m to 1.4m would be set underneath the core wall bottom. Considering local bedrock permeability, seepage control for the dam foundation at two abutments would apply cement curtain grouting.

(3) Dam embanking intensity and duration: For high rockfill dam projects with either an underground powerhouse or a bank side surface powerhouse, the dam embanking intensity and rising speed would be two major factors governing the reservoir initial impoundment and power generation milestones. In the last ten years, construction procedures and equipment for building high rockfill dams, from the dam building materials preparation plants (exploitation, screening for gradation, etc.), transporation vehicles for dam embankment, heavy compacting machines, quality control means to monitoring instruments, which would control dam embankment intensity, have experienced a fast upgrading period based on the late 20th century level. Details refer to Table 8.4-2.

(4) Construction diversion for high rockfill dam projects: Main restriction on the construction diversion of high rockfill dam projects is that, when the dam embankment goes up to a higher level, it would be inappropriate to allow water flowing on the embankment surface, and it is also uneasy to arrange diversion bottom outlets in the dam body. Therefore, besides relying on large-diameter diversion tunnels, deep outlet tunnels or diversion tunnels at different levels must be provided to meet the river diversion and staged flood prevention demands.

Construction diversion for high rockfill dam projects is designed generally in a 50 years flood. The river channel would be closed once for all, and construction can be carried out in the whole year. For control of the diversion tunnel size at a canyon dam site, the upstream cofferdam would be designed with a greater height, and its construction should be completed within one dry period for retaining water and preventing flood. At a dam site with fairly thick overburden in the riverbed, the cut-off wall construction and dam embankment for its upstream cofferdam after river closure would usually turn out to be two important factors governing the overall construction schedule.

Table 8.4-2 Embankment Intensity of High Earth Rockfill Dams

Name of project	Max dam height (m)	Dam type	Dam embankment volume (10^4 m^3)	Concrete cut-off wall		Monthly max. embankment intensity (10^4 m^3/month)	Yearly max. embankment intensity (10^4 m^3/month)	Monthly dam whole section rising speed (m/month)
				Depth (m)	Thickness (m)			
Shuibuya	233	CFRD	1,677	Max depth of impervious curtain 260m	—	90	673	5.1
Sanbanxi	185.5	CFRD	961		—	70	480	8.4
Hongjiadu	179.5	CFRD	920	Max depth of impervious curtain 190m	—	34	367	

Continued

Name of project	Max dam height (m)	Dam type	Dam embankment volume (10^4 m³)	Concrete cut-off wall		Monthly max. embankment intensity (10^4 m³/month)	Yearly max. embankment intensity (10^4 m³/month)	Monthly dam whole section rising speed (m/month)
				Depth (m)	Thickness (m)			
Nuozhadu	261.5	CWRD	3,360		—	127 (94)	1,184 (870)	6.87
Pubugou	186	CWRD	2,372	78	1.2			
Jiudianxia	136.5	CFRD	328	30–35	1.2	30	100	15
Chahanwusu	110	CFGD	410	41.8	1.2	73.5	320	11.6

Note: The monthly/yearly max. embankment intensity and the monthly dam whole section rising speed are the design values, while values in the brackets are actually achieved ones.

(5) Tendency of safety appraisal for high rockfill dams: Many dam experts are of the opinion that the safety of high concrete arch dams is superior to that of high gravity dams, and the safety of high rockfill dams is not inferior to that of arch dams; furthermore, from the geological point of view, a rockfill dam has better adaptability to local geological conditions since an arch dam would have very stringent requirements on the engineering geology at the dam site. The dam building practices in the first ten years of the 21st century show that, in terms of dam construction, it has become difficult to build a 250m to 300m high CFRD, while it is rather rare to find a site suitable for building a 250m to 300m high arch dam due to its austere requirements on geology. Therefore, at the preliminary engineering stage, for the sake of safety, a core wall rockfill dam scheme would be usually proposed as the representative dam type.

References

[1] Pan Jiazheng, He Jing. *Large Dams in China—A Fifty-Year Review*. Beijing: China Water & Power Press, 2000.

[2] China Society for Hydropower Engineering. *Hydropower in China—A Sixty-Year Review*. Beijing: China Electric Power Press, 2009.

[3] Rural Hydropower & Electrification Development Buearu, Ministry of Water Resources. *Small Hydropower in China—A Sixty-Year Review*. Beijing: China Water & Power Press, 2009.

[4] National Energy Administration. *Hydropower in China (1910-2010)—A 100-Year Review*. Beijing: China Electric Power Press, 2010.

[5] China Society for Hydropower Engineering. *Hydropower in China—A Thirty-Year Review after Reform and Opening-up*. Beijing: China Electric Power Press, 2009.

[6] *Technical Development and Engineering Examples of Large Dams in China*. China Water & Power Press, Beijing, 2007.

[7] Zhou Jianping, Niu Xinqiang, Jia Jinsheng. *Design of Gravity Dams—A Twenty-Year Review*. China Water & Power Press, Beijing, 2008.

[8] Zhu Jingxiang, Shi Ruifang. *Hydropower Engineering in China: Hydraulic Engineering Volume*. Beijing: China Electric Power Press, 2000.

[9] Project Group for Hydropower Development, Department of Energy and Mining Industries, Chinese Academy of Engineering. *Strategic Study on Developing Hydroenergy Resources in China*. Beijing: China Electric Power Press, 2005.

[10] Li Zan, Chen Fei and Zheng Jianbo, etc.. *Major Issues of Extra-High Arch Dam Projects*. Beijing: China Electric Power Press, 2004.

[11] *Review of Water Power Resources of the People's Repulic of China (2003)*. Beijing: China Electric Power Press, 2004.

Chapter 9

Design and Construction Technologies of Concrete Gravity Dams

Zheng Shouren[1], Zhou Jianping[2]

9.1 Development and Current Situation of Gravity Dams

Construction of concrete gravity dams in China with modern technologies began during the World War II. In order to plunder the resources rapaciously and to serve their war of aggression, the Japanese invaders built a series of large-medium hydropower projects with gravity dams on rivers in the northeast China. As early as 1937 and 1938, the three hydropower stations, namely, Shuifeng on the Yalu River, Fengman on the Second Songhua River and Jingpohu on the upper reaches of the Mudanjiang River were constructed one after another, and first power unit of the two power stations was put into operation in 1941 and 1943 successively. When Japanese invaders were forced to unconditionally surrender in August 1945, construction of Yizhou and Yunfeng on the Yalu River, Huanren on Hunhe River and Wodu on Liaohe River were started, among them, Shuifeng, Fengman and Yunfeng Concrete Gravity Dams were of 100m-high level.

Independent design and construction of modern gravity dams in China began after the founding of new China. In the 1950s, in order to meet the requirement of river management and economic development for cheap labor force, two concrete slotted gravity dams of Xin'anjiang and Gutian I were constructed; concrete slotted gravity dams of Yunfeng and Danjiangkou, concrete solid gravity dams of Liujiaxia and Sanmenxia were constructed in the 1960s; Hunanzhen Trapezoidal Gravity Dam was constructed in the 1970s; Wujiangdu Arch Gravity Dam was constructed in the early 1980s. By the 1980s, more than 20 gravity dams with height over 70m were constructed in China. The gravity dam construction in this period was featured for efforts on finding various ways to reduce dam quantities and project costs, and a lot of new and light gravity dams were developed and constructed. In the design and construction of Xin'anjiang, Danjiangkou, Wujiangdu and Fengtan Gravity Dams, large amount of research work was done on shape optimization, stress analysis and construction methods etc., rapid development of gravity dam engineering technologies was promoted and gap between China and the world was greatly narrowed. Compared with the concrete solid gravity dams, concrete slotted gravity dams could reduce the concrete quantities by 10%-30%.

Along with the policy of reform and opening-up, due to profound changes in construction management system of hydropower projects and establishment of socialist market economic system in China, river management and hydropower development have entered a new age, damming technologies including construction technologies of gravity dams have also made great progress. After construction of gravity dams of Shuikou, Geheyan, Manwan, Wuqiangxi and Yantan in the 1980s, a series of high gravity dams such as Three Gorges, Longtan, Guangzhao, Jinanqiao and Xiangjiaba were built successively at the turn of the century, and the records of dam height and project scale were surpassed again and again. Longtan Gravity Dam on Hongshui River is the highest gravity dam in China, the design height is 216.5m and concrete quantity is 7.5 million m^3. The characteristics of gravity dam construction in this period were: the dam height increased from 100m or 150m to 200m-high level; the number of solid gravity dam increased as the solid gravity dam is more adaptable to mechanized construction work which favours a higher construction efficiency. A series of new types of energy dissipater were developed and energy dissipation problems of high head and large flood discharge were solved; more and more roller compacted concrete (RCC) gravity dams were constructed, and showing a tendency of gradually replacing the conventional concrete gravity dams.

---

[1] Zheng Shouren, Academician of Chinese Academy of Science and Chief Engineer of Changjiang Water Resources Commission of the Ministry of Water Resources, Professor Senior Engineer.

[2] Zhou Jianping, Chief Engineer of PowerChina Group Corporation, Ltd., Professor Senior Engineer.

At the beginning period of the 21st century, the implementation of the strategy of intensive development in the western China has triggered off the hydropower development in the southwestern China and a new upsurge in hydropower development is expected. Outstanding achievements of dam construction, including design and construction of concrete gravity dams are attained, and Chinese technologies of dam construction have surpassed the world advanced level, and the technologies with Chinese characteristics have attracted worldwide attention. Table 9.1 - 1 lists the concrete gravity dams with height over 100m in China.

Table 9.1 - 1 Concrete Gravity Dams with Height over 100m in China

No.	Project name	River	Dam height (m)	Concrete volume of dam ($10^4 m^3$)	Capacity (MW)	Year of completion
1	Longtan	Hongshui River	192/(Stage-I)	659	4,900/(Stage-I)	2009 (Stage-I)
2	Guangzhao	Beipan River	200.5	280	1,040	2007
3	Three Geroges	Yangtze River	181	1,610	22,500	2010
4	Longyangxia	Yellow River	178	157	1,280	1986
5	Guandi	Yalong River	168	357	2,400	2012
6	Wujiangdu	Wujiang River	165	193.2	1,250	1982
7	Xiangjiaba	Jinsha River	162	1,300	6,400	2013
8	Jin'anqiao	Jinsha River	160	360	2,400	2011
9	Geheyan	Qingjiang River	151	237	1,200	1995
10	Liujiaxia	Yellow River	147	182	1,350	1974
11	Baozhusi	Bailong River	132	200	700	2000
12	Manwan	Lancang River	132	153	1,670	1995
13	Jiangya	Loushui River	131	137	300	2000
14	Baise	Youjiang River	130	260	540	2006
15	Ankang	Hanjiang River	128	321	800	1995
16	Hongkou	Huotongxi River	130	72.7	200	2008
17	Guxian	Luohe River	125	110	60	1995
18	Wudu Reservoir	Fujiang River	120.34	161	150	2012
19	Danjiangkou	Hanjiang River	117 (heightened)	292.8	900	1974
20	Silin	Wujiang River	117	110	1,050	2009
21	Pengshui	Wujiang River	113.5	93.3	1,750	2007
22	Suofengying	Wujiang River	115.95	62	600	2006
23	Yunfeng	Yalujiang River	113.75	250	400	1965
24	Mianhuatan	Tingjiang River	113	61.5	600	2001
25	Gelantan	Lixian River	113	120	450	2009
26	Yantan	Hongshui River	110	172	1,210	1995
27	Dachaoshan	Lancang River	111	130	1,350	2002
28	Jinghong	Lancang River	108	320	1,750	2010
29	Panjiakou	Luanhe River	107.5	280	420	1985
30	Huanglongtan	Duhe River	107	98	510	1976
31	Shuifeng	Yalu River	106.4	340	900	1943
32	Sanmenxia	Yellow River	106	163	410	1960
33	Shatuo	Wujiang River	101	190	1,120	2012
34	Wanjiazhai	Yellow River	105	180	1,080	2000
35	Xin'anjiang	Xinan River	105	138	850	1965
36	Shuikou	Minjiang River	101	181	1,400	1996

9.2 Research on Engineering Geological Conditions

Engineering geologic investigation is very important and basic work for construction of hydropower projects. Special attention should be paid to the investigation and research on topography, geomorphology, lithology of formations, geologic structures, hydrogeologic conditions, and geophysical phenomenon of dam area. And attention should also be paid to complex engineering geology phenomenon such as: active faults, complex and weak rock masses, high slopes and large-scale landslides, high geostress, high ground water, karst leakage and deep overburdens. Engineering geologic investigation should be carried out according to the requirements of dam site selection, general layout, structure design, construction and operation.

In design and construction of gravity dams, on the one hand, detailed investigation, geotechnical testing and field mapping are carried out to study basic geological conditions, rock structure, physical and mechanical properties of the rock masses. Geological quality sorting is carried out correspondingly. On the other hand, careful analysis is needed on influence of engineering geological defects on dam layout, structure design, deformation under stress, and overall stability of the dam body. In so doing, basic data for dam design shall be provided. Suggestions on and solutions to the key engineering geological problems shall be put forward.

9.2.1 Weak Structural Planes and Their Mechanical Parameters

Rock mass is the compound of rock material and rock discontinuities. Due to the low mechanical strength and poor characteristics of weak structural planes and weak intercalation; overall stability of the dam, deformation and seepage stability of the dam foundation are affected. Therefore special investigation and research is needed.

Weak structural plane or weak intercalation at dam foundation constitutes boundaries of deep sliding. Therefore, distribution, connectivity, thickness, properties of structural plane or intercalation, undulation, zoning, integrity of overlying and underlying rocks should be identified; and values for strength, deformation and permeability under different circumstance should also be detected. Continuous weak structural plane and weak intercalation at low dip angle is apt to form sliding surface and is the most harmful to stability of foundations. Structural plane at high angle and other weak structural plane could constitute lateral sliding surface and might affect the overall stability of the dam sections near the banks.

According to research on structural plane and weak intercalations of gravity dam foundation, several structural plane classification methods are put forward by Chinese scholars. According to cause of formation, structural planes are classified as original type, secondary type and tectonic type. And according to composition, there are rock block and, debris type, debris filled with clay type, clay filled with debris type and clay type. Different types of discontinuities and weak intercalations have obviously different engineering properties. Physical and mechanical parameters of discontinuities and weak intercalations are given according to the existing research results, which has profound significance.

9.2.2 Engineering Geological Classification of Dam Foundation Rock Mass

At present, geological classification method of rock mass for evaluating geological conditions and adaptability of foundation becomes more and more popular. In the recent twenty years, research on geological classification of rock mass is still very active and some new progress has been made.

Standard for *Engineering Classification of Rock Masses* (GB 50218—1994) is the first national standard of China for general classification of engineering rock mass. This standard adopts both qualitative and quantitative methods to rate rock mass. The rock mass basic quality (BQ) is firstly defined according to hardness and intactness of the rock mass, and then modified basic quality ([BQ]) of rock mass is defined considering groundwater and geostress situation. This method is not applied widely due to its poor practicality in hydroelectric projects.

In the process of revision of *Code for Hydropower Engineering Geological Investigation* (GB 50287), experience of engineering geological classification of dam foundations in many hydropower projects and research achievements of domestic and foreign experts are summarized. A systematical geological classification method with Chinese characteristics for dam foundation rock mass is developed and applied. Geological classification for dam foundation rock mass plays an important role in aspects such as: rational selection of physical and mechanical parameters, objective assessment of dam foundation stability, design for dam foundation excavation and treatment, etc.

The main factors affecting engineering geological conditions of gravity dam foundation are: compressive strength of saturated rock mass, structure types of rock mass, intactness of rock mass, development and combination of structural planes. In

Code for Hydropower Engineering Geological Investigation (GB 50287), geological classification of rock mass in dam foundation addresses mainly the factors of strength, structure and structural planes properties and the dam foundation rock mass is divided into Grades I-V. Rock mass of Grades I-II has intact structure, high strength and high resistance to sliding and deformation. It can be used as dam foundation without special treatment. Rock mass of Grade V is weak or fractured which can not be used as foundation for gravity dam directly. Special treatment is needed when this kind of rock mass exists locally in part of dam foundation. Rock mass Grades III and IV may be used for gravity dam foundation if proper foundation treatment is provided. In the recent years, more and more attention is paid to geological defects treatment and satisfying progress is obtained, the gravity dam is made more adaptable to foundation conditions.

9.2.3 Selection of Foundation Interface

Rock mass in gravity dam foundation should have enough compressive strength, adequate resistance against sliding and deformation, high impermeability; competent durability is also needed to avoid change of rock properties and structural change under persistent action of high pressure water.

From engineering geological aspect, the factors affecting selection of foundation interface are as follows: lithology, structure, intactness, weathering and unloading degree, hydrogeological conditions and geostress of the rock mass. The aforementioned factors should be considered and evaluated comprehensively for geological classification, and then the criteria for foundation rock mass to be used may be determined, taking working conditions of the specific gravity dam into consideration.

For concrete gravity dams built on hard and intact rock foundation, the selection of foundation interface is mainly controlled by degree of weathering and relaxation of rock mass and the development of weak structural planes. When the height of gravity dam is over 150m, the foundation interface should be in lightly weathered to fresh rock; When the height of gravity dam is 100-150m, the foundation interface should be in slightly weathered or weakly weathered rock; when the height of gravity dam is 50-100m, the foundation interface should be in lower to or middle portion of the weakly weathered rock; when the height of gravity dam is less than 50m, the foundation interface should be in upper or middle portion of weakly weathered rock. For dam sections on both sides of the higher elevation, requirements for the foundation interface could be reduced appropriately. If rock mass in dam foundation is of gentle layered structure with problems of deep sliding, seepage stability, compressive deformation; special analysis should be conducted for the selection of foundation interface.

Before the 1980s, the criteria for gravity dam foundation rock mass to be utilized was determined only based on degree of weathering of rock mass, and the criteria was too high, resulting in deep excavation. Afterwards, beside degree of rock weathering, the quantitative data such as intactness, physical and mechanical parameters and wave velocities are taken into consideration in evaluating the bedrock and the limitations of qualitative judgment based on the experience are overcome.

Beside engineering geological conditions, the blasting damage, relaxation and rebound and treatment to the foundation are also needed to be considered in selecting interface of foundation. Controlled blasting techniques are adopted to avoid damage to the bedrock intactness; pre-reinforcement measures or protective excavation measures are adopted to limit relaxation. According to the requirements of design, the following measures can guarantee rock quality of the dam foundation: controlled blasting techniques, pre-reinforcement measures, protective excavation measures, covering foundation with concrete as soon as possible, consolidation grouting and so on.

Treatments of dam foundation include: curtain grouting, consolidation grouting, chemical grouting, the prestressed anchorage, concrete plug, deep cut-off wall and reinforced concrete pile, etc. When the excavation is deep and construction is difficult and costly, deep excavation should be best avoided, but the strength and stability of foundation could be enhanced through foundation treatment in order to meet the requirements of gravity dam in the aspects of bearing capacity, deformation, seepage control, resistance against sliding and slope stability.

9.2.4 Analysis of Stability Against Sliding and Safety Assessment

Stability against sliding of dam foundation is the reliability of dam to stand the shear sliding failure under various design situations; it is a key issue for gravity dams. Due to different geological structures of dam foundation, the sliding patterns are different and fall into three categories, namely: surface sliding, shallow sliding and deep sliding. In order to find out which sliding pattern controls foundation stability, judgment should be made through computation according to specific geological conditions.

Surface sliding is shearing slip along contact surface between dam concrete and bedrock. The boundary condition of this kind of sliding is simple and safety requirements are easily met owing to high shear strength of the contact surface. At the stage of site selection for gravity dam, only surface sliding is analyzed while shallow sliding and deep sliding in the dam foundation are normally best avoided.

Shallow sliding is shearing slip along weak rock mass, horizontal weak plane or structural plane of fractured rocks. The focus of geological research and stability analysis is to determine the potential track of shallow sliding surface and shear strength parameters of each structural plane so as to perform stability analysis of sliding resistance and make necessary proposals for treatment accordingly.

The problem of deep sliding is complex. Firstly, boundaries of the sliding are complex; including bedding planes, faults and fissures can all serve as the boundaries. Secondly, the abovementioned boundaries might be continuous or discontinuous; so the connectivity rate of structural planes should be studied. Thirdly, the natures of structural planes which form the boundaries are different, including rigid structural plane and weak structural plane. The physical and mechanical parameters and deformation characteristics of the structural planes differ greatly. Fourthly, various composite deep sliding patterns are formed with different sliding boundaries. Therefore, the mechanism of deep sliding is very complex and considerable difficulty exists in design and research work. Based on results of geological investigation, each of the potential sliding patterns should be identified and the engineering geological characteristics and physical and mechanical parameters of different structural planes which form the boundaries should be determined, and engineering solutions be recommended.

9.3 Layout of Hydropower Structure and Energy Dissipation Mode

9.3.1 Characteristics of Hydroelectric Structures Layout

"Spread-eagle" is the first choice for layout of a hydroelectric project with gravity dam. The most striking feature of the dam is that various openings and outlets can be arranged in dam body to meet the requirements of flood discharge, sediment discharge, power generation, water supply and diversion during construction. Moreover, the navigation and fish pass facilities can be integrated into dam body. Generally, flood discharge through dam body is a basic option for layout of gravity dam. Only if there is a narrow mountain pass suitable for spillways, or if the riverbed area is not wide enough for layout of both power plants and discharge facilities, could other arrangement alternatives including spillways on banks or spillway tunnels be considered.

In design of general layout for gravity dams, relationships among flood discharge structures, diversion structures and navigation facilities are the most important item to be reckoned. Special attention should be paid to downstream flow regime, scouring and deposition during flood discharge; so that negative impact on navigation and generating is avoided. River diversion during construction period is an important factor worthy of consideration in general layout. Stage diversion is usually adopted in the broad river channel while tunnel diversion in a narrow one. If flood discharge structures are not arranged in the main channel of the river, river regime will be much different from that of the natual river flow. Therefore, the consequences of changes in flow regime should be studied, and necessary measures should be taken to avoid negative impact.

9.3.2 Typical Layouts of Hydroelectric Projects

Six typical general layouts for gravity dams are given below based on the relative relationships among dams, discharge structures and diversion systems:

(1) Flood discharge structures are in the main riverbed, with underground power house or surface power house on one or both banks.

(2) Flood discharge structures are built at one side of river channel and power house at dam-toe at the other side.

(3) Flood discharge structures and power house at dam-toe are arranged with one overlapping the other in the main riverbed, so that water will overflow or flyover the power houses.

(4) Power house at dam-toe is at main riverbed, while flood discharge structures are at banks.

(5) Flood discharge structures are at main riverbed, while power house at dam-toe is at banks.

(6) Low head gravity barrage dams and water retaining power house.

Overflow open surface outlets on dam crest are usually used as flood discharge structures. Combination of crest overflow outlets, middle and bottom outlets is used in some projects, in order to increase the flood discharge capacity and to meet the requirement of limited total length of overflow section. For projects with difficult discharge and energy dissipation problems, flood discharge sections should be arranged at the main channel of the river which is the most favorable site for flood discharge. Through adjustment of layout of flood discharge outlets and energy dissipaters, the discharge flow could be made to run in the same direction as the original river flow and to join downstream flow smoothly. For the low head projects, when necessary, engineering measures should also be taken to minimize as much as possible the effect of lateral convergence and to increase discharge capacity. If possible, width of water passage of discharge structures should be increased to reduce unit discharge and to facilitate smooth flow of the discharge water into river channel downstream. Discharge flow should be properly controlled to avoid direct scouring of the two banks or too large space between the main stream and the banks, and to avoid strong backflow due to concentration of main stream to scour bank slopes and riverbed so as to reduce the load on energy dissipation work.

Power house at dam-toe is usually adopted in hydroelectric power scheme with gravity dams, while water retaining power house is adopted in low head hydroelectric project. Underground power house, power house within dam or over-flow type power house is also adopted in projects in narrow valleys where layout of structures is difficult, and in an effort to reduce the width of upstream face of the discharge and diversion structures.

The layout of navigation structures should satisfy the requirements of navigation safety such as water depth, flow velocity and direction, length of approach channel and turning radius. Navigation structures are usually arranged at one side of the river. Navigation lock and vertical or inclined ship-lift are common types. Navigation lock is usually used in low head projects, while vertical ship-lift is usually adopted in project with high water head. Because of strict requirements of topography and flow condition at upstream and downstream, model tests are usually carried out to verify feasibility of the navigation structures.

Sediment accumulation usually affects the structures layout in the sediment-laden rivers. Research on sediment transport, sedimentation processes of the reservoir and sediment sluicing is needed; engineering measures and reasonable reservoir operation scheduling is required to avoid the adverse effects of sedimentation.

9.3.3 Innovation in Hydroelectric Project Layout

1. Unit-width discharge and energy dissipater types

Increase of unit-width discharge is good for reduction of engineering quantities and investment; hence it is an important way to optimize flood discharge structures in medium and large-size hydroelectric projects.

Maximum unit-width discharge of the surface outlets in Ankang Dam Project is $254m^3/s$. The energy dissipater design is very difficult as the dam is located at a river bend and the anti-scouring ability of rock mass in energy dissipating area is low while length of stilling basin is difficult to increase. In order to solve this specific problem, we have invented and developed a hybrid dissipater of flared piers and stilling basin. By the action of transversal convergence of flared piers, the flow is forced to form a three dimensional hydraulic jump in stilling basin, in this manner, the coefficient of energy dissipation is increased, especially in the case of a small Froude number.

Maximum unit-width discharge of the surface outlets in Yantan Dam Project is $308m^3/s$. A hybrid dissipater of 'Y' type flared piers and bucket type stilling basin is adopted to cope with the Project's features of "high head, large unit-width discharge and small Froude number". Since the project was put into operation in 1992, the flood discharge fluctuates in the range of $3,000 - 15,000m^3/s$; all of the seven crest overflowing outlets are operating and it shows that flow regime and energy dissipaters are functioning normally at small and medium flood discharge.

Flood discharge and energy dissipation of Xiangjiaba Dam Project have the characteristics of "high head, large unit-width discharge, high sediment concentration". The water level difference between upstream and downstream at check flood level is 85m; the maximum discharge power is 40,000MW; maximum unit-width discharge in stilling basin is $225m^3/s$ and flow velocity into the stilling basin is 35m/s; mean sediment concentration at dam site is $1.72kg/m^3$. The energy dissipation area is close to Shuifu County Seat and the Natural Gas Chemical Plant of Yunnan Province; hence effort was made to minimize the environmental influence of energy dissipation as much as possible. A bucket bottom-flow dissipation scheme is studied in view of the large downstream water depth. A flip bucket is provided at the end of overflow dam. It allows the high velocity zone of flow to be away from the bottom floor of the stilling basin and reduces the impact of pulsating pressure, erosion and sand abrasion, and ensures stability and structural safety of stilling basin floor. Through a series of model tests, the scheme have proved out and the results is quite satisfactory.

2. Large-sized deep and bottom discharge outlets

In order to meet the requirements of flood control, sediment flushing and diversion, large-sized deep and bottom discharge outlets are necessary for controlling pool level or emptying the reservoir, like we did in the projects such as Yantan, Shuikou, Wuqiangxi and Three Gorges Projects, among which, the Three Gorges Project has the largest outlets in the dam in terms of size.

The Three Gorges Project is featured for its large diversion flow, large flood discharge, high head, large variation in water level, complex operating conditions, sediments and debris control. Therefore, based on a lot of research work, overflow monoliths of the Three Gorges Dam are arranged in the middle of riverbed to avoid scouring of bank slopes and adverse impact to navigation. Bottom outlets can satisfy the need of "storing clear and releasing muddy" operation, and sediment accumulation in upstream of dam and in the reservoir is alleviated. Alternative arrangement of surface outlets and bottom outlets in plane are built in flood discharge monoliths, and the bottom diversion outlets are arranged across transversal joints. Taking the advantages of hard and dense rock mass and deep water at downstream, energy dissipation by trajectory jet is adopted in surface, bottom outlets and bottom diversion outlets. Flip buckets of surface outlets and bottom outlets are arranged spatially in a staggered way to minimize unit width flow from the outlets and to disperse the points of falling nappe, and to dissipate the energy in different zones. With this arrangement, the scouring of the downstream area is alleviated and the protection work is reduced.

3. Flyover type rather than Over-flow type

Overflow monoliths of gravity dam are usually arranged in the middle of riverbed and the water flow has vibratory effect to power house when overflowing. The flood discharge of Manwan Project is 22,300m^3/s, with surface outlets of 13m×20m, and maximum unit-width discharge of 268m^3/s. In order to divert the spilling water from the surface outlets over the top of the powerhouse at the toe, a flyover type scheme is adopted, which has a large slotted flip bucket, with nappe spreading laterally and plunging in different roles into a plunge pool. The project has successfully withstood the test of a thirty-year flood. The transversal joints and water stops of the dam are located in the mid-portion of the overflow section. The thickened piers enhance the supporting frame of large sized radial gates on the surface outlets. The emergency gates and service gates at the intakes of penstocks with internal diameter of 7.5m are installed in the piers of the surface outlets. The main transformer house is arranged under the flip bucket. The main and auxiliary power houses are both enclosed structures. On both sides of the intakes of generating units, two of 5m×8m discharge bottom outlets and 3.5m×3.5m silt flushing sluices are arranged respectively.

4. Underground power house at narrow valley

In construction of gravity dams in deep-cut valleys, the contradiction between the arrangements of energy dissipation structures and water diversion structures is very sharp as the valleys are very narrow. In order to solve contradiction of structures layout, underground power house is usually the first choice for projects at narrow valleys. With the use of underground power house, the structure of dam can be simplified for the structure of dam body. RCC dam construction technologies are used in some projects to guarantee the construction schedule, such as Dachaoshan and Longtan Projects. Although some projects have wider riverbeds at dam site, it is still difficult to arrange all the generating units in power house at dam toe due to their large numbers of units and installed capacities. So, a multiple powerhouse scheme is adopted, in which, generating units are arranged in power houses both at dam-toe and underground, such as in Three Georges and Xiangjiaba Projects.

9.4 Shape and Structure Design of Gravity Dams

9.4.1 Shape Structure Design of Gravity Dams

Overall stability of gravity dams depends on its weight, while the weight of the dam depends on its shape and dimensions. In shape design of gravity dams, in the first place, a basic profile and profile parameters are selected and the overall stability of unit width is analyzed; then, a practical profile is obtained by modifying the basic profile according to the need of the configuration; and then stress and stability analysis is carried out and the practical profile is adjusted to meet requirements of safety, economy and easy construction. So shape design of gravity dams can not accomplish in one operation, but needs gradual analysis and repeated modification. Efficiency of design can be increased with the help of "computer aided design", while the procedure remains the same.

When gravity dam body is large, with a long dam axis and many monoliths, the optimization of profile and global shape can reduce concrete quantities effectively. Optimization of profile design is to determine the most economical profile for dam body of unit width, while the stability and stress requirements should be satisfied at arbitrary horizontal section. Optimization design of global shape turns to a mathematical model through mathematical programming method, optimization design objective and constraint conditions are treated and solved with mathematical programming method.

In the design of gravity dam profiles, volume V of dam section is the objective of optimization, which is called objective function. It is known from sensitivity analysis that upstream and downstream slope ratio and elevation of slope rising point have predominant influence to dam volume V; therefore, they are design variables. All the n design variables of gravity dam form a vector X, then $X = [x_1 \ x_2 \ x_3 \ \cdots \ x_n]^T$.

Dam profile is usually subject to some restrictions such as stress and stability requirements of design specifications. These restrictions are constraint conditions for dam profile design. The constraint conditions can be expressed by the following inequality equation:

$$g_j(X) \geqslant 0, j=1,2,3,\cdots,m$$

And optimization design of dam profile can be expressed as follows with mathematical programming method.

$$\begin{cases} \min V(X)_i & X \in R^n \\ \text{s. t. } g_j(X) \geqslant 0, j=1,2,3,\cdots,m \end{cases}$$

where design variables $X = [x_1 \ x_2 \ x_3 \ \cdots \ x_n]^T$; and 's. t.' stands for 'subject to'.

All constraint conditions $g_j(X) = 0$ appear in geometrical form in the design space, and initial design should be within a feasible region. So before optimization design of dam shape, stress and stability analysis should be done to ensure the initial design in feasible region.

Optimization design of dam profile is optimization of a single dam profile, but it is impossible to make each dam monolith into different shapes, considering the requirement of loading, water discharging and vision. Based on optimization design of dam profile, the purpose of global shape optimization design is to select economical and practical shape with concrete volume as objective function. A program for profile design and shape optimization is developed for Longtan Project; overflow section, non-overflow section at riverbed and abutments are included in the program, Fig. 9.4-1 shows typical overflow section and non-overflow section at riverbed.

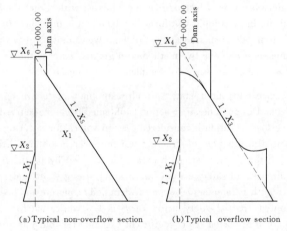

(a) Typical non-overflow section   (b) Typical overflow section

Fig. 9.4-1 Typical Profiles of Dam Section

In global shape optimization design, same upstream slope ratio and elevation of slope rising points are chosen for both overflow sections and water retaining sections. And eight design variables are considered and upper and lower limit is given to each variable, the eight design variables are: upstream slope ratio, upstream elevation of slope rising point, downstream slope ratio and elevation of slope rising point in overflow section, downstream slope ratio and elevation of slope rising point in non-overflow section at riverbed, downstream slope ratio and elevation of slope rising point in non-overflow section at abutment.

Based on shape design and further stress and stability analysis, structure design including detailed design of sub-zoning, concrete materials, structural features, structural jointing, orifices and pipeline structures, gallery system in dam is carried on.

Due to the influence of geological conditions, project layout and function, shape or structure of some special dam sections might be greatly different with conventional dam sections, special design and analysis should be done for these special dam sections. Some commonly used special sections are: sections with complex geological conditions, navigation sections, sections on steep bank slopes, extra-long sections, sections with large orifices at different levels, curved sections and so on.

9.4.2 Assessment Method of the Dam stress and Stability Safety

Single safety factor method and partial safety factor method based on probability theory are used for design and safety evaluation of gravity dams in China.

Single safety factor method is given in *Design Code for Concrete Gravity Dams* (SL 319—2005), an industrial standard for water conservation. This method is a traditional design method in which dam safety classification and load combinations are described and corresponding safety factors are defined. Load action on the dam can be divided into basic and special loads; and based on the possibility of actual loading, the most severe combinations are taken and computed for basic and special load combinations respectively. In the design specifications, stresses in dam body are computed with material mechanics method, and stability against sliding is analyzed using limit equilibrium method for rigid block and shearing strength of rock mass. It also states in the Code that stress in the dam body should be less than or equal to design strength of concrete, and safety factor against sliding should be greater than or equal to the minimum safety factor Analysis methods, strength parameters, loads combinations and safety factors for stress and stability analysis of dams are well matched with each other and following strict rules. However, the safety factor against sliding (value K) is an aggregative indicator for safety of gravity dam; it is not the safety degree in real sense because of that potential overload and potential changes of material strength and other uncertain or unknown factors are included in the factor.

Partial safety factor method based on probability theory is utilized in *Design Code for Concrete Gravity Dams* (DL 5108—1999), an industrial standard for power. Partial safety factor method introduces random variables and reliability indexes which are different from single safety factor method, it is a significant breakthrough of design theory and represents the development direction of engineering structure design theory. At the present stage, reliability index of partial safety factor method is transplanted from single safety factor method, and two methods have not too much difference in this sense.

On review conference of Unified Design Standard for Reliability of Hydraulic Engineering Structures (GB 50199—94), Panjiazheng proposed the guiding ideology of "active and cautious, transition and transplant", which is still an effective shortcut to improve the method of profile design and stability analysis. In the transition of partial factor ultimate state, the foremost issue is how to treat appropriately the conventional design method. It is not advisable to negate or to reject the safety factor based conventional method, while adopting the Reliability method. As the uncertainties in the stochastic variables such as the load action on the dam and the material strength parameters can be quantified by Reliability method, the overall safety factor may be modified and adjusted with target reliability index. At the present stage, reliability index is not a substitute but a complement for safety factor. It is more beneficial to calculate both safety factor and probability of failure rather than one of them separately. Although we cannot identify the factor of safety and probability of failure correctly, the simultaneous use of the two methods can greatly increase the credibility of the design and calculation results. So instead of rejecting each other, they are complementary to each other.

Serviceability limit state and ultimate limit state are analyzed and calibrated in reliability design of concrete gravity dams. Design value of an action is obtained from characteristic values multiplied by partial safety factors, indicating that the overload situation is considered. Design values for shearing strength of dam body concrete and foundation rock masses required in functions of structure resistance can be obtained from the values which are derived from characteristic values f'_k and C'_k of the shearing strength parameters divided by partial factors of the relevant materials, which is indicated that the requirement of safety margin for shearing strength of concrete and bedrock is met. Different partial safety factor of shearing strength parameters between the dam body concrete and bedrock masses (friction coefficient f'_k and cohesive force c'_k) are adopted, partial safety factor for f'_k is 1.3 while that for c'_k is 3.0. The present probabilistic method is limit states design method of partial safety coefficient based on reliability theory, while the partial safety factor can be obtained with safety objectives in deterministic method. Therefore, design results of the specific project from probabilistic and deterministic method is in substantial agreement. Limit states design method of partial safety factor is also widely used in other structures of hydroelectric projects.

However, lots of limitations exist when utilizing the probabilistic limit state design method, such as: dependence on engineering analogue, lack of knowledge on failure mechanism, insufficient statistic samples of material properties which are the basis of probabilistic limit state method, designer's strangeness to the method, imperfection of the related theory and standards etc. Hence special caution should be made when standards and specifications based on reliability theory and method are compiled.

Limit states design method of partial safety factor is widely used in both reliability design and deterministic methods. With the partial safety factor method, the possible stochastic differences and the action effect in the action and material performance in the safety factor K, which contains general uncertainties of different natures, are distinguished from the uncertainties mainly of unknown and fuzzy nature indicated in the resistance analysis and limit states models, so that the designers will have a better understanding of the natures of all uncertainties and their impacts on the structure safety assessment in the design and to have a unified criteria for assessment of the stochastic effect of different structures by reliability index. This "shift of track" will improve the concept of traditional safety coefficient, enhance the collection of statistic information and create positive environment for application of Reliability approach.

For complex dam structures, there are unknown uncertainties in both action effect and resistance computation methods or in models for ultimate limit state, so it is difficult to describe them with probability theory and a structural factor r_d is introduced to express the ultimate limit state with the following partial factor.

$$\gamma_0 \psi S(\gamma_F F_K) \leqslant \frac{1}{\gamma_d} R\left(\frac{f_K}{\gamma_m}, a_K\right)$$

Considering the stochastic nature of the representative values and characteristic values of the action and resistance force, the probability concept is used. Compared with most of the other actions in water and hydroelectric structures, earthquake actions including time of occurrence, localities, intensity, frequency, duration and waveforms are more stochastic and more unexpected, so probabilistic method is more necessary to study actions of earthquakes.

9.4.3 Analysis of Stability Against Deep Sliding

Analysis method of stability against deep sliding for gravity dam is still semi-empirical and semi-theoretical. In recent years, study on stability against deep sliding is still important work for gravity dam design. Further knowledge about mechanism of deep sliding mode is achieved from existing research results; analysis method and the safety evaluation system are established.

Deep sliding mode of gravity dam foundation can be roughly divided into the following categories: single sliding plane mode, double sliding planes mode, triple sliding planes mode, multiple sliding planes mode and 'cutting neck' sliding mode. The abovementioned sliding modes might appear in different sections of the dam. In analysis of stability against deep sliding for Xiangjiaba Gravity Dam, there are three sliding modes as double sliding planes mode, triple sliding planes mode and 'cutting neck' sliding mode.

Double sliding planes mode is the most common instability mode for gravity dam. There are also several kinds of analysis methods and two of them are commonly used. The first one is passive resisting force method which is adopted in design specification for concrete gravity dams. When using the first method, stability factor of the passive sliding block is supposed to be unit and then inter-slice forces of driving sliding block is calculated. The second one is K method, the method and Sarma method is equivalent for double sliding planes mode, assuming that the safety factors of the two blocks are equal and it is the most widely used method for gravity dam design.

In recent years, new methods are put forward based on the development of numerical computing methods, the following are three of them: ①Calculation and analysis are made using design loads and design physical mechanics parameters, and then computation for stability factors of the slips. ②Overload method. Water pressure is multiplied by K, and gradually increase K until instability or failure occurs. ③Strength reduction method. Let strength parameters of weak structural planes decrease gradually until instability or failure occurs in order to obtain strength reserve coefficient.

In some cases, the trend of weak structural planes in gravity dam foundation is different from the sliding direction. And the development of some weak discontinuities in the dam axial direction varies greatly. The number and distribution of weak discontinuities in foundations are different for different sections, and the safety coefficients for stability against sliding for corresponding dam sections are different, so three dimensional analysis of stability against deep sliding is more suitable. Although specific stability coefficient for three dimensional analyses is not provided, the difference between stability coefficient of two and three dimensional analysis plays an important role to judge stability against sliding and to determine the corresponding countermeasures. Three dimensional analysis of stability against sliding was carried out in the Three Gorges, Xiangjiaba and Baise Projects. The analysis results have showed that stability coefficient increased to some extent when three dimensional distributions of weak structural planes are considered.

Taking the Three Gorges Project as an example, the dam bedrock is hard and intact granite rock mass, interface of foundation elevation of power house section at left bank (No. 1 – 5 monoliths) is 90.0m, the bedrock is slightly weathered to fresh, while low-angle joints tending to downstream are relatively developed. And interface of foundation elevation of power house at dam toe is 22.2m, which forms 67.8m high slope with gradient of 54°, the 90m high gravity dam is on top of the slope and the low-angle joints tending to downstream forms sliding plane. The stability against sliding here becomes an important engineering problem.

Therefore, kinds of analysis methods were used to study stability against deep sliding problem of power house dam section at left bank (No. 1 – 5 monoliths). The methods utilized include limit equilibrium method for rigid block, finite element method (including nonlinear finite element method), geomechanics structure model test and so on. With analytical model (No. 3 monolith) and generalized mode of dam foundation stability against sliding under the most unfavorable conditions and parameters, using the rigid body limit equilibrium method (K method), The results from the abovementioned three methods were compared and analyzed. It is shown that the stability meet the requirement of $K \geqslant 3.0$. Due to importance of the Three Gorges Project, complexity of geological conditions and approximation of analysis method, reinforcement, drainage and monitoring measures were taken.

9.5 Design and Construction Technologies of Foundation Treatment

Safety of foundation is prerequisite for safety of gravity dam, especially for high dams. Firstly, overall instability of foundation must be avoided and sufficient safety margin should be provided. For gravity dams, efforts must be made to avoid the sliding of some bedrock along weak discontinuities in the foundation caused by the movement of dam body. The damage of the foundation in this case shall obviously lead to the complete failure of the dam. Secondly, stress and deformation in the foundation should be controlled in the permissive range, and large stress concentration and uneven deformation should be avoided, so as to ensure the safety and normal operation of the foundation and dam. In fact, the two requirements mentioned above should not be separated. Thirdly, mechanical and chemical stability of foundation should be well maintained under long-term seepage effect, and leakage amount and seepage pressure should be controlled in the permissive range.

In design of high dams in China, the foundation conditions of dams and their treatment have been paid great attention. With continuous development of grouting equipments, grouting materials and technologies, more methods are available for treatment of geological defects in foundation. Finite element method and geo-mechnical model test provide effective ways to evaluate the effect of foundation treatment; engineering experience and in-situ monitoring information are strong evidence of effectiveness of foundation treatment.

9.5.1 Excavation of Dam Foundations and Construction Technologies

Rock mass of gravity dam foundation should have competent strength and bearing capacity to keep stress and strain, seepage gradient and pressure in the dam body and foundation in the admissible range. Excavation depth of dam foundation should be determined based on the following factors: physical and mechanical properties of rock mass; restrictions of deformation and stability against sliding; dam structure requirements; technologies and effect of foundation treatment, construction period and investment. If the height of gravity dam is above 100m, the interface of foundation should be in fresh, slightly weathered to lower portion of weakly weathered rock; if the height of gravity dam is 50 – 100m, the interface of foundation could be in lower to middle portion of weakly weathered rock; and if the height of gravity dam is less than 50m, the interface of foundation could be in upper to middle portion of weakly weathered rock.

In selecting of interface of foundations, foundation rock mass quality is defined according to factors of weathered degree, physical and mechanical parameters, and wave velocities and so on. The effect of foundation treatment should also be considered. Consequently the excavation thickness and site of foundation interfaces are established.

Bedrock of the Three Georges Dam is mainly slightly weathered amphibole-plagioclase granite rock with trifle of lower portion of weakly weathered rock in some regions. The foundation rock mass is excellent, 77.1% of rock mass is of very high quality while 16.7% is high quality. Saturated compressive strength of the rock mass is 90 – 110MPa, elastic modulus is 50 –90GPa and compressive wave velocity is 5,000 – 5,400m/s. Only 5.9% and 0.3% of rock mass are medium and relatively low quality respectively, which also meet the requirements for foundation with backfill concrete plug and deeper consolidation grouting.

Technologies of deep hole bench blasting, controlled blasting, pre-split blasting, smooth blasting and cushion blasting are wildly used in foundation excavation for gravity dams. Intactness of rock mass should be protected in blasting during excavation to avoid too many cracks and opening of structural fissures, hence reliable measures should be taken to control excavation quality. Protective layer is reserved to protect foundation rock mass from the damage of excavation blasting. Thickness of reserved protective layer is determined by in-situ blasting experiment, and thickness of 1.5 – 3.0m is usually adopted, occasionally 0.5m and 3.5m are adopted.

Procedure and cost of excavation work is increased due to reserved protective layers, so the protective layer is eliminated in some projects and successful experience is obtained. Step excavation with protective layer was replaced by full excavation in one operation; and horizontal pre-split blasting (or controlled blasting) and blasting with flexible cushion at bottom of the hole. Foundation excavation of Wuqiangxi, Dachaoshan, Baozhusi, Three Gorges and Longtan Gravity Dams all adopt horizontal pre-split blasting once and excavation quality and construction progress are ensured.

9.5.2 Design and Construction Technologies of Grouting

1. Consolidation grouting

The purpose of consolidation grouting is to improve the integrity and uniformity of the bedrock, to improve physical and mechanical properties of shallow rock mass and geological defects in foundation, to improve impermeability of the shallow rock mass. Consolidation grouting should be enhanced in geological defected area, high stress area and areas of complex excavation in terms of shape. Consolidation grouting is also needed in excavation blasting affecting area, fissure concentration area caused by unloading rebound and shallow rock mass and areas upstream and downstream the dams where impermeability of shallow rock masses should be improve.

In the Three Gorges Project, consolidation grouting is done in one fourth width of dam foundation at both dam heel and dam toe where the stress is relatively high. In order to improve impermeability of the shallow rock mass in dam foundation, two rows of twenty meters deep holes and one row of ten meters deep holes are arranged at upstream of main curtain and downstream of the enclosing curtain respectively, the holes are for consolidation grouting and auxiliary curtain grouting. Besides the backfill concrete plug, consolidation grouting is carried out at following regions: large rupture structural belt with poor characteristics, fault zone connections, fissures concentration and strong weathered regions. Necessary support should be done before consolidation grouting at the regions of steep slope, large stress gradient and poor slope stability.

Foundation rock mass of Jinghong Gravity Dam is mainly Early Yanshanian diorite, the rock is hard but with poor integrity and slightly alteration. The foundation rock mass is mainly of Grades Ⅲ-Ⅳ with joints of low inclination angle distributed. Consolidation grouting is designed based on the specific geological conditions and stress distribution characteristics. Shallow rock mass in whole foundation is treated, grouting holes is 5-8m deep, spacing between holes and rows is 3m×3m with plum-shaped arrangement. Consolidation grouting holes are also arranged at upstream of curtain in order to increase thickness of the impermeable curtain. Due to high water level at downstream, two to three rows of consolidation grouting deep holes are set at toe of dam and power house, the consolidation grouting is also partly used as downstream impermeable curtain. The deep holes are 15m deep, distance between holes and rows is 3m and 1.5m respectively with plum-shaped arrangement. Consolidation grouting is enhanced at the regions of upper weak weathered rock mass, fault and compressed zone.

2. Grouting of dam foundation

Diversity of grouting materials could be selected, while cement paste is common grouting paste which can be used for both consolidation and curtain grouting. Wet grinding cement is produced from cement paste by grinding equipment. Curtain grouting and part of consolidation grouting for the Three Gorges Dam adopts wet grinding cement paste, 4.25 level ordinary Portland cement is grinded with three grinding equipments continuously and particle diameter achieves $D_{95} \leqslant 40\mu m$. Fly ash cement paste is used for grouting in Suofengying RCC Dam, cement with 30% Ⅰ-level fly-ash is utilized; and water to binder ratio is 0.7 and 0.5 for consolidation and curtain grouting respectively. In the recent years, new progress is made in chemical grouting for fractured rock mass, Sodium silicate, epoxy resin, polyurethane and acrylate are commonly used chemical grouting materials.

Setting plugs in grouting hole is traditional grouting method and grouting pressure could not exceed 3MPa due to restrains of the plugs in the hole. Orifice-closed grouting method succeeded in curtain grouting for Wujiangdu Dam, the method has the characteristics of high grouting pressure, fast construction and reliable quality. Orifice-closed grouting method is commonly used in curtain grouting or deep-hole consolidation grouting with high pressure.

Generally speaking, weight covering is needed for consolidation grouting of dam foundation, but grouting with leveling concrete was carried out very well in some projects including Three Gorges Project. Leveling concrete could be poured in high temperature seasons and construction interference to dam concrete pouring could be reduced. Leveling concrete is to use concrete to fill up concave places in the foundation. Individual projecting rock mass may be left exposed and the thickness of filling concrete is from 30cm to 50cm. Leveling concrete is mainly used to seal the cracks in bedrock of foundation and to prevent paste leakage during grouting process, and to ensure the quality of consolidation grouting. Weight covering should be adopted when consolidation grouting at regions of faults, fissures concentration area and steep slopes. Galleries in dam or inclined holes are utilized for grouting sometimes to alleviate disturbance to construction schedule.

3. Foundation seepage control

Seepage control design of gravity dam is mainly focused on the dam foundation, and its purpose is to reduce uplift pressure and leakage, to increase long-term permeability stability of weak fillings in structural planes, and to ensure safety operation of the dam.

Uplift pressure including buoyancy and seepage pressure is one of the main loads to gravity dam. The dam safety could be improved and dam volume could be reduced with dropping uplift pressure in dam foundation. Engineering practice and monitoring has showed that curtain grouting with deep drainage holes behind could reduce seepage pressure effectively, and water collecting wells and drainage holes with pumping and draining devices could reduce buoyancy effectively. Pumping and draining measures could also be utilized with inverted filter when soft intercalated layer exists. Effect of pumping and draining measures is more significant for dams with high downstream water level. Results from theoretical analysis and engineering practice prove that fully enclosed pumping and draining could reduce seepage pressure and uplift pressure significantly.

4. Geological defects treatment in dam foundation

Stress, deformation and seepage state of the dam are influenced by geological defects, such as faults, crushed zones, concentrated joint zones, weak rock and Karst caves. The stress distribution, deformation and seepage condition and overall stability of the dam is controlled by unfavorable structural planes with soft intercalated layer of low inclination angle, they must be treated.

Trenching and backfilling concrete plug are usually used to treat the exposed faults, weak rock mass and crushed zone. Anchor bars should be arranged as appropriate after trenching of faults; consolidation grouting is also needed after backfilling concrete. Trenching method could also be used in weathered interlayer and weathered rock mass, the lower weak weathered rock mass could be utilized as dam foundation after clearing strong weathered and upper weak weathered rock mass. Chemical grouting could be used if excavation of weathered interlayer is very difficult. At regions with joints of middle to big inclination angles or concentrated joints band, narrower spacing and larger depth of grouting holes or reinforcing bar mesh are adopted.

Deep cut-off trench, digging backfill concrete, large diameter concrete pile and pre-stressed cable are effective treatment measures for deep soft intercalation in dam foundation. Fault f_{10} is at the foundation of middle orifices dam section of Ankang Gravity Dam, the burial depth of f_{10} is 5-35m and the dip angle is 16°, maximum thickness of f_{10} is 10cm and the fault is filled with fault mud. Stability of the dam section is controlled by combination of fault f_{10} and fault F_{18}. Five main shear resistant tunnels with size of 4m by 5m parallel to dam axis and four auxiliary tunnels with size of 2m by 3m vertical to dam axis are arranged along f_{10}. A fault belt of several centimeters thick is at twenty meters depth in foundation of Yitaipu Dam. It is filled with fault mud ($\varphi=20°$, $c\approx0$). A lattice shear resistant tunnels is adopted. Eight vertical and eight horizontal tunnels with size of 3.5m by 2.5m are used to improve the stability of the foundation. Deep cut-off walls at dam toe and/or heel are also utilized in some projects.

9.6 Stress and Deformation Analysis for Gravity Dams

9.6.1 *Traditional Stress Calculation Method*

Material mechanics method and elastic theory method are introduced in the book of *Gravity Dam Design* by Panjiazheng. When using material mechanics method, vertical normal stress is supposed to be with linear distribution without consideration of the influence of dam deformation and different properties of concrete in different zones, so the calculation result is relatively rough. Stress function of elastic theory method is used to analyze the influence of foundation deformation to dam stress, but the calculation process is very complicated and hence elastic theory solution is not widely applied.

Finite Analysis Method (FEM) is used to analyze stress and strain of gravity dams since the 1960s. After more than fifty years, FEM has already developed into an independent discipline and is widely applied in many fields. Stress analysis, thermal analysis and seepage analysis for complex structures can be effectively simulated using FEM, so FEM become an important tool for quantitative calculation in the current dam design.

In the design specifications for concrete gravity dams, stress and deformation finite element analysis is required for medium to high dams and especially high dams on complex foundations. FEM might also be used for reinforcement design of complex structural part such as orifices inside the dam. If there are weak structural planes and joints with low inclination angle in the foundation, appropriate foundation treatment measures should be determined with limit equilibrium method for rigid block complementary with FEM and geo-mechanical model test.

9.6.2 *New Progresses of Stress Analysis Method*

In design specifications for gravity dams, material mechanics method is the basic method of design calculation and corresponding stress criterion is specified, FEM (mostly linear elastic FEM) is used to check results of material mechanics method. However, foundation rock mass and dam body concrete is complex multiphase material whose stress-strain relationship is nonlinear. Concrete strength and bearing capacity of foundation become important constraint to the performance of high dams, so nonlinear stress analysis and crack analysis on high concrete gravity dam become key issues in design and research work.

Some scholars have introduced the constitutive relations in the solid mechanics to the concrete material. These models are based on kinds of theories, from linear elastic theory, nonlinear elastic and elastic-plastic theory; to endochronic theory, fracture mechanics, damage mechanics which developed in resent years. All of these theory have played an important role in the development of concrete mechanical model and its description of constitutive relations. Fracture mechanics and damage mechanics are two related theories. The expansion of fracture could be regarded as reflection of damage accumulation in local region. Fracture mechanics is mainly applied to analyze macroscopic development of crack, while damage mechanics focus on cracking process of materials and structure caused by microscopic defects of the structure like microcracks or microvoids.

In recent years, damage theory is used to analyze mechanical properties of concrete and rock mass, various damage models are proposed and applied in practice. Fracture mechanical model could be divided to linear elastic fracture model and nonlinear fracture model. Linear elastic fracture model utilizes stress intensity factor and fracture toughness as criterion of crack stability. Nonlinear fracture model could be divided to fictitious crack model and crack band model, softening properties of concrete are considered in both models. Crack band model uses fraction energy as criterion of cracking, and the absorbed energy of complete cracking is fraction energy of concrete material. FEM with smeared crack model according to fracture theory is widely applied in concrete dams. It should be noted that although big progress is made in crack analysis of concrete, existing models and analysis methods still have some limitations and further research work need to be further conducted.

9.6.3 Finite Element Method for Deformation and Stability Analysis

Finite element method (FEM) is widely used in stress, deformation and stability analysis for design of high concrete gravity dams. Nonlinear FEM is usually utilized for analyzing stability against deep sliding; dam body and foundation with weak structural planes (joints) are treated as nonlinear materials in the analysis. With the action of loads and seepage pressure, some elements in local region reach strength limit and equilibrium condition is broken, the residual stress transfers to the nearby elements, new failure region occurs if the nearby elements reach strength limit. Iteration continues until final equilibrium condition is achieved and failure region no longer increases. If failure region increases continually, through crack appears and final equilibrium condition could not be achieved which means instability of the dam. Hence, FEM can simulate the progressive failure of dam foundation, stress and strain in foundation can be observed during failure progress.

Two methods are usually used to evaluate ultimate bearing capacity of dam and foundation using nonlinear FEM. One is to gradually reduce mechanical strength parameters of rock mass until the dam instability occurs, safety factor which is called as strength reserve coefficient is derived through this method. The other method is to gradually increase the load until the dam instability occurs, safety factor which is called as overload coefficient is obtained in this way.

So far, the use of FEM still lacks of mature experience in calculating the dam stress, deformation, and the stabilities of dam foundation, and supporting of corresponding standard provisions in the sliding instability, criteria of damage and its matching safety factors. But FEM has many obvious advantages, such as nonlinear properties of joint structural planes could be considered; progressive failure of structural planes and failure region could be simulated after the structural planes response to force; stress and strain in different parts could be acquired; interaction force between structure faces and blocks could also be got. Therefore, nonlinear FEM, especially three dimensional nonlinear FEM is the developing direction of analysis method for gravity dams.

9.6.4 Coupling Analysis on Seepage and Its Stress

1. Seepage analysis for dam foundation

Dam foundation rock is a complex of rock and joint fractures, which shows the various abilities of deformation, density and seepage under a fixed pressure, water environment and temperature. Though the research of seepage of fractured rock started late, importance and urgency of study on seepage in fractured rock mass were not realized until the failure of Mallpasset Arch Dam in France at 1959, and an upsurge of interest in studying seepage in fractured rock mass was raised in the developed countries since then. Research on seepage in fractured rock mass was widely carried out since the 1980s in China; lots of research achievements have been made in numerical models and methods to meet the requirement of engineering design so far.

There are four common calculation models for fractured rock mass seepage: equivalent continuum seepage model, discrete fracture network seepage model, continuum-discrete coupling seepage model and partition mixture medium model.

(1) Equivalent continuum seepage model takes rock mass as a kind of continuum medium with isotropic seepage, but neglects the interchange of water between fractures and porosities.

(2) Discrete fracture network seepage model takes rock mass as simple fissured medium whose seepage characteristics depend on width, direction and connectivity rate of fissures.

(3) Continuum-discrete coupling seepage model takes rock mass as dual media with fractures and porous media, water head in fractures and water head in porous media are formed separately.

(4) Partition mixture medium model treats the rock mass as media consist of small and medium to large fissures; the medium-sized to large fissures are simulated by discrete fissure network model while small fissures are simulated by equivalent

continuum model, coupled equations can be established based on the consistency principle at the interface of two kinds of media.

Free surface flow and effect of drainage holes are two difficult problems in the numerical simulation. Simulation of free surface could be dealt with fixed grid method or variable grid method; and drainage holes could be simulated with substructure method or bar element.

2. Seepage analysis for RCC dams

Along with the development of RCC dams, systematic study on seepage analysis for RCC dams were carried out both in China and abroad. Experimental study and theoretical analysis were carried out based on Tongjiezi Project in the period of "The Seventh Five-year Plan" and Longtan Project in the period of "The Eighth Five-year Plan" in China. As the structure of RCC layers is similar to fractured rock mass, seepage analysis methods for fractured rock mass are usually adopted in seepage analysis for RCC dams. The four numerical models for the seepage analysis mentioned above have all been tried in the seepage analysis of RCC dams. Some scholars also attempt to use uncorrelated stochastic model to simulate the seepage characteristics of RCC dam, but it is very difficult due to non-uniformity and complexity of the seepage.

3. Seepage stress coupling analysis for gravity dams

Since the 1970s, research on seepage stress coupling analysis becomes one of hot topics. The research could be divided to two aspects; one is research on seepage stress coupling mechanism, the other is research on seepage stress coupling models. Research on seepage stress coupling mechanism takes its emphasis on relations between seepage and stress in single fracture, and there are three research approaches: ①seeking empirical equations between seepage properties and stress condition through physical tests directly. ②deducing the relations from seepage and deformation characteristics indirectly. ③proposing theoretical models. Relevant models for seepage stress coupling analysis are as follows: ①equivalent continuum coupling model. ②discrete fissure network coupling model. ③fracture-porosity media coupling model.

9.7 Study on Dynamic Analysis and Seismic Design

9.7.1 *Study on Seismic Damage and Seismic Fortification Criterion*

Severe damage of structures and huge loss of lives and properties in and around epicenter area might be caused by strong earthquake. Failure of water retaining structures, especially the high dams with large reservoirs, might cause unacceptable consequences. Therefore, great emphasis must be placed on the design of high dams in strong earthquake area, and enough seismic safety of dams must be ensured.

Some high gravity dams at home and abroad as well have withstood the earthquake shock, but few of them suffered from strong earthquake motion. Baozhusi Gravity Dam of 132m high withstood the shock of Wenchuan Earthquake. The dam locates at lower reaches of the Bailong River in China, and the earthquake intensity is 8 while damage of the dam is slight. Other typical cases are Koyna Gravity Dam in Indian, Xinfengjiang Head Buttress Dam in China and Sefid Rud Head Buttress Dam in Iran. After the shock of earthquake, the phenomena of dam concrete cracks, uplift pressure rise and increase of leakage were observed in some low to medium high gravity dams. Through investigation and analysis, seismic damage characteristics of gravity dams could be concluded as follows: ①acceleration amplification effect at crest of gravity dam is remarkable. ②upper part of gravity dam, especially dam slope changing point is the weak point in terms of seismic resistance. ③upper part of high gravity dams is prone to horizontal through cracks when suffering strong earthquake. ④for gravity dams built on fault, dam break is easy to occur if the dam can not sustain the fault dislocation during earthquake.

Unlike civil constructions mainly sustaining vertical loads, the gravity dam is designed mainly to sustain horizontal loads such as water pressure. Horizontal inertia force of earthquake is comparably low and safety margin for stability against sliding is high, so seismic capacity of gravity dam is very high. None of gravity dam has failed because of earthquake in the world so far. Investigation of damage in Wenchuan Earthquake shows that: the dams withstood seismic shock beyond seismic fortification criterion and some dams suffered a certain degree of damage, none of the failed. Investigation of Baozhusi Gravity Dam after earthquake shows that: cracks occur in local region of the dam, but the overall stability was not affected and operation restored soon. It is sure that gravity dams have the ability to resist design earthquake, so long as the dam is appropriately sited with reasonable design and good construction quality.

Seismic design is to ensure the dam meeting seismic functional requirements under particular seismic fortification criteria, which means seismic fortification criterion should match its performance objective. *Specification for Seismic Design of Hydraulic Structures* (DL 5073—2000) provides that: performance objective of dam under "Grade I design earthquake" is that local damage is allowable and normal operation would regain after general repair. Grading fortification criterion is the most widely used method in the design specifications or design guides in some counties, while seismic prevention level and corresponding performance objective of different countries is different.

Based on experience and analysis from Wenchuan Earthquake, it is convinced that seismic fortification criterion in current specification for seismic design of dams in China is generally appropriate, and it is mainly equivalent to that of developed countries including Europe, US and Japan, even more strict in some aspects in China. Technologies of seismic study and design for hydraulic structures in China is among the advanced world level. Fast development of technologies plays an important role to ensure the seismic safety of hydraulic structures. For high dams with large reservoirs which would cause serious damage after failure, fortification of "checking earthquake" is needed besides "design earthquake". Seismic safety is checked with improved fortification criteria or maximum credible earthquake (MCE), and ultimate seismic resistance and failure mechanism of dam are studied.

In order to improve standardization system of seismic fortification criterion for hydraulic structures, Hydropower and Water Resources Planning & Design General Institute proposed to establish two levels of design specifications. The first level is comprehensive and principled regulations, such as *Specifications for Seismic Design of Hydropower Projects*. The second level is specific and compulsory or recommendatory specifications for structures and facilities, such as *Specifications for Seismic Design of Hydraulic Structures*, *Specifications for Seismic Design of Foundations and Slopes*, *Specifications for Seismic Design of Metallic Structures and Facilities*, *Specifications for Emergency Management in Hydropower Projects*. First draft of *Specifications for Seismic Design of Hydraulic Structures* has been accomplished so far. Revision of *Specifications for Seismic Design of Hydraulic Structures'* (DL 5073—2000) is being prepared. A lot of research works based on engineering projects under construction are carried out, such as prevention techniques of seismo-geological disasters, seismic risk analysis for dams, emergency response plan and emergency management for destructive earthquakes, fault tolerant design and early warning preventive technique for disaster.

9.7.2 Seismic Design Method and Evaluation Indexes

Seismic fortification criterion, analysis method and evaluation indexes should match with each other, in order to ensure consistency of seismic fortification criterion and safety objectives. If the dam suffers from a design earthquake, local damage is allowed and normal operation should return through simple recovery. If the dam suffers checking earthquake, wider damage is allowed while dam failure is not permitted, operation could resume after repair. The abovementioned safety objectives are qualitative; evaluation indexes might be different for different analysis method while safety objectives remain the same.

Traditional and modern methods are both used for seismic safety evaluation of high dams. Conventional method includes pseudo-static method and dynamic method based on material mechanics and structural mechanics; modern method includes linear and nonlinear dynamic finite element method. Evaluation indexes system should be determined based on analysis methods respectively. Mature control standards for traditional method have been established based on abundant engineering experiences; therefore, traditional method is still basic method for seismic analysis of high dams. Requirements of bearing capacity should be met under design earthquake condition, while requirements could be appropriately more flexible under check earthquake condition.

Numerical analysis method plays an important role in seismic design of dams. All factors such as complex structures of high dam, nonlinearity of joints, damage and fracture of material, radiation damping, could be simulated in numerical method. Modern method and traditional method are not contradictory. Computation model and method and parameters are still needed to be specified, corresponding indexes system is also needed to be established for design and check earthquake conditions.

1. Studies on dam site ground motion input

Ground motion input is basis for dynamic analysis. In linear elastic dynamic analysis, peak acceleration and normalized design response spectrum are main parameters for ground motion input. In nonlinear dynamic analysis, joints in dam body and foundation, material nonlinearity and time history of acceleration are important parameters of input.

The mode of dam ground motion input is very complex and full of uncertainties as it involves site seismic characteristics and methods of dam response analysis and its numerical models. The accuracy of ground motion input, material strength parameters and dynamic response analysis of dam should match with each other. Sensitivity analysis is needed for input factors that have decisive influence in the input mechanism.

The following methods are used for ground motion input: direct input of peak acceleration with assumption of non-mass foundation; input of peak acceleration with rigid boundary for near-field foundation; input of peak ground acceleration;

setting damper and spring at boundaries of substructure. All the above mentioned methods have certain limitations and assumptions. The results obtained from different methods should be analyzed with comparison and checking to facilitate overall decision making in future. Common view and standardization for ground motion input is needed on the aspects of the basic concept and method of input site seismic studies and ground response analysis.

2. Dynamic analysis method

Instead of the conventional pseudo-static method, linear elastic dynamic analysis method is specified in current seismic design specifications, nonlinear dynamic analysis FEM and dynamic reliability analysis are adopted in some important projects to study dynamic properties and response of the dam.

Dam structure has three types of nonlinear problems, including material nonlinearity, geometrical nonlinearity and contacting nonlinearity. Large strain and deformation is not allowed in dam structure, so material and contacting nonlinearity is main objective of nonlinear dynamic analysis.

Based on the existing research achievements and seismic properties of high dams, *China Institute of Water Resources and Hydropower Research* (Chen Houqun, 2007) establish a nonlinear dynamic analysis program. The program could simulate complex topographical and geological conditions, radiation damping of foundation, joints among dam body and foundation, and the program has been applied for dynamic analysis of several high concrete dams.

Dynamic reliability analysis method is introduced in current seismic design specifications. Peak acceleration and normalized design response spectrum are treated as variables in dynamic reliability analysis, and the first-order second-moment method is used for solution based on full probability. Substantially, the method belongs to static reliability analysis, because seismic action is not only a random variable but random process varying with time as well. Seismic reliability of dam is very complex to obtain if random characters of both acceleration amplitude and frequency are considered. So far, dynamic reliability analysis method is still being explored.

3. Dynamic properties of concrete

Dynamic properties of concrete are the basis of dynamic analysis and seismic safety evaluation. Elastic modulus, tensile strength and strain softening curve of concrete would affect dynamic response and seismic capability of concrete dams. Therefore, dynamic characteristic parameters of material are the basis for seismic design of high dams. Ratio of dynamic strength to static strength is used to reflect dynamic properties of concrete; ratio of dynamic tensile strength to static tensile strength usually adopts 1.15, while ratio of dynamic compressive strength to static compressive strength adopts 1.30. It is reported that dynamic strength of concrete increases with loading rate and the strength increment depends on concrete properties, such as mix proportion, aggregate characters and work conditions. When gravity dam is subjected to earthquake excitation, safety margin of compressive strength, stability against sliding, and tensile strength is from high to low. Tensile strength is the control factor for choice of concrete strength level. Combined action of seismic and static actions should be considered in strength checking of gravity dams, especially when the static stress is high.

As a national key tackling item in The Ninth Five-year Plan of China, dynamic properties test was carried out with complete size grading concrete under different static preloading at IWHR. This is very important to the seismic design of concrete gravity dams and there are limited numbers of study results in China and abroad. Further research should be carried out with concrete dam construction on dynamic properties of concrete.

4. Dynamic stability analysis for dam and foundation

In seismic design and research of dams, seismic strength of dam body is the first concern; however there is less study on seismic stability of foundation which is actually more important for seismic safety of dam. The issue of the aseismic stability of concrete dam is very complicated which is not only because of the dynamical interaction between dam body and foundation rock but also the difficulty to simulation and analysis on discontinuity of foundation rock. The dam body and foundation rock constitute an organic whole, however, in current engineering design, they are calculated separately and evaluate respectively according to individual criterion. Dam body is treated as continuum and dam foundation is a part of it, safety evaluation is done with strength criterion. Foundation is treated as discontinuous medium with considering of structural planes; action of dam is treated as pseudo-static external load, stability evaluation is done according to Mohr-Coulomb criterion. Both of two is incompatible with each other. To evaluate dam and foundation as a whole, combined method of continuum and discontinuous medium should be used considering the coupling relationship between them and according to the unified standard, which is the key research subject in the future.

5. Dynamic model test for gravity dam

The Engineering Seismic Research Center of IWHR established a three dimensional shaking table with six degree of freedom in 1987. The shaking table size is 5m by 5m and maximum loading capacity is 20 tons, maximum horizontal and vertical acceleration is 1.0g and 0.7g respectively, maximum horizontal and vertical displacement is ±4cm and ±3cm respectively, vibration frequency is 0 – 120Hz. The shaking table tests for gravity dams of Three Gorges, Baise, Xiangjiaba were carried out. Acceleration, displacement, strain and dynamic water pressure were measured in the tests. Natural frequency, mode of vibration and damping ratio are available from acceleration response of the structure. Non-contact measurement is realized based on laser displacement measurement techniques recently; accurate measurement of displacement is achieved for joints in dam body and interface between abutment and dam body. Dynamic data acquisition and processing software system has been built and dynamic response signals could be acquired and processed automatically.

Natural frequency of the dam from dynamic tests is close to that from numerical analysis, dynamic interaction between dam and water has significant effect on natural frequency of the dam. Acceleration response of the dam in dynamic test agrees well with that from numerical analysis, especially at the top of the dam. Dynamic interaction between dam and water is considered in FEM with additive mass method, and dynamic water pressure obtained from numerical analysis agrees well with the model tests. By comparison with the calculating results of dam dynamic stress and simulating values, maximum value of stress usually appears at dam heel and middle or upper parts of downstream slope. Due to assumption and simplification are introduced in numerical model and parameters, main conclusion of dynamic analysis should be verified with dynamic model test and the feedback analysis should be conducted by the test of seismic damage examples and strong motion observation records, which will make the results more robust.

9.7.3 *Aseismic Safety Review under Checking Earthquake Condition*

The dam aseismic safety has drawn great attention from the public after Wenchuan earthquake. Hydropower and Water Resources Planning & Design General Institute promulgated *Tentative Regulations for Preparation of Seismic Research & Design and Special Reports of Hydropower Projects* (File No. 24 in 2008) based on the experience from home and abroad. Seismic safety review under checking earthquake is specified to be carried out for very important dams. Checking earthquake is more destructive than design earthquake. Seismic safety review under checking earthquake is based on performance-based seismic design, and emergency management and the prevention of secondary disasters are taken into consideration, which is the developing direction for seismic resistance of major new infrastructure. However, the requirement for checking the aseismic safety review is put forward for the first time in China; hence it is necessary to do some research on analysis method, quantitative and qualitative evaluation indexes.

It is noted that *seismic fortification criterion of dam in the abovementioned Regulations* (File No. 24 in 2008) is not simply tightened the criteria of the dam seismic fortification and the calculation methods and criteria for safety evaluation under design seismic condition also remain unchanged but seismic safety review under checking earthquake is required for only very important dams and "nonoccurrence of dam break" is emphasized to be ensured. Checking earthquake condition is rare event for the dam, so other damages except dam breaking is allowed under that condition. Therefore, seismic review should not focus on material strength but overall stability. "Nonoccurrence of dam break" means nonoccurrence of instantaneous sliding of whole dam and overall stability could be sustained. Reservoir level could be lowered rapidly with emergency disposals and serious threat to downstream area could be dispelled.

Seismic safety review under check earthquake condition should follow the following principles: ①Review methods could not conflict with methods specified in current seismic design specifications. ②Seismic review should be carried out progressively from linear to non-linear according to the objectives. ③As much uncertainties are in the results of computation for check seismic conditions, different methods should be applied and verified with each other, so as to establish standardized calculation methods, values and criteria for evaluation.

Several methods for seismic safety review are briefly introduced as follows.

1. Dynamic method of material mechanics

In stress analysis of gravity dams, assuming that cracks are in the tensile strength area of the dam body, the sectional area and area moment of inertia are adjusted correspondingly. Then calculate them by interaction till convergence is reached. It is in fact a non-linear method. Concrete steps of the method are: ①Cracking region is determined using material mechanics method, the region with tensile stress larger than tensile strength is supposed to be cracking region. ②The cracking region is assumed to be complete opening, and new cracking region with stress redistribution is calculating for the section. ③Repeating step ②until converge or through crack in the section appears.

Variation and dynamic effect of seepage load is not included during calculation in consideration of short dynamic process of earthquake. Evaluation standards for the method are: ①If the calculation completes with convergence, the seismic safety is ensured; ②If the calculation completes without convergence, through crack appears and stability analysis is needed for the isolated dam body. Upper parts of gravity dam, especially dam slope changing point, are part with weak seismic resistance and critical part for seismic review.

2. Dynamic nonlinear FEM

Dynamic nonlinear FEM is a basic method for seismic review calculation, computing method and model should be simple and calculation convergence is used as evaluation standard. Dynamic elastic-plastic finite element method based on Drucker-Prager yield criteria could be also adopted, and yield region expanding degree is used as evaluation standard. Layered element is used to simulate structure with definite layers like RCC dam, and through crack or overall sliding is used as evaluation standard.

Variation and dynamic effect of seepage load is not included during calculation in consideration of short dynamic process of earthquake. The evaluation criteria are: ①If there is not any through cracks, the seismic safety is ensured and there will be no overall instability; ②If through cracks appear and stability analysis is needed for the isolated dam body.

3. Dynamic limit equilibrium method

The method could be combined with material mechanics method and dynamic FEM. Sliding force and anti-sliding force on potential sliding surface are computed, and safety factor is obtained. If safety factor is always larger than 1.0 during earthquake, dynamic instability would not occur. If safety factor during earthquake smaller than 1.0 appears, through cracks might occur, and stability analysis is needed for the isolated dam body.

4. Seismic review methods of stability against sliding for isolated dam body

When through cracks occur in the dam, dam body above crack plane is called isolated dam body. There are two stability review methods for the isolated dam body, namely, dynamic stability review and post-earthquake static stability review. The lower sliding plane of the isolated dam body in the dynamic stability review can be assumed as a plane in the direction of principle stress from the superposition of static and dynamic stresses occurred in the ground motion, or assumed according to the actual seismic hazards, for example, 30° toward upstream (conservatively generalized from the cracks of Koyna Earthquake). The lower sliding plane of the isolated dam body in the static stability review should be better assumed as a horizontal plane. Along with the review of displacement, the stability against toppling should be reviewed too.

Seismic review methods includes: ①Newmark sliding block method. Sliding distances of isolated dam body is calculated and then stability is evaluated accordingly. ②Discontinuous deformation analysis (DDA) and distinct element method (DEM). Movement process of isolated dam body could be computed, and stability is evaluated correspondingly.

Due to the assumption of through cracks for stability review of isolated dam body, cohesion intercept c is negligible, and not f' but f is adopted as internal friction. If overall instability of isolated dam body occurs, consequences of uncontrolled flood should be assessed when the through crack is in upper part of the dam; dam failure would happen when the through crack is at foundation or bottom of the dam.

9.8 Concrete Design and Temperature Control

9.8.1 Concrete Design Fundamentals

Before the 1960s, strength requirement of concrete was focused in concrete design and high water cement ratio was adopted to enhance concrete strength. Many cracks, even transverse vertical cracks at upstream, leakage dissolution and freeze-thaw damage appeared in some projects. Since the 1970s, failure of concrete structures happened due to concrete degradation and environment factors, therefore, high performance concrete with both long durability and high strength are regarded as the direction of dam concrete development.

Durability of concrete is the ability against environment factors, and ability of maintaining appearance integrity and long-term usage. Durability of concrete is mainly reflected in impermeability, frost resistance, crack resistance, abrasive resistance, carbonation resistance, erosion resistance and resistance against alkaline-aggregate reaction. Lots of experiments and research work has been carried out on dam concrete and achievements have been acquired; empirical equation between performance parameters of concrete and mix proportion has been established.

Lots of experimental research work has been carried out for concrete design of the Three Gorges Dam. In order to meet the requirements of the concrete in durability, and on the basis of selection of materials for the Three Gorges dam, comprehensive tests in the fields of mechanics, thermal and deformation as well as economic and technical analysis were conducted. The best concrete mix ratio for the dam was selected. Main characteristics of the concrete design are: ①Fly ash Grade-

I is used to reduce water content and increase workability, to save cement dosage and reduce temperature rise and shrinkage. ②No. 525 cement is selected and content of MgO is increased appropriately. Volume contraction and temperature cracks are reduced due to micro-swelling properties of the cement. ③High efficiency water reducing agent with water reducing ratio higher than 18% is chosen. ④Air entraining agent is used to improve frost resistance of the concrete. ⑤Reducing water to binder ratio (Internal 0.55; external above and below water 0.50; water level fluctuating area 0.45 and foundation 0.50), while increasing fly ash content (Internal 40%, external 30%, foundation 35%, structures 10%–20%). The concrete design for the Three Georges Dam proved in practice reasonable.

(1) Frost resistance of concrete. Frost resistance of concrete is main index of durability. The requirements for the Three Gorges Dams are: F_{100} for concrete in dam body, F_{150} for concrete at dam foundation and F_{250} for concrete at surface and other parts of the dam. Such measures as decreasing water to binder ratio, admixing air entraining agent and Grade I fly ash, reducing water content are adopted to improve frost resistance of the concrete. Laboratory tests and practical application have showed that frost resistance of concrete meets and goes beyond the requirements specified in design standards.

(2) Impermeability of concrete. It is also an indicator of concrete intensity. It has a very important role in durability of concrete, because it controls the seepage rate and reaction of the concrete under heat and icing. Such measures as decreasing water to binder ratio, reducing water content and proper vibration and curing are used to improve impermeability of concrete in the Three Gorges Dam. The inspection results have showed that impermeability of concrete meets the design requirements.

(3) The carbonation resistance of concrete. Reinforced concrete is used in some parts of the dam. Concrete carbonation and corrosion of steel bars should be considered. The protective function of concrete to the steel bars would be lost if carbonation depth is larger than thickness of protective concrete layers. Steel bars would rust and cause concrete to crack and scale off. The safety and durability of structures would be affected. The water and binder ratio of the structural concrete in the Three Gorges Project is greater than 0.45 and flyash admixture is less than 20%. The results of carbonation tests showed that 28 days' carbonization depth is only 1.26cm, which is equivalent to fifty years carbonization depth in natural environments. The protective concrete layer in the dam is normally thicker than 6cm which would take at least five hundred years for carbonization process to reach the depth.

(4) Crack resistance of concrete. Cracks usually lead to leakage, corrosion and thawing damages in concrete and corrosion of steel bars. Serious cracks could also affect the integrity, stability and even safe operation of the structures. So crack resistance is one of important factors for concrete durability. Concrete of the Three Gorges Dam acquires good performance through admixing Class I fly ash and high performance admixture and using moderate heat cement with 4% of MgO. Test results show that: ninety day's ultimate tensile strain is higher than 85×10^{-6}, ninety day's shrinkage strain is smaller than 382×10^{-6}, adiabatic temperature rise is 19.9–22.9℃, grown volume deformation is $(15-45) \times 10^{-6}$, and crack resistance requirements are met.

(5) Alkali aggregate reaction in concrete. Alkali aggregate reaction would cause cracks in concrete and even structural damage to the dam. Granite artificial aggregate is used in the Three Gorges Dam. Although granite aggregate is judged as non active aggregate based on various test methods, strict restrictions is imposed for alkali content in concrete materials, such as cement, fly ash and admixture. Field inspection has showed that alkali content in concrete materials is far below the control index, alkali content in cement and fly ash is 0.4% and 1.0% respectively. Under the most unfavorable situation with maximum alkali content in materials and maximum design strength of concrete ($R_{28}450$), the total alkali content in concrete is 2.3kg/m³ which is smaller than the control index of 2.5kg/m³.

Using of fly ash in dam could reduce water content, increase workability and durability. Generally, early age strength of fly ash concrete is relatively low; flyash mainly acts on long-term strength of concrete. Fly ash could improve internal lubrication and density of concrete, so the impermeability is increased correspondingly. Adding appropriate quantity of fly ash could save cement dosage, reduce temperature rise and increase crack resistance of concrete. Alkali aggregate reaction could also be effectively restrained with adding fly ash. In the recent years, content of fly ash are increasing with constant improvement of the quality, especially with mass production of Grade-I flyash. Flyash content in conventional concrete and roller compacted concrete reaches 30%–40% and 50%–70% respectively.

9.8.2 Thermal Stress in Dam and Control Measures

Gravity dam is a structure of massive concrete. Temperature in dam would rise rapidly due to hydration heat of cement after concreting. When the temperature goes down, large tensile stress would occur in dam body due to restraints and large elastic modulus. Moreover, the mass concrete exposes in atmosphere or contact with water, temperature variation in atmosphere or water would cause tensile stress in the mass concrete. Tensile strength and ultimate tensile strain of concrete are small, cracks would occur in concrete if tensile stress exceeds tensile strength or tensile strain exceeds the limit. The purpose of temperature control for gravity dam is to prevent serious cracks especially through cracks, and the objective is to control temperature difference between dam body and foundation, inside and outside of dam body, upper and lower layers.

1. Study on temperature field and thermal stress

Simple empirical calculating method of thermal stress is not enough for modern gravity dam construction, and it will be an inevitable to master the development and distribution of thermal stress.

Temperature field is required to be analyzed to make clear development and distribution of the thermal stress in concrete dam, and the problem comes down to an equation about heat conduction with given boundary and initial conditions. The heat conduction equation could be solved with different methods to obtain temperature field, analytic solutions could be got for simple conditions while numerical methods are used for complex conditions. Finite difference method and finite element method are two kinds of commonly used methods.

Elastic modulus and creep strain are varying with concrete age, so thermal-creep-stress in concrete is very complex and numerical method is usually utilized for analysis. Finite element method is a mature and effective method and used wildly for simulating temperature field of mass concrete. Lots of valuable research achievements have been achieved for thermal-creep-stress simulation in China, such as zone-merged algorithm, zoned different-step-length algorithm, relocating mesh method and heterogeneous element method are used to increase computation speed. The equivalent and iteration algorithm methods are proposed for simulating the cooling pipes. The elastic modulus considering temperature history and model for adiabatic temperature rise are also proposed for concrete material.

2. Measures for temperature control and its experience

(1) Preferred concrete material and mix proportion are used to ensure that the concrete have the following properties: enough crack resistance, small adiabatic temperature rise, high tensile strength and ultimate tensile strain, low thermal expansion coefficient, delayed micro swelling or low shrinkage.

(2) Reasonable structural jointing and blocking and pouring period is chosen. Thermal stress is sensitive to size of concrete block. Thermal stress would be higher with larger blocks and cracks might be inclined to occur. Concrete is poured without longitudinal joint in many dams with the development of temperature control technologies. It is also beneficial to reduction of thermal stress if concrete pouring is made in cool seasons.

(3) Pre-cooling aggregate, ice mixing and other measures are used to reduce concrete temperature in mixing or transporting process. And then cooling by means of water flow is utilized after placement.

(4) Cooling by means of water flow is utilized to minimize differences in temperatures of concrete poured between inside and outside of dam body before cool seasons set in. The river water is used in the initial period, and then river water is applied in combination with refrigerated water so that the temperature of the concrete placed will meet the requirement of the joint grouting zoning and staged grouting of the dam.

(5) It is preferred to pour concrete in thin layer with short intermission and the dam section should rise evenly. Concrete pouring at foundation should make use of most of the cool reasons, and construction management and technologies should be enhanced and improved to ensure concreting quality.

(6) Surface protection for concrete is one of main measures to prevent surface cracks. Specific surface protection measures should be chosen according to locations and conditions of the concrete. Protection should be emphasized for concrete at upstream faces, constraint area near foundation as well as surfaces of discharge orifices. Concrete curing is also an important measure to ensure concrete strength and prevent shrinkage cracks.

9.9 Diversion and River Closure Construction Technology

9.9.1 Construction Diversion

The aim of construction diversion is to create the condition of dry land construction. The construction of water conservancy and hydropower engineering practices has promoted the development of closure design and construction technology. When the dam site in a wide open valley, the diversion work of gravity dam project uses stage or channel diversion, such as in Wuqiangxi, Yantan, Shuikou, the Three Gorges and Jinghong Projects; When the dam site in a narrow river valley, the river is closed and tunnel diversion is adopted, such as in Dachaoshan, Manwan, Longtan, Guangzhao, Jin'anqiao and Guandi Projects.

The Three Gorges Project's construction diversion was divided into three stages: the first stage is to dry the right side of the main river. Under the protection of the cofferdam, diversion channel is excavated, and concrete longitudinal cofferdam is constructed, while the temporary lock on the left bank and the permanent ship lock are constructed too. The Yangtze River flow is still running in the main channel, and the navigation is normal. The second stage is to cut off the main river. The foundations of upstream and downstream earth-rock cofferdam and concrete longitudinal cofferdam were excavated. The construction of flood discharge dam section for construction period, dam sections for left bank power house and dam-toe power house and non-overflow section of the Left Bank, as well as head works of ship locks continues and construction of the permanent ship locks is completed. The river water is discharged from the diversion channel. Ships are sailing in the diversion channel and through the temporary locks. The third stage is to block diversion channel for construction of upstream and downstream earth-rock and RCC cofferdams of Stage III is constructed. The dam section for the right bank power house, the dam section for dam-toe power house and non-overflow section of the dam are constructed in the Stage III foundation pit. River flow goes through bottom and deep outlets of the flood discharge dam section and ships go through the permanent ship locks. With the Stage III RCC cofferdam in operation, the dam-toe power station on left bank is put into operation.

The world's largest diversion channel was built in the Three Gorges Project in 1998. The total length of the Channel is 3,950m, with a composite profile: high right and low left. The minimum bottom width is 350m, with water depth of 20 - 35m, and design flow of 79,000m^3/s. The open channel also serves as navigation channel during construction period. Diversion channel of the Three Gorges Project in 5 years' operation has experienced 8 time flood peaks with the maximum flood peak discharge of 62,000m^3/s. It has protected the safety of diversion and navigation of the Three Gorges Project.

Among the concrete gravity dams built with aid of diversion tunnels, the Longtan, Jin'anqiao and Manwan's diversion tunnels have the lager sectional size. The right and left bank of Longtan Hydropower Station each has a diversion tunnel. Both of them are of horseshoe shape, 16m wide, and 21m high, fully lined with reinforced concrete. The design diversion flow is 14,700m^3/s, and the height of upstream and downstream cofferdams is 82.7m and 48.5m respectively. Jinanqiao Hydropower Station has two diversion tunnels in the right bank, they are of horseshoe shape too, 16m wide, 19m high, fully lined with reinforced concrete, and the design diversion flow is 10,600m^3/s, and the height of upstream and downstream cofferdam is 61m and 34m respectively. The Manwan hydropower station's right and left bank each has one diversion tunnel, in horseshoe-shape, 15m wide, 18m high, a design diversion flow of 9,500m^3/s. the height of upstream and downstream cofferdam is 64.3m and 39m.

The design and application of diversion bottom outlets has also gained successful experience in the gravity dam construction, such as the 22 diversion bottom outlets and 23 deep flood discharge outlets on the discharge dam sections in the Third Stage of the Three Gorges Project. It has run for 5 years in good condition. Diversion bottom outlets have a sectional area of 6m×8.5m (width×height), with a design head up to 80m, and design diversion flow of 72,300m^3/s (including diversion bottom outlet discharge of 36,100m^3/s), the average flow velocity at the exit of diversion bottom outlet of 32m/s. The Shuikou Hydropower Station, using open channel for diversion purpose in the Third Phase, constructed ten of 8m×15m rectangular diversion outlets and made use of reserve notch on overflow dam section. Design flow of diversion bottom outlets is 25,200m^3/s, and they are operating normally.

The projects mentioned above have demonstrated that China has built the world largest diversion and navigation open channel in terms of discharge capacity and the largest diversion tunnels and the high head bottom outlets in terms of flow section. It has proved that the diversion works of these projects have been built with intelligent designs, advanced technologies, qualified workmanships and reliable operation.

9.9.2 River Closure

River closure is the key object and important milestone in the hydropower projects. In engineering practice, the commonly used closure methods are closure dike method, instantaneous closure method, and closure dike-free method. The most commonly used method is closure dike. Closure dike method is divided into horizontal closure, vertical closure and vertical-horizontal closure. Before the 1940s, horizontal closure method was used around the world. Since the 1940s, the vertical closure method with heavy dump trucks was adopted. The vertical closure method is simple and easy with handy supplies of local materials and mechanical construction, and limited preparation work. Moreover, it is quite flexible in taking any supporting measures according to the specific conditions, such as the "leveled riprap on riverbed", and the "construction of baffle sills" and so on, to prevent river erosion and to reduce the water depth at the closure gap and to prevent the embankment collapse. With the development of closure theory and technology, to the present, the "vertical closure method" has become the main method of closure of major rivers.

The successful closure in the Third Stage of the Three Gorges Project set a series of world records. The design closure flow of the Third Stage is 12,200m³/s, and the total head of the river flow is 5.77m The indicators such as the velocity at closure gap, the maximum unit width energy, and advancing intensity were better than other closure projects, which made it the most difficult closure project in the world. To reduce the difficulties of closure operation, the method of double-dike-sharing gap closure in the upstream and downstream was used; and the steel wire rock cages were dumped into the river bed to protect the base of dike. The steel gabions were placed as baffle sills in the closure gap; with large rock strings and concrete tetrahedron strings; widening of top width of closure dike, increase of the dumping intensity, and the complete removal of upstream and downstream cofferdams of the Second Stage, to meet the design requirements of diversion through bottom outlets. The river is closed by diversion through open channel successfully in one act. The actual closure discharge was 10,300~8,600m³/s, with a total head of 2.85m.

During the river closure, the navigation conditions of channel must be taken into account, and shipping in the Yangtze River should not be interrupted. Through the research and application of the closure hydraulics, navigation hydraulics, hydraulic model test, ship model test hydrological forecast, prototype hydrological and hydraulics observations, river closure engineering technologies and so on, the success of closure of the Project was guaranteed, while the navigation in the Yangtze River was secured as well. It has set a world record of a daily dumping intensity of 194,000m³ in river closure.

9.9.3　*Cofferdam*

In the development of large hydropower projects on the rivers, the cofferdam design and construction have the following three trends: Firstly, the geological conditions of cofferdam are becoming more complex and the height and quantities of the works are increasing. Secondly, highly intensive construction work with a faster cofferdam construction can ensure completion of project in a dry season and close of the river in the current year. Thirdly, cofferdams have to be constructed in the water or even in deep water, but also to complete the foundation treatment in the deep overburdens. Take the Three Gorges Project for an example, its second Stage earth-rock cofferdams both in upstream-downstream were built in 60m deep water. The maximum height of the upstream cofferdam is 82.5m with a filling volume of 5.9 million m³. Thick overburden was treated with double roll of 1m thick concrete cutoff walls, and the maximum height was 74m. The cofferdam is built in just one dry season.

Since the 1980s, the RCC cofferdam is widely used in the water conservancy and hydropower engineering works. Some engineering works use the RCC which is featured for its fast construction and permission for overtopping, and adoption of lower criteria for diversion flow in dry seasons. It serves the purpose of shortening the construction period and cutting down project cost. The maximum height of the upstream and downstream RCC cofferdams in Yantan Project is 54.5m and 39m respectively. The upstream cofferdam was constructed by a method of joint-free, continuous placement, in thin lift and non-stop lift. In 103 days, 18.5 million m³ of RCC were pourwith the raise of cofferdams by 54.5m. The Project met a flood exceeding the set standard in flood season in 1988, with the discharge of 19,100m³/s, and an overtopping flow up to 4,000m³/s, but the arched cofferdam operated normally. The maximum height of Stage III RCC cofferdam of the Three Gorges Project is 121m, with a storage capacity of 14.7 billion m³. The total volume of RCC is 1.6736 million m³, which is done in two stages. The rising height of the cofferdam reaches 25m and the monthly maximum of placement is 476,000m³, intensity of pouring operation of 21,000m³ in the dry seasons of 2002/2003, which sets a world record.

9.10　Concreting Technology in Dam Construction

The construction of the Three Gorges, Longtan, Jinghong and other high concrete dams in China has greatly enhanced the construction management, improved the technology, construction equipment, level of mechanization and quality control. The technology of concreting in dam construction in China has reached the world advanced level.

9.10.1　*Concrete Production*

1. Raw materials for concrete

The cements for concrete gravity dam include ordinary Portland cement, moderate heat Portland cement, low heat Portland cement and low heat Portland slag cement. The Manwan and Dachaoshan projects used ordinary Portland cement; Gezhouba and Guxian projects used low heat Portland slag cement; Ankang, Wuqiangxi, Wanjiazhai, Longtan, the Three Gorges and other projects used moderate heat Portland cement.

In the past, Dam concrete aggregate was natural sandy gravels, but recently, the artificial sand aggregate has been widely used. For example, limestone has been used in Yantan, Jiangya and Longtan Hydropower project. Amphibole plagioclase granite has been used in the Three Gorges Project, basalt has been used in Dachaoshan and quartz sandstone used in the Wuqiangxi project. Technology of making artificial sand aggregate is more perfect.

Flyash as admixture is widely used in concrete construction. In the absence of flyash, other admixtures are studied. For example, a new-type admixture (PT admixture) by mixed phosphorus slag and tuff was used in Dachaoshan RCC gravity dam; Some new admixtures were used effectively in Jinghong and Jin'anqiao projects, such as admixtures with fly ash and ferromanganese slag (FM), tuff and ferromanganese slag (NM), iron slag and limestone (FS), flyash and phosphorus slag (FP). The mixture with phosphorous slag powder (P) and lime stone (L) is researched as the feasibility of concrete admixtures in Ludila project.

Due to the superplasticizer, air-entraining agent and retarders were used in the concrete gravity dam, and the performance of concrete with various indicators is improved greatly.

2. Concrete production

The large size hydropower projects are featured for their demand of large quantity of concrete, intensive workforce, and strict quality requirement. They can be built only with a high degree of mechanization and automation of the concrete mixing plant. The following concrete mixing plants have been produced in China: HL50-2F1000, HL115-3F1500, HL240-4F3000, HL360-4F4500-type and other specifications. The control methods are semi-automatic, automatic and computer controlled types. Most mixing plant can adapt to a variety of mixed admixtures and flyash and pre-cooling of concrete production. For the production of low-temperature concrete, concrete mixing system is usually equipped with refrigeration system such as cooling wind, chilled water and ice equipments.

The Three Gorges Dam concrete production system is the world's largest concrete mixing system, which has nine mixing plants based on site conditions and construction requirements. There were two mixing plants which volume was $4 \times 4.5 m^3$ with a cooling system of $2,150 \times 10^4 kcal/h$, two mixing plants whose capacity was $4 \times 6 m^3$ and $4 \times 3 m^3$ with a cooling system of $1,600 \times 10^4 kcal/h$, two mixing plants with capacity of $4 \times 3 m^3$ and a cooling system of $1,375 \times 10^4 kcal/h$, two mixing plants with capacity of $4 \times 3 m^3$ and $4 \times 4.5 m^3$ and a cooling system of $1,250 \times 10^4 kcal/h$, 1 mixing plant which volume was $4 \times 3 m^3$ and $750 \times 10^4 kcal/h$, respectively. The total production capacity was $1,720 m^3/h$ and $1,360 m^3/h$ of low-temperature (7℃) concrete in summer.

9.10.2 Concrete Pouring

1. Placement of concrete

The concrete was poured by horizontal transport and vertical transport to the job site in the gravity dam construction. Concrete placement can be made in many different ways to suit various conditions of the construction requirements, including cars, rail locomotives, gantry cranes, belt conveyors, cable cranes, tower belt cranes, etc.

The Three Gorges Project adopted a construction program by six tower belt cranes, nine portal tower cranes and two cable cranes and a well coordinated process which allows the concrete to go smoothly from mixing plants to job sites and vibration. Thanks to the arrangement, the concreting work was greatly intensified. The monthly peak of the concrete placement was as much as $49,100 m^3$ per tower belt crane. The $6 m^3$ tower cranes can pour $20,500 m^3$ of concrete per month, and the six tower belt cranes can pour 2.21 million m^3 of concrete per year, accounting for 65.6% of annual amount of concrete placement. Tower belt crane played a major role in concrete placement of the Three Gorges Dam Project.

2. Formworks and re-bars

Some formworks are employed in concrete gravity dams, including composite steel forms, sliding forms, cantilever forms, self-climbing forms and forms with movable cribs and so on. Forms with movable cribs are applied in the Shuikou Dam Project. The integral climbing forms used in construction of the haunches and the shafts of the Three Gorges Dam, while the typified form was used in construction of the galleries and drainages. The development and application of the new formwork technology not only reduces the labor intensity, but also improves efficiency and ensures the concrete quality, and the finish of concrete structures.

Traditionally reinforcement is erected by manual overlapping and welding. Some projects use couplers of cold press linear screw thread sleeve technology. The application of this technology is not affected by the welder's skill and climate. The technology has also some advantages such as reliable coupling, stable quality, fast coupling, and good bonding of weldable and non-weldable re-bars.

3. Concrete spreading and vibrating

The commonly used concrete paving methods after placement are tile method, step method and pouring in thin lift without longitudinal joints. Generally, the thickness of paving blank is 50cm. In recent years, the China's approved Model YPZ-40 vibrating machine has been used in high concrete gravity dam construction. In order to simplify the construction, pouring in thin lift without longitudinal joints method has been used in Wuqiangxi Project and gets good results.

In construction of the Three Gorges dam, concrete were transported from the mixing plant to the job site in the dam directly by belt conveyor and tower belt cranes, while in construction of concrete dam section for right bank power house, the tile method was used with a lift of 3m in high temperature season, and concrete was hauled by double lines of tower belt cranes. In order to ensure the quality of concrete construction, the tile concreting method was used with the aid of tower belt cranes in cold season. When the pouring area is more than 500m^2 in summer, the step method is used. When the pouring area is less than 500m^2, the tile method is used. In order to control the concrete temperature, the pouring time should be properly selected and thin lift (40cm) flat pouring method was adopted in hot season. When the concreting began, the concrete should be cooled with pipe cooling water and the sprayer was started to control the concrete temperature and keep it wet.

9.11 Safety Monitoring of Gravity Dam

Safety monitoring has become an indispensable and important work in engineering construction, operation management, and scientific research. The dam safety monitoring design is an important part of the Gravity Dam Design. The purpose of dam safety monitoring on the one hand is to have a command of the dam performance and make an assessment of dam safety, and on the other hand is to verify the design and test results of scientific research. The construction safety monitoring can guide safe construction and improve the level of construction technology.

In the 21st century, dam safety monitoring management and monitoring technology have been greatly developed. National laws and regulations system have been gradually established and perfect, while the relevant regulations and technical standards have also been promulgated. The automatic monitoring system of the old dam has been updated, while the new dam has more comprehensive and intelligent monitoring system. In addition to conventional deformation, seepage, dam foundation, stresses and strain monitoring program, dam safety monitoring also contains some special contents including the stress and strain of local structures, hydraulic slope stability and deformation near the dam area, structural seismic response and so on.

With electronic and network technology, the dam safety monitoring has been modernized. Sensing performance and automation of monitoring instrument has been greatly improved. A number of new monitoring equipments have been applied in dam projects, particularly deformation monitoring instruments, such as smart CCD, tensile wire telemetry device, laser alignment systems, IBIS and so on. Using smart sensor in the automated system greatly expand flexibility and adaptability of system building. Besides installation laying technology and quality of monitoring instruments is also rising. The intact rate of embedded devices in general is more than 95%. Current dam safety monitoring automation system reliability and utility have a significant improvement.

The development direction of deformation monitoring have Digital photogrammetry, GPS global positioning system, and system for deformation monitoring based on GPS, GIS and RS technology. Measuring robots is becoming major equipment of automated measurement. At present, dam safety monitoring automation system can realize variable data collection, transmission, management, online analysis, which is being to the integration, automation, digital and intelligent direction.

In Three Gorges Dam safety monitoring program, automated data collection system was used, including the measurement control, operation management and communication network. The displacement is measured by wire alignment, borehole inclinometer, multipoint extensometer etc. The dynamics monitoring instrument is also set in dam to monitor the strong earthquake acceleration, dynamic strain of structures for earthquake intensity, vibration and other items.

Monitoring data analysis and information processing has also been developing rapidly. Although the single-point model, including the statistical model, hybrid model and the deterministic model are the still main models used in monitoring, but for the limitations of single-point model, the distribution model has been introduced by Chinese scholars to deal with multiple sensor information about the same monitor point. This model has been systematically studied, and applied. In addition to measuring multi-point model, the country experts and scholars have improved the traditional model to perfect and received many research achievements.

With the development of the management information system (MIS), decision support system (DSS) and the basic theories and technologies, dam safety monitoring management information system has also been developed greatly. Many new technologies or new theoretical methods are continuously introduced and absorbed. For example, application of neural network in comprehensive analysis; using the data mining methods to process and analyze the information, using the internet technology to communicate and remotely manage information and so on.

The development trend of dam safety monitoring and research in the future are as followings:

① In addition to the dam and its ancillary buildings, the scope of monitoring should be expanded to other regions including complex geological conditions and combined with hydrological monitoring to some extent; ② With the development of observation instruments with high accuracy, stability and automation, monitoring tools will be more advanced; ③ Automatic monitoring system will be more perfect, not only to realize the automation of the remote and centralized control, but also to apply GPS, GIS and RS based deformation monitoring system widely in dam construction; ④ The development of data processing from offline centralized processing to online real-time processing, which can provide more timely and accurate information for management in making their policies; ⑤ Feedback analysis has much research results, which can promote the development of dam design and construction technology; ⑥ Mathematical models of monitoring analysis are more and more diversified, in addition to statistical model, deterministic model and the hybrid model, the time series, gray theory, fuzzy mathematics, neural network, stochastic finite element, spectral analysis and other methods were introduced in positive and negative analysis of dam safety monitoring and structural feature; ⑦ The dam safety evaluation studies develop from the single point independent analysis to multiple points and more comprehensive analysis; the risk and reliability analysis will also be launched in depth, and risk control and emergency management are universally applied in the future.

9.12 Planning of Gravity Dam Construction and Key Research Subjects

9.12.1 Dam Projects in the Planning

In order to meet the need of the sustainable economic and social development in 21st century in China, and to guarantee a reliable supply of water and energy resources, hydropower construction will develop actively and orderly based on the principle of protecting the ecological environment and dealing well the resettlement work. Hydropower development in China mainly concentrated in southwest and northwest regions which have rich water recourses. With the implementation of West Power to East Policy and the start of the preliminary work of Power Transfer from Tibet to Outside, the planning work for Hydropower Development of middle and upper reaches of Jinsha River, Lancang River, Yalong River, Dadu River and in the upper reaches of the Yellow River and the Nu River has been basically completed, and the pre-construction work for power projects has been fully spreading.

The hydropower projects located in the western region of China are featured for large scale with a large quantity of corresponding dam construction, and complex engineering geological and hydrological conditions. Meanwhile, transportation conditions are difficult because of that remoteness, high transportation cost of foreign goods, and low support ration of material supply. So in design of projects, priority are normally given to arch dam which use less concrete or embankment dams using local materials. In reality, in the design of the planning stage of the rivers mentioned above, the most of the dams with critical reservoirs are basically arch dam or earth-rock dam considering topographic and geological conditions. When topographic and geologic conditions at the dam site are favorable, the arch dam is preferred. When the topographic and geologic conditions at the dam site are not ideal, the earth-rock dam program is selected. Earth dam is less dependent on imported construction materials, which has a competitive merit.

Some high dams built in Nu River, Lancang, Jinsha, Yalong River and other rivers selected the concrete arch dam alternatives, such as the projects Xiaowan, Jinping I, Xiluodu, Baihetan and Wudongde. Other high dam projects use the earth scheme, such as the projects Lianghekou, Shuangjiangkou, Nuozhadu, Pubugou, Gushui and so on. Compared with the projects mentioned above, concrete gravity dams are less in number. Thus, the technical development of concrete gravity dam faces the new challenge. Only constant innovation can make concrete gravity dam develop further.

Table 9.12-1 lists a number of concrete gravity dams for hydropower projects under planning. The table shows that: the height of designed dam in planning is between 100m and 150m, and the construction of these projects can rely on the mature design and construction experience; Construction of dams over 200m will face great challenge; All of these projects are in high mountains and canyons with very complex regional tectonic conditions and geological phenomenon, so dam construction with complex geological conditions faces substantial technical problems such as fortification of earthquake zone, energy dissipation of the flood flow and reinforcement of the high slope and so on. The design of gravity dams will be competitive only if the construction experience of existing projects is employed, constant innovation is carried out, and solutions to key technical issues are worked out with good construction features and economy.

Table 9.12 - 1 The Planned Concrete Gravity Dams in China

No.	Project name	River	Dam type	Dam height (m)	Storage capacity ($10^8 m^3$)	Capacity (MW)
1	Guanyingyan	Middle reaches of Jinsha River	Rockfill with clay core on the right bank & RCC gravity dam on the river channel and left bank	159	22.50	3,000
2	Ludila	Middle reaches of Jinsha River	RCCPG	140	17.18	2,160
3	Ahai	Middle reaches of Jinsha River	RCCPG	138	8.85	2,000
4	Longkaikou	Middle reaches of Jinsha River	RCCPG	119	5.58	1,800
5	Jinsha	Middle reaches of Jinsha River	Concrete dam	71	1.12	520
6	Kagong	Upper reaches of Lancang River	RCCPG	97	0.03	240
7	Wunonglong	Upper reaches of Lancang River	PG	136.5	2.72	960
8	Tuoba	Upper reaches of Lancang River	RCCPG	158	10.5	1,400
9	Huangdeng	Upper reaches of Lancang River	PG	203	15.49	1,900
10	Dahuaqiao	Upper reaches of Lancang River	PG	107	2.62	900
11	Lumadeng	Middle and Lower reaches of Nujiang River	PG	165	7.35	2,000
12	Bijiang	Middle and Lower reaches of Nujiang River	PG	118	2.80	1,500
13	Yabiluo	Middle and Lower reaches of Nujiang River	PG	134	4.49	1,950
14	Lushui	Middle and Lower reaches of Nujiang River	RCCPG	165	12.88	2,400

Note: RCCPG, RCC gravity dam; PG, gravity dam.

9.12.2 The Development Direction of Gravity Dam Technology

According to statistics, among the dams over 70m in height in China, the gravity dams, arch dams and rock filled dams each is 1/3 in number. But among the dams above 200m, the gravity dams are few in number, less than arch and rock filled dam. The reason is that when the height of gravity dam exceeds 200m, the requirement of ground is high and the volume of dam concrete is large. Compared with the arch dams and the rock-fill dams, the gravity dams have poor adaptability, and lack of competitiveness in project duration and investment.

However, the structure of concrete gravity dams is simple and their loading conditions is clear, with more combinations of discharge structures and high ability against flood, flexibility in power house arrangement, and high adaptability to topographic and geologic conditions. So in appropriate building conditions, the concrete gravity dam is still the main dam type in high dam selection.

Specifically, the concrete gravity dam below 150m (inclusive) has accumulated rich experience in building and design in the world. It is a strong competitive dam type with mature technology. For the dams over 200m in height, there is little experience in design and construction, but more technical difficulties, special study is needed. So far, there is not yet any dam with a height over 300m in the world, and therefore its technical risk is high. Engineering studies have shown that in selection of extra-high dam types, the competitiveness of gravity dam is obviously not as good as arch dam and rock-fill dam. It must be very cautious to choose extra-high gravity dams.

It is worth mentioning that the vitality of gravity dam lies in that, it is necessary to carry out the innovative research according to the specific conditions for project layout, structural design and construction method, so as to suit topographic and geologic conditions better and reliability of the structures and economic rationality of projects. The most important is to focus our studies on foundation stability against sliding, deformation and seepage, energy dissipation and cavitations and erosions caused by high velocity of flow in the narrow valleys, on selection of concrete raw materials and mix ratios and temperature control of massive concrete, river diversion during dam construction, flood control and rapid construction.

When the concrete gravity dam is selected as an alternative dam type, the following factors need to be considered: ①Volume of concrete work. If the work is large, the demand of imported materials such as cement, fly ash and other materials will be huge, then the cost of procurement and transportation of these materials will be high, and higher requirement of transport conditions. ②The stability against deep sliding. If there is soft rock or low-angle weak structural plane in foundation, the problem of stability against sliding of dam becomes an important factor to dam safety. And the treatment of the dam foundation is difficult and expensive. ③Roller compacted concrete dam. Although RCC dams can reduce the amount of cement, simplified temperature control measures and construction costs, the number of lifts is large and high-intensity continuous construction will be required. The key factor of quality control is the reliability of seepage control in upstream face of dam and the quality of binding of each contact plane, where the risk of the dam lies. ④Hydraulic fracturing of high head. It is more complex in high gravity dam heel. The dam will not be "safe and reliable enough", according to the criteria that there is no tensile stress in the upstream face of dam which should be followed in the design. There exist the problems of cracking in the dam heel and cracking and seepage control in upstream face of dam. Taking into account the high head and maintenance difficulties, there should be compressive stress in gravity dam heel and necessary measures should be taken to prevent cracking and seepage.

9.12.3 Key Issues of Technological Innovation

1. RCC dam construction technology

The solutions to the rapid construction, temperature control, cracking prevention of concrete, and reduction of cement consumption, lie in further innovation and development of RCC dam construction technology. The lean cementing material roller compacted concrete dam using the construction technology of rock-fill dam shows good construction features, such as rapid construction reduction of cement consumption and simplified temperature control measures. It has strong economic and competitive advantage. If the RCC dam technology gives vitality to the gravity dam, then, roller compacted concrete dam with lean cementing material will be a new motive force for development.

Drawing on the experience of RCC dam construction both in China and abroad, and assuming an impervious structure is provided in the upstream face of dam and using lean cementing material to build the gravity dam which may be the representative body of a gravity dam with height less than 100m. So it is necessary to carry out in-depth studies on dam impervious system, the performance and mix of RCC dam with lean cementing materials, stability against sliding, stress and deformation of the foundation, dam shape and material function. It needs to get new breakthroughs by industrial test, which is used to increase the adaptability of RCC dam construction technology and to improve its competitiveness.

2. Key technology of extra-high gravity dams

Currently, the planned dams with height of 200m and above are mostly rock-fill or concrete arch dam, virtually no concrete gravity dams. The concrete gravity dam over 200m is few abroad. In view of dam safety and reliability, the ability to resist floods over standard and easy maintenance, concrete gravity dam is still the choice of dam type for high dam construction.

It's no doubt that there will be unprecedented technical difficulties in hydropower development in the Western region in China. The existing domestic and international experience in engineering can not cover a series of challenges in future hydropower development. Many technical problems are related with the engineering geological conditions, such as the complex geological structure, deep overburdens, strong earthquake, karsts geological problems, deep sliding and so on. Besides quality evaluation and quality standards of dam foundation rock, seepage drainage techniques and reinforcement of complex foundation are still the key issues need further study. In addition, zoning, material properties, stress and deformation of the super high gravity dam, ultimate bearing capacity and damage mechanism of the dam require in-depth study, and it's necessary to establish more complete safety evaluation criteria of design.

The following is the key technology issues of high gravity dam: ①research on design of foundation treatment and construction technology; ②research on dam shape, material zoning and structural design; ③studies on dam stress, deformation and stability of safety; ④high-speed flow and energy dissipation technology research, including size of discharge buildings, joint energy dissipater, technology of cavitation and erosion prevention, water pulsation and vibration, abrasion resistance, atomization of flood discharge and so on; ⑤studies on measures of construction technology and processes, including temperature and crack control of massive concrete; ⑥technology research on large-scale supporting construction machinery and equipment, etc.

3. Gravity dam seismic fortification technology

Western region of China is a strong earthquake-prone area. So the multipurpose engineering works with high dams and large storage capacity will inevitably encounter the issue of seismic fortification. In order to prevent failure of dams in earthquake induced disastrous consequences, it should be paid attention to dam seismic work to ensure seismic fortification of the dams.

Key issues of dam seismic technology include: ①seismic safety evaluation and selection of dam site; ②fortification standards and ground motion input studies; ③response of high dam ground motion; ④dam materials and dynamic properties research; ⑤earthquake monitoring; ⑥risk assessment, early warning and emergency plan studies.

Along with the sustainable and stable economic and social development, the infrastructures are becoming more and more complete, and the damage caused by extreme disaster is increasing. Dam failure would be a catastrophic event to the downstream area. Therefore, it is necessary to study emergency plans and countermeasures for a possible earthquake and the damage of the dam, and to ensure the safety of life and property downstream.

4. Key technology of new materials

Materials with good performance, not only ensure the safety of the structure, but also help shorten the construction period and reduce the engineering investment. Different dam site, dam type and head conditions will require different building materials and even special requirements. In addition to active studies on application of new type cement, steel and polymer organic materials, key technologies of new materials to be studied in hydropower project research are: ①new concrete materials and their properties; ②the development of new abrasion resistant materials; ③dam material evolution, creep characteristics and long-term strength characteristics; ④other materials, including lean cementing materials of RCC, cementing sand and gravel or rubble concrete, new and efficient grouting materials, high-performance concrete admixtures, fiber concrete and so on.

5. Ultimate bearing capacity of gravity dams

Massive volume and high consumption of cement, while the strength of the material was not given full play, is the main reason why the gravity dam is lack of competitive advantage. So ideas, principles and methods of design optimization for gravity dams are worthy of further study. In the design of normal concrete gravity dams and RCC gravity dams, zoning of materials must be made on the typical dam sections and its profiles according to the stress condition and stability against sliding, so as to give full play to the concrete strength and to cut down project cost. At present, Design of RCC gravity dam follows the design ideas and principles for the normal concrete gravity dams. In selection and optimization of the cross-sections, the unit concrete volume of the section was taken as the objective function, where the strength of typical portions and the stability against sliding of interface of foundation area was taken into consideration only, without considering the zoning of materials. In the optimization design of concrete gravity dams, the nonlinear optimization program was used, taking the cost of entire dam as the objective function, and constraint function should not only consider the stability against sliding and stress intensity of interface of foundation and upper slope section, but also the stability against sliding and stress intensity of each layer of the roller compacted zones of RCC, and the influence of the shape of the valley to the objective function should be taken into consideration, using a combination of more variables to calculate and analyze, to find the dam shape and the profiles which will meet the requirements of stress and stability, easy construction and least cost. All of above can lead to obtain reasonable parameters of dam body, zoning of materials and properties parameters of materials.

In addition to the traditional material mechanics method and the rigid body limit equilibrium method, but for high concrete gravity dams, finite element method must also be used to calculate state and possible variations of stress and strain between dam and dam foundation, and to review the safety of the possible sliding surface and the weakest part of the dam. Based on the latest developments in elastic-plastic mechanics and nonlinear fracture mechanics, using the nonlinear elastic theory, elastic-plastic incremental theory and nonlinear fracture theory to study the stress state and bearing capacity of gravity dam has become an important research subject.

For the gravity dams with the low-angle weak intercalations in the foundation, because of their geological defects, the safety evaluation and treatment of dam foundation is very difficult. So the stability against sliding of deep foundation has been an important research subject in the dam design. As the failure mechanism of deep sliding of gravity dams and the safety evaluation criteria of stability against sliding based on nonlinear finite element is not involved in current design standards, in-depth further studies should be made. The limit state expression or formula for the stability against sliding should be presented to cope with the current design standards, and the accompanying index of partial factor (including structural factor) or design safety factor evaluation should be established too.

References

[1] Pan Jiazheng. *Gravity Dam Design*. Beijing: China Water & Power Press, 1987, 12.
[2] Zhu Jingxiang, Shi Ruifang. *Hydropower Engineering In China Hydraulic Volume*. Beijing: China Electric Power Press, 2000, 8.

[3] Tan Jingyi. *Hydropower Engineering In China (Construction Volume)*. Beijing: China Electric Power Press, 2000, 8.
[4] Pan Jiazheng, He Jing. *Large Dams in China Fifty Year Review*. Beijing: China Water Power Press, 2000, 9.
[5] Zhou Jianping, Niu Xinqiang, Jia Jinsheng. *Design of Concrete Gravity Dams-20Years*. Beijing: China Water Power Press, 2008, 3.
[6] Ministry of Water Resources International Cooperation about Science and Technology. *Contemporary Water Front*. Beijing: China Water Power Press, 2006, 1.
[7] Sun Gongyao, Wang Sanyi, Feng Shurong. *High Roller Compacted Concrete Gravity Dam*. Beijing: China Electric Power Press, 2004, 3.
[8] Cao Zesheng, Xujinhua. *MgO Concrete Dam Construction Technology*. Beijing: China Electric Power Press, 2003, 6.
[9] Zhang Yanming, Wang, Shengpei, Pan Luosheng. *Roller Compacted Concrete Dams in China-20 Years Perspective*. Beijing: China Water Power Press, 2006, 5.
[10] Mei Jinyu, Zheng Guibin. *Summary on the RCC Dam Construction Techniques in China*. RCC DAMS technology exchange meeting, 2003.
[11] Zhen Shouren, Wang Shihua, Xia Zhongping. *Closure and Diversion Cofferdam*. Beijing: China Water Power Press, 2005, 1.
[12] Zhang Chaoran. *Water Resources and Hydropower Engineering Concrete Construction Manual*. Beijing: China Water Power Press, 2002, 12.
[13] Zhang ye, Yin Longsheng. *Water Resources and Hydropower Engineering Construction Manual-Metal Structure Fabrication and Mechanical Electrical Installation*. Beijing: China Water Power Press, 2004, 9.
[14] Three Gorges Project Corporation. *Three Gorges Water Control Project Phase II (135m) and Navigation Lock and Acceptance Test Summary Document Technical Design Review*, 2003, 5.
[15] Changjiang Water Resources Commission. *Ministry of Water Resources*, Three Gorges Project Technical Design Report, Volume I, 1994, 11.
[16] Three Gorges Project Corporation. *Construction Report on Water Storage and Lock Trial Navigation Inspection of Three Gorges Phase II Project*, 2003, 5.
[17] Three Gorges Project Corporation. *Dam Project Construction Report on Three Gorges water control project (135m) and Lock Trial Navigation Acceptance Inspection File*, 2003, 5.
[18] Three Gorges Project Corporation. *Construction Report on Yangtze River Three Gorges Water Control Project Phase III (156m) Acceptance Documents*, 2006, 8.
[19] Three Gorges Project Corporation. *Construction Report on Three Gorges Water Control Project (156m) Inspection File*, 2006, 8.
[20] Zhu Jianye. *The Development and Prospects of the Hydropower Engineering Geology in China*. Water Power, 2004, 12.
[21] Chen Deji, Cai Yaojun. *China WaterResources and Hydropower Engineering Geology*. Resources Environment and Engineering, 2004, 9.
[22] Wang Sijing. *China of Rock Mechanics and Engineering Achievements*. Nanjing: Hohai University Press, 2004, 9.
[23] Engineering Geology Professional Committee of China Geological Society. *Engineering Geology in China-50 Years Review*. Beijing: Earthquake Press, 2000, 9.
[24] Wu Faquan. *The Summary of Western Hydropower Development and Karst Hydrogeology Engineering Geology*. Journal of Engineering Geology, 2004, 12.
[25] Fan Zhongyuan. *Application Summary of Engineering Geologic Reconnaissance Method and Technology for Water Conservancy and Hydropower Engineering*. Yangtze River, 2005, 3.
[26] Wang Zigao. *Application of New Technologies and Development Programming for Engineering Investigation*. Water Power, 2006, 11.
[27] Lu Youmei, Jia Jinsheng. *Large Dam Construction in China State of the Art and Case Histories*. Beijing: China Water Power Press, 2007, 12.
[28] Zhu Bofang. *Theory and Technology Progress of Concrete Dam*. Beijing: China Water Power Press, 2009, 4.

Chapter 10

Arch Dam Construction Technique

Zhu Bofang[1]

10.1 Introduction

After the founding of new China in 1949, with the rapid development of water conservancy and hydropower industry, the first batch of high concrete arch dams such as Xianghongdian Arch Dam (87.5m high) and Liuxihe Arch Dam (78m high) were built in the 1950s. Afterwards, the construction of arch dams was developing rapidly. According to statistics of *World Register of Dams* (1988), there had been a total of 1592 arch dams with height above 15m built all over the world by 1988, in which 753 arch dams were built in China, accounting for 47.3%; and according to statistics of Chinese National Committee on Large Dams, currently (by September 2010), China had a total of 857 arch dams with height over 30m, and 5 arch dams higher than 200m, namely Xiaowan (294.5m), Laxiwa (250m), Ertan (240m), Goupitan (232.5m) and Dagangshan (210m), wherein Xiaowan Arch Dam is the world's highest arch dam. A large number of arch dams are

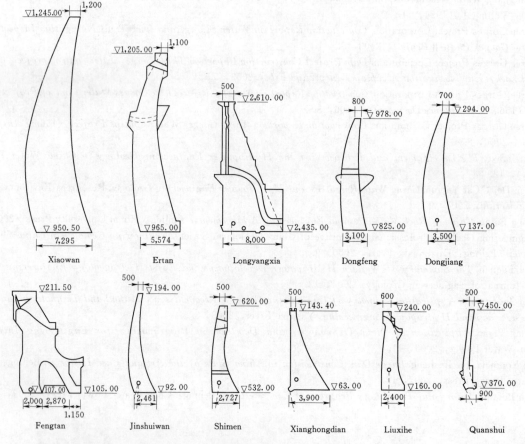

Fig. 10.1-1 Typical Profiles of Some High Arch Dams Built in China (Elevation Unit: m, Size Unit: cm)

---

[1] Zhu Bofang, Academician of the Chinese Academy of Engineering, China Institute of Water Resources and Hydropower Research, Professor Senior Engineer.

Chapter 10 Arch Dam Construction Technique

Table 10.1-1 Characteristics Table of Part of China's Arch Dams Constructed and Under Construction

Dam name	Location		Dam shape	Dam height (m)	Arc length at dam Crest (m)	Thickness of arch at crown (m)		T/H	L/H	Central angle (Degree)	Maximum stress (MPa) (Load distribution method)		Dam foundation of geological Condition	Year of building
	River	Place				Top	Bottom T				Tensile Stress	Compressive Stress		
Jinping I	Yalong River	Sichuan	Parabolic Double Curvature Arch	305	552.25	14	63	0.207	1.81	93.55	1.19	7.77	Marble, sandy slate	2014
Xiaowan	Lancang River	Yunnan	Parabolic Double Curvature Arch	294.5	892.79	12	72.912	0.25	3.058	90.1	1.18	10.37	Hornblende plagioclase gneiss, biotite granitic gneiss	2010
Xiluodu	Jinsha River	Sichuan/Yunnan	Parabolic Double Curvature Arch	285.5	681.51	14	60	0.216	2.451	95.6			Basalt	Under construction
Laxiwa	Yellow River	Qinghai	Logarithmic Spiral Double Curvature Arch	250	475.8	10.0	49	0.196	1.90	92.4	0.82	7.32	Granite	2010
Ertan	Yalong River	Panzhihua	Parabolic Double Curvature Arch	240	769.00	11.0	55.75	0.232	3.21	91.5	0.99	8.82	Basalt, syenite	2000
Goupitan	Wujiang River	Yuqing, Guizhou	Parabolic Double Curvature Arch	232.5	552.55	10.25	50.28	0.216	2.38	88.0	1.15	7.46	Limestone	2009
Dagangshan	Dadu River	Shimian, Sichuan	Parabolic Double Curvature Arch	210	622.42	10	52	0.248	2.964	93.4	1.02	6.41	Granite	Under construction
Longyangxia	Yellow River	Gonghe, Qinghai	Gravity Arch	178	393.34	15	80	0.45	2.21	85.1	1.76	6.36	Limestone	1989
Lijiaxia	Yellow River	Jianzha, Qinghai	Three-center Circle Double Curvature Arch	155	407.2	8.00	47	0.285	2.47	95.4	1.24	7.52	Schist migmatite	1997
Dongjiang	Laishui	Hunan	Double curvature Arch	157	442.66	7.0	35	0.223	2.82	95.0	0.98	6.86	Granite	1992
Dongfeng	Wujiang River	Qingxi, Guizhou	Double curvature Arch	162	254	8.0	25	0.163	1.66	95.0	1.63	4.94	Limestone	1995
Geheyan	Qingjiang River	Changyang, Hubei	Three-center Gravity Arch	151	654	8.0	75.5	0.5	4.33	82.3	1.04	5.41	Acendrada In shilongdong	1995

Continued

Dam name	Location		Dam shape	Dam height (m)	Arc length at dam Crest (m)	Thickness of arch at crown (m)		T/H	L/H	Central angle (Degree)	Maximum stress (MPa) (Load distribution method)		Dam foundation of geological Condition	Year of building
	River	Place				Top	Bottom T				Tensile Stress	Compressive Stress		
Baishan	Second Songhua River	Huadian, Jilin	Three-center Gravity Arch	149.5	676.5	9.0	63.7	0.426	4.53	80.3	0.98	6.57	Migmatite	1986
Jiangkou	Furongjiang River	Chongqing	Elliptic Double Curvature Arch Dam	140	155.61	6	15	0.232	2.41	92.8	1.01	6.34	Limestone	2004
Tengzigou	Longhe River	Shizhu, Chongqing	Elliptic Double Curvature Arch Dam	124	335.44	5	20.01	0.171	2.87	90.5	1.17	5.79	Sandstone mudstone	2006
Jinshuitan	Longquan River	Yunhe, Zhejiang	Three-center Double Curvature Arch	102	350.6	5.0	24.6	0.241	3.44		1.84	6.13	Granite porphyry	1988
Shimen	Baohe River	Hanzhong, Shaanxi	Double curvature Arch	88	260	5.0	27.27	0.31	2.95	129	0.98	3.92	Quartzite, mica	1973
Dayikeng		Qingtian, Zhejiang	Mixed Curve Arch	86.8	155	4.96	10.24	0.118	1.78	92.8				2001
Xianghongdian	Xipi River	Jinzhai, Anhui	Gravity Arch	87.5	367.6	6.0	39.0	0.45	4.2	117	0.74	2.94	Syenite, tuff	1961
Quanshui	Tangpenshui River	Ruyuan, Guangdong	Double curvature Arch	80	148.34	3.0	9.0	0.112	1.75	76.4	0.78	5.93	Granite	1976
Liuxihe	Liuxi River	Conghua, Guangdong	Single Curvature Arch	78	255.52	2.0	24	0.31	3.275	120	0.92	3.18	Granite	1959
Xiwei		Ningde, Fujian	Unified Quadratic Curve Arch	65	130	3.0	10.43	0.16	2.00	98				1999
Zhaixiangkou	Maotiao River	Xiuwen, Guizhou	Double curvature Arch	54.77	151.9	3	8.72	0.159	2.78	113	1.12	8.60	Limestone	1970
Fengle	Fengleshui River	Anhui	Double curvature Arch	54	214.58	2.5	12.5	0.23	3.91	136	0.73	2.84	Sandstone shale	1977

still under construction, including Jinping First-stage (305m), Xiluodu (285.5m), etc. The characteristics of part of China's arch dams are shown in Table 10.1 - 1. The typical profiles of some high arch dams are shown in Fig. 10.1 - 1.

After 60 years of efforts, China has made brilliant achievement in the construction of arch dams, not only ranking first in the number and height of arch dams in the world, but also solving many complex problems in the construction process, and making great achievements in damming technique. The shape of arch dams has many types: single curvature, double-curvature, single-center circle, multi-center circle, parabola, ellipse, logarithmic spiral and so on, and two new arch forms of unified quadratic curve and mixed curve are proposed; and shape design has developed to optimization design from manual design. China has developed linear and nonlinear, static and dynamic, whole-process simulation stress analysis program including the finite element method as well as the trial load method and proposed a finite element equivalent stress method, so as to successfully solve temperature control and crack prevention, high-flow flood discharge and energy dissipation in narrow valley, seepage control for high arch dams in karst regions, reinforcement treatment for complex foundation and etc. China has also built a large number of masonry arch dams, accumulating rich experience in the design and construction of masonry arch dams. Some problems, such as temperature control and crack prevention, flood discharge and energy dissipation, etc. will be elaborated in separate chapters of the book; therefore, this chapter will not repeat.

10.2 Shape Design and Optimization for Arch Dam

Arch dam is a plain concrete shell structure with variable thickness and curvature. Water load acting on the dam is transferred to bedrock at both sides mainly through the arch. By fully utilizing the compressive strength of concrete, dam body has relatively thin cross-section. As long as bedrock on both sides is stable and reliable, the safety of arch dam is higher than various other types of dams, and its earthquake resistance is also better. However, the requirements on topography and geological conditions are higher; the design method is also more complicated.

Experience has shown that the design of shape has an important impact on safety and economy of arch dam. Traditional manual design method relies mainly on experience of engineers and repeated modifications. With the use of optimization method and electronic computer, the optimal shape under the given conditions can be obtained by utilizing optimization program developed in recent years, so as to improve quality and efficiency of design.

The core problem in the design for shape of arch dam is to find the most economical and practical shape under the premise that both stress and anti-sliding stability of dam body meet stipulated requirements.

China's early construction focuses on single-center circle arch dam with uniform thickness, and the shape of arch dam is diversified in later stage, with extensive use of the arch axis of variable three-center circle, parabola, ellipse, logarithmic spiral, unified quadratic curve and mixed curve with variable thickness.

10.2.1 *Diversification of Arch Dam Shape*

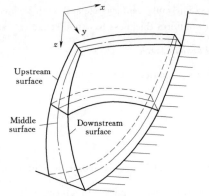

Fig. 10.2 - 1 Shape of Arch Dam

In theory, the shape of arch dam can be determined by two functions for expressing the middle surface and the thickness of the arch dam respectively, for example (Fig. 10.2 - 1):

The middle surface: $y_{mid} = b_1 x + b_2 x^2 + b_3 x^2 z + b_4 z + b_5 z^2$
The thickness: $t = b_6 + b_7 x^2 z + b_8 z$

(10.2 - 1)

In the above formula, b_1-b_8 are eight coefficients on which the shape of arch dam entirely depends. It is obvious that the above formula has the following characteristics: ①in the vertical plane (x=constant), the cantilever axis is a quadratic curve, and there is a linear variation in thickness; ②in the horizontal plane (z=constant), the arch axis is a quadratic curve, and the thickness is also a quadratic curve; and ③the dam axis does not shift, but can be rotated (via b_1 item). The order of variations in middle surface and thickness can also be improved by appropriately adding some items.

In practical engineering design, the shape of arch dam is determined through the description of crown cantilever (vertical) profile and all layers of horizontal arch rings (horizontal profile). Vertical curvature of the arch dam has greater impact on stress, while horizontal curvature shows more influence on anti-sliding stability. This method has the advantage of separating the vertical curvature from the horizontal curvature to facilitate separate analysis and adjustment, and also facilitate comparison between different projects and different schemes, thus being used widely.

10.2.1.1 Profile of Crown Cantilever

In the U-shaped valley, sometimes single-curvature arch dam can be adopted, the upstream surface of dam is cylindrical-shaped. In non-U-shaped valley, double-curvature arch dam is generally adopted, the horizontal coordinate y_u of upstream surface and the thickness t_c of the crown cantilever are respectively expressed as follows:

$$y_u = \sum_{i=0}^{n} b_i z^i, t_c = \sum_{i=0}^{m} c_i z^i \tag{10.2-2}$$

where b_i and c_i are coefficients. The coordinate of the centerline of crown cantilever y_c can be calculated by:

$$y_c = y_u + t_c/2$$

10.2.1.2 Horizontal Arch Ring

The horizontal arch in early stage is generally single-center circle. Afterwards, to improve stress state and stability conditions, these forms such as double-center circle, three-center circle, parabola, ellipse and logarithmic spiral are gradually adopted, with variations in radius and thickness in the horizontal direction.

In the design of arch dam, the central angle of arch ring is an important factor. If only stress constraint is considered, the optimum central angle is 133.6 degrees where the arch is calculated according to circular hoop theory, and 140 degrees or so where the arch is calculated based on elastic arch, concrete can be saved to utmost extent; but if stress and anti-sliding stability constraints are considered simultaneously, the optimal central angle is generally the maximum central angle allowed under anti-sliding stability conditions, which is generally controlled by the anti-sliding stability conditions. Thus, the maximum central angle of arch dam constructed in early stage is generally 110 degrees to 130 degrees. The anti-sliding stability conditions attracts more attention after failure of Malpasset Arch Dam in France, and the maximum central angle of the arch dam is generally reduced from 90 degrees to 110 degrees, which is the so-called "flattening of arch dam".

The forms of various horizontal arches are described as follows.

1. Circular arch

The forms of circular arch dam include single-center circle, double-center circle, multi-center circle, etc. The upstream and downstream surfaces of single-center circle arch are two concentric circles with equal thickness and curvature in the horizontal direction. There are two types of double-center circle arches, namely arch with uniform thickness and double-center arch, wherein the radii of left and right sides of the arch with uniform thickness are different from each other to adapt to asymmetric valley; and the radius of the left side of the double-center arch is the same as that of the right side of the double-center arch, but the radius of the upstream surface is greater than that of the downstream surface, with variable arch thickness. The multi-center circle arch with 3-5 centers of circle is generally divided into three sections, wherein the middle arch with uniform thickness has smaller radius, and side arch with variable thickness has greater radius.

2. Parabolic arch

Arch axis can be expressed as follows with parabola:

$$\left. \begin{array}{ll} \text{Right semiarch} & y = \dfrac{x^2}{2R_R} \\ \text{Left semiarch} & y = \dfrac{x^2}{2R_L} \end{array} \right\} \tag{10.2-3}$$

where R_R and R_L are the radius of right semiarch and left semiarch at $x=0$ respectively.

Generally, the thickness of arch ring is gradually thickened from arch crown to abutment, which can be expressed with one of the following formulas:

$$\left. \begin{array}{l} t(x) = t_C + (t_A - t_C)(x/x_A)^\gamma \\ t(\varphi) = t_C + (t_A - t_C)(\varphi/\varphi_A)^\gamma \\ t(s) = t_C + (t_A - t_C)(s/s_A)^\gamma \end{array} \right\} \tag{10.2-4}$$

where, t_C and t_A are the thickness at the crown and abutment respectively; x_A, φ_A and s_A are horizontal coordinate, central angle and arc length of the abutment respectively; γ is a constant ranging from 1.5 to 3. Fig. 10.2-2 is the shape of Xiaowan Arch Dam.

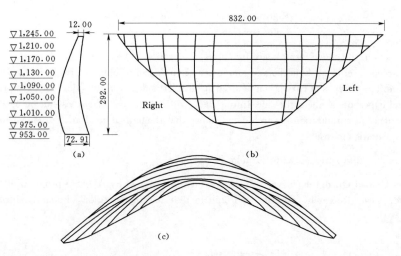

Fig. 10.2 - 2 Shape of Xiaowan Parabolic Arch Dam (Elevation Unit: m, Size Unit: m)

3. Elliptical arch and hyperbolic arch

The centerline of arch can be expressed as follows:

$$\left.\begin{array}{ll} \text{Right semiarch} & x^2 \pm c_R y^2 = b_R y \\ \text{Left semiarch} & x^2 \pm c_C y^2 = b_C y \end{array}\right\} \quad (10.2-5)$$

In the above formula, "+" refers to ellipse, and "−" refers to hyperbola. The thickness of the arch is still expressed with formula (10.2-4). The shape of Jiangkou Elliptical Arch Dam is shown in Fig. 10.2-3.

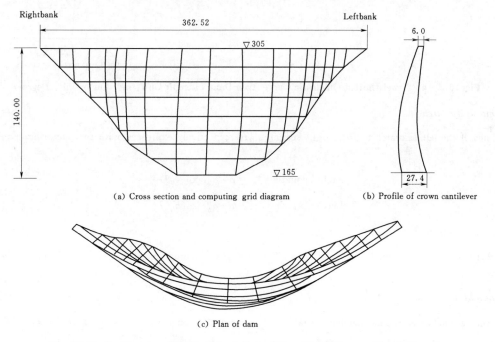

Fig. 10.2 - 3 Jiangkou Elliptical Arch Dam (Elevation Unit: m, Size Unit: m)

4. Unified quadratic curve arch dam

Single-center circle, parabola, ellipse and hyperbola are quadratic curves which can be expressed with the following unified formula:

$$x^2 = ay^2 + by \quad (10.2-6)$$

The curve is a hyperbola, an ellipse and a circle respectively when $a > 0$, $a < 0$ and $a = -1$. By taking unified quadratic curve as the centerline of arch, the left semiarch and the right semiarch can be respectively expressed as follows:

$$\left.\begin{array}{ll}\text{Right semiarch} & x^2+a_R y^2+b_R y \\ \text{Left semiarch} & x^2+a_L y^2+b_L y\end{array}\right\} \quad (10.2-7)$$

All the four parameters in the above formula are functions of elevation. When the value of a_R and a_L varies along with changes in elevation, parabola, ellipse, hyperbola or single-center circle can be obtained respectively. Due to the adoption of different parameters, different types of curves can be used at both left side and right side, for example, ellipse can be adopted at the left side, and hyperbola at the right side. In one dam, different types of curves can be used at different part of the dam. Practical experience in optimization for arch dams shows that the unified quadratic curve is generally the most economical among various quadratic curves.

The thickness of arch is still expressed with formula (10.2-4).

The shape of Xiwei Unified Quadratic Curve Arch Dam which is designed by ADASO program of China Institute of Water Resources and Hydropower Research (IWHR) and built in 1999 in Ningde region, Fujian is shown in Fig. 10.2-4.

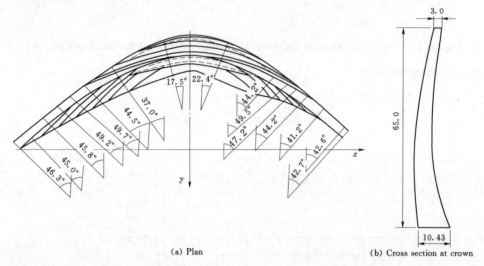

Fig. 10.2-4 Xiwei Unified Quadratic Curve Arch Dam (Length Unit: m, Angle Unit: Degree)

5. Logarithmic spiral arch

The centerlines of the left semiarch and the right semiarch are respectively expressed with two logarithmic spirals as follows:

$$\left.\begin{array}{ll}\text{Right semiarch} & \rho=\rho_{OR}\exp(k_R\theta) \\ \text{Left semiarch} & \rho=\rho_{OL}\exp(k_L\theta)\end{array}\right\} \quad (10.2-8)$$

where, ρ is the radius of curvature.

Laxiwa Arch Dam at the upstream of Yellow River is 250m high, and adopts logarithmic spiral arch. The shape of the arch dam is shown in Fig. 10.2-5.

6. Mixed curve arch

Taking the central angle φ as the independent variable, all kinds of the above curves can be expressed as follows:

Circular arc	$R=$ constant
Logarithmic spiral	$R=R_0 e^{k\varphi}$
Ellipse	$R=R_0/(\cos^2\phi+\alpha\sin^2\phi)^{3/2}$
Parabola	$R=R_0/\cos^2\phi$
Catenary	$R=R_0/\cos^2\phi$
Hyperbola	$R=R_0/(\cos^2\phi-\alpha\sin^2\phi)^{3/2}$

The radius of curvature of a mixed type arch ring can be summarized as follows:

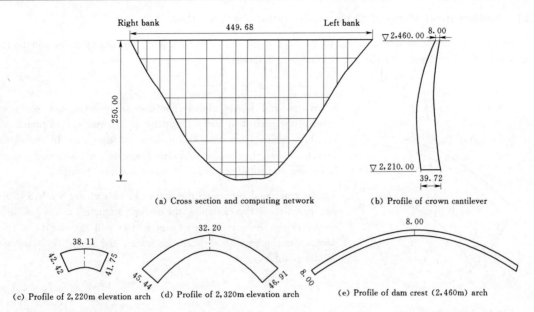

Fig. 10.2-5 Optimized Shape of Laxiwa Logarithmic Spiral Arch Dam (Elevation Unit: m, Size Unit: m)

$$R = R_0 e^{k\phi} / (\cos^2\phi + \alpha \sin^2\phi)^\beta \tag{10.2-9}$$

where, α, β, k, R_0 are parameters to be selected with different values. R_0 is the radius of curvature of arch ring at the crown cantilever. When the two banks of valley are asymmetrical, different radius of curvature should be adopted at both sides of arch dam. The Dayikeng Mixed Curve Arch Dam in Qingtian, Zhejiang designed by Zhejiang University is shown in Fig. 10.2-6.

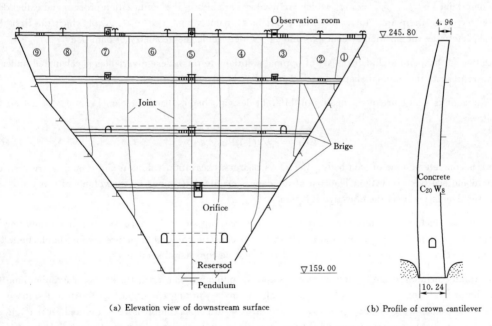

Fig. 10.2-6 Dayikeng Mixed Linear Arch Dam (Elevation Unit: m, Size Unit: m)

10.2.2 Shape Optimization of Arch Dam

In the traditional design for shape of arch dam, a feasible program (meeting the design requirements) is obtained by method of repeated modifications, and thus is generally not an optimal solution. In the optimization shape of arch dam, the optimal design shape under given conditions is obtained by mathematical programming method, which is a new technology developed during the last 20 years in China.

10.2.2.1 Mathematical Model of Shape Optimization for Arch Dam

The valley shape is arbitrary, as shown in Fig. 10.2-7, C point is an upstream point (or central point) at the top of the crown cantilever, and horizontal coordinate of C Point is as follows:

$$x' = x_1, y' = x_2$$

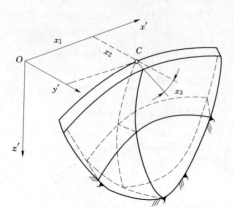

Fig. 10.2-7 Arch Dam

where, x' and y' are global coordinates. The angle between a reference plane of the arch dam and $y'Oz'$ plane is x_3. And x_1, x_2 and x_3 are three variables which determine the position of dam axis. In the optimization process, when they are continuously changing, the dam axis is continuously moving and rotating to find the most favorable position.

The parameters determining the shape of arch dam are referred to as the design parameters. For example, the design parameters of the profile of crown cantilever are the curve of upstream surface and the thickness. The design parameters change with vertical coordinate z and can be expressed with polynomial as follows:

$$\left. \begin{array}{l} f = a_0 + a_1 \eta + a_2 \eta^2 + \cdots + a_m \eta^m \\ \eta = z/H \end{array} \right\} \quad (10.2-10)$$

where, f is the design parameter; z is the vertical coordinate; H is the height of dam; and $a_0, a_1, \cdots, a_m$ are coefficients.

In the formula (10.2-10), let in seguence $\eta=0$, $\eta=\eta_1$, $\eta=\eta_2$, $\cdots$, $\eta=\eta_m$, we get

$$f = f_0, f_1, f_2, \cdots, f_n$$

Thus $a_0, a_1, \cdots, a_m$ can be obtained through inversion of the above equation. It is evident that the design parameter $f(z)$ is completely determined by $f_0, f_1, \cdots, f_m$ which are used as variables in the optimization process and called design variables. For the whole dam, there are a total of n design variables, namely $x_1, x_2, \cdots, x_n$, of which the first three design variables $x_1 - x_3$ determine the position of dam axis and $x_4 - x_n$ determine the geometry of dam body.

In the optimization for shape of arch dam, a set of optimal solutions to the n design variables is calculated under given conditions, which represents the optimal shape of arch dam under the given conditions.

The objective function is a measure of the quality of different design schemes; and the dam's cost can be selected as it, expressed as follows:

$$C(x) = c_1 V_1(x) + c_2 V_2(x) \quad (10.2-11)$$

where, $V_1(x)$ is concrete volume of dam body; $V_2(x)$ is bedrock excavation volume; and c_1 and c_2 are respectively unit price of concrete and bedrock excavation. The cost of the dam is mainly determined by the volume of concrete, so generally the volume of the dam is taken as the objective function.

Constraint conditions, including geometric constraint, stress constraint and stability constraint, must fully meet the provisions of design specifications for arch dam, the requirements of construction and structure layout should be taken into account, and some special requirements of the projects also should be considered sometimes.

Geometric constraints include: range of movement of dam axis, minimum dam crest thickness, maximum dam base thickness, and maximum overhang of dam body surface. For all the main working conditions specified in design specifications, under the action of various design loads, principal tensile stress, principal compressive stress and anti-sliding stability of the dam body must meet the requirements; and the thrust angle of abutment and the central angle of arch ring meet the requirements.

In summary, the optimization for shape of arch dam is to find design variables, which satisfies all the constraint conditions, and leads to the minimum cost of the dam.

Expressed in mathematical language, the shape optimization of arch dam is to find the solution x such that to subjecte:

$$\left. \begin{array}{l} C(x) \rightarrow \min \\ g_j(x) \leqslant 1, j = 1 - m \end{array} \right\} \quad (10.2-12)$$

where, min is the minimum cost of dam; and m is the number of constraint conditions.

This is a mathematical programming problem.

10.2.2.2 Internal Force Expansion Method

In the arch dam optimization process, stress analysis needs to be conducted for thousands of design schemes for shape, and most of the computing time is spent in stress analysis in the optimization process. How to reduce time for stress analysis is a key issue in the optimization for arch dam. Previously, stress expansion method proposed by Schmit was mainly adopted, with very slow convergence, and currently, internal force expansion method proposed by China Institute of Water Resources and Hydropower Research (IWHR) is adopted, with very fast convergence.

Internal force is balanced with the load. When the structure size changes, the load basically remains the same, so the internal force varies within a small range. Based on this idea, IWHR puts forward the internal force expansion method, expanding the internal forces of structural control points (including axial force, bending moment, shearing force, torque, etc.) to be first-order Taylor series as follows:

$$F(x) = F(x^k) + \sum_{i=1}^{n} \frac{\partial F}{\partial x_i}(x_i - x_i^k) \qquad (10.2-13)$$

where $F(x)$ refers to the internal force.

In the optimization process, for any new design program, the internal forces of control points are calculated with the above formula instead of conducting routine stress analysis, and then the principal stress of all the control points is calculated by the formulas of material mechanics.

Experience has shown that the calculating efficiency by internal force expansion method is very high, and Fig. 10.2-8 shows the convergence rate of the two methods, in which the convergence rate of internal force expansion method is much faster. The internal force expansion method is a common method, not only applicable to arch dam, but also can be used in the optimization for other structures. In the process of optimization for arch dam, the anti-sliding stability factor of dam body also can be subjected to Taylor expansion. For optimization methods, penalty function method and sequential quadratic programming method are mainly adopted at present.

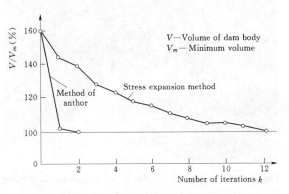

Fig. 10.2-8 Convergence Rate of Optimization for Arch Dam

10.2.2.3 Engineering Application of Optimization for Arch Dam

Many foreign countries such as Britain, Russia, Germany and Portugal are conducting research on optimization for arch dam. But so far, no research has been applied in practical engineering. The main reason is that the mathematical model is not practical.

IWHR takes the lead in optimization for arch dam in China, paying special attention to establish a reasonable and practical mathematical model, and applying optimization for arch dam in practical engineering step by step, from small to medium-sized, then to large and extra large, with continuous improvement of mathematical models and calculation methods in the application. So far, the optimization has been used in more than 100 practical projects, generally saving concrete of dam body by 5% to 30%, thus obtaining significant economic and social benefits.

Ruiyang Arch Dam in Longquan County, Zhejiang Province has maximum height of 54.5m, and length of dam crest chord being 140m. The volume of originally designed dam body is 38,200m³, which is reduced to 26,500m³ after being optimized by IWHR, saving by 30.6%. The dam was constructed in full accordance with optimally designed shape, and completed in 1987 for water storage with normal operation. The Ruiyang Arch Dam is named Outstanding Engineering in Zhejiang Province, as shown in Fig. 10.2-9.

Fig. 10.2-9 Ruiyang Arch Dam - the First Arch Dam Designed by Optimization Method in the World (Completed in 1987 and Fine Operation)

Ruiyang Arch Dam is the world's first arch dam designed practically by the optimization method, and its success leads to rapid popularization of the optimization method for arch dam in China, gradually from the small and medium-sized arch dams to high arch dams.

For example, Laxiwa Arch Dam at the Yellow River is 250m high. Optimal shapes of 6 kinds of arches such as double-center circle, three-center circle, parabola, hyperbola, ellipse and logarithmic spiral are obtained in a short time by the optimization method, and the logarithmic spiral arch dam is selected. According to the calculation of the Northwest China Electric Power Design Institute, compared with the traditional design, concrete was saved by 200,000m³, and bedrock excavation by 130,000m³.

Located on the outskirts of Jiangkou Town, Wulong County, Chongqing City, Jiangkou Hydropower Station has installed capacity of 300MW, and dam height is 140m. The dam was originally designed as a parabolic arch dam, with concrete of dam body of 798,700m³. Commissioned by the Northeast Design Institute, IWHR obtained the optimum shape of six kind of arch dam by self-compiled ADASO program: concrete of unified quadratic curve arch dam is 528,400m³, concrete of elliptic arch dam is 573,400m³, and concrete of parabolic arch dam is 610,000m³. The proprietor selected the optimal shape of elliptical arch dam. Compared with the original design shape, the concrete was saved by 28% (225,700m³), the construction period was shortened by six months, and the Jiangkou Hydropower Station had been built in 2003 for water storage, as shown in Fig. 10.2-10.

Fig. 10.2 - 10   Jiangkou Elliptic Arch Dam in Chongqing Designed by Optimization Method
(Built in 2003 with Height of 140m)

In the traditional dam design, generally one arch form is firstly chosen depending on experience of designers; then some comparison schemes are worked out for this arch form; and finally one scheme program is selected. In the optimization for arch dam shape, a variety of arch forms are optimized simultaneously to seek respective optimal shape, and then the best one is selected. Therefore, both working efficiency and design quality have been greatly improved.

In the planned economy era, although the investment was saved due to the optimization for arch dam, design institutes, especially large design institutes did not benefit from the optimization, thus limiting its application. Under the current market economic conditions, many engineering proprietors request for optimization, with the Jiangkou Arch Dam as one example. Therefore, the application of optimization for arch dam in engineering has become more active in recent years.

10.3   Stress Analysis of Arch Dam

Stress of arch dam is an important factor determining the thickness and shape of arch dam, which is different from gravity dam; therefore, for arch dam, analysis of stress is very important. In early stage, load distribution method of crown cantilever was mainly adopted. After 1970, due to the application of computers, load distribution method of multi-arch cantilever was mainly adopted, which had the advantages of simple computation, long application history, and more mature stress control standards, and the drawbacks of low precision caused by calculation of foundation deformation using Vogt coefficient, and inconvenience in simulation analysis and nonlinear analysis.

After 1970, Finite Element Method (FEM) was gradually used for analysis of stress of arch dam, due to its powerful computing capabilities. In the calculation, complex foundation, construction process of dam body, material nonlinearity and many other factors can be considered. However, the problem of stress concentration exists in finite element calculation, and tensile stress at dam heel is often several times greater than tensile strength of concrete, making it difficult to determine the design shape of arch dam. The finite element equivalent stress method and corresponding stress control standards proposed by IWHR were adopted in the newly compiled design specifications for arch dam which created the condition for adopting the finite element method for arch dam design.

10.3.1   *Impact of Various Factors on Stress of Arch Dam*

In accordance with current design specifications for arch dams, design loads of arch dams mainly include: water load, self-weight of dam body, and changes in average temperature of dam body T_m and equivalent linear temperature difference T_d after arch closure. There are some differences between the actual situations and the loads and calculation methods in the design specifications. The impact of these differences on the stress of dam will be described in this section.

10.3.1.1 Self-weight stress

In the calculation by load distribution method of arch and cantilever, it is generally assumed that all the self-weight is borne by the cantilever; and in the calculation with the finite element method, in the past, it is generally assumed that the self-weight is applied to the whole dam at one time after the whole dam body is formed. Both of the two calculation methods are not realistic. In fact, the generation of self-weight stress is closely related to the construction process and joint grouting. The self-weight stress of dam body can be calculated using incremental method taking the construction process and joint grouting into consideration. The whole dam height is divided into n layers, and the self-weight stress is calculated layer by layer from bottom to top. In the first layer ΔH_1, all the self-weight is borne by the cantilever and then joint grouting is conducted layer by layer. When grouting has been carried out below H_i, the part below the H_i is taken as a whole with the self-weight of zero. In the $(i+1)$ the layer, self-weight of ΔH_{i+1} exists at each block, standing on the arch dam with a height of H_i, causing stress increment $\Delta \sigma_{i+1}$ at this moment. Transverse joints are simulated with a joint element.

Li et al (2000) have conducted analysis on the self-weight stress of the Xiaowan Arch Dam during the construction period. The transverse joints are simulated with a joint element. The dam body is divided into 16 layers in the height direction, and calculation results are shown in Table 10.3-1 and Table 10.3-2. Table 10.3-1 shows all the tensile stress values and tensile stress ranges are significantly reduced in the distributed calculation by comparison with overall calculation.

Table 10.3-1 Comparison between Stresses of Overall Calculation and Distributed Computing of Self-Weight of Xiaowan Arch Dam by FEM (MPa, Normal Tensile Stress)

Scenario	Upstream surface		Downstream surface		Cross section of crown cantilever				Range of tensile stress σ_1	Range of tensile stress σ_z	w_{max} (mm)	Calculation methods	Loads combination
	σ_{1max} (MPa)	σ_{zmax} (MPa)	σ_{zmin} (MPa)	σ_{3min} (MPa)	σ_{1max} (MPa)	σ_{zmax} (MPa)	σ_{zmin} (MPa)	σ_{3min} (MPa)					
2	10.28	7.760	-13.8	-22.9	7.425	5.974	-11.4	-16.6	26.5m deep	14.5m deep	199.44	Overall	Self-weight + water pressure + temperature drop
7	7.202	4.133	-13.0	-21.4	4.212	2.456	-10.8	-15.9	23.1m deep	8.3m deep	247.13	distributed	
3	10.18	7.698	-13.9	-22.9	7.356	5.923	-11.4	-16.6	26.3m deep	14.5m deep	190.56	Overall	Self-weight + water pressure + temperature rise
8	7.105	4.071	-13.0	-21.4	4.141	2.406	-10.8	-15.9	22.8m deep	8.2m deep	238.94	distributed	

Note: σ_{1max} is maximum principal stress on the dam surface; σ_{zmax} is maximum vertical normal stress; σ_{3min} is minimum principal stress on the dam surface (compressive stress with maximum absolute value); and w_{max} is maximum displacement (mm), and stress unit is MPa, tensile stress is positive.

Table 10.3-2 gives the maximum value of the finite element equivalent stress. In overall calculation, the maximum principal tensile stress on the upstream surface is 3.80MPa, which is reduced to 1.54MPa in distributed computing. Assuming that all the self-weight is borne by the cantilever, the maximum principal tensile stress is 1.18MPa in calculation by load distribution method of arch cantilever.

Table 10.3-2 Maximum Finite Element Equivalent Tensile Stress of Xiaowan Arch Dam

Calculation of self-weight	Loads	Principal tensile stress σ_1 (MPa)	Horizontal tensile stress σ_t (MPa)	Vertical tensile stress σ_z (MPa)
Overall calculation	Self-weight + water pressure + temperature drop	3.80	0.58	3.19
Distributed calculation	Self-weight + water pressure + temperature drop	1.54	0.16	1.12

10.3.1.2 Impact of Water Load of Surface of Foundation on Stress of Dam Body

In the calculation of stress of arch dam using the load distribution method of arch cantilever, only the water pressure on the dam surface is taken into account without considering the reservoir water pressure, as shown in Fig. 10.3-1 (a). In fact, as the dam foundation and dam body are integrated, when water pressure of the reservoir causes a larger deformation of the dam foundation, stress of arch dam body will be brought about. It has shown that the reservoir water pressure can greatly increase the tension range and tensile stress values of the arch dam body (Zhu et al, 2002). The reservoir water pressure is applied in two ways, namely a surface force directly applied on the surface of rock in the reservoir basin as shown in Fig. 10.3-1 (b) and a seepage volume force applied on the dam foundation as shown in Fig. 10.3-1 (c); if impermeable silt layer exist on the reservoir bottom, it can be assumed that reservoir water acts on the rock surface; otherwise, calculation based on the seepage volume force is preferable.

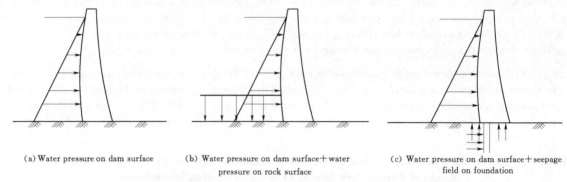

(a) Water pressure on dam surface (b) Water pressure on dam surface + water pressure on rock surface (c) Water pressure on dam surface + seepage field on foundation

Fig. 10.3-1 Scenario Comparison with different Ways of Water Load on Dam Foundation

How to consider the influence of drainage holes is a difficult point in the calculation of three-dimensional seepage field. The substructure method (Wang et al, 1992) is feasible to analyze the seepage field of single dam section, but still has practical difficulties in the analysis of seepage field of whole arch dam foundation due to too many drainage holes. Replacing the curtain of drainage holes by the equivalent drainage interlayer proposed by the author, the three-dimensional seepage field with draining holes can be analyzed by a common three-dimensional seepage program. The permeability coefficient of interlayer is determined according to the condition of equal average water heads; therefore the calculation is convenient and has higher calculation accuracy (Zhu et al, 2007).

The influence of method of application of water load on dam foundation on the stress of Xiluodu Arch Dam body calculated (Zhu et al, 2010) is shown in Table 10.3-3. After taking reservoir water pressure on the rock surface into consideration, maximum values of the principal tensile stress and vertical tensile stress increase by 10% to 20%, and the range of tensile stress on the dam foundation surface also increases; values of the principal tensile stress and vertical normal stress caused by reservoir water pressure which are calculated based on the surface force are greater than those calculated based on the seepage volume force within 10%. As the water permeability of rock mass on the Xiluodu Dam site in

Table 10.3-3 Influence of Method of Application of Water Load on Dam Foundation on the Stress of Xiluodu Arch Dam (Normal Tensile Stress)

	Loads	Upstream of profile of crown cantilever				Upstream of foundation surface	
		Maximum principal tensile stress (MPa)	Maximum vertical stress (MPa)	Range of principal tensile stress (m)	Range of vertical direct stress (m)	Maximum principal tensile stress (MPa)	Maximum vertical stress (MPa)
Normal water storage level	Water pressure on dam surface	10.01	8.70	37.2	32.2	18.82	10.43
	Water pressure on dam surface + water pressure on rock surface	12.09	11.08	59.5	29.8	15.94	12.49
	Water pressure on dam surface + seepage field on foundation	11.53	10.04	59.5	32.2	14.52	11.70

the vertical direction is weak, influences of the reservoir water pressure on the stress of dam body calculated based on the surface force or volume force are smaller, but the influence impact of the reservoir water pressure on the stress of arch dam should not be ignored.

10.3.2 Finite Element Equivalent Stress Method

The computing function of finite element method is far more powerful than the method of structural mechanics. As the calculation software is becoming more perfect, the present calculation is also very convenient, but in the calculation of arch dam stress with the use of three-dimensional elastic finite element method, significant stress concentration exists at positions near the foundation, and stress values increase sharply with the refinement of computing mesh, especially tensile stress calculated by the finite element method is far more than the tensile strength of concrete sometimes, making it difficult to directly use the results of the finite element method to determine the shape of arch dam. For an ideal elastic body, the aforementioned stress concentration phenomenon is understandable, but in the actual project, because of cracks of various sizes existing in rock mass, stress concentration will be relaxed. In the arch dam that has been built, except for individual special circumstances, cracks parallel to the foundation surface are not too much, and a large number of arch dams designed with the method of structural mechanics have been in normal operation, so serious stress concentration reflected in the calculation of arch dam with the finite element method does not necessarily conform to reality. As the impact of large orifice, complex foundation, gravity pier, irregular shape and many other factors can be considered and simulation calculations can be conducted due to the powerful functions of the finite element method, as long as the issue of stress control standards is solved, the application of the finite element method in the arch dam are bound to have good prospects.

At the International Symposium on Arch Dam held in Coimbra, Portugal in April 1987, Zhu (1988) proposed three solutions to the problem of stress concentration on foundation in finite element analysis: ①the finite element equivalent stress method, firstly the internal forces are calculated with the finite element method, and then the stress is calculated with the method of material mechanics; ②a thick shell element is used for the dam body, and a common element is used for the foundation; and ③ element reduced integration. These three methods can solve the problem of stress concentration, but the thick shell element is not conducive to the simulation calculation of temperature creep stress during construction period, and the reduced integration method may cause lower calculation accuracy, so that the equivalent stress method is the best among the three methods. Firstly the stress of arch dam is calculated with the finite element method, and then the internal forces of cantilever and arch are obtained through numerical integration. The stress of the dam body is calculated under the assumption of plane-section, which eliminates the stress concentration. The stress distribution on the foundation surface of crown cantilever of the Dahuashui Arch Dam is shown in Fig. 10.3-2. Some scholars have proposed that the stress distribution can be expressed with quadratic curve, its coefficient can be determined by taking the stress at three points, and then the internal forces of the arch and cantilever are obtained through integration. The quadratic curve has no inflection point; the typical stress distribution arch dam is shown in Fig. 10.3-2, which is a high-order curve with inflection points. The above algorithm is clearly impractical, with unnecessary error, and the numerical integration method is simple and accurate in the direct calculation of the internal forces.

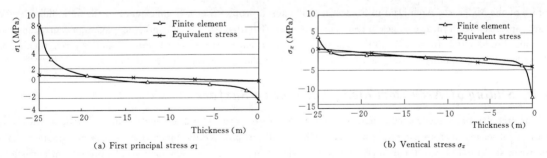

(a) First principal stress σ_1 (b) Ventical stress σ_z

Fig. 10.3-2 Comparison of Stress on the Foundation Surface of Crown Cantilever of the Dahuashui Arch Dam

In combination with the *Research on Stress Control Standards for High Arch Dams* of "9th Five-Year" National Science and Technology Research Plan, we have conducted calculation for more than a dozen arch dams both at home and abroad with multi-arch cantilever method and the finite element method simultaneously, and the calculation results are shown in Table 10.3-4. It can be seen from the above calculation results: ①the finite element equivalent stress method eliminates the impact of stress concentration, thus obtaining the maximum principal stress much smaller than the principal stress value directly obtained by finite element; ②the value of the maximum principal compression stress obtained with the finite element equivalent stress method is comparatively close to that of the maximum principal compression stress obtained with the multi-arch cantilever method; and③the maximum principal tensile stress derived with the finite element equivalent stress method is slightly greater than the principal tensile stress obtained with the multi-arch cantilever method.

Table 10.3 - 4   Analysis Results of Principal Tensile Stress of Arch Dams

Arch dam name	Maximum dam height (m)	Principal tensile stress of load distribution method of arch cantilever		Finite element method	Principal tensile stress of equivalent stress		Ratio of maximum principal stress of equivalent stress and load distribution method of arch cantilever	
		Upstream surface (MPa)	Downstream surface (MPa)	Principal tensile stress (MPa)	Upstream surface (MPa)	Downstream surface (MPa)	Maximum principal tensile stress (MPa)	Maximum principal compression stress (MPa)
Longyangxia	178	−1.09	−1.25	−3.62	−1.37	−1.26	1.10	0.91
Dongfeng	162	−1.03	−0.91	−3.97	−1.14	−0.48	1.11	0.81
Dongjiang	157	−1.17	−1.41	−6.82	−2.63	−1.69	1.87	1.06
Naichuandu	155	−1.71	−1.66	−4.87	−1.24	−1.37	0.80	1.13
Lijiaxia	155	−1.64	−1.28	−4.78	−1.82	−1.08	1.11	0.84
Baishan	149.5	−1.40	−1.25	−4.85	−1.49	−0.86	1.06	1.20
Limenbodeng	146	−0.77	−2.15	−2.92	−0.69	−1.36	0.63	0.73
Pulasimulin	143	−0.73	−1.58	−7.35	−2.21	−1.43	1.40	1.15
Shangzhuiye	110	−0.13	−1.75	−2.33	−0.81	−0.47	0.46	0.85
Ruiyang	50.5	−1.80	−1.82	−4.55	−2.51	−1.62	1.38	0.95

Note: Scenario is self-weight of dam body+water pressure+temperature drop, wherein "−" represents tensile stress.

At the review conference of the newly compiled Design Specifications for Concrete Arch Dams (SL 282—2003), the author systematically introduced the finite element method equivalent stress and the above calculation results, recommended that the finite element equivalent stress method should be brought into the new specifications, and proposed allowing the tensile stress to increase from 1.20MPa (normal load) to 1.50MPa (special load) in the load distribution method of arch cantilever to 1.50MPa (normal load) to 1.80MPa (special load). The author's proposals were adopted in the new standards, creating the conditions for the application of the finite element equivalent stress method in practical engineering.

10.4   Anti-sliding Stability of Arch Dam

The failure of the Malpasset Arch Dam in France in 1959 caused people to attach importance to the anti-sliding stability of arch dam, and resulted in three changes in arch dam projects: ①flattening of arch dam, the maximum central angle of arch dam is reduced to 80°- 100° from 110°- 130° in the past; ②enhancement of geological survey; and ③ strengthening of analysis on the anti-sliding stability of arch dam to improve the anti-sliding stability at dam abutment.

10.4.1   Sliding of Arch Dam along Foundation Surface

Most of the arch dams are constructed in narrow valleys. As the arch dam is an integral structure, as long as the strength of dam body is sufficient, rock mass at skewback is reliable, will not produce too much compressive deformation, or slip along the weak structural plane, the dam is impossible to slide along the foundation surface. The problem of anti-sliding stability of arch dam is essentially the problem of anti-sliding stability of rock mass at skewback. However, under the conditions of wider valley, comparatively moderate bank slope and thin dam body, the arch dam may also slide along the foundation surface or shallow seams under the foundation surface and become instable.

Chengdu Institute of Water Conservancy and Hydro-electric Power Design and Survey has conducted analysis on anti-sliding stability of 12 arch dams in China along the foundation surface, and the results have shown that under the action of normal loads, generally the arch dam will not slide along the foundation surface, and partial sliding along the foundation surface is possible to occur under the action of overload in particular topographic and geologic conditions. Therefore, the sliding stability of arch dam is practically controlled by the sliding stability of rock mass on abutment.

10.4.2 Anti-sliding Stability of Rock Mass on Dam Abutment

Many structural planes exist in rock mass, including joints, soft interlayers, faults, etc. In order to analysis the stability of skewback, firstly the main structural planes in rock mass should be investigated, and then unstable wedge may be formed by a group of structural planes and may slide under the action of arch thrust, shearing force, gravity and other external forces.

As shown in Fig. 10.4 – 1, if three structural planes P_1, P_2 and P_3 existing in the rock mass cut the rock mass at the skewback to form a wedge $OABC$, the wedge may slide under the action of surface forces U_1, U_2 and U_3, weight W and arch thrust Q.

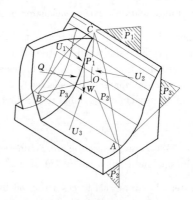

Fig. 10.4 – 1 Unstable Wedge at Skewback

$OABC$—rock wedge; P_1, P_2 and P_3—discontinuous planes in rock mass; U_1, U_2 and U_3—water pressure; W—weight; and Q—arch thrust

As shown in Fig. 10.4 – 2, the most possible sliding mode of the wedge is that the upstream surface OBC is pulled and the wedge slides to the downstream side along the cross ridge OA of side sliding plane P_2 and bottom sliding plane P_3. Theoretically, there are still some other sliding modes; for example, both the side sliding plane and the upstream surface are bulled open, and the wedge slides on the bottom sliding plane P_3 toward a certain downstream direction.

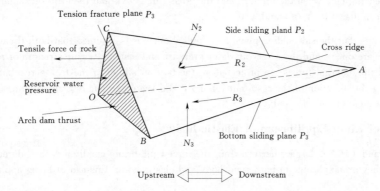

Fig. 10.4 – 2 Schematic Diagram of Stability Analysis on Wedge on Skewback

10.4.3 Analysis Method for Anti-sliding Stability of Arch Dam

10.4.3.1 Method of Limit Equilibrium of Rigid Body

The method of limit equilibrium of rigid body is generally adopted in the calculation of anti-sliding stability of dam. As this method has introduced some assumptions, the results are a bit rough, and cannot correctly reflect the stress state of rock mass, so that only an approximate estimated value of degree of safety can be obtained. However, due to simple calculation and long-term experience of engineering application, and its matching with design specifications, it is still the main method for analysis on sliding stability of arch dam at the present stage. The calculation formula is as follows:

Shear friction formula
$$K_1 = \frac{\sum(Nf_1 + c_1 A)}{\sum T} \tag{10.4-1}$$

Pure friction formula
$$K_2 = \frac{\sum Nf_2}{\sum T} \tag{10.4-2}$$

where, K_1 and K_2 are anti-sliding stability safety coefficients; $\sum N$ is the normal force on the sliding plane; $\sum T$ is the sliding force on the sliding plane; f_1 and c_1 are respectively friction coefficient and cohesion, and should be used in accordance with peak intensity of corresponding materials; f_2 is friction coefficient which should be used in accordance with characteristic value of corresponding materials; for materials in brittle failure, the proportional limit is adopted; for plastic materials or materials in elastic-plastic failure, yield strength is adopted; and for materials which have been already cut off and broken, residual strength is adopted.

The load distribution method of arch cantilever can be used for calculating arch dam thrust Q applied on the wedge, but cannot give the internal stress of foundation, and thus forces U_1, U_2 and U_3 applied on P_1, P_2 and P_3 cannot be calculated with the method, and only can be calculated in accordance with equilibrium conditions of the forces.

As shown in Fig. 10.4-2, it is assumed that the plane P_1 is pulled, the reservoir water pressure Q_1 applied on the crack surface P_1, the residual tensile strength of rock mass Q_2, the arch dam thrust Q_3 and self-weight W are known numbers, it is assumed that the effective normal forces applied on the side sliding plane P_2 and the bottom sliding plane P_3 (after deduction of a seepage force) are respectively N_2 and N_3, sliding occurs along the direction of the cross ridge OA of P_2 and P_3, and sliding resistance is parallel to the direction of OA and can be calculated with the following formula.

$$R_2 = c_2 A_2 + f_2 N_2, R_3 = c_3 A_3 + f_3 N_3 \qquad (10.4-3)$$

where, A_2, c_2 and f_2 are the area, adhesion and friction coefficient of the side sliding plane P_2; A_3, c_3 and f_3 are the area, adhesion and friction coefficient of the bottom sliding plane P_3.

All external and internal forces acting on the wedge are projected in the normal direction of the side sliding plane and the bottom sliding plane, and then we can obtain two equations:

$$\sum F_i \cos\phi_{i2} = 0, \sum F_i \cos\phi_{i3} = 0 \qquad (10.4-4)$$

where, ϕ_{i2} and ϕ_{i3} are respectively the angle of intersection between the force F_i and the side sliding plane P_2 and the bottom sliding plane P_3, the above formula contains two unknowns N_2 and N_3 which can obtained with the equation, and then the sliding force along the direction of the cross ridge can be calculated with the following formula:

$$\sum T = \sum F_i \cos\phi_{it} \qquad (10.4-5)$$

where, ϕ_{it} is the included angle between the force F_i and the cross ridge OA, and then the anti-sliding safety coefficient can be calculated with the formulas (10.4-1) and (10.4-2).

If N_2 is negative, it means that the side sliding plane P_2 is pulled open. It is assumed that both the upstream surface P_1 and the side sliding plane P_2 are open, and the reservoir water pressure and residual tensile strength of rock are given as known external forces; and all the external forces are projected on the bottom sliding plane P_3, a resultant force T is obtained, then sliding occurs along the direction of the resultant force T, and the corresponding anti-sliding safety coefficient can be calculated with the equations (10.4-1) and (10.4-2).

10.4.3.2 Method of Limit Equilibrium of Elastic Body

Equations (10.4-1) and (10.4-2) are derived from the force equilibrium conditions, and do not involve the deformation of the dam, thus they are applicable to both rigid body and elastic body, instead of being applicable to the rigid body only.

For the wedge shown in Fig. 10.4-2, even assuming that the upstream surface is completely separated, under the action of the external forces and self-weight, the magnitude of reaction forces U_2 and U_3 of the side sliding plane P_2 and the bottom sliding plane P_3 of the wedge is closely related to the deformation of rock mass, and the calculation of them is essentially a statically indeterminate problem, so that only an approximate value can be obtained by calculation only through the equilibrium conditions.

In the calculation with the finite element method, the internal stress of foundation can be given, and the normal forces N_2 and N_3 acting on the side sliding plane and the bottom sliding plane can be calculated through integration. As the deformation of the dam body and foundation has been considered, values are more realistic. According to the limit equilibrium conditions, the anti-sliding safety coefficient of the foundation can be calculated with the equation (10.4-1) and equation (10.4-2). This calculation method can be called elastic body limit equilibrium method.

Some scholars have proposed the rigid body spring element method. It is assumed that the wedge is a rigid body, the surrounding rock mass and the dam bodies are elastic bodies. The actual data indicates that the deformation modulus of the rock mass is generally lower than the elastic modulus of concrete, so the calculation results cannot reflect the actual situation. Of course, the original author's intent may mainly aim at being closer to the rigid body limit equilibrium method, but by using conventional finite element program, let the elastic modulus of the wedge large enough can also achieve the same purpose.

10.4.3.3 Tability Analysis Method Based on Upper Limit Theorem of Plastic Mechanics

The abovementioned sliding stability analysis is a simplified method, when the rock mass is cut into several blocks by the structural surface, or the shape of wedge is more complex, it is difficult to use a simple formula to determine the safety coefficient. Chen et al (2005) developed anti-sliding stability three-dimensional limit analysis method for sliding stability analysis which is based on the upper limit theorem of plasticity. Cutting the rock mass into finite strips, the anti-sliding coefficient for a given scheme of computing net is calculated by the principle of virtual work on the assumption of rigid-plastic constitutive relation and Mohr-Coulomb yield criterion, then the relevant parameters of the calculation grid are continuously modified to find the minimum safety coefficient by optimization method.

10.5 Three-dimensional Finite Element Elastic-pastic Analysis of Arch Dam

As the dam body and foundation of arch dam are actually a whole, when the stress of arch dam is calculated with the load distribution method of arch cantilever, and the anti-sliding stability of foundation is calculated with the limit equilibrium method of rigid body the interaction between the dam body and foundation is ignored, and thus there is a difference between the calculation results and the actual situation.

Since 1970, the finite element method has been tremendously developed in China, from 2D to 3D, from linear to nonlinear, from conventional calculation to analysis of arch dam. Currently, the finite element method can be used to conduct linear and nonlinear analysis of arch dam by simulating the construction process of large dams, and considering the temperature field and construction reality. A variety of material/structure and construction factors can be fully reflected in the calculation. China's three-dimensional finite element elastic-plastic analysis and simulation analysis on arch dams have reached international advanced level.

Since the construction of the Xianghongdian Arch Dam in the 1950s, China has also conducted a lot of structural model tests for arch dams, although they can provide an intuitive understanding of the failure process, it is difficult to simulate seepage, temperature and other loads, and fully meet similarity conditions, and due to long cycle and high cost, the structural model tests have been gradually declined with the rapid development of the finite element method.

10.5.1 *Constitutive Model and Yield Criterion*

The material constitutive models generally have elastic constitutive model, ideal elastic-plastic model and brittleness softening model. In nonlinear finite element analysis of arch dams, in general, the brittleness softening model is adopted.

The Mohr-Coulomb criterion with tensile strength is commonly used for Joints and soft interlayer in the rock mass and dam body joints.

$$\left.\begin{array}{l} \tau = c - f\sigma \\ \sigma \leqslant R_t \end{array}\right\} \quad (10.5-1)$$

where, c, f and R_t are cohesion, friction coefficient and tensile strength in sequence; τ is shearing stress; σ is normal stress, and tensile stress is positive.

For the rock mass itself, generally the Drucker-Prager criterion with tensile and compressive strength is used.

$$\left.\begin{array}{l} \alpha I_1 + \sqrt{J_2} - k = 0 \\ \sigma \leqslant R_t, \sigma \geqslant R_c \end{array}\right\} \quad (10.5-2)$$

where, $I_1 = \sigma_1 + \sigma_2 + \sigma_3$, $J_2 = [(\sigma_1 - \sigma_2)^2 + (\sigma_2 - \sigma_3)^2 + (\sigma_3 - \sigma_1)^2]/6$; R_t is tensile strength; R_c is compressive strength; and α and k are constants.

For the concrete dam body, the Ottosen four-parameter criterion is commonly used.

$$\left.\begin{array}{l} a\dfrac{J_2}{R_c^2} + \lambda(\theta)\dfrac{\sqrt{J_2}}{Rc} + b\dfrac{I_1}{Rc} - 1 = 0 \\ \lambda(\theta) = k_1 \cos\left[\dfrac{1}{3}\cos^{-1}(-k_2 \cos 3\theta)\right], \text{if } \cos 3\theta \geqslant 0 \\ \lambda(\theta) = k_1 \cos\left[\dfrac{\pi}{3} - \dfrac{1}{3}\cos^{-1}(-k_2 \cos 3\theta)\right], \text{if } \cos 3\theta \leqslant 0 \end{array}\right\} \quad (10.5-3)$$

where, a, b, k_1 and k_2 are four constants determined by the material testing. For the concrete dam body, the Willam-Warnke five-parameter criterion also can be adopted.

10.5.2 *Loading Modes*

There are several loading modes:

(1) Non-simulation loading. The large dam is poured without simulating the construction process. The self-weight and normal water level are applied at one time, and the overloading modes are as follows: ①increase of water level, with constant self-weight; ②increase of density of water with constant self-weight and water level; ③increase of density of water and dam concrete with same ratio; and ④material strength parameters decrease with same ratio and constant load.

(2) Self-weight simulation loading. The large dam rises up layer by layer, the impacts of stage construction and grouting may be considered by increments of self-weight and water pressure and overloading is conducted after pouring to top. The overloading modes include water level, water density, water and concrete with same ratio, and decrease of material strength.

(3) Overall simulation loading. The construction process of large dam is simulated, all the temperature field, self-weight and water pressure are calculated using the incremental method, and then overloading is conducted or material strength decreases after pouring to top.

Some high arch dams in China have been analyzed for over loading by nonlinear finite element method, but the construction process was not simulated.

Zhang Guoxin, Zhou Qiujing et al have conducted more detailed nonlinear three-dimensional finite element analysis of the Xiluodu High Arch Dam. When the D-P criterion with tensile strength (without regard to crushing damage) is adopted for both the dam body and the bedrock, the calculation results of the three different overloading modes are shown in Fig. 10.5-1, which has showed that the overloading modes have a great influence on the calculation results.

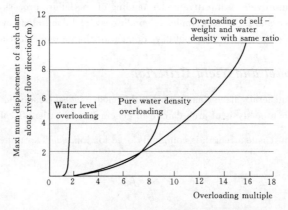

Fig. 10.5-1 Relation Curve of Maximum Displacement Along River Flow Direction and Overloading Multiple with Different Overloading Ways

Failure process of arch dam roughly includes the following steps: firstly, local tension fractures, then local shear failure, and finally compression failure in case that the bedrock is normal. Therefore, if the D-P criterion with tensile strength is adopted for the dam body while ignoring the compression failure, the calculation results will be inevitably greater than the true values.

Zhang Guoxin and Zhou Qiujing have conducted nonlinear calculation for several arch dams, wherein the calculation results of pure water density overloading are shown in Table 10.5-1. It is evident that whether the compressive strength has a great impact on the safety coefficient, and K_1, K_2 and K_3 in the table are respectively the overloading coefficients, when the crack of the dam heel reaches the impervious curtain the displacement load curve has an inflection point and the calculation does not converge in sequence.

Table 10.5-1 Overloading Safety Coefficients of 12 Arch Dams

Projects	Only tensile strength and shear strength are taken into account without considering compressive strength			Compressive strength, tensile strength and shear strength are all taken into account		
	K_1	K_2	K_3	K_1	K_2	K_3
Xiaowan	1.0	4.0	6.25	1.0	2.0	2.25
Xiluodu (without attachment to dam heel)	2.2	4.5	5.75	2.0	2.75	3.0

Continued

Projects	Only tensile strength and shear strength are taken into account without considering compressive strength			Compressive strength, tensile strength and shear strength are all taken into account		
	K_1	K_2	K_3	K_1	K_2	K_3
Xiluodu (with attachment to dam heel)	2.0	5.0	7.0	2.0	3.5	4.0
Sayan	<1.0	3.5	3.75	<1.0	2.5	2.75
Kolnbrein (before reinforcement)	0.75	2.56	5.0	0.75	1.5	1.75
Kolnbrein (after reinforcement)	1.2	2.75	5.0	1.25	1.75	2.0

10.6 Selection of Foundation Surface and Foundation Treatment for Arch Dam

10.6.1 Selection of Foundation Surface

As the stress level of arch dam is high, the selection of foundation surface is an important issue in arch dam engineering. As specified in *Design Specifications for Concrete Arch Dams* (SL 282—2003), the fresh or slightly weathered bedrock is adopted as the foundation surface; and as specified in the *Design Specifications for Concrete Arch Dams* (DL/T 5346—2006), for the high dams, the excavation should be conducted to the rock mass of class II, even to the rock mass of class III at local areas, and requirements on medium and low dams may be appropriately reduced.

Experiences of selection of foundation surfaces for high arch dams such as Ertan, Laxiwa, Goupitan and Xiluodu have shown that, based on the use of appropriate foundation treatment measures, reasonable utilization of slightly weathered rock mass as foundation of high arch dam can greatly reduce foundation excavation and the amount of dam concrete and have significant economic benefits.

Foundation surfaces of abutments are preferably in full-radial direction on the plane, so as to facilitate dam abutment stability. If the skewbacks are very thick, full radial excavation will make slope excavation too high and excessive work amount, in this case radial excavation at half of the downstream and non-radial excavation at half of the upstream side also can be adopted.

10.6.2 Treatment for Foundation Defects

After bedrock excavation, the rock mass at the surface layer still has more cracks, generally consolidation grouting should be carried out, with the hole depth of 5 - 8m. When the bank slope is very steep, contact grouting also should be conducted. Superficial fault and fracture zones in the foundation, of course, can be completely excavated and then concrete is backfilled. However, a wide range of deeper fault and fracture zones exists in the foundation, and the treatment for this is often a difficult point in the project. Due to complex foundation conditions of the Longyangxia Arch Dam, a series of measures including concrete replacement, force transmission tunnel, force transmission plug, mesh replacement and high-pressure consolidation grouting have been adopted, and the tunnels with concrete replacement at F_{18} fault of the dam are shown in Fig. 10.6-1.

10.6.3 Seepage Prevention and Drainage for Dam Foundation

The curtain grouting and drainage holes are used to reduce the uplift pressure and seepage flow of dam foundation. In the traditional grouting process, grouting stopping plugs are adopted for closing the upper part and the lower part of the grouting section, and grouting is conducted step by step from thin slurry to thick slurry. This traditional process has made a major breakthrough in China, in which the grouting stopping plugs are used for plugging the orifices instead of plugging several sections. Full-hole grouting is repeated with each extension of drilling, so as to simplify the process, and also improve the quality of grouting.

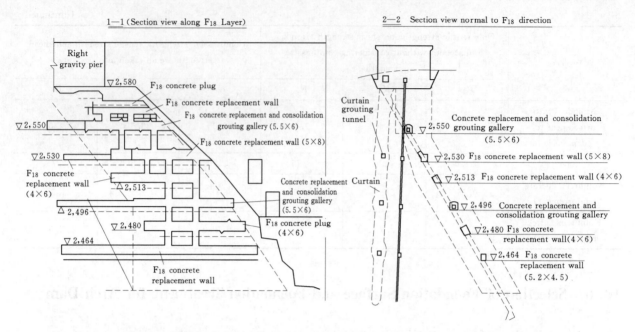

Fig. 10.6 - 1 Treatment Diagram of F_{18} Fault of Longyangxia Arch Dam (Elevation Unit: m, Size Unit: m)

10.6.4 *Construction of Arch Dam in Karst Region*

Located in the karst region, the Wujiangdu Gravity Arch Dam 165m high has adopted a seepage prevention mode of connecting the suspension type cement grouting curtain with upstream shale impermeable layer for the first time, concrete is backfilled for karst caves above the river bed after excavation, and high-pressure grouting is conducted for deep karst caves. With the use of these methods, the problem of seepage prevention in the construction of concrete high dams in the karst region is successfully solved (Eighth Engineering Bureau of the Ministry of Water Conservancy and Electric Power, 1987).

The Dongfeng Arch Dam is another success of construction of high arch dam in the karst region through the grouting treatment after Wujiangdu Dam (Zheng, 1998).

10.7 Empirical Assessment of Shape of Arch Dam

10.7.1 *Flexibility Coefficient of Arch Dam*

Lombardi proposes to conduct empirical assessment of the shape of arch dam with flexibility coefficient C (Lombardi, 1991):

$$C = \frac{A^2}{VH} \qquad (10.7-1)$$

where, A is the area of profile in the arch dam; V is the volume of the arch dam; and H is the maximum dam height.

Flexibility coefficients of some arch dams are shown in Fig. 10.7 - 1. For the same dam height H, the greater the volume V is, the smaller C will be; and the smaller the area of middle surface A is, the smaller C is. Therefore, for the same dam height, the smaller C is, the safer the dam will be. Kölnbrein Arch Dam had serious cracks after impounding, as shown in Fig. 10.7 - 1. For the same dam height, the maximum flexibility coefficient of the dam is 17.7. However, Fig. 10.7 - 1 shows that many lower dams with flexibility coefficients greater than 17.7 are operating normally, showing that the dam height should be considered when judging the degree of safety of the dam body according to the flexibility coefficient.
Consider Dam 1 and Dam 2, they have completely similar geometry, same flexibility coefficient, dam heights of H_1 and H_2 respectively, and maximum stress in the dam of σ_1 and σ_1 respectively, and bear water loads (or self-weight of concrete and rock mass) of with densities γ_1 and γ_2 respectively. Dimensions of γ, H and σ are t/m^3, m and t/m^2, respectively, and $\gamma H/\sigma$ is a dimensionless number. Therefore:

$$\frac{\gamma_1 H_1}{\sigma_1} = \frac{\gamma_1 H_2}{\sigma_2} \qquad (10.7-2)$$

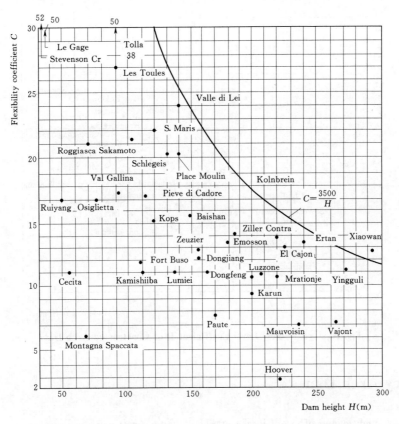

Fig. 10.7-1 Height and Flexibility Coefficient of Arch Dams

However, the density of the water load is the same, and the densities of concrete or rock mass in different projects is also very close, so that it can be made that $\gamma_1 = \gamma_2$, through the formula (10.7-2), the following formula can be obtained:

$$\frac{\sigma_1}{\sigma_2} = \frac{H_1}{H_2} \tag{10.7-3}$$

It can be seen that for two arch dams with completely similar geometry, although their flexibility coefficients are the same, the internal stress of the dams is different and proportional to the dam height. The stress level of low dam is lower that of high dam, so that the degree of safety of low dam is higher than that of high dam with same flexibility coefficient. Therefore, the degree of safety of arch dam cannot be measured only with the flexibility coefficient apart from the dam height.

10.7.2 Stress Level Coefficient of Arch Dam

Based on the above analysis, the stress levels in the dam body and foundation of arch dam are not only related to the flexibility coefficient C, but also proportional to the dam height H. IWHR proposed to take the coefficient D as follows (Zhu, 2000):

$$D = CH = \frac{A^2}{V} \tag{10.7-4}$$

Coefficient D represents the stress level of arch dam (including the dam body and dam foundation), so it can be called stress level coefficient. The stress level coefficients of some constructed arch dams are shown in Fig. 10.7-2.
If D_0 is taken as the upper limit of stress level coefficient, with the formula (10.7-4), then

$$C \leqslant \frac{D_0}{H} \tag{10.7-5}$$

The above formula represents a hyperbolic curve. For example, a hyperbolic curve as shown in Fig. 10.7-1 can be obtained by taking the stress level coefficient D, if $D_0 = 3,500$ according to Kölnbrein Arch Dam.

In short, the stress level coefficient is more meaningful than the flexibility coefficient.

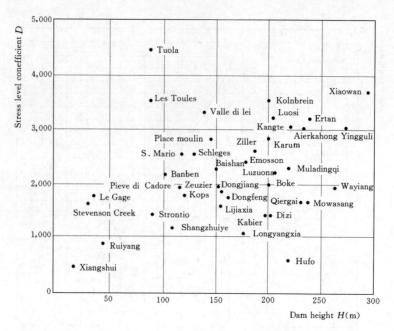

Fig. 10.7-2   Stress Level Coefficient *D* of Constructed Arch Dam

10.8   Masonry Arch Dam

Masonry arch dam has the advantages of use of local materials, simple construction, low cost, etc., and compared with the earth dam, it has high safety, and is convenient in flood discharge and river diversion. Thus, the masonry arch dam has been widely used in irrigation and water conservancy, and construction of small hydropower stations in China. In low and medium arch dams lower than 70m in China, the masonry arch dam accounts for about 90%; and two masonry arch dams 100m high have been completed, namely Yanjinqiao Arch Dam in Huairen, Guizhou Province and Xiahuikeng Arch Dam in Shangrao, Jiangxi Province.

With the development of arch dam technology in China, the design and construction levels of masonry arch dam are increasingly improved: the shape of dam is developed from single-curvature to double-curvature, and the arch form is developed from single-center circle to multi-center circle, ellipse, parabola and logarithmic spiral; and in stress analysis, the crown cantilever method is developed to multi-arch cantilever method, and a lot of masonry arch dams also have been subjected to shape optimization.

The construction technology of masonry arch dam has been greatly improved. Cement mortar was used as cementing material in the past, and now except for the cement mortar used for building the surface of the dam body, fine aggregate concrete are used as cementing material for building all other parts, and fly ash and additives are generally added in the concrete to reduce the amount of cement and heat of hydration, and improve workability. In order to reduce masonry gaps, save cementing material, increase the amount of stones and enhance the integrity of the dam, the masonry stones are erected instead of being flatly paved as before, with masonry height of about 1m each time. By means of these methods, the stone content of the masonry dam reaches 50% to 55%, the amount of cement of unit dam body is about 100kg, and the density of the dam body is up to $2.35t/m^3$.

All manual operations of construction methods in the past are gradually transformed into mechanization, automobiles and belt conveyors are adopted for horizontal transportation, cranes and gantry cranes are adopted for vertical transportation, and vibrators are used in the building of the dam body.

10.9   Conclusion

(1) China has ranked first in the number and height of constructed arch dams in the world, and taken the lead in the scale of arch dams under construction in the world. In the construction process, China has solved a series of complex technical problems. During the past 60 years, China has made brilliant achievements in the construction of arch dams, and attracted

the worldwide attentions.

(2) In order to adapt to different topographic and geologic conditions, China uses a variety of arch shapes, developing from single-center circle arch and three-center circle arch in early stage to parabola, ellipse, logarithmic spiral and many other types, and proposes two new arch forms, namely unified quadratic curve and mixed curve, with the development of shape design from manual design to optimization design.

(3) In the analysis on stress of arch dam, a number of powerful static and dynamic calculation programs of Arc-Beam Load Distribution Method, and linear and nonlinear finite element programs are developed. The finite element equivalent stress method and corresponding stress control standards are proposed, creating conditions for applying the finite element method for shape design for arch dam.

(4) A lot of work have done in anti-sliding stability of arch dams, treatment for complex foundation and other aspects, and Wujiangdu Gravity Arch Dam and Dongfeng Arch Dam are successfully completed in the karst region.

(5) China has constructed a large number of masonry arch dams, accumulating rich experiences in design and construction of masonry arch dams.

References

[1] Pan Jiazheng. *Arch Dam*. Shanghai: Shanghai Science and Technology Press, 1959.

[2] Wang Jingqi. *Design and Calculation of Arch Dam*. Beijing: China Industry Press, 1965.

[3] Li Zhanmei. *Arch Dam*. Beijing: Water Conservancy and Electric Power Press, 1982.

[4] U. S. Bureau of Reclamation. *Design for Arch Dam*. Beijing: Water Conservancy and Electric Power Press, 1984.

[5] Li Zan, Chen Xinghua, Zheng Jianbo, et al. *Design for Concrete Arch Dam*. Beijing: China Electric Power Press, 2000.

[6] Zhu Bofang, Gao Jizhang, Chen Zuyu, et al. *Design and Research of Arch Dam*. Beijing: China Water Resources and Hydropower Press, 2002.

[7] Pan Jiazheng. *Analysis of Sliding Stability of Buildings and Landslide*. Beijing: Water Conservancy and Electric Power Press, 1980.

[8] Eighth Engineering Bureau of the Ministry of Water Conservancy and Electric Power. *Wujiangdu Engineering Construction Technology*. Beijing: Water Conservancy and Electric Power Press, 1987.

[9] Zhu Bofang and Zhang Chaoran. *Research on Key Technology of Structural Safety of High Arch Dam*. Beijing: China Water Power Press, 2010.

[10] Chen Zuyu, Wang Xiaogang, Yang Jian, et al. *Analysis of Rocky Slope Stability-Program of Principles and Methods*. Beijing: China Water Power Press, 2005.

[11] Zhou Weiyuan et al. *Geomechanical Model Test Method for High Arch Dam and Its Application*. Beijing: China Water Power Press, 2008.

[12] Li Yisheng. Quadratic Curve Arch Dam. *Journal of Hydraulic Engineering*, July, 1998.

[13] Liu Guohua and Wang Shuyu. Research on Optimization Model of Mixed Type of Arch Dam and Arch Closure Conditions. *Engineering Mechanics*, Supplement, 1994.

[14] Zhu Bofang, Rao Bin, Jia Jinsheng, et al. Mathematical Model of Optimization for Shape of Arch Dam. *Journal of Hydraulic Engineering*, 1992 (3), 23-32.

[15] Zhu Bofang, Rao Bin and Jia Jinsheng. Calculation Method for Optimization for Shape of Arch Dam Under the Action of Static and Dynamic Loads. *Journal of Hydraulic Engineering*, 1992 (5), 20-26.

[16] Zhu Bofang. Monograph Review of International Symposium on Arch Dam. *Concrete Dam Technology*, February, 1987 and Hydroelectric Power, August, 1988.

[17] Zhu Bofang. Stress Level Factor and Safety Level Factor of Concrete Arch Dam. *Water Resources and Hydropower Technology*, August, 2000.

[18] Lombardi. Kolnbrein Dam: An unusual solution for an unsual problem. *Water Power and Dam Construction*, 1991.6, 31-34.

[19] Chengdu Survey and Design Institute of State Power Corporation. *Safety Evaluation on Sliding Stability of Constructed Arch Dam Along Skewback Basal Plane*, March, 2000.

[20] Wang Lei, Liu Zhong and Zhang Youtian. Analysis on Curtain of Drainage Holes. *Journal of Hydraulic Engineer-*

ing, April, 1992.
[21] Zhu Bofang, Li Yue, Xu Ping, et al. Substitution of the Curtain of Drainage Holes by a Seeping Layer in the Analysis of Seepage Field. *Water Resources and Hydropower Technology*, October, 2007.
[22] Zheng Zhi. Anti-seepage Treatment for Dongfeng Reservoir. *Academic Exchange Proceedings on Water Resources and Hydropower Foundation and Foundation Engineering* 1998. Tianjin: Tianjin Science and Technology Press.
[23] Li Yisheng and Fan Xiuqi. Optimization and Design for Ruiyang Arch Dam. *Water Resources and Hydropower Technology*, August, 1985.
[24] *Design Specifications for Concrete Arch Dam* (SD 145—85). Beijing: China Water Power Press, 1985.
[25] *Design Specifications for Concrete Arch Dam* (SL 282—2003). Beijing: China Water Power Press, 2003.
[26] *Design Specifications for Concrete Arch Dam* (DL/T 5346—2006). Beijing: China Electrical Power Press, 2007.
[27] Li Xuechun, Zhu Bofang, Xu Ping, et al. *Impact of Construction Process, Rock Surface Reservoir Water and Transverse Joints on Stress State of Xiaowan Arch Dam*. Bejjing: China Institute of Water Resources and Hydropower Research, April, 2000.
[28] Yang Bo, Dong Fupin and Zhu Bofang. *Research on Stress Control Standards for High Arch Dams-Analysis on Stress of Arch Dams Constructed at Home and Abroad*. Beijing: China Institute of Water Resources and Hydropower Research, June, 2000.

Chapter 11

Rolled Compacted Concrete (RCC) Damming Technology

Zhang Guoxin[1], Chen Gaixin[2]

11.1 Overview

Since the landmark completion of Kengkou RCC gravity dam in 1989, Roller Compacted Concrete (RCC) dam construction has had a history of above twenty years in China. Through the joint efforts of designers, researchers, engineers, and constructers, China's RCC dam construction has made great strides and a series of advanced technological achievements, and ranked the world's forefront in number, size, height, technical difficulty, and construction technology and other aspects.

So far, a rapid increasing number of dams have been built (Fig. 11.1-1). Apart from a few special high dams still using normal concrete, the vast majority of the concrete dams currently rely on RCC dam construction technology. According to incomplete statistics, till December 2011, China had possessed 123 RCC dams (built +under construction), including 91 gravity dams and 32 arch dams. Table 11.1-1 and Table 11.3-1 list all the RCC gravity dams and arch dams (built, under construction and being designed). Over-100m-level RCC dams have occupied a large proportion in recent years, including Longtan Gravity Dam, 216m high and running for 4 years; Guangzhao Gravity Dam, 200.5m high, completed and put into operation at the end of 2008; Shapai RCC Arch Dam, 130m high, completed and put into operation in May 2003, enduring 2008 Great Wenchuan Shock and kept intact, which proves the safety and reliability of the RCC thin arch dam. Represented by Shapai RCC Arch Dam and Longtan Gravity Dam, a large number of original research results regarding independent intellectual property rights in all aspects of design, research and construction have been obtained: China has accumulated a wealth of experience in the construction of RCC dam.

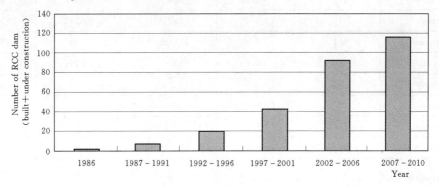

Fig. 11.1-1 Growth Trends of RCC Dam in Number in China

Table 11.1-1 RCC Gravity Dams in China (built & under construction)

No.	Name	Height of dam (m)	Gradient of downstream	Dam crest length (m)	Joint spacing (m)	Joint forming
1	Longtan	192	1:0.68−1:0.7	761	20−44	Regular joint
2	Huangdeng	203	1:0.75	469.6	20−27	

---

[1] Zhang Guoxin, Director, China Institute of Water Resources and Hydropower Research, Professor Senior Engineer.
[2] Chen Gaixin, Vice Director, China Institute of Water Resources and Hydropower Research, Professor Senior Engineer.

Countinued

No.	Name	Height of dam (m)	Gradient of downstream	Dam crest length (m)	Joint spacing (m)	Joint forming
3	Guangzhao	200.5	1:0.75	410	17-25	Regular joint+ Cutting joint
4	Guandi	168	1:0.75	516	20-26	Regular joint
5	Guanyinyan	159	1:0.75	1158	18-35	
6	Yabiluo	164	1:0.70	374	18-25	
7	Jin'anqiao	160	1:0.75	640	30-34	Cutting joint
8	Liyuan	155	1:0.75	525.33	18-35	
9	A'hai	132		482		
10	Ludila	140	1:0.75	622	20-35	Cutting joint
11	Shiyazi	134.5		217.86		
12	Jiangya	131	1:0.80	368	22.5-35	Cutting joint
13	Hongkou	130	1:0.75	340.15	14.5-25	Cutting joint+ Induced joint
14	Baise	130	1:0.75	720	33-35	Regular joint
15	Kelasuke	121.5	1:0.75	1570	15-20	Regular joint
16	Pengshui	113.5	1:0.75	309.53	16-21	Cutting joint+ Induced joint
17	Wudu	120.34	1:0.75	736	14-26	
18	Geliqiao	124		103.9		
19	Longkaikou	116	1:0.75	768	20-25	
20	Silin	117	1:0.70	316	18-25	Induced joint
21	Suofengying	115.95	1:0.70	164.58	19-31	Induced joint
22	Mianhuatan	113	1:0.75	308	33-70	Regular joint
23	Gelantan	113	1:0.75	466	21-38	Cutting joint
24	Dachaoshan	111	1:0.70	460.39	18-36	Regular joint
25	Yantan	110	1:0.65	525	20-46	Induced joint
26	Jinghong	108	1:0.65	704.5	20-34.6	Cutting joint
27	Shangtuo	101	1:0.75	631		
28	Dahuaqiao	106	1:0.70	231.5	14-28	
29	Madushan	107.5	1:0.75	352.96		
30	Gongguoqiao	105	1:0.70	356	14-27	Cutting joint+ Induced joint
31	Linjiang	104	1:0.72	531	15	Regular joint
32	Shuikou	101		783	40	Induced joint
33	Nansha	85		240.22		
34	Yulongyan	100				
35	Nanyi	96.8		193.4		
36	Kuokou	88.4				
37	Fenhe 2nd Reservoir	88	1:0.75	228	66, 45, 90	Regular joint
38	Zaoshi	88	1:0.75	351	20	

Countinued

No.	Name	Height of dam (m)	Gradient of downstream	Dam crest length (m)	Joint spacing (m)	Joint forming
39	Chongxi Ⅲ	86.5	1 : 0.73	190	56, 37, 42, 52	Induced joint
40	Huatan	85.3	1 : 0.90	173.2	49 – 71	Drilled joint
41	Shibanshui	84.6	1 : 0.70	445	41 – 70	Regular joint
42	Wanyao	79	1 : 0.70	390	34.50	Regular joint
43	Leidatan	83	1 : 0.70	219.5		
44	Tongjiezi	82	1 : 0.75	1084.59	16 – 21	Cutting joint
45	Guanyinge	82	1 : 0.70	1040	16	Cutting joint
46	Taolinkou	74.5	1 : 0.63	500	12.5 – 18.2, 17 – 30	Regular joint
47	Yangxishui Ⅰ	82	1 : 0.75	244	45, 54.5, 52.5, 47	Regular joint
48	Tongkou	76	1 : 0.70	220.7		
49	Baisha	75	1 : 0.75	171.8	18 – 29	
50	Zhouba	72.5	—	162.5	18 – 30	Cutting joint
51	Chonghuer	74	1 : 0.70	545.5	15 – 23	
52	Zhouning	73.4	1 : 0.72	201	45, 66, 90	Regular joint
53	Mayandong	73	1 : 0.75	155.8		
54	Xixi	71	1 : 0.70	243	20	Cutting joint
55	Kala	126	1 : 0.75	258	15 – 26	
56	Wan'an	68.51		1097.5	10 – 18	Induced joints
57	Shanzai	64.6		265.86	None	Upstream induced joints
58	Tianshengqiao Ⅱ	61	1 : 0.75	470	68 – 73	Induced joints
59	Jinjiang	62.45	1 : 0.75	229	15	Induced joints
60	Longmentan	58		139	None	
61	Kengkou	57	1 : 0.75	123	None	
62	Daguangba	55	1 : 0.73	719	20 – 40	Induced joints
63	Shuidong	63	1 : 0.65	196.62	None	Induced joints
64	Gaobazhou	57		439.5	18.2 – 25	Induced joints
65	Baishi	49.3	1 : 0.76	513	20 – 24	Cutting joint
66	Rongdi	50	1 : 0.75/1 : 0.65	136	None	
67	Guangxuxia Reservoir	43.5		101.5	None	11 – 16m (normal concrete)
68	Bailongtan	70	1 : 0.70	274	22 – 52	Regular joint
69	Yanwangbizi	34.5		383	15 – 35	Cutting joint
70	Tuoba	158				

11.2 RCC Gravity Dam

11.2.1 *Interlaminar Bonding and Shear Resistance*

Stability of a gravity dam is dominated by shear resistance between concrete and foundation surface as well as that between each layer of concrete. Generally, the interlaminar shear capacity of RCC is lower than that of normal concrete. Therefore, it is critical to take measures to enhance the shear strength between roller compacted layers for the RCC gravity dam. Usually, great attentionh as been focused on the interlaminar bonding quality in design and construction.

In order to study the factors affecting interlaminar bonding capacity and measures for enhancement, Longtan RCC high dam project, from October, 1990 to June, 1993, first carried out three large-scale field compaction tests and on-site and in-situ shearing tests in Yantan Hydropower Station with similar climatic conditions, basically the same binding materials and similar fine aggregates. Then in 2004, three large in-situ tests were conducted at Longtan construction site to study the impacts of interlaminar constructional time intervals, layer surface treatment, temperature conditions on the interlaminar shear strength, with peak value method, single point method, proportional limit method and yield limit method, respectively. The data concluded that: ①most of the specimens failed along the layer surface, stating the RCC layer surface as the weak point; ②interlaminar constructional time intervals coupling with temperature conditions significantly influenced the shear resistance. The interlaminar constructional time intervals for Longtan RCC dam should be no more than 8h, 6h, 4h, at cold, normal and hot seasons respectively; ③cement and flyash slurry strength paved on layer surface were another important factor.

By a large number of tests, recommended parameters of interlaminar shear resistance for Longtan RCC dam are shown in Table 11.2-1. According to Longtan's data, the shear strength can be converted from core-sample test to in-situ test: 1.1 for f', and 1.75 for c'. Generally, in-situ tests have been done for most domestic large-and medium-scale projects, and part of them also did core-sample tests. Table 11.2-2 has showed parts of the test results.

Table 11.2-1 The Suggested Values of Interlaminar Shear Resistance Parameters for Longtan RCC (Pan Luosheng, 2007)

Dam height (m)	Grade level of Coarse Agg.	Regular statistical method				Comprehensive value analytical method				Suggested values for these parameters			
		Comprehensive value		Minimum average value		Guaranteed rate 80%		Deviation coefficient v		$C'_{vf}=0.20$ $C'_{vc}=0.30$		$C'_{vf}=0.20$ $C'_{vc}=0.35$	
		f'	c'	f'	c'	f'	c'	f'	c'	f'	c'	f'	c'
210	3	1.29	2.80	1.13	1.92	1.19	2.59	0.09	0.09	1.07	2.09	1.07	1.97
156	3	1.17	2.10	1.11	1.89	1.08	1.93	0.09	0.10	0.97	1.57	0.97	1.48

After years of engineering exploration and practice, the RCC interlaminar binding quality has been effectively improved. Numbers of in-situ and core samples test results have showed that shear performance of the RCC layer surface by continuous paving method can fulfill the stability requirement of 100m-high RCC dams against sliding; even for 200m-high Longtan Dam and Guangzhao Dam, real interlaminar shear strength are far larger than the design value. Continuously roller compacted hot joint surface no longer impairs the safety and quality of the high RCC dams. Comparatively, construction and cold joint surface, with surface toughing and slurry paving treatment, get slightly weaker shear resistancethan hot joint layer, but still enough to secure stability of high dams against sliding.

11.2.2 Profile Design of Gravity Dams

Gravity dam section design consists of two parts, namely, design for sectional body and materials partition. Size of dam section, i.e. gradients of upstream and downstream slope, is governed by the stability of a dam against sliding. As the shear capacity of the RCC layer surface is less than the normal concrete, the height and width of a RCC gravity dam are usually larger than that of normal concrete gravity dam. However, due to the progress of material mix portion and construction technologies, shear resistance of RCC layer surface continuously improves. As a result, the RCC gravity dam section can be gradually thinned. As a matter of fact, current sectional design of 200m-high RCC gravity dam, like Longtan, is almost the same as normal concrete gravity dam. Table 11.1-1 lists the upstream and downstream slopes of some RCC gravity dams in China.

Material partition of RCC gravity dam section generally follows structural characteristics against mechanical loads, seepage and freeze-thaw cycles. According to the optimization, Longtan dam section was divided into 4 RCC zones and 1 normal concrete zone, as shown in Fig. 11.2-1 and Table 11.2-3. Similarly design strategy is applied for most domestic RCC gravity dams that has completed and been in construction.

Table 11.2-2 Table of Shear Parameters of RCC Layer (joint) Surface in Some Domestic Projects

Dam	Concrete mark	Concrete grading	The dosage of cementation materials (kg/m³)		Testing method	Shear strength		Remark
			Cement	Flyash		f'	c' (MPa)	
Kengkou	$R_{90}100$ (layer surface)	3	60	80	Core sample test	1.12	1.17	Peak shear strength, Pit 2
Tongjiezi	$R_{90}100$ (layer surface)	3	65	85	On-site in situ test	1.54	1.23	Covered before initial setting
Yantan	$R_{90}150$ (layer surface)	3	55	104	On-site in situ test	1.17	1.36	Covered before initial setting
Puding	$R_{90}150$ (layer surface)	3	54	99	Core sample test	1.82	2.75	
Gaobazhou	$R_{90}150$ (layer surface)					1.70	1.58	
	$R_{90}150$ (slurry paved layer surface)	3	88	88	On-site in situ test	1.22	1.78	
	$R_{90}150$ (joint surface)					0.92	2.28	
	$R_{90}150$ (joint surface)					0.97	0.90	
Jiangya	$R_{90}150$ (layer surface)					0.97	0.93	Core sample data in 1997
	$R_{90}150$ (slurry paved layer surface)	3	64	96	Core sample test	1.17	0.99	
	$R_{90}150$ (flat paved layer surface)					1.40	1.03	Core sample data in 1998
	$R_{90}150$ (inclined paved layer surface)					1.27	1.15	
Dachaoshan	$R_{90}150$ (layer surface)	3	67	101 (PT)	Core sample test	2.14	4.00	Concrete age>90d
	$R_{90}150$ (joint surface)					1.88	3.50	
Mianhuatan	$R_{180}150$ (layer surface)		64	96	Core sample test	1.20	2.80	1st dry season
	$R_{180}150$ (layer surface)	3	51	96		1.37	2.55	
	$R_{180}150$ (layer surface)					1.26	2.06	2nd & 3rd dry season
	$R_{180}100$ (layer surface)		48	88		1.24	1.58	
Guangzhao	$R_{90}20$ (layer surface)	2	71.2	87.1	On-site in situ test	1.27	1.35	Hot joint
	$R_{90}20$ (layer surface)	3	60.8	91.2	On-site in situ test	1.02–1.20	1.02–1.40	Hot joint
	$R_{90}20$ (joint surface)				On-site in situ test	1.17–1.26	1.26–1.19	Cold joint+Roughing+Slurry paved

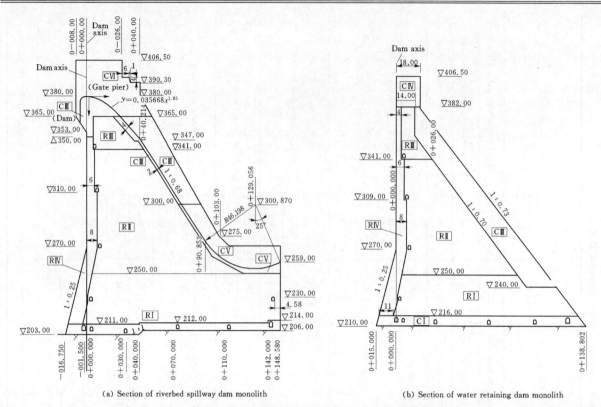

(a) Section of riverbed spillway dam monolith   (b) Section of water retaining dam monolith

Fig. 11.2-1 Section Partition of Longtan Dam (Length/Height Unit: m)

Table 11.2-3 Concrete Partition of Longtan Dam[4]

Partition No.	Roller compacted concrete				Normal concrete
	R I	R II	R III	R IV	C III
Design strength level	$C_{90}25$	$C_{90}20$	$C_{90}15$	$C_{90}25$	25
Seepage resistance level	W6	W6	W4	W12	W12
Freeze-thaw resistance level	F100	F100	F50	F150	F150

11.2.3 Joints in RCC Gravity Dam

As a measure for temperature control and crack prevention, gravity dam joints are designed to reduce axial thermal stress and avoid cracks.

Early RCC technology stated that temperature control measures could be substantially simplified owing to less cement and hydration heat and consequently lower temperature rise. Regarding that joint may influence the large-scale concrete compaction and hinder the rapid construction advantage of RCC, few joints or even no joints were recommended for RCC gravity dam. Early RCC gravity dams, such as Kengkou, Longmentan, Rongdi, etc., had no transverse joints and induced joints, even though length of dam crest reached 120-150m; the spacing of transverse joints of Tianshengqiao II, Mianhuatan and Fenhe 2nd Reservoir were set as large as 70-90m (Table 11.1-1).

In 1999, Zhu Bofang academician pointed out that "similar spacing of joints in RCC gravity dam should be set when compared with normal concrete monolith". In the early stages, engineering practice also proved that no transverse joints or too large joint spacing would result in contraction of the dam under temperature drop, and transverse joint like cracks, penetrating from upstream surface to downstream surface. Currently, joint spacing in RCC gravity dam is set close to that of normal concrete dam, mostly 20-25m. Wider spacing as 35-40m can be used for individual cases due to the layout requirement of water diversion hole, flood discharge equipment, with analytical demonstration and appropriate temperature control measures. Joints in RCC gravity dam are set a combination of transverse joints and induced joints (Table 11.1-1).

Longitudinal joints affect the RCC construction efficiency. Most importantly, they can hardly be grouted as the natural cooling process of the RCC dam, without 2nd phase cooling measure in general, will last for tens of years or even over one hundred years till stable temperature. So longitudinal joints have not been used in RCC gravity dams by far but adopted by full-silo casting method.

11.3 RCC Arch Dam

China's RCC arch dam technology has improved continuously and engineering applications become wider and wider. According to incomplete statistics, till the end of 2010, the total number of RCC arch dams that have been built and under construction rose to 32; many other projects were at the planning and design stage (Table 11.3-1).

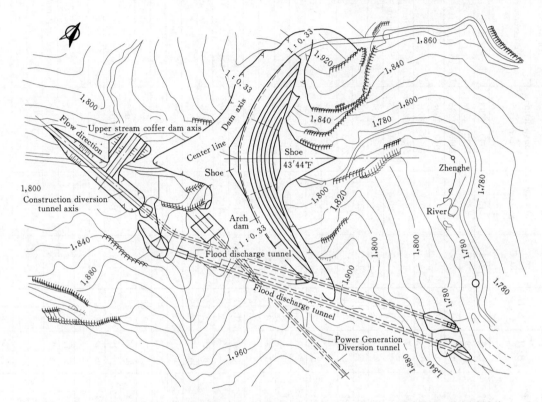

Fig. 11.3-1 Layout of Shapai Station Arch Hub (Seen in Liu Yanshen and Huangwei, 2004)

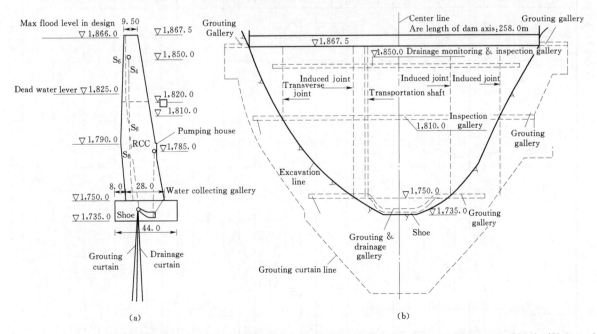

Fig. 11.3-2 Profile and Upstream Diagram of Shapai Arch Dam (Seen in Liu Yanshen and Huangwei, 2004) (Unit: m)

Part II Construction Technologies for Dams and Reservoirs

Table 11.3-1 RCC Arch Dams Built and nder Construction in China

No.	Dam	Type	Dam height H (m)	Dam top length L (m)	Dam top width b (m)	Dam bottom thickness B (m)	B/H	L/H	Amount of RCC ($10^3 m^3$)	Amount of total concrete ($10^3 m^3$)	Construction periods	Divided joints (bar) Transverse joints	Divided joints (bar) Induced joints	maximum joint spacing (m)
1	Puding	S.C	75	195.17	6.3	22	0.37	2.2	125	150	1990–1993	1	2	81
2	Wenquanbao	S.C	46	187	5	14	0.29	3.2	55	62	1992–1995	2	2	34
3	Xibing	S.C	64	101	5	12	0.19	1.2	26	29	1995–1996	Upstream short joints	3	45
4	Hongpo	S.C	55	244	4.5	26	0.48	4.4	70	75	1996–1998	2	2	80
5	Shangpai	S.C	132	250	9.6	28	0.24	1.8	365	392	1998–2001	2	2	45
6	Longshou	D.C	80	140	5	14	0.17	1.76	176	208	2000–2002	Surrounding joints: 2	2	111
7	Shimenzi	D.C	109	177	5	30	0.27	1.3	190	205	2001–2003	Short joint		88
8	Linhekou	S.C	100	311	6	27	0.27	3.11	220	295	2001–2004	1	6	
9	Zhaolaihe	D.C	105	198	6	18.5	0.18	1.89	180	204	2002–2004	2	3	75
10	Xiaqiao	S.C	68	212	6	22	0.32	3.14	67	90	2003–2005		2	
11	Chinglianxi	S.C	96	178				1.86		122	2003–2005	2	2	
12	Dahuashui	D.C	134.5	199	7	25	0.19	2.14	475	629	2004–2006	Surrounding joints: 2	2	
13	Yujianhe	D.C	81	180	4	16.5	0.2	2.22	105	110	2004–2006	2	2	
14	Liubo	S.C	70	258	6	21	0.3	3.68	140	168	2004–2006	2	3	
15	Xuanmiaoguan	S.C	80	242	4	16	0.25	3.04	75	95	2004–2006	1	2	
16	Chilinguan	S.C	77	140	4.5	13	0.16	1.82	45	55	2004–2006	2	2	
17	Longqiao	D.C	95	156	6	22	0.23	1.64	135	160	2005–2007	1	2	
18	Maobaguan	S.C	66	120	5			1.8	80	106	2004–2006		2	
19	Wuchuansha	D.C	87	149	6	20	0.23	1.71	86	92	2004–2006		2	
20	Bailianya	D.C	104.6	421.86	8	25	0.25	4.05	136	71.94	2004–2007	5	4	96
21	Huanghuazai	D.C	110	287.62	6	25.5	0.21	2.1	276	310	2005–2008	2	3	
22	Yunlonghe III	D.C	135	119	5.5	18	0.134	0.88	175	183	2007–2009		2	
23	Luopo	D.C	114	191	6	20	0.174	1.68	182	207	2007–2009		4	
24	Yunkou	D.C	119	152	5	18	0.155	1.27	205	205	2007–2010	1	2	
25	Tianhuaban	D.C	113	160	6	24.2	0.215	1.42	175	182	2007–2011	1	2	
26	Sazhu	S.C	72	160	7	14	0.194	1.6	90	100	2007–2009		2	
27	Sanliping	D.C	133	284.62	5.5	22.7	0.17	2.14	420		2006–	2	3	
28	Maduhe	D.C	99	250	6	17	0.177	2.51	210		2007–2011	2	3	
29	Wehou	D.C	82	271	6.7	18	0.22	3.3	22		2007–2010	Surrounding joints: 2	4	
30	Lizhou	D.C	132.5	175.86						380	2007–			
31	Shannipo		119.4											
32	Wanjiakouzi	D.C	167.5	413.16	9.0	36.0	0.23	2.47	960	980	2009–			

Note: Type, S. C = single arch dam; D. C = hyperbolic arch dam.

As the representative of RCC arch dams in China, Shapai Dam and Dahuashui Dam are 132m and 135m high, respectively. The former is located in the epicentral area of the Great Wenchuan earthquake, the seismic intensity which was beyond IX degrees. No cracks were found after earthquake, which indicates that RCC arch dams are of high safety and China's RCC high arch dam research and construction technology has been raised to a new level. Layout of Shapai Station arch hub is shown is Fig. 11.3 - 1, profile and upstream diogram of Shapai Arc Dam is shown in Fig. 11.3 - 2.

For rapid construction of large rolling equipment, most of the early RCC arch dam bodies belonged to the type of simple linear and thick arch, recently with a gradual shift to the hyperbolic type. The vast majority of RCC arch dams in completion and under construction in China nowadays are the hyperbolic ones. Little difference of thickness can be found between the normal concrete arch dam and RCC arch dam. Thickness-height ratios of more than a dozen recently constructed RCC arch dams are mostly below 0.25.

The biggest difference between RCC arch dams and normal concrete arch dams are the way of concrete casting and arch formation. For normal concrete dams, concrete is poured to form dentate monoliths which are separated by longitudinal joints, while RCC arch dams are constructed with few longitudinal joints. Before impoundment, temperature has to drop in normal concrete case through 2nd phrase water-cooling method, and then normal concrete monoliths are grouted as a whole. However, this process is hardly used for RCC dams as only a few of them have transverse joints or induced joints. Temperature drop in post-cooling stage relies on environmental conditions, and transverse joints or induced joints will be grouted prior to impoundment.

Transverse joints and induced joints were first successfully used in Puding arch dam. And the majority of China's RCC arch dams followed this method. Shapai arch dam developed new joint forming technologies for transverse joints and induced joints. Transverse joint formation took advantage of L-shaped gravity precast member (Fig. 11.3 - 3). Induced joints laid gravity precast members in pairs at intervals of 1 to 2 RCC layers (Fig. 11.3 - 4). Grouting holes were reserved in the pair of precast member for grouting after the opening of the joints.

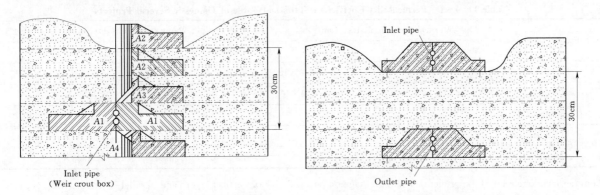

Fig. 11.3 - 3 Schematic of a Transverse Joint   Fig. 11.3 - 4 Schematic of an Induced Joint

Positions and intervals of transverse joints and induced joints in a RCC arch dam shall be determined based the simulation results, in accordance with shape, material, meteorology, hydrology and construction conditions. Joints are generally set in the place with maximum tensile stress and ensure that stress outside the joints is less than the tolerable tensile stress of concrete, in an attempt to avoid cracks. As thermal stress of the arch dam during construction period is closely related with its construction time, when the actual construction schedule is inconsistent with design, it is necessary to make appropriate adjustment to the positions of joints, based on the actual thermal stress, just like the arch dams of Yujianhe, Huanghuazai in Guizhou Province. Joint forming method of existing RCC arch dams are shown in Table 11.3 - 1.

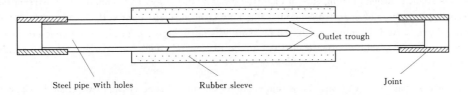

Fig. 11.3 - 5 Sketch up of Repeated Grouting Box

As RCC arch dams generally do not go through two-phase water-cooled process before impoundment, the temperature of dam has not yet dropped to a stable one when the dam is grouted. Therefore, transverse joints may be opened again in the subsequent natural cooling process. In order to grout re-opened the transverse joints and induced joints, China Institute of

Water Resources and Hydropower Research (IWHR) developed repeated grouting system for Shapai arch dam. The key component of this system is repeated grouting box with perforated tube valve (Fig. 11.3-5). The grouting box hinders the entry of external slurry and water. Pressure will be added to open seal boot and slurry will come out in event of re-grouting. This system has been installed in more than a dozen dams and has been successfully used for grouting in Shapai and Linhekou arch dam.

11.4 RCC Seepage Resistance and Drainage System

Because of the RCC interlayer weakness, improper treatment of layer surfaces will be prone to cold joints or concrete aggregate segregation, leading to concentration of large aggregates, poor consolidation and impermeability of concrete. Impermeable layers are currently set in RCC dam on the upstream face, which were 1.5m to 3.0m wide normal concretes in early engineering practice, namely "wrapping silver in gold" method, such as Guanyinge, Tianshengqiao II, Tongjiezi, Daguangba, Yantan, Wan'an and Taolinkou etc. For this type of impermeable layer, a mixture of two sorts of concretes that require separated transports had to be used, resulting in construction inconvenience. What's more, on account of different thermal mechanical properties of the two kinds of concretes, rises of adiabatic temperature of normal concrete are apt to produce large temperature differences, which are more likely to cause upstream surface cracks, thus affecting the seepage resistance. This "wrapping silver in gold" method is basically no longer used.

There were some domestic projects that studied asphalt mortar impermeable layer, reinforced concrete panels, precast concrete panels, etc., but were unable to promote due to complicated structure and apt to crack.

Such projects as Tongjiezi, Yantan, Jiangya and Longtan have done a slew of experimental researches on RCC impermeability. Table 11.4-1 shows permeability coefficient from core samples test and on-site water pressure test of several projects.

Table 11.4-1 Permeability Coefficient Statistical Table of Domestic Several Projects

Project name		Vertical core sample test (cm/s)			Water pressure test (cm/s)		
		Max	Min	Average	Max	Min	Average
Tongjiezi				$4.3 \times 10^{-9} - 5.0 \times 10^{-9}$			$1.96 \times 10^{-8} - 4.20 \times 10^{-9}$
Yantan				$1.3 \times 10^{-8} - 8.6 \times 10^{-10}$			
Tianshengqiao II							$3.95 \times 10^{-6} - 5.0 \times 10^{-5}$
Puding	2 grades				1.99×10^{-6}	1.73×10^{-7}	8.49×10^{-7}
	3 grades				9.54×10^{-6}	1.30×10^{-6}	3.35×10^{-6}

Longtan has done a lot of experiments at design stage on permeability of vertical layers and parallel layers of RCC core samples using field test blocks, with different gradations and in different construction conditions. Combined with test results of these two projects, through summarizing the test data, it can be found that under the premise of guaranteed construction quality, two-grade RCC already has a good impermeability, and can play an effective impermeable role.

The distorted concrete is a roller compacted concrete mixed with a certain amount of slurry, so that it has a certain slump, using plug-in vibrators in construction. The initial purpose is to solve the problem of inconvenient roller compaction of concrete close to the template, so as to get the desired appearance and to improve the binding properties of a variety of concretes. The construction process of distorted concrete can improve interlaminar bonding. Therefore, distorted concrete can be a part of concrete anti-seepage structure.

Impermeability was studied through testing core of distorted concrete in Jiangya RCC dam on the upstream face. The results showed that seepage coefficients of distorted concrete containing layer surface, joint surface and core samples of body concrete all reached 10^{-1} cm/s. And the initial infiltration pressure was much higher than two-grade concrete, reaching the level of normal concrete. Its impermeability outperformed two-grade RCC, and could meet the anti-seepage requirements of the 200m-level RCC dam.

Longtan RCC adopted seepage control plan, i.e. a combination of distorted concrete and two-grade RCC. Distorted concrete layer reaching 1.0-1.5m were set on the dam upstream face, followed by 4.0-10.0m of two-grade RCC dam as an anti-seepage body. And a thicker distorted RCC was used along transverse joints to warp sealing up and drainage pipes. To improve the impermeability, every layer surface had to be treated with slurry or mortar. Reinforcement was used in the distorted concrete against the crack propagation, with 20mm diameter and spacing 200mm×200mm. A 1mm thick cementitious capillary crystalline waterproofing material was brushed on the distorted concrete surface to further improve the impermeability and self-healing the surface cracks.

Longtan has been put into service for 4 years with little leakage observed, which indicates the success of impermeability design. Many in-situ water pressure tests have been done and the data showed that the permeability of RCC with rich binding material was close to that of normal concrete, the scattering of which were similar for two cases. The permeability of RCC decreases with the increase of binding material in mix portion, while such trend becomes slower when the certain amount (170 – 190kg/m^3) is reached. Thus, there is an economic amount between 170kg/m^3 and 190kg/m^3, which could lead to both high shear resistance and low permeability. Later, this research results were applied in Guangzhao and Wudu projects.

Similar drainage system design was used in RCC dam. The main upstream drainage curtain of Longtan dam is set 1.2m to the downstream side of two-graded RCC. Diameter of weep hole was 150mm spacing will be 2.0m if the dam height is below 270m and be 3.0m if the height above 270m. Auxiliary drainage system is comprised of auxiliary drainage holes with height reaching up to 230m in basic longitudinal auxiliary drainage gallery. Auxiliary drainage holes are designed to prevent internal uplift pressure caused by seepage around the main drainage holes arising from scattered seepage of layer surface, and to improve anti-sliding stability of downward layer surface of dam and durability of concrete. The auxiliary drainage hole spacing is 4m. The main drainage holes and auxiliary drainage holes are formed by the drilling method.

Several RCC gravity dam that have been built like Mianhuatan, Jiangya, Dachaoshan and Longtan measured leakage amount in the drainage gallery. And it is usually ranged between 0.4L/s and 0.7L/s, showing a decreasing trend, which indicates that metamorphosis with the combination of distorted concrete and two-grade RCC with rich binding materials is

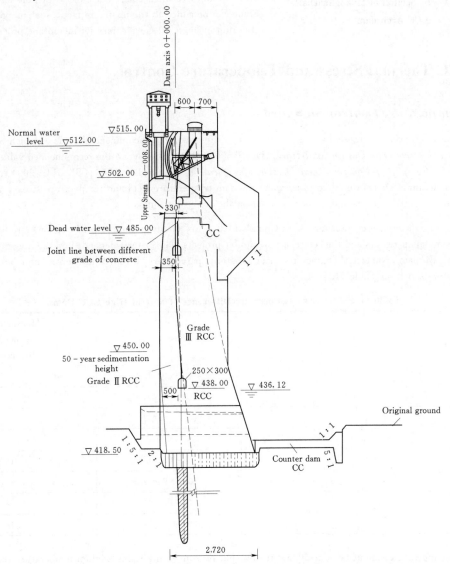

Fig. 11.4 – 1 Section of Linhekou RCC Ach Dam (Elevation Unit: m, Size Unit: cm)

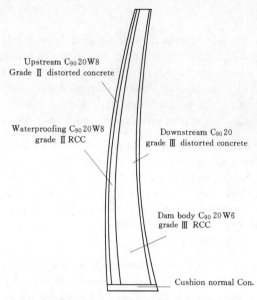

Fig. 11.4 - 2   Section of Huanghuazhai RCC Arch dam

successful and able to fulfill anti-seepage requirements of the 200m-level RCC dams.

Similar with gravity dam, this combination are also applied against leakage of RCC arch dam, e.g. Shapai, Zhaolaihe, Longshou, Huanghuazhai Village and Linhekou. See Fig. 11.4 - 1 and Fig. 11.4 - 2 are typical profiles of RCC arch dams in Linhekou and Huanghuazhai. The distorted concrete on the upstream face is 0.5m thick, with 2m to 5m thick two-grade RCC. The thickness is related to dam height. Practice has proved impermeable structure of this combination is a success in RCC arch dams.

It should be noted that there so far have been failure examples of anti-seepage system in RCC arch dams. Some dams suffer from severe leakage in downstream body and gallery after impoundment, and some dams are subject to jet-like leakage in downstream face, which so seriously affect safety and use of the dams that it has to reduce the water level for seepage strengthening. And the main reason is the poor quality of concrete construction. For thin RCC arch dams, in order to improve the seepage resistance, flexible impermeable membrane can be added to the upstream face, such as polyurea impermeable film by high pressure rotary jet membrane process.

11.5   RCC Thermal Stress and Temperature Control

11.5.1   *Temperature Control Standard*

Temperature control criteria are targeted as base temperature difference, the recommended value of which was once given in *Roller Compacted Concrete Dam Design Guidelines* (DL/T 5005—92). However, the recommended value was canceled in the updated *Roller Compacted Concrete Dam Design Specifications* released in 2003 (SL 314—2004); instead, it is required using of the finite element method to compute the dam temperature field and temperature stress, and then coming up with temperature control criteria and measures based on the analysis results.

The base temperature difference criteria for recent China's RCC dams (built and under construction) are basically specified on the basis of simulation results and the references to similar projects. The criteria depended on the linear expansion coefficient of concrete, the dam geometrical characteristics and material parameters. Temperature control standards of China's existing high dams are shown in Table 11.5 - 1.

Table 11.5 - 1   Base Temperature Difference of Several High RCC Dams

Project name	Coefficient of linear expansion ($10^{-6}/℃$)	Position	Allowable base temperature difference (℃)
Longtan	5.5	$(0-0.2)L$	16
		$(0.2-0.4)L$	19
Guangzhao	5.0	$(0.2-0.4)L$	16
			18
Jinghong	9.5 - 10.0	$(0-0.2)L$	14
		$(0.2-0.4)L$	17
Wudu	8.0	$(0-0.2)L$	16
		$(0.2-0.4)L$	18

In reality, base temperature cannot be controlled directly, but by controlling the maximum temperature that is determined in accordance with stable temperature field of dams on the basis of temperature difference.

11.5.2 Temperature Control Measures

RCC dams have been successively built from South China, the Yunnan border, to Northeast and Northwest China. South China has both high temperature and rainfall, along with dry and hot valley, where local temperature in hot seasons can reach 45℃. The northern winter is extremely cold, with the minimum temperature down to −40℃. Even in such grim regional and seasonal climatic conditions, China also realized uninterrupted RCC construction and has accumulated a large amount of research results and experiences as to thermal stress, temperature control and crack prevention.

Temperature control measures currently used in normal concrete dam construction, such as control of the casting temperature, water cooling, surface insulation and the storehouse spray in hot seasons, have been widely applied for RCC temperature control.

Lower casting temperature depends mainly on air-cooled aggregate, mixing with ice or cold water to reduce inlet temperature of the machine, and it still needs to take measures to reduce the temperature rise amid transport and casting process. Longtan Dam has summed up successful experience in practice, which has been applied in many dams.

1. Air-cooled aggregate mixed with ice or cold water to reduce inlet temperature of the machine

Because of low water-cement ratio and little usage of ice, the inlet temperature could not be much dropped, and pre-cooling the aggregate became the key measure. Using a one-time or twice air-cooled coarse aggregate mixed with pieces of ice or cold water, outlet temperature of the mixing plant couldbe controlled below 12℃.

2. Reducing the temperature rise amid the concrete transport

In the event of dump truck entering the storehouse, a spray device is supposed to be set at the entrance of the mixing plant, so as to reduce the temperature of small environment, and make the truck cool and humid; awning shall be placed at the top of dump truck in course of transportation, with insulation panels affixed to outside; when transporting through high-speed supply line, belt conveyor shall be cooled via cold air before loading materials and discharging them uniformly and continuous. It is necessary to guarantee a certain thickness of the concrete material in the supply line belt, and avoid the thin layer and idling. Shading rainproof plate cover and insulation facilities should be set on the top of the belt machine.

High-speed belt conveyor has been successfully applied in the Longtan Hydropower Project, with actual maximum transport intensity of a single line reaching 320m^3/h and beyond and the average intensity up to 260m^3/h. Through actual measurement, when belt conveyor loads materials from the mixing plant and unload them at storehouse, on conditions of air temperature between 25℃ and 35℃, and concrete temperature at 10℃ to 12℃, shading rainproof plate cover and insulation facilities placed on the top of supply line belt, as well as transport distance ranged between 410m and 585m, the temperature rise will be ranged between 3℃ and 5℃.

In event of dump truck, if temperature is at 25℃ to 35℃, the concrete temperature at 10℃ to 12℃, equipped with awning, the transport distance reaching 2.3km to 5.9km, the temperature rise will be between 0.5℃ and 2.2℃.

3. Controlling the temperature rise in the concrete casting process

In the process of RCC paving and rolling on storehouse surface, improper processes and construction management are likely to cause temperature rise, the Longtan Hydropower Project measures that prevent the temperature rise in the process of pouring: ①Reasonable planning the number of storehouses, determine the size of storehouse surface, and finish the paving and rolling of each storehouse in the shortest possible time. ②After concrete material warehousing, timely paving and timely rolling, it shall be controlled within 1.5h (the shortest is only 1h) from taking materials from the mixing plant to finishing rolling. Immediately after the completion of rolling, insulation quilts shall cover the storehouse surface, and then open them in turn until paving the next layer of material. ③In the process of RCC paving and rolling on storehouse surface, it is necessary to spray to form a small artificial environment. The Longtan project has successfully developed a new type of storehouse surface sprayer, applied to cooling and moisturizing storehouse surface, with good effect. The new sprayer is equipped with a 120° swing device, with spray distance up to 25m. And its droplet size is as small as about 20μm to 60μm, and control range of each sprayer reaches up to 700m^2. The measured results show that the full spray can reduce the temperature of storehouse surface by 3℃ to 6℃.

4. Water cooling

Due to the disturbance of laying cooling pipes with RCC construction, early RCC does without water cooling. The Dachaoshan RCC cofferdam started PE cooling pipe test and make a success, leading to full-scale application in Shapai Arch Dam, Longtan foundation restraint area and upstream face two with District and the hot season pouring part of the laying of PE pipes. Practice has proved that water cooling measures is still effective measures in RCC temperature control. At present, the majority of RCC arch dams, such as Dahuashui, Huanghuazhai, Bailianya, Dahua Bridge, all take water cooling measures. And such gravity dam as Guangzhao, Wudu, Jin'anqiao, Longkaikou have taken water cooling measures in part of dams.

Water cooling measures of RCC dams aim chiefly to reduce the peak temperature. Due to the slow heating of the RCC, this peak clipping effect is more obvious. It is indispensable to pay attention to the control over the temperature difference between the peak temperature and water temperature, which is generally less than 20℃. Meanwhile, it is necessary to control cooling rate. The cooling rate after the peak should be controlled at less than 0.5℃ per day. In practice, when cooling using river water with a small amount of water and without water pipe insulation, the cooling effect will be less satisfactory.

When cooling RCC dams through PE pipes, a special issue requiring attention is to avoid the water pipes burst caused by rolling. The leakage of pipes will lead to bad binding between RCC layers, even and affect safety of dams.

5. Surface insulation

The statistical results show that most of cracks on the surface of concrete dams are caused by temperature difference between inside and outside. Therefore, RCC dams existing or under construction largely adopt the surface insulation measures. The insulation for storehouse surface is a necessary measure regardless of the hot season or cold season. It is able to prevent heat intrusion in the hot season, and to avoid temperature drop caused by surface cracks in cold season. The insulation for the long exposed surface can effectively prevent cracks caused by the temperature difference between inside and outside. Temporary insulation materials currently used mostly are flexible polyethylene insulation quilts, color-stripes, as well as polystyrene board for permanent insulation. The thicknesses of insulation quilts and insulation boards need to be determined through analysis, which generally are 3 - 5cm.

6. MgO micro-expansive concrete crack control technology

Practice shows that concrete mixed with a small amount of light burned magnesium oxide will lead to controllable micro-expansive strain. Therefore, concrete containing or mixed with magnesia, coupled with an appropriate increase in the magnesium oxide content is used to prevent cracks. The early mixture with MgO is only used to fill pond and cast orifice. Since 2000, China has begun to attempt to use MgO concrete to build a whole dam, and to replace some or all temperature control measures for rapid construction. Guizhou completed the first dam, the concrete of which entirely mixed MgO in 2004, equipped with two induced joints and two transverse joints, and with abolition of RCC arch dams that entirely cancelled temperature control measures-Yujianhe, RCC Arch Dam, as shown in Fig. 11.5 - 1. The dam is 81.0m high, with arc length of top arch reaching 179.76m and the volume of RCC reaching 110,000m^3, mixed with MgO that accounted for 3.5% to 4.5% of cementitious materials. After the completion of the dam, only one induced joint opened, and the other one induced joint and two transverse joints were not open before impoundment. Test and its results indicate that, the concrete mixed with 3.5% to 4.5% magnesium oxide can produce the (70 - 100)$\mu\varepsilon$ expansive strain, reducing the tensile stress of concrete dam by 0.6 - 1.0MPa. The maximum expansive strain of dam concrete, through actual measure, reaches 110$\mu\varepsilon$, with stable deformation. The dam has been running for 5 to 6 years with no cracks, indicating that MgO expansive concrete is effective against crack initiation. Huanghuazhai RCC arch dam (dam height is 110m), after design and construction with the same ideas, was completed in October 2010, which helps accumulate experience of technology with respect to using MgO concrete in the whole dam. MgO concrete, as an auxiliary crack prevention measures, is also applied in the Longshou RCC Arch Dam.

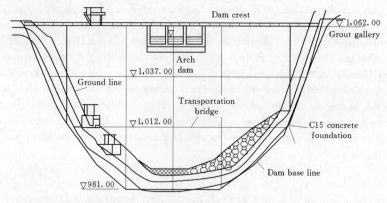

Fig. 11.5 - 1　Yujianhe RCC Arch Dam (Unit: m)

7. RCC temperature control in alpine regions

China's North has cold winters, with the lowest temperature up to $-30°C$ to $-40°C$, and China's Northwest has dry and hot summer, with annual range of temperature between 70°C and 80°C. Such rigorous climate conditions lead to enormous difficulties in the RCC construction and controls over temperature and cracks. Longshou RCC summed up a series of approaches on RCC construction in alpine region in the process of design and construction, which afterwards have been used successfully in other projects.

Longshou Hydropower Project is located in Zhangye City, Gansu Province. The annual average temperature of dam site is 8.5°C, absolute maximum temperature 37.2°C, minimum temperature $-33.0°C$, and the average daily temperature difference exceeds 20°C. RCC Construction began in March 11, 2000 and was completed in April 2001. Concrete pouring is normally performed whether in the hot season when maximum temperature is approaching 38°C or in cold seasons when minimum temperature is below $-30°C$, which achieve year-long RCC construction in dry alpine regions. Temperature and crack control measures of Longshou Arch Dam in cold season are as follows:

(1) Thermal storage construction. Thermal Storage of raw materials refers to heating of raw materials, in an attempt to improve the temperature at machine inlet, by means of aggregate heating and hot water mixing. Warm shed heating in pouring storehouse surface: a warm shed is erected in the pouring storehouse surface, using stove to heat up until storehouse temperature rise above 0°C, so as to meet the growing demand of concrete strength in cold season. Through these two measures, the temperature of mixtures is controlled at about 12°C. In course of transport, mixed concretes using closed transport equipment with insulation measures are carried to storehouse for paving and rolling as soon as possible.

(2) Adding antifreeze agent. When the temperature is at $-3°C$ to $-10°C$, RCC concrete is mixed with 4% DH_8 antifreeze to prevent RCC early frost damage, to improve performance of antifreeze, to lower the freezing point of concrete mixing water, and to keep concrete liquid not frozen in a negative temperature range. And the RCC thus does not suffer damage, ensuring that the cement hydration reaction can proceed, so that the concrete continues to harden.

(3) Thermal insulation. The insulation of RCC dams in alpine regions is the most important measures for temperature control and crack prevention. After discharge and spreading of fresh concrete, it is critical to immediately cover the concrete with insulation quilts; the pads is only uncovered while it has to be rolled along the rolling strip, and then prompt re-cover of the pads is needed. A poured layer shall be immediately covered with multi-layer insulation quilts once rolling is completed. The thickness and amount of insulation layers are determined based on statistical analysis of specific project situation. Before pouring the upper concrete, insulation quilts cannot be open at a time, but be uncovered layer by layer, exposing a layer over a while for adaption of concrete and then proceed, so as to avoid cold shock to concrete layer caused by uncovering all of a sudden.

Insulation on the upstream and downstream surfaces shall be performed in construction period and operation period. The insulation of construction in the alpine region may well use as inside-paste method, aiming to avoid the cold hit by form removal prior to pasting insulation boards. Permanent insulation may well be as combined with that of the construction period. If the combination is impossible, it is necessary to remove the construction of thermal insulation and permanent insulation in the right season. Concrete construction season shall be calculated based on analysis of specific circumstances of projects to avoid the cold hit caused by removal of insulation boards during the construction period and stress-induced cracks caused by temperature changes before pasting permanent insulation boards.

By virtue of thermal storage construction, adding antifreeze agent and flexible use of thermal insulation, Longshou Arch Dam has successfully realized RCC construction in alpine regions in extreme cold seasons, effectively controlled the temperature of dam body, with no hazardous cracks. A RCC gravity dam in the cold region adopted a similar approach, and also successfully finished construction of concrete in the harsh conditions in cold winter, with good results.

11.6 Materials and Mixed Proportion of RCC

The RCC materials in China are featured by medium cementitious content, low cement content, high volume of fly ash (pozzolan), high stone powder content, low water-cementitious material ratio, and low *VC* value. The low content of cement has effectively reduced the temperature rise of the heat hydration of the RCC. The widespread use of superplasticizer has reduced water content and water-cementitious material ratio, while maintaining low *VC* value of RCC, and improved impermeability. Medium cementitious content and the high stone powder content have ensured the compact ability, the density of compaction and the bond between layers of RCC. Adding air-entraining agent has overcome the poor durability of the RCC against frost. The development and application of limestone powder, water quenched slag and other mineral admixtures have got the RCC rid of dependence on fly ash, offering a greater space for the RCC dam construction technology.

11.6.1 General Technical Requirements

The RCC dam in China is generally made from the concrete having two-graded aggregates and three-graded aggregates with maximum sizes of 40mm and 80mm, respectively. The concrete having two-graded aggregates (40mm) is used for the dam upstream face, playing the role of impermeability. The concrete having three-graded aggregates (80mm) is used for the dam inside and downstream surface. For 100m high RCC dams, the interior concrete can be divided into upper and lower parts; for RCC dams with a height of 200m, the interior concrete can be divided into upper, middle and lower parts. Grout enriched vibrated concrete (GEVC) is used for improved the impermeability of upstream face, at the location near formworks, also in the zone surrounding galleries and contact zone with abutment, etc. The RCC materials must firstly meet the design strength requirements, then the workability and temperature control requirements during construction period, and then the durability requirements in the lifetime of dam.

1. Strength

Based on dam height and stress level, the compressive strength of RCC is generally in the class of C10 – C25 at the ages of 90d or 180d. For Kengkou dam, China's first RCC dam, the design compressive strength was $C_{90}10$. For the interior concrete of upper, middle and bottom parts of Longtan RCC dam, the design strength classes were $C_{90}15$, $C_{90}20$ and $C_{90}25$, respectively; in the zone of upstream face, $C_{90}25$ RCC with a maximum aggregate size of 40mm was adopted. For Shapai RCC dam, the design strength of RCC in upstream face zone and interior zone were both $C_{90}20$.

There are specific requirements for the tensile strength and shear strength of the joints between lifts of RCC for high dams and dams with seismic loads.

2. Workability

The workability of fresh mixed RCC mainly includes the VC value and setting times. In China, RCC dams built at early ages adopted RCC with high VC value, and the value then decreased gradually. For example, the VC value of RCC at Kengkou dam was 15 – 25s, then at Yantan RCC dam was 5 – 15s, at Jiangya 3 – 12s, at Suofengying 3 – 8s, at Longtan 5 – 7s, and at Guangzhao 3 – 5s, respectively. According to a large amount of engineering practice, the range of VC value suitable for compaction is 3 – 12s, and the lower values are preferable in normal case.

The RCC setting time under the construction temperature determines the allowable exposure time for placing the next layer without treatments. The allowable time interval for placing the next layer is about 60% to 70% of the initial setting time of fresh RCC, or when penetration resistance does not exceed 5MPa. According to the site temperature conditions, the initial setting time of RCC is generally controlled at the range of 4 – 12h.

3. Limit tensile strain before crack

The limit tensile strain of RCC before crackis generally in the range of $0.65 \times 10^{-4} - 0.85 \times 10^{-4}$.

4. Durability

According to the exposure environment of RCC dams, there are specific requirements for durability on almost all RCC dams. If the dam is located in severe environment, the RCC must first of all meet the durability requirements. If the dam inevitably adopted potentially harmful alkali reactive aggregates, then there must be comprehensive test results to prove that the combination of materials or suppression measures can effectively inhibit the risk of alkali-aggregate reaction. The frost resistance is a representative index to measure the durability of concrete. The class of frost resistance of the exterior RCC should be F50 – F300, and that of interior RCC is F50 – F100. Another important durability index for RCC is impermeability, which is determined by water head and hydraulic gradient of different parts in RCC dams. The impermeability of the RCC on upstream face is generally W8 to W12, and that of the interior RCC is generally W4 – W8.

Table 11.6 – 1 shows the technical requirements of RCC at Longtan; and Table 11.6 – 2 shows the technical requirements of RCC at Dahuashui.

Table 11.6 – 1 Technical Requirementsof RCC at Longtan Gravity Dam

Item	R I (Lower part of dam)	R II (Middle part of dam)	R III (Upper part of dam)	R IV (Upstream face of dam)
Compressive Strength (MPa)	$C_{90}25$	$C_{90}20$	$C_{90}15$	$C_{90}25$
Tensile strength (MPa)	≥2.0	≥1.8	≥1.4	≥2.0

Countinued

Item		R I (Lower part of dam)	R II (Middle part of dam)	R III (Upper part of dam)	R IV (Upstream face of dam)
Tensile Strain Capacity (10^{-6})		≥80	≥75	≥70	≥80
Frost resistance		F100	F100	F50	F150
Impermeability		W6	W6	W4	W12
VC value (s)		5-7	5-7	5-7	5-7
Maximum aggregate size (mm)		80	80	80	40
Degree of compaction (%)		≥98.5	≥98.5	≥98.5	≥98.5
Bulk density (kg/m³)		>2,400	>2,400	>2,400	>2,400
Shear strength	f'	1.0-1.1	1.0-1.1	0.9-1.0	1.0
	c' (MPa)	1.9-1.7	1.4-1.2	1.0	2.0

Table 11.6-2 Technical Requirements of RCC at Dahuashui

The zone of RCC	Compressive strength (MPa)	Tensile strength (MPa)	Tensile Strain Capacity (10^{-6})	Frost resistance	Impermeability	VC value (s)	Maximum aggregate size (mm)	Degree of compaction (%)	Bulk density (kg/m³)
Upstream face	C20(90d)	≥2.2	≥75(28d)	F100	W8	4-7	40	>98	>2,400
Interior dam	C20(90d)	≥2.2	≥75(28d)	F50	W6	4-7	80	>98	>2,400

11.6.2 Raw Materials

11.6.2.1 Cement

For producing RCC, there are no special requirements for the cement, and any type of cement that can be used for producing conventional concrete can also be used for producing RCC. The cement commonly used for construction of RCC dams in China includes ordinary Portland cement and moderate heat Portland cement. Moderate-heat portland cement has no content of blending material, and has stable quality with low heat generation characteristics at early ages. If there is a cement plant around the dam site and the price is acceptable, then moderate-heat Portland cement should be preferred and used in many projects, such as Yantan, Guanyinge, Jiangya, Mianhuatan, Linhekou, Longtan, Baise, Xinjiang Shankou, Pengshui, Jin'anqiao. In Puding project, ordinary portland cement was used. Nevertheless, ordinary portland cement is more easily available and cheaper in most regions, so the great majority of RCC dams in China selected ordinary portland cement, including Shapai, Guangzhao, Dahuashui, and other projects.

It is very important for the quality control of RCC to maintain cement quality uniform and stable. The abnormal fluctuation of gypsum types and content in cement may lead to abnormal setting of the RCC, so they should be strictly controlled. In order to reduce and postpone the temperature rise of RCC in 1-3d, the specific surface area (Blaine) of ordinary Portland cement should not be more than 350m²/kg.

11.6.2.2 Supplementary Cementitious Material (SCM)

The supplementary cementitious material (SCM) used in RCC dams include fly ash, water quenched slag, phosphorus slag, tuff, natural pozzolan, limestone powder, etc. In the majority of RCC dams in China, class I and class II low calcium fly ash (class F) were selected as SCM, but there is no project cases where high calcium fly ash (class C) were used in the construction of RCC dams. The reason why high calcium fly ash (class C) has not been accepted in projects was that its high f-CaO content might cause harmful expansion, and difficult to control. According to fineness, loss on ignition, water requirement ratio, fly ash can be classified into class I, class II and class III, as shown in Table 11.6-3. The quality of class III fly ash is unstable. With high ignition loss, it will strongly adsorb air-entraining agent used as SCM, consequently affecting the VC value and compactibility of RCC. Thus special attention must be paid for using fly ash of class III, the fly ash with low ignition loss would be the best choice. For large and extremely large size RCC dam, during the peak period of dam construction, they often need several thermal power plants to supply fly ash. For example, the total volume of Longtan RCC dam was 6.59 million m³, of which 4.35 million m³ was RCC. During its peak time of construction, over ten thermal power plants supplied fly ash, thus the classification, storage and use of fly ash became also issues to be solved.

Table 11.6-3 Quality Requirements for Fly Ash

Class	Class I	Class II	Class III
Fineness (Amount retained on 45μm sieve) (%)	≤12	≤25	≤45
Loss on ignition (%)	≤5	≤8	≤15
Water requirement ratio (%)	≤95	≤105	≤115
Sulfur trioxide (SO_3) (%)	≤3	≤3	≤3
Moisture content (%)	≤1	≤1	≤1

For the areas in shortage of fly ash, long-distance transporting fly ash would make it very uneconomical to construct RCC dams, so research and application have been conducted on a variety of new SCM ssubstituting for flyash. The Dachaoshan project adopted PT material, made up of 50% phosphorus slag and 50% tuff, and 1.1 million m^3 of dam concrete were placed. Jinghong project adopted the combination of water-quenched iron-slag and limestone powder as SCM at the ratio of 50 : 50 by mass, and over 1 million m^3 of RCC were placed. Compared with tuff powder, limestone powder is easier in production, and it can get along well with portland cement to reduce water content. Limestone has rich reserves in China, and is widely distributed and available easily. Test results have shown that only using limestone powder as SCM, RCC can also be proportioned to meet the design requirements of workability, compactibility, strength and durability. After the construction of Jinghong project, Tukahe, Jufudu and Gelantan RCC dams also adopted the combination of ground water-quenched iron-slag and limestone powder as SCM. The project of Shankou hydropower station located at Tekesi River in Xinjiang province has adopted fly ash and limestone powder as SCM at the ratio of 50: 50 by mass. Nongling hydropower station located at Longjiang River in Yunnan province adopted natural pozzolan as SCMs in RCC and achieved good results.

11.6.2.3 Aggregates

Natural aggregates and crushed aggregates both can be used as aggregates in RCC. RCC using crushed aggregates has better performance in reducing the segregation and higher shear strength between lifts. Among RCC dams built in China, Guanyinge, Shimenzi, Longshou, Jinghong and other projects have adopted natural aggregates; Longtan, Guangzhao, Yantan, Puding, Jiangya, Linhekou, Dahuashui and other projects have adopted crushed limestone as aggregates; Mianhuatan and Shapai projects have adopted crushed granite; Dachaoshan and Jin'anqiao projects have adopted crushed basalt aggregate; Longkaikou and Gelantan projects have adopted crushed dolomite aggregate; and Baise adopted crushed diabase rock aggregate. Limestone can be crushed easily in production, and crushed limestone aggregates have good particle shapes, thus the water content is low for RCC. On the other hand, the thermal expansion coefficient of RCC is small and it has good crack resistance, hence many RCC dams have adopted crushed limestone as aggregates. In Shapai, the RCC using metamorphic granite aggregates is featured with high tensile strain capacity and low elastic modulus, so it has shown excellent crack resistance during the construction period, and no cracking has occurred.

The fineness modulus of crushed sand is generally controlled at the range of 2.2 to 2.9, and the fineness modulus of natural sand at 2.0 to 3.0. When manufacturing crushed sand by wet process, the corresponding dewatering facilities and silos are necessary in order to control the water content of finished sand below 6%. In order to obtain good compatibility, the fine aggregate for RCC should contain sufficient stone powder (fines passing 0.16mm). The content of stone powder in crushed sand should be controlled at the range of 12% to 20% in general, in which the fine particles (passing 0.08mm) should be no less than 5%. The content of stone powder in natural river sand is generally low, so usually fly ash, stone powder or other supplementary fines are added to increase the fines content to more than 8%. At Longtan project, impact-type crusher and rod mill were combined to manufacture sand, and the content of stone powder in the finished sand was still less than 16% to 18%. Later, the stone powder recycling system was equipped to make up for the shortage of stone powder in crushed sand. The content of stone powder in river sand at Jinghong project was 3%-6%, and it was increased to about 16% by adding limestone powder. At Jin'anqiao project, the stone powder content in crushed sand made from basalt was about 13%, and it was increased to 18%-20% by adding limestone powder.

11.6.2.4 Chemical Admixtures

RCC dams have usually mixed chemical admixtures such as slushing agent for concrete and air entraining agent, and so on. Water-reducing and air entraining admixture are diluted in water solution to appropriate concentration, and incorporated in RCC at mixing machine. The type of air-entraining and water-reducing admixture is no longer used in RCC.

At present in China, the water-reducing rate of superplasticizer used in RCC is generally 15% to 20%, and when it is used together with air entraining admixture, the water reducing rate can reach 20%-25%. By changing the type and dosage of retarding or accelerating ingredient in superplasticizers, special types of superplasticizer suitable for different ambient temperature can be made, such as high temperature type (daily average temperature above 25℃), low temperature type (daily average temperature is lower than 5℃) and normal temperature type. Shankou RCC dam is located in Guangdong Province, where the average temperature in summer (June to September) is 28.5℃. During construction, high temperature type superplasticizer was used, and when RCC was placed under the temperature of 28 - 34℃, the initial setting time were controlled at more than 10h. Hence, RCC construction at the temperature of around 35℃ was successfully realized.

It is more difficult to entrain air bubbles in RCC than in conventional concrete. In order to entrain sufficient air content in RCC, the dosage of air entraining admixture should be at least 5 to 10 times more than that in conventional concrete. Sometimes, when fly ash with high ignition loss is used, even if the dosage of air entraining admixture is raisedsubstantially, it is still difficult to increase air content of RCC to more than 3%.

11.6.3 Mixed Proportion

In China, the mix proportion of RCC is designed according to the "water-cement ratio law" and "water content stability" principle, and the steps are the same as conventional concrete. Table 11.6-4 lists the typical mixture proportions of RCC dams in China, showing the following features:

(1) The content of cementitious materials ranges from 140kg/m³ to 190kg/m³, the dosage of fly ash (SCM) is in the range of 50%- 65%, and cement content is about 50 - 100kg/m³.

(2) Water-cementitious materials ratio (W/CM) is in the range of 0.40 - 0.65, and water contentin the range of 75 - 100kg/m³.

(3) RCC mixtures with low VC values (3 - 10s) are more often used.

(4) Sand ratio in the range of 30% to 39%, and paste/mortar ratio above 0.4.

(5) Superplasticizer and air entraining admixture were incorporated in RCC of all dams.

(6) At the upstream of dams, the RCC having two-sizes aggregates were used, with the maximum aggregate size of 40mm, and RCC having three-graded aggregates were used in the interior and downstream of dams, with the maximum aggregate size of 80mm.

11.7 RCC Construction

After more than 20 years of exploration and practice, China has accumulated a lot of experience in the construction of RCC, and has formed a complete set of construction techniques, including production and transportation of aggregates, mixing and transportation of concrete, dumping, spreading, compacting, quality inspection and control, etc. In the arrangements of construction, equipment configuration, site layout, etc., China has developed a set of optimization methods based on simulation analysis, which has been utilized in such large and medium-sized projects as Longtan, Guangzhao, Pengshui projects.

11.7.1 Production of Aggregates

For large and medium-scaled project, the natural aggregates are generally difficult to meet the demand and requirements, so crushed aggregates are more often used. For aggregate production, China mainly adopts wet crushing process, such as the Dafaping crushed aggregate system in Longtan, which can meet the placing rate of more than 325,000m³ per month. The Sinohydro Bureau 8 Co., Ltd developed the semi-dry aggregate production process, and utilized it in the project of Dahua Hydroelectric Power Plant.

11.7.2 Mixing, Transportation and Storing of RCC

Concrete mixing system is mainly made up of concrete mixing plant, aggregate transport system, cement and fly ash transport system, secondary screening system, concrete pre-cooling system, wastewater treatment facilities and other ancillary facilities. Concrete mixing plant can be classified into two types, namely forced mixer and tilting drum mixer. In the late 1990s, the Sinohydro Bureau 8 Co., Ltd collaborated with research institutes to develop the first set of 200m³/h continuous mixer, and successfully applied it in Shapai RCC arch dam, Suofengying RCC gravity dam, and so on. The Longtan concrete production system is equipped with three 2×6.0m³ twin-shaft batch mixing plantand one 4×3.0m³ tilting drum mixing plant, and the total production capacity was up to 1,080m³/h. For Pengshui Hydropower Station, two 4×4.5m³ mixing plant and one 2×3.0m³ mixing plant were utilized, with a total production capacity of 960m³/h.

Table 11.6-4 Typical Mixture Ratio of RCC Dams in China

No.	Dam	Dam type	Mixed portion							VC value (s)	Year of completion	Remarks
			W (kg/m³)	C (kg/m³)	SCM (kg/m³)	W/CM	S (%)	Fine Agg. (kg/m³)	Coarse Agg. (kg/m³)			
1	Kengkou	G	94	65	85 (FA)	0.63	36	782	1,410	5 – 25	1986	$D_{max}=80mm$, interior
2	Yantan	G	90	55	104 (FA)	0.57	30	759	1,490	5 – 15	1992	$D_{max}=80mm$, interior
3	Puding	A	84	54	99 (FA)	0.55	34	768	1,512	5 – 15	1993	$D_{max}=80mm$, dam body
			94	85	103 (FA)	0.50	38	836	1,396			$D_{max}=40mm$, upstream
4	Jiangya	G	93	64	96 (FA)	0.58	33	738	1,520	3 – 11	2000	$D_{max}=80mm$, upper part
			93	46	107 (FA)	0.61	34	761	1,500			$D_{max}=80mm$, bottom part
			103	87	107 (FA)	0.53	36	783	1,413			$D_{max}=40mm$, upstream
5	Mianhuatan	G	88	51	95 (FA)	0.60	35	765	1,460	3 – 8	2001	$D_{max}=80mm$, bottom part
			88	48	88 (FA)	0.65	35	769	1,467			$D_{max}=80mm$, upper part
6	Shapai	A	100	90	110 (FA)	0.50	38	812	1,330	5 – 15	2001	$D_{max}=80mm$
			93	93	93 (FA)	0.50	33	730	1,470			$D_{max}=80mm$, dam body
			102	115	77 (FA)	0.53	37	810	1,378			$D_{max}=40mm$, upstream
7	Dachaoshan	G	84	67	101 (PT)	0.50	35	798	1,521	5 – 10	2002	$D_{max}=80mm$, interior
			94	94	94 (PT)	0.50	38	850	1,423			$D_{max}=40mm$, upstream
8	Longshou	A	85	60	117 (FA)	0.48	30	650	1,518	5 – 7	2002	$D_{max}=80mm$, interior
			91	99	112 (FA)	0.43	32	687	1,461			$D_{max}=40mm$, upstream
9	Linhekou	A	81	66	106 (FA)	0.47	34	754	1,464	3 – 9	2004	$D_{max}=80mm$, dam body
			87	74	111 (FA)	0.47	38	823	1,361			$D_{max}=40mm$, upstream

Continued

No.	Dam	Dam type	Mixed portion							VC value (s)	Year of completion	Remarks
			W (kg/m^3)	C (kg/m^3)	SCM (kg/m^3)	W/CM	S (%)	Fine Agg. (kg/m^3)	Coarse Agg. (kg/m^3)			
10	Suofengying	G	88	64	96 (FA)	0.55	32	680	1,526	3–8	2005	$D_{max}=80$mm, interior
			94	94	94 (FA)	0.50	38	815	1,365			$D_{max}=40$mm, upstream
			94	129	129 (FA)	0.50	38	822	1,372			$D_{max}=40$mm, downstream
11	Jinghong	G	88	64	96 (SL)	0.55	35	795	1,476	3–10	2006	$D_{max}=80$mm, interior
			98	98	98 (SL)	0.50	40	879	1,321			$D_{max}=40$mm, upstream
12	Baise	G	96	59	101 (FA)	0.60	34	814	1,579	3–8	2006	$D_{max}=60$mm, interior
			108	91	125 (FA)	0.50	38	864	1,410			$D_{max}=40$mm, upstream
13	Longtan	G	79	86	111 (FA)	0.40	33	720	1,477	5–7	2007	$D_{max}=80$mm, lower part
			78	70	105 (FA)	0.45	33	727	1,493			$D_{max}=80$mm, middle part
			77	56	104 (FA)	0.48	34	755	1,483			$D_{max}=80$mm, upper part
			87	99	121 (FA)	0.40	38	812	1,340			$D_{max}=40$mm, upstream
14	Guangzhao	G	75	83	83 (FA)	0.45	34	732	1,496	3–5	2008	$D_{max}=80$mm, lower part
			75	68	82 (FA)	0.50	35	755	1,488			$D_{max}=80$mm, middle part
			75	55	82 (FA)	0.55	35	768	1,513			$D_{max}=80$mm, upper part
			83	92	92 (FA)	0.45	39	811	1,348			$D_{max}=40$mm, upstream, bottom part
			83	75	91 (FA)	0.50	39	822	1,366			$D_{max}=40$mm, upstream, middle part
15	Shankou in Xinjiang	G	74	62	99 (FL)	0.46	30	660	1,556	3–8	2008	$D_{max}=80$mm, interior
16	Nongling	G	80	64	126 (NP)	0.50	32	668	1,483	3–9	2009	$D_{max}=80$mm, interior
			93	93	129 (NP)	0.45	35	727	1,352			$D_{max}=40$mm, upstream

Note: Dam type: G=Gravity dam; A=Arch dam;
Mixed portion: W=Water; C=Cement; FA=Fly Ash; PT=a combination of 50% phosphorus slag and 50% tuff by mass; SL=a combination of 50% ground water-quenched slag and 50% limestone powder by mass; FL=a combination of 50% fly ash and 50% limestone powder; NP=ground natural pozzolan; S=sand raito; Fine Agg.=fine aggregate; Coarse Agg.=coarse aggregate.

The transportation of the RCC is divided into horizontal transport and vertical transport, requiring continuity, fastness, large transport capacity, and no segregation. For transporting RCC with pre-cooling aggregates, it is required that temperature recovery should be minimal. When it has long distance between aggregate production system and concrete mixing system, there should be an aggregate transportation system. Large and medium-sized projects usually adopt multi-belt conveyors. In Longtan, finished aggregates were transported using single-hole, and single-belt transport system; the length of belt conveyor was 4km, belt width was 1.2m, speed was 4.0m/s, and designed transport capacity was 3,000t/h.

Both vertical and horizontal transportation are integrated. In recent years, due to the import, development and applications of tower-belt (top machines) and creter crane the horizontal and vertical transport of concrete has been combined, changing the traditional way of concrete transport. The belt conveyer is widely used. And mobile fabric machines and telescopic cantilever distributor have been successfully used in a number of projects.

With construction of high dams in canyon sites, steep-slope and vertical transportation equipment have been developed and applied, and vacuum chute has seen new development; in Dachaoshan and Shapai the 100m vacuum chutes have been adopted. In Dachaoshan hydropower station, two vacuum chutes have been deployed along the two banks of the river. On the left bank of the river, the maximum height difference of the vacuum chute was 86.6m, and it was 120m in length, with a transport capacity of $220m^3/h$. The pipe chute is a highly efficient and new-type vertical RCC transportation equipment that appeared in recent years, the section can be rectangular or circular, with a size of 800mm × 800mm or $\phi 600 - 1,000mm$. When in operation, it is always filled with concrete, so it is both a large-diameter pipeline, and a giant storage box. The installation angle of pipe chute can be $45° - 90°$, height difference is 10 - 100m, and transport capacity is much larger than traditional vacuum chutes. The transport capacity of pipe chute (800mm×800mm) at Guangzhao (gravity dam, 200.5m, completed in 2008) was over $500m^3/h$ (Fig. 11.7-1).

At present, the medium and low dams, as well as the bottom part of high dams, are generally adopting dump trucks to transport RCC to point of placement, while the middle and upper parts of high dams adopt a combination of tower machines, belt conveyors and chutes. For some locations where it is difficult to reach, transport by cableway has to be adopted. For vacuum chute, due to technical constraints, it cannot meet the transport requirements of steep slopes, so the Sinohydro Bureau 8 Co., Ltd developed a new type of vertical transport mode. The technology makes use of vacuum principle, and adopts a combination of horizontal belt conveyor and vacuum pipe in order to vertically transport RCC. It has successfully resolved the problem of high-intensity and vertical transportation of RCC in alpine valley regions, high drop (up to 70m) and steep dip ($60° - 90°$), and it has realized continuous rise of the main body of arch dams by 34.5m. At present, the technology, adopted by Dahuashui, Silin, Geliqiao projects, has been further improved, showing a wide range of application value.

In the practical application, the above mentioned transport methods will be combined if necessary. For example, the transport of RCC at Longtan Dam adopted two tower machines, three lines of high-speed belt conveyors, two 20/25t - 9.5m translational cable crane, vacuum chutes and dump trucks.

Fig. 11.7-1   Box-type Pipe Chute (Guangzhao)

11.7.3 Construction Technologies of RCC

In the construction of RCC, the process of continuous rise of thin layer RCC was commonly adopted. The levelling machines, joint cutter, vibrating rollers, deck cranes, sprayers, materials and methods of embedded cooling pipes, and construction technology of embedded parts have developed along with the development of RCC. The performance of devices can guarantee continuous and rapid RCC construction.

1. Spreading, levelling and compacting of RCC

For concrete spreading, dump trucks are generally used, and bulldozers are used to level the dumped RCC mixes. In order to reduce the segregation of RCC mix, the discharge in layers and stringed spreading are utilized. Segregated RCC is often required to remove or remix by hand labor.

The spreading of concrete should be continuous and uniform, and the pile height is generally less than 1m, and the materials should be unloaded on inclined planes compacted or leveled; the unloading is conducted on sub-bands, and is perpendicular to the spreading direction. After spreading, the concrete should be leveled and compacted immediately. Special levelling machines are used, for example, in the Longtan project, where 3 units of CATD3GLGP leveling machine and 5 SD16L levelling machines were used. When levelling, the thickness of materials should be controlled, with the thickness of each layer reduced, until it reaches the required thickness. Each compacted layer is generally 30cm in thickness, and it is determined according to the mixture proportion and on-site test, usually 33 – 36cm. Thickness deviation in the spreading process is generally controlled within ±3cm. The direction of placing should be perpendicular to waterflow direction as much as possible.

The RCC is compacted by vibratory rollers. The Longtan project was equipped with the 13BW202AD – 2 vibratory roller, and Suofengying project with two BW202AD and one QY12 vibratory rollers. At the same time, they were equipped with small vibratory rollers for compacting RCC at the corner sites. The compacting direction should be perpendicular to the waterflow direction, in order to improve impermeability. The required number of roller passes should be determined according to full-scale trial tests and generally after the spreading, concrete should be rolled without vibration twice, followed by 6 to 8 passes of vibratory compaction, and then 1 to 2 passes if needed. For corner parts, 16 to 24 passes should be rolled with small hand-guided vibratory roller. During RCC placement, the width of overlap between the bands should be not less than 20cm, the walking speed should be controlled in the range of 1.0 – 1.5km/h, the time interval between layers of RCC should be 1 – 2h less than the initial setting time, and the total time lapse between the start of mixing and completion of compaction should be controlled at less than 1.5h. During continuous placement of RCC, allowable time interval for placing the next layer is different according to regions and seasons, generally it is 2 – 4h in summer and 6h in other seasons.

2. Sloped layer method of RCC

For gravity dams, as the construction deck is large, the construction sites are restricted by the intervals between layers, so the RCC can be paved by oblique layer shoving. The oblique layer shoving can be in two directions: one is perpendicular to the dam axis, i. e. the RCC level is oriented to the upstream, and concrete pouring advances from downstream to upstream; the other is parallel to the dam axis, i. e. the RCC level is oriented from one shore to the other. For dam bottom where width is large, the first pouring method is conducive to construction and the tendency from downstream to upstream can contribute to the impermeability and shear. So it is best to have the concrete pouring perpendicular to the dam axis, the RCC level tending to the upstream, the slope of oblique layer at 1 : 15 to 1 : 20, and height at 1.5 – 3m (with minimum value in summer). The Dachaoshan, Longtan, Guangzhao and other projects adopted this method.

3. Treatment of layer surface (joint surface)

For treatment of construction joints and cold joints, surface finishing and other methods can be used to clear the rough surface and loose aggregates on the concrete. After layer surface processing is completed and cleaned, and after inspection and acceptation, mortar in thickness of 1.5 – 2cm can be paved, on top of which the RCC is paved. The requirements for joint surface treatment are as follows: ① The construction joints are cleaned using low pressure water, with water pressure of generally 0.2 – 0.5MPa; surface finishing is conducted after the initial setting and before final setting, usually in 16 – 24h after flattening the concrete, with the minimum value adopted in summer and the maximum value adopted in winter. ② For surface finishing with high pressure water, the water pressure is generally 20 – 50MPa; the surface finishing must be conducted after the final setting of concrete, generally in the 20 – 36h after flattening the concrete, with the minimum value adopted in summer and the maximum value adopted in winter. When high-pressure water surface finishing is practiced, the gun port is 10 – 15cm away from the joint surface, and the angle is around 75℃. ③ Before the RCC pouring, the construction joints must be washed clean and free from water, dirt, etc. ④ In order to facilitate the even distribution of mortar and to ensure construction quality, the consistency of mortar should be between 140 – 180mm. ⑤ On each RCC layer in the upstream impermeable area (RCC concrete in second tier zone), concrete ash paste of 2mm thick should be paved, in order to improve the layer connection and anti-seepage ability.

4. Construction of Grout enriched vibrated concrete (GEVC)

The metamorphic concrete is a technology developed by China independently and is mainly used for areas which are hard or cannot be reached by rolling equipment in upstream and downstream surface and corners; it can also play an anti-seepage role on the surface. After using metamorphic concrete, the whole sections of the dam can get rid of normal concrete, thus speeding up construction and ensuring quality. In China now, almost all the RCC dams have adopted metamorphic concrete.

The slurry amount in metamorphosis concrete needs to be determined experimentally, and is generally 4% to 6%; in Longtan dam it was 5%. Slurry mixing is a crucial process of construction, and is directly related to the quality of metamorphic concrete. The main control processes are divided into the following two aspects. ①Methods of slurry mixing, which are mainly "double mixing" and "groove mixing", in order to achieve the uniform distribution of the slurry; the slurry mixing methods can determine the uniformity of parallel mixing and elevation slurry infiltration. ②Quantitative slurry mixing; right now, the main method is using "containers" to artificially mix slurry, so the shortcomings are human impact factors and ineffective control. In future mechanical slurry mixing should be developed. 10 to 15 minutes after cement and fly ash slurry is mixed with RCC, high-power vibrators should be utilized, controlling the process from mixing to end of vibration in less than 40 minutes.

11.7.4 Transverse Joint and Inducing Joint Technologies

The joint-forming technology for precast concrete developed at Shapai Dam was described above; the precast technology is mainly used for the transverse joints or inducing joints requiring grouting. For transverse joints and inducing joints of gravity dams, they generally do not require grouting, so vibration cutting and joint-forming machines are normally used.

In RCC dam construction, hydraulic vibration joint cutter is utilized; under vibration, plastic deformed blades are developed and embedded in concrete joints, and sealing materials are embedded in the joints together with the blades. Generally, rolling is done before compaction, the fillers are 1 – 2cm from the compacted surface, and after joint-cutting a vibration roller is used 1 to 2 times.

11.7.5 Automatic Rise Template and Continuous Rise of Thin-layer RCC

The pouring layer of RCC is generally 1.5 – 3.0m, and the intermittent layer is usually the weak link in a dam. Without other intermittent needs, the construction method of continuous rise of thin-layer RCC can enhance the speed of building dams, and is beneficial to the anti-seepage and stability of dams. Many projects in China have adopted this technique. For example, when the upstream RCC arch cofferdam of Dachaoshan Hydropower Station was in construction, the continuous rise technique was adopted, so the maximum pouring layer rise was up to 21m, and the arch cofferdam rose by 40.5m in two months during the construction period, to meeting the needs of weathering the flood season safely. In the Suofengying project, the continuous rise technique in blocks was adopted, and continuous rise templates and downstream step templates were designed and supplied in accordance with the features of RCC continuous pouring. The method of continuous rise in blocks was adopted for the RCC dam pouring, setting a record of 31m continuous rise in the main dam. Subsequently the arch dam construction of Dahuashui project hit a new record for continuous rise of 34.5m, and indicated that, with proper template work, concrete placing, temperature control technology and construction measures, the continuous rise of RCC concrete construction can be adopted to ensure construction quality, shorten construction cycle, and save investment.

Fig. 11.7 – 2   Installation of Cantilever Rise Steel Template
(Photo source: Sinohydro Bureau 7 Co., Ltd)

Template installation is an important part of the rapid construction of RCC. At Longtan Dam, according to the structural characteristics and construction needs of continuous thin-layer pouring in RCC dams, 3m × 3.1m continuous rise steel templates were adopted on vertical surface in the upstream and downstream; 3m × 3.1m continuous rise WISA templates were adopted on slope surface in the downstream; 3m × 1.82m continuous rise steel templates were adopted on horizontal transverse surface; cantilever templates were adopted in the pier and side walls; full-face precast concrete templates were adopted for the central corridor of the dam; combination of steel templates were adopted for side corridors in the downstream; special circular steel templates were adopted for the top arch, and the dimensions were all 0.6m × 1.5m; and special cylinder molds were adopted for the shaft. By using the above templates, construction efficiency was greatly improved, and the removal and erection of templates were fast, ensuring rapid construction, appearance and quality of RCC dams (Fig. 11.7 – 2).

For RCC arch dams, due to different curvatures on different dam surfaces in the upstream and downstream, the mature automatic rise templates for gravity dams cannot be applied directly. Sinohydro Bureau 11 Co., Ltd and China Gezhouba Group Corporation, as well as a number of other template companies, tailored to Zhaolaihe RCC arch dam, and developed continuous rise templates suitable for RCC arch dams which can be adapted to dams with large horizontal curvature and double curvatures, and they can be adapted to dam face with the minimum curvature radius of 30m.

11.7.6 Simulation and Optimization Analysis in Dam Construction

For such large-scale construction projects as Longtan Hydropower Station, the organization is extremely complex, so different construction organizations, equipment deployment and pouring procedures will have a great deal of influence on construction efficiency, duration, and cost, so it is difficult to obtain the optimum methods of construction organization through manual analysis. The research and development of methods and procedures through computer simulation, optimization and analysis on construction projects in recent years have provided good tools to optimize the construction organization.

It was adopted principles and methods of system engineering, considered concrete construction technology and regulatory requirements, closely combined the construction features of RCC high dams, and taken full account of the mix of RCC and normal concrete in hydropower station construction processes, various mechanical combinations, and different construction features of dam segments (retaining, spillway, and power generation). It also was integrated digital simulation techniques and database technology with 3D animation technology, and offered simulated calculation, analysis and selection on the placement schemes prepared according to the design. Statistical tables and charts can show the technical indicators of the entire simulation system, and 3D animation technology is applied to draw out the images and construction information of the dam in various stages, offering the model of dam pouring. Through model conversion and simulation of procedures for concrete construction of the dam, the computer simulation application can be written to calculate the concrete construction process so as to accelerate construction process, and improve the accuracy and reliability of concrete construction intensity. Both 2D and 3D graphics are used to visually display the images of the dam in various stages; through analysis of construction intensity, images and analysis on machinery usage, the characteristics of different construction schemes can be grasped, providing basis for optimization decisions. Up till now, most projects in China have adopted computer simulation technology.

11.7.7 Attempts to Thick Layer RCC

In order to speed up the construction progress by adopting advanced construction equipment and processes, under the premise of ensuring the quality of the project, GuizhouZhongshui Construction Management Co., Ltd., the management institution of Huanghuazhai hydropower project, organized experts, design institutions and construction institutions to conduct on-site comparison tests on 100cm, 75cm and 50cm RCC at a site on Luojiao River similar to Huanghuazhai project. Through the field test, the first-hand data were obtained to explore the construction methods for breaking the existing limit of 30cm thick RCC. The test adopted Japanese Sakai SD451 vibro roller, and conducted 3 layers of 100cm thick layer compaction tests on September 27 and 29, 2006 respectively. Due to constraints of mixing capacity, each 100cm-thick RCC layer was made in 4 layers, and each was 27cm in thickness. In contrast, another test was conducted on 50cm and 75cm thickness on October 6,2006, adopting Germany's BMW BW202AD. For the compaction density test apparatus, domestic-produced nuclear moisture density gauge was adopted, and Japanese high RI densimeter was adopted. The results showed that for RCC thickness of 100cm in four layers of 27cm each, paved 10 to 12 times by SD451 vibro roller, can reach the regulatory requirement on relative density of 97%.

The other test adopting BW202AD vibro roller on 50cm and 75cm RCC showed that the vibro roller was not fit for 50 – 75cm RCC layer.

The results of the tests have showed that if the RCC layer is increased from 30cm to 100cm, the effectiveness will increase by twice, and construction speed will be faster, shortening construction period. For RCC arch dams at the level of 100 – 130m, the RCC construction can be completed in three to four months. But it is requested that the capacity of gravel system, mixing system, transportation system, etc. also be increased accordingly; otherwise the efficiency will be difficult to improve.

11.8 Conclusion

After 20 years of engineering practices, China has accumulated a wealth of experience in the construction of RCC dams, developed technologies for building 200m gravity dams, and laid the foundation for the construction of 200m high arch dams. For further development of the construction technologies of RCC dams, some issues still need to be further studied and explored. For example, the characteristics of temperature and stress, as well as temperature control standards, for high RCC gravity dams, temperature control standards for high RCC arch dams, timing of seam grouting and real temperature load, quality and temperature control in prevention of cracking during RCC construction in dry and hot river valleys, management and digital surveillance for RCC construction based on information technology, alternative admixture for RCC in times of fly ash shortage, anti-seepage in high arch dams, long-term operating characteristics of high RCC dams, RCC dam construction techniques under special conditions, etc. Based on the summary of all aspects of successful experience, constant improvement of RCC dam construction technologies, and consummation of dam construction theory, the technologies will be continually enhanced and developed, and will offer better economic and social benefits.

References

[1] Shen Chonggang. Development and Achievements of RCC dams in China.//50 *Years of Large Dams in China*. Beijing: China Waterpower Press, 2009: 482-523.

[2] Liu Yansheng and Huang Wei. New Technologies and Achievements in the Construction of Shapai RCC Dam//*Proceedings of* 2004 *Symposium on High RCC Arch Dam Construction Technology*. Hubei, China. April 2004: 178-182.

[3] Wang Shengpei. *Development of RCC Dam Construction Technology in China. Proceedings of the* 5th *International Symposium on RCC Dams*. Guiyang, China. November 2007: 58-77.

[4] HydroChina Zhongnan Engineering Corporation. *Research on the Key Technologies of 200m RCC Gravity Dams and its application in Longtan* Project. May 2008.

[5] Zhu Bofang, Xu Ping. Temperature Stress and Temperature Control of RCC Gravity Dams. *Water Resources and Hydropower Engineering*, 1996 (4): 18-25.

[6] Zhang Guoxin. Temperature Stress and Temperature Control of RCC Gravity Dams. *China Water Resources*, 2007 (21) 4-6.

[7] Ministry of Water Resource of P.R.C. *Design Specification for Roller Compacted Concrete Dams* (SL 314-2004), January 2000.

[8] Pan Luosheng. Key Technologies for Temperature Control and Anti-Cracking for RCC Construction in Longtan Dam//*Proceedings of the* 5th *International Symposium on RCC Dams*. Guiyang, China. November 2007: 107-111.

[9] Feng Shurong, Luo Junjun, Shi Qingchun, et al. Summary of Key Technologies for the Design of Longtan High RCC Dam//*Proceedings of the* 5th *International Symposium on RCC Dams*. Guiyang, China. November 2007: 211-215.

[10] Sun Gengning. RCC Construction Techniques of the Double-curvature Arch Dam for the *Linhekou* Hydropower Station. *Water Power*, 2004, 30 (2): 48-50.

[11] Zhang Guoxin, Zhao Shijie, Liang Jianwen. Analysis of the Temperature and Stress of the Concrete Cased in High Temperature Seasons in *Longtan* RCC Dam. Proceedings of 2004 Symposium on High RCC Arch Dam Construction Technology. *Water Power*, 2005, 31 (3): 39-41.

[12] Zhu Bofang, Xu Ping. Thermal Stresses in Roller Compacted Concrete Gravity Dams, Dam Engineering, 1995, 6 (3), 199-220.

[13] Zhang Guoxin, Dai Yihua, et al. Research on Simulation of Temperature Stress and Temperature Control in High Temperature Seasons for Jinghong RCC Dam// 20 *years of RCC Dam Construction*. Proceedings of 2006 Symposium on High RCC Arch Dam Construction Technology. July 2006.

[14] Zhu Bofang. On Construction of Dams by Concrete With Gentle Volume Expansion. *Journal of Hydroelectric Engineering*. 2000 (3): 4-16.

[15] Zhang Guoxin, Luo Heng, Luo Jian. Simulation of Thermal Stressin *Yujianhe* MgO Micro-expansion RCC Dams and Discussion of Temperature Compensation Effects//*Proceedings of the* 5th *International Symposium on RCC Dams*. Guiyang, China. 2007: 656-663.

[16] Tian Yugong, Xi Xiangjun. Research and Application of RCC for *Longshou* Arch Dam in Cold and Drought Environment//*Proceedingsof the* 5th *International Symposium on RCC Dams*. Guiyang, China. 2007: 857-863.

[17] Wu Xu. Fast Construction Technologies for RCC Dam Construction Projects at *Longtan* Hydropower Station//*Proceedings of the* 5th *International Symposium on RCC Dams*. Guiyang, China. 2007: 785-799.

[18] Liu Yansheng, Huang Wei. Adhere to Technological Innovation, and Ceaselessly Climbing over the Summits in RCC Technologies—Summary of RCC Dam Building Technologies by Sinohydro Bureau 8 Co., Ltd//*Proceedings*

of the 5th International Symposium on RCC Dams. Guiyang, China. 2007: 107-111.
[19] Chen Gaixin. Research on Materials for RCC Dam Construction. // 20 Years of RCC Dam Buiding in China—Astride from Kengkou Dam to Longtan Dam. Beijing: China Waterpower Press, 2006.
[20] Shi Qingchun, Zhou Huifen et al. Research on Fast Construction Technologies for High RCC Dam at Longtan//Proceedings of 2004 Symposium on High RCC Arch Dam Construction Technology. Hubei, China. 2004: 86-90.
[21] Wang Rongjing, Fu Xing'an. Design and Application of Continuous Rise Templates for Zhaolaihe Double-curvature Arch Dam//Proceedings of 2004 Symposium on High RCC Arch Dam Construction Technology. Hubei, China. 2004: 42-47.
[22] Long Dehai, Gong Yongsheng, Yin Daqiu. RCC Dam Construction Technologies at Dachaoshan Hydropower Station. Water Resources and Hydropower Engineering, 2000 (11): 15-19.
[23] Chen Gaixin. Limestone Powder, A New Type of Supplementary Mineral Fines for RCC//New Progress on Roller Compacted Concrete Dams. Beijing: China Waterpower Press, 2007.
[24] He Hualin, Miu Chengxun. Experience and Lessons of Technical Layout of Sand and Aggregate Processing System for Longtan Hydropower Station. Water Power, 2007 (4): 23-24.
[25] Xiao Zuowen. Measures for RCC Dam Construction under High Temperature at Shankou Hydropower Station in hot season. Journal of Northwest Hydroelectric Power, 2003 (4): 51-54.
[26] Li Chunmin, Liu Shukun. In-situ Test Results and Discussion of 100cm, 75cm, and 50cm thick RCC layer at Huanghuazhai Hydropower Station//Proceedings of the 5th International Symposium on RCC Dams. Guiyang, China. 2007: 743-758.
[27] Zhang Guoxin, Zhang Jinghua, Yang Bo. Research on Arch Closure Temperature and Real Temperature Load of the RCC Arch Dam//Proceedings of the 5th International Symposium on RCC Dams. Guiyang, China. November 2007: 112-119.

Chapter 12

Construction of High Rockfill Dams

Jiang Guocheng[1], Xu Zeping[2]

12.1 Introduction

According to the statement of American consulting engineer J. B. Cooke[1], the construction of rockfill dams was started in late 19th century in America. The replacement of the construction method of dumped rockfill to the thin layers rockfill compaction by vibratory roller in 1956 indicated the beginning of modern rockfill dams construction. Concrete faced rockfill dam (CFRD) and earth core rockfill dam (ECRD) became the two main types of rockfill dam in this period. Besides, asphalt core rockfill dam (ACRD) or rockfill dam with asphalt concrete lining were also developed at the same time.

The construction of rockfill dams in China was started in the middle of 20th century. In its early stage, due to the limitation of construction technology and equipment, rockfill dams were mainly constructed by manpower with the methods of damped rockfill and directional explosion. In general, most of those dams are below 50m in height. The highest rockfill dam constructed by dumped rockfill is Nangudong dam, with the height of 78.5m. Rehabilitation of the dam was conducted several times due to the large deformation of rockfill and the failure of inclined earth core. Nanshui dam is a rockfill dam with height of 82.2m and was constructed by directional explosion. The inclined earth impervious zone was constructed 7 years later after the explosion, when the deformation of the rockfill was almost stopped. The dam performed normally. Shibianyu dam is another rockfill dam constructed by directional explosion and with asphalt concrete face. The dam height is 82.5m. Due to the large deformation of rockfill, the asphalt concrete face and perimetric joints was damaged. Rehabilitation of the dam was conducted by covering geomembrane on the surface of the asphalt concrete face slabs.

Snce 1980s, roller compacted earth core rockfill dam (ECRD) with the height over 100m had been built in China, which include Shitouhe (114m), Lubuge (103.8m), etc. The 154m high Xiaolangdi earth core rockfill dam was built in 1990s. It is constructed on alluvium foundation with the maximum depth up to 80m. Concrete diaphragm wall is used to cutoff seepage of alluvium foundation. This project is considered as the milestone in the development of high ECRD construction in China[2].

Modern technologies for building CFRD were introduced to China in 1980s. With careful studies and series laboratory and in-situ tests, Xibeikou CFRD, as the first testing CFRD in China, with the height of 95m, was successfully built. After that, CFRD is well accepted and rapidly developed in China. Due to its obvious advantages in safety, economy and adaptability, CFRD has soon become a very competitive dam type in dam construction. With 10 years development, Chinese engineers have already managed the modern techniques for building CFRD with the height of 100m. In 1990s, the construction of Tianshengqiao-1 (TSQ-1) CFRD was started. Its dam height is 178m. Along with Aguamilpa CFRD (187m in height) in Mexico, the two projects are the milestone in the development of CFRD from the height of 100m to 200m.

During this time, a number of asphalt concrete faced (or core) rockfill dams had been built, including Maopingxi roller compacted asphalt concrete core rockfill dam with height of 104m, Tianhuangping upper reservoir asphalt concrete faced rockfill dam with a height of 72m, etc. At the same time, composite geomembrane was also used in the construction of rockfill dams.

Experiences and lessons of rockfill dams construction in China before year 2000 had been reported systematically in a series of publications.[3-5]

In the first decade of 21st century, all rockfill dam types, including CFRD, ECRD and ACRD, have been rapidly devel-

---

[1] Jiang Guocheng, China Institute of Water Resources and Hydropower Research, Professor Senior Engineer.
[2] Xu Zeping, China Institute of Water Resources and Hydropower Research, Professor Senior Engineer.

oped in China. Most of the high dams and large-scale projects are located in southwest and northwest region of China, where the natural conditions are quite challenge for dam construction.

1. Earth core rockfill dam (ECRD)

ECRD is one of the main rockfill dam types. In the early stage of China's dam construction, most of the high dams were concrete dam. Before year 2000, only 3 ECRD with the height over 100m, were built in China. From 1980s to the beginning of 21st century, few high ECRD was constructed due to the strong competition and rapid development of CFRD. In contrast, the 7 rockfill dams with the height above 200m in the world are all ECRD. In the first decade of 21st century, due to the requirement for adapting unfavorable natural conditions and the experiences gained from the construction of Xiaolangdi ECRD, several high dam projects accepted ECRD dam type, which include Nuozhadu ECRD (261.5m), Shuanjingkou ECRD (314m), etc. Some high ECRD in China are listed in Table 12.1-1.

Table 12.1-1 High ECRD in China in the First Decade of the 21st Century (with Height Higher than 100m)

No.	Name	Location	River	Height (m)	Volume ($10^6 m^3$)	Reservoir capacity ($10^6 m^3$)	Installed capacity (MW)	Year of completion
1	Nuozhadu	Yunnan	Lancan River	261.5	28.0	23,700	5,850	UC
2	Pubugou	Sichuan	Dadu River	186	24.0	5,390	3,600	2009
3	Xiaolangdi	Henan	Yellow River	154	51.8	12,650	1,800	2001
4	Mao'ergai	Sichuan	Heishui River	147	11.4	535	420	2011
5	Miaowei	Yunnan	Lancan River	139.8		720	1,400	UC
6	Sizhiping	Sichuan	Zagunao River	136	5.8	130	195	2007
7	Jinpen	Shaanxi	Heihe River	130	7.7	200	20	2002
8	Qiaoqi	Sichuan	Baoxing River	125.5	7.2	210	240	2007
9	Shuiniujia	Sichuan	Huoxi River	108	4.8	140	70	2007
10	Qiafuqihai	Sinjiang	Tekesi River	105		1,690	320	2005
11	Shuangjiangkou	Sichuan	Dadu River	314	40.8	2,730	2,000	UC
12	Changheba	Sichuan	Dadu River	240	34.3	1,075	2,600	UC
13	Lianghekou	Sichuan	Yalong River	295	40	10,150	3,000	UC

2. Concrete faced rockfill dams (CFRD)

CFRD dam have been rapidly developed in China. At the end of 20th century, large numbers of CFRD with the height over 100m have been built and the construction of high CFRDs with the height of 200m was started. At the beginning of 21st century, several high CFRD with dam height around 200m were constructed. The highest dam is Shuibuya CFRD (completed) with the height of 233m. Jiangpinghe CFRD, 219m in dam height, is under construction. Besides, several CFRD projects with dam height range from 250m to 320m are in the stage of feasibility studies. For those extra high CFRD projects, their performance cannot be fully predicted by the existing experiences. Further studies and scientific researches shall be conducted. High CFRD with the height over 150m (include completed project and project under construction) in China in the first decade of 21st Century are listed in Table 12.1-2[6-8].

Table 12.1-2 CFRD Projects Constructed in the 1st Decade of the 21st Century (Dam Height Higher than 150m)

No.	Name	Location	River	Height (m)	Volume ($10^6 m^3$)	Reservoir capacity ($10^6 m^3$)	Installed Capacity (MW)	Year of completion
1	Shuibuya	Hubei	Qingjiang River	233	15.3	4,580	1,840	2009
2	Sanbanxi	Guizhou	Qingshui River	185.5	9.6	4,100	1,000	2006
3	Hongjiadu	Guizhou	Liuchong River	179.5	9.2	4,950	600	2005
4	Tianshengqiao I	Guizhou	Nanben River	178	18.0	10,260	1,200	2000
5	Tankeng	Zhejiang	Xiaoxi River	162	9.5	4,190	600	2008

Countinued

No.	Name	Location	River	Height (m)	Volume ($10^6 m^3$)	Reservoir capacity ($10^6 m^3$)	Installed Capacity (MW)	Year of completion
6	Jilintai 1	Xinjiang	Kashi River	157	8.4	2,440	500	2005
7	Zipingpu	Sichuan	Minjiang River	156	11.2	1,110	760	2006
8	Jiangpinghe	Hubei	Loushui River	219	7.0	1,370	450	UC
9	Kajiwa	Sichuan	Muli River	171	5.9	390	440	UC
10	Bashan	Sichuan	Renhe River	155		320	140	2009
11	Liyuan	Yunnan	Jinsha River	155	7.8	800	2,400	UC
12	Malutang	Yunnan	Penlong River	154	6.9	550	240	2009
13	Dongqing	Guizhou	Beipan River	150	9.6	955	880	2010
14	Houziyan	Sichuan	Dadu River	223	11.8	700	1,700	UC

3. Asphalt concrete rockfill dams (ACRD)

The construction of asphalt concrete rockfill dams in China, including roller compacted and poured asphalt concrete, was started in 1950s. To the end of 20th century, more than 50 such dams had been built. But most of them are small or medium scale project. Due to the poor quality of asphalt material, less engineering experiences and no special construction machinery, the development of asphalt concrete rockfill dam lagged far behind the development of CFRD before the end of 21st century. Only few high dams or major projects adopted this dam type at that time. The typical projects are Maopingxi dam and Tianhuangping dam. Maopingxi dam, with the height of 104m, is located at the dam site of Three Gorges project. It is a compacted asphalt concrete central core rockfill dam with the construction works started in 1997. Tianhuangping dam, with the height of 72m, is the main dam of the upper reservoir of Tianhuangping pumped storage plant. It is an asphalt concrete faced rockfill dam and the construction works were completed in 1997. After the beginning of 21st century, with the introduction of advanced technologies abroad and the summarization of engineering practices at home, asphalt concrete rockfill dam has gained a significant development in China. A lot of asphalt concrete rockfill dams have been built with the application of modern techniques. The major projects are Yele and Nierji asphalt concrete core rockfill dams, Baoquan, Zhanghewan and Xilongchi asphalt concrete faced rockfill dams. Quxue asphalt concrete core rockfill dam, with the height of 164.2m, is under construction. It is the highest ACRD in the world. With the progresses of construction technologies, ACRD has been more and more accepted in China. Some recently completed ACRD projects or projects under construction are listed in Table 12.1-3[6,9].

Table 12.1-3 Asphalt Concrete Face (or Core) Rockfill Dams in China Constructed recently

No.	Name	Location	River	Height (m)	Dam type	Reservoir capacity ($10^6 m^3$)	Installed capacity (MW)	Year of completion
1	Yele	Sichuan	Nanya River	124.5	core	298	240	2006
2	Maopingxi	Hubei	Maopingxi River	104	core			2003
3	Xilongchi upper	Shanxi	Longchigou River	50	face	49	1,200	2007
4	Baoquan upper	Henan	Yuhe River	92.5	face	82	1,200	2007
5	Tianhuangping upper	Zhejiang		70	face	9	1,800	1997
6	Zhanghewan upper	Hebei	Gantao River	57	face	8	1,000	2008
7	Xilongchi lower	Shanxi	Hutuo River	97	face	48	1,200	2007
8	Ni'erji	Neimenggu	Nenjiang River	40.55	core	8,610	250	2006

4. Rockfill dams with geomembrane impervious element

Formerly, in the engineering practices, geomembrane was only used as waterproof element of temporary structures or for repairing purpose. Recently it has been used as impervious element of permanent structures. The earth dam of Tiancun reservoir in Guangxi Province, with the height of 40m, used composite geomembrane as the impervious core of the dam. The saddle dam of Wangfuzhou hydroproject on Hanjiang River, 21m in height, used the composite geomembrane as the sloping impervious layer and the upstream blanket, with the area of $1.20 \times 10^6 m^2$ and $0.31 \times 10^6 m^2$ respectively. The project was completed in 1999 and operated normally till now. Xixiayuan rockfill dam on Yellow River, 21m in dam height and $162 \times 10^6 m^3$ of reservoir capacity, faced by composite geomembrane with the area of $160 \times 10^3 m^2$. The upper reservoir of Tai'an pumped storage plant used composite geomembrane as the impervious layer of the bottom of reservoir. The maximum water head is 40m. These engineering practices indicate the good perspective of the application of geomembrane in rockfill dam.

This chapter only focus on the two widely applied rockfill dam types in China's dam construction, i.e. earth core rockfill dam (ECRD) and concrete faced rockfill dam (CFRD).

12.2 Earth Core Rockfill Dam

In the development of China's high earth core rockfill dam, Xiaolangdi ECRD is a milestone project. From the beginning of the 21st century, the construction of ECRD has been rapidly developed. Pubugou, Changheba, Shuangjiangkou are ECRD built on sandy gravel alluvium foundation. Nuozhadu is the high ECRD built on rock foundation. The dam height of shuangjiankou ECRD has reached to the level of 300m, which will be the highest ECRD in the world.

12.2.1 Typical Projects

1. Xiaolangdi multi-purpose water control project [2,6,10]

Xiaolangdi multi-purpose water control project is located on the main stream of Yellow River, 40km north of Luoyang City, Henan Province. The controlled river basin area is 694,200km², about 92.3% of the total area of Yellow River Basin. The project controls 91.2% of the water flow, 100% of the sediment of the Yellow River. The total storage of the reservoir is $126.5 \times 10^8 m^3$ and the installed capacity is 1,800MW. The main function of the project is flood control and sediment reduction. Besides, it also has the benefits of water supply, irrigation and power generation. The basic earthquake intensity of dam site is 7 degree and the designed earthquake intensity for dam is 8 degree.

The project is mainly composed by three parts, which include main dam, flood and sediment discharge structures and power generation structures. The main dam is an inclined earth core rockfill dam, with the height of 154m, crest length of 1,167m. The total volume of filling materials is 51,850,000m³. The dam is built on sandy gravel alluvium foundation with the depth of 80m. Inside the alluvium foundation, there is a liquefiable sand layer. The left abutment is a thin mountain and the right abutment has a large fault that cross to the foundation of the dam. The general layout of the project and the typical section of the dam are shown in Fig. 12.2-1, Fig. 12.2-2.

The distinguishing feature of Xiaolangdi dam is the combination of vertical and horizontal seepage control measures. The vertical seepage control is conducted by concrete diaphragm wall to cutoff the seepage in alluvium foundation. The diaphragm wall is connected by internal blanket to the impervious part of the main dam and cofferdam and also the upstream natural blanket formed by sediment deposit. The design has fully utilized the character of the sediment-laden river. The selection of inclined earth core avoided the adverse impacts of the steep rock slope in the deep river channel on the construction of diaphragm wall. For increasing the strength of the internal blanket and to improve stability of the dam, the mixture of earth core material and sandy gravel material was applied. In dam construction, heavy compaction equipment was used. With the application of advanced construction technologies and management, good construction quality was achieved and the construction was ahead of schedule. Dam construction started in Feb. 1993 and completed in Nov. 2000. Reservoir impoundment started in Oct. 1999.

The construction of 154m high ECRD on alluvium foundation with the depth of 80m is a challenge both in China and world. The successful construction of Xiaolangdi ECRD promoted the high ECRD construction in China. The project was awarded as the International Milestone Project by ICOLD in 2009.

During the operation, some problems were discovered, which include the large leakage, longitudinal crack (along dam axis) on dam crest, large deformation of dam slope, and slow dissipation of pore pressure in the earth core. But those problems have no serious impacts on dam safety and dam operation.

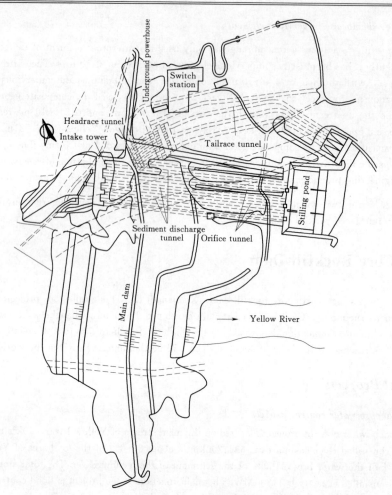

Fig. 12.2-1 General Layout of Xiaolangdi Multi-purpose Water Control Project

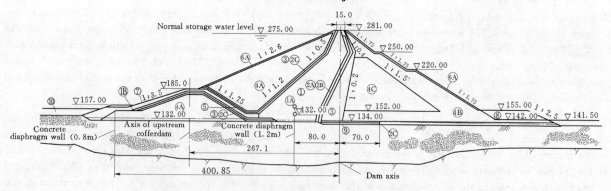

Fig. 12.2-2 Typical Section of Xiaolangdi dam (Unit: m)

① ①B—Clay; ②A—High plastic clay; ②A ②B ②C—Filter; ③—Transition; ④A ④B ④C—Rockfill; ⑤—Admixture; ⑥A—Slope protection block stone; ⑦—Riprap; ⑧—Rock ballast; ⑨—Backfilled sandy gravel; ⑩—Upstream blanket

2. Pubugou hydropower project

Pubugou hydropower project is located at the middle reach of Daduhe River, Sichuan Province. The main purpose of the project is power generation, and it also has the benefits of flood control and sediment block. The total storage of reservoir is $53.9 \times 10^8 \text{m}^3$ and the installed capacity is 3,300MW. The main structures of the project include rockfill dam, spillway, spillway tunnel, bottom outlet and power generation structures. The basic earthquake intensity of dam site is 7 degree and the designed earthquake intensity is 8 degree. Construction of the project started on March of 2004. Reservoir impoundment started in Nov. 2009. At the end of 2009, the first unit was put into operation.

The rockfill dam is a central earth core rockfill dam, with the core material of gravelly soil. The dam height is 186m and the crest length is 573m. Total volume of the fill material is 24,000,000m³. The maximum depth of alluvium foundation is 75.36m. Two concrete diaphragm walls were constructed in alluvium foundation for cutting off the alluvium foundation seepage. Under the bottom of earth core, the depth of diaphragm wall is 78m. The main diaphragm wall is located at the position of dam axis. The top of the wall is connected with earth core by grouting gallery. The bottom of the wall is connected to the grouting curtain of bedrock. The wall inserted 5m into the relatively impervious bedrock with the permeability less than 3Lu and connected with the grouting curtains of abutments. The auxiliary diaphragm wall is located 14m upstream of the main diaphragm wall. The wall directly inserted into earth core with the length of 15m. Under the bottom of the auxiliary wall, there are grouting curtains with the depth of 20m. Blanket consolidation grouting was conducted for the alluvium under the bottom of the earth core. In abutments area, the connecting grouting curtain with the depth of 20m was arranged between the two walls. The connecting section with the length of 8–20m was arranged between the riverbed gallery and the abutment horizontal grouting tunnels. At the two ends of the connecting section, structure joints were arranged. The joints were installed two layers waterstop and filled with asphalt sackings and plastic filler. In the connecting area of the diaphragm walls and the earth core, soft clay was filled. The typical section of the dam is shown in Fig. 12.2-3 and the detailed design for the connection of diaphragm wall and earth core was shown in Fig. 12.2-4.[11]

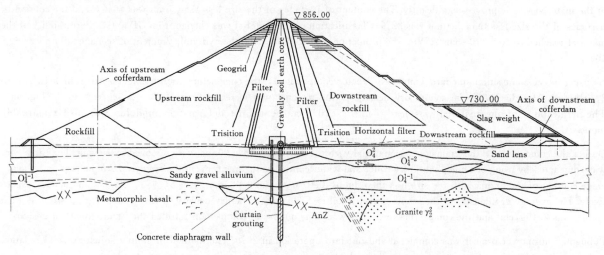

Fig. 12.2-3 Typical Dam Section of Pubugou ECRD (Unit: m)

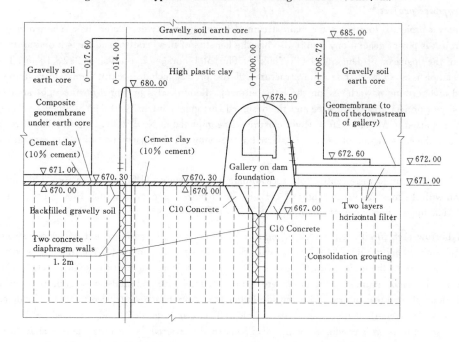

Fig. 12.2-4 Detailed Design of Diaphragm Wall and Earth Core (Unit: m)

For the purpose of bedrock grouting under the alluvium foundation, Pubugou ECRD connect grouting gallery on the top of the main diaphragm wall with earth core and directly insert the auxiliary diaphragm wall into the earth core. This design is not common in the world. After reservoir impoundment, the operation of diaphragm wall and the dam was normal. It is proved that the design of this seepage control system is feasible and reliable. The successful construction of Pubugou dam has accumulated valuable experiences for high ECRD built on deep alluvium foundation. The subsequent project, Changheba ECRD, with the dam height of 240m and alluvium foundation depth of 50m, accepted the same design.

Another feature of Pubugou dam is the use of gravelly soil as the material of earth core [7,12,13]. The Heima borrow area for earth core material is 16km away from dam site. It has good topographic condition and convenient for excavation and transportation, but the soil is gravelly soil with wide range of gradation. The coarse grain (grain size larger than 5mm) content is 50%–65% and the content of particles with the size less than 0.1mm is 8.8%–20%. In soil classification, the material is GP. The permeability of the material after compaction is $10^{-4} - 10^{-5}$ cm/s, which is not satisfied to the requirement of impervious material of high dam. To solve the problem, three measures were proposed: ①Adjust gradation by removing large grains (with the size larger than 80mm or 60mm). ②Increase soil density by using heavy compaction equipment. ③Add clay to increase the content of fine grains. With series studies, the first two measures were employed.

After removing the particles of the size larger than 80mm form the nature wide range gradation gravelly soil, the gradation of the material was improved significantly. The content of particles of the size less than 5mm was 50%, and the content of particles of the size less than 0.1mm was 22%. Classification of the material was change from GP to GC. Permeability of the material reached to $10^{-5} - 10^{-6}$ cm/s. With the protection of filter material, the hydraulic gradient of seepage failure was 60 – 100.

By using heavy compaction standard (modified Proctor compaction), the compaction energy is increased from 604kJ/m^3 to 2,704kJ/m^3. Accordingly, the maximum dry density of the material was increased from 2.23 – 2.32g/cm^3 to 2.375g/cm^3. The permeability of the material was less than 1×10^{-5} cm/s and the deformation modulus was also remarkably increased.

During the construction, the grains of core material with the size larger than 80mm were removed by sieving. The earth core was compacted by using 25t bump vibratory roller and water content adjustment. From the results of site inspection, the average content of the grains with the size less than 5mm, 0.075mm and 0.005mm were 51.6%, 22.6% and 6.5% respectively. The average dry density of the whole gravelly soil is 2.39g/cm^3. The dry density of fine grains (grain size less than 5mm) was 2.15g/cm^3 and the compactness was 101.4%. The quality of compaction fulfilled the requirements of design.

Pubugou hydropower project was completed and put into operation in 2010. The method of simply adjusting soil gradation and using heavy compaction criteria has provided valuable experience to the following projects.

3. Nuozhadu hydropower project[6,14]

Nuozhadu hydropower project is located on the main stream of Lancangjiang River, Pu'er city, Yunnan province. The main purpose of the project is power generation, and it also has the benefits of flood control and navigation improvement. The area of river basin of the upstream of dam site is 144,700km^2. The total storage of reservoir is 237×10^8 m^3. The installed capacity of power station is 5,850MW. The main structures of the project include rockfill dam, open spillway, flood discharge tunnel and underground powerhouse on the left abutment, flood discharge tunnel on the right abutment. The basic earthquake intensity of project area is 7 degree and the designed earthquake intensity of dam is 8 degree. The preparation of project construction started in March of 2004. River closure was completed in Nov. 2007. Construction of rockfill zone started in Aug. 2008 and the construction of earth core started in November. The construction of whole rockfill dam will be finished at the end of 2012.

The dam type of Nuozhadu project is a central core rockfill dam with gravelly soil for the earth core. The dam is constructed on rock foundation with the maximum dam height of 261.5m. The length of dam crest is 608.16m and the volume of the dam is $3,432 \times 10^4$ m^3. The upstream slope of the dam is 1:1.9 and the downstream slope is 1:1.8.

The materials of borrow area for the earth core is mixture of slope washed, residual soil and some strongly weathered rocks. The average grain composition is: 24% gravels with the size larger than 5mm, 44.3% fine grains with the size smaller than 0.074mm, 21.7% of the grains with the size smaller than 0.005mm. Most of the soils are classified as clay sand, low liquid limit clay with sand. As most of the grains are weathered sandstone and mudstone, grain particles are easily broken. After compaction, the content of grins with the size larger than 5mm could be reduced to 10%. The density, deformation parameters and shear strength of the material are very low. Thus, it is decided to add crushed hard rock to the nature borrow material. A series tests were conducted before dam construction started, which include the laboratory tests and site tests for the different proportions of added crushed rocks. Laboratory tests include the test on basic physical properties, mechanical properties under complicate stress paths, dynamic properties, creep deformation, seepage, hydraulic fracture, contact properties of interfaces, etc.

Chapter 12 Construction of High Rockfill Dams

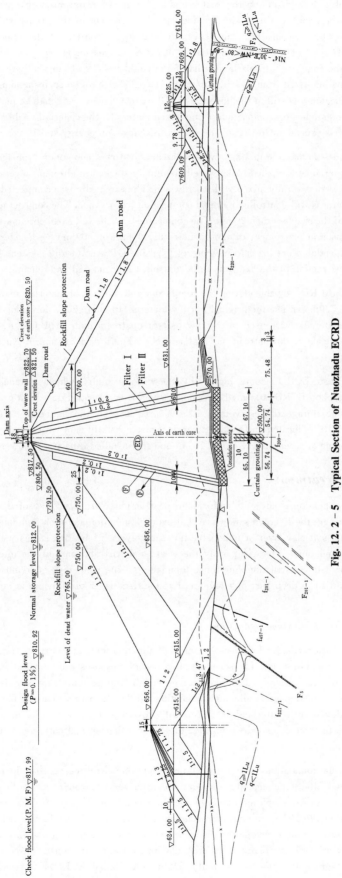

Fig. 12.2－5　Typical Section of Nuozhadu ECRD

The size range of crushed rock to be added in borrow material is 5 – 60mm. After optimization, proportion of the adding material is 35%. From the research results, after adding coarse particles, content of the grains with size larger than 5mm is 50%, content of the grains with size smaller than 0.074mm is 23.6%, content of the grains with size smaller than 0.005mm is 10%. The classification of the mixed material is GC. It is an idea impervious soil for high ECRD. Due to the breakage after compaction, the content of grains larger than 5mm could be 36%. Compare with the original material, the maximum dry density could be increased from 1.7 – 1.8g/m^3 to 1.9 – 2.0g/m^3. The corresponding best water content is about 10% – 15% and the permeability is still in the quantities of 10^{-6}cm/s. The overall engineering properties of the material are greatly improved. From the studies on compaction of the mixed material, the criteria based on heavy compaction energy with the compactness of 95% is equivalent to that of the standard compaction energy and 100% compactness.

For the site test, the techniques on material mixing, water content adjustment and compaction control were tested. For material mixing, the borrow material and crushed rocks were horizontally stacked in alternant layers. By excavating in vertical direction, the crushed rocks were naturally mixed with borrow soil. The earth core was compacted by using 25t bump vibratory roller. Earth core compactness was controlled by rapid compaction test method. The compactness of fine grains was controlled by the test on grains of the size less than 20mm. The compactness of the full earth core material was controlled by the laboratory test with the compaction instrument of the diameter of 600mm or 300mm[15]. Besides, the procedures of dam construction and rockfill compaction were in real-time control by using GPS monitoring system. With the configuration of digital-dam platform, the construction of the dam was fully controlled in visualized and dynamic management.

In section design of Nuozhadu ECRD, by considering the complexity and the cost of materials mixing, the material of earth core with the elevation above 720m can use the original borrow material. In dam shell, for using the low quality excavated material, some soft rockfill materials were used in the low part of upstream dam shell (in the zone near the core) and downstream shell (the dry zone above downstream tail water level). The typical section of Nuozhadu ECRD is shown in Fig. 12.2 – 5.

After construction of Nuozhadu ECRD, some higher ECRD will be built, which include Lianghekou ECRD and Shuangjiangkou ECRD. Lianghekou ECRD is located on Yalong River. It is a rockfill dam with gravelly soil earth core. The dam height is 295m. The total storage of reservoir is 101.54 $\times 10^8 m^3$ and the installed capacity of power station is 3,000MW. Shuangjiangkou ECRD is located on Daduhe River. It is also a rockfill dam with gravelly soil earth core. The dam height is 314m. The total storage of reservoir is 27.32$\times 10^8 m^3$ and the installed capacity of power station is 2,000MW.

12.2.2 Foundation Treatment

For high rockfill dam, soft foundation is not suitable for dam construction. Normally, the soft foundation layers should be excavated. More common cases are the riverbed sandy gravel alluvium foundation and rock foundation. For sandy gravel alluvium, if the foundation has no soft clay layer or liquefiable fine sand layer, its bearing capacity and stability is enough for dam construction. For rock foundation, after clear away the weathered part and part of the intense weathered part, the foundation is also suitable for dam construction. For those foundations, the main task of foundation treatment is seepage control. When sand layer is buried in deep part of alluvium or the surface layer of alluvium is not dense, other treatment measures should be studied.

1. Seepage control of sandy gravel alluvium foundation

In engineering practices, the basic measures for seepage control of alluvium foundation are upstream blanket, various vertical cutoff and downstream filter and drainage. The selection of treatment measures will be based on the requirements of project and the characteristics of the foundation[18]. For high dams, vertical seepage control measure is often used. By using vertical seepage control measures, the seepage in alluvium foundation will be effectively blocked. Therefore, the vertical seepage control, combined with downstream filter and drainage, will provide good protection for foundation and dam free from the damage of seepage failure. In some cases, the utilization of sediment deposit and nature clay layer as horizontal blanket could also be an auxiliary measures.

For high dam with deep alluvium foundation, one of the vertical seepage control measures is the excavation of the alluvium layers below the earth core and filter zones and put the core and filter zones on bedrock. Another vertical seepage control measure is the construction of concrete diaphragm wall. In the engineering practices of China, concrete diaphragm wall is widely applied in various projects. In 1958, column type concrete diaphragm wall was introduced into China. In 1959, the column type diaphragm wall was further developed to the trench plate concrete diaphragm wall. According to the incomplete statistic data, By the end of year 2000, there are 44 concrete diaphragm walls were constructed, with the total area of 54.2 $\times 10^4 m^2$. Among these diaphragm walls, the deepest wall is 81.9m (Xiaolangdi ECRD), the thickest wall is 1.3m (Bikou ECRD) and the wall with largest area is 72$\times 10^4 m^2$ (Huangbizhuang embankment dam). Besides, there also the case of using two diaphragm walls to share water head (cofferdam of Three-gorges project). After 2000, the technologies of building concrete diaphragm wall are further developed. Some engineering cases are briefly introduced below:

(1) Shiziping hydropower station: It is ECRD with gravelly soil earth core, located in Sichuan Province. The dam height is 136m and the maximum depth of concrete diaphragm wall is 101.8m. In the diaphragm wall, two pre-installed grouting pipes are used for bedrock curtain grouting. The construction of diaphragm wall was completed in 2006[2].

(2) Yele hydropower station: The dam is ACRD with dam height of 124.5m. The right bank terrace alluvium has the depth of 420m. In the initial design, the seepage control measure was a concrete diaphragm wall with the depth of 100m plus 100m grouting curtain under the wall. In the implementation stage, the upper 15m was excavated and constructed rockfill dam with concrete core. On the platform of the bottom of the dam, a concrete diaphragm wall with the average depth of 70m was constructed. Under the wall, a construction gallery was built. From the gallery, another diaphragm wall was constructed further down to the alluvium, with the average depth of 66mm and the maximum depth of 84m. The bottom of upper diaphragm wall and the top of construction gallery was connected by grouting curtain. The total area of the lower diaphragm wall constructed from gallery is 21,000m^2. By using this method, the difficulties of constructing deep diaphragm wall were perfectly solved. For the construction in gallery, the low clearance milling and low mast percussion reverse circulation drill were accepted. The project was completed several years ago and is in normal operation at present.

(3) Pubugou hydropower station: The dam is ECRD with gravelly soil earth core. The dam height is 186m and the depth of alluvium is 70m. Due to the high dam height, deep alluvium foundation and the requirement of bedrock grouting, two diaphragm walls with the thickness of 1.3m constructed. The first diaphragm wall was connected to the earth core with the gallery on its top. Another diaphragm wall was directed inserted into the earth core. In the connection of grouting gallery and abutments, joints were arranged to adapt the differential deformation. Bedrock curtain grouting was conducted from gallery. Now the reservoir was impounded. Some leakage was discovered at the connection of grouting gallery and abutments. After treatment, it is in normal operation.

In China, grouting curtain in sandy gravel alluvium is seldom used. In 1958, the test of cement and clay grouting by using sleeve valve drive pipe was successful in Maojiachun hydropower station in Yunnan. In 1959, it is applied in the Baihe main dam of Miyun reservoir in Beijing. The deepest alluvium was treated by upper part wall connected with lower part curtain. The project has been normal operated for 50 years. For Yuecheng reservoir project, part of sandy gravel foundation was treated by circulation drill grouting. As concrete diaphragm wall has more technical and economy advantages, the method of curtain grouting was not widely accepted in high dam project. It is only used in dike engineering, cofferdam, defect small dams and some special cases. One example of the special case is Xiabandi Project in Xinjiang Autonomous area. The alluvium foundation has the depth of 150m. The treatment was conducted by upper part wall and low part grouting curtain.

2. Seepage control of rock foundation

When rock foundation has permeable layers, erodible fault fissures or caves, it is necessary to conduct curtain grouting or curtain grouting combined with consolidation grouting.

Criteria of grouting: the grouting depth should reach to the relatively impermeable layer. According to the present design standard, the permeability requirement for high-class dams or high dams is 3-5Lu. For low-class or low dams, the requirement will be 5-10Lu[23]. For the high dams recently constructed in China, 3Lu is set to be the criteria for the relatively impermeable layer. Those projects are Pubugou, Shiziping, Jinpen, Shuiniujia, Changheba, etc. The only exception is Xiaolangdi project, which used 5Lu as the criteria. For Nuozhadu and Qiafuqihai project, as the low permeability of the bedrock, the criteria for grouting curtain is 1Lu. As the requirement of foundation seepage control is lower than concrete dam, the criteria of 3-5Lu is widely accepted by most of the projects.

In the construction of recent high earth core rockfill dams, the workload of foundation grouting is quite heavy. For avoiding the interference of foundation grouting and rockfill compaction, gallery on the bottom of the dam is widely accepted. Normally the gallery has multiple purposes, include grouting, monitoring and inspection, such as Nuozhadu ECRD. For the high ECRD with concrete diaphragm wall in alluvium foundation, different treatment measures can be used with the consideration of the specific site conditions. For Xiaolangdi ECRD, the bedrock of river channel section is low permeability clay rock, thus there is no grouting curtain under the diaphragm wall. From river channel to the left bank, as the bedrock is permeable rock, the foundation was grouted through the pipes in diaphragm wall. For the following ECRD projects after Xiaolangdi dam, curtain grouting was conducted under the diaphragm wall. By considering of no delay for the construction of dam body, gallery was arranged on the top of the diaphragm wall and the gallery can also be used for observation and inspection. For the case of two diaphragm walls (such as Pubugou ECRD), the gallery is arranged on the top of one diaphragm wall, and another diaphragm wall is directly inserted into earth core. For this arrangement, two projects (Qiaoqi ECRD and Pubugou ECRD) were discovered with leakage at the connection of gallery and abutments. It was due to the large differential displacement at the connection. After treatment, the two projects are in normal operation.

The control of by-pass seepage in abutments request for the grouting curtain extend to a certain depth in the mountain. The area of grouting curtain is determined by three principles: ①The curtain will extend to the intersection of the reservoir normal storage water level and the ground water level before reservoir impoundment. ②The curtain will extend to the intersection of the reservoir normal storage water level and the relatively impermeable layer of the abutments. ③The curtain area is determined by seepage analysis with the requirement of seepage control. For grouting curtain determined by the first two principles, the curtain is considered as closed curtain. But when the closed curtain will be too long or too deep, the area can be determined by the third principle. For the curtain grouting under diaphragm wall, the same principles can be applied.

In grouting techniques and grouting materials, there are also some progresses. The common practice in China is to use high pressure and thick liquid grouting. The grouting pressure is 1-2 times of the water head. The dense and pre-stressed grouting curtain will have good performance in seepage control and durability. For the specific geologic conditions, there are also some special grouting techniques developed, such as chemical grouting, fine cement grouting, etc.

3. Sand layer liquefaction and seepage stability

In alluvium foundation, there are often sand layers or sand lens, which will be liquefied during earthquake or have seepage stability problems. If the sand layers are shallow or locally distributed, it can be simply excavated. But if the sand layers are deeply buried, it is difficult to be excavated. The treatment of this issue will be the difficulties and challenges of a project.

In the case of Xiaolangdi ECRD, the alluvium foundation was excavated to the elevation of 130m. The remained alluvium has 30-37m upper sandy gravel layer, 10-20m sand layer and 5-30m lower sandy gravel layer. In the interface of upper sandy gravel layer and sand layer, there are some local overhanging areas. Seepage control of alluvium foundation was conducted by building concrete diaphragm wall. Part of the bedrock under the wall was not grouted. In operation, the leakage observed by downstream measuring wire was 30,000m^3/d (not include the leakage through alluvium). The leakage through riverbed foundation was account for 60%-70% of the total leakage. From observation data, the piezometer data behind diaphragm wall was reduced 90% of the piezometer in front of the wall. That means the good performance of the diaphragm wall. For the upper sandy gravel layer, from the position behind the diaphragm wall to the downstream measuring wire, the measured hydraulic gradient was 0.001-0.05. This value was far less than the allowable gradient of the sandy gravel soil and also less than the allowable gradient of contact erosion on the surface of sandy gravel layer and sand layer. Besides, the leakage water was clear. It is considered that the leakage through riverbed sandy gravel alluvium has no effects on dam safety. From this case, it is demonstrated that the large leakage does not means seepage instability. The safety of foundation seepage should be analyzed by comprehensive studies.

The average depth of the alluvium foundation of Pubugou ECRD is 40-60m and the maximum depth is 75m. On the bottom of Q_4^{1-2} gravel layer, there are two sand lenses, which are located at upstream and downstream foundation respectively. The average grain size of the upstream sand lens is 0.145-0.32mm, the relative density is 0.71-0.72 and the buried depth on the top is 40-80m. The average grain size of the downstream sand lens is 0.095-0.36mm, the relative density is 0.64-0.72 and the buried depth on the top is 26-40m. The designed earthquake intensity is 8. From the analysis by dynamic FEM and simplified Seed's method, the two sand lenses may have the possibility of liquefaction during earthquake. After the construction of the dam, the stress status of the sand lenses will be improved by the dead weight of dam body and the possibility of liquefaction will be reduced. For safety considerations, 60m long counter weight fill was put at upstream toe and downstream heel to increase the safety margin[6].

According to the Specifications for Seismic Design of Hydraulic Structures, when dam foundation has liquefied soils, FEM dynamic analysis should be conducted for dam and foundation. The safety of dam and foundation upon earthquake will be assessed by the analysis and other studies. Then, the engineering measures will be applied according to the results of the assessment. For foundation treatment, the purpose of aseismic measures is to eliminate or reduce the additional pore pressure of the liquefied soil layer and to increase the strength and stability of the soil. Those measures include: increase soil density and confined stress to reduce pore pressure upon earthquake; arrange drainage to speed up pore pressure dissipation; seal the liquefied soil to avoid flow away of the soil; etc. Density increasing is the ultimate measure for liquefied soil treatment. Using the weight of dam body and the counter weight fill are also the effective and commonly used method for liquefied soil treatment.

12.2.3 Construction Material

The construction materials of rockfill dam include impervious material, filter and transition materials, dam sell materials, drainage materials and slope protection materials. The properties of different materials and its application are always one of the focuses of the studies of rockfill dam. In the design standard published in 1984 (SDJ 218—84), the selection and design of construction material was systematic described. In this design standard, it was clearly mentioned that except boggy soil, bentonite, surface soil and soil with not fully decomposed organics, all the other soils can be used as construction materials of earth and rockfill dam, or be used into different part of the dam by proper treatment. In 2000, the summarization of China's earth and rockfill dam construction also include the experiences and research results on construction materials[2-4]. In 21st century, some important rockfill dams projects were constructed in western part of China, the studies and applications of construction materials were further developed.

12.2.3.1 Impervious Material

The studies on impervious materials of high rockfill dam are to determine the applicability and design index of the material, and also the compaction criteria, gradation requirements, water content adjustment measures, etc. Especially, if the so-called "unsuitable materials" can be used for dam construction through careful studies, it will be important in reducing the cost and improving the safety of the project.

Besides the ordinary clay materials, the studies on impervious materials are mainly focus on two kinds soils: ①Some special soils such as red clay in southern part of China, expansive soil, collapsed loess, etc. ②Gravelly soils. The studies on special soils were mainly conducted in late 20th century. It was mainly applied for the dams with the height around 100m. The studies on weathered gravelly soil conducted in 1980s were applied in the construction of Lubuge ECRD. After that, gravelly soil was gradually become the main selection for the impervious material of high rockfill dams. For the 13 high rockfill dams listed in Table 12.1-1, there are 10 dams used gravelly soil as impervious material. With large amount testing research works of the real projects, engineering properties of gravelly soil were deeply studied.

Table 12.2-1 Provides the Statistic Data of the Gravelly Soils of some High Rockfill Dam in China

Project name	Dam height (m)	$P_{>5}$ (%)	$P_{0.075}$ (%)	$P_{0.005}$ (%)	Dry density (g/cm³)	Gradient	Permeability (cm/s)
Nuozhadu	261.5	38.3	33.2		1.94	25-120 average 67	10^{-6}
Pubugou	186	51.6	22.6	6.5	2.39	7-11 (56)	$<10^{-5}$
Xiaolangdi (inner blanket)	154	37.1			2.05-2.40		
Shiziping	136	57.7	19.7	6.7			
Qiaoqi	125.5	37.1	31	9.9	2.10		0.4×10^{-5}
Shuiniujia	108	41.7			2.07	13.2	10^{-6}
Lubuge	103.8	32	43	19	1.51		$1.5 \times 10^{-5} - 7.9 \times 10^{-6}$

Note: Numbers in the table are average values. $P_{>5}$, $P_{0.075}$, $P_{0.005}$ are grain contents of larger than 5mm, less than 0.075 (0.1) mm, less than 0.005mm respectively. They are all values after compaction. The value of gradient does not consider filter protection. Gradient values in bracket have considered filter protection.

1. Indices of the basic properties of gravelly soil

In the design standard of embankment dam and rockfill dam, specification of site investigation of nature construction materials, there are some regulations on the use of impervious soil and gravelly soil. Take the example of permeability of impervious material, for homogenous dam, it is required to be not larger than 1×10^{-4} cm/s, for central or inclined core rockfill dam, it is required to be not larger than 1×10^{-5} cm/s; Furthermore, the clay content shall be 15%-40% and the plasticity index shall be 10-20; For gravelly soil impervious material, the grains content with the size larger than 5mm is required not more than 50% and the fine grains content with the size less than 0.075mm is required not less than 15%. For the clay content, the specification of site investigation requires the value to be 15%-40% of the content of the grains with the size less than 5mm; the design standard of hydropower project requires the value not less than 8%. According to the laboratory tests and relevant studies, the requirement on gravelly soil for the content of grains with the size larger than 5mm not more than 50% can let the fine grains be filled in the void of the coarse grains. The requirement on gravelly soil for the content of grains with the size less than 0.075mm not less than 15% will guarantee the permeability and the internal stability of the soil. In the design of rockfill dam, the ultimate principle is to meet the requirements on safety of seepage, stability and deformation, and also the requirement on cost reduction. Physic properties indices of materials such as gradation, plasticity index have certain relationship with its engineering properties and are easy to be determined by laboratory test. It can be used for determining the applicability of the materials. But they are not the decisive factors. For some special soils, the relationship of physic properties and engineering properties may different from normal soils. Therefore, the final decision will depend on the studies of real situations of the project. As for the requirement of permeability of impervious material, when permeability reach to the quantity of 10^{-5} cm/s, the real seepage through core is quite small. Thus, it is not necessary to strictly control the permeability to the value of 1×10^{-5} cm/s. For seepage stability, normally, the wide gradation gravelly soil has less clay grains content. It presents some properties in seepage deformation of non-cohesive soils. If seepage deformation mode is judged by considering whether the voids of coarse grains are filled by fine grains [27], the seepage failure mode of the gravelly soils with its gradation satisfies the requirement of design standard will be mass flow or transition type. Besides, the critical seepage gradient of gravelly soil is mainly depended on the filter layer at seepage exit. With the protection of filter layer, the critical seepage gradient of gravelly soil will be greatly increased. From the testing data of the 4 gravelly soils of Pubugou ECRD[27], the content of grains with the size less than 1mm is 17%-48% and the clay content is 4%-12%. Under the density of normal compaction, the maximum hydraulic gradient from the test for the soil with filter layer is 90-140. For this values, even the factor of safety is 5-10, the allowable gradients are still quite high. In the design standard of hydropower projects, it is stated "when requirement for permeability of impervious materials is not satisfied, further studies and assessments shall be conducted in the aspects of seepage, stability and deformation". For the requirement of clay content of gravelly soil, the standard stated, "When the content of fine grains with the size less than 0.005mm is smaller than 8%, further studies and assessments shall be conducted". These statements in the standard have provided the instruction for the flexible application of the regulations in real situations.

2. Gradation and water content adjustment of gravelly soils

Generally, soils with 20% of the grains of the size larger than 5mm are classified as gravelly soil, which include various soils with gravels, clay gravel and weathered rocks and artificially mixed gravelly soil, etc. In 21st century, most of the high ECRD projects in China are located in southwest part of the country. In this area, there are plenty of gravelly soils for impervious core. Usually, the gravelly soils are mixture of rubble, pebble, gravel and fine soils of residual, pluvial and ice accretion deposition. Sometimes, it also includes some totally weathered rocks. Due to the wide gradation and non-homogeneous properties of the natural gravelly soils, its gradation and water content shall be adjusted during application.

For some wide gradation gravelly soils, under the condition of the soil is basically meet the requirements of impervious core, the gradation of the soil can be simply adjusted by removing the large size particles to increase the content of fine grains. The cases of this adjustment are Pubugou ECRD and Changheba ECRD.

For some projects, the soils in borrow area are mainly composed by fine grains. It cannot satisfy the stability and deformation requirements of high dam. In these cases, the soil shall be mixed with crushed rocks or pebbles and gravels to increase the content of coarse grains. Traditional method is conducting laboratory test for the mixed soil to determine an optimal mix proportion. Then, it will be verified by site compaction test. In construction, this determined mix proportion is converted into the ratio of volume. The materials to be mixed are stacked layer by layer at the site and then be mixed by vertically excavation. The construction of the earth core of Fierza ECRD in Republic of Albania, which was design and constructed by Chinese engineer, had use this method. The construction of Nuozhadu ECRD was also used the method (as mentioned above). The same techniques will be applied in the construction of Shuangjiankou ECRD and Lianghekou ECRD.

3. Quality control of the compaction of gravelly soils

There are two kinds of quality control method for soil compaction, i.e. dry density control and compactness (degree of compaction). According to the design standard, the compaction quality of cohesive soil is controlled by compactness. For light compaction and low class dams, compactness should larger than 98%-100%. For heavy compaction, compactness could be slightly reduced, but it should not be less than 95%. For gravelly soil, it is requested to calculate the maximum dry density of the whole material and also to check the dry density of fine grains.

Due to the variation of the classification and composition of soils, its compaction properties are also different. For gravelly soil, it is not possible to only use average dry density as the control criteria. The well recognized method is to use the "three points rapid compaction method" to measure the compactness of the soil. In the construction of Lubuge ECRD, the construction material of central core is weathered soil. After compaction, the content of gravels (coarse grains) had a great change. The three points rapid compaction method was firstly used in compaction quality control. The measured maximum dry density was 14.03-16.48kN/m^3, where the middle value was 15.11kN/m^3. If it is calculated with the middle value of the maximum dry density, the compactness was 90%-106%, which has big difference with the required value of 98%. With the application of the three points rapid compaction method, the compactness of the soil is the same, but the dry densities are different. It reflects the different compaction properties of the soil. After Lubuge project, the method was applied in Daguangba project in Hainan Province. Now the three points rapid compaction method has been widely applied and accepted in the specification of rockfill dam construction.

Gravelly soils contain both the coarse grains and fine grains. In compaction quality control, besides the requirement on compactness of the full material, it also has the requirement on fine grains. The seepage properties and mechanical properties of gravelly soil are mainly depended on the properties and compactness of fine grains. When compaction quality is controlled by the compactness of fine grains, it is unnecessary to conduct large-scale compaction test for full material. When content of coarse grains is below 60%–70%, the dry density of full material will be increased with the increasing of coarse grains content. When content of fine grains is below 20%–30%, the coarse grains have not yet take the function of soil skeleton. The fine grains are fully compacted. Its dry density keeps unchanged. When content of fine grains reaches 30%, the coarse grains start to take function of skeleton. The more content of coarse grains, the stronger is its skeleton function. Thus, the fine grains inside void cannot be fully compacted. The dry density of soil will be reduced with the increase of coarse grains. When content of coarse grains is 60%–70%, the skeleton function of coarse grain is fully realized. The dry density of full material and fine grains reduced synchronously. All its mechanical properties are dropped in big scope and the permeability of soil are increased rapidly. Therefore, in the application of compaction degree control of fine grains, when content of coarse grains is below 25%, the compaction degree for fine grains could be controlled with 100%; when content of coarse grains is 25%–50%, the compaction degree control for fine grains could be reduced to 97%–98%. In conducting the test of the three points rapid compaction, if it is difficult to separate the fine grains of the size less than 5mm, the test can also be conducted for the grains of the size less than 20mm. The compaction energy of the test should be light (correspond to standard Proctor) or heavy (correspond to modified Proctor) compaction as specified in the testing specification. Thus, the test results of different projects and different soils can be compared.

In the construction of Pubugou and Nuozhadu ECRD, systematic studies on compaction properties and control method were conducted for the gravelly soils. Engineering practices and test results proved the applicability of the three points rapid compaction method. For Nuozhadu project, the tests of using different compaction cylinders with the diameter of 600mm, 300mm and 152mm were conducted for the gravelly soils of different gravel content. From the tests, it concluded that the results of 300mm cylinder are the same with the results of 600mm cylinder. The test with 300mm cylinder is enough for the full material compactness control.

The material of earth core of Xiaolangdi Dam is fully composed with fine grains. By considering the difference of the material, the average value of the maximum dry density by periodical multi-points moving Proctor test was used as control criteria. With this method, the real material for dam construction was coordinate with the control parameters. It is also the improvement of the single dry density control method.

12.2.3.2 Filter Material

Normally, the procedure of seepage failure of soil always starts from seepage exit. If the failure is not controlled at the exit, it will further develop into the soil and finally lead to the collapse of the soil. Filter layer can effectively prevent the erosion of fine grains at seepage exit, and it also has the function of drainage. Therefore, the using of filter layer to protect seepage exit is an effective measure of seepage control.

For cohesionless soil, if the soil is homogeneous with the uniformity coefficient less than 5–8, the design of filter layer follows Terzaghi's principle. If the soil is non-homogeneous soil with the uniformity coefficient larger than 5–8, the design of filter will follow the principle of no erosion of the fine grains with the size less than 5mm of the protected soil. The coefficient between layers is calculated based on the character grain size of the fine grains of the protected soil. It can also be determined by Terzaghi's principle. But it should be checked by laboratory test. For cohesive soil, the first layer filter is determined by the content of grains with the size less than 0.075mm. All the above-mentioned design methods were listed in the design standard[23,30]. For gravelly soil with wide gradation, there are no standard methods for filter layer design. It should be studied case by case.

Gravelly soil is composed by several gradations. Although it has some clay grains, as for the full material, it still belongs to non-plastic or low-plastic soil. Therefore, the design of filter will also follow the principle of protecting the fine grains of the protected soil. By taking 2mm as the boundary size of coarse grains and fine grains, 70% of fine grains as the control size (d_k), Liu Jie[27] got the following conclusions from the statistics of laboratory tests of IWHR (China Institute of Water Resources and Hydropower Research) and other testing data:

For mass flow and transition type soil: $D_{20} = (7-13)d_k$
For piping type soil: $D_{20} = (5-10)d_k$

And the corresponding principle for filter:

For mass flow and transition type soil: $D_{20} \leqslant 7d_k$
For piping type soil: $D_{20} \leqslant 5d_k$

The principles follow the same ideas of protecting fine grains of gravelly soil, which is widely accepted in the world. But the boundary size of fine grains and the control size are different.

There are many factors affect the seepage deformation properties of gravelly soils. Besides gradation, density, stress status and the protection of seepage exit will also play an important role on seepage deformation properties of gravelly soils. The mixture of sand, gravel and fine soil with good gradation will have good ability for erosion resistance. If the content of grains with the size larger than 5mm not more than 50% and the fine grains with the size less than 0.075mm less than 15%, and the soil is well compacted (include full material and fine grains), the failure mode of the gravelly soil will be mass flow or transition type. It will have relatively high critical hydraulic gradient. As mentioned above, the laboratory test of the gravelly soil of Pubugou ECRD showed that the maximum hydraulic gradient of the soil was 90 – 140[27] under the condition of exit filter protection, which means the sufficient reliability of the filter protection. As the variability of gravelly soils, the filter design cannot fully rely on calculation. It must be checked by filter tests.

Under the condition of the impervious body have cracks, the hydraulic gradient at seepage exit will be very high. It is suggested that the design of filter should consider the condition of cracks existence.

12.2.3.3 Rockfill

The requirements on deformation control and dam shell materials of ECRD are relatively low than that of CFRD. It may benefit to use more excavated materials. For ECRD, there are no strict requirements on the rockfill in the dry area of downstream dam shell. In Nuozhadu ECRD, with careful test studies and analysis, some excavated soft rockfill were used at the bottom of the upstream of the earth core.

12.2.4 Test Studies and Numerical Analysis

Facing the key technical problems of high rockfill dam construction, China has organized series national research works to conduct scientific studies on technologies of high rockfill dam construction from 1980s. With continuous input and systematic researches, great progresses were achieved. At present, scientific tests and numerical analysis have played an important role for instructing design, construction and operation of high rockfill dam projects.

12.2.4.1 Test Studies

Scientific tests of rockfill dams include laboratory material tests, model test and site prototype tests. The different tests are complement each other and they are also the different scales for verifying results of numerical analysis.

1. Laboratory material tests

Soil and rockfill materials present obvious non-linear and inhomogeneous characteristics. Its properties are related with density, water content and stress status, and also related with stress history and stress path. In addition, it is affected by time factors such as creep, wetting and degradation, etc. By using modern technologies, laboratory test can well control the different factors, to test the soil samples according to the specified conditions, such as density, water content, stress conditions and stress paths, degree of saturation, drainage condition, test duration, static and dynamic loads, etc. It is an indispensable tool in studying constitutive models, getting parameters for numerical analysis and investigating the relationship of material properties.

In recent high rockfill dams construction, most of the construction materials are composed with large particles, such as rockfill and sandy gravel. The impervious materials of earth core are also gravelly soils. Therefore, the laboratory test should be conducted with large-scale testing equipment. With several years' studies, Chinese engineers have developed series large-scale testing equipment, with the features of high pressure, stress status and stress paths controllable, drainage manner and testing duration controllable. These equipment include static and dynamic tri-axial testing machine (the diameter of testing sample is 300mm), plane strain testing machine, true tri-axial testing machine, compression testing machine, compaction testing machine, seepage testing equipment, creep deformation testing equipment, etc. Most of the testing equipment can conduct the tests for the testing materials with maximum grain size of 60mm. It can arrange testing procedures by the real situation of the project, and to verify various mathematic models, provide material parameters and the data for the design and construction of rockfill dams. These testing equipment have been well accepted and widely applied in the high rockfill dam construction in China.

Due to the limitation of the size of testing equipment, grain size of construction materials from site must be reduced in laboratory test. Thus, size effects of testing samples and methods for determine gradation of testing materials was studied. At present, the maximum grain size of soil samples of the large-scale laboratory tests in China is 60mm. The commonly adopted methods for reducing grain size are Particles Eliminate Method (PEM), Equal Quantity Replacing Method (EQRM) and the Similar Particle Distribution Method (SPDM). PEM will increase the content of fine grains and change the shape of gradation curve. It is only used for the soil with few particles exceed the limitation of the maximum particle size. EQRM keeps the contents of small grain size and change the coarse grain size by replacing the particles exceed the maximum allowance with the coarse particle within the allowance according to the equal quantity proportion (Normally, 5mm is considered as the boundary size for fine grains and coarse grains). This method keeps the ratio of fine grains and coarse grains, but change the coefficient of uniformity of the soil. SPDM move the gradation curve in a certain scale. It keeps the coefficient of uniformity (C_u) and coefficient of curvature (C_c) of the original soil, but increase the content of the fine grains. As the above methods may cause some changes in engineering properties of the materials, a new method with the combination of EQRM and SPDM by considering the real situation of the material gradation was developed. From research, when particles that exceed the maximum size allowance are less than 40% of the total amount, EQRM is applicable. When the proportion is large than 40%, SPDM should first be applied by keeping the fine grain ($\phi < 5$mm) proportion not exceed 15%, then EQRM will be applied. The size effect of coarse granular materials is a complex problem. The similarity of laboratory testing material with reduced size and the prototype material is not fully verified. Further studies should be conducted.

2. Centrifuge modeling test

In the studies of rockfill dam structures, besides numerical analysis methods, centrifuge modeling test is often used as a complementary tool for the studies, especially for the important project or complicated structures. The dimension of prototype structure of rockfill dam is huge. In the small size model test, the dead weight stress of model dam is much less than the corresponding stress of prototype. For soil material, its constitutive relationship and strength are closely related with stress level and stress status. Therefore, the small size model dam cannot present the physic phenomena of the prototype. But if the model dam is put in a simulated Ng field of gravity (here N is the scale of model, g is acceleration of gravity), the dead weight stress of the model dam will be the same as the prototype dam, which will make the model dam to have the same mechanical characters as the prototype dam. The field of centrifugal force created by centrifuge in high speed rotation could provide a controllable artificial equivalent gravitational field. Therefore, the model dam in centrifuge operation will reproduce the behaviors of the prototype dam.

The studies of using centrifuge to research geotechnical problems started in 1950s. At that time, it is only limited in research planning, information collection and feasibility studies. In 1970, a 150g·t centrifuge was manufactured and put into operation in 1983. In 1980s, several small and medium scale centrifuges were built, which have the capacity of 20g·t to 100g·t. After 1980s, China has built several large-scale geotechnical centrifuges. The centrifuge in China Institute of Water Resources and Hydropower Research (IWHR) was put into operation in 1991. The maximum acceleration of the centrifuge is 300g and the capacity is 450g·t. At the same time, Nanjing Hydraulic Research Institute has also built a centrifuge with the maximum acceleration of 200g and the capacity of 400g·t. In 1993, Tsinghua University built a centrifuge with the maximum acceleration of 150g and capacity of 50g·t. In 2001, Hong Kong University of Science and Technology (HKUST) built a centrifuge with the maximum acceleration of 150g and capacity of 450g·t. All the above-mentioned large-scale centrifuges have installed shaking table for earthquakesimulation. Besides centrifuge building, Chinese scholars and engineers have also conducted series studies in the basic theories, testing methods and engineering applications of centrifuge modeling tests and achieved some important results.

As an important tool for studying complicated engineering problems, with the combination of numerical analysis, centrifuge model test has been widely applied in studying technical solutions of rockfill dam engineering. Most of the important rockfill dam projects in China have conducted centrifuge model test during design stage, which include Xiaolangdi ECRD, Pubugou ECRD, Tianshenqiao-1 CFRD, Hongjiadu CFRD, Shuibuya CFRD, etc.

Centrifuge model test can also be used to verify numerical analysis method.

3. Dynamic model test on shaking table

Model test on ground shaking table was started in China in 1980s. In recent years, with the progress of high dam construction, technologies of dynamic model test for rockfill dams by using ground shaking table have been rapidly developed. China Institute of Water Resources and Hydropower Research installed a three directions 6m×6m shaking table with six degree of freedom and conducted fundamental researches on dynamic model test of rockfill dam. Several high rockfill dam projects have conducted model test on the shaking table, which include Heiquan CFRD (dam height 123.5m), Lianghekou ECRD (dam height 295m), etc. With the model test on ground shaking table, the procedure and failure mode of rockfill dam can be observed intuitively and the regularity knowledge of the performance of rockfill dam upon earthquake can also be obtained.

Model test uses similar model structure to simulate the prototype to get the mechanical behaviors of the prototype under static and dynamic conditions. By measuring the physic quantities such as displacement, strain, acceleration, etc. of the model structure, the corresponding physic quantities of prototype structure will be obtained by the similarity criteria. Therefore, the similarity criteria of model test not only determine the correctness of model data to prototype data, but also the necessary conditions for establishing the simulating connection of model structure and prototype structure. Normally, for model test on ground shaking table, the test should meet the requirements of thesimilarity on geometrical, kinematical, physical and dynamical conditions, and also the boundary conditions. But for the model test of rockfill dam, the important factor to be considered is the similarity on materials. The construction materials of rockfill dam are granular material, which are highly nonlinear and closely related with stress level and stress paths. Besides, the size of the materials for model test will be reduced to fit the scale of the model. Therefore, the selection of model material is one of the key technologies of ground shaking table model test. At present, most of the model tests on ground shaking table use the same material of prototype dam, with the grains size reduced. The methods for grains size reduction are the same as laboratory tri-axial test. Due to the fact that the stress status of the model on ground shaking table is different with the prototype dam, the model test does not meet the requirement of similarity on stress status.

With the dynamic parameters measured from laboratory test under the condition of low stress level and the earthquake waves measured from shaking table, dynamic analysis could be conducted for the different model dams and testing schemes. By comparison of the testing results and the numerical analysis results, numerical models can be verified. It is a useful tool for improving numerical analysis methods.

4. Site test

As the size effects of testing materials cannot be assessed from laboratory tests, full size site tests are necessary. During the test of compaction test of the construction materials at dam site, the full size site test could be conducted with the real materials and real construction methods. The tests are compression test, direct shear test, load test, seepage test, relative density test, etc. With the data obtained from site tests, back analysis could be conducted based on parameters from laboratory tests. The parameters obtained from this procedure will be more accurate to the real situations. In engineering practices, the first site compression test was conducted at the dam site of Longtan project. After that, the same test was conducted at the dam site of Shuibuya CFRD, Sanbanxi CFRD and jiangpinghe CFRD. The large-scale load test of the full size rockfill material was conducted at the dam site of Gongboxia CFRD. The test of relative density for the sandy gravel materials was conducted at the dam site of Jingpen ECRD, Gongboxia CFRD and Jishixia CFRD. For the application of site tests, more research works should be done in the application of test results on numerical analysis.

12.2.4.2 Numerical Analysis

Finite element method (FEM) has prominent advantages in dealing with nonlinear constitutive relationship, nonhomogeneous materials and complicate boundary conditions. In the analysis of rockfill dam, except stability analysis is conducted by limit equilibrium method, seepage analysis, stress and deformation analysis and dynamic analysis are conducted by FEM. With the application of FEM, some numerical analysis models of earth and rockfill materials are developed recently.

1. Seepage analysis

In the seepage analysis of rockfill dam, due to the difficulty of unknown position of phreatic surface and seepage exit, the solving process is quite complicate. In the early stage, seepage analysis of rockfill dam was conducted by variable mesh method, i.e. the computation field is changed with the procedures of iterative computation of the phreatic surface position. Therefore, the computation mesh is also changed with the iteration. At present, most of the seepage analysis of rockfill dam is conducted by fixed mesh method, which is convenient in finite element mesh arrangement. Technologies of seepage analysis is further progressed with the researches on determine of control equation, treatment of virtual elements and transition elements, management of the possible seepage exit surface, etc.

2. Slope stability analysis

For slope stability analysis of rockfill dam, the commonly used method is limit equilibrium method. It is an approximate analysis method based on several assumptions. It cannot get the stress and deformation of the dam, but it can provide the factor of safety by calculating the ratio of the shear strength to the shear stress of the sliding surface. With finite element method, the distribution of stress level of each element can be obtained. By observing the position of the area with stress level equal to 1, the stability status of the dam can be qualitatively determined. But it cannot get the factor of safety for the design. Recently the strength reduction method by FEM was introduced into China, but its application is not popular in engineering practices.

Slices method is the most commonly used method for slope stability analysis. The critical sliding surface and the corresponding minimum factor of safety are obtained by trial calculation and automatic search. The methods of slices method for slope stability analysis can be classified in different types, which include: method of considering or not considering the acting forces of slices, methods with circular or non-circular sliding surface, methods that satisfying only the main balance conditions (simplified method) or satisfying all the balance conditions (strict method), methods of using effective stress analysis and total stress analysis. In the engineering practices in China, the widely accepted method is simplified Bishop method, which considered the actions between slices and the calculation result is similar with the strict method.

In the early stage of dam slope stability analysis, the linear strength parameters of soil were used. With the increasing of dam height and the use of coarse granular materials, the non-linear features of the material under condition of high stress level were more and more concerned. In the slope stability analysis by linear strength formula, for non-cohesive soil with $c = 0$, the searched critical sliding surface is a very shallow circle. It is not the real critical sliding surface. On the other hand, in the laboratory test, it is discovered that the shear strength of coarse granular material present non-linear characters. The internal friction angle of the materials is reduced with the increasing of stress level. On the condition of low stress level, the friction angle could be higher than 50°, on the condition of high stress level, the friction angle could be lower than 40°. Therefore, slope stability analysis of high dam should use non-linear shear strength formula. This is regulated in the design standard of rockfill dam in China.

In China, the well accepted non-linear shear strength principle is the formula suggested by Duncan. It is a logarithmic form expression, i. e.

$$\phi = \phi_0 - \Delta\phi \lg(\sigma_3/p_a)$$

$$\tau_f = \sigma_n \tan\phi$$

where σ_3 is minor principle stress, i. e. the cell pressure in triaxial test; ϕ_0 and $\Delta\phi$ are parameters; σ_n is the normal stress on the sliding surface.

3. Stress and deformation analysis

Stress and deformation analysis of rockfill dam is mainly conducted by non-linear finite element method. The accuracy of computation depends on computation method, constitutive model and parameters of materials and whether the computation procedures well follow the real situations.

The main methods for studying constitutive models of geotechnical materials are based on the theory of continuum mechanics. From the analysis of apparent features of soil, with the stress-strain relationship obtained by laboratory tests, the constitutive models can be established by the theory of elasticity and plasticity.

Present theories on constitutive relationship include non-linear elastic theory, elastic-plastic theory and visco elastic plastic theory. In the numerical analysis of rockfill dam, the commonly used models are non-linear elastic model and elastic-plastic model.

(1) Non-linear elastic model. For the present applied non-linear elastic models in numerical analysis of rockfill dam, most of the models are belong to the Cauchy elastic model (variable elastic model). It employs the incremental form of the generalized Hook's law:

$$\{\Delta\varepsilon\} = [C]\{\Delta\sigma\}$$

and assume the elastic constants (E, ν, K, G) in the flexibility matrix are only the function of stress status, nothing related to the stress history.

Non-elastic models proposed by Konder (1963), Duncan and Chang (1970), Naylor (1978), Duncan (1980), etc. could simulate the main features of soil and relatively simple in numerical analysis applications. Therefore, these models are widely used in geotechnical engineering. In China, the hyperbola model proposed by Duncan and Chang is the most popular model used in the numerical analysis of rockfill dam. Besides, Chinese scholars have also developed some practical non-linear elastic models, such as the modified K-G model proposed by Sichuan University and the uncoupled non-linear K-G model proposed by Tsinghua University.

(2) Elasto-plactic model. In the elasto-plastic model, the total strain (increment) is decomposed into the elastic and the plastic components, e. g.

$$\{\Delta\varepsilon\} = \{\Delta\varepsilon^e\} + \{\Delta\varepsilon^p\}$$

Accordingly, the elasto-plastic stress-strain relationship can be expressed as:

$$\{\Delta\sigma\} = [D](\{\Delta\varepsilon\} - \{\Delta\varepsilon^p\})$$

where, the elastic strain $\{\Delta\varepsilon^e\}$ can be calculated by generalized Hook's law and the plastic strain $\{\Delta\varepsilon^p\}$ is calculated by following formula:

$$\{\Delta\varepsilon^p\} = \Delta\lambda\{n\}$$

where $\Delta\lambda$ is positive scalar of proportionality depends on the state of stress and load history. It represents the magnitude of the incremental plastic strain and is determined by hardening rule; $\{n\}$ represents the direction of the plastic strain increment vector and is determined by flow rule. The boundary between the elastic strain and the plastic strain is defined by yield surface.

In geotechnical engineering, by considering the anisotropic hardening of the plastic deformation and the dependent of the incremental plastic strain to the direction of the incremental stress, the multiple yield surfaces model will be more applicable. At present, the more popular models used in the numerical analysis are double yield surface model, which include the Lade-Duncan's model and the Shen Zhujing's model (China).

The stress-strain property of the materials of rockfill dam is very complicate and is determined by many factors. Laboratory tests can only measure less data under the relatively simple conditions. Constitutive model requires generalized representation and adaptation. Its application situation is far more complicate than the conditions in laboratory test. Therefore, any constitutive model is simplification and approximate treat of the real situation. The above-mentioned non-linear elastic models are two parameters model with variable elastic modular. They are direct extension of elastic model. For isotropic materials, only two of the four elastic parameters E, ν, B (K), G are independent. Therefore, there are no significant differences between the different models. Theoretically, elasto-plastic model can take into account of more comprehensive characteristics on the stress-strain relationship of rockfill materials. It can consider the volumetric strain (include contraction and dilatancy) caused by shear stress and simulate the development of plastic strain. Compared with non-linear elastic model, elastoplastic model is relatively complicate. In the configuration of the model, assumptions on yield surface and flow rule should be introduced. The accuracy of the assumptions will have important impacts on the applicability of the model. Besides, the determination of parameters of elastoplastic model is also relatively difficult.

At present, the most commonly used models for numerical analysis of rockfill in China are Duncan's hyperbola non-linear elastic model and the double yield surfaces elastoplastic model proposed by Shen Zhujiang. Normally, parameters of model may have more sensitive impacts on the computation results than the model itself. Besides, the accuracy of test results and the computation method also have the inference on the results of analysis.

(3) Dynamic analysis. Dynamic analysis of rockfill dam under the action of earthquake is conducted by finite element method. The key factor of the analysis is the accurate description of the dynamic properties of soil and rockfill materials. From laboratory tests, soil and rockfill material will present the features of non-linear, hysteresis and deformation accumulation under the action of dynamic loads. Besides, it also has the characteristics of stress history dependence, secondary anisotropy caused by unrecoverable plastic deformation, shear stiffness attenuation and energy dissipation. Similar with the configuration of static constitutive model, the establishment of dynamic constitutive model is also based on the theories of elasticity and plasticity.

Dynamic constitutive models of soil based on non-linear elasticity theory mainly include two types: physical model and empirical model. Physical models mainly use the combination of elastic element, viscous element and plastic element to represent the dynamic features of soil. Empirical models are based on laboratory test, which include equivalent linear model (non-linear equivalent visco-elastic model) and "true" non-linear model (Masing model). Dynamic elasto-plastic models mainly include multiple yield surfaces model, boundary surface model and auxiliary elasto-plastic model. At present, the commonly used dynamic analysis models for rockfill dam in China are equivalent linear model and non-linear model.

In the equivalent linear model, stiffness of soil is expressed by secant modulus (equivalent shear modulus) and the hysteresis features of soil are represented by the equivalent damp ration with is varied with the change of shear strain. Soil is considered as a visco-elastic body. The maximum dynamic shear modulus of soil is:

$$G_{\max} = KP_a \left(\frac{\sigma'_m}{P_a}\right)^n$$

where K, n are determined by laboratory tests; $\sigma'_m = (\sigma'_1 + \sigma'_2 + \sigma'_3)/3$, σ'_1, σ'_2, σ'_3 are effective principle stress act on testing sample; P_a is air pressure.

With recursive analysis of the testing data of $G/G_{\max}$ and damp ratio λ with dynamic shear strain γ, the relationship of dynamic shear modulus $G/G_{\max}$ and dynamic shear strain γ, damp ratio λ and dynamic shear strain γ can be obtained. Non-linear model use function expressions to simulate thestress-strain and hysteresis curves of soil and define the corresponding loading/unloading principles. It follows the true loading paths to describe the dynamic properties of soil. The main features of the improved non-linear model by China Institute of Water Resources and Hydropower Research (IWHR) are: ①Compared with the equivalent linear model, it has better simulate the residual strain. Thus it can be directly used for residual deformation calculation. In dynamic analysis, the model can calculate tangent modulus at any moment and to conduct non-linear computation. Therefore, dynamic response obtained by this method will be closer to the real situation. ②Compared with Masing model, it considered the initial loading curve, which provides a seasonable description on shear stress-strain relationship in the case of shear stress ratio beyond the yield shear stress ratio. The hysteresis circle of the model is open so that it can calculate residual shear strain. It considers the impacts of shaking numbers and initial shear stress ratio on the deformation properties of soil.

(4) Post-construction deformation analysis. From observation, in the operation period after rockfill construction completed, dam deformation still in development. In addition, the post-construction deformation will become more and more significant with the increase of dam height. Therefore, it is necessary to study the post-construction deformation properties of high rockfill dam by using the method of laboratory test and numerical analysis. From the analysis of mechanism of rockfill dam deformation, many factors could lead to the development of rockfill dam deformation during its operation. Among these, creep deformation of rockfill is one of the mainfactors. Previous studies on creep deformation of soil mainly focus on soft clay. Recently, with the development of high rockfill dams, the studies on creep deformation properties are very active in China. Several research institutes and universities have installed large scale testing machines for testing creep deformation properties of rockfill or sandy gravel materials. Some numerical analysis models are also developed.

Numerical analysis models for creep deformation analysis of rockfill material use mathematic method to describe the relationship of creep deformation and time. The expressions are simple continuous function. The present methods for establishing creep deformation models are empirical methods. With the relationship between deformation and time obtained from laboratory tests, the functions are selected to simulate the testing curves. Normally, the functions are individual functions with the expression of exponential-type, power function type, hyperbolictype and logarithmic type.

(5) Coupled analysis of different quantities. Soil or rockfill material contains solid particles, liquid and air. In the construction of rockfill dam and reservoir impoundment, the deformation of soil has complicate-coupled relationship with water seepage. To consider the coupled actions of solid particles, water and air, consolidation analysis for considering deformation and seepage should be conducted. The commonly used method for the analysis at present is the effective stress analysis method based on Biot's consolidation theory.

12.2.5 Operation

The operation statuses of completed projects, which include Xiaolangdi, Pubugou, Qiaoqi, Shuiniujia, etc. are normal. All of these high ECRD projects have established instrumentation system and accumulated rich observation data. Some observation data are presented below.

Xiaolangdi Dam: dam height is 156m, built on alluvium foundation. To 2008, the vertical displacement at dam crest was 1,362.7mm, about 0.9% of dam height. The horizontal displacement of dam crest was 934.1mm. But the measured horizontal displacement at the upstream and downstream side of dam crest had certain difference. The leakage through dam and rock foundation (not include the leakage through alluvium) was 30,000m^3/d.

Qiaoqi Dam: dam height is 125.5m, built on alluvium foundation. To 2009, the maximum settlement inside the dam was 1,338.1mm, about 1.09% of dam height. The maximum horizontal displacement was 434.5mm. The maximum leakage was 29.7L/s.

Shuiniujia Dam: dam height is 108m, built on alluvium foundation. To 2008, the maximum settlement inside the dam was 591mm and the maximum horizontal displacement was 92.24mm. Settlement of the dam was trend to stable, but the horizontal displacement was still in development. The piezometers in dam and foundation showed that the whole watertight system performed well.

Some ECRD projects are located at high seismicity area and subjected strong earthquake in 2008, which include Bikou, Shuiniujia, etc. Bikou dam was subjected earthquake with Richtermagnitude of 7.3 in 1976. The dam was not seriously damaged. In the earthquake in 2008, dam site is located in the area with earthquake intensity of 9 degree. Main damages of the dam are rupture and opening of the joints of wave wall, horizontal crack (through left abutment to right abutment) between wave wall and upstream slope protection, significant deformation on downstream slope, etc. The maximum settlement on dam crest due to the earthquake was 24.2cm. Upstream slope and downstream slope moved to upstream side and downstream side respectively. At the connection of dam crest and abutments, some cracks were developed from upstream side to downstream side. There were no significant changes of leakage and seepage pressure before and after earthquake. Shuiniujia ECRD is located in the boundary area with earthquake intensity of 8 degree of the 2008 earthquake. No obvious damages were found. The performance of these high ECRD project during the 2008 earthquake has proved the good capacity of rockfill dam for resisting the action of earthquake.

12.3 Concrete Faced Rockfill Dam

The construction of concrete faced rockfill dams in China has been rapidly developed in the 1st decade of 21st century. Based on the experiences gained from Tianshengqiao I CFRD, several 200m high CFRD had been successfully constructed. Besides, some major projects with the dam height about 300m will select CFRD or take CFRD as the main alternative for dam type selection. As the achievements of CFRD construction in China before year 2000 has been fully reported in the General Report of the 1st International Symposium on Rockfill Dams (2009, Chengdu, China) and other documents, only the latest developments of high CFRD (in the first decade of 21st century) will be presented here.

12.3.1 Typical Cases

1. Tianshengqiao I hydropower station

Tianshengqiao I is the first high CFRD in China with the height close to 200m and constructed on the main stream of large river. The dam height is 178m and the crest length is 1,104m. The total volume of rockfill dam body is $18 \times 10^6 m^3$ and the area of concrete face slab is $17,270 \times 10^3 m^2$. The project has reservoir capacity of $10,257 \times 10^6 m$, installed capacity of 1,200MW and the maximum spillway discharge capacity of $21,750 m^3/s$. The construction started in 1991 and the first impounding started in 1998. The project was completed in 2000. Fig. 12.3-1 and Fig. 12.3-2 is the general layout of the project and the typical cross section of the dam.

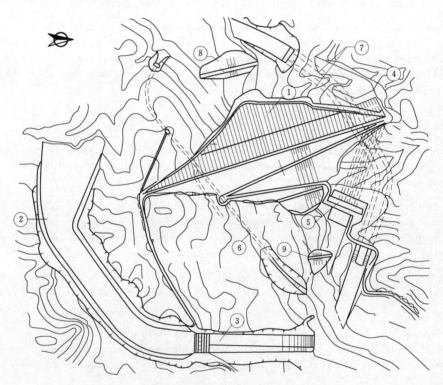

Fig. 12.3-1 Generaln Layout of Tianshengqiao I Project

①—CFRD; ②—Diversion channel of spillway; ③—Discharge chute of spillway; ④—Inlet of power station;
⑤—Power house; ⑥—Discharge and empty tunnel; ⑦—Diversion tunnel; ⑧—Upstream cofferdam;
⑨—Downstream cofferdam

(1) The main features of Tianshengqiao I CFRD can be summarized as follows:

1) Project layout. Open spillway was arranged at a karst trench far from the right abutment to avoid the unfavorable geological conditions of the original spillway arrangement. The modified arrangement can provide better operation condition for the spillway. Although the amount of rock excavation increased, the excavated rockfill can be used as the filling materials of rockfill dam and the raw materials for concrete aggregate. Thus, it is unnecessary to excavate quarry. By keeping the balance of excavation and filling, the maximum economic and ecological benefit had been achieved.

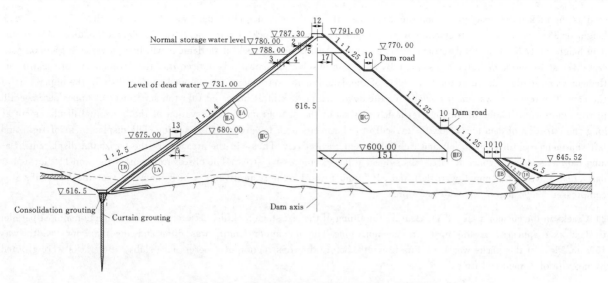

Fig. 12.3-2 Typical Cross Section of the Tianshengqiao Ⅰ CFRD (Unit: m)
Ⅳ, ⅠA—Cohesive soil; ⅠB—Random fill; ⅡA, ⅡB—Cushion zone; ⅡA—Transaction zone;
ⅡB—Main rockfill; ⅢC, ⅢD—Downstream rockfill

2) Zoning and dam materials. The cross section of the dam is shown in Fig. 12.3-2. The boundary of the upstream rockfill and downstream rockfill located at the position of dam axis. Upstream zone is hard rockfill of fresh or weakly weathered limestone. The downstream rockfill in the area below downstream water level and the area near downstream slope is the same as upstream rockfill. Other part of the downstream zone is weak rockfill with the mixture of sandstone, mudstone and thin-layered limestone over the downstream water level. Due to the difference of deformation modulus of unstream and downstream rockfill, large differential deformation occurred between two rockfill zones. Besides, the vertical boundary of upstream and downstream rockfill zone cannot provide a smooth transition of the differential deformation.

3) River division and construction stages. River is closed for one time by upstream and downstream cofferdams and it is diverted through two diversion tunnels at left bank. During the first flood season, water was discharged through diversion tunnels and the original river channel. The maximum flow discharge was 4,750cm³/s where 3,430cm³/s passed through river channel (overflow the cofferdam). In the second flood season, water was discharged through diversion tunnel and the pre-set opening on rockfill dam. The maximum flow discharge was 3,790cm³/s with 1,290cm³/s discharged through the dam. Rockfill filling at abutments area continued in flood season. For the following flood season, the priority section of rockfill dam was used for retaining water. The rockfill dam was constructed in 7 stages and the concrete face slab was constructed in 3 stages. In the last stage of dam filling, the height difference between the top of upstream and downstream rockfill was 122m. The last stage rockfill filling was completed in 4 months with an average raising rate of 1m/d. Concrete face slabs were constructed immediately after rockfill filling reached to the top elevation of face slab construction. Due to the difference of the modulus and deformation time of the upstream and downstream rockfill, the upper part of concrete face slabs produced displacement towards downstream direction, which lead to the structural cracks developed in concrete face slabs. For the purpose of power generation, reservoir impoundment started before the completion of rockfill dam filling. This preloading action is benefit for rockfill deformation control according to the strengthening action of the embankment.

4) Operation performance and lessons. The normal storage water level of Tianshengqiao Ⅰ reservoir is 780m. The initial impoundment of the reservoir started on April of 1998 and the highest reservoir water level in 1998 is 740.36m. The highest reservoir water level in 1999 is 767.40m. The reservoir water level reached to 779.96m in 2000. After that, the project put into normal operation. The highest reservoir water level varies from 773.29m to 779.96m. The measured initial leakage at downstream measuring weir was 183L/s (after deducting the impact of rainfall, the actual leakage value is 150L/s). The initial leakage was reduced gradually in following years. Since 2006, the leakage corresponding to the normal storage water level has been kept around 80L/s.

(2) Main problems occurred during operation are as follows:

1) Large deformation of rockfill dam. By the end of 2007, the observed maximum settlement at riverbed section was 335.17cm, about 1.99% of the dam height. The horizontal displacement toward downstream direction was 113.8cm. The deformation values were much larger than that of the similar high CFRD projects, especially, for the deformation in the stage of operation. If the rockfill deformation occurred before the completion of the third stage concrete face slabs is consid-

ered as the rockfil deformation in construction stage, the settlement of rockfill dam was 295.23cm, about 1.66% of dam height or 83% of the total settlement. Thus, the post-construction rockfill settlement was 59.68cm, about 0.34% of the dam height or 17% of the total settlement. The large post-construction rockfill settlement will have adverse impacts on concrete face slabs and joint waterstop system. By analyzing observation data, it is noticed that the development of rockfill settlement was rapidly reduced at the lower part of the dam. But the settlement is developed continuously in the upper part of the dam. The tendency was more obvious for the downstream rockfill. It indicated that the settlement of upper part rockfill provide more contribution of the long-term deformation of the dam. From the observation of the horizontal displacement along the direction of dam axis, rockfill movedfrom abutments area toward riverbed section. The displacement of right and left abutment rockfill point to left and right abutment respectively. Thus, in the area where the horizontal displacement along dam axis is zero, concrete face slabs subjected large compressive stress. The observation points 10# and 11# located at dam crest moved 28.24mm toward left and 31.70mm toward right respectively, which indicated that the face slabs was compressed 59.94 mm in the width of 48m.

2) Cracks in the cushion zone (2A). During the filling of the last stage rockfill, 37 cracks with its direction almost parallel to dam axis appeared at the upstream cushion zone. The maximum length was 60m and the maximum width was 150mm. Most of the cracks was 1-15mm in width. Before the construction of concrete face slabs, the cracks were grouted by mixture of cement and fly ash.

3) Gaps between concrete face slabs and upstream rockfill. During construction, after the completion of concrete face slab of each stage, a gapbetween concrete face slab and upstream rockfill was discovered at the upper part of face slab. Before the construction of the next stage concrete face slab, grouting of the mixture of cement and fly ash was conducted to seal the gap. For the possible gaps between face slabs and upstream rockfill occurred during operation, grouting treatment was conducted from the tubes installed in parapet wall.

4) Cracks on concrete face slabs. According to the inspection during construction stage, 1,296 cracks were discovered on concrete face slabs. The cracks were basically distributed in horizontal direction (parallel to dam axis). After reservoir impoundment, 4,357 cracks had beendiscovered (mainly on the third stage face slabs, include the old cracks) by detailed inspection during low reservoir water level. Among those cracks, there are 80 cracks with the width larger than 0.3mm. Most of the cracks have their depth less than 10cm. The maximum depth of the cracks was 42cm. The fact indicated that the cracks on concrete face slabs were greatly increased after reservoir impoundment. Most of the cracks were structural cracks caused by large rockfill deformation and the differential deformation of upstream and downstream rockfill.

5) Concrete face slab rupture at the vertical joint of riverbed section. Face slabs concrete rupture was occurred at the vertical joint of the riverbed face slab L3 and L4 in July 18, 2003 and May 29, 2004. The position of the joint was located at the point where the horizontal displacement of face slabs in the direction along dam axis is zero. As the concrete face slabs at the abutments moved toward the center of river valley, the face slabs at the sides of the joint were subjected large compressive stress, which led to the rupture of concrete rupture. Besides, as the rupture was occurred in summer season, thermal stress may also added some supplementary impacts on the event. From the observation, the compressive strain of concrete at the rupture area was over 800μ. The value was reduced below 600μ after the occurrence of concrete rupture. In repairing of the damaged concrete face slabs, the upper part of several vertical joints in riverbed area were cut and filled with 5cm width deformable wood plate. The function of the filling wood plates is to reduce the large compressive stress of concrete face slabs.

Although the concrete face slabs of Tianshengqiao I CFRD has some defects and damages, the leakage through the dam has no significant changes.

The main problems in the operation of Tianshengqiao I CFRD are the large post construction rockfil deformation and the differential deformation of upstream and downstream zones. The main reasons for the problems could be summarized as relative low compaction density of rockfill, improper section zoning and rockfill construction sequence, no sufficient settlement time for rockfill before the construction of concrete face slabs. The construction of Tianshengqiao I CFRD provided valuable experiences and lessons for the following high CFRD construction in China.

2. Hongjiadu CFRD

Hongjiadu CFRD is located in an unsymmetrical narrow valley of the lower reach of Liuchonghe River, tributary of Wujiang River. The dam height is 179.50m, and crest length is 427.79m. The total volume of the rockfill dam is $9.2 \times 10^6 m^3$. The storage capacity of reservoiris $4,947 \times 10^6 m^3$ and the installed capacity is 600MW. The project construction started in 2000 and completed in 2005. The dam section illustrated at Fig. 12.3-3.

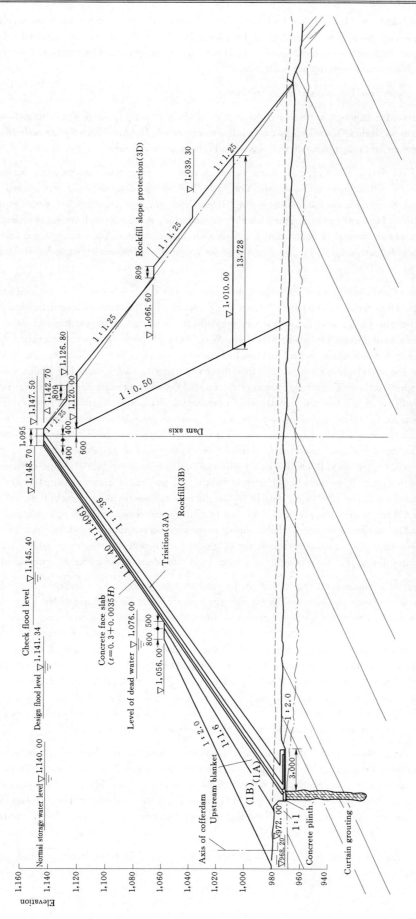

Fig. 12.3-3 Cross Section of Hongjiadu CFRD (Elevation Unit: m; Size Unit: cm)

The construction of Hongjiadu CFRD faced unfavorable conditions. Based on the experiences and lessons gained from other high CFRD projects, especially the lessons learned from Tianshengqiao I, and also the comprehensive scientific researches, advanced technologies and innovative techniques were applied in the design and construction. The project was built successfully and presented very good performance until now.

The main features of the project can be summarized as follows:

(1) As the dam site is located in an unsymmetrical narrow valley, in the plan layout, large-scale tunnels were selected as the flood discharge structures, include tunnel spillway and discharge tunnel. The arrangement has well adapted the topographical and geological condition of dam site, and also avoided high slope excavation.

(2) The left abutment of the dam is a vertical high slope. Technical measures were taken for reducing the large amount excavation of the high slope, which include: ①The application of narrow plinth, i.e. the plinth has a fixed width of 4.5m from dam crest to riverbed. To compensate the insufficient width of plinth, the internal plinth was added at the downstream part of the external plinth. ②The application of special high modulus zone, i.e. set a high modulus rockfill zone (by increasing fine grains and applying intensified compaction) between abutment and dam body to reduce deformation gradient between abutment and rockfill dam body. ③The application of poor concrete to improve (smooth) the local unfavorable slope of the abutment.

(3) Integrated application of techniques on rockfill deformation control. For materials selection, the hard limestone rockfill had been used for the dam. High rockfill compaction density was accepted with the porosity less than 20%, which is to achieve high deformation modulus of the rockfill. Downstream rockfill zone used the same rockfill materials as that in the upstream zone. The lift thickness of downstream rockfill is 1.6m. With the application of impact compaction (polygonal roller), the compaction density of downstream rockfill had reached the same values as that of the upstream rockfill. In the contacting area between dam and abutments, a special high modulus compaction belt with more fine grains was arranged. The function of this zone is to reduce the deformation gradient of the rockfill in abutment area. Except the priority section for the first flood season, the rest part of the dam was constructed in horizontal lift, i.e. no significant height difference for construction layers from upstream to downstream. The construction of concrete face slabs was started with a certain period of time after the finish of rockfill construction, which could avoid the adverse impacts of the initial large deformation of rockfill on concrete face slabs. The top elevation of concrete face slabs (except for the last stage face slab) was controlled lower than the top elevation of constructed rockfill, which could eliminate the gap between face slab and upstream rockfill. The application of the deformation control measures has effectively reduced the post construction deformation and differential deformation of the dam. According to the monitoring data in 2008, the maximum settlement of rockfill dam at normal storage water level is account for 136cm, approximately 0.76% of dam height. The settlement only increased 9cm after the completion of dam construction. The maximum horizontal displacement in the direction of river flow is 24.4cm. The horizontal displacement in the direction of dam axis mainly occurred during dam construction with the maximum value of 1.75cm. The value has no significant changes during the rising of reservoir water level after dam construction was completed.

(4) Integrated application of techniques on improving cracks resistance of concrete face slabs. To control the shrinkage cracks, the shrink compensation concrete was used by adjusting concrete mixture with the use of MgO. To control the initial cracks of concrete, the concrete was mixed with polypropylene fiber. To control the structural cracks due to face slab deformation, double layer reinforcement was arranged. Immediately after the pouring of concrete face slabs, the concrete was cured by temperature control and constantly keeping wet. With the application of those integrated measures and combined with rockfill deformation control, concrete face slabs presented good performance during construction and operation. For the whole concrete face slabs with the total area of $72 \times 10^3 m^2$, there are only 33 cracks and only 1 crack has its width larger than 0.25mm. In the construction of plinth, no construction joints were arranged. The whole plinth was constructed with the normal blocks and the late pouring blocks (1-2m). After concrete shrinkage of the two normal blocks was finished, then the late pouring block was constructed by using the shrink compensation concrete.

(5) For instrumentation of the dam, besides the normal arrangement, a longitudinal observation line was arranged along dam axis in order to identifying the arching effect of high CFRD in a narrow valley. Inside dam body, extra-long extensionmeters were installed to measure the settlement and horizontal displacements of rockfill dam. To distinguish the leakage through dam and abutments, cutoff ditches were arranged in the dam. From the observation data, the leakage through the dam was 130L/s when reservoir water level reached to the highest level. For low reservoir water level, the leakage was 20-30L/s.

(6) During construction, the compaction density of rockfill was double checked by two different methods in order to guarantee the quality of rockfill compaction.

3. Sanbanxi CFRD

Sanbanxi CFRD is located at Qingshuijiang River in Guanzhou Provence. The maximum dam height is 185.5m, with the elevation of dam crest of 482.5m. The crest length is 423.8m and the volume of rockfill is $9,620 \times 10^3 m^3$. The storage capacity of the reservoir is $4,094 \times 10^3 m^3$, while the installing capacity is 1,000MW. Construction of the project was started in 2001 and completed in 2007.

Reservoir impoundment of Sanbanxi project started at the beginning of 2006. For the initial stage of operation, the reservoir water was kept in low level (El. 430m – El. 440m), with the measured leakage at downstream measuring wire of 20 – 40L/s. In June of 2007, when flood season started, the reservoir water level was rapidly raised up. To the end of July, the water level reached to 472m, only 3m below the normal storage water level. From the observation, the leakage suddenly increased to 255L/s and the maximum leakage value was 301L/s. After the flood season, with the drop of reservoir water level, the measured leakage was also decreased. In Jan. of 2008, when water level dropped to the dead water level, El. 425m, the measured leakage was reduced to 150L/s, and further reduced to 100L/s afterward.

In Aug. 2007, it was discovered that the instruments at El. 385m and El. 379m of face slab right MB5 were out of order. Then, a lot of internal measuring instruments of concrete face slabs stopped working. The failure of those instruments was caused by broken of the cables. In Jan. 2008, reservoir water level dropped to the dead water level. With underwater inspection by divers, face slab concrete damage was discovered at the horizontal construction joint with elevation of 385m of the first and second stage concrete face slab. The length of damaged face slabs was 184, involved 12 slabs (from Left MB3 to Right MB9). In the damaged area, concrete was ruptured and rebar was exposed and bended. In some area, waterstop of vertical joints was also damaged. The maximum width of the ruptured concrete was 4m and the maximum depth was 41cm. From 2008, underwater repairing was conducted three times. But the leakage of the dam was still 270L/s when reservoir water level was in 469m.

Detailed back analysis of observation data revealed that the maximum settlement was occurred at the upstream part of the dam and the compression stresses of concrete face slabs in the direction of dam slope was larger than that along the dam axis direction, with the maximum value about 24MPa. Deflection of face slab and the compression stresses in the direction of dam sloping reached its maximum values near elevation 385m, approximately the same place of the rapture zone. Thus the mechanism of the damage could be identified as compression rupture in the direction along dam slope.

As the quantity of rockfill deformation is still in a reasonable scope, with the maximum settlement is 0.95% of dam height, stress distribution of rockfill and concrete face slabs was basically normal, and leakage water was clear, the overall safety of the dam can be guaranteed. The damage of concrete face slab can be repaired when deformation of rockfill is trend to stable.

4. Shuibuya CFRD[40]

Shuibuya CFRD is located on the main stream of the Qingjiang River of Hubei Province. The dam height is 233m (the highest CFRD in the world) and crest length is 660m. The reservoir storage capacity is $4,580 \times 10^6 m^3$ and theinstalled capacity is 160MW. Construction works started in 2002 and the project was completed in 2008. The plan layout and the dam cross section are shown in Fig. 12.3 – 4 and Fig. 12.3 – 5.

Due to its unprecedented dam height, large project scale and complicate topographical and geological conditions, in the design and construction of Shuibuya CFRD, series theoretical and testing studies had been carried out with the participation of design institutes, construction corporations, research institutes and universities. The application of research results in engineering practices had contributed a lot for the successful construction of the Project.

The adopted new techniques are summarized as follows:

1) The use of external plinth and internal plinth was applied. The standard width of external plinth is 8m, which satisfy the requirement of grouting construction. Internal plinth was arranged at the downstream side of the external plinth. The width is calculated with 1/15 of water head. No structural joints were set up along plinth. Several post pouring gaps were arranged with the interval of 12 – 16m. The gaps were filled with contraction compensating concrete 90 days after pouring of plinth concrete at both sides.

2) Sand gravel alluvium within the area of 150m downstream from plinth was excavated. The remaining part of was treated by dynamic compaction. Dry density of alluvium was increased from $2.07 g/cm^3$ to $2.18 g/cm^3$ after treatment.

3) Rockfill materials mainly come from structures excavation. Insufficient part used materials excavated from borrow area. Hard limestone was used in 3B zone. Limestone with thin layer soft rock was used in 3C zone. Rockfill parameters, include gradation, compaction properties, deformation properties, etc., had been determined by in-situ explosion test, compac-

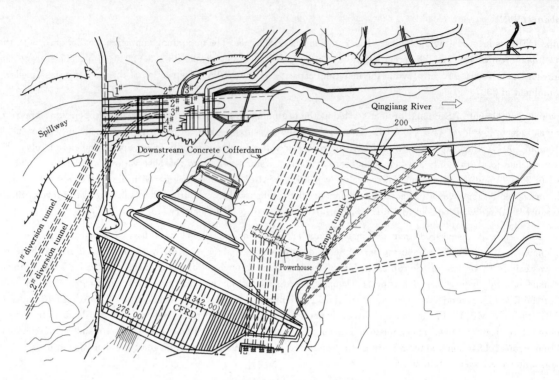

Fig. 12. 3 - 4   Plan Layout of Shuibuya Project

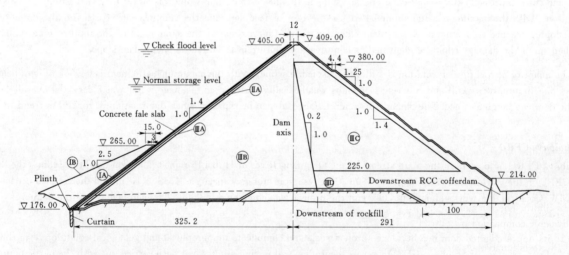

Fig. 12. 3 - 5   Cross Section of Shuibuya CFRD (Unit: m)

ⅠA—Cohesive soil; ⅠB—Randon fill; ⅡA—Cushion zone; ⅡA—Transition zone; ⅡB—Main rockfill;
ⅢC—Downstream rockfill; Ⅲ—Downstream rockfill (free drainage)

tion test, as well as large scale laboratory tests.

4) Technology of extrude curb was used for upstream slope protection. Before the construction of concrete face slab, emulsified asphalt was shoot on the surface of extrude curb to be a bond breaker, and extrude curb was cut at the position of vertical joints to reduce the restraint on face slabs.

5) Concrete face slabs were constructed in stages. Between the slabs of different stages, there are horizontal structure joints. Measures of double layers reinforcement, polyacrylonitrilene fibre concrete, cement-based crystal type coating on the face slab were used to improve the crack resistance properties of concrete face slabs.

6) In the construction, series deformation control measures were applied, which include heavy roller compaction for increasing rockfill density and modulus, control height difference of upstream construction layer and downstream construction

layer, control downstream rockfill higher than upstream rockfill before the construction of face slabs, arrange a certain period of time for settlement of upstream rockfill before the construction of face slabs.

7) A new real-time GPS system with high accuracy was developed to control rockfill compaction quality. With this system, the moving trace, speed and passes of vibratory rollers could be monitored.

8) New waterstop system was developed to adapt large deformation of perimetric joint. The designed capacity of the waterstop system can bear the movement with 5cm opening, 10cm settlement, 5cm shearing and 2.7MPa water pressure.

9) New instrumentation techniques for high dams had been developed, including: extra-long horizontal and vertical displacement measuring set, optical fiber seepage measurement set, optical gyroscope face slab deflection measurement set.

From observation, the maximum settlement of rockfill was 2,121mm (at the end of dam construction completed in 2006, while it increased to 2,495mm in February 2008, approximately 1.07% of the dam height. The observation values and its distribution of dam embankment deformation, stress and strain of the face slab concrete, deflection of the face slab was reasonable. The leakage at the reservoir water level 397.44m (the normal water level 400.00m) in September 2008 was 46.3L/s.

The completed CFRD dams in China with the height exceed 150m are generally in normal operation. Based on engineering practices and systematical researches, the main problems of high CFRD construction were studied and the corresponding solutions were applied. Based on previous practices, Studies on the construction of CFRD with the height about 300m are conducting (see Chapter 23 in this book).

12.3.2 CFRD at Unfavorable Natural Condition

The distribution of CFRD projects covered whole country in China. Many of them were constructed under unfavorable natural conditions, which include: Severe cold areas with the lowest temperature of $-40^\circ C$: Lianhua, Shankou, Zhalong, etc. High altitude areas with elevation over 4,000m: Zhalong, Chusong, etc. High seismicity areas with design earthquake intensity of Ⅸ degree: Jilintai Ⅰ Dam site in narrow valley: Meitai, Longshou Ⅱ, Baiyun, Jiudianxia, etc.

Some project use high concrete retaining wall to improve unfavorable topographical conditions, which include: Xiaogangou, Longshou Ⅱ, Gongboxia, etc.

In China, a lot of dam sites of CFRD projects have deep alluvium. Normally, there are three methodsfor alluvium foundation treatment:

(1) Excavate all alluvium andlet dam be built on rock foundation. For example of Tianshengqiao Ⅰ CFRD, as there was a soft mud deposit located at the bottom of alluvium, all alluvium was excavated and the dam was founded on rock foundation. For alluvium foundation treatment, the common practice is to excavate unfavorable silt and fine sand layers, cohesive soil or any loose material layers.

(2) Plinth is founded on rock foundation by excavating alluvium materials in plinth area and a certain distance downstream of the plinth. The remained part of alluvium is taken as dam foundation. If necessary, the remained alluvium can be treated by dynamic compaction or vibroflotation.

(3) Plinth is founded on alluvium foundation and concrete diaphragm wall is used for seepage control of alluvium. The complete seepage control system from alluvium to dam body is established by using the concrete connecting slabs to connect diaphragm wall and plinth. Table 12.3-1 presents the list of CFRD projects of this type in China. From the list, it could besummarized that: ①3 projects have the dam height above 100m and the highest one is Jiudianxia CFRD, with the dam height of 136.5m. ②The depth of alluvium of most of the projects is more than 50m, with the deepest alluvium is 71m (Tongjiezi CFRD). ③The maximum depth of diaphragm wall is 72.26m (Hengshan CFRD). ④The thickness of the diaphragm wall is 0.8-1.2m. ⑤Most of the projects used one row diaphragm wall, expect Tongjiezi CFRD, which used two row diaphragm walls to form a fame structure, served as high retaining wall and the side wall of diversion channel. The connecting plate is normally constructed after the completion of dam and diaphragm wall to avoid the negative effects of the deformation during construction stage.

12.3.3 CFRD in High Seismicity Area

Modern CFRD possess of good capacity of earthquake resistance. But there are less cases of CFRD that experienced strong earthquake. On May 12, 2008, strong earthquake with 8.0 Richter magnitude hit Wenchuan County in Sichuan Province of China. Zipingpu CFRD, which is 17km away from the epicenter and 5-7km away from the major fault, has experienced the strong earthquake with intensity of Ⅸ-Ⅹ degree. It is the only modern high CFRD in the world that sustained such a strong earthquake.

Table 12.3 – 1 Completed CFRD Projects with Plinth on Alluvium (China)

Name	Location	Year of completion	Dam height (m)	Depth of alluvium (m)	Materials of alluvimun	Treatment measure
Kekeya	Xinjiang	1982	41.5	37.5	Sand-gravel	0.8m diaphragm wall
Tongjiezi sattle	Sichuan	1992	48	48	Sand-gravel with silt layer	2 rows 1m diaphragm wall
Hengshan raied	Zhejiang	1994	70.2	70.2	Laterite, sand, weathering rock	0.8m diaphragm wall
Caoyutai	Zhejiang	1995	16	22	Sand-gravel	0.8m diaphragm wall
Meixi	Zhejiang	1997	40	30	Sand-gravel	0.8m diaphragm wall
Lianghui	Zhejiang	1997	35.4	39	Sand-gravel	0.8m diaphragm wall
Chengang	Zhejiang	1998	27.6	39.5	Sand-gravel	Diaphragm wall
Tasite	Xinjiang	1999	43	28	Sand-gravel	Diaphragm wall
Chusong	Tibet	1998	39.7	35	Sand-gravel	hand excavating and pouring diaphragm wall
Tangpu	Zhejiang	1999	29.6/36.6	18	Muddy sand-gravel	0.8m diaphragm wall
Hanpingzui	Gansu	2006	57	30	Sand-gravel	Diaphragm wall
Nalan	Yunna	2005	109	9 – 24	Sand-gravel	0.8m diaphragm wall
Zhahanwusu	Xinjiang	2008	110	46.7	Sand-gravel	1.2m diaphragm wall
Jiudianxia	Gansu	2009	136.5	56	Sand-gravel	1.2m diaphragm wall

Note: Projects in Hong Kong, Macao and Taiwan are not included.

Zipingpu CFRD is 156m in height. The total storage of its reservoir is $11 \times 10^8 \mathrm{m}^3$ and the installed capacity is 760MW. The designed earthquake intensity is Ⅷ degree with the peak acceleration of 0.26g. The main construction material is limestone rockfill with some sand-gravel materials at the central part of the dam. Plinthis founded on bedrock. Upstream slope of the dam is 1 : 1.4. Downstream slope is designed with two slopes to ensure the stability of dam during earthquake. The upper part of the downstream slope is 1 : 1.5 and the lower part of the downstream slope is 1 : 1.4. The elevation of dam crest is 884m and the width of the dam crest is 12m with a 5m high parapet wall at the upstream side. The normal reservoir water level is 877.00m, while the reservoir water level at the time of earthquake is about 828.74m. According to the information presented by Earthquake Department, earthquake intensity at dam site is Ⅸ-Ⅹ degree and the horizontal earthquake accelerationsare 0.392 and 0.485 respectively, corresponding to the 1% and 2% of 100 years exceedance probability, which is greatly exceed the designed value. From the inspection after earthquake, the dam is stable and safe. It also suffered certain damages on rockfill dam and concrete face slabs, but the watertight system was not failed. Allearthquake damages can be repaired.

1. Damages of concrete face slabs and joints

Along the vertical joint of face slabs 23# and 24# (riverbed), concrete face slabs were ruptured by the action of large compression stress along dam axis. The rupture extended from the top of face slab to the place below water level. The maximum width of rupture zone is 1.7m. Rupture depth not goes through the whole thickness of face slabs. Steel reinforcement was exposed and bent into "Z" shape. The concrete in both sides of vertical joint between face slabs 5# and 6# (left abutment) also ruptured lightly.

On El. 845m, where is the horizontal construction joint between the second and third stage face slabs, the face slabs are overlapped. The maximum height of the overlap is 170mm and the reinforcement across the joint is bent into "S" shape. Surface waterstop at the crossing of the horizontal and vertical joint was upheaved. The rupture area is extended from face slab 5# to 24# and 35# to 38#.

Surface waterstop at the horizontal joint between parapet wall and face slabs was almost totally damaged. The cover plates of surface were also seriously damaged. Surface waterstop in some area presented corrugated shape and its tightening boltswere pull out.

By drilling detection and open inspection at joints, concrete face slabs at the area above elevation 845.00m were separated with rockfill. The maximum gaps between face slabs and rockfill is 230mm.

There are only some small fissures appeared at the surface of the face slabs. The perimetric jointsabove reservoir water level were basically intact.

2. Damages of crest structure and downstream slope

The parapet wall was not collapsed after the earthquake, but the joints near both abutments were opened, and concrete at the two sides of joints in riverbed section were ruptured. Parts of the balustrade at the downstream side of dam crest were collapsed during earthquake. Concrete pavement at dam crest was basically intact, but the downstream concrete sidewalk moved incline to downstream direction, which led to the opening of dam crest pavement and downstream sidewalk. The max opening occurred at riverbed section. At the connection of right part of dam crest and the sidewall of spillway, adifferential settlement of 15 – 20cm was observed.

The top line of the parapet wall was nearly a level line before the earthquake. After the earthquake, it was a curved line with the maximum settlement occurred at riverbed sections.

During the earthquake, the downstream part of rockfill dam has a relatively large displacement. In riverbed section, the dry-laid stone slope protection above the first berm was turned in loose condition, but the cement laid stone slope protection near dam crest is still in good condition.

3. Monitoring data related to the CFRD dam

From the measurement of the mark points of dam crest, the maximum settlement is 744.3mm, about 0.47% of dam height. The maximum displacement of downstream slope to the downstream direction is 270.8mm, at the elevation of 840m. The maximum settlement inside rockfill dam is 810.3mm, whereas the maximum settlement before the earthquake is 683.9mm.

During the earthquake, perimetric joints have produced relatively large displacement. According to the available data, the measured displacements of the perimetric joint at left abutment (Z2, El. 833m) are: 92.85 mm of settlement, 57.85mm of opening and 13.42mm of shearing. At right abutment (Z9, El. 745m, near the bottom of the river bed), the displacement are: settlement of 53.65mm, opening of 26.97mm and shearing of 103.77mm.

The leakage is measured by measuring weir at the dam toe. Before the earthquake, the leakage of the dam is 10.38L/s. After the earthquake, the leakage had increased to 17.38L/s at similar water level. It is obviously increased as compared with that before earthquake, but the total amount was a relatively small value. The first two days after earthquake, seepage water was turbid, and then become clear.

A configuration surveying had expressed that the dam embankment was contracted after earthquake. Most part of the rockfill becomes denser by earthquake action except for the surface part.

4. Repairing measures

The repairing works mainly focus on concrete face slab and waterstop system, include the vertical joint between slabs 23# – 24# and slabs 5# – 6#, as well as the horizontal construction joint at elevation 845.00m between slabs 5# – 24# and slabs 35# –38#. The ruptured concrete at both sides of vertical joint, bent steel reinforcement and damaged waterstop were removed and reinstalled. Compressible material is filled into the vertical joints. The gaps between face slab and rockfill had been grouted with cement and fly ash. After repairing, the reservoir was put into operation and sustained flood of 2008 with normal operation.

Owing to the low reservoir water level during earthquake and the relatively small intersection angle of earthquake direction to the dam axis, it is advantageous to earthquake resistance of the dam. The water discharge facilities were put into operation quickly with emergent repairing, which effectively controlled reservoir water level and relieve the danger on dam safety.

12.3.4 Section Zoning and Materials

Section zoning of CFRD in China is basically similar to the international practices, but it pay more attentionsonthe details.

(1) More rigorous seepage control. With the lessons learned from breaching of Gouhou CFRD and serious damage of face slabs of Zhushuqiao CFRD, the section zoning of CFRD in China emphasizes the importance of safety on seepage. It requir-

es that the seepage stability of dam should be guaranteed even without concrete face slabs. In section zoning, material permeability should be increased from upstream to downstream. Materials of cushion zone (2A) should be semi-permeable, with continuous gradation and inherent seepage stability, and voids of course grains particles should be fully filled by fine grains. Materials of transition zone (3A) should be filter for cushion zone. A special cushion zone (2B) is arranged under perimetric joints, which could be filter for the silt bring by leakage water and to seal the possible passages of leakage. For dam with sandy gravel material or soft rock materials, vertical and horizontal drainage zone should be arranged. Besides, fine particles should be limited in order to control its permeability and property of erosion resistance.

(2) Attentions on deformation control. From engineering practices of CFRD construction and the studies on stress and deformation properties of CFRD, it is noticed that deformation control of rockfill, especially the deformation after face slab construction, is the most important issue for the safety of CFRD. It includes both the total deformation control and differential deformation control.

For section zoning and construction material of high dams, high modulus rock material should be selected and the area of zone 3B should be enlarged. Higher compaction requirement should be applied to increase density of rockfill. The difference of modulus of zone 3B and zone 3C should be reduced. The boundary of zone 3B and 3C should incline to downstream side. In construction, the height difference of upstream and downstream construction layer should be minimized. Before the construction of face slab, the top level of upstream rockfill should higher than the top of face slab. Furthermore, the construction of face slab should behind a certain period of time of the completion of upstream rockfill construction. Compacted sandy gravel material has higher modulus than rockfill and its shearing strength is similar to rockfill under the condition of high stress status. It is good construction material for CFRD construction. It can be used in 3B zone provided that the adequate seepage control measures are applied.

By using high quality compaction to increase rockfill density and modulus is the foundation of deformation control. In addition, the properties of creep deformation, wetting deformation and rockfill degradation by time are all related with compaction density. In the construction of recent high CFRD in China, porosity of compacted rockfill was controlled less than 20%. The controlled compaction quality is great improved than that of Tianshengqiao I. Due to the non-linear relationship between dam deformation and dam height, high dams will have more difficulties in deformation control. Therefore, the requirement on compaction density will also be increased.

(3) Dam materials. Wide ranges of different rocks can be used in CFRD construction. In China, the construction materials for most of CFRD with the height above 100m are hard rocks such as limestone, dolomite, granite, tuff, sandstone etc., as well as sandy gravels. For dams with height over 150m, no dams used soft rocks except Tianshengqiao I. But some projects used combination of hard and soft rocks, with limitation on the position of the soft rockfill zone. The main indexes for determine rock properties are saturated compression strength and coefficient of softing of original rocks. There are requirements on gradation of rockfills, but the variation range is wide. Compaction parameters can be determined by site compaction test.

Physical and mechanical properties of rockfill or sandy gravel materials can be determined by laboratory tests. Therefore, series large-scale test apparatus were developed, including static and dynamic triaxial test machine, compression test apparatus, permeability test equipment, creep deformation test machine, etc., as well as large capacity geotechnical centrifuge and shaking table. With these advanced test apparatus, lots of valuable research results have been obtained.

12.3.5 Plinth, Face Slab and Joint Waterstop

1. Application of external plinth and internal plinth

Based on the previous experiences, the design idea of using internal plinth to meet the requirement on seepage length of plinth is well accepted in China. It is similar to the "4m+x" design idea suggested by J. B. Cooke. With this design, large amount of slope excavation of steep abutment could be saved. In the practices in China, the "4m" width of external plinth is a changeable value, which is determined according to requirement of grouting construction, such as 4.5m for Hongjiadu CFRD and 8m for Shuibuya CFRD.

2. Control cracks and rupture of concrete face slab

The shrinkage cracks and fissures on concrete face slabs appeared in construction stage will be controlled by optimizing row materials selection and mixing ratio, improving construction techniques and careful curing. Structural cracks and concrete rupture are related with rockfill deformation and structural design of face slabs. The deformation control techniques discussed above are effective for avoiding or reducing the structural cracks. After reservoir impoundment, most part of concrete face slabs are subject compression stresses both in the directions of dam axis and dam slope. For Tianshengqiao I CFRD, the observed compression strain of concrete near the rupture area of vertical joints was over 800μ, but reduced to less than 600μ after rupture occurred. For Sanbanxi CFRD, after the rupture of concrete at the horizontal construction joint, back analysis based on monitoring data presented that the compression stress was 24MPa. For avoiding structure cracks and rupture of concrete face slabs, measures could be applied in the detailed design of slab structure, which include: vertical joints filled with soft compressible materials, using double layer reinforcement and anti-spalling reinforcement, etc.

3. Gaps between concrete face slabs and rockfill

The gaps between face slab and cushion zone (2A) are caused by incoordination of rockfill deformation and face slab deformation. It is inevitable, but can be controlled.

During construction stage, the gaps always occur at the top of concrete face slabs. The gaps are related to the settlement and contraction on the top of upstream rockfill. It should be controlled by constructing higher upstream rockfill than the top of face slabs, leaving a certain time for sufficient settlement of upstream rockfill and filling the gaps by fly ash plus cement grouting.

During operation stage, besides rockfill deformation, the gaps are also related to the change of water level. It will open as water level drop down and close as the water level rise up. For Shuibuya CFRD, measured gaps during operation indicated that 85% of the gaps had openings within 15mm, while the remained 15% gaps had openings within 15-30mm. For Hongjiadu CFRD, the monitoring data showed that there were no gaps in abutments area. The gaps in riverbed section were related with reservoir water level. For the first stage face slabs, the opening of gaps was reduced from 13mm at lower water level to 1mm at normal water level. For the third stage face slabs, the opening of gaps was reduced from 65mm at low water level to 18mm at normal water level. For Dongqing CFRD, the maximum opening of the gaps was only 5mm before and after reservoir impoundment.

It can be concluded that if the dam was well compacted, the open gaps between face slabs and upstream rockfill are not significant and it has no serious impacts on normal operation of the dam. The relatively large gaps observed at Tianshenqiao-1 CFRD were mainly caused by large deformation of rockfill. The gaps were treated in 2002 by cement fly ash grouting. After that, no obvious gaps had been found. For Zipingpu CFRD, the gaps enlarged suddenly during earthquake with an intensity of $IX-X$ degree. It is caused by large displacement of the face slabs during earthquake. The treatment measure was cement fly ash grouting.

4. Waterstop structure and materials

Compared with traditional method, the joint waterstop in Chinese CFRD construction was much improved both in structure and material. Normally, the waterstop structure is composed with bottom copper waterstop and a surface waterstop. Surface waterstop include plastic filler and (or) self-healing type waterstop. At present the main waterstop materials used in Chinese CFRD construction are GB series and SR series materials. According to experimental research for Shuibuya CFRD, the waterstop structure and materials can bear three dimensional deformation with opening of 100mm, settlement of 50mm and shearing of 50mm, as well as high water pressure of 2.7MPa. Besides, a series of new techniques are developed to improve the effectiveness of waterstop, which include: strengthening by-pass seepage resistance of copper waterstop by the composition with GB plate; corrugated rubber waterstop and composite cover plate on the top of joint; binding agent for wet concrete surface; installation of plastic filler for surface waterstop by special designed machine; wetting expansion water proof materials. Latest researches indicated that the above-mentioned waterstop system, including its structure and materials, could also be used for CFRD with a height about 300m. The structure could be simplified for CFRD with height less than 100m.

12.3.6 Scientific Research and Analytical Calculation

Scientific researches which include laboratory test on dam materials, geotechnical centrifuge model test, model test on shaking table, numerical analytical, etc. are similar to earth cored rockfill dams. Relevant contents are in section 12.2.4.

12.3.7 Construction Technologies

1. Compaction of rockfill

Due to the huge volume of rockfill dam, heavy construction machinery is necessary for completing dam filling within certain period of time. Normally, the filling speed of modern CFRD is $(500-700) \times 10^3 m^3$/month and sometime the speed could be over $1,000 \times 10^3 m^3$/month. With the requirement of higher density of high CFRD, traditional criterion for the compaction with 4 passes of 10t vibratory roller is not applicable. In high CFRD construction in China, the use of 25t vibratory roller is very popular. Some projects use 32t vibratory roller. Impact compaction roller, which was introduced from South Africa, is a more effective compaction machine with low frequency and high amplitude. It was firstly used in limestone rockfill compaction of Hongjiadu CFRD project. In the compaction, lift thickness was 160cm and passes were 22 – 27passes. The porosity of rockfill compacted by impact compaction was 20% and the corresponding dry density was 2.18g/cm^3. It achieved the same effect of the upstream rockfill compact with 80cm lift thickness and 25t ordinary vibratory roller. After Hongjiadu CFRD, the impact compaction techniques were also applied in Dongqing CFRD.

2. Techniques on upstream rockfill slope protection

In the early stage of CFRD construction, the compacted cement-sand mortar, shooting asphalt emulsion, shooting concrete or mortar had been used for upstream slope protection of upstream cushion zone. Since 2000, techniques of extruded curb were introduced and applied in many CFRD projects in China. But there are also different views on the constraint effects of extrude curb. In recent years, Chinese engineers have developed some new slope protection methods, which include the turn up formwork technique and movable sidewallt echnique.

3. In-situ control and management of construction quality

For construction management, an automatic in-situ real time controlling system for compaction quality management based on GPS system was developed in Shuibuya CFRD project. After that, it was further improved and applied in Pubugou ECRD, Nuozhadu ECRD.

12.3.8 Performance of CFRD

In engineering practice in China, much attention is paid on safety monitoring of CFRD. Almost all the major CFRD projects are installed comprehensive observation instruments. Besides traditional monitoring instruments, some advanced devices have been developed for the requirement of high CFRD construction and automatic data acquisition. At the same time, data processing and back analysis techniques have also been rapidly applied.

(1) The main items of safety monitoring of CFRD dams include stress and deformation of rockfill, displacement of joints, stress, strain and deflection of face slabs, gaps between face slab and upstream rockfill, leakage and seepage pressure, etc. Observation data had indicated that most of the high CFRD with the height over 100m present satisfactory performance. It is noticed that the deformation of rockfill is the key factor on safety of CFRD. There are still some differences between the real behaviors of dam operation and the prediction by numerical analysis. Thus, the techniques for predicting dam behaviors should be further improved.

(2) During the Wenchuan Earthquake in 2008, 28 hydropower stations with installed capacity lager than 30MW are located in the area with earthquake intensity of Ⅵ-Ⅺ degree. Dam types included earth cored rockfill dam, concrete face rockfill dam, roller compacted concrete arch dam, concrete gravity dam and sluice. Site inspection indicated that the instrumentation system of 17 projects located in area with earthquake intensity less than Ⅸ degree werelightly damaged, except one was seriously damaged one. The instrumentation system of Zipingbu CFRD, which is located in area with earthquake intensity of Ⅸ-Ⅹ degree, was also slightly damaged. 3 of 4 projects located in area with earthquake intensity of Ⅴ-Ⅺ degree had seriously damaged. This is valuable data for engineering reference.

(3) It is important to develop automatic information management and safety evaluation system for safety management of CFRD projects. Such system was developed for Tianshengqiao Ⅰ CFRD and had been put into operation. It is a tendency to introduce modern advanced techniques into dam instrumentation system and dam safety evaluation system.

(4) The back analysis of the monitoring data had widely used to verify calculation parameters and numerical analysis method, as well as prediction of dam behaviors. It will play an important role in the field of dam safety assessment, optimization of project operation, verifying the design principles and calculation method, etc.

12.4 Conclusion

The construction of rockfill dams in China began in 1950s. It is rapidly developed, especially in 21st century. For present dam construction in China, project scale, dam height as well as the technical difficulties are unprecedented. Now, the dam height of rockfill dam projects, which are in the stage of construction or planning, has reached to 300m. Technical difficulties of these projects should be solved by combination of theory and practice, as well as engineer's experienced judgment and scientific studies.

References

[1] Cooke J. B. Progress in rockfill dams. The 18th Terzaghi Lecture. *Journal of Geotechnical Engineering*, ASCE, 1984 (110), 10.

[2] Pan Jiazheng, He Jing. *Large Dams in China, A Fifty-years Review*. Beijing: China Water Power Press, 2000.

[3] Modern Rockfill Dams, Proceedings of The 1st International Symposium on Rockfill Dams, Chendu, China. Beijing: China water Power Press, 2009.

[4] Technology for Earth-Rockfill Dams, Proceedings, 2005-2011. Beijing: China Electric Power Press.

[5] Practice and Achievement of Safety Monitoring Techniques for Concrete Face Rockfill Dams, Proceedings of The Symposium on Safety Monitoring Techniques, 2010, Dalian, China, Beijing: China Water Power Press, 2010.

[6] Jiang Guocheng. High Earth-Rockfill Dams//*Hydropower Engineering in China, Hydraulic Structure*. Beijing: China Electric Power Press, 2000.

[7] Cao Keming, Jiang Guocheng. Earth-Rockfill Dams//*Hydropower Engineering in China, Construction*. Beijing: China Electric Press, 2000.

[8] Xu Zeping. Impact of Sichuan Earthquake. Water Power & Dam Construction, 2008.

[9] Xu Zeping. Overview of CFRD Construction in China. *International Journal on Hydropower & Dams*, 2008, 15 (4).

[10] Xu Zeping. Performance of the Zipingpu CFRD during the Wenchuan earthquake. *International Journal on Hydropower & Dams*, 2009, 16 (3).

[11] Xu Zeping, Wei Yingqi, Shao Yu. Seepage Control of the Concrete Faced Sandy Gravel Dam with Deep Alluvium Foundation. *Proceedings of Twenty-second Congress on Large Dams*, Volume III.

[12] Z. Xu, Y. J. Hou, J. H. Liang & L. B. Han. Centrifuge modeling of concrete faced rockfill dam built on deep alluvium//Physical Modelling in geotechnics -6th ICPMG'06, C. W. W. Ng, L. M. Zhang & Y. H. Wang (eds), Vol. 1 Aug. 2006.

[13] Xu Zeping. Technical Progress in China for CFRD Constructed on Deep Alluvium. *International Journal on Hydropower & Dams*, 2010, 16 (1).

[14] Xu Zeping, Deng gang, Zhao Chun. Analysis on the operating performance of Sanbanxi CFRD, Dam and Reservoir under Changing Challenge. CRC Press, Taylor & Francis Group, 2011.

[15] Xu Zeping, Lu Shengxi. Design, construction and operation of China's Jiudianxia CFRD. *International Journal on Hydropower and Dams*, Issue two, 2011.

Chapter 13

New Progress of Cemented Material Dam

Jia Jinsheng[1], Zheng Cuiying[2], Ma Fengling[3], Du Zhenkun[4], Zhao Chun[5]

13.1 Principles for Design and Construction of Cemented Material Dam

In recent years, CSG (Cemented Sand and Gravel) dam[1,2] and Hardfill dam[3] have been widely used in countries including France, Algeria, Morocco, Turkey, Japan, Greece, Philippine, Dominica, USA, etc. Technologies on CSGR (Cemented Sand, Gravel and Rock) dam[4] and RFC (Rockfill Concrete) dam[5] have been investigated and used for new dam construction and old dam rehabilitation in China. The above dams, as well as cemented soil dam (developed in France), represent a dam linking between concrete dams and embankment dams, which was defined by the authors as a new type of dam namely cemented material dams (CMD)[6,7].

CMD emphasizes the new concept of optimizing dam structure to make better use of local materials with much less processed materials compared with concrete. Local materials, including sand, gravel, excavated rock, artificial sand, artificial rock, etc, are mixed with cementing materials including cement, mortar, SCC (self-compacting concrete), etc. to built a dam. Dam structure is designed according to properties of cemented materials produced. In case of clayey material, the concept may include a pretreatment with lime to neutralize clay minerals, before the cement stabilization.

The main principles of CMD was summarized as follows.

(1) Basic Principle.

1) Build a dam with cementing material based on better use of local materials by very simply processing, screening, grading or mixing procedure.

2) Guarantee dam safety, especially in case of overtopping and earthquake.

3) To decrease content of cementing material such as cement, fly ash obviously compared with concrete or RCC in order to decrease the price.

4) To well design the seepage control part with cemented material, RCC, concrete or other material. The water-stop should be well designed based on CFRD technique for block joints if there are.

(2) Structure design Principle.

1) The dam cross section is basically determined between concrete or RCC gravity dam and embankment dam. CMD can be constructed with mix of several types of materials and structures. Dam cross section is enlarged compared with gravity dam to decrease stress level, so as to guarantee strength retention of materials.

2) The stress and stability requirements for CMD is similar to those of gravity dam.

3) Parts of the dam body constructed with cemented materials should be under compressive stress status for any load cases. The other parts of dam with possible tensile stress during construction or operation, can be constructed with material such as enriched cemented material, grouting vibrated cemented material, concrete, reinforced concrete, or RCC.

---

[1] Jia Jinsheng, Vice president, China Institute of Water Resources & Hydropower Research, Professor Senior Engineer
[2] Zheng Cuiying, China Institute of Water Resources & Hydropower Research, Professor Senior Engineer
[3] Ma Fengling, China Institute of Water Resources & Hydropower Research, Professor Senior Engineer
[4] Du Zhenkun, China Institute of Water Resources & Hydropower Research, Professor Senior Engineer
[5] Zhao Chun, China Institute of Water Resources & Hydropower Research, Professor Senior Engineer

4) It is possible to build a CMD lower than 50m for some non-rock foundation if well designed.

(3) Material selection principle. Try the best to make full use of local materials to build CMD in order to reduce the material waste and to be environment-friendly and economic. Soil can be used to construct cemented soil dam. Material with diameter between 0 - 80mm can be used to constructed Hardfill or CSG dam. Material with diameter between 0 - 150mm can be used for CSGR dams. Material with diameter between 0 - 300mm can be used for CSGR cofferdams. Material with diameter larger than 300mm can be used for RFC.

(4) Construction principle.

1) Simplify construction procedure and use modern technology to improve the quality of material and construction.

2) Material should be mixed by reliable equipment, which is much simplified compared with that for concrete or RCC. Backhoe can be used to mix CSGR for cofferdam.

3) Digital system combined with GPS and other IT techniques could be used to improve the construction quality.

CMD consists of CSGR dam (including Hardfill dam, CSG Dam, Cemented sand and rock dam), RFC dam, cemented soil dam and related types of dam. With new development, the research results and application on projects show that CMD is quite different compared with RCC, especially in material selection and its processing, screening, grading and mix design. The concept is proposed in 2012[6,7] and ICOLD Committee on CMD was set up in 2013 to promote this technology worldwide.

13.2 Studies on CSGR Dams and its Application

In reference to experience of France, Japan, Turkey, Greece and other countries on Hardfill dams and CSG dams, CSGR technology including Cemented sand, gravel, rock dam and cemented sand and rock dam, has been studied from 1990s in China. Progress has been achieved in following aspects.

(1) Maximum diameter of aggregate is increased from 80mm to 150mm for dams and 300mm for cofferdams, which causes great change in material properties, construction techniques and quality control.

(2) Sand and gravel from riverbed, excavated rock, artificial aggregate, or mix of above can all be used as aggregate, which extends the usage of local materials.

(3) New anti-seepage materials and structures have been developed.

(4) New construction equipments such as continuous mixer and digital automatic quality control system have been developed to fulfill the requirements of quick construction and quality control with large-gradient-size material.

(5) CSGR dam can be built on some non-rock foundation if well designed.

In 2004, the first CSGR cofferdam—the downstream cofferdam of Jiemian Hydropower project, with the height of 16.3m was constructed in Fujian Province[8,9]. Other applications include Hongkou cofferdam ($H=35.5$m) in Fujian province, Gongguoqiao cofferdam ($H=56$m) in Yunnan province, Shatuo cofferdam ($H=14$m) in Guizhou province, Feixianguan cofferdam ($H=12$m) in Sichuan province, and etc. CSGR dams, such as Shoukoubao Dam ($H=60.4$m) and Naheng Dam ($H=74.5$m), other applications, such as Qianwei dike ($H=14.1$m, $L=2,780$m), are under design and to be built recently[10-12]. A technical guide on CMDs[13] has been issued by Ministry of Water Resources of China in 2014, which will be used as reference for design and construction of CSGR dams and RFC dams.

13.2.1 *Mix Proportion and Performance*

CSGR differs from concrete and RCC. Raw materials of CSGR include sand and gravel in riverbed, or excavated rock, only those with diameter lager than 150mm (for cofferdam, 300mm) should be removed or crushed, while grade classification is not needed. Aggregate is processed very simply, which results in discreteness. Moreover, content of cementing material (cement and flyash) required is much less compared with RCC and concrete. Such difference results in the different material mix proportion design and performance.

In order to guarantee the reliability of strength which has large discreteness usually, the mix design of CSGR is proposed as following.

(1) Conduct screening test for the raw material, concerning the diameter so as to obtain the coarsest gradation, the finest

gradation and average gradation of raw materials.

(2) Raw material used for the mix proportion test, removing those with the diameter larger than 150mm (300mm for cofferdam), should be screened into four grades of coarse aggregate, i. e. diameter in 150 – 80mm, 80 – 40mm, 40 – 20mm and 20 – 5mm, and sand, whose diameter is less than 5mm. They are separately weighed and prepared for testing.

(3) Several cementing material (cement and flyash) contents are selected according to the strength requirement. For each cementing material content, different water contents are determined according to ratio of sand to coarse aggregate for the coarsest gradation, the finest gradation and the average gradation respectively. Set up the relationship between compression strength (28d and design age, normally 180d) and water content and determine the range of appropriate water content and corresponding appropriate strength range required by the construction VC values, i. e. "Mix proportion control range". The water content can fluctuate within the control range. Water content and strength vary with the gradation (see Fig. 13. 2 – 1). The cementing material content and proper range of water content are determined when they both fulfill the requirements that the preparation strength is no less than configuring sthength $f_{cu,k}$ and the minimum strength is no less than the design strength $f_{cu,o}$. $f_{cu,o} = f_{cu,k} + t\sigma$, in which t is probability coefficient, σ is standard deviation of compressive strength.

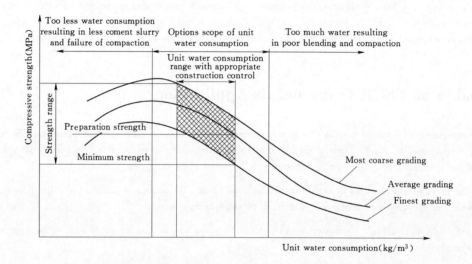

Fig. 13. 2 – 1 Relationship between Unit Water Content and Compressive Strength at Design Age

(4) 450mm cube specimen (full graded) CSGR strength test should be conducted to obtain the ratio between strength of full graded specimen and specimen of 150mm cube at different age.

According to a large amount of experiments and research, main mix design parameters are recommended as following.

(1) The content of cementing material is no less than 80kg/m³, in which cement is no less than 40kg/m³.

(2) When Portland cement, ordinary Portland cement, medium-heat or low-heat Portland cement is used, the total blending amount of fly ash and other admixture could be 40%–60%. When the Portland slag cement, Portland pozzolana cement, Portland fly-ash cement, composite Portland cement are used, the total mixing amount of fly ash and other mixture should be less than 30%.

(3) The proper sand proportion of CSGR is 18%–32%. If the requirement is not met for nature raw material, the gradation can be adjusted by mixing sand or gravel or increasing the content of cementing material.

(4) The water-binder ratio is 0. 7 – 1. 3, which should be determined according to design strength and the characteristics of raw material.

(5) The slurry composed of water, cement and admixture should fill all gaps of sand and envelop all sand. The mortar composed of slurry and sand should fill all gaps among aggregate and envelop all aggregate. That is, cement slurry enveloping rate α and mortar enveloping rate β should be greater than 1.

The permeable dissolution of CSGR under the long-term permeation of pressured water is studied. The test adopts the multi-function pressure concrete dissolution equipment, and the test age of specimen is 28d. The permeable dissolution test results indicate: there are plenty of factors to affect the dissolution of CSGR, in which the main influence factor is the compactness. For a CSGR dam, anti-seepage should be paid special attention to prevent CSGR from long-term penetrable dissolution. According to laboratory tests, Grout Enriched Vibrated CSGR and rich-mix CSGR have anti-permeability grade over W10, permeability coefficient of which is equivalence of about 0.177×10^{-8} cm/s, and frost resisting grade of F300 (at least 300 times of freeze-thaw cycle), which can be used as the upstream/downstream impermeable layer for CSGR dams.

13.2.2 Structure Design

According to material tests, CSGR is a kind of elasto-plastic material with the similar properties to RCC but relatively low strength, elastic modulus and other performance parameters. Due to lower strength, dam cross section of a CSGR dam should be enlarged compared to a gravity dam to decrease stress level, so as to guarantee strength retention of material. The cross section adopted usually is trapezoid section, symmetrical trapezoid in particular. Compared with a gravity dam, stress distribution of a CSGR dam is lower and more uniform, at any load case (Fig. 13.2-2). Stability against sliding of dam body or interlayer is also enhanced compared with a RCC gravity dam.

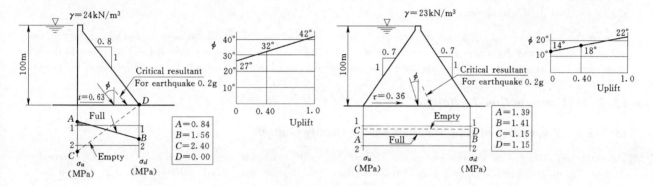

Fig. 13.2-2 Stress Distribution of a Gravity Dam and a CSGR Dam

A CSGR dam utilizes the weight of the material to resist the horizontal pressure of water pushing against it, which is similar to a concrete gravity dam. Therefore, analysis of stability against sliding for a CSGR dam can be done with the method for a gravity dam.

Stability analysis against sliding of a CSGR dam should consider the dam foundation surface, internal dam layer surface and sliding surface of the deep foundation layer. Stability against sliding of dam foundation surface can be calculated with the formula (13.2-1) or formula (13.2-2) below, and stability against sliding of internal dam layer surface shall be calculated with formula (13.2-1).

$$K' = \frac{f' \Sigma W + cA}{\Sigma P} \tag{13.2-1}$$

$$K = \frac{f \Sigma W}{\Sigma P} \tag{13.2-2}$$

Safety factor of stability against sliding (K') calculated by formula (13.2-1) shall be no less than the values in Table 13.2-1. The share safety factor (K) calculated by formula (13.2-2) shall be no less than the values in Table 13.2-2.

Table 13.2-1 Safety Factor of Stability Against Sliding (K') of Foundation Surface and Layer (Joint) Surface

Load combination			K'
Basic combination			3.0
Special combinations		(1) Extreme flood	2.5
		(2) Earthquake	2.3

Table 13.2-2 Shear Safety Factor (K) of Foundation Surface

Load combination		Dam level 1	2	3	4	5
Basic combination		1.10	1.05	1.05	1.05	1.05
Special combinations	(1) Extreme flood	1.05	1.00	1.00	1.00	1.00
	(2) Earthquake	1.00	1.00	1.00	1.00	1.00

As strength of CSGR is relatively low, CSGR is used in dam body where the stress is compressive at any load cases and maximum primary compressive stress of the dam should be less than allowable material compressive stress. The allowable compressive stress of CSGR is determined through the ultimate compressive strength divided by safety factor. Full consideration is given to various uncertain factors, including the difference between laboratorial strength test value and prototype material strength, the possibility of load surpassing the design assumption, non-uniformity in construction, and etc. To ensure the safety and long-term durability, compression safety factor of CSGR dam is set as 4.0 for normal loading case and 3.5 for extreme flood loading case.

To meet the requirements of rapid construction, CSGR dam usually adopts CSGR with the same grade and temperature control is not needed generally. Longitudinal joints may not be set and the transverse joints can be reduced according to the dam foundation. For dams lower than 70m, the number of galleries may be reduced according to the real condition of the project.

13.2.3 Construction Technique

13.2.3.1 Preparation of Raw Materials and its Quality Control

The maximum grain size of the raw material is 150mm. Aggregates with size larger than 150mm should be removed or crushed. The gradation and some properties of raw material at construction stage may be different with that obtained during the stockyard research and survey stage. Therefore, area of reclaiming, storage and gradation of raw material, etc. of the quarry should be re-surveyed before construction.

The reserve of raw material piles shall take into account of the time needed to modify the mix proportion when significant changes happen to the gradation and features of the material. It is recommended to prepare the construction materials which can be used for at least one month. Primary storage pile shall be set at the production site for the raw materials. And secondary storage pile shall be set nearby the belt feed inlet of the mixing station. When carried from the primary storage pile to the secondary storage pile, the raw material should be fully mixed to reduce discreteness.

The main difference between the quality control of raw material of CSGR and that of normal concrete lies in the quality control of sand and aggregate. The primary storage pile of sand and gravel shall be detected every day for its surface dry density, water absorption, gradation, surface moisture content and silt content, etc. As the strength of CSGR is determined by the gradation of raw material and unit water content, the gradation and surface moisture content shall also be measured at the secondary storage pile on the day of construction to confirm the range of gradation variation and unit water content of CSGR. The mix proportion shall be modified when the properties, including color, shape, gradation, surface dry density, and water absorption, etc. of the material is found to have significant changes.

13.2.3.2 Mixing

Modern technology and equipments can be used to simplify construction procedure and guarantee the quality. Mixing of CSGR is typically important and high-yield and efficient continuous mixing equipment is required.

Based on the characteristics of CSGR, a continuous rotary drum-type mixer and mixing syetem are developed, as shown in Fig. 13.2-3 and Fig. 13.2-4, which has the following characters:

(1) Being capable of mixing aggregates whose largest gravel diameter is 200mm.

(2) Having a largest mixing capacity of 200m^3/h, which is adjustable.

(3) Having a sensitive, accurate and reliable weighing system.

(4) Having an adjustable mixing time and volume.

(5) Being capable of mixing the CSGR uniformly.

The mixing sequence, capacity and time of CSGR should be determined by production test at field.

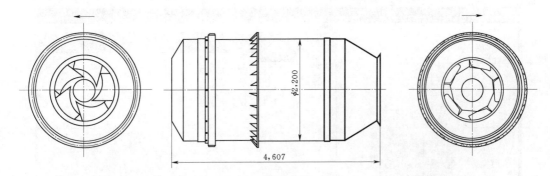

Fig. 13.2-3 Structure of the Drum-type Mixer for CSGR (Unit: mm)

Fig. 13.2-4 Mixing System of CSGR

13.2.3.3 CSGR Construction

Transportation, unloading, spreading and rolling of CSGR, and treatment of placed layer are similar with those of RCC. Placing thickness, rolling thickness and rolling times shall be decided through on-site production tests.

13.2.3.4 Quality Control

The quality control of CSGR mainly includes testing and controlling the gradation range of raw material, VC value of the mixture, density and compressive strength of CSGR after rolling. Rolling times, paving thickness, time interval between the layers, and etc. are all important items to be controlled during construction. Apparent density of CSGR can be detected by adopting nuclear moisture density meter in combination with the water replacement method. The relative compactness of CSGR shall be no less than the design value. For evaluating the production quality of CSGR, sampling should be taken at the mixer outlet, and the compressive strength of 150mm reference cube specimens with standard curing of 28d is adopted as the criterion. Meanwhile, 450mm cube specimen (full graded) CSGR strength tests are suggested to be conducted to review. When necessary, such means as in-hole TV camera, infrared imager, and in-hole sound wave test can be adopted for evaluating the quality of CSGR.

Based on the technology of information and automation, a quality control system, which is used to monitoring construction and control the quality of raw material, mixture and CSGR is being investigated. whose user interface is shown in Fig. 13.2-5.

13.2.4 *Applications*

13.2.4.1 Shoukoubao CSGR Dam [11,12]

In the area of Shoukoubao Reservoir of Shanxi province, there is a large distribution of sand and gravel, which are sufficient for a CSGR dam. After comparison studies with other dam types, CSGR dam has apparent advantages in economy and environment protection. For example, compared with the RCC dam, the total cost is reduced by 13%. According to material tests and structure analysis, symmetrical trapezoidal cross section (see Fig. 13.2-6) with upstream and downstream

Part Ⅱ Construction Technologies for Dams and Reservoirs

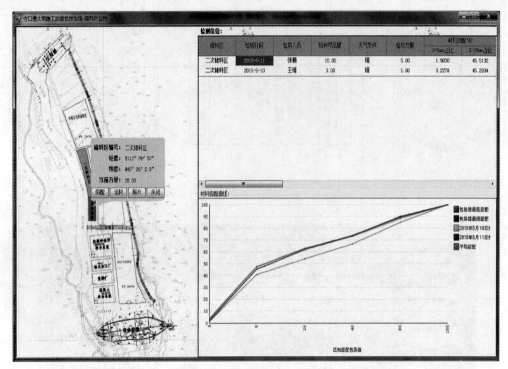

Fig. 13.2-5 Quality Control System for Shoukoubao CSGR Dam

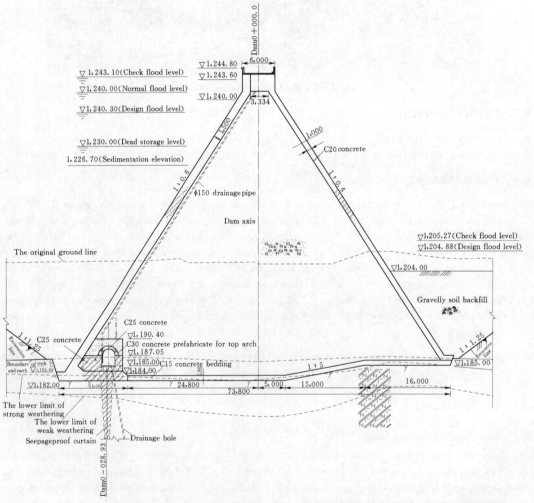

Fig. 13.2-6 Typical Cross Section of Shoukoubao CSGR Dam (Elevation Unit: m, Size Unit: mm)

slope ratio of 1 : 0.6 has been decided for Shoukoubao CSGR Dam. 1.5-meter-thick C20 concrete is used at the upstream face and 1.0-meter-thick C20 concrete is used at the downstream face for seepage control and protection. For any load combination, the safety factor of stability against sliding of foundation surface is larger than 1.31 and the shear safety factor of foundation surface is larger than 4.08, which indicate safety of stability is enough. For safety factors of stability against sliding of interlayers, they are larger. Numerical analysis with the method of material mechanics and finite element method, the stress of main dam body with CSGR is compressive stress at any load case with the maximum compressive stress of 1.293MPa. The design strength of CSGR is 6MPa for 180d.

13.2.4.2 Qianwei CSGR Dike on Non-rock Foundation

Due to the symmetrical trapezoid cross section and larger base compared with a gravity dam, the requirements for strength of material and foundation of a CSGR dam are relatively lower than a gravity dam, which make it possible to build a CSGR dam on non-rock foundation.

Qianwei dike is located along Minjiang River in Sichuan province of China. One part of it is 14.1m high and 2,775m long. The foundation of the dike distributes sand and gravel, which has a thickness of 8.3m to 10.5m. Compared with an embankment dike, a CSGR dike has a higher safety, due to low flood standard of the dike. The cross section for Qianwei dike is designed as shown in Fig. 13.2-7.

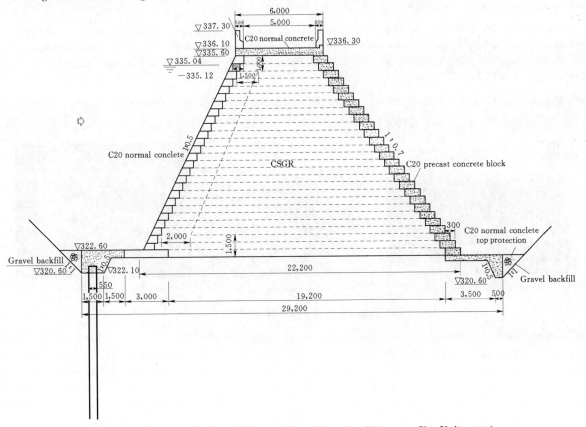

Fig. 13.2-7 Cross Section of Qianwei Dike (Elevation Unit: m, Size Unit: mm)

13.3 Studies on RFC Dam and its Application

Rock Filled Concrete (RFC) dam is another type of CMD.

Rock-Filled Concrete (RFC) construction technology is a new type of concrete technology for massive concrete, which is developed by authors based on the technology of Self-Compacting Concrete (SCC). In the RFC technology, SCC is employed to fill the void space among rock block heaps. As shown in Fig. 13.3.1, RFC is composed of rock block and SCC with coarse aggregate.

A RFC dam makes use of a large deal of rock and the rock proportion in the concrete can reach 55% - 60% in general. Massive rock excavated can be fully used, thus reducing the consumption of cementing material to the largest extent. In

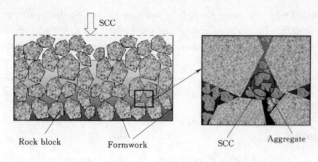

Fig. 13.3-1 Composite of RFC

RFC mass, only about 40%-45% of the volume needs to be filled with SCC, which greatly reduces the cement content in RFC. RFC performs satisfactorily in lowering the cost as well as the heat of hydration, because the unit cement content of RFC with a strength grade of C15 is only $80 - 90 kg/m^3$ [15]. Since RFC was developed in 2003, it has rapidly grown and already been used in a number of hydraulic structures in China, including dams. Projects include auxiliary dam of Henan Baoquan Pumped Storage Power Station, Changkeng Reservoir Rehabilitation Project, Qingyu Project, Shankou Hydropower project, etc. which have achieved the successful experience and remarkable benefit.

Rockfill materials for RFC should be fresh, intact and hard with a shape of pebble or rubble and with a size larger than 300mm. The SCC for RFC has the characteristics of low cement content, low hydration heat, capability of uniform flowing and filling rockfilled gaps. Experimental studies have shown that, RFC has higher strength and modulus of elasticity, lower adiabatic temperature rise, and relatively small volume deformation.

As the strength of RFC is 30% higher than that of the SCC, the compressive strength of standard specimen for SCC can be used during construction for quality assessment. Compressive strength is determined in accordance with standard value of compressive strength of self-compacting concrete cube at 90d age, which is divided into 6 grades, namely, C10, C15, C20, C25, C30 and C35.

For a dam construction with RFC, the relevant design and construction standards of a concrete gravity dam can be referred.

Material test of RFC is shown in Fig. 13.3-2.

(a) Rockfill into the test mold   (b) Specimen molding   (c) Compressive strength test

Fig. 13.3-2 RFC Material Test

RFC technology was implemented in Hengshan arch dam reinforcement project. [16]

Hengshan arch dam was built in 1960s. The maximum height of the dam is 69m. Dam axis is 146.2m long. Dam crest is 2.5m wide and foundation is 15m wide. Since it's built, several problems such as insufficient anti-sliding stability of the dam abutment at the left bank, cracks in the dam body and etc. emerged. In 1964, it was reinforced. Since then, the reservoir impounding water level has been limited under normal water level. In 2002, it was evaluated as a dangerous dam and comprehensive reinforcement should be done.

The reinforcement needed to thicken the dam at the downstream side by 11.04m at dam toe and 1.0m at dam crest. The thickened part was constructed with RFC (see Fig. 13.3-3).

RFC has a low adiabatic temperature rise, 10℃ lower than that of the normal concrete and thus needs no water-cooling temperature control measures and the corresponding stress level is relatively low. The comprehensive cost can be decreased by about 10%.

13.4 Conclusion

CSGR technologies have been used in several cofferdams in China, including Jiemian, Hongkou, Shatuo and Gongguoqiao. 59 million RMB investment (similar to 25% of their total cost) has been saved. The construction period has been decreased by 40%-50% in average. RFC dam technique has been used to more than 20 projects and about 20% investment can

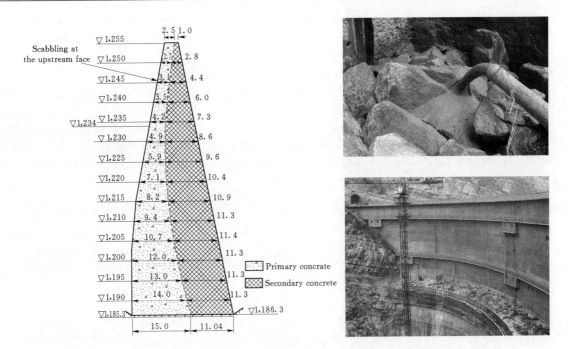

Fig. 13.3 – 3　Hengshan Arch Dam Thickening Section（Unit：m）

be saved compared with the normal concrete under the same conditions.

CMD is a new type of dam, with advantages of saving cost and environmental friendly, which will be an option for dams in planning, especially for the large number of small or medium-sized projects to be built in the world.

References

[1]　Raphael J. M. The Optimum Gravity Dam, Rapid Construction of Concrete Dam, ASCE, New York, 1970.

[2]　Raphael J. M. Construction Methods for Soil-Cement Dam Economical Construction of Concrete Dam. ASCE, New York, 1972.

[3]　P. Londe, M. lino. The Faced Symmetrical Hard-fill Dam: A New Concept for RCC. *Water Power & Dam Construction*, 1992.

[4]　Jinsheng, J., et al. CSGR Dam: Material Property Studies and Engineering Application, *Journal of Hydraulic Engineering*, 2006, 37 (5): 578 – 582.

[5]　Huang, M., An, X., Zhou, H., et al. Rock – Filled Concrete-Development, Investigations and Applications, *International Water Power & Dam Construction*, 2008, 4.

[6]　Jinsheng, J., Cuiying, Z., Fengling, M., et al. Studies on Cemented Material Dam and Its Application in China//*Proceedings of 6th International Symposium on Roller Compacted Dams*, 2012. 10.

[7]　Jinsheng, J., Cuiying, Z., Fengling, M., et al. New type of Dam-Cemented Material Dam and Current Research Progress Related//*Studies on Modern Technology of Rockfill Dam Construction and Hydropower Development*, 2013: 63 – 72.

[8]　Jinsheng, J., et al. Application of CSG dam in China and lab test on CSG material dissolution//*Proceedings of the 2nd EADC Symposium*, 2005.

[9]　Jinsheng, J., et al. Application of CSG dam in China and lab test on CSG material dissolution//*Proceedings of the 22nd ICOLD Congress on Large Dams*, 2006.

[10]　*Studies on CSG Dam Material Properties*. Beijing: China Institute of Water Resources and Hydropower Research, 2005.

[11]　*Research Report of Cemented Material Dam of Shoukoubao Reservoir*. Shanxi Institute of Water Resources and Hydropower Survey and Design, and China Institute of Water Resources and Hydropower Research, 2012.

[12]　*Analysis of Cemented Material Dam Structure of Shoukoubao Reservoir*. China Institute of Water Resources and

Hydropower Research, and Shanxi Institute of Water Resources and Hydropower Survey and Design, 2012.
[13] *Technical Guideline for Cemented Granular Material Dams* (SL 678—2014). Ministry of Water Resources, P. R. China, 2014.
[14] *Research Report of Rockfilled Concrete Damming Technology*. Beijing: Tsinghua University, 2012.
[15] Jin,F., An, X., Shi, J., et al. Study on Rock-filled Concrete Dam. *Journal of Hydraulic Engineering*, 2005, 36 (11): 1348-1351.
[16] *Preliminary Design Report of Rick Removal and Remedy for Hengshan Reservoir at Hunyuan County, Shanxi Province*. Beijing: China Institute of Water Resources and Hydropower Research, 2007.
[17] *Engineering Manual for Construction and Quality Control of Trapezoidal CSG Dam*. Japan Dam Engineering Center, 2009.

Chapter 14

Temperature Control Technology for Dam Concrete

Zhu Bofang[1]

14.1 Introduction

Concrete crack has long plagued dam engineers. Since the 1930s, few studies have conducted thermal stress of the concrete dam in spite of introduction of technical control measures in foreign countries, such as jointing and blocking, water pipe cooling and aggregate precooling. The *Cooling of Concrete Dams*, published by the U.S. Bureau of Reclamation, is so far a foreign representative work concerning concrete temperature control, only presenting temperature computational methods of flat plates and cylinders without any introduction of thermal stress such as the methods for computing thermal stress and law of variation of thermal stresses in various hydraulic structures. It should be said that the driving force of cracks comes from the thermal stress rather than the temperature. For example, under the same temperature difference, cracks perhaps happen to a concrete slab on the batholite but maybe not to a concrete slab on the soil subgrade, this is because stresses produced under the same temperature are different in diverse constraint conditions. Actually the thermal stress of concrete structures is much more complex than that generated by water pressure and dead weight. It can be said that the absence of guidance in the thermal stress results in the problem of "no dam without crack". Since the 1950s, Chinese scholars have always attached importance to exploring the concrete thermal stress of hydraulic structures and conducting lots of pioneering studies through practices in water conservancy and hydropower projects so as to build a more complete theory system for temperature control and crack prevention of concrete dams, including basis theories and computational methods for the temperature field and the visco-elastic thermal stress of concrete, and computational methods, change laws of variation and main features of the temperature field and the thermal stress of various mass concrete structures such as arch dams, gravity dams, buttress dams, dam blocks, sluices, docks, foundation beams and pipelines, as well as technical measures for temperature control and crack prevention in different conditions.

Since the construction of the Foziling Multi-Arch Dam in the 1950s, in order to prevent cracks, China has made great efforts in optimization of concrete materials, mixture of composite materials and admixture agent. In terms of the temperature control technology, China, on the basis of learning foreign technologies, has made a great progress in water pipe cooling, concrete precooling, surface protection and the like by carrying out lots of work.

In terms of the temperature control and the crack prevention of mass concrete, Chinese scholars have put forward a series of new theories and methods, including MgO mixture, semi-mature age of concrete, reservoir water temperature, temperature loads of the arch dam, new ways of water pipe cooling from earlier age with small temperature difference and longer time, equation of equivalent heat conduction for water pipe cooling, fundamental theorems of concrete creep, composite elements, zoned different-step-length algorithms, upstream cracks of the gravity dam, new methods for heightening algorithms and the gravity dam, concept of comprehensive temperature control and long-term superficial thermal insulation, etc.

Until the late 20th century, the concrete dam at home and abroad was suffering the problem of "no dam without crack", but the history is ended with China's first construction of no-crack concrete dams such as the phase-Ⅲ Three Gorges Project and the Jiangkou Arch Dam through long-term and persistent efforts by Chinese scholars and engineers in recent years. It is a significant achievement in dam construction technique and demonstrates China's leading position worldwide in both the theory and the practice of thermal stress and temperature control of mass concrete.

---

[1] Zhu Bofang, Academician of the Chinese Academy of Engineering, China Institute of Water Resources and Hydropower Research, Professor Senior Engineer.

14.2 Basic Theory and Analytical Method of Thermal Stresses of Mass Concrete

The visco-elastic thermal stress of hydraulic concrete structures is actually very complex. Before the 1950s, the theoretical analysis is only conducted to one-dimensional temperature field, but after the 1950s Chinese scholars are vigorously engaged in pioneering studies on the visco-elastic thermal stress of hydraulic concrete structures: firstly, the plane elastic thermal stress is analyzed through influence line and photoelastic experiments, then an analysis on two-dimensional and three-dimensional temperature fields and the visco-elastic thermal stress is made by a finite element method in an attempt to gradually realize a whole-process three-dimensional finite element stimulating calculation on the concrete dam, thus fully reflecting the real situation of construction and operation and proving a strong support to the policy on the temperature control of a concrete structure.

14.2.1 Finite Element Solution for the Concrete Temperature Field

A theoretical solution or a difference method is available to a simple one-dimensional temperature field, while the finite element method is generally applied to two-dimensional or three-dimensional temperature field to get the solution in real projects. According to the variational principle, the discretization is conducted by the finite element in the spatial domain and by the difference method in the time domain so as to get the following system of equations:

$$\left([H]+\frac{1}{s\Delta\tau_n}[R]\right)\{T_{n+1}\}+\left(\frac{1-s}{s}[H]\frac{1}{s\Delta\tau_n}[R]\right)\{T_n\}+\frac{1-s}{s}\{F_n\}+\{F_{n+1}\}=0 \qquad (14.2-1)$$

where, $\{T_{n+1}\}$ is nodal temperature vector when $t=t_{n+1}$; $\{T_n\}$ is nodal temperature vector when $t=t_n$, and $\{F_n\}$ and $\{F_{n+1}\}$ are vectors related to adiabatic temperature rise θ and boundary conditions; wherein $\{T_n\}$, $\{F_n\}$ and $\{F_{n+1}\}$ are known while $\{T_{n+1}\}$ is unknown, that is to say, the above system of equations is $\{T_{n+1}\}$ linear system of equations. After solving, the temperature $\{T_{n+1}\}$ at each node in $t=t_{n+1}$ can be obtained. $s=0$, $s=1$ and $s=1/2$ are equivalent to forward difference, backward difference and midpoint difference respectively, but based on experience, a backward difference method ($s=1$) has a better effect.

14.2.2 Impact of Concrete Creep on Structural Deformation and Stress

Concrete creep has a significant impact on the structural deformation and stress. The predecessor only studies the impact of the creep on a homogeneous structure under a simple boundary condition, while the China Institute of Water Resources and Hydropower Research (hereafter referred to as IWHR) has engaged in studies on the affect of the creep on a non-homogeneous structure under a mix boundary condition, proving two theorems.

In view of non-homogeneous elastic creep bodies made of two different materials, the elastic modulus is set $E_1(\tau)$ and the unit creep is $C_1(t,\tau)$ in the domain Ⅰ, while in the domain Ⅱ the elastic modulus is $E_2(\tau)$ and the unit creep is $C_2(t,\tau)$. Supposing material deformation characteristics of the two domains are in line with the following relation

$$C_1(t,\tau):C_2(t,\tau)=\frac{1}{E_1(\tau)}:\frac{1}{E_2(\tau)} \qquad (14.2-2)$$

It can be seen from the formula above that the creep deformation and its corresponding elastic deformation of the two domains form a proportion, named "proportional deformation". The two following theorems are proved by IWHR.

[Theorem 1] As to a non-homogeneous viscoelastic body whose poisson ratio is constant and satisfies the conditional expression (14.2-2), if the temperature is zero, some boundaries have given external force and the displacement of some boundary is zero, its stress is the same as that of an elastomer under the action of the volume force and the boundary force, and the strain ε_x at any point can be computed as the following equation according to the elastic strain $\varepsilon_x^e(\tau_0)$ at this point:

$$\varepsilon_x(t)=\varepsilon_x^e(\tau_0)[1+E(\tau_0)C(t,\tau_0)] \qquad (14.2-3)$$

[Theorem 2] For a non-homogeneous visco-elastic body whose Poisson ratio is constant and satisfies the conditional expression (14.2-2), if the volume force is zero, some boundaries have given displacement and the external force of some boundary is zero, its displacement is the same as that of an elastomer under the action of temperature and imposed displacement of the boundary, and the stress σ_x at any point can be computed as the following equation by relaxation coefficient method according to the elastic stress at this point:

$$\sigma_x(t) = \sigma_x^e(\tau_0) K(t, \tau_0) \qquad (14.2-4)$$

where, $\sigma_x^e(\tau_0)$ is elastic stress and $K(t, \tau_0)$ is relaxation coefficient.

14.2.3 Finite Element Solution to Visco-Elastic Thermal Stress of Concrete Structure

By incremental method for solving, the general balance equation is as follows:

$$[K]\{\Delta\delta_n\} = \{\Delta P_n\}^L + \{\Delta P_n\}^C + \{\Delta P_n\}^T + \{\Delta P_n\}^0 \qquad (14.2-5)$$

where, $[K]$ is the assembled stiffness matrix; $\{\Delta\delta_n\}$ is node displacement incremental array; and $\{\Delta P_n\}^L$, $\{\Delta P_n\}^C$, $\{\Delta P_n\}^T$ and $\{\Delta P_n\}^0$ are nodal load increments respectively produced by external load, creep, temperature and autogenous volume deformation.

By implicit method, the load increment produced by concrete creep is computed as follows:

$$\{\Delta P_n\}_e^c = \iiint [B]^T [\overline{D}] \{\eta_n\} \, dx\,dy\,dz \qquad (14.2-6)$$

$$\{\eta_n\} = \sum (1 - e^{-r_s \Delta\tau_n}) \{\omega_{sn}\} \qquad (14.2-7)$$

$$\{\omega_{sn}\} = \{\omega_{s,n-1}\} e^{-r_s \Delta\tau_{n-1}} + [Q] \{\Delta\sigma_{n-1}\} \psi_s(\overline{\tau}_{n-1}) e^{-0.5 r_s \Delta\tau_{n-1}} \qquad (14.2-8)$$

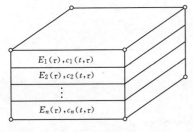

Fig. 14.2-1 Composite Element

14.2.4 Composite Element

The dam block is divided into three zones R_1, R_2 and R_3 from top to bottom. In the upper new-pouring zone R_1, each pouring layer shall be partitioned into multiple layers of finite elements; in the zone R_2 below the zone R_1, multiple layers of elements originated in each pouring layer can be combined into one layer of elements; and in the zone R_3 under the zone R_2, a plurality of pouring layers can be merged into a composite element in which all pouring layers maintain their respective mechanical characteristics and thermal features, as shown in the Fig. 14.2-1.

14.2.5 Zoned Different-Step-Length Algorithm

Requirement on the time step varies with different material properties in different zones. At present only the minimum step length is adopted as the unified time step length for computation. At the early age, it must adopt short-time step length owing to rapid change of material properties. A concrete dam is constructed in a layering manner with constant pouring of fresh concrete at its upper part and the time step length of the whole dam is under the control of fresh concrete, thus the short-time step length must be applied from start to completion, which makes the computation inconvenient. The IWHR raises a zoned different-step-length algorithm, that is, the calculation is based on short-time step length for new pouring concrete and long-time step length for hardened concrete, thus greatly simplifying the computation.

14.3 Thermal Stress of the Concrete Dam Block

The thermal stress of a concrete dam during construction is mainly the thermal stress of a dam block.

14.3.1 Stress Produced by Uniform Temperature Difference and Risk of Long Downtime of a Thin Block

The horizontal stress at the central profile of the dam block under uniform temperature difference is as Fig. 14.3-1. When $H/L \geqslant 0.25$, H/L ratio has less impact on the maximum stress, and when $H/L < 0.25$, the tensile stress creases dramatically with the reduction of H/L ratio and is more uniformly distributed. From this, it can be seen that the "long downtime of a thin block", namely, long-term cease after the pouring of thin and long concrete blocks, is harmful, reflecting big foundation restraint effect and tension of the whole section. In this case surface cracks caused by cold waves are easy to develop into deep cracks.

14.3.2 Thermal Stress of the Dam Block Under Uniform Temperature Effects

During the construction of a concrete dam, the thickness of the pouring layer is much smaller than the length of the dam

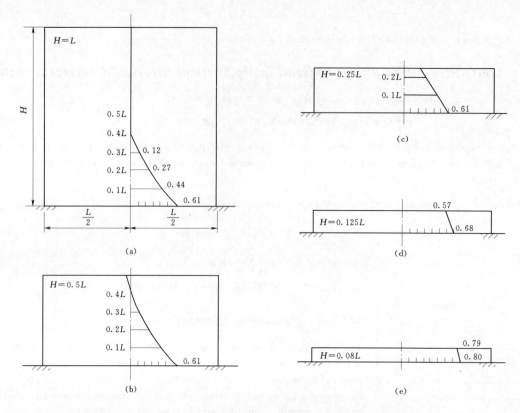

Fig. 14.3-1 Results of Photoelastic Experiment on Restraint Stress of
Dam Blocks due to Uniform Cooling (Unit: $-E\alpha T/(1-\mu)$)

block and heat is mainly dissipated in the vertical direction. When internal thermal stress of the dam block is calculated, it can be deemed that the temperature field $T(y)$ varies only in the direction y. The horizontal stress at the height y on the central vertical section of the dam block can be computed as follows by the influence line raised by the IWHR:

$$\sigma_x(y) = -\frac{E\alpha T(y)}{1-\mu} + \frac{E\alpha}{(1-\mu)L}\int_0^H T(\xi)A_y d\xi \qquad (14.3-1)$$

where, H is the height of the dam block and A_y is stress influence coefficient, as shown in the Fig. 14.3-2.

14.3.3 Remarkable Impact of Temperature Gradient on Thermal Stress

Fig. 14.3-3 shows the distribution of the horizontal stress on the central section of the dam block under the action of stepped temperature (five steps) when $H=L$ and $E_c=E_R$, and the maximum stress restraint coefficient is 0.12, while the maximum stress restraint coefficient under uniform temperature difference is 0.60 (as shown in the Fig. 14.3-4). It can be seen that when the temperature difference is divided into five steps, the tensile stress is reduced to around 1/5.

Fig. 14.3-4 represents the distribution of the horizontal stress on the central section of the dam block under the action of different temperature distribution when $H=L$ and $E_c=E_R$. According to the figure, the temperature gradient has great impact on the tensile stress.

Conclusions above have a significant and practical meaning to temperature control of the concrete dam. It is much difficult to control the temperature of the whole dam block and less difficult to control local temperature, for example, the spacing of cooling pipes may be reduced in the local restraint zone. Computed results above indicate that control over the temperature gradient can reduce the thermal stress remarkably.

14.3.4 Impact of the Height of Cooling Zone on Thermal Stress

Local artificial cooling before joint grouting of a dam and double constraint by the lower base and the upper concrete probably generates larger tensile stress when the cooling zone is not too high.

As shown in the Fig. 14.3-5 (b), the height H of the concrete block is $3L$ and E_C equals to E_R; and when $y=0-b$,

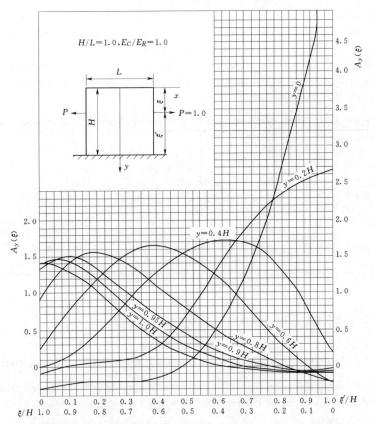

Fig. 14.3-2 Thermal Stress Influence Line of Concrete Block on Rock Foundation ($E_C/E_R = 1$)

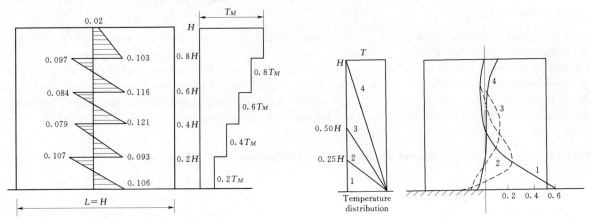

Fig. 14.3-3 Stress of the Dam Block Produced by Stepped Temperature Difference (Unit: $-E\alpha T_m/(1-\mu)$)

Fig. 14.3-4 Impact of Temperature Gradient on Thermal Stress of the Dam Block (Unit: $-E\alpha T/(1-\mu)$)

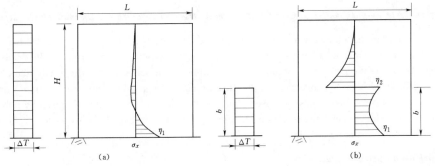

Fig. 14.3-5 Comparison of Overall Cooling and Local Cooling

there exists the temperature difference ΔT. The stress coefficient η of center line of the dam block, calculated by finite elements, is $\eta=(1-\mu)\sigma_x/E_a\Delta T$ (as shown in the Table 14.3 - 1). It indicates that above the bedrock, the ratio b/L of the first cooling zone has a great impact on the thermal stress; for a high dam block with uniform distribution of initial temperature, the impact of the height increment at the subsequent cooling zone on the maximum thermal stress is smaller, but for a low dam block, the impact is still large. It is suggested that the ratio b/L of the cooling zone shall be not less than 0.40 during the construction of the concrete dam.

Table 14.3 - 1 Stress Coefficient η with Different Ratios b/L of the Cooling Zone

Ratio b/L of the cooling zone	0.1	0.2	0.3	0.4	0.5	0.6	0.7	0.8	0.9	1.0	2.0
Stress Coefficient η_2 at the top of the cooling zone	0.85	0.72	0.61	0.53	0.51	0.50	0.49	0.49	0.49	0.49	0.50
Stress Coefficient η_1 at the bottom of the cooling zone	0.86	0.73	0.64	0.57	0.56	0.56	0.56	0.56	0.56	0.56	0.56

14.4 Thermal Stress of the Gravity Dam

14.4.1 Features of Thermal Stress of the Gravity Dam without Longitudinal Joint

The IWHR, based on theoretical analysis and years of experience, has presented the following features of thermal stresses of the gravity dam without longitudinal joint; understanding of which will help one to master the law of the thermal stress of the gravity dam.

1. Internal temperature decreases slowly, small foundation restraint stress and large difference between internal and external temperature

A gravity dam with longitudinal joints is cooled to the steady temperature by pipe cooling before grouting of joints, the internal and external temperature difference is small, but faster temperature reduction during the post cooling make the creep fail to play its role and the tensile stress larger. Natural cooling of the gravity dam without longitudinal joint is very slow, which costs several decades or even centuries to reduce the internal temperature of the dam to the steady temperature, can make the creep fully play its role and the foundation constraint temperature stress smaller, but high internal temperature and large difference between internal and external temperature easily causes deep cracks on the upstream and downstream surfaces of dam.

2. Larger tensile stress caused by temperature difference between the upper and the lower parts

Fig. 14.4 - 1 represents a computing example. The temperature and the stress are computed by finite element method according to local weather conditions and construction progress. The figure indicates that the longer the concrete block is, the larger the temperature stress is. For a block with length of 20m, the thermal stress at the foundation constraint part is large while that at the rest part is small, but for a long block, the temperature difference of the upper and the lower parts may generate bigger tensile stress. In the gravity dam without longitudinal joint, the temperature difference of its upper and lower parts may produce a large tensile stress.

3. Superposition of concrete weight, water pressure and thermal stress will reduce the thermal tensile stress

For a gravity dam with longitudinal joints, when the lower part of concrete block is artificially cooled before joint grouting, concrete is low and there is no water load before the dam, thus it can ignore the stress produced by dead weight and water pressure and only computes the thermal stress. Secondly stage cooling is not necessary for the gravity dam without longitudinal joint. The dam is completed before its internal temperature drops to the steady temperature, thus three loads, namely weight, pressure and temperature, shall be superposed which will reduce the tensile stress caused by foundation constraint inside the dam. Fig. 14.4 - 2 shows calculated results of the gravity dam. In the case of low water level, dead weight and water load can reduce the tensile stress by 0.5MPa.

4. Easy formation of deep upstream cracks

At the early stage of reservoir impoundment of the gravity dam without longitudinal joint, tensile stress is produced on the

Chapter 14 Temperature Control Technology for Dam Concrete

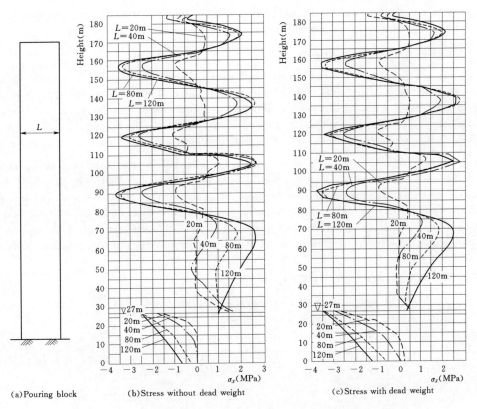

(a) Pouring block (b) Stress without dead weight (c) Stress with dead weight

Fig. 14.4-1 Thermal Stresses in Concrete Blocks with Different-Width

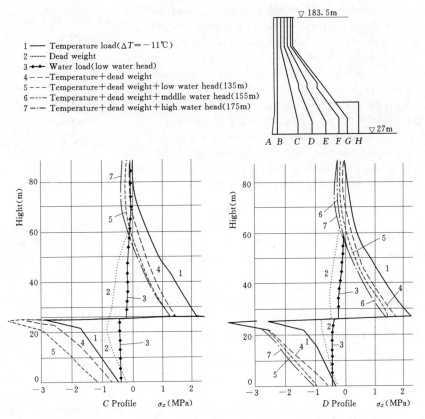

1 —— Temperature load ($\Delta T = -11°C$)
2 ······ Dead weight
3 —•— Water load (low water head)
4 ——— Temperature + dead weight
5 ——— Temperature + dead weight + low water head (135m)
6 ———— Temperature + dead weight + mddlle water head (155m)
7 —··— Temperature + dead weight + high water head (175m)

Fig. 14.4-2 Horizontal Stress σ_x (Tensile Stress is Positive) under the Action of Temperature, Dead Weight and Water Load

upstream face because the temperature of the upstream face is very low while the temperature inside the dam is very high. If surface cracks exist during the construction period, they easily develop into deeper large cracks.

5. Ultra-cooling caused by orifices inside the dam

Orifice edges are in touch with air or water and in winter both the air temperature and water temperature are lower than the steady temperature of the dam, thus "ultra-cooling" phenomenon emerges, probably producing a larger thermal stress nearby the orifices. Orifices of a gravity dam with longitudinal joint are affected less due to smaller size of the pouring block. If cracks appear in the gravity dam without longitudinal joint, they may easily develop into large cracks and endanger the safety of the dam owing to larger influences of orifices, big dam block and wide foundation constraint range.

The development law of the thermal stress of a roller compacted concrete gravity dam is the same as that of a conventional concrete gravity dam without longitudinal joint, but their material properties are a bit different. The elastic modulus E and the linear expansion coefficient α of roller compacted concrete are similar to those of conventional concrete, adiabatic temperature rise of roller compacted concrete is lower, the tensile strength and the extensibility of roller compacted concrete are slightly lower, and the unit creep value of roller compacted concrete is much lower than that of conventional concrete. In a word, the crack-resisting capacity of roller compacted concrete is slightly lower than that of conventional concrete.

14.4.2 Upstream Crack of the Gravity Dam

Upstream crack is a kind of high-risk crack. Surface cracks emerge on the upstream face of the dam during the construction period, of which some extensively expand in sudden and develop into deep upstream cracks after water reserve. Fig. 14.4 – 3 shows the horizontal profile of a dam section nearby the upstream face.

The condition to avoid the extension of surface cracks is as follows:

$$K_I = 1.988(\sigma_T + p - \sigma_b) \leqslant K_{Ic} \qquad (14.4-1)$$

where, K_I is the stress intensity factor at the crack tip; K_{Ic} is the fracture toughness of the first type concrete crack, σ_T is the superficial thermal stress of the dam; p is the water pressure inside cracks; and σ_b is superficial compressive stress of the dam produced by the water pressure p between transverse joint seal and the dam surface. According to the conditional expression (14.4 – 1), big water head (p is large) before the dam, long crack and big difference between the internal and external temperature (σ_T is large) can make surface cracks develop into deep cracks easily. As the gravity dam without longitudinal joint is not subject to second-stage cooling owing to the absence of longitudinal joints, it still has

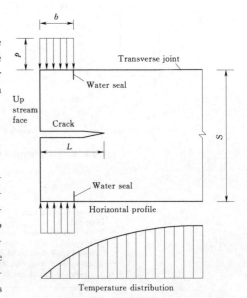

Fig. 14.4 – 3 Abutment Crack

very high internal temperature when water is reserved; and big difference between the internal and external temperature, larger superficial tensile stress produced by temperature and splitting action of fissure water inside cracks easily enable surface cracks to grow into deep cracks.

The following measures shall be taken to prevent deep upstream cracks: ①Effective superficial thermal insulation measures on the upstream face of the dam. ②Moving downstream of the transverse joint seal to increase the compressive stress produced by water pressure between the seal and the surface at the upstream of the dam. ③Partial second-stage cooling except for first-stage cooling when cooling pipes are laid in the dam, so as to maximally reduce the internal and external temperature difference. ④Overall examination on the dam upstream face and anti-seepage handling of all surface cracks before water reserve in order to prevent pressure existing in cracks after water reserve.

14.4.3 Temperature Control of Gravity Dams in Severe Cold Regions

There are two problems for construction of gravity dam without longitudinal joint in severe cold regions: ①long-term stop of pouring concrete in winter between the upper and the lower parts of dam produce big temperature difference; ②concrete poured in summer forms huge internal and external temperature difference due to very low outside temperature in winter. Under the combined action of big temperature difference between the upper and the lower parts and between inside and outside, huge vertical tensile stress on the horizontal joints and large horizontal tensile stress near the upstream and downstream faces may bring on horizontal and vertical cracks.

In the Guanyinge RCC Gravity Dam in Liaoning province, with maximum height of 82m, the pouring of concrete is stopped in five months in the winter because the average temperature is below 0℃. Thermal insulation is conducted on the upstream face, the downstream face and the horizontal top surface respectively through polystyrene foam boards ranging from 3 to 7cm, a layer of straw mattress clamping three layers of polystyrene tarpaulins or 5cm-thick polystyrene foam boards, and a layer of polystyrene paulin covering on three layers of straw mattresses. In real construction, the temperature difference ranges from 14.25℃ to 20.65℃ between the upper and the lower layers and 15.0℃ to 29.0℃ between inside and outside. According to results of finite element stimulation calculation, the vertical tensile stress of the upstream and the downstream faces nearby the overwintering layer is mostly above 2.5MPa and 4.0MPa at maximum, much more than the vertical tensile strength of concrete next to the layer. There emerge 53 cracks on the upstream face of the Guanyinge Dam, of which 51 are horizontal cracks 0.5 - 1.2mm in width and 3.0 - 6.0mm in depth and are mainly at these places nearby three overwintering joint surfaces in 1991 - 1994, and their lengths almost cover the whole dam block. 79 cracks appear on the dam downstream face, of which 60 are horizontal cracks and 19 are vertical cracks.

The Tamagawa Dam in the north of Japan's Honshu Island, 100m in height and constructed in 1987, has emerged lot of vertical cracks on the upstream and the downstream dam surfaces.

Thus the following temperature control measures must be taken to avoid serious crack of the gravity dam in severe cold areas:

(1) Intensifying the superficial thermal insulation. According to finite element stimulation calculation, if superficial thermal insulation is carried out on the overwintering top surface, the upstream and downstream face of the Guanyinge Dam by 17cm-thick polystyrene foam boards, the maximum vertical tensile stresses of the upstream and the downstream faces can be reduced to1.8MPa and 1.30MPa in winter, which basically can prevent crack.

(2) Pipe cooling. Cooling pipes are laid in fresh concrete in order to lower the maximum temperature of concrete and to reduce the temperature difference between the upper and the lower parts.

(3) Presetting artificial crack. Stimulation calculation has showed that as to the Liaoning Baishi Roller Compacted Concrete Gravity Dam 49.3m in height, under the condition of taking scheduled superficial thermal insulation measures, maximum tensile stresses of the upstream and the downstream faces of a water retaining dam are 2.24MPa and 4.59MPa respectively, and 6.06MPa at the anti-arch section of the weir surface of a spillway dam. After study, decision is made to set up horizontal surface artificial stress release cracks on the upstream and the downstream sides of the overwintering layer of the water retaining dam section and vertical surface artificial stress release cracks on the anti-arc section of the spillway dam, and they are 1.0m in depth. After the completion of the dam, except for expansion of preset cracks in winter, there are no cracks nearby them.

As long as strict temperature control measures are taken, dams in severe cold regions can be prevented from cracking. The first step is to intensify temperature control to avoid cracks, and the setting of preset artificial cracks on the upstream face is the second defense line.

14.4.4 Thermal Stress Caused by Heightening of the Gravity Dam

When the gravity dam is heightened, full cooling of old dam blocks forms temperature difference between new and old dam blocks, which will produce the tensile stress not only inside new dam blocks but also on the upstream faces of old dam blocks. It can be computed by the method provided by the IWHR, as shown in the Fig. 14.4 - 4.

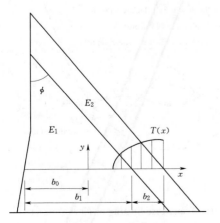
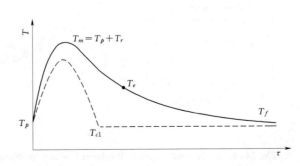

Fig. 14.4 - 4 Heightening of the Gravity Dam   Fig. 14.4 - 5 Temperature Change of New Placing Concrete

By taking a horizontal profile for computation and locating the origin of coordinate on the E-based weighted centroid, the vertical stress on the profile can be computed as follows:

$$\sigma_y = E_i\left(\frac{N}{D} + \frac{M}{F}x\right) - E_i\alpha T(x)\cos^2\phi \tag{14.4-2}$$

where, $N = \cos^2\phi \int E_i\alpha T(x)\mathrm{d}x$ and $M = \cos^2\phi \int E_i\alpha T(x)x\mathrm{d}x$.

As shown in the Fig. 14.4-5, after pouring of fresh concrete, the temperature rises to the maximum temperature $T_m = T_p + T_r$ from the placing temperature T_p, and then drops to the steady temperature after gradual cooling. The temperature difference above the foundation is generally computed as follows: $\Delta T_1 = T_p + T_r - T_f$, while the temperature difference caused by heightening of the gravity dam is:

$$\Delta T = T_c - T_f \tag{14.4-3}$$

where, T_c is the top temperature, that is, the temperature of fresh concrete when heightening is extended to the top, and T_f is the steady temperature that is close to annual average air temperature plus solar radiation.

Why the temperature difference caused by heightening of the gravity dam is not calculated based on the maximum temperature? Because when the temperature of fresh concrete drops from the maximum temperature, the top surface of the new concrete is free and certain stress will appear inside fresh concrete and on the contact surface of fresh and old concrete, but it is a self-balancing force system in a narrow range. According to the Saint-Venant principle, its influence is local and the impact on the overall stress of the dam is very small. Based on the principle above, the temperature control criterion for heightening of the gravity dam is obtained as follows:

$$T_c \leqslant T_f \tag{14.4-4}$$

When $T_c = T_f$, the dam heel does not generate the tensile stress and when $T_c < T_f$, it produces the compressive stress.

Actually, it is impossible to control the maximum temperature T_m below the steady temperature T_f during concrete construction in China's climatic condition, but controlling the top temperature T_c below the steady temperature can be realized. During heightening of the gravity dam in the past, adoption of sliding joints and longitudinal vertical joints in new concrete in most projects makes the construction more complex. According to the expression (14.4-4), conventional temperature control measures including water pipe cooling, aggregate precooling and external superficial thermal insulation can settle the problem of dam stress deterioration and considerably simplify construction measures[59].

Results of simulation calculation and Danjiangkou field test by Changjiang Water Resources Committee indicate that after dam heightening, the vast majority of joint surfaces of new and old concrete are pulled apart, reducing the integrity and the safety of the dam. The method of transverse slit in new concrete and permanent superficial thermal insulation on the downstream dam surface, which is put forward by the IWHR for the first time, can ensure that the joint surfaces of new and old concrete basically do not split. It can be seen in Fig. 14.4-6 that the slitting in new concrete can greatly reduce the tensile stress on joint surfaces in winter.

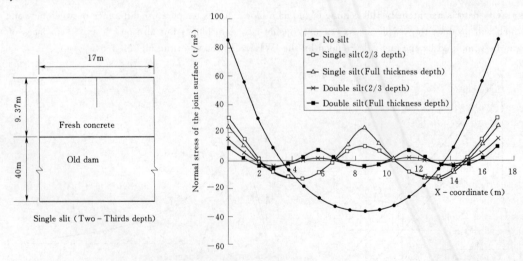

Fig. 14.4-6 **Normal Stress of the Joint Surface in Winter after Transverse Slitting of New Concrete**

14.5 Thermal Stress of the Arch Dam

This section represents the thermal stress of an arch dam after grouting of joints.

14.5.1 *Reservoir Water Temperature*

In the past, there was no formula to compute reservoir water temperature which was a boundary condition of the temperature field of an arch dam. In order to calculate the temperature load of the arch dam, in 1985 the IWHR put forward the following formula to compute reservoir water temperature at any depth y:

$$T(y,\tau) = T_m(y) + A(y)\cos\omega(\tau - \tau_0 - \varepsilon) \tag{14.5-1}$$

where, $T_m(y)$ is the annual average water temperature at the depth y; $A(y)$ is the amplitude of annual variation of water temperature; ε is the phase difference of water temperature (month); y is the water depth; τ is time (month); $\omega = 2\pi/P$ is the circular frequency of water temperature change and $P = 12$ months is the period of water temperature variation.

$$T_m(y) = c + (T_s - c)e^{-\alpha y} \tag{14.5-2}$$

$$A(y) = A_0 e^{-\beta y} \tag{14.5-3}$$

$$\varepsilon = d - f e^{-\gamma y} \tag{14.5-4}$$

where, T_s is annual average surface water temperature; A_0 represents the amplitude of annual change of surface water temperature; and $\alpha, \beta, \gamma, e, d, f$ are constant.

14.5.2 *Temperature Load of the Arch Dam*

Before 1985, China adopted the following empirical formula provided by the U.S. Bureau of Reclamation to calculate the temperature load of an arch dam in the design of arch dam:

$$T_m = \pm 57.57/(L + 2.44) \tag{14.5-5}$$

where, T_m is dam average temperature (℃) and L is dam thickness (m).

The formula above ignores the impact of the following factors: ①local weather conditions, ②dam joint grouting temperature, ③temperature difference between the upstream and the downstream, and④reservoir water temperature.

After 1985, defects mentioned above are overcome by adopting the formula of the IWHR.

The IWHR, based on analysis, points out three characteristic temperature fields of an arch dam during the operation period:

(1) Joint closure temperature $T_0(x)$, that is, dam temperature during joint grouting, a function of the horizontal coordinate x in the thickness direction.

(2) Annual average temperature field during the operation period $T_1(x)$, that is, annual average temperature at x point.

(3) Annual-change temperature field during the operation period $T_2(x)$, that is, annual variation of the temperature at x point.

Based on the three characteristic temperature fields above, the IWHR raises the following formula to compute the temperature load of an arch dam:

$$\left. \begin{array}{l} T_m = T_{m1} + T_{m2} - T_{m0} \\ T_d = T_{d1} + T_{d2} - T_{d0} \end{array} \right\} \tag{14.5-6}$$

where, T_m and T_d represents the mean temperature and equivalent temperature difference of an arch dam; T_{m0} and T_{d0} are mean temperature and equivalent temperature difference of the joint closure temperature field; T_{m1} and T_{d1} are mean temperature and equivalent temperature difference of the annual mean temperature field in the thickness direction during the operation period; and T_{m2} and T_{d2} are mean temperature and equivalent temperature difference of the variable temperature field in the thickness direction during the operation period.

14.5.2.1 Temperature Load of the Arch Dam below Constant Water Level

The boundary condition of the temperature field of an arch dam during the operation period is shown in the Fig. 14.5 –

1. The temperature of the arch dam in the downstream is in touch with air, and it can be expressed as follows:

$$T_d = T_{dm} + A_d \cos\omega(\tau - \tau_0) \qquad (14.5-7)$$

The temperature of the dam upstream face above water equals to air temperature and can be computed according to the formula (14.5-7), and the temperature of the upstream dam below water is equivalent to water temperature. Supposing the water level is constant, the monthly temperature of the dam surface can be expressed by a cosine function as follows:

$$T_u = T_{um} + A_u \cos\omega(\tau - \tau_0 - \varepsilon) \qquad (14.5-8)$$

According to the boundary condition above, we get

$$\left. \begin{array}{l} T_{m1} = \dfrac{1}{2}(T_{UM} + T_{DM}) \\ T_{d1} = T_{DM} - T_{UM} \end{array} \right\} \qquad (14.5-9)$$

$$\left. \begin{array}{l} T_{m2} = \dfrac{\rho_1}{2}[A_D \cos\omega(\tau - \tau_0 - \theta_1) + A_U \cos\omega(\tau - \tau_0 - \varepsilon - \theta_1)] \\ T_{d2} = \rho_2 [A_D \cos\omega(\tau - \tau_0 - \theta_2) + A_U \cos\omega(\tau - \tau_0 - \varepsilon - \theta_2)] \end{array} \right\} \qquad (14.5-10)$$

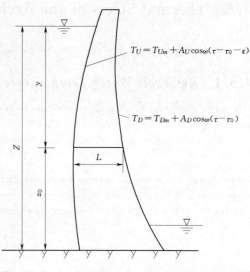

Fig. 14.5-1 Boundary Condition of Temperature Field of the Arch Dam

where,

$$\left. \begin{array}{l} \rho_1 = \dfrac{1}{\eta}\sqrt{\dfrac{2(\mathrm{ch}\eta - \cos\eta)}{\mathrm{ch}\eta + \cos\eta}}, \rho_2 = \sqrt{a_1^2 + b_1^2}, \theta_1 = \dfrac{1}{\omega}\left[\dfrac{\pi}{4} - \tan^{-1}\left(\dfrac{\sin\eta}{\mathrm{sh}\eta}\right)\right] \\ \theta_1 = \dfrac{1}{\omega}\tan^{-1}\left(\dfrac{b_1}{a_1}\right), a_1 = \dfrac{6}{\rho_1 \eta^2}\sin\omega\theta_1, b_1 = \dfrac{6}{\eta^2}\left(\dfrac{1}{\rho_1}\cos\omega\theta_1 - 1\right) \\ \eta = \sqrt{\dfrac{\pi}{aP}}L, \omega = \dfrac{2\pi}{P} \end{array} \right\} \qquad (14.5-11)$$

14.5.2.2 Temperature Load of the Arch Dam below Variable Water Level

In Fig. 14.5-1, water depth y, $y = z - z_0$, wherein z is reservoir water level and z_0 is calculated section height. The temperature of the upstream dam surface can be computed as follows:

$$\left. \begin{array}{l} T_u(\tau) = T_w(z - z_0, \tau), \text{when } z - z_0 \geqslant 0 \\ T_u(\tau) = T_a(\tau), \text{when } z < z_0 \end{array} \right\} \qquad (14.5-12)$$

where, T_w is water temperature and T_a is air temperature.

Under the condition of variable water level, the temperature of the upstream dam surface is not a simple cosine function and but can be expressed by Fourier series as follows:

$$T_u(\tau) = T_{un} + \sum_{n=1}^{\infty} A_{un} \cos\omega_n(\tau - \tau_0 - \varepsilon_{un}) \qquad (14.5-13)$$

The temperature of the downstream dam surface also can be expressed as follows:

$$T_d(\tau) = T_{dm} + \sum_{n=1}^{\infty} A_{dn} \cos\omega_n(\tau - \tau - \varepsilon_{dm}) \qquad (14.5-14)$$

when $\omega_n = 2n\pi/P$, T_{m1} and T_{d1} also can be computed by the formula (14.5-9) and T_{m2} and T_{d2} are calculated by the following formulas:

$$T_{m2} = \sum_{n=1}^{\infty} \dfrac{\rho_{1n}}{2}\{A_{dn}\cos[\omega_n(\tau - \tau_0 - \xi_{dn} - \theta_{1n})] + A_{un}\cos[\omega_n(\tau - \tau_0 - \varepsilon - \theta_{1n})]\} \qquad (14.5-15)$$

$$T_{d2} = \sum_{n=1}^{\infty} \rho_{2n}\{A_{dn}\cos[\omega_n(\tau - \tau_0 - \xi_{dn} - \theta_{2n})] + A_{un}\cos[\omega_n(\tau - \tau_0 - \varepsilon - \theta_{2n})]\} \qquad (14.5-16)$$

where,

$$\rho_{1n} = \frac{1}{\eta}\sqrt{\frac{2(\mathrm{ch}\eta_n - \cos\eta_n)}{\mathrm{ch}\eta_n + \cos\eta_n}}, \rho_{2n} = \sqrt{a_n^2 + b_n^2}, \theta_{1n} = \frac{1}{\omega_n}\left[\frac{\pi}{4} - \tan^{-1}\left(\frac{\sin\eta_n}{\mathrm{sh}\eta_n}\right)\right]$$
$$\theta_{2n} = \frac{1}{\omega_n}\sin(\omega\theta), a_n = \frac{6}{\rho_{1n}\eta_n^2}\sin(\omega_n\theta_{1n}), b_n = \frac{6}{\eta_n^2}\left[\frac{1}{\rho_{1n}}\cos(\omega_n\theta_{1n}) - 1\right]$$
$$\eta_n = \sqrt{\frac{n\pi}{aP}L}, \omega_n = \frac{2n\pi}{P}$$
(14.5-17)

Comparison of two algorithms: ①precise algorithm: water temperature is computed with a one-dimensional numerical model and dam surface temperature is calculated by the formula (14.5-12), with T_u as the Fourier series taking six terms. ②Simplified algorithm: water temperature is counted with the simplified formula (14.5-1), and only one term is taken in the series in the formulas (14.5-14) - (14.5-16). Calculation results show that: ①there exist bigger differences between temperature loads respectively computed by actual variable water level and fixed water level. ②Both based on the computation by variable water level, results of precise algorithm are close to those of simplified algorithm.

The precise algorithm can be applied to an important project while the simplified algorithm to a common project. As for a high-dam with large reservoir focusing on power generation, change of its water level only happens at the upper part of the dam and the maximum tensile stress and compressive stress of the dam are at the lower part of the dam, thus change of temperature loads caused by the change of water level has a limited impact on the maximum stress of the dam, while as to small and medium-sized reservoirs mainly for water supply, big change in its water level has a greater effect on the stress of the dam compared with a high-dam with large reservoir.

14.5.2.3 Temperature Load of the Arch Dam with Thermal Insulating Layers in Cold Regions

1. Calculating T_{m1} and T_{d1}

As shown in the Fig. 14.5-2, supposing the thickness and the coefficient of heat conductivity of the dam are respectively L and λ, the thickness and the coefficient of heat conductivity of the upstream thermal insulating layer are respectively h_U and λ_U, and the thickness and the coefficient of heat conductivity of the downstream thermal insulating layer are respectively h_D and λ_D. The distribution of the annual average temperature field is analyzed by three layers of flat slabs. Both the annual average temperature T_{Um} of the upstream face and the annual average temperature T_{Dm} of the downstream face are constant (not change with time), and the annual average temperature at any point inside the three layer of flat slabs is only a linear function of the horizontal coordinate x without variation with time.

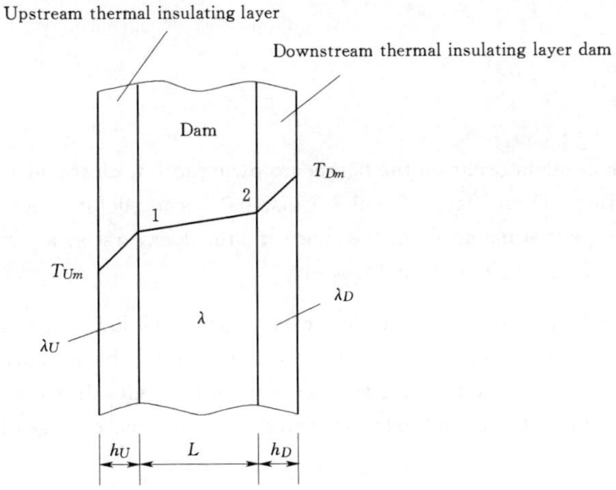

Temperature T_1 and temperature T_2 on the contact surface 1 and the contact surface 2 are:

$$T_1 = \frac{(1+\rho_1\rho_2)T_{Um}}{1+\rho_1+\rho_1\rho_2} + \frac{\rho_1 T_{Dm}}{1+\rho_1+\rho_1\rho_2} \quad (14.5-18)$$

$$T_2 = \frac{T_{Um}}{1+\rho_1+\rho_1\rho_2} + \frac{(\rho_1+\rho_1\rho_2)T_{Dm}}{1+\rho_1+\rho_1\rho_2} \quad (14.5-19)$$

where,
$$\rho_1 = \frac{\lambda_D h_U}{\lambda_U h_D}, \quad \rho_2 = \frac{\lambda_U L}{\lambda h_U}$$

The temperature of the dam upstream face is T_1, the temperature of the downstream face is T_2 and the temperature in the dam varies linearly with the coordinate, thus

$$T_{m1} = \frac{T_1 + T_2}{2}, \quad T_{d1} = T_2 - T_1 \quad (14.5-20)$$

Fig. 14.5-2 Annual Average Temperature

2. Calculating T_{m2} and T_{d2}

The IWHR has raised the exact solution and the approximate solution of T_{m2} and T_{d2}, and the precise computing method is a little complex. Actually the approximate solution may be used which can be obtained as follows: the temperature outside the upstream thermal insulating layer of the dam is computed with the formula (14.5-8) and the temperature of the dam upstream face (one point) in touch with the thermal insulating layer is

$$T = T_1 + k_u A_u \cos\omega(\tau - \tau_0 - \varepsilon - \xi_u) \quad (14.5-21)$$

$$k_u = [1 + 2q\lambda_u/\beta_u + 2(q\lambda_u/\beta_u)^2]^{-1/2} \quad (14.5-22)$$

$$\xi_u = \frac{1}{\omega}\tan^{-1}\left(\frac{l}{1+\beta_u/\lambda_u q}\right) \tag{14.5-23}$$

where,

$$q=\sqrt{\pi/aP}, \omega=2\pi/P$$

The surface temperature of the downstream dam is

$$T = T_2 + k_d A_d \cos\omega(\tau - \tau_0 - \xi_d) \tag{14.5-24}$$

k_d and ξ_d can be got by replacing λ_u and β_u with λ_D and β_D in the formulas (14.5-22) and (14.5-23). T_{m2} and T_{d2} can be computed with the boundary temperature in the two formulas (14.5-21) and (14.5-24) according to the formula (14.5-10).

14.5.3 Temperature Control and Joint Design of the Roller Compacted Concrete Arch Dam

The temperature load of the roller compacted concrete arch dam also can be computed by the formula (14.5-6). If there is no longitudinal joint, the joint closure temperature is actually the maximum temperature of dam concrete which had appeared in the construction period and can be calculated as follows:

$$T_{m0} = T_p + k_r T_r \tag{14.5-25}$$

where, T_p is concrete placing temperature, T_r is temperature rise of concrete due to hydration heat and k_r is reduction coefficient considering early warming, and $k_r = 1.0$ at preliminary calculation.

At the early construction of roller compacted concrete arch dams, only inducing joints are set inside the dam without transverse joints. Research results by the IWHR indicate that: ①As for smaller arch dams in the south of China such as Puding Arch Dam, it is feasible to construct arch dams with only inducing joints if concrete is poured in low-temperature seasons. ②As to high arch dams in general areas, if construction shall go through the whole year, establishment of inducing joints is not enough, transverse joints must be set in order to effectively prevent cracks. ③In order to ensure timely transverse joint grouting, cooling water pipes shall also be laid inside the roller compacted concrete arch dam. The Chengdu Design Institute puts forward the concept of adopting gravity concrete joint forming boards in the construction of Shapai Arch Dam, thus effectively settling the problem of arch joint formation during the construction of the roller compacted concrete arch dam.

14.6 Thermal Stress of the Sluice

The sluice is usually built on the soft foundation. Through cracks seldom occur on the baseboard owing to low elastic modulus and very small constraint of the subgrade, while the pier often emerges large cracks because its thermal deformation is restrained by the baseboard. The IWHR, based on the study on the thermal stress of the sluice and the dock, raises a series solution and a simplified solution (T-shaped beam method) for the first time in the literature.

Cracks of the pier are aroused by the following three factors: ①Heat of hydration during the construction period. The baseboard is poured and cooled first and the pier is poured later so its thermal deformation is restrained by the baseboard. ②Annual change of air temperature. The internal temperature of the pier featuring thinness and exposure of two sides drops more in winter, thus the tensile stress produces therein. ③Cold wave. Big internal and external temperature difference caused by cold wave generates tensile stress on the surface of pier.

Computing Example: please refer to Fig. 14.6-1, the dimension of the sluice has been shown. The pier length × height × thickness = 10.5m × 6.0m × 1.0m and the baseboard length × height × thickness = 10.5m × 9.0m × 1.30m. Calculation is conducted by 1/4 owing to symmetry, concrete adiabatic temperature rise $\theta(\tau) = 30\tau/(1.70+\tau)$°C, and elastic modulus $E(\tau) = 3500\tau/(3.30+\tau)$MPa and $\mu = 0.167$, the concrete creep is considered in computation. The subgrade has no creep and no heat of hydration, $E_f = 50$MPa and $\mu = 0.25$.

The pier is poured after 14 days of baseboard pouring. The variations of the temperature and stress at internal point A and B are shown in the Fig. 14.6-2. And it is evident that the baseboard has little stress in spite of large temperature change, and the pier has less change in the temperature but big tensile stress due to the restraint of the baseboard.

The actual stress is the combined thermal stress of heat of hydration, annual change and cold wave. Fig. 14.6-3 indicates the combined stresses of the lateral surface and the center line of the central cross section of the pier. ① Heat of hydration+

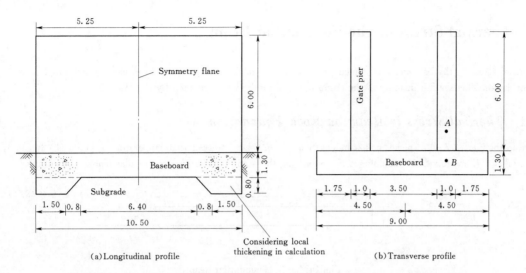

(a) Longitudinal profile (b) Transverse profile

Fig. 14.6 − 1 Example of Computation: the Dimension of the Sluice (Unit: m)

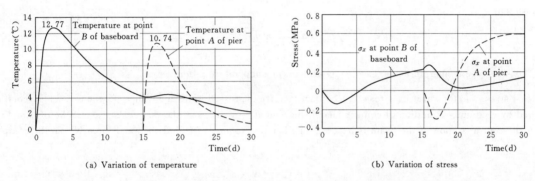

(a) Variation of temperature (b) Variation of stress

Fig. 14.6 − 2 Variations of Temperature and Stress at Point A and Point B inside the Sluice

annual temperature change, the maximum tensile stress of the lateral surface and the center line are respectively 0.43MPa and 0.97MPa. ② Heat of hydration+annual temperature change+cold wave, the thermal surface has the maximum tensile stress of 1.99MPa and the center line 1.49MPa. ③ Heat of hydration+annual temperature change+cold wave+surface hot air heating during baseboard construction period (14 − 16 days), the maximum tensile stress of the lateral surface and the center line are respectively 1.85MPa and 1.18MPa.

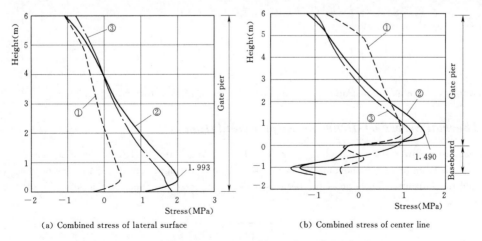

(a) Combined stress of lateral surface (b) Combined stress of center line

Fig. 14.6 − 3 Combined Thermal Stress of the Pier of a Sluice

①—Heat of hydration+annual change; ②—Heat of hydration+annual change+cold wave;
③—Heat hydration+annual change+cold wave+baseboard heating up

14.7 Thermal Stress of the Foundation Beam

The Thermal stress in the baseboard of sluice, ship lock and dock and spillway apron can be computed according to the foundation beam. This section introduces the computational method[7,26] provided by the IWHR.

14.7.1 Thermal Stress in Beam on Rock Foundation

As shown in the Fig. 14.7-1, the beam is $2l$ in length and $2h$ in height, the internal temperature is $T(y)$, a y function, concrete elastic modulus is E and foundation modulus is E_f. Supposing the beam is cut from a rectangular plate, its width is l.

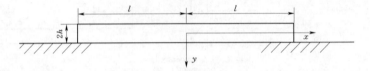

Fig. 14.7-1 Foundation Beam

First the beam is cut along the surface of foundation. The self-generating thermal stress of the beam is computed as the following formula:

$$\sigma_{x1} = \frac{E\alpha}{1-\mu}\left[T_m + \frac{T_d}{2h}y - T(y)\right] \tag{14.7-1}$$

where, T_m is average temperature and T_d is equivalent linear temperature difference.

The normal stress and the shearing stress on a contact face are expressed with Chebyshev polynomial whose coefficient is determined by continuity condition of contact face displacement, in this way the contact stress can be obtained and the stress on the central section of the beam is biggest. The bending moment M_0 and the axial force N_0 can be computed as follows:

$$M_0 = l^2\left(B_1 r - \frac{A_2}{3}\right), \quad N_0 = -B_1 l \tag{14.7-2}$$

where,

$$B_1 = \left(\frac{22.5}{\eta r^3} + 180\right)\frac{E\alpha T_m}{(1-\mu)\Delta} + \frac{18 E\alpha T_d}{(1-\mu)r\Delta}$$

$$A_2 = \left(\frac{67.5}{\eta r^2} - 36\right)\frac{E\alpha T_m}{(1-\mu)\Delta} - \left(\frac{11.25}{\eta r^2} + \frac{45}{r}\right)\frac{E\alpha T_d}{(1-\mu)\Delta}$$

$$\Delta = 331.2\eta + \frac{90}{r} + \frac{54}{r^2} + \frac{45}{r^2} + \frac{11.25}{\eta r^4}$$

$$r = h/l, \eta = E/E_f, \mu = \mu_f = 1/6$$

The restraint stress is calculated as follows:

$$\left.\begin{array}{c}\sigma_{x2\text{up}}\\ \sigma_{x2\text{down}}\end{array}\right\} = \frac{N_0}{2h} \pm \frac{3M_0}{2h^2} \tag{14.7-3}$$

The self-generating stress plus the restraint stress equals to the actual stress, that is, $\sigma_x = \sigma_{x1} + \sigma_{x2}$.

14.7.2 Thermal Stress of Beam on Soft Foundation

The soft foundation has lower elastic modulus, with $E_f = 10 - 50$ MPa. The IWHR provides a complete algorithm of the thermal stress of the beam on soft foundation. When the beam length to height ratio $l/h > 40$, the bending formation is in full restraint and the horizontal deflection of the beam is under elastic restraint. The thermal stress inside the beam can be computed as follows:

$$\sigma_x = \frac{E\alpha}{1-\mu}\left[\frac{T_m}{\text{ch}\lambda l} - T(y)\right] \tag{14.7-4}$$

where,

$$\lambda l = 0.705\sqrt{\frac{E_f(1-\mu^2)l}{E(1-\mu_f^2)h}}$$

14.8 Pipe Cooling of Concrete

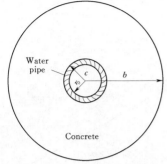

Fig. 14.8 – 1 Computation Model for Non-Metallic Water Pipe Cooling

The pipe cooling method is raised by the U. S. Bureau of Reclamation in the 1930s, and Chinese scholars have done much work in temperature field stress field and method of pipe cooling.

14.8.1 Temperature Field of Pipe Cooling

14.8.1.1 Exact Solution to the Plane Problem of Pipe Cooling

It shall give priority to the problem of postcooling without heat source. Fig. 14.8 – 1 shows a hollow cylinder featuring external thermal insulation and internal water pipe cooling, the initial temperature T_0 is and water temperature is 0℃. The IWHR obtains the exact solution as follows:

$$T(r,t) = T_0 \sum_{n=1}^{\infty} \frac{2e^{-aa_n^2 \tau}}{\alpha_n b} \frac{J_1(\alpha_n b) Y_0(\alpha_n r) - Y_1(\alpha_n b) J_0(\alpha_n r)}{R_1(\alpha_n b)} \tag{14.8-1}$$

$$R_1(\alpha_n b) = -\frac{\lambda}{k_5 b} \alpha_n b \left\{ \frac{c}{b} [J_1(\alpha_n b) Y_0(\alpha_n c) - Y_1(\alpha_n b) J_0(\alpha_n c)] + [J_0(\alpha_n b) Y_1(\alpha_n c) - J_1(\alpha_n c) Y_0(\alpha_n b)] \right\}$$
$$+ \frac{c}{b} [J_1(\alpha_n b) Y_1(\alpha_n c) - Y_1(\alpha_n b) J_1(\alpha_n c)] + [J_0(\alpha_n c) Y_0(\alpha_n b) - Y_0(\alpha_n c) J_0(\alpha_n b)] \tag{14.8-2}$$

where, J_0, J_1, Y_0 and Y_1 are respectively the zero-order, first-order Bessel function, of the first and second kind. $\alpha_n b$ is the root of the following characteristic equation:

$$-\frac{\lambda}{k_5 b} \alpha_n [J_1(\alpha_n c) Y_1(\alpha_n b) - J_1(\alpha_n b) Y_1(\alpha_n c)] + [J_1(\alpha_n b) Y_0(\alpha_n c) - J_0(\alpha_n c) Y_1(\alpha_n b)] = 0 \tag{14.8-3}$$

$$k_5 = \frac{\lambda_1}{c l_n(c/r_0)} \tag{14.8-4}$$

where, λ thermal conductivity coefficient of concrete, λ_1 thermal conductivity coefficient of water pipe, c outer radius of water pipe, r_0 inside radius of water pipe, a thermal conductivity coefficient of concrete. Actually it only takes one item due to fast convergence of the formula. The average temperature is computed as follows:

$$T_m(t) = T_0 e^{-s_1 t}, s_1 = a_1^2 b^2 a/b^2 \tag{14.8-5}$$

The characteristic value α_1, b can be calculated as follows:

$$\alpha_1 b = 0.926 \exp\left\{-0.0314 \left[\frac{b}{c}\left(\frac{c}{r_0}\right)^\eta - 20\right]^{0.48}\right\}, \eta = \frac{\lambda}{\lambda_1} \tag{14.8-6}$$

Supposing adiabatic temperature rise of concrete is $\theta(\tau)$ and water temperature is 0℃, the average temperature at the early stage of water pipe cooling is:

$$T_m(t) = \int_0^t e^{-s_1(t-\tau)} \frac{\partial \theta}{\partial \tau} d\tau \tag{14.8-7}$$

For instance, supposing $\theta(\tau) = \theta_0 (1 - e^{-m\tau})$, thus:

$$T_m(t) = \frac{\theta_0 m}{s_1 - m} (e^{-mt} - e^{-s_1 t}) \tag{14.8-8}$$

Metallic water pipe cooling is a special case, making $\lambda_1 = \infty$.

Fig. 14.8 – 2 shows an computing example of pipe cooling at the early stage, $b = 0.845$m, $c = 1.6$cm, $r_0 = 1.4$cm, $T_0 = 0$℃, $T_w = 0$℃, adiabatic temperature rise. $\theta(\tau) = 25(1 - e^{-0.35\tau})$, $\lambda = 8.37$kJ/(m·h·℃), $\lambda_1 = 1.66$kJ/(m·h·℃). Calculated results of the plastic pipe and the steel pipe are shown in the figure.

14.8.1.2 Approximate Solution to Spatial Problem of Pipe Coding

Fig. 14.8 – 3 represents a hollow cylinder featured by external thermal insulation and internal water pipe cooling, in which water temperature gradually rises due to the absorption of heat released by concrete. The approximate solution to space problem can be obtained by establishing an integral equation according to heat balance. By supposing the initial temperature of concrete is T_0, the adiabatic temperature rise is $\theta(\tau) = \theta_0 f(\tau)$, the water temperature at an inlet is T_w, the water temperature at the point L of the pipe length is T_{Lw}, the average temperature of the concrete section at the point L of the pipe is T_{Lm} and the average temperature of concrete within the length L, the IWHR has put forward the following computational formula:

The overall average temperature of concrete within the whole length L is:

$$T_m(t) = T_w + (T_0 - T_w)\phi(t) + \theta_0 \psi(t) \quad (14.8-9)$$

where,
$$\phi(t) = e^{-pt} \quad (14.8-10)$$

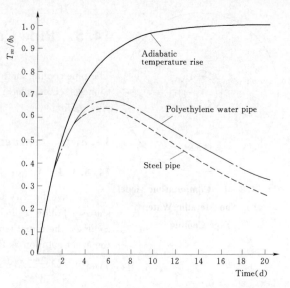

Fig. 14.8-2 Computing Example of Plane Problem on Concrete Water Pipe Cooling with Heat Source

$$\psi(t) = \int_0^t e^{-p(t-\tau)} \frac{\partial f}{\partial \tau} d\tau \quad (14.8-11)$$

$$p = k_2 h, \quad h = \frac{ga}{D^2} = \frac{0.734 ga}{s_1 s_2} \quad (14.8-12)$$

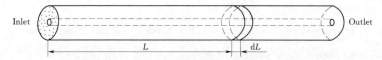

Fig. 14.8-3 Water Pipe Cooling Space Problem

$$k_2 = 2.09 - 1.35\xi + 0.32\xi^2 \quad (14.8-13)$$

$$g = 1.67 \exp\left\{-0.0628\left[\frac{b}{c}\left(\frac{c}{r_0}\right)^n - 20\right]^{0.48}\right\} \quad (14.8-14)$$

$$\xi = \frac{\lambda L}{c_w \rho_w q_w}, \quad \eta = \frac{\lambda}{\lambda_1} \quad (14.8-15)$$

[Computing Example] For the concrete diffusivity $a = 0.10 \text{m}^2/\text{d}$, thermal conductivity $\lambda = 8.37 \text{kJ}/(\text{m} \cdot \text{h} \cdot \text{°C})$, for the polyethylene water pipe outer radius, $c = 1.6\text{cm}$, inside radius $r_0 = 1.4\text{cm}$, thermal conductivity coefficient $\lambda_1 = 1.66 \text{kJ}/(\text{m} \cdot \text{h} \cdot \text{°C})$, $\xi = \lambda L/(c_w \rho_w q_w) = 0.50$. The water cooling function $\phi(t) = \exp(-pt)$ is shown in Fig. 14.8-4. The figure shows that the spacing between water pipes has great impact on $\phi(t)$.

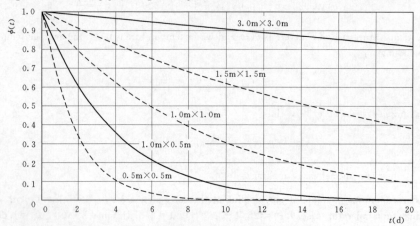

Fig. 14.8-4 Water Cooling Function $\phi(t) = e^{-pt}$

As for hyperbolic adiabatic temperature rise $\theta(\tau) = \theta_0 \tau/(n+\tau)$, when $n = 2.0\text{d}$, temperature rise function $\psi(\tau)$ of pipe cooling can be figured out with the formula (14.8-11), as shown in the Fig. 14.8-5.

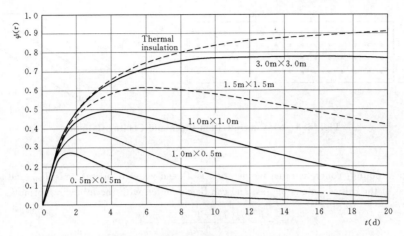

Fig. 14.8 – 5 Water Cooling Function $\psi(\tau)$, Hyperbolic Adiabatic Temperature Rise ($n=2.0\text{d}$)

The water temperature at the point L of the pipe length:

$$T_{Lw} = T_w + (T_0 - T_w)w(\xi,ht) + \theta_0 \int_0^t w[\xi,h(t-\tau)] \frac{\partial f}{\partial \tau}\mathrm{d}\tau \qquad (14.8-16)$$

$$w(\xi,ht) = [1-(1-g)\mathrm{e}^{-\xi}](1-\mathrm{e}^{-2.70\xi})\exp[-2.40(ht)^{0.50}\mathrm{e}^{-\xi}] \qquad (14.8-17)$$

14.8.2 *Equivalent Heat Conduction Equation for Pipe Cooling*

Under the joint action of adiabatic temperature rise and water pipe cooling, the average temperature of the cylinder with external thermal insulation is as the formula (14.8-8), if the adiabatic temperature rise $\theta(t)$ in the general heat conduction equation (14.2-1) is replaced with $T_m(t)$ in formula (14.8-8), the equivalent heat conduction equation considering water pipe cooling can be got as follows[11]:

$$\frac{\partial T}{\partial t} = a\left(\frac{\partial^2 T}{\partial x^2} + \frac{\partial^2 T}{\partial y^2} + \frac{\partial^2 T}{\partial z^2}\right) + (T_0 - T_w)\frac{\partial \phi}{\partial t} + \theta_0 \frac{\partial \psi}{\partial t} \qquad (14.8-18)$$

In calculation by the formula above, the influence of pipe cooling is considered in the functions ϕ and ψ, and the location and the dimension of the water pipe do not take into account in computing net, thus greatly simplifying the computation.

14.8.3 *Stress Field due to Pipe Cooling*

14.8.3.1 Equivalent Algorithm for Computing Stress Field due to Pipe Cooling

If the thermal stress in the real project is directly computed with three-dimensional finite element, there exist two problems: ① the radius of the water pipe is only 10 – 15mm and the computing net must be very intensive; ② existence of stress singularity nearby the water pipe and very large local stress makes the analysis inconvenient; currently the equivalent method provided by the IWHR is in extensive use. As shown in the Fig. 14.8 – 6, a submodel, featuring external thermal insulation and internal arrangement of cooling water pipe, is cut along the middle symmetry plane of each water pipe and its temperature field and stress field are respectively $T_1(x,y,z,t)$ and $\sigma_1(x,y,z,t)$, the equivalent heat conduction equation (14.8-18) can be obtained by replacing the adiabatic temperature rise $\theta(\tau)$ in the heat conduction equation with the average temperature $T_m(t)$ of the submodel. The temperature field $T_2(x,y,z,t)$ and the stress field $\sigma_2(x,y,z,t)$ of the original model can be computed by using the equivalent heat conduction equation and boundary conditions of the original model, and the real temperature field $T(x,y,z,t)$ and the real stress field $\sigma(x,y,z,t)$ are the sum of the temperature field and the stress field of the original model and the submodel:

$$\left.\begin{aligned}T(x,y,z,t) &= T_1(x,y,z,t) + T_2(x,y,z,t) \\ \sigma(x,y,z,t) &= \sigma_1(x,y,z,t) + \sigma_2(x,y,z,t)\end{aligned}\right\} \qquad (14.8-19)$$

14.8.3.2 Theoretical Solution to Self-Generating Local Stress of Pipe Cooling

The IWHR puts forward the theoretical solution to the self-generating temperature creep stress of pipe cooling based on the

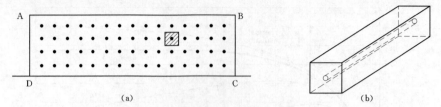

Fig. 14.8-6 Concrete Dam Block and Cooling Cylinder

Fig. 14.8-1.

Local stress of pipe postcooling: concrete initial temperature is T_0 and water temperature is T_w; and as for postcooling $\tau >$ 20d, the tensile stress at the orifice edge is biggest and can be computed with the following formula:

$$\sigma^*(t) = -\frac{E(\tau)\alpha(T_w-T_0)(1-m_1)k_4}{1-\mu}e^{-p_2 s(t-\tau)} \quad (14.8-20)$$

In which concrete creep impact is taken into account with coefficients k_4 and s. As for the polyethylene water pipe with outer radius of 16mm and inside radius of 14mm, parameters K_4, m_1 and s are respectively 0.81, 0.20 and 1.30; and as to the steel water pipe, parameters K_4, m_1 and s can be 1.00, 0 and 1.30 respectively. In case of water temperature changed in steps, $T_w - T_0$ in the formula can be replaced with water temperature difference ΔT_w through superposition method.

Local stress of first stage pipe cooling: concrete adiabatic temperature rise is $\theta(\tau)$, both concrete initial temperature and water temperature are zero, and visco-elastic stress at the orifice edge is

$$\sigma(t) = -\sum_i \frac{(1-m_1)\alpha\Delta\theta(\tau_i)}{1-\mu}\sum_j E(t_j)(e^{-p_1\Delta t_j}-1)e^{-p_1(t_j-t_i)}K(t,t_j) \quad (14.8-21)$$

14.8.3.3 Numerical Analysis of Self-Generating Local Visco-Elastic Thermal Stress of Pipe Cooling

As shown in the Fig. 14.8-7, the computation model is a prismoid with the height S_1, the width S_2 and the length L, and both S_1 and S_2 equal to the spacing of water pipes which are located in the middle of the prismoid. The transverse section is shown in the Fig. 14.8-8.

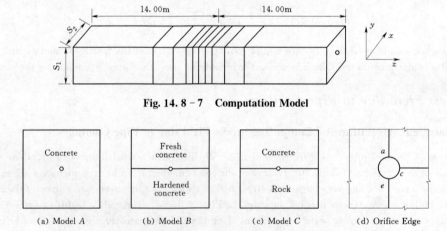

Fig. 14.8-7 Computation Model

(a) Model A (b) Model B (c) Model C (d) Orifice Edge

Fig. 14.8-8 Profile of Computation Model

[Computing Example] the spacing of water pipes is 1.5m×1.5m, $\theta(\tau)=25\tau/(2.1+\tau)$°C, $E(\tau)=35000\tau/(5.0+\tau)$MPa, the initial temperature $T_0=0$, the water temperature $T_w=0$ and the first stage cooling is 20 days. Fig. 14.8-9 presents the variation of tangential stress at the point e of the orifice edge. At the early stage it is the tensile stress, when $\tau=5-7d$, it is biggest and then gradually reduces, and finally there exists the compressive stress due to temperature homogenization. Fig. 14.8-10 indicates the distribution of tangential stress along the thickness direction.

In the late stage of the cooling, owing to similarity of elastic modulus of fresh and hardened concrete, the submodel is similar to the homogeneous structure and the stress calculated by finite element is similar to that computed through the theoretical solution. As Fig. 14.8-10 has shown, since the self-generating thermal stress is very small in the wide range a little far from the pipe edge, the stress calculated by the equivalent heat conduction equation is close to the real stress.

If it is only subject to cooling, the self-generating tensile stress at the orifice edge can be bigger, if the new mode of pipe cooling from earlier age with small temperature difference and longer time is adopted, it is smaller and generally can be ignored.

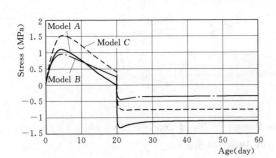
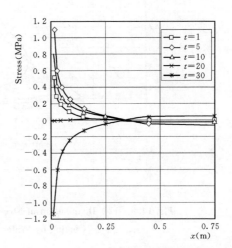

Fig. 14.8-9 First stage Cooling 20d, Variation of Tangential Visco-Elastic stress σ_x at Point e (Spacing between Pipes is 1.5m×1.5m, $\theta_0 = 25℃$)

Fig. 14.8-10 First stage Cooling 20d, Distribution of Tangential Stress σ_x at Cross-section a of Model A (Spacing between Pipes is 1.5m×1.5m, $\theta_0 = 25℃$)

14.8.4 Impact of the Height of Cooling Zone and Cooling Rate on the Stress

As for concrete blocks on the rock foundation, both L and H are 60m, initial temperature T_0 of concrete and foundation is 30℃, cooling water temperature is 8℃, and the spacing between water pipes is 1.5m×1.5m. When $\tau = 90d$, water pipes start cooling until the temperature reaches. $T_f = 10℃$. It is computed by finite element. Fig. 14.8-11 presents the stress on the central profile at the end of cooling, from this we can see that the lower the height of the cooling zone is, the bigger the stress is, and it is recommended that b/L is more than or equal to 0.40 in real construction.

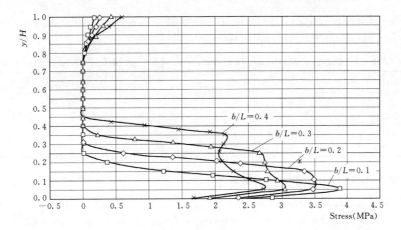

Fig. 14.8-11 Visco-Elastic Thermal Stress on the Central Profile of Pouring Block after Cooling at Different Cooling Heights

Cooling rate also has an impact on the stress. More intensive water pipes and more quick temperature reduction make the thermal stress become larger without fully playing the role of concrete creep.

Fig. 14.8-12 shows an computing example. Both H and L are 60m, b/L is 0.30 and the clock is cooled from the initial temperature $T_0 = 30℃$ to $T_f = 10℃$. It can be seen from the figure that the spacing between water pipes has a great effect on the thermal stress, which is actually very important but ignored before. In order to meet the requirement for allowable tensile stress, suitable measures must be taken, such as adjusting cooling water temperature or closing part of water pipes after the highest temperature of concrete has appeared.

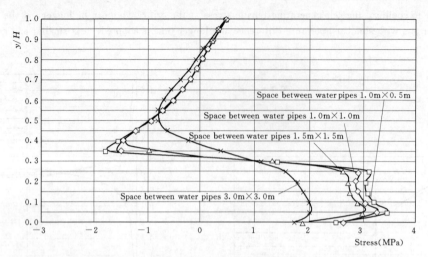

Fig. 14.8-12 Distribution of Stress on the Central Profile of Concrete Block after Cooling by Water Pipes with Different Spacing

14.8.5 Local Intensification of Pipe Cooling

14.8.5.1 Big Latent Capacity of Temperature Control by Closely Laid Pipes

The plastic pipes featuring few joints and flexibility can be laid not only on the surface of old concrete but also during concrete pouring and its vertical and horizontal spacing can be adjusted as needed. Full use of such advantage of the plastic pipe as small spacing can remarkably intensify concrete cooling and reduce concrete temperature stress in order to prevent cracks and to quicken the construction speed of the concrete dam.

Computing Example: Concrete initial temperature $T_0 = 0°C$ and $\theta(\tau) = 25(1 - e^{-0.40\tau})$, concrete maximum temperature T_1 can be computed by the formula (14.8-8), as shown in the Table 14.8-1. When the water temperature T_w is 7°C, and the spacing between water pipes are 1.0m×1.0m, 1.0m×0.5m and 0.5m×0.5m, the maximum temperature of concrete are respectively 25.1°C, 20.7°C and 16.4°C.

When the water pipes are closely laid, the maximum temperature of concrete can be controlled between 16°C and 21°C. It is thus clear that closely laid pipes have very big cooling potential.

Table 14.8-1 Concrete Maximum Temperature T_1 (°C) at Different Water Pipe Spacing

Water pipe spacing(m×m)		0.5×0.5	1.0×0.5	1.0×1.0	1.5×1.5	3.0×3.0
T_w (°C)	12°C	19.66	23.28	27.00	30.72	34.67
	7°C	16.44	20.68	25.07	29.51	34.22
	2°C	13.86	18.36	23.32	28.37	33.78

14.8.5.2 Effect of Closely Laid Pipes in Local Region

[Computing Example] The concrete dam block is 60m in length and 60m in height, and its pouring layer is 3m in thickness. The time interval is 7 days, and both the initial temperature T_0 and the air temperature are 25°C. Within the range 6m above the bedrock surface, the space between water pipes is 1.0m×0.5m and the other part is 1.5m×1.5m. Cooling water temperature is respectively 20°C, 15°C and 11.5°C at the early stage, the intermediate stage and the late stage; and according to finite element stimulation computation, thermal stress is less than the allowable value, as shown in Fig. 14.8-13 and Fig. 14.8-14. Local intensification of pipe cooling costs less but the effect is remarkable.

14.8.6 Pipe Cooling of Concrete Dam from Earlier Age with Smaller Temperature Difference and Longer Time

The maximum temperature of concrete is $T_p + T_r$ during the construction which reduced to the steady thermal T_f before joint grouting, with the temperature difference $\Delta T = T_p + T_r - T_f$. One way to reduce the thermal stress is to lower the maximum temperature, so as to make the temperature difference ΔT decline, and another way is to disperse the temperature difference in order to achieve multiple small temperature differences and prolong the cooling Time. Cooling can be carried out earlier due to small temperature difference. The mode of earlier cooling with small temperature difference and longer time can make the creep fully play its role so as to reduce the thermal stress without affecting the construction progress.

As shown in the formula (14.8-19), the thermal stress consists of the self-generating stress $\sigma_1(t)$ and the restraint stress $\sigma_2(t)$ which will be introduced hereinafter.

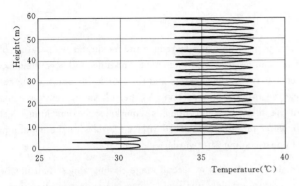

Fig. 14.8-13 Envelope of Temperature at the Central Profile of the Concrete Block

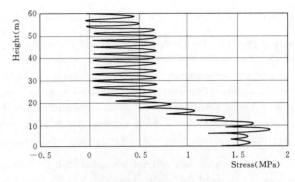

Fig. 14.8-14 Envelope of Thermal Stress at the Central Section of the Concrete Block (Allowable Tensile Stress is 2.1MPa)

14.8.6.1 Reduction of Self-Generating Local Thermal Stress of Pipe Cooling Through Dispersion of Temperature Difference

Computing Example: Supposing the initial temperature of concrete T_0 is 30℃, the steady temperature of the dam T_f is 10℃, the spacing between water pipes is 1.5m×1.5m, $a=0.10\text{m}^2/\text{d}$, and the inner radius and the outer radius of the polyethylene water pipe are 14mm and 16mm respectively, there are four computational schemes and the grades of water temperature is 1, 2, 3 and 6. The maximum Visco-elastic creep stress at the orifice edge is presented in the Table 14.8-2, from which we can see that 6 grades of water temperature enable the maximum elastic stress to drop to 0.97MPa from 4.82MPa.

Table 14.8-2 Maximum Elastic Creep Stress at the Orifice Edge   Unit: MPa

Grades of water temperature	1	2	3	4
Stress at orifice edge (MPa)	4.82	2.50	1.97	0.97

After initial stage of cooling, the compressive stress ranging from 0.5MPa to 1.0MPa will be left at the edge of the water pipe, which can reduce the local tensile stress during the late stage of cooling but can not eliminate it. If the water temperature at late stage is not graded, very big local tensile stress will lead to cracks.

14.8.6.2 Reduction of Total Thermal Stress of Pipe Cooling through Dispersion of Temperature Difference

Computing Example: The dam block is 60m both in length and in height, its pouring layer is 3.0m in thickness, the time interval between layers is 7 days, the adiabatic temperature rise $\theta(\tau)=25\tau/(1.7+\tau)$℃, all the pouring temperature T_p, the air temperature T_a and the bedrock initial temperature are 25℃. The water pipe spacing are 1.0m×0.5m when $y=0-6\text{m}$ and 1.5m×1.5m when $y=6-60\text{m}$. The modes of cooling parts are: ①Two-stage cooling: during the first-stage cooling, $\tau=0-20\text{d}$ and $T_w=20$℃; and cooling at the second stage starts from $\tau=240\text{d}$ until the temperature of concrete is reduced to $T_f=12$℃, with the water temperature $T_w=10$℃. ②Three-stage cooling: the water temperature is respectively 20℃, 15℃ and 10℃. ③Four-stage cooling: the water temperature is respectively 20℃, 15℃, 13℃ and 11℃. Calculated results got by finite element stimulation computation is given in the Table 14.8-3, from which it can be seen that the total thermal stress can be reduced to 1.72MPa from 2.73MPa after temperature difference dispersion.

The medium-stage cooling can be conducted earlier owing to small temperature difference, thus temperature difference dispersion has no impact on the construction progress.

Table 14.8-3 Cooling Modes and Maximum Tensile Stress in the late stage   Unit: MPa

Cooling modes	Two-stage cooling	Three-stage cooling	Four-stage cooling
Maximum tensile stress	2.73	1.90	1.72

In the past, concrete initial temperature and water temperature difference $T_0 - T_w$ is less than or equal to the value ranging from 20℃ to 25℃, which is too big and is actually equivalent to a cold shock. Water temperature shall be divided into several grades in each cooling period to realize gradual reduction. It is recommended to control $T_0 - T_w \leqslant 8 - 10$℃.

14.8.7 Three Principles of Pipe Cooling

The IWHR raised three water pipe cooling principles as follows:

The first principle: Initial cooling immediately starts after concrete pouring in order to reduce the maximum temperature of concrete; the flexibility of the plastic pipe and intensive water pipes can be used in strong constraint zones to lower the maximum temperature of concrete and reduce the temperature difference between the concrete and the foundation and the one between the upper and the lower parts of concrete so as to increase the anti-cracking safety; As the volume of concrete using intensive cooling pipe is small, the total cost of pipe is low, but the intensive water pipe has remarkable anti-cracking effect despite narrow distribution and less expense. A part of intensive water pipes can be closed after 2 - 3 days of the maximum temperature of concrete to prevent bigger tensile stress caused by fast cooling afterwards.

The second principle: the constraint condition shall be improved. The height of the second stage cooling zones shall not be less than 0.4 times of the length of dam block and a certain temperature gradient shall be formed in the vertical direction.

The third principle: The new mode of cooling from earlier age with small temperature difference and longer time is adopted to disperse the temperature difference and reduce the tensile stress. Generally there shall be 1 - 2 intermediate-stages cooling which happens before the post-stage cooling and after the early-stage cooling and the difference between concrete initial temperature and cooling water temperature shall not exceed 10℃.

14.9 Comprehensive Temperature Control and Crack Control Measures

14.9.1 Improvement of Concrete Crack Resistance through Material Optimization

As early as the 1950s China had built its first concrete dam, Foziling Multi-Arch Dam, where a concrete laboratory was established under the guidance of the academician Wang Huzhen and consultation of the academician Wu Zhongwei, making a significant contribution to concrete material optimization and mixture of composite materials and admixture; Thereafter, material studies were conducted in almost all large and medium-sized concrete dams with remarkable effects. The application of fly ash and water reducing agent results in a significant decline in the amount of cement. Now the content of mixture of fly ash is around 30% in normal concrete and 50%-60% in roller compacted concrete. These measures drive remarkable decline of adiabatic temperature rise of concrete.

The technology of concrete admixed with MgO, which is developed by China independently, is applied to the medium-and small-sized concrete dams.

The new concept of semi-mature age of concrete, provided by the IWHR, develops a new way to alleviate the thermal stress of concrete.

14.9.2 Superficial Thermal Insulation of Concrete

Cracks of concrete dams are mostly surface cracks, of which some may develop into deep cracks or penetrating cracks under a certain condition, affecting the structural integrity and durability and resulting in very big danger. The main way to prevent surface cracks is superficial thermal insulation.

Before the 1980s, China mainly adopts straw bags, mats and the like as thermal insulation materials, but they will decay when becoming damp and have such disadvantages as difficult fixation, bad thermal insulation effect and short thermal insulation time, thus there exist lots of surface cracks in real projects. In 1985, it is the first time to successfully apply polystyrene foam boards in the construction of the Jinshuitan Arch Dam for surface protection, hereafter large and mediatesized water conservancy and hydropower projects mainly adopt polystyrene foam boards for surface thermal insulation, realizing good thermal insulation effect, convenient construction, low cost and integration of thermal insulation and moisture preservation functions. Foam boards include of polystyrene foam board, polythene foam board, polyurethane foam board, polrvinyl chloride foam board, phenolic aldehyde foam board, urea formaldehyde foam board, urinary nitrogen foam board, etc., and the first three kinds are mainly used for concrete thermal insulation and their property indexes are presented in the Table 14.9-1.

Chapter 14 Temperature Control Technology for Dam Concrete

Table 14.9 – 1 Physical and Mechanical Properties of Foamed Plastics

Variety	Density (kg/m³)	Coefficient of heat conductivity (kJ/(m·h·℃))	Water absorption (%)	Compression strength (kPa)	Tensile strength (kPa)
Expansive PolyStyrene (EPS)	15 – 30	0.148	2 – 6	60 – 280	130 – 140
Extruding PolyStyrene (XPS)	42 – 44	0.108	1	300	500
PolyEthylene (PE)	22 – 40	0.160	2	33	190
PolyUrethane fat (PUF)	35 – 55	0.080 – 0.108	1	150 – 300	500

Supposing a layer of foamed plastic insulation board is arranged outside concrete and the coefficient of heat conductivity $\lambda =$ 0.1256kJ/(m·h·℃), the surface heat insulation effect can be computed with the heat conduction theory, as shown in the Table 14.9 – 2. It is thus clear that the surface heat insulation effect of concrete is obvious. For example, when the thickness of the foamed plastic board is 3cm, yearly surface temperature range of concrete can reduce by 44% and daily range by 95%, and surface temperature range of concrete during cold wave can cut by 75% – 87% (affected by cooling duration Q).

Table 14.9 – 2 Thickness and Surface Thermal Insulation Effect of Insulation Board

Thickness of foamed plastic board (cm)	Equivalent surface heat transfer coefficient β_s (kJ/(m·h·℃))	Virtual concrete thickness (m)	Ratio A_0/A of concrete surface temperature range to air temperature change			Ratio of amplitude of variation of temperature of concrete surface to that of air temperature during cold wave				
			Daily change	Half-month change	Annual change	$Q=1d$	$Q=2d$	$Q=3d$	$Q=4d$	$Q=5d$
0	82.2	0.11	0.580	0.856	0.968	0.774	0.830	0.857	0.874	0.886
1	10.89	0.83	0.137	0.400	0.790	0.298	0.377	0.427	0.464	0.493
2	5.83	1.54	0.077	0.254	0.657	0.183	0.242	0.282	0.313	0.338
3	3.98	2.26	0.054	0.186	0.558	0.132	0.177	0.210	0.235	0.256
5	2.44	3.69	0.033	0.1206	0.426	0.0845	0.116	0.139	0.157	0.173
8	1.54	5.84	0.021	0.079	0.312	0.0549	0.0762	0.0919	0.1047	0.1158
10	1.24	7.28	0.017	0.064	0.263	0.0445	0.0619	0.1749	0.0856	0.0949
15	0.829	10.86	0.012	0.043	0.190	0.0302	0.0423	0.0513	0.0589	0.0654
20	0.623	14.45	0.0086	0.0330	0.148	0.0229	0.0321	0.0390	0.0448	0.0499

It is advisable to use polystyrene foamed plastic boards to realize thermal insulation of the upstream and downstream surfaces of the concrete dam. Insulation boards are nailed inside the forms before placing of new concrete and can automatically attach to new concrete surface (internal paste method) during template removal. They can also be pasted on concrete surface (external paste method) with plastering agent after template removal. Polystyrene foamed plastic soft quilt can be set on the horizontal construction layer and the side surface with key, and rainproof color striped clothes are pasted to the surface. The Three Gorge Second-stage Project used to adopt such plastic soft quilt which is 1.5m×2.0m in dimension generally and internally arranged with two layers of 1cm-thick polyethylene boards. The price of the polystyrene foamed plastic board 3cm in thickness ranges from 20 (internal paste method) to 38 (external paste method) yuan/m².

Thermal insulation of a constructed concrete dam can be realized by the polystyrene board and the polyurethane foam coating which is achieved by painting method.

An addition of a protective layer outside a polystyrene board can turn it into a permanent insulation board. From structure point the protective layer comprises: ① the polymer mortar protective layer as shown in the Fig.14.9 – 1 (a) is made in the way of mixing crylic acid, cement and sand and is 5mm in thickness; ② the cement mortar protective layer as shown in the Fig.14.9 – 1 (b) is produced in the manner of mixing ordinary portland cement, water, sand and admixture without adding fly ash, and the water cement ratio ranges from 0.35 to 0.40; and in construction the current polystyrene board is externally coated with a layer of interfacial treatment agent to enhance the bonding effect and then is applied with cement mortar with the thickness of 20mm, and after curing a layer of acrylic emulsion is painted on the surface so as to strengthen waterproof and anti-carbonation effects.

The permanent insulation antiseepage board (as shown in the Fig. 14.9 - 2) developed by the IWHR integrates thermal insulation and antiseepage functions and is composed of plastering agent, polymer cement mortar, antiseepage film, plastering agent, XPS board and protective layer in structure. It is advisable to adopt external paste method in construction, the thickness of the XPS board is calculated and determined by weather conditions, which can be applied to dam heels of the upstream faces of the roller compacted concrete dam and the ultra-high arch dam.

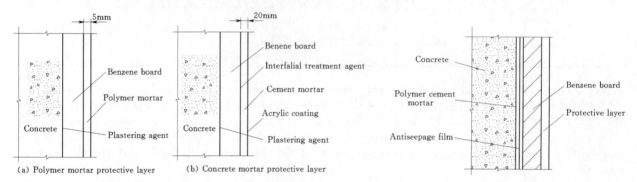

Fig. 14.9 - 1   Permanent Insulation Board and Its Protective Layer

Fig. 14.9 - 2   Heat-Insulation Antiseepage Composite Board

14.9.3   Concrete Precooling

One way to reduce the temperature difference of concrete is to low concrete initial temperature. As for normal concrete, supposing contents of stones, sand, cement and water in concrete are respectively 1,750kg/m³, 480kg/m³, 130kg/m³ and 110kg/m³ and water contents of stones and sand stand at 1% and 3% respectively, according to heat balance, the mixing temperature of concrete at the outlet of concrete mixer:

$$T_0 = 0.631T_g + 0.1896T_s + 0.0446T_c + 0.1342(1-p)T_w - 10.74\eta p \qquad (14.9-1)$$

where, T_g, T_s, T_c and T_w are respectively the temperature of stone, sand, cement and water, p represents the ice mixture ratio, η is the effective coefficient of the ice mixture, ranging from 0.75 to 0.85. It can be seen from the formula (14.9 - 1) that the most effective way to low T_0 is to reduce stone temperature T_g and secondarily sand temperature T_s, and cement temperature has small impact. If the ice mixture ratio p is 0.40 and the effective coefficient η is 0.80, ice mixture of ice can make the temperature reduce by 6.0℃. Thus reduction of concrete mixing temperature at the batch plant mainly relies on stone precooling and ice mixture, sand precooling is inconvenient due to its small size and it is seldom used in real projects.

In the late 1940s, U.S. started developing concrete precooling technology, which was successfully applied to the Gezhouba Hydropower Station in China for the first time. At present concrete precooling is used in large and medium-sized projects. For Conventional concrete, the mixing temperature at the outlet of batch plant can be controlled to 6 - 7℃, while concrete placing temperature will around 12℃ due to temperature rise during concrete transportation and placing. As RCC temperature rise is a bit high in construction and less ice mixture amount, and currently concrete placing temperature can be controlled about 17℃ after precooling.

14.10   Long Time Superfical Thermal Insulation with Comprehensive Temperature Control and End of the History of "No Dam without Crack"

People has attached great importance to temperature control and anti-cracking of the concrete dam starting from the 1930s, but actually at the beginning of this century the problem is "no dam without crack". One of main reasons is that lots of cracks always emerge during the construction of the concrete dam after a big cold wave, thus for quite long time, people only pay attention to early surface heat insulation of concrete and ignore post-stage surface protection. The *Specifications of Hydraulic Concrete Construction* stipulates (DL/T 5144—2001) that concrete within the age of 28d shall be subject to surface protection prior to temperature sudden drop. It gives people an illusion that generally there is no need to carry out surface thermal insulation to concrete after the age of 28d except for some special situations. But that is not true. For instance, lots of cracks occurring on the upstream and downstream faces of a gravity dam are generated in winter of the first and the second years after concrete pouring rather than within the age of 28d. Simulation computation results of such dam are seen in the Table 14.10 - 1. If no thermal insulation measure is taken, both the horizontal and the vertical tensile stress in the first and the second winters exceeds the allowable stress.

Table 14.10 - 1 Maximum Tensile Stress of the Upstream Face of a Gravity Dam in Winter

Thickness of polystyrene boards at upstream and downstream faces (cm)	0	5	3	Allowable tensile stress (post Stage) (MPa)
Thickness of polyethylene insulation quilt on the horizontal plane (cm)	0	2	1	
Horizontal tensile stress (MPa)	4.2	1.6	1.9	2.2
Vertical tensile stress (MPa)	2.6	−0.1	0.1	1.33

In order to completely prevent concrete cracks, long-term surface heat insulation of concrete must be carried out, and the application of foamed plastic realizes not only technical feasibleness but also less expense.

According to practical experience, the theory of thermal stress is correct. Comprehensive strict control over the thermal stress and long-term heat insulation can avoid cracks, ending the history of "no dam without crack". Of course, as for different dam types and dam parts, anti-crack requirements are diverse.

14.11 Anti-crack Safety Factor

As for concrete anti-cracking, the allowable tensile stress can be computed with the following two formula

$$\sigma \leqslant \frac{E\varepsilon_p}{K_1} \quad (14.11-1)$$

or

$$\sigma \leqslant \frac{R_t}{K_2} \quad (14.11-2)$$

where, ε_p is extensibility and R_t is tensile strength of concrete. Since test results of concrete extensibility is not very stable due to its smaller deformation while test result of concrete tensile strength is more stable, the formula (14.11 - 2) is more suitable, but at present design criterion adopts the formula (14.11 - 1).

Lower anti-crack safety factor is one of important causes of concrete cracks. In recent years the adoption of plastic cooling water pipes featuring small spacing can effectively reduce the temperature difference between the concrete foundation and the upper and the lower parts of the dam block and surface heat insulation with foamed plastic can realize effective lowering of the internal and the external temperature difference. Hence, at present it is possible to increase the anti-crack safety factor properly and reduce the possibility of concrete crack maximally.

The author presents a set of methods to determine concrete anti-crack safety factor in the literature. It is recommended to use the following anti-crack safety factors at the preliminary design:

$$\begin{aligned} &\text{Calculated by the formula}(14.1-1): K_1=1.6-2.2 \\ &\text{Calculated by the formula}(14.1-2): K_2=1.4-1.9 \end{aligned} \quad (14.11-3)$$

Compared with the past criterion in which K_1 is from 1.3 to 1.8, the above safety factor is greatly increased and according to our experience, it can remarkably reduce the number of cracks, and its corresponding temperature control measures are achievable in real projects and take both demands and possibility into account in current condition. According to the suggestion of the author on properly increasing the anti-crack safety factor, $K_1 = 1.5 - 20$ is adopted in the new-edition design specification for the concrete gravity dam. From the perspective of development, in future it shall be in gradual transition to adopt the formula (14.11 - 2) for anti-crack computation, which has two advantages: ①tensile strength test result is more stable; and ②in the period of construction generally there is no extensibility test and only tensile strength test, the difference between the real tensile strength and the design adoption value can be got in construction period, thus through the formula (14.11 - 2), temperature control measures can be adjusted in necessity.

The author suggests that the vertical stress on the horizontal construction joint surface is controlled as the following formula:

$$\sigma_y \leqslant \frac{cR_t}{K_2} \quad (14.11-4)$$

c is the tensile strength reduction factor on the joint surface, ranging from 0.5 to 0.7.

14.12 Conclusion

Chinese scholars have established a set of complete theoretical systems for concrete temperature control and anti-crack, including basic theories and computational methods of concrete temperature field and visco-elastic thermal stress, laws of variation and features of thermal stress of various mass concrete structures, as well as technical measures needed in different conditions.

New technical measures such as surface protection with foamed plastic and small spacing of plastic water pipes remarkably enhance temperature control and anti-crack effects.

It is possible to prevent concrete cracks in current conditions, but thermal stress is complex, so we shall not take it lightly and must meticulously undertake design and construction to effectively prevent cracks from occurring.

References

[1] Zhu Bofang. Temperature Calculation of the Concrete Dam. *China Water Resources*, 1956 (11): 8-20. (12): 43-60.

[2] Zhu Bofang. Calculation of Cooling of Mass Concrete with Internal Heat Source by Laying Water Pipes. *Journal of Hydraulic Engineering*, 1957 (4): 87-106.

[3] Compiled by U.S. Department of the Interior, Bureau of Reclamation, Translated by Hou Jiangong. *Cooling of Concrete Dams*, China WaterPower Press, 1958.

[4] Pan Jiazheng. *Calculation for the Concrete Dam Temperature Control*, Shanghai: Shanghai Scientific & Technical Publishers, 1959.

[5] Zhu Bofang, Wang Tongsheng, Ding Baoying. Thermal Stress of the Gravity Dam and the Concrete Pouring Block. *Journal of Hydraulic Engineering*, 1964 (1): 30-34.

[6] Zhu Bofang, Song Jingting. Finite Element Analysis on Concrete Temperature Field and Visco-Elastic Thermal Stress. Selected Anthology of Electronic Computer of Water Resources and Hydropower Engineering Applications. Beijing: WaterPower Press, 1977.

[7] Zhu Bofang. Temperature Stress of the Foundation Beam. *Chinese Journal of Theoretical and Applied Mechanics*, 1979 (3): 200-205.

[8] Zhu Bofang. Thermal Stress of the Dock and the Sluice on the Soft Foundation. *Journal of Hydraulic Engineering*, 1980 (6): 23-33.

[9] Zhu Bofang. On Temperature Loads of the Arch Dam. *Water Power*, 1984 (2): 23-29.

[10] Zhu Bofang. Estimation of Reservoir Water Temperature. *Journal of Hydraulic Engineering*, 1985 (2): 12-21.

[11] Zhu Bofang. The Equivalent Heat Conduction Equation of Concrete Considering Effect of Pipe Cooling. *Journal of Hydraulic Engineering*, 1991 (3): 28-34.

[12] Zhu Bofang. Temperature Control and Joint Design of the Roller Compacted Concrete Arch Dam. *Hydroelectric Power*, 1992 (9): 11-17.

[13] Zhu Bofang. Method Using Different Time Increments in Different Regions for solving Unstable Temperature Field. *Journal of Hydraulic Engineering*, 1985 (8): 46-52.

[14] Zhu Bofang, Xu Ping. Some Problems of Trasverse Cracks and Utra-cooling of Bottom Outlet in Concrete Gravity Dams without Longitudinal Joint, *Water Resources and Hydropower Engineering*, 1998 (10): 14-18.

[15] Zhu Bofang. Effect of Superficial Thermal Insulation for Pipe Cooling in Hot Seasons. *Water Resources and Hydropower Engineering*, 1997 (4): 10-13.

[16] Zhu Bofang. Calculation of Nonmetal Water Pipe Cooling of Mass Concrete. *Water Resources and Hydropower Engineering*, 1997 (6): 30-34.

[17] Zhu Bofang. Stress and Displacement of the Non-homogeneous Viscoelastic Body in Mixed Boundary Conditions. *Chinese Journal of Theoretical and Applied Mechanics*, 1964 (2): 162-167.

[18] Zhu Bofang. Implicit Method for the Stress Analysis of Concrete Structures Considering Effect of Creep. *Journal of Hydraulic Engineering*, 1983 (5): 40-46.

[19] Zhu Bofang. A Numerical Method Using Different Time Increments in Different Regions for Analysing Stresses in Elasto-Creeping Solids. *Journal of Hydraulic Engineering*, 1995 (7): 24-27.

[20] Zhu Bofang. Compound Layer Method for Stimulation Stress Analysis of Multilayer of Concrete Structure. *Journal of Hydroelectric Engineerin*, 1993 (3): 21-30.

[21] Zhu Bofang. On Construction of Dams by Concrete with Gentle Volume Expansion. *Journal of Hydroelectric Engineerin*, 2000 (3): 1-12.

[22] Zhu Bofang, Zhang Guoxin, Xu Linxiang, et al. New Concept and New Techniques for Solving the Thermal Stress Problem in the Heightening of Concrete Gravity Dam. *Water Power*, 2003 (11): 29-33.

[23] Zhu Bofang. Temperature Loads of the Arch Dam with Insulating Layers in Cold Region. *Water Resources and Hydropower Engineering*, 2003 (11): 46-49.

[24] Zhu Bofang. Deep Upstream Cracks of the Gravity Dam. *Journal of Hydroelectric Engineering*, 1997 (4): 86-94.

[25] Zhang Guoxin. Nonlinear Simulation of the Crack of the MgO Gentle-Expansive Concrete Arch Dam. *Journal of Hydroelectric Engineering*, 2004 (3): 51-55.

[26] Zhu Bofang. Thermal Stress of the Concrete Beam on the Subgrade. *China Civil Engineering Journal*, 2006 (8): 99-104.

[27] Zhu Bofang. Two Guiding Ideologies and Two Practice Results of Applying MgO Concrete to Dam Construction. *Water Resources and Hydropower Engineering*, 2005 (6): 42-45.

[28] Zhu Bofang, Mai Shufang. Composite Permanent Thermal Insulation and Antiseepage Board of the Concrete Dam. *Water Resources and Hydropower Engineering*, 2006 (4): 16-21.

[29] Zhu Bofang. Improvement of Calculation Method of Arch Dam Temperature Load. *Water Resources and Hydropower Engineering*, 2006 (12): 22-25.

[30] Zhu Bofang, Xu Ping. Strengthen Superficial Insulation of Concrete Dam to Terminate the History of "No Dam without Crack". *Water Power*, 2004 (3): 28-31.

[31] Zhang Guoxin. Temperature Control of the Roller Compacted Concrete Dam. *Water Resources and Hydropower Engineering*, 2007 (6): 41-46.

[32] Zhang Guoxin, Yang Weizhong, Luo Heng, et al. Study and Application of Temperature Drop Compensation of MgO Micro-expansion Concrete for the Sanjianghe Arch Dam. *Water Resources and Hydropower Engineering*, 2006 (8): 20-23.

[33] Zhang Guoxin, Chen Xianmin, Du Lihui. Reaction Kinetic Model for MgO Concrete Expansion. *Water Resources and Hydropower Engineering*, 2004 (9): 88-91.

[34] Zhu Bofang, Zhang Guoxin, Xu Ping, et al. Decision Support System for Temperature and Stress Control in the Construction of the High Concrete Dam. *Journal of Hydraulic Engineering*, 2008 (1): 1-6.

[35] Yang Bo, Xu Ping, Zhang Guoxin, et al. Study on Temperature Control Measures of the Jiangkou Arch Dam. *Journal of China Institute of Water Resources and Hydropower Researchs*, 2003, 1 (2): 127-133.

[36] Zhu Bofang, Yang Ping, Wu Longkun, et al. Strengthen the Cooling of Concrete by Polyethylene Pipes with Small Spacing. *Water Resources and Hydropower Engineering*, 2008, 39 (5): 36-39.

[37] Zhu Bofang, Yang Ping. Semi-Mature Age of Concrete—a New Method for Improving the Crack Resistance of Mass Concrete. *Water Resources and Hydropower Engineering*, 2008, 39 (5): 30-35.

[38] Ding Baoying, Hu Ping, Huang Shuping. Approximate Analysis on Reservoir Water Temperature. *Journal of Hydroelectric Engineering*, 1987 (4): 17-33.

[39] Yue Yaozhen. Thermal Stress Analysis Considering the Impact of Temperature on Concrete Property. *Water Resources and Hydropower Engineering*, 1993 (1).

[40] Li Zhanmei. Calculation of Arch Dam Temperature Load, *Proceedings of High Arch Dam Symposium*, Beijing: China WaterPower Press, 1982.

[41] Huang Shuping, Hu Ping. Study on Simulation Calculation of Thermal Stress of Guanyinge Roller Compacted Concrete Dam. *Water Power*, 1996 (10).

[42] Zhu Bofang, Wang Tongsheng, Ding Baoying, Guo Zhizhang. *Thermal Stress and Temperature Control of Hydraulic Concrete Structures*. Beijing: China WaterPower Press, 1976.

[43] Zhu Bofang. *Temperature Stress and Temperature Control of Mass Concrete*. Beijing: China Electric Power Press, 1999.

[44] Gong Zhaoxiong, et al. *Temperature Control and Crack Prevention of Hydaulic Concrete*. Beijing: China WaterPower Press, 1999.

[45] Guo Zhizhang, Fu Hua. *Thermal Control of Hydraulic Concrete*. Beijing: China WaterPower Press, 1990.

[46] Zhu Bofang. *New Developments in the Theory and Technology of Concrete Dams*. China WaterPower Press, 2009.

[47] Zhu Bofang, Zhang Chaoran, et al. *Study on Key Technology of Structural Safety of High Arch Dams*. Beijing: China WaterPower Press, 2010.

[48] Zhu Bofang. On Pipe Cooling of Concrete Dams. *Journal of Hydraulic Engineering*, 2010, 5: 505-513.

[49] Hu Ping, Yang Ping, Zhang Guoxin. Study on the Temperature Control and Crack Prevention of Double-curvature Concrete Arch Dam for Laxiwa Hydropower Station. *Journal of Hydroelectric Engineering*, 2007 (11): 51-54.

[50] Hu Ping, Yang Ping, Zhang Guoxin. Studies on Temperature Control and Crack Prevention of RCC Gravity Dam of Longtan Hydropower Station//*Proceedings of 2007 RCC International Conference*. 2007.

[51] Zhu Bofang. *Academican Zhu Bofang's Selected Works*. Beijing: China Electric Power Press, 1997.

[52] Cao Zesheng, Xu Jinhua. *MgO concrete Dam Construction Technology*. Beijing: China Electric Power Press, 2003.

[53] Xing Deyong, Xu Sanxia. Overeview of the Technology of Temperature Control of Massive Body of the Third Phase Work in Three Gorges Pproject. *Journal of Hydroelectric Engineering*, 2005 (10): 14-16.

[54] Zhu Bofang, Wu Longkun, Yang Ping, Zhang Guoxin. Planning of pipe cooling of concrete dams in later age. *Water Resources and Hydropower Engineering*, 2008 (7): 27-31.

[55] Zhu Bofang. Pipe Cooling of Concrete Dam from Earlier Age with Smaller Temperature Difference and Longer Time. *Water Resources and Hydropower Engineering*, 2009 (1): 44-50.

[56] Zhu Bofang, Wu Longkun, Zhang Guoxin. Numerical Analysis of Thermal Stresses around Cooling Pipe in Concrete Dams. *Water Resources and Hydropower Engineering*, 2009 (2).

[57] Zhu Bofang. On Coefficients of Safety for Crack Prevention of Concrete Dams. *Water Resources and Hydropower Engineering*, 2005 (7): 33-37.

[58] Zou Guangqi, Li Guizhi, Wang Chenshan, Du Zhiyuan. Causes and Treatment Measures of Cracking of the Horizontal Construction Joint on the Overwintering Surface of the Guanyinge Concrete Dam. *Water Resources and Hydropower Engineering*, 1995, 8: 49-53.

[59] Wang Chengshan, Han Guocheng, Lv Hexiang. Study and Application of Preset Cracks in Baishi RCC Gravity Dam. *Journal of Hydraulic Engineering*, 2003, 9: 107-111.

[60] Zhu Bofang, Zhang Guoxin, Xu Linxiang, et al. New Concept and New Techniques for Solving the Thermal Stress Problem in the Heightening of Concrete Gravity Dam. *Journal of Hydroelectric Engineering*, 2003 (11): 26-30.

[61] Zhu Bofang, Zhang Guoxin, Wu Longkun. Study for Reducing the Cracking of the Binding Interface between Fresh and Old concrete in Heightening of the Gravity Dam. *Journal of Hydraulic Engineering*, 2007 (6): 639-645.

[62] *Design specification for Concrete Gravity Dams* (SL 319—2005). Beijing: China WaterPower Press, 2005.

[63] *Design specification for Concrete Gravity Dams* (DL 5108—1999). Beijing: China Electric Power Press, 2000.

Chapter 15

Slope Engineering in Dam Construction

Chen Zuyu[1], Wang Yujie[2]

15.1 Introduction

The rapid development of China's hydropower has resulted in the constructions of a large number of high dams one after another, which in turn created many high slopes. Fig. 15.1-1 shows the 295m high hyperbolic arch dam is located in No.6 mountain ridge slope of Xiaowan hydropower station's left bank. Excavations of the access road at the dam crest and the flood discharge tunnel near the river made this slope up to 530m. Slope stability has been an important concern during the feasibility study, construction and operation of these projects. And reservoirs impoundment creates inundated slopes which have important impacts on environment and resettlement. For example, the Three Gorges reservoir has submerged 600km² bank where more than one million people lived. A large number of unstable slopes have been identified, for which careful reinforcement and monitoring programs have been carried out.

Fig. 15.1-1 The Valley No.6 Slope of Xiaowan Dam

A comprehensive review on the advances in the area of water resources slopes in China has been given in the book "*Large dams in China, A Fifty-Year Review*" (Pan et al, 2000). Entering the new century, a number of high dams, including the Three Gorges, Xiaolangdi, Longtan, Hongjiadu, Shuibuya, etc. have been brought to operation. These large projects have witnessed the successful planning, construction and operation of their involved high slopes. In order to provide guidance to the routine design procedures for engineered slopes, the Ministry of Water Resources and the State Planning Council have issued design codes numbered SL 386—2007 and DL/T 5353—2006 for water resources and hydropower based engineered slopes respectively. This Chapter will mainly focus on the achievements and experience obtained in the slope engineering during the past 10 years.

15.2 Engineering Geology

High slopes always involve complicated geological conditions. Different geological features may cause different types of slope failures. It is of prime importance to carry out detailed geological exploration, to understand the particular geological characteristics and to provide appropriate geotechnical parameters for slope design.

Adits and bore holes are normal tools for geological exploration. Geological exploratory adits are extensively used for hydropower projects. They provide more reliable information of rock quality and the structural discontinuities that control stability of slopes.

---

[1] Chen Zuyu, Academician of Chinese Academy of Science, China Institute of Water Resources and Hydropower Research, Professor Senior Engineer.

[2] Wang Yujie, China Institute of Water Resources and Hydropower Research, Professor Senior Engineer.

China's hydro community normally adopts a rock mass classifications system that involves 5 grades. Based on the rating provided by this system, a range of shear strength parameters are recommended. DL/T 5353—2006 also recommendsthe sheer strength indexes from Hoek-Brown's criterion and GSI system (Hoek and Bray, 1977; Hoek and Brown, 1980, 1988; Marinos and Hoek, 2000).

Both SL 386—2007 and DL/T 5353—2006 emphasize the importance of identifying the potential failure types of a slope, based on the orientations of the slope surface and the discontinuities involved in the rock mass. A simple judgment method that allows a quick identification of possible planar, circular, wedge sliding and toppling failure mode has been recommended by Hong Kong Geotechnical Office. SL 386—2007 specifies the application of this approach. Fig. 15.2-1 shows its typical illustrative example (GEO, 1981; Chen et al, 2005).

It has been found that discovering creeping alluvium deposits and fossil landslides during the reconnaissance of geological surveying is also very important. Landslides were triggered right after the excavation of slopes due to the existence of weak seams underlying the alluvium deposit and fossil landslide. Fig. 15.2-2, taken from the Yongdingqiao Project, shows a tree that was unfortunately located at the point where the creeping alluvium deposit just started moving away. It tore the tree into two pieces.

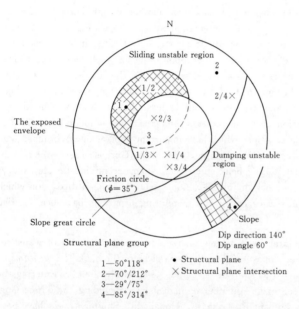

Fig. 15.2-1 Identification of Possible Types of Slope Failures

Fig. 15.2-2 A Tree at the Yongdingqiao Project that was Torn due to the Creeping Alluvium Deposit

The authors would like to share the following experience in identifying a fossil landslide.

The topography of land settlement may still exist in some fossil landslides. Fig. 15.2-3 shows a landslide on the left abut-

Fig. 15.2-3 The Left Abutment of Hongmei Reservoir

ment of Hongmei Reservoir of Yunnan Province. The designers neglected the obvious trace of ground settlement developed by an old landslide, which was reactivated due to the reservoir filling.

If a slope is suspected to be a fossil landslide, geological exploratory adit sare strongly suggested as the most direct and most important mean. In the adits, geologists should carefully map the orientations of the joints and bedding planes. In a fossil landslide, these orientations may exhibit severe randomness, compared to those in the nearby intact rocks. It is very common that in the adits, slicken seams at the slip surface appear. Fig. 15.2-4 shows the muddy sheared seams found in the adits of the Hongmei Reservoir.

In the fossil landslide of Baise Reservoir, the bore hole loggingsor the walls of excavated piles show a distinct change in color and quality at the location of slip surface (Fig. 15.2-5). The underneath parent rock is fresh while that in the overlying fossil landslide is yellow and weathered. This may be an important indication of a fossil landslide.

Fig. 15.2-4 The Slicken Side Taken from the Sheared Seams of the Hongmei Reservoir

Fig. 15.2-5 The Distinct Change in Color and Quality Discovered in Exploratory Work of the Fossil Landslide of Baise Reservoir

It is advantageous to install some field monitoring facilities during the geological reconnaissance stage, which may provide important information on possible movement of slopes. Fig. 15.2-6 shows the slope indicator readings of the alluvium of theYongdingqiao Project. In only 3 months, about 30mm horizontal displacement has been discovered. This information greatly helps for the decision-making.

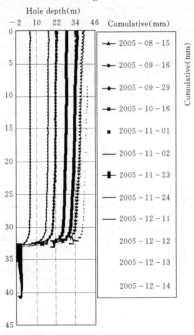
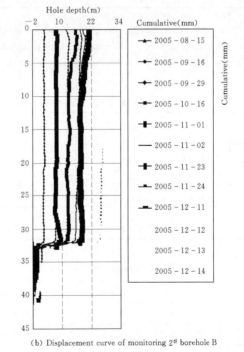

(a) Displacement curve of monitoring 2# borehole A   (b) Displacement curve of monitoring 2# borehole B

Fig. 15.2-6 The Slope Indicator Readings of the Alluvium Deposit of theYongdingqiao Project

15.3 Slope Stability Analysis and Safety Assessment

15.3.1 Slopes Classifications and Allowable Safety Factor

Quantitative analysis for an engineered slope to ensure its sufficient safety provision, normally symbolized by the factor of safety, is a design routine. SL 386—2007 and DL/T 5353—2006 have specified guidelines for this work.

For a particular slope in hydraulic structures, the first step of this work is to define the class of this slope and find its allowable factor of safety. Table 15.3-1 shows the classification specified by SL 386—2007 based on the importance of the hydraulic structure, also symbolized by its grade, and the damages if the slope ever fails. When the grade of a slope is defined, the safety allowable factor can be found in Table 15.3-2 based on the working conditions the slope may possibly encounter. The detailed explanations for these working conditions can be found in SL 386—2007.

Table 15.3-1 The Comparison between the Slope Grading and the Degree of the Hydraulic Structure

The degree of the hydraulic structure	Grading consequence of failure			
	Very serious	Serious	Not serious	Slight
	Slope stability levels			
1	1	2	3	4, 5
2	2	3	4	5
3	3	4	5	5
4	4	5	5	5

Note: 1. Very serious: The related hydraulic structure is completely demolished or totally lost the function.
2. Serious: The related hydraulic structure is demolished or lost the functions to some extent, and special remedial works are required to be recovered.
3. Not serious: The safety of related hydraulic structure is affected but can be repaired.
4. Slight: The safety of related hydraulic structure is sslightlyor indirectly affected.

Table 15.3-2 The allowable safety factors against sliding, SL 386—2007

Operating Condition	Grade of the slope				
	1	2	3	4	5
Normal	1.30-1.25	1.25-1.20	1.20-1.15	1.15-1.10	1.10-1.05
Unusual I	1.25-1.20	1.20-1.15	1.15-1.10	1.10-1.05	1.10-1.05
Unusual II	1.15-1.10	1.10-1.05	1.10-1.05	1.05-1.00	1.05-1.00

DL/T 5353—2006 requires identifying a slope in a hydropower project whether it belongs to a structure slope or a reservoir impounded one, as shown in Table 15.3-3. The three grades listed in Table 15.3-3 are determined based on their impacts on the hydraulic structure with their individual grades. The required minimum factors of safety are specified as shown in Table 15.3-4.

Table 15.3-3 Classifications of Slopes by DL/T 5353—2006

Grade \ Class	Class A — Hydraulic structure slope	Class B — Reservoir slope
I	Affect Class 1 hydraulic structures	Possible to induce reservoir surge and landslide disasters, and slope reinforcement have to be required
II	Affect Class 2 and Class 3 hydraulic structures	Possible to induce landslide but not endanger lives or structures
III	Affect Class 4 and Class 5 hydraulic structures	Partial instability or slow movement can be tolerated

Table 15.3-4 Minimum Factor of Safety Specified by DL/T 5353—2006

Class \ Class and case	Class A — Hydraulic structure slope			Class B — Reservoir slope		
	Normal	Exceptional	Special	Normal	Exceptional	Special
I	1.30-1.25	1.20-1.15	1.10-1.05	1.25-1.15	1.15-1.05	1.05
II	1.25-1.15	1.15-1.05	1.05	1.15-1.05	1.10-1.05	1.05-1.00
III	1.15-1.05	1.10-1.05	1.00	1.10-1.00	1.05-1.00	≤1.00

15.3.2 Slope Stability Analysis Methods

A variety of slope stability analysis methods are available in literatures. SL 386—2007 and DL/T 5353—2006 have agreed to provide unified specifications for hydro-project slopes, which can be highlighted as follows.

(1) For soil slopes without controlled structural continuities plane, Bishop's simplified method (Bishop, 1955) with circular slip surfaces is recommended.

(2) The generalized method of slices proposed by Morgenstern and Price (Morgenstern and Price, 1965) has been specified for slopes with non-circular slip surfaces. Its updated formations by Chen and Morgtenstern (1983) are presented in the codes (See Fig. 15.3 – 1). The force and moment equilibrium equations are respectively:

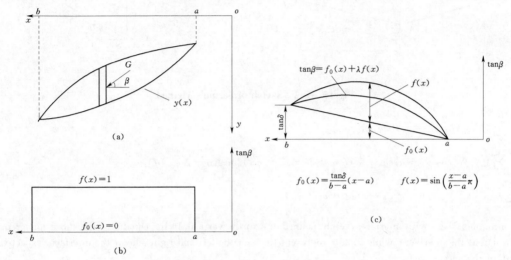

Fig. 15.3 – 1 Sketch of the Generalized Method of Slices Proposed by Morgenstern and Price

$$\int_a^b p(x)s(x)\mathrm{d}x = 0 \qquad (15.3-1)$$

$$\int_a^b p(x)s(x)t(x)\mathrm{d}x - M_e = 0 \qquad (15.3-2)$$

in which

$$p(x) = \left(\frac{\mathrm{d}W}{\mathrm{d}x}+\frac{\mathrm{d}V}{\mathrm{d}x}\right)\sin(\tilde{\varphi}'-\alpha) - u\sec\alpha\sin\tilde{\varphi}' + \tilde{c}'\sec\alpha\cos\tilde{\varphi}' - \frac{\mathrm{d}Q}{\mathrm{d}x}\cos(\tilde{\varphi}'-\alpha) \qquad (15.3-3)$$

$$s(x) = \sec(\tilde{\varphi}'-\alpha+\beta)\exp\left[-\int_a^x \tan(\tilde{\varphi}'-\alpha+\beta)\frac{\mathrm{d}\beta}{\mathrm{d}\zeta}\mathrm{d}\zeta\right] \qquad (15.3-4)$$

$$t(x) = \int_a^x (\sin\beta - \cos\beta\tan\alpha)\exp\left[\int_a^\xi \tan(\tilde{\varphi}'-\alpha+\beta)\frac{\mathrm{d}\beta}{\mathrm{d}\zeta}\mathrm{d}\zeta\right]\mathrm{d}\xi \qquad (15.3-5)$$

$$M_e = \int_a^b \frac{\mathrm{d}Q}{\mathrm{d}x}h_e\,\mathrm{d}x \qquad (15.3-6)$$

$$\tilde{c}' = \frac{c'}{K} \qquad (15.3-7)$$

$$\tan\tilde{\varphi} = \frac{\tan\varphi'}{K} \qquad (15.3-8)$$

$$\tan\beta = \lambda f(x) \qquad (15.3-9)$$

where, K is factor of safety; α is inclination of the slice base; β is the inclination of the inter-slice force. $\mathrm{d}W/\mathrm{d}x$ is weight of the slice per unit width; Q is vertical surface load; $\mathrm{d}Q/\mathrm{d}x$, $\mathrm{d}V/\mathrm{d}x$ is horizontal and vertical force per unit width respectively; h_e is the distance between the horizontal seismic force and base of the slice; u is pore pressure applied on the slice base; $f(x)$ is a function that describes shape of $\tan\beta$, leaving a coefficient to be determined together with K by analytical expressions of (15.3 – 1) and (15.3 – 2).

(3) Sarma's method with non-vertical slices is recommended for rock slopes with sub-vertical joints. The original approach

by Sarma (1979) is based on the force equilibrium conditions (Fig. 15.3 – 2), which has complex recurrence formulations. The codes also presented the formulation proposed by Donald and Chen (1997) that is identical to Sarma's approach but with a simple and explicit equation as the following one.

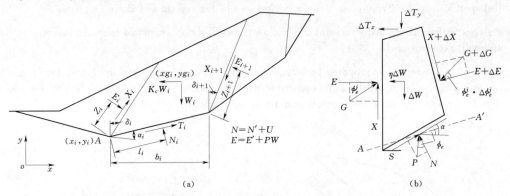

Fig. 15.3 – 2 Sketch of Sarma's Method

$$\sum_{i=1}^{n} \lambda_i [(c_{bi} \cos \widetilde{\varphi}'_{bi} - u_{bi} \sin \widetilde{\varphi}'_{bi}) b_i \sec \alpha_i]$$
$$+ \sum_{i=1}^{n-1} \lambda_{i+1} [(c_{si} \cos \widetilde{\varphi}'_{si} - u_{si} \sin \widetilde{\varphi}'_{si}) \sec(\alpha_i + \delta_i - \widetilde{\varphi}'_{bi} - \widetilde{\varphi}'_{si}) \sin(\Delta \alpha_i - \Delta \widetilde{\varphi}'_{bi}) d_i]$$
$$= \sum_{i=1}^{n} \lambda_i [(W_i + V_i) \sin(\alpha_i - \widetilde{\varphi}'_{bi}) + Q_i \cos(\alpha_i - \widetilde{\varphi}'_{bi})] \qquad (15.3-10)$$

There are n number slices with interfaces, each inclined at δ to the vertical. In the equation, the subscript j represents the referred symbol at the interface, while l and r represent those at the left and right side of the interface respectively. And d is the width of the slice.

$$\lambda_i = \begin{cases} 1 & i = 1 \\ \prod_{k=2}^{i} \dfrac{\cos(\alpha^l_j + \delta_j - \widetilde{\varphi}^l_{bj} - \widetilde{\varphi}_{sj})}{\cos(\alpha^r_j + \delta_j - \widetilde{\varphi}^r_{bj} - \widetilde{\varphi}_{sj})} & i = 2,3,\cdots,n-1 \end{cases} \qquad (15.3-11)$$

$$\tan \widetilde{\varphi}'_{bi} = \frac{\tan \varphi'_{bi}}{K} \qquad (15.3-12)$$

$$\widetilde{c}'_{bi} = \frac{c'_{bi}}{K} \qquad (15.3-13)$$

$$\tan \widetilde{\varphi}'_{si} = \frac{\tan \varphi'_{si}}{K} \qquad (15.3-14)$$

$$\widetilde{c}'_{si} = \frac{c'_{si}}{K} \qquad (15.3-15)$$

$$\tan \widetilde{\varphi}^l_{bj} = \frac{\tan \varphi'^l_{bj}}{K} \qquad (15.3-16)$$

$$\tan \widetilde{\varphi}^r_{bj} = \frac{\tan \varphi'^r_{bj}}{K} \qquad (15.3-17)$$

$$\tan \widetilde{\varphi}_{sj} = \frac{\tan \varphi'_{sj}}{K} \qquad (15.3-18)$$

The factor of safety is implicitly involved in equation (15.3 – 10) and will be determined by solving equations through iterative calculation.

(4) The conventional method based on force equilibrium (Pan et al, 2010) is suggested for wedge failure analysis.

15.4 Engineering Approaches for Slope Stabilization

15.4.1 *Unloading and Surcharge Berms*

The excavation at the top of an unstable slope to reduce its loading can be one of effective and economical measures of slope reinforcement. Meanwhile, backfilling at the toe of a slope is also of great advantage in improving a potentially unstable slope. In the process of engineering application, another optimized approach is to use the excavated soils at the crest as the berm material at the toe.

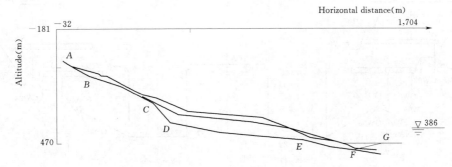

Fig. 15.4 – 1 Slope Reinforcement Located at the Left Bank of Zipingpu Reservoir

Fig. 15.4 – 2 A Landslide Happened at the Left Abutment of Yunqiao Dam

Fig. 15.4 – 1 shows a slope deposit located at the left bank of the Zipingpu Reservoir. About 10 million m³ material has been removed at the elevation and transferred to the toe.

During the construction period, a landslide happened at the left abutment just in vicinity to the right side of Yunqiao Dam (Fig. 15.4 – 2). This landslide is triggered along the bedding plane of the muddy limestone (Song et al, 2011), and the stability of the remaining slope should be drawn a concern. The designers used a berm that sandwiched the slope and the upstream face of the dam (Fig. 15.4 – 3). This approach is fairly cheap with simple construction, compared to the original proposal than involves several hundred cables.

15.4.2 Drainage

Using drainage facility to lower groundwater level is common method of improving the slope stability in China's hydropower projects. Fig. 15.4 – 4 has showed the measured groundwater level in the ship lock slope of the Three Gorges Project is just a few meters higher over the 7 levels of drainage tunnels, which illustrated that drainage tunnel has an important role in lowering the groundwater level. Fig. 15.4 – 5 is an example of the 5 levels of drainage tunnels that will be described in details in this section.

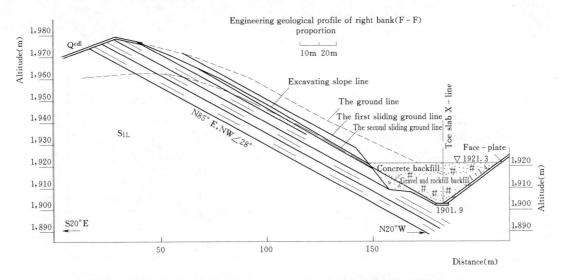

Fig. 15.4 – 3 The Support of Presser Foot at the Left Landslide of Yunqiao Dam

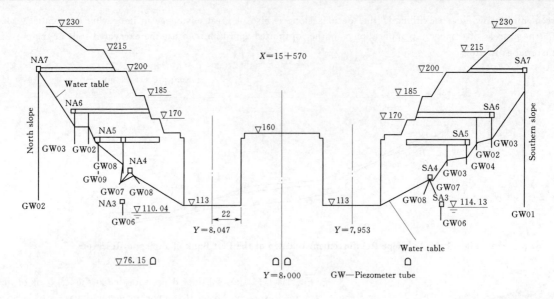

(a) Ground water level distribution map of permanent shiplock's 15 - 15 section

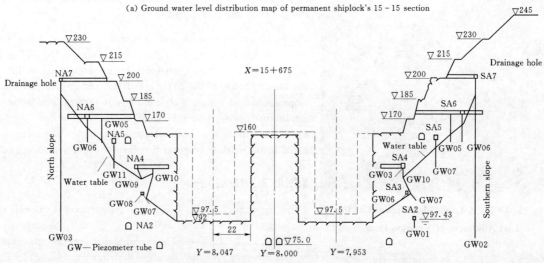

(b) Ground water level distribution map of permanent shiplock's 17 - 17 section

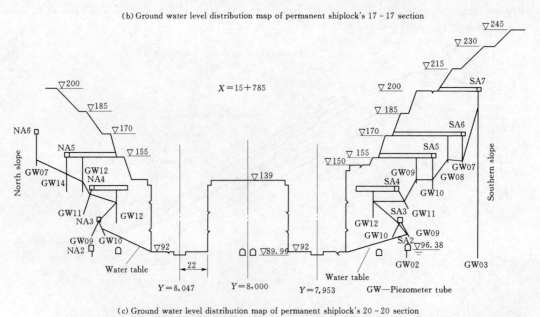

(c) Ground water level distribution map of permanent shiplock's 20 - 20 section

Fig. 15. 4 - 4　The Measured Ground Water Level of the Three Gorges Ship Lock Slope （Unit： m）

Chapter 15 Slope Engineering in Dam Construction

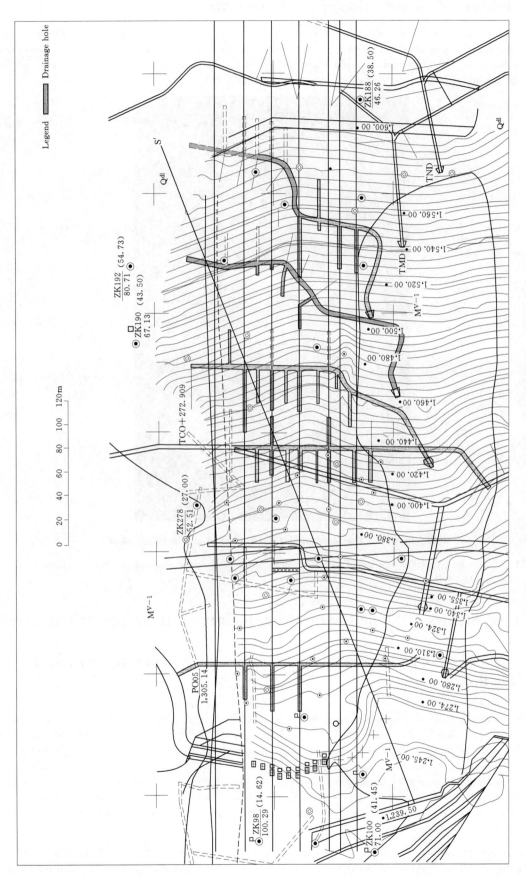

Fig. 15.4-5 The 5 Levels of Drainage Tunnels of the Xiaowan Project

15.4.3 Prestressed Anchor Cables and Anti-slide Piles

Mechanical stabilizing measures are commonly used in hydropower projects if an engineering slope is considered insufficient to perform safely. Prestressed anchor cables are widely used throughout the world. In China, stabilizing piles are also in common use. Recently, these two approaches have been jointly used in many slopes, i.e., applying anchor cables at the top or middle of a pile to reduce cantilever effects.

Fig. 15.4-6   The Accumulation Reinforcement of Yinshuigou Slope

In Fig. 15.4-6, the prestressed anchor cables and anti-slide pile are used in the accumulation reinforcement of Yinshuigou slope to reduce cantilever effects.

15.5   Case Studies of Slope Engineering of China's Hydropower Engineering

15.5.1   Colluviums Deposit Slope at the Left Bank of Xiaowan Hydropower Project in the Lancang River

The Xiaowan hydropower station is a controlling reservoir in the middle and lower reach of the Lancang River that is located between Fengqing county of Lincang and Nanjian countys of Dali. The total reservoir storage is 15 billion m^3 and the installed capacity is 4,200MW. The average annual energy output is 18.853 billion kW·h. This multipurpose project is consisted of a 294.5m high concrete double-curvature arch dam, an underground water supply and power generation system at the right bank, and flood release tunnels at the left bank. The Yinshuigou Colluviums deposit slope is created at the left bank in front of the dam as a result of design layout of this project (Fig. 15.5-1). It is of a tongue shape in plane in a large scale, affecting the safety of dam and the crane system for the concrete conveying platform during dam construction. It has presenteda key technical problem in the construction of Xiaowan hydropower project and is the first time in the international engineering field (Zou et al, 2006; Zhou et al, 2008).

The natural Colluviums have a slope of 32°-35° with elevations of 1,130m at the toe and 1,590m at the crest, the total height of the slope is 470m. The average thickness of deposit is 30-37m with a maximum of 60.63m. The total length is 80 -200m in the north-south direction, and 745-830m in the west-east direction. The total volume is 5.4 million m^3

(Fig. 15.5-1). The deposit is composed of rubbles, boulders in extremely large size, mixed up with gravels, soils, and crushed rocks, underlain by a weak seam consisting of fine soils. The underlying bedrock is of complex structure from toppling and stress release. The geological plan and cross sections of the deposit are shown in Fig. 15.5-2, Fig. 15.5-3 and Fig. 15.5-4.

Fig. 15.5-1 Overview of the Left Bank Slope of Xiaowan Hydropower Station

The preliminary design of deposit is arranged according to arrangement of the nearby hydraulic structures: the excavation started at elevation of 1,645m with a slope of 1:1.3-1:1.5, following a subsequent cut with a slope of 1:1.15-1:1.2 between elevation of 1,500m and 1,380m, and 1:0.8-1:1.2 between elevation of 1,380m and 1,245m. Berms are set per elevation of 20m. The slope excavation started in March 2002. Creep deformation was observed in December 2003 as a result of slope excavation, blasting, and rainfall. By late January 2004, the boundary of the unstable body was clearly identified between elevation 1,245m to 1,460m. The deformation was clearly observed around the contact of the deposit and the bedrock (Fig. 15.5-5 and Fig. 15.5-6). Site investigation and the monitoring data (Fig. 15.5-7) was analysied which showed that the landslide exhibited a translational mode along the contact of deposit and bedrock, and a circular slippage at the top. This is a progressive and translational failure started from a local instability that was extended during construction.

In consideration of scale, geology and environment, consequence of failure, project quantity and investment, reinforcement of Yinshuigou accumulation at left bank was designed as the second class slope. Stability control indexes are as below: after completion, $F=1.20$ during normal running, $F=1.15$ during sudden drawdown of reservoir water level, and $F=1.05$ during Ⅷ degree earthquake; in construction, $F=1.15$ during rainy season, and $F=1.05$ during dry season.

In light of importance of deformation instability mode and stability of Yinshuigou accumulation, the dynamic treatment based on the information is conducted. Reinforcement measures are drainage strengthening, prestressing cable, stabilizing pile and retaining back pressure. The slope of accumulation is comprehensively treated by phases and sections. According to own characteristics of accumulation such as hole collapse, some transnormal technologies are conducted for treatment:

(1) Pipe drilling prestressing cable. The hole collapse during prestressing construction is conquered (Fig. 15.5-8 and Fig. 15.5-9).

(2) Stereo drainage system. To manage the slope, the first step is the water control. In order to lower groundwater in accumulation and perched groundwater on underlying bedrock contact area, 9 layer drainage holes are set above elevation of 1,245m in bedrock lie under the accumulation. Length of master cave is 3,750m with 45 aclinicadits deep down in accumulation (Fig. 15.5-5).

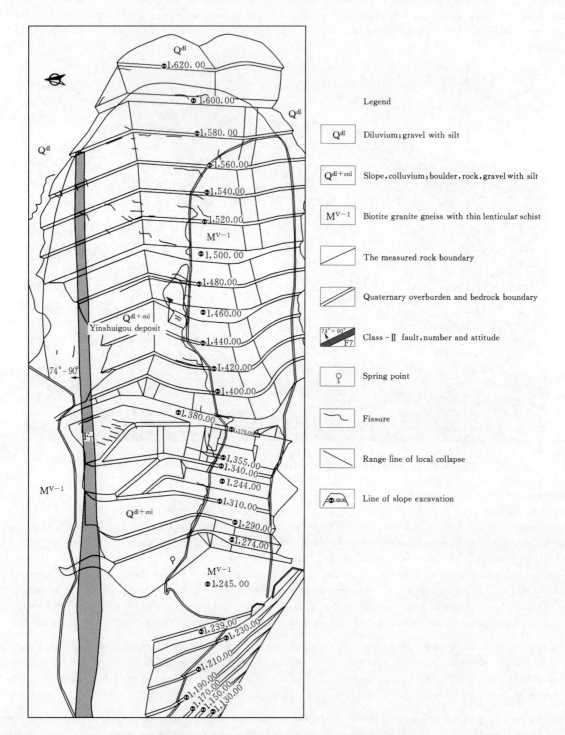

Fig. 15.5-2 The Geological Sketch of Yinshuigou Deposit

(3) Cantilever anti-slide pile. 14 Cantilever anti-slide piles are set at elevation of 1,244m. The pile deepened into bedrock, which makes into a system with concrete wall at cantilever section of elevation of 1,274m (Fig. 15.5-6 and Fig. 15.5-10).

By means of strengthening drainage measures, prestressing cable, stabilizing pile and retaining back pressure by phases and sections, the positive effect of measurement is obvious observed from deformation, seepage monitoring results of 3 typical monitoring sections. The slope is in stable state, which meet design requirement (Fig. 15.5-11).

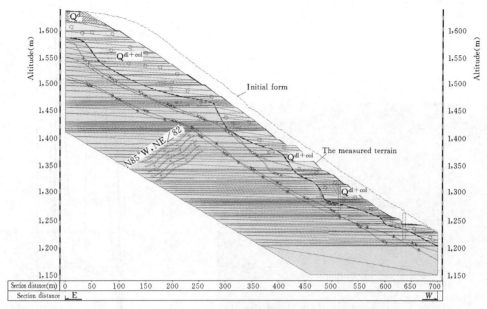

Fig. 15.5-3 Longitudinal Geological Section of the Deposit

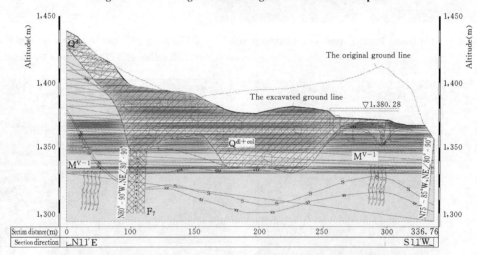

Fig. 15.5-4 Geological Cross Section of the Deposit

Fig. 15.5-5 The Crack Zone Developed by Excavation of Colluviums Slope

Fig. 15.5-6 The Shear Front Exposed by Upstream Side

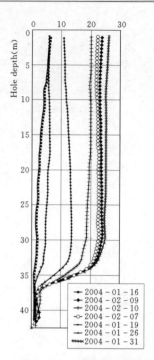
Fig. 15.5-7 Sliding Processes and Surface at Elevation of 1,250m Shown by Gradiograph

Fig. 15.5-8 Prestress Anchoring in Deep Accumulation

Fig. 15.5-9 Pipe Drilling-New Technology for Loose Accumulation Slope Measurement

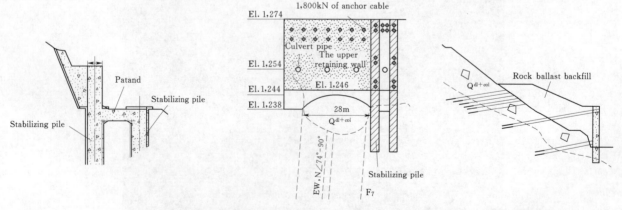
Fig. 15.5-10 "Anchor-pile-slab wall" Combined Bearing and Back Pressure System

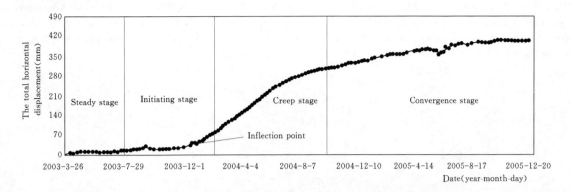

Fig. 15.5-11 Displacement Inflection Point-elevation-time Curve of Typical Point

15.5.2 *Left Bank Dam Abutment High Slope of Yalong River Jingping Ⅰ Hydropower Station*

Jinping Ⅰ Hydropower Station Dam is located at Pusiluogou canyon section between Muli county and Yanyuan county in Sichuan Province, which is the controlling reservoir of the lower reaches of the Yalong River. The dam is a concrete hyperbolic arch dam, which the maximum height is 304m, the highest arch dam under construction in world. The engineering grade is Ⅰ, either is the main hydraulic structures. Due to the construction needs of the abutment excavation and the cable machine platform, the left bank slope was cut with 1 : 0.5 to 1 : 0.3 slope from elevation 2,110m to elevation 1,580m, forming a steep slope up to 530m, which the stability face unprecedented challenges (Fig. 15.5-12). Left Slope has the following characteristics:

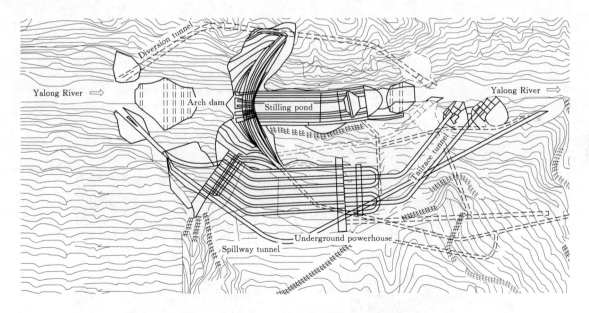

Fig. 15.5-12 Layout of Jinping I Hydropower Project

(1) The vally of Jinping I hydropower station pivotal project is the typical deep V-shaped canyon (Fig. 15.5-13), which relative height difference of cross-strait valley slope reaches 1,500-1,700m. Tectonically it belongs to the three beaches inverted syncline, rock occurrence N15°-60° E/SE∠35°-45°, strike along the river. Bank slope bedrock is mainly middle and upper Triassic Zagunao (T_{2-3z}) metamorphic rocks, the first paragraph (T_{2-3z}^1) green schist, and the second (T_{2-3z}^2) griotte, the third (T_{2-3z}^3) sandy slate. The left slope is inverse slope, elevation 1,820-1,900m below for marble, slope 55°-70°, above for sandy slate, slope 40°-50°, which has micro-morphological features alternate with the ridge and shallow trench. All structural surface characteristics are shown in Fig. 15.5-14.

(2) The left bank of dam site has steep slope, exposed bedrock, cliffs towering, elevation 1,900m below for 60°-90° slope. Unloading rock mass intense, unloading depth is bigger in the horizontal direction, the fault of the dam site area and extruded between layers was relatively developed, the left bank mainly has f_5, f_8, f_{42-9} faults, which f_{42-9} has strike EW,

the dip S∠40°–60°, layer extruded main occurrence is strike N15°–40°E, dip NW∠15°–45°. Advantage occurrence of the deep fissures in the bedrock is strike N30°–50°E, dip SE∠50°–60°. Combined with faults, inter-layer extrusion, former deep unloaded cracks as well as well developed fissures in the slope rock mass each other constitute the controlling structure surface of the left bank (Fig. 15.5 – 13).

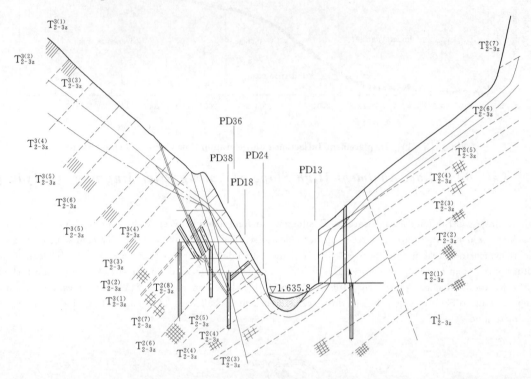

Fig. 15.5 – 13 Section of Left Bank Slope

According to the project layout, the geological conditions and construction technical requirements, the excavation height of left bank abutment slope is about 530m (elevation 2,110 – 1,580m), elevation 1,885m above for one bridle way every 30m, elevation 1,885m below for one bridle way every 15m, which the width is 2 – 3m. Excavation slope is 1 : 0.2 – 1 : 0.15 for grade II, 1 : 0.35 – 1 : 0.2 for grade III, 1 : 0.5 – 1 : 0.4 for grade IV and above, the overburden slope is 1 : 1 – 1 : 0.75 (Fig. 15.5 – 15).

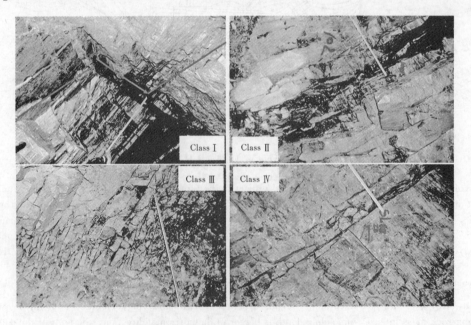

Fig. 15.5 – 14 Stratum Structure Surface

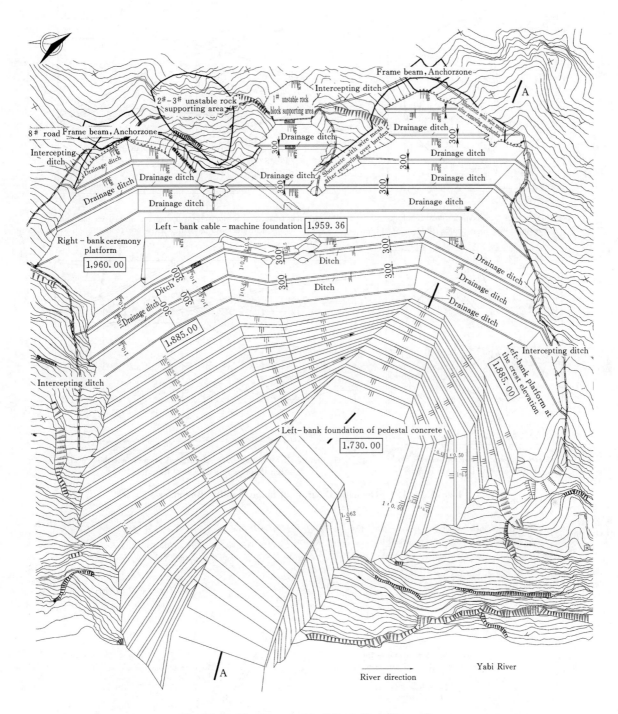

Fig. 15.5-15 Layout of Left Bank Slope Excavation

According to left abutment fault, the inter-layer extrusion, deep developed unloading fissures which are the controlling structure face for slope stability, as well as cracking deformation rock mass constituted by the f_{42-9} fault, the lamprophyre X, SL_{44-1} crack band and KL_{30-10} and KL_{30-12} on the left bank abutment (Fig. 15.5-16), the left bank slope reinforcement measures are (Song et al, 2011):

(1) Extensive using large tonnage, 80-100m long pressure dispersion prestressed anchor cable, more than 4,000 for the left bank slope (Fig. 15.5-17);

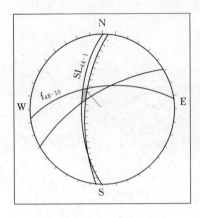
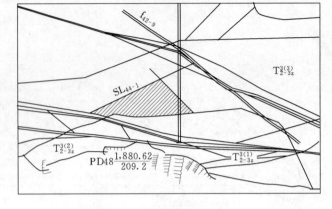

(a) Stereographic projection for unstable blocks constituted with f_{48-10} and SL_{44-1}

(b) Elevation 1,885.00m section schematic diagram for unstable blocks constituted with f_{48-10} and SL_{44-1}

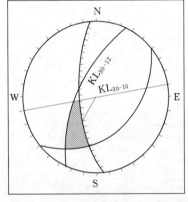

(c) Stereographic projection for unstable blocks constituted with KL_{30-10} and KL_{30-12}

(d) Elevation 1,730.00m section schematic diagram for unstable blocks constituted with KL_{30-10} and KL_{30-12}

Fig. 15.5 - 16 Schematic Diagram of Slope Stability Boundary

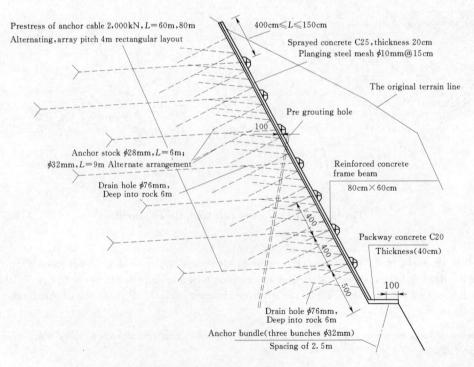

Fig. 15.5 - 17 Section Schematic Diagram for Shallow (Deep) Layer of Reinforcement

(2) For the secondary potential unstable blocks of slope, deformation zone of rock unloading and relaxation, comprehensive reinforcement measures such as slope concrete box grid beams, sprayed concrete, anchor, anchor bundle and prestressed anchor cable were used;

(3) For big blocks constituted by f_{42-9} fault, lamprophyre X, SL_{44-1} tensile fissure band, in addition to using prestressed anchor, in elevation 1,883m, 1,860m and 1,834m along the f_{42-9} strike set anti-shear hole with 9m×10m section, and in the middle of anti-shear hole 4m×5m cross-shaped keyway was set to increase the sliding role of anti-shear hole (see Fig. 15.5-18).

According to the different geological conditions and failure modes of various parts of the dam left bank slope, the slope surface and deep deformation, stress was monitored continuously and comprehensively (Fig. 15.5-19). Up to now, monitoring data shows that the left slope has local excavation unloading cracks, but no signs of continuing development, the slope does not appear destructive deformation and instability, the normal anchor cable force, left bank slope stability during construction.

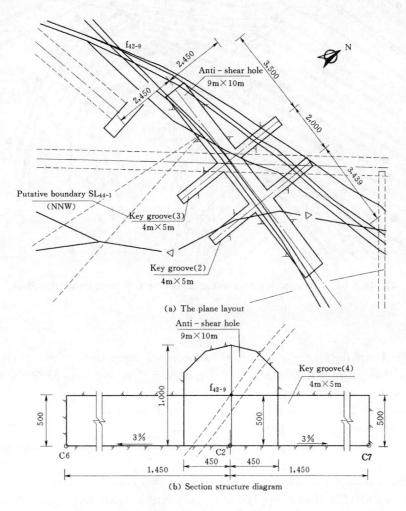

Fig. 15.5-18 The Layout for Anti-shear Hole and Schematic Diagram for the Cross Section Structure

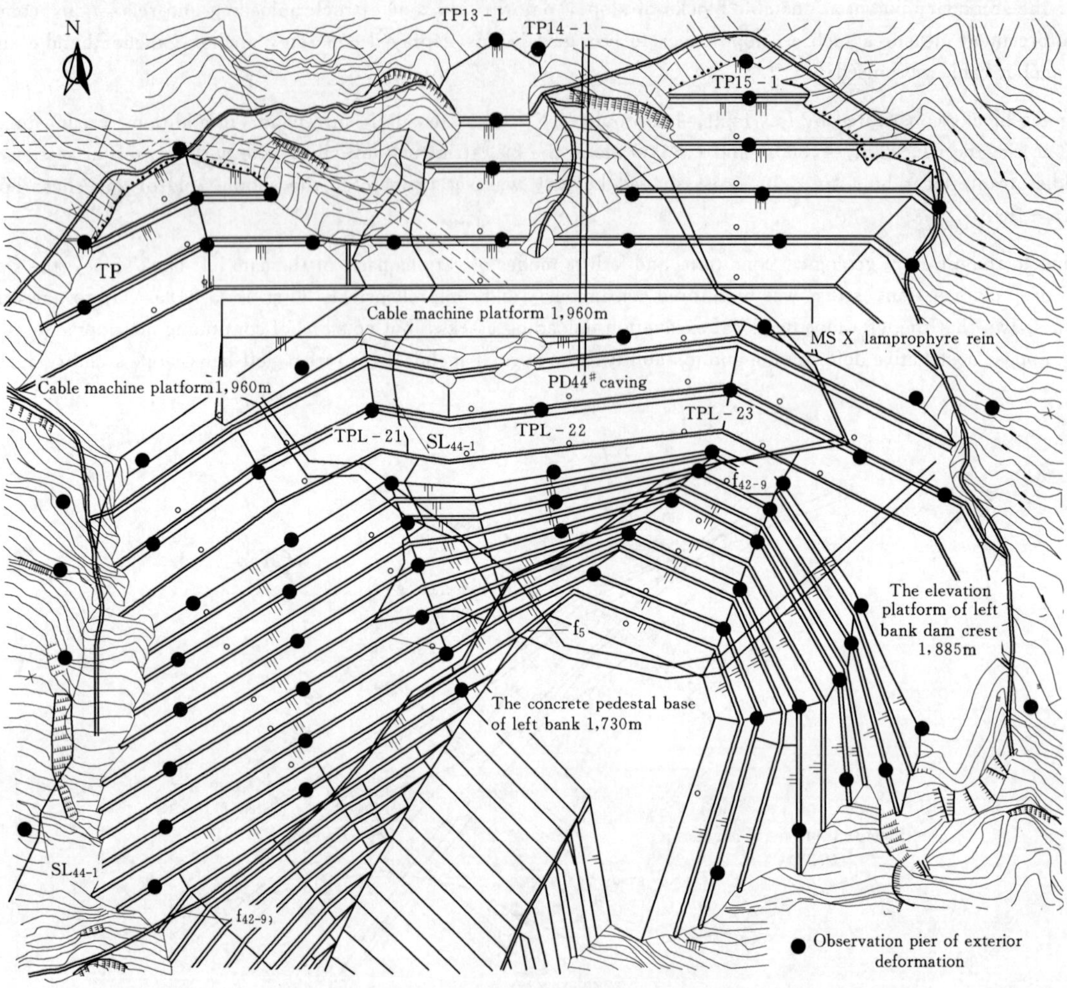

Fig. 15.5-19 Deformation Monitoring Layout for Left Bank Excavation Slope

References

[1] Bishop AW. The Use of the Slip Circle in the Stability Analysis of Slopes. *Geotechnique*, 1955, 5 (1): 7-17, 26.
[2] Chen Z Y and Morgenstern N R. Extensions to the Generalized Method of Slices for Stability Analysis. *Canadian Geotechnical Journal*, 1983, 20 (1): 104-119.
[3] Donald I B and Chen Z Y. Slope Stability Analysis by the Upper Bound Approach: Fundamentals and Methods. *Canadian Geotechnical Journal*, 1997, 34: 853-862.
[4] Geotechnical Control Office, Engineering Development Department, *Geotechnical Manual for slopes*. Hong Kong: 1981.
[5] Hoek E and Bray J. *Rock Slope Engineering*. The Institute of Mining and Metallurgy. 1977.
[6] Hoek E and Brown E T. The Hoek-Brown Failure Criterion-a 1988 Update. In: Curran J C. ed. *Proceedings of the 15th Canadian Rock Mechanics*. Symposium (Torton, Canada, 1988). Toronto: Department of. Civil Engineering, University of Toronto, 1988, 31-38.
[7] Hoek E and Brown E T. Empirical Strength Criterion for Rock Masses. A. S. C. E. *Journal of the Geotechnical Engineering Division*. 1980a (9): 1013-1035.
[8] Marinos P and Hoek E. GSI: a Geologically Friendly Tool for Rock Mass Strength Estimation. In: *The Proceedings of GeoEng* 2000: An International Conference on Geotechnical & Geological Engineering, 19-24 November, Melbourne, Australia. 2000.
[9] Morgenstern N R and Price V. The Analysis of the Stability of General Slip Surface. *Geotechnique*, 1965, 15 (1): 79-93.

[10] Sarma S K. Stability Analysis of Embankments and Slopes. ASCE *Journal of the Geotechnical Engineering Division*, 1979, 105 (GT12): 1511-1524.
[11] Chen Zuyu, Wang Xiaogang, Yan Jian, Jia Zhixin, Wang Yujie. *Rock Slope Stability Analysis -Principles, Methods, Procedures*, Beijing: China Water Power Press, 2005.
[12] Pan Jiazheng, He Jing. *Fifty Years of Chinese Dam* . Beijing: China Water Power Press, 2010.
[13] Song Shengwu, Feng Xuemin, Xiang Baiyu, Xing Wanbo, Zeng Yong. The Key Technology for High and Steep Slope of Hydropower Engineering in Southwest China. *Journal of Rock Mechanics and Engineering*, 2011, 30 (1): 1-22.
[14] The electric power industry standard of the people's Republic of China. DL/T 5353—2006, *the Design Specifiction for Slopes of Hydropower and Water Conservancy Projects*, 2007.
[15] The water conservancy industry standard of the people's Republic of China. SL 386—2007, *the Design Code for Engineered Slopes in Water Resources and Hydropower Projects*, 2007.
[16] Zhou Jianping, Yang Zeyan, Weng Xinxiong. *Chinese typical engineering slope-water conservancy and Hydropower Engineering* Vol. Beijing: China Water Power Press, 2008.
[17] Zou Lichun, Wang Guojin, Tang Xianliang, et al. *The Treatment Theory and Engineering Practice for Complex High Slope*. Beijing: China Water Power Press, 2006.

Chapter 16

Seismic Safety of Dams

16.1 Seismic Safety of High Concrete Dams
Chen Houqun[1]

16.1.1 Introduction

In China large dams are playing a very important role in water resources management, for flood mitigation and clean energy supply, thereby promoting the sustainable development of society and economy. By the end of 2007, there were 130 dams with the height more than 100m in China, in which more than 1/2 are concrete dam. In the high dams of 150m and 200m, the ratio of concrete dam reaches 58% and 78% respectively (Jia et al, 2007). China is a country with frequent seismicity especially in the western region where was concentrated about 80% of the national total hydropower resources. Recently, a series of critical hydropower projects preferable with arch dam of about 300m high are constructed and will be constructed in western China. Such as: Xiaowan arch dam (294.5m) is the highest dam in the world; the Jinping (305m) arch dam is the highest maximum arch dam in the world; the Dagangshan arch dam (210m) has a design seismic acceleration up to $0.577g$ and the Xiluodu arch dam (285.5m) has a design seismic acceleration of $0.325g$. As any accident of serious damage of high dam with huge reservoir during strong earthquake can inflict grave secondary catastrophe upon surrounding communities, to prevent any collapse of high dam under the maximum credible earthquake (MCE) is deeply concerned by our government and social circles. It becomes the strategic priority in the study on seismic safety of high dams and the inevitable challenge has to be faced by dam engineers in China. The flourish of dam construction and the frequency of strong earthquake occurrence give great impetus to the progress of earthquake engineering of dams in the country. Scientific research activities in this field have been conducted on an extensive national scope, including basic and applied researches. Significant advances have been made in study on seismic safety of concrete dams along with their construction.

The basic idea of seismic safety evaluation of high concrete dams should include seismic input at dam site, seismic responses of dam system and dynamic properties of dam concrete as three indispensable and interdependent components. With the deepening of the research in these three aspects, a series of critical problems for dynamic analysis of dams have been solved. Some major progresses achieved in all these aspectsare briefly introduced as follows.

16.1.2 Site-Specific Seismic Input at Dam Site

During the past more than 60 years, mainly in recent decades, the integrated, systematic and thorough researches on the questions of ground motion input of high concrete dams have been conducted. In close combination with China national conditions and high dam project features, a complete set of more reasonable and practicable new ideas and methods in this aspect are provided, including reasonably determining fortification level framework, correctly selecting of ground motion parameters and understanding of seismic input mechanism and reservoir triggered earthquakes. The results have been used in seismic design of many high dam projects in China and accepted by China seismic design code of hydraulic structures.

16.1.2.1 Selecting the Site-specific Input Parameters for Maximum Design Earthquake

(1) As commonly recognized, the pulse peak of acceleration with high frequencies is of little engineering significance to the responses of high dams. So, it is recommended that the seismic hazard evaluation for dam site is based on the response spectrum related "effective peak acceleration (EPA)" instead of the current used "peak ground acceleration (PGA)". By analyzing 145 accelerograms with $M \geqslant 4.5$ recorded at rock sites of events with different magnitude and epicenter distance inter-

---

[1] Chen Houqun, Academician of Chinese Academy of Engineering, China Institute of Water Resources and Hydropower Research, Professor Senior Engineer.

vals in the western United States, an average normalized spectral with its peak at period 0.2 second and an amplification factor of 2.5 is revealed. So, the EPA defined as the spectral acceleration at period 0.2 second divided by an average amplification factor of 2.5 might be accepted (Chen et al, 2004).

(2) The "equal-hazard spectra" usually used as design response spectra is actually an enveloping curve of many earthquakes with different magnitudes and locations. It does not reflect the physical characteristics of response spectrum for a real earthquake associated with the actual magnitude and distance. Therefore, a hybrid method combining both probabilistic and deterministic approaches of selecting specific scenario earthquake to determine the site-specific design response spectrum was recommended. As an important premise, the peak ground acceleration caused by the scenario earthquake must be consistent with the peak acceleration of design exceedance frequency at dam site. Actually, it is always only a few potential seismic sources defined by seismic active faults having contribution to the rather high design peak acceleration. Among all earthquakes with the acceleration consistent with the design EPA at dam site, keeping to the principle of maximum probability of occurrence, the distance R usually nearest to the dam site along the major fault within the potential seismic source and the magnitude M of the scenario earthquake can be selected. Then the spectrum can be defined from the M and R by using an appropriate average attenuation law of ground motion (Chen et al, 2004). Fig. 16.1-1 shows the comparison of the spectra for Xiaowan project. Obviously, the equal-hazard spectrum gives an unreasonable higher spectrum value for the dam with a fundamental period of about 1 second.

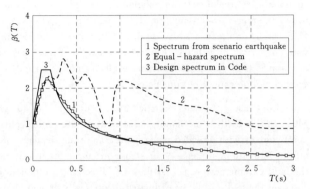

Fig. 16.1-1 Comparison of the Spectra for Xiaowan Project

(3) Earthquake ground motions are strong stochastic processes, non-stationary both in time and frequency domains. However, the influence of the non-stationary in frequency content is usually overlooked, but it may significantly effect on nonlinear responses of the structures during strong earthquakes. Therefore, based on the Priestley theory of evolutionary power spectra (Priestley, 1965), an approach of generating artificial accelerograms non-stationary both in amplitude and frequency and consistent with target evolutionary spectrum is proposed (Zhang, 2009; Zhang et al, 2007). In this method the evolutionary power spectra are calculated by Nakayama method (Nakayama et al, 1994). The target evolutionary spectra are developed based on 80 accelerograms at rock foundation from western United States with magnitudes $M \geqslant 6.4$ and epicenter distances $R < 45$km, from which the statistical prediction model for given M and R was established using regression equations similar to that proposed by Hisao Goto (Goto et al, 1984).

In comparison with the approach developed by Kameda based on evolutionary spectrum calculated by using multi-filter technique (Kameda, 1975), its problem of the damping factor selection can be avoided and the time-dependent characteristic of phase angle can also be considered in the proposed approach. Fig. 16.1-2 shows the regressed target evolutionary spectrum for $M=7.0$ and $R=15$km and the generated artificial accelerogram related to it (Zhang et al, 2007).

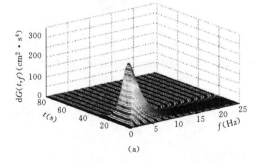 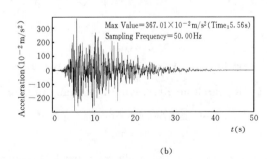

(a) (b)

Fig. 16.1-2 Target Evolutionary Spectrum for $M=7.0$ and $R=15$km and Generated Artificial Accelerogram (Zhang et al, 2007)

16.1.2.2 Determining the Maximum Credible Earthquake for Critical Projects

It is recommended that the critical dams should be checked to prevent uncontrolled release of the reservoir water during the largest reasonable conceivable earthquake defined as the maximum credible earthquake (MCE) at dam site. The MCE is preferably to be defined deterministically as a scenario earthquake located nearest to the dam along the causative fault within the critical potential seismic source with its ultimate magnitude. However, by using the current approach there are two fatal weaknesses significantly affecting on the seismic input of MCE. The first one is that the MCE even with a magnitude more than 7.0 and distance less than 10km has still been regarded as a point seismic source, while, actually, a 3-dimensional spatial area source of near fault should be used. The second one is that due to lack of strong motion records in China we have to use the regressive attenuation relationships mainly based on the limited accelerograms recorded at rock sites in Western U.S. based on the data with magnitudes less than 7 and distances between 20km to 100km. Their extrapolation to the events with very small probability of exceedance like MCE becomes rather uncertain. Moreover, in the relationships the definition of distance R often remains certain perplexity and the rupture directivity and the hanging wall effect for near field strong ground motion cannot be considered.

Therefore, using the deterministic "finite faults method" to synthesize directly the design accelerograms is recommended for MCE. In this method the major active fault of the critical potential seismic source is divided into a series of sub-faults as point sources with provided mode, rate, and time sequence of the rupture. By accumulating the effects of each point source on the dam site in sequence, the strong ground motion near faults at the dam site can be directly predicted. Of course, all the fault parameters, e.g., strike, dip directivity, dip angle, rupture area and its length, width, as well as the average slip over the fault should be estimated theoretically or semi-empirically. The further study should be focused on the slip model describing the heterogeneity of slip on the fault surface with asperities and the calibration of all the fault parameters related with seismic moment based on the data base of the interpolate earthquakes in China. Fig. 16.1-3 shows the synthesized accelerograms for a hydropower project by the stochastic finite faults method and their response spectra which are quite similar to that from the probabilistic method (Shi et al, 2005).

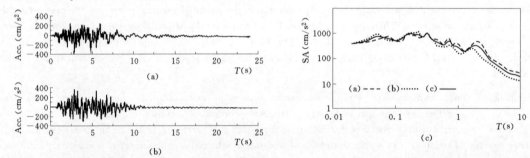

Fig. 16.1-3 Synthesized Accelerograms by Stochastic Finite Faults Method and Their Response Spectra

As no ground motion input has been recorded at Shapai RCC arch dam site during the Wenchuan earthquake in 2008, it has to be reestablished by using the "stochastic finite fault method" with model and parameters identified by the accelerograms recorded at 7 adjacent stations during the earthquake. Fig. 16.1-4 is the reestablished artificial accelerograms at Shapai dam site with PGA of 0.262g and long duration of more than 40 minutes.

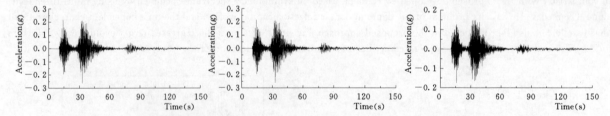

Fig. 16.1-4 Reestablished Artificial Accelerograms at Shapai Dam Site (from Zhang et al, 2007)

16.1.2.3 Research on Reservoir Earthquake in China

Since the occurrence of the reservoir earthquake of Xinfengjiang reservoir in 1960, great efforts have been devoted to the problems of reservoir earthquake in China.

(1) Engineering assessment and monitoring network of reservoir earthquake.

1) On the basis of investigation on the seismo-geological and hydro-geological conditions some major influence factors on the reservoir earthquake have been identified, such as: volume and depth of reservoir, regional stress state, fault activity, lithological characters including karst developing, seismicity background etc. Different engineering assessment approaches have been appliedfor critical projects, such as: experienced discrimination, statistical hazard evaluation, grey clustering, fuzzy estimation andartificial neural network etc. Also, the GIS technique platform has been used (Fig. 16.1-5).

2) As required in the *Seismic Design Code for Hydraulic Structures* (DL 5073—2000) of China, for reservoirs with a

dam higher than 100m and a storage capacity over 500 million m³, if reservoir earthquakes of intensity ⩾ Ⅵ are predicted, a monitoring network as well as an emergency warning and response plan scenario shall be completed before impounding. For the Three Gorges project, a digital remote monitoring network for observing reservoir earthquake has been set up before the impounding of the reservoir. It includes 24 fixed stations, 3 relay stations, 1 network center, 8 movable stations, and 2 non-remote stations. Recently, in the reservoir areas of the 4 cascade hydropower projects at the lower reaches of Jinshajiang River, a microseismic monitoring network with unprecedented scale is under construction. It includes 62 fixed stations and 8 floating stations in the reservoir area. The monitoring area is shown in Fig. 16. 1 - 6.

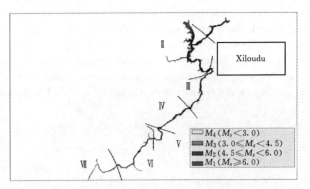

Fig. 16. 1 - 5 Gray Clustering Prediction of RE Induced by Karst for Xiluodu Arch Dam by Using GIS (Wang et al, 2000)

For Zipingpu reservoir a reservoir earthquake monitoring system has been set up in Aug. 2004, and the reservoir impounding started in Oct. 2005. After the Wenchuan earthquake in 2008, the variation of the recorded seismic activity including the occurrence frequency and magnitude before and after the reservoir impounding (Fig. 16. 1 - 7) appeared in a normal range. It mighty demonstrated that the impounding of Zipingpu reservoirs did not trigger the Wenchuan Earthquake.

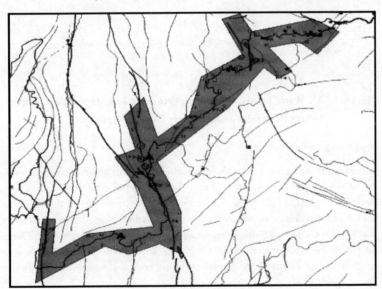

Fig. 16. 1 - 6 The Monitoring Area of RE Network at the Lower Reaches of Jinshajiang River (China Seismic Administration)

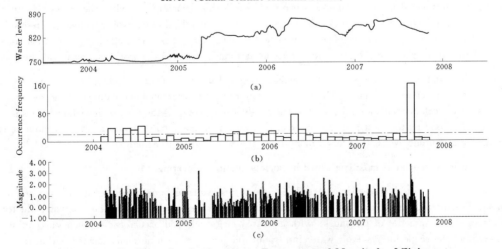

Fig. 16. 1 - 7 Water Level, Occurrence Frequency and Magnitude of Zipingpu Reservoirs (Sichuan Seismic Administration)

For the Three Gorges project the reservoir impounding started in 2003. The occurrence frequency of reservoir earthquake increased with the water level, especially, when the water lever increased from 135m to 156m and from 156m to 175m. All the reservoir earthquakes were located within the predicted area, and most of them were microearthquakes and urtra-microearthquakes. After 2009 the annual average occurrence frequency gradually becomes close to the background level (Fig. 16.1-8). The earthquakes with maximum magnitude of $M=4.1$ in Nov. 2008 and of $M=4.8$ in Dec. 2013, however, did not exceed the maximum magnitude predicted in design.

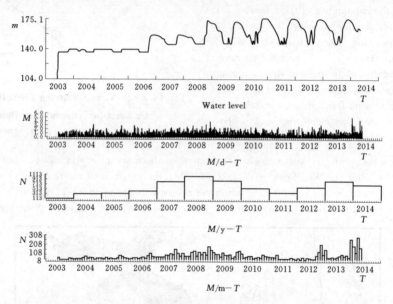

Fig. 16.1-8 Water Level, Occurrence Frequency and Magnitude of Three Gorges Reservoirs (China Three Gorges Corporation)

(2) Some understanding commonly recognized.

1) Reservoir earthquake mechanism. On a global scale the cases of earthquakes by reservoir impounding are only account for very little proportion of the total reservoirs in the world. From about 120 events of more or less recognized reservoir earthquakes in 29 countries 22 events occurred in China. Due to the very limited knowledge of the rheology of crustal material and groundwater movement under high pressures and high temperature conditions in the hypocenter region, the complicated mechanisms of reservoir earthquake are not well understood and may differ from case to case. The more commonly conception is that the pore pressure of the reservoir water permeated into rock mass reduces its effective stresses and cause the drop of shear resistance. The effect of added weight of impounded water on crustal stresses field in epicenter depth is negligible in comparison with the weight of rock mass.

2) Classification of reservoir earthquake. Two kinds of reservoir earthquake can be classified by genetic mechanism.

The first kind of non-tectonic shallow earthquake is featured as follows: ①mainly related to the stress adjustment of karst caves, mining tunnels and surface relaxed rock mass; ②often occurred shortly after reservoir impounding orreservoir water level change, normally related with the variation of reservoir water level; ③usually has magnitude not higher than 3.0 or 4.0 and not harmful on dam and reservoir area. Most of the reservoir earthquake cases in the world belong to this kind so far.

The features of other kind of tectonic reservoir earthquake triggered by reservoir impounding are: ①there is seismogenic fault existed within the reservoir area and has been close to failure before reservoir impounding; ②the reservoir water can be infiltrated into deep rock stratum along the fault to trigger the earthquake; ③the triggered earthquake has no substantial difference from that caused by the fault and cannot exceed its maximum magnitude; ④altogether only four reservoir triggered events have a magnitude over 6.0 so far, among them the highest magnitude is 6.3.

This kind of reservoir earthquake is most concerned by society and in engineering.

3) Specification of reservoir earthquake. Some special features of reservoir earthquake can be summarized as follows: ①in time domain small earthquakes usually appear shortly after the reservoir impounding around the reservoir; the tectonic main shock often occurred after a series of foreshocks with increased magnitude and occurrence frequency; ②in space domain with the limitation of the scope for reservoir water infiltration the epicenters are normally located within the range of 5-10km from the edge of reservoir; the relatively shallow hypocenter depth is around 5km; the ground motions attenuate rap-

idly with the distance from the seismic source but have stronger surface shaking; ③in geological environment within reservoir area there are either seismogenic structures, hydro-geological conditions for reservoir water infiltrating into deep rock stratum or the fractured rock mass, karst cave and mining tunnels; more triggered events are linked to normal and strike slip faulting than to thrust faulting; ④ the reservoir triggered tectonic earthquake normally presents as foreshock-main shock-aftershock type, while the non-tectonic one often as swarm type; ⑤all the ratios of the main shock to the maximum aftershock and the vertical component to horizontal component as well as the b value in the scale-frequency relationship ($\lg N = a - bM$) are relatively higher than these of natural earthquakes.

16.1.3 *Seismic Responses of Dam-foundation-reservoir System*

The high dam system is a comprehensive system including dynamic interaction of dam-foundation-reservoir system. The study on seismic response is the key problem for seismic safety evaluation of concrete dams. As above-mentioned, to prevent any collapse of high dam under MCE is the strategic priority in the study on seismic safety of high dams in China. Except to define the MCE usually with characteristics of large near-site earthquake as above mentioned, the quantitative evaluation of the criterion of the limit state of dam-breach for designer becomes the other obstacle has to be overcame. The significant progresses accumulated in the course of construction of a series of about 300m high concrete dams in the country have provided the preconditions to solve the problem and can be briefly described as follows: ① the analysis method evolves from the "pseudo-static method" to dynamic finite element method; ② the model of the system has been established from a closed linear elasticity vibration system to an opened non-linear wave propagation system with conditions to be as close to reality as possible; ③ the criterion of instability of the system has been evaluated from traditional stability safety factor of separated dam abutment using "rigid limit equilibrium method" to a new concept of being identified by the turning point of displacement response of the whole system to increasing seismic input; ④ the seismic safety of high concrete dam system has been investigated which is not only dependent on the numerical analysis but also assisted by dynamic model test on shaking table and field test; ⑤ the computation program has been developed from routine approach to high performance parallel technique.

16.1.3.1 Seismic Response Analysis Methods

In past 60 years, the seismic response analysis of high concrete dam in China experienced a development process from following foreign analysis method to breakthrough and independent research. Especially in recent 10 years, the development is speeded up under the impulsion by construction of a series of 200m to 300m height high concrete dams.

Up to now, the methods based on structural mechanics with hypotheses of linear distributed deformation at flat section has still been used in static analysis of concrete dams and accepted by design codes both in China and most other countries. Accordingly, the corresponding trial-load method for arch dams is still used in design worldwide. As the seismic safety evaluation must be based on the synthesized results of static and dynamic analyses, the model both in static and seismic analyses must be consistent with each other. Hence the pseudo-static loading has to be applied in seismic analysis of concrete dams for a long period. However the pseudo-static method cannot reflect the real characters neither of ground motion nor of the structure. Until the early 1980s the dynamic methods and computer programs based on structural mechanics including the trial-load method for arch dam has been developed in the country.

The high concrete dam system is a comprehensive system including dam body, foundation, reservoir water and their interaction, which are hard to be considered in structure mechanics model. Along with the application of advanced numerical discretization process and the use of powerful computers, the finite element method has developed rapidly and widely been used for dynamic analyses of concrete dams in China. The first 2D finite element method has been applied for the Xinfengjiang buttress dam in the early 1970s. Later through the China-U.S. cooperation the linear finite element program ADAP with massless foundation was introduced for dynamic analysis of arch dam.

The dynamic analysis results of gravity and arch dams by the finite element methods and the developed structural mechanics methods have been compared. It has shown in Fig. 16.1-9 that the mode shapes and frequencies of lower modes were in quite good agreement.

16.1.3.2 Finite Element Model of Dam-foundation-reservoir System

At present a more realistic 3-D dam dynamic analysis program of finite element method for arch dam with conditions to be as close to reality as possible has been developed and widely used in China as shown in Fig. 16.1-10.

In the continually improved model all following conditions are considered simultaneously, such as: ①dynamic interaction of dam-foundation-reservoir system; ②opening of dam contraction joints during earthquake; ③topography features and geological disturbances including critical potential sliding blocks of near-field foundation rock masses; ④radiation damping of far-field foundation rock masses by using artificial transmitting boundaries or viscous damping boundaries; ⑤ spatial variation of seismic input along the dam-foundation interfaces.

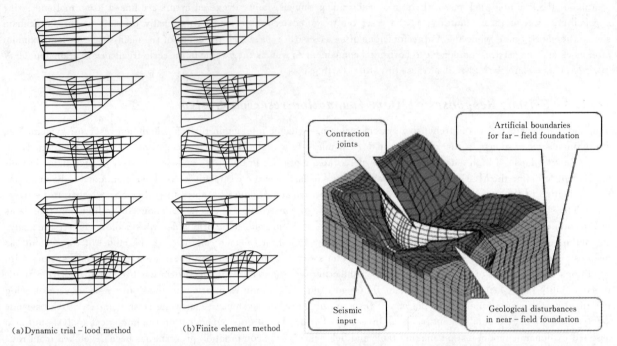

Fig. 16.1-9 The First 5 Modes of the Dongjiang Arch Dam Using Dynamic Trial-load Method and Finite Element Method

Fig. 16.1-10 3-D finite Element Method for Arch Dam

The whole system is discrete in space by finite elements and in time by central finite-differences. It is solved as a boundary non-linear wave propagation problem by explicit integration in time domain.

The factors having major influence on seismic responses have been investigated as following.

1. Hydrodynamic pressure

The key problem for hydrodynamic pressure is the compressibility of the water. A more conclusive view is now commonly recognized that the compressibility effect becomes essential only in the case if the natural frequency of dam (f_d), the first natural frequency of reservoir (f_r) and the predominant frequency of ground acceleration (f_a) are closed to each other. For the arch dams higher than 100m, both the natural frequencies of some first orders of the dam and the effective resonance frequency of the reservoir water are always lower than the predominant frequency of rock foundation. Besides, the reservoir boundaries are not rigid and regular but energy absorbed and with complicate topography, especially for rivers with heavy sedimentation in China. The resonance of reservoir water during earthquake seems hardly could to be induced and actually never be verified in practice so far. Furthermore, when the compressibility of water is considered, a reflection coefficient defined as the ratio of reflected to incident wave amplitudes should be involved. However, the field experimental results have shown that actually this coefficient is not a constant but frequency dependent and varied with position at reservoir boundaries (Yusof et al, 1999). Therefore the compressibility might be neglected and to treat the hydrodynamic pressure in terms of added mass at upstream dam surface in dam engineering practice.

A diagonal mass matrix must be taken for solving the system using explicit integration in time domain. But the added mass from finite element modeling of the hydrodynamic pressure leads to a full matrix at upstream dam surface. Consequently, the Westergaard's added mass determined based on the tributary area around the node and the depth of water at that location is usually used (U. S. Department of the Interior Bureau of Reclamation, 2006). However, for arch dam this rough estimate leads the result overestimated. Therefore, in our practice to use half of the Westergaard's added mass at the upstream dam surface might be more appropriate. Because in this case, the hydrodynamic pressure in terms of added mass will be in good agreement with the results of finite element modeling as shown in Fig. 16.1-11 (Chen et al, 1987).

2. Dam-foundation interaction

The importance of foundation interaction on the responses from loading a concrete dam has been commonly recognized, particularly for dynamic load. In trial-load method analyses the flexibility of the homogeneous foundation is roughly assumed by

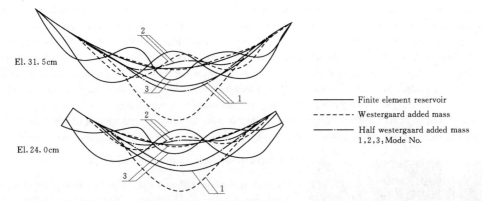

Fig. 16.1-11 Comparison of Hydrodynamic Pressure in Terms of Added Mass and of Finite Modeling

means of Vogt coefficients of the equivalent springs supporting the end of the arch and cantilever members forming the trial-load model. In the finite element model of dam-foundation system only limited foundation block can be included by fix boundaries. In this case the deferent properties of the foundation rock masses can be considered. However, in order to avoid the artificial amplification due to the reflect wave from fixed boundaries and the uncertain vibration properties due to the inertial effect of the arbitrarily chosen volume of foundation block; the foundation has to be assumed as massless and to consider also merely its flexibility. Of course, neglecting the inertial effect and energy dispersion (radiation damping) may have major influence on the seismic responses of the system. Obviously, as a closed vibration problem the dam foundation interaction with all the assumptions in those two methods cannot conform to the objective reality. It should be solved as an opened wave propagation problem to take the seismic energy dispersion into account.

Practically, dam foundation can be divided into two parts. The first one is the near-field foundation with a limited volume formed by extending the boundaries to two or three times of the dam height far away from the dam base along each direction. In this volume the topography features and geological disturbances including some critical potential sliding blocks are contained. The other one is the far-field foundation mainly reflecting the energy dispersion. So it can be replaced by artificial boundaries. The artificial boundaries are added around the perimeter of the near-field foundation adjacent the dam. The total wave field of the system can be divided into the free wave field and the outgoing scattering wave field. The former can be obtained from the theoretical solution of a homogeneous elastic semi-infinite space including both the incident and the reflected waves. The remained latter propagates to the far field with some seismic energy dissipated like radiation damping.

The two kinds of artificial boundaries as transmitting boundary or viscous damping boundaryare more popularly used in engineering practice.

In transmitting boundary based on the conception of a direct simulation of the one-way wave motion, the displacements of the node at boundary can be derived by using multi-transmitting formula from the displacements of outgoing waves of adjacent nodes along the propagation direction at previous time interval (Liao and Liu, 1992).

The viscoelastic damping boundary is represented by a series of damping and spring elements. The interaction force at the artificial boundaries can be defined using the free field responses as

$$\{P_b^f\} = [K_b]\{u_b^f\} + [C_b]\{\dot{u}_b^f\} + [\sigma_b^f][n]\{A_b\} \qquad (16.1-1)$$

where $\{u_b^f\}$, $\{\dot{u}_b^f\}$ and $\{\sigma_b^f\}$ are the free field displacement, velocity and stress vectors induced by the seismic input; A_b and n are the area and normal direction of the boundary; $[K_b]$, $[C_b]$ are the damping and spring matrices. For normal dispersive wave the damping coefficient $C_n = \rho v_p$ and the spring coefficient $K_n = E/2r_b$, while for shear dispersive wave $C_s = \rho v_s$ and $K_s = G/2r_b$. The ρ, E and G represent density, elastic and shear moduli of the foundation rock masses, respectively. $v_p = \sqrt{E/\rho}$, $v_s = \sqrt{G/\rho}$, are the velocities of compressive and shear waves. r_b as an average distance from wave source to the concerned pointis assumed to be the distance from the center of dam base to the concerned artificial boundary (Liu et al, 2006).

Theoretically, the accuracy of a 2-order transmitting boundary will be better than the viscous damping boundary. But it appears less efficient in parallel computation.

Another important problem is the seismic input mechanism for the dam-foundation-reservoir system. As well known, during an earthquake, strong ground motion energy will typically derive from a wide variety of azimuths, incident angles and consist of shear waves, surface waves, and compressional waves. However, the strongest ground shaking at period less than approximately 1 second consists primarily of shear waves. The fundamental period of high concrete dams even of 300m high located on high-velocity materials (e. g., hard-rock sites) will be less than 1 second. As the impedance in terms of the product of density and velocity of the foundation rock masses is increased with the depth, the incident wave tends gradually close to be vertical. For sites of high concrete dams most shear waves are nearly vertically incident into the shallow portion of the foundation. Consequently, the uniform seismic input is assumed to be vertically propagated to the base of the near-field foundation. Based on the theory of plane wave propagating at a homogeneous continuum, the three components of the input acceleration are assumed to be half of their design value defined usually at a flat ground surface of homogeneous rock foundation. In this case the spatial variation of seismic input along the dam foundation interfaces can consequently be reflected.

3. Contraction joints and sliding surfaces

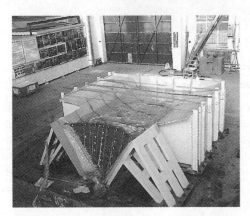

Fig. 16.1 - 12 Arch Dam Shaking Table Model Test for Arch Dam ($H=250$m) with Joints

The dam is constructed as a system of blocks separated by vertical joints at regular interval due to the temperature consideration. However, in traditional analysis both the concrete gravity and arch dams are treated as a 2 - D and 3 - D integral structure without joints. In this case for high arch dam during strong earthquake the tensile stresses in arch direction can be up to a non-accepted lever of 5 - 6MPa. Actually, the grouted contraction joints in arch dam can develop no or only a few of tensile stresses. The opening of the contraction joints releases the dynamic arch stresses and redistribute the stress state of the dam at all. Obviously, the geometrically non-linear features of contraction joints significantly effect on the seismic responses of arch dam and must be considered in the analysis. At first based on the program ADAP - 88 from U. S. was used. In this program the nonlinear joint elements without mass, damping and shear slipping as well as the substructure solution procedure have been used (Gregory et al, 1989). In order to verify the analytical results, the dynamic dam-reservoir model test for an arch dam of 250m high with rigid foundation has been carried out on the triaxial earthquake simulation shaking table of IWHR as shown in Fig. 16.1 - 12. In the test model the contraction joints were simulated and measured by transducers specially developed in laboratory.

By comparing the experimental and calculated results, the opening behavior has basically verified as shown in Fig. 16.1 - 13.

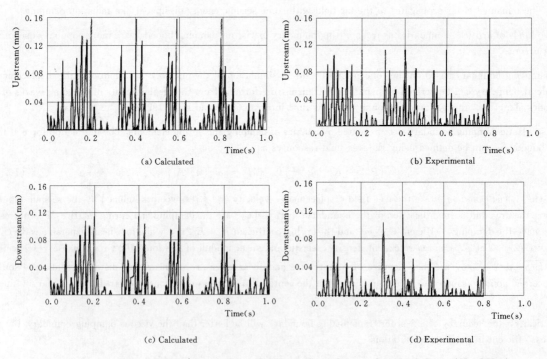

Fig. 16.1 - 13 Joint Opening/Closing Time History at Crest Crown of an Arch Dam ($H=250$m)

However, some problems in the ADAP-88 must be improved, such as:

(1) The non-penetration between contact surfaces and the convergence of algorithm cannot be ensured due to the arbitrary defined stiffness coefficients of the joint elements.

(2) The shear slipping cannot be considered. Actually the slipping is existed even for the contraction joints with shear key while the joint opening is occurred during strong earthquake.

(3) The substructure solution procedure is rather complicated.

Therefore, the effect of the contraction joints of the dam is considered as a dynamic contact problem solved by using a single-step method (Liu et al, 1993). In this method, a two-level algorithm is employed for dealing with the nonlinear boundary conditions caused by the dynamic contact of joint interfaces. At the first lever, an explicit approach is adopted to calculate the nodal displacement $\bar{u}_{ij}^{t+\Delta t}$ of the system without considering the effect of dynamic contact of interfaces

$$\bar{u}_{ij}^{t+\Delta t} = \frac{\Delta t^2}{2m_i}\left[F_{ij}^t - \sum_{l=1}^{n_e}\sum_{k=1}^{n} K_{ijlk} u_{lk}^t - \sum_{l=1}^{n_e}\sum_{k=1}^{n} C_{ijlk} u_{lk}^t \right] + u_{ij}^t + \Delta t\, \dot{u}_{ij}^t \qquad (16.1-2)$$

where $M_i = 2m_i/\Delta t^2$. The additional displacements introduced by the normal and tangential contact forces N_{ij}^t and τ_{ij}^t can be determined by the displacement and stress contact conditions. In this method, both the normal and shear contact forces can be considered and the contact conditions of non-penetrating and with Mohr-Coulomb shear strength in tangential slipping can be satisfied in a single step without iteration. Also, any assumption of the coefficients of stiffness and the penetration of the joint interfaces can be avoided. Besides, the width of any initial opening of the joint can be easy taken into account as well.

16.1.3.3 Criterion of Instability of the System

It is well known that the stability of foundation rock at abutments is a decisive factor for design of concrete dams especially during strong earthquake. For concrete dam, as a massive and redundant structure, being locally overstressed may not be broken because of stress readjustment.

Usually some geological disturbances of the bedrock like fault, cracks and local deterioration of the bedrock etc., almost unavoidably exist at dam site. So, particularly for arch dam some potential sliding blocks cut by these disturbance interfaces within the foundation rock at the abutments of both canyon banks can always be determined by geologic engineers.

Up to now, in conventional design the stability against sliding of such blocks is checked by the traditional limit equilibrium method for rigid body. However, this method is used only as a design approach based on engineering experience, and its main limitation can be summarized as follows:

(1) The maximum thrust from the dam abutment applied to a potential sliding block is determined without consideration of the dynamic interaction of the dam-foundation system during earthquake.

(2) Both the size and direction of the inertial force of the foundation rock at abutment is determined with more uncertainties.

(3) All the forces applied to the block are taken as time-independent, and the stability analysis during earthquake is essentially completed statically.

However, both the thrust from abutment and the inertial force of the rock are time-dependent. Hence both the size and direction of the resultant applied to the block as well as the sliding pattern of the block actually always vary with time. In order to improve the situation, at first an improved approach was adopted for the dynamic stability analysis of arch dam. In this approach, at each time step the thrusts from the abutments were calculated through the dynamic analysis of the dam-foundation system in time domain. Also, the resultant of all other forces applied to the block including the dynamic response of the foundation rock can be calculated by integration of the stresses along all interfaces of the potential sliding block. Then the sliding pattern and the safety factor of stability against sliding at each time step can be calculated. Finally, for each block during the earthquake the time-history of the safety factor of stability against sliding can be obtained as shown in Fig. 16.1-14 for Xiaowan arch dam ($H=295.5$m). From these curves the minimum value of the safety factor for each block can be found (Clough et al, 1984).

Although this approach is more reasonable in comparison with the conventional method, it still is based on the traditional limit equilibrium method for rigid body. In reality, any loss of stability is a process of gradual accumulation of deformation with stress readjustment due to local slippage and opening of the interfaces of the block. Neglecting the dynamic interaction in terms of deformation coupling is the fatal drawback of the traditional limit equilibrium method of stability analysis. Actually, instantaneously exceeding the limit equilibrium does not mean instable, but dam may be severely dam-

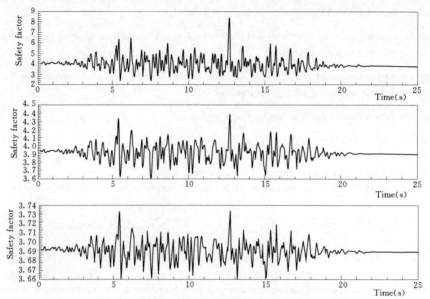

Fig. 16.1-14 Time-history of Safety Factor of Block No. 1, 2 and 3 at Left Bank of Xiaowan Arch Dam

aged due to deformation even the abutments. Particularly, an arch dam is very susceptible to cracking whenever abutment deformation develops. Owing to the existence of contraction joints in dam, the more feasible cracking place due to the abutment deformations probably will be the dam-foundation interfaces near crest during strong earthquake, as it is learned from the lessons of Pacoima dam during San Fernando (1971) and Northridge earthquakes. Obviously, the stability analysis of concrete dam must be based on the dam-foundation-reservoir system interaction and take the overall displacement as the deformation controlled criterion of final instability. However, it is impossible to give a unique threshold displacement as the criterion of overall instability of the system due to the variety of topography and geological disturbances at dam site as well as the type of dam. Therefore, it seems more reasonable to identify the stability of the dam-foundation-reservoir system by the turning points of its gradually increased displacement responses including the opening and slipping of the potential sliding blocks during the progressively enlarged seismic input. As an example, Fig. 16.1-15 shows the overload limit state with suddenly change of displacement of the Xiaowan arch dam of 294.5m (Chen et al, 2003).

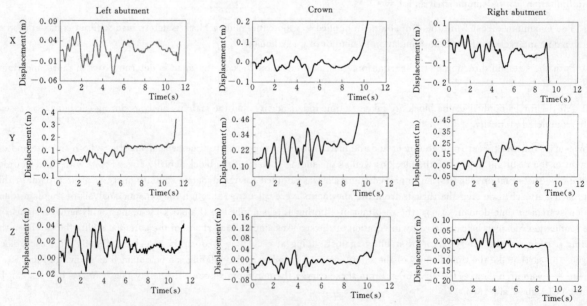

Fig. 16.1-15 Overload Limit State with Break Point of Displacement Response of the Xiaowan Arch Dam ($H=294.5$m)

16.1.3.4 Dynamic Model Test

Since some idealizations and assumptions as well as uncertainties in selecting parameters are hard to avoid in dynamic analysis of such sophisticated system like dam, special emphasis has been placed on the dynamic model test to improve and verify the analytical procedure and its results. For this purpose a large triaxial earthquake simulator with 6 degrees of freedom have been installed in IWHR. It was indicated by famous international experts in this field as the world's best facility for testing concrete dam models (U. S. National Research Council, 1990). A series of dynamic model tests for high arch dams have been completed on the shaking table.

1. Similarity requirements of high concrete dam dynamic model test

As all major quantities governing the behavior of the system in the model must be similar to those in the prototype, the simulated conditions from the theory of similarity specified for concrete dam has been investigated in detail. The ratio between the physical quantities of prototype and model is defined as similitude constant C. The similitude constants of geometry C_l, mass density C_ρ and elastic modulus of material C_E, are usually selected as the basic similitude constants. All the other quantities can be derived from them. As well known, for full similitude it is required to satisfy the relationship $C_E = C_l C_\rho$. It means that the similitude constants C_E, C_l, C_ρ cannot be determined arbitrarily in their own way. In practice, this requirement is hardly to be satisfied. Because the design of scale model is always restricted by a series of factors, such as test facilities, model materials and simulated conditions. However, the study on concrete dam usually limited to small deformation condition and the geometry nonlinearity can practically be neglected. In this case, the relationship $C_E = C_l C_\rho$ is not necessary to be satisfied. Then the similitude constant of displacement C_u is not needed to equal the geometry similitude constant C_L.

Generally, when determining the similitude constant of model in shaking table test, firstly it must be considered whether the dead weight influence is counted according to main content of test and requirements, then whether the geometric nonlinear large deformation issued is considered. Finally the determination of similitude constants for physical quantities need corresponds with the test conditons. Table 16.1-1 summarizes the similitude conditions of various physical quantities under different test conditions which are expressed in power function of basic similitude constants.

Table 16.1-1 Power Function of Physical Quantities for Different Test Conditions

Physical quantity	Considering dead weight influence $c_a = c_g = 1$						Not considering dead weight influence $c_a \neq c_g \neq 1$						
	Large deformation $c_u = c_l$			Small deformation $c_u \neq c_l$			Large deformation $c_u = c_l$			Small deformation $c_u \neq c_l$			
	c_l	c_ρ	c_E	c_l	c_ρ	c_E	c_l	c_ρ	c_E	c_l	c_ρ	c_E	c_a
Geometric dimensions c_l	1	0	0	1	0	0	1	0	0	1	0	0	0
Density c	0	1	0	0	1	0	0	1	0	0	1	0	0
Elastic modulus c_E	0	0	1	0	0	1	0	0	1	0	0	1	0
Displacement c_u	1	0	0	2	1	−1	1	0	0	2	1	−1	1
Acceleration c_a	0	0	0	0	0	0	−1	−1	1	0	0	0	1
Time c_t	1/2	0	0	1	1/2	−1/2	1	1/2	−1/2	1	1/2	−1/2	0
Frequency c_f	−1/2	0	0	−1	−1/2	1/2	−1	−1/2	1/2	−1	−1/2	1/2	0
Stress c	0	0	1	1	1	0	0	0	1	1	1	0	1
Strain c	0	0	0	1	1	−1	0	0	0	1	1	−1	1
Speed c_v	1/2	0	0	1	1/2	−1/2	0	−1/2	1/2	1	1/2	−1/2	1
Deflection c	0	0	0	1	1	−1	0	0	0	1	1	−1	1
Force c_P	2	0	1	3	1	0	2	0	1	3	1	0	1
Pressure c_p	0	0	1	1	1	0	0	0	1	1	1	0	1
Torque c_M	3	0	1	4	1	0	3	0	1	4	1	0	1
Poisson ratio c_μ	0	0	0	0	0	0	0	0	0	0	0	0	0
Damping ratio c	0	0	0	0	0	0	0	0	0	0	0	0	0

2. Dynamic shaking table model test of high concrete dams

Nevertheless, in dynamic test of concrete arch dam considering the dam-foundation-reservoir interaction some special problems still have to be solved as follows:

(1) To simulate the effect of dead load, the similitude constant of acceleration C_a should be taken as 1.0 as the gravity acceleration cannot be changed in model.

(2) To simulate the hydrodynamic pressure of reservoir water, practically only the natural water can be used. Hence the model material for dam and foundation must be specially weighted to have the same density as in prototype due to the requirement of $C_\rho = 1.0$.

(3) To simulate the dam, the tensile/compressive strength ratio of the model material should be the same as that of the dam concrete; to satisfy this requirement, the special brittle pressing block has been developed to be used as the model material of dam. Besides, the contraction joints in the dam should also be modeled.

(4) To simulate the hydrodynamic pressure, the reservoir with a length of no less than three times of the dam height has to be modeled.

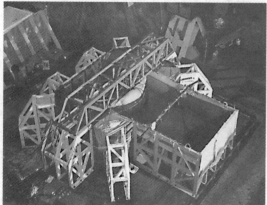

Fig. 16.1-16 Model of Arch Dam System on the Shaking Table

(5) To simulate some near-field foundation including the most critical potential sliding abutment blocks, the shear resistance properties of the possible sliding interfaces must be simulated and also the seepage force should be applied through some small jacks embedded in the foundation.

(6) To simulate the artificial boundaries, rigid shallow groove is needed to set up around the foundation model and pouring in high viscosity fluid to approximately reflect the radiation damping effect of the far-field foundation.

(7) To simulate the seismic input applied at the shaking table, the average uniform acceleration at the location of near-foundation base should be calculated by means of the corresponding analytical results. Of course, in the test the frequency of the acceleration must be converted according to the similitude law.

The acceleration, displacement, strain, hydrodynamic pressure, opening of contraction joints and overload safety factor of concrete cracking are measured and acquired automatically during the tests. Fig. 16.1-16 shows the model of an arch dam-foundation-reservoir system on the shaking table.

16.1.3.5 Field Experiment Investigation

As abovementioned, the structure dynamic model test is still limited by test facilities, model materials and simulated conditions. It is mainly used to verify the analytical results under specific conditions, and cannot fully reflect the real conditions. While the high dam seismic hazard example is fairly in shortage, so the field vibration measurement attracts more attentions to understand and verify the dynamic behaviors of concrete dams. In China, the measurement and research of dam kinetic characteristics such as Fengman and Xinfengjiang dam with the methods including micro-tremor, minor rocker, blasting in water with less dosage and microearthquake records since 50s last century. Till end of 70s, a 20 years a China-US cooperation in seismic research on field vibration experiment of arch dams have been carried out between University of California at Berkeley and IWHR supervised by professors R. W. Clough and K. T. Zhang.

In the first stages the research effort involved in performing harmonic forced vibration tests of two arch dams in China, Xianghongdian (gravity arch dam of 80m high) and Quanshui (thin shell double curvature arch dam of 80m high) using 4 synchronized rotating mass shakers manufactured in IWHR (Clough et al, 1984). It can be seen in Fig. 16.1-17. The test results successfully verified the natural vibration characteristics of arch dam calculated by using finite element program with the assumptions of massless foundation and incompressible reservoir water, if the modulus ratio of rock to concrete is chosen appropriately.

Fig. 16.1-17 Four Synchronized Eccentric Exciter, Xianghongdian Gravity Arch Dam and Quanshuishuangqu Arch Dam

The purpose of the second stage was to evaluate the effects of the compressibility of reservoir water by measuring the hydrodynamic pressure and acoustic reflection coefficients of the soil deposit at the bottom of the reservoir, and to verify the analytical results of the influence of spatial variation of seismic input along the dam-foundation interfaces of arch dams. Two arch dams in China were selected as test structures.

In the first test a reusable excitation approach by detonating explosive charges in an array of boreholes with water sealing in the foundation rock at a distance of about 800m away from the dam toe along the direction approximately parallel to the dam axis was used (Ghanaat, et al, 1993). It can be seen in Fig. 16. 1 - 18.

Fig. 16. 1 - 18   Stimulation by Deep Drilling Water-filling Row Shooting at Downstream of Dongjiang Arch Dam

In the second test due to the limitation of very steep banks at downstream a different approach by underwater detonating in upstream shallow rock surface at a distance of about 1200m away from the dam heel was used (Ghannaat et al, 1999). See in Fig. 16. 1 - 19.

Fig. 16. 1 - 19   Underwater Blasting in Upstream Shallow Rock Surface of Longyangxia Gravity Arch Dam

Both two new excitation approaches developed in the tests were successful to excite the entire dam-reservoir-foundation system by blasting-generated waves travelling through the foundation rock. In the first test from the basically identical records during the two blasting cases, the possibility of reusing the boreholes excited with water sealing has been verified.

In the second test vertical array of blasting caps were fired sequentially in the forebay region in order to measure the reflection of the reservoir boundary using the acoustic reverberation concept. The results indicated that reflection coefficient of the reservoir boundaries actually appears rather complicated. It is not only frequency dependent and but also varied with position at reservoir boundaries. It might be very difficult to be measured. Simply defining it as a constant might always be rather arbitrary and questionable.

The Long-term China-U. S. cooperative research program has successfully completed and significantly promoted the seismic study on arch dam in both countries. There is no question that the results of these cooperative studies were much more than that obtained by any single one of the participating sides.

16. 1. 3. 6   Damage-rupture Process

As mentioned above, with the rapid development of high dam construction in severe seismic region of Western China, it becomes the strategic priority in the study on seismic safety of high dams and the inevitable challenge to prevent any collapse of high dam under the largest reasonably conceivable earthquake which has to be faced by dam engineers in China. However, neither engineering practice all over the world nor lessons from past earthquake cases can be used for reference to ensure seismic safety of such extra high arch dams in severe seismic areas so long. Except clearly defining the MCE and reasonably selecting its site-specific seismic input parameters, the other obstacle is the quantitatively operable criterion of evaluating the limit state of dam-breach for designer. However, in the adopted model even though the boundary nonlinearity has been considered, but both dam concrete and foundation rock masses were still taken as elastic materials. Linear elastic analysis does not account for damage or stress redistribution as damage occurs, and gives the illusion of stress-carrying capacity with stress reversals that not be available. Consequently, the damage-rupture process of the system as the key problem of quantitatively evaluating the criterion of the limit state of dam-breach cannot be reflected. In order to solve the problem, the material nonlinearity of modeling crack initiation and propagation in dam concrete and foundation rock masses must be involved in the seismic response analysis of the system.

As well known, the damage of concrete and foundation rock masses as quasi-brittle materials is a result of continuous developing of random distributed microcracks. There have been many attempts to simulate the cracking process in concrete by using the discrete or smeared models as well as the plasticity-based model (Jeeho 1996; Radin et al, 2003). However, some commonly existed problems as follows in such models must be pointed:

(1) There is no interaction assumed neither between the elastic and crack stiffness matrixes nor among normal and shear components in crack matrix.

(2) It uses arbitrarily defined shear retention factor β and threshold angle α between new crack and former crack (s) directions as well as the parameters of the softening function under reciprocal loading like earthquake action.

(3) The stiffness degradation is not considered in the idealized plasticity-based model.

In general, it seems that to analyze damage-rupture process of concrete dams during strong earthquake by using the abovementioned models appears to be more complicated and might give relative poor accuracy due to some arbitrarily defined parameters.

Therefore, from an engineering point of view, it could be more reasonable to model the experimental phenomenological behavior of quasi-brittle materials like concrete and jointed foundation rock masses at macroscopic level by using damage mechanics based on the basic hypotheses of isotropic damage and strain equivalence.

1. Damage behaviors of dam concrete

As the cracking process in concrete is a continuous forming and connecting of microcracks, the formation of microcracks is represented macroscopically as strength softening and stiffness degradation which are related to a scale damage variable D. It can be taken from zero to one.

In principle, any proposed model and approach should be calibrated with experimental data. So far, only the response behaviors of concrete to uniaxial compressive and tensile loading have been implemented. Fig. 16.1 - 20 shows the typical experimental and idealized damage evolution functions of concrete under uniaxial cyclic tensile loading.

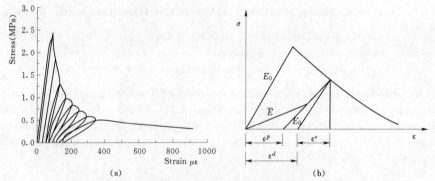

Fig. 16.1 - 20 Experimental and Idealized Damage Evolution Functions of Concrete Under uniaxial Cyclic Tensile Loading

The experimental results of concrete indicate the following special features:

(1) The stress softening and stiffness degradation clearly reveal when the stress exceeds the strength of material.

(2) The unloading and reloading within the cycle under reciprocal loading are elastic with a degraded secant modulus but without further damage. However, an irreversible strain is revealed.

(3) The damage evolution rules of tensile and compressive loading are quite different.

(4) The compressive properties are recovered upon crack closure as the load change from tension to compression.

(5) As a material property of concrete, the fracture energy G is the product of the area under the damaged stress-strain curve 'g' and a characteristic length 'l_{cr}'.

The finite element method is commonly used for dynamic analysis of concrete dams. Usually the characteristic length 'l_{cr}' is assumed equal to the side length of an equivalent tributary area or volume at Gaussian point of an iso-parametric 2 - D or 3 - D finite element.

It should be emphasized that although tensile strength is only one parameter which governs the failure of concrete, it is a dominant one. Particularly for dam engineering the cracking due to tensile stress is more concerned. As both the strength and the design safety factor in compression case are much higher than those in tension case, in engineering practice the concrete dam is hardly to be damaged due to compressive stresses during earthquake.

Considerable work has been performed in dynamic analysis of concrete dam. Most of them are limited in elastic continuum damage (Faouze and Rene, 1995; Miguel and Jene, 1995; Hariri-Ardebili and Mirzabozorg, 2013). However, cracking in concrete and foundation rock masses is always attendant with an inelastic irreversible deformation which cannot be neglected in analysis. So, the plastic-damage model for representing general concrete damage behavior proposed by Lubliner et al (1989) and by Lee and Fenves (1989 and 1998) and also brought into the commercial program ABAQUS (2009) has been often used. The plastic-damage model has provided a general capability for modeling the damage behavior of concrete and other quasi-brittle materials.

But it is worthy to be noted, that the inelastic residual deformation caused mainly by inner friction in concrete is essentially distinguished from the plastic deformation due to crystal dislocation in homogeneous metal material. There is no experimental data to validate that the inelastic residual deformation in concrete is complied with the flow rule of plastic theory. Actually, the effect of inelastic residual deformation has already been embodied in the stiffness degradation in terms of the reduction of elastic modulus as the major damage feature of concrete. Therefore, it is not necessary to involve the anisotropic plastic deformation inconsistent with the basic isotropic damage assumption in damage theory with scale damage variable. Besides, the calculation of the plastic potential has to be defined in effective stress space and an additional damage variable has to be involved for determining the hardening variable in yield surface and the damage evolution function. Also, the non-associated plastic flow has to be used for concrete damaged plasticity. It results in the uncertainties of selecting the dilatancy angle and in the non-symmetry of the material stiffness matrix. Furthermore, the plastic-damage coupling cannot but lead to a rather complicated numerical algorithm.

2. Damage behaviors of foundation rock masses

Up to now, during the dynamic analysis of damage-rupture process of concrete dam suffering strong earthquake, the damage behavior of foundation rock masses has been neglected at all. However, as the damage of jointed foundation rock masses is usually more fragile than dam concrete, it does significantly effect on the damage-rupture process of the whole system and must be considered in the analysis.

Regarding this aspect, some basic concepts of the foundation rock masses have to be confronted. First of all, the damage-rupture process of foundation rock masses is quite complicated. At present there is no generally acknowledged modeling for foundation rock masses. But it is common recognized that the jointed foundation rock masses is always discontinuous with microcracks. So, it might be suggested that both the tensile strength and fracture energy of the rock masses might be less than those of dam concrete. During strong earthquake the damage of jointed foundation rock masses is mainly manifested by the cracking development caused by tensile stress. It is unrealistic to establish theoretical or experimental damage evolution function for rock masses at present. Therefore, it has to be assumed that the evolution function of rock masses and dam concrete are somewhat simulated, as both of them are quasi-brittle non-homogeneous materials. Owing to the fact that no better approach can be found so far, the evolution function of rock masses has to be temporarily borrowed from that of dam concrete only by adjusting the strength.

In view of the above-mentioned facts, for engineering practice, a new approach is proposed to analyze the inelastic damage-rupture process of concrete dams, with the consideration of the irreversible residual deformation under alternated loading of earthquake based directly on the experimental data, avoiding the plastic-damage coupling.

As well known, the strength of concrete is relatively dependent. Numerous researchers reported that both the dynamic tensile and compressive strengths are higher than their static strengths (U. S. Department of the Interior Bureau of Reclamation, 2006). However, in experimental tests on plain concrete (Reinhardt and Weerheijm, 1991), it has been shown that the normalized damage evolution function is little affected by loading rate. It then might be possible to determine the damage evolution function of various loading rates only by measuring the damage evolution function at static loading with adjusted tensile strength.

3. Implementation of the proposed approach for seismic analysis of damage-rupture process of concrete dams

The dam-foundation-reservoir system is multiaxial. In the proposed approach the Barcelona model is used as damage surface function of dam concrete and foundation rock masses (Wang et al, 2000).

$$F=\frac{1}{1-\alpha}[\alpha I_1 + \sqrt{3J_2}+\beta\langle\hat{\sigma}_1\rangle]-c_c=0 \qquad (16.1-3)$$

where the I_1 and J_2 are the first invariant of stress and second invariant of stress deviator respectively, and the α and β are the dimensionless constants evaluated by the initial shape of damage surface which are related to the ratio of equibiaxial to uniaxial compressive strengths and the ratio of compressive to tensile strengths. α and β for concrete can usually be taken as 0.12 and 7.68 (Wang et al, 2000).

As only the uniaxial compressive and tensile loading tests can be implemented, the evolution functions in uniaxial tests must be extended for the multiaxial conditions. The transition from the multiaxial case to uniaxial case is carried out by means of the maximum and minimum principal strains (ε_1, ε_3) as the equivalent strains. Also, in order to combine the tensile and compressive damage variables deduced separately from the uniaxial tests into an equivalent damage variable $\widetilde{D}$ utilized for analyzing the multiaxial system, a weight factor is involved and defined as: $r(\hat{\sigma}) = \sum_{i=1}^{3}\langle\hat{\sigma}_i\rangle / \sum_{i=1}^{3}|\hat{\sigma}_i|$, where $\hat{\sigma}_i$ ($i=1, 2, 3$) are the principal stress components and $\langle x \rangle = \frac{1}{2}(|x|+x)$.

The finite element implementation of the proposed approach can be described as follows:

When the finite element method is used to solve the boundary problem of non-linear seismic responses, the equilibrium equations of the dam-foundation-reservoir system are

$$[M]\{\ddot{u}\}+[C]\{\dot{u}\}+[\widetilde{K}]\{u\}=\{P\} \qquad (16.1-4)$$

where, $[M]$ is the mass matrix including the added mass of hydrodynamic pressure; $[C]$ is the damping matrix; $[\widetilde{K}]$ is the stiffness matrix with the degradation moduli including the effect of the irreversible strains if some part of the system are damaged; $\{P\}$ is the load vector including both the static and dynamic loads; $\{\ddot{u}\}$, $\{\dot{u}\}$, $\{u\}$ are the acceleration, velocity and displacement responses vectors, respectively. The equations are solved by explicit integration in time domain.

During the ith time interval for the jth Gaussian integration point within the kth element, the global strain and stress states as well as the weight factor r can be calculated. Then the damage condition can be checked from the Barcelona damage surface function $F=0$.

If $F<0$, the concerned point is not damaged and its elastic modulus can be taken as the original value E_0, otherwise, the point is damaged and the threshold strength should be calculated relying on the principal stress vector and the damage surface function $F=0$. It will be in terms of tensile strength f_t while $r>0$ and of compressive strength f_c while $r=0$.

The uniaxial tested damage evolution functions like in Fig. 16.1-1 (for tension loading) can be adjusted by the threshold strengths. From them the functions of the cracking displacement to degraded strength w-f and to the increased damage variable w-D can be directly derived. Then, from the experimental tensile or compressive evolution function the apparent modulus $\overline{E}=(1-\overline{D})E_0=\frac{\sigma}{\varepsilon}$ for loading case and $\overline{E}=(1-\overline{D})E_0=(1-D)E_0\left(1-\frac{\varepsilon^p}{\varepsilon}\right)=\frac{\sigma}{\varepsilon}$ for unloading and reloading case can be obtained. Where $\overline{D}$ is the apparent damage variable, D and ε^p are the damage variable and the irreversible strain corresponding to the state of starting unloading of the cycle, respectively.

During the unloading and reloading cases of reciprocal loading, the effect of the irreversible strain has been embodied in the degraded apparent modulus.

The loading/unloading conditions can simply be checked by comparing the values of $r\varepsilon_t$ (for tensile case) or of $(1-r)\varepsilon_c$ (for compressive case) at the current i^{th} time interval with those values at the previous $(i-1)^{th}$ time interval.

Finally, the equivalent damage variable for the multiaxial dam-foundation-reservoir system is assumed as $\widetilde{D}=1-(1-(1-r)\overline{D}_c)(1-r\overline{D})$, and the corresponding equivalent degraded elastic modulus for the multiaxial system is taken as $\widetilde{E}=(1-\widetilde{D})E_0$.

Within the loading and unloading cycle concerned in uniaxial damage evolution function, the irreversible strain in terms of $\varepsilon^p=\varepsilon-\frac{\sigma}{(1-D)E_0}$ can be calculated.

In tension unloading case when ε is close to ε^p as the load tends to be changed from tension to compression, the irreversible strain state of the multiaxial system can be calculated from the formula (16.1-2) with consideration of the equivalent degraded elastic modulus $\widetilde{E}=(1-\widetilde{D})E_0=\{[1-(1-r)\overline{D}_c](1-r\overline{D}_t)\}E_0$ of the damaged point, where the r is the weight factor when the calculated maximum principal strain ε_1 of the concerned Gaussian point equals to the corresponding irreversible strain ε^p of the uniaxial evolution function.

Generally speaking, it's the same in compression loading case, only the $1-r$ instead of r will be taken, and both the irreversible strain states of tension and compression loading cases will be accumulated and remained under reciprocal loading.

However, as above-mentioned; in engineering practice, the concrete dam is hardly to be damaged due to compressive stresses even during strong earthquake. Hence for the time being it might be acceptable with no view of the compressive damage in seismic analysis of concrete dams. Then, the degraded equivalent damage variable and elastic modulus accordingly becomes $\widetilde{D}=1-(1-r\overline{D}_t)$ and $\widetilde{E}=(1-\widetilde{D})E_0$.

Once the irreversible strain state is considered in the multiaxial dam-foundation-reservoir system, in the equilibrium equation (16.1-2) of the system the displacement vector $\{u\}$ will be replaced by $\{u-u^p\}$.

It must be additionally noted that in analysis of the dam-foundation-reservoir system for foundation rock masses the following factors must be considered:

(1) The radiation damping of the far-field foundation must be taken into account.

(2) The vertical and horizontal components of the initial stresses of foundation must be included, which are usually more than 1.2 and 1.0 times the dead load of rock mass respectively, according to the experience of geological engineers.

(3) The seepage stresses must be calculated with consideration of the effects of grout curtain and drainage system in the foundation.

(4) The tensile and compressive strengths can be derived from Mohr-Coulomb formula by the friction coefficient f and viscous coefficient c as well as the deformation modulus E of the rock masses. Those are the only parameters can be provided by geological engineers for dam engineering.

Obviously, the proposed approach seems more consistent to the basic assumption of smeared isotropic damage, and greatly simplifies the analytical algorithm in comparison with the plastic-damage coupling method.

4. Validation of the proposed approach by the behaviors of dams subjected to strong Earthquakes

In order to verify the proposed approach of damage-rupture process of high concrete dams, the validation of dynamic behaviors of the Koyna gravity dam in India and the Shapai RCC arch dam in China both subjected to strong earthquakes by using the recommended damage model and the proposed approach has been implemented.

(1) Validation of the seismic damage of concrete gravity Koyna Dam.

On Dec. 11, 1967 the Koyna gravity dam made of rubble concrete ($H=104$m) above massive basalt foundation in India was subjected to a strong earthquake of $M=6.5$ with a PGA of $0.51g$ at the dam site. Impenetrate cracks were discovered near the abrupt change of downstream slope.

After earthquake the leakage increased from 500L/min to 1500L/min, but the uplift pressures did not indicate any sudden increase and the bore core indicated that the dam-foundation interface was well connected. Obviously, the grout curtain was not damaged (Gupta, 1976; Pender et al, 1998; Murti et al, 1970).

The analysis of dam-foundation system is carried out to consider damage both in dam and foundation rock masses.

With reference to the example of ABAQUS, the same damage evolution function and the material parameters waere adopted The material parameters for foundation were taken as: tensile strength $f_t=2c\cdot\cos\varphi/(1+\sin\varphi)=1.28$MPa($\varphi=54.46°$, $c=2.0$MPa) original elastic modulus $E_0=20$GPa, stiffness damping parameter of Rayleigh type of damping $\beta=0.00323$, Poisson ratio $\mu=0.2$, density $\rho=2,700$kg/m, $G_F=88.4$N/m. In the analysis the viscoelastic damping boundaries were used.

The FEM mesh of dam-foundation system including only the potential damage upper central area with non-linear material properties and the calculated cracking pattern in dam and foundation are shown in Fig. 16.1-21.

To consider non-linear material properties both in dam and foundation rock masses, the analytical results indicate that the cracks appear not along the dam-foundation interface as happened in elastic analysis, but extended vertically in the foundation rock masses near the dam heel. So, both the dam-foundation interface and the grout curtain are not damaged, as it was verified by the investigation after 1967 earthquake in Koyna Dam.

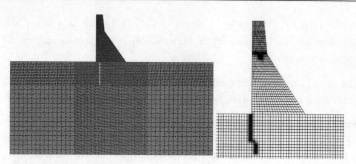

Fig. 16.1-21 FEM Mesh of Dam-foundation System and the Damage Pattern

It is worthy to be noted that the stress concentration mechanism in elastic analysis is not revealed. Actually, the problem of stress concentration at dam heelis largely fictitious due to the fact that the jointed to a significant degree foundation rock masses cannot develop fairly large tensile stress. Probably, the long-standing perplex corner effect, asa major obstacle to apply the FEM in dam design, might be expected to overcome by using the proposed approach.

(2) Verifying the seismic behavior of Shapai RCC arch Dam.

On May 12, 2008 the Shapai RCC arch dam of 132m high in China was subjected to strong Wenchuan earthquake of $M=$ 8.0 with an epicentral distance of about 36km. The reservoir water level is near the normal design level. No damage has been discovered during thorough inspection at dam site after the Wenchuan earthquake. The seismic design of the dam has been carried out with a design peak acceleration of 0.138g corresponding to a return period of about 500 years based on a special seismic hazard analysis for the dam site. As the dam subjected to an earthquake much higher than its design earthquake was not damaged, it can be necessary to be thorough explained. For this purpose it was checked by the proposed approach.

As no ground motion input has been recorded at Shapai RCC arch dam site during the Wenchuan earthquake, it has to be reestablished by using the "stochastic finite fault method" with model and parameters identified by the accelerograms recorded at 7 stations during the earthquake. The reestablished acceleration at Shapai dam site is 0.262g with a long duration of more than 40 minutes. All static loads such as water, seepage and silt pressure, temperature load, dead load and initial ground stresses are considered consistently with the conditions during the Wenchuan earthquake. Fig. 16.1-22 is the FEM mesh of the whole dam-foundation system and the FEM mesh of predicted damage area of dam and adjacent foundation. The viscoelastic artificial boundaries were added. The concrete strength is determined by the testing with the cylinder specimens drilled directly from the Shapai arch dam body after the Wenchuan Earthquake. The splitting tensile test was implemented. The design tensile strength was taken as the apparent strength of the splitting tensile test which is approximately equal to the strength of modulus of rupture test (Test code for hydraulic concrete, 2006). The total number of nodes, elements, and D.O.F are 425,568, 404,090, and 1,276,704, respectively.

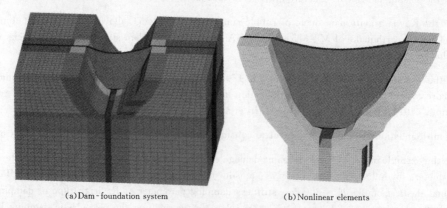

(a) Dam-foundation system   (b) Nonlinear elements

Fig. 16.1-22 FEM Mesh of Dam-foundation Systemand the Mesh of Nonlinear Elements

The calculated results in Fig. 16.1-23 has showed that the dam is basically not damaged even near the bottom pedestal, but the jointed foundation rock body is damaged, as the tensile strength and fracture energy of cracked foundation rock body are less than those of dam concrete.

16.1.3.7 Parallel Calculation Technique

The storage capacity and the computation time of dynamic analysis for seismic responses of concrete dam system are enormous. As a matter of course, the use of high-performance computation for seismic analysis of high concrete dams must be

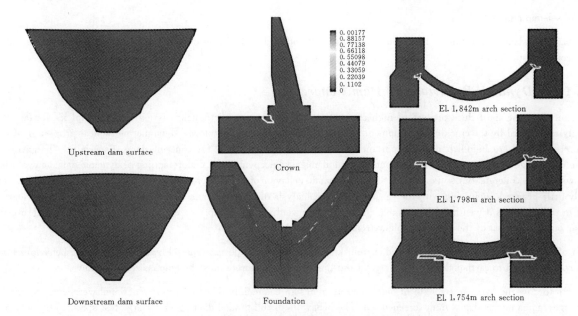

Fig. 16.1-23 Seismic Damage of Jointed Foundation Rock Masses in Shapai RCC Arch Dam Subjected to 0.262g PGA

enforced. Although there are a number of commercial programs like MSC, MARC, ANSYS, FLUENT have been parallelized, right now it seems difficult to satisfy the constantly deepening seismic analysis of high concrete dams. At first, the opened developing environment of parallel finite element programs based on the Parallel Finite Element Program Generator (PFEPG) developed by Liang Guoping (1990) was adopted to solving the problems in seismic analysis of high concrete dams and meso-mechanics analysis of full graded dam concrete specimens.

Through some practical examples of applying the PFEPG based on a LENOVO 1,800 PC cluster parallel computation device of 6 nodes each with double CPU in IWHR, the correctness, reliability and efficiency of the parallel computational programs have preliminary been illustrated and verified.

Fig. 16.1-24 shows the finite element mesh of the Xiluodu Arch Dam ($H=284.5$m) with total number of nodes, elements are 146,769 and 118,828 respectively.

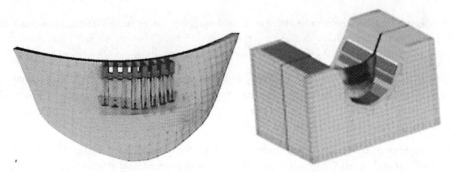

Fig. 16.1-24 Finite Element Mesh of the Xiluodu Arch Dam ($H=285.5$m)

However, the user of the Parallel Finite Element Program Generator (PFEPG) has to know well and master the special finite element programing language, which seems a little difficult and not being popular throughout the dam engineering circles. Besides, the PFEPG cannot include the gradually developed contents of seismic analysis of high concrete dams.

At present, the directly developed parallel program for dynamic analysis of concrete dam-foundation-reservoir system considering all conditions to be as close to reality as possible as well because both the boundary and material non-linearity has been developed. It has successfully used for the dam-foundation-reservoir system with millions degrees of freedom. As one of the example of application, the finite element mesh of Shapai arch dam (132m) with 1,276,724 D.O.F. has already shown in Fig. 16.1-22.

All these parallel programs have been favorably used in the hyper-computer Tianhe No.1 which was being the fastest computer worldwide. Its remarkable efficiency is of extreme importance for the studies on seismic safety of high concrete dams in severe seismic region in China.

The parallel calculation technique has been also successfully applied in 3-D meso-mechanics dynamic non-linear analysis of dam concrete specimens.

16.1.4 Dynamic Behaviors of Dam Concrete

Dynamic response and failure process of high concrete dam under strong earthquakes is extremely complex; its failure is mainly manifested by the cracking development caused by tensile stress. Therefore, dynamic mechanical properties of dam concrete material are an indispensible important part to ensure seismic safety for concrete dams. Especially, the extremely complex damage-rupture process of high concrete arch dams asks deepening the understanding of dynamic damage evolutional laws of dam concrete. But so far the research on dynamic behavior of dam concrete has somewhat overlooked and has relatively little progress compering with the seismic response analysis of concrete dam. Recently the impetus to pay more attention to this aspect is driven by the inevitable challenge of preventing any collapse of extra-high arch dams in western China under potential threat of the unpredictable extreme strong earthquake.

First of all, with the commonly recognized current knowledge, some important special features of dynamic behavior of dam concrete subjected to earthquake quite different from the structure concrete must be emphasized as follows.

(1) Up to now the dynamic strength of dam concrete in design has mainly relied on the test of small wet sieving specimens with aggregates bigger than 40mm screened out. The behaviors of full graded dam concrete large test specimen usually having large size coarse aggregate up to 150mm are more inhomogeneous and significantly different from that of the test with small wet-screened specimen due to the size effect and the change of mix proportion.

(2) During strong earthquake, the damage of concrete dam is mainly manifested by the cracking development caused by tensile stress. However the tensile strength is dependent upon the stress state in structure. The modulus of rupture tensile strengths are more appropriate for stress in a concrete dam particularly during earthquake where the major principal stresses are compressive and the major principal stresses are tensile (U.S. Department of the Interior Bureau of Reclamation, 2006). The conventional uniaxial tensile strength is more appropriate for stress in a case where the major principal stresses are tensile and the minor principal stress is small.

(3) The concrete mechanical properties are highly sensitive to loading type and strain rate. The dynamic strengths are higher than the static strengths. The tensile strengths under cyclic loading are somewhat lower in comparison with the impact loading usually used for dynamic test probably due to the lower-frequency fatigue effect. Under seismic loading the dynamic rate-dependent strength in concrete dam mainly dependent on the major response frequency of dam and having alternating character within a strain rate range of about $10^{-3}-10^{-2}$.

(4) Earthquake mostly occurs when the dam has been in normal operation under static loading. However, the dynamic tests always have been carried out under unitary dynamic loading without static preloads so far. Therefore, the dam engineers have concerned more about the overlooked effect of static preloading on the dynamic strength of concrete for a long time.

(5) For the analysis of damage-rupture process of concrete dams, the experimental damage evolutional functions are the important basis. But, at present, there are only a fewdata of the tests for damage evolutional functions of concrete, and what's more, the tests were still limited in uniaxial compression and tension cases. As a matter of fact, there is a shortage of such data especal for the dam concrete can be used now.

Considering all above-mentioned factors, recently under support of National Natural Science Foundation of China and the cooperation between IWHR and Hohai University, recently a series of experimental studies on static and dynamic behaviors of dam concrete have been completed (Chen et al, 2012). The main ideas are to investigate and explain results of the tests. It includs①dynamic mechanical property tests of high full graded dam concrete; ②manifestation and explanation of the test results based on interior structure of the material by using dynamic 3-D "meso-mechanics" analysis of full-graded dam concrete specimen and X-ray computed topography (CT) technique as well as dynamic acoustic emission technology.

16.1.4.1 Dynamic Mechanical Property Tests of Full Graded Dam Concrete

The test research mainly aims at the principal feature of dynamic mechanical properties for high dam concrete different from general structural concrete, especially at the flexural-tension test of full graded dam concrete specimen integrated with Xiaowan and Dagangshan high arch dams. The specimens were prepared with all the consisted materials transported from the site of those two dams and according to their real mix-proportions.

The test program was mainly focused on the modulus of rupture tension tests as it is the key factor affecting seismic safety of the dam. The modulus of rupture tensile strengths of dam concrete of 180d design age were studied by testing concrete beams simply supported with tri-point loading both in static state and in dynamic state with impact and triangle cyclic waves under different strain rate and static preloads. The specimen sizes of three and four graded large specimens are 300mm× 300mm×1,100mm and 450mm×450mm×1,700mm respectively while the size of wet screened specimens is 150mm× 150mm×550mm according to the *Test code for hydraulic concrete* (SL 352—2006). The test facility and failure shape of full graded dam concrete specimen in the flexural-tension test are shown in Fig. 16. 1 - 25.

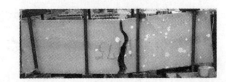

Fig. 16. 1 - 25 The Test Facility and Failure Shape of Full Graded Dam Concrete Specimen in the Flexural-tension Test

Some major test results can be summarized as follows.

(1) The ratios of cube compressive strength of full graded specimens to wet screened specimens for Xiaowan and Dagangshan projects are 0. 87 and 0. 78 respectively. The static cube compressive strength of full graded concrete specimen has commenly been taken as the standard value of dam concrete. However, probably because of taking into account the differences between the specimen and the in-situ concrete, the ratio has usually been inclined to be safe and taken as 0. 67 in dam design.

(2) The ratios of static modulus of rupture tensile strength of full graded and wet screened concrete for the two projects are 0. 58 and 0. 46 respectively. The ratio of three graded concrete is slightly higher than that of four graded concrete. The average value is about 0. 52.

(3) The ratios of static modulus of rupture tensile strength and cube compression strength of dam concrete for two projects are 0. 129 and 0. 125 respectively, of which the average mean is 0. 127.

(4) In dynamic test, the cyclic loading method approximates to the seismic effort and the modulus of rupture tensile strength is lower than impact loading due to low cycle fatigue. Under cyclic loading the ratios of dynamic modulus of rupture tensile strength and static strength of full graded specimens and wet screened specimens of Xiaowan arch dam concrete are 1. 17 and 1. 26, while the ratios of Dagangshan arch dam are 1. 18 and 1. 11. In general, the dynamic modulus of rupture tensile strength is about 20% higher than static modulus of rupture tensile strength.

Therefrom the ratio of standard values of modulus of rupture tensile strength f_{mr} and cube compression strength f_c can be calculated as:

$$\frac{f_{mr}^l}{f_c^l} = \frac{f_{mr}^l}{f_{mr}^s} \frac{f_{mr}^s}{f_c^s} \frac{f_c^s}{f_c^l} = 0.52 \times 0.127 \div 0.67 = 0.0986 \qquad (16.1-5)$$

where, the superscripts l and s indicate the strength f of full graded and wet screened specimens respectively.

Accordingly, the standard valueof modulus of rupture tensilestrength is taken as 10% of cube compression strength standard value in seismic design of concrete dam.

(5) In all test cases for Xiaowan and Dagangshan projects the results show that the dynamic elastic modulus is only 6% higher than the static values. Considering creep effect of dam concrete under the long-term static loading, in dam design its static elastic modulus is only about 2/3 of the value measured in a few minutesin laboratory. According to the many test results, the U. S. bureau of reclamation consider that, in the statistical average sense, dynamic and static elastic modulus determined by testing results can take for equal. Obviously, in comparison with static case the increase of dynamic elastic modulus is essentially different from the increase of dynamic strength due to strain ratio effect, but reflects the difference between long time static creeping effect and instantaneous loading in laboratory.

(6) It was unexpected that if the influence of initial static loading less than 80% of the tensile strength, in general, the increment ratio of dynamic to static strengths both for full graded and wet screened specimens under impact and cyclic loading

of the two project might be somewhat amplified as shown in Fig. 16.1-26.

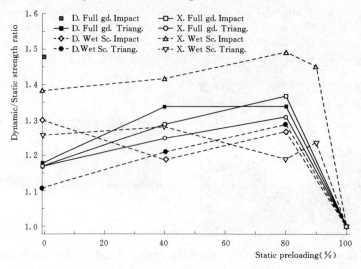

Fig. 16.1-26 Influence of Initial Static Loading on the Increase of the Ratio of Dynamic to Static Strengths

Usually the strain rate effect of weak material will be more conspicuous than strong material. The strain rate effect might be intensified by the material degradation due to development and generation of microcracks caused by static preloading. The results of dynamic test with static preloading might be possible to explain as fellows. If the intensified strain rate effect exceeds the degradation due to static preloading, the dynamic strength might be increased under the following dynamic loading. But when the initial static load exceeds about 80% of tensile strength, the strain rate effect intensified by static preloading could outweigh its degradation effect and the quickly developed and combined microcracks would reduce the tensile strength of concrete. However, under dynamic loading, the damage process and strain rate effect mechanism of concrete are very complex and hard to form common understanding until now. So far as engineering purposes are concerned, it might be inclined to be safe to overlook the effect of static preloading in seismic design of concrete dams. Because the high concrete dam usually has enough safety margin, and the normal static load is far less than 80% ultimate load.

Of course, the proposed explanation of test results has to be further explained and verified through analytical or experimental investigation on the interior structures of the concrete.

(7) For further research on dynamic damage-rupture process of high concrete dams, it is very crucial to establish uniaxial tensile damage evolutional function of dam concrete by direct tension tests. As well known, such kinds of static and dynamic direct tension tests for quasi-brittle materials like concrete are rather difficult. After finishing all preparation like to ensure stiffness of loading system, to improve centering adjustment and clamping device etc., the dynamic tensile damage evolutional function under cyclic loading with cylinder specimens drilled directly from the dam body

Fig. 16.1-27 15,000kN Dynamic Test Machine in IWHR

has been completed as shown in Fig. 16.1-20. Now a 15,000kN dynamic test machine for fully graded specimens of dam concrete has been installed in IWHR (Fig. 16.1-27). The systematic tests for dynamic tensile damage evolutional function with fully graded specimens of dam concrete under cyclic loading of different strain rate are being carried out using this equipment.

16.1.4.2 Manifestation and Explanation of the Test Results Based on Interior Structure of Concrete

In order to manifest and explain the effect of strain rate and verify the influence of the static preloading on dynamic behaviors of dam concrete, the dynamic 3-D "meso-mechanics" analysis on full-graded dam concrete specimen and X-ray computed topography (CT) technique as well as dynamic acoustic emission technology have been used.

(1) Dynamic 3-D "meso-mechanics" analysis.

In order to verify the influence of the static preloading on dynamic behaviors of dam concrete, the non-linear dynamic three-dimensional meso-mechanics analysis under static loading and dynamic loading with different static preloads has been carried out for F. E. model of dam concrete specimens as the composite material consisting of differently sized aggregates, mortar, and their interfacial transition zones satisfying the actual mix proportion requirement of dam concrete (Ma, 2005).

Model is discretized of the system in space by finite elements and in time by Newmark's formula and solved by dynamic analysis. Fig. 16.1-28 is the sketch of modulus of rupture tensile strength test specimen and its finite element mesh of aggregates, mortar, and their interfacial transition zones (Chen et al, 2008).

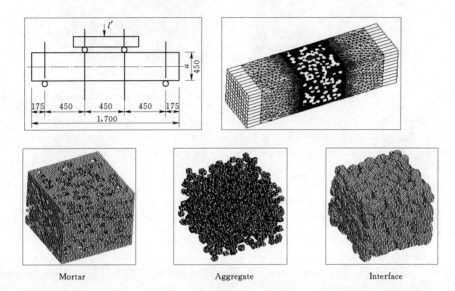

Fig. 16.1-28 Sketch of Test Specimen and its Finite Element Mesh of Aggregates, Mortar, and Their Interfacial Transition Zones

In the analysis following factors are considered:

1) Based on existed experimental data the strain rate effect for tensile stress and modulus of elasticity are taken as exponent functions.

2) A double-broken-line elastic damage evolutional function is assumed.

3) Not only the multi-graded aggregates but also the properties of aggregate, mortar and their interface are random distributed in space using Monte-Carlo method.

4) The interfaces between mortar and aggregate are treated as contact surfaces.

5) In mix ratio of full graded dam concrete, the content of various aggregates amounts to 60% - 70%. The actual concrete specimens are compacted by vibration during pouring. However, the compacting effect cannot be reflected during dosing random distributed aggregates to the specimen space by Monte Carlo Method. So it is hard to meet the requirement of the aggregate content. For this purpose, a "get rid of occupied space" approach has been developed to considerably reduce the time for random dosing aggregates.

6) For concrete specimen there are numerous finest aggregates difficult to be simulated in the meso-mechanics model. So, the fine aggregate and cement mortar have to be considered as two-scale composite materials and to figure out its equivalent mechanics parameters using two-scale analysis theory to significantly reduce the amount of degree of freedom (Cui and Shan, 2000). In the two-scale theory, the composite structures are logically decomposed into a series of assembles with the size of far larger than the maximum grains as single cell of statistical screen. the particle probability distribution of each single cell is same, so it equals to the probability distribution of particles for entire structure.

The results of meso-mechanics analysis also revealed that the dynamic strength increased with the static preload as indicated in Fig. 16.1-29. The analytical and experimental results are mutually in good agreement and the abovementioned preliminary explanation for them probably might be tenable.

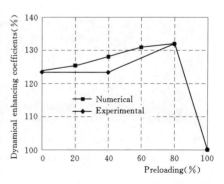

Fig. 16.1-29 Influence of Static Preloading on Dynamical Modulus of Rupture Tensile Strength

To explore the mechanism of low strain rate effect on dynamic tensile strength of concrete, a two-dimensional meso-mechanics analysis of concrete both for static and dynamic loadings was carried out. Different failure patterns for static and dynamic loadings can be clearly displayed as shown in Fig. 16. 1 - 30.

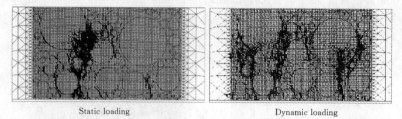

Fig 16. 1 - 30 Different Failure Patterns for Static and Dynamic Loadings

Obviously, under static loading relative fully developed crack propagation mainly follow the weakest monotonous path, while under dynamic loading some partially developed and gentle cracks spread along several different routes something like network simultaneously resulting in motivating a higher dynamic strength with increasing strain rate.

(2) X-ray computed topography (CT) technique.

X-ray computed topography (CT) technique is a useful tool to detect the process of the initiation, propagation and continuous accumulation of interior micro damage leading to macro crack and finally to the complete failure of concrete by non-destructive way (Ding, 2007).

1) To investigate the interior failure structure of concrete under dynamic loading needs the CT machine with rapid scanning function and high image resolution. Right now only the medical electron beam CT (EBCT) machine can be used. Then a potable dynamic real-time loading device specially designed for CT test of concrete specimen has been manufactured as shown in Fig. 16. 1 - 31. The maximum output of the device is 100kN and the allowable maximum diameter of test sample is 150mm. The dynamic loading process can be controlled by the displacement of impulse wave, sine wave or any given waveform with the maximum frequency of 5Hz. by manner and switch over during loading; All the load-time curve, displacement-time curve and load-displacement curve can output automatically.

2) Cracks in CT images were distinguished and their morphology characters were analyzed based on CT physic theory and image processing technique. The difference of cracking process between the static and dynamic compression loading conditions was investigated in detail. Fig. 16. 1 - 32 shows the comparison among the CT images of concrete before cracking and with low and high strain rates. It also reveals the different failure patterns for static and dynamic loadings quite similar to the results of meso-mechanics analyses.

Fig. 16. 1 - 31 Dynamic Real-time Loading Device Specially Designed for Concrete CT Test

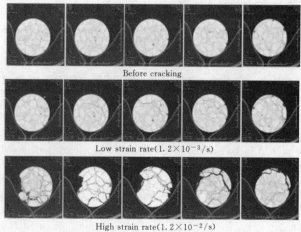

Fig. 16. 1 - 32 CT Images of Concrete before Cracking and with Low and High Strain Rates

3) In order to intuitively demonstrate the crack morphology character and crack evolution process during the loading, the formation of 3-D images and the animated cartoons displaying technique of the crack development by using the CT test results have been completed using 3DMAX software. The movable spatial crack morphology much intensifies the visualized

effect of the damage evolution process of the specimen during the loading and will lead to further deepening understanding of the failure mechanism of concrete. Fig. 16.1 - 33 shows the different failure patterns for static and dynamic loadings in 3-D images of CT test.

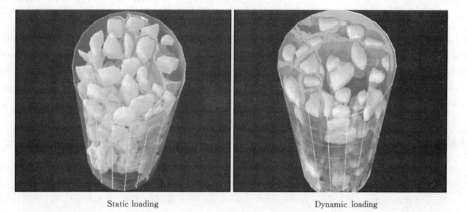

Fig. 16.1 - 33 Different Failure Patterns for Static and Dynamic Loadings in 3-D Images of CT Test

(3) Dynamic acoustic emission technology.

As an effective approach for non-destructive and real-time detection of concrete interior damage status, the acoustic emission technology is applied in research on dynamic damage process of dam concrete (Chen et al, 2012). In the modulus of rupture tensile tests of Xiaowan arch dam concrete subjected to impact with different loading speeds of 0.25kN/s, 2.5kN/s, 25kN/s and 250kN/s, its modulus of rupture tensile strength increases while the accumulative hit number of acoustic emission decreases as shown in Fig.16.1 - 34. The strain rate effect reveals obviously. Fig. 16.1 - 35 shows the axial distribution of acoustic emission hits along the specimen when loading at 0.25kN/s and 2.5kN/s. It is clear that in high speed loading case more spatial dispersion occurs. Because there is no enough time for microcrack propagation to trace the weak part while the loading speed increasing. It also reflects the different failure patterns for static and dynamic loadings.

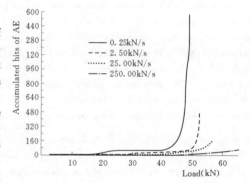

Fig. 16.1 - 34 Accumulative Hits of AE vs. Loading Speed

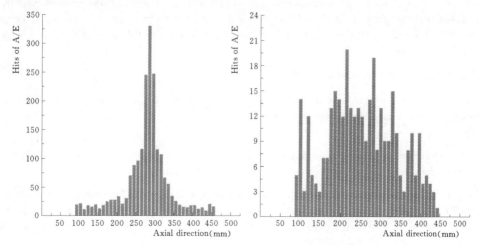

Fig. 16.1 - 35 Distribution of Hits along Axial Direction

Generally speaking, the strain rate effect of dynamic loading seems might be explained by the different failure pattern between dynamic and static loadings, which has identically manifested by all three different approaches abovementioned.

16.1.5 Seismic Design Code

The seismic design code plays an important role in seismic safety guarantee including damage mitigation and secondary disaster prevention. In China the seismic design code of hydraulic structures is of mandatory technical regulation applicable in the professional circle for whole country.

In a long period of past 60 years, the high dam seismic design referred in China was basically related seismic design standards and guidelines in America and former Soviet. Till 1978, the *Seismic design code of hydraulic structures* (SDJ 10—78) is firstly issued as mandatory technical regulation in hydraulic structure construction and put into effect in China. After dam construction advances, its first revision is fulfilled in 1997. At present, in order to meet the seismic safety demand of high dam construction in severe seismic zone in western China and based on the inspection and summary of dam behaviors during the Wenchuan earthquake in 2008, the code is being further revised and upgraded as national standard.

16.1.5.1 Limitation of the First Code

As a common accepted principle, it is well known that for seismic design of a structure, the input ground motion should be consistent with the method of analysis of seismic effects and dynamic resistances of materials. According to the economic and construction situation at that time in China, the first version of *Seismic Design Code of Hydraulic Structures* (SDJ 10—78) issued in 1978 has following major limitations.

(1) It contained only embankment dam, gravity dam, arch dam and slices.

(2) The basic seismic intensity from the *Seismic intensity zonation map of China* compiled using deterministic method was taken as the design intensity of single seismic fortification level. The design intensity in general is equal to the basic intensity. But for important water-retaining structures, the design intensity shall be one degree higher than basic intensity. The intensity is determined according to the "*China earthquake intensity scale*" approximately similar to the modified Mercalli intensity (MMI). The peak accelerations of ground motion corresponding to intensity of Ⅶ, Ⅷ, Ⅸ were taken as $0.1g$, $0.2g$, $0.4g$, respectively. They are the statistical mean values of the actual accelerations recorded in regions with different intensities. The performance object of the seismic design is that when the dams designed based on the code are subjected to the influence of earthquakes with an intensity of less than the design intensity, the dams will not be, or will be only locally damaged and will continue to be service with only ordinary repair.

(3) In the 1970s, there was no dynamic method available for analyzing rather complex seismic responses of dams yet in China. The "pseudo-static method" analysis method with single safety factor criterion has to be used in seismic design. However, in this case the adopted design acceleration was much less than the statistical peak ground accelerations corresponding to the intensity. Even though, among the dams subjected to strong earthquakes, most of them have been survived with only light damages. This fact reveals that dams designed by the conventional methods and safety criteria are capable to resist earthquake actions to certain extent. Probably the resistance capacity of the dam under static loading with large enough safety factors can be instantaneously mobilized to bear the seismic actions with relative short duration during earthquake. In order to fill in this gap, a "comprehensive influence coefficient" $C_Z=0.25$ was involved, and the design peak acceleration was taken as the product of the C_Z and the statistical peak acceleration of ground motion. Furthermore, the dynamic behavior of the structure can only be very roughly reflected by simplified distributions of the design acceleration along the height of various structures.

16.1.5.2 Improvement of the Revised Code

The code had been revised in 1997 based on the development of the dam construction, related research achievement, and lessons learnt from occurred strong earthquakes both in China and abroad. Significant improvement has been achieved in the revision of the code.

(1) The content was extended to involve embankment dam, gravity dam, arch dam, slice, underground hydraulic structures, intake tower, penstock, and surface power house of hydropower station.

(2) During this period a dual performance criteria to address the safety and serviceability of dams through Maximum Design Earthquake (MDE) and Operating Basis Earthquake (OBE) respectively is prevalent in many countries. However, the single performance criterion has still been accepted by the revised code. Because it is in fact very unlikely that a dam that cannot operate any more after the OBE will fulfill the MDE requirements.

(3) The seismic fortification level of hydraulic structures was classified depending upon both the classes of structures related mainly to the probable failure consequences and the intensity of ground motion of the site. The design acceleration was directly taken according to *Seismic ground motion parameter zonation map of China* which was compiled on the basis of probabilistic method of seismic hazard evaluation and providing acceleration at surface corresponding to a 10% probability of ex-

ceedance in 50 years at plain free ground surface of stiff soil site. However, for dam of earthquake fortification class 1 a site-specific seismic hazard evaluation shall be carried out and the curves of probability of exceedance both to peak ground acceleration and intensity must be provided. Fig. 16.1 - 36 shows the curves of probability of exceedance to peak ground acceleration for reference period of 100 years at sites of some importance hydraulic projects. Based on the probability curves of peak acceleration and intensity of a series of important hydropower projects, the probability of exceedance in 100 years for water-retaining structures class 1 with the design intensity one degree higher than the basic intensity can be calibrated as 2% (with a recurrence period of 4,950 years).

(4) Obviously, the traditional "pseudo-static method" can neither adequately reflect the dynamic characteristics of dam body and ground motion nor be able to reasonably explain the earthquake damages. The application of advanced numerical discretization process and the use of powerful computers have increased in a dramatic manner which make it possible today to use dynamic analysis in the seismic design of hydraulic structures. Great advances have been made in dynamic analysis of dam-foundation-reservoir system along with the construction of many large dams located in seismic regions in China during this period. Therefore, it was basically required to use dynamic analysis method in seismic design. However, the pseudo-static method of seismic analyses was still allowed to be used for small and medium-sized concrete dams with height of less than 70m and for all embankment dams. Up to now, there is no commonly accepted approach and criteria of seismic analysis for embankment dams due to the complicated non-linear constitute law of the dam materials.

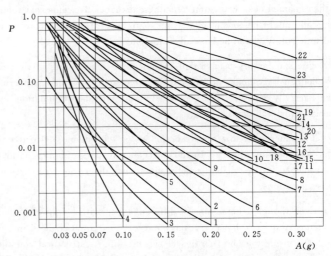

Fig. 16.1 - 36 The Curves of Probability of Exceedance to PGA for 100 Years of some Important Hydraulic Projects

(5) The method of limit state method with partial factors was applied instead of the conventional single safety factor method. The essential advantage of the limit state method is that the different degrees of random uncertainties for actions and resistances can be distinguished by the partial factors based on probabilities of occurrence. However, the single safety factor method involves not only all random uncertainties but also the non-random factor of engineering experience which is still very important and cannot be neglected for hydraulic structures, especially for damsso far. Therefore, a structural factor γ_d has to be added to the equation of ultimate limit state with partial factors to reflect the engineering experience and to be expressed as:

$$\gamma_0 \psi S(\gamma_F F_K) \leqslant \frac{1}{\gamma_d} R\left(\frac{f_K}{\gamma_m}, a_K\right) \tag{16.1-6}$$

where, γ_0 is the importance coefficient of structure; ψ is the design condition coefficient; $S(\cdot)$ is the action effect function of structure; γ_F and F_K are action partial factor and representative value respectively; $R(\cdot)$ is the resistance function of structure; γ_m and f_K are the partial factor and standard value of material performance respectively.

At present, all the partial factors have still to be determined in accordance with the traditional safety factor in order to keep the continuous of the design codes. So, the concerned method is actually a multi-safety factor method having no essential difference with the conventional single safety factor method concerning the setting of safety level as a whole. However, it is entirely different from the reliability analysis method, as a structural factor γ_d is introduced and all partial factors both of action effects and resistances are determined independently from the reliability index. So, neither complex theory and unfamiliar terms nor insufficient statistical data must be worried about by the dam designers.

The statistic average response spectra based on accelerograms recorded in rock sites of different countries was involved as design response spectrum for dynamic analysis in seismic design code in rock soil condition as shown in Fig. 16.1 - 37, where $\beta_{max}=2.5$ for damping ratio of 0.05; the characteristic period $T_g=0.2s$; exponential coefficient $\gamma=0.9$.

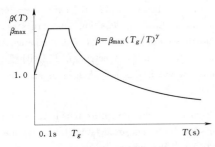

Fig. 16.1 - 37 Standard Response Spectrum

16.1.5.3 Special Features of the Code Being Revised

The Wenchuan earthquake in Sichuan province in 2008 is the most severe earthquake occurred in past 60 years in China. Some major lessons can be learned from the Wenchuan Earthquake.

1) Despite the fact that subjected to strong earthquake over the design level, none of 1,803 dams in earthquake region was collapsed and the integral structural stability and water retaining function of 4 high dams of different types with height more than 100m, especially of the Zipingpu concrete faced rockfill dam and Saipai RCC arch dam located near the epicenter, were still remained even with some repairable damages. It seems that the current seismic design code experienced practical inspection and might be concluded that for dams well designed according to the current Cord and good constructed, their desired seismic safety would be basically anticipated.

2) The satisfactory behavior of the anchored high slopes in the areas of Zipingpu and Saipai dams abutments during the Wenchuan Earthquake manifested the effectiveness of the engineering strengthening measures; even the neighboring slopes were heavily damaged. The geological stability of dam foundation under strong seismic action cannot be overemphasized especially for arch dams.

3) The emergency lowering reservoir water was the most effective measure to prevent the secondary disaster of dam during the Wenchuan earthquake. In the emergency response plan scenario, for destructive earthquake, a great importance should be attached to the reliability of the emergency discharge structures and facilities as well as their energy supply under strong seismic actions.

4) Even though 4 high dams of different types with height of more than 100m have stood up to rigorous tests of the Wenchuan earthquake, however, there might be essential changes in dynamic behaviors while dam height upgrade to a level of 300m. Considering uncertainties of seismic input, structural complexities and serious catastrophic aftermath of secondary disaster of the unprecedented extra high dam located in severe seismic region of the western China, the problem of preventing any collapse of high dam under extremely strong earthquake is deeply concerned by society.

The lessons learnt from the Wenchuan earthquake have given an urgent impetus to revise the current seismic design code of hydraulic structures; and to prevent any collapse of high dam under MCE as the strategic priority for seismic safety of high dams was emphasized. Some special features of the revised code are briefly summarized as follows.

(1) To use dual fortification level for water-retaining structures class 1. An additional fortification level of MCE without uncontrollable release of reservoir water shall be required with following stipulations and a special report of seismic safety evaluation of the dam must be provided for water-retaining structures of class 1.

1) The PGA of MCE shall be selected for the larger one between the values provided deterministically or with a recurrence period of 10,000 years.

2) A site-specific design response spectrum shall be applied based on the scenario earthquake with maximum probability of occurrence.

3) The effect of hanging wall shall be considered while the major concerned seismogenic fault having a dip angle $\theta < 70°$ and the epicentral distance $R \leqslant 30$km.

4) For non-linear analysis of dams the frequency non-stationary feature of artificial input accelerogram shall be considered.

5) The seismogenic fault with $M \geqslant 7$ and $R \leqslant 5$km shall be conducted as an area source to consider the rate, mode, and time sequence of rupture.

(2) To modify the design response spectrum. The design response spectrum was updated based on the US Next Generation of Attenuation (NGA) using 3,551 records of 173 earthquakes, as the crustal and tectonic composition, modern stress condition, seismic genesis and activity features are similar between China and North America continents. The fundamental periods T of concrete dams are usually less than 1s even for dams of 300m high, and the response spectrum $\beta(T)$ with $\beta_{max}=2.5$, $T_g=0.20$s, $\gamma=0.6$ is closer to the results from Abrahamson attenuation low when T is less than 1s as shown in Fig. 16.1-38.

(3) To update tensile strength and elastic modulus of dam concrete. Based on a series of modulus rupture tensile tests with fully-graded dam concrete specimen under earthquake simulated cyclic loading for some high dam projects in China, the material properties are updated as: the tensile to compression strength ratio $f_t/f_c=0.1$ and the dynamic to static strength ratio $f_d/f_c=1.2$ are stipulated. Also, the dynamic to static elastic modulus ratio E_d/E_s is 1.5.

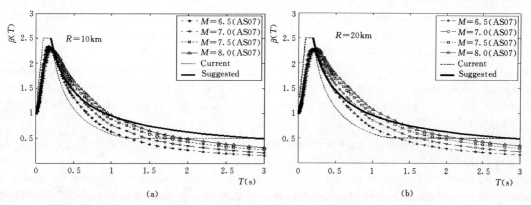

Fig. 16.1-38 Response Spectra from Abrahamson Attenuation Low in NGA

16.1.6 Conclusions

In the past 60 years, during long time exploration process of seismic design and research on high dam in China, especially in recent 10 years, a series of progresses in seismic research on concrete dams has been achieved under the impetus and support given by dam construction practice. Some traditional design ideas and methods have been broken through in combination with China national conditions and high dam features. Nevertheless, to prevent any serious catastrophic aftermath of secondary disaster of the high dams during severe earthquakes, the task is still arduous and the road ahead is rather long. With the unprecedentedly rapid development of construction of high dam with large reservoir in high seismic region of western China, dam engineers still have to face severe challenge to solving a series of key technical problems. Relying on continuous innovation and interdisciplinary study as well as International Corporation and exchange is of extreme importance. This is the purpose of the paper.

References

[1] Jia Jinsheng, Yuan Yulan, Ma Zhongli. *Statistics of Large Dams in China and Introduction to State of World Large Dams*. China Water Power Press, 2007.

[2] Chen Houqun, Li Min, Zhang Baiyan. Input Ground Motion Selection for Xiaowan High Arch Dam. 13th World Conference on Earthquake Engineering, Vancouver, B.C., Canada, August 1-6, 2004, Paper No. 2633.

[3] Priestley M. B. Evolutionary Spectra and Non-stationary Processes. *Journal of the Royal Statistical Society*, Series B Methodological, 1965, 27 (2): 204-237.

[4] Zhang Cuiran. Study on Earthquake Ground Motion Input for Important Hydro-electric Structures. Beijing: China Institute of Water Resources and Hydropower Research, 2009.

[5] Zhang Cuiran, Chen Houqun, Li Min. Earthquake Acceleration Simulation with Statistical Law of Evolutional Power Spectrum, Acta Seismilogic Sinica, 2007, 20 (4): 435-446.

[6] Nakayama T, Fujiwara H, Komatsu S, Sumida N, *Non-stationary Response and Reliability of Linear Systems under Seismic Loadings*. Schueller, Shinozuka & Yao. Structural Safety & Reliability, Balkema, Rotterdam, 1994: 2179-2186.

[7] Goto H, Sugito M, Kameda H, et al. *Prediction of Non-stationary Earthquake Motions for Moderate and Great Earthquakes on Rock Surface*. Annuals, Disaster prevention research institute, Kyoto University, 1984, 27B-2: 19-48.

[8] Kameda H. Evolutionary Spectra of Seismogram by Multifilter. *Journal of the Engineering Mechanics Division*, ASCE, 1975, 101 (6): 787-801.

[9] Shi Yucheng, Chen Houqun, Li Min, et al. The Study and Application of Stochastic Finite Faults Method in Ground Motion Synthesizing. *Earthquake Engineering and Engineering vibration* 2005, 25 (4): 18-23.

[10] Wang Yongxi, Li Min, et al. *Using GIS Technique to Research and Assess the Risk of Reservoir-Induced Seismicity*. Beijing: China Institute of Water Resources and Hydropower Research, 2000.

[11] Chen Houqun, Li Deyu, Guo Shengshan. *Damage-rupture Process of Concrete Dams during Strong Earthquake*. Bijing: China Institute of Water Resources and Hydropower Research, 2014.

[12] Chen Houqun. *Seismic study on high concrete dams*. Collection of the Chinese Academy of Engineering, Higher Education Press Ltd. Co., 2011.

[13] Yusof Ghanaat, Chen Houyqun, Bruce B. Redpath, et al. *Measurement and Prediction of dam-water-foundation*

[14] *State-of-Practice for the Nonlinear Analysis of Concrete Dams at the Bureau of Reclamation*. U. S. Department of the Interior Bureau of Reclamation, January 2006.

[15] Chen Houqun, Hou Shunzha, Yang Dawei. Study on Arch Dam-reservoir Interaction. *Proceeding of International Symposium on Earthquakes and Dams*, May 1987, Beijing Cnina, Vol. 1: 1-24.

[16] Liao Z. P, Liu J. P.. Fundamental Problem of Finite Element Simulation of Wave Motion. *Scientia Sinica (Series B)*, 1992, 8

[17] Liu Jinbao, Du Yixin, Du Xiuli, et al. 3D viscous-spring artificial boundary in time domain, *Earthquake Engineering and Engineering Vibration* 2006, 5 (1): 93-102.

[18] Gregory L. Fenves, Soheil Mojtahedi, Richard B. Reimer. *ADAP-88: A Computer Program for Nonlinear Earthquake Analysis of Concrete Arch Dam*, University of California at Berkeley, Report No. UBC/EERC - 89/12, 1989.

[19] Chen Houqun, Yeh Changhua. Joint Opening of Arch Dam During Earthquake: Experimental and Analytical Results//*Proceedings of the 4th ICOLD Benchmark Workshop on Numerical Analysis of Dams*, Madrid, Spain. 1996.

[20] Jinbo Liu, Wang duo, Yao Ling. A Contact Force Model in the Dynamic Analysis of Contactable Crack. *Acta Mechanica Solida Sinica*, 1993, 6 (4).

[21] Chen Houqun, Tu Jin, Zhang Baoyan. Study on Seismic Behavior of Xiaowan High Arch Dam//*International Commission on Large Dams* Q. 83, R. 68, Montreal, Canada, 2003: 1183-1194.

[22] Panel on Earthquake Engineering for concrete Dams of the U. S. National Research Council. *Earthquake Engineering for Concrete Dams: Design, Performance, and Research Needs*, U. S. National Academy Press, Washington, D. C. 1990

[23] R. W. Clough, K. T. Chang, H. Q. Chen, et al. *Dynamic Response Behaviour of Quan Shui dam*, Report of University of California, Berkeley, U. S. A, No. UCB/EERC - 84/20. 1984.

[24] Yusof Ghanaat, Houqun Chen et al. *Experimental study of Dongjiang Dam for Dam-Water-Foundation Interaction*. QUEST Structures, Emeryville, California. U. S. A. 1993.

[25] Jeeho Lee. *Theory and Implementation of Plastic-Damage Method for Concrete Structures under Cyclic and dynamic Loading*. University of California, Berkeley, 1996.

[26] Radin Espandar, Vahid Lotfi. Comparison of Non-orsogonal Smeared Crack and Plasticity Models for Dynamic Analysis of Concrete Arch Dam. *Computers and Structures*, 2003, 81 (2003): 1461-1474.

[27] Faouze Ghrib, Rene Tinawi. An Application of Damage Mechanics for Seismic Analysis of Concrete Gravity Dams, *Earthquake Engineering and Structural Dynamics*, 1995, 24: 157-173.

[28] Miguel Cervera, Javier Oliver. Sesmic Evaluation of Concrete Dams via Continuum Damage Model, *Earthquake Engineering and Structural Dynamics*, 1995, 24: 1225-1245.

[29] M. A. Hariri-Ardebili, H. Mirzabozorg. A Comparative Study of Seismic Stability of Coupled Arch Dam-foundation-reservoir System Using Infinite Element and Viscous Boundary Models. *International Journal of Strctural Stability and Dynamics*, Vol. 13, No. 6 (2013) 1350032 (24 pages).

[30] J. Lubliner, J. Oliver, S. Oller et al. A Plastic-damage Model for Concrete, *International Journal of Solids and Structures*, 25 (3) (1989) 299-326.

[31] J. Lee, G. L. Fenves. A Plastic-damage Concrete Model for Earthquake Analysis of Dams, *Journal of Earthquake Engineering And Structural Dynamics*, 1998, 27: 937-956.

[32] ABAQUS THEORY MANUAL. ABAQUS, INC, 2009.

[33] *State-of-Practice for the Nonlinear Analysis of Concrete Dams at the Bureau of Reclamation (2006)*. U. S. Department of the Interior Bureau of Reclamation, January 2006.

[34] Reinhardt H W, Weerheijm J. Tensile Fracture of Concrete at High Loading Rates Taking Account of Inertia and Crack Velocity Effects. *International Journal of Fracture*, 1991: 51 (1991): 31-47.

[35] Gupta H K. *Dams and Earthquakes*. Elsevier ScientificPublishing Co. , 1976.

[36] M. D. Pendes, S. N. Hordda, S. Y. Korkarni. Rehabilitation of Koyna Dam-A Case Study//*4th Symposuim on Rehabilitation of Dams*, Noremberg, 1998.

[37] N. G. K. Murti, P. M. Mane, V. R. Deuskar. Koyna Earthquake Remedial Measeres to Crack in Dam. *International*

Commission on Large Dams Q. 39, R. 31, Montreal, Canada, 1970: 573-597.

[38] *Test code for hydraulic concrete* (SL 352—2006), China Ministry of Water Resources, 2006.

[39] Houqun Chen, Shengxing Wu, Faning Dang et al. *Seismic Safety of High Arch Dams*. (in Chinese, will be printed by ELSEVIER SCIENCE B. V. in English) Beijing: China Electric Press, 2012.

[40] Ma Huaifa. *Study on Dynamic Behaviors of Fully-graded Dam Concrete Based on Meso-mechanics*. Beijing: China institute of water resources and hydropower research. 2005.

[41] Chen Houqun, Ma Huaifa, Tu Jin, et al. Parallel Computation of Seismic Analysis of High Arch Dam, *Earthquake Engineering and Engineering Vibration* (English Edition), 7 (1). 2008: 1-11.

[42] Cui J. Z., Shan Y. J. The Two-scale Analysis Algorithms for the Structure with Several Configuration of Small Periodicity, Computational Technique for materials. Composites and Composite Structures, Edited by B. H. V. Topping, CIVIL-COMP PRESS, 2000.

[43] Ding Weihua. *Study on Dynamic Damage Process of Concrete Based on X-ray CT Technique*. Beijing: China institute of water resources and hydropower research, 2007.

[44] Liang Guoping. *Finite Element Program Generator and Finite Element Language*, Advances in mechanics, 1990, 20 (2): 199-204.

[45] Editorial Group of Earthquake Resistant Design Code for hydraulic structures. Earthquake loads for hydraulic structures, Proceeding of 13th Congress of ICOLD, New Delhi, 1979: 1295-1312.

[46] *Seismic design code of hydraulic structures* (SDJ 10—78). Beijing: China Water Power Press.

[47] *Seismic design code of hydraulic structures* (DL 5073—2000). Beijing: China Water Power Press.

16.2 Seismic Safety Evaluation of High Concrete Dams Part Ⅰ: State of the Art Design and Research

Zhang Chuhan[1], Jin Feng[2]

16.2.1 Introduction

In China, the growing demand for electricity and increasing pressure for carbon emission reduction due to global temperature change have accelerated the development of hydroelectric and high dam projects to meet new national goals to increase usage of clean and renewable energy. Nowadays, only 7 percent of the total energy consumption is supplied by hydro-power while 70 percent comes from burning coal, causing serious environmental problems. To reduce use of coal for power generation, the central government plans to invest US $ 265 billion to help obtain 15 percent of its installed capacity from renewable sources including hydro-power by 2020. With total hydropower potential capacity of 540×10^3 MW, the current developed capacity in 2007 is only 26 percent (145×10^3 MW). In the coming two decades, twelve hydropower bases, mainly in southwest China, including the upper reaches of the Yangtze and Lancang Rivers and their branches (Jinsha, Yalong and Dadu Rivers) are being developed with a total capacity of 290×10^3 MW by 2020, 54 percent of the total hydro-power potential of the nation.

To accomplish the goals of hydro-power development, construction of a series of high dams up to 250-300m in height with huge reservoir and flood overflow requirements is being implemented. These high dam projects are mostly located in seismically active region in southwest China, with design peak ground accelerations (*PGA*) of 0.20-0.56g in terms of occurrence probability 0.02 within the period of 100 years. Therefore, seismic safety evaluation of the high dam projects is one of the key issues in the design and construction, together with the other crucial safety checks, such as foundation and abutment stability, high speed flood overflow through the dam and underground powerhouse stability. The main indexes of characteristic parameters of China's high concrete arch dams are listed in Table 16.2-1.

Table 16.2-1 Characteristic Parameters of Main Arch Dams in China

Dam	River	Dam height (m)	Reservoir ($10^8 m^3$)	Installation (MW)	Design flood (m^3/s)	Design *PGA* (g)
Longyangxia	Yellow River	178	247	1,280	7,040	0.237

---

[1] Zhang Chuhan, Academician of Chinese Academy of Seciences, State Key Laboratory Hydroscience and Engineering, Tsinghua University, Professor of Tsinghua University.

[2] Jin Feng, Professor of Tsinghua University.

Countinued

Dam	River	Dam height (m)	Reservoir ($10^8 m^3$)	Installation (MW)	Design flood (m^3/s)	Design PGA (g)
Ertan	Yalong River	240	58	3,300	23,900	0.20
Laxiwa	Yellow River	250	11	5,600	6,000	0.20
Xiaowan	Lancang River	292	151	4,200	20,600	0.308
Xiluodu	Jinsha River	273	140	14,400	50,000	0.321
Jinping I	Yalong River	305	78	3,300	6,900	0.20
Goupitan	Wujiang River	233	65	3,000	27,500	—
Baihetan	Jinsha River	275	180	12,000	40,000	0.325
Dagangshan	Dadu River	210	7.4	2,400	8,320	0.56

Over the past 20 years, with the world-record-height dams in design and construction and extremely strong earthquakes to be considered, the dam engineers and scientists have developed considerable knowledge regarding the seismic safety evaluation for concrete dams especially for arch dams. This paper reports the following aspects of state-of-the-art design and research on seismic safety analysis of high concrete dams:

(1) A brief introduction of the conventional design practice in pre-feasibility study stage.

(2) A summary of advanced design and research in preliminary and final design stage, especially for important projects exceeding 150m high and large index of reservoir and power capacity. These include effects of the five aspects listed in the abstract.

16.2.2 Conventional Seismic Design Practice

(1) According to the Chinese specifications for seismic design of hydraulic structures (DL 5073—2000), the linear elastic gravity method and multiple arch-cantilever method, both with pseudo-static design earthquake loadings, are being used for gravity dams and arch dams respectively. However, when the dam (arch and gravity) is categorized into Grade I or Grade II (related to dam height, reservoir and power capacity, etc.) or the design intensity of earthquakes equals or exceeds 8 degrees (equivalent to $PGA=0.2g$), it is recommended to use the dynamic finite element method (both response spectrum and time history analysis) as a supplemental check.

(2) In pseudo-static analysis of concrete dams, sectional linear prescribed and amplified factor curves are used as the inertial loads acting on the dam, e.g. the amplified curves for arch dams have a coefficient of 3.0 at the crest and 1.0 at the bottom with a uniform distribution along the arch axis. A reduction coefficient of 0.25 is applied for PGA used.

(3) In dynamic response spectrum analysis, the maximum amplification factor for design response spectrum is $\beta_{max}=2.0$ for gravity dams and 2.5 for arch dams respectively with predominant periods of 0.1s to T_g, where T_g equals to 0.2s, 0.3s or 0.4s depending on the category of the rock foundation. In time history analysis, usually, either a time history produced from the design response spectrum of the specification or an artificial time history provided from seismic risk analysis at the site may be used. In dynamic analysis, the damping ratio of 3%-5% for arch dams and 5%-10% for gravity dams are assumed; the assumption of massless rock foundation with 1.0-1.5 times the dam height and added mass method derived from the Westergaard incompressible reservoir are usually considered. The earthquake input is acting on the truncated rock with uniform distribution along the truncated boundary. Dynamic modulus and strengths of concrete may have 30% increase compared with its static counterparts in safety check.

(4) In dynamic stability analysis for foundation and abutment safety, the rigid body limit equilibrium methods with actions of pseudo-static loading (acting both at the dam and abutment rock mass) + reservoir loads + gravity + temperature loads are still being used in design practice. The amplification factor α_d is 3.0 at the dam crest as mentioned before, but the factor for abutment rock α_r is assumed to be 1.0 (no amplification); the criterion of safety factor for abutment rock is 1.20.

16.2.3 Advanced Seismic Design and Research

16.2.3.1 Effects of Earthquake Input Mechanism and Radiation Damping Due To Infinite Canyon

The modeling of infinite canyon and earthquake input mechanism are two crucial aspects for analysis of dam-foundation interaction. Within the past two decades, many sophisticated numerical models were developed for simulation of infinite canyons. These include but are not limited to: finite elements+infinite elements (FE+IE)[1,2], boundary elements (BE)[3,4], finite elements+boundary elements+infinite boundary elements (FE+BE+IBE)[5], finite elements+scaled boundary finite elements (FE+SBFE)[6], finite elements+transmitting boundaries[7] and finite elements+discrete parameter models including viscous dampers-springs-masses[8] (usually matrices defined at the interface). All these methods are proved to be applicable and effective for modeling infinite canyons of dams although each has its own merits and disadvantages in terms of computational efforts.

Regarding the earthquake input mechanism, several procedures have been proposed to study its effects and those of radiation damping on concrete dams, especially on arch dams. Among those, the most commonly used in current practice is the massless foundation model with uniform input acting at the truncated boundaries which was proposed by Clough[9]. The model considers the foundation flexibility but ignores the dam-foundation interaction including radiation damping and non-uniform input. Another input method is the so-called deconvolution approach by which the earthquake input can be deconvoluted using analytical solution or 1-D half-space model from specified free-field motions at the ground surface. This model is usually incorporated with viscous-spring or transmitting boundaries[10]. The third effective earthquake input method including rock mass for dam-canyon interaction is the free-field input method[9,11]. Earthquake excitations are imposed as the free-field motions at the dam-foundation interface directly; thereby a spatially non-uniform free-field input could be specified if it is available.

The research studying the effects of radiation damping due to infinite canyon and earthquake input mechanism include using frequency domain procedures, e.g. Zhang et al[2], Chopra et al[4] and Dominguez et al[3] and time domain procedures, e.g. Zhang & Jin et al[8,11] and Chen & Du et al[7].

From analysis of the Xiaowan, Xiluodu, Laxiwa, Jinping and Dagangshan arch dams, the conclusion of considering radiation effects of infinite canyon is that the reduction of the dam response may reach 25%–40% compared with the results by massless foundation model. Nowadays, there are two differing opinions in China concerning the reduction factor due to the radiation damping. One suggests that it may be considered as a potential safety factor but not to be considered in the design because the damping ratios of structures and foundations are complicated factors, and usually they are measured through prototype tests or seismic measurements in which the material and radiation damping are combined together. The other opinion suggests that we should use this benefit in the design by arguing that the measured responses of several arch dams in the world during moderate earthquake events were much smaller than that from numerical analysis provided the damping ratios of the structures are assumed to be around 0.05[12]. Furthermore, it was also found that a much higher damping ratio as 8% –15% was needed for approximately matching the field response records of the dams. However, the structural damping ratios obtained from forced vibration field tests and ambient measurements were quite low, only about 1%–4%[13] (lower than 0.05). What causes this discrepancy? A rational comparison is of significance for understanding the effects of different earthquake input mechanisms on the response of arch dam-foundation-reservoir systems.

16.2.3.2 Effects of Nonlinearity of Contraction Joint Opening

It is now commonly recognized that high arch dams will experience significant contraction joint opening during strong earthquakes. The experience of contraction joint opening in the Pacoima arch dam during the 1971 San Fernando and 1994 Northridge earthquakes[14,15] is a real example. Because of the contraction joint opening a drastic but instantaneous release of the arch action and substantial increase of the tensile stresses in the cantilever are evident. Due to this non-linear behavior, weakening of dam integrity causing cracking damage of cantilevers and breakage of waterstops between joints is a major concern to dam engineers.

The nonlinear behavior of arch dams due to contraction joint opening during strong earthquakes was first raised by Clough[16,17]. Fenves et al[18,19] presented a discrete 3-D joint element as a direct approach for simulation of the contraction joints. By using a time domain and discrete parameter procedure for modeling dam canyons, Zhang et al[20,21] combined effects of dam-canyon interaction with contraction joint opening. Meanwhile, a contact boundary approach presented by Bathe[22] is also used to analyze contraction joint behavior. Lua et al[23] developed a joint model to simulate both the opening-closing and shear slippage behavior considering shear keys at the contraction joints. Recently, Du et al[10,24] combined explicit FE method with transmitting boundary to study the effects of contraction joint opening on the response of Xiaowan arch dam, but only five joints were simulated; In addition, Chen et al[25], Sheng et al[26] and Wang et al[27] studied contraction joint opening behavior of arch dams by shaking table tests. Lin et al[28] also studied this nonlinear problem using the non-smooth Newton algorithm.

All the analyses and model tests mentioned above reveal that significant contraction joint opening is observed. From the results of Tsinghua research group, the maximum joint opening reaches 16.5mm in Xiaowan Dam, 16.7mm in Dagangshan

Dam, and 11.6mm in Xiluodu Dam. In these analyses, the concrete is assumed to be linear elastic. The massless foundation assumption for canyon rock is usually applied with a few exceptions, where the infinite mass canyon and non-uniform free-field or viscous-spring boundary input models are also considered together with nonlinear joint opening[29,56].

16.2.3.3 Effects of Damage and Fracture Behavior in Dams and Strengthening

Due to the high intensity of design earthquake in certain dam sites such as Jin'anqiao, Xiaowan and Dagangshan, whose PGA reach $0.399g$, $0.308g$ and $0.557g$, respectively, large tensile stresses will inevitably occur at upper portions of gravity dams and cantilevers of arch dams. Dynamic damage and nonlinear fracture analysis have been performed to understand the cracking behavior of dams. In this aspect, both linear and nonlinear fracture mechanics models have been used in analysis of seismic cracking of concrete dams. In linear analysis, Pekau, Feng and Zhang[30] used the boundary element technique to calculate dynamic stress intensity factors and to reproduce the cracking profile of Koyna Dam due to the 1967 earthquake. Meanwhile, a shaking table test was performed for verification of the model. In nonlinear analysis, Wang et al[31] used finite element method and crack band theory to analyze the cracking behavior of Koyna Dam considering strain softening of concrete. Recently, in analysis of damage-cracking of 3-D arch dams, the plasticity-damage model of concrete presented by Lubliner et al[32] and later extended in analysis of damage cracking of Koyna Dam by Lee and Fenves[33] is used and the crack band theory by Bazant[34] is incorporated to ensure the uniqueness of fracture energy without introducing mesh sensitivity of the finite elements. The complete constitutive relationship for concrete including softening behavior can be obtained from rigid machine tests in static case. Thirty percent increase of the tensile strength is adopted for dynamic case. The average fracture energy is assumed to be around 300N/m for dam concrete. The results show evident damage areas occur at the upper portion of the cantilevers. The damage indices computed can then be converted into the equivalent crack width of the dam. Details of analysis and findings are described in Part Ⅱ.

In order to improve earthquake-resistant capacity, several high arch dams under construction in China such as Xiaowan and Dagangshan Dams have been proposed to study some strengthening measures including cantilever reinforcements for alleviating the crack extension in dams under the design earthquakes, joint reinforcements or joint dampers for reducing the joint openings and reinforcement arch-belts for improving the integrity of dams. After a detailed comparison, the owners and designers of the Xiaowan and Dagangshan projects have decided to adopt cantilever reinforcements as the major strengthening alternative and joint dampers as a supplemental measure. In analysis of the response of arch dams with cantilever reinforcement strengthening, the influence of reinforced steel and tension stiffening effect due to steel-concrete interaction are considered. For analysis of cantilever reinforcements, a modified embedded reinforcement model is used. By stiffening the reinforced steel, the model is capable of considering the concrete-reinforcement interaction for lightly reinforced cantilever members (Compared with concrete dam body). In addition, a weighted zoning scheme for dividing plain and reinforced concrete sub-region by An et al[35] is adopted for simplification. Procedure of reinforcement or concrete stiffening[36,37] may be used to reflect the interaction between concrete and steel bars. The aforementioned procedures have preliminarily been used in analysis of strengthening design of the Jin'anqiao gravity dam and the Dagangshan arch dam. It is concluded that the strengthening measures have noticeable effects in controlling the damage index (equivalent to crack width) and extension of cracking. In addition, numerical models for other strengthening measures, such as joint reinforcements, joint dampers and reinforcement belts have been developed for analysis of their effects[21,38]. From the analysis, it is suggested that the Jin'anqiao gravity dam and Xiaowan and Dagangshan arch dams need reinforcement strengthening for crack control. The final designs require 20,000t and 9,400t of reinforced steel for Xiaowan and Dagangshan, respectively.

16.2.3.4 Effects of Dam-Reservoir Interaction

It is recognized that dam-reservoir interaction is an important factor affecting response of concrete dams. Since high curvature arch dams are usually more flexible than gravity dams, a greater influence of arch dam-reservoir interaction than that of gravity dams in evident. The effects of water compressibility on arch dam response were comprehensively studied by Chopra and colleagues[39-41]. Under the frame work of Sino-US scientific cooperation in earthquake engineering, Clough et al[42-44] conducted a series of forced vibration field tests for investigation of arch dam-reservoir-foundation interaction. Valuable data related to dam-reservoir interaction were obtained. Yet the final conclusions concerning the practical significance of the compressibility effect of reservoir on dam response were not conclusive. The important finding is that among the three components of ground motions, the vertical component may have profound effects on the response of arch dams when the fundamental resonant period of the compressible reservoir is in coincidence with the period of vertical ground motion and a rigid reservoir bottom is also assumed. Considering that the effects of reservoir sediment and bottom rock absorption on the reduction of dam response are significant and that most important contributory earthquake components to the dam response are in stream and cross-stream directions rather than the vertical component, the compressibility of reservoir is usually neglected in most of our current design and research practice. Herein, the added mass method is commonly used for arch dam-reservoir interaction due to its simplicity and somewhat conservative nature. The finite element method with incompressible fluid assumption and three times of the dam height for reservoir length are usually used for obtaining the added mass coefficients. However, the compressibility of the water and the sediment effects in the reservoir are also studied in research level[45-47] and deserve further investigation, especially for 300m-high level arch dams.

16.2.3.5 Earthquake Stability of Dam-Abutment System

The traditional seismic design and analysis of concrete dams focus on the response of dam structures from linear elastic to nonlinear damage and fracture states. Little, if any, research has even touched on the stability problems of dam-abutment system. Even in the static case, the dam and foundation or abutment are treated as a completely separate system using entirely different procedures, i.e., finite element models with different assumptions are usually used for the dam and rigid body limit equilibrium method for abutment. In reality, the dam-foundation behaves as an indivisible system during loading and earthquake process. Nowadays, some numerical models and computer codes in rock mechanics field become available. 3-D Discrete Element Code (3DEC)[48,49], Discontinuous Deformation Analysis code (DDA)[50], Particle Flow Code (PFC)[51] and Rigid Body Spring Elements (RBSE)[52] etc. have appeared to be promising and powerful tools for predicting the deformation and collapse response of jointed rock mass. So, it is possible now to combine the dam and the foundation (abutment) into one system using a continuous-discontinuous coupling procedure. The simulation of complete process of the structure-foundation rock starting from linear elastic state to complete collapse becomes possible. As a starting point of exploration, stability studies of post-cracking behavior of the Koyna gravity dam, and abutment failure of the Malpasset arch dam, static and earthquake stabilities of the Xiaowan Arch Dam-abutment and Xiluoduo Arch Dam-abutment System are examples of using DEM and RBSE for static and dynamic studies. Some interesting phenomena have been observed from the study. The details of the results may be found elsewhere[53-55].

16.2.4 Case Study

16.2.4.1 Introduction

1. General description

The Dagangshan arch dam with a height of 210 meters and a dam crest arc length of 609.8 meters is under construction on the Dadu River of Southwest China. The thicknesses of the crown cantilever are 52 meters at the bottom and 10 meters at the crest. The arc length-height ratio and thickness-height ratio are 2.90 and 0.248 respectively. The total number of contraction joints in the dam is twenty-eight. The normal depth of reservoir water is 205 meters and the lowest reservoir depth in operation is 195 meters, and the depth of silt sedimentation during operation is 125 meters. The dam is located in an extremely strong earthquake region with the design $PGA=0.557g$. Safety evaluation of the dam subjected to the design earthquake is a crucial factor for the project. The linear elastic and nonlinear dynamic behavior analyses including contraction joint opening and damage-cracking of concrete are performed to study the response of the dam under the design earthquake. In terms of earthquake input mechanism and foundation modeling, both the massless foundation and mass-viscous-spring boundary model are used for comparison.

2. Material parameters, loading conditions and design earthquake

The values of material parameters adopted in the analysis are obtained from material tests carried out by the Chengdu Hydroelectric Investigation & Design Institute. The dynamic elastic modulus of concrete and rock is multiplied by 1.3 of their static counterparts respectively. The parameters are summarized in Table 16.2-2. The material damping is considered via Rayleigh damping assumption in the analyses.

The normal static loads including the dam gravity, reservoir water, silt pressures, and design temperature loads are imposed first. The dynamic loads include earthquakes in the three directions and the hydrodynamic pressures. The design peak ground accelerations are $0.557g$ in the stream and cross-stream directions, and $0.371g$ in the vertical direction. The response spectra of the design earthquake are shown in Fig. 16.2-1, which are stipulated in the specification DL 5073—2000. The hydrodynamic interaction is modeled via the added mass assumption, based on finite element method with incompressible reservoir fluid. Finite element analysis provides the fundamental periods for full and lowest reservoir elevations are 0.6s and 0.56s respectively.

Table 16.2-2 Material Properties of the Dagangshan Arch Dam

	Concrete	Rock
Elastic modulus (10^4 MPa)	3.12	1.885
Poisson ratio ν	0.17	0.258
Unit weight (kN/m^3)	24.0	26.5

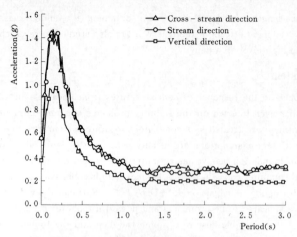

Fig. 16.2 – 1 Design Acceleration Response Spectra

16.2.4.2 Computational Results

1. Comparison study on canyon response by different foundation input models

Since the viscous-spring boundary is efficient and convenient to incorporate with current FE code and has sufficient accuracy without much increasing computational effort, the model is used herein.

In the viscous-spring boundary input model, three pairs of dashpots and springs are installed in each node of artificial boundaries as shown in Fig. 16.2 – 2.

The parameters of springs and dashpots of node l on the artificial boundary are given as follows

$$K_{ln} = a\frac{\lambda + 2G}{r}; C_{ln} = b\rho c_p \qquad (16.2-1)$$

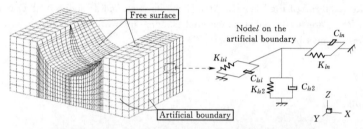

Fig. 16.2 – 2 Sketch of the Viscous-spring Boundary

$$K_{ls} = a\frac{G}{r}; C_{ls} = b\rho c_s \qquad (16.2-2)$$

where subscript l is the node number on the artificial boundary; n and s refer to the normal and tangential direction of the boundary plane; K is the elastic stiffness of the spring; C is the viscous damping; λ and G are the Lame's constants; c_p and c_s denote the P and S-wave velocity respectively; ρ is the mass density; r is the distance from the wave source to the node l; a and b are modification coefficients, which may be determined from parameter analysis.

Herein, earthquake motions are transformed into nodal dynamic loads and exerted directly on the artificial boundaries. To satisfy the force equilibrium conditions at the artificial boundaries, the equivalent traction exerted on the node l can be expressed as

$$f_l(t) = K_l u_0(x_l, y_l, z_l, t) + C_l \dot{u}_0(x_l, y_l, z_l, t) + \sigma_0(x_l, y_l, z_l, t) \qquad (16.2-3)$$

where, $u_0(x_l, y_l, z_l, t)$ is the displacement of the free field at node l; $\dot{u}_0(x_l, y_l, z_l, t)$ and $\sigma_0(x_l, y_l, z_l, t)$ are determined by; $u_0(x_l, y_l, z_l, t)$, K_l and C_l denote the elastic stiffness of the spring and viscous damping of the dashpot at node l respectively. The first two terms on the right-hand side of equation (16.2 – 3) are respectively the added elastic and damping forces to counteract those exerted on the artificial boundary by springs and dashpots. Also, these springs and dashpots may be viewed as radiation energy absorbers for scattering waves reflected by the canyon. The third term, σ_0, represents tractions at the truncated boundaries due to the input waves which usually take one half values of the free-field due to 1 – D deconvolution. Therefore, this model may be categorized as a deconvolution method.

To study the influence of the different earthquake input mechanisms on the canyon response, several dynamic analyses of the Dagangshan canyon without the dam are performed. With verification of accuracy by different mesh size of elements and different range of foundation cut, the canyon cut profiles with viscous-spring boundary shown in Fig. 16.2 – 3 is adequate to represent the canyon cut in a half space. The details of accuracy verification may be found elsewhere[56].

Fig. 16.2 – 4 shows the distribution of Peak Ground Acceleration (PGA) along the canyon surface and the corresponding acceleration response spectra respectively. The PGAs along the canyon surface in massless foundation model are completely uniform and exactly the same as the input. For mass-viscous-spring boundary model, the PGA distributes a spatially non-uniform pattern along the canyon surface due to wave scattering effects. In the two horizontal directions, compared with the massless foundation model, the most PGAs of the canyon surface are significantly reduced except the canyon terrace Point D, where a much larger PGA is observed. However, it is noteworthy that the amplification effects at Point D are simply due to the assumption of the leveling surface of the terrace while it seldom exists in a practical canyon for dam

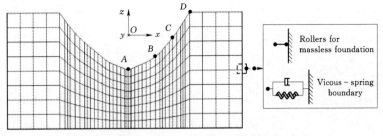

Fig. 16.2-3 Canyon Profile and FE Mesh

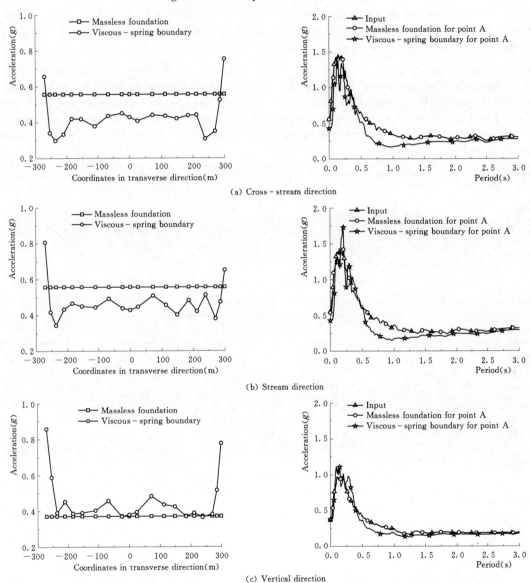

(a) Cross-stream direction

(b) Stream direction

(c) Vertical direction

Fig. 16.2-4 Acceleration Response of the Canyon

projects. Therefore, effects of different slopes of the canyon above the dam crest on the response of crest abutment need further study. In the vertical direction, larger *PGA*s are found in the viscous-spring boundary model. Furthermore, it can be found that smaller values of acceleration response spectra, especially in the period range of 0.5 - 1.0s, are vividly seen implying a significant reduction of the arch dam response can be expected.

2. Comparison study on linear elastic response of the dam by different foundation input models

Finite element discretization for arch dam-foundation system with contraction joints layout is shown in Fig. 16.2-5. When the linear elastic dynamic analysis is performed, the joints are removed from the FE model and the whole dam-foundation system returns to a continuum.

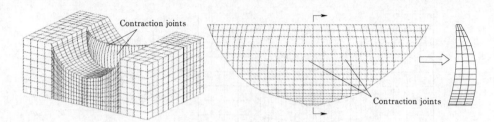

Fig. 16.2-5 FE Discretization of Dagangshan Arch Dam

Three cases of seismic analyses of the Dagangshan arch dam are performed including linear analysis using massless foundation input model with 5% and 10% damping ratio respectively, and viscous-spring boundary input model with 5% damping ratio for dam.

The distributions of maximum principal stresses on dam surfaces and crest displacements are shown in Fig. 16.2-6 and Fig. 16.2-7. Significant difference in stress and displacement response between the two models is evident when both use 5% damping ratios. However, when increasing the damping ratio to 10% for massless foundation input model, the response of the maximum tensile stresses and displacements are similar to those obtained from the viscous-spring boundary input model except that the location of the maximum stress point of the upstream face shifts from the left to right quarter region while retaining a similar value of maximum stress.

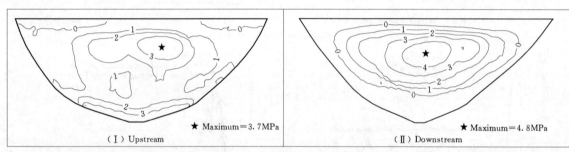

(a) Massless foundation input model with 5% damping ratio for dam

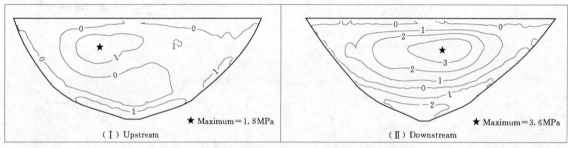

(b) Massless foundation input model with 10% damping ratio for dam

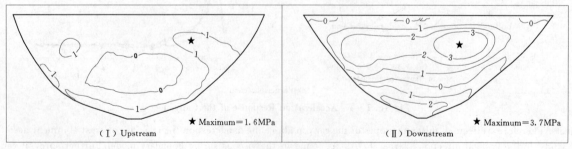

(c) Viscous-spring boundary model with 5% damping ratio for dam

Fig. 16.2-6 Maximum Principal Stresses from Linear Elastic Analysis

It is concluded that the massless foundation input model overestimates the dam response because of ignoring the radiation effects; in this case, the reduction may reach 25%-40% in terms of peak tensile stresses and crest displacements. These results strongly support conclusions of earlier research by Chopra et al[4], Dominguez et al[3] and Zhang et al[2,8]. Interestingly, in the Dagangshan case, if the damping ratio for the massless foundation is increased to 10%, the response will be

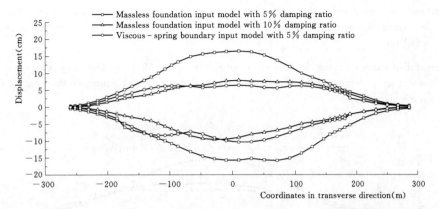

Fig. 16. 2 – 7 Maximum Displacements along the Dam Crest from Linear Elastic Analysis

approximately comparable to considering the effects of radiation damping due to the infinite foundation. Conceptually, for other dam canyons, this increased damping value may change depending on frequency closeness between the dam-canyon system and the input motions, as well as the flexibility of the foundation.

3. Comparison study on contraction joint opening of the dam by different foundation input models

In nonlinear seismic analysis, the contact boundary model[22] with tangential springs is introduced to simulate the behavior of contraction joints. All the twenty-eight joints are simulated in the study. Herein, the same three cases as linear elastic analysis are performed.

Comparisons of stress distributions are shown in Fig. 16. 2 – 8. The maximum tensile stresses from all three analyses are

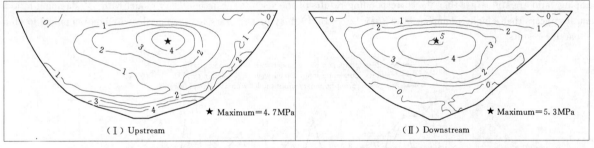

(a) Massless foundation input model with 5% damping ratio for dam

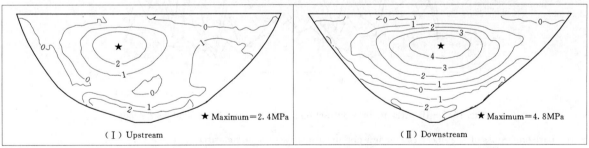

(b) Massless foundation input model with 10% damping ratio for dam

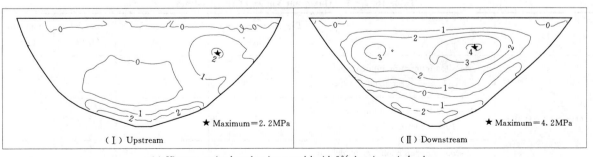

(c) Viscous – spring boundary input model with 5% damping ratio for dam

Fig. 16. 2 – 8 Maximum Principal Stresses from Nonlinear Analysis

10%–25% larger than the corresponding linear elastic cases due to the release of the arch action. The maximum tensile stresses from massless foundation with 10% damping ratio are still higher than the results from the viscous-spring boundary model with 5% damping ratio.

The maximum stream displacements along the dam crest are shown in Fig. 16.2-9. The displacements in the analysis with 10% damping ratio are close (toward upstream) or larger (toward downstream) compared with that of the viscous-spring boundary model with 5% damping ratio.

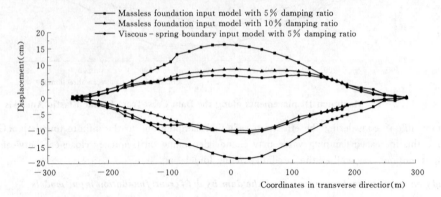

Fig. 16.2-9 Maximum Displacements along the Dam Crest from Nonlinear Analysis

The maximum opening of contraction joints is shown in Fig. 16.2-10. The peak opening is 16.7mm and occurs at the right joint in the massless foundation model with 5% damping ratio, while in the other two cases the peak openings of 13.8mm and 9.6mm occur at the left joint respectively. The joint openings along the entire crest in massless foundation with 10% damping ratio are still larger than the corresponding results of the viscous-spring boundary model with 5% damping ratio for dam. It appears that a higher damping ratio (say 12%–15%) may be necessary for the massless foundation model to obtain comparable results.

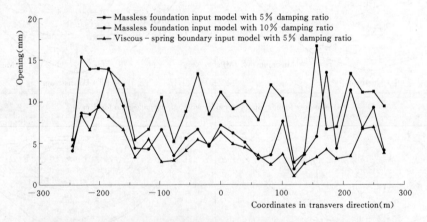

Fig. 16.2-10 Maximum Contraction Joint Openings from Nonlinear Analysis

The comparison results of the linear elastic and nonlinear analysis are summarized in Table 16.2-3.

Table 16.2-3 Maximum Values of the Analyses

	Cases	Displacement (+) (cm)	Displacement (−) (cm)	Stress (MPa)	Opening (mm)
Linear elastic analysis	Massless foundation input model (5% structural damping)	16.4	−15.8	4.8	—
	Massless foundation input model (10% structural damping)	7.8	−9.6	3.6	—
	Viscous-spring boundary model (5% structural damping)	6.5	−10.3	3.7	—

Countinued

	Cases	Displacement (+) (cm)	Displacement (−) (cm)	Stress (MPa)	Opening (mm)
Nonlinear analysis	Massless foundation input model (5% structural damping)	16.2	−18.5	5.3	16.7
	Massless foundation input model (10% structural damping)	8.7	−10.1	4.8	13.8
	Viscous – spring boundary model (5% structural damping)	7.1	−10.7	4.2	9.6

4. Comparison study on damage-cracking of the dam by different foundation input models

For studying damage-cracking of the dam, a plastic-damage model developed by Lee and Fenves[33] is used to model stiffness degradation under cyclic loading conditions. Considering tensile cracking is the most common failure phenomenon while the compressive stresses usually have sufficient safety factors in concrete dams, the compressive damage of crushing is not considered herein. The constitutive relations of the plastic-damage model under uniaxial cyclic loading condition is shown in Fig. 16.2 - 15 (a), where d_t denotes the damage variable in tension state, hence, $(1-d_t)E_0$ represents the residual stiffness after degradation; w_t and w_c are respectively weighting factors for controlling the stiffness recovery when loading changes from one state to another. Herein, the tensile strength of concrete is assumed to be 3.25MPa and other parameters are the same as listed in Table 16.2 - 2. For analysis of damage-cracking behavior, the two-third of the dam body is discretized into refined elements with mesh size of about 2m. The details of modeling may be found elsewhere[57]. The comparison study includes two seismic input models, i.e. massless foundation and viscous-spring boundary model, two assumptions for material modeling, i.e. linear elastic and plastic-damage model; different structural damping ratios of 0.05, 0.10, 0.15 for massless foundation model and 0.05 for viscous-spring boundary model. Contraction joint opening is considered for all cases of study.

The distribution of damage variables of the dam is shown in Fig. 16.2 - 11. It is seen that a region of maximum damage variable d_{max} of 0.9 penetrates through the whole section in the upper portion of the dam when massless foundation is assumed. However, when the viscous-spring boundary is used considering radiation damping, the damage region is limited to a localized area near the down-stream face. The value of d_{max} is also greatly reduced from 0.9 to 0.6. Interestingly, the contours of damage variables obtained from massless foundation model with 10% structural damping ratio are again close to that from viscous-spring boundary model implying a conclusion similar to those obtained in previous sections.

The crest displacements and joint opening are shown in Fig. 16.2 - 12 and Fig. 16.2 - 13. Several important findings are obtained: ①In all cases, the damage cracking and massless foundation input model with 5% structural damping ratio gives the maximum response envelopes of displacements with a peak value of 30cm (Fig. 16.2 - 12) and joint openings of 3.5cm, see Fig. 16.2 - 13 (a); ②When using the massless foundation input model with 10%-15% structural damping ratio, the envelopes of displacements and joint openings are greatly reduced and close to that from viscous-spring boundary model. ③Great difference of the displacements and joint openings are obtained between the two material models when massless foundation input model with 5% damping ratio is used, see Fig. 16.2 - 13 (b). The reason for that is because, in this case, the serious damage region in the upper portion of the dam causes a significant degradation of structural stiffness; however, by considering the radiation damping effects or by increasing the structural damping ratio to 10%-15% for massless foundation, the effects of different material models (linear or damage) on the displacement and joint opening response become much smaller, see Fig. 16.2 - 13 (c) and Fig. 16.2 - 13 (d). In this case, the joint openings at the central portion of the dam by the damage model are still larger than that by the linear elastic model, while the joint openings at the quarter or side portion of the dam behave as an opposite tendency due to different capability of load transfer in arch direction between the two material models.

5. Strengthening of the dam to resist the design earthquakes

Cantilever reinforcements have been selected as a major strengthening measure for the Dagangshan arch dam. The modeling of the reinforcements in massive concrete in arch dams is briefly introduced in the preceding section and its details may be found elsewhere[58].

The layout of reinforcements is shown in Fig. 16.2 - 14, where three layers of reinforcements are equivalently smeared into the membrane elements. The amounts of dam concrete, cantilever steel and arch steel used in the design are $3.14 \times 10^6 m^3$, 7,210 tons and 2,140 tons with a steel-concrete volume ratio of 0.029%. Concrete properties are the same as listed in Table 16.2 - 2. The tensile strength and fracture energy of concrete are $f_t = 3.25$MPa and $G_F = 250$N/m respectively. The steel

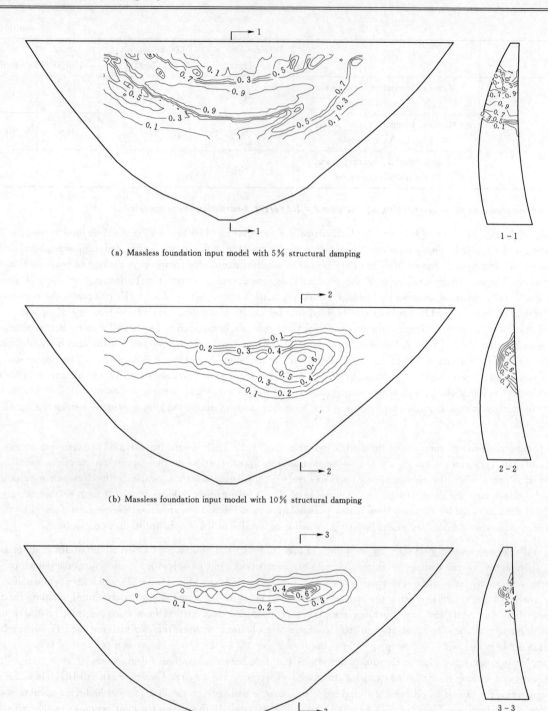

(a) Massless foundation input model with 5% structural damping

(b) Massless foundation input model with 10% structural damping

(c) Viscous-spring boundary model with 5% structural damping

Fig. 16.2-11 Distribution of Damage Variable d on Downstream Face

properties are: $E_s = 260\text{GPa}$; $\nu_s = 0.3$; f_y (yield strength) $= 650\text{MPa}$ and 560MPa for cantilever and arch steel respectively; $\rho_s = 7800\text{kg/m}^3$. The constitutive relations of concrete and reinforced steel are show in Fig. 16.2-15.

The modified embedded-steel model has been implemented in ABAUQS and used in the analysis.

Fig. 16.2-16 shows the envelopes of contraction joint openings. The maximum opening of joints decreases from 35.6mm without reinforcement to 28.7mm with cantilever reinforcement. The openings of joints in the middle region decrease noticeably while those of the other joints near abutments appear little difference.

It is noteworthy that although the peak acceleration of the ground motion beyond 15s are gradually decreased (not shown),

Chapter 16 Seismic Safety of Dams

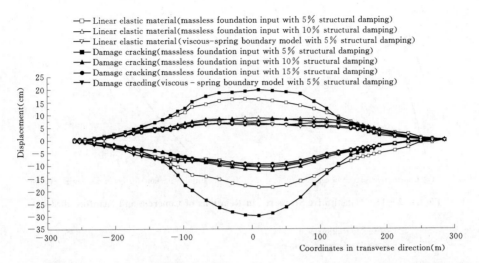

Fig. 16.2-12 Comparison of Maximum Displacements at Dam Crest between Different Material and Input Models

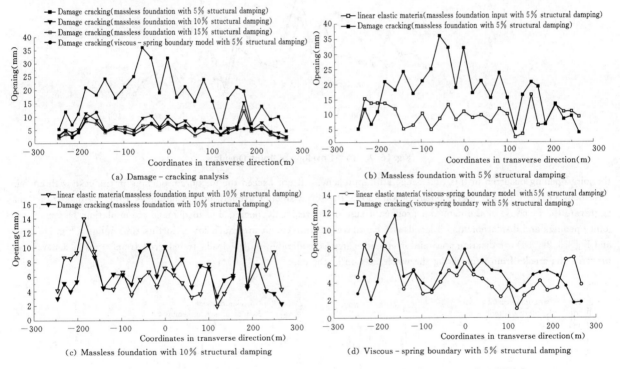

(a) Damage-cracking analysis

(b) Massless foundation with 5% structural damping

(c) Massless foundation with 10% structural damping

(d) Viscous-spring boundary with 5% structural damping

Fig. 16.2-13 Comparison of Joint Opening between Different Material and Input Models

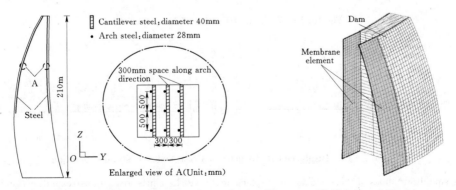

Fig. 16.2-14 Layout of the Strengthening Measure with Reinforced Steel

543

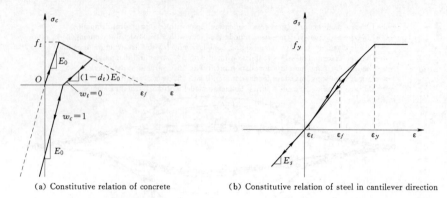

Fig. 16.2 - 15 Constitutive Stress-strain Relations of Concrete and Reinforced Steel

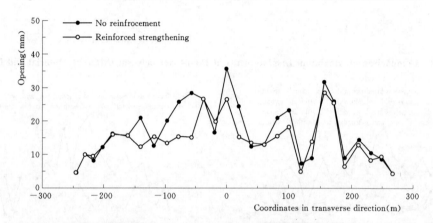

Fig. 16.2 - 16 Envelope of Joint Opening

the joint opening occurred in this time range still approaches, if not exceed the maximum opening in the case without reinforcement (Fig. 16.2 - 17). This phenomenon shows that even smaller ground motion impulses after the major ones may aggravate the response of the dam and produce a significant residual opening due to damage accumulation. However, the joint openings and displacements obtained with cantilever reinforcement are much lower in this time range (Fig. 16.2 - 17 and Fig. 16.2 - 18), indicating the reinforcements carry considerable tensile loads transferred from cracked concrete and prevents the cracks from penetrating the whole section of the dam.

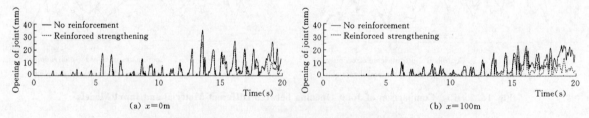

Fig. 16.2 - 17 Time Histories of Joint opening

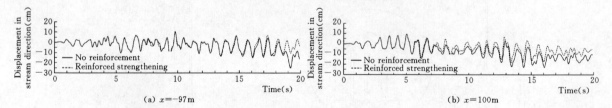

Fig. 16.2 - 18 Displacement Histories of Crest Nodes in Stream Direction

In addition, the equilibrium lines of these oscillating displacement histories shifts toward upstream direction in both analyses (Fig. 16.2 - 18), which are caused by the accumulations of tensile plastic strain after cracking. This phenomenon indicates that there will be a remarkable residual deformation toward the upstream direction after the earthquake.

The evolution of concrete cracking at the downstream face is shown in Fig. 16.2-19, where the cracking strain e_{cr} denotes the maximum cracking deformation occurred during the earthquake. The evolution process in these two analyses shows the difference at all instances listed in the figure; the cracking strain values obtained with cantilever reinforcement are smaller than those obtained without strengthening. The results from Fig. 16.2-20 (a) indicate that the concrete cracks will penetrate the monoliths if reinforcements are not used.

The cracking evolution of monolith at $x = -56\text{m}$ is shown in Fig. 16.2-20. The results show that concrete damage in these two analyses occurs in the downstream zone at the same instance and extends to the upstream face with similar evolution behavior. The results also show cantilever reinforcement significantly limits the extension of concrete crack and prevents the monolith from penetration of entire cross section. Moreover, the maximum value of cracking strain decreases from 3.6×10^{-3} to 2.4×10^{-3} (a decrease of 33.3%).

All the aforementioned results show that cantilever reinforcements can not prevent the dam from cracking because of its low ratio of reinforcement. However, the reinforced steel can carry considerable tensile load transferred from the cracked concrete and improve the stiffness of monolith after concrete damage occurs in the dam. Therefore, the reinforced steel has effects on decreasing the responses of joint opening and displacements as well as limiting crack extension. In addition, the accumulation of tensile plastic strain caused by the concrete damage occurred in the downstream zone may cause residual deformation toward upstream direction. This is even more severe in the case without cantilever reinforcement, which may weaken the load-carrying capacity of the dam.

Consequently, the strengthening measure of cantilever reinforcement has a noticeable benefit for improving earthquake-resistant capacity of the Dagangshan arch dam. Moreover, this study shows that the modified embedded-steel model is applicable to evaluating the effectiveness of reinforcement strengthening and optimizing the design of strengthening measures.

However, the effect of current reinforcement measure is not sufficient to reduce the damage-cracking development of the dam. Therefore, extensive studies of being conducted to investigate the effect of combining seismic-resistant measures are cantilever reinforcement with other strengthening methods, including joint dampers. Moreover, the influence of canyon radiation may be considered in the further investigation.

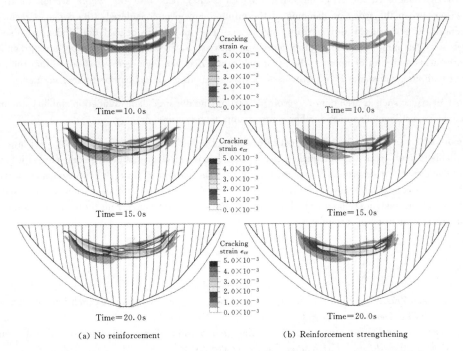

Fig. 16.2-19 Evolution of Cracking Strain e_{cr} at the Downstream Face (without Reinforcement)

16.2.5 Conclusions

(1) Several key issues regarding seismic safety evaluation of high concrete dams, especially arch dams have been studied. Among those, the earthquake input mechanism and foundation modeling may be one of the most important issues in this regard. It is found that compared with the conventional massless foundation model and uniform input of ground motion, the

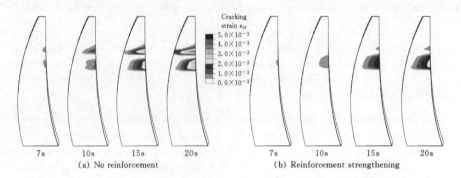

Fig. 16.2 - 20 Evolution of Cracking of the Monolith at $x = -56m$

radiation damping and non-uniform motions due to infinite canyon have important effects leading to a 25%–40% reduction in dam response. Equivalently, use of a larger damping ratio for structures such as 10%–15% may also obtain comparable results of dam response to that of considering radiation damping. This benefit may be (at least partially) considered in the design or as a potential safety factor for resisting strong earthquakes.

(2) Contraction joint opening is another important factor needed to be considered in safety evaluation. The main concern of this nonlinearity is the maintenance of integrity of the dam body and prevention of water stops between joints from breakage. Studies from several arch dams of 200 – 300 meters in height including Xiaowan, Xiluodu and Dagangshan Dams reveal that the maximum joint openings are in the level of 10 – 15mm when the linear elastic model of concrete material is applied, or 20 – 35mm when damage-cracking model for concrete is used. With a special design for withstanding 50 – 80mm in deformation for water stops, the influence of such opening to safety of water stops can be manageable.

(3) Tensile stresses in cantilevers will remarkably be increased due to the response amplification of the dam and joint opening. Thus, damage-cracking will inevitably occur in the upper middle portion of the dam. In the massless foundation analysis for the Dagangshan case, a serious damage area with degradation index of 0.9 is observed from downstream to upstream faces for most of the middle cantilevers, indicating the cracks may penetrate the whole section of the most middle cantilevers. However, when the radiation damping of the infinite canyon is considered, the damage-cracking area will be significantly reduced to a localized downstream region and the maximum degradation index is also decreased to 0.6. Nevertheless, for retaining a sufficient safety margin in the design, it was decided to adopt strengthening measures of cantilever reinforcements and joint dampers in the Dagangshan and Xiaowan arch dams. It is hoped that the predicted earthquake behavior of high arch dams and the effectiveness of strengthening measures can be further verified.

(4) Other important issues such as arch dam-reservoir interaction and dynamic dam-foundation stability are only briefly touched in Part I of the paper. They also need further investigation in the future.

(5) Although more advanced knowledge and sophisticated numerical methods for earthquake analysis of large concrete dams appear to be available to date, realistic earthquake behavior and damage mechanisms of large dams are still far from states of complete clarity. The conventional seismic design practice is that the pseudo-static arch-cantilever method is still used for dam design and the rigid body limit equilibrium for foundation. However, when earthquake input mechanisms and nonlinear interaction behaviors of the system are involved, how to link up the interpretations from these two categories of methods is a crucial issue in safety evaluation of large dams.

References

[1] Zhang Chuhan and Zhao Chongbin. Effects of Canyon Topography and Geological Conditions on Strong Ground Motion. *Earthquake Engineering and Structural Dynamic*, 1998; 16: 1, 81 - 97.

[2] Zhang Chuhan, Wang Guanglun and Zhao Chongbin. Seismic Wave Propagation Effects on Arch Dam Response. *Proc. 9th World Conf. Earthquake Eng.*, Tokyo-Kyoto, 1988, VI, 367 - 372.

[3] Dominguez and Maeso O. Model for the Seismic Analysis of Arch Dams including Interaction Dffects. *Proc. 10th World Conf. Earthquake Eng.*, Madrid, 1992, 8, 4601 - 4606.

[4] Chopra A K and Tan H. Modeling Dam-foundation Interaction in Analysis of Arch Dams. *Proc. 10th World Conf. Earthquake Eng.*, Madrid, 1992, 8, 4623 - 4626.

[5] Zhang Chuhan, Pekau O A and Jin Feng. Application of FE-BE-IBE Coupling to Dynamic Interaction between Alluvial Soil and Rock Canyons. *Earthquake Engineering and Structural Dynamic*, 1992, 21: 367 - 385.

[6] Yan Junyi, Zhang Chuhan and Jin Feng. A Coupling Procedure of FE and SBFE for Soil-structure Interaction in the Time Domain. *International Journal for Numerical Methods in Engineering*, 2004, 59: 11, 1453-1471.

[7] Chen Houqun, Du Xiuli and Hou Shunzai. Application of Transmitting Boundaries to Nonlinear Dynamic Analysis of an Arch Dam-foundation-reservoir System. *Dynamic soil-structure interaction, current research in China and Switzerland* 1997 (Zhang Chuhan and Wolf JP, eds.), Beijing, International academic publisher, 1997: 115-124.

[8] Zhang Chuhan, Jin Feng and Pekau OA. Time Domain Procedure of FE-BE-IBE Coupling for Seismic Interaction of Arch Dams and Canyon. *Earthquake Engineering and Structural Dynamic*, 1995, 24: 1651-1666.

[9] Clough R W. Non-linear Mechanisms in the Seismic Response of Arch Dams. *Proc. Int. Res. Conf. Earthquake Eng*. Skopje, Yugoslavia, 1980, 669-684.

[10] Xiuli Du, Yanhong Zhang and Boyan Zhang. Nonlinear Seismic Response Analysis of Arch Dam-foundation Systems-part I Dam-foundation Rock Interaction. *Bulletin of Earthquake Engineering*, 2007, 5: 105-119.

[11] Zhang Chuhan, Jin Feng and Wang Guanglun. Seismic Interaction between Arch Dam and Rock Canyon//11th *World Conf. Earthquake Eng.*, Mexico, 1996: 595.

[12] Proulxl J, Darbre G R and Kamileris N. Analytical and Experimental Investigation of Damping in Arch Dams Based on Recorded Earthquakes//13th *World Conf. Earthquake Eng.*, Canada, 2004: 68.

[13] Hall J F. The Dynamic and Earthquake Behaviour of Concrete Dams: Review of Experimental Behaviour and Observation Evidence. *Soil Dynamic and Earthquake Engineering*, 1988: 7: 2, 58-121.

[14] Swanson A A and Sharma R P. Effects of the 1971 San Fernando Earthquake on Pacoima Arch Dam. *Proc.* 13th *ICOLD*, New Delhi, Q. 51., R. 3, 1979.

[15] Hall JF. Northridge Earthquake of January 17, 1994 reconnaissance report, volume 1. *Earthquake Spectra* 11 (Supplement C): EERI 95-03, 1995.

[16] Clough RW. Non-linear Mechanisms in the Seismic Response of Arch Dams. *Proc. Int. Res. Conf. Earthquake Eng*. Skopje, Yugoslavia, 1980: 669-684.

[17] Niwa A and Clough RW. Non-linear Seismic Response of Arch Dams. *Earthquake Engineering and Structural Dynamic* 1982, 10: 267-281.

[18] Fenves G L, Mojtahedi S and Reimer RB. ADAP88: a Computer Program for Nonlinear Earthquake Analysis of Concrete Arch Dams. *Report No. EERC* 89-12, Earthquake Engineering Research Center, University of California, Berkeley, CA, 1989.

[19] Fenves G L, Mojtahedi S and Reimer RB. Effect of Contraction Joints on Earthquake Response of an Arch Dam. *Journal of Structural Engineering* (ASCE) 1992, 118: 1039-1055.

[20] Zhang Chuhan, Xu Yanjie and Jin Feng. Effects of Soil-structure Interaction on Nonlinear Response of Arch Dams. *Dynamic soil-structure interaction*, International Academic Publishers: Beijing, 1997: 95-114.

[21] Zhang Chuhan, Xu Yanjie, Wang Guanglun and Jin Feng. Non-linear Seismic Response of Arch Dams with Contraction Joint Opening and Joint Reinforcements. *Earthquake Engineering and Structural Dynamics*, 2000, 29: 1547-1566.

[22] Bathe K J and Chaudhary A. A Solution Method for Planar and Axisymmetric Contact Problems. *International Journal for Numerical Methods in Engineering*, 1985, 21: 1, 65-88.

[23] Lua D T, Boruziaan B and Razaqpur AG. Modeling of Contraction Joint and Shear Sliding Effects on Earthquake Response of Arch Dams. *Earthquake Engineering and Structural Dynamics*, 1998, 27: 1013-1029.

[24] Xiuli Du and Jin Tu. Nonlinear Seismic Response Analysis of Arch Dam-foundation Systems-Part II Opening and Closing Contact Joints. *Bulletin of Earthquake Engineering*, 2007, 5: 121-133.

[25] Chen Houqun. Model Test and Program Verification on Dynamic Behavior of Arch Dam with Contraction Joints. *Report No. SVL-94/2 IWHR*, July, 1994.

[26] Sheng Zhigang, Zhang Chuhan, Wang Guanglun et al. Shaking Table Tests and Numerical Analysis for Nonlinear Seismic Behavior of an Arch Dam Model with Contraction *Joints. Journal of Hydroelectric Engineering*, 2003, 80: 1, 34-43.

[27] Wang Haibo and Li Deyu. Experimental Study of Seismic Overloading of Large Arch Dam. *Earthquake Engineering and Structural Dynamics*, 2006, 35: 199-216.

[28] Lin Gao and Hu Zhiqiang. Earthquake Safety Assessment of Concrete Arch and Gravity Dams. *Earthquake Engineering and Engineering Vibration*, 2005, 4: 2, 251-264.

[29] Liu Xinjia, Xu Yanjie, Wang Guanglun, et al. Seismic Response of Arch Dams Considering Infinite Radiation Damping and Joint Opening Effects. *Earthquake Engineering and Engineering Vibration*, 2002, 1: 1, 65-73.

[30] Pekau O A, Feng Lingmin and Zhang Chuhan. Seismic Fracture of Koyna Dam: Case Study. *Earthquake Engineering and Structural Dynamic*, 1995, 24: 15-33.

[31] Wang Guanglun, Pekau O A, Zhang Chuhan, et al. Seismic Fracture Analysis of Concrete Gravity Dams Based on Nonlinear Fracture Mechanics. *Engineering Fracture Mechanics*, 2000, 65, 67-87.

[32] Lubliner J, Oliver J Oller S et al. A Plastic-damage Model for Concrete. *International Journal of Solids and Structures*, 1989, 25: 3, 299-326.

[33] Lee J and Fenves LG. A Plastic-damage Concrete Model for Earthquake Analysis of Dams. *Earthquake Engineering and Structural Dynamics*, 1998, 27, 937-956.

[34] Bazant Z P and Oh B H. Crack Band Theory for Fracture of Concrete. *Materiaux Constructions*, 1983, 16: 93, 155-177.

[35] An X, Maekawa K, Okamura H. Numerical Simulation of Size Effect in Shear Strength of RC Beams. *Journal of Materials, Concrete Structures and Pavements*, 1997, JSCE, 1997, 35: 564, 297-316.

[36] Gilbert R I, Warner R F. Tension Stiffening in Reinforced Concrete Slabs. *Journal of Structural Division*, ASCE, 1987, 104 (ST12): 1885-1901.

[37] Kwak H G, Filippou F C. Finite Element Analysis of Reinforced Concrete Structures under Monotonic loads. Report No. UCB/SEMM-90/14. Department of Civil Engineering, University of California, Berkeley, California, 1990.

[38] Luo Bingyan, Xu Yanjie and Wang Guanglun. Nonlinear Dynamic Analysis of Dagangshan Arch Dam with Considering Damper Reinforcement. *Journal of Hydroelectric Engineering*, 2007, 26: 1, 67-70.

[39] Chakraparti P and Chopra AK. Earthquake Analysis of Gravity Dams including Hydrodynamic Interaction. *Earthquake Engineering and Structural Dynamics*, 1973, 2: 143-160.

[40] Chakraparti P and Chopra AK. Hydrodynamic Effects in Earthquake Response of Gravity Dams. *Journ. Am. Soc. Civil. Engrs*, 1974, 100 (ST-6): 1211-1224.

[41] Hall JF and Chopra AK. Dynamic Response of Embankment, Concrete-gravity and Arch Dams including Hydrodynamic Interaction. *Report No. UCB/EERC* 80/39, Earthquake Engineering Research Center, University of California, Berkeley, 1984.

[42] Clough R W, Chang KT, Chen HQ, et al. Dynamic Response Behavior of Xiang Hong Dian Dam. *Report No. UCB/EERC* 84/02, Earthquake Engineering Research Center, University of California, Berkeley, 1984.

[43] Clough R W and Chang KT. Dynamic Response Behavior of Quan Shui Dam. *Report No. UCB/EERC* 84/20, Earthquake Engineering Research Center, University of California, Berkeley, 1984.

[44] Clough R W and Ghanaat Y. Experimental study of arch dam-reservoir interaction. *Proceedings of Joint China-US workshop on Earthquake Behavior of Arch Dams*, Beijing, 1987.

[45] Zhang Chuhan, Yan Chengda, Wang Guanglun. Numerical Simulation of Reservoir Sediment and Effects on Hydrodynamic Response of Arch Dams. *Earthquake Engineering and Structural Dynamics*, 2001, 30: 1817-1837.

[46] Du Xiuli, Wang Jinting and Hung Tinkan. Effects of Sediment on the Dynamic Pressure of Water and Sediment on Dams. *Chinese Science Bulletin*, 2001, 46: 6, 521-524.

[47] Qiu Liuchao, Jin Feng and Wang Jinting. Application of Sponge Layer Method for Radiation Damping in Analysis of Dam-reservoir Interaction. *Journal of Hydraulic Engineering*, 2004, 1: 46-51.

[48] Cundall P A. A Computer Model for Simulating Progressive Large Scale Movement in Blocky Systems. *Proc. Symp. Int. Soci. Rock Mech.* 1, Paper No. II-8, 1971.

[49] Cundall P A. Formulation of Three-dimensional Distinct Element Model, Part I, A Scheme to Detect and Represent Contact in System Composed of Many Polyhedral Blocks, *Int. J. Rock Mech. Min. Geomech. Abstr.* 1988, 25: 107-116.

[50] Shi G H, Goodman R E. Two Dimensional Discontinuous Analysis. *Int. J. numer. anal. meth. geomech.* 1985, 9: 541-556.

[51] Cundall PA. PFC^{2d} User's Manual (version 2.0). Minnesota: Itasca Consulting Group Inc, 1990.

[52] Kawai T and Toi Y. New Element Models in Discrete Structural Analysis. *Journal of the Society of Naval Architects and Ocean Engineers*, 1978, 16: 97-110.

[53] Hou Yanli. A Coupling Rupture Model of Distinct Element and Fracture Mechanics for Concrete Dam-Foundations. Ph. D. dissertation. Beijing: Tsinghua University, 2005.

[54] Cui Yuzhu, Zhang Chuhan, Jin Feng, et al. Numerical Modeling of Dam-foundation Failure and Simulation of Arch Dam Collapse. *Journal of Hydraulic Endineering*, 2002, 6: 1-8.

[55] Zhang Chong, Jin Feng and Xu Yanjie. Dynamic Analysis of Arch Dam-abutment Stability. *Journal of Hydroelectric Engineering*, 2007, 26: 2, 27-36.

[56] Zhang Chuhan, Pan Jianwen and Wang Jinting. Influence of Seismic Input Mechanisms and Radiation Damping on Arch Dam Response. *Earthquake Engineering and Structural Dynamics* (submitted).

[57] Pan Jianwen, Wang Jinting and Zhang Chuhan. Analysis of Damage and Cracking in Arch Dams Subjected to Extremely Strong Earthquake. *Journal of Hydraulic Engineering*, 2007, 38: 2, 143-149.

[58] Long Yuchuang, Zhang Chuhan and Jin Feng. Numerical Simulation of Reinforcement Strengthening for High Arch Dams to Resist Strong Earthquakes. *Earthquake Engineering and Structural Dynamics* (will appear in 2008).

16.3 History and Progress of Seismic Analysis and Safety Evaluation of High Dams
Lin Gao[1]

16.3.1 Introduction

The development of hydraulic and hydropower engineering as well as dam technology in China has gone through three stages.

(1) In the 1950s-1960s, with the rapid progress of the Chinese economy, there was a great upsurge in water conservancy works and dam construction. Profiting from the spirit of self-dependence and independent innovation, a batch of dams with a maximum height over 100m have been designed and constructed. Great success was achieved in the modern dam technology during this period.

(2) In the 1980s, the Chinese economy saw another round of rapid growth due to the government policy of reform and opening up. The dam construction went into a new great upsurge. Relevant technology advanced towards maturity and reached a new level. A batch of high dams with the height between 100m to 200m were built on a great number of large rivers.

(3) Around the beginning of the 21st century, an unprecedented golden age appeared for the dam construction in China. In order to meet the need for hydroelectric energy development, a batch of high dams with heights close to or greater than 300m, which is a height exceeding the highest dam all over the world, are being built in the upper reaches of the Yangtse River and the Yellow River. The engineers and researchers in China resolved a series of difficult problems, and pushed the dam technology forward.

The development of dam construction promoted the rapid growth of scientific research on dams. The main objective of the presented paper is to briefly review the progress that has been made towards the research on seismic safety of dams in the past 60 years. Emphasis is placed on the state of the art of seismic response analysis of dams and its application to seismic design and seismic safety evaluation of dams.

China lies between two major world earthquake active zones, i.e., the Circum-Pacific seismic zone and the Eurasia seismic zone. In the west mountainous area of the upper reaches of the Yangtse River and the Yellow River, which are rich in water power resources, moreover, the earthquake activity is highly intensive. Hence, the seismic safety of dams is of great concern for the designers and researchers in dam construction. They have put in a great deal of efforts to enhance the level of seismic analysis and seismic safety evaluation of dams. Drawing inspiration from trial-load method, they presented approaches based on the vibration modes of arch-cantilevers for seismic analysis of arch dams. They investigated the influence of dynamic water pressure on the seismic response of dams taking into consideration of the effect of water compressibility and reservoir boundary absorption. They proposed a variety of methods to study the dynamic interaction between dams (arch dam or gravity dam) and unbounded foundation, including the multi-transmitting boundary method, the approximate visco-elastic boundary method, the coupled finite element-infinite element method, and the scaled boundary finite element method (SBFEM). They analyzed in-depth the effect of contraction joints on the seismic response of arch dams, and extended several effective methods to this problem, such as the Penalty function method and the Lagrangian multiplier method, and obtained the solution through the trial-error method and B-differential equation method. They also performed a great number of dynamic model tests to examine the seismic response of various dams. However, no attempt is made to give a comprehensive discussion on the above-mentioned methods that goes into all the details. Instead, based on the research work carried out in the Dalian University of Technology (DUT), this paper provides most important early technical developments and introduces new progress in seismic response analysis of dam.

16.3.2 A Brief Review of Early Research Work

16.3.2.1 The First Dynamic Model Test of Arch Dam in China

In 1956, a 78m high double curvature arch dam was planned to be built on the Liuxi River in Guangdong Province. By discharging the excess flood water through the crest of the dam, the investment on the project could be greatly reduced. In order to minimize the danger that scouring of the downstream river bed by the impact of the falling water could undermine the dam's toe, a jet bucket was made on the dam crest to let the water jet falling distantly from the dam's

---

[1] Lin Gao, Academician of Chinese Academy of Sciences Professorof Dalian University of Technology.

toe. However, the fluctuating water pressure during flood discharge could induce detrimental vibration of the dam body, the effect on the dam safety was a great concern to the designers. Aiming at resolving this problem, dynamic model tests were carried out at the Talian Engineering Institute (TEI) (The predecessor of DUT) and a set of testing system was established.

The model tests which shed some new light on the following problems[1] were completed within one and a half year.

(1) The rules of dynamic similitude for hydraulic structure were identified[2]:

rule of elastic similarity

$$\tau = \lambda \sqrt{\frac{E_m}{E} \cdot \frac{\rho}{\rho_m}}, \frac{\gamma}{\gamma_m} = \frac{\rho}{\rho_m} \qquad (16.3-1)$$

rule of gravitational similarity

$$\tau = \sqrt{\lambda}, \frac{\gamma}{\gamma_m} = \frac{\rho}{\rho_m} \qquad (16.3-2)$$

rule of elastic-gravitational similarity

$$\tau = \lambda \sqrt{\frac{E_m}{E} \cdot \frac{\rho}{\rho_m}}, \tau = \sqrt{\lambda}, \frac{\gamma}{\gamma_m} = \frac{\rho}{\rho_m} \qquad (16.3-3)$$

where λ denotes geometric scale; τ denotes time scale; E denotes modulus of elasticity; ρ denotes mass density of the dam materials; γ denotes mass density of the reservoirs water. Symbols with subscript 'm' denote those value of the model and symbols without subscript denote those values of the prototype.

(2) A mechanical shaking table with the size of 3m×1.2m and frequency range of 0.83-7.5Hz was designed and manufactured.

(3) The dam model is made of stamped rubber with volume weight of $2.4N/cm^3$, which meets the requirement of similarity for hydrodynamic pressure induced by reservoir water. The geometric scale used for the model is 1:200.

(4) The resonant frequencies of the dam model were measured by the principle of Lissajous figure.

(5) The acceleration and displacement transducers. The transducers for measuring vibration amplitudes of the model dam were also designed and manufactured by ourselves.

(6) Hydraulic experiment was performed to measure the fluctuating water pressure acting on the dam crest during the flood discharge.

The dynamic dam response caused by the flow fluctuation had been studied. It was concluded that the dam can safe withstand such fluctuating load.

16.3.2.2　The First Dynamic Model Test of Embarkment Dam in China

In 1958 a 80m high Maojiacun Earth Dam was planned to be built on the Yili River in Yunnan Province, the dam site is seismic active. The dynamic model test was carried out at the TEI [5]. The structure is designed as a roller compacted central core earth dam. The core and blanket were made of mixed material of clay and gravels, and the shell was composed of sand and gravels. To satisfy the rule of similarity, the sand was used for the model shell and the model core was similated by artificial material with slight cohesive. The dynamic behavior of the dam materials and model materials was tested by dynamic triaxial apparatus mounted on the small shaking table, moving separately in horizontal and vertical direction.

Vibration frequencies and vibration modes of the model dam had been measured. Seismic deformation and settlement were tested and examined.

The stream direction of the reservoir for Maojiacun Earth Dam is nearly parallel to the dam axis, as a result, the opposite river bank stands at a distance no far away from the upstream face (about 600m) of the dam. The safety of the earth dam against overtopping during strong earthquake shaking was a problem of great concern to the designers. The experiment for predicting possible wave height during earthquakes was performed.

16.3.2.3　Experimental Study on the Stability of Buttress Against Lateral Buckling

In 1958 a project of buttress dam was designed for the Baishan Hydropower Station. The height of the buttress monolith is

136.5m, whereas the thickness at the middle is about 10.2m. The lateral stability against buckling was studied in TEI[6], No theoretical solution for the triangular plate against lateral buckling existed. Розанов Н. П. only proposed an approximate energy method, which gives excessive by large factor of safety. At that time a survey of the literature showed that the methods of experiment were rather rough, no critical load can be evaluated. Two approaches were adopted. In the first approach, either the hydraulic water pressure or the gravity load was simulated by three concentrated loads. Through experiment the predicted critical load (resultant of the hydraulic force and the gravitational load) was 260kg. In the second approach, the hydraulic water pressure acting on the upstream face of the buttress and the gravity load were transforming into a concentrated load applied at the crest of the buttress. The tested critical load at the crest of the buttress was 38kg, the equivalent resultant of hydraulic force and the gravitational load was determined as 247kg. The deviation from 260kg is about 5%, which validated the effectiveness of the experimental approaches.

16.3.2.4 The Direct Photographic Method for Model Vibration Test of Dams

In 1971, the 144-meter-high (the actual height reached after construction was 149.5m). Baishan Arch Dam was planned to be built on a tributary of Songhua River. The dam site is located in earthquake active zone. At that time, it was the highest of arch dam in China. The study of earthquake resistant behavior of it has great significance. The experiment carried out at TEI just occurred in the time of Cultural Revolution, most instruments and testing facilities were not available. So the experiment should be designed as simple as possible. For this reason, a new testing approach, i.e. the direct photographic method was presented[7]. The model material was chosen as a glue mixture made of agar-agar, glycerine and water. It has low modulus of elasticity (20-35MPa), light material density (1.2g/cm^3), high strength and very good dynamic restoring elastic behavior. The model dam can be casted in the wooden mold when heated. And it solidifies at room temperature. Under the action of dead weight, the model dam deforms slightly. Before testing, the model dam face was painted into black colour, only measuring points were marked with white colour. When subjected to various resonant frequencies, steady state vibration of the model dam was excited, the vibration model shapes could be clearly seen. All measuring points moved along elliptical trajectories. By taking long time and instant exposures, motion trajectories and phase difference of the measuring points could be recorded by the camera, from which the vibration modes of the model dam were obtained. The direct photographic technique is simple and audio-visual. Through this technique, the testing instruments and devices could be relatively simple. Only low frequency-shaking table is required, making its manufacturing less expensive. The time elapse required to performing the experiments was relatively short. In the meantime, more vibration modes could be captured and the number of recording measuring points could be increased to as many as 50 to 60 for small height of model dam (about 60cm). As a result, the vibration modes could be obtained with higher accuracy. By using the direct photographic technique, the first nine vibration modes (5 symmetric modes and 4 anti-symmetric modes) of Baishan three center circular arch dam was acquired. At that time a maximum of three to four vibration modes has been obtained in Japan by using electromagnetic exciters in the laboratory[8] or by vibration field tests of actual dams[9]. As an alterative, numerical simulation based on finite element method was used for the study[10], but less vibration modes of arch dam could be obtained due to coarse mesh used. The direct photographic technique had also been used to study the dynamic properties of 91.7m high Fengman Gravity Dam in Jilin Province. The first four vibration modes (3 in the horizontal direction and 1 in the vertical direction) were obtained.

16.3.2.5 Computer Method and Computer Program for Studying Hydrodynamic Pressure on Dams

Some preparation work was carried out in TEI in 1973 for the development of the *Specifications for Seismic design of hydraulic structures* in China. A computer method and computer program based on finite difference scheme for the study on the hydrodynamic pressure on the gravity dam as well as on the arch dam was presented[11]. An over-relaxation factor was introduced to accelerate the convergence rate of the computation. An optimum over-relaxation factor was recommended and the induced error was estimated. The research results were adopted as part of the guidelines in the issued *Specification for seismic design of hydraulic structures* in China. (the trying out edition issued in 1978, the 1st edition issued in 1997, etc.)

16.3.2.6 Dynamic Model Failure Test of Concrete Gravity Dams and Arch Dams

In spite of the advances in the theory and numerical procedures in dealing with seismic response of dams subjected to strong earthquake shocks, these analyses have failed to predict the response of real dam structures particularly at their ultimate states, which are important for evaluation of seismic dam safety. For this purpose, experimental investigation was carried out[12,13] to study the seismic damage development and failure process of concrete dams. Based on the understanding that the seismic failure of concrete dams initiates with tensile fracture at the potential vulnerable zones, it is important that the model material simulates the basic characteristics of mass concrete. First, to reproduce the final damage patterns of high concrete dams, the model material is required to be elastic and brittle. Second, modulus of elasticity of the material must be low enough to make the fundamental frequencies of the models lie in frequency range of the shaking table. Third, the tensile strength of the model material must be as small as possible to enable final failure of the model dams to occur within the nominal power of the shaking table. Additionally, it must be possible to mold and demold the models without causing any damage. After hundreds of trial mixes, a special material consisting of cement, river sand, heavy quartz sand, heavy quartz powder, iron powder, and water was developed for the dam models. The elasticity and brittleness of this new material are similar to those of mass concrete, whereas its tensile strength and modulus of elasticity are much lower than of mass concrete.

The model are intended to be as simple as possible to reproduce the essential aspects of the seismic response of the structure under consideration. In view of this, experiments were conducted under rather ideal consideration. The model dams were excited harmonically in their fundamental mode of vibration, which plays dominant role in most cases, until failure. The critical instant at which tensile cracks appeared was defined as the beginning of the seismic failure of the dam. The instantaneous input acceleration of the model, called the first tension crack acceleration, could conservatively be used as the most important parameter needed to evaluate the seismic safety of the real dam. Modification of the results to take into consideration the effect of the input wave form, the loading rate, loading history, and the vertical component were suggested correction factors due to the influence of reservoir water, higher vibration modes, scale effects and discrete distribution of the dam material behavior were also discussed. Some of these factors must be obtained from laboratory experiments and field surveys. Some of the factors need extensive additional research, in particular, the dam-reservoir-foundation interaction is a separate and major research topic. Despite associating with some deficiencies and limitations, dynamic failure tests help better understanding the nonlinear dynamic response of concrete dam and its overloading capacity beyond the design basis.

16.3.3 New Progress in Seismic Response Analysis and Safety Assessment of Dams

The dynamic strength of concrete, the dam-reservoir interaction, the soil-structure interaction, the dynamic stress analysis method etc. are important factors to be considered for seismic design and safety assessment of dam structures. Recent research achievements in these fields are described.

16.3.3.1 Dynamic Properties of Concrete Subjected to Earthquake Loading

In the current design practice, the key factor that determines the capacity of concrete dams to withstand earthquakes is the tensile strength of concrete. However, the acceptable criterion of tensile strength for concrete dams is still a problem which requires engineering judgment.

Concrete is a material that is sensitive to the rate of loading. The material strength, stiffness and ductility (or brittleness) are rate dependent. The wide range of strain rates typically expected in practice for various loading conditions is shown in Fig. 16.3-1. A lot of efforts have been dedicated to the research in this field. However, the great majority of experimental results were emphasized on the compressive behavior under uniaxial monotonic loading conditions. In order to obtain a better understanding of the concrete behavior subjected to earthquake loading, dynamic experiments on about 2,000 specimens were performed in the Dalian University of Technology (DUT).

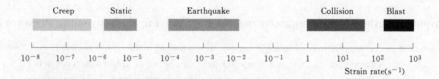

Fig. 16.3-1 Magnitude of Strain Rates Expected for Different Loading Rates

1. Strain rate effect of concrete in direct tension[14]

Dynamic tests were performed using an MTS810 testing machine with dumbbell-shaped specimens (Fig. 16.3-2). The strain-rates covered a range of $10^{-5}/s$ to $10^0/s$, larger than that typical for earthquake loading. The effect of strain rate on the strength of concrete is represented by a dynamic strength increasing factor (DIF), defined as the ratio of dynamic to static strength versus strain rate on a semi-log or log-log scale.

$$DIF = 1 + \alpha \lg(\dot{\varepsilon}_t / \dot{\varepsilon}_{ts}) \qquad (16.3-4)$$

where $DIF = f_t / f_{ts}$; f_t denotes dynamic tensile strength at strain rate, $\dot{\varepsilon}_t$, f_{ts} represents the quasi-static tensile strength at quasi-static strain rate $\dot{\varepsilon}_{ts}$ assumed as $10^{-5}/s$.

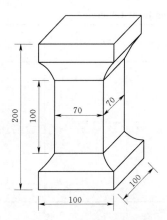

Fig. 16.3-2 Dumbbell-shaped Specimen (Unit: mm)

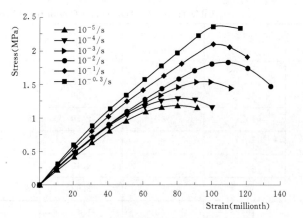

Fig. 16.3-3 Typical Stress-strain Curves of Saturated Concrete (Group LWR)

In the study, effect of environmental factor on the strain rate enhancement was examined, i.e. specimens of different strength were tested under various moisture content and temperature conditions. They were divided into five groups: HDR, LDR, HWR, HDC and HWC, wherein the first letter L or H denotes low or high strength concrete, the second D or W denotes dry or wet environment, and the third R or C stands for room or cold temperature, respectively (Table 16.3-1).

Table 16.3-1 Characteristics of the Specimens

Group	w/c ratio	Moisture (%)	Temperature (℃)
HDR	0.69	0.3	20
LDR	1.02	0.3	20
HWR	0.69	4.8	20
HDC	0.69	0.3	−30
HWC	0.69	4.8	−30

All test results are presented in Table 16.3-2 in which α was a regression coefficient. Note that the general trend of strength enhancement with strain rate is similar to that given by Malvar and Ross[15], which was derived from the test results of relatively small specimens. It can be observed, due to presence of free water the viscous resistance on the strain rate effect appears significantly, the strain rate enhancement coefficient α for fully saturated concrete 0.265 was much greater than that for concrete with normal moisture content 0.134. This phenomenon differs substantially from what under quasi-static loading condition. The static strength for fully saturated concrete 1.30MPa is lower than that for concrete with normal moisture content 2.21MPa (see Table 16.3-2). This is because the presence of moisture forced the gel particles apart and reduced the van der Waals forces[16]. Table 16.3-2 also shows that the static strength has minor effect on the strain rate enhancement. This is different from the results obtained by Cowell[17]. Cowell explained, that lower strength concrete is generally less dense and possesses more micropores, thus higher viscous resistance may be expected. However, this is not always the case. In our experiment, the mix proportion had been optimized, the lower strength concrete was nearly as dense as that of higher strength one. As a result, the strength dependence of the coefficient DIF is insignificant. Under low temperature environment (below 0℃) the free water presenting in micropores of concrete freezes, which contributes to higher strength of concrete, however, it is less sensitive to the strain rate.

The test results on Group HDR specimens have been compared with those from various sources, it is seen that the CEB model code gives a rather conservative estimate of the dynamic strength of concrete at high strain rates.

In the experiments, dynamic enhancement of the modulus of elasticity has also been studied. The results are summarized in Table 16.3-4 where the dynamic modulus enhanced factor β has the meaning similar to that of the strength enhanced factor α. The strain-rate enhancement for modulus is less pronounced than that for tensile strength. The experimental results also show that the critical tensile strain, i.e., the strain corresponding to peak tensile strength, enhances with increasing strain rate. For specimens of Group HDR, LDR and HWR the dynamic critical strain enhanced factors are 0.203, 0.109 and 0.067 respectively. During experiments no definite trend for change of the Poisson's ratio with the variation of strain rates was observed. It was found that the energy absorption capacity of concrete significantly increases with strain rate rising for both saturated concrete and concrete with normal water content.

Table 16.3 – 2 Tensile and Compressive Strengths of Concrete

Group	Strain rate (s^{-1})	Number of specimens	f_t (MPa)	f_c (MPa)	f_s (MPa)	α
HDR	10^{-5}	5	2.21	32.8	3.21	0.134
	10^{-4}	3	2.39			
	10^{-3}	3	2.79			
	10^{-2}	6	2.87			
	10^{-1}	4	3.40			
	$10^{-0.3}$	3	3.93			
LDR	10^{-5}	4	1.18	17.9	1.94	0.135
	10^{-4}	3	1.36			
	10^{-3}	3	1.44			
	10^{-2}	3	1.54			
	10^{-1}	5	1.82			
	$10^{-0.3}$	4	1.88			
HWR	10^{-5}	3	1.30	24.46	2.45	0.265
	10^{-4}	3	1.59			
	10^{-3}	3	1.82			
	10^{-2}	3	2.20			
	10^{-1}	3	2.70			
	$10^{-0.3}$	3	3.07			
HDC	10^{-5}	3	2.53	33.3	3.76	0.115
	10^{-3}	4	2.93			
	10^{-1}	4	3.79			
HWC	10^{-5}	4	6.32	40.2	6.47	

Table 16.3 – 3 Comparison of the Dynamic Strength Enhancement of Concrete from Various Sources

Strain rate (s^{-1})	10^{-5}	10^{-4}	10^{-3}	10^{-2}	10^{-1}	1
CEB model code (1993)	1.0	1.082	1.171	1.267	1.371	1.483
Malvar and Ross (1998)	1.0	1.088	1.184	1.289	1.402	1.526
Eq. (16.3 – 1) for Group HDR	1.0	1.134	1.268	1.402	1.536	1.67

Table 16.3 – 4 Strain Rate Enhancement in Modulus of Elasticity

Group	Strain rate (s^{-1})					Enhanced factor β
	10^{-4}	10^{-3}	10^{-2}	10^{-1}	$10^{-0.3}$	
HDR	4.0%	9.0%	10.1%	11.4%	12.1%	0.023
LDR	1.3%	4.9%	9.3%	12.0%	18.0%	0.037
HWR	28.3%	30.8%	45.8%	70.7%	100.1%	0.188

To account for the mechanism of strain-rate enhancement, various assumptions have been proposed. Some researchers suggested this attributes to the viscous resistance, similar to the Stefan effect[20-22], or the inertia effect for strain-rate higher than $(1-10)$ s^{-1}. From our viewpoint, the microstructure and property of concrete may also be an important factor to be considered. During experiment it was observed, that the fractured surfaces of the failed specimens became flatter as the strain rate rises (Fig. 16.3 – 4) and they cut through an increasing number of coarse aggregates. The representative models of failed specimens are depicted in Fig. 16.3 – 5, where the cut through aggregates are shown by dark colour.

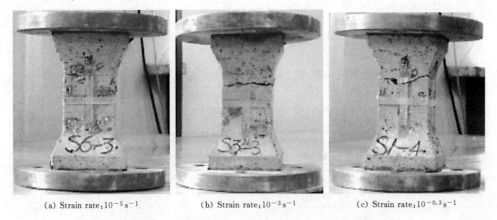

(a) Strain rate: $10^{-5} s^{-1}$　　　(b) Strain rate: $10^{-3} s^{-1}$　　　(c) Strain rate: $10^{-0.3} s^{-1}$

Fig. 16.3 – 4 Typical Failure Modes of Specimens

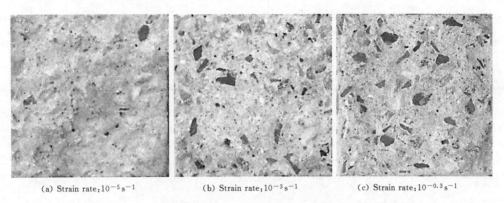

(a) Strain rate: $10^{-5} s^{-1}$　　　(b) Strain rate: $10^{-3} s^{-1}$　　　(c) Strain rate: $10^{-0.3} s^{-1}$

Fig. 16.3 – 5 Typical Fracture Surfaces at Different Strain Rates

The observed phenomena during experiments can be physically explained. Current research shows that concrete is a composite material, the aggregate-mortar interface constitutes the weakest link of it, a large number of bonding microcracks exist. Upon loading, some of these microcracks extend and progressively bridge together to form larger cracks. At low strain rates, the failure surface of the specimen mainly passes through the mortar-aggregate interfaces or the mortar, leaving behind a rough surface as can be seen in Fig. 16.3 – 20 (a). In this case, the crack development takes time to seek a path of least resistance. Whereas under rapid loading, the creation of new cracks is forced to propagate through the region of high resistance, such that a certain part of coarse aggregates are broken and the failed surface becomes flatter. As a consequence, the strength of concrete is enhanced with the rising of strain rate.

2. Dynamic behavior of concrete under biaxial and triaxial loading conditions[21,22]

Lateral confining pressure plays an important role in the dynamic behavior of concrete under multi-axial stress state.

Table 16.3 – 5 presents the experimental results of dynamic compressive strength of concrete subjected to proportional biaxial loading. The dynamic strength enhancement of concrete is a function of both the strain rate and the lateral confining pressure. The relationship may be approximated by the following expression

$$\frac{f_{bd}}{f_{us}} = c_1 + c_2 \lg(\dot{\varepsilon}/\dot{\varepsilon}_s) + \frac{c_3}{(1+\alpha)^2} + \frac{c_4 \alpha}{(1+\alpha)^2} \qquad (16.3-5)$$

where α denotes the stress ratio σ_2/σ_1 of the biaxial stresses; f_{us} denotes the uniaxial quasi-static strength of concrete; $\dot{\varepsilon}$ is the dynamic strain rate; $\dot{\varepsilon}_s$ is the quasi-static strain rate; coefficients c_1, c_2, c_3 and c_4 are determined by fitting to the test data, they are equal to -0.446, 0.0875, 1.43 and 6.42, respectively. The maximum value of the dynamic biaxial strength occurs at a stress ratio in the range between 0.5 and 0.75.

The tested dynamic triaxial compressive strengths of concrete at various confining pressure levels are collected in Table 16.3 – 6.

Table 16.3-5 Dynamic Compressive Strength of Concrete Subjected
to Proportional Biaxial Stress State

Strain rate (s^{-1})	Stress ratios ($\sigma_1 : \sigma_2$)				
	1:0	1:0.25	1:0.5	1:0.75	1:1
10^{-5}	9.84	14.86	16.13	16.39	14.00
10^{-4}	10.63	15.48	16.68	16.75	15.32
10^{-3}	11.38	16.17	17.36	17.54	16.66
10^{-2}	12.32	17.15	18.24	18.66	18.01

Table 16.3-6 Dynamic Triaxial Compressive Strengths of Concrete   Unit: MPa

Confining pressure (MPa)	Strain rate (s^{-1})		
	10^{-5}	10^{-4}	10^{-3}
0	9.84	10.63	11.38
4	30.05	32.11	33.70
8	46.27	48.08	49.39
12	61.21	61.42	61.16
16	72.14	75.34	74.08

Confinement on concrete specimens leads to quite distinct shape of stress-strain relationship. At low confining pressure (e.g. 4MPa), the stress-strain curves exhibit a well-defined ultimate strength with a smooth descending tail part. As the confining pressure close to or greater than the uniaxial static strength (about 10MPa), the stress-strain curves appear in a plateau of concrete strength and exhibit little strength reduction over a large range of strains. In these cases, the dynamic strength of concrete is insensitive to the strain rate.

3. Cyclic loading effect[23]

Earthquake loading is characterized by its cyclic behavior with varying amplitude. In this regard, little is known in the current literature. This research aims to increase the almost nonexistent data base of concrete behavior subjected to varying amplitude cyclic tension and this will be of important significance for designing structures in seismic active area.

To study such an effect, 76 concrete specimens were tested on the MTS810 electro-fluid servo-universal machine at our laboratory to investigate the effect of initial static load, exiting frequency and rate of amplitude increment of cyclic loading on the dynamic properties of concrete.

$$F(t) = A(t)\sin(2\pi f t) + F_0 \qquad (16.3-6)$$

where F_0 is the initial static loading; A is the time varying amplitude and f is the cyclic frequency.

As $\sigma = E\varepsilon$, the expression of strain rate is

$$\frac{d\varepsilon}{dt} = \frac{1}{E}\left(\frac{d\sigma}{dt} - \frac{\sigma}{E}\frac{dE}{dt}\right) \approx \frac{1}{E}\frac{d\sigma}{dt} = \frac{1}{ES}\frac{dF}{dt} \qquad (16.3-7)$$

where E is the elastic modulus and S is the cross section area of the specimen. It is observed that the variation of E with time is fairly slight, and the second term of Eq. (16.3-7) can be neglected. Substituting Eq. (16.3-6) into Eq. (16.3-7), it is seen that the strain rate effects consist of two parts $\dot{\varepsilon}_1$ and $\dot{\varepsilon}_2$, the phase difference between them is 90 and the former is dominant.

$$\frac{d\varepsilon}{dt} = \dot{\varepsilon}_1 + \dot{\varepsilon}_2 = \frac{1}{ES}\left[2\pi f A\cos(2\pi f t) + \frac{dA}{dt}\sin(2\pi f t)\right] \qquad (16.3-8)$$

where $dA/dt = \Delta A \cdot f$, ΔA is the increment of amplitude per cycle. Eq. (16.3-8) clearly shows that the dynamic strength of concrete subjected to time-varying amplitude cyclic loading is mainly determined by the maximum loading rate of the individual cycle $\dot{\varepsilon}_1$, and the effect of the rate of amplitude increment $\dot{\varepsilon}_2$ is insignificant. Hence, the dynamic strength of concrete under cyclic loading conditions can be predicted from the test data of monotonic loading conditions. The tested tensile strength of concrete under monotonic loading conditions is given in Table 16.3-2. The results obtained under cyclic loading conditions are given in Table 16.3-7, where σ_{cp} is the average tensile strength measured by cyclic loading test and $\sigma \pm \Delta\sigma$ is the predicted tensile strength calculated from the result of monotonic loading test (σ is the mean value, and $\Delta\sigma$ is the standard deviation).

These findings may be extended to predict the dynamic strength of concrete subjected to random cyclic loading, such as earthquake excitation. The prediction may be proceeded in the same manner as before, that is, at each critical loading cycle, first, the maximum strain rate is determined and then, based on Eq. (16.3-4), the dynamic strength is estimated.

The experimental results also show that as the number of cycles increase, micro-cracks in the specimen develop, and eventually the irreversible deformation of concrete also increases. The loading history and the corresponding strain history of a typical test are shown in Fig. 16.3-6. Referring to this figure, it can be observed that although the applied cyclic load stayed symmetrically with respect to the central axis along the time (Fig. 16.3-6 (a)), the central axis of the cyclic strains gradually shifted upwards with increasing number of cycles (Fig. 16.3-6 (b)), and the centering position of the cyclic stress-strain curve also shifted with time. This phenomenon suggests that under higher initial strain level (more than approximately 70% of the failure strain in this investigation), unrecoverable plastic deformation grows with increasing number of cycles due to the development of micro-cracks inside the specimens.

The magnitude of the initial static loading F_0 affects the strength of concrete under cyclic loading conditions. Unfortunately, only limited data were obtained in the experiment, because the specimens need to remain in tension, thus reducing the possible range of initial static loading. Two cases of the initial static loading corresponding to 72.4% and 90% of the static strength under monotonic loading were tested (Table 16.3-8). The limited results reveal that with the rising of initial static loading intensity, the dynamic tensile strength of concrete decreases accordingly. This may be explained by the presence of unrecoverable plastic deformation and microcracks induced by the initial static load.

Table 16.3-7 Tensile Strength of Concrete under Cyclic Loading Conditions

Pre-load F_0 (MPa)	Frequency f (Hz)	Increment amplitude (MPa)	Measured Strength σ_{cp} (MPa)	$\dot{\varepsilon}_1$ ($10^{-4}s^{-1}$)	$\dot{\varepsilon}_2$ ($10^{-6}s^{-1}$)	Predicted strength $\sigma \pm \Delta\sigma$ (MPa)
1.6	0.5	0.08	2.49	0.943	1.35	2.38±0.153
	2	0.08	2.74	4.39	4.86	2.54±0.143
	10	0.08	2.88	27.3	27.1	2.81±0.141
	20	0.08	2.86	54.1	54.7	2.83±0.156
	30	0.08	2.87	82.5	82.7	2.85±0.173
1.6	2	0.016	2.49	3.40	0.96	2.38±0.145
	2	0.034	2.82	4.65	2.07	2.55±0.152
	2	0.08	2.74	4.35	4.86	2.54±0.150
	2	0.32	2.73	4.31	19.4	2.54±0.150
2.0	2	0.08	2.51	1.95	4.86	2.43±0.136

Table 16.3-8 Dynamic Strengths of Concrete Versus Initial Static Loads σ_0

No.	Tensile Strength (MPa)	
	$\sigma_0 = 72.4\% f_{ts}$	$\sigma_0 = 90.5\% f_{ts}$
1	2.75	2.43
2	2.73	2.61
3	2.79	2.50
4	2.70	—
Average	2.74	2.51

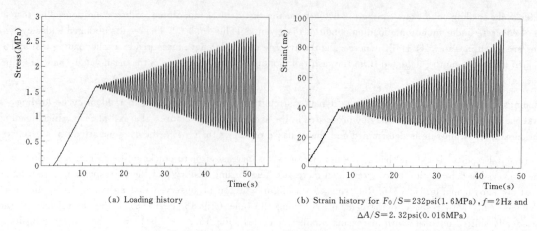

(a) Loading history (b) Strain history for $F_0/S=232\text{psi}(1.6\text{MPa})$, $f=2\text{Hz}$ and $\Delta A/S=2.32\text{psi}(0.016\text{MPa})$

Fig. 16.3-6 Loading History and the Corresponding Strain History of a Typical Test

4. Influence of initial static stress on the dynamic properties of concrete[24]

Majority of concrete structures experienced static load prior to earthquake shock, however, as described above, little is known concerning the influence of initial static load on the dynamic properties of concrete in the current literature. Experiment has been carried out in our laboratory to study the amplitude of initial static stress on the dynamic compressive strength of concrete in order to help better understanding this problem. The designed loading patterns are illustrated in Fig. 16.3-7. Implementation of it was automatically under the control of the computer program.

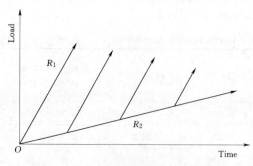

Fig. 16.3-7 Loading Pattern with Different Initial Static Stress

The amplitude of initial static stress equal to 0, 4.72MPa, 9.44MPa and 13.66MPa, which account for 0, 28%, 56% and 80% respectively, of the static compressive strength of concrete, were used for this study. The load was initially applied on the specimen at a very low rate, that is, the loading platen moved at a constant speed toward the specimen ($R_2=10^{-5}\text{s}^{-1}$ in this investigation) and then the dynamic load was applied at a higher strain-rate ($R_1=10^{-2}\text{s}^{-1}$ in this investigation) up to failure of the specimen.

Table 16.3-7 shows the test results, in which $\bar{f}$ denotes the averaged value of tested strength; f_0 corresponds to the initial static stress, at which the strain rate changes from R_2 to R_1; f_d denotes the dynamic strength of concrete loaded monotonically to failure at the strain-rate of R_1 (loading case 1), while f_s denotes such strength at the strain-rate R_2 (loading case 5).

Table 16.3-9 Dynamic Compressive Strength of Conrete Subjected to Various Initial Static Stresses

No.	Loading pattern	f_0 (MPa)	f (MPa)				$\bar{f}$ (MPa)	$\bar{f}/\bar{f}_d$
			1	2	3	4		
1	R_1	—	20.9	19.6	20.6	21.6	20.7	1.00
2	R_2-R_1	4.7	20.1	20.8	20.5	19.6	20.3	0.98
3	R_2-R_1	9.4	19.9	18.9	20.6	20.2	19.9	0.96
4	R_2-R_1	13.7	17.7	19.5	19.0	18.5	18.7	0.90
5	R_2	—	16.3	16.9	17.4	17.5	17.0	0.82

Based on the experimental results, a formulas describing the relationship between initial stress and the ultimate dynamic strength is suggested as follows:

$$\frac{f}{f_d}=\alpha+(1.0-\alpha)e^{(f_0/f_s)^2} \qquad (16.3-9)$$

wherein α is a parameter depending on the material properties and the strain-rate. By fitting to the test data, α is obtained as 1.11 and $R_2=0.81$. The test data and the curve of Eq. (16.3-9) are depicted in Fig. 16.3-8. From this phenomenon it can be concluded that the dynamic strength of concrete is not only affected by the strain-rate at the peak stress point on the stress-deformation curve, but also closely related to the loading history it experiences. The higher the initial static stresses, the lower the dynamic strength.

5. Theoretical study of the strain rate effect on the response of concrete structure[25]

An energy-dissipation based viscoplastic consistency model to describe the yield surface and plastic flow is presented to study the performance of concrete structure subjected to dynamic excitation. Satisfying the natural principles of thermodynamics the model relies on a sound theoretical background. In the process, the least hypotheses are placed on the dissipation function, while none of the assumptions is made on the plastic potential and the flow rule etc. The friction behavior of concrete is inseparately linked to the occurence of non-normal flow rule and the flow rule is straightfowardly derived from the yield surface. The concept of viscoplastic consistency model presented by Wang[27] was adopted, it can relatively easily be implemented in place of classical rate-independent plasticity models. According to Wang, during viscoplastic flow the actual stress state must remain on the yield surface. In this way, in the presented model, the yield surface is rate dependent. In addition, the hardening and softening and the rate dependency are described separately for tension and compression so as to better reflect the concrete behavior. A modified implicit backward Euler integration scheme is used for the numerical computation.

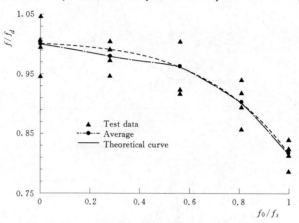

Fig. 16.3-8 **Dynamic Strength of Concrete with Various Initial Static Stress**

Numerical examples are provided to verify the effectiveness of the proposed model. The yield surface and the failure surface are shown in Fig. 16.3-9. The direction of plastic flow is also indicated. The surfaces are perfectly smooth, which ensures the surface gradient is uniquely defined. The rate independent behavior depicted in Fig. 16.3-10 are compared with the experiment data of Kupfer et al[28]. The rate dependent behavior depicted in Fig. 16.3-11, are compared with the experimental results in tension presented in section 3.1-1 and those in compression obtained by Suaris and Shah[29]. The fairly good agreement in most cases reveals that the model is able to describe adequately the behavior of concrete, such as modulus, peak stress strength, critical strain and the strain-rate sensitivity. The somewhat deviation occurring for the cases of complex stress state reminds us about further refinement of the model.

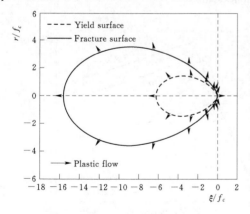

Fig. 16.3-9 **The Yield Surface and the Failure Surface**

In order to get a better understanding of the rate dependent behavior on the dynamic response of concrete structures, the dynamic response of a half-circle plain concrete arch subjected to uniform redial impact is studied. The outer and inner radiuses of the arch are 30m and 20m respectively. The impact loading history is shown in Fig. 16.3-12, and the thickness of the arch is 10m. The material properties are chosen based on the experimental data of Kupfer et al[28]. The structure is discretized with 180 20-node 3D solid elements. Three analyses were performed to study the dynamic response of the arch, namely the linear elastic analysis, the rate-independent plastic analysis and the rate-dependent plastic analysis. The maximum displacements calculated at the crown point A were compared as shown in Fig. 16.3-13. In the elastic stage the displacement curve of three analyses are identical with each other; whereas at the plastic stage the displacement curves for these three analyses start diversified. They reach the maximum value of 47.8mm, 53.8mm and 49.8mm separately at the time instant of 0.112s, 0.120s and 0.116s respectively. General speaking, plastic model leads to most intense response of the structure; linear elastic model overestimates the stiffness of the structure, it results in excessively small response and rate-dependent model produces results lying in the middle of the first two models. The conclusion are somewhat similar to that obtained by employing a viscoplastic Hoffman consistency model. A viscoplastic William-Warnke consistency model was also employed by the authors[26] to carry out seismic response analysis of a 278m high double-curvature arch dam in China. The maximum values of the first and the third principal stresses of the dam are listed in Table 16.3-10 and are also compared with the results obtained by employing the linear elastic model and the rate-independent plastic model. Accordingly, the maximum compressive stresses achieved by these three different models are nearly identical, because most parts of the dam

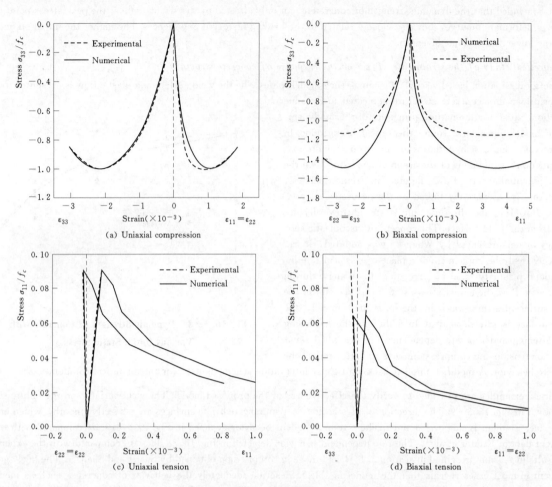

Fig. 16.3-10 Comparison of Rate Independent Stress-strain Relationship between Computed and Experimental Results

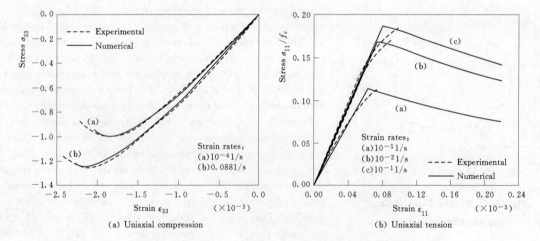

Fig. 16.3-11 Comparison of Rate Dependent Stress-strain Relationship between Computed and Experimental Results

are working in the elastic stage. As far as the maximum tensile stresses are concerned, marked differences can be observed, and the maximum value of tensile stress computed by the rate-dependent model lies within the range obtained by other two models. It may be preliminarily concluded that taking into consideration the rate-dependent behavior of concrete, the dynamic displacements and stresses of concrete structures are lying within the range predicted by the linear elastic model and the rate-independent elasto-plastic model.

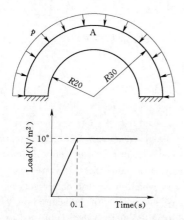

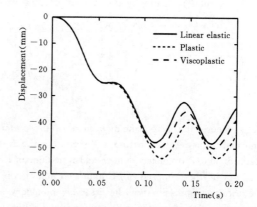

Fig. 16.3 – 12   The Concrete Arch and the Impact Loading History

Fig. 16.3 – 13   Time History of Vertical Displacement at Crown Point of Arch Analyzed by Three Different Models

Table 16.3 – 10   Dynamic Stress Response of a 278m High Arch Dam   Unit: MPa

Model	Max. compressive Stress	Max. tensile stress
Elastic	−12.5	5.17
Rate-independent plastic	−12.5	2.87
Rate-dependent plastic	−12.5	3.22

16.3.3.2   Dam – reservoir Interaction Analysis

1. A SBFE approach for frequency – domain and time – domain hydrodynamic analysis of dam – reservoir system[34,35]

The scaled boundary finite element method (SBFEM) originally put forward by Wolf and Song [32,33] is a promising numerical tool for scientific and engineering analyses. It offers more than combining the advantages of FEM and BEM, the spatial dimension of the problem is reduced by one, but no fundamental solution is required. It is well suited to deal with the unbounded domain problems, the radiation condition at infinity is satisfied rigorously.

A SBFEM approach is developed for hydrodynamic analysis of dam-reservoir system. The reservoir water is assumed as compressive inviscid fluid. The earthquake induced hydrodynamic pressure p satisfies the wave equation

$$\nabla^2 p - \frac{1}{c^2} \ddot{p} = 0 \tag{16.3-10}$$

where c is the velocity of compression waves in water. Assuming the effect of surface wave may be neglected, the boundary conditions at the dam-fluid interface and at the reservoir bottom and sides respectively are defined as

$$p_{,n} = -\rho \ddot{u}_n \tag{16.3-11}$$

$$p_{,n} = -\rho \ddot{v}_n - q \dot{p} \tag{16.3-12}$$

where the subscript following the comma in $p_{,n}$ denotes partial differentiation of p with respect to space; over-dot of p denotes the partial differentiation of p with respect to time; $\ddot{u}_n$ and $\ddot{v}_n$ are normal components of acceleration at the upstream dam face and that at the reservoir bottom and sides respectively; q is damping coefficient characterizing the absorption of pressure waves at the reservoir boundary; α denotes the wave reflection coefficient defined as the ratio of the amplitudes between the reflected and incident pressure waves; q and α have the relationship

$$q = \frac{1-\alpha}{c(1+\alpha)} \tag{16.3-13}$$

A 3 – D arch dam-reseivoir interaction analysis is addressed. In case of the reservoir being idealised ectending to infinity in the stream direction with uniform cross-section, based on SBFEM formulation the hydvodynamic pressure acting on the dam face due to earthquake excitation can be evaluated by the following formulae:

earthquake ground motion exciting in the stream direction

$$\{p\} = -[\Phi_{12}][\Phi_{22}][M^1]^{-1}\{\ddot{u}_n\} \tag{16.3-14}$$

earthquake ground motion exciting in the vertical or cross-stream direction

$$\{p\} = -[\Phi_{12}][\Phi_{22}][M^1]^{-1}\{\ddot{u}_n\} - ([\Phi_{12}][\Phi_{22}]^{-1})[B_1] - [B_2]\rho[C^0]\{\ddot{v}_n\} \quad (16.3-15)$$

with

$$[B_1] = [\Phi_{21}][-\lambda_1^{-1}][A_{12}] + [\Phi_{22}][\lambda_1^{-1}][A_{22}]$$

$$[B_2] = [\Phi_{11}][-\lambda_1^{-1}][A_{12}] + [\Phi_{12}][\lambda_1^{-1}][A_{22}] \quad (16.3-16)$$

where, $\{\ddot{u}_n\}$ denotes the earthquake induced vibration acceleration normal to the upstream dam face; $\{\ddot{v}_n\}$ earthquake acceleration normal to the reseivior bottom and sides. For the evaluation of all the coefficient matrices, only the dam face needs to be discretized. Thus, frequency domain and time domain hydrodynamic analysis of dam-reservoir system can be conveniently carried out by the proposed SBFE approach and all the important factors which affect the dam-reservoir interaction such as the water compressibility, the reservoir boundary absorption etc. can be taken into cousideration with relative ease. Results of complex frequency response functions of the 141.7m high Morrow Point arch dam subjected to earthquake excitation in the stream direction, vertical direction and cross-stream direction, respectively, have been compared with the results obtained by Fok and Chopra, excellent agreement has been reached[35]. However, by employing proposed SBFE approach, the computational effort can be reduced to a great extent.

Several cases of earthquake hydrodynamic analyses of arch dam-reservoir as well as gravity dam-reservoir systems have been carried out. As a result, the following conclusious can be drawn:

(1) The reservoir water compressibility and the reservoir boundary absorption have important influence on the earthquake induced hydrodynamic pressure acting on the dam face, which should be taken into consideration for the earthquake safety evalution of dam structures[31].

(2) Including the effect of water compressibility and the reservoir bottom absorption, phase difference of hydvodynamic pressure appears and varies from point to point over the dam face, in other words, under the action of earthquake excitation, the maximum value of hydrodynamic pressure at different point over the dam face occurs at different instant of the time. This leads to the reduction of hydrodynamic effect.

(3) Reservoir boundary absorption induces additional damping on the dam-resrvoir-foundation system, as a consequences, the hydrodynamic pressure acting on the dam face also tends to be reduced.

(4) The Westergaard's added mass model may considerably overestimate the hydrodynamic force acting on the dam face.

2. Nonreflecting boundary condition for infinite reservoir[42]

For reservoir of arbitrary geometry, to carry out hydrodynamic analysis of the dam-reservoir system, the reservoir is usually split into two parts: a finite near-field region to be discretized by FEM, BEM or SBFEM, and a far-field region to take into account the infiniteness of the reservoir. In this case, a nonreflecting boundary condition (NBC) has to be applied at the interface between the near field and the far field to take account of energy radiation at infinity. This special boundary condition and the truncation distance from the dam face should be defined appropriately to prevent reflection of spurious wave back towards the dam, which would commit significant error in the prediction of the hydrodynamic pressure on the dam face. Among the many NBCs proposed in the literature, the Sommerfeld radiation boundary condition (Sommer B. C.) and the Sharan's the refined boundary condition (Sharan B. C.) have been widely used [43]. The former is simple, however, it performs adequately only when placed at a large distance from the dam face[43]. The later performs better. Both NBCs are generally approximate, absorbing only a portion of impinging wave, and thus introducing some error into the solution.

In order to improve the accuracy of the NBC, a new NBC on SBFEM is proposed[42]. wherein a stress balance condition is introduced at the truncation boundary. The method takes into consideration the effect of water compressibility and the reservoir boundary absorption. The proposed boundary applies to a wide range of exciting frequency, and it is effective for earthquake excitation in the stream direction as well as in the vertical/cross-stream direction. The governing equations of the proposed NBC is expressed as

$$[S(\omega)]\{p\} = [S^\infty(\omega)]\{p\} - \{r_F(\omega)\} \quad (16.3-17)$$

and

$$\{p(t)\} = \frac{1}{2\pi}\int_{-\infty}^{\infty}\{p(\omega)\}e^{i\omega t}d\omega \quad (16.3-18)$$

The readers may refer to[42] in detail.

The accuracy and efficiency of the Sommer B. C., the Sharan B. C. and the proposed B. C. are examined and compared between each other. A rigid gravity dam with vertical upstream face is chosen for this study. The depth of water in the reservoir is assumed to be 100m and the truncated surface is placed at a distance 100m from the dam face. The results evaluated by applying different models of NBC are shown in Fig. 16.3 – 14 (a) – (c). the external excitation is assumed to act in the stream direction only with unit amplitude of $e^{i\omega t}$ in the frequency domain. The absorption of the reservoir boundary is considered, the corresponding reflection coefficient is assigned as 0.925. The closed form solution of Bouaanai et al[43] is regarded as the exact value for reference. In the figure, the abscissa denotes the dimensionless frequency ω/ω_1 and the ordinate denotes the hydrodynamic pressure at the dam heel normalized by the static pressure $\rho g H$. As can be observed, the proposed BC is far superior to the other two NBCs. Adjacent to the resonant peaks the Sommer B. C. deviates greatly from the exact value, the discrepancy between the Sharan B. C. and the analytical solution is still large. On the contrary, the proposed NBC is in full agreement with the analytical solution.

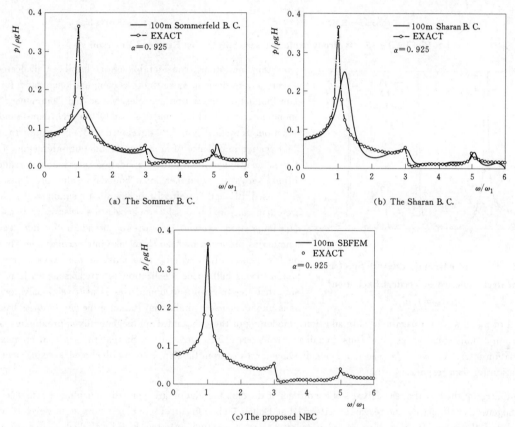

Fig. 16.3 – 14 Hydrodynamic Pressure Spectra Obtained by Various Models of NBC

The Sommer B. C. and the Sharan B. C., apply only for NBC being placed sufficient far away from the dam face. When the discretized reservoir is extended to 1800m away from the dam face, then agreement with the reference solution can be reached for both NBCs (Fig. 16.3 – 15), although there appears a slight oscillation from the analytical solution after the first resonant frequency of the reservoir. Moreover, for exciting frequency high than the third resonant frequency of the reservoir, the results deviate from the exact solution (Fig. 16.3 – 15).

The applicability of the proposed NBC for earthquake excitation in the vertical direction has also been checked. The results shown in Fig. 16.3 – 16 agree entirely with the analytical solution proposed by Chopra[39] for wave reflection coefficient $\alpha = 1.0$. Results by the proposed NBC for sloping bed agree excellently with the BEM solution by Hanna and Humar[37].

16.3.3.3 Dam-foundation Interaction Analysis

Dynamic soil-structure interaction (SSI) effects have always been important in an earthquake-prone environment. Dynamic interaction of dam with unbounded foundation affects earthquake response of dam structure to a great extent because it causes energy dissipation into infinite foundation medium and changes the dynamic behavior (vibration frequencies and mode shapes) of the dam structure. However, carrying out SSI analysis is rather complicated and associates with great computational effort. In the engineering practice, usually a simplified massless foundation was employed to deal with such kind of problem. The radiation damping as well as the inertial effects of the foundation media were ignored. Until recently, to over-

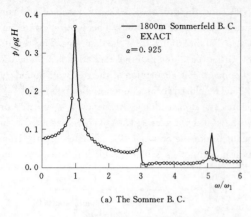
(a) The Sommer B. C.

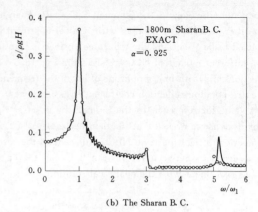
(b) The Sharan B. C.

Fig. 16.3 - 15 Hydrodynamic Pressure Spectra for Extended Reservoir

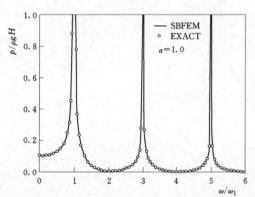

Fig. 16.3 - 16 Hydrodynamic Pressure Spectrum at Dam Heel Subjected to Vertical Excitation for $\alpha=1.0$

come this limitation, considerable effort has been dedicated by the researchers to develop more rigorous computational model for dynamic dam-foundation interaction analyses. Maeso and Dominguesz[44] used a boundary element (BE) model to study the arch dam-foundation interaction. Tan and Chopra[45] developed a BE procedure to determine the dynamic impedance of arch dam-foundation interaction. The three-dimensional dynamic structure-foundation interaction is reduced to an infinite series of two-dimensional boundary value problems. Zhang et al[46] used BE and infinite BE elements to simulate the unbounded arch-dam canyon. Du et al[47] analyzed the seismic response of arch dam-foundation system by introducing non-reflecting transmitting boundaries proposed by Liao[48] to take into account the effects of far-field foundation. By and large, in most of the previous researches a homogeneous half-space idealization is often assumed. It is worth to note that the transmitting boundaries is also valid only for homogenous unbounded soil conditions, because the derivation is based on the assumption of plane wave propagation[48]. In addition, modelling of the unbounded soil is generally approximate, absorbing only a portion of impinging waves, and thus introducing some error into the solution. So that the artificial boundary or the dam-foundation interface have to be placed at some distance far away from the dam face in order to ensure sufficient accuracy of the solution for dam response.

Taking into account the fact that the actual soil conditions of the dam foundation are generally much more complex than an ideally homogeneous half-space, the research performed at Dalian University of Technology aims at to develop efficient as well as accurate algorithms for modelling complex soil-structure interactions effects, based on which systematic investigations were carried out to assess their influence on the earthquake response of arch and gravity dams. Two types of approaches for modelling foundation inhomogeneity were addressed, the first is the scaled boundary finite element (SBFE) approach, and the second is the damping solvent stepwise extraction (DSSE) approach. The algorithms to be used for time domain analysis for soil-structure interaction analysis associated with foundation inhomogeneity are explained in the following.

1. The scaled boundary finite element (SBFE) approach[49]

The SBFE approach is capable to deal with foundation inhomogeneities possessing different shear modulus G, poisson's ratio ν and mass density ρ in different parts as shown in Fig. 16.3 - 17 (a) or the material properties of the foundation medium varying as the power function of the depth along the radial direction or the combination of these two cases.

$$G(r)=G_0 \left(\frac{r}{r_0}\right)^g, \rho(r)=\rho_0 \left(\frac{r}{r_0}\right)^m \qquad (16.3-19)$$

By applying SBFE approach, the governing equations of the displacement dynamic impedance of unbounded foundation in the frequency domain $[S^\infty(\omega)]$ are found to satisfy the following ordinary differential equations[32]

$$([S^\infty(\omega)]+[E^1])[E^0]^{-1}([S^\infty(\omega)]+[E^1]^T)-(s-2)[S^\infty(\omega)]$$
$$-\omega[S^\infty(\omega)]_\omega-[E^2]+\omega^2[M^0]=0 \qquad (16.3-20)$$

where S is the spatial dimension of the problem; $[E^0]$, $[E^1]$, $[E^2]$ and $[M^0]$ are coefficient matrices. For the time domain

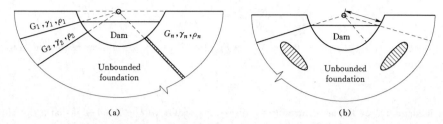

Fig. 16.3 – 17 Various Cases of Foundation Inhomogeneity

analysis it is convenient to transform $[S^\infty(\omega)]$ into the acceleration unit-impulse response matrix $[M^\infty(\omega)]=[S^\infty(\omega)]/(i\omega)^{2[32]}$, and the dynamic interaction force of the dam with unbounded foundation is evaluated as

$$[R(t)] = \int_0^t [M^\infty(t-\tau)]\{\ddot{u}_b(\tau)\}d\tau \qquad (16.3-21)$$

Then the equations for the dynamic response analysis of the general dam-foundation system subjected to earthquake excitation are formulated as

$$\begin{bmatrix} M_{ss} & M_{sb} \\ M_{bs} & M_{bb} \end{bmatrix}\begin{Bmatrix} \ddot{u}_s^t(t) \\ \ddot{u}_b^t(t) \end{Bmatrix} + \begin{bmatrix} C_{ss} & C_{sb} \\ C_{bs} & C_{bb} \end{bmatrix}\begin{Bmatrix} \dot{u}_s^t(t) \\ \dot{u}_b^t(t) \end{Bmatrix} + \begin{bmatrix} K_{ss} & K_{sb} \\ K_{bs} & K_{bb}+K_{bb}^f \end{bmatrix}\begin{Bmatrix} u_s^t(t) \\ u_b^t(t) \end{Bmatrix} = \begin{Bmatrix} 0(t) \\ R^f(t) \end{Bmatrix} \qquad (16.3-22)$$

where $[M]$, $[C]$, $[K]$ denote the mass, damping and stiffness matrices of the dam structure, $\{u^t(t)\}$ denotes absolute displacements from a fixed reference, $\{u_b^t(t)\}$ represents the part of displacements at the dam-foundation interface, $\{u_s^t(t)\}$ the part of the dam but excluding, $\{u_b^t(t)\}$, $\{R^f(t)\}$ represents the part of force amplitudes due to retarded effect of dam-foundation interaction depending on historical displacement response $\{u_b^t(t), [K_{bb}^f]\}\{u_b^t(t)\}$, represents part of the interaction force amplitudes depending on current displacement response $\{u_b^t(t)\}$, $[K_{bb}^f]$ represents a modification matrix related to dam-foundation interaction, which keeps unchanged during the whole time history analysis. A detailed derivation can be found in ref[51].

Based on SBFE approach, the effects of foundation inhomogeneity on the seismic response of arch and gravity dams has been studied[49].

2. The damping solvent stepwise extraction (DSSE) approach[50]

For complex soil conditions of the foundation with irregular inclusions or intercalations in the vicinity of the dam (Fig. 16.3 – 17 (b)), an artificial boundary is introduced which divides the foundation into two parts: the finite near-field and the infinite far field. A transmitting boundary condition must be imposed at the interface such that no energy being reflected towards the dam structure. However, as mentioned above most boundary models currently widely used in dam design are based on homogeneous idealization of the unbounded far-field, as a result abrupt changes of the material properties take place at the artificial boundary, thus often exhibit inadequate accuracy. The damping solvent extraction (DSE) approach firstly proposed by Wolf and Song[32] and later refined by the Dalian University of Technology (DUT) defined as DSSE approach[50,52,53] is more suitable and promising for the study of near field foundations inhomogeneity. The DSE approach is based on the notion that introducing artificial damping ζ to a finite soil domain adjacent to the dam, which may be modeled with FEs or SBFEMs, to attenuate both outgoing and reflected waves, results in negligible amplitudes at the dam-foundation interface. It is reasonable to assume that the dynamic stiffness of the damped finite domain $[S_\zeta(\omega)]$ approaches that of the damped infinite domain $[S_\zeta^\infty(\omega)]$. Hence, after extracting the effects of artificial damping, the actual dynamic stiffness $[S^\infty(\omega)]$ at the soil-structure interface for a real infinite domain is reached.

The governing equations of the DSE approach for the evaluation of the dynamic impedance at the dam-foundation interface in the frequency domain is expressed as[32] (Fig. 16.3 – 18)

$$[S^\infty(\omega)] = \frac{G}{G^*}\left([S_\zeta(\omega)] + [S_\zeta(\omega)]_{,\omega} \frac{a_0 - a_0^*}{a_{0,\omega}^*}\right) \qquad (16.3-23)$$

where superscript " * " denotes the material constants for damped foundation media; G and G^* denotes the shear modulus; a and a^* denote the dimensionless frequencies defined as

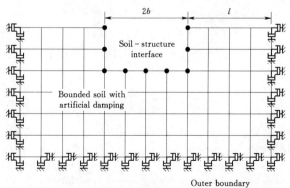

Fig. 16.3 – 18 Finite Soil Domain for DSE Model with Artificial Damping

$$a = \frac{\omega r_0}{c_s}, a^* = \frac{(\omega - i\xi) r_0}{c_s} \quad (16.3-24)$$

c_s is the shear wave velocity; r_0 is a characteristic length (in Fig. 16.3-18 $r_0 = b$). Linear hysteretic material damping with damping ratio ζ is assumed

$$G^* = G(1 + 2i\xi) \quad (16.3-25)$$

Although the artificial damping can be easily introduced, disastrous results will be obtained if that is not appropriately or incompletely extracted. In the standard DSE approach, the extraction of the effects of artificial damping is approximated by the first two terms of Taylor expansion

$$[\overline{S}^\infty(a)] = [\overline{S}^\infty(a^*)] + [\overline{S}^\infty(a^*)]_{,a^*}(a - a^*) \quad (16.3-26)$$

where superscript bar "—" denotes the dimensionless value of the dynamic impedance $[S^\infty(\omega)]$. Note that Eq. (16.3-26) is a conditional equality. To evaluate $[\overline{S}^\infty(a)]$ from $[\overline{S}^\infty(a^*)]$ the following convergent condition should be satisfied.

$$|a - a^*| \leqslant r_c \quad (16.3-27)$$

in which r_c is the radius of convergence. That is to say, in order to assure computational stability and a better convergence for a broad range of frequencies, the artificial damping ζ should be sufficiently small, which leads to larger domain size l/b and more computational effort. The DSSE approach proposed by DUT aims to alleviate the contradiction in the choice of appropriate value of ζ and l/b. The introduced artificial damping is divided into n smaller parts $\Delta\zeta$ which satisfy the requirement of Eq. (16.3-27) and extracted step by step. For each step of extraction the Eq. (16.3-26) is modified as

$$[\overline{S}^\infty(a^*_{k+1})] = [\overline{S}^\infty(a^*_k)] + [\overline{S}^\infty(a^*_k)]_{,a^*_k}(a^*_{k+1} - a^*_k) \quad (16.3-28)$$

with
$$\xi_{k+1} = \xi_k - \Delta\xi \quad (k = 1 \text{ to } n)$$

In this way, a relatively larger artificial damping can be introduced that allows a convenient size of the finite domain to be selected to ensure the desired accuracy and to reduce the computational effort.

In the time domain, the displacement unit-impulse response matrix and the corresponding interaction force at the dam-foundation interface are formulated as[50]

$$[S^\infty(t)] = [S^b_{\xi_{n+1}}(t)] = (1 + \Delta\xi t)^n [S^b_{\xi_1}(t)]$$

$$[S^\infty(t)] = [R_b(t)] = \int_0^t [S^\infty(s-\tau)]\{U_b(t)\}dt = \sum_{l=1}^{n+1} P_{bL}(\Delta\xi, t) R_{bL}(t) \quad (16.3-29)$$

with

$$\{R_b(t)\} = \int_0^t [S^b_{\xi_1}(t-\tau)]\{U_{bL}(\tau)\}d\tau \quad (16.3-30a)$$

$$P_{bL}(\Delta\xi, t) = (-1)^{L-1} c_n^{L-1} \Delta\xi^{L-1}(1 + \Delta\xi t)^{n-L+1} \quad (16.3-30b)$$

where $S^b_\xi(t)$ denote the unit-impulse response matrix of the damped bounded soil domain and $\{R_b(t)\}$ the corresponding interaction force at the dam-foundation interface.

Since convolution integral cost much computational effort, $\{R_{bL}(t)\}$ is suggested to evalueate from solving the equation of motion of the finite damped soil domain as follows[32,50]

$$\begin{bmatrix} \hat{M}_{mn} & 0 \\ 0 & \hat{M}_{bb} \end{bmatrix} \begin{Bmatrix} \ddot{U}_{mL} \\ \ddot{U}_{bL} \end{Bmatrix} + \begin{bmatrix} \hat{C}_{mn} & \hat{C}_{mb} \\ \hat{C}_{bm} & \hat{C}_{bb} \end{bmatrix} \begin{Bmatrix} \dot{U}_{mL} \\ \dot{U}_{bL} \end{Bmatrix} + \begin{bmatrix} \hat{K}_{mn} & \hat{K}_{mb} \\ \hat{K}_{bm} & \hat{K}_{bb} \end{bmatrix} \begin{Bmatrix} U_{mL} \\ U_{bL} \end{Bmatrix} = \begin{pmatrix} 0 \\ R_{bL}(t) \end{pmatrix} \quad (16.3-31)$$

with

$$[\hat{M}] = [M] \quad [\hat{C}] = 2\xi[M] \quad [\hat{K}] = [K] + \xi^2[M] \quad (16.3-32)$$

where $[M]$, $[\hat{C}]$, $[K]$ are the mass, damping and stiffness matrices of the nearfield finite soil domain adjacent to the dam strcture itself; the subscript b and m denote the nodes on the foundation interface and those inside the domain, respectively. For details readers may refer to[50]. By employing the proposed approach nearfield foundation inhomogeneity with inclusions or intercalations can be considered in performing the dynamic soil-strcture interaction analysis with least computational effort, and in the meantime desired accuracy can be ensured.

An approach for performing dynamic interaction analysis of structure with multi-layered half-space is also developed[70].

16.3.3.4 Refined Numerical Methods for Seismic Deformation and Stress Analysis of Dam Structures

With the increasing development of the hydropower energy construction in China, many and many high dams impounding large amount of reservoir water have been and will be built in western mountainous earthquake active area. Higher requirements are imposed to assess the safety and vulnerability of dam structures to withstand strong earthquake shock. Therefore, a versatile, efficient and accurate dynamic analysis technique is of critical important.

Currently the Finite Element Method (FEM) and Boundary Element Method (BEM) have become powerful and robust tools for scientific computation and solution of practical engineering problems, they have also played important role in the seismic response analysis of dams. Either of the two has certain features and advantages of its own. However, when applied to seismic deformation and stress analysis of dams, both methods still exhibit some deficiencies, which are discussed below.

To simulate the mechanic behavior of structures, water and soil, deal with any complicated geometry and material non-linearities, anisotropies and inhomogeneities, the Finite Element (FE) method is a typically versatile in nature and well-established computing method in civil engineering practice. However, this method offers the inherent difficulty in analyzing media of infinite extent since the discretized domain must be large enough in consideration of the radiation damping. This requirement tends to be a severe restriction and burdensome for dynamic analysis of large 3-D structures. To alleviate the computational burden associated with the simulation of infiniteness of the solution domain, various non-reflecting boundary models have been proposed. However, most non-reflecting boundary models are approximate in nature, in which the boundary should be placed far enough from the dam-reservoir or dam foundation interface. Another disadvantage of FE analysis appears in dealing with problems involving stress singularities. Standard FE schemes yield comparatively poor results when encountering geometrical discontinuity and bimaterial interface. The solution converges very slowly near the singular point. In order to improve the results, refining the mesh locally, modifying the shape functions of elements in this region, or introducing super elements are the alternatives. But all these measures increase the complexity of the computation.

The advantage of the BEM is that only the domain boundary needs to be discretized, thus reducing the spatial dimension of the problem by one and saving computational effort to a great extent. In addition, BEM satisfies automatically the radiation damping at infinity, so it has superiority in solving the dynamic interaction problems between the dam and reservoir as well as between the dam and the unbounded foundation. However, a fundamental solution is needed in BEM, which makes the solution procedure complicated and introduces extra singularities. What's more, the fundamental solution is difficulty to find in most cases. What comes last but not least important is that the basic matrices obtained by BEM are full and asymmetric, which further brings trouble in the analysis.

Presently with the development of computational technology, many novel numerical computation methods have been emerged. Based on the research experience gained at Dalian University of Technology, the Scaled Boundary Finite Element Method (SBFEM) and Isogeometric Analysis (IGA) are recommended as two of the most promising alternative. Both have been implemented for the seismic response analyses of dam-reservoir-unbounded foundation interaction system. By employing these models, the accuracy of deformation and stress analysis can be greatly improved, whereas the computational effort is reduced in the meantime. Their features are briefly introduced as below.

The SBFEM was originally presented by Wolf and Song[32,35]. It combines the advantages of both FEM and BEM, and possess its own unique features. SBFEM is semi-analytical, providing analytical solution in the radial direction, so discretization is required only in the circumferential direction. The spatial dimension of the problem is reduced by one, but no fundamental solution is required. The radiation condition at infinity can be satisfied rigorously, and owing to the analytical solution in the radial direction, dealing with problems involving unbounded domain and singularities turns out to be very convenient and simple, yet retaining high accuracy. SBFEM and FEM can be coupled seamlessly. Up to now SBFEM has been successfully applied to the analyses of structural statics and dynamics, soil-structure interaction, structure-fluid interaction, fracture problem, heat transfer problem as well as electro-magnetic wave problem. In the solution of dam-reservoir interaction problem presented above, the advantages of SBFEM have been shown through examples.

What comes next is the IGA. For numerical methods constructed in the framework of FEM, e.g. FEM and SBFEM, low-order Lagrange polynomials are used as basis functions, which introduces error for discretization of domain boundaries with irregular geometry. In addition, only C^0 continuity can be achieved between adjacent elements, and the order of interpolation functions within elements is very low as well, both factors contribute to the poor accuracy of nodal deformation and particularly the stress obtained. In order to improve the accuracy, refined mesh can be used, but the refining process is very involved. So the seismic safety evaluation of dams based on FEM is not reliable enough, and this has become a relatively remarkable problem in seismic analysis of high dams. The emergence of IGA as a new numerical computational method has brought a new dawn for scientific and engineering computation which requires high precision and high performance.

The IGA was presented by Hughes et al[56]. The Non-Uniform Rational B-Splines (NURBS) are employed to generate basis functions for constructing precise geometrical model. The structure analysis and configuration design are integrated, such that the inconsistency between the computational model and the designed geometric model is eliminated. In other words, its basic idea lies in that the computational model and mesh discretization for structural analysis and the geometric model for structural design are formulated in the same frame work of basis functions.

By doing so, the IGA possesses the following advatanges. ①Owing to the high accuracy of NURBS basis functions, relatively coarse mesh with least parameters can be used to accurately discribe the geometrical configuration for structural design. ②For purpose of structure analysis, the mesh is progressively refined and/or the order of continuity elevated, while the exact geometry remains unchanged. That is, in order to meet the precision requirement for the deformation and stress analyses, self-adaptive shape-preserving mesh refinement can be automatically conducted, thus relevant computational effort can be greatly saved. ③In the context of structural dynamics, it is established that the basis functions are complete with respect to affine transformation, meaning that all rigid body motions and constant strain states are exactly represented. Standard patch tests are likewise satisfied. ④By utilizing the features of NURBS functions, the order increase and derivation operation are relatively convenient, which makes it easy to construct high-order continuous conforming elements. This improves the accuracy and efficiency in seismic deformation analysis and stress analysis of dams.

In the research carried out at Dalian University of Technology (DUT), a coupled scaled boundary isogeometric approach (SBIGA) is proposed for the time-domain seismic response analysis of dam-reservoir-foundation system. This new approach (SBIGA) offers more than combining the advantages of SBFEM and IGA[57,58]. SBIGA fully inherits the merits of IGA, including easy mesh refinement, high computational accuracy, thus enabling convenient preprocessing. Concurrently, it possesses the advantages of SBFEM including the following aspects: only the boundary surfaces need to be discretized resulting in the spatial dimension of the problem reduced by one; solution in the radial direction is obtained analytically; problem with unbounded domain or with singularities can be well handled. In the following numerical examples, SBIGA is developed for modeling the semi-infinite fluid domain of reservoir and unbounded elastic half-space of the foundation; whereas IGA is employed for idealization of the dam structure itself. It is worth to note that the dam structure can also be modeled by SBIGA. Both the examples of gravity dam-reservoir-foundation system and the arch dam-reservoir-foundation system validate that the proposed approach is highly accurate and consumes less degrees-of-freedom (DOFs) than conventional widely used FEM and BEM.

1. Basic concept of NURBS based structural analysis

As mentioned above, the analysis framework is based on NURBS, which are built from B-splines. Unlike in standard finite element analysis (FEA), the B-spline parametric space is local to patches rather than elements. That is, the parametric space in FEA is mapped into a single element in the physical space, and each element has its own such mapping. Alternatively, the B-spline mapping takes a patch of multiple elements in the parametric space into the physical space. Each element in the physical space is the image of a corresponding element in the parametric space, but the mapping itself is global to the whole patch, rather than to the elements themselves. Patches play the role of subdomains within which element types and material models are assumed to be uniform.

B-spline curves are constructed by taking a linear combination of B-spline basis functions. The vector-valued coefficients of basis functions are referred to as control points. A piecewise-polynomial B-spline curve is given by

$$C(\xi) = \sum_{i=1}^{n} N_{i,p}(\xi) B_i \qquad (16.3-33)$$

where, $N_{i,p}$ denotes the basis function, i denotes the i-th knot, p is the polynomial order, and n is the number of basis functions used to construct the B-spline curve; B_i ($i=1, 2, \cdots, n$) denotes the corresponding coefficients of control polygon.

Fig. 16.3-19 shows the B-spline curve built from the quadratic basis functions.

In analogous fashion, a tensor product B-spline surface and that B-spline solid are defined by

$$R(\xi,\eta) = \sum_{i=1}^{n} \sum_{j=1}^{m} N_{i,p}(\xi) M_{j,q}(\eta) B_{i,j}$$

$$(16.3-34)$$

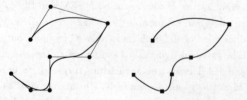

(a) B-spline curve is described by the control points denoted by "●"   (b) The curve is partloued into elements through knots denoted by "■"

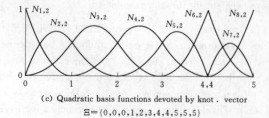

(c) Quadratic basis functions devoted by knot vector
Ξ = {0,0,0,1,2,3,4,4,5,5,5}

Fig. 16.3-19 B-spline Quadratic Curves in R^2

$$S(\xi,\eta,\zeta) = \sum_{i=1}^{n}\sum_{j=1}^{m}\sum_{k=1}^{l} N_{i,p}(\xi)M_{j,q}(\eta)L_{k,r}(\zeta)B_{i,j,k} \qquad (16.3-35)$$

where $B_{i,j}$ and $B_{i,j,k}$ denote control net and control lattice respectively; p, q and r are polynomial orders; the meaning of other symbols can be inferred by analogy.

Transformation from the non-rational B-splines to NURBS opens up the possibility to exactly represent a wide array of geometric entities that cannot be exactly represented by polynomials, many of which are ubiquitous in engineering design. The rational basis functions and NURBS curve are given as follows:

$$R_i^p(\xi) = \frac{N_{i,p}(\xi)w_i}{\sum_{i=1}^{n} N_{i,p}(\xi)w_i} \qquad (16.3-36)$$

$$C(\xi) = \sum_{i=1}^{n} R_i^p(\xi)B_i \qquad (16.3-37)$$

where, w_i is referred to as the ith weight. The expression for a NURBS curve is identical to that for B-splines. Rational surfaces and solids are defined analogously in term of the rational basis functions.

$$R_{i,j}^{p,q}(\xi,\eta) = \frac{N_{i,p}(\xi)M_{j,q}(\eta)w_{i,j}}{\sum_{i=1}^{n}\sum_{j=1}^{m} N_{i,p}(\xi)M_{j,q}(\eta)w_{i,j}} \qquad (16.3-38)$$

$$R_{i,j,k}^{p,q,r}(\xi,\eta,\zeta) = \frac{N_{i,p}(\xi)M_{j,q}(\eta)L_{k,r}(\zeta)w_{i,j,k}}{\sum_{i=1}^{n}\sum_{j=1}^{m}\sum_{k=1}^{l} N_{i,p}(\xi)M_{j,q}(\eta)L_{k,r}(\eta)w_{i,j,k}} \qquad (16.3-39)$$

Fig. 16.3-20 shows the NURBS curve, NURBS surface and NURBS solid, and the corresponding control net and control lattice respectively. Note that the weights play an important role in defining the basis, but they are divorced from any explicit geometric interpretation in the setting. The control points can be chosen freely and independently from their associated weights.

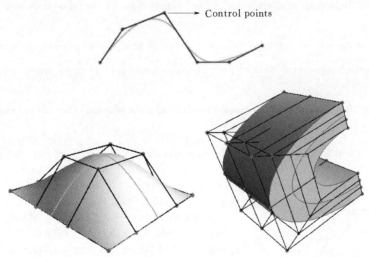

Fig. 16.3-20 NURBS Curve, NURBS Surface and NURBS Solid

Fig. 16.3-20 demonstrate that any desired geometric entities can be described exactly by not many control points. However, for purposes of structural analysis accurate evaluation of the displacements and stresses fields is necessary. For this purpose basis is refined and/or its order elevated. B-spline offer myriad ways in which the basis may be enriched while leaving the underlying geometry and its parameterization intact. Fig. 16.3-21 depicts the refinement by knot insertion, each element has been evenly split in the parametric domain.

To carry out structural analysis based on NURBS, some points and features are addressed:

(1) The isoparametric philosophy is invoked, the solution space for displacement fields is represented in terms of the same basis functions as those for geometry. That is, the coefficients of the basis functions, i.e. the displacements at the control

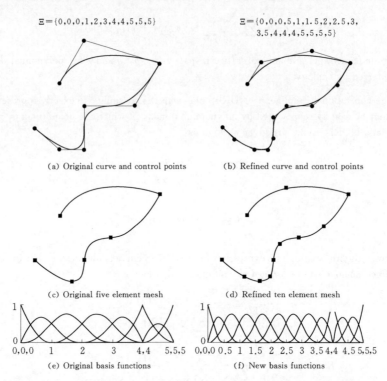

Fig. 16.3 – 21 Refinement by Knot Insertion

points are the degrees-of-freedom (DOFs) of the problem.

(2) A mesh for a NURBS patch is defined by the tensor product of knot vectors. For example, in three dimensions, a mesh is given by $\Xi \times H \times L$, where $\Xi = \{\xi_1, \xi_2, \cdots, \xi_{n+p+1}\}$, $H = \{\eta_1, \eta_2, \cdots, \eta_{m+q+1}\}$, $L = \{\zeta_1, \zeta_2, \cdots, \zeta_{l+r+1}\}$。

(3) Knot spaces subdivide the domain into "elements". The support of each basis function consists a small number of "elements".

(4) Mesh requirement strategies are developed from a combination of knot insertion and order elevation.

(5) Arrays constructed from isoparametric NURBS patches can be assembled into global arrays in the same way as finite elements.

2. Governing equations for time domain seismic response analysis of dam-reservoir-foundation system based on SBIGA and IGA[57,58,57]

The substructure method is employed to carry out seismic response analysis of dam-reservoir-foundation system. Note that in the framwork of NURBS, the DOF of the problem is represented by the displacement field at the control points. Substructuring of the dam-reservoir-foundation system is adopted as shown in Fig. 16.3-22. The dam, the near-field foundation and far-field foundation are taken as substructures. Let n_b represents the number of DOF at the interface between the dam and the near-field, n_d the number of DOF at the interface between the near-field and far-field, n_a the number of DOF in the dam-structure (including dam-reservoir interface) but excluding the n_b DOF, and n_c the number of DOF in the near-field but excluding the n_b and n_d DOF.

Fig. 16.3 – 22 Substructuring of Dam-reservoir-foundation System

The dynamic equation of motion for the dam structure with near-field foundation subjected to the earthquake ground motion is formulated as follows.

$$\begin{bmatrix} m_{aa} & m_{ab} & 0 & 0 \\ m_{ba} & m_{bb} & m_{bc} & 0 \\ 0 & m_{cb} & m_{cc} & m_{cd} \\ 0 & 0 & m_{dc} & m_{dd} \end{bmatrix} \begin{Bmatrix} \ddot{u}_a^t \\ \ddot{u}_b^t \\ \ddot{u}_c^t \\ \ddot{u}_d^t \end{Bmatrix} + \begin{bmatrix} c_{aa} & c_{ab} & 0 & 0 \\ c_{ba} & c_{bb} & c_{bc} & 0 \\ 0 & c_{cb} & c_{cc} & c_{cd} \\ 0 & 0 & c_{dc} & c_{dd} \end{bmatrix} \begin{Bmatrix} \dot{u}_a^t \\ \dot{u}_b^t \\ \dot{u}_c^t \\ \dot{u}_d^t \end{Bmatrix} + \begin{bmatrix} k_{aa} & k_{ab} & 0 & 0 \\ k_{ba} & k_{bb} & k_{bc} & 0 \\ 0 & k_{cb} & k_{cc} & k_{cd} \\ 0 & 0 & k_{dc} & k_{dd} \end{bmatrix} \begin{Bmatrix} u_a^t \\ u_b^t \\ u_c^t \\ u_d^t \end{Bmatrix} = \begin{Bmatrix} R_w^t \\ 0 \\ 0 \\ R_d^t \end{Bmatrix} \quad (16.3-40)$$

where the displacement vector u^t represents total displacement relative to a fixed reference; R_w^t the total nodal forces developed due to dam-reservoir interaction; T_d^t the total nodal forces developed due to dynamic interaction with the unbounded foundation. Partitioning the equation (16.3-40) into two parts as indicated, then it can be written in the compact form:

$$\begin{bmatrix} m & m_g \\ m_g^T & m_{dd} \end{bmatrix} \begin{Bmatrix} \ddot{u}^t \\ \ddot{u}_d^t \end{Bmatrix} + \begin{bmatrix} c & c_g \\ c_g^T & c_{dd} \end{bmatrix} \begin{Bmatrix} \dot{u}^t \\ \dot{u}_d^t \end{Bmatrix} + \begin{bmatrix} k & k_g \\ k_g^T & k_{dd} \end{bmatrix} \begin{Bmatrix} u^t \\ u_d^t \end{Bmatrix} = \begin{Bmatrix} \overline{R}_w \\ R_d^t \end{Bmatrix} \quad (16.3-41)$$

where

$$u^t = \{u_a^t \quad u_b^t \quad u_c^t\}^T \quad (16.3-42)$$

Solution of Eq. (16.3-41) refers to[59] with some modification. The displacement vector is separated into two quasi-static vectors and a dynamic vector as follows:

$$\begin{Bmatrix} u^t \\ u_d^t \end{Bmatrix} = \begin{Bmatrix} \tilde{u}^s \\ \tilde{u}_d^s \end{Bmatrix} + \begin{Bmatrix} \hat{u}^s \\ \hat{u}_d^s \end{Bmatrix} + \begin{Bmatrix} u^d \\ u_d^d \end{Bmatrix} \quad (16.3-43)$$

The first component $\tilde{u}^s$ indicates the value induced by the free-field ground motion u_{bg} at the interface b; whereas the second component $\hat{u}^s$ is produced by the restraint force $\tilde{R}_b^s$ imposed at the interface b to maintain the free-field ground motion u_{bg}. These components are evaluated by the following expressions:

$$\tilde{u}^s = \begin{Bmatrix} -k_{aa}^{-1} k_{ab} \\ I \\ 0 \end{Bmatrix} u_{bg} = \tilde{\Phi}^s u_{bg} \quad (16.3-44)$$

with

$$\tilde{\Phi}^s = [-k_{aa}^{-1} k_{ab}, I, 0]^T$$
$$\tilde{R}_b^s = [k_{bb}^{(2)} - k_{ba} k_{aa}^{-1} k_{ab}] u_{bg} \quad (16.3-45)$$

$$\begin{bmatrix} k & k_g \\ k_g^T & k_{dd} \end{bmatrix} \begin{Bmatrix} \hat{u}^s \\ \hat{u}_d^s \end{Bmatrix} = \begin{Bmatrix} -\tilde{R}^s \\ \tilde{R}_d^s \end{Bmatrix} \quad (16.3-46)$$

In Eq. (16.3-44) I is an $n_b \times n_b$ identity matrix; in Eq. (16.3-45) the superscript (2) indicates that the matrix only contains elements from the dam a; and in Eq. (16.3-46)

$$\tilde{R}^s = [0 \quad \tilde{R}_b^s \quad 0]^T = [0 \quad k_{bb}^{(2)} - k_{ba} k_{aa}^{-1} k_{ab} \quad 0]^T u_{bg} = \hat{\Phi}^s u_{bg} \quad (16.3-47)$$

with

$$\hat{\Phi}^s = [0 \quad k_{bb}^{(2)} - k_{ba} k_{aa}^{-1} k_{ab} \quad 0]^T$$
$$\tilde{R}_d^s = S_{dd}^s \hat{u}_d^s \quad (16.3-48)$$

in which S_{dd}^s is the static impedance of the unbounded far-field foundation at the interface d. Substituting Eq. (16.3-48) into Eq. (16.3-47) yields

$$\begin{bmatrix} k & k_g \\ k_g^T & k_{dd} + S_{dd}^s \end{bmatrix} \begin{Bmatrix} \hat{u}^s \\ \hat{u}_d^s \end{Bmatrix} = \begin{Bmatrix} -\tilde{R}^s \\ 0 \end{Bmatrix} \quad (16.3-49)$$

Substituting Eq. (16.3-43) into Eq. (16.3-41) results in the equation of motion for the evaluation of dynamic components of the displacements

$$\begin{bmatrix} m & m_g \\ m_g^T & m_{dd} \end{bmatrix} \begin{Bmatrix} \ddot{u}^d \\ \ddot{u}_d^d \end{Bmatrix} + \begin{bmatrix} c & c_g \\ c_g^T & c_{dd} \end{bmatrix} \begin{Bmatrix} \dot{u}^d \\ \dot{u}_d^d \end{Bmatrix} + \begin{bmatrix} k & k_g \\ k_g^T & k_{dd} \end{bmatrix} \begin{Bmatrix} u^d \\ u_d^d \end{Bmatrix} = -\begin{bmatrix} m & m_g \\ m_g^T & m_{dd} \end{bmatrix} \begin{Bmatrix} \ddot{\tilde{u}}^s + \ddot{\hat{u}}^s \\ \ddot{\tilde{u}}_d^s + \ddot{\hat{u}}_d^s \end{Bmatrix} - \begin{bmatrix} k & k_g \\ k_g^T & k_{dd} \end{bmatrix} \begin{Bmatrix} \tilde{u}^s + \hat{u}^s \\ \tilde{u}_d^s + \hat{u}_d^s \end{Bmatrix} + \begin{Bmatrix} \overline{R}_w \\ R_d^t \end{Bmatrix} \quad (16.3-50)$$

where the damping terms on the right-hand side of the equation have been neglected, because such terms contribute little to the effective load of the system. Eq. (16.3-50) can be further simplified by incorporating into it the free-field response of the foundation soil before construction of the dam. Removing the contribution of the dam part in Eq. (16.3-41) and denoting the interface forces by R_{dg} yield (for the same reason, the damping terms are neglected)

$$\begin{bmatrix} m^{(1)} & m_g \\ m_g^T & m_{dd} \end{bmatrix} \begin{Bmatrix} \ddot{\tilde{u}}^s \\ \ddot{u}_{dg} \end{Bmatrix} + \begin{bmatrix} k^{(1)} & k_g \\ k_g^T & k_{dd} \end{bmatrix} \begin{Bmatrix} \tilde{u}^s \\ u_{dg} \end{Bmatrix} = \begin{Bmatrix} 0 \\ R_{dg} \end{Bmatrix} \quad (16.3-51)$$

in which the superscript (1) indicates that the matrix contains only those elements contributed by the foundation soil. The static force-displacement relationship for the added dam structure subjected to the free-field ground motion that was used previously in deriving Eq. (16.3-47) and Eq. (16.3-49) may be expressed here as

$$\begin{bmatrix} k^{(2)} & 0 \\ 0 & 0 \end{bmatrix} \begin{Bmatrix} \tilde{u}^s \\ u_{dg} \end{Bmatrix} = \begin{Bmatrix} \tilde{R}^s \\ R_{dg} \end{Bmatrix} \quad (16.3-52)$$

Adding Eq. (16.3-51), Eq. (16.3-52) and Eq. (16.3-46) and then substituting it into the right-hand side of Eq. (16.3-50) leads to

$$\begin{bmatrix} m & m_g \\ m_g^T & m_{dd} \end{bmatrix} \begin{Bmatrix} \ddot{u}^d \\ \ddot{u}_d^d \end{Bmatrix} + \begin{bmatrix} c & c_g \\ c_g^T & c_{dd} \end{bmatrix} \begin{Bmatrix} \dot{u}^d \\ \dot{u}_d^d \end{Bmatrix} + \begin{bmatrix} k & k_g \\ k_g^T & k_{dd} \end{bmatrix} \begin{Bmatrix} u^d \\ u_d^d \end{Bmatrix} = -\begin{bmatrix} m^{(2)} & 0 \\ 0 & 0 \end{bmatrix} \begin{Bmatrix} \ddot{\tilde{u}}^s \\ \ddot{u}_{dg} \end{Bmatrix} - \begin{bmatrix} m & m_g \\ m_g^T & m_{dd} \end{bmatrix} \begin{Bmatrix} \ddot{u}^s \\ \ddot{u}_d^s \end{Bmatrix} + \begin{Bmatrix} \overline{R}_w \\ R_d^d \end{Bmatrix}$$

$$(16.3-53)$$

where

$$R_d^d = R_d^t - R_{dg} - \dot{R}_d^d \quad (16.3-54)$$

Substituting Eq. (16.3-44) and Eq. (16.3-49) into Eq. (16.3-53), finally the governing equations for the dynamic component of displacement field of dam-reservoir-foundation system subjected to earthquake ground motion $u_{bg}(t)$ at the dam-foundation interface b are formulated as follows

$$\begin{bmatrix} m & m_g \\ m_g^T & m_{dd} \end{bmatrix} \begin{Bmatrix} \ddot{u}^d \\ \ddot{u}_d^d \end{Bmatrix} + \begin{bmatrix} c & c_g \\ c_g^T & c_{dd} \end{bmatrix} \begin{Bmatrix} \dot{u}^d \\ \dot{u}_d^d \end{Bmatrix} + \begin{bmatrix} k & k_g \\ k_g^T & k_{dd} \end{bmatrix} \begin{Bmatrix} u^d \\ u_d^d \end{Bmatrix} = [\kappa]\{\ddot{u}_{bg}\} + \begin{Bmatrix} \overline{R}_w \\ R_d^d \end{Bmatrix} \quad (16.3-55)$$

with

$$[\kappa] = -\begin{bmatrix} m^{(2)} & 0 \\ 0 & 0 \end{bmatrix} \begin{Bmatrix} \tilde{\Phi}^s \\ 0 \end{Bmatrix} - \begin{bmatrix} m & m_g \\ m_g^T & m_{dd} \end{bmatrix} \begin{bmatrix} k & k_g \\ k_g^T & k_{dd} + S_{dd}^s \end{bmatrix}^{-1} \begin{Bmatrix} \dot{\Phi}^s \\ 0 \end{Bmatrix} \quad (16.3-56)$$

where $\tilde{\Phi}^s$ and $\dot{\Phi}^s$ are found from Eq. (16.3-44) and Eq. (16.3-47).

The total displacement response of the dam structure is evaluated by combining the dynamic component u^d with two quasi-static components $\tilde{u}^s$ and $\dot{u}^s$, then the internal stresses and the corresponding deformation are determined accordingly.

As we are most interested in the stresses and deformations of the dam structure itself, for the numerical analysis, the discretization model of the dam is constructed by IGA, whereas the discretization model of the reservoir and that of the far-field unbounded foundation respectively are constructed by SBIGA. All the basis functions, either of IGA, or of SBIGA are NURBS based. By employing SBIGA, for the evaluation of $\overline{R}_w(t)$, only the interface DOFs between the dam and the reservoir need to be discretized; while for the evaluation of $R_d^d(t)$, only the interface DOFs between the near-field and the far-field foundation need to be discretized.

Since the dynamic water pressure induced by dam-reservoir interaction $\overline{R}_w$ and the dynamic interaction force with unbounded foundation R_d^d are frequency dependent, calculating the time-domain response needs some special treatment. The solution procedure is detailed in[51], only the basis equations are given here.

Let the number of DOFs at the dam-reservoir interface be denoted by n_r. Similar to that described in section 16.3.3.2, the dynamic water pressure is expressed by complex frequency response function $[\overline{S}(\omega)]$ as (see Eq. (16.3-14) and Eq. (16.3-15))

$$\{p\} = -[\overline{S}(\omega)]\{\ddot{u}\} \quad (16.3-57)$$

After introducing unit impulse response function and transforming pressure intensity into nodal force yields

$$\{R_p(t)\} = -[T]\int_0^t [S(t-\tau)]\{\dot{u}_w(\tau)\}d\tau \quad (16.3-58)$$

where $[T]$ is the matrix which transforms hydrodynamic pressure intensity into nodal forces at the control points. Carrying out discretization with respect to time, the hydrodynamic pressure at nth time step takes the form

$$\{R_p(t_n)\} = -[T]\sum_{j=0}^{n}[S(t_j)]\{\ddot{u}(t_{n-j})\} = -[M_{wp}]\{\ddot{u}(t_n)\} - \{R_{w,n}^{his}\} \qquad (16.3-59)$$

In Eq. (16.3-59), the term $\{R_p(t_n)\}$ is split into two parts: the instantaneous part depending only on the current response $\ddot{u}(t_n)$ and the historic part depending on the retarded response $\ddot{u}(t_j)(j=0,1,2\cdots,n-1)$. Note that the number of DOFs at the dam-reservoir interface n_r is less than that in the dam n_a and n_b, for brevity in the expression, $\{R_p(t_n)\}$ (vector of the size n_r) is generalized in Eq. (16.3-55) as $\overline{R}_w(t)$ (vector of the size $n_a+n_b+n_c$) and written in the form as

$$\overline{R}_w(t_n) = -[M_w]\{\ddot{u}(t_n)\} - \{R_{w,n}^{his}\} \qquad (16.3-60)$$

Analogously, the discretized interaction force with unbounded far field foundation at the n th time step $\{R_d^d(t_n)\}$ can also be split into the instantaneous part and the retarded part and written as

$$\{R_d^d(t_n)\} = -[S_{dd}^d]\{u(t_n)\} - \{R_{d,n}^{d,his}\} \qquad (16.3-61)$$

Note that the retarded part depends not only on $u(t_j)$ but also on $\dot{u}(t_j)$ and $\ddot{u}(t_j)$ (from $j=0$ to $j=n-1$). Substituting Eqs. (16.3-60) and (16.3-61) into Eq. (16.3-55), the nth time step discretization form for the dynamic response component of the dam-reservoir-foundation system is expressed as

$$\begin{bmatrix} m+m_W & m_g \\ m_g^T & m_{dd} \end{bmatrix}\begin{Bmatrix} \ddot{u}^d(t_n) \\ \ddot{u}_d^d(t_n) \end{Bmatrix} + \begin{bmatrix} c & c_g \\ c_g^T & c_{dd} \end{bmatrix}\begin{Bmatrix} \dot{u}^d(t_n) \\ \dot{u}_d^d(t_n) \end{Bmatrix} + \begin{bmatrix} k & k_g \\ k_g^T & k_{dd}+S_{dd}^d \end{bmatrix}\begin{Bmatrix} u^d(t_n) \\ u_d^d(t_n) \end{Bmatrix} = [\kappa]\{\ddot{u}_{bg}(t_n)\} - \begin{Bmatrix} R_{w,n}^{his} \\ R_{d,n}^{d,his} \end{Bmatrix}$$
$$(16.3-62)$$

In this way, seismic analyses of dam-resecvir-foundation system including 2D gravity dam and 3D arch dam structure have been conducted[51,58]. It helps better understanding of the real seismic stress distribution of the dam body.

16.3.3.5 Others

1. An exploratory research on damage simulation, vulnerable analysis and seismic risk assessment of concrete dams[60-62]

Seismic analysis of dams considering its damage evolution and failure modes, leading to more realistic and comprehensive insight into the seismic safety of dams, has become one important frontier in dam design[63]. Various theories and methods have been proposed in the literature. The technique is still under development. So exploratory research in this field has been carried out at Dalian University of Technology.

(1) Numerical simulation of seismic damage evolution and failure process of concrete dams [60].

Seismic failure process analysis of high concrete dams is a very complicated problem. Subjected to strong earthquake shocks, the dam undergoes whole process of elastic deformation, initiation and propagation of cracks, severe damage and ultimate failure. The numerical model should adequately reproduce this strong nonlinear process and has the potential to incorporate various factors including complex geometry of the dam, partitioning of concrete monolith, complex soil condition, dam-reservoir interaction, opening/closing of contraction joints and so on. In addition, time marching process including thousands of time steps also poses a harsh challenge to computational capacity. Existing research fell short in certain aspects. Furthermore, it's worthwhile to point out that concrete has always been assumed as homogeneous in seismic analysis of concrete dams, which may be valid for response analysis in the elastic stage, but may not valid to represent the behavior in the nonlinear range.

Concrete is heterogeneous in nature. In mesoscale concrete is considered to be composed of three phases—the coarse and fine aggregates, the mortar and the interfaces between them.

Tang successfully studied the process of rock failure[64]. He considered that rock is a natural material characterized by discontinuous inhomogeneous, anisotropic and inelastic, he developed a numerical model for rock failure process analysis (RFPA) based on heterogeneity simulation. He also introduced his model to analyze the concrete specimens thereafter considering heterogeneity of each phase in mesoscale. Complicated failure mode of concrete specimens in the shape of cubes and beams resembled those in laboratory tests have been obtained[65].

Tang's model has been extended to seismic failure modeling of high concrete dams. However, as far as a real concrete dam is concerned, it is far from realistic to model concrete exactly in mesoscale in seismic analysis. Yet it is also unnecessary. So some simplification are introduced. Most important of all, the dam is discretized with finite elements and the three phases, i.e. matrices, aggregate and interfaces, may coexist in each single element, in which they are blended. This results in an equivalent homogeneous element with its material properties decided by the original composites. If element sizes are small enough, this process would result in different properties for elements containing different content of substructures. If the dam is discretized with plenty of finite elements, the material properties of elements can be assumed to conform to some specific distribution, the Weibull random distribution law is employed here. In this way, the heterogeneity of concrete can be reflected to some degree. Keeping in mind, the size of elements employed should be small enough. More details refer to[60].

In most cases of seismic response of concrete dams, the nonlinearity of concrete is closely related to initiation and propagation of microcracks, instead of plasticity. So, concrete is simulated as the one exhibiting the feature of elasto-brittle material. In addition, since the size of elements is small, anisotropy and nonlinearity of concrete can be neglected, simple elastic damage constitutive relations (Fig. 16.3 - 23) may be assumed. Damage of element is classified into two categories: the tensile damage and the shear damage. The former occurs when the maximum tensile strain reaches its threshold value; whereas the Mohr-Coulomb criterion is employed to judge whether the later case does appear. Considering the fact that rather larger factor of safety is assigned the compressive strength of concrete in the dam design, and the seismic damage is represented mainly in the form of cracking instead of crushing, the assessment is conducted giving priority to tensile damage. Both tensile damage and shear damage are irrecoverable. The tensile damage does not affect the bearing capacity of the element in compression; on the contrary, compressive damage is accompanied by losing tensile strength of the element; and crushed element can sustain neither compression nor tension any more.

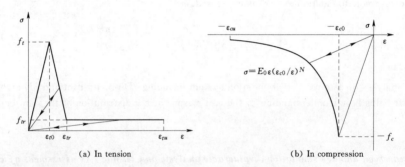

Fig. 16.3 - 23 Stress-strain Curve for Simulating Damage of Concrete Structure

The proposed model has the advantages that no knowledge on the cracking route needs to be known a priori and no remeshing is required in the successive implementation. It enables to simulate the whole process of earthquake damage development from early stage of elastic deformation, subsequent initiation and propagation of microcracks, increasingly severe damage development, until ultimate failure of concrete structure with only a unified model. It is highly simple and efficient. The proposed numerical model has been validated by comparsion with the results of shaking table test performed for a 210m high Dagangshan Arch Dam model[61]. The simulated failure mode agrees well with that of the experimental results.

The proposed model has the advantages that no knowledge on the cracking route needs to be known a priori and no remeshing is required in the successive implementation. It enables to simulate the whole process of earthquake damage development from early stage of elastic deformation, subsequent initiation and propagation of microcracks, increasing severe damage development, until ultimate failure of concrete structure with only a unified model. It is highly simple and efficient.

Then the proposed modal is employed to investigate the failure process and failure modes of a practical gravity dam, some deeper insight into the seismic failure of gravity dam is gained. The project considered is the 160m high Jin'anqiao roller compacted concrete gravity dam. The height of the studied non-overflow section is about 114m. The design PGA is $0.399g$. Six levels of acceleration amplitudes are employed for the earthquake input, i.e. $0.2g$, $0.3g$, $0.399g$, $0.5g$, $0.6g$ and $0.8g$ respectively. A typical scenario showing damage development until complete failure of the dam monolith subjected to earthquake excitation with acceleration amplitude of $0.6g$, about 1.5 times the design PGA, is plotted is Fig. 16.3 - 44.

Four typical failure modes may be distinguished (Fig. 16.3 - 25): cracking and damage at dam neck; cracking at dam heel; cracking at the position of abrupt slope change and further propagating; complete failure and loss of water impounding function; the crack initiates at the dam heel and propagate along the dam-foundation interface going deep into dam body, but does not strictly follow this path all the way, it then turns to downstream face with inclination.

Earthquake damage simulation has also been performed for concrete arch dam. The damage process of the dam subjected to earthquake excitation of the magnitude equal to 1.5 times the design PGA is depicted in Fig. 16.3 - 26 and the typical damage pattern is shown in Fig. 16.3 - 27[66].

(2) Seismic vulnerability and fragility analysis of concrete dams[61].

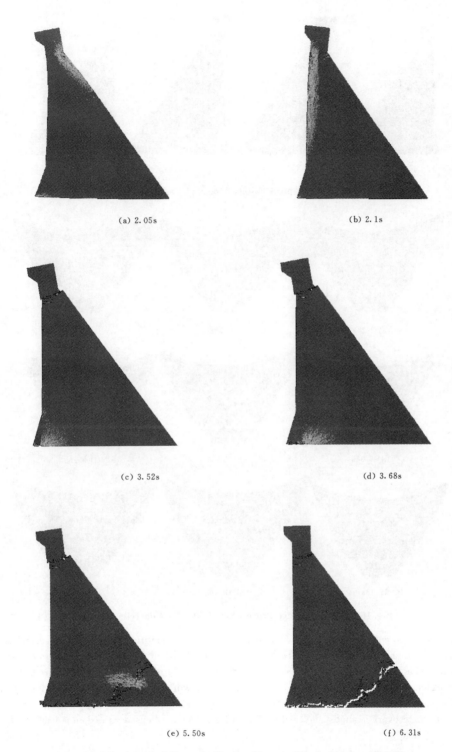

Fig. 16.3-24 A typical Scenario Slowing Damage Development until Failure

Vulnerability analysis of a structure is an integrated process that includes consideration of the uncertainty and randomness including the seismic hazard, structural response and material capacity parameters to give a probabilistic assessment of risk.

The failure modes based Monte Carlo simulation method is applied for this analysis, which has the advantages to deal with large-scale problem like the dam structure and large number of random variables. It makes possible to obtain realistic estimate of the probability of failure and the fragility curves of concrete dams for various predefined seismic damage levels. The uncertainty in material capacity parameters as well as heterogeneity of concrete are considered.

Based on the result of seismic damage simulation described in the previous subsection, as well as considering the seriousness

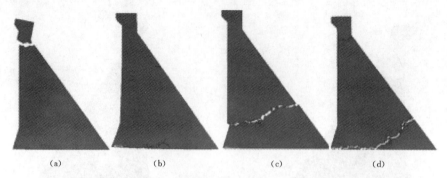

Fig. 16.3-25 Typical Failure Modes of Concrete Gravity Dam

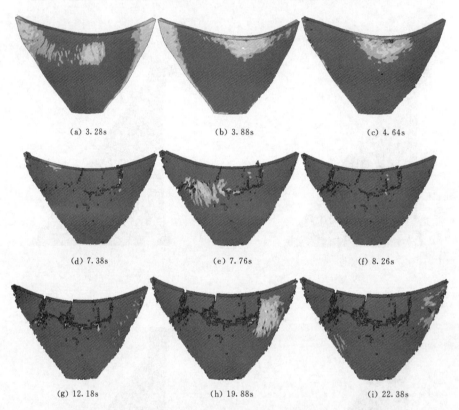

Fig. 16.3-26 Damage Process Simulation for Concrete Arch Dam

of possible consequences and cost for retrofitting, five damage levels are identified: ①Level Ⅰ: surviving almost intact—dam still works in good condition, only fine cracks appears. ②Level Ⅱ: slight damage—some micro cracks occur with limited length, which does not influence normal functioning of the dam, only minor repairing is required. ③Level Ⅲ: medium damage—more cracks spread over the dam body, considerable repair work is needed to restore its normal function. ④Level Ⅳ: severe damage—cracks penetrate deep through the dam, dam head is seriously damaged, long-term extensively repairing work is needed, or recovery is almost impossible. ⑤Level Ⅴ: nearly collapsed—the dam is basically failed to retain its water impounding function.

A sketch showing suggested identification of damage levels for gravity dam as well as for arch dam is given in Fig. 16.3-28. It is expected that with experience broadened for earthquake damage simulation of concrete dams and data accumulated on actual dams subjected to strong earthquake shocks, this work can be further improved.

Probability distribution of various material capacity parameters including the material damping ratio, the modulus of elasticity, the dynamic strength of concrete, the Poisson's ratio and the mass density are referred to the ASCE manual[67] as shown in Fig. 16.3-29. In the figure, the abscissa R denotes the ratio of random value to the deterministic value, and the numbers in parentheses give the deviation of the 10 and 90 percentile from the medium value. To construct the fragility curves ensembles of 6 ground accelerations (0.2g, 0.3g, 0.399g, 0.5g, 0.6g and 0.8g) are used for the gravity dams case and 4 ground accelerations (0.3g, 0.408g, 0.612g, 0.8g) are used for the arch dam case. Where 0.399g and

Chapter 16　Seismic Safety of Dams

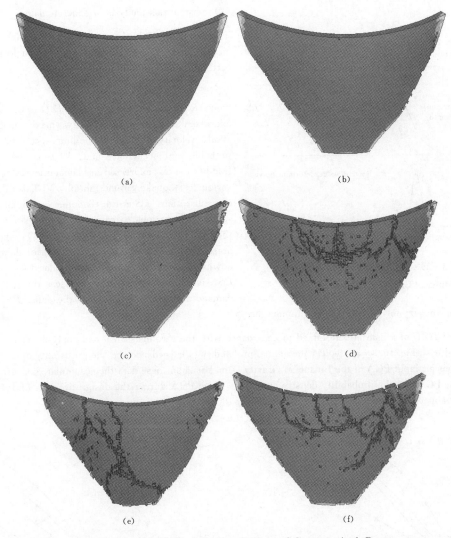

Fig. 16.3-27　Typical Damage Pattern of Concrete Arch Dam

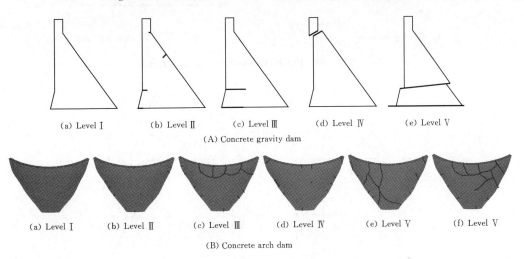

Fig. 16.3-28　Identification of Damage Levels

0.408g are the design PGA for the gravity dam and the arch dam respectively. Thirty randomized dam analyses are carried out for each ground acceleration sample. The resulting fragility curves related to levels of exciting earthquake acceleration for the gravity dam case and for the arch dam case are shown in Fig. 16.3-30 (a) and Fig. 16.3-30 (b) respectively.

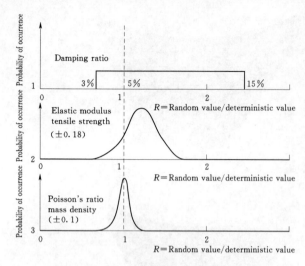

Fig. 16.3-29 Probability Distribution of Material Capacity Parameters

(3) Seismic risk analysis of concrete dams.

The estimation of seismic risk of an engineering project is close related to the economic and human losses induced by earthquake events. It is formulated as

$$RSK = HZD \cdot VUL \cdot LS \qquad (16.3-63)$$

where RSK is the seismic risk; HZD is the seismic hazard, expressed as earthquake intensity, maximum ground acceleration, velocity, or corresponding spectrum value and its probability of occurrence and; VUL is the vulnerability or fragility curve, expressed as damage levels corresponding to various earthquake ground motion, VUL is taken from previous subsection; LS is the economic and human losses corresponding to various damage levels. This study is confined to the first half of the formula (16.3-63), and does not address economic or social consequences of dam failure to the downstream population at risk. Under these circumstances, RSK is defined as an index to assess the risk of earthquake damage during the life time of the dam; RSK may also be used for predicting the amount of economic and human losses.

The seismic hazard HZD of a dam is determined in accordance with the reconnaissance and analysis report on the seismic risk for the dam site[68]. Fig. 16.3-31 depicts the annual probability of exceedance and the probability of exceedance during life time for the seismic input PGA of the Jin'anqiao Gravity Dam. Based on these data the accumulated occurrence probability curve of seismic PGA and the probability density curve of seismic PGA during the design life time of 100 years for the dam-site can be evaluated as shown in Fig. 16.3-32 and Fig. 16.3-33 respectively.

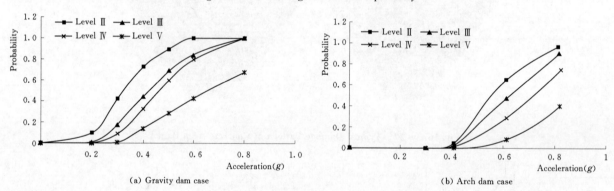

Fig. 16.3-30 Fragility Curves of Concrete Dams

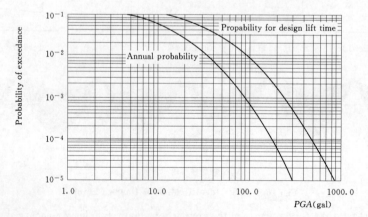

Fig. 16.3-31 Probability of Exceedance Curve of Seismic PGA for Jin'anqiao Dam-site

The probability of occurrence of jth level seismic damage during the design life time of the Jin'anqiao gravity dam is found as

$$P[D_j] = \int_0^\infty f(R)P[D_j \mid PGA]dR \tag{16.3-64}$$

where $f(R)$ denotes the probability density of PGA during the design life time of the dam (see Fig. 16.3-33); $P[D_j \mid PGA]$ denotes the probability of occurrence of jth level of damage for the dam, evaluated from the fragility curve.

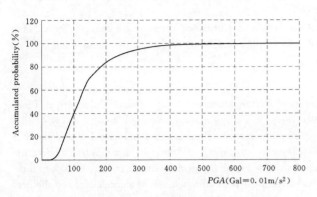

Fig. 16.3-32 Accumulated Occurrence Probability Curve of PGA for the Dam-site

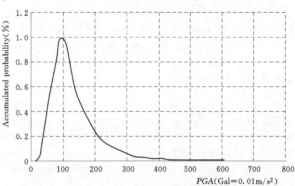

Fig. 16.3-33 Probability Density Curve of PGA for the Dam-site

The probability of occurrentce for various level of seismic damage of Jin'anqiao concrete gravity dam is shown in Table 16.3-10, whereas that of the designed new Fengman concrete gravity dam is given in Table 16.3-11 and that of the arch dam is given in Table 16.3-12.

Table 16.3-11 Probability of Occurrence for Various Level of Damage of the Jin'anqiao Gravity Dam

Damage Level	Level I Surviving intact	Level II Slight damage	Level III Medium damage	Level IV Severe damage	Level V Nearly collapsed
Probability	0.935640	0.035520	0.010411	0.009702	0.008727

Table 16.3-12 Probability of Occurrence for Various Level of Damage of the Designed New Fengman Gravity Dam

Damage Level	Level I Surviving intact	Level II Slight damage	Level III Medium damage	Level IV Severe damage	Level V Nearly collapsed
Probability	0.999362	0.000577	0.000059	0.000002	0.000001

Table 16.3-13 Probability of Occurrence for Various Level of Damage of an Arch Dam

Damage Level	Level I Surviving intact	Level II Slight damage	Level III Medium damage	Level IV Severe damage	Level V Nearly collapsed
Probability	0.9826	0.006	0.0041	0.0025	0.0005

The seismic risk index is given by

$$RSK = \sum_j P[D_j] \tag{16.3-65}$$

The direct economic loss due to seismic events is

$$RSK - E = \sum_j r_{je} \cdot P[D_j] \tag{16.3-66}$$

where r_{je} denotes the direct economic loss due to jth level of seismic damage of the dam. And the loss of human lives is

$$RSK - H = \sum_j d_{je} \cdot P[D_j] \tag{16.3-67}$$

where d_{je} denotes the lost human lives due to the jth level of seismic damage of the dam.

2. Engineering defensive measures

In combination with the study of earthquake response analysis and earthquake safety evaluation of dams, the research on the defensive measures also plays an important role to increase the capacity of dams to withstand strong earthquake shocks. The application of suitable defensive measures will help the structure to perform satisfactorily even under the worse conditions. With regard to the concrete dams, we have conducted preliminary investigation to improve the impermeability and the strength of concrete with some kind of new materials, such as the polyurea, fiber inreforced polymer (FRP)[69] and engineered cementitious compoiste (ECC) etc.

Much additional research is still needed to deepen our knowledge of seismic performance of concrete dams, to improve the reliability and efficiency of present methods for the seismic analyses, designs and safety evaluation of concrete dams.

16.3.4 Conclusion

Since the fifties in the last century, the researchers of the Institute of Earthquake Engineering, Dalian University of Technology have been making constant efforts to promote the scientific research and engineering practice of seismic safety evaluation of dams in China. Benefiting from the spirit of self-reliance, all on their own, they designed and manufactured the first mechanical shaking table in China as well as relevant vibration measuring instruments including acceleration transducers and displacement transducers in 1956. They also autonomously studied suitable materials for model tests and methods for vibration measurement. Between the years from 1956 to 1958, they performed the first vibration test of an arch dam and the first vibration test of an embankment dam in the new China, thereby solved relevant key engineering problems in the design of these two dams. In 1973, by using a self-developed set of test instruments and test method, which was relatively simple and easy to use, they measured the first nine vibration frequencies and vibration modes of the Baishan Arch Dam, as well as the first three vibration modes in the horizontal direction and the first vibration mode in the vertical direction of the Fengman Gravity Dam.

For the purpose to deepen knowledge on seismic safety evaluation of dams, based on model tests and numerical simulation, we kept steadily working to improve the accuracy and efficiency of the seismic response analysis of dams. We performed tests on more than two thousand concrete specimens, and studied the rate sensitivity of concrete material so as to realistically find the strength of concrete in tension when subjected to earthquake loading. We developed procedures based on the scaled boundary finite element method, a promising alternative of finite element method, to accurately and effectively solve the dam-reservoir interaction problems to include the effects of water compressibility and reservoir boundary absorption. Some new numerical models have been developed to deal with the dynamic interaction analyses of arch dam/gravity dam with complex unbounded soil containing inclusions or intercalations in the near-field and certain inhomogeneities in the far-field and to include variation of elastic moduli as a function of depth. We proposed a coupled method based on the SBIGA and the Isogeometric analysis, which advanced the accuracy and efficiency of seismic response analysis of dam-reservoir-foundation system to a new level. We also explored the numerical models to simulate the seismic damage process of arch dams and gravity dams. We are confident that these research will be highly beneficial to the seismic safety evaluation of dams in the sense of better reflecting the reality. Comments from colleagues are most welcome.

It's also worth noting that the presented paper covers only part of our research, the research on the seismic hazard analysis of dam site and the seismic aspect of embankment dams and other hydraulic structures will be presented in due course.

References

[1] Lin G and Chen D. Dynamic Model Test of an arch dam. *Acta Mechanics Sinica*, 1959, 3 (3): 217 - 235.

[2] Lin Gao. Dynamic Similarity Criterion for Vibration Study of Arch Dams, *Journal of Hydraulic Engineering Society of China*, 1958 (1): 79 - 104.

[3] Карцивадзе Г. Н., Медведев С. В., Напетвбридзе Щ. Г. Сейсмостойкое *Стройтелъство за Рубежом*, Госсмройизедам, Москва, 1962.

[4] Зристова В. С., *Арочное Плотиностроение*, Издателъство 《Знергия》, Москва-Ленинград, 1965.

[5] Lin Gao, et al. Earthquake Resistant Model Test of Embankment Dam for 602 - 1 Project. *Jounal of Talian Engineering Institute* (the predecessor of the Journal of Dalian University of Technology), 1959, 5: 71 - 82.

[6] Chen D, Lin G et al. Model Test of the Buttress Stability of a Buttress Dam against Lateral Buckling, *Journal of Talian Engineering Institute*, 1959, 5: 83 - 106.

[7] Lin G and Chen D. The Direct Photographic Method for the Vibration Model Test of Dam Structure. *Journal of Talian Engineering Institute*, 1973, 2: 17 - 37.

[8] Okamoto S and Kato K. A Method of Dynamic Model Test of Arch Dams. *Proc. of the 4th World Conference on Earthquake Engineering*, B-1, 1969: 87 - 97.

[9] Takahashi T. Vibration Studies of an Arch Dam, *Proc. of the 4th World Conference on Earthquake Engineering*, 1969, B-1: 61-71.

[10] Seven R T and Taylor P R. Earthquake Effects on Arch Dam by Response Spectra Methods, *Vibration in Civil Engineering*, session Ⅵ: *Current Practice in Structural Design*, 1966, Paper 19, 221-227.

[11] Lin G and Chen D, Studies on the Computer Method and Computer Program for the Analysis of Hydro-dynamic Pressure on Dams, *Journal of Talian Engineering Institute*, 1975, 4: 72-100.

[12] Lin G, Zhou J and Fan C. Dynamic Model Rupture Test and Safety Evaluation of Concrete Gravity Dam. *Dam Engineering*, 1993, 4 (8): 173-186.

[13] Zhou J, Lin G, Zhu T, Experimental Investigations into Seismic Failure of High Arch Dams. *Journal of Structural Engineering*, 2000, 126 (8): 926-935.

[14] Yan D and Lin G. Dynamic Proportion of Concrete in Direct Tension, *Cement and Concrete Research*, 2006, 36: 1371-1378.

[15] Malvar L J and Ross C A. Review of Strain Rate Effects for Concrete in Tension. *ACI Materials Journal*, 1998, 95 (6): 435-439.

[16] Ross C A, Jerome D M, Tedesco J W, et al. Moisture and Strain Rate Effects on Concrete Strength, *ACI Materials Journal*, 1996, 93 (3): 293-300.

[17] Cowell W L., Dynamic Properties of Plain Portland Cement Concrete. *Technical Report R447*, Naval Civil Engineering Laboratory, Port Hueneme, CA, 1966 (June): 46.

[18] CEB-FIP Model Code 1990: Design Code/*Comité Euro-International du Béton*. Wiltshire, Redwood Books, 1993.

[19] Rossi P, Toutlemonde F. Effect of Loading Rate on the Tensile Behavior of Concrete: Description of the Physical Mechanisms. *Materials and Structures*, 1996, 29 (186): 116-118.

[20] Sercombe J, Ulm F.J and Toutlemonde F. Viscous Hardening Plasticity for Concrete in High-rate Dynamics. *Journal of Engineering Mechanics*, ASCE, 1998, 124 (9): 1050-1057.

[21] Yan D and Lin G. Dynamic Behavior of Concrete in Biaxial Compression. *Magazine of Concrete Research*, 2007, 59 (1): 45-52.

[22] Yan D, Lin G, and Chen G Dynamic Properties of Plain Concrete in Triaxial Stress States. *ACI Materiasl Journal*, 2009, 106 (1): 89-94.

[23] Lin G, Yan D and Yuan Y. Response of Concrete to Elevated-amplitude Cyclic Tension. *ACI Materials Journal*, 2007, 104 (6): 561-566.

[24] Yan D and Lin G. Influence of Initial Static Stress on the Dynamic Properties of Concrete, *Cement and Concrete Composites*, 2008, 30: 327-333.

[25] Leng F and Lin G. Dissipation-based Consistent Rate-dependent Model for Concrete. *Acta Mechanica Solida Sinica*, 2010, 23 (2): 147-155.

[26] Lin G and Xiao S Y. Seismic Response of Arch Dams including Strain-rate Effects. *Proc. of the International Conference on Advance and New Challenges in Earthquake Engineering Research*, Edited by Ko J M and Xu Y L, 2002, Harbin and Hong Kong, 331-338.

[27] Wang M. Stationary and Propagative Instabilities in Metals -a Computational Point of View, PhD dissertation, 1997, TU Delft, Netherlands.

[28] Kupfuer H B, Hilsdor H K and Rusch H. Behavior of Concrete under Biaxial Stresses, *Journal of the American Concrete Institute*, 1969, 66 (8): 656-666.

[29] Suaris W and Shah S P. Constitutive Model for Dynamic Loading of Concrete. *Journal of Structural Engineering*, ASCE, 1985, 111 (3): 563-576.

[30] Winnicki A Pearce C J and Bicanic N. Viscoplastic Hoffman Consistency Model for Concrete. *Computers and Structures*, 2001, 79 (1): 7-19.

[31] Chopra A K. Earthquake Analysis of Arch Dams: Factors to be Considered. *Proc. 14th World Conference on Earthquake Engineering*, 2008, Beijing, China.

[32] Wolf J P and Song C. *Finite-Element Modelling of Unbounded Media*, 1996, John Wiley & Sons, New York.

[33] Song C and Wolf J P. The Scaled Boundary Finite Element Method-alias Consistent Infinitesimal Finite-element Cell Method for Elastodynamics. *Comput. Methods Appl. Mech. Eng.*, 1997, 147 (3-4): 329-355.

[34] Lin G, Du J and Hu Z. Dynamic Dam-reservoir Interaction Analysis including Effects of Reservoir Boundary Absorp-

tion. *Science of China*, *Seria E*: Technological Science, 2007, 50 (Supp, 1): 1-10.

[35] Lin G, Wang Y and Hu, Z. An Efficient Approach for Frequency-domain and Time-domain Hydrodynamic Analysis of Dam-reservoir Systems. *Earthquake Eng. Struc. Dynam.*, 2012, 41 (13): 1725-1749.

[36] Fenves G, Chopra A K. Earthquake Analysis and Response of Concrete Gravity Dams. *Report No. UCB/EERC-84/10*, 1984.

[37] Hanna Y G, Humar J L. Boundary Element Analysis of Fluid Domain. *Journal of Engineering Mechanics Division*, (ASCE) 1982, 108 (2): 436-450.

[38] Sharan S K. Finite Element Modeling of Infinite Reservoirs. *Journal of Engineering Mechanics* (ASCE), 1985, 111 (12): 1457-1469.

[39] Chopra A K. Hydrodynamic Pressures on Dams during Earthquakes. *Journal of the Engineering Mechanics Division* (ASCE), 1967, 93 (6): 205-223.

[40] Fok K L, Chopra A K. Frequency Response Functions for Arch Dams: Hydrodynamic and Foundation Flexibility Effects. *Earthquake Engineering and Structural Dynamics*, 1986, 14 (5): 769-795.

[41] Lin G, Hu Z Q, Du J G, et al. Effects of Reservoir Boundary absorption on the Earthquake Response of Arch Dams. *Proc. 14th World Conference on Earthquake Engineering*, 2008, Beijing, China.

[42] Wang Y, Lin G and Hu Z. Novel Nonreflecting Boundary Condition for an Infinite Reservoir Based on the Scaled Boundary Finite Element Method. *Journal of Engineering Mechanics* (ASCE), 2015. 141 (5: 04014150).

[43] Bouaanani N and Miquel B. A New Formulation and Error Analysis for Vibrating Dam-reservoir Systems with Upstream Transmitting Boundary Conditions. *Journal of Sound and Vibration*, 2010, 329: 1924-1953.

[44] Maeso O, Dominguez J. Earthquake Analysis of Arch Dam I. Dam-Foundation Interaction. Journal of Engineering Mechanics, ASCE 1993, 119: 496-512.

[45] Tan H, Chopra A K. Earthquake Analysis of Arch Dams including Dam-water-foundation Rock Interaction. *Earthquake Engineering and Structural Dynamics*, 1995, 24 (11): 1453-1474.

[46] Zhang C, Jing F and Pekau O A. Time Domain Procedure of FE-BE-IBE Coupling for Seismic Interaction of Arch Dams and Canyons. *Earthquake Engineering and Structural Dynamic*, 1995, 24 (12): 1651-1666.

[47] Du X, Zhang Y and Zhang B. Nonlinear Seismic Response Analysis of Arch Dam-foundation Systems. *New Development in Dam Engineering*, 2004, Edited by Wieland M., Ren Q. and Tan S. Y., Taylor & Francis Group plc, London, U. K.

[48] Liao Z, Wong H L. A Transmitting Boundary for the Numerical Simulation of Elastic Wave Propagation. *Soil Dynamics and Earthquake Engineering*. 1984, 3 (4): 174-183.

[49] Lin G, Du J G and Hu Z Q. Earthquake Analysis of Arch and Gravity dams including the Effects of Foundation Inhomogeneity. *Frontiers of Architecture and Civil Engineering in China*, 2007, 1: 41-50.

[50] Yin X, Li J, Wu C et al. ANSYS Implementation of Damping Solvent Stepwise Extraction Method for Nonlinear Seismic Analysis of Large 3-D Structures. *Soil Dynamic and Eartqhuake Engineering*, 2013, 44: 139-152.

[51] Zhang Y. *Research on Isogeometric Analysis and Scaled Boundary Isogeometric Analysis and Their Engineering Application*, PhD dissertation, Dalian, China: Dalian University of Technology, 2013.

[52] Zhong H, Lin G, Li J B, et al. An Efficient Time-domain Damping Solvent Extraction Algorithm and its Application to Arch Dam-foundation Interaction Analysis. *Communications in Numerical Methods in Engineering*, 2008, 24 (9): 727-48.

[53] Li J B, Yang J, Lin G. A Stepwise Damping-solvent Extraction Method for Large-scale Dynamic Soil-structure Interaction Analysis in Time Domain. *International Journal for Numerical and Analytical Methods in Geomechanics*, 2008, 32 (4): 415-36.

[54] Wong H L, Luco J E. Dynamic Response of Rigid Foundations of Arbitrary Shape. *Earthquake Engineering and Structural Dynamics*, 1976, 4: 579-587.

[55] Song C, Wolf JP. The Scaled Boundary Finite-element Method-Alias Consistent Infinitesimal Finite-element Cell Method-for Elastodynamics. *Computer Methods in Applied Mechanics and Engineering*, 1997, 45: 329-355.

[56] Hughes T J R, Cottrell J A and Bazilevs Y. Isogeometric Analysis: CAD, Finite Elements, NURBS, Exact Geometry and Mesh Refinement. *Computer Methods in Applied Mechanics and Engineering*, 2005, 194: 4135-4195.

[57] Lin G, Zhang Y, Wang Y, et al. Coupled Isogeometric and Scaled Boundary Isogeometric Approach for Earthquake Response Analysis of Dam-Reservoir - Foundation System. *Proc. 15th World Conference of Earthquake Engineer-

ing, 2012, Lisbon, Portugal.

[58] Lin G, Zhang Y, Hu Z. Scaled Boundary Isogeometric Analysis in 2D Elastostatics. *SCIENCE CHINA Physics, Mechanics & Astronomy*, 2013, 56 (8): 1-15.

[59] Clough R and Penzien J. Dynamic of Structures, Second Edition, *Computers and Structures*, Inc., 1995.

[60] Zhong H, Lin G., Li X Y. et al. Seismic Failure Modeling of Concrete Dams Considering Heterogeneity of Concrete. *Soil Dynamics and Earthquake Engineering*, 2011, 31: 1678-1689.

[61] Zhong H, Lin G, Li H J et al. Seismic Vulunerability Analysis of a Gravity Dam Based on Typical Failure Modes, Paper No. 383, *Proc. 15th World Conference on Earthquake Engineeing*, 2012, Lisbon, Portugal.

[62] Li X Y. *Seismic Risk Analysis of Concrete Gravity Dams*, Master thesis, Dalian, china: Dalian University of Technology, 2011.

[63] ICOLD Bulletin 130, *Risk Assessment in Dam Safety*, 2005.

[64] Tang C A and Hudson J A. *Rock Failure Mechanics*. Explained and Illustrated, 2010, Taylor and Francis Group, London, UK.

[65] Zhu W, Tang C. Numerical Simulation on Shear Fracture Process of Concrete Using Mesoscopic Mechanical Model. *Construction and Building Materials*, 2002; 16 (8): 453-63.

[66] Technical Report. *Study on Earthquake Response Analysis of High Arch Dam and the Related Engineering Measures for Disaster Mitigation*. Dalian University of Technology and Central and South China Exploration Design and Research Institute, Nov. 2012.

[67] *Structural Analysis and Design of Nuclear Plant Facilities*. American Society of Civil Engineers, Manuals and Reports on Engineering Practice, No. 58, 1980.

[68] Hu Y X. *Instruction on Seismic Safety Assessment*. Beijing: Seismic Press, 1999.

[69] Zhong H, Wang N L, Lin G. Seismic Response of Concrete Gravity Dam Reinforced with FRP Sheets on Dam Surface. *Water Science and Engineering*, 2013, 6 (4): 409-422.

[70] Lin G, Han Z, Li J. General Formulation and Solution Procedure for Harmonic Response of Foundation on Isotropic as well as Anisotropic Multilayered Half-space. *Soil Dynamics and Earthquake Engineering*, 2015, 70: 48-59.

Chapter 17

Energy Dissipation and High Speed Flow in High Dams

Gao Jizhang[1], Liu Zhiping[2], Guo Jun[3]

17.1 Introduction of Technological Development of High Head and Large Discharge Flow and Energy Dissipation

The past 60 years, particularly the last 30 years of reform and opening up, have witnessed a fast development of water resources and hydropower industries in China, and a large number of large hydropower projects have been constructed. To meet the needs of design and construction, hydraulic research in China has been closely linked to the practical requirements of projects, acquisition of technologies combined with innovation, to provide advanced, safe and economical large flood discharge and energy dissipation schemes for constructions, and develop relevant technologies with Chinese characteristics.

Large flood discharge structures with high head and large flood constructed, being constructed or planned in China are shown in Table 17.1-1, Table 17.1-2 and Table 17.1-3.

China has experienced three stages for its high head, large discharge flow and energy dissipation technology. From 1950 to 1978, 100m to 150m high dams had been constructed with experience learned from the former Soviet Union as well as self-development. Starting from 1978 when China launched its reform and opening up policy, self-innovation had been emphasized to build a lot of 150m to 200m high dams. Since 2000, China has built, or been constructing, a batch of world-level dams, such as Ertan, Xiaolangdi, Three Gorges, Longtan, Xiaowan, Xiluodu, Goupitan and Xiangjiaba, which are 150m to 300m high and have a maximum discharge capacity of more than 100,000m³/s, making China's technology in this field reaching the world's advanced level. The prominent characteristics include high head, large discharge flow and large discharge per unit width, huge discharge power, narrow valley and complex geological conditions. Large outlets in dam body are not only widely used in gravity dams, but also in gravity arch dams and other types of dams. The flood discharge and energy dissipation solution that combines separate discharge and zoned energy dissipation is widely adopted. Methods such as trajectory flow, stilling basin and flip buckets are adopted according to the local situations to dissipate energy. A large amount of new technologies, including differential bucket and heterotypic flip bucket, slit flip bucket, flaring gate pier and the series of dissipation technologies (including stepped spillway that combines with RCC construction technology), flow nappe collision dissipator, plunge pool dissipator, and energy dissipation through vortex shafts reconstructed from diversion tunnels, have been successfully applied to projects construction. In-depth research has been conducted on different types of high speed flow issues, and the fruitful achievements applied to projects.

Currently, a bench of projects with dam heights of 200m to 300m, discharge capacity 20,000m³/s to 50,000m³/s and discharge per unit width 200m³/(s·m) to 300m³/(s·m), are under construction, including Nuozhadu, Jinping I, Baihetan, Wudongde, Shuangjiangkou and Lianghekou dams, etc. There are more difficulties in flood discharge and energy dissipation of the projects, thus challenging hydraulics of large dams and restrains on geology and environment, which has made China's research on high-waterhead and large-flow flood discharge and energy dissipation the advanced level of the world.

---

[1] Gao Jizhang, Former President of China Institute of Water Resources and Hydropower Research, Professor Senior Engineer.
[2] Liu Zhiping, Vice President of China Institute of Water Resources and Hydropower Research, Professor Senior Engineer.
[3] Guo Jun, Vice Chief Engineer of China Institute of Water Resources and Hydropower Research, Professor Senior Engineer.

Chapter 17 Energy Dissipation and High Speed Flow in High Dams

Table 17.1-1 Typical Flood Discharge Structures with High Head and Large Discharge in China (Dam Height Over 100m, Sorted by Discharge)

Year	No.	Name	Type	Completion year	Dam height (m)	Discharge capacity (m³/s)	Surface spillway	Middle outlet	Deep outlet	Chute spillways	Spillway tunnels	Discharge per unit width (m³/(s·m))	Energy dissipation method
2000–2010	1	Three Gorges	PG	2009	181	102,500	22–8×17	2–10×12 discharging debris	23–7×9			134/254/302	Trajectory bucket
	2	Longtan I	RCCPG	2009	192	35,500	7–15×20		2–5×8			223	Trajectory bucket
	3	Ertan	VA	2000	240	23,900	7–11×11.5	6–6×5	4–3×5		2–13×13	125/186/285	trajectory bucket Plunge pool
	4	Dachaoshan	RCCPG	2003	111	23,800	5–14×17.8		3–7.5×10			193.6/267	Flaring gate pier + stepped spillway
	5	Tianshengqiao I	CFRD	2000	178	28,500							Trajectory bucket
	6	Shuibuya	CFRD	2009	233.2	24,400			2–6×7	5–14×21.8		267/233	Slit, trajectory bucket
	7	Xiaolangdi	TE	2001	160	13,990				3–11.5×17	3–D14.5, 3–D6.5; 1–10×12; 1–10×11.5; 1–10.5×13	Orifice tunnel 161.4/179.9 Free flow tunnel 268/246.6/224.5 Spillway 109.1	Trajectory bucket, orifice tunnel, plunge pool
	8	Sanbanxi	CFRD	2006	185.5	27,000				3–20×19	2–5×9	172/288	Trajectory bucket
	9	Gongboxia	CFRD	2006	132.2	7,860			1–7.5×6	2–12×18	1–7×9, spiral-flow tunnel 1–10×12	147/164/103	Trajectory bucket, vortex shaft flow
1990–1999	1	Wuqiangxi	PG	1994	85.83	63,900	9–19×23	1–9×13	5–3.5×7.0			268/287/172	Flaring gate pier + stilling basin
	2	Panjiakou	PG	1992	107.5	54,500	18–15×15		4–4×6			142	Flaring gate pier + trajectory bucket

Part II Construction Technologies for Dams and Reservoirs

Continued

Year	No.	Name	Basic data				Discharge structures (number-width(m)×height(m))					Discharge per unit width (m³/(s·m))	Energy dissipation method
			Type	Completion year	Dam height (m)	Discharge capacity (m³/s)	Surface spillway	Middle outlet	Deep outlet	Chute spillways	Spillway tunnels		
1990–1999	3	Ankang	PG	1995	128	45,000	5–15×17	5–11×12	4–5×8			209.3	Flaring gate pier + stilling basin
	4	Yantan	PG	1994	110	33,400	7–15×21		1–5×8			308/210	Flaring gate pier stilling basin
	5	Geheyan	GV	1995	151	23,458	7–12×18.2		4–4.5×6.5			231/225	Flaring gate pier, plunge pool
	6	Manwan	PG	1995	132	20,910	5–13×20	2–5×6	2–5×8		1–12×12	262–225	Plunge pool
	7	Dongfeng	VA	1995	162	12,580	3–11×7	1–3.5×4.5		Left 1–15×20	1–12×17.5	Spillway and spillway tunnel 281, middle outlets 210/158, surface spillways 67	Trajectory bucket
1980–1989	1	Wujiangdu	GV	1983	165	24,400	4–13×19	2–4×4		Left 1–13×19, right 1–13×19	Left 1–9×10, right 1–9×10	Spillway tunnels 240, spillways 201, surface spillways 144	Overflow/flyover powerhouse + trajectory bucket
	2	Baishan	GV	1986	149.5	19,100	4–12×12		3–6×7			183.0/228	Trajectory bucket
	3	Longyangxia	GV	1989	178	6,000		1–8×9	1–5×7, bottom 1–5×7	2–12×17		268/275/300/187	Trajectory bucket
1970–1979	1	Liujiaxia	PG	1974	147	9,220			2–3×8	3–10×8.5	1–8×9.5	126.3/268.8/250	Trajectory bucket
1960–1969	1	Xin'anjiang	PG	1965	105	10,000	9–13×10.5					76	Flyover powerhouse

Note: PG=Gravity Dam; VA=Gravity Arch Dam; CFRD=Concrete Face Rockfill Dam; TE=Earth Dam; RCCPG=RCC Gravity Dam.

Chapter 17 Energy Dissipation and High Speed Flow in High Dams

Table 17.1-2 Flood Discharge Structures with High Head and Large Discharge in China (under Construction, Dam Height over 100m, Sorted by Discharge)

No.	Basic data				Discharge structures (number-width×height, m)					Discharge per unit width (m³/(s·m))	Dissipator
	Name	Type	Dam height (m)	Discharge capacity (m³/s)	Surface spillway	Middle outlet	Deep outlet	Chute spillway	spillway tunnels		
1	Xiluodu	VA	285.5	50,900	7—12.5×13.5		8—6×6.7		4—14×12	207/267/283	Deflecting flow + plunge pool
2	Xiangjiaba	PG	162	41,200	12—8×26	10—6×9.6				300/331	Flaring gate pier + stilling basin
3	Nuozhadu	ER	261.5	39,500				8—15×20	Left 1:2—5×8.5, right 1:2—5×8.5	162/308/393	Deflecting flow + plunge pool
4	Goupitan	VA	232.5	35,600	6—12×13	7—6×8/7×6			1—11×12	129/228/254	Deflecting flow + plunge pool
5	Xiaowan	VA	294.5	23,600	5—11×15	6—6×6.5			1—15×16.5	146/223/238	Deflecting flow + plunge pool
6	Jinping I	VA	305	15,400	4—11.5×10	5—5×6			1—14×12	65/219/261	Deflecting flow + plunge pool
7	Pubugou	ER	186	15,250				3—12×16	1—12×15/1D9 (emptying tunnel)	201/284/159	Deflecting flow
8	Laxiwa	VA	250	6,310	2—13×9.5		2—5.5×6			106/207	Deflecting flow + plunge pool

Note: VA=Arch Dam; PG=Gravity Dam; ER=Rockfill Dam.

Table 17.1-3 Flood Discharge Structures with High Head and Large Discharge in China (Proposed, Sorted by Discharge)

No.	Basic data				Discharge structures (number-width×height, m)				Discharge per unit width (m³/(s·m))	Dissipator
	Name	Type	Dam height (m)	Discharge capacity (m³/s)	Surface spillways	Middle outlets	Deep holes	Chute spillways	Spillway tunnels		
1	Baihetan	VA	289	50,153	7—12.5×13.5	8—6×7			4—14×12	235/307/267	Deflecting flow plunge pool
2	Wudongde	VA	265	37,500	5—12×18	6—6×8			2—14×10	220/273/259	Deflecting flow + plunge pool
3	Shuangjiangkou	ER	312	8,086				1—16×22	1—9×10.5,1—6×8	260/261	Deflecting flow
4	Lianghekou	ER	295	8,186				1—16×21	1—8×10.5,1—9×7	257/238	Deflecting flow

Note: VA=Arch Dam; ER=Rockfill Dam.

17.2 Technologies and Facilities of High Head, Large Discharge Flow and Energy Dissipation

17.2.1 Layout Characteristics and Scales of Discharge Structures

17.2.1.1 Classification of Layouts of Discharge Structures

The types of flood discharge structures in high dams can be categorized into dam body discharge, bank-run discharge and joint discharge. Bank discharge (chute spillway, spillway tunnel) is suitable for common gravity dams (including RCC dams), arch dams (including gravity dams), earth and rockfill dams (CFRD); while dam body discharge (layered discharge through outlets in dam body) is not suitable for earth and rockfill dams.

17.2.1.2 Characteristics and Scales of Discharge Structures

Influenced by the hydrological characteristics and geological conditions of rivers in China, the characteristics of flood discharge and energy dissipation structures for large hydraulic projects are: ①High head. The height of Er'tan Dam is 240m, Xiaowan and Xiluodu Dams are nearly 300m, and Jinping I Dam is 305m high. The water head of flood discharge is usually 0.5 to 0.8 times of the dam height, and flow speed arranging from 30m/s to 50m/s, making it prominent for problems of fluctuation, vibration, cavitation, atomization, erosion, and river bed scour. ②Large discharge flow. Most high dams in China have a design discharge capacity over 10,000m³/s, with Er'tan larger than 20,000m³/s, Xiluodu over 40,000m³/s, and Three Gorges nearly 80,000m³/s. ③Large discharge flow per unit width. It is generally larger than 200m³/(s·m), with several dams close to 300m³/(s·m). ④Huge flood discharge power. It could be as high as tens or hundreds of gigawatts, making it a difficult task to safely discharge and dissipate such large amount of energy, and one of the key and difficult issues for dam engineering. ⑤Complex geological conditions in the areas for flood discharge and energy dissipation. For example, Ankang and Manwan dams are located on the curve of the rivers; there are landslide bodies in the energy dissipation area of Longyangxia Dam; discharge tunnels of Xiaowan Project run through possibly active faults; bedrocks of Ankang Dam energy dissipation area have a low capacity of scour resistance; while Goupitan Dam energy dissipation area is located on both soft and hard rocks. As a result, more and more large hydraulic projects are adopting combined flood discharge methods.

17.2.2 Flood Discharge Structures with High Head and Large Discharge Flow

17.2.2.1 Types of Flood Discharge and Energy Dissipation

1. Flood discharge and energy dissipation through dam body

With advantages of smooth discharge, easy design of energy dissipation and scour prevention, and reduced investment, flood discharge structures are usually situated in the middle or on one side of the river channel, making it suitable for high arch dams and high gravity dams (including gravity arch dams) in the valley. Flood discharge structures are overlaid on the top of the powerhouse that is a most compact and economical layout; however, it is rarely used now in modern projects. When the flood discharge power is particularly large and the geological conditions make it unsuitable, a scour plunge pool can be adopted. When ski-jump energy dissipation is not suitable and unfavorable geological conditions downstream or other reasons, bottom flow or bucket flow is usually adopted for energy dissipation. To improve dissipation, new dissipators can be adopted for all the energy dissipation methods.

2. Bank-run flood discharge and energy dissipation

Bank-run flood discharge and energy dissipation methods (chute spillways and spillway tunnels) have characteristics of far outflows. When using trajectory energy dissipations, the scour appears at the opposite bank, protective measures should be considered. Plunge pool can also be applied for special flood discharge, landscape and geological conditions.

3. Combined flood discharge and energy dissipation

When discharged flow is large and any of the above-mentioned discharge methods is insufficient, or out of safety or economical considerations, a combined flood discharge structures in and out of dam body will usually be adopted, namely, separate discharge and zoned dissipation. This type of layout has been adopted in projects of Wujiangdu, Er'tan, Xiaowan and Xiluodu, and become a trend for high dams to discharge large discharge flow and dissipate energy. Specifically, this method combines three different types of structures (surface spillway, middle or deep outlet in dam body, spillway tunnels or chute spillway). With dam height of no higher than 150m and wide valley, a lot of projects, mainly gravity dam, adopt a combination of flaring piers gate and stilling basin. With dam height of higher than 150m, narrow valley, relatively large discharge flow, many projects will adopt arch dam as the dam type, and a combination of surface spillway, middle (or deep) outlet and spillway tunnel (or chute spillway) as the flood discharge and energy dissipation facilities, with large plunge pool downstream of the dam. A combination of spillway tunnel or chute spillway is usually adopted for high earth and rockfill dams. When adopting combined discharge methods, energy dissipation zones should be reasonably designed. For narrow valleys, the energy dissipation zones are usually arranged longitudinally along the river course.

17.2.2.2 Characteristics and Applicable Conditions of Flood Discharge via Dam Body Structures

For concrete gravity dams on wide valley with large discharge, flood is usually discharged via dam body structures that are situated on the main river channel. This method makes use of the main river course to discharge flood, it is smooth for flow to fall into river channel, and investment can be reduced, and thus is widely adopted in China's projects. The Three Gorges Project is China's, as well as the world's largest dam for flood discharge. Because the discharge is enormous, the main river channel is designed to discharge flood. Dam sections used to discharge flood are 483m long in total, with 22 surface spillways alternating with 23 deep outlets for the task of flood discharge when the project is in operation. There are also 22 diversion bottom outlets below the surface spillways, and will be used, together with the 23 deep outlets, for the Phase III diversion (Fig. 17.2-1). When the flood control water level is 145m, 56,700m^3/s of flow can be discharged. When the water level is 166.9m (a 100-year event), 69,800m^3/s of flow can be discharged. When it is at the check water level of 180.4m with 80% of units in operation, the total discharge capacity will be 102,500m^3/s. The Three Gorges Project adopts a flood discharge method that mainly depends on deep outlets because of the operational principle of storing clean water and discharging muddy water.

Since the impounding level reached 135m at the first time in June 2003, after several years of operation, both flood control water level and post-flood impounding level of the Three Gorges reservoir have been gradually adjusted. The post-flood impounding level in 2006 was 156m, the flood control level in 2007 was 144m, and the post-flood impounding level in 2008 was 172.7m, entering the experimental impounding period. The experimental impounding period then began. The deep outlets and surface spillways have been checked through the large floods in years of 2004, 2007, 2008 and 2010. After the flood in 2008, part of the deep outlets and surface spillways were operating at a reservoir level of 172.7m, close to the design water level, and the maximum discharge via dam body structures were as large as 33,000m^3/s. The maximum discharge water head of deep outlets was 82.7m, maximum discharge for a single outlet was 2,000m^3/s, maximum speed of bottom flow at flip bucket was 30m/s, and the maximum operational waterhead of surface spillway was 14.7m. In order to understand the hydraulic characteristics and requirements of project safety assessment of bottom outlets, deep outlets and surface spillways when in operation, hydraulic prototype observation of these outlets were conducted while another four times of observations of dam toe scours were conducted during 2004 and 2009. Data of deep outlets and surface spillways were collected at the water level of 172.7m (design water level 175m). Flow pattern, pressure distribution, flow cavitation noise, aerated flow, and aeration concentration distribution on the bottom slab. The results of hydraulic prototype observation show that all hydraulic indicators of the deep outlets and surface spillways are working well, downstream energy dissipation sufficient, and there is no unfavorable flow regimen. Aeration facility downstream radial gates of the deep outlet works well. Although some flow cavitation signals can be detected at the bottom and sidewalls of the discharge chutes, the minimum air concentration that is as high as 2.2% at the bottom of the discharge chute has the effect of cavitation mitigation. Inspection of deep outlets after discharge shows no cavitation erosion.

Because the arrangement of flood discharge structures in Three Gorges Project are so concentrated, scouring downstream in the flood discharge has always been a concern in the process of design and operation. Since the project was put into operation, underwater survey of downstream scour holes had been conducted in 2004, 2008 and 2009 respectively, and the results are basically similar: the distance between the lowest scour hole and the dam toe is over 100m, and the depths of scour holes are between 12m to 18m, close to the figures gained in the physical model experiment. The two horizontal partition dikes for the longitudinal guide wall (longitudinal cofferdam built during the construction phase) near the right-bank power plant, as well as the partial concrete protective slab near dam toe and between No.18 deep outlet and longitudinal guide wall, have good protection of guide wall foundation and dam toe against erosions caused by back flow.

Xiangjiaba gravity dam is the last of the four cascade dams on the lower reach of Jinsha River, its dam height 162m, and the maximum discharge flow 48,680m^3/s. The reservoir adopts a combined discharge facilities (Fig. 17.2-2), with 12 surface spillways (8m×26m) and 10 middle outlets (6m×9.6m). To avoid impacts of discharge atomization on the chemical plants downstream the dam, bottom flow with large volume of the work is adopted for energy dissipation instead of deflecting flow, with 2 stilling pool.

As for Xiangjiaba Dam the discharge per unit width of surface spillways and middle outlets 300m^3/(s·m) and 331m^3/(s·m)

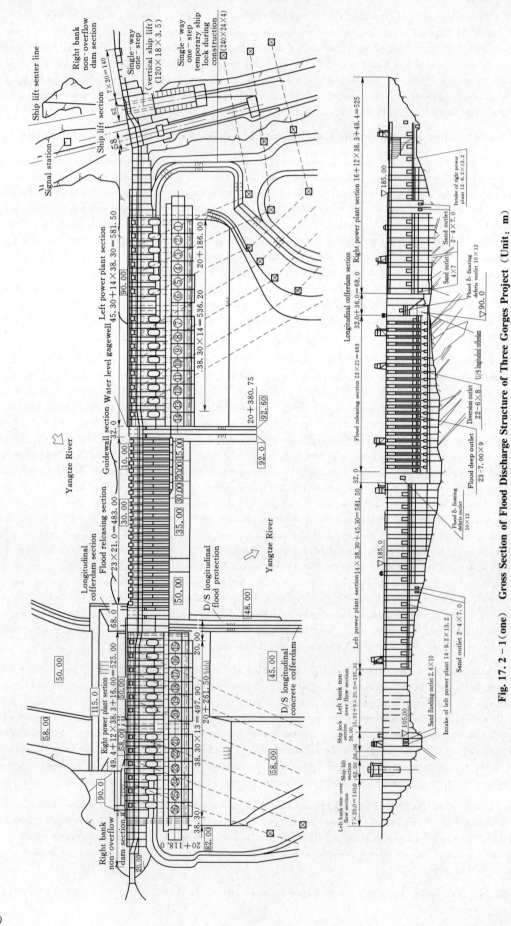

Fig. 17.2-1(one)　Gross Section of Flood Discharge Structure of Three Gorges Project　(Unit: m)

Chapter 17 Energy Dissipation and High Speed Flow in High Dams

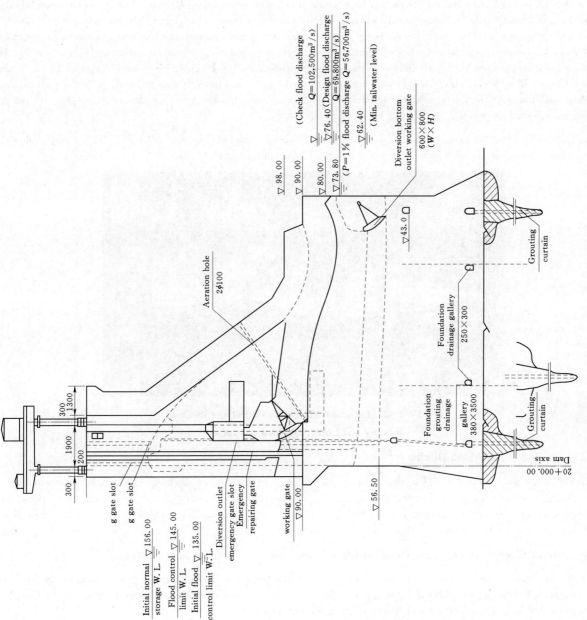

Fig. 17.2−1 (two) Cross Section of Flood Discharge Structure of Three Gorges Project (Unit: m)

respectively, and maximum flow speed into the stilling basin close to 40m/s. The discharge waterhead, discharge per unit width and flow speed into the pool are all at the highest level among other projects, making the safety of stilling basin the most important research task in the design process. After preliminary experimental research, a decision is made to give up on the traditional stilling basin because ① the high dynamic water pressure on the bottom slab in the front of the basin, making the stability of bottom plate not satisfying the design requirement; and ② middle outlets will be used to discharge sediments, while high flow speed and sedimentary erosion increase the difficulty of designing the stilling basin. The research later on focuses on two solutions: contracted middle outlets combined with consecutive flip bucket, and large differential sills of surface spillways and middle outlets. A large off-set is combined with in both solutions. The advantages of first solution is that surface spillways and middle outlets have sound flow patterns in stilling basin in their operational conditions, regulation and operation are both convenient, energy is sufficiently dissipated, discharged flow has less scour on downstream river channel, and less protection is needed although high fluctuating pressure exists in certain areas downstream of the off-set. The advantages of second solution, because surface spillways and middle outlets are separated by guide walls, is that water discharged from surface spillways and middle outlets will suddenly be expended when flowing out of the guide walls into the stilling basin, energy will be sufficiently dissipated, and sound flow patterns in basin can be gained in all operational conditions. As the guide walls are submerged underwater, flow-induced vibration and cavitation flow caused at high speed are both concerned.

Dams that adopt flood discharge methods through dam body also include Xin'anjiang, Manwan, Wuqiangxi, Ankang, Dachaoshan, and Longtan, with different types of energy dissipation methods.

Fig. 17.2 – 2 Layout of Xiangjiaba Project and its Flood Discharge and Energy Dissipation Structures

17.2.2.3 Bank-run Flood Discharge

Earth-rock dams and rockfill dams (including CFRD) usually adopt bank-run flood discharge methods. For concrete dams, when geological condition allows, such methods can also be adopted. Bank-run flood discharge includes chute spillway (or ski-jump spillway) and spillway tunnel. This method diverts water away from the dams, and the areas with poor geological conditions and low scour resistance capability can also be avoided.

1. Characteristics, applicable requirements and shapes of chute spillway (and skip-jump spillway)

To discharge flood via bank-run discharge structures, water is diverted away from the dams, thus imposes no impacts on dam safety, and their relation with tail water of power plants is easy to handle. When the bank slopes are steep, the excavation volume will be large, and there will also be the problems of high slopes.

Spillways with high head often bend at the approach channels, and different shapes of flip buckets are adopted at the end to adjust the direction of the nappe. Hydraulic design of spillway should be able to make sure the nappe falls into the river channel, which means large discharge does not scour the opposite bank while small discharge does not scour the near bank. However, it is inevitable that the nappe flow will scour the near bank in a short period of time at the beginning and ending of flipping, making it necessary to give protection based on the geological conditions.

Shuibuya, 233.2m high, is the highest CFRD built in China. Its chute spillway, situated on the left bank, is China's largest one in the late 20th century. The approach channel is connected to ogee section and to the chute (Fig. 17.2 – 3), where the flow speed exceeds 30m/s. Aeration devices are applied to mitigate the cavitation erosion. Main characteristics of flood discharge and energy dissipation of this project include: ① Slits buckets are adopted at the end of spillway so that the flow jets out longitudinally. On one hand, the energy dissipation efficiency is increased while scour of river channel reduced. On the other hand, the flow drops into the center of the river, causing atomization that has little impacts on left bank slope. ②The application of slit bucket energy dissipator reduces the scour of riverbed. Given the scour resistance capacity of downstream riverbed and the requirement of construction progress, only side wall (instead of the bottom) of plunge pool are protected, the first large plunge pool of this type in China.

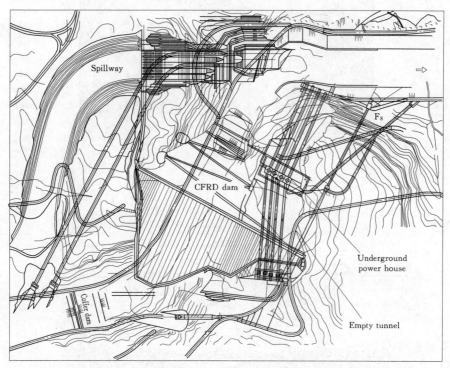

Fig. 17.2 - 3 Layout of Shuibuya CFRD Dam and Chute Spillway

Nuozhadu, a 261.5m high earth core rockfill dam, has spillways as the main flood discharge facility, with a spillway tunnel on both banks (Fig. 17.2 - 4). The layout characteristics of this project include: ①Spillway is designed on the left bank according to the river direction and geological conditions. ②It is the largest spillway among dams with a height of more than 260m in China. The total horizontal length of the spillway is 1,445m, its width 151.5m. With 8 surface spillways, the design and check maximum discharges are 19,398m³/s and 31,318m³/s respectively. ③The spillway is divided into 3 sections, with a large plunge pool at the downstream. Because the water cushion is deep enough for energy dissipation, only side wall are protected instead of the bottom, thus reducing the engineering works. ④Flows are dropped one by one from left to right into the plunge pool. The excavated rocks are used to fill the dam body, thus reducing project investment. ⑤The rear section of No.5 diversion tunnel is the part of the left spillway tunnel. The two spillway tunnels on both banks have a discharge capacity of 6,000m³/s, and are used at the initial stage of dam operation for flood discharge and emptying.

2. Characteristics, applicable requirements and shapes of spillway tunnels

When bank-run flood discharge facility is adopted while chute spillways are not suitable because of the topographic conditions, spillway tunnels might be adopted.

There are a handful of shapes of spillway tunnels with high waterhead and large discharge according to the layout of their longitudinal sections. There are mainly five types: ①horizontal/oblique straight line; ②chute; ③ogee section in front of tunnel; ④ogee section in rear of tunnel; and⑤vortex shaft spillway (Fig. 17.2 - 5).

For the spillway tunnels with deep or bottom outlets, because the elevation difference between the inlet and the outlet is small, they are often in the shape of a straight line on the longitudinal section with small gradient. They are used not only for flood discharge, but also for late-stage diversion. Their operational gate has a high waterhead and small size. Flow in the discharge has a high speed, so aeration and cavitation mitigation facilities are needed. It could be difficult to make sure the bottom cavity of the aeration slot on the slight slope is unobstructed, and careful study is thus needed.

For the spillway tunnels with surface or shallow submerge as their inlet, because the elevation difference between the inlet and the outlet is large, their longitudinal sections could be in the shapes of chute, ogee and inclined section both in the front

(a) Diagram of Nuozhadu Project

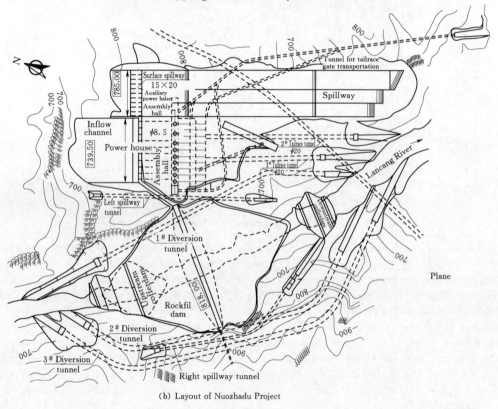

(b) Layout of Nuozhadu Project

Fig. 17.2-4 Nuozhadu Project

and rear of the tunnel, or shaft.

The chute spillway is simple. Flow accelerates along the chute. The size of the sections can remain unchanged considering the influence of aeration. It is relatively easy to install aeration and keep the bottom cavity unobstructed. When rebuilding a diversion tunnel into a spillway tunnel, an ogee section and inclined tunnel is built in the front part of the spillway tunnel. The lower part of the inclined tunnel has a prominent problem of cavitation because of high speed flow. Several projects in China and abroad have ever been damaged by cavitation, and thus aeration and cavitation mitigation measures are very important in the design stage. The ogee part built in the rear part of the tunnel is adopted to shorten the tunnel length with high speed flow and to concentrate high speed flow at the rear of the spillway tunnel. Flow speed of the pressurized section of the front part of the tunnel will not exceed 25m/s to avoid cavitation. The plane bend of spillway tunnel is within the pressurized section to adjust the flow direction and make sure the flow falls into the river reach properly. The rear section with high speed flow is short and strait, and so it is easy and effective to deploy aerators. The spillway tunnels of Xiluodu and Jinping I Projects, designed and constructed in recent years, both adopt this type of tunnel. The reconstruction of diversion tunnels into vortex shaft spillway tunnel is the way widely adopted in China. The large horizontal vortex spillway tunnel of Gongboxia Project has been built and put into operation. Its inlet is directly connected to the shaft to reduce the inlet excavation and avoid high slope excavation. Shaft and horizontal vortex chamber are used to change flow direction, and the tunnel line can be minimized. Energy dissipation efficiency inside the shaft, horizontal vortex chamber and inner plunge pool can be as high as about 80%, and so the speed in the spillway tunnel is low, reducing the impacts of scour and atomization.

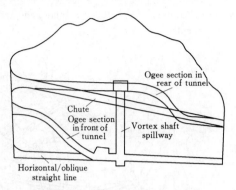

Fig. 17.2 - 5 **Typical Arrangements of Tunnel Spillways**

Xiaolangdi Project mainly adopts spillway tunnels to discharge flood and sediment, and the discharge structures are all on the left bank (Fig. 17.2 - 6). Altogether there are one chute spillway, three spillway tunnels, and three orifice spillway tunnels (all reconstruction from diversion tunnels), with a downstream plunge pool to dissipate energy collectively. The three orifice spillway tunnels are another type of structure to dissipate energy inside the tunnels.

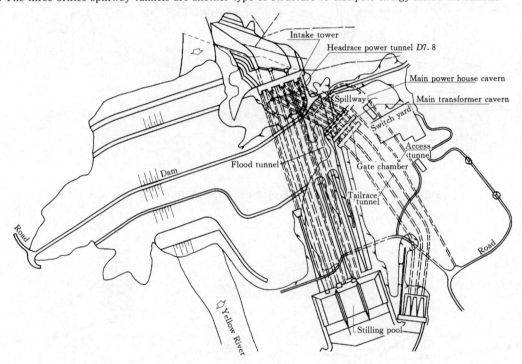

Fig. 17.2 - 6 Layout of Flood and Sediment Discharge Structures of Xiaolangdi Project

17.2.2.4 Combined Flood Discharge

Ertan Project has the typical combined flood discharge arrangement, with 7 surface spillways and 6 middle outlets in the dam body, and 2 spillway tunnels on right bank (Fig. 17.2 - 7). Flow is divided about one third in general and discharged via the surface spillways, middle outlets, and the spillway tunnels. Each of the flood discharge facilities can meet the requirement of discharging normal year flood (about 5 year return period), their operations are flexible, and it is easy to maintenance.

Xiluodu high arch dam, with its maximum discharge flow exceeding $55,000 m^3/s$, adopts combined flood discharge facilities and energy dissipation. With 7 surface spillways and 8 middle outlets in the dam body that discharge more than $31,000 m^3/s$, it also has a large plunge pool at dam toe (Fig. 17.2 - 8). Two large spillway tunnels are designed on each of the two banks, the ogee part is arranged on the rear of the tunnel, the upstream section is low-pressurized tunnel with plane bend, and there are gates in the middle. Water flows out of the gates, along a short horizontal tunnel, into an inclined part, then a short lower horizontal tunnel, and flips into river channel. Flows discharged from surface spillways and middle outlets collide vertically to dissipate energy, while flows from the left and right bank-run spillway tunnels collide horizontally to dissipate energy.

Xiaowan, Jinping I, Baihetan, Wudongde and other high arch dams also adopted such type of combined flood discharge arrangements.

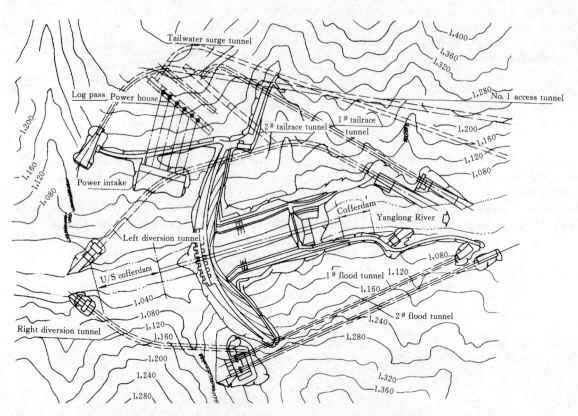

Fig. 17.2-7 Layout of Ertan Project and its Flood Discharge Structures

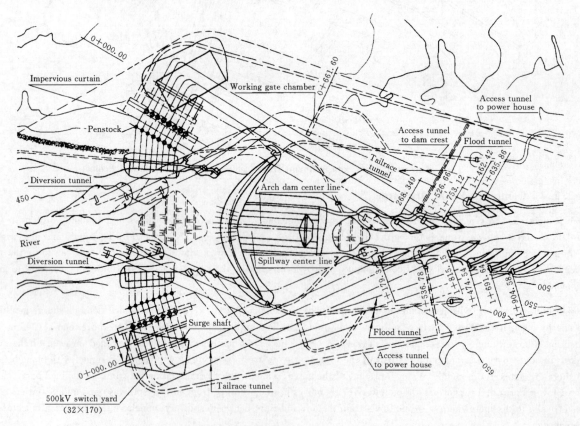

Fig. 17.2-8 Layout of Xiluodu Project and its Flood Discharge Structures

Discharge zones need to be well handled when adopting combined discharge arrangements. For dams on narrow valleys, discharged flood usually falls into the riverbed along the direction of the river. Proper distance needs to be kept between each of the energy dissipation zones. If the distance is too short, overlapping scours will cause instability to bank slopes. If the distance is too long, the discharge structure will have to be extended, increasing the quantity of protection project.

17.2.2.5 Discharge Structures Overlapped with Powerhouse

Overflow and flyover the powerhouses are the types with flood discharge structures overlapping with the powerhouse. When the valley is extremely narrow while the powerhouse needs to be arranged at the dam toe, the overflow and flyover types will be adopted. Xin'anjiang, Wujiangdu and Manwan projects, which were built during the 1960s to the 1990s, adopt such type of flood discharge and energy dissipation layout. However, it is seldom applied in the past two decades.

Xin'anjiang Project, completed in 1965, is China's first high and large-discharge dam that adopts overflow powerhouse. The dam, 105m high with 9 surface spillways to discharge 13,200m^3/s, has been in good operation for the past 40 years.

Wujiangdu Project adopts the flyover type of overlapping powerhouse, with the flow nappe of 4 surface spillways flying over the powerhouse into the riverbed. On both sides are ski-jump spillways (the left one goes above the auxiliary powerhouse), with spillway tunnels in both banks. The flip buckets of the discharge structures alternate with each other on a plane with different elevations. The nappes disperse in the air, fall into the riverbed along the longitudinal direction of the river, and cover most of the riverbed on different layers. From 1982 to 1989, the structures discharge 7,829m^3/s with the operational water level up to the design level, the maximum discharge per unit width 236m^3/(s · m), and cumulative discharge duration 10,143h. The safe operation proves that its layout of flood discharge structures is reasonable.

Manwan Project, completed in the 1990s, is another successful example. It adopts the flyover layout with large differential buckets. The nappes disperse horizontally, and fall into the plunge pool in longitudinal layers to dissipate energy. The structures have been tested by 30-year flood event. Especially in 1993 when the some gates had not been installed to the surface spillways, engineering measures were made, including proper other gates regulation, adding anti-crack reinforcement bars to the roof of powerhouse and using high grade concrete. The powerhouse withstood long-time impact of the jet nappes, and the non-normal discharge was safely conducted.

17.2.3 Reconstruction of Diversion Facilities into Permanent Flood Discharge Structures

17.2.3.1 General Introduction

Diversion facilities can be divided into two types: diversion tunnels and open channels. It is flexible for reconstructing open channels into flood discharge outlets (surface spillways, middle or deep outlets), silt sluices, or shiplocks. Diversion tunnels can be reconstructed into tailrace tunnels of powerhouse or spillway tunnels.

When the site topographic and geological conditions are sound, it is not only for reducing investment to rebuild diversion tunnels into flood discharge tunnels, but also for making the layout of the project more compact. In early time when diversion tunnels were reconstructed into discharge tunnels, the ogee section was usually adopted in front of tunnel. Later on, orifice plate becomes practically. In recent years, several vortex shaft spillway tunnels have been constructed with good results. Characteristics of diversion tunnels have a low elevation inlet and small gradient in tunnel. Thus when reconstructed into spillway tunnels, the high waterhead and flow speed shall be handled properly. In addition, it is difficult to arrange aerators; the outlet elevation is usually below the tailrace elevation or within the fluctuation range, so there would be hydraulic jumps in the tunnel if free flow pattern is adopted; if the outlet is underwater even in dry season, maintenance will be difficult. As a result, if there are diversion tunnels at different elevations, those at higher elevations will be prioritized to be reconstructed into spillway tunnels.

17.2.3.2 Reconstructing Diversion Tunnels into Spillway Tunnels

(1) Ogee section in the front of tunnel, such as Lubuge and Zipingpu projects. Aerators have been widely used since the 1980s. Basically the cavitation problems have been addressed for ogee-type discharge tunnels with waterhead less than 100m, but further research is needed for those with higher waterhead.

(2) Orifice tunnels, such as Xiaolangdi Project. Three orifice plates are arranged in the front of the horizontal section in each tunnel, with gate chamber in the middle, and followed by the open flow tunnel. When flowing through the orifice plate, water is first compressed abruptly, and then dispersed abruptly to cause strong rolling, shearing, and turbulence. Large amount of energy is dissipated, and flow speed reduced to an average of about 15m/s. The major problems for using orifice plates to dissipate energy are cavitation and sediment erosion, and the solutions are to arrange downward slopes on the roof of gate chamber to increase the pressure inside tunnel, optimize the shape of the orifice plate, and use

anti-erosion materials (Fig. 17.2-9). Since the orifice tunnels of Xiaolangdi Project were built, they have been tested under waterheads of 70m and 100m, and the hydraulic prototype observation results show that, the energy dissipation efficiency and pressure distribution characteristics of these three orifice plates are close to the results gained in physical modeling experiment, and the results of energy dissipation are good. Cavitation characteristics of the orifice plates have also been observed. When under an operational waterhead of 100m and the gates are fully open, there is an incipient cavitation noise observed in the No. 2 orifice tunnel. Inspection conducted after operation shows no cavitation damage, and the cavitation characteristics under a design operational waterhead of 120m are yet to be verified.

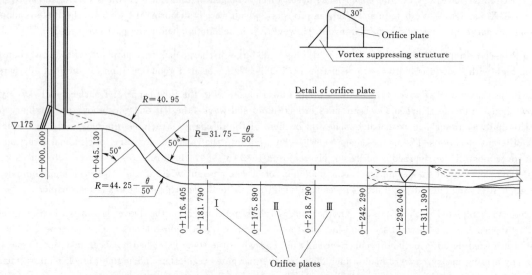

(a) Illustration of the orifice tunnels of Xiaolangdi Project reconstructed from diversion tunnels (Unit:m)

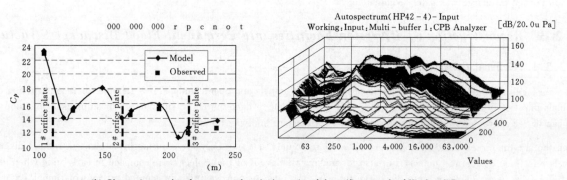

(b) Observation results of pressure and cavitation noise of the orifice tunnels of Xiaolangdi Project

Fig. 17.2-9 Xiaolangdi Project Reconstructed from Diversion Tunnels to Orifice Tunnels

(3) Vortex shaft spillway tunnel. A solution of "pressurized inlet and a short connection tunnel followed by vortex chamber, shaft and open flow tunnel" was proposed and put into practice for the reconstruction of diversion tunnels of Shapai Project in the 1990s. After Wenchuan Earthquake in 2008, this tunnel played its role. Although the operation flow is small, they played an important role in controlling reservoir water level and ensuring dam safety after the earthquake. In the early 21st century, a revised solution of "horizontal vortex spillway tunnel" (Fig. 17.2-10) was proposed, considerations given to the geological conditions, for the reconstruction of diversion tunnels of Gongboxia Project.

When reconstructing diversion tunnels of Gongboxia Project into discharge or emptying tunnels, the ogee section in the front of tunnel was initially adopted. However, after a collapse of tunnel roof during excavating, it was changed into vortex energy dissipation. Two solutions were compared: vortex chamber followed by shaft, and shaft followed by horizontal vortex chamber. Given the weak rock body at the inlet, the second solution was finally selected. A ring-shaped aerator was installed at the lower part of shaft, a vent shaft connected to the horizontal vortex chamber, and an aerator at the joint of the shaft and the horizontal section. These measures have greatly increased the aeration and protection of the shaft and horizontal vortex sections. To increase energy dissipation and adjust the flow direction after dissipation, a plunge pool was arranged inside the tunnel downstream the vortex chamber to make sure water flowing out of the pool smoothly through the open flow tunnel. The project was put into operation in 2006. Hydraulic prototype observation shows sufficient aeration in the shaft and the horizontal vortex section, and sufficient energy dissipation in the horizontal vortex chamber and the plunge pool, with more than 80% of energy dissipated. The video recorded by the cameras on top of the downstream tunnel of the plunge pool show that, in the process of opening and closing the gates, as well as during normal operation, water flowing out of the plunge pool is smooth without unfavorable flow pattern.

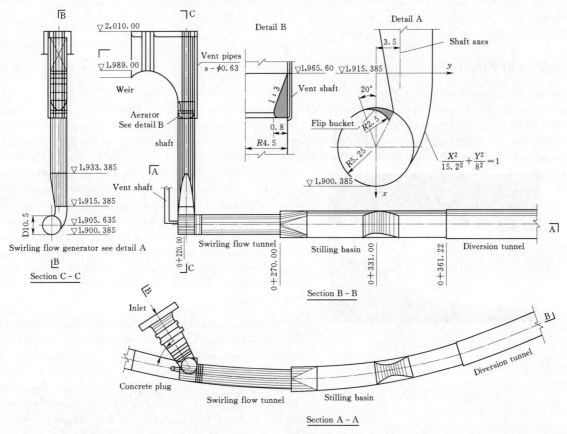

Fig. 17.2 – 10   Illustration of the Shaft, Horizontal Vortex Chamber and Open Flow Tunnel of Gongboxia Project

The orifice spillway tunnel of Xiaolangdi Project and the shaft and horizontal vortex spillway tunnel of Gongboxia Project are two successful examples that are reconstructed from diversion tunnels and dissipate energy inside the tunnels. In addition to increasing flood discharge and emptying flow (1,100m³/s to 1,700m³/s for each tunnel), because most of the kinetic energy has been converted inside the tunnel, the outflow has a low speed, the impacts of scour and atomization greatly reduced, and so it is an environment-friendly method of flood discharge and energy dissipation. Discharge for each tunnel is 1,100m³/s to 1,700m³/s, small for projects with a total discharge of 20,000m³/s, but considerable for projects with a total discharge of lower than 10,000m³/s. Even in China, projects constructed after 2020 will mostly have a total discharge of less than 10,000m³/s, so it is worth further study to adopt this technology.

For high earthfill or rockfill dams, out of safety concerns, emptying operation needs to be considered. Based on the experience gained from Wenchuan Earthquake, it is necessary to arrange emptying tunnels with inlets at low elevation. This can be done by reconstructing diversion tunnels into multi-purpose spillway tunnels for diversion, flood discharge and emptying. Two bottom outlet tunnels for flood discharge and silt sluice of Zipingpu Project have been reconstructed from diversion tunnels. After Wenchuan Earthquake, one of the tunnels was put into operation for flood discharge to control reservoir water level, ensure dam safety and supply water for Chengdu City downstream. Because the opening of gates is controllable, while the reservoir water level controlled does not increase, the reservoir served as an important water transport line for emergency rescue and relief of the reservoir area and the severely hit areas upstream the dam.

17.3 New Types of Energy Dissipator

17.3.1 *General Introduction*

To seek safe and economical solutions for flood discharge and energy dissipation, research and design institutes have worked out a variety of new energy dissipators. Because the research is done based on the requirements of project design, the technologies have been widely and quickly adopted. Up to now, a large number of projects have been completed with new energy dissipators, with prototype observation conducted for a part of them, and some of them tested by large or design floods.

New energy dissipators can be divided into contracted (slit flip bucket and flaring gate pier), dispersed (large differential flip bucket, upward and downward rams and tongue-shaped flip bucket, short side wall separate flip bucket), colliding nappes, nappes falling in layers, and plunge pool.

17.3.2 Slit Flip Bucket

The energy dissipation characteristics of slit flip bucket is that, by contracting water section abruptly at the tip of flood discharge structures, the streamline directions of nappes differ according to their elevations, the outflow angles change consecutively from about $-10°$ to about $45°$, the lower jets has a short throw distance while the upper nappes have the longest throw distance, the whole nappe is sufficiently dispersed in longitudinal and vertical direction, and so this technology fits well on narrow river valleys. Because jets of the nappe have different outflow directions, the nappe is "torn" in the air and aerated sufficiently, reducing the energy per unit cushion area of inflow. Compared to conventional constant-width bucket (with a bucket angle of $30°$) in the same hydraulic conditions, the depth of downstream scour hole is reduced by 50% to 67% compared with normal flip bucket. The shape of slit flip bucket should be selected according to the hydraulic, topographic and geological conditions. Shapes that have been adopted in projects include: rectangle, Y-shaped, and symmetric or asymmetric V-shaped (curved surface angle).

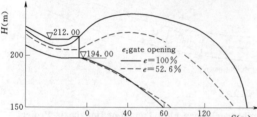

Fig. 17.3-1 Ski-jump Spillway and Flood Discharge in Dongjiang Arch Dam

Slit flip bucket is suitable for projects with a high or medium waterhead, particularly for dams on narrow valleys. It can not only be used for chute spillway and spillway tunnel, but also for surface spillway and deep (bottom) holes in dam body. The main requirement for adopting slit flip bucket is that the Froude number of inflow at the end of chute should be larger than four. When the Froude Number is too large and there is severe shock wave in the flip bucket, the chute can be contracted along the channel to make the Froude number of inflow drop to a proper level. The contraction ratio of slit flip bucket is normally between 1/6 and 1/3. The bucket angle should not be too large, normally between $-10°$ and $+10°$. A typical project is the ski-jump spillway on the right bank of Dongjiang Hydropower Project (Fig. 17.3-1). Prototype observation was conducted in October 1992 after the project was completed. The left-bank ski-jump spillway, left and right ski-jump spillways on the right bank were observed separately and jointly, with the opening of gates between 0.5 and 1.0, and maximum discharge on each $1,045 m^3/s$. The shape of nappe, aeration concentration, downstream scour and silting and other hydraulic factors of the prototype are basically consistent to modeling experiment. Results of modeling experiment can be used as reference for design.

In recent years, slit flip bucket is also adopted by the large spillway tunnels of Xiaowan Project with a discharge waterhead of more than 200m. The results show that it has many advantages compared to tongue-shaped or upward and downward ramps. The nappe falls well into the river reach, the energy dissipation result is good, and the depth of scour is remarkably reduced.

17.3.3 Flaring Gate Pier (Deflecting Flow, Plunge Pool, Bucket, Stilling Basin)

Flaring gate pier is another type new dissipator developed by China. A lot of research has been done since the 1970s, and has been widely used. Flaring gate pier can be divided into two types, i.e. combined with deflecting flow and bottom flow (or bucket flow).

Hydropower projects in China have the characteristics of large discharge flow. Due to narrow river valleys, the discharge per unit width is large. When the Froude Number is small, the energy dissipation efficiency is low, and so long structures are needed for energy dissipation. Several of projects built in China in the 1980s, such as Ankang, Yantan, and Wuqiangxi, were faced with this problem. To tackle this technical challenge, in-depth research has been conducted in Flaring gate pier dissipators, which have been widely adopted. So far, most of the projects have been completed and put into operation, with some having been tested by design flood.

Flaring gate pier with stilling basin. Take as an example the joint operation of flaring gate pier (surface spillway) and stilling basin. The width of nappe is contracted abruptly at the rear part of the pier, forcing the flow to disperse along the dam surface into a narrow and high contracted jet. When the downstream water level is proper, the jet falls into the stilling basin to form hydraulic jumps or submerged jumps, generating 3D strong turbulence with high aeration, which is different from the uniaxial spinning roller of the conventional 2D hydraulic jumps. Because the kinetic energy of flow is largely consumed through the increase of turbulent shearing, the bottom flow speed in the basin decreases dramatically, water surface is lifted and smooth, and the dissipation efficiency is greatly increased. If the water-free zone behind the flaring gate pier is used to arrange bottom outlets, a combined dissipator of "flaring gate pier (surface spillway) - bottom outlet (deflecting flow) - stilling basin" will be formed (Fig. 17.3 - 2). If the position, angle and flow quantity of the deflecting flow nappe falling into the hydraulic jumps are properly selected, both discharge capacity and the dissipation efficiency can be increased.

Flaring gate pier and deflecting flow joint dissipator is first adopted in Panjiakou Project, tested with the three surface spillways near the left bank, and has not been experienced by discharge. Geheyan, a gravity arch dam, adopts flaring gate pier on the surface spillway (the contraction ratio is 0.25, flow connection is deflecting flow, so this can also be seemed as a type of slit flip bucket), with downstream plunge pool for energy dissipation. After put into operation in 1996, prototype observation data of surface spillways was gained.

Flaring gate pier and plunge pool is first adopted in Ankang Project with the following characteristics: ①3 - D hydraulic jumps are formed in the plunge pool, dissipation efficiency increased particularly when the Froude Number is small; ②the second conjugate water depth and jump length of flaring gate pier and plunge pool joint dissipator are dramatically decreased compared to conventional 2D hydraulic jump plunge pool. The operation shows good dissipation result and low atomization impact.

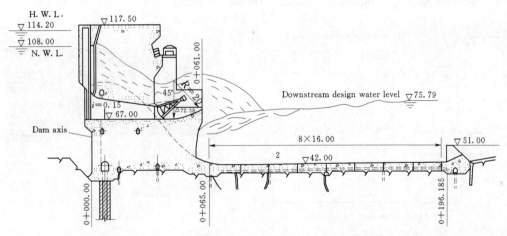

Fig. 17.3 - 2 Flaring Gate Pier (Surface Spillway) -Bottom Outlet (deflecting flow) and Stilling Joint Dissipator

Flaring gate pier and bottom outlet combined with stilling basin joint dissipator. Wuqiangxi Project adopts a joint dissipation solution of "flaring gate pier on surface spillway combined with stilling basin" (three outlets on the right side) and "flaring gate pier on surface spillway and bottom outlet combined with and stilling basin" (six outlets on the left side). The overflow dam has 9 surface spillways. The middle outlet flows are separated into three groups on the right side and six on the left side, with a total width (including the middle outlet) of 244.5m. the radial gate of surface spillway is 19m×23m (width× height). The pier is 5.5m thick, the widening angle of gate pier 16.7°, and the contraction ratio of overflow orifice width 0.368. The downstream stilling basin is 120m long, and the rear part of the pool is Rehbock bucket with different elevations, joined by concrete apron downstream. Below the 5 middle piers of the overflow dam are 5 discharge bottom outlets, the size of their inlets 2.5m×16m (width×height), and controlled by radial gates in the same size.

The project was experienced by unusual floods consecutively in 1995 and 1996 when the construction was still not completed, and the bottom slab of the stilling basin was severely damaged caused by high fluctuating pressure. Attentions need to be paid to the experience.

Flaring gate pier and stilling basin joint dissipator. The surface spillways of Yantan Hydropower Project adopt the Y-shaped flaring gate pier, and connected by downstream stilling basin. The surface spillways have been put into operation since May 1992 with discharge ranging from 3,000m³/s to 15,000m³/s. All the 7 surface spillways have been operated for flood discharge, and the Number 2 to 5 surface spillways have the longest operational duration of about 3,000 hours. The dissipator has been tested by small and medium discharge (10-year event, with discharge of surface spillways 14,760m³/s) with good flow pattern and energy dissipation.

Joint application of flaring gate pier and stepped spillway. In the recent 20 years, while RCC construction technology is fast and widely adopted in dam engineering, energy dissipation using the stepped spillway formed by roller construction has been widely adopted. Its biggest advantage is that, when water overflows the steps, energy dissipation efficiency is increased while the second phase of concrete construction can be canceled, accelerating construction progress. As a result, aeration and energy dissipation mechanisms of water flowing over steps have become a hot topic for research. Compared to smooth dam surface, the research priority rests on the identification of the critical aeration condition of flow so as to ensure the safety of the bench.

Shuidong is China's first project that adopts a joint application of flaring gate pier and stepped spillway, with the maximum discharge per unit width $120m^3/(s \cdot m)$. Because the dam is only 57m high, high-speed flow is not a prominent problem. Dachaoshan RCC gravity dam is one successful example adopting this joint dissipator. The project has the maximum dam height of 111m, its 5 surface spillways adopting the Y-shaped flaring gate pier, downstream dam surface being RCC steps, and the maximum discharge per unit width at the crest being $230m^3/(s \cdot m)$. Because its dam height greatly exceeds that of Shuidong Project, its discharge per unit width is almost twice as large as that of Shuidong. To avoid cavitation damage to the stepped spillway, the first step of the outlet of flaring gate pier is raised to be twice as high as conventional steps. This flips the flow away from the dam surface to form a stable cavity between the flow nappe and the steps and ensure there is sufficient aeration for steps to avoid cavitation damage. During the first flood season (in 2002) after the project is completed, the hydraulic prototype observation on discharge is conducted under the design water level with the discharge per unit width of the surface spillways as large as $150m^3/(s \cdot m)$. The result shows that such type of energy dissipator is successful. The aeration concentration measured on the steps is as high as 30% to 50%, effectively protecting the steps (Fig. 17.3 - 3). After Dachaoshan Project, Suofengying RCC Dam also adopts the X-shaped flaring gate pier. Together with the downstream stepped steps, the energy dissipation efficiency is good.

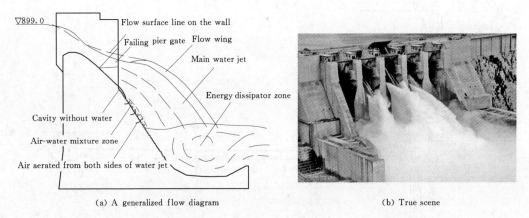

(a) A generalized flow diagram   (b) True scene

Fig. 17.3 - 3   Flaring Pier Gate and Stepped Spillway in Dachaoshan

17.3.4   *Dispersed Flow (Large Differential Flip Bucket, Upward and Downward Ramps and Tongue-shaped Flip Bucket, Short Side Wall Separate Flip Bucket)*

Large differential flip bucket and heterotypic flip bucket are developed based on the conventional continuous flip bucket and differential bucket. By using the elevation, position and shape of differential bucket, the flow nappes are separated, dispersed and falls into the proper area of downstream riverbed while avoiding cavitation. Large differential flip bucket and heterotypic flip bucket can be divided into the following types: large differential flip bucket, upward and downward ramps and tongue-shaped flip bucket, and short side wall separate flip bucket.

Conventional continuous bucket (constant-width flip bucket). When the discharge per unit width is large, the scour holes in the downstream riverbed will be deep. Differential flip bucket increases the dispersion of the nappe flows. However, if its shape is not properly designed, cavitation damages usually occur on flip bucket itself. Large differential flow between flip buckets is widely adopted because it overcomes this shortcoming.

Dispersion at outlet. By changing the shape of flip bucket, the flow nappe is dispersed and falls into the proper area of the riverbed. Shapes commonly adopted include separate flip bucket, upward and downward ramps, tongue-shaped flip bucket, upward and downward tongue-shaped ramps, and the short side wall used in the above-mentioned flip buckets (Fig. 17.3 - 4).

Dispersion at multiple outlets and large differential flip bucket. When the spillway has more than two (or channels), by changing the elevation, position and shape of differential bucket, the flow is dispersed and falls into the proper area of

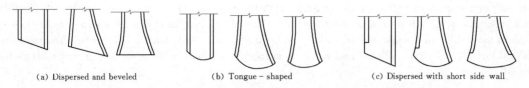

(a) Dispersed and beveled   (b) Tongue-shaped   (c) Dispersed with short side wall

Fig. 17.3-4   Illustration of Various Types of Dispersed Flip Buckets

downstream riverbed while avoiding cavitation.

The five surface spillways and six middle outlets of Ertan Arch Dam have different elevations at the outlets. With different bucket angles, flip buckets are largely differentiated, while deflecting flow is also diffused in different degrees on the plane, to make the flipped flow enclose the riverbed and cut off backflow. The structures were put into operation in 1999 with good dispersion of the nappes and energy dissipation result.

17.3.5   *Joint Application of Colliding Flow Nappes, Falling in Layers and Plunge Pool*

This new type of energy dissipation is developed for high concrete dams, particularly high arch dams, which have large quantity of flood discharged through dam body, on valleys. The dissipation mechanism is by properly arranging the outlets on different layers in the dam body, the flow nappes discharged are layered with large differentials, outflows on different layers collide with each other in the air, and the collision angle is close to 90° to maximize the energy dissipation. At the same time, the shapes of flip buckets are optimized to maximize the longitudinal dispersion of jet flows and reduce the impact hydrodynamic pressure on the plunge pool bottom slab; the secondary dam forms a downstream plunge pool to provide sufficient room for turbulence, dispersion, collision and energy dissipation of the jet flows, so that the effective energy of the jet flow is dissipated through strong turbulent shearing, dispersion and rotation of the mainstream jet. The length of the pool is usually determined by the farthest distance of the nappe of middle outlets, while for the depth of pool and the design of bottom slab, an average impact pressure less than 15×9.8 kPa is adopted in China as the major controlling indicator. Compared to the conventional and common dissipators, the characteristics of jet flow are highlighted by the flow structure in the plunge pool, while the effect of fluctuating pressure highlighted by the pressure on the bottom slab. Methods to protect plunge pools include comprehensive lining, and protection of the banks instead of the bottom (partial reinforcement might be done according to the geological conditions of the bottom slab). When the pool is deep and the base rock is sound and has high anti-scour capacity, project quantity can be reduced to protect the banks instead of the bottom. Practices show that, for high concrete dams on valleys, because the project quantity of bank protection of energy dissipation area is close to that of pool, this type of dissipator is safe, reliable and economical. However, due to the collision of jet flows discharged from surface spillways and middle outlets, splashing and atomization in the downstream areas are heavy when discharging.

These projects have characteristics in common: they are all situated in deep valleys, their dam height 240m to 300m, discharge capacity more than 10,000m³/s, discharge waterhead more than 200m, and the river valley 100m wide under frequent flood. To avoid severe scour of the river channel downstream the dam site when, large plunge pool is arranged downstream of the dam, while secondary dam is used to form large cushion for energy dissipation. One of the protective methods for Wudongde high arch dam being designed is to protect the banks instead of the bottom, given the downstream water cushion is more than 100m deep. So far multiple solutions are being compared and studied.

Proper optimization of the layout of surface and middle (deep) outlets and the outflow bucket angle can reduce the impact hydrodynamic pressure on the bottom slab of the plunge pool, shorten the length of pool, increase energy dissipation rate, and enhance the discharge capacity of dam body. Such optimization has been made for Goupitan, Xiluodu, Baihetan and other projects to increase dam body discharge capacity by about 10%, reduce the scale of bank-run spillway tunnels, and cut the project investment.

The flood discharge structures of Ertan Arch Dam consist of 7 overflow surface spillways (11.5m×11m) and 6 middle outlets (5m×6m) (Fig. 17.3-5 (a)). Downstream the dam is the plunge pool with entire concrete lining. The profile is an inverse trapezium compound cross-section, with a bottom width of 40m and a length of 320m. A Secondary dam is arranged at the end of the plunge pool, with a maximum height of 35m, a maximum dam body discharge of 16,300m³/s, and corresponding discharge power of 26,500MW. Because the surface spillways in the dam body adopt "large differential and dentate drop flow" arrangement while the middle outlets are upwards, the flow nappes of surface spillways and middle outlets collide in the air and diffuse, and so the energy dissipation result is good. The length of plunge pool is controlled by "the distance of nappe of middle outlet discharging only" plus "the length of submerged hydraulic jump when the nappe falls into the downstream water body," while the maximum time-averaged impact pressure on the bottom slab of the pool controlled by the behavior of a single discharging surface spillway, with the maximum impact pressure being 10×9.8 kPa. Prototype observation conducted in 1999 at the design water level of 1,200m shows that the maximum average impact pressure on the bottom slab of the pool when a single surface spillway is discharging is 5.9×9.8 kPa, smaller than the modeling experimental value (Fig. 17.3 - 5 (b)). Plunge pool has good energy dissipation result, while flow downstream the secondary dam is smooth.

Since the successful application of flood discharge and energy dissipation method in dam body in Ertan Arch Dam, this method has been widely adopted in a large number of high arch dams with large discharge flow in China, and the dam body discharge capacity is accounting for 60% to 80% of total discharge.

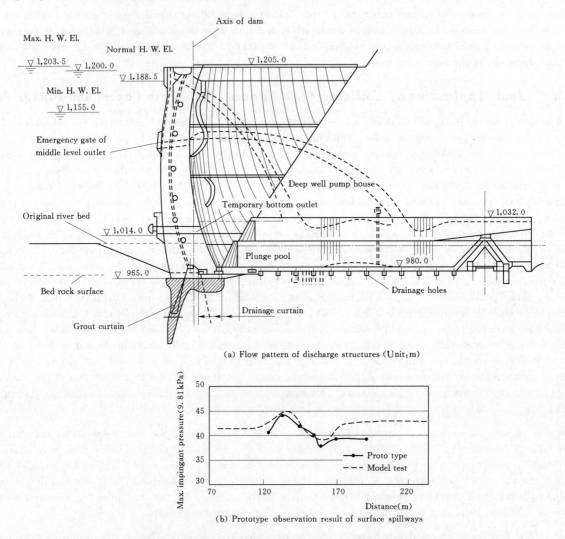

Fig. 17.3 - 5 **The Flow Pattern of Flood Discharge and Energy Dissipation of Ertan Dam**

The dam body flood discharge and energy dissipation scale of Xiluodu Arch Dam is the largest among the arch dams built and under construction in China and in the world (Fig. 17.3 - 6). Table 17.3 - 1 gives a list of high arch dams built and under construction in China in recent years that adopt the "deflecting and drop flow combined with plunge pool" type of energy dissipation via dam body, as well as the major technical specifications. In terms of discharge waterhead, Xiluodu, Xiaowan and Laxiwa are on the same level of 200m or high. However, in terms of dam body discharge under check floods, the maximum dam body discharge of Xiluodu Arch Dam exceeds 30,000m³/s, with the corresponding discharge power exceeding 58,000MW.

Research on the flood discharge and energy dissipation for Xiluodu Arch Dam includes the following main aspects:

(1) To optimize the layout of surface spillways and middle outlets. While the hydrodynamic pressure on the bottom slab of the plunge pool is not large than 15×9.8 kPa, discharge via dam body is maximized to reduce the number of spillway tunnels. After optimization, the maximum dam body discharge reaches 31,500m³/s, an increase of about 10% than the feasi-

bility design value, and reducing one bank-run spillway tunnel.

Fig. 17.3-6 Layout of Dam Body Flood Discharge and Energy Dissipation Structures of Xiluodu Arch Dam

(2) To optimize the bucket shape and angle of both surface and middle outlets, so that collision of nappes in the air can bring better energy dissipation results and falling zones while the impact pressure on the bottom slab can be reduced. After optimization, while the surface spillways adopt the "large spatial differential and tongue-shaped separate bucket", the middle outlets have optimizations of the plane and the outlet bucket angle so that energy dissipation results from nappe collision can be improved, and that flows from the surface and middle outlets can be divided and fall into three zones in the plunge pool.

(3) To optimize the length of the plunge pool and the height of the secondary dam. While the need of flood discharge and energy dissipation is fulfilled, the length of plunge pool, as well as the project quantity, is minimized, and scour of plunge pool outflow on downstream river channel is reduced. After optimization, the length of the plunge pool is 400m, the method of comprehensive lining is adopted to protect the bottom slab, and the height of the secondary dam is 45m.

After comprehensive research and optimization, although Xiluodu high arch dam has the largest scale of dam body discharge and highest technical specifications, good energy dissipation results are achieved, and the maximum hydrodynamic pressure on the bottom slab is under the design controlling value.

Table 17.3-1 Comparison of Dam Body Discharge and Energy Dissipation Solutions for High Arch Dams

Project name	Dam height (m)	Working condition	Discharge waterhead (m)	Total discharge (m³/s)	Dam body discharge (m³/s)	Dam body discharge power (MW)	Energy dissipation rate per unit water body (kW/m³)	Maximum hydrodynamic pressure (×9.8kPa)	Size of plunge pool (m)		
									L	t	B/b
Ertan	240	Check	166.3	23,900	16,300	26,600	13.5	14.1	330	57	126
		design	166.3	20,600	13,200	21,500	11.5			54	40
Xiaowan	294.5	Check	226.6	23,600	15,260	33,900	12.3	11.3	400	48	180
		design	221.8	16,700	9,060	19,700	8.2			42	
Laxiwa	232.5	Check	213	6,310	6,000	12,500	20.5	11.5	217	36	104
		design	213	4,250	4,000	8,350	16.8			30	60
Goupitan	232	Check	148.3	35,600	26,420	38,400	15.3	14.5	311	77	140
		design	150.4	27,900	21,470	31,600	13.4			72	70
Xiluodu	285.5	Check	188.6	50,900	31,496	58,750	10.3	14.2	400	80.8	225
		design	188.1	43,700	23,686	44,540	9.8			76.2	107

Note: discharge power $N=\gamma QH$; $L=$ length; $t=$ dam height of secondary dam; $B/b=$ top/base width ratio.

17.4 Issues of High-speed Flow

With rapid hydropower development in China, dams are higher and higher, and the issues of high-speed flow are becoming more and more prominent, causing special flow phenomena such as aeration, atomization, fluctuation, induced vibration, and cavitation damage. As a result, the issues of high-speed flow attract great attentions.

17.4.1 *Free Flow Aeration, Freeboard of Chutes and Spillway Tunnels*

When flows are moving at high speed, natural aeration will happen because the bottom turbulent flow develops from the boundary layer to free surface and vise verse. Aeration causes changes to the physical characteristics of the flow. For example, to aerate forcibly between flow and solid boundary can reduce cavitation, and so it is widely adopted. However, the volume of water body expends when flow is aerated, increasing water depth. As a result, the free board of 0.5m to 1.5m needs to be added to the elevation of the chute side wall according to curve of the water surface after fluctuation and aeration. For complex regional flow, hydraulic modeling experiments need to be conducted to determine the curve of water surface.

The operation of Ertan spillway tunnels has provided precious experience for proper design of the freeboard of open tunnel. According to China's hydraulic tunnel design code, the clear freeboard of high-speed tunnel above aerated water surface should be 15% to 25% of tunnel area. For long tunnels, except for using the gate shaft for aeration, more air should be added at proper positions. According to the operation of Ertan spillway tunnels and the prototype observation, to ensure the operational safety of such extraordinarily long and large spillway tunnels, there should be sufficient space above water surface, and the upper limit value should at least be adopted.

17.4.2 *Atomization of Discharge, Prevention and Treatment*

In recent years, deflecting and drop flow methods of energy dissipation have been widely adopted for high arch and gravity dams. However, these two dissipators can cause serious problems of atomization, and sometimes even damages. As a result, the atomization issue of discharge structures has attracting increasing attentions. By conducting prototype observation and analysis of discharge atomization on several large projects, discharge atomization is defined as the whole process of flows breaking away the boundary under the effects of inertia and gravity, flying in the air at high speed, diffusing, dropping into water to cause oversplash that rises into the air and diffuse, mixes with the air and form water mist. The strength of atomization caused by the nappes is low, and thus not the major source of the mist. When the nappes drop into the water cushion at high speed, the force of impact or oversplash is enormous. Part of the splashed water body rises into the air and form water mist, whose height is usually close to the maximum height of the nappes. This is the major source of the mist.

For the convenience of research, the atomization area is commonly divided into three zones according to the patterns of atomization: area of mist from splash of nappes, atomized precipitation area, diffusion area of mist. In addition, the water wings breaking away from the main nappe have strong scour force when dropping onto the bank slopes, but it is out of the research scope of atomization.

Since the 1960s, China has begun the research on discharge atomization. Limited by means and methods, the research is mainly qualitative analysis, combined with some prototype observation. In the recent 20 years, the major research methods of discharge atomization are mathematic modeling, physical modeling and prototype observation.

At present stage, prototype observation is the major method to research on atomization in China. In the recent 30 years, systematic prototype observation of discharge atomization has been conducted for Huanglongtan, Fengtan, Baishan, Zhexi, Liujiaxia, Wujiangdu, Lubuge, Lijiaxia, Dongfeng, Ankang, Manwan, Ertan and other hydropower projects to observe and investigate the atomization scope, precipitation strength and the damage.

Through research and prototype observation, some common understanding has been gained on discharge atomization:

(1) The area, shape and strength of atomization areas are affected by discharge capacity, waterhead, energy dissipation method, depth of water cushion (these factors determine the source and size of atomization), topography of downstream valley, and meteorological conditions (wind direction and speed, atmospheric pressure). When gully develops downstream and the oversplash area of inflow nappe is near the gully, water mist will go up along the gully to form a runoff concentration, and the stability of accumulation body and rock mass should be paid attention to. In the same discharge condition at the same observation spot downstream the nappe, when the natural wide has the same direction as the nappe wind, the precipitation is strong, and vice versa. As a result, the precipitation of atomization area is a non-constant field, and attentions

should be paid to the impact of atomization when designing the layout of a project.

(2) The mist is mainly from the oversplash of the nappe dropping into water cushion. The splashed water body (water mist) can reach the highest point of the nappe, while the mist can go up above the dam elevation. Strong precipitation occurs in the area downstream and on both sides of the splashed water body (water mist), and the distance extending toward the downstream area is longer than that diffusing toward both sides.

(3) Generally speaking, as the discharge capacity, waterhead and discharge increase, the maximum precipitation strength in atomization area will also increase.

To reduce the impact of discharge atomization, the nappe area (including water wings) should be arranged in the cushion of the riverbed as much as possible; considering that aeration and dispersion of nappes in prototypes are sufficient, enough margin should be put aside when designing; the dropping area of starting and ending nappe flows should be on hard and sound rock or concrete protective zone, and there should be good drainage for water easily back to river channel; there should not be any power transmission lines or telecommunication lines in the space crossed by nappes; when it is impossible for roads to avoid strong atomization areas, transport gallery with mist protection should be installed; for projects that are sure to have environmental impacts upon downstream residential areas, the discharge method should be changed. For example, because there is a chemical enterprise and large residential area less than 1,000m downstream Xiangjiaba Project on the right bank, in order to reduce the impacts of discharge atomization on this area, the original flip bucket is changed into stilling basin.

Fig. 17.4-1 is the distribution of atomized precipitation strength observed in 1999 when Ertan Arch Dam was discharging. When in operation, the reservoir water level was close to the design level, with 4 surface spillways and 4 middle outlets to perform a joint discharge, and the discharge was 7,757m³/s. The hydraulic and atomization characteristics include: ①it is the highest dam (240m) built and in operation in China; ②it has the largest discharge capacity, and the collision of nappes from surface and middle outlets cause discharge atomization; ③strong atomization precipitation area is limited to both sides downstream the dam to a certain scope downstream the plunge pool, asking for proper protection based on the geological conditions of both sides; ④there should not be important facilities on both sides of the plunge pool and within a certain scope downstream the pool; ⑤this area has strong precipitation and high-speed nappe wind, so tunnels should be used as the transport facilities. The observation results of discharge atomization of Ertan dam body provide helpful technical support for other high arch dams to analyze the impact of discharge atomization and study the prevention measures, particularly for Xiaowan, Xiluodu, Baihetan and Wudongde arch dams that have a dam height of about 300m.

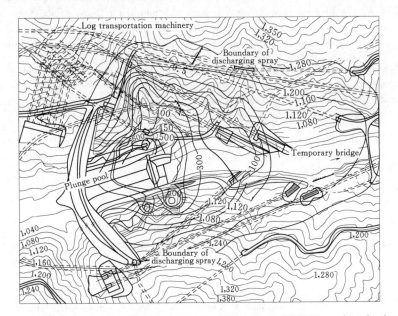

Fig. 17.4-1 Precipitation Intensity Distribution of Discharge Atomization Caused by the Discharge of the 4 Surface Spillways and the 4 Middle Outlets of Ertan Arch Dam in 1999

Apart from the above-mentioned engineering measures to prevent discharge atomization, numerical simulation forecast is also widely conducted. Through mathematic modeling, a large number of prototype observation information is used as the calibration and validation parameters for numerical modeling, and forecast can be made on the distribution and affected area of atomization precipitation for projects in design and under construction stages. So far impressive progress has been made. For example, the forecast of the intensity and distribution of discharge atomization precipitation when Xiaowan Project adopts different combination of flood discharge methods has provided effective technical support for optimal design of preventive and protective measures and operational management of projects. Numerical forecast has become one of the important methods for high dams in valley areas in China to design preventive and protective measures against discharge atomization and operational management.

17.4.3 Cavitation and Aeration

17.4.3.1 Roles and Positions of Aerators and Cavitation Mitigation Measures

Liujiaxia spillway tunnel and Longyangxia bottom outlet tunnel are the typical cases of China's first projects damaged by cavitation in the 1970s and 1980s. Cavitation damage occurs at the lower reach of the ogee section of Liujiaxia spillway tunnel, and there was no aerator arranged due to limited knowledge of cavitation damage. Similar damage occurs at the downstream side wall behind the radial gate of Longyangxia bottom outlet tunnel because of the "rising offset" effect caused by unevenness of construction. Afterward, a lot of research and prototype observation are conducted on aeration and cavitation mitigation, and breakthrough has been made in both theory and practice. Aerators were installed to a large number of spillway tunnels and spillways with good operational results. According to China's design code, when the flow speed exceeds 30m/s and the cavitation number smaller than 0.2, aerators shall be installed.

17.4.3.2 Type and Design of Aerators

The purpose of installing aerators is to increase the aeration concentration of flows near the walls, so that the surface of solid walls can be protected, and cavitation damage mitigated or avoided. As a result, ventilation in aeration facilities should be unobstructed, wind speed in the aeration vent should not be too high (when the wind speed is too high, the sound noise might affect safety), the negative pressure in the cavity should be controlled properly (when it is too high, it means air is not sufficient, and the size of the aeration pipe cross-section is not large enough), and disturbance of flow, while requirements are fulfilled, should be avoided as much as possible to avoid or reduce adverse impacts. The facilities should be matched with the shape of flood discharge structures to fit into different conditions. With aeration facilities, it is easier to fulfill the demand of operation.

Aerators can be divided into different types: ramp, offset, and slot that combines the above-mentioned two types. The most adopted type is the offset or combined with ramp, such as the surface spillway tunnel of gravity dams, chute spillway and spillway tunnels that have large gradient. All the different types of aerators have been adopted in China's flood discharge structures. No matter how they are combined, the offset heights are mostly ranging from 20cm to 160cm. After operational test and prototype observation, it is proved that the aerators play an important role in protecting flood discharge structures.

In terms of the types, overflow surfaces and chute spillways with large gradient can adopt offset and slot type; tunnel or chute slop with small gradient can adopt ramp, differential or ring-shaped ramp; when the flow is large, gradient is steep, and the water depth is large, two-step aerator can be adopted. In order to deal with the curved surface water-stop problems of high-waterhead radial gates, the sudden enlargement type is developed and has been widely adopted.

The sudden enlargement type of aeration facility that fits well with curved surface water-stop issue of radial gates has been widely adopted in lot of projects, including the emptying tunnels of Dongjiang Project, deep outlets of Longyangxia Project, emptying tunnels of Tianshengqiao I Project, and the orifice tunnels of Xiaolangdi Project. The offset on bottom is 80cm to 150cm high, and the side enlargement is about 40cm to 60cm. Most projects have been tested with good aeration results.

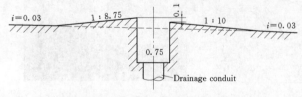

Fig. 17.4-2 Upward and Downward Ramps

When the spillway tunnels or discharge channels have a small gradient, it is not suitable to use the conventional ramp-slot type because slot is easy to be filled by water. By conducting experimental research on the aeration facilities of Longyangxia bottom outlet tunnel, a upward and downward ramp combined with slot is proposed for the chutes with small gradient (Fig. 17.4-2).

Because flow speed is high in the spillway tunnels of Ertan Project, the conventional offset is not effective. A special differential offset aerator is proposed to enhance aeration (Fig. 17.4-3 (a)).

17.4.3.3 Research and Application of 3D Aeration and Erosion Reduction Facilities

In the 1980s to 1990s, aeration in China is mainly bottom slab (i.e. 2D aeration). Then in 2001 cavitation damage occurred

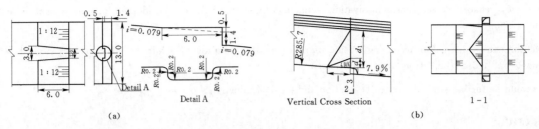

Fig. 17.4-3 Ertan 2D Differential Offset Aerator (a) and 3D Aerator (b)

in the No. 1 spillway tunnel of Ertan Project. Through profound analysis, it is believed that, for large spillway tunnels or spillways of dams higher than 200m, because the flow speed of bottom slab is more than 35m/s, and the water depth is usually more than 8m, it is also important to protect the bottom part as well as the side walls through aeration. Research on 3D aeration is conducted with the repair of Ertan No. 1 spillway tunnel. In the 2D model, the side wall 5m to 20m downstream the rear part of the ogee section and within 7m above the bottom slab is unaerated. The aeration concentration measured there is very low. In the 3D modeling experiment, the side aeration cavity is very stable, the maximum length of side wall bottom cavity is 23m to 25m, and the maximum aeration concentration at the points 15m and 25m from the aerator, and 2.8m above the bottom slab, are 86% and 43% respectively. The 3D aeration type recommended by research institute is adopted by design. After repair, hydraulic prototype observation was conducted in 2005 of the No. 1 spillway tunnel. Take as an experiment the working condition with a water level of 1200m (design water level) and a full opening of the gate (discharge in one tunnel is 3,600m^3/s). The aeration concentration at the point 5m from the end of aerator and 3.5m above the bottom slab is 83%. Hydrophone is installed on the side wall 25m from the aerator and 3m above the bottom slab. The noise level difference measured is 1.0dB to 3.5dB without cavitation. The modeling experiment and prototype observation prove that the 3D aeration and erosion reduction measures are effective, and the research has filled the gap in this field (Fig. 17.4-3 (b)).

When reconstructing diversion tunnels into flood and sediment sluice tunnels at Zipingpu Project, the 3D aerators are developed into slight slope spillway tunnel. And at Gongboxia Hydropower Station, it is adopted to rebuild the division tunnels into the horizontal vortex spillway tunnel and a ring shape aerator installed in the shaft tunnel. Both projects have been tested by flood discharge, and prototype observations also conducted. The results show that these 3D aerators are effective and enlarge its application.

17.5 Conclusion

Design or early-stage researches have been basically completed for the large and medium sized hydropower stations that are to be built before 2020. In addition to further address the technical problems of these projects, basic and applied research needs to be conducted as early as possible for dams to be built from 2020 to 2050. In the view point of flood discharge and energy dissipation, the characteristics of dam construction are:

(1) Dam constructions from 2020 to 2050 will concentrate on Jinsha River, Lancang River, Dadu River, upper reach of Yalong River, Yarlung Zangbo River, and upper and middle reaches of Nujiang River. The more dam heights will mainly be 50m to 150 m but there will also be high dams of more than 200 m so that the high speed flow will also be the key issue in the design.

(2) There will be few projects with their discharge exceeding 20,000m^3/s, but the most common is that the dam site is located in narrow valley. So it should be paid attention to the problem of flood release and energy dissipation.

(3) The altitude in dam sites will be high, with most of the projects being at 2,500m to 3,700m, and some situated on perpetually frozen soil, which has great impacts on flood discharging atomization. And it should be paid more attention.

(4) Some projects are near the natural reserves, and so more stringent requirements are for ecological and environmental protection.

In terms of research on flood discharge and energy dissipation technology, attentions need to be paid to the following aspects:

(1) For high arch dams, effective dissipation method that has less atomization impacts, and that combine plunge pool without collision of nappes, need to be studied.

(2) The safe, reliable and economical dissipation technology by stilling basin can be adopted for gravity dams with a height of 100m to 200m.

(3) Existing energy dissipation technology inside tunnels need to be improved, while new technologies studied, including new tunnels and those rebuilt from diversion tunnels.

(4) It should be further studied on the 3D aeration and erosion reduction.

References

[1] Gao Jizhang, Liu Zhiping, Guo Jun. Energy Dissipation and High Speed flow in High Dams//Editor-in-Chief: Pan Jiazheng and He Jing, *Large Dams in China: A Fifty-Year Review*. Beijing: China WaterPower Press, 2000.

[2] Peng Cheng, Chief Editor, *China's Hydropower Projects in the 21st Century*. Beijing: China WaterPower Press, 2006.

[3] Zhou Jianping, Niu Xinqiang, Jia Jinsheng, Chief Editor. *Two Decades of Designing Gravity Dams*, Beijing: China WaterPower Press, 2008.

[4] Editing Committee. *Development of Dam Engineering in China and Examples of Projects*. Beijing: China WaterPower Press, 2007.

[5] Editing Committee. *Scientific and Technological Frontline of Modern Water Resources*. Beijing: China WaterPower Press, 2006.

[6] Zhou Xiaode, Li Guifen, Wang Lianxiang, Wang Dongsheng, Chief Editor. *Progress of Hydraulics and Hydroinformatics in* 2009. Xi'an: Xi'an Jiaotong University Press, 2009.

[7] Li Guifen, Wang Lianxiang, Li Jia, Chief Editor. *Progress of Hydraulics and Hydroinformatics in* 2005. Chengdu: Sichuan University Press, 2005.

[8] Li Guifen, Wang Lianxiang, Chief Editor. *Progress of Hydraulics and Hydroinformatics in* 2003. Tanjin: Tianjin University Press, 2003.

[9] Guo Jun. Recent Achievements in Hydraulic Research in China//Editor-in-Chief: Ali Sayigh. *Comprehensive Renewable Energy*. Oxford: Elsevier Ltd., 2012.

Chapter 18

Underground Hydraulic Structures

Wang Renkun[1], Chen Zhonghua[2], Chen Ping[3], Cui Wei[4], Liao Chenggang[5], Xing Wanbo[6], Fu Yingru[7]

18.1 Introduction

As a renewable resource, hydropower resource is rich in China, with the hydropower capacity of 541,000MW for economic development ranking first in the world. And China's hydropower resources mainly concentrate in the west region, especially the alpine and gorge valleys in southwest region that is usually an appropriate choice to build high dam, underground hydropower station and long diversion tunnel, as concentrated high head drop is available. These large-and super-large-sized hydropower stations have the characteristics of high design head and large flood discharge, so the flood discharging tunnel with large cross-section or construction diversion tunnel need to be built due to the narrow river valley. Meanwhile, as much as possible to make full use of hydropower resources, small-and medium-sized hydropower stations have been built widely; many of them choose the structure of long diversion tunnel or underground powerhouse. The development of hydraulic underground structures has been effectively promoted in our country for nearly 30 years by the increasing demand for the pumped storage power station. China is a large country with huge population, where water resources are relatively poor and uneven distributed, water shortage is severe in arid regions especially in northern droughty areas and in the industry, agriculture and urban life, so it is necessary to build many of inter-basin water transfer project. The east-route and middle-route parts of national south-to-north water transfer project have commenced at the end of the 20th century, and west-route part has begun to plan, as well as multiple provincial water transfer projects have been building at around the same period, including many long distance water conveyance tunnel engineering through the watershed and the mountainous regions. China's hydraulic underground engineering construction has achieved remarkable development in the last century, since the founding of new China's over 50 years. *Large dams in China—A Fifty-year Review* (Chapter 11) has an overview on the development, practical technology and technological innovation of the hydraulic underground structure in China.

Along with the sustained economic growth since the 21st century, the hydraulic underground engineering construction is also developed by leaps and bounds in China. In the end of 20th century, China's total installed capacity of hydropower has been ranked first in the world, which is more than 200,000MW. Since the construction sites of hydropower stations focus further on high valley in southwest China with abundant water, the underground hydropower stations have significant breakthrough in the number and size. The newly-built Longtan underground hydropower station has a total installed capacity of 6,300 (9×700)MW, far more than the Ertan hydropower station of 3,300 (6×550)MW which used to ranking with first place. There are more than a dozen or more underground powerhouses of large-sized hydropower stations being constructed or proposed excluding the pumped storage power station, and the Xiluodu hydropower station is under construction, with a total installed capacity of 13,860 (18×770)MW ranking 3rd in the world. In these projects, the underground cavern groups with the large underground hydropower stations as the core have large scale and complex structure, some may be staggered arrangement with the diversion tunnel and flood discharging tunnel in the space, which are unprecedented century projects. With the rapid development of small-and medium-sized hydropower station in the recent ten years, the number of un-

---

[1] Wang Renkun, Vice President and Chief Engineer, Power China Chengdu Engineering Corporation Ltd., Professor Senior Engineer.
[2] Chen Zhonghua, China Institute of Water Resources and Hydropower Research, Professor Senior Engineer.
[3] Chen Ping, China Institute of Water Resources and Hydropower Research, Professor Senior Engineer.
[4] Cui Wei, China Institute of Water Resources and Hydropower Research, Senior Engineer.
[5] Liao Chenggang, Power China Chengdu Engineering Corporation Ltd., Professor Senior Engineer.
[6] Xing Wanbo, Power China Chengdu Engineering Corporation Ltd., Senior Engineer.
[7] Fu Yingru, Power China Chengdu Engineering Corporation Ltd., Professor Senior Engineer.

derground powerhouses and power generation water diversion tunnels has been fold increased, according to incomplete statistics, there are more than 60 of diversion tunnels with the length more than 5km both newly-built and under construction, but only 17 similar tunnels have been accounted in the book *Large dams in China—A Fifty-year Review*. There are 22 inter-basin diversion water conveyance tunnel projects newly-built and under construction with length more than 5km, one tunnel with the longest length, which is Dahuofang I, has the total length of 85km. Prophase work of west-route part of water regulation project is still in a positive progress, and the water conveyance tunnel through a high altitude Bayankala mountain with the total length of about 320km, where the geological conditions are extremely complex and the work conditions are extremely tough. 22 hydraulic tunnels, with area more than 140m^2 or span more than 12m, are listed in "Large dams in China—A Fifty-year Review ", however, there are about 60 tunnels newly built or under construction in the past ten years with the similar scale, many of them are the construction diversion, flood discharge and tailrace tunnel of the hydropower station, parts of them are diversion tunnels for new generating unit with large capacity. The scale of hydraulic underground structure has unceasingly expanded in our country, the number has dramatically increased, transferring focus on construction in mountainous and gorge area with complicatedly hydrological, meteorological and geological conditions and severe working conditions. Therefore challenging projects with high water head, large scale, long distance and shorter construction period are emerging unceasingly, some projects still need to arrange a large number of tunnel groups for flood discharge, desilting and power generation in the limited space, thus put forward a series of key technical problems to be solved, which promoted the development of design and construction technology for hydraulic underground structure in China. As the rapid growth of the national comprehensive strength, China's engineering construction of hydraulic underground structure has been rapidly developed since the end of the 20th century till the beginning of this century, some projects have been completed and started to create profits. With the further development of the construction, designing and scientific research personnel, it has been accumulated a wealth of experience. This chapter will briefly introduce the development general situation and new progress of China's underground powerhouse and hydraulic tunnel for hydropower station in recent years. Please see Chapter 21 for the contents of the underground hydraulic structure about a pumped storage power station.

18.2 Development of Underground Hydraulic Structure in China

18.2.1 Introduction to the Development of Underground Powerhouse of Hydropower Station

With the rapid development of the national economy and the urgent demand for energy, China's hydropower construction has been marching in a high speed. As the 4th hydroelectric generating unit of Xiaowan hydropower station was put into operation, the total installed capacity of China had reached 200,000MW ranking first in the world by August 25th, 2009. By the end of 2020, the total installed capacity of hydropower stations under construction and proposed construction will reach above 300,000MW. Due to more than 70% of hydropower projects located in southwest region with rich water energy, but also with high mountains, steep slopes, narrow valleys, ground space is very limited. So making full use of underground space is a practical and effective approach to solve the problem of the general layout and to improve efficiency of water resource. Because of less influence from flood and meteorological factors, the underground powerhouse can be built in all year round so as to shorten construction period and minimize the destruction of the vegetation, as well as avoid high slope problems from ground powerhouse, etc. Therefore, the underground powerhouse is widely used in hydropower project and has made considerable progress.

The underground powerhouses of hydropower projects were constructed in the 1950s with only span of 10m and unit capacity less than 10MW. Among the hydropower stations built in the 1990s, the underground powerhouse of Er'tan hydropower station was built with span of 30.7m and unit capacity of 550MW; the underground powerhouse of Dachaoshan hydropower station with span of 30.92m and unit capacity of 225MW; for hydropower projects all under construction and proposed construction, such as Longtan, Three Gorges, Xiluodu, Xiangjiaba, etc., the underground powerhouses with spans more than 30m and maximum unit capacity of 700-800MW; the proposed Baihetan hydropower station with unit capacity of around 1000MW. It is clear that underground powerhouse of hydropower station is heading to a direction with large unit capacity, wide span of powerhouse cavern and large-scale cavern group.

Table 18.2-1 is a statistics table for 30 underground powerhouses of large and medium-sized hydropower stations built, been building and proposed.

Table 18.2-1 Statistics for Underground Powerhouse of Large-scale
Hydropower Station Built, Planned and under Construction

No.	Hydropower station name	Located river	Installed capacity (MW)	Maximum height of cavern H(m)	Maximum span of cavern B(m)	Condition
1	Ertan	Yalong River	6×550	65.38	30.70/25.50	Built
2	Xiluodu	Jinsha River	18×770	75.10	31.90/28.40	Under construction
3	Jinping I	Yalong River	6×600	68.63	29.60/25.90	Under construction
4	Pubugou	Dadu River	6×600	70.10	30.70/26.80	Built
5	DaGangshan	Dadu River	4×650	73.78	30.80/27.30	Under construction
6	Changheba	Dadu River	4×650	73.35	30.80/27.30	Proposed
7	Guandi	Yalong River	4×600	76.30	31.10/29.00	Under construction
8	Xiaoliangdi	Yellow River	4×300	61.44	26.20/25.00	Built
9	Right bank of Three Gorges	Yangtze River	6×700	87.30	32.60/31.00	Under construction
10	Xiaowan	Langcang River	6×700	79.38	30.60/29.50	Built
11	Longtan	Hongshui River	9×700 (II)	75.10	30.30/28.50	Built
12	Xiangjiaba	Jinsha River	4×800	85.50	33.00/31.00	Under construction
13	Manwan II	Langcang River	1×300	69.40	26.60 /	Under construction
14	Jinping II	Yalong River	8×600	72.20	28.30/25.80	Under construction
15	Laxiwa	Yellow River	6×700	74.84	30.00/27.80	Built
16	Dachaoshan	Lancang River	6×225	67.70	26.80	Built
17	Shuibuya	Qingjiang River	1840	68.00	23.00/21.50	Built
18	Sanbanxi	Qingshui River	4×250	60.01	22.70/21.00	Built
19	Guangxu	(pumped storage)	8×300	44.54	22.00/21.00	Existed
20	Xilongchi	(pumped storage)	4×300	49.00	21.75 /	Under construction
21	Tianhuangping	(pumped storage)	6×300	47.73	22.40/21.00	Built
22	Tongbai	(pumped storage)	4×300	57.25	25.90/24.5	Built
23	Baoquan	(pumped storage)	4×300	47.30	22.90/21.50	Built
24	Tan-an	(pumped storage)	4×250	55.28	25.90/24.50	Built
25	Yixing	(pumped storage)	4×250	52.40	23.40/22.00	Built
26	Langyashan	(pumped storage)	4×150	46.20	21.50 /	Built
27	Suofengying	Wujiang River	3×200	57.91	24.00/21.30	Built
28	Dongfeng	Wujiang River	3×170		/ 20.85	Built
29	Mianhuatan	Branch of Tingjiang River	4×150	52.08	21.90/20.90	Built

Note: The number before and after the symbol of "/" in column of "Maximum span of cavern" refer to the upper span and lower span of the crane bean of the machine hall cavern.

18.2.2 *Introduction to the Development of Hydraulic Tunnel*

This section introduces the development of China's hydraulic tunnel from three aspects, which are diversion pressure tunnel of power generation, hydraulic tunnel for water transfer project, and hydraulic tunnel with large section and high velocity.

Part II *Construction Technologies for Dams and Reservoirs*

1. Pressure hydraulic tunnel

During the period of the construction of large-sized hydropower station, China's small-sized hydropower station also all-round developed, in which adopt the way of pressure tunnel for diversion and power generation are numerous. In Chapter 11 of this book called *Large Dams in China—A Fifty-Year* Review enumerates 17 existed pressure tunnels with span more than 5km and total length of 141km. According to incomplete statistics, the number of newly-built and under construction similar tunnels in the near ten year in China is more than 60, the total length is about 623km, considerably exceed the total number and total length of pressure tunnel built 50 years ago. In Table 18.2-2 only lists 26 pressure tunnels with length more than 10km, of which the total length is about 361km. These projects mostly was distributed in the west region of China, especially the alpine valleys in southwest region, facing all kinds of unfavorable geological conditions such as high ground stress and deep underground water level, as well as some technical challenges like fault fracture zone, which accumulated valuable experience for pressure tunnel design and construction.

Table 18.2-2 Statistics of Pressure Tunnels for Built and under Construction in China with the Length of above 10km

No.	Project name (Province)	Geology	Length (km)	Section Form	Section dimension (m)	Lining form pattern	Flow discharge (m^3/s)	Pressure head (m)
1	Jinping II (Sichuan)*	Marble, limestone	16.7	Round	12.4 (A) 12.0 (B)	A	465.0	321 TBM
2	Jinping II (Sichuan)*	Marble, limestone	16.7	Horseshoe	13.0 (A) 13.6 (B)	A	465.0	321 drilling and blasting
3	Muzuo (Sichuan)*	Slate, sandstone	11.94	City-gate section	4.4×4.4	A, B	43.02	372
4	Sinanjiang (Yunnan)*	Sandstone, mudstone	10.32	Round	5.3	A	75.70	329
5	Raozi (Sichuan)*	Phyllite, sandstone	18.672	Round	4.50	A, B, D	56.20	490
6	Baoxing (Sichuan)*	Dolomite	18.05	Horseshoe (Round)	4×5.4 (5.4)	A, B		305
7	Futang (Sichuan)*	Granite	19.34	Horseshoe (Round)	8×10.4 (9.0)	B (A)	250	159
8	Shiziping (Sichuan)*	Sandy slate	18.71	Round (Horseshoe)	5.5 (4.42×6.0B)	A, B	57.0	390
9	Xuecheng (Sichuan)*	Sandstone, phyllite	15.1	Horseshoe	5.6 (6.8×8.4)	A, B	113.9	160
10	Gucheng (Sichuan)*		16.374	Round	6.0	A, B	148	125
11	Jisha (Yunnan)*	Limestone	14.47	Round	3.3-4.3	A, B		485
12	Lugu (Sichuan)*	Emeishan basalt	15.83	Round (Horseshoe)	5.0-4.0	A, B, C	35	328
13	Qinglong (Sichuan)*	Limestone	13.6	Horseshoe (Round)	6.5×8.5	A, B	132.66	110
14	Zhouning (Fujian)*	Porphyroclastic lava	12.35	Round	6.8	A	69	452
15	Chengzigou (Gansu)*	Phyllite	17.2	Round	10.5	A, B	259.5	50.4
16	Se'ergu (Sichuan)*	Phyllite	10.14	Round	8.5	A	213	93
17	Liuping (Sichuan)*	Phyllite	10.67	Round	9.0	A		
18	Youfanggou (Yunnan)*	Quartz fine	10.9	Round	4.4	A	46.7	
19	Laqi (Yunnan)*	Limestone	10.158	Horseshoe (Round)	3.6×4.6 (3.8)	A, B	26.2	231
20	Dongshuixia (Gansu) △	Quartz schist	10.661	Horseshoe	2.125*×4.25	A	32.4	189
21	Jia'mi (Sichuan	Sandstone, limestone	11.078	Round	6.4		45.45	90*
22	Bobona (Xinjiang)*		11.63	Round	4.6		72.6	238
23	Liuhong (Sichuan)*	Limestone, shale	10.218	Horseshoe	4.5× (6.2-7.0)	A, D	57	358

Continued

No.	Project name (Province)	Geology	Length (km)	Section Form	Section dimension (m)	Lining form pattern	Flow discharge (m³/s)	Pressure head (m)
24	Pingtou (Sichuan)*	Siltstone	12.76	Horseshoe (Round)	5.5×6.2	A, D	72	290
25	Jinkang (Sichuan)*	Limestone, phyllite	16.302	Horseshoe	4.8×5.2 (5.2×5.6)	A, B	37.4	453
26	QinggangheⅡ (Yunnan)	Limestone	11.242	Horseshoe	4.8	A, B, D	54	*

Note A, B, C and D refer to the lining code, A refers to concrete or reinforced concrete, B refers to shotcrete anchor, C refers to unlined, D refers to steel lining.

* Refers to the state of existed, and △ refers to the state of under construction.

The new characteristics of diversion hydropower tunnel construction are increased length of tunnel line, the increased height of internal water pressure head, and the increased dimension of section size. To make full use of the river drop, the way of long tunnel diversion is adopted for many hydropower stations, according to statistics, there are 18 newly-built and under construction high-pressure tunnel projects with length longer than 5km and pressure head more than 300m, such as the lengths of pressure diversion tunnels in Raozi (Baoxing River in Sichuan), Zhouning (Zhouning county in Fujian), Gaoqiao (Xiyu River in Shaotong, Yunnan) and Jinwo (Tianwan River in Sichuan) are 18.67km, 12.35km, 8.86km and 7.57km respectively, the largest pressure heads are 490m, 452m, 550m and 619m respectively. Because the large capacity units was used in several large-sized hydropower stations or demands for larger flow, diversion and power generation tunnel in large section is also designed in some projects under construction. The diversion tunnel in Jinping Ⅱ has unit capacity of 600MW, length of 16.7km and section span of 12.0-13.0m. The diameters of round diversion and power generation tunnels in Pengshui, Three Gorge and Xiajiaba hydropower stations are 14m, 13.5m, and 14.4-13.1m respectively. The diversion tunnels in these projects adopted penstock diversion, and the related technical problems in the process of penstock manufacturing, transportation, hoisting and welding were solved so as to avoid the risks of construction. The diversion tunnel of NO.4 unit in Xiajiaba hydropower station adopted the layout pattern of one tunnel and one unit for the diversion tunnels of 4 units, with unit capacity of 800MW ranking first in the world. The penstock of diversion tunnel has the characteristics of longer distance and much more inflection and kick points. The penstock material is O7MnCrMoVR low-alloy and high strength steel, the largest lifting weight of single unit is 46 tons, the total quantity is 2,796 tons. The biggest diameter is 14.4m, which is ranked first for super-large diversion penstock in the world. The installation plan changed traditional round welding way in the tunnel into the method of prefabricated single section round welding in the underground workgroup, and then lifting into the tunnel to welding together. The designed pressure of diversion penstock with highest head in China is up to 10.5MPa, which was used in Xilongchi pumped storage hydropower station. Because of the reinforced concrete lining is bound to crack under internal water pressure more than 150m, for the high pressure hydraulic tunnel with strict seepage control requirements, besides adopting steel lining, pre-stressed concrete lining has reached practical promotion stage in China, in which, grouting concrete lining applied in Baishan project, post-drawing unbonded concrete lining applied in Xiaolangdi project and post-drawing bonded concrete applied in Geheyan and Tianshengqiao projects are all successfully applied.

In the tunnel design, the general principle of making full use of the self-stability capacity, bearing capacity and impermeability ability is further implemented. Designed static pressure of internal water of diversion tunnel in Gouzhou pumped storage power station is up to 6.1MPa, which adopts high pressure reinforced concrete lining and bifurcated pipe. Since the station has been built and put into operation, then the technology applied in Tianhuangping, Tongbo and Huizhou pumped storage power stations successfully. The concepts, which was designed the high pressure reinforced concrete lining and bifurcated pipe as the permeable structures and the surrounding rock to bear the internal water pressure as the impervious body, have been adopted by the designers. Meanwhile, to use finite element numerical calculation as the auxiliary supporting design method, the design methods of surrounding-rock high-pressure consolidation grouting and lining against crack are improved gradually, which make full use of surrounding rock. The construction technique and technology also get the further process, including a set of routing technique such as inclined shaft and vertical shaft, high-pressure consolidation grouting and tunnel formwork.

2. Hydraulic tunnel in water transfer project

In the recent 30 years, China's national economy lasts rapid growths and the demands for water resources from industry, agriculture and urban residents are increasing day by day, meanwhile, coupled with frequently severe natural disasters, the shortage of water resources has become a bottleneck of restricting the development of national economy. Besides the east-and middle-route parts of south-to-north water transfer project, a series of water diversion projects had to be built in many provinces and regions, especially long-distance inter-basin water transfer projects so as to meet local water demands. Statistics for non-pressure hydraulic tunnels completed or under construction in China (length>5km) were shown in Table 18.2-3.

Table 18.2-3 Statistics for Non-pressure Hydraulic Tunnels Existed and under Construction in China (Length>5km)

No.	Project name	Geology	Length (km)	Section form	Section dimensionn (m)	Lining form	Discharge (m³/s)	Construction method
1	Aojiang river diversion project in Fuzhou of Fujian province*		21.2	Round	3.0	A	10	
2	Fuqing-Duxi diversion project tunnel in Fujian province*		7.053	City-gate section	3.2×3.0	A	6.67	
3	Diversion tunnel in I the first period of Dahuofang project in Liaoning province*	Migmatite	85	Round	7.16	AO	70	TBM
4	Caoyuling tunnel in Yinqin project in Linfen irrigation district of Shanxi province*	Sandy mudstone	19.4	City-gate section	3.4×4.0	A	17-21	
5	Qinling tunnel in water diversion from Qian to Shi in Shaanxi province*	Gneiss	18.04	City-gate section	2.5×3.05	A, B	8	Drilling and blasting
6	Qinling tunnel in water diversion project from Hong to Shi in Shanxi province	Gneiss	19.76	Round	3.90 (3.3×3.55)	AO, A	13.5	TBM drilling
7	Dingshan tunnel in the water diversion project from E to Wu in Xinjiang province*	Sandy conglomeate	15.35	Horseshoe	5.2×5.2	A	47.4-55	Soft rock tunnel
8	Dawushan tunnel in water diversion project from Niulanshan to Dianchi in Yunnan province△		35		5.3×5.75	AO, A	23	
9	Jinkuidi tunnel in water diversion project from Niulanshan to Dianchi in Yunnan province△	Mudstone	15.160	Horseshoe	4.0×4.6	A	23	
10	Shangongshan tunnel in water diversion from Zhangjiuhe to Dianchi in Yunnan province*	Slate dolomite	13.769	Round	3.0	A, AO	8	TBM
11	Changkou tunnel in water diversion from Zhangjiuhe to Dianchi in Yunnan province*	Breccia	11.15	Round	3.0	A	8	
12	Cross Yellow River tunnel in the first-stage of I east route of the south-to-north water transfer project*	Limestone	7.87	Round	7.5	A	50-100	
13	Cross Yellow River tunnel in I the first-stage of middle route of the south-to-north water transfer project (two tunnels)*	Sandstone	4.25	Round	7.0	AO	265-320	Perforated Shield
14	Dabanshan tunnel in water diversion project from Da to Yellow River in Qinghai province	Igneous rock	24.3	Round	5.0	A, AO	35	TBM drilling and blasting
15	Buzihe tunnel in Dongjiang water diversion in Shenzhen city △		6.23	City-gate section	4.6×5.85	A	30	
16	The 4th tunnel in the main canal in I the first-stage of water diversion from Zhao in Gansu province*	Argillaceous siltstone	5.82	Horseshoe	4.72×4.93	A	32-36	drilling and blasting
17	The 7th tunnel in the main canal in I the first-stage of water diversion from Zhao in Gansu province*	Argillaceous siltstone	17.286	Round	4.96	AO	32-36	TBM

Continued

No.	Project name	Geology	Length (km)	Section form	Section dimensionn (m)	Lining form	Discharge (m^3/s)	Construction method
18	The 9th tunnel in the main canal in I the first-stage of water diversion from Zhao in Gansu province*	Argillaceous siltstone	18.275	Round	4.96	AO	32–36	TBM
19	Water delivery tunnel in water diversion from DaLian in Liaoning province *	Granite	14.11	Circular arch and vertical wall	3.2×3.48	A		
20	Zhongjiang shiya water diversion tunnel in Sichuan province*	Sandy mudstone	6.13	City-gate section	4.15×4.0	A	28	
21	Qishan tunnel in water diversion from Yellow River to Luo River△	Sandy conglomerate	20	City-gate section	3.2×4.2	A	9	
22	Hongqi reservoir water diversion project in Sichuan province △		7.97	City-gate section	3.6×4.0	A	6.0	

Note: A, B, AO refer to the lining code, A refers to concrete or reinforced concrete, B refers to shotcrete anchor, AO refers reinforced concrete duct piece.

* Refers to the state of built, and △ refers to the state of under construction.

Since 1991, Shuimogou water-conveyance tunnel in the water transfer project from Datong River to Qinwangchuan firstly adopted the tunnel boring machine (TBM), which has applied in several hydraulic tunnels. Through 20 years' engineering practice, rich experiences have been accumulated in the aspects of advance geological forecast and unfavorable geological section construction in hydraulic tunnel, a set of routing construction method has been formed initially, and the TBM surrounding rock classification method of hydraulic tunnel has preliminary applied.

The east-to-west Dahuofang water transfer project in Liaoning province divert water from Hengren reservoir in the Hunhe River to Dahuofang reservoir in Fushun city, and water is supplied to six water shortage cities which are Fushun, Shenyang, etc. in the Hunhe River and Taizi River, the maximum water diversion discharge is 60m^3/s, the average annual diversion quantity is 1.8 billion m^3. The length of diversion tunnel is 85.3km, cross through 50 mountains and 50 valleys, which is the longest diversion tunnel project in China, and the first phase of the project has been built completely in 2008. The construction mainly adopts TMB and supported by the drilling and blasting. The drilling and blasting method is used for first 24.58km, and TBM (three open-type machines) is used for last 60.73km, in which, the cross-section of TBM construction is round with excavated diameter of 8m and final diameter of. The bottom slope of tunnel is 1/2,380. The most burial depths of caverns in the tunnel are within 100–300m, and surrounding rock is slightly weathered fresh rock with high strength and good stability. However, serious landslides caused by fault fracture zone, close-jointed zone and the compaction of weak structure affect TBM construction schedule, therefore, the countermeasures, such as enhancing advanced geological forecast, dynamic adjustment of support scheme, driving parameter, advancing speed, and timely supporting and closing the excavated surrounding rock, have been taken.

Inter-basin south-to-north water transfer project in Shaanxi province includes water diversion project from Qianyou River to Shifa reservoir in the east route project, water diversion project from Han River to Wei River in the middle route project and water diversion project from Hongyan River to Shitou River in the west route project, all these projects divert water from Yangtze River basin to Weihe River in the Yellow River basin, crossing the diversion tunnel in Qinling. After the completion of these projects, it is expected to ease the problem of water shortage in Guanzhong region, especially in Xi'an, and to protect and improve the ecological environment of Weihe River basin with serious degradation. The water diversion project form Qianyou River to Shifa reservoir has been completed with maximum diversion quantity of 8m^3/s, and annual average available diversion quantity to Xi'an is 49.43 million m^3. The construction of Qinling's diversion tunnel decreases the investment and shortens construction period, with the help of the favorable conditions of Qinling tunnel near Xikang highway. The length of diversion tunnel is 18.04km, which is a small-section tunnel with complex lithology of surrounding rock (mainly are compound gneiss and mixed granite), with maximum buried depth of 1,600m and existing the problems of rock burst. The comprehensive measures are adopted to solve the problem of rock burst under construction, and the useful experiences are summarized. The length of Qinling diversion tunnel in the water transfer project from Hongyan River to Shitou River is 19.76km, with maximum designed diversion flow of 13.5m^3/s. After the storage in Shitou reservoir, the average water supply quantity of the reservoir is up to 0.266 billion m^3, supplying water to Xi'an, Xianyang, Baoji, Yangling and supplement ecological flow to Wei River. The buried depths of caverns are within 150 – 300m in common, the deepest is 450m. For the construction with drilling and blasting method, when the tunnel line goes through fault fracture zone with rich water, it is difficult to excavate, thus the construction technique by combining pipe shed and pre-grouting is adopted to solve the problems of collapse and roof-fall. For TMB construction, part of tunnel section with concentrated fault fracture zone, weak rock layer and weak interlayer, and strong swelling and fracture fissure water in local area, PUR grouting material is adopted under construction to plug water in advance and fasten the loose rock so as to rapidly form the solid water-resisting layer with thickness of several meters, which effectively solves the problems of heading collapse and cutter sticking.

Water transfer project from Hanjiang River to Weihe River is the key project in the south-to-north water transfer project in Shaanxi province, which has two water sources, one is Huangjinxia reservoir in the trunk Stream of Hanjiang River, and another is Sanhekou reservoir in the branch Ziwu River. The way of water diversion is through Huangjinxia pump station to lift the water to 113.5m and transfer it to diversion tunnel from Huangjinxia to Sanhekou, after storage in Sanhekou reservoir, through Qinling tunnel and arrive in Hanzhong region finally. The project is divided into two stages, Sanhekou reservoir and Qinling tunnel are constructed during the first stage with the annual diversion quantity of 0.5 billion m^3; huangjinxia reservoir, water source pump station and water diversion project from Huangjinxia to Sanhekou are constructed during the second period with annual diversion quantity of 1.5 billion m^3. The total length of diversion tunnel is 97.37km, in which, the diversion tunnel crossing Qinling is 81.58km, the designed flow is 64.0m^3/s. Qinling diversion tunnel through the middle part of Qinling with maximum buried depth of around 2,012m, which belongs to the deep and super-long tunnel, and the total length of tunnel, the penetration length of single tunnel and the length of inclined tunnel are all ranked high in the world. The project is proposed to adopt drilling & blasting and TBM method to construct, and the integral section of the tunnel is 40km due to the mountain belt with wild life reserve, so TBM can only be adopted in the manner of digging from the north and the south. The tunnel is 500-600m away from the ground, into the main tunnel with the method of inclined shaft, set up 11 caves totally. The longest No. 3 tunnel is 3.87m; the comprehensive longitudinal gradient is 8.15%; the vertical dropping is 288m. The main project issues of Qinling tunnel are complicated geological conditions, indivisible super-long tunnel section, deepest buried depth, high crustal stress, high ground temperature and severe water and mud bursting. Analysis of exploration actual data and engineering analogy inference indicate that the temperature of original rock in the maximize embedded depth is 42℃, the maximum horizontal crustal stress is expected to more than 50MPa. The extraordinary design and main construction technologies include the prediction and prevention of high crustal stress rock burst, extrusion deformation, stability analysis and supporting design of surrounding rock with high crustal stress; real-time monitoring and support optimization design of surrounding rock under construction; geological disaster forecast and information management system of super-long tunnel with deepest buried depth, etc. TBM driving distance is more than and the longest ventilation distance of drilling and bursting construction is, which are both beyond the requirements of current codes and actual engineering experiences and challenges in the measurement accuracy control, ventilation arrangement and TBM construction technology. The Qinling project makes the design and construction technology in super-long hydraulic tunnel with maximum depth to reach a new height.

The section of across Yellow River in East route of South-to-North Water Diversion Project is composed of maintenance gate of import of buried pipes, buried pipes of bottom land, tunnel crossing the Yellow River, tunnel crossing the Yellow River and breakout gate. The tunnel of cross Yellow River adopts the structure of inverted siphon, including shaft in the south bank, adit across the river, inclined shaft in the north bank and import & export buried pipes. The adit is located in 70m below the riverbed of Yellow River with the total length of 7,870m, including 585.38m inversed siphon tunnel. The tunnel lining adopts reinforced concrete structure with round-shaped section, and diameter is design as 7.50m. Except the inlet/outlet in the rock adopts the way of open-dig and pipes are buried by cast-in-situ reinforced concrete, others all excavated in the Cambrian limestone. According to the structural requirements and crack limit design, the thickness of tunnel lining is determined by taking other similar projects as reference. The lining thickness of the adit is 0.55m; the inlet lining thickness of inclined shaft is 0.75m; the lining thicknesses of the middle part of inclined shaft and vertical shaft are both 0.65m. The first phase of the project was completed and put into operation in 2007, the discharges across the Yellow River are within 50 –100m^3/s.

The tunnel project of the middle route scheme of South-to-North Water Diversion Project across the Yellow River is the structure combination with main canal of the middle route and the Yellow River, which is located in the eastern of Zhengzhou city of Henan province, is the most complex project in the main canal with large-scale, also a route path controlling construction scheme, the designed flow of the first phase is 265m^3/s. The project adopts shield tunnel across Yellow River. The tunnel across Yellow River is composed by Mangshan inclined tunnel and Yellow River tunnel in parallel arrangement, the total length of single tunnel is 4.250km, the inner diameter of tunnel is 7.0m and the external diameter is 8.7m; the maximum internal water pressure is 0.51MPa. The difficulty of tunnel is to cross wandering reach, which is located in sandy soil layer, with complex geological conditions and high external/internal water pressure. According to the requirements of tunnel construction and later operation, across Slurry shield TBM is adopted firstly for the shield tunnel across the Yellow River in China, which designs double lining structure. The outer layer is precast reinforced concrete lining by 7 segments, and the inner layer is routine reinforced concrete or pre-stressed reinforced concrete lining. The tunnel project across the Yellow River commenced in 2005, the driving of two shield tunnels has been completed, and the segment lining has been finished.

3. Hydraulic tunnel and high-velocity flow tunnel of large section

The design and construction of large-section hydraulic tunnel and the problems of scouring abrasion and corrosion caused by high velocity flow have been concerned in the recent 20 years.

Currently, the focus of China's hydropower construction is in the western mountainous and valley areas, the layouts of power plant, diversion and flood discharge buildings for large-and super-large-sized hydropower projects are restricted by the narrow topography, so the underground powerhouse for power generation is the only option. Due to the concentrated rainy season in the river basin and heavy flow in the valley, the cofferdam diversion and discharge by dam body may not meet the overflow, especially for the earth-rockfill dam, thus the only way is to adopt large section for diversion and flood discharge. Moreover, it is difficult to arrange the spillway in both banks of valley with steep slope and high mountains, which is the main reason to adopt hydraulic tunnels for flood discharge. Due to the substantial increase of unit capacity, the sections of diversion and tailrace tunnels are also increased. Table 18.2-4 lists project features of 26 large-section hydraulic tunnels existed and under construction with span of mare than, such as the diversion tunnel of Longtan hydropower station adopts the section pattern of city-gate, with net span of 16m and height of 21m, as well as the designed flow of 14,700m^3/s; and the flood discharge tunnel of Longtoushi hydropower station adopts the section pattern of city-gate, with excavated span of 20m and height of 21m, as well as the designed flow of 9,600m^3/s. The section pattern of city-gate is adopted in the tailrace tunnel of Pubugou hydropower station, with the spans of 20m and height of 24.2m.

Table 18.2 – 4 Statistics for Large-section Hydraulic Tunnels Existed and under Construction (Span is more than 14m after Lining)

No.	Project name	Geology	Length (m)	Section form	Section dimension (m×m)	Purpose	Flow discharge (m^3/s)	Flow velocity (m/s)
1	Hongjiadu	Limestone	754.543	City-gate section	14×21.5	Tunnel-spillway(two)	4,591	37.5
2	Dachaoshan	Basalt	687	City-gate section	15×18	Diversion	6,400	23.7
3	Dachaoshan	Basalt	1,144/1,092	Round	15.0	Tailwater		
4	Longtan	Sandstone	595/874	City-gate section	16×21	Diversion	14,700	
5	Longtan	Sandstone		Round	21.0	Tailwater(three units and one tunnel)		
6	Goupitan	Caly rock	888,674	Horseshoe	15.6×17.7	Diversion	10,415	
7	Longtoushi	Granite	511 – 646	City-gate section	20×21 (Excavated)	Flood discharge	9,600	
8	Pubugou	Granite	1,137/1,075	City-gate section	20.0×24.2	Tailwater(three units and one tunnel)	417 (single unit)	
9	Xiaowan	Gneiss	861/980	City-gate section	16×19	Pressure diversion tunnel	10,300	
10	Xiaowan	Gneiss	958(No.1)	Round	18.0	Tail water(three units and one tunnel)		4.3
11	Sanbanxi	Tuff	734	City-gate section	16×18	Diversion		38
12	Guandi	Basalt	687/1004	City-gate section	16×19	Diversion	9,780	

619

Continued

No.	Project name	Geology	Length (m)	Section form	Section dimension (m×m)	Purpose	Flow discharge (m³/s)	Flow velocity (m/s)
13	Guandi	Basalt	736(915)	City-gate section	16×18 (12×16)	Tail water tunnel	1,100(773)	
14	Xiluodu	Basalt	562/1,050	Round, City-gate section	15.0, 14×18	Flood discharge (two in left bank, three in right bank)	3,860 – 4,127 (single tunnel)	
15	Xiluodu	Basalt		City-gate section	18.0×22.0	Diversion (three in each bank)		
16	Xiluodu	Basalt		City-gate section	(15 – 18)×20	Tail water(No. 1 and No. 6 is 15×20)	423.8(one unit)	(three units and one tunnel)
17	Jinping I	Marble		City-gate section	15×19	Diversion(one for each bank)		
18	Jinping I	Marble	505(360)	City-gate section	15×16.5	Tail water(three units and one tunnel)	337.4(single)	
19	Jinping II	Marble	593	City-gate section	14×15	Diversion	1,690	
20	Wudongde	Marble		City-gate section	16×20	Tail water(two units and one tunnel)		
21	Three Gorges	granite Granite	97 – 138	City-gate section	15×25	Tail water(one units and one tunnel)		
22	Shenxigou	dolomite Dolomite	1,350	City-gate section	15.5×18	The length of No. 2 tunnel is 1,493	7,980	
23	Pankou	Crystalline schist	530	City-gate section	15×18	Diversion	4,350	
24	Tianhuaban	Dolomite	485	City-gate section	12×14	Diversion (with pressure)	2,130	13.66
25	Ludila	Metamorphic sandstone	865	City-gate section	14.5×17.0	Diversion	2,170	
26	Laxiwa	Granite	519/712	Round with pressure	17.5	Tail water(three units and one tunnel)	1,140	
27	Xiaolangdi	Sandstone	1,134/ 1,121/1,121	Round with pressure	14.5	Three for flood discharge	1,727/ 1,549/1,549	Energy dissipation by orifice in tunnel

The location and axis direction of these large-section overflow tunnels are restricted by the layouts of main structures and underground powerhouse and the inlet/outlet hydraulic conditions, thus it is difficult to completely avoid unfavorable conditions caused by geological structures and ground stress direction. The slopes of inlet/outlet have the characteristics of great depth unloading, serious rock weathering and broken rock, which brings numerous difficulties to the design and construction of large-section tunnels, especially for adverse geological section, such as mudstone softening section of diversion tunnel in Zipingpu Hydropower Station, fault fracture zone of diversion tunnel in Nuozhadu Hydropower Station, incompetent bed of diversion tunnel in Goupitan Hydropower Station, strong water section and high stress rock burst zone in Jinpin Hydropower Station, etc. The conventional design methods limited generally to the hydraulic tunnel with span less than 10m, but there are numerous hydraulic tunnels newly built or under construction with the span of more than 10m. For the design of large-section hydraulic tunnel, more advanced numerical methods such as finite element and block analysis methods are utilized to analyze the stability of the surrounding rock and optimize the plans of tunnel excavation, surrounding rock reinforcement and supporting, which accumulates the design and construction experiences in adverse geological section.

The difference between upstream and downstream heads is large when high-velocity flow is formed in the high dam. The lining is abraded seriously by flow carrying silt when the velocity in cavern is more than 15m/s; severe cavitations damage is caused by flow when the velocity in cavern is more than 30m/s. The velocities in many flood discharge tunnels of hydraulic projects in our country are more than 30m/s, such as the velocity of spillway in Hongjiadu hydropower station is up to 37.5m/s, the velocity of flood discharge tunnel in Pubugou hydropower station is up to 37.5m/s, the velocity of flood discharge tunnel in Ertan hydropower station is up to 50m/s.

In the recent years, hydraulics and structure prototype observation, hydraulic model testing and hydraulics numerical simulation in the aspects of cavern against cavitation and downstream fogging have innovated and achieved good results. Dragon-raise-head and dragon-down-tail are the typical arrangements for flood discharge tunnel with high head and large discharge in the world. The spillway tunnel of dragon-raise-head adopt short pressure inlet linking up with steep slope section, then through ogee section jointing with low adit to discharge flood, usually make use of the latter part of the original diversion tunnel, but the spillway tunnel of dragon-down-tail adopt longer small bottom slope inlet section linking up with the ogee section and downward adit, and the lock chamber is located in the middle part of discharge tunnel. The flow property and the risk of cavitation erosion in two arrangements are compared (Guo et al, 2006), which indicates that the early dragon-raise-head has big risk and the improved dragon-down-tail can reduce the risk. Some large-sized flood discharge tunnels, such as Xiluodu, Jinping and Baihetan hydropower stations, all adopt dragon-down-tail in China. The cavitation damages mainly occur in the end of ogee section. The research on the technology of 3D aeration corrosion reduction has made greater progress, which shows that the optimized U-shaped ridge with air can form a real U-shaped cavity from the side-wall to the bottom to greatly improve the cavitation resistance. The research and application of ecological environment friendly energy dissipation technology has made great progress. The advantages of energy dissipation within the flood discharge facilities are becoming more and more obvious in the premise of eco-friendly, which are in the aspects of overcoming the cavitation damage of the structures, preventing the atomization and reducing the project investment, etc. The development of research and application is summarized in the world (Gao et al, 2008). The results of hydraulics and prototype observation indicate that shaft-horizontal vortex flood discharge tunnel is feasible in Gongboxia, the vortex energy dissipation facilities are environment friendly, which is the most effective energy dissipation technology and has important application value. A new trumpet-shaped shaft flood discharge tunnel with diving up spin pier is proposed through design theory and experimental research (Dong et al, 2011), which can produce stable cavity rotation movement under various heads to avoid cavitation erosion and increase the efficiency of flood discharge energy dissipation at the same time. This new pattern has been applied to the design of shaft flood discharge tunnel of Qingyuan pumped storage power station in Guangdong province. There are 3 multistage orifice-plate tunnels rebuilt from the diversion tunnels in Xiaolangdi project, which is also one pattern of energy dissipation within the tunnel. The concrete lining in our country mainly adopts silicon power concrete and HF high strength fly-ash concrete. The flood discharge tunnel of Xiaowan hydropower station is of a pattern varying from pressure to non-pressure, which is arranged in manner of dragon-raise-head. The discharges of flood discharge tunnel are $3,535m^3/s$ and $3,811m^3/s$ respectively under designing and checking conditions, with maximum discharge head of around 212m and maximum velocity of 45m/s. The net width of non-pressure section in lined up is 14m. The thickness of concrete lining is 1.2m. The bottom slab of tunnel and side-wall concrete both adopt silicon powder concrete against abrasion ($C_{90}60W_{90}8F_{90}100$) and concrete against abrasion ($C_{90}50W_{90}8F_{90}100$) respectively added with polypropylene fiber and steel fiber. The effective measures, such as precooling concrete, optimized concrete mix proportion, timely finish and long-term water curing, are utilized to ensure the construction quality of silicon powder concrete under the concrete construction.

18.2.3 Research and Application on Underground Engineering of China's Hydropower Station in China

In the past 20 years, the theories, methods, techniques and measures on design, construction and management have advanced significantly in underground powerhouse of hydropower stations, especially in large underground caverns, such as the continuous innovation and breakthrough on the idea of design and construction, gradual perfection and application on new technology and methods, accumulated experiences on design and construction, and making the significantly economic and social benefits.

The underground geotechnical engineering has the following characteristics: ①the complex geological conditions and structures (the heterogeneity, discontinuity of the 3D space, and the existing weak intercalated layer, fault, fracture and interformational distributed belts); ②pre-stressed structure features (The unloading or loading caused by initial stress field during excavation); ③the dimension of structure and time with a wide range; ④unknown strength and deformation of rock mass (the properties after peak value are important); ⑤coupling effect by fluid and solid, which make the issue more complicated; ⑥the features of variability and uncertainty. The underground engineering is taken as data-limited rock mechanics issue doomed by the above characteristics. Until now, there are not still reasonable and effective methods to analyze and evaluate the stability of surrounding rock in caverns, without universally accepted quantitative safety indicators.

The research methods of underground geotechnical engineering can be summarized as follows: ①laboratory testing of structure, such as rock and joint; ②geo-mechanical model testing; ③measurement and testing on site; ④classification system of experience; ⑤numerical simulation technique. Methods of No. 1 - No. 4 are traditional, nearly 20 years, with the support of computer hardware update, the technology of numerical simulation is developing rapidly; in the recent 20 years, many complex mechanical models and boundary conditions can be simulated by computer with the updated computer hardware and rapid development of numerical simulation technique, thus it is widely applied with the broad prospect.

Nowadays, numerical analysis methods of rock mechanics in underground engineering include: ①Finite Element Method (FEM); ②Finite Difference Method (FDM, FLAC); ③Block Element Method; ④Interface Element Method; ⑤Stochastic Finite Element Method; ⑥Boundary Element Method (BEM); ⑦Discrete Element Method (DEM); ⑧Lagrangian Element Method (LEM); ⑨Discontinuous Deformation Analysis (DDA); ⑩Element Free Method (EFM); ⑪Manifold Element Method (MEM). In spite of the advantages and disadvantages, the basic mechanical mechanisms of all numerical methods are the same. In addition, the numerical models and methods for the analysis of fractured rock seepage have become more and more mature, however, there are still a lot of urgent problems on seepage control optimization, seepage-stress coupling analysis in large-sized underground cavern groups. In recent 10 years, the theory research and application in the above research field have made great progress with the efforts of researcher and designer on rock mechanics and engineering.

In practices of underground engineering, because of the difference between the influencing factors of the stability of surrounding rock masses, it is difficult to establish a uniform criterion to evaluate the stability of underground caverns, thus the multi-criterion or combined-criterion methods are usually adopted. At present, there are many stability analysis methods in the world, such as empiric analogue approach, safety factor method, reliability, degree of stability and failure probability; numerical direct data (displacement, stress, strength and plastic zone of rock masses); disturbing energy method, etc. Numerical stability research of surrounding rock for underground engineering is leading towards refinement simulation, large-sized calculation, and whole-process simulation on elastic, plastic, viscous, crack, damage, softening induced by damage, sudden change, etc. Meanwhile localized problems in project have been concerned considerably by the researchers in the world, which are common failures of rock masses such as the intersections of continuous-and discontinuous-deformations, local bifurcate instability, flow softening of rock interface and local instability.

The studies on construction and excavation sequences of underground caverns have also made some progresses in the world. Both plane analysis and 3D nonlinear analysis are used to quantitatively evaluate the stability of surrounding rocks. Meanwhile the optimized construction organization has been introduced in numerical researches to close the mechanics analysis with the construction system project, which has turned to the developing trend of the study on construction and excavation sequences.

18. 2. 4 *Development Survey of Hydraulic Underground Engineering Construction Technology in China*

The hydraulic underground engineering in our country is rapidly developed, the construction technology of underground engineering is greatly improved by the increased quantities, scales and construction difficulties of diversion tunnels in large and superlarge underground hydropower stations and inter-basin water transfer projects. The integrated innovation technology emerges constantly, and the routing construction technology is becoming more and more perfect. Only the new development of construction technology of hydraulic underground engineering is introduced briefly in this section as space is limited, the contents derived from the book named *Hydraulic Underground Engineering Construction in China* published in 2011. The construction technology and development of hydraulic underground engineering had been comprehensively summarized and reviewed in this book. The new achievements of hydraulic underground engineering construction are mainly presented as follows:

(1) Rapid construction technology of larg underground diversion and power generation system;

(2) Rapid construction technology of long tunnel in large-section;

(3) Construction technology of high-pressure reinforced concrete bifurcation without steel lining;

(4) Rapid construction technology of high-pressure long inclined (vertical) shaft;

(5) Construction technology of concrete formwork of underground engineering.

Drilling and blasting method is widely used for the construction of hydraulic tunnel in China. Drilling and blasting method is the core of the above five technologies, the contents of construction technology of hydraulic underground engineering are summarized in the manners of stablity, efficiency and accuracy. "Stablity" is to ensure that the surrounding rock is stable under the whole process of construction, thus the personnel, equipment and construction can be guaranteed under the favorable conditions. A full set of technology has been summarized in this aspect, which include the shotcrete bolt supporting as core, advanced surveying, pre-stressing or pre-grouting, control blasting, short footage, early closed, strong support and regular measurement. The construction problems of large-section tunnel in the unfavorable geological conditions have been solved effectively. "Efficiency" is to excavate quickly and support timely, which can present the economic effect of underground engineering construction. The construction methods of multiple operations in plane and multilayer in space have been summarized in this aspect. The characteristics of stable and fast are both presented in the excavation and supporting process of "side first, then middle" and "soft first, then hard". "Accuracy" is accuracy of measurement, accuracy of advanced geological forecast, accuracy of monitoring data from buried equipment, accuracy of the result of fine control blasting. Accuracy is the evaluation criterion of the above five techniques.

In the recent 30 years, China's construction technology of underground engineering experiences the process of introduction, digestion, absorption to innovation, and transform into integrated innovation constantly. It covers the each link of construction, including the following aspects: ①the development of high performance blasting equipment and the design of detonating network, which make the higher efficiency of the blasting and meet the requirements of stricter blasting control; ②the development of measurement technology and the improved drilling accuracy and contour blasting technology make the excavation quality nearly perfect; ③the traditional steel fabric shotcrete is replaced by the application of micro fiber and polypropylene fiber shotcrete; ④the stability technology of large cavern excavation in unfavorable geological conditions is more mature because of the development of geotechnical anchoring technique and various kinds of new type anchors, combined with optimization of hierarchical block and pre-reinforcement technology; ⑤the rapid construction of long inclined shaft and deep vertical shaft is achieved by the application of raise-boring machine and open-type TBM and the improvement of slip forming; ⑥the problem of smoke ventilation in complex caverns and long tunnel is solved by the research on the ventilation method and the development of effective ventilator; ⑦the method of multiple operations in plane and multilayer in space is widely used, which greatly improve the construction speed of underground powerhouse; ⑧the combination of various advanced formwork techniques can be adapted to the application of new type structure, which promotes the reform of underground engineering structures; ⑨taking the feedback analysis of engineering safety monitoring as a method, and the information integration technology and digital video monitoring technology are utilized to improve the technology protection for supporting design optimization and safety operation; ⑩the reasons for the differences between calculation results of cavern mathematical model and the actual monitoring results are further analyzed so as to promote the development of geotechnical engineering discipline.

18.3 New Development of Underground Powerhouse of China's Hydropower Station

18.3.1 *Research on Rational Layout of Large Underground Cavern Groups*

18.3.1.1 Geological Conditions of Large Underground Cavern Groups

Cavitation geological conditions, supporting measures and project cost for the large-sized cavern groups are closely related to the engineering geology, hydrogeology, rock mechanics parameters, magnitude of ground stress of rock mass and classification of surrounding rock in the underground powerhouse.

The layout of large-sized underground cavern groups requires a region with good geological structure, hard rock, small groundwater, moderate ground stress, without major fault, development of joint and fracture, fracture zone and large karst caves and underground rivers and etc.

According to a statistical analysis based on more than 20 large-sized underground powerhouses existed and under construction, the proportion of main body of caverns located in section of surrounding rock with grade Ⅰ, Ⅱ, Ⅲ is more than 80%. For the partly located in section of surrounding rock with grade Ⅳ, Ⅴ, the stability of surrounding rock can also be met by the effective support measures.

The initial ground stress is one of the main loads for surrounding rock of underground caverns. The layout of underground caverns, analysis on the stability of surrounding rock, and supporting design all depend on the rational and reliable distribution rules and characteristics of initial ground stress field. The important task of research on underground cavern is to utilize the inversion of actual measurement to approach the real ground stress field. The inversion analysis on initial ground stress field of underground powerhouse shall be based on the actually measured ground stress, in general, which includes the boundary loading method, multiple regression inversion method, 3D finite element method or finite difference method, etc. Difference from ground stresses with various inversion methods may be caused by the actual measured points, inverse regression methods and treatment technology.

The limited data, such as the initial and secondary ground stresses at the actual measured points and the actual displacements of the cavern under excavation, can be obtained at present. The distribution of initial stress field can be studied by the inversion analysis of above information.

18.3.1.2 Location of Large-sized Underground Cavern Groups

Depending upon the location of underground powerhouse in water diversion system, the underground powerhouse can be classified as the prime type, the central type and the tail type. An appropriate location of underground cavern group is chosen by comprehensive analysis and comparison in the layout of main structures, topographical and geological conditions, construction conditions, operation requirements and environmental protection, etc.

1. Depth of the overlying rock mass of underground cavern

There shall be not only suitable geological conditions, but also enough depth of overlying and side rock mass in the position of large underground cavern groups. The depth of overlying and side rock mass is synthetically analyzed and determined by the factors such as the integrity of rock mass, weathering and unloading level, the scale of ground stress, the movement of the ground water and the scale of the caverns, etc. The rock crown with thin cover thickness has the unfavorable effects on the formation of load-bearing arch, full use of bearing capacity of surrounding rock and the support of upper rock mass of side-wall; the ground stress will become higher as the thick cover thickness and deeper buried depth of underground cavern, which is likely to cause the rock burst and unfavorable for the stability of surrounding rock.

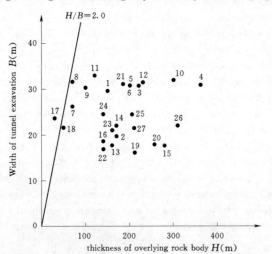

Fig. 18.3-1 The Statistics of Rock Mass Overlying Depth of Underground Powerhouses Existed or under Construction in China

1—Jinping Ⅰ Hydropower Station; 2—Taipingyi Hydropower Station; 3—Pubugou Hydropower Station; 4—Dagangshan Hydropower Station; 5—Changheba Hydropower Station; 6—Ertan Hydropower Station; 7—Xiaolangdi Hydropower Station; 8—Right bank underground power house, Three Gorges Hydropower Station; 9—Longtan Hydropower Station; 10—Xiluodu Hydropower Station; 11—Xiangjiaba Hydropower Station; 12—Laxiwa Hydropower Station; 13—Se'ergu Hydropower Station; 14—Zhile Hydropower Station; 15—Shiziping Hydropower Station; 16—Liuhong Hydropower Station; 17—Dafa Hydropower Station; 18—Renzonghai Hydropower Station; 19—Shuiniujia Hydropower Station; 20—Xiaotiandu Hydropower Station; 21—Guandi Hydropower Station; 22—Erdaoqiao Hydropower Station; 23—Tianhuangping pumped storage power station; 24—Tongbai pumped storage power station; 25—Tai'an pumped storage power station; 26—Yixing pumped storage power station; 27—Baoquan pumped storage power station

Fig. 18.3-1 shows the statistics of the overlying rock mass depth of some underground powerhouses existed or under construction in China. According to the experiences of existed projects in the world, the depth of the top rock of main cavern should be not less than 2 times of the excavated span. For the underground caverns with good geological conditions, the cover thickness of rock mass, which has been demonstrated and treated with reliable reinforcement measures, has been improved.

2. Reasonable distances between caverns

It is difficult to obtain a unified standard for the distance between underground caverns, because of the factors such as the geological conditions and scale of cavern, etc. A larger distance between main caverns is favorable for the stability of surrounding rock with less supporting, however, the lengths of subsidiary caverns need to be longer and more corresponding facilities are needed, such as the length increase of bus tunnel and bus, power loss and increased investment, etc., which may not be efficient way.

In the demonstration of distance between caverns for the stability analysis of surrounding rock, it is generally required that the plastic yield area or relaxation area of surrounding rock mass must not be connected, so as to avoid the decrease of rock strength. However, in fact, the integrity and certain bearing capacity of surrounding rock can be ensured by rigid control measures for the construction blasting technology and other effective reinforcement measures.

According to the engineering experience in the world, the rock thickness of adjacent caverns with surrounding rock of grade Ⅰ, Ⅱ, Ⅲ should be not less than 1.00 - 1.50 times of the average excavated width of adjacent caverns, for the caverns in high ground stress region or with "hard, brittle and broken" rock mass, the thickness should be not less than 1.5 times; and the distance between adjacent caverns should be not less than 0.50 - 0.80 times of the height of larger cavern. The distances between main caverns for large underground powerhouses should be within 40 - 50m.

According to the statistics, the rock thicknesses of main caverns are rarely less than 1.0 times or more than twice of the average excavated widths of adjacent caverns, mostly within 1.30 – 1.80 times, and mainly within 1.50 – 1.80 times.

The distance between the main caverns of underground powerhouse is related to the height of large caverns, and according to the abroad practice, the rock thickness of adjacent caverns shall be no less than 0.60 – 0.80 times of the height of the larger cavern. In terms of the statistics of the existed or under-construction underground caverns in China, the distances between main caverns are mostly within 0.50 – 0.75 times of the heights of larger caverns. For example, the distances in Baishan Hydropower station and Xiangjiaba hydropower station are 0.31 times and 0.46 times respectively.

3. The longitudinal axis directions of main caverns

The longitudinal axis direction of underground powerhouse shall be determined synthetically according to the distribution of initial ground stress field, joint fracture structure and fault fracture zone. In low ground stress region, the longitudinal axis direction of underground powerhouse mainly depends on rock mass structure, but in high ground stress region, it mainly depends on ground stress.

(1) Main structural surface. The included angle between the longitudinal axis direction of main cavern and the direction of main structural surface (such as faults, main joints and layer surfaces, etc.) should be more than 60° so as to enhance the stability of high side-wall of underground powerhouse, meanwhile, the unfavorable effects of secondary structural surface on cavern stability shall be concerned. When main structural surface has the gentle dip less than 35°, which has an obvious effect on the crown stability of cavern, thus the effect of the secondary structural surface on the stability of high side-wall shall be paid attention at same time. In terms of the influence of weak structural surface, not only the number and dip angle, but also the properties, especially the friction, shear strength, and stability against slide, shall be taken into account.

(2) The maximum principal stress. The included angle between the longitudinal axis direction of powerhouse and the direction of maximum principal stress of ground stress should be within the range of 15°– 30°. Following this, the unloading stress is released gently during excavation to reduce the side pressure and deformation and benefit the stability of sidewall. However, the longitudinal axis of underground powerhouses should not be totally parallel to the direction of the maximum principal stress; otherwise it would have the adverse effects on the surrounding rock stability for caverns for inlet and tailrace systems arranged vertically with the longitudinal axis of underground powerhouses, the rock pillar between bus tunnels and high end-wall of main cavern.

4. Layout of main caverns

Main caverns are the main powerhouse, main transformer chamber and main surge chambers (including upstream surge chamber or tailrace surge chamber), which are usually arranged parallel. Then the auxiliary tunnels, such as access and transportation tunnels, bus tunnels, draft tubes, tailrace tunnels, air tunnels, outgoing line tunnels, drainage tunnels, etc., shall be arranged. The distances among three large chambers are determined preliminarily according to the geological conditions and engineering analogy. The common layout patterns and their characteristics are listed in Table 18.3 – 1.

Table 18.3 – 1 The Layout Patterns and Characteristics of Main Caverns

Program	The location of main transformer chamber	Diagram	Advantage	Disadvantage
I	Between the main powerhouse and the tailrace surge chamber		1. Compact layout, convenient operation and maintenance. 2. The separate layout of the main powerhouse and the main transformer chamber can reduce the harm of accidents	The compressible leeway between the main powerhouse and tailrace surge chamber is small, and otherwise it's unfavorable to the surrounding rock stability of caverns
II	In the upstream of the main powerhouse		1. Compact layout, convenient operation and maintenance. 2. Reduce the harm of accidents. 3. Shorten the length of draft tube	1. It's difficult to set up the main transformer chamber and the drainage system for seepage control between the powerhouse and the inclined pressure shaft. 2. The stability of the surrounding rock of the powerhouse is relative worse

Countinue

Program	The location of main transformer chamber	Diagram	Advantage	Disadvantage
III	Within the main powerhouse		1. The shortest bus and the lowest power energy loss. 2. Easy to operate and manage. 3. Shorten the length of draft tube	1. The main transformers are near to the units, so it is more dangerous for fire or explode. 2. The main powerhouse with larger dimensions is unfavorable to the stability of the surrounding rock of powerhouse
IV	Between and above the main powerhouse and the tailrace surge chamber, trilaterally located		1. Shorten the distance between the main powerhouse and the tailrace surge chamber and the length of draft tube. 2. Reduce the harm of accidents	1. Inconvenient to operate and manage. 2. Longer bus, more investment, higher power energy loss and the complex issues on ventilation and heat dispersion. 3. The increase of the facilities for lifting and ventilation and the transportation channels

According to the number of units, hydraulic stability and geological conditions, the patterns of the surge chambers can be gallery-or cylinder-shaped, the former is conducive to the layout and operation of maintenance gates, and the latter is conducive to the surrounding rock stability of caverns. The gallery-shaped surge chambers (or gate chambers) are adopted in Er'tan, Xiluodu, Pubugou, Dagangshan, Changheba and Jinping II hydropower stations, which are arranged parallel to the main powerhouse and transformer chamber. The rock partition walls are set up in surge chambers (or gate chambers), and the top of which forms an interconnected channel. It can only be utilized by three to four units at the same time. On the other hand, Laxiwa, Xiaowan and Jinping I hydropower stations adopt cylinder-shaped surge chambers.

5. Cases: *the layout of large underground cavern group*

(1) Jinping I Hydropower Station. Jinping I Hydropower Station has a large-sized underground powerhouse cavern group which mainly consist of the diversion tunnel, underground powerhouse, bus tunnel, main transformer chamber, tailrace surge chamber and the tailrace tunnel, etc., of which the three main chambers are arranged parallel. The underground powerhouse is located in the mountain on the right bank, 350m downstream from the dam, with a horizontal buried depth of 110 – 300m and a vertical buried depth of 180 – 350m. The total length of the powerhouse is 276.99m. The excavation span above the crane beam is 28.90m, the one below the crane beam is 25.60m and the excavation height is 68.80m. The main transformer chamber is located the downstream of main powerhouse and the distance between the crown center line and the axis of the main powerhouse is 67.35m. The thickness of rock pillar between the powerhouse and main transformer chamber is 45m and the powerhouse is an arch-shaped cave with straight wall. The main transformer chamber has length of 197.10m, width of 19.30m and height of 32.70m. The tailrace surge chamber consists of two surge chambers with circular-shaped, the diameters of which are 37.00m and 41.00m respectively, and the distance between their center lines is 95.1m, and the distance between the center lines of tailrace surge chamber and the crown of main transformer chamber is 77.65m.

The positions of underground powerhouse are determined by the factors as follows:

1) Try to minimize the effects of the fault f_{13} and the above fissure zone (gushing water zone) between NNW and NW on the underground powerhouse, and arrange the powerhouse below the fault f_{13} so that the fault f_{13} only has an impact on the erecting bay.

2) To make sure that the buried depth for underground powerhouse is large enough. The cavern group of underground powerhouse have a vertical buried depth around 160 – 420m and a horizontal buried depth around 100 – 380m, which can meet the requirements of underground cavern group with large span.

3) Evade the effects of the greenschist and marble in the small first layer ($T_{2-3z}^{2(1)}$) of the Zagunao 2nd section on underground powerhouse. The rock mass in this layer has a low strength, its rock category is IV$_1$ and the stability of surrounding rock is poor. The underground powerhouse should be arranged near the adit PD01CZ by comparison so as to meet the above demands.

The axis directions of three main caverns are formulated by the factors as follows.

1) The underground powerhouse is located in the high ground stress area and the direction of the maximum principal stress σ_1 is N48.7°W, thus the axis direction of the three main caverns should be as parallel to σ_1 as possible.

2) The advantageous directions of main structural surfaces are: ①N40°-60°E, NW∠15°-30°; ②N50°-70°E, SE∠50°-80°; ③N50°-70°W, NE (SW) ∠80°-90°. The axis directions of caverns should intersect with the directions of main structural surfaces in a large angle as much as possible. Finally the axis direction of three main caverns is ascertained as N65°W, which has a 15° angle with σ_1 and an angle of 70°-90° with the structural surfaces ① and ② (rock stratums and faults), which are favorable to the stability of the surrounding rock of cavern. The details are shown in Fig. 18.3-2.

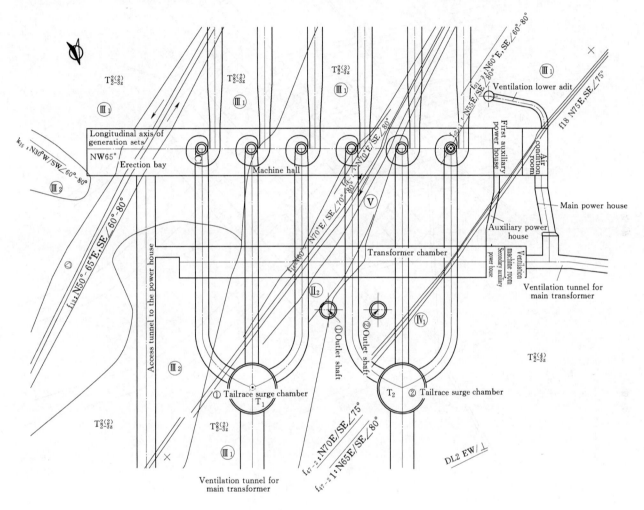

Fig. 18.3-2 Layout for the Underground Powerhouse of Jinping I Hydropower Station

(2) Pubugou Hydropower Station. The underground powerhouse of Pubugou Hydropower Station is arranged in the area with simple-structured, hard and complete rock of grade II, which avoids the lager faults, concentrated joint band, fracture zone and high ground stress areas. The quality of rock in the powerhouse can be classified into 3 main types shown from Fig. 18.3-3: in the west part of fault f_{12} and f_{13}, development of small faults, poor integrity of rock mass, most of them belong to grade III; in the south part of fault f_7, the weak weathered rock mass, most of them belong to grade III; in the east part of fault f_{12} and f_{13} and also in the north part of fault f_7, simple structure and good integrity of rock mass, most of them belong to grade II which is suitable to arrange the large underground caverns. Thus the underground powerhouse should be arranged in the downstream part of the 45# caving and the hole with depth above 260m, with surrounding rock of grade II. According to the overall layout of the project favorable to the safety operation of power station, shortening the construction period and saving the engineering quantity and investment, the longitudinal axis of underground powerhouse is selected in consideration of the directions of the four small faults existed in the main powerhouse and the direction of maximum principal stress. N42°E, which has been chosen as the axial direction, has a larger angle with main structural surface, which is usually more than 50°, and has a smaller angle with maximum principal stress, which is usually within 26.7°-36.7°.

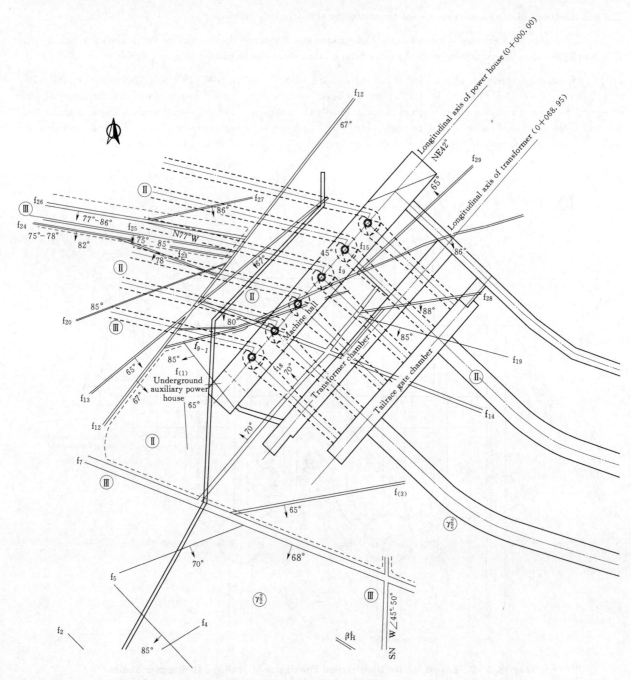

Fig. 18.3-3 Layout for the Underground Powerhouse of Pubugou Hydropower Station

18.3.2 Research and Evaluation Criteria of Surrounding Rock Stability of Underground Caverns

1. Analysis of the affecting factors on the stability of surrounding rock of underground cavern

There are many factors affecting the stability of surrounding rock of underground caverns, such as the geological factors, engineering factors, and construction factors, etc., in which the geological factors are mainly analyzed. The significant factors affecting the stability of surrounding rock include: the regional geology and seism, topography, lithology, geological structure, physical geological actions, hydro-geological conditions (including karsts channels and the distribution of mined areas, etc.), ground stress, rock mass physical and mechanical properties, rock type, structural surface properties, rock structures, surrounding rock types and their spatial distribution, and distribution of stratum containing radioactive minerals and harmful gases. Only the geological factors have been identified and analyzed, the detailed geological basis can be provided for the stability assessment and support design of surrounding rock.

2. The relationship between the structure and characteristics of rock and the form of deformation and damage for the surrounding rock of underground caverns

The stability of surrounding rock mostly depends on the rock structures. Therefore, the study of rock structures is an important basis for stability of surrounding rock. The rock quality can by classified by rock structure. The combination of geological and engineering works is the basis for the joint work carried out by the geological and design engineers. The key point of rock structure research is to determine the level, natural features, distribution rule, development density, status characteristics, mechanical properties, continuity, spatial combination, the scale of spatial structure, geometry, orientation and strong resistance for various structural surfaces The research methods of rock structure include the surveys of rock structural surface and structures, as well as the analysis of rock structure.

Rock structures are one of the main factors determining the rock integrity and rock deformation and failure, and also the main basis to determine the rock category and the supporting type. Therefore, it's particularly important to analyze and evaluate the stability of surrounding rock on the types of rock structure, the deformation and failure and their relationships.

3. Theories and methods for stability analysis of surrounding rock

The main factors affecting the stability of surround rock include the type of surrounding rock, the rock mass stress, underground water and project factors, therefore, qualitative and quantitative analysis of integrity or local stability of various surrounding rocks should be conducted based on above factors so as to adopt the relatively reasonable excavation schemes use the effective supporting measures to fasten the surrounding rock, and play full use of the rock bearing capacity, which can ensure the stability of surrounding rock.

There are two methods for the stability analysis of surrounding rock of underground powerhouse, qualitative and quantitative analysis. The qualitative analysis methods are the geological analysis method and engineering analogy method, and the quantitative analysis methods include the numerical analysis method, model testing method, on-site monitoring method and feedback analysis method.

The engineering analogy method is still the main method for the design of underground project, and the rock classification is the basis of engineering analogy method, which has been widely used in the design by the designers from around the world. The engineering analogy method should be used the whole process of engineering design and construction. For some stability analysis of underground cavern with complex geological conditions or large-span, the engineering analogy method cannot ensure the reliability and rationality of the design, thus the comprehensive evaluation with multiple methods must be made in many ways.

4. Evaluation criteria of surrounding rock stability

There is still lack of a reasonable and effective method of rock stability analysis and assessment in the integrity stability evaluation of surrounding rock of underground cavern. There are also no generally accepted quantitative safety indicators, and there are no uniform standards in the world. The calculation methods of stress state and rock mechanics model are not uniform, but the elastic-plastic theories are commonly used for the geotechnical theory researches and project actual experience. The stability assessment only based on elastic theory is unreasonable. The surrounding rocks are not with the plastic state, which are contrary to the modern theory of supporting, and the self-supporting capacity of rocks cannot be fully used. In fact, if part of the rock turns into plastic state, the surrounding rock will not be unstable for sure, and the rocks can provide fully self-supporting capacity, therefore some of the supporting works can be saved. In practical project, the suitable mechanical model is chosen to analyze based on the lithology and the state of ground stress.

Although there is still no unified conclusion on stress and stability evaluation technology on the surrounding rock of underground powerhouse caverns, but the awareness is in line. Based on the rock force, deformation, damaged or unstable areas, and changes in energy state, etc., the stability of surrounding rock of underground caverns should be evaluated from all sides by the strength criterion, the energy criterion and the critical value of deformation. In the current engineering practice, the stability evaluation criteria and the stability control indicators of surrounding rock of underground projects should be combined with the observation, analogy and analysis, so as to conduct full and comprehensive assessment.

18.3.3 Research on Support Design and Optimization of Large Sized Underground Cavern Group

1. Theories of supporting design of surrounding rock

The theory of supporting design of underground project is closely related to the development of rock mechanics. One of the important subjects of the theory is how to determine the load on the supporting structure. In the early 1960s, the surrounding rock was taken as the load, and the support was taken as the bearing structure, which formed the system of "load-structure". The surrounding rock was taken as loose structure, and its weight was calculated by the theory of collapse arch. The reinforced concrete lining was widely used as supporting structure. After the 1960s, rock mass was regarded as one of the load-bearing structures. The new system of "surrounding rock-supporting" was formed and worked together. The modern bolt-shotcrete supporting theory combined support and surrounding rock was formed by the concept transformation. The development of this theory, not only gives full play to bearing capacity of surrounding rock, also provides a broad space for the supporting design of surrounding rock of underground caverns.

The modern bolt-shotcrete supporting theory has the characteristics of timeliness, adhesiveness, flexibility, deep-going, mobility and closure, etc.

2. Status and development tendency of system supporting

Bolt-shotcrete supporting is adaptable, convenient, flexible and fast, which is suitable for rock mass with various structural types. In the engineering practice, it is used widely not only in rock above class Ⅲ, but also in rock with class Ⅳ and Ⅴ, it is used for the initial safety supporting and combined with the permanent supporting.

The theory of New Austrian Method merges the design, construction and monitoring into one organic whole, which conducts the dynamic design principles. It provides good prospects for the development of supporting design. Practice has proved that the New Austrian Method has been widely used not only in underground engineering but also in ground engineering such as the treatment of high slope, etc.

Over the past decade, the excavated dimensions of the cavern cross-section have increased; anchor construction equipment also has made significant progress, the supporting strengths of system bolts (cables) of underground engineering have increased, and the shotcrete supporting spans of corresponding types of surrounding rocks have also greatly exceeded the original scope.

Statistics show that the lengths of crown system bolts are general 0.25 to 0.30 times of the cavern span for the underground caverns with large spans. When the high-span ratio of cavern is 0.8 - 1.2, system bolt length of side-wall is generally equivalent to the length of crown system bolt. A minimum length of the top bolt is $0.25B$, and the length of bolt below the arch line should be as $0.20H$ stipulated by USACE.

The experiences of domestic underground powerhouses of large-sized hydropower projects have proved that the high-span ratio (H/B) of main chamber can be up to 2.13 - 2.67 (H/B are 2.13 and 2.67 respectively for Er'tan and Three Gorges underground powerhouses), significantly exceeds routine range of high-span ratio (0.8 to 1.2) of original bolt-shotcrete underground caverns. Thus when the high-span ratio is more than 2.0 or cavern group occurred, the stability and deformation of surrounding rock of side-wall are becoming the outstanding issues. Side-wall becomes the main position which needs to be supported and fastened, and its length of long bolt (cable) can be close or even over the cavern span, which can reach 0.25 to 0.45 times of side-wall height.

18.3.4 Research on the Mechanism, Prediction and Prevention of Rock Burst in High Geo-stress Underground Caverns

Rock bursting is a common and complex dynamic failure phenomenon during the construction of underground project with large buried depth, with the continuous development of hydropower project, there are more and more large-span and deep-buried underground cavern groups. The issues of high ground stress become more and more outstanding, and the rock burst occurs frequently. Especially in Western China, the deep-buried underground caverns are generally located in the area with high ground stress due to the special geological structures. The rock burst occurs repeatedly during the excavation of underground caverns for the existed or under-construction hydropower projects, such as Er'tan, Laxiwa, Pubugou, Taipingyi, Jinping Ⅰ, Jinping Ⅱ and Shuangjiangkou hydropower stations (See Fig. 18.3-4 and Fig. 18.3-5).

Some rock burst cases of large underground projects are showed in Table 18.3-2.

High ground stress and rock excavation disturbance are necessary conditions for rock burst generation. Rock bursting is a dynamic phenomenon of geological disasters, usually with catastrophic consequences such as casualties, damage of construction equipment and even project abandonment, which has become a worldwide problem of underground projects.

According to field investigation and indoor model testing, the researches on the occurrence mechanism of rock bursting, prediction, and control measures have been conducted by many scholars. However, due to some inherent characteristics of the rock mass such as anisotropy and heterogeneity, a lot of achievements are restricted to certain aspects of the post-verification without unified understanding. Therefore, the research of rock bursting is far from forming the systematic results.

Fig. 18.3 - 4 Collapse Caused by Rock Burst at Draft Tube of Underground Powerhouse in Ertan Hydropower Station

Fig. 18.3 - 5 Thin Films of Rib Spalling Produced by Rock Burst in Pubugou Hydropower Station

Table 18.3 - 2 Rock Bursting Cases Occurred in some Underground Projects

Project Name	Depth (m)	Lithology	σ_θ	σ_c	σ_θ/σ_c	n_b	Rock burst situation
Ertan hydropower	194	Syenite	90.0	220.0	0.41	29.7	Minor rock burst
Taipingyi underground cavern	400	Granite, diorite	62.6	165.0	0.38	17.6	Different Scale
Pubugou underground cavern	—	Granite	43.4	123.0	0.35	20.5	Occurred rock burst
Laxiwa underground cavern	300	Granite	55.4	176.0	0.31	24.1	Occurred rock burst
Lubuge underground cavern	—	Limestone	34.0	150	0.23	27.8	Pie core
Jinping II diversion tunnel	150	Marble	98.6	115.0	0.82	31.1	Mainly strong and weak rock bursts
Tianshengqiao II diversion tunnel	400	Dolomitic limestone	30.0	88.7	0.34	24.0	Moderate
Yuzixi diversion tunnel	200	Granodiorite and diorite	90.0	170.0	0.53	15.0	Moderate and strong rock bursts

Note: σ_θ shear stress of surrounding rock (MPa); σ_c compressive strength of rock (MPa); brittleness degree of rock, $n_b = \sigma_c/\sigma_t$, σ_t is tensile strength of rock (MPa).

Domestic and abroad scholars have proposed various theories and doctrines of rock burst, which include the energy, stiffness, strength, impact orientation, damage and fracture, instability, nonlinear science (such as mutation, fractals, bifurcation and chaos theory, etc.), rock burst theory based on the prophetic experience (such as fuzzy mathematics, neural networks, etc.).

Combined with the mechanism of rock bursting, three measures can be taken to engineering rock bursting, as follows:

(1) Prediction. A variety of methods are used to implement the comprehensive prediction, such as the prediction method with the criterion, energy-storage analysis and testing of rock bursting, the prediction method of acoustic emission on spot monitoring, low temperature detecting method and geological forecast method, etc.

(2) Prevention. The main prevention measures include the improvements of physical and mechanical properties of surrounding rock (short footage, less frequency, small dose, smooth blasting), stress relief method, adjustment of operation and efficient measures, as well as personnel and equipment protection.

(3) Treatment. The strength can be reduced by the measures such as high pressure of water-jet, deep water injection, etc., and the surrounding rock can be fastened by the measures such as shotcrete, anchor and grid steel frame, etc.

18.3.5 *Rapid Monitoring and Feedback Analysis in Large Underground Cavern Group Under Construction*

There are many uncertainties for the underground projects, thus it is very difficult to accurately determine the basic parameters of rock mass, boundary conditions, initial conditions, optimized excavation plan and supporting parameters before construction. It is necessary to adopt the method of "Dynamic Design". Rapid monitoring and feedback analysis play an important part of "Dynamic Design" during the construction of underground cavern, and they also play an important role in the process of the supporting design. The common methods used for monitoring and feedback analysis are conventional method of analysis (engineering analogy, etc.), numerical analysis and mathematical & physical model analysis.

Nowadays, the monitoring and feedback analysis have been carried out in construction process of large-sized underground powerhouses existed and under construction in China, such as existed underground powerhouses of Ertan, Xiaowan and Laxiwa, under construction projects of Pubugou, Xiluodu and Jinping I, which have achieved great results through monitoring and feedback analysis. They have provided a favorable basis for parameter adjustment of excavation and supporting designs during construction of large underground powerhouse caverns, and ensure the stability and safety of surrounding rock during construction.

Fig. 18.3 - 6 presents the arrangement of equipment and related feedback analysis for the typical monitoring profile of Xiluodu underground powerhouse.

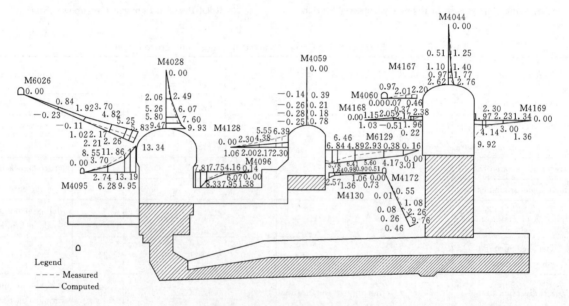

Fig. 18.3 - 6 Arrangement of Multipoint Displacement Meter and Related Feedback Analysis for the Typical Profile of Xiluodu Underground Powerhouse

According to the monitoring and feedback analysis of Xiluodo underground powerhouse during the construction, the stability evaluation of the surrounding rock and the predictive effects of subsequent excavation on the surrounding rock shall be conducted in layers. These works provide the important basis for adjustment of the supporting design parameters and the excavation schemes during the construction.

18.3.6 Rock-bolted Crane Beam of Large Underground Powerhouse

The rock-bolted crane beam has the characteristics as follows: the bearing capacity of the surrounding rock can be fully utilized; the crane can be installed and put into operation ahead of schedule, which is beneficial to speed up the progress of the civil construction and equipment installation, and the span of the main powerhouse can be reduced.

The beam and pillar structure was used mostly in early underground powerhouse as ground powerhouse. In consideration of the characteristics of underground powerhouse, the suspended (hanging) crane beam was first put into usage in the 1950s to the 1960s abroad. Since the 1970s, the rock-bolted crane beam has been utilized gradually with the development of the underground excavation technology and rock mechanics. Over the past decade, the rock-bolted crane beam has been adopted in mostly large-sized existed or under-construction underground powerhouses such as Xiaowan, Laxiwa, Longtan, Pubugou, Xiluodu, Jinping I and Jinping II underground powerhouses. With the increase of scale and lifting weight of the underground powerhouse, the load applied in the rock-bolted crane beam is also increased.

The design methods of rock-bolted crane beam include the rigid limit equilibrium method and finite element method. For the rock-bolted crane beam of large underground powerhouse with complicated geological conditions, the finite element method should be utilized. The finite element computation include: the actions and effects of the excavation from lower part of underground powerhouse on the anchor bolt and beam body of the rock-bolted crane beam and surrounding rock under construction; the actions and effects of the crane loads on the anchor bolt and beam body of the rock-bolted crane beam and surrounding rock under operation; the finite element assessment on safety stability of the rock-bolted crane beam.

The rock quality of rock-bolted crane beam supports and the combination of the unfavorable joints and cracks and poor structural surfaces are the main factors affecting the stability of the crane beam rock foundation. The excavation and construction techniques and the reasonable excavation and construction procedures and technologies of the rock-bolted crane beam are the key points to ensure the slope formation of rock wall and the design requirements. The tensile and compressive bolts of the crane beam are anchored by the efficient measures so as to share the loads by the crane beam and side-wall surrounding rock, as well as ensure the stability of crane beam and rock of support.

With the wide utilization of crane beam structures, the theoretical methods, construction technologies and experiences of the rock-bolted crane beam have been enriched and improved. It has been proven by many engineering practices that the bolt stress and deformation of surrounding rock are influenced greatly by the excavation from lower part of underground powerhouse, especially for the underground powerhouse with large span and complicated geological conditions. The countermeasures are of controlling the timing of the concrete pouring and the track installation of the rock-bolted crane beam. Taking Jinping I underground powerhouse as an example, the concrete pouring of rock-bolted crane beam is conducted after the excavation of bus tunnel. However, the track installations are conducted in batches based on the geological conditions of different sections. Most of them are installed when the excavation reaches the machine stable. In this way, the service schedule of crane is met, and the effects of the excavation from lower part of underground powerhouse on the rock-bolted crane beam are also reduced effectively. In view of this, the favorable basic conditions for the safety operation of rock-bolted crane beam are provided.

18.3.7 *Designs on Seepage Control and Drainage of Large Underground Powerhouses*

The designs on seepage control and water drainage are important parts in the design of the underground powerhouse, which can greatly reduce the adverse effects of underground water on the surrounding rock and the seepage pressure implied on the supporting structures, especially in the cases of the rock mass with developing fractures and large seepage discharge and underground powerhouse near by the reservoir. The operation conditions of underground powerhouse can be improved at the same time.

The basic design principles of seepage control and drainage for the underground powerhouse are as follows:

(1) The design is conducted on the basis of the geological and hydrogeological conditions of the underground powerhouse in the engineering, as well as the relative position relationship between main structures and underground powerhouse (head part, middle part and tail part).

(2) Combination of the drainage and seepage control, mainly for the drainage and seepage control outside of the powerhouse, supporting with the drainage inside of the powerhouse.

The function of the seepage control outside of the powerhouse can be described in two aspects. On one hand, the underground water level and seepage pressure head can be reduced effectively, thus, the seepage pressure of surrounding rock is also reduced simultaneously. In particular, for the rock mass softened in the water, the strength index of the rock below the underground water level will be reduced sharply. Furthermore, higher underground water level is not beneficial to the stability of the surrounding rock. On the other hand, it can reduce the seepage discharge from the surrounding rock into underground powerhouse.

Seepage prevention curtain is generally used as the measures for seepage control outside of the powerhouse. According to the engineering geology, hydrogeological conditions and seepage prevention curtain position in front of the powerhouse, the position of seepage prevention curtain is determined by the engineering analogy method, if necessary, 3D seepage finite element analysis method shall be also used.

In normal conditions of the fracture water and seepage discharge, the drainage system inside of the powerhouse is as follows: For the drainage of the crown and side wall, the boreholes are mostly used. Sequentially, the seepage water and drip water from production water supply are flumed to the collecting well of underground powerhouse through the drainage tubes and ditches. Finally, the water is pumped to the outside of the powerhouse.

The seepage discharge of the underground powerhouse should be determined by the engineering analogy method. It is determined synthetically by the calculation results of 3D seepage finite element method if necessary.

The typical layout of seepage control and drainage system for the underground powerhouse of Er'tan Hydropower Station is shown in Fig. 18.3 - 7.

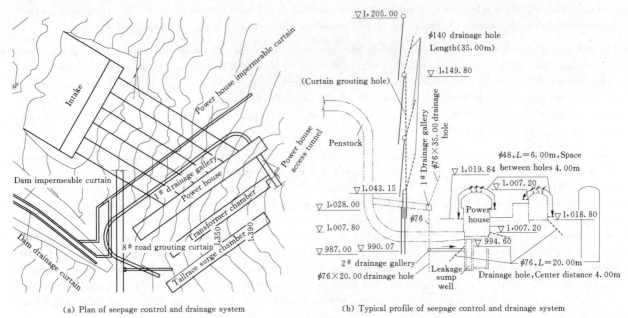

(a) Plan of seepage control and drainage system   (b) Typical profile of seepage control and drainage system

Fig. 18.3 - 7   Typical Layout of Seepage Control and Drainage System for
Underground Powerhouse of Ertan Hydropower Station (Unit: m)

18.4   New Progress of China Hydraulic Tunnel Construction

18.4.1   Introduction to China's Existing Hydraulic Tunnel Norms

From the end of the 20th century to the beginning of the 21st century, in view of that *Specification for design of hydraulic tunnel* SD 134—84 (trial) (hereinafter referred to as SD 134—84) issued by the former Ministry of Water Conservancy and Electric Power in 1985 has been used for many years, in combination with the design and construction experience and research achievements in China hydraulic tunnel, and absorbing foreign advanced technology, the planning and design departments of the Ministry of Water Resources and the former Ministry of Electric Power managed and revised the original SD 134—84 specification, compiled respectively SL 297—2002 and DL/T 5195—2004 *Specification for design of hydraulic tunnel*, the two industry standards for water conservancy industry and electric power industry (hereinafter referred to as SL 297—2002 and DL/T 5195—2004).

The new adjusted, incorporative, revised specification clauses have text normativity and comprehensive contents, which embody the characteristics of hydraulic tunnel. In the clause notes, the project examples have been added, which illustrate the reasons of the continuous use of the original clauses, the inductive, incorporative and new contents, and the guiding ideology for compiling in detail. Based on DL/T 5195—2004 and focusing on the primary and secondary supporting design, China's current hydraulic tunnel specifications were introduced briefly as follows.

(1) The surrounding rock classification stipulated in the current hydraulic tunnel specifications was according to *Code for engineering geological investigation of water resources and hydropower* (GB 50287—1999). The hydraulic tunnel surrounding rock classification adopted predominantly qualitative standard and the surrounding rocks were divided into class I - class V from good to bad.

(2) The rock cover layer of pressure tunnel should satisfy the conditions that the cover layer does not lift and the hydraulic fracturing and the seepage failure do not occur. When the formula is used for judgment, the current specification recommended the Norway criterion and gave the notes to the use of other criteria.

(3) In DL/T 5195—2004, the general principle that the hydraulic tunnel design should make full use of the self-stability, the carrying and permeability capacities of surrounding rock were clearly put forward. SL 297—2002 also made clear this principle in the specific clauses.

(4) DL/T 5195—2004 complied with the design principle of GB 50199 *Unified design standard for reliability of hydraulic engineering structures*, and used the limit state expression of the partial factors to calculate the tunnel supporting. The changed the partial factors maintained the original safety level. SL 297—2002 continued to use the algorithm of the single safety factor.

(5) In the calculation of concrete and reinforced concrete lining, DL/T 5195—2004 used two kinds of design method of crack design and crack limit design, cancelled the calculation method that the crack was not permitted. According to the three kinds of anti-seepage requirements such as "strict", "general" or "no", SL 297—2002 stipulated the three permeability resistance design principles of the corresponding crack resistance design, the crack limit design and the unrestrained crack width design.

(6) In the informative annex of DL/T 5195—2004, according to the classification of surrounding rock, tunnel diameter or span, the shotcrete and rock bolt support types and their reference parameter table were given. In SL 297—2002, according to the structure permeability resistance design principles, whether surrounding rocks meet the requirements of the minimum cover thickness, whether have the ability to bear the internal pressure and classification of surrounding rocks, the caves lining (including shotcrete) type selection table was given.

(7) The current specifications allowed using the method of elastic mechanics and structure mechanics to calculate the normal concrete and reinforced concrete lining, retained part of the calculation method of elasticity mechanics and structural mechanics in the clauses. For the cross area tunnel and bad geological cavern section with diameter or span more than 10m or bifurcation pipe, DL/T 5195—2004 suggested adopting finite element method for the structural analysis. In SL 279—2002, the guidance suggestions on the applicability of the mechanical model of surrounding rock, the construction process simulation and the application of seepage flow stress coupling analysis in finite element calculation were given.

(8) As a kind of information construction method in drilling and blasting excavation, NATM involves many links in the design, construction and construction monitoring and dynamic feedback design. The current specifications had summarized the existing engineering experience and had complemented and enriched correspondingly in the relevant clauses.

(9) In SL 279—2002, on the basis of summarizing the past project experience and lessons, the adverse geological section design chapter was added, the bad geological hole section was defined, and the design principles were given. The contents involved tectonic intersection zone, water, rock burst, harmful gases, lava zone, quicksand or loose soil, high plastic rheological rock, swelling rock and other adverse geological cavern sections.

(10) Both the two specifications supplemented the contents on prestressed concrete lining, high pressure reinforced concrete bifurcation pipe and the tunnel plugging design.

The current *Specification for design of hydraulic tunnel* reflects the design, scientific research and new technology level of China at the beginning of 20th century, which has currently been widely used and has promoted the standardization of the hydraulic tunnel design in the higher level.

18.4.2 New Progress in Structure Design of Hydraulic Tunnel

Structure design of hydraulic tunnel should make full use of the self-stability, carrying and permeability capacities of surrounding rock. The surrounding rock and the supporting should jointly be applied as an integration structure. In the process of carrying out the design principle, China has summarized the existing design experience at home and abroad. With the aid of modern computing, monitoring and information interaction means, China has used modern numerical analysis and feedback analysis method to carry out the dynamic design in the indoor and field test and tunnel construction, and has studied on security and stability of surrounding rock and supporting. A lot of work has been done.

1. No lining and shotcrete and rock bolt support design

China has set up a more mature no lining and shotcrete and rock bolt support design method used in preliminary design stage. Shotcrete and rock bolt support types and their parameter table, and the instructional supporting suggestions on the class I - class V surrounding rock and the cavern with excavation hole diameter or span $D<30m$ were given in DL/T 5195—2004. For example, it does not need to consider supporting for class I surrounding rock with $5m<D<10m$, or class II surrounding rock with $D<5m$. Or, it can use shotcrete and rock bolt support design of 50mm thick. It only does not need to consider supporting for class I surrounding rock with $D<5m$. And for class IV surrounding rock with $10m<D<15m$, it recommends to adopt 200mm thick steel fabric shotcrete and arrange anchor bars with 4.0 - 5.0m long and 1.0 - 1.5m distance. The secondary supporting should be adopted for the part with big deformation in situ monitoring. When necessary, it needs to set steel arch or grid arch. The secondary supporting can be shotcrete and rock bolt support or cast-in-place reinforced concrete support. As another example, no suggestion was given in the table for the cavern in class V surrounding rock with 10m<D<15m. It shows that special argumentation and research must be done for the design. For the cavern section at different geological conditions in general project, currently, it has been possible to adopt experience analogy method and partial block balance method to make a basically reliable stability analysis on surrounding rock, and in the construction stage, make the successful supporting design through the analysis of the geological exploration data to adjust new supporting type and parameters. For important project and the adverse geological cavern section lack of existing engineering experience, the design generally adopts more advanced finite element numerical analysis method to assist supporting design. And we should strengthen the lead geological prediction and dynamic feedback design in the construction and has obtained the successful experience.

With the development of modern computer monitoring, geological exploration technology and engineering monitoring technology, the research and application of the establishment of classification methods and standards for surrounding rock that conform to the real situation of hydraulic tunnel surrounding rock have also been further carried out. For example, the surrounding rock classification experience for diversion tunnels of Rongdi Hydropower Station, Zhangfeng Reservoir and other engineering, the application research of the surrounding rock classification method of Yindajihuang tunnel on the combination of the Q system and China's rock classification surrounding water resources and hydropower, and the discussion on the other for TBM construction, lava area and swelling rock tunnel surrounding rock classification method have made the new foundation work on the hydraulic tunnel surrounding rock supporting design.

2. Concrete and reinforced concrete lining crack limit design and research

For hydraulic tunnel with strict seepage control requirements, the specification recommended to use prestressed concrete or steel lining. Concrete and reinforced concrete lining mainly adopts the crack limit design principle. Since 1930s, international began to research on the theory and calculation method of crack developing of concrete and reinforced concrete component. Chinese scholars started the research on this problem since the 1950s. Although the important research results have been incorporated into a variety of specifications of each country, so far a consensus of crack width calculation theory has not reached on the international. The formula and the crack limit standards in various specifications are also different. At the same time, the work foundation of the calculation formula both at home and abroad mainly aimed at the space builder of the r general reinforced concrete, rarely gave attention to the application of hydraulic tunnel, which did not meet the idea that made the surrounding rock to become the bearing and seepage control body as far as possible. Lining crack limit calculation formula in China derived from SDJ 20—78 *Design code hydraulic reinforced concrete structure* (hereinafter referred to as 78 code) before 1996, and then derived from DL/T 5057—1996 *Design code for hydraulic concrete structures* (hereinafter referred to as 96 code). The formula in these two codes derived from using the theoretical formula derived from the mechanism of crack developing. The 78 code was based on the bond-slip theory, and the 96 code was based on the comprehensive theory of the two theories of integrated bond slip and no slip. Slip theory thought that the bond force between steel and concrete after cracking were destroyed and led the slip between the two, and the width of the crack depended on the differential deformation of the two. No slip theory thought that the destruction and slip of the bond force were negligible, and cracks were caused by the retraction of concrete, and the crack width depended on the back shrank, and the deciding factor was the thickness of the covering layer out of the reinforcement. Guangzhou Pumped Storage Power Station in China used the empirical formula in American ACI318 in the design of the lining. The formula was according to the general crack calculation theory combined with the mathematical statistics method. The actual case result showed that for the same crack width, the amount of reinforcement calculated in the formula of the code (DL/T 5057—1996) was bigger than that of thecode (SDJ 20—78), and with the increase of protective layer, the reinforcement quantity linearly increased. The ACI formula also had the same situation. Because China basically used the 78 code formula since the 1960s, and no design errors caused by the calculation were found, this formula was still used in many designs. The allowable value of China's control crack width was 0.15 - 0.30mm. In fact, this is just a design standard, and can not achieve the goal of the control on real crack width.

Chinese scholars put forward crack calculation formula aiming at hydraulic tunnel lining, more considering the effect of surrounding rock, fissure water pressure and the contact condition of surrounding rock and lining, pushing the applicability of the formula one step further. However, the adoption of relatively simple formula can not reflect the surrounding rock and lining joint work state. The assumption on the way and distribution mode of crack developing had bigger difference with the actual. It was difficult to consider the nonlinear of surrounding rock and reinforced concrete materials, the contact condition of the two and the coupling relationship between fissure water pressure and stress. There was also the same problem on the strength calculation of reinforced concrete lining.

The further research on the joint action of surrounding rock and reinforced concrete lining includes hydraulic tunnel indoor and field water pressure test, seepage, stress and seepage flow stress coupling numerical analysis of the joint action of surrounding rock and lining, and the feedback analysis based on the test data. China's direct pressure water test of hydraulic tunnel lining was introduced (Pan, 2000). The experiment indicated that after the lining cracking, the high-pressure water in the seam made tensile stress into compressive stress, and tensile stress value far from the lining seam outside also decreased accordingly. It was difficult to produce new cracks and the original cracks increased significantly. The law of the crack was few and wide, which reflected the special effect of the high pressure internal water leaking on lining working state. Check the reinforced concrete lining bearing the high internal water pressure in practical project after emptying the tunnel. Generally, only in part of the cavern section and bifurcation pipe appeared relatively and obviously dense cracks. The crack width was not large after emptying, and part close under the effect of outside pressure. In the operation, the reinforcement stress was not enough to reach yield value generally.

Previous domestic indoor pressure water test focused on the research on bearing capacity, using steel or balloon to load. Recent research on the lining crack test (Shen, 2010) first adopted the large scale hydraulic tunnel model of the real water pressure loading. The model was designed according to the geometric and physical similitude. Geometric scale was 1 : 100. The model was 6.50m long, 5.50m deep, 5.70m high. The tunnel diameter was 80cm and the lining was 6cm thick. The concrete grade was C20 and connected by transition section. Single or double reinforcement were simulated by the two segments respectively. On the top and one side of the Lining, there was inducing joint with 1mm wide and 5mm deep. Surrounding rock was made of the C10 concrete mixing with air-entraining agent, correspondingly class IV surrounding rock. Within a certain range of the surrounding rock model, the rock fracture was simulated. The fly ash, lime and cement mix slurry were adopted to backfilling the seam. The water pressure adopted hierarchical pressure to 1.7MPa, and unloaded after the pressure stability. The test simulated better the effect of the internal water leaking. The lining internal water pressure was relatively high of 1.3MPa. Reinforcement stress measured values were small. In addition to the measured value at the bottom of the single-layer reinforcement nearly 100MPa, the rest were 40MPa. Nearly a quarter of the measured values of the measuring points were under 5MPa. The reinforcement stress curve had a tendency of the stress decrease with the load increase. It was similar with the test results of reinforcement stress in the field water pressure test. The calculation results of crack width using the formula in the code of SDJ 20—78 showed that in addition to the crack width of a measured point with the stress measured values nearly 100MPa was about 0.03 mm, the calculated value of the rest each measuring point were less than 0, namely no cracking. The author discussed in this paper on constant 0.7×10^{-4} in the crack width calculation formula (SDJ 20—78), namely the average tensile strain value of the structural concrete and the reasonable value range of steel stress.

Under the encouragement and the suggestion of specifications, results given priority to the finite element numerical simulation and closer to the working state of the reinforced concrete lining tunnel have emerged greatly, which provides the reference value for the design. Here, no introduce. The test and calculation results enrich the understanding of the joint working state of surrounding rock and lining, and the actual engineering numerical calculation results guide the hydraulic tunnel design of the major projects, which greatly improve the design level of China's lining crack limit.

3. Seepage, stress and seepage flow stress coupling analysis of the integrated surrounding rock and lining

The conventional calculation method of concrete and reinforced concrete lining is to separate lining and surrounding rock, appropriately consider the effect of surrounding rock by a simplified method. The external water pressure adopts single factor method to determine. Geostress, external pressure of surrounding rock use a simplified method to determine. Because it does not consider the material nonlinear properties of surrounding rock and lining, it is difficult to reasonably consider the joint action of surrounding rock and lining. In China, according to the different needs of design and research, the integrated seepage, stress and seepage flow stress coupling numerical analysis method of surrounding rock and lining has been selectively and reasonably applied, and has become a powerful tool for hydraulic tunnel design and the first choice means to solve the design difficulties, promoting the development and application of the new design concept. In China, the finite element method is mainly used in the calculation of actual hydraulic tunnel lining. In addition to the early use of elastic and elastoplastic model, the material model also includes the viscoelastic model, which is suitable for soft rock and rheological surrounding rock model, the fracture mechanics model, the reinforced concrete model, etc. In seepage and seepage flow stress coupling calculation, especially for high internal and external water pressure, the surrounding rock and concrete lining act as the permeable structure, and water pressure is imposed by volume force.

To really simulate the initial seepage field and stress field as far as possible is the most important step to know the real working state of the surrounding rock support. For those engineering domains that are able to provide borehole water level, water pressure test and springs emersion materials, engineering geological sliced and profile figures within the calculation scope, generally, it can use the assumed boundary conditions to solve the initial seepage field, or use the feedback analysis to get the relatively reasonable permeability coefficient, free surface of groundwater and a relatively reliable initial seepage field. Although the rock fracture seepage analysis obtained great development in recent years, it is still not enough to be used for seepage analysis of large area. The finite element feedback analysis method includes direct solution trial and error method and more complete back analysis method theoretically. direct solution trial and error method was successfully used in the analysis of lining strength and surrounding rock stability of Yamzho Lake power diversion tunnel external water pressure, the thickness optimization of $7^{\#}$ cavern duct piece lining in Water Diversion Project from Yellow River to Shanxi

Province as well as the evaluation on seepage of mountain in the left bank and leakage amount of underground cavern group of the Xiaolangdi, and mountain settlement analysis, and other projects. Seepage back analysis is seldom used in large area and the whole hydraulic tunnel area. The initial geostress field finite element regression analysis by using geostress measured values has been widely used in the design of the underground cavern group with high internal water pressure and high geostress in large engineering.

The construction process simulation methods, which is based on the initial seepage field and geostress field, the appropriate selection of sub calculation domain, the application of the finite element incremental theory and with the aid of seepage control and reinforcement measures, can obtain in principle osmotic pressure, hydraulic gradient and seepage flow of the key parts, the stress, deformation, hydraulic fracturing of the structure and characteristic physical parameters of other damage of the surrounding rock and lining at any time during construction and running periods. The actual calculation complexity and precision depends on the complexity and importance of the calculation object, the design calculation purposes and needs, and the corresponding geological data, the material parameters and construction procedure and the completeness degree of the plan. The sensitivity analysis of the calculation parameters is often used in the solution comparison of the designers. The simplified calculation results are often used for designers to judge structure safety and compare the solution of the elastic or structural mechanics. Seepage flow stress coupling numerical analysis is mainly used in the concrete lining and the bifurcation pipe of high internal water pressure.

At present, the calculation software of the domestic autonomous development and the secondary development after the introduction from abroad, and computer and network equipments have basically met the needs of the complicated analysis and calculation of hydraulic underground structure.

4. Permeable lining

Many internal water pressure head of waterway system of the underground hydropower station in China are more than 300 m. The highest is more than 600m. A lot of underground cavern groups with high internal water pressure are built in the surrounding rock, giving priority to class I, class II. The layout of waterways and workshop has been reviewed and demonstrated by layer thickness, hydraulic fracturing and the seepage stability. The surrounding rocks have the ability to independently bear the internal load and anti-seepage after high pressure consolidation grouting treatment. Therefore, to use high pressure reinforced concrete lining and bifurcation pipe to replace steel pipe and steel bifurcation pipe can yet be an economic and reasonable choice. When the internal pressure increases to a certain value, the reinforced concrete lining is bound to crack, and the crack directly reach rock face. When internal water seeps into the crack rock, the surrounding rocks become the main body to bear the internal pressure and the anti-seepage. The lining design no more has strength requirement. Therefore, the proper limit of crack developing width to prevent too much the seepage flow often becomes the main design principle. This design can play the role of load-bearing and anti-seepage ability as large as possible, which is widely attached importance by the design and research personnel. They actively carry out the study of high pressure permeable reinforced concrete lining.

In Guangzhou first-stage Pumped Storage Power Plant, the diameter of high pressure tunnel concrete lining and main bifurcation pipe hole is 8.0m. The hydrostatic pressure is 610m. The design personnel combined engineering practice at home and abroad to research the design method of permeable lining, through the analysis of the seepage field of surrounding rock and lining, and the internal and external pressure exerting seepage load by the seepage volume force, analyzed the force of the lining structure under the internal water pressure load. The results showed that under the constraint of the deformation of surrounding rock, the lining reinforcement cannot give full play to the role, and we could boldly reduce the number of the lining reinforcement. The successful use in Guangzhou Pumped Storage Power Plant enabled "permeable lining" to enter the practical stage for the first time. In recent years, the permeable lining design has been used in high water head pumped storage power stations, such as Tianhuangping, Tongbai, Huizhou.

Recently, the design and the research on permeable lining and bifurcation pipe adopted finite element method, and carried out the geostress field, seepage field, seepage flow stress coupling analysis. Combining with the prototype monitoring data, the different deep researches were carried out on load sharing of surrounding rock and lining, the plastic zone developing in surrounding rock, the crack developing in lining, seepage hydraulic gradient and seepage flow. For example, it was set up the relation equation of the strain – crack width – permeability coefficient of reinforced concrete lining (Zhang et al, 2008), used three dimensional elastoplastic damage finite element to build seepage flow stress coupling model, put forward a set of relatively complete permeable lining calculation method. The method was used in Huizhou Pumped Storage Power Station Project. The calculation results coincided well with the monitoring data. It was used a similar method to research on high pressure reinforced concrete lining crack opening and closing, permeable and leakage passages of the surrounding rock and the effect of consolidation grouting of Huizhou Pumped Storage Power Station (Deng et al, 2006), which provided a method for seepage flow calculation. The adoption of the equivalent continuum mechanics model to carry out flow stress coupling analysis considering permeable volume force, the simulation of the joint effect of surrounding rock and lining is a big step in the design of the lining structure. On the other hand, after the surrounding rock and reinforced concrete cracking, the material discontinuous state is very complicated. The use of the characteristic physical quantity, such as strain calculation of crack width, still faces the problem that the crack developing theory and calculation formula is inconsistent at present. In addition, the discontinuous problems, such as crack and slip of lining and surrounding rock interface, deserve attention. It was discussed the contact condition of concrete and surrounding rock in pressure tunnel cracks, and gave the discriminant formula of lining independently work when detaching surrounding rock (Peng and Zhong, 2010).

Some projects that used reinforced concrete lining pressure tunnel and bifurcation pipe had big leakage during the process of the first water filling trial in waterway, which were far beyond the expected value of the designers. In fact, obvious changes have occurred in workshop, high pressure diversion cavern, all kinds of construction cavern, and drainage cavern to form high water head spatial seepage field and high geostress field before and after plant area. Due to the uncertainty of geological defects and the unavoidable undiscovered hazards in the construction, The permeable channels connected in surrounding rock after the high pressure water leakage enables the seepage field to present anisotropy and highly non-uniform state. To limit the value of the hydraulic gradient J (e.g., $J < 10-12$) is an averaging standard. It is extreme possible for local hydraulic gradient to break through the standard. In the secondary stress field after the excavation, the minimum main stress is generally less than the high water head pressure on the edge of the cavern. It is difficult to avoid that the hydraulic fracturing and asymptotic damage within limits occur in local of the surrounding rock reinforcing ring. The experience of Guangzhou, Tianhuangping and Huizhou Pumped Storage Power Station showed that for surrounding rock with the main class I, class II and meeting the conditions of high pressure consolidation grouting, The pressure-bearing and seepage control body is safe, which is composed of high pressure reinforced concrete lining and surrounding rock of 600m level. Although a large amount of seepage appeared in the water filling test, according to the analysis of monitoring data and the result of the human tour inspection, repair the lining concrete cracks and the engineering defect, backfill the useless cavern, and take supplementary measures, such as consolidation grouting, the leakage can be reduced to below the expected value of the designers. High pressure reinforced concrete diversion tunnel and bifurcation pipe have a high requirement on the geological condition. The surrounding rock characteristics of different project are not identical. The successful experience of project should not be simple to paraphrase. When design personnel select reinforced concrete channel to replace steel structure, they should be carefully argument on the feasibility.

18.4.3 Diversion Tunnel in Jinping II Hydropower Station

18.4.3.1 Engineering Layout Overview

The big natural fall 310m of Jinping river bend was utilized in the Jinping II Hydropower Station, and straightened excavation of tunnel and diversion for power generation. The total installed capacity is 4,800MW. Average annual power output is 24.23TW·h. It is the highest water head, the largest capacity and best hydropower station in the twenty-one cascade hydropower stations on Yalung River.

The project area is located in the transit diagonal valley belt of Tibetan plateau to Sichuan basin valley. The main landform belongs to erosive mountains. The terrain belongs to the alpine valleys. The maximum cutting depth is 3,000m. The mountains are abundant, and continuous. The main mountain elevation is above 4,000m. the highest mountain elevation is 4,488m. On tectonic areas in this area, it is located at the southeast of Songpan-Ganzi geosynclines fault-fold belt, in the "Sichuan-Yunnan rhombic block" surrounded by multiple fracture, which is more complex geologic structure fault-fold belt. The lithology along the diversion tunnel engineering section is mainly marble, limestone, sandstone and slate of upper triassic in triassic system, belonging to karst development area.

Jinping II Hydropower Station is composed of a large underground cavern complex. Seven tunnels are arranged across the Jinping Mountain, including about 16.67km long the diversion tunnel, two parallel auxiliary caverns with the length of 17.5km and one construction drainage cavern. On the upstream of the Power plant workshop, the excavation diameter of the four surge-chambers is 26m and the height is 150m. The excavation diameter of eight high pressure shafts is 9.2m and the height is 260m. The three big caverns of the main factory workshop, the main transformer cavern, tailrace surge chamber are parallelly arranged. Among them, the main-side workshop is 352.4m long, 28.3m wide and 71.2m high. The total amount of excavation is more than 13.2 million m^3.

Jinping Secondary Hydropower Station diversion tunnel axis and Jinping Mountain ridge line are nearly orthogonal. The mountains along the line are abundant. The largest tunnel buried depth is about 2,525m. The cavern sections with the buried depth more than 1,500m account for 76.7%. The excavation cavern diameter is 12.4-13.0m. It has the characteristics of deep bury, long cavern line, big cavern diameter, high geostress level, high groundwater pressure, complicated karst hydrogeological conditions and difficult construction layout, and is the world biggest comprehensive overall scale and the most difficult tunnel engineering that had been built and under construction.

18.4.3.2 Key Problems and Solutions in Engineering

Jinping Secondary Hydropower Station diversion tunnel has the following engineering geological problem basically.

1. High geostress and rock burst

The engineering area is located in the southwest high geostress area in China. The measured geostress results showed that geostress value increased with the increase of the buried depth, and the maximum principal stress of the cavern depth between 600m and 3,000m for changed from parallel bank slope into nearly vertical, namely, the geostress shifted from horizontal stress state to give priority to with vertical stress state, and the measured maximum stress value was 42.11 MPa. According to the geostress regression analysis, the maximum principal stress value of the tunnel line elevation was 70.1MPa, and the minimum principal stress value was 30.1MPa. According to the cavern excavation construction and theoretical analysis, it showed that the occurrence of rock burst was controlled by lithology, structure, underground water, wall rock types, and rock mass structure. And it mainly was class II surrounding rock with good integrity rock mass, and hard and dense rock. The following was class III surrounding rock. In class IV surrounding rock, it appeared less. Rock burst was given priority to weak to moderate. The local area would appear strong rock burst. The rock burst form was given priority to continuous and scattered distribution obscission and exfoliation. Through the construction of the auxiliary cavern, the prediction to the development degree of rock burst was correct. In the rock burst of 3[#] diversion cavern TBM tunneling section, there was a strain-based rock burst with exfoliation. The other was caving of tectonic fracture zone of surrounding rock under high geostress.

In the process of tunnel excavation, the rock burst was given priority to medium to medium weak. The local had strong rock burst. In the construction cavern section of borehole-blasting method, the adoption of comprehensive treatment, protective measures of stress relief hole, quick rock bolt, hanging net shotcrete supplemented by grid arch can effectively reduce the impact of rock burst on the construction. In the TBM construction cavern section, in addition to the conventional control method, combining with the characteristics of the construction optimization of supporting design, the quick anchorage surrounding rock temporary support and permanent support way were adopted, such as using the mechanical expanding shell type expansion rock bolt, or water expansion water rock bolt, using nanometer material concrete or steel fiber, fiber reinforced concrete that can quickly close surrounding rock, spraying concrete with ultrafine zeolite powder, mixing accelerator, and using the flexible fence to replace the common reinforcement net and other measures. Due to limited by the characteristics of the structure and properties of TBM and not enough flexible construction, rock burst often had not clear precursor, and strong sudden, and occurred before the surrounding rock revealed on the shield, which was difficult to play the TBM safe driving advantage at a high speed. How to solve the rock burst problems in TBM construction such as large buried depth, high geostresshave been accumulated certain experience at present, however, the response method is not yet mature, it will be further improved from all aspects in the future.

2. Underground water gushing prediction and its processing

The conditions revealed by caving, and the comprehensive study on karst hydrogeological indicated that although the existing karst hydrogeological condition was relatively complex, the overall development degree of karst was relatively weak. It mainly showed that the corrosion along the crack surface, and there was no hall form, rivers form underground karst system. Another key problem of the diversion tunnel construction was the outburst flood forecast of groundwater and its processing method. Hydrogeologic karst fissure water was rich. Caving had produced large water many times, and the major water points had the characteristics of high water head and large flow. The prediction of the gushing water characteristic of tunnel excavation was similar with the caving water situation. For this, before Jinping Secondary Hydropower Station diversion tunnel construction, a drainage construction cavern was firstly excavated with the auxiliary cavern at the same time.

Aiming at the problem that it was possible to appear diversion tunnel water outburst, the geological prediction adopted an integrated forecast method. Combining with engineering geology method, instrument method, and advanced drilling, utilized macro and long distance forecasted results, analyzed key sections in advance, focused on strengthening forecast for lithology change, fault, fold, and other water burst part. Through macro forecast, middle-long distance forecast (30 – 150m in front of the tunnel face), short distance forecast (15 – 30m in front of the tunnel face), the forecasting system of geological (groundwater) early warning near disaster, a set of geological forecast system was established for the construction of the diversion tunnel.

Due to the limitation of operating conditions, the water diversion tunnel can not be arranged as gable drainage bottom slope. The western construction import section can only be reverse slope construction. Therefore, it was prominent in the safety of underground water bursting construction. For this purpose, using deep-buried long tunnel construction experience at home and abroad for reference, the overall design idea includes "preliminary geological exploration work → 5km long caving construction → gable advanced auxiliary caving (Jinping auxiliary cavern) construction → hydraulic tunnel construction", the underground water problems were processed step by step in different stages. Simplify the processing difficulty of diversion tunnel underground water, improve the construction conditions of the hydraulic tunnel, and avoid the construction risk of hydraulic tunnel, so as to reduce the construction difficulty of the whole project. In the construction process of long caving and auxiliary cavern, high pressure underground water was repeatedly encountered, both the concentrated big water outburst, and the more a lot of loose fissure water. The maximum instantaneous flow rate of single point was 4.91 m^3/s, and the stable flow 2 - 3 m^3/s. For this purpose, the construction principle was "first caving then digging", "given priority to block, combining with block and drainage". The loose fissure water can be effectively controlled through the high pressure grouting. For the water bust with large water volume and high water pressure encounted by the tunnel face, the excavation diversion and guide drainage diversion was adopted. The diversion guide tunnel was designed and arranged for plugging according to the condition of surrounding rock and effluent structure. The section size took the meeting of the drilling, deslagging, and convenient steel tube transport as the prerequisite. After pouring the concrete of the main cavern side and grouting, closed the high pressure valve of the diversion guide tunnel and completed the water plugging. The practice showed that this method was feasible.

3. About high ground temperature and harmful gas

Generally speaking, the deep-buried long tunnels will encounter the high ground temperature above the permissible range of the construction. However, it was found in Jinping long caving construction that the highest temperature was 17.68℃ (at 639m cavern depth), and the minimum temperature was 10.80℃ (at 2,949.5m cavern depth) in the process of excavation. Generally, it was between 11 - 12℃. With the increase of cavern depth (bury depth), the ground temperature gradually decreased. The main reason was that the geothermal field of Jinping was controlled by underground water seepage field in low temperature. According to the construction process of auxiliary cavern and drainage cavern, it was forecasted that in the construction of the diversion tunnel, the high temperature problems affecting tunnel construction would not appear.

Through the monitoring of harmful gases, the analysis of rock chemical composition, and analysis of ventilation test, the S content in surrounding rock was very small. In the construction, the H_2S content was commonly below 6‰, not causing harm to personnel. The blasting fume containing CO, CO_2, construction dust and diesel mechanical exhaust were the main pollution sources, particularly serious in the Dutou section. The difficulty of construction ventilation scheme on the arrangement was that the west section of the tunnel had no ground outlet and only led to the ground through construction branch tunnel of the four column tunnel. The construction unit took different ventilation schemes according to the construction characteristics of different times and different parts. With the formation of the construction roadway respectively connected $1^\#$ and $2^\#$, $3^\#$ and $4^\#$ tunnels, the partition tunnel ventilation scheme was adopted.

The four diversion tunnel of Jinping II Hydropower Station has been connected, solving a large number of technical problems of deep-buried long tunnel construction and bringing China's deep-buried long tunnel engineering technology to a new level.

18.4.4 Air-cushioned Surge Chamber

Air-cushioned surge chamber was also called closed surge chamber, pneumatic surge chamber was a kind of superior performance water hammer and surge wave control facilities. Its working principle: part of the indoor was filled with water. The enclosed space above the water was filled with high pressure air, forming a cushion. Utilize the air compression and expansion to reduce the fluctuation amplitude of the pressure regulating indoor water level. Compared with the conventional surge tank, the height of the air-cushioned surge chamber can be greatly reduced. It can be less restricted by terrain conditions and protect the surface environment. It also can be arranged at the location closer to the factory, making the derivation on the profile closer to a straight line, thus shortening the length of the derivation, reducing the engineering quantity and the head loss, and improving the regulation performance of the unit.

Compared with the traditional surge tank, especially in the mountain valley area, using air-cushioned surge chamber can greatly reduce the height of the surge chamber. And it also does not need to build surge chamber road transportation. It not only can greatly reduce the quantity of the open excavation engineering, ensure the construction period, reduce the cost of surge chamber, still can greatly reduce the destruction to the aboveground vegetation, effectively protect the surrounding environment.

The world first air-cushioned surge chamber is Norway Driva Hydropower Station. It was built in 1973. The study of air-cushioned surge chamber in China began in the 1970s. In recent years, based on the introduction of Norway's advanced technology of water and electricity, and according to the needs of engineering, China has successfully designed and built Qinghai Da'gangou, Sichuan Ziyili, Xiaotiandu and other air-cushioned surge chambers. Da'gangou air-cushioned surge chamber (gas tank) adopted the top-closed cylinder structure, which erected on the ground and was not underground surge chamber. China's first underground air-cushioned surge chamber was Ziyili Hydropower Station. Its completion filled the gap on this project in China.

Ziyili Hydropower Station is located on Huoxi River within the territory of Pingwu, Sichuan Province. The power plant has a total installed capacity of 130MW. The design water head is 445m, and quotative discharge is 34m^3/s. The total length of the diversion tunnel is 9,167m. The tunnel section is 4.2m×4.6m (width×height, the same below) of the city gate shape. The design velocity is 1.9m/s. Along the tunnel, the minimum depth is greater than 0.5 times of the internal water pressure. The surrounding rock along the line is two-mica granite. The vast majority of the cavern section adopts no-lining tunnel. Air-cushioned surge chamber horizontal distance with the workshop is about 450m. The vertical cover thickness is 350m (calculated from weak weathering layer, similarly hereinafter). The lateral cover thickness is 280m. Air-cushioned surge chamber is moderate weathering indosinian period two-mica granite. In the rock mass, develop a set of structural plane with occurrence of N60°E/SE∠73°. The fracture is close and nearly orthogonal with cavern. The rock mass is overall-block structure, given priority to class II surrounding rock.

Air-cushioned surge chamber floor elevation is about 150m above the installation elevation. The pressure pipeline is 523m long. According to the hydraulics preliminary calculations, the stability area of the surge chamber is 1,000m^2. Air-cushioned surge chamber is composed of a gas chamber and a water curtain cavern. Chamber is 110m×10m×13m (length×width×height, similarly hereinafter) city gate shape. The floor height is 1,707.00m. The surface height is 1,710.00m. The water bed is 3.0m deep. The design gas pressure is 3.25MPa. The design gas volume is 9,998m^3. The maximum gas pressure is 3.8MPa. The smallest gas pressure is 3.1 MPa. The largest volume is 11,057m^3. The minimum volume is 8,881m^3. The water curtain cavern is 318m×4m×6m city gate shape, which is symmetrical arranged on both sides of the air chamber. The minimum distance from the air chamber is 15m. The floor elevation is 1,707.00m. On the top of the arch, the horizontal angle is 42°45′40″. Two curtain of drainage holes are arranged respectively on the upper end toward gas chamber with 18 holes per row. The aperture is ϕ70mm. The hole depth is 32m. The minimum distance from water curtain hole to gas chamber is 7m. The water curtain overpressure is 0.3MPa, namely, the pressure of water curtain is designed as 4.1MPa. The value is less than the minimum crustal stress. Surge chamber adopts no lining. Based on the geological conditions and the degree of fracture development, locally adopt high pressure consolidation grouting. Utilize geological exploration flat hole to reconstruct surge chamber transportation cavern. The section is 4m×4.5m city gate shape, in which arrange air compressor room, water pump room and other observation and control equipments. Air-cushioned surge chamber operation is according to small air cushion way. All the parameters meet the design requirements. The overall operation of the unit is smooth. Because the air-cushioned surge chamber air leakage greatly exceeds the design requirements, the normal operation of the power plant should have four sets of 15m^3/min air compressor put into operation. The operation and maintenance cost is relatively higher. It is difficult for maintenance and management. Because of this, in the subsequently-built Xiaotiandu and Jinkang Hydropower Stations that adopted air-cushioned surge chamber, in the aspect of preventing gas leakage, the steel lining concrete or steel cover were adopted to prevent gas leakage. The operation of Jinkang Hydropower Station showed that under the condition of normal power generation for two units, only in the case of one 1m^3/min air compensating air compressor operation, the gas chamber can maintain indoor water level. The gas chamber air leakage was about 20m^3/h, which proved that the design was successful. In the process of operation and management of the air-cushioned surge chamber, we had collected a large amount of data and accumulated a certain amount of operating maintenance experience, which settled abundant practical basis for the subsequent power station to adopt the design, operation and maintenance of air-cushion surge chamber.

At the downstream of Ziyili Hydropower Station on Huoxi River, Muzuo Hydropower Station and Yinping Hydropower Station have been built one after another. Both adopted air-cushioned surge chamber. Air-cushioned surge chambers were arranged in front of the workshop in Muzuo and Yinping Power Station where the steel cover structure to prevent gas leakage was adopted. Muzuo Hydropower Station gas chamber adopted sandwich type steel cover structure similar to Xiaotiandu Hydropower Station. Yinping Hydropower Station gas chamber adopted plug-in type steel cover structure. Set flat warehouse outside the steel cover. This structure effectively simplified the process and saved the cost, which was an innovation and breakthrough in the steel cage type air-cushioned surge chamber design. The operation of the air-cushioned surge chamber of Muzuo and Yinping Power Stations completely achieved the design anticipated goal. Yunnan Gangqu River Secondary Hydropower Station also used the air-cushion surge chamber, and succeeded. It can be expected that in the development and construction of hydropower station in western China in the future, there will be more hydropower stations to adopt the air-cushioned surge chamber.

18.5 Prospects of Hydraulic Underground Engineering Construction in China

China has the most rich hydropower resources in the world. However, the current development level is far lower than that of the developed countries. The current development level is about 35%-40%. On August 26, 2010, *China hydropower 100th anniversary conference* was held by the National Energy Administration in Kunming. It was put forward at the meeting that China was committed to develop hydropower stalled capacity of 5 billion MW. By 2010, the hydropower development had reached two billion MW. According to the hydropower development planning, the installed hydropower capacity will reach 3.8 billion MW, and the degree of hydropower development will reach more than 70% by 2020.

According to the incomplete statistics, China has more than twenty hydropower station with the installed capacity greater than 100MW. After the reconnaissance, design, construction, operation and the practice of the scientific research work of underground powerhouse cavern group of so many large scale hydropower stations, China has accumulated relatively rich experiences. In the early 1980s, in combination with Lubuge Hydropower Station construction, national-level research on some key issues of such large span underground caverns as underground powerhouse of hydropower station were carried out. A number of practical meaningful results were obtained. After the completion of Lubuge Hydropower Station, the underground powerhouses of many more large-scale hydropower stations have been built.

The installed capacity of Jinping Secondary Hydropower Station on Ya-lung River was 4,800MW. The four diversion tunnel cavern diameter was 9.5m deep and 18.7km long. The biggest buried depth was 2,525m. There was a series of challenging and complex engineering problems, such as high underground water, high tectonic stress. In the recently-built Han Ji-Wei works cited water transfer project, the length of water conveyance tunnel in Qinling is 81.58km. The maximum buried depth is about 2,012m. The design and construction faces many key technical problems beyond the scope of the rules and regulations and existing engineering experience. South-to-North Water Diversion West Project transfers water from Yalong River main stream and its tributaries, Dadu River tributaries respectively to the upstream main stream of the Yellow River. The project proposal design stage work has almost finished. The water transfer route passes through the Bayan Har Mountains. The total length of the newly-designed cavern line is about 320km. The maximum single section cavern is 72km long. The cavern line elevation is between 3,440m to 3,640m. The project area has cold climate, thin air, and severe engineering conditions, and is the world's first super large plateau inter-basin water transfer project. The construction of these projects will make China super long deep-buried hydraulic tunnel design and construction level to reach a new level.

Xiluodu hydropower station on Jinshajiang of upstream of the Yangtze River is the follow-up project of The Three Gorges Water Conservancy Hub. The installed capacity is 13,860MW (18×770MW), and hyperbolic arch dam is 300m high. The design flood discharge flow is 43,800m³/s, and the total power of flood discharge is close to 100,000MW. Three flood discharge tunnels are arranged on left bank and right bank respectively. The maximum net sectional area is 14m×19m. The underground powerhouse consisting of 18 units of 770MW unit are arranged in the mountain of left and right bank, each bank nine units. The size of the two underground powerhouse is about 420m×31.9m×75m. Supporting by the underground powerhouse, each bank has nine diversion tunnels with inner diameter of 10m, and each bank has three the tailrace tunnels, in which two tailrace tunnels at the outside of the mountain joint with the diversion tunnel. The left and right bank cavern group of Xiluodu Hydropower Station is intensive with large scale, and large quantity. In order to solve problem in the design and construction of the Xiluodu underground cavern, the national science and technology research has been carried out. The underground powerhouses of the built and under construction hydropower stations in China have large scale and quantity. The technical difficulty is also big. It is a magnificent and challenging underground engineering. The successful completion of these projects will create a new situation for human to conquer nature.

Chapter 19

High Dam Construction in China

19.1 Construction of High Earth-rock Dam

Miao Shuying[1], Chu Yuexian[2]

19.1.1 Development History of High Earth-rock Dam Construction

Earth-rock dam is one of the oldest dam types with a long history in China, and the modern technology for dam construction was applied since the 1950s. The development of earth-rock dam construction technology in China is characterized by improvement of geotechnical construction technology and raise of dam height, which can be mainly divided into the following stages.

1. Early construction stage (1950-1970)

The Huaihe River regulation project commended in 1950, and then a batch of earth dams with height of 50m below were constructed consecutively during the next 7 or 8 years. Most of the dams are homogeneous earth dams or sand-gravel dams with clay core. The dam foundations were treated mainly with clay cutoff trench or upstream blanket. The dam construction relies on man power with a few machineries. Limited by construction machinery, construction technology in this period failed to develop. After 1958, number of dams increased a lot, which were mainly homogeneous earth dams, sand-gravel dams with earth core or inclined core. As for construction method, besides roller compaction, other damming technologies were developed, like hydraulic filling, dumping soil into water and directional blasting which a few equipment was required and suitable for mass damming, and a batch of middle-and small-sized earth-rock dams were constructed successfully. In 1958, the column-type concrete cutoff wall was successfully set up in Qingdao, and the concrete cutoff trench was successfully constructed in Beijing, which demonstrated a huge breakthrough in foundation seepage control for deep sand-gravel overburden of earth-rock dam. The representative rolled earth-rock dams in this period included Songtao homogeneous earth dam (dam height 80.1m, completed in 1970), Yuecheng homogeneous earth dam (dam height 53m, completed in 1964), Maojiacun sand-gravel dam with core (dam height 82.5m, completed in 1971), Baihe sand-gravel dam with inclined core (dam height 66m, completed in 1961), and Bikou earth-rock dam with earth core (dam height 101.8m, completed in 1976), etc. The representative directional blasting rockfill dams were Nanshui project with a height of 80.2m, completed in 1971; Shibianyu project with a height of 82.5m, completed in 1973; and Yunnan Yiyi project with a height of 90m, completed in 1978. In this period, some dumping rockfill dams were constructed, such as Maotiaohe II concrete faced rockfill dam with a height of 47.8m, completed in 1966. Construction of bituminous concrete core (inclined core) dam was started in the early 1970s, the dams were normally built by placing limited by construction technologies; Gansu Danghe I bituminous concrete core dam was completed in 1974 with a height of 58.5. Restricted by construction machinery and source of high-quality bitumen, this kind of dams failed to enjoy due growth. In this period, the representative earth-rock dams were Baihe sand-gravel dam with inclined core and Bikou earth-rock dam with earth core, etc.

Bikou earth-rock dam with earth core was completed in 1976 with a height of 101m, a crest length of 297m, and filling amount of 3.97 million m³. The dam was built on a valley, and embankment materials are mainly earth, gravels and rockfill. The embankment materials were basically transported by trucks. In the late construction, the towed vibrating drum roller was introduced to roll gravel and rockfill of dam shell, which was a pioneering work at that time.

In this period, Shimen rockfill dam with earth core (dam height 133m and completed in 1964) and Zengwen rockrill dam with earth core (dam height 133m and completed in 1973) were set up in Taiwan, and Gaodao bituminous concrete core rockfill dam (dam height 107m and completed in 1979) was built in Hong Kong.

---

[1] Miao Shuying, Former Chief Engineer of SinoHydro Corporation Engineering Bureau 15 Co., Ltd.
[2] Chu Yuexian, Deputy Director of Engineering & Science, SinoHydro Group Ltd.

2. New development period (1980 – 1990)

With advancement of large-scale efficient complete construction machinery and construction technologies, as well as improvement of earth-rock mechanics and testing technologies, high earth-rock dam enjoyed rapid growth. After the middle of 1970s, based on extracting advanced foreign experiences, introducing large-scale construction machinery and conducting scientific researches, many construction enterprises were equipped with heavy earth-rock machinery and supporting facilities, thus a new period for development of earth-rock dam begun as our overall national strength improved.

Development of earth-rock dam construction technology in the new period is represented by the successful practice of large-scale machinery and equipment, such as heavy earth-rock machinery and vibrating roller, therefore, rockfill dam with earth core and concrete face rock fill dam became the dominant types of modern high earth-rock dams. Dense rockfill with less deformation was formed by using vibrating roller to compact the rockfill in thin layers, thus the problems of face cracking, joint opening and heavy leakage due to dump rockfill were solved so that this kind of dam are adopted popularly. By vibrating compacting with large-scale machinery, all rocks exploited by blasting may be used as embankment material, including sand and gravel with large grain size. Meanwhile, it can improve the compaction density of weak rock so as to increase rock strength. Along with pad foot vibratory roller, plate vibrator and other compaction tools are gradually applied. More materials are suitable to serve as impervious ones. The use of vibrating roller improves the safety, economic efficiency and applicability of earth-rock dam. For embankment material delivery, the advanced and high efficiency trackless transportation replaced gradually the trace transportation by automobiles, and this method was first used in construction of Bikou earth-rock dam completed in 1976. At the end of 1970s and the beginning of 1980s, the construction of Shitouhe earth-rock dam demonstrated the construction technology of earth-rock dam has made a new advance. In the middle and late period of 1980s, the Lubuge core dam served as a link in the development of construction technology. The construction of Xiaolangdi rockfill dam with inclined core and Hehei Jinpen sand-gravel dam with clay core, both completed in the last years of the 20th century, greatly enriched the practical experience of earth-rock dam construction and fully improved construction technologies.

The Shitouhe core earth-rock dam, with dam height of 114m, crest length of 590m long and filling amount of 8.35 million m^3, was built on a valley and completed in 1982. For the dam construction, embankment materials included clay soil, river bed sand and gravel, and blasted rockfill; the two methods were adopted in exploitation of earth materials, vertical exploitation by bucket-wheel excavator and horizontal exploitation by bulldozer and carry-scraper, and belt conveyor and 12t auto-dumper were used for material delivery; bulldozer was used for material leveling. In the early compaction, 30t pneumatic tyre roller and 8.2t sheep-foot roller were used, and then self-made 8t pad foot vibratory roller was used. For dam shell materials, 4m^3 excavator was used for exploitation, and 18t dumper for transportation, 320HP bulldozer for material leveling, and self-made 14t towed vibrating drum roller for compaction. The thickness of sand and gravel layer placed on shell is 1.5m with the maximum grain size of 100cm, and the thickness of rockfill placed on shell is 1.0m with the maximum grain size of 80cm. The high mechanization was reached in construction of this project, which demonstrated a new era for earth-rock construction in China.

Lubuge rockfill dam with earth core was completed in 1989, with 103.8m high, 216m long and filling amount of 3.96 million m^3. The impervious material for the core is mainly the mixture of weathered sandy shale and residual soil. Weathered soil and shell rockfill materials were delivered by 15t and 20t dumpers, and compacted with pad foot vibratory roller and drum roller. The earth material was filled as the design compactness and the three-pint rapid compacting method was adopted for quality control of earthfill so as to adapt to the features of weathering rock and the changed of gradation. For the construction quality of this project, the compactness and construction parameters were adopted for controlling the earthfill and the construction parameters were used for controlling the rockfill.

The Xiaolangdi rockfill dam with inclined earth core is 154m high and 1,667m long at the crest and filling volume of 53.07 million m^3. The sand-gravel overburden of dam foundation exceeds 80m at the deepest part and the cutoff wall of main dam has area of 21,120m^2. The dam construction was started in 1994, the river closure was made in October 1997, and completed in 2000. It's the highest core rockfill dam at that time in China. The core material is silty loam, shell material is silicon fine sandstone blasted. The construction flow of rockfill is as follows: making boreholes with hydraulic track-mounted drill →loosing rock by bench blasting → loading with 10.3m^3 excavator →delivery to dam with 60t dumper → spreading with bulldozer → rolling with 17t self-propelled vibrating drum roller → construction quality control by construction parameters. The construction flow of core earth material is as follows: exploitation and collection by bulldozer → loading by loader → delivery to dam with 36t and 60t dumpers → rolling with 17t self-propelled pad foot vibratory roller. Due to high mechanization and rigid construction management for dam construction of Xiaolangdi project, high intensity, high level and high efficiency reached, with the following indexes, maximal monthly filling intensity 1.58 million m^3, maximal daily filling intensity 67,000m^3, maximal monthly rise 8m, and monthly uneven coefficient of filling 1.31. The Xiaolangdi dam represents the construction level of earth-rock dam in the 20th century in China, and is rated as the milestone of rockfill dam construction in the world.

The rolled concrete faced rockfill dam is the dam type resurrected in the 1960s. Due to its good safety, economy, adaptation and convenient for construction, it enjoyed rapid development. Since the middle of 1980s, modern construction technologies had been adopted in construction of concrete faced rockfill dams in China. The Xibeikou dam was the first started project and Guanmenshan dam was the first completed project, which represented China's concrete faced rockfill dam development had entered a new stage from the 50m-level Guangmenshan dam and 100m-level Xibeikou dam. After 1990, this type dam developed faster. There had been 39 such dams completed in the period of 1985 to 1998, and 9 dams over 100m high were set up in 2000, among which one is about 178m high. This kind of dams has become a dominant type for high and middle earth-rock dams. Representative projects include Xibeikou, Tianshengqiao, Wuluwati, and Shanxi, etc.

Tianshengqiao concrete faced rockfill dam is 178m high, with filling amount 18 million m^3, and concrete face area 180,000m^2. Its project scale and technical difficulties rank the top in the world. It's also the first concrete faced rockfill dam from 100m level to 200m level in China, and its filling intensity reached a new level to 48,000m^3/d and 1.18 million m^3/month. The technologies, like the finishing cushion slope with laser-guiding backhol, and the excavation of dam materials in spillway by controlled blasting, had played an important role.

3. New period of high earth-rock construction (since 2001)

Since the 21st century, construction achievements of earth-rock dams in China have drawn worldwide attention. A batch of high earth-rock dams and super high earth-rock dams are constructed consecutively, China's construction technologies in earth-rock dams have ranked among the top in the world. During current 10 years, over 30 earth-rock dams over 100m high have been built (100m high dams over 50 in total), among which there are 10 over 150m high (12 in total). There are 21 high earth-rock dams over 150m completed and during construction. There are 21 high earth-rock dams over 150m completed or under construction (refer to table 19.1-1 for earth-rock dams over 150m high completed or under construction). The tonnage of carrier vehicles and capacity of loaders used in construction of high earth-rock dams increased with the embankment enlarges. Most rolling equipment adopt heavy vibrating roller with large exciting force, and impacting compacting equipment are applied.

Selection of construction equipment becomes more rational and normative. Dam material planning and embankment zoning are more scientific and reasonable, and processing technology of dam material has improved significantly with more acquaintance to the features of gravel and earth. Testing measures and burying technology of observation instruments developed simultaneously, focus has been given to collection of observation data in construction stage and same results have been achieved. The technology of upstream slope protection is constantly improved, and several protection types have appeared. Research on crack control of face concrete is deepened and good effects have been achieved. The understanding of dam body settlement before placing face is becoming more rational. Water flow control during filling and deep foundation treatment has got favorable results.

In this period, a batch of high earth-rock dam projects have been built in plateau, cold and arid areas, as well as earthquake-prone areas. All of these have accumulated more experiences for comprehensive development of earth-rock dam.

Bituminous concrete as impervious material used in rockfill dams has enjoyed development in these over ten years since it has the advantages of excellent imperviousness, good adaptability to deformation, few work quantity and accumulated practical experiences. Some dams have been set up with modern construction technology, such as: Tianhuangping bituminous concrete faced rockfill dam (72m high) in Zhejiang, Maopingxi bituminous concrete core rockfill dam (104m high) of Three Gorges project, Yele bituminous concrete core rockfill dam (124.5m high) in Sichuan which have been completed using technologies, as well as Shimen bituminous concrete core rockfill dam (106m high) under construction in Xinjiang and Allah bituminous concrete core sand-gravel dam (105.26m high) under construction in Xinjiang, which have indicated that this dam type has become a competitive one.

In China, the construction and management of earth-rock dams have gained constant improvement in practice and active exploration. The projects like Nuozhadu rockfill dam with earth core, Pubugou rockfill dam with earth core, Shuibuya concrete faced rockfill dam, Zipingpu concrete faced rockfill dam and Yele bituminous concrete core rockfill dam express construction level in this period.

Table 19.1-1 Table of Earth-rock Dams Higher than 150m Completed or Under Construction in China

No.	Name	Location	Type of main dam	Height(m)	Time of completion
1	Shuangjiangkou	Sichuan	Rockfill dam with earth core	314	Construction of diversion tunnel
2	Lianghekou	Sichuan	Rockfill dam with earth core	295	Construction of diversion tunnel
3	Nuozhadu	Yunnan	Rockfill dam with earth core	261.5	Under construction

Countinued

No.	Name	Location	Type of main dam	Height(m)	Time of completion
4	Shuibuya	Hubei	Concrete faced rockfill dam	233.2	Apr. 2008
5	Houziyan	Sichuan	Concrete faced rockfill dam	223.5	Construction of diversion tunnel
6	Jiangpinghe	Hubei	Concrete faced rockfill dam	219	Under construction
7	Sanbanxi	Guizhou	Concrete faced rockfill dam	185.5	2006
8	Pubugou	Sichuan	Rockfill dam with earth core	186	Jul. 2009
9	Hongjiadu	Guizhou	Concrete faced rockfill dam	179.5	2005
10	Tianshengqiao I	Guangxi and Guizhou	Concrete faced rockfill dam	178	1999
11	Kajiwa	Sichuan	Concrete faced rockfill dam	171	Under construction
12	Qianzhong	Guizhou	Concrete faced rockfill dam	162.7	Under construction
13	Tankeng	Zhejiang	Concrete faced rockfill dam	162	2009
14	Liyang	Jiangsu	Concrete faced rockfill dam	161.5	Under construction
15	Longbeiwan	Hubei	Concrete faced rockfill dam	158.3	Under construction
16	Jilintai	Xinjiang	Concrete faced rockfill dam	157	2005
17	Xiaolangdi	Henan	Inclined core rockfill dam	160	1999
18	Zipingpu	Sichuan	Concrete faced rockfill dam	156	2005
19	Malutang II	Yunnan	Concrete faced rockfill dam	154	2010
20	Bashan	Chongqing	Concrete faced rockfill dam	155	2010
21	Dongjing	Guizhou	Concrete faced rockfill dam	150	2012
22	Liyuan	Yunnan	Concrete faced rockfill dam	155	Under construction

19.1.2 Examples of High Earth-Rock Dam Construction

Until 2011, there're 12 earth-rock dams over 150m in China, among which there're 10 completed in the early 10 years of 21st century, and the highest one is 233m, and the maximal height under construction is over 300m. The representatives are described as follows.

1. Zipingpu concrete faced rockfill dam

Zipingpu concrete faced rockfill dam is located in 6km of upstream of the world-known Dujiang Dam, with reservoir capacity of 1.112 billion m^3, maximal dam height of 156m, crest length of 634m and crest of width 12m. The dam body consists of upstream blanket, face, footstep, cushion, transition zone, main rockfill zone, secondary rockfill zone, downstream rockfill zone, and dry masonry slope protection, with the horizontal width of 3m for the cushion and 5m for the transition zone. The total filling amount is 11.8 million m^3. The dam site is located in the big turning of Shajinhe, a tributary of Minjiang River. This area is the south section of Longmenshan fracture tectonic belt, and the basic earthquake intensity is 7 degree.

During filling, PC-650 face shovel and PC-400 backhoe were adopted as the main loaders and assisted with 3m^3 loader for, loading, and 26t and 32t dumpers for delivery of the main and secondary rockfills, and CAT385 backhoe, CAT320 backhoe and 26t dumper were used for the downstream rockfill and transition materials. The cushion material was made of processing, and loaded and delivered by 3m^3 loader and 20t dumper. During rolling, YZ-26C self-propelled vibrating roller was used and sprinkling was made carefully. The slope of cushion was rolled with BW75S2 vibrating roller. To satisfy the need for long-distance transportation (about 10km) and high construction intensity, enough equipment were provided and the peak filling intensity had reached 700,000m^3/month.

To meet the requirements of high construction intensity and concreting quality, the batching plant was arranged near the concreting area, which is of great help to improve cracking of concrete face.

During the earthquake of 8 degree (the epicenter intensity is IX) on May 12, 2008, in Wenchuan, the Zipingpu concrete faced rockfill dam which is 17km away from the epicenter was subjected to the earthquake in excess of the design intensity (dam site intensity is between VIII and IX). After the earthquake, the maximal settlement of dam was 83cm, the middle and right bank of wave wall on the highest dam part was damaged by squeezing with obvious settlement and maximal horizontal deformation 32cm; part of concrete face was unjointed, construction joints between 2nd and 3rd phases opened with offsets, water stop between wave wall and face were damaged so as to increase seepage flow significantly. However, the dam was generally stable, and the damaged parts were able to be repaired and had less influencing on impounding. The main outlet structure returned back to normal operation shortly after the earthquake, and the water level of reservoir can be controlled normally. This project has stood the severe test of the big earthquake and been significant in the field of dam construction in the world.

2. Shuibuya concrete faced rockfill dam

Shuibuya key water control project is located in Badong country of Hubei province, with total storage of 4.58 billion m³. The main structures include concrete faced rockfill dam, side spillway on left bank, emptying tunnel on right bank and underground diversion type power station. The dam is the highest one in the world, with dam height of 233m, maximal effect head of 200m, dam axle length of 670m, crest width of 12m, upstream dam slope 1 : 1.4, average downstream slope 1 : 1.4 and total filling amount of 15.68 million m³. The basic earthquake intensity is Ⅵ, and the design earthquake intensity of the main structures is Ⅶ.

For the project, the construction preparation period was in 2001 - 2002, and the river closure was conducted in the late October of 2002, dam filling started in February of 2003 and was basically completed in October of 2006 with filling duration of 52.5 months. In July 2007, the first generation unit put into power generation, and all 4 units were installed and put into operation in September 2008.

Till the end of 2008, the dam experienced the test of normal impounding level. The maximal settlement of dam body is 2,473mm, the maximal deflection of face is 573mm, and the maximal water seepage amount observed at the measuring weir is 40L/s. Comprehensive analysis from the monitoring data suggests that the performance indexes of the dam are within the design control scope.

The main features for construction of Shuibuya dam are as follows:

(1) Because filling of the dam adopts many measures, including the optimization of earth and rock materials transportation, construction machinery configuration and site management, and sequential filling from downstream to upstream with the filling method of "pre-compacting against being lifted", which have effectively reduced unbalanced settlement, kept balanced filling intensity, and ensured filling quality. In the 45-month filling period, the average monthly intensity is 386,000m³ and the maximal monthly intensity is 751,110m³ with unbalanced coefficient of 1.47, 1.23 and 1.38 for the phases of 2nd, 3rd and 4th, respectively.

(2) In early-stage diversion, cofferdam was adopted to retain water during dry season, diversion tunnel was used to discharge, cofferdam and dam face were allowed to be overflowed during flood season. In midterm diversion, the dam height completed was enough to retain water so that the dam body, diversion tunnel and emptying tunnel were used jointly to discharge. In later stage diversion, the emptying tunnel was used in the 1st dry season after closure of diversion tunnel, and then the spillway and powerhouse were taken to discharge. The upstream cofferdam adopted the type of high weir for water retaining (El. 223m) and low weir for discharge (El. 215m), while the downstream RCC cofferdam heightened with 3m (from El. 211m to El. 214m) to serve the purpose of temporary protection the embankment below El. 210m. The maximal flow speeds at the temporary upstream cofferdam section were reduced to 10.48m/s and 5.43m/s, with smooth flow, so as to reduce the difficulty and risk for protection of cofferdam and dam greatly and shorten the construction duration of protection works.

(3) The sand and gravel overburden of dam foundation was reserved partly and treated by dynamic compaction, which have effectively reduced its settlement and deformation. Before construction, the dynamic compaction test was conducted first to determine the equipment, technique and quality control standards by the indexes including dry density of bed gravel, permeability coefficient, grain composition, bearing capacity and other physical mechanics parameters.

(4) The crack control is very important for the dam since its maximal inclined length of concrete face is 392m, and a series of comprehensive measures have been adopted to prevent the face cracking, such as: balanced and continuous high-intensity filling and rigorous quality control to effectively lessen the deformation of dam body; improving raw material and structures of extruded concrete sidewall technology to effectively reduce negative impacts on the face from deformation of dam body; optimizing the design mix of face concrete by adding the third generation polycarboxylic acid additive and the fourth generation polyacrylonitrile fiber admixture to greatly improve the concrete performances of crack control, seepage control and durability; improving the sliding filling design and the supporting structure of formwork, as well as rigid quality control and careful operation to eliminate the possible weak planes on the face; painting the new protective materials of crystallization type at the surface of concrete under EL270m to enhance the crack control performance. Through vigorous dispatching and

management of the whole process, and construction in low-temperature period, effectively preventing and reducing cracks on the concrete face.

3. Pubugou rockfill dam with soil core

Pubugou hydropower station is located in the middle reach of Dadu River, the water retaining structure is a rockfill dam with earth core. The elevation of the bottom is 670m, and of the top is 857.20m. The dam is 186m high with the crest length of 573m. The maximum width of the dam bottom is 780m. At the elevation of 795m, a berm of 5m wide was set of which the up slope is 1 : 2.0 and the down slope is 1 : 2.25. There is a zigzag road with a slope of 1 : 1.8 on the downstream slope. The dam filling amount is 23 million m^3, in which core gravel and earth required is 2.76 million m^3.

The core of the dam is mainly composed of impervious body and artificial filtering system. The top and bottom of impervious body are at El. 854m and El. 670m, with top width 4m, and upstream and downstream slopes 1 : 0.25. The core material is mainly spreading graded gravel and earth. Highly plastic clay is used at the core bottom, between core and bank slope, and cutoff wall top and concrete gallery periphery. The two filtering layers are set at upstream side and downstream side of the core, respectively, with the thickness of 4m for the two upstream layers and 6m for the two downstream layers. The transition material and shell material are filled outside the filtering layer. Two rows of concrete cutoffs with 1.2m thick and 14m spacing are set on the overburden of core lower part, and the maximal wall depth under the core base is about 78m. The connection of cutoff wall and core is penetrating, and the penetrating depth of cutoff wall is 15m, and concrete gallery is set between walls.

Difficulties for construction of the dam are as follows: ① Total dam filling volume is 23 million m^3, among which gravel and earth of core is 2.76 million m^3, and the distance between to the borrow area to the dam is about 15km. Exploitation, processing, transportation and filling for this spreading graded gravel are challenging. ② Maximal depth of dam bed overburden is 77.9m, with great processing difficulty. ③ A deformable body with 150m long and 450m elevation difference existed nearby the dam bank, bring great difficulty for construction.

(1) Exploitation of Gravel and Earth Material. Total impervious soil material required is 2.76 million m^3. According to the specifications, the bank measure needs 5.64 million m^3, and the reserve of Heima borrow is 6.8 million m^3, which satisfy the requirements. Among that, the reserve only in need of simple gradation adjustment is 5.51 million m^3, and the others need to mix with clays.

Since the grain composition and moisture content of gravel were uneven distribution at plane and elevation in the borrow area, and investigation was conducted to acquaint the changes in moisture content and underground water level in different seasons, at different parts of planes and elevations in advance of one year, with the investigating frequency which could understand the changing principles at different places, elevations and in different seasons. According to the investigation of borrow area and available date, an exploitation scheme was worked out, adopting single-point exploitation at elevation and multiple sections exploitations at plane (at least 4 working faces) to coordinate coarse and fine grains, thick and thin layers, and moisture content, with the graded plan distribution as the main control item and equal emphasis on changes in underground water level and moisture content in different seasons.

(2) Screening and Transportation of Gravel and Earth Material. After comparison and selection, the Heima borrow area is an ideal material source. However, the content of natural grains less than 5mm is low, and the gravels with oversize (some even more than 1m) are too many, so simple adjustment must be conducted to screen out some gravels so as to meet the design requirement. Through field tests, it's decided to adopt the secondary screening scheme with bar screen and square hole vibrating screen.

1) Production flow. Dumper transportation → primary screening by bar screen (screen out the grains over 250mm) → secondary screening by vibrating screen (screen out the grains with over size ones) → delivery by belt conveyor → stockpiling by stacker → dumper transportation to the dam.

2) Screening system, including bar screening system, transfer belt conveying system and vibrating screening system. Through calculation, 3 bar screens and 3 square hole vibrating screens are arranged. The size and interval of bar screen are 10m×4m (length×width) and 250mm, with a slope of 34°, each bar screen equipped with a 20m^3 bin, the vibrating from the vibrator on the bin wall are transferred through feed opening and discharging belt. The actual production capacity of screening system is 750m^3 (1,275t) /h.

3) Belt conveying system, mainly including conveying system, conveyer loading hopper and stoker. The belt outside the tunnel is 2km long and the belt inside the tunnel is 4km long (elevation difference of inlet and outlet is 460m). The speed of belt is 3.15m/s in front of the tunnel, and 4m/s inside the tunnel and 4m/s behind the tunnel. Conveying flow: screening system → belt conveyor downside (with slope no larger than 15°) to tunnel inlet → delivering to tunnel outlet by belt

conveyor → loading hopper → discharging belt conveyor → earth material belt conveyor → stocker. The maximal design conveying capacity is 1000t/h, and the maximal actual production capacity is over 1500t/h. Practice has shown that the downside belt conveying system has high insurance rate and can save power energy and provide vigorous guarantee for filling construction. This production scheme has won the award of scientific and technical achievement at provincial level.

(3) Filling of Gravel and Earth Material. Construction parameters identified through rolling test are as follows: ①25t self-propelled pad foot vibratory roller, placing depth of 40cm and rolling 8 times; ②18t self-propelled pad foot vibratory roller, placing depth of 30cm and rolling 8 times.

The moisture content 1%-2% higher than the optimal is better for compaction.

For concentrated blocks at dump foot due to segregation from stockpiling by stocker, loading was made by backhoe and loader, with mixing coarse material in one bucket and fine material in several buckets, and scattering the coarse material in the dumper; adopting end-dump advance method for unloading, spreading by bulldozer, the self-propelled pad foot vibratory roller run parallel with the dam axle with vibrating rolling speed of 2-3km/h, mainly using the method of staggering forward and backward.

For filling and rolling the adjacent areas of other structures, adopt hand-held vibrating roller, vibrating tamping plate and frog hammer for assistant compaction; adopt pad foot roller at connecting area with filtering zone.

According to construction features of Pubugou, a special temporary road was set up to facilitate cross-core transportation of the filtering material. The road subgrade was made of gravels with top width of 4m and bottom width of 6m, and compacted with vibrating roller. 20mm manganese steel plate was placed as pavement. When filling completed at the two sides of core, the road was dug up, backfilled and compacted according to requirements.

The filling started in April, 2007, with the average monthly rise of around 7m. The filling quality is good and the progress is advanced a little. This has shown that construction of high rockfill dam with imperious core has reached a new height.

4. Nuozhadu core rockfill dam

Nuozhadu hydropower station is located in the downstream mainstream of Lancang River, Pu'er city in Yunnan, its main structures include core rockfill dam, spillway on left bank, discharge tunnels on left and right banks, underground convey and power tunnel as well as diversion works. The dam is at El. 821.5m, with crest length of 627.87m and crest width of 18m. The core of the dam is of central upright, namely central straight core made of gravel and earth, and filtering layers are set at the two sides of the core, outside that is the rockfill shell. The main features are as follows, maximal dam height 261.5m, upstream dam slope 1:1.9, downstream dam slope 1:1.8; core top and foundation elevations El. 820.5m and El. 560m, core top width 10m; total filling volume 32.15 million m^3. For the core materials, earth material blended with gravel is filled at the part below El. 720 and earth mixture without gravel filled at the part above El. 720. The mixing ratio of earth material blended with gravel is 65% clay material and 35% gravel material. The fill moisture contents of the earth mixture and those mixed with gravel are controlled within $-1\%-+3\%$ of optimum one, and the fill moisture content of contact clay material is controlled within $+1\%-+3\%$ of optimum one.

For the project, the engineering features and construction difficulties are as follows, large-sized diversion tunnel (2 diversion tunnels with dimension of 16m×21m each), difficult river closure, deep foundation of cofferdam, short time for construction of impervious body, large filling volume, high filling intensity, and mixing ratio of gravel and earth for the core. Therefore, reasonable scheme for river closure, filling technologies and advanced management system are adopted, including mixing technology of gravel and earth, optimization of dam construction roads, optimization of dam filling stages, filling construction technology and construction quality and safety information management system. This project has successfully conducted closure in November of 2007, and dam filling started in August 2008. It will be completed in 2015, and current construction is smooth.

(1) River closure. The river closure construction has such features as high flow speed, large discharge per unit width, large closure scale, high dumping intensity, and difficulties in closure of narrow river bed and construction road arrangement. During closure construction, the maximal flow is $2,890m^3/s$, maximal flow speed is 9.02m/s, and maximal stream energy per unit width is 528t·m/(s·m).

The roads for closure are independently arranged according to the bank-off situation. The road, with minimal width of no less than 12m and maximal slope of less than 10%, is of clay-bound macadam pavement, and the trunk road is 18-25m wide for temporary stop of the special transportation vehicles.

The excavating and loading equipment is allocated according to the dumping intensity of closure, $2,005m^3/h$. $4.0-6.0m^3$ face and backhoe shovels and loaders are selected, EX1100 and EX1200 face shovels, and EX870 and PC650 backhoe shovels for big blocks and 20 – 50t truck crane for hoisting concrete tetrahedrons and rebar gabions. In total, there're 173 dumpers with 32t or 20t, 10 bulldozers, 19 loaders and 3 truck cranes.

The river closure started on Nov. 3 of 2007. The construction parameters are as follows, closure dike axle 111m long, closure gap 66.6m wide with its water surface width of 50.1m, upstream water level 608.8m, difference of dike 2.2m, upstream flow 2,480m^3/s, diversion tunnel flow 816m^3/s. The method of full section dumping and upstream corner advancing, direct dumping, concentrated accumulating and dumpling are used during closing. The dumping is mainly conducted at right bank assisted by the left bank. The closure lasted for 27 hours, and total quantity included dumping 69,700m^3, among which big blocks 34,000m^3, 1,140 rebar gabions of 3m^3, 60 rebar gabions of 4.5m^3, 286 tetrahedrons and 25 hexahedrons.

The correct design of bank-off provides technical guarantee for successful closure, and the reasonable combination of dumping materials promotes the dike advancing. The upstream and downstream corners were closed by rebar gabion series and tetrahedron series, with rock ballasts follow-up in the middle section. The concentrated dumping of special materials is the effective measure for closure.

(2) Blending Technology of Gravel and Earth Material. 4 storage bins, with total storage capacity of 140,000m^3, are provided at the blending site, with 2 for material storage, 1 for material preparation and 1 for exploitation, which can satisfy the maximal filling intensity for about 15d. The blending technology is as follows:

The earth material blended with gravel is blended and spread in the specific area, with the mixing ratio of 65 : 35 for earth and gravel. Spreading method: spreading one layer of 50cm thick gravel material first, then one layer of 110cm thick earth material, and the material in each bin can be used for 3 earth layers and 3 gravel layers. The gravel material to be blended is unloaded with end-dump advance method, and leveled by bulldozer timely. The earth material is unloaded with retreating method.

When 3 gravel layers and 3 earth layers are spread, blend them with 4m^3 face shovels. After completion of material preparation in each bin, they wereblent uniformly before transportation. Blending method: load at the bottom for face shovel, and lift up the bucket to dump the material free for 3 times. After that, load those qualified blended materials with 4m^3 face shovel, and transport the materials to the work site by 20 – 32t dumper.

(3) Dam Filling. Because of climate influence on the core construction, divide the filling period to dry season and rainy season according to rainfall. For rockfill material, fine material and filtering material, filling construction can be conducted in whole year. For core materials, the effective numbers of working days are determined as blending earth materials and mixtures in the construction period is from September 16 to May 31 next year. The period from June 1 to September 15 is rainy season and belongs to work suspension time with the total suspension time of 3.5 months.

The dam roads are arranged by combining bank slope and dam slope. The main road parameters are as follow, subgrade width of the trunk road 12.5 – 16.5m, pavement width 10 – 15m, maximal longitudinal slope 8%; for the dam road on slopes of upstream and downstream, the pavement of 10 – 12m wide and maximal longitudinal slope below 12%.

At elevations of 760m, 695m and 656m of right bank and 715m and 645m of downstream, the dam roads are arranged; at elevations of 660m of left bank and 670m and 625m of downstream, the dam roads are arranged.

Set the temporary left-bank access tunnel in front of No. 5 diversion tunnel to the upstream platform at EL 656m (top elevation of upstream cofferdam). The excavated materials from the stilling basin of spillway are transported through No. 5 diversion tunnel, temporary left-bank access tunnel, upstream cofferdam and upstream dam road to the filling site.

Coarse and fine rockfill materials are transported by 20 – 42t dumpers, filtering material, highly plastic clay and earth material blende with gravel are transported by 20t dumper. The coarse rockfill material and core earth materials are mainly unloaded with end-dump advance method and levelled with bulldozer. Filtering material and fine rockfill material are mainly unloaded with retreating method and leveled by bulldozer. In the connecting areas of coarse and fine rockfill materials, and fine rockfill material and filtering materials, bulldozer assisted with 1.2m^3 backhoe are used for levelling; in the connection area of filtering zone and earth material zone, bulldozer assisted with labors are used for levelling. For core gravel earth material and earth mixture, 20t pad foot vibrating rollers are used for compacting; for highly plastic clay, wheel loader of over 18t or 20t pad foot vibrating roller are used for compacting; for filtering material, 26t self-propelled drum rollers are used for compacting; for fine and coarse rockfill materials, 26t self-propelled vibrating roller are used for compacting.

(4) Quality and Safety Management System of Digital Dam Project. The quality and safety management system of digital dam project serves the functions of effective managing each link in the whole construction process, online real-time monitoring and feedback control; integrated management and analysis for dam design, quality and safety inspection information; setting the comprehensive digital information platform and 3D model; providing the information service and supporting plat-

form for quality supervision in construction period and the safety diagnosis in operational period.

For the supervision and analysis system of construction material source and transportation, the vehicle-mounted GPS and PDA are utilized to conduct the dynamic supervision and warning for material balance from various borrow areas, summarize the supplying quantities of various material sources and traffic densities of transportation vehicle, optimize vehicle dispatch at working sites, and realize 3D dynamic supervision of transportation.

For the embankment and compaction quality, the GPS supervision system is mainly used for real-time monitoring rolling rack, speed and number of rolling times; monitor compaction thickness and calculate compaction rate; real-time warning through the supervision terminal and mobile PDA SMS; provide the measure of dam construction quality control to realize "double control".

For the dam construction information PDA collection system, it mainly collects on-site testing data (test pits) and on-site photographs; collect the information, such as: added water quantity, exciting force of vibrating roller, vehicle data, sectional calibration; conditions of dam materials, borrow area and vehicles. Transmit the on-site collected and analyzed data to the central system database through PDA wireless transmission.

For the construction schedule system, it mainly realizes the 3D dynamic simulation in whole construction progress (planned progress); conduct real-time simulation and prediction of next construction progress (predicted progress); set 3D dynamic model of general schedule (actual progress) according to the progress plan and actual progress.

For the safety supervision system, it mainly makes a safety supervision 3D model for deformation, settlement and seepage, as well as visual management of safety supervision dynamic information and statistical analysis of observed value at supervision points.

19.1.3 Technical Advancement and Achievements of Earth-Rock Dam Construction

1. Technical advancement of filling construction

(1) The heavy rolling equipment is applied widely, and the compaction quality is gradually improved.

In the current construction of earth-rock dam, attention is mainly paid to selectively prefer the heavy vibrating and rolling equipment with capacity of no less than 10t, and pull-type vibrating rollers with the self weight of 18 to 26 tons and self-propelled vibrating rollers with the weight of 26t are used widely. For the high face dam with 100 to 200m high built in recent years, the porosity of more compacted dams is better than that of the filling standard specified in the design code issued at the end of the 20th century because heavy rollers and vibrating rollers with high excitation force as well as relevant supporting measures are utilized, wherein the porosity of 200m-high dams are controlled within 20%, which is 2 percentages lower than the foundation of the Tianshengqiao dam (178m high). To some high face dams, the secondary rockfill material at 3C zone is compacted by heavy vibrating rollers and impact rollers so as to get the same density (porosity and compression modulus) as that at 3B zone to improve the deformation performance of dams, such as Hongjiadu dam and Dongqing dam, and the effects are favorable. The compacting quality and the uniformity of every dam zone are obviously improved, and the standard deviation of the dry density is commonly lower than that required by the design code.

For the Hongjiadu project, through field rolling tests and careful demonstration, the impact rolling technology is firstly applied to the downstream rockfill, and the compaction amount is 1.85 million m^3. Because of obvious compacting effect, the impact rolling technology is also applied to the main rockfill zone, and the compaction amount is about 290,000m^3. Practice has proven the compaction of impact roller has the advantages of filling thickness and high work efficiency, etc., and has an application prospect in the filling construction of earth-rock dam.

(2) Continuous and uniform filling.

How to effectively control the sedimentation of the dam is a difficult point of the construction of the faced rockfill dam, particularly to prevent the adverse influences on uneven sedimentation on the concrete face, water stop and internal stress distribution of the dam, thus, trying best to realize balanced and continuous dam body raise is the key point of the of construction management. For the high faced dams built in recent years, enough attention has been paid to that, and gets an outstanding effect.

The faced of Gongboxia rockfill dam was filled since August 1,2002, and reached to the dam top (132m high) on October 22,2003, the filling construction lasted for 15 months with 4.5 million m^3 of various dam materials, the mean filling strength of 300,500m^3 per month and the maximum filling strength of 524,000m^3 per month. Since the horizontal and continuous filling flow process was adopted from upstream to downstream, the management of the material storage was strengthened, and the new quality check method and long-face one-time construction were used, thus, the uniformly filling and rolling, and even deformation were obtained, with the total sedimentation of lower than 0.5% in the construction period. The project is also a successful example for construct the faced rockfill dam in plateau cold region in China.

For the Shuibuya faced rockfill dam (233m high), in order to ensure the dam continuous and uniformly raising, besides the optimization of layout, equipment allocation and soil-rock materials deployment, and reinforcing onsite commander, the following measures are adopted, filling the downstream rockfill zone (3C) higher than the upstream rockfill zone advanced so as to make the soft limestone in 3C zone settle as early as possible and reduce the uneven deformation of the dam; adopting periodic intervals between filling layers in stages, reverse-raise filling, high-compaction rolling and continuous filling to control the sedimentation and deformation of the embankment effectively from aspects of time and space. The total embankment of the Shuibuya dam is 15.68 million m^3, and it is filled in six stages with the total time of 45 months, the mean monthly filling intensity of 378,000m^3, the maximum monthly intensity of 751,000m^3, the maximum height difference of upstream and downstream embankments is 32m in various stages. the uniformity of filling is lower than 1.5 and uniformly raising is realized. In October, 2006, the project began to impound water, and put into generating operation in July, 2007. The accumulated sedimentation of the dam is 247.3cm in May, 2009, which is 1.06% of the dam height, the mean leakage 31L/s, and the maximum leakage is 40L/s. Shuibuya project is a significant practice for the whole coordinated deformation of 200m high faced rockfill dams.

Qiaoqi core rockfill dam combines the engineering experiences and the contractor's experiences and adopts the method of balance-raise filling so as to decrease the adverse influences of the climate factors, such as rainfall, increase the utilization rate of the construction equipment, and it also beneficially ensures the construction quality.

(3) New stage of construction technology of gravelly soil material.

The super-high core rockfill dam has high requirement on the performance of impervious soil, besides outstanding impervious performance and impervious stability, and the mechanical property required is higher than that of the common high dams. In the last 20 years, in order to take the weathered granular soil and spreading gradation gravelly soil as impervious soil material, different construction methods for quarrying, processing and rolling are tested and practiced, so as to widen the scope of impervious soil materials and make the new stage for application of the super-high dam core material.

At the late period of 1980s, weathered soil was used to build the Lubuge core dam (101m high), which is a successful beginning. The method of rolling in thin layers with heavy roller was used in rolling and compacting of the weathered soil material, which not only improved the gradation of the material but also met the compaction degree and the impervious requirement.

For the Qiaoqi dam (125.5m high), river closure was carried out in December, 2003 and embankment completed in May, 2007. The exploitation of gravelly soil is made on the principle of stripping first and then quarrying from top to bottom, and the bench elevation quarrying is adopted with bench 4 to 6m high each, and the water content and the gravel distribution are uniform before loading. For the gravelly soil material, its construction parameters are as follows, 40cm thick filling layer compacted by 20.5t heavy pad foot vibrating roller, filling volume of 7.3 million m^3, filling duration 19 months, the mean monthly filling strength 380,000m^3, the maximum monthly filling strength 580,000m^3, the mean monthly raising height 7.05m, and the maximum raising height 13.6m. The construction of Qiaoqi dam achieves the advanced level of core rockfill dam under similar conditions in China.

Both the Pubugou dam and the Nuozhadu dam are gravelly soil core dams, and have achieved many creative foundation works on the aspect of construction of gravelly soil and contribute to the development of super-high gravelly soil core rockfill dams with the height of 200m and 300m.

(4) Development and application of dynamic balance system of embankment materials.

The optimization of excavating and transporting balance for embankment materials is helpful to reduce the construction cost and accelerate the construction progress, and meanwhile, by increasing the rate of directly filling material, reducing unused materials and excavating quantity, it is finally favorable to protect the ecological environment. For rock and soil material excavating and transporting in the past, the plan and management were made on the contractor's experiences, therefore, the construction progress and cost were influenced from thoughtless sometimes, thus, the optimized dispatch of embankment materials cannot be achieved. Although, qualitative models were provided, systematic analysis and the effective scheme have not been worked out.

For construction of the Shuibuya project, the optimized dispatch of the embankment materials is a key path in construction schedule since the total amount of excavating, transporting and filling of embankment materials is more than 50 million m^3. By detailed analysis of dispatch of the embankment materials, and comprehensively considering of the time and space relationship, and the physical performances of the materials, the multi-dimension dynamic model and the optimizing dispatch and management system are developed for excavating and transporting balance of embankment material. Through construction practice, the dispatch of the embankment materials is optimized, the balanced filling construction with a high-strength is ensured, and finally, the ideal effects on the construction schedule, quality, cost, environment protection, etc., are realized, for example, 100% of the useful materials in the excavated material is used for the dam, and the rate of directly filling materials is 86.23%.

For the Liyang pumped storage power station, its upper reservoir comprises a main dam, two secondary dams and the reservoir basin formed by excavating. The main dam is a concrete face rockfill one, with the crest at El. 295m, the maximum dam height 161.5m and the crest length 984.3m. The embankment materials of the main dam include those from the main borrow area and the excavated materials from the upper and lower reservoir basins, and the underground powerhouse, the utilization of the excavated materials and the construction of the dam are contradicted in time and space, thus, the dynamic simulation system of dam construction is developed. This system can realistically reflect the construction process of the dam, and has the functions of dynamic dispatch of soil and rock material, simulated calculation of filling, optimization of machinery allocation, statistics and analysis of results, dynamic displaying of filling process, 3D animate vision, etc.

For construction of the Qiaoqi core rockfill dam, the simulated analysis technology is used for generating the dynamic construction situation and displaying the construction process of the rockfill dam in 3D images as required, which well reflects the dynamic influences of various factors on the whole construction process so as to realize the analysis of the construction scheme and communication of information, and provide decision support for the effective management of the construction.

(5) New technical means for filling quality control.

1) The additional quality method (also called the wave exciting measurement method), it is a rapid non-destructive test method for testing the density of rockfill body, and applicable to the test of embankment materials with different grain diameters. After being applied to the Xiaolangdi inclined core rockfill dam, Hongjiadu faced rockfill dam, Shuibuya faced rockfill dam, Nuozhadu core rockfill dam, etc., the results have shown that the method with competent precision can satisfy the requirement on testing the rockfill density, and can get the whole-process control of the unit works. In the process of testing, if there are oversized blocks or poor coupling contact between the launching vibration exciter and the receiving sensor, the testing results will be influenced.

2) The GPS real-time monitoring system, it monitors the running track and the running speed of the rolling machinery, and enables to proper control the filling construction continuously and automatically, with good visual interface. Besides reducing the workload of construction and monitoring personnel at site, and increasing the construction efficiency, it also effectively ensures the filling quality. This monitoring system is used in Shuibuya dam, Nuozhadu dam and others.

3) The whole quality inspection method (also called the test method of compacting deformation) is used for inspecting the filling quality for Hongjiadu project, Tai'an project, etc., and the effect is favorable. For this method, the inspection of compaction quality is carried out by comparing the settlement (the mean value) of embank material compacted, which is measured at the nodes of the square grid, with the calibrated test data in advance.

4) The K30, K50 method is used for assisting to inspect the filling quality, and the data measured can be correlated with the modulus in Gongboxia faced rockfill dam, which is helpful for measuring the displacement of the dam in the construction period.

(6) Recognition on dam zoning in construction period obviously improved. Construction practice has shown the height difference of embankment lifts at different stages should not be too large, or else, it will be unfavorable to the coordinated deformation of the dam. Thus, the newly-revised *Construction Specification for Concrete Faced Rockfill Dams* (DL/T 5128—2009) requires the longitudinal and transverse height differences of embankment lifts in the rockfill zones should not be higher than 40m. The Gongboya faced rockfill dam and other projects adopted some measures and realized the uniformed raise of the full dam section, the maximum height difference at the difference rockfill zones is 32m for the Shuibuya dam, and is smaller than 40m for the Hongjiadu dam, and is 45m for the Sanbanxi dam. The Shuibuya dam adopts the construction sequence as filling the downstream rockfill zone in advance (also called inverse-raise type), which is helpful for improving the deformation of the dam. Generally, the embankment zones should be filled horizontally and continuously, and be orderly filled from downstream to upstream, and the downstream zones can be higher than the upstream zones.

There are several high dams over 150m adopted the corresponding measures to control the height differences of embankment lifts at different stages within a small scope (about 30m), under the precondition of comprehensive coordinated flood regulation and key paths in the construction schedule. When the height difference cannot be reduced, filling lifts can be added, and the width of construction platform added cannot be narrower than 30m.

In construction of the face, the camber shall be provided at the top of the embankment lift constructed in the early period so as to lessen the adverse influence of the embankment settlement on the face later. The camber height is more than 10m for most projects, such as the Sanbanxi dam, the height differences of the face top and the placement platform are 25m and

38m at the first face placing stage and the second face placing stage, respectively. For the super high dams, the top elevation of the face should be at least 20m lower than the embankment surface.

(7) Innovation and promotion of consolidation technology of upstream surface of concrete face cushion. When the faced rockfill dams were constructed in the first two decades in China, the consolidation of upstream slope surface of the cushion basically adopted the slope cutting method, i. e., spreading and rolling the cushion material toward the upstream with over-width, and then cutting the over-wide part by manual work or back shovel and compacting the slope surface by slope vibrating roller or hydraulic plate vibrator; and using cement mortar or emulsified asphalt to consolidate the cushion slope after the slope was compacted and qualified, finally, placing the concrete face. A more economical and safe construction method is required for consolidating the cushion slope since the slope cutting method was complicated and slope surface was easy to be eroded by water during construction in rainy season. After demonstration, based on the experience in Brazil, the Gongboxia dam adopted the extruded concrete sidewall method to consolidate the cushion slope by the extrusion machine which is independently developed in 2002. The extruded concrete sidewall method is widely recognized and rapidly promoted by its characteristics of high compacting quality, improving the protection of the upstream slope and simple construction, etc. Till December, 2010, according to the primary statistics of the manufacturer of extrusion machines, there are more than 70 faced rockfill dams adopting this construction technology in China, including the super-high dam, Shuibuya. In 2008, the extrusion machines were upgraded and improved, providing the automatic leveling system and the function of adding material at slope corner. The effects of upgraded machines are excellent proven by practices in the Jishixia face slab dam and the Wenquan face slab dam in Xinjiang.

In recent years, in the construction practice, the new technologies for protecting the cushion slope are continuously developed, such as Shuanggou concrete faced rockfill dam (110m high), the upstream slope protection adopts the turning-over formwork. The formwork is a wedge-shaped climbing one with a wedge-shaped plate anchored in the cushion, and the working steps of the formwork are as follows, withdrawing the wedge-shaped plate out after the primary rolling completed, placing a thin-layer mortar, and conducting the final rolling to make the completion of filling cushion material and consolidating the slope one time and get ready for retaining water in flood season. For the Qagan Us faced rockfill dam (110m high), the technology of mobile sidewall is used for slope consolidation, that is to put the prefabricated concrete mobile sidewall in place, spread the cushion material inside the sidewall, make rolling and qualified, and then the second layer sidewall is arranged and the cushion material is filled. Only three layers of sidewalls are set, and the fourth layer one is made of that of the first layer one disassembled. Then the redundant cushion material on the triangle part is cut by back shovel and the mortar for consolidation is placed with the raise speed of the embankment.

2. Achievement of crack control technology for face concrete

(1) Reasonable selection of face pre-settlement duration during construction. The newly-revised *Construction Specification for Concrete Faced rockfill dams* (DL/T 5128—2009) requires the pre-settlement duration of embankment is 3 to 6 months, and a certain camber should be reserved in the construction of face when the face is constructed in stages, which is concluded by the experience and lessons.

Currently, there are also other auxiliary means for controlling the pre-settlement duration: ①It is controlled on the settlement rate of 3mm to 5mm per month at the face top; ②The face is constructed after the main settlement resulting from compressive deformation of the embankment is completed (it can be known from the settlement curve). For the Sanbanxi project, the design pre-settlement duration is 5 months, and the actual one is 6 to 7 months, it is controlled with the settlement rate of the embankment at the face top is lower than 5mm per month before placing the face. For the Gongboxia dam, the pre-settlement duration is more than 4 months, and the embankment is in the secondary settlement period of compressive deformation before placing the face.

The concrete of connection slab between the cut-off wall at the dam foundation and the plinth is placed after the deformation in the construction period is basically completed. Generally, the wave wall on the dam crest is constructed after completion of the face placement.

(2) Technique for improving concrete quality. To control temperate cracks and shrinkage cracks of the face concrete in construction of many projects, besides high-range water reducing agent, air entraining agent, shrinkage reducing agent, thickening agent and other additives, polypropylene fiber or steel fiber, fly ash, silicon powder and other modified measures are used to improve the performance of the concrete. The design level of the mix proportion of the concrete has continuously increased, the curing measures are continuously improved and innovated so as to reduce cracking of the concrete face, particularly, and the effect is obvious in severe cold and drought regions.

The concrete mixing plant arranged near the dam or on the dam surface is a means for improving the concrete quality and reducing cracks. The cement consumption and the water cement ratio are reduced for transportation distance shortened and transportation process simplified, with the same work performance of the concrete placement area.

In the severe cold region, the face concrete is placed by the vacuum method at water level fluctuation area so that the frost-resistance and durability of the face can be improved since the water cement ratio is obviously reduced.

3. Make river flow control scheme according to local conditions

For the high dams built in recent years, much attention is paid to control the river flow during construction. The suitable flow and flood control schemes are made on the local environment and the construction conditions so as to get the best technical economic results.

The Gongboxia project located on the Yellow River sufficiently utilizes the upstream reservoir of regulation the flood and storage, through analysis on the combined regulation of the two reservoirs, adjusts the flood flow in the flood period and create the conditions for safety by the high cofferdam and construction in the foundation pit in the whole year to make horizontal and continuous placement at full sections and the dam deformation uniform. Based on the same conditions, the Jishixia dam (101m high) adopts the mode of flood protection by the high cofferdam in the whole year. It was adopted the mode of retaining water by the cofferdam in the whole year and discharging through the diversion tunnel for the Jilintai dam (157m high), and the upstream and downstream cofferdams are overtopped ones, made of rock and soil materials, with the maximum height of upstream cofferdam 33m. For the Miaojiaba faced rockfill dam (100m high), the river closure was carried out in 2008, the flood protection by the cofferdam (28m high), and construction in the foundation pit and dam placement were all conducted in the whole year. Generally, the flood protection mode of discharging through the diversion tunnel, river closure by the cofferdam and the whole-year construction in the foundation pit is a new feature in the construction of high soil-rock dams in China in recent years.

When the dam cannot be embanked to the elevation of retaining water for flood protection or the high cofferdam cannot be adopted for flood protection in the first dry season, the flood protection is ensured by adopting the overtopped cofferdam and overflowing from embankment is admissible in the first flood season. The projects, such as the Tankeng, Shuibuya, Tianshengqiao, etc., have made corresponding schemes of arrangement and protection. For Tankeng project, the 20-year flood and the peak flow of $10,400m^3/s$ are adopted. The protection of the dam face is made with rebar gabions and large blocks. Six floods occurred in the flood season of 2006, with the maximum flood flow of $3,400m^3/s$, the flow passing the dam surface of $2,100m^3/s$ and the mean flow speed of 3.51m/s at dam axis. After the floods, the cofferdam and the protection of the dam face were completed and only partially wash existed, so that placement can be conducted continually after removing the rebar gabions and placing transition material on the large blocks. The practices have shown the dam type, the flow-pass protecting measures and the flood protection scheme are practical and feasible. For the Shuibuya project, the flood protection scheme is as follows, retaining water by the high soil-rock cofferdam at upstream (El. 223m), discharging by the low cofferdam (El. 215m), raising the RCC cofferdam at downstream (from El. 214m to El. 211m) and the elevation of the embankment not higher than 210m, the maximum speeds of passing flow reduced to 10.48m/s and 5.43m/s at the upstream cofferdam and the temporary embankment surface, respectively, the flow is smooth, so as to reduce the protection difficulty and risk of the cofferdam and the embankment surface greatly. For the Tianshengqiao project, the cofferdam was overtopped in 1995 and the dam was not filled, and in 1996, a gap of 120m was reserved on the embankment surface for discharging. Otherwise, there were some projects no subjected to discharge flood during construction in low water years although the design scheme of the overtopped cofferdam and the overflow embankment surface was worked out.

The Hongjiadu and the Sanbanxi dams adopted the scheme of retaining water by cofferdam in low water season and the flood protection by the embankment body. This scheme of the flood protection is common, that is to retain water by cofferdam in low water season and place the embankment body to the height higher than the specified level for flood protection.

4. Achievements of deep overburden treatment

The impervious treatment upon deep overburden by means of concrete cut-off wall and grouting curtain has conducted for years and the relative experiences have been accumulated in China. Since the 1990s, the level of deep overburden treatment for high dams is rapidly improved, and the adopted technologies include the grouting autographic recorder, the grouting strength method and the slurry stabilizing method, the new-type cutoff wall with improved concrete proportioning or high-quality slurry, the percussive reverse circulation drill, the rapid construction process of cutoff wall, the hydraulic trench cutter with double wheels, etc. The construction machinery for cutoff walls has been developed, including percussive reverse circulation drill, crab bucket, hydraulic trench cutter, etc., and the new construction processes have been applied, such as drilling-grabbing, cutting-grabbing, cutting-grabbing-drilling, etc. New records of cutoff wall depth are continuously created, and the new achievements of cutting off the deep overburden with the mode of upper wall and lower curtain have been obtained in the western regions in China. Therefore, the construction technology of cutoff wall for soil-rock dams in China has achieved the world leading level.

In 1997, the trial concrete cutoff wall was completed in the Yele project. This wall section is 100m deep, 7.8m long and 1.0m thick, and comprises a trench hole, a single hole and a double inverse-arch joint, which provides a condition for depth of cut-off walls over 100m in China.

In 1998, a concrete cutoff wall with a depth of 81.9m was built in the Xiaolangdi inclined core rockfill dam. For this wall, high-pressure jet grouting is applied at the contact section of the partial wall with the rock mass or with the overhang cliff; For the 60m deep overburden of the upstream cofferdam, the plastic concrete cutoff wall is adopted. For dam foundation treatment, multiple new technologies are adopted: such as the two-drilling and one-grabbing; the plate type joint of two wall sections by drilling on the joint part with protection of transverse trench and plastic concrete; and the special processing technique to sand interlayer; the grouting strength method and the slurry stabilizing method for curtain grouting.

In 2001, the grouting curtain for the natural barrier formed by earthquake was constructed in Chongqing, with the maximum grouting depth of 80m. In this works, slurry of cement clay is used to consolidate the barrier, and the slurry is poured by the circulating drilling and pouring method. The consolidation effect is excellent.

For the Xiabandi asphalt concrete core dam (dam height of 78m), the maximum depth of the overburden on the dam foundation is 150m, and the grouting depth is 168m made by the circulating drilling and pouring method.

In 2010, for the Pangduo sand-gravel dam with asphalt concrete core (dam height of 72.3m) at El. 4,000m over in Tibet, the foundation treatment was conducted by heavy percussion drills and new-type grab machines, and the cut-off wall is 158.47m deep and the connecting trough is 158m deep.

5. Effective discharging of upward seepage for concrete face dam

In the construction of face dams in China, some accidents had occurred that the cushion layer, the slope consolidated and the face were damaged by upward seepage resulting from the downstream level higher than the upstream level, or from the discharge pipes frozen and the construction of auxiliary impervious bodies, therefore, the prevention and treatment of upward seepage have attracted the attention. Through continuous practices and experience conclusion for years, the method for solving the problem is mature, and relevant requirements are also incorporated into the relevant Chinese codes.

For the Xibeikou reservoir, a 7m upward water head was formed because of pumping from the upstream foundation pit, resulting in more than 20 springs concentrated in the 60m scope of the cushion material, 6 partial collapses with the maximum depth of 0.5m. This part was removed fully and backfilled on the design requirement.

Before the upstream blanket is not backfilled, the upward water pressure may lift or break the face, thus, when the upstream blanket material is not be filled to a certain elevation, it is forbidden that the upward water pressure is higher than the permeability gradient of the cushion material or can lift the face.

The effective measure for removing the upward water pressure is to be set the free or forced water discharge system in the dam. The discharge capacity of the forced water discharge system shall satisfy the design requirements and ensure the normal operation, and only when the upstream filling elevation is higher than the highest level of upward water, the discharge facility can be blocked.

For the Tianshengqiao dam, 2m×2m rebar gabions are used as upward pressure well, the water is freely discharged to the upstream through communicated steel pipes, and pumping in the well is combined when necessary. For the Hongjiadu dam, the 150mm-diameter perforated steel pipe wrapped with gauze is buried in the 3B zone for discharging freely, and the collector well is cancelled. For the Songshan face dam with 80.8m high, located in the severe cold region, two ϕ150mm pipes are buried at the lowest part of the dam for discharging the upward water, the system is normal in the early period. Before the winter of 2000, the 1-stage face was completed, and covered with thermal insulation. In May of next year, map cracks on the bottom of the face were found when removing the thermal insulation, the face and the plinth were staggered and the face was partially detached, and water flowed out from the peripheral seams of the bottom. Through analysis, it proved the face cracks were caused by the force applied by the frozen of accumulated water on the dam bottom. The treatment was made as following: grouting cement mortar into the detached part, placing concrete of the plinth and the face again at the reinforcing area, and providing other water stoppage system. Before the winter of 2001, the whole face was completed, the discharge pipe of the dam was blocked before filling soil material, but it found that the reinforced face was obviously lifted and had the tendency of deterioration after 10 hours blocked. Inspection immediately made by drilling hoe to discharge water for pressure reduction, the cracks were found on the face but the face was not broken, and GB material was used to coat on the reinforced face.

6. Breakthrough of rapid construction technology of rockfill dam under the conditions of plateau, serve cold and drought

In Western and Northern China, several asphalt concrete rockfill dams, high concrete faced rockfill dams and soil core rockfill dams have completed or under constructed, their successfulness provided experiences for rapid construction of soil-rock dams in plateau and severe cold areas. Under a certain condition of the negative temperature, a lot of test research and practice is carried out, such as placement of cement concrete and asphalt concrete, and construction technique of various embankment materials, and the achievements are fruitful. In the process of construction, the adverse influences on the construction personnel and the equipment efficiency generated by the environment were researched and analyzed, some methods were tried to how to reduce the construction personnel and arrange the construction personnel in balance under bad climate conditions, and plentiful experiences for construction of soil-rock dams in the plateau and severe cold region is accumulated.

For Wuluwati project (dam height of 133m), it is located in Hetian, Xinjiang, where is dry, cold, windy and sandy influenced by the Taklimakan Desert. The climate environment is bad there, with less rainfall is less (annual rainfall 79.1mm), high evaporation (annual evaporation 2,079.7mm), extremely dried, high temperature difference between day and night (measured as 12℃ to 18℃), change of annual air temperature, high windy frequency and high wind velocity, there 200 aerial dust days in a year. To this condition, new technical measures are adopted in the construction of slope consolidation, processing and installation of reinforced bars, transportation of concrete, concrete curing, face protection in winter, etc., so as to balance the allocation of manpower and equipment, accelerate the construction, and the projects have obtained high quality.

The Longshou II project is located on the Heihe River, in the Qilian Mountain, with a height of 146.5m. The dam site belongs to the area of drought, severe cold and plateau, with the mean annual rainfall of 175mm, the mean annual evaporation of 1,378.7mm, the highest annual temperature 37℃, the lowest annual temperature −33℃, and the mean annual temperature 8.5℃. Generally, the Heihe River freezes in mid-November and thaws in the end of March. The project started in 2002 and completed in 2004. Since the construction period was short, the filling of embankment materials and the placing of partial concrete (such as the extruded side wall, concrete high cut-off trench, etc.) had to be carried out in the severe cold season, thus, the construction experiences have be summarized in concrete placing, curing and filling of the embankment materials in the severe cold climate.

The Manla clay core rockfill dam in Tibet is 76.3m high, at El. 4,200m. Under the conditions of severe cold, oxygen insufficiency and capricious climate, the dam is constructed on the high quality standard. The project was completed in July, 2000, and got the state prize of outstanding project construction in 2003. In the last decade, several concrete faced rockfill dams have built in the area higher than 4,000m above sea level in Tibet.

7. Timely prepare and revise relevant construction standards (codes and regulations) to direct the construction

In recent ten years, a multitude of large and super large hydroelectric dams and pumped-storage power plants has been built or started construction, especially for the completion of the Three Gorges Dam on the Yangtze River, Longtan dam, Gongboxia, Xiaowan, etc. and the beginning to construct the hydropower station of Xiangjiaba, Xiluodu, Jinping I and II etc., and China has achieved great success in the hydropower development. Led by Sinohydro Corporation, China's hydropower construction companies have continuously developed new technique, new materials and new methods, the construction techniques and management skills have grown by leaps and bounds to shorten the construction cycle of projects, the construction practice and theory study has made a revolutionary process and has accumulated rich experiences in project construction management, which can be focused on a series of construction standards of the power industry compiled by Sinohydro Corporation and have applied widely in the domestic hydropower development. These standards is not only used in our country but also started to apply in the international countries in recent years, which have changed the former situation of corner standard markets by Europeans and Americans. To formulate, revise and compile the relevant standards (specifications and regulations) of hydropower construction techniques in time, it is of major strategic importance and practical significance to direct the dam construction.

19.1.4 Conclusion

In China, the modern technology for building soil-rock dams has experienced for more than 60 years, starting from low homogeneous soil dams and rockfill dams, up to now, the technology has stepped into the time of high dams and super-high dams including soil core rockfill dam, concrete faced rockfill dam, asphalt concrete impervious rockfill dam, etc., and the construction personnel devote their painstaking, and have got abundant experiences. Currently, the diversion tunnels are constructing for both 300m super-high gravelly soil core rockfill dams, Shuangjiangkou and Lianghekou. The preliminary survey and design are being proceeded for about 250 to 300m high dams, including Gushui, Maji, Qizong, Cihaxia, etc. The construction personnel for hydropower development are expecting the further development and the technical challenge of high soil-rock dams construction, particularly the high concrete faced rockfill dams in the 21st century.

References

[1] Editorial Committee of *Construction Manual of Water Conservancy and Hydropower Engineering*, Volume of Soil and Rock Cube Engineering of Construction Manual of Water Conservancy and Hydropower Engineering. Beijing: China Electric Power Press, 2002.

[2] Editorial Committee of *Construction Manual of Water Conservancy and Hydropower Engineering*, Volume of Foundation and Basic Engineering of Construction Manual of Water Conservancy and Hydropower Engineering. Beijing: China Electric Power Press, 2004.

[3] Wang Bole. *Modern Chinese Soil-Rock Dam Engineering*. Beijing: China Water Power Press. 2004.

[4] Editorial Committee of *Chinese Hydroelectric Engineering*. Construction of Chinese Hydroelectric Engineering. Beijing: China Electric Power Press, 2000.

[5] Hydropower and Water Resources Planning and Design General Institute, etc. Thesis Collection of Soil-Rock Dam Technology in 2008. Beijing: China Electric Power Press, 2000.

[6] Zhou Jianping. et al. Advancement of Modern Rockfill Dam Technology—2009. Beijing: China Water Power Press 2009.

[7] Jiang Guocheng, Xu Zeping, Summarization of Development of International Concrete Face rockfill dam—General Report of 1st Rockfill Dam International Seminar Summarization, Development of Modern Rockfill Dam Technology—2009. 2009.

[8] Zhao Zengkai, Several Reviews and Suggestions on Building High Concrete Face rockfill dams. Progress of Modern Rockfill Dam Technology—2009. 2009.

[9] *Construction Specification for Concrete Face rockfill dams* (DL/T 5128 - 2009), Beijing. China Electric Power Press, 2009.

[10] Wu Guiyao, Huang Zongying, Jiang Jianlin. Construction Technology of High Concrete Face rockfill dam, *Technology of Water Resource*. 2006 (7).

[11] Yu Yang, Zhu Xiangpeng, Huang Zongying, Wang Hongyuan. Construction Technology for Mixing Rock and Soil in Nuozadu Dam. *Technology of Water Resource*, 2010 (5).

[12] Pan Jiazheng, He Jing. Chinese Dams in 50 Years. Beijing: China Water Power Press, 2009.

[13] Xuyong. Enlightenment of Construction of Concrete Face rockfill dam of Grade 1 Hydropower Station in Tianshengqiao and Discussion for Building Higher Dams//*Thesis Collection of Soil-Rock Dam Technology in* 2008, Beijing: China Electric Power Press, 2008.

[14] Zhou Yanjiang. Filling Construction of Concrete Face rockfill dam in Zipingpu//*Thesis Collection of Soil-Rock Dam Technology in* 2008, Beijing: China Electric Power Press, 2008.

[15] Fan Shaoyong, Jiang Changchun, Xing Xinyuan, etc., Quarrying, Transportation and Rolling of Gravelly Soil of Dam in Pubugou. *Hydroelectric Power*, 2010 (6).

[16] Peng Guoyong. Construction of Gravelly Soil of Core Wall of Hydropower Station Dam in Pubugou, Progress of Modern Technology for Rockfill Dam—2009. Beijing: China Water Power Press, 2009.

[17] Li Shiqi, Liu Qiongfang, Jin Guohui. Construction Technology of High Core-Wall Rockfill Dam of Nuozadu Hydropower Station. *Hydroelectric Power*, 2010 (1).

19.2 High Concrete Gravity Dam Construction

Dai Zhiqing[1], Sun Changzhong[2], Huang Jiaquan[3]

19.2.1 *Achievements and Development of High Concrete Gravity Dam Construction in Recent Years*

Since the 1950s, China's water conservancy and hydropower development has made rapid development and been impressive

---

[1] Dai Zhiqing, Deputy Chief Engineer of China Gezhouba Group Corporation, and Chief Engineer of the Three Gorges Branch, Professor Senior Engineer.

[2] Sun Changzhong, Deputy Chief Engineer of the Three Gorges Branch of China Gezhouba Group Corporation, Professor Senior Engineer.

[3] Huang Jiaquan, Deputy Director of Science Department in the Three Gorges Branch of China Gezhouba Group Corporation, Senior Engineer.

for numerous projects, large scale and fast speed. Among them, large and medium-sized hydropower projects completed or under construction with high concrete gravity dam (height above 100m) account for more than 50%, hold a dominant position, and are one the most important part of water conservancy and hydropower project construction.

Review the development history of high concrete gravity dam construction, under the national planning economy system before the 1980s, developing rate is slower; after the reform and opening-up policy, the construction technology has obtained the rapid development, and the engineering construction gained brilliant success. Especially in the twenty-first Century, along with the rapid development of the national economy, water conservancy and hydropower development is entering a high-speed development period, in recent ten years, the construction of high concrete gravity dams has made breakthrough progress in advanced technologies, new techniques, new-type materials and new-type equipment, and has formed a series of the construction techniques and methods to meet large complex structure construction requirements.

Table 19.2 - 1 is the features of concrete gravity dams with the height of more than 100m completed and under construction in recent 10 years. From the table, high concrete gravity dam construction technology development mainly has the following achievements.

Table 19.2 - 1 Features of Concrete Gravity Dams with Height More Than 100m
Completed and Under Construction in Recent 10 Years in China

No.	Name	River	Location (province)	Dam type	Dam high (m)	Reservoir capacity (billion m^3)	Installed capacity (MW)	Completed year
1	Mianhua'tan	Tingjiang River	Fujian	RCC gravity dam	113	2.035	600	2002
2	Dachaoshan	Lancang River	Yunnan	RCC gravity dam	111	0.94	1,350	2003
3	Jiangya	Loushui River	Hunan	RCC gravity dam	131	1.741	300	1999
4	Suofengying	Liuguanghe River	Guizhou	RCC gravity dam	115.8	0.201	600	2005
5	Baise	Youjiang River	Guangxi	RCC gravity dam	130	5.66	540	2006
6	Three Gorges	Yangtze River	Hubei	Concrete gravity dam	181	45.05	22,500	2009
7	Guangzhao	Beipanjiang River	Guizhou	RCC gravity dam	200.5	3.245	1,040	2009
8	Longtan Ⅱ	Hongshui River	Guangxi	RCC gravity dam	216.5	29.92	6,300	under construction
9	Hongkou	Huotongxi River	Fujian	RCC gravity dam	130	0.45	200	2010
10	Wudu	Fujiang River	Sichuan	RCC gravity dam	120.3	0.594	150	2009
11	Gelantan	Lixianjiang River	Yunnan	RCC gravity dam	113	0.409	450	2009
12	Jinghong	Lancang River	Yunnan	RCC gravity dam	108	1.139	1,750	2009
13	Silin	Wujiang River	Guizhou	RCC gravity dam	117	1.593	1,050	2011
14	Jin'anqiao	Jinsha River	Yunnan	RCC gravity dam	160	0.913	2,400	2011
15	Guandi	Yalong River	Sichuan	RCC gravity dam	168	0.76	2,400	2012
16	Shatuo	Wujiang River	Guizhou	RCC gravity dam	101	0.91	1,120	2012
17	Gongguoqiao	Lancang River	Yunnan	RCC gravity dam	105	0.349	900	2011
18	Longkaikou	Jinsha River	Yunnan	RCC gravity dam	116	0.558	1,800	under construction
19	Ludila	Jinsha River	Yunnan	RCC gravity dam	140	1.718	2,160	under construction
20	Xiangjiaba	Jinsha River	Yunnan/Sichuan	Concrete gravity dam	162	5.163	6,400	under construction
21	Guanyin'ge	Jinsha River	Yunnan	RCC gravity dam	159	2.25	3,000	under construction

Firstly, the construction technology for roller compacted concrete (RCC) dams has been rapidly popularized. Among 21 high concrete gravity dams completed and under construction, 19 are RCC dams, taking the absolute dominance, because it has the advantages of less cement consumption, fast construction speed, short duration, low cost, therefore, RCC has generally been adopted in gravity dam construction.

Secondly, large project scale and fast construction speed for RCC dam have ranked first in the world. Although RCC dam construction started late in China, the development history is less than 30 years, but the development speed is very fast. Especially in the past decade the Longtan Hydropower Station with the maximum dam height of 216.5m is a represent of high-concrete gravity dam construction successfully, it has had the height of concrete gravity dam become the 200m level quickly, and set the new records of RCC dam construction at home and abroad, including the highest dam height, the largest dam RCC quantity of 6.59 million m^3 and the short construction duration of less than 9 years.

Finally, the construction technology of high concrete gravity dam has reached the world advanced level. Over the current decade, the construction technology has innovated continuously on the basis of past 20 years, the key technologies have made breakthrough, such as, construction information management, high strength aggregate production, rapid construction, temperature control and crack prevention of mass concrete, high performance concrete application, large hydraulic steel structure installation and concrete quality control, and the achievements include the National Grand Prize of Science and Technology Progress, more than 500 major scientific and technological achievements, 120 national patents, prepared 30 national or industry standards and state-level engineering methods. Longtan Hydropower Station won the "International Honor Roller Compacted Concrete Dam Engineering Award", which indicates that China's high concrete dam construction technology has gone into the world's leading level.

19.2.1.1 High-quality, High-intensity Aggregate Production Technique Improvement

Over the recent decade, water conservancy and hydropower development has entered an unprecedented peak in China. Now, the high concrete gravity dams completed and under construction are mostly located in mountain and valley areas of central and western regions. The artificial aggregation has to be used as the concrete material because of lack of natural aggregate resource there. With the construction technology and equipment continue to upgrade, construction duration shortened, dam concrete production intensity increased, the aggregate production system is faced with new challenges, in terms of large construction scale, short construction duration, high product quality, low processing cost, and harmony with the environment. The artificial aggregate production system is the most important one of hydropower station construction ancillary facilities. In order to adapt to the demand of hydropower development, numerous scientific and technical personnel through unremitting efforts and innovation, created the key production techniques relying on those remarkable projects, such as "natural aggregate milling process" "rock material vertical transport technique at high steep narrow area", in the aspects of production scale, or the aggregate processing mechanization degree and product quality, breakthrough achievements and progress has been obtained, and the first-class level in some scopes has reached in the world, and a strong supporting has given for sustainable development of hydropower projects.

1. *High intensity aggregate production technology for Longtan Hydropower Project*

For Longtan project, the aggregate processing system at Dafaping bears the concrete aggregate production for main dam and cofferdam works and a total aggregate volume of 14.12 million tons is required for RCC and conventional concrete. This system is a comprehensive production system for making artificial sand and gravel, and it has the largest scale, the most comprehensiveness and advanced technology in China, with its design capacity of 2,650t/h and aggregate production capacity of 2,150t/h. The system consists of following processes, quarrying, pre-screening and washing, reduction and fine crushing, secondary and third screening, sand crushing and stone powder recycling, water supply and wastewater treatment, power supply and electrical control components, the main process adopted the crushing method of four steps, that is coarse, medium, fine and superfine crushing. In the whole process, two production lines are adopted, and wet screening, wet crushing sand, dry medium crushing, dry fine crushing, dry superfine crushing are conducted, in which the medium and fine crushing and secondary screening are taken as the semi-closed production, and the superfine crushing and tertiary screening as the closed-circuit production.

In this system, not only the most advanced production equipment in the world, a large mobile crusher and Drake fine material recovery device, are used first time, also uses a large cylinder stone scrubber, long belt conveyer for concrete aggregate (4.0km long) and other large production facilities. The system operation successfully sets the new records of monthly production sand and aggregate 1.24 million tons and supplying of 955,000 tons in China, and has win the praise by the owner, supervisors and industry experts.

(1) Application of advanced equipment for aggregate processing. The Longtan aggregate processing system is equipped with 550 sets of new equipment (sets), worth 135 million Yuan (in which imported equipment nearly 100 million Yuan), with the total installed power 21MW. For the key processing equipment such as crusher, stone washing machine, powder recovery, etc., high-performance imported equipment are selected, with advanced technology, high unit production and reliable quality, including large mobile crusher station, portable belt conveyer, large cylinder stone scrubber, powder recovery device and Pneuma dredging system. The configuration of the system is the first time application in China, and it is a beneficial attempt in this kind of processing system.

1) Application of large mobile crusher station and portable belt conveyor. The quantity of mining and hauling rock material is huge, so that the balance of stone material horizontal and vertical transportation is the key path to control the transportation cost, construction safety and reliability feeding in aggregate processing. The traditional transportation is vehicular mainly, and feed chute is also used in some projects. There are two techniques for chute feeding, that is chuting→breaking and breaking→chuting, the two methods have been applied in Dafaping aggregate processing system.

The key point of technology for the breaking→chuting, large mobile crusher station and mobile belt conveyor are set at the working face and can flexible move crushing and convey to the chute. After researching and demonstration for years, 2 sets of METSO LT140E large mobile crusher station and LL12 mobile belt conveyor are imported. The main characteristic of this large crushing station is that it can be arranged in the quarry, and change its working position with expanding of exploitation and loading operation, thereby rock material is crushed nearby the face, which is hard to do for fixed crusher. While combining with the mobile belt conveyor, subsequent conveying of excavated material is conducted.

Due to the application of large-scale mobile crusher station, stone material blasted at the quarry, is exploited and loaded by excavator in the working face, and crushed by crusher into small size, and then transport by mobile belt conveyor to a chute or semi-finished stockpile.

2) Wastewater treatment by Pneuma dredging system. The wastewater treatment volume of Dafaping aggregate processing system is $4500m^3/h$, according to the site's conditions, a new wastewater treatment system is adopted, which include the following components: fine sand recycling, sedimentation concentration, Pneuma system, pipeline delivery and tailings. The Pneuma system is a pumping station, Italy patent product model-300/600-VS, with 2 air compressors ATLAS-GA25010.

This new system is to collect the wastewater of processing system into the settling tank, have the silt particles in the wastewater deposit on the bottom of the tank by natural precipitation and the clean water at the upper part of settling tank flow to the recycling pool through pipeline gravity. The Pneuma pump is arranged in the dredging ship, and used to pump the silt powder on the tank bottom to the spoil pile through pipe. The filter well is provided at the spoil area and can make the water after standing flow into the drainage culvert. In the whole process does not cause secondary pollution of the water in settling tank and not pollute the discharged into water bodies during water delivery, so this system is called as the environmental dredging system.

The Pneuma system can pump the silt slurry with high concentration to the spoil area 2,000m away, with the slurry concentration of greater than 50% and the highest up to 92%. The turbidity index of recycled water and drainage turbidity indicators of the soil pile have reached the design requirements.

(2) Composite recycling technology of stone powder. The moisture content of artificial sand has a greater influence on concrete properties, in the concrete specification clearly states: the contents of artificial sand and stone powder used for RCC shall be 16% to 22%, in which the content of fine material less than 0.08mm is 6%-9%. When dry or semi-dry method is used in crushing sand, the content of stone powder is generally 8% to 9%, therefore certain measures shall be taken to recover the powder from the waste water.

The artificial sand produced by the Dafaping system are mainly used for RCC, in which the content of stone powder is 16%-22%. Three sets of DER-RICK 2SG48120W4A fine material recovery device were selected in the original design, each recovery device is equipped with 2 groups of cyclone of 100mm diameter, and has the design flow of 600t/h, pressure of $2.6kg/cm^2$ and recycling capacity of 25-30t/h. But it was found in production that the cyclone diameter is small with the outlet diameter of 15mm only, when the discharging particles larger than 2mm or 1mm makes up 25% and more, no powder can be recovered basically. Moreover, when the wastewater flow is uncertain, or fails to meet the flow requirement, the recovery effect is not good.

In order to achieve the ideal recovery effect, the composite recycling technique is developed, that is using cyclone and sand-scraping machine to recover the powder. Besides the powder recovery by cyclone, three sand-scraping machines are adopted, the one is arranged in front of cyclone process, and the others are set behind the cyclone process. Thus, before wastewater into the cyclone, the part fine particles can be precipitated and scraped out. While overflow water from the cyclone or excess water without through the cyclone can be further precipitated, and some powder is recovered by the sand-scraping machines set behind.

The application of composite recycling technique takes the advantages of the cyclone with high efficiency, large output and less cost of sand-scraping machine, convenient setting and less fault, and the technique solves the problem of easy blocking. The powder recovery process is smooth, the amount of powder recovery has greatly increased, and thereby the demand of high content of stone powder is met for RCC.

(3) High-speed deep-trough long distance belt conveyance of aggregate. The transportation mileage is about 7.5km from the Dafaping processing system to the dam concrete production system (straight line distance of about 4.5km), and road conditions are poor there, with road width less than 6m partly, more bends and steep slope, therefore, road transport by vehicle is very inconvenient and cannot meet the traffic requirements, there still exist a great potential safety hazard. Thus, long distance belt conveyor and controllable start transmission (CST) device are selected to effectively solve the problem of vehicle transport, and realize the integration of mining, crushing and transportation. The semi-finished materials are conveyed from the quarry to the semi finished products stockpile by the belt conveyor with long cantilever cable which is set along hillside with two transition bins so that the finished material can be delivered in the case of high height difference; and the materials are conveyed to the mixing plant by the belt conveyor in the long tunnel, the conveyer has the total length of 3.95km, 3.55km of which set inside the tunnel, with the design conveying capacity of 3,000t/h.

The operation of long-distance belt conveyor is driven by the high-voltage motor and the controllable starting transmission (CST) device, and the automatic delivery system is controlled through PLC and computer automatic control system and camera monitoring in the central control room. The control system has the detection and protection functions for major faults and can realize unmanned operation. In the design of long-distance belt conveyor, taking into account the high wear of sand and gravel material delivered, large conveying quantity and long conveying distance, the high-quality rubber steel cord is used as the conveyor belt which is characterized by deep groove, high speed, wear resistance and shock resistance. In addition, the auxiliary devices are provided against deviation, skid, blocking, tear, and emergency stop (rope) switch and removing water. A flapper is set in front of the machine to prevent the material falling or being stuck in the belt and the roller. The two feed points are provided behind the machine, the front guide groove can be lifted and the back guide groove is fixed, when the back guide groove is fed, the front one is leveled off. The belt is driven by friction between transmission drum and the belt. A stepless annular band is formed by the belt rounding the transmission drum and turnover drums, the up and down branches of belt are supported by the up and down rollers respectively, and the tension force is provided by the tension device for belt, the transmission drum is driven the driving device (motor and controllable start transmission CST device). The materials are constantly fed on the belt and then be conveyed through the guide groove.

As compared with other transportation means, the actual operation condition of high-speed deep-trough long distance belt conveyer has the following characteristics: ①transmission speed 4m/s, transmission intensity up to 2,800 – 3,000t/h, high efficiency and low operation cost; ②simple structure, components and parts being welded with section steel and steel plate on the drawings in the factory, and then being jointed with bolts or welded in the working site, quick and convenient assembly and disassembly, easy operation; ③safe and reliable operation, the controllable drive device CST, in addition to meet requirements of tension force and speed rate control, can prevent adverse effect from momentary load impacting and avoid the damage by system sudden power outages; ④flexible arrangement, especially suitable for the working site with poor access and high conveying intensity, low cost due to small occupying land and less temporary works.

The successful application of long distance belt conveyer in Longtan Hydropower Station is a major breakthrough for aggregate transportation and has provided valuable reference for construction of hydropower projects. For example, the artificial aggregate processing system in the Xiangjiaba Hydropower Project adopts the long distance belt conveyor set in the tunnel, with conveying length of 31.07km, tunnel length of 29.31km, transmission speed of 4m/s and capacity of 3,000t/h, the conveyer can meet the concrete placement intensity of 347,000m^3/month and is the longest of the artificial aggregate belt conveyors in the world.

2. Sand crushing technique from natural aggregate in Tingzikou Project on the Jialing River

For high RCC gravity dam construction, volumes of aggregate is required, while RCC have strict requirements to the powder content and water content of artificial sand and may be 20% for the powder content of finished sand, but by traditional rod mills and the wet technique of sand crushing and breaking, a large amount of stone powder will loss and the content of stone powder is hard to meet the design requirements. Therefore, more and more attentions have given to artificial sand for gradation adjustment need and for meeting the quality of finished materials.

The aggregate processing system at left bank of Tingzikou Water Control Project, on Jialing River, is the largest natural aggregate production system in Asia, which supply about 5.20 million m^3 aggregate required for the concrete of main structures, which is constructed and used by the China Gezhouba Group. In the natural material source, undisturbed sand content is insufficient and stone powder content is lower than the requirement of RCC, so that the adjustment of aggregate gradation is needed. The proportion of sand is 31.43%, while the undisturbed sand makes up a half of the required proportion, so it is necessary to increase sand proportion by the processing system. On the demand of finished material, the powder content in mixed undisturbed sand and crushed sand (broken material by crusher and rod mill) is about 11%, which only meet the requirements of powder content for normal concrete, but the powder content is 14%–20% required for RCC, therefore, it has to add 3%–9% the stone powder content to meet the powder content requirement.

In view of the above problems, the new technique research is conducted on how to improve the crushing effect of artificial sand and increase the level of natural undisturbed gravel preparation to get the ideal powder mixture and quality control, and the technique of crushing powder from natural gravel is developed. For this technique, new-type sand mill with median speed is used to make stone powder by crushing natural pebbles. The technique effective expanse the stone powder material source, and can conduct blending timely and control mixing uniformly, so that the powder content of RCC is controlled between 14%–20% successfully and the quality requirements of finished aggregate is met.

(1) Crushing sand by vertical shaft crusher. In the common scheme of vertical shaft crusher application, the mixture with size less than 80mm is put into the vertical shaft crusher to make the material less than 5mm. But in Tingzikou project area, the natural undisturbed pebbles has high compressive strength (246MPa) without cracks and joints, therefore, the quality of crushed material is poor if putting the material directly into the vertical shaft crusher, at the same time, it is uneconomic to blend big, middle and small stones as raw material to making sand for lack of middle and small rock materials there. In order to improve the quality of crushed sand, only natural boulders are taken as raw material to make sand. For this technique, a step "pre-broken" is added, that is putting the boulders first into a cone crusher to make the material cracking so as to facilitate crushing by vertical shaft crusher. Meanwhile, a part of sand can be produced when crushing made by cone crusher, and used as the finished product.

The operation of the system has proven that, taking the "pre-broken" pebble as raw material, the crushed sand rate by the vertical shaft crusher can reach 26%, higher than that of traditional graded gravel processing technique.

(2) Crushing san by rod mill. The sand quantity has been greatly increased due to the vertical shaft crusher is adopted, but the product is characterized by lack of intermediate size, high fineness modulus, unreasonable gradation, lack of intermediate grade for finished sand, therefore, the blend material with undisturbed sand and crushed sand has unreasonable gradation, but the production cost is low; and the artificial sand made by rod mill has well gradation, controlled fineness modulus, but the production cost is high. The two techniques have a good complementary in technology and economy, therefore, the best solution of the adjustment of finished sand gradation is to adopt the traditional rod mill.

Through the above steps, the crushed sand with qualified fineness modulus can be produced, and the stone powder content under the circumstances of the reserves and recovery, only reach about 14%, and cannot fully meet the requirements of stone powder content for RCC, it is need to take measures to make powder.

(3) Crushing and blending powder. The content of sand and powder is generally up to 17% produced by vertical shaft crusher, and that is generally around 11% produced by rod mill, while the powder content in undisturbed sand on the natural floodplain is generally not more than 5%. Moreover, the stone powder of blend sand mixed with undisturbed sand and crushed sand (produced by vertical shaft crusher and rod mill) is about 11%, which can only meet the requirement of normal concrete to the powder content, and the technique of crushing and blending powder is need if the product is used for RCC.

1) Powder Preparation. Since the powder manufactory is far away from the working site, procurement powder is not economic, so the powder preparation is adopted. In view of the operation of common ball mill occupies a large area, with high steel consumption, by studying, comparing and choosing, firstly adopt a new-type sand mill with median speed for crushing powder (see Fig. 19.2-1).

Fig. 19.2-1　MTM160 Sand Mill with Median Speed

MTM160 sand mill is mainly composed of a host machine, fan, separator and dust collecting equipment. Main technical parameters: unit capacity 253kW; production capacity 13 – 22t/h, the content of small than 0.16mm above 90%, adjustable fineness; powder is delivered to the powder tank by pneumatic conveying device.

2) Powder Blending. The fine powder in fine aggregate shall be uniformly dispersed in concrete mixing, in order to have the fine aggregate uniform after powder being blended, the finished sand source should be normal sand which is naturally stockpiled and dewatered qualified (water content less than 6%), therefore, the pre-dewatering yard is set for stacking the crushed sand without dewatering, after their moisture content meet the condition, and then blend them with powder, the process is shown in Fig. 19.2 – 2.

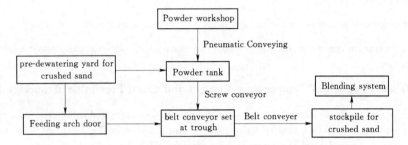

Fig. 19.2 – 2 Process Flow Diagram for Powder Blending

3) Quality control measures of powder blending

a. Control of powder content.

In order to make the powder content of the crushed sand within 14% – 20% and remain stable, the four steps are adopted: Firstly, calculate the delivery volume of the belt conveyor set at trough through testing, and determine the opening of feed door within the corridor for pre-dewatering of crushed sand; Secondly, calculate the powder amount to be added according to the test powder content of normal sand and the delivery volume of the belt conveyor at trough and the target powder content; Thirdly, determine the screw conveyor speed by testing according to the relationship of screw conveyor speed and powder volume to be conveyed; Fourthly, determine the actual powder content of the mixture, compared with the target powder content, and adjust the screw conveyor speed until qualified.

b. Control of uniformity.

For the powder to be blended, it is put on the finished normal sand, and then the powder and the sand are blended through dropping to the belt conveyor each time, finally, the blended material is delivered to the finished crushed sand stockpile. In the process, a mixer (similar to a mortar mixer) is set in front of the belt conveyer for discharge between the two belt conveyers so that the material can be blended once again during dropping and then be delivered into the next belt conveyer, therefore, the powder and the sand are blended uniformly. The fine aggregate blended is loaded by the loader, and transported by dump truck to the feeding point of the mixing system, or directly to the mixing plant by the belt conveyer, thus, the powder in the crushed sand by blended times is distributed uniformly when it is delivered into the mixing plant.

In the above two steps of making sand and powder, the fineness modulus is adjusted through traditional rod mill, the dry method is adopted in crushing so as to retain the powder content of crushed sand as much as possible, and then the powder milled is added, therefore, the high-quality sand is produced for RCC.

19.2.1.2 Significant Achievements for High Intensity Quick Concrete Placement Technology

Over the past decade, as represented by the Three Gorges Project, high-intensity concrete placement technology has made significant progress. Before the Three Gorges Project built, the highest annual, monthly and daily records of concrete placement volume were created by Kuybyshev Dam in the world, with 3.13 million m^3, 0.389 million m^3 and 19,000m^3, respectively. In China, the highest annual and monthly records of concrete placement volume were created by the Ertan project with 2.12 million m^3 and 0.245 million m^3, respectively, and the highest daily intensity was 16,900m^3 created by Gezhouba project. For the Three Gorges Project, the highest annual, monthly and daily records of placement volume are 5.48 million m^3, 0.5535 million m^3 and 22,000m^3 respectively at the second stage works, in which the yearly placement intensity is over 400 million m^3 and the monthly one is up to 0.39 million m^3 from 1999 and 2001, and the world record is broken in the successive three years. During concrete placement at the third stage works of Three Gorges Project, the other construction records were created by the Gezhouba Group, such as the average annual dam rise height is 63.7m for the first placing dam block, and the highest annual dam rising height is 68m, these records have shown that high-intensity rapid concrete construction technology reached a new level. The main construction techniques are as follows:

(1) For the second-stage works of Three Gorges Project, the dam concrete is placed and delivered by the tower belt crane and the computer integrated monitoring system of production and transportation, which has provided the technology and equipment guarantee for high intensity and mass concrete placement.

(2) A series of advanced technology, new-type technique and new kind of materials are used in the construction, such as high range water-reducing agent ZB-1A, JG3, X404.

(3) The successful application of thick concrete lifts, with the thickness of 3.0 – 4.0m.

(4) The design and application of new-type formworks, such as doka form, integral lift form, prefabricated form, dies of surface outlet and trash outlet.

(5) The new-type rebar mechanical connection technology, such as rolling straight thread joint, upsetting straight thread joint.

(6) The integrated temperature control technology, which can ensure the concrete construction quality while accelerating the construction progress.

19.2.1.3 Significant Progress of Temperature Control and Crack Prevention Technology for Mass Concrete

The temperature control and crack prevention for mass concrete is the key point and difficult in dam construction, but the high standard comprehensive temperature control technology is first adopted in the whole construction process and all of concrete structures in the concrete dam of the Three Gorges project, with the technology, harmful penetrating cracks did not appear and the surface cracks were considerably less than those of similar projects at home and abroad, so that the temperature control and crack prevention technology for mass concrete has achieved significant progress and reached a new level in the world.

In the Three Gorges project area, climate is very hot in summer. The concrete placement works is characterized by the largest placing intensity in the world, the large placement blocks and complicated block shape. In dam construction process, a series of comprehensive temperature control measures are adopted, including continuous optimization design, selection of raw materials, mix proportion, aggregate secondary cooling, mixing by adding cold water or iced mixing, shading and spraying in aggregate transportation, spraying over fresh concrete area, specific cooling water system, wet curing, exposed parts isolated by benzene board, orifices and flow passage coated by polyurethane, and a set of mature temperature control technology has developed for mass concrete placement. Among them, the typical concrete temperature control measure is the three warning systems: ①the warning system of concrete temperature control, including the warning of concrete placing temperature, supplying cooling water and concrete internal temperature; ②the warning system of delay between placement lifts; ③the warning system of special weather during placement, including the construction in rainy season, or in high temperature season, sudden drop in air temperature. The implementation of the comprehensive temperature control measures effective controls the temperature rise of dam and successfully prevent cracking in concrete.

19.2.1.4 Concrete Quality Control Level on Upgrade

In recent 10 years, concrete quality control has developed on the principle of completely, full, whole process quality management, and made prevention beforehand, vigorously promote the personnel skills training and quality awareness education, continuous strengthen process control and detail management, sturdy advance institutionalization, fine and standardization of the project quality management, and achieve the major changes of management concept and behavior. Through unremitting efforts, the quality control of concrete placement has upgraded a new level.

1. Strict technical interpretation system

Technical disclosure is an important means in construction management, which can make each constructor clear the project specific construction conditions, construction organization and technical requirements and key technical measures. Before each procedure conducted, the technical disclosure is made by chief engineer or technical department in phases and levels to have the construction personnel clear task and know fairly well. The contents of technical disclosure include: project features, content of construction drawings; technical requirements of main work elements and diversion works, construction method and sequence, quality standards, safety requirements, etc. ; the main structure axis, elevation, structure outline and relative introduction; the technical problems occurring in crossing with working procedure and work type; construction measures in the winter, the rainy season, high temperature season and operation methods and the matters needing attention; technical problems of the key parts and special process; measures and plan of flood prevention and passing. For example, at the third stage works of Three Gorges project, the quality control points of bearing walls upstream and downstream, that of walls and columns of upstream auxiliary plant at El. 82m, construction quality measures of tailwater diffusion, concrete quality measures of scroll case second-stage concrete, concrete quality measures of 6m lift, these made by Gezhouba Group. These quality control points and quality measures provide the guidance and the basis for construction personnel, and technical support for the works becomes "the high-quality product".

2. Strict controlling construction process and enhancing detail management

In order to strengthen the field quality management in refinement, routinization, and standardization and control construction process strictly, the fine management method is formulated to research the potential quality problem prior to construction and strengthen the foreseeability and the ability to resist the quality risk. In the construction process control, the strict quality standard is made for each procedure, which is used in inspection and acceptance of the procedure by construction technicians and quality inspectors. The next procedure cannot be conducted when this procedure does not meet the standard. The special inspection of procedure quality is made every week, including placing preparation, placing technique, appearance quality, curing by cooling water, the inspection is made strictly on the list of quality inspection item by item. The problems found are reported timely and the corrective and preventive measures are worked out to prevent similar quality problems happening again.

(1) Adhere to the "three inspections system" and ensure process controlled. In construction, adhere to implement the internal "three inspections" system that is the initial inspection by construction team, the re-inspection by construction management department and the final inspection by the quality and safety department. When the internal three inspections are all qualified, the application of acceptance is submitted. Moreover, pay special attention to the quality information collection, so as to ensure construction and information feedback synchronously. At the same time, strengthen the collecting and processing of the original records to make the information integrity, accuracy and traceability.

(2) Adhere to the "three certificates system", ensure the quality of excavation. In cavern and foundation excavation, in order to ensure excavation quality and strive for "high-quality products", Gezhoubu Group put innovatively forward the "three certificates" system for excavation management, namely excavation certificate, excavation finish certificate and blasting certificate. Through the strict implementation of the "three certificates" system, drilling quality and powder charging quality are controlled, and the excavation quality is ensured in order to have the excavation works become a "high-quality products".

3. Continuous improving and constantly upgrading construction quality level

(1) Testing and research of concrete bleeding, laitance film, aggregate segregation. In order to solve the common defects in construction, such as concrete bleeding, laitance film and aggregate segregation, conduct the comprehensive performance tests aim at to the concrete mix proportion, and put forward the optimal construction mix proportion so that phenomena of laitance film and aggregate segregation have greatly improved, and bleeding also have certain better.

(2) Application of large integral steel form and shaped steel form to the improvement of concrete construction quality. In nearly 10 years' construction, new-type formworks have developed and constantly improved, this is not only a technical means, but also is a continuous quality control and improvement measure. The successful application of the new-type formworks ensures the concrete structure size and improves the appearance quality. At the second-stage and third-stage of Three Gorges project construction, the contractor-Gezhouba Group comprehensive design and plan the application of formworks, such as, doka forms used for the exposure dam faces upstream and downstream, shaped steel forms for the crown of galleries, shaped large forms for the side walls, uniform shaped steel forms for drainage ditches of galleries and shaped steel forms set between the two ways of water stops. By these standardized construction means, the formworks are not moved during concrete placing, and the appearance quality has improved significantly.

(3) Live observation, learning each other and improving together. The contractor of Three Gorges project has organized the construction teams for different sections to do live observation to some key techniques, including: concrete placing by the tower belt crane, appearance quality of the concrete, gallery construction, construction joint treatment, concrete placing at the part with reinforcing fabric. Through observation and summarization, the key techniques have been continuously improved and enhanced.

(4) "Destroy persistent defects, create first-class works". In recent years, the project owners have organized the labor contest "destroy persistent defects, create first-class works" in the project construction process. The contractors make full use of this good external environment, and actively organize the personnel taking part in all kinds of lessons, training and key subject research. This have a positive progress to improve the quality consciousness of the whole personnel, improve the techniques, promote technology innovation, and destroy the quality defects.

Through the effective combination of quality control measures with technology, the Three Gorges project has been obtained significant achievement. For example, at the $15^{\#} - 20^{\#}$ blocks on the right bank for third-stage works of Three Gorges project constructed by the Gezhouba Group, no crack was found on the 2.8 million m^3 concrete monolith, which has set a precedent of no cracking for the dam concrete placement. By the Academies Pan JiaZheng known as "the Three Gorges Dam is not only the world's most magnificent concrete gravity dam, also is a dam with good quality, safety and reliability. The right bank dam at the third-stage of the Three Gorges three project has not a crack, not only fully demonstrated the level of concrete temperature control technology in the high concrete gravity dam construction, also reflected the concrete quality control upgrade to a new level. The builders of the Three Gorges project wrote the record in the history of dam works, created a miracle in architectural history. Not only the dam is a high quality one, all parts of the Three Gorges project is fine in quality." The following data can prove the above conclusion.

For the $15^\#$ – $20^\#$ blocks at the third-stage works of Three Gorges project, 26 holes were drilled for self-inspection (total drilling footage of 430.48m) in which all were of I-class, and 233 holes for surface inspection (total drilling footage of 461.1m) in which 232 were of I-class and one was of II-class, the concrete with satisfactory density and met the design requirements. The once pass rate of UT detection was over 99.6%, and that of RT detection was over 96.1%. In quality assessment of 7,812 unit works, the qualified rate was 100% and the good rate was 94.57%.

19.2.1.5 Application of New Kind of Concrete Materials with High Performances

1. Abrasion resistance concrete

The high-speed sediment flow will wear and erode the wetted area of concrete surface, which is common disease of hydraulic structures, such as overflow dam, discharge tunnel (slots) and discharge gate. Especially high velocity flow carried with bed load or suspended load, the erosive wear and erosion on the structures are more serious. In China, many rivers are heavily silt-carrying ones. There are 42 rivers with average annual sediment load of more than 10 million tons, and 60 rivers with maximum annual sediment load of more than 10 million tons. The Yellow River with average annual sediment load of 1.64 billion tons is a rare heavily silt-carrying river in the world. For the Sanmenxia Water Control project, the average sediment concentration at the scour bottom outlet is 80 – 100kg/m³ during flood season, with the maximum instantaneous sediment concentration of 911kg/m³. A large sediment load directly inflow into the sea every year from the Yangtze River, the Haihe River, etc. In recent years, with the development of the western regions and the power transmission from west to east, a number of large hydropower projects with high heads were constructed, such as Xiaowan, Longtan, Laxiwa, Goupitan, Xiluodu, Xiangjiaba, Jinping I, Baihetan, Nuozhadu. The velocity of outlet structures had reached 40 – 50m/s, and it put forward higher request to the performances of construction material against cracking, abrasion and quick easy construction (especially the materials for protection and repair).

In view of the current problem of abrasion resistance material in engineering application, combined with the development of the western region and the power transmission from west to east, the scientist and technical personnel make full use of the advanced technology of modern construction materials, and have developed the high range abrasion resistance additive. As compared with the ordinary concrete, the abrasion resistance strength and durability of concrete added with the abrasion resistance additive increased by more than 40%, meanwhile, the workability of the abrasion resistance concrete significantly improved.

2. Interface agent for bonding the interface of new and old concrete

The typical interface agent to bond the interface of new and old concrete is applied in the dam heightening works of Danjiangkou water conservancy project, the successful application of this new type material fill the blank in concrete construction.

The Danjiangkou project was built in the 1960s, and its dam heightening works is a part of middle route of south-to-north water transfer project. The concrete used for new built and existed structures has a larger difference in performance, therefore, the interface treatment of both structures is very important, especially in the wetted areas of gate pier and chamber. To ensure the quality of old and new concrete bonding, the high range interface agent, HTC-1 and HTC-2, are developed. HTC-1 with the ratio of water to material 0.16 : 1 and HTC-2 with the ratio of water to material 0.27 : 1 are mixed uniformly and coated on the wetted surface of the existed structure in 1 – 3mm thick. The mixed agent should be used up in 45 minutes as far as possible, if exceed 45 minute, the agent liquidity has slightly lower, but it has a little influence on the effect and can be used after adding a little water. When the interface is coated by the agent under the conditions of no wind and no sunlight, the fresh concrete can be placed on the interface in 1.0 hour.

This interface agent used in the Danjiangkou project has received high praise and reviewed by the experts and the owner for its favorable effect and low cost, and it has been used to the projects such as the diversion tunnel block of the Pubugou hydropower project and the repair of diversion tunnel bottom board of Xiluodu Hydropower Station.

19.2.1.6 Application of Large Tower Belt Crane

Tower belt crane is the specific machinery for dam concrete placement and serves as concrete horizontal and vertical hauling and spreading, which has the characteristics of feeding continuous and high intensity and has been used in high concrete gravity dam construction, the Three Gorges project and Xiangjiaba project, such as: six tower belt cranes were used in concreting of the dam and powerhouse at the second-stage works of the Three Gorges project, in which 1#、2#、3#、4# tower belt crane, 5# top belt crane and the feeding line were installed and operated by the Gezhouba Group, and 6# top belt crane was installed and operated by the Qingyun Company. The application of grouped tower belt cranes is the first time for concrete placing and hauling system in the world.

The design and fabrication defects existed in the tower belt crane system itself have been eliminated through the practice of operation and management, and the difficult problems resolved successfully, such as: concrete temperature control, aggregate segregation, bonding between concrete lifts. Rich experiences have been accumulated in quality control of raw materials, equipment allocation at the placement area, concreting technique, equipment running management, and an important guarantee was provided for high intensity and mass concrete placement in Three Gorges project. Moreover, "Tower belt crane operation management manual" was prepared and published by Gezhouba Group in order to guide the construction practice, and it provides an important technical support for construction of the Xiangjiaba project.

19.2.1.7 Fabrication and Installation of Large Hydraulic Steel Structures

The technologies of large hydraulic steel structure fabrication and installation have made a great progress in the Three Gorges project, and also is an achievement in high concrete gravity dam construction in nearly 10 years.

In the Three Gorges project construction, the installation quantity is 91,000 tons for steel structures by the Gezhouba Group, which is equivalent to the total installation quantity of steel structures of more than 10 hydropower station with installed capacity of 1,000MW, including 3 bottom emergency gates, 11 bulkhead gates of bottom intakes, 11 bulkhead gates of bottom outlets, 23 bottom service gates and 22 service gates of high level intakes. The all-position automatic welding technique was applied for the first time in fabrication and installation of steel penstocks, and the expansion joint was installed integrally in installation of spiral case and penstock, which are all filling the blank in construction of hydropower projects at home and abroad. In the installation of the bottom gate and gate slot at the discharge section, "reversal installation method" was adopted and successfully solved the schedule contradiction of the gate installation and civil works construction. In deep tunnel with large steel liner construction, welding technique of clad stainless steel was adopted, which is the first time to be used popularly in hydroelectric construction industry. In the second-stage works of the Three Gorges project, the new records of hydroelectric construction were created at the same dam monolith, including daily installation quantity of 425 tons, monthly installation quantity of 2,500 tons and one of 18,200 tons within one year. In 2002, the State Council acceptance expert team inspected and accepted the steel structure installation works of diversion bottom outlet, in unit works 128, the good rate was 100%, and the construction quality was rated as the good level. When gating and impounding after diversion tunnel being blocked, no seepage was found.

19.2.1.8 Ship Lock Stage Construction

The ship lock is the part of high concrete gravity dam, the Three Gorges lock stage construction is the representation of progress of lock construction technology nearly 10 years, especially the perfect combination with the computer information technology, and it is the best example of hydropower construction.

The Three Gorges ship lock was built in two stages to adapt to the different operation water level, that is to say, the lock stills and gates for upstream and downstream lock heads of the first-stage lock. In stage construction, the following major construction difficulties were solved: installation of long steel box girder chamber; lift and hang of 850 tons miter gate with the large ratio of height to width and high center of gravity; putting the miter gate in place accurately; In the existing building space constraints, high drop conveying of concrete and high intensity placement in the space restricted by the existed structure; concrete temperature control of large concrete monolith in the strong restricted zone. The technical difficulty was resolved for adapting to large water level fluctuation in stage construction of large-scale lock.

19.2.2 *Future Development and Prospect of High Concrete Gravity Dam Construction in China*

The water resources and hydropower construction has experienced more than one hundred years in China, and successfully built various types of large, complex technology giant hydropower stations, and obtained the remarkable achievements and progress. In recent years, with rapid development of national economy, the development of water resources and hydropower projects entered a high speed development time in structure size or in construction speed. Especially in high concrete gravity dam construction fields, through the construction practices, such as the Three Gorges, Longtan, Xiangjiaba, Guangzhao hydropower stations, the construction technologies and management have been improved significantly, the rich achievement of science and technology have been obtained, so that a series of construction techniques and methods which can meet the large and complicated structure construction requirements, have been developed, and technical experts and construction teams with innovative ability have been emerged, which means that China's high concrete gravity dam construction has stepped into the advanced rank in the world.

In order to adapt to the demand of the national sustainable development and the "western region development strategy", the construction of water resources and hydropower project is in rapid development. And this development is mainly focus on the southwest regions with rich water resource, Yunnan, Guizhou and Sichuan provinces. With the development of river water resources, such as Lancang River, Nujiang River, Yalong River, Jinsha River and Dadu River, currently, there are more than 10 high concrete gravity dams with height over 100m under construction, otherwise nearly 10 high concrete gravity dams are in planning stage. In which, the retaining dam of the Huangdeng hydropower station, on Lancangjiang River, is as high as 202m and is the second high dam in the world and the highest concrete gravity dam is the dam of Longtan Hydropower Station in the world. For these dams under construction or proposed, the concrete quantity has reached millions of cubic even 10 million m^3, and most of them are RCC dams, since compared with normal concrete, RCC dam not only save cement, but has simple construction procedure so that it can give full play to the advantages of large construction machinery and rapid rolling construction in large placement area so as to save cost and shorten construction duration.

Therefore, in the future, the main development of high concrete gravity dam construction technology will focus on RCC damming technology, involving with new kind of damming materials, large mechanization, informatization, standardization, intelligent, energy conservation and environmental protection. The main subjects include strengthening basic research, developing and using of new kind of damming materials, advanced technology, develop intelligent integrated system, and research on the application of the Internet of Things technology in construction and large-scale construction machinery and complete corollary equipment.

(1) Development of new type of RCC damming material. Damming material is one of the main factors of affecting the structure's resistance, only the damming material with good performances can ensure the safety of the dam structure, adapt to various dam types, dam site conditions and the stress state of dam body. Moreover, the optimal damming material also is benefit to speed up the construction progress, shorten the construction duration and reduce the construction cost. The main topics of RCC damming material research include: ①the development and application of new type cementations material and admixtures, especially new type cement; ②research and application of high performance concrete, such as low hydration heat, high ultimate tensile value, high durability, abrasion resistance; ③research of the other materials, such as poor cementing RCC, bound gravel or rockfill concrete, and new type grout with high range, silica fume concrete, magnesium oxide concrete, fiber concrete, etc.

(2) Research of large construction machinery and complete corollary equipment. Since concrete gravity dam with large monolith size and high placing intensity, only an appropriate placement method can achieve the results of high quality and less construction duration. The research point mainly aims at how to decorate the grouped concrete transportation equipment and make automatic placement continuously for large concrete placement area to meet the requirement of high intensity placement. In addition, how to improve the design and fabrication defects which is existing in the specific placing equipment, to solve the difficulties of temperature control and aggregate segregation during concrete hauling.

(3) Research of mass concrete temperature control and crack resistance. The temperature control and crack resistance are the key and difficulty in dam construction since mass concrete placement is a prominent feature of concrete gravity dam construction. Based on the existed comprehensive temperature control measures, the future research will focus on the cooling technology of dam concrete to gradually realize from artificial control to intelligent control, temperature monitoring standardization and automatic operation so as to improve the management level of cooling by circulating water. In construction practices of the Xiangjiabu Station and Xiluodu hydropower Station, the intelligent integrated control technology of cooling by circulating water for concrete dams has been developed and some results have been obtained.

(4) Research and promotion of construction information and management technology. In recent years, along with the high speed development of computer information technology, engineering construction organization and management has been improved. Various softwares of project management have been used in construction management, for example, computer P3 software (which has already been developed into P6) has commonly been applied to construction schedule arrangement and management. OA system (Office Automation) management software has also brought the office affairs into the automation era. The integrated information system has also developed, which serves as production, management, quality safety and completion technology. However, the information system which is real used for management and control of construction field is still few. Although the construction simulation technology repeatedly tried in different construction fields, it is hard to promote because that boundary conditions of assist operation cannot be accurately foreseen. Therefore, it also is a main issue in future research and development.

(5) Study environmental protection technology in construction. In recent years, the governments have paid more and more attention to environmental protection works, and the requirement is upgraded. The main issue of environment protection is the technical measures of quarrying and processing aggregate in concrete production process for construction of hydropower projects. The key research subjects are the environmental measures of quarrying, dust removal and noise control,

wastewater treatment and recycling utilization, etc.

19.2.3 Achievements and Practices for High Concrete Gravity Dam Construction Technology in Environmental Protection and Sustainable Development

19.2.3.1 Research and Implementation of Cooling Water Recycling

At present the water resources and hydropower project construction has undergone rapid development in China, but the development has led to a certain environmental pollution and ecological system damage, therefore the environmental protection is particularly important in construction. The water recycling is one of main measures taken to protect environment and save water resource. Therefore water recycle in the dam construction has been one of the main means to environmental protection, full exploitation of water resources and water conservation. The control strategy for recycling by the cooling water was studied and applied in the Xiangjiaba Hydropower station.

For Xiangjiaba hydropower project, the cutoff trench of dam foundation is backfilled to the elevation 240m with RCC on the design requirement. According to the general schedule, RCC construction is conducted in hot season, May to September, meanwhile, the concrete placement for cutoff trench of stilling pool, left stilling pool and left guide wall proceeded, so that the refrigerating capacity of 800m^3/h is required based on concrete placing intensity. But the total productive capacity of cooling water is merely 540m^3/h with 9 refrigeration units, which is far from meeting the needs of cooling water in peak construction period. The cost of increasing refrigeration units will be about RMB 3.6 million, which service time is only about 3 months, so the contractor made a special research on the pipeline of cooling water supply and recycle, and the following measures are taken.

1. Layout refrigeration station

In the front of the dam at El. 300m, 1# mobile refrigeration station is set on a platform for supplying cooling water to the center and upper monoliths of overflow section and the upper monoliths of part right non-overflow section. Behind the right non-overflow section El. 260m, 2# refrigeration station is set on a platform for supplying cooling water to the lower monoliths of overflow section and lower monoliths of part right non-overflow section. After completing consolidation grouting and recovering concrete placing for the stilling pool, 3# mobile refrigeration station is set on the bottom of left stilling pool to supply cooling water for the stilling pool. The above three mobile refrigeration stations are equipped with 9 refrigeration units, with production capacity (8 - 10℃) is about 540m^3/h.

2. Layout of recycling water pipe

For the second-stage works of the Xiangjiaba project, the cooling water recycling system is mainly made up of the recycling water tank, pressure pump and return main pipe and return pipe network in the gallery. The return pipes are set on vertical and horizontal side walls of the gallery respectively, and connected each other to form a recycling network. Branch pipes are embedded in the gallery side wall on the return pipe, and then the stand pipe is set up with rising of placement block and used to recycling the cooling water. Specific procedures are as follows.

(1) Layout of recycling water pipe network in gallery. At the overflow sections of ①-⑦ below El. 240.0m, the DN300 main pipe of the recycling pipe network in the gallery is brought from the temporary shaft in front of the overflow section ② to 1# water tank at upstream the overflow section ⑧ along the excavated slope in front of the dam. The return water in the galleries above El. 240m are collected to 1# and 2# tanks through return main pipe set on the outlet of temporary gallery at upstream the overflow section ⑤and the embedded return main pipe on the overflow section (13) (0 + 120.0m). The return main pipe which brought from the temporary gallery lay to 1# tank in front of the overflow section ⑧ along with the platform at EL. 240.0m. The return pipe network of stilling pool in the gallery connect with the DN200 return main water pipes of curtain grouting gallery at the right guide wall and then with the DN300 return main pipe in the curtain grouting gallery at the overflow section (13) downstream and lead to 2# recycling water tank behind the non-overflow section ①. When the main return water pipes in the galleries at each layer connected with the return main pipe in the elevator well at the non-overflow section ①, gate valves are provided for separately control the supplying of cooling water when the height difference of return water is increased. The cooling water of placement area between upper and lower galleries are recycled through the embedded pipe in the dam and brought to the return main pipe at the lower gallery, and finally back to the refrigeration units through the recycling water tank.

The recycling water pipe branch out several branch pipes, and the branch pipes are one-to-one corresponding with the outlets of cooling water pipes embedded in the dam. The branch pipes are bound with rubber tube to ensure seal, the cooling water discharged from the cooling water pipe flow into the recycling water tank or the refrigeration tank by gravity through head difference between the inlet of water pipe and the outlet of return pipe as well as the supplying pressure of refrigeration unit. The outlets of the recycling branch pipes and main pipes and cooling water pipes are all equipped the gate valves so as to control flow flexibly, match the pipe connection and facilitate the plugging of pipes in the late stage.

(2) Recycling of cooling water on the placement surface. The cooling water on the placement surface is recycled through the buried pipes, namely welding DN150 steel pipe on the return water branch pipe in the gallery at each monolith, and embedding the DN150 pipe into the gallery wall along with the placement area, setting the stand pipe with the placement area rising with the exposed length no less than 50cm and closed top end. And branching several branch pipes and connecting the branch pipes with the outlets of cooling water pipes on the placement surface tightly. If necessary, horizontal branch pipes may be put and led to the outlets of cooling water pipes on the placement surface, and then connected with the outlets of cooling water pipes. Through the above ways, the cooling water on the placement surface will be recycled to DN200 return branch pipes in the gallery, finally, led out the gallery into the recycling water tank. When the placement area continues to rise or the circulating water is completed, the connection of return pipes on the placement surface will be cut and the plugging of DN150 return branch pipe is removed, and then the pipes will be led with the rising of placement surface.

(3) Arrangement of steel water tank. Conventional arrangement of steel water tanks is close to the cooling water outlets or return pipes, that is placed in the gallery, but this arrangement needs not only the bump with large capacity and high-lifting in the water tank, but also the returning water can be bumped to the refrigeration station, and has influence on the construction personnel passage in the gallery and an emergency escape from a sudden accident.

In view of this, in the second-stage works of Xiangjiaba project, the arrangement of water tank was optimized based on the actual field conditions. 1# and 2# recycling water tanks are set on the El. 240m platforms in front of the dam and behind the dam, respectively, and 1# water tank recycle mainly the cooling water from 1# refrigeration station and 2# water tank recycle mainly the cooling water from 2# refrigeration station. by the head difference between the inlets of cooling water pipes and the outlet of main return pipes, cooling water will be recycled into the recycling tank freely, and it need not to set pressure pumps between the recycling water tank and the return pipes led from the galleries. The arrangement has not only less influence on the surrounding construction, but also convenient management and maintenance of the tanks, and will not affect the personnel and equipment passing in the gallery.

In the second-stage works of Xiangjiaba project, the "recycling pipe + recycling water tank" or cooling water recycling pipe is adopted to directly recycle the cooling water, by which not only solve the temperature control problem of RCC and guarantee the construction progress, also improve the reuse rate of cooling water and the efficiency of refrigeration unit, and save the cost. At the same time, the technology relieves the drainage pressure in the foundation pit and provides a good construction environment for concrete placement. The successful implementing of cooling water recycling technology has obtained favorable effect, won the consistent high praise of the owner and the supervisor and provided the good practice experiences for subsequent similar project construction.

19. 2. 3. 2　Technology of Using Low Temperature River Water and Groundwater to Cool Concrete

Because most of concrete gravity dams are mass concrete structures, large amounts of hydration heat is not easy to emit in initial hydration period after concreting so that the temperature cracks are easily produced. Through research and engineering practices for years, a set of complete concrete temperature control measures have been developed. The key technology is to control the temperature difference between internal and external concrete, on the one hand, to reduce the concrete internal temperature through cooling water, and on the other hand, to cure and insolate the concrete surface. For cooling by circulating water, large flow cooling water is supplied in 4 - 5 days before placing new concrete specially, supply cooling water during placing instead of circulating water after complete of placement, which are effective measures to drop the early hydration heat of concrete. This can dramatically reduce concrete internal temperature and laid a foundation for circulating water in middle and late periods.

However, a large amount of cooling water is required for mass concrete construction because the period of supplying water often lasts dozens of days, and the temperature of water below 10℃ maintain 10 - 15 d in early time. One of the main measures for energy conservation and environmental protection is to use river water to cool the concrete. For example, in the Three Gorges project construction, nearby river water was used to cool concrete based the concrete grade and change of seasonal water temperature. ①season options: river water temperature is 11 - 15℃ in December to March every year. Namely, using river water as cooling water in placement period at the early of the year, supplying artificial cooling water in the other seasons with air temperature of higher than 15℃. ②determination of water flow: in order to ensure the quality of cooling water, adjust circulating water flow according to the concrete grade, water temperature of inflow and outflow and concrete temperature measured by meters embedded in concrete, and ensure the highest temperature of concrete qualified so as to achieve the dynamic management of placement area cooled by water and customize supplying cooling water. Actual water flow control standard in the construction are as follows: the water flow of 40 to 60 L/min in early 4 days and 30 to 40 L/min in late 6 days after complete of placement for $R_{28}250$ concrete, 35 to 45 L/min in early 4 days and 25 to 35 L/min after complete of placement for $R_{28}200 - R_{90}200$ concrete, and 20 to 30 L/min for $R_{90}150$. In the process of circulating water, change the direction of inflow and outflow every day, control the temperature difference of inflow and outflow above 3 - 5℃, dynamic adjust cooling water flow when temperature difference is less than 3℃ and then reduced the flow to the lower limit of the control standard.

In the future, the technology of using low temperature river water to cool down the dam concrete will be applied in more mass concrete works. Moreover, low temperature groundwater and spring water will developed and utilized further, Since the underground temperature is 5 - 10℃ in summer in north China and there is rich low temperature spring water in high valley area in the Midwest China, these natural low temperature water can be utilized in construction of water resources and hydropower projects.

19.2.3.3 Reasonable Planning of Construction and Life Camps and Permanent Infrastructure

The reasonable planning and construction of infrastructure in water resources and hydropower project field are significance to ensure the project progress smooth and save the cost, therefore, the construction general layout shall be made according to the construction need and site conditions, taking into consideration of combining temporary facilities with permanent infrastructure on the principle of "plan as a whole, reasonable arrangement, appropriately advance, implement step by step", as far as possible to avoid the repetition works, such as remove and rebuild.

(1) On-site access. For large hydropower projects, the work quantity of access roads at the site is large since the construction site has large area, and generally the road construction is completed before commencement of the main structure construction or in early period of construction. The access roads at the site have a high utilization rate, with features of heavy transport volume, high intensity, strong seasonality and overweight or oversize goods transportation, therefore, in the design of on-site access roads, the main attention shall be given to consider the main features above, and the combination of the permanent and temporary works, early and late construction, and reasonable planning and layout of main trunk roads, selecting the corresponding road grade.

(2) Facilities of life camp. According to the existing engineering bidding mode, one large hydropower station project is divided into several construction contracts, there are many contractors participate in project construction at the same time, therefore, it is difficult to plan the temporary construction camps combined with the power plant or permanent facilities. The project practices have demonstrated that the repetition works happened often in construction of many projects due to no full consideration combining the life camps with the permanent facilities in the early planning. Therefore, the planning of life camps, should meet the construction need and not increase the project cost as much as possible, meanwhile try to make temporary facilities combining with permanent facilities, properly consider the urban development plan of local government, save land and save project cost.

(3) Greening in the construction site. In recent years, with development of water resources and hydropower projects, the debate whether or not the dam construction destroying the ecological balance can be heard often. Now, the environmental issues caused by hydropower project construction has drawn the attention of the developers, Huaneng Lancang River Hydropower Co., LTD strictly comply with the state regulations and environmental protection system, always consider the environment protection in the planning, design, construction and operation management, the company has become the model of harmonious enterprises. In construction process, the company adhere to the principle of "prevention first, prevention combined with treatment, plan as a whole, and reasonable arrangement, comprehensive treatment", strengthen environmental protection in periods of construction preparation and construction, prevent damage and pollution natural environment, such as the completed Jinghong hydropower station, the greening area is more than 95%, known as the "ecological type power station" with the features of tropical forests and the Dai ethnic culture, provides valuable experiences for hydropower projects under construction and proposed to build.

19.3 Construction of High Concrete Arch Dam
Cheng Zhihua[1]

19.3.1 *Achievements and Development of High Arch Dam Construction in Previous Decade*

Arch dam is curved upstream in plan so as to transmit the water load to the abutments. This type of dam can give full play to

---

[1] Cheng Zhihua, Deputy Director of Science and Technology Department of China Gezhouba Group Corporation, Senior Engineer.

the dam materials characteristics of high compressive strength, with the features of small size, low cost, superior economy and safety. The quantity of arch dam works is 50% or less than that of a gravity dam.

The development of arch dam has a long history and can be traced to the earliest Romans who had built arch dams on the arch principle. In the early time, the arch dam was made of masonry. The concrete arch dam was developed in the mid-nineteenth century. Since the early 20th century, the height of arch dam had breakthrough. In the 1970s and 1980s, many arch dams were constructed in China. Through statistics from International Commission on Large Dams, China had become the country having the largest numbers of arch dams at the end of 1986. In recent three decades, the construction technology of high concrete arch dam had made a considerable progress in China. A number of high arch dam projects built successfully, such as Longyangxia, Lijiaxia, Wujiangdu and Dongfeng, especially the completion of arch dam Er'tan and safety operation, had shown that the technology of 200m level of high arch dam has matured, and has a condition to construct 300m high arch dam in China. With the ultra-high arch dams, Xiaowan, Xiluodu, Jinping and Laxiwa, were constructed and completed, the construction experiences of 300m high arch dams have basically mastered, and improved rapidly. In previous decade, 100m high arch dams built or under construction are listed in Table 19.3-1.

Table 19.3-1 Arch dams Higher Than 100m Built or Under Construction in Recent Decade

No.	Name	Completed year	River	Location	Height (m)	Length (m)	Volume ($10^4 m^3$)	Capacity ($10^8 m^3$)
1	Jiangkou	2003	Furong	Chongqing	140	380.71	667	5.05
2	Shapai	2002	Caopo	Sichuan	132	250.25	392	0.18
3	Dongping	2006	Zhongjian	Hubei	135	267	403.5	3.36
4	Tengzigou	2006	Longhe	Chongqing	124	339	610	1.93
5	Shimenzi	2002	Taxi	Xinjiang	110	176.5	211	0.8
6	Zhaolaihe	2005	Zhaolai	Hubei	109.5	220.296	22.27	0.7033
7	Huaguangtan II	2005	Fenshui	Zhejiang	104			0.036
8	Xiahuikeng	2002	Huating	Jiangxi	101	264.6	180.8	0.35
9	Linhekou	2003	Lanhe	Shaanxi	100	311	293	1.47
10	Xiaowan	2010	Lancang	Yunnan	294.5	892.8	848	150
11	Laxiwa	2010	Yellow River	Qinghai	250	459.64	258	10.79
12	Goupitan	2009	Wujiang	Guizhou	232.5	520	557.11	64.51
13	Jingping I	under construction	Yalong	Sichuan	305	552.23	252	79.88
14	Xiluodu	under construction	Jinsha	Sichuan/Yunnan	285.5	698.07	665	126.7
15	Dagangshan	under construction	Dadu	Sichuan	210	622.42		7.77

Over the past decade, the high concrete arch dams have been constructed rapidly, which has become one of the main dam types in large hydropower development. The applicability of arch dam has expanded, so this kind of dams could be adopted in the area with complicated topographic and geologic conditions, and many dam shapes have been developed. The technologies of rapid construction for RCC arch dam and high arch dam have been made great breakthrough.

1. Arch dam built in the areas with complicated topographic and geologic conditions

(1) Built on the terrain with larger aspect ratio. For the existed small and medium-sized arch dams the aspect ratio was up to 6, and for the existed high arch dams with height of 100-150m, the aspect ratio was generally 3.2 to 4.5, such as 4.525 for the Baishan arch dam and 3.21 for the Er'tan arch dam.

(2) Built on asymmetric valley. For arch dams built on asymmetric valley, the following measures are taken, concrete gravity pier, filling deep groove, setting concrete pedestal and adjusting the center and radius of arch, such as Jinping and Lijiaxia projects.

(3) Built under complicated geological conditions. For arch dams built in the area with complicated geological conditions, such as heterogeneity rock, more faults and fissures, deep overburden on the riverbed and dam foundation weathered and deep fractured, a lot of foundation treatment have to be conducted. The successful engineering examples include Zhaixiangkou arch dam in Guizhou province, with dam height of 44.5m and overburden thickness of 27m, and Ertan, Xiaowan,

Laxiwa and Xiluodu projects are located in the high seismic intensity zones.

2. Diversification of arch dam profile

Most of the dam shape was of arc arch in the early time. With construction of non-circular arch abroad, the non-circular arch has gradually been adopted in recent twenty years in China, such as Zhaolaihe dam with a logarithmic spiral arch, Jiangkou dam with elliptic arch. The following shapes are developed, parabola arch, ellipse arch, hyperbola, poly-centered circle, logarithmic spiral, uniform quadratic curves, or other variations curvature. A reasonable dam shape will help to improve the thrust direction of arch abutment and reduce the moment at arch end, and are favorable to hung and dam stress conditions. The best case is the arch axis coinciding with the load pressure line.

3. New breakthroughs of RCC dam construction technology and rapid development of RCC arch dams

In China, RCC arch dams with a height of over 100m built or under construction include Shapai (dam height 132m), Shimenzi (dam height 109m), Linhekou (dam height 100m), Dahuashui (height 134.5m), Bailianya (dam height 104.6m), Yunlong Ⅲ (dam height 135m), Tianhuaban (height 110.5m), as well as Longshou (thick-height ratio 0.170), Zhaolai River (thick-height ratio 0.176). China has currently become the country with the largest number, the highest and the thinnest arch dams in the world.

4. Rapid construction technology of arch dam developed

In construction of RCC arch dams, continuous climbing forms and the repeated grouting system for transverse joints were adopted so as to speed up the dam's construction schedule, such as the Zhaolaihe dam, the largest monthly rise of dam placement is up to 27.3m.

For normal concrete arch dams, the technologies of joint grouting in the whole year, thickening dam lifts and increasing placement intensity are adopted so as to speed up the dam's construction schedule, such as Xiaowan dam, the largest annual rising height is more than 70m.

19.3.2 Significant Developments and Achievements of High Arch Dam Construction Technology in Recent Ten Years

Over the past decade in China, high arch dam construction technologies have made significant development, and obtained remarkable achievements, and have ranked in the advanced level in the world. The main development includes the following items.

19.3.2.1 Excavation Control of High Arch Dam Abutment and Foundation with Complicated Topographic and Geologic Conditions

In China, most of existed high arch dams were set in the mountain valley with a high abutment, large ground stress, complicated geological conditions and construction difficult. Over the past decade, through continuous research, it has made a major breakthrough in excavation of abutment and foundation with high ground stress and treatment of abutment and foundation under complicated geological conditions.

1. Excavation technology for dam abutment and foundation with high ground stress

The excavation of high ground stress abutment and foundation may cause the foundation rebound and result in unloading change so as to lower the quality of rock and increase foundation treatment difficult. Now there are three construction technologies to solve the problem of excavating abutment and foundation with high ground stress.

(1) Smooth ogee excavation. The smooth ogee excavation is to make the excavation on a smooth transition curve from the slope to the riverbed. This excavation is not only reducing the amount of excavation, but also lowering the strength of stress concentration and decreasing the rebound deflection and the loose range of relief so as to ensure and improve the quality of bedrock.

For example of Laxiwa hydropower project, the ground stress of dam foundation is 30-70MPa, the smooth ogee excavation technology was adopted in excavation of abutment and foundation. During excavation, the following measures were taken to ensure the excavation made on an ogee curve: ①the dam foundation excavation zones are adjusted according to the distribution of ground stress timely; ②the maximum initiating charge in single shot was controlled strictly based on geological conditions and bursting scale, so that the vibration velocity of each blasting particle is controlled within the safe range to ensure the stability of high abutment slope and foundation; ③after the excavation, the foundation surface and slope are anchored in time to ensure the foundation surface stable, slow down the rock relief; ④strengthen the geophysical acoustic monitoring on the foundation, understand the mechanical parameters of rock and comprehensive quality, measure the range and the thickness of relaxation of foundation surface rock after foundation excavation, and then adjust construction methods timely according to the foundation conditions; ⑤in order to reduce stress concentration as far as possible and meet the design requirements, During the protective layer excavation, the excavation was conducted by cutting a trench at middle section and extending toward the left and right banks, the base surfaces of floor and the slope were of circular shape and the working faces were increased for accelerating construction. According to the monitoring results, because of adopting the smooth ogee excavation technology, the relief rebound deflection at abutment trench is much smaller than the calculated value and the quality of bedrock is maintained, with high foundation geophysical velocity and good rock quality, thereby the elevation of foundation surface is raised 2m.

(2) Stress relief holes and buffer holes. For the Laxiwa project, at the left bank energy dissipation zone, a high slope with a drop of 220m in level will be formed after excavation, with variational lithology and high ground stress in rock mass. To ensure the excavation quality of high stress slope, for the rock mass with high ground stress, strengthen the rock burst prediction, drill a certain amount of stress relief holes in advance and inject water in the holes during excavation. While the buffer holes are drilled between the pre-split holes (depth of 20m to 25m) and the main blasting holes on the excavated bench, the distance is 2 times that of the pre-split holes. Through selecting the rational blasting parameters and strictly controlling drilling quality, to reduce the pressure on the hole wall by filling loess and rock powder into the holes and the blast influence on the excavated slope. The simple support is adopted for the excavated high slope which stress is relieved through blasting, so that the slope remains stable in whole slope excavation. Meanwhile, the pre-split face is smooth and semi-porosity is above 90%, and the high slope keeps stability.

(3) Anchoring succeeded by excavation. The technology of anchoring succeeded by excavation is a means of preventing the relaxation of foundation rock mass from stress relief. That is to embed anchor bolts or steel tendon in advance the excavation of the protective layer in high ground stress area.

For Jinping I project, the abutment is located in high-stress area at El. 1,670 – 1,650m, a protective layer is set during excavation, the protective layer is excavated with the technology of anchoring succeeded by excavation. The excavation steps are as follows, reserve a protective layer with 3 – 5m thick, adopt the pre-split blasting next to the protective layer, and then embed anchor bolts or steel tendon. After acceptance of anchoring construction, the second pre-split blasting is conducted on the slope protection layer, and excavation is made to the design slope so as to ensure the quality of rock mass excavation.

2. Foundation treatment under complicated geological conditions

In the west region in China, the areas of mountains and valleys tend to have higher ground stress and many faults and fracture zones on the slopes, the geological conditions are complicated, therefore, it needs to take corresponding measures to ensure the safety of slope.

(1) Replacement and reinforcement of faults. For Jinping project, the two major faults f13 and f14 were found on the right bank slope, the replacement and reinforcement technology was applied. The procedures of the technology are as follows, set adit, inclined shaft and cutoff inclined shaft, remove fractured rocks in the faults, and then adopt anchoring and mining, grouting, backfilling and concrete liner to make replacement and reinforcement. For examples, adits at three levels, six inclined shafts and one cutoff inclined shafts are provided for f_{13} fault, and adits at three levels, three inclined shafts and one cutoff inclined shafts are provided for f_{14} fault.

Since large quantity of replacement and reinforcement works, tight schedule and high risk of paralleled construction of treatment and placement, the measures of setting the mucking shaft, and taking grouting adits and drainage adits as the transport path of replacement and reinforcement works are taken so as to solve the safe problem of slope treatment and the dam placement and ensure the construction schedule.

(2) Providing shear resisting plug and retaining concrete plug. For Dagangshan project, the shear resisting plug and retaining concrete plug are used mainly as measures for reinforcing the relief cracks on the right bank slope. For the reinforcement works, the following measures are taken, shear resisting plugs at six levels (El. 1,060m, 1,120m, 1,150m, 1,190m, 1,210m, 1,240m), retaining concrete plugs at two levels (El. 1,210m, 1,240m) and two inclined shafts, meanwhile, keys are set on both sides of the shear resisting plugs and inclined shafts to conduct replacing, with replacement depth of 3.00m each side.

19.3.2.2 Complicated Profile Control of High Arch Dam

The shapes of arch dams mainly include of single-curvature, double-curvature arch dam, and variable-curvature, the complex structural profile is difficult to be controlled. Through research and project practices, the technologies of formwork for controlling arch dam profile and control survey have become matured and provided conditions for quick and high quality construction of arch dams.

1. Controlling dam profile by formworks

The controlling technology of arch dam profile can be mainly utilized the large folding flat forms and large curvature turnover forms, the forms can not only meet the design requirement to dam profile and expedite the construction.

(1) D22K folding form construction technology. For the concrete double curvature arch dam of Xiaowan project, its curvature is variable along vertical and horizontal directions, the forms are required to be moved and adjusted within a certain range vertically and horizontally, and can be recycled efficiently. According to the characteristic parameters of dam profile, the D22K tilting form can meet the design requirements. This type of forms is adaptable to the complicated profile of arch dam: ①D22K can match the curvature variation of arch dam upstream and downstream faces since it can be inclined forward with the maximum 30° and backward with the maximum 29°; ②D22K can be continuously adjusted up and down 20cm, and toward left and right 20cm to adapt the curvature variation of dam profile in vertical and horizontal directions. The maximum inclination of Xiaowan dam is 31.8°, so the inclination of the form itself to the dam surface plus its adjustable angle of forward and backward can meet the tilting requirement on the full height of arch dam.

For the Xiaowan arch dam, the dimensions of the form need to be determined when it is selected. In order to facilitate fabrication and construction of the forms, the folding forms were taken instead of curvature ones since placement was made in lifts. By analysis of drawings, the folding flat forms with height 3.75m and width 3.75m are selected, and the maximum deviation of dam profile is controlled within the scope of permissible error, so this type of forms can meet the specifications.

According to the analysis of the form arrangement on upstream dam face, the form openings are set on the back of upstream dam face, so the forms at the first monolith can be tightly arranged. When the gap between forms is enlarged along with the dam height rising, the shim plate is used in the gap, when the gap width is greater than 250mm, the local forms should be replaced by wider forms. For downstream dam face, the opening between adjacent forms is set on the front, a gap between the two adjacent forms at the first monolith is set with a certain spacing (generally 2mm), and the shim plate is used in the gap. When the gap width is greater than 250mm, the local forms should be replaced by wider forms. The shim plate with a width of 20 – 250mm is connected with the two adjacent forms by hook bolts, and used for 20 – 250mm wide gaps, which is easy to be operated.

The Xiaowan project has been built, the dam profile met the design requirements through utilization of the folding flat forms.

(2) Large curvature turnover forms for RCC arch dam construction. For Zhaolaihe hydropower project, its retaining dam is a RCC double-curvature variable-thickness one with logarithmic spiral center, and the dam profile is complicated with large curvature change and large overhang. Considering this type of dam, a turnover form has been developed successfully for RCC concrete arch dam with large curvature. The forms are equipped with adjustable front and tail sliding blocks so that the upper and lower forms can relatively travel, or the shim plates can be added or reduced, or the vertical joints between forms can be filled with the shim plates, and the profile of arch dam may be adapted through adjusting flexible hinge, level and vertical adjustment systems.

For Zhaolaihe project, the application of large curvature turnover forms has met the requirements of RCC arch profile and ensured its rapid construction.

2. Measurement control technology for arch dam construction

Because the arch profile is complicated, the traditional measuring method cannot meet the requirements of form raising quickly and positioning accurately. Therefore, a set of practical and accurate measurement and fast detection software is prepared specifically for each arch dam project based on the profile parameters combined with advanced computer programming technology. The software development principles are as follows:

The software program generally consists of three parts: ①calculate the corresponding elevation parameters base on the measured elevation; ②calculate the maximum central angle and the maximum arc distance at the elevation through corresponding parameters; ③input the plane coordinates of the measured points, and determine the deviation of the measuring point with the design value so as to adjust the forms as the calculated results at the site and then detect the adjusted forms until the forms are qualified. Follow this procedure, detect one point on every other form and adjust to the points to the design position. The software can be used for setting out of an arch at any elevation, rapidly detecting the form edge and construction setting out at any elevation.

In preparation of the software, the traditional method of construction setting out is given up, and the total station is used together with the software developed, the limitation of design drawings on construction setting out is not restricted. When the form needs to be adjusted, measure any coordinate point on the form top and input the measured value to the software, the deviation of the form can be determined in several seconds, and the position and elevation of the form has no influence on the calculation so that this technology provides a guarantee of quick detection for continuous raising of dam and laid a solid foundation for dam profile precise control and form rapid detection.

19.3.2.3 Construction Technology of Cracking Resistance and Temperature Control

Theory and practice have shown that the temperature stress is the main reason for cracking in concrete arch dams. Based on the high-arch dam construction technology and structural features, such as Xiaowan and Jinping projects, the temperature control of high arch dams has the following characteristics: ①Due to a high arch dam restricted by freedom height of itself, the placed concrete bottom need to be grouted to form an entire, but the dam body of arch closured has a large restricted effect to the upper part, so the entire dam is defined as the constraint area. For these dam sections, high temperature stress of foundation existed and a strict temperature control is required on the foundation since steep abutments, more sloped sections and more constraint zones at bedrock and dam openings. ②With a high design strength of dam concrete, large cemented material consumption and high hydration heat, temperature control is difficult. ③Generally, an arch dam is placed without longitudinal joint and the placement area is a narrow strip with large size, so that surface cracks and through cracks are easy to be produced because of slowly concrete temperature drop and long internal constraints period caused by temperature difference inside and outside. It can be seen, the temperature control and crack resistance is difficult in dam construction.

To reduce the temperature stress, prevent and reduce cracks, a set of measures has been taken in the dam construction including adopting high-performance concrete, shortening the concreting interval, enhance cooling by circulating water and insulation curing the placed concrete.

1. Mix proportion of high-performance concrete

The concrete cracking is mainly causes by the tensile stress of concrete exceeding the tensile strength or the tensile deformation coefficient larger than the limit tension. When the deformation of concrete drying shrinkage, cooling shrinkage and autogenous shrinkage are restrained by the foundation and surrounding environment (limit restraint), tensile stress is generated within the concrete and may cause cracks in concrete, so the application of high-performance concrete is particularly important.

Since aggregate, cement, additive and admixture adopted are different in various projects, the dam concrete performances of high-strength, high ultimate tensile value, low hydration heat, slow temperature rise, low elastic modulus and less shrinkage can be obtained through selection of high-quality raw materials and optimization of concrete mix proportion. For the arch dam concrete with high requirements to mechanics, deformation, cracking, durability and temperature control, compound high-quality fly ash, water-reducing admixture and air entraining agent are selected to improve the performance of concrete, and the dosage of fly ash is increased to reduce the concrete temperature rise in hardening process if permitted. Application of the high-performance concrete could help to improve the crack resistance of concrete.

2. Reducing concreting interval

Long concreting interval will increase the cracking risk of concrete caused by being exposed to the weather, or bottom concrete becoming a strong constraints area and having large restrain to the upper concrete to ensure the concrete of dam body rise intermittently and homogenously. But, the consolidation grouting of the dam foundation often requires a long interval, so cracks are produced. In order to reduce or avoid cracks in the dam foundation concrete, a new consolidation grouting technique is developed, which involves without cover-weight, the pipe and cover-weight, that is to conduct grouting 5m below bedrock before concreting, and then do grouting with cover-weight for the parts above 5m bedrock by the pipe to the appropriate positions of the upstream and downstream dam faces when the placed concrete with a certain thickness has met the design strength and the temperature of dam concrete cooled to that of arch closure, thus the long interval of foundation concrete placement is shortened. For the positions cannot be avoided long placing interval, a certain percentage of steel fibers or plastic fibers may be mixed on the top layer of the placement block to increase the surface crack resistance.

3. Cooling by circulating water

Cooling by circulating water is the important measure to control the maximum temperature rise inside the concrete of arch dam and cool down the concrete temperature to meet the requirement of the arch closure. At the same time, different circulating water strategies are adopted at different cooling stages so as to have the concrete temperature cool down slowly and evenly and improve the concrete crack resistance.

(1) Cooling in stages. Generally, cooling is divided into three stages, initial, intermediate and final cooling in accordance

with the cooling purposes and periods. The main purpose of initial cooling is to control the maximum temperature rise inside the concrete, that of intermediate cooling is to prevent the concrete temperature rise again, and that of final cooling is to have the internal temperature of concrete drop to the temperature of joints grouting required.

The initial cooling is divided into two steps of temperature control and temperature drop depends on the purpose of temperature control. The temperature control is to control the temperature of concrete inner does not exceed the maximum design value; and the temperature drop is mainly to have the temperature drop from the highest to a stable one slowly. For the initial cooling, the cooling range is not more than 6℃ and cooling rate is not larger than 0.5℃/d. The purpose of intermediate cooling is to have the temperature drop slowly and not rise again in the period of the initial cooling finished to the final cooling beginning, and the cooling rate is not more than 0.3℃/d. The final cooling is divided into two steps of temperature drop and temperature control, the temperature drop is to lower the concrete temperature to the design arch closure temperature, and the temperature control is to keep the concrete temperature close to that of arch closure (+0.5℃), to meet the requirement of joint grouting, and the cooling rate is no larger than 0.3℃/d.

(2) Requirement of Temperature gradient. The following experiences have drawn from Xiaowan project construction. In addition of cooling in stages, the temperature of dam concrete shall be controlled in grades so as to form a proper temperature gradient for temperature and drop range at different levels and reduce concrete gradient stress and prevent concrete cracking. Currently, the high arch dam is divided into five zones of temperature control, which are the grouted zone, grouting zone, cooling zone, transition zone and cover-weight zone from bottom to up. The suitable gradient of upper zones in joint grouting is shown in Fig. 19.3-1.

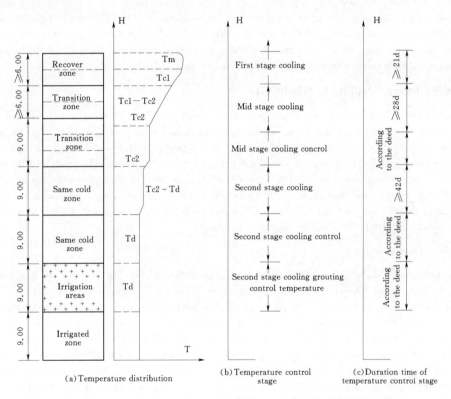

Fig. 19.3-1 Schematic Diagram of Temperature Gradient Control

In order to meet the demand of cooling by circulating water, save cost and decrease loss, the following measures are taken, mobile chiller set at downstream of dam, reasonable arrangement of cooling water pipes, timely adjustment of cooling water flow, according to the structure features of the arch dam and circulating water requirement.

1) When cooling water is used, in order to make cooling uniform, change the water flowing direction of chillers every 24 hours.

2) The cooling water pipes within the dam are generally laid perpendicular to the direction of water flowing.

3) In order to avoid too large temperature difference between dam upstream and downstream, coil the water main in upstream from one side to the other side to make the water inflow from upstream and the upstream flow larger than that of downstream.

4) To form the equivalent temperature difference of closure temperature field at the same level, cooling water with different water temperatures are adopts, that is the lower temperature water for upstream area and the higher temperature water for downstream area.

5) To compare the daily parameters, including the ranges of temperature rise and drop, and the cooling rate so as to control water flow and inflow temperature.

4. Insulation curing

Due to ambient temperature has big effect on thin arch dam body, a "professional" curing shall be conducted for the dam concrete. In summer season, automatic rotary nozzle is set on the placement area for providing continuous wet curing, perforated pipe is put on the transverse joints face and upstream and downstream dam faces for spraying curing, and artificial sprinkling curing is made on local parts. In winter season, benzene board and polyurethane protective coating are adopted as permanent insulation.

For the arch dam of Jinping I project, 5cm benzene boards are put on the permanent concrete dam surface after form removal so as to remain the concrete internal temperature basic unchanged no matter how outside temperature changes. Moreover, 5cm benzene boards are set outside of steel forms to ensure heat preservation during construction, and put on the transverse joints after form removal for heat preservation and windproof. In case of air temperature below 10℃ at the placement area or in hot season, double insulation (polyethylene roll) is covered on the construction joints for thermal insulation.

For Shimenzi RCC arch dam, the new type of polyurethane hard foam insulation is adopted so that the temperature difference between inside and outside was reduced to 14℃ from 40℃ at temperature of -20℃, with reduction rate of 65%, and the temperature difference can reduce about 7℃ in summer. When the ambient humidity was 40%, the concrete humidity did not vary with outside humidity change at 5cm below the polyurethane protective coating and the concrete humidity of 97% is basic remained, with the concrete humidity of 89% at 1cm deep.

19.3.2.4 Arch Closure Grouting in Whole Year

Arch dams become a whole by grouting, thus it can bear loads. Therefore, joint grouting is an important measure in arch dam construction. Transverse joints are usually set on concrete arch dams, with spacing of 15–20m, and grouting is made when the concrete surrounding these joints are cooled to the required temperature. For the lower part of the dam, it can be raised after completion of joint grouting because the dam is high and the cantilever has the requirement to its free height. At the same time, the joint grouting for corresponding parts must be completed before flood season, so that the parts become a whole and can bear loads and meet the requirements of retaining water. If joint grouting can be made in cold season, it will have a great influence on the dam construction schedule. In recent years, with development of temperature control technology, arch closure grouting of high arch dam in whole year is possible.

1. Requirement about joint grouting

Construction Specification of Cement Grouting Used for Hydraulic Structures (SL 62—94) has specified that "for joint grouting, the age of dam block concrete on both sides of grouting zones should be more than six months, and it is not less than 4 months if effective measures taken", this provision often bring a difficulty to the dam construction. In order to speed up construction and make early impounding, the following requirements are considered in arrangement of joint grouting time: ①the dam concrete should have sufficient strength to withstand the principle stress of dam; ②the slurry in joints should have sufficient strength to withstand the normal stress; ③the joint grouting should be conducted when the adiabatic temperature rise of concrete has been finished on both sides; ④the temperature of foundation dam block should meet the requirement of tolerant temperature difference. According to this request, a requirement of temperature gradient has been proposed in dam grouting, and the dam body have been divided into five zones from bottom to up, grouted zone, grouting zone, cooling zone, transition zone and cover-weight zone so as to ensure the upper part temperature of each zone and cooling rate form a suitable gradient, see Fig. 19.3-1 in details.

2. Cooling water circulation by mobile chiller

Cooling water is an important means to control the temperature of arch dam. Mobile chillers are selected according to the influence of main factors on the concrete temperature, such as placement intensity, joint grouting plan, concrete thermal properties, weather conditions, insulation measures, cold water temperature, and the relationship among the temperature of cooling water, cooling time and total refrigerating volume.

Mobile chillers can recycle the cooling water and save water. While mobile chillers are set on the trestle at dam downstream which is raised with the dam placement, they are close to the dam and the cooling water pipes are covered by rubber sponge insulation boards to reduce the temperature loss of cold water. The above measures can make the dam be cooled to the grouting temperature in hot season, and the cost will not increase too much. The use of mobile chillers can provide a guarantee of cooling the dam to the arch closure temperature and make arch closure grouting in whole year.

For the Xiaowan project, the application of mobile chiller technology reduces the installing quantity of cooling water pipes and the amount of insulation materials, and the number of chillers decreased from 8 units to 4 units; while the power consumption of chillers was reduced too, with the consumption of electricity for unit cooling water decreasing by 10% to 20% and the cooling water cost for unit cooling water decreasing by about 20%, supplementary water amount decreasing to 5%-10% from the design 20%, generally, the cost of cooling water drops 15%-20% the bidding cost.

19.3.2.5 Rapid Construction for High Arch Dam

1. Construction technology of form with large lift height

In order to speed up the construction for a high arch dam, the large flat forms with 4.0-4.5m lift height have been used in construction of normal concrete arch dam. Moreover, the specific forms are developed for special, such as, hydraulic lifting forms for bracket, hydraulic slip forms for elevator shaft, and large turnover forms for RCC arch dams with complicated profile. In order to quickly control the dam profile, according to the parameters of the dam profile, the use of modern programming technology and the total station can make measurement and setting out quickly at the site, and control forms in place, so the rapid construction is implemented.

For the XiaoWan project, the dam of 4m lift height was successful, and the form deformation and placing meet the requirement. For Jinping project, 4.5m lift height was adopted at most parts. Through improvement of form support, the profile deviation of placement was less than 20mm, and the concrete placement intensity, temperature control, concrete quality all meet the requirements. The forms with large lift height and placement technology have been mature.

2. High intensity placement

In order to meet the requirements of rapid construction for high arch dam, by the completed aggregate processing and transport system, forced mixing plant, placing with multiple cable machines at the same time as well as providing sufficient concrete spreaders and vibrators, the placement intensity can be reached. This technique has ensured the rapid raising of concrete dam body and the construction quality of concrete.

(1) Forced mixing plant. As the concrete mixed by forced mixer not only have good quality, but also the mixing time is short. In order to meet the requirement of high intensity placing, the forced mixing systems have been adopted in some large projects, such as Longshou, Linhekou and Jinping I.

(2) Quickly placing with cable crane. Arch dam is generally set in a narrow valley, have used large cable cranes have been used widely in some high arch dams existed or under construction. For example, 6 sets of 30 t cable cranes were arranged at two levels in Xiaowan project and 5 sets of 30t cable cranes were set at single level in Jinping I project. The cable crane is not only a tool for hoisting construction materials, but also is an important means for concrete placing. Through the experiences of high arch dam construction, a technology of quick placing by cable crane has developed.

1) Standardization of construction techniques for placing horizontally in single placement area.

Generally, horizontal placing is used in order to ensure the quality of construction, the construction techniques of horizontal placing method are as follows:

The placement areas are zoned on the control axis of cable crane, as the cable crane axis and the placement area have a certain angle, and placing is conducted horizontally in areas and in strips; for placing on the single strip, it can avoid the cable crane frequently travelling so as to eliminate the position difficulty and increase the efficiency.

For placement machinery, one cable crane is generally provided with one spreader and one vibrator. Spreading of concrete is made from upstream to downstream, with the spreader operating on the placing strip and the vibrator working at the downstream side. For the first concrete lift placed with horizontal placing method, since the cooling pipes have been laid within the placement area, a strip with 4m wide and 8m long shall be placed first, using 2 to 3 buckets of concrete material, and then the spreader can operate on the surface of placed concrete and the vibrator can travel in the space of the cooling water pipes downstream. The vibrator will be moved to the upstream side when the first strip has been completed. In case the cooling pipes are added in a middle lift, the placing method is the same as the above. For the second concrete lift, the placing is made by the same method, the spreader operates on the placed strip, and the vibrator works at downstream side. As many machines are required in the placement area, the reasonable zoning and standardized placing process are needed so that the placed concrete surface can be covered timely and the concrete temperature inversion can be prevented in summer.

The thickness of concrete lifts should be strictly controlled on the design. The maintenance of spreaders and vibrators should be cared, and the spreaders and vibrator arms shall be cleaned before lifting into the placement area. In hot season, it must pay close attention to the organization and management the placing in placement area so as to play the efficiency of cable crane and mixing plant, speed up the placing and enhance insulation covering to ensure the quality of concrete placed.

2) Un-intermittent transferring and synchronous placing. In Xiaowan and Jinping I projects, the technique of un-intermittent transferring and synchronous placing of two placement areas (monoliths) was developed.

The technique of un-intermittent transferring between two placement areas in Xiaowan project is as follows: five cable cranes are arranged in sequence, they cannot cross each other generally and their positions are fixed. The sequence placement is as one riverbed monolith and one bank monolith or both bank monoliths, that is the four cable cranes are used to place one placement area, and the fifth cable crane is used to make preparation for next placing. The preparation of next placing shall be completed before end of the last placement so that the four cable cranes can be used to the next placement area in time and the transfer of placing position is un-intermittent. The transferring sequence and time of cable cranes depend on the control of placing complete, which is the key of transferring position. By this technique, the average daily intensity is $6,313.5 m^3/d$ and the maximum daily intensity is $7,019 m^3/d$, it achieves a continuous high-intensity concrete placing in construction peak period.

The technique of synchronous placing two monoliths in Xiaowan project is as follows: with the increase of dam height, the area of placing decreased gradually, and the placing intensity was reduced too since five cable cranes are used to place a single placement area at the same time. In order to take a full use of the cable cranes, it was decided that the two placement area would be placed synchronously through analysis of the relationship between the placement areas and the position of the cable cranes, placing intensity of typical placement area, the necessary placing conditions and influencing factors at different positions. Three schemes for synchronous placing the two areas have been developed: ① the two placement areas are completely staggered; ② the two placement areas are overlapped partly; ③ the two placement areas are completely overlapped and parallel. The first scheme applies to the two areas at the river bed monolith and the bank monolith; the second and third schemes apply for the small placement area at high elevation. The scheme selection shall be made on the actual situations. The practice in Xiaowan project has proved that the placing intensity of two placement areas has higher than that of single placement area, the two placement areas were placed with $4+1$ or $3+2$ of cable cranes.

The technology of quick placing with large cable cranes is widely adopted in high arch dam construction in Xiaowan and Jinping I projects, it has provided a guarantee for the rapid construction of high arch dam.

3. Burying cooling pipe during concreting

In strong constraint area of concrete arch dam, if lift thickness of 3m or over is adopted, cooling pipes shall be set in two layers generally, and the second layer pipes should be laid during concreting. When the concrete are placed to the level where cooling pipes to be set, placing concrete and laying cooling pipes can be conducted at the same time so as not to affect the next layer concreting. In order to speed up the laying of cooling pipes, usually HDPE pipes are used, with diameter of 28mm and outside diameter of 32mm, and watertight expansion joints are used and fixed by $\phi 6 - \phi 8$ round rod. The length of coil pipe is controlled within 300m or less per coil. At the corner, edge and elbow of cooling pipe, the spacing and size shall be controlled strictly to avoid the pipe crossing over the transverse joints. For the needs of the second cooling and the acquainting of temperature gradient at same elevation and along the direction of dam thickness, cooling pipes should be laid perpendicular to the direction of flow. The cooling pipes are arranged in accordance with the positioning coordinates in the design, and strive to overlap in the projection plane, and the layout of all cooling pipes shall be provided to guide the grouting and sampling of drilling hole. Before cooling pipes are buried into the concrete, their internal and external walls should be cleaned and the scale is eliminated, and flushing test shall be made to identify whether the pipes have been blocked or leaked. The laid pipes should be protected carefully from displacement or damage. The pipes out of concrete should be capped or covered for protection.

The branches and other pipes should be connected effectively at any time and can be installed and removed quickly, and the flow of each pipe can be controlled reliably without affecting the normal operation of other cooling pipes.

4. Balanced rising

An arch dam is usually taken the double curvature shape with a large overhang, in order to ensure the safety of dam construction. The maximum free cantilever height is required for each monolith. If the dam body cannot be raised equivalently, the rising speed may be affected and the safety of dam construction period may also be of influence. For balanced rising of dam body and making cooling synchronistically, it can make the grouted joints at the same elevation reach the temperature required at the same time and meet the requirements of arch closure grouting, therefore, it provides conditions for balanced dam raising and retaining water in advance. In order to meet the balanced raising of dam, the construction schedule shall be made reasonably and construction resources shall be prepared suitably to ensure the dam construction can be smoothly conducted without interruption.

19.3.2.6 Rapid Construction Technology

1. Repeated grouting

In order to prevent cracks and reduce tensile stress in concrete, inducing joints and transverse joints are in high RCC arch dams, but the joints need to be grouted so as to make the integrity of the dam. The repeated grouting is required for the dam integrity and retaining water request.

For the transverse joints or inducing joints set in RCC dam construction period, the repeated grouting system has provided in the joints since the temperature of dam body does not meet the stable temperature condition and the retaining water of dam body is required. The repeated grouting technology mainly consists of the repeated grouting system with single return pipeline, grout box with rubber sleeve valve and modified ultrafine cement grout which can be filled into fine cracks.

This construction technique has been successfully applied in the Shapai project. Through grouting 14 grouting zones repeatedly beneath El. 1,930.0m where the two inducing joints set, the expected effect of grouting has been obtained. The grout box with rubber sleeve valve and single return pipeline system are characterized by simple structure, easy to be installed and washed, repeatable use and low cost, the repeated grouting system is more suitable for RCC arch dam construction.

2. Formwork

Formwork for rapid construction of RCC includes mainly the turnover form with large curvature for controlling RCC arch dam profile and the precast concrete gravity form for setting joints.

(1) Precast concrete gravity form for setting joints. In order to ensure ventilation, rolling operation and transverse joints or inducing joints, and rapid construction of RCC arch dam, the construction technique of precast concrete gravity form is adopted, that is to install the gravity concrete form first, and then to place RCC. The cross-section of precast concrete form is of gravity, with length of 1.0m, height of 0.25 - 0.3m and bottom width of 0.3 - 0.35m (plug-in holes set on toe plate), the joint surface is straight, and grouting hole, exhaust pipe and keyway are set in the joint. The form adapts to assemble and work rapidly at construction site and has a good effect on setting joint. It has good results for accelerating joint surface cracking, reducing the strength of inducing joints and playing the role of inducing joints. This precast concrete gravity form has been used in construction of Shapai project.

(2) Turnover form with large curvature. RCC arch dam is featured as complicated profile, changeable curvature and large overhung, requirements of quick construction and high quality. In order to meet the construction requirements of RCC arch dam with complicated profile, a large curvature turnover form is developed, and its work principle is as follows:

The lateral pressure of RCC transfers to the truss system through the facing system, and then to the truss system of next form through vertical adjusting system, and finally to the anchorage system. The lateral pressure is balanced by the grip between anchor embedded in the concrete and concrete. The form adapts the changes of complicated arch dam profile by setting and adjusting the flexible splice, level and vertical control system. The form is provided with a truss and two anchorage systems to share the lateral pressure of RCC, and can increase the grip between concrete and the anchor through the ring-shaped dovetail anchor bar. The vertical gaps between the forms can be eliminated by setting and adjusting the front and tail sliding blocks to have the relative movement between upper form and lower form, or adding or reducing the standard shim plates, or combining joints. Zhaolaihe dam is double-curvature variable-thickness one with logarithmic spiral center, the construction technology of the turnover form with large curvature is used successfully there.

3. Concrete mix proportion

The concrete with good crack resistance should have the high tensile strength, large ultimate tension, low elastic modulus, low dry shrinkage, low adiabatic temperature rise and small temperature deformation coefficient and small autogenous shrinkage. For the RCC used in Shapai arch dam, a special cement of micro-expansion with low-brittle is developed successfully, through comparison and study, and choose of local factories, reducing C_3A and C_3S and increasing the content of CAF and C_2S in clinker cement, and introducing low alkaline slag admixture. For this cement, the index of hydration heat is lower than that of moderate heat cement. The RCC prepared by this cement shows characteristics of low elastic modulus, high ultimate tension and slight expansion. At the same time, the quantitative study of cracks for artificial aggregate reveals the concrete having low elastic modulus, large ultimate tension and creep. Therefore, the concrete mix proportion applicable for dam RCC has been worked out, all technical indicators have reached the design requirements after reviewing of laboratory testing and drilling core.

19.3.2.7 RCC Construction Techniques in Hot Season

1. Cooling by circulating water

As the arch dam has a high height and large scale, so the temperature stress is an outstanding issue. In addition to improving crack resistance of concrete and setting reasonable structure zones, the measure of cooling is needed to reach the temperature requirement. The polyethylene cooling pipe was first used in Shapai RCC arch dam, and then it was promoted to the Shimenzi and Longshou RCC dams after this measure was successful in Shapai project, it has made a new progress for improving hydration temperature rise, joint grouting and contact grouting, and good results have been obtained.

The construction practice has proven, the cooling technique by using high strength polyethylene pipes is applicable to the construction conditions of RCC, with simple burying process and little interference in other construction activities, and the favorable cooling effect.

2. Heat insulation and spraying

In order to control of the maximum temperature of RCC and prevent thermal flow from direct sunlight, the measures of covering the placement surface with polyethylene insulation sheet and spraying on the surface by the sprayer are taken during concreting in summer.

The temperature rising control in the placement area is very important, and covering the placement surface with thermal insulation quilt is a convenient and important measure. As the placement area is large, so thermal insulation quilt should be large enough with light weight and good insulation performance. Generally, two layers of 1cm thick polyethylene insulation sheet winded with plastic woven cloth are used, it has the following performances, low thermal diffusivity, high flexibility, high tensile strength, good waterproofing, low density, good elastic, glossy appearance, good impact resistance, good aging and chemical resistance, non-toxic, odorless, low temperature resistance, oil proof and fire proofing, it can be used to cover uneven placement surface in the placement area and play the effect of heat insulation. Through testing at site, the temperature can be reduced $5-6℃$ when the placement surface is covered with the insulation quit at temperature of $28-35℃$ under direct sunlight.

Spraying is an important measure to prevent concrete temperature rise and keep wetting, it can significantly change the microclimate in placement area and reduce temperature $4-6℃$, it is very favorable for temperature control. In RCC placing, special sprayers are used commonly, the spray radius is 30m with swing angle of $120°$ and swing frequency of 5 times/min.

3. Fast placing

RCC is mainly delivered by dump truck, belt conveying, negative pressure chute (pipe) and special vertical tube, and the cable crane, portal crane and tower crane can also be used as auxiliary delivery equipment.

Because RCC arch dams have large height usually, several placing methods are used depending on the positions and elevations of placement operation so as to speed up placement. Trucks are used mainly for RCC delivery at the lower part of the dam, and the combination with cable crane, vacuum tube, chute, movable distributor and overhang distributor are adopted for vertical delivery at the middle and upper parts.

When RCC is delivered by trucks, the trucks can access to the placement area along the dam roads backfilled with macadam. When the trucks directly go to the placement area, the entrance should be set reasonably, and protective plates shall be put at the entrance in order to protect the structure of dam body, commonly the lifting steel trestle may be used as the protective plates, and it can protect the placement surface from damage and be transferred quickly.

19.3.3 Development Trends and Prospects of High Arch Dam Construction in China

The major trends of high concrete arch dam construction technology development will focus on the construction of RCC dam, and the corresponding new type of dam materials, large mechanization, informatization, standardization, intelligence, energy saving and environmental protection. It is mainly to strengthen basic research, develop and use of new type of dam materials with excellent performance and new technique, intelligent integrated system and Internet of Things of civil works construction, and research large-scale construction machinery and equipment supporting. The main research directions are as follows.

19.3.3.1 Research of High-performance Concrete Materials

Consideration of features of high arch dams, the concrete shall have the following performances, high strength, high ultimate tension, low hydration heat, long time of hydration, low elastic modulus and low shrinkage. In construction of water conservancy and hydropower projects, generally, concrete aggregate is local materials, cement and fly ash and other primary materials rely on the local supply, but some researches can be made, including special admixtures, additives and the design parameters of mix proportion, to improve the concrete performances of high arch dam.

19.3.3.2 Research of High-intensity Construction Technology under Bad Conditions

High arch dam is general laid at the mountain valley, the common lifting means are difficult to be adopted, and therefore, cable crane is the main tool for the arch dam construction. But, the cable crane is often difficult to play the efficiency in bad climate, such as cold, rainy and foggy. So, it needs to research how to use intelligence and information technology to improve the automation and the efficient of machinery in adverse climate conditions, and achieve the purpose of automatic placing concrete, so that the concreting intensity can be improved. In addition, in order to solve the difficulties of concreting for high arch dam, it needs to study the vertical delivery system with high drop for coarse grading concrete delivery. The main research items include concrete aggregate segregation in high drop and long-distance delivery, secondary crushing of aggregate, concrete temperature rise, pipe blockage, transportation of large size aggregate and concrete distributing successively in several placement areas so as to provide more means for concrete placing of high arch dams.

19.3.3.3 Monitoring and Simulation During Construction

With advancing technology, the construction monitoring and simulation technology have developed rapidly, but it still has much room to develop, the future research and application include the following systems:

1. Visual dynamic simulation during construction

The dynamic simulation technology for arch dam construction will be researched to achieve the show of 3D visualization (virtual reality), establish the simulation and management system with numerical simulation and virtual reality interaction, and achieve the simulation and optimization for major construction schemes and the real-time dynamic management and control in the whole construction process, including the development of construction simulation system, 3D visual modeling, simulation database and interactive of 3D visualization and so on.

The core of simulation technology of arch dam construction schedule is to establish the computer mathematical model of the true system on the principles of system engineering, and then use the model to make experiments and researches on a computer instead of the true system. The main purpose is to solve two problems: ①rational allocation of machinery and equipment, proper arrangement of placing sequence, correct forecast of placing period; ②real-time control in construction. The basic idea of placing process simulation is as follows: Firstly, numbers of placing blocks and machinery or equipment were used for placing and transporting. Then, arrange the placing sequence for placing blocks and calculate the construction duration of the placing blocks according to the constraints conditions in the placing scheme, thus the construction duration by the scheme can be worked out. A number of schemes are calculated repeatedly to obtain some results, and the optimum scheme can be selected through comprehensive comparison. The goal of simulation is as follows: in the premise of meeting a variety of constraints, arrange the placing sequence for whole placing blocks and the machinery or equipment so as to make the placing duration shorten, the idle time of machinery or equipment least and reach the purpose of quick construction, low cost, high quality and safety.

2. Intelligent monitoring for temperature control and crack prevention of concrete and real-time simulation system of 3D temperature field

The research of automatic control system for temperature of dam concrete, includes mainly the following items, cooling by circulating water of dam concrete and automatic collection method and data processing system of the internal temperature of concrete, relationship between water cooling and temperature change of dam concrete under different climatic conditions, control measures of water cooling, automatic control system of water cooling so as to improve the feedback and processing efficiency of the temperature data for water cooling and internal temperature of concrete, and achieve truly the specific circulating water and precisely control of internal temperature of concrete in a planned way.

At the same time, by collecting the internal temperature of dam automatically, 3D real-time simulation of the temperature field of dam, it is helpful to take specific measures for temperature control and crack prevention.

3. Information monitoring system for dam construction

In construction of large-scale water resources and hydropower projects, the construction site has a large area and complicated conditions, the construction arrangement and command by personnel are hard work and very dangerous since the personnel should often walk through the several placement areas for inspecting one area, so it is difficult to improve efficiency. Therefore, it needs to establish an integrated computer monitoring system for control of production, transportation and placing concrete. The monitoring system mainly consists of video monitoring subsystem, testing subsystem of concrete production process, management and decision-making subsystem, optimal scheduling subsystem of concrete production and transportation operations, network and database subsystems and so on, it serves the functions of monitor, control and management of machinery and equipment used in critical links of concrete production, transportation and placement, and control quality of concrete, it should ensure the quality in production, transportation, placing process and the safety of various construction machinery and equipment, it can greatly enhance the efficiency of placing concrete and ensure the quality of concrete construction.

19.3.3.4 Research on Building Information Modeling (BIM) System

The combination of current construction technology and information technology is an important direction of technological innovation. Some Chinese construction companies have utilized the BIM technology to improve construction management. It needs to research and develop the BIM technology for hydropower projects and lead the development direction of hydropower industry.

BIM is a new term of construction industry in the last two years. It is a new technology to lead the information technology of construction industry to a higher level. The comprehensive application of BIM will take significant effects for technological advance in construction industry, and can greatly improve the integration of construction. At the same time, it also has brought enormous benefits for development of construction industry, significantly improved the design and the quality and efficiency and reduce the project cost.

BIM is established based on 3D digital technology, it integrates a variety of engineering information data models, and it is an expression of detailed information about the project. BIM is a direct application of digital technology in construction industry and solve the description of construction in the software, it has designers and engineers make the right response for a variety of building information, and provide a solid basis for collaborative design.

BIM is also a digital method used in the design, construction, management, this method supports the integrated management of construction, and can improve the entire construction process efficiency and reduce the risk significantly.

As the BIM needs an integrated management environment of full life-cycle of construction, so the structure of BIM is a composite structure containing the data model and behavior model. In addition to the data model which contains the geometry and data, it also contains the behavioral model related to management, the combination of the two models can be used to simulate the behaviors in real-world.

The application of BIM can support continuous and real-time applications of a variety of information involving with the project, the information are high quality, reliable, integrative and compatible, it can improve the design, quality, and efficiency of the whole project and reduce the costs significantly.

The application of BIM can immediately get the benefits of making construction faster, cheaper and more accurate, it also can cooperate better with the various types of construction work and reduce the errors of drawings, and the long-term benefits has gone beyond the stage of design and construction, it can be a great help to operation, maintenance and facilities management of buildings in future and to save the cost.

19.3.3.5 Research on New-type of Formworks

(1) Research on form materials. The research and analysis have been conducted for materials with preparing forms, including plywood, bamboo plywood, combination of steel frame, small steel form, steel form and large steel form, at the same time, the plastic form is tested, which the technology is not yet mature, for developing environmentally friendly formworks and replacing the traditional steel and wooden forms.

(2) Modular research on hydraulic jack system. The modular design of hydraulic jack system for climbing forms has been studied, and achieved a variety of combinations between hydraulic jack system and the form to make more flexible application of the form.

(3) Development of large-scale cantilever form which is used for placement of bracket. In order to improve the lift thickness of the bracket parts, it is necessary to develop large-scale cantilever forms.

Chapter 20

Hydraulic Generating Equipments

Zhu Yaoquan[1], Zhu Yunfeng[2], Liu Jie[3], Zhang Liangying[4]

20.1 Hydraulic-Turbine and Generator Unit

20.1.1 *The Developing History of Hydraulic Power Unit*

From the first hydraulic turbine made by Ji Yanhong in 1927, the developing history of hydropower equipment in China went across more than eighty years. The growth history of hydropower equipment in China may be classified into the five stages of infancy, early development, development, full development, and facing the world, independence and innovation.

20.1.1.1 Infancy Stage (1927 – Sept. 1949)

In 1927, engineer Ji Yanhong from Fujian Province made a hydraulic turbine of 3kW at Xiadao Hydropower Station in Nanping. The prelude of hydraulic turbine made in China was opened. Before the found of New China, Chinese factories had made 60 sets of hydraulic turbine applied to fifty hydropower stations. The largest capacity of Francis hydraulic turbine made by Chongqing Mingsheng Machinery Works was 735kW and installed at Xiadong Power Station in Changshou County, Sichuan Province. The largest capacity of hydraulic-turbine generator strypped-down from a frequency conversion generator by the 4th National Electric Works (the former of Kunming Electric Machinery Works) was 1,550kW and installed at Xiadong Hydropower Station. The largest capacity of the hydraulic-turbine generator designed and manufactured independently by the Huasheng Electric Works in China was 500 hp for Xiannvdong Power Station in Wan County.

20.1.1.2 Early Development Stage (1949 – 1965)

In 1951, the first Francis type hydraulic-turbine generator unit made in China was 800kW at Xiadong Hydropower Station. And then the Francis type hydraulic-turbine generator units of 10MW of the Guanting Power Station and 12 – 15MW of the Shizitan, Shangyou Power Station, etc and Kaplan type hydraulic-turbine generator unit of 16MW of the Dahuofang Hydropower Station were made. The turbines were mainly imitated from the design of SMS Company of America as the type of E2, H2, G2 and Leningrd Metal Factory of the former Soviet Union as the type of HL211, HL638, HL587 etc. The technology mainly followed the former Soviet Union. In the beginning of 1950s, 4 sets of 72.5MW hydraulic turbine and 5 sets of generator were imported for the Fengman Power Station. Through the installation and operation of the equipments at the Fengman Power station, China had mastered more quickly the technique of design, manufacture, installation and operation from Czechoslovakia and the former Soviet Union. In the early 1960s, the Kaplan type turbine of 3.3m in diameter and the Francis type turbine of 4.1m in diameter were put into operation, especially the units of the Xin'anjiang and Yunfeng Hydropower Station were manufactured entirely by China and the units of Yunfeng won the national price of quality.

The General Administration of Hydropower Construction, the Administration of Hydropower Survey and Design, and Beijing and Shanghai Investigation and Design Institutes were found by the Ministry of Fuel Industry in 1954. These design institutes were reorganized into eight Institutes of Hydro-power Design in 1957. From that time, the hydropower stations with several hundreds of thousand kilowatts could be designed independently, such as Xin'anjiang. In 1954, the large manufacture workshop for hydropower equipments was built in Harbin Electric Machinery Works (HEC for short). It established the first manufacture base of hydro-power equipment in China. And the middle-scale manufacture works of Chongqin, Hangzhou, Tianjin, Kunming were set up early or late. In 1963, the hydraulic turbine model test stand in the Institute of

---

[1] Zhu Yaoquan, China Institute of Water Resources and Hydropower Research, Professor Senior Engineer.
[2] Zhu Yunfeng, China Institute of Water Resources and Hydropower Research, Engineer.
[3] Liu Jie, Mechanic and Electrical Engineering Bureau, China Three Gorges Corporation, Engineer.
[4] Zhang Liangying, Three Gorges Power Station, China Yangtze Power Co., Ltd, Senior Engineer.

Harbin Large Electric Machinery was passed through appraisal and the research and development work of hydraulic generating equipments were begun independently. Some great achievements, such as the hydraulic turbine new model, dual water cooled generator, the cast-welding structure of the melting electro slag welding at the blade and crown at the 9th unit in the Xin'anjiang Power Station had been acquired. At the same time, China Institute of Water Resources and Hydropower Research (IWHR for short) conducted the work of R & D in developed to a certain scale and got more achievements in the research of the field test, cavitation, silt erosion, blade crack, etc, and promoted the development of hydraulic power equipments. Up to the middle of 1960s, the early development stage of hydraulic power equipments had been accomplished in China.

20.1.1.3 Stage of Early Development (1966 – 1980)

By the operation of Liujiaxia Hydropower Station with the installed capacity larger than 1,000MW and the unit capacity of 300MW, the anual increasing amount of generator equipment reached to 1,000MW and entered the second stage of the development of hydraulic power equipments. To 1980, the total installed capacity of hydraulic hydro power equipments reached 20,318MW in China. The capacity increased by 17,305.4MW in fifteen years and the average annual increasing capacity reached 1,153.7MW and firstly exceeded 1,000MW. In this period, the second large hydro power generating equipment manufacture factory, Dongfang Electric Machinery Company (DFEM for short), was set up.

The first batch of research achievements of the hydraulic turbine model, such as A001, A008, A009, A29, A012 in HEC, had been applied to engineering practice. The specific speed coefficients of Francis turbine was increased to 2000 from 1000. The hydraulic turbine generator unit in Liujiaxia was the largest capacity in China at that time, and was one of the largest Francis turbine in the world at that time. The diameter of Kaplan turbine was increased to 5.5m in Qingtongxia Units, and the diameter of Kaplan turbine was increased to 8.0m in Fuchunjiang, Xijin Hydropower Stations from 3.3m in DahuofanSg Power Station. It showed the level of Kaplan turbine had been reached the world level.

In 1964, the first pumping storage unit was imported from Fuji Co. Ltd. for pumping water in the period of low water period and increasing capacity of peak load regulating. This was the first project of technique introduction in hydro-power industry. The unit was the type of the diagonal flow pumping hydro turbine. The speed of turbine was different from the speed of pump. This provided the experience in introducing foreign advanced technique as the earliest pumped storage unit. The unit was later installed at the Gangnan Hydro-power Plant in Hebei Province.

20.1.1.4 The Full Development Stage (1980 – 1990)

The operation of Gezhouba Power Station with the installed capacity of 2,715MW indicated that the key techniques of design, manufacture and installation of the huge hydro-generating unit for the class of the million kilowatts hydro-power station had been mastered. Through the bids inviting and contract signing of the hydro-power equipment of the Lubuge project, it was the first time for China to accumulate the experience of equipments and technology of inviting bids and signing contract with abroad. In these ten years the installed capacity of 15,727MW was increased, especialy 10,000MW in the period of the 7th five-year plan and 2,000MW on the average was put into operation. The main sign in this stage was that the large Francis and Kaplan units in Gezhouba, Wujiangdu, Longyangxia, Baishan, etc. were designed and manufactured independently by China. In Pellton turbine unit (head over 1,000m, 15MW) was made successfully in Tianhu Power Station and the Bulb turbine generator unit (15MW) at the Anju and Duping Hydro-power Station were put into operation. The tide tubular hydraulic generator unit in Jiangxia Power Station, and 4MPa oil pressure equipment and electronic hydraulic governor had been manufactured.

The hydraulic turbine in Gezhouba Power Station adopted 2 units of ZZ560 ($D=11.3m$, $P=170MW$) with four blades and 19 units of ZZ550 ($D=10.2m$, $P=125MW$) with five blades. The blade material of hydraulic turbine of 170MW hydraulic turbine and the blade material of ZZ500 hydraulic turbine adopted 0Cr13Ni6Mo and 0Cr13Ni5Mo stain steel respectively, which was successfully developed by China for the first time. The two type units of Gezhouba Hydro-power Station were manufactured by HEC and DFEM and were the biggest Kaplan turbine in the world. It was appreciated by the experts home and abroad till now, and rewarded the special class of National Science and Technology Progress Prize.

The large Francis turbine units ($D=6m$, $P=320MW$) in Longyangxia Power Station were made by DFEM. The type is HLD06A-LJ-600. The blade material was 0Cr13Ni6Mo and the structure adopted the eccentricity and heterogeneous steel welding structure in halves. The eccentricity of interrupted face is 250mm. The material of spiral casing was HT62CF high intensity steel plate. The pre-heating temperature before welding was 120℃ and the technology of de-hydrogen was adopted after welding. The five units of hydraulic generator ($P=300MW$) of Baishan Power Station were also self-designed and manufactured. It indicated that China had entered the maturity stage of the manufacture of 300MW unit after the manufacturing unit of Liujiaxia Power Station.

The unit of Lubuge project was the first input item of hydro-generator unit after reformation and opening in China. The input principle was preferred bidding, shop around, trade and technology combination and manufacturing cooperation. The twelve companies in America, Germany, Norway, Sweden, France, Japan, etc. nine countries joined the technique exchanging and negotiation, at last the companies of KB and Siemens won the bid and cooperated to manufacture with HEC. It created the experience of the cooperation between the abroad and domestic manufactures. The maximum model efficiency of the high head Francis turbine of Lubuge project reached 94.6%. The runner was made up of 15 long blades and 15 short blades. The efficiency was high and the stability was better in the area of part load operation condition. The main shaft was coupled with runner by friction and the technical supplying water was taken from the top cover. These new techniques were all from KB in Norway.

The input of the pump-storage unit of Panjiakou was an important practice after the input of the pump-storage unit of Miyun Power Station in the 1960s. The unit adopted two rotating speeds in pump operation for solving the problem of the large varying range of operation head. The unit average efficiency may be increased by 13% to 15%.

Besides, the design of hydraulic turbine selecting, workshop arrangement and the calculation of regulation guarantee were widely used with computer, and the progress had been made in the optimization hydraulic design of the flow passage of hydraulic turbine.

20.1.1.5 The Stage of Independence and Innovation Facing the World (1991 - 2010)

The development of the unit was continued toward the big scale, large capacity and high specific speed. The Francis unit was from Longyangxia unit (320MW, $D=6.0$m) in the 1980s to Er'tan unit (550MW, $D=6.247$m) and Yantan unit ($D=8.0$m), the Wuqiangxi hydraulic hydro turbine ($D=8.3$m) in 1998 and these units were also the huge hydro-generating unit in the world. After 2003, the unit capacity of Three Gorges Project (TGP for short) put into operation was 700MW and the runner diameter reached more than 10.0m. The great achievements and successful experience had been mastered in hydraulic performance, manufacture craft and operation technique of hydro-generating unit.

After the operation of the unit of Gezhouba in the 1980s, the Kaplan turbine of Tongjiezhi ($P=150$MW, $D=8.5$m, $H_{max}=40$m) and Shuikou ($P=200$MW, $D=8.0$m, $H_{max}=57.8$m) were put into operation in succession. The Shuikou unit was among the international forefront.

The successful operation of the hydro-generating units in Three Gorges and Longtan Power Station proved that the design and the manufacture of the hydro-generating unit had reached the advance level of the world. The capability had been provided with the Three Gorges Project (TGP) unit (capacity 840MkVA, thrust bearing load 5500 tons, 10.6m in runner diameter), Er'tan unit ($H_{max}=189.2$m, $P_{max}=610$MW, 6.427m in the runner diameter, pole capacity 14.57MVA with air cooling), the Wuqiangxi unit ($P=800$MW, $D=8.3$m) and the Shuikou Kaplan turbine ($P=200$MW, $D=8.0$m, thrust bearing load 4100 tons), and the Yele impulse turbine ($H_{max}=637.2$m, $P=120$MW, $D=2.6$m). Evaporation-cooling technique in Lijiaxia generator ($P=400$MW) had been manufactured. These progresses have shown the level of design and manufacture in China.

The bulb tubular and pumping storage units were developed rapidly. The water resource for tubular generating unit is quite abundant. The installation capacity of the hydo power stations ($H \leqslant 20$m, $P \geqslant 25$MW) in programming item is 13,400MW and the generating electricity is about 60×10^9 kW·h. The small tubular generating unit was input in the 1960s, but up to 1984, the tubular unit of the Baigou project ($P=10$MW, $D=5.5$m) in Guangdong Province began to be manufactured in China. The rapid development was after the 1990s. The tubular units of Bailongtan ($H_{max}=18$m, $D=6.4$m, $P=32$MW) and Wangpuzhou ($D=7.2$m, $P=27.25$MW) had been made by Fuchunjiang Power Generation Equipment Works. The 30MW units in Linjingtan, Guigang, Hongyanzi Power Station, 40MW units in Jingyentai and 45MW units in Hongjiang were developed by HEC and DFEM. The bulb tubular unit in Qiaogong Power Atation ($P=57$MW, $D=7.45$m) has been put into operation. The design and manufacture of this bulb tubular unit has reached the advanced world level.

The design and manufacture technology of large pump-storage unit has been basically mastered and has the preliminary competitive ability with the world strong opponents from the unifying bid and importing technology of the 16 sets of 300MW in Baoquan, Huizhou and Bailianhe Pumping-storage Station to the 12 sets of the 300MW supported by the localized production projects in Pushihe, Heimifeng and Huhehaote, and the research and development of 4 sets of 250MW in Xiangshuijie, 4 sets of 300MW in Xianyou and 6 sets of 250MW in Liyang. At the same time, DFEM and HEC had mastered the design and manufacturing technology of large-scale pumped storage, and preliminarily had strong competitiveness with the developed-country competitors.

20.1.2 *The Advanced Technology of Hydro-turbine Generator Unit*

In the construction of hydro-power project, because the quality and performance of the hydraulic-turbine generator unit are directly related to the economic profit of the Hydro-power Station, the importance is paid more attention. The domestic hydro-power unit manufacture companies develop independently with the transferred technology digesting and absorbing, especially the innovation based on the conveying techniques of cooperation manufacture in TGP, the key techniques of hydraulic design, electric magnetism, ventilation, thrust bearing, insulator, structure optimization, manufacture craftwork of huge Francis type unit have been mastered and realized the span of thirty years of hydro-power manufacture in a short time of ten years. The manufacture levels of the key techniques and important parts have reached the same as those of the abroad. In product varieties, the huge Francis unit (700MW, 10m in diameter), the Kaplan unit (200MW, 11.3m in diameter) and the bulb unit (50MW, 7m in diameter) had been manufactured and the level of research and manufacture had advanced in the world.

20.1.2.1 The Advanced Technology of Hydro-turbine Generator Unit Design

1. Hydraulic design and testing technology

The runner is the key part of hydraulic turbine and its performance directly affects the economic profit and operation stability. In the 1990s, hydraulic turbine model performance of domestic company was lower than that of foreign company. A majority of the runner of hydro-power station adopted foreign technology. Advanced CFD software and analysis method were introduced to domestic company by TGP. With the accumulation of the runner database and the progresses of test method and model manufacture, the hydraulic performance of model turbine by domestic company improved rapidly and arrived at the level of the world. In recent five years, more runners whose efficiencies were higher than 94% were worked out, some models efficiencies had reached 95%.

During the development of new runner for TGP Right Bank Power Station, based on digesting and absorbing the transferred technology, HEC and DFEM applied new design concept and method to develop the new runner whose performance was better than that for TGP Left Bank, the higher efficiency of model and stability was improved much with no high-part-load pressure fluctuation in whole operating range and it solved the difficult problem that puzzled this industry in the world. This achievement was confirmed by identical comparison test on the hydraulic test stand in IWHR. The new runner had been applied in prototype turbine of Right Bank in Three Gorges Project.

In the respect of hydraulic machinery model test, the domestic test stands realized high accuracy test and automatization. It can carry out any performance test according to international standards. The observation system of flow pattern as fiber, endoscope and vidicon was equipped and the happening and developing of the blade vortex and cavitation in all operation range can be observed. The PIV technology has been researched now in the interior flow pattern test of hydraulic turbine. Numerical control machining technology has been applied in the manufacture of runner and model device. Spiral casing and draft tube adopted the material of Tantalum, stain steel or other complex material and installed after digital control machining. The guide vane and stay ring of model also adopted digital control moulding craftwork and the precision increased greatly.

2. Cooling technology of hydro-power generating unit

Water-cooling technology is most popular in huge generator. This technology has completely mastered by co-operation with foreign company and applied in the 700MW generator of TGP which runs well now.

Air-cooling system has the characteristics of simple structure, easy maintenance and stable operation, but for the large load and volume of huge generator, the air-cooling system design has more difficulty. Through the co-operation with foreign companies, HEC has researched more progresses in air-cooling of the huge generator about flow field between poles, heat dissipating coefficient on core surface, axial and radial 3D flow field. The methods of aeration design and analysis has been established and applied in the design of many 700MW generators as TGP Right Bank, Longtan, Xiaowan, Laxiwa. HEC has also set up 1:5 aeration test stand for complete air-cooling generator, and tested more strictly, applied the full air-cooling system in hydro-electric generator of TGP Right Bank, Longtan, Laxiwa, Xiaowan Project.

The evaporation-cooling is an inner cooling technique and is to use the boiling of low boiling point liquid to absorb the latent heat of vaporization for generator cooling system. It has high insulator performance, no erosion to motor part and self-circulation feature without pump. The 400MW generator of Lijiaxia unit adopted the evaporation-cooling.

3. Thrust bearing technology of hydro-power generating unit

For optimizing thrust bearing structure and performance, improving reliability, HEC and DFEM have set up the thrust bearing test stands separately with load up to 3,000 tons and 1,000 tons to research and test bearing performance.

The elastic metal plastic tile has many advantages and developed by Tianjin Institute of Hydroelectric and Power Research, HEC and DFEM. The domestic manufacturing elastic metal plastic tiles have been applied in Tianshenqiao, Yantan, Dahua

Power Stations, etc.

The elastic metal plastic tile thrust bearing for the 4100t thrust bearing of Shuikou, 3,300t of Gezhouba and 3600t of Xiaolangdi were made by HEC, and the 5,600t thrust bearing of TGP is the maximum thrust load in the word and also made by HEC. The research and test of 6,000t thrust bearing for TGP with elastic metal plastic tile and babbitt metal were carried out by HEC cooperated with ABB. The study of 1,000t babbitt metal thrust bearing for pumping-storage station also had been taken. The imported spring used for the support of babbitt metal thrust bearing was replaced by the domestic spring of DFEM. The test results of maximum velocity and special pressure were all satisfactory. The results were the basis for TGP thrust bearing that adopted elastic metal plastic tile.

4. *Stator major insulation technology*

The major insulation of stator coil uses the class F mica paper epoxy resin rich mould press system. The stator bar is continuously wrapped with resin rich mica tape and then covered with corona protective tape on the surface, finally solidified and formed by heat press. It has uniform dimension and good compatibility. The major insulation, corona prevention, winding head and slot structure and insulation material are all made in domestic.

The stator bars of the hydro-generating unit (200kV/700MW) in TGP manufactured with domestic main insulation technology were tested and confirmed strictly by Alstom and Siemens. The appearance quality, dimension, surface resistance, medium dissipation, cool and heat circulation test and electrical aging test, etc. have reached the same level of Alstom and VGS. The products are operated well in the Power Stations.

5. *Structure stiffness-strength analysis for hydro-power generating unit*

The evaluation technology of structure optimizing, fatigue resistance design and mechanical stability were applied in the structure design of main parts of hydro-generating unit.

To compare stiffness-strength of major parts with different manufacturers, the design difference in the same component in various companies was found. From the system analysis on the three aspects of structure topology, geometry and slab thickness, the main factors to determine structure stiffness and strength have been found. The own structure type has been formed and the automatic optimization of structure analysis has been realized by the parameterization-modeling method. The main parameters can be analyzed by variable analysis method and the design curve is formed. The information is the basis of the structure improvement. If it was applied in the unite optimization method of the dimension and the shape, the best structure for both the plate thickness and the geometry shape may be found.

The runner stress analysis, local stress sub-analysis and dynamics characteristic analysis had been programmed in fatigue resistance design of hydraulic turbine. Various factors are considered, such as runner type, blade number, local thickness on blade, crown and ring thickness and strengthen triangle. And then the way to improve strength performance of runner is concluded. Optimization strategy is advanced. Underwater fatigue capability and crack propagation rate of runner material are tested. With steady stress analysis and dynamic stress test data, acceptable defect level and fatigue capability could be estimated, and then main factors of runner fatigue resistance are concluded.

In the aspect of the evaluation of stability of hydro-power unit, by the detail calculation analysis of the axial system critical speed and dynamical response, the systematic quantitative analysis of the factors affected the axial stability of unit, such as support stiffness, mass and the electric unbalance, etc, and the analysis of the stiffness and damp of bearing oil film, the strategy of controlling the axis system stability of hydro-power unit is tabled by the control of support stiffness and the accuracy of rotor balance. On the other hand, the technology of coupling of fluid and solid is adopted for the analysis of the dynamic feature of the main parts of flow passage. The natural frequency of main parts should keep away from the frequency of every energizing frequency and Karman vortex frequency in order to avoid resonance. At the same time, the vibration analysis of top cover, upper bracket and lower bracket is also carried out to guarantee the level of vertical and horizontal vibration to reach the good class.

20.1.2.2 The Manufacture Technology of Hydro-turbine Generator Unit

1. *Digital control processing technology*

With the development of electronic technique, the numerical control equipments are applied greatly in recent twenty years. Large digital control machine tool and special digital control equipments have become the conventional measures in hydraulic power equipment manufacturing.

(1) Numerically controlled cutting material technique: Plasma digital laser cutting machine suitable for 120mm thickness has replaced the method of cutting stainless steel with flame vibration cutting and carbone arc-air gouge together. Geometry shape and welding bevel can be completed at one time. It basically meets the requirements of steel cutting in large hydro-

power equipment and the problem of cracks generated after the welding has been solved. The technique can increase the rate of material using and get more economic benefits by digital control program. The digital cutting machine with three cutting torches of angle variable and infinite turning make it possible for the blanking of steel plate for components such as spiral casing and spherical tank. This equipment is an advanced cutting machine.

(2) Large component with digital control machining: The application of digital control boring and milling machine, digital control vertical lathe and multiple purpose digital control machining center have greatly enhanced the processing ability, as well as the processing accuracy. The domestic backbone enterprises of hydraulic power equipment manufacturing have been equipped with more large digital control machines by technique reformation in recent ten and more years, such as digital control vertical lathe ($\phi 22$m), digital control horizontal lathe ($\phi 4.3$m$\times 18$m), boring and milling machine ($\phi 254$mm) and 5 axis linkage digital control planer type milling machine ($\phi 4.5$m$\times 13$m). The coaxiality of shaft hole of guide vane at top cover and bottom ring can be guaranteed by digital control machining. The relative position of guide vanes can be guaranteed with digital control machining of the surface and sealing face of guide vane and it leads to shorten the pre-installation time of guide vane greatly. With the increasing of coaxiality, the uniform of end and vertical clearance of guide vane, the product quality are greatly enhanced.

(3) Turbine blade with digital control machining: because the runner is the key part of hydraulic turbine, its manufactured quality is directly related to the operation performance and service life and hence the runner manufacture technology is very important. The molded line of blade curved surface can keep the consistency with the value of theoretical design and the waveslope of blade can be guaranteed with digital control machining. At present, the blade of large runner and welding bevel all adopted digital control manufacturing method and the blade of middle and small runner is developing toward the digital control machining.

The largest blade of Francis turbine manufactured with digital control in all positions is the blade of the turbine of Three Gorges Project. The weight of single blade is 16,000 kg after processing and the machining period is seven days. The tolerance of molded line of blade may be controlled within ± 1mm. The weight difference between blades with the maximum and minimum weight is controlled within 2.2% of the theoretical blade weight. The largest blade of Kaplan turbine manufactured with digital control in all positions is the blade of the turbine of Letan Power Station and the diameter of runner is $\phi 10,400$mm. The blade of pump-turbine with digital control machining is the blade of Baisan Power Station and the diameter of runner is $\phi 5,130$mm. The digital control machining technology of the runner of Pelton turbine is highly developed at abroad and the domestic technology is in research.

(4) Other technology of digital control machining: The generator bar was wrapped by handwork before, the bar wrapping quality is more affected by the worker's technique level, working condition and surrounding condition. At present, the use of six axis digital control wrapping machine has realized uniform wrapping with constant tension and promoted the bar wrapping quality. The laser cutting machine is adopted in the processing of magnetic yoke stamped steel. This technology can reach the purpose of reasonable use of material, decreasing cost, raising quality and benefit.

2. Welding technology

Because the more parts of hydro-power generating unit are steel plate welding and casting welding structure, the welding level is highly related with the new product quality. The successively application of such equipments as wide band submerged welder, submerged arc auto-welder, narrow submerged arc welder, micro arc plasma welder, tunnel welder, large displaced machining and roller carrier play a significant role in the selection of process program and in the welding quality and efficiency.

(1) The technique of gas shielded welding: The method of gas shielded welding has been developed and promoted since the 1960s and the advantage of this welding method has been more and more materialized along with the improvement and advancing of welding techniques. In the 1980s, manual arc welding was basically substituted by the high efficient gas shielded welding. The semi automatic gas shielded welding technique has significant technical superiorities. With flexible operation, stable welding and reliable quality, this welding technique meets more the requirements of circular welding such as small lot of single part, various circular shapes, multilayer box type and multi rib plate. The gas shielded welding with carbon dioxide and rich Ar mixed gas has been applied extensively to low alloy steel and alloy steel.

(2) The narrow gap automatic submerged arc welding: For the welding of shafts and ring type parts, the narrow gap automatic submerged arc welding is a mature technology and the application is extensive, such as in the welding of the shafts of Three Gorges Project, Letan and Suofengying Hydro-generator unit. The effective wall thickness of the shaft welding has reached 380mm. High efficiency, low welding material consumption and excellent welding quality are the features of this welding technique.

(3) Wide-band submerged arc welding: To solve the problems of serious cavitation, silt erosion of flow passage parts in

the power station, a layer stainless steel material of anti cavitation and anti wear is welded onto the surface of carbon steel and low alloy steel to increase the ability of anti cavitation and anti wear. This method is used extensively and the effect is better. The hardness after layered surfacing welding with martensite welding materials can reach HB380 - HB400. This welding technique has already been applied in the Power Stations of Lubuge, Tianshengqiao, Yantan, Geheyan and Yingxiuwan. The surfacing materials and fluxes are gradually getting domestically produced, the quality is steady and the effect is better.

(4) The technique of "welding robot": in the welding of large runner, the workpiece needs to warm-up to 100℃ above for increasing the welding quality to avoid the crack. The scurviness of work condition easily affects the welding quality. The application of "welding robot" technique is one approach to solve the problem.

The application of "welding robot" now is a welding work station system in strict meaning. The operation frame of this welding device can be moved in X, Y, Z three degrees of freedom, and the welding torch can be moved in U, V, C three degrees of freedom. So the six coordinates can be programmed in demonstration way to control the striking and ending of arc, the supply and recovery of the welding flux, as well as to control the welding voltage, current flow, welding speed, heat input amount, welding wire parameters, voltage attenuation compensation and system configuration parameters. On-line adjustment can be realized to the welding voltage, current and welding speed. Though the "welding robot" can not carry out the welding of full flow passage, the application will play a strategic role in the domestic manufacturing of power generating equipments.

(5) Other welding technology: The middle frequency welding technology was used in welding of water-cooling winding in order to raising the welding quality of stator winding and conducting bar. To ensure the location accuracy of stator ventilating slot steel and the ventilating property of generator, digital control automatic welding and digital control laser welding have been adopted.

3. Other advanced technique of manufacture

(1) The technique of the blade hot mould pressing: From the consideration of the batch production and raising the material quality of the runner blade, the runner blade would be manufactured by the technique of the blade hot mould pressing. This technique is that the steel plate of the blade out spreading after heated to a certain temperature is put into a mould with the upper and backer shape of the blade and pressed into a blade of the runner.

The research and the using of the technique of the blade hot mould pressing are earlier at abroad. This technology was learned from the cooperation manufacture of the runner of Lubuge Power Station with Kvaerner Brug in Norway. The blade of the runner of Lubuge Power Station is uniform thickness. It is in favor of using this technology. The hot mould pressing technique of nonuniform thickness blade was researched at home. One method is to use the technique of hot mould pressing of uniform thickness and then to use the digital control machining. Another method is to use the technique of hot mould pressing of nonuniform thickness and achieve successfully. Technique of the blade hot mould pressing used now is adopted in the special software to expand the blade approximately, and then to simulate the deformation rule of blade material under pressure field and temperature field by finite element method, to calculate the pressure centre and pressure tonnage, to design the molds and positioning correctly and determine the pressing speed and cooling time.

(2) The techniques of runner static balance: The traditional technique of runner static balance adopts steel ball and runner plate balancing device, however, now the two methods of the hydraulic ball bearing balancing and measuring bar balancing are adopted in large runner balancing. The operating principle of the hydraulic ball bearing balancing is that the non-balancing location and weight is computer calculated by passing a coat of high-pressure oil film through a set of ball bearing friction pair with high precision and with computer synchronous control and electric measurement signal feedback. The balancing precision is far higher than that of traditional steel ball and runner plate structure. The operating principle of measuring bar balancing technique uses the strain caused by the main gauge installed on the measuring bar and the balancing location and weight are calculated by computer. The cost of this technique is lower than that of hydraulic ball bearing balancing.

(3) Integral runner field manufacturing technique: Due to the limitation of transportation facilities, the runner is manufactured by segments in the factory and assembled together in the field. With the hydraulic design and the optimization of rigidity analysis of the runner, the structure of large runner is thinner and lighter than that of traditional runner. And the arising of the requirement for high performance and stability, the integral runner needs to be made. In the recent ten years, the techniques, such as field welding, hoisting and turning over, heat treatment, mechanical machining and balancing of the runner, have been mastered in China. The Yantan runner ($D = 8.0$m) had realized the field integrally manufacture in 2003. This technique has been applied to 7 runner of Longtan Hydro Station including crown, band, blade transported to working site and group welding, finishing machining, balancing and entire annealing on construction site.

(4) The melting technique of extra-low-carbon stainless steel: The melting technique of extra-low-carbon stainless steel is

closely related to the quality of runner cast and new product. A large number of researches on runner material were carried out. With the increasing of the size of the runner, the welding quality of runner blade, sand-erosion resistance and anti-cracking ability have high requirements. Based on the progress of melting technique, the refining technique of extra-low-carbon of AOD/VOD was extensively applied in the refining of the stainless steel. The minimum carbon content may be controlled under 0.03% and carbon content of S can be controlled under 0.015% and less. The liquid steel possesses high cleanliness, low air content and less non-metal inclusion. The stable feature of the alloy leads to stabilization of mechanical feature of heat treatment of product at last. The surface quality and dimensional accuracy are raised by leaps and bounds and the cost is decreased. Synthetically mechanical property of stainless steel and the corrosion resistance are all increased, particularly, the welding feature is visibly improved.

(5) The forming techniques of the resin-rich insulation and VPI forming techniques: the forming techniques of the resin-rich insulation and VPI forming techniques are two types of insulating craftworks and are usually applied to the main insulation of high voltage stator bars in the world. The thermo compression forming technique of the resin-rich insulation is one innovation in China. The prpcess is as follows: The resin-rich mica tape is consecutively wound on windings and wound anticorona tape with outsourcing, and then heated, mold pressed and solidified "once shaping". The dimension of bars is consistent and has good interchangeability. The VPI technique is consecutively wound on windings (bars) with dry mica tape, and then the bars are tied up on framework and put into a pressure tank, vacuum-impregnated, curing and forming with pressure and heating. The thermo compression forming technique of resin-rich insulation is adopted mainly in domestic.

20.1.2.3 Advanced Installation Technology

Back in the early 1950s, with the commencement of the postwar restoration of the Fengman Hydro-Power Station in Northeast China, the Fengman E&M Installation Company was established, who installed the 72.5MW generator set made in the former Soviet Union, marking the beginning of China's own E&M installation industry. Half a century since then and specially in the last twenty years, China's E&M installation industry has covered a course of development and grown from scratch into being, small to big and weak to strong. The total installed capacity in China have reached 200 GW, 97% were installed by China's specialist hydro-power E&M installation companies. With special technologies and expertise, Installation Companies have managed to be self-dependent and innovative, achieved excellent performance and technical advancement of far reaching influence, and obtained world leading construction technologies and installation techniques on the projects, such as Lijiaxia, Longyangxia, Ertan, Xiaolangdi, Three Gorges, Longtan, Shisanling, Guangzhou Pumped Storage Power Station, Tianhuangping Pumped Storage Power Station, etc. China's hydro-power development has created a group of typical installation companies with strong comprehensive capacity, and this is witnessed by the successful installation and subsequent operation of the aforesaid 80 GW large and mid-large generator sets. Chinese professionals of E&M installation have mastered the technology and expertise for installation, adjustment and commissioning of complete sets of hydro-power E&M equipment as summarized below.

(1) Since the 1980s, the technology and workmanship to assemble in-situ of whole lamination of generator stators, to insert in-situ all coils and to lift and install the whole stators have made the assembly and installation of stators the same as those of rotors in large and mid-large turbine generators, i.e., to assemble and install in their entirety on site, thus to improve the technology of manufacturing, transporting, assembling and installing stators.

(2) The use of disc holders (exceeding 6m in diameter) in lieu of supporting arms (either of I shape or box shape) to hold large turbine generators, and the lieu of welding instead of the traditional bolt connection for site assembly have brought fundamental changes to structure and assembling technology, which enhance both dial and axial strength of the rotors.

(3) New material, new structure and new technology have been introduced to the manufacturing of thrust bearing tiles and guide bearing tiles by substituting elastic metal plastic for pure babbitt metal as the surface material of tiles. The applied technology and associated installation adjustment ensure an optimal superposition between the deformation of thrust tiles against hydrodynamic oil film and that due to their own mechanical reasons and heat deformation, as builds an ideal working tile surface for hydrodynamic sliding bearings and a fundamental solution to the working reliability of large thrust bearings.

(4) Now, the industrialized manufacturing method to the large and extra-large embedded parts of turbines, such as draft tube, base ring, spiral case, pit liner, etc., which enables the site cutting and rolling, has provided a solution to the dual problems puzzling manufacturers of these embedded parts with respect to their manufacturing cost and delivering difficulty.

(5) The technology advancement from delivering the runner of large Francis type turbines in halves to site for assembling and welding to assembling in-site the whole runner from spare parts by using gang welding, along with the increasing improvement of site assembly workshop of the large runner crowns in halves, has a practical significance to large-scale hydro-power development in Southwest China.

(6) New internal cooling technology of hydro-generator is applied in China. The assembling workmanship and adjustment test of the evaporation cooled stator has succeeded first in China.

(7) The welding technology of high-strength steel of 588MPa (60kg/mm^2) and 784MPa (80kg/mm^2) and stress-relieving technology are used for the manufacture of penstock pipes, high-pressure bifurcated steel pipes and spiral cases.

(8) The technology to install and commission intellectualized and electromechanically integrated control equipment for speed regulator and excitation system has made the unit operation mode and transient quality totally conform to the requirements of the stability of hydro-power equipment in modern electric power system.

(9) The technology to install and commission the sealed combination electrical equipment (GIS) of high voltage (330kV, 500kV) and the large capacity three-phase compound transformer has been achieved.

(10) The installation and commission technology of the computer monitoring and control system in hydro-power station and the status monitoring and fault diagnosis of hydro-power unit have been achieved.

(11) The technology to start up and test reversible pumped storage units along with the technical definition on the startup, testing and taking over of the units clearly set up the testing items and commissioning schedule prior to the units being put into commercial operation for the first time.

(12) Based on mature construction experience, the establishment of various E&M installation norms provides, for the time being in the world, the richest, most comprehensive and appropriate guidelines and specifications, which have become the most important technologyand quality guarantee for China's E&M installation.

20.1.3　*Examples for Hydro-turbine Generator Unit*

20.1.3.1　Three Gorges Hydraulic-turbine Generator Unit

In Three Gorges Project, the installed gross capacity is 22,500MW, 32 units, unit capacity 700MW and 2 units, unit capacity 50MW. The main parameters of the hydraulic turbine are rated head 80.6m/85.0m, maximum head 113.0m, minimum head 61.0m, speed 75.0/71.4r/min, runner diameter 10.43m/9.88m. The hydro-generator unit is made up of penstock, stay ring, spiral case, distributor, runner, top cover, draft tube, main shaft, upper and lower bracket, rotor, stator, etc. The unit structure adopts semi-umbrella type with three bearings. The load of thrust bearing is 5050/5520t. The main parameters of the generator are rated power 778MVA, power factor 0.9, maximum output of generator 840MVA, rated voltage 20kV, stator with water cooling, rotor with air-cooling. The outer diameter of stator frame is 21.42/20.9m and the inside diameter of stator core is 18.5/18.8m, the height of stator core is 3.13/2.95m. The unit weight is about 7000t.

Thirty-two hydro-generator units are designed to be installed in TGP, in which there are fourteen units in Left Bank Power Station, the design and manufacture are relying mainly upon foreign enterprises and domestic enterprises participate subcontract manufacture. The import substitution rate is about 50%. Eight units in twelve units of Right Bank Power Station and four in six units of undergrounded hydroelectric power plant are domestic units to have self-owned intellectual property rights.

1. Penstock

There is one extra large penstock with diameter of 12.8m poured in concrete dam from the water inlet of the dam to the front of spiral case in each unit. It is made up of seventy-two pipe sections and divided into four parts: upper inclined straight section, lower bend, lower horizontal section and taper pipe transition region. It was manufactured with wall thickness from 26mm to 60mm and high intensity 60MPa steel plate by roller bending machine, and the weight of the simple section was about 20t to 50t.

2. Spiral case and stay ring

The inlet diameter of spiral case is 12.4m and the section dimension of radius is gradually reduced from inlet 6.2m to 2.1m. The dimension and weight of spiral case are all the top level at home. The spiral case made in VGS is made up of thirty-three sections and the total weight is 690 t and the spiral case made in Alstom is composed with thirty sections and the total weight is 739t. It was rolled with wall thickness from 24mm to 120mm and high intensity 60MPa steel plate. The 3mm corrosion allowance was considered in design. The length of welding line above second-class is over 12.6km and the consumed welding electrode is over 185 tons in the unit of the Left Bank Power Station.

The stay ring is located in the inside of the spiral case and is a flat plate welding structure. It is made up of upper and lower ring plates, stay vane, guide plate, transition plate and tongue plate. The stay ring supplied by VGS is divided into six petals for transportation convenience and the maximum weight of simple section is about 70.5 tons, and the total weight is 382 tons. The outer diameter after installation is about 14.5m and the height is 4.625m. The weight of the stay ring supplied by Alstom is 345.5 tons and the maximum weight of simple section is about 65 tons, the outer diameter after installation is about 15m. The stay ring is divided into six petals to be made and annealed in factory and installed in the field. At first, the pieces are screwed with prestress bolt in the turbine pit, then welded as the whole, and at last inspected by Magnetic Particle Test (MT) and Ultrasonic Test (UT). The requirement of dimension accuracy is high and the manufacture and installation are very difficult.

3. Distributor

The distributor is made up of bottom ring, top cover, twenty-four pieces of guide vane, control ring and operating mechanism, etc. and composed of more than one thousand parts and the total weight is nearly 1000 tons. The diameter of bottom ring is 11.6m, the height is 0.7m and the weight is 112 tons. The diameter of the top cover is 13.29m, the height is 2.275m and the weight is about 380t. The bottom and the top covers are all divided into four pieces to be made and the stress relief treatment is taken in factory. The height of the guide vane is 3m and the weight is about 11 tons for each piece.

The electroslag cast technique is adopted for guide vane cast by Shenyang Research Institute of Foundry. The electroslag remelting casting technique is combined with electroslag remelting and cast solidification molding. The key technical problems, such as material, craftwork, tools and equipments, have been resolved in the guide vane electroslag casting. The guide vane casts for ten units of TGP hydraulic turbine have been provided for HBC and DEFM. The product cost is decreased by 30% and the life is increased by 30%. It alters the situation of import-dependence as this high-end component.

4. Runner and manufacture

The runner is made up of crown, band, blades, crown seal, band seal and cone. The main factors of runner supplied by Alstom are 10.427m in diameter, 5.08m in height, 9.8m in throat diameter, 15 blades, 450 tons in net weight. It is the largest cast steel welding part of hydro-power in the world now. The weight of the crown is 112tons, the weight of band is 58 tons, the weight of each blade is 17.49 tons and the dimension is $4,537mm \times 4,951mm \times 2,300mm$. The blade cast adopts VOD liquid steel refined and the inspection class is CCH70-3. The dimension, weight, technology content and manufacture difficulty of the runner in TGP may be rated as the top of the similar products in the world today.

The structure of the runner supplied VGS is small different with that of Alstom. The runner is made of crown, band and thirteen pieces of blade and welded together. The main factors of the runner supplied by VGS are 10.007m in diameter, 5.565m in height, 9.4m in throat diameter, thirteen blades, 473 tons in net weight of the runner.

In the manufacture of the hydraulic turbine in Three Gorges Project, the parts of crown, band and blade have the highest technology content, the most manufacture difficulty and the longest manufacturing cycle. Because the runner material adopts ZG06Cr13Ni4Mo of martensitic stainless steel, and this type of material is very sensitive to the temperature, it is easy to crack at high and low temperature. When multi-clads of steel liquid are casted in the same time, the chemical components of each clad steel liquid must be uniform. In the solidification and heat treatment, the cast is easy to deform. The technical parameters and the kiln loading mode of heat treatment are not easy to be mastered. The blade is a cast of stainless steel plate shaped with 3-D twist variable cross-section structure. Its quality requirement is the highest and the manufacture difficulty is the largest. Its manufacture accuracy directly affects the unit hydraulic performance, such as efficiency, etc.

In the cooperation with foreign manufacturers, HBC invested more than 300 million Yuan to set up the Binhai large part processing factory at Huludao in Liaoning Province. The technique of "welding robot", large scale welding positioner, bell-type annealing furnace and 500-ton class static balance device of runner were adopted in runner manufacture and welding. In DEFM, the workshop carried out technical reformation. The pulse gas protection and multi-angle ultrasonic inspection were applied in runner manufacture. The primary qualified rate of inspection of runner reached 96% and the crack did not happen.

5. Main shaft

The main shaft of hydraulic turbine is the type of inner flange. The main shaft is 4.125m in diameter, 6.3m in length, 116mm in wall thickness and 117 tons in weight. The accuracy of main shaft axis is especial strict, and the vibration of the radial and end plane is less than 0.05mm. Its dimension and the quality requirements are on an unheard of scale in the history of hydro-power equipment manufacture in the world. HEC and DFEM imported a batch of world advanced equipments, such as the digital control vertical lathe ($\phi 20m$ to $\phi 22m$), digital control heavy horizontal lathe ($\phi 4.3m \times 18m$) and five axis digital control machining center, to meet the requirements.

6. The overview of the generator structure

The generators in TGP Left Bank were supplied by Alstom/ABB and VGS combo and in the Right Bank, the manufacture of generators realized to be made at home. The structure is semi-umbrella structure with three guide bearings and the thrust bearing is located on the lower bracket. It is made up of rotor, stator, upper bracket, lower bracket, thrust bearing, air-cooling and magneto, etc. The main factors are: rated capacity 777.8 MVA/700MW, maximum capacity 840MVA, rated voltage 20kV, rated current 22,453 A, rated speed 75 rpm, thrust bearing load 5520 tons, total weight 3343 tons and the cooling mode is water cooling in stator winding and air cooling in rotor. It is the largest hydro-generator in the world.

7. Rotor structure of generator

The rotor is made up of rotor center body, disk arm, magnetic yoke, magnetic pole, torsion block, upper and lower compression plate, permanent bolt, upper and lower back plates and braking vane. The rotor supplied by VGS is 18.43m in maximum diameter, 3.435m high and 1,694.5 tons in weight. And the rotor supplied by Alstom is 18.738m in maximum diameter, 3.639m high and 1,780 tons in weight, which is the heaviest generator in the world. The install loading with sling lifting is nearly 2,000 tons and two units of 1,200 tons bridge crane are used for the lifting or installation.

The magnetic yoke is the key part of rotor and is used to fix magnetic pole. It is installed by more than 13,500 pieces of magnetic yoke punching and its weight is more than 1,300 tons. Each group of magnetic yoke punching has fifty holes arranged uniformly. The error of the distance between holes is 0.05mm. The magnetic yoke punching was punched by punching machine before, the steel plate edge is easily to produce rough edge and in the installation, the magnetic yoke punching is unfairness. In the 1990s, the laser cutting machine came into use for the manufacture of the magnetic yoke punching. In 2002, DFEM brought off the manufacture of the magnetic yoke punching with the laser cutting machine made by Wuhan Huagong Tech. The rotor has eighty magnetic poles and the weight of each pole is 5,467kg.

8. Stator structure of generator

The stator is made up of frame, core and winding, etc. The outer diameter of the frame is 21.42/20.9m, the core is 18.5/18.8m in inner diameter, 3.13/2.95m high and 326.4 tons in weight. The stator current is 22,453 A. The winding is the important and key part to determine the life of the generator. The winding will be taken the comprehensive affection of magnetic force, heat effect and mechanical stress in the operation, starting or closing the machine, particular in the condition of short circuit the vibration and impacting deformation may happen. The cooling mode is water cooling in stator and the air cooling in rotor of the generator supplied by VGS and Alstom. Each unit has 1,020 bars. The dry mica tape vacuum impregnation process is adopted in the main insulator of the bars. By consensus, HEC and DFEM are allowed to adopt the molding process of the resin-rich insulation and successfully completed the manufacture of stator winding, which the working field intensity is 2.51kV/mm.

9. Cooling technique of stator bars

The cooling of bars is one of the key techniques of the large generator. At present, the cooling mode has mainly two types, i.e., semi water-cooling and full air-cooling. The semi water-cooling is that the stator winding is directly cooled by water and the core and the winding of magnetic pole are cooled by air. The full air-cooling is that the windings both stator and rotor are all cooled by air. The generators supplied by foreign manufactures in TGP Left Bank all adopted the mode of semi water-cooling.

To compare with the semi water-cooling, the full air-cooling is more safe and reliable. With the generator load increasing, the technical difficulty is greater and greater. HEC has set up 1 : 5 aeration test stands for complete air-cooling generator and simulation calculation of ventilation system is carried. By four years of research, the stator bars structure and the main insulator material for 20kV full air-cooling are developed and applied to four units at TGP Right Bank Power Station with self-owned intellectual property rights.

10. 6,000-ton class of the thrust bearing

The thrust bearing is one of the nucleus parts of the unit and is one of the most difficult parts of the unit manufacture. The axis load of the thrust bearing of Three Gorges unit is 5,520 tons and is the most in the world. The thrust bearing is arranged above the rotor. It is made up of thrust collar, thrust bearing runner, thrust sector, bearing block and oil groove, etc. the weight of thrust collar and thrust bearing runner is 68 tons. It needs to load 5,520 tons, including the weight of unit rotating parts 2,600 tons and the water thrust force 2,920 tons. The allowance error of the perpendicularity and smooth finish of thrust collar and thrust bearing runner is only 2mm.

The support of thrust bearing is made up of a set of small spiral springs in the units supplied by VGS and the support is made up of a set of cylindrical pins with different diameters in the units supplied by Alstom. Both the units supplied by VGS and Alstom, the bush material are all babbit metal and the outer circulation structure is adopted in immersion oil system. The diameter of the thrust bearing runner is 5.4m in VGS units and is 5.2m in Alstom units. In the beginning of the 1990s, HEC invested 50 million Yuan to build a 3,000-ton class test rig of thrust bearing and 6,000-ton class test rig of thrust bearing of elastic metal plastic bush. Based on the operation conditions of TGP, more than one thousand tests of thrust bearing performance were carried out and more than ten thousand data were obtained. In 1997, HEC was successful to finish the manufacture of the 6000-ton class thrust bearing of elastic metal plastic bush and to finish the full-scale simulation test. The elastic metal plastic bush thrust bearing made at home can meet the requirement of the operation condition of 500 times of starting and closing unit.

20. 1. 3. 2 Longtan Hydro-power Station Unit

Longtan Hydro-power Station will be constructed by two stages. The first stage is designed to have a normal storage level (NSL) of 375m, a minimum water head of 97m, a maximum water head of 154m and 7 hydraulic-turbine generator units (700MW each). The second stage will have a NSL of 400m, a minimum water head of 107m, a maximum water head of 179m and two more units. Through appraisal and approval of related Chinese authorities, DFEM & Voith Siemens have successfully won the package of manufacturing seven sets of turbines and their ancillaries and HEC & ALSTOM have successfully won the package of manufacturing seven sets of generators and their accessories.

1. *Structure characteristics*

(1) Structure type. The hydraulic-turbine generator unit has a semi-umbrella construction consisting of upper guide bearing, lower guide bearing and main guide bearing. The thrust bearing is arranged of the lower bracket. The turbine and generator have their own shafts. The rotation direction of the unit is counterclockwise viewing from top.

(2) Runner. According to the transportation condition and provisions of the bidding document, the field integrity delivery way will be adopted, i. e., the runner crown, runner band and runner blades will be transported to the project site separately and then assembled and welded in the field. The buyer shall provide the assembly and welding site and the seller's related professional will conduct the field assembly and welding. The runner has a diameter of 7.900m and a weight of 260 tons. DFEM has made a case study for the field assembly in his bid and optimization on the basis of the practical scheme of field assembly and welding of runner for Xiaolangdi Hydro-power Station.

(3) Cooling type for Generator. Through investigation, research and technical exchange, the manufacturer has confirmed that the full air-cooling of generator for Longtan Hydro-power Project is appropriate. After consultation, the experts also have agreed that the full air-cooling should be the primary option for the project. In case the full air-cooling can meet the normal operation of generator, the water-cooling for the stator shall not be adopted as far as possible. In addition, the experts have suggested that the manufacturer do simulating ventilation test to prove the rationality and correctness of the ventilation design. Therefore, the full air cooling is adopted for the project. The scale of ventilation model is 1 : 5.

(4) Thrust Bearing filled the Lower Bracket. The thrust bearing can be arranged on the head cover through bearing resisting plate or on the lower bracket of generator. The above two arrangements have well proved experiences at home and abroad and can ensure the safe and stable operation of unit. Generally speaking, the arrangement of thrust bearing on the head cover can reduce the height of the generator, so it is adopted for the large-sized unit installed in underground powerhouse. But for Longtan Project, the two arrangements have little impact on the height of the powerhouse under the condition of satisfying the arrangement of electro-mechanical equipment, the water pressure fluctuation at unit operation can not be presumed correctly due to great change of water head, violent variation of load and great difference of downstream water level. So, the impact of water pressure fluctuation on the thrust bearing can not be predicted. In order to ensure the unit stability, it is better not to adopt the arrangement of thrust bearing on the head cover. In addition, since the project has the features of many sets of units and difficulty in design and manufacture and also based on the project schedule's of putting three units into operation in a year, the arrangement of erecting the thrust bearing on the lower bracket is favorable for reducing the co-operation trouble of turbine and generator and facilitating award the package of turbine and generator separately. Through case study and experts'consultation, it is finally determined to accept the arrangement of erecting the thrust bearing of the lower bracket. The thrust bearing is designed by Alstom and the double-layer tungalloy shoe is supported by small spring pillow. The cooling means of self-circulation with pocket plus outside circulation pump is employed and the thrust load is 3,100 tons.

(5) Stator housing shall satisfy the lifting requirement of stator with windings. The project has an underground powerhouse with nine sets of units arranging on the Left Bank. In order to meet the requirement of putting three units into operation in a year based on the provisions of the bidding documents, the stator housing shall be designed with sufficient rigid strength to meet the needs of assembling in the erection bay or generator pit and endure the stress caused by lifting the stator with its windings from erection bay or one generator pit to the destination generator pit without any deformation.

The stator designed by HEC with a weigh of 760 tons, the outside diameter of stator housing of 17.43m and height of 5.80m is supported by inclined elements. The iron core's concentric expansion is allowable to avoid ellipsoidal deformation and twist as well as void ellipsoidal deformation and vibration caused by magnetic force. The stator housing is welded and assembled into one circular integrity by six splits after being transported to the site and the construction with big tooth ballast, which can increase the rigidity and quality of core assembly, shall be employed. The rigid strength calculation has been carried out for various conditions in the bid. The design of stator housing is considered to lift the integral stator with windings into the destination generator pit from erection bay or generator pit with the help of radial lifting beam and power house bridge crane.

2. Major parameters of turbine

(1) Selection principle of turbine parameters. The selection principle of turbine parameters of the project is to put the stable index of the turbine on the first priority. High weighted average efficiency will be tried to get on the premise of guaranteeing stability, especially the stability at high water head.

(2) Related height of guide vane. The relative height of guide vane is a characteristic dimension of water passage of turbine. It not only directly influences the passage shapes of spiral casing and runner as well as the change rule of corresponding passage area, but also greatly influences the hydraulic characteristics of runner. The selection of b_0 must follow the principles: ①meeting the digit and strength requirements of the components of passage and runner; ②reducing hydraulic loss as much as possible. According to the theoretical analysis and practical experience, the relative height of guide vane $b_0 = 0.23$ is employed.

(3) Specific speed and its factors. The specific speed is a basic characteristic parameter of turbine and it can reflect the comprehensive characteristics such as output, cavitation and efficiency as well as design and manufacture level of turbine. According to statistical analysis and in combination with the practical operation conditions of large-size Francis hydraulic-turbine generator unit at home and abroad and the actual conditions of Longtan Project, the specific speed factor $K = 2,225$ is adopted and the specific speed n_s is 188 m · kW accordingly.

(4) Unit discharge. The unit discharge of turbine is a major energy index. The increment of unit discharge can reduce the diameter of runner, weight of turbine and dimensions of powerhouse. But excessive unit discharge may result in some unfavorable phenomena, such as increment of flow velocity and cavitation factor in the water passage and decrement of cavitation performance. Through computation and overall analysis, the unit discharge of $0.77 m^3/s$ is adopted.

(5) Rated speed of unit. The speed of unit is the basis to determine other parameters of a power plant and is restricted by structural dimensions of powerhouse. It not only influences the parameters and performance of turbine, but also has close relation with slot current and number of branch circuits of stator windings, cooling means of stator winding, dimensions and weight of generator. Through computation, the optional speeds are 111.1r/min and 107.1r/min for the project. After extensive exchange with manufacturers at home and abroad, and analysis and study of CFD on turbine model made by related manufacturers and universities, the rated synchronous speed of the unit is finally determined to be 107.1r/min.

(6) Stability index. The investigation has indicated that although the huge-sized turbines with a runner's diameter over 6m and power of 248-620MW which have put into operation in recent years are designed by advanced software of hydraulic design and modern rigid strength design, most of them have been subjected to pressure fluctuation and vibration more or less and some cracks have occurred on the runner at the beginning of the operation. These problems undoubtedly have affected the economic profit of hydro-power station. Therefore the parameter selection of turbine for the project has always put stability first. The stability of turbine has been strengthened from parameter selection, hydraulic design and structural design and some practical measures have been taken to ensure good stability and reliability of huge turbine of Longtan Project. The final stability indexes are determined as follows.

- The pressure fluctuation limits of draft tube at various water heads and different powers are given in the Table 20.1-1.

Table 20.1-1 Pressure Fluctuation Limits of Draft Tube at Various Water Heads and Different Powers

Water head (m)	P (MW)	$\Delta H/H$ (%)
107-125	70%-100% of predicated power at various water heads	4
125-160	499-714	3
160-179	499-714	5
Other operation conditions		8

- The double amplitudes of vertical vibration and radial oscillation measured on the head cover of prototype turbine at any operation conditions in the specified operation range shall not exceed 0.12mm and 0.15mm respectively. The absolute double amplitudes of main shaft throw shall not exceed 0.25mm.

• Within the operation range of turbine, Karman vortices and the vortex in blade channels shall not be permitted and also the resonance is not allowed.

3. Major parameters of generator

(1) Power factor. The power factor of generator depends on the economic indexes of linking into power grid, operation mode, reactive power compensation, generator cost and operation. The determination of the value shall consider the influence to the cost and system stability and the demand of the reactive power compensation of power grid. The rated power factor of generator may be 0.9 – 0.95. In order to meet the future plan of the power grid and make the generator cost in a rational range, the power factor 0.9 is adopted through comparison.

(2) Rated voltage. The rated voltage of generator shall be determined by many factors, such as the whole scheme of generator, and be a comprehensive parameter. It directly influences the selection of the construction of generator, transformer, bus-bar of big current, voltage switchgear installation and has relation with generator capacity, cooling means, rational slot current and rated speed. The selection ranges of rated generator voltage of the project are 15.75kV, 18kV or 20kV. From the point of view of the generator itself, it has the characteristics that the lower rated voltage, the cheaper and the easier to ensure the insulation level. But in case that 15.75kV is adopted with a rated working current over 28 kA, it results in the difficulty in type selection of voltage switchgear installation and cost increment of electrical equipment. In case that 20kV is adopted with high requirements for the grade of insulation material and the corona-resistant measure, it results in great difficulty in unit manufacture, transportation and installation. From the consideration of economic and rational slot current and increment of spacer factor, the rated voltage of 18kV is comparatively rational. Therefore, the rated voltage is determined to be 18kV.

(3) Direct-axis transient reactance (X'_d). The generator's X'_d is mainly determined by leakage impedance of stator winding and field winding and the value of X'_d influences the stability of power system and the cost of generator. The less the X'_d is, the more the dynamic stability limit is and the less the change rate of transient voltage is. But in case X'_d is reduced, the diameter or length of stator core have to be increased, which results in the increment of generator dimension and weight, and cost. The generator cost has inverse relation with the square root of X'_d. In consideration with the decrement of cost, the X'_d shall be increased and also at the same time the operation demand of power system shall be met. Therefore, in consideration with the 500kV power grid condition and quick exciting measures taken for the units, the bigger value of X'_d can be adopted within the rational parameter range. After comprehensive consideration, X'_d finally is determined to be not more than 0.33.

(4) Vertical-axis sub-transient reactance X''_d. X''_d is mainly determined by leakage impedance of damping winding and also has the relation with the leakage impedance of stator windings and field windings. The value of X''_d mainly influences short circuit current and involves the selection of electrical equipment and design of earthing system. The more the X''_d is, the less the amplitude of short circuit impulse current is. In consideration with selection of electrical equipment and design of earthing system, X''_d shall be higher. But X''_d is determined by leakage impedance of damping winding. It is difficult to take it higher. In consideration with the generator manufacture, it is not good to adopt higher value. After comprehensive consideration, X''_d is finally determined to be not more than 0.24.

(5) Short-circuit ratio (SCR). SCR refers to the ratio of excitation current at rated voltage of no-load and excitation current when taking 3-phase steady state short circuits currents as rated value. SCR shall also be determined by the specific conditions of power system, besides the relation with generator design. Generally speaking, the higher SCR will increase the steady-state stability of generator in the power system operation, but make cost of generator increase. Through comprehensive consideration, it is finally determined that SCR is not less than 1.11.

(6) Flywheel moment (GD^2). The flywheel moment is a product of weight of rotating components of generator and square of its inertia diameter, directly influences rise of unit speed when suddenly rejecting load and pressure rise of power intake system at various working conditions, and transient stability of power system, and has close relation with the unit cost. Generally speaking, the bigger the flywheel moment is, the longer the inertia time constant has and mole profit the transient stability of power system owns. In addition, the big flywheel moment also increases the dimensions or weight of unit and unit cost. Through calculation of regulation guarantee done by university and in consideration with the stability requirement of power system and bidding condition, GD^2 is finally determined to be not less than 220,000t · m².

20.2 The Advanced Technology of Hydraulic Turbine Governor

20.2.1 *The Developing History of Hydraulic Turbine Governor*

The development of hydraulic turbine governor has undergone several stages with mechanical-hydraulic governor, radio tube electro-hydraulic governor, transistor electro-hydraulic governor, integrated circuit electro-hydraulic governor, and micro-governor in China.

1. Mechanical-hydraulic governor

The most of early produced mechanical hydraulic governor was modeled on foreign products as W-900, PO-40 and T-100 in China. In the 1960s – 1970s, the T-100, PO-40 and TT series mechanical hydraulic governor were developed and became main matching products of hydro-turbine generator unit.

2. Analog electro-hydraulic governor

In the beginning of the 1960s, China first DT-100 radio tube electro-hydraulic governor was made by Harbin Electric Machinery Works and put into operation at Liuqihe Power Station in Guangdong Province. At the end of the 1960s, China first BDT-100 semi conductor electro-hydraulic governor was developed by Tianjin Electrified Transmission Institute and Changjiang Water Resources Commission and operated at Lushui Hydro-power Station in Hubei Province. In the 1970s, China first JST-100 integrated circuit electro-hydraulic governor was designed by Changjiang Water Resources Commission and manufactured by Wuhan Thermal Meter works and operated at Hemianshi Power Station in Guangxi Province.

Hydraulic turbine governor was developed greatly in the 1970s – 1980s. The system structure was mostly definitely of buffered system structure and regulation rule was proportional-integral (PI) mode in mechanical-or electro-hydraulic governor. The YT series was the typical products of mechanical hydraulic governor and the JST-100 and YDT-1800A presented the national level of electro-hydraulic governor at that time.

3. Micro-electro-hydraulic governor

Micro-electro-hydraulic governor began to be developed and manufactured in the early of the 1980s and synchronized with the developed countries in the world. Huazhong University of Science & Technology had taken the lead in developing self-adaptive varying parameters micro-electro-hydraulic governor in China and the governor was put into operation at Ouyanghai Power Station in Hunan Province and later the dual micro governor was manufactured by cooperating with other units. In several years, there were a hundred of these type governors to be put into operation. Its function and feature can meet the requirements of computer control in hydro-power station. Its superiority was proved and welcomed by consumer. Micro governor fastly entered the broad market of technological transformation.

In the 1990s, the international famous brand industrial controllers were adopted in China. The STD-BUS industrial controller was adopted at the beginning and later the programmable logic controller (PLC), programmable computer controller (PCC) and industrial controller (IPC) were adopted.

4. Servo motor controlling governor

The research work of stepper motor and DC servo-motor began to be used as electrical-mechanical converter components in the 1980s. Because the electrical-mechanical converter component did not need oil and had the features of simple structure, good reliability, easy maintenance and control, the problems, such as the poor ability of anti-oil and high failure rate of electro-hydraulic conversion parts, were solved. In the 1990s, the governor controlled with servo-motor, including stepper motor, DC servo-motor and AC servo-motor, had become the leading product and the output accounted for 70%~80% of the gross production of large and middle scale governors.

20.2.2 Developing Situation of Hydraulic Turbine Governor

The feature and function of digital (micro) micro-electro-hydraulic governor keep the synchronous regime with the world advanced level. The technique, such as double-unit redundancy, double-unit crossing redundancy, adaptive varying parameter regulation, electro-hydraulic conversion part with servo-motor/stepper-motor, has Chinese characteristics and at a leading position in the international competition.

Adopting international famous brand industrial controller (IPC), programmable logic controller (PLC) and programmable computer controller (PCC) as hardcore of electric cabinet promotes the quality, process level and reliability of electric cabinet to the world advanced level.

All micro-governors adopt the electronic regulators with the mode of electric-hydraulic servo system structure. Technical specification reached the stipulation by the national governor standard and satisfied with the requirements of hydro-electricity industry. Besides the governor inputted with non-technique factor, all types of governor for Francis, Kaplan, tubular, Bellton, pump-storage and tide hydraulic turbine can be manufactured and supplied now in China. The homemade hydraulic turbine governor has been exported to abroad and participated in the international market competition.

In resent years, the industry standard hydraulic parts have been adopted in micro-governor, such as electro-hydraulic proportional valve and digital valve for electric-hydraulic converter component, cartridge valve for combination valve, step-closure device and oil pump shut-off valve, etc. It ends the situation of the regulation technique separated from the outside of the modern hydraulic technology and the backward situation of the new hydraulic technique application. In proportional-servo governor, the signal is controlled by flow and pressure. There is no drive of mechanical shift and the structure is simple, the redundancy and hydraulic integration can be easily realized. The application of this type governor increases greatly in large hydro-power station year by year.

The governors with stepper motor, AC servo-motor and DC servo-motor as electrical-mechanical converter components have formed a new kind of governor possessing proprietary intellectual property rights. This governor has high ability of anti-oil, simple structure and good reliability and is particularly suitable for the power station operation under the condition of the difficult guarantee of the oil quality and the low standard management.

The working oil pressure is generally increased to 4.0MPa and 6.3MPa in large hydraulic turbine. In some of medium and small hydraulic turbines, the bag type accumulator was used and the working oil pressure may be increased to 14–20MPa.

Micro-computer has powerful calculation ability, memory ability, logic judgment and communication function, so it has more advanced functions. At present, the functions of frequency tracking, electrical opening limit, artificial dead zone, fault diagnosis and treatment and communication with upper computer have been regulated to be realized. The functions of hand automatic switching with the un-condition and undisturbance, offline diagnostics, maintenance and computer-aided testing have been arranged in micro-governor. Besides the above mentioned, the functions of fault data recording, error proofing and fault-tolerant feature, dead zone and zero shift compensation are also realized.

In recent years, due to the improvement and perfection of electric-mechanical converter with AC servo-motor and stepper-motor, the recovery function is automatically realized after power failure. This type electric-mechanical converter may be connected directly with pilot valve in the electric-hydraulic service system. System structure is simplified and the reliability is increased. The jam and failure are impossible to happen in this type of electric-mechanical converter. Therefore, the mechanical opening limiter and its system of bars can be canceled and it has better static and dynamic performances. The simple degree of the structure is no less than that in the governor with proportional valve and hydraulic integrated structure.

20.2.3　*Examples for Hydraulic Turbine Governor*

20.2.3.1　WB（L）T Type of PCC Governor

WB (L) T type of PCC governors for Three Gorges and Yantan Hydro-power Station have been developed by Yichang Nengda Co. and put into operation favorably. The characteristics of system structure and hardware layout are as follows.

(1) Controller. The programmable compute control made in Austria B&R Corporation was used for the governor controller. 32-bit Microcontroller MC68 series 68332 control chip made by Motorola was adopted as its inner core.

(2) Electric-hydraulic converter. The double-electric-hydraulic converting unit of the governor was used by WB (L) of PCC governor, which is the two transforming widget with different work by proportional servo valve and stepper conversion device. When the governor operates at automatical mode, the main control pattern is the controller → proportional valve → main distributing valve → main servomotor. When operates in stand-by mode, the controller → stepper-servo system → main distributing valve→ main servomotor. In the condition of power failure or operating in manual model, the main distributing valve may return to the middle position automatically to stabilize the present working condition. At that time, the main distributing valve and main servomotor may be controlled with hand operation system. When the proportional servo-motor is working, the stepper servo-motor is at hot-backup status. At the time of switching into hand operation, the ball valve can quickly switch to stepper servo-motor system in 20 ms and realize the switching without disturbance. The valve core of main distributing valve adopts the mode of three valve discs of the same diameter. Upper chamber is controlling chamber and the lower chamber is constant pressure chamber. The area of thrust surface in upper chamber is two times area of thrust surface in lower chamber. When the upper chamber is opened to connect to pressure oil, the main valve core moves downward and when the upper chamber is opened to connect to discharge of oil, the main valve core moves up. When the upper chamber is closed, the main valve core is fixed steadily.

(3) Power supply. The power supply of governor in Three Gorges Power Station has three power supply ways, thereinto two AC power supplies and one DC power supply. The power supply of governor in Yantan Power Station adopted double power supplies, thereinto one AC and one DC. Each control system of governor used independent power supply without affecting each other.

(4) Frequency measurement. The four modes of the frequency measuring of the hydro-turbine governor are configured. One

mode is from electric network and other three modes are from the unit, including one from the unit potential transformer (PT) and two from the double-probe fluted disc. Frequency measurement circuits are mutual backup. The frequency from PT confirmed correctly by comparing with signal from fluted disc is supplied for governor using in normal operation. When the frequency measurement system from PT is failure or the compared result between the frequency from PT and from fluted disc goes beyond the scope, the signal from fluted disc is as signal of frequency measurement for governor using. The frequency measured from electric grid is also as standby of frequency measurement.

(5) Redundancy control. The governor in the Left Bank of Three Gorges Power Station adopts three sets of controllers, the two sets of NEYRPIC1500 controllers of them are used as redundancy control and control TR10 electric-hydraulic converter with the selecting switching mode. Another NEYRPIC1000 controller is designed to control ED12 electric-hydraulic servo-valve by power or hand control. The governor in the Right Bank of Three Gorges Power Station adopts three units crossing redundancy system and separately controls two sets of proportional valves. The dual-unit redundancy system control was adopted in the governor of Yantan Hydro-power Station. Each controller controls one convert module. It is equivalent to the operation of two governors for redundancy control. There is no switching plate in electric circuit. The control system is simple and the reliability is increased.

(6) The test results. The feature test of WB (L) type of PCC governor was carried out at the No. 4 unit of Yantan Power Station in November, 2006. The test results are shown in Table 20.2-1.

Table 20.2-1   Governor Feature Test Results

Item	A (Proportional servo-motor valve)	B (Stepper motor)
Dead space of rotation speed	$i_s=0.018\%$	$i_s=0.014\%$
Sewing value of rotation speed	$\Delta f \leqslant 0.13\%$	$\Delta f \leqslant 0.124\%$
Non-rotation time of servomotor	$t_q=0.11\mathrm{s}$	$t_q=0.15\mathrm{s}$
Regulating time when 100% of rated load being turned	$t_批 \leqslant 8.0\mathrm{s}$	$t_批 \leqslant 9.7\mathrm{s}$

20.2.3.2　GLT-100 Type Governor for Tubular Turbine

Tubular turbine can be divided into the four types, such as shaft-extension type tubular turbine, pit type tubular turbine, rim-generator turbine and bulb turbine. Among them, the bulb turbine is employed most extensively and is a model of good feature for large and middle tubular generator unit. Its characteristic is horizontal layout and the generator is arranged in the bulb body, which is made of metal under water. The generator of bulb turbine generator unit is arranged inside of bulb body. Due to the inference of water flow condition, the bulb ratio is limited and the dimension and the weight of the generator are also limited, and the moment of inertia and flywheel moment CD^2 are small. Therefore, the regulation rule of bulb turbine has unique characteristic compared with the common unit. At present, more types of governors of tubular unit have been manufactured domestically and the feature is at high level. For example, the governor is introduced as follows, which is developed by Wuhan Yangtze River Institute of Control Equipment and installed at Zhaoshandu Hydro-power Station in Zhejiang Province.

(1) Main features of the governor. The PID control technique of variable structure and parameter is adopted to guarantee steady operation of unit under different conditions. The application of the digital on-cam technique can make the unit operate in the area of high efficiency. The features of water flow and vibration of unit are improved and the surge control is realized. The use of the mechanical closing device in steps decreases the rising of water hammer pressure at the time of the fastly closing guide vane. The new heavy hammer closing control device based on cartridge valve as main part is employed for safely and reliably closing unit at the fault condition. The 6.3MPa oil pressure device is employed with three sets of oil pumps, particularly, one small oil pump is installed to add oil for normal consumption. The noise and fluctuation of service power can be decreased due to the frequent starting of large oil pump. Under the consideration of the characteristics and the demand of tubular generator unit, the electric cabinet is standard cabinet and arranged by the unit. The mechanical-hydraulic converter, oil pump and motor are located on the tank of back oil. The pressure oil tank and backoil tank are located separately at operation layer. The pipe connected between oil pressure tank and back oil tank, control pipe of governor, control cable and power cable are laid in pipe trench at operation layer. There is no need to layout the special pipe trench in power house, which not only saves civil engineering costs, but also the arrangement of main power house is neat and tidy, spacious for easily operation. After the tubular generator unit is connected grid, the water level control mode is put into operation, the reservoir water level is regulating target. Based on the given difference of control water head and practical head, the governor carries out the PI calculation for controlling the opening of guide vane and blade which can control the flow of unit and reach the goal of automatic regulation of reservoir water level.

(2) Governor structure. The governor is made up of electric cabinet, oil pressure device, closing device in steps and heavy hammer closing control device.

(3) Field test results. GLT-100 governor of the tubular generator unit operating in Zhaoshandu Power Station in Zhejiang Province was tested by the Centre of Hydro-electric Power Equipment Quality Test on 26 - 27, June 2001. The field test results are shown below.

Static characteristics and blade servo system: dead space of rotation speed of servomotor $i_x = 0.029\%$ is superior to the national standard 0.04%. The uncertainty of blade servo system is 0.35%, which is superior to the national standard 1.5%.

When 25% rated load being turned, the non-rotatable time of servomotor $T_q = 0.16s$ is superior to the national standard 2s.

When 100% rated load being turned, adjusting time 21s is superior to the national standard 40s. Wave crest of more than 3% of steady-state speed is 2 times, which meets the national standard.

Under the work condition of no-load and automatism the turn sewing of rotational speed is 0.113Hz which is superior to 0.15Hz of the national standard.

When AC and DC power are exchanged, the servomotor does not move. The result meets the national standard.

20.3 Advanced Excitation Technique of Hydraulic-turbine Generator

20.3.1 *Summarization of Excitation of Hydraulic-turbine Generator*

With the advanced progress both of control theory and electronic pieces from the 1950s, the achievements promoted the development of the excitation control technique of the hydraulic-turbine generator. The excitation device was developed from mechanical type to electromagnetic type or electronic type, from analogical mode to digital mode. The exciting mode was developed from exciter excitation or phase compound excitation to static excitation. To compare with other excitation modes, the static excitation system has more merits such as the simple main circuit, excellent regulating performance, high reliability and easy maintenance.

At present, new hydraulic-turbine generator unit is equipped the static excitation system without exception. The old hydraulic-turbine generator unit is gradually out of the exciter exciting or phase compound exciting and adopts the static excitation system. The static excitation system is a high-tech product with strong comprehensiveness. It integrates the software and hardware, integrated theory of electrical engineering, power electronic technology, control theory and system engineering and field bus, etc., advanced technologies. The new product has displayed that the highly interdisciplinary and interaction between the power technique and other subjects is more and more obvious.

The exciting system of hydraulic-turbine generator units basically is now used with self shunt excitation system at present and adopted thyristor rectifier technology. The exciting regulator is mainly based on analog system in small power stations and in large and middle power stations, the exciting regulator is mainly based on digital system. The research and development of the microcomputer excitation controller was carried out earlier. The first set of microcomputer excitation controller was put into operation in 1985. There are many factories to study this work. In the near twenty years, the great successes in microcomputer excitation controller have been achieved and the applications are more and more mature day by day.

In abroad, the microcomputer excitation controller entered the operation is in the 1980s also, TOSHIBA is in 1989, CGE is in May, 1990, ABB developed the UNITROL-P type of microcomputer excitation controller and other companies developed microcomputer excitation controller in succession. These large companies have great ability of research and development. The core of the computer system used high speed programmable controller or high speed microprocessor in general and the controller adopted the self-manufactured special control plate. The products have compact structure and high reliability. The UNITROL-P type regulator of ABB has been applied in Lijiaxia Power Station and the UNITROL-5000 type is operated in Longtan, Dachaoshan Power Station. The excitation controller in TGP was supplied by SIEMENS. The SILCO dual-channel microcomputer excitation controller of CGE was used in Geheyan Power Station. The excitation systems in the Ming Tombs and Tianhuangping Pump-storage Power Station are supplied by ELIN Company. These excitation controllers all adopted the PID+PSS controlling mode and the functions of controlling and limiting were perfect. The manufacture level of the device was high.

On the whole, the control algorithm of microcomputer excitation system in China is at the international forefront and the device function is also very strong, however, the reliability of the components and the manufacture technique has a definite difference.

20.3.2 *Example for Excitation of Hydraulic-turbine Generator*

EXC9000 type of self shunt excitation system adopted in Gongboxia Power Station is made up of excitation controller, power rectifier cabinet, de-excitation circuit, rectifier transformer, voltage transformer and current transformer. The working principle is that the excitation current is taken from generator and the DC excitation voltage is controlled by the phase of trigger pulse of thyristor. After the voltage is decreased by rectifier transformer, the current is rectified by full-bridge controlled rectifier.

The basic configuration is made up of one excitation cabinet, three excitation power cabinets, one incoming cubicle, one de-excitation switch cabinet, one nonlinearity resistor cubicle and one excitation transformer. The main features are:

(1) Full digital system. Function software and system digitization are main features of the system. The system digitization not only shows in regulator, but also in power cabinet and de-excitation cabinet. Each part of the excitation system can realize intelligent detection, intelligent display, intelligent control, intelligent information transmission and intelligent test.

(2) CAN bus technology is applied the interconnecting of excitation system. The application of CAN bus has realized interior information exchanging and control in excitation system. The communication has good real-time, high speed, simple hardware, wiring simplification and convenient maintenance. It has promoted the system technique and operation reliability.

(3) The friendly human-machine interface. Regulator cabinet, power cabinet and de-excitation cabinet are all installed with intelligent touch screen, full Chinese display, and inspecting and recording of operation status and parameters on real time. It can be operated to many types of system control and the fault can be warning and showing at real-time. It can realize the record of the large capacity fault and power-off.

(4) Perfect test software. The test software can modify the regulating parameters on line and carry out some function tests, record the test results automatically, configure the definition of output quantity of switch easily, regulate intelligently power cabinet and de-excitation cabinet.

(5) Excellent electromagnetic compatibility function. Based on the multi-interdiction measures used in components selecting, circuit design, technology and structure of system and giving full play to the functions of shielding, grounding, filtering and wiring in design, the regulator fault is significantly decreased. The electromagnetic compatibility test is passed through by the national laboratory.

(6) The special micro-computer three channels configuration. The regulator is made up of dual microcomputer digital voltage regulating channels and a manu-regulating channel, which forms the entirely independent three channels from the measuring circuit to the pulse output circuit and realizes mutual benefit between digital and analogical regulator and has greatly raised the reliability.

(7) Multi-CPU architecture mode. The excitation adopts collaborative work mode with multi-computer of 486+DSP+FPGA. The wafer of 486 is used to regulation calculation, logical process and communication. DSP is used to take samples and FPGA is used to form trigger pulse. The division of responsibilities of multi-CPU raises greatly the working speed and the data processing function of the regulator.

(8) AC sampling technology. High periodic sampling is realized by DSP. The timeliness and accuracy of sampling data are raised by moving window arithmetic. That is in favor of increasing response time and accuracy of adjustment and convenient to adjust PSS parameters exactly.

(9) Perfect fault inspecting system. The perfect fault inspecting system is entirely independent of the regulator. And the scope of the inspecting system includes the power, hardware and software. With the combination of self-inspection and mutual inspection, it is efficient to prevent the phenomena of sendinerror and omission of fault signals, guarantee the fault inspecting sufficiently and successfully switch the channel during fault.

(10) The residual voltage for excitation build-up. The triggering technology with high frequency pulse is used for the residual voltage for excitation build-up. When the residual voltage is lower, it can be guarantee the excitation only with smaller auxiliary current. The soft start technique can be adopted to realize the zero overcontrol of generator voltage during excitation build-up.

(11) Intelligent dynamic current equality technology. Standing on the method of the dynamic current equalizing, the current equalizing between cabinets and between phases can be auto-adjusted without any other measures as long cable, thyristor option, the current equality factor may reach 98%. When one or multi power cabinets are dropped out of operation, the op-

eration power cabinets can realize the dynamic current to be uniformity.

(12) High-voltage pulse transformer. The AC resistance voltage of pulse transformer is 20kV high and it can guarantee the efficient insulation between main circuit and control circuit.

(13) The powerful operation ability during the air blower drop out. Due to the reasonable design of the structure and selecting of air-cooling heat emitter, it is efficient to raise the output capacity of power cabinet during the air blower stopping. The power cabinets in parallel-running can be continuously running one hour at the rated operation condition of the unit and the air blower off.

(14) The special thyistor cross-border control. The thyistor cross-border control is used for over voltage protection of rotor and de-excitation circuit. There is no mechanical contact. The control is simple and reliable. The bilateral diode is used to set accurately the action value of overvoltage protection and does not need maintenance.

20.4 Computer Monitoring and Control System

20.4.1 Summarization of Computer Monitoring and Control System in China Hydro-power Station

Under the advocacy of the National Electrical Power Company, the automation of hydro-electric power station has experienced scientific research experimental stage, application popularization stage and extensive application popularization stage and got great success in China.

1. Scientific research experimental stage

The application of the hydro-power station comprehensive automation in China started in the beginning of the 1980s. At that time, the China Institute of Water Resources and Hydropower Research (IWHR), Nanjing Automatic Institute of Water Conservancy and Hydrology and Tianjin Institute of Electric Drive carried out separately preliminary study in Fuchunjiang, Gezhouba Erjiang and Yongdinghe Cascade Hydro-power Plants. The control system with multiple microprocessors developed successfully by IWHR was put into operation in December, 1984 and was rewarded the Third Class of National Science and Technology Progress Prize in 1986. By the preliminary study, it tried the superiority and feasibility of the application of the computer technique to the hydro-power station monitoring and control system. Some practical popularization modes were formed. It trained and exercised a crop of engineers and technicians for research and development, design, installation and maintenance and has accumulated valuable experience. But there are more discrepancy with foreign products in scale, function, processing, reliability and software.

2. Application popularization stage

The computer application planning of hydro-power plant automatization in the 7th Five-Year Plan was made in 1987, and the hydro-power application and popularization planning of computer monitoring and control system both in the 8th Five-Year Plan and in the year of 2000 was made in 1993 by the former of Ministry of Water Conservancy and Electric Power. It was planned sixty-seven large and middle hydro-power stations were planned to develop in the two plan periods. By 1993, twenty-seven power stations were adopted early or late the different types of computer monitoring and control system, such as Gezhouba Erjiang, Lubuge, Fuchunjiang, Danjiangkou, Xin'anjiang, Tongjiezi, Ankang, Shiquan, Longyangxia, Dongjiang, Baishan, etc. The standardization of software and hardware was got primary succession, the industrial production was formed basically and a few of the recommended types were developed. Some research institutes have indenpendently undertaken the manufacture of computer monitoring and control system. The level of science and technology was raised greatly and a large number of technology talents were grown sturdily.

Based on the suggestion of the meeting at Taipingwan charged by the former Ministry of Power Industry, "Some Regulations Concerning 'Nobody on Duty' (Few on Duty) of hydro-power station (try out)" was promulgated and carried in 1996 and at the same time "Notice on the Appraisal Standards of First-class Hydro-power Plant" was also promulgated by Ministry of Power Industry. The five hydro-stations as Gezhouba Erjiang, Taipingwan, etc. were confirmed to be the first batch of selected units for "Nobody on Duty" (Few on Duty) on the meeting of Taipingwan. In 1996, the selected units were extended to nine hydro-power stations as Baishan, Jinshuitan, Gongzui, etc for second batch of selected units. The features in the stage were as follows.

(1) The system has the features of completed functions, high standard degree of software and hardware, short development period and high successful rate and the system, such as Baishan Cascade Power Station remote centralized supervisory

system, Gezhouba and Jinshuitan monitoring and control system. The systems met the requirement of "Nobody on Duty" (Few on Duty).

(2) Research and development teams in national level formed and created the independent brand computer monitoring and control systems, such as HP9000 type computer monitoring and control system of IWHR and SSJ type of NARI. These two types of products basically occupied the domestic market and have certain popularity in the world.

3. General popularization and application stage

Entering the 21st century, the hydro-power construction appears good situation of vigorous development. The first unit in Three Gorges Project, which has attracted the world attention, was put into operation in July, 2003 and in September, 2005, 14 units in TGP Left Bank Power Station were all put into operation. The first unit in TGP Right Bank was operated in June, 2007. In a batch of huge hydro-power stations, such as Longtan, Xiaowan, Laxiwa, Goupitan and the power stations located at the upper stream of Jinsha River, was put into operation in succession. It indicated that the hydro-power construction in China had entered the new history stage. The computer monitoring and control systems of IWHR and NARI applied respectively in TGP Right Bank and Longtan hydro-Power Station.

20.4.2 Example for the Computer Monitoring and Control System of Hydro-power Station

20.4.2.1 The Computer Monitoring and Control System in Three Gorges Right Bank Power Station

The unit capacity is large and the unit is more in Three Gorges Power Station. It is in the nucleus position of power system of national cross-region interconnection and its importance is very outstanding. The Three Gorges Right Bank Power Station is an important part of Three Gorges Power Station. The control system not only has high reliability to guarantee the safe operation of the unit, but also has good expandability and maintainability to guarantee the smooth transition from construction period to operation period, because of the long construction space of times, and different construction progress of different manufacturers.

1. System configuration

Based on the practical situation and the principle of hierarchical distribution, the computer monitoring and control system of Three Gorges Right Bank Power Station adopted the configuration of hierarchical distribution open system with three networks and four layers and full system redundancy.

2. Hierarchical distribution

(1) The three nets include station control network, station management network and information publish network. The different kinds of information can be transmitted with different channels by hierarchical distribution configuration. It can avoid the interference and guarantee the real-time, safety and reliability.

The station control network adopts the Gigabit Backbone Industrial Ethernet switch of MACH3002 manufactured by HIRSCHMANN. It is connected with the equipments of local control layer and station control layer. The information related to the real-time supervisory and control in the field is mainly transmitted by station control network, such as ascending information and control order of LCU etc. The main switch ports are directly connected to each LCU with fiber. The LCU is as star distributed. Star network has high network speed and simple arrange of fiber. Dual-redundant fiber network has high reliability and can meet the requirements of real-time, reliability and fast maintenance of huge hydro-power station.

The station management network is made up of dual-redundant hot-backup switches. The backbone network switch uses the type of Cisco®Cata-lyst®4503 and is connected as star type with the equipments of station control layer and station management layer by fiber or twisted pair. The station management network transmits the information related with production management, particular history data, such as data processing information, historical data back-up, reports printing and so on.

The information publish network is mainly connected with the equipments of the information publish layer and adopts CISCO3550 100M type of network switch. The information publish layer is connected with the station management layer by network security devices.

In addition, based on the practical needs and the conditions of equipments, the field bus technology, such as Profibus-DP, S908, MB+, RS485, is adopted in LCU.

(2) Four layers include the local control layer, station layer, station management layer and information publish layer. The four layers possess their own local point and were coordinated to finish the full functions of the computer monitoring and control of hydro-power station.

The local control layer is made up of all the equipments of each local control unit to finish the local monitoring task. It is mainly composed of Unity Quantum programmerable logic controller of Schneider, industrial control computer, AC sample system and Transmitter. Industrial control computer may be as local monitoring window and also as local operating control desk. The two 586 CPU with 32 bits of Unity Quantum PLC are composed of hot-backup system and may be automatically switched with undisturbance. I/O may be hot plugging and the resolution of SOE module is 1 ms. The communication ability of MB+ is high. The temperature acquisition is carried out by RTD module of PLC. I/O terminal uses fast cable system.

The task of station control layer is the real-time information acquisition and processing, monitor and control from all equipments in power station and is composed of data acquisition server, operator station, application server, intra-plant communication server and dispatching gateway server. The hardware adopts Sun Fire440 server or Sun Blade2500 work station and installs the real time data base of H9000/RTDB with redundant distribution for the guarantee of system real-time.

The task of the station management layer is the information management and classification of all equipments operation. It is composed of the history data server, simulator training station, voice fault monitoring and report print server. The hardware adopts Sun Fire490 server, shares mdadm-raid or Sun Blade2500 working station and is installed the Oracle Relational database to form H9000/HistA history data management system.

The function of the information publish layer is information issue and query. It is composed of condition monitoring and trend analysis server, WEB browsing server, WEB browsing data server and browsing terminal and adopts the H9000/WOX system software of browsing and issuing with B/S structure.

3. Distribution

The functions of the control system distribute in different equipments of different layers. Based on the cooperation of the equipments, all functions are realized.

(1) Each LCU is configured according to the control target, such as unit LCU, switch station LCU, station power supply LCU and overall common device LCU. Each LCU completes the data acquisition, monitoring and control for its control target.

(2) The monitoring and control functions of main station distribute in the equipments of the station control layers and the station management layer. The data acquisition and processing server completes the task of data acquisition and processing. The data management server takes charge of the management of the real-time data base. The operator work station is in charge of the function of human-machine relationship of monitoring and control system. The historical data management server completes the data management and so on.

(3) Because there are more units in Three Gorges Right Bank Plant and the task of data acquisition and processing is great heavy, the computer monitoring and control system in Right Bank Plant is configured two sets of data acquisition and processing servers, named the Right 1 and Right 2. According to the requirements, it will install other data acquisition and processing server later.

(4) The overall plant real-time data base is distributed in computer node. Each local unit data base is arranged in each LCU. Each function is configured in each system node and each node implements the appointed task.

(5) The net work equipments are configured the station control net work, station management net work and information publish net work. It is for the equilibria of the network load and guarantee the real-time and reliability of the control.

(6) By the reasonable distribution of the functions, the load of each node in system meets the design requirement and any fault of part equipment can not affect the normal operation of other functions.

(7) The clock system adopts hierarchical distribution system. Each set of LCU is equipped with one set of the second class clock checked by main GPS satellite clock system.

4. Redundant measure

To insure the safe reliability operation of the control system, various effective redundant measures are adopted in the links of the control system.

(1) The hot-standby dual-redundant equipment is configured in each node equipment of main station, such as data acquisition server, data management server, operator workstation, history data management server, high application server, plant communication server, dispatch server and GPS clock. The double modular systems operate the same task at same time. The stand-by machine does not transmit any data in general. They check each other and are mutual standby with each

other. When the fault is detected in the main device, the standby device may be switched smoothly as main device operation.

(2) The dual network redundant structure is used in the station control network and the station management network. Two networks work at the same time and are mutual standby with each other.

(3) The redundant measure is adopted in various aspects of LCU, such as dual-CPU, dual field bus, dual sample power. The important signals of I/O also configure the redundant structure. The synchronizing device adopts the automatic quasi-synchronization and the manual synchronizing device is as standby.

(4) The main station power is configured hot-standby dual-device switched with no disturbance.

(5) Two sets of LCU in Right Bank switch station are connected by the field bus and they can realize the sharing and the exchanging of the information.

(6) The LCU controller is used in the CONVERT manufactured by POWER ONE Company. I/O working power may adopt AC220V or DC220V. When the I/O is far from the controller, each I/O independent working power is set. The power of all PLC is supplied by duplicate power module.

5. *The characteristics of the system configuration*

(1) The overall design of the system is completed one-time. The system is constructed by stage according to the field progress. It has obvious division surface with the follow-up engineering.

(2) High reliable redundant design: besides the I/O module of PLC, all important devices basically adopt redundant technology to ensure high reliability of the system, such as data server, operator working station, net equipment, each kind of power, CPU and power of PLC, bus and so on. All detections and switching of the redundant devices are completed by software or manual operation without hardware switching devices for decreasing new fault point of hardware and increasing the reliability of the system.

(3) Advanced net system design with high reliability: the schema of the system network adopts the mode that the station control network is separated with the management information. Important control device is connected with the control network and the management auxiliary equipment is connected with the information net. It avoids the influence of management information to the control network and ensures real-time, safety and reliability of the control system function. The backbone network of the station control network adopts dual redundant loop fiber ethernet structure with 1,000 MB. It avoids the shortcomes, such as more nodal points, long transmission time in single loop fiber ethernet structure and makes the system possess good reliability and real-time. Main equipments as all computer working stations, PLC in LCU are directly connected with the backbone network and it may increase the ability of communication and resource sharing.

(4) Hierarchical distribution system structure: The function distribution can guarantee that any equipment fault only affects the local function of the system. When the fault of main control class of equipment or other LCU is taken place, the LCU can be independently operated with no influence.

(5) Reasonable distribution of the system load: the system functions between main station and LCU, among every nodal points of main station and among every modules of LCU are allocated reasonably and the communication loads in different nets and nodes are also distributed reasonably.

(6) System adopts modular design and structured design and there leaves extended interface and capacity in hardware and software.

(7) The system has the ability of external communications and interface, the safety measures are confirmed with the newest security specifications related to the automatic system.

(8) The type of the equipments of the system hardware is as consistent as possible. It avoids the poor interchangeability and the difficulty of spare parts due to the more types of hardware.

20.4.2.2 NC2000 Computer Monitoring and Control System for Hydro-power Station

As a new software of computer monitoring and control system for large and middle hydro-power station, based on the summarizing of the years of the developing experience of computer control system and absorbing advanced foreign science and technology, NARI NC2000 adopts object-oriented programming cooperated with the new requirements of hydro-power station integrated automation, NARI NC2000 developed by Nanjing Automation Research Institute is a new computer monitoring and control system with distributed object technology and cross-platform character. NC2000 follows many good advantages of NARI Access and has high reliability, openness and extendibility. It provides more functions and applications and its consumer interface is more friendly.

1. Applicable scope

It can be widely used in large and middle hydro-power stations, pump-storage plants, cascade hydro-electric power stations and water pump stations.

2. Performance characteristics

(1) It has good openness, extendibility and heterogeneous platform adaptability. Its hardware may select the servers of HP, SUN, IBM, UNIX work station, PC work station, industrial controller and other high reliable PC. The operation system may select UNIX system or the heterogeneous system made up of UNIX, Linux and Windows.

(2) It has powerful configurator with object-oriented programming. Customers can easily complete each configurating function through configurator interface whether in UNIX system or in Windows system.

(3) Design of object-oriented database makes each application near the reality.

(4) The high efficient and convenient visualization tool with sequence control function makes the control be more reliable and easy.

(5) The object-oriented graphical interface makes the display be more plentiful and the interface is much friendly.

(6) Perfect warning notice of intelligence is more suitable for the condition of nobody on duty.

(7) The object-oriented history database is convenient for the sharing and application.

(8) Reporting system with high-quality pictures and written explanations has intuitive show and easy use.

(9) Abundant senior application software provides the widely special application.

(10) The perfect browsing function can be used as the same as operator workstation for inquiring about menu, report forms, schedule and curves.

3. Main functions

The computer monitoring and control system can control in real time, accurately and efficiently for the controlling objects. Its main functions include data acquisition and processing, operation monitoring and fault warning, control and regulation, automatic generating control, automatical voltage control, dispatcher's control for the provincial, cascade and net dispatching, record and statistical of operating parameters and production management, human-machine interface, data communication, the balance analysis of the generating quantity and transmitting and dissipation electricity, self-diagnosis system and redundancy switching, development and maintenance of other software, Web browsing in Internet/Intranet and remote diagnosis, etc.

20.5 Key Technology of Hydraulic Turbine Operation

After the hydraulic turbine is put into operation, the problems, such as vibration, cavitation and silt erosion, etc, would be occured. Many problems have been solved in a better way by relevant study, practice and processing.

20.5.1 *The Operation Stability of Hydraulic Turbine*

20.5.1.1 The Common Vibration in Hydro-generating Units

The common vibration is defined as the vibration caused by some unbalance forces. The vibration caused by vortex in draft-tube at the partial load in Francis turbine belongs to this kind of vibration.

(1) The vibration caused by the unbalance forces. The unbalance forces in hydro-power units include three kinds of imbalances, such as mechanical, electrical and hydraulic forces. The mechanical imbalance can be solved by matching the weights. Based on the reason analysis of the electrical and hydraulic imbalance forces, some measures can be adopted to solve or decrease these kinds of vibration. Experience shows that the method of solving the imbalance problem in vertical hydraulic generator unit is distinctly different from that of the horizontal unit. The vibration of the guide bearing or the throw of main shaft caused by the mechanical unbalance may not consist with the rule that centrifugal force is proportional to the power of the rotational speed. When multi-imbalance factors co-exist at the same time, to adopt a single processing method will not have a good result and the mutual effect of the three unbalance factors must be considered. Whether the vibration caused by single or three unbalance factors, the processing or correcting methods based on the gesture of the dynamic axis

line of the rotation part in vertical unit is effective and simple. The obvious result was got by this method in Yuzixi Power Station. The bow-shape involution of the unit rotor would greatly affect the correction of unbalance in vertical turbine unit with a high rotational speed, particularly when the speed is over 500r/min.

(2) The vibration caused by the vortex in draft tube. The vortex in draft tube is a common phenomenon of hydraulic unbalance when Francis turbine is operated at small guide opening. The pressure pulsation of the vortex in draft tube causes the vibration of unit and increases the throw of the main shaft. The phenomenon is significant in large size units such as in Liujiaxia, Geheyan and Wuqiangxi Power Stations. More model and prototype tests indicate that the evaluation of the vortex pressure pulsation should be based on the amplitude of the absolute pressure pulsation of prototype unit. The evaluation of the vortex pressure pulsation may be based on the maximum amplitude of peak to peak in the operation range or at the operation conditions of the design head. The amplitude of peak to peak may be measured at the condition of mixed frequencies or main frequency. The method of the value taken must be identical with the evaluation standard or stipulation in the contract. The amplitude of peak to peak of mixed frequencies is the superposed value including main frequency and other harmonic waves and does not include other components of non-vortex pulsation. The method of pushing air into the area of vortex is an effective measure. The guiding flow equipments fitting in draft tube may decrease the vortex pressure pulsation, but it may be easily broken by flow in large unit.

20.5.1.2 The Abnormal Vibration in Hydro-power Unit

The vibrations of resonance and self-excitation of mechanical parts caused by some factors belong to abnormal vibration in hydro-power units.

(1) The resonance of the shaft system. The resonances of the shaft system happened in Dahuofang Power Station in 1963 and in Gezhouba II Power Station in 1980. The tests showed that the resonance was excited by the several times of frequency of rotation speed. The results of prototype tests also showed the critical rotation speed was much smaller than the value of theory calculation, which was related to the clearance of the guide bearing. For example, the value of the critical rotation speed of theory calculation is 4.6Hz, but the resonance frequency is only 2.7Hz in Gezhouba II unit. It indicates the clearance of guide bearing significantly affects the critical rotation speed, and the reason is the clearance of guide bearing may not keep the fixed value. Meanwhile the force at the guide bearing is different with the supposed conditions in calculation. Therefore, the critical rotation speed is not a fixed value and varies in a wide range. So the calculation method of the critical rotation speed should be further studied for evaluating the effect degree of the clearance of the guide bearing on the critical rotation speed.

(2) The resonance of Karmen vortex of the blade. The Karmen vortex at blade outlet edge may excite the resonance of the blade and cause the cracks of the blades such as in Huangtankou, Hongmen Power Stations. Modifying the shape of the outlet of the blade or adding the piles between the blades can solve the problem.

(3) The self-excitation caused by the water leakage of the labyrinth. Violent vibration happened at No.4 unit in Yuzixi Power Station in 1973. By the field test and analysis, the reason of the vibration was the self-excitation caused by the water leakage of the labyrinth of the unit. The calculation method of the water flow in the labyrinth clearance was studied in IWHR and Dalian Science and Technology University and the suggestions for decreasing the self-exciting vibration were provided.

(4) The resonance of the water body in penstock. The resonances of the water body in penstock happened in succession in Liujiaxia, Danjiangkou, Gongzhui, Fongshuba Power Stations, etc. The powerful pressure pulsation caused violent vibration of the unit. For example, in Liujiaxia Power Station, the amplitude of pressure pulsation in peak to peak was 48m water column, which was about 50% of the operation head, and the vibration amplitude of the upper frame in peak to peak was above 1mm accompanied by strong noise. The rule of the resonance of the water body in penstock was mastered by field tests and theoretical analysis. It happens when the natural vibration frequency of water body in penstock is near or 2 times of the frequency of the rotation speed. The resonance may be solved by pushing the suitable quantity of air into the flow passage.

(5) The resonance of the local water body in turbine passage. The resonance of the local water body was found when analyzing the pulsation of the turbine of Yantan Power Station. There were two positions to occur two kinds of the resonance of the local water body. One position was at the loop between guide vane and the inlet of the runner, the vibration frequency was about 25Hz to 30Hz and the amplitude was about 30m (relative value was about 48.8%). Another position was in the draft tube, the vibration frequency was about 1.66Hz and 6.93Hz and the amplitude was about 27m (relative value was about 45.6%). By calculation and analysis of the resonance of the water body after guide vane was excited by Karmar vortex, the resonance of the water body in draft tube was excited by non-symmetry rotation flow.

The fact illustrates that due to the large dimension of the flow passage and the low natural frequency of the water body, the resonance may be easily excited by the pressure pulsation of low frequency. Therefore, it must be noted in the turbine design. Another opinion about the strong resonance of unit and workshop structure in Yantan Power Station is caused by the blade vortex.

20.5.1.3 The Vibration in Operation Conditions of Turbine

When the operation conditions of hydraulic turbine are far from the optimum operation condition, the vibration of the turbine will be caused by instability flow. This phenomenon is generally called the vibration of the operation condition, including low, high water head and transient operation condition. The vortex vibration at partial load mentioned above also belongs to this kind of vibration.

(1) The vibration at low water head. Generally speaking, the vibration would happen when the turbine operates at low head and far from optimum operation condition. But in many prototype and model tests, the serious vibrations did not happen at the low head operation condition, such as tests in Gangnan Power Station at 58.7% of the design head and in Sanmenxia Power Station at 56% of the design head.

(2) The vibration at high water head. The turbine vibration at high head is different from that at low head operation condition. Firstly, the intensity of vortex pulsation at high head is higher than that of low head. Secondly, the impact and flow separation at high water head are larger than that of low water head. Thirdly, the blade vortex at high head is stronger than that of low head. So it should pay more attention to the vibration at high water head.

(3) The vibration at transient operation condition. At the transient process of load rejection, the rate of speed ascent and the degree of the maximum speed near the critical speed are the most important factors to excite the vibration. In general, the rate of speed ascent is strictly limited in design and the over-speeding protection device is provided. So the maximum value of vibration at the condition of load rejection is often limited if the critical speed of the turbine is high enough.

20.5.1.4 The Breaking Problems of Blades and Other Parts of the Turbine

The breaking problem often appears at the parts of the turbine, such as the blade, draft tube liner, stay vane, releasing load plate and the top cover of the turbine, etc.

Plentiful observation analysis and processing experience have proved that more crack positions have the obvious characteristics of the fatigue failure. It also discovers that the breaking position of the parts appears some manufacture faults, such as casting faults, welding faults, stress concentration, residual stress, material and other structure faults. It shows that the vibration is only one of the factors inducing the breaking of the parts of the turbine. If the intensity of the structure is high enough and there is no fault in the parts, the breaking problem may not happen.

20.5.1.5 The Magnetic Vibration of Hydro-turbine Generator Unit

The magnetic vibration of hydro-turbine generator unit is induced by some electric-magnetic reasons of the stator iron core. In general, there are two kinds of the magnetic vibration of hydro-turbine generator unit. One is the rotation frequency vibration induced by electric-magnetic unbalance and the other is the pole frequency (100Hz) vibration induced by harmonic wave or negative-sequence current. The pole frequency vibration happens at some time. Cold state magnetic vibrations usually take place at the joints of petal stator where have some clearance. The pole frequency vibration is obvious during the iron core being in the state of resonance.

20.5.1.6 The Forecast and Prevention of the Vibration of Hydro-turbine Generator Unit

The key of vibration forecasting technique is to find out the calculation method of the dynamic correspondence and determine the exciting load. The study on this field is carried out now and has achieved certain results; however, it should need further study.

The prevention of the common vibration in hydro-power units is to raise the level of the structure design and improve the processing quality, to reduce various imbalance aroused by manufacture deviation and to improve the hydraulic design of hydraulic turbine so as to decrease the vortex pulsation pressure.

The key of prevention abnormal vibration is to master the inherent frequency of the vibration body and the frequency range of the excited vibration force, and let them not coincide with each other.

20.5.2 The Silt Erosion of Hydraulic Turbine

20.5.2.1 General Introduction

The exploitable hydro-power in China is about 380GW, but one of the features of the river is high silt content. There are 115 rivers transporting sand over 10 million tons, the amount of sand directly entered sea is up to 1.94 billion tons. The hydraulic turbine in hydro-power stations at the Yellow River and its tributaries, Jinshajiang, Dadu River, Mingjiang, Hong River, Yangtze River and its part reaches of the tributaries existed or will meet the problem of silt erosion.

According to primary statistics, 132 units of hydraulic turbine (about 12,000MW and not including TGP) met silt erosion in the large and middle hydro-power stations of China. The total installed capacity of middle and small hydro-power stations is about 22,000MW. 30% of the hydraulic turbines (about 6,600MW, 13,000 units) had silt erosion. In addition, 100,000 units of pump (about 3,400MW) also had silt erosion in different degrees at the station pumping water from Yellow River and this problem puzzled the safe operation in the Yellow River Division Project.

The effect of silt erosion of hydraulic machinery is to shorten the repairing period of unit and increase workload greatly. For example, in Sanmenxia Power Station, after 15,000 h of turbine operation, the unit overhaul must be arranged. Generally speaking, the overhaul period of hydraulic turbine operated in clean water is about five years, but the 4# hydraulic turbine in Sanmenxia Power Station only operated for two years. The flow passage was damaged fatigue failure and the efficiency decreased about 8.7%. In Gezhouba Power Station, the silt erosion rate at the blade outlet of 17# turbine is about 3.5mm/10,000 h and the erosion rate is about 4.3mm/10,000 h at the inlet of turbine blade. The water pumps at the both sides of banks of Yellow river operated 2000 h to 4000 h must be discarded as useless or replaced. The pump operated at high head was rejected only after 1,000 h operation.

20.5.2.2 Protection and Measure for Silt Erosion of Hydraulic Machinery

(1) To decrease the sand passing through the turbine as few as possible in the design of project and the reservoir operation: In the plan and design of hydro-project, it should fully be considered the condition of land configuration and the reasonable project layout, such as the intake elevation of the water passage and bottom outlet, etc. for decreasing the sand, particularly, the coarse sand, passing through the hydraulic turbine. At same time, by using the water and sediment regulation, storing clean water and releasing sand water, the sand content through the turbine may be decreased. For example of Sanmenxia Power Station after reconstruction, before the period of generating electricity with sand water, using the method of discharging flood or decreasing the level of reservoir and using the land configuration in front of the dam, and the configuration of river bed funnel near the dam, the velocity and sand content distribution in the stream line at the front of dam may be changed in favor of decreasing the sand content through the hydraulic turbine. In addition, according to the conditions of the unit operation and the coming water and sand, it tries to open the bottom outlet best to discharge the water for increasing the ratio of emission flow and decreasing the sand content through hydraulic turbine. The sand sluice bottom outlet has been generally arranged in hydraulic project at the more sand-carrying rivers.

(2) To choose reasonable turbine parameters for units and the relative flow control: The elevation of hydraulic turbine centerline is based on the operation requirement and determined by the cavitation characteristics of the turbine. It is not dependable to use the cavitation coefficient σ_m got by model test under the condition of clean water to calculate the elevation of hydraulic turbine centerline operating at sand-carrying River. The σ_m should be modified by $\sigma_{ms} = \sigma_m + \Delta\sigma_s$, $\Delta\sigma_s$ is the increment value affected by sand-content flow. Based on the research of Yellow River Institute of Hydraulic Research and Tianjin Investigation and Design Institute, the modified value of $\Delta\sigma_s$ is suggested to be equal to 0.1 when sand content $s \leqslant 12\text{kg/m}^3$ and 0.15 when sand content is $s > 12\text{kg/m}^3$. In general, the range of $\Delta\sigma_s$ is 0.1 to 0.15.

The reasonable parameters of hydraulic turbine and the control of the flow velocity through the passage of turbine are the key factors in silt-carrying water power stations. The flow velocity is a major contributing factor, it is necessary to decrease the level of the parameter of turbine in more silt-carrying water power station, particularly to control the flow velocity passing through hydraulic turbine. Based on the practice experience and test achievements, in selecting the parameters of the turbine in silt-carrying river, the parameters are controlled as follows: $V_m \leqslant 12\text{m/s}$ to 15m/s, $U_2 \leqslant 30\text{m/s}$, $W_2 \leqslant 32\text{m/s}$ in Kaplan and tubular, and $U_2 \leqslant 38\text{m/s}$ in Francis turbine. Based on the test results of A3 steel, 20SiMn, 0Cr13Ni4Mo, 0Cr13Ni6Mo, Tianjin Investigation and Design Institute suggested: U_2 does not exceed 38-40m/s when $s \leqslant 5\text{kg/m}^3$ and U_2 does not exceed 34m/s when $s \geqslant 12\text{kg/m}$. After the limit of velocity is determined, the velocity pass through hydraulic turbine may be selected according to the sand feature, the operating mode and scope of unit, and it can be as a constraint condition to determine the parameters of water turbine suitably.

(3) It is important to optimize the bulk material of hydraulic turbine. Generally speaking, the material of sand erosion resistance should have good combination as high toughness, high hardness, high stress, high fatigue limit, homogeneous

quality, fine grain size, compact structure, etc. and workability and weldability. The progress and application of the material of sand erosion resistance is achieved with the exploration and practice. In the primary period, the high hardness material of Cr5Cu and high fatigue limit of 0Cr18Ni9Ti stainless steel were adopted in 4# hydraulic turbine of Sanmenxia Power Station. However, the field results proved the sand erosion resistance of Cr5Cu was not better than that of epoxy silicon carbide and the resistance of cavitation erosion of 0Cr18Ni9Ti stainless steel was not good in sand water. Later in the 1970s, the high-intensity stainless steels, such as 0Cr13Ni4Mo, 0Cr13Ni6Mo and 0Cr13Ni5Mo, were adopted extensively. The sand erosion resistance of hydraulic turbine was enhanced to a certain extent. Recently the material of ZG06Cr16Ni5Mo (16-5 for short) was developed from the martensite depositing stainless steel (17-4PH). Its carbon content was lower and weldability was better. The material of 16-5 was applied to the high head turbine of Lubuge Power Station. After the operation of six flood seasons, the erosion of cavitation and sand was light. The operating result in Sanmenxia Power Station showed the sand erosion resistance of 16-5 cast steel was better than that of 13-6 steel.

(4) Coating protection with good materials in the position of sand erosion. The methods of the surface protection with metal are mainly built up welding, surface hardening, etc. The results of field and laboratory test show that the resistance of the sand erosion is proportional to the hardness of metal. In general, the welding electrode of high hardness, as named wearing No. 1 welding electrode, has good feature not only in sand erosion resistance, but also in cavitation erosion resistance. In recent years, the austenitic electrode of GB No. 1 was developed. Its effect is better and the price is lower. In recent years, the technology of spray welding is greatly applied in middle and small hydraulic machinery. The spray welding has been used to the turbine blades of Gezhouba, Sanmenxia Power Stations, etc. Due to the excellent binding power of the spray welding and the parent metal, the spray welding can bear not only sand erosion, but also cavitation erosion.

Due to the simple construction techniques and low price, nonmetal coating, such as epoxy diamond sand, polyesteramide, clad nylon, etc. has been applied extensively in China. These nonmetal coatings are all good abrasion resistance and applied successfully to the blade of the Kaplan turbine, such as the turbines in Sanmenxia, Qingtongxia, Tianqiao, Gezhouba, Power Stations, etc. Even if in the more sand-carrying of Sanmenxia Power Station, the turbine can be operated safely in several flood seasons. The insufficiency of nonmetal coating is that the binding power is not too strong and therefore the nonmetal coating would fall off in the position of cavitation as the back of blade and middle ring of runner chamber.

Supra polymer polyethylene began to apply in hydraulic machinery in recent years as in the inlet ring of pump and in wearing plate of the Francis turbine. Supra polymer polyethylene not only has good feature of abrasion resistance, but also has good feature of cavitation resistance.

(5) Reasonable operation and repairing. The hydraulic turbines are not allowed to operate with super load at more sand content condition in flood period. If the conditions permit, the turbine should avoid to be operated at the period of sand peak. The turbine with sand erosion must be repaired in time and enough spare parts must be prepared for replacing in the overhaul period. In repairing, the shape of water flow passage and the smooth finish of surface must be noted greatly. It is because of the efforts in all respects, the problem of unit sand erosion is greatly improved and the safe operation is guaranteed. But the sand erosion is a quite complicated problem that needs more efforts to further study.

20.5.3 *Hydro-generator Status Monitoring and Fault Diagnosis Technology*

Hydro-power unit is the key equipment in hydro-power plant. Its operation status directly affects the unit safe operation. With the increasing unit capacity and highly demanding for repair, maintenance, operating and management, the status monitoring and fault diagnosis of hydro-power unit play an important role in the guarantee of the unit safe operation for the large and medium sized hydroelectric generating sets. Meanwhile, with the development of monitoring technology and the improvement of management mode, the equipment maintenance mode is transiting to the condition based maintenance from the time based maintenance and the management mode is developing to the mode of "Nobody on Duty" (Few on Duty). Based on the status monitoring and fault diagnosis of hydro-power unit, the status maintenance has become the inevitable developing trend of the unit safe operation technology.

20.5.3.1 Research and Application of the Status Monitoring and Fault Diagnosis Technology

In nearly 10 years, the technology of status monitoring of hydro-power unit was studied by many manufactures and institutes at home and abroad. The products were developed and used in hydro-power plants. In the developed and applied system of monitoring the hydro-generating units, the analysis function is good, but the diagnosis function is weak. The domestic integral network system of hydro-power unit monitoring, diagnosis, management, maintenance of multi-area and multi-unit has not achieved by now. Generally speaking, the domestic system of the status monitoring and fault diagnosis has some features as follows.

(1) The development of the product of unit monitoring system has achieved some progresses. It has monitored the parame-

ters from the single monitoring of vibration and throw in the past to more parameters now, such as vibration, throw, pressure, temperature, air gap, magnetic field intensity, partial discharge, oil and air, etc.

(2) The monitoring technology of unit stability has derived to maturity stage. However, the air gap and partial discharge, etc. are just monitored preliminarily. So the monitoring technology needs to be improved also, such as how to discriminate the noise interference in partial discharge monitoring. The problems of the line-monitoring of low-frequency vibration, cavitation and cavitation erosion, the stress and crackle of key parts, rotor coil temperature are not perfectly solved by now and need to be further perfected.

(3) The technology of expert system and nerve network system is the hot point in the research of unit fault diagnosis in recent years, but the development is only stayed at the stage of method joining and arithmetic improving, and the practical product has not been achieved. The expert system is used relatively better than nerve network system. Some primary expert systems of unit fault diagnosis have been put into trial operation.

(4) The domestic integral network system of hydro-power unit monitoring, diagnosis, management, maintenance of multi-area and multi-unit has not achieved by now. The analysis function of monitoring system of hydro-power unit is not too strong. And the expert diagnosis system is primary and short of the support of profound theory and technique. The condition-based maintenance is almost blank.

20.5.3.2 The Status Monitoring Technology of Hydraulic Turbine

To guarantee high efficient and safe operation of power station, the hydraulic turbine status monitoring mainly includes two aspects:

(1) Stability monitoring. Include the status monitoring of shaft throw, unit structure vibration and hydraulic pressure pulsation. The on-line stability monitoring system of hydraulic turbine is relatively mature. The domestic system of TN8000, HJS and S8000 monitoring system, together with some foreign hydro-power unit systems, such as Swiss Vibro-Meter system, are mature and applied expansively.

(2) The status monitoring. Include the status monitoring of hydraulic turbine energy efficiency, cavitation and silt erosion, stress and crack of key parts of hydro turbine. Energy efficiency monitoring has been studied mainly on unit over flow, working head, active power, pressure of inlet and outlet section of spiral case, and servomotor stroke, etc. The energy character monitoring has realized automatic acquisition and the accuracy of the data is high in large and middle hydro-Power Station. The monitoring technology of cavitation, silt erosion, the stress and crackle of key parts was in blank and needs to be studied further.

20.5.3.3 The Status Monitoring Technology of Generator System

The main items of the status monitoring of generator include stator vibration, stator coil vibration, air-gap and magnetic field intensity, insulator and partial discharge parameters, temperature of stator bars and iron core, temperature of magnetic poles and rotor bars.

The vibration monitoring of stator and stator bars of generator and the stability monitoring of hydraulic turbine belong to the stability monitoring item of main unit. The measures and methods are the same and the technologies are relatively mature.

The monitoring of air-gap and magnetic field intensity, insulator and partial discharge of generator had been studied in domestic and oversea. The related products have applied in hydraulic-turbine generator unit in China, but the air-gap monitoring is only applied in a few power plants as TGP, Gezhouba, Longtan, etc. The domestic monitoring technique and application of air-gap are comparatively dropped behind. The interior signal of partial discharge of generator is transmitted to outside through two routes: the generator bus-bar and earthed neutral point. The signals from these two routes are affected by electric-magnetic interfere. If the problem of magnetic interfere is not solved, the monitoring of partial discharge will not get a good result. Because the rotor is a rotating part, the direct measure means of temperature monitoring of rotor is being studied now.

20.5.3.4 The Status Monitoring Technology of Bearing System

The status monitoring objects of the bearing system includes the temperature of thrust bearing, upper and lower guide bearings, turbine guide bearing, and the oil temperature in oil basin, cooling water discharge and the temperature at inlet and outlet, thrust load, oil level and mixing of oil and water etc. The main inspecting parameters mentioned above can be obtained from the monitoring system, but some inspecting parameters of the operation feature of the thrust bearing will need to be researched further.

20.5.3.5　The Status Monitoring Technology of Auxiliary Equipments

The objects of the status monitoring of auxiliary equipments include cooling system of generator, speed-governoring system, exciting system, main transformer and GIS and other auxiliary system. The main inspecting parameters include the air temperature, water temperature, water amount in unit cooling system; alarming and normal operation in speed governor, the voltage and current of exciting system; the oil temperature of main transformer oil tank, cooling water amount, water temperature, and water pressure of inlet and outlet, fault of main transformer cooler; SF6 density abnormal of phase A, B, C of GIS and density low of circuit breaker; and the water flow amount of main shaft seal, etc. These parameters can be taken from the computer monitoring and control system.

20.5.3.6　The Fault Diagnosis Technology of Hydro-power Unit

The study of the fault diagnosis technology of hydro-power unit was started later in China. In recent years, some progresses are achieved, but it still stays at the stage of theory study and has not successfully applied in project. The application effect of the diagnostic system of artificial intelligence is not ideal and the majority system can not arrive at the function target of the on-line automatic diagnosis.

References

[1] Sha Xilin. Achievement and Prospects of Hydro-power Equipment in China. *Water Power*, 1999, 10.
[2] Wang Chuming. Technical Achievement and Prospects in the 50 Years of China Hydrogenator, Proceedings of the 14th Symposium on Chinese Hydro-power Equipment, 2000.
[3] Pan Jiazheng, He Jing. *Large Dams in China—A Fifty-Year Review*. Beijing, China Water Power Press, 2000.
[4] Wang Zhigao. The Present Situation of Abrasion-cavitation of the Hydraulic Machinery and Advancement of Protection Measure. *Design of Water Resources & Hydroelectric Engineering*, 2002 (3).
[5] Zhong Liding. Development of Electomechanical Technology of Hydro-power Project in China, *Water Power*, 2006, 32 (1).
[6] Liang Weiyan. *Technological Progress on the Large-scale Hydro-power Unit in China*, The 1st International Conference on Hydro-power Technology & Key Equipment 2006, Beijing, China Electrical Power Press, 2006.
[7] Fu Yuanchu. *Hydro-power Electromechanical Installation In China: 50 Years of Development and Technical Innovation*, The 1st International Conference on Hydro-power Technology & Key Equipment 2006, Beijing, China Electrical Power Press, 2006.
[8] Wang Quanlong. *Advanced Technologies of Harbin Electric Machinery Co., Ltd. in the Field of Hydraulic turbines*, The 1st International Conference on Hydro-power Technology & Key Equipment 2006, China Electrical Power Press, 2006.
[9] Tao Xingming. *Development of Key Technology on Hydro-power Generating Unit*, The 1st International Conference on Hydro-power Technology & Key Equipment 2006, China Electrical Power Press, 2006.
[10] Du Jinchen. *Manufacturing Technique of Large Generating Unit*, The 1st International Conference on Hydro-power Technology & Key Equipment 2006, China Electrical Power Press, 2006.
[11] Fu Fengqi, Zhou Wenkai. New Machining Technology of Huge Hydro Turbine Runner. *Dongfang Electric Review*, 2007, 21 (2).
[12] Xu Lijia, He Yinzhi. Structure Characteristics and Performance Parameters of Hydro-generating Units for Longtan Hydro-power Station. *Water Power*, 2007, 33 (4).
[13] Zhu Chen, Shi Chong, Li Bin. Development of Computer Supervision and Control System for Giant Hydro-power Generators. *Hydro-power Automation and Dam Monitoring*, 2008, 32 (2).
[14] Li Youping, Yi Lin, Zheng Yumin, Huang Jian. Analysis of Current Condition-based Maintenance Technology for Domestic Hydro-generator Sets. *Hydro-power Automation and Dam Monitoring*, 2008, 32 (1).
[15] Liang Weiyan, Beng Fengshan, et al. China Electrical Engineering Canon, Volume 5 Hydraulic Power Engineering. Beijing: China Power Press, 2009.
[16] Shi Qinghua. Technical Advancement of Hydraulic Development for Hydro Turbines. *Dongfang Electric Review*, 2009, 23 (2).
[17] Gong Jingjiao. New Development of the Technology for Huge Francis Hydroelectric Generating Units. *Large Electric Machine and Hydraulic Turbine*, 2009 (4).
[18] The Maximum Water Turbine in the World-The Development of 700MW Hydro-power Generating Unit in Three Gorges Project (part 2, 2009.11.05), The List of Super Projects in China, http:/blog.sina.com.cn/Chinatop10/.
[19] Lu Juan, Pan Luoping, Gui Zhonghua, Zhou Ye. The Current Status of On-line State Monitoring and Fault Diagnosis Technologies for Hydrogenating Unit. *Large Electric Machine and Hydraulic Turbine*, 2010 (2).

Chapter 21

Pumped Storage Power Station

Qiu Binru[1], Lv Mingzhi[2], Jiang Zhongjian[3]

21.1 Development of Pumped Storage Power Station in China

The first pumped storage power station in the world was built in Switzerland in 1882, but the more development was started from the 1950s, most of them were built in the developed countries, such as Europe, US and Japan. Since the 1990s, the development of pumped storage power station has stressing on Asia area, especially China.

The development of pumped storage power station in China started late. The first 11MW pumped storage power unit in China was put into operation in 1968. Nevertheless, China's pumped storage power station has been developed rapidly. By the end of 2009, 24 pumped storage power stations (See Table 21.1-1) had been constructed and put into operation with the total installed capacity of 14,545MW, ranking the 3rd in the world after Japan and the United States. However, the installed capacity of pumped storage power station is only 1.66% of the total installed hydropower power capacity in China.

Table 21.1-1 Existing Pumped Storage Power Stations in China (by the end of 2009)

No.	Project name	Location	Development style	Total installed capacity (MW)	Number of unit	Unit capacity (MW)	Rated head (m)	Date of first unit put into operation
1	Gangnan	Hebei	Mixture type	11	1	11	47	May, 1968
2	Miyun	Beijing	Mixture type	22	2	11	70	Nov., 1973.
3	Panjiakou	Hebei	Mixture type	270	3	90	85	Sep., 1991
4	Cuntangkou	Sichuan	Pure pumped	2	2	1	31	Nov., 1992
5	Guangzhou	Guangdong	Pure pumped	2,400	8	300	512	Mar., 1994
6	Shisanling	Beijing	Pure pumped	800	4	200	430	Dec., 1995
7	Yamzhog Yumco	Xizang	Mixture type	90	4	22.5	816	May, 1997
8	Xikou	Zhejiang	Pure pumped	80	2	40	240	Dec., 1997
9	Tianhuangping	Zhejiang	Pure pumped	1,800	6	300	526	Sep., 1998
10	Xianghongdian	Anhui	Mixture type	80	2	40	45	Jan., 2000
11	Tiantang	Hubei	Pure pumped	70	2	35	43	Dec., 2000
12	Shahe	Jiangsu	Pure pumped	100	2	50	97.7	Jun., 2002
13	Huilong	Henan	Pure pumped	120	2	60	379	Sep., 2005
14	Baishan	Jilin	Mixture type	300	2	150	105.8	Nov., 2005
15	Tongbai	Zhejiang	Pure pumped	1,200	4	300	244	Dec., 2005
16	Tai'an	Shandong	Pure pumped	1,000	4	250	225	Jul., 2006
17	Langyashan	Anhui	Pure pumped	600	4	150	126	Sep., 2006
18	Yixing	Jiangsu	Pure pumped	1,000	4	250	363	May, 2008

---

[1] Qiu Binru, Former Chief Engineer of Hydrochina Beijing Engineering Corporation, Professor Senior Engineer.
[2] Lv Mingzhi, Chief Engineer of Hydrochina Beijing Engineering Corporation, Professor Senior Engineer.
[3] Jiang Zhongjian, Vice Chief Engineer of Hydrochina Huadong Engineering Corporation, Professor Senior Engineer.

Continued

No.	Project name	Location	Development style	Total installed capacity (MW)	Number of unit	Unit capacity (MW)	Rated head (m)	Date of first unit put into operation
19	Baoquan	Henan	Pure pumped	1,200	4	300	510	Aug., 2008
20	Zhanghewan	Hebei	Pure pumped	1,000	4	250	305	Dec., 2008
21	Xilongchi	Shanxi	Pure pumped	1,200	4	300	640	Dec., 2008
22	Huizhou	Guangdong	Pure pumped	2,400	8	300	517.4	May, 2009
23	Heimifeng	Hunan	Pure pumped	1,200	4	300	295	Jun., 2009
24	Bailianhe	Hubei	Pure pumped	1,200	4	300	195	Aug., 2009

Note: Minghu and Mingtan pumped storage power stations in Taiwan China have not been taken into account, and the installed capacities are 1,000MW and 1,620MW respectively.

The development process of China pumped storage power station can be generally divided into three stages (Fig. 21.1-1).

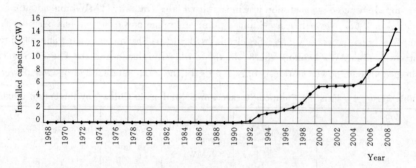

Fig. 21.1-1 Development Process of Pumped Storage Power Station in China

1. The starting stage

The first stage was from 1960s to 1970s, it can be called as the starting stage. In this stage, China began to learn and introduce the technology of foreign pumped storage power project, two small-sized pumped storage power stations, Gangnan (11MW) and Miyun (2×11MW), were built in 1968 and 1975 respectively, then, came a period of stagnating. The main reason was that at that time, the total economic development level was still low and there was not urgent demand for pumped storage station, and it was difficult to raise the fund for importing large pumped storage units and lack of understanding on pumped storage power stations' function and benefit.

2. The first height of construction stage

The second stage from 1980s to 1990s was the first upsurge stage for the construction of pumped storage power stations.

After 1978, China has implemented the policy of reform and opening to the outside world, national economy has developed rapidly with an average GDP growth of 9.8% in thirty years, therefore, the need for power has increased rapidly and the difference between system peak load and low load became bigger and bigger. The contradiction between peak supply and demand became outstanding increasingly in the power grids which are dominant by thermal power generation, such as North China power grid and East China power grid. However, hydropower resource available for peaking was quite rare, the consensus that building pumped storage power station to meet peak demand has gradually been formed. In addition, in later 1980s, Guangdong Dayawan nuclear power station and East China Qinshan nuclear power station began to be constructed, which also need the cooperation with pumped storage power stations to ensure their safe and economic operation. All the above reasons make China pumped storage power stations increased rapidly.

Panjiakou mixture type pumped storage power station (270MW) in Hebei Province was constructed in 1984 and put into operation in 1991. During the 1990s, there were 9 pumped storage power stations successively been constructed, including Guangzhou (2,400MW) in Guangdong Province, Shisanling (800MW) in Beijing and Tianhuangping (1,800MW) in Zhejiang Province, etc. By the end of 2000, the total installed capacity of pumped storage power stations has reached 5,590MW.

3. The second height of construction stage

Before 2003, only Shahe pumped storage power station (100MW) was constructed and put into operation, which was a hard time for construction of pumped storage power station.

In 2003, China's GDP per capita exceeded 1,000 US dollars. From 2003 to 2007, the GDP kept growing rapidly during the consecutive 5 years, so did the power consumption. In 2004, 26 provincial power grids had to implement power-cut because of the expanding gap between peak and valley power loads. Moreover, the first batch of pumped storage power stations put into operation had played a good role in power grids. The necessity and advantages of pumped storage power station had been recognized. Therefore, a new construction upsurge of pumped storage power station had begun. From 2005 to 2009, 12 large and medium-size pumped storage power stations were constructed and put into operation with a total installed capacity of 14,545MW.

By the end of 2009, 9 pumped storage power stations were put into construction, with a total installed capacity of 10,240MW, see Table 21.1-2. It is estimated that all these power stations will be put into operation around 2015, then, there will be 33 pumped storage power stations in mainland of China, in which 22 of them are large-size ones (≥300MW). The total installed capacity will be 28,385MW, in which the large-sizepower stations take 96.5%, with an installed capacity of 27,380MW.

Table 21.1-2 Pumped Storage Power Station Under Construction in China (by the end of 2009)

No.	Project name	Location	Development style	Total installed capacity(MW)	Number of units	Unit capacity (MW)	Rated head (m)
1	Pushihe	Liaoning	Pure Pumped	1,200	4	300	308
2	Xiangshuijian	Anhui	Pure Pumped	1,000	4	250	190
3	Foziling	Anhui	Mixture type	160	2	80	54.2
4	Qingyuan	Guangdong	Pure Pumped	1,280	4	320	470
5	Huhhot	Inner Mongolia	Pure Pumped	1,200	4	300	521
6	Xianyou	Fujian	Pure Pumped	1,200	4	300	430
7	Xianju	Zhejiang	Pure Pumped	1,500	4	375	437
8	Hongping	Jiangxi	Pure Pumped	1,200	4	300	540
9	Liyang	Jiangsu	Pure Pumped	1,500	6	250	259

From the regional distribution, it can be seen that pumped storage power station construction has already been extended to central area from eastern of China. In 2000, only five provinces and cities possessed pumped storage power stations and all of them were located in the most economically developed eastern regions. In the second upsurge stage, the pumped storage power stations were constructed in 16 provinces and cities in eastern and central China.

Viewing from the constitution features of power grids, the first batch of pumped storage power stations were all built in the thermal power dominated grids and the region where nuclear station located. The second batch of pumped storage power stations have already been extended to the regions with rich hydropower resources, such as central China and Fujian province.

Though the construction of large-size pumped storage power stations started late, the starting point was high, and most of them were large-size ones with an installed capacity more than 1,000MW.

The development process for rated head of pump-turbine unitsare shown in Fig. 21.1-2, the rated heads above 300m take 68%, the rated head of Xilongchi pumped storage power station reaches 640m. The unit capacitiesof 250MW or above account for 86% of the total units, the unit capacity of Xianju pumped storage power station under construction is 375MW (rated head 437m), which is the largest one in China.

Based on the experiences in construction of large-size hydropower stations, technology of civil works adopted in construction of pumped storage power stations also reached a

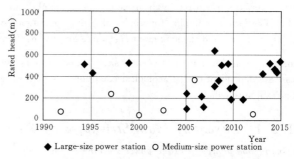

Fig. 21.1-2 Development of Rated Head for Pumped Storage Power Units in China

high level. For examples, little leakage of concrete face covered on whole basin of reservoir; 97m rockfill dam with asphalt concrete face built on deep overburden; reinforced concrete bifurcated pipe with a large HD value, over 6,000m·m (design head H multiplied by diameter of bifurcated pipe D); shotcrete and anchorage support and other flexible supporting technology widely used for the underground powerhouses built within surrounding rocks of Class Ⅲ or Ⅳ.

21.2 Layout of Main Structures of Pumped Storage Power Station

In China, there are 30 large and medium-size pumped storage power stations completed or under construction, in which 83% are pure pumped storage power ones. Pure pumped storage station indicates no river water supply to the upper reservoir. Almostall the large-size pumped storage power stations (with installed capacity≥300MW) are pure ones, except for Baishan station (300MW). In the pure pumped storage power stations, 80% of them have rated head higher than 250m. Almost all the mixture type pumped storage power stations are medium or small size, with rated head less than 100m.

The layout of main structures of pumped storage power station is usually influenced by upper and lower reservoirs. In China, the mixture type pumped storage power stations generally adopt the existing reservoirs as the upper ponds, and 60% of the lower reservoirs are the existing ones. Yamzhog Yumco mixture type pumped storage power station is a special case, taking the Yamzhog Yumco lake as the upper reservoir and the Yarlung Zangbo River as the lower reservoir, no extra dam constructed.

For pure pumped storage power stations, the combination types of upper and lower reservoirs are diversified. 42% of the lower reservoirs (or upper reservoirs) utilize the existing ones. About 58% are newly-built reservoirs, with a tendency of rapid increase nowadays.

There are many reasons for the above mentioned tendency, the usage of existing reservoirs will involve a lot of complicated problems. The utilization of existing reservoirs (as upper and lower reservoirs) may reduce the cost of construction, but it is always required to heighten and strengthen the dams; for multipurpose reservoirs, the water quantity has to be redistributed and the construction cost shall be shared clearly; new management system for reservoir operation hasto be worked out.

The pump turbine with high head requires large suction head, so it is difficult to arrange a ground powerhouse, and 83% of pumped storage power stations have underground powerhouses. There are two semi-underground and two ground powerhouses, which all belong to medium-size power stations with low head, only the powerhouse of Panjiakou station is locatedat dam toe.

There are three types of layout for underground powerhouses, head, middle or tail. The head type is unfavorable for seepage control, operation management and construction. If the topographical and geological conditions permit, the underground powerhouse is not usually arranged with head type. The tail type is good for operation and construction and is often adopted by many stations when the topographical and geological conditions permit. Since the layout of pumped storage power station is mainly influenced by the locations of upper and lower reservoirs, a relatively optimized location of underground powerhouse can only be chosen in the limited range.

Because of the features of powerhouse layout, except part of the high pressure pipes in Yamzhog Yumco (ground powerhouse) and Xikou (semi-underground powerhouse) pumped storage power stations are exposed steel pipes, the underground pressure pipes are used for most of pumped storage power stations. About 2/3 of the high pressure pipelines are lined with concrete, but more and more penstocks are adopted currently in view of safety and reliability for the stations with water head above 500 – 600m.

Penstock can be lined with concrete or steel. One pipe for two units is mostly adopted for high pressure pipes. The internal water pressure of concrete high pressure pipe is mainly transferred to the surrounding rocks, which is not sensitive to HD value. The type of one pipe four units also has been adopted, such as Guangzhou, Huizhou, Qingyuan pumped storage power stations, and the maximum HD value exceeds 6,000m·m.

The ratio of inclined shaft to vertical shaft is about 2:1, which has little relationship with lining types, and the types of shaft are dependent on the geological conditions, unit transient requirement and distance between the powerhouse and the upstream surge shaft.

21.2.1 *Layout of Main Structures of Pure Pumped Storage Power Station*

1. Guangzhou pumped storage power station

For Guangzhou pumped storage power station, the installed capacity is 8×300MW, and construction of the station is divided into two stages. It is the largest pumped storage power station in China for the present. The first unit was put into operation in June, 1993 and all of the units were put into operation in March, 2000.

Guangzhou pumped storage power station is located in the second tributary of Liuxihe River upstream, the upper and the lower reservoirs are formed by damming on the natural topography. The horizontal distance between the two reservoirs is about 3km, with natural drop over 500m, the ratio of distance and height is $L/H \approx 6$. The power station situates in granite area, which is a satisfactory site for pumped storage station, it is also one of pumped storage power stations with the lowest investment.

The layout of main structures of Guangzhou pumped storage power station is shown in Fig. 21.2-1 and Fig. 21.2-2, with main features as follows.

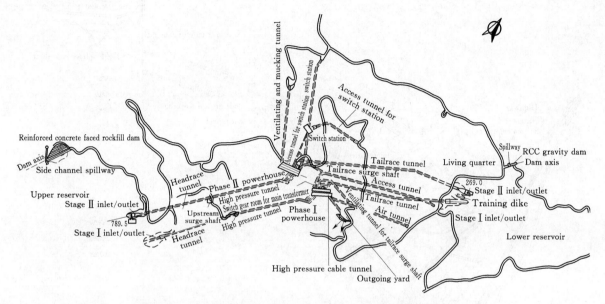

Fig. 21.2-1 Plan Layoutof Guangzhou Pumped Storagepower Station

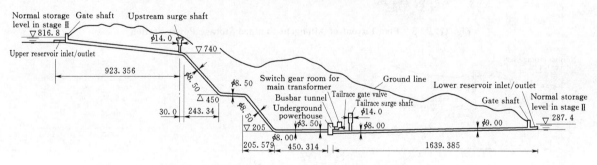

Fig. 21.2-2 Layout of Guangzhou Project (Unit: m)

(1) The upper and lower reservoirs are all formed by damming on the natural valleys. Rockfill dam with concrete face (dam height 68m and crest length 319m) is used for theupper reservoir, while RCC gravity dam (dam height 43.5m and crest length 153m) is adopted for the lower reservoir.

(2) For headrace and tailrace systems, the scheme of one tunnel for four units is adopted, surge shafts are provided for the two systems. High pressure pipes are arranged with two inclined sections with 50° slope, the high pressure tunnel with a diameter of 8.5-8m and bifurcated pipe ($HD=5,800$m·m) are all lined with reinforced concrete.

(3) Underground cavern groups are located in granite area, which are all supported by flexibly bolting and shotcreting. SF_6 high-voltage switch of stage I powerhouse is arranged in the main transformer cavern and that of stage II powerhouse is set on the ground.

(4) 500kV oil-filled cable goes to the ground via a 29° long shaft with 200m drop.

2. Xilongchi pumped storage power station

Xilongchi pumped storage power station is a large-size one with the highest water head (rated head 640m) in China, which is located in Wutai County, Shanxi Province, with the installed capacity of 4×300MW. The construction of main works commenced in August, 2003, and the first unit was put into generation on November 28, 2008. The layout of main structures is shown in Fig. 21.2 - 3 and Fig. 21.2 - 4.

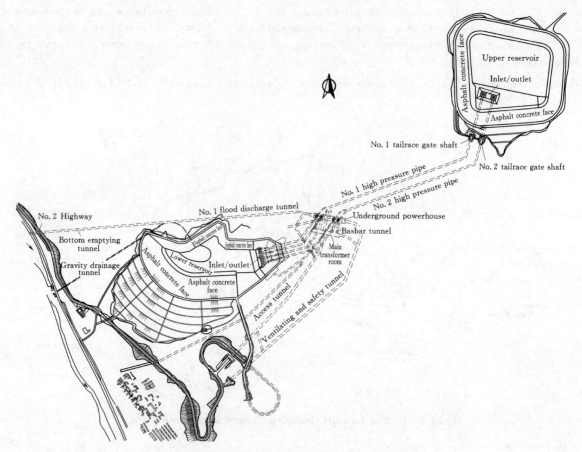

Fig. 21.2 - 3 Plan Layout of Xilongchi Pumped Storage Power Station

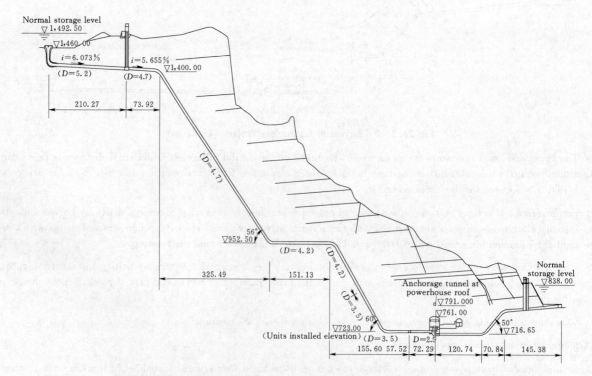

Fig. 21.2 - 4 Profile of Water Conveyance System of Xilongchi Pumped Storage Power Station (Unit: m)

The upper reservoir is located on the summit of the left bank of the Hutuohe River. The foundation rock is limestone and the groundwater table is low. There are faults, cracks and karst development in the rock mass which has strong water permeability. Therefore, asphalt concrete face with total area of 215,700m^2 is adopted for seepage control. The upper reservoir is made up of a main dam and two auxiliary dams. The main dam is 50m high with crest length of 401m, and the reservoir bank and dam upstream slopesare both 1 : 2.

The water conveyance system, with the total length of 1,819 – 1,871m, passes through the stratum composed of limestone, dolomite and sandstone. The shaft type in-outlet is adopted for the upper reservoir, which is the first case in China. The ratio of distance and height for the headrace system is $L/H=2.92$, which is the lowest one in China. The surge chambers are not provided for headrace and tailrace systems. The "two units sharing one pipe" is adopted for the pressure pipe with inner diameter of 5.2 – 3.5m. The inclined pressure pipe enters into the powerhouse at an angle of 65°. Each tailrace tunnel with a diameter of 4.3m is adopted for every unit. The pressure pipes are lined with steel plate with the total lining steel of 13,184 tons. The maximum thickness of steel plate is 60mm (strength class 800MPa), and the exterior surrounding of the tunnel is filled with concrete in thickness of 60cm. The bifurcated pipes are crescent rib reinforced Y-shaped pieces with the HD value of 3,552.5m · m, the maximum thickness of shell is 56mm and the rib plate is 120mm thick.

The surrounding rock of the underground powerhouse are stratified limestone, and supported with flexible reinforcement measures, such as steel fiber shotcrete, anchor bolt and prestressed anchor cable. In view of stability of surrounding rocks, the assembly bay is set up in the center of the machine hall. The main auxiliary powerhouse and switchyard are arranged on the ground. Gravity drain tunnel is adopted.

The lower reservoir is about 150m higher than the Hutuohe River, the rock masses of reservoir bank are mostly limestone and dolomite. The bedrock has strong water permeability, the overburden of the reservoir bottom and the dam foundation is 20 – 40m thick with high permeability. For the lower reservoir, the whole reservoir basin is treated for seepage control, the reservoir bottom and dam slope are lined with asphalt concrete face, and the reservoir bank composed with rocks is lined with reinforcement concrete face slab. The dam features of the lower reservoir are as follows: rockfill dam with asphalt concrete face, maximum height 97m, crest length 537m, upstream dam slope 1 : 2, and downstream dam slope 1 : 1.7, excavated reservoir bank slope 1 : 0.75. The emptying tunnel is arranged at the right bank of the reservoir.

21.2.2 Layout of Main Structures of Mixture Type Pumped Storage Power Station

At present, there are 6 mixture type pumped storage power stations in China, and most of them are medium-and small-sized ones (installed capacity<300MW), only Baishan pumped storage power station with an installed capacity of 300MW belongs to a large-size one.

Usually, the mixture type pumped storage power stations are formed by reconstruction or expansion of conventional hydropower stations, and then pump/turbine units are added. The common features of this kind of stations are as follows: the upper reservoirs are large-or medium-sized multipurpose reservoirs; the operation mode of conventional hydropower stations depends on water volume and cannot meet the requirements of peak shaving. After adding pump/turbine units, the station operation modeis changed to meet the water demand for multipurpose use, and improve the peaking capacity of the station.

Most existing reservoirs are used as the upper and lower reservoirs, except for Panjiakou and Xianghongdian stations, their lower reservoirs are formed by building new dams.

Usually the water head of the mixture type pumped storage power station is not high (≤100m), but Yamzhog Yumco pumped storage power station is a special case, its water head is 816m.

For Panjiakou pumped storage power station with the powerhouse set at dam toe, the pump/turbine units and conventional units are installed in the same machine hall. Since the suction heights of two kinds of units are different, the layout and construction of powerhouse are both complicated. Generally, the pump/turbine units are arranged in another hall for most mixture type pumped storage power stations. Because the water head is low and the suction height is small, newly-built powerhouses are arranged at underground or ground. The water conveyance systems are all arranged in underground.

1. Xianghongdian pumped storage power station

The upper reservoir of Xianghongdian pumped storage power station is the existing Xianghongdian reservoir, and the lower reservoir is 8.8km away from the upper reservoir. The downstram of reservoir is a new gravity dam with the height of 16m and crest length of 260m, see Fig. 21.2 – 5. The conventional hydropower station and pumped storage power station are arranged separately. The existing conventional hydropower station is located on the right bank of the upper reservoir dam. The newly-built pumped storage power station is set up on the left bank of the upper reservoir dam, the water head is formed by using natural drop of the river bay and cutting the curve course. The underground powerhouse is arranged at tail, installed with two 40MW pump/turbine units. One headrace tunnel is shared by two units, and the diameter of main pipe lined by reinforced concrete is 5.5m, the impedance surge shaft is set up at upstream. The one tunnel for one unit is adopted in tailrace system.

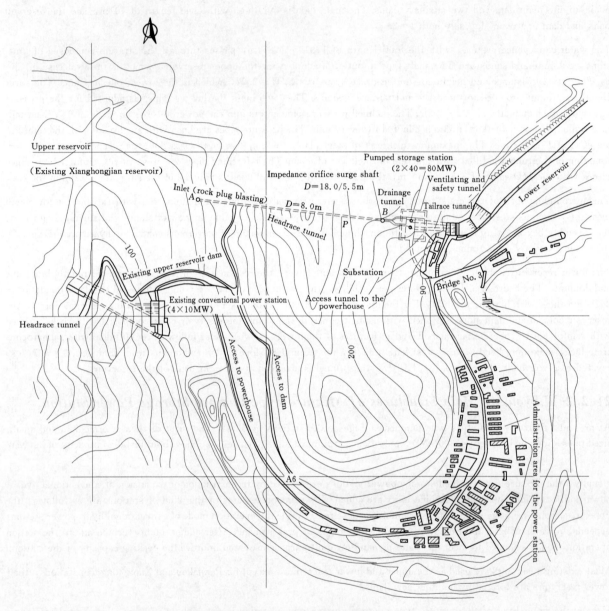

Fig. 21.2-5 Plan Layout of Xianghongdian Pumped Storage Power Station

This expanded pumped storage power station has the following advantages:

(1) By water storage of the upper and lower reservoirs, make the pump/turbine units and conventional units work for power output and peak shaving in non-irrigation period or low water using period, and the original station under half-year working and half-year resting has the peak capacity in a whole year.

(2) Because of the combining operation of conventional units and pump/turbine units, power discharge is increased so that all the irrigation water and abandoned water in flood period can be used for power generation.

(3) The discharge capacity of reservoir is increased, the frequent operation of discharge tunnel for discharging irrigation water is avoided and reservoir operation conditions are improved.

2. Foziling pumped storage power station

The existing Mozitan reservoir locate on the branch stream of Huaihe River is used as the upper reservoir and the existing Foziling reservoir is used as the lower reservoir for the Foziling pumped storage power station, which is arranged on the left bank of Mozitan reservoir dam. The underground powerhouse is located at head, installed with two 80MW pump/turbine units. The arrangement of "one tunnel for one unit" is adopted in headrace tunnel, and the type of "one tunnel for two units" is adopted in tailrace tunnel, see Fig. 21.2-6. Mozitan and Foziling reservoirs are connected head to tail. Foziling reservoir water level has great change. In order to avoid the unfavorable influence of low level of lower reservoir on pumped storage power station operation, a dam is built 3km away from the in-outlet of the lower reservoir to form a transitional reservoir with a storage capacity of 15.5 million m³. The dam is overflow RR sand-gravel rockfill dam with maximum height of 15.6m, which is immersed in water most of the time. When the water level of Foziling reservoir is high, it serves as the lower reservoir, and when the water level is low, the transitional reservoir serves as the lower reservoir, of which the using rate is about 15%.

Features of Foziling pumped storage power station are as follows: ①Utilization of the existing reservoirs as upper and lower reservoirs, less environmental impact and occupied construction land, no reservoir inundation and resettlement, simple project construction conditions and arrangement. ②Combining operation of pump/turbine units and conventional units, using abandoned water for power generation and transferring a part of output into peak power. ③Large storage capacity of upper and lower reservoirs with flexible operation and capability of carrying out alternate tasks for a long time in emergency. ④Rated head 54.2m, large head variation, low conversion efficiency (70% of pumping/generating or 75% after taking abandoned water for power generation).

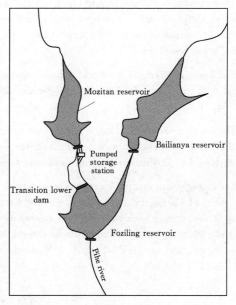

Fig. 21.2-6 Layout of Foziling Project

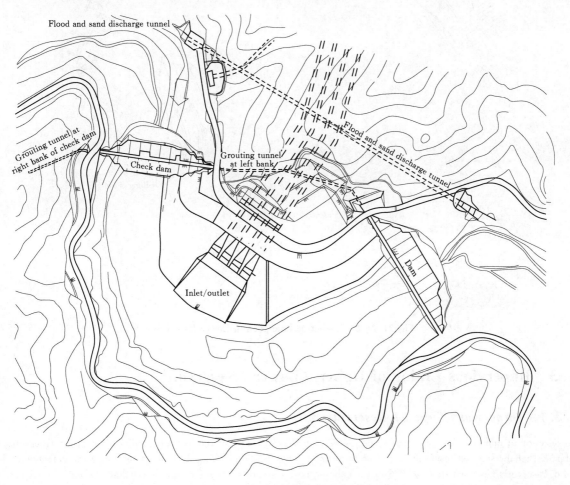

Fig. 21.2-7 Layout of Huhhot Lower Reservoir

21.2.3 Layout of Lower Reservoir in Silt-carrying River

There are many silt-carrying rivers in north China, most of them have the characteristics of "more water more silt, or less water with little silt (or even no silt)", and some of them have the average annual silt content exceeding $10kg/m^3$. Since the silt content through a high head turbine unit is required to be lower than $40g/m^3$, de-silting measures shall be taken to the lower reservoirs.

The lower reservoir of Xilongchi pumped storage power station is arranged in a branch gully which is higher than the original river course, and is away from the main river because the average annual silt content of Hutuohe River is $17.5kg/m^3$ (Fig. 21.2-3).

The lower reservoir of Huhhot pumped storage power station is located in the Halaqin ditch with average annual silt content of $7.95kg/m^3$. Therefore, a sediment control dam is arranged at the upstream of the lower reservoir and a discharge/flushing tunnel is provided to discharge sediment-laden flow to ensure the clearness of water (Fig. 21.2-7).

For Zhanghewan pumped storage power station, the lower reservoir is located in Gantaohe River with an average annual silt content of $11.1kg/m^3$. A sediment control dam is built at 1.8km upstream of the retaining dam, and the inlet/outlet is set up at the concave bank between the retaining dam and sediment control dam. An open silt-releasing channel is excavated at a saddle 200m upstream of the sediment control dam, water with high silt content diverted directly to the front of the retaining dam, and then is discharged to the downstream by discharging and desilting facilities without passing the inlet/outlet (Fig. 21.2-8).

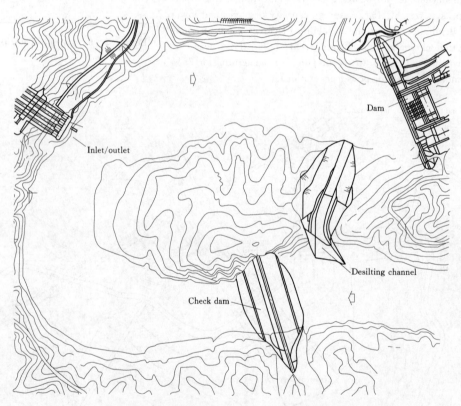

Fig. 21.2-8 Layout of Zhanghewan Lower Reservoir

21.3 Impervious Lining of Upper (Lower) Reservoir

21.3.1 Impervious Type of Reservoir Basin

The selection of pumped storage power station sites is greatly affected by the leakage conditions of upper reservoirs. It is preferred that the selected reservoir only need to be treated by curtain grouting partially. Therefore, the percentage of this kind of reservoirs has risen from 46.7% of the existing power stations to 75% of the power stations under construction.

Among the upper reservoirs of existing pumped storage power stations, 3 basins are fully lined with asphalt concrete face and 3 with reinforced concrete face. For asphalt concrete face, due to its high imperviousness and good adaptability to inhomogeneous deformation of foundation, its competitiveness is improved along with large-scale production of domestic high quality asphalt, and development of specific construction equipment and professional construction team. Among the upper reservoirs of pumped storage power stations under construction, one basin slope is lined with asphalt concrete face and one basin slope is lined with reinforced concrete face. For impervious measures on reservoir bottom, the geomembrane has been put into use successfully in the upper reservoir of Taian pumped storage power station. The geomembrane is cheap in price, easy for construction and good in impervious effect. Clay blanket has once been used in the upper reservoir of Baoquan pumped storage power station, but there are some disadvantages about it, such as farmland occupation for borrow area, significant effectof weather on construction, and relatively weak imperviousness.

Among large and medium size pumped storage power stations completed and under construction, the types of newly-built dams of upper reservoirs are listed in Table 21.3 - 1, in which, earth-rockfill dams account for 87%. The excavated earth and rock in construction of the upper reservoir can be used as material for dam construction and can increase the effective storage capacity of reservoir, but the cement used for concrete dam has to be transported from other places. Moreover, the catchment area of upper reservoirs is usually small and the flood peak discharge is also small, so there is no need to arrange water release structures for the concrete dam. In general, the cost of concrete dam is higher than that of earth-rockfill dam, and concrete faced slab rock-fill dam is the popular type selected, accounting for 56.5%, this tendency accords with the conventional hydropower stations. The asphalt concrete faced dam takes the second with 21.7%, which is mainly adopted when high imperviousness is required for upper reservoir.

Table 21.3 - 1 New Built Dam Types in the Upper and Lower Reservoir

No.	Dam Type	Upper reservoir		Lower reservoir	
		Amoutn	Percentage (%)	Amoutn	Percentage (%)
1	Concrete face rockfill dam	13	56.5	4	23.5
2	Asphalt concrete faced dam	5	21.7	1	5.9
3	Earth rockfill dam with clay core	3	8.7	2	11.8
4	Homogeneous earth dam	0	0	1	5.9
5	RCC gravity dam	2	8.7	7	41.2
6	Concrete gravity dam	1	4.3	2	11.8

The types of newly-built dams of lower reservoirs are also listed in Table 21.3 - 1. In which, concrete dams account for 53%, which are adopted in case of large catchment area of lower reservoir and large flood peak discharge. The concrete faced rockfill dam is the second, taking 1/4. Only the dam of Xilongchi lower reservoir is asphalt concrete faced dam, where high imperiousness and adaptability to foundation deformation are required since the reservoir basin and the dam foundation are built on thick overburden. The dam of Xiangshuijian lower reservoir is homogeneous earth dam, located in marsh and lake depression areas. The earth can be borrowed in local, and reservoir can be formed by closing dike.

21.3.2 *Concrete Face*

The technology of concrete faced rockfill dam is mature, which has been widely used in water conservancy and hydropower projects in China. The upper reservoir of Shisanling pumped storage power station is lined with concrete face and makes great success, afterwards, concrete face slab as impervious measure has been adopted for more and more pumped storage power stations, see Table 21.3 - 2.

The operating conditions of pumped storage power stations are special as follows:

(1) Due to water loss of upper reservoir caused by leaking and evaporating, water is pumped from lower reservoir for hundred meters of lift, and high imperviousness is required for upper reservoir.

(2) Because of the frequent and quickly fluctuation of water level in the reservoirs, the stability, frost-resistance ability and durability are required strictly for the face under the action of reversed water pressure.

(3) Normally, the upper reservoir is located in the mountaintop or platform where water table is low and rock mass is severe weathered with developed fissures, therefore, the whole reservoir basin shall be lined for seepage control. Because of the differences in rock types, weathering degrees, and influence of geological formations, the deformation of reservoir basin is largely varied. The concrete face is rigid structure, its adaptability to foundation uneven deformation is rather weak, which requires corresponding reinforcement measures.

Table 21.3 - 2 Features of Concrete Face in Existing Pumped Storage Power Stations in China

No.	Project name	Location of lining	Lining area ($10^4 m^2$)	Height of main dam (m)	Length of dam crest (m)	Dam slope ratio Upstream	Dam slope ratio Downstream	Face thickness (m)	Year of operation
1	Upper reservoir of Shisanling	Whole reservoir basin	17.5	75	550	1 : 1.5	1 : 1.7	0.3 (uniform)	1995
2	Upper reservoir of Yixing	Whole reservoir basin	18.07	75	495	1 : 1.3	Average 1 : 1.48	1 : 0.3 (uniform)	2008
3	Upper reservoir of Taian	Dam slope, reservoir bank	8.27	99.8	540	1 : 1.5	Average 1 : 1.63	0.3 (uniform)	2006
4	Lower reservoir of Xilongchi	Rock slope, reservoir bank	6.72			Reservoir slope 1 : 0.75		0.4 (uniform)	2008

1. *Layout of upper reservoir with impervious concrete face on the whole basin*

The two upper reservoirs of the pumped storage power stations, Shisanling and Yixing, adopt concrete face to prevent leakage. The layouts are shown in Fig. 21.3 – 1 and Fig. 21.3 – 2.

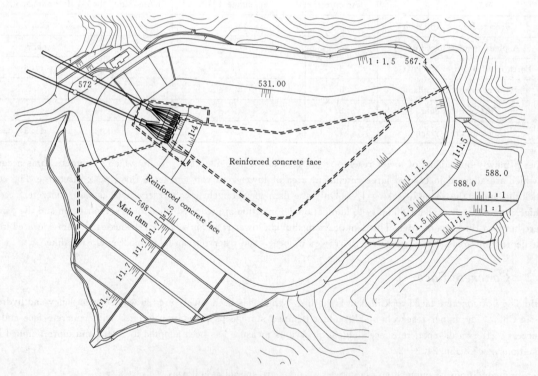

Fig. 21.3 - 1 Plan Layout of Shisanling Upper Reservoir (Unit: m)

The concrete face area of Shisanling upper reservoir is 175,000m², the maximum dam height of main dam is 75m, with upstream slope and excavated reservoir slope 1 : 1.5 and downstream slope 1 : 1.7, the thickness of concrete face is 0.3m. The 30cm-thick no-fines concrete drainage layer is under the concrete face of reservoir slope.

The concrete face area of Yixing upper reservoir is 180,000m², the maximum dam height of the main dam is 75m, with the upstream dam slope of 1 : 1.3, downstream dam slope 1 : 1.48 and excavated reservoir slope 1 : 1.4. The thickness of concrete face is 0.4m. The materials under the concrete face of reservoir slope are as follows: 400g/m² geotextile, 1 – 1.5kg/m² emulsified asphalt, and 30cm thick porous concrete drainage layer. The 70cm thick crushed-stone drainage layer and 28cm filter layer are set at the bottom of reservoir.

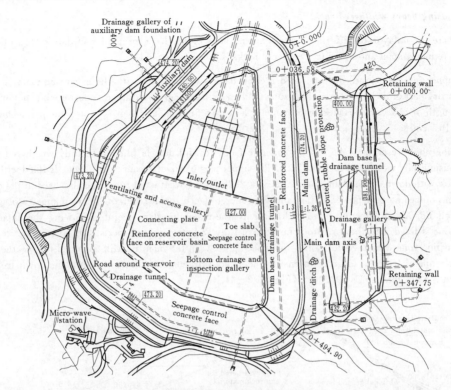

Fig. 21.3-2 Plan Layout of Yixing Upper Reservoir (Unit: m)

2. Layout of the concrete faced rockfill dam on the base of steep valley

It is common to build a dam on the base of steep valley for upper reservoir of a pumped storage power station. If a rockfill dam is built on the foundation of a ramp, the foundation shall be excavated into step-shape.

The foundation surface of Yixing upper reservoir's main dam is inclined toward downstream, with the dip angle more than 20° or partial over 30°. The bedrock is quartz sandstone, argillaceous siltstones and silty mudstone. The base is inclined toward downstream for 10°–15°. The maximum dam height is 75m at dam axis. In the original design, the rockfill dam is built attaching to the steep slope, the height difference is 285m from bottom to top of the dam. The optimized design is a 45.9m high gravity concrete retaining wall 135.5m away from the axis of dam downstream. The height difference is 138.2m from retaining wall toe to the dam top (Fig. 21.3-3). Detailed analysis was made on the dam materials, foundation, massif and drainage system of gravity retaining wall in the design. The operation of the power station is normal after reservoir impoundment.

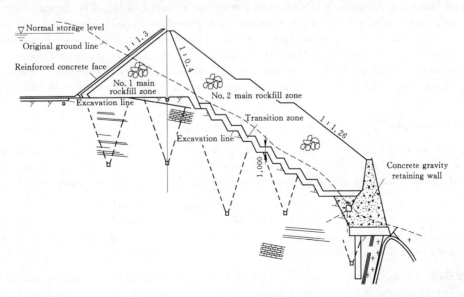

Fig. 21.3-3 Cross-section of Concrete Faced Rockfill Dam for Yixing Upper Reservoir

3. Dam material using heavy weathered rocks

In order to mitigate the unfavorable effects on environment, one of the basic principles in dam design of pumped storage power station is reducing spoil and utilizing weathered rocksas much as possible.

For Shisanling upper reservoir, the concrete faced rockfill dam is constructed with the excavated materials from reservoir basin. The upper dry part of downstream rockfill zone is filled with strongly weathered andesite with the minimum saturated compressive strength of 11MPa, particles with diameter less than 5mm is less than 20% and the permeability coefficient is larger than 1×10^{-2} cm/s.

For Tianhuangping upper reservoir, the asphalt concrete faced rockfill dam is filled with fully weathered rocks, which is the first case of using fully weathered rock in China. The upper part of downstream rockfill zone is filled with fully and heavily weathered rock with downstream dam slope 1 : 2 - 1 : 2.2. Heavy weathered rocks are also used in the clay core rockfill dam of Qingyuan pumped storage power station and the concrete faced rockfill dam of Xianju pumped storage power station under construction. For Xianju concrete faced rockfill dam, the maximum dam height is 86.7m, the upper part of downstream rockfill zone is filled with the earth-rock mixture of fully and strongly weathered tuff lava, andesite and tuff breccias. Drainage transition zone with the width of 1.5m is set up between the main rockfill zone and the weathered rocks, with downstream dam slope of 1 : 2.2. In order to ensure the smooth drainage of the dam body, the downstream dam body is taken as the main rockfill zone at El.597.00m (Fig.21.3-4).

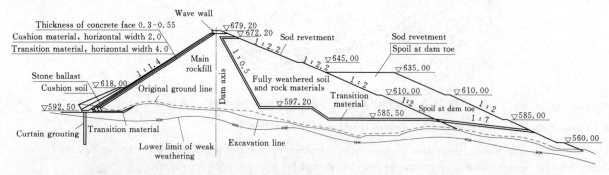

Fig. 21.3-4 Profile of Concrete Face Rockfill Dam of Xianju Upper Reservoir (Unit: m)

4. Joint of concrete face

Concrete face is a rigid structure, proper joints are an effective way to adapt the uneven settlement of foundation and avoid harmful cracking. However, too many joints and irregular divided sections will make it difficult for waterstopping construction, and may create new leaking passages. Therefore, face joints shall achieve a balance between face flexibilityto adapt to the uneven deformation of foundation and the requirement of waterstopping reliability.

The concrete face of Shisanling upper reservoir is divided into 477 blocks, as shown in Fig. 21.3-5. The rocky slope and rockfill section are identical in division of face joint, with a vertical joint spacing of 16m. The spacing of vertical joints is 8-10m on the two end sections of main dam and the maximum length of joint is 65.4m. Peripheral joints are set up at the border of main-and auxiliary-dam rockfill zones and the border of in-outlet where deformations are large. In order to reduce penetrating cracks caused by uneven deformation on face at reservoir bottom, the joints on the bottom face are set up at the positions where foundation medium are different, including in-outlet area, fractured and fault zone, central line of drainage gallery and edge of retaining wall. For the face at reservoir bottom, width is 16m at the normal level and the length is limited to below 35m.

5. Water stop of concrete face

Water stop at joints is a weak link of concrete face, where leakage occurs easily. As to the whole reservoir basin lined with concrete face, the work quantity of water stop at joints is large, for example, the total length of various joints of Yixing upper reservoir is 29,000m and Shisanling upper reservoir is 21,300m. Therefore, the water stop structures are important.

In Shisanling upper reservoir, three courses of water stop are set up in the peripheral joint of facing, two courses of water stop are set up in tensile joint and compressive joint respectively, one course of water stop is set up in the joint of face and wave wall. The water stop structure of peripheral joint is shown in Figure 21.3-6, the improvements of water stop structure, materials and technique are as follows.

(1) To add GB material in rubber strip, it can chemically bind with cast-in-situ concrete to stop any leakage from the joints of rubber strip. The GB material in filling chute on top of the face may be squeezed into the joints under the water pressure,

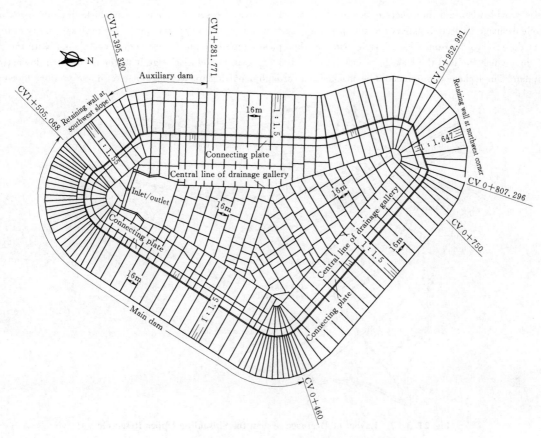

Fig. 21.3 - 5 Concrete Face Joints of Shisanling Upper Reservoir

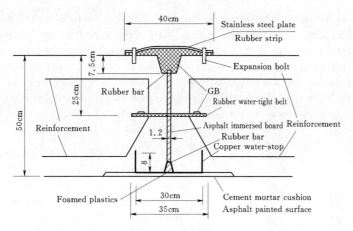

Fig. 21.3 - 6 Arrangement of Peripheral Joint Water Stop

and plays a role of water stop.

(2) Adopted copper water stop, reduce the number of welding joints as far as possible and shape it according to the joint length at site, the stamping forming of X-shaped joint and T-shaped joint are made once in workshop.

(3) Using GB material as flexible seal, it has good water resistance and durability.

6. Drainage system

Drainage system beneath the concrete face is necessary to drain leakage water from the face timely. It is more important to ensure that the drainage is smooth for the gentle slope of reservoir bottom, and the permeability coefficient of drainage blanket is usually larger than 1×10^{-2} cm/s.

The drainage system of Shisanling upper reservoir is shown in Fig. 21.3 - 7. The 30cm-thick drainage blanket with no-fines concrete is provided beneath the concrete face in the rocky slope area of reservoir bank; 50cm-thick stone drainage blanket and a circle drainage inspection gallery are set up at the reservoir bottom. In order to improve the drainage ability of drainage blanket at the reservoir bottom, 15cm-inside diameter rigid plastic pipes are placed in the drainage blanket with the spacing of 20 - 30m, which can lead the leakage of face into the drainage gallery and discharge it to the drain ditch downstream of the main dam. The leakage position could be checked out through departing the drainage blanket into 8 sections by concrete.

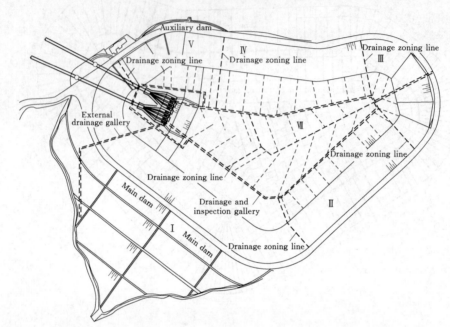

Fig. 21. 3 - 7   Layout of Drainage System for Shisanling Upper Reservoir

7. Cracking of concrete face

In China, lots of successful experiences have been gained in preventing concrete faced rockfill dam from cracking, such as the lower reservoir of Tongbai pumped storage power station. There is almost no crack on the concrete face. However, as to the reservoirs with whole basin lined, due to the complicated operation conditions and strong constraint on rocks of reservoir slope, concrete face cracking of the existing projects is common. It was found that the concrete face slab on reservoir bottom is almost intact and the face on dam slope has less cracking, but more cracks on rocky slope of reservoir bank as shown in Fig. 21. 3 - 8. There are more cracks on the face of turning sections than that of straight sections, and more cracks on no-fines concrete blanket than that on geotextile ones. And there are fewer cracks in concrete face slab casted early than that casted latter.

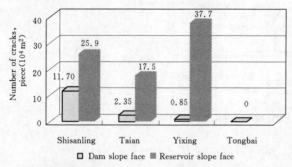

Fig. 21. 3 - 8   Cracks of Reinforced Concrete Face on Dam Slope and Reservoir Bank

The successful experiences and the effective measures of preventing cracking of concrete face have been taken into use in the recently built Yixing upper reservoir, such as mixing polyacrylonitrile fiber, setting double-decked reinforcement, keeping heat and humidity timely, spraying emulsified asphalt and laying geotextile on no-fines concrete blanket. At last, the cracks on rocky slope of reservoir bank are still more than that on dam slope, which can only be explained that the constraint of bedrock on reservoir bank is stronger than it is anticipated. It also means that effective measures of preventing concrete face from cracking for seepage control in a whole reservoir basin need to be studied further.

8. Leakage of concrete face

Fig. 21. 3 - 9 shows the 13-year monitoring records of the leakage volumes in Shisanling upper reservoir. The leakage volumes vary in yearly period, which is larger in winter and smaller in summer. The leakage volume was rather large in initial impoundment of reservoir, and the maximum value was 16. 7L/s. After twice complete emptying of the reservoir and repairing of cracks, the maximum leakage volume reduced to 10L/s below, in the winter of 2002, it dropped to 3. 7L/s. However, in recent years, there is a increasing tendency, with a maximum leakage volume of 11. 61L/s.

The anticipated permeability coefficient of the concrete face is 1×10^{-7} cm/s, the maximum leakage of the upper reservoir is 15.58L/s, which equals to the ratio of daily leakage to total storage capacity 1/3306. The measured maximum leakage is slightly smaller than the design value.

In Fig. 21.3-9, Q_1 and Q_2 are respectively the monitoring value of leakage volume from measuring weirs set up at the dam foot and the values at the drainage gallery outlet at reservoir bottom. Since the leakage from the measuring weir at the dam foot includes mountain leakage, the leakage volume is larger than that measured from the drainage gallery outlet. During early construction, the discharge quantity is increased by construction consumption, so the measured values from the two weirs are quite different. In the operation period, leakage volume is influenced by rainfalls, and the difference between the values of two weirs becomes small.

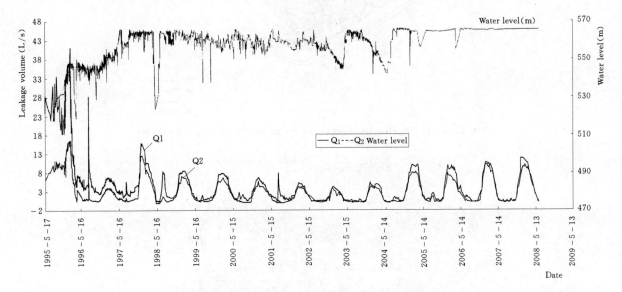

Fig. 21.3-9 Leakage Graph of Shisanling Upper Reservoir

The impoundment of Yixing upper reservoir began in June, 2007. The maximum leakage volume measured from the drainage gallery outlet at reservoir bottom was about 28.5L/s (exclusive of rainfall), which is about 1/2155 of the reservoir's total storage capacity. The leakage is mainly in the inlet/outlet area, which accounts for 70%–83% of the total leakage at reservoir bottom. There is a tendency of decrease in recent years.

It is obvious that the seepage control with concrete face applied to the upper reservoirs of pumped storage power stations is successful in China, and the engineering measures are effective. After treatment of concrete face cracking and water stop at joints, the permeability coefficient of concrete face reaches 1×10^{-7} cm/s, which can meet the requirement of daily leakage less than 1/2000 of the total storage.

9. Ice regime of Pumped-storage Power Station in chilly region

In early operation period of Shisanling station, the icing was monitored during the consecutive 3 years. The lowest air temperature was $-14\,^{\circ}\!\text{C}$, but ice sheet was not formed on the reservoir surface. The icing are as follows: ①when the lowest air temperature is above $-5.0\,^{\circ}\!\text{C}$, no ice is formed in any shape; ②when the lowest temperature drops to $-7--9\,^{\circ}\!\text{C}$, frazil ice occurs on the surface of reservoir; ③when air temperature decreases to $-9--14\,^{\circ}\!\text{C}$, glare ice is formed and bound on the concrete face at drawdown area, frazil ice in water increases, see Fig. 21.3-10.

The water surface area of Shisanling lower reservoir is 2.57 million m². According to the actual measurement, the unfrosted area in front of the inlet/outlet is only 204,000m², which extends along with the flow direction, see Fig. 21.3-11.

In order to meet the requirement of ice control for pumped storage power stations in cold regions, it is necessary to let a certain number of units keep operating so that fluctuated water is less likely to freeze. Moreover, the first unit should not be put into operation in November or December, and the defects elimination after unit trial running should be scheduled in the unfreezing period.

21.3.3 Asphalt Concrete Face

From 1970s to 1980s, some asphalt concrete face dams were built in China, due to the limited quality of domestic crude oil and processing technique, the quality of asphalt was poor and construction equipment and techniques were backward,

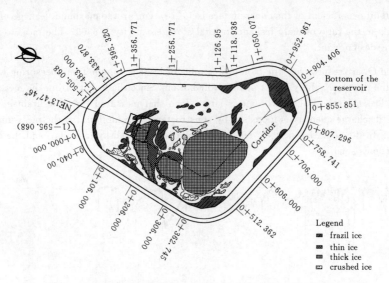

Fig. 21.3 - 10 Spatial Distribution of Ice in Shisanling Upper Reservoir

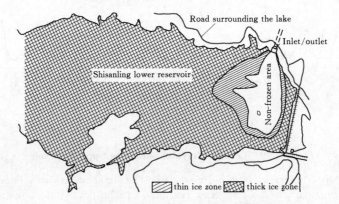

Fig. 21.3 - 11 Spatial Distribution of non-ice Area of Shisanling Lower Reservoir

which caused cracking and leakage of the asphalt concrete face and overland flowing. The construction of this kind of dam was once stopped. With the advancement of technology in the 1990s, asphalt concrete face was used for seepage control in many projects and gained success. The engineering features of asphalt concrete face are shown in Table 21.3 - 3. The area of asphalt concrete face in Zhanghewan pumped storage power station is 337,000m². The asphalt concrete face dam of Xilongchi lower reservoir is built on 40 - 50m thickness overburden, with the maximum dam height of 97m, which has reached a high level.

Table 21.3 - 3 Engineering Features of Asphalt Concrete Face of Existing Pumped Storage Power Stations

Item	Completion year	Storage capacity ($10^4 m^3$)	Dam height (m)	Upstream slope of dam	Area of face ($10^4 m^2$)		Thickness of face (cm)	
					Slope	Bottom	Slope	Bottom
Upper reservoir of Tianhuangping	1997	885	72	1:2.0 - 1:2.4	18.2	10.4	20.2	18.2
Upper reservoir of Zhanghewan	2007	770	57	1:1.75	20	13.7	26.2	28.2
Upper reservoir of Xilongchi	2008	469	50	1:2	10.2	11.4	20.2	20.2
Lower reservoir of Xilongchi	2007	494	97	1:2	6.8	4.0	20.2	20.2
Upper reservoir of Baoquan	2007	730	94	1:1.7	16.6	—	20.2	—

1. Layout of asphalt concrete face for seepage control

The whole basin lined with asphalt concrete face is adopted for upper reservoirs. The upper reservoirs of Tianhuangping, Zhanghewan and Xilongchi are shown in Fig. 21.3 - 12, Fig. 21.3 - 13 and Fig. 21.3 - 14 respectively.

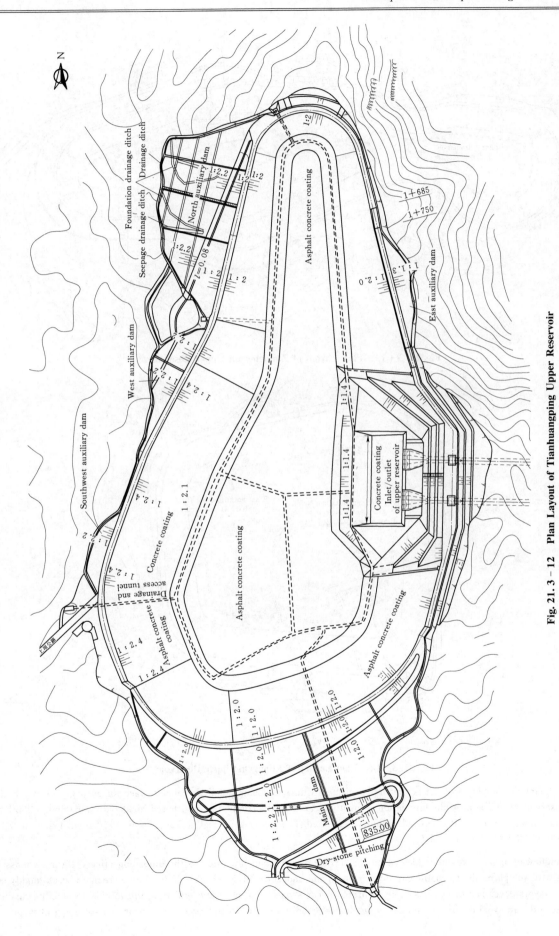

Fig. 21.3-12 Plan Layout of Tianhuangping Upper Reservoir

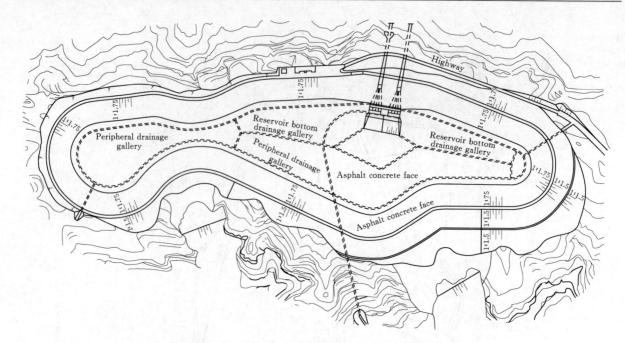

Fig. 21.3-13　Plan Layout of Zhanghewan Upper Reservoir

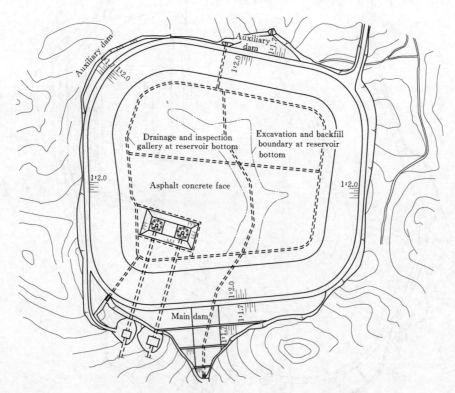

Fig. 21.3-14　Plan Layout of Xilongchi Upper Reservoir

The Tianhuangping upper reservoir is the first project in China that using asphalt concrete face for seepage control in the whole reservoir basin. Because the foundation is mostly located on the completely weathered rocks with a maximum depth of 35m, the asphalt concrete with high adaptability isselected. The impervious area is 286,000m², with dam height of 72m and face slope of 1:2 (partially 1:2.4).

The Zhanghewan upper reservoir is located on the terrace of mountain top, due to the unloading effect, the rock mass is pervious with the permeability coefficient of 0.2-3m/d and underground water table 250m. There are many weak muddy intercalated layers developed in bedrock, the saturated shear strength is quite low which is unfavorable to stability against sliding. In order to avoid deterioration of hydrogeological conditions after completion of reservoir and softening of weak intercalated layer, the asphalt concrete face is applied to line the whole basin. The impervious area is 337,000m^2, with the maximum dam height of 57m and face slope of 1 : 1.75.

The Xilongchi upper reservoir is located in the sulcus on the mountain top, the surrounding mountain body is thin and the outside of reservoir bank are deep dissected gullies. The high dip angle fissures are developed in the rock mass. The underground water is buried deep and leaking is severe. Therefore, the asphalt concrete face is used for seepage control in the whole basin. The impervious area is 215,700m^2, with the maximum dam height of 50m and face slope of 1 : 2.

The Xilongchi lower reservoir is the only project that using both asphalt concrete face and reinforced concrete face, as shown in Fig. 21.3 - 15. The main dam is asphalt concrete faced rockfill dam with the maximum height of 97m (the highest in China), and the depth of overburden is 30 - 50m at dam foundation. In order to meet the demands of storage capacity, the downstream axis of dam is convex toward downstream; the slope of upstream face is 1 : 2. Because of the steep mountain, the excavated slope of rock reservoir bank is taken as 1 : 0.75 to avoid forming high slope, and asphalt concrete face is adopted for seepage control. The foundation of reservoir bottom consists of partial overburden and partial rock mass, in order to adapt large deformation and irregular settlement, the asphalt concrete face is adopted for seepage control. The asphalt concrete and reinforced concrete faces are connected via concrete toe slabs with gallery.

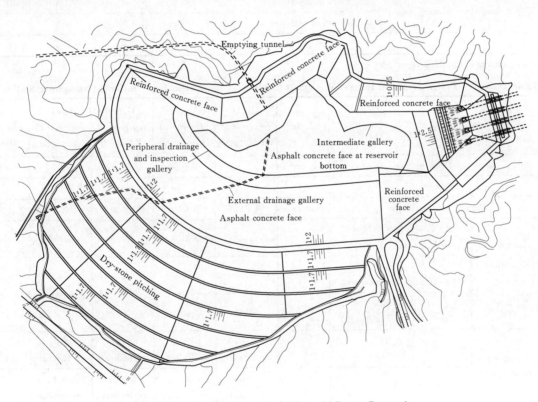

Fig. 21.3 - 15 Plan Layout of Xilongchi Lower Reservoir

2. *Structure design of asphalt concrete face*

(1) Structure and thickness of face. In China, most of the asphalt concrete faces in existing pumped storage power stations are the simple structure with single layer of lining, consisting of impervious layer and leveling cement layer. In Zhanghewan upper reservoir, the asphalt concrete face is a simplified dual structure consisting of upper impervious layer, drainage layer and leveling cement layer, the leveling cement layer and bottom impervious layer are merged into one layer. The thicknesses of layers are listed in Table 21.3 - 4, the seal coats on the surface are all 2mm thick. Each layer is placed by one course.

(2) Design technical indexes of each layer. The impervious layer is the main part of face and shall meet the requirements of seepage control, deformation, thermo-stability at inclined surface, cracking resistance at low temperature, immersion stability and ageing resistance according to all kinds of loads or actions to encounter. During the design of Zhanghewan and Xilongchi stations, many tests and finite element analysis have been carried out. Xilongchi pumped storage power station is located in the chilly area, the extreme lowest air temperatures at upper and lower reservoirs are −34.5℃ and −30.4℃ respectively. Considering crack resistance at low temperature, modified asphalt is used in the impervious layer of face at the slope section of upper reservoir, plain petroleum asphalt is used in other parts, the same as that used in other projects. For the deformation resistance, apart from the requirements on flexibility for disk testing, there are also requirements on ben-

ding strain for beam testing and tensile strain for uniaxial tensile testing. The main technical indexes of impervious layer in asphalt concrete face are listed in Table 21.3-5.

Table 21.3-4 Asphalt Concrete Structure and Slab Thickness of Existing Pumped Storage Power Stations

No.	Project name	Position	Leveling cemented layer (cm)	Drainage layer (cm)	Upper impervious layer (cm)
1	Upper reservoir of Tianhuangping	Slope of reservoir	10	—	10
		Bottom of reservoir	8	—	10
2	Upper reservoir of Xilongchi	Slope of reservoir	10	—	10
		Bottom of reservoir	10	—	10
3	Lower reservoir of Xilongchi	Slope of reservoir	10	—	10
		Bottom of reservoir	10	—	10
4	Upper reservoir of Zhanghewan	Slope of reservoir	8①	8	10
		Bottom of reservoir	8①	10	10
5	Upper reservoir of Baoquan	Dam slope	10	—	10

① leveling cemented and impervious layer

Table 21.3-5 Design Technical Indexes of Asphalt Concrete of Impervious Layer in Built Hydropower Station

No.	Project		Unit	Upper reservoir of Tianhuangping	Upper reservoir of Zhanghewan	Upper reservoir of Xilongchi	Lower reservoir of Xilongchi	Upper reservoir of Baoquan
1	Density		g/cm³		>2.30	>2.35	>2.35	>2.35
2	Porosity		%		≤3.0	≤3.0	≤3.0	≤3.0
3	Permeability coefficient		cm/s	≤1×10^{-8}	≤1×10^{-8}	≤1×10^{-8}	≤1×10^{-8}	≤1×10^{-8}
4	Slope flow value	1:2.0 (or 1:1.75), 70℃, 48h (marshal)	mm		≤2.0	≤0.8	≤0.8	≤0.8
		1:2.0, 70℃ (Van Asbeck test specimen)		≤5.0		≤2.0	≤2.0	
		1:2.0, 60℃ (Van Asbeck test specimen)		≤1.5				
5	Flexibility (disk test)	25℃	%	≥10, no leakage	≥10, no leakage	≥10, no leakage	≥10, no leakage	≥10, no leakage
		2℃ (5℃)		≥2.5, no leakage	≥2.5, no leakage	≥2.5, no leakage	≥2.5, no leakage	≥2.5, no leakage
6	Bending strain	2℃, rate of deformation 0.5mm/min	%		≥2.0	≥3.0	≥2.25	≥2.0
7	Tensile strain	2℃, rate of deformation 0.34mm/min	%		≥0.8	≥1.5	≥1.0	≥0.8
8	Frost break temperature		℃		≤−35	≤−38	≤−35	≤−30
9	Immersion stability				≥0.9	≥0.9	≥0.9	≥0.9
10	Expansion (unit volume)		%	≤1.0	≤1.0	≤1.0	≤1.0	≤1.0
11	Asphalt content		%	6.9%	7.5%	7.5% (modified asphalt)	7.5%	7.0%
12	Maximum particle size of aggregate		mm	16	16	16	16	16
13	Design thickness of layer		cm	10	10	10	10	10

The leveling cemented layer is located beneath the concrete face, which serves the function of binding the face and stone cushion and providing even and sound bed for impervious layer to ensure the spreading quality of the impervious layer. It is not exposed to the reservoir water, air and sunlight, and is not required in seepage control. The design of leveling cemented layer shall meet the following requirements: imperviousness, heat stability and immersion stability. The main technical indexes of leveling cemented layer of asphalt concrete face are listed in Table 21.3-6.

Table 21.3-6 Design Technical Indexes of Asphalt Concrete of Leveling Cemented Layer in Completed Pumped Storage Hydropower Station

No.	Project	Unit	Upper reservoir of Tianhuangping	Upper and lower reservoirs of Xilongchi	Upper reservoir of Baoquan
1	Density (volumetric method)	g/cm³	>2.1	>2.1	>2.20
2	Porosity	%	10-15	10-14	10-14
3	Permeability coefficient	cm/s	$5\times10^{-2}-1\times10^{-4}$	$5\times10^{-3}-1\times10^{-4}$	$1\times10^{-2}-1\times10^{-4}$
4	Immersion stability			≥0.85	≥0.85
5	Heat stability			≤4.5	≤4.5
6	Asphalt content	%	4.3	4.0	4.0
7	Maximum particle size of aggregate	mm	22.4	19	19
8	Design thickness of layer	cm	10 (reservoir slope) 8 (reservoir bottom)	10	10

For Zhanghewan upper reservoir, the leveling cemented layer and the lower impervious layer are merge dinto one layer, which is designed as a relatively impervious layer. It can collect leakage water from the upper impervious layer and lead the leakage into the drainage gallery. The main technical parameters are as follows: porosity no larger than 5%, permeability coefficient no larger than 5×10^{-5} cm/s, asphalt content in the mixture no less than 5% and maximum particle size 16mm.

The function of asphalt concrete drainage layer is to drain leakage water from impervious layer in time, avoid reverse water pressure behind the impervious layer and monitor the working conditions of impervious layer. The design parameters of drainage layer for asphalt concrete face of Zhanghewan upper reservoir are as follows: porosity no less than 16%, permeability coefficient no less than 1×10^{-1} cm/s, asphalt content in the mixture about 4% and maximum particle size 19mm.

(3) Drainage system at underlying bed of asphalt concrete face. The asphalt concrete face almost cannot stand any reversed water pressure, therefore, drainage system shall be provided for the underlying bed. The drainage system of Xilongchi upper reservoir bottom is shown in Fig. 21.3-16. A 0.6m thick crushed stone drainage cushion is set up beneath the asphalt concrete faces on the bank and bottom of reservoir, with the permeability coefficient larger than 8×10^{-3} cm/s. A drainage-inspection gallery is set up at the reservoir bottom, connecting with the crushed stone cushion via drainage pipes. Because the slope of reservoir bottom is gentle ($i=2.84\%$), 100mm-diameter drainage pipes are laid in the drainage cushion with a space of 20m, leading water to the drainage gallery directly.

(4) Treatment of uneven base of asphalt concrete face. The asphalt concrete face is flexible with a high adaptability to deformation. The asphalt concrete face at the southern section of Tianhuangping upper reservoir bottom is located on fully weathered rock mass, but the concrete face does not crack when the maximum settlement is 36.8cm after impounding. However, the face ability of adapting uneven settlement is still quite weak so that effective engineering measures shall be taken, especially in the areas where the deformation modulus of foundation changes rapidly.

At Xilongchi upper reservoir, the elastic modulus of the fully weathered rock mass in stratum O_{2s}^{2-6} is just 2MPa, while the elastic modulus of strongly weathered rock mass is 2GPa. Engineering measure of material substitution is taken, the deformation modulus of substituted cement and stone is 1GPa and the substitution depth 2m, as shown in Fig. 21.3-17. The mix proportion of substituted cement and crushed stones is as follows: cement 40-50kg/m³, water consumption 60-65kg/m³ and crushed stone 2,150-2,200kg/m³. The cushion is constructed by rolling.

For the Xilongchi lower reservoir, the uneven deformations may occur on the foundation of asphalt concrete face at reservoir bottom, between the bedrock and overburden, among the soil lens in overburden and voids of stones. Therefore, a gentle slope (1:3) is excavated at the boundary of bedrock and overburden, the thickness of cushion on the bedrock area is thickened to 1m, the cushion on the overburden is also enlarged to 2m, as shown in Fig. 21.3-18. The measurements are effective since the dam is under good condition after impoundment and operation.

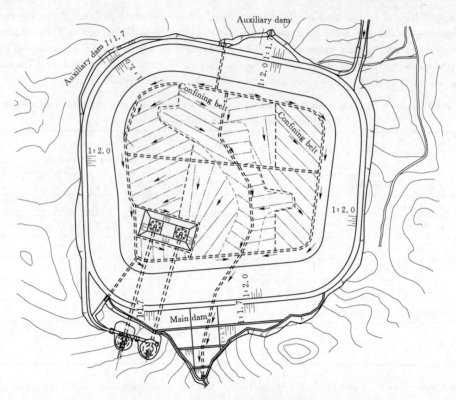

Fig. 21.3-16 Arrangement of Drainage System at the Upper Reservoir Bottom of Xilongchi Pumped Storage Station

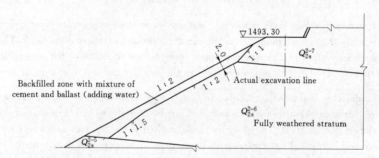

Fig. 21.3-17 Foundation Treatment at Substitution Position of Fully Weathered Stratum of Xilongchi Upper Reservoir (Unit: m)

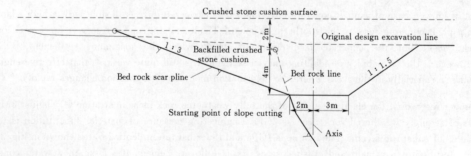

Fig. 21.3-18 Foundation Treatment of Steep Slope at Xilongchi Lower Reservoir Bottom

3. Measured leakage of asphalt concrete face

The asphalt concrete face is good for seepage control. A qualified design and construction of asphalt concrete faces slab shall have no any leakage.

For Xilongchi upper reservoir, the total volume of leakage at the drainage gallery exit of the main dam was 2.44L/s on

Dec. 22, 2009. the leakage volume measured from the measuring weir in the gallery at two shaft type inlets/outlets were 1.51L/s and 0.89L/s respectively, summing up to 2.40L/s. It means that the leakage volume mostly comes from the concrete structure of inlet/outlet, and the leakage from the asphalt concrete face is nearly zero.

The measured results from the Zhanghewan upper reservoir are similar to that of the Xilongchi. No any leakage is found on the asphalt concrete face in the drainage gallery at reservoir bottom.

It is concluded that the seepage control performance of asphalt concrete face is much better than concrete face. The design permeability coefficient of asphalt concrete is 1×10^{-8} cm/s, which is much smaller actually. The asphalt concrete faces of the above two power stations almost have no any leakage. But the leakage from concrete structures (inlet/outlet) and at the connecting zone between the concrete structures and the faces shall be paid attention.

21.3.4 Other Types of Seepage Control

For existing pumped storage power stations in China, the mixture type of seepage control for reservoirs is also utilized, which includes the concrete face, asphalt concrete face, clay blanket and geomembrane (Table 21.3-7).

Table 21.3-7 Samples of Mixture Type Seepage Control Works for Existing Storage Power Stations in China

No.	Project name	Scope of seepage control	Dam slope	Rock slope of reservoir	Reservoir bottom
1	Xilongchi, lower reservoir	Seepage control for whole basin	Asphalt concrete face	Concrete face	Asphalt concrete face
2	Baoquan, upper reservoir	Seepage control for whole basin	Asphalt concrete face	Asphalt concrete face	Clay blanket
3	Langyashan, upper reservoir	Partial seepage control	Concrete face	Curtain grouting in partial zone, concrete backfill after solution cavern excavated above impervious line	Auxiliary impervious with horizontal clay blanket on the reservoir bottom in auxiliary dam zone
4	Taian, upper reservoir	Partial seepage control	Concrete face	Concrete face on the right bank, vertical impervious curtain at the left side of basin	Horizontal geomembrane

1. Geomembrane

Geomembrane has the advantages of low permeability coefficient, easy fit for foundation deformation, fast construction and low cost. It has been applied in seepage control works of dam or dike of low-head hydraulic projects in recent years. Due to the special operation requirements of pumped storage power stations, geomembrane was used on the upper reservoir of Taian pumped storage power station.

In Taian pumped storage station, there is a longitudinal fault F_1 developed along the upper reservoir bottom, which is about 33-52m in width with relative imperviousness horizontally. The fault divides the reservoir left and right banks into 2 different hydrogeological zones. The concentrated joints in the right bank develop with high permeability, so the engineering treatment shall be made for overall seepage control.

The seepage control with concrete face is adopted at upstream face and right bank slope in the upper reservoir. The face has a thickness of 30cm and an 80cm thick crushed stone cushion is laid beneath the face; the geomembrane is placed on the cushion surface at the reservoir bottom with impervious area of 177,000m² and bearing water head of 36m. The geomembrane connects with the seepage control concrete face on the right bank and concrete face of rockfill dam, and the grouting curtain is conducted at the perimeter for sealing. The line of perimeter sealing grouting curtain is arranged along the side wall of observation gallery at the reservoir bottom with a depth of 20-60m and the curtain is deepened in the zones of faults F_1 and F_2. The perimeter sealing curtain and dam foundation curtain on the left bank are connected together with the impervious curtain on the right bank at the end of reservoir, see Fig. 21.3-19 and Fig. 21.3-20 for details.

(1) Structure of geomembrane. The puncture resistance ability of geomembrane is rather poor, therefore, in addition to using geomembrane materials with good tensile and puncture resistance abilities, ensuring meticulous construction and strengthening protection in laying process, necessary protective measures shall also be taken. For instance, taking the advantages of puncture resistance ability and tensile strength of the nonwoven geotextile, use nonwoven geotextile as the pro-

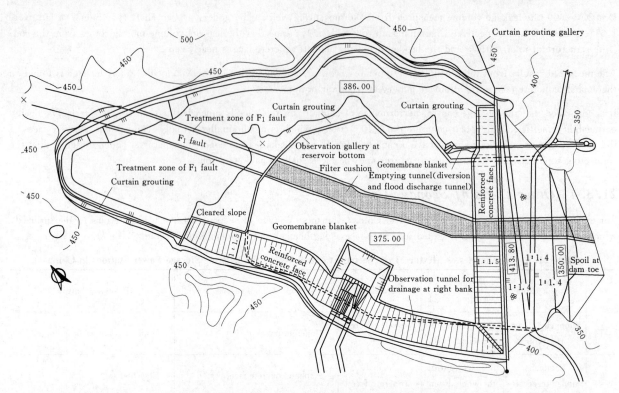

Fig. 21.3-19　Layout of Taian Upper Reservoir （Unit：m）

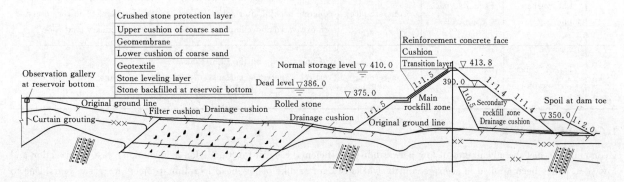

Fig. 21.3-20　Schematic of Typical Section of Taian Upper Reservoir （Unit：m）

tective layer together with geomembrane. The pattern of structure adopted for the seepage control lining layer at the bottom of Taian upper reservoir is given below （from top to bottom）：

1） 250g/m² terylen needle-punched nonwoven geotextile bag packing with 30kg sand aggregate.

2） 500g/m² terylen needle-punched nonwoven geotextile.

3） HDPE geomembrane impervious layer with a thickness of 1.5mm.

4） 500g/m² terylen needle-punched nonwoven geotextile.

5） Geofabric pad with a thickness of 5mm.

6） Crushed stone bed with a thickness of 60cm （20cm thick at upper part with particle size smaller than 2cm, and 40cm thick at lower part with particle size smaller than 4cm）.

7） Rockfill transition layer with a thickness of 120cm.

8） Rockfill zone at reservoir bottom with the compacted layer thickness of 80cm.

In order to strengthen the drainage capacity beneath the geomembrane, a drainage pipe network is provided inside the crushed stone cushion with the spacing of 30m×30m and inner diameter of ϕ90mm. The drain pipes are wrapped with the geotextile and connected with the drainage gallery at reservoir bottom.

(2) Joint Treatment. The key point for seepage control of geomembrane lining is the treatment of joints, including the joints between geomembranes or between geomembrane and surrounding structures. Welding joint, usually with double welds, is adopted between geomembranes to facilitate the vacuum testing and to ensure the reliability of welding. Double water seals are adopted for the joints between geomembrane and surrounding gallery and between geomembrane and concrete of connection plates: one is the join seal between geomembrane and concrete by mechanical anchoring and pressing, another is the connection seal around flexible materials, see Fig. 21. 3 - 21.

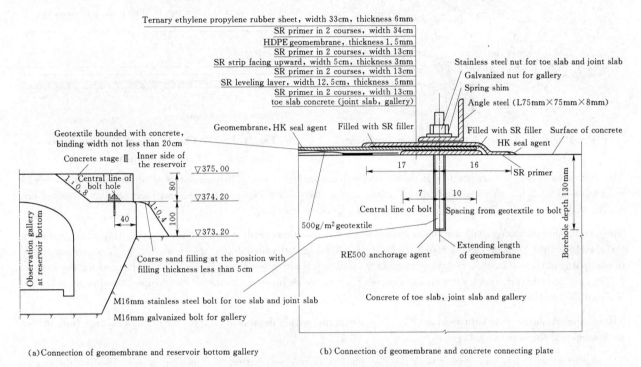

(a) Connection of geomembrane and reservoir bottom gallery   (b) Connection of geomembrane and concrete connecting plate

Fig. 21. 3 - 21 Connection Structure of Geomembrane and Gallery in
Taian Upper Reservoir (Elevation unit: m; Dimension unit: cm)

(3) Design of drainage and air release systems. Two drainage systems are provided in Taian upper reservoir. One is the drainage channel at the lower part of dam connecting the crushed stone cushion, transition layer and rockfill body beneath the geomembrane. It is necessary to set up the auxiliary drainage channel because of the non-uniform permeability coefficient of the embankment material during construction. The other is the drainage gallery added surrounding the geomembrane at the reservoir bottom, which serves the functions of connecting the geomembrane and the perimeter sealing grouting curtain and observing the leakage at the reservoir bottom during operation. In order to release leakage and air from the lower parts of geomembrane, the $\phi 150$mm air release pipes with vertical and horizontal spacing of 30cm are arranged on the top of underlying transition layer of geomembrane and connected with the drain pipes in the drainage gallery at the reservoir bottom.

(4) Effectiveness of geomembrane. The geomembrane placement on the reservoir bottom of Taian upper reservoir took 2 to 3 months. After the station was put into operation for two and a half year, the maximum leakage was 41. 17L/s measured from the measuring weir behind dam and 3. 89L/s from the observation gallery at the reservoir bottom.

Since there are obvious advantages in geomembrane and successful experiences in combined seepage control of concrete face and geomembrane, the scheme of combined seepage control of concrete face and geomembrane is adopted on the Liyang upper reservoir.

2. Clay Blanket

Clay has the advantages of low permeability coefficient and excellent self-healing. Clay is usually taken as impervious material forearth-rockfill dam if enough clay is available near the construction site. Baoquan pumped storage power station is the only case that clay is used as impervious material for upper reservoir. In Baoquan project, the ground water table is over 100m lower than the normal storage level of the reservoir. There are 7 large faults in the reservoir area, 5 of them are tension faults extending from the middle part of reservoir area to the bottom of the dam, with serious seepage problems. If no seepage control measures, the estimated leakage may be 30,000m^3/d to 50,000m^3/d. Therefore, it is decided to take seepage measures for the whole basin. The asphalt concrete faces are used on reservoir slope and dam slope, while clay blanket is set on the reservoir bottom, see Fig. 21.3-22 and Fig. 21.3-23.

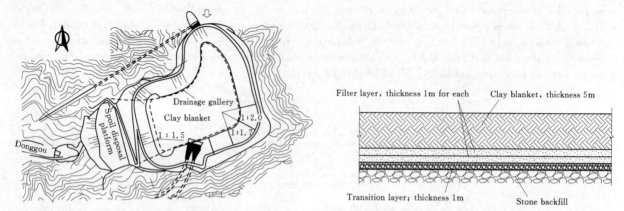

Fig. 21.3 - 22 Layout of Baoquan Upper Reservoir

Fig. 21.3 - 23 Clay blanket of Baoquan Upper Reservoir Bottom

The gross area of asphalt concrete face is 166,000m² with a slope of 1 : 1.7. The face slab is of single lining structure with a thickness of 20.2cm, which is composed of 0.2cm sealing coat, 10cm impervious layer and 10cm leveling cemented layer. The underlying layer is stone drainage cushion with a thickness of 60cm.

The clay blanket on the reservoir bottom is composed of gravel mixture, with an area of 135,000m². Hydraulic gradient is taken as 10 for the blanket. The calculated thickness is 3.99m, and the final design thickness is 4.5m. The compaction ratio required is not less than 98% and the permeability coefficient is not larger than 10^{-6}cm/s. Two filters with a thickness of 0.5cm each and one transition layer with a thickness of 1.0cm are set up at the bottom of clay blanket. Spoil material is filled between the transition layer and reservoir bottom.

The ϕ100mm drainage pipes with spacing of 15 - 25m are buried within the transition layer for draining leakage from the basin bottom into the drainage gallery.

The asphalt concrete face on reservoir bank and clay blanket on reservoir bottom is connected in-line, that is, the asphalt concrete face is supported on the concrete drainage gallery at the reservoir bottom. A connection slope of clay blanket with crest width of 3.0m and slope ratio of 1 : 5.0 is arranged within 20m wide surrounding the reservoir, to extend the contact seepage path of clay and asphalt concrete face. The highly plastic clay is placed on the contact part between clay and facing, with the thickness of 0.5m (See Fig. 21.3-24).

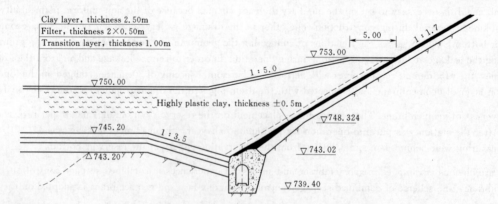

Fig. 21.3 - 24 Connection of Clay Blanket and Face Slab for Baoquan Upper Reservoir (Unit: m)

The connection section between face and clay blanket at the reservoir bottom is 40m higher than the surface of dam foundation. The asphalt concrete face inserts into the clay blanket in ogee type with the radius of 30m and total overlap length of 47m.

The period from completion of clay blanket construction and reservoir impoundment in the next year takes a whole winter. A protective layer with a thickness of 0.3m on the clay blanket is needed to avoid damaging the clay blanket by construction machinery. The protective layer is filled with excavated spoil in the quarry (including reservoir area).

The prospect of clay blanket used on the upper reservoir of pumped storage power stations is unpromising because the clay exploitation usually occupies a large amount of farmland. The clay borrow areas in Baoquan pumped storage power station cover an area of 170,000m², which are mainly farmlands and orchards. Furthermore, the reliability of clay blanket for seepage control, especially in the connection parts, and the complexity of the construction in rainy areas are also the main influencing factors.

21.4 High Pressure Pipe and Powerhouse

21.4.1 *High-head Pressure Pipe*

The feature of water conveyance system of pumped storage power station is that the design head of the pressure pipe is high and the product of design head H and the main pipe's diameter D is large. For the concrete bifurcated pipe of Huizhou pumped storage station, HD is 6,160m·m; In Qingyuan pumped storage station, the bifurcated pipe HD is 6,780m·m, see Table 21.4-1. Crescent rib bifurcation is adopted for all large steel bifurcated pipe and Xilongchi power station has the largest size, $HD=3,552$m·m, for which, 800MPa high strength steel is used. Though the surrounding rock may share partial internal water pressure, the wall thickness of the bifurcated pipe shall be 56mm and the thickness of ribbed plate shall be 120mm, see Table 21.4-2.

Table 21.4-1 Large Concrete Bifurcated Pipe of Pumped Storage Power Stations in China

	Power station	Number × Installed capacity	Bifurcated pipe arrangement	Minimum thickness of rock cover	Rock type	Type of bifurcation pipe	HD	Design head H	Diameter of main pipe D	Diameter of main pipe	Bifurcation angle of bifurcated pipe	Thickness of pipe wall	Grouting pressure	Concrete grade
		MW		m			m·m	m	m	m		cm	MPa	
In operation	Huizhou	8×300	1→4	350	Granite	⊢-type flat bottom	6,160	770	8	3.5	60°	60, bottom 120		C30
	Guangzhou Phase I	4×300	1→4	420	Granite	⊢-type	5,800	725	8	3.5	60°	60	6.5	R_{28}300
	Guangzhou Phase II	4×300	1→4	370	Granite	⊢-type flat bottom	5,800	725	8	3.5	60°	60, bottom 120	6.5	
	Tianhuang-ping	6×300	1→3	330	Tuff	⊢-type flat bottom	5,775	825	7	3.2	60°	60	9	
	Baoquan	4×300	1→2	550	Gneiss	⊢-type	5,200	800	6.5	4.5	55°	60	6.3	
	Pushihe	4×300	1→2	280	Granite	⊢-type flat bottom	3,888	480	8.1	5	60°	80		C25
	Heimifeng	4×300	1→2	215	Granite	Y-type	3,842	452	8.5	5		100		
	Tongbai	4×300	1→2	305	Granite	⊢-type	3,645	405	9	5.5	53°	70	5.2	C30
	Taian	4×250	1→2	260	Granite	Y-type	3,224	403	8	4.8	50°	80	5	C30
	Baishan	2×150	1→2	170	Migmatite	⊢-type flat bottom	1,520	190	8	4.9	45°	80		C25
Under construction	Qingyuan	4×320	1→4	365	Granite	⊢-type flat bottom	6,780	737	9.2	4	60°		6.5	C30

21.4.1.1 Concrete Lined Pressure Pipe

1. Concrete lined pressure pipe of Guangzhou pumped storage station (Phase II)

The maximum static head of high pressure tunnel in Guangzhou pumped-storage station (Phase II) is 610m. In consideration of surge pressure, the maximum design water head is 725m and the diameter of the concrete lined tunnel is 8.5-8.0m. The lining thickness is 60cm, for which concrete No. 300 is used.

Table 21.4-2 Large Crescent Rib Steel Bifurcated Pipe of Existing Pumped Storage Power Stations in China

Power station	HD (m·m)	Design pressure head H (m)	Diameter of main pipe D (m)	Diameter of branch (m)	Bifurcation angle (°)	Steel	Thickness of bifurcated pipe-wall (mm)	Thickness of ribbed plate (mm)
Xilongchi	3,552	1015	3.5	2.5	75	SHY685NS	56	120
Yixing	3,120	650	4.8	3.4	70	P500M	60	100
Zhanghewan	2,704	520	5.2	3.6	70	SHY685NS	52	120
Shisanling	2,599	684	3.8	2.7	74	HT80	62	124

Construction design for high pressure tunnel is carried out according to the principles of permeable lined tunnel. Given that cracking is inevitable for the lining under the force of high head, the internal water exosmose forms a seepage field in the surrounding rocks and the internal water pressure is born by both concrete lining and the surrounding rocks through consolidation grouting. Furthermore, the internal water pressure acting on the tunnel lining is considered as seepage volume force. Employ the heterogeneous, three dimensional anisotropic seepage finite element modal in any main seepage direction, simulate reinforced concrete lining tunnel with large diameter and high internal water pressure, calculate the coupling of concrete lining and reinforcement stresses through stress field and seepage field and determine the width of crack development after concrete cracking.

The calculation results of high pressure seepage lining are adopted for reinforcement. The circumferential reinforcement reduces dramatically comparing with the tunnel section under the same working conditions of Phase I works. For instance, the lower adit bearing the maximum water head decreases from $\phi 32@100$ to $\phi 25@125$ and the longitudinal reinforcement mainly reinforces steel bars according to the structure ($\phi 16@200$) and single-skin reinforcement is adopted.

2. Concrete bifurcated pipe of powerhouse A in Huizhou pumped storage station

The bifurcated pipe in powerhouse A of Huizhou pumped storage station recently built is the largest reinforced concrete bifurcated pipe and the arrangement of one tube for four units is adopted for its water conveyance system. The diameter of the main pipe is 8m, and the branch pipe is 3.5m. The maximum static head is 627m, the design head is 770m and the HD is 6,160m·m.

The bifurcated pipe is composed of 3 paratactic ⊦-type branch pipes of 60° and the arrangement of flat bottom and equivalent pipe diameter is adopted for main pipe (See Fig. 21.4-1). Such arrangement can reduce the head losses, and is convenient for drainage and construction, as well as reduce the investment.

The bifurcated pipe structure is also designed by way of permeable lining tunnel, and finite element analysis is carried out through the three dimensional model on the whole bifurcated pipe with four branch pipes. The outside pressure conditions are calculated with the method of weakening the surrounding rocks, that is, stimulate the incomplete binding function of lower surrounding rocks and the concrete outsider surface under the outer pressure by reducing the element elastic modulus of surrounding rocks (about 1m in thickness) adjoining the concrete lining to 1/50 of the former surrounding rocks' elastic modulus. The final designed bifurcated pipe concrete lining has a thickness of 60cm at the top and 120cm at the bottom. The reinforcement for main pipe in circumferential direction is $\phi 25@100$; for fork in longitudinal direction is $2\times\phi 25@100$ and fork in circumferential direction is $2\times\phi 25@200$.

3. Consolidation grouting pressure

The high pressure consolidation grouting of the high head concrete high pressure pipe and the bifurcated pipe is one of the key factors for ensuring the bearing capacity of surrounding rocks and reducing leakage. Low grouting pressure leads to bad seepage control effect; and high grouting pressure may result in hydraulic fracture of the surrounding rocks. The grouting pressure of the existing projects is usually 1.2-1.5 times of the static pressure of the pipe, see Fig. 21.4-1.

4. Risk of high water head to high pressure reinforced concrete pipe

For some existing pumped storage stations in China, problems such as surrounding rock fracture or large internal water leakage have been encountered when the maximum static water head of concrete high pressure pipe or bifurcated pipe reaches to level 600m. According to the analysis, the thickness of rock coverage of the high pressure pipe with level 600m is usually 400-500m and the minimum ground stress of rock bodies with tiny closed cracks in corresponding development is around 6-7MPa. The safety margin is rather small comparing with the internal water pressure (See Table 21.4-2), so the hydraulic fracture or large leakage may occur in some weak parts, which needs to be highly concerned.

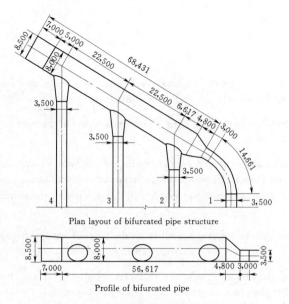
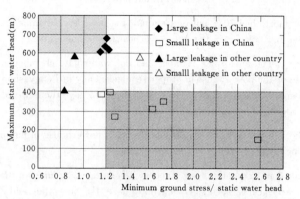

Fig. 21.4-1 Layout of Bifurcated Pipe in Huizhou Powerhouse A (Unit: mm)

Fig. 21.4-2 Relationship Between Leakage Accidents and Minimum Ground Stress/maximum Static Water Head for Pipes Lined with Concrete

21.4.1.2 Penstock

1. The application of high-tensile steel with level 800MPa on penstock

As more and more pumped storage stations with high water head, high-tensile steel plate is widely used in the penstock. Steel with tensile strength at grade 800MPa (HT80) is adopted in the existing pumped storage stations such as Shisanling, Tianhuangping, Zhanghewan, Xilongchi and Baoquan and Hohhot under construction.

The penstock of Xilongchi pumped storage station has the highest design water head (1,015m), the maximum HD value (3,552m·m) and the most 800MPa grade high-tensile steel consumption. The total amount of steel reaches 9,715 tons, in which, the SUMTEN780 steel plate is 5,201 tons.

The Xilongchi penstock is lined totally by steel. C15 micro-expanded concrete is back filled to the thickness of 60cm outside the steel. In the structure design of penstock, it is considered that steel, concrete and surrounding rocks jointly bear the internal water pressure. The sharing ratio of surrounding rocks on the internal water pressure is limited to not larger than 45%. SUMTEN510-TMC steel (with a thickness of 16 - 34mm) is used above elevation 1,203.395m; SUMTEN610-TMC steel (with a thickness of 24 - 38mm) is used above elevation 961.386m; SUMTEN780 800MPa steel (with a thickness of 36 - 60mm) is used below middle horizontal section.

2. Design of embedded bifurcated pipe with surrounding rocks bearing the internal water pressure

See Table 21.4-2 for the large crescent rib bifurcated pipes of pumped-storage stations built in China. The HD value has reached 2,500 - 3,500m·m. Previously, the embedded bifurcated pipes were designed to be open conduits and the surrounding rocks sharing the internal water pressure are only regarded as a safety reservation. However, with the increase of HD value, if the bifurcated pipe is designed as open conduit, the pipe wall is too thick, which will increase not only the investment, but also the difficulty of fabrication, installation and the security risks. Considering that the measured stress of embedded steel bifurcated pipe of the existing projects is much smaller than the design value of open conduit, after deep and comprehensive experimental studies, the pumped storage stations built in recent years, such as Xilongchi, Zhanghewan and Yixing are designed based on the theory that surrounding rocks bear partial internal water pressure.

The joint force-bearing of steel bifurcated pipe and surrounding rocks ispresented in two aspects: ①under the action of internal water pressure, the same as the underground embeddedcircular pipe, and corresponding to the stiffness ratio of surrounding rocks and steel pipe, the surrounding rocks bear partial internal water pressure; ②the bifurcated pipe is restricted by surrounding rocks after deformation, which reduces the peak stress of bifurcated pipe knuckle points, and fully develops the strength of materials. The peak stress of steel bifurcated pipe knuckle points of Xilongchi pumped storage station and the elimination rate of partial membrane stress is basically twice of that of average surrounding rocks sharing ratio.

21.4.2 *Powerhouse*

21.4.2.1 Underground Powerhouse

1. *Layout of underground powerhouse*

Most of the pumped storage stations have been completed or under construction in China, especially the pure pumped storage stations, have the underground powerhouse. See Table 21.4-3 for its layout and main dimensions.

For more than 70% power stations, the headrace penstock axis obliquely intersects the powerhouse axis. On the contrary, over 70% of the draft tubs are vertical to the powerhouse.

The main transformer chamber and tail water gate chamber are usually arranged paralle linside the cavern at downstream side of the powerhouse in proper sequence. Only in Langyashan and Huilong pumped storage stations, the main transformer chambers are arranged at the end of the main powerhouse. And in Langyashan power station, the split-winding transformer is adopted for convenience of layout.

The ball valves are set up inside the main powerhouse in pumped storage stations except Bailianhe pumped storage station, which has a separate ball valve chamber.

Only 4 power stations have their switch stations arranged in the underground transformer chambers. Most of them were built in early years, such as Guangzhou Phase Ⅰ, Shisanling and Zhanghewan pumped storage stations. In Guangzhou Phase Ⅱ pumped storage station, the layout of ground switch station indoor GIS is adopted, which is popular in the layout of most of the pumped storage stations built recently.

Since the draft head of pump-turbine units is rather high and the underground powerhouse is embedded deeply, leakage water in underground powerhouse of previous pumped storage stations is mostly pumped and drained. In recent years, in order to reduce the risk of flooding the powerhouse, gravity-drainage hole is more and more adopted in power stations if the topographical conditions allow, such as Tianpinghuang, Xilongchi, Baoquan and Huizhou pumped-storage stations. The gravity-drainage hole of Qingyuan pumped storage station has length of 4,790m. Although some of the power stations do not have suitable conditions for gravity-drainage hole conducting, the upper gravity-drainage hole is also adopted for draining off leakage water from the upper part of the powerhouse.

Table 21.4-3 Layout of Underground Powerhouses in Large-size Pumped Storage Stations Completed or Under Construction in China

Power station	Number × single unit capacity (MW)	Position of power house	Included angle of water conveyance system and powerhouse		Main powerhouse	Main transformer chamber	Location of auxiliary cavern		Gravity-drainage hole
			Upstream	Downstream	Length × width × height (m×m×m)	Length × width × height (m×m×m)	Draft Tube Gate Chamber	Switch yard	Length (m)
Bailianhe	4×300	Tail	90°	90°	146.7×21.85×50.9	134.4×19.7×19.43	Downstream of transformer chamber	Ground indoor GIS	Upper gravity drainage
Huhhot	4×300	Tail	90°	90°	152×23.5×50	121×17×33.5	Downstream of transformer chamber	Transformer chamber	—
Tianhuangping	6×300	Tail	64°	79°	200×21×46	166×17×21	Downstream of transformer chamber	Ground indoor GIS	1,624
Baishan	2×150	Tail	90°	90°	95×21.7×50.6	65.1×20×15	None	Ground	—
Xilongchi	4×300	Tail	65°	65°	164.5×22.25×49	130.9×16.4×17.5	None	Ground indoor GIS	1,220

Continued

Power station	Number × single unit capacity (MW)	Position of power house	Included angle of water conveyance system and powerhouse		Main powerhouse Length × width × height (m×m×m)	Main transformer chamber Length × width × height (m×m×m)	Location of auxiliary cavern Draft Tube Gate Chamber	Switch yard	Gravity-drainage hole Length (m)
			Upstream	Downstream					
Heimifeng	4×300	Tail	90°	55°	136×25.5×57.2	131×20×19.5	None	Ground indoor GIS	Yes
Tongbo	4×300	Tail	65°	90°	182.7×24.5×53	144.15×18×21.45	None	Ground indoor GIS	Yes
Zhanghewan	4×250	Middle	90°	83°	151.6×23.8×50	117.39×17.8×28.8	None	Transformer chamber	—
Baoquan	4×300	Middle	45°	90°	147×22×47.3	134×18×19.8	Downstream of transformer chamber	Ground indoor GIS	2,295
Guangzhou Phase I	4×300	Middle	65°	90°	146.5×21×44.5	138×17×27.4	Downstream of transformer chamber	Transformer chamber	—
Guangzhou Phase II	4×300	Middle	65°	90°	146.5×21×47.6	138×17×17.6	Downstream of transformer chamber	Ground indoor GIS	—
Shisanling	4×200	Middle	90°	60°	145×23×46.6	136×16.5×25.7	Downstream of transformer chamber	Transformer chamber	—
Huizhou A	4×300	Middle	65°	90°	154.5×21.5×48.25	131.5×18.15×18.95	Downstream of transformer chamber	Ground indoor GIS	4,406
Yixing	4×250	Middle	65°	90°	155.3×22×52.4	134.65×17.5×20.7	Downstream of transformer chamber	Ground indoor GIS	Upper gravity drainage
Xiangshuijian	4×250	Middle	60°	60°	175×25×55.7	167×18×20.8	None	Ground indoor GIS	Upper gravity drainage
Hongping	4×300	Middle	65°	90°	161×22×51.1	150×18.5×20.3	Downstream of transformer chamber	Ground indoor GIS	Upper gravity drainage
Taian	4×250	Head	65°	90°	180×24.5×51.3	164×17.5×18.4	Downstream of transformer chamber	Ground indoor GIS	—
Langyashan	4×150	Head	80°	59°	156.7×21.5×46.17	24.1×21.05×20.07	Combined with tailrace surge	Ground indoor GIS	—

Continued

Power station	Number × single unit capacity (MW)	Position of power house	Included angle of water conveyance system and powerhouse		Main powerhouse	Main transformer chamber	Location of auxiliary cavern		Gravity-drainage hole
			Upstream	Downstream	Length × width × height (m×m×m)	Length × width × height (m×m×m)	Draft Tube Gate Chamber	Switch yard	Length (m)
Pushihe	4×300	Head	60°	90°	161.8×22.7 ×54.1	131.45×20 ×23.7	Downstream of transformer chamber	Ground indoor GIS	—
Qingyuan	4×320	Head	70°	90°	168.5×25.5 ×57.9	155×19.65 ×19.05	Downstream of transformer chamber	Ground indoor GIS	4,790
Xianyou	4×300	Head	90°	90°	161.5×23.5 ×50	160.4×19.7 ×20.9	Downstream of transformer chamber	Ground indoor GIS	—
Xianju	4×375	Head	81°	90°	181×25.5 ×54.5	168×19.5 ×21.5	Downstream of transformer chamber	Ground indoor GIS	1,200
Liyang	6×250	Head	65°	90°	219.9×23.5 ×55.05	193.16×19.7 ×22	Downstream of transformer chamber	Ground indoor GIS	Upper gravity drainage

Note: Not include Minghu and Mingtan pumped storage power stations in Taiwan, China.

2. Flexible supporting measures for underground powerhouse

The design of stability and supporting of surrounding rocks for underground powerhouse of pumped-storage station is the same as normal hydropower stations. Except for the case that reinforced concrete lining is used at the crown of Shisanling pumped-storage station, the flexible supporting measures are taken in almost all the pumped-storage stations, even under poor geological conditions of class IV surrounding rocks.

(1) Xilongchi pumped-storage station-Large-spanning underground powerhouse in horizontal thin-bedded surrounding rocks. The surrounding rock of Xilongchi underground powerhouse is mainly horizontal thin-bedded limestone, which mainly belongs to class $III_b - III_a$. The stability of crown surrounding rocks is the key point in constructing underground powerhouse in horizontal thin-bedded limestone. The design scheme of crown supporting of Xilongchi underground powerhouse is: 20cm thick steel fabric shotcrete with shotcrete reinforcement ribbed arch locally; systematic pre-stressed anchor bolt with the diameter of ϕ32mm, row spacing 1.5m×1.5m and length 5.2m and 7.2m; systematic anchorage cable with 7 rows on the crown, 4 rows of 1,600kN internal anchorage cables, 3 rows of 2,000kN op-thread anchorage cables with the spacing of 4.5m×4.5m, see Fig. 21.4-3. The construction method is: to excavate middle pilot tunnel first, and then to enlarge the excavation from side arch with short proceeding footage and frequent supporting. The key measurement is: firstly excavate the anchor hole expanded from the geological exploration hole at 28.5m above the powerhouse crown and drill the anchor hole in advance. During the excavation of pilot tunnel of the powerhouse, complete the construction of counter strike anchor cables in time, aided by steel fabric shotcrete and anchor bolt.

(2) Yixing pumped storage station-large-size underground powerhouse constructed under adverse geological conditions. The surrounding rock of Yixing underground powerhouse is medium-thick debris sandstone intercalated with argillaceous siltstone, in which, the argillaceous siltstone accounts for 12%–23% with the dip angle of 15°–30°. 10 layers of soft rock strata are distributed with in 40m above the powerhouse crown, and each layer has a thickness of 0.2–0.9m. 7–10 layers of soft strataare also distributed alongthe side wall of the powerhouse. The powerhouse situates between two large faults F_{204} and F_{220}. The width of Fault F_{220} is 5–15m and its minimum distance to the powerhouse is 4.7–8.7m; the width of Fault F_{204} is 1.7–5m and its minimum distance to the powerhouse is 17.3m. Dozens of small faults also develop between the two

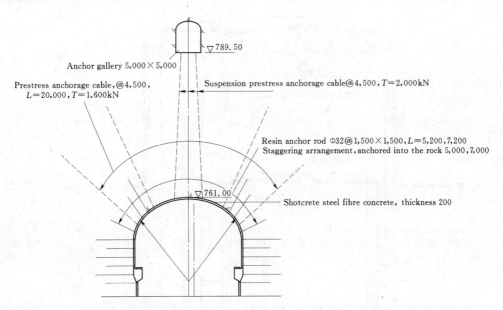

Fig. 21.4-3 Typical Section of Crown Support of Xilongchi Underground Powerhouse (Unit: mm)

faults. The rock mass are mainly moderately broken to broken with the average RQD value of 8.8%–32.6%. Surrounding rock Class IV and V account for a considerable proportion and the seepage during excavation of the underground powerhouse reaches 4,800–6,600m³/d.

Under such poor geological conditions, the flexible permanent supporting is adopted for the main powerhouse. In addition, 15–20cm thick CF30 steel fiber shotcrete, 6–12m long systematic anchor bolts and pre-stressed anchor bolt (300kN), as well as 13.4–40.6m long pre-stressed anchorage cable (1,000kN) are also adopted. Steel arch rib with a thickness of 20cm and permanent drainage holes are set up at the powerhouse crown.

Drainage gallery is constructed in advance to lower down the underground water level and keep the underground powerhouse dry; pre-consolidation grouting is made on the powerhouse crown by drilling downward from the crown drainage gallery before excavation; the underground crown is excavated from pilot tunnels at both sides to the rock pillar in the middle; monitoring methods are fully used and supporting design parameters aretimely revised.

Through the above measures, the excavation of underground powerhouse is completed successfully and the surrounding rocks remain stable, and the maximum displacement of crown is 17.39mm by the end of excavation.

21.4.2.2 Semi-underground Powerhouse

Among the existing pumped-storage stations, only the medium-size pumped storage power stations, Xikou (80MW) and Shahe (100MW), have the semi-underground powerhouse. Shahe Station has a shaft powerhouse installed with two 50MW pumped storage units (see Fig. 21.4-4). The shaft is built in ignimbrite in cylindrical structure. Its inner diameter is 29m and total height is 38m. The shaft wall has the compound supporting structure combining spouting concrete anchorage wall rock and concrete lining with the thickness of 1m. The main powerhouse inside the shaft has four stories: spiral case floor, water pump turbine floor, intermediate floor and generator floor. The generator pier and nose cover are of cylindrical structure. The mass concrete is below the pump turbine floor. There are vent shaft and stair case at upstream side of themain powerhouse. There are auxiliary powerhouse, cable shaft, vent shaft, lift shaft and stairs at the downstream side. In order to prevent the underground water of the surrounding rock bodies from penetrating into the shaft, a circular drainage gallery is set up at 8m outside the shaft and drainage holes are set up to form a drainage curtain. No damp-proofing wall is needed inside the shaft. An aluminum-plastic perforated plate is set up on the surface of the shaft wall, and a water proofing and sound-absorbing cushion is set up between the plate and the wall. Ventilation and air conditioning system is installed, and a dehumidifier is equipped at each of the floors for spiral case and pump turbine.

Frame structure is adopted for powerhouse above the ground, and the steel truss and prefabricated reinforced concrete face are used for roofing. The assembly bay is located at north of the shaft, which is connected with the road access to the station. The auxiliary powerhouse is located at downstream of the shaft, and the substation is built adjacent to the auxiliary powerhouse on the ground.

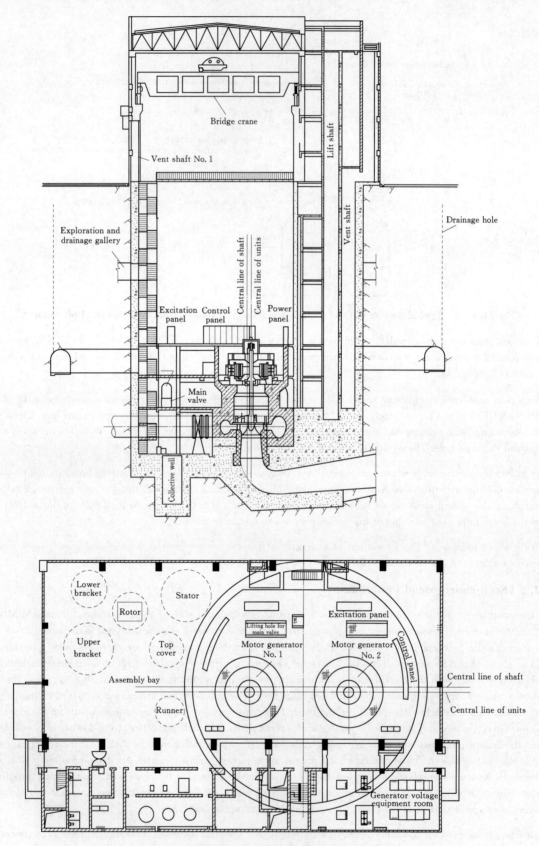

Fig. 21.4-4 Layout of Shahe Shaft Powerhouse

References

[1] Qiu Binru, Liu Lianxi. *Engineering Technologies for pumped-storage stations* [M]. Beijing: China Electric Power Press, 2008.
[2] Peng Cheng. *China Hydropower Engineering in the 21st Century* [M]. Beijing: China Water Power Press, 2006.
[3] Power Grid Peak Shaving and Pumped Storage Station Professional Committee of Chinese Society of Hydroelectric Engineering. *The Thesis compilation of the Annual Conference on Academic Exchanges of Pumped Storage Station Construction* [C]. 2002-2009.

Chapter 22

Typical Projects of High Dam

22.1 Three Gorges Project

Sun Zhiyu[1], Chen Xianming[2]

22.1.1 Overview of Three Gorges Project

22.1.1.1 Project Introduction

The Three Gorges Project (TGP) is the key backbone project for the harnessing and development of Yangtze River, consisting of a hub project, reservoir inundation treatment and immigrant resettlement project, and the power transmission project.

The Three Gorges Project got its name because it is situated in the Three Gorges section of the stem stream of Yangtze River. The dam site of the Three Gorges Project (hereafter Three Gorges Dam Site) is situated in the middle section of Xiling Gorge in Sandouping Town of Yichang City, Hubei Province. It is 38km upstream of the Gezhouba Water Project. The Three Gorges Project is the largest water project in the world, consisting of dam, hydropower plant, navigation structures and guard dam. The dam is a concrete gravity dam, with a dam crest elevation of 185m, a maximum dam height of 181m, and a dam axial length of 2,309.5m. The hydropower stations on the left and right banks just downstream of the dam are installed with 26 sets of 700MW turbine generator units, a total installed capacity of 18,200MW, and an average annual electricity generation of 84.7 billion kW · h. Considering the expanded underground hydropower station on the right bank (6× 700MW) and power supply station (2×50MW), the total installed capacity of the Three Gorges power stations reaches 22,500MW with an average annual electricity generation of 88.2 billion kW · h. The navigation structures include navigation lock and ship lift, where the navigation lock is the double-line five-level continuous navigation lock, and can be opened to the navigation of ten-thousand-ton-level fleet, and the ship lift, a single-line one-level vertical ship lift with a maximum lifting height of 113m, can be opened to the navigation of 3,000 t-class ships, ranking first in the world. Maopingxi Guard Dam is built as an earth rockfill dam with asphalt concrete core, with a dam crest elevation of 185.00m, an axial line length of 889m and a maximum dam height of 104m. A panorama of the Three Gorges project is given in Fig. 22.1-1.

Fig. 22.1-1 Panorama of the Three Gorges Project

---

[1] Sun Zhiyu, Director of Science & Technology and Environmental Protection Department, China Three Gorges Corporation.
[2] Chen Xianming, Chief Engineer of China International Water & Electric Corp.

22.1.1.2 Engineering Construction

The construction of the Three Gorges Project started on December 14, 1994, and all significant milestone progress was completed as in the preliminary design and plan. At the end of 2008, 26 power generator units in the hydropower stations just downstream of the dam on left and right banks of the Three Gorges Project came into operation for power generation one year ahead of schedule, and the Three Gorges Project (excluding the ship lift to be postponed) was completed, and the reservoir possessed the ability to impound water until the normal pool level of 175m is reached. The significant milestones of the Three Gorges Project are shown in 22.1-1.

Table 22.1-1 Significant Milestones in the Construction of the Three Gorges Project

Date	Event	Compared with preliminary design
July 26, 1993	《Report on Preliminary Design of Three Gorges Hydro Complex on Three Gorges of Yangtze River (Hydro Complex Engineering)》approved by Three Gorges Project Construction Commission of the State Council	—
December 14, 1994	Three Gorges Project Hydro Complex Engineering started	—
October 6, 1997	Open diversion channel made accessible to navigation	1 month in advance
November 8, 1997	Success of main watercourse closure on Yangtze River	1 month in advance
May 1, 1998	Temporary navigation lock open to navigation	1 month postponed
October 26, 2002	The dam on the left bank was cast to the design elevation 185.00 m	Conforming
November 6, 2002	Success of the closure for open diversion channel	1 month in advance
April 16, 2003	Upstream roller compacted concrete (RCC) cofferdam in Phase III was cast to the design elevation of 140.00 m	2 months in advance
June 10, 2003	The reservoir impounded till the 135m water level, and the hydro complex was put into the period of power generation with the water retaining by the cofferdam	Conforming
June 18, 2003	Trial navigation of navigation lock (suitable for the 135-139m water level)	Conforming
July 18, 2003	The first 2 power generator units in the hydropower station on the left bank were networked for power generation and put into commercial operation	3 months in advance
July 8, 2004	One year after the trial operation of the navigation lock, it was accepted and put into navigation	—
September 16, 2005	All the 14 power generator units in the hydropower station on the left bank were put into commercial operation	1 year in advance
May 20, 2006	The whole line of the dam on the right bank is cast to the design elevation 185.00m, and the Hydro Complex has achieved the designed flood control capacity	7 months in advance
October 27, 2006	The reservoir impounding level was up to 156m, and the Hydro Complex was put into the initial operation period	1 year in advance
April 30, 2007	The construction of the navigation lock was completed according to the final operation water level (145-175m)	—
June 11, 2007	The first power generator unit in the hydropower station on the right bank was put into commercial operation	1 year in advance
October 30, 2008	All the 12 power generator units in the hydropower station on the right bank were put into commercial operation. Thus all of 26 power generator units were put into operation	1 year in advance
November 4, 2008	The reservoir test impounding level was up to 172.8m	—
August 29, 2009	Passed the acceptance of the State Council Acceptance Committee on normal impounding level up to 175m	—
October 26, 2010	The trial impounding level of the reservoir was up to the normal impounding level 175m	3 years in advance

According to the regulation of Three Gorges Project Construction Commission of the State Council on the approval of the preliminary design, the ship lift should have been put into operation in May, 2003. In May, 1995, Three Gorges Project Construction Commission of the State Council approved the postponement for the construction of ship lift. In September, 2003, Three Gorges Project Construction Commission of the State Council approved the new design proposal of the ship lift, in which the original "wire rope hoisting lift" proposal is replaced by "rack and pinion climbing" proposal. The water-retaining structures on the lock head for the ship lift were completed in 2003, and the upstream diversion navigation channel and auxiliary structures were completed in 2006, meeting the requirements for a gradual increase in the reservoir water level. In September, 2007, the construction of the ship lift resumed. And the ship lift is expected to be put into operation in 2015.

In the preliminary design, the position for the subsequent expansion of underground hydropower station was preserved on the right bank, and it was required that the water intake part for the underground hydropower station shall be firstly constructed, to avoid the difficulty of construction after impounding water in the reservoir. The preconstruction of the water intake was totally completed in 2004, meeting the requirements of a gradual increase in the reservoir water level. Considering high-speed national economic growth and the intense power supply nationwide, the construction for the remaining part of the underground hydropower part with the approval of Three Gorges Project Construction Commission of the State Council got started in March 2005, for which the investment would be included in the engineering budget of the Three Gorges Project. All of the six power generator units in the underground hydropower station are expected to be put into operation for power generation in 2012.

To ensure safe operation and power supply system safety of the Three Gorges Project, and improve the power supply reliability of the Three Gorges Project, Three Gorges Project Construction Commission of the State Council approved the construction of the power station with self-contained power source in September 2003, with the installation of two 50MW water turbine generator units. The power supply station is made as the underground power plant, where two water turbine generator units were put into operation in February 2007.

22.1.1.3 Operation of the Three Gorges Project

Since the water retaining level of 135m with the cofferdam, 156 m-level operation in the initial period and the 175 m-level trial impounding, the Three Gorges Project has met the expected goal with respect to the flood control, power generation, navigation, drought relief and water supply.

With respect to the flood control and drought relief, the Three Gorges Project received the first flood peak of $70,000 m^3/s$ on July 20, 2010, which was the third largest flood peak in the history record of Yichang City Hydrologic Station. It controlled the flood discharge, and played a huge role in alleviating the flood control pressure in the middle and lower reaches of Yangtze River. At the same time, the Three Gorges Project has also played the role of drought relief and water supply in the dry season of the Yangtze River. In the dry season from 2009 to 2010, the Three Gorges Reservoir supplied a water volume of more than 13.0 billion m^3 for the middle and lower reaches of the Yangtze River.

With respect to the power generation, the Three Gorges power stations had generated the power of more than 400.0 billion kW·h by July 2010, and the electric energy has been constantly transmitted from Three Gorges to East China, Central China and Guangdong region, providing a large amount of clean energy for the national economy and daily life of the people, equivalent to saving 134 million tons of standard coal, reducing the emission of 304 million tons carbon dioxide, and reducing the emission of 3.60 million tons sulfur dioxide.

With respect to the navigation, navigation lock of the Three Gorges operates well and the quantity of shipment continuously grows. Since the reservoir started impounding water in 2003, the navigation industry on the Yangtze River has rapidly developed. Take the quantity of shipment in Gezhouba Ship Lock as an example, the annual average shipment has been up to 49.46 million tons after the impoundment, which is four times 9.58 million tons before the impoundment. Because of the improvement of the water flow in the navigation channel, the navigation energy consumption has been reduced and the cost has also been decreased.

During the trial impoundment and drawdown of the Three Gorges Reservoir from 2008 to 2010, comprehensive monitoring and testing were carried out on the sediment accumulation in Chongqing Port, the stability of the bank near the Three Gorges Dam, earthquake in the region near the dam, safety monitoring on the Three Gorges Project, operation of high water head for the generator units and water quality in the reservoir area, and the results have shown that the Three Gorges Project and the reservoir operate normally in various aspects. The operation in recent years has shown that the quality of Three Gorges Project was effectively tested, and the reservoir sediment has been well predicted, having achieved huge social and economic benefit. Successful construction of Three Gorges indicates that the hydropower in China has been at the forefront in the world and that China has grown from a large hydropower country into a strong hydropower country.

22.1.2 Dam Structure and Arrangement

22.1.2.1 Design Standard for the Dam

(1) Engineering grade and building class: Three Gorges Project as the first-grade project; the dam and left guide wall as the first-class structure; right guide wall (downstream longitudinal concrete cofferdam) as the second-class structure; downstream apron as the third-class structure.

(2) Design flood criteria. See Table 22.1-2 for the design flood criteria.

Table 22.1-2 Design Flood Criteria

Project and operation period	Design flood frequency (%)	Check flood frequency (%)	Remarks
Dam	0.1	0.01, being enlarged with 10	
Water retaining and flood protection for the Dam after the blocking of the bottom hole	0.5	0.2	
Scour protection of left and right guide walls	1	0.1	
Downstream scour in the spillway dam section	5	1	

See Table 22.1-3 for the characteristic water level and flow of the project.

Table 22.1-3 Characteristic Water Level and Flow of Three Gorges Project

Item		Unit	Initial	Later period	Power generation period with cofferdam
Normal pool level		m	156	175	135.0
Water level limited for flood control		m	135	145	135.0
Drawdown low level in the dry season		m	140	155	135.0
20-year flood	Reservoir water level	m	150.7	157.5	135.4
	Discharge	m³/s	56,700	56,700	71,750
100-year flood	Reservoir water level	m	162.3	166.9	139.8
	Discharge	m³/s	56,700	56,700	75,250
1,000-year flood	Reservoir water level	m	170	175	
	Discharge	m³/s	71,000	69,800	
	Downstream water level	m	76.6	76.4	
Flood once 5,000 years	Reservoir water level	m		175.4	
	Discharge	m³/s		90,400	
	Downstream water level	m		80.9	
Check flood (once 10,000 years plus 10%)	Reservoir water level	m		180.4	
	Discharge	m³/s		102,500	
	Downstream water level	m		83.1	

(3) Earthquake-resistant fortification standard. Basic seismic intensity is Degree 6; the design seismic intensity of the Dam is Degree 7; the left and right guide walls are non-backwater structures. But considering special significance of the Three Gorges Project and the difficulty in case of any damage, it is rechecked as Degree 7 during the design.

22.1.2.2 Structural Arrangement of the Dam

In the middle of the river course for the Three Gorges Dam, the spillway dam section is designed, the left side is designed with the dam section of left guide wall, the dam section of the plant building on the left bank and the non-overflow dam section on the left bank; while the right side is designed with the dam section of longitudinal cofferdam, the dam section of the plant building on the right bank, and the non-overflow dam section on the right bank. The spillway dam section is arranged with 23 deep holes and 22 surface holes, in which the deep holes are arranged in the middle of every spillway dam section, and the surface holes are arranged by means of joint-crossing in the upper part; in addition, 3 floating trash removal holes, 7 desilting holes (excluding the desilting holes for the underground hydropower station on the right bank) and 2 flushing sluices. See Fig. 22.1-2 for the layout of the dam.

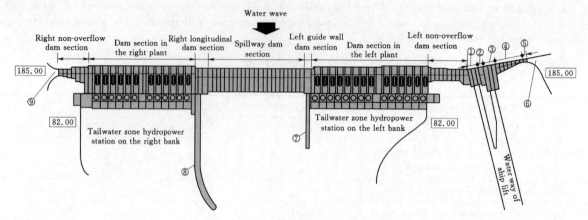

Fig. 22.1-2　Schematic Diagram for the Layout of the Dam (Unit: m)

①—Dam section near the ship (Length 62.00m); ②—Left non-overflow dam section (Length 23.07m); ③—Ship lift dam section (Length 62.00m); ④—Left non-overflow dam section 1#-7# (Length 140.00m); ⑤—Connecting left non-overflow dam section (Length 15.00m); ⑥—Highway leading to the dam on the left bank; ⑦—Left guide wall; ⑧—Longitudinal cofferdam in the downstream of the right guide wall; ⑨—Highway leading to the dam on the right bank

22.1.2.3　Hydraulic Design of Spillway Facilities

1. Arrangement and discharge capacity

Main considerations for the arrangement of the spillway facilities include the reservoir flood regulation as well as the general arrangement, reservoir sand discharge, floating debris removal, engineering protection and construction diversion.

Considering the high water head, large flood discharge and sand discharge capacity of the Dam, and in combination with the requirements in flood control, desilting, project protection, desilting before the plant and floating debris removal, the deep hole and surface hole are alternatively arranged for the spillway facilities of the Dam. At the same time, the 26 generator units in the hydropower station are used for flood discharge. See Fig. 22.1-3 for the typical section of the spillway dam. See Table 22.1-4 for the discharge capacity of the Three Gorges Project.

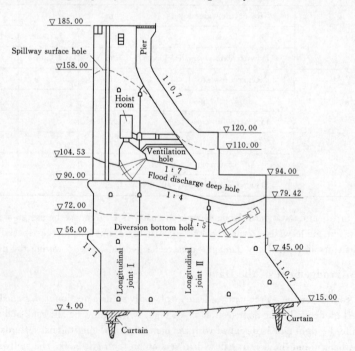

Fig. 22.1-3　Typical section of spillway dam section (Unit: m)

Chapter 22 Typical Projects of High Dam

Table 22.1-4 Discharge Capacity of the Three Gorges Project

Reservoir water level (m)	Flood discharge capacity(m³/s)					Total flood discharge capacity(m³/s)
	Deep outlet	drains for floatage	Desilting bore	Surface outlet	26 generator units	
135	33,510	90	2,230	—	22,360	58,190
140	35,760	570	2,330	—	23,010	61,670
145	37,880	1,290	2,430	—	23,660	65,260
150	39,880	2,170	2,540	—	24,310	69,800
155	41,790	2,930	—	—	25,090	69,810
160	43,620	3,580	—	820	24,050	72,070
165	45,370	4,140	—	5,850	23,010	78,370
170	47,060	4,620	—	13,630	22,100	87,410
175	48,680	5,070	—	23,540	21,450	98,740
180	50,260	5,470	—	35,260	21,060	112,050
183	51,180	5,700	—	43,090	20,670	120,640

During its calculation according to the flood control scheduling plan and the scheduling application, 10% of the sluices out of control are considered for the deep holes, floating debris removal holes and desilting holes for the purpose of allowance, and the flood discharge capacity is calculated as 90%; 70% of the 26 generator units (18 units) are considered with the overflow. At the same time, in case of the super large flood with a return period of 1,000 years, recheck was carried out on the situation that all these generator units in the hydropower station clogged up. See Table 22.1-5 for the calculation results of flood regulation.

Table 22.1-5 Calculation Results of Flood Regulation

Working condition	Discharge and water level	Flood frequency			
		5%	1%	0.1%	0.01%, being enlarged with 10%
Final scale: Water level limited for flood control 145m; Normal pool level 175m	Reservoir water level (m)	157.5	166.9	175.0	180.4
	Discharge (m³/s)	56,700	56,700	69,800	102,500
	Downstream water level (m)	73.8	73.8	76.4	83.1
Initial operation: Water level limited for flood control 135m Normal pool level 156m	Reservoir water level (m)	150.7	162.3	170.0	—
	Discharge (m³/s)	56,700	56,700	71,000	—
	Downstream water level (m)	73.8	73.8	76.6	
Water level of water retaining with cofferdam for power generation during the construction period of 135m	Reservoir water level (m)	135.4	140.0	—	—
	Discharge (m³/s)	71,750	75,250	—	—
	Downstream water level (m)	77.0	77.8	—	—
All generator units in the hydropower station are clogged up	Reservoir water level (m)	—	—	—	182.7
	Discharge (m³/s)	—	—	—	95,500
	Downstream water level (m)	—	—	—	81.8

For the purpose of diversion and flood regulation, 22 diversion bottom holes are designed; the long pressurized pipe with short and open discharge channels is used, where the outlet size of the pressurized pipe is 6m × 8.5m. The diversion bottom holes came into use before the interception of open channels in November 2002. In June 2003, the reservoir storage level was 135m, and in July of the same year, the first generator units in the hydropower station on the left bank were put into operation, and the Three Gorges Project was put into the power generation period with cofferdam for water retaining. During the flood period, 2 generator units were considered with overflow, which would shoulder the task of flood control with 22 diversion bottom holes and 23 deep holes, 3 desilting holes (dam section of the left hydropower station building).

2. Energy dissipation and scouring prtection

(1) Downstream energy dissipation. The surface flow energy dissipation is used for flood discharge deep outlet and surface outlet. According to the hydraulic simulation test, the most unfavorable situation for the downstream scour pits happens during the discharge of a 1,000-year flood, with surface outlet and deep outlet jointly working for flood discharge. The lowest scouring pit is in the middle of the river course, with an elevation of 21.00m, about 142m from the dam toe. Considering that the highest foundation construction surface is 45m for the spillway dam section and the upstream slope of the scour pit is 1 : 5.9, the foundation for the dam toe is safe.

During the period of power generation with the cofferdam for water retaining, the outflow of the bottom hole was mixed with the surface flow, there was reverse eddy under the tongue with a maximum reverse flow rate of about 4 – 5m/s. In addition, since the diversion bottom hole on the right side is about 50m from the right guide wall, the large backflow zone was formed on the right side of the discharge zone, with a maximum backflow rate of up to 10 – 11m/s, so that when the reservoir level was 135m, the elevation for the partial scouring depth on the right side of the dam toe was 24.6m, and the elevation on the left side of the right guide wall was about 25.0m. During the construction period, flood discharge became the controlled working condition for the scouring protection design of the dam toe and right guide wall.

(2) Scouring protection design. The minimum scour pit elevation on the left side of the right guide wall was 31.0m, and the elevation of the construction foundation surface was 45.0m. For this reason, the concrete anti-scour wall was built within the scour pit on the left side of the right guide wall, and the elevation at the wall bottom was 30.0m. The elevation for the construction foundation surface of the left guide wall was lower than the elevation of the scour pit, thus needs no protection.

During the period of power generation with the cofferdam for water retaining, deep holes and surface holes were used for flood discharge, and the backflow in wide range would be generated at the right side under the dam. To eliminate the backflow and ensure the safety, two anti-scour blocking piers were designed on the left side of the right guide wall, with a height of 10m and 18.5m respectively.

22.1.2.4 Design of Deep Sliding Stability in the Dam Section of the Bank Slope

The foundation rock in the dam is porphyritic granite with hard and complete rock body, uniform lithology and high mechanical strength. Most sections of the dam foundation have the low-angle dip and aren't developed with the fissures, which will not become the controlled sliding surface, but the 1 – 5 dam sections in the left plant and in the 24 – 26 dam sections of the right plant, the structural surface with gentle slope in the foundation rock was relatively developed. To meet and improve the safety of the deep sliding stability, the following engineering structure measures are adopted in the 1 – 5 dam sections of the left plant.

(1) Properly reduce the construction foundation surface of the dam, from 98.0m to 90.0m in the elevation, and install the key walls in the upstream with the bottom elevation 85.0m.

(2) Enlarge the width at the dam bottom towards the upstream, use the support corbel of the trash rack at the water inlet of the water diversion pipe in the hydropower station to extend the dam heel upward 17.5m; at the same time, move forward the main curtain and discharge of the dam, and increase a row of main curtains.

(3) The plant building is close to the rock slope of the dam, the contact grouting is used between the concrete and rock slope in the plant building below the 51.0m elevation, to ensure the joint force on the dam at the plant.

(4) At No. 1 – 5 dam sections in the left plant, the horizontal joints are designed with the keys and grouted, to strengthen the overall effect; the tail channel of the hydropower station is consolidated and grouted within certain range to ensure the effect of the resistance body in the downstream.

(5) The enclosed pumping drainage system is designed between the dam the plant building foundation. At the elevations of 73.0m and 50.0m on the upstream of construction foundation rock, two rows of drainage tunnels are arranged, which are interconnected with a drainage hole drilled between them. At the 25.0m elevation on the downstream of the dam foundation, a drainage tunnel is dug (from No. 17 non-overflow dam section in the left to No. 6 dam section in the left plant); the No. 1 – 6 dam sections of the left plant and the plant building have formed the enclosed pumping drainage zone.

(6) Use the controlled blasting technology to ensure the integrity of the foundation rock.

(7) Strengthen the anchor support of the side slope.

(8) Strengthen the consolidation grouting. Apply the ordinary consolidation grouting on the dam foundation, and add the consolidation grouting on the fissures with shallow dip in the shallow layer and on the checked long and large fissure surface within certain depth.

(9) Preset the corridor in the field, to provide the conditions for the consolidation treatment in the subsequent period.

(10) Strengthen the safety monitoring, and list the No. 1 - 5 dam sections in the left plant as the critical monitoring positions.

(11) See Fig. 22. 1 - 4 for the consolidation treatment on the dam foundation for the No. 1 - 5 dam sections in the left plant. From the calculation and analysis, the safety coefficient of the deep sliding stability on the No. 1 - 5 dam sections in the left plant of the Dam is more than 4.0, thus meeting the design requirements, with considerable safety allowance. Therefore, for the No. 1 - 5 dam sections in the left plant, comprehensive engineering structure measures have been taken, thus the deep sliding stability in the dam foundation is ensured.

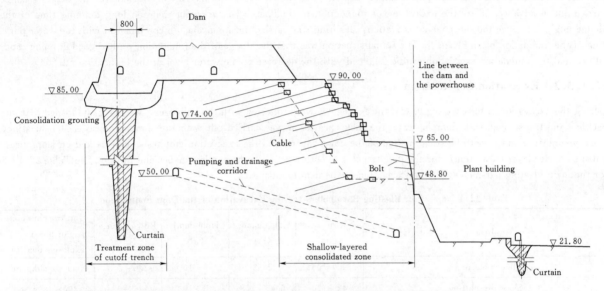

Fig. 22. 1 - 4   Schematic Diagram for Consolidation Treatment on the Dam Foundation for the No. 1 - 5 Dam Sections in the Left Plant (Unit: m)

22. 1. 3 *Dam Foundation Excavation and Foundation Treatment*

22. 1. 3. 1 Rock Excavation at the Dam Foundation

(1) Large engineering scale and high excavation intensity. The total foundation excavation quantity between the dam and the plant building was 28.30 million m^3, which was completed in phases. The construction peak happened in the second phase with an excavation quantity of 10.75 million m^3, an annual excavation intensity of up to 7.25 million m^3, and the excavation construction period of the building foundation was 7 months with the monthly excavation intensity up to 800 thousand m^3. In the third phase, monthly excavation intensity was also above 400 thousand m^3.

(2) Hard foundation rock and large working quantity of the drilling and blasting. Except the loose soil at first several meters from the ground surface, 70% of the foundation excavation was the rock, the compressive strength of the fresh stone was up to 110MPa, thus increasing the excavation intensity.

(3) High quality requirement. The levelness of the foundation surface for the dam shall be controlled at ±20cm; the underbreak isn't allowed in the foundation surface of the plant building and the overbreak capacity shall not be more than 20cm. Pipe trench and cutoff trench in the dam section of the bank slope are complex in their shapes, thus raising high requirements for excavation.

(4) Large excavation height for slope. There are many positions with the excavation slope at the height above 100m, and the excavation height for the slope of the plant building at the bank slope is up to 122.0m, therefore practical and effective measures shall be taken for safety protection and quality control.

22. 1. 3. 2 Excavation Measures

(1) Firstly remove the upper top in advance, thus reduce subsequent excavation quantity, alleviate the excavation intensity, and complete the excavation handover interface in advance. For example, in the excavation for Phase I on the right

bank and Phase I on the left bank, the upper top removal excavation was firstly completed.

(2) The favorable terrain of the bank slope is used for the foundation excavation in the dam section of the bank slope and the foundation excavation for the hydropower station building, the progress was about 6 – 12 months in advance compared with the excavation of the large foundation pits in the same period. Therefore there is considerable progress and economic benefit. For example, the foundation excavation for No. 1 – 6 dam sections in the left plant, No. 24 – 26 dam sections in the right plant and plant building was commenced in advance, and the construction is continued by means of hard water retaining with the preserved stone during the interception and cofferdam construction.

(3) Finish the underwater desilting of the foundation pit in advance according to the conditions. After the evacuation of the foundation pit, it cannot be followed by the sludge of mechanical excavation, and shall be combined with the water drainage of the foundation pit to complete the desilting task and accelerate the excavation progress; for the sludge with good bleeding (one week about after the evacuation of the foundation pit, it can be naturally drained and added with mechanical excavation), the desilting can be carried out in advance. Before the excavation of foundation pit in Phase II, worrying that too long draining time of the sludge might influence the subsequent excavation, desilting was carried out in advance in combination with the water pumping of the foundation pit in Phase II. The actual situation was that the sludge of the foundation pit had good draining capability, and the foundation excavation can be followed with the decrease of the water level in the foundation pit.

22.1.3.3 Excavation Construction

During the excavation of the covering layer, the aggregates were collected with the bulldozer, were dug and loaded with the backhoe and transported with dumping car; the rock excavation was carried out with the combination of bench blasting, slope presplitting and smooth blasting method of the cover layer, and the excavation emphasis shall be laid on slope presplitting, cover layer excavation, and the pipe trench and cutoff trench with complex shapes and profiles. See Table 22.1 – 6 for the main blasting parameters for the excavation of the dam foundation.

Table 22.1 – 6 Main Blasting Parameters for the Excavation of the Dam Foundation

Item		Hole diameter (mm)	Hole distance (m)	Hole depth (m)	Row distance (m)	Unit consumption or linear explosive density
Bench blasting		90 – 100	2 – 3	5 – 10	2.0 – 2.5	0.50 – 0.60kg/m³
Slope presplitting		42 – 100	0.45 – 0.8	3 – 10	—	400 – 480g/m
Excavation of the cover layer	Vertical shallow hole	42	0.80 – 1.0	1.7 – 2.0	0.6 – 1.0	0.50 – 0.55kg/m³
	Horizontal presplitting	42 – 90	0.45 – 0.8	3 – 10	—	380 – 450g/m
	Horizontal smooth blasting	42	0.50	3 – 4	—	200g/m
	Vertical main blasting hole	42	1.20	3 – 4	1	0.45 – 0.49kg/m³
	Horizontal main blasting hole	42	1.50	3 – 4	0.8	0.40 – 0.45kg/m³

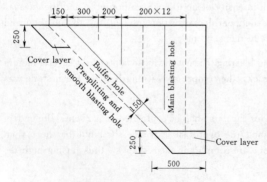

Fig. 22.1 – 5 Arrangement Section for Presplitting and Smooth Blasting Hole (Unit: cm)

(1) Slope excavation. According to the slope profile, the presplitting blasting technology was used, where the bench height was 3 – 10m, while manual pneumatic drill and YQ-100B submerging hole was used for holing. The 2 – 3 blasting holes adjacent to the presplitting surface of the slope are used as the buffering holes, and the hole distance was 1/3 – 1/2 reduced by comparison with the main blasting hole, to control the single dynamite amount less than 100 kg. To create good free face, a presplitting is generally designed in every level of the slope. See Fig. 22.1 – 5 for the arrangement section of presplitting and smooth blasting hole.

(2) Excavation of horizontal cover layer. From the production experience, it was determined that the thickness of the preserved cover layer on the construction foundation surface is 2.5m. There are two methods for the excavation of the cover layer: horizontal smooth blasting hole and vertical shallow hole, horizontal main blasting hole and horizontal smooth blasting hole. Underbreak is not allowed for construction foundation surface of the plant building, and all holes shall be made with manual pneumatic drill, with the circulating depth 3.0 – 4.0m; the construction foundation surface of the Dam was generally made with the vertical holes on the top of manual pneumatic drill, 100-type down-the-hole drill was used to create the horizontal holes, with the

depth of 8 – 10m. The distance from the bottom of vertical hole to the horizontal presplitting surface is controlled at 0.5 – 0.8m, and vertical hole and horizontal hole were blasted with the interval of 75 – 100m.

(3) Pipe trench excavation. In the dam section of bank slope, the trench excavation was started on the water diversion pressurized pipe, and carried out layer by layer according to the excavation method of the cover layer, with the thickness of each layer 2.5m. Every layer was excavated into two zones, and the area of 2.5m at both sides was the excavation zone for the cover layer. The middle part was the trench cutting zone, in which one layer shall be excavated, so that the excavation zone of the cover layer shall be interfaced with the gentle slope of the trench cutting zone, to make full use of the free face in the downstream, and the work surface of steel pipe trench is maintained with the dustpan excavation shape, and 6 degrees slightly inclined towards the downstream, to facilitate the drainage and debris removal on the work surface. See Table 22.1 – 7 for the blasting parameters of steel pipe trench excavation drill.

Table 22.1 – 7 **Blasting Parameters of Steel Pipe Trench Excavation Drill**

Item	Hole diameter (mm)	Hole distance (m)	Hole depth (m)	Row distance (m)	Cartridge diameter (mm)	Unit consumption (kg/m³)	Linear dynamite amount (g/m)	Single-hole dynamite (kg)
Main blasting hole	42	1.2	2.6	1.0	32	0.67	—	2.00
Reinforcing hole	42	1.0	2.6	0.8	32	0.88	—	2.00
Buffering hole	42	1.0	2.6	—	25	0.40	600	1.08
Presplitting hole	42	0.4	2.6	—	25	—	180 – 220	0.72

(4) Blasting vibration control. During the blasting excavation, it is required that the blasting vibration shall not result in the cutoff wall of cofferdam, newly cast concrete, and seepage-proof curtain, and it is required to strictly control the single dynamite amount, thus controlling the safety vibration speed on the mass point for the surrounding buildings. The control indices are given below. See Table 22.1 – 8 for the safety vibration speed on the mass point on the foundation of the newly cast concrete.

Table 22.1 – 8 **Vibration Speed on the Safety of the Mass Point**

Concrete age (d)	Initial setting – 3	3 – 7	7 – 28	>28
Control parameter (cm/s)	1.5 – 2.0	2.0 – 3.0	3.0 – 5.0	5.0 – 8.0

Note: The control shall be based on the above criteria before the formal conclusion for the site blasting shock test hasn't been drawn.

The vibration speed of the safety mass point on the design slope shall not be more than 10cm/s; the vibration speed of the safety mass point at the grouted part shall not be more than 1.2 – 1.5cm/s; the vibration speed of the safety mass point on the foundation or wall surface of the drainage tunnel shall not be more than 10cm/s; the vibration speed of the safety mass point on the foundation or wall surface of the desilting hole and vertical shaft shall not be more than 10cm/s.

22.1.3.4 Excavation Quality

The construction foundation surface is mainly covered with slightly fresh rock. The overbreak and underbreak shall be generally controlled within −20cm to +5cm, and in the sound wave test of the construction foundation surface, the average wave speed shall be above 5,200m/s. See Fig. 22.1 – 6 for the excavation quality about the surface of the construction foundation on the spillway dam section.

Fig. 22.1 – 6 Impression Drawing for Actual Excavation of Construction Foundation Surface on the Spillway Dam Section

22.1.4 Seepage Control

22.1.4.1 Seepage Control Measures

The seepage control for the dam foundation is realized with the proposal combining the seepage-proof drainage and the enclosed pumping drainage; enclosed pumping drainage is used in the river course dam section of relatively low construction foundation surface and large downstream water depth, while the general proposal with curtain seepage prevention and the drainage with drainage tunnel is used in the bank slope dam section at relatively high construction foundation surface. The natural water drainage is used for the dam section of the check level beyond the elevation on the construction foundation surface on the downstream and both ends of the dam, and the seepage-proof curtain under the seepage-proof wall is used on some section between the dam and the ship lock. Considering that the fissure with gentle dip on No. 1 - 5 dam sections of the left plant on the bank slope and No. 24 - 26 dam sections of the right plant is relatively developed, the height difference between the construction foundation surface of the dam and the foundation of the plant building behind the dam is about 68m, thus unbeneficial to the sliding stability in the depth of the dam foundation. The enclosed pumping drainage range for the dam foundation in this position is extended to the foundation of the plant building. The seepage-proof curtain is designed as 1Lu. Fig. 22.1 - 7 shows the foundation seepage-proof curtain and drainage arrangement in typical dam sections.

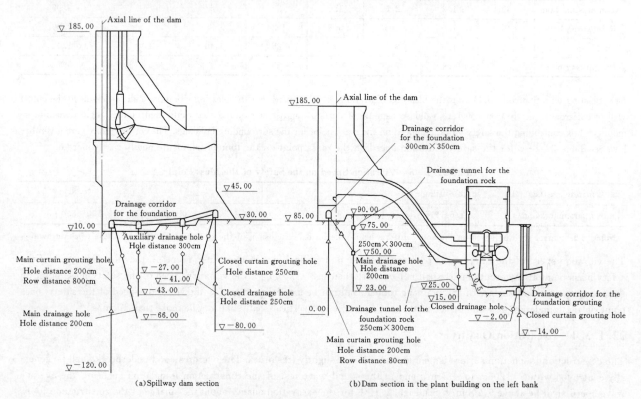

Fig. 22.1 - 7 Arrangement Diagram for Foundation Seepage-proof Curtain and Drainage Arrangement in Typical Dam Sections (Unit: cm)

22.1.4.2 Curtain Grouting Construction

Curtain grouting construction is carried out with the process of "small diameter, enclosed hole, from up to down, circulation in hole and high-pressure grouting" under the rule of sequencing and thickening.

The construction of curtain grouting is generally carried out in three phases, No. 1 non-overflow dam section in the left-No. 11 non-overflow dam section in the left for the construction in Phase I ; the No. 12 non-overflow dam section in the left-No. 2 right longitudinal dam section, non-overflow connecting section in the left and non-overflow connecting section in the right for the construction in Phase II ; right plant drainage-No. 7 non-overflow dam section in the right for the construction in Phase III. General depth is about 200,000m.

The drilling and grouting length of the foundation rock for main curtain of the dam is 130,700m. Average unit cement injection quantities in the I-sequence hole, II-sequence hole and III-sequence hole are 18.42kg/m, 6.90kg/m and 4.49kg/m respectively.

The drilling and grouting length of the foundation rock for enclosed curtain of the dam is 68,100 thousandm. Average unit cement injection quantities in the I-sequence hole, II-sequence hole and III-sequence hole are 28.32kg/m, 9.96kg/m and 6.23kg/m respectively.

After supplementing the chemical grouting, it is required to carry out the after-grouting inspection, with the water permeation ratio 0.01-0.11Lu, which is far below the designed 1Lu, and the fine fissures have been effectively grouted. With the increase of the hole sequences, the water permeation ratio and unit cement consumption have been obviously and progressively decreased.

22.1.4.3 Construction Quality

Quality inspection shall be carried out with water pressure test, sound wave test and drilling core.

The quantity of the holes for quality inspection shall be generally controlled with 10% of grouting holes, and one unit shall be arranged with at least one hole. The inspection holes shall be arranged on the curtain axial line, and shall cover the positions with complex geological conditions, the hole sections and positions with large grouting quantity, the positions with relatively large hole deviation, and the positions with abnormal grouting or with the problems in quality analysis.

In the water pressure test after the grouting, the sections where the water permeation rate of main curtain is not more than 1Lu account for 99.56%; the sections where the water permeation rate of enclosed curtain is no more than 1Lu account for 99.06%. The unqualified hole sections have been treated, and meet the qualified standard after the recheck.

The single-hole and cross-hole sound wave test was applied, which was 5,100-5,400m/s on average before grouting, and was improved to 5,300-5,600m/s, with the improvement percentage about 2.45%-8.40%.

From the large-diameter drilling hole inspection, it is known that the cement stone is dense and hard with good adhesion. Single-axis compressive strength of stone core sample: 26.1MPa at dry state, 23.8MPa in the saturated state; tensile strength 1.0MPa, flexural strength 1.2MPa, and stone anti-permeability test 1.4-2.1MPa.

22.1.5 Concrete Works

22.1.5.1 Construction Characteristics

The concrete construction of the dam was divided into three phases, and the construction peak was in the second phase. The concrete construction of the dam had the main characteristics as follows.

(1) Large engineering quantity, and tight construction period, high-intensity continuous construction required. The total concrete quantity for the dam in preliminary design was 16.08 million m^3, in which the concrete quantity for the dam in the second phase was about 12.00 million m^3, and the controlled construction period was 45 months.

(2) Complex dam structure and large construction difficulty. The dam was designed with 105 tunnels and channels, including the desilting (sand flushing), floating debris removal and diversion pressure pipe, and the three-layer holes and tunnels in the spillway dam section were alternately arranged with densely distributed reinforcements and various grades of concrete, as well as metallic structure and equipment installed and inserted.

(3) Large difficulty in concrete temperature control. The dam concrete was mostly mass concrete, whose temperature shall be strictly controlled; the characteristics of the high-intensity and continuous construction have determined that in the high-temperature season, normal construction was still required, thus raising the requirements of high difficulty for the pre-cooling concrete production and site temperature control.

(4) High quality requirements. Three Gorges Dam was the core structure in the Three Gorges Project, "crucial for generations to come and linking the national fate", where the first-level project shall achieve the first-level quality.

(5) There are many construction interference factors and large difficulty management and coordination. The dam and the plant buildings were subcontracted for the section construction, with many standard sections and construction teams as well as many construction devices, thus increasing the difficulty in organization, coordination and management.

Considering the above construction characteristics, in order to achieve the orderly and high-efficiency construction and ensure the engineering quality, the construction proposal was fully studied, the cable crane proposal, belt tower crane proposal (tower belt crane proposal), elevated portal crane and large tower crane proposal were compared, thus the tower belt crane proposal and large tower crane proposal was compared.

22.1.5.2 Concrete Construction

In the first phase, only one tower belt crane was used in the construction of the right longitudinal dam section, and the portal (tower) cranes were used in other construction positions; in the second and third phases, the concrete construction of the dam was mainly realized with the construction proposal mainly using tower belt crane and supplemented by large portal tower crane and cable crane, where the construction of concrete below the elevation 160.0m was mainly used with the tower belt crane, and the construction of the part above this elevation was realized with the large portal (tower) crane at the 120.0m elevation.

1. Construction proposal

(1) The dam on the left bank in the second phase. The dam construction in the second phase was arranged with 6 tower (top) belt cranes, 9 large portal (tower) cranes, 2 luffing cable cranes as well as 4 creter cranes as the flexible support. In addition, the construction organization also provided small and medium-sized mechanical devices. See Table 22.1-9 for main construction machineries. The spillway dam section and No. 7-14 dam sections of the left plant are situated in the positions with relatively low elevation on the construction foundation surface within the foundation pit, characterized by high dam, large volume and tight construction period. 6 tower (top) belt cranes were arranged in the lower blocks of No. 1, 7, 14 and 21 dam sections of the spillway dam section and in the middle blocks of No. 8 and 12 dam sections in the left plant, covering all the above scope and bringing the high-intensity casting of the tower belt crane into full play. The casting height of the tower belt crane was up to 155.0-160.0m, and the casting quantity was about 7.40 million m^3, accounting for about 60% of total concrete quantity in Phase Ⅱ in the plant dam; the part above 160.0m elevation was the dam top structure with complex structures, for which the construction was realized with portal (tower) crane on the trestle at the 120.0m elevation. See Fig. 22.1-8 for the elevation arrangement of typical construction machinery in the spillway dam section.

Table 22.1-9 Configuration of Main Construction Equipment for the Dam in Phase Ⅱ

Equipment name	Type	Quantity (pcs)	Installed position
Tower belt crane	TC2400	4	Spillway dam section
Top belt conveyor	MD2200-TB30	2	Dam section in the left plant
Elevated portal crane	MQ2000	6	Spillway dam section, dam section in the left plant, No. 12-18 non-overflow dam sections in the left
	SDTQ1800	1	Spillway dam section, dam section in the left plant
Dedicated portal crane for metal structures	MQ6000	1	82m trestle in the plant site
Large tower crane	KROLL-1800	1	Spillway dam section
Luffing cable crane	20t×1416m	2	Covering the spillway section and dam section in the left plant
Creter crane	CC2200	4	Flexible use

The material supply line of the tower belt crane was arranged in front of and behind the dam, with the supply speed about 3.5-4.0m/s, the transportation time of concrete on the belt was about 3.0-5.0 min, and the temperature rise of the pre-cooled concrete in the high-temperature season was about 0.8-1.0℃/100m.

1) Arrangement of portal (tower) crane. It is situated on the construction trestle and construction platform. Three construction trestles are arranged: 45.0 m-elevation trestle at the foundation pit downstream of the spillway dam section, 120.0 m-elevation trestle on the slope surface downstream of the dam (60.5m from the axial line of the dam), and the 82.0 m-elevation trestle between the plant dams. 9 large portal (tower) cranes (seven MQ2000 elevated portal cranes, one K1800 tower crane and one MQ6000 dedicated portal crane for metal structures) are arranged on the above trestles and construction platforms, mainly responsible for the embedded part installation, concrete casting and placing area preparation of metal structures and generator units.

2) Arrangement of cable crane. The cable crane has a span of 1,416.1m, a single lifting weight of 25 tons, and 2 cable cranes can lift 46 tons large piece. 2 luffing cable cranes are arranged at 10m and 40m downstream of axial line of the dam, and the main tower is designed at 185.0m elevation in the No. 8 non-overflow dam section in the left, and the auxiliary tower is designed at 160.0m elevation in the right longitudinal No. 2 dam section. The cable crane crosses the dam in the second phase, and the control range of smooth flow is 15m upstream of the axial line of the dam to 65m downstream of the axial line of the dam. It is mainly responsible for auxiliary work, such as the installation of metal structures, transfer of placing area devices, and material transfer within the range, and also for the casting of a small amount of concrete.

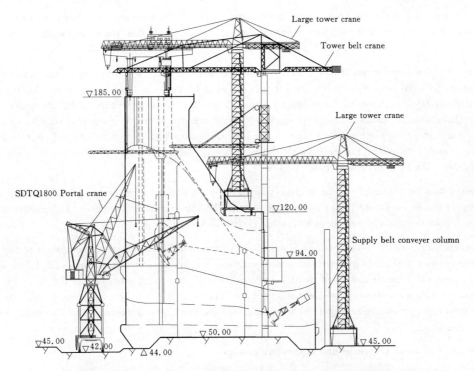

Fig. 22.1-8 Elevation Arrangement of Typical Construction Machinery in the Spillway Dam Section

3) Creter crane. Honored as "walking tower belt crane", it has not only the characteristics of the tower belt crane for continuous high-strength casting, but also the advantages of flexible movement and quick installation. 4 creter cranes are widely used in the positions of the dam foundation, apron and blind zone, thus having played a significant role in the concrete construction.

(2) The dam on the right bank in the third phase. The mechanical devices for the construction of the dam on the right bank in the third phase are similarly arranged as in the second phase, totally 4 tower belt cranes, 6 large portal (tower) cranes and 2 creter cranes. In addition, the construction organization provided several small and medium-sized mechanical devices. See Table 22.1-10 for the configuration of the construction machinery on the dam in the right bank in the third phase.

Table 22.1-10 Configuration of the Construction Machinery on the Dam in the Right Bank in the Third Phase

Equipment name	Type	Quantity (pcs)	Installed position	Description
Tower belt crane	TC2400	4	Dam section in the right plant	No. 16, 19, 21 and 25 solid dam sections in the right plant
Elevated portal crane	MQ2000	2	Dam section in the right plant	On the 120m trestle behind the dam
	SDTQ1800	1	Dam section in the right plant	On the 58m trestle in front of the dam, and then moved to 120m trestle
	MQ1260	1	Dam section in the right plant	On the 120m trestle, then moved to 82m trestle between the plant and the dam
Large tower crane	KROLL-1800	1	Dam section in the right plant	On the 58m trestle in front of the dam, and then moved to 120m trestle
Dedicated portal crane for metal structure	MQ6000	1	82m trestle on the plant and dam	Mainly used for the lifting of large machine units and embedded parts
Creter crane	CC2200	2	Plant and dam on the right bank	Flexible use

1) Arrangement of tower belt crane: 4 tower belt cranes are situated in the middle of the solid dam section of No. 16, 19, 21 and 25 dam sections in the right plant (where the tower belt cranes are numbered as No. 10, 9, 8, 7 respectively),

which are 42m downstream of the axial line of the dam, mainly used for the casting of the large mass concrete below the 160.0m elevation. The concrete above the 160.0m elevation shall be cast till the dam top with the portal tower crane on the 120.0m trestle.

2) Arrangement of portal tower crane. During the initial construction, the portal tower cranes are arranged on the platforms with the 58.0m and 103.0m elevation, mainly used for the initial and intermediate concrete casting, auxiliary placing area work, and the lifting and installation of the embedded pipes in the dam; at the same time, MQ1260 portal crane is added on the 82m trestle between the dam and plant, to support the first-line steel tube lifting and pipe trench concrete casting behind the dam, and support the plant building construction behind the dam. In the intermediate and later period, the portal tower crane was mainly arranged on the trestle with 120m elevation, mainly used for the dam top structure and back pipe behind the dam above the 160.0m construction elevation.

Concrete is mainly used for the supply to the mixing system at the 150.0m elevation. In case of system fault or insufficient supply, the supply will be supported from the mixing system at the 84m elevation. The concrete coarse and fine aggregates shall be produced from the artificial aggregate processing system in Xia'anxi.

3) Construction trestle. The dam construction is designed with three trestles. During the construction of the spillway dam section, one construction trestle will be set at the 45m elevation on the downstream of the dam and at 120m elevation on the slope behind the dam (45m trestle and 120m trestle, same below); during the construction of the dam sections at the left and right plant buildings, a construction trestle is arranged at the platform with the 82m elevation between the plant and the dam, and one construction trestle is arranged on the slope behind the dam, and connected with 120m trestle arranged in the spillway dam section.

2. Casting technology of tower belt crane and supporting process

Tower belt crane integrates large tower crane and belt conveyor, not only having the function of the tower crane, but also integrating the characteristics of the belt conveyor. It integrates concrete horizontal and vertical transportation and placing area material spreading functions, combined with the concrete supply line, realizing the plant-based one-stop concrete construction from the mixing building to the placing area. Its significant characteristics include continuous supply characteristics, high strength, placing area material spreading functions, thus beneficial to the construction with safety and civilization. It has simplified the production links, largely improved the productivity and realized the horizontal casting construction.

The tower belt crane is used under the "one building, one belt, one crane" supporting work mode, that is, one supply line shall be equipped with one mixing building and one tower belt crane. To bring its uniform, continuous and high-strength material supply into play, it has adopted the measures such as optimizing concrete material and mix proportion, making reasonable configuration of the placing area construction materials and improving the casting process and so on, to form a set of methods and processes for the concrete casting with tower belt crane.

(1) Concrete material and mix proportion. To prevent the aggregate segregation and breakage during the transportation of concrete mixtures, the following measures are taken, such as reducing the large-stone percentage, adjusting the particle size and controlling the stone beyond the diameter. Reduce the percentage of super large stone from 30% to 20%-25%, adjust the particle size from 80-150mm to 80-120mm, strictly control the percentage of the stone beyond the diameter at the level no more than 5%, and control the use of total cementing materials for concrete at no less than 160kg/m^3.

To control the workability of concrete, maintain the slump stability of concrete and reduce the bleeding, the modulus for the particle size of the sand shall be controlled at 2.6±0.2, and the water content shall be controlled within 6%.

Class I fly ash is used. The "micro-sphere" in the fly ash serves as the "bearing" in concrete, and can largely improve the workability and improve the blocking phenomenon of belt conveyor during the material charging.

(2) Supporting device for placing area. To bring into play and adapt to the advantages in continuous and high-strength material supply of the tower belt crane, the resources in placing area shall be sufficient and proper.

The tower belt crane has the function of spreading the materials, but when the pouring cylinder has the blind zone within certain range, or when the operation is immature leading to non-uniform materials, it is required to arrange the concrete spreading machine, and generally one spreading machine will be arranged in the silo.

The tower belt crane has high material supply capacity, sufficient vibrating devices shall be equipped on the silo surface, and the vibrating capacity shall be designed as the 1.5-2 times the placing strength; it is required to use the large-power vibrating arm (vibrator with 6-8 heads, 150mm diameter, 850mm height, vibration frequency 7,000-8,000 times/min, and vibration amplitude 2.8mm) according to the actual conditions of the silo, to be combined with the handheld vibrating

Chapter 22　Typical Projects of High Dam

rod; the designated personnel shall be responsible for the concrete around water stop strip, grout stop strip and formwork, embedded parts and corridor, and the handheld vibrating rod shall be used for careful vibration, to ensure the compactness of concrete.

In the steel pipe dam section of the dam, the silo position 40m(L)×25m(W) is generally arranged with 2 concrete spreading machines, 4 vibrating rods of $\phi130$, 2 vibrating rods of $\phi100$; the solid dam section 40m(L)×13.3m(W) will be generally arranged with 1 concrete spreading machine, 4 vibrating rods of $\phi130$, and 2 vibrating rods of $\phi100$. In special positions, it is required to additionally configure the long-handle vibrating rods or soft-tube vibrating rods according to the requirements.

(3) Placing method. The tower belt crane has the capacity of continuous and high-intensity concrete supply, and in principle, the horizontal casting method shall be used whenever possible. The covering time interval for the base layers shall be no more than 4–6 hours in the low-temperature season and no more than 2–4 hours in the high-temperature season, and the average casting intensity shall not be less than 80–120m³/h. The silo has complex structure and densely distributed reinforcement network, so it is improper to be used in the high-intensity construction. The step method shall be chosen for construction. In the step method, it is required to use the relatively large step width (above 8–10m), and the quantity of the steps shall not be more than 4 layers, and the casting intensity shall be consistent with the size of the silo surface so as to ensure no initial setting at the material head.

(4) Placing area in the construction process. The tower belt crane shall be used with the 9–15m length skin tube for material charging. To prevent the separation of concrete in this link, emphasis shall be put on controlling: ①For the placing area without the reinforcement, the distance from the discharge outlet of the skin tube to the placing area shall be no more than 1.5–2m, and the spread materials shall be uniformly moved, without too high buildup; ②The material spreading strip shall be clear with sufficient width, and the strips have the connection of fish scale shape; ③During the material spreading around the formwork, the distance between the unloading point and the formwork shall be within 1–1.5m; ④The height difference between the front and rear blocks of the dam and the adjacent blocks shall be controlled within 6m to create good conditions for material spreading.

Within 8m of the water face of the dam, it is required to use the 20cm thickness secondary-aggregate grading concrete in the same grade, and in other positions, it is required to use 40cm thickness of third-grading rich mix concrete in the same grade.

Concrete mixing system was equipped with the refrigerating system, and was configured with sufficient precooling capacity. Secondary air cooling aggregate, ice addition and cold water addition were used to realize the production of high-intensity precooled concrete in the high-temperature season, and with the support of the temperature control measures throughout the process, it ensured normal continuous and high-intensity concrete construction in high-temperature season. During the transportation of the precooled concrete, the sun shading measures were added above the belt conveyor to prevent the second temperature rise.

3. Large integral and standard steel formwork

To make the dam rise in a quick and good manner, and ensure the concrete solid inside and smooth outside, the following measures were taken on the choice of the formwork.

(1) Promote the use of large integral steel formwork. In the plane positions with the shape change upstream and downstream of the dam as well as the horizontal joint surface, it is required to use the large cantilevered integral steel formwork (Doka formwork). See Fig. 22.1-9.

(2) Optimize the corridor arrangement and unify the formworks. The corridors in the dam are intersected with various specifications. If the formworks were erected one by one, this would limit the quick rise of the dam. To resolve this difficulty, the climbing corridors with different types, sizes and slopes were optimized to make them achieve the consistent shape and slope so as to carry out the formed steel formwork construction with uniform specification. At the same time, to ensure the quality in the details, the water drainage ditches for the corridors were also made with the shaped formworks.

4. Lift concrete placing of 3m thickness

To accelerate the construction, the concrete construction of the dam was used with the 3m lift layer. For the 3m rise layer, the critical problems centered on three aspects: the strength and rigidity of the formwork, temperature control in the summer and placement strength.

The formwork problem is addressed with the Doka formwork of 2m lift layer to height and consolidating the formwork. With respect to the temperature control, mainly 7℃ precooled concrete, densified cooling water pipes and personalized water

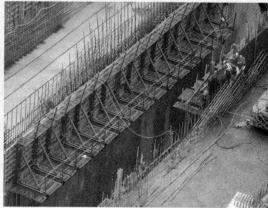

Fig. 22.1 - 9 Large Cantilevered Integral Steel Formwork Used During the Dam Construction

supply was used. The concrete placement strength was mainly addressed with the double-hole supply in the mixing building.

In the high-temperature season, the overtemperature ratio for the concrete placement was controlled below 0.3%; maximum temperature didn't exceed the indices. Average placement strength was 80 - 100m³/h, and maximum placement strength was 180m³/h. The average time for the concrete base layer was 3.2h. This realized an average annual rise height of 60.0m, and a maximum block rise height of 81.5m for the dam.

22.1.5.3 Temperature Control and Crack Control

Since concreting was carried out all year round, effective temperature control and crack control measures shall be used. Main temperature control measures during the dam construction were:

(1) Optimize the mix proportion, reduce the cement use, and reduce the heat of hydration.

(2) Use the aggregates to produce the precooled concrete with the secondary air cooling process. For the foundation restriction zone and in the period other than low-temperature season, the upper mass concrete shall be placed with precooled concrete.

(3) In the large mass concrete, cooling water pipe is embedded into the system to realize initial, intermediary and later artificial cooling. Water supply for cooling in the initial period was mainly used to control or reduce the maximum temperature peak in concrete. The intermediate water supply for cooling was mainly used to reduce the temperature in concrete, reduce the temperature difference between inside and outside, and make preparation for the concrete to pass the winter. In the later period, the water supply for cooling was used to further reduce the temperature in concrete so as to achieve the stable temperature and meet the requirements for joint grouting.

(4) Implement interval control between the layers and realize the thin-layer, uniform and continuous rise.

(5) New insulating materials were used to prevent the risk for crack incurred upon the dam due to the attack from the sudden drop of atmospheric temperature.

(6) Technical and management measures, such as temperature control warning and personalized water supply, were used so as to realize effective implementation of concrete temperature control.

1. Production of precooled concrete

(1) Secondary aggregate air cooling technology. To address the difficulty for the temperature control of large mass concrete, particularly for continuous high-intensity concrete placement in the high-temperature season, the secondary aggregate air cooling technology was developed. This technology was successfully used in the system at the 98.7m elevation on the left bank in the first phase, and was widely promoted and used in the second and third phases. The production of precooled concrete was mainly realized with secondary aggregate air cooling, slice ice adding and cold water adding.

Secondary aggregate air cooling process was to make full use of the aggregate adjustment silo on the ground in the mixing system by taking -6°C to -4°C cold air as the medium, to reduce the initial setting temperature of the coarse aggregates below 8°C. When the coarse aggregates were transported to the material storage silo on the mixing building, the coarse aggregates would be cooled with -12°C to -8°C cold air for secondary cooling, thus achieving the aggregate temperature -1°C to $+1.5$°C after the secondary air cooling. See Table 22.1 - 11 for air temperature and temperature index in secondary aggregate air cooling.

Table 22.1-11 Schedule on ontrol Temperature and Aggregate Temperature Indices for the Secondary Aggregate Air Cooling

Aggregate type	Particle size (mm)	Primary air cooling (℃)		Secondary air cooling (℃)	
		Air temperature	Aggregate temperature	Air temperature	Aggregate temperature
Superlarge stone	80-150	−6−−4	≤8	−12−−8	−1−−1.5
Large stone	40-80	−6−−4	≤8	−12−−8	−1−−1.5
Medium stone	20-40	−6−−4	≤8	−12−−8	0-0.5
Small stone	5-20	−1-1	≤8	−12−−8	1-1.5

When the product aggregates produced by the aggregate processing system were transported to the concrete production system, they were sent for secondary screening before being delivered into the adjustment silo. At this moment, the small stone surface has high water content, which will easily result in the freezing of the silo if air cooling is used. Therefore, it is required to adjust the air cooling temperature to achieve the primary air cooling temperature generally at −1 to +1℃, and after the surface is dried in air, the secondary air cooling may be carried out with other aggregates.

Primary air cooling is used mainly to largely reduce the aggregate temperature. Secondary air cooling is used mainly to make the aggregates totally cool down. The adjustment silo of primary air cooling is larger than the mixing silo and the storage silo upstairs, thus requiring longer air cooling time, so it is critical to control the primary air cooling effect. It is required to design with primary air cooling adjustment silo for aggregates with proper volume, and check the air cooling time of the aggregates according to the production intensity of concrete. Generally speaking, the time for primary air cooling of the aggregates is above 60-90min, and the time for secondary air cooling of the aggregates is above 45-60min. If the air cooling time is not enough, the aggregates will not be thoroughly cooled down. If the air cooling time is too long, it will result in the silo freezing, which will be uneconomic.

(2) Production process of precooled concrete. After secondary air cooling of the aggregates, the slice ices is added at the rate of 30-50kg/m^3 and the refrigerated water at 2-4℃ is added, thus producing the precooled temperature. By adjusting the quantity of added slice ice and refrigerated water, the precooled concrete of different temperatures can be produced. During the initial construction period of the dam, the temperature at the production outlet of precooled concrete is controlled at 7℃, 9℃, 10℃, 12℃ and 14℃ respectively.

2. Outlet temperature control

The outlet temperature control for precooled concrete is mainly realized with the following measures: ①Control the temperature of cement before being delivered into the tank not higher than 60℃, particularly in the summer. ②Strengthen the inspection and control on the air temperature and aggregate temperature in secondary air cooling, adopt the stone breaking and temperature measurement means to check whether the aggregates are fully cooled down. ③Set the ground alley and corridor to get the aggregates in the bottom floor, or set the shading louver so as to avoid the aggregates being exposed to the sunshine in the summer. ④Maintain stable concrete production strength, shorten the shift change time, defrosting time and building washing time, and the time for the building stop shall be generally controlled at no more than 30 min. ⑤Arrange the inspection and maintenance and daily maintenance in the winter so that the refrigeration system can operate under good conditions. ⑥The air cooling adjustment silo level, refrigerating device operation parameters, air temperature and aggregate temperature shall be included in the inspection and control range, and it is required to establish the corresponding inspection and evaluation tables for the dynamic tracking and control.

The outlet temperature is controlled on the basis of T, where the measured outlet temperature $t \leqslant T$ is qualified; $T < t \leqslant T+3℃$ belongs to overtemperature, which can be arranged in some secondary positions; $t > T+3℃$ is treated as the wastes. It is specified that the outlet qualification rate shall not be less than 85%. Take the outlet temperature of 7℃ as an example, $t \leqslant 7℃$ is qualified, $7℃ < t \leqslant 10℃$ belongs to overtemperature, but can still be used, $t > 10℃$ is treated as the wastes.

During the transportation of the precooled concrete, it is required to reduce the temperature rise. The transporting truck shall be covered with the shading louver and the supply line shall be insulation materials and shall be totally insulated. In case of long interruption of the supply line, and under the premise of being allowed by the placing area, the short-time cold water conveying method may be used to reduce the temperature of the belt.

3. Maximum temperature control

In order to control the maximum temperature inside the concrete, in addition to optimizing the concrete mix proportion and choosing the proper outlet temperature, it is required to control the casting temperature and adopt the artificial water supply for cooling. In addition, during the dam construction on the right bank, when the gap between the first line of the pipe trench behind the dam and the No. 2 non-overflow dam section is enlarged, the 42.5 low-heat cement shall be used.

(1) Placing temperature control. According to the casting in different positions, different casting means and different seasons, the counter calculation was made from maximum temperature, to determine the concrete placing temperature and the corresponding outlet temperature. See Table 22.1-12.

Table 22.1-12 Control and Comparison of Concrete Placing Temperature and the Corresponding Outlet Temperature

Item	Gradation	Placing device	December-next February		March, November		April-October	
			Placing temperature	Outlet	Placing temperature	Outlet	Placing temperature	Outlet
Foundation restricted zone	Fourth grade	Portal tower crane	Natural placing	Ambient temperature	12 – 14℃	10℃	12 – 14℃	7℃
		Tower belt crane	Natural placing	Ambient temperature	12 – 14℃	7℃	14 – 16℃	7℃
Apart from foundation restricted zone	Fourth grade	Portal tower crane	Natural placing	Ambient temperature	Natural placing	Ambient temperature	16 – 18℃	10℃
		Tower belt crane	Natural placing	Ambient temperature	Natural placing	Ambient temperature	16 – 18℃	7℃
Second and third grade			Natural placing	Ambient temperature	12 – 14℃	7℃	14 – 16℃	7℃

Main measures for the placing temperature control include:

The materials are supplied from the double inlets in the mixing building, the resources in the placing area are allocated to improve the placing strength, and control the covering time of the base layer in the high-temperature season no more than 2 –4h, and no more than 4 – 6h in the non-high-temperature season.

Choose the low-temperature time section to open the silo, or spray water before opening the silo so as to reduce the temperature in the placing area. The spraying facilities are prepared in the placing area, and during the placing process, it is difficult to control the placing temperature, and when continuous over-temperature happens, it is required to spray water immediately to reduce the ambient temperature.

During the placing process, the position of the material head and the placing area shall be timely covered with self-made insulation quilt with the size of 3.0m×1.0m, which is made by about 2cm PVC insulation quilt enwrapped with tarpaulin.

Implement the over-temperature warning and silo stop system. When the placing temperature is too high, it will send out the warning. When the difference between the measured placing temperature and control criteria is 2℃ or if the covering time of the cover layer is up to 2.5h, it will send out the alarm. The concrete placing temperature shall be generally tested once two hours (three test points) to make more frequent tests on the high-temperature time section, once per 30 minutes. If the placing over-temperature is detected in three consecutive times (with the duration of 1h), the silo shall be stopped under the command of the general director. With comprehensive use of the above measures, the placing temperature of the concrete in the Three Gorges Dam was effectively controlled, and the over-temperature ratio was controlled within 0.80%.

(2) Initial water supply for cooling. During the concrete hardening process, there will be initial heating speed and large heating quantity. As a result, in order to control the maximum temperature rise in the concrete, the measure of artificial water supply for cooling was used.

1) Arrangement of cooling water pipe. In the positions with large mass concrete, steep slope, filling pond, both sides of the joint, plug and other initial, intermediate and later water supply for cooling, it is required to embed the cooling water pipes, and embed the cooling water pipes in the positions for the large mass structure with the size of structural long side of the pier and wall. The cooling water pipe was made with 1 inch (interior diameter 2.54cm) black iron pipe, and may also

be made with high-density PE pipe materials.

The cooling water pipe is arranged as S-shaped, generally at the size of 1.5m×2.0m. During the embedment, it is required that the water pipe shall be 2.0 - 2.5m from the upstream dam surface, 2.5 - 3.0m from the downstream dam surface, and 1.0 - 1.5m from the water pipe to the joint surface and holes in the dam. The length of single water pipe is generally not more than 250m. The S-shaped water pipes in the dam shall be introduced into the corridor from nearby via the water planning according to the range of the joint grouting zone. The water pipes for the introduced corridor shall be in order, and shall concentrate so as to avoid partial overcool of concrete. The horizontal or vertical pipes introducing the corridor shall be generally not less than 1.0m. Before concrete placing, it is required to be filled with water for leakage detection, with the water passing pressure 0.3 - 0.4MPa.

The high-grade concrete is densely arranged with the cooling water pipes in more layers, and shall be provided with the black iron pipe in the bottom layer, and PVC pipe in the upper part. It is zoned according to the concrete grade, and the cooling water pipes are also properly zoned.

2) Water supply requirements. In initial period, the cooled water is generally the refrigerated water at 10 - 12℃ (or 8 - 10℃). In the low-temperature season, when the river water temperature is relatively low (below 15℃), the maximum temperature is controllable after the test, and river water or mixed water can be supplied for artificial cooling. The water supply direction shall be changed once 24 hours. The difference between the water temperature for cooling and the concrete temperature shall not be more than 20℃, and the concrete temperature decrease rate shall be controlled at the level not more than 1℃/d. The water supply time shall be 10 - 15 days, which will start generally within 12h after the concrete placement, with the water flow no less than 18L/min. In the high-temperature season, it is generally clip the peak by 6 - 8℃. In the low-temperature season, the water supply time shall be properly prolonged, until the requirements for the temperature are met for passing the winter. For the high-grade concrete, water shall be supplied for cooling after starting the concrete placement, and it is required to increase the water flow in the initial period, or properly reduce the water supply temperature.

3) Water supply measure. To ensure that water is replaced every 24h, the cooling water pipes are designated with A and B pipe inlets (or with different colors), which are classified into two groups, group A for the odd day and group B for the even day so as to strengthen the supervision and ensure uniform temperature decrease of concrete. Water supply is controlled by means of water flow, thus realizing the control through the combination of total quantity control single pipe census, instrument embedment and temperature measurement pipe, or the stuffing temperature (iron pipe 5 - 7 days, PVC pipe 7 - 10 days) are used to test the change of the temperature in the concrete so as to guide the water supply. The main pipe of the cooling water pipe is insulated with rubber so as to ensure that the temperature of the water along the pipe will be not more than 1 - 1.5℃.

Promote "personalized water supply". For the typical regions with the use of a large amount of cement, such as high-grade zone and flowing concrete, the temperature measurement pipes shall be embedded to track the temperature, and the cooling water pipes are densified and added with more layers, and implement large water flow (25 - 35 L/min) during the initial period, and then after the highest temperature is achieved, change it into the small water flow (18 - 20 L/min) to continue the water cooling, thus effectively controlling the maximum temperature inside concrete, and preventing too quick concrete cooling.

(3) Curing of the placing area. After the final setting of concrete, the placing area shall be cured with flowing water by means of rotary spraying and hanging the embossing pipes on the side.

With the above comprehensive measures, during the dam construction in the second phase, the maximum temperature has been basically controlled. During the dam construction in the third phase, the qualified rate of maximum temperature in 2003 was 82.4%, and above 96.6% in 2004 - 2006. The personalized water supply positions are used to achieve the qualified rate of 100%.

4. Concrete insulation

(1) Insulation proposal. Permanently exposed surface shall be insulated with polystyrene board. The significant positions, such as inlet flow channels (including inlet, desilting hole and floating debris discharge channel) shall be insulated with the foam-in-place materials of the coated polyurethane. The horizontal placing area and horizontal seam need the short-term insulation and shall be insulated with the insulating quilt. The position where the longitudinal face will be exposed in more than two winters shall be insulated with the polystyrene board, and the positions to be exposed for a short term shall be insulated with the insulating quilt. All holes shall be blocked with the integral tarpaulin. At such significant positions as the inlet and outlet (including the desilting holes) of the intake pipe in the hydropower station, not only was the hole blocking implemented, but also the polyurethane materials were used for the insulation in the hole.

For the permanently exposed surfaces, concrete cast during the period from October to the next April, permanent insulation shall be immediately used after the casting and mold stripping. For the concrete casting from May to September, it shall be covered with permanent insulating layer in the early October. Before every autumn (end of September), the inlets and outlets of all water intakes, desilting holes, floating debris removal holes, vertical shafts, corridors and other holes shall be blocked. Permanent insulation during the construction period means from the insulation to the operation of the project.

(2) Insulation measure. 5 cm-thick polystyrene board shall be used in the upstream side, and the time following the placement in the low-temperature season shall not be more than 5 days, and time following the placement in the high-temperature season shall not be more than 7 days. 3cm-thick polystyrene board shall be used in the downstream side, and a casting layer shall be followed once being delayed.

At such positions as intake pipe, the inlet section of the desilting hole and floating debris removal hole in the hydropower station, the polyurethane materials being coated and foamed in the site with 1.5 - 2cm thickness shall be used for insulation.

Horizontal joint and the previously covered longitudinal joint are insulated by fixing the insulation of 2cm thickness with the timber beads, and in case of the key, the insulation shall be implemented with the fluctuation. The longitudinal joints to be exposed for more than two winters shall be insulated with the 3cm thickness polystyrene boards.

The steel pipe trench bottom is insulated with 2cm thickness insulating quilt; and the side wall to be exposed for more than two winters shall be insulated with 3cm-thick polystyrene boards, and others shall be insulated with 3cm-thick insulating quilt.

The insulation of all placing areas and the blocking of all holes shall be completed before the coming of the autumn (end of September). At the non-plane and complex positions such as reinforcement mesh, longitudinal and horizontal keys, and the bottom step of the pipe trenches, the fit insulation shall be completed according to the fluctuation of the positions. During the insulation period, except necessary water spray is required for humidification, it is prohibited to use the cross flow of water during the construction.

(3) Insulation effect. The insulation test results on the upstream and downstream surfaces of No. 17-1 dam sections in the right plant (5cm polystyrene board on the upstream side, 3cm polystyrene board on the downstream side), and No. 24-1 A (3cm polystyrene board) in the right plant have shown that when the change amplitude of atmospheric temperature in 2 - 3 days is about 10 - 20℃, the temperature change in 5cm insulating board is about 1 - 2℃, and the temperature difference between inside and outside is about 5 - 15℃. When the temperature change in the 3cm insulating board is about 5 - 7℃, the temperature difference between inside and outside is equivalent to that for 5cm polystyrene board. Surface temperature of the concrete in polystyrene board is basically constant, thus indicating good insulation effect.

In the spray coating site of polyurethane, the foaming insulation materials are close to concrete, thus achieving water tightness and air tightness, that is, very good insulation and humidification effect. Before being immersed into water, it will be totally stripped for inspection, in which no concrete cracking is detected.

22.1.5.4 Construction Quality

1. Test results

(1) Concrete construction quality of the Dam is generally very good. During the initial construction period of the Dam in the second phase, there were such problems as the concrete below the diversion bottom hole and bottom board not dense, and the cracking in the upstream surface of the spillway dam section. But with the implementation of the refined construction and full-process temperature control and cracking control technology, the subsequent construction quality was gradually and stably improved. The construction quality achieved the very good level. From the site testing:

The guarantee rate for compressive strength of concrete was up to 92.5% - 99.9%, the coefficient of dispersion was 0.040 - 0.194, and the qualified ratio was up to 97.0% - 100%.

Average qualified rate for concrete anti-freezing index was up to 97.1%, average qualified rate for anti-seepage index of concrete was up to 99.3% and average qualified rate for ultimate tension of concrete was up to 95.5%.

(2) Concrete compactness test. Concrete compactness test was mainly realized with drilling and core taking, video in hole and sound wave test. The core taking test covers water pressurizing and water pumping, specific gravity, aggregate distribution, interlayered combination, being overhead or not, and with bubble or not. Total concrete quantity of the dam is 16.08 million m^3, 429 inspection holes were made for concrete compactness test, total advance was 6,248.96m, and the hole rate for inspection was 38.9mm/m^3, including:

Final inspection holes for the concrete mass of the dam in the left bank were arranged at the rate of 20 - 100mm/m³, totally 334 holes were made for core taking, totally 5,250.57m. Average qualified rate of the pressurized water test is 93.5%, the qualified rate of compressive strength and specific gravity of concrete is 100%, the availability rate of core samples was above 97%, and the qualified rate of sound wave test was above 98.2%. The expanded test and cement grouting reinforcement were carried out and all were qualified after the grouting. See Table 22.1 - 13 for the test results about the concrete mass and core taking in the dam on the left bank.

Table 22.1 - 13 Test Results on the Quality of the Concrete in the Dam on the Left Bank

Dam section	Pressurized water qualified ratio (%)	Qualified ratio for compressive strength (%)	Qualified ratio for specific gravity (%)	Availability ratio of core sample (%)	Qualified ratio for video in hole (%)	Qualified ratio for sound wave test (%)
Left plant	98.20	100	100	98 - 100	98.5	98.5
Spillway	93.50	100	100	97 - 100	98.3	98.2

Note: left non-overflow dam section: 23 concrete final test holes, advance 448.09m, 19 Category I holes and 4 Category III holes.

For the dam on the right bank, 95 core holes were made with total advance 998.39m. In the pressurized water test of concrete in the seepage-proof layer, the water permeation ratio was less than 0.1Lu, the water permeation ratio of internal concrete was less than 0.3Lu, and the pressurized water quantity and pumped water quantity met the design requirement. Average availability rate of the core samples was 97.2%, the core samples were characterized by dense cementing, uniform distribution of aggregates, smooth surface and good interlayered combination. From the core samples, it was concluded that all met the requirements in the quality standard of Category I hole in the Three Gorges Project. See Table 22.1 - 14 for the test results about the concrete mass and core taking in the dam on the right bank. The concrete construction of the dam on the right bank was realized with full-process and refined temperature control, the qualified rate of over-temperature control index was overall improved, and there wasn't any cracking in the large mass concrete.

Table 22.1 - 14 Test Results About the Concrete Mass and Core Taking in the Dam on the Right Bank

Core taking position	Hole diameter (mm)	Silos (silo)	Hole number (pcs)	Advance (m)	Water permeation ratio (Lu)		Core sample availability ratio (%)		Longest core sample (m)
					Maximum	Average	Minimum	Average	
Concrete in the seepage-proof layer	76	10	12	159.23	0.038	0.010	90.0	96.4	0.84
	91	10	11	105.73	0.073	0.027	92.4	97.2	0.68
	110	7	7	70.89	0.05	0.011	96.5	98.2	0.96
	219	5	5	73.37	0.03	0.005	90.0	99.8	14.43
	Total	32	35	409.22	0.073	0.015	90.0	97.5	14.43
Quality criteria of Category I hole					≤0.1	—	—	—	—
Internal concrete	76	25	39	387.18	0.079	0.010	80.0	96.5	0.97
	91	9	10	65.40	0.07	0.021	91.3	97.4	0.77
	110	4	4	39.46	0.02	0.005	95.83	97.1	1.66
	219	7	7	97.13	0.027	0.011	97.5	99.4	15.08
	Total	45	60	589.17	0.079	0.012	80.0	97.03	15.08
Quality criteria of Category I hole					≤0.3	—	—	—	—

Note: In this table, the statistic data are the test results from the concrete core taking below the 160.0m elevation, and core taking and test wasn't carried out for the structural concrete above the 160.0m elevation.

2. Cracking problem in the upstream of spillway dam section and the treatment

(1) Cracking situation. In October, 2000, it was firstly discovered that there was the cracking between the bottom holes in the upstream surface of No. 16 spillway dam section, till the middle of December, 7 cracks were discovered in 5 dam sections. During September-October, 2001, preliminary inspection was carried out with telescope once again, in which no new crack was discovered. Since November, 2001, new cracks were generated continuously, and the original cracks were developed, till February, 2002, totally 40 cracks were discovered. In the 23 dam sections of the spillway dam section, there were 1 - 2 cracks in every dam section, some even with 3 cracks. The cracks were mainly distributed in the middle part, and

the positions with the cracks were mostly at the 45.00 – 77.00m elevation. The highest crack reached the root of the corbel in the deep hole, and the lowest crack was extended to the position of the foundation rock. In No. 16 spillway dam section, the longest crack was up to 35m, with the general width 0.1 – 0.3mm, and in the No. 9 spillway dam section, the widest crack was 1.25mm, and the crack depth was generally less than 2m. See Table 22.1 – 15 for the statistics of the cracking situation.

Table 22.1 – 15 Basic Information of the Cracks in the Dam on the Left Bank

Item	No. 1 – 23 spillway dam section	No. 11 – 14 dam sections in the left plant	No. 1 – 10 dam sections in the left plant
Inspection date (year and month)	November, 2001 – April, 2002	November, 2001 – April, 2002	November, 2001 – April, 2002
Dam section with cracks	Totally 23 dam sections, 1-2 cracks in every dam section, some even with 3 crack	Totally 4 dam sections, in which there were the cracks in 3 dam sections, and there wasn't the crack only in the No. 13 dam section on the left plant	Totally 11 dam sections, in which there were the cracks in 7 dam sections, and there wasn't the crack in No. 3.5 and 6 dam sections and the safety Ⅲ dam section in the left plant
Crack number	40	5	15
Direction	Vertical	Vertical	Vertical
Main distributed position	In the middle of the dam	In the middle of the steel pipe dam section	In the middle of the steel pipe dam section
Elevation range (m)	45 – 80	44.33 – 73.84	55.5 – 94.98
Maximum crack length (m)	35.13	10	9
Maximum crack width (mm)	1.25	0.24	0.75
Maximum crack depth (m)	<3	<0.5	<3

(2) Crack reason. In the early February, 2002, China Three Gorges Corporation held the consulting conference by inviting the experts about crack reason and treatment proposal. In the consulting conference, it was thought that the cracks resulted from many factors and belonged to the surface shallow-layered temperature cracks. Because of annual change of atmospheric temperature and sudden drop of the atmospheric temperature in winter, the dam surface wasn't insulated in the winter of 2001, thus the draft in the diversion bottom hole resulted in very large temperature difference inside and outside and large surface temperature gradient, as the main factor causing the cracking in the dam surface; the dam had complex structure, the bottom hole and board were arranged as reinforced concrete boards with fillet joint, and the cross-joint for the embedded parts of metal structure, having certain influence on the cracking in the dam surface.

(3) Crack treatment. The treatment method combining blocking and prevention was adopted:

1) For the cracks, the proposal of chemical grouting + chisel groove (5cm×8cm) for pointing the water stop materials + sticking rubber pieces + SR seepage-proof covering pieces was adopted. The grouting material was epoxy series material LPL, and the pointing material was SR – 2 plastic water-stopping material (plastic water stopping material with butyl rubber and organic silicon as the main material, with the extension ratio 800% – 850%), and the rubber piece was made with 60cm width chloroprene rubber.

2) For the protection at both sides of the cracks and on upper and lower parts, the protection was provided on both sides of the cracks; the upper protection of the crack was extended to the 84.0m elevation; the protection for the cracks in the No. 1 – 4 spillway dam sections was extended to the 24.0m elevation, and the protection for the cracks in No. 5 – 23 spillway dam sections was extended to the foundation rock surface. Protective measures: within 8m in the middle of the dam section, SR seepage-proof pieces were attached, the thickness of 25 – 40cm concrete slabs were cast outside (PVC boards were anchored in the counter-slope, deep and narrow V-grooved backfilling concrete was made in the position of foundation rock) to protect the seepage-proof covering pieces. Within 6.5m at both sides of SR seepage-proof covering pieces, the spray coating of cement-based seepage-proof material KT1 was made, and within 2m on both sides of the reinforced concrete slabs below the 45.0m elevation, it is required to spray 8 – 10cm polypropylene fiber concrete on the KT1 surface.

3) Treatment of the crack in the dam heel. The cracks in No. 18 – 23 spillway dam sections were close to or equivalent to the

foundation rock surface. To prevent the high-pressure water from seeping into the cracks via the foundation rock after the reservoir impounding, after the casting of the mud-jack plate in the position of the dam heel in the dam section, the rock around the crack was consolidated and grouted with acrylate. See Fig. 22.1 - 10 for schematic diagram of the crack treatment proposal in the spillway dam section.

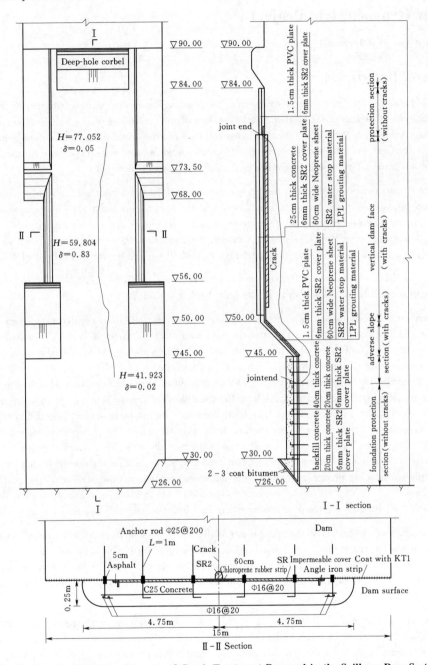

Fig. 22.1 - 10 Schematic Diagram of Crack Treatment Proposal in the Spillway Dam Section

In the end of April 2002, the crack treatment in the spillway dam section was completed and accepted. On the cracking places in the upstream dam surface, 9 test cracks were embedded. In December, 2002, the test results showed that after the cracking treatment, the measured opening was very small, basically within the range of the observation error.

Before the impounding (135m water level), the crack opening on April 21, 2003 was −0.18 - 0.21mm. After the impounding to 135m water level, the crack opening was −0.22 - 0.19mm, suggesting that the crack was in the closed status. After the impounding to 156m water level, the inspection was carried out on the seepage of the drainage pipe in the dam below the 90.0m elevation, and the inspection and analysis were made on the water drainage pipe with large seepage in the hole, and the core taking, video in the hole and sound wave tests were carried out on three dam sections in order to test the depth of the crack. The inspection concluded that in the corridor below the 90.0m elevation, there wasn't abnormal phenomenon, such as new crack and water leakage point. The total seepage amount in the corridor was very small and the drainage pipe in the dam with the crack area was normal in operation. The crack depth monitoring and analysis have shown no new crack development.

22.2 Ertan Hydroelectric Project

Zhu Zhonghua[1], Wang Gang[2]

22.2.1 General Description

Ertan Hydroelectric Project is located on the lower reach of the Yalong River, the biggest tributary of the Jinsha River, in western Sichuan Province. The dam site is located 33km upstream of the confluence of the Yalong River and the Jinsha River. It is approximately 46km from Panzhihua City, an industrial base of steel, vanadium and titanium production. Ertan Hydroelectric Station is the first one constructed among the 21-cascade hydropower stations planned on the Yalong River, and the biggest hydropower project built and in operation in the 20th century in China.

Ertan Hydroelectric Project is developed mainly for power generation. The catchment area of the project is about 116,400 km^2. The normal operation level of the reservoir is 1,200m, and the minimal operation level of the reservoir for power generation is 1,155m. The total storage capacity of the reservoir is 6.1 billion m^3 with an active capacity of 3.37 billion m^3. The reservoir can regulate seasonal river flow. The power plant has 6 generator units, 550MW each, with a total installed capacity of 3,300MW. Its average annual energy production is 17,000GW·h and guaranteed output of 1,000MW.

The main structures of Ertan Hydroelectric Project consist of a concrete parabolic double curvature arch dam, an underground powerhouse in the left bank and spillways in the right bank (Fig. 22.2-1).

The double curvature arch dam has a maximum height of 240m with a crest length of approximate 769m (Fig. 22.2-2 and Fig. 22.2-3). The width of the dam crest is 11m and the base width of arch crown is 55.74m. The hydrostatic load on the dam reaches 9.8 million tons. The Ertan arch dam was the first arch dam with height over 200m built in China. For the sake of flood discharge and lowering the reservoir level, the dam was equipped with 7 surface overflow spillways (11.5m high and 11.0m wide), 6 orifices (5.0m high and 6.0m wide) in the middle, and 4 orifices (5.0m high and 3.0m wide) at the bottom. The surface overflow spillways and the middle orifices are used for flood discharge, with a maximum discharge capacity of 9,800m^3/s and 6,450m^3/s, respectively. The bottom orifices are not designed for flood discharge but for emptying the reservoir when necessary. Furthermore, at the bottom of the dam there are 4 temporary diversion conduits with gates and hoists. The diversion conduits are plugged after the second stage of the river diversion. Four layers of galleries are embedded in the dam to serve for observation, grouting, drainage and transportation. Behind the dam are the energy dissipating and anti-scour facilities including a plunge pool, a plunge pool dam, and an apron located on downstream of the plunge pool dam. The plunge pool is concrete lined, with a trapezoid-shaped cross section, about 300m long and 40m wide. At the downstream end of the plunge pool, a 35m-high plunge pool concrete dam was built to retain plunge pool water.

Besides the overflow spillway at the crest, the middle level orifices and the low level orifices through the dam, there are 2 free flow spillway tunnels located at the right bank for flood discharge. The spillway tunnels are 922.0m and 1,248.0m long, respectively. The tunnel cross section has vertical side walls and circular top with the width of 13m and the height of 13.5m. The design flood discharge capacity of the two tunnels is 7,400m^3/s. The maximum design flow velocity in the tunnels is about 45m/s.

The hydroelectric system is placed in the left bank (Fig. 22.2-4). An intake tower about 83m high was built in the left bank upstream the dam. Water flows through 6 headrace tunnels and 6 turbines about 300m deep under the ground. The underground powerhouse complex includes the machine hall, the transformer chamber and the tailrace surge chamber oriented parallel to each other. The underground powerhouse is 25.5m wide, 280.3m long and 65.4m high, with 6 units of 550MW each. The transformer chamber is 18.3m wide, 215m long and 24.9m high. The tailrace surge chamber is 19.8m wide, 203m in long and 69.8m high, followed with 2 tailrace tunnels (16.5m high and 16.5m wide).

During construction, a two-staged river diversion was adopted. At the first stage, a rockfill cofferdam was built to block the river and protect foundation ditch of the dam. Two diversion tunnels located in each bank were built to divert the river flow during the construction period. The diversion tunnels were 17.5m wide and 23m high, the biggest of its type at that

---

[1] Zhu Zhonghua, Deputy Chief Engineer of Yalong River Hydropower Development Company, Ltd., Professor Senior Engineer.
[2] Wang Gang, Director of Yalong River Hydropower Development Company, Ltd., Senior Engineer.

Chapter 22 Typical Projects of High Dam

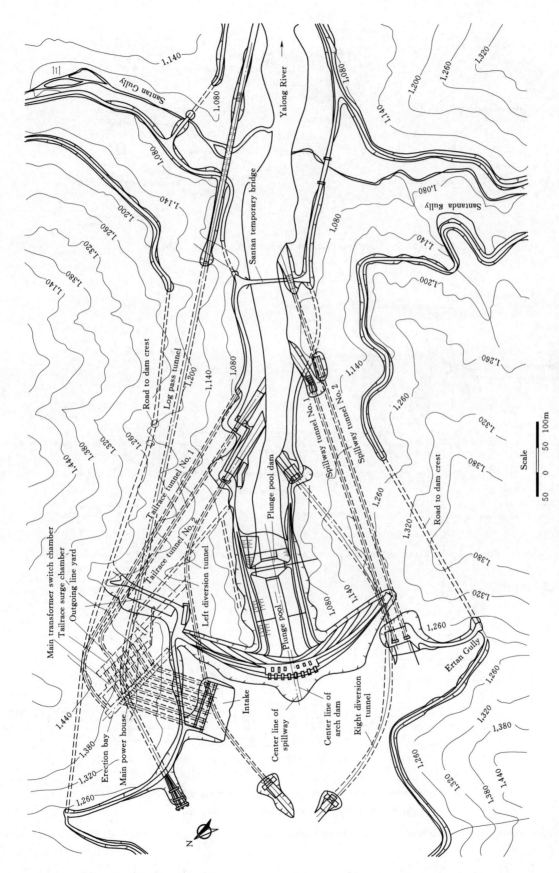

Fig. 22.2 – 1 Layout of the Ertan Hydroelectric Project

Part II Construction Technologies for Dams and Reservoirs

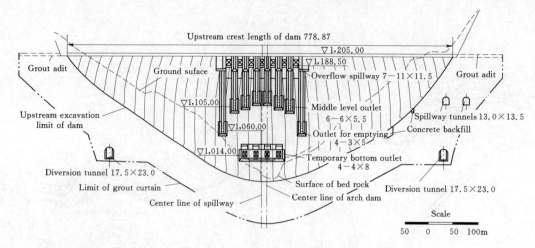

Fig. 22.2-2 Upstream View of the Ertan Arch Dam (Unit: m)

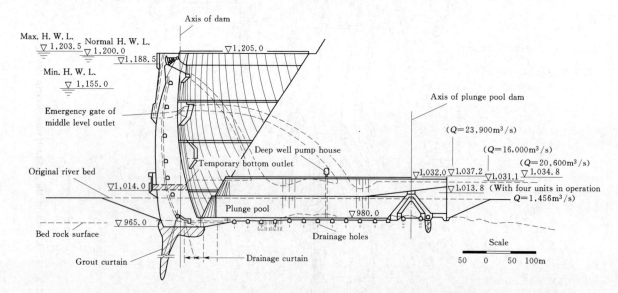

Fig. 22.2-3 Cross Section of Crown Cantilever of the Ertan Arch Dam (Unit: m)

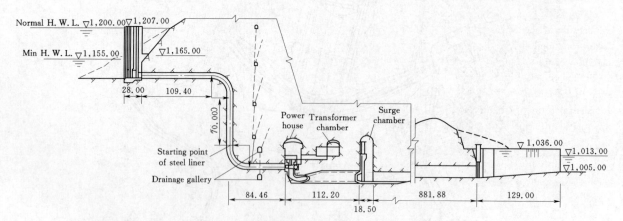

Fig. 22.2-4 Cross Section of the Waterways and the Underground Powerhouse Complex, the Ertan Hydropower Power Station (Unit: m)

time. At the second stage, the diversion tunnels were plugged, 4 temporary diversion conduits in dam were used to divert the river flow.

The main works of the Ertan project include: ①a total volume of open excavation of 9.27 million m³, ②a total volume of underground excavation of 3.368 million m³, ③a total volume of concrete and reinforced concrete of 5.857 million m³, and ④a total quantity of steel structure production and installation of 21,000 tons. The budget of the project was RMB 28.554 billion, and the actual cost of the project was RMB 28.263 billion.

The commencement date of main construction work was September 14,1991. On November 26, 1993, the river was blocked. On May 1, 1998, 4 temporary diversion conduits in the arch dam were closed, reservoir impounding started. In the end of 1998, two units were put into commercial operation. In 1999, the rest four units commenced commercial operation. The whole project was completed in 2000.

The fund used for the project was partly a World Bank loan for Reconstruction and Development. Open global bidding was adopted to select the contractors and supplies. It is the first project in China with large funding from World Bank.

Ertan hydroelectric project is on a grand scale, complex technique, and ranking first in most technical indicators. It was carefully designed on the base of the large amount of scientific tests, and absorbed extensively the suggestions of experts at home and abroad. And it was used the international competitive bidding mechanism in the dam construction, and developed and managed according to international project contract of FIDIC with advanced construction technology, equipment and craft to effectively control the safety, process, quality and investment in the dam construction. In January 2007, Ertan hydroelectric project won the sixth Zhan Tian You Award, which is a top honor in recognition of civil engineering advances in China.

22.2.2 Main Achievements

22.2.2.1 Foundation Selection of the High Arch Dam

The dam is located in hard igneous rock formation consisting of Permian basalt and intrusive syenite. The shallow zone rock at the dam site is mostly moderately weathered. The location of the moderately weathered zone changes with the elevation, and is deeper in the middle of the valley than other places. The maximum depth of the moderately weathered zone of the valley is about 100m in horizontal direction. According to the Chinese *Design specification for concrete arch dams* at that time, the base of arch dam must be laid on sound or slightly weathered rock mass, so the maximum excavation depth at the dam abutment may reach 113.5m at lower parts of the dam axis. When determining the foundation of the Ertan Arch Dam, a comprehensive consideration of dam strength and stability to meet the safety requirement after the rock excavation and treatment is more emphasized instead of the weathering degree of rock. By the stress-strain and stability analysis of dam and foundation system, three criteria are proposed to determine the base of the Ertan Arch Dam as follows:

(1) Strength criterion of the rock mass. The rock mass of moderately weathering at Ertan dam site is of massive or massive-jointed structure, with the saturated compressive strength of the rock mass over 100MPa. The rock mass is strong enough as a foundation of an arch dam.

(2) Deformation criterion of rock mass. The effective deformation modulus is a composite of deformation moduli for all materials within a particular part of the foundation. The deformation modulus of the rock mass varies largely over the entire contact area with the dam foundation. The dam-foundation interaction analysis model is established, and the effect of foundation rock modulus on the dam stress distribution is carefully studied. It is found that the distribution of dam stress is reasonable when assuming the foundation's equivalent deformation modulus is approximately 28GPa and the deformation modulus of dam concrete ranges from 21 to 25GPa, and the following three requirements are satisfied: ①the ratio of the maximum to the minimum value of the equivalent deformation modulus of the rock foundation is less than 4, ②the ratio of the equivalent deformation modulus of adjacent foundation rock masses at the same bank is less than 2, and③the ratio of the dam concrete deformation modulus to the equivalent deformation modulus of the foundation rock mass is less than 3.

(3) Stability criterion of the dam abutment. The insertion depth of the dam abutment into the canyon wall mainly depends on the stability of the dam abutment. The stability of dam abutment can be ensured by foundation treatment to reinforce the rock masses and by adjusting the layout and shape of arch dam. By technical and economical comparison, the methodology of adjusting the layout and shape of the arch dam is adopted. A parabolic curve shape arch ring in horizontal is proposed to optimize the direction of thrust from arch dam to the abutment. Stability analysis is performed to check the safety of the dam abutment.

Based on the above mentioned analyses, a systematic criterion is developed to determine the availability of the rock mass as foundation of the Ertan arch dam with regard to the quality (as illustrated in Fig. 21.2-5) as follows: ①The fresh to slightly weathered rock mass, and lower part of moderately weathered rock mass (i.e. A-C class) can be utilized directly as dam foundation. ②The middle part of moderately weathered rock mass (i.e. D class) can be utilized after properly treatment. ③The upper part of moderately weathered rock mass (i.e. E class) can be used partially depending on the

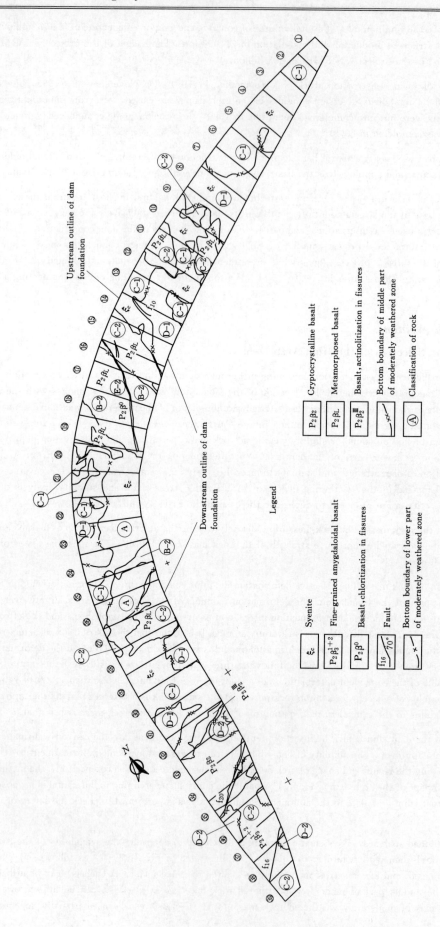

Fig. 22.2-5 Geological Map of Ertan Arch Dam Foundation

situation after treated and reinforced. ④The strongly weathered rock mass cannot be utilized. Most of the moderately weathered rock mass (C and D classes) are utilized as dam foundation; its area is about 67 percent of the total foundation base area.

Consequently, compared with that at the Preliminary Design Stage the average excavation depth of dam foundation decreased about 7.35m, the dam concrete volume reduced about 370,000m³, the open excavation of rock masses decreased about 80,000m³, and the construction time was shortened for 1 year. The design of the Ertan arch dam foundation is beyond the relevant Chinese Specifications that required only fresh to slightly weathered rock masses as a high arch dam foundation at that time, achieved significant economic benefit for the project. This also promotes the philosophy of arch dam foundation selection, and makes it an important reference for the latter arch dam project development.

22.2.2.2 Shape Design of the High Arch Dam

The Ertan arch dam was designed in the 1970s and 1980s, as the first double curvature arch dam higher than 200m in China. At that time, the mathematical programming method had not been developed well enough for the optimization of the arch dam shape. The design parameters such as radius, central angle and thickness were determined by an iterative trial-and-error method, in combination with sensitivity analysis, until a satisfied shape of the Ertan arch dam was finalized. The characteristics of the Ertan arch dam shape are as follows:

(1) The selected shape fits with the conditions of large difference of rock foundation deformation modulus and the asymmetric canyon shape. It makes it possible to optimize the excavation of the foundation. The thickness of the crown cantilever base decreases from 70.34m at the preliminary design stage to 55.74m. This reduced the placement area of dam concrete. No longitudinal construction joints have been set. This simplifies the construction procedures, and speeds up the construction schedule.

(2) The parabolic arch was chosen finally after a comprehensive review. The angle between the thrust direction and the dam axis of the parabolic arch is 5% larger than that between the thrust direction and the dam axis of a three-centered arch, 20.9% larger than a single centered arch. The increase of the angle of thrust is beneficial for stability of the arch dam abutment.

(3) In consideration of the asymmetric characteristics of the foundation deformation, the arch was designed with an optimized asymmetric shape. Four parabolas were adopted to model the upstream and downstream surfaces of the double-curvature arch dam. So the optimization procedure resulted in the dam with a big portrait curvature, and adjusted horizontal arch ring section, and finally reduced the excavation depth of the dam abutment.

(4) The Ertan Arch Dam adopted a variable-thickness horizontal arch ring. The end of the arch is thicker than other parts in order to improve the distribution of the stress. The thickness of the arch terminal is usually 1.1 to 1.5 times of that at the crown.

Table 22.2-1 Geometry Characteristics of the Ertan Arch dam Shape

Crest thickness of crown cantilever (m)	11.0	Chord length of crest arch axis (m)	769.32	The ratio of the thickness and the height	0.232
Base thickness of crown cantilever (m)	55.74	the maximum slope of overhang at the upstream face	0.18	the ratio of chord and the height	3.21
Maximum thickness of arch terminal (m)	58.51	volume of dam concrete ($10^4 m^3$)	391	the ratio of the thickness on the base top and the bottom	0.197
Central angle of crest arch (°)	87.09	dam flexibility $K=F_2/V$	3,120	watar pressure on the dam ($10^4 t$)	980
Maximum central angle of arch (El. 1,130m) (°)	91.49	Flexibility factor $C=K/H$	13.3		

22.2.2.3 Temperature Control Design of the Dam

Stress analysis by the multi-arch-cantilever method shows that the maximum tension stress of the Ertan Dam is about 0.82MPa, and the maximum compression stress of the Ertan Dam is about 8.66MPa. According to the stress distribution of the dam, the dam concrete can be classified into 3 zones by concrete grade of strength: zone A, zone B, zone C. Zone A includes the region adjacent to the foundation and orifices, zone B mainly includes the middle region of the dam, and zone C mainly includes the upper region of the left and right part of the dam. The concrete in the interior or exterior region of the dam are the same and not distinguished. The strength of the concrete at the age of 180 days was adopted for design strength in order to utilize the long term strength of concrete and decrease the amount of cement and flyash. The compression strength of the concrete at the age of 180 days in zone A, zone B and Zone C are 35MPa, 30MPa and 25MPa respectively.

The Ertan concrete double-arch dam is divided into 39 sections of dam,; the spaces between transverse joints are from 20 to 23m; and the largest placing surface is 58m×23m and the maximum casting block has an area of 1,200m². For those reasons, the temperature control required high efficiency and accuracy. Based on the success experience in China and abroad, and the results of tests and numerical analysis, the project of concrete temperature control of Er'tan arch dam was designed reasonably and practically. The project guaranteed the quality of the concrete and well controlled the cracks. Its main characters are described as follows:

(1) The division of temperature control zones was simplified. The temperature control region is divided into restricted zone (Zone II) and unrestricted zone (Zone I). The divided zones of temperature control here didn't follow the suggestion of the *Design Specification for Arch Dam*, which suggest three divided zones as strongly restricted zone, weakly restricted zone and unrestricted zone. The restricted zone means the region within distance from the foundation rock of $t/4$ or from the old concrete of $t/8$, where t is the lift length. The old concrete is the concrete with its age over 14 days.

(2) Criterion for difference of temperature: for the restricted zone (Zone II) $\Delta T \leqslant 14°C$, for the unrestricted zone (Zone I) $\Delta T \leqslant 20°C$, ΔT is the difference between the highest temperature and the stable temperature.

(3) The casting temperature was strictly controlled. Allowable maximum temperature of the concrete in Zone I and Zone II is 10°C, and that in those not very important zones is 14°C.

(4) The interval between treatments of two layer surfaces is controlled strictly form 3 to 14 days. When the interval is beyond 14 days, the overlaying layer concrete will be treated as aged concrete.

(5) Advanced concrete mixed proportion is adopted. The amounts of cement materials, which is consisted by cement and 30% fly ash, used in Zone A, B and C, are 190kg/m³, 182kg/m³ and 175kg/m³ separately. The amount of the water used in those zones is 85kg/m³.

(6) High strength PE type plastic pipes are used as cooling coils. Those pipes can be paved and repaired quickly, and can be paved in any place inside the casting block. The prices of material and field operation for those pipes are lower than steel pipes. During construction, the thickness of the casting layer in restricted zone (Zone II) is changed form 1.5 to 3.0m. This change reduces the time in treating the surface of each casting layer and improves the casting speed.

22.2.2.4 Flood Discharge and Energy Dissipation Design

The double-curvature arch dam is 240m high, with the maximum discharge of 23,900m³/s and the total flood discharge power of 39,000MW. The issues like energy dissipation and scour prevention are serious, with the characteristics of high head, huge discharge flow and large power, narrow river valley and high in-situ stress in the rock mass. There is no former double-curvature arch dam with so big discharge flux at that time. And thus studies on flood discharge and energy dissipation of high concrete dam are carried out to find solutions and key issues are highlighted as follows:

(1) Design flood standard for the Ertan project. As the Ertan hydropower project is a huge size project, the recurrence interval of design flood and check flood is supposed to be 1,000 years and 10,000 years respectively according to the Chinese relevant design specification during the preliminary design stage. Assuming the event of reservoir inflow larger than the design and the event of insufficient flood discharge capacity are independent, further analysis of flood relief risk has been performed. The results show that when the recurrence interval of flood for check is assumed to be 5,000 years, the probability of water flow overtopping the dam is 5.96×10^{-6}. Approved by the authority, the flood recurrence interval during the tender design stage was changed to be 1,000 years for a design flood of 20,600m³/s, and 5,000 years for a check flood of 23,900m³/s.

(2) General arrangement of the structures for flood discharge and energy dissipation. Comparisons and analyses have been carried out for different arrangements of flood discharge structures, different schemes of flood discharge and energy dissipation. Based on the characteristics of high water head, large discharge volume and narrow river valley, a comprehensive measure was proposed for the Ertan project, including dispersion of flow to decrease its erosion capacity, and reinforcement of riverbed to enhance the erosion-resistance. The structures for flood discharge and energy dissipation consist of 7 surface o-

verflow spillways on the dam crest, 6 middle level orifices through dam and 2 discharge tunnels at the right bank. The jets from the surface overflow spillways and from the middle level orifices impact each other, so the energy is absorbed by impact. An auxiliary dam and a plunge pool downstream of the dam were designed to protect riverbed from scour.

(3) Optimization of the structures for flood discharge and energy dissipation. Many researches have been performed on the flood discharge and energy dissipation structures, such as the shape optimization of the surface spillways and the orifices, energy dissipation by the impact between water flows, the erosion resistance of riverbank and riverbed, flow aeration measures to prevent cavitation of spillways etc. There are innovations in designing the surface spillways and the orifices. A new kind of energy dissipater of free fall drop and slotted bucket with diversion teeth was used in the surface spillways. A new kind of energy dissipater of turned on plan, upturned on facade, and a flaring deflector at the end to fan the jet into a thin sheet was used in the orifices.

(4) Model test research on flow-induced dam vibration. The problem of flow-induced structural vibration has attracted word wide interesting for a long time. A model of the Ertan dam with the scale of 1 : 200, which is the largest in China at that time, was built to study the problem of dam vibration resulting from flood discharge. The results show that the vibration of dam due to flood discharge can be perceived by human beings, but the magnitude of the vibration amplification is small and the dam is safe.

The arrangement for flood discharge and energy dissipation in Ertan Hydroelectric Power Station has been proved to be safe, reliable and economic in practice.

22.2.2.5 Support Design of the Large Underground Plant Chambers

The average depth of underground plant chambers in Ertan Hydroelectric Power Station is from 250 to 300m. The geostatic stress of the surrounding rocks is high. It is very difficult to design such underground plants with such large scale excavation in high geostatic stress region. And also there is almost no experience for such project both in China and abroad. At the preliminary design stage, some experts worried about the rock burst, large rock deformation and other difficult issues. In order to solve them, a series of model tests to simulate the real situation of the underground plant was conducted. And also during the construction, the deformations and rock bursts of the surrounding rocks was carefully monitored, and based on the monitor data back analysis to evaluate the status of the supporting system is conducted. The design of supporting system was adjusted as quickly as we could in accordance to the analysis results. Eventually we succeed in design and construction of such large plant chambers. Our experience is very valuable.

22.2.2.6 Large Turbine Generator Sets and Ultra High Voltage Electrical Equipment

Ertan Hydroelectric Power Station has been installed with six Francis turbines and generators. The rated head is 165m and maximum design head is 189.2m. The rated power of each unit is 550MW with maximum power of 612MW. The units (turbine and generator) suitable for the Ertan were selected by combination research on the performance parameter of the units, structure characteristics of the units, and the rotational speed control of the turbines. The main parameters for the turbines were determined by combination research on the specific speed of the turbines, synchronous speed of the generators, parallel circuits of generators, branch current and rated voltage of generators, and the salient parameter of the turbine are as follows: the specific speed is 184m · kW, specific speed coefficient is 2,368, the highest efficiency for single turbine is 96.05%, which represented the most advanced level at that time. The parameters of the turbines are rational and advanced, and harmonize with the parameters of generators. At the same time, 500kV dry-core cable was used as high voltage lead-out of transformers, which reduced the budget on electrical equipment. And also the GIS switching station was moved from underground to ground surface, which help to reduce the underground excavation.

22.2.2.7 Construction Technique and Achievement

The achievement of construction has reached an advanced level in the world at that time. The rate of concrete placement has created a record of 2,260 thousand cubic meters per year and 245,000m³ per month in China. The rate of concrete placement in the arch dam was 1.552 million m³ per month and 163,600m³ per month, which ranked respectively the third and fourth place in the world at that time.

Many new construction techniques have been adopted or developed during the construction of the Ertan project. The prestressed pier anchors, used in the radial service gates of intermediate outlets, were changed from line to ring shape. This is the first time a ring-shaped anchor used in China. When ring-shaped anchor used, the access galleries and anchor's end wells inside the dam or at the gate pier can be cancelled. So small damage was made to the dam body structure and at the same time the stress distribution around the anchor was improved. The stress distribution of arc part serving as end fixture for the anchor is good. Even if the anchorage cable was damaged or not qualified during or after tension, it could be replaced and re-fixed easily.

High density PE (HDPE) type plastic pipes were used as cooling pipes instead of steel pipes in large volume concrete. The use of plastic pipes broke the limitation that the placement space of the traditional steel pipes was limited by the thickness of casting layers. The HDPE pipe can be placed anywhere rapidly, and repaired easily. In the construction of the dam concrete, the depth of casting layers was changed from 1.5 to 3m, and the vertical space of the cooling pipes were kept by placing the HDPE cooling pipes in the middle of the casting layer. This measure reduced the work of surface treatment of casting layers and speeded up the concrete construction.

22.2.3 Environmental protection and running Benefit

22.2.3.1 Environmental Protection during Dam Construction

During the construction process of Ertan Hydroelectric project, the requirements of two authorized files were strictly followed, i.e. *Report on Influence of Ertan Hydropower Station of Yalong River on Environment and Environmental Protection Design for Ertan Hydropower Station of Yalong River*. Following the Principle of Three Simultaneity, environmental protection measures were performed simultaneously with the construction. The investment on the environmental protection of the Er'tan project is more than RMB 31 milliontotally. To guarantee the effect of these environmental measures, specialists from the World Bank were invited to give counsel about environmental protection.

22.2.3.2 Benefits of the Ertan Project

Ertan Hydroelectric Power Station has operated safely for almost 12 years and achieved great benefits of economy, ecology and society.

(1) benefit of power generation. Up to June, 2012, the total energy output of Ertan Hydropower Station reached 180 billion kW·h. As a clean energy provider, Ertan Hydropower Station provides 1,7000GW·h clean energy per year. Comparing to the same scale heat-engine plant, Ertan Hydropower Station saves 6.47 million tons of coal yearly, reduces 310,000 tons emission of SO_2 and 1.81 million tons of slag yearly. These data shows great effect on changing the consuming energy proportion and reducing the emission of waste gas and slag.

(2) benefit of shipping. At normal storage level of the reservoir, the 145km reach of main stream and the 40km reach of tributary in the reservoir area have become the main traffic aisle for the merchant shipping and tour transportation of the five counties nearby.

(3) benefit of flood control. In the year of 1998, the construction of Ertan Dam was still not completed; early impounding stage was still not finished. At that time, a catastrophic flood happened in the Yangtze River. By the precisely controlling the operation of the reservoir, Ertan Reservoir stored flood water for 3.3 billion m^3 in the six flood peak period, playing a very important role in flood control.

(4) benefit of local climate. Before Ertan Reservoir impounding, the dam location had a typical dry-hot valley climate. After impounding, the rainfall of the surrounding areas increased obviously, the air humidity was larger than before. The original local climate with such features as dry, rain lack, windy weather, dust weather, sharp raining and dry season was changed. Taking the observation data of Xiaodeshi Weather Station as an example, it is located away from 12km downstream of the Ertan Dam to reprehensively illustrate the change before and after the impounding. The data from 1978 to 1997 are taken as a representation of the weather before impounding, and the data from 1998 to 2006 are taken as a representation of the weather after impounding. It can be found from the comparisons between these data: the yearly average temperature is 19.6℃ after impounding, which is 0.2℃ lower than before; the yearly average humidity is 68%, which is 1.6% higher than before; annual average wind speed is 0.8m/s, which is 0.4m/s lower than before; the yearly average evaporation is 1,846.2mm, which is 299.4mm lower than before.

(5) benefit of land ecology. The local weather variation has a lot benefits to the vegetation growth in the reservoir area. Ertan people follow the idea of *Retaining Water by Trees and Maintaining Trees by Water*, do their best to optimize the ecology environment by means of forest planting. Along the mountain region of the river, the virescence area is more than 400 thousand mu, and the forest coverage is about 50%, 19% higher than the average of Sichuan Province. The lake area of an elevation from 1,200m to 4,196m, which was dry-hot valley before, now is covered by evergreen broad-leaf forest, coniferous forest, quercus and coniferous mixed forest, mountain coniferous forest, and subalpine coniferous forest. According to the survey, now there are more than 700 kinds of advanced plants and 325 kinds of land and aquatic birds living in the reservoir area. These birds can be classified into 47 families and 153 categories, including five kinds of migratory birds such as White Swan. The Ertan reservoir area has become a natural reserve park with most kinds of bird in Sichan Province. In recent years, black-necked cranes, first-grade protection birds have been found in the Ertan reservoir area. Now the Ertan reservoir area has become a new transfer station for black-necked cranes and other kind of mi-

gratory birds.

(6) benefit of tour. After the Ertan construction, the improvement of the environment of Ertan area brings a large number of tourists. A new tour route, called Tour of Sunshine from Chengdu to Panzhihua, attracts many tourists from Chengdu, Kunming and many other places.

(7) benefit of society and economy. From 1991 to 1999, the construction of the Ertan hydroelectric project had a contribution of 76.8% of the GDP to Panzhihua City of Sichuan Province, increased RMB 1.11 billion industrial output, increased a tax of RMB 3 billion. And during that time, a new county near the dam site, Yanbian, was built up. The local social economy was improved obviously.

In February 2004, the Ertan reservoir area was designated as an Artificial Wetland Natural Reserve of Sichuan Province. In December 2008, China National Tourism Administration approved Ertan reservoir area as a 4A Class Scenic Destination. In June 2006, the Ertan hydroelectric project was awarded with the honor of the National Environmentally Friendly Project, which is the top award in China to praise the environmental protection in dam construction.

22.3 Xiaolangdi Project
Yin Baohe[1]

The Xiaolangdi Project (referred to as the "XLD Project" hereinafter) is located in the mouth of the last gorge on the main stream of the Yellow River, 40km north of Luoyang City, 130km downstream of Sanmenxia Dam and 128km upstream of Huayuankou Hydrological Station. The project is mainly for flood control (including ice-jam prevention) and sediment reduction, and consideration is given to water supply, irrigation and power generation as well, with clear water stored and muddy water discharged. The reservoir is designed to provide a total storage capacity of 12.65 billion m³. The rockfill dam with an inclined loam core has a maximum height of 160m and a total volume of 50.73 million m³ (Fig. 22.3 - 1). The XLD Project is recognized as one of the most challenging projects in the history of dam construction in China and the globe for its complicated geological conditions, special flood/sediment regimes and severe operation demands. The large-scale efficient equipment, up-to-date technology and strict quality control incorporated into construction have produced a dam of high quality.

Fig. 22.3 - 1 Xiaolangdi Dam

During the initial operation period, enormous social, ecological and economic benefits were achieved following the operational principles of safety first, exercising integrated regulation of water resources and prioritizing public benefits with the help of safety inspection and monitoring. The XLD Project has won a series of titles, including the "Prize for International Milestone Rockfill Dams", "Prize for 100 Classic and Fine Projects in Celebration of the 60th Anniversary of the Founding of

---

[1] Yin Baohe, Director general of Xiaolangdi Project Construction & Management Center, Ministry of Water Resources, Professor Senior Engineer.

the People's Republic of China", "China Prize for Excellent Quality of Water Developments (Dayu)" and "China Zhantianyou Prize for Civil Works', as well as other awards for excellence in investigation, design, operation and ecological protection.

22.3.1 Dam Structures and Features

The XLD Dam, a rockfill dam with an inclined loam core, has a maximum height of 160m, a crest elevation of 281m, a crest length of 1,667m, a crest width of 15m, a maximum bottom width of 864m, an upstream slope of 1 : 2.6 and a downstream slope of 1 : 1.75, with 14m and 6m benches arranged at El. 220m and El. 250m respectively as shown in Fig. 22.3-2. The upstream cofferdam is located at the upstream dam heel as a part of the dam. There is a 80 m-wide berm at the downstream dam toe at El. 155m. The lower portion of the core is inclined core with a maximum bottom width of 101.9m, while the portion above El. 250m is changed to a normal core with a crest width of 7.5m. Several characteristics of the project are described in the following sections.

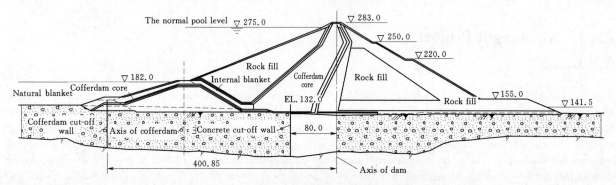

Fig. 22.3-2 Typical Section of the Dam（Unit: m）

(1) Double Seepage Control Consist of a Primary Vertical Seepage and a Secondary Horizontal Seepage Control Systems. The dam contains an inner blanket which is a 6 m-thick artificial gravel structure and integrates the river diversion berm, the starter cofferdam, the main cofferdam and the main dam as a whole and establishes a natural blanket acting as a secondary horizontal seepage control system when the reservoir sedimentation develops. A 1.2m concrete cutoff wall is applied to the bed cofferdam for the purpose of seepage control, with the lower portion inserted into the bedrock to a depth not less than 1m and the upper portion inserted into the clay core by 12m. The wall forms a vertical seepage control system together with the loam clay in the dam and the grouting curtains on both banks. This arrangement enables double seepage control of a primary vertical and a secondary horizontal seepage control systems, and greatly improves the reliability of the dam seepage control.

(2) A Dam Based on a Deep Overburden, with Complicated Outline of Bedrock and High Strength of Concrete Cutoff Wall. The overburden is deeper than 70m to the maximum and contains sand-layer and large boulder rocks. There exist 2 high benches in the bedrock outline from left to right, representing steep sills and inverted slopes. A concrete cutoff wall with a designed thickness of 1.2m and concrete strength of $R_{28}=35$MPa was placed in the overburden for the convenience of seepage control. The construction of the cutoff wall was on the critical path after river closure. The tight schedule together with the steep sills, inverted slopes and high-grade concrete works increased the degree of difficulty in construction.

(3) Core Seepage Control System with a Combination of Normal and Inclined Portions. In order to connect the core base with the foundation seepage control system and enable the upstream slope stability, the lower portion of the core is designed as an inclined core and the upper portion changed to a normal one. From the point of view of construction, the inclined lower portion is helpful for placing downstream shell materials when it is not ready to place soil materials, but also makes it difficult to construct the upper normal core where the upstream and downstream slopes are steep and all zones are narrowed abruptly.

(4) A Large Quantity of Placement Work and a Great Varity of Fill Materials. The valley at the dam site is wide open, causing a long dam axis. The dam height is 160m and the volume is up to 50.73 million m³; the placement volume was the first in China and the third in the world at the time of construction. The dam embankment is made of 17 types of filling materials, each with its own special demands, including Zones 1, 1A, 1B, 5 and 10 as seepage control layers, Zones 2A, 2B and 2C as filter layers, Zone 3 as a transition layer, Zones 4A, 4B and 4C as shell rockfill materials, Zones 6 and 7 as slope protection works, Zone 8 as a downstream berm, Zone 9 as a foundation backfill sand-gravel layer, and Zone 10 as an upstream blanket (Table 22.3-1). Parallel and cross with various types of materials interfered with each other.

Table 22.3-1 Filling Materials for Different Zones of Dam

No.	Zones	Title	Placement Requirements
1	Zone 1	Dam core	Calcareous content be not more than 8%
2	Zone 1A	Zone of high plastic clay	Calcareous content be not more than 8%
3	Zone 1B	Core of upstream cofferdam	Calcareous content be not more than 8%
4	Zone 2A	First filter layer downstream	Grain size 0–20mm
5	Zone 2B	Second filter layer downstream	Grain size 5–60mm
6	Zone 2C	Filter layer upstream	Grain size 0–60mm
7	Zone 3	Transition Zone	Maximum grain size 250mm, less than 5% smaller than 0.1mm
8	Zone 4A	Rockfill A for upstream	Maximum grain size 1m, less than 5% smaller than 0.1mm. Content of siltstone and claystone less than 5%
9	Zone 4B	Rockfill B for downstream	Maximum grain size 1m, less than 5% smaller than 0.1mm. Content of siltstone and claystone less than 10%
10	Zone 4C	Rockfill C for downstream	Maximum grain size 1m, less than 5% smaller than 0.1mm. Content of siltstone and claystone less than 20%
11	Zone 5	Mixed impervious blanket	Soil mixed with sandy gravel, maximum grain size 60mm
12	Zone 6A	Upstream crushed rock revetment	Angular sandstone, not less than 50% of grain size >700mm, not more than 10% of grain size <400mm
13	Zone 6B	Downstream crushed rock revetment	Angular sandstone, not less than 50% of grain size >500mm, not more than 10% of grain size <80mm
14	Zone 7	Upstream backfill rock revetment	Sandstone, maximum grain size 500mm, not less than 50% of grain size >400mm, not more than 10% of grain size <100mm
15	Zone 8	Ballast weight	Maximum grain size 1m, less than 10% smaller than 0.1mm
16	Zone 9	Sandy gravel backfill	Excavated material from riverbed
17	Zone 10	Upstream loess blanket	Loess excavated at right bank above El. 140m

(5) An Inner Blanket Linking the Inclined Cofferdam Core to the Dam Core. There is a large sediment discharge at the dam site. The measured long-term annual average sediment discharge was 1.35 billion tons and the annual average sediment inflow was 1.275 billion tons in the base year. To use the sediment in front of the reservoir as a natural blanket and help seepage control in the main dam, Zone 5 mixed impervious materials were placed to link the inclined cofferdam core to the dam core and establish an inter blanket leaning downstream.

(6) Application of Well-graded Wide and Thick Filter Zones Containing Artificial Materials Suitable for Mechanical Construction. Filter zones were provided prudently between dam core materials and adjacent materials to enable stable permeability. On the downstream side of the inclined core are 6m and 4m wide filter layers with grain sizes ranging from 0.1 to 20mm and 5 to 60mm, respectively. On the upstream side are 4m mixed filter layers with grain size of 0.1 to 60mm. The transition materials with a maximum size not larger than 250mm were placed between filter layers and rockfill embankments. All filter materials were produced artificially and well graded.

(7) Varying and Unsymmetrical Cross Sections of the Valley. The bed deviates to the left bank, with gentle slopes in the open floodplain on the right bank and steep ones on the left bank, where there is a 220m belt-like gentle zone above El. 260m with the foundation surface slightly inclined to the left bank. The dam here is only about 20m high. Such topography is extremely prone to uneven settlement, so it is very important to control the compaction properly.

(8) Left-bank Weak Rock Mass Used as Extension of the Dam. The dam seepage control line was extended to the left bank about 1km, with curtain grouting performed on the surface and the grout galleries at El. 235m and El. 170m to provide a complete impervious curtain. On the downstream side of the impervious curtain are drainage curtains. Also, unsuitable materials were placed on the weak left-bank rock mass to ensure its stability upon reservoir filling.

22.3.2 Construction Technology and Outcome

22.3.2.1 Construction Technology

Large capacity construction equipment, up-to-date technologies, processes and materials have been introduced in the XLD Project and a series of technical difficulties was solved successfully.

1. Mechanized assembly-line process

To use high-efficiency construction equipment with a large capacity and emphasize the joint operation of equipment is one of important measures to speed up the construction of rockfill dams. Imported complete sets of construction equipment in good conditions have been used for the XLD Dam, all are hydraulic, easy to operate, reliable and able to perform lifting, excavating, loading and other reciprocal actions in short time. All items of equipment are very high in efficiency.

In the process of construction, equipment was allocated properly to ensure efficient operation of each item of equipment, enabling mechanized flow process. Rockfill materials were loaded with 10.3m^3 excavators, hauled with 65 tons dump trucks, spread with 42-ton CATD9N bulldozers and compacted with 17t plane vibrating rollers. Filter and transition materials were loaded with 10.3m^3 loaders, fed with 65-ton dump trucks, processed, transported with 65-ton (or 36-ton) dump trucks, spread with 2.2m^3 hoe shovels and compacted with 17-ton plane vibrating rollers. Impervious materials were collected with 37.5-ton CATD8N bulldozers, loaded with 10.7m^3 or 5.9m^3 loaders, transported with 65-ton (or 36-ton) dump trucks, spread with CATD8N bulldozers, raked and leveled with CAT14G graders, and compacted with 17-ton sheep-foot vibrating rollers.

2. Rockfill placement without water

It is stated in the "*Specifications of Rolled Earth Rock Dams*" (SDJ 213—83), Sub-section 8.1.2, that "adequate water shall be added to spread sand-gravel and rockfill materials". In this case, 2 contrast tests have been conducted specially, one with water and the other without water. The test results indicated that, when the quantity of water added to Zone 4A rockfill materials took up 50% of the quantity of fill, the settlement was only 1–3mm higher than the value without water, and the bulk density was only 0.013 t/m^3 higher at most. Thus the construction technology without water was used for construction.

Based on field test results, there was an average bulk density of 2.130g/cm^3 for rolled Zone 4A materials, 2.135g/cm^3 for Zone 4B and 2.169g/cm^3 for Zone 4C, indicating an good outcome of compaction. Control test and recording test results showed that the graduation of Zone 4 rockfill materials complied with the specifications. Placement without water, with the help of severe quality control on site, both the quality and progress of the dam rockfill placement met the requirements of contract and specifications, and it has saved costs and produced huge economic benefits while avoiding complicated contractual issues.

3. Construction of filter materials in winter and wet season

Normally, the natural content of water in Zone 1 soil was 1.3%–3% higher than the optimum value. With the loss in transportation and loading/unloading taken into account, the content of water without moisture adjustment met the required value for placement (with a content of water in the range of −1%–+2% of the optimum value). But special construction technologies were needed for the construction in winter and wet season.

(1) Construction in winter. During construction in winter, priority was given to operating borrow areas with a southern exposure and on the lee side, with frozen topsoil removed as the first step to guarantee a soil temperature ≥ 0℃ at the time of loading. The content of water in the field was controlled within the allowable limit, not higher than 90% of the allowable plastic content of water. Any frozen topsoil was removed. Frozen soil with a suitable content of water and free from wind drying was raked, leveled and then spread quickly when broken with sheep-foot rollers. The temperature of loose soil prior to compaction was not lower than −1℃. When the air temperature is −10℃ or below 0℃ and the wind speed is over 10m/s, construction was suspended. To ensure good contact of layers, wind desiccated soil in a rolled surface was watered in accordance with air temperature, or the dry soil was scraped off and roughened before soil was spread as soon as possible. At the same time, the work face was reduced and generally rolled by 40–60m sections. If and when work was suspended for snow, any snow and unsuitable soil on the dam surface was cleaned and checked before work was resumed.

(2) Construction in wet season. During the placement of impervious materials, a series of construction technologies was adopted to mitigate/offset the effect of rain and the work for post-rain treatment, including but not limited to: ①make each layer of the core slightly lean against the upstream to help storm water to drain into the upstream shell fills; ②roll the surface smooth with plane vibrating rollers prior to a forecast of or an actual heavy rain to facilitate drainage and reduce the soakage of superficial fill materials; ③suspend when the surface failed to meet the required content of water subsequent to

rain; ④drain off any water in low depressions subsequent to rain without walking or operating equipment on the surface before it was dry; ⑤clean unsuitable wet soil prior to the resumption of work to ensure a clean and wet surface; ⑥roughen the treated surface with graders and resume work after the loose soil met the required content of water.

4. Placement of diversified fill materials

Up to 17 types of fill materials have been incorporated into the XLD Dam. Accordingly technologies for placing diversified materials have been developed in the light of the working process and schedule.

(1) Placement in parallel. The progress of construction in different zones was balanced to enable placement in parallel and harmonious rise of adjacent zones. This arrangement is advantageous in that it ①reduced the sequence of work for jointing and sloping; ②ensured the work surface as large as possible to benefit the performance of mechanical construction; ③facilitated the arrangement of access roads necessary for placing impervious materials; ④avoided dump trucks for filter and transit material cutting through the impervious zones; ⑤facilitated to reduce boundary deviation and cross-joint compaction of adjacent zones placed in parallel; ⑥balanced the construction intensity.

(2) Placement in sequence. In practice, the sequence of work was determined according to the slope strike of different zones. Filter materials and impervious materials were placed alternately, first sand and then soil, or vice versa, while other types of materials were placed in the normal sequence, i.e. first fine and then coarse materials. All zones were raised in a balance.

(3) Control of boundary deviation. The boundary of each layer was surveyed and set out, with setting-out pegs arranged at a space of 10 - 20m. When materials were spread with hoe shovels, the boundary was finished against the established pegs to ensure correct sections.

(4) Cross-joint compaction. The method of cross-joint compaction was employed in addition to parallel placement of diversified fill materials. The accuracy of boundaries was guaranteed before this sequence was performed, with large rocks removed from construction joints and boundaries between rockfill and transition materials, joints and boundaries filled and leveled using small stones.

5. Processing and placement of filter materials

(1) Establishment of efficient sand-gravel processing plants. A sand-gravel processing plant was located in the right-bank floodplain, 6km downstream of the dam. This plant is of the following characteristics: ①high automation, central control of all items of equipment, quick, flexible and well-coordinated; ②good quality and high efficiency equipment; ③good quality of products; ④advanced sand production and recovering process and plant, strict separation of coarse and fine grains, prompt and adequate dehydration, stable content of water; ⑤well-designed and well-arranged sewage treatment system and reuse of treated sewage without any environmental pollution.

(2) Placement of filter materials. Filter materials are pervious and resistant to the loss of soil grains in seepage control works. Technical measures taken during construction include: ①piling up finished products at the plant by 2m layers, mix and load with loaders, ensuring uniformity and no segregation during transportation; ②locating a transfer stockpile area in an appropriate position close to the dam, avoiding possible delays due to inadequate supply of filter materials; ③maintaining 0.25m thickness of layers, rolling 2 times with 17-ton plane vibrating rollers, placing soil, Zone 3 transition materials or different filter layers in parallel, and performing cross-joint compaction; ④unloading materials vertically to the dam axis, spreading and mixing them with 2.2m³ hoe shovels to ensure against segregation and achieve even thickness; ⑤controlling the boundary of filter materials strictly to maintain clear boundaries and even thickness.

6. Construction of concrete cutoff wall

The sand-gravel overburden is deeper than 70m and contains clays, large boulder rocks, steep sill and inverted slopes. A concrete cutoff wall was therefore constructed for the purpose of seepage control. A combination of clamshells and hydrofraises (also called hydraulic trench cutter) was used for excavation. The sand-gravel overburden was excavated largely by clamshells, though heavy hammers were used to break large pebbles and boulders. Bedrock excavation was performed predominately by hydrofraises. Hole verticality control, bedrock inspection, hole scrapping and slurry replacement prior to concrete placement were all assisted by hydrofraises.

Efficient excavation equipment made it possible to create a new type of cutoff wall joints. Phases I and II cutoff walls pioneered in applying lateral board joints under the protection of plastic concrete (Fig. 22.3-3). The construction was performed as follows: excavating a 2.8m × 1.2m lateral joint panel in designated positions of primary and secondary panel joints; pouring plastic concrete into the panel; excavating a primary slot 10cm beyond the centerline of the lateral panel; removing the over-cast 10cm concrete in primary I with a hydrofraise at the time of secondary II panel excavation to result in a fresh and smooth rough surface left by close-spaced vertical and uniform hydrofraise cuts and enabling close contact between primary slot concrete and secondary panel concrete. The plastic concrete was placed beyond both ends of the main panel adhering to both ends of the joint for the purpose of protection.

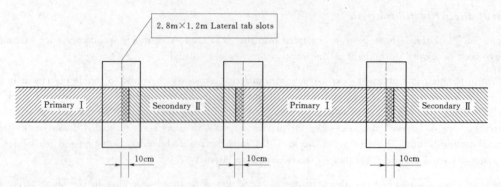

Fig. 22.3 - 3 Lateral Joint Panel

For the construction of panel walls, quality bentonite made in France was purchased to mix slurry, which reduced the possibility of hole failure and thick bottom deposit and joint mud. A complete set of vibrating sand removers capable of repeated use was used, which reduced the cost of slurry. Desirable indexes of slurry were maintained in the whole process of construction, including specific gravity <1.1, viscosity 35 - 50s, maximum content of sand 3.2%, minimum 0.1% and average <1%. These ensured good quality of wall concrete.

7. High-pressure jet grouting

The characteristics of high-pressure jet grouting technology applied to the dam are as follows: ①up-to-date equipment, including rigs with off-balance bits and capable of drilling with casing, crawler grouting machines equipped with multifunctional LUTZ CL88 self-recording instruments to enable automatic recording and real-time monitoring of hole depth, slurry pressure, air pressure, flow, revolution, lifting speed and other necessary parameters at the time of grouting; ②wide application for a number of purposes including foundation seepage control in the bed section of the upstream cofferdam, improving the bearing capacity of the sand-gravel of the core foundation at the foot of the left-bank slope, and closure and reinforcement of the "overhanging rocks" in the main dam cutoff wall. These have proved effective and fully reflected the advantages of the technology.

8. GIN grouting

The GIN method is a grouting technique involving closing boreholes and performing circulating grouting from up to bottom, which is an innovation suitable for conditions in China. This method was applied to curtain grouting in the rock mass on both banks of the dam. On the basis of numerous tests, slurry with a water to cement ratio of 0.7 : 1 and 0.75 : 1 was selected, which enabled good stability and flow ability. Grouting self-recording instruments were introduced as a pioneering practice in China, with data acquisition and processing completed in the course of grouting to enable real-time monitoring of the grouting process and provide process curves of grouting pressures and flow-time variations. According to the geological and overburden conditions, different GIN values were selected, generally 50 - 150MPa · L/m in 20 m-deep holes, 150 - 200MPa · L/m in 20 - 40 m-deep holes, and 200 - 250MPa · L/m in holes deeper than 40m. Also, remote control of a group of 8 grouting machines was exercised for the first time in China to provide real-time process curves, which was helpful for GIN quality control. The GIN method in comparison with conventional grouting is advantageous in good quality, high efficiency and less consumption, quite practical and cost effective.

22.3.2.2 Outcome

1. Dam placement at "High Rate and High Efficiency"

The dam was placed at a peak annual work rate of 16.36 million m^3 (1999), a peak month rate of 1.58 million m^3 (March 1999), a peak daily rate of 67,000m^3 (22nd January 1999), an average month rate of 1.12 million m^3 and an average month progress of 6.33m. The availability factor of major equipment reached 80% - 86%. Human resources were used very efficiently, with about 1,500 people involving in the construction during the peak period. The non-uniformity factor of dam embankment was 1.31, the second in the world.

2. Quality at entry complying with the specifications

Impervious materials in the core had a measured average bulk density $\gamma = 1.720 g/cm^3$, average degree of compaction $\geqslant$

100% and a water content in the specified range. The graduation curve of filter and transition materials were controlled within the envelope, with a measured average bulk density of 1.893 - 2.228g/cm^3 for filter materials, 1.91 - 2.37g/cm^3 for transition materials and 1.850 - 2.298g/cm^3 for rockfill materials. The quality of embankment complied with the specifications.

The core ratio of inspection holes in concrete cutoff wall was over 97%. Joints were in tight contact, with pressure test results under 3Lu. Water test results from curtain grout inspection holes were less than 5Lu, meeting the specifications. When the reservoir was filled, readings of the osmometer behind the curtains were stable, showing normal operation of the seepage control system.

3. Work progress ahead of schedule

Critical items and control points were put under control in the whole process of construction, with the schedule of work approved strictly and achieved ahead of schedule. The upstream cofferdam was raised to El. 185m by April 24, 1998, 68 days ahead of the contractual schedule. The concrete cutoff wall in the main dam was completed on March 10, 1998, 58d ahead of contractual schedule. The dam was raised to El. 236m on October 30, 1999, 8 months ahead of schedule, and totally completed by November 30, 2000, 13 months ahead of schedule.

4. Up-to-date efficient construction equipment and adequate management promoting the course of earth rock dam construction in China

Efficient large-capacity equipment and mechanical flow process were used for dam placement, including clamshells, hydrofraises and lateral joint schemes for concrete cutoff walls and GIN among other up-to-date techniques for curtain grouting works. Also, strict quality control and schedule control were exercised during construction, such that all parameters of work reached the world's best practice at the time. This provided rich experience in high earth and rockfill dam construction in China.

22.3.3 Tour Inspection and Safety Monitoring to Ensure Safe Operation of the Project

The XLD Project has been operated at a maximum pool level of El. 265.69m since it was put into initial operation in October 1999, i.e. 9.31m from the normal pool level of El. 275m. The reservoir has operated at a high level above El. 260m for a total number of 192 days. The maximum daily rise is 3.26m (22nd August 2001) and the maximum drawdown is 3.47m (1st July 2007). Tour inspection and safety monitoring results show that all structures are in stable operation.

22.3.3.1 Adequate Safety Monitoring System

Altogether 3,201 observation points were arranged in the dam, involving 33 types of instruments and facilities. Since 2003, dead instruments have been replaced or updated by the Owner as advised by experts. By the end of August 2008, the instrumentation system contained a total number of 3,380 observation points, including 112 hydraulic instrument bases. Since the initial taking over of completed works, 96.6% instruments have been in good condition.

An adequate safety monitoring system has been established for the XLD Project, with 885 major instruments in critical locations introduced into Geomation 2,300 system to enable automatic data acquisition. Other instruments are read by man according to the specified frequency. Surface deformation observation is carried out by satellite positioning, measuring robots and other advanced means.

Pursuant to the current specifications, observation data were interpreted and analyzed in time, and monthly, quarterly and annual reports were prepared regularly together with special reports to provide reliable information for safe operation of the project.

22.3.3.2 Inspection and Maintenance of Hydraulic Structures

Routine inspection regulations have been formulated in the view of the actual operation conditions of hydraulic structures, with the route and sequence of inspection on the basis of the time, location, item and requirement of inspection. Tour inspections are divided into 3 types, namely, routine inspection, annual inspection and special inspection. Routine tour inspection is regular inspection of all structures included in the project. Annual tour inspection is conducted before and after the flood season each year, together with rating of structures. If and when there is a rainstorm, major flood, sensible earthquake and abrupt rise/drawdown or constant high water level, special tour inspection is arranged. Inspection records are interpreted and kept in file.

Based on tour inspection results and in consideration of defaults identified in inspection, hydraulic structure maintenance schemes have been developed, with maintenance performed by the Owner or external agencies according to the degree of difficulty.

Maintenance represents difficult work due to the high complexity of hydraulic structures and the large quantity of metal works. Procedures of work orders and operation orders have been developed with reference to the work bill and operation bill management regulations in power sector. This has provided the sequence of work to be followed in the work area, enhanced the implementation of regulations and secured the safe operation of equipment and the personal safety of workers.

22.3.3.3 Dam Safety Monitoring

Dam safety monitoring consists of deformation observation, seepage monitoring, stress/strain monitoring, seepage quality monitoring, reservoir sedimentation monitoring and earthquake monitoring. Deformation monitoring results show that vertical and horizontal displacements of the dam generally tends to be convergent and stable. Seepage monitoring results indicate that the recorded hydraulic gradients in all areas are less than the designed allowable value, without affecting safe operation of the project. Seepage quality monitoring results suggest that reservoir seepage will not cause erosion damages to the structures and foundations. Reservoir sediment monitoring results demonstrate that the channel in the range of 4.2km upstream of the dam is generally in balance, with a gentle horizontal slope. Earthquake monitoring results show that the frequency and degree of earthquakes practically fall within the normal range of variation. All monitoring results prove stable operation of structures and all monitoring data are continuous, reliable and complete, with an accuracy meeting the specifications.

22.3.3.4 Termite Control

The XLD Project has a rockfill dam with an inclined loam core, with surroundings suitable for termite to move and live. In this case, termite control is included as an important part of project management and the concept "One ant hole may cause the collapse of a thousand-mile dike" is established, with potential damages checked in the bud. Principal measures for termite control include: ①studying habits and characteristics of termite and formulating action plans; ②conducting a census during the termite monitoring and disposal period of April to October each year, with 2 tour inspections conducted per month; ③eradicating termite by destroying nests, luring to death with insecticide, grouting with insecticide, digging grooves and building isolation walls as instructed by experts; ④making efforts for popular education and dissemination to raise the awareness of termite control, with a termite control library (showing center) established and special films produced to enable a deep understanding of the damages, popularize the general knowledge of termite control, and improve the termite control skills of operators.

These measures have proven effective with termite nests been destroyed. As shown by tour inspection results, the movement of termite has not caused damages to the dam.

22.3.4 *Scientific Management and Proper Operation Has Played an Important Role in Regional Economy and Produced Huge Benefits*

The Yellow River is known for its high silt load. Little water and much sediment distributed unevenly in time and space cause the lower Yellow River "to suspend above ground", which endangers the people living on the river banks at all times. Since the 1990s, the Yellow River has witnessed consecutive dry years, worsening zero-flow events and sharpening water supply-demand contradictions, which has hindered the socioeconomic development in the basin. The XLD Project is the most direct and effective means of water regulation in the lower reach, a critical step of integrated water regulation and management in the basin.

Since the start of initial operation, the XLD Project has kept in mind its strategic importance in harnessing the Yellow River and its significance in exercising united water dispatch and management in the basin. It has generated huge social, economic and ecological benefits by sticking to the operation principle of "putting safety first, exercising integrated water regulation and prioritizing public benefits" and absolutely following the "instruction to water distribution" as a fundamental rule of operation and management.

22.3.4.1 Effective Retention of Sediment from the Middle Reach, Mitigation of Flood Loss and Basic Release of Ice-jam Risks in the Lower Reach

The XLD Project, when put into operation, increases the flood protection level in the lower reach of the Yellow River from 60-year flood to 1,000-year flood, basically releasing the lower reach from ice-jam risks.

The reservoir has operated safely for consecutive 11 years since it was filled, i.e. the start of initial operation. Faced with the serious situation of long-duration and extensive rainfall in the middle and lower reaches in late September 2005, the reservoir retained 2.575 billion m³ water from the upstream reach, with flood flows in the lower reach always controlled within the discharge capacity. The XLD Project, when operated together with other reservoirs, releases a part of base flows into the lower reach during the ice-jam season, avoiding the lower reach from freezing due to small flows and practically releasing it from ice-jam risks. In the winter 2001, air temperature in the lower reach was lower than that in normal years. At that critical period when the river was to be frozen soon, the reservoir replenished water at the rate of 500m³/s continuously, setting a precedent of unfrozen lower reach in severe cold years.

22.3.4.2 Sediment Reduction in the Lower Reach by Means of Sediment Retention and Flood/Sediment Regulation

The XLD Reservoir is designed to have sediment storage of 7.55 billion m³, which means a large capacity to retain sediment in the initial operation period. From the filling of reservoir to the end of 2010, a total volume about 2.73 billion m³ sediment was deposited in the reservoir. Sediment retention and flow/sediment regulation have enabled degradation of the downstream bed, in lieu of aggregation without the XLD Project, with a total volume of 1.2 billion m³ flushed downstream. From 2002 to 2010, water/sediment regulation was carried out at 12 times, with outflows controlled accurately to ensure the achievement of regulation goals and to flush the downstream bed effectively. The discharge capacity of the main channel is enlarged from less than 1,800m³/s in local sections to approximately 4,000m³/s.

22.3.4.3 Improved Availability Factor of Floods for Higher Probability of Irrigation and Municipal Water Supply in the Downstream Area

Operation of the reservoir enables effective flood storage during the flood season, provides water during the non-flood season and alters the uneven distribution of water in time and space. The reservoir, when it was put into operation, has further improved the water regulation capacity of the Yellow River, increased the probability of water supply for the downstream irrigation districts fed by the Yellow River, relaxed the situation of insufficient water supply for production and domestic purposes in the downstream area, and diverted the Yellow River water to Tianjin 5 times, Qingdao 9 times and Baiyangdian 3 times. In the late spring of 2000 and 2001, the reservoir level was lowered below El.195m to ensure water supply for the arid downstream areas, with reservoir nearly emptied and generator units suspended for 4 months.

22.3.4.4 Proper Operation of Project and Reasonable Utilization of Water Resources for Improving Environmental and Ecological Benefits

Proper operation of the XLD project and reasonable utilization of water resources have greatly helped to avoid zero-flow events in the lower reach, maintain the healthy life of the Yellow River, and especially improve the ecosystem in the reservoir area and the estuary. During construction, great importance was attached to environmental aspects, with forceful efforts devoted to soil conservation and environmental protection to improve the ecosystem in the reservoir area. The establishment of large wetlands provides good habitats for aquatic species and effectively ameliorates the ecosystem in the estuary. The XLD Project has been given the title of "100 Top Environmental Projects", "Model of Water and Soil Conservation in Developments", "National AAAA Scenic Spot" and "Top 10 Hot Scenic Spots in Henan".

22.3.4.5 Better Quality of Power Supply for Promoting Regional Economic Development

As the main peak/frequency regulator and emergency power source of the Henan power grid, the XLD Power Plant improves the safety and quality of power supply from the grid, effectively relieves power shortage and promotes regional economic development. Clean energy helps energy reduction, emission reduction and ecological improvement. Equipment sizing in the special flood/sediment conditions and operation under the condition of "power generation depending upon water availability" represents rich experience. By the end of March 2011, a cumulative output of 49.71TW·h had been achieved.

22.4 Xiaowan Hydropower Project
Yu Jianqing[1]

22.4.1 *Project Overview*

Xiaowan Hydropower Project is located on the middle reaches of Lancang River, at the boundary between Nanjian County, Dali Prefecture and Fengqing County, Lincang City, in Yunnan Province. It is the second cascade of the eight cascades planned in the middle and lower reaches of Lancang River, and a critical cascade of the river section. It is 455km away from Kunming by road. Lancang River is originated from Mountain Tanggula in Qinghai-Tibetan Plateau, which flows across

---

[1] Yu Jianqing, Deputy Chief Engineer of Hydro China Kunming Engineering Co., Ltd, Professor Senior Engineer.

Qinghai-Tibetan Plateau, Hengduan Mountain Range and west Yunnan-Guizhou Plateau, via Qinghai and Tibet into Yunnan from north to south. It flows out of the Chinese territory via Xishuangbanna Prefecture, Yunnan Province, after which it is referred to as Mekong River. The length of Lancang River in the Chinese territory is about 2,100km and the basin area is 174,000km^2. The length in Yunnan Province is 1,240km, and the basin area is 91,000km^2.

The controlled basin area U/S of the Xiaowan dam site is 113,300km^2, the perennial average discharge is 1,210m^3/s, the perennial average annual runoff is 38.2 billion m^3, the actually measured maximum discharge is 9,150m^3/s, and the minimum discharge is 273m^3/s. Through hydraulic analysis and calculation, the peak discharges at the dam site upon flood events of 100 years of recurrence period, 500 years of recurrence period, 1,000 years of recurrence period and 10,000 years of recurrence period are 13,100m^3/s, 16,700m^3/s, 18,300m^3/s and 23,600m^3/s, respectively. The perennial average suspend load sediment at the dam site is 46.35 million tons, in which the high-water season (June-October) accounts for 96%. The perennial average sediment concentration is 1.21kg/m^3, and the annual bed load sediment is about 1.5 million tons.

The peripheral seismic geological environment of the Xiaowan dam site is very complicated, with frequent earthquakes in the history. The basic seismic intensity of the dam site area is VIII degree. The non-retaining structures are to be protected against earthquakes exceeding 5% probability in 50 years of recurrence period and the corresponding horizontal acceleration of bedrock peak value is 0.205g. The arch dam is to be protected against earthquakes exceeding 2% probability in 100 years of recurrence period, and the corresponding horizontal acceleration of bedrock peak value is 0.313g.

Xiaowan Hydropower Station is a high dam and a large reservoir. It is a Large (1) Level I project, and the main permanent hydraulic structures are Level 1 structures. The main tasks for project development are power generation, together with such comprehensive utilizations as flood control, irrigation, sediment retention and navigation in the reservoir area. The dam-style concentrated development is applied to the project complex. The main permanent structures comprise a concrete hyperbolic arch dam, an after-dam plunge pool, a secondary dam, and the L/B flood-release tunnel and R/B underground water diversion and power generation system, etc., which is shown in Fig. 22.4-1. With a total storage capacity of about 15 billion m^3 and an effective storage capacity of 9.9 billion m^3, the reservoir is an incomplete perennial regulating reservoir. The installed capacity is 4,200MW; its firm output is 1,778MW; the perennial average annual energy production is 19 billion kW·

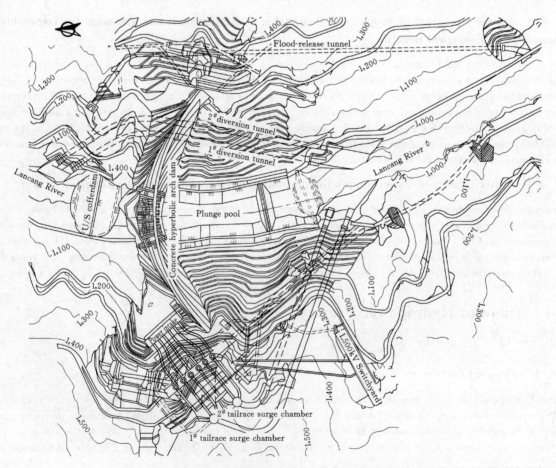

Fig. 22.4-1 Plan Layout of Complex of Xiaowan (Unit: m)

h, and the number of annual utilization hours is 4,520h. Xiaowan is one of the main power supplies for the "West-East Power Transmission" in China, and a strategic project for implementing the "Great West China Strategy".

The evaluations of the project proposal of Xiaowan and the feasibility study report were approved by China International Engineering Consulting Corporation (CIECC) in March 2000 and September 2001 successively. The project was officially commenced in January 2002. The river closure was in October 2004, one year before schedule. The first concrete placement to the arch dam was in December 2005. The first unit was put into power generation in September 2009, and all the six units were put into operation in August 2010. To be specific, operation of the 4th unit enabled the installed capacity of Chinese hydropower to exceed 200 million kW, and it was nominated by National Energy Administration and China Electricity Council as a symbolic unit for Chinese installed capacity of hydropower to exceed 200 million kW · h.

22.4.2 Background and Socio-Economic Benefits of the Project Construction

Since reforms and opening up, the economy of Guangdong Province has been developing very fast, and electric power demand has drastically increased. However, Guangdong has very poor energy sources, and most of the coal and fuel need to be transported from North China or imported from overseas countries. Due to the limitation of transporting coal from North China to South China, Guangdong had to enlarge the quantity of coal to be imported, and arranged a number of large thermal power plants along the coast, which caused pollution to the local environment, and Guangdong has become one of the hard-hit areas of acid rain in China. Therefore, the National Development and Reform Commission and State Power Grid Corporation of China definitely expressed that the development of hydropower in west China and transmitting electric power to Guangdong will play an important role in promoting the economic development of Yunnan Province and Guangdong Province, improving the composition of the Chinese energy structure, alleviating pressure on coal transportation and environment, realizing optimized allocation of resources, and promoting implementation of sustainable development strategy in Guangdong Province. In August 2000, the People's Government of Guangdong Province and the People's Government of Yunnan Province signed an "Agreement on 'Yunnan-Guangdong Power Transmission'", which symbolized that the strategic project of "West-East Power Transmission" has made a solid leap forward. In December 2000, National Development and Reform Commission clearly pointed out in "Examination and Approval of State Planning Commission on Requests for Instructions on Project Proposal of Xiaowan Project on the Lancang River in Yunnan Province" that it was planned that Xiaowan Hydropower would transmit about 2,000MW electric power to Guangdong Province, so that Xiaowan would become one of the critical projects for implementing the strategy of "West-East Power Transmission".

In the power supply source structure of Yunnan Province, the percentage of runoff hydropower is excessively high, with rather poor regulating properties, resulting in very prominent contradictions of high-water season and low-water season in the power grid. At the end of 2009, the installed capacity of hydropower in Yunnan Province was 20,904MW, accounting for 66% of installed capacity in the province. The total installed capacity of hydropower in the provincial power grid was 14,905.6MW, most of which had only monthly or weekly regulating capacity, or even without regulating capacity, and the firm output during the low-water season was very low. Therefore, the task of strategic adjustment of power supply structure in Yunnan Province was very onerous, and it was absolutely necessary and pressing to construct large hydropower stations with good regulating properties. Xiaowan project has perennial regulating properties, and in addition to providing 4,200MW of generation capacity and 1,778MW of firm output for the power grid, the perennial average energy production will be 19 billion kW · h. It also may bring significant compensation benefits for the D/S cascade HPPs, and conduct trans-basin coordinated optimized scheduling through the power grid, which will greatly improve the regulating properties of cascade hydropower in Yunnan Province, and improve the unreasonable structure of power source in Yunnan Province.

Following the completion of Xiaowan Hydropower, the firm output of the three cascade hydropowers in the D/S, i.e., Manwan, Dachaoshan and Jinghong, was increased by some 1,100MW, and the energy production increased by about 2.7 billion kW · h (both being energy production of low-water period). Moreover, it turned about 1.4 billion kW · h of energy production during the flood-season into energy production of the low-water season, and totally increased about 4.1 billion kW · h energy production during the low-water season for the three cascade hydropowers. The ratio of high-water season energy production to low-water season energy production was improved from 1.67 : 1 without Xiaowan Hydropower to 0.93 : 1. The percentage of firm output in annual energy production was improved from 46.6% without Xiaowan hydropower to 86.4%, which was equivalent to building a 1 million kilowatt peak-regulating power plant at no charge. By taking advantage of the huge regulating capacity of Xiaowan Reservoir and through coordinated trans-basin optimized scheduling of the power grid, it also might greatly improve the coordinated regulating properties of hydropower clusters in the Yunnan power grid, increase the percentage of hydropower capacities with regulating capacity in the total installed capacity of hydropower in Yunnan Province, and improve the passive situation of "abandoning water during high-water season and being short of water during low-water season", which had been a long-standing situation in the power grid of Yunnan.

In accordance with the unique conditions of Yunnan, in the course of implementing the strategy of "Great West China Strategy", Yunnan Province would build five national bases, in which the construction of a hydropower base was an important component for the strategic objective of building Yunnan as a powerful province in green economy, and Xiaowan hydropower was one of the critical projects for Yunnan hydropower base construction.

It was estimated that during the dam construction period, about RMB 10 billion of project investments was spent in Yunnan Province. Once Xiaowan was completed, it would annually generate about RMB 100 billion of economic benefits for the entire society.

Xiaowan dam may substitute about 25.2 billion kW·h of coal-fired thermal energy production on an annual average, save about 7.56 million tons of standard coal, so as to greatly reduce emission of SO_2, CO_2 and other waste gases, and discharge of wastewater and solid wastes, improved the environmental quality, and promoted healthy development of electric power industry in Yunnan Province.

Moreover, Xiaowan Reservoir formed some 178km of deep waterway on mainstream, and some 125km deep waterway on Heihui River, its tributary, created conditions for developing navigation industry in the reservoir area. The reservoir operation of storing water during the high-water season to supplement the low-water season may also increase the discharge of the downstream channel of Lancang River during the low-water season, and improve the navigation conditions of the downstream of Lancang-Mekong River.

22.4.3 Environmental Protection

In the course of Xiaowan construction, the following measures were taken for environmental protection and water and soil conservation.

(1) To treat production wastewater and domestic wastewater according to quality of domestic wastewater, ICEAS biochemical treatment + filtration and adsorption + recycling process was adopted to treat domestic wastewater from the Employer's camp, and activated sludge method + MBR + AOP process was adopted to treat wastewater from ground control building in the powerhouse and dam site area. The reclaimed water after treatment was recycled, and realized "zero discharge" of wastewater.

(2) Such measures as transplanting and building botanical garden and animal natural reserve were taken to preserve terrestrial animals and plants. The rare and endangered plants were protected by means of transplanting, collecting seeds for cultivating young plants and protecting seed resources, and establishing Jinguangsi Nature Reserve of Yongping County as animal refuge. And it was to establish Qinghua Green Peafowl Reserve at Weishan County (which has presently been upgraded as a provincial level nature reserve) and Ma'anshan Macaque Reserve at Weishan County, which were to reinforce protection of such rare animals as green peacocks and macaques respectively.

(3) Environmental monitoring of the construction workmanship was enhanced by strictly following EIA, water and soil conservation report and relevant approval document requirements. Professional agencies were employed to conduct environmental monitoring, water and soil conservation monitoring, ecological monitoring, people's health monitoring and meteorological monitoring, etc. Monitoring of noises generated in the course of project construction, air quality, river water quality, construction wastewater, rare fish resources, people's health and meteorological changes was conducted. Regular bulletins and annual reports were prepared to timely feed back and guide subsequent environmental protection through monitoring data.

(4) Special supervision was implemented for environmental protection and water and soil conservation. In order to further improve environmental protection and water and soil conservation, HYDROCHINA Huadong Engineering Corporation was appointed to conduct environmental protection and water and soil conservation supervision of the project.

(5) According to the social and environmental characteristics of Lancang River Basin, several domestic and foreign specialized agencies were appointed to study environmental impacts of hydropower development to the downstream areas, including such fields as hydrology, sediment, agriculture, navigation and fish, etc.

22.4.4 Main Structures of the Project

22.4.4.1 Concrete Hyperbolic Arch Dam

1. Structural layout of arch dam

The type of arch dam was a parabolic thickened hyperbolic arch dam, the dam crest elevation was 1,245.00m, the lowest base surface elevation was 950.50m, the maximum dam height was 294.50m, and the arc length of dam crest centerline was 892.786m. The arch dam was divided into 43 dam sections, the width of the flood-release dam section was 22 - 26m, and the width of the remaining dam section was 20m. The dam crest width from the center to arch end gradually changed from 12 to 16m. The maximum central angle of the arch dam was 92.791°, the beam bottom width of arc crown was 73.124m, the arc height ratio was 3.035, and the thickness height ratio was 0.248. The left bank of the arch dam was established with thrust piers. The bottom elevation of which was 1,210.00m, the height of which was 35m, and the bottom length of which was 48m.

Dam sections $20^\#$ - $25^\#$ were arranged with 5 surface outlets and 6 middle outlets; $19^\#$ and $26^\#$ were each arranged with a reservoir emptying bottom outlet; $20^\#$ and $25^\#$ were each arranged a diversion bottom outlet, which were respectively located at the lower part of flood-release middle outlets $1^\#$, and $6^\#$; $21^\#$ - $23^\#$ were arranged with three diversion middle outlets, which were respectively located at the lower part of $2^\#$, $3^\#$ and $4^\#$ flood-release middle outlets.

The foundation of the arch dam was arranged with a curtain grouting gallery, $1^\#$ foundation drainage gallery, $2^\#$ foundation drainage gallery and a horizontal access gallery. The dam was arranged with five layers of galleries at 1,010.00m, 1,060.00m, 1,100m, 1,150m, and 1,190m for checking and observing the dam body and drainage. All layers of galleries were connected with a curtain grouting tunnel and a drainage tunnel at corresponding elevations of both dam abutments, and the upper five layers are connected with after-dam permanent berms of downstream dam at corresponding elevations. The arch dam was arranged with a 1.5 t elevator each at dam sections $18^\#$ and $27^\#$, from 1,010.00 - 1,245.00m, the operation height reached 235m.

2. Arch dam foundation treatment

(1) Geological defects replacement treatment of dam foundation. Trench excavation and backfilling of concrete was conducted to treat fault F_{11}, alteration zone E_1, alteration zones $E_4 + E_5$, E_9 and L/B f_{64-1}, etc. The unloaded rock masses distributed at low elevations of two sides of dam foundations were also improved by excavation and concrete replacement treatment.

(2) Dam foundation curtain grouting. The dam foundation was arranged with complete curtain grouting, and it was to arrange 1 - 3 rows of grouting holes according to the size of acting heads, and the maximum hole depth of main curtain was 120m. Both banks were arranged grouting tunnels at a spacing of 40 - 50m, in the grouting tunnel, except arranging the same number of rows of curtain grouting holes at the tunnel portal section as the same dam foundation, which were generally arranged with a single row of curtain grouting holes.

(3) Dam foundation consolidation grouting. The dam foundation was entirely treated with consolidation grouting, and all the grouting was conducted with concrete cover weight, and consolidation grouting was divided into six areas. Area A was located at central part of the dam foundation, and the hole depth was 10m; Area B and Area C were located upstream of the dam foundation, and the hole depth was 15m; Area D was located downstream of dam foundation, including consolidation grouting on fillet, and the hole depth was 15m; Area E was located at right side of fault alteration zone, and the hole depth was 25 - 30m; Area F was located near dam abutments on both banks, and the hole depth was 45m.

(4) Dam foundation drainage. The dam foundation was arranged with two rows of drainage galleries, in which drainage holes were arranged, and it was designed as a concealed water pumping and drainage system. It was to arrange drainage tunnels on both banks at the same elevation as curtain grouting galleries, in which it was arranged with reverse drainage holes. The dam foundation along the river developed steep dip angles and fissures along the slope. Therefore, all the downward drainage holes below El. 975.00m were inclined toward the hills, and all the upward drainage holes above El. 975.00m were inclined outward the hills so as to run across more fissures, and enhance drainage effects.

3. Geological defects treatment of resistance body of dam abutment

Geological defects of in the resistance body of the R/B dam abutment mainly include faults F_{11}, F_{10}, f_{10} and f_9, altered rock zones of E_1, $E_4 + E_5$, E_9 and unloaded rock masses; Geological defects of resistance body of L/B dam abutment mainly include faults F_{11}, F_{20}, f_{34} and f_{12}, altered rock zones of E_8, unloaded rock masses of $4^\#$ Ridge and Longtan'gan Gully at the dam toe with low elevation. The following comprehensive treatments were conducted to the above mentioned geological defects.

(1) Cutting surface slope for reducing loads. Cutting slope was conducted to the strongly unloaded rock masses at surface layers of $4^\#$ ridge on left bank, and the slope cutting depth was about 15m.

(2) Underground concrete replacement. The left side dam abutment was arranged with four floors of replacement holes between El. 1,160.00 - El. 1,220.00m. The right side of dam abutment was arranged with 10 layers of replacement holes as per 20m height differences. According to the widths and shapes of different parts, different elevations and different fault al-

teration zones, the dimensions of replacement holes included 4m × 5m, 5m × 5m, 5m × 8m, 6m × 10m and 10m × 10m. Concrete placement of the replacement holes was conducted in two phases. In Phase I lining, it was internally left with 3.5m × 3.5m grouting tunnels, which were to conduct Phase II backfilling after completion of consolidation grouting. Phase I concrete was not precooled, but cooling water pipes were embedded for cooling with running water. The low-heat, micro-expansion concrete for backfilling is applied to Phase II backfilling concrete.

(3) High-pressure consolidation grouting. All the replacement holes were arranged with consolidation grouting, which was simultaneous contact grouting between concrete and bedrock. Grouting pressure included a high-pressure zone and low-pressure zone. Low-pressure grouting with maximum pressure of 2MPa was conducted at the outlets and outside of the unloaded rock masses of 4# Ridge on L/B, which was to form closed areas for high-pressure grouting areas. The 6MPa high-pressure grouting was applied to the remaining parts.

(4) Underground and surface drainage system. After completing consolidation grouting of the replacement holes, it was to sweep open some consolidation grouting holes and deepen as drainage holes. In order to lower the underground water table in the dam abutment and plunge pool slope, six rows of underground drainage holes were constructed along the river on the two sides below El. 1,245.00m, from the dam foundation drainage tunnel to the secondary dam. The average spacing was approximately 40m. Reverse upward drainage holes were arranged in all the drainage holes.

(5) Surface prestressed anchorage. Since the 3# Ridge on the R/B dam abutment and 4# Ridge on the L/B dam abutment were severely unloaded and located at critical stressed positions, prestressed anchor cables were installed to stabilize the critical parts of resistance bodies on both banks. Prestressed anchor cables arranged were mainly 3,000kN level, with a few 6,000kN anchor cables arranged below El. 1,050.00m near the dam, while the prestressed anchor cables far away from the dam were mainly 1,800 kN level, and the total number of anchor cables was about 1,400.

22.4.4.2 Flood Release through Dam Body and After-Dam Energy Dissipation Structures

1. Distribution of flood release capacity of the complex

The combined flood release scheme through five surface outlets and six middle outlets in the dam, plunge pool and secondary dam after dam, and a flood release tunnel on the left bank were applied to the complex.

After cutting the peak discharge of the reservoir, the maximum release discharge of the complex was 20,710m³/s, the maximum head was 225m, and the corresponding release power reached 46,000MW. The flood-release energy dissipation issue of the high-head and large discharge was extraordinary, and the flood-release vibration of the dam body was also a concern.

The gate opening dimensions of all the flood-release structures and flood release distribution are shown in Table 22.4-1.

Table 22.4-1 Portfolio of the Flood-Release Structures

Position of Flood-Release Structures	Inlet Bottom Slab Elevation (m)	Service Gate Dimension (m) Number of openings-Width × Height	Check Flood ($P=0.01\%$) El. 1,242.51m	Design Flood ($P=0.2\%$) El. 1,238.30m	Greatest Flood in 100 years El. 1,236.90m	Greatest Flood in 5 Years El. 1,236.50m
			Flow Release(m³/s) (Percentage in Total Flow Release%)			Independently operating flow release(m³/s)
Surface Outlet of Dam	1,225.00	5 - 11×15	8,625 (41.7%)	5,530 (30.4%)	4,610 (27.0%)	4,355
Middle Outlet of Dam	1,165.00 1,152.50 1,140.00	65 - 6×6.5	8,264 (39.9%)	8,038 (44.2%)	7,962 (46.5%)	7,940
Flood Release Tunnel	1,200.00	1 - 13×13.5	3,811 (18.4%)	3,535 (19.4%)	3,439 (20.1%)	3,410
Flow of Unit	1,140.00	—	0	1,095 (6.0%)	1,095 (6.4%)	5×365
Maximum discharge that may be released from the complex at corresponding levels / flow release of flood regulation			20,700 / 20,710	18,198 / 15,691	17,106 / 12,971	5,678
Natural peak flood discharge of corresponding flood standards			23,600	16,700	13,100	6,330
Peak Cutting Percentage			12.25%	6.0%	1%	—

2. Layout of flood-release structures on the dam body

Flood release of the dam was designed as per the principle of "longitudinal separation by layers, horizontal diffusion by single units, integral entering water and returning to the trench, combined operation of surface and middle outlets for collision in air for energy dissipation". The flood-release structures of the dam are symmetrically arranged, and the overflow centerline and arch dam centerline coincided.

The five overflow surface outlets on the dam body were arranged at $20^\#$ – $25^\#$ dam sections, all being arranged crossing the horizontal joints, and the surface outlets were open. The orifice/opening weir crest elevation was El. 1,225.00m, and the opening dimensions were 11m×15m. The diffusion angle of surface outlet wall was $4°30'$ – $7°42'$, the bucket angle or depression angle was $10°$ – $20°$.

The six flood-release middle outlets in the dam body were deep water-release outlets, $1^\#$ – $6^\#$ flood-release middle outlets were respectively arranged at $20^\#$ – $25^\#$ dam sections, located at lower part of the surface outlet gate piers. In the course of flood release, the plan layout means of symmetrical pairs was adopted; the elevations of inlet and outlet were arranged in three layers; the upstream elevation view was arranged as an inverted "八" shape. The entire outlet was steel plate lined. The inlet sill elevations of $1^\#$ and $6^\#$ middle outlets, $2^\#$ and $5^\#$ middle outlets, and $3^\#$ and $4^\#$ middle outlets were El. 1,165.00m, El. 1,152.50m, and El. 1,140.00m, respectively. The plane diffusion and elevation upwarping are applied to outlet section. The elevations of $1^\#$ and $6^\#$ outlet bottom sills, $2^\#$ and $5^\#$ outlet bottom sill and $3^\#$ and $4^\#$ outlet bottom sill were El. 1,164.249m, El. 1,156.147m, and El. 1,149.580m, respectively. The outlet service gate opening dimensions were 6.0m × 6.5m. In order to tackle the issue of radial concentration of water-release jet of middle outlet, the middle outlet adopted plane deflection angles of $1°$ – $2.5°$.

3. After-dam plunge pool and secondary dam

The plunge pool is closely connected with the back of the dam, the azimuthal angle of centerline of front section was SE178°. It coincides with crest outlet overflow centerline and azimuthal angle of rear section was SE168°. The full length of the plunge pool was 350m, adopting compound trapezoidal sections. The bottom slab elevation was El. 965.00m, and the bottom width of the narrowest place was 70m. The full-section reinforced concrete lining is applied, the thickness of the bottom slab was 3m, the surface was arranged with 0.5 m-thick $C_{90}60$ abrasion-resistant silica fume concrete, and it was to conduct anchorage to the bottom slabs.

The end of the plunge pool was established an secondary dam, the dam crest elevation was El. 1,004.00m, the base surface elevation was El. 960.00m, the maximum dam height was 44m. The full length of the secondary dam was 164.2m. 20m deep curtain grouting was arranged under the secondary dam. The foundation was arranged with 5m deep consolidation grouting. The secondary dam of the plunge pool was arranged with closed drainage system, around which there was no curtain but 30m deep drainage holes. Triangular drainage ditches between the pouring blocks were constructed, and shallow drainage holes 5m deep in the trench were drilled. After the auxiliary dam, there was a 32m long apron.

22.4.4.3 Flood-Release Tunnel

Flood release tunnel was arranged in the left massif of the river channel. The tunnel was arranged to transit from pressurized into non-pressure "dragon head rise", comprising an inlet section, a pressure section (including plane curve section), a service gate chamber, a non-pressure section, an outlet flip bucket, an air replenishment tunnel (well) and other auxiliary tunnels and chambers. The horizontally projected full length of the tunnel axis was about 1,600m. The inlet was located on the downstream of Gouyazi Gully on the L/B in front of the dam, U/S of Yinshui Gully, adopting bank-tower intake. The length of the inlet section was 20m, the azimuthal angle of axis was SE123°0'0″, and the water intake was a trumpet comprising elliptic curve. The elevation of bottom slab was El. 1,200.00m, after which it was connected with emergency bulkhead gate, which was a plain gate, the opening dimensions were 15m×16.5m $(W \times H)$. The full-section reinforced concrete lining is applied to tunnel body of the pressure section of the flood-release tunnel ("Release 0+020.00-0+ 428.444 m"), and the bottom slope of the entire tunnel section was 1.5%. The section was D16.5m circular tunnel section. The service gate chamber of the flood-release tunnel was fully embedded underground chamber, which was installed with a radial service gate, the opening dimensions were 13m×13.5m, the gate chamber bottom slab elevation was El. 1,193.873m, the elevation of maintenance platform was El. 1,211.873m. From the U/S to the D/S, the non-pressure section of the flood-release tunnel included Ogee section, ogee section, straight channel and slope section. The non-pressure tunnel section was circular arch and straight wall type, which was arranged with 7 aeration sills, and the total area of vent shafts on both sides of the aeration sills was 7.20m². The outlet flip bucket of the flood-release tunnel was located at bank slope U/S of Wa'xielu Gully, to which the narrow joint flip bucket to dissipate energy is applied, and the elevation of the lowest point of ogee at the outlet was El. 1,031.00m. Three air replenishment tunnels (shafts) were respectively arranged on right wall of service gate chamber, non-pressure tunnel crowns of "Release 0 + 820.00 m" and "Release 1 + 091.00 m", which were to make up air above the water level of the non-pressure section.

22.4.4.4 Headrace and Power Generation System

Due to the limitation of topographic and geological conditions, the Headrace and Generation System was located at right underground of the complex, occupying an area of some 1.5 km². It comprised a headrace system, the main and auxiliary powerhouses, the main transformer switch house, a tailrace system and power transmission work, etc. The headrace system included power intake and 6 penstocks, while the tailrace system included 2 tailrace surge chambers, 2 tailrace tunnels and outlet structures, and the power transmission works included 2 line-outgoing tunnels and 500kV ground switchyard.

1. Headrace system

The headrace system comprised power intake and penstocks. In combination with topographic and geological conditions and master layout of the complex, the bank-tower arrangement is applied to the power intake, the full length of the intake tower was 160.5m, the width was 32m, the height was 105m, the inlet bottom slab elevation was El. 1,140.00m, and the crest elevation was El. 1,245.00m. The dimensions of the trash rack slot were 0.62m×3.5m, the bulkhead gate opening dimensions were 7.5m×10.5m ($L \times W$), and the emergency bulkhead gate opening dimension was 7m×10m ($W \times H$).

The single penstock for single Unit is applied to the hydropower station. According to calculation and analysis, the overburden depth could satisfy the Norwegian criteria and the Snow mountain criteria, vertical criteria, and minimum main stress criteria, and integrity of the rock masses was good. Therefore, except that the steel liner is applied to the lower horizontal section and the end 1/3 of the lower curve section, the reinforced concrete lining is applied to the remaining pipe sections. The total lengths of six penstocks (1# - 6#) were 300.787m, 298.081m, 296.744m, 296.828m, 298.357m and 301.232m in turn. The centerline elevation of the upper horizontal section was El. 1,145.00m, and the centerline elevation of the lower horizontal section was El. 980.00m. The standard section of the tunnel was circular, the inner diameter was 9m (the lower curve section was 8.5m, the end section was 6.5m), the lining thickness was 1.75m (end section) / 0.8m / 0.75m / 0.5m. The steel liner is applied to inner walls of the end 1/3 of lower curve section and lower horizontal section, the length of single penstock was 51.52m, and the steel liner thickness was 38mm / 42mm / 46mm.

2. Underground main and auxiliary powerhouses

The underground main and auxiliary powerhouses were arranged in a parallel line, the section was square circular, the maximum dimensions were 298.4m×30.6m×79.38m ($L \times W \times H$). The maximum dimension of the main powerhouse was 264.5m×30.6m×79.38m ($L \times W \times H$), the spacing between the units was 33m, and the length of erection bay was 50m. The unit installation elevation was El. 980.00m, which was totally arranged by six floors, in which the elevation of disc valve drainage equipment floor was El. 971.50m, the elevation of the generator water supply equipment floor was El. 975.00m, the spiral case floor elevation was El. 980.00m, the elevation of turbine floor was El. 985.00m, the elevation of the intermediate floor was El. 991.00m, and the elevation of generator floor was El. 998.50m. Auxiliary powerhouses included auxiliary powerhouse at the terminal and auxiliary powerhouse beside the erection bay. The auxiliary powerhouse at the terminal was arranged at right terminal of the main powerhouse, with a dimension of 13.6m×28m×46.92m ($L \times W \times H$), which was arranged in eight floors, and arranged with simplified control room, high-pressure air compressor chamber, low-pressure air compressor chamber, electrical lab, and secondary panel chamber, etc. The auxiliary powerhouse beside the erection bay was arranged at left terminal of the main powerhouse, with a dimensions of 20m×28m×24.55m ($L \times W \times H$), which was arranged in six floors, and arranged with turbine oil tank chamber, oil treatment chamber, lighting panel chamber, air conditioner control room, public transformer chamber, and lighting transformer chamber, etc.

In order to meet the ventilation requirements of the underground powerhouse, it was to arrange an air conditioner chamber at left end of underground powerhouse, the section was square circular, the maximum dimensions were 56.5m×18.6m×10.18m ($L \times W \times H$), and the bottom slab elevation was El. 1,022.00m.

3. Main transformer chamber

The main transformer chamber was arranged in parallel to the main powerhouse on the downstream side, 50m away from the downstream wall of the main powerhouse, which connects with the main powerhouse through six omnibus bar tunnels. The section of the main transformer chamber was square circular, the maximum dimensions were 230.6m×19m×24.05m ($L \times W \times H$), which was totally arranged in three floors, in which the elevation of the transformer floor was El. 998.05m, the elevation of cable floor was El. 1,010.50m, and the elevation of the ventilation floor was El. 1,015.00m. In accordance with layout requirements of E/M equipments, the section of the omnibus bar tunnel was to adopt square circular shape, the dimension was 8.3m×9.946m ($L \times W$), and the length was 50m.

4. Tailrace system

Three generators with three tailrace adits-tailrace surge chamber-tailrace tunnel, comprising tailrace adits, tailrace bulkhead gate chamber, tailrace surge chamber and tailrace tunnel are applied to tailrace system.

Total lengths of six tailrace adits ($1^{\#} - 6^{\#}$) were 133.918m, 118.300m, 133.918m, 143.918m, 128.300m and 143.918m in turn, the standard section was inverted U shape, and the clearance dimension of the section was 10m×15m ($W \times H$).

To facilitate tailrace maintenance of the unit, it was to arrange unit service gate chamber before tailrace surge chamber of the Units. The total length of gate chamber was 206.800m, the width was 11m, the height was 184.8m, the bottom elevation was El. 853.20m, and the top elevation was El. 1,038.00m. The bulkhead gate opening dimension was 12m × 15m ($W \times H$).

According to the calculation results of complex layout of headrace and generation system and hydraulic transitional process, the surge facilities must be arranged to tailrace system. After comparing multiple proposals, the proposal to use a cylinder impedance surge shaft was adopted. The tailrace surge chamber was a double cylinder impedance surge shaft with an upper chamber, and the internal diameter was 32m; the bottom elevation was El. 953.00m, the top elevation was El. 1,039.497m, and the cylinder wall lining thickness was 3m / 1,15m. It connected to the upper chamber bottom elevation of El. 1,013.90m, the top elevation was El. 1,030.173m, the standard section of connecting upper chamber was inverted U shape, and the clearance dimension was 18m×16.273m ($W \times H$). The impedance hole was circular, the diameter was 11m, the bottom elevation was El. 971.00m, and the top elevation was El. 972.50m. Water level in the surge chamber: One generator at El. 992.140m, six generators at El. 1,000.03m, the minimum surge level at El. 975.00m, and the maximum surge level at El. 1,020.50m.

The total length of $1^{\#}$ tailrace tunnel was 932.448m (to tailrace bulkhead gate chamber), while the total length of $2^{\#}$ tailrace tunnel was 718.31m. Standard section of the tunnel was circular, the inner diameter was 18m, and the liner thickness was 1.3 m/0.5m.

5. Power transmission works

Power transmission works comprised line-outgoing tunnel and 500kV GIS switchyard. In order to avoid flow-release atomization and facilitate traffic layout, it was suitable to arrange the power transmission works at El. 1,250.00m, and arrange the 500kV ground switchyard on the U/S side of Xiushanda Gully on the R/B. The foundation was to adopt semi-excavation and semi-filling mode, and its foundation top elevation was El. 1,250.00m. According to the equipment layout requirements of 500kV ground switchyard, its dimensions were 200m×18.5m×26.6m ($L \times W \times H$), arranged in four floors, in which the elevation of the reactor floor was El. 1,250.00m, the elevation of cable floor was El. 1,257.00m, the elevation of the GIS floor was El. 1,262.00m, and elevation of the roof line-outgoing floor was El. 1,276.60m.

22.4.5 Critical Technologies of the Complex Works

22.4.5.1 The 300m-High Arch Dam Geometry and Measures for Crack Control and Earthquake Resistance

The Xiaowan arch dam was built in an area with high seismic intensity. The dam crest arc length was almost 900m, and the total hydraulic thrust reached 180 million kN. The layout and geometry of arch dam was subject to particular topographic and geological conditions, master layout of complex and dam abutment stability. The dam geometry design must simultaneously satisfy the stress control criteria during the construction period, normal operation period and seismic working conditions, which was extremely difficult. Therefore, except adopting conventional multi-crown cantilever methods to optimize the arch dam geometry, such methods as thick shell element method and finite element method were introduced into optimizing the high arch dam geometry. It was proposed not only to focus on maximum stress value meeting control criteria of the spots, but also to control the high tensile and compressive stress areas of dam heels and dam toes, and to optimize the high arch dam geometry of the high dynamic stress area at middle and upper elevations of the dam body, which well coordinated and solved the safety and economic problems in the dam geometry design of the Xiaowan arch dam. Fig. 22.4-2 has shown the Xiaowan Arc Dam after its completion.

In order to standardize the cracking of dam heels of high arch dams which was not involved yet, several analysis methods both at home and abroad were adopted. We conducted thorough studies from different perspectives, based on which we proposed engineering measures for preventing or reducing dam heel cracking of the Xiaowan arch dam-arranging structural induced joints at the dam heel, applying anti-seepage system at the upstream dam surface and arranging clogging materials before dam, etc. As to enhance the earthquake resistance of the arch dam, except considering it as an important element in optimizing the geometry, studying and adopting high-strength concrete in the high dynamic stress area, possible expansion of horizontal joints of middle and upper elevations of the dam body was considered carefully when there is a strong earthquake. We adopted multiple analysis methods and model tests to conduct analysis and study, arranged aseismic steel bar, and conducted thorough value analysis and model test study of its acting mechanism and effects.

Fig. 22.4 - 2 The Completed Xiaowan Arch Dam

22.4.5.2 Stability and Consolidation Measures for Dam Foundation and Dam Abutment under Complicated Geological Conditions

The safety of high arch dam not only depends on the arch dam itself, but also critically depends on stability of the dam abutment and dam foundation. The complex layout, dam line position and arch dam geometry optimization of Xiaowan hydropower were all conducted under the conditions that the resisting rock masses of the dam abutment are good and that the arch ends had sufficient embedded depth. Moreover, extensive analyses, studies and model tests were conducted on the stability of dam foundation and dam abutment rock masses and their consolidation measures, and slope stability, deformation and overload safety of dam abutment rock masses were thoroughly analyzed and studied. Comprehensive measures, such as high-pressure consolidation grouting, concrete tunnel and shaft plug replacement, prestressed anchor cable and rock bolt, surface and underground drainage system, were taken, and in combination with slope treatment of plunge pool, comprehensive consolidation treatment and protection of resisting rock masses of the dam abutments.

As to the swelling and relaxation after excavation arising from high ground stresses in areas of mountains and valleys, advanced geophysical prospecting technologies (borehole digital imaging, acoustic wave, etc.) were adopted to investigate the range and degree of impacts. And refined and complicated 3D non-linear finite element model were adopted to thoroughly study impacts of relaxation of dam foundation rock masses on dam and the dam foundation itself, and a series of oriented engineering measures, such as strengthened anchorage and consolidation grouting, were adopted to ensure safety and stability of the arch dam.

22.4.5.3 Energy Dissipation of High-Head and Large-Discharge Flood Release

The maximum flood-release discharge and flood-release power of Xiaowan HPP reached 20,700m^3/s and 46,000MW respectively. The river valley of the dam site area was narrow, the bank slopes were steep, the flood-release energy dissipation area was located in the range of resisting rock masses of dam abutment, with high dam and great fall of released flood, and there were a series of high-speed flow issue (cavitation and cavitation erosion, vibration and atomization, etc.); As the longitudinal dimensions of the flood-release surface and middle outlets of the dam were restricted, and there was the problem of radial concentration of flood-release flow and great discharge of single width of water, the flood-release energy dissipation problem was prominent and very difficult. Therefore, extensive numerical analyses and model test studies were conducted on such issues, and the flood-release means of combining dam surface and middle outlets with bankside flood-release tunnel to ensure flood-release safety of the complex, and dam flood release solved energy dissipation problem according to the principle and method of being longitudinally layered and separated, horizontally diffused in single units, colliding in air to dissipate energy, and generally spreading out and returning to channel, and realized ideal energy dissipation effects.

22.4.5.4 Ultra-large Underground Caverns and Chambers

The underground headrace and generation system was of huge scale. The main and auxiliary powerhouses, main transformer chamber, tailrace surge chamber and gate chamber were arranged in parallel, in which the length of the main powerhouse was 298.4m, the width 30.6m, the maximum height 79.38m, the diameter of surge chamber 32m, and the height 92m. The surrounding rock stability of underground caverns and chambers had become a prominent issue for the headrace and generation system.

Based on 3D non-linear finite element analysis of the simulated excavation and support steps and back of the construction process, it was to stress the timeliness and orientation of support measures, optimize and adopt economic and effective support measures convenient for construction to maintain stability of surrounding rocks. We conducted the first integrated equipment hydraulic transitional process test and studied both at home and abroad, and solved the stability issue of tailrace system by adopting cylinder impedance surge shaft with upper chamber (Fig. 22.4-3).

Fig. 22.4-3 Xiaowan Underground Powerhouse

22.4.5.5 Turbine-Generator Units with High Heads and Large Capacities

The turbines of Xiaowan HPP had high operating heads, large unit capacity and high rotation speed of units (H_{max} = 251m, N_r = 700MW, n_r = 150r/min), which was the unit with the largest single installed capacity at the said head in the world. Simultaneously, since the dam head had varies in a large range ($\Delta 87$m), the unit served as peak regulation, frequency regulation and accident backup, which was demanding for operation stability of the turbine. In order to improve the operation stability of the turbines, after model tests and studies, we raised the rated head of the turbines, increased the suction height and rated voltage of generators, and adopted fully air cooling means for the generators. The pressure maintaining concrete placement method is applied to the spiral case, with pressure maintaining value up to 190m H_2O, which was currently the hydropower station with the highest pressure maintaining value of 700MW installed capacity.

Each pole capacity of the generator was 19.45MVA, the rated capacity × runaway speed was 222,450 (MVA × r/min), thrust of the thrust bearing × rated rotation speed was 420,500, which was the highest in the world. The length of the stator core was 3.65m, which was among the top ones in the world, and it had exceeded the limits of air cooling Unit of $H \leqslant$ 3.5m.

22.4.5.6 Large Steel Structures with High Heads

Due to large openings, high heads, high velocity and large flow release of flow-release structures, it was highly difficult to design, manufacture and install the steel structures. Structural vibration resistance, water stopping, supporting, gate slot hydraulics, and air cavitation were all global hard nuts to be cracked. Let's take the service gate of emptying bottom outlet of the dam (5m×7m) as an example, the designed head was 160m, the total hydraulic pressure reached 115,000kN, its emergency bulkhead gate (5m×12m) was plane hinge gate bearing 160m head, and the total hydraulic pressure reached 107,000kN. Through thorough analysis and model tests and studies, the relevant design problems was successfully solved.

22.4.5.7 Temperature Control and Crack Prevention of High-Strength Concrete and Mass Concrete

The length of the largest pouring block at middle and lower section of the Xiaowan arch dam was more than 90m, and temperature control and crack prevention was extremely important. It was started from analyzing and determining the temperature control parameters of dam concrete, then thoroughly analyzed and studied steady temperature field, quasi-steady temperature field, temperature difference and temperature stress criteria of typical dam sections, the unsteady temperature field and temperature stress of pouring blocks, grouting temperature of closure dam crest/crown, emulated and analyzed the temperature fields and stress fields and horizontal joint openness of the entire course from starting concrete placement to normal operation of 43 dam sections of the entire arch dam. On such a basis, it was studied and proposed such controlling criteria as reasonable lowest temperature of concrete mixer outlet, pouring temperature, water cooling temperature and time, crown closure grouting temperature, as well as pouring lift thickness and time interval, etc.

22.4.5.8 The 700m Complicated High Slope

The complex area was located in an area with high mountains and deep valleys, with steep slopes on both banks, which were higher than the river level by some 1,000m. In order to satisfy the layout of permanent and temporary structures and needs of the sites, many excavated high and steep slopes were formed. One of the slopes is as high as 700m. These high slopes are of poor stability conditions, with all kinds of different instability modes and complex boundary conditions, which was one of the first few critical technical issues encountered in this project. The technical difficulties were rare in the history of the Chinese and foreign hydropower construction. Treatment of high slopes was still in the exploration stage both at home and abroad and there was no uniform understanding in terms of analytical theories, calculation methods and corresponding safety indicators and control criteria. No highly workable regulations or specifications are in place yet.

Therefore, it was conducted a series of analyses and carefully studied the high and steep slopes of Xiaowan hydropower station. Based on thorough analyses of slope instability mechanisms, it was adopted multiple analytical methods (such as rigid body limit equilibrium method, spatial wedge vector analysis, and finite element method) to evaluate the slope stability; explored a set of methods suitable for treatment of rocky slopes with fissured and soil slopes with uneven components (particularly falling accumulations), and successfully solved the problem of stabilization of 700m complicated high slope.

22.5 Shuibuya Hydropower Project

Sun Yi[1], Chen Runfa[2], Cai Jinyan[3]

22.5.1 Overview

Located in Badong County, Hubei Province, China, the Shuibuya Hydropower Project is the first cascading hydropower project in the middle reach of the Qingjiang River, 117km away from the Enshi City in the upstream and 92km away from the Geheyan Hydropower Project in the downstream. The project is mainly used for power generation, flood control and other benefits. At the normal water level of 400m, the corresponding reservoir capacity is 4.312 billion m^3, while at the check flood level of 404m, the total reservoir capacity reaches 4.58 billion m^3. The total installed capacity of the hydropower project is 1840MW. The project belongs to large-sized Type I and consists of a concrete-faced rockfill (CFR) dam, an underground powerhouse, a chute spillway on the left bank and a sluice tunnel on the right bank, as shown in Fig. 22.5-1 and Fig. 22.5-2. The main engineering properties are listed in Table 22.5-1.

Table 22.5-1 Shuibuya Project Features

Engineering location		Badong County in Hubei Province	Approval starting time	January 2002
The main comprehensive benefits		Power generation and flood control		
Hydrological characteristics	The controlled basin area on the upstream of dam site	10,860km²	Reservoir properties	Total storage capacity	4.58×10⁹m³
	Average annual runoff	9.44×10⁹m³		Normal water storage level	400m
	Average annual discharge	299m³/s		Design flood level	402.2m

---

[1] Sun Yi, General Manager of Beijing Xinheng Hydroelectric Development Co., Ltd., Professor Senior Engineer.
[2] Chen Runfa, Hubei Qingjiang Hydroelectric Development Co., Ltd., Professor Senior Engineer.
[3] Cai Jinyan, Hubei Qingjiang Hydroelectric Development Co., Ltd., Professor Senior Engineer.

Chapter 22 Typical Projects of High Dam

Continued

Engineering location			Badong County in Hubei Province	Approval starting time		January 2002
Characteristics of main buildings	Dam	Dam type	Concrete face rockfill dam	Power station indexes	Total installed capacity	1840MW
		Maximum dam height	233.2m		Guaranteed output	312MW
		The elevation of dam crest	409.00m		Average annual generated energy	3.984×10^9 kW·h
		The length of dam crest	674.66m			
	Discharge structures	Type	Chute spillway	The total quantity of pivotal project	Earth-rock excavation	2.22×10^7 m^3
		The size of sluice gate (width×height), quantity	14m×21.8m, 5			
		Design flood discharge capacity	16300m^3/s		Earth-rock fill	1.76×10^7 m^3
	The emptying tunnel	Type	pressured tunnel to connect free-flow discharge tunnel		Concrete and reinforced concrete	2.05×10^6 m^3
		The size of sluice gate (width×height), quantity	6m×7m, 1		Metal structure installation	22476t
		Maximum flood discharge capacity	1,605m^3/s			
	The main power house	Type	Underground		Curtain grouting	3.176×10^5 m
		Inner dimension (length×width× height)	168.5m×23m× 65.47m		Consolidation grouting	2.063×10^5 m

The dam crest elevation is 409.0m, the crest length is 674.66m, the maximum height is 233m, and the crest width is 12m. The upstream slope of the dam is 1 : 1.4 and the average downstream slope is 1 : 1.46. The dam body is mainly composed of concrete face slab, cushion layer, transition layer, main rockfill zone, secondary rockfill zone and downstream rockfill zone. The thickness of the face slab is 0.3 - 1.1m, the area of the face slab is 138,700m^2, and the total length of the joints is 12,500m.

Cofferdams are used to intercept the riverbed at one time and tunnels are used for diversion. The water is blocked by cofferdams in the period with low water level, while the construction pit is submerged in flood season.

The river closure of the project was finished in October 2002, and water storage was started in October 2006. The first generator was put into operation in July 2007, and all the 4 generators were put into operation and connected with the power grid in September 2008. On November 2, 2008, the reservoir water level reached 399.51m, reaching the normal operation scale.

22.5.2 Research History

Beginning with the pre-feasibility study, some research topics were conducted. Because the topographic and geologic condi-

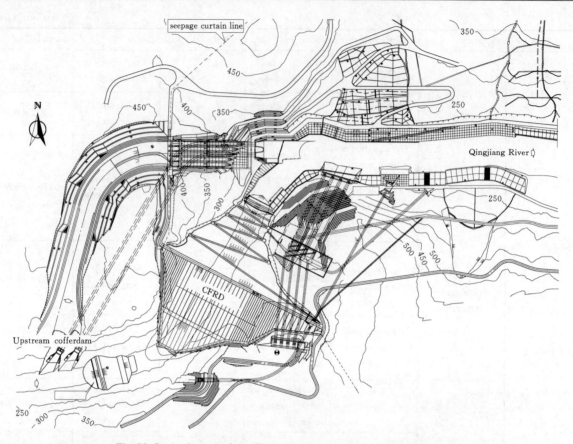

Fig. 22.5-1 Layout of the Shuibuya Hydropower Project (Unit: m)

Fig. 22.5-2 The Holo-image of the Shuibuya Hydropower Project

tions of the dam site are suitable for the construction of a local materials dam, the core rockfill dam and concrete face rockfill dam with the same depth were compared in the pre-feasibility and feasibility study stage. Safety, economical efficiency, environmentally friendly and fast construction progress and other advantages led faced rockfill dam to be selected as the target dam type, with its height of 233m.

In the process of the study up to 15 years, more than twenty organizations participated in the research, including Hubei Qingjiang Hydropower Development Company Limited, Survey, Planning and Design Institute of the Yangtze River, Yangtze River Scientific Research Institute, Hydropower Planning and Design of Water Conservancy Institute, China Institute of Water Resources and Hydropower Research, Nanjing Hydraulic Research Institute, Chinese Academy of Sciences,

Wuhan Institute of Rock and Soil Mechanics, Tsinghua University, Wuhan University, Hohai University, Three Gorges University, Wuhan University of Technology, Huazhong University of Science and Technology, State Power Corporation Kunming Investigation and Design Institute, Mid-South Design and Research Institute, East China Investigation and Design Institute, the China Gezhouba Construction Group, Jiangnan Water Resources and Hydropower company, participated in the research on the Shuibuya project, which lasted for 15 years. A series of direct guidance was provided by many domestic experts and scholars; Besides, internationally renowned companies of the Canadian International Project Management (CIPM), The COPEL Company of Brazil, and other well-known face slab dam experts such as Cook, Pinto, Moree also provided advice. Relying on national science and technology research programs, the National Natural Science Foundation, special research programs, design research, special research and other projects, a series of innovative research achievements has been made in dam construction technology for the superhigh concrete faced rockfill dam, flood discharge and energy dissipation, underground power station, air-raid shelter, seepage control engineering, diversion and flood prevention techniques, project planning and management, and so on.

22.5.3 Construction Technology of High Concrete-faced Rockfill Dam

In the early 1990s, there were still no design/construction specifications to follow for concrete-faced rockfill dams in China, nor the experience in building more than 200m's concrete-faced rockfill dam at home and abroad. At that time, the highest concrete-faced rockfill dam in the world, i.e. the Mexican Aguamilpa concrete-faced rockfill dam (187m), was just built in 1994, and the highest rockfill dam in China, i.e. Tianshengqiao I concrete-faced rockfill dam (178m), just started its construction in 1993. It faces a series of high concrete-faced rockfill dam technical problems: an additional height of about 50m over the maximum dam height already built, the design concept of ultrahigh concrete-faced rockfill dam, the mechanical properties of the dam fillers under high stress condition, the deformation control technology of high concrete-faced rockfill dam, high-performance face concrete, the sealing structure adapting to the large deformation and the new monitoring means of high concrete-faced rockfill dam. The owner of Shuibuya Project, Hubei Qingjiang Hydroelectric Development Co. Ltd and the design organization, The Changjiang River Survey Planning and Design Institute have organized many domestic and foreign organizations to develop a lot of research work about the dam construction technology of concrete-faced rockfill dam. The subproject "The engineering technology research on 200m high concrete-faced rockfill dam" that is the key science and technology project in the national Ninth Five-Year Plan was set up. Relying on the Shuibuya concrete-faced rockfill dam, according to the five technical problems of the Shuibuya concrete-faced rockfill dam, making joint efforts, we obtains a series of innovative research results and forms systematic dam construction key technology of ultrahigh concrete-faced rockfill dam, which changes the status of the concrete-faced rockfill dam as a kind of "an empirism-based dam".

22.5.3.1 Anti-seepage System Technology of High Concrete-faced Rockfill Dam

By a new structure type of toe board with a combination of "standard facing and watertight facing" and the concrete crushing-type side wall fracture, for the first time, it was set up permanent horizontal joints on the concrete facing, and polyacrylonitrile fiber is used in concrete-faced rockfill dam. The conventional sealing structure and sealing material of peripheral joints and vertical joints are systematically improved and the sealing structure of the central sealing type and surface self-healing type, adapting to the large deformation, is developed. Then a complete set of anti-seepage technology system of ultrahigh concrete-faced rockfill dam was formed.

22.5.3.2 The Test Methods for Dam Material (coarse-grained materials) Performance

The physical properties and mechanical properties test system of high concrete-faced rockfill dam materials are proposed; The CT triaxial tests of coarse-grained materials have been conducted. According to the long-period large-scale rheological test of limestone, it was proposed that the rheology of high concrete-faced rockfill dam has an important impact on dam integral deformation and a new rheological constitutive model of limestone.

22.5.3.3 Dam Deformation Control

1. Stress and strain analysis of super-high concrete-faced rockfill dams

Prior to taking account the variations of the contact state of concrete facing and concrete crushing-type side wall, dam body rheology, the temperature change of the environment and other factors in the Shuibuya dam, systematic stress and strain analysis has been conducted to obtain the distributions of displacement and stress in the dam body, concrete facing and joints under different conditions. Under comprehensive argumentations, the stress/deformation for the dam body and concrete facing in the 233m Shuibuya concrete-faced rockfill dam are acceptable and controllable, and the control index of selecting sealing structure was proposed.

2. Dam materials selection and dam body zoning

In order to ensure the safe operation of the dam in the water head of above 200m and make full use of the excavated materials to reduce the project cost, we should design the dam body zoning and propose the rolling parameters of fillers, according to the working condition and stress condition of various parts of dam body, the source of fillers, characteristics and the parameters of the site blasting and compaction test.

The dam body zoning and dam materials design principle of high concrete-faced rockfill dam and it was proposed that controlling the inhomogeneous settlement of dam body zoning and maximizing utilization of the excavate materials (Fig. 22.5-3).

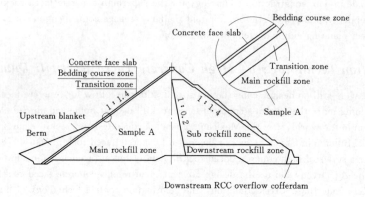

Fig. 22.5-3 Dam Zones

3. Inhomogeneous deformation control

To reduce inhomogeneous deformation, several measures were taken: a smooth toe board to layout line was used; the dividing line between the main and secondary rockfill zones inclines to the downstream from the dam axis; filling and rolling construction parameters for main and secondary rockfill zone are basically the same; the two abutment slopes contacting with the upstream dam were trimmed and the deformation coordination area was set up; the thickness of upstream rockfill materials of the peripheral joints and downstream toe board is in the uniform thickness range of less than 1m and defect topography was transformed.

4. Construction process control

Rational planning filling duration and concrete facing pouring time in the dam construction can allow a pre-settlement period for the "temporary dam" corresponding to the concrete facing. Sufficient height difference between the top of the concrete facing with sub-stage construction and the top of the "temporary dam" is reserved, reducing the impact of later deformation on the poured concrete facing. Filling along the dam axis and rising uniformly can make the rockfill deformation continuous parallel to the dam axis. It was adopted "anti-lift method" to fill the downstream dam first and realize the pre-settlement of dam deformation.

22.5.3.4 Dam Seepage Control

1. Permeability characteristics of dam body

According to the engineering properties of dam materials in the high concrete-faced rockfill dam of Shuibuya hydropower, the seepage stability experiments for the combination of dam cushion, transition material, large specific of main rockfill and high water head are carried out together with the analysis of seepage field. Under comprehensive argumentations for various conditions, the seepage stability of concrete-faced dam can be guaranteed.

2. Seepage control of the foundation

As to the seepage stability of foundation and seepage around abutment, numerical seepage analysis was conducted, and the appropriate depth of anti-seepage curtain was selected. Moreover, the effects of anti-seepage curtain and faults on the seepage field are analyzed.

Considering special circumstances of karst development in dam site and high water head in dam foundation, seepage control system for foundation, which is characterized by strengthening anti-seepage of toe plates, changing fracturing rock from karst rock and dynamic optimization of anti-seepage curtain, is adopted.

3. New structures and materials of water sealing

As shown in Fig. 22.5-4, a structure joint seal system, which is characterized by multiple sealing and limit leakage and can strengthen the surface sealing and has the function of anti-seepage and self-healing, was put forward (Fig. 22.5-4). The surface, middle and bottom sealing structure can play their respective roles. According to the characteristics of high concrete-faced rockfill dam, main surface sealing materials such as GB and SR plastic filler were improved. The improved plastic filler has advantages of large plasticity, seepage resistance, cold and heat resistance, aging resistance, strong cementation with concrete, good capability, convenience of construction and so on.

4. Crack control of face slabs

The cracks of face slabs can be divided into structural cracks and temperature cracks. The maximum dip length of the face slabs is 392m, which is the highest in the world. So the crack control problem is vital of importance. After study, a face slab crack control technology which integrates structural improvements, concrete mix proportion optimization, construction technology improvement and heat and wet preservation has been carried out.

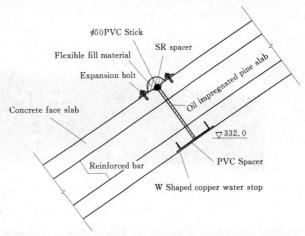

Fig. 22.5-4 Horizontal Joint Sealing Structure of Face Slabs

The main treatments include: setting up permanent horizontal fracture in the height of 332m in the second stage (Fig. 22.5-5); cutting off those corresponding slots in sidewall between supporting and extruded surface as well as the vertical joints; spraying emulsification asphalt on the surface of extruded sidewall; cutting off reinforced frame of face slabs; setting up buffer compartment in vertical joints of panel; setting up reinforced steel near joints. For the construction of the face-concrete, the medium-heat cement, class-I coal ash, high-performance water-reducer and mixing fiber materials were used. During construction, some methods were adopted to control later deformation, such as preparing settlement before deformation and anti-lift method. Face-concrete should be completed during the seasons with low temperature.

Fig. 22.5-5 Horizontal Joints of Concrete-faced Rockfill Dam

22.5.3.5 Construction Technology of High Concrete-faced Rockfill Dam

In the construction, the high-pressure grouting technology of thin-layer toe board (from 0.6m to 1.2m), multi-dimensional dynamic earthwork allocation that improves utilization ratio of earthwork have been applied. Moreover, the new process of one time molding of sealing profiled connector, copper water seal film as a whole roll forming machine and the technology of the upstream slope extruded sidewall fast solid slope has been used to solve the problem of the construction of high concrete-faced dam 233m in height.

22.5.3.6 Quality Control, Monitoring Technology and Feedback Analysis of High Concrete-faced Rockfill Dam

Ultra-large-scale multi-parameter numerical inversion method has been used and a dynamic feedback optimization has been achieved in the whole construction process. GPS precision real-time quality monitoring system, as shown in Fig 22.5 - 6, and additional mass method have been creatively used in rapid detection of the dry density of dam material filling, which changes filling quality from stage control to whole process control. The project used specially 520m long horizontal and vertical displacement. Moreover, fiber optic gyroscope was firstly applied in the monitoring of deflection of face slabs, horizontal and vertical displacement of dam body. In addition, grating measuring temperature system was firstly used in seepage monitoring of the dam.

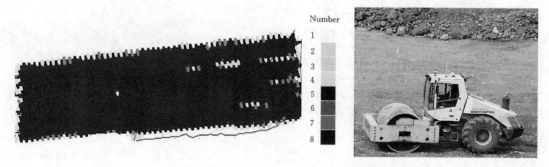

Fig. 22.5 - 6 GPS Construction Quality Monitoring System

During the construction period of the Shuibuya dam, according to the results of deformation monitoring and actual dam filling procedures, it was predicted that final deformation value was no more than 2,700mm through dynamic feedback optimization study on dam.

22.5.4 *Flood Discharge and Energy Dissipation Technology*

(1) As demonstrated in Fig 22.5 - 7, the shore spillway deflector dissipation and anti-scouring wall were for the first time adopted in the hydropower station energy dissipation design, with large flood discharge power and complex soft rock environment. That is, underground reinforced concrete diaphragm wall was introduced in the two sides of energy dissipation for the purpose of preventing the toe of the slope from erosion. Comparing with conventional concrete plunge pool program, the scheme has the advantage of avoiding flood risk, shortening construction period, and saving investment.

Fig. 22.5 - 7 The Structure of Flood Release and Energy Dissipation

(2) Based on a series of model experimental studies, zoned chute slot ladder export was originality adopted in the energy dissipation arrangement. That is, the spillway is divided into five zones by the long piers in the direction of pier, which would shrink into narrows at the end. To better conform to the complex river regime in the downstream export of Shuibuya spillway, export stakes and elevations were arranged into a ladder-like in the buckets. Overall, this new energy dissipation type ensures a smooth flow of tail water on the right bank, and shape of scour holes is left high, with the lowest scour hole elevation of the downstream river bed more than 1,720m.

(3) After a comparative study on energy dissipation mechanism of narrow and different shape parameters, an optimal parameter combination style, which is applicable to the large vent, high drop hydropower station was proposed. This style can ensure to form a typical slit water tongue shape and reduce the reflux strength and bank scouring in Mayayu rockfill dam. Sedimentation is avoided in the export of tail water and right half side of downstream.

(4) After the investigation of water wings attacking bank slope in the condition of small opening operation, fillet slit bucket which let the water wings tend to the river center by using shock wave asymmetric intersection principle was proposed. This method can solve the problem of water wings attacking bank slope, at the same time, water tongue shape is not affected in

the large opening and open discharge condition.

(5) Based on experimental studies on series of atmospheric pressure and vacuum, a method that prevents cavitation by aeration was proposed. This method can improve the scouring river bed in the downstream and enrich the design idea and research result in oversize spillway.

(6) An anti-scouring structure which consists of an anti-scouring wall, pre-stressed anchor cable and drainage, is employed for the first time. This structure can not only meet the structural requirements, but also strengthen the wall rock and speed up the construction schedule. At the same time, the groundwater level of the mountain is dropped, and the stability of anti-scouring wall and Maya high slope is improved.

(7) A combination construction method consisting of flat hole stratified and wide shaft sub-sequence method was applied to anti-scouring wall. That is, a flat hole stratified method was introduced in the soft rock anti-scouring wall excavation and concrete pouring, while wide shaft sub-sequence method was adopted in left bank of the cover layer. Construction problems in the soft rock and cover layer were solved by this combination method, with construction progress, were quality and safety ensured.

To ensure the shallow grouting pressure, auto-recording and alarm system was adopted to monitor and control the whole process of upward deformation.

22.5.5 Underground Hydropower Station Technology

(1) The ring beam style advance soft rock closed support body structure is adopted to deal with the soft rocks (p1q3) in the up-middle part of the sidewalls which control the stability of surrounding rocks of the underground powerhouse. After the treatment, the stiffness and integrity of the surrounding rocks above the powerhouse are guaranteed, the transmission of forces between the upper and lower layers in the soft rocks are ensured, and the soft rock compression and squeezing out deformation is prevented. Meanwhile, take advantage of the closed ring beams to generate confining pressures and basically maintain the primitive stress state of rock layers in the soft rocks and preserve the bearing capacity of soft rocks, achieving the goal of taking the full advantages of the soft rocks.

(2) Measures for preserving the soft rock for support division pier and carrying out advanced composite reinforcement are adopted to deal with the soft rock layers (p1q1, the Maan coal seam and the Huanglong shear zone) in the lower part of the side walls that control the stability of surrounding rocks of the underground powerhouse. The treatments can make the division pier effective in supporting the upstream and downstream side walls of the underground powerhouse, and restricting the deformation of the side walls of the powerhouse. Finally the goals of effectively reducing the whole section excavation height of the powerhouse and restricting the rebound deformation of the lower soft rocks, is achieved.

(3) On the basis of the features that soft and hard rocks distribute alternatively in the surrounding rocks in Shuibuya hydropower project, combining the use of the ring beam style advance soft rock closed support body structure, a new structural style of buttress rock anchor composite crane beam is adopted (the largest wheel pressure capacity is 750kN, the external suspension degree of the track axis is 75cm). This structural style has the benefits to rock anchor beam and wall (pillar) crane beam, explicit structural forces and flexible construction arrangement. Hence it provides a new structural style for the design of high-level wheel pressure underground powerhouse crane beams.

(4) By choosing appropriate stress conditions around section, excavating every other adjacent caves and adopting perfect support design (spray 20cm thick steel fiber reinforced concrete, 20A 'H' style steel arches, advance bolts, advance small duct preliminary grouting and so on), the large-scale tailrace group (4 parallelly arranged, diameter 13.7m, the rock pillars between caves are 1.26 times the cave excavation span) were constructed safely and successfully in the soft rocks (the Maan coal seam, the Huanglong shear zone and the Xiejingsi formation, surrounding rocks are Ⅳ and Ⅴ types) under the condition that the rock pillars between caves do not meet the specification requirements (generally requires 2 times cave excavation span).

(5) Taking advantage of the advanced monitoring, the non-contact surrounding rock 3D quick monitoring technology and the feedback analysis, the information construction of the underground powerhouse were completely realized. Before the underground powerhouse excavation, based on the pre-installed multi-point extensometers, the whole process information of the powerhouse deformation were obtained by using the exploration audits and the drainage tunnels outside the powerhouse; The complete and a wealth of monitoring data can be further obtained by using the total station, carry out the surrounding rock 3D quick deformation monitoring (as a kind of non-contact deformation monitoring technology, this one makes the deformation monitoring very simple and fast, and has the features of high accuracy monitoring results and monitoring process automation) in the underground powerhouse; the technologies of carrying out coefficient back analysis of the underground

powerhouse surrounding rocks and predicting the stability of surrounding rocks based on the back analysis results, is a powerful tools to give some guidance on the optimization design and construction.

(6) According to the junction arrangement characteristics and hydropower station operation features, the high current enclosure-continuous isolated phase bus with a vertical height difference of 118m is adopted. Shuibuya hydropower station is the first engineering application example which adopts high current enclosure-continuous isolated phase bus with vertical height difference exceeding 100m at home. In the engineering design, a series of key technology problems of the high vertical and high current enclosure isolated phase bus are solved, such as heating, force, supporting way, ventilation, installation, maintenance and so on.

22.5.6 Emptying Tunnel Technology

(1) After a series of hydraulics model tests under normal and reduced pressure, an optimal combination of arrangement type for the outlet of pressure tunnel with sudden expansion and sudden drop pattern was proposed, which was suitable for the operation and could work safely with water head ranging from 0 to 100m. It can successfully solve the difficult cavitation problems which frequently occur under high water head and high pressure for the pressure tunnel.

(2) Based on comparison tests of a variety of energy dissipater types, a new type of hyperbolic differential flip bucket was employed. The sidewalls of the bucket outlet asymmetrically spread (left 20.8°, right 8°), longitudinal skewed (the left sidewall was 27.69m, the right sidewall was 37m), and the bucket with bank and groove pattern was big differential (flip angled 33°–66°). Under the condition of 110m large-amplitude water head, the flip distance of water jet varied only 60m. The flow of the flip bucket diffused sufficiently and the scour pool was shallow. The new type of flip bucket successfully solved the hard problem of high water head in empty tunnel, and large amplitude of water head, narrow valley and poor geological conditions in Shuibuya, and it appears in the world for the first time.

(3) The roller wheel pressure value of plane fixed wheel accident bulkhead gate was 5400 kN round in Shuibuya. It was the biggest value in the same type of gates all over the world. Through reasonable choices of wheel and matched track materials, heat treatment and roller structure layout, the technology problem of big roller wheel pressure in the plane fixed wheel accident bulkhead gate under high water head condition could be solved.

(4) The maximum seal water head of the plane accident bulkhead gate was 152.2m, which was the biggest one for the same type of all gates. By using the LD-19-A72 rubber and plastic composite water seal materials and the new type of "double cavity pressurized enhanced" water seal structure pattern, and utilizing the reservoir water to add back pressure before preloading 0–5mm water head on the head of water seal rubber, the technology problem of designing water seal at the plane accident bulkhead gate under high water head condition could be solved.

(5) Casting-forging combined section form was first used in the main rail designing of plane fixed wheel gate with the value of the roller wheel pressure being 5,400kN. The main rail tread manufactured by alloy thick plate forged steel, and the rail seat made by ordinary carbon steel, then combined them as a whole with bolts after the end of their procedures. The problems of the whole cast of the main rail's heat treatment process, and stringent quality requirements, and high scraped rate could be solved.

(6) The maximum water head was 152.2m for the eccentric hinge compacted water seal of the radial gate, which was the biggest value in the same type of gates all over the world. By using the LD-19-A72 rubber and plastic composite water seal materials and the new type of "double cavity pressurized enhanced" water seal structure pattern, the technology problem of designing water seal for the eccentric hinge of the radial gate under high water head condition could be solved.

(7) As the total water pressure reached 89,634N on the radial gate, in order to prevent the gate from deviating from the normal water-retaining position induced by the gate slipped downward under the big compression deformation, the normal gate locking device was set on the hoist machinery. Moreover matched installing roller bearing gate locking device on the gate leaf bottom and buried part's bottom sill of the radial gate with eccentric hinge was first proposed and employed.

22.5.7 Seepage Control Engineering Technology

(1) Applying the principle of "changing karst body into fissure body", and based on the exact definition the position and size of the karst piping system to optimize the line of curtain, the karst cave within the scope of the curtain was first cleaned up and backfilled, and then proceeded with curtain grouting, reducing the difficulty and engineering quantity of curtain grouting. This way of curtain grouting had important theoretical and practical reference value for the design of similar engineering.

(2) With the breakthrough and extension of traditional consolidation grouting concept, "uniform consolidation with cur-

tain" technology was proposed. It emphasized on the performance of the consolidation grouting for "seepage control". By using consolidation grouting to enhance the thickness and compactness of impervious body, the seepage gradient can be reduced, and the grouting pressure of shallow curtain could be increased, then the maximum pressure reached up to 0.5MPa at the consolidation grouting contact segment. At the same time the contact segment's maximum pressure of the curtain grouting was 1.5MPa. It was the highest grouting pressure in contact area of shallow toe slab basis in the current high CFRD.

(3) On the basis of systematic bolts of toe slab, together with the results of in site grouting test, the system process measures of booster grouting was designed in order to ensure the shallow grouting pressure. Some technologies are employed, such as setting the initial grouting pressure, the target grouting pressure, the extreme grouting pressure and grade steady voltage, the limiting injection rate, increasing the consistence of slurry, employing automatic recording and alarm system to monitor and control the whole process of upward deformation, etc.

22.5.8 Diversion and Flood Prevention Technology

(1) High sub-weir cofferdam was adopted in the upstream debris over water cofferdam. That is, the overflow weir elevation is below the elevation of cofferdam, while the height of sub cofferdam is 8.5m. Flow rate is reduced in the surface by adjusting the height difference between upstream and downstream, and the difficulty for protection of filled dam is dropped.

(2) A horseshoe-shaped diversion tunnel with sloping wall and Curved bottom were built in the hard and soft interphase limestone with inclination 7°–15°. Section of this tunnel is the biggest in the world, with the size of 14.89m×15.72m.

(3) In the hard and soft interphase rock range of Qixia group between part 2 and part 10, the fastest record for the large diameter cavern comprehensive footage is 65.6m/m, with the section size 14.89m × 15.72m and the axis length 1,355.22m, which plays a key role to shorten the construction duration by one year.

22.5.9 Project Planning

In order to guarantee the flood control standard of En'shi City and improve the generation benefits, after comparing several schemes and taking advantage of the small flood discharge in summer and autumn and the stage flood design result, the concept of flood season flood-prevention controlling water level (397m) was firstly proposed on the basis of normal water level 400m, limit level for flood control 391.8m. As a result, the goal of the optimum usage of water and increasing annual average generated energy by 70 million kW · h was achieved.

22.5.10 Project Management

Adopting advanced management mechanisms, carrying out the design supervision, the immigration supervision and the quality consultation, the project quality was guaranteed, saving the project investment by nearly 100 million, generating power a year ahead and creating direct economic benefits nearly 1.4 billion. During construction, great attention was paid to the environmental protection, establishing perfect environment protection system and acquiring good social benefits.

22.5.11 Operation Status

In October 2006, the elevation of dam body filling reached to 405m. At the same time, the project was checked and began to store water. Since the reservoir water storage, the dam has been normally operated for four years. Monitoring indicators such as settlement and horizontal displacement of the dam body, stress state of the dam, deformation of the face slabs, stress state of the face slabs, vertical joint deformation of the face slabs, peripheral joints deformation of the face slabs and seepage state of the dam, are all in the control range. Among them, the accumulated largest settlement is 2.5398m and the largest seepage discharge is 66.6L/s. The work state of the dam is safe and good. The Shuibuya Engineering has got the national organization's security identification.

Reference

[1] The Yangtze River Water Resources Commission. *Study on Three Gorges Project Dam and Power Station Plant*. Wuhan: Hubei Science and Technology Press, 1997.
[2] Lu Youmei, Cao Guangjing et al. *Three Gorges Project (Technology)*. Beijing: China Water Power Press, 2010.

Chapter 23

Typical Problems of High Dams around 300m

23.1 Typical CFRD Projects with the Height around 300m and the Main Challenges

Ma Hongqi[❶]

Due to its good adaptability, safety, and economy features, concrete faced rockfill dam (CFRD) is widely adopted in the world. According to statistics, by the end of 2008, about 275 modern CFRD over 30m high and ten CFRD over 150m high had been built in foreign countries, among which, the highest is Bakun CFRD of 203.5m high built in Malaysia. 170 CFRDs over 30m high had been built in China. Seven of those are over 150m high, and the highest is Shuibuya CFRD with the height of 233m, which is currently the highest CFRD in the world.

The western region of China is rich in hydropower resources. To use the resources, high dams and large reservoirs need to be built for runoff regulation. However, most of the high dam projects are located in mountainous and deep valley regions with inconvenient transportation, and complicated seismic and geological conditions. High CFRD, on the other hand, can use local materials to reduce the pressure on materials transportation. Besides, high CFRD have strong adaptability to the complicated topographical conditions and geological conditions. They are also good in seismic performance, and have obvious advantages in economic comparison. Therefore CFRD is often considered to be the first choice for dam type selection. For example, Gushui, Maji, Rumei and Cihaxia Hydropower Station are all with dam height of 250-300m, and CFRD is the main option on dam type selection.

However, both in China and abroad, there are only few construction experiences for building CFRD of 200m in height. Also, technical accumulation for 300m level CFRD is insufficient. Its engineering characteristics, key technologies, and operation characteristics are difficult to be handled. Therefore, China Hydroelectric Power Consultant Group, Huaneng Lancang River Hydropower Co. Ltd., and China Huadian Corporation Nujiang River Hydropower Co. Ltd. took the lead, and together with reconnaissance organization, scientific research institutes, and higher education organization carried out "research on the adaptability and countermeasures for the height of 300m". Based on the experiences of CFRD with the height of 200m, and consulting of Gushui, Maji, Rumei, and Cihaxia Hydropower projects, systematic studies were carried out in aspects of CFRD with the height of 300m such as economic advantages, dam construction conditions, section zoning and dam construction parameters, stability evaluation and deformation control, seepage and water stop structure, construction technology requirements, and quality control etc.. Main research results are introduced as follows.

23.1.1 Experiences and Lessons Learned from CFRD with the Height of 200m

23.1.1.1 Main Problems

Mature experiences and design ideas of CFRD with the height of 100m were used in the design and construction of 200m high CFRD that built before 2000, such as the 187m high Aguamilpa Dam built in Mexico in 1993, and the 178m high Tianshengqiao-1 CFRD built in China in 2000. The section zoning of these dams are shown in Fig. 23.1-1 and Fig. 23.1-2. The problems in these dams including large settlement of rockfill, lots of horizontal structural cracks on face slabs, gaps between face slabs and surface of cushion zone, concrete rupture at the vertical joints of face slabs, and large leakage etc. The reason is that the damming experience of CFRD with the height of 100m is used to the higher dam and its relevant deformation control requirements did not put forward with aspects to the dam zoning, characteristics of dam material, and compactness etc.

According to J. B. Cooke, the horizontal tensile cracks on Aguamilpa Dam are caused by overlarge settlement difference be-

---

[❶] Ma Hongqi, Academician of Chinese Academy of Engineering, Senior Consultant of Huaneng Lancang River Hydropower Co. LTD.

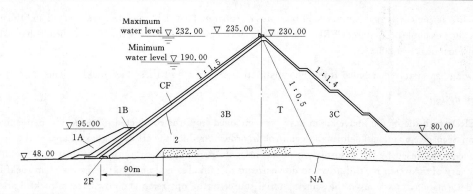

Fig. 23.1-1 Section Zoning of Aguamilpa Dam

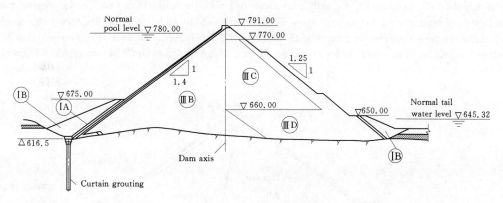

Fig. 23.1-2 Section Zoning for Tianshengqiao I Dam

tween the upstream and the downstream rockfill zones. The compressive modulus of upstream rockfill zone is 5.5 times that of downstream rockfill zone. Overlarge gradient of compressive modulus and part of dam body at upstream inclined to the downstream, caused the cracks on face slabs. Excavation materials of soft rocks such as sandstones and mudstones are used for downstream rockfill zones of Tianshengqiao Ⅰ CFRD, and the compressive modulus ratio between the upstream rockfill and the downstream rockfill is 2.0. Besides, for the requirement on retaining water of flood season, high priority section was constructed, with the height difference of 123m between the top of upstream and downstream rockfills is formed. After the flood season, the upstream and downstream rockfill bodies are filled and leveled up at the speed of 1.0m/d. It can produce overlarge settlement difference between the upstream and the downstream rockfill, and cause tensile deformation of the upstream face and cracking of cushion zone, the gaps between the face slab and the cushion material, and many horizontal cracks on the face slabs. Besides, the study of compaction density of rockfill materials also stays in the stage of experience of dam, for example, the porosity requirements for upstream and downstream rockfill zones of Tianshengqiao I are respectively 22% and 24%. According to analysis above, it's considered that most of horizontal loads are transferred to the foundation through dam body above the dam axis.

According to the analysis above, the idea of "deformation modulus of dam materials decreases from upstream to downstream" is not completely applicable for 150–200m CFRD. And the high dam design need to pay a special attention on deformation control. Professor Li Nenghui (2007) proposed that "zoning design should follow four principles: material source, hydraulic transition, excavation material utilization, and deformation coordination. And the most important one is the principle of deformation coordination, which requires not only deformation coordination of all parts of the dam, but also synchronized coordination between dam deformation and face slab deformation".

23.1.1.2 Main Experience

After 2000, there are 18 CFRD including the ones constructed, under construction and planned to build in the nation with respect to different unfavorable topography, geological conditions and climatic conditions. Engineering design and construction are successful in general and it was therefore accumulated experiences for tackling with different difficulties. CFRDs in China, no matter in terms of quantity, or dam height and scale, are in the front ranks of the world. We have achieved world's leading techniques for CFRD construction by re-innovation on the basis of import, digestion and absorption. For example, Hongjiadu CFRD in Guizhou, which has been completed in 2005, has a dam height of 179.5m. Sanbanxi CFRD in Guizhou, which has been completed in 2007, has a dam height of 185.5m. And Shuiyabu CFRD in Hubei, which has also been completed in 2007, has a dam height of 233m. Although these three CFRDs suffer from varying degrees of structural cracks of face slab, facing cavity and compression fractures of concrete face slab, the seepage is small and the overall operation is favorable. For example, Zipingpu CFRD which has been completed in 2006, with 158m in height has experienced the 5.12 Wenchuan Earthquake in 2008.

The main following aspects are concluded from the construction experience of these face rockfill dams:

1. Section zoning design

These three CFRDs with the height above 200m have been impoved from the completed projects. The dam subarea has been paid attention on Seepage control and deformation control. In order to reduce the effect on rockfill area at the upstream of the dam body and face slab caused by the deformation of downstream rockfill area, the range of upstream rockfill area is enlarged and boundary of upstream rockfill area and downstream rockfill area is dipped to the downstream rockfill. In order to coordinate the deformation, it was used the same material source in the upstream and downstream rockfill areas or tried to reduce compressive modulus of upstream and downstream rockfill area; in addition, special rolling area (modulus increasing area) is arranged on the steep bank slope and upper elevation, which requires higher compaction degree than the main rockfill area. Based on requirements for seepage control, there are vertical and horizontal drainage zones inside the dam body, which can make the seepage water smoothly discharge to the downstream and keep the dryness status of dam body above the water level of the downstream. Refer to section division design of Shuiyabu CFRD for typical section of dam body (Fig. 23. 1 – 3).

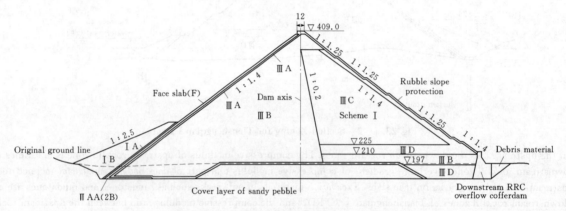

Fig. 23. 1 – 3 Section Division of Shuibuya Dam

2. Standard for compaction

Aggregate materials with good gradation above medium hardness are applied to the rockfill body. Porosity is required to be reduced below 20%. Large-scale vibrating roller with 20 – 30 tons is used to increase the compaction degree of the rockfill body. In addition, rolling compactor, with a striking and vibrating force of 200 – 250 tons, is applied to Hongjiadu dam and Dongjing dam. GPS monitoring technique is adopted to achieve real-time & on-line construction quality control. Methods such as additional mass method, etc. are adopted to inspect the dry density. Porosities in upstream rockfill areas of Hongjiadu, Sanbanxi and Shuiyabu are 19.6%, 17.62% and 19.6% respectively and porosities of downstream rockfill areas are 20.02%, 19.48% and 20.7% respectively, which are 2%–4% lower than Cascade I of Tianshengqiao. The maximum sedimentation rates of these three rockfill dams are 0.74%, 0.78% and 0.96% respectively, while that of Cascade I of Tianshengqiao is 1.94%. The compression modulus ratio of upstream and downstream rockfill body is 1.3 or less (Ma and Cao, 2007).

3. Stage division of dam filling

Having learned from the Cascade I of Tianshengqiao that there is a height difference of 123m between the upstream and downstream rockfill body, the original filling procedure of "high in the front and low in the back" are abandoned. It focuses on combining the stage division of dam body construction and filling with deformation control of the dam body. Except for the temporary section of dam body to be used during the first flood season to retain water for flood control the rest of stage division of the dam body filling shall try to achieve a balance rising of both upstream and downstream. High in the front and low in the back is not allowed. But it can be high in the back and low in the front to reduce settlement difference of upstream and downstream rockfill body. Before stage division of concrete face slab construction, pre-settlement control theory for rockfill body has been proposed creatively. Two indexes for pre-settlement control of dam body have been proposed: i. e. before every stage of face slab construction, the rockfill body shall have about half year of pre-settlement period while the deformation rate of settlement has been inclined to be restrained. Moreover, it is required that the crest elevation of the face slab shall be 15 – 20m lower than the rockfill body. These are important measures to avoid or reduce structural cracks of the face slab.

4. Structure of face slab and joint water stop

The thickness of crest of face slab is added from 30cm to 40cm. For face slabs with a reservoir depth over 120m, new calculation formula for thickness $T = 0.0045H$ will be adopted. To improve the anti-cracking ability of the concrete, it mixes with compound admixture and other fibrous materials. In order to avoid compression damages of the face slab, besides increasing strength grade of the concrete, compression reinforcing steel bars shall be arranged within a definite range of compression seam end and joint sealing materials, with high resilience and absorption and deformation abilities, shall be filled into the compression seam; meanwhile, height of copper water stop nose shall be reduced to reduce section size of the face slab to be pressed.

Types of joint seal structure have developed from signal water stop type to the combination with self-healing water stop. Intermediate water stop has been eliminated by most of projects in China. Surface water stop is strengthened, which makes the structure of water stop more reliable. Materials for water stop shall be able to adapt to larger stretch and shearing deformation and the durability will be enhanced in a large scale. Tri-compounded rubber plate covering layer with favorable aging resistance shall be developed and widely used.

5. Back analysis of typical CFRDs

In order to verify the computation model and computation parameters, inversion analysis is conducted for the five CFRDs with the height of about 200m, including Tianshengqiao I, Hongjiadu, Sanbanxi, Shuibuya and Bakun, on the basis of observation data of prototype. And computation model and its methods are modified and also adjusted to the stress deformation calculation of CFRD with the height of 300m which can be provided theoretical basis for this dam construction. And the results of inversion analysis can be shown below.

(1) Duncan E-B model and double-yield surface model proposed by Mr. Shen Zhujiang are basically suitable for the computation analysis of 200m-grade CFRD. It reflects the sedimentation and horizontal displacement process of representative test points in dam body as well as the distribution regularity of deformation of dam body, and provides referential basis for project design, construction and operation management.

(2) The model parameters obtained from inversion calculation are different from the test parameters due to the difference of scale effect in indoor test and grain breakage mechanism as well as the difference of stress state and stress route, which hardly reflects the mechanical properties of original gradation grains from an objective view.

(3) The sedimentation values obtained from inversion analysis is close to the practical test values but still smaller; the value of horizontal displacement is obviously greater than practical test value and the distribution regularity is also different. Therefore, there are some defects of the horizontal displacement mode for the rockfill stack.

(4) The creep, wetting and degeneration of rockfill stack has an obvious influence on the sedimentation deformation of high CFRD, especially the super-high CFRD. The existing model cannot accurately reflect the creep property of rockfill material.

23.1.2 Engineering Analysis of Typical CFRD with the Height of 300m

The research is based on the projects of Gushui Dam in Yunnan with a dam height of 310m, Cihaxia Dam in Qinghai with a dam height of 253m, Maji Dam on Nujiang River with a dam height of 270m and Rumei Dam in Tibet with a dam height of 315m.

23.1.2.1 Economic Advantages and Dam Building Conditions

The face rockfill dam has a strong economic advantage and wide adaptability when it was compared to other dam types, based on the research results of four projects and the comprehensive analysis of landform and geological conditions, natural building materials, the adjustment of dam foundation, hydropower layout, construction strength, project period and dam building conditions. It also avoid lacks of soil material, and the situation where the quality cannot meet the requirements, as well as the adverse influence on environment and soil and water conservation due to occupation of farmland. The project period can be shortened for one year, and this can save several millions of RMB.

23.1.2.2 Rockfill Zoning, Design Parameters and Deformation Properties

Based on the Gushui Project, performance test of rockfill material and engineering property study under high stress status are conducted and Duncan E-B model parameter are put forward; large permeability test, triaxial and compression test for rockfill zoning with different porosity and under high stress are conducted, mechanical properties and permeability property test for cushion and transitional materials of different levels are conduct, and suitable design parameters for the CFRD material with the height of 300m are proposed; the influence of dam material division, filling process and dam material compaction on the dam body deformation is greatly studied, and design suggestion of dam body division is proposed (See Fig. 23. 1 – 4).

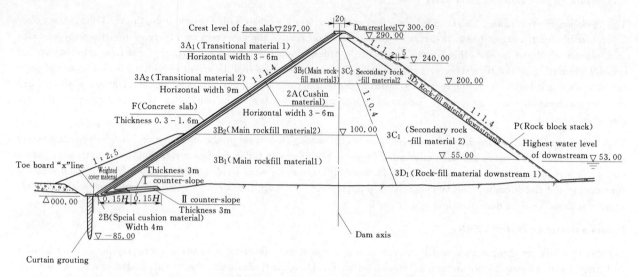

Fig. 23. 1 – 4　Section Division of Face Rockfill Dam of Gushui Station

According to the calculation, there are no obviously abnormal or great changes in the dam body stress and deformation, joint displacement, stress regularity of face slab of these 4 CFRDs when compared to the built CFRD with the height of 200m. For the CFRD with the height of 300m which is made of hard rock, its porosity is less than 20%, and sedimentation of dam body can be controlled within 1% of the dam height.

23. 1. 2. 3　Stability Estimation and Deformation Control Standard

Non-linear strength index shall be adopted for the stability analysis on dam slope of high face rockfill dam. Provided that medium hard rock is adopted for the dam building material and under tight compression condition, the stability of dam slope of rockfill stack will not be the main factor which influences CFRD with the dam height of 300m. Great importance shall be attached to the seepage stabilization of CFRD with the height of 300m. It was studies on the adaptability of existing filtering regulations, the thickness of cushion layer and transitional layer shall be appropriately increased, the design shall be conducted according to filtering regulations, and it was necessary to be verified through the test.

Based on the analysis of deformation regularity and characteristics of rockfill stack in the built projects, the establishment of the deformation relationship between preliminary dry density and axial direction and strength and time under confining pressure as well as the creep deformation influence of rockfill zoning, the modification guiding and measures of constitutive model for CFRD with the height of 300m are proposed, which mainly focus on the harmony of the self-deformation of rockfill and the harmony of the deformation between rockfill and face slab, and mainly stress the deformation property of rockfill for a long term, especially the weathering deformation, wetting deformation and creep deformation.

23. 1. 2. 4　Seepage Control and Water Stop System

According to the practical evidence of CFRD project with the height of 200m, the existing water stop systems of plastic filling type and self-healing filling type are effective. Besides, plastic filling type also has a good anti-seismic property. For CFRD with the height of 300m, it is necessary to design the water control for each pass which can independently stop the water. The test research shows that the water control structures on the crest and bottom made of proper structure and material can meet the joint water control requirements for CFRD with the height of 300m.

The preliminary research study on compression damage of pressure joint face slab shows that the compression damage to the concrete on the both sides of pressure vertical joints is mainly caused by the concentration of pressure stress. And proper design of cushion material for the joints can reduce the concentrative stress.

23. 1. 3　*Key Research Topics of CFRD with the Height of 300m*

Dam deformation control and crack control of concrete face are the key technical issues of the ultrahigh CFRD. Materials and section design, deformation control and crack control technologies of a face rockfill dam with height of 200m can be used for reference for the one of 300m one. However, the pertinence and effects of some measures should be checked if they are applied in the construction of CFRD with the height of 300m. With the increase of dam height, the stress level of the rockfill increases and the creep property becomes distinct, and dam deformation in the later stage will have great impact on working characteristics of the face slab. The recent computation model and test methods of high CRFD with the height of 200m have certain limitations. Theoretical analysis results only tally with monitoring results to some extent. The forecast error due to limitation of computation model and test methods is within the control scope for the face rockfill dam about 200m. As for the dam height of 300m. whether the condition degenerates from quantitative changes to qualitative changes, both computation model and test methods have to be improved, and quantitative analysis should be more accurate and reliable. Researches on basic science and key technology should be carried out further for the face rockfill dam with the height of 300m.

(1) To further carry out researches on key technologies of dam stability, stress deformation and crack control of face slab as well as corresponding construction technologies, basing on CFRD with the height of 300m in good construction conditions.

(2) To carry out researches on rockfill density indexes and dam section zoning is adapted to the dam height, valley shape and protolith characteristics of the rockfill. In accordance with control requirements of dam deformation, it was proposed the design criteria of dam material and zoning design principle to meet the requirements of anti-sliding slope stability and seepage stability.

(3) To discuss the suitability of applying non-linear strength indexes to analyze the slope stability of the face rockfill dam with the height of 300m and the design criteria of the slope stability. It should be carried out the experiment technique researches on seepage failureof large-scale rockfill under the action of ultrahigh water pressure.

(4) To study and develop test equipment to apply on a larger scale and improve the accuracy of large test equipment. It should be conducted on rockfill large-scale test technology as well as in-situ test and numerical analyses to make progress in the test method of complex stress path of the rockfill.

(5) To research on reasonable rockfill constitutive model and methods of determining parameters. More reasonable contact surface model and simulation method should be explored to develop more accurate simulation methods and improve accuracy of numerical simulation. It is also studied the rockfill deformation mechanism in a microscopic manner so as to predict dam deformation reasonably.

(6) To analyze mechanism and causes for squeezing failure of vertical joints of the high CFRD, and to predict the possibility of squeezing failure of CFRD with the height of 300m. Face slab structure and measures about materials should be put forward to control dam deformation and improve the anti-extrusion of face slab.

(7) To study on water control structures and materials which are suitable for large deformations, including opening, sinking, cutting of joints of the face rockfill dam with the height of 300m. The existing water control structures should be improved to strengthen their self-healing and stability under a high head of large deformation.

(8) According to seismic resistance examples of existing face rockfill dams, to summarize all anti-seismic methods and engineering anti-seismic measures. To take research on their pertinence, reliability and working mechanism to put forward corresponding anti-seismic reinforcement measures.

(9) To research adaptability and durability of overlong vertical and horizontal displacement meter for the dam with the height of 750m, and to develop monitoring technologies for new type deformations of the dam. A real-time intelligent feedback and prediction system for CFRD deformations with the height of 300m should be developed to provide technique support for the follow-up construction of CFRD in time and effectively.

(10) With the experimental projects, it should be formulated control standards suitable for a 300m high CFRD's construction diversion, filling stages, height differences between upstream rockfills and downstream ones, and rockfill height at the top of face slab in different stages, and to be prepared proper compacting standard, quality control standard and pre-settlement control standard, and developed large-scale rolling machines to improve compactness of the rockfill.

References

[1] Hydrochina Corporation, Huaneng Lancang River Hydropower Co., Ltd., China Huadian Corporation Nujiang River Hydropower Co., Ltd., Report of Adaptability of 300m-Grade Face Rockfill Dam and the Counterplan Research, August, 2010.
[2] Ma Hongqi, Cao Keming, Key Technical Issues on Ultrahigh CFRD. *Engineering science*, November, 2007 (11).

[3] Li Nenghui. Discussion on Design Concept of High Concrete Face Rockfill Dam, *Chinese Journal of Geotechnical Engineering*, 2007 (8).

23.2 Typical Projects of 300m Earth Core Rockfill Dam and Main Problems

Zhang Jianhua[1], Yao Fuhai[2], Xiao Peiwei[3], Ma Fangping[4]

23.2.1 Introduction

Earth Core Rockfill Dam is a well developed dam type either in theory and in practice. It has been adopted widely in China with its characteristics of local construction material, good anti-seismic performance, good foundation adaptation and quick construction. In recent 30 years, together with the development of geotechnical technology in the world, the rapid growth of the construction mechanical technology and the improved construction management, core Rockfill dams higher than 100m experienced four stages as follows.

(1) 100m-level Earth Core Rockfill Dam stage. In the early 1980s, with the completion of the Bikou and Shitou River Dam, the experience in building earth core rockfill dams with the height of 100m or higher accumulated rich experience in construction. Bikou Dam, completed in 1983 on the Bailong River in Gansu Province, with dam height of 101.8m, overburden layer depth of 34m, and the height of concrete diaphragm wall about 135.8m above the bedrock. After the magnitude 8.0 of Wenchuan Earthquake in 2008, the dam is still under good operation condition.

(2) 150m-level Earth Core Rockfill Dam stage. In August 2000, Xiaolangdi Dam on Yellow River was completed, which is 156m in dam height with overburden layer depth of 80m. It has a concrete diaphragm wall and seepage height above the bedrock is 236m. The safe operation in recent 10 years indicates the success of Xiaolangdi Dam construction.

(3) 200m-level Earth Core Rockfill Dam stage. In August 2009, Pubugou Rockfill Dam was completed and reservoir was impounded in the early December in the same year. The dam height is 186.0m, with overburden layer thickness of 77m and with the height of two concrete diaphragm walls of 263.0m above the bedrock. Pubugou dam is the highest core Rockfill dam today in China built on thick overburden. After one year operation, the dam performance is satisfactory with all the stress-strain and anti-seepage parameters under the original designed allowable range. It was awarded as a milestone project by ICOLD in September, 2011 after the dam withstood three flood seasons. The dam construction has added more experience to core Rockfill dam construction with the height of 300m in our country.

(4) 300m-level Earth Core Rockfill Dam stage. Besides the Pubugou Dam, there are four 300m rockfill dams under construction or to be constructed in recent 3 years, namely the Nuozhadu Earth Core Rockfill Dam on the Lancang River and earth core wall is located on the bedrock with the maximum dam height of 261.5m; the Shuangjiangkou Earth Core Rockfill Dam on the Dadu River and the core wall is located on the bedrock with the maximum dam height of 314m; the Shuanghekou Earth Core Rockfill Dam on the Yalong River and earth core wall is located on the bedrock with the maximum dam height of 295m and the Changhe Earth Core Rockfill Dam on the Dadu River and the core wall is located on the 50m deep overburden, the maximum dam height is 240m with the total anti-seepage height of 290m above the bedrock.

Generally speaking, the dam height can be defined as the distance from the riverbed construction base to the dam crest. For core Rockfill dam located in deep overburden after treatment, the treated overburden is actually the important part of the dam. Such as Changhe River Dam, if it is constructed on the bedrock, the filling dam height will be 298.0m. If it was directly put on the overburden as for Shuangjiangkou Dam, the filling dam height will be only 246m. For the convenient of comparison, the core rockfill dams on thick overburden with the total anti-seepage height of more than 260m are classified as 300m-level dam.

With the successive completion of above five dams, the construction techniques of Earth Core Rockfill Dam will be greatly advanced in China. To summarize the current construction conditions in time, the following contents introduce some projects rockfill and discuss the major problems in construction and operation, by considering Earth Core Rockfill Dam on thick overburden and on bedrock.

---

[1] Zhang Jianhua, Vice President, Tibet Part of China Guodian Group Corporation, Professor Senior Engineer.
[2] Yao Fuhai, China Guodian Dadu River Hydropower Development Co. Ltd., Professor Senior Engineer.
[3] Xiao Peiwei, Sichuan Leather Tie hydropower Development Co. Ltd., Engineer.
[4] Ma Fangping, China Guodian Dadu River Hydropower Development Co. Ltd., Senior Engineer.

23.2.2 Two Earth Core Rockfill Dam on Thick Overburden

23.2.2.1 Pubugou Gravel Soil Earth Core Rockfill Dam

1. Design criteria, topographic and geological conditions

Pubugou Dam is located at the conjunction area between Hanyuan County and Ganluo County in Sichuan Province which is in the lower reach of the Dadu River. The maximum height of the gravel soil Earth Core Rockfill Dam is 186m. Reservoir's normal water level is 850.00m, flood control water level is 841.00, and the dead water level is 790.00m. The total reservoir capacity is 5.337 billion m^3, which is an incomplete annual regulation reservoir. The overburden layer with sand lenses, vacancies, large different particles and complicated geological structures in different layers, had an remarkable thickness variation of 40 - 60m, with maximum thickness of 77.9m, the foundation is lack of 5 - 0.5mm grains, but has strong permeability and poor uniformity with potential geological problems such as leakage, uneven deformation, instability of seepage and earthquake induced sand liquefaction.

The dam's designed flood frequency is $P=0.2\%$, corresponding flood flow of $9460m^3/s$. Its check flood at possibly maximum flood (PMF) has corresponding flood flow of $15,250m^3/s$. The regional basic anti-seismic intensity is VII, basic rock acceleration is 0.21g. The anti-seismic design of the dam adopts intensity of VIII, and the exceeding probability is $P=1\%$ ($ah=268gal$) for benchmark 100-year. The safety factor of the dam slope for stability is 1.5 during the normal operation period (steady seepage period), 1.3 during the unusual operational period I (construction completion period and impounding period), and 1.2 during the unusual operational period II (earthquake during steady seepage period).

2. Dam structure and stress calculation

The gravelly soil Earth Core Rockfill Dam has an upper-reach slope of 1 : 2 - 1 : 2.25, a lower-reach slope of 1 : 1.8, dam crest width of 14m, and is composed mainly of four areas, namely the gravelly soil core, filter layers, transitional layers and rockfill body. The cofferdams are combined with the dam rock body.

The core has crest elevation of 854.00m, with top width of 4m, upper and lower-reach slope ratio of 1 : 0.25, bottom elevation of 670.00m, and bottom width of 96.0m. There are filters on both sides of the core with thickness of 4.0m at upper reach and 6.0m at lower reach. And at the bottom of the core on lower-reach of the diaphragm wall, two layers of filter with total thickness of 1m is placed and connected to the filter on lower reach of the core. There is a transitional layer between the filter layer and the dam shell, with the contact slope ratio of 1 : 0.4.

There are two concrete diaphragm walls constructed in the dam foundation, with the wall thickness of 1.2m and the central interval of 14m between two walls. The maximum depth of upper-reach diaphragm wall below elevation 670.00m is 76.85m, and the maximum depth of lower-reach diaphragm wall below elevation 670.00m is 75.55m. The upper-reach diaphragm wall above elevation 670.00m is directly inserted in the dam core with the inserting depth of 10m. On top of the lower-reach diaphragm wall there is a 3.5m×4m corridor for grouting and monitoring. The lower reach dam toe has two-level back-pressure rockfill body with the top surface elevation of 730.00m and 692.00m respectively, and the lower reach cofferdam is part of the back-pressure body. Between the core wall and the bedrock at abutments, there is a layer of 3 m-thick highly plastic clay which is also adopted at the top of diaphragm wall, corridor surrounding and at the bottom of the core.

Fig. 23.2-1 shows the typical section of the Pubugou Earth Core Rockfill Dam. The construction quantity for the dam is 16,400m^2 for the diaphragm wall, 228,400m for curtained grouting, 1,060,000m^3 for earthwork excavation and 22.37 million m^3 for refill, including 2.65 million m^3 of gravel soil and 13.266 million m^3 of rockfill.

The finite element analysis shows that the maximum settlement of the dam is 191.7 - 234.5cm, the maximum horizontal displacement is 52.9 - 69.9cm. The stress level inside the core is relatively low, with less possibility of horizontal tension cracks and vertical tension cracks at abutments, although obvious arch effect exists in the core. The stress and strain conditions of the two diaphragm wall are similar, with the maximum horizontal displacement of 27.1 - 39.2cm and maximum settlement of 15.5 - 16.9cm for the upper reach diaphragm wall, and the maximum primary stress is 31.59 - 45.47MPa. For lower reach diaphragm wall, the maximum horizontal displacement is 25.5 - 34.7cm, the maximum settlement is 15.9 - 19.4cm, and the maximum primary stress is 35.7 - 43.6MPa. Since the corridor is located on the top of the diaphragm wall, relatively large tension stresses exist in the corridor with maximum tension stress of 3.66 - 3.9MPa.

3. In situ Compaction test and main technique Requirement

(1) Material. The sub-proluvial gravel soil at Heima borrow (grain size above 80mm is discarded) and the sub-proluvial gravel soil on slopes (grain size above 80mm is discarded), and the mixed soils after mixed with Guanjiashan clay at three proportions (weight ratio) of 15%, 20% and 25% respectively, and expressed successively with soil number ①-⑤.

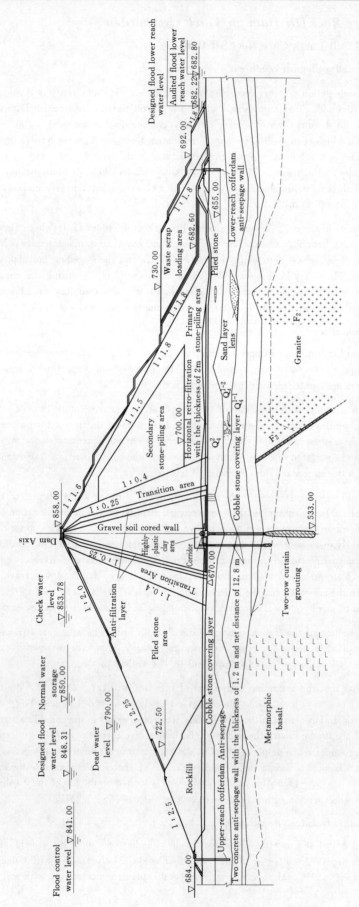

Fig. 23.2-1 Typical Section Plan of Pubugou Dam (Unit: m)

The characteristics of grain sizes. The grain size distribution curve of soil number ①is close to the designed average line, with average $P5$ content of 47%, and grain size less than 0.075mm in average content of 17.5%. The grain size distribution of soil number ②is close to the designed lower line, with average $P5$ content of 61%, and grain size less than 0.075mm in average content of 11.8%. The fine particles of the mixed soil increase with the mixed clay content, and the graduation curves are gradually close to the designed average line.

Seepage. Soil supply ①has the best anti-seepage performance, with the penetration coefficient more than 10^{-6} after the soil tamping, meeting the designed demands ($\leqslant 1 \times 10^{-5}$). Thesoil supply ②has the large penetration coefficient, hard to reach the designed demands and can not be used directly as the core wall anti-seepage material. Soil number ④and ⑤partial penetration coefficients meet the designed demands but have some diffusive character.

Compaction. The water content of soil number ①is slightly higher than the optimum water content. The soil is paved with thickness of 30-40cm per layer, and compacted with vibrating pad foot roller for 6 times or more, and reaches 98% in degree of compaction. By considering of efficiency and evenness of compaction, the construction adopted the water content of 1%-2% over optimum one, pavement thickness of 40cm, and compacted 8 times with vibrating pad foot roller.

After comparison of test results, soil number ①was finally chosen as the core material. Vibrating padfoot 18t-roolers were used for construction, pavement thickness of 40cm which was compacted 8 times.

(2) Technical requirements on dam body construction. For gravel soil: maximum grain size $\leqslant$80mm, grain content less than 5mm $\geqslant$45%, grain content less than 0.075mm $\geqslant$15%, seepage coefficient $<1 \times 10^{-5}$cm/s, fine material compaction degree $\geqslant$100% (Standard Proctor test).

Filter material: For the first layer, the maximum grain size$<$20mm, seepage coefficient $>5 \times 10^{-3}$cm/s, relative density$\geqslant$0.8, and later adjusted to 0.8-0.95. For the second layer, the maximum grain size $<$80mm, seepage coefficient$>8 \times 10^{-3}$cm/s, and relative density is the same with that in the first layer.

Transition material: maximum grain size $\leqslant$ 300mm, grain content less than 5mm $<$20% (later adjusted to $<$25%), seepage coefficient $>1 \times 10^{-2}$cm/s, porosity$\leqslant$20%.

Rockfill material: maximum grain size $\leqslant$ 800mm, grain content less than 5mm $<$10%, seepage coefficient $>5 \times 10^{-2}$cm/s and porosity $\leqslant$22%.

4. Construction Brief

After river closure in November 2005, the construction of diaphragm walls for the upper and lower-reach cofferdams began in December, and the cofferdams completed in May 2006. The diaphragm walls for the main dam finished in December 2006. During the early stage of dam construction, the soil materials were transported with a 4km-long belt and 20-ton-trunks. At the upper and lower reaches of the piling ground, rockfill material was transported to the dam with 20t-self-unloading-trunks. The peak dam filling volume reached 1.5 million m³. The dam construction started in April 2007 and finished in August 2009.

After quality inspection and appraisal on the dam, the gravel soil material, filter material, transition material and their seepage coefficients meet the design requirements, except the water content for small portion of gravel soil was higher than the designed value and the water content for the part of the highly plastic clay was lower than the designed one. The quality of dam compaction also met the design requirement.

5. Appraisal on initial operation

The water level in the reservoir reached dead water level of 790.00m on Dec. 8, 2009, and on Oct. 14, 2010, reached normal water level of 850m. At the dead water level, accumulated seepage was detected in the horizontal tunnels on both abutments and part of the water pressure pipes indicated higher water pressure inside. Therefore, reinforcement grouting was given to the area of accumulated leakage and the weak section of the diaphragm walls. It was found that the seepage flow was obviously reduced, the seepage water pressure was also decreased and the water head at the lower reaches of the diaphragm wall reduced 68.5%. The seepage capacity in the horizontal tunnel on both abutments reduced to normal range after the reinforcement grouting. The pore water pressure and its distribution inside the core were normal, and the dam deformation is also in a rational scope, which indicates the normal dam operation.

During the impounding period from the dead water level to the normal water level, the monitoring results showed that the dam stress strain and settlement met the requirements. At the impounding period, the maximum dam crest settlement was 655.75mm, the maximum deformation at the corridor structural joints was 25.86mm, and seepage through both abutment was 87L/s, which are normal for the dam operation.

23.2.2.2 The Changhe Dam-Gravel Soil Core Rockfill Dam

1. Design standards and topographic and geological conditions

The dam is located in Kangding County, Sichuan Province in the upper reach of the Dadu River. It was designed at 1000-year flood and checked at possible maximum flood (PMF). The earthquake intensity of Ⅸ was adopted for the anti-seismic design, with horizontal peak acceleration 0.359g at bedrock and exceeding probability of 2% for benchmark 100-year. The added height for anti-seismic safety is 1.5m for normal flood and 1.0m for abnormal flood. The safety factor of dam slopes is 1.50 in normal condition, 1.30 for abnormal condition Ⅰ and 1.20 for abnormal condition Ⅱ (seismic condition).

The river valley at dam site is relatively open and wide. The thickness of the overburden layer near dam axis is about 30-40m, and it can reach 50m for local area. The seepage coefficient of the riverbed gravels is about $1.22 \times 10^0 - 4.82 \times 10^{-2}$ cm/s, with high permeability. The water permeability rate is $q \leqslant 3$Lu at 100m deep beneath the river bedrock at the dam axis, with a relatively water-resistant layer in the depth. The upper portion of the weak weathered rocks on both abutments has a strong to extremely strong water permeability, medium to strong water permeability at lower portion, and a little permeability for fresh rocks.

2. Dam structure and calculation of strain

The bottom elevation of the dam is 1,457.00m and the maximum dam height is 240m on the riverbed. Both dam slopes are 1:2.0 in slope ratio. Between the core and the upper and lower-reach dam shell, filters and transition layers are constructed. Horizontal filter layers are implemented at the bottom of the core and lower reaches of the diaphragm wall, and also between the downstream dam shell and the overburden. The top elevation of the core is 1696.4m, with crest width of 6m, upper and lower reach slope ratio 1:0.25. The bottom elevation of the core is 1457.00m, with maximum bottom width of 125.7m. In the core and both abutments, there is a layer of highly plastic clay with the thickness of 3m. The horizontal thickness of the filter at upper and lower core is 8m and 12m respectively. The upper and lower transition layers are both 20m in horizontal thickness.

The diaphragm walls were constructed through the 50m overburden, and curtain groutings were given to the bedrocks beneath the overburden layer and both abutments, with controlled permeability $q \leqslant 3$Lu. There are two diaphragm walls with thickness of 1.4m and 1.2m for upper and lower reaches respectively. The net distance between two walls is 14m. The upper main diaphragm wall is located in the dam axis plane, connected with the core through a grouting corridor. The bottom of the diaphragm wall was embedded into bedrock in 1.5m. The maximum depth of the wall is 50m. T lower reach diaphragm wall is inserted into the core for 15m, and embedded into bedrock for 1.5m, with maximum wall depth of 50m.

Fig. 23.2-2 shows the largest section of the Changhe Dam. The dam construction includes earthwork excavation of 2.131 million m³, curtain grouting of 53,700m, diaphragm wall of 16,200m², dam refill of 36.29 million m³, including rockfill of 22.90 million m³ and core material of 5.66 million m³.

Changhe Dam under construction today is known as the highest Earth Core Rockfill Dam in the world. The gravel soil was taken from the Changba borrow, about 22km away in the upper stream. The natural density of gravel soil is 2.068g/cm³, dry density is 1.86g/cm³, average water content is 10.7%, permeability coefficient $k = 8.67 \times 10^{-7} - 1.05 \times 10^{-7}$ cm/s, compression coefficient $a_v = 0.016$MPa^{-1} and compression modulus $E_s = 76.6$MPa.

Through calculation, the dam's maximum settlement is 272.5cm, maximum horizontal displacement is 92.6cm. The maximum primary stress for upper and lower reach diaphragm walls is 45.2MPa and 42.7MPa respectively, and the maximum horizontal displacement is 42.0cm and 37.4cm respectively. The maximum tension and compression stress in the corridor is 46.2MPa and 60.0MPa respectively. Technical progress in Changhe Dam was based on the construction experience and achievement of the Pubugou Dam.

3. Current construction condition

The river closure was finished in November, 2010, the core construction started in May 2013, and the dam will be completed at crest elevation of 1697.00m in May 2017.

23.2.3 Three Earth Core Rockfill Dams on Bedrocks

23.2.3.1 Nuozhadu Core Rockfill Dam

1. Design brief

The dam is located at the downstream of the Lancang River in Lancang County of Yunnan Province, with normal water level of 812m, the total reservoir capacity of 23.7 billion m³, and regulating capacity of 11.3 billion m³. The maximum dam height is 261.5m.

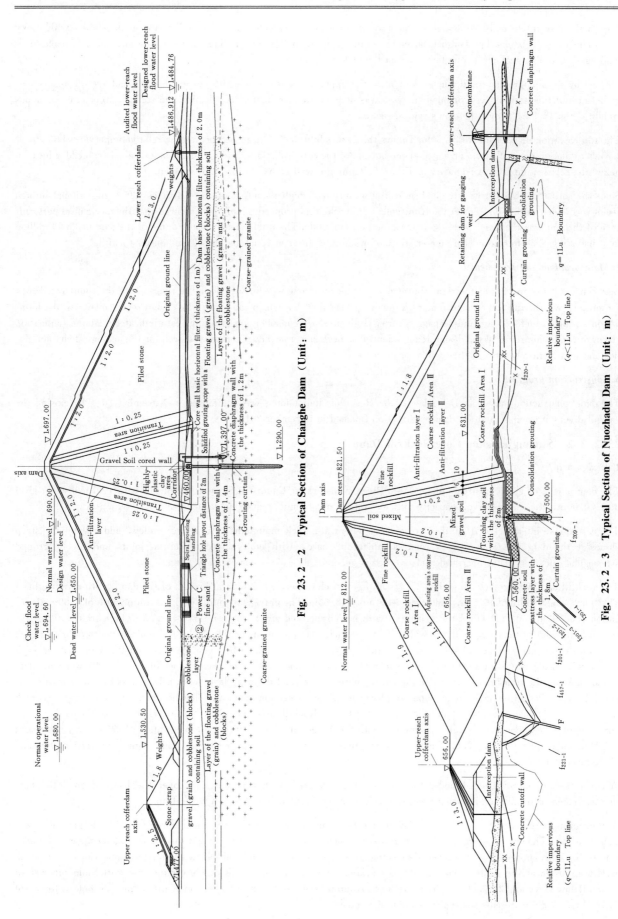

Fig. 23.2-2 Typical Section of Changhe Dam (Unit: m)

Fig. 23.2-3 Typical Section of Nuozhadu Dam (Unit: m)

The earthquake intensity is Ⅶ degree at the project area. The river valley is in a "V" shape at dam site, with rocky mountains on both banks. The bedrock is mainly granite, with shallow weathering rocks at dam site and left bank, less faults and favorable geological condition for dam foundation.

Limited by the topographic conditions at dam site, the upper and lower reaches slopes are 1 : 1.9 and 1 : 1.8 respectively. Because of the steep slopes, detailed analyses was done on the dam materials selection and zoning, to ensure the slopes stability. Fig. 23.2 - 3 shows the dam's typical section.

The top elevation of the core is 821.50m. Taking the permeability $q \leqslant $ 1Lu in bedrock as the relative impervious boundary, 2 rows of diaphragm walls were arranged at the bottom of the core wall. The 1st row of diaphragm walls is embedded into 5m below the relative impervious boundary and the 2nd diaphragm wall is mainly for reinforcement.

Since the soil from the borrow sites had more fine grains with higher content of clay particles, 35% of artificial broken rocks was added to the soil to meet the requirement on mechanical properties. Soil tests showed that the mechanical performance of the soil was improved with less deformation of dam body. The total construction volume of the dam is 33.60 million m^3, including 4.68 million m^3 core material, 2.02 million m^3 filter material and 26.89 million m^3 rockfill.

2. Main problem and treatment

Between Fault F_{12} and F_{13a} at the right abutment, a weak rock zone was formed along the river due to the complicated rock structural, weathering and erosion. The rock quality is classified V, which can't meet the design requirements on the foundation strength, deformation and seepage control. High pressure consolidation grouting tests indicated that the foundation strength can be improved to the design requirement. It is necessary to conduct further research for the geological characteristics and treatment measures of the foundation.

3. Construction progress

The dam construction started in December 2008. Reservoir impounding was committed in November 2011 by closing the gate of diversion tunnel, and the dam construction finished in December 2012.

23.2.3.2 Shuangjiangkou Core Rockfill Dam

1. Design brief

The dam is located at the upstream of the Dadu River in Ma'erkang County of Sichuan Province, with a normal water level of 2500.00m, the total reservoir capacity of 2.897 billion m^3, annual regulating capacity of 1.917 billion m^3.

The earthquake intensity is Ⅶ at the project area. There are thick mountains, deep river valley, and asymmetry steep slopes at the dam site. The exposed rock is intact and hard, with undeveloped structural joints and higher ground stress. The thickness of the overburden at riverbed is generally 48 - 57m, maximum 67.8m.

The rockfill dam with gravel soil core is 2510.00m in crest elevation, 2196.00m at bottom of the dam, with dam height of 314.00m, crest width of 16.00m, and crest length of 648.66m. The gravel soil cutoff wall has a top width of 4.00m at elevation 2508.00m and 2202.00m at the bottom, with both upper and lower slopes ratio of 1 : 0.2. There are two filter layers on both sides of the core, with the width of 4m for each layer on the upper side of the core and the width of 6m for each layer on the downstream side of the core. Transitional layers are constructed between the core and the dam body on both sides, with top elevation of 2504.00m, top width of 10m and upper and lower reach slopes of 1 : 0.3. Between the dam foundation and the transition layer under the core, there is filter and drainage layer with a thickness of 2m at lower reaches. Riprap rocks on the upper dam slope are placed for slope protection above the elevation of 2410.00m, and also rock blocks on the downstream slope at elevation above 2330.00m. Rockfill dregs are piled on the heel and toe of the dam as pressure weight. The upper reach weight is combined with the upper cofferdam, with the top elevation of 2330.00m, top width of 156m and slope ratio of 1 : 2.5. The lower reach pressure weight has a top elevation of 2330.00m, top width of 90m and slope ratio of 1 : 2.5.

A concrete base is constructed under the core on the riverbed, with concrete slab cover the bedrock on both abutments. Cement consolidation grouting was given below the concrete slab with depth of 5 - 10m.

The grouting curtains of this project consist of riverbed foundation curtains, abutment curtains and underground power house curtain. The curtain grouting of the dam was conducted to the foundation through the corridor along the core foundation to the both abutments with 6-layer of grouting adits. The cutoff curtain is 5m below the relative impervious layer which is 1Lu in permeability. In the bedrock above elevation 2460.00m, there are one grouting row with hole interval of 1.5m. Below elevation 2460.00m, there are two grouting raws with the row distance of 1.5m and hole interval of 1.5m. Fig. 23.2 - 4 presents the typical section of the dam.

Chapter 23 Typical Problems of High Dams around 300m

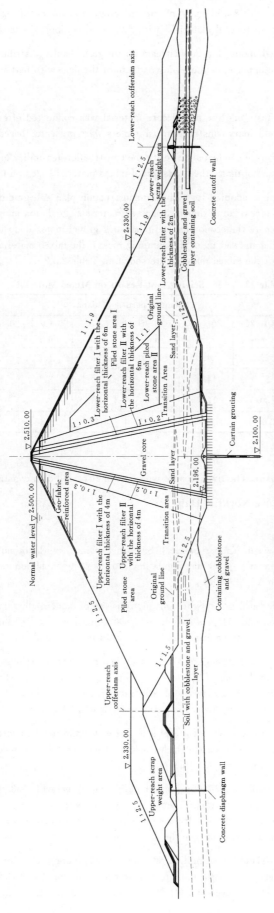

Fig. 23.2-4 Typical Section of Shuangjiangkou Dam (Unit: m)

The construction quantity of the dam is 5.08 million m³ of core material, 153,300m³ of high plactic clay, 2.19 million m³ of filter material and 32.72 million m³ of rockfill material, with a total excavation volume of 5.01 million m³.

The main borrow of the dam is located about 7km downstream on the right bank, distributed 600m higher than the river surface. And it was transported to the upper stream and downstream of the dam with belt convenyors.

2. Field compaction test

In 2006, the initial field compaction tests one the mixed core material was conducted after laboratory tests. The material mixing plan was verified, and the preliminary construction parameters were presented based on the lab tests.

In 2009, more field compaction tests were conducted to verify the construction feasibility of mixing gravel material or mixing with clay and gravel, and adjust and optimize the gradation of mixed material based on tests results as given as follows:

The large scale rock processing system is adopted to study the construction plan and grading selections of rock and gravel from No. 1 diversion tunnel dumping rocks, to optimize the mixed materials' gradation and construction measures through the material mixing tests, and to prepare the economical artificial rock-gravel material up to the design requirements. The test results show that after the crushing and sieving of the dumping rocks, the mixed material can satisfy the design requirements on gradation. The test results of the mixed material are shown in Table 23.2 - 1.

Table 23.2 - 1   Seiving Test Results on Mixed Material

Group	Content of particle gradation (%)									
	100-60 mm	60-40 mm	40-20 mm	20-10 mm	10-5 mm	5-2 mm	2-0.5 mm	0.5-0.25 mm	0.25-0.075 mm	<0.075 mm
1	13.0	12.6	29.7	24.0	12.9	3.0	1.3	0.8	1.4	1.3
2	19.6	21.8	25.2	14.4	9.2	3.3	2.1	1.5	1.9	1.0
3	17.9	21.2	30.5	12.7	8.6	3.5	2.0	1.1	1.4	1.1
Average	16.83	18.53	28.47	17.03	10.23	3.27	1.80	1.13	1.57	1.13

3. General construction plan

Now slope excavation started at dam site. The river closure will start in 2013 and dam construction will be finished in 2018.

23.2.3.3 Lianghekou Earth Core Rockfill Dam

1. Design brief

The dam is located on the Yalong River in Yalong County, Sichuan Province, with the normal water level of 2865.00m, appropriate capacity of 10.154 billion m³, perennial regulating capcity of 6.56 billion m³. The dam crest elevation is 2,875.00m with maximum dam height of 295.00m.

The erosive river valley has overburden at dam site which mainly consisits of modern alluviums and rock collapses and residual on both banks. The dam is built on undeveloped and intense weathering bedrock. The alluvial materials on the riverbed and the unstable rocks on both banks need to be excavated or removed for dam foundation.

The dam structure is similar as that in Shuangjiangkou Dam, with dam body combined with the upper and lower-reach cofferdams. At the bottome of the core with elevation of 2,580.00m, there is a 2m-thick concrete base, with curbtain grouting corridor (3m×4m) underneath. A 1m-thick plastic clay layer is placed on top of the concrest base, before filling the core mateiral. The width of dam crest is 16.00m, and the upper and lower reach dam slope ratio is 1:2.0 and 1:1.9 respectively.

The impervious components of the dam is the type of gravel soil core wall, and the dam core has filter layers and transition layers on both sides. On both abutment, there is a 1m-thick concerete base, and there is a contacting clay layer between the core and the concrest base with 4m in the horizontal thickness.

The rockfill zone I is placed above elevation 2,658.00m, and rockfill zone II below 2,658.00m. Inside the lower-reach dam shell between 2,630.00m and 2,804.13m, the rockfill zone III is placed with the zone I material for the outskirt dam shell. Horizontal steel bars, geo-grids and concrete frames on both dam slopes above elevation 2,820.00m is designed in the filters, transition layers and dam shells to improve the dam's anti-seismic peroformance.

In dam construction, core material is about 4.294 million m³, soil is 164,000m³, clay for cofferdam closure is 8,900m³, the total is 172,900m³; rockfill material is 29.492 million m³, transition material is 4.697 million m³, filter material is 1.903 million m³, riprap rocks for slope protection is 1.5955 million m³ and artificial brocken rocks is 5.27 million m³.

Fig. 23.2-5 is the typical dam section.

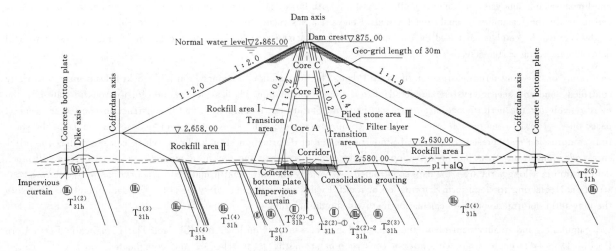

Fig. 23.2-5　Typical Section of Lianghekou Dam（Unit：m）

2. Related researches

(1) Study on the properties of rockfill materials and lab measurment techniques. For the 300m-level rockfill dam as Lianghekou with the huge project scale, the heavey vibrating roller will be widely used, with much improvement of impaction energy and dam filling density. With the increased compaction energy and the size of the rock blocks, the gradation similarity method and density replaceing method in the laboratory tests may not reflect the strength and deformation characteristics of the real rockfill materials and need to be verified. Therefore, it is necessary to do research on the methods for large scale lab tests on rockfill materials, to meet the need of the ultra-high rockfill dams.

(2) Study on the constitutive models and deformation of rockfill materials under high confining pressure and high water head. The Lianghekou Hydropower Station has a dam heigtht of 295m, with the pressure more than 6MPa at the bottom and confining pressure more than 3MPa. The rockfill matetrial has obvious anisotropy, which lead to the deformation of partical structures under the stress path variation due to high pressure and high water head. The constitutive model adopted by numerical analysis of rockfill dam deformation usually does not reflect this phenomenon. Therefore, it shall be conducted to study on the compression and deformation characteristics of rockfill materials under the high confining pressure and high water head on the basis of relevant constitutive model should be established.

(3) Study on the structural contacting surface characteristics of rockfill material and it constitutive model. Because of the different characteristic between core and rockfill materials, there are relatively large shear stress and displacement discontinuity on the contacting surface. This problem is more serious for Lianghekou dam. The characteristics of the contacting area should be studied and clearly understood, for taking appropriate engineering measures to reduce the disparity of contacting materials, minimize the shear stress and displacement discontinuity, and to ensure the construction quality.

(4) Static deformation analysis of rockfill dam and deformation control techniques. With the increased dam height and construction scale, the dam's deformation and settlement have become the key factors for dam safety. The deformation of rockfill dam and its appropriate inflence on the core may not be fully understood and apprized with merely the existing experience on dam construction, simple linear estimation of deformation and the stability analysis method of limit equlibrium. Those traditional methods mentioned above can not meet the need for the optimizing of dam zones and the structures of anti-seepage system. Avanced numberical analysis should be adopted for the dam design. However the current software can not meet the requirments. The constitutive models should be further studied and optimized.

(5) Study on the dynamic deformation of high rockfill dam and its antiseismic measures. There are few highrockfill dams subject to strong earthquake. Less records and documents are availble for the high rockfill dams at strong seismic region. Meanwhile, theories of strong earthquake on rockfill dam are not mature such as in dynamic constitutive model, coupled dynamic interactions, and non-linear dynamic analysis method. Nowadays the actual dynamic response of the high rockfill dam under earthquake can not be obtained with theoretical analysis and numerical calculations, neither for the understanding of mechnism and accurate evaluation of dam safety under earthquake, which undoubtedly affects the reliability, economy and rationnal antiseismic design of high rockfill dams.

It's neceasry for such a high rockfill dam as Lianghekou, to study on its dynamic deformation and anti-seismic measures, to accumulate and analyse the earthquake records and actual seismic damages on rockfill dams, to do research on the seismic mechanism of rockfill dam, to develop numerical methods with the function of multi-factor coupled and three-dimensional non-linear seismic analysis, to conduct model tests both with large scale shaking table and dynamic centrifugal model tests, and to verify the feasibility of analysing methods. Relevant assessment theories and methods for high rockfill dams under earthquake can be established based on the understanding of seismic mechanism and the failure characteristics of rockfill dams under strong earthquakes.

(6) Study on the long-term deformation of rockfill dam. The deformation of Lianghekou high rockfill Dam must be strictly controled, and anti-seepage system should adapt to the dam deformation. The less deformation for the rockfill material, the more reliable for the performance of anti-seepage system. The design indexes for rockfill construction control are general based on the triaxial compression tests and limited compression tests, and stress and strain analysis on the dam. The short time compression tests present the material properties of initial deformation after loading, which can not reflect the influence under long-term stress. The deformation monitoring results on existing rockfill dams indicated that there is still much deformation in a long term after the dam constructed, especially the soft rockfill material or poorly compacted during construction. Neglecting the long-term deformation of rockfill will cause potential safety risks for the dam operation. Therefore the long-term deformation and its calculation methods should be studied.

(7) Simulation and quality control on the construction process of 300m-level rockfill dam. The Lianghekou Earth Core Rockfill Dam is large in scale with intense construction strength, tight schedule and complicated construction conditions. Since the project is lcoated in the upper river valley on the Yalong River, the construction layout is quite difficult. The dam construction is the key project of the Lianghekou Hydropower pivot. Efficient control on the construction process is important to guarantee the construction progress and quality. The simulation of the dam construction process will help to optimize the construction procedures, improve the efficiency of construction and management, and hence ensure the construction. The real-time GPS qualty monitoring system will be introduced for the construction quality control, to watch the compaction process of rollers and to monitor the compaction quality effectively.

3. Construction preparation

Now the Lianghekou Dam is in the design stage for bidding document and the dam construction is schduled to be finished in 2018.

23.2.4 Comments on the Main Problems of Core Rockfill Dam with the height of 300m

23.2.4.1 Design Safety Standards

For the design standard of 300m Earth Core Rockfill Dam, it's a common practice to adopt the upper limit value for design flood, dam compaction parameters, safety factor of dam slope, anti-seismic criteria, and PMF for check flood. Attention should be paid to the cascade effect of 300m-level rockfill dam when the dams on the upstream of the river are collapsed. The instant flow capacity from the upstream collapsed reservoirs may much greater than the PMF capacity for the design, which will be a catastrophe for the dam and downstream projects. Therefore the free-board of the 300m high rockfill dam and the discharge capacity of spillway should be appropriately increased on the basis of existing standards after sufficient study.

23.2.4.2 Corridor Structural on Soft Foundation

The corridor on soft foundation is usually arranged on top of concrete diaphragm wall. The diaphragm wall will deform horizontally after the reservoir impounding, and which causes deformation of the structural joints between the riverbed corridor and corridors on both banks. For the copper water-stop in the corridor structural joint, the joints on the side walls may adapt to the horizontal deformation, but this is not the case for the bottom joints and side joints near the bottom, the copper water-stop may be easily damaged due to the direction change and cause leakage. In recent 3 years, the leakage of structure joints was found in three Earth Core Rockfill Dams with the height of 100m in Sichuan Province, indicating that the actual deformations of those structure joints were larger than the designed values. The same phenomenon was also detected during the construction and initial operation stage of Pubugou rockfill Dam.

It takes a long time for the deformation stabilization on the structure joints, which is always a potential safety risk of the dam. There are three measures for solving this problem ①to increase consolidation grouting in the lower reach of the structural joints during the construction period to limit the wall displacement; ②to install the water-stop in two stages, completing the plastic water stop in first phase, and after the dam deformation, install the copper water stop in the reserved second phase; and ③to construct concrete lining at structural joints and add water stops in the lining after the complete deformation of the dam.

23.2.4.3 Cutoff Curtain Arrangement

In recent years, some domestic rockfill dams have had leakage through abutments larger than the designed value after the impounding, mainly because ①lack of strict management on the curtain grouting construction process; ②lack of reinforced curtain grouting on slopes and developed rock joints; ③wrong construction process on the corridor construction near abutments. It is suggested that for the 300m Earth Core Rockfill Dam with unfavorable geological conditions on the abutments, main grouting curtain and auxiliary curtain should be implemented.

23.2.4.4 Properties of Gravel Soil

There are two types of gravel soils, natural and artificial. Natural gravel soil should be processed for excessive large blocks. There are two methods for producing artificial gravel soil, one is to mix artificial broken rocks and another is to mix natural fine gravels, the later method is recommended. It is suggested that for a 300m Earth Core Rockfill Dam, the gravel soils should be placed at different zones according to their properties of permeability and compaction, and the water head undertaken.

23.2.4.5 Deformation Control in Dam Construction Period

Deformation control is important for the quality management of Earth Core Rockfill Dam construction. The following techniques are general accepted at present: ①For Earth Core Rockfill Dam on overburden, the fine sand in the foundation should be excavated or replaced before the consolidation grouting in overburden, to improve the foundation bearing capacity. ②Reduce the dam uneven settlement by adopting the same compaction standard for the both the upper and the lower reach rockfill materials. ③It usually takes 1-2 years for primary settlement after dam completion, therefore the wave wall on dam crest should be constructed 2-3 year after the dam construction. ④Highly plastic soil on abutments may reduce the arch effect of the core. The five dams mentioned above adopt the design of highly plastic layers.

For the settlement control criteria, it is suggested that in related standards, specific regulation should be made for 300m rockfill dams with different foundation and dam types. For the criteria of anti-hydraulic fracturing of the core, it is suggested that reasonable values should be presented based on the monitoring results of above 5 dams during their operations.

23.2.4.6 Construction Management

Considering the problems in current Earth Core Rockfill Dam construction, attention should be given to the following problems for the design and construction of Earth Core Rockfill Dams with the height of 300m: ①The scope of land requisition are general large with a lot of arable land. Excavation plan should be made before the construction, by collecting the surface arable soil at the beginning and paved back after the excavation. ②The quarry excavation should leave room for later land use. ③The speed of dam construction should be strictly controlled, better more than 4 years for Earth Core Rockfill Dams with the height of 300m. It was found that fast and intensive construction does no good to the dam safety after water storage. ④It is reasonable to introduce third party on the quality inspection of diaphragm wall and curtain grouting construction.

23.2.4.7 Initial-stage Operation

After the reservoir impounding, due to the large stress difference on both sides of the anti-seepage walls, the dam's normal settlement speed is affected to some degree, and longitudinal cracks may appears on dam crest because of rapid change of water levels in the reservoir. Therefore it is suggested that the allowable water level change for Earth Core Rockfill Dam with the height of 300m be controlled according to its own conditions. The monitoring work during dam operation mainly consists of stress-strain and seepage observations. Because of the extremely complicate seepage conditions, it is suggested that the gauge weir for seepage water measurement should be implemented together with the design of the diaphragm wall for the downstream cofferdam.

23.3 Outstanding Issues of High Arch Dams with the Height of 300m

Duan Shaohui[●]

23.3.1 *General*

The 240m arch dam of Ertan Hydropower station on Yalong River was put into service at the end of the last century and since then, a number of arch dams with the height of 300m have been constructed and are in the early stages of impounding, such as the 250m heigh Laxiwa dam on the Yellow River, the 294.5m high Xiaowan dam on the Lancang River, the

---

[●] Duan shaohui, Deputy Chief Engineer, Construction & Management Center, Yalong River Hydropower Development Company Ltd., Professor Senior Engineer.

285.5m heigh Xiluodu dam on the Jinsha River, and the 305m heigh Jinping I dam on the Yalong River, etc. As a result, our national technical level of design, research, construction and management of 300m level arch dams has been greatly enhanced.

In China 300m high arch dams are mostly located in western areas, where the typical characteristics are high mountains and canyons, steep slopes, complex geological conditions, high earth stresses, and high earthquake intensity. As such, the outstanding issues of 300m high arch dams include difficulties of project layout and construction arrangement; stability analysis and treatment of steep artificial slopes; difficult treatment for stabilization due to the steep natural slopes with widespread potential instability of rock mass; new challenges of arch dam geometric design and safety evaluation; issues of high arch dam structure, earthquake resistance and crack prevention; complicated and varied foundation treatment; high requirements for performance of materials and concrete for arch dams; temperature control and crack prevention of arch dam concreting; and new challenges of flood discharge and energy dissipation, as well as prevention and cure of atomization effects.

The following is to introduce some ideas on how these outstanding issues are being addressed, taking Jinping-1 hydropower station as a typical project.

23.3.2 *Stability and Reinforcement Treatment of High and Steep Slope Abutments*

23.3.2.1 Stability and Reinforcement Treatment of Abutments-artificial High-steep Slopes

1. *Stability of abutments under the artificial high-steep slope*

Abutments of high arch dams with the height of 300m are generally high and steep rock slopes, slopes with local overburden, or deformed slopes. The issues of slope stability are mainly mass stability and deformation of the high and steep rock slopes and apart from conditions of geological structure and hydrogeology, they are closely related to the sequence of excavation and stabilization treatment. Failure modes of high-steep slope stability are generally through sliding failure, yield failure, tension fracture, toppling failure, etc. Analyses are carried out to use a variety of methods such as: rigid limit equilibrium, limit state analysis, continuous medium dynamic finite element, block theory of non-continuous medium dynamics, discrete element method, manifold method, etc, all variously applied as analytical methods to establish stability and deformation of the high-steep rock slopes. The stability and treatment of Jinping I project left abutment artificial slope is used as a case study below.

Jinping I project left abutment artificial slope has 3 sections: slope above the cable-crane platform, slope between cable-crane platform and dam crest, and spandrel groove slope below the dam crest. The cable-crane platform of the left abutment is 14m wide, at the elevation of 1,960m, maximum height of slope excavation is 150m, and inclination is 1 : 0.5, with a 3m wide berm every 30-40m. The inclination of the slope between the elevation of 1,960m and dam crest at the elevation of 1,885m is also 1 : 0.5, separated by 3 berms with the width of all 3m, except the berm at the elevation of 1,885m which is 6m. Excavation profile of the spandrel groove slope below the elevation of 1,885m is divided into upstream side slope, downstream side slope and groove slope; maximum length of excavation is 305m; there are 3m wide berms every 30-40m height; every 50m height of fresh, breeze and weak weathering floor rock mass has a berm; the left cushion is concreted between spandrel groove at el 1,730m and dam crest. The maximum height of left abutment slope is designed as 530m.

Exposed strata of Jinping I dam site are the second section and the third section of Zagu, nao petrofabric of the Middle Upper Triassic. The second section comprises 3-8 layers of marble at the left bank below the elevation of 1,820-1,850m, the riverbed and right bank; lithological characteristics are massive overlying marble, onyx marble, ruin marble, and grey to deep grey thin to medium thick banded marble, with local green schist lenses. Above the left bank of 1,820-1,850m is the third section, comprising sandstone and slate.

The slope of the left abutment is a counter-inclined rock slope, the gradient below the elevation of 1,820m is 50°-65°, while the gradient above slow to 45°. The sandstone and slate of the left bank slope have obvious tension fractures and toppling deformation; unloading of the sandstone is strong, the horizontal depth of strong unloading is 50-90m, while the horizontal depth of weak unloading is 100-160m, with the deepest at 200m or more. A dense belt of deep unloaded fissures exists horizontally, some 200-300m deep, and is a peculiar geological phenomenon of the Jin Ping-1 left abutment slope. Local buckling failure modes of the left abutment slope are toppling-slip failure controlled by toppling deformation at the bottom surface or unloading joint, wedge sliding failure combined by small fault and joints or bedding fault zone, and slip-collapse failure controlled by toppling fracture. Overall stability of the left abutment slope is controlled by wedge sliding body with fault f_{42-9} (N80°-90° E, SE $\angle 45°-55°$) as the sliding surface, deep seated rupture zone SL_{44-1} (SN-N20°W, E$\angle 55°-60°$) as the upstream boundary, and the lamprophyre vein X (N45°-55° E, SE $\angle 65°-70°$) as the posterior boundary.

Overall stability is controlled by the wedge sliding body of the left abutment slope; stability analysis can be checked using limit equilibrium method, elastic-plastic finite element method or three-dimensional discrete finite element method analysis. At Jinping I, the 3D limit-equilibrium method from the 3D-SLOPE program was used; the general occurrence of SL_{44-1} relaxation and cracking zone has an important effect on the wedge body; two orientations SN and N20°W respectively were input in the analysis calculation to get wedge stability models A and B. The highest elevations of the sliding bodies model A and model B were both 2,050m, and lowest elevations were respectively 1,700m and 1,750m. The rigid equilibrium limit method analysis was carried out to check the overall stability of the left abutment slope, considering natural conditions, excavation to dam crest, and spandrel groove excavation of the three stages of excavation to completion. In the natural condition, while stability safety coefficients of varying working conditions of the wedge were all greater than 1, they were incompliant with Specification requirements. Stability safety coefficients of the two models both increased for the excavation to dam crest case, and were compliant with Specification requirements. After completion of the spandrel groove excavation, in normal conditions, the safety coefficients of models A and B were respectively 1.13 and 0.995, hence overall stability of the slope was incompliant with the requirements. Left abutment overall stability was checked using the 3D nonlinear FEM strength reduction method, available in the two FEM programs, PHASE^2 and FINAL respectively. Calculation analysis results of these two programs showed the slope was stable in the natural condition and in the case of excavation to dam crest. After the excavation of the spandrel groove, the safety coefficients calculated by programs PHASE^2 and FINAL were respectively 1.05 and 0.95, which was incompliant with requirements. And it was thus required to reinforcement treatment.

During excavation of the left abutment slope, other methods are taken included: safety monitoring and warning and prediction, safety monitoring and feedback analysis, and micro seismic monitoring, tracing stability analysis results of the slope of excavation and reinforcement, and provided important technical support for slope safety excavation.

2. Abutment artificial high-steep slope reinforcement measures

Various measures for slope reinforcement are available, such as water interception and discharge, shotcrete rock bolt support, slope surface protection, anti-shear cavity strengthening, anti-slide pile strengthening, etc.

Actual stability reinforcement measures adopted at Jinping I left abutment slope were: anchor cables in the slide surface block, f_{42-9} anti-shear cavity, SL_{44-1} grouting tunnel, lamprophyre vein (X) drainage tunnel or long drainage hole, f_5 fault local concrete replacement, etc. Seven scenarios combined by these measures were compared, each case being analyzed by the 3D-SLOPE program for stability safety appraisal. The final measures of slope reinforcement were determined as follows:

(1) Interception drain peripheral slope, excavating slope surface drain, slope surface drainage mainly by drainage holes.

(2) Rock mass of slope above left abutment at the elevation of 1,960m was fragmented, treated with pre-grouting first to increase stability of slope excavating.

(3) Surface shotcrete protection.

(4) Supported excavated slope except spandrel groove slope with bolting, mortar bolts length of 6m and 9m, the two diameter is 28mm and 32mm respectively, spacing 2m×2m; the excavation boundary and berm of slope above the left bank at the elevation of 1,960m was arranged with three anchor bars (ϕ32 at 2m spacing on part of the slope, as pre-grouting holes of lower slope excavating.

(5) Spandrel upstream and downstream slopes, cushion groove slope, and slope above dam crest were all supported by anchor cables, except spandrel groove slope; anchor cables were 40 - 80m deep with design capacity 2,000kN, local was 3,000kN, systematically arranged on toppling deformation, high weathered rock mass, with spacing 4m×4m, drilled anchor cable holes and reinforced slope by grouting; arrangement on the III rock slope below the elevation of 1,960m systematic on a 6m×6m grid.

(6) It was applied steel concrete frame beam in the range of toppling deformation above the elevation of 1,960m at the left bank, used steel concrete vertical beam in the range of high weathered and type IV rock, and reinforced toe of the slope by steel concrete block protection on each step of berm downstream line.

(7) 3 lines of anti-shear galleries were excavated along f_{42-9} fault at the elevation of 1,886m, 1,860m and 1,834m; all are 9×10m city-gate section, key slots were dug as an anti-shear cavity by cross way, up and down shafts were dug in rock, the depth of shafts was twice the diameter of the gallery, consolidation grouting was carried out surrounding the anti-shear gallery.

23.3.2.2 Stability and Prevention Measures of Abutment Natural Slope with Potential Unstable Rock Mass

The 300m high arch dam is located in a canyon, with solid mountains on both banks, where height difference between natural slope and riverbed is more than 1,000m, rock foundation is mostly exposed, and natural slope is generally stable. It is the characteristics of geological conditions. However, high-steep slope instability is mostly due to unloading fracture development, local formed unstable and dangerous rock masses after further tension and extension, weathering and erosion. The destabilization of the natural slope would be a gross safety threat during the dam construction and operation because of dangerous rock mass, and any movement must be prevented and the rock mass treated according to stability conditions and risk of unstable destruction.

1. *Destabilization model and stability of dangerous rock mass*

Any dangerous rock mass is either suspended, notched-in, or isolated on a slope. There are three common modes of unstable failure about dangerous rock mass of rocky slope: the first one is collapse and sliding along freeing or steeply inclined structural planes; the second is block toppling cut by an unfavorable combination of negative or steep structural planes, and the third is shearing failure-detachment of the over-hanging rock or rock above a cavity along vertical structural planes. Such seven unstable failure modes of dangerous rock mass in the Jinping Ⅰ dam region are described below.

(1) Compressed-toppling deformation mode. Developing in steep slope of left bank, medium-thick metamorphic sandstone with sandy slate are interbedded; the upper is hard, the lower part soft, and in a state of compressed deformation; slope rock mass is deformed, twisted and toppled along the downward or structural planes to a gully/valley to free-face. See Fig. 23.3 – 1.

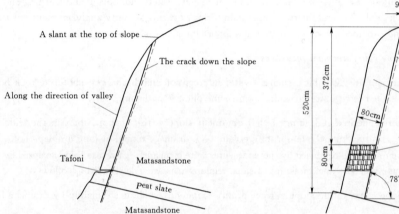
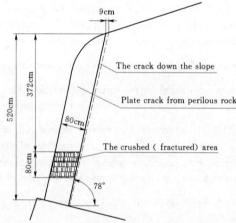

Fig. 23.3 – 1 Compressed-toppling Deformation Mode **Fig. 23.3 – 2 Toppling-cracking Deformation Mode**

(2) Toppling-cracking deformation mode. Developed in sandstone and slate of the left bank steep-inclined slope, it is evident as a laminated rock or rock pillar along the downward/steep-inclined fracture, with overhanging beam deformation under the effect of dead weight or pressure due to surface run-off. The lower part of the flakey rock pillar was bent by pressure, making the dip angle less steep; slope structure relaxation was obvious, falling away under gravity occurred often at the slope front. Once the maximum bending fault plane had been attained, the root of the beam slab would have cracked and been crushed, followed by toppling failure. See Fig. 23.3 – 2.

(3) Eccentric rolling (sliding) mode. Overburden of colluviums and solitary stone piles at the toe of the steep slope were well developed at the left bank, overhanging at the slope edge; the lower part formed a cavity, affected by progressive sliding and gravity; the block gravity center had shifted gradually then lost support, leading to rolling sliding destabilization. See Fig. 23.3 – 3.

(4) Ladder sliding deformation mode. This had developed in the left bank sandstone and slate, and was controlled by two groups of joint fracture, one was downward/steep-inclined, the other was downward/gentle-inclined, with steep-inclined joint opening, gentle-inclined joint sliding; destabilization evidently started from the front, and transferred upward in a stair-stepping way. See Fig. 23.3 – 4.

(5) Block collapse and slide mode. Developed in the left bank, this was a non-growing fracture, comprising shallow unloading metamorphic siltstone and steep wall of right bank marble; there was an unfavorable combination of 2-3 joints, and a free-face; small scale collapse and sliding destabilization developed along the intersection line. See Fig. 23.3 – 5.

(6) Overall sliding mode. A cliff of the whole right bank high level slope had developed, forming a rock bridge of dangerous rock mass, where the parent rock was sheared and isolated under the effect of gravity; overall sliding happened due to zero

support at the lower part of the cliff. See Fig. 23. 3 - 6.

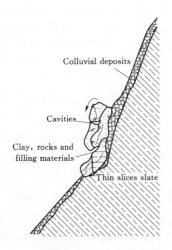

Fig. 23. 3 - 3 Eccentric Rolling (sliding) Mode

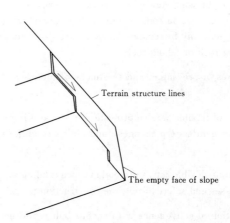

Fig. 23. 3 - 4 Ladder Sliding Deformation Mode

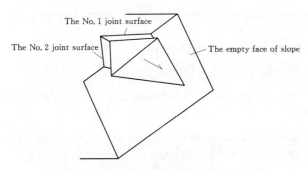

Fig. 23. 3 - 5 Block collapse and Slide Mode

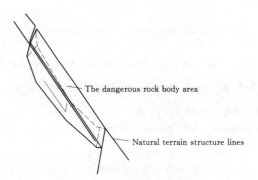

Fig. 23. 3 - 6 Overall Sliding Mode

(7) Faulted failure mode. Where the whole right bank high level slope cliff had developed, under the effect of gravity, the rock bridge of dangerous rock mass and parent rock was sheared and isolated, leading to overall collapse. See Fig. 23. 3 - 7.

Dangerous rock mass developed along clear structure planes leading to deformation in an unstable mode; it acquired reliable physical and mechanical parameters of the structural plane, and the stability safety factor could be quantitatively calculated by identifying the stable state of dangerous rock mass.

At the Jinping I project, dangerous rock mass of two banks mostly developed on the slope or cliff above the elevation of 2,100m, over a relatively short distance and detailed geological investigations could not be carried out. The degree of instability of the dangerous rock mass was therefore difficult to quantify. The stability of the dangerous rock mass was then determined by estimation and macro-geological diagnosis according to regional geological background of the project, the development of geological conditions, and geological characteristics of the dangerous rock mass, and possible deformation form. At the left bank of Jinping I, there were 32 such dangerous rock mass zones of natural slope, 10 out of 32 were unstable, 5 of the 32 were not so unstable, while the rest were generally stable; 34 dangerous rock mass development zones of natural slope at the right bank were identified, 7 of which were unstable, 12 were not so unstable, the rest were generally stable.

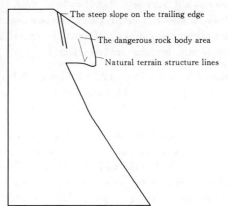

Fig. 23. 3 - 7 Faulted Failure Mode

2. The hazard analysis and prevention measures of dangerous rock mass

Hazard degree of dangerous rock mass is the damage level to hydraulic structures and people caused by dangerous rock mass destabilization. In accordance with the effects of dangerous rock mass on buildings and people, hazard of dangerous rock mass in Jinping I project had 3 degrees: severe, not so severe, and general. The left bank had 10 zones of severe hazard, 22 zones not so severe; while at the right bank, 34 zones of high level dangerous rock masses were identified as severe.

Main prevention measures of dangerous rock mass are: removal, excavation, shotcrete, rock bolting, pre-stress anchor cables, active protection fences, passive protection fences, unstable rock barricade wall, masonry or concrete reinforced buttresses, dental concrete, etc. Removal and excavation are generally applied to small scale and lower elevation dangerous rock mass, the effect on construction of lower part being relatively controllable. Dangerous rock mass above construction slope cannot generally be removed or excavated due to danger of uncontrollable damage to the lower areas of work or buildings by flying rock or falling rock.

Main measures of prevention and treatment of high level dangerous rock mass applied to Jinping I left and right banks were:

(1) Erection of flexible passive protection fences at different elevations and locations according to slope dangerous rock mass distributions, gentle-steep changes, and gullies conditions; about 24,000m^2 of 750kJ and 1,000kJ passive protection fence were used.

(2) Erection of flexible active protection fence totaling about 190,000m^2, depending on degree of fragmentation of dangerous rock masses and safety during the construction period.

(3) Installation of active fence with anchor bolt reinforcement, the anchor bolts connected with active fence as a precaution for "less unstable" dangerous rock masses, and enlarged anchor cable were adopted to strengthen for the bigger volumes.

(4) Other measures included rock barricade walls, masonry or concrete supporting buttresses, dental concrete, etc.

23.3.3 Arch Dam Geometric Design and Integral Safety Degree
23.3.3.1 Arch Dam Geometric Design

Arch dams are statically indeterminate structures; hence the requirements for dam foundations are very high. The 300m high arch dams with particularly high water pressures require:

(1) Load-carrying capacity of dam foundation to be high, the deformation of dam foundation must be small and homogeneous overall, rock body of 300m high dam foundation base must be fresh and complete.

(2) The dam foundation to have a high degree of impermeability, and seepage failure must not occur under high water pressure.

(3) The arch dam geometry to have adequate integral rigidity in order to withstand large water pressures, but stress and deformation must be within allowable limits. Hence, a double curved arch dam structure geometry with good mechanical characteristics is generally used, combined with arch dam foundation base determination, foundation treatment, consideration of dam body stress, deformation, overall stability and safety degree, quantity of work, etc.; the combination of factors need to be compared and selected.

Jinping I dam is located on a 1.5km reach between the Pusiluo and Shoupa gullies; the dam region has ideal topographic conditions for such a high arch dam, but the geological conditions are very complicated especially the left bank with the X, f_5, and f_8 faults and deep unloaded fissures, relaxation zones, etc.

In addressing the unfavorable geology, a 155m high cushion was designed, while grid replacement to the X, and f_5 faults with large scale high pressure consolidation grouting were carried out at the deep unloaded relaxation zone. Referring to the *Concrete arch dam design code*, SD 145—85 (hereinafter referred to as *arch dam design code*) stipulates that the main reference of arch dam stress analysis to evaluate strength and safety is calculation by the arch-cantilever method. After project analogy, Jinping I stress control standard co-ordination with arch-cantilever method was determined. See Table 23.3-1 - Table 23.3-2.

Table 23.3-1　Jin Ping-1 arch Dam Stress Control Standard

Load combination		Allowable compressive stress (MPa)	Allowable tensile stress (MPa)	
			Upstream subface of dam	Downstream subface of dam
Common combination		9.0	1.2	1.5
Special combination	No earthquake	10.0	1.5	1.5
	With earthquake	As per hydraulic strcutures anti-seismic design code, controlled by partical coefficient		

Table 23.3-2 Jin Ping-1 arch Dam Dynamic Stress Control Standard

Concrete partitions	Compression (MPa)	Tensile (MPa)
A (Basic constraint and the orifice area)	17.70	3.30
B (In the middle of dam)	14.70	2.70
C (The upper dam)	12.30	2.30

After comparison and selection of arch dam foundation base and geometry at the feasibility study stage, the shape of the Jinping I dam was determined to be a parabolic double-curved arch dam, with crown cantilever thickness-height ratio of 0.19. However, according to the terrain and geological conditions in the dam region; detailed treatment planning of the foundations and the geo-mechanical model test results, and in order to further decrease the maximum stress of the dam body and increase integral safety degree of the arch dam; the arch dam foundation base and geometry were subsequently adjusted during the tender and construction stages. The crown cantilever thickness-height ratio was increased to 0.207, together with other parameters of dam geometry; please see Table 23.3-3. The maximum compressive stress of the arch dam under normal working conditions has been decreased from 8.40MPa at the feasibility study stage to 7.78MPa, while the maximum tensile stress is still 1.18MPa. Stability against sliding of the left abutment was compliant with Specification requirements, but stability of the right abutment was affected by the 4th layer of a marble downward/low-angle green schist lens body. The dead weight of the surrounding rock mass acted as a major element in the stability analysis, except block R24, where the grout curtain drainage partly failed under working conditions, and shear and wear stability safety coefficients of all other blocks were larger than 2.5. The current domestic 300m high arch dam has the geometric characteristic parameters, please see Table 23.3-3.

Table 23.3-3 The geometric Characteristic Parameters of High Arch Dam with the Height of 300m in China

Items	Unit	Jinping I arch dam		Xiaowan arch dam	Xiluodu arch dam
		Feasibility study stage	Construction stage		
Arch curve form		parabolic double-curved arch		parabolic double-curved arch	
Crest elevation	m	1,885.00	1,885.00	1,245.00	610.00
Dam height	m	305.00	305.00	294.50	285.50
Crown roof thickness	m	13.00	16.00	12.00	14.00
Crown bottom thickness	m	58.00	63.00	73.124	60.00
Spandrel maximum thickness	m	62.00	68.50	73.124	
Dome central line arc length	m	568.62	552.43	892.786	678.65
Maximum center angle	(°)	95.71	93.55	92.791	95.58
Thickness-height ratio		0.19	0.207	0.248	0.216
Arc height ratio		1.864	1.811	3.035	2.441
Flexibility coefficient		9.326	7.99	12.46	
Total volume	$10^4 m^3$	435.59	476.47	761.7	
Total water thrust	$10^4 t$		1200	1700	

23.3.3.2 The Whole Safety for Arch Dam

Generally dam stresses and the degree of integral safety are comprehensively evaluated by numerical simulation analysis methods and geo-mechanical model test methods, as well as engineering analogy.

Degree of integral safety for Jinping I arch dam was evaluated three times by geo-mechanical model tests and three-dimensional finite element analyses using ABAQUS TFINE, and FLAC^{3D} software. Normal, overload and strength reduction conditions were calculated. Through model testing and calculation analysis, after arch dam foundation treatment, and under normal load, above 2.5P_0, the upstream dam heel would crack (non-linear overload multiples $K_2 = 4-5$, $K_3 \geqslant 7.5$). The stability of dam abutment rock masses was adequate, with effective foundation treatment. Compared to practical experience on other large projects (see Table 23.3-4), after foundation treatment at Jinping I arch dam, the integral overload safety coefficient is between Xiaowan's and Er'tan's, which is meant that the whole safety for Jinping I arch dam is higher.

Table 23.3-4 Comparison Table of Typical Dam Geo-mechanical Model Tests for Safety Coefficients (SC)

Project name	Height (m)	Flexible SC K_1	Non-linear SC K_2	Limit state SC K_3
Jinping I in Jan. 2003	305	1.5 – 2	3 – 4	5 – 6
Jinping I (foundation treatment) in Sep. 2003	305	2.0	3.5 – 4	6 – 7
Jinping I (foundation treatment) in May. 2006	305	2.5	4 – 5	7.5
Xiluodu dam (feasibility study XLD824 geometric untreated)	285.5	1.8	5.0	6.5 – 8
Xiluodu dam (optimizing XLD03 geometric untreated)	285.5	1.8 – 2	4.5	8.5
Xiaowan double-curved arch dam	294.5	1.5 – 2	3.0	7.0
Laxiwa in Nov. 2004	250	2.0	3.5 – 4	7 – 8
Ertan double-curved arch dam (rock simulation cohesive value C)	240	2.0	4.0	11 – 12
Ertan double-curved arch dam (non-simulation rock cohesive value C)	240	2.0	3.5	8.0

Note: The data in this table are from the research results of Qing Hua university.

23.3.4 *Foundation Treatment of Complicated Geological Conditions*

Geological conditions for 300m high arch dams located in high mountains and canyons are always complicated, the loads of the 300m high arch dams are especially large, the foundation base must have good homogeneity, anti-deformation and high impermeability; fault and soft fracture zones of the foundation must be replaced and treated by consolidation grouting and substantial curtain grouting, these are common foundation treatments, but the requirements are higher.

The problem of high geostress also occurs during the foundations excavation of 300m high arch dam, and must be treated carefully. To withstand the huge water thrust of super high arch dam, dam abutment forces and resisting rock mass must have good completeness, and sound overall rigidity. As for Xiaowan arch dam, there are large scale fault fracture zones, alteration zones and soft rock, and the left bank of Jinping I arch dam also had the faulting at a larger scale within resisting force of the left bank, lamprophyre veins, deep unloaded fracture zones, weakened rigidity of force resisting rock, and all should be treated in order to increase the integral rigidity of the resisting rock.

23.3.4.1 High Geostress Dam Foundation Excavation and Unloading Relaxation Prevention

Geostress in the valleys, which is located the arch high dam with the height of 300m, is high, and particularly important during the excavation of the foundations. Recently, constructed high arch dams, such as Laxiwa, Xiaowan and Jinping I arch dams for hydropower, all had high geostress and unloading relaxation problems during the excavation of dam foundation.

The maximum main stress of Laxiwa arch dam valley floor was more than 33MPa, and local stress was 42 – 56MPa. The riverbed excavation was designed as a smoothly connecting reverse curve primarily in order to reduce high geostress and unloading relaxation during excavation. The valley stress-concentrated area shifted to the lower part of the foundation base while excavating the dam foundations; stress concentrations at both sides of the dam foundation, together with the range of yielding stresses were decreased. In the event, there was no obvious rock burst, stripping, or high geostress phenomena while excavating the dam foundations. Unloading relaxation was effectively controlled during excavation, with small section bench blasting and timely anchor support.

Plane geostress test results of Xiaowan dam region showed the extent of abutment horizontal depth to be more than 50m, with $\sigma_1' = 8 - 17$MPa; the valley floor elevation was 930 – 960m, and nearby stresses were $\sigma_1' = 22 - 35$MPa, local stress concentrations were $\sigma_1' = 44 - 57$MPa. Following excavation of the dam foundations, strong unloading relaxation occurred at the riverbed below the elevation of 975m; relaxation depth was around 5m; then a second excavation was required to clear away the unloading relaxation rock mass. Then the sequence followed as pre-rock bolting, foundation excavating, immediately more rock bolting, foundation base clean-up, covering with dam concrete rapidly, and so on.

The bank slope of Jinping I dam was high-steep, and unloading of bank slope rock mass was strong. Space geostress tests were carried out during the preliminary exploration period at the dam location side slopes showed: left bank horizontal depth within 150m, $\sigma_1 = 5.84 - 20.72$MPa, fluctuation of maximum main stresses was big, that is 20.72MPa occurring at exploration profile II, the elevation of 1670m, at a depth of 28m; left bank stresses beyond 150m were $\sigma_1 = 21.49 -$

40.40MPa; the right bank horizontal depth was less than 100m, with $\sigma_1 = 12.96 - 15.42$MPa; at horizontal depths between 100m and 250m, stresses were $\sigma_1 = 15.08 - 35.70$MPa.

In order to prevent riverbed dam foundation excavation unloading relaxation, below el 1,670m, foundation reinforcement for protective coverings were planned, together with systematic rock bolting; thickness of protective cover was 3 - 5m; anchor bolts were: ϕ28mm or 32mm, the length of 6m or 9m, on a 2m grid. During construction, comparison tests of fore bolting excavation with protective covering (of excavated rock left in place) and bolting-in-time after the excavation were carried out above the elevation of 1,650m. There was no obvious change in anchor bolt stresses or sliding micrometer displacement in either plan. Therefore, in the left bank above the elevation of 1,630m and right bank above the elevation of 1,610m, the dam foundation was excavated by protective covering and support-in-time measures. The measures below the aforesaid elevations were protective covering reserving and fore bolting excavating.

Geophysical exploration results of foundation base rock mass excavation blasting relaxation showed that the whole change of dam foundation rock mass in blasting unloading relaxation depth of left bank foundation El. 1,730 - 1,580m was significant, between 1 and 3.4m, generally 1 - 2m. At the right bank foundation at the elevation of 1,710 - 1,580m, dam foundation shallow unloading relaxation depth was between 0.8 and 3m, mainly 1 - 2m. Left and right dam foundation rock mass long-term observation well test results showed that in the process of the dam foundation excavation, rock mass relaxation depth had no obvious change, except for a little decrease of sound wave in the locations away 5m from of left bank dam foundation relaxation ring.

23.3.4.2 Resisting Body Weak Rock Replacement and Consolidation Grouting

The rigidity of arch dam force resisting rock masses would be insufficient in areas of extensive weak rock, and deformation could form under the arch dam thrust, especially under the large water loads of 300m high arch dams; the effect on stress of arch dam structures and deformation is very significant. Therefore, certain types of weak rock of 300m high arch dam force resisting rock must be replaced and/or treated with high pressure grouting.

At Jinping I, the left bank abutment showed some unloading crack development; the left bank marble deep unloading base horizontal depth was generally 150 - 200m; left bank sand slate deep unloading base horizontal depth was 200 - 330m; left abutment main geological structure had a lamprophyre vein (X), and f_8, f_5 and f_2 faults, where the rock characteristics were poor, distributed throughout the left bank force resisting rock. The right bank abutment of main geological structure had faults f_{13} and f_{14} fracture zones, but the width of the affected zone was relatively narrow at the dam foundation, and wider on force resisting rock. According to stress deformation calculations and geo-mechanical model test results for arch dam, the left bank deep unloading cracks and main geological structure and the right bank fault all had a significant effect on arch dam stresses, deformation and cracking, and had to be treated with concrete replacement and extensive grouting. A replacement shaft was excavated in the right bank force resisting rock f_{13} fault with the elevation from 1885m to 1601m, and 3 layers of replacement horizontal tunnels and 5 replacement inclined shafts were dug at the f_{14} fault.

Weak rock replacement and consolidation grouting treatment were carried out at Jinping I left bank force resisting rock; 2 layers of horizontal tunnels were arranged between 1,670m and 1,730m, and 3 concrete replacement inclined shafts at upstream side, and one concrete replacement inclined shaft at f_5. See Fig. 23.3-8. Replacement horizontal tunnels were excavated at els 1,829m, 1,785m and 1,730m at the left bank lamprophyre vein (X), and 3 replacement inclined shafts were dug between els 1,785m and 1,730m; an impermeability inclined shaft was dug at the upstream side of a cavern crest elevation between 1,885m and el 1,670m. See Fig. 23.3-9. Five anti-shear transferring tunnels were dug at left bank, the force resisting rock connecting with the cushion, two of which was at the elevation of 1,730m, two at elevation of 1,785m and the fifth was at the elevation of 1,829m; these 5 anti-shear transferring tunnels are connected to each layer of replacement tunnel and internal construction accesses.

The horizontal distance between the inclined shafts is 30 - 35m; the height of the f_5 inclined shaft horizontal section is 15m, while the height of the X inclined replacing shaft horizontal section is 12m; width of horizontal section is based on the width of the fault, lamprophyre vein and fracture zone; height of replacing tunnel is 10m, width was determined by the width of fault and lamprophyre vein; section of anti-shear transferring tunnel was 9m×12m (width×height). Consolidation grouting was mainly carried out at left bank force resisting rock for deep cracks and lower wave speed IV and III$_2$.

For rock masses in a certain range of arch end and cushion, the vertical scope of consolidation grouting was between 1,635 and 1,885m; plane scope of mountain side was inside the lamprophyre vein of 5 - 10m, with VI$_2$ rock masses as the boundary; plane scope of river side was the connecting line along the dam or cushion foundation; consolidation grouting hole base to downstream was almost parallel to slope; depth of consolidation grouting holes on foundation base was 25m.

According to calculation analysis on arch end displacement, extension length from above the elevation of 1,820m to down-

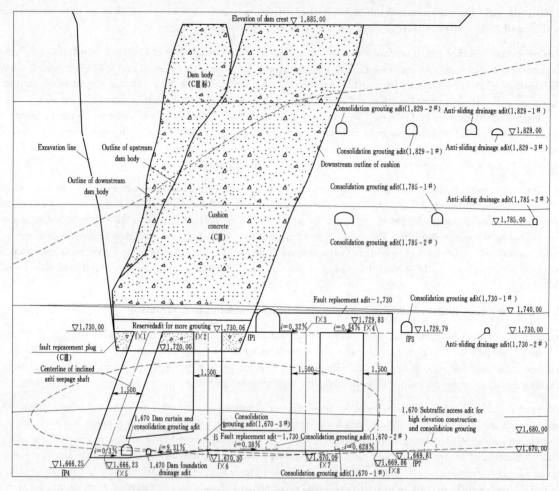

Fig. 23.3-8 JinPing-1 f₅ Fault Replacement Grid Section (Elevation Unit: m; Size Unit: cm)

stream was more than 3 times the arch end thickness; below the 1,820m it was around 2.5 – 2.7 times the arch end thickness, and below the elevation of 1,670m it was 1.5 times the arch end thickness. An arrangement of left bank force resisting consolidation grouting was designed, and combined with each replacing tunnel, grout curtain tunnel, anti-shear transferring tunnel and construction access tunnel; also at the various elevation of 1,670m, 1,730m, 1,785m, 1,829m and 1,885m there are specially arranged consolidation grouting tunnels.

Consolidation grouting holes along the gallery central line, above the elevation of 1,785m spacing is 3m, and below the elevation of 1,785m is 4m; the holes inside the ring are taken downward, drilling angle in a 90° range as main control holes to cover the scope of the grouting body. Upward holes have a drilling angle range of 270° as auxiliary overlap and sealing leakage paths; grouting holes were arranged along the tunnel periphery radially, with the hole spacing at the bottom of 4m. Force resisting rock consolidation grouting had a control grouting area and main grouting area: control grouting area was of side 15m of arch dam or cushion foundation base, the other boundaries were 5 – 7.5m wide, required to be carried out before the main grouting area, with low pressure thick grout. Maximum grouting pressure of control grouting area was 1.5MPa, while max grouting pressure of main grouting area above the elevation of 1,785m was 3MPa, below was 5MPa.

23.3.5 *Arch Dam Structure Cracking Resistance*

In China, 300m high arch dams are located in the western areas; terrain and geological conditions of dam regions are complicated; in addition to huge loads due to high water head and thermal loads, there is a high intensity of earthquake action, and the risk of cracking is higher than in normal high arch dams.

Analysis on dam cracking and integral stability, overall safety evaluation using finite element analysis and geo-mechanical model, combined with arch dam geometric design, foundation treatment design and temperature control design, are all used to determine the weak areas in the arch dam structure liable to cracking. It is necessary to investigate anti-crack measures according to the research results.

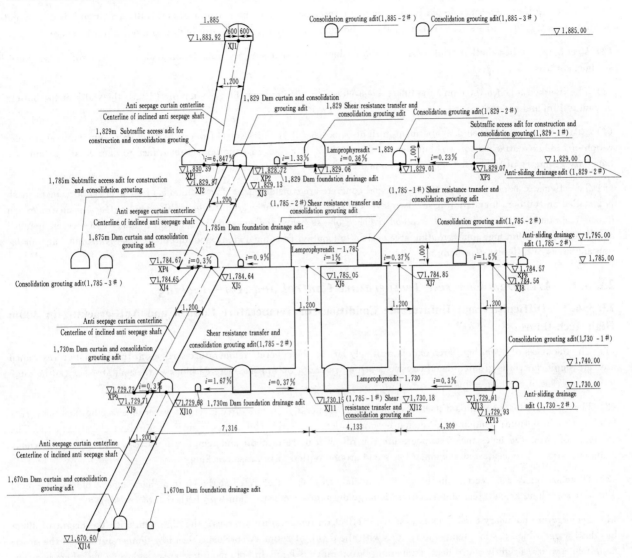

Fig. 23.3-9 Jinping I Lamprophyre Vein Replacement Grid Section (Elevation Unit: m; Size Unit: cm)

The terrain of the Jinping I dam region and hydro-power station, left and right banks are asymmetrical; below the elevation of 1,810m at the right bank is a 70°-90° steep cliff, while above it, it is a 35°-40° gentle slope; the gradient below the elevation of 1,820m is 50°-65°, gradient above is gentler and around 45°. Geological conditions of left bank and right bank are asymmetrical, middle and upper dam foundation above elevation of 1730m of the left bank is relaxation cracking rock mass, the structures comprise the f_5, f_8 faults and lamprophyre vein, most were III_2, IV_2 and even V rock mass. The lower dam foundations, near elevation of 1,660m have the f_2 fault and interlayer pressurizing zone. Deeply below the dam foundation at the right bank riverbed, there are green schist rock strata, the f_{13} fault through dam profile near the top arch and the f_{14} fault through dam profile near elevation of 1,770m. Downstream of the foundation base there are green schist lens inter-layered, NW-W orientated preferred fracture path, and almost S-N orientated steep inclined fracture, through f_{18} fault near elevation of 1,590m.

Measures applied to improve asymmetric geological conditions comprised: concrete cushion, hacking-out and concrete plugging, transferring tunnels, concrete replacement grid, force resisting rock mass unloading relaxation rock mass for consolidation grouting, etc. But these improvements were still of limited value.

Arch dam 3-D finite element value analysis and 3-D geo-mechanical model test results showed that arch dam deformation and stress distribution were still asymmetric, and connecting lines of arch dam downstream, max displacement and tensile stresses were all inclined along lines toward the left bank. Left and right bank dam foundations had some asymmetrical distributed geological defects, which might cause stress in the dam body, and concentrated local cracking. Therefore, after research on Jinping I arch dam, the main anti-crack measures were decided as follows:

(1) The strength of dam concrete partitions; there are three parts including partition A is $C_{180}40$, partition B is $C_{180}35$ and partition C is $C_{180}30$; large stress areas near the foundation base are concreted with high grade partition A concrete.

(2) steel bars installed the left bank cushion; the cushion upstream side and foundation rock connected zone is an unrestrained surface.

(3) The riverbed touched the dam's toe fillet; crest elevation of fillet is at the elevation of 1610m, the width at the bottom of dam is 10m and the top is 5m.

(4) At the upper part of arch dam abutment in right bank has fillet; fillet starts from right bank dam foundation at the elevation of 1,820m, attached to two abutment blocks of No. 25 and No. 26 from downstream face to dam crest, to improve boundary geometry of right bank upper dam foundation and stress of dam body.

(5) The effects of bond and slip between steel and concrete are considered. ϕ 32@300mm centres steel reinforcement mesh is installed on full dam foundation base; two layers of steel reinforcement are installed at the dam surfaces of upstream and downstream on foundation restrained area and riverbed blocks below the elevation of 1,630m; one layer of steel is installed at lift surfaces per 9m high joint grouting area at dam foundation restrained area, and anti-crack steel is installed at foundation and areas of different dam body materials.

23.3.6 Arch Dam Concrete Temperature Control and Anti-cracking

23.3.6.1 Difficulties and Unfavorable Conditions of Temperature Control and Anti-cracking in 300m High Arch Dams

(1) The thickness of 300m high arch dam is relatively large; the concrete temperature stresses at the foundation restrained area are significant; the grouted dam body has a strong restraint on upper concreted blocks; thermal stresses can be quite large, and risk of cracking is high.

(2) The structure of the arch dam itself is complicated; Generally, it has three layers of water discharging facilities: surface, deep and bottom; temporary diversion bottom outlets, galleries and access tunnels at different elevations, etc. The existence of large size holes makes temperature control even more difficult and temperature stress distribution extremely complex; stresses concentrated at thinner parts and around outlets may cause cracking.

(3) There are many dam sections in the steep slope and stress is concentrated at the bottom sharp angle, blocks concreted with a narrow long shape, temperature stress is large during dam construction with high risk of cracking.

(4) Temperature boundary conditions have a major effect on temperature stresses; the high dam with the height of 300m has deep water which has low temperature zone with the depth of 200m. Arch closure has low temperature, and the maximum temperature is tightly controlled, temperature drop range is large. In high mountain areas and canyons, there is a typical mountain climate, with big temperature differences, and frequent sudden temperature drops, so any careless concrete temperature control may cause cracking; winds, especially during the low temperature season, often occur, causing drying-out shrinkage cracking on concrete surfaces.

(5) Concreting the large surface area of a lift in thin layers, which means that temperature stresses may be large during concreting.

(6) Many areas have long intervals between concreting, due to dam foundation consolidation grouting and installation of deep and bottom outlet steel liners; and the time interval is generally longer than the design requirement, so concrete insulation and curing are difficult; risk of cracking is high.

(7) Concrete materials and their performance have a large effect on temperature stress and temperature control for anti-cracking measures, however, the selection of materials is often limited, such as local geological conditions, proximity of cement manufacturers, etc.

23.3.6.2 Concrete Temperature Control and Anti-crack Measures in Arch Dams

Due to difficulties and unfavorable conditions of 300m high arch dam temperature, control of anti-cracking, and the requirements of temperature controlling standard and measures are higher. Recently, concrete temperature control standards of the three arch dams with the height of 300m in process, are all much higher than the Code, and strict temperature control and anti-crack measures are applied. Below, taking Jinping I hydropower station as an example, the chief concrete temperature control standards and measures for current 300m high arch dams are described.

(1) Joint closure temperature required at Jinping I arch dam is 12–15℃; temperature control of the whole dam was de-

signed based on a restrained area; concrete temperature difference of each area is $\Delta T \leqslant 14℃$, peak temperature is controlled between 26℃ and 29℃, related placement temperature is specified at 11℃.

(2) Lift concreting thickness of riverbed blocks at the bottom within the height of 0.4L (where L is length of dam footprint) is 1.50m, the height of remaining lifts was generally 3m, however, after site trials and temperature stress simulation analyses, subsequent lift thickness was increased to 4.5m. Time interval between successive lifts for lift of 3m is 5 – 14 days, 4.5m height lifts is 7 – 14 days.

(3) Cooling water pipe layout spacing in partition A concrete of the arch dam foundations, and dam body outlet areas is 1m × 1.5m; in other areas, spacing is 1.5m × 1.5m; spacing at abutment blocks of 4.5m height lifts in the foundation restrained area is 1m × 1.2m; for local $C_{90}50$ grade 2 concrete areas spacing is 0.8m – 1m × 1m.

(4) Concrete cooling has 3 stages: stage Ⅰ, intermediate stage and stage Ⅱ; target temperatures of each cooling stage are 21 – 23℃, 18℃, 12 – 15℃ respectively; cooling durations should not be less than 20 days, 28 days and 42 days, respectively; cooling amplitudes should not be greater than 6℃, 5℃ and 6℃, respectively; cooling rates are 0.5℃/d, 0.3℃/d and 0.3℃/day respectively; there are two sets of cooling water: water temperature of the first is 14 – 16℃, mainly applied to stage Ⅰ and intermediate stage cooling, while water temperature of the second set is 8 – 10℃, mainly applied to stage Ⅱ post cooling.

(5) Annual heat preservation and the maintenance of flowing water were used the insulation and spray is applied to reduce the temperature in the dam storehouse when the air temperature at the lift is higher than 20℃ or the sun's rays are shining directly; cooling with mister units during placement; the full lift surface is covered with insulation mats as soon as the concreting has finished; 50mm extruded polystyrene foam plates are attached on all vertical surfaces after removal of the forms, either till the next lift preparation (side forms), or water storage (u/s and d/s faces) for generating power; equivalent heat exchange coefficient of insulation mats and polystyrene panels is $\beta \leqslant 10kJ/(m^2 \cdot h \cdot ℃)$; curing using running water or garden sprinkles starts as soon as the concrete in the lift surface has attained its final set.

(6) Temperature gradient controls were carried out based on grouting field, same cool field, transient field and covering field; the height of same cool field was controlled at 0.2 – 0.3 of grouting field dam body thickness. Cooling amplitude and rate were controlled strictly according to requirements of each stage.

(7) Both Jin Ping-1 and Xiluodu projects used concrete with PVA fiber material at the dam foundation and consolidation grouting layers; it also was used at foundation restrained areas and surrounds of outlets in order to increase anti-crack performance of the concrete.

(8) Personalized refining and dynamic management were carried out for temperature control; both Jin Ping-1 and Xiluodu projects utilized informational management to build personalized temperature files of each lift concreted, recorded whole temperature control process of each lift concrete, and applied other measures according to the actual situation.

23.3.7 *Arch Dam Discharge Structure Arrangement and Atomizing Improvements*

23.3.7.1 Arch Dam Discharge Structure Arrangement

Considering flood discharge and energy dissipation; 300m high arch dams have the characteristic of narrow valley, high water head, big stream flows, therefore the typical layout of flood discharge and energy dissipation on dam body is via surface outlets, deep outlets, bottom emptying outlets, diversion bottom outlets, etc. Four level discharge structures are common, with trajectory energy dissipation into downstream plunge pools; the layout of bank slope spillway tunnels is normally *tail inclined to downward*, pressure tunnels give way to non-pressure ones, with trajectory energy dissipation.

The designed discharge capacity of Jin Ping-1 project is 13,308m³/s, (checked at 13,913m³/s), with a maximum discharge water head of 221.68m; discharge power is 33,456MW.

Discharge structures on the dam body are 4 surface outlets, 5 deep outlets and 2 bottom outlet, with 1 spillway tunnel at the right bank. Discharge distribution of each structure, which isthe surface outlets and spillway tunnel, are both capable of discharging the perennial flood independently. This is with ample spare capacity, in order to increase the operational flexibility and safety of discharge facilities, that is any set of discharge facility combined with the powerhouse generator units flowing can discharge the perennial flood. The 4 surface outlet dimensions are 11m × 12m $(W × H)$; 5 deep outlets have dimensions 5m × 6m. Total discharge capacity of surface outlets and deep outlets is 9,068 – 10,074m³/s from the designed flow discharge to the checked; dimensions of the 2 emptying bottom outlets are 5m × 6m.

The mode of right bank spillway is pressure tunnel connecting to non-pressure tunnel, tail inclined downward; bottom elevation of spillway intake is at the elevation of 1,830; emergency gate dimensions are 12m × 15m, outlet dimensions of radial working gates are 13m×10.5m, with trajectory energy dissipation at outlet. Spillway tunnel discharge capacity is 3,229 - 3,320m³/s from the designed flow discharge to the checked.

23.3.7.2 Latest Development of Trajectory Energy Dissipation Forms and Discharge Atomizing

The arch dams with the height of 300m are normally constructed in high mountains and canyon terrain, and trajectory energy dissipation is one of the main energy dissipation forms. Ertan arch dam utilized surface outlets and middle outlets, with stratified flow and jet flow, respectively; the *jets impact in air*, and thus reduce the impact force on the plunge pool by jet flow; it is an effective form of energy dissipation. Trajectory energy dissipation of jet flows from high arch dams was applied in Goupitan, Xiaowan and Xiluodu arch dams for dam body discharge and energy dissipation. However the resulting severe atomization from airborne collision, has a significant effect on the slopes safety for the dam.

Flood water energy dissipation of Jinping I arch dam, in both the feasibility study stage and tender design report, applied the same method as Ertan arch dam; trajectory energy dissipation form from surface and deep outlets was outlet diffused shape; the jets impacted in air to dissipate energy but, when tested in a hydraulics model, the atomizing rainfall intensity was huge. Both the left and right bank slopes of Jin Ping I plunge pool are steep; unloading relaxation was severe, so strong atomizing rainfall would make for a potential safety hazard of the slopes' stability.

Even during dam construction, several optimizing tests were being investigated, and the form of surface outlets was changed to narrow channel energy dissipation, creating a new form of dam body trajectory. Thus energy dissipation is now: surface and deep outlets stratified flow and jet flow, no airborne collision, plunge pool separate energy dissipation, resulting in greatly reduced intensity of atomizing rainfall.

From Table 23.3-5 the optimized construction plan of left bank atomizing rainfall intensity is 19.4%-32.0% of that in the tender planning.

Table 23.3-5 Jinping I Surface Outlets, Deep Outlets Different Geometric Plans of Discharge Atomizing Maximum Rainfall Intensity

Design plan	Right bank slope maximum rainfall intensity at the right bank slope of model				Maximum rainfall intensity at the left bank slope of model			
	Chainage (m)	Elevation (m)	Checked (mm/12h)	Design (mm/12h)	Chainage (m)	Elevation (m)	Checked (mm/12h)	Design (mm/12h)
Tender plan	0+170.1	1677.53	390.95	211.32	0+170.1	1677.53	1653.39	972.58
Optimized plan	0+170.1	1677.53	125.28	61.40	0+135.3	1677.53	321.35	273.26

Jet flow of spillway tunnel outlet is normally applied to an oblique cutting, inverted bucket or narrow channel form of energy dissipation. Due to high water level andnarrow riverbed, any oblique cutting or narrow channel flip bucket form at the spillway tunnel would have a major influence on the riverbed downstream or opposite bank. Through trial and error, a new flip bucket form with dovetail was designed, which greatly reduces scouring in the downstream riverbed and opposite banks and water flow conditions downstream have also improved.

References

[1] Chengdu Hydroelectric Investigation & Design Institute of CHECC. Jinping I hydro-power station dam project design adjustment report, Yalong River, Sichuan (draft for review), 2010.
[2] Prevention of geological hazard and geological environment protection national special laboratory of Chengdu University of technology. High slope dangerous rock mass investigation, risk elimination and countermeasure research for Jinping 1 hydro-power station, Yalong River, Sichuan, 2005.
[3] Chengdu Hydroelectric Investigation & Design Institute of CHECC. Si Chuan, Yalong River Jinping I hydro-power station dam project design adjustment report special subject 16 hydro junction area special research of hazard analysis and protective measures for natural slope dangerous rock mass, 2007.
[4] Chengdu Hydroelectric Investigation & Design Institute of CHECC. Sichuan, Yalong River Jinping I hydropower station dam project bidding design report.
[5] Chengdu Hydroelectric Investigation & Design Institute of CHECC. Yunnan, Nanchang River, Xiaowan hydropower complex project impoundment safety appraisal 1st stage project design self-inspection report (6th album) Hydraulic structure design, 2008.

[6] Chengdu Hydroelectric Investigation & Design Institute of CHECC. Si Chuan, Yalong River, Jinping I hydro-power station dam project design adjustment report, special subject 7 arch dam concrete temperature control design report, 2010.
[7] Chengdu Hydroelectric Investigation & Design Institute of CHECC. Si Chuan, Yalong River, Jinping I hydro-power station dam project design adjustment report, special subject 8 dam body flood discharge and energy dissipation structure deepening design special report, 2010.

Chapter 24

Dams in the World Constructed by China

Wang Ruihua[1], Liang Jian[2]

24.1 The International Brand of Dams Built by China

Electric power industry, as a basic energy industry, is a priority in the development of national economy, and hydropower is a clean and economical way of power generation with large development and production scales. Dam is one of the core areas of hydropower development. Besides, many developing counties with rich water resources are in strong demand for hydropower development and dam construction. Considering such situation, China made the "going out" policy and took efforts in developing overseas markets. Dams built by China showed that China had lead the world in technology of water conservancy and hydropower construction.

Hydropower design, construction technology and equipment manufacture in China has been developing rapidly since 2000 and China has more and more influence and a bigger advantage in the hydropower industry of the world. It is a very good opportunity for China's "going out". In these years, China has made a rapid growth in international market based on its advantages in design, construction, equipment manufacture and financing. Great achievements have been made in the larger international market, bigger construction projects, better project management, and diversified participation patterns. Except traditional EPC contracts, enterprises actively explored and applied high-end and differentiated modes, such as EPC, BOT, BOOT, F+EPC, and broke the monopolization of large western contracting companies. This is a leap for "going out", and China's international brand "SINOHYDRO" is established.

Sinohydro Group Ltd (SINOHYDRO), China Gezhouba Group Corporation (CGGC) and China International Water & Electric Corp. (CWE), already successfully constructed some large or super-large projects, and become famous internationally in the field of hydropower construction. Sinohydro alone takes up 50% of the international market share, and "SINOHYDRO" becomes the first brand and representative of China's hydropower construction industry.

On August 30, 2010, America Engineering News-Record (ENR) issued the list of 2010 top 225 international contractors according to international business income. SINOHYDRO, CGGC and CWE ranked 41st, 84th and 125th respectively in the list.

Before the 21st century, China's international market mainly stayed in Asia and Africa. The situation changed a lot since 2000. Chinese enterprises made breakthroughs in contracting and constructing large and super-large projects and in developing international regional markets. Examples are Thailand Kelongtaidan project, Malaysia Bakun hydropower station (2,400MW), the longest earth-rock dam in the world—Sudan Merowe hydropower station (1,250MW), Ethiopia Tekeze Hydropower Project (300MW), Iran Taleghan water control project, Ecuador Coca-Codo-Sinclair (CCS) hydroelectric project (1,500MW), etc. They also successfully contracted some large or medium hydropower stations in Middle East and Central Asia, such as Iran Rudbar Lorestan Dam (450MW), Mollasadra hydropower station (100MW), Turkey Feke II hydropower station (70MW), Kazakhstan Mayina hydropower station (300MW). What's important is the breakthrough in Latin American market. On October 5, 2009, Sinohydro signed the EPC contract with Ecuador for the CCS project with an installed capacity of 1500MW and a contract value of US $ 2 billion. The project started in July 2010. And in August 2010, CGGC won the tender of Ecuador Sopladora hydropower station for US $ 672 million.

Unremitting endeavor of China hydropower organizations makes foreign customers, especially foreign governments get to know China's construction and contracting strength in hydropower stations. They also see China's advantages in design, consultation, equipment manufacture and installation, river basin planning, development, investment, management and operation. "SINOHYDRO" has become a famous brand in the field of international engineering construction.

---

[1] Wang Ruihua, Senior Executive of International Engineering, China Power Construction Corporation Ltd., Senior Engineer.

[2] Liang Jian, Former Vice President of China International Water & Electric Corporation, Professor Senior Engineer.

24.2 Earth-rock Dams and Concrete Face Rockfill Dams (CFRD)

24.2.1 *Merowe Dam in Sudan*

Merowe Dam in Sudan is located in the city of Merowe, 450km north of Khartoum, capital of Sudan. It is the second large-scale hydropower station built on the main stream of Nile after Aswan Dam in Egypt. This project was invested by Sudan Ministry of Irrigation and Water Resources, with Sudan Dams Implementation Unit as the owner and Germany Lahmeyer International GmbH as the project supervisor.

Merowe Dam is a key water control project for power generation and irrigation. It is 9.80km long with a maximum height of 67m, and the main structures included the dam and the plant. Main dam of Merowe is a combination of three types—concrete face rockfill dam, clay core wall dam, and concrete gravity dam. The total reservoir capacity is 14 billion m³. Ten power units of 125MW each provide a total installed capacity of 1,250MW, which exceeds two times of existed capacity in Sudan, and is named the "Three Gorges" of Sudan (Fig. 24.2-1 and Fig. 24.2-2).

Fig. 24.2-1 Overlook of Merowe Dam

Fig. 24.2-2 Merowe Dam and Spillway

Sinohydro and China International Water & Electric Corp. Won the tender for civil work and metal construction installation (2A, 2B, 2C and 3D), and they entrusted a joint venture of Bureau No. 7 and Bureau No. 5 to carry out the specific construction. The contract value is about 600 million Euros, and contracted commence date was on June 15, 2003 and completion date on May 31, 2009.

Merowe Dam is very large in scale with several dam types and construction technology and condition are complicated. Flood control is crucial in this project, and technology of diversion scheme is highly required. According to the design, in phase I, construction is carried out in sections and longitudinal cofferdam is applied for diversion; and in phase II, diversion is realized through spillover dam preset gullets which are plugged gradually.

The main part of Merowe Dam was completed on May 15, 2009, and the whole project entered the final stage. On the world renowned Nile, a dragon-like 10km dam sprung up, and a lake of 14 billion m^3 appeared in the desert, which is really astonishing. By the end of 2009, accumulated workload was: earth-rock excavation 9.5 million m^3, earth-rock filling 17 million m^3, concrete placement 1.9 million m^3, reinforcing steel fabrication and installation 117,000 tons, metal structure fabrication and installation 18,000t, totaling a settlement value of Euros 790 million.

Sudan is a country with relatively backward infrastructures, and the completion of Merowe Dam made Sudan a country without national power gird before having a national backbone power gird. Gravity irrigation was formed in a scope of 400km at the downstream, and water supply for production and life use of over four million people was solved. Meanwhile, the generated electricity was also enough to build a series of lift irrigation stations at the upstream, which greatly improved Sudan's agriculture distribution. Generally speaking, construction of Merowe would promote Sudan's economy development greatly, change its economic structure fundamentally, and lay a solid foundation for its economic take-off.

During the construction, Chinese staff in field of hydropower made great efforts and achieved successes in construction progress, project quality, construction safety and civilization, environment protection, technology optimization, etc. They got approval and praises from high level leaders and the Dam Implementation Unity (DIU). Merowe is also another monument of friendship between China and Sudan, and a landmark project in foreign countries of Chinese hydropower construction. Construction of Merowe laid a solid foundation for further exploration of Sudan market and even African market.

24.2.2 Bakun Concrete Face Rockfill Dam in Malaysia

Bakun Hydropower Station is located on the Balui River in the middle of Sarawak, Malaysia, 37km from Belaga at the downstream and 180km from the booming port city Bintulu in East Malaysia. Bakun dam is 205m high, the second highest CFRD in the world, and its installed capacity is 2,400MW, granted output is 1,771MW, annual energy output is 15.517 billion kW·h, and is called the "Three Gorges" of Southeast Asia. The general contractor of the civil work CW2 (EPC) is the Sino-Malaysia joint venture (MCH-JV), made up of Sinohydro and six Malaysian companies. Sinohydro takes up 30% of the share as the second largest shareholder and did the specific design and construction work.

The average annual flow at the dam is 1,249m^3/s, reservoir capacity at normal impounded level EL228m is 43.8 billion m^3, and the controlled drainage area is 14,750km^2. The main work includes concrete face rockfill dam with the height of 205m, spillway with four radial gates, each gate height 20m, chute width 50m, length about 610m, and a flip bucket at the end, intake tower with eight bulkhead gates, length 168m, width 58.5m and height 67.7m, eight concrete lining and steel lining diversion tunnels, inner diameters 8.5m and 7m respectively, ground plant stretching for 325m, including eight generator units, each 300MW. three concrete lining diversion tunnels, each 1,400m in length, 12m in inner diameter, one of which was changed into discharge tunnel and for the other two bulkhead sealing was applied.

Bakun Hydropower Station (Fig. 24.2-3) is constructed in a tropical rainy climate, where the annual precipitation is 4,600mm. Under such severe condition, a maximum monthly excavation of 2.08 million m^3 was realized, and in a succession of 11 months, the excavation exceeded 1.5 million m^3. Meanwhile, a maximum monthly filling of 860,000m^3 was realized and a high record of over 600,000m^3 in a succession of 12 months was created.

Bakun project caught high attention in the field of dam construction in the world. Its owner hired a special consulting team of world class experts from America, Brazil, Norway, etc. They made inspection and assessment every half year, and gave high appraise.

The dam of Bakun Hydropower Station is mainly made up of concrete face, cushion (2A & 2B), transition material (3A), main rockfill (3B), secondary rockfill (3C), and downstream rockfill (3D). At the downstream, big rock blocks were used to protect the slopes, while at the upstream, a certain width of rolled silt (cohesiveless soil) was used to cover the middle and lower part of the concrete face and put another layer of any material on the cohesiveless soil layer. Its upstream slope is 1:1.4 and virtual downstream slope is 1:1.3, dam crest is 12m wide and 748.85m long. Total filling is

Fig. 24. 2 - 3　Full-view of Bakun Dam

16.72 million m³, and face area is 130,000m².

Foundation excavation of Bakun dam started in June 2003. Dam body filling started on June 18, 2004, and was filled to the crest on May 31, 2007. Main part filling lasted 35.5 months, and total filling is 16.72 million m³, with an average monthly filling of 471,000m³. And in a succession of nine months from December 2005 to August 2006, monthly filling exceeded 700,000m³ with a maximum monthly filling of 864,000m³.

Filing of Bakun dam was conducted in five phases according to the demands of construction strength, settlement control, concrete face construction phases and flood protection. Phase I is filling of the part below El. 70.00m at the downstream. Phase II is filing the dam to El. 90.00m at the downstream and 125.60m at the upstream, 5m over the designed face elevation of phase I. Phase III is filling the dam to El. 229.00 - El. 230.40m. At the beginning of phase III, construction of the concrete face in phase I is going on. At the end of phase I, altitude difference between main rockfill area and secondary rockfill area is reduced gradually to the same elevation. When dam filling reached El. 190m, bedding face flat layer is filled to the crest. Phase IV is filling stock ground before the dam, and phase V is filling inner side of wave wall to the crest.

Filling of the concrete face is conducted in two phases: phase I, to El. 121m, started in December, 2005 and finished in March, 2006; phase II, to crest, finished in July, 2008. Except some parts was not applicable, the whole concrete face was covered with sliding mode.

Usually, concrete face construction in phase I goes smoothly, but before filling in phase II, cracking happened on main concrete face rockfill dams in the world, especially sever on Campos Novos Dam in Brazil. This brought great concern from owner, consultant, and designer. It is decided to revise the design to prevent similar cracking after impoundment.

24.2.3　Taleghan Water Control Project in Iran

Two thousand years ago, ancient Chinese explored the "silk road" to the Western Regions, and brought there silk and friendship of Chinese people. Two thousand years later, Chinese hydropower goes out to West Asia and built a landmark—Taleghan water control project in Iran, which got renowned in the world for its economic benefit and high quality. It also won the 2009 NCS Reuben award.

Taleghan water control project is an EPC-T project, in which Sinohydro provided 85% of the financing through Export-Import Bank of China supplier credit and Iran Ministry of Energy Tehran Water Organization contributed the other 15%. The construction was carried out by Bureau No. 10. Construction duration for this project is 46 months, and the project mainly includes a 101m high clay core wall rockfill dam with a total filling of 15.3 million m³, a spillway with a maximum discharge quantity of 2,500m³/s and underground powerhouse with installed capacity of 18MW (Fig. 24.2 - 4).

In February, 2001, Sinohydro and the owner, Iran Ministry of Energy Tehran Water Organization, signed the master contract of the EPC/T project, contract value being about 143 million US dollars, and contract scope covering a package

Fig. 24.2-4 Full-view of Taleghan Dam

work of supplier credit financing, design, purchase, construction, installation, completion transfer and defects patching. The contract is settled by unit price of fixed sealed top price and was validated in January, 2002. The construction started on March 15, 2002 and was completed on August 31, 2006.

During the construction, as the workload of the clay core wall dam is too big in a short period of construction, management is complex, and filling is too difficult as a result, a set of advanced and reasonable management and motivation mechanisms was designed so as to fully mobilize the initiative of workers and managerial staff. Meanwhile, optimized measures were applied as many as possible. After 38 months, the project was finished and it was a new record in China's hydropower industry for projects of the same type with a maximum filling intensity of 1.05 million m³ per month.

24.2.4 Vau i Dejës Dam in Albania

Vau i Dejës hydropower station was named as "Mao Zedong Hydropower Station" (Fig. 24.2-5). It was a key project in Albania's fourth "Five-year Plan", and also the largest one at that time. It is located in the major economic zone in the north, Shkoder District, and it is the third cascade power station in the water resources development project on Drin River. The contractor was Ministry of Water Conservancy and Electric Power of China at that time. The major constructor was Bureau No. 12, and the designer was Beijing Survey and Design Institute at that time. The construction started in January 1967 and ended in June 1973.

Fig. 24.2-5 Full-view of Completed Vau i Dejës Hydropower Station

The dam is 60m high with a reservoir capacity of 567 million m³, a total installed capacity of 250MW. Five mixed flow hydroelectric generator units of 50MW were installed, and the average annual power generation at the beginning was 1.07 billion kW·h, and guaranteed output was 46MW. Fierza Hydropower Station at the upstream had an average annual power generation of 1.22 billion kW·h, and the guaranteed output was 110MW.

This key water control project includes: main dam (an earth-rock clay core wall dam), two auxiliary dams (crest width 9m, length 380m), maximum height of 60m for main dam, and 45m for auxiliary dams; concrete gravity dam connecting auxiliary dams for reservoir flood discharge and station diversion (height 52m); 11m × 11m door-like discharge tunnel (length 262m); power generation plants, tailrace, booster substation, etc. A hydropower station of 250,000kW was the biggest foreign aid project at that time. Albania's 30th anniversary of its Labor Party and the Sixth National Party Convention were coming in November 1971, and Albania wanted to have two generator units working at that time. To finish such a tough work during a short period of time, Chinese construction team had to start the survey, design and construction at the same time. On one hand, manufacturing of the five Francis hydro-turbine generator units were listed major task with priority to ensure the installation need abroad; on the other hand, China send out over 200 experts to help in investigation and research, scientific experiments, organization and coordination, and to give technical guidance, so as to speed up the construction with guaranteed quality.

Albanian people were quite satisfied with the qualified construction and the obvious economic benefits.

24.2.5 *Fierze Dam in Albania*

In order to solve the problem of self power supply, Albania was in urgent need of China's aid in building Fierze Dam, and insisted their power generation schedule. To meet their requirement, China paid a lot in organizing working staff, preparing resources, and solving high-end technical problems.

Fierze Hydropower Station was the second large-scale hydropower station China aided on Drin River after the Vau i Dejës hydropower station. It is a clay core wall rockfill dam with a height of 166m. At that time, China did not have any experience of building a dam of this kind over 150m in height. In order to solve the problems, a series of experiments were made in China: hydraulic model experiment for reservoir surge caused by landslide, model test of rockfill over-current, mixture and compaction test of admixture for clay core wall, flood discharge test of large scale tunnel. At that time, China's mechanical and electrical manufacturing level was not so high and properties of some products could not reach the standard. In order to provide highly qualified equipment for Albania, special fund was allocated for technical transformation tests of generator units.

Fierze Hydropower Station has an installed capacity of 500MW and an average annual power generation of 1.8 billion kW·h. Construction of the station started in November 1971, and the first generator began working in May 1978. After operation, the station not only met the electricity demand of the country, but also had surplus for sale to two neighboring countries, which became Albania's one of the most important sources of foreign currency earnings.

24.2.6 *Boukourdane Dam in Algeria*

China International Water & Electricity Corp. (CWE) has constructed a large amount of dam projects since the 1980s. Its works had good effect and won approval from owners and consulting engineers.

Boukourdane Dam is the first large-scale water control project for irritation and water supply in Algeria (Fig. 24.2-6). Its clay core wall rockfill dam is 74.4m high and the reservoir capacity is 120 million m^3. The designer of this project is BG Consulting Engineers of University of Lausanne, Swiss. As the prospecting work was not thoroughly enough, there appeared problems during the construction. After in-depth study in construction, design and scientific research, China proposed a scheme for amendment and got approval from the owner and the design company. This played an important role in the project construction. For example, in the initial design, a concrete cutoff wall will be built in the covering layer of river bed, and a grouting tunnel will be excavated in the weathered rock of dam foundation for upward grouting to connect with the dam foundation concrete cutoff wall; there will also be downward grouting to form an anti-seepage curtain. In fact, there existed sever defects in the design scheme. As rock mass of dam foundation was broken, tunnel excavation becomes very difficult; there existed potential safety hazards in construction of the concrete cutoff wall and excavation of the grouting tunnel; tunnel grouting can not ensure the wholeness of the anti-seepage curtain, and may cause leaking. Considering all these factors, China proposed open cut for dam foundation, advantages of which include: weathered basement can be completely eliminated, part quirk incision is easier to deal with, and safety hazards in tunnel excavation in weathered rock can be avoided. The new construction plan not only simplified the process, avoided excavation hazard, but also guaranteed the quality of dam foundation seepage prevention body and was good for dam safety.

24.2.7 *Imboulou Dam in Congo*

Imboulou hydropower project in Republic of Congo is located at the downstream of Laifeini River, branch of Congo River, 14km from confluence of Congo River, and 215km from capital Brazzaville.

Fig. 24.2-6 Full-view of Boukourdane Dam

Imboulou Hydropower Station is mainly for power generation, performing as the backbone power station, conducting peak load regulation and frequency modulation of the power supply system. Total reservoir capacity is 584 million m^3; four axis Kaplan turbines set in the powerhouse has a single-machine capacity of 30MW, total capacity is 120MW, and annual energy output is 685 million kW·h. The project is classified Grade I. Permanent buildings include dam, sluice gate, and station powerhouse, designed as Grade I building. Minor buildings include inlet diversion canals of station and sluice gate, energy-dissipating buildings at the sluice gate downstream, station tail water channel, etc, designed as Grade III buildings (Fig. 24.2-7).

Fig. 24.2-7 Full-view of Imboulou Dam

Imboulou Hydropower Station includes earth dam, sluice gate, river bed station at the left bank, and earth dam at the right bank. Position of left bank dam is the original river bed. And it is 287m long with crest width 7m, maximum height 32.5m, and a 3m wide packway 8m lower the crest at the downstream slope. Sluice gate is 37m long, the same elevation as the dam. River bed power station dam (including clipping room) is 128.4m long with four generators of single capacity 30MW. Right bank dam is 132.6m long with crest width 7m, and maximum height 32.5m. Both dams are soil texture anti-seepage body zoned dams. The project also includes approach road, house building of operating village, installation of electromechanical equipment, installation work of sluice gate and headstock gear. Commercial operation maturity is one year.

Major real work quantity: open cut of earth work 2.9067 million m^3, open cut of rock work 552,500m^3, earth-rock filling 1.4276 million m^3, concreting 259,800m^3, reinforcing steel fabrication and installation 9,837 tons, steel structure and sluice gate fabrication and installation 2,966 tons, installation of 4×30MW hydroelectric generating sets and installation of fire control and plumbing equipment, 57km approach road, 6,387m² house building of operating village, etc.

Capital source of this project is form loans of Export-Import Bank of China. Project owner: Republic of Congo Délégation Générale des Grands Travaux (DGGT). Project supervisor: Germany Feixitena Consulting Corporation.

The project is contracted by China National Machinery Imp. & Exp. Corp. (CMEC), and sub contractors will be decided through tendering in China. On October 29, 2004, CMEC and Sinohydro signed the subcontract, a price contract totaling RMB 939 million RMB (45,411,672 in US dollar and 563.82622 million in RMB), in Beijing. Contracted construction duration is from November 1, 2004 to April 30, 2009, a total of 54 months, one year's business guidance not included. By the end of August 2010, the civil work was basically finished with some equipment installation and fitment work left; 72 hours' test run of the last generator (1# generator) was over on August 14, and all generators were combined to the gird.

This project was designed according to China's national standard and the industrial standard, which means national regulations and standards as well as International Engineering Consortium (IEC) and relative international standards in technology management should be abided by. Key technologies for the construction are as following:

(1) According to the layout region of the project, the geological condition of dam foundation is infinite strong permeable soft rock, which may cause seepage deformation and influence construction safety and diversion foundation pit dewatering and drainage in phase I. To solve such problems, construction measures of foundation pit earth-rock open cut, foundation pit dewatering and drainage, and foundation grouting were adopted.

(2) Diversion in phase II is rinsed through built sluice gate bottom hole. The flow of Laifei Ni River is 550 – 700m³/s around the year. As the covering layer of river bed resist rushing, and flow rate is low. But flow rate at closure gap becomes high. Besides, there is no closure material around the project site. All of these factors bring difficulty to the closure of the river bed.

(3) There are three difficulties in earth dam filling: First, the project is near the equator in a tropical rainy climate and the rainy season lasts long with a large annual precipitation. Earth dam filling can only be constructed during June and August every year, and so effective construction period is too short. Second, because of the rainfall, nature moisture content of earth is higher that the optimum value. How to make tedding and bakeout to meet the need of moisture content is crucial in the construction. Third, after dam foundation excavation, basement will expose in the air for a long period of time. Long-term seepage of foundation pit may cause crevice corrosion and large amount of seepage water, which makes drainage and grouting at the places of seepage difficult.

(4) Strengthen technical management according to specific circumstances of Congo Imboulou project. There are a series of regulations, such as working drawings examination system, construction organization design (plan) formation and management system, examination and approval system, field operation instruction compilation and its supervision and examination system, technology approval and design alteration management system, technical disclosure and review system, and technology standardization system.

24.2.8 Jatigede Dam in Indonesia

Jatigede Dam (Fig. 24.2-8) was planned to be built on Cimanuk River, at about 25km of Rentang Barrage upstream. The project is in Cijuengjing village, Jatigede subdistrict, Sumedang District, West Java Province, 16km from Cirebon-Sumedang main stem, and about 75km from Cirebon. First purpose of the project is irrigation and then power generation. (In the beginning, power generation was the first purpose; later, the demand changed and irrigation became the first goal.) Jatigede reservoir controls a drainage area of 1,460km², and the dam is a clay core wall rockfill dam with a maximum height of 110m and a total reservoir capacity of 1.063 billion m³. At normal impoundment level of 260.00m, reservoir area is 39.53km², reservoir capacity is 796 million m³; at the dead water level of 230.00m, reservoir area is 12.70km², and reservoir capacity is 183 million m³. Installed capacity is 110MW, and average annul power generation has been 643 million kw·h for years.

It is a general contract. A joint venture (Sinohydro-CIC Jo) of Sinohydro and four Indonesian state-owned construction companies (CIC) was formed, and Sinohydro takes up 67% of the contract share, and the construction is carried out by Sinohydro Bureau No. 10. Co., Ltd. Construction duration: 65 months, date of commence being November 5, 2007.

Capital source: Export-Import Bank of China (EIBC) preferential export BC buyer credit. Total contract value: US $ 239.5 million. The project mainly includes: dam, spillway, irrigation tunnel, water diversion system and powerhouse.

The project was developed in two phases: phase I includes dam, spillway, irrigation tunnel, inlet of diversion tunnel (to the 150m tunnel at the side of the downstream gate shaft); phase II includes tunnel, upstream surge-chamber, pressure

Fig. 24.2-8　Full-view of Jatigede Dam

pipeline, plants, tail water surge-chamber, tail water tunnel, etc. of the water power generation system.

Jatigede Dam is a clay corn wall rockfill dam 1,715m long, crest width 12m, crest elevation 265.00m, dam height 110m, and total filling amount about 6.063 million m^3.

Diversion tunnel is 732.5m in length, and 10m in diameter.

Most of the spillway structure is on the dam, total concrete volume about 433,800m^3, four gates of 13m (width) × 14.5m (height) installed on weir crest.

Water intake tunnel for irrigation is about 372m long, inner diameter 4.5m, including intake structure, access door, cleaning door, etc..

Power generation system: first 150m of the diversion tunnel is 4.5m in diameter, including water inlet and gate shaft.

On April 30, 2007, the signing ceremony of Indonesia Jatigede Dam Project was held in the auditorium of Indonesia Public Works Ministry.

24.2.9　Bougous Dam in Algeria

Bougous Dam (Fig. 24.2-9) is located in the east of Algeria, on the dried up Bougous River, 20km east of Willaya El Taref neighboring Tunisia. Main parts of the project in the original contract included a series of hydraulic structures—clay core wall rockfill dam 71.4m high and 636m long, spillway, grouting gallery, intake tower, diversion corridor, etc. The original contracted construction duration was 34 months, and the contract value was 1,982,963,067 dinars and US $31,370,610 (totally US $59.22 million, converted according to contracted exchange rate 1 US $ = 71.205 dinars). The dam owner, Agence Nationale des Barrages et Transferts (ANBT) announced commencement on January 25, 2005.

Sinohydro won the tender of this project and entrusted Sinohydro Bureau No. 16 Co., Ltd. to carry out the construction. It is the first large-scale earth-rock dam Sinohydro constructed independently in Algeria.

All things are difficult before they are easy. As the first dam project Sinohydro carried out in Algeria, the construction of Bougous Dam witnessed a lot of efforts. At the beginning, there were difficulties of all kinds. However, after experiencing a tough process, the whole staff gradually explored a way out and finally got fully adapted and blended in. On January 24, 2006, the first group of working staff entered the site, and later on July 1, 2005, excavation commenced. Concreting started on January 25, 2006, dam filling started on May 20, 2007, reservoir impoundment started on February 23, 2010, and filling of the main part finished on April 8, 2010. During the construction, Director of ANBT visited and inspected the site several times, and Algeria Minister of Water Resources also paid an inspection visit to Bougous Dam on December 3, 2009. Both of the officials expressed their satisfaction for the construction progress and quality.

On the completion, contract value of Bougous Dam added to US $79.02 million from the original US $59.22 million, excavated volume reached 1.78 million m^3, filling volume reached 4.3 million m^3, reinforcing steel fabrication and installation reached 5,700 tons, and concreting reached 140,000m^3.

Fig. 24. 2 - 9   Full-view of Bougous Dam

Completion of Bougous Dam in Algeria will effectively relieve the water shortage in the surrounding area. The impoundage of 66 million m³ at the normal water level of 139m can meet the demand of drinking water and industrial water with a daily water supply of 300,000m³. There is also a surplus to supply water for Mexenna Reservoir at the downstream.

24. 2. 10   Algeria's Draa Diss Dam

Algeria DRAA DISS Dam (Fig. 24. 2 - 10) locates in Tachouda town, east of EL EULMA City, SETIF province which is in eastern Algeria. The dam is 11. 5km away from the urban area. The project is part of the eastern water supply system. It transfers about 189 million m³ water from the north to the south each year for irrigation of 31,700hm² of land. It also provides EL EULMA with industrial water and drinking water.

Fig. 24. 2 - 10   Draa Diss Dam under Construction

The water-retaining structure consists of a main dam and four auxiliary dam. The main dam is a clay core rockfill dam, crest length of which is 956m; crest width is 10m and a maximum height of 76m. Four auxiliary dams are alluvial material filling homogeneous dam. The four auxiliary dams are at lower altitude, locating on four passes on the left bank of the reservoir. The right bank abutment is covered with clay blanket for seepage control. The clay blanket is of 288,000m², and the blanket thickness of 6.0m.

The project is mainly composed of the retaining structure, the water intake structure, and the emptying tunnel and transport facilities.

The diversion, water pipe, and the emptying tunnel are arranged in a same tunnel, the length of which is 429.17m. On the upstream of the tunnel lies the underwater-style water chamber, the stilling basin and tailrace lie on the downstream. An irrigation pipe with a diameter of 1.6m and a water supply pipe with a diameter of 0.6m are arranged under the tunnel. Above the tunnel is a horseshoe-shaped emptying tunnel, with the height of 3.0m, diameter 4.0m, and bottom width of 3.6m. The tunnel excavation length is 384.71m, the excavation section is horseshoe-shaped, and is with the height of 5.35m, diameter of 4.7m, and bottom width of 3.0m. The tunnel also bears the diversion task during the construction period.

The lower part of the water chamber is cylindrical, with the inner diameter of 15m, the outside diameter of 17m. The top part is a spherical cap, with the inner radius of 7.5m, thickness of 1.0m. The maximum height is 22.05m. In the water chamber, there are intake pipes and control gates for water supply, irrigation and reserved flow. The project has a 162.8m long horizontal transportation corridor and a 50.6m deep shaft as the transport corridors of the water chamber. A control room is arranged on the upper part of the shaft.

The project's contract engineering quantity is: earth rock excavation of 676,500m³, earth rock backfill of 4.9484 million m³, geotextile lying of 288,000m², concrete pouring of 29,100m³ and borehole grouting of 104,000m.

The Algeria Dam Division signed the DRAA DISS Dam contract on May 6, 2008, and issued the Notice to commence on May 27, 2008. The project contract period is 34 months. The project is shared between Sinohydro Bureau 13 Co., Ltd. and Sinohydro Bureau 3 Co., Ltd, and Sinohydro Bureau 13 Co., Ltd is the responsible party for the project.

Since the commencement of the project, we built the living area and office buildings according to local architectural style. We completed the construction of the gradating material processing system using the height difference of 50m of the site terrain, which saved usage of steel structure, reduced power consumption so as to reduce cost for the project.

Local rock developed rock joints, layered thin, and sandwiched with layers of clay, all of which brings great difficulties to the mining and processing of gradating materials. Facing the reality, we organized experts to study the above problems, and finally decided to install bar screen in the mining area to simplify the process. This accumulated a wealth of experience for us on gradating material mining and processing in the case of poor source material.

24.2.11 Botswana Dikgatlhong Dam

Dikgatlhong dam (Fig. 24.2-11 and Fig. 24.2-12) is located on the northeast of Botswana, 3km on the upstream of the confluence reaches of the Border River with the neighboring country Zimbabwe, Shashe River and Ramokgwebana River. It's 450km away from Gaborone. The project is the water source project for the country's North-South Water Diversion Project, is also the country's largest water conservancy project. Its design capacity is about 400 million cubic meters. The project investment is about US $200,000,000. It will meet the country's growing mining industry and domestic water needs after the project is completed. It will supply water for the capital, other major cities and major mining area.

Fig. 24.2-11 Filling Construction for the Downstream of the Riverbed Dam Section Core Wall

Fig. 24.2-12 Construction of Apron for the Spillway Diversion of the Dikgatlhong Dam

Dikgatlhong dam is constructed by Sinohydro Bureau 11 Co. , Ltd. The project started construction on March 9,2008, started water storage on October 1,2011, and is completed on February 8,2012. The contract period is 47 months. The scope of the project include: a 49m high, 4.6km long clay core rockfill dam; a 200m wide spillway with open concrete lining; a 1000m wide, non-lining auxiliary spillway; a 50m high cylindrical reinforced concrete intake tower; a 280m long drain pipe with 3m diameter and buildings for its exports flow control; 43m gravel approach road, among which 33km is adapted from existing roads, 10km is newly built.

The sewage treatment has been impeding the progress of the project. Because of different cultural ideas, the owner insisted that the living sewage should be transported to 60km away, which will raise the cost and is very difficult to accomplish. We actively discussed with the owner and the engineers, continuous optimized the sewage treatment measures. On the other hand, we managed to contact the Minister of Water Resources hoping to get some government support. After one and a half year's negotiation, a result to every party's satisfaction is finally reached. Through the construction of septic tanks, adaption of infiltration pond, construction of wetlands and other measures, the owners are finally satisfied with the sewage treatment and we also reduced the cost.

Through careful analysis and accurate grasp on the terms of the contract, we managed to protect our own interest. The excavation of Dikgatlhong Dam is graded in two levels: class 1 and class 2, which is equivalent to China's earth and rock classification. On the spillway excavation grading, we have a lot of differences with the engineers. The engineers tended to use the actual limit excavation depth of backhoe excavation or bulldozer's limit plow depth as the dividing line, which is a very harsh condition, due to their inability to quantify what is the "limit level". We insisted that use whether single bulldozers plow depth is less than 30cm as the dividing line. After tough negotiation, the two parties finally decided to use the BOQ proportion in the contract, namely class1/class2 is 32.7%/67.3% respectively.

24.2.12 Laos Nam Lik 1-2 Hydropower Project

Laos Nam Lik 1-2 Hydropower Project (Fig. 24.2 - 13) is constructed by Sinohydro. It's a BOT project invested by China International Water & Electric Corp. ; it's also the first BOT project signed by Laos' government with a foreign company. The project is attached great importance by the two governments. The main part of the project is completed ahead of schedule, which greatly improved the competitive advantage of Sinohydro, and has a great significance on the continued sustainable development in the Lao market.

Laos Nam Lik 1-2 Hydropower Project started construction on September 1, 2007, the main part of the project was completed on May 31, 2010, and the project began commercial operation from September 1, 2010. 25 years after the initial commercial operation, the project will be transferred to the Lao government.

Laos Nam Lik 1-2 Hydropower Project is a concrete face rockfill dam, with auxiliary dam, spillway hole, discharge/diversion tunnel and power generation building. The height of the main dam is 103m, its capacity 1.359 billion m³. It's a carryover storage reservoir. The installed capacity is 100,000 kW, and annual average generating capacity is 435 million kW·h. The engineering quantity of the main part is: earth rock excavation of 2.93 million m³, earth rock backfill of 2.1 million

Fig. 24.2 - 13 Upstream View of the Laos Nam Lik 1-2 Hydropower Project

m³, concrete pouring of 250,000m³, reinforcing bar creation and install of 16,000 t, anchor root of 290 million, curtain grouting and consolidation grouting of 46,000m, backfill grouting of 20,000m², metal structure fabrication and installation of 3,265 t.

The implementation procedures for overseas investment projects are very complicated. It can be divided into three periods: project preparation period, construction period and commercial operation period. In which the project preparation is essential, it is fundamental to the success of investment projects. During the preparation period, all the contract negotiations and signing should be completed, all approval documents should be obtained, the project financing and project insurance should be completed.

24.3 Concrete Dam

24.3.1 *Khlong Thadarn Dam Project in Thailand*

On October 28, 1999, the CCVK joint venture which is formed by Sinohydro and other companies officially signed the Khlong Thadarn Dam Project construction contract with Thailand Royal Irrigation Hall (RID) in Bangkok, Thailand. The total investment of the project is 8.27 billion baht (equivalent to about US $ 300 million); the project period is five years.

Khlong Thadarn Dam project (Fig. 24.3 - 1) is located on Thadarn River, Nakhon Nayok House, to the northeast of Bangkok. It's about 150km from Bangkok. The project is large scale project which is designed as flood control, irrigation and tourism. The total reservoir capacity is about 225 million m^3. The project mainly include: the dam, irrigation works, road works, electrical works and camps.

Fig. 24.3 - 1 Khlong Thadarn Dam Project

The project is characterized by the huge size of the RCC dam and its high construction difficulty. The project is a roller compacted concrete (RCC) gravity dam, the dam height is 95 m, dam length is 2,161m, and earth work of the RCC amounted to 5.47 million m^3. In terms of earth work, the project can be described as the world's leading mega-roller compacted concrete dam.

Khlong Thadarn Dam project is Thailand's first RCC dam project. Thai side does not have the technology, or construction experience. According to the contract, Sinohydro sent a technical expert group to guide and monitor the implementation of the project.

24.3.2 *Vaitarna Dam in India*

Vaitarna Water Supply Project is the main project of the Urban Water Supply Project invested by Mumbai Urban Construction Company under the government of India. The dam is a roller compacted concrete (RCC) dam, its height of 105m, its length of 520m, and the total volume of the RCC of 1 million m^3. The main project investment is more than US $ 100 million.

On September 3, 2007, Mumbai Urban Construction Company, the owner of the Urban Water Supply Project issued its bid document. On December 15, 2007, the CSC joint venture formed by China International Water & Electric Corp. and Indian companies formally submitted the Vaitarna RCC dam tender document. On September 9, 2008, the owners noticed CSC that they won the bid. China Institute of Water Resources and Hydropower Research were responsible for technical advice.

24.3.3 *Muruo Hydropower Station in Malaysia*

Muruo Hydropower Station (Fig. 24.3 - 2) is the secondary hydropower station in the four cascade development in the Upper Rajang River; it's about 70km away from the downstream Bakun hydropower station. The power station installed four 236MW mixed flow generating units, with a total installed capacity of 944MW. The dam site controls about 2,750km^2 basin area.

The project is mainly composed by the roller compacted concrete gravity dam, spillway, diversion and power generation system. The crest length is 473m, the maximum dam height is 141m. The normal reservoir water level is 540m, dead water level is 515m, the total capacity is 12.043 billion m^3, the adjust the capacity is 5.475 billion m^3.

Total contract value is 5,262 billion Yuan. The constructor

Fig. 24.3 - 2 Closure of the Muruo Hydropower Station

of the project is the China Three Gorges Corporation. Sinohydro is responsible for civil construction, fabrication and installation of metal structure, and installation of permanent electrical and mechanical equipment. The China National Machinery and Equipment Import and Export Corporation are responsible for the procurement of permanent mechanical and electrical equipment. The Yangtze Surveying and Planning design and Research Institute of the Yangtze Water Resources Commission is responsible for engineering design. The project commenced construction in October 2008, the total duration is five years, the contract amount is US $ 1 billion. The river closure is completed on May 1, 2010.

24.3.4 Kampuchea Kamchay BOT Hydropower Station

Kamchay Hydropower Station (Fig. 24.3 - 3) is located on Kamchay River in Kampot Province, the southwest of the Kingdom of Cambodia. It's about 150km away from the capital Phnom Penh. The power station has an installed capacity of 193,200kW, is by far the largest hydropower project in Cambodia. It's known as Cambodia's "Three Gorges Project".

The Cambodian Ministry of Industry began an international tender for the project in July 2004. Through hard work, Sinohydro finally won the bid with the best technical and business solutions. On February 23, 2006, Sinohydro, the Cambodian Department of Energy and Mining and the Cambodian Power Company officially signed the project implementation and power purchase agreement. According to the agreement, Sinohydro will invest US $ 280 million for the project, and develop the project in BOT (build-operate-transfer) approach. Project financing period is one year, the franchise operation period is 44 years including construction period of four years.

Fig. 24.3 - 3   Kamchay Hydropower Station under Construction

Kamchay Hydropower Station has significant value. It is Sinahydro's first overseas BOT hydropower project; it's also one of the three major works of Chinese aid in Cambodia in 2006. The project is currently the largest introduction of foreign investment in Cambodia, with a typical reference value.

The project is a RCC gravity dam, the total reservoir capacity is 718 million m^3, the dead storage is 354 million m^3, the capacity below the normal water level is 681 million m^3, the effective capacity is 327.1 million m^3, and the capacity factor is 19.3%. It's a partial annual regulation reservoir. The dam crest elevation is 153.00m, dam base elevation 41.00m, maximum height 112.00m, crest width 6.0m. The upper part of the dam upstream surface is vertical. The slope is 1 : 0.3 under the elevation of 84.0 m. The slope for the downstream surface is 1 : 0.75. the elevation of the folded slope point is 145.00 m. The dam has 10 dam sections, the space of transverse joints is generally 42 - 60m; transverse joint adopts straight joints. Dam spillway crest is open spillway, which is arranged in the river. The spillway crest elevation is 135.00m. There're altogether 5 holes, net width of each hole is 12m. The central pier width is 3.0m; side pier width is 3.0m. Five 12m×15m arc shaped steel gate are used, corresponding five winches are arranged to control the gate.

Continuous, rapid construction is not only the construction feature of a roller compacted concrete dam, but also is the way to ensure the quality of the dam. During Kamchay Hydropower Station's construction, we used many new damming technologies and process that were developed in recent years, such as Horizontal Spreading Method in RCC Inclined Layer, metamorphic concrete, Limestone Powder as Roller-Compacted Concrete Additive, etc. This played a crucial role on flood control in the flood season of 2009 and 2010, ensured on time working of the first generating unit and obtained good engineering and economic benefits.

The advantages that the Horizontal Spreading Method in RCC Inclined Layer in Kamchay Hydropower Station include: ①smaller pouring strength can cover a larger pouring area, there by reducing the usage of resources; ②shorten the interval time between layers; ③pouring cube is large thereby to avoid joint spacing; ④because the covering an area is small, it is easy to do spray moisturizing for pouring cube.

24.3.5 Ethiopia Tekeze Dam

Tekezé Dam (Fig. 24.3 - 4and Fig. 24.3 - 5) is a double-curvature arch dam in the Tigray region of northern Ethiopia on the Tekezé River, 140km to Mekelle City. The dam is at present the largest hydropower project in Ethiopia, with the aim of power generation and flood control. It's composed by concrete double-curvature arch dam, underwater water diversion and power generation system on the right bank, diversion project on the right bank, substation on the left bank and other buildings. Four generating units were installed in the underground plant, unit capacity of 75MW, with a total installed capacity of 300MW. The dam is 188m high, is currently the highest concrete hyperbolic arch dam in Africa. Normal concrete volume of the dam is 1.02 million m^3.

Fig. 24.3 - 4 Tekeze Dam under Construction

Fig. 24.3 - 5 Downstream of Tekeze Dam

The owner of the dam is the Ethiopian Electric Power Corporation (EPPCO), designer and consultant (Supervision Unit) is Montgomery the WATSON HARZA.

The project was officially started in June 2002, it's divided into a total of seven tenders, where 1B, 2, 3 tenders are the dam, water diversion and power generation system, civil engineering of the guide and closure works, design, installation of the metal structures and construction of the permanent road. Those tenders are contracted by Sinohydro and Gezhouba Group. The total contract value is US $ 282 million.

During the implementation of the project, we overcome many difficulties, firstly timely completion and approval of design drawings of the gate hoist, which eased the very tight manufacturing schedule, and stopped the situation that the gate devices were delaying the progress of civil construction. At the beginning of the project, project design standards and calculation methods have not been able to get the owners recognition, thus affecting the approval of the drawings.

Project team organized multiple meetings to find out the problems, communicated with the consultant engineers of the owner in a timely manner, and then renewed the drawings. By doing this, we finally won support from all parties so as to make sure the project proceed smoothly. Secondly, from the project leader to the workers, the whole team actively communicated with the owner's consultant engineers, paid special attention to the installation technologies and the quality problem. Through the whole team's hard work, the installation work for 230 tons gate hoist was finally completed in time for the 2007 flood season, which was a miracle. Thirdly, we promptly resolved the major technical problems and made reasonable arrangements for domestic manufacturing, and coordinated the progress of quality acceptance, ensured more than 3550 tons of equipment shipped to the site as planned. By doing all these, we managed to make the dam and intake gate hoist ready for water storage before the flood season of 2008 under extreme tight schedule for equipment manufacturing and installation.

During the dam construction, the consulting engineers made a large number of design changes; added a lot of steel and anchor, which made the schedule even tighter. Because the Tekeze River is very seasonally distinct, if water storage milestone cannot be realized before the flood season of this year, power generation will have to be delayed for nearly a year. And if the water-retaining structure and gate installation cannot be achieved a certain level, the plant is most likely to be destroyed by the flood. To this end, the project continuously adjusted the plan according to the deadline, increased human and material resources into the construction and made sure have the grasp of every aspect of the construction site. After the efforts of all parties, the project team finally achieved the goal that concrete structure reaches EL1103 elevation in the last minute of the 2009 dry season, which laid a solid foundation to achieve the goal of generating electricity by the end of September 2009.

The Ethiopian government attaches great importance to the Tekeze project, regarding it as an important project to develop the economy and to improve the people's livelihood. Government ministers had been to the construction site several times to coordinate and solve problems, and the Chinese Ambassador in Ethiopia also regarded the Tekeze project as an important project for the development of bilateral relations and put forward many good suggestions and opinions for the project's construction.

Chinese hydropower builders strictly performed their duties in the contract, strengthened project management, overcome a number of unfavorable factors such as poor construction conditions, not used to local living environment, etc, overcome the fact that the equipment units came to the construction site late than the schedule, organized the construction carefully, set the goal that one successfully installation, one successful adjustment, and one unit power generation be realized as soon as possible. By doing this, the team got a good grasp of the whole process, later units were installed quicker and quicker, finally Unit 4, 1, and 3 realized power generation on August 15, 2009, September 5, 2009 and September 28, 2009 respectively.

On November 14, 2009, under the assistance of the MWH Global, Inc., we organized an opening ceremony for the power station, and attracted a large number of domestic and foreign government officials, media and business leaders.

The total installed capacity of the Tekeze hydropower station accounts for 40 percent of the country's total power generation capacity of 683MW at that time. Due to the lack of fossil fuels, the power supply in Ethiopia is almost entirely from hydropower. This station laid the foundation for the sustainable development of Ethiopia's society and economy.

24.3.6 *Laos Nam Ngum5 Hydropower Station*

Laos Nam Ngum5 Hydropower station is located on Nam Ting River, an upstream tributary of the Laos Nam Ngum. The Station is about 300km north of the capital, Vientiane, in the mountains of northern Laos. The project is invested by International Engineering department of Sinohydro in BOT form. It's constructed by Sinohydro Bureau 10 Co., Ltd and Sinohydro Bureau 15 Co., Ltd.

The dam's controlled drainage area is 483km^2, 2.9% of the total drainage area; average annual flow is 22.8m^3/s, and annual runoff is 719 million m^3. The dam is mainly composed by: a 104.5m high roller compacted concrete dam, 8.6km water delivery tunnel, 2.8km penstock, 197m high surge shaft, 428m long tailrace tunnel and plant booster station project (Fig. 24.3-6).

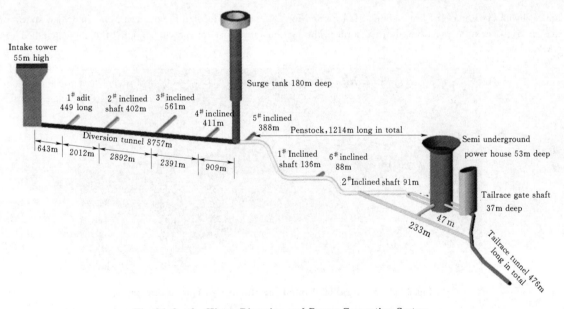

Fig. 24.3 - 6 Water Diversion and Power Generation System

The power station's installed capacity is 2×60,000kW, and guaranteed output is 44.8MW. As an annual regulation power station, the annual generation capacity is 507 million kW·h, the annual operating hours are 4,225h.

The total investment of Laos Nam Ngum5 Hydropower station is approximately US $ 200 million.

The Dam was designed under 500 year flood return period, checked under 10,000 year flood return period. The designed flood level is 1,100.27m, checked flood level is 1,101.84m. The reservoir's normal water storage level is 1,100.27m, crest elevation is 1,103.00m, dam height is 104.50m, crest width is 6.0m, bottom width of the crown cantilever is 42.00m, the thickness high ratio of the crown cantilever is 0.424, centerline arc length of the dam top is 234.85m, arc-height ratio is 2.42, and the top arch central angle is 92.80°.

The dam uses surface hole for flood releasing, the surface holes lie in the middle of the riverbed. The spillway crest elevation is 1,089.00m, a total of 3 holes are arranged, each hole's net width is 12m, the central pier width is 3.0m, and side pier width is 3.0m. At the crest of the spillway, three 12m×11m steel radial gates are installed, with 3 winches controlling the gate. The surface of overflow weir adopts the WES weir, chute slope is 1 : 2. The elevation of the ogee section at the end of the surface hole is 1,082.15m, ogee radius is 15m, angle of emergence of weir is 21°, and the bucket elevation is 1,083.15m. As to the downstream energy dissipation, trajectory energy dissipation is adopted. According to the design standard, the discharged volume is 2,480m³/s, discharged volume per surface hole is 68.89m³/(s·m), the corresponding upstream water level is 1,100.00m, downstream water level is 1,025.82m. According to the checking standard, the discharged volume is 3,231m³/s, discharged volume per surface hole is 89.75m³/(s·m), the corresponding upstream water level is 1,101.84m, downstream water level is 1,027.95m. The access bridge has the same elevation with the crest. Considering the arch gate's opening and closing, the access bridge is arranged at the tail of the spillway; with width of 5m.

Fig. 24.3 - 7 RCC Dam of the NN5 Project under Construction

The Dam's total concrete pouring volume is 332,946m³, RCC peak monthly placing intensity is 36,000m³. The upstream face adopted the cantilever climbing template for construction; the downstream face adopted the same method (Fig. 24.3 - 7).

24.3.7 Myanmar Shweli Hydropower Station

Myanmar Shweli Hydropower Station (Fig. 24.3 - 8 and Fig. 24.3 - 9) is located in northern Myanmar Shan state, on Shweli River, right next to the China-Burma border. The station is a diversion power station; the total installed capacity is 600MW.

Fig. 24.3 - 8 Head of Pivot of the Shweli Hydropower Station

The hydropower station is developed by the joint venture, which is formed by Yunnan Joint Power Development Company (YUPD) and Myanmar government-owned department of hydropower implementation (DHPI), in BOT (build-operate-transfer) way. The main construction is contracted by Sinohydro Bureau 14 Co., Ltd.

The project began construction in February 2004. It completed on May 16, 2009.

The station is a diversion power station; it mainly comprises head of pivot, water diversion system and factory hub.

The head of pivot is mainly the concrete gravity dam (including the spillway section and non spillway section), the maximum height of the dam is 47m, crest length is 143m. Non-spillway section crest width is 6m, and the upstream slope is vertical whereas the downstream slope is 1 : 0.8. The spillway is placed in the central of the dam; its length is 44.0m. There're 3 spillway holes, each width is 10m. The spillway hole crest elevation is 709.00m, 3 arch gates and a flat maintenance door are arranged. Hole length for flood releasing and scouring sand is 20.0m, the hole's bottom elevation is 695.00m; a flat maintenance door and a flat working door are set there. The stilling basin is arranged after the spillway with the length of 60m. The right bank tunnel for flood releasing and scouring sand (diversion) is mainly composed by intake

Fig. 24.3 - 9 Factory Overview of the Shweli Hydropower Station

tower, the pressure hole section, export chamber section and chute section. The total length of the tunnel body is 259.41m, the elevation of the intake floor is 695.00m, and the intake tower is 40.0m high, with a flat maintenance door set. The tunnel cross-sintion shape is circular, the inner diameter $D=10.0$m, concrete lining thickness is 0.8m, and the bottom slope is 1.2%. The export chamber is about 25m long; the elevation of the export floor is 692.57m. At the tunnel exit, two flat work doors are arranged. And then follows the 140m chute section.

Diversion system is arranged in the right bank, and it's mainly composed by the diversion tunnel, surge shaft, and steel penstock (underground pipes). From inlet to the surge shaft, the diversion tunnel is altogether 5,118.30m long; the bottom slope of the tunnel is 5 ‰. Pressure hole with circular section is adopted; the intake flow is 229.14m^3/s. The diversion tunnel is divided into the reinforced concrete lining hole and ejector anchor hole. The diameter of the lining hole is 7.0 - 8.5m, the inner diameter of the ejector anchor hole is 10.0m. The surge shaft is of the reinforced concrete structure, the inner diameter is approximately 16.5m, height is about 96m, the highest water level is approximately 764.10m, and the lowest water level is 700.50m. In the surge tank, there are two flat emergency gate set up, gate hole size is 5.2m × 5.2m. The inner diameter of the steel penstock is 5.2m, the length of 1$^\#$, 2$^\#$ pipe (primary pipe) is about 1077.569, and 1086.422 (to the fork tube center), it's composed by the horizontal section - inclined (inclination of 60°) section - horizontal section - inclined (inclination of 60°) section - horizontal section. From the fork tube, the pipe is divided into three pipes, branch pipe and conical tube, which leading to the main plant. The diameter of the branch pipe and the conical tube is respectively 4.24m and 3.0m. The conical tube's diameter reduced from 3.0m to 1.8m by the upstream wall side of the main plant, and then follows the valve.

The factory hub is formed by the primary powerhouse, auxiliary powerhouse and indoor booster station. In the powerhouse, six 100MW hydroelectric generating sets were installed. The factory's centralized arranged, the primary machine room, the installation room, the upstream auxiliary plant, indoor GIS building, and auxiliary plant at the end and tailrace are arranged together. The main plant's structural dimensions are 85.5m × 36.58m × 37.6m. It adopted the reinforced concrete frame structure; and the roof is a lightweight steel roof.

24.3.8 Burma's Yewa Hydropower Station

Yewa Hydropower Station (Fig. 24.3 - 10) located on Meitege River, is the largest hydropower station in Burma; the dam is constructed on the first tributary-Myitnge of Irrawaddy River, with maximum reservoir capacity of 2.6 billion m^3, the total installed capacity of 790MW, installed with four hydraulic turbine generator units and 197.5MW for each. It contributes 50% power of Burma and is another "Three Gorges Project" in the country. The project is also the largest project with maximum single unit capacity of exported generation units of Sinohydro.

The project includes a RCC dam, with dam crest elevation of 197m, the maximum height of 137m, the length of 690m, crest width of 12m, and the volume is 2.650 million m^3. Spillway is at the center of the dam, and overflow

Fig. 24.3 - 10 Downstream of Yewa Dam

865

water discharge is adopted. One of the two diversion tunnels is functioned as bottom hole of the dam. Its powerhouse is based at dam toe on the left bank. Hydropower system includes four inlets, four steel pressure pipes and four vertical Francis water turbine-generator units with installed capacity of 197.5MW for each.

On 14:30, February 21, 2010, 72-h trial load operation of the first unit (No.1 generation unit) of Yewa Hydropower Station was successfully completed, and commissioning was conducted under conditions of non-stopping operation. During trial operation, power output and operation status were sound, with generated power of 11.15 million kW • h. Premier of Burma made an on-site personal visit to the project and Minister of Power Ministry of Burma extended his congratulation and thanks to Sinohydro through addressing a letter.

Yewa Hydropower Station has a total installed capacity of 790,000kW, and thus the total installed capacity of the country mounted up by 50%, which has greatly improved the power quality of Burma and eased the power shortage status in Mandalay Province, playing an important role in promoting economic and social development in middle region even the whole country-Burma.

24.3.9 Ghana's BUI Hydropower Station

BUI Hydropower Station (Fig. 24.3-11) on boundary of North Ghana and Cote d'ivoire, is based on Volta River, 150km upstream of Volta Reservoir (one of the largest reservoirs in the world). The pivot building mainly consists of a 110m RCC dam and a powerhouse at dam toe; the hydropower station is equipped with 3 generator units, with total installed capacity of 400MW and annual power generation of 1 billion kW • h. After completion, the hydropower station will be listed as the second largest one in Ghana. The project is contracted by Sinohydro on an EPC type and implemented by Sinohydro Bureau 8 Co., Ltd. The contract content includes hydropower station, power transmission & transformation project, environmental emigration and irrigation project (temporarily decided), totaled US $ 597 million.

Fig. 24.3-11 River Closure of BUI Hydropower Station

On August 24, 2008, BUI Hydropower Station commenced construction. Since the first batch of large equipment was put in place in April 2008, Sinohydro devoted an input of about RMB 200 million for construction equipment within seven months. Projects of temporary construction roadway, contractor's encampment, the owner's temporary encampment, temporary sand & stone production and concrete mixing system, affiliated processing factory and warehouse have been finished. Construction of excavation of dam abutment and diversion channels of main body was conducted successively. River closure of the dam was achieved one year ahead of the schedule.

24.3.10 Mongolia's Taixier Hydropower Station

Taixier Hydropower Station (Fig. 24.3-12) is based at Wulanbumu Canyon of Zhabukehan River, border of Gobi-Altai Province and Zavkhan province, 1050km west of Ulan Bator (Capital of Mongolia). The project is contracted by Sinohydro and implemented by Sinohydro Bureau 11 Co., Ltd.

The RCC dam has a height of 55m, dam crest of 190m, and the volume is of 200,000m^3. The water surface area is 50km^2, and storage capacity of 930 million m^3. Stepped chute spillway without gate is set on width of 75m at center of the dam and a ground powerhouse at dam toe is also constructed. The rated total installed capacity is 11MW, installed with three sets of 3.45MW generator units and one 650kW francis type water turbine-generator unit. Width of dam crest is 5m, top elevation of water-retaining dam section is 1708m, and top elevation of overflow dam section is 1704m. It is vertical surface above EL1,668m at upstream surface of the dam, slope surface on the left of intake tower below EL1668m, with slope 1 : 0.85 and vertical surface on the right of intake tower. Downstream surface of the dam adopted step-type, with slope 1 : 0.725.

Winter is extremely cold while summer is fervent in this region, with annual average temperature of 0°C; the extreme temperature in January is −51°C and 39°C in July, indicating large temperature difference between day and night. The regional annual precipitation is 200mm. dominant wind direction is bias Northwest and north here, with heavy frequent wind-sand in Apr. and May.

Dam seepage control utilizes single-row grouting anti-seepage curtain and dam surface anti-permeate film. Curtain grouting is constructed on grouting platform casted on upstream face of the dam, with maximum embedded rock depth of 35m. Bottom

Fig. 24.3-12 Full-view of Taixier Dam

of dam surface anti-permeate film is anchored on grouting platform while its top is fixed on normal concrete (thickness of 0.4m) on RCC; anti-permeate film of intake tower and overflow dam is installed on normal concrete. Installation of anti-permeate film, a adventurous technology innovation, has fundamentally solved the technical problem of concrete seepage control in high and cold area, and is an active measure in application and popularization of advanced construction technology and new materials, and thus thought of hydraulic structure seepage control has changed. In the past, only the way of improvement of concrete performance is available to meet requirements of seepage control; currently, high-quality anti-permeate film can be used to satisfy needs of seepage control of hydraulic structures.

In winter of 2007, water level rose to El. 1,677.1m, but any leakage is not found at the dam. This option can simplify RCC construction, save a lot of cement and lower the hydration heat temperature rise of mass concrete; water seal installation of dam joint is cancelled, and cold joints and cracks require no treatment during dam construction, therefore fast casting of dam concrete can be achieved, which speeds up construction progress, reduces the cost and embodies characteristics of rapidness, economy and beauty.

24.3.11 Belize Chalillo Dam and the Powerhouse

Chalillo Dam and its powerhouse (Fig. 24.3-13 and Fig. 24.3-14) are located in a Central America country, Belize. It's exclusively invested by the Canadian company-Fortis, the owner of the project is Belize Electric Company Ltd. (BECOL), and the engineering consultant is Gilbert-Green & Associates Inc, Canada. Through international tender, International Engineering department of Sinohydro became the contractor and signed the contract with the owner in February 2003.

The project is located on Macal River, the southwest of the country. It's about 12km away from the existing downstream dam, Mollejon Power Plant. This is Sinohydro's first hydropower project in Central America. And the powerhouse, power generator units, and power transmission line are all EPC project.

The local weather is tropical rainforest climate. It has two distinct seasons, dry season (January to May) and rainy season (June to December). The annual rainfall is about 1500mm; the highest temperature reaches 38.9℃, the lowest temperature is 11.7℃.

The project consists of a roller compacted concrete gravity dam, a powerhouse at the toe of the dam, substations and a high voltage transmission line. Its main purpose is to provide water for existing Mollejon Power Plant so as to ensure power generation all the year.

The RCC gravity dam is 425m long, 50m high, crest width is 5-5.75m, the upstream dam surface slope is 1 : 0.1 and downstream dam surface slope is 1 : 0.75. There's designed to be a 1m thick normal concrete impermeable layer on the up-

Fig. 24.3-13   Downstream Overview of the Chalillo Dam

Fig. 24.3-14   Overview of the Chalillo Dam

stream face of the dam. There're a total of three levels of spillways. Dam section L3 and L4 can give vent to regular floods, dam section L1, 2, 5 and 6 can discharge flood with return period of 1000 years, dam section R1-7 deal with flood with return period over 1000 years. Due to its relatively poor foundation, L7 and L12 is non-spillway dam section.

The powerhouse is 25m long, 17.5m wide and 36.8m high. Two 3.65MW Kaplan turbine units are installed. The 115KV transmission line is 17.3km long, American anti-corrosion log poles are selected.

The project officially started in April 2003, completed in October 2005, lasted for 30 months.

The engineering quantity completed in the project include: 570,000m^3 of earth and stone excavation, 200,000m^3 of concrete pouring, 1,400 tons steel and steel structure installation, 13,400 m grouting, equipment and drain hole drilling, two sets of Kaplan unit installation, 17.3km high-voltage transmission line. The actual investment is approximately US $ 29,020,000.

The project introduced large scale of third-country labor, which speeded up the access of workers; the concrete were transported by two ways: negative pressure trough and automobile. Inclined-layer Placing Method was used on larger placement area; because it's very difficult to find suitable admixture, after repeated experiments, we finally chose the soft rock sand at the right bank of the dam to add in the RCC, which on one hand could meet the performance requirement of the RCC, on the other hand could save cement. 21.24 m of sliding mode was used for the spillway construction, which speeded up the progress and ensured the quality of construction.

Belize Chalillo Dam's completed on schedule, which has not only accumulated experience for Sinohydro in the Central American market, but also laid the foundation for its further expansion in the American market.

24.4 Prospect Forecasting

The course of 60 years overseas construction of hydroelectric dams depicted the glorious development history of China's hydropower, from small and medium-sized hydroelectric dam reconstruction in African countries from the 1960s until the 1980s, to global market competition at the end of 20th century, and to the successful implementation of world-class high dam in the early 21st century, and fostered China hydropower's brand on the international market, indicating a better future for China hydropower's overseas expansion and development.

In October 2009, Sinohydro and the Government of Ecuador signed the contract to construct the country's largest hydroelectric project in Quito-the Coca Codo Sinclair Hydroelectric Project (CCS project). The project locates on the Ecuadorian Amazon River. After the completion of the project, the total installed capacity will reach 1500MW. The annual power generation amounts to 8.8 billion kW·h, which will meet 75 percent of the country's electricity demand, and become the country's largest hydropower base.

On March 15, 2011, Sinohydro and Hydropower Development Company of Iran signed the contract for Bakhtiari dam in Tehran. The dam is a 315m high concrete double arch dam. After its completion, it will be the highest arch dam in the world.

Hydropower is clean energy. There are rich water resources in Southeast Asia, Africa and South America, and the utilization rate is not high in these regions. There are a large number of hydropower resources to be developed, which will provide a huge market for the international hydropower industry. In April 2010, China Foreign Aid Work conference was held in Beijing. On the conference, the government gave more support to the "going out" policy, which gave great impetus to China's hydropower development in the overseas market. There will be more and more international landmark projects being implemented by China and the dam height record will be constantly refreshed.

Part III

Operation and Management of Dams and Reservoirs

Chapter 25

Dam Operation and Management

Cai Yuebo[1], Sheng Jinbao[2], Yang Zhenghua[3], Wang Shijun[4], Wu Suhua[5]

25.1 General

25.1.1 *Introduction*

Dams are the important engineering measures to regulate and control the spatial and temporal distribution of water resources and optimize their allocation, and constitute the significant components of flood control engineering system of rivers. These reservoir infrastructures are irreplaceable for economic and social development, an inseparable part of ecological environment improvement, and public welfare and strategic decision making. They play important roles not only in the safety of flood control, water supply and grain production, but also in economic, ecological and national security.

China is one of the countries with the longest dam building history in the world. One of the oldest dams—the Anfengtang Dam in the Huaihe River Basin, built 2,600 years ago, is operating well and playing its benefits. Most of dams have been constructed after the founding of New China. According to *2011 Yearbooks of China Water Resources Press*, China is one of the countries with the most dams in the world. China (excluding Hongkong, Macao and Taiwan) has 87,873 dams which consist of 552 large dams, 3,269 medium-sized dams and 84,052 small dams, with a total reservoir storage capacity of 716.2 billion m^3. From the dam distribution in space, 61.3% of China's dams were constructed in seven provinces including Hunan, Jiangxi, Guangdong, Sichuan, Shandong, Hubei and Yunnan (Table 25.1-1). Among all provinces, Hunan has the largest number of dams, which is 12,092, accounts for 13.8% of the national total. According to the dam type, 93% of China's dams are earth-rockfill dams and 26,300 dams have a height of more than 15m on the base of statistical results of dam height in China.

25.1.2 *Dam Management System*

In China, dams (including hydropower station) are under the jurisdiction of various government sectors including water resources, energy, construction, transportation and agriculture, etc. According to *Regulations for Dam Safety Management* (RDSM), dam safety management is implemented under a responsibility system from the Central Government to local governments. The Water Resources Department and other relevant Departments supervise all the dams in the country. The Water Resources Departments of local governments at or above county level in conjunction with relevant departments supervise the dams in its jurisdictional area. All levels of water resources, energy, construction, transportation, agriculture and other relevant departments take responsibility for dams in their jurisdictional areas. Administrative leadership responsibility system is implemented at all levels of local governments and Administrative Department.

The administrative agencies for dam management include the Ministry of Water Resources of the State Council and its branch agencies in major river basins and local Water Administrations at all levels. The Office of State Flood Control and Drought Relief Headquarters and the Flood Control and Drought Relief Headquarters at all governmental levels are responsible for the reservoir flood control work and issue reservoir regulation commands based on the approved plans.

---

[1] Cai Yuebo, Vice President of Nanjing Hydraulic Research Institute, Vice President of Dam Safety Management Center of the Ministry of Water Resources, Professor Senior Engineer.

[2] Sheng Jinbao, President of Dam Safety Management Department, Nanjing Hydraulic Research Institute and Deputy chief engineer of Dam Safety Management Center of the Ministry of Water Resources, Professor Senior Engineer.

[3] Yang Zhenghua, Vice Director of the regulation & supervision department in Dam Safety Management Center of the Ministry of Water Resources, Professor Senior Engineer.

[4] Wang Shijun, Vice President of Dam Safety Management Department, Nanjing Hydraulic Research Institute, Professor Senior Engineer.

[5] Wu Suhua, Nanjing Hydraulic Research Institute, Professor Senior Engineer.

Part III Operation and Management of Dams and Reservoirs

Table 25.1-1 Number Distribution of Dams Built in China

Region	Reservoir size			Total No.
	Large	Medium-sized	Small	
China	552	3,269	84,052	87,873
Beijing	4	17	61	82
Tianjin	3	11	14	28
Hebei	22	42	1,002	1,066
Shanxi	8	58	667	733
Inner Mongolia	12	81	404	497
Liaoning	33	74	844	951
Jilin	16	99	1,528	1,643
Heilongjiang	26	97	790	913
Jiangsu	8	42	860	910
Shanghai	0	0	0	0
Zhejiang	31	152	4,034	4,217
Anhui	14	105	4,700	4,819
Fujian	20	154	3,051	3,225
Jiangxi	26	240	9,543	9,809
Shandong	37	214	6,040	6,291
Henan	23	108	2,221	2,352
Hubei	63	252	5,533	5,848
Hunan	26	281	11,785	12,092
Guangdong	34	306	7,097	7,437
Guangxi	38	186	4,143	4,367
Hainan	7	72	917	996
Sichuan	12	109	6,638	6,759
Chongqing	9	64	2,767	2,840
Yunnan	12	186	5,360	5,558
Guizhou	16	62	1,995	2,073
Tibet	3	6	66	75
Shaanxi	9	58	954	1,021
Gansu	8	38	267	313
Ningxia	1	29	196	226
Qinghai	7	10	140	157
Xinjiang	24	116	435	575

In addition to the water administrations at the central and provincial levels, there are professional management agencies responsible for dam safety management. These agencies entrusted by governments, are responsible for dam safety management as an extension of governmental functions, including the Dam Safety Management Center, the Center for Construction Management of Water Works of the Ministry of Water Resources, and Dam Safety Management Centers and Water Resources Project Administrations at the provincial level, etc.

For multi-purpose reservoirs for flood control and irrigation, each of the water administrations at all levels functions as owners on behalf of the government at the same level. Most of the small dams are owned by rural collective economic entities. In the power sector, the owners of hydropower dams are the regional or provincial power companies affiliated to the

State Power Corporation. A small number of dams in other industries are owned by their respective superior administrations. Dam owners are fully responsible for the dam operation and safety, and set up reservoir operation management entities in charge of the daily operation and management.

25.1.3 Laws, Regulations and Standards for Dam Safety Management

Before the 1980s, the safety management for dams in China mainly depended on administrative means, and various types of administrative documents to direct and implement reservoir management. Since the 1980s, technical standards such as *General Rule for Reservoir Project Management* (SLJ 702—81) were issued and dam safety management entered a new period of standardized management. These standards laid the foundations for the existing technical standards for reservoir management. The promulgation and enforcement of the *Water Law of the People's Republic of China* in January 1988 launched a new era of systematic development of dam safety management laws, regulations and technical standards. The issuance and enforcement of the *Regulation of Dams Safety Management* (RDSM), in March 1991, based on the *Water Law of the People's Republic of China*, laid the legal foundations for dam safety management in China. Subsequently, a series of relevant regulations, specifications and technical standards have been issued and a standardized system has been established for dam safety management laws, regulations and standards. (Fig. 25.1-1).

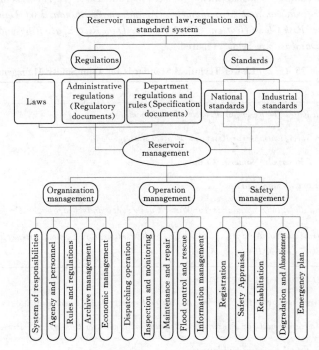

Fig. 25.1-1 Structure Charts of Reservoir Management Regulation and Standard System in China

1. Management regulations

At legislative levels, the safety management laws and regulations for dams in China include the national laws, administrative regulations and codes of the State Council, department regulations and specification documents, and local laws and regulations. After years of development, an improved system of dam safety laws and regulations has been established.

National laws on dam safety management are based on the *Water Law of the People's Republic of China* and *Flood Control Law of the People's Republic of China*, and include *Water and Soil Conservation Law*, *Water Pollution Control Act*, *Environmental Protection Law*, *Environmental Impact Assessment Law*, *Law on Protecting Against and Mitigating Earthquake Disasters*, *Work Safety Law*, *Law on Emergency Response*, *Law on Land Management* and other relevant laws. Among them, the "*Water Law*", promulgated on Jan. 21, 1988 and revised on Aug. 29, 2002 is the highest level law for dammanagement, and the *Flood Control Law* promulgated on Aug. 28, 1997 is the law that must be obeyed when reservoir flood regulation is conducted.

Regulations and laws specifically for dam safety management in China consists of RDSM, *Flood Control Regulations*, and relevant laws and detailed rules for implementation. The main regulatory documents are *Implementation Opinions of Water Resources Project Management System Reform*, *State plan for rapid response to public emergencies*, and National E-

mergency Plan for Flood Control and Drought Relief. Among them, as RDSM issued on Mar. 22, 1991 and put forward by State Council 77th Document, which is a special law on dam safety management and the core of legal system for dam safety management.

The Department regulations mainly consist of *Registration for Dams*, *Safety Appraisal Rule for Dams*, *Management for Reservoir Degradation and Abandonment* (*for trial implementation*), and include some important specification documents promulgated in recent years, such as *Announcement on Strengthening Safety Management Work for Reservoirs* ([2002] No. 188), *Announcement on Strengthening Safety Management Work for Reservoirs* ([2006] No. 131), *Guidelines on Preparation of Emergency Preparedness Plan for Dams* (*for trial implementation*) ([2007] No. 164), and "*Safety Management Measures for Small Reservoirs* ([2010] No. 200), and so on.

2. Technical standards

In addition to national standards, the standards of the Ministry of Water Resources are complete and most widely used for dam safety management. In the power industry, the dam safety standards are systematically applied to hydropower in the industry.

Technical standards issued for dam safety management are listed in Table 25.1-2. A series of new technical standards such as *Standard for Reservoir Degradation and Abandonment*, *Abandonment Standard of Dam Safety Monitoring Instruments*, *Standard for Reservoir Risk Classification*, *Guideline for Dam Risk Evaluation* and *Standard for Structure and Identifier of Water Project Database* are being prepared.

Table 25.1-2 List of Technical Standards for Dam Safety Management

Category	Title	Number
Comprehensive management	General Rules for Reservoir Project Management	SLJ 702—81
	Design Specifications for Reservoir Project Management	SL 106—96
	Codes of Reservoir Names in China	SL 259—2000
	Flood Control Standards	GB 50201—94
	Grading and Flood Standards for Water Conservancy and Hydropower Projects	SL 252—2000
	Specifications for Analysis, Calculation and Assessment of Economic Benefits of Completed Flood Control Projects	SL 206—98
Organization management	Code of Conduct for Water Sector	SL 301—93
	Tentative Standards for Staffing of Water Project Management Entities	SL J705—81
	Basic Information Code Preparation Regulation for Water Conservancy Projects	SL 213—98
Safety management	Dam Safety Assessment Guidelines	SL 258—2000
	Standard of Decommission for Metal Structures in Water Conservancy and Hydropower Projects	SL 226—98
Operation management	Rules on Assessment of Reservoir Flood Regulation	SL 224—98
	Standards for Acceptance of Reserve Materials for Flood Control	SL 297—2004
	Regulation for Preparing Quota of Reserve Materials for Flood Control	SL 298—2004
	Specifications on Operation of Reservoirs for Large and Medium-sized Hydropower Stations	GB 17621—1998
	Technical Specifications for Safety Monitoring of Earth and Rockfill Dams	SL 60—94
	Technical Specifications for Safety Monitoring of Concrete Dams	SDJ 336—89
	Specifications for Compilation and Preparation of Safety Monitoring Data for Earth and Rockfill Dams	SL 169—96
	Basic Technical Conditions for Dam Safety Automatic Monitoring Systems and Equipment	SL 268—2001

Continued

Category	Title	Number
Maintenance	Specifications for Maintenance and Repair of Earth and Rockfill Dams	SL 210—98
	Specifications for Maintenance and Repair of Concrete Dams	SL 230—98
	Technical Regulation for Safety Inspection and Testing of Steel Gates and Hoisting Equipment	SL 101—94
	Standards for Grading for Gates, Hoisting Equipment and Boat Lifts in Water Conservancy and Hydropower Projects	SL 240—1999
	Specifications for Anti-corrosion of Metal Structures in Water Conservancy Works	SL 105—95

25.1.4 Dam Operation Management System

Based on RDSM, and in conjunction with the practical and development needs in reservoir management, China has already established a basic operation management system covering the whole life cycle of dams (Fig. 25.1-2).

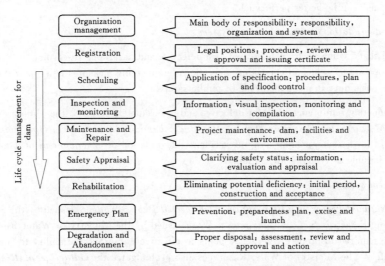

Fig. 25.1-2 **Dam Operation Management System**

1. Registration system

Dam registration is one of the basic systems in reservoir management, and is the fundamental means to verify the existence of the reservoir project, conduct dam safety supervision, define relevant management responsibilities and share the basic information of the project.

In *Registration Measures for Dams* issued by the Ministry of Water Resources in 1995, the registration institutions and registration procedures, and cancellation, certification and file management were clearly defined. By 2009, interim results had been obtained in dam registration, all existing reservoirs had been registered, and the information update mechanism was formalized.

2. Operation, maintenance and monitoring system

In 1993, the Ministry of Water Resources issued the *"Operation Rules for Multi-purpose Reservoirs"*, and subsequently, the relevant departments issued technical standards such as *"Specifications for Operation of Large and Medium-sized Hydropower Stations"* (GB 17621—1998) and *"Rules on Assessment of Reservoir Flood Regulation"* (SL 224—98). With the development of social economy and sciences and technologies, constant advancement has been made on methodology for reservoir operation. At present, the important subjects of reservoir operation are mainly on the utilization of floodwater as resources which includes flood regulation staging, scheduling control on dynamic limited water level, dam emergency dispatch of rescues and disaster relief work, and uniform regulation of water resources in a basin or a region, etc.

Maintenance includes daily maintenance, repair before and after a flood, major and annual repair, which is an important work to mitigate aging affects, improve project safety and appearance. As promoted by the institutional reform in water project management organizations, maintenance for reservoir works has been improved. The relevant technical standards include "*Specifications for Maintenance and Repair of Earth and Rockfill Dams*", "*Specifications for Maintenance and Repair of Concrete Dams*" (SL 230—98) and "*Specifications for Anti-corrosion of Metal Structures in Water Works*" (SL 105—95).

Inspection and monitoring is a measure of dam safety management based on provisions and requirements in RDSM. Inspection and monitoring can be performed in the form of manual patrol inspection, and instruments monitoring. The relevant technical standards include "*Technical Specification for Safety Monitoring of Earth and Rockfill Dams*" (SL 60—94), "*Specifications for Compilation and Preparation of Safety Monitoring Data for Earth and Rockfill Dams*" (SL 169—96) and "*Technical Specification for Safety Monitoring of Concrete Dams*" (SDJ 336—89).

3. Dam safety appraisal system

Based on "*Safety Appraisal Measures for Dams*" (promulgated in 1995 and revised in 2003), a regular safety appraisal system shall be implemented for dams, the first safety appraisal shall be performed within 5 years following project completion acceptance, and every 6 – 10 years afterwards. If a serious dam safety hazard occurs or a major change has been made during reservoir operation, an overall dam safety appraisal should be carried out. In 2000, the Ministry of Water Resources published the "*Dam Safety Assessment Guidelines*" (SL 258—2000), as an associated technical standard of "*Safety Appraisal Measures for Dams*". It defines the specific technical requirements on dam safety assessment, so that "Safety Appraisal Measures for Dams" can be better performed.

4. Dam rehabilitation, degradation and abandonment system

It is specified in RDSM that, when a dam is identified as a defective and hazardous reservoir, or a serious dam incident or hazard has occurred in operation, which affects safety operation, engineering measures must be taken to remove the risk. It is the reservoir owner's basic duty to meet the needs of public safety.

While energetically conducting reinforcing work for defective and hazardous reservoirs, the Ministry of Water Resources issued the *Management Measures for Reservoir Degradation and Abandonment (for trial implementation)* in May 2003 which provided measures for decision making on reinforcement, degradation or abandonment of dams if the dam reinforcement is not technically feasible or economically rational.

5. Emergency management system

According to *General National Emergency Plan against Unexpected Public incidents* issued by the State Council and *Notification on Strengthening Reservoir Safety Management Work* issued by the Ministry of Water Resources in Apral 2006, all reservoir owners are required to prepare emergency plans to effectively improve the ability to respond to dam incidents and to ensure public safety. The Ministry of Water Resources also promulgated the *Guidelines on Preparation of Emergency Preparedness Plan for Dams (for trial implementation)* in May 2007, which indicates that emergency management for dam incidents has become an important part in dam safety management, and emergency management system has become one of the important systems in modern dam management.

25.1.5 Achievements in Dam Safety Management

The Chinese Government has paid great attention to dam safety and management. Especially, since China's policy to reform internally and open internationally, a series of steadfast efforts have been made such as the implementation of responsibility system on dam safety, actively pushing forward institutional reform on reservoir management system, establishing and completing laws and regulations and standard system on dam safety management, energetically reinforcing defective and hazardous reservoirs, and strengthening non-structural measures. As a result, the safety conditions of dams have been substantially improved and the management level continuously upgraded.

1. Continuously enhancing and implementing responsibility system on dam safety

The responsibility system for dam safety consists of administrative heads as the core, and the persons in charge of the governments, water resources authorities and management units for dam safety have been defined. According to the principle of responsibility at different levels, the Ministry of Water Resources publishes to the public a list of personnel responsible for the safety of large reservoirs in China every year, and the local authorities also published a list of personnel responsible for other types of reservoirs, to strengthen social supervision, enhance the implementation of responsibilities, and ensure safe operation of dams.

2. Steadily advancing reservoir management system reform

To change the outstanding problems in reservoir management that have been accumulated over a long time, such as non-flexible mechanism, shortage of fund, overstaffing, extensive management and low social security, the General Office of the State Council issued *Opinions on Implementation of Reform to Water Project Management System* ([2002] No. 45) on Sept. 17, 2002 by the general office of the state council, stating a water conservancy project management system and operation mechanism to be built in 3 – 5 years which shall be suitable for China and meet the requirements of water resources and socialist market economy.

In recent years, the Ministry of Water Resources has urged and guided the local authorities to energetically advance the reform in reservoir management system with a series of measures, and reservoir management units are classified, to provide budget for staffing and expenditure. Great efforts have been made in local authorities to push ahead this reform with appointed jobs and quota-based budget and basic personnel expenditure and maintenance expenditure for public welfare projects as the core, and interim results have been obtained, promoting the establishment of a favorable long-term operation mechanism for reservoirs.

3. Laws, regulations and standard system for dam safety management have been basically established, and operation management system has been gradually improved

After over 30 years of efforts, a regulation and standard system for dam management has been established with the *Water Law* and *Flood Control Law* as the foundation, RDSM as the core, associated with department regulations and specification documents such as *Registration Measures for Dams*, *Safety Appraisal Measures for Dams* and *Safety Management Measures for Small Reservoirs*, and supported by technical standards such as *Management and Design Specifications for Reservoir Project* (SL 106—96), *Dam Safety Assessment Guidelines* (SL 258—2000), *Specifications for Maintenance and Repair of Earth and Rockfill Dams* (SL 210—98), *Specifications for Maintenance and Repair of Concrete Dams* (SL 230—98), and it suits the Chinese conditions and satisfies the needs in economic and social development.

As required by RDSM, Department regulations and specifications in conjunction with the practical and development needs in dam operation management, the dam operation management systems including registration, regulation and application, flood control and disaster rescue, inspection and monitoring, maintenance, safety appraisal, rehabilitation, degradation and abandonment, and emergency management have been gradually improved, which have effectively promoted and standardized the dam safety management.

4. Reinforcing defective and hazardous reservoirs on a large scale

Following the severe flood in Aug. 1975, strengthening and eliminating dangers of 65 large-sized defective and hazardous reservoirs nationwide were conducted to enhance flood control standard in China. During 1986 – 1992, seepage prevention and reinforcement for unsafe reservoirs were 43 in the first group and 38 in the second group. Following the big flood in 1998, the Central Government speeded up the pace of reinforcing defective and hazardous reservoirs by incorporating 3458 defective and hazardous reservoirs in the Central Government 10-year enforcement plan. At the Rural Area Working Conference of the Central Government held in Dec. 2007, the CPC Central Committee and State Council determined to complete the reinforcement of all large and medium-sized reservoirs and important small reservoirs in the whole country in about 3 years, in addition to 2,300 defective and hazardous reservoirs, which were reinforced in the first two groups. 6,240 defective and hazardous reservoirs were selected for reinforcement by 2010.

To further solve the outstanding problems associated with small-sized defective and hazardous reservoirs, the Ministry of Water Resources and Ministry of Finance jointly formulated *Rehabilitation Program for Small-sized Defective and hazardous Reservoirs in China* in 2010, planning to reinforce 5,400 small-sized defective and hazardous reservoirs by 2013. In Dec. 2010, the *Reinforcing Program for Important Small Type Ⅱ Defective and Hazardous Reservoirs in China* was approved, with a plan to reinforce over 40,000 small type Ⅱ defective and hazardous reservoirs by 2015.

Dam reinforcement has significantly improved dam safety in China and brought huge economic and social benefits. Firstly, it has eliminated potential dam safety hazards, ensuring dam safety and protecting people's lives and properties downstream; secondly, it has increased the capacity of flood control and water resources regulation of reservoirs, and brought substantial benefit from flood control and drought relief; and thirdly, the reservoir management facilities have been greatly improved, which has promoted the development of water undertakings. And it can change the dam situation completely, improve the management facility, further consolidate the basis for water infrastructure and improve the capacity of water development and management.

5. National dam database has been developed, substantially upgrading information management level

With special funding from the Central Finance, the Ministry of Water Resources developed the "National Dam Database" in 2010. It includes the basic information of over 85,000 dams, provides easy and convenient access to these data and information for decision-making support by the Ministry of Water Resources. It also provides an information service platform for management and response to major water resources emergencies, and greatly enhances information management capacity in China.

In addition, with the advancement of standardization, regulation and modernization process in dam safety management and in conjunction with the implementation of rehabilitation for defective and hazardous reservoirs, automatic dam safety monitoring system has been installed in many large and medium-sized reservoirs, and provides real-time dam performance information. These systems, together with the national dam database, make it possible to achieve information-based management of dams in China.

6. Strengthening development of non-structural measures, and improving management system and innovation management concept

With the economic and social development in China, the tasks of reservoirs for flood control, drought relief and disasters mitigation are growing with social and public expectations for dam safety getting higher. In recent years, in addition to upgraded efforts to reinforce defective and hazardous dams, the Ministry of Water Resources has also paid special attention to developing non-structural measures for dam management, including: ①management system reform, establishing staff, enhancing specialized management, carrying out staff performance evaluation and optimizing staffing management structure; ②rationally defining reservoir storage capacity for generating benefits and for flood control and implementing rational reservoir regulation; ③strengthening emergency warning, enhancing Emergency Surveillance and early Warning, and raising risk concept and disaster prevention awareness of the public; ④degrading or decommissioning seriously defected dams; and⑤introducing and promoting dam risk management concepts and approaches to create a comprehensive safety management system covering dam safety mechanism and risk control mechanism for downstream communities.

25.1.6 Existing Problems in Dam Safety Management

Dam safety management in China has made considerable progress, but there are still many severe problems due to congenital defect, structure aging, poor operation and maintenance, and frequent extreme events, etc. China has faced great challenges due to higher expectations from the economic and social development.

1. Problems concerning defective and hazardous dams remain serious and the dam rehabilitation is still a long-term arduous task

In China, over 90% of dams were built in the 1950s–1970s. Most of these dams were built with low technical standards and comparatively poor construction quality, and lack of proper maintenance and management, causing various defects and hazards in many dam projects. In recent years, rehabilitation for defective and hazardous dams has become the major task in water sector, however, up to the end of 2010, only the large and medium-sized reservoirs and important small type I reservoirs has been reinforced, and there are tens and thousands defective and hazardous dams still need to be reinforced. In the meantime, new dam safety deficiencies and hazards are expected to occur every year in many dams due to ageing, washout, earthquake or poor maintenance. Therefore, it will be quite difficult to completely solve the severe problems of deficiencies and hazards of dams in China in a short period of time. This means that not only the task of dam rehabilitation is quite arduous, but also poses challenges to the safety management of dams.

2. Dam classification system needs to be improved and "Regulations for Dam Safety Management" needs to be revised as early as possible

In China, classification of dams and their design standards are mainly based on reservoir storage capacity. Therefore, some high dams with small reservoir storage capacity but high potential consequences are presently classified as low class dams. This approach has resulted in dysfunctional dam classification system.

As the first national regulation on Dam Safety Management in China, the RDSM was issued in March 1991. It has played an important role in ensuring the safety operation of dams. However, after more than 20 years, this Regulation no longer meets the current needs. Firstly, great changes have taken place in the legal background which is the basis of this regulation. Secondly, the dam management system has been reformed from the single state-owned and collective-owned to the coexistence of a number of forms such as state-owned, collective-owned, private-owned and shareholding system. Thirdly, with the economic and social development, new expectations have been raised on the management of water resources, water areas and reservoir projects, and great changes have also taken place in the functional demand and management scope for dams. Fourthly, advanced technologies in dam safety management field have been made. Therefore, RDSM should be revised as early as possible.

3. It is urgent to strengthen the safety management for small dams due to weak management

A high percentage of the small dams are identified as defective and hazardous ones and lack of proper operation and management. Firstly, poor management systems, inadequate management personnel and many small type II reservoirs are not attended with personnel. Secondly, normal maintenance is seriously inadequate for dams, and deterioration of dams due to aging and lack of repair. Thirdly, poor management facilities with incomplete safety monitoring, flood control and communication infrastructure result in big concerns in flood control and safety protection system.

4. Improve emergency management capacity in response to increasing safety adverse impact caused by extreme events

In recent years, extreme events have caused significant adverse impacts on dam safety. Climate change has caused more flood disasters, extreme storms, extraordinary droughts and more frequent super-strong typhoons. Meanwhile, earthquake and geological disasters also occur more frequently. These extreme events are abrupt, hazardous and increase the risks of dam safety, causing many accidents or even dam failures every year. The severe Wenchuan earthquake on May 12, 2008 caused hazard to over 2,400 reservoirs dams, including 379 high hazard dams in Sichuan Province. In 2008, typhoons and rainstorms in Guangdong Province caused damage to nearly 150 dams of different types. In 2009, rainstorms and flood in South China caused high hazard to Kama Reservoir in Guangxi province. In early 2010, the extreme droughts in the southwest of China made many reservoirs dried up, and cracks occurred on dam and reservoir bottom, affecting safety impoundment of reservoirs. During the flood season in 2010, 11 dams were at the risk of bursting in China burst. Among them, Dahe dam failure in Jinlin Province resulted in the significant casualties. It is urgent to strengthen dam risk management, enhance monitoring and warning of emergencies, improve the effectiveness and operability of Emergency Preparedness Plan (EPP), and raise the public awareness of dam failure risk through propaganda, training and rehearsals.

25.2 Dam Operation and Management Technologies

At present, it is widely recognized that China has a wealth of experience and state-of-art technologies in construction of different types of dams, especially high dams. However, the dam operation and management technologies are relatively lagging behind due to the concept of "emphasizing construction while overlooking management" over a long time.

With the rapid development of economy and society, dam safety is drawing increasing concerns in our society and the public. The Chinese government had been paying increasing attention to dam safety and management, pushing ahead the constant progress in dam safety management technologies. Especially, since the promulgation and implementation of RDSM in early 1990s, active efforts had been made in the institutional reform of water project management system, implementing the safety responsibilities system in reservoir management, enhancing safety supervision and inspection, and establishing and completing dam safety management system. In the meantime, many key technical issues to improve the safety management level and establish sustainable operation management system and mechanism for dams in China have been solved through research efforts on various programs such as the National Science & Technology Program (scientific and technological support program), the "973" Program, Natural Science Foundation, Scientific and Technological Innovation Program and social benefit special projects in the water sector, and social benefit dedicated research program of the Ministry of Science and Technology, making impressive progresses and obtaining rich results in the fields of dam potential deficiencies detection, dam safety assessment and deficiency diagnosis, dam safety monitoring and emergency event forecasting and pre-warning, dam risk analysis and risk management, defective and hazardous reservoir rehabilitation and decision-making, and reservoir operation, etc.

25.2.1 *Technology for Detecting Dam Potential Deficiencies*

In China, researches and application of dam potential deficiencies detection technology started in the 1980s, when the Hydraulic Research Institute of Shandong Province first developed an electrical instrument to detect potential deficiencies of dams. In the National Science & Technology Program during the 7th Five-year Plan period (1986-1990), China Institute of Water Resources and Hydropower Research (IWHR) developed an instrument to detect concrete dam cracks using pulse surface waves. In 1992, "Technical Research on Potential Deficiencies Detection of Embankments" was included in the 8th Five-Year Plan to solve key scientific and technological issues, and then in the national program to promote key scientific and technological research results in the 9th Five-year Plan period. In Nov. 1999, a major scientific and technological program of the Ministry of Water Resources "Development of Instrument for Dam Potential Deficiencies Detection" was initiated. In 1999 and 2000, The Office of State Flood Control and Drought Relief Headquarters implemented two test sites in Yiyang, Hunan, located in the southern China and in Daxing, Beijing, which is in the north, and conducted geophysical exploration instrument testing and potential deficiencies detection technology field tests, giving powerful push to the advancement of dam potential deficiencies detection technologies, obtained first-hand data for all types of testing instruments. In Sept. 2000, the Office of State Flood Control and Drought Relief Headquarters held the "Seminar on Detection Technologies for Embankment Potential Deficiencies and Leakage" in Zhengzhou. In Nov. 2005, the Ministry of Water Resources held the

"International Symposium on Dam Safety and Detection of Potential Deficiencies in Dams and Dikes" in Xi'an, and these activities promoted the academic discussions and development on embankment potential deficiency detection technologies.

The current widely applied dam potential deficiencies detection technologies in China include high density electrical resistivity method, ground penetrating radar method, transient electromagnetic method, surface wave method and computerized tomography (CT) method.

25.2.1.1 High Density Electrical Resistivity Method

This method was first proposed in UK in the 1970s, marketed by Japan Geological Co., Ltd. in the 1980s and introduced to China in late 1980s. Based on the principle that electrical resistivity in rock and soil are different, this method studies the change and distribution of underground conducting current when an electrical field is applied, similar to conventional electrical resistivity method, with high density of measuring points and with equal polar spacing in the arithmetic coordinates system. It is a combination of electrical profile method and electrical sounding method, can complete the process of vertical and horizontal (two-dimensional) exploration in one time, with high observation precision, reliable data acquisition, and able to provide images of electrical structure in the ground. So it can be used to obtain rich geological information. Cracks in dams, caves, uneven mass, soft layer can be clearly and visually reflected on detection diagrams. It is the main method for detection of potential dam deficiencies.

25.2.1.2 Ground Penetrating Radar (GPR)

This method is also referred to as geological radar method. By GPR technology, a high frequency electromagnetic wave beam is used to detect objects with medium loss, and the beam in the form of electromagnetic wave with wide band and short pulses, is injected from ground with transmitting antenna, and then returned to the ground as reflected by geological strata or object mass with electromagnetic difference, and received by another antenna. The received signals are analyzed to detect potential defects in dam. It is suitable for dry soils at shallow depth.

Up to now, commercially available GPR radar are the MK Ⅰ and Ⅱ of Microwave Associates of US, the Pulse EKKO series of SSI Company of Canada, the SIR series of the GSSI of US, the RAMAC boring radar system of SGAB of Sweden, the XADAR system of the XADAR Inc. of Russia, and the radar instrument of ERA Engineering Technology of UK.

Despite of the short term application of GPR technology in China, a lot of data and results have been obtained. It has been applied in bedrock detection, groundwater investigation, geological stratification, Karst mapping, river and lake bed profiling, Karst and cavity exploration, dam fill investigations, landslide investigations, dam mass quality testing, etc.

25.2.1.3 Transient Electromagnetic Method

Transient Electromagnetic Method or TEM is also referred to as time domain electromagnetic method, and it detects geological anomaly by using eddy current of different strength produced by geological strata at different locations and depths. The higher conductivity of the stratum, the greater eddy current produced and the stronger secondary magnetic field. The transient electromagnetic system normally consists of the transmitter transmitting coil, receiving oil, receiver and microcomputer data acquisition and plotting system. This method is not affected by topography and grounding resistance, featuring a great detection depth, therefore it is mostly used in detection of dam leakage locations in soft and loose layers.

Fang et al. (2010) firstly applied this method to detecting leakage and potential deficiencies at earth dams and dykes in China, and successfully developed type SDCO2 dam leakage detector. This instrument can detect leakage and potential defects in dykes and earth dams, locate leakage channels, with maximum detection depth of 60m, location resolution of 1-5m (with transversal detection spacing set arbitrarily), depth resolution of 1-2m, and relative resolution (diameter to depth ratio) about 8%. It uses a small coil sized $70cm \times 70cm \times 5.6cm$ with no need to insert the electrode into the ground, so it is not subject to variation of grounding resistance. The operation of this method is easy and quick, requiring less than 0.5min for each point measurement. This system was used in potential defects detection in Yueyang dikes in Hunan Province, with a measuring line of 17.4km and a detection depth of 0-60m. The detected potential piping areas was verified in the flood season; it was also successfully applied at seven dams including Nishan, Miyun, Yuecheng and Zhangze for detection of seepage channel, foundation leakage, by pass leakage, a ground water detection and foundation grouting quality checking.

25.2.1.4 Surface Wave Method

This method is also referred to as elastic wave frequency detection, and is a new shallow seismic exploration method that has developed at home and abroad in recent years. The surface wave is divided into Rayleigh wave (R wave) and Love wave (L wave), as R wave features the highest energy, maximum magnitude and lowest frequency in the vibration wave set, easy to identify and measure, so normally R wave is used in the surface wave exploration.

In performing field detection tests, a hammer or weight dropping on the ground surface produces a transient exciting force, which generates wide band waves. Waves at different frequencies superpose and transmit to the outside in the form of pulse signal. After the pulse vibration signal is received by the multi-channel low frequency detector, waves at different frequencies are separated through data acquisition and spectrum analysis to obtain the corresponding values, and then the surface wave frequency dispersion curve is plotted.

In performing dam potential deficiencies detection with this method, an impact vibrating source is used to produce seismic waves, the ground motion is acquainted and recorded via multi-channels, and the surface wave frequency dispersion characteristics are drawn by the surface wave analysis software, to analyze the dam structure and physical properties. This method can produce apparent effect in laminated layers with different wave speeds. Therefore, the various zones in the inhomogeneous dam can be understood from the frequency dispersion curve of the surface wave.

The R surface wave has been widely applied in engineering geological survey and non-destructive detection in China since the 1990s with its low attenuation, high S/N ratio, strong resistance to interference and frequency dispersion characteristics in laminated media.

25.2.1.5 CT Method

In the CT (Computerized Tomography) method, a 2D image on specific layer of the object is re-built by computer with some numerical computation processing on the basis of physical quantities such as wave speed or X-ray strength obtained around the object, without damaging the structure of the object, and then a 3D image of the object is re-built using a series of 2D images, to determine if there is any internal quality defect in that object. It is mainly used to detect potential defects in concrete dams and buildings.

25.2.2 Dam Safety Assessment and Deficiency Diagnosis Technologies

In China, periodical dam safety appraisal is conducted according to RDSM and *Safety Appraisal Rule for Dams*. Dam safety assessment and deficiency diagnosis are the main tasks in dam safety appraisal. An overall analysis and assessment of dam safety is based on field inspection, review of project geological investigation information, design document, construction record, reservoir operation record, dam safety monitoring information, supplementary geology exploration and testing. Dam safety appraisal includes assessment of dam construction quality and operation management, review of flood control standard, analysis of dam structure stability, dam seepage stability and seismic stability, assessment of metal structure safety, and analysis of potential deficiencies as well as their causes and hazards, so as to provide scientific basis for dam operation, maintenance and safety management, and reinforcement design for the defective and hazardous dams.

In the 1980s, after massive testing and research work, the safety appraisal was first performed at Yangmaowan Dam in Shaanxi, and then at Wang Yao Dam in Shaanxi Province and Jiubujiang Dam in Hunan Province. In the 1990s, the two key scientific and technological projects in water resources, "Research on Comprehensive Safety Assessment Methods for Earth and Rockfill Dam" and "Research on Failure Mechanism and Process of Gouhou Concrete Faced Sandy Gravel Fill Dam" were completed, and the latter was the first time to study on dam failure mechanism in China. On the basis of the above researches, the Ministry of Water Resources issued the *Dam Safety Assessment Guidelines* (SL 258—2000) in 2000, introducing the safety assessment system (Fig. 25.2 - 1) and assessment standards for dams in China, and a comprehensive safety analysis and assessment theories and methods have been established.

25.2.2.1 Construction Quality Assessment

Construction quality assessment is conducted mainly by field inspection, analysis of historical data and supplementary boring tests, including assessment of engineering geology, hydrological geology conditions, review of construction quality (including quality of foundation treatment, structure form and materials) in line with current specification requirements, and checking quality changes after dam construction for the safe operation.

25.2.2.2 Operation and Management Assessment

Operation and management assessment is to review the reservoir water diversion according to the approved operation procedures (or plan), check the adequacy of flood control and communication facilities, the existence of rules and regulations and emergency preparedness plan, documents, review the structure integrity and status of maintenance and repair, check the adequacy of hydrological forecasting and dam safety monitoring facilities, examine the implementation of safety monitoring and the application of safety monitoring information to serve the operation of reservoir.

25.2.2.3 Flood Control Standard Review

Review of flood control standard is to examine whether the dam flood control capability can meet the requirements in current

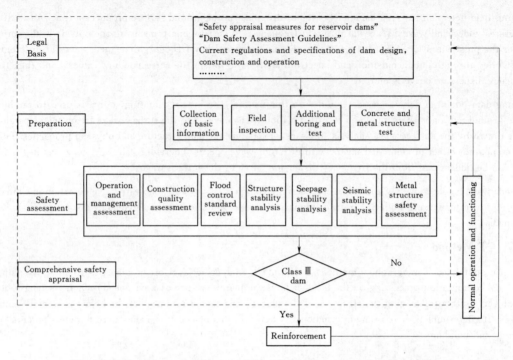

Fig. 25.2-1 Schematic Diagram of Dam Safety Assessment System in China

standards by rechecking of design flood and calculation of flood routing based on hydrological data used for dam design, and the extended hydrological data during the operation period, also taking into account the effect of human activities in upstream area since the completion of the dam and the status of the dam project.

For reservoirs with large storage capacity in natural river channel, the re-check calculation should be based on reservoir inflow flood. If the design flood was based on dam site flood, it should be changed to reservoir inflow design flood whenever possible, or to estimate the adverse impact if design flood is not changed. For medium-sized and small reservoirs whose inflow data are not available, design flood can be calculated from rainfall records or using empirical formula. The calculation results should be checked for its rationality.

25.2.2.4 Structure Stability Analysis

Based on current standards, the analysis includes deformation, strength and stability of dam, water discharging and conveying structures under static loading, via field inspection, analysis of monitoring data and calculation. The emphasis will be placed on deformation and stability analysis for earth and rockfill dam, and on strength and stability analysis for concrete dam and water discharging and conveying structures.

25.2.2.5 Seepage Stability Analysis

This analysis includes checking whether the seepage control and drainage facilities of the project are adequate and whether the design and construction (including foundation treatment) meet the requirements in current standards via field inspection, analysis of monitoring data, calculation and analysis, model testing and empirical comparison, investigating if any abnormal seepage occurred during the operation and judging its effect on dam safety, assessing the performance and status of various seepage control and drainage facilities under the existing conditions of the project, assessing dam safety assuming seepage induced by potential higher reservoir water levels in future operation, and analyzing the causes and possible hazards to the dams induced by potential adverse seepage.

25.2.2.6 Seismic Stability Analysis

This includes checking whether the dam meets the seismic stability requirements of current standards, including the design seismic intensity or ground accelerations, seismic design criteria and corresponding seismic calculation methods, analyzing the seismic stability safety of dam, dam foundation, reservoir banks near the dam, and evaluating liquefaction potential and remedial measures for layers susceptible to liquefaction of earth and rockfill dam.

25.2.2.7 Metal Structure Safety Assessment

This includes checking whether the gates, hoisting equipment and penstocks of the water discharging and conveying structures can be operated in a safe and reliable manner, based on field inspection, testing, calculation and analysis. The mental structure safety assessment includes calculation of strength, rigidity and stability of gates, checking closing and opening ability and power supply reliability of hoisting equipment, and calculation of strength and stability of penstocks under external loadings.

25.2.2.8 Comprehensive Dam Safety Appraisal

On the basis of the above analyses and assessment, the dams can be classified as Class Ⅰ, Class Ⅱ and Class Ⅲ. A "Class Ⅰ" dam is one with flood control capability meeting the requirements in *Flood Control Standard* (GB 50201—94), no major construction quality defects, and its structure stability, seepage safety, seismic safety and metal structure safety are meeting the specification requirements, with adequate facilities for safety monitoring and operation management, perfect operation and management rules, being properly maintained, operation normally and producing benefits according to the design. A "Class Ⅱ" dam is one with flood control capability lower than the requirement in GB 50201—94 but higher than the short term operation flood standard for rehabilitation projects issued by the Ministry of Water Resources ([1989] No. 21), or the construction quality defects and potential structure deficiencies not expect to result in serious consequences, operation is basically normal, and able to operate safely under certain constraints. A "Class Ⅲ" dam is one with flood control capability lower than "Class Ⅱ" dam, or with serious construction quality defects, high hazards, and unable to operate normally as designed, and it is also referred to as defective and hazardous reservoir, requiring rehabilitation.

The former Power State Cooperation classified hydropower station dams as normal dam, defective dam and hazardous dam according to the safety conditions of the projects.

A "normal dam" is one with design standards meeting the current specifications no operation accident, the dam and foundation in good conditions or with local deficiencies not affecting the overall structure safety, and reservoir banks near the dam free of geological problem that may endanger the safety of the dam. A "normal dam" is equivalent to "Class Ⅰ" dam.

A "defective" dam is the one with design standards not complying with current specifications, but the dam operations have been restricted. The dam foundation has local potential defects but does not pose a safety threat to the dam. The dam is stable and structure safety meets the specifications, but when the dam has local deficiencies, the dam still functions normally to retain water. Local deficiencies in the energy dissipation facilities will not affect the safety of water retaining structure of the dam. Potential slope failures or landslide in the reservoir area near the dam will not pose a threat to the water retain structure of the dam. So a "defective" dam is equivalent to "Class Ⅱ" dam.

A "hazardous" dam is the one with design standard below the requirements of current specifications and the dam has not been rehabilitated or renovated or the operation condition has not been changed. The dam foundation potential deficiencies could endanger the safety of the dam. The dam stability or structure safety does not meet the current specifications, and no structural or non-structural remediation has been taken. The dam has signs of problems (including local break or failure); a serious collapse of slope or a landslide in the reservoir area near the dam could endanger the safety of the dam. A "hazardous dam" is equivalent to "Class Ⅲ" dam.

25.2.3 *Dam Safety Monitoring Technologies*

There has been a long history of dam building, but dam safety monitoring technologies were first developed and applied in foreign countries in the beginning of last century. Dam safety monitoring in China started in early 1950s, monitoring of earth and rockfill dam surface transversal and horizontal displacement, settlement and internal stratified settlement (consolidation) was performed for Guanting Reservoir in Beijing (1951 - 1954), Nanwan Reservoir (1952 - 1955) and Boshan Reservoir (1952 - 1954) in Henan, and Dahuofang Reservoir in Liaoning (1954 - 1958), and large size open standpipes were used to observe the phreatic line for earth dams; during the same period, temperature and strain monitoring started for concrete dams of Fengman on the No. 2 Songhua River, Foziling and Meishan on the Huaihe River.

In early 1960s, monitoring of pore water pressure, pressure relief well water level and seepage quantity in large earth and rockfill dam were conducted in China. Since then, various new types of imported and domestic sensors were used. The technology of dam safety monitoring instruments were largely improved with the support of the National Science & Technology Program during the 6th Five-Year Plan (1981 - 1985)" and the 7th Five-Year Plan (1986 - 1990). There were different types of vibrating wire instruments, such as pore water pressure meter, movement meter and earth pressure meter, gradually forming series of vibrating wire instrument; strain meters and three-dimensional displacement sensors gradually came into use, with their performance developing to wide range and high precision; the successful development of instruments such as water tube settlement sensor, tension wire horizontal displacement sensors, electromagnetic settlement sensors, resistance strain chip and servo acceleration inclinometer enabled monitoring deep level displacements of dams.

In the 1980s, China started to introduce and independent develop domestic dam safety monitoring automation technologies. In 1989, supported by one of the key projects in the National Science & Technology during the 7th Five-Year Plan, which is "Development of dam safety automatic monitoring microcomputer system and instruments", the first concrete dam safety monitoring automation system in China was put in service at Canwo Reservoir in Liaoning; In 1993, in conjunction with the Water Resources Science and Technology pilot project "Establishment of Real-Time Dam Safety Monitoring System", the first safety monitoring automation system for earth and rockfill dam in China was built at Dongwushi Reservoir in Hebei, marking the start of practical application phase of dam safety monitoring automation systems in China. At the beginning of the 21st century, dam safety monitoring automation technologies in China have reached a more mature stage, and technical standards such as *Fundamental Specification of Equipment of Automation System for Dam Safety Monitoring* (SL 268—2001) and *Technical Specification for Dam Safety Monitoring Automation* (DL/T 5211—2005) were issued. In recent years, with the support by the Ministry of Science and Technology of "Research on Real-Time Safety Regulation Technology for Large Scale Water Projects", one of the Key Projects in the National Science & Technology Program during the 11th Five-Year Plan of "Research on Dam Safety Information Monitoring and Prediction and Warning Technologies" and "948" project "Application of High Precision GPS System in Dam Safety Monitoring" and other projects, the dam safety monitoring automation system integrating structure performance, hydrology information and gate operation information was developed and applied in projects, which can provide decision-making support for real time safety reservoir regulation for dams.

Over the past 60 years, the dam safety monitoring technologies have gradually developed from nothing into a fairly complete dam safety monitoring and management system that integrates instrument research and development, design and installation, data compilation and analysis, and codes and standards preparation and is able to conduct monitoring and automatic data acquisition of all kinds of items. Generally, the dam safety monitoring technology in China has reached the international level, meeting the needs of dam safety monitoring in China. Meanwhile, great efforts are being made on technology foreland such as development of new type of optical fiber sensors, improvement of automation monitoring system, construction of networked dam safety monitoring and management information system, and development of expert system for monitoring dam safety with data analysis.

25.2.3.1 Basic Principles of Dam Safety Monitoring

(1) Define the purpose of monitoring, and determine the temporary and long-term monitoring items by taking into account various monitoring requirements in construction period, reservoir impounding period and operation period.

(2) Based on relevant technical specifications, the selection of monitoring items and location of monitoring points should consider project characteristics, project class and size, structure type, service period, topological and geological conditions and geographic environment.

(3) To emphasize the key elements with due consideration of overall situation, pay attention to critical area reflecting structure performance and consider requirements on safety monitoring data analysis, safety monitoring calculation model establishment and monitoring indicator comparison and verification, and monitoring items related to each other should be coordinated in arrangement for comprehensive analysis.

(4) Safety monitoring items are different for different types of dams. Emphasis is on seepage, deformation, water level and rainfall monitoring for earth and rockfill dam, while on deformation, stress and strain, temperature, seepage and water level monitoring for concrete dams.

(5) Safety monitoring for new dams should be based on theoretical design, taking into account monitoring in construction period for direct project construction, and safety monitoring during reservoir impounding period, to verify design and monitor dam operation performance; for existing dams, especially defective and hazardous dams, monitoring should be focused on abnormal and critical area of the project.

25.2.3.2 Dam Safety Monitoring Items

Dam safety monitoring consists of field inspections and instrumentation monitoring. The latter includes environmental parameter monitoring, deformation monitoring, seepage monitoring, stress, strain and temperature monitoring, and seismic response monitoring.

(1) Environmental parameter monitoring. It includes upstream and downstream water levels, rainfall, water and air temperature, and atmospheric pressure.

(2) Deformation monitoring. It includes dam external deformation, internal deformation, foundation deformation, cracks and joints, concrete face deformation, reservoir bank displacement.

(3) Seepage monitoring. It includes seepage pressure, seepage quantity, and seepage water turbidity.

(4) Pressure monitoring. It includes pore water pressure, uplift pressure, earth pressure and contact earth pressure.

(5) Stress, strain and temperature monitoring. It includes concrete stress and strain, anchoring rod (cable) stress, rebar stress, steel plate stress, bedrock displacement and temperature monitoring.

(6) Seismic response monitoring. It includes strong motion acceleration and dynamic pore water pressure monitoring.

25.2.3.3 Dam Safety Automated Monitoring

Dam safety automated monitoringsystem consists of sensors, measuring and control units, central control unit and communication links (Fig. 25.2 - 2). The measuring and control units are installed at project site for automatic data acquisition and control of acquisition process, with the functions of data measurement, communication and storage, self-check and diagnosis, protection against lightning and electromagnetic interference, automatic power supply and protection against power failure. The central control unit is the equipment and software to perform centralized management of the distributed data acquisition system, with the functions of automatic control, manual control, self-check, remote control, and monitoring data management and analysis.

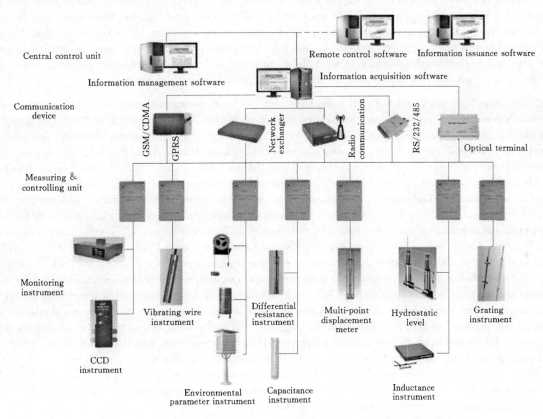

Fig. 25.2 - 2 Structure Diagram of Dam Safety Automated Monitoring System

Monitoring instruments and measuring & controlling devices usually communicate via specific cables, and relevant wireless or optical cable communication can also be used for different types of monitoring instruments; the bus approach is normally adopted between measuring & controlling devices and central controlling unit, and the communication media can be twisted pairs cable, optical fiber, radio and GPRS. For the communication between central controlling unit and outside, a number of media such as wire, radio and public network are used to realize remote monitoring.

25.2.3.4 Analysis on Dam Safety Monitoring Data

Analysis methods for monitoring data include comparison of measured values, hydrograph analysis, correlation analysis, distribution map analysis, characteristic value statistics, and analysis of statistic numerical models, to find out changing law of measured quantities, and then to assess the operation performance and status of dams.

Methods for making numerical models for monitoring data include regression analysis, fuzzy mathematics, gray system theory and neural network theory. As statistic numerical models usually produce poor precision in extended calculation, mixed numerical models can be established by combining statistic numerical models with deterministic ones, i. e. the analysis results from observation data can be used to invert calculation parameters in the deterministic numerical model, and after rationality verification, it is believed that this deterministic numerical model is closer to the practical conditions of the project, so it can be used to deduce the dam safety at high water level, to demonstrate the prediction, forecasting and monitoring functions of the system.

Today, information technology is being applied more extensively to the analysis of dam safety monitoring data. The storage, retrieval, statistics and chart processing of monitoring data has become more convenient and rapid, and many advanced and practical monitoring data analysis software and information management systems have been developed and promoted for application.

25.2.4 Dam Risk Analysis and Risk Management Technologies

Risk analysis technology started in United States in the late 1970s and the early 1980s. In 1991, BC Hydro of Canada started to apply the risk management technology in the field of dam safety management, which drew attention and was generally accepted in the dam engineering field in the world, and it has been applied and developed in many countries. In 1997, a seminar on dam risk evaluation was held in Norway. In 2000, at the 20th International Conference on Large Dams held in Beijing, dam risk analysis technology was taken as a key topic for the first time, and Bulletin "Risk Assessment-Assist Dam Safety Management" was issued. In 2003, the International Commission on Large Dam (ICOLD) issued "Risk Assessment in Dam Safety Management", and risk management has gradually become the developing trend in dam safety management. In developed countries such as Canada, United States and Australia, dam risk management technologies have formed a discipline system, and dam risk management has come into the practical application phase.

Early in this century, the researches and application about dam risk analysis and risk management technologies started in China through international cooperation and technical exchangement. In 2003, with the support of "948" program of "Reservoir Risk Evaluation and Intelligent Dam Safety Monitoring Technology", the experts from China and Australia jointly conducted a systematic risk analysis before and after reinforcement of Shaheji Dam in Anhui Province, and the reservoir risk standard was studied and discussed. It was the first case of dam risk evaluation in China. In 2004, the framework and standards for dam risk evaluation in China were initially established based on the researches on "Risk Judgment Standard System for Defective and Hazardous Reservoirs". In 2005, the dam safety assessment methods based on risks were proposed in conjunction with research on "Key Technologies for Risk Analysis and Emergency Response for Existing Water Projects" as a key scientific and technological innovation project of the Ministry of Water Resources, and portfolio risk assessment and pilot application was conducted in five medium-sized reservoirs in Ganzhou of Jiangxi Province. This was the first application case of dam risk assessment technology into the actual operation management in China; in 2006, the book "Dam Risk Assessment and Risk Management" was published as a mark that dam risk analysis and risk management technology research has initially become a system in China; during 2006 - 2007, with the support of Sino-Canada cooperation project "Dam Safety Management and Reservoir Risk Analysis", China selected and sent a large number of engineers and reservoir operation management personnel to Canada to systematically study dam risk analysis and risk management technologies and conduct case study for the east auxiliary dam of Shahe Reservoir in Jiangsu in cooperation with Canadian experts, which further promoted the technical progress of dam risk analysis and risk management and the popularization of risk concept in China; in May 2007, the promulgation and implementation of the regulation document *Guidelines for Preparation of Dam Emergency Preparedness Plan* of the Ministry of Water Resources based on risk management concept indicated that dam risk management technologies have come into the practical application phase in China.

In recent years, in conjunction with researches in many key scientific and technological projects such as Public Benefit Sector Scientific and Research Project "Research on Dam Failure Life Loss Evaluation, Warning and Emergency Plan" funded by the Ministry of Science and Technology, and several projects of the national key scientific and technical supporting program of the 11th Five-Year Plan, "Study on Dam Breach Test and Simulation Techniques", "Research on Risk-based Dam Safety Assessment Methodology System", "Research on Non-structural Measures for Dam Risk Control", and "Research on Dam Risk Standards", a number of field dam breach tests and laboratory dam failure model tests have been conducted, the 2D numerical models of dam failure and flood routing results for earth and rockfill dam have been built, the dam failure flood numerical simulation and GIS integration realized, the risk-based dam safety assessment system established (Fig. 25.2-3), and dam risk management model suitable to conditions in China put forth, so that the dam risk analysis and risk management technologies in China have been further developed and improved. In the meantime, relevant technical standards such as *Risk Evaluation Guidelines for Dams* and *Risk Classification Standard for Dams* are being prepared. It can be foreseen that in the near future, dam risk analysis and risk management technologies will be broadly applied to the dam safety management in China.

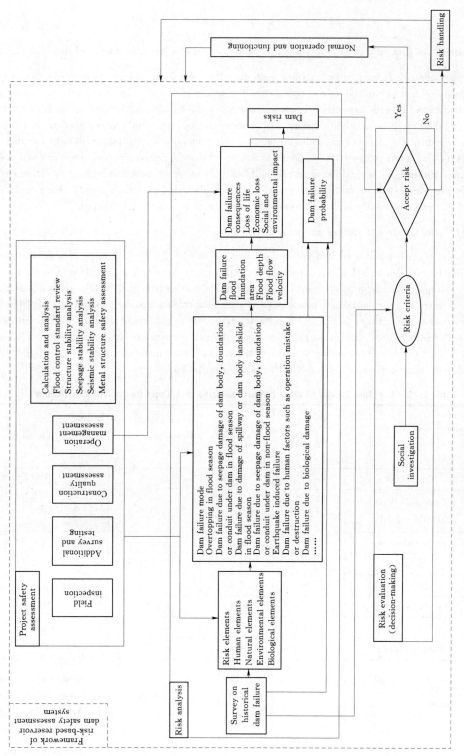

Fig. 25.2-3　Schematic Diagram of Risk-based Reservoir Dam Safety Assessment System

25.2.4.1 Dam Risk Analysis

Dam risks are inherent from the effect of dam failure on downstream areas, as a measure of the probability dam failure and severity of an adverse effect to life, health, property or the environment. It can be expressed as the product of dam failure probability and dam failure consequence, i.e.:

$$R = P_f \times L \qquad (25.2-1)$$

where R is the risk of the dam, P_f is the dam failure probability, and L is the consequence of dam failure.

Dam risk analysis methods include qualitative analysis, semi-quantitative analysis and quantitative analysis. In qualitative analysis, the possibility of dam failure and its consequences are described in words or by numerical range based on historical dam failure statistic data and experts' experience. In quantitative analysis, the dam failure probability is calculated using numerical calculation method based on reliability theory. In the initial period, the risk analysis method was applied to dam safety management, quite a few experts and scholars preferred using quantitative analysis method to calculate dam failure probability, however, due to the complexity, randomness and uncertainty in dam engineering structure and operation environment, up to now, the analysis method based on reliability has not become a practical method to determine dam failure probability, and further studies and exploration are required. Today, dam risk analysis is mainly used as a decision-making tool, mostly using the qualitative analysis based on expert experience with quantitative analysis.

1. Identification of failure mode and failure pathway

According to the latest statistic analysis of dam failures in China (Table 25.2-1), the main dam failure modes include: ①overtopping in flood season due to lack of flood discharge facilities or insufficient discharge flow, insufficient freeboard or gate malfunction; ②seepage damage in flood season; ③damage of spillway or dam slope failure in flood season; ④seepage damage in non-flood season; ⑤earthquake; ⑥human factors such as operation mistake or sabotage; and ⑦biological damage. Failure pathway means the pathway leading to dam failure event due to damage of water retaining, conveying or discharge structures under certain load, and can be expressed as load-structure-damage-dam failure, such as "flood-gate malfunction -reservoir water level rising-overtopping-scouring dam body-ineffective intervention-dam failure", or "flood-concentrated seepage in dam foundation-piping -rescue-ineffective rescue-dam failure".

Table 25.2-1 Main Causes of Dam Failure in China and Their Respective Proportions

Classification of dam failure causes		No. of dams	Proportion (%)	Annual average dam failure probability (10^{-4})	Remarks
Overtopping	Extra flood	435	12.6	1.0996	1,737 dams, accounting for 50.2% of total
	Inadequate flood discharge capability	1302	37.6	3.2912	
Dam quality	Dam or foundation seepage	701	20.2	1.7720	1,205 dams, accounting for 34.8% of total
	Dam slopes	110	3.2	0.2781	
	Spillway	208	6.0	0.5258	
	Flood discharge tunnel	5	0.1	0.0126	
	Culvert	168	4.9	0.4247	
	Dam settlement	13	0.4	0.0329	
Improper management		185	5.3	0.4676	Including unattended, excessive impoundment, improper operation and weir building in spillway
Others		212	6.1	0.5359	Human destruction, landslide on reservoir slopes, blocking of spillway and improper project layout, etc.
Total		3339		8.75	

2. Dam failure probability calculation

After the failure mode and pathway have been identified, experts can assign value P_{ij} for event probability with reference to the qualitative description of the failure event and the corresponding probability (Table 25.2-2), and the product of occurring probability of all events under each dam failure pathway is the dam failure probability P_m of that failure mode. The sum of dam failure probabilities of all possible dam failure pathways is the dam failure probability P_f of that dam.

Table 25.2-2 Qualitative Description of Dam Failure Occurrence and Corresponding Probability

Qualitative description	Corresponding probability	Judgment criterion
The event is extremely unlikely to occur	$1\times10^{-6} - 1\times10^{-4}$	For a specific event and based on historical data and dam safety appraisal results, and in conjunction with dam safety management and long-term operation conditions
The event is basically unlikely to occur	$1\times10^{-4} - 1\times10^{-2}$	
The event is possible to occur	$1\times10^{-2} - 1\times10^{-1}$	
The event is quite likely to occur	$0.1 - 0.5$	
The event is extremely likely to occur	$0.5 - 1.0$	

Assuming there are k experts, and their weight in determining the dam failure probability can be determined by the factors of their experiences and ability. If it is considered at equal weight, the occurring probability $\overline{P_i}$ at the ith link is:

$$\overline{P_i} = \frac{\sum_{j=1}^{k} P_{ij}}{k} \qquad (25.2-2)$$

Where P_{ij} is the event occurring probability assigned by the jth expert for the ith link.

Assuming there are m links in the dam failure pathway under the ith failure mode, the dam failure probability P_m is:

$$P_m = \prod_{i=1}^{m} \overline{P_i} \qquad (25.2-3)$$

Where m is the total number of links in a given failure mode.

The total failure probability P_f of the dam is:

$$P_f = \sum_{m=1}^{n} P_m \qquad (25.2-4)$$

Where n is the total number of failure modes.

3. Dam-break flood analysis

Firstly, Dam-break flood was calculated to obtain the hydrograph of the flow and flow velocity at the dam site, and the hydrograph of the reservoir water level and other hydraulic elements and breach shape parameters, to provide basic data for downstream flood routing calculation. For earth and rockfill dam, normally the gradual dam break mode is selected, with commonly used calculation models as BEED, BREACH and FLDWAY; for concrete dams, normally instantaneous dam break mode is selected.

Dam break flood routing calculation is mainly used to determine the inundated area and water depth, water flow velocity and duration of the dam break flood and the empirical formula method, simplified analysis method and numerical physical model method are used in calculation. The most commonly used at present is the numerical model method, for which many mature commercial software versions are available, such as DAMBRK and FLDWAY developed by NWS, HEC-RAS developed by USACE, and the MIKE series developed by the Water Science Research Institute of Denmark. In recent years, the GIS technology has been found applicable extensively to dam failure flood routing analysis, so that the dam failure flood routing computation software has more powerful functions and can provide more visual results (Fig. 25.2-4).

4. Dam failure consequences calculation

Dam failure consequences mainly consider life and economic losses.

(1) Loss of life. The factors considered in loss of life due to dam failure include number and distribution of population under risk, time of dam failure, alarming time, water depth and flow velocity, flood water rising rate and evacuation conditions. Simple and practical estimation methods are Dekay & McClelland method (Formula 24.2-5) and Graham method (USBR, 1999), both taking into account the effect of alarm time, population under risk and flood intensity. The Dekay &

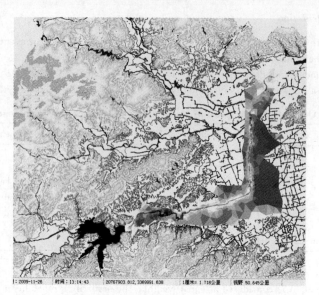

Fig. 25.2-4　Water Depth Distribution Map from 2D Flood Routing Calculation Downstream of a Reservoir ($t=20h$)

McClelland method is based on historical statistic data, and the Graham method also takes into account the effect of public awareness of flood severity.

$$L_{OL}=0.075P_{AR}0.56\exp[-0.759W_T+(3.790-2.223W_T)F_C] \quad (25.2-5)$$

Where L_{OL} is the potential loss of life due to dam failure; W_T the alarm time; P_{AR} the population under risk; F_C the flood intensity, which is taken as $F_C=1$ for high flood risk areas such as large reservoirs, high dams and mountainous areas, and $F_C=0$ for low flood risk areas such as medium-sized and small reservoirs, low dams and plain areas.

(2) Economic Losses in Case of Dam Failure.

Economic losses resulting from dam failure include direct and indirect losses. The direct economic losses include damage to reservoir project and various direct losses that can be measured by currency. The latter includes losses that can be measured by currency other than direct losses, such as additional cost in taking countermeasures, reduction and stoppage of industrial and agricultural production resulting from dam failure, human resources in rescue, and expenditure on restoration of production after the disaster. As the indirect losses involve a wide range of areas and have difficulty defining the boundaries for calculation, direct estimation method or coefficient method is used to estimate the loss.

25.2.4.2　Risk Management for Dams

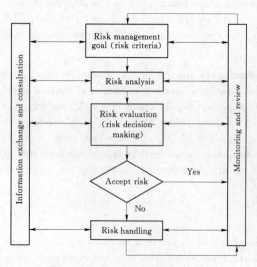

Fig. 25.2-5　Block Diagram for Dam Risk Management

Risk management for dams is a management mechanism based on risk measurement to accept, refuse, reduce or transfer risks through a full-process management system (Fig. 25.2-5). In a risk management system, a "safe" dam means that first of all, its risks are acceptable to the public and its designed functions can be fulfilled. Risk management for dams is the expansion of existing dam safety management mode, which links the dam safety with public safety, incorporates dam safety management into social public safety management, and thus sets more clear management goals for the dam safety management department, suggesting a major change in management concept. Risk management can go through the whole life cycle of the dam. In all phases of planning, design, construction, operation, degradation, abandonment and removal, the risk analysis technology can be used to perform risk analysis, the risk standards can be used to assess level of risks, and risk control strategy and decision-making technology to mitigate and control risks as well as ensure realization of the overall goal with this series of operations.

(1) Risk standard. Risk standard is also the goal of dam risk manage-

ment, involving all aspects of social traditional culture background, values, management system and insurance system. The development of and changes in society and economy can directly or indirectly affect the determination of the risk management goal. A rational risk standard can achieve the overall management goal of coordinating project safety, safety of downstream areas and sustainable development. China is now studying and formulating dam risk standard.

(2) Risk evaluation (decision-making). Risk evaluation is a process of decision-making. On the basis of the risk analysis result, it is determined whether the existing risks are tolerable, the risk control measures are appropriate, and how the risks are reduced with structural or non-structural measures.

(3) Risk handling. Risks that are not tolerable must be handled, including mitigating, transferring, avoiding and retaining such risks.

(4) Information exchange and consultation. This involves information exchange with internal and external risk bearers at appropriate time in all phases of risk management, to ensure that risk management decision-makers understand and decide whether special measures should be taken. It ensures that risk management can respond to the effect of continuously changing external information on risk assessment.

(5) Monitoring and review. Monitoring and review are a process of performing monitoring and review of the risk management system to ensure the normal functioning of this process.

25.2.5 *Defective and Hazardous Reservoir Reinforcement and Decision-making Technologies*

Defective and hazardous reservoirs are not only unable to perform their normal functions, but also have comparatively high risks of dam failure, posing serious threats to the safety of lives, properties and infrastructures downstream. As a result, the governments and water resources authorities have always paid high attention to reinforcement of defective and hazardous reservoirs. After the severe flood in Aug. 1975, China reinforced 65 large reservoirs mainly by upgrading the flood control standard during 1976—1985. Main defective and hazardous reservoirs were designated for reinforcement with 43 in the first stage from 1986 and 38 in the second stage from 1992, including 69 large reservoirs and 12 medium-sized reservoirs. After the big flood in 1998, the Central Government speeded up the pace of reinforcing defective and hazardous reservoirs, and over 2300 defective and hazardous reservoirs specifically planned in two phases were reinforced during a period of 10 years. At the Rural Area Working Conference by the Central Government in Dec. 2007, the State Council determined to basically complete the reinforcement of all large and medium-sized reservoirs and major small reservoirs in the whole country within 3 years. The Ministry of Water Resources attached great importance to this task, and 6,240 defective and hazardous reservoirs were selected for rehabilitation and scheduled to be completed by the end of 2010, including 86 large reservoirs, 1096 medium-sized reservoirs and 5,058 small reservoirs. During the "12th Five-year Plan" period, the State Council will continue to invest a huge sum of money in reinforcing defective and hazardous dams.

In recent years, scientific research and construction units have developed many new technologies, new processes and methods of defective and hazardous reservoirs reinforcement.

25.2.5.1 Reinforcing Technologies for Defective and Hazardous Reservoirs

1. *Reinforcement for flood control*

To raise the flood control ability of reservoirs, there are 2 main methods of reinforcement: one is to increase the height of the dam and improve the storage and regulation capacity of the reservoir, and the other is to add or expand flood discharge facilities to increase the flood discharge ability.

(1) Raising dam height.

1) Adding a parapet wall. It is suitable for dams without any parapet wall on the crest. The dam height can be raised by adding a parapet wall on the upstream side of the crest. It involves minimum work in reinforcing. The parapet wall is normally 1-1.2m high, and should not exceed 3m.

2) Raising the crest elevation. The dam crest elevation can be increased from the existing level to the elevation that meets the requirement of flood control. While the dam height increased, the width of dam section also increased. Therefore the work can be done in conjunction with the dam slope reinforcement.

(2) Adding or expanding flood discharge facilities. In case raising dam crest is not feasible due to geological and topological features, reservoir inundation or resettlement conditions, it can be considered to add or expand the existing flood discharge facilities, to lower the design and check flood water level by increasing the flood discharge capacity while maintaining the

normal water storage level, thus ensuring that the existing dam crest elevation meets the flood control standard requirements.

As small reservoirs are poorly managed and even unattended, a new spillway should be considered, and be provided with an open overflow weir without gate control. In case of expansion, it is also suggested to rebuild it as an open overflow weir without gate control.

2. Reinforcing to control seepage

Seepage is one of the main problems in defective and hazardous reservoirs, and one of the main causes of dam failure. Safety problems in dam structure (such as cracks and loss of stability on dam slope) are also frequently caused by seepage. Therefore, to control seepage is often a key to the success of reinforcing defective and hazardous reservoirs. In recent years, a number of reinforced dams have burst because of improper treatment in the reinforcement against seepage.

(1) Treatment of foundation seepage.

1) Weight ballasting. In case piping or flowing soil appears downstream of the dam, filter weight ballasting method can be used.

2) Drainage to reduce pressure. When the low permeable layer in the upper foundation is thin, a pressure relief ditch can be provided at downstream toe and parallel to the dam axis, to drain water and reduce pressure. The ditch shall be excavated into the high permeable layer and filter protection shall be provided. If it is difficult to dig a ditch where the low permeable layer is thick, pressure relief wells can be provided to drain water and reduce pressure.

3) Cutoff trench. A clay cutoff trench can be dug near the upstream toe of the dam, parallel to the dam axis, down to the weak weathered bedrock, to cutoff seepage in dam foundation. It is suitable for dams where dam foundation high permeable overburden and severely weathered bedrock layer are not deep.

4) Blanket. A layer of impervious material (clay or geomembrane) can be provided at the upstream toe of the dam to reduce and control foundation seepage.

5) Grout curtain. A curtain can be formed in the dam foundation by grouting to control seepage through the pervious layer and reduce seepage flow, and protect dam foundation against erosion from concentrated seepage flow and scouring.

6) Karst grouting. In Karst developed areas, in case of concentrated seepage channels in the dam foundation, filter can be formed by boring and backfilling graded materials and then cement mortar can be grouted to block such channels.

7) Seepage cutoff wall. It is the presently the most common method for controlling foundation seepage to construct a concrete wall in the highly pervious dam foundation to thoroughly cut off the seepage channel. The impervious wall can be constructed by various methods such as underground concrete, Saw-type engineering, hydro-cutting, interlocking piles and jet grouting. The selection of method is based on the thickness and gradation of overburden and the local construction methods and conditions.

8) Vertical plastic seepage cutoff sheet. It is suitable to small and long dams in plain areas and the dam foundation is silt soil or silty sand and fine sand.

(2) Treatment of abutment seepage.

1) Blanket. Impervious materials (clay or geomembrane) can be provided to cover the upstream slope to form a blanket layer, and cutoff the abutment by-pass seepage. It is suitable for dams with inclined core and homogeneous dams with flat slopes.

2) Grout curtain. For abutment slopes with thick weathered layers and developed fractures, and without any cutoff trench or with a cutoff trench that is not deep enough, curtain grouting can be used. The grout curtain should be reliably connected to dam and foundation seepage cutoff works.

3) Contact grouting. For rock at abutment slope is intact and contact seepage between the dam and rock slope is the only concern, grouting can be done only for the contact surface.

4) Karst grouting. In areas with developed karst, concentrated seepage channels in the abutments can be cut off by grouting.

5) Downstream filter protection. A filter layer can be constructed at the seepage exit to prevent seepage erosion.

(3) Treatment of dam seepage.

1) Split grouting. It was extensively applied in China during the 1970s and 1980s. However, based on results of investigations, the actual effect was not satisfactory, and seepage gradually increased and became serious again in many treated dams after years of operation. It can be used for low and "fat" dams, but is not suitable for high and "slim" dams.

2) Inclined wall. A clay inclined wall or geomembrane is provided on the dam upstream face to cut off seepage, with its lower end connected to the foundation cutoff trench or clay blanket upstream of the dam.

3) Vertical plastic seepage cutoff sheet. It can be constructed together with that for the dam foundation, and is suitable for dams in plain areas.

4) Seepage cutoff wall. It is now in extensive application. There are many construction methods or types, such as excavated trench backfilled with clay, jet grouting, panels filled with plastic concrete, and the construction is often performed concurrently with the seepage cutoff treatment for dam foundation.

5) Impervious seepage cutoff connected to geomembrane. This consists of a seepage cutoff wall below a given elevation upstream (such as dead water level), and geomembrane above that elevation. It is commonly used today.

6) Seepage diverting downstream. Filter protection along downstream slope is provided below the possible highest seepage exiting point, to protect the seepage exiting areas against erosion and deformation. It is normally used as an auxiliary or emergency treatment means.

(4) Seepage treatment at contact of dam and buildings.

1) Excavation and backfilling with impervious key wall. The junction location is excavated to add 1 - 2 concrete key walls, and then it is backfilled with clay.

2) Split grouting. It is done upstream embankment with a grouting pressure about 3 times the water head. Cement mortar or cement mortar mixed with clay is used at the junction of earth dam and side wall, and clay mortar can be used in areas far away from the wall.

3. Structure reinforcement

Structural defects commonly with earth and rockfill dam are cracks and loss of dam slope stability.

(1) Treatment of cracks.

1) Providing a protection layer. Dry shrinking cracks and freezing/thawing cracks often occur on surface of homogeneous clay dams and on top of clay core dams. Slope and crest protection can be provided with riprap, crushed rocks, sandy soil and concrete, with a thickness greater than the depth of the frozen soil.

2) Excavation and backfilling. This involves removing all cracks, and then backfilling and compacting with materials similar to the surrounding soil. It is the most commonly used method to treat cracks, and is suitable to all types of shallow cracks.

3) Backfill grout. For deep and big cracks, the shallow part can be treated by excavation and backfilling, and deeper part can be treated by grouting. But grouting for cracks should be used with caution to prevent slope failures.

4) Split grouting. This can be used to treat deep and concealed cracks. It is suitable mainly to wide homogeneous dams and dams with wide core, but is not suitable to cracks resulting from slope failure, and may possibly result in new cracks in the impervious zone of the dam.

(2) Dam slope stabilization and reinforcement.

1) Lowering the phreatic line in the dam. Lowering the phreatic line will increase the effective stress of dam fill and the stability of the dam downstream slope.

2) Improving drainage. A ditch filled with reverse filter or a wedge of toe drain can be provided in downstream slope areas to increase its stability.

3) Flattening dam slopes. The dam slopes can be flattened as appropriate in conjunction with slope treatment to meet the stability requirements.

4) Combination of load reduction and slope cutting. For dams with excessive freeboard, dam fill can be removed from the crest to reduce loading, and then the slopes can be cut accordingly.

5) Reinforcing dam toes. Masonry stone can be placed on the upstream toe or a masonry retaining wall can be built at the downstream toe to reinforce the toes.

6) Weight ballasting at dam toes. Ripraps can be placed on the upstream toe to increase resistance to sliding, and pervious materials can be filled at the downstream toe to increase resistance to sliding, provide drainage and reduce pressure.

4. Seismic reinforcement

Earthquakes can do great harm to earth and rockfill dam, and the common harmful effects include cracking, settlement, loss of slope stability and soil liquefaction. The reinforcing treatment methods for cracks and loss of slope stability are the same as above.

If the freeboard is insufficient due to settlement after an earthquake, the dam height should be increased, or a parapet wall should be built.

Reinforcing methods for sandy soil against seismic liquefaction can be basically grouped into three categories: replacement, densification and adding weight. By replacement, the liquefiable sandy soil is removed and replaced with non-liquefiable materials such as excavate rockfill spoil. By various densification methods, such as vibration, vibrating compaction, and blasting are applied to increase the density of the soil to prevent seismic liquefaction. By adding weight, weight berm is placed above the sandy soil to increase its effective stress and resistance to seismic liquefaction.

5. Reinforcing dam culverts and tunnels

(1) Replacing culvert with tunnels. In many existing reservoirs in China, buried dam culverts were used as the water conveying facilities. Due to dam deformation, poor quality of culvert and improper structure design, water leakage has occurred in many culverts due to breaking or damage of water stops, resulting in contact scouring damage and seriously endangering the dam safety. Therefore, masonry culverts on soil foundation should preferably be blocked and scrapped after break and leakage have been observed and it shall be replaced by building tunnels at river banks. If it is rebuilt at the original location, special attention should be paid to preventing the piping and erosion of soil placed at contact between the culvert and dam fill.

(2) Reinforced concrete lining. A reinforced concrete lining can be applied to the existing tunnel section to meet the structure safety and seepage control requirements. But this will reduce the cross section of the tunnel, affecting the water conveying capacity.

(3) Steel lining. A steel lining can be applied in the existing culvert or tunnel to meet the structure safety and seepage control requirements. The gap between the steel lining and wall can be grouted with cement mortar to join the lining with the existing wall. As the steel lining roughness is low, the water conveying flow will normally not be reduced after adding the steel lining.

(4) Applying high strength carbon fiber fabric liner. For tunnel concrete liner with insufficient anti-cracking ability, or the liner has defects such as cracks and cavities, applying 1 - 3 layers of high strength carbon fiber fabrics on the wall can increase its strength and achieve the purpose of seepage control.

25.2.5.2 Risk-based Decision-making Technology for Reinforcing Defective and Hazardous Reservoirs

Presently in China, dams are classified by their safety levels, which are used as basis to determine whether reinforcement is required. This method is obviously unreasonable as it does not link with the consequences of dam failure. For example, some dams, although in poor safety conditions as "class Ⅲ" dams, do not have high hazards because of low dam failure consequences; but some dams, although in slightly better safety conditions as "class Ⅱ" dams, may have extremely high hazards because their consequences of dam failure are high, with significant impact on the downstream area, and some other dams, although in similar engineering safety conditions, may differ greatly in failure risks due to different conditions such as reservoir volume, elevation, dam height and impact on the downstream area. None of these factors can be taken into quantitative consideration when dams are classified by safety level of projects. Therefore, we often make some unreasonable decisions that dams with low risks are reinforced before those dams with higher risks. Especially when risk mitigation and reinforcement are required for a large number of "class Ⅲ dams" at the same time with shortage of fund, it would be quite difficult to make a scientific decision to ensure that defective and risky dams with high hazards can be reinforced first. Therefore, great attention has been paid to the application of risk ranking technology for defective and hazardous dams.

1. Risk ranking for group of dams

Comprehensive assessment of dam failure consequences. In the dam failure risk calculation, the dam failure consequences include loss of life, economic loss and social and environmental impact. The three items are in different units and cannot be added directly. Therefore, it is necessary to take into account the effect of them in a comprehensive way. So the dam failure consequences comprehensive assessment function L is introduced to establish the dam failure consequences comprehensive assessment method.

The dam failure consequences comprehensive assessment function L reflects the overall effect of dam failure on loss of life, economic loss and social and environmental impact, and it is composed of the following linear weighing method:

$$L = \sum_{1}^{3} S_i F_i = S_1 F_1 + S_2 F_2 + S_3 F_3 \quad (25.2-6)$$

where S_1, S_2 and S_3 are respectively the weighing coefficients of loss of life, economic loss and social and environmental impact; and F_1, F_2 and F_3 are respectively the seriousness coefficient of loss of life, economic loss and social and environmental impact.

Studies have shown that S_1, S_2 and S_3 taken respectively as 0.737, 0.105 and 0.158 are suitable for the conditions and social and economic development level in the present stage in China. The dam failure consequences seriousness coefficient F_i is determined using the normalized function model as shown in Fig. 25.2-6, where

$$F_1 = r_1 = \frac{1}{5^{0.1}} (\lg x)^{0.1} \quad (25.2-7)$$

where x is loss of life (people), and F_1 the seriousness coefficient for loss of life.

$$F_2 = r_2 = \frac{1}{5^b} \left(\lg \frac{x}{10}\right)^b, b=0.1 \text{ or } b=0.2 \quad (25.2-8)$$

where x is economic loss (10000RMB), and F_2 the seriousness coefficient for economic loss.

$$F_3 = \frac{1}{4} \lg f \quad (25.2-9)$$

where f is the social and environmental impact coefficient, and F_3 the seriousness coefficient for social and environmental impact.

After the dam failure consequences comprehensive assessment function L is obtained, the risk index R of the reservoir can be calculated using the following formula:

$$R = 1000 P_f L \quad (25.2-10)$$

where P_f is dam failure probability.

For a series of defective and hazardous reservoirs requiring reinforcement, their risk index R is calculated respectively, and these risk indices are used for prioritization of reinforcing dams.

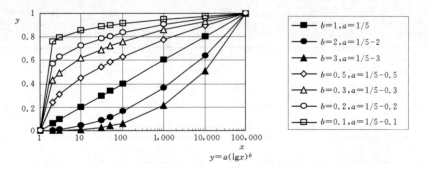

Fig. 25.2-6 Normalized Function Model

2. Ranking technology of risk elements for a single reservoir

For a single reservoir, the dam failure consequences are certain. It is only necessary to rank the qualitative or quantitative analysis results of different events (elements) in the dam failure probability analysis and then the importance of different risk elements can be determined. This can ensure that higher risk elements can be treated in priority when there is no sufficient funding available to reinforce the dam.

FMECA (Failure Mode, Effects and Criticality Analysis) is a qualitative analysis method suitable for risk element ranking for a single reservoir, and is a combined analysis method of failure mode effect analysis (FMEA) and Criticality Analysis (CA).

(1) Identification of risk elements. The dam is divided as a system into a number of sub-systems, which are further decomposed into a number of elements, and elements are the minimum units constituting certain functions of the reservoir. For a dam system, the system is finally divided into many elements, each with many different functions.

Preliminary evaluation is performed on the effect of each element on the system performance after the element fails. The elements with minor effect on system performance are removed, and only those with key functions to system performance are retained for further analysis and studies.

For all selected elements, their failure mode should be identified. The damage of a certain element may not have direct effect on the whole system, but it may be a certain link in a series of elements and the damage of this series of elements may have serious effect on the system. Therefore, in identifying the failure mode of elements, it is necessary take into account the interaction between elements. Usually, an element tree or fault tree can be used to analyze the interaction between elements and the failure sequence of a series of elements.

To determine the extent of failure of elements, it is necessary to determine the direct effect and final effect (consequences) of element failure. Usually, failure of one element may not directly affect the functions of other elements or sub-systems of such elements, but may affect other elements indirectly, or directly affect the operation of the system, or directly affect objects outside the system. Some elements produce apparent direct consequences when failing, such as power supply lines and communication equipment. However, due to the interaction between elements, it would not be easy to determine the failure consequences of some elements. For the failure mode of each element, the emphasis should be placed on direct effect of the element failure mode, while taking into account the final effect as far as possible.

(2) Analysis of elements hazard extent. Hazard extent is a qualitative measurement of risks when it is difficult to calculate the dam failure probability, and is related to the likelihood of element failure mode, seriousness of failure consequences and probability of successful human intervention, being the product of above three factors.

The likelihood of element failure mode is determined by experts according to their experience, and the determination criteria are as shown in Table 25.2 – 3.

Table 25.2 – 3 **Failure Mode Likelihood Values**

Failure mode likelihood	Annual occurrence probability	Determination criteria
Extremely impossible	Below 0.2‰	Extremely impossible to occur during the life cycle of project, such as maximun credible earthquake (MCE) or probable maximum flood (PMF)
Impossible	2‰ – 0.2‰	Quite impossible to occur during the life cycle of project
Basically impossible	2% – 2‰	Possible but not expected to occur during the life cycle of project
Possible	20% – 2%	May occur during a certain phase in the life cycle of project
Extremely possible	Greater than 20%	Frequently occur, or may occur within recent 5 years if not treated

Different criteria can be used to evaluate the seriousness of failure mode consequences of each element and its effect to the system. As it is a qualitative evaluation, it is just to take into account the relative importance of consequences. If a 5-level criterion is adopted, the determination criterion is shown in Table 25.2 – 4.

Table 25.2 – 4 **Consequence Classification**

Consequence category	Determination criteria
Minor	Economic loss is not exceeding RMB 50,000, without casualty, without environmental impact or external impact
Fairly little	Project economic loss between RMB 50,000 – 1 million, with no casualty, or downstream property loss of RMB 25,000 – 500,000, or pollutant with permanent effect discharged and no apparent impact on agriculture, or without environmental impact, or with no serious external impact, or reinforcing expenditure of RMB 20,000 – 200,000, or the combination of the above
Serious	Project economic loss between RMB 1 – 10 million, with a number of serious or fatal casualties, or downstream property loss of RMB 500,000 – 5 million, or pollutant with permanent effect and discharged causing long-term harm to environment or agriculture, or the combination of the above

Continued

Consequence category	Determination criteria
Very serious	Project economic loss between RMB 10 – 100 million, with apparent casualties, or downstream property loss of RMB 5 – 50 million, or causing large area harm to environment or agriculture, or the combination of the above
Extremely serious	Project economic loss exceeding RMB 100 million, with heavy casualties, or downstream property loss exceeding RMB 50 million, or causing major and long-term harm to environment or agriculture, or the combination of the above

Between the element failure mode and final serious consequences caused, it is necessary to measure the possibility of occurrence. When the 5-level criterion is used, the determination criteria of occurrence are as shown in Table 25.2 – 5. It includes human intervention. If timely human intervention is obtained during the event, the seriousness of the consequences will be reduced.

Table 25.2 – 5 **Values of Possibility of Consequences**

Consequences possibility factor	Evaluation of possibility	Determination criteria
Extremely impossible	Below 5%	The failure mode may produce effect, but the consequence is very unlikely to occur
Basically impossible	5% – 25%	The failure mode may produce effect or consequence, but it is expected not to occur
Possible	25% – 75%	The failure mode is expected to produce effect or consequence, with equal opportunities to occur or not to occur
Very possible	75% – 100%	The failure mode is expected to produce effect or consequence
Surely to occur	100%	The failure mode will surely produce effect or consequence

The hazard indicator of failure mode of each element is the combination of possibility of failure mode occurrence, seriousness of failure consequences and possibility of failure consequence occurrence of the element, and the value assignment criteria are as given in Table 25.2 – 6. The variation range of the hazard indicator is 1 – 20, 1 means extremely low failure possibility and failure consequences, and the failure consequences are unlikely to occur and 20 means extreme possibility of element failure, and the failure consequences are extremely serious and likely to occur.

Table 25.2 – 6 **Values of Hazard Indicators**

Consequence		Possibility of element damage				
Severity	Possibility	Extremely impossible	Highly impossible	Basically impossible	Possible	Extremely possible
Very low	Extremely impossible	1	2	4	5	7
	Basically impossible	2	3	5	7	8
	Possible	3	5	7	8	9
	Highly possible	4	5	7	9	10
	Surely to occur	4	5	7	9	10
Fairly low	Extremely impossible	3	5	7	8	9
	Basically impossible	5	6	8	9	11
	Possible	6	8	9	11	12
	Highly possible	6	8	10	11	13
	Surely to occur	7	8	10	11	13

Continued

Consequence		Possibility of element damage				
Serious	Extremely impossible	6	8	10	11	12
	Basically impossible	8	9	11	13	14
	Possible	9	11	12	14	15
	Highly possible	9	11	13	14	16
	Surely to occur	10	11	13	15	16
Very serious	Extremely impossible	9	11	13	14	15
	Basically impossible	11	12	14	16	17
	Possible	13	14	16	17	19
	Highly possible	13	14	16	17	19
	Surely to occur	13	14	16	18	19
Extremely serious	Extremely impossible	11	13	14	16	17
	Basically impossible	12	14	16	17	18
	Possible	14	15	17	19	20
	Highly possible	14	16	18	19	20
	Surely to occur	14	16	18	19	20

After the hazard indicator of each element failure mode is calculated, the hazard extent of each element and the proportion in the sub-system and in the system are counted, and the hazard extent of each sub-system and its proportion in the system are counted. The hazard extent of each element is a simple addition of the hazard extent of all failure modes of that element, and the hazard extent of a sub-system is a simple addition of the hazard extent of hazard extent of all elements of that sub-system.

Based on the extent of hazard extent, all failure modes, elements and sub-systems are ranked. The higher the hazard extent is, the greater the risks are. By this means, the main failure modes and main risk elements of the dam are determined, and this can ensure that reinforcement for main risk elements can be prioritized.

25.2.6 Reservoir Operation Technologies

In developed countries, the reservoir management system, meteorological and hydrological characteristics, population density and its distribution, water resources supply and demand and social insurance mechanism differ greatly from those in China and their reservoir regulation is focused on flood control safety and ecological environmental efficiency, and the traditional planned and design values are usually used as flood limit water level, with little study conducted on dynamic control method of flood limit water level.

In the early 1950s, China completed a number of reservoirs, and the reservoir flood control and operation system was established with reference to the relevant experience and theories of the former USSR. Due to low prediction and short series of reservoir flood sample data at that time, it was not able to meet the fundamental condition to study stage control of flood limit water level and therefore, a fixed flood limit water level was designed for the whole flood season. In the 1960s, with accumulation of flood data, which demonstrated staged characteristics of flood, and the development of national economy and increased demand for water resources, therefore study of staged control method of flood limit water level was started. The first trial was conducted at Fengman Reservoir in Jilin province, and then it was revised and upgraded step by step and was incorporated into specification.

The key to staged control of flood limit water level is rational phasing of flood season. Before the 1980s, the phasing of flood season was largely based on mathematical statistic method of hydrological elements. In the 1980s, flood season phasing was analyzed from the viewpoint of meteorological origin. Later on fuzzy theory was introduced in flood season phasing, and the phasing periods were gradually shortened, and the concept of flood limit water level variation curve and its deducing method were proposed. In the 1990s, with the development of hydrological and meteorological sciences and computer sciences, flood forecast technologies and precision were improved continually, providing favorable conditions for the development, study and application of reservoir flood control and regulation technologies. In the 21st century, with the progress in dam safety monitoring technology and the extensive application of modern technologies such as satellite, optical fiber, network and GIS, dynamic flood limit water level regulation, emergency regulation, real time safety regulation and decision-making support technologies have been developed and applied rapidly to meet the actual demand in high efficiency utilization of water resources and reservoir emergency management.

25.2.6.1 Flood Regulation

1. Flood regulation calculation

The objectives of building dams are primarily in two areas. One is to the effective utilization of water to meet the water demand in different industries, and the other is flood control to mitigate flood disasters in the downstream areas.

(1) Basic principles and methods in flood regulation calculation. Flood flows in a reservoir as unsteady flow in open channel, satisfying the following Saint-Venant equations:

Continuity equation

Motion equation

$$\left.\begin{array}{l}\dfrac{\partial \omega}{\partial t}+\dfrac{\partial Q}{\partial S}=0 \\ -\dfrac{\partial Z}{\partial S}=\dfrac{1}{g}\dfrac{\partial v}{\partial t}+\dfrac{v}{g}\dfrac{\partial v}{\partial S}+\dfrac{Q^2}{K^2}\end{array}\right\} \quad (25.2-11)$$

where ω is discharge section area; t is time; S is distance in flow direction; Q is flow; v is average flow velocity at section; Z is water level; K is flow modulus; g is gravitational acceleration.

Usually, it is quite difficult to get precise analytical solutions to the above equations, and normally they are solved with simplified method. The water quantity balance can be expressed by the reservoir water quantity balance equation (Formula (25.2-12) and Fig. 25.2-7, and dynamic balance can be reflected by reservoir storage and discharge equation (Formula (25.2-13)).

$$\frac{Q_1+Q_2}{2}\Delta t - \frac{q_1+q_2}{2}\Delta t = V_2 - V_1 \quad (25.2-12)$$

$$q = f(V) \quad (25.2-13)$$

where Q_1 and Q_2 are inflow at the beginning and end of time period Δt; q_1 and q_2 are discharge flow at the beginning and end of time period Δt; V_1 and V_2 are amount of water stored in reservoir at the beginning and end of time period Δt; Δt is calculation time period.

Flood regulation calculation is to obtain simultaneous solutions to formulas (25.2-12) and (25.2-13), to get the hydrograph of the reservoir discharge flow, maximum discharge flow q_m, flood regulation volume V_{flood} and reservoir maximum flood water level Z_{flood}. The commonly used methods are trial method, semi-graphical method and simplified triangle method.

(2) Flood control regulation mode. Reservoir flood regulation mode is formulated normally for two types: flood control and no flood control.

1) Reservoir regulation mode with flood control downstream of the reservoir. Most reservoirs in China undertake the task of flood control for downstream areas. They should ensure not only the safety of their own, but also the flood control safety in downstream areas. According to the distance away from the reservoir downstream control point, flood control regulation mode can be formulated by taking into account incoming water in the interval and without taking into account incoming water in the interval.

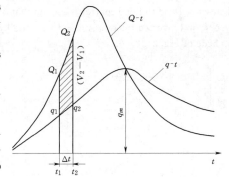

Fig. 25.2-7 Schematic Diagram of Reservoir Flood Regulation Calculation

a. Regulation mode without taking into account incoming water in the interval. When the reservoir is close to the downstream flood control point, the incoming water flow in the interval is low, and can be neglected, so the regulation mode of fixed discharge flow can be adopted. The fixed discharge flow velocity should be determined according to the importance of the protection object downstream and the flood control ability. For downstream protection objects with different flood control ability, and different losses when subjected to flood, flood regulation by discharge under staged control is adopted.

By flood regulation with discharge flow under staged control, in principle, it is regulated from small flood to big flood by stages. If the downstream area of a reservoir is under control by two stages of flood control, the flood hydrograph of flood control standard is as shown in Fig. 25.2-8, and the safe discharge flow for different flood control standards is respectively $q_{safe\ 1}$ and $q_{safe\ 2}$. In flood regulation, regulation calculation is performed first for the minimum level of flood, as shown in

Fig. 25.2-8 (a), with the controlled discharge flow as $q_{safe\,1}$, to obtain the corresponding $V_{control\,1}$ and the flood control high water level $Z_{control\,1}$. Then, regulation calculation is performed first for the next level of flood, as shown in Fig. 25.2-8 (b), first, the discharge flow is controlled as $q_{safe\,1}$, and at the time t_2, when the reservoir water storage has reached $V_{control\,1}$, discharge is started at $q_{safe\,2}$, and the flood control volume $V_{control\,2}$ and the corresponding flood control high water level $Z_{control\,2}$ are obtained. After that, regulation calculation is performed for flood of higher level of standard with the above method. When the incoming water exceeds the reservoir flood control standard, as shown in Fig. 25.2-8 (c), the discharge flow will be increased up to full discharge, with emphasis on protecting the dam.

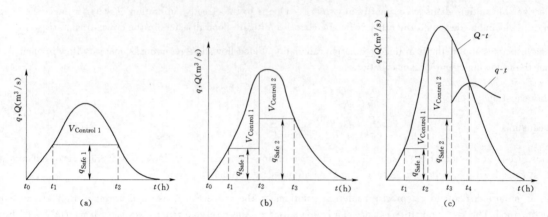

Fig. 25.2-8 Schematic Diagram of Staged Flood Regulation

b. Regulation mode by taking into account incoming water in the interval. When the reservoir is far from the downstream control points and there is a large catchment area between them, the incoming water in this interval cannot be neglected. It is necessary to adopt the regulation method of compensated regulation or regulation by shifting peaks, to make full use of flood control reservoir volume.

Compensated regulation means that the reservoir discharge flow plus the incoming water in the interval should be less than (equal to) the permissible safe passage flow q_{safe} at the downstream flood control point, and this requires that the reservoir discharge flow be reduced when the interval flood passes through the control point. As shown in Fig. 25.2-9, $Q_{interv}-t$ is the hydrograph of interval flood, $Q-t$ is the hydrograph of reservoir incoming flood, and the difference between travel time of interval flood to flood control point and that of reservoir discharged flow to the control point is Δt, $Q_{interv}-t$ is moved back by Δt, and inverted under q_{safe}, and then the hydrograph of discharge flow from the reservoir at different time is as shown in abcde of Fig. 25.2-9.

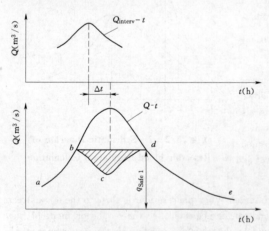

Fig. 25.2-9 Schematic Diagram of Compensated Regulation

When the interval flood collecting time is too short and it is not possible for the reservoir to discharge water by time period according to the hydrograph of the interval flood, to ensure that the sum of reservoir discharge flow and interval incoming water does not exceed the safe discharge flow of the downstream flood control standard, the reservoir will control the discharge flow in a period of time according to the time and process of interval flood, to shift the flood peak.

2) Reservoir regulation mode without flood control in downstream. Reservoir flood regulation without flood control in the downstream reaches is mainly to ensure the safety of hydraulic buildings of the reservoir, and the discharge flow is not restricted and can be at maximum discharge capacity.

For reservoir spillway without gate, normally the normal storage water level of the reservoir is at the crest of spillway weir, and water can discharge freely when the water level exceeds the spillway weir crest. The discharge flow depends on the variation of reservoir water level.

For spillway with gate control, the gate is opened gradually when flood comes to control the discharge flow equal to the incoming flow, and maintain the reservoir water level constant. With the increase of incoming water, the gate will be opened more widely until it is fully open to realize free discharge, and the reservoir water level rises and discharge flow increases. When the incoming flood peak arrives, the reservoir water level reaches the maximum. After that, the incoming flow gradually decreases and is less than the discharge flow, so the control gate is gradually closed to maintain the reservoir water level at the flood limit water level.

(3) Flood forecast and regulation. With the development and application of new technologies and theories such as electronic technology, remote sensing and measuring technology, hydrological simulation technology and information theory, system theory and control theory, the flood forecast time is shortened with increasing precision. Therefore, pre-discharge can be done before the arrival of flood to make volume for better protection against flood. After the flood peak passes, the flood will be stored as much as possible for water utilization.

1) Pre-discharge based on forecast. The purpose of pre-discharge based on flood forecast is to get some reservoir volume below the flood limit water level before the arrival of flood to regulate and store the coming flood. After obtaining the rainfall information in the catchment area of the reservoir, the inflow process can be obtained according to the flood runoff forecast. As shown in Fig. 25.2-10, at time t_0, the water level in the reservoir is the flood limit water level. According to the flood forecast period, hydrograph of the incoming flood $Q-t$ and downstream safe discharge flow q_{safe}, the reservoir starts to increase the discharge flow up to time t_1, to get available some reservoir volume V_1, then the inflow is greater than q_{safe}, so the water in the reservoir increases gradually, reaching the maximum at time t_2. Without pre-discharge, the flood control volume required for the reservoir is V_2, but after pre-discharge, the actually required flood control volume is V_2-V_1. This shows that by pre-discharge based on flood forecast, the dedicated flood control volume can be reduced to increase the overall utilization efficiency of the reservoir.

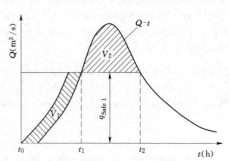

Fig. 25.2-10 Schematic Diagram of Flood Discharge in Advance

2) Storing floodwater based on forecast. Whether the reservoir can store the floodwater at the end of flood period to increase the water level up to the normal storage water level is related to whether it can ensure the water supply demand in the coming dry season.

The time to close the gate to store water at the end of flood period should be determined by analyzing the historical hydrological information, the flood characteristics of the catchment area, the weather situation in current year, and the medium and short term flood forecast, in conjunction with the reservoir regulation experience and comprehensive analysis. By analyzing the flood receding law, the correlation of receding flow Q and flood tail flow W, and flood probability $P(\%)$ can be plotted as shown in Fig. 25.2-11. With the flood magnitude and required water storage amount of reservoir known, the corresponding flood receding flow can be found on the diagram to get the corresponding time to close the gate to store water.

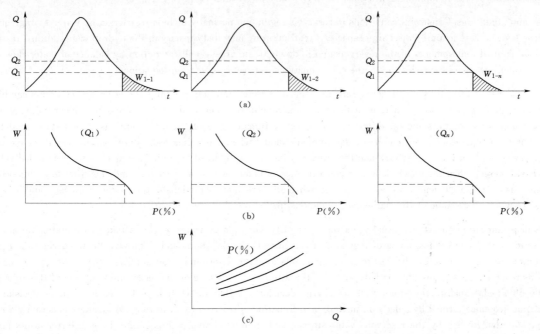

Fig. 25.2-11 $Q-W-P$ Correlation Diagram

2. Optimization technologies for reservoir operation

The purpose of optimizing reservoir operation is to make full use of reservoir volume to regulate flood and display the comprehensive benefit of water resources according to the reservoir design requirements and incoming hydrological condition. As mentioned above, the conventional flood regulation method is by plotting regulation diagram based on measured data to direct the reservoir operation, which is simple and visual, and fairly reliable. However, as regulation diagram is restricted by experience to a certain extent, the regulation result is normally a feasible solution instead of the optimal one.

In optimization of reservoir operation, the system analysis method is used to study how to operate the reservoir to achieve the maximum efficiency with minimum detrimental effect with given engineering parameters of the reservoir project. Starting with the system engineering theory, optimizing reservoir flood regulation can mainly be summarized as two aspects: ①reservoir regulation numerical model, including system input model, target function and restricting equation; ②optimization technology.

(1) Optimization regulation model for reservoir. The numerical model for optimization reservoir regulation usually consists of the optimal objective function and constraint conditions. The former is usually expressed in the form of efficiency or cost, related to optimal criteria. The constraint conditions reflect the capacity of various equipment and constraint conditions in operation.

1) Optimal criteria. Optimal criteria are designed to measure if the reservoir operation mode has reached the best level. For a reservoir with a single target or with a target as the main, the optimal objective and criterion are relatively simple. For example, with a reservoir mainly for power generation, it can be the minimum outlay for power system while reasonably meeting the water demand from other departments or the minimum total consumption by the power system or the maximum power generation from the hydropower station. For a reservoir mainly for flood control purpose, the optimal criterion can be, while taking into reasonable consideration the demand for other comprehensive utilization, the minimum disaster caused from discharged flow after leveling the flood peak or shortest weighed duration when the safe discharge flow is exceeded. For a reservoir with multiple targets or for a complicated water resources system, it should be the optimal comprehensive indicator, such as maximum efficiency in national economy or minimum expense from national economy.

2) Objective function.

The general expression of an objective function is:

$$Z = \max f(x_i, S_j, P_k) \tag{25.2-14}$$

where x_i is decision-making variable, S_j status variable and P_k system parameter.

The objective function is taken as maximization or minimization depending on the set criterion. It should be maximized when based on efficiency, and be minimized when based on cost or expenditure.

3) Constraint conditions. Constraint conditions in reservoir regulation normally include restriction by water storage quantity (or storage level), restriction by discharge capacity, restriction by installed capacity in the hydropower station, restriction by the flood control requirements of the reservoir and in downstream areas, and restriction by balance of water flow and electricity. Usually they are expressed by mathematic equations as a set of constraint equations.

The numerical model formed by the objective function and restricting condition equations for reservoir regulation can be classified as static and dynamic model, deterministic model and stochastic model, linear model and non-linear model according to different inputs and outputs and the differences in objective function and restriction conditions. It is a static model when system variation is independent of time. It is a deterministic model when determined values are used for variables and parameters in a given range of time and space, and the efficiency indicators obtained by optimization are also determined values. It is a stochastic model when the uncertainty of some variables is taken as random variables and the efficiency obtained from optimization is expected efficiency. It is a linear model when all mathematic equations in the model are linear; and it is a non-linear model when all or some of the mathematic equations in the model are non-linear.

In reservoir optimization operation, usually the water quantity (or storage water level) is taken as a status variable, the storage water quantity at the beginning of regulation is taken as the initial status and is known. When a given time period is taken along the time coordinate, the task of reservoir regulation is to determine water supply, water storage and discharge flow of the reservoir during this time period, and also the water storage status of the reservoir at the end of this time period should be obtained. Usually, the storage and discharge decision adopted for each period is referred to as decision-making. The time sequence formed by a decision-making in each period is referred to as strategy. A strategy is actually a regulation plan. The plan formed by the reservoir water storage and time as shown in Fig. 25.2 – 12 is referred to as strategy space.

The storage and discharge decision adopted by the reservoir in each period is not only directly related to the output of this period, but also to the output in other periods via status transfer, therefore, reservoir regulation is actually a multi-period decision-making process. The basic content in reservoir optimization regulation is to solve mathematic models of reservoir regulation to find the optimal control application plan according to the incoming flow process of the reservoir using the optimizing method. To operate the reservoir in water storage and discharge, maximized total efficiency can be achieved in flood control, irrigation and power generation as a whole in the total calculation period, with minimum detrimental effect. From the mathematic viewpoint, finding the optimal reservoir regulation plan is a question of optimization to solve a multi-period decision-making process including the time factor.

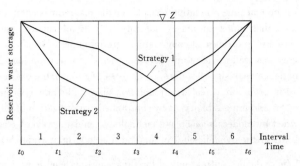

Fig. 25.2 - 12 Reservoir Regulation Strategy Space Schematic Diagram

(2) Optimization technology. After determining the reservoir regulation model, it can be solved using the optimization technology. The present commonly used methods include linear planning (LP), non-linear planning (NP) and dynamic planning (DP). And among these methods, DP is the most suitable for solving optimization regulation of reservoir. It is a recurrence optimization approach to study multi-period decision-making processes, by rendering a complicated high dimensional question into a series of simpler low dimensional sub-questions, so it is an optimization technology applied frequently with good effect in reservoir optimization regulation.

25.2.6.2 Dynamic Reservoir Regulation with Flood Limit

With the social and economic development, flood disasters and shortage of water resources coexist, imposing higher requirements on flood control safety and assurance of water supply for reservoirs, and further sharpening the contradictions between flood control and water utilization of reservoirs. Scientific design and application of flood limit water level and rational utilization of the "overlapped" of flood control and water utilization volume of reservoirs while ensuring flood control safety of reservoirs, are one of the effective means to make use of comprehensive efficiency of reservoirs and mitigate the contradictions between scarce water resources and flood control and disaster mitigation in our country.

Traditional methods of reservoir flood control level are based on planning and design requirements without taking into account forecast, and the discharge flow is determined by water level to keep the water level in the reservoir below the flood limiting level, or to have it drop back below the flood limiting level when it is exceeded. The design concept believes that flood is stochastic, and flood in different years are independent without correlation, and the probability of flood is the same in every year. This guiding ideology focused the attention of people on low probability events to guard against check flood at all times, resulting in such a situation as "water is discarded in regulation due to restriction by flood limit water level, but no water could be stored after the flood" and leading to waste of water resources. In recent years, research and application have been performed on dynamic control methods of flood limit water level to cope with this situation, such as flood limiting level staged control, pre-discharge capacity restriction, and flood control breakthroughs regulation. These methods can use some hydrological statistic information and forecast information, but are still based on the above-mentioned design concept. With the development of hydrological and meteorological forecast theory, the construction of modern observation and monitoring system projects, and improvement in the precision of flood and precipitation forecast, much more information is available in the practical operation than in design phase for reservoir flood limit water level, and regulation personnel can know in advance the magnitude of the coming inflow flood according to the forecast information, without the need of taking into account low probability event of check flood. In this new situation, it is quite necessary to implement dynamic flood limiting level regulation by using all information in the real time regulation phase.

The existing methods of flood limiting level control can be divided into two categories: one is based on the most detrimental design flood from the viewpoint of planning and design, mainly including the traditional fixed flood control level, phased flood control level based on hydrological statistic information, regulation of flood control level based on flood control forecast regulation method with forecast information and pre-discharging capacity restriction method, and cascade reservoir flood control level based on hydrological characteristics and reservoir volume difference compensation. The other is to control the flood limiting level from the viewpoint of regulation implementation, based on real time water, rain and operation conditions and comprehensive information of hydrological and meteorological forecast. The core of dynamic control of reservoir flood limiting level is an issue in the real time regulation phase, with the upper and lower envelope of the planned and designed flood limiting level as the constraint domain.

In the real time regulation phase after the reservoir is completed, much more information is available for use in flood control level than in design phase, not only statistic information based on random theory, but also deterministic information from actual monitoring and comparative deterministic information from theoretical analysis and calculation based on physical origins, as well as fuzzy information provided to decision-makers based on the above-mentioned information and regulation experience analysis. Therefore, it is necessary to establish a new concept of real time application of flood control level by using all information in real time regulation phase. The new concept of real time flood control level is to predict the most probable event and impossible event by comprehensive utilization of all useful information of telemetering, forecast and statistics provided by modern science and technology, and to prevent emergent event using remedial measures. This new concept is the basis of the dynamic control method for flood limit water level.

The dynamic control method for flood limiting level is to determine the specific values in the predicted period within the domain of upper and lower limits permitted by design values based on the reservoir catchment area weather forecast information (including short time and medium trend forecast) at real time regulation time, precipitation runoff forecast information, and the hydrological information, operation and disaster information at current time, while meeting the requirements of reservoir water storage, discharge capacity and flood control and water conservancy.

Existing dynamic control methods for flood limiting level are mainly the following.

(1) Real time pre-storage and pre-discharge method. With this method, the control is based on the reservoir discharge capacity and taking into account the flood or precipitation forecast information. The basic idea is to raise the flood limiting level up to the limit of the discharge capacity within the forecast flood period, and the influence factors include inflow hydrological forecast information at current time, discharge capacity of discharge equipment, permissible pre-discharge flow of downstream river channel, stability and speed of decision-making information transfer, and gate operation time. The safety of this method and the flood resource utilization rate mainly depend on such key factors as the precision and effective prediction period of flood or precipitation forecast, and reservoir discharge capacity.

Real time pre-storage and pre-discharge method is based on about the same calculation principle as the pre-storage and pre-discharge capacity restricting method in planning phase, with the difference that the inflow hydrological forecast information, discharge equipment capacity, permissible pre-discharge flow of downstream river channel used by the former are real time information at the current time, while the latter used the generalized safest value, with more ideal results.

The real time pre-storage and pre-discharge method features complete theoretical basis and simple calculation, and is suitable for reservoir with highly reliable hydrological forecast information, especially those with fairly high pre-discharge capacity.

(2) Comprehensive information deducing model method. This is a macroscopic decision-making method. It establishes a deducing model with historical data of all factors influencing the dynamic control of reservoir flood limiting level on the basis of analyzing all factors influencing the control of flood limit water level, referred to as "grand premises", and then the comprehensive information at the current time of reservoir real time regulation, referred to as "minor premises", is used in the deducing, to obtain a satisfactory plan for flood control levelin the predicted period, to guide the storage and discharge of the reservoir. For any reservoir, the comprehensive information for dynamic flood control level is mainly the actual reservoir water level at the current time; the forecast incoming water from the current time to the next precipitation; the future precipitation information forecast by the meteorological station; the normal water supply and discharge capacity; and the flood control level experience and lessons of regulation experts and decision-makers. There are three key issues with this method: one is to analyze the relationship between the flood control level value and its influence factors; the second is to establish a logic deducing relationship; and the third is to study and select the deducing method. The key factors include flood and precipitation forecast information, real time water, rain and operation information, statistic results of historical flood, and multi-year experience of regulation personnel and decision-makers.

This method is highly logic and demanding of rich regulation experience, in addition, it requires preparing application software for dynamic flood control system which is easy to operation and convenient for decision-making and procedure to ensure implementation of dynamic flood control level. It is suitable for reservoirs that have operated for years and accumulated rich experience of flood control and water utilization regulation, with high flood forecast precision and available regional precipitation forecast information.

(3) Optimization method of dynamic flood control level coupled with flood control real time forecast and regulation system. This method integrates the interacted decision-making sub-system of comprehensive information deducing model with the current flood control real time forecast and regulation system of reservoirs. The flood control level and decision-making plan is optimized and upgraded repeatedly with the change of time sequence and refreshing of comprehensive information. The key factors in this method is the integration of key factors of the two sub-systems of interaction decision-making

and real time forecast and regulation in which the key factors of interaction decision-making system are hydrological forecast information, real time water, rain and operation conditions, statistic results of historical floods, and years of control experience and preference of different decision-making factors of regulation personnel and decision-makers. The key factor of flood control real time forecast and regulation system is remote sensed precipitation information, which is the basis of flood process forecast, and its precision and steady transfer is the foundation for the normal operation of the whole coupled system.

This method is in line with the trend of extensive application of flood control regulation decision-making support system, with high degree of intelligence and convenient for operation. It can be used at all reservoirs with flood control real time forecast and regulation system, which is operated reliably and steadily with high forecast precision, and has summarized comprehensive information deducing model after years of operation.

(4) Dynamic flood control level for table-decision support form by comprehensive information. This method enables decision-makers to quickly check and calculate the table plan for real time dynamic flood control level. This form is prepared on the basis of the flood regulation calculation principle for conventional reservoirs, and the permissible flood control level at the current time is calculated using the reverse deducing method based on the inflow process and permissible discharge process in the effective prediction period and the flood control level domain at the end of the prediction period. With this form, the permissible flood control value and controlled discharge flow can be selected from the comprehensive information at the current time. As the inputs are characteristic of scattered data, and the integration of calculated results, there is no unique key factor in the decision support form.

This method features a simple and clear calculation process, makes the decision-maker clear about the adopted plan, and is highly operable. It can be used at all reservoirs as a macroscopic decision-making tool.

(5) Assessment decision-making method. This method relies on hydrological forecast, and determines the dynamic flood limit water level plan by introducing the opinions of decision-maker on the basis of analyzing and balancing the benefits and risks possibly resulting from various feedback messages at the current time in the regulation process. The assessment of dynamic flood control level mainly includes benefit and risk assessment on flood control, water supply and power generation and ecology. Key factors are the indicators and their influence factors selected in building up the assessment system.

This method incorporates the preference of decision-makers, and includes the possible effect of regulation plan in the control method, therefore covering comprehensive factors. It is suitable for all reservoirs, and can produce better results when merged with other control methods.

(6) Compensation regulation method. Compensation regulation method is suitable to dynamic flood control level of a reservoir group jointly undertaking a flood control objective in the downstream area. In the real time regulation, it dynamically controls the flood limiting level of all reservoirs according to the difference between upstream and downstream reservoirs volume at the current time or the hydrological non-synchronism of forecast. The key factors are the volume conditions of reservoirs at the current time and the forecast future inflow flood information.

25.2.6.3 Real Time Safety Regulation

In recent years, hydrological automatic measurement, dam safety automated monitoring and gate monitoring system have been established at many large and medium-sized reservoirs, making it possible to perform dam safety regulation in real time based on the monitoring outcomes.

1. Rule-based regulation mode

It is the basic regulation mode for reservoir real time safety regulation, and is built on the approved reservoir scheduling procedures, with long-term application experience.

The rule-based regulation mode flow is as shown in Fig. 25.2-13. Based on the protection standard of different flood control protection objects upstream and downstream of the reservoir, a series of controlled discharge flow (Q_{K1}, Q_{K2}, ⋯ and Q_{KE}) is set corresponding to design flood levels of given probability (H_{1K1}, H_{1K2}, ⋯ and H_{1KE}), and during regulation, gates are opened to discharge water according to this water level-discharge flow correlation, until all flood discharge facilities are open and the reservoir discharges freely at the maximum capacity. If the downstream river channel water level (H_{2K1}, H_{2K2}, ⋯) should be controlled concurrently, compensation regulation can be performed as per $Q \leqslant Q_{Kn}$ and $H_2 \leqslant H_{2Kn}$.

2. Water level control regulation mode

This mode is adopted in case of serious danger with the dam and water discharge or conveying building; or in case there is a serious flood disaster upstream of the reservoir but the flood control pressure in the downstream area is relatively low,

Part III Operation and Management of Dams and Reservoirs

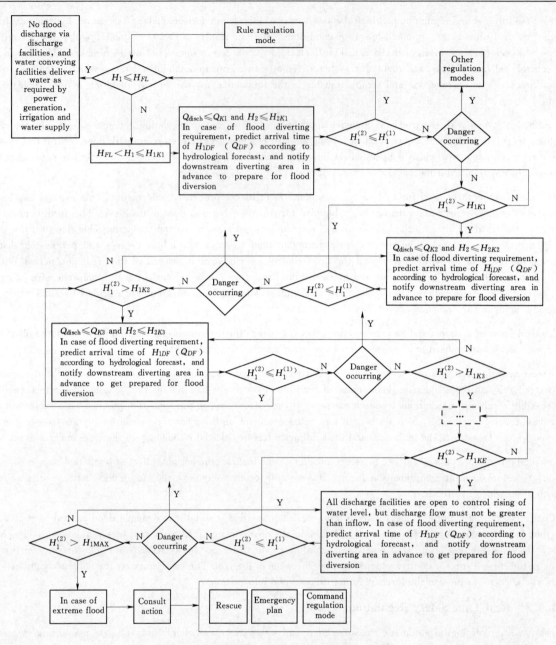

Fig. 25. 2 – 13 Rule Regulation Mode Flow Chart

and the reservoir should be run at controlled water level.

The water level control regulation mode flow process is as shown in Fig. 25. 2 – 14. At the current water level of $H_1^{(1)} \leqslant H_C$ (safety control water level), the rule regulation mode can be applied when it can still satisfy this condition. Otherwise, flood regulation should be performed with $H_1 = H_C$ as control condition using the water quantity balance equation according to the real time flood forecast result. At a current water level $H_1^{(1)} > H_C$, discharge should be at the maximum discharge capacity of the reservoir to lower the reservoir level below the safety control level H_C as quickly as possible. The safety control level H_C can be increased if the danger is mitigated after emergency rescue during the regulation and it should be further lowered if the danger keeps on developing. When new danger occurs at the project, it should be decided if other regulation mode should be implemented according to the category of danger. If the discharge flow ($Q_{disch.}$) required by flood regulation exceeds the discharge capacity Q_{max} of the facilities, it indicates that the reservoir water level will exceed safety control level H_C, so actions should be taken to increase the discharge capacity of the reservoir, and an alarm should be sent in advance. When necessary, the emergency plan for dam sudden incident should be launched.

3. Discharge flow control regulation mode

When it is required to control the reservoir drawdown rate in case of landslide accident on dam upstream slope and reservoir

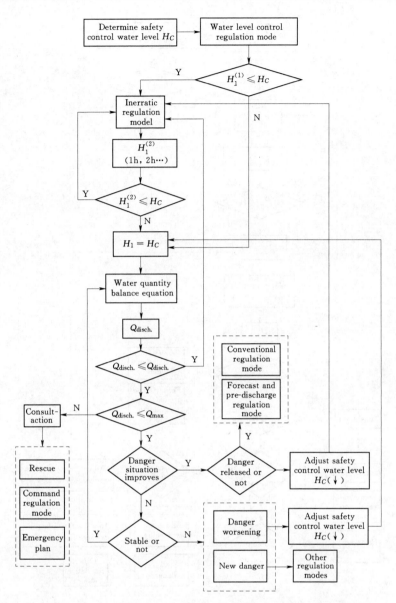

Fig. 25.2-14 Water Level Control Regulation Mode Flow Chart

bank near the dam or in case serious danger has occurred at discharge facility structure and gate or the reservoir discharge flow should be control when there is a serious flood disaster and serious danger with embankment downstream of the reservoir while the reservoir still has sufficient flood regulation capacity, the discharge flow control regulation mode is adopted.

The discharge flow control regulation mode flow process is shown in Fig. 25.2-15. At the current discharge flow of $Q_{disch.}^{(1)}$ $\leqslant Q_C$ (safety control discharge flow), the rule regulation mode can be applied when it can still satisfy this condition, otherwise flood regulation should be performed with $Q_{disch.} = Q_C$ as control condition using the water quantity balance equation according to the real time flood forecast result. During the regulation, the safety control discharge flow Q_C can be increased if the danger is mitigated after emergency rescue and it should be further lowered if the danger keeps on developing. When new danger occurs at the project, it should be decided if other regulation mode should be implemented according to the category of danger. If the reservoir water level $H_1^{(2)}$ obtained by flood regulation exceeds the maximum flood water level H_{1max} of the reservoir, it indicates that there is the danger of dam overtopping, so the water level control regulation mode should be adopted with H_{1max} as the safe control water level.

4. Compensation regulation mode

The compensation regulation mode flow process is as shown in Fig. 25.2-16. At the current water level of $H_1^{(1)} \leqslant H_C$ (safety control water level) and discharge flow $Q_{disch.}^{(1)} \leqslant Q_C$ (safety control discharge flow), the rule regulation mode can be

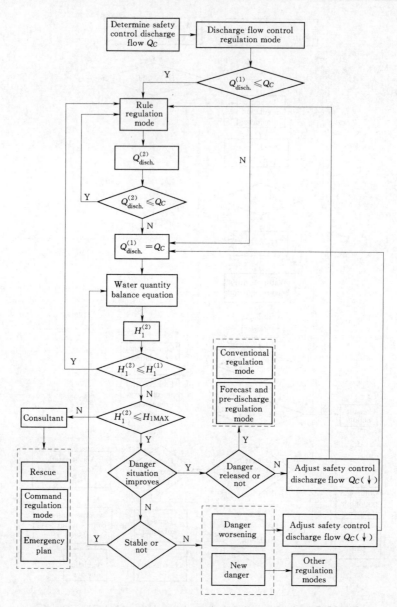

Fig. 25.2-15 Discharge Flow Control Regulation Mode Flow Chart

applied when it can still satisfy this condition, otherwise regulation should be performed by control of discharge flow when $H_1^{(2)} \leqslant H_C$ and $Q_{disch.}^{(2)} > Q_C$ and regulation should be performed by control of water level when $Q_{disch.}^{(2)} \leqslant Q_C$ and $H_1^{(2)} > H_C$. During the regulation, the safety control water level H_C and safety control discharge flow Q_C can be increased if the danger is mitigated after emergency rescue and the safety control water level H_C and safety control discharge flow Q_C should be further lowered if the danger keeps on developing. If the reservoir water level $H_1^{(2)}$ and discharge flow $Q_{disch.}^{(2)}$ obtained by flood regulation both exceed the control indicator, an alarm should be sent in advance. When necessary, the emergency plan for dam sudden incident should be launched.

5. Command regulation mode

If the danger cannot be brought under control by implementing the above-mentioned emergency regulation mode, and the dam may possibly fail, especially after the emergency plan of dam has been launched, members of the emergency commanding organ, expert group and rescue people usually gather at the reservoir site to perform real time regulation according to field commands on the basis of analyzing the situation of danger, water and disaster and real time safety monitoring (Fig. 25.2-17).

25.2.6.4 Reservoir Regulation Decision Support Technology

Reservoir regulation, especially in case of emergency, usually faces instantly changing conditions, demands judgment and

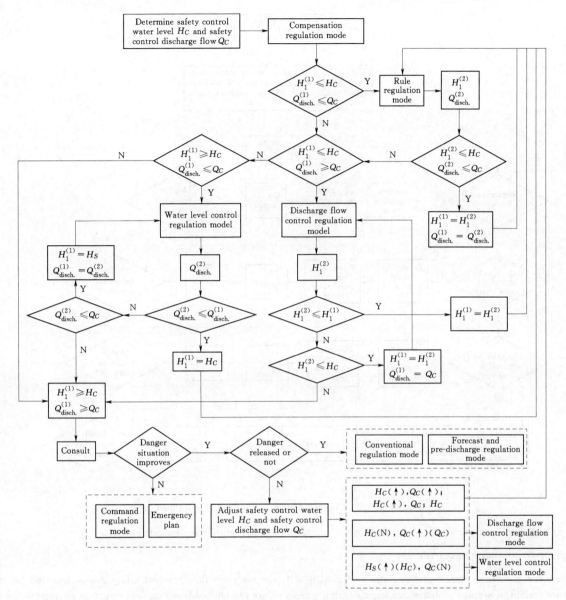

Fig. 25.2 – 16 Compensation Regulation Mode Flow Diagram

decision in the first time and selection of a suitable emergency regulation mode, which should also be adjusted according to changes in water, hazard and disaster conditions. This process involves information collection and processing, determination of disaster factors and hazard category, real time analysis of dam safety status, consultation, decision-making, regulation mode selection and implementation, rescue, relief and risk evaluation, in which all uncertainties interact, and mistake in any link may result in serious consequences.

In recent years, the reservoir emergency regulation decision support system has been developed using modern information technology and means and in conjunction with experts' experience to provide a digital platform with functions of information support, flood forecast, dam safety real time assessment, flood risk evaluation and emergency regulation decision support, so that a number of emergency regulation plans can be generated following reservoir emergency for selection by decision-makers.

1. *Flood forecast*

After continual development, reservoir flood forecast has developed from experience-based approaches into conceptual or precipitation and runoff models with physical foundation, and is developing in the direction of real time flood forecast. Flood forecast systems with rich functions have been established at many reservoirs.

Flood forecast models are established according to the runoff characteristics in the reservoir catchment area, to forecast

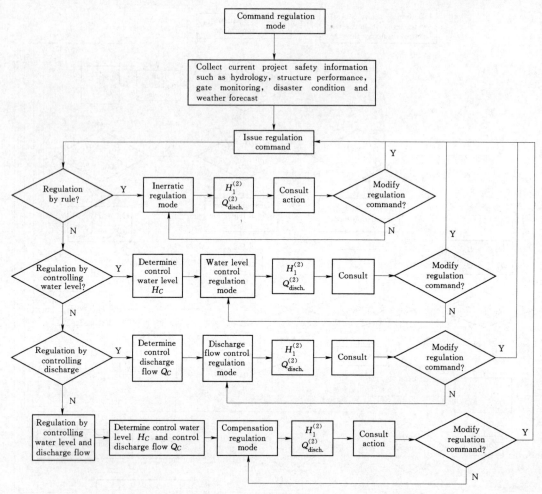

Fig. 25.2 – 17 Command Regulation Mode Flow Chart

flood in real time, and provide flood information for emergency regulation in case of sudden events.

2. Dam safety assessment and prediction

Deterministic numerical models and statistic numerical models for reservoir flood control safety assessment and for dam structure stability, seepage stability and seismic stability analysis are established based on dam structure characteristics and dam safety monitoring data, to analyze or predict dam safety performance in real time, and determine the dam safety control water level and the falling rate of water level, as emergency management of water resources scheduling problems with the growing emergency.

3. Generation of emergency regulation plans

A number of possible emergency regulation plans for rule regulation, water level control regulation, discharge flow control regulation and compensation regulation are generated via flood regulation on the basis of daily regulation rules of the reservoir, and by taking into overall consideration various possible influence factors and restriction conditions such as flood forecast results, dam safety control water level and the falling rate of water level, for selection by decision-makers.

4. Downstream flood risk analysis

Suitable dam failure models and downstream flood routing calculation methods are selected on the basis of dam structure characteristics and downstream topographic data to simulate flooding process, including flooding depth, flooding scope and flooding velocity and other flooding elements, and the superposition analysis method in GIS is used to calculate inundation loss (Fig. 25.2 – 18).

5. Assessment of emergency regulation plans

Indicator systems for assessment of emergency regulation plans are established and corresponding plan simulation and assessment platforms are developed to assess the above-mentioned reservoir emergency regulation plans for different target re-

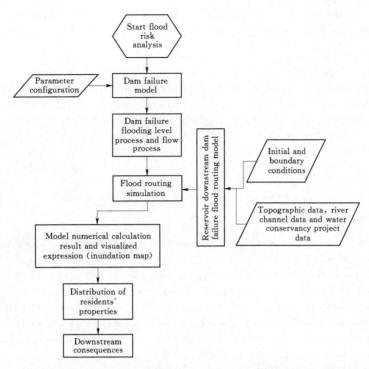

Fig. 25.2 – 18 Reservoir Downstream Flood Risk Analysis Flow Chart

sults by arousing the participating awareness of people to select the optimal regulation plan under specific restriction conditions.

A fuzzy comprehensive assessment model is used as the design tool of indicator systems for assessment of emergency regulation plans, to express the human judgment in quantities, the influence objects and classified and graded starting from the after effect of implementing the emergency regulation plans and based on the downstream flood risk analysis results, and scores and weights are assigned based on possible extent of influence, to enable decision-makers to determine the plan finally.

The reservoir regulation decision support system is an auxiliary decision support system with a number of comprehensive management functions such as information receiving and processing, commanding and regulation, disaster visualizing and time and space analysis. It first integrates the information monitoring sub-systems of all types and professional models of flood forecast, dam safety status analysis, flood routing and risk analysis, then conveys all information via the data collecting platform and computer communication network to the multi-purpose database, to select the optimum emergency regulation plan with the support by the database server and application server, via application interfaces for emergency regulation information service, flood forecast, dam safety and safety control water level analysis, emergency regulation, plan assessment and plan management, and make real-time adjustment to it according to the changing conditions, so as to provide decision support for dam emergency management. The specific functions are:

(1) Assist the reservoir management department in scientific management of reservoir regulation basic information, flood risk information, emergency management information and emergency regulation plans in daily operation management.

(2) Assist the relevant department in scientific decision-making according to the existing emergency regulation plan, to implement emergency regulation, avoid or alleviate dam failure, minimize loss, and win time and get prepared for personnel evacuation.

(3) Collect information on emergency regulation and consequences of emergency regulation after completion of emergency disposal, and summarize experience to improve and complete the emergency regulation plans.

25.3 Dam Management Informatization

Dam management informatization is an important part in IT application to water resources, and it also is an important mark of dam management modernization, and will play an important technical supportive role in raising the dam safety management efficiency and performance, optimizing water resources configuration, upgrading the dam safety prediction and warning ability and its ability to cope with emergencies, and effectively ensuring dam safety, drinking water safety and public safety.

With respect to the characteristics in dam safety management and demand in water management informatization, dam management informatization mainly includes dam information acquisition, information transmission management and information service and application. Information acquisition covers basic engineering information and real time monitoring information, as the infrastructure for dam management informatization. Information transmission and management covers transmission of commands and collected information, network platform building and data integration management, as the basic condition to realize information sharing and information service and application are to provide the information service and exchange platform to provide decision support for management, being the ultimate goal of management informatization.

25.3.1 *Information Acquisition*

In early days, dam information acquisition was basically done manually, mainly for dam's basic characteristic parameters, such as water level, precipitation, dam saturation line and surface deformation. For example, in the 1950s, dam surface deformation and internal vertical displacement observation, and saturation line measurement with standpipes were started for earth and rockfill dam, and measurement of temperature and strain were performed on concrete dams. In the 1960s, measurement of pore water pressure, pressure relief well water level and seepage flow was started in dams. The reservoir hydrological monitoring was started in the 1970s. In the 1980s-1990s, great attention was paid to safety automated monitoring for dam safety, and relevant technical standards were issued to formalize information acquisition, such as *Technical Specification for hydrological data acquisition system* (SD 159—85), *Tipping bucket raingauge* (GB 11832—89), *Float type staff gauge* (GB 11828—89), *Technical specifications for safety monitoring of concrete dams* (SDJ 336—89), *Technical specifications for safety monitoring of earth and rockfill dam* (SL 60—94), *Specifications for Compilation and Preparation of Safety Monitoring Data for Earth and Rockfill Dams* (SL 169—96), and *Water quality monitoring specification* (SL 219—98), and so on.

With the development of management informatization demand and relevant technologies, the scope of information acquisition was gradually expanded, and information includes relevant laws and regulations, technical standards, management information on dam safety and static information about dam basic information and dam operation and maintenance information, and real time monitoring information such as reservoir hydrological, operation conditions, gate opening, water supply flow and water quality, improving the formalization and completeness of basic information, and providing complete information resources for information services and applications.

25.3.2 *Information Transmission and Management*

Before the 1980s, except for hydrological information that was transmitted by radio and automatically collected by USW means, a large amount of information was collected and transmitted manually. From the 1980s, with the development and demand of automation, communication, network and system integration technologies, automation was gradually accepted for hydrological measurement and dam safety monitoring. A number of communication means such as optical fiber, wireless and GPRS and the safe and efficient network technology based on Internet/Intranet became more mature, realizing transmission and sharing of collected data. Relevant technical standards were issued, such as *Basic Technical Conditions for Dam Safety Automatic Monitoring Systems and Equipment* (SL 268—2001), *Specifications for Managing Radio Technology in Water Conservancy System* (SL 305—2004), *Guidelines for Communication Operation in Water Conservancy System* (SL 292—2004), *Communication Operation Regulation in Water Conservancy System* (SL 306—2004), and *Technical Specifications for Dam Safety Monitoring Automation* (DL/T 5211—2005).

Information management and standardization is an important guarantee for management informatization efficiency and performance, the real time data, historical data, analysis result and management data of various information acquisition systems are integrated, isolated and backed up according to the specified datasheet structure and identifier standard to reduce data redundancy and effectively ensure information security and sharing. Early in this century, information management and standardization was carried out step by step, and *Standards for Basic Hydrological Datasheet Structure and Identifiers* (SL 324—2005), *Standards for real time rain and hydrological database sheet structure and identifiers* (SL 323—2005), and *Specifications for Water Quality Datasheet Structure and Identifiers* (SL 325—2005) were issued, and *Specifications for Dam Basic Datasheet Structure and Identifiers* are being prepared to provide the basic guarantee for efficient information service and sharing.

25.3.3 Information Service and Application

In the 21st century, the construction of major Service Application Systems has been speeded up while completing the water conservancy infrastructures according to the idea of IT application in the water sector. Websites were available to reservoir management units at all levels and authorities, development and application have been strengthened for associated service and decision-making systems, such as dam safety analysis, assessment and warning system, reservoir risk evaluation system, water quality evaluation system and information visualization simulation system. Modern information technology, model technology, artificial intelligence technology and virtual simulation technology are used to achieve intelligent and visualized services in information application.

25.3.4 Cases of Dam Management Informatization

25.3.4.1 Management Websites of Dams

Dam management website is an important part in reservoir management informatization, and they can provide important platforms for the government, water authorities, reservoir management units, relevant technical personnel and the public to understand the dam management activities, retrieve relevant information, perform technical exchange and propagate dam management. The dam safety management website of the Ministry of Water Resources is an industrial service platform oriented to the public, covering management information, safety management, laws, regulations and standards, industry services, academic exchange and overview of dams, as shown in Fig. 25.3 – 1.

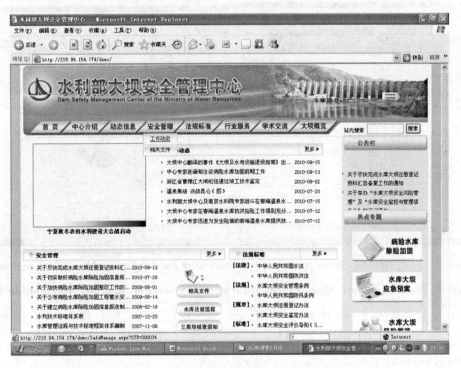

Fig. 25.3 – 1 Dam Management Website Interface

25.3.4.2 National Dam Basic Data and Information System

To formalize and share national dam basic information, the Ministry of Water Resources has developed the "National dam basic data and information system", built a platform for the dam basic data and information service, and realized storage, inquiry, statistics, updating and maintenance services of basic data for dam engineering characteristics, hydrological characteristics, registration, operation and reinforcement and downstream consequences in the whole country. The system is now providing decision support and information services for water administration authorities and industrial management departments, and providing information support for project safety and emergency management. Fig. 25.3 – 2 shows block diagram for national dam basic data and information system. Fig. 25.3 – 3 shows interface between national dam basic data and information management system.

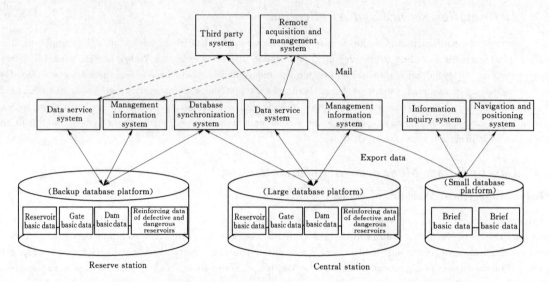

Fig. 25.3 - 2 Block Diagram of National Dam Basic Data and Information System

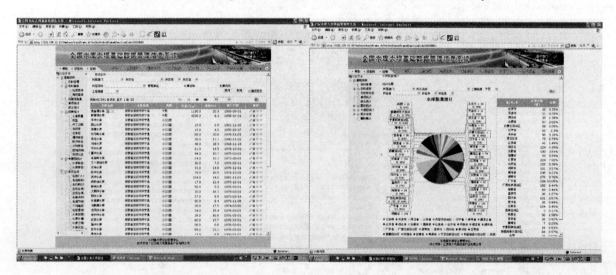

Fig. 25.3 - 3 Interface Between National Dam Basic Data and Information Management System

25.3.4.3 Safety Monitoring and Management Information System of Shanmei Reservoir in Fujian

Shanmei Reservoir is located in the upstream reach of Dongxi tributary in Jinjiang River Basin, with a catchment area of $1,023km^2$ above the dam site, and a total reservoir storage capacity of 655 million m^3. It is a large multi-purpose water project for flood control, water supply, irrigation and power generation, etc. The reservoir project consists of main and auxiliary dams, spillway and power station. To ensure the reservoir safety and its comprehensive efficiency, the reservoir safety monitoring and management information system was set up in 2005. The system consists of 15 sub-systems of real time information acquisition, computer communication network, comprehensive database and application services (Fig. 25.3 - 4).

(1) Dam safety monitoring system. After studying the safety status of Shanmei Dam, the monitoring items, measuring point arrangement was arranged for key variables and locations influencing and controlling the safety performance of the project according to relevant technical specifications. The monitoring items include main dam deformation, main dam seepage flow, auxiliary dam seepage flow and environmental parameters. The acquisition, storage and management of dam safety monitoring information are established.

(2) Hydrological data acquisition system. It consists of 1 central station, 1 sub-central station, 1 relay station and 17 remote measuring stations (12 for rain gauge, 3 for rain guage for water level, 1 for gauging station and 1 for evaporation plant), and dual-channel communication mode is provided with USW as the main and GSM mobile phone message as auxiliary to realize real time acquisition, reporting and processing of precipitation, water level, evaporation and flow in the river basin.

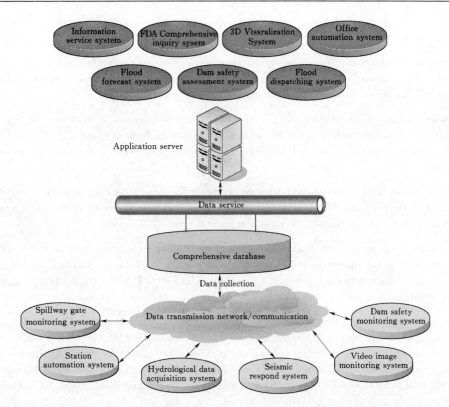

Fig. 25.3-4 Block Diagram of Safety Monitoring and Management Information System of Shanmei Reservoir

(3) Spillway gate monitoring system. It consists of 1 monitoring host computer, 1 PLC field monitoring unit, 6 manual control cabinets, 6 gate opening meters, 6 limit meters, and the UPS power pack. Computer is used for automatic control of the opening and closing of 6 spillway gates, the gate position and status are controlled by both field control and remote control, so that the gate opening can be adjusted automatically according to the fluctuations of reservoir water level, to maintain a steady discharge flow. The actual gate opening and closing operation and energy dissipation at the gates can be monitored by the monitoring system.

(4) Seismic response monitoring system. The system consists of a strong motion observation matrix of 7 measuring points on the dam structure and free bedrock ground and a monitoring center. It monitors the seismic response and response to normal environmental vibrations in the dam structure, identifies earthquakes intelligently and can trigger automatically to obtain real time seismic response data of dam structure and trigger earthquake alarms.

(5) Power station automation system. It is in a hierarchic and distributed structure to perform real time acquisition, processing, monitoring and storage of various operation parameters of the hydropower station, and control objects in the station with diversified manners to realize remote measuring, telecommunication and monitoring of the power station by higher authority regulation automation system.

(6) Video image monitoring system. The system can monitor remotely the operation and performance conditions of the dam, gates and generator sets in the power station, and can monitor in real-time operation of main parts of the reservoir by video switchover, to visually display the external status and remote operation status of equipment.

(7) Computer network communication system. The network communication system includes two local networks in the Shanmei Reservoir Administration at reservoir site and in Shanmei Official Mansion in Quanzhou City, and dedicated optical cables are used for communication between them to ensure the system transmission quality, and the reliability, safety and confidentiality of data communication.

(8) Comprehensive database system. According to the unified standards, comprehensive database collects the real time data, historical data, system performance data and management data from different information acquisition systems, for data integration, segregation and backup to reduce data redundancy and effectively ensure information security and information sharing.

(9) Office automation system. It consists of the individual office, archive and file management, public service, document circulation and information automatic inquiry systems, to integrate information management, business handling and pro-

duction management. It can process and inquire information according to the authorization limit set at any computer in the network. It has raised the management level and work efficiency at all levels of department and achieved office automation.

(10) PDA comprehensive information inquiry system. It combines the wireless communication, PPC and embedded GIS technology with the actual applications in the water sector, achieves flood condition monitoring and information inquiry, and provides basic information for mobile-based office work and decision-making.

(11) Information service system. By using browsers, it displays the meteorological, rain, water and real time flood conditions in a clear and simple manner with both graphs and text in multimedia form of graphs, tables and text to reflect the power generation, regulation and management information in real time. It has created a good platform to use reservoir management information.

(12) Flood forecast system. It has realized automation by computer for flood forecast with flood forecast software in the form of browser. The main functions include: automatic forecast at preset time intervals, manual intervened interaction forecast, follow-up forecast and forecast of total rainstorm precipitation.

(13) Flood regulation system. In the form of browser, it takes into overall account the flood control contradiction between upstream and downstream areas of reservoir and that between flood control and water utilization, to handle multi-objective relations correctly, make decisions in a scientific and optimized manner for regulation, and perform dynamic analysis of real time flood control situation.

(14) Dam safety analysis and assessment system. It uses automatic real-time monitoring data by the dam safety monitoring system, manual observation data and visual inspection records, and establishes the dam safety analysis and assessment model based on dam safety assessment guideline and experts' experience and using rational mathematic models and judgment criteria, to analyze dam safety operation status in real time, assess dam safety classification and provide warning on potential deficiency in the dam so as to provide support for decision-making for reservoir safe operation and dynamic control of flood limit water level.

(15) 3D visualization system. It reproduces the 3D visualization of Shanmei Reservoir with 3D GIS technology, 3D modeling technology, virtual simulation technology and database technology to display in real time the reservoir inundation area, dam safety monitoring real time acquainted information and safety status, dynamically simulate gate opening and closing conditions, and realize dynamic and real time performance between 3D scene and water analysis models, and the 3D visualization of dam scene and reservoir safety analysis technology.

The safety monitoring system of Shanmei Reservoir is based on the concept of unified network platform, database platform and application platform, and a cross-platform system structure is adopted to form an integrated digital platform covering the reservoir flood monitoring, flood control regulation, dam safety management, station management, spillway gate control and office information management. It has realized automatic information acquisition, information transmission via network, standardized information processing and scientific flood control and regulation, basically achieving modernization in reservoir management.

25.3.4.4 Dam Safety Real Time Monitoring and Warning System of Yanshan Reservoir in Henan

Yanshan Reservoir is located on Ganjiang River and upstream of Lihe River, a main tributary of Shaying River in Huaihe River Basin. It is a multi-purpose water project designed mainly for flood control, and also for water supply, irrigation and power generation. It is one of the 19 key projects of the State Council for controlling floods in the Huaihe River. To ensure project safety and raise safety management modernization level, the safety monitoring and warning system was established at Yanshan Reservoir in 2008 (Fig. 25.3-6 and Fig. 25.3-7). This system consists of remote data transmission, dam safety diagnosis, dam safety remote warning, dam information management and system information management. The system is based on B/S mode, and it has realized intelligent and visualized project safety warning with a cross-platform J2EE three-level system structure.

25.3.5 *Prospects for Dam Management Informatization*

(1) The management informatization standard system should be further improved. Dam information includes static information on project characteristics, operation and maintenance and relevant standards and regulations and dynamic supervision and monitoring information such as hydrological, structure performance, gate opening, water supply flow and quality. Establishment of information standardizing system is a prerequisite to realize information sharing. Presently in China, information sharing standard system has not been completed for dam safety management, and the communication and database interface standards and the associated policies and regulations need to be improved.

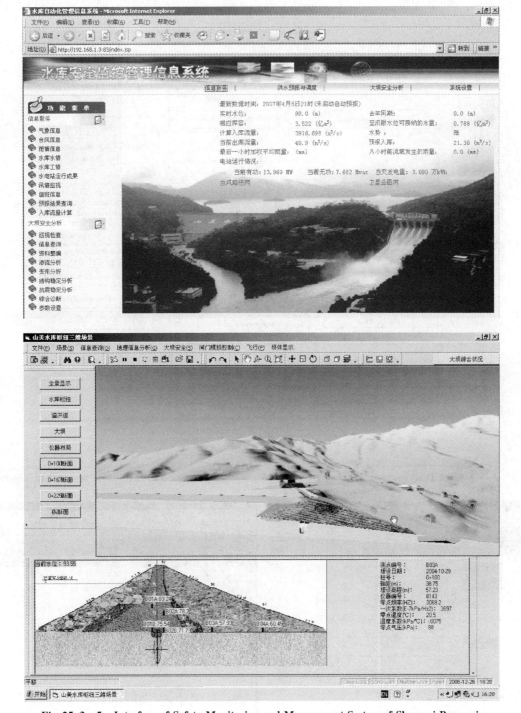

Fig. 25.3-5 Interface of Safety Monitoring and Management System of Shanmei Reservoir

(2) Information acquisition resources should be further expanded and integrated. With the shift from traditional dam safety management to risk-based dam safety management, the scope and objective of dam safety management have been expanded to include the population, social, economic and environmental and ecological factors in the areas downstream of reservoirs, in addition to dam project, and the management objects, technologies and tasks have also been expanded. Therefore, the management information acquisition resources should also be expanded. Meanwhile, integration and sharing of information resources should be further strengthened to avoid insufficient connection between planning and implementation and repeated poor quality construction. Continual expansion and integration of information resources is a development trend in dam management informatization.

(3) Information application service ability should be strengthened. With the expansion of information resources, the increasing public concern on dam safety and occurrence of reservoir sudden incidents, we should establish dam management

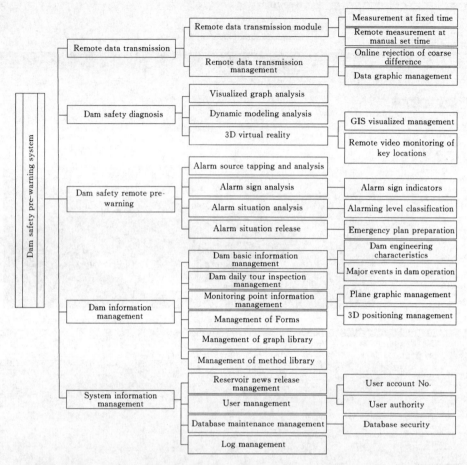

Fig. 25.3-6 Block Diagram of Dam Safety Real Time Monitoring and Pre-warning System of Yanshan Reservoir

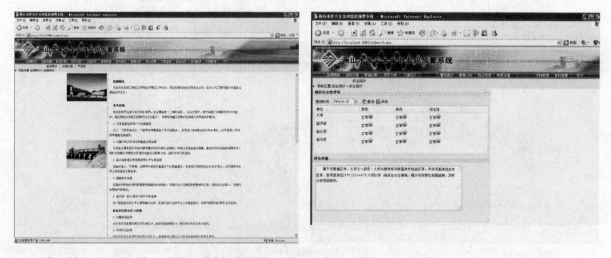

Fig. 25.3-7 Interface of Dam Safety Real Time Monitoring and Pre-warning System of Yanshan Reservoir

service and emergency platforms with rich resources available to the public. It strengthen information tapping and merging technology, upgrade the resources utilization rate and remote application ability, make the business information more intelligentized and visualized, and provide dam safety information service and exchange platforms for the government, water authorities, owners, engineers and the public. It will improve the decision-making and emergency response capacity of the government, the management ability of the owners, and the participation ability of the public.

(4) Safety informatization management mechanism should be established. Establishing operation and maintenance mechanism for an efficient dam management information system is an important guarantee for sustainable development of management informatization. We should formulate operation and maintenance quotas for dam management information systems, and

incorporate the information system operation and maintenance fund into department budget. We should establish operation and maintenance mechanism for complete information system, implement system of responsibilities and improve the information system maintenance and management level. We should upgrade the information system monitoring and emergency response ability and means so as to ensure safe, steady and efficient operation of the system.

(5) Information safety system should be completed. Dams are infrastructures of the State, and their safety is related to the safety of the public, drinking water and grain production in the country. Therefore, dam safety is a big issue concerning the national stability and social development. Dam safety is an integral part of the national security system. We should establish and complete dam safety monitoring system and improve the ability to cope with and prevent network safety accidents. We should pay great attention to emergency response capacity for information security, perfect and complete the ability of information safety emergency handling and safety notification system, constantly improve contingency plan for information safety, establish the information system disaster backup system, and enhance the ability to prevent information from destruction and the ability to restore the information after disasters.

References

[1] *2011 Yearbook of China Water Resources*, Beijing: China Water Power Press, 2011.
[2] Sun Jichuan, Dam Safety Management in China [J], *Journal of China Water Resources*, 2008, (20).
[3] Special Rehabilitation Program for Defective and hazardous Reservoir Dams in China, Ministry of Water Resources, 2007.
[4] Fang Cungang, Yao Chenlin, Jia Yongmei. *Non-destructive Testing Technology and Instruments for Detecting Leakage and Potential Deficiencies at Embankments*. Beijing: China Water Power Press, 2010.
[5] Li Lei, Wang Renzhong, Sheng Jinbao et al. *Risk Assessment and Management for Reservoir Dams*. Beijing: China Water Power Press, 2006.
[6] Sheng Jinbao, He jian, Wang Zhaosheng. Risk-based Dam Rehabilitation Decision Making Technology, *Advances in Science and Technology of Water Resources*, 2008 (4).
[7] Sheng Jinbao, Shen Dengle, Fu Zhongyou. Classification and Rehabilitation of Defective Dams in China, *Hydro-Science and Engineering*, 2009 (12).
[8] Sheng Jinbao, Wang Zhaosheng, Dam Safety Assessment, *Journal of China Water Resources*, 2010 (2).
[9] Sheng Jinbao, Li Lei, Wang Zhaosheng, et al. Real-time Safety Regulation Technology for Reservoir Dams, International Symposium on Detection of Embankment Potential Deficiencies, 2005.
[10] Sheng Jinbao, Li Lei, et al. *Non-structural Measure to Control Reservoir Risks*. Nanjing: Hydraulic Research Institute, 2010.
[11] He Yongjun, Liu Chendong, Xiang Yang, et al. *Dam Safety Monitoring and Automation Technology*, Beijing: China Water Power Press, 2008.

Chapter 26

Dam Operation and Management

Zhang Xiuli[1], Xie Xiaoyi[2], Du Dejin[3], Zhao Huacheng[4],
Huang Shiqiang[5], Xu Chuangui[6]

26.1 Dam Safety Supervision and Administration Organizations

26.1.1 Development of Dam Safety Supervision Management

Along with the changes and development of administration institutions, the dam safety administration has experienced three stages in China.

The first is on the groping stage of management. Since 1949, China started infrastructure construction on a large scale, including reservoirs and hydropower stations. At that time, China had a poor industrial foundation and weak national power, and the technical level was low, so all activities were carried out in exploring with in practice. Therefore, the so-called 'three same-time performed projects' occurred, which means their investigation, design and construction, were carried out at the same time. Defects and hidden dangers remained in the projects. The primary task at that time was to build. The regular electrical power safety production sequence was destroyed due to the Cultural Revolution, the dam safety administration of hydropower stations in operation was as blank as a white paper. Thus some serious accidents happened. The former Ministry of Water Resources and Electric Power (hereinafter MWREP) approved the establishment of Dam Safety Supervision Center (hereinafter LDSSC) in 1985, which is the professional organization responsible for the dam safety management of hydropower stations. Since its establishment, LDSSC conducted general survey of the basic conditions of hydropower dams, mastered the dam safety conditions, developed a series of legislations about dam safety, and laid foundation for dam safety management in operation.

The second is the vertical integration management stage under the planned economic system. Since the 3^d Plenary Session of the 11th Central Committee of the Communist Party of China, China has entered into a new stage of reform and opening up, as well as the construction of socialist modernization. Hydropower industry stepped into a new developing stage. In this stage, the dam safety management mainly relied upon administration. China issued *Reservoir Dam Safety Management Regulations*, *Flood Control regulations of People's Republic of China* and *Flood Control Law of People's Republic of China*, etc. The administrative departments in charge of hydropower stations, such as Ministry of Water Resources and Electric Power, Ministry of Power Industry, Ministry of Energy and State Power Corporation, issued a series of rules and regulations in succession. The safety production regulatory framework of enterprises was built up step by step. The safety objective management and check were advocated in the power enterprises. The safety responsibility system was conducted in response. These measures changed the situation of no regulations to follow and no following the regulations. The enterprises extensively performed the activities of Safety Production Month, and took measures to avoid accidents. The organization system of flood control was strengthened. Inspection of safety and professional training were preformed. The first and second rounds of regular dam safety inspections were carried out. The dam safety registration was carried out. Meanwhile the dam

---

[1] Zhang Xiuli, Chief Engineer and Director of Committee of Experts, Large Dam Safety Supervision Center, State Electricity Regulatory Commission, Professor Senior Engineer.

[2] Xie Xiaoyi, Deputy Chief Engineer and Vice Director of Committee of Expert, Large Dam Safety Supervision Center, State Electricity Regulatory Commission, Professor Senior Engineer.

[3] Du Dejin, Office Manager of Large Dam Safety Supervision Center, State Electricity Regulatory Commission, Professor Senior Engineer.

[4] Zhao Huacheng, Secretary General of Committee of Expert, Large Dam Safety Supervision Center, State Electricity Regulatory Commission, Professor Senior Engineer.

[5] Huang Shiqiang, Vice President of Huadong Engineering Safety Technical (Zhejiang) Co., Ltd, Professor Senior Engineer.

[6] Xu Chuangui, Large Dam Safety Supervision Center, State Electricity Regulatory Commission, Executive Vice Chief Editor and associate Senior Translator of the Journal of *dam and safety*.

safety registration and the dam safety grade appraisal were regarded as the checking basis for the managing capacities of the enterprises. Theses measures greatly pushed management level of dam safety forward and improved the dam safety conditions in quality.

The third is the supervision management stage under the market economic system with Chinese characteristics. Since 2002, the marketing reform of "the separation of hydropower plants from the grids and the separation of the sideline enterprises from the main production bodies" was conducted in the Chinese electric power system, and the electric power industry began a new framework from planning to market, from monopolization to competition and from centralization to dispersion. The main body of the market, the interest pattern of the enterprises, as well as the survival and development environment for the enterprises changed greatly. It brought challenges to the dam safety management. The State Council issued *Notice on Strengthening the Electric power Safety Production* in 2003, and authorized State Electricity Regulatory Commission (hereinafter SERC) to be in charge of the electric power safety administration. SERC established the Department of Safety Administration in 2004, and the Office of Central Institutional Organization Commission LDSSC into SERC. It guaranteed the settlement of the dam safety administration function in organization. From then on, a new stage of the dam safety management appeared of *"the enterprise responsibility, government administration and social supervision"* with Chinese characteristics under socialist market economic condition. After taking over the function of hydropower dam safety administration, SERC revised and issued *Dam Operation Safety Management Regulation for Hydropower Stations*, *Dam safety Registration Rules for Hydropower Stations*, *Regular dam safety Inspection Rules for Hydropower Stations*, *Dam Safety Information Report Rules for Hydropower Stations*, *Dam Safety Monitoring management Rules for Hydropower Stations*, *Dam Safety Rehabilitation Rules for Hydropower Stations*, and etc. These legislations and regulations defined the dam safety responsibilities, requirements and programs under the market economy conditions, and laid the foundation of the dam safety administration in new situation.

Currently the dam safety management mechanism is gradually set up, and the stipulations and regulations are increasingly perfect. The situations of dam safety are steadily improved. The theory of the dam safety risk management and the emergency management capability are substantially developed. The dam safety managements have been transferred from the accident management to potential deficiency management, from traditional administrative approach, economic method, conventional supervision and inspection into the combination of modern legislative approach, scientific method and cultural approach. The dam safety management has been updated to a new stage of the systemic management from basic standardization and normalization to scientific management with humanism.

26.1.2 *Dam Safety Supervision and Administration Organizations and Their Functions*

Followed by the Chinese electric institution reform, the General Office of the State Council declared *Notice on Strengthening Electric power Safety Production* in Dec. 5, 2003, demanding to strengthen the electric power safety production. SERC was authorized to be in charge of the electric power safety supervision administration and the State Administration of Work Safety was authorized to be in charge of the comprehensive administration. In order to meet the requirements of electric power safety supervision, SERC established the Department of Safety Administration to be in charge of the electric power safety supervision. The Office of Central Institutional Organization Commission LDSSC into SERC, as the institution directly under SERC, to be in charge of the regular supervision of dam safety administration. SERC established hereafter six regional supervision bureaus, twelve provincial offices responsible for the regional or provincial electric power safety supervision.

The dam safety supervision administrative responsibilities of SERC are as follows:

(1) Supervise and check the operation organizations of hydropower stations on their execution of safety production legislations, administrative regulations and technical rules and regulations.

(2) Supervise and inspect the dam safety conditions, if the hidden dangers were found, the operation agencies should be enforced to take measures at once.

(3) Participate in the construction project completion acceptance of the hydropower stations.

(4) Promote and check the operation agencies to conduct the dam strengthening and the deficiency treatment of hydropower stations, as well as the reinforcement for the risky dams and the dams with defects.

(5) Other responsibilities stipulated by the legislation and administration.

LDSSC, as the institution directly under SERC, is responsible for dam safety technical supervision and support according to *Operation Safety Management Regulation of Hydropower Dams* (Decree 3 of SERC). The responsibilities of LDSSC are as follows:

(1) Handle the dam safety registration of hydropower stations.

(2) Instruct the operation agencies to perform the dam safety inspection.

(3) Instruct the operation agencies to strengthen the risky dams and the dams with defects, and timely eliminate the potential possibilities of dams emergency.

(4) Organize the regular inspection and special inspection of the dams, propose the review comments on the regular inspection and the special inspection reports, and report to SERC for the record.

(5) Be responsible for training the dam safety personnel of the operation agencies.

(6) Establish and manage the dam safety supervision database and archives.

(7) Participate in the design review and completion acceptance of dam reinforcement, participate in the design review and completion acceptance of the ancillary facilities of the dams, and participate in the impounding and completion approval of the hydropower projects.

26.1.3 Introduction of Large Dam Safety Supervision Center

26.1.3.1 Historical Evolution

The former MWREP declared an urgent document and approved the establishment of LDSSC in 1985. The document demanded to set up LDSSC in East China Investigation Design Institute. The management mode is one administration group with two names. East China Investigation Design Institute provides technical and management supports. The task of LDSSC is responsible for the dam safety supervision of hydropower stations under jurisdiction of MWREP. LDSSC is the institution with the department/bureau level. It is headed by the president of East China Investigation Design Institute. The party organization, personnel management and logistics are managed by East China Investigation Design Institute. There are related departments and personnel in LDSSC in charge of related business.

Along with the development of state government organs, LDSSC was under jurisdiction of MWREP, the Ministry of Power Industry and Ministry of Energy in different periods. With the electric power institution reform and separation of enterprise from administration, LDSSC was under jurisdiction of State Power Corporation in administrative relation, but its business was under the guidance of the Safety Production Supervision Administration of State Economic and Trade Commission and thereafter State Administration of Work Safety.

In June, 2004, the Office of Central Institutional Organization Commission declared the document No. 75, [2004] and LDSSC into SERC, changed the name as LDSSC of SERC. LDSSC became the institution directly under SERC.

26.1.3.2 Main Tasks

LDSSC is responsible for the hydropower dam safety technical supervision support. The effective dam safety management mechanism is formed to guarantee the hydropower dam operation safety through dam safety registration, regular dam safety inspection, dam safety monitoring management and informatization construction, dam rehabilitation management and service, dam safety emergency management, personnel training and technical exchange, etc.

The working principals of LDSSC are as follows: promoting the dam safety management informatization construction, forming new pattern of dam operation safety management with the combination remote management and site inspection, finding and eliminating the hidden dangers timely, the risk forecast, the emergency preparation plan, the abolished dam removal and the ecological environment restoration, to realize the dam safety management during the whole process.

1. Dam safety registration

Through the site inspection and spot check in the dam safety registration, the hydropower enterprises are urged to carry out the related national stipulation and requirements with regard to safety responsibilities, the personnel equipment, the department establishment, the safety inputs, establishing rules and regulations, the routine monitoring and maintenance, etc. The registration grades are defined as A, B and C depending on the safety grades of the regular dam safety inspection. The licenses intervals respectively are 5 years, 3 years and 3 years.

By Dec. 2010, the number of registered dams in LDSSC is 211. Although it is a small portion of over 80,000 dams in China, the 211 hydropower stations are generally with high dams, huge reservoir storages and large installed capacities. The installed capacity of the registered hydropower stations reached 30.7% of the total nationwide, and the reservoir storages reached 33.5% of the total. Among them there are 5 reservoirs with the storage over 10 billion m^3 and there are 18 hydropower stations with the installed capacity over 1,200MW. The registered hydropower stations include many important large-scale ones, such as Longyangxia, Xin'anjiang, Fengman, Gezhouba, Ertan, etc.

Table 26.1 – 1   Dam Characteristic Parameters of the Registered Hydropower Stations and Its Percentage in China

Items	Number in China	Registered hydropower dams	
		Number	Percentage in China (%)
Dam number	87,873	211	0.24
Installed capacity (MW)	213,400	65,460	30.7
Total storage ($10^8 m^3$)	7,162	2,401	33.5

Notes: 1. The installed capacity in China means the total hydropower units put into operation by the end of 2009.

2. Data comes from Statistic Bulletin on National Water Conservancy Development in 2009.

2. Regular dam safety inspection

Regular dam safety inspection is the safety physical examination of the dams with the interval of 5 years. The expert team comprehensively assessed the dam safety grade through site inspection, the systemic examination to the hidden dangers, the special subject analysis and study. The safety grades are divided into three levels: normal dam, the dam with defects and risky dam.

Since the establishment, LDSSC has completed the general survey, the first round regular inspection, the second round regular inspection and the third round regular inspection. Now the fourth round regular inspection is under way from 2011 to 2016.

Seven dams were detected with defects during the first round regular inspection, they were Qingtongxia, Luodong, Foziling, Lvshuihe, Majitang, Yilihe Ⅳ and Baiyutan, and two dangerous dams were Xiuwen and Tianqiao. Eight dams with defects were detected during the second round regular inspection, and they were Meishan, Liguan, Shuidong, Yilihe II, Lvshuihe, Huanglongtan, Nangao and Fengman, and there was no risky dam. By the end of 2010, 135 dams had undergone the regular inspection in the third round, Fushi was assessed as the dam with defects due to the low flood control capacity, and others were in normal condition.

Based on the experiences summarized from the first, second and third round regular inspections, LDSSC will carry out classified management during regular inspection, with regard of special conditions of different dams. The regular inspection mode will be optimized through the application of the information construction results and the new technical methods to find out the hidden dangers and improve the dam safety.

3. Dam safety monitoring management and information construction

The main tasks of the dam safety monitoring management are as follows: compilation of monitoring technical standard, evaluation on monitoring system, review and completion acceptance of improvement plan for monitoring devices, monitoring data analysis and evaluation of dam behavior, assessment and approval of monitoring devices sealing and disuse, and the development of monitoring software, etc.

Since 2006, LDSSC performed dam safety information construction according to *Dam Safety Information Construction Program of Hydropower Stations* issued by SERC, in order to realize remote acquisition of dam safety monitoring information, remote management and diagnosis, remote warning and feedback, to realize real time monitoring of the major components of the dam and key items step by step, and to find out and trace the hidden dangers timely. The dam safety dynamic management should be realized by modern means to enforce the dam safety risk and emergency management.

4. Dam reinforcement

The dam safety registration, regular dam safety inspection, monitoring management and information construction are the means of dam safety management. The goal is to detect the hidden dangers and take necessary measures to eliminate deficiencies and ensure the dam safety. So the dam deficiency elimination, the dam reinforcement and the modification are always the important tasks of LDSSC. The hidden dangers of dams in operation were treated in the bud through the reinforcement management and services by LDSSC.

5. Dam safety emergency management

The emergency management is currently an important component of the safety production administration. In accordance with the arrangement of SERC, LDSSC takes advantage of technology to enforce the development of dam safety emergency preparation plan, and the exercise of the emergency reaction plan and the construction of Hydropower Dam Safety Emergency Technical Support Center. Entrusted by the World Bank in 2001, LDSSC developed Dam Safety Plan for Jiangya Dam financed by the World Bank. LDSSC and Xin'anjiang hydropower Plant developed Emergency Preparation Plan of Xin'anjiang Dam in 2006. LDSSC completed the flood re-check and the dam breach analysis of cascade hydropower station group in Lixianjiang basin, as well as the development of the emergency preparation plan. LDSSC instructed their EPP exercises of some hydropower stations on invitation. During every flood season, especially the emergency events of the heavy disasters, such as 5.12 Wenchuan Earthquake in 2008, LDSSC timely analyzed and evaluated the information submitted from these hydropower stations in disaster area and provided necessary support and assistance for their operation and disaster relief.

6. Technical training and exchange of Dam safety personnel

Since the establishment, LDSSC held 28 training classes on the dam safety management and monitoring personnel carrier with more than 1000 participants. There were more than 300 leaders and chief engineers from the owners taking part in the senior dam safety management training. In recent years, LDSSC annually held the dam safety management training class and the dam safety technical specification training class together with the related universities. The lecture contents were plentiful, task-focused and received feedback from trainees who learnt a lot.

Meanwhile, in order to improve the dam safety management, LDSSC carried out technical exchange both home and abroad on dam safety extensively, and promoted new techniques application; published the journal of *Dam and Safety*; opened the LDSSC website. These activities popularized the newest dam safety management theory and means home and abroad, and improved the professional quality of the dam safety personnel as well.

26.2 Dam Safety Administrative Legislations

26.2.1 *The Importance of Managing the Dams by Laws*

The dam safety involves human life and property safety upstream and downstream of the dams, and the sustainable development of the national economic. It is an issue of public safety.

The dam operation is a long term and complex process which will encounter all kinds of problems and relate to many social respects. It is a complex systemic project. For instance, a synthetic development project will be related with departments of the generation, navigation, flood control, irrigation, fishery, entertainment and water supply, also related with departments of communication, mining, building materials, land management, soil and water conservation, earthquake and civil air defense. Different dams probably belong to the different departments, such as conservancy, electric power, urban construction, agricultural reclamation, army reclamation and metallurgy, etc. In complex relationship mentioned above, each party probably holds its different viewpoint with its own unique position and the contradiction is difficult to avoid. Only the legislation can take account of each party's interest, regulate and coordinate the relationships between the parties to seek for the benefit maximization of all. Thus, it is important and necessary to manage the dams by laws.

In order to ensure the orderly development of the dam safety management, China has built up complete dam safety management legislation system to standardize each party' activities and regulate the interest relationship of each party. For instance, the flood control issues should be dealt with *Flood Control Law of the People's Republic of China* and *Flood Control regulations of the People's Republic of China*. With regard to dam management, these issues should be dealt with *Dam Safety Management Regulation* and *Operation Safety Management Regulation*.

26.2.2 *Current Dam Safety Administration Legislations and Departmental Regulations*

1. Legislations

(1) *Safety Production Law of the People's Republic of China*. It was approved in the 28th Standing Committee meeting of the 9th National People's Congress in June 29, 2002 and enforced from Nov.1, 2002.

(2) *Water Law of the People's Republic of China*. It was approved in the 29th Standing Committee meeting of the 9th National People's Congress on Aug. 29, 2002 and enforced from Oct. 1, 2002.

(3) *Flood Control Law of the People's Republic of China*. It was approved in the 27th Standing Committee meeting of the 8th National People's Congress in August 29, 1997, issued by the 88th Chairman decree of the People's Republic of China in August 29, 1997, and enforced from Jan. 1, 1998.

(4) *Flood Control Regulation f the People's Republic of China*. It was approved in the 87th Standing meeting of the State Council in June 28, 1991, issued by the 86th State Council decree of the People's Republic of China and enforced in Jul. 2, 1991, and revised in 2005.

(5) *Reservoir Dam Safety Management Regulation of People's Republic of China*. It was issued by the 77th State Council decree of the People's Republic of China and enforced in March 22, 1991 and enforced from this day.

2. Departmental regulations

(1) *Dam Operation Safety Management Regulation for Hydropower Stations* approved in the Chairman meeting of SERC, issued in Dec. 1, 2004, and enforced in Jan. 1, 2005.

(2) *Dam safety Registration Rules for Hydropower Stations*, issued by SERC in Nov. 10, 2005.

(3) *Regular dam safety Inspection Rules for Hydropower Stations*, issued by SERC in Nov. 10, 2005.

(4) *Dam Safety Information Report Rules for Hydropower Stations*, issued by SERC in Sep. 29, 2006.

(5) *Dam Safety Monitoring Management Rules for Hydropower Stations*, issued by SERC in the Feb. 24, 2009.

(6) *Dam Safety Reinforcement Rules for Hydropower Stations*, issued by SERC in Oct. 21, 2010.

The above legislations and regulations formed current legislation system of the dam safety management for hydropower stations in operation in order to guarantee all activities in smooth performance. In our routine activities, *Reservoir Dam Safety Management Regulation of People's Republic of China* and *Dam Operation Safety Management Regulation for Hydropower Stations* are closely involved with the dam safety management; therefore the following is the brief analysis on them.

26.2.3 Brief Analysis on Dam Safety Management Regulation of People's Republic of China

1. Scope of application

The scope of application covers the dams with height over 15m or the storage over 1 million m^3. The dam structures include the permanent impounding structure, as well as spillways, flood discharge and water supply structures and ship locks, etc.

From the safety point of view, if the impounding structure, flood discharge and water supply structure of the dam projects were damaged, it would result in a heavy disaster with a large quantity release of the reservoir water. Among them, the safety of the impounding structure is of major importance. So the dam indicated in this *Regulation* includes the impounding, flood discharge and water supply structures, this does not only conform to the Chinese custom and grade classification, but also matches the international generic terms.

2. Responsibilities of dam safety management

According to the flood control experience of many years, responsibilities should be implemented to ensure the dam safety. The *Regulation* clearly defines the duties and responsibilities of the dam safety management in order to combine the dam safety supervision system with the current dam safety management system. Under the principle of management by different departments and at different levels, the water administration department of the State Council is responsible for the flood control of the whole country and the routine works of State Flood Control and Drought Relief Headquarters, as well as undertakes the supervision of the national dam safety. As for the dams under jurisdiction of water conservancy, electric power, construction, metallurgy, land reclamation and army reclamation departments, their safety are directly managed by their own authorities. The local governments at different levels are responsible for the dam safety under their jurisdiction and undertake the administrative head responsibility institution. The chief executive should undertake the corresponding responsibilities to the dam safety accidents due to misgoverning and improper operation.

Dam safety involves the investigation, design, construction and operation management. If any part among them was ignored, it will result in safety problem. The *Regulation* clearly defines that the authorities at different levels should strictly enforce the design review, the completion acceptance and operation management, and defines the responsibilities of each party. The purpose is to make clear of the dam safety responsibilities of the related parties during dam construction and management, in order to consciously put the dam safety in the most important position in their work and to avoid unnecessary loss.

3. Dam construction

Dam construction should be paid high attention to because its construction quality is related with dam safety. The *Regulation* defines that the dam construction should comply with the dam safety specifications developed by the water administration of the State Council, jointly with the related dam authorities.

Dam investigation and design are the important basis for dam construction. The quality of dam investigation and design involve directly dam safety. The units undertaking the dam design should pass the qualification check from the authority, hold the qualification license corresponding to the magnitude of the dam designed. The hydrological and geological design should meet with the in-depth requirements of design phase. Dam safety is the prerequisite for the comparison and selection of the design schemes. The safety demonstration is necessary for the application of new technique concerning dam safety. The design documents and drawings should be strictly examined to ensure the project quality.

Biding and tendering system shall be adopted for dam construction. The project owners should conduct qualification examination of the potential contractors, notarization is required if necessary. The contractors should be qualified with the construction ability and construction experiences corresponding to the magnitude of the dam project. The dam construction should be strict in accordance with the documents, drawings and the related specifications stipulated in the construction contracts in order to ensure the project quality. Strict quality check system and quality control procedures are required for dam construction. The project owners and designers should assign their representatives to the site, closely coordinate with the contractors, strengthen the project quality inspection and implement the supervision system. When the construction quality was found not to meet the requirements, construction should be done over again for the poorly done project. If disputable, it should be reported to the higher project development unit for adjudication. If the remedial measures are necessary, it must be reliable and taken without decreasing the dam safety level.

The completion acceptance is the last control to guarantee the project quality, so completion acceptance procedure shall be followed strictly. Due to historical reasons, many dam projects transferred to the operation management units without going through the acceptance procedure so that a large quantity of closing works and project quality deficiencies remained. These seriously affected the implementation of dam safety management due to the dealing with the above remained works. Therefore, the project owners should apply for the project acceptance from the dam administrative department after the dam's completion. The completion acceptance, based on the acceptance of the partitioned projects, the sub-divisional works and the periodic acceptance, is taken by the authority together with the project acceptance committee composed of the related departments from project development, design, construction, management and the related departments of the local government. The following criterions should be met in addition to the acceptance rules for infrastructure:

(1) The project has been completed according to the requirements stipulated in the design documents; the quality meets the requirements; the project operates normally and has gone through the acceptance of the partitioned projects, the sub-divisional works and the periodic acceptance.

(2) The deficiencies detected in the each periodic acceptance have been treated and settled.

(3) The operation and management conditions stipulated in the design documents have been achieved.

(4) The relocation compensation and the project management land requisitioned, etc. have been completed and well-handled.

If the above mentioned conditions are not completely obtained, the remained problems will not affect dam safety operation, the completion acceptance can be performed. The measures and expenses should be settled down for the remained works and closing works, the responsibilities should be defined and the works must be completed within the stipulated period. The completion acceptance report should be submitted when the project passed the completion acceptance, and take effect as soon as the members of the completion committee signed on it. It will be the evidence of the project transferred to the management unit. If the remained works are important and the acceptance committee can not come up with a consensus, it should be reported to higher authority for discussion and decision. The management unit can refuse to handle the acceptance procedures.

The purpose to build a dam is to make the project producing more benefits through the dam safety management. In order to create necessary conditions for dam safety management, the design of the managing facilities, such as engineering monitoring, communication, power supply, lightening, transportation, fire control, buildings, should be incorporated in the project budget in a new dam design (enlargement, modification and strengthening). The contractors should have full and complete installations according to the design requirements. Dedicated staff should be designated for the dam safety monitoring instruments. The instruments should be embedded carefully to ensure the effectiveness, monitored and managed during dam construction according to the requirements of the related specifications. The contractors should transfer the monitoring instruments and their data of the construction phase to the management units after the project completion acceptance.

Dam management scope and protection scope should be defined by inviting the government above the county level for approval, according to the management requirements, with regard to the natural geographical conditions, historical conditions and social, economical conditions. With regards to the land and its appurtenance, its ownership belongs to the whole people and the operating right belongs to the dam management units. Activities, such as quarry, fetching earth, well drilling, reclamation, which may endanger the dam safety, should be restricted to ensure dam safety.

In order to strengthen the dam safety management, dam administrative department should set up the dam operation management unit at the beginning of dam construction as soon as possible. During the construction, the management units should take part in the construction quality check and the acceptance of the partitioned projects, the sub-divisional works and the reservoir impounding acceptance. The purpose is that the management unit can master the project development and project quality, and lay a solid foundation for thereafter dam safety management.

4. Dam management

All kinds of dams are the valuable treasures of the country and people, and the significant infrastructure to guarantee the economic construction and safety of people's lives and properties. The *Regulation* elucidates the state protection right and the legislative evidence to the dam project and its related facilities, and prohibits the activities which can endanger the dam safety, regardless of the activities from the organization or individual. The main duties of the dam management are the followings:

(1) Dam safety monitoring. The monitoring is the ear and eye of dam safety management, also an important approach to master the dam operation situation. The *Regulation* stipulates that the management units should enhance the dam monitoring according to the related standards and specifications.

(2) Dam maintenance and repair. The in-time maintenance and repair are the important means to ensure the dam safety, and the hoist devices in good operation. The *Regulation* clearly prohibits the activities of the random lumbering and reclaiming abrupt slope in the dam watershed. Reservoir sedimentation issues are serious in Northwest China, so the activities which can endanger the mountain mass, such as land reclamation, quarry and fetching earth are forbidden in the reservoir area. It is forbidden to build wharf, channel in the dam body and to accumulate materials and to air-dry the grains, as well as to build fishponds at the toe of downstream slope.

(3) Reservoir regulation. On the premise of ensuring dam safety, reservoirs should be managed by scientific regulation to fully produce the comprehensive benefits. When coming across the contradictions of safety *vs* benefit, integrity benefit *vs* partial benefit, reservoir upstream benefit *vs* downstream benefit, the *Regulation* stipulates that the dam management units should regulate the reservoir according to the approved plan and the instructions from the dam administrative department. No unlawful interference is allowed for reservoir regulation. Dangerous situation will probably appear at the high reservoir water level during flood control regulation and thus the dam safety is especially important. The dam management department should closely contact with the meteorological department to strengthen the meteorological and hydrological forecast, auto-monitoring of water regime, and improve the planning and foresee ability of the reservoir regulation and operation. According to the rainfall, the reservoir can be reasonably pre-discharged to reduce the storage for the coming flood when the heavy flood is foreseen from the upstream and the flood discharge is permitted to the downstream channel. When it is necessary for flood peak shifting in the downstream and it is possible for the reservoir, adequate discharge is allowed from the reservoir. If there is no enough storage for flood peak shifting in the reservoir, dam safety shall be the priority and immediate discharge shall be carried out. Blind over-storage is not permitted in order to prevent the disaster of dam failure.

(4) Dam safety inspection. There are different terms adopted in the world regarding dam safety inspection, and their types, frequencies and forms of organization vary, while their objectives and inspection items, and scope all focus on unstable factors, aiming at evaluating the dam status and controlling the operation. In China, dam safety inspection means the regular examination of the dam safety and integrity. Before and after the floods, after storm, extreme flood and heavy earthquake, it is necessary to carry out regular inspection or immediate inspection. It is one of the important approaches to detect dam safety deficiencies, together with the instruments monitoring. Regular inspection of hydropower dam safety is enforced and formed based on historical lessons and experience from dam operation home and abroad. Some dam safety issues cover extensive aspects, deep research and difficulties, so the conclusions shall be of higher authoritativeness. Thus, in general, management units can not complete the dam safety inspection themselves. According to the practical experience, after 3 years to 5 years of dam completion, it is necessary to organize expert team to undertake the safety inspection. This kind of inspection is of decisive meaning to the evaluation of project flood control standard, project characteristics and afterwards operation modes. For the old dams with operation period over 40 to 50 years, it is necessary to carry out the comprehensive review and inspection because the natural conditions and operation modes probably have changed greatly, and degradation of engineering components and materials are obvious.

(5) Flood control. The *Regulation* stipulates that dam management unit should prepare flood control materials and reinforce meteorological and hydrological forecast, and ensure the information exchange of water regimen and warning, as well as the communication between the dam management units and the authority and the higher flood control headquarters. As the sign of the danger appears, the dam management unit should report at once and take immediate measures. The reservoir flood control headquarter can take exceptional measures to ensure dam safety according to the existing plan under the circumstance of extreme rainstorm, when the water level of reservoir rises rapidly, and rainfall continues upstream, overtopping would happen, and the dam will be in danger, and there is no time to ask for instructions. As the dam is in great danger or the exceptional measures are to be taken, the reservoir flood control headquarter should send out warning through all

possible means to inform rapidly the related units downstream of the dam to guarantee the public's safe evacuation.

5. Treatment of dangerous dams and dams with defects

There are many dams with hidden dangers due to reasons of design, construction, operation and dam degradation, directly relating to the safety of the downstream human life and property, as well as the development of national economy; these dangers should be settled as soon as possible. This is the reason why treatment of dangerous dam is individually stipulated in the *Regulation*. At present the definition of a dangerous dam is that the dam cannot meet the standard of designed floods, with insufficient seismic fortification standard, or with serious quality deficiencies, at the same time with great potential threats to the downstream. The *Regulation* stipulates that dam administrative department should prepare strengthening plan for the dangerous dam under jurisdiction of its own, give priority in expense arrangement and eliminate the danger within the defined time limit. The design and contractors undertaking the strengthening project of the dangerous dam should have relevant qualifications and make good preparatory work. Confined by the existing structure, dangerous dam treatment is more difficult than to build a new one. The dangerous dam strengthening scheme can not be implemented until the review approval are available. After completion, its acceptance should be organized by the dam administrative department. Before being listed to the strengthening plan, the dam management unit should prepare the emergency methods to guarantee the safety of the dangerous dam. If the original design mode needs to be changed by demonstration, it should be reported to the dam authorities level by level for approval according to the stipulation. At the same time, it should make pre-evaluation for possible dam failure and flooding range, develop emergency preparation plan and report to the higher flood control headquarter for approval.

26.2.4 Brief Analysis on Dam Operation Safety Management Regulation for Hydropower Stations

26.2.4.1 Scope of Application

The *Regulation* is applicable to large and medium-sized dams in operation under jurisdiction of Chinese electric power industry. The dams include all permanent impounding structures, flood discharge structure and the impounding structure of the water conveyance structure and the ship lock facilities crossing the riverbed and saddles around the reservoir, as well as the structure foundations, adjacent dam, slope and auxiliary facilities.

26.2.4.2 Guiding Ideology and Basic Principles

1. Guiding Ideology

The policy of "precaution crucial, safety first and comprehensive administration" should be carried out to enhance the dam safety management of hydropower stations and to guarantee safety production. The enterprise self-regulation is emphasized, and simultaneously the government should enforce the supervision on the safety of hydropower dams.

2. Basic Principles

(1) Precaution first and overall supervision. When the dams utilize natural resources, change the natural environment, promote the beneficial and abolish the harmful, as well as create huge benefits for the human beings, they also have potential safety risk. Any oversight in the management will possibly result in dam failure and significant impacts on the safety of people's life and property, national economic construction, ecological environment and social stability. So we must enhance the supervision and management in every aspect to prevent the possible accidents.

(2) Power and responsibility well defined and taken. Hydropower dam safety not only relates to the enterprise safety production and operation, but also to the public safety. Thus, as the enterprise is responsible for the hydropower dam safety, the government must enhance the safety supervision of those dams. The *Regulation* clearly defines the main responsibilities of SERC, LDSSC and the enterprises. It also defines the investigation, design, construction, supervision, and operation parties participated in dam construction to implement their works in accordance with the legislation, rules and regulations, specification and contracts, and take corresponding responsibilities.

(3) Uniform standards and easy implementation. The *Regulation* defines four dam safety inspection patterns: routine examination, annual detail inspection, regular inspection and special inspection, and also defines the standards for the dam safety grade assessment and safety registration so that the enterprises can undertake dam safety responsibilities.

3. Main contents

The *Regulation* has 7 chapters, totally 43 articles, which define the enterprise safety management responsibilities, the contents, procedures and standards of the hydropower dam safety inspection and assessment, the dam registration organization and registration requirements, as well as the dam safety supervision administrative organizations, their responsibilities and the punishment for the infringement. The *Regulation* emphasizes on the dam safety inspection and assessment, dam safety registration and government supervision administration.

(1) Dam safety inspection and grade assessment. The dam safety inspections are divided into routine examination, annual detail inspection, regular inspection and special inspection. In order to enforce the hydropower dam safety supervision administration, to timely detect and treat the hidden dangers, as well as to strengthen the seriousness and authoritative of the dam safety grade assessment, the assessment organizations and the assessment procedures for the above mentioned inspections are appointed in the *Regulation*, and it is stipulated that the safety grades assessment should be organized and carried out by LDSSC.

(2) Dam safety registration. Dam safety registration license is the certificate of dam operation, also the warrant of hydropower station in normal operation and the acceptance of government supervision. The *Regulation* defines that dam operation should pass the safety registration procedure. Dams must apply for the safety registration and operate with the license. The dam will be regarded illegal, if it is put into operation without application of the safety registration.

(3) Dam safety supervision administration. The *Regulation* defines that the supervision administrative department of SERC is responsible for the government supervision of the hydropower dams, and LDSSC is responsible for the technological supervision and support of dam safety.

4. Main works of hydropower enterprise self-regulation

The dam operation agencies should undertake the following works during the dam operation safety management.

(1) Establishing and improving the safety production responsibilities, allocating adequate professionals qualified for the practical requirements, and guaranteeing the necessity of the safety production funds. The head of the enterprises is the first responsible person of its own enterprise safety production and responsible for the whole enterprise safety production. The safety production involves all aspects of routine production and management, only if the safety production responsibilities are defined, can the effective safety management system be set up to create the good safety environment for safety production. Thus, to establish the dam safety production responsibility system of the hydropower station operation agencies is the key point of implementing dam safety management. Moreover, well-defined dam safety production responsibility system is the premise of personnel allocation and safety production funds input.

(2) Implementing flood control. Viewing from the whole world, flood is the significant reason causing dam failures. So flood control is the focal point of the dam safety management. Before every flood season, hydropower station operation agencies have to implement flood control responsibilities, improve the rules and regulations, develop annual scheme for reservoir flood, enforce flood control safety inspection and inspect during flood season to ensure a safe flood season.

(3) Establishing and improving dam safety inspection institution, enhancing dam safety inspections. Although dam failure is unexpected incident, the dam will represent quantitative change process during qualitative change. The implementation of dam safety inspection helps to detect dam anomaly during the degradation so remedial work can be carried out in time to ensure dam safety. Dam safety inspections are divided into routine examination, annual detail inspection, regular inspection and special inspection. Among them, routine examination and annual detail inspection are the responsibilities of the enterprise, and are organized and completed by hydropower station operation agencies. The regular inspection and special inspections are the behaviors of the government technological supervision, organized and completed by LDSSC. Hydropower station operation agencies should establish inspection institution to carry out routine examination and annual detail inspection according to the stipulations, and support LDSSC to implement the regular inspection and special inspection to ensure the dam safety.

(4) Strengthening the monitoring of the dam safety, analyzing the monitoring data and keeping track of the dam behavior. Dam safety monitoring data is the significant evidence for dam safety evaluation, and it can represent the actual responses of the design variables of each dam. On the one hand, the data analysis results can be applied into the safety evaluation and forecasting; on the other hand, it can be used as the feedback and help to verify the design.

(5) Applying for the dam safety registration to ensure legal operation with the license. Dam safety registration license is the certificate of dam operation, as well as the legal document for hydropower station to be put into normal operation and under the government supervision. Dam registration according to the stipulation means that dam safety is under the government supervision and administration. Furthermore, it shows that the organization of a hydropower operation unit is healthy, the management is perfect, the staff is qualified, the fund is guaranteed, the maintenance is performed and the dam safety can be ensured.

(6) Enforcing professional training. Dam safety management is closely related with the quality of the operation professionals and their sense of responsibility. The hydropower station operation agencies should be allocated with capable professionals according to the position requirement, and training on safety legislation and technical skills for the operation and management should be held regularly.

(7) Responsible for dam repair and rehabilitation. As soon as the dam deficiencies are found, the dam operation agencies should take actions to repair the dam to normal operation condition to ensure the normal operation.

(8) Developing the dam emergency forecast and preparation plan, deploying necessary devices and facilities for emergency treatment. The dam emergency forecast and preparation plan are the key points of the dam safety management, and the important safeguard to rescue and reduce losses of life and property in case of dam failure. Thus, the dam operation agencies should explore the possible modes of dam failure, develop, improve and revise the emergency preparation plan according to its own dam characteristics, and exercise the emergency actions.

(9) When dam emergency omen appear, hydropower station operation agencies should timely take effective measures to rescue, and report to the related departments of the local government and higher authority, to ensure the effective rescue and to prevent the spread of the incident.

5. Supervision administration of the government

While the *Regulation* emphasizes on the enterprise self-regulation to strengthen the dam safety management, it also defines the specific requirements for the government supervision on hydropower dam safety.

(1) The dam operates with the safety registration institution. According to related requirements of the *Regulation*, SERC is the authority of the dam safety registration, and LDSSC is responsible for the routine work of hydropower dam safety registration. LDSSC annually reports the previous year's dam safety registration summary to SERC. SERC regularly declares the hydropower dam safety registration list.

The *Regulation* stipulates explicitly the requirements for registration: "dam operation agencies should apply for the dam safety registration to LDSSC within one year after the completion assessment of a new dam, or within half year after the first regular inspection completed. The hydropower station operation which has not applied for safety registration by the stipulated deadline is not permitted."

(2) Carrying out regular and special inspection of dam safety to track the dam safety. Regular inspection of dam safety is an effective system to ensure normal dam operation, also an important measure to implement government supervision. It should be undertaken by the authorized organizations which regularly carry out the comprehensive inspections for the dam design, operation, management, monitoring and etc. according to the related stipulations. The *Regulation* defines the frequency of the regular inspection is generally with interval of 5 years. Since the first round dam safety regular inspection was performed in China in 1987, three rounds of regular inspections have been completed until now. The above mentioned regular inspections have detected some dam deficiencies, and some significant problems affecting the dam safety had been found out, providing evidences to strengthen dams and making sure the comprehensive safety conditions of the hydropower dams. The regular inspections made dam safety significantly improved.

When extreme floods, strong earthquakes or abnormal conditions that can probably affect the dam safety occur, hydropower station operation agencies shall apply LDSSC for special inspections, and LDSSC should organize expert teams to implement the special inspections.

(3) The supervision and administration on the hydropower stations safety monitoring system and management. The dam safety monitoring system is an important facility to ensure dam safety operation and to detect the dam deficiencies, also the ears and eyes of routine dam operation management. The *Regulation* defines as follows: as for the dam with height over 70m, or the low or medium-sized dams with complicated monitoring systems, the *Regulation* defines that the legal person of the project development organization should organize the related units to carry out special design and special review for the dam safety monitoring system according to the related stipulations. When the project is completed, specific inspection and completion acceptance shall be carried out.

As for the dam safety monitoring system in operation period, regular inspection, assessment and evaluation are required. As for the modification of the monitoring systems, the hydropower station operation agencies should organize specific design and specific review. When the modification project is completed, they should apply for the specific inspection and completion acceptance from LDSSC.

(4) Supervision and administration on the enlargement and modification of the hydropower stations.

The enlargement or modification of the existing facilities will have some impacts on existing structures. Improper handling will bring about disastrous consequences to the dam. Due to the inherent characteristics of the dam in operation, the *Regulation* stipulates that the enlargement or modification shall not endanger dam safety. The hydropower station operation agencies should send structure enlargement or modification design (including the construction procedures and construction methods) to the original design unit for examination and report to LDSSC for safety evaluation.

(5) Supervision and administration on the reinforcement and modification work of hydropower stations. Design examination should be implemented for the reinforcement and strengthening of the dam. For the projects involving dam foundation or concealed work, specific acceptance should be carried out according to concealed work criterion, and completion acceptance should be performed after completion.

(6) Supervision and administration on emergency mechanism for dam in danger. The hydropower station operation agencies should develop dam dangerous situation forecast and preparation plan according to related stipulations, and report to the higher flood control headquarter for approval and report to local government (county, city and province) for record.

(7) Supervision and administration on the corresponding technical support and service organizations undertaking the dam regular inspections and special inspections. The technical support and service organizations are responsible for undertaking the specific subject inspection in hydropower regular dam safety inspection and specific inspection. The report of the specific subject inspection is the primary evidence of assessing the dam safety grade and developing regular and specific inspection report. Thus, specific inspection is of significant importance. The *Regulation* defines that these technical support and service organizations should have the corresponding qualifications and good achievements. Meanwhile, LDSSC should regularly announce the list of technical services organizations meeting the requirements, helpful for the supervision and administration of these organizations, and provide convenience for the operation agencies.

26.3 Dam Safety Technology Supervision Administration

26.3.1 *Dam Safety Registration, Regular Inspection and Technology Supervision Administration*

26.3.1.1 Dam Safety Registration Administration

In order to strengthen dam safety supervision and administration, to push forward the improvement of dam safety management and to step gradually on the track of normalization, institutionalization and standardization, the former Ministry of Electric Power Industry began to enforce the dam safety registration regime in 1996, issued *Dam Safety Registration Rules* in Oct. 1996 and stipulated the dams managed by the electric grid enterprises (such as the electric authorities, the provincial electric bureau and river development companies, etc.) to apply for the dam safety registration. The registration grades of dams are classified into grade A, B and C according to the dam safety conditions and management standards. The former LDSSC of MWREP was responsible for processing the applications, assessing the grades and issuing the registration licenses.

SERC issued *Dam Operation Safety Management Regulation for Hydropower Stations and Dam Operation Safety Registration Management Rules* in succession in 2004 and 2005, further strengthening the dam safety registration institution. The *Rules* defines that the dam authorities should apply for the original registration within one year after completing the project acceptance safety assessment. The registration grade depended upon the site examination result on dam safety management conditions. The examination mainly covers the following six aspects.

(1) Implementing the dam safety legislations, rules and regulations. Establishing clear dam safety responsibilities, and implementing duties of each departments and important positions.

(2) Dam safety regulation specifications institution. Establishing and improving the technical regulations and management institutions, including water management, hydraulic inspection and maintenance, hydraulic monitoring, hydraulic machinery operation overhaul, flood control, emergency treatment etc.

(3) Dam safety professional qualification and training.

(4) Dam safety inspection, monitoring, the implementation of the dam inspections and monitoring institution and the quality of their work.

(5) Dam safety information and archives. Dam design, construction, scientific research, and operation information are kept on files and archives management.

(6) Dam maintenance, deficiencies treatment and safety funds. The dam leftover problems and defects treatment, dam safety funds assurance.

The dam registration grades are verified by LDSSC and reported to SERC for approval. The registration grades are managed dynamically, re-registration and change shall be made according to the results of the regular dam safety inspection and the annual dam registration spot check.

From the year 1996 to Dec. 2000, there are 218 dams of large or medium-sized hydropower stations registered in LDSSC. Through long-term and dynamic management, most dam owners paid high attention to the dam safety management, established dam safety position responsibilities institution, arranged monitoring professional training, strictly carried out the dam inspection institution, timely enforced repair and reinforcement, so that these dams were continuously registered as grade A. The dams assessed with defects were temporarily registered as grade C according to the stipulation, some of them obtained grade A when they were re-assessed as normal dams after the defect treatment and reinforcement. The management standards of few dams should be improved, especially some of the new dams, the dam safety responsibilities system were not fully implemented, the project site were short of hydraulic staffs, the technical specifications were short of pertinence and practicability, so rectification and improvement shall be conducted or they shall be reiterated as grade B.

26. 3. 1. 2 Regular Inspection of Dam Safety

In 1987, former MWREP issued *Dam Safety Management Interim Regulation for Hydropower Stations*, the large and medium-sized stations belonging electric power system implemented the regular dam safety inspection institution so as to ensure the safety reliability of dam structure and operation, timely detect dam abnormal phenomenon, hidden dangers and existing defects, and perform the deficiency and defect treatment and reinforcement. Beginning from Gutian I as the experimental regular dam safety inspection in 1987, three-round regular dam safety inspections have been carried out. Especially since *Dam Operation Safety Management Regulation for Hydropower Stations* issued by SERC in Dec. 2004 and the third round regular dam safety inspection initialized in 2005, regular dam safety inspections have been fulfilled comprehensively as the important measures of the government supervision for dam safety.

1. First round regular dam safety inspection and the treatment for the dangerous dams and the dams with defects

The first round regular dam safety inspection began from Gutian I dam in 1987, and ended in 1998, regular inspections of 96 dams were completed, including some high dams and huge reservoirs such as Fengman, Xin'anjiang, Liujiaxia, Gezhouba. Through design examination, construction recheck and dynamic operation performance analysis, the first round inspection basically figured out the dam safety conditions which were put into operation before 1980s, detected out some significant defects and hidden dangers threatening the dam safety, and solved some safety puzzles existing for a long term. In accordance with the statistics, 350 deficiencies and issues were detected out in 96 dams. The deficiencies involved the flood control standard, dam foundation treatment, dam structure and strength, seepage safety, discharge facility erosion prevention, hydraulic machinery, reservoir rim collapse, monitoring instruments, etc. Along with the regular inspections going into depth, most deficiencies were settled during the regular inspections, such as the dam heightening, structure strengthening, adding or modifying the discharge facilities, anti-seepage grouting and drainage pressure relief, energy dissipater and bank slope repair, monitoring facilities modification. By assessment, 86 dams are assessed as normal dams, 7 dams were assessed as with defects, for instance, Qingtongxia, Luodong, Foziling, Yilihe IV, Lvshuihe, Baiyutan and Majitang. 2 dams were regarded as risky dams, for instance, Xiuwen, Tianqiao. The safety grade of Huanglongtan dam has not been approved up to now. More than 10 dams among the normal dams should be classified into dangerous dams or the dams with defects due to their original deficiency property and extend, the related organizations took strengthening measures during the regular inspections and timely eliminated the unsafe hidden dangers.

Assessed as the dangerous dams or the dam with defects in the first round regular inspection, dam heightening were carried out in Luodong, Baiyutan and Xiuwen dams, additional spillway were constructed in Lvshuihe dam, and structure strengthening were implemented in Yilihe IV, Qingtongxia and Majitang dams. The above mentioned dangerous dams and the dams with defects are reassessed as the normal dams Foziling was possessed by local government. Risk-removal reinforcement of Tianqiao dam is scheduled to be completed by 2012.

2. The second round regular dam safety inspection and the treatment of risky dams and the dams with defects

The second round regular inspections began from the trial of Qingtongxia and Fengshuba dams in 1997, ended in 2005 by completing Fengman regular dam safety inspection, totally 124 dam safety inspections were completed. Focuses were on gate metal structure detection and review, underwater examination in the second round inspection Through inspections, 8 dams, such as Liguan, Shuidong, Meishan, Lvshuihe, Yilihe II, Nangao, Huanglongtan and Fengman, were assessed as dams with defects, others were normal dams and no risky dams.

Liguan and HUanglongtan dams were assessed as dams with defects due to low flood control standard. Liguan dam reached the normal dam standard by dam heightening. Along with construction of Pankou reservoir, the problem of flood control standard of Huanglongtan dam will be thoroughly solved due to the flood regulation function of Pankou reservoir in the future.

Serious defects in anti-seepage system Shuidong and Nangao dams were found, with terrible seepage calcareous segregation. Cement grouting of Shuidong dam was successfully carried out against seepage, as well as underwater anti-seepage treatment on the upstream face cracks, and the problem of the seepage calcareous segregation was thoroughly solved. The original wet downstream dam face was already in dry condition. So Shuidong dam timely recovered to normal condition and was re-assessed as normal dam. Nangao dam was charged by local government; the strengthening work started in July, 2010, and was scheduled to be finished in July, 2012. After strengthening, reassessment will be done regarding dam safety grade.

Many defects and hidden dangers existed in the gates and hoists in Meishan, Lvshuihe and Yilihe II dams such as inadequate material quality, serious corrosion, equipment aging, so the normal operation could not be realized. After updating the gate and hoist system, Lvshuihe and Yilihe II dams were reassessed as normal dams.

Fengman dam has inborn deficiencies and operated safely up to now through continuous construction, modification, long term strengthening and reinforcement. But the inborn deficiencies could not be solved thoroughly; e. g. only continuous reinforcement could be done concerning poor concrete construction quality of the dam body. Surface concrete frost heaving damage resulted from leakage of dam body endanger discharge safety of overflow dam and dam durability, a large scale of damaged overflow face resulted from the discharge in 1986. During the first regular inspection of Fengman dam, it was assessed as normal dam just when the dam had undergone a new round of reinforcement and experienced the discharge of a 100 years flood in 1995. In the second regular inspection, considering the lack of anti-seepage system in dam body caused by inborn concrete quality defects, dam body leakage had not been effectively controlled, dam body uplift pressure was high, and frost-heaving was serious, especially the concrete frost heaving cracking on the overflow face would greatly affect the discharge safety, Fengman dam was assessed as the dam with defects. From 2008 to 2009, the dam administrative department took the grouting measures of "thick mud, close holes, high pressures, superfine cement" to construct the anti-seepage system of dam body and its mass reached 0. 15Lu. The seepage quantities from new drilled drainage holes of the dam body were rather small. At present, according to capital construction program, the authority of Fengman dam is pushing forward the research on comprehensive treatment scheme of Fengman dam.

3. *The third round regular dam safety inspection fulfillment*

On the background of *Operation Safety Management Regulation of Dams* issued by SERC, the third round regular inspection was performed abiding by the principle "systematical investigation, overall evaluation", examination and evaluation were carried out on the aspects of engineering flood control, structure safety, seepage control, discharge safety, operation performance, gate and hoist reliability, and monitoring reliability. By the end of Dec. 2010, 182 hydropower dams received the regular inspection. At the same time, based on 5.12 Wenchuan Earthquake damages, Bikou, Yingxiuwan, Tongtou, Baozhusi, Taipingyi, Yuzixi dams received special inspections. Among 135 dams which had undergone regular inspections, only Fushi dam in Guangxi Autonomous Region was assessed as the dam with defects due to inadequate flood control capacity, other 118 dam all were assessed as the normal dams. The major dams receiving the first regular inspections, such as Ertan, Longyangxia, Lijiaxia, Tianhuangping, Tianshengqiao I, Mianhuatan and Dachaoshan dams were all assessed as the normal dams.

Through systematic investigation, some important issues and their influences were made clear further during the third round regular inspection, such as the downstream face cracks of Ertan arch dam, the instable mode of reservoir bank landslide of Longyangxia dam and higher level of ground water in its right abutment, the erosion in front of the gates of Majitang dam, the cracks of Chencun dam, the deformation anomaly of Lijiaxia arch dam, the shear zone operational behavior of Gezhouba gate dam, the foundation deficiencies and lateral stability of the dam blocks at the right bank of Xinanjiang dam. According to statistics, there were few significant deficiencies need to handled, except that the flood control capacity of Fushi dam needs to be improved, the deficiencies are the followings: the heightening of the spillway crest of Baizhangji I dam, the maintenance function and recovery of discharge tunnel (lower tunnel) gates and hoists of Dongjiang, Yilihe I and, the treatment of penetrating crack of Taipingshao dam pier, the heightening of wave wall of Xueshanhu dam, the continuing construction of spillway of Habaheshankou dam, the spillway gates and hoists replacement of Qushui dam.

Through the fulfillment of the regular inspection institution for more than 20 years and the rehabilitation based on the regular inspection, the dam safety conditions belonging to the hydropower system have been improved continuously and laid a good foundation for the safety of the society and enterprises, as well as their healthy development.

26.3.2 Dam Safety Monitoring Technology and Supervision Management

26.3.2.1 Dam Safety Monitoring Technology and Supervision Legislation

Along with the improvement of awareness about dam safety and the strengthening supervision of the dam safety, the governments at different levels in China issued in succession a series of the dam safety monitoring technology supervision administration legislations and department regulations, such as the *Dam Safety Management Regulation of People's Republic of China*, *Dam Operation Safety Management Regulation for Hydropower Stations*, *Dam Safety Information Report Rules*, *Dam Safety Monitoring Management Rules*. Those legislations laid a solid foundation for Chinese dam safety monitoring technology, supervision administration and the monitoring activities following the legislation and rules.

The 19th article of the *Dam Safety Management Regulation of People's Republic of China* defines that the dam operation management units should carry out the dam safety monitoring and inspection, compile and analyze the monitoring data according to the related technical specification to keep track with the dam conditions. If abnormal phenomenon and unsafe factors were found, the dam operation management units should report to the dam administrative department at once and take actions in time.

Regarding newly built high dams or low and medium-sized dams with complex monitoring systems, *Dam Operation Safety Management Regulation* stipulates that, specific design, review and acceptance should be performed; while with regard to the hydropower stations with inadequate monitoring facilities or that with the monitoring items not meeting dam safety operation requirements, modification should be taken within the specified time. Specific design and review should be performed for modification schemes. Specific acceptance would be carried after completion.

Dam Safety Monitoring Management Rules defines that, after a dam is in operation, the operation and administration of monitoring system shall be the responsibility of hydropower station units, dam safety monitoring rules and operation instruction should be the established, technical files of dam safety monitoring should be kept; dam safety monitoring should be performed in compliance with related requirements, monitoring items, observation points, frequencies and time limits can not be altered without authorization; routine examination, annual detail inspection and regular inspection should be strengthened in daily work. The monitoring data should be timely compiled and preliminarily analyzed and the annual compilation should be completed by the end of every March. In combination with regular dam safety inspections, the synthetic analysis of the monitoring data should be performed, emphasizing on the trend analysis and abnormal phenomenon diagnosis. As for the key monitoring items, the operation warning value should be proposed. The dam safety monitoring information should be reported to the superior authority according to the *Dam Safety Information Report Rules*. If earthquakes, extreme floods or abnormal situation happens which would affect dam safety occur, the following should be done, strengthening routine examination, increasing monitoring frequencies (add the monitoring items if necessary), analyzing the monitoring data, evaluating the dam operation performance, and reporting the related information to higher level. The dam safety monitoring professionals should be trained and qualified to their positions.

Dam Safety Monitoring Management Rules defines that, LDSSC of SERC is responsible for the synthetic management of dam operation safety information in the hydroelectric station system, and for the establishment of the dam operation safety information management system; the dispatching agencies of SERC are responsible for the supervision of the dam operation safety information submission, the dam safety information subsystem construction and operation of the dams belonging to the owners and the hydropower station operation agencies. The hydropower stations operation agencies should collect, disposal and keep the dam operation safety information in time, realize automatic acquirement, management and report of the dam operation safety monitoring information step by step. The dam operation safety information reporting should be timely, accurate and complete. The contents should include dam safety monitoring information, annual dam safety detail inspection report, annual summary of dam safety work, special reports of design, acceptance and operation of dam safety monitoring system upgrading and modification, as well as the dam danger anticipation analysis report.

26.3.2.2 Dam Safety Monitoring Information Construction Practice

The thoughts of Chinese dam operation safety technological supervision management are as follows: vigorously pushing the dam safety management informational construction, forming new frame of the dam safety management which will be the combination of remote management with field inspection, timely detecting and eliminating hidden dangers, making preparations of anticipated danger and emergency plan, and performing abandoned dam removal and ecological rehabilitation, to realize modernization of dam safety management throughout the whole process. Therefore, the leaders at each level pay high attention to informational construction, and consider it as the important content of hydropower modernization construction and technological supervision management.

The principles of the dam safety management information construction are based on the modern communication and information processing technology, making full use of available hardware resources such as the internet, hydropower net, mobile communication net, monitoring automatic system, applying flexibly advanced technologies such as Web Services, XML, intelligent client, combining Chinese dam safety management practice, and establishing unified platform of dam safety management and technical exchange with remarkable characteristics of authority, integrity and timeliness. At present this system has been put into operation, significantly improving the whole level and working efficiency of the Chinese dam safety management, and providing strong technological guarantee for the new dam safety management frame which combines remote management with field inspection, scientific reservoir operation regulation and timely keeping track of the dam operation performance.

1. System frame

The system frame is made according to the function requirements, customer conditions and communication conditions. The functions mainly include the information inquiry and information analysis, among them, dam safety monitoring information analysis is the common function. The monitoring information analysis will use the monitoring records up to ten thousands, and generate various graphs and charts rapidly, requiring a high degree of flexibility of human-computer interaction and system processing speed. As for the function requirements, the most favorable Client/Service structure is selected in the system. As for the clients, besides the dam safety engineers, there are leaders at each level and a large quantity of dam safety management staff. The later often use information inquiry and statistics. Therefore, the software will be easily set up and manipulated. It is best to directly inquire the dam safety information through browser instead of loading the specific software. The system uses the B/S structure for this kind of function requirements. As for the communication conditions, it is best to use the available communication devices such as common net and electricpower net in order to reduce the costs of system construction and maintenance. The technology based on web was adopted, such as http and web service. Therefore the software structure of the system has adopted the combination of B/S and C/S structures. The B/S structure is used for the information inquire function. The C/S structure is used for the frequent monitoring information graphical processing and monitoring quantity calculations. In addition, the connection among main system, subsystem and branch subsystem has adopted the Web Services and the background information automatic processing has adopted Windows Services.

2. System composition

The whole system is composed of the main dam safety information system, established in LDSSC of SERC, the dam safety information branch system, established in every administrative department, and the dam safety information subsystem, established in each hydropower station operation agency. The clients can not only easily inquire the related management legislations, rules and regulations and the dam safety archives, compile and analyze the monitoring data, interpret the monitoring information in time and order the information services, but also acquire the data in time from the monitoring points connecting monitoring automatic system through internet or mobile phone, so that the clients can keep track of the dam safety conditions timely when they are far away from the field.

No matter which one, the main system, branch system and subsystem are all equipped with independent database. The three systems can work independently, and they are connected each other regularly or in the fixed time in order to keep the integrity and consistency of information. The information communication among the three systems adopts the active sending mode by one-way direction from the bottom to the up. That means information submission is actively uploaded from the subsystem to branch system, then from branch system to main system. If there is no branch system, the information should be uploaded directly from subsystem to the main system.

3. System functions

(1) Main system. The main system is divided into two parts: website information inquiry and information processing according to the function type.

1) Main functions of the website: propaganda of dam safety legislations and policies, dam safety information and technical exchange, dynamitic news about dam safety, dam safety legislation and specifications inquiry and download, dam safety specialist information inquiry, dam safety technical service organization information inquiry, each dam safety detail information inquiry, each dam safety operation behavior inquiry, every dam safety management condition inquiry, dam safety information issue in special period.

2) Main functions of information processing: remote acquisition of dam safety monitoring data; uploading monitoring data, reception, processing, storing and reply; dam safety monitoring data processing and analysis, synthetic evaluation for dam safety condition; synthetically processing the information submitted such as annual report on dam safety, dam safety registration, regular dam safety inspection, dam safety field patrol examination, emergency preparation plan; automatic reception, processing and reply for message, email; processing and submitting information ordered by user; back stage

management of dam safety website.

Fig. 26.3-1 shows the interface of users' login into the main system website. Fig. 26.3-2 shows the interface of hydropower station information inquiry of the main system website. Fig. 26.3-3 shows the monitoring data inquiry interface of main system website (monitoring point layout).

Fig. 26.3-1 User Login Interface of Main System Website

Fig. 26.3-2 Inquiry Interface of Hydropower Station Information Inquiry of Main System Website

(2) Branch system. The branch system can satisfy the requirements of the new mode of dam safety centralized management. It is used to process and manage safety information of the dams under jurisdiction of each dam authorities, for instance the river development companies, power generation corporations. The main functions include the followings: dam safety technical archives inquiry; dam safety registration information inquiry and statistic; dam safety annual report inquiry

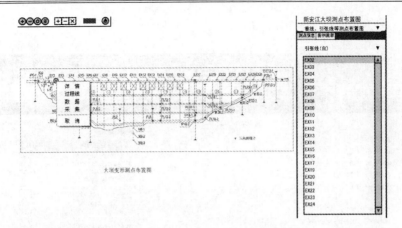

Fig. 26.3-3 Monitoring Data Inquiry Interface of Main System Website

and statistic; regular dam safety inspection inquiry and statistic; dam safety monitoring information inquiry and statistic, monitoring graph and chart diagramming, abnormal condition warning; dam safety field patrol examination information inquiry; reservoir regulation condition inquiry; dam safety information reception or automatic uploading and sending, etc.

Fig. 26.3-4 shows the hydropower station browser interface of the Yellow River Hydropower Corporation subsystem. Fig. 26.3-5 shows the monitoring information inquiry interface of the Yellow River Hydropower Corporation branch system.

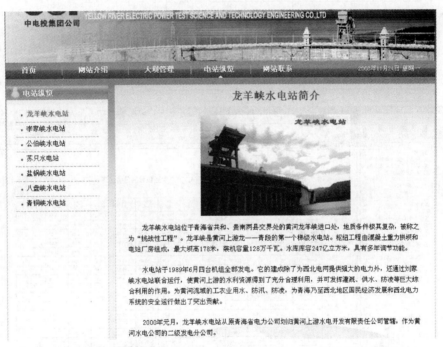

Fig. 26.3-4 Browsing Interface of Hydropower Station in the Yellow River
Hydropower Corporation Branch System

(3) Subsystem. The subsystem is used to process and manage the safety information of the dams managed by hydropower station operation organizations. The main functions include the followings: automatic acquisition of dam safety monitoring data; dam safety monitoring information management; dam safety monitoring information compilation, chart diagramming and management; dam safety information automatic uploading and sending; dam safety monitoring data analysis; dam safety operation performance evaluation; dam safety information management such as annual report, registration, regular inspection, reservoir regulation, patrol inspection; dam safety technical archives management and inquiry.

Fig. 26.3-6 shows diagramming compiling interface of the subsystem with the monitoring data. Fig. 26.3-7 shows the monitoring data processing and analysis interface of the subsystem. Fig. 26.3-8 shows the monitoring data analysis-typical hydrograph of the subsystem. Fig. 26.3-9 shows the monitoring data analysis-component hydrograph of the subsystem. Fig. 26.3-10 shows the dam safety supervision and synthetic evaluation interface of the subsystem.

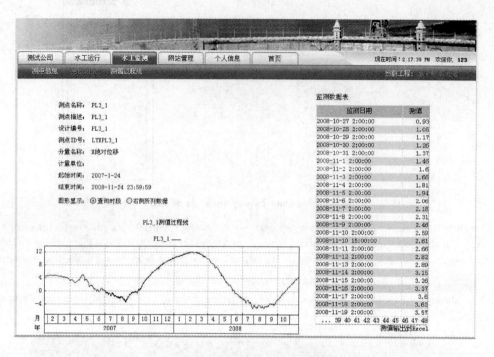

Fig. 26.3-5 Monitoring Information Inquiry of the Yellow River Hydropower Corporation Branch System

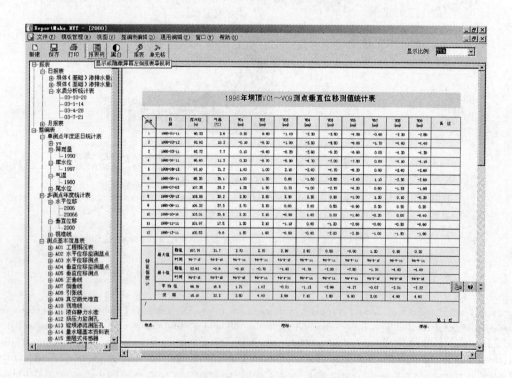

Fig. 26.3-6 Diagramming Compiling Interface of the Subsystem with Monitoring Data

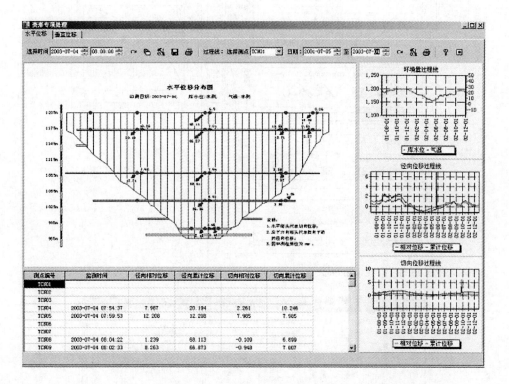

Fig. 26.3 - 7 Monitoring Data Processing and Analysis Interface of the Subsystem

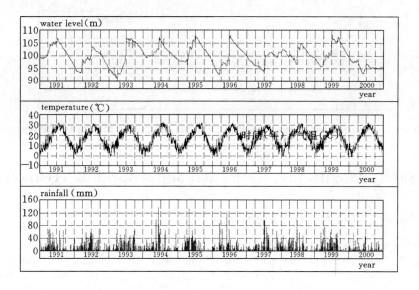

Fig. 26.3 - 8 Typical Hydrograph of the Subsystem with Monitoring Data

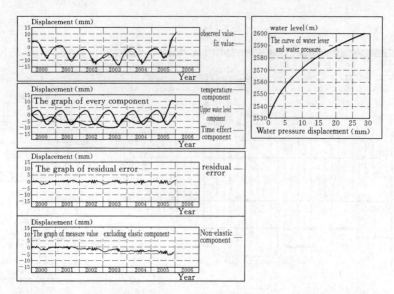

Fig. 26.3 - 9 Component Hydrograph of the Subsystem with Monitoring Data

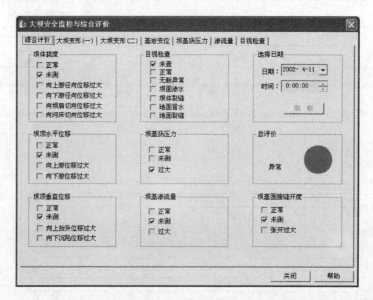

Fig. 26.3 - 10 Dam Safety Supervision and Synthetic Evaluation Interface of the Subsystem

26.4 Dam Safety Monitoring, Detection and Appraisal Technology

26.4.1 *Dam Safety Monitoring Technology*

The dam safety monitoring (early stage called prototype observation) in China began in the 1950s. The related department of Ministry of Water Resources started to develop the draft of hydraulic structure observation interim regulation in the 1960s and published *Hydraulic Structure Observation Technology Manual* in 1964. Since the 1970s, with the development of research on monitoring item identification, instrument selection, instrument arrangement, instrument embedment, observation method and monitoring data analysis, great achievements have been made. Since the 1980s, the dam prototype observation which originally mainly served for research, design and construction, developed to serve monitoring dam safety operation concerning social public security and was renamed safety monitoring, at the same time, a series of safety monitoring design, construction, acceptance criteria; monitoring instrument specifications and monitoring data compilation specifications were issued in secession. Since the 1990s, along with the research going into depth and engineering practical experiences accumulating, the improvement of monitoring design concept, the development and the application of monitoring instrument and its acquisition system, the continuous improvement of monitoring data analysis and information feedback method, now dam safety monitoring in China has already been in a leading position in the world.

Coming into 21 century, China has ushered in a new opportunity of dam engineering development. The monitoring theory and method of dam safety were continuously improved. The dam safety monitoring and management automation have been gradually perfected and improved. With monitoring data automatic acquisition system, the dam safety information management system and real time dam safety monitoring and forecasting realized by existing individual monitoring models, safety monitoring has fully played a more and more important role in the dam safety management.

26.4.1.1 Monitoring Instruments

1. Deformation monitoring instruments

Dam external deformation observation in China started in the 1950s. The instruments were mainly optical level and transit. Now the water levels with high precision and total stations have been used in practice, for instance the full automatic total station called "georobot". The dam deformation monitoring system with GPS technology developed by China itself has been widely used.

The inverse pendulum and reversed pendulum systems were widely used in dam deformation monitoring since the 1950s and performed effectively. The vertical plumb coordinator of mechanical contact type was used in the early stage, and the optical vertical plumb coordinator was used since the 1960s. It is still a main artificial measurement instruments. Since the 1980s, along with the development of science and technology, different kinds of the telemetric vertical plumb coordinators were sequentially developed. Typical instruments are mainly the followings: the stepped motor type of vertical plumb coordinator, capacitive vertical plumb coordinator and CCD vertical plumb coordinator. The inverse pendulum and reversed pendulum systems have the characteristics of simple use, easy maintenance, reliability and high precision. They are not only flexible in artificial reading, but also easy to achieve automatic monitoring, as well as function as checking datum marks for wire alignment and laser collimation measurement methods.

The wire alignment system for measuring the horizontal displacement of dam body is of simple devices, direct and reliable measured values, weatherproof and higher precision. It is not only suitable for artificial measurement, but also easy to achieve monitoring automation. The research and development of the wire alignment telemetry instruments were in the same pace with vertical plumb coordinator.

China began to develop and use the atmospheric laser collimation system and vacuum laser collimation system in the 1970s. The application and technology of the vacuum laser collimation system gained further development in the 1990s. The sealed laser point source and the charge coupled devices (CCD) were adopted as the sensor, the new approach and new methods such as the new-type zone plate and automatic circling water cooling device with vacuum pump were used in, further improving the reliability of the vacuum laser collimation system.

The multi-point extensometer and inclinometer for measuring the rock and soil deformation were used since the 1960s. The sensors used for the multi-point extensometer were mainly as follows: vibrating wire gauge, differential resistor sensor, inductance sensor, etc. The inclinometers were mainly servo accelerometer sliding inclinometer and fixed inclinometers of leveler type, vibrating wire type and strain gauge type.

Along with the rapid development of damming technology in respect of concrete slab rock fill dams in China, the instruments monitoring the internal horizontal displacement and settlement of the rock fill dams have been developed and widely used, for instance wire alignment horizontal displacement gauge and water level gauge, as well as a series of instruments, suitable for large scale displacement monitoring for rock fill dams, such as potentiometer displacement gauge and resistor strain gauges.

2. Seepage monitoring instruments

The piezometer was the earliest instruments used to measure the seepage pressure in the dam body, and dam foundation and the saturation line in the rock fill dams. Sounding lead was used to measure the level in the tube of the piezometer. Afterwards, electric fluviograph and pressure sensors of vibrating wire type, differential resistor type, strain gauge type and differential transformer type were adopted to measure the level of the piezometer, or directly embedded into the dam body to monitor the change of the seepage pressure.

The volumetric method and the measuring weir are the basic methods to measure the seepage quantity. Originally the data was recorded by artificial observation. Along with the improvement of the science and technology, as well as the development and application of the instruments with high precision, the seepage quantity monitoring instruments are mainly the followings: electromagnetic flow-meter and Ultrasonic flow-meter to measure the flow quantity in the tube, and various kinds of micro-pressure sensors, floating type level meter, Ultrasonic level meter and needle water level gauge, etc. to measure the water head over measuring weir.

3. Stress, strain and temperature monitoring instruments

The resistant thermometer was the earliest instruments embedded in the concrete dams. At present the embedded type thermometers have developed into various kinds, such as resistor type, vibrating wire type, thermistor type and semi-conductive type. The resistance thermometers are widely used in the hydropower projects in China.

The stress strain monitoring instruments are mainly the differential resistant type and vibrating wire type instruments. China began to develop differential resistant instruments in 1956. The earliest application was in large concrete dams built in 1950s, the typical project is Xin'anjiang hydropower station. Through efforts for half century, many key technical difficulties had been solved, for example, the 5 cores measurement method eliminating the resistant effect of the core wires, and the hydraulic balanced type, the differential resistant type strain gauges and joint meters suitable for high dams and the projects with high water heads were all firstly developed in success and used in all large and medium sized hydropower projects in China.

Through the continuous efforts of decades, especially in the recent ten years, the dam safety monitoring instruments have gained great achievements in the instrument principles, varieties, properties and automation, and meeting the practical engineering requirement of safety monitoring. Now there are more than 10 types of monitoring instruments manufactured with more than 10 principles are widely used in the safety monitoring of the hydraulic structures, such as the differential resistant type, the vibrating wire type, the capacitive type, the resistant strain gauge type, the inductive type, the electromagnetic type, the slide wire resistant type.

Now tremendous amount of work have been done in developing new sensors with the application of optical fiber technology. This variety is with the advantages against the influences of corrosion, humidity, lightening and strong magnetic field. Some developed products have been used in engineering. As for monitoring the seepage from dam body and dam foundation by "seepage thermal monitoring "technology, and the integration and synthetic application of GIS, RS, GPS and Expert System (ES) into the monitoring and management for the dams in a whole valley or into the safety monitoring and supervision for single dam, systematical research and practices have been carried out and some results have been made.

26.4.1.2 Monitoring Data Automatic Acquisition System

The research on Chinese dam safety monitoring data acquisition system began in the 1970s and was further improved in the 1980s. After entering the mid-1990s, with the development of the electric technology, computer technology and communication technology, various types of dam safety monitoring data acquisition systems had been developed, significantly improving the dam safety real-time monitoring, practicability and reliability, and realizing remote monitoring and supervision. Around 70% hydropower stations have successfully adopted the dam safety monitoring data automation acquisition systems.

The dam safety monitoring data automation acquisition system in China experienced the development process from the centralized type to the distributed type. The centralized dam safety monitoring data acquisition system was the product developed in the early stage. Limited by micro-electric technology development level and electric element cost, centralized type was mostly adopted in the monitoring system developed in early stage, in order to adapt many varieties and large quantities of dam monitoring instruments, remote cross box around the instruments were placed. Along with the development of micro-electric technology development and communication technology, electric element cost decreased significantly, data acquisition unit replaced remote cross box, and the occurrence of the distributed dam safety monitoring data automation acquisition system appeared. The data acquisition unit can select points to measure analog signal and digital signal of various instruments and transfer them with the standard output, with the functions of data store and data communication. Thus, the effect resulted from the analog signal of long distance transportation on the monitoring result can be eliminated, and the precision can be improved. Due to distributed measurement risk of instruments, reliability increases. In addition, the anti-lightening property of the distributed system is significantly better than that of the centralized system.

26.4.1.3 Monitoring Data Analysis and Information Management System

The monitoring data analysis and information processing are important components of dam safety monitoring activities. Since the 1980s, especially since the 1990s, monitoring data analysis and information processing technology had gained rapid development along with the fast development of the computer and information processing technologies, as well as high attentions to the dam safety management paid by the governments and leaders at various levels.

1. Monitoring data analysis mathematic models

By years of study and practice, statistic model, deterministic model and hybrid model concerning the measured values of a single point have been used widely during the quantitative analysis of the monitoring data and safety monitoring. The model of multiple point and multiple directional were studied and developed on this basis. The concept of "distributed mathematical model" was proposed to deal with the monitoring data from multiple points of one monitoring variable. Breakthrough improvements have been made in using gray system, neural network model and fuzzy mathematical model to perform the dam safety monitoring data analysis.

2. Synthetic analysis evaluation methods

It is a trend to apply the modern mathematical theory and information processing technology into the synthetic analysis and evaluation of dam monitoring. Now there are mainly analytic hierarchy process and the comprehensive analysis inference methods. In addition, Chinese scholars studied the monitoring performance by the analytic hierarchy process from the multiple angles and multiple approaches, including the method of fuzzy evaluation combined with analytic hierarchy process, fuzzy pattern recognition approach, fuzzy integral estimation method, hierarchical gray correlation analysis, catastrophe theory analysis, attribute recognition theoretical model, etc. Systematic engineering approach in modern mathematical field was adopted in these methods and researches, bringing many valuable research results. These applications are helpful to the solution for the synthetic analysis evaluation of complex dam safety monitoring performance from many respects.

3. Inverse analysis

The establishment of conventional hybrid model for single point and deterministic model included the contents of the inverse analysis. At present, hybrid model or deterministic model for multiple points are widely used in deformation inverse analysis, and gained reasonable application and successful cases. In addition to the research of displacement spatio-temporal model and the fuzzy mathematical inversion, satisfactory application results have also been gained.

4. Monitoring index identification

At present the main methods of developing the monitoring index for operation period are the followings: monitoring index is determined through mathematical model of monitoring quantity with consideration of mathematical expression composed by certain confidence interval; threshold value is taken as the monitoring index which is derived according to mathematical model substituted by the most adverse combinations, taking into account of errors factors; the allowable value is the monitoring index, which is inversely operated by means of the criticality safety and reliability meeting the stability and strength conditions; concerning practical engineering issues, by identifying the grades and calculating physical model, with the mechanics parameter through the inverse analysis of the measured deformation data, then specific deformation monitoring index are finalized.

5. Dam safety monitoring information management system

The research and development of the dam safety monitoring information management system began in the 1950s. Along with the improvement of the theory and methods, the developing of MIS (management information system) and DSS (decision support system), as well as dam monitoring technology, great progress has been made in the development of the systems. Many new practical techniques and new theoretical approaches were continuously introduced and absorbed. For instance, the neural network method and the theoretical approaches of the data warehousing and data-mining were adopted for monitoring information analysis and processing in synthetic analysis evaluation, and the communication management of dam monitoring information was carried out by internet. All these researches and applications have greatly improved the Chinese dam safety monitoring information processing technology.

The software of *"Engineering Safety Supervision and Monitoring Information Management System"* developed by LDSSC has been widely used in Tianhuangping Pumped Storage Power Station, Guangzhou Pumped Storage Power Station, Shisanling Pumped Storage Power Station, Er'tan hydropower station and Longyangxia Hydropower Station. This system software is an engineering safety supervision information management software package with powerful function, friendly interface and easy operation, including rapid online safety evaluation, offline analysis, supervision model-analytical model-forecasting model management, engineering documentation and chart/graph diagramming management, database management, network system management, remote monitoring and ancillary services, supporting system, demo system and etc.

26.4.1.4 Monitoring Technical Specifications

Since the 1980s, along with key science and technology program going into depth and continuous accumulation and summarization of the engineering practical experiences, some issues in the safety monitoring have been gradually solved. The monitoring design and monitoring data analysis feedback method has been continuously improved. The related authorities of Chinese government developed and issued a series of dam safety monitoring specifications, which are continuously revised and improved in practice. The current specifications are mainly the followings: *Safety Monitoring Technology Specification* for Concrete Dams (DL/T 5178—2003), Safety Monitoring Data Compilation Specification for Concrete Dams (DL/T 5209—2005), Dam Safety Monitoring Automation Technical Criteria (DL/T 5211—2005), Dam Safety Monitoring System Completion Acceptance Criteria (GB/T 22385—2008), as well as "Differential Resistant Instrument Specifications", "Capacitive Instrument Specifications", "Vibrating wire Instrument Specifications", and etc. (see the table 26.4-1). The issues and fulfillment of those specifications filled in the blanks of the related technology specifications of Chinese dam safety monitoring field, gradually improved the dam safety technology specification institution and laid a foundation for the standardization and normalization of dam safety monitoring activities.

Table 26.4-1 Current Main Dam Safety Monitoring Specifications

No.	Name of specifications	Serial No. of specification
1	Safety Monitoring Technology Specification for Concrete Dams	DL/T 5178—2003
2	Serial Types for Monitoring Instruments of Concrete Dams	DL/T 948—2005
3	Serial Types for Monitoring Instruments of Rock fill dams	DL/T 947—2005
4	Dam Safety Monitoring Automation Technical Criteria	DL/T 5211—2005
5	Safety Monitoring Data Compilation Specification for Concrete Dams	DL/T 5209—2005
6	Capacitive Wire Alignment Gauge	DL/T 1016—2006
7	Capacitive Displacement Gauge	DL/T 1017—2006
8	Capacitive Joint Gauge	DL/T 1018—2006
9	Capacitive Plumb Coordinator	DL/T 1019—2006
10	Capacitive Hydrostatic Leveling Gauge	DL/T 1020—2006
11	Capacitive Weir Level Gauge	DL/T 1021—2006
12	Vibrating Wire Joint Meter	DL/T 1043—2007
13	Vibrating Wire Strain Gauge	DL/T 1044—2007
14	Vibrating Wire Piezometer	DL/T 1045—2007
15	Wire Alignment Horizontal Displacement Gauge	DL/T 1046—2007
16	Observation Tube Settlement Gauge	DL/T 1047—2007
17	CCD Plumb Coordinator	DL/T 1061—2007
18	CCD Wire Alignment Coordinator	DL/T 1062—2007
19	Differential Resistant Displacement Gauge	DL/T 1063—2007
20	Differential Resistant Anchor Dynameters	DL/T 1064—2007
21	Differential Resistant Anchor Stress Meter	DL/T 1065—2007
22	Construction Supervision Criteria for Dam Safety Monitoring System	DL/T 5385—2007
23	CCD Hydrostatic Leveling	DL/T 1086—2008
24	Dam Safety Monitoring System Completion Acceptance Criteria	GB/T 22385—2008
25	Specification of Strong Earthquake Safety Monitoring For Hydraulic Structures	DL/T 5416—2009
26	Vibrating Wire Instrument Specifications	DL/T 1133—2009
27	Instrumentations For Vibrating Wire Instruments	DL/T 1134—2009
28	Potentiometers Type Displacement Gauge	DL/T 1135—2009
29	Vibrating Wire Steel Bar Stress Meter	DL/T 1136—2009
30	Vibrating Wire Type Soil Pressure Gauge	DL/T 1137—2009
31	Potentiometer Type Displacement Gauge Measuring Instrumentation	DL/T 1104—2009
32	Stepping Plumb Coordinator	DL/T 327—2010
33	Stepping Wire Alignment Gauge	DL/T 326—2010
34	Communication Protocol of Dam Safety Monitoring Automation System	DL/T 324—2010

Continued

No.	Name of Specifications	Serial No. of specification
35	*Vacuum Laser Collimation Displacement Measurement Devices*	DL/T 328—2010
36	*Safety Monitoring Technology Specification for Rock fill dams*	DL/T 2659—2010
37	*Safety Monitoring Data Compilation Specification for Embankment Dams*	DL/T 2656—2010
38	*Differential Resistant Strain Meter*	GB/T 3408.1—2008
39	*Differential Resistant Steel Bar Meter*	GB/T 3409.1—2008
40	*Differential Resistant Joint Meter*	GB/T 3410.1—2008
41	*Differential Resistant Piezometer*	GB/T 3411—1994
42	*Resistant Bridge*	GB/T 3412—1994
43	*Embedded Copper Resistant Thermometers*	GB/T 3413—1994
44	*Electromagnetic Type Settlement Gauge*	GB/T21440.2—2008
45	*Liquid Settlement Systems*	GB/T21440.3—2008
46	*National Grades 1 and 2 Leveling Survey Criteria*	GB/T12897—2006
47	*National Grades 3 and 4 Leveling Survey Criteria*	GB/T 12898—2009
48	*Optoelectronic Ranging Code for Medium and Short Ranges*	GB/T 16818—1997
49	*National Triangulation Survey Standard*	GB/T 17942—2000

26.4.2 *Dam Safety Detection Technology*

During long-term operation, due to adverse effect of concrete aging, reinforcement corrosion, geology, underground water, earthquake, artificial and biological destruction, the dam safety level may turn down gradually. Therefore, it is necessary to implement safety detections for the important parts of dam. Furthermore, the detection results are the significant basis for dam safety evaluation and rehabilitation.

From the1970s, more importance is attached to dam patrol inspection and dam safety detection worldwide. In the1980s, China started systematic research on geophysical methods to detect dam safety, including electrical resistivity method, acoustic method, seismic wave method, method of ground probing radar, transient surface wave method, transient electromagnetic method, impact-echo method and acoustic CT method. These methods are playing an important role in dam defects detection, safety inspection and rehabilitation. The example of safety inspection and research on Fengman dam is representative, witnessing the development process of dam safety detection technology in China.

Fengman dam was constructed in 1937 and completed in 1953. It has been in operation for more than 70 years. Because of historical reasons and poor technology then, quality defects existed ever since. In the recent 50 years, many detections, inspections and rehabilitations were carried out. From 1958 to 1990, by core-drilling method, borehole acoustic method, borehole television, etc., several corporations (namely Hydro China Beijing Engineering Corporation, China Water Northeastern Investigation, Design & Research Corporation, Beijing Building Materials Academy, the Yangtze River Scientific Research Institute, Shanghai Investigation Design & Research Institute and Fengman Hydropower Plant) undertook the detection of dam defects such as low strength concrete, cracks, honeycombed concrete, segregation. From 1991 to 1992, by application of acoustic method, rebound method, core-drilling method, borehole television and acoustic CT method, Japan International Cooperation Agency carried out the detection on Fengman dam. In 2005, Dalian Saiwei Information Corporation and Fengman Hydropower Plant detected the surface defects on underwater concrete of the upstream dam. From 2006 to 2008, by use of exterior investigation, core-drilling method, single hole acoustic method, acoustic plane test, acoustic transmission method, acoustic-echo method, acoustic CT, borehole deforming modulus method, borehole high-density resistivity method, ground penetrating radar, strength tests, etc., Hydro China Huadong Engineering Corporation implemented systematic inspection and evaluation on concrete surface cracks and inner defects, as well as grouting effect.

After 40-year's development, the detection system for dam safety has been established in China, with various methods and approaches, reaching international advanced level.

Dam safety detection includes: exterior patrol examination, concrete performance test, concrete cracks detection, concrete defects tests, seepage inspection, sedimentation investigation and metal structures tests. For face-slab rock-fill dam and earth-rock dam, detection of cracks on the face slab, inspection of separation between face slab and concrete cushion, dam body seepage inspection and dam body density test should be added into the detection.

26.4.2.1 Exterior Patrol Examination

According to statistics, more than 70% of the safety hazards were found by experienced technicians during patrol examination. Exterior patrol examination is one of main means for dam safety detection.

1. Artificial exterior inspection

The inspection is mainly depending on visual observation, with the assistance of tape, plug gauge, crack width gauge, magnifier, telescope, camera, video camera, aiming at performing necessary inspection and measurement for different dam parts; for underwater structures, underwater camera, underwater robot and diver will be mobilized.

2. Exterior scan detection

The principle of exterior scan detection is that the CCD camera equipped with zooming long-focus lens, inclinometer and laser range finder and controlled by high-precision servo motor, with observation points set around the structure, the structures surfaces within 300m can be scanned by at full range by automatic control telemetry system, forming 3D model of structures surfaces. The space positions, physical properties (length, width, and state) and images of the defects on structures' surface are measured with high-precision, a complete database is established. Based on periodic scan results, the dynamic development and trend of the defects can be analyzed.

26.4.2.2 Concrete Performance Test

The concrete performance test contains carbonation depth test, strength test, reinforcement distribution and protective layer thickness test.

1. Method of concrete carbonation depth test

As concrete is exposed to the air, carbon dioxide would penetrate the concrete slowly and react with the calcium hydroxide. The reaction product is calcium carbonate, and from external to internal, the concrete turn from alkaline to neutral gradually. Concrete carbonation depth can be determined by application of phenolphthalein alcohol, which will become red in alkalinity. After holes drilling and cleaning on the concrete surface, a 1% phenolphthalein alcohol solution is applied to concrete. If the concrete has undergone carbonation (neutral), no color change will be observed. Otherwise, it will turn bright pink. In this way, we can find out the carbonation depth.

2. Methods of concrete compressive strength tests

The methods of compressive strength tests mainly include rebound method, Ultrasonic-rebound method and core-drilling method.

(1) Rebound method. By test of concrete surface hardness, compressive strength can be calculated accordingly. When the plunger of rebound hammer driven by spring hit the surface of concrete, the distance (rebound value) of the hammer rebound is measured. Based on the empirical relation between concrete strength and Rebound Value, concrete compressive strength can be calculated or found by looking up the table.

(2) Ultrasonic-rebound method. Rebound Value depends on surface hardness of concrete. The acoustic velocity in concrete depends on concrete elastic property. With Ultrasonic-rebound method, the Rebound Value of the elastic mass and the acoustic velocity in concrete are measured, the compressive strength of concrete can be determined according to their correlations, and evaluation of compressive strength and homogeneity of concrete can be assessed.

(3) Core-drilling method. Core-drilling method is semi-destructive. The cylinder core samples obtained are taken for compression test and splitting tensile strength test to check and verify the strength.

3. Tests of reinforcement distribution and protective layer thickness

Main methods adopted are electromagnetic induction method and radar method.

(1) Electromagnetic induction method: electromagnetic field is radiated by the coil supplying alternating electric current, and a secondary magnetic field is radiated by induced current generated by exciting reinforcement bar, and the secondary magnetic field will excite the coil to produce induced electromotive force. Different distances between the coil and bar will produce different induced voltages. The maximum induced voltage would appear while the reinforcement bar is right beneath the coil. Thus, by measuring the induced voltage and its changing trend, the location of reinforcement and the protective layer thickness can be detected with electromagnetic induction method.

(2) Radar method: high-frequency electromagnetic wave is emitted towards the concrete with transmitting antenna, the electromagnetic wave will produce strong reflected electromagnetic wave, based on the signal and image of reflected electromagnetic wave, the location of reinforcement bar can be determined, the protective layer thickness of reinforcement bar can be calculated according to the return travel time of the reflection and the electromagnetic wave velocity in concrete.

26.4.2.3 Concrete Crack Detection

Concrete is a kind of porous, cementitious, and artificial stone material, which is of high compressive strength, low tensile strength and elongation. Concrete is subject to cracking, and dam cracks are common. The concrete shrinkage, temperature deformation, load and uneven settlement of foundation may cause cracking. Distribution, direction, length, width, depth are the main description indexes for crack.

1. Crack investigation

There are several kinds of harmful cracks: leaky cracks, cracks wider than 0.2mm which cannot self-heal, penetrating cracks, cold joint leaky cracks, cracks reaching deep to structural reinforcement, and mesh pattern distributed cracks, cracks shortening the duration of concrete structure, the cracks harmful to the safety and stability of concrete structures.

Tape is adopted for length measurement of cracks on the concrete surface, plug gauge and optical micrometer is used for the width, the location, length, width, direction of cracks, seepage and calcium-segregation are recorded, cracks distribution is mapped. Investigation mainly contains: cracks location, direction (vertical, horizontal, oblique), length, width, depth and their extending direction, cracking time, cracks development, cracks leakage, materials inside the cracks and calcium-segregation.

2. Crack depth investigation

Core-drilling method and Ultrasonic method are widely used in crack depth investigation. As regard with the types of Ultrasonic method, Ultrasonic methods are divided into single plane method, double plane Ultrasonic transmission method and borehole Ultrasonic transmission method.

(1) Core-drilling method. Core-drilling is semi-destructive. In the middle of the crack, along the predicted direction of depth development, drilling straddling the crack will reach deeper than the predicted crack depth. By observing the pinch location of the crack from the intact core sample, the crack depth will be obtained by measuring the sample length between the surface and the pinch. B can also be used to the crack depth by observing the pinch location of crack inside the core hole.

(2) Single plane Ultrasonic method. Single plane Ultrasonic method is applicable for shallow cracks when there is on one surface available for detection, suitable for the cracks deep less than 500mm.

Wit this method, Ultrasonic waves are transmitted by powerful transmission energy converter from one side of the crack and are received on both sides of the crack. The crack depth can be calculated based on the time that Ultrasonic waves travel around the end of the cracks and acoustic velocity in concrete, as shown in Fig. 26.4-1.

(3) Double plane Ultrasonic transmission method. Double plane Ultrasonic transmission method is applicable when two sides of concrete structures are available for detection. On the concrete surface on which cracks are visible, transmitters are placed in a certain space vertical to the crack on both sides of the crack, receivers are placed on the other side on the direction of cracks extending, with diagonal method, the acoustic wave signals are recorded in each point be moving from one side of the crack to the other side, the amplitude of the first acoustic wave of each ray is read, and the acoustic wave ray corresponding saltation point in the first acoustic amplitude is considered as the ray passing the crack tip. With the same method, reverse detection is done to locate the ray on the other end, the intersection point of the two rays is the crack tip. The line connecting this point and the surface cracks is the extending direction of the cracks, and the depth of this point is crack depth.

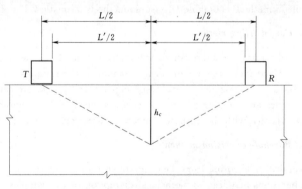

Fig. 26.4-1 The Schematic Diagram of Single Plane Ultrasonic Method

(4) Borehole Ultrasonic transmission method. Double plane Ultrasonic transmission method is not applicable in mass concrete. If the cracks are deeper than 500mm by estimation, borehole Ultrasonic transmission method is adopted. Its principle is similar to that of double plane Ultrasonic transmission method, to drill a hole on each side of the crack, and the depth is

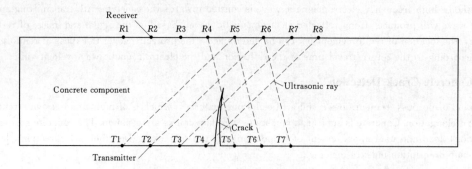

Fig. 26. 4 – 2 Sketches of Ultrasonic Transmission Method for Crack Detection

500mm deeper than the estimated crack depth with its diameter larger than 40mm. The distance between crack and the holes will preferably be 50 – 150cm, if the crack is too deep and inclined, the maximum value shall be adopted, in order to make sure the crack staying in the middle of the two holes. Prior to detection, the transmitter and receiver are placed in the two holes filled with water, and are moved from top down synchronously respectively at the same elevation and at equal intervals, the depth of each detection point and the amplitude of the first acoustic wave are recorded to make a curve of acoustic amplitude-depth, and the acoustic wave ray corresponding saltation point in the first acoustic amplitude is considered as the crack tip. The depth of this point equals to the depth of the crack, as shown in Fig. 26. 4 – 3.

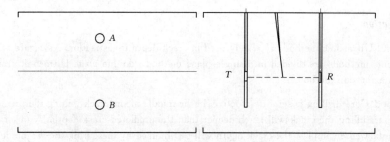

Fig 26. 4 – 3 Illustration of Borehole Ultrasonic Transmission Method for Crack
Detection（Left is the Plane, and the Right is Profile）

26. 4. 2. 4 Concrete Defects Detection

Concrete defects refer to the honeycomb, pores, segregation, crack, poor cementation, low strength part and inclusions such as sand and clay which would destroy the continuity and integrity of the mass concrete. Meanwhile, the defects would reduce the concrete strength and durability. Defects detection methods include core-drilling method, borehole televisiontelevision method, borehole Ultrasonic method, acoustic CT and ground penetrating radar (GPR) method.

1. Core-drilling method

By observing the aggregate distribution, cementation, density, solid cement in cracks, honeycomb, pores, segregation, crack and inclusions on the drilled samples, the samples are described and appraise, and their appearance quality is assessed. Compressive strength test, impermeability test are carried out with interception of core samples, and core-drilling samples are used for acoustic inspection, acoustic CT, borehole deformation modulus, water pressure test and borehole television, with the purpose of overall inspection.

2. Borehole television method

As drill hole coring from the concrete with low strength or from defected section of the concrete is very difficult, improper operations may lead to mechanical disruption of the samples and affecting the description and appraisal. Television may be used to observe the real state of concrete in original position in the hole wall.

Borehole television is a high resolution digital color video taping system, with camera, compass and in-depth counter to obtain continuous, detailed and orientated 360° true color images of borehole wall and record the orientation and depth of the borehole wall. The borehole images is edited, analyzed and processed to obtain the sketch view and drill column so as to present the original borehole wall. In consideration of the location, extent and feature of the defect concrete with borehole samples, synthetic assessment of concrete appearance quality will be made.

3. Acoustic detection method

Concrete is a kind of inhomogeneous, anisotropic and elastic-plastic material. The high non-homogeneity causes a considerable sound attenuation and low sound velocity in concrete with defects. By observing the velocity and amplitude changes, concrete defects are determined. As there are many factors affecting the velocity and amplitude which vary in a wide range, it is impossible to establish a uniform standard of evaluation. So in general, concrete defects are evaluated based on acoustic velocity and acoustic amplitude and their corresponding variations. With different detection methods, there are surface wave detection, single borehole acoustic wave and penetrating acoustic wave detection.

(1) Surface wave detection. On the concrete surface, one transmitter and several receivers are set in a straight line in a certain intervals. The signal traveling time between transmitter and receivers are measured and recorded respectively. The acoustic wave velocity can be calculated based on the space between two receivers and the time difference of acoustic wave velocity traveling. According to acoustic wave velocity and its variation, surface concrete quality can be assessed and the location and extent of defect concrete can be determined.

(2) Single borehole acoustic detection. With single borehole acoustic wave method, on the concrete surface, one transmitter and two receivers are set inside the borehole, and acoustic wave is transmitted from the transmitter and received by the two receivers. The acoustic wave traveling time differences between the two receivers are read. The acoustic wave velocity in the concrete in the borehole wall where the receivers are set can be calculated by distance divided by time. The acoustic probes are moved from top down point by point, the acoustic wave velocity in borehole concrete of each detection point and the amplitude of the first acoustic wave are recorded to make a curve of acoustic wave velocity-borehole depth, and a curve of amplitude-borehole depth. According to acoustic wave velocity, amplitude and their variation trends, and concrete quality can be assessed and the location, extent and degree of the defect concrete in the borehole can be determined.

(3) Penetrating acoustic wave detection. In the case that acoustic wave can penetrate through the part between two surfaces/ two boreholes/ one borehole and surface, the penetrating acoustic wave detection can be employed. In accordance with different surfaces positions, boreholes dips, the spatial location of the range detected, and the defect property and scope, horizontal, oblique and fan-shape penetration are selectable. The velocity of each acoustic wave ray, the amplitude of the first wave etc are calculated, according to the acoustic velocity, amplitude and their variations, the locations, extents and scales of the defect concrete between surface/borehole and surface/borehole can be determined. Then the defect property and degree can be inferred.

4. Acoustic CT method

Acoustic computerized tomography (acoustic CT for short) is based on the principle of wave transmission. With two boreholes in a plane or on concrete structure surface, acoustic wave is transmitted from on borehole (side) and is received in another borehole (side), large amount of intersectant acoustic wave rays are obtained by fan-shaped observation, and the first arrival time of each ray is read. Based on the start-stop space coordinates of each ray and the traveling time acoustic wave, following the principles of Fermat and Hyngens to track the ray with minimum traveling time and the shortest route, and with Radon Transform, the acoustic velocity in each unit of the model is inverted and fitted to obtain the profile acoustic wave velocity image in the section. According to the distribution characteristic of acoustic wave velocity in the section to estimate its internal structure and structural feature, defect concrete exists possibly in the relatively low velocity zone, thus, the distribution and location of defect concrete is determined.

Acoustic CT method includes dam acoustic CT detection and borehole CT detection. In dam acoustic CT detection, detection section is perpendicular to dam axis. Transmitters are set on the upstream dam surface while receivers are set on the downstream or dam crest, and a unified coordinate system is established. The excited acoustic on the upstream dam surface (as the propagation distance is usually longer than 100m, high-energy sources such as electric spark, electric detonator and hammering are recommended) are revived at the detection points at the dam rest or downstream, and the first arrival time of each acoustic wave read. Based on the start-stop space coordinates of each ray and the traveling time acoustic wave, the acoustic velocity distribution can be obtained, furthermore, the dam concrete quality can be judged and defect concrete can be located. Due to trapezoidal cross section of the dam body, there is a long distance between the upstream face and downstream face, and the dam becomes larger from the top to the bottom, and due to the occupation of equipment inside the dam and the layout of the galleries etc, the acoustic waves are not evenly distributed. Especially at the middle and lower part of the dam, the acoustic waves become scattered, even decrease to 0 at some parts, bringing adverse influence on the resolution and accuracy of acoustic CT detection (depend upon wave length and distribution density). In addition, lower-frequency source adopted is another influencing factor in detection of small defects. In this case, boreholes are required in the dam, acoustic CT detection is carried out between the boreholes or between the boreholes and the dam face, in this way, the dam cross section can be effectively reduced to improve acoustic CT detection efficiency.

Borehole acoustic CT detection is applicable enables at any part of the dam. Firstly, boreholes are drilled near the parts to be detected, and usually the hole's depth shall be larger than the space between boreholes. Acoustic CT detection with high frequency source between two boreholes will produce high detection resolution and effects. In addition, acoustic CT detection can also be applied between the boreholes on the dam body and the upstream and downstream faces.

5. Ground penetrating radar method

Ground penetrating radar method is a technique that uses electromagnetic waves with frequencies from 1 MHz to 1 GHz to detect the inner distribution of medium. According to electromagnetic waves theory, reflected waves occur when they meet different electrical interfaces during electromagnetic propagation. In general, the inner structure and moisture rate in defect concrete differ with those in normal concrete, their electrical conductivity, dielectric constant etc differ distinctively those of normal concrete. Reflected waves will occur when electromagnetic waves encounter defect concrete. The location and range of defect concrete are determined by analyzing the strength, phase, frequency, traveling time and event shape of the electromagnetic wave signals.

The effective detection depth and resolution depend on the center frequency of ground penetrating radar. The higher the center frequency is, the faster the signals will attenuate, the smaller the effective detection depth will be, and the higher the resolution be obtained. The size, embedded depth of defect concrete and resolution and other factors are considered when selecting the center frequency of an antenna. High frequency antenna is applicable for the shallow and narrow distributed concrete, while low frequency antenna for deep and wide distributed defects concrete.

26.4.2.5 Detection of Voids Beneath Face Slab

Stage construction of concrete face embankment dam and repeated load would cause dam deformation which would lead to voids beneath face slab. Separation between concrete face slab and cushion layer means losing support from embankment, which would finally result in cracks. To detect voids beneath face slab, the impact-echo method and infrared thermography (thermal imaging) are adopted respectively or combined.

1. Impact-echo method

The method is also called vertical reflection of acoustic wave method. With the theory of acoustic wave vertical reflection, when an acoustic wave signal is exited on the concrete face slab, reflection will occur when the signal propagate at the bottom boundary of concrete face slab. Because the air wave impedance is far small than that of rockfill cushion, the acoustic reflection coefficient in the voids area (nearly 1.0) is greater than that in normal concrete (generally 0.2 - 0.5). The location, range and degree can be determined according to the number and distribution of acoustic reflection coefficient in each part of face slab.

2. Infrared thermography

Because the concrete face slab is rather thin, under the sunshine, the heat the concrete face slab absorbed will radiates through the slab. And because the thermal conductivity of rockfill body is higher than that of air, the face slab temperature rises fast for the poor thermal conductivity of the air in voids area under the sunshine, and the face slab temperature rises slowly for the good thermal conductivity of the cushion in non voids area. At night, it is the reverse, the heat accumulated in the bottom will radiate through the slabs, slab temperature is low in voids area due to poor thermal conductivity of the air, while the slab temperature is high in non voids area due to good thermal conductivity of the cushion. With infrared imaging devices to measure the heat radiation intensity of concrete face slabs in suitable time, the voids defects can be determined by the temperature distribution on the slab surface.

26.4.2.6 Detection of Embankment Density

Owing to compaction in layers, the embankment density is an important index in quality evaluation for embankment dam. The density of embankment mass can be determined indirectly by the nuclear density methods, Rayleigh wave method and added mass method.

1. Nuclear density method

Nuclear density method determines the concrete density by the attenuation of gamma rays. The ability of concrete to absorb gamma radiation is directly proportional to its density. Gamma radiation is absorbed by means of Compton scattering. A definite relation which is fixed by sand cone method exists between the number of gamma photons that penetrate the concrete and the density of concrete. The gauge measures the gamma photons and it is inversely proportional to density. When performing site inspection, inserting a density measuring tube into the rock fill dam, by measuring the gamma rays penetrated to the surface, the density of rock fill dam can be indirectly determined.

2. Rayleigh wave method

On the basis of the theory of Rayleigh wave propagation in layer medium, Rayleigh wave method is developed to detect rock fill density. The propagation velocity directly reflects the rock fill density. By measuring velocity and dispersion curves, according to the relation between velocity and density that was confirmed before, rock fill density can be calculated. Between the transient-state detection and steady-state detection, transient-state detection is usually preferred.

3. Added mass method

In added mass method, the vibration state of rock fill is modeled and the rock fill is equivalent to a linear elastic system of single degree of freedom. The vibration mass of rockfill is obtained by adding different added mass. Measuring the dynamic stiffness and length of the elastic longitudinal wave, with known relation among dynamic stiffness, vibration mass and rock fill density, its density can be determined.

26.4.2.7 Dam Seepage Detection

As the difference of water levels between the upstream and downstream of the dam, the reservoir water starts to percolate through cracks and uncompacted area of dam as well as joints, faults and karsts in foundation. Tracer method, pressure-water method and flow field method are applicable to detect the water inlet of seepage and the connection between upstream and downstream ends.

1. Tracer method

Tracer method can detect the sources that the ground water appeared in the drainage holes downstream of the dam and /or leakage to the downstream through dam body foundation. Determination of seepage passage and water inlet plays a guide part in further treatment and strengthening. Artificial tracer method and natural tracer method are addressed to solve this problem. Natural tracer consists of temperature tracer, electrical conductivity tracer and dissolved oxygen tracer. Artificial tracer comprises dye tracer, saline tracer and isotope tracer.

Natural tracer method measures the temperature, electrical conductivity and dissolved oxygen of seepage water downstream. Then, by comparing them with the water upstream, the seepage source can be determined by similar rule.

Artificial tracer is to put the dyed tracer, saline tracer or isotope tracer into boreholes or specific point of reservoir and its concentration is detected. Then, based on dilution method, the seepage velocity is calculated and the leakage source is determined. With velocity change, seepage field can be analyzed.

2. Pressure-water method

Empty the upstream reservoir first. By use of connection of seepage passage, inject dyed water (for example potassium permanganate solution) from downstream water outlet and exert pressure on it. Keep the pressure on until dyed water flow out upstream. In this way, the water intake upstream is located.

3. Flow field method

The method to examine seepage is on the basis of the relation between pseudorandom electrical field and seepage flow field. It begins from producing artificial electrical field and detecting its distribution. By analysis on the spatial and temporal distribution of electrical field and abnormal flow field, the inlet of seepage water is figured out.

26.4.2.8 Sedimentation Detection in Front of the Dam

As time goes by, the sedimentation accumulated in front of the dam will reduce the reservoir capacity and affect its operation. In sedimentation monitoring, measurement of initial conditions is necessary to establish a monitoring baseline for comparison. It is recommended to carry out periodical hydrographic surveying and mapping to calculate sediment quantity change, scope and thickness. The method mainly includes measurement of fixed cross sections and multi-beam complete coverage measurement.

1. Measurement on fixed cross sections

Set some cross sections and perform periodical measurement. By data comparison, the changes of sediment scope and thickness for the cross sections and time intervals can be obtained, and total amount can be inferred. Sounding rod, sounding lead and acoustic depth finder are used to get the water depth values of different points, and theodolite, level, total station and GPS are used to measure elevation and location.

2. Multi-beam complete coverage measurement

A multi-beam sonar system provides fan-shaped coverage of the underwater terrain with the output data in the form of 2D depths. The system will measure and record the time of the acoustic signal traveling from the transmitter to the sediment surface and back to the receiver. The patterns usually run as a series of parallel lines that ensure overlapping coverage of the multi-beam sonar. The locators involved are GPS, motion sensors, gyro-compass, sounding velocity profiler etc.

26.4.2.9 Metal Structure Detection

1. Patrol examination

Patrol examination aims to check the integrity of gates and related structures and status of the flow. The examination objects include: the operation of gate and hoist; the flow patterns in the water way of gate and in front/back of the gate slot during discharging; the leakage when the gate closed; cracks, denudation and aging of pier, breast wall and bracket; cavitation, scouring and erosion of gate slot; opening/closure and offset of expansion joints on pier and bottom plate, as well as their influences on gate and hoist; collapse, block of air hole; cracks and leakage of hoist chamber; effectiveness of anti-freezing measures in cold areas; integrity of hydraulic system and its protection and controlling system; standby power supply.

2. Appearance inspection of the gate

Visual inspection is mainly adopted with the assistance of some instruments. Before inspection, the staff should know well about the manufacture, installation, operation, maintenance and overhaul of the gate and hoist. The components for appearance inspection consists of the gate body, water stop, supporting component, running gear, gate slot, lock-up devices, pressure balance and connecting parts.

3. Corrosion inspection

Usually, thickness gauge and other professional instruments are necessary. Inspection items are as follows: corrosion location and distribution; depth, size and density of corrosion pits; the ratio of corroded area to gate and hoist surface; the size of the part without corrosion. According to the resulting data, the corrosion severity is classified into 4 grades: A is mild corrosion; B is general corrosion; C is serious corrosion; and D is extreme corrosion.

4. Detection of materials

Each element has a characteristic line spectrum that can be used to identify the element. The absorption or emission of radiation can occur by "jumps" of the electron from one stationary orbit to another, which produces the characteristic line spectrum for element identification. The variable measured is often the intensity which is used to determine the content of the element and assess the hardness of the material. The material designation is determined by synthetic analysis. The detection methods of materials selected as follows:

(1) If sample-taking is feasible, designation of the material can be identified by chemical analysis and mechanical test.

(2) If sample-taking is impossible, extract some crumbs at force-free parts for chemical analysis and hardness measurement. And its designation can be determined.

(3) In-situ determination of the designation by spectrometric analysis.

5. Detection of material effective thickness and rust

Ultrasonic thickness gauge is employed to measure the effective thickness and rust of the material. As the Ultrasonic pulse reaches the boundary between two materials, it is reflected to receiver. By application of propagation time and velocity, the effective thickness can be inferred. Comparing them with the design parameter or the former detection data, the rust quantity and rust velocity can be calculated.

6. Non-destructive detection on weld

Welding quality means a lot for structural safety. The detection on weld is a key process. Appearance detection and non-destructive detection should be carried out one by one. Inspection on weld's appearance goes first, and then comes the detection and assessment on the non-destructive detection on weld.

Visual inspection, with the help of 5 fold magnifier is usually preferred. Cracks, incomplete fusion, slag inclusions, crater and other defects are not acceptable. The size of weld should be measured by gauge or caliper.

Methods for non-destructive detection on weld mainly are ultrasonic testing, X-ray testing, magnetic particle inspection and penetrant testing.

(1) Ultrasonic non-destructive testing. As soon as the ultrasonic wave comes to a change in material characteristic, e. g. cracks, inclusions, porosity, et al., the propagation will change, such as reflection, refraction and conversion of

waveform. Transverse wave testing is most often used to detect defects. Based on with the transmitter position and echo amplitude, the depth, location, size and pattern of defects can be determined.

(2) X-ray testing. The penetrability of X-ray has a direct relationship with the wave length of X-ray, as well as the density and thickness of the testing material. The typical welding defects, lack of fusion, cracks or porosity inside, would cause variations in weld density, namely variations of wave impedance. The image on film displays the defects inside the weld. X-ray testing is to record the information with the film, mainly used to test the marco-geometry defects inside the components tested and welded. .

(3) Magnetic particle inspection. Magnetic particle inspection (MPI) is used for the detection of surface and near-surface flaws in ferromagnetic materials. A magnetic field is applied to the specimen, either locally or overall. If the material is sound, most of the magnetic flux is concentrated below the material's surface. However, if a flaw is present, so that it interacts with the magnetic field, the flux is distorted locally and "leaks" from the surface of the specimen in the region of the flaw. Fine magnetic particles, applied to the surface of the specimen, are attracted to the area of flux leakage, creating a visible indication of the flaw.

(4) Penetrant testing. Penetrant testing is based upon capillary action, where low surface tension fluid penetrates into clean and dry surface-breaking discontinuities. Penetrant may be applied to the tested component by dipping, spraying, or brushing. When adequate penetration time is up, the excess penetrant is removed, a developer is applied. The developer helps to draw penetrant out of the flaw where a visible indication becomes visible to the inspector.

7. Stress measurement of the gate

In structural strength assessment of gate, its static stress and dynamic stress should be measured. The tested components includes main beam, edge beam, supporting arm, panel and lifting lugs of gate, as well as gantry, supporting beam, frame and other force-bearing parts of the hoist. Before measurement, based on the material characteristic, structural parameters and load condition, the stress analysis and calculation are performed to understand the stress distribution and arrange the measuring points. For the important, huge gate and hoist with serious defects, residual stresses on the main components should be measured and their influence on structural safety should be analyzed.

8. Measurement of stress concentration of the gate

Fatigue attributes to more than 95% failure of steel structures, which is caused by stress concentration. Large gates often encounters stress concentration. Effective measurement of stress concentration for gate is basic to prevent its failure.

Metal magnetic memory testing is primarily based on magnetomechanical effects. Ferromagnetic materials have magnetic domain structures and a self-magnetization property. When an external load is applied to this type of material, the residual magnetic induction and the spontaneous magnetization increase at stress concentration zones. This generates fixed nodes of magnetic domains, which appear on the surface in the form of a leakage field. Metal magnetic memory testing can effectively detect the area of stress concentration, therefore, it can be used to diagnose premature defects for metal structures and show the development of the cracks.

9. Measurement of residual stress

Residual stresses in a structural material or component are those stresses which exist in the object without the application of any services or other external loads. Generally, two kinds of residual stresses are defined: the macro stresses and micro stresses. The macro residual stresses vary within the body of the component over a range much larger than the grain size, while the micro residual stresses result from differences within the microstructure of a material. They vary on the scale on an individual grain or exist within a grain. Commonly, the residual stress refers to the macro residual stress. Residual stress in the material causes the interplanar spacing of the material to change. In the X-ray diffraction testing, the changes in the interplanar spacing can be used with the Bragg's equation to detect elastic strain through a change in the Bragg scattering angle. At present the methods to measure the residual stress are various, and the X-ray stress diffraction testing is most typical in them. By analyzing the diffracted peak changes, the residual stress values and directions in the materials can be detected.

10. Inspection of the hoist performance

The tested components are the hoist, the controlling system, protection devices and the electric circuits. Through the inspection, we may know better about the operation and performance of the hoist.

11. Inspection of lifting capacity of the gate

Several different methods have been developed to evaluate the lifting capacity of the gate, mainly sensor method, strain gauge method and other methods. For strain gauge method, the monitoring points should be arranged at the uniformly forced parts of the spreader, suspender and lifting lug and more than 4 monitoring points are required. The whole hoisting process should be monitored. Before inspection, theoretical calculation on lifting capacity is necessary. Process graph of lifting capacity is drawn to find out maximum value. For emergency gate, monitoring of holding force in hydrodynamic condition is required.

12. *Inspection of structural vibration*

Strong vibration of gate is dangerous to operation safety. It even may initiate resonant vibration and dynamic instability under certain circumstances. Inspection of structural vibration aims to test the parameters of vibration amplitude, natural frequency and damping characteristics. Thus, a foundation is laid for later analysis on vibration characteristics, counter-measure design and optimization of structure design.

Displacement sensors, speed sensors, accelerometers and resistant strain gauges are set on the parts with strong vibration. The measurement is performed in condition of normal operation, electromagnetic excitation or shock excitation. The measurement items are vibration momentum, structural free-vibration characteristics.

13. *Inspection of the mechanical and electrical safety of the hoist*

Inspection of the mechanical and electrical safety of the hoist mainly includes the appearance inspection and mechanical and electrical parameters test of the hoists. Main components such as open gear, reducer gear box, drum and their wearing and rusting are observation objects in appearance inspection, as well as other abnormal operation conditions directly discovered by appearance inspection. Electrical parameters to be inspected are insulation resistance, current, voltage and so on.

14. *Recheck of gate structural safety*

Based on the testing data, we can recheck the strength, stiffness and stability of the gate by the application of finite element method.

26.4.2.10 Technical Standards for Detection

The main technical standards of detection in China are presented as follows.

Table 26.4-2 Main Technical Standards of Dam Safety Detection in China

No.	Name of specifications	Serial No. of specification
1	*Method for manual ultrasonic testing and classification of testing results for steel welds*	GB 11345—1989
2	*Standard method to of corrosion data statistics analysis*	GB/T 12336—1990
3	*Test code for hydraulic concrete*	DL/T 5150—2001
4	*Code of operation in drilling for hydropower and hydraulic engineering*	DL/T 5013—2005
5	*Code for engineering geophysical exploration of water resources and hydropower*	DL/T 5010—2005
6	*Code of water pressure test in borehole for hydropower and water resources engineering*	DL/T 5331—2005
7	*Technical criterion of safety inspection for hydraulic steel gate and hoist machinery*	DL/T 835—2003
8	*Technical specification for inspection of concrete compressive strength by rebound method*	JGJ/T 23—2001
9	*Technical specification for detecting strength of concrete by ultrasonic-rebound combined method*	CECS 02:2005
10	*Technical specification for inspection of concrete defects by ultrasonic method*	CECS21:2000

26.4.3 *Dam Safety Evaluation Technology*

The regular dam safety inspection and the special inspection are a main carrier for dam safety evaluation. The regular inspection is implemented every 5 years. Experts would examine the dam carefully in aspects below: recheck of original design data, method and degree of safety, according to current codes; discussing on construction method, quality, problems encountered and their treatment; recheck of flood, reservoir capacity and flood discharge capacity; review of dam operation records and analysis on monitoring data; implementing in-situ inspection (including underwater inspection); classifying the dam safety grades. Special inspection is only for problems in special operation condition, usually after floods, storm, earthquake or accident which caused possible hazards for dam safety. The contents of special inspection are determined by the kinds of hazard and its seriousness. Some examples are given here.

26.4.3.1 Case study: Technical Appraisal of Key Issues in the First Regular Inspection for Ertan Hydropower Station in Sichuan Province

1. Influence of downstream cracks on structural safety

3 cracks were found in December 2000, on the right bank of the dam section 33 and 34 and lying near foundation of Ertan dam. The lengths were respectively 3m, 6m and 7m, their horizontal angles were 40°-45°, extending upwards. After then, more and more cracks were found. Until December 2005, there were 127 cracks on the right bank and 52 cracks on the left bank. Some cracks never stopped developing.

Since the cracks were found, Ertan Hydropower Development Corporation entrusted the experts from HydroChina Chengdu Engineering Corporation and Tsinghua University to do research on the causes, stability and treatment measures of the cracks.

(1) Calculation and analysis result of HydroChina Chengdu Engineering Corporation.

1) According to monitoring data, it can be concluded that the increment of dam horizontal displacement was convergent, manifested as the effect of aging and residual deformation. Vertical displacement reflected the aging and residual deformation of dam foundation. To arch dams, it is an inevitable self-adjustment process to loads and foundation deformation during first impoundment. As to Ertan dam, the process was convergent, combined with seepage monitoring data, the analysis showed that the arch dam and foundation were in normal operation condition.

2) Through analysis on temperature effect and simulation, especially the effect of monitored temperature, water temperature and temperature field change on temperature load of dam body, and they performed calculation on temperature stress. The analysis result was presented below: from first impoundment, the dam experienced 2 cold snaps and 1 steep temperature drop which caused the tensile stress downstream reached 2-4MPa. Taking the influence of nonlinear temperature load on thinner cross sections of upper dam part into consideration, the tensile stress downstream would increase another 0.02-6MPa. Nonlinear temperature difference was the main contributing factor for cracks.

3) As known, deformation modulus of dam foundation is one factor greatly affecting arch dam body stress. By hole coring, water pressure test and acoustic wave test at the foundation of dam sections 33 and 34, the experts rechecked the geology condition and deformation modulus. Furthermore, they studied its sensitivity. The calculation showed that: because of low deformation modulus of the weak zone (10-15m wide) of chlorite-actinolite basalt, tensile stress still existed at the cracked part, and it had increased as the modulus decreased. As time moves on, with load of storage water, the creep and residual deformation generated at dam foundation (especially the weak zone) were greater than syenite rock foundation which decreased the deformation modulus of dam foundation. It prompted the stress adjustment of dam body. Parts where tensile stress exceeded the tensile strength finally suffered cracks.

(2) Calculation and analysis result of Tsinghua University. Combined with the mechanical parameters of dam and foundation material reflected by monitoring dam deformation, experts from Tsinghua University used linear finite element and nonlinear finite element methods to simulate water level and dam temperature. Meanwhile, they studied the distribution of stress and its changing trend. In current arch dam design standards, nonlinear temperature difference and steep temperature drop is not considered in temperature difference load analysis. But in operation, it did exert adverse influence on dam surface. According to calculation result, as the effort of asymmetrical surface temperature caused by uneven sunshine, the maximum principal stress downstream was not symmetrical on the left and right bank. On the left bank, the stress was in the form of compressive stress. On the contrary, the stress on the right bank was in the form of tensile stress, and the maximum value exceeded 2MPa. In the first 10 days of January 2000, the maximum tensile stress was 2.73MPa and caused cracks (in the zone of first crack appeared, the maximum principal stress then was 2.32MPa). Considering steep temperature drop (simulation was performed in the unit of day), the maximum tensile stress on the downstream right bank would increase.

About crack stability, analysis results were as follows:

1) Tensile stress concentration was apparent at the ends of the crack. If the crack ends were in high stress zone, the more serious tensile stress concentration encountered. And, stress was released in the small zone in normal direction of crack. It demonstrated that influence of crack on stress was shallow and little.

2) While the crack was long and shallow, as the crack depth increased, the tensile stress zone of dam was increasing, but convergent. In the case that cracks were 6-12m long and 4-7m deep, the cracks may develop to greater depths, but would be convergent.

2. Stability appraisal of Jinlong mountain slope

Jinlong mountain slope locates on the left bank, 600 – 1,300m far away from the dam. Jinlong gully and Abulangdang gully cut through the slope respectively on the east and west side. The slope is projecting into the air from the mainland without support. With different geological condition and unstable mode, the slope is divided into 3 zones, marked Ⅰ, Ⅱ and Ⅲ. Among the three zones, the slope compaction by slag and the local slope protection had been implemented for the zone Ⅰ and Ⅱ. The total volume of slag is about 700,000m³.

Before impoundment, surface displacement was small, except the monitoring point 11 at the slope of the clay mine on the elevation 1,490m. The slope was stable. Early deformation and landslide were local and shallow which didn't threaten the project construction.

By comparison and analysis on the geological stability, deformation characteristics, influencing factors, deformation trend and slope stability before and after impoundment, experts reached the conclusion as follows.

The sliding energy of zone Ⅰ was completely released. With the help of anti slide measures, after long-term stress adjustment, its deformation tended towards stable. Zone Ⅱ was stable, although great deformation occurred at the stronger creep-the creep-slide-mass (including shallow layer of V$^#$ adit) and the mined-out area which showed the lower level of stability. Due to the following: ①as for the stronger creep – the creep-slide-mass, no sliding surface formed; ②deformation of middle layer and deep layer of V$^#$ adit was much less than shallow layer; ③the geological structure of zone Ⅱ stayed unchanged and the soft stratum was deeply buried as well, the toe of slope was without free surface; ④middle-section rock was of good and comprehensive quality which can afford great deformation. Even deformation occurred at rocks and soft stratum, the middle-section rock would hold the zone back from structural failure; ⑤no more mining at the clay mine diminished the adverse effect on the slope stability; ⑥the anti slide measures carried out at the toe of slope enhanced the stability in the zone Ⅱ, generally speaking, zone Ⅱ was stable. Concerning zone Ⅲ, it was of lower stability after impoundment. Fortunately, sliding surface was not generated and no deformation acceleration was found. All these demonstrated its stability. To sum up, except the hazard of deformation mass at zone Ⅲ to the highway 6$^#$, the slope was stable, exerting no impact on the operation of hydropower station.

26.4.3.2 Case Study: Technical Appraisal of Key Issues in the First Regular Inspection of Longyangxia Dam in Qinghai Province

1. Stability appraisal of the left bank abutment

The stability against-sliding of left abutment lies mainly on the combination type of F_{73} fault and low-angle structural plane as well as morphology of scour holes on river bed. In the upper rock mass above F_{73} fault, low-angle silted fractures such as F_{215}, T_{12} and T_{25} had developed beneath elevation 2,450.00m, which may form the bottom slide surface combined with F_{73} fault. There were 2 steeply dipping joints on the side sliding surface and ground sliding surface of the sliding mass, namely NWW and NE. Their sliding directions were SE 110° and NE78°.

In the project completion safety appraisal report, the NE78° joint which was regarded as the side sliding surface in stability calculation for left bank abutment was described as below: after excavation of abutment and chambers, an investigation and analysis on connectivity rate and fluctuation difference of fracture surface were carried out. For NE-direction fracture, when its buried depth <30m, the connectivity rate was assumed to be 100%; when its buried depth was 30 – 50m, the rate was assumed to be 85%; when buried depth was >50m, the rate was assumed to be 70%. Experts all approved the principles adopted by HydroChina Xibei Engineering Corporation in determination of shear strength of abutment rock. In the design, the shear strength and connectivity rate adopted were carefully considered for an enough margin.

During the recheck of abutment stability in 2003, to obtain the data of underground water level, the data of 2 newly added holes No. 47 and No. 49 at the left bank were used. The reservoir had been in operation with low water level for long time after impounding, so the underground water level in condition of normal storage only can be deduced based on monitoring data with low water level. Recheck result showed that the friction coefficient and shear friction coefficient based on the measured stable underground water level both had met the requirements on stability. However, some deduced indicators based on the deduced underground water level couldn't meet the stability control requirements.

From July 18th to November 18th, 2005, the water level of Longyangxia reservoir gradually rose from 2,570.29m to highest water level 2,597.62m, then fell to 2,592.29m on March 14th 2006. During this process, the underground water level remained stable, except for a few occasions affected by rainfall or other factors (for example water consumption for afforestation), and nothing to do with reservoir level. As a result, the stability coefficient of left bank abutment approximated to the figure in recheck based on the measured stable underground water level in 2003 and met the stability controlling requirements.

In fact, the water level values of holes No. 47 and No. 49 adopted in stability calculation were conservative. Firstly, the two holes are located on the inner side of slope which was not within the scope of the rock mass rechecked. For the two holes, the longitudinal drainage gallery on the left bank didn't play an obvious role in reduction of pressure. Secondly, the seepage pressure on F_{73} structural surface was a main index in recheck. As the effect of drainage system set nearby and the seepage prevention effect of rock, the underground water level was not representative of seepage pressure on F_{73} structural surface. Thus, during the first regular inspection, 4 observation holes were set in March 2006. Among the four, only hole A_2 had a water level of 2.8m while other three were all 0. Then, it came to a conclusion that the coefficient of the stability against sliding was high in practice, meeting the stability controlling requirements.

There were 2 range finders set on F_{73} fault. The observation result during July 4th to December 5th showed that the scope's vertical change was 0.57 – 0.68mm. No clear trend was found. The stretching value of the period was 0.036mm, while maximum stretching value was 0.11mm (on November 14th). The oblique change scope was −0.01 – 0.06mm, with apparent stationary trend.

Of the two inverted perpendiculars on elevation 2,463m, IP II-2 was anchored on the bottom of F_{73} while IP II-1 was anchored on the upper part of F_{73}. During the period of water level rising, the measured value of IP II-2 changed smoothly. Till February 6th 2006, the deformation toward upstream was 0.31mm; deformation toward left bank was 0.2mm. The increment went to the opposite direction of failure mode. The measured value of IP II-1 was stable. The above evidence proved the stability of F_{73} fault.

By analysis, we may conclude that the abutment of Longyangxia arch dam was stable.

2. Stability of reservoir bank near dam

The steep slope on the right bank near dam is 300 – 500m. The middle and lower parts mainly consist of compacted, overconsolidated, half-diagenesis and nearly horizontal attitude clay with inclusion of thin layer sandy soil. Under saturated environment, the clay is softened and its permeability is poor. On the contrary, the permeability of sandy soil is much better. Experts of HydroChina Northwest Engineering Corporation are in the opinion that the mechanisms of landslide of slope near dam after impoundment are mainly divided to two types: one is that as the wave erosion at slope bottom causes the lose of support, the shallow layer rock collapses as tension crack occurs. Most of the accidents of collapse happened so far are of this kind. Since the rock volume is not great, it doesn't exert threats on dam safety. The second type is that slope bottom is softened and turns to creeping zone. Under gravitational force, it leads to tensile cracks on the slope top. As the development of crack, it would finally cause instability and failure. The great volume would exert threats on dam safety undoubtedly. The landslide happened at Longxi slides No. 2 and No. 3, Nongchang slides No. 3 and No. 4 caused surge higher than 10m.

As the volume of landslide is associated with the location of tensile crack, the closer it is to the mountain body, the greater volume of crack will be. In this occasion, deeper and wider cracks provide possibility of landslide prediction. The main monitoring items for Longxi and Nongchang slopes are crack width, underground water level, surface displacement (by application of GPS technology). Furthermore, a monitoring cave is set on mountain side (before impoundment) to monitor the change of crack inside the cave.

The long-term inspection found that there were some transverse (parallel to slope) and longitudinal (vertical to slope) cracks both on slope surface and in cave of Nongchang and Longxi slopes. Deep cracks of Longxi slope were J8, J21, J5, J4, Jzl and so on. Their corresponding landslide volume is more than 3 million m^3. The crack widths were 1 – 3mm. J8 crack was of major concern. J8 crack was discovered in May 1993. The long-term observation showed that there was no development, and no obvious cracks were detected in the corresponding part on hill top. A longitudinal crack in Nongchang cave was paid close attention to in the patrol inspection. The crack width was 1 – 2mm. The glass strip used in observation suffered tensile fracture.

The observation of two cracks on high slope began in 1994 and more monitoring instruments were put into operation in the early of 2006. The observation data showed that the crack width was within 2mm and no development, which was identical with the patrol inspection results.

According to long-term monitoring data, the rising velocity of underground water level was much lower than that of reservoir water level. The reservoir water level reached 2,597.62m in November 2005. Correspondingly, the underground water level of Nongchang slope and Longxi slope were lower than 2,555m and 2,560m respectively. Water pressure outside was much larger than that inside. With unchanged mechanical parameters, the slope stability was improved comparing to natural state.

Operation with high water level would bring down the stability coefficient of landslide mass. Whether large-scale sliding would start depended on the scope of creeping zone at bottom of slope, and on the crack locations reflected on the mountain body. As a matter of fact, there were some cracks deep inside the slope and the boundary conditions existed for landslide with volume more than 3 million m³. But it remained stable for over ten years. Some cracks were formed before impoundment and they didn't widen during high water level operation. So far, the slope was safe. Considering the poor permeability of clay rock and short permeation time in condition of high water level, the observation and patrol inspection of deep cracks should last for a very long time to find out whether large-scale creeping zone is formed which may cause giant landslide.

After impoundment, slope failures occurred were in the form of shallow layer collapse. Collapses happened in succession 28m in front of Longxi monitoring cave. On October 16, 2005, landslide of Chadong slope happened, with volume of about 30,000m³, leading to no big threat to dam safety. After slope collapse, the front of the left slope was vertical, worsening the shallow layer collapse condition. It was proved by the widening of cracks 5m in front of monitoring cave of Longxi slope during high reservoir water level operation. It was estimated that, in condition of lower underground water level than reservoir water level, the intact stratigraphic slope would fail in the form of collapse and shallow-layer sliding and pile up underwater. As the accumulation masses rise to reservoir water level, it would protect the intact stratigraphic slope and slow down its failure process. At last, dynamic equilibrium of underwater corrosion and collapse of slope can be achieved.

26. 4. 3. 3   Case Study: Technical Appraisal in the Second Regular Inspection of Fengman Dam in Jilin Province

Fengman hydraulic project is made up of concrete gravity dam, flood discharge structures, power house at dam toe and third-stage power house on the left bank. The maximum dam height is 91.7m and the length of dam crest is 1080m. The elevation of dam crest is 267.70m. Flood discharge structures consist of 11 spillways at dam body with the top elevation at 252.5m and hole size of 12 m×6m, spillway tunnel on the left bank with diameter of 9.2m and length of 682.9m. The project construction began in 1937. The first power generation unit was put into operation in 1943. During 1951 - 1953, extension and reconstruction were carried out and the whole project was completed in 1953.

In the first regular dam safety inspection for Fengman dam which was implemented during 1995 - 1998, the dam was assessed as normal dam. The second regular inspection started in September 2003 and ended in December 2005, the appraisal result was as below:

(1) Storage capacity of Fengman dam is 10.988 billion m³. Its installed capacity is 1002.5MW. The hydropower station is classified as Grade I and structures are designed as first-grade, which satisfies the requirement of *Standard for flood control* (GB 50201—94) and *Classification & Design Safety Standard of Hydropower Project* (DL 5180—2003). According to the classification and considering the important Jilin city and railways downstream, it is proper to have the flood control standard as 1000 years for flood design and 10,000 years for flood check.

(2) The seismic safety appraisal of Fengman dam given by Engineering Seismology Research Center of Jilin province is that: F_{67} fault at dam foundation belongs to the second Songhua river fracture which is Quaternary late Pleistocene fracture. No activity is detected after late Pleistocene (Q_3), so the fracture is inactive. The seismic peak acceleration with a 2% 100-year exceeding probability is 0.131g which is smaller than design fortification intensity 8° and review value 0.161g.

(3) In the second regular inspection, flood recheck adopted the 68-year flood series from 1933 to 2000. The calculation showed that the original flood figure was reasonable. The curve of reservoir capacity and curve of flood discharge used in flood regulation calculation was newly approved. The calculation principle was based on current flood regulation, i.e. the flood control water level was set on 260.5m. By application of alternative peak reservoir capacities of Baishan reservoir, and the joint regulation of Fengman and Baishan reservoir is feasible. In case that Fengman reservoir encounters a flood less than 500 years one, the discharge volume can be controlled step by step to meet the requirements of downstream flood control. Under the check flood condition ($P=0.01\%$) and taking generating flow 1300m³/s into account, the reservoir check flood water flood level is 267.44m, lower than dam crest elevation (267.70m). Required by related standards, flood regulation discharge dose not include the generating flow in the case check flood. Since the water regime forecasting system is rather perfect with high flood forecast accuracy, there is 24 - 48 h in advance for discharge forecast. The calculation result demonstrated that with 24-hour prediction, flood discharge without generating flow was satisfactory.

(4) Except that sections 34-36 of dam foundation are located on F_{67} fault, other dam foundation rock bodies are composed of hard metamorphic conglomerate. The conglomerate was intact and of Grade II. Some local low-angle joints didn't contribute to the stability against sliding of dam foundation because of poor continuity. In the second regular inspection, the recheck standard adopted was *Design Specification for Concrete Gravity Dam* (DL 5108—1999). The stability of anti-sliding met the standard requirement.

F_{67} fault fracture zone was made of some small high dip angle fractures along the river. Its width was 40m, the central width was about 0.5 - 0.7m. The main content was light-green gouge, and the rest was cataclasite, belonging Grade III-IV. The gouge was of good compaction. As long as good quality of seepage prevention measures is ensured, seepage failure of fracture zone and uneven settlement can be avoided. In-situ shear test showed that, according to current design specification for concrete gravity dam, the shear strength parameters used in the first regular inspection was still applicable in recheck. The recheck result showed that, the stability of anti-sliding of dam sections 34, 35 and 36 met the requirements of the specification.

There were some problems with the interlayer bonding of horizontal construction joints at upper dam body (elevation 240 - 266.5m). In 1986, the reinforcement by prestressed anchor cable was used to solve these problems. After then, the requirements were met.

(5) No keyway was set on the three longitudinal joints and no grouting, so the integrity of dam was weakened. The dam had experienced long-term static condition, but its inner stress would be worsened under seismic load. In the modification in 1953 and rehabilitation of overflow surface in 1987, measures such as concrete pouring upstream and downstream as well as connection of AB joint by steel bar were employed. It enhanced the dam integrity at that time, but it was not designed for joints, the connection measures needed to be improved.

(6) The dam has been under supervision for more than 20 years. Monitoring data showed that the dam was in normal condition. The annual changes of horizontal and vertical displacements stayed stable. Time-effect displacement of dam crest toward upstream still can be found but not great, within 5mm in recent 20 years. Dam sections No. 33 - 37 were located in poor geology. The foundation deformation monitoring points were relatively stable and no variation trend was found. The longitudinal joints' opening and closing changed with temperature and water level changes.

(7) Uplift pressure of dam foundation was smaller than designed value and there was a little seepage (about 0.5L/s). Recent curtain grouting at foundation of dam section No. 22 reminded us of a lack of depth in former grouting, but it was totally of good seepage prevention effect on foundation. At the same time, slow rising of uplift pressure of some dam sections were observed, including dam sections No. 34 - 36 in poor geology, calling for close attention.

(8) Being limited by poor construction and management at the time when Fengman dam was built, the quality defects of concrete were quite common, such as low strength, weak cementation and low compactness, as well as poor treatment of construction joints, longitudinal joints and transverse joints, huge seepage took place at the beginning of operation. After treatment for many times according to the Design Document 366, the seepage prevention was improved but the seepage problem still remained. According to monitoring data of uplift pressure at dam foundation in 1996 and 1997, the uplift pressure coefficient behind drainage curtain was higher than the value (0.15 - 0.3) stipulated in the standard. In 2006, grouting of dam section No. 14 proved the existence of large-consumption holes which reflected the defects in seepage-prevention system. Space between drain holes was 4.5m and was almost blocked by calcareous precipitation. So the drainage system was basically failed and seepage pressure remained at high level. The seepage not only damaged the durability of concrete, but also caused the frost heave and cracks on overflow surface.

(9) Because of seepage of dam body, the concrete beneath buffer layer of overflow surface was lower than saturation line. The frost heave cracked the concrete buffer layer. It couldn't function any more. In 1986, quite a large area of overflow surface was damaged during flood discharge. Fortunately, no accidents happened. The comprehensive rehabilitation was carried out in 1988 and the flood discharge experience in 1995 demonstrated its good effect. Regardless of the positive effects, the concrete on overflow surface got worse as a result of frost heave, more and more leakage points and cracks appeared, crack became wider, most of the cracks distributed on the area of low elevation and high flow velocity. By the second regular inspection, the crack distribution was of high density in some parts of the overflow surface which destroyed its anti-scour resistance. So it was difficult to guarantee the safety discharge. Before flood season in 2006, surface treatment was implemented for high-density crack zones, but frost heave cannot be eliminated. It is recommended that engineering measures to be taken to improve the drainage system.

(10) After the first regular inspection, the innovation of overflow equipments on dam crest and emergency gate at the discharge tunnel were completed. Meanwhile, penstocks were reinforced and passed the completion acceptance.

The water intake gate has been in operation for more than 60 years, but its stress, stiffness can basically meet the requirements. The maximum lifting capacity of the hoist can meet the need of the normal operation.

(11) By inspection, there are no stability issues that will endanger the dam safety at the reservoir banks near the dam.

Conclusions for the second regular inspection of Fengman dam:

With non-engineering measures added, the flood control capacity of Fengman dam could meet the standards requirements; dam stability and its stress in normal operation condition could meet standards requirements well; dam deformation was regular; there was no abnormal displacement of dam foundation, seepage control was of good effect; the flood discharge gate and hoist were applicable after innovation; slopes near dam were stable; problems with seepage prevention system caused by innate poor concrete quality exerted adverse influence on durability and seepage was not well controlled; the drainage system lost its effectiveness and saturation line was high; serious frost heave cracked the overflow surface concrete, threatening the safety of flood discharge. According to the 24th Article of *Regulation of Operation Safety of Dam*, Fengman dam was assessed as the dam with defects.

26.4.3.4 Case Study: Technical Appraisal of Special Safety Inspection of Tianshengqiao-I Dam in Guizhou Province

Extrusion damage happened twice at longitudinal joint between the face slabs No. L3 and L4 of Tianshengqiao I dam. To make clear of the causes and its effect, special safety inspection was launched in September 2004.

By analysis on dam deformation, stress and strain of face slab and its damage patterns, compression and shear failure on surface layer was regarded as the main damage form. Main influencing factors were:

(1) Great settlement of upper dam combined with deformation toward river center caused the extrusion of face slab.

After impoundment (especially after to normal water level), great deformation occurred at the upper dam and was not stable yet. Station No. 0+630−0+822m on dam crest experienced huge settlement. From the beginning of 2001 to July 2003, the deformation was 110−130mm; by the end of 2004, the deformation was 150−170mm. It was decreasing every year but its convergence rate was small.

The rockfill embankment deformed from two banks toward river bed. With friction, the face slab deformed together with the embankment. Due to the different properties of the two materials and confinement of face in horizontal direction, the face slab displacement falls behind that of rockfill. Thus, horizontal strain energy accumulated. Before extrusion damage, the monitored maximum horizontal average strain value was 948×10^{-6}.

The upper dam embankment in the left bank was carried out later than that in the right bank, it was reflected in the monitoring dada that the large-strain zone in horizontal direction was distributed on the left river bed, namely the Station No. 0+600−0+900m. The maximum point was at station No. 0+725m, on elevation 746m.

(2) Temperature stress on face slab during high temperature and low water level operation intensified the horizontal extrusion between face slabs.

As an over-year regulation reservoir, the reservoir operated with low water level, leaving the face slab exposed to the sunshine before flood. The temperature on surface of face slab would reach as high as 50℃ which intensified the extrusion. According to the calculation result of Hydro China Kunming Engineering Corporation, the temperature stress was 4.5MPa as the temperature difference between outside and inside the face slab was 15℃. In the first repair of the damaged face slab, thermometer and compression stress meters were embedded. The monitoring data showed that the compression stress reached 4−6MPa as the face slab temperature was 30−50℃. So, temperature stress of face slab during high temperature and low water level every May to July was one of the causes for face slab extrusion.

(3) Adverse combination of biaxial stresses at some parts of face slab.

Based on monitoring data of section at station No. 0+630−0+726m, the upper parts of stage 2 and stage 3 face slabs suffered transition from tensile stress to low compression along slope. The tensile strain on section 0+630m near dam crest exceeded 500×10^{-6}, and the face slab was in state of biaxial "compression-tension". In fact, the tensile stress along slope on the face slob of stage 3 exceeded its tensile strength at early stage, which caused many cracks. Highest compression stress existed at the interfaces of the face slabs of stage 2 and stage 3 and compression strain along slope was small. The compression strain along slope before the first damage at the monitoring point on section 0+630m and elevation 745m was 158.92×10^{-6}. The compression strain at the monitoring point on section 0+726m and elevation 745m was 55×10^{-6}. The combination of high horizontal compression stress and tensile stress along slope/ low compression stress increased the possibility of horizontal extrusion damage.

(4) Strength difference between concretes of face slabs L3 and L4.

The core-drilling test showed that the compressive strength of face slab L4 was lower than face slab L3. This may explain why the damage mainly happened on face slab L4.

(5) The V-type groove on the surface of longitudinal joints and copper water-stop on the bottom reduced the contact area of upper face slab and increased the compression stress.

To prevent extrusion damage, 5 longitudinal joints in high-stress area were chosen to be reinforced (including the damaged joints L3/L4). Rubber plate of 2mm thick was embedded in the joints to alleviate the accumulation of strain energy. After rehabilitation, several longitudinal joints were damaged, but the damage area was small.

26.5 Reinforcement Technology for Dams

26.5.1 Introduction

The dam would continually afford the water pressure, seepage pressure and other kinds of loads inevitably in operation and would encounter corrosion, scouring, freeze-thawing damage which exerts adverse influences on dam structural stability and strength. As time goes by, the long-term effects would finally bring impact on dam safety and durability.

In China, regular inspections are implemented for dams, thus the problems found can be solved in time. Through regular dam inspections, the dam reinforcement is promoted.

Dam reinforcement technology is developing in high speed and covers many technical areas. With the emerging of new materials, new technology and innovation, not only traditional measures, but also the most advanced technologies are introduced and applied to dam reinforcement, such as the underwater floating gate blocking technology, carbon fiber composite.

In aspect of foundation and abutment treatment, we have cases of Dashankou dam in Xinjiang Uygur Autonomous Region and Hongmen dam in Jiangxi province. Rehabilitation of gravity piers and anti-seepage treatment were carried out at Dashankou hydropower station and concrete grouting was adopted at Hongmen hydropower station.

In aspect of seepage treatment for dam body, we have cases of Luowan dam in Jiangxi province, Shuidong dam in Fujian province and Fengman dam in Jilin province. Luowan reservoir was emptied and concrete face slabs were poured again to improve its seepage prevention performance. Concrete grouting was employed at Shuidong dam and Fengman dam, which was of good effect.

In aspect of underwater treatment and rehabilitation, advanced construction technology and new material were used in treatment for cracks of Zhexi dam, the emergency gate slot of discharge tunnel of Zhelin dam, sand sluicing bottom hole No. 10 of Gongzui dam in Sichuan province, discharge tunnel gate and gate slot of Liujiaxia dam in Gansu province etc.

In practice of dam reinforcement, new technology and materials were invented, employed and popularized. Some typical cases were introduced here to show the achievements in China.

26.5.2 Underwater Blocking Gate

For hydropower stations constructed in 1960s and 1970s, they have been in operation for more than 40 years. Bottom sills of water discharge gate and emergency gate of some stations suffered serious scouring. Once the gate is open for water discharge, the gate slot will be damaged seriously by water. This is a big threat to dam safety. Regarding these problems, underwater blocking gate was developed to create dry construction environment and it proved to be successful in practice.

26.5.2.1 Engineering Application and Innovation of Underwater Blocking Gate

The underwater blocking gate is to place a water tight floating gate at the intake of the discharge tunnel before dam to create dry construction environment. In the past, the installation of floating gate was mainly accomplished by hoisting machinery on shore, with the assistance of divers. For example, this kind of underwater blockage was used in the rehabilitation of emergency gate of discharge tunnel of Zhelin dam (the bell mouth was of 3.5m wide and 8.0m high; designed water head was 21.0m; the project was carried out in 1999) and rehabilitation of outlet gate and gate slot of Liujiaxia dam (the bell mouth was of 7.0m wide and 15.3m high; designed water head was 62.3m; the project was carried out in 2000). But this construction method was subject to the environmental condition, especially when the water depth reached 60m (the maximum safe limit for breathing air is about 60m. It is the limit depth of conventional air diving). Divers breathe a mixture of oxygen, helium and nitrogen for deep dives and this kind of mixed gas is very expensive. To realize deepwater blocking, automatic positioning floating gate with automatic remote control was developed and applied in the reinforcement of bottom plate of emergency gate at Gongzui dam.

Till now, the underwater blocking gate technology is widely used in the defect treatments for a great number of reservoir dams and hydraulic structures, especially when cofferdam is impractical. In the foreseeable future, the blocking gate construction will gain applause and be well popularized.

26.5.2.2 Technical Advantages of the Underwater Blocking Gate Construction

Underwater blocking gate has advantages as follows:

(1) It could supersede the earth-rock cofferdam, especially in the case of deep water. It may make the construction project easier in the dry environment, shorten the construction period and guarantee the quality.

(2) The gate may be installed on spot by application of portal crane, so large civil machinery is not necessary. The blocking and locating will be instructed by the diver. It is convenient to install.

(3) The dry construction condition solves the problem of poor welding quality underwater and saves special materials and professionals, which improves the efficiency.

(4) The gate may be reused in different positions and different holes. It is economical.

26.5.2.3 Case Study: Application of Underwater Floating Gate Blockage for Bottom Outlet of Shuikou Hydropower Station in Fujian Province

There are 2 bottom outlets at Shuikou hydropower station, set on dam blocks No. 22 and No. 36 respectively. In 2003, the concrete between main-rails of emergency gate at dam block No. 2 was found cracked and scoured. And it worsened in 2005. Underwater reinforcement for emergency gate was carried out in 2005 but it was found damaged during the inspection in 2006. During May to December in 2008, floating gate was applied in underwater blocking for outlet bottom hole No. 2.

After accomplishing the rehabilitation of bottom outlet No. 2, the blocking gate was reused to reform the original floating blocking gate and rubber water seal, fitting the structure of outlet No. 1 from Nov. 2009 to Feb. 2010. Then, the underwater blocking was implemented. In January 2010, floating gate (8.7m wide and 15m high, total weight was about 173t, total buoyancy was about 230t) was used in blockage of the bottom outlet No. 1. Before blockage, debugging and positioning were done and the floating gate was emplaced into water by $2\times1,600$kN portal crane. The divers opened the water-filling chambers No. 1 - 8 in the floating gate, connected the chamber mouth with the filling pipes pre-connected in turn, adjusted the position of the floating gate, turned the gate vertical, started filling water and made the gate downward, then placed it to the installation position. Four depth sounders were installed on the floating gate and set to 0. The gate was pulled to the selected position. At the same time the underwater monitoring system was installed in the corresponding location. Finally, the divers slowly opened the working gate to discharge. The floating gate was pressed to the inlet of the bottom hole and blocked the water with a water of pressure 7,186 t on it. The adjustment and emplacement of the floating gate were showed as Fig. 26.5-1 and Fig. 26.5-2.

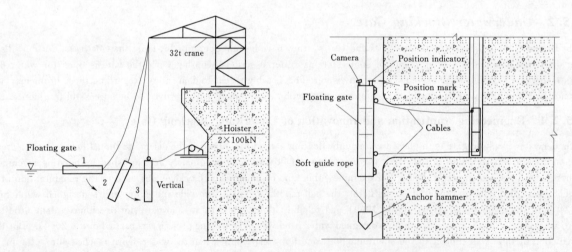

Fig. 26.5-1 Floating Gate Turning to Vertical   Fig. 26.5-2 Emplacement of the Floating Gate

The floating gate was reused in Shuikou project in Fujian Province. It was economical and effective, and the experience can be learnt by other similar projects.

26.5.3 Seepage Prevention Technology by Concrete Grouting at Dam Body

Seepage was commonly believed as a systematic defect. Seepage treatment is difficult and costly. In the past few years, the consolidation grouting with the characteristics of as high pressure, thick slurry, close-drilling and fine cement was applied in the treatment projects at Shuidong dam in Fujian province and Fengman dam in Jilin province, and good effect was achieved.

26.5.3.1 Case Study: Reinforcement by Grouting of Shuidong Dam in Fujian Province

Shuidong dam is integral roller compacted concrete gravity dam. It is located in Youxi County in the middle of Fujian province, lies on the Youxi river which is the first tributary of the middle reaches of Min river. The maximum dam height is 63.0m and the length of dam crest is 196.6m. Since the poor binding between upstream anti-seepage layer and roller compacted concrete layers, seepage was great in cold weather and calcium-precipitation was serious. So concrete grouting was carried out for reinforcement.

1. Main technical parameters of dam reinforcement

The space between grouting holes was determined by grouting test: the vertical grouting holes on the left and right water retaining dam blocks were single-rowed and spaced 0.5m. They were drilled 2-3m deep into dam foundation; the vertical grouting holes on sluice section were single-rowed and spaced 0.5m. They were drilled deep to the location of 1m to the top of gallery. The radicalized holes on upstream side of gallery were drilled by ϕ40mm air drilling with the row space of 0.5m. The bottom vertical holes were single-rowed and spaced 0.5m. They were drilled 2-3m deep into foundation.

In the slurry test, it was found that the lower density of slurry, the higher bleeding rate and the lower strength of the hardened cement. Sometimes, it may cause damage to dam structure. By application of the cement slurry with water cement ratio of 0.6 : 1, this problem can be solved. But its workability, flowability and pourability decreased. To compensate the insufficiency, wet milled cement with some additives was used to grouting holes whose absorption was less than 40L/min.

The test result showed that, for the 0.6 : 1 cement slurry, after two wet milling, the specific surface area of cement increased from 3,680cm^2/g to greater than 8,200cm^2/g. Non-air-entraining efficient retarding water-reducing agent weighing 0.5% of the cement weight was added to the cement slurry after wet milling, the slurry would increase its flowability and workability. Thus, it would spread to micro-cracks and fill them. Further, compared with Portland cement, it had better bleeding stability. The lower water cement ratio, the higher stability after wet milling would be obtained. Moreover, the wet milled cement usually produce a pressured bleeding process due to greater grouting pressure added in construction practice. As a result, the compactness, strength of hardened concrete and impermeability were improved significantly.

As thick slurry was adopted, to guarantee the grouting quality, high grouting pressure was required. The grouting pressures for different sections were determined by their locations.

2. Reinforcement effect

Judging from slurry absorption amount, core-drilling test, water pressure test, acoustic detection and video record, the curtain grouting was of good effect. The ratio of grouting pressure to water pressure was larger than 3, and the thickness of curtain met the design requirement. Besides, no harm was caused to dam body during construction.

The 1-year operation after grouting showed that the seepage of gallery decreased from 3L/s to less than 1L/s now; the calcium-precipitation was lessening; the elevation of wet part downstream lowered by 4m. The rehabilitation had achieved the anticipated results and was a success.

6 years after reinforcement, a detailed inspection was carried out. The wet area downstream disappeared. The monitoring data showed that the temperature field was stable; the average dam temperature increment was about 3℃ and the amplitude reduced by 50% than before.

26.5.3.2 Case study: Rehabilitation by Grouting at Overflow Sections of Fengman Dam in Jilin Province

Fengman hydropower station lies on the middle reaches of main stream of secondary Songhua River in Jilin province. The dam is concrete gravity dam with maximum dam height of 91.7m and length of 1,080m. The dam is divided into 60 dam blocks, and the length of each block is 18.0m. The dam was constructed in 1937 and has been in operation for more than 70 years. Because of quality defects after the dam construction, some dam blocks encountered the problems of high uplift pressure, seepage and corrosion. In 2008, grouting was carried out on dam blocks No. 8-20.

1. Arrangement of dam grouting and main technical parameters

The anti-seepage grouting for overflow dam blocks was implemented on overflow weir crest. The grouting holes were vertical and double-rowed. The first row was basic grouting holes, the axis stations was No. 0+002.7m downstream of the dam axis. The second row was strengthening grouting holes, positioning downstream the dam axis station No. 0+003.0m. The space between holes was 1.0m and the holes were arranged interleaving. The basic grouting holes consisted of 18 holes and the strengthening grouting holes consisted of 17 holes. The water cement ratio of slurry was 0.6 : 1. The average particle diameter of the wet milled cement was no greater than 10μm. The usual grouting pressure was 1.5MPa. For densely distributed grouting holes, the grouting pressure could be raised to 2MPa. The criterion of acceptability for grouting was: the dam permeability should be no greater than 0.1Lu. If the dam permeability was less than 0.1Lu after grouting of basic grouting holes, then the grouting of strengthening holes was not needed. If the permeability was greater than 0.1Lu after grouting of basic grouting holes, then the grouting of strengthening holes was necessary.

2. Technology of anti-seepage grouting

The orifice-closed grouting method was adopted. Its sketch drawing was presented below (the Fig. 26.5 - 3).

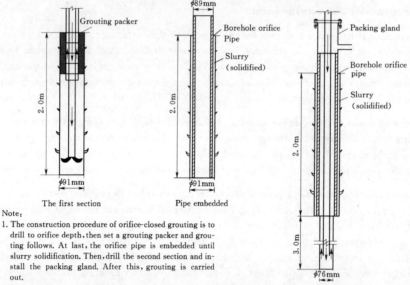

Note:
1. The construction procedure of orifice-closed grouting is to drill to orifice depth, then set a grouting packer and grouting follows. At last, the orifice pipe is embedded until slurry solidification. Then, drill the second section and install the packing gland. After this, grouting is carried out.
2. The construction procedure and method of the third section and following sections are absolute the same with the second section.
3. Orifice-closed grouting method is simeple, effective and can be repeated.

Fig. 26.5 - 3 Sketch of Orifice-close Grouting Construction Technique

When grouting thick slurry from top to bottom, grout backflows from outside of the packer is quite common. When the grout backflow happened, interval, pressure reduction and cease are not beneficial for grouting quality. The orifice-closed grouting can be repeated. At the same time, as the drill pipe functions as a slurry jetting pipe, during the grouting process, it may spin at regular time and lift up or go down at will. Compared with application of packer, the accident of forming "inner grouting pipe" can be avoided, and grouting quality will be guaranteed.

3. Quality inspection result

The project acceptance conclusions given by the construction party was as follows: the grouting parameters and technology satisfied the design requirements; the average particle diameter was no greater than $10\mu m$ after wet milling; the slurry stability should meet the requirements stipulated in standard. Borehole bottom inclining measurement showed that all holes inclinations were less than the distance between holes. The permeability of all hole sections were no greater than 0.5Lu, and the sections whose permeability were less than 0.15Lu accounted for 95.5%. According to acoustic CT test result, the wave velocity before grouting was 4,000 - 4,700m/s, after grouting was 4,200 - 4,800m/s, increased 3% - 10%. At the parts where wave velocity before grouting was 3,500 - 3,800m/s, the velocity was enhanced to 40,000m/s after grouting. Borehole TV showed that 92.1% fractures and cracks were fully filled, 4.9% fractures and cracks were partly filled and 3.0% were unfilled.

By analysis on the results of water pressure test, acoustic CT and borehole TV inspection, conclusion can be achieved that the permeability of grouting concrete got an obvious decrease; the wave velocity in concrete enhanced; the fractures and cracks were filled up well; the grouting quality was satisfactory. The project proved the advantages of orifice-closed grouting such as efficient and practical, which was worth of popularization.

26.5.4 *Reinforcement by Carbon Fiber Composite*

26.5.4.1 Advantages of Carbon Fiber Composite

Reinforcement by carbon fiber composite is a newly developed method. The main material used includes two types: carbon fiber material and matching resin. The tensile strength of carbon fiber is 8 times than that of the steel material, and its elastic modulus is almost the same as steel. Carbon fiber is of good workability and durability which makes it an excellent reinforcing material. The matching resin includes primer resin, leveling resin and luting resin. The function of the former two is to enhance the bond quality of carbon fiber while the function of luting resin is to composite with concrete structure. Thus the carbon fiber may work together with the structure, which enhances the shear strength and flexural capacity of structure.

Comparing with traditional reinforcement method, carbon fiber has advantages as below:

(1) High strength. The carbon fiber has excellent physical and mechanical properties. Its tensile strength is above 4,000MPa which is 8 times of steel material and its elastic modulus approximates to steel. It is applicable in reinforcement of structures.

(2) Stable performance. It has good durability and resistance to chemical attack, namely acid-resistant, alkali-resistant, salt-tolerant and weather corrosion resistant. No periodical maintenance is needed. It not only protects the inner concrete structure from corrosion, but also enhances the physical and mechanical properties of the structure. It is dual-functional.

(3) Light-weighted. The regular densities of carbon fiber composite are $200g/m^2$ and $300g/m^2$, the thicknesses respectively are 0.111mm and 0.167mm. It is light-weighted, without additional load and sectional size to the structure.

(4) Convenient and swift construction. In the construction process, no wet work, no large mechanism or fixed facility is needed and the carbon fiber sheet can be cut in any way at will. The construction procedure is simple and short, no dust or noise pollution will be released.

(5) High construction quality. The carbon fiber composite is quite flexible. Even the structure surface is irregular, and the material can be bonded with the surface 100%.

26.5.4.2 Case study: Application of Carbon Fiber Composite into Rehabilitation of Reinforced Concrete Structure of Er'tan Hydropower Station in Sichuan Province

Er'tan hydropower station lies on the mainstream of Yalong river in Panzhihua city, Sichuan province. The station is constructed mainly for power generation. Its installed capacity is 3,300MW. In the inspection, it was found that there were some cracks on most roof beams of GIS building, approximates to 0.3 - 0.6mm wide. The reinforcement was carried out and carbon fiber composite was applied into the reinforcement.

The UT70-30 carbon fiber sheet was adopted, whose density was $300g/m^2$ and thickness was 0.167mm. the matching binder was WSX carbon fiber resin.

According to cracks widths and depths, U-groove was cut on surface, sealing the cracks with IFS-II inorganic instant sealing material or WEP structural perfusion glue.

On the two sides of roof beam of No. 29 axis and away from 400mm of beam and surface, 300mm-wide carbon fiber sheet was bonded in the direction of length to overcome the stress caused by concrete shrinkage. Then, to compensate the loss of bearing capacity due to cracks and enhance its shear strength and flexural capacity, 500mm-wide carbon fiber sheet was bonded to the bottom of beam. At last, 300mm-wide U type carbon fiber hoops, spacing 200mm, were bonded to the beam to enhance its shear strength.

26.5.5 Patent Sealing Structure in SR Seepage Prevention System

26.5.5.1 Research on the Sealing Structure in new SR Seepage Prevention System

The working principle of plastic sealing material is that it flows to expansion joints under huge water pressure and seals the cracks or seams on the water stop. Then it integrates into the water stop and plays as a whole part. After years of research by Chinese scientific institutes, the technology of flexible sealing for expansion joints is leading in the world.

The structure of SR seepage prevention system is that copper seal on the bottom, steel-edged hollow rubber tube seal in the middle and plastic sealing material flexible water stop on the top. The steel-edged hollow rubber tube is of great stiffness which can support itself. In the concrete pouring and vibration process, it keeps straight and doesn't interfere with the pouring which promises a good seepage prevention effect and high strength of concrete face slab. The steel edge binds together with the concrete and its thermal expansion coefficient approximates to the figure of concrete, this enhances its by-pass seepage prevention ability. The L-type and U-type steel-edged water stop developed by bending the steel edge as well as SR sealing material binding to the steel edge can give further improvement in by-pass seepage prevention ability. Fig. 26.5-4 is the sealing structure of SR seepage prevention system.

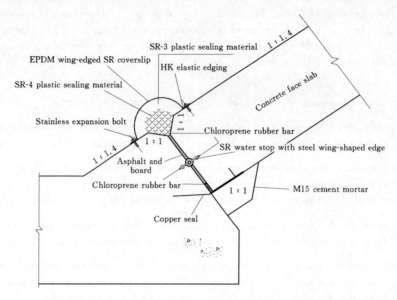

Fig. 26.5 – 4 Sealing Structure of SR Seepage Prevention System

26.5.5.2 Case Study: Application of SR Seepage Prevention System in Lianhua Hydropower Station in Heilongjiang Province

As for the RCC dams built in the cold region, such as Lianhua hydropower station in Heilonjiang province, expansion bolts being pulling out, angle steel being bended, anti-seepage face membrane being torn off etc are common phenomena occurred in operation. The reason is that the common surface sealing structure can not effectively withhold the cold weather condition in North China. The fact is, in the cold winter, the ice layer combines the expansion bolts, angle steels and face membrane together as a whole. The wind and waves push forward the ice layer to strike the surface water stop and the settlement of ice layer caused by lowering water level due to power generation would result in relatively strong drag force on the water stop which destroys the surface water stop. For dam safety, reinforcement is necessary.

Taking into consideration of the practical destroyed surface water-stopping on face-slab rock-fill dams in cold area and combining with the anti-freezing design requirements during operation of the face-slab rock-fill dams in the pumped storage power station in cold region, the improvement of seepage prevention performance for joints were carried out in three aspects:

(1) Anchoring method. Due to the disadvantages of expansion bolts (mechanical fatigue, the seam between the bolt and the hole providing the passage for seepage water, damage of concrete by the frost heave of water in the seam, the freezing of outcropped bolt head), the anchoring method by combination of countersunk expansion bolt and HK 98 epoxy anchor agent replaced the application of common expansion bolts. At the beginning of installation of countersunk expansion bolts, the bolts were dipped into HK 98 epoxy anchor agent to enhance its anti-freezing performance and pullout resistance. At the same time, the outcrop height of the bolt head was lowered.

(2) Less protruded parts of the seepage prevention structure. Grooves were dug to make the SR flexible seepage prevention system go down and the stainless flat steel instead of stainless angle steel. The SR rubber cover slip was widened and the radian of SR flexible filling bulge was decreased. However, the deformation capacity of joints remained.

(3) Sealing of parts. The matching HK elastic edging agent specially developed was used to seal the metal strips and bolt heads, filled the grooves. Therefore, it prolonged the seepage path, avoided the protruded parts from ice layer and enhanced the anti-freezing capacity. HK elastic edging agent is two-component sealing material. It is of advantages such as good brushability and adhesion property, high strength and elastic extension, great anti-seepage and anti-freezing performance. Meanwhile, it suits for deformation caused by thermal expansion and contraction. As the special edging for SR seepage prevention system, it plays an important role. Furthermore, chemical anchor agent can be used to seal the seam between bolt and hole, and the SR leveling layer can be used to seal the seams among flat steel, SR cover slip and bolt.

For Lianhua hydropower station built in the coldest area in China, SR-3 plastic sealing material was developed which was still practicable in $-50℃$ environment. The test showed that from $-50℃$ to room temperature and soaked in the medium, it was bonded to concrete and was of excellent extension, water resistance and medium resistance, anti-freeze-thawing, heat resistance and anti-freezing properties, as well as the simply cool construction performance. It can meet the requirements of engineering sealing below $-50℃$. SR seepage prevention structure was set in the joints beneath surface of the concrete face slab. It is a kind of the hydraulic concrete deformation joint sealing structure to effective decrease the damage of frost-heaving and freeze-thawing. Fig 26. 5 - 5 is the water-stopping structure of SR-3 cold resistance.

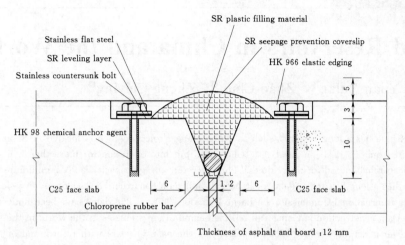

Fig. 26. 5 - 5 Water-stopping Structure of SR-3 Cold Resistance

Appendix

Dams and Reservoirs in China and the World

Jia Jinsheng[1], Yuan Yulan[2], Zhao Chun[3], Zheng Cuiying[4]

Dam construction in China has a long history, such as Anfengtang Project, which is located in Shou xian County, Anhui Province and was built from 598 B. C. to 591 B. C. It has been put into operation for more than 2,600 years. The dam is 6.5m high and has been rehabilitated for many times. The reservoir storage capacity is 90.7 million m^3 and the surface area covers $34km^2$. China has a long history of dam construction, but dam construction developed very slowly in the early stages. According to the International Commission on Large Dams in 1950, 5,268 dams have been built in the world, only 21 dams were in China. Dam construction in China still fell far behind other countries in the world in the total storage capacity and the amount of power generation. Since 1950, especially in the past 30 years with reform and opening-up policy, dam construction and technology has been developed quickly in China, and guaranteed the safety for flood control, irrigation, water supply and energy.

Among dams built in China before 1990, most were middle and lower dams. Of them only 3 dams were higher than 90m. By the end of 2009, there were 5,443 dams higher than 30m, 1 of them is higher than 300m, 11 dams are between 200 - 300m, 31 are between 150 - 200m, 124 are between 100 - 150m, 509dams are between 60 - 100m and 4,767dams are below 60m. Among them, 862dams are arch dams including 40 Rolled Compacted Concrete (RCC) arch dams, 635 gravity dams including 89 RCC gravity dams, 536 rockfill dams including 240 concrete faced rockfill dams (CFRD), 3,065 earth dams, Number of others are 345. The projects such as the Three Gorges, Ertan and Xiaolangdi, etc. have been put into operation, which has marked the major achievements in dam construction in China, and is now taking a leading position in many aspects of dam development. Many reservoirs and dams had withstood severe challenges such as the 1998's flood and the 2008 Wenchuan earthquake. The achievements in construction of reservoirs and dams in this stage have won the worldwide recognition for high quality design, rapid construction speed, high construction quality and good comprehensive benefits.

After 60 years of development, China has made great achievements in dam construction. However, to look forward to the future and meet the requirement of national development, China should be emphasized on a series of the technical challenges. In this article, projects in the world with the top ranking in dam height, reservoir storage capacities and installed capacity has been listed for references. Typical data can be found in Table 1 to Table 13.

For better understanding of the dam classification in China, Table 14 has illustrated the dam classification by reservoir capacity and installed capacity.

The data in these tables are from China's dam statistical database in 2011 by Secretariat of Chinese National Committee on Large Dams (CHINCOLD) and dam statistical database in 2003 by International Commission on Large Dams (ICOLD). It was also referred to profile information about the reservoirs and dams from Dam Safety Management Center of the Ministry of Water Resources, and Large Dams Safety Supervision Center of National Energy Administration of the People's Republic of China.

The codes of different dam types in the tables are from ICOLD's code. Dam type can be identified by the following list:

PG: Gravity Dam

VA: Arch Dam

ER: Rock Fill Dam

TE: Earth Dam

RCCPG: Rolled Compacted Concrete Gravity Dam

RCCVA: Rolled Compacted Concrete Arch Dam

CFRD: Concrete Faced Rockfill Dam

MV: Multiple Arch Dam

CB: Buttress Dam

BM: Barrage

---

[1] Jia Jinsheng, Vice President of China Institute of Water Resources and Hydropower Research, Vice President of Chinese National Committee on Large Dams, Professor Senior Engineer.

[2] Yuan Yulan, Office Director, Secretariat of Chinese National Committee on Large Dams.

[3] Zhao Chun, China Institute of Water Resources and Hydropower Research, Professor Senior Engineer.

[4] Zheng Cuiying, China Institute of Water Resources and Hydropower Research, Professor Senior Engineer.

Appendix Dams and Reservoirs in China and the World

Table 1 The Top 100 Dams by Storage Capacity in China

No.	Dam name	River	Location	Dam type	Dam height (m)	Total reservoir storage ($10^8 m^3$)	Installed capacity (MW)	Years of construction
1	Three Gorges	Yangtze River	Hubei	PG	181	450.5	22,500	1994-2010
2	Danjiangkou	Hanjiang River	Hubei	PG	117	339.1	900	Constructed from 1958 to 1961; heightened from 2005 to now
3	Longtan	Hongshui River	Guangxi	RCCPG	192 (Phase I); 216.5 (Phase II)	188 (Phase I); 299.2 (Phase II)	4,900 (Phase I); 6,300 (Phase II)	2001-2009 (Phase I); To be constructed (Phase II)
4	Longyangxia	Yellow River	Qinghai	VA	178	274	1,280	1977-1989
5	Nuozhadu	Lancang River	Yunnan	ER	261.5	237.03	5,850	2006-2015
6	Xin'anjiang	Xin'an River	Zhejiang	PG	105	216.26	850	1957-1965
7	Xiaowan	Lancang River	Yunnan	VA	294.5	150	4,200	2002-2012
8	Shuifeng	Yalvj River	Liaoning	PG	106.4	146.66	900	1937-1943
9	Xinfengjiang	Xinfeng River	Guangdong	PG	105	138.96	355	1958-1977
10	Xiaolangdi	Yellow River	Henan	ER	160	126.5	1,800	1994-2001
11	Fengman	Songhua River	Jilin	PG	91.7	109.88	1,002.5	1937-1953
12	Tianshengqiao I	Nanpan River	Guizhou/Guangxi	CFRD	178	102.57	1,200	1991-2000
13	Sanmenxia	Yellow River	Henan/Shanxi	PG	106	96	410	1957-1960
14	Xiluodu	Jinsha River	Sichuan	VA	285.5	126.7	13,860	2005-2015
15	Nierji	Nenjiang River	Heilongjiang/Inner Mongolia	TE	40.55	86.1	250	2001-2006
16	Jinping I	Yalong River	Sichuan	VA	305	79.88	3,600	2005-2014
17	Zhelin	Beixiu River	Jiangxi	ER	63.5	79.2	420	1958-1975
18	Goupitan	Wujiang River	Guizhou	VA	232.5	64.54	3,000	2003-2009
19	Liujiaxia	Yellow River	Gansu	PG	147	64	1,350	1958-1974
20	Ertan	Yalong River	Sichuan	VA	240	61	3,300	1991-2000
21	Dongjiang	Xiangzhileishui River	Hunan	VA	157	92.7	500	1978-1992

Appendix Dams and Reservoirs in China and the World

No.	Dam name	River	Location	Dam type	Dam height (m)	Total reservoir storage (10^8m^3)	Installed capacity (MW)	Years of construction
22	Baise	Youjiang River	Guangxi	RCCPG	130	56.6	540	2001–2006
23	Baishan	No. 2 Songhua River	Jilin	VA	149.5	59.21	1,800	1975–1986
24	Pubugou	Dadu River	Sichuan	ER	186	53.37	3,600	2004–2009
25	Xiangjiaba	Jinsha River	Yunnan/Sichuan	PG	162	51.63	6,400	2006–2015
26	Hongjiadu	Mainstream of Wujiang River	Guizhou	CFRD	179.5	49.47	600	2006–2015
27	Shuibuya	Qingjiang River	Hubei	CFRD	233.2	45.8	1,840	2002–2009
28	Miyun Reservoir (Main dam on chaohe River)	Chaobai River	Beijing	TE	66.4	43.75	91.5	1958–1960
29	Wuqiangxi	Yuanshui River	Hunan	PG	85.83	42.9	1,200	1986–1998
30	Changzhou	Xunjiang River	Guangxi	PG	56	56	630	2004–2009
31	Tankeng	Oujiang River	Zhejiang	CFRD	162	41.9	604	2004–2008
32	Guanting	Yongding River	Beijing	TE	52	41.6	30	1951–1954
33	Sanbanxi	Mainstream of Yuanshui River	Guizhou	CFRD	185.5	40.94	1,000	2002–2006
34	Tingzikou	Jialing River	Sichuan	RCCPG	116	40.67	1,100	2009–2014
35	Yunfeng	Yalu River	Jilin	PG	113.75	38.95	400	1959–1965
36	Zhexi	Zishui River	Hunan	PG	104	35.7	947.5	1958–1975
37	Huanren	Hunjiang River	Liaoning	CB	75	34.62	246.5	1958–1972
38	Yantan	Hongshui River	Guangxi	RCCPG	110	33.8	1,210	1985–1995
39	Songtao (Nanfeng)	Nandu River	Hainan	TE	80.1	33.45	24.9	1958–1967
40	Guangzhao	Beipan River	Guizhou	RCCPG	200.5	32.45	1,040	2003–2007
41	Ankang	Hanjiang River	Shaanxi	PG	128	32.03	800	1978–1995
42	Xijin	Yujiang River	Guangxi	PG	41	30	242.2	1958–1964
43	Panjiakou	Luanhe River	Hebei	PG	107.5	29.3	420	1975–1985
44	Bilahe	Bila River	Inner Mongolia	PG/ER	83.3	28.76	255	
45	Geheyan	Qingjiang River	Hubei	VA	151	27.06	180	1987–1995
46	Chencun	Qingyi River	Anhui	VA	76.3	27.06	180	1958–1971

Appendix Dams and Reservoirs in China and the World

Continued

No.	Dam name	River	Location	Dam type	Dam height (m)	Total reservoir storage ($10^8 m^3$)	Installed capacity (MW)	Years of construction
47	Xianghongdian	Xipi River	Anhui	VA	87.5	26.32	80	1956–1961
48	Shuikou	Minjiang River	Fujian	PG	101	26	1,400	1987–1996
49	Hongshan	Xiliao River	Inner Mongolia	TE	31.4	25.6	8.72	1958–1969
50	Jilintai I	Kashi River	Xinjiang	CFRD	157	25.3	500	2001–2005
51	Kalasuke	Eerqisi River	Xinjiang	RCCPG	121.5	24.19	140	2007
52	Hualiangting	Changhe River	Anhui	ER	58	23.66	40	1958–1976
53	Pankou	Duhe River Mainstream	Hubei	CFRD	114	23.38	500	2007
54	Wujiangdu	Wujiang Tributary	Guizhou	VR	165	23	1,250	1970–1982
55	Dahuofang	Liaohe River	Liaoning	ER	49.8	22.68	40	1953–1958
56	Baozhusi	Bailong River	Sichuan	PG	132	25.5	700	1984–2000
57	Meishan	Shihe River	Anhui	PG	88.24	22.63	40	1954–1956
58	A'ertashi	Yeerqiang River	Xinjiang	CFRD	164.8	22.51	730	2011–2017
59	Caojie	Jialing River	Chongqing	PG	83	22.12	500	2004–2010
60	Guanyin'ge	Taizi River	Liaoning	RCCPG	82	21.68	20.75	1990–1995
61	Zhanghe	Juzhang River	Hubei	TE	66.5	21.13	9.22	1958–1965
62	Hu'nanzhen	Qiantang River	Zhejiang	CB	129	20.67	320	1958–1980
63	Mianhuatan	Tingjiang River	Fujian	RCCPG	113	20.35	600	1998–2001
64	Fengshuba	Dongjiang	Guangdong	PG	95.3	19.32	200	1970–1974
65	Feilaixia	Beijiang River	Guangdong	PG/TE	52.3	19.04	140	1994–1999
66	Jiemian	Youxi River	Fujian	CFRD	126	18.24	300	2002–2007
67	Shanxi	Feiyun River	Zhejiang	CFRD	132.8	18.24	200	1997–2001
68	Jingpohu	Mudan River	Heilongjiang	PG	10.9	18.2	96	1938–1978
69	Erlongshan	Dongliao River	Jilin	TE	32.16	17.92	8.36	1943–1966
70	Qiafuqihai	Tekesi River	Xinjiang	ER	105	17.7	320	2002–2005
71	Lijiaxia	Yellow River	Qinghai	VA	155	17.5	1,600	1988–1997
72	Fengtan	Youshui River	Hunan	VA	112.5	17.4	815	1970–1979

Appendix Dams and Reservoirs in China and the World

Continued

No.	Dam name	River	Location	Dam type	Dam height (m)	Total reservoir storage ($10^8 m^3$)	Installed capacity (MW)	Years of construction
73	Jiangya	Lizhi River	Hunan	RCCPG	131	17.41	300	1995-2000
74	Huanghuazhai	Getu River	Guizhou	RCCVA	110	1.748	60	2005-2008
75	Ludila	Jinsha River	Yunnan	RCCPG	140	17.18	2,160	2007
76	Daguangba	Changhua River	Hainan	RCCPG	55	17.1	240	1989-1995
77	Gangnan upper Reservoir	Haihe River	Hebei	TE	64.5	17.04	4.1	1958-1968
78	Fushui	Fuhe River	Hubei	TE	46.8	16.65	40	1958-1966
79	Suyahu	Ruhe River	Henan	TE	16.2	16.56		1958
80	Baishi	Daling River	Liaoning	RCCPG	49.3	16.45	9.6	1995-2001
81	Nanwan	Huaihe River	Henan	TE	38.3	16.3	5.92	1952-1955
82	Silin	Wujiang River	Guizhou	RCCPG	117	15.93	1,050	2006-2009
83	Yuqiao	Jiyun River	Tianjin	TE	24	15.59	5	1950-1965
84	Centianhe (kuo)	Xiaoshui River Tributary	Hunan	CFRD	113	15.1	200	1966-1970
85	Pengshui	Wujiang River	Chongqing	RCCPG	113.5	14.65	1,750	2003-2007
86	Zaoshi	Lishui River	Hunan	RCCPG	88	14.39	120	2004-2007
87	Xiashan	Yellow River	Shandong	TE	21	14.05	4.125	1958-1960
88	Wangkuai	Haihe River	Hebei	TE	62.5	13.89	23.5	1958-1960
89	Jinshuitan	Oujiang River	Zhejiang	RCCPG	102	13.93	300	1982-1988
90	Tuokou	Ruanshui River	Hunan	RCCPG	82	13.84	830	2009-2011
91	Jiangpinghe	Loushui River	Hubei	CFRD	219	13.66	450	2007
92	Yahekou	Hanjiang	Henan	TE	34.6	13.39	12.8	1958-1959
93	Luhun	Yellow River	Henan	TE	55	13.2	10.45	1959-1965
94	Bailianhe Lower Reservoir	Xishui River	Hubei	ER	69	13.2	1,200	2005-2009
95	Yuecheng	Zhanghe River	Hebei	TE	55.5	13	17	1959-1970
96	Chaersen	Taoerhe River	Inner Mongolia	TE	39.7	12.53	12.8	1973-1990
97	Shengzhong	Yangtze River-Jialing River-Xihe	Sichuan	ER	79	13.39	4.4	1977-1987
98	Longjiang	Longjiang River	Yunnan	VA	110	12.17	240	2006-2010
99	Guxian	Yellow River	Henan	PG	125	11.75	60	1978-1995
100	Jinghong	Lancang River	Yunnan	RCCPG	108	11.39	1,750	2003-2008

Table 2 The Top 100 Dams by Height in China

No.	Dam name	River	Location	Dam type	Dam height (m)	Total reservoir storage ($10^8 m^3$)	Installed capacity (MW)	Years of construction
1	Jinping I	Yalong River	Sichuan	VA	305	79.88	3,600	2005–2014
2	Xiaowan	Lancang River	Yunnan	VA	294.5	150	4,200	2002–2012
3	Xiluodu	Jinsha River	Sichuan	VA	285.5	126.7	13,860	2005–2015
4	Nuozhadu	Lancang River	Yunnan	ER	261.5	237.03	5,850	2006–2015
5	Laxiwa	Yellow River	Qinghai	VA	250	10.79	4,200	2001–2010
6	Ertan	Yalong River	Sichuan	VA	240	61	3,300	1991–2000
7	Changheba	Dadu River Mainstream	Sichuan	ER	240	10.75	2,600	2005–2013
8	Shuibuya	Qingjiang River	Hubei	CFER	233.2	45.8	1,840	2002–2009
9	Goupitan	Wujiang River	Guizhou	VA	232.5	64.54	3,000	2003–2009
10	Houziyan	Dadu River	Sichuan	CFRD	223.5	7.06	1,700	2011–2017
11	Jiangpinghe	Loushui River	Hubei	CFRD	219	13.66	450	2007
12	Longtan	Hongshui River	Guangxi	RCCPG	192 (Phase I); 216.5 (Phase II)	188 (Phase I); 299.2 (Phase II)	4,900 (Phase I); 6,300 (Phase II)	2001–2009 (Phase I); No construction (Phase II)
13	Dagangshan	Dadu River	Sichuan	VA	210	7.77	2,600	2005–2015
14	Guangzhao	Beipan River	Guizhou	RCCPG	200.5	32.45	1,040	2003–2007
15	Pubugou	Dadu River	Sichuan	ER	186	53.37	3,600	2004–2009
16	Sanbanxi	Ruanshui River Mainstream	Guizhou	CFRD	185.5	40.94	1,000	2002–2006
17	Three Gorges	Yangtze River	Hubei	PG	181	450.5	22,500	1994–2010
18	Deji	Dajiaxi	Taiwan	VA	181	2.32	234	1969–1974
19	Hongjiadu	Wujiang River Mainstream	Guizhou	CFRD	179.5	49.47	600	2000–2005
20	Longyangxia	Yellow River	Qinghai	VA	178	274	1,280	1977–1989
21	Tianshengqiao I	Nanpan River Mainstream	Guizhou/Guangxi	CFRD	178	102.57	1,200	1991–2000
22	Kajiwa	Muli River	Sichuan	CFRD	171	3.745	440	2008–2014
23	Guandi	Yalong River	Sichuan	RCCPG	168	7.597	2,400	2010–2012
24	Wujiangdu	Wujiang River	Guizhou	VA	165	23	1,250	1970–1982

Appendix Dams and Reservoirs in China and the World

Continued

No.	Dam name	River	Location	Dam type	Dam height (m)	Total reservoir storage ($10^8 m^3$)	Installed capacity (MW)	Years of construction
25	Aertashi	Yeerqiang River	Xinjiang	CFRD	164.8	22.51	730	2011 – 2017
26	Pingzhai	Sancha River	Guizhou	CFRD	162.7	10.89	140.2	2010 –
27	Xiangjiaba	Jinsha River	Yunnan/Sichuan	PG	162	51.63	6,400	2006 – 2015
28	Tankeng	Oujiang River	Zhejiang	CFRD	162	41.9	604	2004 – 2008
29	Dongfeng	Wujiang River	Guizhou	VA	162	10.25	695	1984 – 1995
30	Liyang Pumped Storage Power Station Upper Reservoir	Zhongtianshe River	Jiangsu	CFRD	161.5	0.14	1,500	2008 – 2009
31	Wanjiakouzi	Gexiang River	Yunnan	RCCVA	167.5	2.793	180	2009 –
32	Xiaolangdi	Yellow River	Henan	ER	160	126.5	1,800	1994 – 2001
33	Jin'anqiao	Jinsha River	Yunnan	RCCPG	160	9.13	2,400	2003 –
34	Gaixiaba	Changtan River	Chongqing	VA	160	3.54	132	2009 –
35	Tuoba	Lancang River	Yunnan	RCCPG	158	10.394	1,250	2010 – 2016
36	Dongjiang	lei shui River	Hunan	VA	157	92.7	500	1978 – 1992
37	Jilintai I	Kashi River	Xinjiang	CFRD	157	25.3	500	2001 – 2005
38	Zipingpu	Minjiang River	Sichuan	CFRD	156	11.12	760	2001 – 2006
39	Lijiaxia	Yellow River	Qinghai	VA	155	17.5	1,600	1988 – 1997
40	Liyuan	Jinsha River	Yunnan	CFRD	155	8.05	2,400	2008 – 2014
41	Bashan	Renhe River	Chongqing	CFRD	155	3.154	140	2005 – 2009
42	Malutang II	Panlong River	Yunnan	CFRD	154	5.46	240	2005 – 2009
43	Geheyan	Qingjiang River	Hubei	VA	151	27.06	180	1987 – 1995
44	Dongqing	Beipan River	Guizhou	CFRD	150	9.55	880	2005 – 2010
45	Baishan	No. 2 Songhua River	Jilin	VA	149.5	59.21	1,800	1975 – 1986
46	Liujiaxia	Yellow River	Gansu	PG	147	64	1,350	1958 – 1974
47	Maoergai	Heishui River	Sichuan	ER	147	5.35	420	2008 – 2011
48	Jilebulake	Irtysh River	Xinjiang	CFRD	147	2.32	160	2009 – 2013
49	Longshou II	Heihe River Mainstream	Gansu	CFRD	146.5	0.862	157	2001 – 2005

Appendix Dams and Reservoirs in China and the World

No.	Dam name	River	Location	Dam type	Dam height (m)	Total reservoir storage ($10^8 m^3$)	Installed capacity (MW)	Years of construction
50	Shiziguan	Hongjia River	Hubei	PG	145	1.41	10	1999 – 2004
51	Xigu	Jiulong River	Sichuan	CFRD	144	0.9986	249	2009 – 2013
52	Deze	Niulan River	Yunnan	CFRD	142	4.48	30	2008 – 2011
53	Ludila	Jinsha River	Yunnan	RCCPG	140	17.18	2,160	2007 –
54	Jiangkou	Wujiang River	Chongqing	VA	140	5.05	300	1999 – 2004
55	Qinglong	Maweigou River	Hubei	RCCVA	139.7	0.283	40	2009 – 2011
56	Wawushan	Zhougong River	Sichuan	CFRD	138.76	5.843	240	2003 – 2007
57	Shiziping	Zagu'nao River	Sichuan	ER	136	1.35	195	2003 – 2007
58	Buxi	Yazui River	Sichuan	CFRD	135.8	2.52	20	2007 – 2011
59	Longma	Lixian River-Babian River	Yunnan	CFRD	135	5.904	240	2003 – 3007
60	Dongping	Qingjiang	Hubei	VA	135	3.43	110	2002 – 2006
61	Yunlonghe Ⅲ	Yunlong River	Hubei	RCCVA	135	0.44	40	2006 – 2008
62	Jiyin	Keliya River	Xinjiang	CFRD	134.7	0.83	240	2008 – 2013
63	Dahuashui	Wujiang River tributary	Guizhou	RCCVA	134.5	2.765	200	2005 – 2007
64	Sujiahekou	Binglang River	Yunnan	CFRD	133.75	2.25	315	2005 – 2010
65	Shimen	Dahanxi River	Taiwan	TE	133.1	3.0912	90	1956 – 1964
66	Jiudianxia	Taohe River	Gansu	CFRD	133	9.43	300	2005 – 2008
67	Zengwen	Zengwenxi River	Taiwan	TE	133	7.127	50	1967 – 1973
68	Sanliping	Nanhe River	Hubei	RCCVA	133	4.99	70	2006 –
69	Wuluwati	KalaKashi River	Xinjiang	CFRD	133	3.47	60	1995 – 2001
70	Shanxi	Feiyun River	Zhejiang	CFRD	132.8	18.24	200	1997 – 2001
71	Gongboxia	Yellow River Mainstream	Qinghai	CFRD	132.2	6.2	1,500	2001 – 2006
72	Baozhusi	Bailong River	Sichuan	PG	132	25.5	700	1984 – 2000
73	Manwan	Lancang River	Yunnan	PG	132	10.06	1,670	1986 – 1995
74	Ahai	Jinsha River	Yunnan	RCCPG	132	8.85	2,000	2008 – 2013
75	Shapai	Caopo River	Sichuan	RCCVA	132	0.18	36	1997 – 2004

Continued

Appendix Dams and Reservoirs in China and the World

Continued

No.	Dam name	River	Location	Dam type	Dam height (m)	Total reservoir storage ($10^8 \mathrm{m}^3$)	Installed capacity (MW)	Years of construction
76	Jiangya	Lizhi River	Hunan	RCCPG	131	17.41	300	1995–2000
77	Baise	Youjiang River	Guangxi	RCCPG	130	56.6	540	2001–2006
78	Hongkou	Huotongxi River Mainstream	Fujian	RCCPG	130	4.497	200	2005–2008
79	Jinpen	Weihe River Tributary	Shaanxi	ER	130	2	20	1996–2002
80	Yinzidu	Yangtze River-Wujiang River-Sancha River	Guizhou	CFRD	129.5	5.31	360	2000–2003
81	Kensiwate	Manasi River	Xinjiang	CFRD	129.4	1.88	100	2009–2013
82	Hunanzhen	Qiantang River	Zhejiang	CB	129	20.67	320	1958–1980
83	Ankang	Hanjiang River	Shaanxi	PG	128	32.03	800	1978–1995
84	Hongqigang	Qiantang River	Zhejiang	TE	128	0.0378	0.4	1966–1970
85	Shankouyan	Yuanhe River	Jiangxi	RCCVA	126.71	1.05	12	2007–
86	Jiemian	Youxi River	Fujian	CFRD	126	18.24	300	2002–2007
87	Eping	Duhe-Huiwan	Hubei	CFRD	125.6	3.027	114	2002–2006
88	Qiaoqi	Qiyi River	Sichuan	ER	125.5	2.12	240	2002–2008
89	Guxian	Yellow River	Henan	PG	125	11.75	60	1978–1995
90	Yele	Dadu Rive	Sichu River	ER	124.5	2.98	240	2000–2006
91	Baixi	Baixi River Mainstream	Zhejiang	CFRD	124.4	1.684	18	1998–2001
92	Shimenkan	Lixian River	Yunnan	VA	111	1.97	130	2007–2010
93	Tengzigou	Longhe River	Chongqing	VA	124	1.93	70	2003–2006
94	Geliqiao	Qingshui River	Guizhou	RCCPG	124	0.774	150	2007–2010
95	Heiquan	Huangshui River	Qinghai	CFRD	123.5	1.82	12	1996–2000
96	Jintan	Huangdong River	Guangdong	VA	123.3	2.49	47	2007
97	Shitouxia	Datong River	Qinghai	CFRD	123.1	9.85	90	2011
98	Feicui	Beishixi River	Taiwan	VA	122.5	3.9	70	1981–1987
99	Hekoucun		Henan	CFRD	122.5	3.17	11.6	2011–2015
100	Kalasuke	Irtysh River	Xinjiang	RCCPG	121.5	24.19	140	2007

Appendix　Dams and Reservoirs in China and the World

Table 3　The Top 100 Dams by Installed Capacity in China

No.	Dam name	River	Location	Dam type	Dam height (m)	Total reservoir storage ($10^8 m^3$)	Installed capacity (MW)	Years of construction
1	Three Gorges	Yangtze River	Hubei	PG	181	45.05	22,500	1994–2010
2	Xiluodu	Jinsha River	Sichuan	VA	285.5	126.7	13,860	2005–2015
3	Xiangjiaba	Jinsha River	Yunnan/Sichuan	PG	162	51.63	6,400	2006–2015
4	Longtan	Hongshui River	Guangxi	RCCPG	192 (Phase I); 216.5 (Phase II)	188 (Phase I); 299.2 (Phase II)	4,900 (Phase I); 6,300 (Phase II)	2001–2009 (Phase I); No construction (Phase II)
5	Nuozhadu	Lancang River	Yunnan	ER	261.5	237.03	5,850	2006–2015
6	Jinping II	Yalong River	Sichuan	BM	34	0.192	4,800	2007–2015
7	Xiaowan	Lancang River	Yunnan	VA	294.5	150	4,200	2002–2012
8	Laxiwa	Yellow River	Qinghai	VA	250	10.79	4,200	2001–2010
9	Jinping I	Yalong River	Sichuan	VA	305	79.88	3,600	2005–2014
10	Pubugou	Dadu River	Sichuan	ER	186	53.37	3,600	2004–2009
11	Ertan	Yalong River	Sichuan	VA	240	61	3,300	1991–2000
12	Goupitan	Wujiang River	Guizhou	VA	232.5	64.54	3,000	2003–2009
13	Gezhouba	Yangtze River	Hubei	PG	53.8	7.11	2,715	1970–1996
14	Changheba	Dadu River	Sichuan	ER	240	10.75	2,600	2005–2013
15	Dagangshan	Dadu River	Sichuan	VA	210	7.77	2,600	2005–2015
16	Guandi	Yalong River	Sichuan	RCCPG	168	7.597	2,400	2010–2012
17	Jin'anqiao	Jinsha River	Yunnan	RCCPG	160	9.13	2,400	2003
18	Liyuan	Jinsha River	Yunnan	CFRD	155	8.05	2,400	2008–2014
19	Guangzhou Pumped Storage Power Station Upper Reservoir	Zhaodashui River	Guangdong	CFRD	68	0.26	2,400	1988–1994
20	Huizhou Pumped Storage Power Station Upper Reservoir	Xiaojin	Guangdong	RCCPG	56.7	0.317	2,400	2003–2011
21	Ludila	Jinsha River	Yunnan	RCCPG	140	17.18	2,160	2007

Appendix Dams and Reservoirs in China and the World

Continued

No.	Dam name	River	Location	Dam type	Dam height (m)	Total reservoir storage ($10^8 m^3$)	Installed capacity (MW)	Years of construction
22	Ahai	Jinsha River	Yunnan	RCCPG	132	8.85	2,000	2008－2013
23	Shuibuya	Qingjiang River	Hubei	CFRD	233.2	45.8	1,840	2002－2009
24	Xiaolangdi	Yellow River	Henan	ER	160	126.5	1,800	1994－2001
25	Dongjiang	Leishui River	Hunan	VA	157	92.7	500	1978－1992
26	Baishan	No. 2 Songhua River	Jilin	VA	149.5	59.21	1,800	1975－1986
27	Longkaikou	Jinsha River	Yunnan	RCCPG	116	5.58	1,800	2007－2013
28	Tianhuangping Pumped Storage Power Station Upper Reservoir	Shanhegang River	Zhejiang	ER	70	0.0885	1,800	1993－1997
29	Pengshui	Wujiang River	Chongqing	RCCPG	113.5	14.65	1,750	2003－2007
30	Jinghong	Lancang River	Yunnan	RCCPG	108	11.39	1,750	2003－Under Construction
31	Houziyan	Dadu River	Sichuan	CFRD	223.5	7.06	1,700	2011－2017
32	Manwan	Lancang River	Yunnan	PG	132	10.06	1,670	1986－1995
33	Lijiaxia	Yellow River	Qinghai	VA	155	17.5	1,600	1988－1997
34	Fengtan	Youshui River	Hunan	VA	112.5	17.4	815	1970－1979
35	Mingtan Pumped Storage Power Station Upper Reservoir	Shuilixi River	Taiwan	RCCPG	61.5	0.14	1,600	1987－1995
36	Liyang Pumped Storage Power Station Upper Reservoir	Zhongtianshe River	Jiangsu	CFRD	161.5	0.14	1,500	2008－2009
37	Gongboxia	Yellow River Mainstream	Qinghai	CFRD	132.2	6.2	1,500	2000－2006
38	Shuikou	Minjiang River	Fujian	PG	101	26	1,400	1987－1996
39	Liujiaxia	Yellow River	Gansu	PG	147	64	1,350	1958－1974
40	Dachaoshan	Lancang River	Yunnan	RCCPG	111	9.4	1,350	1997－2002
41	Tianshengqiao II	Nanpan River	Guangxi	RCCPG	60.7	1.1569	1,320	1984－1992
42	Longyangxia	Yellow River	Qinghai	VA	178	247	1,280	1977－1989

Appendix Dams and Reservoirs in China and the World

Continued

No.	Dam name	River	Location	Dam type	Dam height (m)	Total reservoir storage (10^8m^3)	Installed capacity (MW)	Years of construction
43	Qingyuan Pumped Storage Power Station Upper Reservoir	Qinhuang River	Guangdong	ER	54	0.12	1,280	2010–2015
44	Wujiangdu	Wujiang River tributary	Guizhou	VA	165	23	1,250	1970–1982
45	Tuoba	Lancang River	Yunnan	RCCPG	158	10.394	1,250	2010–2016
46	Yantan	Hongshui River	Guangxi	RCCPG	110	33.8	1,210	1985–1995
47	Tianshengqiao I	Nanpan River Mainstream	Guizhou/Guangxi	CFRD	178	102.57	1,200	1991–2000
48	Xilongchi Pumped Storage Power Station Lower Reservoir	Longchigou River	Shanxi	CFRD	97	0.0494	1,200	2002–2007
49	Tianchi Pumped Storage Power Station Upper Reservoir	Huangya River	Henan	CFRD	95.4	0.0881	1,200	2010–2015
50	Baoquan Pumped Storage Power Station Upper Reservoir	Yuhe River	Henan	ACFRD	92.5	0.07984	1,200	2002–2007
51	Wuqiangxi	Ruanshui River	Hunan	PG	85.83	42.9	1,200	1986–1998
52	Pushihe Pumped Storage Power Station Upper Reservoir	Pushi River Mainstream	Liaoning	PG	78.5	0.1351	1,200	2007–2010
53	Xianyou Pumped Storage Power Station Upper Reservoir	Mulanxi River	Fujian	CFRD	72.6	0.1735	1,200	2009–2014
54	Tongbai Pumped Storage Power Station Lower Reservoir	Baizhangxi River	Zhejiang	CFRD	71.4	0.13	1,200	2002–2006
55	Heimifeng Pumped Storage Power Station Upper Reservoir	Xiangjiang River	Hunan	CFRD	69.5	0.1035	1,200	2005–2010
56	Bailianhe Lower Reservoir	Xishui River	Hubei	ER	69	13.2	1,200	2005–2009
57	Huhehaote Pumped Storage Power Station	Daqingshan River System	Inner Mongolia	CFRD	62.5	0.066	1,200	2007–2012
58	Lianhua Pumped Storage Power Station Upper Reservoir	Mudan River Mainstream	Heilongjiang	CFRD	49	0.0954	1,200	1992–1998
59	Shatuo	Wujiang River	Guizhou	RCCPG	101	9.1	1,120	2009–2012
60	Tingzikou	Jialing River	Sichuan	RCCPG	116	40.67	1,100	2009–2014

Appendix Dams and Reservoirs in China and the World

Continued

No.	Dam name	River	Location	Dam type	Dam height (m)	Total reservoir storage ($10^8 m^3$)	Installed capacity (MW)	Years of construction
61	Wanjiazhai	Yellow River	Shanxi	PG	105	8.96	1,080	1994－2000
62	Silin	Wujiang River	Guizhou	RCCPG	117	15.93	1,050	2006－2009
63	Guangzhao	Beipan River	Guizhou	RCCPG	200.5	32.45	1,040	2003－2007
64	Jishixia	Yellow River	Qinghai	CFRD	103	2.94	1,020	2005－
65	Fengman	Songhua River	Jilin	PG	91.7	109.88	1,002.5	1937－1953
66	Sanbanxi	Ruanshui River Mainstream	Guizhou	CFRD	185.5	40.95	1,000	2002－2006
67	Taian Pumped Storage Power Station Upper Reservoir	Yingtaoyuan gou River	Shandong	CFRD	99.8	0.1097	1,000	2001－2005
68	Xiangshuijian Pumped Storage Power Station	Yangtze River	Anhui	CFRD	87	0.1748	1,000	2006－2011
69	Yixing Pumped Storage Power Station Upper Reservoir	Taihujingxi River	Jiangsu	CFRD	75	0.053	1,000	2001－2006
70	Banqiaoyu Pumped Storage Power Station Lower Reservoir	Baihe River	Beijing	VA	70	—	1,000	2002－2009
71	Songta	Songta River	Shanxi	X	62.6	0.974	1,000	2009－2011
72	Minghu Pumped Storage Power Station	Shuilixi River	Taiwan	PG	57.5	0.097	1,000	1981－1985
73	Zhanghewan Upper Reservoir	Gantao River	Hebei	ACFRD	57	0.0785	1,000	2002－2008
74	Zhexi	Zishui River	Hunan	CB	104	35.7	947.5	1958－1975
75	Danjiangkou	Hanjiang River	Hubei	PG	117 (added)	339.1	900	1958－1974(First) 2005－now(Added)
76	Shuifeng	Yalu River	Liaoning	PG	106.4	146.66	900	1937－1943
77	Gongguoqiao	Langcang River	Yunnan	RCCPG	105	3.49	900	2007
78	Dongqing	Beipan River	Guizhou	CFRD	150	9.55	880	2005－2010
79	Xin'anjiang	Xin'an River	Zhejiang	PG	105	216.26	850	1957－1965

Continued

No.	Dam name	River	Location	Dam type	Dam height (m)	Total reservoir storage ($10^8 m^3$)	Installed capacity (MW)	Years of construction
80	Tuokou	Ruanshui River	Hunan	RCCPG	82	13.84	830	2009-2011
81	Huanghuazhai	Getu River	Guizhou	RCCVA	110	1.748	60	2005-2008
82	Ankang	Hanjiang River	Shaanxi	PG	128	32.03	800	1978-1995
83	Shisanling Pumped Storage Power Station	Shangsigou River	Beijing	CFRD	75	0.0445	800	1992-1997
84	Zipingpu	Minjiang River	Sichuan	CFRD	156	11.12	760	2001-2006
85	Aertashi	Yeerqiang River	Xinjiang	CFRD	164.8	22.51	730	2011-2017
86	Badi	Dadu River	Sichuan	—	112	4.65	700	2009-2014
87	Gongzui	Dadu River	Sichuan	PG	85	3.73	700	1966-1979
88	Longtoushi	Dadu River	Sichuan	ER	58.5	1.392	700	2005-2009
89	Dongfeng	Wujiang River	Guizhou	VA	162	10.25	695	1984-1995
90	Shenxigou	Dadu River	Sichuan	PG	49.5	0.33	660	2007-2012
91	Changzhou	Xunjiang River	Guangxi	PG	56	56	630	2004-2009
92	Tankeng	Oujiang River	Zhejiang	CFRD	162	41.9	604	2004-2008
93	Hongjiadu	Wujiang River Mainstream	Guizhou	CFRD	179.5	49.47	600	2000-2005
94	Suofengying	Wujiang River Tributary	Guizhou	RCCPG	115.95	2.012	600	2002-2006
95	Mianhuatan	Tingjiang River	Fujian	RCCRG	113	20.35	600	1998-2001
96	Lubuge	Huangni River	Yunnan	ER	103.8	1.224	600	1982-1992
97	Letan	Hongshui River	Guangxi	PG	82	9.5	600	2001-2005
98	Tongjiezi	Dadu River	Sichuan	PG	82	2.6	600	1985-1992
99	Tongzilin	Yalong River	Sichuan	PG	71.3	0.912	600	2010-2016
100	Langyashan Pumped Storage Power Station Upper Reservoir	Huangshangou River	Anhui	CFRD	64	0.18	600	2001-2005

Appendix Dams and Reservoirs in China and the World

Table 4 The Top 30 Earth Dams by Height in China

No.	Dam name	River	Location	Dam type	Dam height (m)	Total reservoir storage (10^8m^3)	Installed capacity (MW)	Years of construction
1	Nuozhadu	Lancang River	Yunnan	ER	261.5	237.03	5,850	2006–2015
2	Changheba	Dadu River Mainstream	Sichuan	ER	240	10.75	2,600	2005–2013
3	Shuibuya	Qingjiang River	Hubei	CFRD	233.2	45.8	1,840	2002–2009
4	Houziyan	Dadu River	Sichuan	CFRD	223.5	7.06	1,700	2011–2017
5	Jiangpinghe	Loushui River	Hubei	CFRD	219	13.66	450	2007
6	Pubugou	Dadu River	Sichuan	ER	186	53.37	3,600	2004–2009
7	Sanbanxi	Ruanshui River Mainstream	Guizhou	CFRD	185.5	40.94	1,000	2002–2006
8	Hongjiadu	Wujiang River Mainstream	Guizhou	CFRD	179.5	49.47	600	2000–2005
9	Tianshengqiao I	Nanpan River Mainstream	Guizhou/Guangxi	CFRD	178	102.57	1,200	1991–2000
10	Kajiwa	Muli River	Sichuan	CFRD	171	3.745	440	2008–2014
11	Aertashi	Yeerqiang River	Xinjiang	CFRD	164.8	22.51	730	2011–2017
12	Pingzhai	Sancha River	Guizhou	CFRD	162.7	10.89	140.2	2010
13	Tankeng	Oujiang River	Zhejiang	CFRD	162	41.9	604	2004–2008
14	Liyang Pumped Storage Power Station Upper Reservoir	Zhongtianshe River	Jiangsu	CFRD	161.5	0.14	1,500	2008–2009
15	Xiaolangdi	Yellow River Mainstream	Henan	ER	160	126.5	1,800	1994–2001
16	Jilintai I	Kashi River	Xinjiang	CFRD	157	25.3	500	2001–2005
17	Zipingpu	Minjiang River	Sichuan	CFRD	156	11.12	760	2001–2006
18	Liyuan	Jinsha River	Yunnan	CFRD	155	8.05	2,400	2008–2014
19	Bashan	Renhe River	Chongqing	CFRD	155	3.154	140	2005–2009
20	Malutang II	Panlong River	Yunnan	CFRD	154	5.46	240	2005–2009
21	Dongqing	Beipan River	Guizhou	CFRD	150	9.55	880	2005–2010
22	Jilebulake	Irtysh River	Xinjiang	CFRD	147	2.32	160	2009–2013
23	Maoergai	Heishui River	Sichuan	ER	147	5.35	420	2008–2011
24	Longshou II	Heihe River Mainstream	Gansu	CFRD	146.5	0.862	157	2001–2005
25	Xigu	Jiulong River	Sichuan	CFRD	144	0.9986	249	2009–2013
26	Deze	Niulan River	Yunnan	CFRD	142	4.48	30	2008–2011
27	Wawushan	Zhougong River	Sichuan	CFRD	138.76	5.843	240	2003–2007
28	Shiziping	Zagu'nao River	Sichuan	ER	136	1.35	195	2003
29	Buxi	Yazui River	Sichuan	CFRD	135.8	2.52	20	2007–2011
30	Longma	Lixian River-Babian River	Yunnan	CFRD	135	5.904	240	2003–2007

Appendix Dams and Reservoirs in China and the World

Table 5 The Top 30 Arch Dams by Height in China

No.	Dam name	River	Location	Dam type	Dam height (m)	Total reservoir storage ($10^8 m^3$)	Installed capacity (MW)	Years of construction
1	Jinping I	Yalong River	Sichuan	VA	305	79.88	3,600	2005 – 2014
2	Xiaowan	Lancang River	Yunnan	VA	294.5	150	4,200	2002 – 2012
3	Xiluodu	Jinsha River	Sichuan	VA	285.5	126.7	13,860	2005
4	Laxiwa	Yellow River	Qinghai	VA	250	10.79	4,200	2001 – 2010
5	Ertan	Yalong River	Sichuan	VA	240	61	3,300	1991 – 2000
6	Goupitan	Wujiang River	Guizhou	VA	232.5	64.54	3,000	2003 – 2009
7	Danggangshan	Dadu River	Sichuan	VA	210	7.77	2,600	2005
8	Deji	Dajiaxi River	Taiwan	VA	181	2.32	234	1969 – 1974
9	Longyangxia	Yellow River	Qinghai	VA	178	274	1,280	1977 – 1989
10	Wujiangdu	Wujiang River Tributary	Guizhou	VA	165	23	1,250	1970 – 1982
11	Dongfeng	Wujiang River	Guizhou	VA	162	10.25	695	1984 – 1995
12	Wanjiakouzi	Gexiang River	Yunnan	RCCVA	167.5	2.793	180	2009
13	Gaixiaba	Changtan River	Chongqing	VA	160	3.54	132	2009
14	Dongjiang	Leishui River	Hunan	VA	157	92.7	500	1978 – 1992
15	Lijiaxia	Yellow River	Qinghai	VA	155	17.5	1,600	1988 – 1997
16	Geheyan	Qingjiang River	Hubei	VA	151	27.06	180	1987 – 1995
17	Baishan	No. 2 Songhua River	Jilin	VA	149.5	59.21	1,800	1975 – 1986
18	Jiangkou	Wujiang River Tributary	Chongqing	VA	140	5.05	300	1999 – 2004
19	Qinglong	Maweigou River	Hubei	RCCVA	139.7	0.283	40	2009 – 2011
20	Yunlonghe III	Yunlong River	Hubei	RCCVA	135	0.44	40	2006 – 2008
21	Dongping	Qingjiang River	Hubei	VA	135	3.43	110	2002 – 2006
22	Dahuashui	Wujiang River	Guizhou	RCCVA	134.5	2.765	200	2005 – 2007
23	Sanliping	Nanhe River	Hubei	RCCVA	133	4.99	70	2006 –
24	Shapai	Caopo River	Sichuan	RCCVA	132	0.18	36	1997 – 2004
25	Shankouyan	Yuanhe River	Jiangxi	RCCVA	126.71	1.05	12	2007 –
26	Tengzigou	Longhe River	Chongqing	VA	124	1.97	70	2003 – 2006
27	Shimenkan	Lixian River	Yunnan	VA	111	1.97	130	2007 – 2010
28	Jintan	Huangdong River	Guangdong	VA	123.3	2.49	47	2007
29	Feicui	Beishixi River	Taiwan	VA	122.5	3.9	70	1981 – 1987
30	Shanmipo	Beipan River	Guizhou	RCCVA	119.4	0.85	185.5	2009

Appendix Dams and Reservoirs in China and the World

Table 6 The Top 30 Gravity Dams by Height in China

No.	Dam name	River	Location	Dam type	Dam height (m)	Total reservoir storage ($10^8 m^3$)	Installed capacity (MW)	Years of construction
1	Longtan	Hongshui River	Guangxi	RCCPG	192 (Phase I); 216.5 (Phase II)	188 (Phase I); 299.2 (Phase II)	4,900 (Phase I); 6,300 (Phase II)	2,001 – 2009
2	Guangzhao	Beipan River	Guizhou	RCCPG	200.5	32.45	1,040	2003 – 2007
3	Three Gorges	Yangtze River	Hubei	PG	181	450.5	22,500	1994 – 2010
4	Guandi	Yalong River	Sichuan	RCCPG	168	7.597	2,400	2010 – 2012
5	Xiangjiaba	Jinsha River	Yunnan/Sichuan	PG	162	51.63	6,400	2006 – 2015
6	Jin'anqiao	Jinsha River	Yunnan	RCCPG	160	9.13	2,400	2003
7	Tuoba	Lancang River	Yunnan	RCCPG	158	10.394	1,250	2010 – 2016
8	Liujiaxia	Yellow River	Gansu	PG	147	64	1,350	1958 – 1974
9	Shiziguan	Hongjia River	Hubei	PG	145	1.41	10	1999 – 2004
10	Ludila	Jinsha River	Yunnan	RCCPG	140	17.18	2,160	2007 –
11	Manwan	Lancang River	Yunnan	PG	132	10.06	1,670	1986 – 1995
12	Ahai	Jinsha River	Yunnan	RCCPG	132	8.85	2,000	2008 – 2013
13	Jiangya	Lizhi River	Hunan	RCCPG	131	17.4	300	1995 – 2000
14	Baise	Youjiang River	Guangxi	RCCPG	130	56.6	540	2001 – 2006
15	Hongkou	Huotongxi River Mainstream	Fujian	RCCPG	130	4.497	200	2005 – 2008
16	Ankang	Hanjiang River	Shaanxi	PG	128	32.03	800	1978 – 1995
17	Guxian	Yellow River	Henan	PG	125	11.75	60	1987 – 1995
18	Geliqiao	Qingshui River	Guizhou	RCCPG	124	0.774	150	2007 – 2010
19	Kalasuke	Eerqisi River	Xinjiang	RCCPG	121.5	24.19	140	2007
20	Wudu	Peijiang River	Sichuan	RCCPG	120.34	5.72	150	2004
21	Silin	Wujiang River	Guizhou	RCCPG	117	15.93	1,050	2006 – 2009
22	Danjiangkou	Hanjiang River	Hubei	PG	117	339.1	900	1958 – 1974 (First) / 2005 – now (Added)
23	Zangmu	Yaluzangbu River	Tibet	PG	116	0.886	510	2010
24	Longkaikou	Jinsha River	Yunnan	RCCPG	116	5.58	1,800	2007 – 2013
25	Tingzikou	Jialing River	Sichuan	RCCPG	116	40.67	1,100	2009 – 2014
26	Suofengying	Wujiang River Tributary	Guizhou	RCCPG	115.95	2.012	600	2002 – 2006
27	Yunfeng	Yalu River	Jilin	PG	113.75	38.95	400	1959 – 1965
28	Pengshui	Wujiang River	Chongqing	RCCPG	113.5	14.65	1,750	2003 – 2007
29	Mianhuatan	Tingjiang River	Fujian	RCCPG	113	20.35	600	1998 – 2001
30	Gelantan	Lixian River	Yunnan	RCCPG	113	4.09	450	2003 – 2009

Appendix Dams and Reservoirs in China and the World

Table 7 The 30 Concrete Faced Rockfill Dams by Height in China

No.	Dam name	River	Location	Dam type	Dam height (m)	Total reservoir storage ($10^8 m^3$)	Installed capacity (MW)	Years of construction
1	Shuibuya	Qingjiang River	Hubei	CFRD	233.2	45.8	1,840	2002－2009
2	Houziyan	Dadu River	Sichuan	CFRD	223.5	7.06	1,700	2011－2017
3	Jiangpinghe	Loushui River	Hubei	CFRD	219	13.66	450	2007
4	Sanbanxi	Ruanxi River Mainstream	Guizhou	CFRD	185.5	40.94	1,000	2002－2006
5	Hongjiadu	Wujiang River Mainstream	Guizhou	CFRD	179.5	49.47	600	2000－2005
6	Tianshengqiao I	Nanpan River Mainstream	Guizhou/Guangxi	CFRD	178	102.57	1,200	1991－2000
7	Kajiwa	Muli River	Sichuan	CFRD	171	3.745	440	2008－2014
8	Aertashi	Yeerqiang River	Xinjiang	CFRD	164.8	22.51	730	2011－2017
9	Pingzhai	Sancha River	Guizhou	CFRD	162.7	10.89	140.2	2010
10	Tankeng	Oujiang River	Zhejiang	CFRD	162	41.9	604	2004－2008
11	Liyang Pumped Storage Power Station Upper Reservoir	Zhongtianshe River	Jiangsu	CFRD	161.5	0.14	1,500	2008－2009
12	Jilintai I	Kashi River	Xinjiang	CFRD	157	25.3	500	2001－2005
13	Zipingpu	Minjiang River	Sichuan	CFRD	156	11.12	760	2001－2006
14	Liyuan	Jinsha River	Yunnan	CFRD	155	8.05	2,400	2008－2014
15	Bashan	Renhe River	Chongqing	CFRD	155	3.154	140	2005－2009
16	Malutang II	Panlong River	Yunnan	CFRD	154	5.46	240	2005－2009
17	Dongqing	Beipan River	Guizhou	CFRD	150	9.55	880	2005－2010
18	Jilebulake	Eerqisi River	Xinjiang	CFRD	147	2.32	160	2009－2013
19	Longshou II	Heihe River Mainstream	Gansu	CFRD	146.5	0.862	157	2001－2005
20	Xigu	Jiulong River	Sichuan	CFRD	144	0.9986	249	2009－2013
21	Deze	Niulan River	Yunnan	CFRD	142	4.48	30	2008－2011
22	Wawushan	Zhougong River	Sichuan	CFRD	138.76	5.843	240	2003－2007
23	Buxi	Yazui River	Sichuan	CFRD	135.8	2.52	20	2007－2011
24	Longma	Lixian River-Babian River	Yunnan	CFRD	135	5.904	240	2003－2007
25	Jiyin	Keliya River	Xinjiang	CFRD	134.7	0.83	240	2008－2013
26	Sujiahekou	Binglang River	Yunnan	CFRD	133.75	2.25	315	2005－2010
27	Jiudianxia	Taohe River	Gansu	CFRD	133	9.43	300	2005－2008
28	Wuluwati	Karakash River	Xinjiang	CFRD	133	3.47	60	1995－2001
29	Shanxi	Feiyun River	Zhejiang	CFRD	132.8	18.24	200	1997－2001
30	Gongboxia	Yellow River	Qinghai	CFRD	132.2	6.2	1,500	2000－2006

Table 8 The 30 Roller Compacted Concrete dams by Height in China

No.	Dam name	River	Location	Dam type	Dam height (m)	Total reservoir storage ($10^8 m^3$)	Installed capacity (MW)	Years of construction
1	Longtan	Hongshui River	Guangxi	RCCPG	192 (Phase I); 216.5 (Phase II)	188 (Phase I); 299.2 (Phase II)	4,900 (Phase I); 6,300 (Phase II)	2001–2009 (Phase I); No construction (Phase II)
2	Guangzhao	Beipan River	Guizhou	RCCPG	200.5	32.45	1,040	2003–2007
3	Guandi	Yalong River	Sichuan	RCCPG	168	7.597	2,400	2010–2012
4	Wanjiakouzi	Gexiang River	Yunnan	RCCVA	167.5	2.793	180	2009
5	Jinganqiao	Jinsha River	Yunnan	RCCPG	160	9.13	2,400	2003–
6	Tuoba	Lancang River	Yunnan	RCCPG	158	10.394	1,250	2010–2016
7	Ludila	Jinsha River	Yunnan	RCCPG	140	17.18	2,160	2007
8	Qinglong	Maweigou River	Hubei	RCCVA	139.7	0.283	40	2009–2011
9	Yunlonghe III	Yunlong River	Hubei	RCCVA	135	0.44	40	2006–2008
10	Dahuashui	Wujiang River Tributary	Guizhou	RCCVA	134.5	2.765	200	2005–2007
11	Sanliping	Nanhe River	Hubei	RCCVA	133	4.99	70	2006
12	Ahai	Jinsha River	Yunnan	RCCPG	132	8.85	2,000	2008–2013
13	Shapai	Caopo River	Sichuan	RCCVA	132	0.18	36	1997–2004
14	Jiangya	Lizhi River	Hunan	RCCPG	131	17.4	300	1995–2000
15	Baise	Youjiang River	Guangxi	RCCPG	130	56.6	540	2001–2006
16	Hongkou	Huotongxi River Mainstream	Fujian	RCCPG	130	4.497	200	2005–2008
17	Shankouyan	Yuanhe River	Jiangxi	RCCVA	126.71	1.05	12	2007
18	Geliqiao	Qingshui River	Guizhou	RCCPG	124	0.774	150	2007–2010
19	Kalasuke	Eerqisi River	Xinjiang	RCCPG	121.5	24.19	140	2007
20	Shannipo	Beipan River	Guizhou	RCCVA	119.4	0.850	185.5	2009
21	Wudu	Fujiang River	Sichuan	RCCPG	120.34	5.72	150	2009
22	Silin	Wujiang River	Guizhou	RCCPG	117	15.93	1,050	2006–2009
23	Longkaikou	Jinsha River	Yunnan	RCCPG	116	5.58	1,800	2007–2013
24	Tingzikou	Jialing River	Sichuan	RCCPG	116	40.67	1,100	2009–2014
25	Suofengying	Wujiang River Tributary	Guizhou	RCCPG	115.95	2.012	600	2002–2006
26	Luoboba	Lengshui River	Hubei	RCCVA	114	—	30	2007–2010
27	Pengshui	Wujiang River	Chongqing	RCCPG	113.5	14.65	1,750	2003–2007
28	Mianhuatan	Tingjiang River	Fujian	RCCPG	113	20.35	600	1998–2001
29	Gelantan	Lixian River	Yunnan	RCCPG	113	4.09	450	2003–2009
30	Tianhuaban	Niulan River	Yunnan	RCCVA	113	0.657	180	2006–2010

Appendix   Dams and Reservoirs in China and the World

Table 9   The Top 100 Dams by Storage Capacity in the World

No.	Dam name	Country	Dam type	Total reservoir storage ($10^8 \mathrm{m}^3$)	Dam height (m)	Installed capacity (MW)	Year of completion
1	Owen Falls Reservoir	Uganda	PG	2,048	31	180	1954
2	Kariba	Zambia/Zimbabwe	VA	1,806	128	1,500	1976
3	Bratsk	Russia	PG	1,690	125	4,500	1964
4	Haswan High Dam	Egypt	ER	1,620	111	2,100	1970
5	Akosombo Dam	Ghana	ER	1,500	134	1,020	1965
6	Daniel Johnson Dam	Canada	VA	1,418.5	214	2,656	1968
7	Guri	Venezuela	PG	1,350	162	10,235	1986
8	W. A. C. Bennett	Canada	TE	743	183	2,730	1967
9	Krasnoyarsk	Russia	PG	733	124	6,000	1972
10	Zeya Dam	Russia	PG	684	115	1,330	1978
11	La Gande II	Canada	ER	617.2	168	7,722	1992
12	La Gande III	Canada	ER	600.2	93	2,418	1984
13	Ust-Ilim	Russia	PG	593	102	3,840	1979
14	Boguchany	Russia	TE	582	87	4,000	2010
15	Kuibyshev	Russia	PG	580	45	2,320	1957
16	Serra Da Mesa	Brazil	TE	544	154	1,275	1998
17	Caniapiscau	Canada	ER	537.9	54	712	1981
18	Cahora Bassa	Mozambique	VA	520	171	4,150	1987
19	Upper Wainganga	India	TE	507	43	600	1998
20	Bukhtarma	Kazakhstan	PG	498	90	675	1966
21	Ataturk	Turkey	ER	487	169	2,400	1991
22	Irkutsk	Russia	TE	481	44	660	1958
23	Tucurui	Brazil	PG	455.4	98	8,370	2002
24	Three Gorges	China	PG	450.5	181	22,500	2010
25	Bakun	Malaysia	CFRD	438	205	2,400	Under Construction
26	Cerros Colorados	Argentina	TE	430	35	450	1978

Appendix Dams and Reservoirs in China and the World

Continued

No.	Dam name	Country	Dam type	Total reservoir storage (10^8 m^3)	Dam height (m)	Installed capacity (MW)	Year of completion
27	Hoover Dam	United States	VA	373	221.4	2,080	1936
28	Vilyui Reservoir	Russia	ER	359	75	650	1976
29	Kouilou Dam	Congo	VA	350	137	1,050	1992
30	Sobadinho	Brazil	TE	341	41	1,050	1979
31	Danjiangkou	China	PG	339.1	117	900	1968 (Completed); 2005 (Added)
32	Glen Canyon Dam	United States	VA	333	216	1,021	1966
33	Churchill Falls	Canada	TE	323.2	32	5,428	1974
34	Skins Lake I	Canada	TE	322	25	4,600	1953
35	Jenpeg/Kiskitto	Canada	TE	317.9	15	135	1979
36	Volgograd	Russia	TE	315	47	2,563	1962
37	Sayano-Shushenskaya	Russia	VA	313	245	6,400	1989
38	Keban	Turkey	PG	310	210	1,330	1975
39	Garrison Dam	United States	TE	302.2	64	583.3	1953
40	Iroquois	Canada	PG	299.6	20	1,880	1958
41	Longtan	China	RCCPG	192 (Phase I); 216.5 (Phase II)	188 (Phase I); 299.2 (Phase II)	4900 (Phase I); 6300 (Phase II)	2001–2009 (Phase I); No construction (Phase II)
42	Oahe Dam	United States	TE	291.1	74.7	595	1962
43	Itaipu	Brazil/Paraguay	PG	290	196	14,000	1991
44	Missi Falls	Canada	TE	283.7	18	—	1976
45	Kapchagay	Kazakhstan	TE	281	56	432	1972
46	Kossou	Ivory Coast	TE	276.8	58	175.5	1973
47	Longyangxia	China	VA	274	178	1,280	1989
48	Rybinsk Reservoir	Russia	TE	254	33	346	1945
49	Mica	Canada	TE	250	243	1,740	1973

Appendix Dams and Reservoirs in China and the World

Continued

No.	Dam name	Country	Dam type	Total reservoir storage ($10^8 m^3$)	Dam height (m)	Installed capacity (MW)	Year of completion
50	Brokopondo	Suriname	PG/ER	240	66	120	1965
51	Tsimliansk	Russia	PG/TE	240	41	160	1952
52	Kenney Dam	Canada	ER	238	104	1,670	1954
53	Nuozhadu	China	ER	237.03	261.5	5,850	2015
54	Fort Peck	United States	TE	235.6	78	185	1940
55	Ust-Khantaika	Russia	ER	235	65	441	1972
56	Furnas	Brazil	ER	229.5	127	1,216	1963
57	Rajghat Dam	India	PG/TE	217.2	44	45	2006
58	Xin'anjiang	China	PG	216.26	105	850	1965
59	Ilha Solteira	Brazil	PG	211.7	74	3,444	1978
60	Tres Marias	Brazil	TE	210	75	387.6	1969
61	Yacyreta	Argentina/Paraguay	PG	210	43	3,100	2010
62	Bureya Dam	Russia	PG	209	139	2,010	1998
63	El Chocon Dam	Argentina	TE	202	86	1,200	1972
64	Porto Primavera	Brazil	TE	199	38	1,540	1999
65	La Grande Ⅳ	Canada	ER	195.3	128	2,779	1985
66	Toktogul	Kyrgyzstan	PG	195	215	1,200	1978
67	Kakhov	Ukraine	PG	181.8	37	351	1958
68	Emborcasao	Brazil	ERPG	175.9	158	1,192	1982
69	Balbina Reservoir	Brazil	PG	175.4	33	250	1987
70	Verkhne-Svirskaya	Russia	PG	175	32	160	1952
71	Itumbiara	Brazil	PG	170	106	2,082	1981
72	Mingechaur	Azerbaijian	TE	160	80	370	1953
73	Xiaowan	China	VA	150	294.5	4,200	2012
74	Gordon Dam	Australia	VA	150	140	750	1974
75	Kainji	Nigeria	PG	150	80	960	1968

991

Appendix Dams and Reservoirs in China and the World

Continued

No.	Dam name	Country	Dam type	Total reservoir storage ($10^8 m^3$)	Dam height (m)	Installed capacity (MW)	Year of completion
76	Boruca	Costa Rica	TE	149.6	267	1,400	1990
77	Reindeer	Canada	PG	148.6	12	—	1942
78	Shuifeng	China/North Korea	PG	146.66	106.4	900	1943
79	Tabqa	Syria	TE	140	60	824	1976
80	Gibe III	Ethiopia	RCC	140	243	1,870	Under Construction
81	Pamouscachiou I	Canada	PG	139	20	—	1955
82	Xinfengjiang	China	PG	138.96	105	355	1977
83	Nizhne-Kamsk	Russia	PG	138	36	1,248	1979
84	Tarbela Dam	Pakistan	TE	136.9	143	3,478	1976
85	Kenyir	Malaysia	ER	136	155	400	1985
86	Kremenchug	Ukraine	TE	135.2	33	625	1960
87	Bhumibol	Thailand	VA	134.6	154	535	1964
88	Tres Irmaos	Brazil	PG	134.5	90	1,292	1991
89	Saratov	Russia	TE	129	40	1,360	1971
90	Nova Ponte Dam	Brazil	ER	128	142	519	1994
91	Cheboksary	Russia	PG	128	45	1,404	1986
92	Xiaolangdi	China	ER	126.5	160	1,800	2001
93	Saint Maguerite-3	Canada	CFRD	126.34	171	882	2003
94	Sao Simao	Brazil	ER	125.4	127	1,710	1978
95	Mosul Dam	Iraq	TE	125	131	750	1984
96	Piedra Del Aguila	Argentina	PG	124	170	1,400	1992
97	Kama	Russia	TE	122	37	504	1958
98	Grand Goulee	United States	PG	118	168	6,494	1980
99	Nagarjuna Sagar	India	PG/TE	125	115.6	815.6	1974
100	Verkhne-Tulom	Russia	PG/TE	115.2	46.5	268	1966

Appendix Dams and Reservoirs in China and the World

Table 10 The Top 100 by Height Dams in the World

No.	Dam name	Country	Dam type	Dam height (m)	Total reservoir storage ($10^8 m^3$)	Installed capacity (MW)	Years of construction
1	Jinping I	China	VA	305	79.88	3,600	2014
2	Nurek	Tajikistan	TE	300	105	2,700	1980
3	Xiaowan	China	VA	294.5	150	4,200	2012
4	Xiluodu	China	VA	285.5	126.7	13,860	2015
5	Grande Dixence	Switzerland	PG	285	4	2,069	1962
6	Kambarazin I	Kyrgyzstan	TE	275	36	1,900	1996
7	Inguri	Georgia	VA	271.5	11	1,320	1980
8	Boruca Dam	Costa Rica	TE	267	149.6	1,400	1990
9	Vajont	Italy	VA	262	1.69		1961
10	Nuozhadu	China	ER	261.5	237.03	5,850	2015
11	Chicoasen	Mexico	TE	261	16.8	2,430	1981
12	Tehri	India	TE	261	35.4	1,000	1990
13	Alvaro Obregon	Mexico	PG	260	4	86.4	1952
14	Mauvoisin	Switzerland	VA	250.5	2.12	114	1991 (Added)
15	Laxiwa	China	VA	250	10.79	4,200	2010
16	Guario	Colombia	TE	247	9	1,600	1992
17	Deriner	Turkey	VA	247	19.7	670	2004
18	Alberto Lleras	Colombia	ER	243	9.7	1,150	1989
19	Mica	Canada	TE	243	250	2,104	1972
20	Gibe III	Ethiopia	RCC	243	140	1,870	2013
21	Sayano-Shushenkaya	Russia	VA/PG	242	313	6,800	1989
22	Changheba	China	ER	240	10.75	2,600	2013

993

Appendix   Dams and Reservoirs in China and the World

Continued

No.	Dam name	Country	Dam type	Dam height (m)	Total reservoir storage ($10^8 m^3$)	Installed capacity (MW)	Years of construction
23	Ertan	China	VA	240	61	3,300	2000
24	La Esmeralda (Chivor)	Colombia	ER	237	8.2	1,000	1975
25	Kishau	India	PG	236	18.1	600	1995
26	Oroville	United States	TE/ER	235	43.67	762	1967
27	El Cajon	Honduras	VA	234	70.85	300	1985
28	Chirkey	Georgia	VA	233	27.8	1,000	1977
29	Shuibuya	China	CFRD	233.2	45.8	1,840	2009
30	Goupitan	China	VA	232.5	64.54	3,000	2009
31	Karun Ⅳ	Iran	VA	230	21.9	1,000	2010
32	Contra	Switzerland	VA	230	1.05	105	1965
33	Bekhme	Iraq	TE	230	170	1,560	Under Construction
34	Tasang	Myanmar	RCCPG	227.5	—	—	Under Construction
35	Bhakra Dam	India	PG	226	96.2	1,325	1963
36	Luzzone Dam	Switzerland	VA	225	1.1	418	1963
37	Houziyan	China	CFRD	223.5	7.06	1,700	2017
38	El Platanal	Peru	CFRD	221	—	—	Under Construction
39	Hoover Dam	United States	VA	221.4	373	2,080	1936
40	Nam Ngum Ⅲ	Laos	CFRD	220	13.2	440	2002
41	Contra Dam	Switzerland	VA	220	1.1	105	1965
42	Mratinje	Yugoslavia	VA	220	8.9	86	1976
43	Jiangpinghe	China	CFRD	219	13.66	450	Under Construction
44	Dworshak	United States	PG	219	42.8	400	1973

Appendix Dams and Reservoirs in China and the World

Continued

No.	Dam name	Country	Dam type	Dam height (m)	Total reservoir storage ($10^8 m^3$)	Installed capacity (MW)	Years of construction
45	Longtan	China	RCCPG	192 (Phase I); 216.5 (Phase II)	188 (Phase I); 299.2 (Phase II)	4,900 (Phase I); 6,300 (Phase II)	2001–2009 (Phase I); To be constructed (Phase II)
46	Glen Canyon	United States	VA	216	333	900	1966
47	Toktogul	Kyrgyzstan	PG	215	195	1,200	1978
48	Daniel Johnson	Canada	VA	214	1418.5	2,656	1968
49	Keban	Turkey	PG	210	310	1,330	1975
50	Ermenek	Turkey	VA	210	45.8	300	2007
51	Dagangshan	China	VA	210	7.77	2,600	2015
52	La Yesca	Mexico	CFRD	210	—	—	Under Construction
53	Portugues	Puerto Rico	RCCVA	210	—	—	Under Construction
54	Auburn	United States	VA	209	31	750	1975
55	Irape Dam	Brazil	ER	208	59.6	360	2006
56	Zimapan	Mexico	VA	207	9.96	400	1994
57	Bakun	Malaysia	CFRD	205	438	2,400	Under Construction
58	Karun III	Iran	VA	205	29.7	2,280	2005
59	Lakhwar	India	PG	204	5.8	300	1996
60	Ross	United States	VA	204	17.4	400	1949
61	Dez	Iran	VA	203	33.4	520	1962
62	Almendra	Spain	VA	202	26.5	828	1970
63	Campos Novos	Brazil	CFRD	202	16.5	880	2006
64	Berke	Turkey	VA	201	4.3	512	1996
65	Khudoni	Georgia	VA	201	3.7	2,100	1991
66	Guangzhao	China	RCCPG	200.5	32.45	1,040	2007
67	Karun I	Iran	VA	200	31.4	2,000	1976
68	San Roque	Philippines	TE	200	8.4	345	2003
69	Koelnbrein	Austria	VA	200	2	881	1977
70	Kabir	Iran	VA	200	29	1,000	1977
71	New Bullards Bar	United States	VA	197	11.9	284.4	1969

Appendix Dams and Reservoirs in China and the World

Continued

No.	Dam name	Country	Dam type	Dam height (m)	Total reservoir storage (10^8m^3)	Installed capacity (MW)	Years of construction
72	Itaipu	Brazil/Paraguay	PG	196	290	14,000	1991
73	Altinkaya	Turkey	ER	195	57.6	700	1988
74	Boyabat	Turkey	PG	195	35.57	513	2014
75	Seven Oaks	United States	TE	193	1.79	—	1999
76	Karahnjukar	Iceland	CFRD	193	21	690	2008
77	New Melones	United States	ER	191	35.4	300	1979
78	Sogamoso	Colombia	CFRD	190	—	—	2005
79	Miel I	Colombia	RCCPG	188	5.7	375	2002
80	Aguamilpa	Mexico	CFRD	187	69.5	960	1994
81	El Cajon Dam	Mexico	CFRD	187	50	750	2007
82	Pubugou	China	ER	186	53.37	3,600	2009
83	Kurobe 4	Japan	VA	186	2	335	1963
84	Zillergrundl	Austria	VA	186	0.9	360	1986
85	Sanbanxi	China	CFRD	185.5	40.94	1,000	2006
86	Barra Grande	Brazil	CFRD	185	50	708	2005
87	Mossyrock	United States	VA	185	20.8	300	1968
88	Katse	Lesotho	VA	185	19.5	—	1996
89	Oymapinar	Turkey	VA	185	3	540	1984
90	Arachtos	Greece	PG	185	—	—	Under Construction
91	W. A. C. Bennett	Canada	TE	183	743	2,730	1967
92	Shasta	United States	VA	183	56.2	676	1945
93	Three Gorges	China	PG	181	450.5	22,500	2010
94	Deji	China	VA	181	2.32	234	1974
95	Nam Ngum 2	Laos	CFRD	181	—	615	2010
96	Dartmouth	Australia	TE	180	40	150	1979
97	Karaj	Iran	VA	180	2.5	120	1961
98	Tignes	France	VA	180	2.4	93	1952
99	Emosson	Switzerland	VA	180	2.3	357	1974
100	Hongjiadu	China	CFRD	179.5	49.47	600	2005

Table 11 The Top 30 Earth Dams by Height in the World

No.	Dam name	Dam height (m)	Country	Year of completion
1	Nurek	300	Tajikistan	1980
2	Kambarazin I	275	Kyrgyzstan	1996
3	Boruca Dam	267	Costa Rica	1990
4	Nuozhadu	261.5	China	2015
5	Manuel M. Torres	261	Mexico	1981
6	Tehri	261	India	1990
7	Guavio	247	Colombia	1992
8	Alberto Llerasc	243	Colombia	1989
9	Mica	243	Canada	1972
10	Changheba	240	China	2013
11	La Esmeralda (Chivor)	237	Colombia	1975
12	Oroville	235	United States	1967
13	Shuibuya	233.2	China	2009
14	Bekhme	230	Iraq	Under Construction
15	Houziyan	223.5	China	2017
16	El Platanal	221	Peru	Under Construction
17	Nam Ngum III	220	Laos	2002
18	Jiangpinghe	219	China	Under Construction
19	La Yesca	210	Mexico	Under Construction
20	San Roque	210	Philippines	2003
21	Irape	208	Brazil	2006
22	Bakun	205	Malaysia	2003
23	Campos Novos	202	Brazil	2006
24	Karahnjukar	198	Iceland	2008
25	Altinkaya	195	Turkey	1988
26	Seven Oaks	193	United States	1999
27	New Melones	191	United States	1979
28	Sogamoso	190	Colombia	2005
29	Aguamilpa	187	Mexico	1994
30	El Cajon	187	Mexico	2007

Appendix Dams and Reservoirs in China and the World

Table 12 The Top 30 Arch Dams by Height in the World

No.	Dam name	Dam height (m)	Country	Year of completion
1	Jinping I	305	China	2014
2	Xiaowan	294.5	China	2012
3	Xiluodu	285.5	China	2015
4	Inguri	271.5	Georgia	1980
5	Vajont	262	Italy	1961
6	Mauvoisin	250	Switzerland	1958 (Completed) 1991 (heightened)
7	Laxiwa	250	China	2010
8	Deriner	247	Turkey	2004
9	Sayano-Shushenskaya	242	Russia	1989
10	Ertan	240	China	2000
11	El Cajon	234	Honduras	1985
12	Chirkey	233	Georgia	1977
13	Goupitan	232.5	China	2009
14	Kaun 4	230	Iran	2010
15	Contra	230	Switzerland	1965
16	Luzzone	225	Switzerland	1963
17	Hoover	223	United States	1936
18	Contra	220	Switzerland	1965
19	Mratinje	220	Yugoslavia	1976
20	Glen Canyon	216	United States	1966
21	Daniel Johnson	214	Canada	1968
22	Dagangshan	210	China	2015
23	Ermenek	210	Turkey	2007
24	Portugues	210	Puerto Rico	Under Construction
25	Auburn	209	United States	1975
26	Zimapan	207	Mexico	1994
27	Karun III	205	Iran	1980
28	Dez	203	Iran	1962
29	Almendra	202	Spain	1970
30	Berk	201	Turkey	1996

Table 13　The Top 30 Gravity Dams by Height in the World

No.	Dam name	Dam height (m)	Country	Year of completion
1	Grande Dixence	285	Switzerland	1962
2	Alvaro Obregon	260	Mexico	1952
3	Kishau	236	India	1995
4	TaSang	227.5	Burma	Under Construction
5	Bhakra	226	India	1963
6	Dworshak	219	United States	1973
7	Longtan	192 (Phase I); 216.5 (Phase II)	China	2009 (Phase I); To be constructed (Phase II)
8	Toktogul	215	Kyrgyzstan	1978
9	Keban	210	Turkey	1975
10	Lakhwar	204	India	1996
11	Guangzhao	200.5	China	2009
12	Itaipu	196	Brazil	1991
13	Boyabat	195	Turkey	—
14	Miel I	188	Columbia	1999
15	Arachtos	185	Greece	—
16	Three Gorges	181	China	2010
17	Revelstoke	175	Canada	1984
18	Alpe Gera	174	Italy	—
19	Piedra Del Aguila	170	Argentina	1992
20	Guandi	168	China	2012
21	Grand Coulee	168	United States	1934-1951; 1967-1980
22	Ingapata	166	Ecuador	—
23	Wujiangdu	165	China	1982
24	Guri	162	Venezuela	1986
25	Xiangjiaba	162	China	2015
26	Songwon	160	North Korea	1995
27	Jin'anqiao	160	China	Under Construction
28	Shafarud	159	Iran	Under Construction
29	Tuoba	158	China	2016
30	Okugadami	157	Japan	1961

Appendix   Dams and Reservoirs in China and the World

Table 14   **Classification of Hydropower Projects in china**

Classification	Scale	Total reservoir storage ($10^8 m^3$)	Installed Capacity (MW)
1	Large (Ⅰ)	≥10	≥1,200
2	Large (Ⅱ)	<10 ≥1	<1,200 ≥300
3	Medium-sized	<1.00 ≥0.10	<300 ≥50
4	Small (Ⅰ)	<0.10 ≥0.01	<50 ≥10
5	Small (Ⅱ)	<0.01	<10

Reference

[1] Zhao Chunhou, Zhu Zhenhong and Zhou Duanzhuang. *Worldwide Rivers & Dams*. Beijing: China Water & Power Press, 2000.
[2] *A collection of High Arch Dams*, Northwest Institute of Hydro-electric Design and Survey, 1992.
[3] *Overview of Arch dams in the world*, The third Engineering Bureau of Water Resources and Power Ministry, 1978.